COMPENDIUM OF METHODS FOR THE

MICROBIOLOGICAL EXAMINATION OF FOODS

FOURTH EDITION

EDITED BY
FRANCES POUCH DOWNES
KEITH ITO

AMERICAN PUBLIC HEALTH ASSOCIATION

The American Public Health Association is an Association of individuals and organizations working to improve the public's health. It promotes the scientific and professional foundation of public health practice and policy, advocates the conditions for a healthy global society, emphasizes prevention, and enhances the ability of members to promote and protect environmental and community health.

American Public Health Association
800 I St., NW
Washington, DC 20001-3710

Mohammad N. Akhter, MD, MPH
Executive Vice President

10 M 3/01
Library of Congress Catalog Card Number: 99-073697

ISBN: 0-87553-175-x

Printed and bound in the United States of America.
Cover Design: Sam Dixon Design
Typesetting: Susan Westrate, Joseph R. Loehle
Set in: Palatino and Universe
Printing and Binding: Sheridan Books, Inc.

Contents

About the Editors

Frances Pouch Downes is the director of the Michigan Department of Community Health (MDCH) Bureau of Laboratories. Until July 1999, she was director of the Infectious Diseases Division in the Bureau of Laboratories. She has served as director of the Molecular Biology and Virology laboratories at MDCH. She earned doctorate and masters degrees in laboratory practice from the University of North Carolina School of Public Health. She has a bachelor of science degree in medical technology from Indiana University. Dr. Downes has worked internationally in the laboratory field in Africa, Asia, and the Carribean. She is adjunct clinical professor in the Medical Technology Program of Michigan State University.

Keith A. Ito is Senior Vice President of the National Food Processors Association's Technical Assistance Center in Dublin, California. In this capacity, he oversees the Association's technical consulting activity at the Center. He has a Bachelor of Arts degree in Bacteriology from the University of California, Berkeley. He is also Director of the University of California Laboratory for Research in Food Preservation, which is administered by the Food Science and Technology Department at University of California, Davis. He has been associated with the *Compendium* as an author, reviewer, and member of the editorial committee since the first edition.

Editorial Committee

Gary R. Acuff, Texas A&M University, Dept. of Animal Science, College Station, TX 77843-2471

Stephanie Doores, Dept. of Food Science, 1078 Borland Lab., Pennsylvania State University, University Park, PA 16802

Frances Pouch Downes (Co-Chair), Michigan Dept. of Community Health, 3500 N. Martin Luther King Jr. Blvd., Lansing, MI 48909

Russell S. Flowers, Silliker Laboratories Group, Inc., 900 Maple Road, Homewood, IL 60430

Keith Ito (Co-Chair), National Food Processors Association, 6363 Clark Ave., Dublin, CA 94568

George Jackson, Office of Special Research Skills, HFS-500, U.S. Food and Drug Administration, Washington, DC 20204

John H. Silliker (Retired), 12321 Williams Ct., Crown Point, IN 46307.

Don F. Splittstoesser (Retired), Cornell University, Dept. of Food Science & Technology, New York State Agricultural Experiment Station, Geneva, New York 14456

Mary Lou Tortorello, U.S. Food and Drug Administration, National Center for Food Safety and Technology, 6502 South Archer Rd., Summit-Argo, IL 60501

Joan B. Rose, University of South Florida, 13301 Bruce B. Downs Blvd., Tampa, FL 33612

Jacqueline S. Scott (APHA Representative), Michigan Dept. of Community Health Laboratories, P.O. Box 30035, Lansing, MI 48909

Ellen T. Meyer (APHA Staff Liaison), American Public Health Association, 800 I St. NW, Washington, DC 20001

Authors

Carlos Abeyta, U.S. Food and Drug Administration, Seafood Products Research Center, 22201 23rd Dr. SE, Bothell, WA 98021-4421

Gary R. Acuff, Texas A&M University, Dept. of Animal Science, 2471 TAMU, College Station, TX 77843-2471

Jean E. Anderson, Pillsbury Technology East, 737 Pelham Blvd., St. Paul, MN 55114

Wallace H. Andrews, U.S. Food and Drug Administration, Center for Food Safety and Applied Nutrition, 200 C St., SW, Washington, DC 20204

Stephen S. Arnon, California Dept. of Health Services, Room 506, 2151 Berkeley Way, Berkeley, CA 94704

David H. Ashton (Retired), ConAgra,Grocery Products Inc., 1967 Quiet Ranch Rd., Fallbrook, CA 92028

J. Stan Bailey, Russell Research Center, U.S. Dept. of Agriculture, P. O. Box 5677, Athens, GA 30613-5677

John A. Baross, University of Washington, School of Oceanography, Box 357940, Seattle, WA

Timothy J. Barrett, Centers for Disease Control and Prevention, 1600 Clifton Road, Mailstop C03, Atlanta, GA 30333

Reginald W. Bennett, U.S. Food and Drug Administration, Center for Food Safety and Applied Nutrition, HFS-516, 200 C St., SW, Washington, DC 20204

Negash Belay, Dept. of Health and Human Services, U.S. Food and Drug Administration, Public Health Service, HFS-516, 200 C St. SW, Washington DC 20204

Dane T. Bernard, National Food Processors Association, 1350 I St. NW Suite 300, Washington, DC 20005

Larry R. Beuchat, University of Georgia, Dept. of Food Science and Technology, 1109 Experiment St., Griffin, GA 30223

Jeffrey W. Bier (Retired), U.S. Food and Drug Administration, Center for Food Safety and Applied Nutrition, 200 C St., SW, Washington, DC 20204

Sarah G. Birkhold, Texas A&M University, Dept. of Poultry Science, 2472 TAMU, College Station, TX 77843-2472

Patrick Boerlin, University of Guelph, Ontario Veterinary College, Dept. of Pathobiology, Guelph, ON Canada N1G 2W1

Robert E. Brackett, U.S. Food and Drug Administration, Center for Food Safety and Applied Nutrition, HFS-300, 200 C St. SW, Washington DC 20204

Fred Breidt, Jr., U.S. Dept. of Agriculture, Agricultural Research Service, North Carolina State University, Dept. of Food Science, Box 7624, Raleigh NC 27695

Raymond G. Bryant, Microbial Diseases Laboratory, California Dept. of Health Services, 2151 Berkeley Way, Berkeley, CA 94704

Lloyd B. Bullerman, University of Nebraska, Dept. of Food Science & Technology, 349 FIB, East Campus, Lincoln, NB 68583

William Burkhardt, III, U.S. Food and Drug Administration, Gulf Coast Seafood Laboratory, P.O. Box 158, Dauphin Island, AL 36528-0158

Kevin R. Calci, U.S. Food and Drug Administration, Gulf Coast Seafood Laboratory, P.O. Box 158, Dauphin Island, AL 36528-0158

Hui-Cheng Chen, Food Industry Research and Development Institute, P.O. Box 246, Hsinchu, 300 Taiwan

Michael C. Cirigliano, Unilever U.S. Inc., Thomas J. Lipton Co., 23 Seventh St., Cresskill, NJ 07626

Rocelle S. Clavero, National Food Processors Association, 1350 I St. NW Suite 300, Washington, DC 20005

Dean O. Cliver, University of California Davis, Dept. of Population Health and Reproduction, School of Veterinary Medicine, One Shields Avenue, Davis, CA 95616

David L. Collins-Thompson, Nestle R&D Center, 201 Housatonic Ave., New Milford, CT 06776

David W. Cook, U.S. Food and Drug Administration, Gulf Coast Seafood Laboratory, P.O. Box 158, Dauphin Island, AL 36528-0158

Oliver D. Cook, U.S. Food and Drug Administration, Office of Regulatory Affairs, HFC 130, 5600 Fischer Lane, Rockville, MD 20857

Maribeth A. Cousin, Purdue University, Dept. of Food Science, 1160 Food Science Bldg., West Lafayette, IN 47907-1160

Kurt E. Deibel, Tropicana North America, 1001 13th Ave. E., Bradenton, FL 34208

Robert H. Deibel, Deibel Labs, 7165 Curtiss Ave., Sarasota, FL 34231

Cleve B. Denny (Retired), National Food Processors Association, 6230 Valley Rd., Bethesda, MD 20817-3253

Angelo DePaola, Jr., U.S. Food and Drug Administration, Center for Food Safety and Applied Nutrition, Gulf Coast Seafood Laboratory, P.O. Box 158, 1 Iberville Dr., Dauphin Island, AL 36528-0158

Ralph DiGiacomo, Pepsico Research and Tech Services, 100 Stevens Avenue, Valhalla, NY 10595

Catherine W. Donnelly, University of Vermont, Dept. of Nutrition and Food Sciences, Room 216 Carrigan Hall, Burlington, VT 04505

L. Scott Donnelly, Wyeth Nutritionals, Inc., P.O. Box 2109, Georgia, VT 05468

Michael P. Doyle, University of Georgia, Center for Food Safety and Quality Enhancement, Georgia Station, Griffen, GA 30223-1797

Phyllis Entis, QA Life Sciences, Inc., 6645 Nancy Ridge Dr., San Diego, CA 92121

John P. Erickson, Best Foods Technical Center, 150 Pierce St., Somerset, NJ 08873-6710

George M. Evancho, Campbell Soup Co., Group Director, Food Safety, World Headquarters, 1 Campbell Place, Camden, NJ 08103-1799

Jeffrey M. Farber, Health Canada, Microbiology Research Division, Banting Research Centre, Postal Locator 2204A2, Ottawa, ON, Canada, K1A 0L2

Ronald Fayer, U.S. Dept. of Agriculture, Agricultural Research Service, Animal and Natural Resources Institute, Bldg. 1040, Beltsville, MD 20705-2732

Aamir M. Fazil, Decisionalysis Risk Consultants, Inc., 515 Brookstone Place, Newmarket, ON, Canada L3X 2H3

Peter Feng, U.S. Food and Drug Administration, Center for Food Safety and Applied Nutrition, HFS-516, 200 C St., SW, Washington, DC 20204

Joseph L. Ferreira, U.S. Food and Drug Administration, Office of Regulatory Affairs, Southeast Region, 60 8th St., Atlanta, GA 30309

Gunnar Finne, RR3 Box 9K, LeSueur, MN 56058

Henry P. Fleming, U.S. Dept. of Agriculture, Agricultural Research Service, North Carolina State University, Dept. of Food Science, Box 7624, Raleigh, NC 27695

Russell S. Flowers, Silliker Laboratories Group, Inc., 900 Maple Road, Homewood, IL 60430

Joseph F. Frank, University of Georgia, Dept. of Food Science and Technology, 204 Dairy Science Bldg., Athens, GA 30632

Sharon K. Franklin, U.S. Dept. of Agriculture, Enteric Diseases and Food Safety Research Unit, 2300 Dayton Road, Ames, IA 50010

T. A. Freier, Silliker Laboratories Group, Inc., 2725 61st Ln. 4, Vinton, IA 52349

Scott J. Fritschel, DuPont/Qualicon, P.O. Box 80375, Wilmington, DE 19880

Daniel Y.C. Fung, Kansas State University, Dept. of Animal Sciences & Industry, 139 Call Hall, Manhattan, KS 66506-1600

Phyllis Gallagher, Pepsico Research and Tech Services, 100 Stevens Avenue, Valhalla, NY 10595

H. Ray Gamble, U.S. Dept. of Agriculture, Agricultural Research Service, Animal and Natural Resources Institute, Bldg. 1040, Beltsville, MD 20705-2732

Richard K. Gast, U.S. Dept. of Agriculture, Agricultural Research Service, SE Poultry Research Lab, Athens, GA 30605-2720

Steven M. Gendel, U.S. Food & Drug Administration, National Center for Food Safety & Technology, 6502 S. Archer Road, Summit-Argo, IL 60501

Hassan Gourama, Penn State University Berks Campus, Food Science, Tultehocken Road, Box 7009, Reading, PA 19610-6009

Rodney J. H. Gray, Hercules, Inc., 11333 Southeast Hercules Plaza, Wilmington, DE 19711

Mansel W. Griffiths, University of Guelph, Dept. of Food Science, Guelph, ONT N1G 2W1 Canada

Michael J. Haas, USDA, Eastern Regional Research Center , 600 East Mermaid Lane, Wyndmoor, PA 19038

Paul A. Hall, Kraft Foods Inc., 801 Waukegan Rd., Glenview, IL 60025

Lester Hankin (Retired), Dept. of Biochemistry and Genetics, The Connecticut Agricultural Experiment Station, 123 Huntington St., Box 1106, New Haven, CT 06504

John H. Hanlin, The McCormick Company, Research and Development, 18 Loveton Circle, Sparks, MD 21152

Todd Hannah, Siligan Containers Mfg. Corp., 1190 Corporate Center Drive, Oconomowoc, WI 53066

Paul A. Hartman (Deceased), Iowa State University, Dept. of Microbiology, Ames, IA 50011

W. S. Hatcher, Jr., The Minute Maid Co., P. O. Box 2079, Houston, TX 77252

Jennifer A. Hudnall, Texas A&M University, Dept. of Animal Science, 2471 TAMU, College Station, TX 77843-2471

Timothy C. Jackson, Nestle USA Quality Assurance Laboratory, 6625 Eiterman Road, Columbus, OH 43017-6516

Michael Jantschke, National Food Processors Association, 6363 Clark Ave, Dublin, CA 94568

James M. Jay, University of Nevada at Las Vegas, Dept. of Biological Science, P.O. Box 454004, Las Vegas, NV 89154-4004

Eric A. Johnson, University of Wisconsin, Food Research Institute, 1925 Willow Dr., Madison, WI 53706

Jennifer L. Johnson, Emmpak Foods, 4700 North 132nd St., Butler, WI 57002

Thomas Jones, The Dried Fruit Association of California, 1855 South Van Ness Avenue, Fresno, CA 93721

Charles A. Kaysner, Seafood Products Research Center, U.S. Food and Drug Administration, 22201 23rd Drive SE, Bothell, WA 98021

Henry Kim, U.S. Food and Drug Administration, Center for Food Safety and Applied Nutrition, HFS-306, 200 C St., SW, Washington, DC 20204

A. Douglas King, Jr., U.S. Dept. of Agriculture, Western Regional Research Lab, P. O. Box 361, Monte Rio, CA 95462

John A. Koberger, University of Florida, Dept. of Food Science & Human Nutrition, P.O. Box 110370, Gainsville, FL 32611-0370

Patrick J. Konkel, Nestle USA, N 57 W 37773 Country Lane, Oconomowoc, WI 53066

Jeff L. Kornacki, Silliker Laboratories Group, Inc., 3688 Kinsman Blvd., Madison, WI 53704

Ronald C. Labbe, University of Massachusetts, Dept. of Food Science and Nutrition, Amherst, MA 01003

G. H. Lacy, Laboratory for Molecular Biology of Plant Stress, Virginia Polytechnic Institute and State University, Blacksburg, VA 24061-0330

Anna M. Lammerding, Health Canada, Health Protection Branch, 110 Stone Road West, Guelph Ontario N1G3W4 Canada

Keith A. Lampel, U.S. Food and Drug Administration, Center for Food Safety and Applied Nutrition, HFS-237, 200 C St., SW, Washington, DC 20204

Gayle A. Lancette, U.S. Food and Drug Administration, Office of Regulatory Affairs, HFR-SE600, 60 8th St., NE, Atlanta, GA 30309

Warren L. Landry, U.S. Food and Drug Administration, Office of Regulatory Affairs, Southwest Region, 3032 Bryan St., Dallas, TX 75204

Loralyn Ledenbach, Kraft Foods Inc., 801 Waukegan Rd., Glenview, IL 60025

J. Ralph Lichtenfels, U.S. Dept. of Agriculture, Agricultural Research Service, Animal and Natural Resources Institute, Bldg. 1040, Beltsville, MD 20705-2732

J. Eric Line, U.S. Dept. of Agriculture, Agricultural Research Service, Poultry Microbiological Safety Research Unit, Richard B. Russell Agricultural Research Center, 950 College Station Rd., Athens, GA 30604-5677

John A. Marcy, University of Arkansas, Dept. of Poultry Science, Center of Excellence for Poultry Science, O-203 Poultry Science Center, Fayetteville, AR 72701

Jeffrey P. Massey, Michigan Dept. of Community Health, Molecular Biology Section, 3423 N. Martin Luther King, Jr. Blvd., Lansing, MI 48909

Susan A. McCarthy, U.S. Food and Drug Administration, Gulf Coast Seafood Laboratory, P.O. Box 158, Dauphin Island, AL 36528-0158

Roger F. McFeeters, U.S. Dept. of Agriculture, Agricultural Research Service, North Carolina State University, Dept. of Food Science, P.O. Box 7624, Raleigh, NC 27695

Lynn McIntyre, Massey University, Institute of Food, Nutrition and Human Health, Albany, New Zealand

Ann Marie McNamara, Sara Lee Foods, 8000 Centerview Pkwy., Suite 300, Cordova, TN 38018

Jianghong Meng, University of Maryland, College Park, MD

Thaddeus F. Midura (Retired), Microbial Diseases Laboratory, California Dept. of Health Services, 162 Ardith Dr., Orlando, CA 94563

Lloyd J. Moberg, Borden Foods Corporation, 2001 Polaris Pkwy., Columbus, OH 43240

Mark A. Moorman, (T) Kellogg Company, 2 Hamblin Ave. East, Battle Creek, MI 49016-3232

R. Dale Morton, Quaker Oats, 617 W. Main St., Barrington, IL 60018

Elsa A. Murano, Texas A&M University, Dept. of Animal Science, Food Science and Technology, 2471 TAMU, College Station, TX 77843-2471

Ranzell Nickelson II, Foodbrands America, Inc., 7401 Will Rogers Blvd., Fort Worth, TX 76140

Anita J. Okrend, U.S. Dept. of Agriculture, Agricultural Marketing Service, 10322 Inwood Ave., Silver Spring, MD 20902

Karl E. Olson, The Dial Corp., Consumer Product Group, 15101 N. Scottdale Rd, Scottdale, AZ 85254

Samuel Palumbo, U.S. Dept. of Agriculture, Eastern Regional Research Center, 600 E. Mermaid Lane, Wyndmoor, PA 19038

Greg M. Paoli, Decisionalysis Risk Consultants, Inc., 1831 Yale Avenue, Ottawa, Ontario K1H 6S3

M. E. Parish, University of Florida, 200 Experiment Station Rd., Lake Alfred, FL 33850

Nina Gritzai Parkinson, National Food Processors Association, 6363 Clark Ave., Dublin, CA 94568

Ruth L. Petran, The Pillsbury Company, 200 South 6th St., Mail Stop 23B1, Minneapolis, MN 55402

Joan M. Pinkas, Mc Cormick & Co. Inc., Corp. Analytical Science, 1104 McCormick Rd., Hunt Valley MD 21031

J. I. Pitt, CSIRO, Division of Food Processing, P.O. Box 52, North Ryde NSW, Australia 2113

Wayne Payton Pruett, Jr., Silliker Laboratories Group, Inc., Director of Microbiology, Analytical Services, 1304 Halsted St., Chicago Heights, IL 60411

Gary P. Richards, USDA Agricultural Research Service, Microbial Food Safety Research Unit , Delaware State University, 1200 N. DuPont Highway, Dover, Delaware 19901-2277

Ronald L. Richter, Texas A&M University, Dept. of Animal Science, 2471 TAMU, College Station, TX 77843-2471

Steven C. Ricke, Texas A&M University, Dept. of Poultry Science, 2472 TAMU, College Station, TX 77843-2472

Elliot T. Ryser, Michigan State University, Dept. of Food Science and Human Nutrition, 2108 S. Anthony Hall, East Lansing, MI 48824

Linda Schroeder-Tucker, U.S. Dept. of Agriculture, National Veterinary Services Laboratory, Animal Plant Health Inspection Service, Ames, IA 50010

Virginia N. Scott, National Food Processors Association, 1350 I St. NW Suite 300, Washington, DC 20005

Wayne P. Segner (Retired), Crown Cork and Seal, 6340 Americana Dr., Unit 1002, Clarendon Hill, IL 60514

Anthony N. Sharpe, Bureau of Microbial Hazards, Health Protection Branch, Health Canada, Tunney's Pasture, Ottawa ONT K1A OL2 Canada

Linda M. Sieverding, Iowa State Iniversity, Dept. of Microbiology, Ames, IA 50011

John Silliker (Retired), 12321 Williams Ct., Crown Point, IN 46307

Richard B. Smittle, Silliker Laboratories Group, Inc., 420 Tyler Dr., Clear Brook, VA 22624

Haim M. Solomon, U.S. Food and Drug Administration, Center for Food Safety and Applied Nutrition, HFS-516, 200 C St., SW, Washington, DC 20204

Kent M. Sorrells (Retired), National Steak and Poultry, 32249 River Island Drive, Springville, CA 93265

William A. Sperber, Cargill Citro-America, 2301 Crosby Rd., Wayzata, MN 55391

Don F. Splittstoesser (Retired), Cornell University, Dept. of Food Science & Technology, New York State Agricultural Experiment Station, Geneva, New York 14456

Gerard Stelma, U.S. Environmental Protection Agency, EMSL Microbiology, Research Division, 26 ML King Dr., Cincinnati, OH 45268

Norman J. Stern, U.S. Dept. of Agriculture, Agricultural Research Service, Poultry Microbiological Safety Research Unit, Richard B. Russell Agricultural Research Center, 950 College Station Road, Athens, GA 30604-5677

Kenneth E. Stevenson, National Food Processors Association, 6363 Clark Ave., Dublin, CA 94568

William H. Sveum, Oscar Meyer, Oscar Meyer Food Division, 910 Mayer Ave., Madison, WI 53704-4256

Katherine M. J. Swanson, The Pillsbury Co., Director, Microbiology and Food Safety, 737 Pelham Blvd., Saint Paul, MN 55118

R. Bruce Tompkin, ConAgra Refrigerated Prepared Foods, 3131 Woodcreek Dr., Downers Grove, IL 60515

Mary Lou Tortorello, U.S. Food and Drug Administration, National Center for Food Safety and Technology, 6502 South Archer Rd., Summit-Argo, IL 60501

John A. Troller, John A. Troller Incorporated, 314 Ritchie Ave., Cincinnati, OH 45215

Kevin D. Tyler, Health Canada, HPB, P.L. 22042A2, Tunneys Pasture, Ottawa, ON Canada, K1A 0L2

P. C. Vasavada, University of Wisconsin, Dept. of Animal and Food Science , 410 3rd St., River Falls, WI 54022-5001

Ebenezer R. Vedamuthu (Retired), Quest International, 2402 7th St., NW, Rochester, MN 55901

Isabel T. Walls, International Life Sciences Institute, Senior Scientist, Risk Science Institute, 1126 16th St. NW, Washington DC 20036-4810

Guodong Wang, Kraft Foods Inc., Mail Box TD 2-1, 555 South Broadway, Tarrytown, NY 10591

Stephen D. Weagant, U.S. Food and Drug Administration, Office of Regulatory Affairs, HFR-PA360, 22201 23rd Dr. SE, Bothell, WA 98041-3012

Cheng i Wei, Auburn University, Nutrition and Food Science Dept., 328 Spindle Hall, Auburn University, AL 36849-5605

John L. Weihe, The Minute Maid Co., 2000 St. James Place, Houston, TX 77056

Irving Weitzman, U.S. Food and Drug Administration, Office of Regulatory Affairs, HFC 130, 5600 Fischer Lane, Rockville, MD 20857

Irene V. Wesley, U.S. Dept. of Agriculture, National Animal Disease Center, Enteric Diseases and Food Safety Research Unit, 2300 Dayton Road, Ames, IA 50010

B. B. Woodward, Florida Dept. of Agriculture, 1238 Sedgefield Rd., Tallahassee, FL 32311

Reviewers

The editorial committee gratefully acknowledges the assistance of the following colleagues who reviewed one or more chapters of this book.

Wallace H. Andrews, Division of Microbiological Studies, Office of Special Research Skills, Center for Food Safety and Applied Nutrition, U.S. Food and Drug Administration, Washington, DC 20204

Negash Belay, Division of Microbiological Studies, Office of Special Research Skills, Center for Food Safety and Applied Nutrition, U.S. Food and Drug Administration, Washington, DC 20204

Reginald W. Bennett, Division of Microbiological Studies, Office of Special Research Skills, Center for Food Safety and Applied Nutrition, U.S. Food and Drug Administration, Washington, DC 20204

Jeffrey W. Bier, Office of Seafood, Center for Food Safety and Applied Nutrition, U.S. Food and Drug Administration, Washington, DC 20204

Cheryl Bopp, Foodborne and Diarrheal Diseases Laboratory Section, Centers for Disease Control and Prevention, Atlanta, GA 30333

William Burkhardt, III, Division of Science and Applied Technology, Office of Seafood, Center for Food Safety and Applied Nutrition, U.S. Food and Drug Administration, P.O. Box 158, Dauphin Island, AL 36528-0158

Kevin R. Calci, Division of Science and Applied Technology, Office of Seafood, Center for Food Safety and Applied Nutrition, U.S. Food and Drug Administration, P.O. Box 158, Dauphin Island, AL 36528-0158

David W. Cook, Division of Science and Applied Technology, Office of Seafood, Center for Food Safety and Applied Nutrition, U.S. Food and Drug Administration, P.O. Box 158, Dauphin Island, AL 36528-0158

J.-Y. D'Aoust, PhD, Microbiology Research Division, Sir F.G. Banting Research Centre, Health Canada, Ottawa, ON, K1A OL2, Canada

Angelo DePaola, Division of Science and Applied Technology, Office of Seafood, Center for Food Safety and Applied Nutrition, U.S. Food and Drug Administration, P.O. Box 158, Dauphin Island, AL 36528-0158

Lawrence Elliott, MD, Department of Community Health Sciences, University of Manitoba, 770 Bannatyne Ave, Rm T162, Winnipeg, MB R3E OW3, Canada

Peter Feng, Division of Microbiological Studies, Office of Special Research Skills, Center for Food Safety and Applied Nutrition, U.S. Food and Drug Administration, Washington, DC 20204

Steven M. Gendel, Division of Food Processing and Packaging, Office of Plant and Dairy Foods and Beverages, Center for Food Safety and Applied Nutrition, U.S. Food and Drug Administration, 6502 S. Archer Avenue, Summit-Argo, IL 60501

Todd Hannah, Siligan Containers Manufacturing Corporation, 1190 Corporate Center Drive, Oconomowoc, WI 53066

Robert Holland, DVM, National Food Safety and Toxicology Center, Michigan State University, East Lansing, MI 48823

George J. Jackson, Office of Special Research Skills, Center for Food Safety and Applied Nutrition, U.S. Food and Drug Administration, Washington, DC 20204

Henry Kim, Division of Plant Product Safety, Office of Plant and Dairy Foods and Beverages, center for Food Safety and Applied Nutrition, U.S. Food and Drug Administration, Washington, DC 20204

Keith Lampel, Division of Virulence Assessment, Office of Plant and Dairy Foods and Beverages, Center for Food Safety and Applied Nutrition, U.S. Food and Drug Administration, Washington, DC 20204

Gayle Lancette, Southeast Regional Laboratory, Office of Regulatory Affairs, U.S. Food and Drug Administration, 60 8th Street NE, Atlanta, GA 30309

Susan A. McCarthy, Division of Science and Applied Technology, Office of Seafood, Center for Food Safety and Applied Nutrition, U.S. Food and Drug Administration, P.O. Box 158, Dauphin Island, AL 36528-0158

Joan B. Rose, University of South Florida, 13301 Bruce B. Downs Blvd., Tampa, FL 33612

Haim M. Solomon, Division of Microbiological Studies, Office of Special Research Skills, center for Food Safety and Applied Nutrition, U.S. Food and Drug Administration, Washington, DC 20204

Mary Grace Stobierski, DVM, MPH, Communicable Disease Epidemiology Section, Michigan Department Community Health, 3423 N. Martin Luther King Jr. Blvd., Lansing, MI 48909

Nancy Strockbine, PhD, Foodborne and Diarrheal Diseases Laboratory Section, Centers for Disease Control and Prevention, Atlanta, GA 30333

Sterling S. Thompson PhD, Hershey Foods Corporation, 1025 Reese Avenue, Hershey, PA 17033-0805

Mary Lou Tortorello, Division of Food Processing and Packaging, Office of Plant and Dairy Foods and Beverages, Center for Food Safety and Applied Nutrition, U.S. Food and Drug Administration, 6502 S. Archer Avenue, Summit-Argo, IL 60501

Irving Weitzman, Division of Emergency and Investigational Operations, Office of Regulatory Affairs, U.S. Food and Drug Administration, 5600 Fishers Lane, Rockville, MD 20857

Preface

The Centers for Disease Control and Prevention listed safe foods as one of the 10 significant public health accomplishments of the past 100 years. Development of industry standards and microbiology-based quality assurance programs have contributed significantly to the abundant and wholesome food supply currently available in developed countries.

Since the publication of the 3rd edition of this Compendium, consumer expectations have focused attention on food safety as an emerging issue with both public health and international trade implications. Foodborne disease outbreaks draw public and media attention to the continued need for vigilance to protect the public's health. Not only does food safety concern health, but the expanding global trading of foods means that foods have an increasing role to play in global economics and supply. Globalization also means that food pathogens can be rapidly and widely distributed to susceptible populations. The evolution of the science and practice of food microbiology also continues to be challenged by developments in food packaging, consumer preferences and food choices, and personal and retail practices in food preparation.

Healthy People 2010 identified challenges and trends that will make control of foodborne disease a viable public policy issue, including emerging pathogens, improper food preparation, storage and distribution practices, insufficient training of retail and food service employees, an increasingly global food supply, and an increase in the number of people at risk of foodborne disease because of immunosuppression. In this 4th edition of the Compendium, we have attempted to address the evolving technologies in microbiology and molecular biology that have been applied to or promise to play an important role in meeting the above challenges. Recognizing the role of emerging pathogens in the practice of food microbiology, several chapters have been added or expanded to reflect new technologies available for detection, identification, and appreciation of the role of these microbes. For example, a new chapter on pathogenic *Escherichia coli* (Chapter 35) and revised chapters on *Listeria* (Chapter 36) and Parasites (Chapter 42) are included in this edition. The new chapter on Risk Assessment (Chapter 29) introduces a strategy to target safety intervention at key steps from production to consumption. Several chapters have been updated and revised to include newer technologies, procedures, and techniques that have been developed to detect pathogens and spoilage organisms, including Methods for Detection, Identification, and Enumeration (Chapter 10), Identification and Differential Methods (Chapter 11), and Labor Saving and Automation (Chapter 12). The placement at the beginning of the book of the chapter on Quality Assurance (Chapter 1) reflects the continued importance of this topic in the operation of food microbiology laboratories in all settings. Many chapters have increased in length because of the new information and recognition of the role of emerging technologies. The references in all chapters have been updated and the information incorporated into the text. Because of the need to incorporate all this information into one volume, a new format was chosen. We believe that the new dimension of this edition will make the book more useful in the laboratory and will facilitate its continued use as a reference source consulted daily.

The editors wish to thank the editorial committee; the contributing authors and reviewers; Ellen Meyer and Susan Westrate of APHA; the project liaisons from CLaSP, Jacqueline Scott and Joan Rose; and our assistants Jill Molina, Karen Herndon, Dona Dewey, and Janet Braatz. We also wish to acknowledge the support and commitment of employers and families without whom this project would not have been completed.

Chapter 1

Laboratory Quality Assurance

Richard B. Smittle and Anita J. Okrend

1.1 INTRODUCTION

For a compendium of microbiological methods to begin with a chapter on laboratory quality assurance appropriately reflects that subject's importance. In recent years it has become common practice for domestic and foreign food manufacturers to rely on data generated from laboratories outside of their control. In addition, data generated by non-governmental laboratories has become increasingly important to regulatory efforts. Consequently, a documented and functional quality system with national and international recognition has become critical to the operation of a food testing laboratory.

The objective of laboratory quality assurance is to verify the accuracy and precision of information obtained from analyses and to ensure that data obtained from analyses are suitable for use in decision making. The accuracy and precision required may vary, depending on the use to be made of the information, but however strong the effort for precision may be, some variation in results from biological analyses is to be expected.

Generally, food microbiology laboratories—governmental, industrial, academic, or commercial—are active in one or more of the following areas:

1. Basic research determines the characteristics of microorganisms important in foods and the factors that affect their growth and survival.
2. Applied research uses basic information to develop and test new food products and microbiological methods.
3. Quality control monitors food production to detect deviations from good manufacturing practices, to assure conformance to criteria, and to detect contaminants.
4. Investigative activities seek to determine the causes of foodborne illness and spoilage problems.

Although the principal objective of a laboratory quality assurance program is to ensure the correctness of data, the systems and procedures required for an effective program of this sort provide additional benefits. For example, using good microbiological techniques not only prevents cross-contamination of the samples being examined, but also protects personnel against infection and their working environment against contamination. Monitoring and maintaining equipment to ensure proper functioning decreases the risk of electrical and fire hazards and reduces the possibility of equipment failure that leads to lost samples and analytical down time. Another indirect benefit of a laboratory quality assurance system is the standardization of analytical methods, resulting in decreased intra- and inter-laboratory variation. Finally, the record keeping activities required for laboratory quality assurance provide information that helps management and individual benchworkers evaluate proficiency. The ability of management and workers to monitor the quality of work promotes confidence and pride in the laboratory.

This chapter introduces the principal concepts in quality assurance for the food microbiology laboratory. It is not possible to present here specific quality assurance programs to meet the needs of all food microbiology laboratories. Rather, this chapter is concerned with the particular laboratory functions that must be controlled and the probable sources of laboratory error. The selection of appropriate methods for quality assurance will be addressed, but specific methods will be referenced rather than discussed in detail. Thus, the reader should be mindful that this chapter is intended to serve as a guide for the design and implementation of laboratory quality assurance programs, not as a detailed statement of the system itself.

1.2 MANAGEMENT'S ROLE IN LABORATORY QUALITY ASSURANCE

Direct responsibility for the design and implementation of a laboratory quality assurance program lies within the management of the laboratory. Management must evaluate the risks associated with laboratory error and the costs and benefits of reducing error through systematic quality assurance. Measurement of the cost of these activities requires a system to account for material, labor, and overhead costs. The likelihood and significance of errors arising from each particular activity of the laboratory and the costs associated with controlling and monitoring a particular activity must be considered in the development and operation of a quality assurance program. The principles of Hazard Analysis Critical Control Points (HACCP)[20] are useful for the development of concepts for a laboratory quality assurance system.

The following steps describe a general approach that should be followed in establishing a laboratory quality assurance program[38]:

1. Formalize objectives and policies that specify the accuracy of work required, including the selection of appropriate methods and sample handling procedures.
2. Select appropriate personnel to perform laboratory functions to maintain the desired level of quality.

3. Provide facilities and equipment necessary for the performance of laboratory functions.
4. Establish monitoring methods and record keeping protocols to verify the accuracy and consistency of the laboratory work.
5. Establish procedures for the initiation of corrective measures when unacceptable quality is discovered.

1.3 GENERAL LABORATORY OPERATIONS

The first step in implementing a quality assurance program for the laboratory is to make a formal statement of the objectives and scope of activities for the laboratory, after which general operating procedures are developed to meet the objectives. Specific standard operating procedures are installed to ensure that approved sample preparation and analytical methods are chosen and followed. Within any analytical method, procedural steps may be subject to interpretation that can result in minor modifications, which have a pronounced influence on the outcome of an analysis. Standard operating procedures (SOP) should be sufficiently described to prevent minor modifications or deviations that could arise from misinterpretation of a method and to provide an accurate record of each stage of a procedure. When changing methodology, the new method should be validated against the previously used method before the change is instituted to show that the laboratory is proficient in the new method. This should be done even when the new method is considered to be a standard method. An SOP for this validation process is necessary to provide consistency for methodology changes. For research laboratories, general standard operating procedures may be established for development of a research proposal, approval of the project, and periodic evaluation of the progress made on the project. Specific operating procedures, such as method validation and execution, may need to be modified during a project as new data are obtained and interpreted.

1.31 Sample Management

Chapter 2 offers specific information on the collection, shipment, and preparation of samples for microbiological analysis. This section discusses the criteria used to determine the acceptability of samples received by the laboratory and the proper handling of samples accepted for analysis.

1.311 Criteria for acceptance of a sample

1. Adequate documentation and identification must accompany samples, including a description of the place of collection, manufacturer, date and time of collection (especially for perishable samples), reason for collection (such as, compliance with legal standards or routine surveillance), sampling plan followed, analysis requested, and storage conditions.
2. The original condition of samples and the integrity of sample containers must be maintained from collection until receipt at the laboratory. The manner of shipment must be appropriate for the type of sample.
3. Other considerations for acceptance of samples are (1) the samples must be representative of the lot of product and processes sampled, and (2) the laboratory must have the capability to do the requested analyses.

1.312 Handling of samples in the laboratory

1. After receipt, samples must be stored to maintain their original condition until analyzed. Samples should be tested as soon as possible after receipt. Facilities should be available for both short-term storage before and during analyses and, when required for forensic reasons, long-term storage after analyses have been completed.
2. It may be necessary for an individual in the laboratory to have the responsibility as sample custodian to maintain accountability records for samples in the laboratory. This individual may (1) receive samples, (2) record the date and time received, (3) initially verify the identity of samples, (4) store according to instructions accompanying samples, (5) record the date and time when samples are delivered to analyst(s) for examination and the date and time when they are returned to storage following analyses, (6) maintain a long-term sample storage system, and (7) dispose of samples as necessary.

1.32 Analytical Methods

The need for standardized methods to promote consistency of results nationwide by reducing inter- and intra-laboratory variation was recognized early by a committee of the American Public Health Association, whose report in 1910[3] was the beginning of "Standard Methods for the Examination of Dairy Products" (SMEDP), which is now in its 16th edition. SMEDP contains methods assembled by a committee of experts who have selected the analytical procedures most appropriate to examine samples for groups of microorganisms and for specific pathogens.

The Association of Official Agricultural Chemists was organized in 1884 to provide to governmental regulatory and research agencies methods that perform with the required accuracy and precision under usual laboratory conditions.[1] Since its founding, the Association, now known as *AOAC International*, has provided a mechanism to select methods of analysis from published literature or to develop, approve, and publish new methods.[2]

Other organizations have published useful collections of microbiological methods. The American Public Health Association published methods for the microbiological analysis of foods in 1958[4], and it published the first edition of this book in 1976. The United States Food and Drug Administration (FDA) has published the procedures it uses to examine foods for regulatory compliance.[15] The International Commission on Microbiological Specifications for Foods (ICMSF)[19] has been active since 1962. ICMSF efforts have been directed to the selection and recommendation of suitable methods for microbiological evaluation of foods moving in international commerce. ICMSF has published a number of collaborative studies and books.

Less laborious and more rapid methods are always needed, and many adequate methods are available for analytical food microbiologists. Unfortunately, the methods available to be used for specific analyses vary considerably both internationally and nationally. To reduce variation, laboratory management should try to select standard or approved methods. Modifications of standard methods should not be used unless collaborative or comparative studies have shown that the modified methods are at least as reliable as the reference methods.

1.33 Laboratory Standard Operating Procedures

Each laboratory should have a standard operations manual that describes (1) organization and management, (2) quality systems for media and reagent quality control, audit and review, (3) personnel requirements and training, (4) accommodation and environment critical to the integrity of test results such as cleaning and sanitation (5) equipment and reference materials use and maintenance, (6) calibration of equipment and test materials suitable for the tests being performed traceable to national and/or international sources, (7) calibration procedures and validated test methods, (8) handling of calibration items (thermometers, weights, reference cultures) and samples, (9) records, (10) certificates and/or reports, (11) sub-contracting or testing, (12) outside support and supplies, and finally (13) complaints concerning the quality of the data.

The manual should provide instructions to cover most of the normal situations that the laboratory will encounter. It will inform personnel about appropriate methodology, ensure uniformity of sample analysis, and promote quality assurance programs. All employees should read a copy of the manual and sign a statement attesting that they have done so.

The selection of methods to be included in the manual will depend to a certain extent on the type of laboratory, i.e., governmental, commercial, or industrial. The description of each method in the manual should be complete and detailed enough that reference to other publications is unnecessary. Included should be all the necessary controls and checks on materials, media, reagents, positive and negative controls, the desired response from each control, and corrective measures that should be taken if a control is not correct.

Limitations of each test should be included, where known, as well as a list of precautions to be taken. Possible interference should be described, such as natural inhibitory substances in foods that must be diluted out before growth of any organism can occur.

The manual should fully describe the quality assurance program in use. Included should be each quality assurance procedure, the frequency of use, specific analytical methods that may be required in each procedure, applicable tolerances for each procedure and, if possible, remedial steps to correct out-of-control procedures, and the names or titles of personnel to be notified if out-of-control procedures are found.

1.4 PERSONNEL

Quality results require quality personnel. The personnel selected must have the education, experience, and motivation necessary to perform their jobs and to carry out the requirements of the quality assurance program.[38] The number of persons and their technical training may vary considerably from laboratory to laboratory, depending on the types and numbers of analyses to be performed. However, general guidelines for selection of laboratory personnel are available. The U.S. Environmental Protection Agency (EPA) has published guidelines for the selection of personnel for water bacteriology[12] laboratories, and some states have specified personnel requirements for certification of analytical laboratories.[21] These guidelines should help develop personnel requirements, but they may be too tightly framed to fit the needs of food microbiology laboratories.

Successful management of laboratory personnel through motivation, training, supervision, and workload direction is as important as selection of the personnel. Workers must be properly trained to perform their duties, and the first step in training is to specify as completely as possible the duties of the position and the importance of these duties to the quality of results generated by the whole laboratory. The goals of training are that the workers know the exact duties they are to perform, with sufficient instruction and time to learn how to perform these duties so as to obtain results of the highest quality. Employees must be made fully aware of their quality assurance responsibilities and especially of the adverse consequences that will arise from' failure to carry out their duties carefully. All training of employees should be documented and filed appropriately.

Motivation of laboratory personnel to do high-quality work is essential for a successful quality assurance program. To achieve this requires a safe, efficiently designed facility, sufficient supplies and equipment, workloads that are not excessive, suitable compensation, and top management committed to high quality performance.

1.41 Evaluation of Personnel

Uniform application of laboratory procedures by all analysts is very important for consistent results. The routine evaluation of the accuracy of each analyst's performance is necessary for such consistency. The supervisor is the best person to evaluate worker performance on a day-to-day basis. An appropriate outside agency or individual may occasionally check on-site worker performance. However, experience has shown that on-site evaluations alone are not enough to minimize variation among analysts.[34]

The proficiency of each analyst should be rated by having the laboratory participate in split/check sample programs in which the performance of each analyst can be compared to the performance of other analysts in the same laboratory or other laboratories. Split or check samples to verify analyst performance can be generated internally or supplied by external sources. Whenever possible, the split or check samples obtained should be composed of the product most commonly analyzed in the laboratory. AOAC International has recently initiated a proficiency check sample program. To ensure a quality product, AOAC International has specified that the providers must conform to ILAC (Jan. 1998), ISO/IEC Guide 43-1, and 43-2 requirements.[18,24,25] Comparison of results obtained from laboratories participating in an external split/check sample program provides a valid evaluation of the proficiency of the laboratory as a whole. Statistically acceptable performances by individual analysts and the laboratory as a whole give the best indication that all elements associated with diagnostic procedures (personnel, media, reagents, and equipment) are satisfactory. Critics have asserted that split/check sample proficiency testing programs in clinical laboratories do not truly measure day-to-day capabilities because the laboratory staff usually is aware of the source of these samples and tends to be more careful than usual. A more realistic measure of performance is the introduction of internal "blind" unknown samples.[39]

Split samples can be naturally or artificially contaminated. Naturally contaminated samples have the advantage that they represent the kinds of samples routinely received and examined by analysts. However, their microbiological character may not be known, and it may be difficult to obtain large enough quantities of such samples to distribute to several analysts or laboratories. A discussion of the preparation and distribution of split samples and the statistical procedures for evaluation of split sample results have been presented by Olson et al.[31] and Donnelly et al.[11]

Studies of clinical laboratories suggest that continuous participation in proficiency surveys results in improved analyst performance.[35] Participation in the national milk laboratory proficiency testing program has resulted in marked improvement in analyst performance. Although ongoing proficiency studies have not been conducted in food microbiology laboratories, it is likely that improvement in analyst and laboratory performance can be obtained.

1.5 FACILITIES

The safety of workers is of utmost importance in design and construction of laboratory facilities. The food microbiology laboratory should be designed and built to meet this priority and also to provide for the convenience of the workers and operations. It should be adequately equipped to carry out the stated objectives of the laboratory. The following points should be considered when designing a laboratory.[12,31]

1.51 Laboratory Design

1.511 Ventilation, temperature, and humidity

Laboratories should be well ventilated, preferably by use of central air-conditioning, to reduce the amount of particulates in the air and minimize temperature variation. Temperature and relative humidity should be comfortable for workers and suitable to the requirements of the laboratory equipment. Normally, an ambient temperature of 21° to 23°C and a relative humidity of 45% to 50% are recommended. Consideration should be given to appropriate air flow and air filters.

1.512 Lighting

Laboratory lighting should be maintained at an average intensity of at least 50, and preferably 100, footcandles. Dependence upon natural sunlight during the day should be discouraged because of high variability in its intensity. Because direct sunlight is known to have deleterious effects on media, reagents, and specimens, preparation or storage of these items in direct sunlight should be avoided.

1.513 Laboratory space and bench areas

Laboratory space should be organized to maximize usefulness. Where possible, media preparation and glassware cleaning areas should be separated from the analytical areas. Equipment and materials should be positioned to make the maximum amount of bench space available. For most routine work, 6 linear feet is the recommended minimum work area for each analyst. The ideal bench top height is 36 to 38 in with a depth of 28 to 30 in. The walls and ceiling of the laboratory should be covered with good-grade enamel or epoxy paint, or other material that provides a smooth, impervious surface that is easily disinfected. Floors should be covered with high-quality tile or other impenetrable material. Cracks and crevices should be minimized, as they provide an opportunity for the buildup of debris that may contribute to cross-contamination of samples. Open floor drains are undesirable since they frequently provide direct access to a sewer system. If present, care must be taken to assure that the trap always contains water. Unnecessary traffic through the laboratory should be prohibited. Eating or smoking should never be permitted in a microbiology laboratory.

1.514 Storage areas

Storage space for equipment, materials, and samples should be sufficient for needed media, reagents, glassware, and plasticware. The use of cabinets with doors and drawers will minimize dust buildup and allow easier cleaning and disinfecting of laboratory surfaces. Test samples must be stored under conditions outlined in the particular analytical procedure being followed. Samples stored at room temperature should be placed in sealed containers to prevent the proliferation of pests. The laboratory should have a written policy for rodent and insect control. A written standard policy should outline the conditions and length of storage time for samples. Storage areas should be routinely inventoried and superfluous samples, and outdated media and reagents should be disposed of according to an established, documented policy. All media and reagents should be labeled with the date received and date opened to facilitate the disposal of out-of-date products. Hazardous chemicals should be stored properly as dictated by their reactivity and quantity.

1.515 Other utilities

Every laboratory should be equipped with enough electrical outlets of the appropriate voltage and amperage, enough natural gas jets for bunsen burners, a waste disposal system, and laboratory-grade water. The laboratory also should have an adequate number of sinks with hot and cold tap water and, preferably, lines for deionized or distilled water. Sinks with foot-operated handwashing taps are recommended.

1.516 Laboratory-grade water

Laboratory-grade water, which should be available in the food microbiology laboratory, is defined as water that has been treated to free it from nutritive and toxic materials. Laboratory-grade water may be produced by distillation, reverse osmosis, ion exchange, filtration, or a combination of these. Viable microorganisms have been shown to accumulate in laboratory water systems such as ion exchange systems, and they should be monitored routinely for microbial growth according to a written standard operating procedure. In addition to this microbiological monitoring, the following physical-chemical elements should be measured monthly and documented to have met the indicated parameters:

1. Trace metals, a single metal not greater than 0.05 mg/L
2. Total metals, equal to or less than 1.0 mg/L
3. Specific conductance greater than 1.0 megohm-cm resistivity.
4. pH 5.5 to 7.5
5. Residual chlorine less than 0.1 mg/L

Standard Methods for the Examination of Water and Wastewater, 19th ed., describes a test for the bacteriological suitability of laboratory water.[7] This is a sensitive test for the determination of toxic or stimulatory substances in laboratory water. *Standard Methods for the Examination of Dairy Products*, 17th ed., describes a procedure for the toxicity testing of phosphate-buffered dilution water.[8] Both of these tests should be performed annually to validate the quality of laboratory water.

1.517 Personnel safety

To ensure the safety of personnel, all facilities should be designed according to established federal, state, and local building and safety codes. All laboratories should be equipped with fire extinguishers and alarms, fire blankets, sprinkler systems, eyewash stations, and safety showers. Approved safety glasses should be available to laboratory workers and visitors, and autoclave gloves and appropriate protective clothing should be available for the staff. A comprehensive safety program, including worker training and routine inspection of safety equipment, should be a vital part of laboratory procedures. Laboratories should not permit employees to work alone in the facility, all doors should have locks, there should be at least two exits from the laboratory, and there should be phones available in the laboratory that can contact rescue services. Material Safety Data Sheets should be available to all employees, and the laboratory should conform to all Occupational Safety and Health Administration (OSHA) regulations. The OSHA regulations can be found in the Code of Federal Regulations (CFR), 29, Part 1910. Specific regulations for Occupational Exposure to Hazardous Chemicals in Laboratories can be found in 29 CFR Part 1910.1450.

1.518 Animal facilities

Some food microbiology procedures, such as botulinal toxin screening, require the use of laboratory animals. Laboratory animals should be maintained in separate areas other than those where routine analytical tests are performed. Animal rooms should have all air discharged to the outside without recirculation. A minimum of 15 air changes per hour is recommended. A specific written operating procedure should be designed outlining the details of animal maintenance.

Those who contemplate the use of animals for laboratory work should refer to the Public Health Service Policy of Humane Care and Use of Animals, Public Law 99-158, November 20, 1985, "Animals in Research," which can be obtained from the Office for Protection from Research Risks, National Institutes of Health (NIH).

1.52 Housekeeping

A routine cleaning and disinfecting schedule for the entire laboratory should be established, documented, and monitored for effectiveness. Disinfectants such as iodophor, quaternary ammonium compounds, or phenolic disinfectants should be employed. All laboratory benches and equipment should be disinfected before and after each use. If a cleaning service is used for floors etc., they should receive specific instructions concerning possible hazards. The cleaning personnel should have an emergency phone number to call if something is broken during the cleaning operation.

Laboratory materials should be stored after use in order to maintain a clutter-free work area. Unneeded and outdated materials should be discarded according to a written procedure that includes a description of the methods of disposal, safety precautions, and frequency of inventory measurement.

Dust and soil should not be allowed to build up in a microbiology laboratory. Close attention should be paid to corners and hard-to-clean areas. Floors should be wet-mopped, preferably with a suitable disinfectant-detergent solution, not dry-mopped or swept with a broom because these practices will contribute to airborne contamination.

1.53 Environmental Monitoring

To assess the efficacy of the established laboratory disinfecting schedule and to determine the microbial profile of the laboratory, a written operating procedure on environmental monitoring should be followed. This operating procedure should contain a description of the environmental sampling procedure and statements of locations to be sampled, tolerance limits, and frequency of monitoring. Chapter 3 describes procedures for environmental sampling of food plant environments, including equipment, containers, air, and water. These same procedures are appropriate for sampling of the laboratory environment.

1.6 EQUIPMENT AND INSTRUMENTATION

The reliability of an analytical procedure is only as good as the reliability of the equipment and instruments used for the procedure. A protocol should be established to verify the reliability of the equipment and instruments. **Only properly trained personnel should use equipment.** All equipment (balances, pH meters, etc.) should be cleaned before and after use.

1. ***Thermometers and temperature recorders.*** Thermometers should meet the minimum specifications outlined in Standard Methods for the Examination of Dairy Products.[8] Their accuracy should be checked at least annually with a thermometer certified by the National Bureau of Standards (NBS). Mechanical windup temperature-recording devices are preferred for incubators and water baths that are used continuously. Such mechanically driven temperature recorders are recommended over electrical plug-in recorders in order to measure temperature fluctuations during power failures. This is especially important during non-working hours. The recorders should be validated annually against a certified or traceable NIST thermometer and after any repair work. Written specifications should be established for these recorders, and at least daily, if not continuously, temperatures should be monitored in all active incubators, water baths, refrigerators, freezers, and ambient laboratory environments. Results of such monitoring should be placed in the permanent quality assurance records.
2. ***Balances.*** Laboratory balances should be sensitive to 0.1 g with a 200 g load. An analytical balance having a sensitivity of 1 mg with a 10 g load should be used for weighing small quantities. Single pan balances are preferred. The accuracy of laboratory balances should be checked routinely, preferably daily, by standard reference weights that are calibrated annually against a certified or traceable NIST set of weights. Generally, the balance should be checked with several different weights. Written documentation should be maintained on the calibration of the standard reference set of weights as well as on routine accuracy checks of balances. Balances should be certified yearly by an outside expert contractor.
3. ***pH meters.*** pH meters should be standardized with a minimum of two standard buffers (pH 4.0, pH 7.0, or pH 10.0) before use. Aliquots of buffer solution should be used once and discarded. Standard buffer solutions should be dated upon receipt and an expiration date established after opening the container. The pH meter should be accurate within 0. 1 pH unit. The life of pH electrodes will vary by type, brand, and frequency of use. Manufacturers' directions should be followed for servicing pH electrodes. Each standardization of the pH meter should be written in the permanent record. See Chapter 64 for further details on pH.
4. ***Autoclaves.*** Autoclaves should be equipped with accurate pressure and temperature gauges. They should be equipped with a calibrated thermometer located properly on the exhaust line to register the minimum temperature within the sterilizing chamber. It is preferable that the autoclave be equipped with a temperature recorder in order to provide a permanent record for each sterilization cycle. A permanent, record keeping system should be established to document each sterilization cycle. This consists of a daily chart for each cycle listing such items as (1) temperature and time settings, (2) materials in the chambers, (3) pressure and temperature readings once the autoclave has reached the sterilizing region of the cycle, and (4) date and time that the sterilizing cycle is started and finished, followed by the signature or initials of the operator.

   Ensuring the proper functioning of autoclaves is essential. This can be done through the use of biological indicators and through physical measurements. Biological indicators, such as *Bacillus stearothermophilus* spore ampules or strips, are available from several commercial sources and should be used at least monthly. Physical measurements can be made with thermocouples and maximum registering thermometers. A combination of both biological indicators and physical measurements should be used to validate sterilization processes. Physical validations employing thermocouples located in various areas of the autoclave chamber should be performed annually. Temperature sensitive tape and maximum registry thermometers should be employed with each use.[12]

5. ***Hot air sterilizing ovens.*** Each sterilizing oven should be equipped with a thermometer and, preferably, a temperature recorder, both calibrated against a NIST or NIST traceable thermometer. A time and temperature record should be maintained for each sterilization cycle. In addition, periodic physical and biological validations are suggested.
6. ***Other equipment.*** Equipment such as water baths, incubators, and refrigerators should be equipped with thermometers or temperature recorders, or both, calibrated against a NIST or NIST traceable thermometer. Temperature specifications for these items are listed in specific chapters on analytical methods. Operating procedures should be established so that proper written records are maintained for each piece of equipment. All instrument malfunctions and repairs should be documented.

   Laminar flow hoods should be checked with a particle counter on a routine basis by the dioctyphthalate test.[10] Specifications should be established on the filtering efficiency of the hoods. They should be routinely disinfected and monitored through the use of RODAC plates. Airflow rates should be monitored periodically with a certified-flow meter.

   Anaerobic chamber and glove boxes should contain appropriate anaerobiosis indicators. Indicator strips, which are commercially available, should be changed daily. Written records should be maintained to daily document the existence of anaerobic conditions in the chamber.

1.61 Preventive Maintenance of Equipment

It is important that the food microbiology laboratory have a formal, written preventive maintenance program. If the laboratory is small and does not have a maintenance department, preventive maintenance agreements with the manufacturers or dealers from whom the equipment was purchased should be arranged. Reputable independent maintenance firms also can provide preventive maintenance services. Regardless of laboratory size, the following points, as outlined in "Quality Assurance Practices for Health Laboratories," should be considered when establishing a preventive maintenance program[6]:

1. *Inventory.* Each piece of equipment, with its location, complete name, age, and description, and the appropriate supervisor or person responsible for the item, should be listed in an inventory. The pertinent information should be outlined on a separate page or card for each item of equipment. This record constitutes a portion of the inventory control.
2. *Definition of service tasks.* For each item of equipment, the tasks needed to keep the equipment calibrated, operating, and clean must be defined. Supportive information may be obtained from manufacturers' brochures, product guides, and journal reprints. These specific tasks should be included in the laboratory's written statement of standard operating procedures.
3. *Interval establishment.* The frequency with which the defined tasks should be performed must be determined. This will vary with the item of equipment, the type of installation, and the workload of the particular item of equipment. However, even equipment used infrequently must have minimum standards of preventive maintenance.
4. *Personnel.* Those individuals should be listed who are immediately available, or may be ultimately available, for function verification tasks, cleaning, preventive maintenance, troubleshooting, and repair. This assignment should be customized to the laboratory's own situation. In general, it is preferable to depend upon in-house personnel who can perform maintenance activities economically and efficiently. However, for some instruments, one must depend on a manufacturer's services or on an independent service company.
5. *Job assignments.* If the program is to succeed, responsibilities for the tasks outlined above must be assigned so that each person will know the responsibilities. Job assignments should be matched with training, experience, and aptitude.
6. *Training.* Laboratories having a large number of well-trained personnel who are familiar with laboratory equipment are indeed fortunate. Most laboratories must carry out some in-service training to teach personnel the use of special monitoring devices and the performance of some of the more difficult service tasks.
7. *Special instruments.* The monitoring devices, techniques, materials, and types of special equipment used to check each type of instrument in the laboratory should be listed. If the laboratory does not have the necessary monitoring equipment, it must be acquired.
8. *Setting up the system.* Once the effort has been made to inventory the equipment, define the tasks, and train the personnel on special instruments, the program should continue year after year. A record for each item of equipment should be established in which all entries are made. In addition, it may be necessary to develop a reminder system so that appropriate personnel are notified when certain tasks are to be performed.
9. *Records and documentation.* Documentation is needed to record that the appropriate service tasks have been accomplished. This may be the appropriate place to incorporate a system of reminders to ensure that the tasks are performed on time. Index card systems are good for this purpose and are inexpensive. Some laboratories use a computer to remind technologists of these tasks. The documentation scheme must be tailored to the laboratory's specific needs.
10. *Surveillance.* After setting up a program, periodic surveillance should be carried out to ensure that the records are legible and complete.

1.7 LABORATORY GLASSWARE AND PLASTICWARE

Specifications of laboratory glassware and plasticware should be established and followed. For example, the calibration of newly purchased glass or plastic pipettes should be checked upon receipt in the laboratory. The calibration marks on dilution bottles should be checked with NIST certified volumetric glassware.

Glassware should be made of high-quality, low-alkali borosilicate glass. Glassware composed of soft glass presents problems because of leaching of components and the presence of surface alkali, which may interfere with some analytical procedures.

Etched or chipped glassware should be discarded. Plasticware should be free of defects and toxic residues.

Procedures must be established to sterilize and wash microbiologically contaminated reusable glassware or plasticware. Microbiologically contaminated reusable labware must be sterilized by autoclaving or other suitable means prior to being washed.

Reusable glassware and plasticware should be washed manually or mechanically with hot water containing a suitable detergent. Stubborn residues may be removed by soaking in a potassium dichromate cleaning solution on glassware before washing. Screw caps, test tube caps, and other reusable closures also should be washed in a detergent solution and rinsed thoroughly. Many detergents have a high affinity for glassware and plasticware and some are highly bacteriostatic; it is imperative to ensure their removal after washing. After washing and rinsing, all glassware should be rinsed in laboratory grade water. Glassware and plasticware should be checked routinely for alkaline or acidic residues by applying a few drops of bromthymol blue pH indicator. This indicator is useful because it displays color changes from yellow to blue-green to blue in a pH range of 6.5 to 7.3. Since most cleaning solutions are either acidic or alkaline, this simple test assures proper rinsing.

1.71 Toxicity Testing

Disposable glassware and plasticware may be sterilized by ethylene oxide gas. If these items, pipettes, petri dishes, etc. are not properly rinsed after the sterilization treatment, toxic residues may remain. Therefore, it is important to check these items periodically for toxic residues and to request certification from the supplier that no toxic residues are present. Similarly, glass items washed and sterilized in the laboratory may contain toxic detergent residues not detected by the bromthymol blue pH test. These items should be checked periodically for toxic detergent residues. Procedures for toxicity testing are detailed in several publications.[7,8,12] The washing procedures should be checked at least annually by performing toxicity tests and should be modified if necessary.

1.72 Sterility Testing as a Quality Assurance Tool

The sterility of sterilized supplies and equipment must be ensured. A sterility test may be performed on a portion of the sterilized items. Sterility control tests on petri dishes may be performed by simply pouring a nonselective medium such as Standard Methods Agar into several randomly selected plates from a case. Upon solidification, the plates are then incubated aerobically or anaerobically and examined for growth. Sterility controls on sampling containers, utensils, and dilution bottles may be performed by the rinse filtration technique. According to this technique, the items are aseptically rinsed with sterile phosphate buffer that is filtered through a membrane filter. The filter is placed on the surface of a nonselective agar and incubated under prescribed conditions.

Test tubes may be checked for sterility by adding a broth such as fluid thioglycollate to the tubes and observing for turbidity after incubation. Also, microbial growth controls should be performed for each plating group by placing 1 to 2 mL of dilution fluid from bottles into the agar medium being used. This procedure is a sterility check for the agar medium, diluent, petri dishes, and pipettes. However, this sterility monitoring method has limited reliability. A statistically valid sterility testing system is available that permits determination of the probability that n negative samples will occur at a specified concentration of organisms per test unit (dish, pipette, etc.).[31]

1.8 MEDIA AND REAGENTS

Food microbiology laboratories use many different media and reagents to detect and enumerate microorganisms, and most of them are purchased already prepared or in dehydrated form. Reagents and media should be tested before using to validate their efficacy.[27,33] Those media and reagents formulated in the laboratory also should be prepared carefully and validated for performance.

Common errors that occur in preparation of media and reagents are listed here:

1. Incorrect weighing of dry material.
2. Use of dry material that has deteriorated as a result of exposure to heat, moisture, oxidation, or other environmental factors.
3. Incorrect measurement of water volume, or use of tap water or water from a malfunctioning still or deionizing resin column. Water must meet the requirements for laboratory pure water and be proven to be microbiologically suitable.
4. Use of containers and glassware that are contaminated with detergent or other chemicals.
5. Incomplete mixing or solubilization of ingredients during preparation of media or solutions. This may result in excessive or insufficient gel strength of the medium and uneven concentrations of constituents among aliquots.
6. Overheating during preparation and sterilization, or holding too long in the molten state before dispensing into plates, tubes, or bottles. Overheating can result in loss of a medium's productivity through hydrolysis of the agar, caramelization of carbohydrates, lowering of pH, increase or decrease in inhibitory action because of the loss of dye content in selective or differential media, and the formation of inhibitory precipitates.
7. Improper determination of pH, resulting in the addition of too much acid or alkali.
8. Improper addition or incorporation of unsatisfactory supplements or enrichments, or addition of supplements at the wrong temperature, possibly causing chemical changes in the supplements if the temperature is too high, or solidification of media before proper mixing if too cold.
9. Failure of the laboratory to subject samples of finished media to quality control procedures before the media are used.
10. Failure of the laboratory to test samples of dehydrated media purchased from suppliers to ensure that the media are productive.
11. Failure of the laboratory to allow for evaporation of liquid in dilution blanks during autoclaving, resulting in dilution blanks containing less than the required diluent.

1.81 Receipt of Media, Reagents, and Ingredients

Containers of media and reagents should be dated upon receipt. The laboratory should maintain a media/reagent control file where the following information is recorded for each shipment of media or reagents received:

1. Manufacturer and manufacturer's code.
2. Quantity received, i.e., size and number of containers.
3. Date received.
4. Date opened.

5. Location where medium/reagent is to be stored.
6. Initials of person receiving and placing the item into stock.
7. Results of productivity and selectivity testing, if performed (see Section 1.83).

Each lot of medium/reagent should be inspected before use for volume, tightness of closure, clarity, color, consistency, and completeness of label.

1.82 Storage of Dehydrated Media, Reagents, and Ingredients

Directions for the storage of most media and ingredients are generally listed by the supplier on the label of each container. In addition, the supplier will often indicate an expiration date after which the item should not be used. When available, the supplier's directions should be followed. Some general guidelines are listed below.

1. Store dehydrated media in tightly capped bottles or tightly closed plastic liners in a cool, dry place protected from light. If specified, keep under refrigeration and in the dark.
2. Keep no more than 6 months' to a year's supply on hand, being sure to use older stocks first. Do not exceed supplier's expiration date.
3. Dehydrated media and reagents should be free-flowing powders or crystals. If a change is noted in this property or in the color, the item in question should be discarded.
4. Media containing dyes should be protected from light by storage in a dark room or a dark glass bottle or by wrapping the container with foil or brown paper.

1.83 Productivity/Selectivity Testing of Media and Reagents

Historically, clinical microbiologists have been concerned with performance testing of media and reagents. "Cumitech 3" recommends the verification of media and reagents' performance for each new lot number or shipment received. This publication provides a detailed outline of performance tests on various media, recommended control organisms, and expected results. No guidelines exist for performance testing of media and reagents employed specifically in food microbiology but a CDC Lab Manual titled "Quality Control in Microbiology" is a general microbiology laboratory quality control reference, and may be of value for a food laboratory.[27] Studies of commercially available media and reagents for use in clinical laboratories have suggested that these items generally perform as expected.[30] However, the studies provide several examples of media and reagent failure.[29] Considering the consequences of such failure, i.e., the generation of faulty data, it is appropriate that each new lot of medium or reagent be subjected to performance testing.

1.831 Media

Standard Methods for Examination of Dairy Products[8] gives a procedure for evaluating the performance of new lots of standard methods agar by comparing them to a standard control lot. This procedure can be used for testing any nonselective general purpose agar medium. Liquid nonselective media can be tested similarly using dilution-to-extinction techniques to compare new lots of media to standard lots. Careful consideration must be given to the selection of samples and cultures to be evaluated in these tests. If test cultures are to be employed in the evaluation of nonselective media, organisms should be chosen that are at least as fastidious as those for which the medium will be used routinely. Furthermore, it is suggested that more than one culture be used to evaluate a given medium.

In the examination of selective or differential media, cultures must be chosen that test both the productivity and selective/differential characteristics of the medium. Liquid media can be evaluated by inoculation with cultures expected to grow in the medium and with cultures expected to be suppressed. Following suitable incubation, titers on each culture must be determined. An acceptable medium should show high titers of organisms to be detected and low titers of organisms to be suppressed.[19] Performance of new lots should be similar to control or standard lots. The evaluation of solid media is performed in a similar manner. Organisms of the group for which the medium was designed should be recovered almost quantitatively compared to nonselective media. Organisms that should be suppressed by the medium should not develop or should be greatly reduced when compared to enumeration on nonselective media. Furthermore, the differential characteristics of the medium should be the same as expected for the known cultures.

Unfortunately, the variety of media employed in most food microbiology laboratories, and the number of control cultures and lengthy procedures required for the evaluation of these media, make the routine evaluation of each new lot a burden.

Mossel and coworkers[29] developed a relatively simple "ecometric" technique to evaluate liquid and solid selective media. In this method, cultures of test strains of the groups of cultures to be detected and those that are to be suppressed, are streaked in parallel lines onto a solid medium or a liquid medium that has been solidified by the addition of about 15 g of agar/L, and poured into each section of a quadrant petri plate. Organisms for which the medium was designed should develop in all quadrants, whereas "background" flora should develop in only the first or second quadrants streaked. Based upon the pattern of growth, an Absolute Growth Index (AGI)[29] is calculated for each test organism. The AGIs for new lots of media should be similar to those of standard or control lots.[29]

1.832 Reagents

A committee, part of a Centers for Disease Control and Prevention (CDC) task force, was created to develop standards and create product classes for diagnostic reagents.[30] This committee divided diagnostic reagents into three general classes: immunodiagnostic reagents, microbiological media, and miscellaneous reagents. The miscellaneous reagents class consists of stains and chemical reagents, most of which are available from commercial sources. Although the manufacturer is charged with ensuring that the reagents meet minimum product standards, the committee suggested that the recipient perform quality control tests as well. Reagents routinely employed should be checked at least weekly with positive and negative control cultures. Less frequently used reagents should be checked by including positive and negative controls with each use. In addition, some commercial systems are available for evaluation of reagents (Analytab Products, Plainview, N.Y.), and recently a simple technique for testing a variety of reagents has been described.

Antisera to a number of species are being commonly used in food microbiology laboratories. On occasion, it has been noted that somatic polyvalent and grouping sera have failed to react or reacted weakly.[36] Problems have occurred with the labeling of antisera. Often containers and brochures did not adequately identify agglutinins contained in the sera or labeled them alphabetically without regard to the antisera groups contained. Based upon this information, it appears essential that both positive and negative controls be run on all antisera received. A control containing only culture in saline should also be performed to test for autoagglutination.

Such tests should be performed before initial use and monthly thereafter, depending on the frequency of use. Recommendations of the manufacturer for preparation, storage, and use of antisera should be followed closely.

The maintenance of stock cultures for productivity/selectivity testing of media and reagents is important. The stock culture collection must consist of strains of microorganisms that will reflect the productive, selective, and differential characteristics of each medium and reagent employed in the laboratory. Stock cultures may be maintained in the laboratory, usually by one of three methods: (1) lyophilization, (2) ultrafreezing, and (3) maintenance in appropriate media with frequent transferring. Microorganisms used for quality control also may be obtained from several commercial sources.

1.84 Performance and Sterility Testing of Prepared Media and Reagents

The tests described above have been designed to test the performance of new lots of media and reagents and should be applied to each batch of prepared medium or reagents in the laboratory. For accreditation to AALAC requirements this is essential. The decision of whether to test only new lots, or to test each batch of medium or reagent must be made by the laboratory management based upon the cost of such tests and confidence in the quality of the laboratory personnel, equipment, and facilities to prepare each batch properly (*see section 1.8*).

Each batch of media and reagents, whether prepared in the laboratory from dehydrated ingredients or purchased in prepared form, should be checked for hydration, clarity, surface condition, color, thickness and sterility. Selected plates or tubed media, representative of the batch, should be incubated prior to use or along with inoculated media. Generally, incubation of sterility controls at the temperature normally used for these media is an adequate check of sterility. Selective media may present some problems, having been designed to inhibit a variety of microorganisms. Gross contamination may not be evident visually in the form of turbidity or colony formation. This problem can be overcome by transferring a portion of liquid selective media to a nonselective medium or by swabbing the surface of an agar plate and then incubating the swabs in a nonselective medium.

Procedures for sterility testing of reagents are presented in "The United States Pharmacopeia XXIII".[37]

1.9 RECORD KEEPING

It is essential for a food microbiology laboratory to maintain accurate and permanent records of sample analyses and quality assurance programs. The nature of the records can vary from individual worksheets and data books to entire electronic data processing systems. The length of time and the manner in which records are retained will depend on the laboratory's objectives, the nature of the records, the scope of the work performed, and the space available for storage.

1.91 Sample Analytical Data

1. Records should be kept documenting the care and disposition of samples during their time in the laboratory. These should show the storage conditions, the personnel with custody of samples, and the final disposition of samples when no longer required.
2. Records of all aspects of sample analyses, including sample descriptions, storage conditions and reverse sample retention, descriptions of analytical methods, all raw data, and observations, calculations, and conclusions, are required. The analyst(s) responsible for each segment of the procedure should be identified in the record.

These records may be in the form of worksheets that become a part of the entire record for each sample or a notebook that can be referred to in the sample records and correspondence but regardless of form, all records should be recorded in ink. When an error is made, it should be corrected by drawing a single line through the error, writing the correct information above it, and dating and initialing the change. Never use white out.

3. Analytical records should be reviewed for completeness and accuracy before the results are reported. This review should be at least a two-step operation, with the first review done by another analyst in the laboratory and a second by the supervisor.

1.92 Research Data

Analytical laboratories are occasionally faced with the need to develop a new method of analysis or to modify an existing method. The details of these research activities must be properly documented. One should also consider adopting applicable elements of the following recommendations to other types of record keeping in the laboratory.

The standard, bound laboratory notebook is the preferred medium for the recording of research data. At times, however, other media may replace or be used in conjunction with the laboratory notebook. Record maintenance procedures are as follows:

1. Each research project should have its own set of notebooks.
2. The first pages should be reserved for a brief table of contents that should include dates, type of information, and page numbers.
3. Notebook entries are to be made only in black or blue-black ink, with a fountain pen or ballpoint pen. Pencil or pen with water-soluble ink is not to be used. All entries must be dark and clear enough to be photocopied.
4. Illustrations such as charts, graphs, photographs, etc., may be pasted securely in the notebooks if they approximate the size of the page. Voluminous printouts, photographs, charts, etc. may be maintained in supplemental files and referenced in the notebook.
5. All entries made by someone other than the notebook owner should be initialed.
6. Entries are to be neat and legible. No erasures are to be made; errors will be marked through with a single line, initialed, and dated.
7. Experimental results will be summarized.
8. All unused notebook pages will be cancelled with diagonal lines.
9. Once a study has been completed, the researcher will maintain the related notebook(s) and other research records. The first-line supervisor is responsible for assuring access and will maintain a log of all such materials under his or her responsibility.
10. A notebook assigned to an employee who is separating from an organization will be returned to the first-line supervisor, who will be responsible for maintaining the notebook and all related information.

1.93 Quality Assurance Data

Records should be kept of all quality assurance and control testing. These records can be kept on analytical worksheets as a part of the analytical routine or on separate log sheets. The analysts responsible for each check should be indicated, and the steps taken to bring back into control any procedures or functions out of tolerance should be recorded. Quality assurance records should be maintained for the following:

1. Analytical split or check sample results.
2. Purity and authenticity data on biological standards such as bacterial or fungal stock strains.
3. Calibration records on non-precalibrated volumetric implements.
4. Calibration/standardization records on analytical instruments such as gas chromatographs, spectrophotometers, pH meters, etc.
5. Annual calibration data and daily use check weighing on analytical balances.
6. Temperature records for freezers, refrigerators, incubators, waterbaths, etc.
7. Moisture level test results in incubators.
8. Time-temperature-pressure records for autoclaves.[13]
9. Thermometer calibration results.
10. Animal room hygrothermograph records.

Periodic review of these records should be carried out by supervisory personnel to assess the effectiveness of the quality assurance program. This review can be valuable in revealing potential problem areas and making corrections before serious problems occur. The reviewing supervisor should sign and date all records at the completion of the review.

1.94 Storage and Retrieval of Data

A system that provides storage and ready retrieval of all the data generated in the laboratory is necessary. The type of system will depend on the type of laboratory and the analyses performed. The length of time that these records should be kept can vary greatly. In the case of regulatory agency laboratories, they may need to be available for several years.

1.95 Issuance of Results

It is very important that the laboratory analytical results report be accurate and understandable. A clerical or transcription error can cause a great deal of damage to the client and the laboratory. Each laboratory should develop a system to review final reports when they are signed and before they are submitted to the client. The reports should contain the definition of any symbols used and definitions of abbreviations. If an error has been made, it should be corrected by drawing one line through the error, making the correction, and then initialing and dating the correction. White out must never be used on an analytical result report.

1.10 ACCREDITATION OF TESTING LABORATORIES

Accreditation of quality systems is becoming increasingly important both nationally and internationally. With the increased emphasis on food safety domestically and the globalization of the food market place, the need for accurate and reliable microbiological test results has become an essential part of public safety and commerce. The pressure for food microbiology laboratories to demonstrate competency in testing has accelerated in recent years.

The ultimate goal of an accreditation system is to recognize laboratories with a demonstrated ability to produce accurate, reliable and consistent results using validated methods. It is desirable that any system to be recognized by government and the food industry be consistent with international standards such as ISO Guide 25, 43 and 58.[23-26] In the United States the ISO/IEC Guide 25, 1990, "General Requirements for the Competence of Calibration and Testing Laboratories," has been adopted by the AOAC International as the guide for operating a testing laboratory.[23] The quality systems found in the ISO/IEC Guide 25 are general and can be applied to many types of laboratories.[23]

It was written specifically for laboratories and addresses testing competency to a specific method. Because the Guide 25 is general, specific criteria have been developed for food microbiology laboratories. In Europe, quality systems have been developed for microbiology laboratories and are found in "Eurachem/EAL, Accreditation for Laboratories Performing Microbiological Testing," and ISO 7218.[13,22] In the United States these criteria have been delineated by the Food Laboratory Accreditation Working Group (FLAWG), "Accreditation Criteria for United States Laboratories Performing Food Microbiological Testing," Feb. 10, 1998, which was presented to the AOAC International for further development and maintenance.[16] The new working group in AOAC International is called Analytical Laboratory Accreditation Criteria Committee (ALACC) and is a part of the Technical Division Laboratory Management.

1.11 REFERENCES

1. AOAC. 1920. Official methods of analysis, 1st ed. Assoc. Off. Anal. Chem., Washington, D. C.
2. AOAC. 1995. Official methods of analysis, 16th ed. Assoc. Off. Anal. Chem., Arlington, Va.
3. APHA. 1910. Report of the committee on standard methods of bacterial milk analysis. Am. Public Health Assoc. Am. J. Pub. Hyg. 6:315.
4. APHA. 1958. Recommended methods for the microbiological examination of foods, 1st ed. American Public Health Association, Washington, D. C.
5. APHA. 1976. Compendium of methods for the microbiological examination of foods, 1st ed., ed. M. L. Speck (ed.). American Public Health Association, Washington. D. C.
6. APHA. 1978. Quality assurance practices for health laboratories. Amican Public Health Association, Washington, D. C.
7. APHA. 1995. Standard methods for the examination of water and wastewater, 19th ed. American Public Health Association, Washington, D. C.
8. APHA. 2001 (In press). Standard methods for the examination of dairy products, 17th ed. American Public Health Association, Washington, D. C.
9. Blazevic, D. J., C. T. Hall, and M. E. Wilson. 1976. Cumitech 3. Practical quality control procedures for the clinical microbiology laboratory, American Society for Microbiology, Washington, D. C.
10. Chatigny, M. A. 1986. Primary barriers. *In* B. M. Miller (ed.), Laboratory safety: principles and practices, American Society for Microbiology, Washington. D. C.
11. Donnelly, D. B., E. K. Harris, L. A. Black, and K. H. Lewis. 1960. Statistical analysis of standard plate counts of milk samples split with state laboratories. J. Milk Food Technol. 21:315.
12. EPA. 1975. Handbook for evaluating water bacteriological laboratories, 2nd ed. EPA070/9-75-006. U.S. Environmental Protection Agency, Cincinnati, Ohio.
13. Eurachem/EAL. 1996. Accreditation for laboratories performing microbiological testing. 1st ed. LGC (Teddington) Ltd.
14. FDA. 1982. Bureau of foods laboratory quality assurance manual. Food and Drug Administration, Washington, D. C.

15. FDA. 1995. Bacteriological analytical manual, 8th ed. Food and Drug Administration Association, Office Analytical Chemistry, Arlington, Va.
16. Food Laboratory Accreditation Working Group (FLAWG). 1998. Accreditation criteria for United States laboratories performing food microbiological testing. Feb. 10, 1998, AOAC International, Gaithersburg, Md.
17. Hicock, P. 1. and Marshall, K. E. 1981. Reagent quality control in bacteriology: cost- effectiveness, easy-to-use methodology. J. Clin. Microbiol. 14: 119.
18. ILAC. 1998. ILAC requirements for accreditation of providers of proficiency testing schemes, 13 January 1998.
19. ICMSF. 1980. Appendix 1. *In* Microbial ecology of foods, vol. 11, p. 945. Intern. Comm. on Microbiol. Spec. for Foods. Academic Press, New York.
20. ICMSF. 1988. Microorganisms in foods 4: application of the Hazard Analysis Critical Control Point (HACCP) System to Ensure Microbiological Safety and Quality. Intern. Comm. on Microbiol. Spec. for Foods. Blackwell Scientific, Boston. Mass.
21. IDPH and IEPA. 1980. Certification and operation of environmental laboratories. Department of Public Health and Illnois Environmental Protection Agency, Springfield, Ill.
22. ISO 72i8. 1985. (E) Microbiology—General rules for microbiological examinations. 1st ed. International Organization of Standardization, Geneva, Switzerland.
23. ISO/IEC Guide 25. 1990. General requirements for the competence of calibration and test in laboratories. 3rd ed. International Organization of Standardization, Geneva, Switzerland.
24. ISO/IEC Guide 43-1. 1997. Proficiency testing by interlaboratory comparisons—Part. 1: Development and operation of proficiency testing schemes. International Organization of Standardization, Geneva, Switzerland.
25. ISO/IEC Guide 43-2. 1997. Proficiency testing by interlaboratory comparisons—Part 2: Selection and use of proficiency testing schemes by laboratory accreditation bodies. International Organization of Standardization, Geneva,
26. ISO Guide 58. 1993. Calibration and testing laboratory accreditation systems—General requirements for operation and recognition. 1st ed. International Organization of Standardization, Geneva, Switzerland.
27. Miller, J. M. 1990. Quality control in microbiology. Revised November 1987 reprinted August 1990. Centers for Disease Control, Atlanta, Ga.
28. Mossel, D. A. A. 1980. Food microbiology, how it used to be in the 1950s and what it may become in the 1980s. Culture 1:1. Oxoid Ltd., Hampshire. England.
29. Mossel, D. A. A., F. Van Rossem, M. Koopmans, M. Hendricks, M. Verdouden, M., and Eelderink, 1. 1980. A comparison of the classic and the so-called ecometric technique. J. Appl. Bacterial. 49:439.
30. Nagel, J. G., and L. J. Kunz. 1973. Needless retesting of quality assured commercially prepared culture media. Appl. Microbiol. 26:31.
31. Olson, J. C., R. A. Belknap, A. R. Brazis, J. T. Peeler, and D. J. Pusch. 1978. Food microbiology. *In* Quality assurance practices for health laboratories. American Public Health Association, Washington, D. C.
32. Peddecord, K. M., and R. L. Cada. 1980. Clinical laboratory proficiency test performance. Its relationship to structural, process, and environmental variables. Am. J. Clin. Pathol. 73:380.
33. Power, D. A. 1975. Quality control of commercially prepared bacteriological media. *In* J. E. Prier, J. T. Bartola, and H. Friedman (ed.), Quality control in microbiology, University Park Press, Baltimore, Md.
34. Prier, J. E., J. T. Bartola, and H. Friedman. 1975. Quality control in microbiology. University Park Press, Baltimore, Md.
35. Snyder, J. W. 1981. Quality control in clinical microbiology. API Spores 5:13. Analytab Products, Plainview, N. Y.
36. Suggs, M. T. 1975. Product class standards (specifications) and evaluation of microbiological in vitro diagnostic reagents, p. 87. *In* J. E. Prier, J. T. Bartola, and J. T. Friedman (ed.), Quality control in microbiology, University Park Press, Baltimore, Md.
37. US Pharmacopeial Convention, Inc., 1995. The United States Pharmacopeia XXIII. Rockville, Md.
38. Wilcox, K. R., T. E. Baynes Jr., J. V. Crable, J. K. Duckworth, R. H. Huffaker, R. E. Martin, W. L. Scott, M. V. Stevens, and M. Winstead. 1978. Laboratory management, p. 3. *In* S. L. Inhorn (ed.), Quality assurance practices for health laboratories. American Public Health Association, Washington, D. C.
39. Wilson, M. E. 1975. Microbiological proficiency: what basis for confidence, p. 119. *In* J. E. Prier, J. T. Bartola, and H. Friedman (ed.), Quality control in microbiology, University Park Press, Baltimore, Md.

Sampling Plans, Sample Collection, Shipment, and Preparation for Analysis

Thaddeus F. Midura and Raymond G. Bryant

2.1 INTRODUCTION

The objective of this chapter is to enable the user to obtain representative samples of a food lot, submit the samples to the laboratory in a condition that is unchanged microbially from the time of sampling, and prepare the samples for analysis. The procedures described in this chapter apply generally to collecting, labeling, transporting, storing, and preparing samples for analysis. For specific information and discussion about sampling and analytical procedures for canned foods, see Chapters 61 and 62. Other foods also may require special sampling and preparation procedures depending upon the specific microorganisms involved. For such procedures, see the chapters covering the relevant food types and specific microorganisms.

2.2 GENERAL CONSIDERATIONS

A representative sample that is appropriately transported to the laboratory, and prepared for examination, is the first priority in the microbiological examination of any food product. Laboratory results and their interpretation are valid only when appropriate samples are examined. Samples must be representative of the entire lot of material under evaluation, must be the proper type for the determination to be made, and must be protected against extraneous contamination and improper handling, especially at temperatures that may significantly alter the microflora. Refrigeration often must be provided to prevent destruction or growth of organisms in a sample. Unfrozen samples must be refrigerated, preferably at 0° to 4.4°C, from the time of collection until receipt at the laboratory. To avoid contact of the product with melting ice, a sealed eutectic coolant is preferable for use in the shipment container. Samples collected while frozen should be kept solidly frozen. When dry ice is used, the containers should have tight closures to prevent pH changes in the sample caused by the absorption of carbon dioxide. As a general rule, samples should be examined within 36 hr after sampling. Perishable items that cannot be analyzed within 36 hr should be frozen or retained at refrigerated temperatures, depending upon the type of product, reason for analysis, and type of microorganisms sought. Nonfrozen samples of shellfish should be examined within 6 hr after collection; they cannot be frozen.[3]

Samples must be clearly and completely identified with the following information: sample description, collector's name, name and address of the manufacturer, lot number, dealer or distributor, and date, place, and time of collection. Frequently the temperature at the time of collection is also useful to the laboratory for the interpretation of results. Further, it is often desirable that the reason for testing be given, e.g., samples may be collected as part of a quality control or surveillance program, as official samples to determine conformity to regulatory standards, or as part of a foodborne disease investigation.

2.3 EQUIPMENT, MATERIALS, AND REAGENTS

1. **Instruments for opening containers.** Sterile scissors, knives, scalpels, can openers, or other hand tools as required.
2. **Sample transfer instruments.** Sterile multiple- or single-use spatulas, scoops, spoons, triers, forceps, knives, scissors, tongue depressors, drills and auger bits, corers, dippers, metal tubes, and swabs as required.
3. **Sample containers.** Sterile multiple-use containers, either large or small-mouthed designs, nontoxic, leakproof, and presterilized polyethylene bags or other suitable sterile, nontoxic containers as appropriate.

   Nonsterile, nontoxic, single-service vials, polyethylene bags, or bottles that are clean and dry and do not have a viable bacterial count in rinse tests in excess of one organism per mL of capacity are acceptable transport containers. Sterile, evacuated sampling equipment also can be used. Sterile glass containers usually are not desirable because of possible breakage and consequential glass contamination of the sampling environment.
4. **Thermometers.** Thermometers that measure –20 to 100°C with graduation intervals not exceeding 1°C. A metal dial type or digital electronic unit is preferred due to the risk of glass thermometers breaking. Thermometers should be sanitized by dipping in a hypochlorite solution not less than 100 mg/L, or other equivalent microbicide, for at least 30 sec before being inserted into foods.
5. **Microbicide.** Medium strength (100 mg/L) hypochlorite solution or other approved disinfectant.
6. **Labeling supplies.** Pressure-sensitive tapes and labels, tags of adequate size to hold sample information, indelible marking pens.
7. **Sample shipping containers.** For frozen or refrigerated samples, rigid metal or plastic containers that are insulated and equipped with a tight-fitting cover. Each container

should have ample space for the refrigerant so that samples will remain at the desired temperature until arrival at the laboratory.

Containers for nonperishable samples should be made of sturdy corrugated cardboard or other material capable of withstanding abusive shipping conditions.

Refrigerant or dry ice is to be added as needed for perishable samples.

8. **Balance.** A balance with 2000 g capacity having a sensitivity of 0.1 g with a 200 g load is acceptable.
9. **Blenders and mixers.** Mechanical blenders with several operating speeds or rheostat speed control with sterile glass or metal blending jars and covers or a Stomacher (Tekmar Company, Cincinnati, Ohio) are examples of acceptable devices.
10. **Diluents.** Sterile Butterfield's phosphate buffer (Chapter 63). Sterile 0.1% peptone water (Chapter 63). Sterile sodium chloride solutions (Chapter 63). Special diluents required for specific microorganisms and special analytical conditions (specific chapters).

2.4 PRECAUTIONS

Adequate precautions should be taken to prevent microbial contamination of samples from external sources, including the person taking the samples, air, sample containers, sampling devices, and the shipping vehicle. When foods are packaged in small, sealed containers, collect the unopened containers rather than portions from each container.

The sampling operation should be organized in advance with all the needed equipment and sterile containers at hand. For collecting samples, use an instrument appropriate to the physical state of the food.

Protect sampling instruments from exposure and contamination before and during use. When removing sampling instruments from the food container to the sample container, do not pass them over the remaining presterilized instruments. When opening the sterile sampling container, open it sufficiently to admit the sample, then close and seal it immediately. Do not touch the inside of the sterile container lip or lid. Do not allow the open lid to become contaminated. Do not hold or fill a sampling container over the top of a bulk food container when transferring a sample. Fill the sample container not more than 3/4 full to prevent overflow and to allow proper mixing of sample in the laboratory. Do not expel air when folding or whirling plastic sample bags. Submit an empty sterile sampling container similarly opened and closed as a control.

The sample collector must keep their hands away from the mouth, nose, eyes, and face. Hands must be washed immediately before beginning the sampling and during sampling if the hands become contaminated. Sterile plastic gloves may be useful to limit contamination during the procedure.

Contaminated sampling equipment must be placed into proper containers for ultimate disposal and/or sterilization. Labels must never be moistened with the tongue. Use direct labeling of the container or pressure-sensitive labels.

Chapter 1 gives additional safety precautions to ensure proper handling and preparation of samples for microbiological analysis. Food samples may contain infectious microorganisms or toxic materials that may be potentially hazardous. The best protection against hazards is the use of good sampling techniques and treating each sample as if it were contaminated.

2.5 PROCEDURES

2.51 Sampling Plans

Sampling plans were first developed in 1923 by engineers at the Western Electric Company, but were not immediately adopted.[21] After World War II the Department of Defense developed Military Standards for attribute and variables sampling plans[21,22,23,24] (definitions in Section 2.511). These plans considered single and multiple sampling based on the history of the producer performance. Levels of sampling (i.e., tightened, normal, and reduced) that reflected prior producer performance were established for these plans. Special categories of sampling plans have been adapted to agricultural products.[5,6,14,19,20,25]

A comprehensive set of single sampling attribute plans has been published for food microbiology.[7,12] Typical points of sampling where sampling plans may be applied are shown in the following chart:

Points of Sampling
raw materials
↓
production line
↓
producer's warehouse
↓
retail storage or sales outlet
↓
international port—export or import

This discussion of sampling plans will be clarified if certain basic concepts are stated at the outset. First, the lot to be sampled must be defined. When a given quantity of product is surveyed and units or portions of this lot are taken for examination (including microbiological analyses), the procedure is called "sampling." Sampling can be applied in a way that ensures that results obtained will be statistically valid. In this case, a "sampling plan" is developed, where selection of samples taken from a lot is carried out in a manner that ensures that each sample has the same chance of being chosen. For example, assume that a lot defined as "1000 packages in a warehouse" is to be sampled. Each package is assigned a number. If the 1000 items are stacked equally in 10 rows, then 1 to 100 would be assigned to the units in the first row, 101 to 200 to the second, and so on until all packages are numbered. A random number table is used to choose the required numbered containers for sampling.[12,17] In this example, if we select five as the required number of packages, then five typical random numbers (lot size being 1000) could be 586, 973, 99, 838, and 737. In addition to random sample selection, a sampling plan will also contain instructions that specify the number of packages to be taken and the basis for accepting or rejecting the lot. Thus, a sampling plan as employed in this context states the number of units required to be randomly collected from a lot and lists the acceptance and rejection criteria. Before using a sampling plan, it is usually prudent to consult a professional statistician to ascertain that the lot of food to be sampled meets the criteria required by that particular sampling plan.

2.511 Definitions Used in Sampling Plans

The following statistical terms are frequently used in the sampling literature.[9,12,13,24,25]

1. **Acceptance number(s)**—The maximum number(s) of defectives in an attribute sampling plan for which a lot will be accepted.

2. **Acceptance quality level (AQL)**—The maximum percent defective (or the maximum number of defects per 100 units) that for purposes of sampling inspection can be considered satisfactory as a process average.[23]
3. **Analytical unit**—The amount of sample actually analyzed.
4. **Attribute**—A qualitative characteristic of a sample unit, e.g., the results of an analysis are positive or negative for *Salmonella.*
5. **Average outgoing quality limit (AOQL)**—The maximum possible percent defective that will result after a rectifying inspection of the lot has been performed. In such an inspection, 100% of rejected lots are screened to remove defective items, provided such screening is 100% effective.[24] Typically, AOQL values do not exceed 5%.
6. **Binomial distribution**—Distribution of a population in which a proportion (p) of the units in the population has a certain characteristic, and a proportion (q = 1–p) does not. Thus, each individual unit falls into one of only two classes, either p or q.
7. **Consumer protection**—The ability of a sampling plan to identify unacceptable lots. This is measured as the complement of the probability of acceptance for Limiting Quality (LQ) lots.[25] The consumer protection is often set at 90%.
8. **Consumer's risk**—The risk a consumer takes that a lot will be accepted by a sampling plan even though the lot does not conform to requirements.[25] This risk is known as a Type I risk, and in many standards[8,21,23,25] this risk is nominally set at 10%. A second type is a consumer beta risk, which is the risk that the lot will be rejected even though it is actually satisfactory.
9. **Continuous distribution**—A distribution of a population of measures that take on a continuum of values. A variables sampling (Section 2.515) plan may be applicable.
10. **Control chart**—A graphic device that can be used to monitor repeated sampling from a manufacturing or measurement process.
11. **Destructive testing**—A testing process that results in destruction of the unit or sample under test.
12. **Discrete variate**—A random variable consisting of isolated points that can have a finite or infinite number of values. These values can be used in an attribute sampling plan. (Sections 2.512, 2.513, 2.514).
13. **Estimate**—Any value computed from sample data and used to infer a corresponding population (lot) value, e.g., the sample mean (average).
14. **Estimated process average**—The average percent defective or average number of defects per 100 units of product found at the time of original inspection.[24]
15. **Frequency distribution**—The mathematical description of the way members of a population are distributed. The information about the distribution is used to calculate the probability of lot acceptance or rejection. The discrete random variable takes on a countable number of values, and the probability distribution is defined by a probability mass function. A continuous random variable is defined by a density function.
16. **Homogeneity of variances**—The equality of variances among populations, which must sometimes be tested.
17. **Homogeneous**—A product having a uniform texture or content.
18. **Limiting Quality (LQ)**—Percent defective or defects per 100 units.[25] A lot having a 10% probability of acceptance is referred to under many common standards[8,21,23,25] as a lot having a quality level equal to LQ.
19. **Lot**—The number of sample units produced in one batch or in some specified period of time such that the units will be about the same quality. Each lot or batch should consist of units of product of a single type, grade, class, size, and composition. The lot should be manufactured under the same conditions and at essentially the same time.
20. **Lot inspection by attributes**—Inspection whereby either the sample unit is classified as defective or nondefective with respect to a requirement or set of requirements (when on a "defective" basis), or inspection whereby defects in each sample unit are counted with respect to a requirement or set of requirements (when on a "defect" basis).[25]
21. **Lot quality**—A measure of the characteristic being controlled. The results of lot inspection are often expressed as percent defective units (e.g., a unit containing *Salmonella* is defective). Less frequently, the quality will be expressed in terms of the variable measured (weight/unit, coliform/g).
22. **Operating characteristic (OC) curve**—A graphical representation of the relation of the probability of lot acceptance to lot quality (usually expressed as percent defective units). This curve will depend on the number of units required in the sampling plan and the acceptance number. The curve also shows the lot quality associated with the consumer's risk and the producer's risk. It thus describes the consequences of the sampling plan (decision rule) for accepting lots of different quality.
23. **Population**—Any finite or infinite collection of individuals (samples or units) on which decisions are to be made.
24. **Probability**—An estimate of the frequency of occurrence of an event, e.g., probability of n sample units out of a population being positive for *Salmonella,* expressed as a value from 0 to 1. For a sampling plan, assumption of a particular probability distribution (i.e., binomial or Poisson) allows the estimation of the computation of the probability of lot acceptance versus lot quality in an operating characteristic curve.
25. **Producer's risk**—The risk that a producer takes that a lot will be rejected by a sampling plan even though the lot conforms to requirements.[25] This is called an alpha risk, and in many plans[8,21,23,25] this risk is set at 5%. Another type is the producer's Type I risk, which is the risk that the lot will be accepted even though it does not conform to requirements.
26. **Proportion defective units (P)**—The number of defective units divided by total units in a lot. Proportion defectives P or percent defectives (100P) are often plotted as the abscissa on an OC curve.
27. **Random sample**—A sample that was chosen in such a way that every sample or unit in the lot had equal chance of being selected. This is often achieved with the aid of a random number table.
28. **Representative sample**—In the widest sense, a sample which is representative of a population, and not merely a portion of it. Some confusion arises if "representative" is regarded as meaning "selected by some process that gives all samples an equal chance of appearing to represent the population." In fact, a representative sample is one that is considered typical of a population in respect of certain characteristics, however chosen. To obtain a truly representative sample, there are four requirements: (a) determine the location of sampling points critical to the population; (b) establish a sampling method representing the population characteristics; (c) select the sample size; and (d) specify the frequency of sampling.

29. **Sample unit**—The smallest definable part of a lot, also called a unit. For example, a singleton or a package is a sample unit. When lots are bulk packaged in bins, barrels, bags, etc., then the sample unit is arbitrary and may depend on the sampling device. The use of the word in this context should be differentiated from the analytical sample unit specified by the analytical method.
30. **Sampling plan**—A design that indicates the number of units to be collected from each lot and the criteria to be applied in accepting or rejecting the lot. A *single sampling plan* requires the lot to be judged on the basis of one set of sample units (Section 2.512). A *double sampling plan* (Section 2.513) is a sampling inspection in which the inspection of the first sample leads to a decision to accept, to reject, or to take a second sample. The inspection of a second sample, when required, then leads to a decision to accept or reject. Two or more such collections would mean that a multiple sampling plan was being used. Another procedure where units are drawn one by one (or in groups) and drawing at any stage determines whether to accept, reject, or continue sampling is called a sequential sampling plan.
31. **Sample size**—The number of samples selected from a lot. The sample size should accurately describe the population, and should be the most economical possible to reach a certain level of accuracy. If the testing method is nondestructive, it may be possible to conduct sampling of 100% of the units in a lot. Typically, microbiological testing destroys product, and thus 100% sampling is not possible. Sample size can be determined by using sampling plan master tables, such as those found in Military Standard 105D, published by the American National Standards Institute.[23]
32. **Stratified random sampling**—A procedure for sampling where the lot is divided into parts or strata that differ with respect to the characteristics under study. In some cases, this is a way of improving the estimate of lot quality.
33. **Zero defective tolerance**—An evaluation system implying that a lot must be free of the undesirable characteristic or defect. All units in the lot must be tested in order to ensure zero defectives. This can be performed only where the test is nondestructive. The requirement of zero defects is definitely not possible in microbiological sampling plans since these cannot be applied to all units in a test lot.

2.512 Single Sampling (Two-Class) Attribute Plans

Single sampling procedures are useful in food inspection since, in many production processes, the lot can be sampled and tested only once. In addition, the results of microbiological tests are clearly defined as attributes. For example, an attribute such as the presence or absence of a microorganism, e.g. *Salmonella*, is frequently reported. In other instances, a certain level of organisms may be acceptable. For example, a unit may be acceptable if it has less than 3 *E. coli* colonies /g of sample.

Single sampling attribute plans also have the advantage that the true distribution of the variable in question (e.g., *E. coli*/g) is not required. Single sample plans and the multiple sample plans in the next section (2.513) can be evaluated by using the hypergeometric, binomial, or Poisson distributions. The choice of distributions used to compute the probability relationships[8,9] depends on the number of units (N) in a lot. When N is large relative to the sample size (n), or when the number of defective samples in a lot is small, a Poisson distribution can be assumed, serving as an approximation of the binomial distribution. The type of lots acceptable for single sample procedures are assumed to be large, of homogeneous quality, and to satisfy the conditions above.

Steps for choosing and applying an attribute sampling plan are as follows:

- Select the measurements of interest.
- Define the sampling units that constitute a lot.
- Determine a value of consumer's and/or producer's risk to ensure the lot quality (LQ or AQL) desired.
- Obtain an estimate of process average.
- Compute or select a plan that meets the risk and lot quality requirements.
- Calculate the OC curve.
- Apply the plan on a group of randomly selected units from a lot.
- Maintain records on the process average (if you are a producer) and make changes in the plan as necessary.

Figure 1 presents the operating characteristics (OC) curves for seven single sampling plans. These OC curves for the sampling plans will be used to help illustrate the process of choosing a plan. The reader should refer to Dodge and Romig,[8] Duncan,[9] and International Commission on Microbiological Specifications for Foods[12] for presentations of complete sets of plans where OC curves for additional values of n or c are given. Figure 1 has an acceptance number (c) of zero, or zero tolerance, which means that any positive results on a test lot will result in the rejection of that specific lot.

As an example, consider that a lot N > 1000 units is to be analyzed for coliform MPN/g and a unit is to be called defective if it

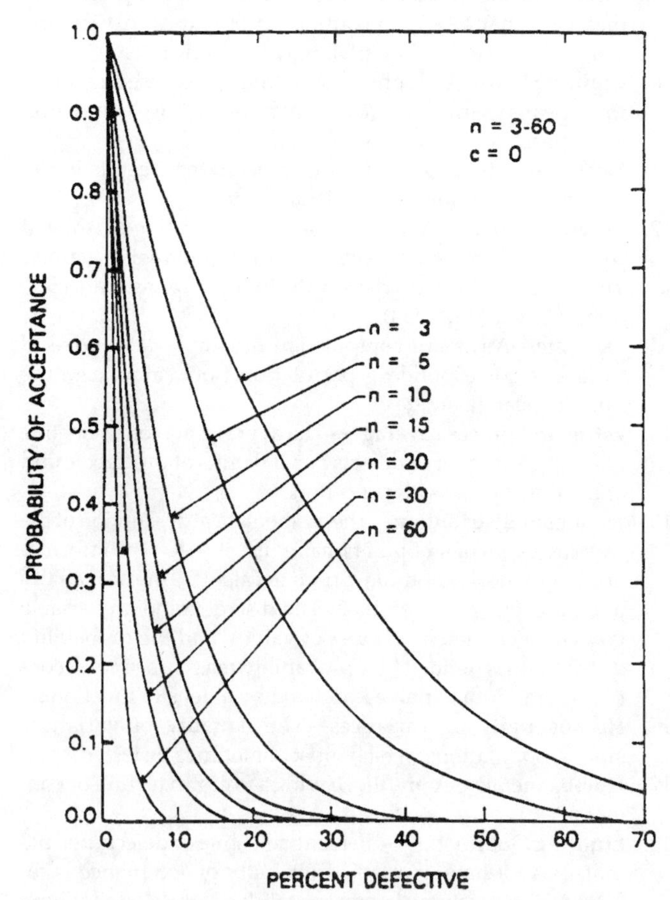

Figure 1. *Single sampling (two-class) attribute plans for sample sizes n = 3, 5, 10, 15, 20, 30, 60, and c = 0.*

has MPN ≥ 100/g, and to be called acceptable or nondefective if it has an MPN < 100/g. A sampling plan is wanted to define the number of units (n) to sample where the number of defectives (c) equals zero and the probability (consumer's risk) of accepting lots with 8% or more defective units is 10%. Using these criteria, the sampling plan to be chosen (Figure 1) is n = 30 and c = 0.

Since microbiological tests are destructive (the sample unit cannot be used after testing), cost of sampling may be balanced with the cost of the risk. For example, a plan may be designed to inspect incoming raw materials and, on the other hand, may be designed to sample consumer products before releasing lots for sale. The measurement (i.e., coliform MPN/g) may be the same, but the choice of plan will be affected by other factors such as (a) processing conditions, (b) potential health hazard of the products, (c) persistence of organisms under different storage conditions, and (d) type of plan used by regulatory agencies to inspect the same lots. The n = 15, 20, 30, and 60 with c = 0 plans in Figure 1 are in common use[11,12] to test for *Salmonella.* The n = 3 and 5 with c = 0 plans could be used for screening raw materials.

Some analytical procedures are sufficiently sensitive to measure the presence of a single organism when the sampling units are pooled. A positive test indicates that one or more of the units was positive from n pooled units. This produces the same decision for a plan (e.g., n = 5, c = 0) as if the units had been analyzed separately. Pooling of sampling units cannot be done when c > 0 or when a positive sample is defined quantitatively (e.g., *E. coli* colonies/g). Calculations with OC curves are found in the references.[8,9]

It should be noted that as sample size n increases, the OC curve becomes idealized, with a probability of acceptance being 100% with 0 defects, and falling to a probability of 0% when the fraction of defects is > 0. Also, as the acceptance number c decreases, the OC curve shifts to the left, evidence that with a given number of defective units, the probability of accepting the lot decreases.

2.513 Multiple Sampling (2-Class) Attribute Plans

Sampling plans of this type are designed to inspect lots based on multiple samplings. The lot is accepted or rejected based on sequential decisions. As an example, consider a double sampling plan (Figure 2) of $n_1 = 10$, $c_1 = 0$ and $n_2 = 6$, $c_2 = 1$. This plan requires that 10 units be analyzed and the lot be rejected if any unit is defective. If only 1 of the 10 units sampled from the same lot is defective, then 6 additional units are tested. If one or more of the 6 units are defective, then the lot is rejected; otherwise it is accepted. The operating characteristic curve for this plan is given in Figure 2. In this example, if there were 24% defectives in the lot, the plan would accept 10 of 100 lots and the average outgoing quality would be 5% defectives.

Figure 2 shows the operating characteristic curve for the double sampling plan and one for a single sampling plan (n = 7, c = 0) that have the same average (5%) outgoing quality limit (AOQL). Although the AOQLs are equal in the two plans, it can be seen that the probability of accepting a defective lot is lower for the single sampling plan. In addition, the choice of taking 15 units (c = 0) for a single sampling results in a lowering of the AOQL from 5% to 2.5%. Thus, in choosing sampling plans, one must consider the probability of acceptance of a defective lot that will be tolerated.

A double or other multiple sampling plan requires that the lot of units be available if extended sampling is necessary. This is not practical in many types of foods. However, the advantage of multiple sampling plans is that lots of good quality (i.e., low coliform count) will on average require fewer samples to be tested than if tested by a single sample plan giving equal protection. For example, a two-stage or double sample plan of $n_1 = 10$, $n_2 = 6$ might require that just the first 10 units be analyzed. Thus, the average sample number tested would be decreased.

Further information on the design of multiple sampling plans can be found in Dodge and Romig[8] and Duncan.[9] A plan can be devised if the desired average quality limit (LQ), consumer (AQL) and/or producer's risk has been selected. Dodge and Romig[8] present the operating characteristic curves for double sampling plans and several outgoing quality limits (0. 1 to 10%, p. 112). The figures allow a quick assessment of risks for each plan. The diagram shown in Figure 3 is an interpretation of double sampling plans.[8]

2.514 Three-Class Attribute Plans

Three-class attribute plans were developed by Bray et al.[7] in conjunction with the ICMSF[12] to be used with methods recommended by Thatcher and Clark[17] who stated[12] that the "test is concerned primarily with plans that may be applied to lots presented for acceptance at ports or similar points of entry." It was assumed that very little, if any, information would be known about the quality of lots and, thus, attribute plans, two-class plans (Section 2.513), and three-class plans would be applicable. These plans are

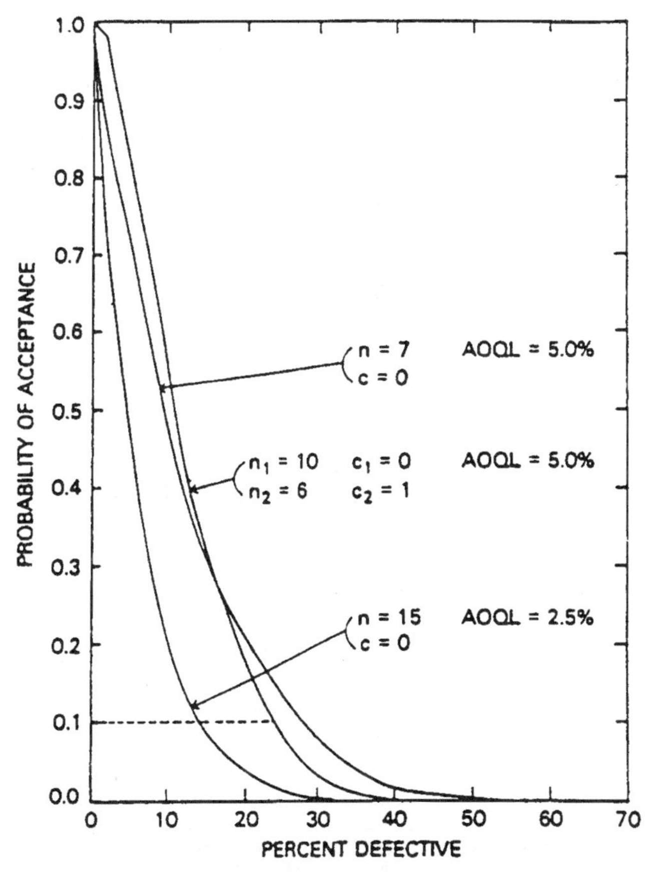

Figure 2. *Operating characteristic curves for two single and one double sampling plan with average outgoing lot quality limit 2.5 and 5.0%. The c = 0 for a single sampling plan indicates that a lot will be rejected if a single unit of the sample is positive. Rejection of the lot will occur if two or more units of the first or second sample are positive for the double sampling plan.*

also useful for inspecting lots within a country or corporation where more information is known about a lot.

Three-class attribute plans differ from those described in Section 2.513 by having two microbiological limits that create three classes of product. Bray et al.[7] noted that the choice of limits (e.g., for coliform MPN, a quality or sanitation indicator) is difficult. Nonetheless they state that "most scientists can provide two numbers: one below which they have little or no concern and a higher value above which they clearly begin to have critical concern. If we denote the lower level by m and the larger one by M, then the set of values in the range [m, M] can be considered safe." Observations falling in the regions defined by the two limits are called acceptable, marginal, and unacceptable. The acceptable results have values equal to or less than m ($\leq$ m). Marginal observations fall between greater than m and equal to or less than M (e.g., [m, M]). Unacceptable results are greater than M. Three-class attribute plans can be specified by a sample size n, the number of units allowed (c_1) between limits $\geq$ m and $\leq$ M, and the number of units allowed (c_2) equal to or above M. It is assumed that all units $\leq$ m are acceptable. However, as stated above, the value of M is the decision point for this type of sampling plan. The value of $c_2 = 0$ was set in all plans discussed in *Microorganisms in Foods 2*[12] and in the discussion below. Thus, the sampling plans are noted as n, c where c is the number of marginal samples allowed. The three-class attribute sampling plans found in *Microorganisms in Foods 2*[12] present a detailed description of the type of microbiological methods to be recommended, potential risk of contamination, and subsequent hazard for a wide group of products and the choice of limits m, M for these products. Suggested sampling plans are presented for fish, fishery products, vegetables, dried foods, frozen foods, milk, milk products, raw meats, processed meats, shelf-stable canned foods, and fresh or frozen raw shellfish. Different measurements are of interest for each product and include aerobic plate count, coliforms, fecal coliforms, *Salmonella*, *V. parahaemolyticus*, *Staphylococcus aureus*, *B. cereus*, *C. perfringens*, and *C. botulinum*. Limits [m, M] were chosen for product-measurement combination and plans were established based on risk.

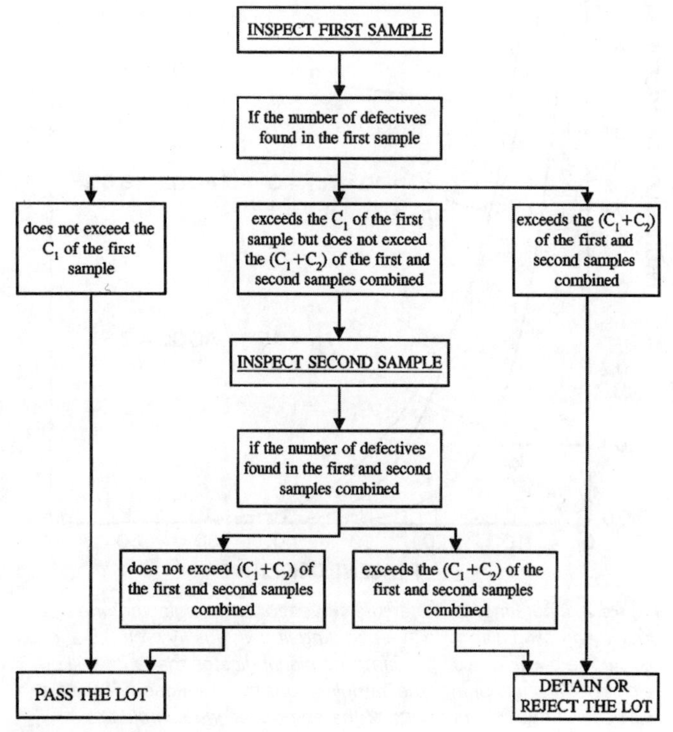

Figure 3. *Interpretation of a double sampling plan.*

The ICMSF[12] categorized the types of microbiological hazards and conditions to which a lot of food would be exposed. This categorization is an aid to the microbiologist when considering a choice of sampling plan. These hazards were defined as follows:

1. No direct health hazard, utility (e.g., general contamination, reduced shelf life, and spoilage).
2. Health hazard low, indirect (indicator).
3. Health hazard moderate, direct, limited spread.
4. Health hazard moderate, direct, potentially extensive spread.
5. Health hazard severe, direct.

These hazards are linked with three conditions of risk (reduction, no change, or increase) that reflect how a food was expected to be handled and consumed after sampling. The combination of the types of hazards and conditions of risk yields 15 cases where suggestions of sampling plans and limits could be presented for products, although only a few cases would be realistically applicable for a given product and measurement.

Sampling plans for hazards 1, 2, and 3 are three-class plans in which n = 10, c = 1; n = 5, c = 3; n = 5, c = 2; and n = 5, c = 1. In these cases (of the 15 defined cases) limits [m, M] are set to define marginal quality. When the hazards are more severe (i.e., 4 and 5) two-class plans in which n = 60, c = 0; n = 30, c = 0; n = 20, c = 0; n = 15, c = 0; n = 10, c = 0; and n =5, c = 0 are suggested. As noted in Section 2.513, the two-class plans above mean that n units are analyzed and the lot is accepted only if all units are negative for some characteristic (e.g., *Salmonella*). The operating characteristic curves are given for the two-class plans above in Figure 1.

The approach of the ICMSF[12] on three-class plans was to define risks, suggest limits for a wide variety of specific cases, and suggest plans to be employed. This approach differs from other texts on sampling plans in that the mathematical tools for computing and comparing plans are presented, and the reader must make a choice of various conditions. Those desiring to compute their own three-class plan should consult Bray et al.[7]

Operating characteristic contours presented as two-way tables are used rather than operating characteristic curves for three-class attribute plans. These contours reflect the fact that the true marginal percent can vary from 0 to 100% at the same time the proportion of defective units (value $\geq$ M) varies from 0 to 100% with the obvious restriction that the sum of percent good units, percent marginal units and percent bad units must be 100%.

The choice of an ICMSF sampling plan is straightforward. First, specify the food and measurement (e.g., freshwater fish and fecal coliforms). Second, specify the risk. ICMSF[12] suggests case 4 in this example, which reflects low indirect health hazard at reduced risk. The sampling plan is n = 5, c = 3. The last step is to choose limits m and M for the product to be sampled. ICMSF [12] recommends m = 4 and M = 400 for freshwater fish. The complete plan is listed (e.g., n = 5, c = 3; m = 4, M = 400), once the product and measurement are specified.

To derive a three-class plan, one must consider (a) the assignment of risk to the product, (b) the choice of sampling plan based on a comparison of probability of acceptance versus percent marginal units (assuming we will not accept any bad units), and (c) the values of m and M for the product. The choice of risk and the setting of m and M may differ depending on the purpose of the sampling. One set of FDA sampling plans for *Salmonella* has values of n = 15, c = 0; n = 30, c = 0; n = 60, c = 0.[11]

2.515 Variables Sample Plans

A variables plan can be used when the probability density function of a measurement is known, and a majority of published plans are computed when the distribution of the variable or its transformation is normal (Gaussian). An advantage of the variables plans over single sampling plans is that fewer samples are required, resulting in a lower cost to achieve the same protection as a single attribute plan. Disadvantages include the calculations involved in evaluating a lot, the different calculations required for each variable (coliform, *E. coli*, etc.), and the requirement that the probability distribution must be known or assumed for each measurement. The last two requirements may prove to be the most inconvenient. A series of three or four variable (e.g., aerobic plate count, coliform, *E. coli, and S. aureus*) measurements may be required on a product. The transformation that ensures a normal distribution may not be known, and the estimates of variance are often different for each measurement. For these reasons, variables sampling plans are not widely used in the food industry for microbiological measurements. For further information about derivation of variables sampling plans, refer to Duncan,[9] Kramer,[14] and MIL-STD 414.[22]

2.516 Sampling Procedures for Low Contamination Levels

The microbiologist often encounters a situation where production lots of a product with low incidence of a pathogen (low number of contaminated units) must be sampled (e.g., 1 positive sample unit in 500). How many sample units should be taken to have a high probability of detecting the pathogen? Suppose 200 sample units have been analyzed and all are negative. What can be reported about the lot? Clearly, the whole lot may not be free of the pathogen. The same situation also arises in sterility testing. In both of these situations, it is assumed that the pathogen or contaminant can be detected if present in an analytical sample (e.g., 25 g) and the results will be reported as positive or negative. Other necessary assumptions are given in the discussion of single sampling attribute plans (Section 2.512).

Table 1 lists the number of sample units (n) needed to detect a positive result at a given level of fraction positive analytical units in a lot. This table can be used to determine how many units should be analyzed in order to detect a positive at a given fraction positive level. Suppose a regulator knows that the fraction positive of a certain pathogen is 0.04. How many sample units should be obtained in order to find a positive unit with probability 0.95? In Table 1, we go down the left hand (fraction positive samples) to 0.04 and over two columns to (1-Pr) = 0.95. Pr is the probability that all n units are negative. The required sample size is 75.

Let us examine the situation from another point of view. Suppose a producer knows that pathogens can be found in his product with a frequency of 6 in 1,000 units. The regulator takes 30 sample units per lot. What is the probability (Pr) that all 30 samples will be negative? The answer is $Pr = e^{-(30)\ (0.006)} = 0.84$. Thus, the regulator has a 0.16 chance of detecting a positive from a lot with a frequency of defective units of 0.006. Simply stated, 84% of the time that 30 samples are collected from this defective lot, no defects will be found, while defective units will be found 16% of the time.

Another situation arises when a sample of n units is examined and all are negative for the pathogen in question. What can be said about the rest of the lot in terms of any units being defec-

Table 1. Number of Samples Needed to Detect a Fraction Positive[a] with Probability 0.90, 0.95, and 0.99 Where at Least One Positive Result Occurs

Fraction positive[b] samples (P)	Number of analytical units to be tested (n) Probability (1-Pr) 0.90	0.95	0.99
1.0	3	4	4
0.9	3	4	5
0.8	3	4	6
0.7	4	5	7
0.6	4	5	8
0.5	5	6	9
0.4	6	8	12
0.3	10	10	16
0.2	12	15	23
0.1	23	30	46
0.09	26	34	51
0.08	29	38	58
0.07	33	43	66
0.06	39	50	77
0.05	46	60	92
0.04	58	75	115
0.03	77	100	154
0.02	115	150	230
0.01	230	299	461
0.001	2,303	2,996	4,605
0.0001	23,026	29,963	46,052

[a] Can be computed from Thorndike Table, p. 35 of Dodge and Romig.[8]

[b] The fraction of positive units (e.g., 90 positives in 100 analytical units).

Table 2. Fraction Positive Samples When the Probability[a] Is That All n Samples Are Negative

Number of analytical negative units (n)	P = Fraction positive samples Probability (Pr) 0.10	0.05	0.01
3	0.77[b]	1.00	1.50
5	0.46	0.60	0.92
10	0.23	0.30	0.46
15	0.15	0.20	0.31
20	0.12	0.15	0.23
25	0.092	0.12	0.18
30	0.077	0.10	0.15
35	0.066	0.086	0.13
40	0.058	0.075	0.12
45	0.051	0.067	0.10
50	0.046	0.060	0.092
100	0.023	0.030	0.046
200	0.012	0.015	0.023
300	0.0077	0.012	0.015
400	0.0058	0.0075	0.012
500	0.0046	0.0060	0.0092
1000	0.0023	0.0030	0.0046

[a] Can be computed from Thorndike Table, p. 35 of Dodge and Romig.[8]

[b] Rounded to two significant digits.

tive? Table 2 shows one way of expressing the result. The fraction positive (defective units) per lot can be related to the probability, Pr, when all n sample units examined are negative. If 100 sample units were negative, then there is only 1 chance in 10 (Pr = 0.10) that the fraction positive exceeds 0.023 per lot. Conversely, the probability is 9 chances in 10 that the fraction positive is 0.023 or lower. Stated yet another way, there is a probability of 10% (at most) that 230 sample units in 10,000 have a pathogen. So even in a lot size of 10,000, there is some chance (although only 10%) that a positive unit will reach a consumer. There is an additional, though small, probability that the consumers will receive the unit. This is because the distribution of microbial contaminants in the lot is not uniform. The fact that contaminants are not evenly distributed throughout the lot increases the chances of not detecting them, and thus increases the probability of accepting the defective lot.

If the samples were homogeneous and the organisms could be assumed to be randomly distributed throughout the lot, then an estimate of concentration could be obtained (Chapter 6). For example, assuming that one *Listeria monocytogenes* organism occurs in each 10 mL of milk, if n samples of milk are taken from a well-agitated bulk milk tank, then the concentration of pathogen could be estimated. If the analytical sample is 1 mL, how many sample units are required to obtain at least one positive sample with probability 0.95? From Table 1 we see that 30 samples should be examined.

Let us further suppose that 2 of the 30 results are positive upon examination. Based on the results of a single dilution, the estimate of concentration of *L. monocytogenes*/mL = ln(n/# negatives) = ln(30/28) = 0.069. Thus, sampling plans are useful not only for lot inspection, but in determining risk and in predicting a specified level of organisms associated with a given probability, although this is limited to lots where the organism is evenly distributed. In addition, the concentration can also be estimated in some restricted cases.

2.517 Summary

The discussion of sampling plans in this section has been modified for application to microbiological measurements. A general survey of plans is presented, but some of these procedures do not lend themselves directly to microbiological sampling in foods. The brief discussion of multiple sampling plans does not mean that these procedures may not be found useful. The sampling for microorganisms is destructive, the results may be delayed for several days because of incubation requirements, and the foods may be perishable. Thus, single sampling plans are generally best suited to most situations.

Variables sampling plans are subject to additional disadvantages beyond those stated in Section 2.515. The general level of statistical knowledge in the typical food facility may prevent the use of these plans. Calculations needed to choose and apply the variables plans may prove burdensome. The two- and three-class attribute plans are easy to apply and thus should aid in the spread of statistics in microbiological sampling of foods. Additionally, the ICMSF[12] has presented extensive suggestions of specific plans and microbial criteria to be used on a wide variety of products. This reduces the amount of work necessary in calculations and time spent in plan selection. Thus, the single sampling two-class and three-class attribute plans have the most utility in sampling of foods for microbiological analysis. Individuals who wish to derive their own sampling plans are advised to use Dodge and Romig,[8] Duncan,[9] and MIL-STD 105.[23]

2.52 Sampling Procedures

It is necessary to consider the physical state of the product to be sampled (e.g., dry, semisolid, viscous) and the reason for sampling and testing before taking samples so that the number of units will be representative and/or statistically significant for the intended use.[12] The sampling procedures described in this section are those that generally apply. Unusual sampling procedures that may be required for specific foods are given in the chapters dealing with particular microorganisms or commodities.

2.521 Finished Products

Consumer packages of foods should be sampled from original unopened containers of the same processing lot. Record processing information and product code of samples on forms submitted with samples. The practice of submitting unopened containers prevents contamination that might be introduced by opening at the sampling location. In addition, it allows laboratory examinations to be performed on products and packages or containers as they are offered to the public.

2.522 Bulk Liquid Material

If the products are in bulk form or in containers of a size impractical for submission, aseptically transfer a representative sample portion to a sterile, leakproof container. Before drawing a sample, aseptically mix the food mass to ensure that the sample is as homogeneous as possible. If adequate mixing or agitation of the bulk product is not possible, multiple samples should be drawn from the bulk container. Do not fill sample containers over bulk containers of food. Care must be taken to select sample containers with enough capacity to accommodate the needed sample volume when the container is 3/4 full. Avoid the use of glass sample containers. Thermometers used in bulk food containers should be sanitized before use.[4] Cool perishable samples to 0° to 4.4°C quickly if they are not already refrigerated. Metal thermometers are preferred since breakage of a mercury thermometer would contaminate the product.

When appropriate, a temperature control sample should be prepared and submitted with the samples to be tested. If a temperature control sample is needed for frozen samples, use a container of ethylene glycol. The sample should be sealed in its container so that the container will not break or leak and introduce extraneous contamination. If the sample is to be examined for a regulatory purpose, the sample container must be sealed so that it cannot be opened without breaking the seal. In addition, an empty sterile sample container should be submitted as a container control.

2.523 Bulk Solids or Semisolids

Sampling of dry or semisolid foods should be done with sterile triers, spoons, or spatulas. Sterile tongue depressors may be substituted for spatulas. Aliquots from several areas of the food under examination should be taken to ensure a representative sample. Care should be taken to protect this type of sample from excess humidity.

2.524 Frozen Bulk Materials

Frozen bulk foods may be sampled with sterile corers, auger bits, and other sharp sampling instruments. A presterilized auger bit or hollow tube may be used to obtain enough material for analysis. Frozen samples should be kept frozen until arrival at the laboratory. Thawing and refreezing of samples must be avoided.

A suitable procedure for obtaining test units of frozen foods (particularly from larger samples) is to use an electric drill combined with a funnel.[1] The sterile auger bit is inserted through a sterile plastic funnel (which has been cut off so the hole is just slightly larger than the bit) held against the frozen sample (Figure 4). The frozen shavings are conveyed to the surface and collected in the funnel. The shavings then can be placed in a sterile sample container. For larger solid food samples—frozen or unfrozen, test units should be taken aseptically from several areas using sterile knives and forceps. These portions should be mixed as a composite to provide a sample representative of the food to be evaluated.

2.525 Line samples (in-process samples)

1. **Liquids.** Sterile metal tubes or dippers may be suitable sampling instruments at certain plant locations for liquid food samples. Disposable plastic transfer pipets can also be used. A special line sampling technique involves the use of a disposable sterile hypodermic needle and syringe. The needle is inserted into a rubber closure of a stainless steel nipple. The nipple can be clamped on or permanently located at any desired spot.[10]

   Sampling cocks on holding tanks and product pipelines may be used, but disinfection and material flow-through of the sampling cock must be assured before collecting the sample.
2. **Solids.** Sampling of solid line samples may be accomplished using the same equipment and procedures as would be used for bulk solid or semisolid products. Automatic sampling devices are available for powdered products and other solid products not requiring refrigeration. When automatic samplers are used, the manufacturer's directions must be strictly followed.
3. **Special purpose samples (foodborne outbreaks, consumer complaints).** In some instances samples are tested as part of a foodborne disease outbreak investigation or on the basis of a consumer complaint. Consult a clinical or public health laboratory. Such situations may be the subject of legal proceedings, and laboratory personnel may be required to testify concerning the results of their examinations. If the record of sample collection and handling is incomplete, or if samples are received in a partially decomposed state or having been temperature abused, the laboratory results may be of little or no value.

   In outbreak situations, collect all suspect foods. If there are no leftover foods, try to get samples of items prepared in a similar manner subsequent to the suspected food. Collect ingredients or raw items used in the suspect food, if available. All of these foods should be held under suitable conditions until an analysis of the attack rate data and other facts can define more accurately the suspect food(s).[12] The original containers in which the foods were found also should be collected, labeled, and submitted for examination. Other specimens as well as foods from outbreak investigations are essential. Human specimens may include stools, vomitus, and serum, depending upon the suspected cause of illness. These specimens should be collected in sterile leakproof containers and should be properly identified as to the patient's name, type of specimen, and date of collection. Transport media is sometimes indicated. Aseptic techniques always should be used to obtain samples even if the foods have been mishandled grossly.
4. **Samples for water activity measurements.** Collect samples intended for water activity in sealable, vapor tight containers. Obtain small, unopened, retail-size packages if possible. When sampling from bulk containers, work quickly to minimize changes in water content.
5. **Samples for pH measurements.** Samples to be tested for their pH and/or titratable acidity should be collected into tightly sealed containers. If the material to be tested is undergoing fermentation or some other gas producing reaction, use a vented container or flexible plastic bag with ample space for expansion.

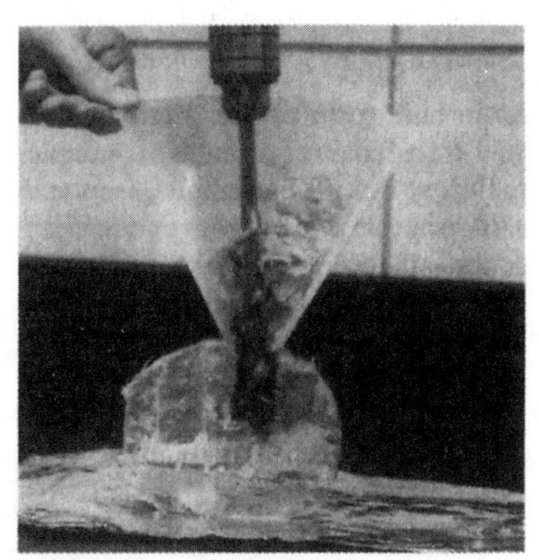

Figure 4. *Funnel collection apparatus.*

2.526 Storage and Shipment of Samples

When it is necessary to store samples prior to shipment, a storage area for frozen (–20°C) and for refrigerated (0° to 4.4°C) samples should be available. Waterproof labels should be used on sample containers to prevent loss of labels. Direct labeling of containers is preferred.

Whenever possible, samples should be submitted to the laboratory in original unopened containers.

Samples should be delivered to the laboratory as rapidly as possible. The condition, time, and date of arrival at the laboratory should be recorded. The samples should be packed to prevent breakage, spillage, or change in temperature. Since laboratory examination of food samples requires preparatory work, the laboratory should be given advance notice, if possible, of the number and types of samples being submitted. If the product is in a dry condition or is canned (flat or normal), it need not be refrigerated for shipment. Swollen containers should be shipped with coolant. The product label may also indicate whether refrigeration is required. Samples not requiring refrigeration or freezing may be packed in a cardboard box using appropriate packing material to prevent breakage. Refrigerated products must be transported in an insulated shipping container with sufficient refrigerant to maintain the samples at 0° to 4.4°C until arrival at the laboratory. Water frozen in plastic containers or cold packs serves well for 0° to 4.4°C shipping and should last 48 hr under most conditions. Do not use loose ice as this may cause product contamination if the container breaks or leaks. Dry ice may be used for longer transit times if the sample is separated from the dry ice packing material to avoid freezing. Refrigerated products should not be frozen, as destruction of certain microorganisms will occur. Frozen samples can be kept frozen by ensuring that the samples are shipped with dry ice. Frozen samples collected in plastic bags, however, must not come in direct contact with dry ice as the plastic bag will become brittle and subject to rupture from the extreme temperature. Use paper or another suitable material to protect the sample. Such samples should be transported to the laboratory by the fastest possible means. Mark the shipment of samples as "Perishable," "Packed in Dry Ice," "Refrigerated Biologic Material," or "Fragile," as appropriate. Label the shipment according to federal postal service rules and Department of Transportation regulations.[26,27]

2.527 Preparation of Homogenates

Use aseptic technique. Prior to opening, swab the exterior area of the container with 70% ethanol to destroy microorganisms that might contaminate the sample.

Samples should be examined promptly. When initiation of analysis must be postponed, store frozen samples at –20°C until they are to be examined. Refrigerate unfrozen perishable samples at 0° to 4.4°C for not more than 36 hr.

Store nonperishable, canned (normal, flat), or low-moisture foods at room temperature until ready for analysis. Frozen samples should be thawed at refrigeration temperatures (≤4.4°C) for no longer than 18 hr in the original container in which received.[4] If the sample must be removed from the original container, it should be done aseptically. Alternatively, higher temperatures may be used for a short period of time, but the temperature must remain low to prevent destruction of microorganisms (<40°C for ≤15 min). Frequent shaking of samples is necessary when samples are thawed by the alternate procedure. A thermostatically controlled water bath with agitator is recommended for the rapid thawing of samples.

Liquid or semiliquid samples in containers that have an airspace can be mixed by rapidly inverting the sample container 25 times. Sample containers that are 2/3 to 3/4 full should be shaken 25 times in 7 seconds over a 30 cm arc. The interval between mixing and removing the test portion should not exceed 3 min. To ensure a homogeneous sample when no airspace is present, aseptically open the container and pour the product from the filled container back and forth into a sterile container three times.

Dry samples should be aseptically stirred with a sterile spoon, spatula, or other utensil to ensure a homogeneous sample. Test portions of nonviscous liquid products (i.e., viscosity not greater than milk) may be measured volumetrically using a sterile pipette (11 mL into 99 mL, or 10 mL into 90 mL, or 50 mL into 450 mL). If the pipette becomes contaminated before completing the transfer, replace it with a sterile pipette. Do not insert the pipette more than 2.5 cm below the surface of the sample. The pipette should be emptied into the diluent by letting the column drain from the graduation mark to the rest point of the liquid in the tip of the pipette within 2 to 4 sec, then touch the lower edge of the pipette tip against the inside of the neck of the dilution container. Do not blow out the last drop or rinse the pipette in the dilution fluid.[4] When measuring products having a viscosity similar to milk, the last drop should be blown from the pipette. For viscous liquid products, the test portion for the initial dilution should be aseptically weighed (11 ± 0.1 g) into a sterile 99 mL dilution blank (or 10 ± 0.1 g into 90 mL, or 50 ± 0.1 g into 450 mL). This provides a 1:10 dilution.

Test portions of solid or semisolid foods should be 50 ± 0.1 g. The 50 ± 0.1 g test portion should be weighed aseptically (using sterile forceps or spatulas) into a sterile tared blender cup[11,12]; then add 450 mL of sterile diluent. A variety of diluents may be used depending upon the nature of the product. Those most commonly used are Butterfield's phosphate buffer and 0.1% peptone water. When analyzing for specific organisms, other diluents may be appropriate, e.g., 3% NaCl for *Vibrio parahaemolyticus*. (Refer to the chapters on specific groups of organisms for other diluents.) When analyzing fatty foods or lump-forming powder, wetting agents such as Tergitol Anionic-7 (1% weight/volume) may be included in the diluent to promote emulsification. Blend for 2 min at low speed (approximately 8000 rpm) to disperse the material.[11] The blending time may vary depending on the type of food.[12] Some blenders may operate at speeds lower than 8000 rpm. It is preferable to use a higher speed for a few sec initially. No more than 15 min should elapse from the time the sample is blended until all dilutions are in appropriate media.

Optionally, if the entire food sample is less than 50 g, weigh to the nearest 0.1 g portion approximately one-half of the sample into a sterile tared blender cup. Add sufficient sterile diluent to make a 1:10 dilution (i.e., add an amount of diluent equal to nine times the weight of the test portion in the blender cup). The total volume in the blender cup must cover the blades completely. Blend as described above.

If the sample is not homogeneous, weigh 50 g from a representative portion of the package into a sterile, tared blender cup or analyze each portion of food separately Proceed as described above. Caution should be exercised in the blending step to prevent excessive heating. The amount of heating may vary with foods of different consistencies and may be expected to increase if blending times greater than 2 min are required. Chilled diluent (i.e., tempered in an ice-water bath) may be employed to decrease the chances of excessive heating.

Stomaching is acceptable as an alternative to blending in preparing the food sample homogenate.[16] In this procedure, the food sample with diluent is placed in a clean, preferably sterile plastic bag. The plastic bag is positioned within the stomacher and pummeled for 1 to 2 min. Because the sample is contained in a plastic bag, the developers recommend that samples with bones or other sharp or protruding objects not be prepared by stomaching. Thirty foods were evaluated using this procedure to determine the usefulness in a regulatory agency's laboratory. Results indicated that only certain food homogenates should be prepared using this procedure.[2]

In some solid food products, the microbial flora is restricted primarily to the surface area. More accurate enumeration of these microorganisms may be obtained by rinsing the sample with sterile diluent rather than by blending. This can be accomplished by placing the sample in a suitable sterile container (plastic bag or sealable bottle) and adding a volume of sterile diluent equal to the weight of the sample. The container is then shaken in a manner similar to that used for preparing an initial dilution of a liquid food sample. Each mL of "rinse" thus prepared represents 1 g of sample.

See Chapter 6 for preparation of further dilutions and plate count procedures. Use media recommended in the specific chapter for the organisms of interest.

2.528 Preparation of Samples for Water Activity Testing

The ideal preparation of samples for water activity measurements is to grind the material to a fine consistency before testing. It is important to avoid heat build-up in the sample during the grinding process. Certain emulsions such as oil/water emulsions may be difficult to measure unless the water phase can be separated by low temperature cycling or centrifugation. Judgement must be used in evaluating the accuracy of measurements done on the water phase of an emulsion. In any case, the prepared sample should be added to the test chamber of the water activity meter quickly to avoid an exchange of moisture with the air. If prepared samples will not be tested promptly, store them in vapor tight, sealed containers. Avoid storage in high humidity environments.

When performing the analysis, transfer sample portions to the test instrument sample holder and follow the manufacturers instructions. A reliable reading may take from 5 minutes to several hours depending on the type of instrument in use. Proper maintenance and calibration of water activity instruments requires some skill and experience. High quality training and support is available from the instrument manufacturers. A complete description of water activity testing can be found in the work of Troller and Christian,[18] and the symposium series by Rockland and Beuchat.[15] See Chapter 64 for a further discussion of water activity measurement.

2.529 Preparation of Samples for pH Determinations

Many types of liquid samples require very little preparation for pH determination. Semisolid samples, mixtures of solids and liquids, emulsions, and various types of marinated foods in oil, require special preparation steps. Semisolid samples can be blended to a thick paste and a small amount (no more than 20 mL/100 g) of distilled water added to provide a more fluid test portion. Mixtures of solids and liquids can be tested either by blending the mixture to a paste and measuring directly or by separating the solid and liquid portions using a U.S. standard #8 sieve. For marinated products in oil, separate the oil from the product, blend the solids (adding a small amount of distilled water if necessary) and test the paste. When attempting to determine the pH of emulsions, it may be necessary to separate the water phase for testing by low temperature cycling or centrifugation.

Temperature effects on the pH electrodes and the actual hydrogen ion activity will modify the pH readings from electronic pH meters. Instruments that have temperature compensation adjust the response of the electrodes. The ionic activity of the sample, however, cannot be corrected for. For accurate results, the standardization and actual determinations should all be done at the same temperature and within the range of 20° to 30°C. See Chapter 64 for a further discussion of pH measurement.

2.6 REFERENCES

1. Adams, D. M., and F. F. Busta. 1970. Simple method for collection of samples from a frozen food. Appl. Microbiol. 19:878.
2. Andrews, W. H., C. R. Wilson, P. L. Poelma, A. Romero, R. A. Rude, A. P. Duran, F. D. McClure, and D. E. Gentile. 1978. Usefulness of the stomacher in a microbiological regulatory laboratory. Appl. Environ. Microbiol. 35:89-93.
3. APHA. 1970. Recommended procedures for the examination of seawater and shellfish, 4th ed. Am. Public Health Assoc., Washington, D. C.
4. APHA. 1985. Standard methods for the examination of dairy products, 15th ed. Am. Public Health Assoc., Washington, D. C.
5. Bartlett, R. P., and H. H. Weatherspoon. 1964. USDA variables control chart plan applied to fill weights. Food Technol.18:40-42,44-45,48.
6. Bartlett, R. P., and J. B. Wegener. 1957. Sampling plans developed by United States Department of Agriculture for inspection of processed fruits and vegetables. Food Technol.11:526-532.
7. Bray, D. F., D. A. Lyon, and I. W. Burr. 1973. Three-class attributes plans in acceptance sampling. Technometrics 85:575.
8. Dodge, H. F., and H. G. Romig. 1959. Sampling inspection tables, 2nd ed. John Wiley & Sons, Inc., New York.
9. Duncan, A. J. 1959. Quality control and industrial statistics, rev. ed. Richard D. Irwin, Inc., Homewood, Ill.
10. Elliker, P. R., E. L. Sing, L. J. Christensen, and W. E. Sandine. 1964. Psychrophilic bacteria and keeping quality of pasteurized milk. J. Milk Food Technol. 27:69-75.
11. FDA. 1984. Bacteriological analytical manual, 6th ed. AOAC, Arlington, Va.
12. ICMSF. 1986. Microorganisms in foods, vol. 2. Sampling for microbiological analysis: Principles and specific applications, 2nd ed. University of Toronto Press, Toronto, Canada.
13. Kendall, M. G., and W. R. Buckland. 1960. A dictionary of statistical terms, 2nd ed. Hafner Publ. Co., New York.
14. Kramer, A., and B. A. Twigg. 1966. Fundamentals of quality control for the food industry, 2nd ed. AVI Publishing Company, Westport, Conn.
15. Rockland, L. B., and L. R. Beuchat. 1987. Water activity: Theory and applications to food. IFT Basic Symposium Series. Marcel Dekker, Inc., New York.
16. Sharpe, A. N., and A. K. Jackson. 1972. Stomaching: A new concept in bacteriological sample preparation. Appl. Microbiol. 24:175-178.
17. Thatcher, F. S., and D. S. Clark (ed.). 1968. Microorganisms in foods: Their significance and methods of enumeration. University of Toronto Press, Toronto, Canada.
18. Troller, J. A., and J. H. B. Christian. 1978. Water activity and food. Academic Press, New York.
19. U. S. Department of Agriculture, ARS. 1964. United States standards for sampling plans for inspection by attributes single and double sampling plans. U.S. Government Printing. Office, Washington, D. C.
20. U.S. Department of Agriculture. 1966. Accuracy of attribute sampling: A guide for inspection personnel. USDA Consumer and Market Service, Washington, D. C.
21. U.S. Department of Defense. 1954. Administration of sampling procedures for acceptance inspection (H-105). U.S. Government Printing Office, Washington, D. C.
22. U.S. Department of Defense. 1957. MIL-STD 414. Military standard, sampling procedures, and tables for inspection by variables for percent defective. U.S. Government Printing Office, Washington, D. C.
23. U.S. Department of Defense. 1963. MIL-STD 105D. Military standard, sampling procedures, and tables for inspection by attributes. U.S. Government Printing Office, Washington, D. C.
24. U.S. Department of Defense. 1969. MIL-STD 109B. Military standard quality assurance terms and definitions. U.S. Government Printing Office, Washington, D. C.
25. U.S. Government. National Archives and Records Administration. 1988. CFR. Title 7. Part 43 - Standards for sampling plans, p. 320.
26. U.S. Government. National Archives and Records Administration. 1997. CFR. Title 49. Part 172. Hazardous materials table, special provisions, hazardous materials communications, emergency response information, and training requirements. Subparts C-F, p. 286-343.
27. U.S. Government. National Archives and Records Administration. 1997. CFR. Title 49. Part 173. Section 217 B Carbon dioxide, solid (dry ice), p. 487-488.

Chapter 3

Microbiological Monitoring of the Food Processing Environment

George M. Evancho, William H. Sveum, Lloyd J. Moberg, and Joseph F. Frank

3.1 INTRODUCTION

The survival and growth of microorganisms in a food processing environment may lead to contamination of the finished product that may, in turn, result in a reduction of microbiological safety and quality. Sources of environmental microbial contamination include raw materials, processing equipment, manufacturing activities, sanitation and maintenance practices, workers, waste, animal and insect pests, and microbial growth niches embedded in equipment and in structural components of the building.

Most food plants have locations that can promote the growth of pathogens and spoilage microorganisms that may be transferred directly onto product or carried into additional niches. The origins or these growth habitats are mainly unhygienic design, construction, and maintenance and repair activities that prevent easy cleaning and disinfection. Both water and nutrients (food product) are required to form a microbial growth niche, and the chemical composition of the food and conditions of water activity, pH, temperature, etc., will select the "normal" organisms that can grow there.

The pathogen *Listeria monocytogenes* has been isolated from the environments of dairy plants[10] and ready-to-eat meat plants.[4] In a potato processing plant, *Listeria* spp. were isolated from floors and drains, condensed and stagnant water, process equipment, conveyor belts, and wiping cloths.[15] In addition, *Listeria* spp. have been isolated from dehumidifiers, air handling systems, wet insulation, cracks and crevices of floors, milk case conveyor belts, and crevices of many types of processing equipment in dairy[24] and meat processing plants.[4] Examples have been found of *Salmonella* contamination of dry foods, such as milk-based products, eggs, soybean meal, chocolate, and peanut butter, which resulted from environmental contamination.

A biofilm is formed when microorganisms colonize a surface. Most food industry biofilms consist of microorganisms and their exocellular polymers intermingled with food residues or mineral deposits. Any surface within a processing facility that is exposed to water or moist food will support biofilm formation if it is not effectively cleaned at regular intervals.[6] Environmental surfaces such as drains, floor mats, and equipment exteriors may support growth of pathogen-containing biofilms.[29] *Listeria monocytogenes* survives well in multispecies biofilms that might accumulate in such environments.[40] Worn, abraded or corroded food contact surfaces tend to accumulate biofilms because they are difficult to clean and sanitize.[28,38] In addition, poor equipment design or ineffective cleaning regimens will ultimately lead to biofilm formation. Microorganisms, including *L. monocytogenes*, are highly resistant to chemical sanitizers when growing in biofilms.[27,45] However, if the biofilm is completely disrupted and dissolved by the cleaning process, these microorganisms are readily inactivated by commonly used chemical sanitizers.[27] If equipment design does not allow this degree of effective cleaning, then sanitizers especially suited for biofilm inactivation may need to be used.

Microbial growth niches may be established when water is used to clean dry processing environments not designed for wet cleaning and not all points in the equipment are promptly and completely dried.

Adherence to good manufacturing practices, such as hygienic design, construction, and maintenance of the factory; hygienic operation and maintenance of the processes and equipment; and application of appropriate (e.g., dry vs. wet) cleaning and disinfection procedures constitute the principal effective approach for control of microbial contamination and growth. In order to suppress the establishment of niches, the environment must be designed and fabricated to resist microbial growth or be made easily cleanable.

3.2 ENVIRONMENTAL SAMPLING STRATEGIES

Microbiological monitoring of the food processing environment may be conducted to meet one or more of the following objectives: (1) verification of the effectiveness of cleaning and disinfection practices; (2) determination of the frequency required for cleaning and disinfection; (3) determination of the presence of foodborne pathogens in the environment; (4) discovery of environmental sources of spoilage organisms; (5) determination of the frequency required for special maintenance procedures, e.g., changing of air filters to reduce airborne mold contamination; (6) evaluation of hygienic design and fabrication of food processing equipment and facilities. ATP bioluminescence may be used to measure the efficacy of cleaning and sanitation of food plant environments and equipment, and for providing a means to quickly validate that effective cleaning and sanitation have occurred.

A sampling program may be established to verify control of environmental critical control points (CCPs) within the context of a Hazard Analysis Critical Control Point (HACCP) system,[39] after the microbiological hazards and risks have been determined and appropriate critical control points have been identified. Cleaning and disinfecting procedures for processing, conveying, and packaging equipment may be critical control points for prevention of post-processing recontamination, and monitoring can be best accomplished through sensory inspection (the plant and equipment look, smell and feel clean), and chemical (measurement of sanitizer concentration and pH), physical (measurement of temperature), and microbiological tests of the equipment and environment. Microbial criteria for acceptance of cleanliness of equipment and the environment can be developed using a data base derived from repeated, routine sampling and testing of specific sites.[39] However, the time required to obtain microbiological results is usually too long to make sampling and testing an effective tool for day-to-day monitoring. To be effective, monitoring systems must provide information promptly, if a critical point is to be controlled. Of all the rapid tests currently available, the only one that potentially offers real-time results is ATP bioluminescence.

Four approaches may be used to verify microbiological acceptability of food processing equipment and environments: (1) sampling and testing equipment; (2) measuring microbial loads in food products after all processing, packaging, and handling are completed; (3) collecting and testing in-process samples; (4) collecting and testing samples from the food processing environment.[39]

Environmental sampling and testing can be an early warning system to detect and eliminate niches of undesirable microorganisms before the risk of product contamination increases significantly. Collection of microbiological samples should not be limited to sites that are easily cleaned and sanitized, because the results from only these points may not reveal critical hazards and risks. If cleaning and sanitizing is effective, such easily disinfected sites should yield satisfactory results. The verification procedures should include collections of food or other organic residue samples from inaccessible or neglected niches.

To avoid errors of judgment and interpretation in quantitating microbial hazards and risks associated with the equipment and environment, and to help establish which microorganisms should be sought in a monitoring or verification program, it is important to understand the microbial ecology of the specific food and its process. The ecological pressures of extrinsic and intrinsic factors, including heat processing steps, ambient or storage temperatures, packaging atmospheres, acidity, and water activity, should determine which organisms are important in a particular environment. In plants that process dry foods not intended to be cooked before consumption., e.g., nonfat dry milk, chocolate, and peanut butter, *Salmonella* may be a significant hazard, especially when high-moisture-containing niches are present.

The environmental monitoring program should be designed to measure the occurrence and numbers of the normal spoilage flora and the pathogens that present the greatest risk to the product.

If a food is processed to be an ingredient for another product, it is prudent to monitor the ingredient's production environment for those organisms that will be a hazard to the ultimate finished product. To cite one example, dried milk products are used to make chocolate and confectionery products and are a potential source of salmonellae; therefore, the milk processing environment should be monitored for salmonellae in order to reduce risks further up the processing chain. This is consistent with the HACCP approach in that if a control step does not exist in the manufacture of the finished product, control must be exercised up stream at the ingredient stage.

An environmental monitoring or verification program that includes tests for indicators, such as the aerobic plate count, family *Enterobacteriaceae*, and fungi, as well as tests for the plant's unique microflora, will permit a more accurate assessment of microbial contamination of equipment and the plant than pathogen testing alone will. Negative pathogen test results may be misinterpreted as meaning that the site is microbiologically inert. Such pathogen test results merely indicate that the pathogen of interest was not detected in that site at the time of sampling; they do not provide useful information about the general microbiological risks associated with that site.[6]

3.3 SAMPLING OF SURFACES—EQUIPMENT AND PHYSICAL PLANT

Sampling sites on equipment should be selected to include all points that are liable to harbor microorganisms that may directly or indirectly contaminate the product. Sampling sites should not necessarily be limited to direct product contact zones because microbial contamination can also be transferred indirectly into product from condensation, aerosols, lubricants, packaging materials, line workers' garments, and so on.

The distinction between what is and is not a product contact zone is not always easy to make, especially in open systems where the product is exposed to the processing environment and not continuously protected by enclosure in a pipeline or vessel.[39] Direct product contact surfaces include pipeline interiors, conveyors, product storage vessels, fillers, utensils, work tables, mixers, grinders, and so forth. Nonproduct contact sites include structural components of machinery; the exterior of equipment, pipelines, and vessels; walls; motors; bearings; floors and floor drains of buildings; heating, ventilation, and air-conditioning equipment; forklifts; workers' garments and footwear; mechanics' tools; and cleaning tools.

Microorganisms can be transferred from nonproduct contact surfaces to direct product contact surfaces during production and between cleaning and sanitation cycles. Failure to clean and disinfect all sites that harbor microbial growth will increase the risks of contamination of finished product. Verification that sites are microbiologically acceptable is best accomplished by sampling and testing. Sensory evaluation is useful to detect environmental conditions that may lead to microbial growth and survival, but visually clean sites will still harbor microorganisms. Therefore verification of cleanliness and microbiological acceptability requires sampling and testing. Sampling and testing merely provide a rough estimate of the quantity of food debris and microbial populations on equipment, but accumulation of data from repeated tests will permit development of criteria by which to judge the hygienic condition of specific pieces of equipment.[20]

Preprocess sampling and microbiological testing of equipment by conventional microbiological methods may not be useful because of the time required after sampling to obtain results. Generally, production will begin before sites with unacceptable numbers or types of microorganisms are identified. Application of ATP bioluminescence for monitoring sanitation and hygiene has recently gained some acceptance in the food industry and may offer some evidence of surface cleanliness and hygiene. However, microbiological analysis of preoperative samples taken of equipment surfaces may be a useful technique to verify historically the efficacy of the cleaning and sanitizing procedures in order to evaluate the performance of the cleaning and sanitizing crew or the cleanability of a particular piece of equipment.

3.4 RINSE SOLUTION METHOD FOR SAMPLING CONTAINERS AND PROCESSING EQUIPMENT SYSTEMS[51]

3.41 Equipment, Supplies, and Solutions

1. Sterile pipettes
2. Sterile petri dishes
3. Stock phosphate buffer solution
4. Sterile buffered rinse solution
5. Sodium thiosulfate solution
6. Standard methods agar
7. Violet red bile agar
8. m-Endo broth MF
9. Nutrient broth (May be used as a substitute for buffered rinse solution since it effectively neutralizes chlorine and quaternary ammonium compound.)

3.42 Packaging Containers

Remove containers from the conveyor line or container cartons.

For 1-L or smaller containers, aseptically add 20 mL of sterile buffered rinse solution into each container; for 1.89-L containers, use 50 mL; and for 3.78-L or larger containers, use 100 mL. After addition of the rinse solution, recap the container.

Holding the container firmly with its long axis in a horizontal position, shake vigorously 10 times through a 20-cm arc. Turn the container 90 degrees and repeat the horizontal shaking treatment. Turn the containers 90 degrees twice more and repeat the horizontal shaking. Swirl the container vigorously 20 times in a small circle with the long axis in the vertical position, then invert and repeat. Stand the container upright before removing the sample.

For small containers, determine the number of bacteria in the rinse solution by distributing 5 mL equally between two sterile petri dishes. For larger containers, place 2 mL of the rinse solution into a single plate. Determine the number of coliforms in the rinse solution by dividing a total of 10 mL of the rinse solution among three sterile petri dishes. Pour 15 to 20 mL of the desired medium, e.g., plate count agar or violet red bile agar, using the appropriate incubation conditions for each. Yeasts and molds, proteolytic bacteria, and other specific microorganisms may be determined by the use of appropriate differential media and incubation temperatures and times as described in their respective chapters. To calculate the residual bacteria count per container capacity, multiply the total number of colonies by the volume of the rinse solution divided by the volume of the sample plated.

When using 100 mL or greater portions of rinse solutions, follow membrane filtration procedures for analysis, particularly if low levels of contamination are expected. Membrane filtration may also be used for analysis of the 20-mL rinse samples.

Interpretation of results from container rinse samples should take into consideration both numbers and types of microorganisms. The types of microorganisms may be important in terms of their potential to cause spoilage. Normally, the number of microorganisms should be very low. In most cases, the number of organisms added to the product from the container will be much lower than that indigenous to the product. However, under certain circumstances, such as with aseptic packaging, the microbiological condition of containers is a critical control point. Application of microbiological guidelines and standards is not common for the myriad of containers in use today. Nevertheless, bacterial standards for multiuse and single-serve containers for packaging pasteurized milk and milk products[23] and bottled water[65] have been published. These standards require that such containers have a residual bacteria count of one colony or less per mL of capacity or not over one colony per square cm of contact surface. No coliform organisms may be present. For pasteurized dairy product containers, four containers from any given day are sampled, and three out of four samples must meet this standard. The standard for bottled drinking water containers states that at least once each 3 months, bacteriological swab or rinse samples are to be taken from at least four containers and closures selected just prior to filling and sealing.

3.43 Processing Equipment Systems (Tanks, Pipelines, Fillers, etc.)[51]

Water for large-volume rinse-sampling of equipment should be heat sterilized or may be treated by chlorinating to a residual concentration of 25 mg per L, holding for 10 min, and then neutralizing by adding an excess of sterile 10% sodium thiosulfate solution. Tap water may be used after sterilizing by membrane filtration followed by the addition of sterile 10% weight per volume (wt/vol) sodium thiosulfate to inactivate residual disinfectant.

A sufficient volume of treated rinse water is added to the system at the upstream end of the assembly, then pumped or allowed to flow by gravity through the assembly. A control sample (about 1 L) of the treated rinse water is taken before using the water for rinsing.

Samples of rinse water are collected from the discharge end of the assembly from the first, middle, and final portions of the rinse water. Samples may also be collected at various points throughout the assembly.

The membrane filtration procedure (Chapter 7) may be used to analyze large volumes of rinse water. Analyses of rinse solutions from CIP processing assemblies and control samples require use of membrane filtration procedures. Average the number of colonies obtained from rinse samples taken at the beginning, middle and end of drainage, and subtract the number of colonies (if any) obtained from the control samples. Calculate the ratio of sample volume to rinse volume, and multiply by the corrected yield to obtain an indication of numbers of organisms present in the entire system. The presence of specific types of organisms may be determined by employing appropriate differential media and incubation temperatures. Refer to the specific chapters covering these organisms.

3.5 SURFACE CONTACT METHODS

Meaningful microbiological examination of surfaces requires selection of an appropriate method. The replicate organism direct agar contact (RODAC) procedure, the Petrifilm™ aerobic count plate procedure, and the swab procedure are the usual methods of choice for sampling of surfaces. Swab techniques should be used for surfaces with cracks, corners, or crevices, i.e., areas of such dimensions that the swab will be more effective in recovering organisms from them. Swab procedures should also be used for sampling utensils, tableware, and kitchenware. Sponge/swab procedures are useful for sampling large areas of food processing equipment and environmental surfaces. The RODAC and Petrifilm procedures should be used only on flat, impervious surfaces that are relatively easy to clean and disinfect. Selection of the proper technique is essential to obtain meaningful results.

3.51 Swab Contact Method[8,64]

3.511 Equipment and Supplies

1. Sterile nonabsorbent cotton swabs with the head firmly twisted to approximately 0.5 cm in diameter by 2 cm long

on a wooden applicator stick 12 to 15 cm long may be used. Swabs should be packaged in individual or multiple convenient protective containers with the swab heads away from the closure. Calcium alginate, dacron, and rayon swabs may also be used. Presterilized swabs may be purchased or the swabs may be sterilized in the laboratory. A commercially available test system that includes a swab sampler and various agar recovery media (Millipore Corp., Bedford, Mass.) has been shown to be comparable to conventional swab sampling procedures.[18,61]

2. Swabs made of calcium alginate fibers are soluble in aqueous solutions (rinse, culture media, etc.) containing 1% of sodium hexametaphosphate, sodium glycerophosphate, or sodium citrate. All organisms captured on the swab will be liberated from the calcium alginate swab. Presterilized calcium alginate swabs contained in various transport media are commercially available. The transport medium maintains microbial viability while inhibiting multiplication.
3. Vials, small, screw-capped, 7 to 10 cm long, are to be prepared to contain 5 mL (4.5 mL if calcium alginate swabs are used) of buffered rinse solution after autoclaving.

When sampling is to be carried out on surfaces previously subjected to chemical disinfection, appropriate neutralizers should be incorporated into the rinse solution. A commonly used neutralizer is 0.5% polysorbate (Tween 80) plus 0.07% soy lecithin. Dehydrated media for preparing neutralizing solutions are commercially available. Polysorbate 80 neutralizes some substituted phenolic disinfectants, and soy lecithin neutralizes quaternary ammonium compounds. The efficacy of any disinfectant neutralizer should be validated under actual use conditions.

3.512 Sampling Procedure

To sample equipment surfaces, open the sterile swab container, grasp the end of a stick, being careful not to touch any portion that might be inserted into the vial, and remove the swab aseptically.

Open a vial of buffered rinse solution, moisten the swab head, and press out the excess solution against the interior wall of the vial with a rotating motion.

Hold the swab handle to make a 30°-angle contact with the surface. Rub the swab head slowly and thoroughly over approximately 50 cm² of surface three times, reversing direction between strokes. Move the swab on a path 2 cm wide by 25 cm long or other dimensions to cover an equivalent area. Return the swab head to the solution vial, rinse briefly in the solution, then press out the excess. Swab four more 50-cm² areas of the surface being sampled, as above, rinsing the swab in the solution after each swabbing, and removing the excess.

After the areas have been swabbed, position the swab head in the vial, and break or cut it with sterile scissors or other device,[9] leaving the swab head in the vial. Replace the screw cap, put the vial in a waterproof container packed in a suitable refrigerant, and deliver to the laboratory. Analyze the sample within 24 hr after collection.

When sampling utensils such as knives and ladles, moisten the swab with dilution fluid and then run the swab slowly and firmly three times over the significant surfaces of the utensil, reversing the direction each time. After the utensil has been swabbed, return the swab to the buffered rinse solution using the procedure described above.

When unmeasured surface areas such as pump impellers, gaskets, rings, valve seats, and filler nozzles have been swabbed, the results may be reported on the basis of the entire sampling site instead of a measured area.

3.513 Plating Swab Rinse Solutions

At the laboratory, remove the vial from refrigerated storage. Shake it vigorously, making 50 complete cycles of 15 cm in 10 sec, striking the palm of the other hand at the end of each cycle. Groups of vials may be shaken together to save time.

Plate 1- and 0.1-mL portions of rinse solution plus additional dilutions if deemed necessary. Pour plates with SMA, or other appropriate media depending on the organisms of interest, incubate, count colonies, and then calculate the number of colonies recovered from 50 cm² (equivalent to 1 mL of rinse). When groups of microorganisms other than the aerobic plate count are sought, plate with appropriate selective/differential media and incubate as required.

3.514 Interpretation

As a guide, the U.S. Public Health Service recommends that adequately cleaned and sanitized food service equipment have not more than 100 colonies per utensil or surface area of equipment sampled.[66] Interpretation of results obtained from unmeasured surface areas such as utensils, gaskets, and pump impellers should be based on knowledge of historical data obtained when the surfaces had been documented to be thoroughly cleaned and sanitized. Generally, the levels of microorganisms should not exceed more than a few colonies per sampling site. *In many cases, the types of microorganisms may be more significant than the numbers alone.* For example, the presence of even very low numbers of *Saccharomyces bailii* and/or *Lactobacillus fructivorans* on salad dressing processing equipment may be highly significant with respect to potential spoilage of the finished product. Thus, with spoilage organisms for specific foods, the standards for evaluating sanitation may be much more stringent than when only the total numbers are considered. When swabbing is done for purposes other than evaluation of sanitation procedures, interpretation of results must be based on knowledge of the product, process, and equipment in order to determine the significance of data. In addition, the objectives of sampling may govern the interpretation of results.

3.52 Sponge Contact Method[60]

3.521 Equipment and Supplies

1. Cellulose or polyurethane[47] sponges free of antimicrobial preservatives should be cut into approximately 5-cm × 5-cm pieces, placed in individual kraft paper bags, and autoclaved. Alternatively, commercially sterile cotton gauze surgical swabs (ca. 10.2 cm × 10.2 cm) may be used. Sterile plastic bags are suitable to contain the sponges after sampling.
2. Sterile buffered rinse solution, nutrient broth, or 0.1% peptone water may be used as the rinse solution. If the surface to be sampled contains fatty materials, 0.5 to 1.0% Tween 80 or other noninhibitory surfactant solution may be used. For sampling equipment that may contain residual disinfectants, the use of neutralizers in the buffered rinse solution is recommended (see Section 3.511). It is prudent to incorporate neutralizers in all fluids used to collect samples from equipment and the plant. Neutralizing buffers and transport media are commercially available.
3. Sterile crucible tongs, sterile rubber or plastic gloves, or other means may be used to hold the sponge aseptically during sampling.

3.522 Sampling Procedure

Moisten the sponge with approximately 10 mL of the appropriate sampling fluid. While holding the sponge aseptically with tongs or sterile gloves, swab the surface to be sampled by vigorously rubbing the sponge over the designated area. If the surface is flat, the rinse solution may be applied directly to the surface and then taken up into the sponge by the rubbing action. An area of several meters may be effectively swabbed.

After sampling, place the sponge aseptically in a sterile plastic bag and transport it to the laboratory under refrigeration.

3.523 Plating and Analysis

Because large areas may be sampled with the cellulose sponge, this technique is particularly useful for detecting pathogens *(e.g., Salmonella* or *Listeria)* or spoilage microorganisms in the food plant environment. For *Salmonella or Listeria* analyses, the sponge is introduced directly into an enrichment broth, incubated, and then tested by approved methods for *Salmonella* (Chapter 37), and *Listeria* (Chapter 36).

The sponge sample may be subjected to a variety of microbiological analyses in the same fashion as fabric-tipped swabs. For quantitative analyses, 50 to 100 mL of diluent is added to the bag containing the sponge. The sponge is then vigorously massaged with diluent for 1 min or more to release the microorganisms. Aliquots of the diluent are removed from the bag, further diluted if required, and plated into the desired media for the microorganism(s) in question.

After incubation, the numbers of microorganisms per unit surface can be calculated on the basis of the area swabbed, the amount of diluent used, and the size of aliquot plated. For example, if 50 colonies are obtained from a 1-mL aliquot derived from a sponge in 100 mL of diluent that swabbed 1 m^2 the count per m^2 will be 5,000.

3.524 Interpretation

Interpretation of results from sponge samples taken from cleaned and sanitized equipment is essentially the same as for results obtained from fabric-tipped swabs.

Historically, the sponge technique has been useful in sampling the environment for *Salmonella*. Recent experience shows it also to be useful for *Listeria*.[30] This technique can be used to evaluate the efficacy of cleaning and sanitizing programs for the environment, particularly for foodborne pathogens. Obviously, results should always be negative after the application of appropriate cleaning and sanitizing procedures. Sponge swabs can be taken to identify areas that harbor pathogens, and the results can be used to develop a program to control the organisms. Evaluation of results from samples taken of cleaned and sanitized floors and other areas where relatively high residual microbial levels are expected depends on the history and experience relating to those particular sites in a given plant. As a rule of thumb a 4- to 5-log cycle reduction in the residual microbial level should be obtained on most floor surfaces after cleaning and sanitizing.

3.53 RODAC Plate (Agar Contact) Method[5,35,67]

The RODAC plate method (agar contact) provides a simple, valuable contact technique for estimating the sanitary quality of surfaces. The method is recommended particularly when quantitative data are sought from flat, impervious surfaces. It is not intended to be used for crevices or irregular surfaces, although the RODAC plate may be useful even if its only purpose is to demonstrate the presence or absence of a specific microorganism. Ideally, the RODAC plate method should be used on previously cleaned and sanitized surfaces. Samples taken from heavily contaminated areas will result in overgrowth on the plates. If accurate colony counts are desired, the plates should have fewer than 200 colonies. A sufficient number of sites should be sampled to yield representative data. Randomization of site selection may permit additional comparisons and inferences.

3.531 Equipment and Supplies

1. Disposable plastic RODAC plates may be purchased prefilled with test medium, or they may be filled in the laboratory. When prepared in the laboratory, the plates should be filled with 15.5 to 16.5 mL of the appropriate medium. The meniscus of the agar should rise above the rim of the plate to give a slightly convex surface. This is necessary so that the agar makes proper contact with the surface to be sampled.
2. Normally, plate count agar is used for aerobic plate counts. However, if qualitative data for specific microorganisms are desired, selective or differential media may be used, e.g., LBS agar for lactic acid bacteria, violet red bile agar for coliforms, or Baird-Parker agar for *Staphylococcus aureus.*
3. Dey-Engley neutralizing medium may be used in place of plate count agar. This medium incorporates a variety of ingredients capable of neutralizing any of the germicidal chemicals likely to be encountered on surfaces.[21] Following preparation, the plates should be incubated at 32°C for 18 to 24 hr as a sterility check. They should be used within 12 hr after preparation unless wrapped and refrigerated.
4. In lieu of RODAC plates, two commercially available systems, PetriFilm™ (3M Medical-Surgical Division, St. Paul, Minn.)[53] (Section 3.54) or Con-Tact-It (Birko Chemical Corp., Denver, Colo.),[61] can be used as a medium contact method. Another method is to use mylar adhesive tape (Dynatech Laboratory, Inc., Alexandria, Va.), which is transferred to the surface of an appropriate agar plate after being pressed to the surface of the sample site.[13]

3.532 Sampling Procedure

Remove the plastic cover from the RODAC plate and carefully press the agar surface to the surface being sampled. Make certain that the entire agar meniscus contacts the surface, using a rolling uniform pressure on the back of the plate to effect contact.

3.533 Incubation and Colony-Counting Procedure

Replace the cover and incubate in an inverted position under the appropriate time and temperature conditions for the microorganism(s) in question.

Colonies should be counted using a Bactronic or Quebec colony counter and recorded as the number of colonies per RODAC plate or number of colonies per cm^2.

3.54 Petrifilm Aerobic Count Method

The Petrifilm direct-contact method provides a simple means of detecting bacterial contamination on both flat and curved surfaces. The procedure should not be used for surfaces with cracks or crevices.

3.541 Equipment and Supplies

Petrifilm plates are provided by the manufacturer (3M Company, St. Paul, MN 55144) in sealed foiled pouches. Sealed pouches may be stored at 2° to 4°C until the specified expiration date. Plates must be prehydrated before use.

3.542 Sampling Procedure

Prehydrate plates by dispensing 1 mL of sterile dilution water onto the center of the bottom film. If the surface has been treated with sanitizer, incorporate an appropriate neutralizer into the sterile dilution water. Replace the top film down onto the diluent. Distribute the diluent by exerting downward pressure on the center of the plastic spreader. Do not slide the spreader across the film. Remove the spreader and allow 30 minutes for the gel to solidify.

To sample the test surface, lift the top film of the prehydrated plate without touching the growth surface. The gel should adhere to the top of the film. Allow the gel and the top of the film to contact the test surface. Firmly rub fingers over the entire film side of the gelled area to ensure good contact with the surface. Lift the film from the surface and rejoin the top and bottom sheets of the plate.

3.543 Incubation and Colony-Counting Procedure

Incubate the plates in a horizontal position with the clear side up at 32°C for 48 hours. Count all the red colonies in the 20-cm^2 circular growth area. When very high concentrations of colonies on the plate cause the entire growth area to become red or pink, record the plate results as greater than 250.

3.6 MICROBIOLOGICAL AIR-SAMPLING STRATEGIES

With many nonperishable foods, the quality of the air in a food plant does not directly affect the microbiological safety or keeping quality. On the other hand, some perishable products, such as dairy products and baked goods, are particularly sensitive to airborne contaminants. Environmental air quality, especially in the packaging areas, is a critical control point for these foods. Aseptically packaged foods may require that the air supplies in packaging rooms have very low microbial loads such as that supplied by air filtered through laminar flow systems. Measurement of the microbial quality of air is useful for assessing the effectiveness of disinfection procedures for air-handling equipment.

Microorganisms occur in air as aerosols consisting of single unattached cells or cells in clumps. They can become airborne from environmental sources such as worker activity, sink and floor drains, water spraying, air-conditioning systems, dust generated from raw material, and specific food-processing systems. Microorganisms may adhere to a dust particle or may exist as a free-floating particle surrounded by a film of dried organic or inorganic material. Particulates in microbial aerosols may range in size from <1 μm to 50 μm. Particle size is the major factor influencing aerodynamic behavior. Vegetative bacteria may be present in lesser numbers in air than bacterial and mold spores. Many vegetative bacterial cells ordinarily will not survive for long in air unless the relative humidity and other factors are favorable or unless the organism is enclosed in some protective matrix. As a rule of thumb, microbial aerosols generated from the environment will be primarily bacterial spores, molds, and yeasts. When personnel are the source of microbial contamination, the primary types are vegetative bacteria, especially staphylococci, streptococci, micrococci, and other organisms associated with the human respiratory tract, hair, and skin.

Quantitative and qualitative guidelines should be established that relate numbers and types of microorganisms per volume of air to critical levels of product contamination. These guidelines must be established for each plant or process so that data collected in an air-sampling program can be used to make decisions relating possible sources, such as air flow patterns, filtration systems, or personnel density and activity, to product contamination. Significant increases above an established guideline may indicate a breakdown of standard contamination control barriers. National Aeronautics and Space Administration (NASA) air cleanliness standards[54] (Table 1) may be used as a reference point, but their suitability for application in a particular processing environment will have to be determined experimentally.

3.7 AIR-SAMPLING METHODS[1,16,19,32,36,52]

Viable airborne microorganisms can be determined quantitatively by a variety of methods, including sedimentation,[34,48] impaction on solid surfaces,[31,44,49] filtration,[25] centrifugation,[55] electrostatic precipitation, impingement in liquids,[46] and thermal precipitation. Of these, sedimentation and impaction on solid surfaces are most frequently used. Aerosol-sampling methods have been reviewed recently by Kang and Frank.[41,42,43]

Many collecting and culturing media are available for biological aerosol sampling. The selection of nutrient medium will depend on the nutritional requirements of the organism(s) under study, the type of information desired from the study, the sampling method, and the sampling conditions. When initial collection is in a liquid medium, the microorganisms must remain viable without growth until aliquots are taken for culture. Some of the more common liquid media used are tryptose saline, buffered gelatin, peptone water, buffered gelatin enriched with brain-heart infusion, buffered saline, and buffered water. These media are used also as diluting fluids to obtain suspensions suitable for plating. Buffered saline and buffered water are used only for collecting spores and other resistant microbial forms.

When collection is made directly on solid nutrient medium, a sufficient concentration of agar (1.5% to 2.0%) to produce a stable medium capable of withstanding the action from a rapidly flow-

Table 1. NASA Air Cleanliness Classes*

	Class, English (Metric) System		
Test	100 (3.5)	10,000 (350)	100,000 (3500)
Max. no. of 0.5 μm and larger particles per ft^3 (per L)	100 (3.5)	10,000 (350)	100,000 (3500)
Max. no. of 0.5 μm and larger particles per ft^3 (per L)	†	65 (2.3)	700 (25)
Max. no. of viable particles per ft^3 (per L)	0.1 (0.0035)	0.5 (0.0176)	2.5 (0.0884)
Avg. no. of viable particles per ft^3 (per M^2) per week	1,200 (12,900)	6,000 (64,600)	30,000 (323,000)

* NASA standards for clean rooms and work stations for the microbially controlled environment.[25]

† Statistically unreliable except when a large number of samplings is taken.

ing airstream should be used. Some of the more common solid nutrient media employed for general bacterial air sampling are blood agar, tryptose agar, trypticase soy agar, proteose extract agar, and nutrient agar. These media are also employed for culturing the liquid collecting media by surface-plating methods, the pour-plate method, and the membrane filter method.

Under certain sampling conditions, incorporation of selective agents into the medium to inhibit interfering contaminants is desirable. Some commonly used inhibitory agents are crystal violet, brilliant green, potassium tellurite, and cycloheximide. Chemicals should not be used unless preliminary screening has demonstrated that they do inhibit the target organism.

Air samplers should be sanitized or sterilized prior to use. Sieve samplers and filtration-type samplers that have been wrapped in kraft paper, as well as liquid impingers with cotton plugs inserted in the intake and exhaust ports, can be conveniently autoclaved. In actual use, it will be found that swabbing the sampler with disinfectant prior to each sampling period is adequate and convenient. Gaseous sterilization techniques can be used to sterilize all of the samplers.

The following air sampling methods are commonly used in environmental microbiology, and six types of commercially available aerosol samplers are listed in Table 2. It is important that the manufacturer's directions be followed for each sampler and that the limitations of each be understood. The methods listed below for air sampling are by no means comprehensive. The laboratory worker should review Public Health Monograph 60 for a detailed discussion of air-sampling principles.[37]

3.71 Sedimentation Methods

Sedimentation methods are easy to use, inexpensive, and collect particles in their original state. The exposure agar plate and microscopic slide exposure method rely on the force of gravity and air currents to deposit airborne particles on a nonselective or selective agar surface. Results are obtained as colony forming units (CFU) or particles per min. Particle size distribution may be obtained by direct microscopic observation. The 16th edition of "Standard Methods for the Examination of Dairy Products"[51] recommends 15-min exposure of standard size (90-mm diam) petri plates containing standard methods agar or a selective medium. After exposure, plates are incubated according to the appropriate procedure. In addition, microscope slides coated with agar can be exposed and particles counted using a microscope. This technique is only used for total particulate counts.

Sedimentation methods have several disadvantages, including their measure of airborne microorganisms quantitatively, i.e., the number of viable particles per cubic unit of air, and their weak correlation with counts obtained by other quantitative methods.[57] They are useful only when fallout onto a particular surface is of interest, and they require a relatively long sampling time. Air movement will influence the deposition of the particles. Thus, these methods are heavily biased toward large particles, which would settle more rapidly than smaller particles.

Samples may be taken at (a) openings in equipment subject to potential contamination from organisms transported by air currents, (b) selected points for testing general room air, (c) areas of employee concentration, and (d) process air passages where air is incorporated into products. Because of air turbulence during operating hours, sampling by volumetric methods will be more effective and dependable than sedimentation samples.[51]

3.72 Impaction Methods

Impaction usually involves collecting microbial aerosols on an agar surface, but dry or coated surfaces may be used for special purposes such as particle size determination. An impactor consists of an air jet that is directed over the impaction plate so that the particles collide with and stick to the surface. Impaction methods give higher particle recovery than other methods.[26,62,63] Impac-

Table 2. Commercial Sources of Aerosol Samplers

IMPINGERS

All-Glass Impinger 30 and Pre-Impinger; Ace Glass, Inc., P.O. Box 688, Vineland, NJ 08360

Midget Impinger with Personal Air Sampler; Supleco Inc., Supleco Park, Bellefonte, PA 16823-0048

May 3-stage Glass Impinger; A.W. Dixon Co., 30 Anerly Station Road, London, S.E.20, England

IMPACTORS (SLIT TYPE)

Casella Single-slit and Four-slit Sampler; BGI Incorporated, Air Sampling Instructions, 58 Guinan Street, Waltham, MA 02154

Mattson-Garvin Air Sampler; Mattson-Garvin Company, 130 Atlantic Drive, Maitland, FL 32751

New Brunswick STA Air Sampler; New Brunswick Scientific Company, Inc., P.O. Box 986, 44 Talmadge Road, Edison, NJ 08817

IMPACTORS (SIEVE TYPE)

Andersen 6-stage, and 2 -stage Samplers; Andersen Samplers, Inc., 4215-C Wendell Drive, Atlanta, GA 30336

Ross-Microban Sieve Air Sampler; Ross Industries, Midland, VA 22728

Personal Particulate, Dust Aerosol Collector; SKC Inc., 334 Valley View Road, Eighty Four, PA 15330

FILTRATION SAMPLERS

Millipore Membrane Filterfiled Monitor; Millipore Corporation, Bedford, MA 01730

Gelman Membrane Filter Air Sampler; Gelman Sciences Inc., 600 S. Wagner Road, Ann Arbor, MI 48106

MSF 37 Monitor; Micro Filtration Systems, 6800 Sierra Court, Dublin, CA 94568

Satorius MD8 Air Sampler; Satorius Filters Inc., 30940 San Clemente St., Bldg. D, Hayward, CA 94544

CENTRIFUGAL SAMPLERS

RCS Centrifugal Sampler; Folex-Biotest-Schluessner, Inc., 6 Daniel Road East, Fairfileld, NJ 07006

ELECTROSTATIC PRECIPITATION SAMPLERS

LVS Sampler; Sci-Med Environmental Systems Inc., 8050 Wallace Road, Eden Prairie, MN 55344

General Electric Electrostatic Air Sampler; General Electric Co., Lamp Components & Technical Products Div., 21800 Tungsten Road, Cleveland, OH 44117

tion results in low sampling stresses after collection, and sample manipulation is not required. Impactors are of two types: slit samplers (e.g., Casella slit sampler) and sieve samplers (e.g., Andersen multistage sieve sampler).

3.721 Slit Sampler

Slit samplers are usually cylindrical and have a slit tube that produces a jet stream when the air is sampled by vacuum. Beneath the slit is a platform that accepts a culture plate and that is rotated by a clock mechanism. The rate of the plate rotation may be varied. These samplers require a vacuum source sufficient to draw a constant flow of air through the sampler—usually, 28.3 L per min, although the air flow may be changed by altering the dimensions of the slit. Some of the common characteristics of slit samplers are relatively high collection efficiency, fabrication from metal, ruggedness, portability, simplicity of operation, and relatively high sampling volume. Some slit samplers cannot be sterilized by autoclaving. While gaseous sterilization is desirable, swabbing with disinfectant is often sufficient. Samplers employing agar are limited to use in temperatures above 0°C unless some method of heating is provided to avoid freezing the medium. Slit samplers do not discriminate for size of airborne particles, and can be used to detect bursts of contamination associated with specific activities at certain times.

3.722 Sieve Samplers

Sieve samplers are operated by drawing air through a large number of small, evenly placed holes drilled in a metal plate (sieve). The suspended particles impact on an agar surface located a few millimeters below the perforated plate. There are single stage (e.g., Ross Microban) and multistage (e.g., Andersen) sieve samplers. A multistage sieve sampler consists of a series of two, six or eight stacked sieves and plates, each with successively smaller holes. This arrangement causes increased particle velocity as air flows through the apparatus. Large particles impact at the initial stage, and small particles follow the air flow until accelerated sufficiently to impact at a later stage. The commonly used Andersen six-stage sampler consists of sieves with holes ranging from 1.81 mm to 0.25 mm. The distance of the agar collecting surface from the sieve, which is critical, is controlled by utilizing a special petri dish containing 27 mL of medium. However, with newer designs, conventional prefilled disposable petri plates can be used. Air is drawn successively through each of the sieves at increasing velocities so that larger airborne particles (>7 mm) impact in the first stage and smaller particles, depending on their sizes and inertia, impact on the later stages. The optimum flow rate is 28.3 L per min. After sampling, the plates are removed and incubated. Some models have only two stages, which are designed to differentiate nonrespirable particles (>5 µm) from respirable particles (<5 µm). Some units have a single stage, which does not differentiate particle size. The multistage sieve samplers are used not only to detect the number of viable particles per unit volume of air during a prescribed sampling time but also to yield a size profile of the particulate in the microbial aerosol. This information is usually much more important in health care settings than in food processing environments. As with the slit sampler, no diluting or plating procedures are required. Final assay results are expressed as particles per unit volume.

Associated with sieve samplers are the following limitations: multistage sieve samplers are cumbersome to handle and are expensive; the exact volume of agar must be poured into all plates so that the gap between the sieve and the agar surface meets the manufacturer's specifications; and the inside of the sampler and the outside of the pre-poured agar plates should be maintained sterile until sampling, as they can contribute to contamination.

3.73 Centrifugal Samplers

Centrifugal force can be used to propel aerosol particles onto a collection surface. When the aerosol is spun in a circular path at high velocity, the suspended particles impact on the collecting surface with a force proportional to the particle's velocity and mass. Centrifugal samplers do not generate high-velocity jet flow during sampling, so less stress is imposed on airborne microbes as compared to impingement and impaction methods. Centrifugal samplers are simple and easy to operate and may be less expensive than impactor types. Generally, centrifugal samplers can rapidly sample a high volume of air, resulting in a more representative sampling. Assay results are expressed as particles per unit volume of air, e.g., CFU per L.

Limitations of some centrifugal samplers are associated with failure to generate sufficient centrifugal force to propel small particles onto the collection surface. The recovery efficiency of these samplers depends on the particle size being sampled and the amount of centrifugal force generated.

The Reuter centrifugal air sampler (RCS sampler, Biotest Diagnostics Co.) is battery operated, portable, lightweight (2.5 lb.), and convenient to use. A plastic strip containing a culture medium lines the impeller drum. Air from a distance of at least 40 cm is sucked into the sampler by an impeller. Air enters the impeller drum concentrically from a conical sampling area. The impeller is set in rotation and the aerosol is impacted by centrifugal force onto the agar surface. Air then leaves the sampling drum in a spiral outside the cone of entering air. After the sample has been taken, the agar strips are incubated and the colonies counted. The sampler has a self-timer for sampling from 30 sec to 88 min. The actual sampling rate is 280 L per min. However, the manufacturer has published an effective sampling rate or separation volume of 40 L per min for 4 µm particles, a value derived from an attempt to reconcile the actual number of viable particles collected from an air sample with measurements involving airflow direction, air velocity, and available collecting surface area. Clark and Lidwell indicated that the effective sampling volume of the RCS sampler will vary widely depending on the aerosol particle size.[12] Consequently, the results obtained by using this sampler must be interpreted with considerable caution. Macher and First measured the collection efficiency with increasing particle size.[50] Particles larger than 15 µm are almost 100% collected, those in the 4 to 6 µm range are collected at 55% to 75% efficiency, and particles smaller than 1 µm pass through the sampler without significant retention. Although the RCS sampler does not accurately estimate total viable particle concentration, Placencia and Oxborrow recommended this sampler for good manufacturing practices investigations.[55] These investigators found that the RCS sampler will collect more viable particles than a slit sampler, and it could detect the difference in the environmental quality of each medical device manufacturing facility tested. In addition, the RCS sampler effectively detects various types of microorganisms.[55]

3.74 Filtration Methods

Filters are widely used for aerosol sampling because of their low cost and simplicity of operation. The air filtration apparatus consists of cellulose fiber, sodium alginate, glass fiber, gelatin membrane filter (GMF) (pore size 3 µm) or synthetic membrane filters (pore size 0.45 µm or 0.22 µm) mounted in an appropriate holder and connected to a vacuum source through a flow rate controller (e.g., critical orifice). After a fiber filter is used, the whole filter or a section of it is agitated in a suitable liquid until the particles are uniformly dispersed.

Aliquots of the suspension are then assayed by appropriate microbiological techniques. Membrane filters can be either treated similarly to fiber filters or placed directly on an agar surface and incubated.

The gelatin membrane is water soluble so that it can easily be diluted for plating or be solubilized on top of a nutrient medium, resulting in microbial colonies that are easily counted. The hygroscopicity of the gelatin membranes causes difficulty in sampling because of the swelling of the membrane when the relative humidity exceeds 90%.[58] The large number of pores in these membranes allows a large volume of air to be sampled during a short time (2.7 L of air per min per cm^2 per 500-mm water column).

The technique has been shown to be effective in certain types of environments,[26] although some investigators have cautioned against drying of vegetative bacteria on the membrane filter and the consequent difficulty of recovery. Fields and co-workers have shown that recovery rates between membrane filter techniques and slit samplers are comparable for naturally occurring airborne microorganisms that have already survived drying.[25,26]

Filtration methods are good for enumerating mold or bacterial spores, but they may not be effective for counting vegetative cells because of the stress of dehydration produced during sampling.[22] The shorter sampling times used in gelatin membrane filtration may reduce this stress. Filtration methods do not discriminate particle size.

3.75 Impingement Methods

Impingement methods use a liquid to collect microorganisms from air. When air is dispersed through the liquid, particles in the air are trapped. Quantification of airborne microorganisms is accomplished by plating the collection fluid or by using a membrane filtration plating technique when the expected microbial level load is low.

Liquid impingers can be either low-velocity or high-velocity samplers. Low-velocity samplers utilize the air washing principle: airborne particles entering the sampler at low velocity through a large jet, fitted glass dish, or perforated tube are bubbled through and trapped in the liquid collecting medium. Small particles (<5µm) are not efficiently trapped in low-velocity samplers, remaining in air bubbles and being carried out with discharged air.

High-velocity samplers draw air through a small jet, directing it against a liquid surface. While these samplers efficiently collect all particle sizes greater than 1 µm diameter, the high velocity tends to destroy some vegetative cells. High-velocity collection disperses clumps of cells, producing counts that may be higher than those of gentler collection methods.

A suitable collecting medium for liquid impingement samplers must preserve the viability of the microorganism while inhibiting its multiplication. The more common collecting media include buffered gelatin, tryptose saline, peptone water, and nutrient broth. Use of an antifoam agent in the collecting medium is suggested if excessive foaming occurs. Acceptable agents are Dow Corning Anti-Foam A and B, General Electric Anti-Foam 60, and olive oil.

With extended sampling, air impact has a cooling effect on the liquid. If the ambient temperature is 40°F, the collecting liquid is likely to freeze. Use of a low-freezing-point diluent such as glycerol or some means of temperature control is necessary in such a situation.

After sampling, an aliquot of the collecting liquid is plated and incubated in a growth medium to obtain a viable count. In quantitative studies, the total air flow must be measured to calculate microorganisms per volume of air. The volume of collecting fluid must also be measured to determine the number of cells collected. This method is not suited to low concentrations of airborne microorganisms.

The All Glass Impinger (AGI-30, Ace Glass, Inc.) sampler is a high velocity impinger widely used for air sample collection. The jet is held 30 mm above the impinger base and consists of a short piece of capillary tube designed to reduce cell injury. The AGI-30 sampler operates by drawing aerosols through an inlet tube curved to simulate the nasal passage.[14] This makes it especially useful for studying the respiratory infection potential of airborne microorganisms. The usual sampling rate is 12.5 L per min. When it is used for recovering total airborne microorganisms from the environment, the curved inlet tube should be washed with a known amount of collecting fluid after sampling since larger particles (i.e., over 15 µm in diameter) are collected on the tube wall by inertial force.

The glass impinger is relatively inexpensive and simple to operate, but viability loss may result from the amount of shear force involved in the collection. The air stream approaches sonic velocity when particulates impinge on the collection fluid, resulting in almost complete collection of suspended particles; however, this condition tends to cause the destruction of vegetative cells[3] or may result in overestimation because of the dispersion of dust particles and the breaking up of clumps of bacteria.[57] Another constraint is that the glassware should be sterilized before each sampling. Also, the apparatus is easily broken.

3.76 Electrostatic Precipitation

Electrostatic precipitation samplers impart a uniform electrostatic charge to incoming airborne particles, which are then collected on an oppositely charged rotating disc. A known volume of air at a given rate is sampled. Electrostatic precipitators may employ a variety of solid collecting surfaces, such as glass or agar. A liquid collecting medium with added wetting agent, to aid in uniform distribution, can also be used to wash the collected particles centrifugally into a collecting vessel. Although these precipitators can sample at a relatively high rate (up to 1,000 L per min) with high collection efficiency and low resistance to air flow, they are complex and must be handled carefully. Furthermore, little is known about the effect on viability and clumping of electrostatically charged particles. During ionization of the air sample, oxides of nitrogen and ozone are produced that may be toxic to microorganisms. Although several electrostatic precipitators are manufactured specifically for sampling microbial aerosols, they are not widely used for this purpose.[59]

3.77 Comparison Studies on Aerosol Samplers

Comparison studies of air-sampling devices indicate that the choice of the correct sampler to use is seldom obvious. A multistage sieve sampler such as the Andersen may be most efficient at viable particle recovery, but it may not be suitable for routine sampling, and it requires a vacuum source. Filter samplers work well for quality control monitoring of molds[59] and bacterial spores, but bacterial recovery efficiency may be less, depending on the extent of dehydration that occurs during sampling.[11] In addition, a vacuum source is required. The RCS sampler is convenient to use, creates its own air flow, and recovers bacteria as well as molds. Even though the RCS sampler does not recover the smallest viable particles, it is useful for determining relative air quality on a routine basis.[17,55,56] Slit samplers may not be as convenient to use as the RCS sampler, especially if a vacuum source is required. However, slit samplers are more efficient at recovering small particles.

3.8 AEROSOL SAMPLING AND MEASUREMENT GUIDELINES

3.81 Standard Methods for Examining Dairy Products

The 16th edition of the "Standard Methods for the Examination of Dairy Products"[51] lists no Class A standard method for testing the microbiological quality of air and dairy environments, though there are methods designated as Class D and B. Favero et al. introduced air-sampling strategies and various air-sampling methods in the previous edition of this compendium.[22] They pointed out that the first and most important decision is whether air sampling at any level is required. If it is, then quantitative and qualitative guidelines should be established that relate numbers and types of microorganisms per volume of air to critical levels of product contamination.

3.82 NASA Air Cleanliness Standards

Favero et al. also suggested that the NASA air cleanliness standards be used as a reference point after experiments to determine suitability.[22] The "NASA Standards for Clean Rooms and Work Stations for the Microbially Controlled Environment"[54] defines three air cleanliness classes (Table 1). According to the standards, the collection methods must conform to "Standard Procedures for the Microbiological Examination of Space Hardware (NHB 5340.1 or revisions thereof)," which specifies use of a slit sampler.

3.83 Federal Standard 209C[33]

Federal standard 209C for "Clean Rooms and Work Station Requirements, Controlled Environment" establishes standard classes of air cleanliness for airborne particulate levels in clean rooms and clean zones. These classes are based only on particle enumeration and place more emphasis on small particles that are not necessarily viable.[33] Consequently, this standard is not useful for food plant applications.

3.84 Standard Reference Samplers

Brachman et al. recommended the AGI-30 sampler as a standard reference sampler because of historical use, economics, availability, and simple design.[7] On the other hand, the American Conference of Governmental Industrial Hygienists Committee on Bioaerosols used the Andersen multistage air sampler as the reference sampler for its committee activities and reports.[2] In the pharmaceutical industry, the slit sampler is the most widely used device for monitoring sterile manufacturing and quality control environments.[3]

3.9 REFERENCES

1. ACGIH. 1978. Air sampling instruments for evaluation of atmospheric contaminants, 5th ed. Amer. Conf. of Govern. Indust. Hyg., Cincinatti, Ohio.
2. ACGIH. 1986. Committee on Bioaerosols. ACGIH hygienists committee activities and reports. Appl. Ind. Hyg. 1:R19.
3. Akers, M. J. 1985. Sterility testing, p. 1. *In* Parenteral quality control. Marcel Dekker, New York.
4. AMI. 1988. Interim guideline: Microbial control during production of ready-to-eat meat products. Controlling the incidence of Listeria monocytogenes, 2nd ed. Am. Meat Inst., Washington, D. C.
5. Angelotti, R. J., L. Wilson, W. Litsky, and W. G. Walter. 1964. Comparative evaluation of the cotton swab and RODAC methods for the recovery of *Bacillus subtilis* spore contamination from stainless steel surfaces. Health Lab. Sci. 1:289.
6. Blackman, I. C., and J. F. Frank. 1996. Growth of *Listeria monocytogenes* as a biofilm on various food-processing surfaces. J. Food Prot. 59:827-831.
7. Brachman, P. S., R. Ehrlich, H. R. Eichenwald, V. J. Gabelli, T. W. Kethley, S. H. Madin, J. R. Maltman, G. Middlebrook, J. D. Morton, I. H. Silver, and E. K. Wolfe. 1964. Standard sampler for assay of airborne microorganisms. Science 144:1295.
8. Buchbinder, L., T. C. Buck, Jr., P. M. Phelps, R. V. Stone, and W. D. Tiedeman. 1947. Investigations of the swab rinse technic for examining eating and drinking utensils. Am. J. Public Health 37:373-378.
9. Buck, T. C., Jr., and E. A. Kaplan. 1944. A sterile cutting device for swab vial outfits utilizing wood applicators. J. Milk Technol. 7:141-142.
10. Charlton, B. R., H. Kinde, and L. H. Jensen. 1990. Environmental survey for *Listeria* species in California milk processing plants. J. Food Prot. 53:198-201.
11. Chatigny, M. A. 1978. Sampling airborne microorganisms, p. E1. *In* Air sampling instruments for evaluation of atmospheric contaminants, 5th ed. Am. Conf. of Govern. Indust. Hyg., Cincinatti, Ohio.
12. Clark, S., and O. M. Lidwell. 1981. The performance of the Biotest RCS centrifugal air sampler. J. Hosp. Infect. 2:181.
13. Cordray, J. C., and D. L. Huffman. 1985. Comparison of three methods for estimating surface bacteria on pork carcasses. J. Food Prot. 48:582-584.
14. Cox, C. S. 1987. The aerobiological pathway of microorganisms. John Wiley & Sons, New York.
15. Cox, L. J., T. Kleiss, J. L. Cordier, C. Cordellana, P. Konkel, C. Pedrazzini, R. Beumer, and A. Siebenga. 1989. *Listeria* spp. in food processing, non-food and domestic environments. Food Microbiol. 6:49-61.
16. Curtis, S. E., R. K. Balsbaugh, and J. G. Drummond. 1978. Comparison of Andersen eight-stage and two-stage viable air samplers. Appl. Environ. Microbiol. 35:208-209.
17. Delmore, R. P., and W. N. Thompson. 1981. A comparison of air-sampler efficiencies. Med. Device Diagn. Ind. 3:45.
18. Devenish, J. A., B. W. Ciebin, and M. H. Brodsky. 1985. Evaluation of Millipore swab-membrane filter kits. J. Food Prot. 48:870-874, 878.
19. Dimmick, R. L., and A. B. Akers. 1969. An introduction to experimental aerobiology. Wiley-Interscience, New York.
20. Elliott, R. P. 1980. The microbiology of sanitation, p. 35-60. *In* A. M. Katsuyama and J. P. Strachan (eds.), Principles of food processing sanitation. The Food Processors Institute, Washington, D. C.
21. Engley, F. B., and B. P. Dey. 1970. A universal neutralizing medium for antimicrobial chemicals. *In* Proceedings of the 56th Meeting of the Chemical Specialties Manufacturers Association, New York.
22. Favero, M. S., D. A. Gabis, and D. Vesley. 1984. Environmental monitoring procedures, p. 47. *In* M. L. Speck (ed.), Compendium of methods for the microbiological examination of foods, 2nd ed. Am. Public Health Assoc., Washington, D. C.
23. FDA. 1985. Grade A Pasteurized Milk Ordinance—Recommendations of the Public Health Serv. Transmittal No. 87-3 IMS-a-25, Public Health Serv. Publ. No. 229. Dep. Health, Human Serv. U.S. Government Printing Office, Washington, D. C.
24. FDA and Milk Industry Foundation/International Ice Cream Association. 1988. Recommended guidelines for controlling environmental contamination in dairy plants. Dairy Food Sanit. 8:52-56.
25. Fields, N. D., G. S. Oxborrow, C. M. Herring, and J. R. Puleo. 1973. An evaluation of two microbiological air samplers, abstr. E11, p. 2. *In* Abstracts of the Annual Meeting of the Am. Soc. Microbiol., American Society for Microbiology, Washington, D. C.

26. Fields, N. D., G. S. Oxborrow, J. R. Puleo, and C. M. Herring. 1974. Evaluation of membrane filter field monitors for microbiological air sampling. Appl. Microbiol. 127:517-520.
27. Frank, J. F., and R. A. Koffi. 1990. Surface-adherent growth of *Listeria monocytogenes* is associated with increased resistance to surfactant sanitizers and heat. J. Food Prot. 53:550-554.
28. Frank, J. F., and R. A. N. Chmielewski. 1997. Effectiveness of sanitation with quaternary ammonium compound or chlorine on stainless steel and other domestic food-preparation surfaces. J. Food Prot. 60:43-47.
29. Gabis, D., and R. E. Faust. 1988. Controlling microbial growth in food processing environments. Food Technol. 42:81-82, 89.
30. Gabis, D. A., R. S. Flowers, D. Evanson, and R. E. Faust. 1989. A survey of 18 dry dairy product processing plant environments for *Salmonella, Listeria* and *Yersinia*. J. Food Prot. 52:122-124.
31. Greene, V. W., D. Vesley, R. G. Bond, and G. S. Michaelsen. 1962. Microbiological contamination of hospital air. I. Quantative studies. Appl. Microbiol. 10:561-566.
32. Gregory, P. H. 1973. Air sampling technique, p. 126. *In* The microbiology of the atmosphere, 2nd ed. John Wiley & Sons, New York.
33. GSA. 1987. Federal standard 209C. Clean room and work station requirements, controlled environment. Gen. Serv. Admin., Federal Supply Service, U.S. Government Printing Office Washington, D. C.
34. Hall, L. B., and H. M. Decker. 1960. IV. Procedures applicable to sampling of the environment for hospital use. Am. J. Public Health 50:491-496.
35. Hall, L. B., and M. J. Hartnett. 1964. Measurement of the bacterial contamination on surfaces in hospitals. Public Health Rep. 79:1021-1024.
36. Heldman, D. R., and T. I. Hedrick. 1971. Air-borne contamination control in food processing plants. Res. Bull. 33. Mich. State Univ. Agric. Exp. Sta., East Lansing, Mich.
37. HEW. 1959. Sampling microbiological aerosols. Public Health Monogr. 60, Public Health Serv. Publ. No. 686. Dep. Health Ed., Welfare, U.S. Government Printing Office, Washington, D. C.
38. Holah, J. T., and R. H. Thorpe. 1990. Cleanability in relation to bacterial retention on unused and abraded domestic sink materials. J. Appl. Bacteriol. 69:599-608.
39. ICMSF. 1988. Cleaning and disinfecting, p. 93-116. *In* ICMSF, Microorganisms in foods 4: Application of the hazard analysis critical control point (HACCP) system to ensure microbiological safety and quality. Blackwell Scientific Publications, Palo Alto, Calif.
40. Jeong, D. K., and J. F. Frank. 1994. Growth of *Listeria monocytogenes* at 10°C in biofilms with microorganisms isolated from meat and dairy processing environments. J. Food Prot. 57:576-586.
41. Kang, Y. J., and J. F. Frank. 1989. Biological aerosols: A review of airborne contamination and its measurement in dairy processing plants. J. Food Prot. 52:512-524.
42. Kang, Y. J., and J. F. Frank. 1989. Comparison of airborne microflora collected by the Andersen sieve sampler and the RCS sampler in a dairy processing plant. J. Food Prot. 52:877-880.
43. Kang, Y. J., and J. F. Frank. 1989. Evaluation of air samplers for recovery of biological aerosols in dairy processing plants. J. Food Prot. 52:655-659.
44. Kraidman, G. 1975. The microbiology of airborne contamination and air sampling. Drug Cosmet. Ind. 116:40.
45. Lee, S-H., and J. F. Frank. 1991. Inactivation of surface-adherent *Listeria monocytogenes* hypochlorite and heat. J. Food Prot. 54:4-6, 11.
46. Lembke, L. L., R. N. Kniseley, R. C. Van Nostrand, and M. D. Hale. 1981. Precision of the all-glass impinger and the Andersen microbial impactor for air sampling in solid-waste handling facilities. Appl. Environ. Microbiol. 42: 222-225.
47. Llabres, C. M., and B. E. Rose. 1989. Antibacterial properties of retail sponges. J. Food Prot. 52:49-50, 54.
48. Loughhead, H. O, and J. A. Moffett. 1971. Air-sampling techniques for monitoring microbiological contamination. Bull. Parenter. Drug Assoc. 25:261.
49. Lundholm, I. M. 1982. Comparison of methods for quantitative determinations of airborne bacteria and evaluation of total viable counts. Appl. Environ. Microbiol. 44:179-183.
50. Macher, J. M., and M. W. First. 1983. Reuter centrifugal air sampler: Measurement of effective air flow rate and collection efficiency. Appl. Environ. Microbiol. 45:1960-1962.
51. Marshall, R. T. (ed.). 1993. Standard methods for the examination of dairy products, 16th ed. Am. Public Health Assoc., Washington, D. C.
52. May, K. R. 1967. Physical aspects of sampling airborne microbes, p. 60-80. *In* Airborne Microbes, 17th Symposium of the Society for General Microbiology. Cambridge Univ. Press, New York.
53. McGoldrick, K. F., T. L. Fox, and J. S. McAllister. 1986. Evaluation of a dry medium for detecting contamination on surfaces. Food Technol. 40:77-80.
54. NASA. 1967. NASA standards for clean rooms and work stations for the microbially controlled environment. NHB 5340.2. Natl. Aeronautics and Space Admin. U.S. Government Printing Office, Washington, D. C.
55. Placencia, A. M., and G. S. Oxborrow. 1984. Technical Report. Use of the Reuter centrifugal air sampler in good manufacturing practices investigations. U.S. Food and Drug Admin., Sterility Research Center, Minneapolis Center for Microbiol. Invest., Minneapolis, Minn.
56. Placencia, A. M., J. T. Peeler, G. S. Oxborrow, and J. W. Danielson. 1982. Comparison of bacterial recovery by Reuter centrifugal air sampler and slit-to-agar sampler. Appl. Environ. Microbiol. 44:512-513.
57. Radmore, K., and H. Luck. 1984. Microbial contamination of dairy factory air. S. Afr. J. Dairy Technol. 16:119.
58. Scheurrman, E. A. 1972. The gelatin membrane filter method for the determination of airborne bacteria. Pharm. Ind. 34:756.
59. Silas, J. C., M. A. Harrison, J. A. Carpenter, and J. B. Floyd. 1986. Comparison of particulate air samplers for detection of airborne *Aspergillus flavus* spores. J. Food Prot. 49:236-238.
60. Silliker, J. H., and D. A. Gabis. 1975. A cellulose sponge sampling technique for surfaces. J. Milk Food Technol. 38:504.
61. Stinson, C. G., and N. P. Tiwari. 1978. Evaluation of quick bacterial count methods for assessment of food plant sanitation. J. Food Prot. 41:269-271.
62. Sullivan, J. J. 1979. Air microbiology and dairy processing. Aust. J. Dairy Technol. 34:133-138.
63. Sunga, F. C. A., D. R. Heldman, and T. I. Hedrick. 1966. Characteristics of airborne microorganism populations in packaging areas of a dairy plant. Mich. State Univ. Agric. Exp. Sta. Q. Bull. 49:155-163.
64. Tiedman, W. D. (chair). 1948. Technic for the bacteriological examination of food utensils. Committee report, p. 68. *In* American Journal of Public Health Yearbook 1947-48 (Part 2). Washington, D. C.
65. U.S. Government. National Archives and Records Administration. 1998. CFR. Title 21. Part 192, Processing and bottling of bottled drinking water. p. 264-269.
66. U.S. HEW. 1967. Procedure for the Bacteriological Examination of Food Utensils and/or Food Equipment Surfaces. Pub. Health Serv. Publ. No. 1631, Tech Info. Bull. No. 1. Dep. Health, Ed., Welfare, Food and Drug Admin., Washington, D. C.
67. Walter, W. G., and J. Potter. 1963. Bacteriological field studies on eating utensils and flat surfaces. J. Environ. Health 26:187.

Chapter 4

Microscopic Methods

J. W. Bier, D. F. Splittstoesser, and Mary Lou Tortorello

4.1 INTRODUCTION

The microscope is often regarded as the definitive tool of the microbiologist, perhaps because the microbial world was discovered through its use. If there is one instrument that connects all of the subdisciplines of microbiology, it is the microscope. It is the only instrument that allows us to observe microorganisms directly.

Although the microscope has very old origins, it is by no means an obsolete device. In the development of more rapid microbiological methods, we should be reminded of the attractiveness of the microscope for its speed of analysis. When used in combination with concentration methods and diagnostic staining techniques, the microscope can provide specific detection and quantitation of microbial cells in short order.

Microscopy has been featured in a number of analytical methods for the microbiological examination of foods. Bright field microscopy is necessary, for example, in Gram staining, the first step in microbial identification[50]; for indicating the presence of endospores of *Bacillus* and *Clostridium*[60,61,71]; in differentiation of parasite eggs and protozoan parasites[13]; and for determining the invasiveness of enteropathogens by observing their entry into mammalian tissue culture cells.[6] Dark field and phase contrast microscopy are recommended for differentiating *Bacillus* species by motility,[61] for observing the characteristic corkscrew motility of *Campylobacter*[34] and the tumbling motility of *Listeria*.[30] Fluorescence microscopy is useful for fluorescent antibody identification of pathogens,[2] and for visualization of the autofluorescent oocysts of *Cyclospora*.[11]

Microscopy can be a quantitative technique. Direct microscopic counts are among the standard methods for microbiological examination of eggs[7] and for grading of milk.[52] Mold[8] and yeast[10] contamination of foods may be quantified by bright field microscopy and fluorescence microscopy, respectively. The direct epifluorescent filter technique (DEFT) combines membrane filtration with epifluorescence microscopy for sample concentration and determination of total microbial cell counts.[53]

This chapter will be concerned with the types of microscopy mentioned above. However, there are other types that can be useful in food microbiology, but which are more appropriate in research investigations because of applicability, cost and the technical expertise required. Among them are the confocal laser scanning microscope, an instrument which eliminates blur from out-of-focus sections of an image, thus allowing resolution of detail in specimens thicker than possible with conventional optics[78]; and electron microscopy, which provides even greater resolution, with detection limits in the nanometer range. Transmission electron microscopy allows detection of virus particles and can provide details of subcellular structures of bacteria and their hosts. Scanning electron microscopy can provide a three-dimensional impression of surfaces, for example in studying pathogen-host interactions and biofilm structure.[81]

4.2 GENERAL PROCEDURES

Cleanliness is an essential practice for successful microscopy. Microscope lenses, slides, and cover slips should be clean and free from dust and residues. Because oil immersion lenses are commonly used in microbiological examination, it is important to clean the lenses after every use and to avoid leaving residues of oil which can become dry and obscure visibility later on. It is also important to avoid oily contamination of lenses not meant to be used with oil. Dried oil residues can be removed from lenses with small amounts of lens cleaner such as xylene.

Proper illumination is also critical. The position of the condenser lens and the field diaphragm (if present) should be adjusted to achieve optimal lighting and specimen contrast (Koehler illumination). Procedures for optimizing illumination vary with the type of microscope and are available from the manufacturer. Room lighting should also be considered, especially in fluorescence and dark field microscopy, where a darkened room is best for viewing.

Finally, it is important to set up the microscope for the physical comfort of the microscopist. The interocular distance of a binocular microscope must be adjusted for the distance between the eyes so that a single field of view results, and the adjustable ocular lens should be focused for the sightedness of the microscopist. Chair height also should be positioned so that the neck and back muscles are in a relaxed state.

Practical tips on cleanliness and instructions for illumination and set-up are usually available from the microscope manufacturer, and general principles pertaining to these topics have been published.[16,49]

4.3 FLUORESCENCE MICROSCOPY

The major advantage of fluorescence microscopy is that its high sensitivity makes possible the visualization of substances in low concentrations or particles smaller than the resolution allowed by the transmitted light microscope. The numerous applications of fluorescence microscopy are quite specific in their details and beyond the scope of this general discussion.

Fluorochromes are substances that absorb light energy (often in the ultraviolet range) and when activated dissipate the absorbed energy in the form of longer wavelength radiation upon return to the resting state. The microscopist is interested in that the dissipated energy occurs in the visible spectrum. To observe this light the microscopist must visualize it on a dark field for contrast. The direct fluorescence microscope consists of an excitation filter which limits the light emitted from the source to a specific range that activates a substance of interest in the specimen. A barrier filter transmits light in the range emitted by the specimen when it returns to the resting state. Most fluorescence microscopes today are of the epifluorescence type which allow visualization of fluorescence reflected from the fluorochrome. The epifluoresence microscope includes a dichroic mirror in its light path in addition to the excitation and barrier filter.

Various light sources are utilized for fluorescence microscopes. Xenon or mercury vapor lamps are common sources of ultraviolet light; some confocal microscopes use lasers for their light source. The important feature of any light source is that it produce an adequate amount of emissions in the range necessary for activating the fluorescent substance of interest.

There are several types of fluorescence used by microscopists. Autofluoresence is the natural fluorescence found in tissues. Induced fluorescence is the result of a chemical reaction with a component in the tissue. Numerous dyes of various specificities are available that fluoresce. These flurochromes may be linked to an immune component to give immunofluoresence. Numerous fluorochromes and many antibody-linked fluorochromes are available from chemical supply houses. Microscope manufacturers produce a limited number of filter blocks that contain the excitation filter, barrier filter and dichroic mirror combination necessary for use with the most common dyes, but specialty manufacturers produce interchangeable filter cubes for more exotic dyes.

4.4 STAINS, PROBES, AND INDICATORS (see Chapter 63)

4.41 Direct Staining

Most microbial cells do not have sufficient contrast for visualization by bright field microscopy (notable exceptions being cells which contain intracellular pigments such as chlorophyll). Chemical stains are used to provide the contrast needed for observing most microbial cells by bright field microscopy. Staining destroys the viability of the cells, however. If it is necessary to observe viable cells, e.g. in determining motility, then the instrument of choice is the phase contrast microscope, which increases the contrast of the cells against the background so that they become visible without staining.

Direct staining can help to generally classify microbial cells, e.g. bacteria, yeasts, molds, protozoa, as well as to determine cellular morphology: rods, cocci, spirals, filamentous forms. Bacterial endospores can be discerned because they are not readily stained, but appear as refractile bodies within or separate from stained vegetative cells.

Heat fixation is commonly done prior to direct staining to promote cell adherence to the microscope slide. A thin film of the cell suspension is smeared onto the slide, allowed to dry, and the underside of the slide is quickly passed over a Bunsen burner flame several times to fix the cells onto the slide. The stain is then applied to cover the entire area of the smear. Commonly used stains are methylene blue, crystal violet and carbol fuchsin, all of which can provide adequate staining in one minute or less. After gentle rinsing with a stream of water and drying between sheets of blotting paper, the slide is ready for examination.

4.42 Differential Staining

Differential stains allow microbial cells to be distinguished based on a particular property or characteristic. The most important differential stain for the food microbiologist is the Gram stain, which categorizes most bacteria as either gram positive or gram negative depending on their reaction to the staining procedure. The primary reason for the distinction lies within the structure of the cell wall, which determines whether the procedure's decolorization step will cause the stain to be washed out of the cells. It is best to use freshly-grown cultures, because older cultures, especially gram positive bacteria, may show variable reactivity in the staining procedure.

Gram staining reagents are available commercially or may be prepared from component dyes and solvents. Although a number of variations exist, the staining protocol generally consists of 4 steps: the initial stain (crystal violet), the mordant (iodine), decolorization (alcohol) and the counterstain (safranin). A typical Gram staining protocol involves crystal violet staining of a heat-fixed smear for 30 seconds, application of the iodine mordant (30 seconds), alcohol decolorization (15 seconds) and safranin counterstaining (30 seconds), with gentle water rinsing after each step.

4.43 Special Structural Stains

A number of stains for special structures are of use to the food microbiologist. These include staining for bacterial endospores, storage materials, protozoan cysts and parasite eggs. They will be briefly described here, but are also covered in relevant chapters of this *Compendium*.

Bacterial endospores are not easily stained because of their impermeable nature. Heat must be applied to weaken the endospore wall and allow uptake of the stain. The Schaeffer-Fulton procedure involves staining of a fixed smear with malachite green on a rack above a boiling water bath for 5 minutes. After rinsing with water, a counterstain such as safranin, is applied for 30 seconds to stain the vegetative portion of the cell red and provide contrast for the green-stained endospore. Endospore staining helps in speciation, because the location and shape of the endospores within the cell are often diagnostic; for example, *Bacillus cereus* endospores appear ellipsoidal and are located in a central to subterminal position,[61] while *Clostridium botulinum* produces a characteristic "tennis racket" appearance.[71] Before testing a culture for the presence of *C. perfringens* enterotoxin, which is produced during sporulation, the concentration of endospores should be determined microscopically, so that adequate levels of the enterotoxin are ensured for the assay.[60]

Certain bacterial species accumulate polymers as storage materials, e.g. glycogen, poly-β-hydroxybutyrate and polyphosphate. These storage materials can be observed microscopically through the use of special stains.[50] Food microbiologists are generally not concerned with storage material staining, except in the case of *Bacillus* speciation, where it becomes important in distinguishing between the foodborne pathogen *B. cereus* and the closely related insect pathogen *B. thuringiensis.* The two species are identical for many phenotypic characteristics, but a primary distinguishing trait is the production of insecticidal protein toxin by *B. thuringiensis*.[61] The protein toxin is visible as diamond-shaped crystals after sporulation of the species. Heat-fixed smears are treated with methanol for 30 seconds, then air-dried. The smear is covered with carbol fuchsin or basic fuchsin, then heated gently over a flame until steaming of the dye is observed. After cooling the slide for 1–2 minutes, the heating is repeated, followed by rinsing of the slide with water.

Protozoan cysts and parasite eggs may be identified by their microscopic morphology. After separation and concentration from the food matrix, the cysts and eggs may be stained with Lugol's iodine,[13] or by fluorescent antibody reaction.

4.44 Specific Probes

Microbial cells may be distinguished by the use of specific probes, notably antibodies and oligonucleotides. Probes may be designed for recognition of various levels of specificity, e.g. genus, species, strain. Labeled with fluorescent dyes and used in conjunction with the fluorescence microscope, antibodies and oligonucleotides can be direct, specific identifiers of microbial cells.

Antibodies are protein molecules produced as part of the immunological response to foreign substances (antigens) in animals. The proteins, polysaccharides, and their derivatives which make up microbial cell structure are antigenic and promote production of antibodies, which, in turn, can be used as specific probes of the microbial cells. Identification of a "target" antigen that is unique to a microbial group (genus, species, strain, etc.) is the first step in the design of an antibody probe. Antigens usually are present in the cell in amounts sufficient for detection by microscopy; however, certain antigens may be influenced by environmental conditions and may not be expressed at all by the cell. In fact, antibody probes have been used to study regulation of expression by tracking the appearance of antigens microscopically.[28]

Fluorescent antibodies have been used for microscopic identification of *Salmonella* in foods for many years, and the Fluorescent Antibody Screening Method for *Salmonella* has undergone the extensive testing required of an official method.[2] Its use is limited to screening, i.e. presumptive identification rather than confirmation, because of cross-reactivity of the commercial antibody preparation to some species of the *Enterobacteriaceae* other than *Salmonella*. Immunofluorescence microscopy has been used for identification of many other microbial species, including bacterial foodborne pathogens, e.g. *Listeria monocytogenes*,[41] *E. coli* O157:H7,[73] and protozoa, e.g. *Cryptosporidium* and *Cyclospora*.[51] As is true for *Salmonella* identification, the method is often limited by the specificity of the fluorescent antibody probe.

Oligonucleotides are short nucleotide polymers which bind to specific segments of nucleic acids by complementary base pairing. As with the antigens, identification of a unique target sequence (DNA or RNA) is the first step in oligonucleotide probing. Unlike the antigens, however, genetic sequences are usually present in the cell in amounts too low for detection by light microscopy. An exception is the relatively abundant amounts of RNA present in the ribosomes of growing cells. Fluorescent oligonucleotides complementary to ribosomal RNA sequences have been used as "phylogenetic stains" for bacterial cell detection.[22] For low-copy number sequences, amplification of the sequence[32] or of the probe signal[1,46] has been used to increase detectability. The procedure for oligonucleotide probing is slightly more involved than the one-step reaction usually needed with antibody probes. Microbial cells may need to be pre-treated to permit entry of the oligonucleotide to the interior of the cell. Heat and chemical treatment are required for hybridization of the probe to the appropriate nucleic acid sequence. The use of oligonucleotide probes for direct microscopic analysis has not been as popular among food microbiologists as it has been among microbial ecologists.

4.45 Indicators of Viability

The microscope is a good tool for determining cell viability, because it allows a direct determination to be made which does not depend on the ability of a cell to form a colony. Although the generally accepted definition of a viable cell is based on this ability, the definition is flawed, because not all viable cells can be grown in culture.[1,47,68] Of practical importance to food microbiologists are cells which have been sublethally injured by food processing treatments.[18,36,58,59] Although much work has been done on methods for resuscitation of such cells,[25,37,57] uncertainty regarding their detection remains. The ability to directly determine viability of cells with the microscope may help to improve detection.

Acridine orange is the stain most commonly used in the direct epifluorescent filter technique[53] for quantitation of the total microbial population in a food sample (section 4.55). Binding of this fluorescent dye to single-stranded RNA or double stranded DNA results in red-orange or green fluorescence, respectively, and because growing cells have an abundance of RNA, the characteristic red-orange fluorescence was associated with cellular viability.[31] However, the differentiation depends on the matrix in which the bacterial cells are suspended and does not hold under a wide variety of conditions.[40] Acridine orange is used more appropriately as a general stain rather than a viability indicator.

The development of microscopic methods for distinguishing living and dead bacteria has been pursued for over a century, and a history of such methods has been published.[67] Several viability indicators which measure different aspects of viability and which may be used in microscopy are described in this section.

The earliest microscopic viability methods described were based on dye exclusion: viable cells with impermeable membranes excluded certain dyes while dead cells were stained due to dye uptake. A variety of dyes are available and appropriate for use with bacteria, yeast or other eukaryotic cells.[67] For example, permeability to the fluorescent dye propidium iodide has been used to identify dead *E. coli* cells.[39] Because only the dead cells take up certain dyes, it is useful to apply an appropriate counterstain to the population for differentiation of viable cells.[9,38,48]

Enzyme activity is also indicative of viability. An example is the enzymatic activity associated with respiration, i.e. functional electron transport system components. Redox indicators which accumulate in the cell as insoluble chromogens[80] or fluorescent derivatives[66] upon reduction have been used to indicate electron transport activity in direct microscopic assays. Viable microbial populations in bottled water[24] and in milk[12] have been enumerated with redox indicators. Fluorescent antibody staining has been coupled with the redox indicator CTC (cyanoditolyl tetrazolium chloride) to achieve counts of specific viable *E. coli* O157:H7.[56]

A useful fluorescent staining reagent that depends on both the presence of an intact membrane and enzyme activity is fluorescein diacetate (FDA) and derivatives.[77] FDA is a hydrophobic compound that accumulates slowly in viable cells due to the low permeability of the membrane to hydrophobic compounds. Dead cells do not accumulate the compound due to their leaky membranes. Inside the cell, the FDA molecule is cleaved by esterase activity to produce the fluorescent product fluorescein.

Cells that are capable of undergoing cell division increase in length but do not complete the formation of new cell walls in the presence of nalidixic acid and nutrients; thus elongated cells indicate viability in the direct viable count (DVC) assay.[44] The assay compares 2% formaldehyde-fixed cells with unfixed cells suspended in yeast extract (0.025%) and nalidixic acid (0.002%). The cells are incubated for 6 hours to allow growth processes to proceed, collected on a membrane filter, stained with acridine orange and examined by epifluorescence microscopy. The use of inhibitors other than nalidixic acid has extended the technique to a variety of bacteria, including gram positive cells.[17,27,69] The DVC assay also has been combined with fluorescent antibody staining to specifically enumerate viable *Vibrio* and *Salmonella*.[14,23] Much of the information on the viable but nonculturable state in bacteria has been obtained by using the DVC assay.

Viable cells incubated in the presence of nutrients can form microcolonies which are detectable microscopically.[63, 64] The microcolony assay involves membrane filtration of a sample, then incubation of the filter on selective media for several hours to allow bacterial cell division and formation of microcolonies on the filter surface. A stain is then applied, and the filter is examined by epifluorescence microscopy. Acridine orange[63, 64] or fluorescent antibody[65, 70] staining may be used to provide total and specific viable counts, respectively. The microcolony assay has the same limitation as plate counting, i.e., it presumes provision of appropriate growth conditions for the cells. If the cells are injured, their ability to form colonies (even microcolonies) on selective media is compromised. Detergent and enzyme treatments that are routinely used to allow filtration of the food sample were found to inhibit microcolony formation,[63] and resuscitation measures have been incorporated into the procedure to overcome the inhibition.[64]

Active transcription of messenger RNA has also been used as an indicator of bacterial viability in the reverse transcriptase-polymerase chain reaction (RT-PCR),[42] and *in situ* hybridization techniques have been developed for microscopic detection of cells possessing specific genes or gene products by PCR and RT-PCR, respectively.[32] Even low levels of gene product expression, e.g. *lac* in *Salmonella typhimurium* cells, have been visualized microscopically by *in situ* RT-PCR.[74]

4.5 QUANTITATION

4.51 Introduction

Rapid, direct quantitation of microbial populations may be obtained by microscopy. The sensitivity or limit of detection often cannot match that of agar plate counts, but depends on the procedure, especially the preparative steps. Counting chambers, dried films on microscope slides, and membrane filters are some of the implements of quantitation by microscopy.

4.52 Counting chambers for bacteria

Numerous counting chambers are available,[43] including the Helber, the Hawksley, the Petroff-Hausser, and the hemocytometer. All are similar: each consists of a grid of etched squares of a given area and is covered with a glass slip that is positioned a fixed distance from the etched surface. Counts are usually made at about 400X magnification, although some chambers such as the Hawksley permit the use of an oil immersion lens.

The volume of liquid within the etched square equals the area of the square times the depth of the film. The average cell count per square multiplied by the reciprocal of the volume in milliliters (i.e., the chamber factor) will equal the concentration of the microorganisms in the diluent. Chamber factors commonly range from 4×10^6 to 2×10^7; thus the procedure is most applicable to foods that contain large microbial populations.

Major sources of error are the difficulty of accurate filling of chambers and the adsorption of cells to glassware surfaces.[43] When using a microscope in which the focus dial reads directly in micrometers, the exact chamber thickness can be measured by focusing on cells attached to the cover slip and to the bottom of the slide. Adsorption can be reduced by using anionic detergent diluents and plastic-tipped pipettes.

The material to be analyzed should be prepared in diluent, e.g. 0.1% peptone water containing 0.1% lauryl sulfate, so that the concentration of bacteria will equal 5–15 cells per small square of the counting chamber grid. The diluted material is added to fill the counting chamber, and allowed to settle for about 5 minutes. Using phase contrast, a sufficient number of squares is counted to give a total count of about 600 cells. The number of microbial cells per gram or milliliter is calculated by multiplying the average count per small square by the chamber factor by the dilution factor.

4.52 Howard Mold Count

The Howard mold count was established many years ago[33] to ensure that catsup would be made from tomatoes that were relatively free of visible rots. Although most widely applied to tomato products (catsup, juice, paste, sauce, canned tomatoes, soup, pizza sauce, etc.), the Howard mold count has also been used to assure the quality of other foods such as frozen berries, cranberry sauce, citrus and pineapple juices, fruit nectars and purees, and pureed infant food.

The mold count is a standardized procedure[4] to determine the percentage of microscopic fields containing mold filaments whose combined lengths exceed one-sixth the diameter of the field. The Food and Drug Administration has established for many food products regulatory action guidances that include Howard mold count criteria.[26]

The analyst must be familiar with the microscopic appearance of sound food tissue and with the morphology of the more common molds in order to distinguish mold filaments from other fibers. Methodologies for preparation of individual food products for Howard mold counting have been established.[4] The food sample is placed on the center of a Howard slide and spread evenly with a scalpel. The cover slip is lowered rapidly so that the material is distributed evenly over the center of the slide but not drawn across the moat. Proper contact between the cover slip and the slide can be confirmed by observation of colored bands known as Newton's rings.[4] Twenty-five fields from two or more mounts should be counted. Positive fields are those in which (a) a single filament exceeds one-sixth of the field diameter, (b) a filament plus the length of its branches exceeds one-sixth of the field diameter, or (c) an aggregate of not more than three filaments exceeds one-sixth of the field. Results are calculated as percentage of positive fields.

4.53 Geotrichum Count

The incidence of filaments of the mold *Geotrichum candidum* in canned and frozen fruit and vegetable products is an indicator of the hygienic condition of the food processing equipment. *Geotrichum* grows on the surfaces of processing equipment,[79] and thus has been termed machinery mold. The method involves microscopic counting of typical filaments using a rot fragment slide at 30–45× magnification.[3,20] Low filament counts, under 1 per g, often do not correlate well with the aerobic plate counts on frozen vegetables.[72]

Specific procedures for various types of foods have been established.[3] The following procedure describes the determination of *Geotrichum* mold in canned vegetables, fruits and juices.[5] The net weight of the can contents is determined, then drained on a No. 8 sieve positioned over a pan. The food is removed from the sieve and discarded, and the container and sieve are washed with about 300 mL water. The liquid and washings are transferred to a 5″ No. 16 sieve that rests on a 2 L beaker. The residue is washed with 50 mL water, and the residue is discarded. The liquid and washings are transferred to a 5″ No. 230 sieve, tilted at a 30° angle. The solids are flushed to the sieve edge with a wash bottle, transferred to a 50 mL graduated centrifuge tube, and diluted to 10 mL. If the initial volume of the residue exceeds 10 mL, it should be concentrated by centrifugation. One drop of crystal violet is added to the 10 mL of residue, the tube is mixed, and then 10 mL of stabilizer solution is added to bring the total volume to 20 mL. After thorough mixing, 0.5 mL is transferred as a streak approximately 4 cm long to a rot fragment counting slide. Duplicate slides should be prepared and counted. The entire surface of the slide is examined at 30–45×, using transmitted light. *Geotrichum* fragments that contain three or more hyphal branches at 45° angles from the main filament are counted. The *Geotrichum* count per 500 g of food is equal to $(S/V) \times (500/W) \times 20$, where S = average number of fragments per slide; V is the total volume counted (0.5 mL per slide multiplied by the number of slides counted); W = net weight of sample in g.

4.54 Dried Films

The microscopic examination of a thin film of food dried onto a slide is one of the simplest microbiological techniques available. The method can be used to determine the morphological types of bacteria within a food sample, e.g. staphylococci[16] or endosporeformers,[45] but it can also be used for quantitation of microbial populations in a food. Although a quantitative dried film procedure was originally developed[15] and widely used for grading of milk,[52] similar methods have been applied for the examination of liquid eggs and other foods.[7,16] The general principle is that a known quantity of food is spread over a prescribed area of a microscope slide. After the film is dried and stained, the average number of organisms per microscope field is determined. This count can be converted into numbers per g or mL of food based on the area of the microscope field. Advantages of the method are that it is rapid and that the slides may be retained for later reference. A disadvantage is that it is applicable only to foods containing large populations of microorganisms.

The microscope slides used for dried films have one or more circular 1-cm² (diameter 11.28 mm) areas circumscribed by either painted or etched rings, which are used to contain the material for analysis in a defined area. The material (0.01 mL) is transferred into the circle with a pipette, spread uniformly over the area with a bent needle, and air-dried at 40–45°C. The film is fixed by immersion in 95% ethanol. If the material contains a high fat content, the film may be defatted by immersing in xylene for 1–2 min; washed in methanol and dried. After fixation, the film is stained. North's aniline oil methylene blue, which is recommended for liquid eggs, may be used to stain the film for 10–20 min. The number of microbial cells in 10–100 fields is counted, and the average count per field is determined.

The next task in the calculation of microbial cell concentration in the food is to determine the microscope factor (MF). Each microscope lens will have a different MF; for typical bacterial counts, the 100× lens is used. The MF is the number of microscope fields in the 1 cm²-slide area, divided by the volume of material (0.01 mL) applied to the slide. The area of the microscope field is determined by viewing a stage micrometer through the microscope (100× objective lens in place), and measuring the diameter of the field in mm to the third decimal place. The field area in cm² is calculated by applying the formula for the area of a circle (πr^2); i.e., dividing the field diameter by 2 to get the radius, squaring the radius and multiplying by 3.1416; then dividing by 100 to convert mm² to cm². The MF is obtained by dividing the slide area (1 cm²) by the field area to obtain the number of microscope fields in the slide area; and then dividing this quantity by 0.01, which is the volume in mL spread over the slide area. Therefore, in condensed form,

$$MF = 10{,}000 \text{ divided by } \pi r^2$$

where r = radius of the microscope field in mm. Finally, the microbial cell concentration in the food is determined by multiplying the average count per field by the MF. If a dilution of the food material was made prior to analysis, the calculated concentration should be multiplied by the reciprocal of the sample dilution. Results may be expressed in number of microbial cells per mL of food.

4.55 Direct Epifluorescent Filter Technique

Membrane filtration may be used to concentrate material for analysis by collecting it on the filter surface. Concentrating the material, including the microbial cells present within it, provides an effective way to lower the limit of detection of an analytical method, thus increasing its sensitivity. Membrane filtration has been combined with microscopy in the direct epifluorescent filter technique (DEFT)[53] to increase sensitivity by several orders of magnitude over other microscopic methods. The membrane filter not only concentrates the cells, but also provides the surface upon which the microscopic examination is made. Incident light as in epifluorescence, rather than transmitted light, is the type of illumination required for examination of the membrane filter surface.

Procedures have been developed for analysis of beverages,[10,62] diluted food homogenates[54] and surface rinses[35] by DEFT. Although DEFT has primarily been used for enumeration of single cells and cell clumps, the transfer of membranes to selective agar media following filtration also permits the counting of microcolonies.[23] A fluorescent dye such as acridine orange is used to stain the cells. Fluorescent antibodies have also been applied for specific enumeration of microbial cells.[75,76]

The methods differ depending upon the food type, the microorganisms of interest, and whether cells or microcolonies are to be enumerated. Various food suspensions, for example, will require different pre-treatments or enzymatic digestions to allow passage through the membrane filter.

The following procedure describes the enumeration of microbial cells in raw milk.[19,52,55] All reagents should be filter-sterilized through 0.22-µm pore size membrane filters before use in the DEFT. Somatic cells and lipid micelles in milk are lysed by adding 0.5 mL rehydrated Bacto-trypsin (Difco) and 2 mL of 0.5% Triton X-100 to 2 mL of milk. The mixture is incubated at 50°C for 10 min, then added to a previously warmed filter assembly (Millipore) holding a 25 mm-diameter, 0.6-µm pore size black Nucleopore polycarbonate membrane (shiny side up). A vacuum is applied to filter the digested milk, and the assembly is rinsed with 5 mL pre-warmed 0.1% Triton X-100. The vacuum is disconnected, and the membrane is overlayed with 2 mL of acridine orange stain for 2 minutes. The membrane is rinsed under vacuum with 2.5 mL 0.1 M citrate-NaOH buffer, pH 3, followed by 2.5 mL 95% ethyl alcohol. After air drying, the filter is mounted on a slide in a drop of nonfluorescent immersion oil, and a cover slip is applied. It is examined using an epifluorescence microscope fitted with an appropriate fluorescence filter combination for acridine orange.

Orange fluorescent microbial cells are counted in randomly selected microscope fields around the filter. The number of fields that should be counted depends on the microbial cell density per field.[52] For fields with 0–10 cells, 15 fields are counted; for 11–25 cells, 10 fields; for 26–50 cells, 6 fields; for 51–75 cells, 3 fields; for 76–100 cells, 2 fields. If there are more than 100 cells per field, the sample should be diluted before analysis. As in the dried film procedure (section 4.54), the number of cells per mL is obtained by multiplying the average number of cells per field by the MF; in the DEFT, the MF calculation is based on the area of the membrane filter. The membrane filter microscope factor (MFMF) is calculated by determining (A) the area of the membrane through which the sample was filtered, in mm^2 and (B) the area of the microscope field of view, in mm^2 (see section 4.54). The MFMF is determined by A divided by B. The microbial cell concentration is calculated by multiplying the average number of cells per field by the MFMF, and dividing by the volume of material filtered. If a dilution of the original material was made prior to filtration, the reciprocal of this dilution should be multiplied into the calculation.

4.6 REFERENCES

1. Amann, R. I., W. Ludwig, and K.-H. Schleifer. 1995. Phylogenetic identification and *in situ* detection of individual microbial cells without cultivation. Microbiol. Rev. 59:143-169.
2. AOAC International. 1997. Official methods of analysis, 16th ed., 3rd rev. *Salmonella* in foods, fluorescent antibody screening method. AOAC International, Gaithersburg, Md. Method 975.54.
3. AOAC International. 1997. Official methods of analysis, 16th ed., 3rd rev. *Geotrichum* mold counting. AOAC International, Gaithersburg, Md. Method 984.30.
4. AOAC International. 1997. Official methods of analysis, 16th ed., 3rd rev. Howard mold counting, general instructions. AOAC International, Gaithersburg, Md. Method 984.29.
5. AOAC International. 1997. Official methods of analysis, 16th ed., 3rd rev. Mold in vegetables, fruits and juices (canned), *Geotrichum* mold count. AOAC International, Gaithersburg, Md. Method 974.34.
6. AOAC International. 1997. Official methods of analysis, 16th ed., 3rd rev. Invasiveness of mammalian cells by *Escherichia coli*, microbiological method. AOAC International, Gaithersburg, Md. Method 982.36.
7. AOAC International. 1997. Official methods of analysis, 16th ed., 3rd rev. Techniques for eggs and egg products, microbiological methods. AOAC International, Gaithersburg, Md. Method 940.37.
8. AOAC International. 1997. Official methods of analysis, 16th ed., 3rd rev. Howard mold counting, general instructions. AOAC International, Gaithersburg, Md. Method 984.29.
9. Autio, K., and T. Mattila-Sandholm. 1992. Detection of active yeast cells (*Saccharomyces cerevisiae*) in frozen dough sections. Appl. Environ. Microbiol. 58:2153-2157.
10. Bandler, R., M. E. Stack, H. A. Koch, V. H. Tournas, and P. B. Mislivec. 1995. Yeasts, molds and mycotoxins, p. 18.01-18.08. *In* Bacteriological analytical manual, 8th ed. AOAC International, Gaithersburg, Md.
11. Berlin, O. G. W., J. B. Peter, C. Gagne, C. N. Conteas, and L. R. Ash. 1998. Autofluorescence and the detection of *Cyclospora* oocysts. Emerg. Infect. Dis. 4:127-128.
12. Betts, R. P., P. Bankes, and J. G. Banks. 1989. Rapid enumeration of viable microorganisms by staining and direct microscopy. Lett. Appl. Microbiol. 9:199-202.
13. Bier, J. W., G. J. Jackson, A. M. Adams, and R. A. Rude. 1995. Parasitic animals in foods, p. 19.01-19.09. *In* Bacteriological analytical manual, 8th ed. AOAC International, Gaithersburg, Md.
14. Brayton, P. R., M. L. Tamplin, A. Huq, and R. R. Colwell. 1987. Enumeration of *Vibrio cholerae* O1 in Bangladesh waters by fluorescent antibody direct viable count. Appl. Environ. Microbiol. 53:2862-2865.
15. Breed, R. S. 1911. The determination of bacteria in milk by direct microscopic examination. Zentralbl. Bakteriol. II. Abt. 30: 337-340.
16. Bryce, J. R., and P. L. Poelma. 1995. Microscopic examination of foods and care and use of the microscope, p. 2.01-2.06. *In* Bacteriological analytical manual, 8th ed. AOAC International, Gaithersburg, Md.
17. Buchrieser, C., and C. W. Kaspar. 1993. An improved direct viable count for the enumeration of bacteria in milk. Int. J. Food Microbiol. 20:227-236.
18. Busta, F. F. 1976. Practical implications of injured microorganisms in food. J. Milk Food Technol. 39:138-145.
19. Champagne, C. P., N. J. Gardner, J. Fontaine, and J. Richard. 1997. Determination of viable bacterial populations in raw milk within 20 minutes by using a direct epifluorescent filter technique. J. Food Prot. 60:874-876.
20. Cichowicz, S. M., and W. V. Eisenberg. 1974. Collaborative study of the determination of *Geotrichum* mold in selected canned fruits and vegetables. J. Assoc. Off. Anal. Chem. 57:957-960.
21. Colwell, R. R. 1993. Nonculturable but still viable and potentially pathogenic. Zentralbl. Bakteriol. 279:154-156.
22. DeLong, E. F., G. S. Wickham, and N. R. Pace. 1989. Phylogenetic stains: ribosomal RNA-based probes for the identification of single cells. Science 243:1360-1363.
23. Desmonts, C., J. Minet, R. Colwell, and M. Cormier. 1990. Fluorescent antibody method useful for detecting viable but nonculturable *Salmonella spp.* in chlorinated wastewater. Appl. Environ. Microbiol. 56:1448-1452.
24. Ferreira, A.-C., P. Vasconcellos Morais, and M. S. da Costa. 1994. Alterations in total bacteria, iodonitrophenyltetrazolium (INT)-positive bacteria, and heterotrophic plate counts of bottle mineral water. Can. J. Microbiol. 40:72-77.
25. Foegeding, P. M., and B. Ray. 1992. Repair and detection of injured microorganisms, p. 121-134. *In* C. Vanderzant and D. F. Splittstoesser (ed.), Compendium of methods for the microbiological examination of foods, 3rd ed. American Public Health Association, Washington, D. C.
26. Food and Drug Administration. 1996. Foods, colors and cosmetics, p. 165-338. *In* Compliance policy guides. Food and Drug Administration, Rockville, Md.
27. Frank, J. F., M. A. Gassem, and R. A. N. Gillett. 1992. A direct viable count method suitable for use with *Listeria monocytogenes.* J. Food Prot. 55:697-700.
28. Harry, E. J., K. Pogliano, and R. Losick. 1995. Use of immunofluorescence to visualize cell-specific gene expression during sporulation in *Bacillus subtilis.* J. Bacteriol. 177:3386-3393.
29. Haugland, R. P. 1996. Assays for cell viability, proliferation and function, p. 365-398. *In* Handbook of fluorescent probes and research chemicals, 6th ed. Molecular Probes, Inc., Eugene, Ore.
30. Hitchins, A. D. 1995. *Listeria monocytogenes,* p. 10.01-10.13. *In* Bacteriological analytical manual, 8th ed. AOAC International, Gaithersburg, Md.
31. Hobbie, J. E., R. J. Daley, and S. Jasper. 1977. Use of nucleopore filters for counting bacteria by fluorescence microscopy. Appl. Environ. Microbiol. 33:1225-1228.
32. Hodson, R. E., W. A. Dustman, R. P. Garg, and M. A. Moran. 1995. *In situ* PCR for visualization of microscale distribution of specific genes and gene products in prokaryotic communities. Appl. Environ. Microbiol. 61:4074-4082.

33. Howard, B. J. 1911. Tomato catsup under the microscope with practical suggestions to insure a cleanly product. U.S. Department of Agriculture Bureau of Chemistry. Circular No. 68.
34. Hunt, J. M., and C. Abeyta. 1995. *Campylobacter*, p. 7.01-7.27. *In* Bacteriological analytical manual, 8th ed. AOAC International, Gaithersburg, Md.
35. Hunter, A. C., and R. M. McCorquodale. 1983. Evaluation of the direct epifluorescent filter technique for assessing the hygienic condition of milking equipment. J. Dairy Res. 50:9-16.
36. International Commission on Microbiological Specifications for Foods. 1980. Injury and its effect on survival and recovery, p. 205-214. *In* Microbial ecology of foods, vol. 1. Factors affecting life and death of microorganisms. Academic Press, New York.
37. Jay, J. M. 1996. Metabolically injured organisms, p. 207-213. *In* Modern food microbiology, 5th ed. Chapman and Hall, New York.
38. Jenkins, M. B., L. J. Anguish, D. D. Bowman, M. J. Walker, and W. C. Ghiorse. 1997. Assessment of a dye permeability assay for determination of inactivation rates of *Cryptosporidium parvum* oocysts. Appl. Environ. Microbiol. 63:3844-3850.
39. Jorgensen, F., and C. G. Kurland. 1987. Death rates of bacterial mutants. FEMS Microbiol. Lett. 40:43-46.
40. Kepner, R. L., Jr., and J. R. Pratt. 1994. Use of fluorochromes for direct enumeration of total bacteria in environmental samples: past and present. Microbiol. Rev. 58:603-615.
41. Khan, M. A., A. Seaman, and M. Woodbine. 1977. Immunofluorescent identification of *Listeria monocytogenes*. Zentralbl. Bakteriol. Org. A 239:62-69.
42. Klein, P. G., and V. K. Juneja. 1997. Sensitive detection of viable *Listeria monocytogenes* by reverse transcription-PCR. Appl. Environ. Microbiol. 63: 4441-4448.
43. Koch, A. L. 1994. Growth measurement, p. 248-277. *In* P. Gerhardt, R. G. E. Murray, W. A. Wood, and N. R. Krieg (ed.), Methods for general and molecular bacteriology. American Society for Microbiology, Washington, D. C.
44. Kogure, K., U. Simidu, and N. Taga. 1979. A tentative direct microscopic method for counting living marine bacteria. Can. J. Microbiol. 25:415-420.
45. Landry, W. L., A. H. Schwab, and G. A. Lancette. 1995. Examination of canned foods, p. 21.01-21.22. *In* Bacteriological analytical manual, 8th ed. AOAC International, Gaithersburg, Md.
46. Matsuhisa, A., Y. Saito, H. Ueyama, Y. Aikawa and T. Ohono. 1994. Detection of staphylococci in mouse phagocytic cells by *in situ* hybridization using biotinylated DNA probes. Biotechnic Histochem. 69:31-37.
47. McKay, A. M. 1992. Viable but non-culturable forms of potentially pathogenic bacteria in water. Lett. Appl. Microbiol. 14:129-135.
48. Molecular Probes, Inc. 1996. Live/Dead *Bac*Light™ Bacterial Viability Kit, package insert MP 7007. Molecular Probes, Inc., Eugene, Ore.
49. Murray, R. G. E., and C. F. Robinow. 1994. Light microscopy, p. 7-20. *In* P. Gerhardt, R. G. E. Murray, W. A. Wood,a nd N. R. Krieg (ed.), Methods for general and molecular bacteriology. American Society for Microbiology, Washington, D. C.
50. Murray, R. G. E, R. N. Doetsch, and C. F. Robinow. 1994. Determinative and cytological light microscopy, p. 21-41. *In* P. Gerhardt, R. G. E. Murray, W. A. Wood, and N. R. Krieg (ed.), Methods for general and molecular bacteriology. American Society for Microbiology, Washington, D. C.
51. Ortega, Y. R., C. R. Roxas, R. H. Gilman, N. J. Miller, L. Cabrera, C. Taquiri, and C. R. Sterling. 1997. Isolation of *Cryptosporidium parvum* and *Cyclospora cayetanensis* from vegetables collected in markets of an endemic region in Peru. Am. J. Trop. Med. Hyg. 57:683-686.
52. Packard, V. S., Jr., S. Tatini, R. Fugua, J. Heady, and C. Gilman. 1992. Direct microscopic methods for bacteria or somatic cells, p. 309-325. *In* Standard methods for the examination of dairy products, 16th ed. American Public Health Association, Washington, D. C.
53. Pettipher, G. L. 1986. Review: the direct epifluorescent filter technique. J. Food Technol. 21:535-546.
54. Pettipher, G. L., and U. M. Rodrigues. 1982. Rapid enumeration of microorganisms in foods by the direct epifluorescent filter technique. Appl. Environ. Microbiol. 44:809-813.
55. Pettipher, G. L., R. Mansell, C. H. McKinnon, and C. M. Cousins. 1980. Rapid membrane filtration - epifluorescent microscopy technique for direct enumeration of bacteria in raw milk. Appl. Environ. Microbiol. 39:423-429.
56. Pyle, B. H., S. C. Broadaway, and G. A. McFeters. 1995. A rapid, direct method for enumerating respiring enterohemorrhagic *Escherichia coli* O157:H7 in water. Appl. Environ. Microbiol. 61:2614-2619.
57. Ray, B. 1979. Methods to detect stressed microorganisms. J. Food Prot. 42:346-355.
58. Ray, B. 1993. Sublethal injury, bacteriocins and food microbiology. ASM News 59:285-291.
59. Read, Jr., R. B. 1979. Detection of stressed microorganisms—implications for regulatory monitoring. J. Food Prot. 42:368-369.
60. Rhodehamel, E. J., and S. M. Harmon. 1995. *Clostridium perfringens*, p. 16.01-16.06. *In* Bacteriological analytical manual, 8th ed. AOAC International, Gaithersburg, Md.
61. Rhodehamel, E. J., and S. M. Harmon. 1995. *Bacillus cereus*, p. 14.01-14.08. *In* Bacteriological analytical manual, 8th ed. AOAC International, Gaithersburg, Md.
62. Rodrigues, U. M., and R. G. Kroll. 1985. The direct epifluorescent filter technique (DEFT): increased selectivity, sensitivity and rapidity. J. Appl. Bacteriol. 59:493-499.
63. Rodrigues, U. M., and R. G. Kroll. 1988. Rapid selective enumeration of bacteria in foods using a microcolony epifluorescence microscopy technique. J. Appl. Bacteriol. 64:65-78.
64. Rodrigues, U. M., and R. G. Kroll. 1989. Microcolony epifluorescence microscopy for selective enumeration of injured bacteria in frozen and heat-treated foods. Appl. Environ. Microbiol. 55: 778-787.
65. Rodrigues, U. M., and R. G. Kroll. 1990. Rapid detection of salmonellas in raw meats using a fluorescent antibody-microcolony technique. J. Appl. Bacteriol. 68:213-223.
66. Rodriquez, G. G, D. Phipps, K. Ishiguro, and H. F. Ridgway. 1992. Use of a fluorescent redox probe for direct visualization of actively respiring bacteria. Appl. Environ. Microbiol. 58:1801-1808.
67. Roszak, D. B., and R. R. Colwell. 1987. Survival strategies of bacteria in the natural environment. Microbiol. Rev. 51:365-379.
68. Russell, A. D., J. R. Furr, and M. J. Day. 1993. Enumeration of damaged or otherwise stressed bacteria. Lett. Appl. Microbiol. 16:237-238.
69. Servis, N. A., S. Nichols, and J. C. Adams. 1995. Development of a direct viable count procedure for some Gram positive bacteria. Lett. Appl. Microbiol. 20:237-239.
70. Sheridan, J. J., I. Walls, J. McLauchlin, D. McDowell, and R. Welch. 1991. Use of a microcolony technique combined with an indirect immunofluorescence test for the rapid detection of *Listeria* in raw meat. Lett. Appl. Microbiol. 13:140-144.
71. Solomon, H. M., E. J. Rhodehamel, and D. A. Kautter. 1995. *Clostridium botulinum*, p. 17.01-17.10. *In* Bacteriological analytical manual, 8th ed. AOAC International, Gaithersburg, Md.
72. Splittstoesser, D. F., M. Groll, D. L. Downing, and J. Kaminski. 1977. Viable counts versus the incidence of machinery mold (*Geotrichum*) on processed fruits and vegetables. J. Food Prot. 40: 402-405.
73. Tison, D. L. 1990. Culture confirmation of *Escherichia coli* serotype O157:H7 by direct immunofluorescence. J. Clin. Microbiol. 28:612-613.
74. Tolker-Nielsen, T., K. Holmstrom, and S. Molin. 1997. Visualization of specific gene expression in individual *Salmonella typhimurium* cells by *in situ* PCR. Appl. Environ. Microbiol. 63: 4196-4203.
75. Tortorello, M. L., K. F. Reineke, and D. S. Stewart. 1997. Comparison of antibody-direct epifluorescent filter technique with the most probable number procedure for rapid enumeration of *Listeria* in fresh vegetables. J. AOAC Int. 80:1208-1214.
76. Tortorello, M. L., and D. S. Stewart. 1994. Antibody-direct epifluorescent filter technique for rapid, direct enumeration of *Escherichia coli* O157:H7 in beef. Appl. Environ. Microbiol. 60:3553-3559.

77. Tsuji, T., Y. Kawasaki, S. Takeshima, T. Sekiya, and S. Tanaka. 1995. A new fluorescence staining assay for visualizing living microorganisms in soil. Appl. Environ. Microbiol. 61:3415-3421.
78. Vodovotz, Y., E. Vittadini, J. Coupland, D. J. McClements, and P. Chinachoti. 1996. Bridging the gap: use of confocal microscopy in food research. Food Technol. 50 (6):74-82.
79. Wildman, J. D., and P. B. Clark. 1947. Some examples of the occurrence of machinery slime in canning factories. J. Assoc. Off. Agric. Chem. 30:582-585.
80. Zimmermann, R., R. Iturriaga, and J. Becker-Birck. 1978. Simultaneous determination of the total number of aquatic bacteria and the number thereof involved in respiration. Appl. Environ. Microbiol. 36:926-935.
81. Zottola, E. A. 1997. Special techniques for studying microbial biofilms in food systems, p. 315-346. *In* M. L. Tortorello and S. M. Gendel (ed.), Food microbiological analysis: new technologies. Marcel Dekker, New York.

Chapter 5

Cultural Methods for the Enrichment and Isolation of Microorganisms

W. A. Sperber, M.A. Moorman, and T. A. Freier

5.1 INTRODUCTION

This chapter describes the general principles and methods for the enrichment and isolation of microorganisms. It does not describe the detailed requirements for the enrichment and isolation of specific microorganisms. These are presented in other chapters of the *Compendium* addressing each microorganism. The principles and methods described here are important for the isolation of *Salmonella, Listeria*, other foodborne pathogens, and indicator microorganisms.

There are several reasons for the use of enrichment methods. Primary among these is the need to grow a detectable population of cells from a very low initial level. It is not unusual for the target microorganism to be present in foods at levels of about one cell per 100 g of food. With enrichment methods as little as one cell per 500 g of food have been detected.[13] Enrichment techniques are also used for the recovery of injured microorganisms. Microorganisms in foods are typically in a stressed condition. They often lack the optimal nutrients for growth, they may be in an environment (e.g. pH, temperature) that will not support their growth, or they may be damaged by the food process. These microorganisms will require a period of time in the appropriate conditions so that cellular damage can be repaired and metabolic pathways can be activated. Last, enrichment methods permit the proliferation of the target microorganism to detectable levels while repressing the growth of competing nontarget microorganisms.[15]

Unlike quantitative microbial recovery, as described in Chapter 7 and Chapter 10, enrichment methods are qualitative. They will indicate the presence or absence of the target microorganism, but not its numbers. Enrichment methods can, however, yield quantitative results when they are used in conjunction with the MPN technique described in Chapter 6.

The isolation methods described in this chapter are necessary to obtain pure cultures for the biochemical, serological, and genomic identification of the target microorganisms.

5.2 ENRICHMENT METHODS

Enrichment methods determine the presence or absence of a target organism. These methods are not conducted to determine the level or quantity of the organism. Many direct plating or quantitative methods exist for the recovery and enumeration of target microorganisms[8] (Chapter 6). However, quantitative methods are not appropriate in several situations.

1. The permissible level of the organism is less than the maximum sensitivity of the quantitative procedure.
2. The organism is surrounded by large numbers of competing microorganisms.
3. The suspending food is inhibitory to the target organism.

The goal of the enrichment method is to permit the growth of the target microorganism, and if a selective enrichment, to suppress or inhibit the growth of competing microorganisms. There are various types of enrichment protocols which either individually or combined, permit the growth of the target organism to levels necessary for detection or recovery by diagnostic or selective plating procedures respectively.

5.3 PREENRICHMENT

The purpose of preenrichment is to allow the stressed target microorganism to resuscitate in either a non-selective or moderately selective environment. While the microorganism may resuscitate, very little growth may occur during the preenrichment step (Figure 1). These preenrichment media are either nonselective or are designed to be moderately selective against competing microorganisms. If the media contain any selective components, these must be balanced to permit the growth of the target organ-

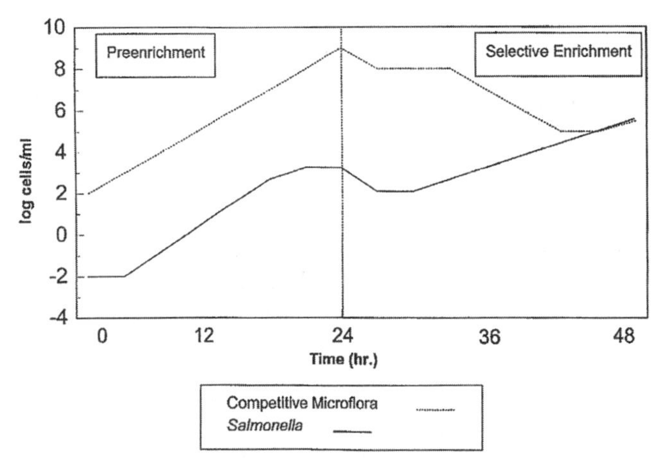

Figure 1. *Growth of Salmonella and competitive microflora during preenrichment and selective enrichment.*
Modified from Silliker Laboratories Group, Inc.[10]

ism and repress the growth of competing microorganisms in the food sample. A preenrichment procedure may not be necessary when the enrichment medium has been validated to support the resuscitation and growth of the target microorganism. The liquid version of Baird Parker Agar (Liquid Baird Parker), without agar, has been successfully used to recover *Staphylococcus aureus* in foods.[16]

The formulation of the preenrichment media will depend upon the level of competing microorganisms and the ability of the food to inhibit microorganisms. The Food and Drug Administration Bacteriological Analytical Manual describes 10 different preenrichment media for the recovery of *Salmonella* in foods.[8] These media range from sterile deionized water when recovering *Salmonella* from non-fat dry milk to Brilliant Green-supplemented Milk for the recovery of *Salmonella* from chocolate. These media use either the nutrients from the food or contain ingredients (e.g., casein) to neutralize the inhibitory nature of the food (e.g., chocolate).[3] Lactose is commonly used for the preenrichment medium for *Salmonella,* even though few species of this genus are capable of metabolizing this carbohydrate. The presence of lactose is not directly essential for the recovery of *Salmonella.*[5,15] It has been suggested that some selectivity of preenrichment media containing lactose is generated by the reduction in media pH when the lactose is fermented by the mixed competitor flora.[12]

5.31 Preenrichment Considerations

5.311 Sample Weight

The amount of sample analyzed is either defined in the analytical procedure or based upon a sampling plan. The sample weight for preenrichment can vary from as little as one gram to as much as 1500 g. Typically these food samples are diluted in the preenrichment media at a 1:9 food/media ratio. The preenrichment may be prepared by weighing the sample or by conducting a rinse, such as that for poultry or other meat carcasses.

5.312 Compositing

The enrichment protocol is intended to recover viable microorganisms in the tested sample. The sample analyzed may be a single unit, or represent a composite of multiple units dispensed into the sample preenrichment. Composite analysis enables the laboratory to examine many units in one preenrichment, thereby reducing the number of samples to be analyzed in the laboratory. For example, category 3 analysis of samples following the *FDA Salmonella* composite protocol requires the analysis of fifteen 25-g samples in one 375-g sample.[8] Composite analysis for *Salmonella* has been shown to not reduce the method's sensitivity.[13] The increase in sample weight will increase the volume of the preenrichment media often necessitating a larger incubator depending on the number of samples to be tested.

5.313 Preenrichment Temperature

Incubation temperatures used for preenrichment or enrichment are typically near the ideal temperature for the growth of the target microorganism. Incubation temperatures may be employed which provide a selective advantage for the target organism. *Listeria monocytogenes,* for example, a psychrotrophic pathogen, was initially enriched at refrigeration temperatures.

The temperature of the preenrichment media prior to inoculation will have an effect on the recovery of the target microorganism. Large volume preenrichments require extended time to equilibrate at the appropriate incubation temperature following sample addition. For preenrichments at mesophilic temperatures these large volumes of media may require equilibration at room temperature or higher prior to inoculation.

5.314 Mechanics of Testing

The objective of any microbiological assay is to recover the microorganisms present in the food sample, and not that from the laboratory environment. The food sample must be handled aseptically with the test portion dispensed into the enrichment container without introducing external contaminants. Good laboratory practices require that neither the test container, nor the dispensing utensil, fall below the plane of the preenrichment container. After dispensing, the preenrichment container can be shaken to dissolve powders or homogenized by other means to adequately release those microorganisms trapped within the food matrix. Typically this homogenization can last one to two minutes. Surfactants may be added to aid the release of the trapped microorganisms. After homogenization, the preenrichment pH should be verified to be within the appropriate range. Many foods will affect media pH requiring addition of acid or alkali to reestablish the appropriate pH for the preenrichment.

5.315 Incubation Time

The preenrichment period must be long enough for effective resuscitation of stressed or injured microorganisms. If this time is too short, the target microorganism may not have recovered resulting in a false negative test result. Theoretically, if the preenrichment time is excessive, overgrowth by competing organisms may occur again resulting in a false negative test result. D'Aoust found that increased recovery of *Salmonella* with prolonged incubation was not due to *Salmonella* growth but rather to the higher rate of death of competing microorganisms due to their greater sensitivity to elevated temperatures and toxicity of selective media. The appropriate incubation period must be validated based upon the unique nature of the food sample, the enrichment medium, and the incubation time.

5.4 SELECTIVE ENRICHMENT

The preenrichment will result in resuscitation of the target microorganism and moderate levels of proliferation. The selective enrichment furthers the growth of the target microorganism while suppressing or inhibiting growth of competing microorganisms (Figure 1). The selectivity of the media is provided by agents or conditions which are antagonistic or inhibitory to competing microorganisms. These selective agents include temperature, antimicrobials, salts, acids and metals. The preenrichment also permits the analysis of food samples that could not be analyzed directly using the selective enrichment. These food samples will obviate the selective agents of this enrichment thereby reducing the efficacy of the medium. Taylor and Silliker determined that lactose preenrichment prior to selective enrichment in tetrathionate and selenite cystine increased the recovery of *Salmonella* from albumin.[15] Furthermore, direct inoculation of food samples with large numbers of competing microorganisms into selective enrichments may result in false negative results due to the reduction in the media selectivity.[5]

Conversely, food material in a preenrichment transferred to a selective enrichment will positively affect media efficacy. Abbiss demonstrated that the presence of food material in buffered peptone water preenrichment enhanced the recovery of *Salmonella typhimurium* from selective enrichments. This improved recovery was due to amelioration of the initial inhibitory environment *Salmonella* encounters when transferred to the selective medium. In the case of *Salmonella*, neither tetrathionate nor selenite cystine enrichment broths will support the growth of all strains of *Salmonella*.[1] To reduce the risk of a false negative, many selective enrichment protocols will employ more than one medium following the preenrichment.

To ensure availability of nutrients to the target microorganism, enrichments may be shaken during incubation. Facultatively anaerobic microorganisms may experience shortened lag phase and generation times when oxygen is added during the enrichment. The oxygen will support an aerobic metabolism that yields higher energy than anaerobic metabolism. Duffy, et al, however studied the growth kinetics of *Listeria monocytogenes* and found that aeration of the selective enrichment did not alter the length of lag phase or growth rate of *Listeria monocytogenes*.[6]

5.41 Mechanics of Transfer

The transfer of the preenrichment to selective enrichment requires the aseptic transfer of an aliquot of preenrichment to the selective enrichment. This step is highly prone to laboratory contamination and must be done with great care. Typically, ratios of 1:10 preenrichment inoculum to selective enrichment are attained. The preenrichments should be gently shaken or stirred prior to transfer.

5.42 Motility Enrichment

Motility enrichment media support the growth of the target organism while immobilizing this strain through the use of antisera specific to the target organism. This protocol has been successfully used as a selective agent and diagnostic aid in the recovery and identification of *Salmonella*.[14]

5.43 Selective Enrichment

Some media favor the growth of the target microorganism while possessing none of the inhibitory effects of selective agents. An example of such a media is M Broth, which is termed elective enrichment.[14] Originally formulated to promote antigen development after selective enrichment, M Broth contains sodium citrate and D-mannose as the only fermentable carbohydrates. Both compounds are fermented by salmonellae, thereby serving as anaerobic energy sources and providing a further competitive advantage over microorganisms that can not ferment either carbohydrate. It is conceivable that other applications of the elective enrichment phenomenon could be developed to improve the sensitivity of procedures for the detection of the target microorganism.

5.44 Selective and Differential Isolation Methods

While enrichment methods increase the proportion of the target microorganisms, the cultures still contain similar competing microorganisms that often must be eliminated before the target microorganism can be isolated and identified. These results are accomplished by the use of agar plating media and biochemical tests. Many selective and differential agents are used in these tests.[2]

5.441 Selective agents

Antibiotics. Many antibiotics are available for the selective isolation of microorganisms from foods. Those commonly used are: polymyxin B, ampicillin, moxalactam, novobiocin, oxytetracycline, D-cycloserine, vancomycin, trimethiprim, and cycloheximide.

Other chemicals. It is also possible to inhibit the growth of nontarget microorganisms in selective plating media by using chemicals other than antibiotics. Those commonly used are: dyes such as brilliant green, sodium selenite, bile salts, potassium tellurite, and sodium lauryl sulfate.

Anaerobiosis. The exclusion of oxygen can sometimes provide a selective advantage for the target microorganism. This is usually accomplished by the physical or chemical removal of oxygen inside a sealed incubation chamber. A similar selective effect can be achieved by the use of respiratory inhibitors such as sodium azide and potassium cyanide in the culture medium. In general, these prevent the growth of catalase-positive microorganisms, while permitting the growth of catalase-negative microorganisms. Oxygen can be removed from tubed media by boiling and tempering these media just before inoculation. Diffusion of oxygen into these media can be retarded by the use of sterile 3% agar or mineral oil overlays.

pH. Acidified media are commonly used to select particular groups of microorganisms. Media acidified to pH 3.5 are used for the isolation of yeasts, molds, and alicyclobacilli. Similar use of pH control can be used for the selection of other aciduric bacteria.

Water Activity. The water activity of selective media can be lowered by the addition of numerous solutes such as sodium chloride, glucose, ethanol, and propylene glycol. This approach is effective when the target microorganism can tolerate the chemical properties of the solute and the reduced water activity.

Temperature. As described above for enrichment media, the incubation temperature can be a very effective means to select particular target microorganisms.

5.442 Differential agents

Several differential agents are useful for the screening and presumptive identification of the target microorganism.

pH indicators. The indication of pH change is one of the most common differential methods for the determination of a particular metabolic activity. The fermentation of carbohydrates is accompanied by the production of acids which lower the medium's pH value. The decarboxylation of amino acids and the hydrolysis of urea are accompanied by the production of ammonia or amines which raise the medium's pH value. Commonly used pH indicators in microbiological media are phenol red, methyl red, and brom cresol purple.

H2S indicators. Some microorganisms produce hydrogen sulfide as a by-product of sulfur-containing amino acid metabolism. The production of hydrogen sulfide (H_2S) can be detected by the use of iron salts such as ferrous citrate, ferric ammonium citrate, or ferric ammonium sulfate. These combine with H_2S to form ferrous sulfide, FeS, a black compound that is produced anaerobically in the deep portion of tubed media, or under colonies on agar surfaces. Because of its solubility, FeS is able to diffuse throughout the medium.

Egg yolk reaction. Egg yolk is added to some microbiological media to assist in the recovery of injured microorganisms. Some of these produce a characteristic reaction that consists of a zone of clearing and/or a light precipitate around the colony, depending on the type of lipolytic enzymes excreted by the microorganism.

Blood hemolysis reactions. Plates of various species of blood agar can be used to differentiate pathogens such as *Staphyloccocas aureus, Streptococcus pyogenes, and Listeria monocytogenes.* The hemolytic reaction will vary depending on the particular pathogen and the species of blood that is used.

5.443 Agar plates

Enrichment cultures are streaked onto the surface of agar in Petri dishes to obtain isolated colonies so that pure cultures of the target microorganism will be available for identification tests. It is essential that this step be done well so that the isolated colonies originate from single cells of the target microorganism

Media handling and preparation. While commercially-prepared media are often available, many of the media used for the selective and differential isolation of microorganisms are dehydrated media that are reconstituted in the laboratory. It is important that these media be handled and prepared in accordance with the good laboratory practices described in Chapter 1.

Plates of differential media should be poured deep enough to permit observation of the differential characteristic. This can be accomplished by pouring 20 mL of molten agar into a 15 × 100 mm petri dish. This technique tends to reduce the diffusion of excreted metabolic products and the masking of a particular colony's reaction by neighboring colonies.

The shelf lives of prepared media vary depending on the stability of the components. Manufacturers' instructions or other reference materials should be consulted. In general, prepared media should not be stored longer than one month. Some media require refrigeration and/or storage in the dark. The degree of hydration of plating media is often critical. If agar surfaces are too wet, spreading and swarming will prevent the isolation of a pure culture. If the surface is too dry, the microorganisms may grow poorly, or not at all. Dehydration of prepared media can be prevented by storage in sealed bags or containers.

Streaking technique. To assure the isolation of a pure culture of the target microorganism, it is essential that the streaking technique will provide isolated colonies. The streaking technique can be varied by skilled technicians because highly-selective media are more "forgiving" than less-selective media. For general purposes, however, the following streaking technique should be used (Figure 2).

1. Using a sterile inoculating loop, transfer a loopful of the enrichment culture to the surface of the agar plate near the edge of the plate. Streak the loop back and forth over the top quarter of the plate about five to ten times in a tight "Z" fashion. The streaking lines should not cross each other.
2. Resterilize the inoculating loop and allow to cool. Streak the right quarter of the plate by passing the loop through the original area of inoculation (top quarter of plate) and streaking in a tight "Z" about five to ten times. The loop should not contact the original area of inoculation after the first pass.
3. Resterilize the inoculating loop, allow to cool. and streak the remainder of the plate in the same fashion beginning with a single pass through the second streaked area (right quarter of plate).

If presterilized inoculating loops are used, a separate loop must be used for each step of this procedure.

If isolated colonies are not obtained after incubation of the plates, or if the purity of a culture is in doubt, a portion of the growth on the plate can be restreaked onto a fresh agar plate.

It is sometimes necessary to streak a culture several times in order to obtain pure isolated colonies.

Picking colonies. The proper colony-picking technique can help assure the isolation of a pure culture (Figure 3). A sterile inoculating wire should be used. Never use a loop to pick a colony. The tip of the wire should be touched only to the top and center of the colony so that a miniscule amount of the colony is picked. This technique is important when colonies are being picked from mixed cultures on selective plates. The selective plating medium will permit the survival, and probably some growth, of nontarget microorganisms. Some of these may be on the agar surface beneath growing colonies of the target microorganism. The use of this picking technique may permit the isolation of a pure culture from a mixed colony.

Tubed media. Picked colonies are usually inoculated into one or more tubed solid or liquid media for the determination of biochemical characteristics and the amplification of antigens or genetic material for serological or genetic testing.

Media handling and preparation. All media should be handled and prepared in accordance with the good laboratory practices described in Chapter 1.

The tubed media should be prepared deep enough so that anaerobic reactions can occur and be observed. In particular, agar slants should be prepared with 10 to 12 mL of agar in each tube so that after solidification the butt portion of the tube will contain at least 5 to 6 mL of agar.

Inoculation of tubed media. A sterile inoculating wire is used to pick a colony as described above. The tubed media are inoculated by stabbing the butt and streaking the surface of agar slants and by gentle twirling in liquid media.

It is possible to inoculate multiple tubes without returning to the original colony to obtain more cells. The need to inoculate multiple tubes sometimes induces the inexperienced technician to take too much growth from the colony, even using the needle

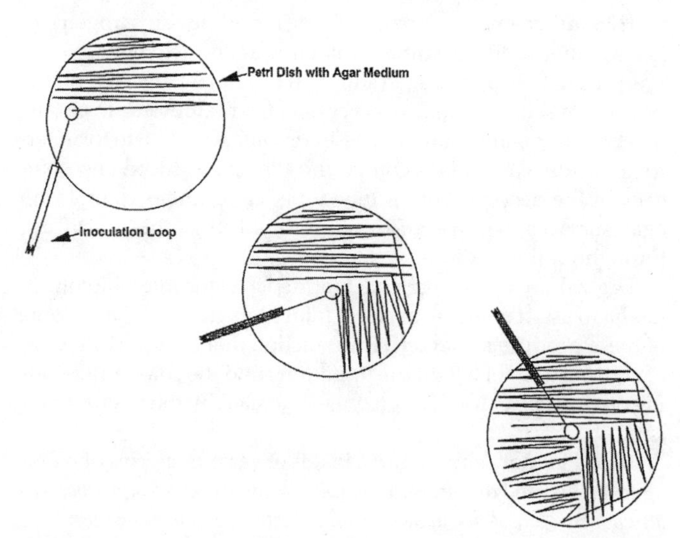

Figure 2. Recommended streaking technique to obtain isolated colonies.

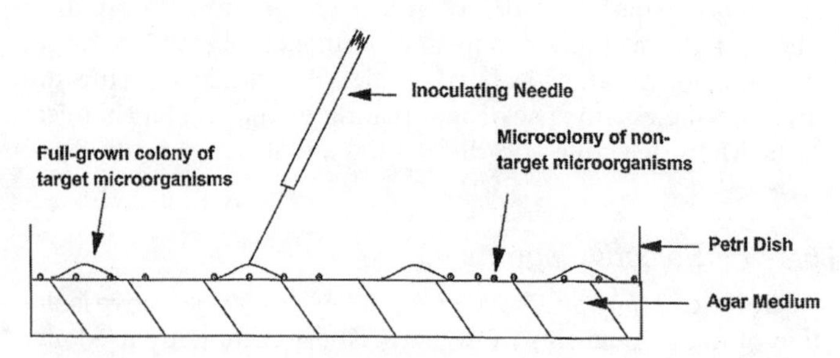

Figure 3. Recomended technique for picking colonies to obtain pure cultures.

or an inoculating loop to scoop the entire colony from the agar surface. This practice increases the chance of getting an impure culture. The minuscule, even invisible, amount of growth obtained by the proper colony picking technique described above is sufficient to inoculate many tubes. (One of us (WHS) inoculated 144 successive agar tubes with the minuscule growth on a single inoculating wire. All of the tubes showed good growth of the test microorganism.)

5.5 QUALITY ASSURANCE OF ENRICHMENT AND ISOLATION METHOD

5.51 Temperature Control

Temperature control is a critical selective or elective component of many microbiological enrichment protocols. The first step in the quality assurance process of temperature control is the purchase of the appropriate equipment. Most inexpensive gravity flow connection air incubators can maintain incubation temperatures within about ± 3° C of the desired temperature. For protocols requiring more stringent control, forced air or water-jacketed incubators, or shaking or circulating water baths should be used.

Incubators and water baths are usually equipped with temperature controls and indicators. These "built-in" devices should never be relied upon as the sole means of temperature verification. Incubators can have hot and cold spots, so a calibration should be performed annually to establish a relationship between the temperature reading of the internal device and the true temperature at key locations within the chamber. This can be done using a recorder equipped with thermocouples that have been calibrated against a reference thermometer.

Thermometers of the proper type (partial, total or complete immersion) and of sufficient accuracy and precision should be permanently placed within incubators and water baths. These thermometers should be calibrated at least once per year against a reference thermometer whose accuracy is certified to be traceable to an NIST thermometer. Large incubators should have at least two thermometers, one located towards the top, and one towards the bottom. If calibration indicates that a correction factor is necessary, this correction factor must be used each time the temperature is noted.

A temperature monitoring log form should be developed that includes the manufacturer and model number or serial number of the incubator or water bath being monitored, the date and time of day when the temperature was taken, the location of the thermometer (top/bottom), the allowable temperature range for that piece of equipment, the initials of the person recording the temperature, and any corrective action that was taken.

5.52 Media

Each lot of media used for enrichment and isolation, whether prepared from individual ingredients, purchased as a dehydrated blend, or ready to use, should be subjected to quality assurance before use. All incoming media should have the received date written on the package along with the opened date. An inventory system should be developed that documents the type of medium, supplier, supplier's lot number, internal lot number, expiration date, size or weight, date received, date opened, date discarded, and the initials of the person checking in the medium. First in/first out stock rotation should be practiced. If the manufacturer supplies a certificate of analysis, this should be kept on file. If no certificate is available, the medium should be subjected to quality assurance testing before it is used for analysis. These tests should include tests for performance for intended use, selectivity (if applicable), sterility, appearance and pH (Chapter 1).

If media are prepared in the lab, a media preparation log book should be developed. This log should include the date made, supplier or internal lot number, the final pH after sterilization, batch size or quantity, autoclave used, autoclave load number, fill volume, pre-sterilization pH (where applicable; i.e. buffers), the initial of the person making the media, and an approval signature. Nonselective/nondifferential media should be tested for sterility and performace for intended use and/or differentiation.

5.521 Sterility

All new batches of media should be tested for sterility. This testing should be performed on all media immediately after sterilizing. A randomly chosen tube, agar plate, or bottle is set aside and allowed to cool. The media are then placed at the appropriate temperature and incubated for the appropriate amount of time. After the desired incubation time is completed, the medium is checked for growth or turbidity. Results should be documented and any necessary corrective action taken. Ideally, media should be "quarantined" until results of quality assurance testing are known. If media must be used before results are known, a trace-back system must be developed in case the quality assurance testing indicates problems.

5.522 Performance for Intended Use

Media used for enrichment and isolation need to be tested for performance, to help insure against false-negative results. Productivity analysis verifies that the medium, as formulated and prepared, will support the growth of the target microorganism(s). Unfortunately, it is very difficult to detect batches of media that are slightly more inhibitory than normal. Inoculation with freshly passaged laboratory adapted cultures may result in growth, while slight increases in inhibitory properties may not allow growth of injured target organisms from food or environmental samples. This problem can be partially overcome by using very low level inoculums in liquid media and by using the ecometric technique for agar plate media.[11] Appropriate stock cultures should be chosen, media should be incubated at the appropriate temperatures and times, then checked for typical reactions.

5.523 Selectivity

Selectivity testing needs only to be done on media that have selective or differential properties. Selectivity analysis verifies that the medium, as formulated and prepared, will prevent the growth of competing microorganisms. Microorganisms that produce a "negative" reaction (do not grow, or do not produce typical reactions) should be chosen. The newly prepared media should be inoculated with the selectivity control, then incubated at the appropriate temperature. and time. The growth or reaction should be documented. Many enrichments require the addition of sterile components after the base medium has been sterilized. Great care should be taken to prevent contamination during this step. Reagents should be tested for sterility by transferring a small amount (usually 1 mL) to a nonselective medium such as Standard Methods Agar or Typticase Soy Broth, then incubated for sufficient time (usually 48 hrs.) to prove sterility. Reagents should be divided into containers in amounts sufficient for a single use. The use of large containers of reagents that are used multiple times should be avoided, as each use increases the chances for contamination.

5.53 Laboratory Environment

Enrichment techniques are designed to be as efficient as possible at detecting extremely low levels of target microorganisms. This means that accidental contamination from the laboratory environment is a potential threat, particularly for laboratories located within food manufacturing facilities. The Food Laboratory Accreditation Work Group states, "Laboratories located in facilities where products or ingredients are manufactured must not test for pathogens (such as *Listeria monocytogenes, Salmonella, Escherichia coli* O157:H7, *Shigella, Campylobacter, Vibrio cholera*) unless the laboratory is physically separated with limited access, negative air flow, and supervised by a qualified microbiologist."[7] Laboratories handling infectious agents should meet the general requirements of "Biosafety Level 2."[4]

Laboratories should be designed with physical separation between critical areas such as sample check-in, storage, preenrichment set-up area, enriched culture transfer areas and media preparation and sterilization areas. Hand wash stations should be conveniently located and stocked with antimicrobial soap. Disposable paper towels or warm air dryers should be available.

Analysts who work with enriched cultures or highly contaminated samples should do this only in designated areas that are separated from other areas, and should wash their hands and change their lab coats before entering other areas of the lab. In short, many procedures for limiting cross-contamination that are considered Good Manufacturing Practices in the food manufacturing plant should also be applied to the food microbiology laboratory.

The air supply for food laboratories conducting enrichments and isolations should reduce the levels of contamination, lower humidity, and control temperature. Airborne microbiological contamination should be controlled by using filters and the air quality should be verified by doing microbiological monitoring (air sampling devices, air settling plates, surface swabs). Typically, total bacteria or yeast and mold are monitored, but monitoring for specific target organisms may be appropriate.

Critical work surfaces should be routinely monitored for the presence of the target organisms being enriched or isolated (Chapter 2). Another important area that should be monitored is the analysts' hands. Samples from each analyst who is involved in performing the test should be taken before and after various critical steps in the procedure. Results can be used for verification of the hand washing procedures, and can serve as a valuable source of information when investigating possible cross contamination issues. Should the target organism be detected on the analyst's hands prior to enrichment or transfer of the sample, and the sample tests positive, this result should be considered suspect due to environmental/analyst contamination.

Great care needs to be taken when transferring preenrichment or enrichment cultures. Analysts should be trained in the elimination of aerosols and microdroplets, and in all aspects of aseptic technique. A useful technique for training new analysts is to place brown paper towels on the lab bench during transfer practice exercises. After transfers are complete, the towels can be checked for droplets. Micropipettors should be used only with extreme caution, and the use of micropipet tips containing a filter should be considered. Mixing test tube contents by using a vortex mixer before transfer is not necessary in most cases, and can be a source of cross contamination. Receptacles for contaminated pipets and micropipettor tips should be located as close as possible to the operation being performed to reduce the potential for dripping.

If testing indicates that results may have been compromised by the laboratory environment, policies and procedures must be in place that allow for interpreting, evaluating and reporting equivocal results. Strict documentation at every step provides valuable information for the investigation of equivocal results. This documentation may include analyst, enrichment time, transfer time, sample order and rack order.

5.54 Positive Control Cultures

The use of positive control cultures is absolutely necessary to verify that the analysis will detect the target organism. The presence of large numbers of healthy target organisms in the positive control can be a source of laboratory cross-contamination. One way to lessen the consequences of this occurrence is to chose specific organisms for the positive control that are easily distinguishable from typical sample isolates. This can be accomplished by choosing "rare" organisms. An example is the use of *Salmonella abettatube* for the positive control in the salmonella assay. Another approach is to use a control strain that has been genetically altered with an easy-to-detect characteristic, such as antibiotic resistance, luminescence, or fluorescence.[9]

5.6 REFERENCES

1. Abbiss, J. S. 1986. A study of the dynamics of selective enrichment of *Salmonella*. The British food manufacturing industries research association. Number 565. 1-27.
2. Atlas, R. M. 1993. Handbook for microbiological media. CRC Press, Inc. Boca Raton, Fla.
3. Busta, F. F., and M. L. Speck. 1968. Antimicrobial effect of cocoa on salmonellae. Appl. Microbiol. 16:424-425.
4. CDC. 1988. "Biosafety in microbiological and biomedical laboratories, 2nd ed. U.S. Government Printing Office, Washington, D. C.
5. D'Aoust, J-Y. 1981. Update on preenrichment and selective enrichment conditions for detection of *Salmonella* in foods. J. Food Prot. 44:369-374.
6. Duffy, G. J. L. Sheridan, R.L. Buchanan, D.A. McDowell, and I.S. Blair. 1994. The effect of aeration, initial inoculum and meat microflora on the growth kinetics of *Listeila monocytogenes* in selective enrichment broths. Food Microbiol. 11:429-438.
7. Food Laboratory Accreditation Work Group. Accreditation Criteria for United States Laboratories Performing Food Microbiological Testing. Draft July 21, 1997, available at www.aoac.org.
8. Food and Drug Administration. 1998. Bacteriological Analytical Manual. 8th ed. AOAC International, Washington, D.C.
9. Fratamico, P. M., M. Y. Deng, T. P.Strobaugh, and S. A. Palumbo. 1997. Construction and characterization of *Escherichia coli* O157:H7 strains expressing firefly luciferase and green fluorescent protein and their use in survival studies. J. Food Prot. 60:1167-1173.
10. Laboratory Methods in Food Microbiology. Short Course. Chap. 5, *Salmonella*. Silliker Laboratories Group, Inc., Chicago, Ill.
11. Mossell, D. A. A., F. Van Rossem, M. Koopmans, M. Hendricks, M. Verdouden, and I. Eeldrink. A comparison of the classic and the so-called ecometric technique. J. Appl. Bacteriol. 49:439-454.
12. North, W.R. 1961. Lactose preenrichment method for isolation of Salmonella from dried cgg albumin. Its use in a surve of commercially produced albumen. Appl. Microbiol. 9:188-195.
13. Silliker, J.H., and D.A. Gabis. 1973. ICMSF methods studies. I. Comparison of analytical schemes for detection of *Salmonella* in dried foods. Can. J. Microbiol. 19:475-479.

14. Sperber, W. H,. and R. H. Deibel. 1969. Accelerated procedure for *Salmonella* detection in dried foods and feeds involving only broth cultures and serological reactions. Appl. Microbiol. 17:533-539.
15. Taylor, W. I., and J. H. Silliker. 1961. Isolation of *Salmonellae* from food samples. IV. Comparison of methods of enrichment. Appl. Microbiol. 9:484-486.
16. Van Doorne., H., R. M. Baird, D. T. Hendricks, D. Margaretha, D. M. Van der Kreek, H. P. Pauwels. 1981. Liquid modification of Baird-Parker's medium for the selective enrichment of *Staphylococcus aureus.* Antonie van Leeuwenhoek.. 47:267-278.

Chapter 6

Culture Methods for Enumeration of Microorganisms

Katherine M.J. Swanson, Ruth L. Petran, and John H. Hanlin

6.1 INTRODUCTION

Many analyses performed in food microbiology laboratories involve enumeration of microorganisms present in a sample. Although light microscopy can be used to enumerate microorganisms, the technique suffers from three significant limitations. First, it is difficult to differentiate live from dead cells. Second, it is almost impossible to observe bacteria under light microscopy at cell densities less than 10^6 per mL. Third, solid materials such as food particles cannot be viewed without mechanical disruption under high-power light microscopy.

Three general methods employed to estimate the number of viable microorganisms present in the samples are plate count procedures (PC), most probable number procedures (MPN), and membrane filtration (MF). The initial stages of these procedures are the same. A portion of sample is measured by weight or volume, a series of dilutions is prepared, and aliquots are added to either an agar medium (PC), tubes of liquid media (MPN), or passed through a membrane filter that retains microorganisms and is then placed on the surface of a growth medium. The basic principle of these methods is that following incubation, the population of microorganisms originally present can be estimated by counting the number of colonies or number of tubes showing evidence of growth and multiplying by the dilution which was added to the plates, tubes, or membrane filter. By varying the growth medium and incubation conditions, different microorganisms can be enumerated by any of these basic methods. The optimum medium and conditions for determining the colony count may vary from one food to another. However, once a procedure for a given food is determined, it can be very useful for routine microbial analysis of the food. Since minor variations in procedures can alter the results obtained with the colony count,[13] the competency and accuracy of the analysts are very important.

This chapter describes the basic principles and techniques used to perform cultural methods for enumeration of microorganisms. Specific details on media, equipment used, and incubation conditions and interpretation of results are contained in subsequent chapters for specific tasks. Procedures described in Chapter 1—Laboratory Quality Assurance and Chapter 63—Media, Reagents, and Stains should be used for accurate results.

6.2 DILUTIONS

6.21 Basic principles

Enumeration of microorganisms requires dilution of samples to achieve a population that is countable by the chosen method. Generally, decimal or ten-fold dilutions are used for ease of calculation of final results. A variety of diluents are available including phosphate buffer, and 0.1% peptone water. Distilled water should be used in the preparation of diluents. However, use of plain distilled or deionized water as a diluent is inappropriate due to the potential for osmotic stress on diluted cells. In fact, use of diluents with high levels of sugar is needed for osmophilic yeast tests. Refer to the appropriate chapter to determine the correct diluent for the organism under consideration.

6.22 Liquids

Test portions of nonviscous (i.e., viscosity not greater than milk) liquid products or homogenates may be measured volumetrically using a sterile pipette. Do not insert the pipette more than 2.5 cm below the surface of the sample. Empty the pipette into the diluent e.g., phosphate buffered water or 0.1% peptone water, by letting the column drain from the graduation mark to the rest point of the liquid in the tip of the pipette within 2 to 4 sec. Promptly and gently expel the last drop when pipetting the undiluted sample[18] or when using a pipette designed to be blown out. Do not rinse the pipette in the dilution water.

If the pipette becomes contaminated before completing transfers, replace it with a sterile pipette. Use a separate sterile pipette for transfers from each dilution. Dilution blanks should be at room temperature (15° to 25°C) when used. Caution: Do not prepare or dispense dilutions or pour plates in direct sunlight. When removing sterile pipettes from the container, do not drag tips over exposed exteriors of the pipettes remaining in the case because exposed ends of such pipettes are subject to contamination. Do not wipe or drag the pipette across the lips and necks of dilution bottles. Draw test portions above the pipette graduation, then raise the pipette tip above the liquid level. Adjust to the mark by allowing the pipette tip to contact the inside of the container in such a manner that drainage is complete and excess liquid does not adhere when pipettes are removed from sample or dilution bottles.[5] Do not flame pipettes.

Pipetter aids, assists, or automatic pipetters that are accurately calibrated and comply with pipette standards can be used instead of traditional pipetters. Apply all precautions identified for routine pipetting when using automatic pipetters or pipetting devices.

6.23 Solid Sample Homogenates

For viscous liquid products or food homogenates, the test portion for the initial dilution should be aseptically weighed (11 ± 0.1 g into a sterile 99 mL dilution blank, or 10 ± 0.1 g into 90 mL, or 50 ± 0.1 g into 450 mL). This provides a 1:10 dilution. Vigorously shake all dilutions 25 times in a 30-cm arc in 7 sec.[14] Optionally, a mechanical shaker may be used to shake the dilution blanks for 15 sec.[18]

6.3 PLATING TECHNIQUES

6.31 Basic Principles

The introduction of agar media in the late 1800s allowed the development of methods to enumerate microorganisms by colony count. Such methods are used extensively for determining approximate viable microbial populations in foods. These procedures are based on the assumption that each microbial cell in a sample will form a visible, separate colony when mixed with an agar or other solid medium and permitted to grow. Since microorganisms in foods often represent a number of populations with many different growth requirements, some organisms may not be capable of growth under conditions used in colony count methods. Additionally, not all microorganisms exist as single cells and closely associated clumps or chains of organisms will appear as a single colony. Consequently, the counts are at best an estimate and should not be reported as absolute.

The aerobic plate count (see Chapter 7) is the major, but not the only application of the colony count method. A more descriptive evaluation of the microorganisms present in the food sample may be obtained by using several nonselective media and incubating under more than one set of conditions; i.e., temperature, atmosphere, etc. Also, specific microorganisms can be enumerated using selective media, conditions, or both.

6.32 Precautions and Limitations

Colony count methods provide an estimate of the number of viable microorganisms in food according to the medium employed and the time and temperature of incubation. Microbial cells often occur as clumps or groups in foods. Whereas shaking samples and dilutions may uniformly distribute the clumps of bacteria, this may not completely disrupt the clumps themselves. Mixing the initial dilution in a mechanical blender may provide better breakdown of the clumps. However, this does not ensure that the microorganisms will be distributed as single cells. Consequently, each colony that appears on the agar plates can arise from a clump of cells or from a single cell and should be referred to as a colony forming unit (CFU).

Precision is defined as the likelihood of obtaining similar results when the same person or other analysts make repetitive counts. Accuracy is the minimizing of the difference between the count obtained and the "true" count. When considering the entire procedure and the results obtained, both are important.

The failure of some microorganisms to form visible colonies on the agar medium limits the accuracy of a colony count method. This failure can result from nutritional deficiencies of the medium, unfavorable oxygen tension, unfavorable incubation temperature, or cell injury. Incubation time and temperature also may be factors. The presence of inhibitory substances on glassware or in diluents or produced by competitive microorganisms in the agar may adversely affect some microbial cells and limit their ability to form colonies.

Another factor that affects apparent counts is the analyst's ability to see colonies distinctly. This depends on colony separation and morphology. Procedures that enhance colony growth and improve size, shape, contrast, and distribution should be used. The analyst's eyesight and fatigue may reduce the reliability of the count.

Other factors that may influence the accuracy of the colony count include:

1. improper sterilization and protection of sterilized diluents, media, and equipment;
2. inaccurate measurement of sample and dilutions;
3. improper distribution of the sample in or on the agar medium;
4. unsatisfactory working areas that permit contamination;
5. erratic mixing or shaking of sample or dilution;
6. inaccurate determination of colonies because of the presence of artifacts such as food particles in low dilutions and scratches on plates;
7. improper evaluation of spreaders or pinpoint colonies; and
8. other errors in counting and in computing counts.

Although there are some inherent limitations in enumerating microorganisms by the colony count method, many of the errors can be minimized if the analyst follows directions carefully and exercises extreme care in making all measurements.[5] Consistently accurate and meaningful results can be obtained from the routine examination of a food only if the same procedures are used to analyze each sample of that food. This includes sampling procedures, sample preparation, preparation of dilutions, plating medium, incubation conditions, and counting procedures.

6.33 Procedures

Bacteria, yeast and mold can grow either on or within a nutrient-rich substrate. As such both a pour plate method and a spread plate method can be used to enumerate microorganisms.

6.331 Pour Plate

6.3311 Sample preparation. Refer to Chapter 2 for complete details on sample preparation. For viscous or solid foods an initial 1:10 dilution is usually prepared. High-fat foods such as butter may require use of warm (40°C) diluent to facilitate mixing.

6.3312 Labeling. Label all petri plates, and tubes and bottles where necessary, with the sample number, dilution, date, and any other desired information. The bench area should be cleaned and sanitized, and all possible sources of contamination removed or reduced to a minimal level.

6.3313 Dilutions. For an accurate count, dilutions should be selected to ensure that plates containing the appropriate number of colonies will be produced. Different ranges for the appropriate number of colonies on plates may be applicable for certain procedures because of the crowding of colonies and other factors. For many methods such as the aerobic colony count, plates should contain between 25 and 250 colonies for accurate counts.[26] If the count is expected to be in the range of 2500 to 250,000 per mL or g, prepare plates containing 1:100 and 1:1000 dilutions. Figure 1 shows a schematic drawing of examples for preparing dilutions using a single plate for each dilution. For increased accuracy, two or more plates per dilution should be employed.

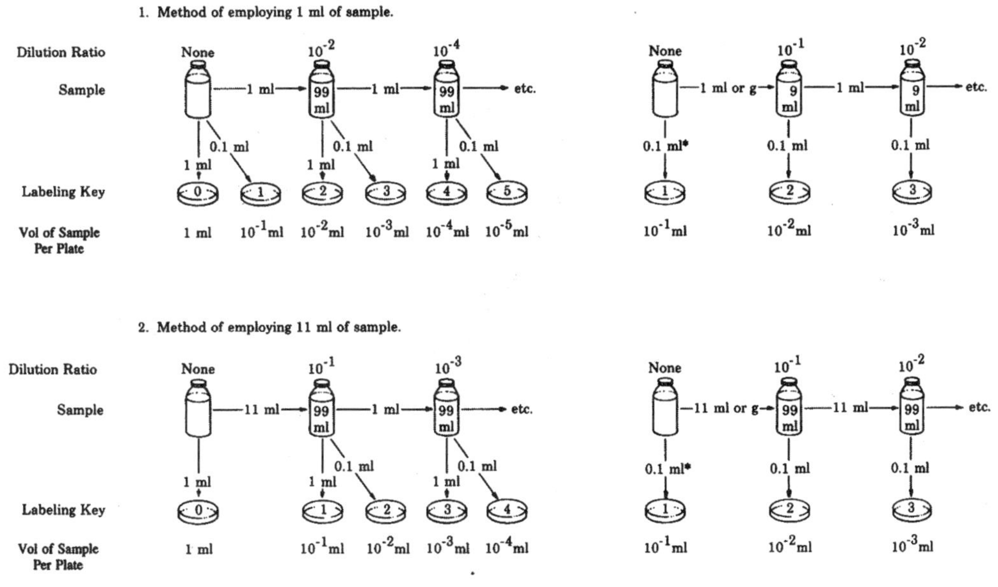

Figure 1. Preparation of dilutions from a nonviscous liquid food sample.

6.3314 Melting and tempering media. Melt agar media in flowing steam or boiling water, avoiding prolonged exposure to high temperatures. Temper melted media promptly and maintain between 44°C and 46°C until used. Set a thermometer into water or medium in a separate container similar to that used for the test medium; this temperature control medium must have been exposed to the same heating and cooling as the test medium. Do not depend upon the sense of touch to indicate the proper temperature of the medium when pouring agar. Cold gelling agents may be substituted for agar if previously shown to be equivalent.

6.3315 Plating. When measuring diluted samples of a food into petri plates, lift the cover of the petri plate just high enough to insert the pipette. Hold the pipette at about a 45° angle with the tip touching the inside bottom of the petri plate. Deposit the sample away from the center of the plate to aid in mixing. If a 1.0 or 1.1 mL pipette is used, allow 2 to 4 sec for the sample to drain from the 1 mL graduation mark to the rest point in the tip of the pipette. Then, holding the pipette in a vertical position, touch the tip once against a dry spot on the plate. Do not blow out. When 0.1 mL quantities are measured, hold the pipette as directed and let the diluted sample drain from one 0.1 mL graduation point down to the next 0.1 mL mark. Do not retouch the pipette to the plate when only 0.1 mL is delivered.[18] Replicate plates may be prepared for each dilution plated. Roll tubes, screw-cap tubes, bottles, or other containers may be used as alternatives to petri plates if all appropriate standardization is made and precautions are considered to assure equivalency.

6.3316. Pouring agar. After removing tempered agar medium from waterbath, blot the bottle dry with clean towels to prevent water from contaminating the plates. Pour 12 to 15 mL of liquefied medium at 44° to 46°C into each plate by lifting the cover of the petri plate just high enough to pour the medium. Avoid spilling the medium on the outside of the container or on the inside of the plate lid when pouring. This may require holding the bottle in a near horizontal position or refraining from setting down the bottle between pouring steps. As each plate is poured, thoroughly mix the medium with the test portions in the petri plate, taking care not to splash the mixture over the edge. This can be accomplished by rotating the plate first in one direction and then in the opposite direction, by tilting and rotating the plate, or by using mechanical rotators. Allow agar to solidify (no longer than 10 min) on a level surface.

Select the number of samples to be plated in any one series so that not more than 20 min (and preferably 10 min) elapse between diluting the first sample and pouring the last plate in the series.[2,15] Should a continuous plating operation be conducted by a team, plan the work so that the time between the initial measurement of a test portion into the diluent or directly into a dish and the pouring of the last plate for that sample is not more than 20 min. Avoid stack pouring unless the plates are distributed singly on a cooling surface immediately after mixing.[17]

Note: To obtain countable plates for foods having low colony counts, low dilutions must be used. For some foods this results in the presence of many food particles in the plate, which makes it difficult to distinguish the colonies easily for accurate counting. This problem often can be overcome by adding 1 mL of 0.5% (wt/vol) 2, 3, 5-triphenyltetrazolium chloride (TTC) per 100 mL of melted agar medium just prior to pouring the plates. Most bacteria form red colonies on an agar medium containing TTC. Counts should be made initially with and without TTC to determine if the TTC has any deleterious effect on the count. The TTC should be prepared as an aqueous solution and sterilized by passage through a sterilizing filter. To avoid decomposition, the solution must be protected from light and must not be exposed to excessive heat.

Sterility controls of medium, diluents, and equipment are recommended. Pour control plates for each lot of dilution blanks, medium, petri plates and pipettes.

6.3317 Incubation. After solidification, invert the plates to prevent spreaders, and promptly place them in the incubator. Incubation conditions for specific methods and commodities are presented in appropriate chapters. Agar within the plates should equilibrate to incubation temperature within 2 hr. Slower equilibration caused by excessive height of stacked plates or crowded incubators must be avoided.

Avoid excessive humidity in the incubator to reduce the tendency for spreader formation, but prevent excessive drying of the medium by controlling ventilation and air circulation. Agar in plates should not lose weight by more than 15% during 48 hr of incubation. Under some conditions, humidity control may become essential.

6.3318 Counting colonies. Count colonies with the aid of magnification under uniform and properly controlled artificial illumination, using a tally. Routinely use a colony counter[23] equipped with a guide plate ruled in square centimeters. Examine plates in subdued light. Try to avoid mistaking particles of undissolved medium, sample, oil droplets, or precipitated matter in plates for pinpoint colonies. Examine doubtful objects carefully, using higher magnification, if necessary, to distinguish colonies from foreign matter. A stereo microscope or magnifying glass may be useful for this examination. Carelessness, impaired vision, or failure to recognize colonies can lead to erroneous results. It is generally suggested that laboratory workers who cannot duplicate their own counts on the same plate within 5%, and the counts of other analysts within 10%, should discover the cause(s) and correct such factors.[8] However, others indicate that these percentages should be 7.7% and 18.2%, respectively.[10] Schedules of the laboratory analyst should be arranged to prevent eye fatigue and the inaccuracies that inevitably result from this.

Count all colonies on selected plates containing the appropriate number of colonies promptly after the incubation period.[26] Refer to Item 6.3319, Computing and reporting, for guidelines on selecting plates and computing counts. If impossible to count at once, the plates may be stored, after the required incubation, at approximately 0° to 4.4°C for under 24 hr.[10] This should not be a routine practice.

6.3319 Computing and reporting. To compute colony counts, multiply the total number of colonies (or the average number if replicate plates of the same dilution are used) per plate by the reciprocal of the dilution used. Record the dilution used and the number of colonies counted or estimated on each plate. To avoid giving false ideas of precision and accuracy when computing colony counts, record only the first two left-hand digits. Raise the second digit to the next highest number only when the third digit from the left is 5, 6, 7, 8, or 9; use zeros for each successive digit to the right of the second digit (Table 1). Report counts (or estimates thereof) as CFU per g or mL, as applicable.

When counts on duplicate plates or consecutive dilutions are averaged, round off counts to two significant figures only at the time of conversion to the CFU per g (Table 1, Sample No. 1117).

The appropriate number of colonies to count on a plate is a function of colony size, plate size, and size of differential properties produced on the medium. Typically, 25 to 250 colonies per plate yield reliable results. Use this as a guide unless an alternate range is indicated for specific methods. The following guidelines or "rules" should be used for selecting plates and calculating the CFU per g or mL, as applicable:

1. *One plate with 25 to 250 colonies.* Select a plate with 25 to 250 colonies unless excluded by spreaders or lab accidents (see Item 8). Count all colonies, including those of pinpoint size and record the dilution used and the total number of colonies counted (Table 1, Sample Nos. 1001, 1004, 1011, and 1012).
2. *Duplicate plates.* Count plates with 25 to 250 colonies and average the counts to obtain the colony count (Table 1, Sample No. 1112). If only one plate of a duplicate pair yields 25 to 250 colonies, count both plates,

Table 1. Examples for Computing Colony Count per Gram or Milliliter

	Colonies counted				
	Dilution				
Sample no.	1:100	1:1000	Count ratio[a]	Colony count[b] (CFU/g or mL)	Rule
Common application, one plate from each of two dilutions					
1001	234[c]	23	—	23,000	1
1002	243	34	1.4	29,000	3
1003	140	32	2.3	14,000	3
1004	Spr[d]	31	—	31,000	1, 8
1005	0	0	—	<100 est.	6
1006	TNTC	7150	—	>5,600,000 est.	7
1007	18	2	—	1800 est	5
1008	Spr	Spr	—	Spr	8
1009	325	20	—	33,000 est.	4, 7
1010	27	215	—	LA[e]	8
1011	305	42	—	42,000	1
1012	243	LA	—	24,000	1, 8
1013	TNTC	840	—	840,000 est.	7
Procedure where two plates per dilution are poured					
1111	228	28	1.2	25,000	2, 3
	240	26			
1112	175	16	—	19,000	2
	208	17			
1113	239	16	—	28,000	2
	328	19			
1114	275	24	—	30,000	2
	280	35			
1115	138	42	2.4	15,000	2, 3
	162	30			
1116	228	28	1.1	24,000	2, 3
	240	23			
1117	224	28	1.4	24,000	2, 3
	180	Spr			
1118	287	23	—	28,000 est.	4, 7
	263	19			
1119	18	2	—	1700 est.	5
	16	0			
1120	0	0	—	<100 est.	6
	0	0			

a Count ratio is the ratio of the greater to the lesser plate count, as applied to plates from consecutive dilutions having between 25 and 250 colonies.
b All counts should be made in accordance with instructions in Section 6.631, No. 8, as well as any other rules listed or given in the text.
c Underlined figures used to calculate count.
d Spr = Spreader and adjoining area of repressed growth covering more than one-half of the plate
e LA = Laboratory accident.

unless excluded by spreaders, and average the counts (Table 1, Sample Nos. 1113 and 1114). When counting duplicate plates from consecutive decimal dilutions, compute the count per g for each dilution and proceed as in Item 3 (Table 1, Sample Nos. 1111, 1115, 1116, and 1117).

3. *Consecutive dilutions with 25 to 250 colonies.* If plates from two consecutive decimal dilutions yield 25 to 250 colonies each, compute the count per g for each dilution and report the arithmetic average as the CFU per g, unless the higher computed count is more than twice the lower one. In that case, report the lower computed count as the CFU per g (Table 1, Sample Nos. 1002, 1003, 1111, 1115, 1116, and 1117).
4. *No plate with 25 to 250 colonies.* If there is no plate with 25 to 250 colonies and one or more plates have more than 250 colonies, select plate(s) having nearest to 250 colonies and count as in Item g for crowded plates. Report count as the estimated (est.) CFU per g (Table 1, Sample Nos. 1009 and 1118).
5. *All plates have fewer than 25 colonies.* If plates from all dilutions yield fewer than 25 colonies, record the actual number of colonies on the lowest dilution (unless excluded by spreaders) and report count as est. CFU per g (Table 1, Sample Nos. 1007 and 1119).
6. *Plates with no colonies.* If plates from all dilutions have no colonies and inhibitory substances have not been detected, report the estimated count as less than (<) one times the corresponding lowest dilution (Table 1, Sample Nos. 1005 and 1120).
7. *Crowded plates (more than 250 colonies).* If the number of colonies per plate exceeds 250, count colonies in portions of the plate that are representative of colony distribution to estimate the aerobic colony count. If there are fewer than 10 colonies per cm^2 count the colonies in 12 cm^2, selecting six consecutive squares horizontally across the plate and six consecutive squares at right angles, being careful not to count a square more than once. When there are more than 10 colonies per cm^2, count the colonies in four representative squares. In both instances, multiply the average number of colonies per cm^2 by the area of the plate to determine the estimated number of colonies per plate. Individual laboratories should determine the area of the plate and the proper factor for multiplication; however, the area of a standard 15 × 100 mm plastic petri plate is approximately 56 cm^2 and therefore the appropriate factor is 56. For an example using an average count of 15 colonies per cm^2 on a 56 cm^2 plate, see Table 1, Sample No. 1013. Do not report counts on crowded plates from the highest dilution as "too numerous to count" (TNTC).

   Where bacterial counts on crowded plates are greater than 100 colonies per cm^2, report as greater than (>) the plate area multiplied by 100, multiplied by the highest dilution plated. For example, for a 56 cm^2 plate, the count would be 5600 times the highest dilution plated. Report as est. CFU per g (Table 1, Sample No. 1006).

   When all colonies on a plate are accurately counted and the number exceeds 250, report as est. CFU per g (Table 1, Sample Nos. 1009 and 1118).
8. *Spreaders.* There are three distinct types of spreaders. The first type is a chain of colonies, not too distinctly separated, that appears to be caused by disintegration of a bacterial clump when the inoculum is dispersed in or on the plating medium. If one or more chains appear to originate from separate sources, count each as one colony. Do not count each individual colony in such chain(s) as separate colonies.

   The second type of spreading colony develops in a film of water between the agar and the plate.[8] The third type forms in a film of water at the edge or over the surface of the agar. These two types develop mainly because of moisture accumulation at the point from which the spreader originates, and these spreaders may repress the growth of individual colonies. When dilution water is uniformly distributed throughout the medium, bacteria rarely develop into spreading colonies. Steps to eliminate spreaders of this type should be taken if 5% of a laboratory's plates have spreaders covering 25% of the plate.

   If spreaders occur on the plate(s) selected, count colonies on representative portions thereof only when colonies are well distributed in spreader free areas and the area covered by spreader(s), including the total repressed growth area if any, does not exceed 50% of the plate area. Calculate the estimated count by multiplying the average count per cm^2 by the area of the plate. Where the repressed growth area alone exceeds 25% of the total area, report as "spreaders" (spr) or "laboratory accident" (LA) (Table 1, Sample Nos. 1008 and 1010).

   Inhibitory substances in a sample may be responsible for the lack of colony formation. The analyst may suspect the presence of inhibitory substances in the sample under examination when plates show no growth or show proportionately less growth in lower dilutions. Such developments cannot, however, always be interpreted as evidence of inhibition, and unless inhibition is demonstrated, should be reported as LA.

6.332 Surface or Spread Plate Method

Methods of plating designed to produce all surface colonies on agar plates have certain advantages over the pour plate method.[16] The use of translucent media is not essential with a surface or spread plate but is necessary with the pour plate to facilitate location of colonies. The colonial morphology of surface colonies is easily observed, improving the analyst's ability to distinguish between different types of colonies.[21] Organisms are not exposed to the heat of the melted agar medium, so higher counts may be observed in some situations.[3,8,22,28] On the other hand, since relatively small volumes (0.1 to 0.5 mL) of the sample must be used, the method may lack precision for samples containing few microorganisms.

1. Prepare sample (see Chapter 2).
2. Pour approximately 15 mL of the agar into sterile petri plates. To facilitate uniform spreading, the surface of the solidified agar should be dried by holding the plates at 50°C for 1.5 to 2 hr. Plates may also be dried at a lower temperature (25° to 35°C) for longer periods (18 to 24 hr), or in a laminar flow hood, with covers ajar, for 0.5 to 1 hr.
3. Label all petri plates as in pour plate method.
4. Prepare ten-fold serial dilutions following general procedures described previously.
5. Measure 0.1 mL of diluted sample onto agar surface using a sterile pipette (graduated into 0.1 mL divisions). Larger volumes may be appropriate under certain situations, but take precautions to ensure that the liquid does not remain on the agar surface to promote spreaders. For example, 0.2 mL of a 1:10 dilution can be delivered to each of five plates to get the equivalent of 1 g of food for CFU determinations.
6. Spread the diluted sample on the surface of the agar medium with a sterile bent glass rod (hockey stick), or equivalent, as quickly and carefully as possible. Use a separate glass rod

for each plate or spread the plates for a given sample starting with the most dilute plate and proceed to the least dilute plate in series with aseptic technique throughout. Allow the plates to dry at least 15 min prior to inversion.
7. Incubate plates.
8. Compute and record colony counts as in pour plate method.

6.4 MOST PROBABLE NUMBER TECHNIQUES

6.41 Basic Principles

The most probable number (MPN) dilution technique uses results that are reported as positive or negative in one or more decimal dilutions of the sample to estimate the number of organisms present. Thus, unlike the aerobic plate count, the MPN does not provide a direct measure of the bacterial count. In addition, the MPN is more variable than the plate count.[19]

Although the MPN is not a precise measure, a specific value can be computed for a single dilution[11] or for multiple dilutions,[12] provided the results are not all positive or negative for all dilutions used, and assuming[4] that the organisms to be measured are distributed randomly throughout the sample and that growth will occur when one or more organisms are present in a tube. Halvorson and Ziegler[12] demonstrated that for a multiple tube MPN, precision depends only on the number of tubes per dilution; for single dilution tests,[11] it depends on bacterial numbers and number of tubes. Eisenhart and Wilson[9] and Oblinger and Koburger[20] discuss the early history of dilution techniques. The latter article is useful for training students in understanding the test.

The composition of many food products and ingredients makes it difficult to use standard plating procedures, particularly when the microbial concentration of the sample is less than 10 CFU per gram. Ziegler and Halvorson[30] showed that in these low-count situations, the MPN technique gave higher values for bacterial populations than did the plate count method. The direct microscopic count gave the same value as the plating and MPN method only when it was used on cultures that had not entered the death phase. McCarthy et al.[19] also demonstrated a considerable positive mathematical bias in MPN values relative to plate counts.

As a sample is serially diluted, some of the aliquots eventually contain such small amounts of sample that they will contain no microorganisms. The MPN method is based on diluting out of the microorganisms, and therefore may be described as the "multiple tube dilution to extinction method." The most satisfactory information is obtained when all of the tubes with the large sample portions show growth and the tubes with the smaller portions show no growth.

Applications of the MPN method are numerous. Use of the method is particularly important in the standard coliform procedure used for water and wastewater testing and in testing foods in general. The method is also used in the isolation and enumeration of staphylococci, streptococci, *Vibrio parahaemolyticus,* and salmonellae when quantitative rather than qualitative analysis is necessary.

The method also can be applied when a single sample dilution is used in several tubes, e.g., five 0.1-g samples for enumeration of very low numbers of organisms. For this type of application, special tables are required.[1,20]

Because the method uses liquid media, it offers the user considerable flexibility as to sample size. If allowances for appropriate dilutions of sample and ratios of medium to sample are made, sample volumes can be quite large. Increasing the number of tubes within each effective dilution improves precision. At low population levels, sensitivity is generally greater with the MPN than with the plate count;[18] however, this is not always the case. The "bathing" aspect of nutrient availability in a liquid medium may enhance recovery of organisms. Subsequent transfer of samples to a more inhibitory environment is possible after a period of resuscitation.

Variation among replicate aliquots is by far the most important source of error.[27] Extreme care is needed in preparing dilutions and transferring aliquots within the same dilutions. Other important factors that contribute to spurious results include difficulty in obtaining truly representative samples from a given lot and the possibility of an uneven distribution of microorganisms within the sample units selected.[24]

If the sample contains inhibitory substances, or the product itself is inhibitory (e.g., sodium chloride), growth in the tubes with high concentrations of sample may be inhibited. The possibility of injured cells that cannot grow out should not be overlooked. Nutrient in the sample may also interfere with the selectivity of the medium. For example, sucrose in a food will lead to a false indication of the presence of coliforms.

One set of tubes from each batch of medium prepared should be used as an uninoculated control. If, for example, the five-tube MPN method is being used, a set of five tubes should be incubated as uninoculated controls to ensure that the medium was properly sterilized.

6.42 Procedures

1. Prepare sample (see Chapter 2).
2. Label all tubes.
3. *Dilutions.* Unless previous experience with a sample indicates the appropriate number of dilutions needed, a minimum of five 10-fold dilutions should be prepared.
4. *Inoculating tubes.* Usually three or five tubes are inoculated for each dilution used within 20 min of preparation of the initial dilution. Typically one part of sample to ten parts of medium should be maintained; e.g., 0.1-g sample should be dispersed in 1 mL of medium, or 1-mL aliquots into 10 mL of broth. The strength of the medium can be adjusted so that the concentration of medium after the sample is added equals single strength medium. For example, in water analysis, one frequently uses double strength broth with an equal sample volume to avoid excessive nutrient and inhibitor dilution.
5. *Incubation.* Incubation conditions for specific methods and commodities are presented in appropriate chapters for each analysis. An air incubator or a water bath may be used.
6. Detection of positive tubes.
   a. *Turbidity.* When using samples that do not cloud the medium in the tubes, the development of turbidity after incubation indicates growth (positive tubes). When the sample causes turbidity, other methods must be used to determine positive tubes.
   b. Metabolic end products.
      (1) Detection of gas production. Gases produced by developing microorganisms can be captured and observed with gas traps or inverted vials that are placed in the medium in the growth vessels before sterilization. A positive reaction is recorded when gas bubbles are observed in traps at the end of the incubation period. Other methods used to capture and observe the gases produced include overlay with vaspar or agar. Obviously these are useful only when the microorganisms to be enumerated are known to produce gas

under the conditions of the test. If tubed media are stored at low temperature, small bubbles may accumulate in the inverted fermentation tubes when media is brought to incubation temperatures due to air dissolving in the cold medium. Steaming or boiling the tubed media before use removes these bubbles; however, one must consider the possibility of denaturing sensitive medium components (depending on the medium being used).

(2) *Detection of acid or base.* Acid or base production can be determined after incubation by measuring the pH or titratable acidity in each tube or by using a medium containing a pH-indicating dye. Detecting positive tubes by this method requires that the microorganisms being enumerated produce a pH change from a defined substrate.

(3) *Detection with reduction methods.* Electron acceptors (e.g., resazurin, methylene blue or 2, 3, 5-triphenyltetrazolium chloride) that change color upon reduction can be incorporated into the medium. Reduction of any of these compounds by microbial action also indicates growth.

(4) *Other.* Specific media can be developed to assay for certain metabolic activities, e.g., NO_3 reduction, indole production, starch hydrolysis, and H_2S production, depending on the information desired.

7. Confirming inconclusive tests

a. *Direct microscopic examination.* Microscopic examination is done by placing a loopful of the medium from the tube on a slide, drying, heat fixing, and staining. The slide is examined for the specific microorganism, using oil immersion at 1000 × magnification. Care should be taken if the original sample contains high numbers of killed or inactivated microorganisms to distinguish live from dead cells using procedures such as dye techniques.

b. *Subculturing.* To confirm growth in questionable tubes, medium from the tube is transferred to a nonselective medium and incubated for an appropriate additional period of time. Growth in this medium confirms the presence of viable microorganisms in the tube. To confirm a questionable growth reaction (such as acid) in tubes of selective media with heavy food turbidity, media from the incubated tube is transferred to a tube of the identical medium and similarly incubated. Growth reactions in such subcultures can be readily observed since they are free of the color and turbidity.

Aliquots may also be streaked from MPN tubes onto either selective or nonselective media to ascertain appropriate growth.

6.43 Use of MPN Tables

Since the MPN procedure provides an estimate of the count present, confidence intervals are used to indicate the precision of the MPN estimates. If we are considering a 95% confidence interval, then the true, but unknown, number of organisms in the sampled population lies within the limits 95% of the time. Tables in this book follow the Dairy Products Standard Methods.[18] Halvorson and Ziegler[12] presented the formulae for computing the probability of a combination (i.e., 3-2-0) for given dilutions and organism concentrations.

Tables 2 and 3 provide combinations of positive results that occur frequently enough to be statistically significant. Combinations that occur less than 1% of the time are omitted. When compared to other references[1,20,27] reported confidence limits are slightly different, a situation attributed to assumptions made and computational methods used to derive the values. Review the original work if further insight into these tabular differences or the methods used to compute the values is needed.[6,7] If the appropriate combination of positive results is not found, the analysis should be repeated or more complete tables should be consulted.

When the multiple tube method is used, results are usually reported as "the most probable number of microorganisms per gram (or mL)." When specific groups of microorganisms have been estimated, results can be reported as a "presumptive MPN estimate" for that specific group until appropriate confirmatory tests have been completed. Tables 2 and 3 show the most probable numbers of microorganisms corresponding to the frequency of positive tubes obtained from three 1:10 dilution series beginning with 0.1-g test portions. Results for both three tubes and five tubes per dilution are given along with the 95% confidence limits.

Table 2. Selected MPN Estimates and 95% Confidence Limits[7] for Fermentation Tube Tests When Three Tubes with 0.1-g, 0.01-g, and 0.001-g sizes Are Used[a]

No. of positive tubes/3 tubes				95% confidence limits	
0.1 g	0.01 g	0.001 g	MPN/g[b]	Lower	Upper
0	0	0	<3	—	—
0	1	0	3+	<1	17
1	0	0	4	<1	21
1	0	1	7+	2	27
1	1	0	7	2	28
1	2	0	11+	4	35
2	0	0	9	2	38
2	0	1	14+	5	48
2	1	0	15	5	50
2	1	1	20+	7	60
2	2	0	21	8	62
3	0	0	23	9	130
3	0	1	39	10	180
3	1	0	43	10	210
3	1	1	75	20	280
3	2	0	93	30	380
3	2	1	150	50	500
3	2	2	210+	80	640
3	3	0	240	90	1400
3	3	1	460	100	2400
3	3	2	1100	300	4800
3	3	3	>1100	—	—

[a] Normal results, obtained in 95% of tests, are *not* followed by a plus (+). Less likely results, obtained in only 4% of tests, are followed by a plus (+). Combinations of positive tubes not shown occur in less than 1% of tests, and their frequent occurrence indicates that technique is faulty or that assumptions underlying the MPN estimate are not being fulfilled. MPN estimates for combinations that are not shown may be obtained by extrapolation (or by Thomas' formula, Section 6.44) to the next highest combination that is shown in the table. For example, a result of 2-0-2 has an MPN of approximately 20, which is the MPN for a more likely result of 2-1-1.

[b] All figures under "MPN/g" in this table may be multiplied by 100 for reporting "MPN/100 g."

Table 3. Selected MPN Estimates and 95% Confidence Limits[7] for Fermentation Tube Tests When Five Tubes with 0.1-g, 0.01-g, and 0.001-g volumes Are Used[a]

No. of positive tubes/5 tubes				95% confidence limits	
0.1 g	0.01 g	0.001 g	MPN/g[b]	Lower	Upper
0	0	0	<2	—	—
0	0	1	2+	<1	10
0	1	0	2	<1	10
1	0	0	2	<1	11
1	0	1	4+	1	15
1	1	0	4	1	15
1	2	0	6+	2	18
2	0	0	4	1	17
2	0	1	7+	2	20
2	1	0	7	2	21
2	1	1	9+	3	25
2	2	0	9	3	25
3	0	0	8	3	24
3	0	1	11	4	29
3	1	0	11	4	30
3	1	1	14+	6	35
3	2	0	14	6	35
3	2	1	17+	7	40
3	3	0	17+	7	41
4	0	0	13	5	38
4	0	1	17	7	45
4	1	0	17	7	46
4	1	1	21	9	55
4	2	0	22	9	56
4	2	1	26+	12	65
4	3	0	27	12	67
4	3	1	33+	15	77
4	4	0	34+	16	80
5	0	0	23	9	68
5	0	1	31	13	110
5	1	0	33	14	120
5	1	1	46	20	150
5	1	2	63+	22	180
5	2	0	49	21	170
5	2	1	70	30	210
5	2	2	94+	40	250
5	3	0	79	30	250
5	3	1	110	40	300
5	3	2	140	60	360
5	4	0	130	50	390
5	4	1	170	70	480
5	4	2	220	100	580
5	4	3	280+	120	690
5	4	4	350+	160	820
5	5	0	240	100	940
5	5	1	350	100	1300
5	5	2	540	220	2000
5	5	3	920	300	2900
5	5	4	1600	600	5300
5	5	5	>1600	—	—

[a] Normal results, obtained in 95% of tests, are *not* followed by a plus (+). Less likely results, obtained in only 4% of tests, are followed by a plus (+). Combinations of positive tubes not shown occur in less than 1% of tests, and their frequent occurrence indicates that technique is faulty or that assumptions underlying the MPN estimate are not being fulfilled. MPN estimates for combinations that are not shown may be obtained by extrapolation (or by Thomas' formula, Section 6.44) to the next highest combination that is shown in the table. For example, a result of 4-0-2 has an MPN of approximately 21, which is the MPN for a more likely result of 4-1-1.

[b] All figures under "MPN/g" in this table may be multiplied by 100 for reporting "MPN/100 g."

Tables 4 and 5 give examples for determining MPN estimates for three-tube and five-tube MPN series, respectively, when 1-mL sample aliquots from serial dilutions are planted. Note that the tabular values are treated in terms of the actual sample volumes planted in these dilutions.

When more than three dilutions of a sample are prepared, the results from only three consecutive dilutions are used in determining the MPN. First, for all dilutions having all tubes positive, select the highest dilution (smallest sample size). Second, use the next two higher dilutions (smaller sample sizes), as shown in examples a and b of Tables 4 and 5. When none of the tested dilutions yield all tubes positive, select the first three consecutive dilutions for which the middle dilution contains the positive result, as in example c of Tables 4 and 5. If a positive result occurs in a higher dilution (smaller sample size) than the three selected, add the number of positive tubes in this dilution to the highest dilution (smallest sample size) of the three selected, as in example d of Tables 4 and 5. When all dilutions tested have all tubes positive, select the three highest dilutions (smallest sample sizes), as in example e of Tables 4 and 5.

Often it is necessary to calculate the MPN from initial sample sizes other than those listed in Tables 2 and 3. If the largest (greatest) sample size used for the table reference is 0.01 g, multiply the MPN index listed in the table by 10. Thus, results of a five-tube MPN determination showing 3 positive 0.01-g portions, 2 positive 0.001-g portions, and 1 positive 0.0001-g portion (3-2-1) are read from Table 3 as 17, and multiplied by 10 to arrive at 170 as the actual MPN/g for the sample. Similarly, if the greatest portion used for the table reference is 1 g rather than 0.1 g, divide the MPN derived from the table by 10. Thus, the result of a three-tube MPN determination for salmonellae showing 3 positive 1-g portions, 1 positive 0.1-g portion, and 0 positive 0.01-g portions (3-1-0), is read from Table 2 as 43 and divided by 10, or 4.3 as the presumptive MPN/g for the sample.

An alternative approach to obtain the MPN per gram uses the following formula:[24]

[(MPN/g from Table/100)] × dilution factor of the middle tube = MPN/g.

To obtain an MPN/100 g, multiply by 100.

MPN estimates are often credited with unfounded precision. The tabular MPN estimate represents a range and not an absolute value. Most MPN tables[18,27] include 95% confidence limits for the tabular MPN estimates. The true number of organisms lies between these limits 95% of the time. One must be able to read each table properly and understand the significance of the results. For a three-tube test, the 95% confidence limits cover a 33-fold range from approximately 14% to 458% of the actual tabular MPN estimate and for a five-tube dilution test, the 95% confidence limits cover a 13-fold range from approximately 24% to 324% of the MPN.[29]

Results should be recorded as "Number of microorganisms per quantity (g or mL) of sample by the MPN method," e.g., coliform MPN/g = 11. With the report of microbiological counts by the MPN method, the number of tubes used in each dilution is included, i.e., five-tube MPN or three-tube MPN, and the particular method used.

6.44 Approximate MPN and 95% Confidence Limits

Thomas[25] published a simple formula to approximate MPNs. The results may not exactly agree with those in tables; however, deviations are usually slight and of no practical consequence. Additionally, the formula is not restricted to the number of tubes and

Table 4. Examples of Determining MPN Estimates: Three-tube Series (1-mL sample aliquot per tube)

	Sample volume (g)					Reported	MPN
Example	0.1	0.01	0.001	0.0001	0.00001	positive values	estimate/g
a	3/3[a]	3/3	2/3	0/3	0/3	3-2-0	930
b	3/3	3/3	3/3	2/3	0/3	3-2-0	9300
c	0/3	0/3	1/3	0/3	0/3	0-1-0	30
d	3/3	3/3	2/3	1/3	1/3	3-2-2	2100
e	3/3	3/3	3/3	3/3	3/3	3-3-3	>110,000

[a] Numerator / Denominator = No. positive tubes / No. tubes inoculated.

dilutions used and may be applied to all types of tests. Thomas' approximation is given by the following:

$MPN/g = P/\sqrt{NT}$

P = the number of positive tubes

N = the total quantity of sample (g) in all negative tubes

T = the total quantity of sample (g) in all tubes.

For example, consider the two-fold dilution series given below:

g of sample	*number of tubes*	*number of positive tubes*
8	5	5
4	5	4
2	5	2
1	5	0
0.5	5	1
0.25	5	0

$P = (5 + 4 + 2 + 1) = 12$

$N = (8 \times 0) + (4 \times 1) + (2 \times 3) + (1 \times 5) + (0.5 \times 4) + (0.25 \times 5) = 18.25$

$T = 5(8 + 4 + 2 + 1 + 0.5 + 0.25) = 78.75$

$MPN/g = 12/\sqrt{(18.25)(78.75)} = 0.32/g$

Estimates of the 95% confidence limits can be obtained as follows:[4]

$\log (MPN/g) \pm (1.96)(0.55)\sqrt{(\log a)/n}$

a = the dilution ratio

n = the number of tubes per dilution

For the above MPN example, the approximate 95% confidence limits are:

$\log 0.32 \pm (1.96)(0.55)\sqrt{(\log 2)/5} = -0.495 \pm 0.265$,

then the lower limit is antilog (–0.76) = 0.17/g and the upper limit is antilog (–0.23) = 0.59/g.

6.5 MEMBRANE FILTRATION

6.51 *Basic Principles*

For certain foods or food ingredients, the ability to test relatively large samples will improve the accuracy of quantitative microbiological analyses. Large volumes of liquid foods or solutions of dry foods that can be dissolved and passed through a bacteriological membrane filter (pore size 0.45 µm) may be analyzed for microbial content by the membrane filter method. The method is especially useful for samples that contain low numbers of bacteria. Additional details on rapid and commercial kits using membrane filtration methods are described in Chapter 10.

6.6 OTHER METHODS

A number of innovative and convenient methods are commercially available for enumeration of microorganisms. These include Petrifilm™ Plate method, pectin gel method, calibrated loop method, drop plate method, Hydrophobic Grid-Membrane Filter method (HGMF), Spiral Plate method, SimPlate™, impedance, and luminescence. Refer to Chapter 7 for general procedures and information.

6.7 ANAEROBIC OR OTHER ATMOSPHERES

The choice of atmospheric conditions to which the plates will be exposed is vital to the successful enumeration of microorganisms. The three most commonly used atmospheres for the growth of microorganisms are aerobic, anaerobic and microaerophilic. If the preferred atmosphere is aerobic, incubation is conducted in an incubator under conventional atmosphere.

An atmosphere devoid of measurable oxygen is preferred for the manipulation and enumeration of anaerobic microorganisms such as *Clostridium* spp. There are a number of methods that enable a researcher to achieve an anaerobic atmosphere. For example this can be achieved with an incubator in a sealed hood with an atmosphere of approximately 85% nitrogen and 15% hydrogen. Alternatively Brewer jars or candle extinction jars are lower cost methods to achieve a similar end result. Under conditions of anaeobiosis it is important to verify the absence of oxygen within the system. A number of testing methods are available, ranging

Table 5. Examples of Determining MPN Estimates: Five-tube Series (1-mL sample aliquot per tube)

	Sample volume (g)					Reported	MPN
Example	0.1	0.01	0.001	0.0001	0.00001	positive values	estimate/g
a	5/5[a]	5/5	2/5	0/5	0/5	5-2-0	490
b	5/5	5/5	5/5	2/5	0/5	5-2-0	4900
c	0/5	0/5	1/5	0/5	0/5	0-1-0	20
d	5/5	5/5	3/5	1/5	1/5	5-3-2	1400
e	5/5	5/5	5/5	5/5	5/5	5-5-5	>160,000

[a] Numerator / Denominator = No. positive tubes / No. tubes inoculated.

from sophisticated electrical sensors to rudimentary indicator strips.

Methods for determining total anaerobic colony counts are similar to those outlined in Chapter 23 for anaerobic sporeformers except that steps to eliminate vegetative cells are omitted. If the analyst is concerned with enumerating strict or sensitive anaerobes, the methodology developed at the Anaerobe Laboratory, Virginia Polytechnic Institute and State University, may be of help. The Hungate method with pre-reduced, anaerobically sterilized media in tubes with butyl rubber stoppers is suggested for analysis of food samples for obligatory anaerobic bacteria.

The specific procedure can follow individual steps of the pour plate or spread plate techniques (6.321 and 6.322). The media may include plate count agar, Andersen's pork pea agar, blood agar, or similar complex media, and may be overlaid with thioglycollate agar. Incubation is under anaerobic conditions in an anaerobic culture jar, an anaerobic incubator, or some similar container that can contain an atmosphere free of oxygen and maintain an appropriate atmosphere. Such atmospheres can be achieved using an incubator connected to a free-flowing cylinder of an appropriate gas mixture. Another method to achieve a microaerophilic atmosphere is through the use of a Brewer jar with an appropriate catalyst system.

6.8 REFERENCES

1. APHA. 1989. Standard methods for the rxamination of water and wastewater, 17th ed. American Public Health Association, Washington, D. C.
2. Berry, J. M., D. A. McNeill, and L. D. Witter. 1969. Effect of delays in pour plating on bacterial counts. J. Dairy Sci. 52:1456.
3. Clark, D. S. 1967. Comparison of pour and surface plate methods for determination of bacterial counts. Can. J. Microbiol. 13:1409.
4. Cochran, W. G. 1950. Estimation of bacterial densities by means of the "Most Probable Number." Biometrics 6:105.
5. Courtney, J. L. 1956. The relationship of average standard plate count ratios to employee proficiency in plating dairy products. J. Milk Food Technol. 19:336.
6. deMan, J. C. 1975. The probability of most probable numbers. Eur. J. Appl. Microbiol. 1:67.
7. deMan, J. C. 1977. MPN tables for more than one test. Eur. J. Appl. Microbiol 4:307
8. Donnelly, C.B., E. K. Harris, L. A. Black, and K. H. Lewis. 1960. Statistical analysis of standard plate counts of milk samples split with state laboratories. J. Milk Food Technol. 23:315.
9. Eisenhart, C., and P. W. Wilson. 1943. Statistical methods and control in bacteriology. Bacteriol. Rev. 7:57.
10. Fowler, J. L., W. S. Clark Jr., J. F. Foster, and A. Hopkins. 1978. Analyst variation in doing the standard plate count as described in *Standard Methods for the Examination of Dairy Products*. J. Food Prot. 41:4.
11. Halvorson, H. O., and N. R. Ziegler. 1933a. Application of statistics to problems in bacteriology. II. A consideration of the accuracy of dilution data obtained by using a single dilution. J. Bacteriol. 26:331.
12. Halvorson, H. O., and N. R. Ziegler. 1933b. Application of statistics to problems in bacteriology. III. A consideration of the accuracy of dilution data obtained by using several dilutions. J. Bacteriol. 26:559.
13. Hartman, P. A., and D. V. Huntsberger. 1961. Influence of subtle differences in plating procedure on bacterial counts of prepared foods. Appl. Microbiol. 9:32.
14. Huhtanen, C. N., A. R. Brazis, W. L. Arledge, E. W. Cook, C. B. Donnelly, R. E. Ginn, J. N. Murphy, H. E. Randolph, E. L. Sing, and D. I. Thompson. 1970. Effect of dilution bottle mixing methods on plate counts of raw-milk bacteria. J. Milk Food Technol. 33:269.
15. Huhtanen, C.N., Brazis, A.R., Arledge, W.L., E. W. Cook, C. B. Donnelly, R. E. Gin, J. J. Jezeski, D. Pusch, H. E. Randolph, and E. L. Sing. 1972. Effects of time and holding dilutions on counts of bacteria from raw milk. J. Milk Food Technol. 35:126.
16. ICMSF. 1978. Microorganisms in foods: their significance and methods of enumeration, 2nd ed. Intern. Comm. On Microbiol. Spec. Foods. Univ. of Toronto, Canada.
17. Koburger, J. A. 1980. Stack-pouring of petri plates: A potential source of error. J. Food Prot. 43:561.
18. Marshall, R. T. (ed.). 1993. Standard methods for the examination of dairy products, 16th ed. American Public Health Association, Washington, D. C.
19. McCarthy, J. A., H. A. Thomas Jr., and J. D. Delaney. 1958. Evaluation of the reliability of coliform density tests. Am. J. Public Health 48:1628-1635.
20. Oblinger, J. L., and J. A. Koburger. 1975. Understanding and teaching the most probable number technique. J. Milk Food Technol. 38:540-545.
21. Punch, J. D., and J. C. Olson Jr. 1964. Comparison between standard methods procedure and a surface plate method for estimating psychrophilic bacteria in milk. J. Milk Food Technol. 27:43.
22. Ray, B., and M. L. Speck. 1973. Discrepancies in the enumeration of *Escherichia coli*. Appl. Microbiol. 25:494.
23. Richards, O. W., and P. C. Heijn. 1945. An improved darkfield Quebec colony counter. J. Milk Technol. 8:253.
24. Silliker, J. H., D. A. Gabis, and A. May. 1979. ICMSF methods studies XI. Collaborative/comparative studies on determination of coliforms using the most probable number procedure. J. Food Prot. 42:638-644.
25. Thomas, H. A. 1942. Bacterial densities from fermentation tube tests. J. Am. Water Work Assoc. 34:572-576.
26. Tomasiewicz, D. M., D. K. Hotchkiss, G. W. Reinbold, R. B. Read Jr., and P. A. Hartman. 1980. The most suitable number of colonies on plates for counting. J. Food Prot. 43:282.
27. U.S. Department of Health, Education & Welfare. 1984. Bacteriological analytical manual, 6th ed. U.S. Food and Drug Administration. U.S. Dept. Health, Education and Welfare, Washington, D. C.
28. Vanderzant, C., and A. W. Matthys. 1965. Effect of temperature of the plating medium on the viable count of psychrophilic bacteria. J. Milk Food Technol. 28:383.
29. Woodward, R. L. 1957. How probable is the most probable number? J. Am. Water Works Assoc. 49:1060-1068.
30. Ziegler, N. R., and H. O. Halvorson. 1935. Application of statistics to problems in bacteriology. IV. Experimental comparison of the dilution method, the plate count, and the direct count for the determination of bacterial populations. J. Bact. 29:609-634.

Chapter 7

Aerobic Plate Count

R. Dale Morton

7.1 INTRODUCTION

The aerobic plate count (APC) is used as an indicator of bacterial populations on a sample. It is also called the aerobic colony count, standard plate count, mesophilic count or total plate count. The test is based on an assumption that each cell will form a visible colony when mixed with agar containing the appropriate nutrients. It is not a measure of the entire bacterial population, but as its name implies, it is a generic test for organisms that grow aerobically at mesophilic temperatures. APC does not differentiate types of bacteria. Alterations in agar nutrient content, incubation temperature, etc. will change the type of organisms that will grow and be counted.

While there are limitations, the aerobic plate count can be used successfully to gauge sanitary quality, organoleptic acceptability, adherence to good manufacturing practices and, to a lesser extent as an indicator of safety. The aerobic plate count can give a food processor information regarding raw materials, processing conditions, storage conditions and handling of product. This test may also provide information regarding shelf life or impending organoleptic change in a food.[23]

Aerobic plate counts are poor indicators of safety in most instances, since they do not directly correlate to the presence of pathogens or toxins. The existence of low APC counts does not mean the product or ingredient is pathogen free. Nevertheless, some products or ingredients showing excessively or unusually high APCs may reasonably be assumed to be potential health hazards, pending pathogen screening results. Interpretation of the APC results must take into consideration knowledge of the product and whether a high APC is expected.

Depending on the situation, APC can be valuable in evaluating food quality. Large numbers of bacteria may be an indication of poor sanitation or problems with process control or ingredients. Certain products, such as those produced through fermentation, naturally have a high APC. Again, low APC numbers do not equate to an absence of pathogens. Often, it is necessary to assay foods for specific pathogens or spoilage organisms before ruling on food safety or food quality.

7.2 MICROBIOLOGICAL SPECIFICATIONS

Quality and safety guidelines or specifications are often applied to raw materials and finished goods. Using APC for ingredients may or may not be appropriate as a quality indicator. A food manufacturer's decision to apply APC guidelines on ingredients must be based on the ingredient's effect on the finished product. For instance, in dried foods, APC can be used as a means of assessing the adequacy of moisture control during the drying process. For meats, APC can be used to check the condition of incoming carcasses to potentially identify suppliers who provide those with excessively high counts. APC can also be used to evaluate sanitary conditions of equipment and utensils. This can be done during processing to monitor buildup and after sanitation to gauge its effectiveness. Tables 1 and 2 offer some general guidelines for expected aerobic plate counts on ingredients and finished products. Bear in mind, APC specifications are not always appropriate. For example, raw agricultural commodities can have widely fluctuating plate counts. In these situations, APC can provide meaningful data to the processor who has a better understanding of factors that may influence the count, but they provide little value in relation to acceptance criteria.

7.3 GENERAL CONSIDERATIONS

7.31 *Other Tests on Same Sample*

If additional tests are to be performed on the sample, first aseptically remove the portions for microbiological analysis.

7.32 *Preparation*

Equipment and supplies should meet the specifications described in Chapter 2 unless otherwise specified in this book. All media and materials to be sterilized in the autoclave should be steril-

Table 1. FAO/WHO Microbiological Specifications for Foods[18]

Product Category	Maximum APC per gram
Dried and frozen whole egg	50,000
Dried instant products	1,000
Dried products requiring heating before consumption	10,000
Precooked frozen shrimp	100,000
Ice mixes	25,000
Edible ices	50,000
Dried milk	50,000
Caseins	30,000

Table 2. Typical Commodity Aerobic Plate Counts

Commodity	Aerobic Plate Count per gram
Wild Rice (before hulling)[11]	1,800,000
Wild rice (after hulling)[11]	1,400
Almonds[16]	3,000–7,000
Walnuts[15]	31,000–2,000,000
Refrigerated and frozen doughs[15]	100–1,000,000
Baked goods[15]	10–1,000
Soy protein[15]	100–100,0000
Pasta[15]	1,000–100,000
Dry cereal mixes[15]	100–100,000
Breakfast cereals[15]	0–100
Cocoa[15]	10,000
Frozen vegetables[24]	1,000–1,000,000
Raw Milk[20]	800–630,000
Deli salads (chicken, egg, potato, shrimp)[19]	10,000–100,000
Fresh ground beef[2]	100,000
Prepackaged cut chicken[2]	100,000
Frozen potatoes[7]	1,000–100,000

ized as described either by the author, the manufacturer, or for the specific method employed.

7.33 Controls

Sterility tests of media, plates and diluents should be conducted for each lot. Laboratory quality assurance procedures should include periodic checks of old vs. new lots of media with known cell numbers to verify that recovery of organisms is optimum (Chapter 1, Chapter 6).

7.4 EQUIPMENT, MATERIALS AND REAGENTS

7.41 General

Oven
Autoclave
Balance
Quebec colony counter
Dilution bottles 150 mL
Incubator
Mechanical shaker or vortex
Petri plates 100 × 15 mm
Pipettes
Pipette aids
Refrigerator
Thermometers
Work area—clean, level bench or table
Water bath
Media—plate count agar or equivalent
Bent glass rods

7.42 Additional Equipment for Alternative Methods

Membrane Filter Method

Membrane filters, 0.45 μm
Media—plate count agar or nutrient pads
Filter holders
Filtering flask
Vacuum tubing
Vacuum source
Forceps

Dry Rehydratable Film Method

Petrifilm™ AC plates (3M)
Plastic spreader

Pectin Gel Method

Redigel™ pretreated dish (3M)
Redigel™ aerobic count pectin gel bottles

Hydrophobic Grid—Membrane Filter (HGMF) Method

HGMF membrane filter
Filtration units
Vacuum manifold
Peptone / Tween 80 diluent
Tryptic soy-fast green agar
Tris buffer 1.0M
Stock solutions

Spiral Plate Method

Spiral plater
Spiral colony counter
Vacuum trap
Disposable beakers 5 mL
5% NaOCl

Calibrated Loop Method

Volumetrically calibrated loop

Drop Plate Method

Calibrated pipette

SimPlate™

SimPlate™ pretreated dish (Idexx Laboratories)

7.5 PRECAUTIONS AND LIMITATIONS

Procedures described in Chapter 1, Chapter 6, and Chapter 63 should be used for accurate results. A delay can result in lower counts for several reasons, including possible diluent toxicity or adherence of the bacteria to the dish itself. Berry et al.[4] demonstrated the risk of a significantly lower plate count if the plates are not poured within 10 minutes (Table 3).

Another potential source of error in plate counts is the stack-pouring of petri plates. Koburger[17] found that in a stack of three plates, the middle and top plate took longer to cool resulting in lower counts.

Hutanen et al.[14] demonstrated that for raw milk, increased holding time of dilutions in buffer leads to higher counts. Holding dilutions for 20 minutes resulted in a count that was 22% higher than the control, possibly due to bacterial growth or breaking up of clumps. The closest approximation to the true counts was at a five minute holding time.

7.6 STANDARD PROCEDURES

7.61 Dilutions

Refer to Chapter 6 for complete details on dilutions.

1. **Basic Principles.** Enumeration of microorganisms requires dilution of samples to achieve a population that is countable by the chosen method. Generally decimal or ten-fold dilu-

Table 3. Effect of Delay in Pouring Plates on Total Counts

Delay (minutes)	Sample 1 (cfu)	Sample 2 (cfu)	Sample 3 (cfu)
0	173	137	138
5	156	102	123
10		90	101
15	120	68	98
20			93
30	98	69	84
45	81		
60	51	28	28

tions are used for ease of calculation of final results. A variety of diluents are available including phosphate buffer and 0.1% peptone water. Distilled water should be used in the preparation of diluents. However, use of plain distilled or deionized water as a diluent is inappropriate due to the potential for osmotic stress on diluted cells.

2. **Liquids.** For non-viscous liquid samples (viscosity not greater than milk), pipette 11 mL of well-mixed sample into a 99 mL dilution blank using aseptic technique. For a viscous liquid sample, weigh 11 g of a well-mixed sample into a 99 mL dilution blank to assure accuracy of the sample size.
3. **Solid Sample Homogenates.** Refer to Chapter 2 for initial preparation of sample homegenates. For fine granular or powdered samples, mix samples thoroughly, weigh 11 g into a sterile sample container, and add 99 mL of diluent. If a larger sample is desired, other sample and diluent quantities can be used to arrive at a 1:10 dilution; e.g., a 50 g sample can be added to 450 mL of diluent. For solid and particulate material samples, prepare a 1:10 dilution selecting a sample size that assures a representative sample is tested. The appropriate diluent volume for various sample sizes may be determined by multiplying the sample weight by nine. Dilutions may also be determined by weight.
4. **Blending.** Refer to Chapter 2.
5. **Stomaching.** Refer to Chapter 2.
6. **Swabs and sponges.** Refer to Chapter 3.

7.62 Pour Plate Techniques

Although there are some inherent limitations in enumerating microorganisms by the colony count method, many of the errors can be minimized if the analyst follows directions carefully and exercises extreme care in making all measurements. Consistently accurate and meaningful results can be obtained from the routine examination of a food only if the same procedures are used to analyze each sample of that food. This includes sampling procedures, sample preparation, preparation of dilutions, plating medium, incubation conditions, and counting procedures.[3,8,21] Refer to Chapter 2 for sample preparation and Chapter 6 for the pour plate method.

7.63 Surface or Spread Plate Method

Methods of plating designed to produce all surface colonies on agar plates have certain advantages over the pour plate method.[5] The use of translucent media is not essential with a surface or spread plate but is necessary with the pour plate to facilitate location of colonies. The morphology of surface colonies is easily observed, improving the analyst's ability to distinguish different types of colonies. Organisms are not exposed to the heat of the melted agar medium, so higher counts may be observed in some situations. However, since relatively small volumes (0.1 to 0.5 mL) of the sample must be used, the method may lack precision for samples containing few microorganisms. Refer to Chapter 2 for sample preparation and Chapter 6 for the spread plate method.

7.7 ALTERNATIVE METHODS

7.71 Membrane Filtration

For certain foods or food ingredients, the ability to test relatively large samples will improve the accuracy of quantitative microbiological analysis. Large volumes of liquid foods or solutions of dry foods that can be dissolved and passed through a bacteriological membrane filter (pore size 0.45 µm) may be analyzed for microbial content via the membrane filter method. This method is especially useful for samples that contain low numbers of bacteria. Additional details on rapid and commercial kits using membrane filtration methods are described in Chapter 10.

1. Aseptically assemble the membrane filter apparatus following the manufacturer's directions and connect to the vacuum system.
2. Prepare sample as previously discussed (Chapter 2).
3. Introduce appropriate amount of sample into the funnel with a sterile pipette or graduated cylinder. For sample volumes of less than 10 mL, aseptically pour approximately 20 mL of sterile diluent in the funnel prior to adding the sample. If a graduated cylinder is used, rinse the cylinder with approximately 50 mL of sterile diluent and add the rinse to the funnel.
4. Apply vacuum to the filtering apparatus and allow the liquid to pass completely through the filter into the flask. Do not turn off vacuum.
5. Rinse inside of funnel with sterile diluent using an amount at least equal to the volume of liquid just filtered. After the rinse has passed completely through the filter, turn off the vacuum.
6. Carefully and aseptically disassemble the part of the apparatus containing the filter. Using sterile smooth-tipped forceps, remove the filter and carefully place it, avoiding air bubbles, onto the surface of the chosen saturated nutrient pad or plate count agar medium.
7. Incubate plates as in pour plate method.
8. Count the colonies with the aid of low power magnification. An acceptable range of colonies per filter is 20 to 200.
9. Compute counts and report as membrane filter colony count per milliliter or per gram based on the amount of sample filtered.

7.72 Dry Rehydratable Film Method

Petrifilm™ Plate Method

The dry rehydratable film method (3M™ Petrifilm available from Medical-Surgical Division, 3M Center, St. Paul, MN 55144) is a ready-made culture medium system which consists of two plastic films coated with standard methods nutrients, a cold-water-soluble gelling agent, and a tetrazolium indicator that facilitates colony enumeration. The Petrifilm AC plate has been collaboratively studied with milk, dairy products and foods and has been found to be not significantly different from the APC method.[6,10] The plate is inoculated with 1 mL of undiluted or diluted sample by pipet, pipettor or plate loop continuous pipeting syringe. The sample is spread over approximately a 20 cm^2 growth area. The gelling agent is allowed to solidify, the plates are incubated and then enumerated.

1. Prepare sample (see Chapter 2).
2. Place the labeled Petrifilm plate on a flat surface.
3. Prepare ten-fold decimal dilutions following general procedures.
4. Lift the top film and inoculate 1 mL of sample onto the center of the bottom film with either a pipet, pipettor or plate loop continuous pipetting syringe.
5. Release the top film down onto the inoculum. Place the supplied plastic spreader on the top film over the inoculum following manufacturer's instructions. Distribute the sample with a downward pressure on the center of the plastic spreader.
6. Remove spreader and leave the plate undisturbed for 1 min to permit the gel to solidify.
7. Incubate plates as in pour plate method, with the clear side up in stacks not exceeding 20 plates.
8. Count all red colonies regardless of size or intensity and report results. The circular growth area is approximately 20 cm^2. Estimates can be made on plates containing greater than 250 colonies by counting the number of colonies in one or more representative squares and determining the average number per square. Multiply the average number per square by 20 to determine the estimated count per plate.

7.73 Pectin Gel Method

Redi-Gel™ plates (available from Medical-Surgical Division/3M, 3M Center, St. Paul, MN 55144) rely on a temperature independent liquid pectin formulation that gels in the presence of calcium ions. All components are pre-sterilized. A pretreated petri dish is coated with calcium ions. An appropriately diluted sample is poured into the dish and the calcium ions diffuse upward producing a calcium-pectate gel within 30–40 minutes. Because this process is not temperature dependent, the amount of condensate is markedly decreased. Redigel Aerobic Count test is an AOAC Official Method (AOAC Official Method 988.18, Aerobic Plate Count–Pectin Gel Method for Food and Dairy Products). Collaborative tests against conventional methods indicate that Redi-Gel performs favorably.[22]

1. Prepare sample (Chapter 2).
2. Mix 1 mL portions of appropriate dilutions with enough Redigel bottles as are needed to properly estimate population.
3. Swirl bottle to mix.
4. Pour contents of bottle in pretreated Redigel petri dish.
5. Place the unsolidified Redigel dish upright in the incubator. If desired, the dishes may be inverted anytime after they are solidified. Continue normal incubation. Alternately, the Redigel dish may be allowed to solidify on a level surface for 40 minutes before being incubated.
6. Count the colonies as described previously for pour plates.

7.74 Hydrophobic Grid—Membrane Filter (HGMF) Method

The HGMF method represents another use of membrane filtration. Although colonies form, enumeration by this technique is based upon most probable number (MPN) determinations. The aerobic count HGMF method uses a tryptic soy agar medium containing the stain Fast Green FCF, which eliminates the need for post-incubation staining of the growth. Colonies develop varying degrees of intensity of green stain. At the recommended concentration the stain produces no evidence of toxicity. The method has Official First Action status accorded by the AOAC International.[1] Refer to HGMF General Procedures in Chapter 10 on Rapid Methods for general information on carrying out HGMF counts.

7.75 Spiral Plate Method

The spiral plate method (Spiral Systems Instruments, Inc., Bethesda, MD 20814) for enumerating microorganisms has been collaboratively tested with milk and milk products, foods, and cosmetics, and was found equivalent to the standard pour plating procedure.[9] A known volume of sample is dispensed onto a rotating agar plate in an Archimedes spiral. The amount of sample decreases as the spiral moves out toward the edge of the plate. A modified counting grid, which relates the area of the plate to sample volume, is used to count the colonies on an appropriate area of the plate. With this information the colony count for the sample can be computed.

1. Prepare sample (see Chapter 2). If necessary, let diluted solid samples settle a few minutes before removing test portions since particles may clog tubing.
2. Check stylus tip angle daily by using the vacuum to hold a cover slip against the face of the stylus. The cover slip should be parallel to and 1 mm from the surface of the agar. Adjust, if necessary.
3. Decontaminate stylus tip and tubing by pulling household bleach and then sterile water through the system before pulling the liquid sample into the stylus.
4. Place a pre-poured plate count agar Petri plate on turntable, and lower stylus. The sample is differentially dispersed as the stylus tip rides on the surface of the rotating agar plate. Remove the inoculated plate while returning the stylus to the starting position. Decontaminate the stylus and load for the inoculation of another sample.
5. Incubate plates as in pour plate method.
6. Count colonies and report results as described by the manufacturer.

7.76 Calibrated Loop Method

The plate loop method[26] and the oval tube (or bottle culture) method,[13] uses volumetrically calibrated loops (0.01 or 0.001 mL) for transferring samples in lieu of decimal dilutions. These methods are useful only for nonviscous liquids having greater than 2500 CFU/mL or for viscous and solid foods having greater than 25,000 CFU/mL. For the plate loop method, fit a calibrated loop at the end of a Cornwall continuous pipetting device. After dipping the loop into the sample, rinse the measured volume in the charged loop into a petri plate by depressing the Cornwall plunger. The oval tube method transfers a calibrated loopful of sample, or diluted sample, directly to agar in a tube. After solidification, incubate tube as desired.

7.77 Drop Plate Method

The drop plate method of enumerating microorganisms is similar in principle to the spread plate method except that bent glass rods are not used to spread the diluted sample on the agar surface. The diluted samples are measured onto the surface of pre-poured plates by adding a predetermined number of drops from a specially calibrated pipette. The drops are allowed to spread and dry over an area, usually 1.5 to 3 cm in diameter, of the agar surface. The plates are incubated for the required temperature and time. The colonies are counted and the computation of the colony count is based on the number of drops per plate, the number of drops per mL, and the dilution factor. The method is not recommended for food samples having counts of less than 3000 per g. For details, see the literature.

7.78 SimPlate™

SimPlate™ (IDEXX Laboratories, One IDEXX Drive, Westbrook, Maine 04092) for total plate count is used to quantify the total bacterial concentration in food. This method has been collaboratively tested with a variety of food products and compared to the conventional plate count methods, Petrifilm, and Redigel. It relies on the correlation of enzyme activity to the presence of viable bacteria in food. The media provided contain multiple enzyme substrates. These substrates are used by enzymes commonly found within many foodborne bacteria in various biochemical pathways key for growth and survival. A food sample is mixed with the medium and distributed into a SimPlate that contains a fixed number of wells. When the substrates are hydrolyzed by bacterial enzymes, a blue fluorescent molecule (4-methylumbelliferone) is produced and is readily visible when the media is exposed to UV light. The number of visible wells is correlated to a number of bacteria via an MPN table.

7.8 REFERENCES

1. AOAC. 1986. Aerobic plate count in foods hydrophobic grid membrane filter method first action. J. Assoc. Off. Anal. Chem. 69:376.
2. Ayres, J. C. 1951. Some bacteriological aspects of spoilage of self-service meats. Iowa State Coll. J. Sci. 26:631.
3. Babel, F. J., E. B. Collins, J. C. Olson, I. I. Peters, G. H. Watrous, and M. L. Speck. 1955. The standard plate count of milk as affected by the temperature of incubation. J. Dairy Sci. 38:499.
4. Berry, J. M., D. A. McNeill, and L. D. Witter. 1969. Effect of delays in pour plating on bacterial counts. J. Dairy Sci. 52:1456.
5. Clark, D. S. 1967. Comparison of pour and surface plate methods for determination of bacterial counts. Can. J. Microbiol. 13:1409.
6. Curiale, M. S., P. Fahey, T. L. Fox, and J. S. McAllister. 1989. Dry rehydratable films for enumeration of coliforms and aerobic bacteria in dairy products: collaborative study. J. Assoc. Off. Anal. Chem. 72:312.
7. Duran, A. P., A. Swartzentruber, J. M. Lanier, B. A. Wentz, A. H. Schwab, R. J. Barnard, and R. B. Read. 1982. Microbiological quality of five potato products obtained at retail markets. App. Env. Microbiol. 44:1076.
8. Fowler, J. L., W. S. Clark, J. F. Foster, and A. Hopkins. 1978. Analyst variation in doing the standard plate count as described in standard methods for the examination of dairy products. J. Food Prot. 41:4.
9. Gilchrist, J. E., C. B. Donelley, J. T. Peeler, and J. E. Campbell. 1977. Collaborative study comparing the spiral plate and aerobic plate count methods. J. Assoc. Off. Anal. Chem. 60: 807.
10. Ginn, R. E., V. S. Packard, and T. L. Fox. 1986. Enumeration of total bacteria and coliforms in milk by dry rehydratable film methods: Collaborative study. J. Assoc. Off. Anal. Chem. 69:572.
11. Goel, M. C., B. L. Gaddi, E. H. Marth, D. A. Stuiber, R. C. Lund, R.C. Lindsey, and E. Brickbauer. 1970. Microbiology of raw and processed wild rice. J. Milk Food Technol. 33:571.
12. Hartman, P. A., and D. V. Huntsberger. 1961. Influence of subtle differences in plating procedure on bacterial counts of prepared frozen foods. Appl. Microbiol. 9:32.
13. Heinemann, B., and M. R. Rohr. 1953. A bottle agar method for bacterial estimates. J. Milk Food Technol. 16:133.
14. Huhtanen, C. N., A. R. Brazis, W. L. Arledge, E. W. Cook, C. B. Donnelly, R. E. Ginn, J. J. Jezeski, D. Pusch, H. E. Randolph, and E. L. Sing. 1972. Effects of time of holding dilutions on counts of bacteria from raw milk. J. Milk Food Technol. 35:126.
15. ICMSF, 1980. Microbial ecology of foods, vol. 2, Food commodities. Academic Press, New York.
16. King, A. D., Jr., M. J. Miller, and L. C. Eldridge. 1970. Almond harvesting, processing and microbial flora. Appl. Microbiol. 20:208-214.
17. Kohburger, J. A. 1979. Stack pouring of Petri plates: a potential source of error. J. Food Prot. 43:561.
18. NRC, Subcommittee on Microbiological Criteria for Foods and Food Ingredients, 1985. An evaluation of the role of microbiological criteria for foods and food ingredients. National Academy Press, Washington, D. C.
19. Pace, P. J. 1975. Bacteriological quality of delicatessen foods: Are standards needed? J. Milk Food Technol. 38:347.
20. Randolph, H. E., B. K. Chakraborty, O. Hampton, and D. L. Bogart. 1973. Microbial counts of individual producer and commingled grade A raw milk. J. Milk Food Technol. 36:146.
21. Randolph, H. E., B. K. Chakraborty, O. Hampton, and D. L. Bogart. 1973. Effect of plate incubation temperature on bacterial counts of grade A raw milk. J. Milk Food Technol. 36:152.
22. Roth, J. N. 1988. A temperature-independent pectin gel method for aerobic plate count in dairy and nondairy food products; collaborative study. J. Assoc. Off. Anal. Chem. 71:343.
23. Silliker, J. H. 1963. Total counts as indexes of food quality. *In* Microbiological quality of foods, Academic Press, New York.
24. Splittstoesser, D. F., 1973. The microbiology of frozen vegetables. Food Technol. 27:54-56, 60.
26. Thompson, D. I., C. B. Donnelly, and L. A. Black. 1960. A plate loop method for determining viable counts of raw milk. J. Milk Food Technol. 23:167.

Chapter 8

Enterobacteriaceae, Coliforms, and *Escherichia coli* as Quality and Safety Indicators

J. L. Kornacki and J. L. Johnson

8.1 INTRODUCTION

Escherich observed the ubiquity of the organism now known as *Escherichia coli* in human stools in 1887. Schardinger and, later, Smith independently introduced the use of *E. coli* and related organisms as indicators of the potential presence of enteric pathogens (e.g. *Salmonella typhi*) in water[55] in 1892 (Australia) and 1895 (USA), repectively. The premise was that *E. coli* in water indicated fecal contamination because *E. coli* and enteric pathogens were found in feces of warm blooded animals. The U.S. Public Health Service standard for water was changed from *E. coli* to coliforms around 1915, based on the questionable assumption that all coliforms possess equal value as sanitary indicators. The practice of testing for *E. coli* and coliforms spread from the original application in assessing water contamination, first to pasteurized milk and dairy products, then to other foods.

Little thought seems to have been given to the validity of using indicator organisms in widely disparate matrices with divergent microbial ecologies. Papers cautioned, as early as 1924, that some of the analytical methods which worked well with water were not well suited to the analysis of milk.[64] Breed et al. (1937)[10] stated that, "the usefulness of tests of organisms of the *Escherichia-Aerobacter* types and so-called intermediates in dairy products has been complicated by a tendency on the part of many workers to carry over the interpretations of results from the water sanitation field into the dairy field."

These historical paradigms originating in the field of water sanitation continue to cause confusion in the area of food testing. The assumption that the presence or high numbers of *E. coli*, coliforms, fecal coliforms, or *Enterobacteriaceae* in a food means that fecal contamination has occurred is invalid for a number of reasons. These are as follows: 1) *E. coli*, coliforms and other *Enterobacteriaceae* are not obligate inhabitants of the intestinal tract of warm blooded animals; 2) environmental reservoirs of these organisms are known to exist; 3) these organisms are common in food manufacturing environments and may become part of the resident microflora of the facility, especially when sanitation is inadequate; 4) growth of *E. coli* can occur on some foods when temperatures exceed approximately 7–8°C; and 5) growth of some coliforms, fecal coliforms, and other *Enterobacteriaceae* can occur on some foods under refrigeration.[14,16,44]

An attempt was made in the 1970s to differentiate between the use of *E. coli*, the *Enterobacteriaceae*, and coliforms as a marker or ***index*** of the potential presence of pathogens (i.e. food safety) and the use of these organisms as ***indicators*** of overall food quality.[52] This differentiation in function is critical because it is rare that an organism or a group of organisms can be used to simultaneously address both food safety and food quality.

Index organisms signal the increased likelihood of a pathogen originating from the same source as the index organism and thus serve a predictive function.[12] The original application of *E. coli* to assess water safety, for example, used that organism as an index of *Salmonella* contamination. Like most other historical uses of the *Enterobacteriaceae* group as index organisms, this application assumed that the common source of these bacteria was the intestinal tract. There is no reason to assume that the *Enterobacteriaceae* group would have any value in predicting the presence of pathogens such as *Listeria monocytogenes* which typically originate from extra-intestinal sources. An unpublished study of raw hamburger patties from 16 producers indicated that as the level of *E. coli* increased in ground beef patties the incidence of *Salmonella* also increased (Table 1).

However, numerous studies have determined that *E. coli*, coliforms, fecal coliforms and *Enterobacteriaceae* are unreliable when used as an index of pathogen contamination of foods.[50,68,70] Higher levels of index organisms may (in certain circumstances—see Table 1), but often do not, correlate with a greater probability of an enteric pathogen(s) being present and the absence of the index organism does not always mean that the food is free from enteric pathogens. National and international advisory committees such as FAO/WHO and the U.S. National Research Council's Subcommittee on Microbiological Criteria have concluded, on this basis, that it is invalid to attempt to predict the safety of food products based on levels of coliforms, fecal coliforms, *Enterobacteriaceae* or *E. coli*.[24,57] Data from the 1994 outbreak of *Salmonella enteritidis* in ice cream support this view. Sufficient *S. enteritidis* cells were present in the ice cream to cause nearly a quarter million human infections despite very low coliform and *E. coli* counts (< 1 CFU/g).[78]

The greatest application of coliform, *Enterobacteriaceae*, and *E. coli* testing is in assessment of the overall quality of a food and the hygienic conditions present during food processing. Enumeration of these organisms in heat-pasteurized foods, for example, can be used to assess the adequacy of a heating process designed to inactivate vegetative bacteria. Dairy microbiologists used *E. coli* as a true indicator organism to assess post-pasteurization contamination of milk as early as 1927, especially that coming from improperly cleaned bottles.[75] The presence of *E. coli* in pasteurized milk may indicate inadequate pasteurization, poor hygienic conditions in the processing plant, and/or post-processing contamination because proper pasteurization inactivates levels of *E. coli* anticipated in raw milk. A number of factors must be considered before testing for a particular indicator organism or group of organisms: the physico-chemical nature of the food; the native microflora of the food; the extent to which the food has been processed; the effect that processing would be expected to have on the indicator organism(s); the physiology of the indicator organism(s) chosen; and the accuracy with which the intended testing method can identify the indicator organism(s).

8.2 DEFINITIONS

The term "*Enterobacteriaceae* group" will be used in this chapter to refer collectively to coliforms, fecal coliforms, *E. coli*, and other bacteria in the Family *Enterobacteriaceae*. The definitions used for coliforms and fecal coliforms are best described as working concepts since these groups have no taxonomic validity. "Coliforms" and "fecal coliforms," practically speaking, are those microorganisms that are detected by the "coliform test" and the "fecal coliform test," respectively.

8.21 *Enterobacteriaceae* Family

The taxonomically defined family, *Enterobacteriaceae*, includes those facultatively anaerobic gram-negative straight bacilli which ferment glucose to acid, are oxidase-negative, usually catalase-positive, usually nitrate-reducing, and motile by peritrichous flagella or nonmotile. Common foodborne genera of the Family *Enterobacteriaceae* include *Citrobacter*, *Enterobacter*, *Erwinia*, *Escherichia*, *Hafnia*, *Klebsiella*, *Proteus*, *Providencia*, *Salmonella*, *Serratia*, *Shigella*, and *Yersinia*. Psychrotrophic members of this family are not uncommon,[44,54] although the *Enterobacteriaceae* are widely regarded as being mesophilic. Psychrotrophic strains of *Enterobacter*, *Hafnia*, and *Serratia* may grow at temperatures as low as 0°C.[61]

The *Enterobacteriaceae* have been used for years in Europe as indicators of food quality and as indices of food safety. The use of coliforms as indicators of food quality or insanitation in food processing environments is based upon tradition in the United States. This practice arbitrarily bases judgments on food quality or manufacturing plant insanitation upon recovery of those members of the *Enterobacteriaceae* group (i.e. coliforms) which ferment lactose, thus ignoring the presence of non-lactose fermenting members. Mossel et al.[56] recommended the examination of food products for *Enterobacteriaceae* rather than coliforms in an attempt to better assess glucose-positive, lactose-negative members (e.g. *Salmonella*) of the food microflora.

8.22 Coliforms

The term "coliform" appears to have been coined by Blachstein[8] in 1893 to refer to those bacilli which resembled *E. coli* and yielded similar colonies in culture (specifically on plates). The coliform group is defined on the basis of biochemical reactions, not genetic relationships, and thus the term "coliform" has no taxonomic validity. Coliforms are aerobic and facultatively anaerobic, gram negative, non-sporeforming rods that ferment lactose, forming acid and gas within 48 hr at 35°C.[32] An incubation temperature of 32°C is usually used for dairy products.[15,33] Test media include a variety of lactose-containing liquid or solid media supplemented with dyes, and/or surface-active agents.

The coliform group may contain organisms not included, or only provisionally included, in the *Enterobacteriaceae*, e.g., *Aeromonas* species. Representatives of 20 or more species may be classified as coliforms, on the basis of available evidence, when lactose fermentation to gas is used as the defining criterion. It is also possible to detect coliforms by assaying cultures for their β-galactosidase activity with synthetic substrates rather than by lactose fermentation. Detection may be based either upon the formation of colored end-products or volatile o-nitrophenol.[19,25,74] However, assays using synthetic substrates for β-galactosidase may detect *Enterobacteriaceae* that do not ferment lactose to gas in the traditional MPN assay and hence would not be classified as coliforms.

8.23 Fecal Coliforms

"Fecal coliforms" are defined as those coliforms which can ferment lactose to acid and gas within 48 hr at 44.5 to 45.5°C, usually in EC broth.[35] Strains of *E. coli*, *Klebsiella pneumoniae*, *Enterobacter* spp. (including *agglomerans*, *aerogenes*, and *cloacae*), and *Citrobacter freundii* may be recovered by the fecal coliform test, depending on the food product and incubation temperature.[72] The term "fecal coliform", as with coliforms, has no taxonomic validity. The term "thermo-tolerant coliforms" is sometimes used to refer to these organisms and is, perhaps, a more accurate description than is "fecal coliform." The practice of incubating coliform tests at elevated temperatures to separate organisms thought to be of fecal origin from other coliform organisms began with Eijkman in 1904 and was originally used to assess fecal contamination in water. High-temperature incubation and gas production from lactose will not, however, reliably select for organisms originating in the intestinal tract or in feces.[6,27]

The fecal coliform test is essentially a shortened version of the *E. coli* MPN test which does not involve isolation or the lengthy and laborious IMViC tests traditionally used to confirm *E. coli*. However, a 48 h IMViC assay was developed which offers the advantage of contact with differential chemicals in solid media of the much higher bacterial populations in a colony as compared with the traditional assays using broth culture. Reactions can be read on a single quad plate.[62] A variety of incubation temperatures are used for the detection of fecal coliforms. A temperature of 45.5 ± 0.2°C is widely used for foods while 44.5°C is recommended for shellfish.[35] Data indicate that incubation of EC broth

Table 1. *Salmonella* Incidence in Relationship to *E. coli* Most Probable Number (MPN) in Raw Preformed Meat Patties

E. coli MPN/g	Samples within MPN Range	Samples Positive for *Salmonella* within MPN Range	% Positive within MPN Range
<3	270	2	0.7
3-5	406	20	4.9
51-100	54	3	5.6
101-240	96	4	4.1
241-1,100	65	3	4.6
1,101-11,000	56	9	16.1
>11,000	25	5	20.0

at 45.5°C may be more specific for *E. coli* while incubation at 44.5°C may yield slightly higher numbers of other fecal coliforms. Evidence indicates that an incubation temperature of 45.0° ± 0.2°C for all tests would be a logical compromise.[79] Well-regulated circulating water baths are preferred to air incubators, given the narrow temperature range (typically ± 0.2°C) allowed for incubation of fecal coliform and *E. coli* assay tubes. It should be noted that the elevated incubation temperatures used for the recovery of fecal coliforms and *E. coli* from foods are unsuitable for the growth of some pathogenic *E. coli*. Strains of *E. coli* O157:H7, for example, are typically incapable of growth at 44 to 45.5°C.[18] Enteropathogenic strains of *E. coli* will be described in Chapter 35.

8.24 *Escherichia coli*

The identification and enumeration of *E. coli* of sanitary significance relies upon isolate conformance to the coliform and fecal coliform group definitions. *E. coli* isolates were traditionally identified by their IMViC pattern: + + – – (Type I) and – + – – (Type II). In this scheme "I" refers to the ability of the organism to produce indole from the metabolism of tryptophane; "M" indicates the ability of the organism to ferment glucose to "high" acid as detected by Methyl Red pH indicator dye in the medium; "Vi" stands for the production of the neutral products 2,3 butanediol and/or acetoin from glucose metabolism, otherwise known as the "Vogues-Proskauer" reaction, whereas "C" represents the ability of the bacterium to use citrate as a sole carbon source. Recent data indicate that defining *E. coli* by IMViC profile is inadequate for identification of the species (see Section 8.37). There are, for example, *E. coli* strains which do not give IMViC reactions corresponding to either Biotype I or Biotype II.[39] The relatively high incidence of Type II "*E. coli*" in some specimens is at least partly explained by the fact that many isolates require 48 hr to produce a detectable amount of indole, and additional tests are essential for speciation.

8.3 PRECAUTIONS

8.31 Cultures

Stock cultures of *E. coli* (IMViC pattern + + – –) and *Enterobacter aerogenes* (IMViC pattern – – + +) should be maintained for quality control testing IMViC media and reagents. Check the performance of all media. The control strain of *E. coli* should produce gas in EC medium at 45.5°C within 24 hr. The control strain of *E. aerogenes* should produce a negative reaction.

8.32 Dilutions

Prepare only as many test dilutions as can be inoculated within 15 min.

8.33 Media

Exhausting autoclaves too rapidly can result in air bubbles which form inside Durham tubes used in MPN assays for *Enterobacteriaceae*, coliforms, fecal coliforms or *E. coli*. These bubbles can confound interpretation of the assay(s). Liquid media also absorb air during cold storage and should be allowed to reach laboratory temperature gradually after removal from refrigeration. Tubes should be inspected before use and those with gas bubbles discarded. Include uninoculated medium as a negative control.

8.34 General Limitations of the MPN Liquid Enrichment Test

MPN determinations offer the potential for enhanced recovery due to enrichment of injured cells, absence of heat injury of cells from molten "solid" media (as can occur with direct plating), lack of confusion from particulates on solid media, and adjustment of the sensitivity of the assay by changing the quantity and/or dilution tested (e.g. one could enrich a 100 g or a larger quantity at the lowest dilutions for increased sensitivity).

However, the most probable number (MPN) technique indicates only the most likely number of organisms present in a sample (see Chapter 6). The convention, among food microbiologists, has been to state only the MPN value rather than both the MPN and its associated confidence interval. This reporting method is unfortunate because it does not convey the idea that the MPN is purely a statistical approximation; the true number of organisms in the sample is an unknown value occurring within the MPN confidence interval. Regrettably, confusion about the MPN value is not uncommon.

A glance at a 3 dilution (9 tube) MPN table indicates that the lower limit of detection is accompanied by quite a wide confidence interval (see Chapter 6). For example, an MPN of <3.6 results when quantities of 0.1 g, 0.01 g, and 0.001 g are tested in triplicate and none of the tubes is positive (0-0-0 result). However, the upper limit of the 95% confidence interval is 9.5 MPN per g or mL. The confidence intervals increase with greater numbers of positive tubes. Consequently, an MPN value of 1100 (from a 3-3-2 result) has a 95% confidence interval from 180 to 4100 MPN/g. One may question whether the labor needed to perform the MPN is justified, given this inexactitude, or whether the simpler, less laborious plating methods may be used. It is possible to achieve a lower limit of 1 CFU per g or mL by pour-plating 1 mL of the 1 to 10 dilution onto each of 10 plates. In practical terms, a result of < 1 CFU per g or mL by a direct count method may be just as useful as an MPN result of < 0.36 MPN per g or mL. The decision to use an MPN versus a direct count approach should take into consideration the food matrix, the level of required sensitivity and expected accuracy, labor and time considerations and the analyst's experience with the product given considerations described above.

8.35 Minimal Number of Fecal Coliforms

The fecal coliform assay is identical to the assay used for detection of presumptive *E. coli*. However, the confirmation step to determine what proportion of the fecal coliform count (if any) are *E. coli* is not performed. Some food manufacturers stop at this point and do not undertake the time consuming and laborious IMViC testing typically done to confirm *E. coli*. A fecal coliform count is occasionally wrongly interpreted as being equivalent to the *E. coli* count. All *E. coli* are fecal coliforms, but all fecal coliforms are not *E. coli*. It is impossible to extrapolate the *E. coli* population from a fecal coliform count unless confirmatory testing is done or unless a particular food manufacturer has a comprehensive collection of in-house data indicating a preponderance of *E. coli* amongst fecal coliforms for a specific food product. The number of fecal coliforms in a food corresponds to maximum number of *E. coli* present if *E. coli* confirmations are not performed.

8.36 General Limitations of L-EMB Agar

L-EMB is not especially selective for *E. coli*, and physical and subjective limitations are encountered in the use of this agar.[73] The ability to discern one distinctive colony among many, and specifically to recognize an *E. coli*-like colony among many coliform colonies, is critical to the success of the entire analytical procedure. Unfortunately, some biotypes of *E. coli* do not produce a typical colony with green sheen; slow or non-lactose-fermenters produce colorless colonies. Some non-*E. coli* colonies (e.g. *Klebsiella pneumoniae*) may also exhibit typical *E. coli*-like morphology on L-EMB.[77] These authors reported a 28% *E. coli* confirmation rate for "typical" colonies taken from L-EMB versus a confirmation rate of 86% for MUG-positive colonies taken from VRBA-MUG. Finally, inoculum taken from a "typical" *E. coli* colony on an L-EMB plate may represent a mixed population and mixed cultures confound IMViC and other confirmatory tests.[39] Selection of a pure culture is therefore essential.

8.37 Limitations of the IMViC Tests for E. coli Confirmation

The use of the biochemical tests now referred to as the IMViC series (Indole, Methyl Red, Voges Proskauer, and Citrate) to characterize lactose-fermenting bacilli dates back to the 1920s. Originally, IMViC tests were used in an attempt to differentiate coliforms thought to be of intestinal origin from those believed to be soil-born. The IMViC tests were incapable of differentiating coliforms on the basis of habitat, even when supplemented with other biochemical characterization tests, however.[46]

The routine use of the IMViC tests for confirmation of *E. coli* was questioned as early as 1934.[7] Advances in bacterial taxonomy and improvements in biochemical and genetic confirmation methods for *E. coli* also cast doubt on the wisdom of relying on the IMViC tests for *E. coli* confirmation. Various references indicate that the following members of the *Enterobacteriaceae* may give IMViC reactions indicative of Biotype I *E. coli*: *Escherichia* spp. (*E. fergusonii, E. hermannii, E. blattae,* and non-gas producing strains of *E. coli*); *Enterobacter agglomerans, Edwardsiella* spp. (*hoshinae, tarda*); *Leclercia* (*Escherichia adecarboxylata*); *Kluyvera cryocrescens; Morganella morganii; Proteus* spp. (*mirabilis, morganii, vulgaris*), *Providencia rustigianii; Shigella* spp. (*boydii, dysenteriae, flexneri*); and *Yersinia* spp. (*enterocolitica, frederiksenii, intermedia, kristensenii*). Biotype II *E. coli* reactions are also shared with some strains of the following *Enterobacteriaceae*: *Budvicia aquatica; Edwardsiella hoshinae; Enterobacter* spp. (*agglomerans, hafniae*); *Escherichia* spp. (*blattae, vulneris*); *Klebsiella pneumoniae; Moellerella wisconsensis; Pragia fontium; Proteus penneri, Providencia heimbachae; Salmonella choleraesuis* subsp. *indica; Shigella* spp. (*boydii, dysenteriae, flexneri, sonnei*); and *Yersinia* spp. (*aldovae, bercovieri, enterocolitica, kristensenii, mollaretti, pestis, pseudotuberculosis, ruckeri*).[20,23,37,39] Silliker Laboratories of Wisconsin (unpublished results) examined 16 isolates from a freeze-dried meat extract recovered from a modified USP enrichment procedure[76] which displayed IMViC reactions typical of *E. coli* Biotype I and II organisms. Further biochemical characterization (with the Biolog identification system) indicated that none of the isolates was *E. coli*. Isolates were instead identified as *Escherichia* spp. (*vulneris, hermanii*), *Citrobacter freundii* (lactose and non-lactose fermenting biotypes), *Leclericia adecarboxylata*, and *Enterobacter* spp. (*agglomerans* and *cloacae*, with isolates corresponding to both Biotype I and Biotype II). Selected strains did not produce gas after 48 hr incubation in EC broth at 45.5°C. This indicates the importance of elevated incubation temperature (e.g. 44.5–45.5°C) in the "Completed" MPN Test for *E. coli* to reduce the number of false positives (see section 8.91).

It is important to note that most of the classical methods for coliforms, fecal coliforms, Enterobacteriaceae, and *E. coli* are targeted at "typical" organisms. The MPN assay for *E. coli*, for example, quantifies only typical, gas-producing organisms. Anaerogenic strains which do not produce gas from lactose within the required 48 hr or strains which are incapable of growth at 44.5 to 45.5°C would not be detected. Furthermore, classical detection methods for coliforms and fecal coliforms are based on biochemical and physiological responses to enrichment; isolation of pure cultures and confirmatory testing is not done. A similar situation exists with the enumeration of *Enterobacteriaceae* on Violet Red Bile agar with glucose (VRBGA). These tests all sacrifice some accuracy in order to obtain (relatively) rapid results. It remains to be seen whether these limitations will continue to be ignored in favor of expediency or whether more sophisticated testing procedures (e.g. genetic or extensive biochemical confirmation) will be utilized to detect *E. coli* from foods. These approaches may become more necessary in the face of an increasing litigious climate in the food industry.

8.38 Interferences

8.381 False-Negative Results

The importance of false-negative reactions in the recovery of the *Enterobacteriaceae* group from water supplies has been described.[22,45] Similarly, false-negative results can be obtained when analyzing food samples or food-plant environmental samples. Difficulty in recovering these organisms is partially attributable to sub-lethal cellular damage or stress introduced by food preservation practices or environmental conditions such as drying, refrigeration (including freezing), heating, acidification/fermentation, and/or the use of bactericidal or bacteriostatic agents.[12,26,36,63] Sub-lethally damaged or stressed cells are unable to tolerate inhibitory agents (e.g. bile salts, desoxycholate) present in selective media, especially if high incubation temperatures are used.[73]

Recovery of sub-lethally damaged or stressed members of the *Enterobacteriaceae* group can often be improved by the use of resuscitation procedures.[66,71] Resuscitation can occur in either diluent or in non-selective agar or broth media (e.g. before addition of selective agents). Periods of 2–6 hr at room temperature in non-selective liquid media are usually sufficient for repair of injured cells but may permit some growth of healthy and repaired cells.[47] This type of approach has been used to improve recovery of coliforms with VRB agar and on Petrifilm™ Coliform Count Plates[58] and recovery of *Enterobacteriaceae* on VRBGA.[42,54] Resuscitation in solid (agar) media appears to be as effective as in liquid media and minimizes problems caused by growth during resuscitation.[47,71] Speck et al. [71] determined that a 1-2 hr resuscitation period in Tryptic Soy Agar was sufficient to allow the repair of injured coliforms in frozen foods. The value of spread-plating over pour-plating on the non-selective agar has been debated with some researchers advocating spread-plating[71] and others finding no advantage in this procedure.

A second type of false-negative result is obtained when an analytical procedure fails to recover or enumerate strains which exhibit atypical behavior. The recovery method can be modified so as to include these organisms when past experience indicates that a particular food product commonly contains atypical strains. Anaerogenic strains of *E. coli* which do not produce gas in Lauryl Sulfate Tryptose (LST) broth, for example, are known to occur in foods. In muscle food, anaerogenic strains of *E. coli* may constitute 3 to 74% of the *E. coli* population.[4] Detection of these anaerogenic *E. coli* can be accomplished by streaking turbid but not gassy LST tubes (see Section 8.71) onto a solid medium containing 4-methylumbelliferyl-β-D-glucuronic acid (MUG) and enumeration of fluorescent (e.g. MUG positive) colonies under ultraviolet light or by testing colonies for indole production.

A false-negative result may also be obtained because of microbial competition during enrichment. Growth of *Proteus vulgaris*, for example, may suppress gas production by *E. coli* in LST when both organisms are present.[26] Slow growth and/or enzyme production is an additional cause of false-negative results. A 48 hr incubation period is usually recommended for MUG-containing liquid media in order to detect *E. coli* which are slow producers of β-glucuronidase (GUD). One study indicated that 34% of *E. coli* strains isolated from human fecal samples were MUG negative in LST broth containing MUG.[13] In some cases, a false-negative β-glucuronidase test result is obtained because the gene responsible for GUD production is present but not expressed due to catabolite repression by lactose in the media.[77] Transfer to a lactose-free minimal medium restored GUD activity. Indole-based *E. coli* detection tests (specific for Biotype I only) may give false-negative results for *E. coli* Biotype I when dairy foods are tested directly since indole production is inhibited by high levels of carbohydrate.[36]

False-negative results may also occur because of inappropriate pH. MUG hydrolysis, for example, is optimal under alkaline conditions.[21] False-negative MUG results may occur in agars containing high levels of fermentable carbohydrates but lacking good buffering capacity.

8.382 False-Positive Results

Various false-positive reactions are known to occur in the analysis of foods for members of the *Enterobacteriaceae* group. False-positive reactions are especially common in MUG-based assays. Some batches of LST/MUG reportedly auto-fluoresce[21] as do glass test tubes containing cerium oxide.[32]

False-positive reactions and auto-fluorescence may occur when fish (e.g. salmon, tuna), Crustacea (e.g. shrimp), or shellfish (e.g. oysters, clams) homogenates are added to MUG-containing broths. The interference is apparently caused by an enzyme native to the muscle tissue[34] and can usually be removed by centrifugation.[77] Raw liver also contains endogenous β-D-glucuronidase which may interfere with MUG-based assays.[59] False-positive results due to food matrices can usually be ruled out by sub-culturing MUG-positive primary enrichment broths (containing relatively high levels of the potentially interfering food) into a secondary MUG-containing medium. Fluorescence in the secondary medium is nearly always representative of bacterial GUD activity since sub-culturing effectively dilutes out the interfering food matrix. Non-*Enterobacteriaceae* present in a food sample may also produce β-D-glucuronidase and interfere with MUG-based *E. coli* tests. Strains of *Staphylococcus warneri* and *S. xylosus* may exhibit fluorescence when grown in LST-MUG at 37°C but no fluorescence is observed when the LST-MUG enrichments are sub-cultured to EC-MUG and incubated at 44.5°C.[34] Incubation periods for MUG-containing agars are typically limited to 24 hr or less since MUG diffusion into the agar surrounding GUD-positive colonies will occur during prolonged incubations and give false-positive results.[26]

False-positive test reactions due to food matrices may also occur when the test relies on pH indicators. Tests which rely upon the detection of acid production, for example, may yield false-positive results when an acidic food is tested without neutralization. Infrequently, individual food particles may contain sufficient acid to change the color of the pH indicator in the agar immediately surrounding the food particle. False positives (e.g. red appearing particulates) have been observed with swine hair and flour products on VRB agar (Kornacki, unpublished observations).

8.4 EQUIPMENT, MATERIALS, AND REAGENTS

This section lists equipment, media, reagents and stains used in many of the common assays for the *Enterobacteriaceae*; media and reagents specified for use with various commercially-available assay kits are not included. Chapter 63 discusses composition and preparation of media, reagents, and stains.

8.41 Equipment

Air incubators: 32° and 35°C
Glass-rod or plastic spreader, sterile
Microscope with illumination capable of 1000× magnification.
Petri dishes, 15 × 150 mm, glass or plastic, sterile
Pipette bulb or filler
Pipettes: 1 mL, 2 mL, 5 mL, 10 mL sterile serological; Pasteur
Test tube racks (stainless steel, epoxy-coated, or plastic) for various-sized tubes
Total immersion thermometer, approximately 45 to 55 cm long, range 1°C to 55°C, and standardized against a NIST-certified thermometer or equivalent
Ultraviolet lamp, 365 nm wavelength (long-wave)
Water-bath for tempering agar (45 ± 2°C)
Water baths with plastic or metal gable covers and mechanical circulation systems capable of maintaining temperatures of 44.5, 45, or 45.5 ± 0.2°C

8.42 Materials

Brilliant green bile (BGB) broth
E. coli (EC) broth
Koser's citrate medium or Simmons' citrate slants
Lauryl sulfate Tryptose (LST) broth: single-strength and double-strength
Levine eosin methylene blue (L-EMB) agar
Methyl red-Voges Proskauer (MR-VP) broth
Tryptic or Trypticase Soy agar
Tryptone (tryptophane) broth
Violet Red Bile (VRB) agar
Violet Red Bile Glucose (VRBG) agar

8.43 Reagents

Butterfield's phosphate-buffered dilution water
Gram stain reagents
Indole reagent, Kovac's formulation
Methyl red indicator
4-Methylumbelliferyl-β-D-glucuronic acid (MUG)
0.1% Peptone water diluent
Voges-Proskauer reagents

8.5 SAMPLE PREPARATION

8.51 Preparation of Food Test Samples

Chapter 2 provides information on sample collection and preparation before analysis while Chapter 6 deals with enumeration of foodborne microorganisms. If the food is frozen and must be thawed, refrigerate at 2–5°C for approximately 18 hr before analysis. Aseptically weigh 25 g of unfrozen food into a sterile weighing jar, sterile blender jar, or sterile Stomacher bag. Add 225 mL of sterile Butterfield's phosphate buffer diluent or sterile 0.1% peptone to the jar or Stomacher bag and homogenize for 2 min. If a larger sample size is desired (e.g. 50 g), maintain the 1:10 ratio between sample weight and diluent volume.

A typical dilution series used for analysis of the *Enterobacteriaceae* group is 10^{-1} through 10^{-3} but additional dilutions may be required for the analysis of samples suspected of having higher counts. FDA recommends in the "Bacteriological Analytical Manual"[35] that dilutions be prepared by adding 10 mL of the previous (lower) dilution to a fresh 90 mL diluent blank.[5] Dilutions prepared in this manner are mixed by shaking 25 times in a 1-ft (30 cm) arc for 7 sec. Many food microbiology laboratories use 9 mL diluent blanks and prepare serial 1:10 dilutions by adding 1 mL of the previous dilution to a test tube containing 9 mL of sterile diluent. The contents of tubes are then mixed by using a vortex mixer.

8.6 THE *ENTEROBACTERIACEAE*

Traditionally, analyses for *Enterobacteriaceae* have been conducted at 35–37°C. Lower incubation temperatures (4, 10, 25, or 30°C) may be more effective when analyzing a refrigerated food which might be expected to contain psychrotrophic *Enterobacteriaceae*. Psychrotrophic organisms are often incapable of growth at 35°C but may grow at 30°C and will grow at lower temperatures.[54]

8.61 Enrichment Method

In general, enrichment for *Enterobacteriaceae* is done less frequently than is enumeration via plating with VRBGA (section 8.62). Enrichment procedures for this family typically use *Enterobacteriaceae* Enrichment (EE) Broth, a modification of BGB broth in which the lactose has been replaced by glucose. EE broth may be used in either a straight enrichment procedure or in an MPN assay. For the isolation of *Enterobacteriaceae*, Mossel et al.[56] originally described the incubation of this broth at 37°C for 20–24 hr followed by streaking on VRBGA and incubation at 37°C. In North America, food microbiologists rarely use the 37°C incubation temperature, preferring 35°C instead. If desired, differentiation between psychrotrophic *Enterobacteriaceae* and "regular" mesophilic types can be achieved by incubation at 43°C for 18 hr; psychrotrophic types are unable to grow at the elevated temperature.[56]

8.62 Plating with VRBG Agar

Pour approximately 10 mL of VRBGA tempered to 48°C into plates containing 1.0 mL portions of diluted sample. Swirl plates to mix well and allow media to solidify. Overlay each plate with 5–8 mL VRBGA. After solidification, invert plates and incubate 18 to 24 hr at 35°C. Examine plates with illumination under a magnifying lens. Count purple-red colonies, 0.5 mm in diameter or larger, surrounded by a zone of precipitated bile acids. Optimally, plates should have 15 to 150 colonies; *Enterobacteriaceae* colonies on more crowded plates may remain small and fail to reach 0.5 mm in diameter. Multiply the number of typical *Enterobacteriaceae* colonies by the reciprocal of the dilution used and report results as *Enterobacteriaceae* count (CFU/g or mL). If desired, *Enterobacteriaceae* colonies on VRBGA may be isolated and speciated by conventional or miniaturized biochemical tests.

8.63 Petrifilm™ Methods

All of the Petrifilm methods employ a "dry dehydratable film" medium which is applied to a card. Petrifilm plates used for the detection of *Enterobacteriaceae*, coliforms, and *E. coli* use a film containing selective and/or differential agents together with a cold-water soluble gelling agent. The plating medium is hydrated when 1 mL of a diluted or undiluted sample is added to a 20 sq. cm Petrifilm or 5 mL of sample is added to a High-Sensitivity Petrifilm (60 sq. cm). The plastic overlay film is then carefully lowered onto the inoculated plate so as not to trap small gas bubbles. Pressure applied to a plastic spreader placed on the overlay film distributes test portions over either 20 sq. cm (for the *Enterobacteriaceae*, coliform, and *E. coli* count plates) or 60 sq. cm (for the High-sensitivity coliform plate). The sample-containing Petrifilm plate is allowed to stand at room temperature for several minutes to allow gelation to occur. Petrifilm plates are then stacked upright (in stacks of 20 or less) and placed in the incubator. Incorporation of triphenyltetrazolium dye in the Petrifilm facilitates colony counting; a magnified illuminator (e.g. Quebec colony counter) may be also be used when counting Petrifilm plates. Uptake of triphenyltetrazolium dye results in red colonies. However, red colonies are not indicative of *Enterobacteriaceae*, coliforms, or *E. coli* specifically but only of microbial metabolism in general. Gas production is another characteristic detected by the Petrifilm plates methods, since members of the *Enterobacteriaceae* group typically ferment carbohydrates (lactose in the case of coliforms and *E. coli*; glucose in the case of the family *Enterobacteriaceae*) to gas. These organisms typically produce colonies on Petrifilm plates which have a gas bubble either adjacent to the colony or within 1 colony-diameter of the colony or exhibit a ring of gas bubbles around the colony. The preferred counting range for standard, High-sensitivity, and Series 2000 Petrifilm plates is typically 15–150 colonies for coliforms and *E. coli* and 15–100 colonies for *Enterobacteriaceae*. Colonies on the "white foam dam" (outside the gridded well of the Petrifilm) should not be counted. If desired, colonies may be isolated from the Petrifilm gel and subjected to confirmation procedures.

Difficulties with Petrifilm interpretation may be encountered when sample particulates are dark and present little contrast to the background medium. Low dilutions of chocolate milk, cocoa powder, and dried herbs have been reported to be problematic. Buffers containing citrate, bisulfite or thiosulfate should not be used with Petrifilm or growth inhibition may result. Differential characteristics for colony counting are discussed below.

8.631 *Petrifilm* Enterobacteriaceae *Method*

The Petrifilm *Enterobacteriaceae* plate is one of the very few "rapid" methods directed at the enumeration of this family of bacteria. The *Enterobacteriaceae* Petrifilm method is an alternative to the standard Enterobacteriaceae count method using VRBGA. First ensure that the pH of the diluted sample is between 6.5 and 7.5, then Petrifilm plates (20 sq. cm) are inoculated with 1 mL of sample as described above (Section 8.64) then incubated aerobically at 35°C for 24 ± 2 hr. All colonies visible on the plate will be red in color after incubation. *Enterobacteriaceae* colonies are defined as those colonies which are associated with one or more gas bubbles (within one colony diameter of the colony) and/or are surrounded by a yellow zone indicative of acid production.

8.7 COLIFORMS

It should not be assumed that the analysis of a single food sample by several coliform enumeration methods will recover the same types of organisms or yield the same quantitative results. For example, organisms classified as coliforms on solid media (e.g. VRBA or Petrifilm) may be incapable of producing gas when tested by the standard LST/BGB method and therefore not be counted as coliforms by the MPN procedure.[49,67] The proportion of true coliforms among VRBA isolates varies markedly with the food product under analysis.[49]

This section deals only with methods for the enumeration of coliforms; methods for the simultaneous enumeration of coliforms and *E. coli* or coliforms, fecal coliforms, and *E. coli* are discussed in Section 8.9 (below).

8.71 Presumptive Test for Coliform MPN

In general, a 3-replicate, 3 dilution tube MPN procedure is used for the analysis of foods. Certain procedures do, however, specify a 5-replicate tube MPN format. The MPN for fecal coliforms in shellfish and shellfish meats, for instance, specifies that 5 replicate tubes be prepared at each dilution.

Inoculate 3 replicate tubes of LST broth per dilution with 1 mL of the previously prepared 1:10, 1:100, and 1:1,000 dilutions. Using the current 3-tube MPN table (Table 1 of the 1998 BAM),[30] this dilution range will cover an MPN range of <3.6 MPN/g or mL to >1,100 MPN/g or mL. It is permissible to add 10 mL of the original 1:10 sample dilution to tubes containing 10 mL of double-strength LST, in addition to the aforementioned dilutions, if the food is expected to contain low levels of coliforms.[41] This adaptation lowers the limit of detection of the 3-tube MPN to <0.36 MPN/g or mL. Tubes are incubated at 35° ± 0.5°C for 24 and 48 ± 2 hr, after inoculation. Tubes are examined for evidence of gas production at the end of 24 hr incubation. Gas production is measured either by gas displacement in the inverted vial (e.g. Durham tube) or by effervescence produced when the tube is gently shaken. Negative tubes are re-incubated after recording the results for an additional 24 hr. Tubes are again examined for gas production (Chapter 6). After reference to the MPN rules and an appropriate MPN table (see Chapter 6), report results as the presumptive MPN of coliform bacteria per g or mL. Tubes giving presumptive-positive coliform results are confirmed as described in Section 8.72 (below).

8.72 Confirmed Test for Coliforms

Gently agitate all LST tubes exhibiting gas production within 48 ± 2 hr (see Section 8.71) then subculture into BGB broth by means of a 3-mm loop or other appropriate transfer device. Some laboratories utilize pre-sterilized wooden sticks (about 120 mm × 3 mm) as a convenient and inexpensive transfer device. Avoid the pellicle (if present) when transferring. Incubate all BGB tubes at 35° ± 0.5°C for 48 ± 2 hr. Gas production in BGB at 35 ± 2 °C is considered confirmation of coliform presence. Examine tubes for gas production then record results and refer to the appropriate MPN table (See Chapter 6). Report results as confirmed MPN of coliform bacteria per g (or mL).

8.73 VRBA Method for Coliforms Not Expected to Be Stressed or Damaged

Transfer 1-mL aliquots of each dilution to separate labeled petri dishes. Pour 10 mL of VRBA (boiled after hydration, not autoclaved, per manufacturer's recommendations) tempered at not more than 48°C and adjusted to pH 7.0–7.2 into the plates. (VRBA at pH pH 6.9 or lower should not be used as these values are indicative of flaws in media preparation and/or storage).[33] Swirl plates to mix well, and let solidify. Overlay each plate with about 5 mL VRBA and let solidify. Invert plates after solidification and incubate 18 to 24 hr at 35°C. Incubate plates at 32°C for dairy products. Examine plates with illumination under a magnifying lens. Count purple-red colonies, 0.5 mm in diameter or larger, surrounded by a zone of precipitated bile acids. Plates should ideally have 15 to 150 colonies; coliform colonies on more crowded plates may remain small and fail to reach 0.5 mm in diameter. Multiply the number of presumptive-coliform colonies by the reciprocal of the dilution used and report results as presumptive VRBA count (CFU/g or mL).

The coliform count obtained on VRBA can be confirmed by selecting representative colonies and testing them for gas production in BGB (see Section 8.72). Colonies producing gas from lactose in BGB are confirmed as coliform organisms. Perform a Gram stain on a test portion from any tube showing a pellicle in order to exclude gram-positive, lactose-fermenting bacilli. Determine the confirmed number of coliforms per g (or mL) by multiplying the percentage of BGB tubes confirmed as positive by the presumptive VRBA count. Report as "estimated" counts derived from plates outside the 15–150 colony per plate range.

8.74 VRBA-Overlay Method for Coliforms Expected to Be Stressed or Damaged

Sub-lethally damaged or stressed coliforms may be unable to grow and form typical colonies on selective agars like VRBA, as noted above (Section 8.391). This limitation is commonly overcome by plating the sample in a non-selective agar (e.g. Tryptic Soy agar, TSA), allowing several hours of resuscitation at room temperature, then overlaying the plate with VRBA.[71] The FDA Bacteriological Analytical Manual[35] describes an overlay of single-strength VRBA while Standard Methods for the Examination of Dairy Products[15] specifies overlaying TSA with double-strength VRBA (VRBA-2). Studies have indicated that the VRBA-2 method consistently yields higher numbers of coliforms than non-resuscitative methods (e.g. VRBA or Petrifilm) when analyzing foods expected to contain sublethally-damaged or stressed cells. Differences in coliform levels obtained with the VRBA-2 method versus non-resuscitative methods may exceed 1 $\log_{10}$ CFU/g.[67]

Transfer 1-mL aliquots of each dilution to separate labeled petri dishes. Pour approximately 10 mL of TSA tempered to 48°C into the plates. Swirl plates to mix well, and let solidify. Allow TSA plates to incubate at room temperature for 2.0 ± 0.5 hr. Then overlay plates with 8 to 10 mL of melted, cooled VRBA or VRBA-2 and allow to solidify. Invert plates after solidification and incubate 18 to 24 hr at 32°C (for dairy products) or at 35°C (for other food products). Count colonies as described in Section 8.73 (above).

8.75 Petrifilm Methods for Coliforms

8.751 Petrifilm Coliform Count Plate

This coliform method is an alternative to the coliform plate count method using VRBA. Petrifilm plates (20 sq. cm) are inoculated with 1 mL of sample as described above (Section 8.63) then incubated aerobically at 32°C (for dairy samples) or at 35°C (for other food samples) for 24 ± 2 hr. After incubation, all colonies visible on the plate will be red in color. Coliform colonies are defined by the method as those colonies which are associated with one or more gas bubbles (within one colony diameter of the colony). No additional confirmation is necessary because gas production from lactose fermentation by bile salt-resistant colonies is a characteristic of coliforms and is assumed to result from fermentation of lactose in the medium (as opposed to fermentation of other carbohydrates carried over in the food matrix). The Petrifilm Coliform Count Plate is an AOAC Official Method[2] for the analysis of coliforms in milk (method #986.33) and in other dairy products in general (method #989.10).

8.752 Petrifilm High-Sensitivity Coliform Count Plate

The High-sensitivity Petrifilm Coliform Plate was developed to analyze sample volumes of 5 mL rather than the standard 1 mL analyzed with the Petrifilm Coliform Count Plate. Analysis of larger sample volumes on 60 sq. cm Petrifilm plates improves sensitivity when recovering low levels of coliforms. Analysis of samples with the High-sensitivity coliform plates is identical to the procedure described in Section 8.751 except that the pH of the diluted sample must be adjusted (if necessary) to pH 6.5 to 7.5 and 5 mL of sample is used. High-sensitivity coliform count plates should be incubated at 32°C (for dairy products) or 35°C (for other food products). The Petrifilm High-sensitivity Coliform Count Plate is an AOAC Official Method[2] for the analysis of coliforms in dairy products (method #996.02).

8.753 Petrifilm Series 2000 Rapid Coliform Count Plate

The Petrifilm Series 2000 Rapid Coliform Count plate incorporates the usual features of the Petrifilm Coliform Count plate plus a pH indicator which permits more rapid identification of presumptive Coliform colonies. Analysis of samples with the Series 2000 Rapid Coliform Count plate is identical to the procedure described in Section 8.751 except that the pH of the diluted sample must be adjusted (if necessary) to pH 6.5 to 7.5 and the counting is done differently. Plates may be examined for yellow zones indicative of acid production and presumptive coliform growth, as early as 6 to 14 hr after incubation. The final coliform count is usually read after 24 hr of incubation and is done in a manner identical to that used for other Petrifilm™ Coliform plates.

8.76 Pectin Gel Method (Redigel™ Violet Red Bile Test)

Instead of using a standard VRB agar-based medium for preparing pour plates, the Pectin Gel method uses an agar-free medium which gels when poured into specially coated petri plates. Typically, the sample homogenate (or dilutions thereof) is added to a single-use bottle of Redigel VRB medium, then the bottle is mixed gently by inversion, and the inoculated medium is added to a Redigel plate. The AOAC procedure calls for adding the Redigel medium to a Redigel plate then adding the inoculum to the medium in the plate. The plates are swirled to mix the medium and sample and then left at room temperature to solidify. The inoculated medium in the plate should be overlaid with sterile Redigel medium, within 15 min of pouring the plate. The plates are allowed a solidification period of approximately 40 min and then incubated aerobically at 32°C (for dairy samples) or at 35°C (for other food samples) for 24 ± 2 hr. All pink or red colonies are counted to obtain the coliform count, after incubation. The Pectin Gel method (e.g. with Redigel Violet Red Bile test) is an AOAC Official Method[2] for the analysis of coliforms in dairy products (method #989.11).

8.77 HGMF Method for Coliforms

The Hydrophobic Grid Membrane Filter (HGMF) method involves filtering a diluted food sample through a membrane filter then placing the filter on a plate of selective/differential agar, incubating, and counting colonies possessing certain color characteristics. To obtain a coliform count, the inoculated membrane filter is placed right-side up on the surface of a pre-dried plate of m-FC agar (without rosolic acid). The plate is then incubated at 35°C for 24 ± 2 hr. Grid-cells containing one or more blue colonies (any shade of blue) are counted. This value is then plugged into a standard formula to calculate the MPN. The MPN is multiplied by the reciprocal of the dilution used and this number is reported as the MPN coliforms per g or per mL of food. Additional confirmation is not required. The HGMF[1] is an AOAC Official Method[2] for the analysis of coliforms in foods (method #983.25).

8.8 FECAL COLIFORM GROUP

8.81 EC Broth Method for Fecal Coliform MPN

The first stage of the EC broth MPN method is the presumptive coliform MPN test, as noted above. Subculture all LST tubes exhibiting gas within 48 ± 2 hr (Section 8.71) to *E. coli* (EC) broth by means of a standard 3-mm loop or other appropriate transfer device. Incubate EC tubes 24 ± 2 hr at 44.5° ± 0.2°C for waters and shellfish or 45.5 ± 0.2°C for foods, preferably in a circulating water bath. If a variety of food types are to be examined, a single incubation temperature of 45.0° ± 0.2°C should suffice." Examine tubes for gas which is indicative of a positive result. Report results as fecal coliform MPN per g or mL, after reference to an appropriate MPN table (see Chapter 6). Fecal coliforms are defined as those organisms giving positive results by this procedure.

8.82 Modified MacConkey Procedure for Shellfish and Shellfish Meats

The official Food and Drug Administration (FDA) procedure for the bacteriological analysis of domestic and imported shellfish is fully described elsewhere.[3] Methods are described for examining shellfish, fresh-shucked frozen shellfish, and shellfish frozen on the half shell. This procedure does not apply to the examination of crustaceans (crabs, lobsters, and shrimp) or to processed shellfish meats such as breaded, shucked, precooked, and heat-processed products.

8.9 ESCHERICHIA COLI

8.91 "Completed" MPN Test for E. coli

The "completed" test for *E. coli* begins with LST cultures generated during the presumptive test for coliforms (section 8.71). Subculture all positive LST tubes into EC broth and incubate the EC tubes in a circulating water bath at 45.5° ± 0.2°C. All EC tubes that show gas within 48 ± 2 hr should be sub-cultured by streaking on L-EMB agar plates and incubating the plates aerobically for 18 to 24 hr at 35°C. Examine plates for typical nucleated, dark-centered colonies with or without a metallic sheen which are indicative of *E. coli*.

8.92 Confirmation of E. coli (Including IMViC Tests)

If typical *E. coli*-like colonies are present on the L-EMB plates, select two colonies from each L-EMB plate by touching an inoculating needle to the top center of the colony and transferring each isolate to a Plate Count Agar (PCA) slant. If there are no typical colonies on the L-EMB plates, pick two or more atypical colonies and transfer to PCA slants. Incubate slants at 35°C for 18 to 24 hr. Transfer growth from PCA slants into the following broths for identification by biochemical tests:

- Tryptone broth. Incubate 24 ± 2 hr at 35°C and test for indole using Kovac's indole reagent.
- MR-VP medium. Incubate 48 ± 2 hr at 35°C. Remove 1 mL

into a small glass test tube and test for acetylmethycarbinol by adding Voges-Proskauer reagents (α-napthol solution then 40% KOH solution). Incubate remainder of MR-VP culture an additional 48 hr and test for methyl red reaction by adding methyl red indicator.

- Koser's citrate broth. Incubate 96 hr at 35°C and record growth (as evidenced by turbidity). Alternatively, Simmon's citrate slants (a modification of Koser's formulation) may be used. The advantage of Simmon's slants is that citrate utilization is signaled by a green to bright blue color change rather than turbidity.[69] Incubate Simmon's slants at 35°C for 96 hr.
- Lauryl sulfate Tryptose (LST) broth. Incubate 48 ± 2 hr at 35°C. Examine tubes for gas formation from lactose.
- Gram stain. Perform Gram stain on a smear prepared from 18 to 24 hr PCA slant. Coliforms are non-sporeforming bacilli which stain red (gram negative); gram positive organisms stain purple.

Compute MPN *E. coli* per g (or mL), considering gram-negative, nonsporeforming rods producing gas in lactose and producing + + – – (Biotype I) or – + – – (Biotype II) IMViC patterns as *E. coli*. Note: this procedure will not enumerate anaerogenic strains. If desired, a miniaturized biochemical identification system may be used to confirm the identity of isolates suspected of being *E. coli*.

8.93 β-Glucuronidase-based tests for E. coli

Presumptive *E. coli* colonies may be identified by the production of β-glucuronidase (GUD) and testing suspect colonies on L-EMB plates for GUD production has been proposed as an alternative to the IMViC tests.[39] GUD is commonly produced by *E. coli* and has been utilized as a differential characteristic in coliform recovery media containing various β-D-glucuronic acid substrates. For example, 4-methylumbelliferyl-β-D-glucuronic acid (MUG is a fluorogenic substrate). A fluorescent product, 4-methylumbelliferone, is generated when non-fluorescent MUG is cleaved by GUD. The 4-methylumbelliferone exhibits a bluish fluorescence when exposed to long-wave (365 nm) ultraviolet light in a darkened room. A level of 50 mg MUG per mL of broth is usually used for liquid media while agar media is supplemented at 100 mg MUG per mL agar. Chromogenic GUD substrates such as 5-bromo-4-chloro-3-indolyl-β-D-glucuronide (BCIG, sometimes called X-GLUC) may also be incorporated into coliform-selective agars. Enzymatic cleavage of BCIG gives *E. coli* colonies on the agar plate a dark blue color. Levels of 50 to 125 mg BCIG per liter of agar reportedly give optimal differentiation of *E. coli* colonies and the blue chromophore does not diffuse into the agar like 4-methylumbelliferone.[28,60]

Reports indicate that 92–99% of *E. coli* isolates, including many anaerogenic strains, produce GUD.[17,26,31,32] Complete reliance on GUD production to indicate *E. coli* is not recommended, however. Some pathogenic serotypes of *E. coli* (principally the Enterohemorrhagic strains) do not produce GUD[18] and GUD production has been observed in various non-*E. coli* organisms. GUD-producing *Enterobacteriaceae* include strains of *Shigella* (especially *S. sonnei*), *Salmonella* (including *S. choleraesuis*), *Escherichia vulneris*, *Citrobacter*, *Enterobacter*, *Proteus* (including *P. mirabilis*), *Klebsiella* (including *K. ozaenae*), *Serratia*, and *Yersinia enterocolitica*.[38,39] Non-*Enterobacteriaceae* known to produce GUD include *Flavobacterium* spp., *Pseudomonas* spp., *Clostridium* spp., *Micrococcus*, and *Staphylococcus* spp.[38] GUD-production by gram-positive organisms is especially problematical when minimally-selective media (e.g. Peptone Turgitol Glucuronide agar) are incubated for 48 hr or more.[17]

8.931 LST-MUG MPN for E. coli and Coliforms

The procedure used for the simultaneous presumptive-MPN determination of coliforms and *E. coli* is the same as that outlined in Section 8.71, with two exceptions.[51] First, LST broth that contains 50 mg MUG/mL broth is used. Second, incubation is usually terminated after 24 ± 2 hr, which will identify 83 to 95 % of *E. coli*-positive tubes, depending on product. Incubation for 48 hr will identify 96 to 100% of *E. coli*-positive tubes.[51] The LST-MUG tubes are examined under a long-wave (365 nm) ultraviolet light in a darkened area for fluorescence. A 6-watt, hand-held UV lamp is satisfactory for this purpose. More powerful UV sources (e.g. a 15-watt fluorescent tube type of lamp) may be used but the user should be aware of the potential for false-positive fluorescence (AOAC Chapter 17, p. 21) and the need for protective safety gear (protective glasses or goggles and gloves). Fluorescence results are used to obtain an MPN value by consulting a standard MPN table. Fluorescent-positive tubes are then streaked out onto L-EMB plates which are incubated at 35°C for 24 ± 2 hr. Confirmation of L-EMB isolates with *E. coli*-like morphology is done using the IMViC tests (Section 8.92) or another confirmation method. The LST/MUG assay is an AOAC Official Method[2] for the analysis of *E. coli* in chilled or frozen foods (AOAC method #988.19).

When analyzing foods known to have endogenous GUD activity (e.g. shellfish or fin fish), all growth-positive LST tubes should be transferred to EC broth containing 50 mg MUG/mL broth. EC/MUG tubes exhibiting fluorescence should then be struck to L-EMB and presumptive colonies confirmed as described immediately above.

8.932 ColiComplete™ MPN Method for E. coli and Coliforms

The ColiComplete method combines the principles used in the LST-MUG test with an enzymatic assay for coliforms. ColiComplete uses a substrate supporting disc containing MUG and 5-bromo-4-chloro-3-indolyl-β-D-galactopyranoside (a colorimetric indicator for β-galactosidase activity, a trait common to most coliforms). Discs are added to inoculated tubes of LST then the LST tubes are incubated at 35°C as usual (Section 8.71). After 24 and 48 hr, tubes are examined for an insoluble blue precipitate indicative of the presence of coliforms. After 30 ± 2 hr of incubation, tubes are examined under long-wave UV light for fluorescence resulting from MUG hydrolysis; the presence of fluorescence indicates that *E. coli* is present. The manufacturers of ColiComplete recommend that a known *E. coli* be used as a positive control together with two negative controls (uninoculated media and media inoculated with a non-*E. coli* coliform such as *Klebsiella* or *Enterobacter*). The ColiComplete method yields confirmed results for *E. coli* and coliforms; no additional confirmation is necessary. The ColiComplete assay is an AOAC Official Method[2] for the analysis of *E. coli* and coliforms in all food products (AOAC method #992.30).

8.933 VRBA/MUG Method for E. coli and Coliforms

Depending on the nature of the sample and the expected physiological state of the target organisms, one of several VRBA/MUG plating methods may be used. Pour-plating typically uses 1.0 mL portions of sample dilutions plated in standard-sized 10 cm diameter/1.5 cm depth petri plates. Larger sample aliquots (up to

4 mL) may be pour-plated when larger and deeper plates are used and approximately 15 to 20 mL VRBA are added to the plate.[35]

- For unprocessed foods expected to contain healthy cells, the VRBA/MUG method follows the VRBA method outlined in Section 8.73 except that MUG is added to the VRBA at a level of 100 mg MUG/mL agar or to VRBA-2 agar at a level of 200 mg MUG/mL agar.
- For processed foods or environmental samples which may be expected to contain sublethally-damaged or stressed *E. coli* or coliforms, a procedure similar to that described in Section 8.74 is used except that MUG is added at a level of 100 mg MUG/mL VRBA or 200 mg MUG/mL VRBA-2. The procedure for preparing the basal TSA plate is not well standardized and several alternative procedures exist:
    - Pour-plate 1 mL of the diluted sample with TSA then proceed with resuscitation.
    - Spread-plate 0.1 mL of the diluted sample onto a TSA plate then proceed with resuscitation.
    - Pour-plate 1 mL of the diluted sample with TSA contain ing MUG at a level of 100 mg MUG/mL agar then pro ceed with resuscitation.
- Spread-plate 0.1 mL of the diluted sample onto a TSA plate containing MUG at a level of 100 mg MUG/mL agar then proceed with resuscitation.

Coliform colonies are counted as usual after incubation (see Section 8.73). Plates are then exposed to long-wave UV light and fluorescent colonies indicative of *E. coli* are counted. The coliform and *E. coli* counts are then multiplied by the reciprocal of the dilution used and results reported. Confirmation is typically not done when using the VRBA/MUG method or the VRBA-2/MUG method.

8.934 HGMF with MUG Method for E. coli and Coliforms

The HGMF-MUG method involves filtering a diluted food sample through a membrane filter then placing the filter on a plate of selective/differential agar, incubating, and counting colonies possessing certain characteristics (color or fluorescence; see discussion in section 8.93). To obtain a coliform count, the filter is placed on a pre-dried plate of Lactose Monensin Glucuronate (LMG) agar then incubated at 35°C for 24 ± 2 hr. Squares containing one or more blue colonies are counted then the number of positive (blue) squares is converted to a MPN using a mathematical formula. The coliform MPN/g or mL is then calculated by multiplying the MPN by the inverse of the dilution factor of the filtered sample homogenate. If coliforms are present, an *E. coli* count can be obtained on the same filter by transferring it to the surface of a pre-dried Buffered MUG Agar (BMA) plate. The filter is incubated at 35°C for 2 hr and then examined under long-wave UV light in a darkened room. Squares containing one or more large blue-white fluorescent colonies are considered positive. The number of positive squares is then converted to a MPN using the mathematical formula and the *E. coli* MPN/g or mL is calculated by multiplying the MPN by the inverse of the dilution factor. The HGMF MUG method yields confirmed results for *E. coli* and coliforms; no additional confirmation is necessary. The HGMF/MUG ISO-GRID method is an AOAC Official Method[2] for the analysis of *E. coli* and coliforms in foods (AOAC method #990.11).

8.935 Petrifilm for E. coli and Coliforms

The Petrifilm *E. coli* count plate allows simultaneous enumeration of coliforms and *E. coli* on a single Petrifilm. Differentiation of *E. coli* is accomplished by the addition of BCIG to a dehydrated film similar to that used in the Coliform Count Plate (Section 8.751). Plates are inoculated as described above (Section 8.63) then incubated aerobically at 32°C (for dairy samples) or at 35°C (for other food samples) for 24 ± 2 hr. Non-*E. coli* coliform colonies (red colonies associated with one or more gas bubbles within one colony diameter of the colony) and *E. coli* colonies (blue colonies associated with one or more gas bubbles within one colony diameter of the colony) are counted after 24 hr incubation according to the Interpretation Guide provided with the kit. The total number of coliforms equals the number of red gas-producing colonies plus the number of blue gas-producing colonies. Petrifilm *E. coli* count plates are re-incubated for an additional 24 ± 2 hr after counting in order to recover slow GUD producers and *E. coli* colonies are again counted. The number of *E. coli* equals the number of blue gas-producing colonies. The Petrifilm *E. coli* Count Plate yields confirmed results for *E. coli* and coliforms; no additional confirmation is necessary.

The Petrifilm *E. coli* Count Plate is an AOAC Official Method[2] (AOAC method #991.14) for the analysis of *E. coli* and coliforms in foods. The AOAC procedure specifies that only blue colonies with gas are counted as *E. coli* but this practice may result in the omission of anaerogenic *E. coli*. Some researchers[9,48] have recommended that blue, non-gas producing colonies be confirmed in order to determine whether they represent *E. coli*.

8.936 The Redigel™ ColiChrome™ 2 Test for E. coli and Coliforms

The Redigel ColiChrome method follows the same procedure as the Redigel Violet Red Bile test (Section 8.76) only a different gel formulation is used. Like the Petrifilm *E. coli* count plate, the ColiChrome medium incorporates indicator systems for both β-galactosidase (for detection of coliforms) and β-glucuronidase (for detection of *E. coli*). Redigel ColiChrome plates are incubated at 32°C (for dairy products) or at 35°C (for other foods) for 24 to 48 hours. Coliform colonies are pink to red, *E. coli* colonies are purple, and non-coliform colonies are white or cream in color. The countable range for the Redigel ColiChrome plate is 15 to 150 CFU per plate. No additional confirmation is necessary.

8.937 The IDEXX SimPlate™ Coliforms/E. coli test and the Reveal® Bio-Plate for E. coli and Total Coliforms

The general principle behind both the IDEXX SimPlate Coliforms/*E. coli* test and the Reveal® Bio-Plate for *E. coli* and total coliforms is very similar to that outlined in section 8.76 and 8.936 in that the sample is added to a special selective/differential medium then plated onto a special plate. Media enters the wells on the dimpled SimPlate or Bio-Plate then the remainder of the sample-containing medium is poured off the plate and discarded. The plates are incubated at 35 or 37°C for 24 hr. The number of colored wells is counted and this number is entered into a special MPN table. The MPN value is then multiplied by the reciprocal of the dilution used to obtain the MPN coliforms per g or per mL. The plate is next examined under long-wave ultraviolet light and fluorescent wells are counted. The number of fluorescent wells is then entered into the special MPN table and this value is multiplied by the reciprocal of the dilution factor to determine the MPN *E. coli* per g or per mL. No confirmation is necessary for either the IDEXX SimPlate Coliforms/*E. coli* test or the Reveal® Bio-Plate for *E. coli* and total coliforms.

8.94 *Indole-based Methods for* E. coli *and Coliforms*

Indole production at 44°C has been used for the detection and identification of *E. coli* since 1948.[46] Type I (indole-positive) *E. coli* strains are regarded as "typical" *E. coli*. Some 95% of *E. coli* strains recovered from foods are indole-positive.[4] Bacteria other than *E. coli* which are indole-positive include *Klebsiella* spp. (*pneumoniae* and *oxytoca*), *Citrobacter diversus*, and *Providencia* spp. Indole-positive bacteria other than *E. coli* may comprise 3–5% of the indole positive isolates on foods.[64] Indole-based detection methods, like MUG-based assays, offer a means of detecting anaerogenic strains of *E. coli* which might be missed by tests predicated upon the production of gas from lactose. Combining indole and GUD assays improves the specificity of "rapid" *E. coli* determinations.

8.941 *Modified Baird-Parker Procedure for* E. coli *and Coliforms*

The modified Baird-Parker procedure involves inoculating a membrane with the diluted food sample then successively transferring the membrane from a non-selective plated medium (for resuscitation) to a selective/differential medium. Subsequently, indole production is determined for identification of *E. coli* colonies.

Portions (0.5 to 1.0 mL) of a diluted food sample are inoculated onto a membrane placed on a plate of Nutrient agar or Minerals Modified Glutamate Medium. Plates are incubated (right-side up) at 35°C for 4 hr after spreading with a sterile spreader to allow resuscitation to occur. Food carbohydrates which might interfere with indole production also diffuse into the non-selective medium during the resuscitation period. Membranes are then transferred to the surface of a Tryptone Bile Agar (TBA) plate. TBA plates are incubated at 44.5 ± 0.2°C for 18 hr. The membrane is removed from the surface of the TBA plate after incubation, and placed in Vracko-Sherris indole reagent for 5 min. The membrane is then removed and drained of excess reagent before being dried under a long-wave UV light for about 20 min. Exposure to UV light "fixes" the color of the pink-stained, indole-positive colonies and helps to minimize color fading. Pink-stained, indole-positive colonies representative of *E. coli* are then counted. *E. coli* CFU/g or mL is then calculated by multiplying the number of indole-positive colonies by the reciprocal of the dilution used. Confirmation is not possible with this approach, since the indole reagent kills the *E. coli* during the staining process.

8.942 *HGMF Method for* E. coli *and/or Fecal Coliforms*

The method is similar to the modified Baird-Parker procedure except that it uses filtration and an oxidant-accelerated indole reagent; the same filter can be used to obtain both an *E. coli* count and a fecal coliform count. A diluted food sample is first filtered through a membrane filter as with other HGMF procedures. The inoculated filter is then placed on the surface of a pre-dried plate of either non-selective agar or selective/differential agar depending on the nature of the sample and the expected physiological state of the target organisms. A pre-dried plate of TSA with $MgSO_4$ (TSAM) is used if the food or environmental sample is expected to contain sub-lethally damaged or stressed organisms. The TSAM plate is then inverted and incubated at 25°C for 4–5 hr (if analyzing dry foods) or at 35°C for 4–5 hr (for all other foods) to allow for resuscitation. The filter is transferred to the surface of a pre-dried plate of TBA agar, after filtration or resuscitation. The plate is then inverted and incubated at 44.5° ± 0.5°C for 18 to 24 hr in an incubator with good temperature control. The number of grid squares containing one or more blue colonies (any shade of blue) is counted after incubation and this count is used to calculate the MPN of fecal coliforms with a standard formula. The MPN value is then multiplied by the reciprocal of the dilution used and the MPN fecal coliforms per g or per mL of food reported. Additional confirmation is not required.

The filter can then be stained with an oxidant-accelerated indole reagent to detect indole-positive colonies, if an *E. coli* count is desired. The grids should be stained at room temperature for 10–15 minutes afterwards the HGMF grids containing one or more pink-red (indole-positive) colonies are counted as *E. coli*. This count is used to calculate the *E. coli* MPN with a standard formula. The MPN value is then multiplied by the reciprocal of the dilution used and the MPN *E. coli* per g or per mL of food reported. Additional confirmation is not required. The HGMF[1] is an AOAC Official Method[2] for the analysis of *E. coli* and fecal coliforms in foods (method #983.25).

8.95 *The Gene-Trak®* E. coli *DNA Hybridization Test*

The Gene-Trak® *E. coli* DNA hybridization test is a semi-quantitative assay for *E. coli* which can be used to identify food samples containing ≥ 3 *E. coli* per g or mL of food. The 24-hr protocol uses a 2-stage enrichment procedure including enrichment of the diluted sample in Brain Heart Infusion broth for 4 hr at 35°C followed by transfer to Tryptose Phosphate Broth and incubation at 42°C for 22 ± 2 hr. The 48-hr protocol specifies the addition of the diluted sample to LST broth and incubation at 35°C for 24 ± 2 hr followed by a transfer to fresh LST broth and another incubation identical to the first. A portion of the enriched culture is tested with an *E. coli*-specific probe using the Gene-Trak® colorimetric DNA hybridization format. Sample enrichment cultures which give positive results with the DNA hybridization test can be culturally and biochemically confirmed.

8.10 INTERPRETATION OF DATA: THE VALUE OF *ENTEROBACTERIACEAE*, COLIFORMS, AND *E. COLI* AS INDICATORS OF QUALITY AND INDEXES OF PATHOGENS

The enumeration of *Enterobacteriaceae*, coliforms (including fecal coliforms), and *E. coli* in foods is far from an exact science, as has already been discussed. Considerable effort has been expended over the past 100 years in attempts to improve the specificity of assays for *Enterobacteriaceae*, coliforms, and *E. coli* but these efforts have achieved relatively little. However, most food manufacturers seem willing to sacrifice some specificity in exchange for more rapid results. Results obtained from assays for *Enterobacteriaceae*, coliforms, and *E. coli* (including those obtained by direct count methods) are best regarded as estimates and data interpretation must take this limitation into account.

8.101 *Use of* Enterobacteriaceae*, Coliform, or* E. coli *Counts in Microbiological Criteria*

Regulatory agencies, for the most part, have recognized the futility of assuming that index counts reflect the safety of a particular food product. This is particularly true with processed (as opposed to raw) food products. The assumption that index counts reflect the presence or absence of a particular pathogen in a processed (as opposed to raw commodity) is not supported by the scientific literature (see Section 8.1). Consequently, regulatory criteria for index/indicator organisms, when issued, typically take the form of "guidelines" which are relatively lenient rather than "standards" which are the more restrictive in that they are legally enforceable.

The presence of high levels of coliforms, *Enterobacteriaceae*, or *E. coli*, in foods processed for safety, may indicate one of the following possibilities: 1) inadequate processing and/or post-processing contamination; and/or 2) microbial growth. The presence of high levels of any of these organisms is not *de facto* evidence that fecal contamination has occurred.[40] In raw foods, the presence of members of the *Enterobacteriaceae* group is to be expected since these organisms are wide spread in the natural environment.

End-product specifications describing acceptable levels of indicator organisms in various food products and food ingredients are commonly issued by purchasers, often to the consternation of the would-be supplier. Specifications are, too often, scientifically unsupportable and arbitrary. There is a common tendency among food manufacturers to assume that a specification is "universal" and can be applied to a variety of foods despite diverse origins, composition, physical/chemical characteristics, processing, and storage. In order to be effective, a microbiological specification must be: 1) food or ingredient specific; 2) supported by organoleptic and/or microbiological data relating the specification to the quality and/or safety of the product in question; and 3) capable of being met when the particular product/ingredient is produced under "optimal" hygienic conditions. An end-product specification should also detail both the sampling procedure and the microbiological method used to obtain the results. Analysis of a single food sample by two different microbiological assays, as discussed earlier in this chapter, often yields different numerical results which may be reflective of different sampled microbial populations. Confirmation tests such as the IMViC tests for *E. coli* may also be unreliable (see section 8.37).

The point in the production and distribution process at which a product is sampled has a tremendous impact on the recovery of coliforms, *Enterobacteriaceae*, or *E. coli* from that food. Levels of these organisms on raw foods (e.g., vegetables or raw meats) are not indicative of product quality unless processing procedures (e.g., washing) have been applied to reduce levels of naturally-occurring microorganisms. Analytical results obtained from product in refrigerated distribution will not be reflective of production conditions since some coliforms and *Enterobacteriaceae* are psychrotrophic[44] and can, therefore, grow during refrigerated storage. Similarly, coliforms and *E. coli* levels may decrease over time in cultured dairy products like yogurt, buttermilk, and sour cream.

8.102 The Enterobacteriaceae

Despite their name, members of the family *Enterobacteriaceae* are not confined to the intestinal tract and may be isolated from a variety of non-intestinal sources.[16] *Enterobacteriaceae* are relatively heat-sensitive and easily killed during cooking or milk pasteurization. Their resistance to freezing and other processing procedures varies between members of this Family. *Enterobacteriaceae* are good indicators of environmental hygiene because they are readily inactivated by sanitizers and capable of colonizing a variety of niches in the processing plant when sanitation is inadequate.[16,29]

8.103 The Coliform Group

Recovery of the coliform group from foods has less interpretive impact than the single index organism, *E. coli*, or the fecal coliform group because the coliform group may contain non-enteric members such as *Serratia* and *Aeromonas*. The specificity of the coliform group as an index of fecal contamination is diminished by the anonymity of its individual members and the diversity of this group.[53] It is possible to use coliforms as indicators of inadequate sanitation on pre-operational equipment contact surfaces since these organisms are not resistant to sanitizers. The presence of coliforms on ready-to-eat heat-processed foods can serve as an indication of either inadequate heat-processing or post-pasteurization contamination. The inability of many coliforms to survive freezing makes them of questionable use when analyzing frozen foods.[80]

Coliform counts can differ significantly, depending on food tested, medium used, and other testing conditions. Various conditions in food processing establishments (such as drying, acidification, heating, sanitation, etc.) may cause cell injury and assays must be capable of recovering injured cells. Specification of the medium and temperature used to obtain coliform counts is critical to interpretation of data.

8.104 The Fecal Coliform Group

The term "fecal" coliform is a misnomer since organisms enumerated by the fecal coliform assay may or may not have originated in the intestinal tract. Organisms such as *Klebsiella* spp., *Enterobacter* spp., and *Citrobacter freundii*, for example, may grow outside of the intestinal tract.[16,43,72] The presence of these organisms within the fecal coliform group compromises the group's specificity and represents a deficiency in methodology or nomenclature (see section 8.23). Since the proportion of *E. coli* within the fecal coliform population varies between samples, there is little reason to stop at the fecal coliform test when *E. coli* is really the object of interest.

8.105 Escherichia coli

E. coli is regarded as being the most valid indicator of fecal contamination of raw foods. This is not to say, however, that *E. coli* is a good indicator of fecal contamination of processed foods. Contemporary data indicate that *E. coli* can grow in a variety of extra-intestinal niches, including in the processing plant environment.[16] Recovery of *E. coli* from heat-processed foods is indicative of either inadequate processing or subsequent contamination. Differences in sensitivity to various food processing technologies (e.g. drying) between *E. coli* and enteric pathogens also limits the usefulness of *E. coli* as an index organism.

8.11 ACKNOWLEDGMENTS

This chapter is based on the text prepared by the late Ira J. Mehlman and later modified by A. D. Hitchins, the late P. A. Hartman and E. C. D. Todd.

8.12 REFERENCES

1. AOAC. 1985. Total coliform, fecal coliform and *Escherichia coli* in foods, Hydrophobic Grid Membrane Filter method. J. Assoc. Off. Anal. Chem. 68:404
2. AOAC. 1998. Microbiological methods, chap. 17, p. 12c. *In* Official methods of analysis, 16th ed., 4th revision. p. 12c. AOAC International, Gaithersburg, Md.
3. APHA. 1985a. Laboratory procedures for the examination of seawater and shellfish, 5th ed. American Public Health Association, Washington, D. C.
4. Anderson, J. M., and A. C. Baird-Parker. 1975. A rapid and direct plate method for enumerating *Escherichia coli* biotype I in food. J. Appl. Bacteriol. 39:111.

5. Andrews, W. H., and G. A. June. 1998. Food sampling and preparation of sample homogenate. Chapter 1. *In FDA bacteriological analytical manual*, Rev. A, 8th ed., AOAC International, Gaithersburg, Md.
6. Bagley, S. T., and R. J. Seidler. 1977. Significance of fecal coliform-positive *Klebsiella*. Appl. Environ. Microbiol. 33:1141.
7. Bardsley, D. A. 1934. The distribution and sanitary significance of *B. coli, B. lactis aerogenes*, and intermediate types of coliform bacilli in water, soil, faeces, and ice-cream. J. Hyg. 34:38.
8. Blachstein. 1893. Contribution a l'étude microbique de l'eau. Ann. Inst. Pasteur 10:689.
9. Bloch, N., H. Sidjabat-Tambunan, T. Tratt, K. Lea, and A. J. Frost. 1996. The enumeration of coliforms and *E. coli* on naturally contaminated beef: a comparison of the Petrifilm™ method with the Australian standard. Meat Sci. 43:187.
10. Breed, R. S., and N. F. Norton. 1937. Nomenclature for the colon group. Am. J. Public Health 27:560.
11. Brodsky, M. H. 1995. The benefits and limitations of using index and indicator microorganisms in verifying food safety. Presented at the FSIS meeting on The Role of Microbiological Testing in Verifying Food Safety, May 1-2, 1995, Philadelphia, Pa.
12. Brodsky, M. H., P. Boleszczuk, and P. Entis. 1982. Effect of stress and resuscitation on recovery of indicator bacteria from foods using Hydrophobic Grid-Membrane Filtration. J. Food Prot. 45:1326.
13. Chang, G. W., J. Brill, and R. Lum. 1989. Proportion of β-D-glucuronidase-negative *Escherichia coli* in human fecal samples. Appl. Environ. Microbiol. 55:335.
14. Chordash, R. A., and N. F. Insalata. 1978. Incidence and pathological significance of *Escherichia coli* and other sanitary indicator organisms in food and water. Food Technol. 32:54.
15. Christen, G. L., P. M. Davidson, J. S. McAllister, and L. A. Roth. 1992. Coliform and other indicator bacteria, chap. 7, p. 246. *In* R.T. Marshall (ed.), Standard methods for the examination of dairy products, 16th ed. American Public Health Association, Washington, D.C.
16. Cox, L. J., N. Keller, and M. van Schothorst. 1988. The use and misuse of quantitative determinations of *Enterobacteriaceae* in food microbiology. J. Appl. Bacteriol. Symp. Suppl. 237S.
17. Damaré, J. M., D. F. Campbell, and R. W. Johnston. 1985. Simplified direct plating method for enhanced recovery of *Escherichia coli* in food. J. Food Sci. 50:1736.
18. Doyle, M. P., and J. L. Schoeni. 1984. Survival and growth characteristics of *Escherichia coli* associated with hemorrhagic colitis. Appl. Environ. Microbiol. 48:855.
19. Edberg, S.C., M. J. Allen, and D. B. Smith. 1991. Defined substrate technology method for rapid and specific simultaneous enumeration of total coliforms and *Escherichia coli* from water: collaborative study. J. Assoc. Off. Anal. Chem. 74:526-529.
20. Edwards, P. R., and W. H. Ewing. 1972. Identification of *Enterobacteriaceae*, 3rd ed. Burgess Publishing Company, Minneapolis, Minn.
21. Entis, P., and P. Boleszczuk. 1990. Direct enumeration of coliforms and *Escherichia coli* by hydrophobic grid membrane filter in 24 hours using MUG. J. Food Prot. 53:948.
22. Evans, T. M., R. J. Seidler, and M. W. LeChevallier. 1981. Impact of verification media and resuscitation on accuracy of the membrane filter total coliform enumeration technique. Appl. Environ. Microbiol. 41:1144.
23. Ewing, W. H. 1986. Edwards and Ewing's identification of *Enterobacteriaceae*, 4th ed. Elsevier, New York, N.Y.
24. FAO/WHO (Food and Agricultural Organization of the United Nations and World Health Organization). 1979. Report of a joint FAO/WHO Working Group on Microbiological Criteria for Foods, Geneva, 20-26 February, 1979. WHO, Geneva. Document WG/Microbiol; 79/1.
25. Feldsine, P. T., M. T. Falbo-Nelson, and D. L. Hustad. 1993. Substrate supporting disc method for confirmed detection of total coliforms and *E. coli* in all foods: collaborative study. J. AOAC Int. 76:988.
26. Feng, P. C. S., and P. A. Hartman. 1982. Fluorogenic assays for immediate confirmation of *Escherichia coli*. Appl. Environ. Microbiol. 43:1320.
27. Fishbein, M. 1962. The aerogenic response of *Escherichia coli* and strains of *Aerobacter* in EC broth and selected sugar broths at elevated temperatures. Appl. Microbiol. 10:79.
28. Frampton, E. W., L. Restaino, and N. Blaszko. 1988. Evaluation of the β-glucuronidase substrate 5-bromo-4-chloro-3-indolyl-β-D-glucuronide (X-GLUC) in a 24-hour direct plating method for *Escherichia coli*. J. Food Prot. 51:402.
29. Gabis, D. A., and R. E. Faust. 1988. Controlling microbial growth in food processing environments. Food Technol. 12:81.
30. Garthright, W. E. 1998. Most probable number from serial dilutions. Appendix 2. *In* FDA bacteriological analytical manual, app. 2.01-2.09. AOAC International, Gaithersburg, Md.
31. Hansen, W., and E. Yourassowsky. 1984. Detection of β-glucuronidase in lactose-fermenting members of the family *Enterobacteriaceae* and its presence in bacterial urine cultures. J. Clin. Microbiol. 20:1177.
32. Hartman, P. A. 1989. MUG (β-glucuronidase) test for *Escherichia coli* in food and water, p. 290. *In* A. Ballows, R. C. Tilton, and A. Turano (eds.), Rapid methods and automation in microbiology and immunology. Brixia Academic Press, Brescia, Italy.
33. Hartman, P. A., and P. S. Hartman. 1976. Coliform analyses at 30°C. J. Milk Food Technol. 39:763.
34. Himelbloom, B. H., and R. C. Pfutzenreuter. 1998. False-positive fluorescence by pink Salmon tissue and Staphylococci in a rapid test for *Escherichia coli*. J. Food Prot. 61:1119.
35. Hitchins, A. D., P. Feng, W. D. Watkins, S. R. Rippey, and L. A. Chandler. 1998. *Escherichia coli* and the coliform bacteria, chap. 4, p. 4.01. *In* FDA bacteriological analytical manual, Rev. A, 8th ed. AOAC International, Gaithersburg, Md.
36. Holbrook, R., J. M. Anderson, and A. C. Baird-Parker. 1980. Modified direct plate method for counting *Escherichia coli* in foods. Food Technol. Aust. 32:78.
37. Holt, J. G., N. R. Kreig, P. H. A. Sneath, J. T. Staley, and S. T. Williams (eds.). 1994. Bergey's manual of determinative bacteriology, 9th ed. Williams and Wilkins, Baltimore, Md. p. 787.
38. Holt, S. M., P. A. Hartman, and C. W. Kaspar. 1989. Enzyme-capture assay for rapid detection of *Escherichia coli* in oysters. Appl. Environ. Microbiol. 55:229.
39. Huang, S. W., C. H. Chang, T. F. Tai, and T. C. Chang. 1997. Comparison of the β- glucuronidase assay and the conventional method for identification of *Escherichia coli* on Eosin-Methylene Blue agar. J. Food Prot. 60:6.
40. ICMSF (International Commission on Microbiological Specifications for Foods of the International Union of Biological Societies). 1978. Microorganisms in foods 1. Their significance and methods of enumeration. University of Toronto Press, Toronto. p. 10.
41. International Standardization Organization. 1991. International Standard 4831, Microbiology—General guidance for the enumeration of coliforms—Most Probable Number technique. ISO, Geneva.
42. International Standardization Organization. 1991. International Standard 8523, Microbiology—General guidance for the detection of *Enterobacteriaceae* with pre-enrichment. ISO, Geneva.
43. Knittel, M. D., R. J. Seidler, C. Eby, and L. M. Cabe. 1977. Colonization of the botanical environment by *Klebsiella* isolates of pathogenic origin. Appl. Environ. Microbiol. 34:557.
44. Kornacki, J. L., and D. A. Gabis. 1990. Microorganisms and refrigeration temperatures. Dairy Food Environ. Sanit. 10:192.
45. LeChevallier, M. W., and G. A. McFeters. 1984. Recent advances in coliform methodology for water analysis. J. Environ. Health 47:5.
46. Mackenzie, E. F. W., E. W. Taylor, and W. E. Gilbert. 1948. Recent experiences in the rapid identification of *Bacterium coli* type I. J. Gen. Microbiol. 2:197.
47. Mackey, B. M., C. M. Derrick, and J. A. Thomas. 1980. The recovery of sublethally injured *Escherichia coli* from frozen meat. J. Appl. Bacteriol. 48:315.
48. Matner, R. R., T. L. Fox, D. E. McIver, and M. S. Curiale. 1990. Efficacy of Petrifilm™ *E. coli* count Plates for *E. coli* and coliform enumeration. J. Food Prot. 53:145.

49. Mercuri, A. J., and N. A. Cox. 1979. Coliforms and *Enterobacteriaceae* isolates from selected foods. J. Food Prot. 42:712.
50. Miskimin, D. K., K. A. Berkowitz, M. Solberg, W. E. Riha Jr., W. C. Franke, R. L. Buchanan, and V. O'Leary. 1976. Relationships between indicator organisms and specific pathogens in potentially hazardous foods. J. Food Sci. 41:1001.
51. Moberg, L. J., M. K. Wagner, and L. A. Kellen. 1988. Fluorogenic assay for rapid detection of *Escherichia coli* in chilled and frozen foods: collaborative study. J. Assoc. Off. Anal. Chem. 71: 589.
52. Mossel, D. A. A. 1978. Index and indicator organisms—a current assessment of their usefulness and significance. Food Technol. Aust. 30:212.
53. Mossel, D. A. A. 1985. Media for *Enterobacteriaceae*. Int. J. Food Microbiol. 2:27.
54. Mossel, D. A. A., I. Eelderink, M. Koopmans, and F. van Rossem. 1979. Influence of carbon source, bile salts, and incubation temperature on recovery of *Enterobacteriaceae* from foods using MacConkey-type agars. J. Food Prot. 42:470.
55. Mossel, D. A. A., and P. A. Van Netten. 1991. Microbiological reference values for foods: a European perspective. J. Assoc. Off. Anal. Chem. 74:420.
56. Mossel, D. A. A., M. Visser, and A. M. R. Cornelissen. 1963. The examination of foods for *Enterobacteriaceae* using a test of the type generally adopted for the detection of Salmonellae. J. Appl. Bacteriol. 3:444.
57. National Research Council (NRC), Food and Nutrition Board, Committee on Food Protection, Subcommittee on Microbiological Criteria. 1985. An evaluation of the role of microbiological criteria for foods and food ingredients. p. 436. National Academy Press. Washington, D. C.
58. Nelson, C. L., T. L. Fox, and F. F. Busta. 1984. Evaluation of dry medium film (Petrifilm™ VRB) for coliform enumeration. J. Food Prot. 47:520-525.
59. Ogden, I. D., and N. J. C. Strachan. 1993. Enumeration of *Escherichia coli* in cooked and raw meats by ion mobility spectrometry. J. Appl. Bacteriol. 74:402.
60. Ogden, I. D., and A. J. Watt. 1991. An evaluation of fluorogenic and chromogenic assays for the direct enumeration of *Escherichia coli*. Lett. Appl. Microbiol. 13:212.
61. Patterson, J. T., and P. A. Gibbs. 1976. Some growth characteristics of psychrotrophic *Enterobacteriaceae* isolated from meat. J. Appl. Bacteriol. 41:xxii.
62. Powers, E. M., and T. G. Latt. 1977. Simplified 48-hr IMViC test: an agar plate method. Appl. Environ. Microbiol. 34:274.
63. Ray, B. 1986. Impact of bacterial injury and repair in food microbiology: its past, present and future. J. Food Prot. 49:651.
64. Rayman, M. K., G. A. Jarvis, C. M. Davidson, S. Long, J. M. Allen, T. Tong, P. Dodsworth, S. McLaughlin, S. Greenberg, B. G. Shaw, H. J. Beckers, S. Qvist, P. M. Nottingham, and B. J. Stewart. 1979. ICMSF methods studies. XIII. An international comparative study of the MPN procedure and the Anderson-Baird-Parker direct plating method for the enumeration of *Escherichia coli* biotype I in raw meats. Can. J. Microbiol. 25:1321.
65. Rochaix, A. 1924. La recherche du colibacille dans l'eau et dans le lait, au moyen des milieux a l'esculine. Lait 4:541-544.
66. Sharpe, A. N., and L. J. Parrington. 1998. Membrane filter method based on FC-5-Bromo-4- Chloro-3-Indolyl-β-D-Glucuronide medium facilitates enumeration of *Escherichia coli* in foods and poultry carcass rinses. J. Food Prot. 61:360.
67. Silk, T. M., E. T. Ryser, and C. W. Donnelly. 1997. Comparison of methods for determining coliform and *Escherichia coli* levels in apple cider. J. Food Prot. 60:1302.
68. Silliker, J. H., and D. A. Gabis. 1976. ICMSF methods studies. VII. Indicator tests as substitutes for direct testing of dried foods and feeds for *Salmonella*. Can. J. Microbiol. 22:971.
69. Simmons, J. S. 1924. A culture medium for differentiating organisms of typhoid-colon aerogenes groups and for isolation of certain fungi. J. Bacteriol. 9:59.
70. Solberg, M., D. K. Miskimin, B. A. Martin, G. Page, S. Goldner, and M. Libfeld. 1977. Indicator organisms, foodborne pathogens and food safety. Assoc. Food Drug Off. Quart. Bull. 41:9.
71. Speck, M. L., B. Ray, and R. B. Read Jr. 1975. Repair and enumeration of injured coliforms by a plating procedure. Appl. Microbiol. 29:549.
72. Splittstoesser, D. F., D. T. Queale, J. L. Bowers, and M. Wilkison. 1980. Coliform content of frozen blanched vegetables packed in the United States. J. Food Safety 2:1.
73. Stiles, M. E., and L.-K. Ng. 1980. Estimation of *Escherichia coli* in raw ground beef. Appl. Environ. Microbiol. 40:346.
74. Strachan, N. J. C., and I. D. Ogden. 1993. A rapid method for the enumeration of coliforms in processed foods by ion mobility spectrometry. Lett. Appl. Microbiol. 17:228.
75. Swenarton, J. C. 1927. Can *B. coli* be used as an index of the proper pasteurization of milk? J. Bacteriol. 13:419-429.
76. United States Pharmacopeal Convention. 1995. Microbial limits test, chap. 61, p. 1681. *In* U.S. Pharmacopeia: National Formulary, 23rd ed. U.S. Pharmacopeal Convention, Rockville, Md.
77. Venkateswaran, K., A. Murakoshi, and M. Satake. 1996. Comparison of commercially available kits with standard methods for the detection of coliforms and *Escherichia coli* in foods. Appl. Environ. Microbiol. 62:2236.
78. Vought, K. J., and S. R. Tatini. 1998. *Salmonella enteritidis* contamination of ice cream associated with a 1994 multistate outbreak. J. Food Prot. 61:5.
79. Weiss, K. F., N. Chopra, P. Stotland, G. W. Reidel, and S. Malcolm. 1983. Recovery of fecal coliforms and of *Escherichia coli* at 44.5, 45.0, 45.5°C. J. Food Prot. 46:172.
80. Wilderson, W. B., J. C. Ayres, and A. A. Kraft. 1961. Occurrence of Enterococci and Coliform organisms on fresh and stored poultry. Food Technol.

Chapter 9

Enterococci

Paul A. Hartman, Robert H. Deibel, and Linda M. Sieverding

9.1 INTRODUCTION

Classification of the enterococci is in a state of flux. In the second edition of this book,[10] all streptococci of fecal origin that produce group D antigen were considered enterococci. These included *Streptococcus avium, S. bovis, S. faecalis* (and its varieties *liquefaciens* and *zymogenes*), and *S. faecium* (and its varieties *casseliflavus* and *durans*). Mundt (1986) included only *S. avium, S. faecalis, S. faecium,* and *S. gallinarum* in the most recent edition of *Bergey's Manual*[34]; *S. bovis* and *S. equinus* were placed in a group of "Other Streptococci."[23] Recent molecular biology studies (including oligonucleotide cataloging of 16S rRNA and DNA-DNA and DNA-rRNA hybridization), combined with physiological studies, resulted in a more elaborate classification,[39] wherein a new genus, *Enterococcus,* was established. Members of this genus are *E. avium, E. casseliflavus, E. durans, E. faecalis, E. faecium, E. gallinarum, E. hirae, E. malodoratus,* and *E. mundtii* (Table 1). All these bacteria usually grow at 45°C, in 6.5% NaCl, and at pH 9.6; most grow at 10°C. *S. bovis* and *S. equinus,* which are negative in two or more of these properties, were assigned to a miscellaneous group of "Other Streptococci."[39] Almost all (99%) are susceptible to vancomycin, and very few (less than 1%) produce gas from glycerol.[1]

The enterococci have conventionally been identified by physiological as well as serological methods. When the former are employed, a spectrum of characteristics (Table 1) must be examined because no single, two, or three traits will establish a definitive identification.[9] Identification medium formulations and interpretations of tests have been described by Facklam and Collins.[15] All enterococci, as well as *S. bovis* and *S. equinus,* produce group D antigen, although presence of this antigen is difficult to demonstrate with some isolates.

Generally, the *Enterococcus* habitat is the intestinal contents of both warm- and cold-blooded animals, including insects.[33] Some enterococci have adapted to an epiphytic relationship with growing vegetation. None of the enterococci can be considered as absolutely host specific, although some species evidence a degree of host specificity.

Enterococcus faecalis and *E. faecium* are relatively heat resistant and characteristically may survive traditional milk pasteurization procedures. *Enterococcus faecium* is markedly heat tolerant and is a spoilage agent in marginally processed canned hams. Most of the enterococci are relatively resistant to freezing, and, unlike *Escherichia coli,* they readily survive this treatment.

In the past, some investigators have associated food poisoning outbreaks with these bacteria,[24] but definitive experiments with unequivocal positive results were lacking until recently, when several enterococci were implicated as causes of diarrheal diseases in neonatal animals.[12] Other enterococci, especially those that are resistant to chemotherapeutic agents, can cause serious illness in humans.[26]

9.2 GENERAL CONSIDERATIONS

An abundance of media has been advocated for the selective isolation and/or quantitation of enterococci.[25] Many selective agents, incubation conditions, and combinations of these have been described, but all have one or more shortcomings. The media and methods that are available presently lack selectivity, differential ability, quantitative recovery, relative ease of use, or a combination of these to various degrees. Some strains of fecal streptococci from anaerobic environments, for example, initiate growth only in the presence of elevated levels of CO_2 until the cultures have been adapted to an aerobic environment.[30] Therefore, a compromise must be reached in the selection of a general-purpose medium for the recovery of enterococci from foods. In food microbiology, *E. faecalis* and *E. faecium* are the most common enterococci encountered. This undoubtedly influences the rationale of employing KF streptococcal agar for the estimation of enterococci in foods.[28] The selectivity of KF streptococcal agar is not absolute,[13] quantitative recovery is less than ideal, and preparation of the medium necessitates an aseptic addition of an indicator. Nevertheless, many industry and regulatory agencies have accepted KF agar for the quantitative estimation of enterococci in nondairy foods. For dairy products, a more selective medium or higher incubation temperature (45°C) may be necessary to reduce background growth of lactobacilli and lactic streptococci.

KF streptococcal agar[28] is a selective differential medium that employs sodium azide as the chief selective agent and triphenyltetrazolium chloride (TTC) for differential purposes. The medium contains a relatively high concentration of maltose (2.0%) and a small amount of lactose (0.1%). Most, but not all, enterococci and streptococci ferment these sugars. The intensity of TTC reduction varies. *E. faecalis* reduces the compound to its formazan derivative, imparting a deep red color to the colony. Other group D enterococci and streptococci, if they grow on KF

Table 1. Some Characteristics of the Enterococci and Group D Fecal Streptococci[a]

Species	G + C content (mole %)	Serological Group	PYR (see text)	Growth, 10°C	Growth, 45°C	Growth, 6.5% NaCl	Growth, pH 9.6	Hydrol. of arginine	Hydrol. of esculin	Hydrol. of hippurate
E. avium	39–40	D + Q	+	+	+	+	+	−	+	−
E. casseliflavus	41–45	D	+	+	+	+	+	v	+	v
E. durans	38–40	D	+	+	+	+	+	+	+	v
E. faecalis	37–40	D	+	+	+	+	+	+	(+)	v
E. faecium	37–40	D	+	+	+	+	+	+	+	(−)
E. gallinarum	39–40	D	+	+	+	+	+	+	+	(+)
E. hirae	37–38	D	+	+	+	+	+	+	+	(−)
E. malodoratus	40–41	D	+	+	+	+	+	−	+	−
E. mundtii	38–39	D	+	+	+	+	+	+	+	−
S. bovis	36–38	D	−	−	+	−	−	−	+	−
S. equinus	36–39	D	−	−	+	−	−	−	+	−

Species	Acid from arabinose	Acid from glycerol	Acid from melezitose	Acid from melibiose	Acid from sorbitol	Acid from sorbose	Acid from tagatose	Motility	Yellow pigment	H_2S produced
E. avium	+	v	+	v	+	+	+	−	−	+
E. casseliflavus	+	v	−	+	v	−	(−)	+	+	−
E. durans	−	−	−	v	−	−	−	−	−	−
E. faecalis	−	(+)	(+)	−	(+)	−	+	−	−	−
E. faecium	+	−	−	v	v	−	(−)	−	−	−
E. gallinarum	+	−	(−)	+	−	−	(+)	+	−	−
E. hirae	−	−	v	(+)	−	−	v	−	−	nd
E. malodoratus	−	−	−	+	+	+	−	−	−	+
E. mundtii	+	−	−	+	v	−	−	−	+	nd
S. bovis	v	−	nd	+	v	−	nd	−	−	−
S. equinus	−	−	nd	nd	−	nd	nd	−	−	−

[a] All are gram-positive, catalase-negative, facultatively anaerobic cocci or coccobacilli. Note that "false-positive" and "false-negative" test reactions can be obtained, depending on the sensitivity of the assay that is used.[19]

+ = positive; (+) = most strains positive; (−) = most strains negative; − = negative; v = variable; nd = not determined (but probably would be H_2S-negative. Differential characteristics of *Enterococcus* spp. are listed in references [3,6,7,12,14,15,16,18,20] and by L. M. Sieverding (unpublished data). See Facklam and Collins (1989) for abbreviated identification schemes, including those for three new species (*E. pseudoavium, E. raffinosus,* and *E. solitarius*).

agar, are feebly reductive and the colonies appear light pink. (Tetrazolium reactions of *E. hirae*[18] and *E. malodoratus*[6] have not been described.) Most other lactic acid bacteria are partially or completely inhibited; however, some strains of *Pediococcus, Lactobacillus,* and *Aerococcus* may grow, producing light pink colonies. A "repair-detection" procedure[22] should be considered when the enterococcal population of a food may contain a large proportion of injured cells (Chapter 5).

KF Streptococcal medium is available commercially with or without agar. A broth is available for the most probable number (MPN) procedure to detect low numbers of enterococci, but the MPN procedure is rarely used for foods.

KF streptococcal agar contains azide, which is inhibitory to many strains of *S. bovis* and *S. equinus* and possibly some of the newly named *Enterococcus* spp. Therefore, an alternative procedure[32] that permits the recovery of a wider variety of enterococci from foods is included in this section. This alternative procedure, fluorogenic gentamicin-thallous-carbonate (fGTC) agar, utilizes inhibitors other than azide. Dyed starch and a fluorogenic substrate are included to impart differential qualities to the medium. Enterococcal counts from foods may be two or more orders of magnitude higher of fGTC agar than on KF agar. Further, the incubation period for fGTC agar is only 18 to 24 hr whereas it is 48 hr for KF agar. Neither KF nor fGTC agars have been tested for recoveries of some of the newly named enterococci, and the performance of both media should be reevaluated in the light of recent advances in enterococcal classification.

As in other microbiological plating procedures, sample preparation is important. For example, dried foods are often reconstituted and immediately diluted and plated. In one study,[41] however, the optimum procedure of sample preparation involved the addition of 25 mL of 0.1% peptone water diluent to 25 g of dry food in a sterile pint jar. The jar was "swirled" and allowed to remain at 4°C for 60 min. Then 200 mL of sterile peptone water were added to the jar and mixed to obtain a final 1:10 dilution. Enterococcal counts of dried soup mix were increased by 42% by using the "swirl-hold-dilute" method.[41]

9.3 EQUIPMENT, MATERIALS, AND REAGENTS (CHAPTER 63)

Bile-esculin agar[16]
Brain Heart Infusion broth (BHI)
Filter-sterilized 1% aqueous triphenyltetrazolium chloride (TTC)
Fluorogenic gentamicin-thallous-carbonate (fGTC) agar[32]
Hydrogen peroxide (3%)
KF streptococcus (KF streptococcal) agar[28]
Long-wave (365-nm) ultraviolet light
6.5% salt medium (BHI + 6.0% NaCl)

9.4 PRECAUTIONS

Many foods contain from small to large numbers of enterococci, especially *E. faecalis* and *E. faecium*. Certain varieties of cheese and occasionally, fermented sausage may contain more than 10^6 organisms per g. Relatively low levels, 10^1 to 10^3 per g, are common in a wide variety of other foods. The shelf life of sliced, prepackaged ham (and sometimes other similarly prepared cured meats) may be dictated by controlling the initial numbers of contaminating enterococci.

Many investigators have reported a lack of correlation between *Enterococcus* and *E. coli* counts, and the unreliability of *Enterococcus* counts as a reflection of fecal contamination is established. The ability of enterococci to grow in food processing plants, and possibly other environments, long after their introduction, as well as the observation that enterococci can establish extraintestinal epiphytic relationships, reinforces these observations.

No acceptable levels of enterococci can be stated because *Enterococcus* counts vary with product, holding conditions, time of storage, and other factors. In general, enterococci serve as a good index of sanitation and proper holding conditions. However, the entire history of each product must be established and the culture medium and conditions must be standardized before setting specific criteria.

9.5 ENUMERATION OF ENTEROCOCCI

9.51 KF Streptococcal Agar[28]

Prepare the sample for culturing by the pour plate method as directed in Chapter 7 or Section 9.2 above. Dispense 1 mL of decimal dilutions into duplicate petri plates. If a low count is expected, the accuracy and sensitivity may be increased by plating 1 mL of a 1:10 dilution into each of 10 petri plates, in which case the total number of colonies on the 10 plates represents the count per g of food. Add 12 to 15 mL of KF agar cooled to 45°C, and allow to solidify. Incubate the plates for 48 ± 2 hr at 35° ± 1°C. Using a dissecting microscope with a magnification of 15 diameters or a colony counter, count all red and pink colonies. Report this number as the KF enterococcal count.

9.52 fGTC Agar[32]

Prepare plates as directed in Section 9.51 above. Add 12 to 15 mL of fGTC agar[32] cooled to 45°C, and allow to solidify. Incubate the plates for 18 to 24 hr at 35° ± 1°C. Observe for starch hydrolysis (a zone of clearing around a colony under visible light) and fluorescence (a zone of bright bluish fluorescence when the opened plate is held under a long-wave ultraviolet [UV] lamp). Three phenotypic groups are identifiable: (1) starch hydrolysis and fluorescence, indicative of *S. bovis*, (2) no starch hydrolysis but fluorescence, indicative of *E. faecium* and related biotypes, and (3) no starch hydrolysis or fluorescence, indicative of *E. faecalis*, *E. avium*, *S. equinus*, and other streptococci.[32] Use all colonies to calculate the fGTC enterococcal count, which can be divided, if desired, into subgroups based on starch hydrolysis and fluorescence.

9.6 CONFIRMATION OF ENTEROCOCCI

9.61 Conventional Procedures

If confirmation is desired, pick 5 to 10 typical colonies and transfer each into a separate tube of brain heart infusion (BHI) broth. Incubate at 35°C for 18 to 24 hr. Prepare gram-stained smears of the BHI cultures and observe for typical enterococcal morphology (gram-positive cocci, elongated, in pairs and occasionally short chains). Test for catalase activity by adding 1 mL of 3% hydrogen peroxide to a culture and observe for the generation of oxygen bubbles. Enterococci are catalase-negative, and no reaction should occur. Caution: Do not test for catalase activity directly on azide-containing media such as KF streptococcal agar. Observe for growth and blackening on bile-esculin agar[13] after incubation for 24 hr at 35°C. Examine for growth in BHI broth containing 6.5% NaCl after incubation for 72 hr at 35°C. Test for growth at 45°C in BHI broth that has been tempered to 45°C prior to incubation. Note: If growth in the salt-containing medium and growth at 45°C are to be determined, subcultures must be inoculated before testing for catalase.

S. equinus and *S. bovis* are not enterococci, but they can be of fecal origin. Most do not grow at 10°C, in media containing 6.5% NaCl, or at pH 9.6, but all should grow at 45°C. An excellent confirmatory test for enterococci/fecal streptococci is the ability of an isolate to grow on bile-esculin agar. Enterococci and group D streptococci tolerate bile (grow on bile esculin agar) and hydrolyze esculin.[13] Esculin is 6,7-dihydroxycoumarin-β-D-glucoside. Some bacteria produce an "esculinase" (β-D-glucosidase) that hydrolyzes esculin and releases esculetin (6,7-dihydroxycoumarin); the esculetin reacts with Fe^{+3} in the medium to form a dark brown or black complex.

9.62 Rapid Methods

A 15-min "esculinase" test was devised using *p*-nitrophenyl-β-D-glucopyranoside as the substrate for β-D-glucosidase ("esculinase") determination,[42] and a 4-hr combined sodium chloride tolerance-esculin hydrolysis test also has been described.[37]

Another rapid confirmatory test is the PYR test (Table 1), which detects the ability of a culture to hydrolyze pyrrolidonyl-β-naphthylamide (L-pyroglutamic acid-β-naphthylamide). Hydrolysis of this aminopeptidase substrate is detected by formation of a reddish color within 2 min of addition of PYR reagent. Of the streptococci, only *S. pyogenes* (group A) and the enterococci are positive; *S. bovis* and *S. equinus* are negative.[16,20] Prepackaged PYR test reagents are available[38] (Strep-A-Chek, E-Y Laboratories, Inc., San Mateo, CA 94401; Identicult-AE, Scott Laboratories, Inc., West Warwick, RI 02893; Roscoe Diagnostica, 2630 Taastrup, Denmark).

Convenient tri-plates, quad-plates, and tubed[29] media for key identification tests are available from many suppliers of prepared media. A RapID STR kit (Vitek Systems, Hazelwood, MO 63042)[2,45] may also be time-saving. This kit is similar to the API-20S (Analytab Products, Div. Sherwood Medical, Plainview, NY 11803).[1,8,17,19,21,27,36] The efficacy of this kit and the data bank used with it are based on isolates from human clinical material; the efficiency may differ when isolates from other animals or food are studied.[36] (See Chapter 10).

9.63 Automated Identification

Both the Vitek AutoMicrobic gram-positive identification system[1,17] and the General Diagnostics Autobac system[4] can be used. The latter system includes only bile-esculin agar and salt broth to confirm whether an isolate is or is not an *Enterococcus* or group D *Streptococcus*; the Autobac system makes no attempt to speciate. An impedance method that will detect 10 to 100 "fecal streptococci" in dry milk within 18 hr has been described.[43] (See Chapter 10).

9.64 Serological tests

If serological confirmation is deemed necessary, commercial grouping sera are available from BBL Microbiology Systems (Becton Dickinson Co., Cockeysville, MD 21030) and Wellcome

Reagents Div. (Burroughs Wellcome Co., Research Triangle Park, NC 27709). A variety of serological kits is available: BBL Strep typing kit; Bacto Strep Grouping Kit, Difco Laboratories, Detroit, MI 48232; Streptex, Wellcome Reagents[5,20,35,38,40]; SeroSTAT, Scott Laboratories, Inc., West Warwick, RI 02893[5,31,38,44]; Strepslide, NCS Diagnostics, Inc., Buffalo, NY 14206; and Phadebact, Pharmacia Diagnostics, Piscataway, NJ 08854.[5,20,35,38] These kits vary in efficacy, and false-negative group D reactions are common. Consult the references listed before using these kits.

It is often difficult to demonstrate the presence of the group D antigen in some strains; only 77% of 188 *Enterococcus* strains tested were positive.[15] The method of group antigen preparation is important.[11,35]

9.7 REFERENCES

1. Appelbaum, P. C., Jacobs, M. R., Heald, J. I., Palko, W. M., Duffett, A., Crist, R., and Naugle, P. A. 1984. Comparative evaluation of the API 20S system and the AutoMicrobic system gram-positive identification card for species identification of streptococci. J. Clin. Microbiol. 19:164.
2. Appelbaum, P. C., Jacobs, M. R., Palko, W. M., Frauenhoffer, E. E., and Duffett, A. 1986. Accuracy and reproducibility of the IDS RapID STR system for species identification of streptococci. J. Clin. Microbiol. 23:843.
3. Bridge, P. D. and Sneath, P. H. A. 1982. *Streptococcus gallinarium* sp. nov. and *Streptococcus oralis* sp. nov. Intl. J. Syst. Bacteriol. 32:410.
4. Brown, L. H., Peterson, E. M., and de la Maza, L. M. 1983. Rapid identification of enterococci. J. Clin. Microbiol. 17:369.
5. Chang, G. T. and Ellner, P. D. 1983. Evaluation of slide agglutination methods for identifying group D streptococci. J. Clin. Microbiol. 17:804.
6. Collins, M. D., Jones, D., Farrow, J. A. E., Kilpper-Bälz, R., and Schleifer, K. H. 1984. *Enterococcus avium* nom. rev., comb. nov.; *E. casseliflavus* nom. rev., comb. nov.; *E. durans* nom. rev., comb. nov.; *E. gallinarium* comb. nov.; and *E. malodoratus* sp. nov. Intl. J. Syst. Bacteriol. 34:220.
7. Collins, M. D., Farrow, J. A. E., and Jones, D. 1986. *Enterococcus mundtii* sp. nov. Intl. J. Syst. Bacteriol. 36:8.
8. Colman, G. and Ball, L. C. 1984. Identification of streptococci in a medical laboratory. J. Appl. Bacteriol. 57:1.
9. Deibel, R. H. 1964. The group D streptococci. Bacteriol. Rev. 28:330.
10. Deibel, R. H. and Hartman, P. A. 1984. The enterococci. In "Compendium of Methods for the Microbiological Examination of Foods," 2nd ed., ed. M. L. Speck, p. 405. American Public Health Assn., Washington, D.C.
11. Elliott, S. D., McCarty, M., and Lancefield, R. C. 1977. Teichoic acids of group D streptococci with special reference to strains from pig meningitis (*Streptococcus suis*). J. Exper. Med. 145:490.
12. Etheridge, M. E., Yolken, R. H., and Vonderfecht, S. L. 1988. *Enterococcus hirae* implicated as a cause of diarrhea in suckling rats. J. Clin. Microbiol. 26:1741.
13. Facklam, R. R. and Moody, M. D. 1970. Presumptive identification of group D streptococci: The bile-esculin test. Appl. Microbiol. 20:245.
14. Facklam, R. R. and Carey, R. D. 1985. The streptococci and aerococci. In "Manual of Clinical Microbiology," 4th ed., ed. E. A. Lennette, A. Balows, W. J. Hausler, Jr., and H. J. Shadomy, p. 154. Amer. Soc. for Microbiol., Washington, D.C.
15. Facklam, R. R. and Collins, M. D. 1989. Identification of *Enterococcus* species isolated from human infections by a conventional test scheme. J. Clin. Microbiol. 27:731.
16. Facklam, R. R., Thacker, L. G., Fox, B., and Eriquez, L. 1982. Presumptive identification of streptococci with a new test system. J. Clin. Microbiol. 15:987.
17. Facklam, R., Bosley, G. S., Rhoden, D., Franklin, A. R., Weaver, N., and Schulman, R. 1985. Comparative evaluation of the API 20S and AutoMicrobic gram-positive identification systems for non-beta-hemolytic streptococci and aerococci. J. Clin. Microbiol. 21:535.
18. Farrow, J. A. E. and Collins, M. D. 1985. *Enterococcus hirae*, a new species that includes amino acid assay strain NCDO 1258 and strains causing growth depression in young chickens. Intl. J. Syst. Bacteriol. 35:73.
19. Fertally, S. S. and Facklam, R. 1987. Comparison of physiologic tests used to identify non-beta-hemolytic aerococci, enterococci, and streptococci. J. Clin. Microbiol. 25:1845.
20. Gordon, L. P., Damm, M. A. S., and Anderson, J. D. 1987. Rapid presumptive identification of streptococci directly from blood cultures by serologic tests and the L-pyrrolidonyl-β-naphthylamide reaction. J. Clin. Microbiol. 25:238.
21. Groothuis, D. G., Elzenaar, C. P., and Van Silfhout, A. 1986. An evaluation of the API-20 Strep system (Rapid-Strep system). Syst. Appl. Microbiol. 8:137.
22. Hackney, C. R., Ray, B., and Speck, M. L. 1979. Repair detection procdure for enumeration of fecal coliforms and enterococci from seafoods and marine environments. Appl. Environ. Microbiol. 37:947.
23. Hardie, J. M. 1986. Other streptococci. In "Bergey's Manual of Systematic Bacteriology," ed. P. H. A. Sneath, N. S. Mair, M. E. Sharpe, and J. G. Holt, Vol. 2, p. 1068. Williams and Wilkins, Baltimore, Md.
24. Hartman, P. A., Reinbold, G. W., and Saraswat, D. S. 1966. Indicator organisms—A review. II. The role of enterococci in food poisoning. J. Milk Food Technol. 28:344.
25. Hartman, P. A., Petzel, J. P., and Kaspar, C. W. 1986. New methods for indicator organisms. In "Foodborne Microorganisms and Their Toxins: Developing Methodology," ed. M. D. Pierson and N. J. Stern, p. 175. Marcel Dekker, Inc., New York, N.Y.
26. Hoffman, S. A. and Moellering, R. C. Jr. 1987. The enterococcus: "Putting the bug in our ears." Ann. Intern. Med. 106:757.
27. Jorgensen, J. H., Crawford, S. A., and Alexander, G. A. 1983. Rapid identification of group D streptococci with the API 20S system. J. Clin. Microbiol. 17:1096.
28. Kenner, B. A., Clark, H. F., and Kabler, P. W. 1961. Fecal streptococci. I. Cultivation and enumeration of streptococci in surface waters. Appl. Microbiol. 9:15.
29. Kim, M. J., Weiser, M., Gottschall, S., and Randall, E. L. 1987. Identification of *Streptococcus faecalis* and *Streptococcus faecium* and susceptibility studies with newly developed antimicrobial agents. J. Clin. Microbiol. 25:787.
30. Latham, M. J., Sharpe, M. E., and Weiss, N. 1979. Anaerobic cocci from the bovine alimentary tract, the amino acids of their cell wall peptidoglycans and those of various species of anaerobic *Streptococcus*. J. Appl. Bacteriol. 47:209.
31. Levchak, M. E. and Ellner, P. D. 1982. Identification of group D streptococci by SeroSTAT. J. Clin. Microbiol. 15:58.
32. Littel, K. J. and Hartman, P. A. 1983. Fluorogenic selective and differential medium for isolation of fecal streptococci. Appl. Environ. Microbiol. 45:622.
33. Mundt, J. O. 1982. The ecology of the streptococci. Microb. Ecol. 8:355.
34. Mundt, J. O. 1986. Enterococci. In "Bergey's Manual of Systematic Bacteriology," ed. P. H. A. Sneath, N. S. Mair, M. E. Sharpe, and J. G. Holt, Vol. 2, p. 1063. Williams and Wilkins, Baltimore, Md.
35. Poutrel, B. 1983. Comparative evaluation of commercial latex agglutination and coagglutination reagents for groups B, C, and D mastitis streptococci. Amer. J. Vet. Res. 44:490.
36. Poutrel, B. and Ryniewicz, H. Z. 1984. Evaluation of the API 20 Strep system for species identification of streptococci isolated from bovine mastitis. J. Clin. Microbiol. 19:213.
37. Qadri, S. M. H., Flournoy, D. J., and Qadri, S. G. M. 1987. Sodium chloride-esculin hydrolysis test for rapid identification of enterococci. J. Clin. Microbiol. 25:1107.
38. Rappaport, T., Sawyer, K. P., and Nachamkin, I. 1988. Evaluation of several commercial biochemical and immunologic methods for

rapid identification of gram-positive cocci directly from blood cultures. J. Clin. Microbiol. 26:1335.

39. Schleifer, K. H. and Kilpper-Bälz, R. 1987. Molecular and chemotaxonomic approaches to the classification of streptococci, enterococci and lactococci: A review. Syst. Appl. Microbiol. 10:1.
40. Shlaes, D. M., Toossi, Z., and Patel, A. 1984. Comparison of latex agglutination and immunofluorescence for direct Lancefield grouping of streptococci from blood cultures. J. Clin. Microbiol. 20:195.
41. Ting, W.-T. and Banwart, G. J. 1985. Enumeration of enterococci and aerobic mesophilic plate count in dried soup using three reconstitution methods. J. Food Prot. 48:770.
42. Trepeta, R. W. and Edberg, S. C. 1987. Esculinase (β-glucosidase for the rapid estimation of activity in bacteria utilizing a hydrolyzable substrate, *p*-nitrophenyl-β-D-glucopyranoside. Antonie van Leeuwenhoek 53:273.
43. Tsang, N., Firstenberg-Eden, R., and Lamb, M. 1987. Assessment of the microbial quality of dairy powder using the impedance technique. Dairy Food Sanit. 7:516. (abstract)
44. Vanzo, S. J. and Washington, J. A. II. 1984. Evaluation of a rapid latex agglutination test for identification of group D streptococci. J. Clin. Microbiol. 20:575.
45. You, M. S. and Facklam, R. R. 1986. New test system for identification of *Aerococcus*, *Enterococcus*, and *Streptococcus* species. J. Clin. Microbiol. 24:607.

Chapter 10

Rapid Methods for Detection, Identification, and Enumeration

Phyllis Entis, Daniel Y.C. Fung, Mansel W. Griffiths, Lynn McIntyre, Scott Russell, Anthony N. Sharpe, and Mary Lou Tortorello

10.1 INTRODUCTION

Microbiological testing is both time- and labor-intensive, and there is a need to develop easier, faster methods for detecting, identifying and enumerating microorganisms in foods. The reasons for the difficult nature of microbiological testing are well known. First, the microorganism of interest (MOI) is often present at low levels relative to the mixed community of species in which it exists; i.e., finding and recognizing it are difficult tasks. Second, food is a complex matrix, which causes interference with analytical methods; i.e., some degree of separation or purification of the MOI is required. Most standard microbiological methods employ measures that address both problems. Selective enrichment allows the smaller MOI population to increase in number while keeping in check the growth of other species. The enrichment culture is then plated onto a differential medium to obtain isolated colonies, and in doing so, the MOI is effectively purified from the food matrix. Detection of typical colonies is followed by subculture and a variety of tests for species identification. Enumeration of the MOI is an additional task that is often needed, and in most cases it is even more difficult than detection and identification.

Rapid methods for detection, identification and enumeration of microorganisms offer savings in both time and labor. While many rapid detection and identification methods have been developed and brought to market, the choices available in rapid methods of enumeration are much more limited. The world of rapid methods is expanding and changing, however, and innovations undoubtedly will continue to be made.

This chapter covers selected rapid procedures for detection, identification and enumeration of microorganisms in foods. The various sections of the chapter cover methods based on antibody and nucleic acid recognition; biochemical and enzymatic methods; phage probes; membrane filtration; and impedance. Many of these rapid methodologies have been commercialized and/or have undergone collaborative testing. Details for performing rapid assays for specific microorganisms or groups of microorganisms appear in many other chapters in this book.

Because of its dynamic nature, it is difficult to portray the field of rapid methods with the currency to last until the next edition of the *Compendium*. Updates may be obtained from more frequently published information sources, such as the trade journals, the publications of professional organizations, newsletters and catalogs of commercial suppliers, and the Internet.

10.2 ANTIBODY-BASED METHODS

10.21 Introduction

Antibodies are protein molecules which bind to specific cellular structures or products, generally called antigens. The identification of an antigen that is unique to a particular microbial group (genus, species, strain, etc.) is the first step in the design of an antibody-based method. Antibody-based methods are available for a wide variety of microbial groups, and in fact, have been used for microbial identification for decades. Improved assay design and the use of monoclonal antibodies have significantly reduced the specificity problems which plagued many of the first-generation antibody-based methods. Now these methods are among the most commonly used techniques in food microbiology for rapid detection and identification. Antibodies are also used in affinity separation applications, in which microbial cells or their products are purified from the food matrix for use in subsequent assays. The applications of immunoassay technology to food analysis are broad; however, their most significant impact in food microbiology, both commercially and from a food safety perspective, has been in the detection of foodborne pathogens or their toxins.[7,30,42] A list of antibody-based rapid methods for foodborne pathogen detection has been published.[18] Because so many antibody-based methods have been commercialized, their individual descriptions will not be attempted in this chapter. Many of the subsequent chapters in this book provide details of the methods as they pertain to particular microorganisms.

10.22 Immunofluorescence

Immunofluorescence methods utilize fluorescent antibodies, i.e., antibodies coupled to fluorescent dyes, for specific identification of microorganisms. Fluorescent antibody techniques were among the first rapid methods available in food microbiology.[2] In the traditional immunofluorescence assay, microbial cells from an enrichment culture are fixed to a microscope slide, and the fixed cells are treated with fluorescent antibody. After excess reagent is removed, the slide is observed using a fluorescence microscope. The presence of the specific microorganisms is visualized by fluorescent staining of the cells. Sensitive quantitation of specific bacteria may be achieved by including membrane filtration to concentrate the sample, prior to performing the fluorescent antibody reaction on the surface of the filter.[40] Immunofluorescence is the basis for flow cytometry, a technology used for many years in clinical applications, e.g. recognition of lymphocyte subpopulations based on antigen expression. Flow cytometers have been modified so that detection of cell sizes on the order of microorganisms is possible. The instrumentation involves detection of fluorescent cells moving in a fluid stream past an optical sensor. Cell sorting capability also provides the possibility of isolating the fluorescent cells from a mixture. Applications to the microbiological analysis of foods, including pathogen detection, have been limited but promising.[9,14,26,28,41]

10.23 Latex Agglutination

Antibodies coupled to latex microspheres are the reagents used in latex agglutination assays for microbial identification. When in the presence of reactive microbial cells, the microspheres agglutinate due to formation of molecular cross-bridges, resulting in a visible "clumping" reaction. The microbial cells are generally obtained from an isolated colony, although a centifuged sample of an enrichment culture may also be used. The sensitivity of the latex agglutination test is generally in the range of 10^7–10^8 cells per mL of sample. Latex agglutination assays have been described for several foodborne pathogens, including *Salmonella*,[13] *Shigella*,[29] *Campylobacter*,[22] and *E. coli*.[27] The assays are often used to confirm the identity of colonies from selective agar. It is important to perform appropriate controls each time to ensure that the microspheres have retained reactivity and to detect false-positives caused by spontaneous agglutination of the microspheres.[6]

In reverse passive latex agglutination (RPLA), antibodies attached to latex microspheres react with soluble antigens (such as toxins) in test tubes or wells of microtiter plates. If an antigen-antibody reaction occurs, a diffuse pattern is observed at the bottom of the tube or well. In the absence of a reaction, a ring or "button" is observed.[20] RPLA assays have been commercially developed for detection of enterotoxins, including those produced by *Staphylococcus*,[31] *Bacillus cereus*,[3] *Clostridium perfringens*,[21] *E. coli*,[4] and *Vibrio cholerae*.[1]

10.24 Immunoimmobilization

Antibodies directed against cellular flagella may be used in immobilization assays to detect specific motile microorganisms. Motile microbial cells traverse a semisolid medium until they meet the antiflagellar antibodies diffusing from the opposite direction. A visible arc, which forms at the antibody/microbial cell interface, is the indication that target microorganisms are present.[12,15,36] Non-motile variants of the microorganisms, or those expressing different flagellar antigens, are not detected by the assay.

10.25 Enzyme immunoassay

Enzyme immunoassay (EIA), also known as enzyme-linked immunosorbent assay (ELISA), is a common technique for antibody-based detection of microorganisms or their products. Since its origin in the 1970's, the EIA has become a standard tool in many food microbiology laboratories. Many different EIA formats have been developed.[8,37,39] However, the common element in all of the formats is the involvement of antibody recognition of a target antigen, with some type of linkage to an enzyme-substrate reporter system. The enzyme's activity upon its substrate produces a measurable product, which indicates whether the antigen-antibody reaction took place and, therefore, whether the target antigen was present in the sample. In food microbiology, the "sandwich" assay is a familiar format and is illustrated in Figure 1. In the sandwich assay, antibody that is immobilized onto a solid phase, such as the well of a microtiter plate, captures the specific antigen, e.g. the target microbial cells present in an enrichment culture. After rinsing to remove unbound substances, a second antibody that is conjugated to an enzyme, such as horseradish peroxidase or alkaline phosphatase, is added. Another wash is then performed. After addition of an appropriate substrate for the enzyme, a visible product of the enzyme-substrate reaction is formed, which indicates the presence of the target microbial cells in the original sample. Most often, the reaction product causes a visible color change. The intensity of the colored product formed is directly proportional to the concentration of the target antigen, within a certain range, and may be quantified by optical instrumentation. Variations on the EIA theme employ fluorescent or chemiluminescent reaction signals instead of color development.[19,25,45,47]

An essential element in EIA interpretation is the numerical value, obtained from the optical instrumentation, that is used to discriminate between a positive test and a negative test, i.e., the "cut-off" value. The terms "positive" and "negative" are used to indicate the presence or absence of target antigen, respectively. The cut-off value is often set to be 2–3 times the mean of several negative control results.

Perhaps the most widely used support for the EIA is the microtiter plate, but test tubes, blotting membranes, and even micropipet tips have been used.[38,46] EIAs have become so popular that fully automated systems have been developed for their performance.[11,24]

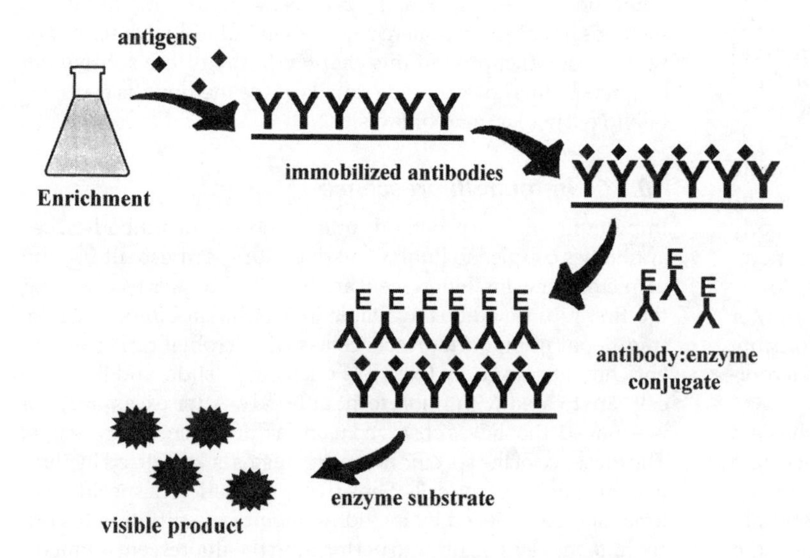

Figure 1. *The "sandwich" enzyme immunoassay (EIA) format*

10.26 Immunoprecipitate Detection

Recent entries to the world of rapid assays are the immunoprecipitate detection devices (Figure 2), in which an aliquot of enrichment culture is added to the sample port on a "paddle," and the sample is wicked through a chromatographic matrix to a reaction area containing antibodies conjugated to some type of precipitable material, such as colored latex particles or colloidal gold. If antigen is present,

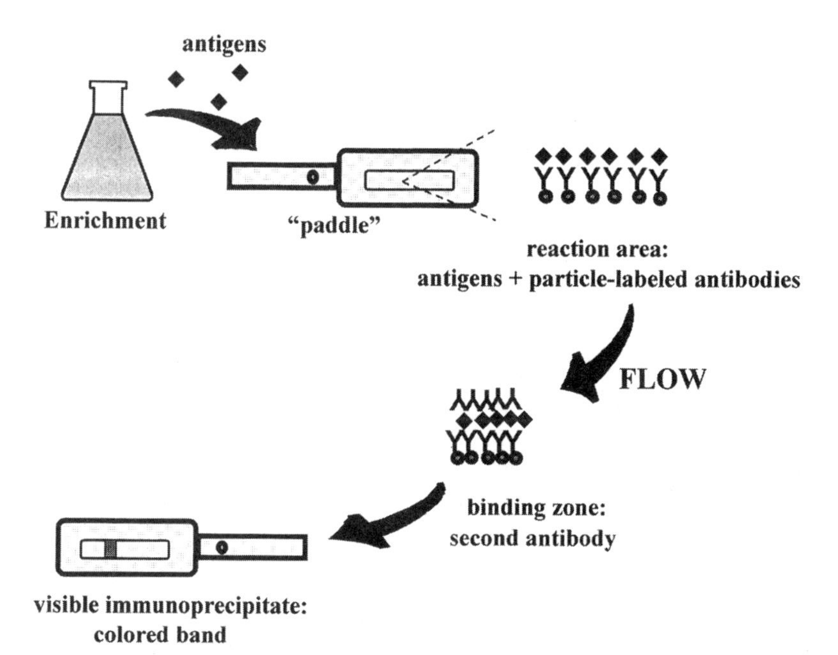

Figure 2. *Generalized design of an immunoprecipitate detection device.*

the antigen-antibody complex flows through the matrix to a binding zone, where it binds to an immobilized second antibody. The localized concentration of the antibody-antigen-antibody complex results in a visible immunoprecipitate, indicated by a colored band on the matrix. A control reaction area also exists on the paddle to provide visible assurance that the detection device is working correctly. A band in both reaction areas is a positive test for the presence of the target antigen. A band only in the control area is a negative test. Immunoprecipitate detection devices have been described for several food pathogens.[5,16,17] The reaction is extremely rapid, with an assay time as short as 5 minutes.

10.27 Immunoaffinity Chromatography and Immunomagnetic Separation

In addition to detection and identification of target antigens, antibodies have been used to achieve separation of antigens from complex matrices by their specific binding characteristics. Immunoaffinity chromatography is a quick way to separate and purify soluble antigens, and immunomagnetic separation allows isolation of particulate antigens, i.e., microbial cells, from a mixture in one step.

Antibodies coupled to a solid inert matrix are used in immunoaffinity chromatography. The sample containing the target antigen is allowed to flow through the immunoaffinity matrix, which is often packed in a column for ease of use. The target antigen is retained in the immunoaffinity matrix while other materials are removed by extensive washing of the column with buffer. The target antigen is then eluted by changing the composition of the buffer appropriately. Immunoaffinity columns have been used for detection of toxins in foods.[34]

Immunomagnetic separation (IMS) is performed by reaction of microbial cells, usually from an enrichment culture, with antibody-coated magnetic microspheres, or beads (Figure 3). The antibody molecules "capture" the microbial cells through their specific binding capability. After a magnet is applied to the side of the test tube, the magnetic beads are drawn to the wall of the tube, effectively concentrating and purifying the captured cells from the mixture. Non-target cells and food debris are washed away, leaving a purified suspension of target cells, which may then be used in subsequent assays. They may be plated onto selective agar to achieve isolation of the microorganism,[35,44] or used as a purification step prior to use in other assays, e.g., immunoassays,[10] nucleic acid assays,[23] or optical methods such as microscopy, flow cytometry and electrochemiluminescence detection.[32,43,48] IMS improves the efficiency of colony isolation and confirmation so that shorter enrichment times may be possible.[33] Applications for the use of these handy reagents will no doubt continue to increase in number in the future.

10.28 References

1. Almeida, R. J., F.W. Hickman-Brenner, E.G. Sowers, N. D. Puhr, J. J. Farmer 3rd, and I. K. Wachsmuth. 1990. Comparison of a latex agglutination assay and an enzyme-linked immunosorbent assay for detecting cholera toxin. J. Clin. Microbiol. 28: 128.
2. Ayres, J. C. 1967. Use of fluorescent antibody for the rapid detection of enteric organisms in egg, poultry and meat products. Food Technol. 21: 115.
3. Beecher, D. J. and A. C. L. Wong. 1994. Identification and analysis of the antigens detected by two commercial Bacillus cereus diarrheal enterotoxin immunoassay kits. Appl.Environ. Microbiol. 60: 4614.
4. Beutin, L., S. Zimmermann and K. Gleier. 1996. Rapid detection and isolation of shiga-like toxin (verocytotoxin)-producing *Escherichia coli* by direct testing of individual enterohemolytic colonies from washed sheep blood agar plates in the VTEC-RPLA assay. J. Clin. Microbiol. 34: 2812.

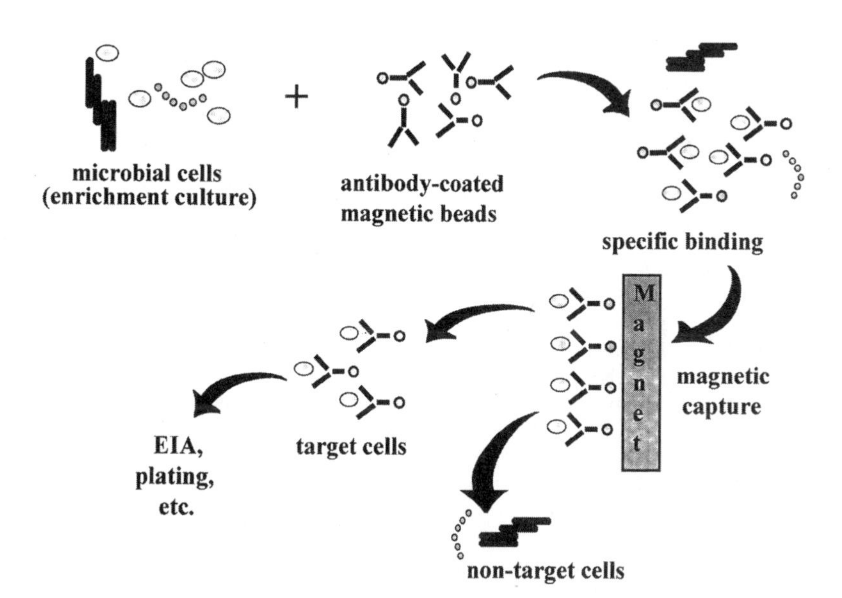

Figure 3. *Immunomagnetic separation of target microbial cells from food matrix and other microorganisms.*

5. Bird, C. B., R. L. Miller and B. M. Miller. 1999. Reveal for *Salmonella* test system. J. AOAC Int. 82: 625.
6. Borczyk, A. A., N. Harnett, M. Lombos and H. Lior. 1990. False-positive identification of *Escherichia coli* O157 by commecial latex agglutination tests. Lancet 336: 946.
7. Candish, A. A. G. 1991. Immunological methods in food microbiology. Food Microbiol. 8: 1.
8. Chan, D. W. and M. T. Perlstein. 1987. Immunoassay: a practical guide. Academic Press, Inc., Orlando, FL.
9. Chang, T. C. and S. H. C. Ding. 1992. Rapid detection of *Staphylococus aureus* in food by flow cytometry. J. Rap. Meth. Automat. Microbiol. 1: 133.
10. Chapman, P. A. and C.A. Siddons. 1996. Evaluation of a commercial enzyme immunoassay (EHEC-Tek) for detecting *Escherichia coli* O157 in beef and beef products. Food Microbiol. 13: 175.
11. Curiale, M. S., V. Gangar and C. Gravens. 1997. VIDAS enzyme-linked fluorescent immunoassay for detection of *Salmonella* in foods: collaborative study. J. AOAC Int. 80:491.
12. D'Aoust, J.-Y. and A. M. Sewell. 1988. Reliability of the immunodiffusion 1-2 Test™ system for detection of *Salmonella* in foods. J. Food Prot. 51: 853.
13. D'Aoust, J.-Y., A. M. Sewell and P. Greco. 1991. Commercial latex agglutination kits for the detection of foodborne *Salmonella.* J. Food Prot. 54: 725.
14. Donnelly, C. W., G. J. Baigent and E. H. Briggs. 1988. Flow cytometry for automated analysis of milk containing *Listeria monocytogenes*. J. Assoc. Off. Anal. Chem. 71: 655-658.
15. Farmer, 3rd, J.J. and B. R. Davis. 1985. H7 antiserum-sorbitol fermentation medium: a single tube screening medium for detecting *Escherichia coli* O157:H7 associated with hemorrhagic colitis. J. Clin. Microbiol. 22: 620.
16. Feldsine, P. T., R. L. Forgey, M. T. Falbo-nelson, and S. L. Brunelle. 1997a. *Escherichia coli* O157:H7 Visual Immunoprecipitate Assay: a comparative validation study. J. AOAC Int. 80: 43.
17. Feldsine, P. T., A. H. Lienau, R. L. Forgey and R. D. Calhoon. 1997b. Visual Immunoprecipitate Assay (VIP) for *Listeria monocytogenes* and related *Listeria* species detection in selected foods: collaborative study. J. AOAC Int. 80: 791.
18. Feng, P. 1998. Rapid methods for detecting foodborne pathogens. In: Bacteriological Analytical Manual, 8th Ed., Rev. A. AOAC International, Gaithersburg,Md. Appendix I.
19. Flowers, R. S., M. J. Klatt, S. L. Keelan, B. Swaminathan, W.D. Gehle, and H. E. Chandonnet. 1989. Fluorescent enzyme immunoassay for rapid screening of *Salmonella* in foods: collaborative study. J. Assoc. Off. Anal. Chem. 72: 318.
20. Gribnau, T. C. J., J. H. W. Leuvering and H. van Hell. 1986. Particle-labeled immunoassays: a review. J. Chromatogr. 376: 175.
21. Harmon, S. M. and D. A. Kautter. 1986. Evaluation of a reversed passive latex agglutination test kit for *Clostridium perfringens* enterotoxin. J. Food Prot. 49: 523.
22. Hazeleger, W. C., R. R. Beumer and F. M. Rombouts. 1992. The use of latex agglutination tests for determining *Campylobacter* species. Lett. Appl. Microbiol. 14: 181.
23. Jinneman, K. C., P. A. Trost, W.E. Hill, S. D. Weagant, J. L. Bryant, C.A.Kaysner and M. M. Wekell. 1995. Comparison of template preparation methods from foods for amplification of *Escherichia coli* O157 shiga-like toxins type I and II DNA by multiplex polymerase chain reaction. J. Food Prot. 58:722.
24. Kerdahi, K. F. and P.F. Istafanos. 1997. Comparative study of colorimetric and fully automated enzyme-linked immunoassay system for rapid screening of *Listeria spp.* in foods. J.AOAC Int. 80:1139.
25. Kricka, L. J. 1998. Prospects for chemiluminescent and bioluminescent immunoassay and nucleic acid assays in food testing and the pharmaceutical industry. J. Biolumin. Chemilumin. 13:189.
26. Laplace-Builhe, C., K. Hahne, W. Hunger, Y. Tirilly and J. L. Drocourt. 1993. Application of flow cytometry to rapid analysis in food and drinks industries. Biol. Cell 78:123.
27. March, S. B. and S. Ratnam. 1989. Latex agglutination test for detection of *Escherichia coli* serotype O157. J. Clin. Microbiol. 27:1675.
28. McClelland, R. G. and A. C. Pinder. 1994. Detection of *Salmonella typhimurium* in dairy products with flow cytometry and monoclonal antibodies. Appl. Environ. Microbiol. 60:4255.
29. Metzler, J. and I. Nachamkin. 1988. Evaluation of a latex agglutination test for the detection of *Salmonella* and *Shigella* by using broth enrichment. J. Clin. Microbiol. 26:2501.
30. Notermans, S. and K. Wernars. 1991. Immunological methods for detection of foodborne pathogens and their toxins. Int. J. Food Microbiol. 12:91.
31. Park, E. E. and R. Szabo. 1986. Evaluation of the reversed passive latex agglutination (RPLA) test kit for detection of staphylococcal enterotoxins A, B, C, and D in food. Can. J. Microbiol. 32:723.
32. Pyle, B.H., S.C. Broadway and G.A. McFeters. 1999. Sensitive detection of *Escherichia coli* O157:H7 in food and water by immunomagnetic separation and solid-phase laser cytometry. Appl. Environ. Microbiol. 65:1966.
33. Restaino, L., E.W. Frampton, R.M. Irbe and D. R. K. Allison. 1997. A 5-h screening and 24-h confirmation procedure for detection of *Escherichia coli* O157:H7 in beef using direct epifluorescent microscopy and immunomagnetic separation. Lett. Appl. Microbiol. 24:401.
34. Scott, P. M. and M. W. Trucksess. 1997. Application of immunoaffinity columns to mycotoxin analysis. J. AOAC Int. 80:941.
35. Skjerve, E., L. M. Rorvik and O. Olsvik. 1990. Detection of *Listeria monocytogenes* in foods by immunomagnetic separation. Appl. Environ. Microbiol. 56: 3478.
36. Swaminathan, B. and R. L. Konger. 1986. Immunoassays for detecting foodborne bacteria and microbial toxins. In: Foodborne Microorganisms and Their Toxins: Developing Methodology. M. D. Pierson and N. J. Stern, eds. Marcel Dekker, New York, NY.
37. Swaminathan, B., J. M. Denner and J.C. Ayres. 1978. Rapid detection of salmonellae in foods by membrane filter-disc immunoimmobilazation technique. J. Food Sci. 43: 1444.
38. Szabo, R., E. Todd, J. Mackenzie, L. Parrington, and A. Armstrong. 1990. Increased sensitivity of the rapid hydrophobic grid membrane filter enzyme-labeled antibody procedure for *Escherichia coli* O157 detection in foods and bovine feces. Appl. Environ. Microbiol. 56: 3546.
39. Tijssen, P. 1985. Practice and theory of immunoassays. Elsevier, Amsterdam, The Netherlands.
40. Tortorello, M. L. and S. M. Gendel. 1993. Fluorescent antibodies applied to direct epifluorescent filter technique for microscopic enumeration of *Escherichia coli* O157:H7 in milk and juice. J. Food Prot. 56:672.
41. Tortorello, M. L., D. S. Stewart and R. B. Raybourne. 1998. Quantitative analysis and isolation of *Escherichia coli* O157:H7 in a food matrix using flow cytometry and cell sorting. FEMS Immunol. Med. Microbiol. 19:267.
42. Van Der Zee, H. and J. H. J. Huis in 'T Veld. 1997. Rapid and alternative screening methods for microbiological analysis. J. AOAC Int. 80: 934.
43. Wang, X. and M. F. Slavik. 1999. Rapid detection of *Salmonella* in chicken washes by immunomagnetic separation and flow cytometry. J. Food Prot. 62: 717.
44. Weagant, S. D., J. L. Bryant and K. G. Jinneman. 1995. An improved rapid technique for isolation of *Escherichia coli* O157:H7 from foods. J. Food Prot. 58:7.
45. Weeks, I., M. L. Sturgess and J. S. Woodhead. 1986. Chemiluminescence immunoassay: an overview. Clin. Sci. 70:403.
46. Woody, J. M., J. A. Stevenson, R. A. Wilson and S. J. Knabel. 1998. Comparison of the Difco EZ Coli rapid detection system and Petrifilm test kit-HEC for detection of *Escherichia coli* O157:H7 in fresh and frozen ground beef. J. Food Prot. 61: 110.
47. Yolken, R. H. and F. J. Leister. 1982. Comparison of fluorescent and colorigenic substrates for enzyme immunoassay. J. Clin. Microbiol. 15: 757.

48. Yu, H. and J. G. Bruno. 1996. Immunomagnetic-electrochemiluminescent detection of *Escherichia coli* O157 and *Salmonella typhimurium* in foods and environmental water samples. Appl. Environ. Microbiol. 62: 587.

10.3 NUCLEIC ACID-BASED METHODS

10.31 Introduction

The sequence of nucleotide bases in the DNA of a microorganism is ultimately its most defining feature. The entire DNA molecule, however, contains too much information to process; therefore, shorter nucleotide sequences that are unique but more readily analyzed are the targets of nucleic acid-based detection methods. Hybridization and the polymerase chain reaction (PCR) are the nucleic acid-based methods that are most commonly used in food microbiology,[33] and they will be described in this section. Both methods depend on the rules of complementary nucleotide base pairing, i.e: adenine binds to thymine, and cytosine binds to guanine. In hybridization, the target sequence is recognized by a complementary nucleotide probe that is labeled in some way for direct detection of the sequence. In PCR, two nucleotide primers bind to complementary regions that flank the target sequence, which is then enzymatically amplified for detection. In either method, the DNA molecule must first be denatured to separate the double-stranded helix into single strands so that the probe or primers can bind to the appropriate sequence.

Whereas hybridization requires approximately 10^5–10^6 copies of the target DNA for detection, PCR is much more sensitive, producing millions of copies of a single gene target in a few hours.[21] Hybridization can be quantitative, if performed as a colony hybridization using the same format as agar plate counting. Quantitation is possible but difficult with PCR, which is most commonly used as a qualitative or presence/absence assay. Compared to antibody-based methods, there are relatively few commercialized versions of nucleic acid-based methods, which will be described in this section.

10.32 Hybridization

Hybridization is the process by which a DNA probe binds to a complementary region of single-stranded DNA or RNA. A DNA probe is usually a 20–2000 nucleotide base sequence of DNA that is unique to a particular microbial group. Probes can identify specific genes, e.g. toxin genes; regions of ribosomal RNA (rRNA); or DNA sequences that have unknown function but are common to particular microbial groups.

DNA probes may be prepared synthetically or biologically. Short oligonucleotide probes (20–30 bases in length) are easily prepared with the aid of a DNA synthesizer, and the proliferation of commercial suppliers offering such services has made oligonucleotide probes both widely available and affordable. Biological preparation of probes involves cloning of a nucleotide sequence into a plasmid that replicates many copies of the sequence. The entire plasmid may be used as a probe, but better specificity is likely if the nucleotide sequence is isolated by restriction endonuclease digestion and subsequent gel electrophoresis. In either case the probe sequence must be labeled so that it provides a signal that can be detected in the hybridization assay.

The radioactive isotope ^{32}P was the first probe label to be used. Although highly sensitive, ^{32}P as a probe label has all the disadvantages of handling and disposal of radioactive material, as well as a short shelf-life. There has been a trend toward development of non-isotopic methods of probe detection, such as fluorescent labels. Non-isotopic methods often involve enzymes, e.g., horseradish peroxidase or alkaline phosphatase, which can act upon substrates, producing color or chemiluminescence. Most commonly the enzyme is conjugated to a protein that binds to a specific ligand on the DNA probe to form an enzyme-binding protein-ligand-DNA complex. Many variations on the theme have been applied. A commonly used ligand-binding protein pair is biotin-streptavidin. Antibodies have also been used as binding proteins for ligands on the DNA probe.

There are two basic hybridization formats: solid-phase and liquid-phase. The phase refers to the location of the target sequence during hybridization. In solid-phase hybridization, the target sequence is bound to a solid support, e.g. a membrane filter. In liquid-phase hybridization, the target sequence is suspended in liquid, and after the hybridization reaction has occurred, it is removed from solution for detection.

Solid-phase hybridization may be performed as colony blots or dot-blots. Colony blotting involves transfer of individual colonies that have grown up on agar plates to a membrane filter. In dot-blotting, microbial growth from liquid or solid culture is transferred, e.g. with a loop or other device, as a "dot" onto a localized area of a membrane. In either case, the membrane is then treated with reagents that lyse the cells and denature the DNA. The released single strands of target DNA stick in place to the membrane, and then the probe is added. After the probe hybridizes with the matching DNA sequence on the membrane, unhybridized DNA is washed away. The composition of the hybridization buffer, temperature, and the conditions of washing are all critical to the specificity of the reaction and may be varied to affect the accuracy of the probe binding. After hybridization, an appropriate method is used for detection of the probe-target hybrid. If the probe is labeled with ^{32}P, a piece of X-ray film is placed on the membrane. The radioactive label exposes the film and develops spots where the probe hybridized to the DNA. If a non-isotopic probe is used, the reagents are added to develop visible spots directly on the membrane.

Colony blots may be used to enumerate target microbes in a food, if the food is plated directly. After development of the probe signal on the membrane, the reactive spots are counted and may be related directly to colony forming units of the target microbe. Dot blots are not quantitative. A generalized illustration of a colony hybridization is shown in Figure 4. Detailed procedures have been published for performing colony hybridizations for major foodborne pathogens, including *E. coli, Shigella, Listeria, Staphylococcus, Vibrio,* and *Yersinia*.[20] Hydrophobic grid membrane filtration (section 10.5) has been combined with probe hybridization to provide improved identification of *Listeria* and *E. coli* colonies.[35,39]

In liquid-phase hybridization, both probe and target DNA are suspended in solution, where the hybridization proceeds. Following hybridization, the target DNA with its hybridized probe is separated from the solution through capture of the hybrids on a solid support. Capture may be effected on hydoxylapatite, because of the relatively high affinity to hydroxylapatite that double-stranded DNA shows compared to single-stranded DNA. A streptavidin-coated matrix may be used to capture DNA that has been labeled with biotin. A solid support coated with polydeoxythymidylic acid (poly-dT) may be used in conjunction with a "capture probe," which contains a polydeoxyadenylic acid (poly-dA) tail at the end of a sequence complementary to the target nucleic acid. The probe can bind to the target nucleic acid and to the poly-dT by complementary base pairing, resulting in removal of the probe with its hybridized target nucleic acid from solution.

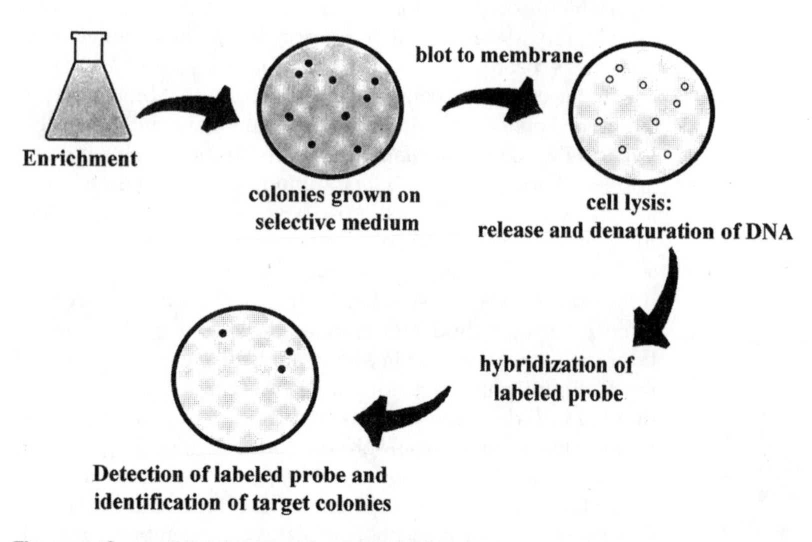

Figure 4. *Generalized diagram of colony hybridization.*

Liquid-phase hybridization has been commercialized in the GENE-TRAK® Assays, which are dual-probe systems based on 16S ribosomal RNA detection in enrichment culture samples.[13,14] The target ribosomal RNA hybridizes on one end to a poly-dA-tailed probe and on the other end to a fluorescein-labeled detector probe. The hybrids are captured on a poly-dT-coated "dipstick." After binding of an enzyme-conjugated anti-fluorescein antibody, detection is accomplished colorimetrically. GENE-TRAK® Assays have been developed for various food pathogens, including *Salmonella, Listeria, E. coli, Staphylococcus, Campylobacter* and *Yersinia.*[18] Another commercial assay based on liquid-phase hybridization is the AccuPROBE® test, which uses a chemiluminescence detection system. Aimed primarily at the clinical market, the AccuPROBE® test is designed for colony identification, and is not useful for screening of enrichment cultures.[34] *Campylobacter, Staphylococcus* and *Listeria* are the food pathogen species represented in the AccuPROBE® line.[18]

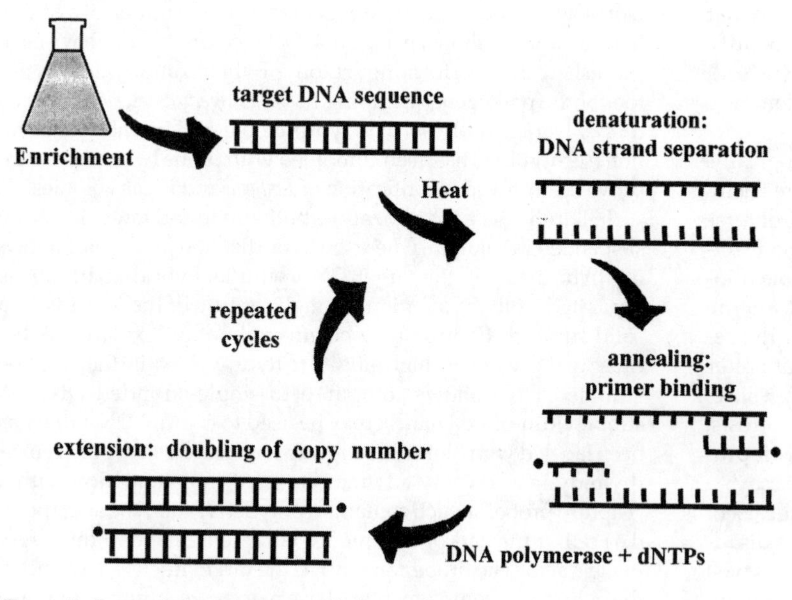

Figure 5. *Steps of the polymerase chain reaction*

10.33 PCR

Since its development in 1983,[30] PCR has become an essential tool in molecular biology, and its use is increasing in food microbiology. PCR uses enzymatic amplification to produce, within a few hours, millions of copies of a target nucleic acid sequence. The enzyme DNA polymerase copies the target sequence, which is refered to as the template. Oligonucleotide primers are needed to provide starting points for the DNA polymerase, which adds nucleotide bases to the primers in the correct order according to the sequence of the template. The primers are designed to flank the target sequence; i.e., they define the ends or boundaries of the target sequence, and they are oriented in such a way that the DNA polymerase copies the template in the correct direction.

The PCR procedure consists of repeating cycles of alternating high and low temperatures that bring about the amplification. Thermostable DNA polymerase enzymes are used because of their ability to withstand the temperature fluctuations. Each cycle consists of three steps: denaturation, annealing and extension (Figure 5). Denaturation at high temperature causes the double stranded target DNA to separate into single strands. In the annealing step, the temperature is lowered, and the primers bind to their complementary sequences on opposite strands of the target DNA. The temperature is raised again, and DNA polymerase adds nucleotide bases onto the ends of both primers (extension), building two new polynucleotide chains according to the sequences within the target region. Thus, the original target is copied. The newly made copies then serve as templates for additional rounds. Because it is doubled in each successive cycle, the amount of target DNA is increased exponentially. The cycle is repeated until an amount of product sufficient for detection is made. Detailed protocols for PCR may be found in Chapter 11 of this book, where its use in several microbial typing methods is described.

Gel electrophoresis is typically used to detect the amplified product. Because the primers bind to specific sites surrounding the target DNA, the size of the amplified product can be anticipated and detected as a band of known size on the gel. Proof of the band's identity may be obtained by blotting of the gel and DNA probe hybridization to the band.

The appeal of PCR technology lies in the amplification of small quantities of DNA to detectable levels in a short amount of time. Theoretically, a single gene target can be amplified a million-fold for detection in a few hours. Thus, direct detection, i.e., without the need for a lengthy enrichment step, is possible. In practice, however, direct detection of a single gene target is difficult to achieve. The small sample volumes used in PCR (microliter-scale) limit the sensitivity of the procedure for examining foods directly. In addition, inhibitory substances that interfere with the PCR are present in many foods. Thus, a preliminary enrichment culture step to increase target levels is common in PCR detection protocols. Enrichment culture also provides some assurance that the DNA detected is derived from viable cells. The superior sensitivity of the

PCR compared to conventional culture methods may allow shorter enrichment times, however.[19,40]

Sample preparation procedures to remove inhibitory substances are critical to the success of the PCR, whether it is to be used as a direct assay or coupled with an enrichment culture step. PCR inhibitors may be removed to some extent by optimizing DNA extraction protocols or including purification steps such as immunoprecipitation or membrane filtration.[6,15,23,27,32] DNA polymerases from various sources show different degrees of susceptibility to interference from inhibitors.[1]

The intricacies of preparing materials for performing the PCR have been eased by the availability of commercialized systems that supply many of the reagents in ready-to-use form. The BAX™ systems supply tableted PCR reagents in prepackaged tubes, buffers and other materials for detection of the amplified product by gel electrophoresis using pre-cast gels. BAX™ systems have been used for screening food samples for *Salmonella*,[3] *E. coli*,[24] and *Listeria*.[38] Other commercialized systems have incorporated alternatives to gel electrophoresis detection. Probelia™ utilizes colorimetric detection of amplified product in microtiter plate wells. After PCR, the amplified product is hybridized in solution to a capture probe linked to the microplate well surface and to an enzyme-labeled detector probe. Detection is then achieved by adding the enzyme substrate and colorimetric development. Probelia™ has been used for *Salmonella* detection,[12] and a system for *Listeria* is also available. TaqMan™ is based on fluorescence detection of the PCR product, with the amplification and detection performed in a single tube. The unique TaqMan™ probe is labeled with two different fluorescent dyes, which function as reporter and quencher dyes. If no product is present, the fluorescent signal of the quencher dye is detected. As amplification proceeds, the probe hybridizes to the product, and the quenching effect is relieved; thus, the fluorescence of the reporter dye is detected. The ratio of reporter to quencher dye is measured in an automated system and indicates whether the PCR product was present. TaqMan™ detection of *Salmonella*,[25] *Listeria*[5] and *E. coli*[31] has been tested.

PCR procedures have been described for detecting many of the bacterial foodborne pathogens in food matrices, including *Salmonella*,[7,11,28] *E. coli*,[29] *Clostridium perfringens*,[17,43] *C. botulinum*,[2,16] *Listeria monocytogenes*,[4] *Staphylococcus aureus*,[44] *Shigella*,[26] and *Campylobacter*.[42,45] Multiplex PCR protocols, which are designed to test for more than one gene target simultaneously, can be used to detect the presence of more than one microoganism or to identify a microorganism based on multiple gene markers.[8,9,23,37]

The detection of foodborne viruses is an area in which PCR may have its most significant impact. Because of the extreme difficulty or inability to culture many viruses in the laboratory, PCR provides an opportunity for sensitive detection of viral nucleic acid in food. Sample preparation, as discussed above, is crucial, because an enrichment step for increasing the amount of target nucleic acid is not feasible. Procedures have addressed the need for both purification and concentration of virus particles from food matrices.[22] The viruses of importance in foodborne illness have RNA, rather than DNA, as genomic material.[10] For their detection, RT-PCR has been applied, a procedure which uses the enzyme reverse transcriptase (RT) to make a DNA copy of the RNA genome before proceeding with the standard PCR protocol. Combining purification/concentration procedures with RT-PCR has provided improved detection of viral disease agents in foods.[36,41]

10.34 References

1. Abu Al-Soud, W. and P. Radstrom. 1998. Capacity of nine thermostable DNA polymerase to mediate DNA amplification in the presence of PCR-inhibiting samples. Appl. Environ. Microbiol. 64: 3748.
2. Aranda, E., M. M. Rodriguez, M. A. Asensio and J. J. Cordoba. 1997. Detection of *Clostridium botulinum* types A, B, E and F in foods by PCR and DNA probe. Lett. Appl. Microbiol. 25: 186.
3. Bailey, J. S. 1998. Detection of *Salmonella* cells within 24 to 26 hours in poultry samples with the polymerase chain reaction BAX system. J. Food Prot. 61:792.
4. Bansal, N. S. 1996. Development of a polymerase chain reaction assay for the detection of *Listeria monocytogenes* in foods. Lett. Appl. Microbiol. 22:353.
5. Bassler, H. A., S. J. Flood, K. J. Livak, J. Marmaro, R. Knorr and C. A. Batt. 1995. Use of a fluorogenic probe in a PCR-based assay for the detection of *Listeria monocytogenes*. Appl. Environ. Microbiol. 61:3724.
6. Bej, A. K., M. H. Mahbubani, J. L. Dicesare, and R. M. Atlas. 1991. Polymerase chain reaction-gene probe detection of microorganisms by using filter-concentrated samples. Appl. Environ. Microbiol. 57:3529.
7. Bej, A.K., M. H. Mahbubani, M. J. Boyce and R. M.Atlas. 1994. Detection of *Salmonella spp.* in oysters by PCR. Appl. Environ. Microbiol. 60:368.
8. Brasher, C. W., A. DePaola, D. D. Jones and A.K. Bej. 1998. Detection of microbial pathogens in shellfish with multiplex PCR. Curr. Microbiol. 37: 101.
9. Bubert, A., I. Hein, M. Rauch, A. Lehner, B. Yoon, W. Goebel and M. Wagner. 1999. Detection and differentitation of *Listeria spp.* by a single reaction based on multiplex PCR. Appl. Environ. Microbiol. 65: 4688.
10. Cliver, D. O. 1997. Foodborne viruses. In: Food Microbiology, Fundmentals and Frontiers. M. P. Doyle, L. R. Beuchat and T.J. Montville, eds. ASM Press, Washington, D.C., Chapter 24.
11. Cohen, H. J., S. M. Mechanda, and W. Lin. 1996. PCR amplification of the *fimA* gene sequence of *Salmonella typhimurium*, a specific method for detection of *Salmonella spp.* Appl. Environ. Microbiol. 62: 4303.
12. Coquard, D., A. Exinger and J.-M. Jeltsch. 1999. Routine detection of *Salmonella* species in water: comparative evaluation of the ISO and PROBELIA™ polymerase chain reaction methods. J. AOAC Int. 82:871.
13. Curiale, M. S., T. Sons, L. Fanning, W. Lepper, D. McIver, S. Garramone and M. Mozola. 1994. Deoxyribonucleic acid hybridization method for the detection of *Listeria* in dairy products, seafoods and meats: collaborative study. J. AOAC Int. 77: 602.
14. D'Aoust, J.-Y., A. M. Sewell, P. Greco, M. A. Mozola and R. E. Colvin. 1995. Performance assessment of the GENE-TRAK® colorimetric probe assay for the detection of foodborne *Salmonella spp.* J. Food Prot. 58: 1069.
15. Dickinson, J. H., R. G. Kroll and K. A. Grant. 1995. The direct application of the polymerase chain reaction to DNA extracted from foods. Lett. Appl. Microbiol. 20: 212.
16. Fach, P., M. Gibert, R. Griffais, J. P.Guillou and M. R. Popoff. 1995. PCR and gene probe identification of botulinum neurotoxin A-, B-, E-, F-, and G-producing *Clostridium spp.* and evaluation in food samples. Appl. Environ. Microbiol. 61:389.
17. Fach, P. and M. R. Popoff. 1997. Detection of enterotoxigenic *Clostridium perfringens* in food and fecal samples with a duplex PCR and the slide latex agglutination test. Appl. Environ. Microbiol. 63: 4232.
18. Feng, P. 1998. Rapid methods for detecting foodborne pathogens. In: Bacteriological Analytical Manual, 8th Ed., Rev. A. AOAC International, Gaithersburg, Md. Appendix I.
19. Gouws, P. A., M. Visser and V. S. Brozel. 1998. A polymerase chain reaction procedure for the detection of *Salmonella spp.* within 24 hours. J. Food Prot. 61: 1039.

20. Hill, W. E., A. R. Datta, P. Feng, K. A. Lampel and W. L. Payne. 1998. Identification of foodborne bacterial pathogens by gene probes. In: Bacteriological Analytical Manual, 8th Ed., Rev. A. AOAC International, Gaithersburg,Md., Chapter 24.
21. Hill, W. E. and S. P. Keasler. 1991. Identification of foodborne pathogens by nucleic acid hybridization. Int. J. Food Microbiol. 12: 67.
22. Jaykus, L.-A., R. de Leon and M. D.Sobsey. 1996. A virion concentration method for detection of human enteric viruses in oysters by PCR and oligoprobe hybridization. Appl. Environ. Microbiol. 62: 2074.
23. Jinneman, K. C., P. A. Trost, W. E. Hill, S. D. Weagant, J. L.Bryant, C. A. Kaysner and M. M. Wekell. 1995. Comparison of template preparation methods from foods for amplification of *Escherichia coli* O157 shiga-like toxins type I and II DNA by multiplex polymerase chain reaction. J. Food Prot. 58: 722.
24. Johnson, J. L., C. L. Brooke and S. J. Fritschel. 1998. Comparison of the BAX for screening/*E. coli* O157:H7 method with conventional methods for detection of extremely low levels of *Escherichia coli* O157:H7 in ground beef. Appl. Environ. Microbiol. 64: 4390.
25. Kimura, B., S. Kawasaki, T. Fujii, J. Kusunoki, T. Itoh and S. J. Flood. 1999. Evaluation of TaqMan PCR assay for detecting *Salmonella* in raw meat and shrimp. J. Food Prot. 62: 329.
26. Lampel, K. A., J.A. Jagow, M. Trucksess and W. E. Hill. 1990. Polymerase chain reaction for detection of invasive *Shigella flexneri* in food. Appl. Environ. Microbiol. 56: 1536.
27. Lantz, P. G., F. Tjerneld, B. Hahn-Hagerdal, and P. Radstrom. 1996. Use of aqueous two-phase systems in sample preparation for polymerase chain reaction-based detection of microorganisms. J. Chromatogr. B. Biomed. Appl. 680: 165.
28. Lin, C. K. and H. Y. Tsen. 1996. Use of two 16S DNA targeted oligonucleotides as PCR primers for the specific detection of *Salmonella* in foods. J. Appl. Bacteriol. 80: 659.
29. Maurer, J. J., D. Schmidt, P. Petrosko, S. Sanches, L. Bolton and M. D. Lee. 1999. Development of primers to O-antigen biosynthesis genes for specific detection of *Escherichia coli* O157 by PCR. Appl. Environ.Microbiol. 65: 2954.
30. Mullis, K.B. 1990. The unusual origin of the polymerase chain reaction. Sci.Am. 240: 56.
31. Oberst, R. D., M. P. Hays, L. K. Bohra, R. K. Phebus, C. T. Yamashiro, C. Paszko-Kolva, S. J. A. Flood, J. M. Sargeant, and J. R. Gillespie. 1998. PCR-based DNA amplification and presumptive detection of *Escherichia coli* O157:H7 with an internal fluorogenic probe and the 5′ Nuclease (TaqMan) assay. Appl. Environ. Microbiol. 64:3389.
32. Ogunjimi, A. A. and P. V. Choudary. 1999. Adsorption of endogenous polyphenols relieves the inhibition by fruit juices and fresh produce of immuno-PCR detection of *Escherichia coli* O157:H7. FEMS Immunol. Med. Microbiol. 23: 213.
33. Olsen, J. E. , S. Aabo, W. Hill, S. Notermans, K. Wernars, P. E. Granum, T. Popovic, H. N. Rasmussen and O. Olsvik. 1995. Probes and polymerase chain reaction for detection of foodborne bacterial pathogens. Int. J. Food Microbiol. 28: 1-78.
34. Partis, L., K. Newton, J. Murby and R. J. Wells. 1994. Inhibitory effects of enrichment media on the Accuprobe test for *Listeria monocytogenes*. Appl. Environ. Microbiol. 60:1693.
35. Peterkin, P. I., E. S. Idziak and A. N. Sharpe. 1991. Detection of *Listeria monocytogenes* by direct colony hybridization on hydrophobic grid membrane filters by using a chromogen-labeled DNA probe. Appl. Environ. Microbiol. 57: 586.
36. Shieh, Y.-S. C., K. R. Calci and R. S. Baric. 1999. A method to detect low levels of enteric viruses in contaminated oysters. Appl. Environ. Microbiol. 65: 4709.
37. Soumet, C., g. Ermel, N. Rose, V. Rose, P. Drouin, G. Salvat and P. Colin. 1999. Evaluation of a multiplex PCR assay for simultaneous identification of *Salmonella sp., Salmonella enteritidis* and *Salmonella typhimurium* from environmental swabs of poultry houses. Lett. Appl. Microbiol. 28: 113.
38. Stewart, D. and S. M. Gendel. 1998. Specificity of the BAX polymerase chain reaction system for detection of the foodborne pathogen *Listeria monocytogenes*. J. AOAC Int. 81: 817.
39. Todd, E. C. D., R. A. Szabo, J. M. MacKenzie, A. Martin, K. Rahn, C. Gyles, A. Gao, D. Alves and A. J. Yee. 1999. Application of a DNA hybridization-hydrophobic grid membrane filter method for detection and isolation of verotoxigenic *Escherichia coli*. Appl. Environ. Microbiol. 65: 4775.
40. Tortorello, M. L., K. F. Reineke, D. S. Stewart and R. B. Raybourne. 1998. Comparison of methods for determining the presence of *Escherichia coli* O157:H7 in apple juice. J. Food Prot. 61: 1425.
41. Traore, O., C. Arnal, B. Mignotte, A. Maul, H. Laveran, S. Billaudel and L. Schwartzbrod. 1998. Reverse transcriptase PCR detection of astrovirus, hepatitis A virus and poliovirus in experimentally contaminated mussels: comparison of several extraction and concentration methods. Appl. Environ. Microbiol. 64: 3118.
42. Waage, A. S., T. Vardund, V. Lund and G. Kapperud. 1999. Detection of small numbers of *Campylobacter jejuni* and *Campylobacter coli* cells in environmental water, sewage and food samples by a seminested PCR assay. Appl. Environ. Microbiol. 65: 1636.
43. Wang, R. F., W. W. Cao, W. Franklin, W. Campbell and C. E. Cerniglia. 1994. A 16S rDNA-based PCR method for rapid and specific detection of *Clostridium perfringens* in food. Mol. Cell. Probes 8: 131.
44. Wilson, I. G., J. E. Cooper and A. Gilmour. 1991. Detection of enterotoxigenic *Staphylococcus aureus* in dried skimmed milk: use of the polymerase chain reaction for amplification and detection of staphylococcal enterotoxin genes *entB* and *entC1* and the thermonuclease gene *nuc*. Appl.Environ. Microbiol. 57: 1793.
45. Winters, D. K., A. E. O'Leary and M. F. Slavik. 1997. Rapid PCR with nested primers for direct detecton of *Campylobacter jejuni* in chicken washes. Mol. Cell. Probes. 11: 267.

10.4 BIOCHEMICAL AND ENZYMATIC METHODS

10.41 Introduction

Biochemical and enzymatic methods are among the oldest methods in microbiological analysis. They have been used since the dawn of bacteriology to study the metabolic activities of microorganisms for identification purposes. These methods eventually became standard tests in bacteriology laboratories. As time passed, improvements were made to obtain results more rapidly or easily. Fung[5] made a historical account of the development of rapid bacteriological procedures, including miniaturized and automated systems. A comprehensive treatment of rapid testing is provided in the *Handbook for Rapid Methods and Automation in Microbiology*.[6]

10.42 Miniaturized Microbiological Methods

Hartman[15] compiled over 1,200 citations of rapid miniaturized techniques for microbiological testing in his classical book *Miniaturization of Microbiological Methods*. A significant advance in miniaturization came with the development of the microtiter plate format, which was introduced in the 1960s. Ninety-six reactions could be performed in the wells of the microtiter plate, each holding approximately 0.2 mL of media or reagents. The reduced volume compared to conventional test tubes, which held 5–10 mL, and the multiple tests which could be performed in a single plate provided a great savings in time, materials, space, and labor.

Miniaturized methods have been used to study large numbers of isolates from foods,[7,17,18] including yeasts and molds.[3,8,13,14,19,20] Useful microbiological media were developed using the miniaturized format,[11,23] which is also ideal for doing challenge testing of chemicals against many types of microorganisms.

10.43 Commercial Miniaturized Diagnostic Kits ("Minikits")

A host of commercial diagnostic kits were marketed beginning in the early 1970s, as a result of the development of miniaturized methods. Some of the systems survived the test of time. Many fell by the wayside due to various reasons, e.g. poor performance or design. The following is a synopsis of currently available commercial minikits. A recent review on this subject was made by Russell et al. [21]

10.431 Agar Media Minikits

10.4311 Enterotube (Roche Diagnostics, Nutley, NJ). The only agar-based diagnostic kit system still widely used is the Enterotube II, a self-contained plastic tube containing twelve different conventional media in separate compartments. The tube is inoculated with a wire threaded through each compartment along the length of the tube. The unit permits assay of 15 standard biochemical tests from a single bacterial colony. Reagents are added into the Indole test and Voges-Proskauer test compartments before recording of the color reactions and gas formation. Table 1 describes the color reactions of the tests, which are recorded on data sheets (Figure 6). A five-digit number is generated from the test results and matched to the manufacturer's code book, which provides the identification of the isolate.

Similar color reaction charts, data recording sheets and code book formats are available for the other minikits described in this section. Some systems utilize a seven- or ten-digit code number. The code book of one system cannot be used to interpret data from another minikit. Sometimes an isolate cannot be identified from the code generated. An impure or insufficient inoculum may be the cause, or perhaps an incorrectly prepared diluent. After exhaustive testing, if the isolate is still not identifiable, a culture may be sent to the manufacturer for further testing, and the results may be added to the data base. Nevertheless, it is very unusual to "discover" a new organism in this way.

The Enterotube II was developed to identify species of the Enterobacteriaceae only. A similar unit called Oxi/Ferm Tube was designed for gram-negative nonfermenters.

The accuracy of the Enterotube II was 94% compared to the conventional method.[20] A 90–95% accuracy of a minikit against the standard conventional methods is considered good to excellent. When accuracy drops to 85%, the system is considered marginally acceptable, and any value below that is not acceptable.

Advantages of the Enterotube II include speed and ease of inoculation. There is no need to prepare an inoculum suspension, and a single colony can be used for identification. Disadvantages include difficulty in stacking of the units in the incubator, short shelf-life and a limited data base for food isolates.

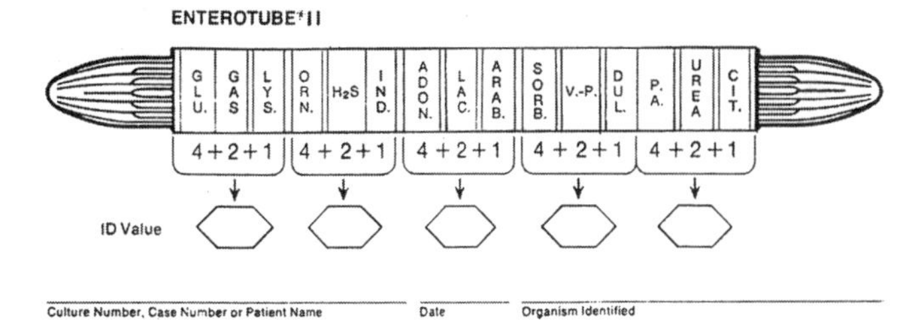

Figure 6. Enterotube II data sheet

10.432 Dehydrated Media Minikits

A variety of minikits utilize dehydrated media, which become hydrated upon inoculation. Longer shelf-life, minimal contamination, and lighter weight of the units are major advantages of dehydrated versus agar media kits.

10.4321 API (bioMerieux Vitek, Inc., Hazelwood, MO). The API is probably the most widely used diagnostic system in the world for the *Enterobacteriaceae*. Originally, the data base was very strong for clinical isolates, but in recent years, it has been expanded to cover food, environmental and industrial isolates as well. A flow sheet on the operation of the API system is shown in Figure 7.

The API 20E has 20 small elongated wells housed in a plastic panel, each containing a dehydrated medium, and is designed to perform 23 miniaturized biochemical tests from a pure agar plate culture. Colony growth is mixed with buffer to make a suspension of cells, which is used to inoculate the wells of the API strip using a small pipet. The unit is then placed into a plastic chamber, into which water is added to avoid moisture loss during incubation. After incubation for 4 hr or 18 hr at 35–37°C, the unit is removed from the incubator, and reagents are added into certain wells. After the color reactions are recorded, a 7-digit code is generated. The unit can also be read and interpreted by a computerized system. The API has been tested extensively, with reported accuracies ranging from 90.2%[12] to 93%.[18]

Advantages of the API include an excellent data base, a proven record of testing over the past 25 years, and long shelf-life. Disadvantages include a difficult and time-consuming inoculation into the units, especially if many are to be prepared simultaneously. Handling and stacking of the flexible trays and lids are inconvenient. Reading and interpretation of reactions require experience and training.

The following systems are available in formats similar to the API 20E:

API NFT, for gram negative non-*Enterobacteriaceae*

API CAMP, for *Campylobacter* spp.

API Staph-IDENT, for staphylococci and micrococci

API 20A, for anaerobes

Table 1. Enterotube II Color Reactions

Abbreviation	Test	Positive Reaction	Negative Reaction
GLU	Glucose utilization	Yellow	Red
Gas	Gas production	Wax lifted	Wax not lifted
LYS	Lysine decarboxylase	Purple	Yellow
ORN	Ornithine decarboxylase	Purple	Yellow
H_2S	H_2S production	Black	Beige
IND	Indole formation	Pink-Red	Colorless
ADON	Adonitol fermentation	Yellow	Red
LAC	Lactose fermentation	Yellow	Red
ARAB	Arabinose fermentation	Yellow	Red
SORB	Sorbitol fermentation	Yellow	Red
VP	Voges-Proskauer	Red in 20 minutes	No red color
DUL	Dulcitol fermentation	Yellow	Green
PA	Phenylalanine deaminase	Black to smoky gray	Green
UREA	Urease	Red-Purple	Yellow
CIT	Citrate utilization	Deep blue	Green

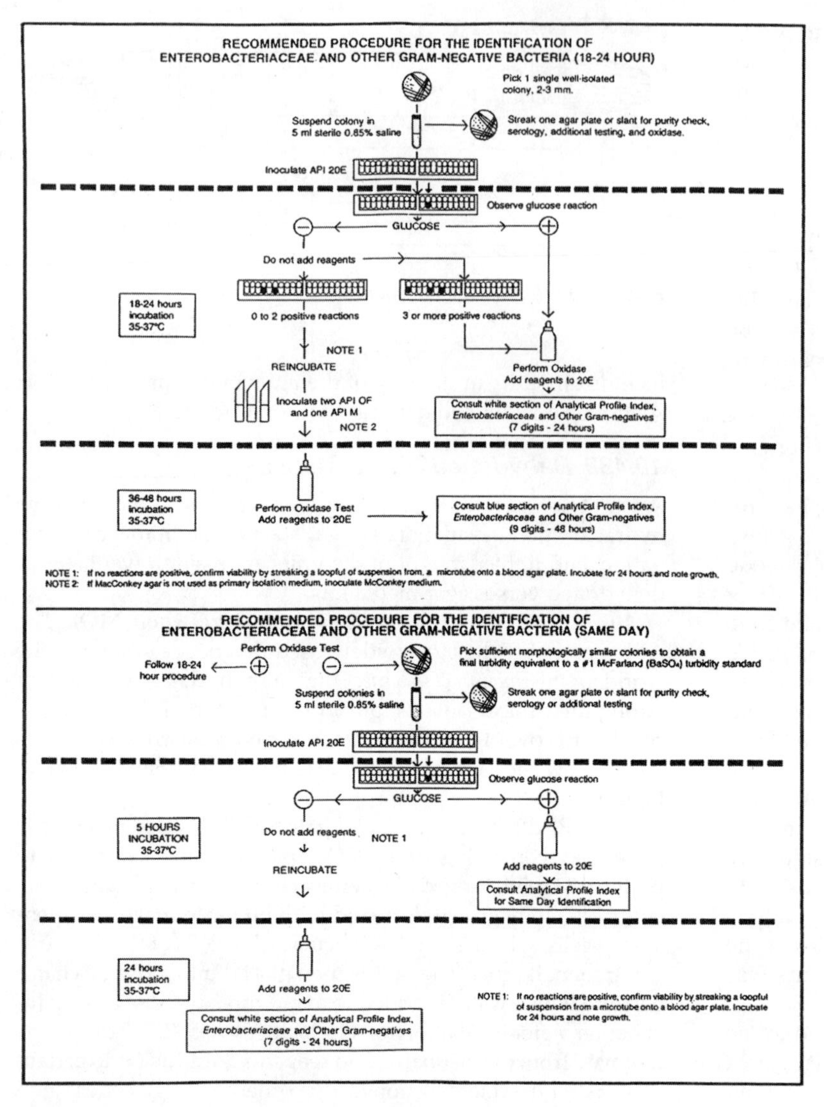

Figure 7. *Procedure for isolate identification using API 20E system.*

API 50 CH, for *Lactobacillus*
API 20C, for yeasts
Rapid STREP, for streptococci
Rapid CORYN, for *Corynebacterium*

10.4322 ATB (bioMerieux Vitek, Inc., Hazelwood, MO). The ATB is a system for testing assimilation of 32 different substrates by a microbial isolate. Cells are suspended in liquid and introduced into the unit. After incubation for 4–24 hr, depending on the culture, the tests are read manually or automatically. Test strips are available for anaerobes, staphylococci, micrococci, yeast, *Enterobacteriaceae*, streptococci, and gram negative bacilli.

10.4323 Minitek (Becton Dickinson Microbiology, Franklin Lakes, NJ). The Minitek system consists of a plastic unit of ten wells into which are placed paper discs impregnated with different substrates. A suspension of bacterial cells is injected into the wells by a calibrated "gun." Mineral oil is overlayed in some anaerobe tests, and the units are incubated at 35–37°C for 18 hr before reactions are read. A code is generated and the identification is made using the Minitek code book.

Thirty-six different substrates are available for use with the system. A special dispensing chamber may be used to allow 10 types of discs to be dispensed into the 10 test wells of the unit. Two units may be used for the 20 common tests for identification of the *Enterobacteriaceae*.

Fung et al.[10] rated the accuracy of the Minitek at 97%. The system has the flexibility to allow the analyst to use different substrate discs for characterization of genera and species other than the *Enterobacteriaceae*. The many steps involved in setting up the units and the need for several instruments for inoculation of the wells are disadvantages of the system.

10.4324 BBL Crystal (Becton Dickinson Microbiology, Franklin Lakes, NJ). The two-part unit design of the BBL Crystal system provides convenience in use and reading of results. The bottom chamber of the unit is for holding one mL of a liquid suspension of the test culture. The top unit contains 30 small rods with dehydrated substrate housed at the tip of each rod. The rods are clipped into the bottom unit and in one motion the test culture comes into contact with 30 different substrates. The unit is then incubated overnight at 37°C, and the results are read either manually or by using the Crystal light box. The 10-digit code generated by the results can be interpreted using the Crystal code book or by computer.

The Enteric/Nonfermentor ID Kit with 30 tests is designed for the identification of both *Enterobacteriaceae* and nonfermentative gram negative organisms, and the gram Positive ID kit allows identification of 121 gram positive cocci and bacilli. Advantages include sturdiness of the units, ease of operation and computer-assisted identification. There are no significant disadvantages.

10.4325 MicroID (Organon Teknika, Durham, NC). The first minikit capable of gram negative isolate identification in four hours was the MicroID (Figure 8). The system consists of a molded, styrene tray containing 15 reaction chambers and a hinged cover. The first five chambers contain a substrate disc and a detection disc in the lower and upper portions of the chamber, respectively, and the other ten chambers contain combination substrate/detection discs. Except for the Voges-Proskauer test, the discs contain all the substrates and detection reagents required to perform the biochemical tests. Bacterial cells are suspended into liquid, then 0.2 mL is inoculated into the top of each chamber. The inoculum flows to the bottom of the chamber, where it hydrates the disc. After incubation for 4 hr at 35°C, two drops of 20% KOH are added to the Voges-Proskauer chamber. The unit then is rotated 90 degrees, so that the substrate and detection reagents of the first five chambers are mixed. The reactions of the 15 chambers are read, the first five from the upper portion of the chamber. The MicroID code book is used to interpret the code number generated from the results. The accuracy of the MicroID has been reported as 93.5%[12] and 97%.[20] A similar format with different substrates is available to identify *Listeria spp.*

Advantages of the MicroID are its accuracy, speed of reaction, convenience in operation, addition of only one reagent, and long shelf-life. Disadvantages include having to individually inoculate each of the 15 chambers and difficulty in reading the color of marginal reactions.

10.4326 RapID One System (Remel, Lenexa, KS). The RapID One System consists of a chamber with an inoculum trough con-

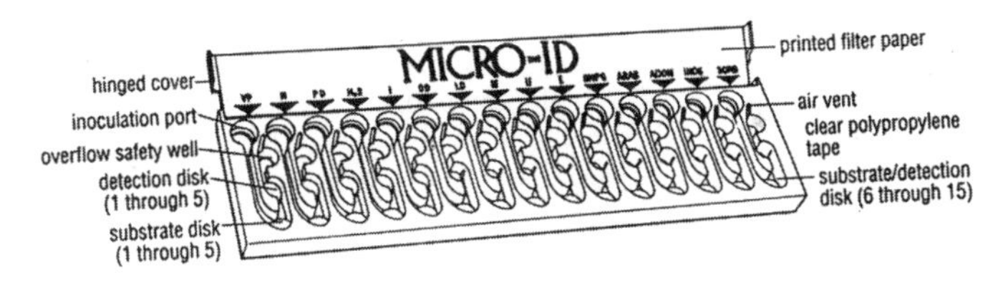

Figure 8. *Micro ID system.*

nected to 20 separate wells, each housing a different dehydrated substrate. A bacterial cell suspension is introduced into the trough, then the unit is tilted so that the inoculum flows into the 20 test wells in one motion. The reactions can be read after a 4-hr incubation. Remel markets test kits for *Enterobacteriaceae, Leuconostoc, Pediococcus, Listeria, Neisseria, Haemophilus*, nonfermentors, yeast, anaerobes, streptococci, and urinary tract pathogens. Identification accuracies of various anaerobes to the genus level using the anaerobic RapID system range from 83–97%, and to the species level, 76%–97%.[18]

Advantages of the RapID One System include the reaction speed and one-step inoculation. A disadvantage is the skill required in reading the color changes.

10.44 Chromogenic Media

Microorganisms growing in the presence of certain dyes and organic compounds may fluoresce under UV light or produce distinctive colors. Such differential color reactions have been exploited in the development of growth media for rapid screening and identification of microorganisms.

Although not a new concept,[19] recently there has been interest in developing novel differential media that utilize a new generation of chromogenic substrates. The new chromogens and media ingredients allow differential coloring of different bacterial groups on the same agar plate. BBL CHROMagar (Becton Dickinson, Franklin Lakes, NJ), BCM (Biosynth, Naperville, IL), Chromocult and Fluorocult Agars (Merck, Darmstadt, Germany) and Rainbow Agar (Biolog, Hayward, CA) are representative. *Salmonella, Listeria, E. coli* O157:H7 are a few of the microorganisms for which chromogenic media have been developed.

10.45 Quantitative Enzymatic Methods

The previous sections described the use of biochemical and enzymatic activities for the identification of microorganisms to the genus and species levels in rapid assays. These metabolic activities may also be used as rapid indicators of biomass, providing an estimation of the size of the microbial population in foods.

Metabolic enzyme activity can be measured directly or indirectly. For example, a direct enzymatic test is the widely-used MUG test for *E. coli*, which detects action of the enzyme β-glucuronidase on the compound 4-methylumbelliferyl-β-D-glucuronide (MUG), resulting in production of a fluorescent compound visible under UV light. Theoretically, the more MUG-positive *E. coli* cells present in the liquid, the more fluorescence will be generated in the presence of the MUG reagent. An example of an indirect test is pH change, caused by acid formation by the microbial population in a food. The amount of acid formed, and therefore the magnitude of the pH drop, would be indicative of the microbial load in a food.

Enzyme activity tests are valuable because of the speed with which a biomass estimate can be obtained. There are drawbacks, however. (1) Food components may interfere with the measurement of enzyme activity. (2) The food may not provide physiologically optimal conditions; therefore, enzyme activity may be suboptimal and may not directly reflect the size of the microbial population. The following sections describe several enzyme activity tests used for rapid evaluation of biomass in foods and the food production environment.

10.451 ATP Bioluminescence

Bioluminescence is the emission of light from viable cells and has been used as an estimation of microbial load, as well as an index of general cleanliness of the food production environment. Bioluminescence has been coupled to an ATP-dependent enzymatic reaction. Because ATP is a marker for viable cells, its measurement in a bioluminescence reaction indicates the presence of biomass. In the presence of luciferin-luciferase, oxygen and magnesium, ATP promotes the light-generating reaction. The amount of light generated is proportional to the amount of ATP, and therefore biomass, in the sample.

Many attempts have been made to correlate ATP levels with the total microbial count, with mixed success. The amount of light generated does not always reflect microbial numbers for various reasons: biological food components (e.g. meat or vegetable tissue) may have an intrinsic ATP content; the microbial population may be metabolically inactive or slow; ATP levels may vary among different groups of a mixed microbial population. Siragusa[22] reviewed applications of ATP bioluminescence technology for microbial monitoring in a wide variety of materials. Recently, ATP research and application work has been focused on sanitation monitoring in the food plant environment. ATP indicates the presence of biomass, and after cleaning, ATP levels drop dramatically.

A number of companies currently manufacture portable handheld instruments for measuring ATP levels from environmental swabs. After the swab is reacted with the enzyme reagents in the self-contained instrument, results are obtained in a matter of minutes. The technology allows sanitation monitoring in "real-time." Some of the instruments can detect as little as 100 fg ATP.

The following companies market bioluminescence-based instruments for ATP measurement:

Biotrace-Ecolab Inc. (St. Paul, MN)
Celsis Inc. (Evanston, IL)
Charm Sciences Inc. (Malden, MA)
Cogent Technologies (Cincinnati, OH)
EM Science (Gibbstown, NJ)
GEM Biomedical Inc. (Hamden, CT)
IDEXX Laboratories Inc. (Westbrook, ME)
Lumac (Silver Spring, MD)
Millipore Corp. (Bedford, MA)
New Horizons Diagnostics (Columbia, MD)

10.452 Limulus Amebocyte Lysate Assay

All gram negative organisms produce endotoxin as a structural part of the cell wall. In the presence of endotoxin, amebocytes from the horseshoe crab *Limulus* form a gel. The gel formation is the basis of the endotoxin measurement system known as the Limulus Amebocyte Lysate (LAL) Assay. Jay[16] used this simple and sensitive assay to estimate Gram negative cell number in ground beef, but it has also been used to monitor the presence of endotoxin in the hospital environment, pharmaceuticals, air, water and other materials.

10.453 Catalase Test

Catalase is an enzyme that breaks down hydrogen peroxide (H_2O_2) into water and oxygen gas (O_2). The O_2 liberated in the reaction can be detected as bubbles in the solution of H_2O_2, and the test can be performed in a simple capillary tube.[4] Not all microorganisms produce the catalase enzyme, but many of the food spoilage species do, e.g. *Pseudomonas spp., Micrococcus, Staphylococcus*, as well as a number of enterics. The test may be used as a rapid method for estimating levels of these microbial populations in certain applications. Because catalase is also present in other biological materials, e.g. meat, the test may be useful in sanitation monitoring in the same way that ATP bioluminescence is used (see section 10.451). The catalase test has also been used to monitor the end-point heating temperature of food. Ang et al.[1] showed that heating poultry meat to 71°C destroys both bacterial and animal catalase. The test is 99% accurate and is simple and easy to perform.

10.454 CO_2 Measurement

Many organisms produce CO_2 as a metabolic by-product, and its measurement may be used to estimate microbial population levels. The Bactec system (Becton Dickinson, Franklin Lakes, NJ) uses radiolabeled substrates in selective media to monitor radiolabeled CO_2 production, and therefore measure population sizes of the microbial groups capable of growth in the media.

10.46 References

1. Ang, C. Y. W., F. Liu, W. E. Townsend, and D. Y. C. Fung, 1994. Sensitive catalase test for end-point temperature of heated chicken meat. J. Food Sci. 59:494-497.
2. Celig, D. M., and P. C. Schreckenberger. 1991. Clinical evaluation of the RapID-ANA II panel for identification of anaerobic bacteria. J. Clin. Microbiol. 29(3):457-462.
3. Chein, S. P., and D. Y. C. Fung. 1991. Acriflavin violet red bile agar for the isolation and enumeration of *Klebsiella pneumoniae*. Food Microbiol. 7:73.
4. Fung, D. Y. C. 1985. Procedures and methods for one-day analysis of microbial loads in foods, p. 656-664. *In*: Rapid methods and automation in microbiology and immunology. K.-O. Habermehl (ed.). Springer-Verlag, Berlin.
5. Fung, D. Y. C. 1992. Historical development of rapid methods and automation in microbiology. J. Rapid Methods Automation Microbiol. 1(1):1.
6. Fung, D. Y. C. 1998. Handbook for Rapid Methods and Automation in Microbiology. Workshop, Dep. of Animal Sci. & Ind., Kans. State Univ., Manhattan, Kans.
7. Fung, D. Y. C., and C. Liang. 1989. A new fluorescent agar for the isolation of *Candida albicans*. Bulletin d'information des Laboratores des Service Veterinaries (France). No. 29/30. p. 1-2.
8. Fung, D. Y. C., and P. A. Hartman. 1975. Miniaturized microbiology techniques for rapid characterization of bacteria, pp. 347–370. In C. G. Heden and T. Illeni (ed.), New approaches to the identification of microorganisms. John Wiley, New York.
9. Fung, D. Y. C., and R. D. Miller. 1973. Effects of dye on bacterial growth. Appl. Microbiol. 25:498-500.
10. Fung, D. Y. C., M. C. Goldschmidt, and N. A. Cox. 1984. Evaluation of bacterial diagnostic kits and systems at an instructional workshop. J. Food Prot. 47:68.
11. Goldschmidt, M.C., D.Y.C. Fung, R. Grant, J. White, and T. Brown. 1991. New aniline blue dye medium for rapid identification and isolation of *Candida albicans*. J. Clin. Microbiol. 29(6):1098-1099.
12. Goosh, W. M., III, and G. A. Hill. 1982. Comparison of MicroID and API 20E in rapid identification of *Enterobacteriaceae*. J. Clin. Microbiol. 15(5):885-890.
13. Hart, R. A., and D. Y. C. Fung. 1990. Evaluation of dye media selective for *Aspergillus* and/or *Penicillium*. In Proceedings of Annual Meeting. Am. Soc. Microbiol. Anaheim, Calif., May 13-17, p. 279.
14. Hart, R. A., O. Mo, F. Borius, and D. Y. C. Fung. 1991. Comparative analysis of Trypan blue agar and Congo red agar for the enumeration of yeast and mold using HGMF system. J. Food Safety, 11:227.
15. Hartman, P. A. 1998. Miniaturized microbiological methods. Academic Press, New York.
16. Jay, J. M. 1996. Modern food microbiology, 4th ed. Van Nostrand, Reinhold, New York.
17. Lin, C. C. S., and D. Y. C. Fung. 1985. Effect of dyes on growth of food yeast. J. Food Sci. 47:770.
18. Lee, C. Y., D.Y.C. Fung, and C. L. Kastner. 1982. Computer-assisted identification of bacteria on hot-boned and conventionally processed beef. J. Food Sci. 47:363.
19. Lee, C. Y., D. Y. C. Fung, and C. L. Kastner. 1985. Computer-assisted identification of microflora on hot-boned and conventionally processed beef: Effect of moderate and slow initial chilling rate. J. Food Sci. 50:53.
20. Lin, C. C. S., and D. Y. C. Fung. 1987. Critical review of conventional and rapid methods for yeast identification. Crit. Rev. Microbiol. 14(4):273.
21. Russell, S. M., N. A. Cox, J. S. Bailey, and D. Y. C. Fung. 1997. Miniaturized biochemical procedures for identification of bacteria. J. Rapid Methods Automation Microbiol. 5(3):169-178.
22. Siragusa, G. R. 1996. Primer on ATP bioluminescence for microbial detection and hygiene monitoring. Proceeding. 16th Food Microbiology Symposium, University of Wisconsin-River Falls. October 1996.
23. Thakur, R. A. H., and D. Y. C. Fung. 1991. Effects of dyes on the growth of food molds. J. Rapid Methods Automation Microbiol. 4(1):1-25.

10.5 MEMBRANE FILTRATION

10.51 Hydrophobic Grid Membrane Filter

The hydrophobic grid membrane filter (HGMF) was first introduced to extend the counting range of conventional membrane filters, thus reducing the number of dilutions required to analyze a sample.[37,38] At first, its applicability was limited to water samples and other suspensions that did not contain a significant amount of debris.[22] Later, prefilters were introduced to eliminate particles from the food homogenates prior to filtration,[21,32] and a variety of diluents and enzyme digestions techniques were introduced to enable more efficient filtration of the prefiltered homogenates.[21,31,40]

The HGMF has numerous advantages over other food microbiology methods. In most cases, its counting range eliminates the need for multiple dilutions[4,38]; the process of filtration separates the microorganisms from the food matrix, thus eliminating interference with growth as well as with interpretation of differential reactions[18]; confining colonies inside the individual growth squares on the HGMF usually results in taller, denser colonies which are easier to detect[4,37] and restrains fast-growing organisms from spreading and overgrowing other colonies; the non-staining nature of the membrane filter allows introduction of non-toxic dyes into the culture medium, enhancing visibility of the colonies[17,19,27]; the ordered arrangement of colonies on the filter reduces the need for subjective interpretation during counting, improving between-analyst reproducibility[37,41]; and the filter can be transported from medium to medium (or staining reagent) without disturbing the colonies, allowing for preliminary incubation of an injured population on a repair medium prior to selective incubation[2,16] or for sequential reading of two or more differential reactions on the same filter.[6,11,18,24]

The disadvantages of the HGMF are the need to clean and resterilize the filtration apparatus, the difficulty encountered in filtering certain types of foods, and the eye fatigue involved in counting up to 1600 individual grid squares. The labor required to clean and resterilize filtration units is largely offset, in most cases, by the elimination of multiple dilutions. Handling protocols to facilitate rapid filtration must be established for each food product prior to incorporating that product into the test routine, but, once established, they remain consistent unless the formulation of the food product changes. Finally, both manual (ISO-GRID Line Counter, see Figure 9) and automated[42] counting aids have been developed to assist in interpreting growth patterns on HGMF.

Soon after the introduction of food filtration techniques, several procedures were developed for the detection or enumeration of microorganisms of interest to the food microbiologist, including coliforms and *Escherichia coli*,[3,39] aerobic plate count,[4] yeast and mold count[4], *Staphylococcus aureus*,[3] enterococci[3] and *Salmonella*.[22] Later, HGMF-based tests for *Vibrio parahaemolyticus*,[6,16] fluorescent pseudomonads,[26] *Zygosaccharomyces bailii*,[23] *Aeromonas hydrophila*,[45] lactic acid bacteria,[5,24,28] *Listeria monocytogenes*,[35] and *E. coli* O157:H7[7,20] were also described. Of these, the coliform, *E. coli*, aerobic plate count, yeast and mold, *Salmonella* and *E. coli* O157:H7 procedures have been accepted as Official Methods by AOAC International.[8-15]

The hydrophobic grid membrane filter, with its ordered growth patterns, has been used by several researchers as the basis for enzyme-labeled antibody (ELA) enumeration and presence/absence tests for organisms of interest in food microbiology. Early efforts in this area were directed towards *S. aureus*[33] and *E. coli* O157.[7,43] ELA-HGMF methods have also been described for verotoxigenic *E. coli*,[29] *Clostridium botulinum*[30] and *Salmonella*.[44] The ELA procedure is carried out either directly on the HGMF or on a "blot" or replica of HGMF colonies on another membrane filter. When a replica is used, the colonies on the original filter remain viable and can be subcultured for further detailed investigation, if desired.

Colony hybridization methods have also been developed for use with HGMF. A chromogen-labeled DNA probe was developed and used to detect colonies of *Listeria monocytogenes* directly on the filter.[34,35] Also, a colony hybridization technique has been developed to enumerate and differentiate *Vibrio parahaemolyticus* and *V. vulnificus*.[25] In this last procedure, the HGMF was used for primary isolation of suspect colonies from samples and a blot replica of the filter was probed with a chromogen-labeled DNA probe.

10.511 Equipment Needed

ISO-GRID Hydrophobic Grid Membrane Filters (QA Life Sciences, Inc.)

ISO-GRID Filtration Units with built-in 5μm pore size prefilter (QA Life Sciences, Inc.)

ISO-GRID stainless steel clamp (QA Life Sciences, Inc.)

Membrane Filter forceps (blunt-tipped, no gripping ridges)

Vacuum manifold or 1 liter vacuum flask

Vacuum pump (1/6 hp. minimum for use with vacuum flask; 1/3 hp. minimum for use with manifold

Vacuum tubing with Y-connector and tubing clamp

Counting instrument (ISO-GRID Line Counter; QA Life Sciences, Inc.). Quebec Colony Counter or other illuminated magnifiers are acceptable alternatives

10.512 Sample Preparation for Quantitative Analyses

Prepare a 1:10 sample homogenate. Ideally, a minimum 10 g sample should be used. The choice of diluent will depend on the nature of the sample. For example, most meat and dairy products filter best when prepared in a diluent containing 1% Tween 80, whereas most other products are best prepared in a surfactant-free diluent such as 0.1% peptone or Butterfield's Phosphate Buffer. Homogenates can be prepared by using a conventional lab blender, by shaking, or by using a "stomacher" or equivalent instrument. Some foods produce viscous homogenates which are difficult to filter. These can often be handled simply by diluting the homogenate to 1:100 and filtering 10 mL of the diluted homogenate. Dilution reduces the viscosity of the sample homogenate, allowing it to pass through both the prefilter and membrane filter more easily, while filtration reconcentrates the sample, allowing the analyst to obtain a sensitivity level of 10 MPN/g. A number of foods require digestion with one or a combination of enzymes to enable 1 mL of a 1:10 dilution of the homogenate to filter. Detailed recommendations on choice of diluent, enzyme digestion and sample dilution for optimum filtration, together with instructions on enzyme preparation and use, can be found elsewhere.[1,36]

10.513 General Procedure for Quantitative Analysis

The process of setting up samples for analysis by HGMF is very consistent, regardless of the analysis. It is comprised of the following steps:

1. Prepare the sample homogenate for filtration.
2. Place a sterile filtration unit on the vacuum flask or manifold and rotate the hinged funnel to the open (horizontal) position.
3. Use sterile forceps to aseptically position a filter on the base of the filtration unit. Rotate the hinged funnel to the shut (vertical) position.
4. Slide the stainless steel clamp over the flanges protruding from both sides of the filtration funnel and base and rotate the clamp arm to the shut (horizontal) position (see Figure 10).

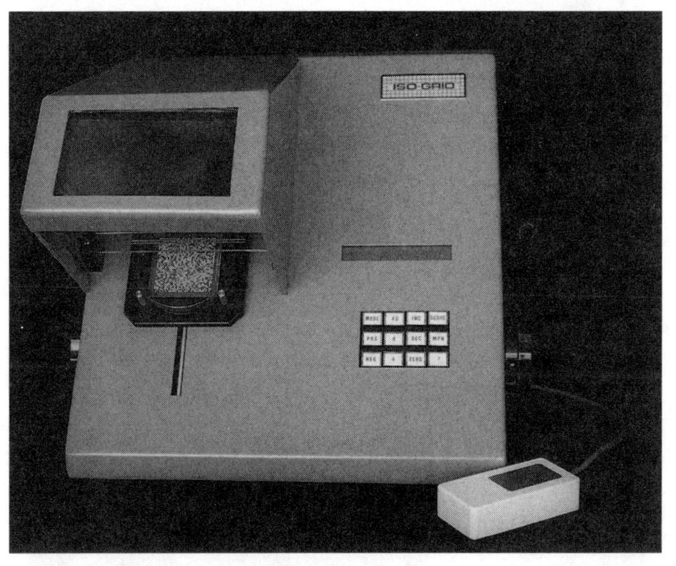

Figure 9. *ISO-GRID Line Counter*

5. Dispense approximately 10–15 mL of sterile water or diluent into the top of the funnel. This will begin to drip through the prefilter onto the membrane filter, preventing the sample from sticking to either filter.
6. Without waiting for the sterile liquid to drip through the prefilter, measure the desired volume of sample homogenate or diluted homogenate (typically, 1 mL of 10^{-1} homogenate to obtain 10 MPN/g sensitivity) into the funnel. Dispense an additional 10-15 mL of sterile water or diluent into the funnel to ensure uniform dispersal of the sample.
7. Using a Y-branched vacuum tubing, gently apply vacuum to the filtration funnel just below the prefilter to draw liquid quickly into the lower chamber of the funnel. Vacuum should also be applied simultaneously to the filtration unit base below the HGMF to prevent the membrane filter from being dislodged during the prefiltration step.
8. Apply vacuum to the base of the filtration unit to draw the sample liquid through the membrane filter.
9. Remove the membrane filter from the filtration unit, place it on the desired culture medium and incubate as appropriate for the test.
10. Following incubation, examine the membrane filter for squares containing target colonies, count the number of positive squares, convert to the corresponding most probable number (MPN) index and multiply by the appropriate dilution factor to obtain the MPN per gram or milliliter. MPN conversions can be obtained by using the formula MPN = $1600 \times \log_e (1600/[1600 - x])$, where x represents the number of positive squares.

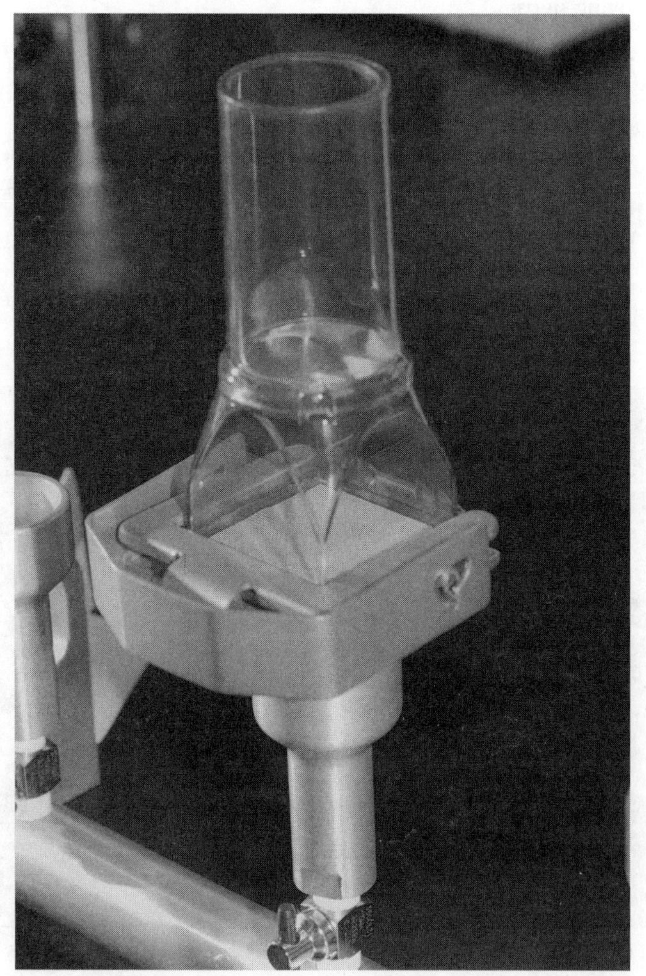

Figure 10. *ISO-GRID Filtration Unit clamped shut with filter in place.*

If multiple quantitative analyses are to be run on a sample, a single filtration unit can usually be used for all those tests. After the first filtration is complete and the membrane filter has been placed on the appropriate medium, a second membrane filter can be inserted into the filtration unit and a second filtration carried out. The number of filtrations that can be run for a single sample on a single filtration unit without washing is limited only by the ease with which the sample homogenate passes through the 5 µm stainless steel prefilter. Table 2 summarizes culture media, incubation conditions (duration and temperature) and target colony colors for several common quantitative analyses.

10.514 Counting and Dilution Strategies for Quantitative Analyses

The HGMF counting range spans nearly 4 $\log_{10}$ cycles, although counts obtained at the upper end of the range should be treated with caution. When 95% or more of the squares are positive, the filter produces an underestimate of the true population. Therefore, any result exceeding 1520 positive squares should be reported as an estimated count. In the vast majority of cases, filtering a single dilution will suffice to cover the desired counting range (as compared to running three consecutive $\log_{10}$ dilutions as is done with most conventional procedures). There are two approaches to deciding what sample dilution should be filtered for any analysis. The simplest is to choose the lowest dilution that would be run if a conventional method was being used. For example, if a coliform test is usually run on 1:10, 1:100 and 1:1,000 sample dilutions, then 1 mL of the 1:10 dilution should be filtered for the HGMF analysis. Similarly, if an Aerobic Plate Count on a particular product is usually carried out on the 1:1,000, 1:10,000 and 1:100,000 dilutions, then the filtration should be run on the 1:1,000 dilution.

A second method of determining dilutions is based on the quantitative product specification for the analysis in question. For example, if an Aerobic Plate Count result is considered unsatisfactory at levels above 10^5 CFU/g, then a dilution should be chosen to enable accurate and convenient counting in the vicinity of that count level. Filtering 1 mL of a 1:1,000 dilution from a sample with an Aerobic Plate Count of 10^5 CFU/g would produce approximately 100 positive squares. This level of growth is relatively easy to count, is well within the accurate counting range of the HGMF, and allows for easy and accurate counting both above and below the critical target level, giving the analyst a clear picture of how "clean" or, conversely, how far out of specification the sample is. Alternatively, if only an occasional sample is expected to be out of specification and the analyst wishes more information at the lower end of the count range, 1 mL of a 1:100 dilution can be filtered instead.

Occasionally, almost all of the squares will contain target colonies. When nearly all squares are positive, one should count the number of negative squares and subtract from 1600 to determine the number of positive squares. On other occasions, a filter may contain several hundred positive squares. If the growth is randomly distributed over the entire 1600 squares, and if at least 25% of the squares (400 squares) appear to be positive, the number of positive squares over the entire filter can be estimated by counting the number of positive squares contained in eight contiguous rows and multiplying by five to determine the estimated number of positive squares on the entire filter.[26]

Table 2. Hydrophobic Grid Membrane Filter Applications for Quantitative Analysis

Analysis	Ref.	AOAC Official Method	Culture Media	Incubation Conditions	Interpretation
Aerobic plate count	10	986.32	TSAF Agar	48 ± 3 hours; 35–37°C	Count all squares containing growth (usually appear as green colonies)
Total coliform count	8, 17	990.11	LMG Agar	24 ± 2 hours; 35–37°C	Count all squares containing blue colonies
E. coli count	8, 17	990.11	BMA Agar	2–5 hours[a]; 35–37°C	Count all squares containing fluorescent colonies under long wave UV light
Fecal coliform count	31	983.25	m-FC Agar	24 ± 2 hours; 44.5 ± 0.5°C	Count all squares containing blue colonies
Yeast & mold count	34	995.21	YM-11 Agar	50 ± 2 hours; 25 ± 1°C	Count all squares containing growth (usually appear as blue colonies)
E. coli O157:H7	36	997.11	SD-39 Agar	24 ± 2 hours; 44.0–44.5°C	Count all squares containing pink colonies; subculture and confirm serologically
Enterococci	20		m-Enterococcus Agar	48 ± 3 hours; 35–37°C	Count all squares containing pink or red colonies; confirm using catalase test
S. aureus count	20		Baird-Parker Agar	48 ± 3 hours; 35–37°C	Count all squares containing black or dark gray colonies; confirm using coagulase test
Lactic acid bacteria	15		MRSD Agar	48 hours (anaerobic); 25°C	Count all squares containing growth
Zygosaccharomyces bailii	23		ZBM Agar	2–3 days; 30°C	Count all squares containing smooth, moist, round-convex colonies (usually appear blue)
Fluorescent pseudomonads	22		TSA Agar, then S-1Agar	4 + 48 hours[b]; 25°C	Count all squares containing colonies that fluoresce under long wave UV light

[a] After counting coliform colonies on LMG Agar, transfer filter to BMA and incubate 2–5 hours.

[b] First incubate on TSA for 4 hours, then transfer to S-1 Agar and incubate 48 hours.

Regardless of the dilution filtered, whether positive or negative squares are counted, and how many rows are examined, calculations must always proceed in the following sequence:

a) Convert any negative square count or estimated count to obtain a total number of positive squares on the filter ("Score");
b) Convert the Score to the corresponding MPN either by application of the formula given above or by using a look-up table[36]; and
c) Multiply the resultant MPN by the dilution factor to obtain the MPN/g.

While the reproducibility of HGMF results is usually sufficient to avoid the need for duplicate filtrations,[9-11,14,15] greater precision might occasionally be desired. When calculating results obtained from replicate filters, determine the Score and MPN separately for each filter, take the average of the MPN values and multiply the resultant average MPN by the dilution factor to obtain the MPN/g.

10.515 Procedure for Salmonella Presence/Absence Testing

The details of this procedure are given elsewhere.[1] The *Salmonella* method can be summarized in the following steps:

1. Suspend a minimum 25 g sample in 9 volumes of nonselective enrichment broth. Mix as appropriate, adjust the pH, if necessary, to 7.0 ± 0.2, and incubate for 18–24 hours at 35–37°C.
2. Transfer 0.1 mL (100 µL) of overnight nonselective enrichment into a 10 mL volume of Mueller-Kaufmann Tetrathionate Broth. Incubate 6–8 hours at 35–37°C in a water bath.
3. Using the filtration technique described in the quantitative analysis procedure (steps 2–9), filter 0.1 mL (100 µL) of Tetrathionate Broth culture. Place the membrane filter onto EF-18 Agar and incubate 18–24 hours at 42 ± 0.5°C.
4. Examine the membrane filter for the presence of green colonies. These are presumptive *Salmonella* colonies. If no growth is evident on the membrane filter, or if only yellow colonies are present, the sample may be reported as "*Salmonella* Not Detected."
5. If presumptive positive colonies are present, subculture a representative number of these (three is usually sufficient) for biochemical and, if appropriate, serological confirmation.

10.52 References

1. AOAC International. Official Methods of Analysis of AOAC International, 16th ed., vol I. AOAC International; 3rd rev., 1997.

2. Brodsky, M. H., Boleszczuk, P., and Entis, P. 1982. Effect of stress and resuscitation on recovery of indicator bacteria from foods using hydrophobic grid-membrane filtration. J. Food Prot. 45: 1326.
3. Brodsky, M. H., Entis, P., Entis, M. P., Sharpe, A. N., and Jarvis, G. A. 1982. Determination of aerobic plate and yeast and mold counts in foods using an automated hydrophobic grid-membrane filter technique. J. Food Prot. 45: 301.
4. Brodsky, M. H., Entis, P., Sharpe, A. N., and Jarvis, G. A. 1982. Enumeration of indicator organisms in foods using the automated hydrophobic grid-membrane filter technique. J. Food Prot. 45: 292.
5. Chu, J. L., Elo, A., Maissin, R., and Decallonne, J. R. 1987. La quantification des populations de bacteries lactiques par la methode HGMF. Belgian J. Food Chem. Biotechnol. 42: 65.
6. DePaola, A., Hopkins, L. H., and McPhearson, R. M. 1988. Evaluation of four methods for enumeration of *Vibrio parahaemolyticus*. Appl. Environ. Microbiol. 54: 617.
7. Doyle, M. P., and Schoeni, J. L. 1987. Isolation of *Escherichia coli* O157:H7 from retail fresh meats and poultry. Appl. Environ. Microbiol. 53: 2394.
8. Entis, P. 1984. Enumeration of total coliforms, fecal coliforms, and *Escherichia coli* in foods by hydrophobic grid membrane filter: collaborative study. J. AOAC. 67: 812.
9. Entis, P. 1984. Enumeration of total coliforms, fecal coliforms, and *Escherichia coli* in foods by hydrophobic grid membrane filter: supplementary report. J. AOAC. 67: 811.
10. Entis, P. 1986. Hydrophobic grid membrane filter method for aerobic plate count in foods: collaborative study. J. AOAC. 69: 671.
11. Entis, P. 1989. Hydrophobic grid membrane filter/MUG method for total coliform and Escherichia coli enumeration in foods: collaborative study. J. AOAC. 72: 936.
12. Entis, P. 1990. Improved hydrophobic grid membrane filter method, using EF-18 agar, for detection of Salmonella in foods: collaborative study. J. AOAC. 73: 734.
13. Entis, P. 1996. A research note: Validation of the ISO-GRID 2-day rapid screening method for detection of Salmonella spp. in egg products. J. Food Prot. 59: 555.
14. Entis, P. 1996. Two-day hydrophobic grid membrane filter method for yeast and mold enumeration in foods using YM-11 agar: collaborative study. J. AOAC Int. 79: 1069.
15. Entis, P. 1998. Direct 24-hour presumptive enumeration of Escherichia coli O157:H7 in foods using hydrophobic grid membrane filter followed by serological confirmation: collaborative study. J. AOAC Int. 81: 403.
16. Entis, P., and Boleszczuk, P. 1983. Overnight enumeration of Vibrio parahaemolyticus in seafood by hydrophobic grid membrane filtration. J. Food Prot. 46: 783.
17. Entis, P., and Boleszczuk, P. 1986. A research note: Use of fast green FCF with tryptic soy agar for aerobic plate count by the hydrophobic grid membrane filter. J. Food Prot. 49: 278.
18. Entis, P., and Boleszczuk, P. 1990. Direct enumeration of coliforms and Escherichia coli by hydrophobic grid membrane filter in 24 hours using MUG. J. Food Prot. 53: 948.
19. Entis, P., Brodsky, M. H., and Sharpe, A. N. 1982. Effect of pre-filtration and enzyme treatment on membrane filtration of foods. J. Food Prot. 45: 8.
20. Entis, P., Brodsky, M. H., Sharpe, A. N., and Jarvis, G. A. 1982. Rapid detection of Salmonella spp. in food by use of the ISO-GRID hydrophobic grid membrane filter. Appl. Environ. Microbiol. 43: 261.
21. Entis, P., and Lerner, I. 1996. Two-day yeast and mold enumeration using the ISO-GRID-membrane filtration system in conjunction with YM-11 agar. J. Food Prot. 59: 416.
22. Entis, P., and Lerner, I. 1997. 24-Hour presumptive enumeration of Escherichia coli O157:H7 in foods using the ISO-GRID method with SD-39 agar. J. Food Prot. 60: 883.
23. Erickson, J. P. 1993. Hydrophobic membrane filtration method for the selective recovery and differentiation of Zygosaccharomyces bailii in acidified ingredients. J. Food Prot. 56: 234.
24. Holley, R. A., and Millard, G. E. 1988. Use of MRSD medium and the hydrophobic grid membrane filter technique to differentiate between pediococci and lactobacilli in fermented meat and starter cultures. Int. J. Food Microbiol. 7: 87.
25. Kaysner, C. A., Abeyta, Jr., C., Jinneman, K. C., and Hill, W. E. 1994. A research note: Enumeration and differentiation of Vibrio parahaemolyticus and Vibrio vulnificus by DNA-DNA colony hybridization using the hydrophobic grid membrane filtration technique for isolation. J. Food Prot. 57: 163.
26. Knabel, S. J., Walker, H. W., and Kraft, A. A. 1987. Enumeration of fluorescent pseudomonads on poultry by using the hydrophobic-grid membrane filter method. J. Food Sci. 52: 837.
27. Lin, C. C. S., Fung, D. Y. C., and Entis, P. 1984. Growth of yeast and mold on trypan blue agar in conjunction with the ISO-GRID system. Can. J. Microbiol. 30: 1405.
28. Millard, G. E., McKellar, R. C., and Holley, R. A. 1990. Simultaneous enumeration of the characteristic microorganisms in yogurt using the hydrophobic grid membrane filter system. J. Food Prot. 53: 64.
29. Milley, D. G., and Sekla, L. H. 1993. An enzyme-linked immunosorbent assay-based isolation procedure for verotoxigenic Escherichia coli. Appl. Environ. Microbiol. 59: 4223.
30. Nakano, H., and Sakaguchi, G. 1991. An unusually heavy contamination of honey products by Clostridium botulinum type F and Bacillus alvei. FEMS Microbiol. Lett. 79: 171.
31. Peterkin, P. I., Idziak, E. S., and Sharpe, A. N. 1991. Detection of Listeria monocytogenes by direct colony hybridization on hydrophobic grid-membrane filters by using a chromogen-labeled DNA probe. Appl. Environ. Microbiol. 57: 586.
32. Peterkin, P. I., Idziak, E. S., and Sharpe, A. N. 1992. Use of a hydrophobic grid- membrane filter DNA probe method to detect Listeria monocytogenes in artificially- contaminated foods. Food Microbiol. 9: 155.
33. Peterkin, P. I., and Sharpe, A. N. 1980. Membrane filtration of dairy products for microbiological analysis. Appl. Environ. Microbiol. 39: 1138.
34. Peterkin, P. I., and Sharpe, A. N. 1981. Filtering out food debris before microbiological analysis. Appl. Environ. Microbiol. 42: 63.
35. Peterkin, P. I., and Sharpe, A. N. 1984. Rapid enumeration of Staphylococcus aureus in foods by direct demonstration of enterotoxigenic colonies on membrane filters by enzyme immunoassay. Appl. Environ. Microbiol. 47: 1047.
36. QA Life Sciences, Inc. ISO-GRID- Methods Manual, 3rd ed. QA Life Sciences; 1989 (revised 1995).
37. Sharpe, A. N., Diotte, M. P., Dudas, I., Malcolm, S., and Peterkin, P. I. 1983. Colony counting on hydrophobic grid-membrane filters. Can. J. Microbiol. 29: 797.
38. Sharpe, A. N., Diotte, M. P., Peterkin, P. I., and Dudas, I. 1986. Towards the truly automated colony counter. Food Microbiol. 3: 247.
39. Sharpe, A. N., and Michaud, G. L. 1974. Hydrophobic grid-membrane filters: new approach to microbiological enumeration. Appl. Microbiol. 28: 223.
40. Sharpe, A. N., and Michaud, G. L. 1975. Enumeration of high numbers of bacteria using hydrophobic grid-membrane filters. Appl Microbiol. 30: 519.
41. Sharpe, A. N., Peterkin, P. I., and Dudas, I. 1979. Membrane filtration of food suspensions. Appl. Environ. Microbiol. 37: 21.
42. Sharpe, A. N., Peterkin, P. I., and Malik, N. 1979. Improved detection of coliforms and Escherichia coli in foods by a membrane filter method. Appl. Environ. Microbiol. 38: 431.
43. Todd, E. C. D., MacKenzie, J. M., and Peterkin, P. I. 1993. Development of an enzyme- linked antibody hydrophobic grid membrane filter method for the detection of Salmonella in foods. Food Microbiol. 10: 87.
44. Todd, E. C. D., Szabo, R. A., Peterkin, P., Sharpe, A. N., Parrington, L., Bundle, D., Gidney, M. A. J., and Perry, M. B. 1988. Rapid hydrophobic grid membrane filter-enzyme-labeled antibody procedure for identification and enumeration of Escherichia coli O157 in foods. Appl. Environ. Microbiol. 54: 2536.

45. Warburton, D. W., McCormick, J. K., and Bowen, B. 1994. Survival and recovery of Aeromonas hydrophila in water: development of methodology for testing bottled water in Canada. Can. J. Microbiol. 40: 145.

10.53 Commercial References

Commercial Supplier for ISO-GRID Hydrophobic Grid Membrane Filters and Related Products
QA Life Sciences, Inc., 6645 Nancy Ridge Drive, San Diego, CA 92121

10.6 IMPEDANCE, CONDUCTANCE, AND CAPACITANCE

10.61 Theory

The use of electrical methods to determine bacterial growth dates back to the late 19th century.[60] In the early 1900s, conductivity was found to be useful for measuring ammonia produced by clostridia in various environments,[61] and was used for measuring bacterial proteolysis.[1] Although electrical methods for enumerating bacteria were evaluated throughout the 1900s, it was not until the early 1980s that these methods became automated and were used to estimate bacterial populations on a variety of foods.[17,18,19,21,23,26,34,64]

Impedance is defined as the resistance to the flow of a sinusoidal alternating current through a conducting material. Determination of impedance is complex in that it consists of a resistive component (R) and a reactive component (X), which, for the purposes of monitoring bacterial growth, is measured as capacitive resistance (X_c). Impedance (Z) is calculated using the equation $Z = \sqrt{(R^2 + X_c^2)}$ where $X_c = 1 / 2 \pi fC$ (f = frequency; C = capacitance).[9] The conductance (G) of a bacteriological growth medium is a function of the resistance (R) of the medium, where $G = 1/R$ and is dependant upon the ionic strength of the solution.[19,28] Capacitance changes in a growth medium occur very close to the electrode. The electrode serves as one side of the capacitor and a layer of ions serves as the opposite side. Smaller ions produce a thinner layer and this is reflected as a change in capacitance.[13] Capacitance has been proven to be useful in high conductivity media used for estimation of yeasts and molds.[28]

As bacteria multiply in a growth medium, they convert large molecules into smaller, more mobile ionic metabolites, which change the impedance of the medium.[18] These metabolites increase the conductance and decrease the impedance of the medium. Once microbial populations reach a level of 10^6 to 10^7 cells/mL, the impedance, conductance, or capacitance (depending on which parameter is being measured) of the medium shifts to the extent that the instrument can detect the change.[18] At this point, a detection time (DT) in hours is automatically reported and recorded by the instrument. Hence, DT is a function of the temperature at which the assay is conducted, the medium, the generation time of the bacterium or group of bacteria able to grow in the medium, the extent of repair the organisms need to begin multiplying, and the inhibitors in the sample or medium.

A common misconception about impedance microbiology is that the method is incapable of enumerating low numbers of bacteria from samples. This confusion is due to the misunderstanding about the sensitivity of the instrument versus the sensitivity of the assay. The sensitivity of the instrument is 10^6 to 10^7 cells, because there must be sufficient metabolite accumulation to enable a large impedance shift. The sensitivity of the assay is one viable cell because, as one cell multiplies to 10^6 to 10^7, its presence may be detected.

The time required for the monitoring instrument to detect the impedance shift depends upon the number of microorganisms in the sample initially. For example, if a sample contains bacterial populations at a level of 10^5 cells/mL, the detection time would be reached rapidly because only four to eight cell divisions would be required to exceed the detection threshold level.[28] However, for samples that contain only 1 microorganism/mL, more time would be required for microbial populations to multiply to the detection threshold. Hence, detection times are inversely proportional to initial bacterial numbers.

Most commercial systems (section 10.62) can be programmed such that samples that detect rapidly and are "highly contaminated" may appear red on the report screen of the computer. This allows for immediate recognition by the technician that a sample is highly contaminated and the product should be held. Likewise, detection time ranges that would be considered "caution zones" can be programmed to appear yellow on the screen and ranges that would be considered "acceptable" would be programmed to appear green. Using this technique, a technician can determine the disposition of many products being tested by glancing at the computer screen.

10.62 Commercial Systems

10.621 The Bactometer

The Bactometer Microbial Monitoring System is a computer controlled impedance, conductance, and capacitance monitoring device. The Bactometer can be used to: 1) perform total microbial, coliform, yeast and mold, lactic acid bacteria, *Escherichia coli*, and *Pseudomonas fluorescens* counts, 2) detect the presence of *Salmonella* and *Listeria*, 3) predict shelf-life, 4) conduct challenge and functional sterility testing, and 5) monitor environmental samples.

The Bactometer includes the following components: a computer, printer, and Bactometer Processing Unit (BPU). The BPU is comprised of two incubators that can be set independently of one another with a temperature range from 10°C below ambient temperature to 55°C. Each BPU can simultaneously monitor from 1 to 128 samples. Three additional BPU's (each containing two independent incubators) can be added to bring the total capacity of the system to 512 samples.

Test samples (total of 2.0 mL) are loaded into the Bactometer disposable module wells (16 wells per module) which are inserted into the BPU. The system is capable of monitoring conductance, capacitance, or total impedance for each sample. Readings on each sample are taken every six minutes.

The Bactometer software utilizes an algorithm that determines the onset of acceleration in the sample's impedance curve. The DT's are displayed as soon as they are available. Samples exceeding the microbial specification level appear in red, borderline or marginal samples appear in yellow, and samples that are acceptable and below the specification level appear in green. The software allows the user to print reports, view and plot curves, and perform statistical analysis required for the creation of calibration curves.

10.622 The RABIT

RABIT for Windows™ is a compact and versatile system for rapid enumeration of bacteria. The basic system is composed of a PC with Windows™ software and a 32-channel incubation block. The test cells appear to be similar to test tubes; however, the bottom of the cell contains an electrode assembly that may be removed and disassembled for cleaning, autoclaving, and reuse. Thus, the cost per test is very low because the cells are reusable rather than disposable. A working volume of 2–10 mL of medium may be added to each test cell. The design of the test cell allows the user to conduct direct or indirect impedance assays (section 10.648).

The 32-channel incubation block is composed of two sides which may be individually controlled with regard to temperature. The system can be expanded to provide a total of 512 test channels by increasing the number of incubation modules. The user has full access to information about each individual cell throughout incubation, allowing each test to be set up or terminated on an individual basis.

The Windows™ based software facilitates sample entry and analysis of results. Impedance data may be exported to spreadsheets and database programs for manipulation. Statistical Process Control (SPC) software is available for companies that utilize SPC programs. Final impedance readings may be viewed on screen or printed.

10.623 The Malthus

The Malthus-V can be used to detect a wide range of microorganisms. Up to 256 samples can be tested simultaneously for the presence of aerobic, microaerophilic, and anaerobic microorganisms.[28] The computer-based instrument is designed for continuous automated analysis. Liquid, colloidal, and solid samples require no special preparation. The samples are analyzed using reusable electrodes over a temperature range of 4°C to 47°C. Color-coded pass/fail/suspect results are automatically displayed on an integral HP-Vectra microcomputer, and programmed checks ensure precise analysis. The Malthus Analyzer has the capability for on- and off-line communications with other computers.[28]

10.624 The BacTrac

The BacTrac may be used to detect the proliferation of various types of bacteria, yeast, and mold. Included with the system are a 386 or 486 computer, an incubation block that can contain 20 or 40 measuring cells with working volumes of 10 or 100 mL. The test cells are similar to the RABIT in that they contain an electrode assembly that may be removed and disassembled for cleaning, autoclaving, and reuse. The design of the test cell allows the user to conduct direct or indirect impedance assays (section 10.648). Up to six incubators may be linked to allow for a maximum system capacity of 240 samples.

As with other impedance methods, the impedance of the medium (M-value) may be monitored. However, this system is also able to simultaneously monitor the impedance associated with the area around the electrode (E-value). While the Bactometer is able to monitor impedance, conductance, and capacitance (signal around the electrode), it is unable to separate these signals and interpret them individually for each individual sample. Monitoring these two signals independently is called the Impedance-Splitting method. This method is especially useful for monitoring bacteria that are grown in high salt media, such as *Staphylococcus aureus*, because, at high salt concentrations, the M-value does not significantly decrease as the bacteria multiply; however, the E-value decreases drastically.[44]

10.63 General Considerations and Precautions

The total aerobic plate count (APC) method is different than impedance, conductance, and capacitance with regard to how estimations of bacterial populations are performed. With the APC method, each individual bacterium on the surface of a plate containing bacteriological growth medium with agar must multiply to the extent that a visible biomass (colony) is reached. The colonies are then counted and an estimate of the original number of bacteria in the sample is determined. Impedance, conductance, and capacitance, are based on the electrical changes that occur in medium as bacteria break down large molecules into smaller, more mobile metabolites. Thus, preparation procedures, media, and growth conditions for conducting electrical assays are different than those used for the APC method. Detailed reviews of approaches to the development electrical methodologies are presented elsewhere.[19,53,54]

10.631 Media

The growth medium used to conduct electrical assays greatly influences the electrical shift that must be interpreted by the instrument. If medium components are not designed to optimize the electrical shift during bacterial growth, impedance curves will be inferior. The following circumstances may lead to curves that are not able to be interpreted by the instrument: 1) the baseline may drift, 2) the percent change in impedance may reverse direction during the growth phase of the organism, 3) the acceleration of the impedance change may be weak, 4) a lower maximum impedance change value may be obtained, 5) the curve may be bi-phasic which cannot be interpreted, 6) erratic curves may develop if bacteria are able to produce gas while growing in the medium.[19] Thus, media must be developed to optimize the signal by controlling the use of ingredients that contribute to erratic curves such as certain types of salts and carbohydrates.

10.632 Incubation Temperature

Incubation temperature is one of the most important considerations when conducting an impedance assay. Temperature is used as a selective component of the assay. For example, if a sample containing 1 coryneform and 10^5 pseudomonads were placed into an impedance instrument and allowed to incubate at 30°C, the coryneform would be the organism that is detected and enumerated.[19] Although the pseudomonads in the initial sample exceeded the coryneform by a factor of 100,000 in the initial sample, they did not contribute significantly to the impedance change of the medium and, thus, were not enumerated. This example illustrates how temperature may be used to select for the growth of a particular species of bacteria to be enumerated in samples containing mixed populations. Detailed methods for developing and optimizing media have been reported.[47,53]

10.633 Sample Preparation and Handling

Because time to impedance shift or "detection time" is the parameter being measured, sample preparation and handling must be extremely strict when conducting impedance assays. Samples should always be placed into growth medium that is a consistent temperature. The amount of time to inoculate the medium, place the sample into the module well, place the module well into the incubation chamber, and initiate the impedance readings must be held constant from sample to sample. For example, if a sample is placed into cold growth medium, and held in a refrigerator for 1 hr before being placed into the module and then incubated, this sample will have a much different DT than an identical sample that is placed into a warm medium and incubated immediately. Thus, every effort must be made to standardize sample preparation and handling procedures.

10.634 Generation Times

When using impedance to estimate bacterial numbers in samples containing mixed populations, it is essential that the media and temperature be selected so that differences in generation times between the different species of bacteria in the sample are minimized.[18] Firstenberg-Eden and Tricarico[21] revealed that 18°C is the appropriate temperature for monitoring samples that contain mixed flora (mesophilic and psychrotrophic) when enumerating total populations of bacteria using impedance. This is due to the fact that 18°C is the temperature at which generation times for the mesophiles and psychrotrophs are most similar at 1.5 and 1.2 hr, respectively.[21] Minimization of differences in generation times allows most of the genera in the sample to multiply at a similar rate and hence, gives a more accurate indication of the total populations present.

10.635 Inhibitors

Some foods contain natural compounds that are inhibitory to the growth of bacteria. Other foods such as processed products, may contain preservatives. Natural bacterial inhibitors or preservatives may prolong the lag time of the bacteria being enumerated and delay detection, giving a false estimation of the number of organisms in the original sample. Interference by inhibitors or preservatives may be detected by conducting assays on serial dilutions of the sample. If the diluted samples have similar detection times or shorter detection times than concentrated samples, then the presence of an inhibitor may be suspected.[19]

Ions in foods may interfere with the impedance signal and cause excessive drift by increasing or decreasing the impedance readings, or by reacting with components in the medium to produce compounds which mimic bacterial metabolites. Specifically, chloride (Cl^-) ions cause excessive drift and destroy the oxide layer coating on electrodes in the Bactometer module.[19] Bisulfite (HSO_3^-) ions can form $SO_3^=$ in neutral medium, which changes the conductivity component of the medium.

Inhibitory substances should be neutralized using dilution or separation procedures. Dilutions should only be conducted on those samples that contain a sufficient number of bacteria initially.

10.636 Selective Components

Selective components in commercial media may act as inhibitors when conducting impedance assays. A general knowledge of the components of the medium to be used is necessary for determining those species of bacteria that will be inhibited if selective components are present during impedance monitoring.

10.637 Damaged Cells

Sublethal injury to microorganisms may be caused by heating, cooling, freezing, or exposure to inhibitory or toxic substances. Sublethally injured cells would not be detected using a standard APC that is incubated for 48 hr, because in that time interval the bacteria would not have sufficient time to recover, reproduce, and produce a visible biomass. In an electrical assay, the accumulation of metabolites could be detected for cells which have been injured and must repair themselves before initiating replication. For processed foods that contain injured cells, it is useful to develop separate calibration curves.

10.638 Particulate Matter

Some foods contain significant quantities of particulate matter such as spices, that interfere with the electrical impulse as it passes between electrodes. For these foods, the sample should be diluted (if there are enough organisms to allow dilution) or filtered to remove the particulate matter. Another technique involves covering the electrodes with hot sterile plate count agar and then layering the inoculated medium over the plate count agar. As bacteria multiply and produce metabolic by-products, the metabolites diffuse into the plate count agar and change the impedance between the electrodes.

10.64 Typical Applications

In general, impedance assays require much less time and effort to conduct than traditional plate counts. With traditional plate counts, the sample must be diluted using serial dilutions, plated using duplicate plates, incubated, and then counted. For impedance assays, samples can be analyzed without using serial dilutions, provided they contain at least one viable bacterial cell and no more than 10^7 cells. Samples that contain greater than 10^7 cells must be diluted because the impedance threshold will be reached within 1 hr and the instrument will not be able to interpret such rapid impedance changes. In most cases, with impedance, the sample is diluted 1:10 into a microbiological growth medium and 1 mL of the growth medium is then placed into the Bactometer module well (Bactometer) or tube (RABIT, Malthus, and BacTrac). The module or tube is then incubated and impedance is read automatically by the instrument.

10.641 Estimation of Aerobic Plate Counts

To estimate APC using impedance, differences in the generation times of the species in the sample must be minimized by choosing the proper medium and temperature of incubation.[19] By minimizing differences in generation times of species in a sample, the microorganisms, although different, will multiply at a similar rate and will reach the detection threshold of 10^6–10^7 at similar times. Using this method, correlations between impedance and plate counts will be highest. By correlating DT to APC, future APC can be estimated based on impedance readings.

A commonly used procedure to estimate APC from foods of animal origin was described by Firstenburg-Eden.[17] A 10 g sample is placed into 90 mL of brain heart infusion broth (BHI) and stomached. One milliliter of the strained homogenate is then placed into a module well (Bactometer) or tube (RABIT, Malthus, and BacTrac). The module or tube is then incubated in the instrument at 30°C for 48 hr. Depending upon the initial microbial load in the sample, results may usually be obtained in 1 to 12 hours.

10.642 Estimation of Selected Groups of Microorganisms

Impedance has been used as a means of enumerating coliforms,[20,22,31,53,61,62] fecal coliforms,[33,46,58] mesophiles,[49] and *Enterobacteriaceae*[11] from samples of wastewater, dairy products, poultry, and meat. The growth of lactic acid bacterial starter cultures have been monitored using impedance.[3] Other researchers used impedance as a means of monitoring the metabolic activity of lactic acid bacteria in milk.[30] Direct and indirect impedance techniques have been used to estimate yeast populations in frozen fruit juices.[13,14] Using specific media and temperature, the growth of these groups of bacteria or fungi may be enhanced over competitors in mixed samples. By enhancing their growth, these organisms are able to multiply, produce enough metabolites to change the impedance of the medium, and be responsible for the detection time.

10.643 Estimation of Specific Bacteria

Using media and temperature that are exclusively selective for a particular microorganism, impedance may be used as a means of detecting and enumerating the organism from samples containing mixed populations. Media have been developed specifically for detecting *Salmonella* in foods using impedance.[2,8,12,15,24,35,36,38,41,42,59] Because bacterial reduction of trimethylamine oxide to trimethylamine is generally repressed under aerobic conditions, it was originally thought that detection of *Salmonella* using impedance would be optimal using anaerobic medium.[57] However, another researcher found that for *Salmonella* detection, impedance changes increased and time to detection decreased when impedance was conducted in aerobic conditions.[37]

Immunomagnetic separation procedures were combined with impedance as a means of capturing and concentrating *Salmonella* from preenriched broth medium.[40] The medium used was selenite cysteine/trimethylamine oxide/mannitol medium (SC/T/M). The authors demonstrated that *Salmonella* detection was enhanced by eliminating competition from other bacteria in dried milk samples using selective medium.[40]

Impedance methods for detecting *Salmonella* have been compared to conventional methods and have been found to be suitable for screening, since negative results can be obtained in less than 38 hr.[44] Others found that impedance outperformed rapid microbiological techniques, such as the Gene-Trak System and *Salmonella*-Tek.[45] Impedance as a means of detecting *Salmonella* has been given AOAC first action approval following a 17-laboratory collaborative trial.[25]

Listeria spp. have been detected by conducting impedance assays using a modified growth medium containing acriflavine, ceftazidime, nalidixic acid and esculin.[43] Other species of bacteria were able to grow in the medium; however, *Listeria* spp. could be differentiated from the others by the shape of the impedance curve. Glucose-enriched nutrient broth supplemented with proflavine and moxalactam has been used to detect *Listeria* spp. in spiked (10^3 cfu/gm) cheese.[27] Only Bolton and Gibson[7] have described a method for detecting *Listeria* spp. from unspiked foods. For further information regarding the use of impedance as a means of detecting pathogenic bacteria, please refer to an excellent review by Silley and Forsythe.[57]

A procedure to enumerate *Escherichia coli* (*E. coli*) from foods of animal origin in as little as 1 to 7 hours has been developed.[16] A significant linear correlation ($r^2 = 0.92$) was observed between $\log_{10}$ Petrifilm™ *E. coli* Count Plates (AOAC International, approved methodology number 991.14) and *E. coli* detection times (DT) using CM Medium. *E. coli* were enumerated from broiler chicken carcasses in 1 to 7.6 hours, which would allow processors to analyze products prior to shipment.

A capacitance method for enumerating *Pseudomonas fluorescens* from broiler carcass rinses was described by Russell.[47] Brain heart infusion broth (BHI) containing the following additives: nitrofurantoin (4 µg), carbenicillin (120 µg), and Irgasan (25 µg) per mL (BHI-PSA-patent accepted and pending) was able to effectively select for the growth of *Pseudomonas fluorescens* in samples containing mixed microflora, such as a broiler chicken carcass rinse. The time required to enumerate *P. fluorescens* was <22.4 hr. This method may be useful for enumerating *P. fluorescens* on chicken as a means of predicting the potential shelf-life of fresh products.

10.644 Estimation of Psychrotrophic Plate Counts

Psychrotrophic plate counts at 7°C for milk samples containing high numbers of psychrotrophs were highly correlated ($r = -0.95$) to impedance detection times, when samples were incubated at 18°C.[21] The authors found that by reducing the incubation temperature to 18°C, while conducting impedance measurements, the generation time of the psychrotrophic bacteria was similar to the mesophilic bacteria in the sample. By reducing the generation time, the psychrotrophs were able to multiply and change the impedance of the medium, which resulted in a detection time that was not influenced by the presence of mesophilic bacteria.[21]

A selective medium was developed for rapidly enumerating *Pseudomonas fluorescens* from broiler chicken carcasses using capacitance.[47] Log_{10} detection times were highly correlated ($r^2 = 0.96$) to psychrotrophic plate counts conducted at 7°C. The author concluded that capacitance, in combination with selective media, can be used to enumerate psychrotrophs from fresh chicken in less than 18 hours, as opposed to 10 days, as is required by the standard method.[48]

10.645 Estimation of Shelf-life

Estimation of shelf-life of fresh foods is based on enumerating the organism or group of organisms responsible for producing spoilage defects (odor, slime, discoloration) on the food. Mead and Adams[32] described the challenge of rapidly enumerating small numbers of psychrotrophs from samples containing much higher numbers of mesophiles by stating "the rapid isolation of psychrophilic spoilage bacteria is of interest in relation to various foods and numerous attempts have been made previously to develop selective media which would permit more rapid growth of the organisms at higher incubation temperatures whilst inhibiting the growth of mesophiles." The traditional method for enumerating psychrotrophs to predict shelf-life is described in section (Chapter 13) and requires 10 d at 7°C. After 10 d, most foods would have already been purchased and consumed before results could be obtained.

A rapid impedance based method for determining the potential shelf-life of milk has been reported.[5] In this study, milk samples were preincubated in plate count broth at 18 or 21°C for 18 hr. After preincubation, the samples were placed in module wells containing modified plate count agar (MPCA) and were incubated at 18° or 21°C. In another study, high correlation coefficients ($r^2 = 0.87$ and 0.88) were observed between the shelf-life of milk and impedance readings taken at 18 and 21°C, respectively.[4]

To rapidly predict the shelf-life of fresh fish, samples have been analyzed using conductance at 25°C in trimethylamine oxide nitrogen medium (TMAO).[29] Using this method, H_2S producing bacteria enumerated using conductance were highly correlated ($r^2 = 0.96$) to the shelf-life of fresh fish.

The potential shelf-life of fresh poultry can be predicted in less than 18 hr.[48] In this study, $\log_{10}$ capacitance detection times were highly correlated to day of storage ($r^2 = 0.86$).

10.646 Determination of Temperature Abuse

A series of studies were conducted to develop a method to rapidly determine the occurrence of temperature abuse of fresh poultry.[49–54] By incubating samples at 42°C, increases in populations of mesophilic bacteria as a result of temperature abuse were detected. To make the procedure more sensitive to small fluctuations in temperature abuse, coliforms were monitored at 43°C, such that significant increases of coliforms on groups of chicken carcasses were considered to be due to temperature abuse because, under refrigeration, populations of coliforms remain constant or decrease in number.[53]

10.647 *Sterility Testing of Ultra-High Temperature (UHT) Products*

Many shelf-stable foods are sterilized using ultra-high temperature processing (UHT). The impedance procedure is often used as a screen to insure that all viable cells or spores have been destroyed during the sterilization process. These assays are conducted to determine if any microbial growth can be detected, not for enumeration purposes; thus, equilibration of generation times is not an issue. Food samples may be pre-incubated with or without growth medium to allow any viable organisms or spore-forming bacteria to proliferate. Samples are then placed in one or more growth media and incubated at one or more temperatures (often 35° and 55°C are used to allow mesophiles and thermophiles to grow). Procedures for screening dairy products,[55] fruit juices,[56] and low acid canned foods[10] have been reported.

10.648 *Indirect Methods*

A method for indirectly monitoring yeast growth using conductance has been described.[63] As yeast proliferate, they evolve CO_2. The evolution of CO_2 causes the conductance of an alkaline solution, which is contained in an insert containing electrodes, to change. The indirect method is advantageous for measuring the growth of yeasts because the electrical signal is greatly magnified.[13] Using indirect methods, culture media containing high concentrations of salt or other inhibitory agents, that negatively influence the electrical signal when conducting direct impedance measurements, can be used for the detection of microorganisms without interference.[6] An indirect impedance procedure has also been reported for detecting the presence of *Escherichia coli* in potable water.[63] The authors stated that when compared with the Colilert® method, indirect impedance eliminated the need for confirmation and was found to be more rapid than Colilert®. It is also possible for the method to be used with samples containing high solids/low volume where membrane filtration would be inappropriate.

10.65 *Interpretation of Results*

10.651 *Calibration Curve Method*

Before using electrical methods for estimation of bacterial numbers in foods samples, a calibration curve should be developed for each "group" of products that will be assayed in the future. The calibration curve is used to define the relationship between the electrical method and a standard method (usually APC) or the electrical method and some other parameter, such as shelf-life.

To develop a calibration curve, 80 to 100 samples of a food product with varying levels (3–5 $\log_{10}$ cycles) of the organism(s) to be assayed are analyzed using the electrical and conventional methodologies. For example, to enumerate *E. coli* from milk, 100 samples of milk should be obtained that contain *E. coli* at levels of 10^1 to 10^5 and should be tested using electrical and conventional techniques.[19] These data are then analyzed using linear or quadratic regressions. Because there is error associated with using electrical methods and conventional methods, and the two methods are not measuring identical things (metabolites versus visible colonies), the data points will generally be scattered about the regression line. The correlation coefficient (**r**) is used to describe the correlation between the two methods. The closer **r** approaches –1.0, the higher the level of agreement between the two methods. The ideal **r** is negative as opposed to positive 1.0, because DTs are inversely proportional to APC or other bacterial counts. The correlation between electrical and conventional methods will vary based on the number of samples tested and the range of the concentration of microorganisms on the samples. In general, on samples where the generation time differences of the target organisms can be minimized, a correlation coefficient of **r** = 0.85 or higher is achievable if 80 to 100 samples containing levels of 10^1 to 10^5 are tested.[28]

When developing a calibration curve, several factors must be considered. Samples that contain few (< 10) microorganisms are subject to sampling error. Curves generated from these samples contain more scatter and lower correlation coefficients than samples containing higher numbers of microorganisms. In addition, when a sample contains few microorganisms, extended lag times may result in long and variable DTs.[28] For samples that contain more than 10^7 microorganisms/mL, the detection threshold is reached so rapidly, that the number of microorganisms cannot be distinguished. It is often necessary to remove extreme data points (low or high) from the calibration curve equation, so as to reduce the amount of error associated with the curve. Any data point that lies far outside the standard error should be evaluated to determine if 1) an error was made in calculating or recording the colony count, 2) an error was made in entering the data into the computer, 3) an inaccuracy of DT resulted from an abnormal curve, and 4) the data point was derived by assay of an unusual sample source.[28]

Most food products do not contain a naturally broad range of microorganisms to be used for calibration curve purposes. To increase counts artificially, food samples are temperature abused. If samples must be temperature abused to develop a calibration curve, every effort should be made to mimic actual abuse situations (temperature, moisture, and oxygen conditions) that would occur naturally, so as not to accidentally select for the growth of a microorganism or group of microorganisms that will interfere with the desired counts. Any alteration of the natural flora may result in inaccurate DTs.[28]

Calibration curves may also be influenced by the type of food being analyzed. A separate calibration curve must be developed for foods that contain ingredients that increase or decrease the growth of bacteria, i.e., sugars or preservatives. Foods should be grouped according to these factors and a calibration curve should be developed for each group.

10.652 *Sterility Method*

Foods that contain low levels of microorganisms may be tested using an "industry sterility" protocol. This protocol was developed such that DTs are achieved only if the product is contaminated with more microorganisms than the specified level. An appropriate dilution scheme is used to dilute out the normal background flora and allow the detection of higher numbers of organisms. Because dilutions are made, any detection means that the sample is contaminated with the organism at a level that exceeds the specified level. For example if samples normally contain 1 to 50 organisms and the specified level is 1000 cells/mL, a 1:100 dilution would be used. This method may be used for various foods and different levels of contamination, depending on the dilution scheme chosen. It is effective for foods in which a calibration curve cannot be developed because of low- or high-count samples, inability to calibrate, presence of inhibitors, etc. For the sterility technique to be effective, however, a large difference in counts must exist between acceptable and unacceptable samples.[28]

10.66 *References*

1. Allison, J. B., Anderson, J. A., and Cole, W. H. 1938. The method of electrical conductivity in studies on bacterial metabolism. J. Bacteriol. 36:571-586.

2. Arnott, M. L., Gutteridge, C. S., Pugh, S. J., and Griffiths, J. L. 1988. Detection of salmonellas in confectionary products by conductance. J. Appl. Bacteriol. 64:409-420.
3. Asperger, H., and Pless, P. 1994. Salmonella detection in cheese-comparison of methods in regard to the competitive microorganisms. Wiener Tierarztlichie Monatsschrift 81, 12-17.
4. Bishop, J. R., and White, C. H. 1985. Estimation of the potential shelf-life of pasteurized fluid milk utilizing bacterial numbers and metabolites. J. Food Prot. 48:663-667.
5. Bishop, J. R., White, C. H., and Firstenberg-Eden, R. 1984. Rapid impedimetric method for determining the potential shelf-life of pasteurized whole milk. J. Food Prot. 47(6):471-475.
6. Bolton, F. J. 1990. An investigation of indirect conductimetry for detection of some food-borne bacteria. J. Appl. Bacteriol. 69:655-661.
7. Bolton, F. J., and Gibson, D. M. 1994. Automated electrical techniques in microbiological analysis. In: Rapid Analysis Techniques in Food Microbiology ed. Patel, P. pp. 131-136. Glasgow:Blackie Academic and Professional.
8. Bullock, R. D., and Frodsham, D. 1989. Rapid impedance detection of salmonellas in confectionary using modified LICNR broth. J. Appl. Bacteriol. 66:385-391.
9. Cady, P. 1978. Progress in impedance measurements in microbiology. Pages 199-204 in: Mechanizing Microbiology. Sharpe, A. N., and Clark, D. S., eds. Charles C. Thomas, Springfield, Illinois.
10. Coppola, K., and Firstenburg-Eden, R. 1988. Impedance based rapid method for detection of spoilage organisms in UHT low acid foods. J. Food Sci. 53:1521-1524, 1527.
11. Cousins, D. L., and Marlatt, F. 1989. The use of conductance microbiology to monitor Enterobacteriaceae levels. Dairy, Food, Environ. Sanit. 9(10):599.
12. Davda, C., and Pugh, S. J. 1991. An improved protocol for the detection and rapid confirmation within 48 h of salmonellas in confectionary products. Lett. Appl. Microbiol. 13:287-290.
13. Deak, T., and Beuchat, L. R. 1993. Comparison of conductimetric and traditional plating techniques for detecting yeasts in fruit juices. J. Appl. Bact. 75:546-550.
14. Deak, T., and Beuchat, L. R. 1993. Evaluation of the indirect conductance method for the detection of yeast in laboratory media and apple juice. Food Microbiol. 10:255-262.
15. Easter, M. C., and Gibson, D. M. 1985. Rapid and automated detection of Salmonella by electrical measurements. J. Hyg. 94: 245-262.
16. Edmiston, A. L., and Russell, S. M. 1998. A rapid conductance method for enumerating Escherichia coli from broiler chicken carcasses. Submitted to J. Food Prot.
17. Firstenberg-Eden, R. 1983. Rapid estimation of the number of microorganisms in raw meat by impedance measurement. Food Technol. 37: 64-70.
18. Firstenberg-Eden, R. 1985. Electrical impedance method for determining microbial quality of foods. in: Rapid Methods and Automation in Microbiology and Immunology. Habermehl, K. O., ed. Springer-Verlag, Berlin. p. 679-687.
19. Firstenberg-Eden, R., and Eden, G. 1984. Impedance Microbiology. Research Studies Press LTD. Letchworth, Hertfordshire, England.
20. Firstenberg-Eden, R., and Klein, C. S. 1983. Evaluation of a rapid impedimetric procedure for the quantitative estimation of coliforms. J. Food Sci. 48:1307-1311.
21. Firstenberg-Eden, R., and Tricarico, M. K. 1983. Impedimetric determination of total, mesophilic and psychrotrophic counts in raw milk. J. Food Sci. 48:1750-1754.
22. Firstenberg-Eden, R., Van Sise, M. L., Zindulis, J., and Kahn, P. 1984. Impedimetric estimation of coliforms in dairy products. J. Food Sci. 49:1449-1452.
23. Gibson, D. M. 1985. Predicting the shelf-life of packaged fish from conductance measurements. J. Appl. Bacteriol. 58:465-470.
24. Gibson, D. M. 1987. Use of conductance measurements to detect growth of Clostridium botulinum in a selective medium. Lett. Appl. Microbiol. 5:19-21.
25. Gibson, D. M., Coombs, P., and Pimbley, D. W. 1992. Automated conductance method for the detection of Salmonella in foods-collaborative study. J. AOAC Int. 75:293-302.
26. Gibson, D. M., Ogden, I. D., and Hobbs, G. 1984. Estimation of the bacteriological quality of fish by automated conductance measurements. Int. J. Food Microbiol. 1:127-134.
27. Hancock, I., Bointon, B. M., and McAthey, P. 1993. Rapid detection of Listeria species by selective impedimetric assay. Lett. Appl. Microbiol. 16:311-314.
28. Hartman, P. A., Swaminathan, B., Curiale, M. S., Firstenburg-Eden, R., Sharpe, A. N., Cox, N. A., Fung, D. Y. C., and Goldschmidt, M. C. 1992. Rapid methods and automation. *In*: Compendium of Methods for the Microbiological Examination of Foods. 3rd ed. Vanderzant, C., and Splittstoesser, D. F., eds. American Public Health Association, Washington, D. C. pp. 688-693.
29. Jørgenson, B. R., Gibson, D. M., and Huss, H. H. 1988. Microbiological quality and shelf-life prediction of chilled fish. Int. J. Food Microbiol. 6:295-307.
30. Lanzanova, M., Mucchetti, G., and Neviani, E. 1993. Analysis of conductance changes as a growth index of lactic acid bacteria in milk. J. Dairy Sci. 76:20-28.
31. Martins, S. B., and Selby, M. J. 1980. Evaluation of a rapid method for the quantitative estimation of coliforms in meat by impedimetric procedures. Appl. Environ. Microbiol. 39:518-524.
32. Mead, G. C., and Adams, B. W. 1977. A selective medium for the rapid isolation of pseudomonads associated with poultry meat spoilage. Br. Poultry Sci. 18:661-670.
33. Mischak, R. P., Shaw, J., and Cady, P. 1976. Specific detection of fecal coli by impedance measurement. Abst. Ann. Meet. Am. Soc. Microbiol. P3, p. 187.
34. Ogden, I. D. 1986. Use of conductance methods to predict bacterial counts in fish. J. Appl. Bacteriol. 61:263-268.
35. Ogden, I. D. 1988. A comductance medium to distinguish between Salmonella and Citrobacter spp. Int. J. Food Microbiol. 7:287-297.
36. Ogden, I. D. 1990a. Salmonella detection by a modified lysine conductance medium. Lett. Appl. Microbiol. 11:69-72.
37. Ogden, I. D. 1990b. The effect of air on the response of salmonellas in conductance media. Lett. Appl. Microbiol. 11:193-196.
38. Ogden, I. D., and Cann, D. C. 1987. Modified conductance medium for the detection of Salmonella spp. J. Appl. Bacteriol. 63:459-464.
39. Owens, J. D., Thomas, D. S., Thompson, P. S., and Timmerman, J. W. 1989. Indirect conductimetry: a novel approach to the conductimetric enumeration of microbial populations. Lett. Appl. Microbiol. 9:245-249.
40. Parmar, N., Easter, M. C., and Forsythe, S. J. 1992. The detection of Salmonella enteritidis and Salmonella typhimurium using immunomagnetic separation and conductance microbiology. Lett. App. Microbiol. 15:175-178.
41. Pettipher, G. L., and Watts, Y. B. 1989a. Effect of carbohydrate source in selenite cystine trimethylamine oxide broth on the detection of salmonellas using the Bactometer. Lett. App. Microbiol. 9:241-242.
42. Pettipher, G. L., and Watts, Y. B. 1989b. Evaluation of lysine-iron-cystine-neutral red for the detection of salmonellas in the Bactometer. Lett. App. Microbiol. 9:243-244.
43. Phillips, J. D., and Griffiths, M. W. 1989. An electrical method for detecting Listeria spp. Lett. Appl. Microbiol. 9:129-132.
44. Pless, P., Futschik, K., and Schopf, E. 1994. Rapid detection of salmonellae by means of a new impedance-splitting method. J. Food. Prot. 57:369-376.
45. Quinn, C., Ward, J., Griffin, M., Yearsley, D., and Egan, J. 1995. A comparison of conventional culture and three rapid methods for the detection of Salmonella in poultry feeds and environmental samples. Lett. App. Microbiol. 20:89-91.
46. Rowley, D. B., Vandemark, P., Johnson, D., and Shattuck, E. 1979. Resuscitation of stressed fecal coliforms and their subsequent detection by radiometric and impedance techniques. J. Food Prot. 42:335-341.

47. Russell, S. M. 1997a. A rapid method for enumeration of Pseudomonas fluorescens from broiler chicken carcasses. Journal of Food Protection 60(4):385-390.
48. Russell, S. M. 1997b. A rapid method for determination of shelf-life of broiler chicken carcasses. Journal of Food Protection 60(2):148-152.
49. Russell, S.M., Fletcher, D. L., and Cox, N. A. 1992a. A model for determining differential growth at 18 and 42 C of bacteria removed from broiler chicken carcasses. J. Food Prot. 55:167-170.
50. Russell, S.M., Fletcher, D. L., and Cox, N. A. 1992b. A rapid method for the determination of temperature abuse of fresh broiler chicken. Poultry Science 71 (8):1391-1395.
51. Russell, S. M., Fletcher, D. L., and Cox, N. A. 1993a. Effect of freezing on recovery of mesophilic microorganisms from broiler chicken carcasses subjected to temperature abuse. Poultry Sci. 73(5):739-743.
52. Russell, S. M., Fletcher, D. L., and Cox, N. A. 1993b. The effect of incubation temperature on recovery of mesophilic bacteria from broiler chicken carcasses subjected to temperature abuse. Poultry Sci. 73(7):1144-1148.
53. Russell, S. M., Fletcher, D. L., and Cox, N. A. 1995. Comparison of media for determining temperature abuse of fresh broiler carcasses using impedance microbiology. Journal of Food Protection 58(10):1124-1128.
54. Russell, S. M., Fletcher, D. L., and Cox, N. A. 1996. The effect of temperature mishandling at various times during storage on detection of temperature abuse of fresh broiler chicken carcasses. Poultry Sci. 75:261-264.
55. Schaertel, B. J., and Firstenburg-Eden, R. 1986. Impedimetric determination of low level microbial contamination in UHT processed dairy products. Abstr., 46th Ann. Mtg., Inst. Food Technol., abstr. 85.
56. Schaertel, B. J., and Firstenburg-Eden, R. 1986. Impedimetric measurement of microbial contamination in aseptically packaged fruit juices. Abstr., 46th Ann. Mtg., Inst. Food Technol., abstr. 130.
57. Silley, P., and Forsythe, S. 1996. Impedance microbiology-a rapid change for microbiologists. J. Appl. Bacteriol. 80:233-243.
58. Silverman, M. P., and Munoz, E. F. 1979. Automated electrical impedance technique for rapid enumeration of fecal coliforms in effluents from sewage treatment plants. Appl. Environ. Microbiol. 37:521-526.
59. Smith, P. J., Bolton, F. J., Gayner, V. E., and Eccles, A. 1990. Improvements to a lysine medium for detection of salmonellas by electrical conductance. Lett. Appl. Microbiol. 11:84-86.
60. Stewart, G. N. 1899. The changes produced by the growth of bacteria in the molecular concentration and electrical conductivity of culture media. J. Exp. Med. 4:235-247.
61. Strauss, W. M., Malaney, G. W., and Tanner, R. D. 1984. The impedance method for monitoring total coliforms in wastewaters. Folia Microbiol. 29:162-169.
62. Tenpenny, J. R., Tanner, R. D., and Malaney, G. W. 1984. The impedance method for monitoring total coliforms in wastewaters. Folia Microbiol. 29:170-180.
63. Timms, S. K., Colquhoun, O., and Fricker, C. R. 1996. Detection of Escherichia coli in potable water using indirect impedance technology. J. Microbiol. Meth. 26:125-132.
64. van Spreekins, K. J. A., and Stekelenburg, F. K. 1986. Rapid estimation of the bacteriological quality of fresh fish by impedance measurements. Appl. Microbiol. Biotechnol. 24:95-96.

10.67 Commercial References

Bactometer Microbial Monitoring System M128 (BioMérieux Vitek, Inc., Hazelwood, MO)
Rapid Automated Bacterial Impedance Technique (RABIT-Don Whitley Scientific Limited, UK)
Malthus V (Malthus Diagnostics, Inc., North Ridgeville, OH)
BacTrac 4100 (Sy-Lab, Vertriebsges.m.b.H., Purkersdorf, Austria)
Gene-Trak Systems (Hopkinton, MS)
Salmonella Tek™ (Organon Teknika Corporation, Durham, NC)
Petrifilm E. coli Count Plates (3M Microbiology Products, St. Paul, MN)
CM Medium (bioMérieux Vitek, Inc., Hazelwood, MO)
Colilert® (IDEXX Laboratories, Inc., Westbrook, ME)

10.7 PHAGE PROBES

10.71 Introduction to Bacterial Viruses

10.711 History of Discovery

The existence of bacterial viruses known as bacteriophages—*eaters of bacteria*—was first reported by the British scientist F.W. Twort in 1915, and somewhat controversially in 1917 by the Canadian bacteriologist F. d'Herelle. Twort's publication in the Lancet documented the phenomenon of "glassy transformation" observed in colonies of micrococci grown on agar in the presence of small pox vaccine. He postulated that the "transparent dissolving material" might be caused by: i) an enzyme; ii) could be part of the life cycle of the bacterium; or iii) was caused by a *virus*, a term first used vaguely by Jenner in 1796 and used more specifically in 1898 by the Dutch botanist Martinus Beijerinck in his study of Tobacco Mosaic Virus.[114]

d'Herelle, however, was much more certain of the viral nature of this bacteriolytic agent, subsequently published a book in 1921 entitled *Le bactériophage: son rôle dans immunité*. However, he mistakenly assumed that a single bacteriophage existed,[40] and that it could act against a number of different species of bacteria due to adaptive variation.[88] This is, in hindsight, not a surprising assumption when one considers that virus taxonomy and classification was still many years away from the generally accepted system in use today.[39] Australian scientist Burnett, between 1924 and 1934, disproved d'Herelle's theory of a single bacteriophage, and went on to identify a myriad of viruses, which while exhibiting different physical and biological properties, were able to maintain these characteristics despite repeated subculturing.[88]

Significant scientific advances were to be made in the study of bacteriophages during the late 1930s to 50s. In particular, the formation of the Phage Group—a group of scientists to include Max Delbruck, Alfred D. Hershey, Salvador E. Luria, and James D. Watson—played an enormous role in both the beginnings of molecular biology and the elucidation of the structure of DNA. During this period, researchers showed that viruses were composed of either DNA or RNA with a protein component, and in 1946, Hershey demonstrated that bacterial viruses could undergo genetic recombination. In 1952, Zinder and Lederberg first demonstrated host transduction (the transfer of genetic information via a virus particle) by bacteriophage P22, while Brenner, Jacob and Meselson discovered messenger RNA in 1961.[40] The presence within the Phage Group of the electron microscopist Anderson enabled the identification of viral architecture, a feature now used by the International Committee on Taxonomy of Viruses (ICTV) to aid in identification and classification of viruses along with nucleic acid information.[39] The attachment of bacteriophages to bacterial cells was also demonstrated by electron microscopy. Ultimately this led to the elucidation of replication of these parasites within the host cell.

10.712 Phage Replication

Phages, in common with other viruses, do not possess the biochemical machinery required to replicate themselves, and thus rely on the assistance of an actively metabolizing bacterial cell to do the job for them. As phages have no means of actively seeking out replicative hosts, coming into contact with one is a completely random event, relying on their interaction during movement through a medium.[112] Replication will only follow if the cell contains specific receptor sites for the phage in question, such as cell wall lipopolysaccharides, proteins, teichoic acids, pili or flagella. Variations within receptor groups further increase the phage's specificity for a particular host, with the result that bacteriophages are typically specific in their host requirements to at least the genus level, and usually beyond.[35] The replicative cycle can differ depending on the phage, but there are sufficient commonalities to use T4[100] and Lambda[22] phages as general examples of the lytic and lysogenic replicative cycles respectively.

10.7121 Attachment and infection. Attachment, as described above, occurs when a phage encounters a bacterial cell possessing a particular receptor specific for that particular phage. In the case of the extensively studied DNA bacteriophage T4, fibers associated with its contractile tail mediate attachment, and as more fibers make contact with the receptor, the tail plate of the phage settles down onto the surface of the cell. Thus attachment is generally a two-step event involving first a reversible step, followed by irreversible binding. For each phage-host combination, various salt concentrations (in particular calcium and magnesium ions) govern the rate of attachment, influencing the formation of electrostatic bonds between tail fibers and cell receptors. Either reduction or removal of these ions, or receptor mutations will effectively prevent adsorption and subsequent phage multiplication. After attachment the next step of phage replication follows: infection of the host.

First, conformational changes occur to the adsorbed tail base plate and sheath, and the tail sheath contracts. The central tube of the phage tail then enters through the bacterial wall and DNA is pushed out of the phage head, similar to the action of a syringe, and into the host cell. This constitutes infection of the cell, although it can also occur by other means, such as attachment to pili and transfer of DNA either through or along the pilus. Shortly after introduction of phage DNA into the bacterium, synthesis of host DNA and protein is suppressed. Two events govern the subsequent synthesis of phage DNA: first, the degradation of host DNA into its nucleotide components occurs, providing building blocks for viral DNA synthesis and preventing further host gene expression. Second, modification of bacterial RNA polymerase occurs, with the result that it now recognises viral promoters, thus promoting the transcription of viral genes rather than host genes. Viral DNA synthesis commences within 5 minutes of infection, followed by synthesis of proteins responsible for phage structure, assembly, and cell lysis and phage release. It takes approximately 15 minutes for the first complete T4 particles to appear following infection, and another 7 minutes for cell lysis.[112]

10.7122 The lytic versus lysogenic cycle Two modes of replication have been observed for bacteriophages: i) the lytic cycle, where replicated DNA is packaged into phage heads and subsequently released from the bacterial host by cell lysis; such phages (e.g. T4) are described as virulent, and ii) the lysogenic cycle, where phage DNA is incorporated into and replicated within the bacterial genome; in this case phages are termed temperate, the best characterized being the double-stranded DNA phage lambda. Virulent phages will typically lyse and destroy susceptible hosts (although some filamentous phages are secreted without cell lysis), while temperate phages can adopt either mode. The choice between the lytic and lysogenic cycle depends on the relative expression rates of phage repressor encoded by the *cII* gene (promoting lysogeny) and the *cro* protein, capable of turning off repressor gene expression and starting the lytic pathway.[22] Following infection of susceptible cells with lambda phage, a very small proportion will be lysed but the majority will survive to become lysogens. These infected cells continue to replicate as normal, creating a clone of cells containing phage DNA. The latent form of the phage genome remaining in the host is termed the prophage, and its presence prevents superinfection of the host with the same virus. Temperate phages can be induced, in particular by agents causing DNA damage, such as UV light or chemical mutagens, and once induced will enter the lytic cycle, resulting in production of phage at the expense of the bacterial host. Two classes of genes are involved in lysogeny and induction; those establishing the lysogenic state, and those maintaining the prophage in a repressed state within the bacterial genome. Mutations in either group of genes will result in loss of lysogenic function, and depending on the class involved mutants can either resort to lytic activity, lose the prophage completely, or be unable to release viable phage particles.[45]

10.7123 Host Implications. The lysogenic relationship between host and phage can be mutually beneficial. The incorporation of phage DNA into the bacterial genome is considered a survival tactic of the virus, and typically occurs under adverse conditions such as host nutrient deprivation or a high multiplicity of infection where the host cells are in danger of being completely destroyed, exposing phage to environmental hazards. One of the most significant properties conferred to host bacteria following lysogenization by a temperate phage is that of toxin production. Toxins including Shiga-like toxins in *Escherichia coli*, botulinum toxins in *Clostridium botulinum*, enterotoxins in *Staphylococcus aureus*, diphtheria toxin in *Corynebacterium diphtheriae*, and pyrogenic toxins in *Streptococcus pyogenes* have all been identified as proteins produced as a result of "phage conversion" of the bacterial host.[63] Another benefit of lysogenic infection has been reported in the literature as a type of "suicide bombing" where a small proportion of *Bacillus subtilis* cells infected with a phage release virus, thus killing off any susceptible strains present and giving the remaining bacteria (those which can safely harbor the phage) a competitive edge.[3]

Although it is mentioned above that phage infection can be an advantage to the host, a proportion of the bacterial population is going to be lysed as the virus takes over the replicative machinery of the cell and puts it to the task of manufacturing more viral particles. This has been problematic in fermentation processes, where starter culture failures can occur as a result of bacteriophage infection, and control is difficult particularly when a large number of variable phages are continually introduced into the processing environment via raw materials such as raw milk,[20] and even from the starter cultures themselves.[124]

10.713 Microbiological Applications of Bacteriophages

10.7131 Bacterial typing. The use of characterized bacteriophages to type bacteria beyond the species and serotype level is well known, and perhaps the best recognized system is the International Typing System for *Listeria monocytogenes* defined in 1985 by Rocourt *et al.*[119] developed following the successful application of phage typing to other bacteria such as *Staphylococcus aureus*.[122] Identification of new bacteriophages has resulted in an optimized typing system,[90] as well as the development of alternate bacteriophage typing sets[78] and a reversed phage typing procedure.[38,79] Bacteriophage typing has also proven useful as an epidemiological tool in the study of food poisoning outbreaks associated with *L. monocytogenes*, *Bacillus cereus*,[1] *Salmonella enteritidis*,[17,155] and *E. coli* O157:H7.[8,49,123] This allows the epidemiologist to confirm the relationship between outbreak strains isolated from infected patients and those cultured from suspected food samples. In addition, environmental sources of contamination can be identified, their movement within food processing plants monitored, and persistent problems distinguished from sporadic contaminants.[80]

10.7132 Biological indicators. The most frequently used phages as biological indicators are those of the coliphage group. Their use as indicators of water quality was proposed based on their occurrence in waters along with their bacterial hosts, with numbers of phage theoretically reflecting the numbers of host bacteria present. Several studies have utilized these phages instead of the host strains as an indication of fecal pollution,[18,103] and evidence suggests that they are as good as, or indeed better indicators than bacterial systems currently in use. A coliphage-based water assay is described in the Standard Methods for the Examination of Water and Wastewater,[2] and using *E. coli* strain C as the host bacterium, five or more phages/100 mL of water can be detected within 4–6 hours. Problems have been encountered with naturally occurring coliphages in environmental waters, which affect the correlation between phages and fecal pollution. However, viruses tend to survive better than coliforms in water, and have been shown to exhibit a higher resistance to chlorine. Their application as indicators for coliforms in foods has seen limited study.[68, 69] Overall this method may be useful as an alternative to *E. coli* or coliform determinations, or as an indicator of enterovirus survival.[67]

10.7133 Cloning vectors. Extensive genetic characterization of certain bacteriophages has seen their extensive application as cloning vectors for amplification of DNA in *E. coli.* Indeed, phage Lambda was the prototype vector long before recombinant molecules could be made *in vitro.* Lambda phage offers a number of attractive features for cloning: recombinant genomes can be efficiently recovered in packaged phage heads, plaques provide a large amount of DNA for hybridization studies, and cloned genes can be easily analyzed. Phage M13 has been particularly useful in cases where single-stranded DNA is required, such as in site-directed mutagenesis, while phage P1 offers a large cloning capacity of up to 100 kilobases. P1 has also been demonstrated to have a wide host range, which could facilitate the transfer of DNA to bacteria other than *E. coli.*[101]

10.7134 Phage therapy. A prophetic aspect of the discovery of bacteriophages was the suggestion by d'Herelle that bacteriophage could be used as a form of therapy for the treatments of 'intestinal disturbances', bacillary dysentery, staphylococcal infections, and bubonic plague.[33] While this idea was viewed, certainly by d'Herelle himself, as perhaps the universal cure for all bacterial infections, it was never explored to any great extent, and indeed with the advent of antibiotics was all but discarded. However, with the increasing incidence of bacterial resistance to certain antibiotics, bacteriophage therapy has been revisited in the 1990s.[75,98,115,132]

While scientists such as Lederberg[73] have been reluctant to completely rule out the usefulness of phage therapy, a number of pertinent questions remain unanswered in the wake of work by Merril et al.[98] to isolate and utilize long-circulating mutants of *Escherichia coli* phage and *Salmonella typhimurium* phage P22. An identified alteration to the phage head protein was found to be responsible for the mutants' success in avoiding elimination by the reticuloendothelial system of test mice, one of the more important problems to be addressed in phage therapy. Despite this work, scepticism remains for many reasons such as the potential for bacteria to develop resistance to phages (a phenomenon which may be slowed by the use of phage combinations) and a lack of hard data despite claims that phage therapy has been successfully used in parts of Europe for many years.[115]

10.7135 Bacterial detection. As a result of the usefulness of phages in the typing of bacteria, and as cloning vectors, their ability to identify microorganisms to at least the genus level has seen their increased application in techniques for the detection of important food-associated bacteria. Ultimately, their specificity dictates the success of any phage-based system, and thus identification of useful phages for species- or strain-specific detection of bacteria is of the utmost importance. The next section of this chapter will focus on the use of phages in a variety of molecular, biochemical and physical methods applicable to the microbiological analysis of foods, some of which are in use, and their imminent application in more recently developed technologies.

10.72 Bacteriophage-Based Detection Methods

10.721 Phage-Based Reporter Genes

10.7211 *lux*-based bioluminescence. The use of bacterial bioluminescence, in particular the application of *lux* genes, and aspects of its molecular organization and regulation have been extensively reviewed,[7,59,60,96,97,127,136-139] and readers are referred to these publications for a comprehensive introduction. Briefly, bacterial bioluminescence is a light-emitting reaction involving the oxidation of reduced flavin mononucleotide ($FMNH_2$) and a long chain aliphatic aldehyde, such as tetradecanal (considered the natural substrate of the reaction) or dodecanal, by molecular oxygen. The reaction is catalysed by intracellular luciferase according to the following equation:

$$FMNH_2 + RCHO + O_2 \; FMN + RCOOH + H_2O + LIGHT$$

where light emission is in the form of a glow with an estimated quantum yield (photons of light produced per molecule of substrate) of 0.1.[19] Given that the production of $FMNH_2$ from flavin mononucleotide is essential to the reaction, only living cells capable of electron transport will be able to produce this blue-green light (490nm emission), and bioluminescence is thus an excellent means of establishing the metabolic status of an organism. Light can be measured by instrumentation including luminometers, scintillation counters and sensitive photon counting imagers such as the BIQ Bioview Image Quantifier (Cambridge Imaging, Cambridge, U.K.) which is capable of detecting single bioluminescing cells.[64] Naturally bioluminescent bacteria have been isolated from marine, freshwater, and terrestrial environments, and the majority belong to one of four genera: *Vibrio*, *Alteromonas*, *Photobacterium*, or *Xenorhabdus*.[7] In symbiotic relationships, bacterial bioluminescence is thought to serve as a means of communication for the host, but its purpose is less clear for free-living bacteria.[97]

A heterodimeric enzyme, bacterial luciferase is composed of two subunits, (40kDa) and (36kDa), respectively encoded by the *luxA* and *luxB* genes. These genes are flanked by *lux* genes *C*, *D*, and *E*, respectively coding for reductase, transferase, and synthetase proteins responsible for the production of aldehydes from fatty acids. These five genes compose the *lux* operon, transcribed in the order *luxCDABE*, and these genes have been cloned and sequenced for some luminescent *Vibrio*, *Photobacterium*, and *Xenorhabdus* spp.[97] Other *lux* genes have since been identified in *V. fischeri*, including a second operon consisting of regulatory genes *luxI* and *luxR* located immediately upstream from the *lux* operon. These genes code for an autoinducer and receptor required for activation and expression of luminescence, in conjunction with another gene, *luxG*, encoding a reductase involved in the production of $FMNH_2$.[97]

Transfer of *lux* genes has been approached from a number of angles, depending on both the organisms to be transformed, and the application of the bioluminescent phenotype. Plasmids, containing either the complete *luxCDABE* operon, *luxAB* luciferase genes, or fused *luxAB* genes, have been constructed and employed to transfer a bioluminescent phenotype to a desired bacterial recipient.[60] The complete operon encodes substrate in addition to luciferase, thus providing a continuously luminescent organism. Innate substrate production provides a non-destructive means of real-time monitoring of luminescent bacteria in mixed culture and within eukaryotic cells, but is only useful for gram-negative organisms capable of recognizing transcription and translation signals originating from bioluminescent bacteria. Additionally, continuous expression of substrate places energy constraints on the cell, and is therefore not suitable for physiological studies.

Alternatively, only the genes for luciferase are required if substrate is added exogenously, and this does not significantly influence cell metabolism. Bacteria expressing *luxAB* are therefore non-luminescent before substrate is added, and this less complicated procedure is favored for use with gram-positive bacteria where a promoter also has to be engineered for expression. Another option when working with gram-positive bacteria is the use of *luxAB* fusions. A monomeric fusion product was reported to increase bioluminescence levels in *Bacillus subtilis* about 100 times more than the use of unfused *luxAB* genes. Furthermore, this increased light output was comparable to that obtained from a control *E. coli* strain demonstrating optimal bioluminescence levels.[60] Transposon vectors have also been engineered to contain *lux* genes without promoters, such that the genes are only expressed when the transposon inserts into the bacterial chromosome beside a promoter region.[60] In addition, the source of luciferase genes influences the temperature stability of the resulting enzyme, and should be carefully considered based on the growth temperature of the microorganism of interest and the application of the bioluminescent phenotype. It is worth noting that fusion genes produce products with temperature optima 7–10°C lower than that of the wild type.[60]

Such a variety of alternatives has resulted in the transformation of a number of both gram-negative and gram-positive bacteria, and these now bioluminescent microorganisms have most commonly been used as reporters of, for example, bacterial injury and repair, biocide efficacy,[154] regulation of gene expression[36,107,127] and survival of pathogens such as *Salmonella enteritidis*[27] and *Escherichia coli* O157:H7 in food processing.[65,146]

Bioluminescence-based detection methods for bacteria have also been developed by the use of a third gene vector: bacteriophage engineered to carry *lux* genes within their genome. Depending on the host-specificity range of these phage vectors, generic or specific detection of microorganisms can be achieved by bacterial expression of bioluminescence within 30–50 minutes of infection and transduction by *lux*+ phage unable to express the bioluminescent phenotype themselves.[71] This approach to bacterial detection, first described by Ulitzer and Kuhn,[149] has led to a number of microbiological applications, in particular the detection of various foodborne and clinically important pathogens.

A near on-line method for detecting enteric bacteria using recombinant *lux* bacteriophage was subsequently proposed by Kodikara et al.[71] Using a slaughterhouse processing pig carcasses as a test environment, the authors reported the ability of recombinant *lux*+ Enterobacteriaceae-specific phages to convert enteric bacteria sampled from carcasses, meat-contact factory surfaces, and various abattoir areas into bioluminescent bacteria within one hour of sampling. The limit of detection of this assay was around 10^4 target cells per g or cm^2, but was substantially improved by the incorporation of a 4 hour enrichment in Luria broth at 37°C, reducing the detection limit to less than 10 enterics per g or cm^2 with a required total assay time of 5 hours. Despite these promising results and the applicability of such a rapid detection system in a Hazard Analysis Critical Control Points (HACCP) system, industrial application of such on-line analysis has not been adopted.[136] Regardless, this research prompted significant research activities in the area of bacterial detection, and alternative applications have been forthcoming.

In 1993, Turpin et al.[147] reported the development of a phage-based luminescent most-probable-number (MPN) method for the detection of *Salmonella typhimurium*, employing a commercially available *lux*+ P22 *Salmonella* phage (Amersham International). A 15-tube MPN test protocol was established, with each of 0.1, 1, and 10 mL volumes of samples (including soils, sewage sludge and water) artificially inoculated with *S. typhimurium* pipetted into five tubes of sterile buffered peptone water. After overnight enrichment, samples were tested by both the MPN method (inoculated into Rappaport-Vassiliadis broth) and the *lux*-based protocol where samples were inoculated into Luria broth containing 6.9×10^{10} plaque forming units (PFU) of *lux*-P22 phage. MPN samples were incubated at 43°C overnight, streaked onto xylose lysine desoxycholate and brilliant green agars after 24 and 48 hour incubations, and presumptive colonies were biochemically confirmed using the API system of identification. *lux* samples were incubated for 90 minutes at 30°C and aldehyde substrate was then added. Subsequent luminescence was measured using a luminometer, with positive tubes producing relative light outputs 100 to 1000 times greater than negative samples. When compared with the MPN method, the *lux*-based method had an efficiency of 100% with no false positive or negative results. A *S. typhimurium* detection limit of 10 colony forming units (CFU) per gram of soil or per 100 mL of lake water was demonstrated in the presence of large numbers of competing bacteria, thereby reducing the requirements for selective enrichment and decreasing the duration of the protocol to less than 24 hours.

Three recombinant *lux* phages, containing either *luxAB* or the entire *lux* operon, were later constructed by Chen and Griffiths[26] to expand the detection range to *Salmonella* isolates belonging to different serogroups. One hundred percent of the *Salmonella* strains belonging to groups B and D, and selected group C isolates tested were identified using a recombinant phage plaque assay, while all of the 22 non-*Salmonella* isolates tested negative. The optimum adsorption time and temperature for subsequent phage infection and luminescence was determined to be 30 minutes at 42°C, producing detectable luminescence in less than one hour for *Salmonella* concentrations of 10^8 CFU/mL. Employing a 6 hour pre-incubation step, as few as 10 CFU/mL of *Salmonella* could be reliably detected, which was reported to agree well with previous estimates using a similar detection system. The method was then evaluated in a food system by directly inoculating *S. enteritidis* into the white of eggs followed by recombinant phage addition. After a 16 hour incubation period, luminescence was detected in eggs inoculated at an initial level of about 10^3 CFU, but lower concentrations of 100 and 10 CFU of *S. enteritidis* per egg went undetected. However, increasing the incubation period to 24 hours resulted in luminescence detection at all levels, with the opportunity to observe the internal location of luminescent organisms by CCD imaging of the whole egg. Additional protocols using both i) *E. coli* Petrifilm and X-ray autoradiography, and ii) liquid media and luminometry were evaluated and demonstrated to be useful alternatives to the original methodology, offering a less expensive means of light detection and a more portable assay system.

Similarly, Loessner et al.[83] reported the construction and use of a *lux*$^+$ reporter bacteriophage, based on the specificity of *Listeria* phage A511 for about 95% of *Listeria monocytogenes* isolates belonging to serogroups 1/2 and 4. A *V. harveyii luxAB* gene fusion was inserted just downstream of the strong promoter for the major capsid protein gene *cps*. Using the recombinant phage A511::*luxAB*, as few as 500 CFU/mL of tested isolates could be detected after phage infection and a two hour incubation at 20°C. To increase the sensitivity even further, *Listeria* selective enrichment broth was utilized to enrich artificially contaminated lettuce samples over a 16 hour period at 30°C before phage-based testing. The sensitivity of detection was subsequently increased to less than one cell of *L. monocytogenes* Scott A per gram of salad, with a total time investment of 22–24 hours. While the bacteriophage used was reported to be highly specific for certain *L. monocytogenes* serogroups, it has a very broad host range with demonstrable activity against *L. ivanovii* and *L. seeligeri* isolates. Therefore, this particular phage will only gain application as a generic detection method for *Listeria* spp., assuming that it is also capable of infecting other species within the genus such as *L. innocua*, which has been proposed as a useful indicator for the presence of *L. monocytogenes*.

The detection of *L. monocytogenes* Scott A in various artificially contaminated foods by the above method was further examined by Loessner's research group,[85] and a MPN-based method was also evaluated. The method was shown to reliably detect *L. monocytogenes* at a level of 100 CFU/g or less in all tested foods following a 20 hour selective incubation, with even lower detection limits for liverwurst, shrimp, pasteurized milk, cottage cheese, ricotta cheese, chocolate pudding, and cabbage. Detection of lower levels of target cells was improved for minced meat, liverwurst, shredded Swiss cheese, and cabbage samples following an increase in enrichment time to 44 hours, although detection sensitivity never improved beyond 10 CFU/g for minced meat and Camembert, likely due to interference by a large microbial background flora. Naturally bioluminescent bacteria were also responsible for non-specific light emission by a shrimp sample, but this was easily identified due to light emission by a negative sample not containing *lux* phage. Naturally contaminated food samples were also tested, with 57 determined to be *Listeria*-positive by selective plating, and 55 positive by the *lux* phage method. Differences were observed in the number of positive results each method identified, with the phage-based method identifying more positive dairy products and dairy plant environmental isolates, while the plating method identified 20% more positive poultry samples. Failure to isolate *Listeria* cells by plating was proposed to be related to sensitivity of some strains to selective agents used in Oxford agar, while failure to detect poultry isolates by phage-mediated luminescence was due to the fact that phage A511::*luxAB* was determined not to infect those isolates.

A modified *lux*-MPN method was developed and evaluated, using *Listeria* selective broth as the growth medium in a 3 × 3 tube method, with artificially contaminated samples of minced meat, shrimp, ricotta cheese, and cabbage.[85] Samples were removed from positive tubes after 20 and 44 hours of enrichment at 30°C, and tested using the *lux*$^+$ phage method. Numbers calculated from the luminescent MPN method agreed well with the levels of *L. monocytogenes* Scott A initially inoculated into food samples following both 20 and 44 hour enrichment, with only minced meat samples showing slightly higher counts after 44 hours, demonstrating the rapidity and reliability of such an assay. Furthermore, it was demonstrated to be easily modifiable to a microtitre plate format, enabling a large number of samples to be processed simultaneously.

10.7212 *luc*-based bioluminescence. The natural bioluminescence of the North American firefly *Photinus pyralis* has been similarly exploited in bacterial systems, following the successful cDNA-cloning and expression of the *luc* gene, encoding a single polypeptide, 62 kDa luciferase, in *E. coli*.[32] The production of light by this means relies on the interaction of ATP, present in all living cells as an energy transfer molecule, with substrate, heterocyclic carboxylic acid luciferin, in the presence of molecular oxygen and magnesium, and is catalysed by the firefly luciferase enzyme according to the following equations:

Luciferase + luciferin + ATP luciferase–luciferyl–AMP + PPi (Eq. 1)

Luciferase–luciferyl–AMP + O_2 luciferase + oxyluciferin + AMP + CO_2 + light (Eq. 2)

This reaction results in the highly efficient production of light, in the form of a flash, with an estimated quantum yield of 0.88, substantially more efficient than that generated by bacterial luciferase.[19] The luciferase of *P. pyralis* emits yellow-green light with peak emission at 560 nm, but other firefly and insect species (including click beetles) have been shown to emit different colors of light, ranging from green (547 nm) to red (604 nm).[102] This reporter system is particularly attractive for use with bacteria because they are not capable of natural *luc*-encoded luminescence, and therefore will not interfere with detection of *luc*-transformed bacteria. Any natural bioluminescence will be of *lux* origin, and can be differentiated from *luc* by their differing wavelengths of light emission. Given that only one gene is involved in encoding insect luciferases, they also offer the advantage over *lux* genes of being somewhat simpler to work with. *luc*-encoded bioluminescence can be measured by the same instrumentation used for *lux*$^+$ detection, or a customized apparatus such as the system reported by Gailey et al.,[43] who developed an imaging system based on a single-photon counting photomultiplier tube. This system is capable of real-time monitoring of luciferase gene expression in cancer cells but at a substantially reduced cost.

There have, however, been problems associated with the use of firefly luciferase relating to its pH and temperature stabilities, sensitivity to proteolysis, and the rapid decay of light output during the bioluminescence reaction. Given that different species of insects can produce different light emissions while sharing a common substrate, differences have been attributed to enzyme structure.[102] This, along with the recent publication of the structure of firefly luciferase, has been the impetus for genetic manipulation studies of the structure-function relationship in an attempt to create enzyme derivatives with improved light emission, and more applicable pH and thermal stabilities. For example, substitution of the amino acid serine at position 286 in the firefly luciferase sequence by either leucine, valine, glutamine, tyrosine, lysine or phenylalanine results in a shift in the luciferase emission spectrum producing red-orange light as opposed to the wild-type yellow-green.[5,87] Additionally, amino acid substitutions can decrease or improve thermal stability, as demonstrated by White et al.[159] who generated a series of mutant enzymes by replacing glutamate at position 354 with all other possible amino acid residues. With the exception of glycine, proline and aspartate, any amino acid replacing the glutamate residue produced an enzyme with enhanced thermostability as compared to the wild-type firefly luciferase. The flash kinetics of the reaction, caused by the interaction of luciferase and oxyluciferin, have been addressed to some degree by the use of chemicals such as coenzyme A, reported to prolong the half-life of light to ten minutes. Another system of reagents, known as LucLite™, has been developed and it has been estimated to improve the half-life of light output to over 5 hours, with the production of a glow-type light signal.[121] In addition, Thompson et al.[144] reported that a single serine to threonine substitution at position 198 resulted in an enzyme with a 150-fold increase in half-time for light emission decay, as well as a higher pH optimum.

The permutations are numerous, and offer an approach which will be able to produce enzymes with more desirable properties for a particular application, along with the simultaneous differentiation of bacteria based on the emission spectra of the *luc* genes used. The use of dual bio- and chemi-luminescence reporter genes for luciferase and β-galactosidase has also been reported,[91] and current advances in characterisation of luciferases from other species of insects[152,163] will no doubt advance the application of *luc*-based bioluminescence even further.

As with the *lux* genes, the firefly *luc* gene has been engineered into both plasmid and bacteriophage delivery vectors, facilitating gene cloning of many eukaryotic and prokaryotic cells, including *E. coli* O157:H7[41,106] and *Staphylococcus aureus*.[134]

10.7213 Ice nucleation activity. Ice nucleation, i.e., formation of ice nuclei in pure water supercooled below its freezing point, by bacteria was first reported in 1973 by Schnell and Vali, and this frost damage phenomenon was subsequently demonstrated to be caused by the plant pathogen *Pseudomonas syringae*. Since then, a number of strains of *P. syringae* as well as strains of *Xanthomonas campestris*, *Erwinia ananas*, *E. herbicola*, *P. fluorescens* and *P. viridiflava* have been identified as ice nucleation active (INA) bacteria.[89] Bacterial ice nucleation has been shown to occur at threshold temperatures ranging from –2 to –10°C, higher than those associated with the ice nucleation activity of lichens, freeze-tolerant amphibians and insects[156] and these threshold values have been used to further classify bacterial ice nuclei into three groups.[128]

Although it is not known what advantage this phenotype conveys to INA gram-negative bacteria, a single chromosomally-encoded gene has been determined to be responsible for ice nucleation activity, and a variety have already been sequenced, including *inaA* (*E. ananas*), *iceE* (*E. herbicola*), *inaU* (*E. uredovora*), *inaW* (*P. fluorescens*), *inaX* (*X. campestris* pathovar *translucens*) and *inaZ* (*P. syringae*), ranging in size from 3.4kb to 7.5kb.[77] Recently, a new ice nucleation gene, *inaV*, of about 3.5kb in size was isolated from *P. syringae*.[128] This gene encodes a 1200 amino acid residue protein, InaV, and shares 92.4% homology with *inaZ*. Based on 77% sequence conservation of their protein products,[128] it is thought that these individual *ina* genes from different bacteria evolved from a common gene. Also, their action seems to be host independent in nature, since antibodies raised against the inaW protein of *P. fluorescens* will recognize and react with the inaZ protein of *P. syringae* expressed in *E. coli*.[89] Ina proteins are located in the outer membrane of bacterial cells, and have a functional requirement for lipids.[50] It has been proposed that ina proteins act as templates for ice formation, and that differing aggregation of these proteins dictates variations in threshold temperatures.[156]

Ice nucleation genes cloned into *Escherichia coli* have been shown to convey ice nucleating properties to their host,[105] and this has led to the application of bacterial ice nucleation in a variety of ways, ranging from freezing processes in the food industry,[156] artificial snow production,[50] a reporter for gene expression in moderately halophilic bacteria[6] and promoter analysis in *Zymomonas mobilis*,[34] to a reporter for the detection of bacteria such as *Salmonella* spp. in food.[160] It has been estimated that the use of *ina* gene fusions for the detection of gene expression is 100- to 1000-fold more sensitive than the *lux* detection system.[50] With a low number of proteins (300 or less) required for ice nucleation to occur, such activity could be a useful proposition for the detection of injured cells still capable of minimal cellular activity.

By cloning an ice nucleation gene into a delivery vector such as a bacteriophage, which cannot express the INA phenotype on its own, this phage can be used to transduce susceptible cells. An INA phenotype can thus be passed on to infected cells, with ice nuclei synthesized within 15 minutes of *ina* gene transcription. This principle was employed to develop a sensitive and rapid method for the detection of *Salmonella* spp., known as the BIND™ (Bacterial Ice Nucleation Diagnostic) assay, marketed by IDEXX Laboratories, Inc., located in Westbrook, Maine, USA. This test employs a proprietary phage mixture, engineered to contain *ina* genes, to detect ice nucleation in susceptible *Salmonella* cells. Phage and samples being tested for the presence of *Salmonella* are mixed in the presence of a freezing-indicator dye, carboxyfluorescein, and incubated at 37°C to allow phage attachment and transduction of susceptible bacteria. Samples are then incubated at 23°C to optimize expression of the ice nucleation protein and formation of ice nuclei. Ice nucleation activity is subsequently detected by chilling treated samples and controls to a temperature of –9.5°C. Freezing of cells is indicated by an orange, non-fluorescent appearance, while negative samples appear fluorescent green. The level of bacterial contamination can be calculated, based on the concentration of ice nuclei determined by the droplet freezing assay.[160]

The performance of the BIND™ prototype assay in foods such as milk, meat, gelatin and eggs was most encouraging, with little or no interference from the food samples themselves or from background ice nuclei. In an optimized prototype assay using an *ina*-transduced phage constructed from P22 (with a similar host specificity to the parent phage), a sensitivity of 10 cells per mL or less was obtained in test samples containing either *S. enteritidis*, *S. typhimurium*, *S. dublin*, *S. paratyphi* B, or *S. gallinarium*. A similar detection limit was achieved for *S. Dublin* in the presence of high numbers of non-*Salmonella* bacteria. The test is also capable of detecting sublethally injured cells which are still able to produce ice nuclei. Any enrichment step can thus be relatively short, enhancing the rapidity of the method, which, at around two hours for a positive result is a considerable time saver as compared to other detection methods for *Salmonella* spp.[160]

Information regarding the BIND™ assay and the agreement of assay-positive isolates with identification by the API 20E biochemical system was presented in 1995 by Wolber at the 95th general meeting of the ASM.[161] Ten food matrices were spiked at low (20 to 50 cells/25g) and high (10^2 to 10^3 cells/g) levels with 20 of the commonest *Salmonella* serotypes isolated in the U.S., and after a non-selective pre-enrichment using buffered peptone water, samples were tested for ice nucleation ability. Positive isolates were then identified by the API 20E system, and agreed 100% with the results obtained by the BIND™ assay.

More recent information from IDEXX regarding the BIND™ assay claims an assay sensitivity of one viable *Salmonella* cell per 25g of sample, following a single overnight enrichment step. Handling time for the assay is estimated at less than one minute per sample, and including enrichment, results are obtained within 22 hours. The test can now detect 500 different *Salmonella* spp. without producing false-positive results.[4]

10.7214 Green Fluorescent Protein. Another reporter system of tracking gene expression and protein localization in a variety of organisms and cells is the green fluorescent protein (GFP), naturally occurring in the bioluminescent jellyfish *Aequorea victoria*. GFP is activated in *A. victoria* by the absorption of blue chemiluminescence generated from the calcium binding activation of photoprotein aequorin, and produces a stable, strong green fluorescence.[25] It has been proposed that GFP evolved as a means of amplifying the relatively low levels of chemiluminescence produced by photoproteins into higher intensity fluorescence with no extra energy expenditure on the part of the cell.[31] The 27-kDa GFP, encoded by the *gfp* gene, is 238 amino acid residues in size and absorbs light at 395nm (with a smaller peak at 475nm). The formation of GFP is favored by lower temperatures (~22°C) and takes about four hours in *E. coli*.[54] Once GFP is formed, it is stable at temperatures up to 65°C, at alkaline pH, and is resistant to proteases. Its crystal structure has been determined to consist of 11 beta-sheets forming a barrel-like arrangement around a central alpha-helix containing the fluorescence center.[99] Its activity requires the presence of oxygen, but does not need additional genes, co-factors or substrates for fluorescence to occur. This allows for the detection of sub-lethally injured cells, making it an attractive alternative to the use of substrate-dependent marker systems such as *lux* and *luc*.

Research interest in GFP soared following the demonstration by Chalfie et al.[25] that the *gfp* gene engineered into a plasmid vector could produce fluorescence in both *E. coli* and eukaryotic cells when the transformed cells were excited by near-U.V. or blue light. Additionally, the *gfp* gene can be fused to genes of interest in living cells without interfering with cell function and growth, although it has been suggested that the functionality of some fusion proteins may be affected by the size of the GFP tag.[99] This makes it a potentially useful marker system in the study of, for example, bacterial biofilm formation. Such breakthroughs, together with the availability of fluorescence detection methods such as flow cytometry, epifluorescent microscopy, spectrofluorimetry, and confocal laser microscopy, have led to the application of GFP technology in a number of research areas.

In a special issue of the journal *Gene*, dedicated to GFP, some bacterial applications were reported. For example, host-pathogen interactions following infection of live mammalian cells with either *Salmonella typhimurium*, *Yersinia pseudotuberculosis*, or *Mycobacterium marinum*, were monitored by bacterial expression of the GFP phenotype.[150] Bacterial movement through environmental materials such as quartz sand could be tracked by fluorescence[21] and plasmid transfer was studied using GFP as a marker of genetic transfer between bacterial cells.[28] Recent publications have focused on GFP as a successful marker system for evaluating movement[145] and survival of genetically engineered microorganisms in the environment,[74] the survival of *E.coli* O157:H7 in orange juice and apple cider,[41] and detection of individual *P. aeruginosa* cells in a mixed biofilm population.[15]

Despite the obvious attractiveness of this marker system, it has yet to be reported as a bacteriophage-based detection method. Promising results have been published reporting the use of virus-based *gfp* vectors such as potato virus X,[16] tobacco mosaic virus,[24] and retroviruses,[70] and it is anticipated that bacteriophage-mediated delivery of GFP will be available in the near future. Inevitably, problems associated with this new technology have arisen, which have in part already been addressed by researchers. One of the most crucial aspects is the fluorescence intensity of the protein, and efforts have been made to improve fluorescence by single point mutations targeting primarily the serine65-tyrosine66-glycine67 chromophore. Replacement of Ser65 with alanine, valine, leucine, cysteine, and threonine residues resulted in loss of the 395nm excitation emission peak, and moved the more photostable 475nm peak slightly towards a longer wavelength, producing spectra similar to that of the sea pansy *Renilla* GFP. When Ser65 was replaced by threonine, a six-fold increase in brightness of GFP was observed as compared to wild-type GFP, with decreased photobleaching.[55] Cormack et al.[30] reported the generation of GFP variants, by random amino acid substitution, which when expressed in *E. coli*, showed a 100-fold increase in fluorescence intensity, attributed to shifts in excitation maxima and more efficient protein folding than that of wild-type GFP.

The color emission of GFP has also been studied, and it was demonstrated that replacement of Tyr66 and Tyr145 with histidine and phenylalanine, respectively, produced isoforms emitting blue instead of green light.[54,99] A red-shifted variant, RSGFP4, has also been created by mutagenesis,[162] and similarly, photoactivation of GFP in low oxygen environments produced red instead of green fluorescence.[37] This provides exciting advantages to the use of GFP and its variants in simultaneous detection of cells expressing different fluorescence emissions as demonstrated by Yang et al.,[162] who used dual colour microscopy to visualize cells expressing red and green fluorescence. While the use of GFP has been praised due to the absence of interfering background fluorescence, care should be taken in the application of its variants, as reported by Lewis and Errington.[76] These researchers found that a blue *gfp* variant was not useful in the study of sporulation of *Bacillus subtilis*, due to the innate fluorescence of the spore-former in the blue portion of the spectrum.

Little information is currently available regarding the detection limits of GFP. While it has been estimated as being 10^5 bacterial cells using a derivative Tn5GFP1 transposon vector[21] and as low as 1 CFU/mL using a commercial cDNA vector,[41] the use of a strong promoter and modified *gfp* genes[140] should improve sensitivity greatly.

10.722 Phage-Based ATP Bioluminescence

The use of ATP (adenosine-5′-triphosphate) bioluminescence as a rapid method for the detection of bacterial biomass and its applications in food microbiology, hygiene monitoring and HACCP programs have been extensively reviewed.[46,47,52,148] The ATP bioluminescence reaction has already been discussed with regard to *luc*-encoded bioluminescence and readers are directed to this section for details of the reaction. Protocols for the analysis of food samples typically include primary chemical treatments, filtration and washing steps to remove background and/or non-microbial ATP. Bacterial cells are concentrated on the filter and are then chemically lysed to release microbial ATP which can be quantified by the addition of the luciferin-luciferase complex. This assay can be performed either in liquid form using a luminometer, or by direct application of ATP extractants and luminescent reagents onto filters, followed by image processing.[142] Light output, quantified in terms of relative light units, can thus be used to estimate biomass[48] based on the generally accepted fact that the ATP content of each bacterial cell is reasonably constant. Alternatively, in hygiene monitoring, both microbial ATP and ATP present in food debris are of relevance and thus a total ATP estimate is desirable to evaluate hygiene levels relating to microbial contamination and overall cleanliness.[129]

While ATP bioluminescence has been widely accepted as a rapid and convenient method for the analysis of overall bacterial load, the lack of specificity of the procedure is a major drawback in its application to many areas of food microbiology. Chemical extraction of microbial ATP is inherently non-specific, and while measurement of total ATP levels has been shown to be sufficient for the purposes of quality assessment of food products and factory hygiene,[9] the detection of specific groups of microorganisms in mixed cultures, for example, foodborne pathogens, starter cultures, and spoilage bacteria, has been impossible to achieve. Two approaches have been studied in an attempt to improve the specificity of the ATP bioluminescence reaction, both of which exploit the specificity of lytic bacteriophages. These breakthroughs could make ATP bioluminescence a more complete tool for the microbiological analysis of foods.

10.7221 Lytic phages. The first approach has focused on the replacement of chemical extractants with bacteriophages capable of specifically lysing groups of microorganisms. Very simply, the lysis of bacterial cells by phage will result in release of intracellular ATP which can be detected by bioluminescence. A sample containing added bacteriophage and bacterial cells susceptible to phage infection will show an increased amount of ATP in the system, as compared to a control sample containing no phage. This approach was first published in 1995 by Sanders,[125] who reported the use of a bacteriophage for the specific detection of *Listeria monocytogenes* even in the presence of the closely related species *Listeria innocua*. In a broth system, the presence of *L. monocytogenes* ATCC 23074 was confirmed within 80 minutes of addition of bacteriophage ATCC 23074-B1, when the light output was at its maximum. This method was also capable of detecting 10^5 CFU/g of *L. monocytogenes* in artificially contaminated lettuce within 15 hours, incorporating a selective enrichment step.

Despite the specificity of the assay, the lack of sensitivity has become an overriding concern. The bioluminescence assay itself has a detection limit of around 10^3–10^4 bacterial cells, and is unreliable for cell numbers below this level. For the phage assay, even in pure culture, at least 10^5–10^6 cells are required to produce a detectable increase in ATP as compared with a control in which no phage lysis has occurred. For optimum results an even higher cell count is required. As an alternative to pre-enrichment techniques, suggestions to improve sensitivity have focused on i) the use of adenylate kinase (AK) and excess amounts of ADP as a means of driving the equilibrium of AMP, ADP and ATP towards the anabolism of ATP,[133] and ii) recycling of ATP by the use of specific substrate-enzyme mixtures to effectively increase the net production of ATP.[53] The AK assay, employing lytic phage to release intracellular AK, currently has a detection limit of less than 10^4 cells/mL for both *E. coli* and *S. newport*. Results were achievable in less than one hour for *E. coli* while the *Salmonella* assay took up to two hours.[13] The ATP recycling method was demonstrated to improve the sensitivity of the standard ATP assay from 550nM to 250pM ATP.[53] However, this was achieved using pure ATP standards and not by ATP extraction from microorganisms, as the publication title suggests.

10.7222 Bacteriolysins. A second approach to improving the ATP bioluminescence assay has also exploited the activity of lytic bacteriophages, more specifically the endolysins they produced as late gene products during phage replication. Endolysins have been demonstrated to facilitate the release of progeny phage from the bacterial cell; this is known as 'lysis from within'.[143] These lysins fall into one of at least four categories of enzymes: lysozymes (as is the case with phage T4), transglycosylases, endopeptidases, and amidases. The enzymes gain access to the peptidoglycan layer by the formation of non-specific membrane lesions or holes, produced by a group of membrane proteins known as holins.[165]

The genes responsible for endolysin production have been isolated and cloned for a number of phages,[164] and subsequent information gained from characterization of these lytic enzymes has led the way towards more applied uses for phage endolysins. More recently, Loessner et al.[81] reported the use of a recombinant phage L-alanoyl-D-glutamate peptidase endolysin ply118 of *L. monocytogenes*, overexpressed in *E. coli* to very high levels, as an efficient means of gently releasing DNA, RNA and proteins from *Listeria* cells. This procedure was subsequently modified to incorporate a 12 amino acid purification tag without adversely affecting the lytic activity of the protein, and enabled the production of lysins purified to more than 90% homogeneity.[84] This endolysin was shown to be specific for *Listeria* cells, but did not lyse other gram-positive and gram-negative bacteria. It thus provides an efficient and rapid means of lysing specific cells. Analysis of gene sequences upstream of endolysin genes *ply*118 and *ply*500 revealed highly conserved holin (*hol*) genes, which the endolysins require for transportation across the cell membrane.[82]

The continued characterization of phage lytic enzymes[58,86,130] will greatly enhance our knowledge of the molecular action of such enzymes, but as well, will increase their applicability in other areas of microbiology. For example, such recombinant lysins are currently being evaluated as ATP extractants for use in bioluminescence assays,[84] in which intact target cells could be selectively lysed and rapidly detected by firefly luciferase. In addition, they offer the potential of being used as antimicrobial agents to eliminate pathogens in the food industry and in medical applications.

10.723 Phage-Based Fluorescence

10.7231 Labeling of intracellular phage nucleic acid. Fluorescent stains have been reported as a useful accessory to the use of phage probes, as a means of labeling phage nucleic acid within infected cells. Sanders et al.[126] reported the use of two fluorescent dyes, chromomycin A3 (CA3) and Hoechst 33258 (H33258), to bind and label respectively G-C- and A-T-rich regions of phage T4 DNA present inside infected *E. coli* cells. Following infection of *E. coli* ATCC 11303 with phage T4, samples were placed on ice to interrupt viral replication, washed and ethanol-fixed in preparation for fluorescence staining. Cells were then analyzed by dual-laser flow cytometry, and within 35 minutes of phage infection, T4-infected cells could be differentiated from uninfected cells based on fluorescent detection of phage DNA. While this technique was developed to study viral DNA within the infected bacterial host, it has also provided an excellent means of fluorescent labeling of bacterial cells infected by phage. Thus it could be used in conjunction with phages of known host specificity as a detection method for particular groups of microorganisms.

10.7232 Labeling of virus-packaged nucleic acid. Fluorescent labeling of nucleic acid packaged within the virus head has also been used to directly count viruses in samples from freshwater and marine environments using epifluorescence microscopy.[56] This labeling technique was employed to create differential phage probes using fluorescent stains with different excitation wavelengths for the identification and enumeration of specific groups of marine bacteria.[57] More recently, labeling of phage DNA has been adapted to enable fluorescent detection of *E. coli* O157:H7 in ground beef, in conjunction with immunomagnetic separation and flow cytometry.[44] Prior to sampling, phage DNA was fluorescently labeled with the benzoxazolium-4-quinolinium fluorescent dye YOYO. Ground beef and milk samples were artificially inoculated with low levels of *E. coli* O157:H7 and pre-enriched in non-selective media for 6 hours. Target bacteria were subsequently removed from the food matrices by immunomagnetic separation with *E. coli* O157:H7 Dynabeads, and the fluorescent phage preparation was then used to label the recovered cells. Using both confocal laser scanning and epifluorescent microscopy techniques, in addition to a more automated flow cytometric method, target *E. coli* O157:H7 cells were counted based on their fluorescence. The detection limits achieved were approximately 2.2 CFU/g in ground beef and 10^3 CFU/mL in milk. The potential application of this method for detection of other food-borne pathogens is enormous, given the large number of fluorescent stains commercially available for phage labeling and the adaptability of magnetic separation techniques for the detection of a wide variety of bacteria.[104]

10.724 Metabolic Inhibition Assays

A number of methodologies, including electrical impedance, turbidimetry and colorimetry, that rely on metabolic activity and microbial growth as a means of investigating microbial populations are available. Similar to cultural methods, total counts of microorganisms can be determined by using non-selective media, while specific detection of microorganisms has involved the development and use of suitable differential or selective culture media. By incorporation of pH indicators, fermentable carbohydrates, organic acids and alcohols, or other inhibitors such as antibiotics, preferential growth conditions for the microorganism of interest are provided at the expense of the background microbial flora. Such technologies have found many applications in food microbiology, particularly with advancement in the automation of such systems.[151] They are particularly promising for the study and application of bacteriophages, where metabolically active bacterial cells are required for viral replication.[159]

10.7241 Electrical impedance. Of the metabolic methods to be considered in this section, electrical impedance, though more complicated in theory, has been the most widely studied and developed in terms of its microbiological applications. Detection by impedance is based on changes to the growth medium associated with the increase in biomass of metabolically active microorganisms. The components of growth media, such as proteins, carbohydrates and lipids, are uncharged or weakly charged substrates, but as microorganisms metabolize these compounds, they are converted into more highly charged molecules such as amino acids, lactate, and acetate. These metabolic products increase the conductivity of the medium. However, this increase is small, therefore requiring a microbial level of at least 10^6 cells/mL to be reached before significant changes can be detected. The period between initiation of the test and detection is known as the detection time (D.T.), the value of which is inversely proportional to the number of cells initially present in the test sample.

A large number of reports regarding the electrical detection of specific microorganisms have been published, particularly since the introduction of automated systems such as the Vitek Bactometer® Microbial Monitoring System and the Malthus Microbial Growth Analyzer. There are a number of differences between the two systems, but the most pertinent is their electrical signal measurement. The Malthus Analyzer measures only conductance, while the Vitek system offers more flexibility by measuring either conductance, capacitance, or total impedance. Applications utilizing these systems have focused primarily on microbial enumeration, detection of contaminants[66,72] and pathogens, particularly Enterobacteriaceae[11,29,42,108,111] and *Listeria* spp.,[51,110,120] shelf-life studies,[12,109] sterility testing of UHT products,[141] and detection of bacteriophages implicated in cheese starter culture failures.[23,153]

Despite its many uses, the direct impedance technique has disadvantages, the most crucial being that it does not operate under high salt concentrations found in some growth media, and this has prompted the development of alternative methods. An indirect approach, where metabolic production of CO_2 is considered as an indicator of microbial growth rather than biomass, prompted the development of the RABIT (Section 10.6) automated bacterial impedance technique.

10.7242 Turbidimetry. Turbidimetry is a relatively simple means of estimating bacterial generation times. The use of turbidity measurements takes advantage of the increase in optical density of sterile microbiological growth media associated with growth of inoculated bacteria. Although this method is used to estimate the rate of growth of microorganisms, generated optical density data can also be used to estimate cell populations when compared to total plate counts. The automation of optical density readers has been particularly useful in increasing the number of samples capable of being analyzed at one time, and computer interfacing has provided an easier means of graphing and analyzing generated growth curves. One particular automated system, the Bioscreen C microbiology plate reader (Labsystems, Helsinki, Finland), has provided food microbiologists with a useful tool for the evaluation of turbidity as a means of estimating bacterial levels in food samples. Capable of incubating up to 200 samples at various temperatures, the Bioscreen C can be programmed to read the optical density of each sample well a particular number of times over a defined incubation period. In the case of food samples, dilutions are often required to reduce the background turbidity to a minimum, and the resulting growth curves can be analyzed using a number of parameters appropriate to the application. These include the time taken for optical density to increase by a defined amount, maximum optical density, slope of the logarithmic growth phase, or turning point time (the middle of the logarithmic phase).

The Bioscreen C has been used by researchers to, for example, predict colony forming units in raw milk[93] and other foods,[92] and assess the effects of antimicrobials such as bacteriocins[131] on bacterial growth. In addition it has proved very useful in the development of data for predictive modeling,[94] and most recently has been used to investigate recovery of single heat-injured *Salmonella* cells by measuring the lag phase of cells pre-enriched in non-selective media.[135] To date bacteriophage-based turbidimetric methods for the detection of foodborne pathogens have not been reported.

10.7243 Colorimetry. The suitability of color changes to indicate the status of biochemical reactions is evident by their extensive application in numerous methodologies used in microbiology, such as differential and selective culture media, biochemical tests for identification of isolates, and enzyme-linked immunosorbent assays (ELISA). This approach has also been used in the development of media for both electrical impedance, where the incorporation of a color change is typically used in conjunction with detection times and curve characteristics to confirm a particular microorganism, and colorimetric assays, where the color of growth media containing pH indicators is monitored over time for changes corresponding to, for example, fermentation of carbohydrates. This has been applied particularly in dairy microbiology where colorimetric tests have been successfully developed to evaluate the presence of antibiotics, coliforms, pathogens or psychrotrophs, enzyme activity, stale flavor development, lactic culture activity, and shelf life of dairy products. Instrumentation includes the the Omnispec™ 4000 bioactivity monitor[117] and the versatile LabSMART™ color scanner and software package.[116] Both instruments operate on the basis of reflectance color readings, which eliminates any problems associated with background turbidity of food samples.[118] Phage-based colorimetric assays have not been reported to date.

10.7244 Phage-based applications. What these methods have in common (regardless of their mode of detection), and what has primarily limited their application in pathogen detection in foods, is their requirement for differential or selective media. The use of such media can increase testing time considerably, and does not always allow for the detection of stressed and injured bacterial cells. Generally, they do not differentiate to the species level, and can also be affected by the growth of contaminating microorganisms associated with the background microflora of foods, resulting in false-positive detections. One approach to these problems is the use of bacteriophages, specifically their effect on the metabolic activity of their hosts, which can offer an alternative means of detecting specific microorganisms in the presence of the background flora of foods. Based on the target population's inability to replicate (in the presence of specific phage) to similar cell concentrations achieved by a negative control not exposed to phage, positive detection of pathogens can be confirmed. If target cells are present, they will be infected and lysed by the phage, resulting in delayed detection in samples with added phage. Control samples without phage will have normal metabolism and growth, resulting in a shorter detection time.

This approach was evaluated by Pugh and Arnott[113] in their development of impedimetric media for the detection of *Salmonella* spp. in confectionery products. A combination of two impedimetric media, selenite-cystine-trimethylamine oxide-dulcitol medium (SC/T/D) and modified lysine decarboxylase broth (MLD), was used in an attempt to take into account those *Salmonella* strains not capable of fermenting dulcitol, but able to decarboxylate lysine. In addition, phage Felix O-1, estimated to be lytic for 96.1% of tested *Salmonella* strains, and phage G47, included to compensate for the tendency of Felix O-1 not to infect O groups E1 to E4, were employed to confirm the positive genus identification made by impedance microbiology. Of 81 *Salmonella* strains tested in the Bactometer system by both media and the phage test, all strains except one strain of *S.* Houten, were positively identified by at least two of the three protocols, while false-positive reactions by *Citrobacter freundii, E. coli, Proteus vulgaris*, and *Pseudomonas aeruginosa* were obtained in either SC/T/D or MLD, but never both broths. The non-*Salmonella* strains were all negative according to the phage test. A pre-enrichment step using brilliant green-milk broth was incorporated into the assay when artificially inoculated samples were analysed. In all cases but one, where *Salmonella* isolates were identified by conventional selective enrichment and plating, positive identifications were made by at least two of the three test protocols. The one exception was an isolate of *S. typhimurium* which was positively identified by the phage test but neither by the conventional nor the Bactometer methods. Despite the apparent usefulness of this phage-based impedimetric method, confirmation of presumptive positives made the method more time consuming, and no further application of bacteriophages to impedance has been reported.

Recently, these metabolic methods were reevaluated by McIntyre and Griffiths[95] in an effort to i) develop specific pathogen detection methods using bacteriophages specific for *Listeria* spp., *E. coli*, and *Salmonella* spp., ii) to demonstrate the ability of particular phages to differentiate particular isolates, and iii) to investigate whether decreased selectivity of media could be counter-balanced by the incorporation of specific bacteriophages for the bacteria of interest. This attempted to reduce the amount of time required for testing, and take into account the deleterious effect of selective media on stressed or injured cells, while the phage test would indicate if target cells were present.

Impedimetric and colorimetric methods were successfully developed for the detection of bovine isolate *E. coli* G2-2 in raw milk and ground beef using Vitek Coliform Medium (CM), a lactose-containing medium designed to change color from purple to yellow in the presence of lactose-fermenting coliforms, and phage AT20, demonstrated to be lytic for isolate G2-2. Less than 10 CFU of *E.coli* per mL of raw milk could be detected by both phage-based methods within 10 hours, while a detection limit of less than 1000 CFU/g of ground beef was achieved using a test-tube based phage colorimetric assay in 10.25 hours. Using phage AT20, specific for non-shiga toxin (stx) *E. coli* but not stx-positive isolates, the presence of non-stx-producing *E. coli* G2-2 in raw milk and ground beef was subsequently confirmed by an increase in the time required to reach the detection threshold in the presence of phage as compared to identical samples without added phage. Equivalent impedimetric detection limits were achieved for skim milk powders artificially contaminated with various *Salmonella* spp. using phage Felix 0-1 and CM (able to support the growth of *Salmonella* but with no lactose fermentation), but the specificity of the assay was greatly enhanced by the use of a second bacteriophage, SJ2, with a host range narrow enough to positively detect only *S. typhimurium* and *S. enteritidis*.

The effect of using less selective media for the impedimetric detection of *L. monocytogenes* ATCC 23074 in combination with specific phage ATCC 23074-B1 gave rather interesting results. In pure culture using both Listeria selective impedimetric broth and less selective Demi-Fraser broth, large differences in detection times for *L. monocytogenes* were observed in the presence and absence of phage, with, as expected, slower detection occurring in the presence of phage. In addition, detection was substantially faster in the less selective medium. However, when raw milk was artificially contaminated with *L. monocytogenes* and tested in a similar fashion, the reverse result was obtained in Demi-Fraser broth where slightly faster detection times were achieved in the presence of phage. It is likely, in this case, that the lower selectivity of the medium plus the lowering of the numbers of *Listeria* present due to the lytic action of the phage actually allowed faster growing members of the background microflora to dominate. These microorganisms were thus able to reach a detectable level faster than was the case in the absence of phage, where original *Listeria* levels were competing with the raw milk microflora. Interpretation of results by this method should thus be considered carefully, depending on the selective nature of the medium, the food material being tested, and the growth rates of the microorganisms in question. Phage-based turbidimetric detection of pathogens has also been demonstrated using similar media and phage combinations, but the need to dilute food samples at least 100–1000 times detracts from the sensitivity and overall usefulness of this type of assay. However, the Bioscreen C has been used effectively in pure culture work to assess the susceptibility of bacterial strains to lytic phages, and has been useful in establishing optimum phage concentrations for use in other assays.

10.725 Miscellaneous Phage Probe Applications

10.7251 High-performance liquid chromatography. A high-performance liquid chromatography (HPLC) method for the detection of *Salmonella* spp. utilising the Felix O-1 phage was reported in 1983.[61] Using an E-1000 m-Bondagel gel permeation column, it was demonstrated that HPLC could be used to distinguish and detect *Salmonella* spp. based on the detectable increase in bacteriophage titres associated with phage replication and cell lysis. Susceptible host cells and Felix O-1 phage were mixed and allowed to interact for two hours during which replication of phages would occur. The aqueous phase was removed, treated with chloroform, and centrifuged to remove cell debris. This bacteriophage supernatant was then injected into the HPLC column and the resulting chromatograph was analysed for a specific peak corresponding to the retention of bacteriophage previously determined to occur 3.2 minutes after injection. It was determined from the results obtained that a minimum bacteriophage concentration of approximately 10^9 PFU/mL was required to give a reliable detection (response of at least 10mm over background), corresponding to around $3x10^6$ bacteria per mL required in the reaction mixture. This peak occurred only when *Salmonella* spp. were present in the reaction mixture, and a positive detection was demonstrated even when non-*Salmonella* spp. were also present. The high levels of bacteria required for detection was a definite shortcoming of this method, and was subsequently addressed by the use of filtration and enrichment in follow-up research outlined below.

These modifications were incorporated in order to detect *Salmonella* spp. in milk using this HPLC method.[62] Large pore electropositive filters were employed to remove bacteria from artificially-inoculated milk samples, and subsequently treated with alkaline broth media to elute the trapped microorganisms from the filter. This method was estimated to recover about 40% of the bacteria originally added to the milk, and subsequent overnight enrichment in the presence of brilliant green dye was adequate to ensure that 75-100% of the total bacterial population consisted of salmonellae. Following enrichment, the cells were centrifuged and washed, resuspended in $CaCl_2$-supplemented nutrient broth, and bacteriophage was added to defined cell densities. Following preparation of bacteriophage supernatants (as previously described), 25 mL volumes were injected into the HPLC column, and the presence of a 3.2 minute bacteriophage peak was indicative of the presence of *Salmonella* in the original sample. Less than five salmonellae per mL of milk were detectable by this method, resulting in a positive detection result within 24 hours of sampling. However, as previously reported,[61] problems arose when certain *Salmonella* serotypes were not infected by this phage. The false-negative results were explained by the potential presence of either prophages, known to limit or inhibit replication of Felix O-1, or bacteriophage-restricting plasmids. Despite the improved detection limit achieved in this research, the use of this assay has not been more recently reported.

10.7252 Immobilized-phage capture assay. More recently, the 'Sapphire' lytic phage (Amersham International) has been evaluated in a novel immobilized capture method for the separation and concentration of *Salmonella* strains by Bennett et al.[10] Using both a microtitre plate and dipstick method, phage particles were allowed to adhere to the solid phase, washed several times to remove unbound virus, and treated with bovine serum albumin to minimize non-specific binding of cells. The phage-coated solid phase was then exposed to several *Salmonella* serotypes including *S. typhimurium* and *S. enteritidis*, and non-salmonellae including *E. coli*, *Proteus mirabilis*, and *Hafnia alvei* in liquid suspensions. Using the polymerase chain reaction (PCR), the method was evaluated for i) its ability to remove cells from the suspensions and ii) specifically distinguish and remove *Salmonella* cells from mixed cultures of Enterobacteriaceae. A commercially available *Salmonella*-specific BAX PCR system was chosen, but used in a procedure slightly different from that suggested by the manufacturer, previously demonstrated to significantly increase the sensitivity of the system. Data obtained showed that capture of *Salmonella* spp. was possible using both solid phases mentioned above, but, based on the detection limits of the PCR assay, high numbers of cells (around 10^7 CFU/mL) were required to subsequently obtain a positive end result. Specific capture of *S. typhimurium* in the presence of equal numbers of competing Enterobacteriaceae using the microtitre plate test format was also a viable option, but again, high numbers of cells were required for a positive result. While these results were encouraging, a number of problems were encountered in this assay. Most importantly, *S. ealing* and *S. arizonae* were not detected by this capture method, and the 'Sapphire' phage was subsequently shown not to infect these particular organisms. Second, phage capture was very inefficient and it is likely that some phage particles were binding to the solid phase via the tail rather than the head, thus rendering them useless in the cell capture process. The small surface area of the microtitre plates may also have contributed to the poor capture rate. In addition, unusually high cell counts obtained using control (uncoated) dipsticks demonstrated that phage or cell clumping may have arisen as a result of uneven surface chemistry. A solution to the surface area issue is the use of magnetic spheres as a solid phase, but further work in this area was not reported by these or other authors.

10.7253 Phage-linked immunosorbant assays. The use of phages as reporter molecules has also been extended to the development of phage-linked immunosorbant assays (PHALISA), similar to the ELISA method, where the amount of phage bound by immobilised antigen-antibody complexes can be related back to the number of target cells present by its subsequent infection of susceptible bacterial cells. Block et al.[14] employed bacteriophage M13, engineered to contain the *E. coli lacZ* gene, as a 'bio-amplifiable tag' for the detection of herpes simplex virus (HSV). HSV-infected cells were incubated with HSV-specific monoclonal antibodies raised in mice, followed by rabbit anti-mouse serum and mouse anti-M13 serum. The resulting complexes were incubated with phage M13. The bound phage was eluted and then quantified by plaque assay and acquired beta-galactosidase activity in phage-infected *E. coli* cells. This PHALISA assay was more sensitive than an ELISA method under similar test conditions and demonstrated the potential for detecting bacterial and viral antigens in foods.

10.7254 Phage amplification assays. A similar role for phage has been exploited in the development of methods to detect *Pseudomonas aeruginosa* and *Mycobacterium tuberculosis*.[136] Known as the 'phage amplification assay', it exploits the lytic cycle of bacteriophage to indicate the presence of target bacterial cells. Following infection of cells, samples are first treated with a special virucidal reagent to destroy external virus particles not already involved in the infection process, and then the reagent is neutralized to prevent later destruction of phage progeny. If susceptible target cells are not present in the sample, then all bacteriophage will be destroyed. Otherwise, the bacteriophage replicates normally within the host cell and eventually lyses the cell wall, releasing its progeny into the external medium. Resulting virus particles can then be amplified further using a non-pathogenic host and detected by plaque assay. This method has been reported to have a detection limit of about 100 mycobacteria per mL of sample, detectable within 10 hours, and offers the advantages of destroying virulent *M. tuberculosis* cells during testing if they are indeed present in the clinical sample, simplicity of use with no expensive instrumentation required, and a less expensive application of phages as opposed to more costly cloning techniques mentioned previously. As with other phage-based detection methods, resistance to a particular bacteriophage is the most likely problem associated with this approach. However the use of phage cocktails could be a means of overcoming this shortcoming.

10.73 Conclusions

Bacteriophages are versatile biological particles, capable of identifying host bacteria to levels well beyond the capabilities of many conventional microbiological methods. Their demonstrated suitability for bacterial typing and cloning has resulted in their successful application in a variety of molecular-based and/or 'rapid' methodologies, particularly for the detection of pathogenic bacteria. Regardless, the limiting factor in the advancement and application of phage-based detection systems will always be the bacteriophages themselves. For example, the generic detection of *Salmonella* spp. using the broad-spectrum phage Felix O-1 has been extensively documented, but it is not a viable option for more specific detection to the serovar level. Continuing phage isolation and characterisation still remains the best way of identifying potentially useful bacterial viruses, and this avenue, in combination with more novel molecular approaches to the question of phage specificity, should ensure continued research interest in bacteriophage-based detection methods.

10.74 References

1. Ahmed, R., P. Sankar-Mistry, S. Jackson, H.-W. Ackermann, and S. S. Kasatiya. 1995. *Bacillus cereus* phage typing as an epidemiological tool in outbreaks of food poisoning. J. Clin. Microbiol. 33:636-640.
2. American Public Health Association. 1998. Standard methods for the examination of water and wastewater. 20th ed. American Public Health Association, Washington, D. C.
3. Anonymous. Suicide bombing, Bacillus style. Discover 1995(12):28.
4. Anonymous. 1997. Microbiological hazards. Food Safety Security 1997(June):8-11.
5. Arslan, T., S. V. Mamaev, N. V. Mamaeva, and S. M. Hecht. 1997. Structurally modified firefly luciferase. Effects of amino acid substitution at position 286. J. Am. Chem. Soc. 119:10877-10887.
6. Arvanitis, N., C. Vargas, G. Tegos, A. Perysinakis, J. J. Nieto, A. Ventosa, and C. Drainas. 1995. Development of a gene reporter system in moderately halophilic bacteria by employing the ice nucleation gene of *Pseudomonas syringae*. Appl. Environ. Microbiol. 61:3821-3825.
7. Baker, J. M., M. W. Griffiths, and D. L. Collins-Thompson. 1992. Bacterial bioluminescence: applications in food microbiology. J. Food Prot. 55:62-70.
8. Barrett, T. J., H. Lior, J. H. Green, R. Khakhria, J. G. Wells, B. P. Bell, K. D. Greene, J. Lewis, and P. M. Griffin. 1994. Laboratory investigation of a multistate food-borne outbreak of *Escherichia coli* O157:H7 by using pulsed-field gel electrophoresis and phage typing. J. Clin. Microbiol. 32:3013-3017.
9. Bautista, D. A., L. McIntyre, L. Laleye, and M. W. Griffiths. 1992. The application of ATP bioluminescence for the assessment of milk quality and factory hygiene. J. Rapid Meth. Automation Microbiol. 1:179-193.
10. Bennett, A. R., F. G. C. Davids, S. Vlahodimou, J. G. Banks, and R. P. Betts. 1997. The use of bacteriophage-based systems for the separation and concentration of *Salmonella*. J. Appl. Microbiol. 83:259-265.
11. Bird, J. A., M. C. Easter, S. Gaye Hadfield, E. May, and M. F. Stringer. 1989. Rapid *Salmonella* detection by a combination of conductance and immunological techniques, p. 165-183. *In* C. J. Stannard, S. B. Petitt, and F. A. Skinner (ed.), Rapid microbiological methods for foods, beverages and pharmaceuticals. Society for Applied Bacteriology Technical Series, Blackwell Scientific Publications, Oxford, England.
12. Bishop, J. R., C. H. White, and R. Firstenberg-Eden. 1984. A rapid impedimetric method for determining the potential shelf-life of pasteurized whole milk. J. Food Prot. 47:471-475.
13. Blasco, R., M. J. Murphy, M. F. Sanders, and D. J. Squirrell. 1998. Specific assays for bacteria using phage mediated release of adenylate kinase. J. Appl. Microbiol. 84:661-666.
14. Block, T., R. Miller, R. Korngold, and D. Jungkind. 1989. A phage-linked immunosorbant system for the detection of pathologically relevant antigens. Biotechniques 7:756-761.
15. Bloemberg, G. V., G. A. O'Toole, B. J. J. Lugtenberg, and R. Kolter. 1997. Green fluorescent protein as a marker for *Pseudomonas* spp. Appl. Environ. Microbiol. 63:4543-4551.
16. Boevink, P., S. Santa Cruz, C. Hawes, N. Harris, and K. J. Oparka. 1996. Virus-mediated delivery of the green fluorescent protein to the endoplasmic reticulum of plant cells. Plant J. 10:935-941.
17. Boonmar, S., A. Bangtrakulnonth, P. Pornrunangwong, J. Teragima, H. Watanabe, K.-I. Kaneko, and O. Masuo. 1998. Epidemiological analysis of *Salmonella enteritidis* isolates from humans and broiler chickens in Thailand by phage typing and pulsed-field gel electrophoresis. J. Clin. Microbiol. 36:971-974.
18. Borrego, J. J. M. M. A., A. de Vicente, R. Córnax, and P. Romero. 1987. Coliphages as an indicator of faecal pollution in water, its relationship with indicator and pathogenic microorganisms. Water Resour. 21:1473-1480.
19. Bronstein, I., J. Fortin, P. E. Stanley, G. S. A. B. Stewart, and L. J. Kricka. 1994. Chemiluminescent and bioluminescenet reporter gene assays. Anal. Biochem. 219:169-181.
20. Bruttin, A., F. Desiere, N. d'Amico, J.-P. Guerin, J. Sidoti, B. Huni, S. Lucchini, and H. Brussow. 1997. Molecular ecology of *Streptococcus thermophilus* bacteriophage infections in a cheese factory. Appl. Environ. Microbiol. 63:3144-3150.
21. Burlage, R. S., Z. K.Yang, and T. Mehlhorn. 1996. A transposon for green fluorescent protein transcriptional fusions: application for bacterial transport experiments. Gene 173:53-58.
22. Campbell, A. M. 1994. Lambda bacteriophage, p. 772-776. *In* R. G. Webster and A. Granoff (ed.). Encyclopedia of virology. Academic Press, London.
23. Carminati, D., and E. Neviani. 1991. Application of the conductance measurement technique for detection of *Streptococcus salivarius* ssp. *thermophilus* phages. J. Dairy Sci. 74:1472-1476.
24. Casper, S. J., and C. A. Holt. 1996. Expression of the green fluorescent protein-encoding gene from a tobacco mosaic virus-based vector. Gene 173:69-73.
25. Chalfie, M., Y. Tu, G. Euskirchen, W. W. Ward, and D. C. Prasher. 1994. Green fluorescent protein as a marker for gene expression. Science 263:802-805.
26. Chen, J., and M. W. Griffiths. 1996. *Salmonella* detection in eggs using *Lux*[+] Bacteriophages. J. Food Prot. 59:908-914.
27. Chen, J., R. C. Clarke, and M. W. Griffiths. 1996. Use of luminescent strains of *Salmonella enteritidis* to monitor contamination and survival in eggs. J. Food Prot. 59:915-921.
28. Christensen, B. B., C. Sternberg, and S. Molin. 1996. Bacterial plasmid conjugation on semi-solid surfaces monitored with the green fluorescent protein (GFP) from *Auquorea victoria* as a marker. Gene 173:59-65.
29. Colquhoun, K. O., S. Timms, and C. R. Fricker. 1995. Detection of *Escherichia coli* in potable water using direct impedance technology. J. Appl. Bacteriol. 79:635-639.

30. Cormack, B. P., R. H. Valdivia, S. Falkow. 1996. FACS-optimized mutants of the green fluorescent protein (GFP). Gene 173:33-38.
31. Cubitt, A. B., R. Heim, S. R. Adams, A. E. Boyd, L. A. Gross, and R. Y. Tsien. 1995. Understanding, improving and using green fluorescent proteins. Trends Biochem. Sci. 20:448-455.
32. de Wet, J. R., K. V. Wood, D. R. Helinski, and M. DeLuca. 1985. Cloning of firefly luciferase cDNA and the expression of active luciferase in *Escherichia coli*. Proc. Natl. Acad. Sci. USA 82:7870-7873.
33. d'Herelle, F. 1926. The bacteriophage and its behaviour, G. H. Smith (translator). Williams & Wilkins, Baltimore, Md.
34. Drainas, C., G. Vartholomatos, and N. J. Panopoulos. 1995. The ice nucleation gene from *Pseudomonas syringae* as a sensitive gene reporter for promoter analysis in *Zymomonas mobilis*. Appl. Environ. Microbiol. 61:273-277.
35. Duckworth, D. H. 1987. History and basic properties of bacterial viruses, p. 1-43. *In* S. M. Goyal, C. P. Gerba, and G. Bitton (ed.). Phage ecology. John Wiley, New York.
36. Eaton, T. J., C. A. Shearman, and M. J. Gasson. 1993. The use of bacterial luciferase genes as reporter genes in *Lactococcus*: regulation of the *Lactococcus lactis* subsp. *lactis* lactose genes. J. Gen. Microbiol. 139:1495-1501.
37. Elowitz, M. B., M. G. Surette, P-E Wolf, J. Stock, and S. Leibler. 1997. Photoactivation turns green fluorescent protein red. Curr. Biol. 7:809-812.
38. Estela, L. A., and J. N. Sofos. 1993. Comparison of conventional and reversed phage typing procedures for identification of *Listeria* spp. Appl. Environ. Microbiol. 59:617-619.
39. Fauquet, C. M. Taxonomy and classification—general, p. 1396-1410. *In* R. G. Webster and A. Granoff (ed.). Encyclopedia of virology. Academic Press, London.
40. Fenner, F. 1994. History of virology, p. 627-644. *In* R. G. Webster and A. Granoff (eds.). Encyclopedia of virology. Academic Press, London.
41. Fratamico, P. M., M. Y. Deng, T. P. Strobaugh, and S. A. Palumbo 1997. Construction and characterization of *Escherichia coli* 0157:H7 strains expressing firefly Luciferase and Green fluorescent protein and their use in survival studies. J. Food Prot. 60:1167-1173.
42. Fryer, S. M., and K. Forde. 1989. Electrical screening of powdered dairy products, p. 143-153.*In* C. J. Stannard, S. B. Petitt, and F. A. Skinner (ed.), Rapid microbiological methods for foods, beverages and pharmaceuticals. Society for Applied Bacteriology Technical Series, Blackwell Scientific Publications, Oxford, England.
43. Gailey, P. C., E. J. Miller, and G. D. Griffin. 1997. Low-cost system for real-time monitoring of luciferase gene expression. Biotechniques 22:528-534.
44. Goodridge, L. D. A fluorescent bacteriophage assay for detection of *Escherichia coli* O157:H7 in ground beef and raw milk. University of Guelph, Guelph, Ontario, Canada.
45. Gottesman, M. and A. Oppenheim. 1994. Lysogeny and prophage, p. 814-823. *In* R. G. Webster and A. Granoff (ed.). Encyclopedia of virology. Academic Press, London.
46. Griffiths, M. W. 1993. Applications of bioluminescence in the dairy industry. J. Dairy Sci. 76:3118-3125.
47. Griffiths, M. W. 1996. The role of ATP bioluminescence in the food industry. Food Technol. 50:62-73.
48. Griffiths, M. W., L. McIntyre, M. Sully, and I. Johnson. 1991. Enumeration of bacteria in milk. *In* P. E. Stanley and L. J. Kricka (ed.). Bioluminescence and chemiluminescence: current status. John Wiley, New York.
49. Grimm, L. M., M. Goldoft, J. Kobayashi, J. H. Lewis, D. Alfi, A. M. Perdichizzi, P. I. Tarr, J. E. Ongerth, S. I. Moseley, and M. Samadpour. 1995. Molecular epidemiology of a fast-food restaurant-asociated outbreak of *Escherichia coli* O157:H7 in Washington State. J. Clin. Microbiol. 33:2155-2158.
50. Gurian-Sherman, D., and S. E. Lindow. 1993. Bacterial ice nucleation: significance and molecular basis. FASEB J. 7:1338-1343 .
51. Hancock, I., B. M. Bointon, and P. McAthey. 1993. Rapid detection of *Listeria* species by selective impedimetric assay. Lett. Appl. Microbiol. 16:311-314.
52. Hawronskyj, J-M, and J. Holah. 1997. ATP: A universal hygiene monitor. Trends Food Sci.Technol. 8:79-84.
53. Hawronskyj, J.-M., R. S. Chittock, C. W. Wharton, and J. Holah. 1995. Low level bacterial contamination measured using a novel bioluminescent assay, p. 411-414. *In* A. K. Campbell, L. J. Kricka, and P. E. Stanley (ed.). Bioluminescence and chemiluminescence: fundamentals and applied aspects. John Wiley, New York.
54. Heim, R., D. C. Prasher, and R. Y. Tsien. 1994. Wavelength mutations and posttranslational autoxidation of green fluorescent protein. Proc. Natl. Acad. Sci. USA. 91:12501-12504.
55. Heim, R., A. B. Cubitt, and R. Y. Tsien. 1995. Improved green fluorescence. Nature 373:663-664.
56. Hennes, K. P., and C. A. Suttle. 1995. Direct counts of viruses in natural waters and laboratory cultures by epifluorescence microscopy. Limnol. Oceanogr. 40:1050-1055.
57. Hennes, K. P., C. A. Suttle, and A. M. Chan. 1995. Fluorescently labeled virus probes show that natural virus populations can control the structure of marine microbial communities. Appl. Environ. Microbiol. 61:3623-3627.
58. Hertwig, S., W. Bockelmann, and M. Teuber. 1997. Purification and characterisation of the lytic activity induced by the prolate-headed bacteriophage P001 in *Lactococcus lactis*. J. Appl. Microbiol. 82:233-239.
59. Hill, P. J., and G. S. A. B. Stewart. 1994. Use of *lux* genes in applied biochemistry. J. Biolum. Chemilum. 9:211-215.
60. Hill, P. J., C. E. D. Rees, M. K. Winson, and G. S. A. B. Stewart. 1993. The application of *lux* genes. Biotechnol. Appl. Biochem. 17:3-14.
61. Hirsh, D. C., and L. D. Martin. 1983a. Rapid detection of *Salmonella* spp. by using Felix O-1 bacteriophage and high-performance liquid chromatography. Appl. Environ. Microbiol. 45:260-264.
62. Hirsh, D. C., and L. D. Martin. 1983b. Detection of *Salmonella* spp. in milk by using Felix O-1 bacteriophage and high-performance liquid chromatography. Appl. Environ. Microbiol. 46:1243-1245.
63. Holmes, R. K., and M. P. Schmitt. 1994. Bacteriophage toxins and disease, p. 101-106. *In* R. G. Webster and A. Granoff (ed.). Encyclopedia of virology. Academic Press, London.
64. Hooper, C. E., R. E. Ansorge, H. M. Browne, and P. Tomkins. 1990. CCD imaging of Luciferase gene expression in single mammalian cells. J. Biolum. Chemilum. 5:123-130.
65. Hudson, L. M., J. Chen, A. R. Hill, and M. W. Griffiths. 1997. Bioluminescence: a rapid indicator of *Escherichia coli* 0157:H7 in selected yoghurt and cheese varieties. J. Food Prot. 60:891-897.
66. Irving, T. E., G. Stanfield, B. W. T. Hepburn. 1989. Electrical methods for water quality testing, p. 119-130. *In* C. J. Stannard, S. B. Petitt, and F. A. Skinner (ed.), Rapid microbiological methods for foods, beverages and pharmaceuticals. Society for Applied Bacteriology Technical Series, Blackwell Scientific Publications, Oxford, England.
67. Jay, J. M. 1992. Modern food microbiology, 4th ed. Van Nostrand Reinhold, New York.
68. Kennedy, J. E., Jr., J. L. Oblinger, and G. Bitton. 1984. Recovery of coliphages from chicken, pork sausage and delicatessen meats. J. Food Prot. 47:623-626.
69. Kennedy, J. E , Jr., WCI, and J. L. Oblinger. 1986. Distribution of coliphages in various foods. J. Food Prot. 49:944-951.
70. Klein, D., S. Indraccolo, K. von Rombs, A. Amadori, B. Salmons, Gunzburg. 1997. Rapid identification of viable retrovirus-transduced cells using the green fluorescent protein as a marker. Gene Ther. 4:1256-1260.
71. Kodikara, C. P., H. H. Crew, and G. S. A. B. Stewart. 1991. Near online detection of enteric bacteria using *lux* recombinant bacteriophage. FEMS Microbiol. Lett. 83:261-266.
72. Kyriakides, A. L., and P. A. Thurston. 1989. Conductance techniques for the detection of contaminants in beer, p. 101-117.*In* C. J. Stannard, S. B. Petitt, and F. A. Skinner (ed.), Rapid microbiological methods for foods, beverages and pharmaceuticals. Society for Applied Bacteriology Technical Series, Blackwell Scientific Publications, Oxford, England.

73. Lederberg, J. 1996. Smaller fleas ... ad infinitum: therapeutic bacteriophage redux. Proc. Natl. Acad. Sci. USA 93:3167-3168.
74. Leff, L. G., and A. A. Leff. 1996. Use of green fluorescent protein to monitor survival of genetically engineered bacteria in aquatic environments. Appl. Environ. Microbiol. 62:3486-3488.
75. Levin, B. R., and J. J. Bull. 1996. Phage therapy revisited: the population biology of a bacterial infection and its treatment with bacteriophage and antibiotics. Am. Nat. 147:881-898.
76. Lewis, P. J., and J. Errington. 1996. Use of green fluorescent protein for detection of cell-specific gene expression and subcellular protein localization during sporulation in *Bacillus subtilis*. Microbiology 142:733-740.
77. Li, J., and T.-C. Lee. 1995. Bacterial ice nucleation and its potential application in the food industry. Trends Food Sci. Technol. 6:259-265.
78. Loessner, M. J. 1994. Taxonomical classification of 20 newly isolated *Listeria* bacteriophages by electron microscopy and protein analysis. Intervirology 37:31-35.
79. Loessner, M. J. 1991. Improved procedure for bacteriophage typing of *Listeria* strains and evaluation of new phages. Appl. Environ. Microbiol. 57:882-884.
80. Loessner, M. J., and M. Busse. 1990. Bacteriophage typing of *Listeria* species. Appl. Environ. Microbiol. 56:1912-1918.
81. Loessner, M. J., A. Schneider, and S. Scherer. 1995a. A new procedure for efficient recovery of DNA, RNA, and proteins from *Listeria* cells by rapid Lysis with a recombinant bacteriophage endolysin. Appl. Environ. Microbiol. 61:1150-1152.
82. Loessner, M. J., G Wendlinger, and S. Scherer. 1995b. Heterogeneous endolysins in *Listeria monocytogenes* bacteriophages: a new class of enzymes and evidence for conserved holin genes within the siphoviral lysis cassettes. Mol. Microbiol. 16:1231-1241.
83. Loessner, M. J., C. E. Rees, G. S. A. B. Stewart, and S. Scherer. 1996a. Construction of Luciferase reporter bacteriophage A511::*luxAB* for rapid and sensitive detection of viable *Listeria* cells. Appl. Environ. Microbiol. 62:1133-1140.
84. Loessner, M. J., A. Schneider, and S. Scherer. 1996b. Modified *Listeria* bacteriophage lysin genes (*ply*) allow efficient overexpression and one-step purification of biochemically active fusion proteins. Appl. Environ. Microbiol.62:3057-3060.
85. Loessner, M. J., M. Rudolf, and S. Scherer. 1997a. Evaluation of Luciferase reporter bacteriophage A511::*luxAB* for detection of *Listeria monocytogenes* in contaminated foods. Appl. Environ. Microbiol. 63:2961-2965.
86. Loessner, M. J., S. K. Maier, H. Daubek-Puza, G. Wendlinger, and S. Scherer. 1997b. Three *Bacillus cereus* bacteriophage endolysins are unrelated but reveal high homology to cell wall hydrolases from different bacilli. J. Bacteriol. 179:2845-2851.
87. Mamaev, S. V., A. L. Laikhter, T. Arslan, and S. M. Hecht. 1996. Firefly luciferase: Alteration of the color of emitted light resulting from substitutions at position 286. J. Am. Chem. Soc. 118:7243-7244.
88. Maniloff, J. H.-W. Ackerman, and A. Jarvis. 1994. Bacteriophage taxonomy and classification, p. 93-100. *In* R. G. Webster and A. Granoff (ed.). Encyclopedia of virology: Academic Press, London.
89. Margaritis, A., and A. S. Bassi. 1991. Principles and biotechnological applications of bacterial ice nucleation. Crit. Rev. Biotechnol. 11:277-295.
90. Marquet-Van Der Mee, N., and A. Audurier.1995. Proposals for optimization of the international phage typing system for *Listeria monocytogenes*: Combined analysis of phage lytic spectrum and variability of typing results. Appl. Environ. Microbiol. 61:303-309.
91. Martin, C. S., P. A. Wight, A. Dobretsova, and I. Bronstein. 1996. Dual luminescence-based reporter gene assay for luciferase and b-galactosidase. Biotechniques 21:520-524.
92. Mattila, T. 1987. Automated turbidometry—a method for enumeration of bacteria in food samples. J. Food Prot. 50:640-642.
93. Mattila, T., and T. Alivehmas. 1987. Automated turbidometry for prediciting colony forming units in raw milk. Int. J. Food Microbiol. 4:157-160.
94. McClure, P. J., M. B. Cole, K. W. Davies, and W. A. Anderson. 1993. The use of automated turbidimetric data for the construction of kinetic models. J. Ind. Microbiol. 12:277-285.
95. McIntyre, L. Application and evaluation of bacterial viruses in rapid methodologies for the detection of food-borne pathogens: University of Guelph, Guelph, Ontario, Canada.
96. Meighen, E. A. 1991. Molecular biology of bacterial bioluminescence. Microbiol. Rev. 55:123-142.
97. Meighen, E. A. 1993. Bacterial bioluminescence: organization, regulation, and application of the *lux* genes. FASEB J. 7:1016-1022.
98. Merril, C. R., B. Biswas, R. Carlton, N. C. Jensen, G. J. Creed, S. Zullo, and S. Adhya. 1996. Long-circulating bacteriophage as antibacterial agents. Proc. Natl. Acad. Sci. USA 93:3188-3192.
99. Misteli, T., and D. L. Spector. 1997. Applications of the green fluorescent protein in cell biology and biotechnology. Nat. Biotechnol. 15:961-964.
100. Mosig, G. 1994. T4 bacteriophage and related bacteriophages, p. 1376-1383. *In* R. G. Webster and A. Granoff (ed.). Encyclopedia of virology. Academic Press, London.
101. Murray, N. E. Bacteriophages as cloning vehicles, p. 113-116. *In* R. G. Webster and A. Granoff (ed.). Encyclopedia of virology. Academic Press, London.
102. Ohmiya, Y., T. Hirano, and M. Ohashi. 1996. The structural origin of the color differences in the bioluminescence of firefly luciferase. FEBS Lett. 384:83-86.
103. O'Keefe, B., and J. Green. 1989. Coliphages as indicators of faecal pollution at three recreational beaches on the Firth of Forth. Water Resour. 23:1027-1030.
104. Olsvik, Ø, T. Popovic, E. Skjerve, K. S. Cudjoe, E. Hornes, J. Ugelstad, and M. Uhlén. 1994. Magnetic separation techniques in diagnostic microbiology. Clin. Microbiol. Rev. 7:43-54.
105. Orser, C., B. J. Staskawicz, N. J. Panopoulos, D. Dahlbeck, and S. E. Lindow. 1985. Cloning and expression of bacterial ice nucleation genes in *Escherichia coli*. J. Bacteriol. 164:359-366.
106. Pagotto, F., L. Brovko, and M. W. Griffiths. 1996. Phage-mediated detection of *Staphylococcus aureus* and *Escherichia coli* O157:H7 using bioluminescence. Bacteriol. Qual. Raw Milk IDF Special Issue 9601:152-156.
107. Park, S. F., G. S. A. B. Stewart, and R. G. Kroll. 1992. The use of bacterial luciferase for monitoring the environmental regulation of expression of genes encoding virulence factors in *Listeria monocytogenes*. J. Gen. Microbiol. 138:2619-2627.
108. Petitt, S. B. 1989. A conductance screen for Enterobacteriaceae in foods, p. 131-141.*In* C. J. Stannard, S. B. Petitt, and F. A. Skinner (ed.), Rapid microbiological methods for foods, beverages and pharmaceuticals. Society for Applied Bacteriology Technical Series, Blackwell Scientific Publications, Oxford, England.
109. Phillips, J. D., and M. W. Griffiths. 1985. Bioluminescence and impedimetric methods for assessing shelf-life of pasteurized milk and cream. Food Microbiol. 2:39-51.
110. Phillips, J. D., and M. W. Griffiths. 1989. An electrical method for detecting *Listeria* spp. Lett. Appl. Microbiol. 9:129-132.
111. Prentice, G. A., P. Neaves, D. I. Jervis, and M. C. Easter. 1989. An inter-laboratory evaluation of an electrometric method for detection of Salmonellas in milk powders, p. 155-164.*In* C. J. Stannard, S. B. Petitt, and F. A. Skinner (ed.), Rapid microbiological methods for foods, beverages and pharmaceuticals. Society for Applied Bacteriology Technical Series, Blackwell Scientific Publications, Oxford, England.
112. Prescott, L. M. H. J. P.; Klein, D. A. 1993. Microbiology, 2nd ed. Wm. C. Brown Publishers, Dubuque, Iowa. pp. 346-366.
113. Pugh, S. J., and M. L. Arnott. 1989. Automated conductimetric detection of Salmonellas in confectionery products, p. 185-201.*In* C. J. Stannard, S. B. Petitt, and F. A. Skinner (ed.), Rapid microbiological methods for foods, beverages and pharmaceuticals. Society for Applied Bacteriology Technical Series, Blackwell Scientific Publications, Oxford, England.

114. Radetsky, P. 1994. The invisible invaders: viruses and the scientists who pursue them. Little, Brown and Company, Boston, Mass.
115. Radetsky P. 1996. The good virus. Discover 17(11):50-58.
116. Richardson, G. 1997. LabSMART™ software user manual. Version 2.0.
117. Richardson, G. H., J. T. C. Yuan, and D. V. Sisson. 1994. Enumeration of total bacteria in raw and pasteurized milk by reflectance colorimetry: collaborative study. J. AOAC Int. 77:623-627.
118. Richardson, G. H., R. Grappin, and T. C. Yuan. 1988. A reflectance colorimeter instrument for measurement of microbial and enzymatic activities in milk and dairy products. J. Food Prot. 51:778-785.
119. Rocourt, J., B. Catimel, and A. Schrettenbrunner. 1985. Isolation of *Listeria seeligeri* and *L. welshimeri* phages: Phage-typing of *L. monocytogenes*, *L. ivanovii*, *L. innocua*, *L. seeligeri* and *L. welshimeri*. Zentralbl. Bakteriol. Hyg. A 259:341-350.
120. Rodrigues, M. J. R. M., C. J. Capell, and R. M. Kirby. 1995. A new capacitance medium for presumptive detection of *Listeria* spp. from cheese samples. J. Microbiol. Meth. 23:291-296.
121. Roelant, C. H., D. A. Burns, and W. Scheirer. 1996. Accelerating the pace of luciferase reporter gene assays. Biotechniques 20:914-917.
122. Ryser, E. T., and E. H. Marth. 1991. Listeria, listeriosis and food safety. Marcel Dekker, New York.
123. Samadpour, M., L. M. Grimm, B. Desai, D. Alfi, J. Ongerth, and P. I. Tarr. 1993. Molecular epidemiology of *Escherichia coli* O157:H7 strains by bacteriophage lambda restriction fragment length polymorphism analysis: Application to a multistate foodborne outbreak and a day-care center cluster. J. Clin. Microbiol. 31:3179-3183.
124. Sanders, M. E. 1994. Bacteriophages in industrial fermentations, p. 116-121. *In* R. G. Webster and A. Granoff (ed.). Encyclopedia of virology. Academic Press, London.
125. Sanders, M. F. 1995. A rapid bioluminescent technique for the detection and identification of *Listeria monocytogenes* in the presence of *Listeria innocua*. *In* A. K. Campbell, L. J. Kricka, and P. E. Stanley (ed.). Bioluminescence and chemiluminescence: fundamentals and applied aspects. John Wiley, New York.
126. Sanders, C. A., D. M. Yajko, P. S. Nassos, W. C. Hyun, M. J. Fulwyler, and W. K. Hadley. 1991. Detection and analysis by dual-laser flow cytometry of bacteriophage T4 DNA inside *Escherichia coli*. Cytometry 12:167-171.
127. Schauer, A. T. 1988. Visualizing gene expression with luciferase fusions. TIBTECH 6:23-27.
128. Schmid, D., D. Pridmore, G. Capitani, R. Battistutta, J.-R. Neeser, and A. Jann. 1997. Molecular organisation of the ice nucleation protein InaV from *Pseudomonas syringae*. FEBS Lett. 414:590-594.
129. Seeger, K., and M. W. Griffiths. 1994. Adenosine triphosphate bioluminescence for hygiene monitoring in health care institutions. J. Food Prot. 57:509-512.
130. Sheehan, M. M., J. L. Garcia, R. Lopez, and P. Garcia. 1997. The lytic enzyme of the pneumococcal phage Dp-1: a chimeric lysin of intergeneric origin. Molecular Microbiol. 25:717-725.
131. Skyttä, E., and T. Mattila-Sandholm. 1991. A quantitative method for assessing bacteriocins and other food antimicrobials by automated turbidometry. J. Microbiol. Meth. 14:77-88.
132. Soothill, J. S. 1992. Treatment of experimental infections of mice with bacteriophages. J. Med. Microbiol. 37:258-261.
133. Squirrell, D. J. and M. J. Murphy. 1995. Adenylate kinase as a cell marker in bioluminescent assays, p. 486-489. *In* A. K. Campbell, L. J. Kricka, and P. E. Stanley (ed.). Bioluminescence and chemiluminescence: fundamentals and applied Aspects. John Wiley & Sons, New York.
134. Steidler, L., W. Yu, W. Fiers, and E. Remaut. 1996. The expression of the *Photinus pyralis* luciferase gene in *Staphylococcus aureus* Cowan I allows the development of a live amplifiable tool for immunodetection. Appl. Environ. Microbiol. 62:2356-2359.
135. Stephens, P. J., J. A. Joynson, K. W. Davies, R. Holbrook, H. M. Lappin-Scott, and T. J. Humphrey. 1997. The use of an automated growth analyser to measure recovery times of single heat-injured *Salmonella* cells. J. Appl. Microbiol. 83:445-455.
136. Stewart G. S. A. B. 1997. Challenging food microbiology from a molecular perspective. Microbiology 143:2099-2108.
137. Stewart, G. S. A. B., and P. Williams. 1992. *lux* genes and the applications of bacterial bioluminescence. J. Gen. Microbiol. 138:1289-1300.
138. Stewart, G. S. A. B., and P. Williams. 1993. Shedding new light on food microbiology. ASM News 59:241-246.
139. Stewart, G. S. A. B., T. Smith, and S. Denyer. 1989. Genetic engineering for bioluminescent bacteria. Food Sci. Technol. Today 3:19-22.
140. Suarez, A., A. Güttler, M. Strätz, L. H. Staendner, K. N. Timmis, and C. A. Guzmán. 1997. Green fluorescent protein-based reporter systems for genetic analysis of bacteria including monocopy applications. Gene 196:69-74.
141. Suhren, G., and W. Heeschen. 1987. Impedance assays and the bacteriological testing of milk and milk products. Milchwissenschaft 42:619-627.
142. Tanaka, H., T. Shinji (Iwano), K. Sawada, Y. Monji, S. Seto, M. Yajima, O. Yagi. 1997. Development and application of a bioluminescence ATP assay method for rapid detection of coliform bacteria. Water Res. 31:1913-1918.
143. Tarahovsky, Y. S., G. R. Ivanitsky, and A. A. Khusainov. 1994. Lysis of *Escherichia coli* cells induced by bacteriophage T4. FEMS Microbiol. Lett. 122:195-200.
144. Thompson, J. F., K. F. Geoghegan, D. B. Lloyd, A. J. Lanzetti, R. A. Magyar, S. M. Anderson, and B. R. Branchini. 1997. Mutation of a protease-sensitive region in firefly luciferase alters light emission properties. J. Biol. Chem. 272:18766-18771.
145. Tombolini, R., A. Unge, M. E. Davey, F. J. de Bruijn, and J. K. Jansson. 1997. Flow cytometric and microscopic analysis of GFP-tagged *Pseudomonas fluorescens* bacteria. FEMS Microbiol. Ecol. 22:17-28.
146. Tomicka, A. T., J. Chen, S. Barbut, and M. W. Griffiths. 1997. Survival of bioluminescent *Escherichia coli* 0157:H7 in a model system representing fermented sausage production. J. Food Prot. 60:1487-1492.
147. Turpin, P. E., K. A. Maycroft, J. Bedford, C. L. Rowlands, and E. M. H. Wellington. 1993. A rapid luminescent-phage based MPN method for the enumeration of *Salmonella typhimurium* in environmental samples. Lett. Appl. Microbiol. 16:24-27.
148. Ugarova, N. N. 1993. Bioanalytical applications of firefly luciferase (review). Appl. Biochem. Microbiol. 29:135-144.
149. Ulitzer, S., and J. Kuhn. 1987. Introduction of *lux* genes into bacteria, a new approach for specific determination of bacteria and their antiviotic susceptibility, p. 463-472. *In* J. Scholmerich, R. Andreesen, A. Kapp, et al. (ed.), Bioluminescence and chemiluminescence, new perspectives. John Wiley, New York.
150. Valdivia, R. H., A. E. Hromockyj, D. Monack, L. Ramakrishnan, and S. Falkow applications for green fluorescent protein (GFP) in the study of host-pathogen interactions. Gene 173:47-52.
151. Vasavada, P. C. 1993. Rapid methods and automation in dairy microbiology. J. Dairy Sci. 76:3101-3113.
152. Viviani, V. R., and E. J. H. Bechara. 1995. Bioluminescence of Brazilian fireflies (Coleoptera: Lampyridae): spectral distribution and pH effect on luciferase-elicited colors. Comparison with elaterid and phengodid luciferases. Photochem. Photobiol. 62:490-495.
153. Waes, G. M., and R. G. Bossuyt. 1984. Impedance measurements to detect bacteriophage problems in cheddar cheesemaking. J. Food Prot. 47:349-351.
154. Walker, A. J., S. A. A. Jassim, J. T. Holah, S. P. Denyer, and G. S. A. B. Stewart. 1992. Bioluminescent *Listeria monocytogenes* provide a rapid assay for measuring biocide efficacy. FEMS Microbiol. Lett. 91:251-256.
155. Ward, I. R, J. De Sa, and B. Rowe. 1987A phage typing scheme for *Salmonella enteritidis*. Epidemiol. Infec. 99:291-294.
156. Warren, G., and P. Wolber. 1991. Molecular aspects of microbial ice nucleation. Molecular Microbiol. 5:239-243.
157. Watanabe, M., and S. Arai. 1994. Bacterial ice-nucleation activity and its application to freeze concentration of fresh foods for modification of their properties. J. Food Eng. 22:453-473.

158. White, P. J., D. J. Squirrell, P. Arnaud, C. R. Lowe, and J. A. H. Murray. 1996. Improved thermostability of the North American firefly luciferase: saturation mutagenesis at position 354. Biochem. J. 319:343-350.
159. Wiggins, B. A., and M. Alexander. 1985. Minimum bacterial density for bacteriophage replication: implications for significance of bacteriophages in natural ecosystems. Appl. Environ. Microbiol. 49:19-23.
160. Wolber, P. K., and R. L. Green. 1990. Detection of bacteria by transduction of ice nucleation genes. Trends Biotechnol. 8:276-279.
161. Wolber, P. K., and R. L. Green. 1995. Rapid detection of *Salmonella* contamination of foods by transduction of bacterial ice nucleation genes. Presented at 95th General Meeting of American Society for Microbiology, Washington, D. C.
162. Yang, T-T, S. R. Kain, P. Kitts, A. Kondepudi, M. M. Yang, and D. C.Youvan. 1996. Dual color microscopic imagery of cells expressing the green fluorescent protein and a red-shifted variant. Gene 173:19-23.
163. Ye, L., L. M. Buck, H. J. Schaeffer, and F. R. Leach. 1997. Cloning and sequencing of a cDNA for firefly luciferase from *Photuris pennsylvanica*. Biochim. Biophys. Acta 1339:39-52.
164. Young, R. 1992. Bacteriophage lysis: mechanism and regulation. Microbiol. Rev. 56:430-481.
165. Young, R., and U. Bläsi. 1995. Holins: form and function in bacteriophage lysis. FEMS Microbiol. Rev. 17:191-205.

Chapter 11

Molecular Typing and Differentiation

Jeffrey M. Farber, Steven M. Gendel, Kevin D. Tyler, Patrick Boerlin, Warren L. Landry, Scott J. Fritschel, and Timothy J. Barrett

11.1 INTRODUCTION

The identification of bacterial isolates to the species level is often sufficient to determine whether a food or product is contaminated with a human pathogen and should be withheld or withdrawn from the market. In many cases, however, bacterial isolates must be grouped below the species level. Bacterial subtyping (grouping below the species level) has become a critical part of epidemiologic investigations to match case isolates with suspected sources of infection. Examples of bacterial subtyping include phenotypic methods such as serotyping, phage typing, biotyping, and antimicrobial resistance typing, as well as the newer molecular methods.

In an environment where simultaneous circulation of many bacterial clones occur, epidemiological studies can only be done with reliable, reproducible and highly discriminatory typing techniques. There are five major criteria one can use to evaluate any typing system: 1) How universal the method is with respect to strains that it can type (typeability); 2) How reproducible are results upon repeated trials (reproducibility); 3) How adequately does the method discriminate between closely related strains of the same organisms (discriminatory power); 4) How straight forward are the results to interpret (ease of interpretation); 5) How rapidly can the results be obtained, at what expense and with how much technical expertise (speed of performance, expense and technically simplicity).[11]

Conventional phenotypic typing methods are usually only relevant to the organism for which they were developed and are not universally applicable. They are also subject to variation in expression within the same isolate due to environmental pressures and acquisition of extra chromosomal DNA carrying genes that may alter phenotypic traits. This can result in i) vastly distant bacteria displaying similar phenotypic patterns leading to a false conclusion that they are related or ii) similar strains exhibiting different phenotypic patterns resulting in the false conclusion that they are different. Further, phenotypic methods are usually not sensitive enough to study bacterial population genetics. Due to the close relationship or clonal origin of some bacteria, there is a requirement for more sensitive typing methods that can pick up subtle differences in the chromosome due to point mutations, small insertions, deletions or rearrangements. For these reasons a variety of genetic approaches have evolved.

Genotypic methods are in theory considered to have 100% typeability because all organisms contain nucleic acid and it is the genome that is being targeted with these methods. Genotypic methods are generally rapid, inexpensive and simple once established but require an initial investment in finances, time and technical skills in order to evaluate, interpret and standardize. An excellent review of genomic typing methods can be found in the "Rap on REP" by Kerr.[7] Table 1 provides a listing of the advantages and disadvantages of traditional vs contemporary typing method with consideration of the above criteria.

11.2 CELLULAR FATTY ACID ANALYSIS

11.21 Introduction

The Food and Drug Administration (FDA) as a regulatory agency has a need for rapid automated microbial identification procedures. The classical procedures for microbial identification outlined in the Bacteriological Analytical Manual[4] are time consuming and for some identifications can take up to 2 weeks. For FDA to protect the consumers and regulate industry in a timely fashion, more rapid methods of identification need to be developed and implemented.

The only commercially available gas-liquid chromatography (GLC) system dedicated to the identification of bacteria by cellular fatty acids (CFA) analysis is the Microbial Identification System (MIS). It was co-developed by the Hewlett-Packard Co. and Microbial ID, Inc., Newark, DE, and is now marketed by Microbial ID. The original data base for identification of aerobic bacteria was developed by Sasser.[14]

The MIS employs a gas-liquid chromatograph to analyze and identify microorganisms isolated in pure culture on culture media. The MIS uses sophisticated software and fatty acid libraries to compare whole-cell fatty acid extracts of unknown microorganisms and determine the identity of the unknown based on similarity indices. The MIS has been developed for use by microbiologists, and it does not require extensive experience with gas chromatography in order to operate. The software is designed to provide constant calibration and monitoring of the system in order to ensure proper functioning.

Upon injection of a sample into the column containing a specified flow of hydrogen (carrier) gas, the fatty acids are separated because of different retention times under conditions of increas-

Table 1. Advantages and Disadvantages of Typing Methods

Method	Advantages	Disadvantages
Phenotypic		
Biotyping	• availability of commercial kits and automated devices • relatively low cost to maintain stocks	• results can be highly variable even within isolates of same strain due to enviromental factors causing variation in gene expression; thus reproducibility suffers • poor discriminatory power where biochemical diversity is uncommon • may falsely relate unrelated isolates or vice versa when only a narrow panel of biochemical tests are employed
Serotyping	• has long standing history • association with foodborne and invasive disease maintain its value • relatively quick and easy to perform requiring little technical ability • well suited for testing large number of isolates	• surface proteins may be hard to type due to capsules • expense of producing and maintaining a complete inventory of reagents—some organisms have diverse serotypes (e.g., *Salmonella* has over 2200 antiseras) • antiseras may cross react and some strains may not be typeable due to lack of available antiseras • need to create new sera for untypeable strains • many strains may belong to only a few commonly occuring serotypes thus lacking discriminatory power of other methods that can differentiate within the same serotype (e.g., PFGE and RAPD) • may only be suitable for reference laboratories
Phagetyping	• stability allows for long term surveillance of large populations • large numbers of isolates can be readily processed	• constant QC and replenishment of bacteriophage stocks is time consuming, labor intensive and expensive • some organisms may be untypeable due to lack of available phages • need to isolate new phages for untypeable strains • primarily applicable to only a limited group of organisms (e.g., *S. aueus*) • requires significant expertise to perform and interpret and can yield ambiguous results • may only be suitable for reference laboratories
Antimicrobial susceptibility	• easy to use, score and interpret • technically simple and low cost • good reproducibiltiy	• relatively poor discriminatory power because antimicrobial resistance is under tremendous selective pressure • may falsely relate unrelated isolates or vice versa when plasmids carrying resistance genes are picked up or lost
Multi-Locus Enzyme Electrophoresis (MLEE)	• virtually all strains are typeable	• technically very demanding and only moderately discriminatory
Electrophoretic protein typing & Immunoblotting	• virtually all strains are typeable	• large number of bands leads to difficult and subjective interpretations • requires stocks of antisera to be maintained
Genotypic		
Plasmid typing	• relatively quick and easy • results can be standardized using known markers • plasmids can be readily isolated with simple cell lysis procedures	• many strains contain no plasmids (or only 1 or 2) producing limited useful information • other strains may readily aquire or lose plasmids, making the method an unstable and unreliable marker for typing • different plasmids can appear to be the same size, or concatemers can form, giving the false presence of a larger plasmid.
Restriction Enzyme Analysis (REA)	• universally applicable • rapid, inexpensive and relatively easy to perform • very reproducible	• have to screen a number of restriction endonuclease (RE) which may be expensive in time and money • usually produces far too complex a banding pattern and is thus subject to interpretation due to the numerous overlapping and unresolved fragments (approx. 0.5–50 Kb) • a single fragment may contain fragments of similar size from many different areas of chromosome and larger fragments tend not to migrate from top of gel, therefore not all potentially useful bands are resolved

Table 1. Advantages and Disadvantages of Typing Methods—*Continued*

Method	Advantages	Disadvantages
PFGE	• a tool for both taxonomic and epidemiological studies • highly reproducible and discriminatory • produces around 5–30 easily visible bands approx. 10–800 Kb so interpretation is straightforward	• may require extensive screening of expensive RE • equipment is expensive and method is technically complex • more tedious and time consuming than other methods (1 to 3 days) • chromosomal DNA of some bacteria is degraded due to endogenous nucleases before it can be cleaved properly (e.g., *Clostridium difficile*)
Southern based hybridization with chomosomal DNA (IS typing and other hypervariable probe based typing schemes)	• single probe can be used on all isolates of a particular species • good discriminatory power and reproducibility in important pathogens (*S. aureus, M. tuberculosis*) • some applications are well standardized (e.g., *M. tuberculosis*)	• may require radioactive tags • searching for suitable locus is time consuming and hypervariable probes are usually restricted to a single species • may be poorly discriminatory when organism contains fewer than 5 copies of element • method is tedious and time consuming
Ribotyping	• single probe can be used to subtype all eubacteria due to highly conserved sequences • reproducible patterns obtained with reasonable number of fragments (10-15) • can be automated	• more tedious and time consuming than other methods due to multiple steps • only moderately discriminating when compared to some newer methods (rrn operon covers only about 0.1% of chromosomal DNA) • may not be useful when only 1 or 2 copies of rRNA gene are present within the genome of the organism (e.g., *Mycobacterium*)
RAPD	• universally applicable to all eubacteria, rapid and easy to perform • easily adjustable levels of discrimination through choice of primers • does not require isotopic labeling nor use of restriction endonucleases • does not require prior knowledge of DNA sequence • amenible to rapid exraction technics (i.e., crude lysates and whole cell)	• necessary to screen possibly hundreds of potential primers in order to find those which are discriminatory and reproducible • even so, method is subject to technical variation thus inter-run and inter-lab reproducibility is problematic • requires strict standardization to achieve even modest reproducibility • only looking at possible differences from a small % of total genome when one primer used • comparison of different intensity bands of the same size may be troublesome
rep-PCR	• very reproducible and moderately discriminatory • use of conserved primers are widely applicable	• may not be as discriminatory or as widely applicable as other methods (e.g., PFGE)
PCR-ribotyping	• universal bacterial primers can be used • produces stable, easily detectable amplification patterns in a rapid manner • potential to be widely useful in epidemiology	• has been shown to be of lower discriminatory power when compared to PFGE and possibly RAPD
PCR-RFLP & PCR-SSCP	• speed, simplicity and reproducibility	• prior knowledge of the DNA in the region of interest is required • discriminatory power of the method tends to vary substantially depending on the different species, loci, and restriction enzyme used • may require large amounts of purified DNA • inefficient cutting by REs may occur due to methylation of nucleotides at the restriction site

ing temperature. A computer-controlled temperature program begins at 170°C and is gradually increased to 270°C at 5°C/min. When each separated component reaches the end of the column, a signal from the flame ionization detector is recorded as a peak by the integrator. The area under the peak reflects the relative amount of individual fatty acid. The retention times for known fatty acids are used by the computer to calculate equivalent chain lengths for the molecules. The equivalent chain length is equal to the number of carbon atoms of a straight-chain saturated fatty acid or to a number that can be calculated by interpolation with a mathematical formula for other fatty acids. The amounts of CFAs detected are calculated as a percentage of the total amount, and a summary is printed at the end of each run to show the names and amounts of the CFAs and the most likely identification according to similarity to entries in the library. The multi-variate statistical method of principal-component analysis is used by the computer as the basis for interpreting data and matching an unknown with library entries. Each CFA quantitatively represents the objects of the principal-component analysis, resulting in pattern recognition as the basic means of identifying an isolate.[9]

Fatty acids extracted from unknown cultures are automatically quantified and identified by the MIS to determine the fatty acid composition. The fatty acid profile is then compared by pattern recognition to a library of reference strains stored in the computer to determine the identity of the unknown. Unknown isolates may be identified to the genus, species, and sub-species level. The MIS is capable of identifying a wide range of microorganisms.[15] Screening isolates for physiological or biochemical grouping prior to analysis is not required.

11.22 Principle

Isolates are grown under controlled conditions, then harvested into Teflon-lined screw-cap glass tubes for derivation to fatty acid methyl esters (FAME). The FAMEs are analyzed by a computer automated GLC instrument equipped with a fused silica capillary column and flame ionization detector. The MIS quantifies and identifies the fatty acid composition and compares it to the computer library of organisms.

11.23 Apparatus

a. Gas chromatograph: Hewlett Packard 5898A equipped with flame ionization detector, automatic sampler, integrator, and a computer. Operating conditions: temperatures injector port 250°C; detector 300°C. Gases (purity 99.995%): hydrogen and nitrogen and air (best quality available).
b. GC column: Fused silica, 25 m × 0.2 mm methyl phenyl silicone capillary.
c. Water bath: Dual chamber 80/100 ± 1°C.
d. Glass tubes: 13 × 100 mm. With leak-tight, Teflon-lined screw caps.
e. Vials: 2 mL, with 11 mm crimp caps.
f. Pipetters: Brinkman adjustable volume 1 to 5 mL.
g. Rotator: Hematology/chemical mixer.
h. Vortex shaker: Variable speed.
i. Analytical balance: Accurate to ± 0.01 g.
j. Glassware: Volumetric flasks, 100, 250, and 1000 mL; Pasteur pipettes.
k. Incubator: 28 ± 1°C.

11.24 Reagents

Alkaline Peptone Water (APW)
Trypticase Soy Broth Agar (TSBA)
Saponification Solution
Methylation Solution
Extraction Solution
Base Wash
Solvents: Hexane, methanol, methyl tert-butyl ether
Sodium hydroxide

11.25 General Instructions

All glassware or screw caps should be new or cleaned in a high quality biological cleaner prior to use. Only Teflon and glass should come in contact with reagents. Tips of disposable Pasteur pipettes packaged with foam pads should be heated or flamed prior to use to eliminate contaminating peaks in the chromatographic analysis. Extinguish all flames and heat sources before use of reagents. Handle reagents in a chemical fume hood. Excess time or temperature in extract procedure can degrade cyclopropane compounds, altering fatty acid profiles.

11.251 Delay During Sample Preparations

a. Samples can be held in the 28°C incubator for 1 to 2 h or quickly frozen following the harvesting step (before the addition of saponification solution).
b. Do not hold samples during the saponification step. Once the saponification step is complete, it is acceptable to stop, but not longer than 1 h, before continuing with the next step.
c. Following methylation, proceed immediately with extraction. The extraction step cannot be delayed.
d. After extraction and removal of the aqueous phase, the organic phase can be held refrigerated overnight.
e. Do not hold the samples in the base wash step. Remove the organic layer immediately.
f. The completed extract can be refrigerated in the crimped-top GC vial for several days before GC analysis.
g. Following GC analysis, the extracts can be stored refrigerated for several weeks. Replace the pierced crimped-top septum with a new one to prevent contamination from septa bleed or loss through evaporation.

11.252 Preparation of Isolates

All isolates are inoculated into APW and grown at 28 ± 1°C for 24 ± 2 h. The isolates are checked for purity by streaking onto TSBA plates and incubation at 28 ± 1°C for 24 ± 2 h.

11.253 Preparation of Cellular Fatty Acids

One large loopful of cells is added to 1 mL Saponification Solution in a 13 mm screw-cap test tube with a Teflon-lined cap and incubated for 5 min at 100°C to lyse the cells and liberate the fatty acids from the cellular lipids. The cells are vortexed and incubated for an additional 25 min at 100°C, and cooled. Fatty acids are converted to methyl esters by addition of 2 mL Methylation Solution and incubation for 10 min at 80°C. After rapid cooling, the fatty acid methyl esters are extracted with 1.25 mL Extraction Solution. After removal of the aqueous phase, the organic phase is washed with 3 mL of base wash solution for 5 min. The organic layer is then transferred to a GC vial for GLC analysis.

11.254 Calibration Standard

The calibration standard is controlled by computer software incorporated into the capillary chromatographic system. The computer monitors the calibration data and uses it to identify the fatty acid peaks in each extract. The analysis sequence commences with one acceptable calibration injection. The calibration standard is reanalyzed every 11th injection. The computer checks the results of each calibration injection against the peak naming table, which contains the expected retention time and the amount for each peak in the calibration analysis. The straight chain C9:0 to C20:0 fatty acid methyl esters are used by the system to quantitatively calibrate and compensate for peak area discrimination between the low and high boiling point fatty acids. Five hydroxy acids are added to the mixture to detect injection port liner or column degradation, which can result in poor peak shape (tailed hydroxy peaks) or an actual loss of hydroxy acid peak area. Deviations from the expected values result in a failure to calibrate, and a warning message to the user.

A second function of the calibration standard is to provide accurate retention times for the straight chain saturated fatty acid methyl esters C9:0 to C20:0. These retention times are used to calculate the Equivalent Chain Length (ECL) values by which peaks in subsequent analysis are named. The system calculates how much the calibration analysis has deviated from the expected retention times and reports the Root Mean Square (RMS) fit error. If a calibration run is invalid due to a high RMS fit error, the MIS prints a message to warn the user and then repeats the calibration analysis. If the system fails to calibrate after two consecutive attempts, the error message is repeated and the sample sequence aborted.

As the performance of the MIS degrades to an unacceptable level, a small peak at 10.914 ECL usually appears, indicating the breakdown of the 14:0-3OH peak to a 12:0 aldehyde. Also the 14:0-3OH peak may drop below 1.00%, or the 16:0-2OH peak may drop below 2.05%, indicating poor performance. If any of the above cases are observed, corrective action should be taken. The computer's ability to properly identify each acid is essential for successful matching of profiles.

11.255 Column Calibration

The head pressure on the column and the oven calibration temperature can be adjusted to compensate for the small variation between new columns and also for long term drift. Using the procedure outlined in the HP manual, the system can be calibrated so that the ECLs of the polar hydroxy fatty acids will be centered on the expected ECLs as listed in the peak-naming table. This will make the system appear identical to those that generated the libraries, and will allow for maximum drift while still correctly naming the peaks.

The ECL of the 2OH fatty acids varies +0.0012 per +1°C change in the elution temperature of the peak. The first hydroxy peak in the run (10:0-2OH) is the one most influenced by the starting temperature and can be adjusted by the oven temperature calibration. The late-eluting hydroxy peak (16:0-2OH) is most influenced by the linear velocity (flow) of the carrier gas and can be adjusted by the column head pressure. The two adjustments are interrelated. The procedure and the table in the HP manual describe the simultaneous adjustment of both the pressure and temperature to correctly calibrate the column.

If during normal operation of the system, the average ECL of the 10:0-2OH or the 16:0-2OH drifts more than ± 0.002 from the target values of 11.156 and 17.234 respectively, one should make the minor adjustments needed to bring them back to the target values. If large corrections must be made, it may be necessary to repeat the procedure until convergence. If the ECL target values for the hydroxy acids are 0.001 below the peak-naming table values, it means that the large straight chain reference peaks in the calibration standard are slightly overloaded, causing the hydroxy acids to elute early.

11.26 Limitations

Deviation from the recommended growth medium, growth conditions (temperature or time), or extract preparation procedure will alter the fatty acid profiles, resulting in poor similarity matches. Several investigators have demonstrated that the same bacterial strain grown and analyzed repeatedly will yield highly reproducible cellular fatty acid profiles.[3,12] Differences in physiological age of the cell population at the time of analysis may result in variable results. Several investigators have demonstrated a variation, both quantitative and qualitative, between the fatty acid composition and the culture age of bacteria.[1,6,10] The recommended culture conditions for the MIS, i.e., 28°C on TSBA for 22 to 26 h, does offer the potential for variation in physiological age differences as a consequence of the relatively low incubation temperature and the resulting increase in generation time.[13] The influence of temperature can be observed mainly in the relative proportions of CFAs rather than a loss or gain of major components. The CFA composition of cells varies, at least quantitatively, with culture conditions.[2,8,10] The most stable fatty acid compositions are from cultures in the late log phase.

There are slight but significant differences in the chemical composition of the media from different suppliers, which may result in quantitative shifts in the fatty acid profiles.[5] Consequently, one must follow preparation procedures of these media carefully.

The quadrant streak pattern is recommended for culturing cells on plates for identification by the MIS. This streaking pattern affords ample material for analysis, while confirming the presence of a single colony type or pure culture. The organisms found in the third quadrant are in the late log phase and have the most stable fatty acid compositions.

Vibrio spp. occasionally demonstrate variations in colony morphology. In some cases, these different morphological types produce different fatty acid profiles, also resulting in failure to identify or poor similarity matches.

The total area count under all the peaks, in units of microvolt-seconds, reflects the relative amount of fatty acids . The total area of a sample should be between 80,000 and 300,000. If the total area count is below 80,000, the sample should be rerun. If the total area count is above 300,000, the sample should be diluted and rerun. Too few cells may not yield enough fatty acids for a reliable comparison to the data base. Too many cells will result in excessively high area counts and column overload, with subsequent shifting of peak retention times. Such peaks may not be named properly, and comparison to the data base will be invalid.

It is important to follow all steps of the procedure exactly as described. Any deviation may result in misidentification, no identification, or a poor similarity index.

Completed extracts can be stored in the refrigerator for several weeks in crimped-top GC vials. Replace pierced crimped-top septa with new ones to prevent contamination from septum bleed or loss through evaporation.

11.27 Library

Analysis and identification of samples are completely automatic. The chromatograph is quantitatively and qualitatively calibrated, ECL values calculated, and a pattern recognition program compares each unknown to a standard library developed by Dr.

Myron Sasser of MIDI, Newark, DE. The library was developed with well-characterized strains of reference cultures from microbiologists specializing in areas such as clinical, environmental, plant and soil, industrial, and food microbiology. Each entry was developed by analyzing at least 10, but typically 20, different strains. The library is periodically updated as new strains and species are included. The pattern recognition program compares the profile of an unknown to every library entry. It compares expected ranges and considers the expected relationship of each fatty acid to all others. The bacteria will be identified if present in the library. If some similarity to other bacteria exists, the next best fit at the species and genera levels will be reported, with the probability of misidentification calculated. If the unknown is not closely related to any entry, it will be reported as no match.

11.271 Interpretation of Library Search

If the search results in more than one possible match, the suggested identities are listed in descending probability. The most probable genus identification is printed first. The next line identifies the species match, preceded by an abbreviation of the genus. Other likely matches, at each level, are listed by decreasing similarity index values.

11.28 Similarity Index

The similarity index (SI) is a numerical value which expresses how closely the composition of an unknown isolate compared with the fatty acid composition of the library matches. This index value is a calculation of the unknown's distance in n-dimensional space, from the mean profiles of the closest library entries.

A similarity index of 0.6 to 1.0 is an excellent match, with 1.0 being the highest possible match. An identification giving only one match with a SI greater than 0.500 has a strong likelihood of being correct. Single-match identifications with SI's less than 0.300 may indicate an organism not in the library, but related to the match. Low SI's may also be a result of not following the procedure or a system malfunction.

11.29 References

1. Cullen, J., M. Phillips, and G. Shipley. 1971. The effects of temperature on the composition and physical properties of the lipids of Pseudomonas fluorescens. Biochem. J. 125:733-742.
2. Drucker, D., and F. Veazey. 1977. Fatty acid fingerprints of Streptococcus mutans NCTC 10832 grown at various temperatures. Appl. Environ. Microbiol. 33:221-226.
3. Eerola, E., and O. Lehtonen. 1988. Optimal data processing procedure for automatic bacterial identification by gas-liquid chromatography of cellular fatty acids. J Clin Microbiol. 26:1745-1753.
4. FDA. 1998. Bacteriological analytical manual, 8th ed., Revision A. AOAC, Gaithersburg, Md.
5. Hewlett-Packard. 1984. 5895A microbial identification system operation manual. Hewlett-Packard Co., Avondale, Pa.
6. Hunter, M., and D. Thirkell. 1971. Variation in fatty acid composition of Sarcina flava membrane lipid with the age of the bacterial culture. J. Gen. Microbiol. 65:115-118.
7. Kerr, K. G. 1994. The rap on REP-PCR-based typing systems. Rev. Med. Microbiol. 5:233-244.
8. Knivett, V., and J. Cullen. 1965. Some Factors affecting cyclopropane acid formation in *Escherichia coli*. Biochem. J. 96:771-776.
9. Maliwan, N., R. Reid, S. Plisha, T. Bird, and J. Zvetina.1988. Identifying Mycobacteriumtuberculosis cultures by gas-liquid chromatography and a computer-aided pattern recognition model. J. Clin. Microbiol. 26:182-187.
10. Marr, A., and J. Ingraham. 1962. Effect of temperature on the composition of fatty acids in *Escherichia coli*. J. Bacteriol. 84:1260-1267.
11. Maslow, J. N., M. E. Mulligan, and R. D. Arbeit. 1993. Molecular epidemiology: Application of contemporary techniques to the typing of microorganisms. Clin. Infect. Dis. 17:153-162.
12. Mukwaya, G., and D. Welch. 1989. Subgrouping of Pseudomonas cepacia by cellular fatty acid composition. J. Clin. Microbiol. 27:2640-2646.
13. Osterhout, G., V. Shull, and J. Dick. 1991. Identification of clinical isolates of gram-negative nonfermentative bacteria by an automated cellular fatty acid identification system. J. Clin. Microbiol. 29:1822-1830.
14. Sasser, M. 1990. Identification of bacteria through fatty acid analysis. *In* Z. Klement, K. Rudolph, and D. Sands (eds.), Methods in phytobacteriology. Akademiai Kiado, Budapest.
15. Welch, D. 1991. Applications of cellular fatty acid analysis. Clin. Microbiol. Rev. 4:422-438.

11.3 MULTILOCUS ENZYME ELECTROPHORESIS

11.31 Introduction

Multilocus enzyme electrophoresis (MEE or MLEE) was introduced to the field of bacteriology in the early 1980s mainly by the pioneering works of Selander and coworkers.[7] This technique has been subsequently used to study the epidemiology of numerous human and animal pathogens.[4]

The principle of MEE is based on the separation of intracellular enzymes by electrophoresis under non-denaturing conditions. Subtle differences in electrical charge, conformation, and size of the enzyme molecules result in changes in migration rates, which can be detected after specific enzyme staining (Figure 1). Differences in migration patterns for the variants of an enzyme are directly dependent on its amino acid sequence and therefore on the nucleic acid sequence at the corresponding genetic locus. In practice, the genotype of an organism is determined by analysis of several enzyme loci (usually 15 to 25), thus allowing the differentiation of many subtypes within a species and the subsequent assessment of the genomic relatedness among isolates. MEE may be more cumbersome and less discriminatory than DNA-based

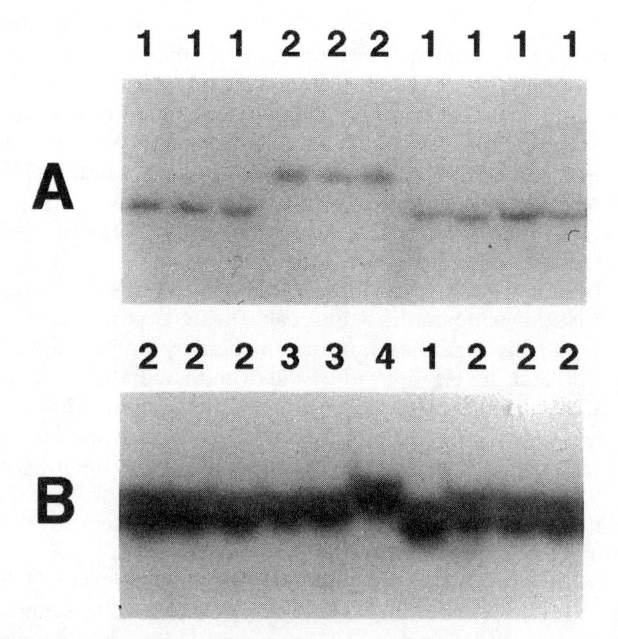

Figure 1. MEE gels for ALDH (A) and PEP (Phenylalanyl-Leucine, B) obtained with Listeria monocytogenes. Numbers above each lane represent electromorphs. Note that electromorphs for PEP appear as double bands but are interpreted as a whole, and not each band separately.

typing techniques. However, none of these has been as successful as MEE in large-scale epidemiology and in bacterial population genetics.

Electrophoresis can be performed in starch, in cellulose acetate, in agarose, in polyacrylamide, or in gels made of a mixture of the two latter components. Differences exist between these variants in terms of discriminatory power. However, for simplicity and uniformity, only the widely used electrophoresis in starch gels[1] will be described here. With this technique several slices are obtained from a single gel and the mobility of several enzymes can be determined with a single electrophoretic run. Since one of the major interests of MEE is in the study of global epidemiology, a certain degree of standardization would be welcome in its applications, particularly with regards to enzymes, electrophoresis conditions and the use of common collections of reference strains with known enzyme mobility patterns.

11.32 General Description of the Method

11.321 Preparation of Bacterial Lysates

Pure cultures of the organism are grown under optimal conditions to late log phase and harvested by centrifugation or scraping. Depending on the organism and laboratory settings, 100 mL to 1 L of broth culture or 1 to 20 plates may be needed per strain. The bacteria are resuspended in 2 mL of cold lysis buffer (10 mM Tris-HCl, 1 mM EDTA, 0.5 mM NADP, pH 6.8). From this point on, and in order to avoid denaturation of enzymes and loss of activity, all the manipulations are made on ice. Cell lysis is most efficiently obtained by sonication with a microtip in an ice-water cooling bath (optimal conditions strongly depend on organisms, container, and sonicator). Vortexing the bacterial suspension together with glass beads (100 mm and less) for a few minutes with intermittent cooling on ice may be a good alternative for some microorganisms. With both techniques, beware of aerosol formation and take preventive measures for safety of the experimenter. After lysis, the samples are centrifuged in a high-speed refrigerated centrifuge (20,000 g, 4°C, 20 min). The supernatants are sterile filtered through cooled 0.22 mm filter and aliquoted into cold containers. Lysates can be used immediately or frozen at –80°C for later use. Lysates can usually be kept at –80°C for several months without significant loss of activity.

11.322 Gel Preparation

An 11% mixture of hydrolyzed starch (Connaught Lab. Toronto, Ontario, Canada) in gel buffer (see Electrophoresis section below) is prepared and heated over a Bunsen burner in a large thick-walled Erlenmeyer flask under constant vigorous swirling until the mixture becomes homogeneous and starts to boil. The mixture is then degassed for 1 min and rapidly poured into a heat resistant plastic mold to obtain an approximately 1 cm thick gel (400 mL are optimal for an 18 cm × 20 cm × 1 cm gel). The gel is cooled to room temperature for 1 to 2 h, wrapped in a plastic foil, and kept at 4°C until the next day.

11.323 Electrophoresis

Small filter papers (Whatman filter paper #3, Whatman International Ltd., Maidstone, Kent, UK) are soaked into thawed lysates, the excess liquid is blotted away, and the filters are inserted regularly into a slit cut into the gel perpendicularly to the direction of migration, a few centimeters from its cathodic end. One of the filter papers is loaded with a marker dye (amaranth red) instead of a sample to allow control of the progression of the migration. The gel is then loaded onto a horizontal electrophoresis unit and constant voltage is applied until the amaranth dye has migrated approximately 8 to 10 cm. The gel is cooled during the electrophoresis using ice or a cold water circulating system. An example of a homemade electrophoresis assembly for MEE is illustrated in Figure 2. Beware of the danger of high voltage during electrophoresis. The buffer systems for starch electrophoresis[1] that have been used for foodborne pathogens are indicated in Table 2 and described in Media and Reagents. (Buffer pH is adjusted with NaOH, HCl, or the acid used in the buffer, where needed).

11.324 Gel Slicing

After electrophoresis, the gel is cut horizontally in 1 mm to 1.5 mm thick slices using a slicing tray as illustrated in Figure 3. A thin wire (a violin or guitar string) spanned in a small saw can be used for this purpose. The wire is regularly drawn through the gel and the slices placed into plastic trays for subsequent staining. In general, a maximum of 4 to 5 slices can be obtained for each gel (the side of the gel close to the cooling system usually presents an uneven migration profile and cannot be used).

11.325 Staining Procedures

A list of enzymes commonly examined for foodborne pathogens is presented in Table 2, and the staining solutions most frequently used in our laboratory are presented in Media and Reagents. Methods for enzymes not described here can be found elsewhere.[1,8,23] An agarose overlay is recommended for certain enzymes to avoid diffusion of the reaction products and to obtain sharper bands (see staining solutions in Media and Reagents). For this purpose, 25 mL of a 2% agarose (electrophoresis grade) so-

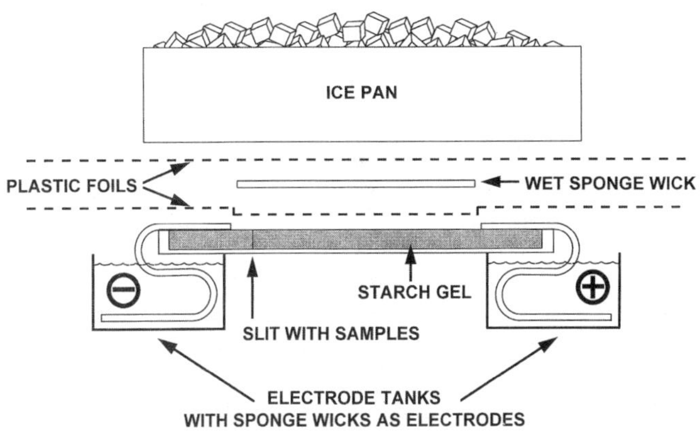

Figure 2. *Example of a homemade electrophoresis unit for horizontal electrophoresis of starch gels.*

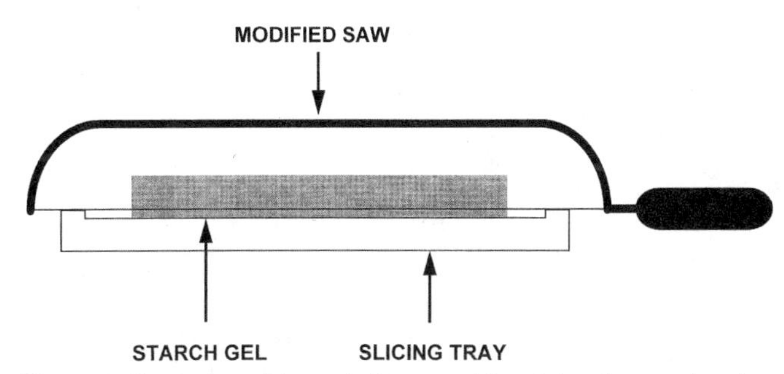

Figure 3. *Illustration of the technique used for slicing the starch gels after electrophoresis (applying a steady pressure with the hand on top of the gel is recommended during slicing).*

Table 2. Enzymes and Foodborne Pathogens Studied by MEE

Enzyme	*E. coli* and *Shigella*[a]	*Salmonella* spp.[b]	*Yersinia enterocolitica*[c]	*Aeromonas* spp.[d]	*Vibrio cholerae*[e]	*Campylobacter* spp.[f]	*Listeria monocytogenes*[g]	*Staphylococcus aureus*[h]	*Clostridium* spp.[i]	*Bacillus cereus*[j]
Acid phosphatase (ACP)	F	A					A,B,F			
Aconitase (ACO)	B	A	+	A		B	A	+		
Adenosine deaminase (ADA)		A								
Adenylate kinase (ADK)	A	A	+	A	A	A	A,F		A	
Alanine dehydrogenase (ALDH)		+		A	A		A			+
Alcohol dehydrogenase	C	A,C						+		
Aldolase (ALD)				A			B			
Alkaline phosphatase (ALP)	F	A,D	+			F	+			
Beta-galactosidase (BGA)	A									
Catalase (CAT)		A	+			F	A,F	+		
Citrate synthase (CST)		+								
Carbamate kinase (CAK)	+	A	+					+		
Creatine kinase (CK)								+		
Diaphorase (DA1, DA2)		A			A	+		+		
Esterases (EST)	D	A	+	A			A,B,D,F	+	+	+
Fructose-1,6-diphosphatase or biphosphatase (FDP or FBP)	+									
Fumarase (FUM)		A,D	F			F	A,B,F			
Glucose 6-phosphate dehydrogenase (G6P)	C	A	D		G		A,B	+		+
Glucose dehydrogenase (GDH)		+								
Glutamate dehydrogenase (GD1 and GD2)	+	A,D	+	A			A,F	+	A	
Glutamic-oxalacetic transaminase (GOT)	D	A,C	+	A	G		A		A	
Glutamic-pyruvic transaminase (GPT)					G					
Glyceraldehyde-phosphate dehydrogenase (GP1 and GP2)	B	+	+	A			A,B,F		+	
Hexokinase (HEX)	+	+								
Hydroxybutyrate dehydrogenase (HBD)								+		
Indophenol oxidase (IPO or SOD)	B	A,C	+	A		+	A,F	+	A	
Isocitrate dehydrogenase (IDH)	B	A	+	A	A	B	A,B			
Lactate dehydrogenase (LDH)	B	A			A		C,F	+		
Leucine aminopeptidase (LAP)	A	A	+	A	G					
Leucine dehydrogenase (LED)										+
Malate dehydrogenase (MDH)	A	A	+	A	G	A		+		+
Malic enzyme (ME)	+	A	+	A	G	F				+
Maltose dehydrogenase (MAD)								+		
Mannitol-1-phosphate dehydrogenase (M1P)	B	A	+					+		
Mannose phosphate isomerase (MPI)	C	A,C	+				A,B	+		
Nucleoside phosphorylase (NSP)	A	A	+	A	A		A	+	A	+
Peptidase (Glycyl-Glycine, PEP)		+								
Peptidase (Leucyl-Leucyl-Alanine, PEP)										+
Peptidase (Leucyl-Leucyl-Glycine, PEP)	A									
Peptidase (Leucyl-Proline, PEP)										+
Peptidase (Leucyl-Tyrosine, PEP)	+	+								
Peptidase (Leucyl-Glycyl-Glycine, PEP)	+	A	+				A,B	+		
Peptidase (Phenylalanine-Leucine, PEP)	A	A	+	F		A	A,B	+		
Phosphoglucomutase (PGM)	F	A,D	+	F			A,F		+	+
6-phosphoglucose isomerase (6PG)	A	A	+				A	+		+
Phosphoglucose isomerase (PGI)	B	A	+	A,F	G		A	+	+	+
Shikimate dehydrogenase (SKD)	+	+						+		
Threonine dehydrogenase (THD)	+			A	A	F			+	

Letters indicate electrophoresis buffer systems described in the literature. Only the major references covering all the enzymes and buffer systems are mentioned in the footnote. [a] see references 11, 21, 24, 26, 31, 32; [b] see references 5, 8, 13, 16, 22, 23, 27; [c] see references 9, 14; [d] see references 3, 29; [e] see reference 12; [f] see references 1, 19; [g] see references 7, 10, 30; [h] see references 18, 28; [i] see references 2, 20; [j] see reference 33.

lution in the staining buffer is cooled to 50 to 60°C, mixed with 25 mL of the staining solution and quickly poured onto the gel slice. Except for a few particular cases (see staining solutions in Media and Reagents), the gels are incubated at 37°C in the staining solution in the dark until clearly visible bands appear (shaking recommended). Bands may be dark on a light background (most staining procedures), or light on a dark background (for instance CAT and IPO). The reactions are stopped by rinsing the gels in water and soaking in fixing solution (acetic acid, methanol, water, 1:5:5, v/v/v). Many components of the staining solutions are potent toxic or corrosive chemicals, carcinogens, or mutagens, and must be handled with great care. Protective measures should be taken to avoid exposure of the experimenter. Special care also must be taken for discarding the residual solutions and the gels after staining, following local regulations for toxic waste disposal.

11.33 Interpretation of the Gels

In general, one band appears on the stained gels for each isolate (Figure 1A). Mobility patterns or electromorphs are recorded, relative to those of reference strains. By convention, electromorphs are numbered in order of decreasing mobility and the absence of an *E. coli* O157:H7 2band is recorded as zero. In the latter case, fresh bacterial lysates should be prepared again with the same strain to confirm the lack of activity for the specific enzyme. Because of possible gel deformations, experience shows that electromorph determination should not be based on relative mobility (i.e., relative to the dye migration front), but only by putting the isolates to be compared side by side on the same gel. This is of particular importance for electromorphs with only minor differences in mobility (Figure 1B) or for enzymes with a high degree of diversity. Several electrophoretic runs may be needed until the electromorph of an isolate is accurately determined. For some enzymes, several bands may appear for each isolate on a gel (Figure 1B). In this case, the pattern is interpreted as a whole and not each band separately, except when strong support is present for allocation of bands to different loci (i.e., different affinities for specific substrates as for EST and PEP or clear lack of linkage between the migration profiles of the different bands). Beware that staining procedures for an enzyme may also reveal other enzymes because of common components in the biochemical reactions used for detection of enzyme activities. For instance, staining for GPT often reveals ALDH and not GPT. Some organisms also present non-specific dehydrogenase activities not associated with specific substrates. These may be mistaken as specific activities or may give rise to additional bands on gels.

The electromorphs are recorded over all the enzymes studied, which results in an allelic pattern for each isolate (electrophoretic type or ET). For simple epidemiological typing, the comparison of allelic patterns among isolates over a limited number of enzymes with a high degree of diversity is usually adequate (for instance esterases). However, in most instances, the assessment of genomic relationships between isolates with different allelic makeup is highly desirable. Estimates of genetic distances between isolates are obtained by calculating the proportion of enzyme loci with dissimilar alleles among all the enzyme loci tested for each pair of isolates (in this case, the enzyme loci should be randomly selected). This results in a matrix of pairwise genetic distances, which can subsequently be used as a basis for a more understandable graphical representation of genomic relationships. Among the numerous phenetic and cladistic analysis methods available, only the unweighted pair group method with arithmetic average (UPGMA) and principal component analysis have been widely used for this purpose and good illustrations of these techniques can be found elsewhere.[1,17] Whatever the analysis method, it should be born in mind that graphical representations are only simplified pictures of complex genomic and evolutionary relationships and should not be over-interpreted.

11.34 References

1. Aeschbacher M, Piffaretti J-C. Population genetics of human and animal enteric Campylobacter strains. Infect Immun 1989;57:1432-7.
2. Altwegg M, Hatheway CL. Multilocus enzyme electrophoresis of Clostridium argentinense (Clostridium botulinum toxin type G) and phenotypically similar asaccharolytic clostridia. J Clin Microbiol 1988;26:2447-9.
3. Altwegg M, Reeves MW, Altwegg-Bissig R, Brenner DJ. Multilocus enzyme analysis of the genus Aeromonas and its use for species identification. Zbl Bakt 1991;275:28-45.
4. Bibb WF, Schwartz B, Gellin BG, Plikaytis BD, Weaver RE. Analysis of Listeria monocytogenes by multilocus enzyme electrophoresis and application of the method to epidemiologic investigations. Intern J Food Microbiol 1989;8:233-239.
5. Beltran P, Musser JM, Helmuth R, Farmer JJ, Frerichs WM, Wachsmuth IK, et al. Toward a population genetic analysis of Salmonella: genetic diversity and relationships among strains of serotypes S. choleraesuis, S. derby, S. dublin, S. enteritidis, S. heidelberg, S. infantis, S. newport, and S. typhimurium. Proc Natl Acad Sci USA 1988;85:7753-7.
6. Boerlin P. Applications of multilocus enzyme electrophoresis in medical microbiology. J Microbiol Meth 1997;28:221-31.
7. Boerlin P, Piffaretti J-C. Typing of human, animal, food, and environmental isolates of Listeria monocytogenes by multilocus enzyme electrophoresis. Appl Environ Microbiol 1991;57:1624-9.
8. Boyd EF, Wang F-S, Whittam TS, Selander RK. Molecular genetic relationships of the Salmonellae. Appl Environ Microbiol 1996;62:804-8.
9. Caugant DA, Aleksic S, Mollaret HH, Selander RK, Kapperud G. Clonal diversity and relationships among strains of Yersinia enterocolitica. J Clin Microbiol 1989;27:2678-83.
10. Caugant DA, Ashton FE, Bibb WF, Boerlin P, Donachie W, Low C, et al. Multilocus enzyme electrophoresis for characterization of *Listeria monocytogenes* isolates: results of an international comparative study. Intern J Food Microbiol 1996;32:301-11.
11. Caugant DA, Levin BR, Ørskov I, Ørskov F, Svanborg Eden C, Selander RK. Genetic diversity in relation to serotype in *Escherichia coli*. Infect Immun 1985;49:407-13.
12. Chen F, Evins GM, Cook WL, Almeida R, Hargrett-Bean N, Wachsmuth K. Genetic diversity among toxigenic and nontoxigenic *Vibrio cholerae* O1 isolated from the Western hemisphere. Epidemiol Infect 1991;107:225-33.
13. Cox JM, Story L, Bowles R, Woolcock JB. Multilocus enzyme electrophoretic (MEE) analysis of Australian isolates of *Salmonella enteritidis*. Intern J Food Microbiol 1996;31:273- 82.
14. Dolina M, Peduzzi R. Population genetics of human, animal, and environmental *Yersinia* strains. Appl Environ Microbiol 1993;59:442-50.
15. Harris H, Hopkinson DA. Handbook of enzyme electrophoresis in human genetics. New York: American Elsevier Publishing Co; 1976.
16. Kapperud G, Lassen J, Dommarsnes K, Kristiansen B-E, Caugant DA, Ask E, et al. Comparison of epidemiological marker methods for identification of *Salmonella typhimurium* isolates from an outbreak caused by contaminated chocolate. J Clin Microbiol 1989;27:2019-24.
17. Murphy RW, Sites JW, Buth DG, Hauler CH. Proteins: Isoenzyme electrophoresis. In: Hillis DM, Moritz C, Mable BK, editors. Molecular systematics. 2nd ed. Sunderland (Mass): Sinauer Associates, Inc.; 1996. P. 51-120.

18. Musser JM, Schlievert PM, Chow AW, Ewan P, Kreiswirth BN, Rosdahl VT, et al. A single clone of *Staphylococcus aureus* causes the majority of cases of toxic shock syndrome. Proc Natl Acad Sci USA 1990;87:225-9.
19. Patton CM, Wachsmuth IK, Evins GM, Kiehlbauch JA, Plikaytis BD, Troup N, et al. Evaluation of 10 methods to distinguish epidemic-associated *Campylobacter* strains. J Clin Microbiol 1991;29:680-8.
20. Pons J-L, Combe M-L, Leluan G. Multilocus enzyme typing of human and animal strains of *Clostridium perfringens*. FEMS Microbiol Letters 1994;121:25-30.
21. Pupo GM, Karaolis DKR, Lan R, Reeves PR. Evolutionary relationships among pathogenic and nonpathogenic *Escherichia coli* strains inferred from multilocus enzyme electrophoresis and mdh sequence studies. Infect Immun 1997;65:2685-92.
22. Reeves MW, Evins GM, Heiba AA, Plikaytis BD, Farmer JJ. Clonal nature of *Salmonella typhi* and its genetic relatedness to other *Salmonellae* as shown by multilocus enzyme electrophoresis, and proposal of *Salmonella bongori* comb. nov. J Clin Microbiol 1989;27:313-20.
23. Selander RK, Beltran P, Smith NH, Helmuth R, Rubin FA, Kopecko DJ, et al. Evolutionary genetic relationships of clones of *Salmonella* serovars that cause human typhoid and other enteric fevers. Infect Immun 1990;58:2262-75.
24. Selander RK, Caugant DA, Ochman H, Musser JM, Gilmour MN, Whittam TS. Methods of multilocus enzyme electrophoresis for bacterial population genetics and systematics. Appl Environ Microbiol 1986;51: 873-84.
25. Selander RK, Levin BR. Genetic diversity and structure in *Escherichia coli* populations. Science 1980;210:545-7.
26. Selander RK, Korhonen TK, Väisänen-Rhen V, Williams PH, Pattison PE, Caugant DA. Genetic relationships and clonal structure of strains of *Escherichia coli* causing neonatal septicemia and meningitis. Infect Immun 1986;52:213-22.
27. Seltmann G, Voigt W, Beer W. Application of physico-chemical typing methods for the epidemiological analysis of *Salmonella enteritidis* strains of phage type 25/17. Epidemiol Infect 1994;113:411-24.
28. Tenover FC, Arbeit R, Archer G, Biddle J, Byrne S, Goering R, et al. Comparison of traditional and molecular methods of typing isolates of *Staphylococcus aureus*. J Clin Microbiol 1994;32:407-15.
29. Tonolla M, Demarta A, Peduzzi R. Multilocus genetic relationships between clinical and environmental *Aeromonas* strains. FEMS Microbiol Letters 1991;81:193-200.
30. Trott DJ, Robertson ID, Hampson DJ. Genetic characterisation of isolates of *Listeria monocytogenes* from man, animals and food. J Med Microbiol 1993;38:122-8.
31. Whittam TS, Wolfe ML, Wachsmuth IK, Ørskov F, Ørskov I, Wilson RA. Clonal relationships among *Escherichia coli* strains that cause hemorrhagic colitis and infantile diarrhea. Infect Immun 1993;61: 1619-29.
32. Woodward JM, Connaughton ID, Fahy VA, Lymbery AJ, Hampson DJ. Clonal analysis of *Escherichia coli* of serogroups O9, O20, and O101 isolated from Australian pigs with neonatal diarrhea. J Clin Microbiol 1993;31:1185-8.
33. Zahner V, Momen H, Salles CA, Rabinovitch L. A comparative study of enzyme variation in *Bacillus cereus* and *Bacillus thuringiensis*. J Appl Bacteriol 1989;67:275-82.

11.4 RIBOTYPING

11.41 Introduction

Ribotyping refers to a Southern blot analysis in which strains are characterized for the restriction fragment length polymorphisms associated with the ribosomal operon(s). In general, this analysis consists of digesting bacterial DNA with a restriction endonuclease, separating the restriction fragments by agarose gel electrophoresis, and then transferring the fragments onto a nylon or nitrocellulose membrane ("blotting"). The fragments containing specific information from the ribosomal RNA loci are detected by using a labeled piece of homologous DNA as a probe. The probe binds to those restriction fragments which contain genetic information complementary to the probe sequences. The positions of fragments which hybridize with the probe are determined relative to known molecular weight standards on the blot. Different strains carry a variable number of copies of the rRNA operon and also vary in the position of restriction sites within and in flanking regions to the operon.

Ribotyping has been carried out in a manual format using a variety of labeling chemistries. The following method describes manual Ribotyping using a chemiluminescent label.[2,3]

11.42 Reagent Preparation

The *rrnB* rRNA operon from *E. coli*, inserted and replicated in pGEM, is digested with *Eco*RI before labeling. After digestion, the DNA is precipitated, dissolved in water to a concentration of 0.8–1.0 mg/mL, denatured by immersion in a boiling water bath, and chilled on ice. To label the DNA by sulfonation, a volume of 2.0 M sodium bisulfite solution (pH 5.6) equivalent to one-half of the DNA-solution volume and a volume of 1.0 M methoxylamine hydrochloride solution (pH 6.0) equivalent to one-eighth of the DNA-solution volume are added. The samples are mixed, and the pH of the solution is adjusted to 6.0 or less with HCl before incubating overnight at 30°C. Labeling reagents are removed by Sephadex G-25 column chromatography.

A conjugate of anti-sulfonated DNA monoclonal antibody (Orgenics, Yavne, Israel) and alkaline phosphatase (AP) is prepared by adding 15 times molar excess of N-succinimidyl-4-(N-maleimidemethyl)-cyclohexane-1-carboxylate at 10 mg/mL in dimethyl sulfoxide to 50 mg of dialyzed antibody in 10 mM sodium phosphate/300 mM NaCl, pH 7.0. The mixture is incubated in the dark in a 25°C water bath for 30 min and then placed on ice to stop the reaction. Unreacted cross-linking reagent is removed by Sephadex G-50 column chromatography. Fractions containing activated monoclonal antibody are pooled, and the molar concentration is determined.

N-Succinimidyl-S-acetylthioacetate at 10 mg/mL in dimethyl sulfoxide is added to an AP solution at a 15 times molar excess. The amount of AP (10 mg/mL) used is that required to produce a 1:1.6 (wt/wt) ratio of monoclonal antibody to AP. The mixture is incubated in the dark in a 25°C water bath for 30 min and the reaction is stopped by placing the mixture on ice. The sulfhydryl groups are deprotected by adding 500 µL of 1.0 M hydroxylamine for every 10 mg of AP and placing the solution in the dark in a 25°C water bath for 30 min. This reaction is stopped by placing the mixture on ice and is followed by the removal of the unreacted cross-linking agent by Sephadex G-50 column chromatography. Fractions containing activated AP are pooled, and the molar concentration is determined.

AP containing sulfhydryl groups is conjugated to the maleimide-activated monoclonal antibody by mixing the two solutions at a ratio of 1 mg of monoclonal antibody: 1.3 mg of AP and incubating the mixture in the dark in a 25°C water bath for 2 h. This reaction is stopped by adding 18 µL of 0.1 M N-ethylmaleimide for every 10 mg of monoclonal antibody and incubating the mixture in a 25°C water bath for 30 min. The conjugate is concentrated by using an Amicon ultrafiltration stirred cell, model 8050, with a YM100 Diaflo membrane and purified by Sephacryl S300 column chromatography using 50 mM Tris/100 mM NaCl, pH 8.0. The purified conjugate is stored at –20°C after 1:1 (vol/vol) dilution with Storage Buffer.

11.43 Lysis of Bacteria and DNA Extraction

Strains grown overnight in 3 mL of brain heart infusion broth are collected by centrifugation in a 1.5-mL tube, resuspended in 200 µL of 10 mM Tris.HCl, pH 8.0/10 mM NaCl/50 mM EDTA, pH 8.0 and heated at 75°C for 10 min. Cells are treated with 30 µL of N-acetylmuramidase at 1 mg/mL, 30 µL of lysozyme at 20 mg/mL, 5 µL of lysostaphin at 5000 units/mL, and 5 µL of Rnase, Dnase free at 2000 units/mL at 37°C for 15 min, followed by addition of 40 µL of crude achromopeptidase at 20 mg/mL and 15 min of additional incubation at 37°C. After the addition of 100 µL of 10% SDS and 126 µL of proteinase K at 10 mg/mL, the solution is incubated at 65°C for 30 min. The cell lysate is transferred to a 1.5-mL phase-lock gel I light centrifuge tube, extracted with phenol/chloroform and the DNA is ethanol-precipitated with 3 M NaOAc. The precipitated DNA is collected by centrifugation, and the ethanol is removed. The pellet is washed with ethanol and air-dried for at least 15 min. The DNA is resuspended in 500 µL of 1X TE Buffer and 5 µL of Dnase-free Rnase, incubated at 37°C for 4 h and stored at 4°C.

11.44 DNA Digestion and Electrophoresis

After determination of nucleic acid concentration by absorbance at 260 nm, 5 µg of DNA is diluted to 158 µL with water, 40 µL of 5X *Eco*RI Buffer and 2 µL of *Eco*RI at 50 units/µL are added, and the DNA is digested overnight at 37°C. Gel Loading Solution 2.5 µL) is added to 17.5 µL of digested DNA. Four microliters of the resulting solution (87.5 ng) is applied per lane to a 0.8% agarose gel prepared with TTNE Buffer and electrophoresed at 40 V for 2.75 h. To ensure accurate data extraction, size standards consisting of pooled *Bgl* I, *Cla* I, and *Ssp* I digests of pKK3535 (10) are electrophoresed in lanes adjacent to samples yielding eight samples and five standards per gel.

11.45 Electrophoretic Transfer, Denaturation, and UV Cross-Linking

After electrophoresis, the DNA is transferred to a nylon membrane with a commercial transfer unit via a 1-h electrophoresis at 1.0 A in TTNE buffer at 4°C. After transfer, the membrane is rinsed in TTNE buffer to remove residual agarose and placed, DNA-side up, on a sheet of 3MM paper saturated with Denaturing Solution. After 2.5 min, the membrane is transferred to a second blotter paper saturated with TTNE buffer for 15-20 s. The membrane is dried for 30 min under a heat lamp, DNA-side up, on a piece of 3MM paper. The DNA is cross-linked to the membrane with an auto-crosslinker at an automatically timed dose of 0.6 J/cm^2.

11.46 Hybridization and Detection

Each membrane is pre-hybridized with 5 mL of Hybridization Solution for 10 min at 66°C in a roller bottle apparatus. This solution is then replaced with 6 mL of probe solution (Hybridization Solution containing 1.5 µg of heat-denatured sulfonated DNA). After overnight hybridization at 66°C, the probe solution was decanted, and 20 mL of 66°C Hybridization Wash Buffer is added. The bottle is returned to the hybridization oven for 15 min. This treatment with wash buffer is repeated for a total of four washes. The membrane is removed from the bottle, placed on a blotter, DNA-side up, and dried for 10 min at 30°C.

Before application of the conjugate, each hybridized membrane is submerged in 20 mL of freshly prepared Blocking Buffer and placed on a rocker in a 30°C incubator. After 30 min, the buffer is decanted and replaced with an amount of conjugate (based on titer, typically 1:100 to 1:300) in blocking buffer. The tray is returned to the rocker for an additional hour. This incubation is followed by three successive 5-min washes in 20 mL Membrane Washing Solution at room temperature. The membrane is then washed three times for 5 min each with Assay Buffer. After decanting the final assay-buffer wash, 20 mL of assay buffer containing 220 µL of phosphate phenyl dioxetane (10 mg/mL) is added for each membrane. The tray is covered and held at room temperature for 5 min with rocking, followed by 10 min without rocking. The membrane is taped DNA-side up on blotter paper and dried in a 30°C incubator for a minimum of 30 min. The membrane is then removed from the blotter paper and heated in a 700-W microwave oven on the high setting for 10 s.

The chemiluminescent images are recorded electronically by using a high-sensitivity, super-cooled Star One camera (Photometrics, Tucson, Ariz.) in a dark environment. The images are stored on computer disk.

11.47 Patterns: Data Processing

For each membrane image, the software locates the lane positions, reduces the background and noise, scales each lane's image intensity, and uses the data from the lanes containing DNA standards of known sizes to normalize the band positions. The normalized position and intensity profile for each lane, referred to as a pattern, is then stored as an individual record consisting of 512 bytes in a data base.

Additional custom software based on the method of Hubner[3] is used to analyze the levels of correlation between pairs of patterns. This software uses the 512 intensity values for each lane as coordinates in a Euclidean, 512-dimensional space. Each pattern represents a single point in the 512-dimensional space. Each pair of patterns is compared by measuring the angle between the pair of lines constructed from the origin of the 512-dimensional space to each of the two points created from the two patterns being compared. Similar patterns have angles approaching zero degrees and cosines approaching unity (one).

11.48 Automated Ribotyping

Recently, the RiboPrinter® System has been introduced as an instrument which automates all steps of the Ribotyping process (Qualicon™, Inc., Wilmington, DE).[1,4] (Figure 4). This instrument, together with pre-packaged reagents, allows for all steps of the process to be carried out in just 8 h with no operator intervention.

Sample preparation for the automated instrument involves picking a colony from the lawn of an overnight culture of the organism to be typed and suspending the inoculum in the sup-

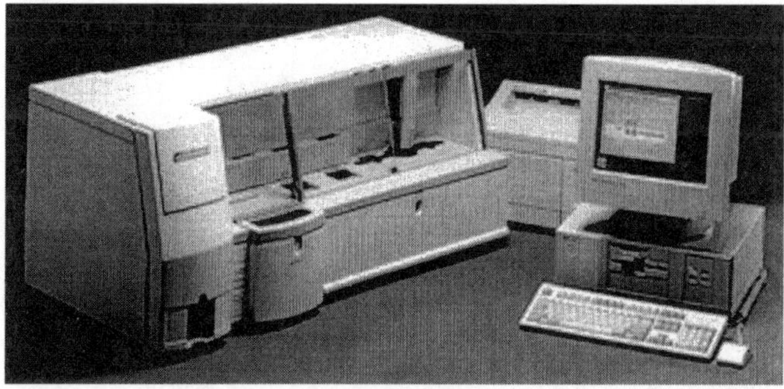

Figure 4. *The RiboPrinter® System (Qualicon™, Inc., Wilmington, DE)*

plied lysis buffer. Gram-positive organisms are prepared with two picks in 40 µL of buffer while gram negatives are prepared with a single pick into 700 µL of buffer. Up to 8 individual cultures can be processed in a single batch. The sample carrier is then placed into a heat treatment station which brings the samples to 80°C for 10 min. This treatment reduces the viability of the organisms and inactivates nucleases.

The sample carrier is then loaded into the instrument along with the required consumables for that batch. Once the operator enters strain-tracking and related sample information, the instrument will automatically process the strains. The instrument carries out the cell lysis and restriction digestion and loads the restricted DNA on an agarose gel. Electrophoresis and direct blotting onto a nylon membrane are carried out, followed by hybridization of the membrane with a labeled probe derived from the ribosomal operon of an *E. coli*. Following development of the image by use of a chemiluminescent substrate, the image is digitized using a CCD camera. This image data is stored in the system computer's hard drive memory.

Software extracts information from the image. The software recognizes data lanes on the image and distinguishes between reference marker and sample lanes. The position and intensity of well-characterized marker fragments, run simultaneously with the unknown sample, allows the system algorithms to normalize the resulting output data.

After the automated system processes a batch of samples, it generates a pattern for each sample and marker lane using proprietary algorithms. The pattern for each lane consists of a series of light and dark bands which can also be represented as waveforms (Figure 5).

The system statistically compares the output pattern to patterns obtained previously. This is a two-part operation. For identification, statistical analysis allows the conclusion that the unknown sample can or cannot be matched with known standards stored in the identification database and a classical taxonomic name applied. Those samples with matches above a fixed similarity threshold are identified; those below the threshold are not. For classification, the sample patterns are grouped with all existing pattern types run in the system to form RiboGroups. A separate similarity threshold is used to determine grouping. If a sample cannot join an existing RiboGroup because it falls below the threshold, the system creates a new RiboGroup for that sample.

The standardization and automation brought by the RiboPrinter system removes most of the labor required to generate ribotype patterns and brings an outstanding level of accuracy and reproducibility to the technique.

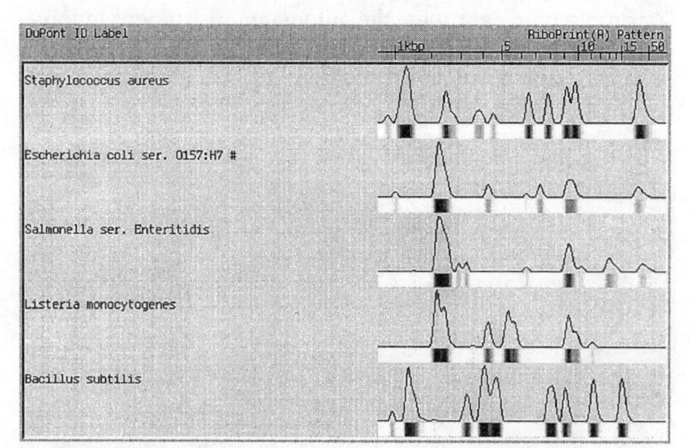

Figure 5. *Typical patterns generated by RiboPrinter® System (Qualicon™, Inc., Wilmington, DE)*

11.49 References

1. Bruce, J. L. 1996. Automated system rapidly identifies and characterizes microorganisms in food. Food Technol. Jan;77-81.
2. Bruce, J. L., Hubner, R. J., Cole, E. M., McDowell, C. I., Webster, J. A. 1995. Sets of *Eco*RI fragments containing ribosomal RNA sequences are conserved among different strains of *Listeria monocytogenes*. Proc. Natl. Acad. Sci. 92:5229-5233.
3. Hubner, R. J., Cole, E. M., Bruce, J. L., McDowell, C. I., Webster, J. A. Predicted types of *Listeria monocytogenes* created by the positions of *Eco*RI cleavage sites relative to rRNA sequences. Proc. Natl. Acad. Sci. 1995;92:5234-5238.
4. Sethi, M. R. 1997. Fully automated microbial characterization and identification for industrial microbiologists. Am. Lab. May.

11.5 RESTRICTION FRAGMENT LENGTH POLYMORPHISM FINGERPRINTING

11.51 Introduction

Restriction Fragment Length Polymorphis (RFLP) typing is similar to ribotyping in that a labeled probe is used to identify a specific set of fragments in restricted genomic DNA. The basic procedure is straightforward—genomic DNA is isolated and cut with a restriction enzyme, the DNA fragments are separated by gel electrophoresis and transferred to a membrane support, and the resulting Southern blot is hybridized with a labeled probe. The pattern of labeled bands is detected, recorded, and compared to related patterns.

RFLP fingerprinting has been used with a wide variety of foodborne microbes, both pathogens and non-pathogens. Table 3 presents several examples intended to show the variety of both the organisms and the probes that have been used for RFLP fingerprinting. The chief advantage of RFLP fingerprinting over ribotyping lies in the ability to target genomic regions that evolve faster than the ribosomal operons or that carry specific genes of interest. The chief disadvantage lies in the fact that new probes need to be developed for each species and target combination.

Because methods for producing and probing Southern blots are well known[1,15] and are the same as those used for ribotyping, they will not be discussed here (see Section 11.4). This discussion focuses on issues related to the choice of genomic target, the choice of restriction enzyme, and on the efficacy of combining data from multiple assays. It should be noted that, in recent years, the use of Southern blots has been increasingly supplanted by PCR amplification of the genomic target, followed by restriction digestion. PCR-RFLP techniques are discussed in Section 11.614.

11.52 Genomic Target

The major advantage of RFLP fingerprinting is flexibility in choice of genomic target. Probes can be derived from specific genes, random cloned fragments, or various repetitive elements. Repetitive elements can include insertion sequences (IS), non-coding sequences such as enterobacterial repetitive intergenic consensus (ERIC) sequences, or coding regions such as tRNAs. The RFLP patterns produced with probes for repetitive sequences reflect heterogeneity throughout the genome, and thus have a greater probability of detecting differences between closely related strains.[20]

Probes that target known coding regions can focus on genes that are functionally important, such as toxin or virulence genes. This may be important for species such as *E. coli* that consist of

Table 3. Examples of RFLP Fingerprinting of Foodborne Microbes

Species	Probe Target	Restriction Enzyme(s)	Comments	Reference
Bifidobacterium spp.	Cloned random fragments	*Bam* HI, *Xho* I	Probe and enzyme varied by species	9
Campylobacter	Flagellin, ATPase, elongation factor	*Bgl* II, *Cla* I, *Pst* I, *Pvu* II	Combined data from all probes and enzymes	21
Campylobacter fetus	Surface array protein	*Eco* RV, *Hae* III, *Hind* III, Xba I	Recommend using multiple enzymes	5
Campylobacter jejuni	Commercial probe	*Hae* III		6
Escherichia coli	Toxins (2)	*Eco* RI, *Pvu* II	Combined data from both enzymes & both probes	8
Escherichia coli	Toxins (2), λ Phage	*Pvu* II	Combined data from all probes	7
Escherichia coli	Toxins (2)	*Bam* HI, *Eco* RI, *Hind* III, *Pvu* II	Combined data from all enzymes and both probes	13
Escherichia coli	λ Phage	*Pvu* II		14
Lactobacillus helveticus	Plasmid fragment	*Eco* RI		11
Lactobacillus spp.	M13 Phage	*Bgl* II, *Eco* RI, *Hind* III, *Pst* I	Combined data from multiple enzymes	10
Listeria monocytogenes	Low temperature growth protein	*Hind* III	Single Band	22
Listeria monocytogenes	Cloned random fragments	*Nci* I	Combined two probes in one hybridization	12
Salmonella heidelberg	Insertion sequence	*Pst* I		18
Salmonella spp.	Cloned random fragment	*Hind* III		19
Salmonella typhimurium	Insertion sequence	*Pst* I		17
Shigella dysenteriae	Toxin	*Cla* I, *Eco* RV, *Hind* III	Combined data from all enzymes	3
Shigella spp.	Insertion sequence	*Eco* RI		16
Streptococcus salivarius	Cloned fragment	*Hae* III, *Nde* II	Combined data from both enzymes	4
Vibrio cholerae	Toxin	*Ava* I, *Bgl* I, *Eco* RV, *Pst* I		2

both pathogenic and non-pathogenic strains. Gene-specific probes can also target essential metabolic enzymes, ensuring homologous sequences will be present in all organisms of interest.

If possible, RFLP fingerprinting should be carried out using well characterized probes to facilitate data comparison and exchange between laboratories. If such probes are not available for a particular system, repetitive sequence probes are the best choice for initial fingerprinting because they are most likely to detect polymorphisms. Probes for specific (non-repetitive) coding sequences should be used only in situations where information about those genes is important for distinguishing strains of concern, because of the lower probability of finding sequence heterogeneity in the limited genomic regions involved.

11.53 Restriction Enzyme

It is difficult to predict which restrictions enzyme will generate the most informative pattern for RFLP fingerprinting. The most useful enzyme(s) vary not only with species, but with target sequence within a single genome. This is particularly true for target sequences that occur within pathogenicity islands, mobilizable IS elements, or other regions that may be subject to horizontal transfer. In general, it is usually not necessary to use exotic or expensive enzymes. The enzymes used should produce DNA fragments in the range of about 0.5–50 kbp (kilobase pairs) (depending on the gel system used), and the reaction conditions should be tested to insure that the genomic DNA is completely digested. It is generally not necessary to consider whether the enzyme cuts within the target sequence as long as it is not cut into too many fragments.

11.54 Combining Assays

It is common practice to characterize strains by using two or more restriction enzymes and/or probes to produce RFLP fingerprints (see Table 3 for examples). An overall pattern type is assigned to each strain based on these multiple patterns. However, there have been no studies carried out to determine whether the efficacy of using multiple assays differs between target sequences or microbial species. It appears that the use of multiple fingerprints is most helpful when using gene-specific probes, because these probes sample a limited region of the genome.

11.55 Other Considerations

In the past, fingerprint patterns were classified by visual observation and manual comparison. Increasingly, these activities are being carried out with scanners and pattern recognition software. When this technology is used, it is important to standardize all sample treatments and electrophoretic procedures as much as possible, and to include appropriate markers and controls on each gel. In some cases, automated systems are capable of considering both band intensity and position in comparing fingerprint patterns. Human observers tend to place the most emphasis on the location of major bands when making classifications, while pattern recognition systems may give greater weight to faint bands that vary between samples. Care must be taken to ensure that differences in sample loading, probe labeling, or signal development do not cause such variations (See also section 11.48).

11.56 References

1. Ausubel, F., Brent, R., Kingston, R., Moore, D., Siedman, J., Struhl, K., editors. 1993. Current protocols in molecular biology. New York: Wiley.
2. Bhadra R, Roychoudhury S, Banerjee R, Kar S, Mujumdar R, Sengupta S, et al. Cholera toxin (CTX) genetic element in Vibrio cholerae O139. Microbiology 1995;141:1977-1983.
3. Blaser M, Miotto K, Hopkins J. Molecular probe analysis of *Shigella dysenteriae* type 1 isolates from 1940 to 1987. Int J Epidemiol 1992; 21:594-598.
4. Colmin C, Pebay M, Simonet J, Decaris B. A species-specific DNA probe obtained from *Streptococcus salivarius* subsp. thermophilus detects strain restriction polymorphism. FEMS Microbiol Lett 1991; 65:123-128.
5. Denes A, Lutze-Wallace C, Cormier M, Garcia M. DNA fingeprinting of *Campylobacter fetus* using cloned constructs of ribosomal RNA and surface array protein genes. Vet Microbiol 1997;54:185-193.
6. Geilhausen B, Mauff F, Vlaes L, Goossens H, Butzler J. Restriction fragment length polymorphism for the identification of *Campylobacter jejuni* isolates. Zbl Bakt 1990;274:366- 371.
7. Grimm L, Goldoft M, Kobayashi J, Lewis J, Alfi D, Perdichizzi A, et al. Molecular epidemiology of a fast-food restaurant-associated outbreak of *Escherichia coli* O157:H7 in Washington state. J Clin Microbiol 1995;33:2155-2158.
8. Kudva I, Hatfield P, Hovde C. Characterization of *Escherichia coli* O157:H7 and other shiga toxin-producing *E. coli* serotypes isolated from sheep. J Clin Microbiol 1997;35:892-899.
9. Mangin I, Bourget N, Simonet J, Decaris B. Selection of species-specific DNA probes which detect strain restriction fragment polymorphisms in four *Bifidobacterium* species. Res Microbiol 1995;146:59-71.
10. Miteva V, Abadjieva A, Stefanova T. M13 DNA fingerprinting, A new tool for classification and identification of *Lactobacillus spp.* J Appl Bacteriol 1992;73:349-354.
11. Reyes-Gavilan C, Limsowtin G, Tailliez P, Sechaud L, Accolas J. A *Lactobacillus helveticus*-specific DNA probe detects restriction fragment length polymorphisms in this species. Appl Environ Microbiol 1992;58:3429-3432.
12. Ridley A. Evaluation of a restriction fragment length polymorphism typing method for *Listeria monocytogenes*. Res Microbiol 1995;146:21-34.
13. Samadpour M. Molecular epidemiology of *Escherichia coli* O157:H7 by restriction fragment length polymorphism using shiga-like toxin genes. J Clin Microbiol 1995;33:2150-2154.
14. Samadpour M, Grimm L, Desai B, Alfi D, Ongerth J, Tarr P. Molecular epidemiology of *Escherichia coli* O157:H7 strains by bacteriophage restriction fragment length polymorphism analysis: Application to a multistate foodborne outbreak and a day-care center cluster. J Clin Microbiol 1993;31:3179-3183.
15. Sambrook, J., Fritsch, E., Maniatis, T. 1989. Molecular cloning: A laboratory manual. Cold Spring Harbor (NY): Cold Spring Harbor Press.
16. Soldati L, Piffaretti J. Molecular typing of *Shigella* strains using pulsed field gel electrophoresis and genome hybridization with insertion sequences. Res Microbiol 1991;142:489-498.
17. Stanley J, Baquar N, Threlfall E. Genotypes and phylogenetic relationships of *Salmonella typhimurium* are defined by molecular fingerprinting of IS200 and 16S *rrn* loci. J Gen Microbiol 1993;139:1133-1140.
18. Stanley J, Burnens A, Powell N, Chowdry N, Jones C. The insertion sequence IS200 fingerprints chromosomal genotypes and epidemiological relationships in *Salmonella heidelberg*. J Gen Microbiol 1992;138:2329-2336.
19. Tompkins L, Troup N, Labigne-Roussel A, Cohen M. Cloned, random chromosomal sequences as probes to identify *Salmonella* species. J Infect Dis 1986;154:156-162.
20. Versalovic, J., Koeuth, T., Lupski, J. 1991. Distribution of repetitive DNA sequences in eubacteria and application to fingerprinting of bacterial genomes. Nucleic Acids Res. 19:6823-6831.
21. Weijtens M., van der Plas J, Bijker P, Urlings H, Koster D, van Logtestijn J et al. The transmission of *Campylobacter* in piggeries; An epidemiological study. J App Microbiol 1997;83:693-698.
22. Zheng W, Kathariou S. Differentiation of epidemic-associated strains of *Listeria monocytogenes* by restriction fragment length polymorphism in a gene region essential for growth at low temperature (4o C). Appl Environ Microbiol 1995;61:4310-4314.

11.6 RAPID PCR-BASED METHODS

11.61 Introduction

The recent wave of rapid amplification methods along with their variations and combinations has had a great impact on how modern investigations into the epidemiology of microorganisms are conducted. It is difficult to find a journal that does not describe some use of these techniques. Researchers have been tempted by these relatively easy and inexpensive approaches because they require limited or no previous knowledge of the genomic sequence under consideration. The following should serve as an introduction to these frequently pursued methods.

11.611 RAPD

Arbitrary amplification of polymorphic DNA sequences has been increasingly reported as a method for genetic characterization of microorganisms and there are many variations of this technique. Arbitrarily primed polymerase chain reaction (AP-PCR), first described in 1990, employed arbitrarily synthesized oligonucleotide primers approximately 20 nucleotides in length and made use of initial non-specific and subsequent specific amplification. The primer was allowed to anneal non-specifically only during the first two cycles at an annealing temperature of 40°C. The annealing temperature was then raised to 60°C for 10 cycles to allow products generated in the first phase to amplify specifically. This was followed by incorporation of ^{32}P at 60°C for a further 20–30 cycles. The relatively few amplicons generated were visualized with polyacrylamide gel electrophoresis and autoradiography.[95]

Random Amplified Polymorphic DNA (RAPD), described in 1990 as a companion method to AP-PCR, used arbitrarily synthesized oligonucleotide primers of 10-12 bases at a 36°C annealing temperature for 45 cycles. DNA fragments of between 300–3000-bp were generated and visualized after agarose gel electrophoresis and ethidium bromide staining.[100]

DNA Amplification Fingerprinting (DAF), described in 1991, used arbitrarily synthesized oligonucleotide primers as short as 5 bases at an annealing temperature of 30°C for 30–45 cycles. Amplicons of up to 500 bp were non-specifically generated and visualized after polyacrylamide gel electrophoresis and silver staining.[11]

The use of AP-PCR and DAF methods have fallen off in favor of RAPD and thus will not be considered further. Although not commonly found to be as discriminatory as some genotypic methods, such as PFGE, RAPD is continually found to be superior to phenotypic methods such as serology and phagetyping. RAPD (now used synonymously with AP-PCR) has gained widespread acceptance as a rapid method for epidemiological typing.

In RAPD analysis, single arbitrary DNA sequences target an unspecified region of the genome and generate a genetic profile. Amplification is conducted at low annealing temperatures which allows for mismatches, and thus permits arbitrary primer sequences to bind non-specifically as well as specifically to the DNA template. Amplimers are generated whenever two correctly orientated copies of the primer are close enough for the PCR to operate efficiently. In his review, Powers provides some insight into RAPD typing in microbiology and the mismatch annealing of primers.[73] A review by Caetano-Anollés highlights these methods and encompasses them under the global term multiple arbitrary amplicon profiling (MAAP).[12]

Unlike regular PCR, where an increase in DNA and/or primer concentration is expected to increase existing products, arbitrary methods may amplify new targets or reduce amplification of previous ones.[64] The resulting profile is thus a combination of artifactual variation mixed with true polymorphism.

Theoretically, arbitrary primers will generate a consistent amplification pattern in related strains of a species and it has been commonly accepted that any polymorphism observed between related individuals or strains is due to loss of priming sites, either by mutation, deletion, or insertion of genetic elements.[6,100] However, although these methods have proven valuable in rapid discrimination between individual isolates and in identification of outbreak strains, the inter-laboratory reproducibility of arbitrary amplification protocols can be problematic. The issue of reproducibility has been the focus of several publications since the inception of this arbitrary method and variation in several parameters have been cited as influencing the stability of banding patterns upon successive trials. Included in these are: DNA extraction method, concentration and secondary structure of the DNA; primer synthesis, length, GC content and concentration; primer/template ratio; concentration of magnesium ions; source and concentration of DNA polymerase; thermocycling profile, model and heating strategy; type of detection system used (e.g., percentage of polyacrylamide vs agarose) and RNA contamination.[5,8,12,15,20,22,26,32,37,43,50,60,61,70,71,77,91,101,103,104] Factors which do not influence reproducibility include: inclusion of various additives in the reaction buffer, such as gelatin and DMSO, the pH of the reaction mix, strain subculturing and plasmid content.[21,26,60] For these reasons caution must be taken when comparing and interpreting like-data between labs.

A growing acceptance of the inherent lack of reproducibility among laboratories has lead to much focus on standardization. Several independent, multi-center studies have been designed and carried out in order to improve inter-laboratory reproducibility by ascertaining where the variability lies and how best to approach solving the problems associated with various organisms (*Staphylococcus aureus*,[89] plant pathogens,[71] *Listeria monocytogenes* [WHO study],[97] *Clostridium difficile*[75]). With the recent push for standardized methods in other areas that would well be served with all typing methods,[90] and with the emergence of rigorously tested species specific primers, commercially available 10-mer kits and RAPD tablets which may contain any number of reaction components including primers and DNA polymerase, RAPD may well prove to be the ideal quick solution in ascertaining relationships among individual isolates.

Since DNA consists of only 4 bases in an ordered sequence, any random primer of 10 base pairs in length should occur once every 410 or 1,000,000 bp. Bacterial genomes are on the order of 5 MB or 5,000,000 bp and so would be expected to carry at least five copies of any random 10-mer. For the PCR to efficiently generate a single intervening product, two such binding sites need to occur on opposite strands of the target DNA within 2000 bp of each other. Larger fragments of up to 4000 bp are still possible but may not be efficiently amplified. However, with the introduction of long PCR, which utilizes a more efficient polymerase, products up to 35 Kb may be generated. This would not be advantageous for RAPD as only smaller fragments are desired. Since products formed by RAPD are rarely the result of homologous annealing events due to the use of low stringency annealing temperatures that favour non-specific binding, one can envision the numerous banding possibilities. The use of higher levels of primer, dNTPs and DNA polymerase will favour the stabilization and extension of mismatched annealing events thus creating an authentic primer molecule for further amplification. The closer the similarity between the primer and the mismatch site, the more likely the chances are that this product will be amplified and contribute to the spectrum of PCR products that appear in the reaction. Subsequent amplification will preferentially favor those products that have been extended most efficiently (short fragments tend to accumulate faster and are preferentially amplified) and it is these that will contribute to the final RAPD profile.

Because, in practice, many primers may give little or no amplification, the choice of primers for RAPD analysis is one of the most critical factors. Several may first have to be screened in order to find ones that are suitable. It appears that some arbitrary primers may work better than others and be more reproducible.[71] The fact that some arbitrary primers give little or no amplification may be due to the presence of extremely rare oligonucleotide sequences as demonstrated in mammals and other organisms.[10]

RAPD primers are generally used as single oligos that arbitrarily anneal in outwardly facing directions to amplify the region inbetween. Combinations of primers can be used individually to generate separate profiles which are then considered together in assigning types or may provide added information when used as combinations in the same reaction to generate a single profile with improved scanning of the genome.[18,33,62,96] It is advisable to use another typing method in combination with RAPD analysis and to employ at least two useful primers, either individually or in pairwise combinations, before considering the likelihood of relatedness. This is often reported to alleviate some of the concerns with reproducibility.

Since RAPD primers vary widely among species due to variations in their respective genomes, there are no generic primers that are frequently used. However, many have been applied and reported to work successfully on a number of separate occasions with a variety of organisms. Table 4 provides a variety of nucleotide sequences that have been used as primers for the various methods and Table 5 provides information on the successful application of these primers to the various organisms by method.

11.612 REP & ERIC

As an alternative to the above arbitrary approach, known conserved regions can be amplified using a single DNA primer set in a way which gives rise to polymorphic DNA fingerprints. Repetitive DNA motifs are particularly amenable to this approach and several highly conserved intergenic repetitive consensus nucleotide sequences have been reported in the literature that exploit this principle.

Initially identified in enteric bacteria, similar sequences have now been revealed in many diverse eubacterial species.[94] The 33-40-bp Repetitive Extragenic Palindromic (REP) elements, first recognized in 1981,[14] but not formally reported until 1982,[35] have been discovered in the genomes of *Escherichia coli* and *Salmonella typhimurium* and are present in approximately 500–1000 copies occupying up to 1 % of the bacterial genome.[84] These sequences have also been referred to as Palindromic Units (PU).[29,30,35] Primers targeting these REP sequences have been used at annealing tem-

Table 4. Primers Shown to be of Value with Multiple Organisms for the Various Methods (see also Table 5)

Primer name & sequence	Annealing temp. (°C)	RAPD	rep-PCR		PCR-ribotyping on 16S-23S rRNA spacer	PCR-RFLP or SSCP on 16S rRNA gene for various species	PCR-RFLP on coagulase gene for *S. aureus* digested with *Alu*I	PCR-RFLP on flagellin gene for *Campylobacter spp.* (flaA) digested with *Dde*I
			Gram-negative species	Gram-positive species				
1254 (5′- AAC CGA CGC C-3′)	36	✔						
HLWL74 (5′-ACG TAT CTG C-3′)	36	✔						
REP1R-I (5′III ICG ICG ICA TCI GGC-3′)								
REP2-I (5′-ICG ICT TAT CIG GCC TAC-3′)	40		✔	✔				
REP1R-Dt (5′-III NCG NCG NCA TCN GGC-3′)								
REP2-Dt (5′-NCG NCT TAT CNG GCC TAC-3′)	40		✔	✔				
ERIC1R (5′-ATG TAA GCT CCT GGG GAT TCA C-3′)								
ERIC2 (5′-AAG TAA GTG ACT GGG GTG AGC G-3′)	52		✔	✔				
BOXA1R (5′-CTA CGG CAA GGC GAC GCT GAC G-3′)	52		✔	✔				
BOXA2R (5′-ACG TGG TTT GAA GAG ATT TTC G-3′)	40		✔	✔				
G1 (5′-GAA GTC GTA ACA AGG-3′)								
L1 (5′-CAA GGC ATC CAC CGT-3′)	55				✔			
P11P (5′-GAG GAA GGT GGG GAT GAC GT-3′)								
P13P (5′-AGG CCC GGG AAC GTA TTC AC-3′)	55					✔		
COAG2 (5′-CGA GAC CAA GAT TCA ACA AG-3′)								
COAG3 (5′-AAA GAA AAC CAC TCA CAT CA-3′)	55							✔
flaA FORWARD (5′-GGA TTT CGT ATT AAC ACA AAT GGT GC-3′)								
flaA REVERSE (5′-CTG TAG TAA TCT TAA AAC ATT TTG-3′)	55							✔

Table 5. Methods Used in Typing Various Organisms Listing Primers or Targets Most Commonly Employed if Applicable

Organism	RAPD	rep-PCR (REP, ERIC, BOX)	PCR-RFLP	PCR-SSCP	PCR-ribotyping
Staphylococcus spp.	with primer 1254[47] & ERIC2[89]	with primers sets REP1R-I & REP2-I, REP1R-Dt & REP2-Dt, ERIC1R & ERIC2[102]	on coagulase gene with primer set COAG2 & COAG3, digested with *Alu*I on 16S rDNA with *Eco*RI or *Hind* III[78,44,31]	on 16S rRNA gene[98,99] combined with Cleavase I digestion[9]	with primer set L1 & G1[78,42]
Streptococcus spp.	✔	with primer sets REP1R-I & REP2-I, REP1R-Dt & REP2-Dt, ERIC1R & ERIC2[102] and single primers BOXA1R, BOXA2R[45]	on 16S rDNA with *Hha*I, *Rsa*I or *Msp*I (16S rRNA gene lacks heterogeneity) on virulence genes (eg. M-protein (emm) genes with *Hae*III)	on 16S rRNA gene[76,98,99]	✔ (Group A lacks heterogeneity in 16S-23S spacer region) combined with SSCP
Enterococcus spp.	✔	with primer set REP1R-Dt & REP2-Dt and single primer BOXA2R[53]	on 16S rDNA using primer set P11P & P13P digested with *Alu*I, *Rsa*I, *Bst*UI, *Cfo*I, *Tru*9I, HhaI or *Msp*I[92] on 16S-23S spacer[99]	on 16S rRNA gene[92,99]	with primer set L1 & G1 (may not be helpful with *E. faecium*)[92]
Lactococcus lactis	✔	—	—	—	—
Listeria spp.	with primer HLWL74[97,56] and ERIC1 & ERIC2[80]	✔	on 16S rDNA using primer set P11P & P13P digested with *Alu*I, *Rsa*I, *Bst*UI, *Cfo*I, *Tru*9I[92,99] on 16S-23S rRNA spacer region with enzyme pairs (*Pvu*II + *Sau*3A, *Sau*3A + *Taq*I, *Taq*I + *Alu*I) on virulence genes (hlyA, iap, mpl, prfA, actA, inlA, inlB)	on 16S rRNA gene using primer set P11P & P13P[92,98,99]	with primer set L1 & G1 (has been combined with RFLP using enzyme pairs - *Pvu*II + *Sau*3A, *Sau*3A + *Taq*I, *Taq*I + *Alu*I[92,42]

Continued

Table 5. Methods Used in Typing Various Organisms Listing Primers or Targets Most Commonly Employed if Applicable—*Continued*

Organism	RAPD	rep-PCR (REP, ERIC, BOX)	PCR-RFLP	PCR-SSCP	PCR-ribotyping
Bacillus spp.	with primer HLWL74[60]	with primers sets REP1R-I & REP2-I, REP1R-Dt & REP2-Dt, ERIC1R & ERIC2[34,102]	on 16S rDNA using primer set P11P & P13P digested with *Alu*I, *Rsa*I, *Bst*UI, *Cfo*I, *Tru*9I[92] (*Alu*I & *Rsa*I digestion has also been comined with SSCP[99] on 16S-23S rRNA spacer on virulence gene (cerolysin AB)	on 16S-23S rRNA spacer[76,17] on 16S rRNA gene[98] (*B. cereus* combined with RFLP on 16S rDNA with *Alu*I or *Rsa*I	with primer set L1 & G1[92] (has been combined with RFLP & SSCP)
Lactobacillus spp.	✔	—	—	—	—
Clostridium spp.	✔	—	on 16S rRNA gene[92,99] on virulence genes (eg. toxA & toxB)	on 16S rRNA gene using primer set P11P & P13P[98,99] on 16S-23S rRNA spacer[76]	with SSCP & RFLP
Camplylobacter spp.	with primer HLWL74[57]	✔	on flagellin gene (flaA & flaB digested with *Alu*I, *Dde*I or *Hinf* I)[65,51,83] on 16S rDNA (not digested by *Eco*RI, *Hind* III or *Mlu*I)[44]	on 16S rRNA gene (has been combined with Cleavase I digestion)[9]	—
Pseudomonas spp.	✔	with primers sets REP1R-I & REP2-I, REP1R-Dt & REP2-Dt, ERIC1R & ERIC2[102,48] and single primers BOXA1R, BOXA2R[80,48]	on flagellin gene on 16S rDNA[92,99] (not digested by *Eco*RI, *Hind* III or *Mlu*I)[44]	on 16S rRNA gene using primer set P11P & P13P[98,99]	✔
Escherichia spp.	with primer 1254[52,47] & HLWL74[60]	with primers sets REP1R-I & REP2-I, REP1R-Dt & REP2-Dt, ERIC1R & ERIC2[102] and single primers BOXA1R, BOXA2R[45]	on 16S rDNA with *Eco*RI, *Hind* III or *Mlu*I[92,99,44] on virulence genes (fimA, gyrA, fliC, G adhesions, PaPE, VT2B)	on 16S-23S rRNA spacer[76] on 16S rRNA gene using primer set P11P & P13P[98,99] (has been combined with Cleavase I digestion)[9]	with primer set L1 & G1 (has been combined with SSCP)[42]
Shigella spp.	—	with primers sets REP1R-I & REP2-I, REP1R-Dt & REP2-Dt, ERIC1R & ERIC2[102]	on 16S rDNA with *Eco*RI, *Hind* III or *Mlu*I[44]	on 16S rRNA gene[98,99] (has been combined with Cleavase I digestion[9])	—
Salmomella spp.	with primer 1254[37,47]	with primers sets REP1R-I & REP2-I, REP1R-Dt & REP2-Dt, ERIC1R & ERIC2[102,7]and single primers BOXA1R, BOXA2R[45]	on rRNA	on 16S rRNA gene[98,99] (has been combined with Cleavase I digestion[9])	with primer set L1 & G1[42]
Klebsiella spp.	with primer 1254[47]	with primers sets REP1R-I & REP2-I, REP1R-Dt & REP2-Dt, ERIC1R & ERIC2[102] and single primers BOXA1R, BOXA2R[45]	on TEM beta-lactamases	on 16S rRNA gene	—
Enterobacter spp.	✔	with primers sets REP1R-I & REP2-I, REP1R-Dt & REP2-Dt, ERIC1R & ERIC2[102] and single primers BOXA1R, BOXA2R[45]	—	on 16S rRNA gene using primer set P11P & P13P[98,99]	with primer set L1 & G1[42]
Yersinia spp.	✔	with single primers BOXA1R, BOXA2R[45]	on virulence genes (eg. ailA & ailNA)	on 16S rRNA gene[98] (has been combined with Cleavase I digestion[9]	with primer set L1 & G1[42]
Vibrio spp.	✔	✔	on 16S rDNA with MluI[44]	—	✔
Aeromonas spp.	✔	—	on 16S rDNA with EcoRI or MluI & virulence gene (aroA)[44]	—	—

peratures of 40 and 44°C for 30 cycles and the resulting amplicons visualized after agarose gel electrophoresis and ethidium bromide staining.[48,94] The 124 to 127 bp Enterobacterial Repetitive Intergenic Consensus (ERIC) elements, described in 1990, have also been discovered in the genomes of the aforementioned organisms as well as other gram-negative species, and are present in approximately 30–150 copies.[40] These sequences have also been called Intergenic Repetitive Units (IRU).[81] Primers aimed at these ERIC sequences have been used at an annealing temperature of 52°C for 30 cycles and the resulting amplicon visualized as with REP-PCR.[48,94]

Microorganisms contain a variety of interspersed repetitive DNA sequences in their genome which by definition occur in multiple copies.[49] Unlike rRNA, which by nature is also a repetitive element but is expected to be fixed within a given species, rep elements vary by both location and copy number within the chromosome of a particular species, making them an attractive target for DNA typing.[40] Rep primer sets have been designed from each half of the conserved stem of the palindromic sequence such that the 3' ends are outward facing and can therefore amplify the hypervariable region (HVR) between these elements. The principle on which all rep-PCR is based is inverse PCR.[68] Inosine is sometimes added to REP primers in order to balance out the number of nucleotides. Inosine is capable of forming weak but stable Watson-Crick base pairs with all four nucleotides.[55] Inosine has also been placed at specific positions to make these positions partially degenerate. REP primers have been shown to be most effective when used as primer sets whereas ERIC1R and ERIC2 primers provided the same information when used together or when ERIC2 is used alone.[94] This single use of a primer may appear the same as RAPD, but unlike RAPD, is directed at a specific repetitive element under high stringency conditions.

Although there are several published sequences for REP and ERIC primers,[93] very few are getting widespread repeat usage. Although originally described in gram-negative bacteria, REP and ERIC sequences have been equally successful in typing several gram-positive species (see Table 4 and Table 5). An excellent introduction to rep-PCR can be found in the paper by Lupski and Weinstock.[49]

11.613 BOX

The 154 bp BOX element, described in 1992, was discovered in the genome of *Streptococcus pneumoniae* and is present in approximately 25 copies. From 5′–3′, BOX elements are composed of 3 sub-units: the 59 nucleotides of box a, the 45 nucleotides of box b and the 50 nucleotides of box c.[54] They appear to be the gram-positive equivalent of the REP and ERIC sequences, and are the first such example of repetitive elements to be described in gram-positive organisms. Primers aimed at the 'box a' sub-unit have been subject to an annealing temperature of 53°C for 30 cycles and visualized as with REP and ERIC-PCR.[48]

As with REP and ERIC sequences, these motifs are genetically stable and differ only in their copy number and chromosomal locations between species, making them a desirable target for strain differentiation.[40] Collectively, primers targeting repetitive elements have been referred to as rep-PCR.[48] An excellent description of the BOX element and overview of the conservation of the 'box a' subunit among several bacteria can be found in the paper by Koeuth et al.[45]

11.614 PCR-RFLP

In order for PCR-RFLP to be feasible, one needs to find a target containing suitable polymorphism to allow discrimination within the species. The primer set required is therefore dependant on the DNA target. The amplified DNA fragment is then subjected to digestion by a specific restriction endonuclease or combination of nucleases and run on an agarose gel to visualize the resulting polymorphism. The digestion should result in a banding pattern containing at least 7 to 10 fragments. There are a variety of preferred targets that have been shown to have adequate polymorphisms to make them useful as targets for typing. The majority of these involve virulence factor genes, due to the fact that they are most commonly found among pathogenic strains of a particular species and have the potential for rapid rates of recombination and change due to environmental pressures. Some examples of targets are the flagellin gene (*flaA*) in *Campylobacter spp.* amplified with *flaA* forward and *flaA* reverse primers and digested with *DdeI*[51,65,83] and the coagulase gene in *Staphylococcus aureus* amplified with the COAG2 and COAG3 primer set and digested with *AluI*.[31,78] Schwarzkopf et al.[79] found the coagulase gene in general not to be discriminatory enough on its own and recommended it be used in combination with other methods. The Protein A (*spa*) gene in *S. aureus* has also been used as a target for RFLP, however this technique has been found to be of no value for clonal analysis.[38]

A somewhat more universal approach has been shown to be successful by Kobayashi[44] and Vaneechoutte et al.[92] who applied the use of the 16S rRNA gene as a target for amplification and restriction digest to several organisms.

A multi-centre study was recently carried out on *Pseudomonas aeruginosa* with various typing methods. Of these, PCR-RFLP of the 16S rRNA gene was found to be the most discriminatory.[1]

Universal primers that have seen repeat usage for amplifying the 16S rRNA gene in various microorganisms can be seen in Table 4 and Table 5.

11.615 PCR-SSCP

SSCP (single strand conformation polymorphism), as originally described, involves restriction digest of whole genomic DNA, denaturing an aliquot and running it out on a neutral (non-denaturing) polyacrylamide gel.[69] When applied to PCR, this allows further discrimination of isolates based on the fact that the structure formed by ssDNA under non-denaturing conditions displays mobility shifts due to conformational changes as a result of nucleotide substitutions. The advantage of SSCP over RFLP is that it can detect DNA polymophisms and point mutations at a variety of positions in the ssDNA fragment, as opposed to using restriction enzymes which only recognize one specific mutation. Two bands from different isolates may run at the same position, indicating an identical fragment, when in fact, the underlying sequences are different. SSCP helps to resolve this concern by detecting differences in the electrophoretic mobility of ssDNA as a function of the structures that will form.

Operating on the same principle as PCR-RFLP, PCR-SSCP requires a primer set to amplify a specific target which contains adequate polymorphisms to make the target useful. Therefore the same targets and primers which are useful for PCR-RFLP can be used for SSCP analysis and the technique has been extensively applied to the popular 16S rRNA gene[92,99] (see Table 4 and Table 5).

11.616 PCR-Ribotyping

Another technique called PCR-ribotyping is rivalling the traditional ribotyping method which involves far more time and effort. The ribosomal RNA operon, consisting of the 5s, 16s and 23s subunits, has been extensively characterized and determined to be quite stable (conserved) among individual species. However, the intergenic spacer regions between these subunits appear to evolve at a much higher frequency and can therefore be used as a target for interspecies differentiation.[42] By designing primers that target the flanking regions of the well characterized and conserved subunits, the variable spacer regions in between can be amplified and the resulting products resolved by traditional electrophoresis on an agarose gel.

PCR-ribotyping requires a primer set to amplify across the specific rRNA target and since oligonucleotide primers are complementary to highly conserved regions, a single primer set may be adequate for typing a broad spectrum of species. In addition, primers and probes can be sought that are species specific within this region in order to develop a single step diagnostic and typing method.[4,6,25,66,86]

The polymorphisms detected are a function of the size of the 16S–23S spacer region and its copy number in an organism, which is a direct function of the number of copies of rRNA genes present. This can be anywhere from 2 to 11 copies per bacterial cell.

It has been demonstrated that the majority of variability lies within the HVR between the 16s and 23S ribosomal subunits (VH1) as opposed to the spacer region between the 23S and 5S subunits (VH2). This region encodes several tRNAs as well as containing several direct repeat sequences.[46] The VH2 region may in fact be of little discriminatory value without the application of restriction enzymes.[46,82] However, in some cases it may be advantageous to amplify the entire 23S rRNA gene along with its adjacent HVRs. With this strategy, even a small variation in sequence length of either of the spacer regions could generate useful polymorphisms for differentiating isolates.[82] Evidence to support the continued use of this strategy has been published recently by Sallen et al.,[74] who found several signature regions within the 23S rRNA gene of *Listeria*.

Alternatively, organisms that are poorly discriminated by amplifying the 16S–23S region may have discriminatory power increased by applying RFLP to the few resulting fragments in order to increase chances of hitting a polymorphism and thus increasing discriminatory power.[17,46,82] Sontakke and Farber[82] applied a combined enzyme strategy and performed a double digest to this purpose. Universal primers that have seen repeat usage are listed in Tables 4 and 5.

11.617 Recommended Controls

In order to evaluate any method, a suitably comprehensive panel of organisms must be selected for testing. One needs a diverse group of isolates that are representative of the environment for which the technique will be applied. Their epidemiology should already be established and well documented by virtue of previous characterization with more traditional and established methodologies. Naturally occurring and disease (outbreak) strains as well as geographically and epidemiologically related and distinct strains are all essential components of a good study group. Pitt[72] has also suggested that one use colonial variants of the same strain, pairs of organisms from the same and different isolation sites, multiple colonies from primary plating of specimen and subsequent passages and antibiotic sensitive and resistant strains from the same source.

Not all targets will be suitable to discriminate every organism. Primers must first be screened and carefully evaluated against a properly selected collection of previously characterized strains. The selection of a proper control study test group is critical to future statements regarding applications of the method.

Ideally, multiple parties should participate and try to achieve the same results in different locations and times with their own cultures of the same isolate. For one lab to retest their own work is not wholly adequate, especially given that what some labs claim as reproducibility amounts to re-running the same sample at a different time. The same isolate should be grown from original culture and all aspects of the procedure repeated at another time preferably by another experienced researcher in the same lab to see if the same patterns are achieved, or at least the same groupings appear.

Alm et al.[2] have developed an oligonucleotide database (OPD) that researchers are beginning to discover and draw upon to find previously tested primers and probes for their applications. Addressing the numerous difficulties when relying on the literature for information regarding primers and probes, this database centralizes and supplies details on design and characterization, including a standardized name, probe sequence, nucleotide position within the target gene, optimal hybridization or annealing conditions, intended target group and original citations, as well as citations that experimentally validate target group specificity. The majority of probes in the current data set target rRNA, however, this database is still in its infancy and it is envisioned that it will accommodate probes and primers for several gene families. Online submissions are welcomed and encouraged. The database can be accessed through the World Wide Web at http://www.cme.msu.edu/OPD or the authors contacted at OPD@uiuc.edu.

11.62 Preparation of Genomic DNA from Bacteria for Use with Rapid Methods

11.621 Introduction

The application of molecular biology techniques to the analysis of bacterial genomes has previously depended on the ability to prepare pure, high-molecular-weight DNA. This generally involved gently lysing the cells, solubilizing the DNA and removing contaminating proteins, RNA, and other macromolecules by one or more enzymatic or chemical steps. Depending on the application, further purification and concentration of nucleic acids may have been required to separate plasmid from genomic DNA as well as separate genomic DNA from residual debris in the cell lysate. With the advent of rapid methods for DNA analysis has also come rapid methods for DNA extraction and the time consuming purification steps of the past have been shown to not always be necessary. Several researchers now appear to be taking a whole cell or crude lysate approach to the preparation of DNA for the PCR template rather then purifying and claiming no comparable differences in the profiles generated. However, it should be noted that variations in the quality of DNA preparations may affect reproducibility[61] and that unpurified DNA preparations are not stable and PCR products can be degraded by residual thermostable nuclease activity.[28] These facts might argue against a whole cell or crude lysate approach. However the effects of nuclease activity may be reduced or inhibited by the addition of EDTA (60 mM, pH 8.0) or storage at –20°C.[28] At any rate, whole cell preparations should only be used with caution and standardization. Everyone applying a PCR protocol using these practices would have to be standardized on the same method in order for it to be of value with respect to interlab comparison of results. For a further discussion on whole cell approaches see the end of this section.

The next section presents a crude approach for the aquisition of DNA from gram-positive and gram-negative organisms that uses whole cells or boiled/heated cell lysates directly as template in the PCR reaction mix.

11.622 Special Equipment, Supplies and Media

Variable Temperature Incubator (aerobic, CO_2, or anaerobic). The exact specifications will depend on the organism under investigation.

Selective growth media (agar based) and enrichment media (broth based). Again the choice would depend on the organism's requirements.

1.5 mL centrifuge tubes
High speed centrifuge with rotor for 1.5 mL tubes
Vortex
Set of micro-pipettors and barrier tips (for 0.2 µL–1 mL)
Spectrophotometer and micro-cuvettes (holding maximum of 500 µL)
Freezer (–20°C)
0.9% (w/v) NaCl
TE buffer

11.623 Procedure

11.6231 Preparation of culture.

1. Inoculate the organism onto an appropriate growth medium (e.g., LB or Brain Heart Infusion (BHI) agar seems to get good all round usage) and incubate accordingly.
2. Pick a single colony and use this to inoculate a tube of appropriate enrichment growth broth (e.g., 5 mL of LB or BHI broth). Incubate accordingly with or without agitation till stationary phase is reached.

   Note: Suitable incubation conditions will vary with the organism being investigated. Allow culture to grow until the growth is adequate. This may take several hours to several days, depending on the growth rate. When using O/N plate growth as a starting material, there is a significant opportunity for variability in the amount of culture taken. Particularly with whole cell preparations, there is the possibility of using too many cells (overloading the PCR) and there is a chance of contamination with the agar material itself, thus leading to a general inhibition or decline in efficiency of the PCR.[63] For this reason, it is recommended that one use liquid culture as starting material for DNA preparations and standardized OD values for preparation of working stocks.

11.6232 Preparation of the cells.

1. Transfer 1 mL of broth culture to a 1.5 mL centrifuge tube and centrifuge until a compact pellet forms (3 min at 15,000 × g). Carefully pour off/remove the supernatant.
2. Wash pellet in 1 mL of a suitable sterile bacterial buffer (e.g., TE buffer) and vortex vials to resuspend pellet. Centrifuge again to obtain compact pellet. Repeat step 4 if desired.

   Note: As a precaution, with some organisms such as *Salmonella* spp. where nucleases are known to be present,[28] researchers have used an initial pre-heating step to inactivate the intrinsic DNases (resuspend in 300 µL TE buffer, heat at 75°C for 10 min, re-spin to pellet).[36]

11.6233 Standardizing whole cells. Following this whole cell approach allows the flexibility of having a standardized whole cell stock on hand to be used directly as template or in the preparation of crude lysates. Alternatively, one can skip directly to the boiling method below. Please see notes on whole cell preparations at end of section before proceeding.

1. Carefully pour off/remove the supernatant and add 50 µL sterile saline. Resuspend the cells using a micro pipette, until a homogeneous suspension is obtained. This is called the stock suspension.
2. The optical density at 600 nm (OD_{600}) of the stock suspension is determined using a spectrophotometer, according to the following procedure. Take an aliquot of the stock suspension and prepare a dilution in an appropriate diluent (e.g., water, saline). The dilution should be made in such a way that the OD_{600} reading is in the range 0.2–0.8 (usually this means a 100 fold dilution of the stock suspension—5 µL in 495 µL sterile dH_2O and use micro cuvette that holds maximum of 500 µL—may only need 200–300 µL for assay depending on the spec). If the OD_{600} is outside the given range, the dilution should be adjusted accordingly. From this, the OD_{600} of the stock suspension is calculated (see Example below).
3. If required, the stock suspension can now be stored at –20°C for up to 3 months in convenient aliquots. Stock suspension should not be thawed more than once.
4. Dilute a fresh aliquot of the stock suspension in sterile distilled water (or other suitable bacterial buffer) to prepare the working suspension. This suspension should have a concentration of 10^6 cells/µL which translates to an OD_{600} of 1.8.* The appropriate dilution ratio is calculated from the value determined in step 6 (see Example below). Divide the working suspension into small aliquots (20-50 µL) and freeze individually. Again, the same aliquot should not be thawed more than once. This can be used directly as template for whole cell PCR.

   *Note: This value of 1.8 has been empirically determined for *Listeria*. The concentration of cells will vary with the organism (i.e., the value may not be transferable between organisms as they all have their particular growth characteristics). It may be necessary to first empirically determine the growth curve and calculate the log phase number of cells as related to the OD_{600} value for the particular organism. However, as long as one is consistent, the problems with reproducibility should be lessened.

Example: In step 6, 6 µL of the stock suspension is mixed with 594 µL saline (100-fold dilution). In the spectrophotometer, this results in an approximate OD_{600} = 0.36. This means that the OD_{600} of the stock suspension is 0.36 x 100 = 36. Thus, to obtain the working suspension in step 8 of OD_{600} = 1.8, the stock suspension must be diluted 36/1.8 = 20 fold (i.e. 10 µL stock diluted with 190 µL water).

11.6234 Boiling method. Note: If boiling of cells is required, continuing from the above whole cell approach gives the advantage of being able to process a standardized amount of cells each time. Alternatively one can start from the end of step 4.

1. If using whole cell preparations in combination with boiling, use up to 500 µL of the OD_{600} 1.8 standard in a 1.5 mL eppendorf tube. If continuing from step 4 resuspend pellet in 0.5 mL sterile dH_2O (or other suitable bacterial buffer).
2. Secure tubes in a floating rack and boil 10 min (piercing top with syringe may avoid tops from popping open). Alternatively, tubes may be heated in a heating block or PCR machine at 100°C for 10 min to lyse cells.

   Note: Given that some organisms may be harder to lyse than others, researchers have proposed addition steps to improve on this method and inactivate some of the contaminants which may have an effect on the PCR efficiency or reproducibility. A pre-incubation with lysozyme followed by proteinase K and heat inactivation preceding the boiling step has been found useful for *Listeria*[16] and *Lactococcus* lactis[23] [250 μL standardized whole cells, spin down, resuspend in 100 μL PCR buffer with $MgCl_2$, add 4 μL lysozyme (10 mg/mL) and incubate 37°C for 30 min, add 1 μL proteinase K (20 mg/mL) and incubate 55°C for 1 h, boil 10 min]. The use of lysostaphin followed by a lysing buffer containing proteinase K has been used on *S. aureus*[31] [resuspend pellet in 500 μL TE with 7.5 U lysostaphin (15 U/mL), incubate 1 h at 37°C, add 1 mL lysing buffer (0.45% Nonidet P-40, 0.45% Tween 20, 0.6 μg/ mL proteinase K in PCR buffer with $MgCl_2$, incubate 1 h at 56°C, inactivate enzymes at 95°C 10 min] as well as direct lysing in an NaOH/SDS solution[78] [resuspend pellet in 500 μL 0.02 M NaOH/0.1% SDS, heat at 90°C for 15 min, make 10 fold dilution in H_2O, centrifuge and transfer]. For *Pseudomonas cepacia* pre-incubation steps with lysozyme and proteinase K prior to boiling as well as a post-treatment with DNase free RNase have been used[46] [resuspend pellet in 500 μL TE and boil 5 min, add 5 μL lysozyme (10 mg/mL) (approx. 100 μg/ mL final), incubate 15 min on ice, add 2.5 μL proteinase K (20 mg/mL) (approx. 100 μg/mL final), incubate at 55°C 10 min, heat inactivate enzymes at 95°C for 15 min, add 0.5 μL DNase-free RNase (10 mg/mL) (approx. 10 μg/mL final) incubate 37°C for 15 min]. *Salmonella* has been found to benefit from a sonication step prior to boiling[7,102] [resuspend pellet in 500 μL TE in 1.5 mL eppendorf tube, place in floating rack in H_2O and place sonication probe in water surrounding tubes at 40 W for 5 min; heat tube at 100°C for 5 min].

3. Centrifuge 2 min at high speed and transfer supernatant to clean tube. Store lysates at –20°C in convenient aliquots if not used directly. Storage at 4°C is not recommended due to the possibility of residual nucleases as mentioned in the introduction.

   Note: If one does not use the standardized whole cells for boiling application, then one can try and estimate the DNA concentration in the lysate directly by taking the OD_{260} reading of a diluted aliquot. In this way, the amount of template taken for the PCR can be standardized. Adjust the concentration of lysate to 0.25 in sterile dH_2O (μg/mL = $OD_{260} \times$ dilution factor (df) × 50, where 50 is a constant used for double stranded DNA molecules) and use 2 μL directly in the PCR (~25 ng). Not all of this is DNA because the contaminating RNA, protein and other extracts contribute to the OD_{260} reading, but at least it should be a standardized amount each time. An OD readout may not be possible by standard spectrophotometry due to the small amount of DNA that may be present or may not be relevant due to the large amount of contributing contaminants. Thus, an alternative is to use a fluorometer with a fluorescent DNA binding dye as mentioned in the notes on whole cell preparations at end of this section. The use of fluorometry may be more applicable to boiling methods where a small yield of highly contaminated DNA may be present, but can also be useful with purification techniques for the same reason.

11.63 Notes

11.631 Notes on Whole Cell Preparations and Crude Lysates

Rapid boiling or heating methods have been the most popular extraction method when using rapid typing methods, however they may not always be suitable. A whole cell or boiling approach may not work with all organisms and additional steps may be required or a standard purification method may be in order. Mazurier et al.[57] found that the whole cell method previously shown to be successful for *Listeria* did not work when applied to *Campylobacter*, however a step up to the boiling method worked fine. Hejazi et al.[33] found that separate boiling methods previously successful for *E. coli* and *Serratia marcescens* were inadequate for their work with RAPD typing of *S. marcescens*. Only purified DNA provided an acceptable level of reproducibility in banding patterns. Since *S. marcescens* was the organism under investigation in each case, this questions the reproducibility of whole cell extracts in the hands of others. Powers[73] also found that fingerprint patterns of *Candida albicans* using unpurified DNA obtained from simple boiling procedures were quite unstable. This is due to the difficulty in estimating the DNA concentration due to many impurities that affect the OD readings of traditional spectrophotometry. However, this can be compensated for by using a fluorometer in conjunction with a fluorescent DNA binding dye such as Hoechst 33258 or pico green that binds specifically to the base paired nucleotides.[19] Also inhibitors may be present that affect the efficiency of PCR; this can be addressed by the addition of proteinase K during the boiling of cell lysates. Another problem with crude preparations is that the DNA may be sheared resulting in templates of various sizes and structures, adding to potential reproducibility problems. Woods et al.[102] have tested whole cell and boiling/heating protocols on a variety of enteric and pathogenic gram-positive and negative organisms and offer their recommendations for the best organism/method combinations. Some papers describe boiling methods whereby a few colonies are picked or a loopful scraped from a plate, resuspended, boiled and an aliquot of the lysate used directly without further standardization.[15,19,47,78,80,92] Although this may be fine for single run assays, it is not ideal for reproducibility and should be avoided with rapid PCR typing techniques, especially RAPD. The protocols presented above, in our opinion gives the most flexibility for standardization and reproducibility.

11.64 Preparation, Amplification, Detection and Analysis of PCR for Rapid Methods

11.641 Introduction

This protocol is based on a set of parameters most commonly used for RAPD. Ranges and values for other methods are shown in Table 6 and again represent those most often cited or successfully applied to a number of organisms. These values can be readily substituted into any protocol. Ranges and suitable values presented have been compiled from a variety of sources and represent the values that have been claimed to be reproducible over time. This protocol is only intended as a guide or starting point based on the majority of use with a wide variety of organisms. It can be applied equally well to purified templates, whole cell preparations or crude lysates. However, many whole cell preparations require an additional pre-heating step e.g., *Bacillus*, Klebsiella, *Pseudomonas*, *Staphylococcus* and *Streptococcus*[102] or sonication e.g., *Salmonella*[7,102] (see Note 2 in protocol). The protocol given below can be followed for all amplification methods discussed, taking into account the PCR parameters listed in Table 6.

11.642 Additional Special Equipment, Supplies and Media

Ice maker
Small sterile bottles for making master mixes (5–10 mL)
deionized, distilled H_2O
10X PCR buffer without $MgCl_2$

Table 6. PCR Parameters—Ranges and Most Common Values[a]

Parameter to insert	Rapid PCR Typing Method						
		rep-PCR Method					
	RAPD	REP-PCR	ERIC-PCR	BOX-PCR	PCR-RIBO Typing	PCR-RFLP of 16S rRNA gene	PCR-SSCP of 16S rRNA gene
Reaction Volume (µl)	20–**25**–100	20–**25**–50	20–**25**–50	**25**	20–**50**–100	40–**50**	**50**–100
Template (ng/µl)	0.4–**1**–2	1–2–**4**	1–2–**4**	2–**4**	0.5	1–2	0.25
Primer (µM)[b]	**0.2**–2.4	1–**2**	1–**2**	**2**	**0.2**–1	**0.2**–1–1.5	**0.2**–0.4
Length (bp)	8–**10**–20	**18**	**22**	**22**	**15**–25	**23–24**	**20**
GC content (%)	45–**65**–100	**50**	**50–55**	32–**41–68**	**45–60**–68	**48–62**	**60**
$MgCl_2$ (mM)	1.5–**2**–4	**1.5**–3.5	**1.5**–3.5	**1.5**	**1.5**–2.5	1.5–**2.5**–3.0	1.5–**2.5**–7
Tris-HCl (mM)	**10**–20	**10**	**10**	**10**	7.5–**10**	**7.5**–10	**7.5**–20–50
pH**8.3**–9.0	**8.3**–9.0	**8.3**–9.0	**8.3**	8.3 -**8.8**–9.0	**9**	8.5–**9.0**	
KCl (mM)	**50**–75	**50**	**50**	**50**	**50**	**0**–50	**0**–50
$(NH_4)_2SO_4$ (mM)[c]	0–16	0	0	0	20	**20**	16–**20**
Gelatin (%)	0–0.01	0	0	0	0–0.1	0.1	0
Triton X100 (%)	0–0.1	0.1	0.1	0.1	0–0.1	0	0
BSA (%)	0–0.01–0.15	0	0	0	0–0.01–0.2	0	0.2
Tween 20 (%)	0	0	0	0	0–0.1	0–0.1	0–0.1–0.5
dNTPs (mM)	0.08–**0.2**–0.25	0.2–0.625–**1.25**	0.2–0.625–**1.25**	**1.25**	0.1–**0.2**	0.05–**0.1**–0.2	0.05–**0.1**
Polymerase (U/µl)	0.002–**0.025**–0.1	0.014–0.05–**0.08**	0.014–0.05–**0.08**	**0.08**	0.005–**0.025**–0.125	0.005–**0.025**–0.25	0.002–**0.025**
Cycles (#)	30–**35**–45	**30**–35	**30**–35	**30**	25–**35**–45	30–**35**–40	25–**35**–30
Initial (min/°C)	0/0–0.5/94–**5/94**	2/94–**7/95**	2/94–**7/95**	**7/95**	0/0–**2/94**–5/95	0.6/94–**5/95**	**5/95**
Denature (min/°C)	.08/94–**1/94**	**0.5/90**–1/94	**0.5/90**–1/94	**1/94**	0.5/95–**1/94**	0.3/94–**0.75/95**–0.5/95	**0.3/95**–1/94
Anneal (min/°C)	1/25–0.5/36–**1/36**–2/35–1/45	**1/40**–1/44	**1/52**–1/55	**1/52**–1/53	0.5/50–**1/55**–2/55–7/55	0.25/57–**0.75/55**–2/55	**1/55**
Extend (min/°C)	1/72–**2/72**–4/74	**8/65**–2/72	**8/65**–2/72	**8/65**	1/72–**2/72**–2.5/72	0.25/70–**1/72**–4/72	0.16/72–**1/72**
Ramp speed between temps (min)	0–1.5	6	6	6	2	-	-
Final (min/°C)	0/0–3/72 -**10/72**	**16/65**–10/72	**16/65**–10/72	**16/65**	2/72–**5/72**–10/72	2/72–**7/72**	**7/72**

[a] Values in bold are ones most commonly used.

[b] Primer value is for each primer (ie. If have two primers, then total in reaction mix is double).

[c] The use of $(NH_4)_2SO_4$ is generally as a replacement for KCl.

$MgCl_2$ (25 µM)
dNTP Mix (500 µM of each = 2 mM total)
DNA Taq polymerase (1 U/µL)
DNA template (10^6 cells/µL) (see Notes on Template, section 11.652)
Appropriate primers (25 µL) (see Notes on Primers, section 11.653)

Positive control DNA (A standardized commercial preparation of *E. coli* DNA should be adequate for most PCR based methods as they have all been tried successfully with this organism—see Note in section 11.6231)

Thermocycler

Mineral oil if required by thermocycler

Powdered high purity agarose or pre cast gels

10X TBE buffer

Gel loading dye

Molecular weight standards (e.g., 100 bp or 123 bp DNA ladders are suitable to size most amplified fragments)

Electrophoresis apparatus. Mini-gel systems (8 × 10 cm) are convenient for preliminary work whereas large gel systems (15 × 20 cm or greater) are good for larger scale analysis.

Ethidium bromide stain

Transilluminator (302 nm wavelength)

Photo documentation system. Digital systems are preferred to allow subsequent analysis of DNA profiles with computer software.

Computer and DNA analysis software (optional).

The majority of these items are commercially available or require little time to prepare.

11.643 General PCR Reaction Mix for a 25 µL Assay

Prepare the reaction tubes and keep on ice until the vials are placed in the thermal cycler. If many samples are analysed in one run, prepare a master reaction mixture in bulk. One can distribute this master mix into 23 µL aliquots in PCR tubes ready for use and store at –20°C for several months. When subsequently used, only the 2 µL of template need be added. This assumes that working stocks of DNA have been diluted to the recommended standards where purified DNA is at a concentration of 12.5 µg/mL (or ng/µL), whole cell preparations are approximately 10^6 cells/µL ± one log and crude lysates are approximately A_{260} = 0.25 or prepared from standardized whole cell preparations containing DNA from 10^6 cells/µL. The bulk preparation may not be suitable if a pre-heating step is required prior to the PCR (see Note, section 11.644).

Each reaction tube will contain:

2.5 µL 10X PCR buffer without $MgCl_2$
2.0 µL $MgCl_2$ (25 µM)
2.5 µL dNTP mixture (dATP, dCTP, dGTP, dTTP, 2 mM)
1 µL primer (5 µM)
0.625 µL *Taq* DNA polymerase (1 U/µL)
2 µL template (10^6 cells/µL, or 12.5 ng/µL DNA)
14 µL H_2O

Add 40 µL of paraffin oil to tops of tubes if using older generation machines and spin at 15,000 × g for 30 sec. Newer generation machines with heated lids do not require oil.

Note: Always include a negative control blank for each run where the template is replaced by ddH2O, as well as a positive control using some commercially standardized DNA (e.g., *E. coli*) and dilute it to a working stock of 12.5 ng/μL, thus using 25 ng for assay (2 μL). If concentrations of the various stocks differ from that assumed or multiple primers one being used, then adjustments should be made accordingly.

11.644 Thermal Cycler program

Place the vials into the PCR thermal cycler and run the following program:

4 to 8 min at 94°C (time required for efficient cell lysis may be organism dependant), followed by:

(1 min at 94°C—1 min at 36°C—2 min at 72°C) for 35 cycles, followed by:

10 min at 72°C followed by:

4°C (until gel analysis).

Note: If using whole cell templates, a pre-heating step cell was found to be necessary for species of *Bacillus, Klebsiella, Pseudomonas, Staphylococcus* and *Streptococcus* but not for enteric organisms.[102] Genomic typing of *Listeria* has been equally successful with or without this step.[56,82] If required, mix 2 μL DNA sample with PCR buffer and the H_2O and heat at 80°C for 15 min prior to adding rest of PCR reaction components in a hot start fashion (i.e., leave tubes in the PCR machine at 80°C while adding in the rest of the components and then set the standard protocol to run. However, one must be careful as there is a great opportunity for contamination of the PCR reaction when using this hot start procedure).

11.645 Gel Analysis

1. After the PCR run is complete, mix each amplicon with 1 to 2 μL loading buffer.
2. Load the sample on a horizontal 1.5% agarose gel made with 1X TBE buffer and submerged in the same buffer. The gel may also contain ethidium bromide stain or post-staining can be done subsequent to the run. Load between 5–20 μL per well depending on the depth and the yield of PCR product.
3. Load a molecular weight marker preparation in the first and last lanes of the gel. If certain DNA analysis software packages are to be used it is recommend that an additional centrally located marker be run in order to achieve more accurate normalization of the gel during the analysis procedure. The marker chosen should have adequate range to size all expected fragments. Some convenient multipurpose markers are the 100 and 123 bp ladders. Loading between 0.8–1.2 ng usually gives good visibility of the marker.
4. Electrophorese the samples at 5v/cm (between 80–150 volts depending on the system used) until the tracking dye has migrated approximately 1 cm from the bottom of the gel for mini gels and 10 cm from the top for larger format gels. The time required will therefore depend on the size of the gel being used.
5. Place the gel on a transilluminator (302 nm) and photograph. If using a digital camera or fluorescent scanner, save the image to disk for further analysis.

11.646 Interpretation and Quality Assurance

1. If DNA analysis software is available, analyse gels with that. If not, compare the profiles by eye. Cutting up a photo into separate lanes can be very helpful. For RAPD profiles obtained from duplicate amplifications there should be virtually indistinguishable patterns. If, however, the duplicate profiles show minor variations, i.e., very faint bands, these can be discounted. If major differences are observed (prominent bands), this means that problems have occurred with respect to the reproducibility of the assay. In the latter case, no conclusions can be drawn from the results.
2. If doubt exists as to whether the profiles of two strains are identical or different, the analysis should be repeated from step 8) or even step 1) of the DNA extraction protocol. If, in the second assay, the difference is found again or the duplicate profiles are indistinguishable, then the RAPD profiles are regarded as different.
3. With RAPD, groups of strains from the same collection should be analyzed at different times and reproducibility between separate experiments monitored. For this, one or two reliable strains from the collection should be analyzed on each occasion with these RAPD profiles serving as internal standards.

11.65 Notes

11.651 General Notes

The main difference between standard PCR based gene detection and DNA fingerprinting is the generation of multiple differently sized DNA fragments. Since smaller amplicons are preferentially amplified over larger ones, some modifications to the PCR reaction can be made in order to efficiently amplify a wide spectrum of fragments. Excess dNTPs and primers, greater initial amounts of DNA template and longer elongation times during each cycle will all help achieve this goal.

Storage of components becomes a critical factor as deterioration may affect profiles, especially with whole cell DNA preparations (i.e., purified templates would theoretically be more stable than crude extracts). Also the activity of DNA polymerase may decrease with improper or long term storage, therefore it is recommended to freeze components in small one use aliquots and prepare master mixes where possible to cut down on variability (i.e., a large volume 10X frozen stocks dispensed into usable aliquots (also see specific suggestions in notes below).

11.652 Notes on Templates

For most PCR protocols there is a wide range that can be used for the DNA template concentration in order to get adequate amplification. However with RAPD methods in particular, it is essential to standardize the amount of DNA used in order to reduce the problems associated with reproducibility (See Table 7). As a general rule, template DNA ranging from 10 pg/μL to 10 ng/μL should provide for adequate reproducibility.[19,87] The essential thing is to ensure your concentration is standardized so that the same amount is taken each time.

11.653 Notes on Primers

Primers used for these methods can either be made by the participant in their own laboratory or obtained from an outside supplier. If purchasing from outside, the amount of primer is usually written on the tube as an nmol amount making calculations of concentration somewhat easier. If preparing primers in house, one can use a convenient approximation for determining concentration: 1 OD_{260} = 10 μM or 33 μg/mL (for single stranded DNA only). Primers are generally reconstituted in sterile double distilled water (ddH_2O) and allowed to reconstitute for several hours at room temperature or overnight at 4°C. For a detailed calculation see example below. Primers may be fluorescently labelled by incorporation of fluorescently labelled nucleotides at the time of synthesis or tagged at the 5′ end subsequent to synthesis, with commercially available kits.

Example: If the primer to be used has 110 nmol of oligonucleotide in the tube and 200 μL of sterile water is added in order to reconstitute then the concentration is 550 μM (110 nmol in 200 μL is equal to 0.55 nmol/μL or 550 pmol/μL or 550 μM). This is the stock concentration. If you then desire a convenient working concentration of 5 μM, simply do a 1:110 dilution of stock (550/5 = 110) (e.g., dilute 2 μL of reconstituted primer in 218 μL water).

Alternatively one can prepare a working concentration based on the OD_{260} of the stock. Make a dilution of the reconstituted primer to achieve an OD_{260} reading of between 0.2–0.8 (usually a 50–100 fold dilution). If adding 5 μL of reconstituted primer to 495 μL ddH_2O (1:100 dilution) and the OD_{260} = 0.25, then the OD_{260} of the stock is 25 (100 × 0.25). Applying the formula 1 OD_{260} approximately equals 10 μM or 33 μg/mL, the stock becomes 250 μM (25 × 10) or 825 μg/mL (25 × 33). Therefore a 1:50 dilution is required for a 5 μM working concentration (e.g., 2 μL of stock in 98 μL ddH_2O).

There is a more accurate alternative to the above that requires one to know the molecular weight of the stock primer, as well as its concentration. The molecular weight is calculated by adding up the number of like bases, multiplying them by their individual molecular weight (in μg/μmole) and taking the sum of all bases where A = 347.2, C = 323.2, T = 322.2 and G = 363.2 (e.g., the 10-mer 5′-ACG TAT CTG C- 3′ would equal 3357 μg/μmole (2 × 347.2 + 3 × 323.2 + 3 × 322.2 + 2 × 363.2)). The concentration is calculated based on the estimate that 1 OD_{260} = 33 μg/mL for single stranded DNA (50 μg/mL for double stranded). So if OD_{260} = 25 as in above example, then the concentration is 825 μg/mL (25 × 33). The molar concentration then becomes 825,000 μg/L divided by 3357 μg/μmole = 245.75 μM.

11.654 Notes on $MgCl_2$ and DNA Polymerase

Increasing the concentration of magnesium ions will generate more visible bands.[26,43] Higher concentrations of polymerase in the reaction mix tend to generate DNA fingerprint profiles consisting of high molecular weight fragments while lower concentrations generate profiles with lower molecular weight fragments. The concentration of polymerase used is dependant on activity and this is a direct factor of the brand being used. There are some 'super' polymerases available as well as ones derived from sources other than *Thermusaquaticus* that may require lower concentrations due to higher activity.

Since it has been shown that enzyme activity variations from one suppliers brand to the next, or between tubes of the same supplier, may affect the DNA profile (see Table 7), it is recommended to pool a large supply of DNA polymerase of the same lot number and store it in small aliquots to achieve the best reproducibility.

Due to extraneous contamination of *Taq* DNA polymerase, background amplification may occur when tested in control blanks containing no experimentally introduced DNA template. Frequent contamination of *Taq* DNA polymerase was first recognized in 1990 with the advent of universal rDNA primers[8] and methods for eliminating this have now been suggested.[39,59] With the realization of the potential for background amplification, one should always run the appropriate negative controls. Exogenous contamination of *Taq* DNA polymerase has been observed by those who have used the appropriate controls, but it has been reported that this background does not affect amplification of the DNA sample.[60,100] Tyler et al.[87] found this not to be so clear cut and suggest that the background may be a function of the primer and *Taq* being used, especially with arbitrary primers (i.e., some primers may not amplify the contaminating DNA in the *Taq* whereas others will). It is therefore very important to run control blanks that look for primer/template interaction in the *Taq* and account for any background bands. The fact that different patterns result from using different tubes of *Taq* DNA polymerase, even within the same lot number,[77,87] indicates that this contamination may arise from various sources. Hughes et al.[39] attempted to sequence the contaminating DNA from the *Taq* polymerase and found that it appeared to represent more than one strain or species of eubacterial DNA. However, it could not be identified as coming from the host bacteria from which it was extracted, *Thermus aquaticus*, nor the cloning vector, *Escherichia coli*.

11.655 Notes on PCR Machines

Parameters for PCR machines are highly dependent on their age and thickness of PCR tubes used. Many researchers are now using the next generation of thermocyclers which use heated lids and a Peltier system for faster temperature ramping (e.g., MJ Research DNA Engine). This, in combination with thin walled PCR tubes, allows for more rapid heat exchange and a great reduction in ramping times (i.e., 10–30 sec. as opposed to 1–2 min).

It has been cited that fluctuations in the number of cycles or cycle time and the use of different brands of thermocycler will often result in different amplification products and variations between models is well documented.[20,32,50,60] It has been further shown that the temperature across the block of some machines may vary as much as 5°C with the majority of variations occurring in outside wells.[32,91] It is therefore recommended that only the inside core of older models be used and empty spaces be filled up with water blanks.[91]

11.656 Notes on Agarose Gels and Detection Methods

Co-migration of RAPD generated amplicons has recently been reported as a problem in the typing of *Aeromonas hydrophila*.[67] When amplification primers produce several clustering bands rather than well separated and distinct bands, a group of closely sized fragments may appear as one and lead to a false assumption of similarity. In the extreme, two apparently identical bands may in one case consist of three closely sized fragments and in the other case consist of two closely sized fragments which may share no sequence similarity to the first. This would lead to a false positive assumption of similarity. Conversely, false negative assumptions of similarity may result when an identical band in another sample has been inefficiently amplified, rendering it only faintly detectable or undetectable on the ethidium bromide stained agarose gel.[88] In the extreme case these two patterns would be called different on the basis of a missing band, even though they contain the identical genetic sequence. The percentage of agarose used may also affect reproducibility by suppressing separation of some bands or encouraging others. In general, larger-pore (lower percentage) agarose gels are used to resolve fragments larger than 500 to 1000 bp and smaller pore (higher percentage) acrylamide or sieving gels are used for fragments smaller than 1000 bp. Although polyacrylamide gels with ethidium bromide or silver staining and high performance agaroses may perform better, the same problems of reliably resolving all fragments generated still exist.

Many researchers are beginning to use fluorescently labelled primers in their PCR reaction mixtures which allows detection of amplicons by way of automated DNA sequencers. Oligonucleotide primers are 5′ end labelled with a fluorescent tag. As the fluorescently tagged DNA fragments pass by a sensor on their way out of the gel, a peak is detected and recorded. A profile is generated as a function of peaks vs time. Polyacrylamide gels containing high concentrations of urea as a denaturant provide a powerful system for resolution of shorter fragments (<500 nucleotides) and sequences differing by only a single nucleotide in length can be detected. Resolution of larger fragments can be accomplished using denaturing agarose gels.

As another alternative to conventional agarose gels, some researcher have drawn on techniques that for years have been applied to proteins and other macromolecules. Capillary zone electrophoresis (CZE) avoids polymorphism misinterpretation with respect to resolving fragments due to its superior resolving capabilities. It has been reported on as a general approach for resolving RAPD fragments and specifically applied to the RAPD typing of *Listeria*.[80,88] Capillary Electrophoresis (CE) has also been applied to the rapid detection and analysis of SSCP profiles.[3] It has the ability to detect a 1 bp difference in amplicons and still generate a peak (indicating presence of a band) when the concentration of a particular fragment is too low to be visualized on regular agarose. The draw back of CE is that it requires expensive equipment and is best suited for bands in the 200–800 bp range. Although this is adequate for the majority of rapid typing methods, it may not be suitable for all applications. Although these limits can be increased by varying the parameters,[41] larger size fragments above 1000 bp are still more reliably detected by traditional means. CE and fluorescently labelled amplicon can now be combined with the development of Fluorescent CE. This approach has been extended to the analysis of tRNA-spacer-PCR typing in *Listeria*.[92]

11.657 Notes on Analysis Procedures

There is still the issue of what constitutes sameness in two related patterns. For PFGE analysis, Tenover et al.[85] suggest a two to three band difference is likely to result from one genetic event (i.e., point mutation) or genetic rearrangement (i.e., small insertion or deletion) and such strains can be considered very similar. Two genetic events may give rise to a four to six band difference and now the organisms may be related but are not similar. Seven or more band differences indicates definitively that two organisms are completely unrelated.

As a general rule of thumb, polymorphic DNA bands should only be scored as such if they are observed in repeat amplifications involving different DNA preparations and if their presence or absence is not affected when the amount of target DNA is doubled.[24]

There has been a move towards the digital imaging of gels via fluorescent scanning or use of digital cameras for photographing (i.e., Molecular Dynamics fluorImager SI and Bio Image Workstation) in order that banding patterns can be analysed and compared with computer software (i.e., Molecular analyst (GelCompare), WinCam, Whole Band analysis). It has generally been the norm to compare isolates pair-wise and present strain relationships in rank order of similarity in the form of dendograms. What was once a complex art form is now technically simple due to the advent of user friendly, menu driven computer software aimed at DNA analysis and the computer aided generation of dendograms. These software packages have now become common place in everyday analysis of genetic profiles, because they help alleviate some of the subjective interpretations with respect to calling bands and help manage the more complex banding patterns. However, these advancements are not without their concerns. Different packages handle analysis in different ways and combined with the fact that there is still room for user interpretation in the final outcome, the reproducibility of dendograms is still an issue.[27]

11.66 References

1. [No authors listed]. 1994. A multicenter comparison of methods for typing strains of Pseudomonas aeruginosa predominantly from patients with cystic fibrosis. The International Pseudomonas aeruginosa Typing Study Group. J. Infect. Dis. 169:134-142.
2. Alm, E. W., D. B. Oerther, N. Larsen, D A. Stahl, and L. Raskin. 1996. The oligonucleotide probe database. Appl. Environ. Microbiol. 62:3557-3559.
3. Arakawa, H., S. Nakashiro, M. Maeda, and A. Tsuji. 1996. Analysis of single-strand DNA conformation polymorphism by capillary electrophoresis. J. Chromatogr. A 722:359-368.
4. Barry, T., G. Colleran, M. Glennon, L. K. Dunican, and F. Gannon. 1991. The 16s/23s ribosomal spacer region as a target for DNA probes to identify eubacteria. PCR Meth. Appl. 1:51-6.
5. Berg, D. E., N. S. Akopyants, and D. Kersulyte. 1994. Fingerprinting microbial genomes using the RAPD or AP-PCR method. Methods Mol. Cell Biol. 5:13-24.
6. Berthier, F., and S. D. Ehrlich. 1998. Rapid species identification within two groups of closely related lactobacilli using PCR primers that target the 16S/23S rRNA spacer region. FEMS Microbiol. Lett. 161:97-106.
7. Beyer W, Mukendi FM, Kimmig P, Bohm R. Suitability of repetitive-DNA-sequence-based PCR fingerprinting for characterizing epidemic isolates of Salmonella enterica serovar Saintpaul. J. Clin. Microbiol. 1998;36:1549-1554.
8. Bottger, E. C. 1990. Frequent contamination of Taq polymerase with DNA. Clin. Chem. 36:1258125-9.
9. Brow, M. A., M. C. Oldenburg, V. Lyamichev, L. M. Heisler, N. Lyamicheva, J. G. Hall, N. J. Eagan, D. M. Olive, L. M. Smith, L. Fors, and J. E. Dahlberg. 1996. Differentiation of bacterial 16S rRNA genes and intergenic regions and Mycobacteriumtuberculosis katG genes by structure-specific endonuclease cleavage. J. Clin. Microbiol. 34:3129-3137.
10. Burge, C., A. M. Campbell, and S. Karlin. 1992. Over- and under-representation of short oligonucleotides in DNA sequences. Proc. Natl. Acad. Sci. USA 89:1358-1362.
11. Caetano-Anolles, G, B. J. Bassam, and P. M. Gresshoff. 1991. DNA amplification fingerprinting using very short arbitrary oligonucleotide primers. Biotechnology 9:553-557.
12. Caetano-Anolles, G. 1993. Amplifying DNA with arbitrary oligonucleotide primers. PCR Meth. Appl. 3:85-94.
13. Caetano-Anolles, G., B. J. Bassam, and P. M. Gresshoff. 1992. Primer-template interactions during DNA amplification fingerprinting with single arbitrary oligonucleotides. Mol. Gen. Genet. 235:157-165.
14. Clement, J. M., and M. Hofnung. 1981. Gene sequence of the lambda receptor, an outer membrane protein of E. coli K12. Cell 27(3 Pt 2):507-514.
15. Cocconcelli, P. S., D. Porro, S. Galandini, and L. Senini. 1995. Development of RAPD protocol for typing of strains of lactic acid bacteria and enterococci. Lett. Appl. Microbiol. 21:376-379.
16. Czajka, J., and C. A. Batt. 1994. Verification of causal relationships between Listeria monocytogenes isolates implicated in food-borne outbreaks of listeriosis by randomly amplified polymorphic DNA patterns. J. Clin. Microbiol. 32:1280-1287.
17. Daffonchio, D., S. Borin, A. Consolandi, D. Mora, P. L. Manachini, and C. Sorlini. 1998. 16S-23S rRNA internal transcribed spacers as molecular markers for the species of the 16S rRNA group I of the genus Bacillus. FEMS Microbiol. Lett. 163:229-236.

18. Daud Khaled, A. K., B. A. Neilan, A. Henriksson, and P. L. Conway. 1997. Identification and phylogenetic analysis of Lactobacillus using multiplex RAPD-PCR. FEMS Microbiol. Lett. 153:191-197.
19. Davin-Regli, A., Y. Abed, R. Charrel, C. Bollet, P. de Micco. 1995. Variations in DNA concentrations significantly affect the reproducibility of RAPD fingerprint patterns. Res. Microbiol. 146:561-8.
20. del Tufo, J. P., and S. V. Tingey. 1994. RAPD assay. A novel technique for genetic diagnostics. Meth. Mol. Biol. 28:237-241.
21. Elaichouni, A, van Emmelo J, Claeys G, Verschraegen G, Verhelst R, Vaneechoutte M. Study of the influence of plasmids on the arbitrary primer polymerase chain reaction fingerprint of Escherichia coli strains. FEMS
22. Ellsworth, D. L., K. D.Rittenhouse, R. L. Honeycutt. 1993. Artifactual variation in randomly amplified polymorphic DNA banding patterns. Biotechniques 14:214-217.
23. Erlandson, K., and C. A. Batt. 1997. Strain-specific differentiation of lactococci in mixed starter culture populations using randomly amplified polymorphic DNA-derived probes. Appl. Environ. Microbiol. 63:2702-2707.
24. Farber, J. M. 1996. An introduction to the hows and whys of molecular typing. J. Food Prot. 59:1091-1101.
25. Forsman, P., A. Tilsala-Timisjarvi, and T. Alatossava. 1997. Identification of staphylococcal and streptococcal causes of bovine mastitis using 16S-23S rRNA spacer regions. Microbiology 143 (Pt 11):3491-3500.
26. Gao, Z., K. M. Jackson, and D. E. Leslie. 1996. Pitfalls in the use of random amplified polymorphic DNA (RAPD) for fingerprinting of gram negative organisms. Pathology 28:173-177.
27. Gerner-Smidt, P., L. M. Graves, S. Hunter, and B. Swaminathan. 1998. Computerized analysis of restriction fragment length polymorphism patterns: Comparative evaluation of two commercial software packages. J Clin Microbiol. 36:1318-1323.
28. Gibson, J. R., and R. A. McKee. 1993. PCR products generated from unpurified Salmonella DNA are degraded by thermostable nuclease activity. Lett. Appl. Microbiol. 16:59-61.
29. Gilson, E., J. Clement, D. Perrin, and M. Hofnung. 1987. Palindromic units: A case of highly repetitive DNA sequences in bacteria. Trends Genet. 3:226-230.
30. Gilson, E., J. M. Clement, D. Brutlag, and M. Hofnung. 1984. A family of dispersed repetitive extragenic palindromic DNA sequences in E. coli. EMBO J. 3:1417-1421.
31. Goh, S. H., S. K. Byrne, J. L. Zhang, and A. W. Chow. 1992. Molecular typing of Staphylococcus aureus on the basis of coagulase gene polymorphisms. J. Clin. Microbiol. 30:1642-1645.
32. He, Q., M. K. and J. Viljanen. 1994. Mertsola effects of thermocyclers and primers on the reproducibility of banding patterns in randomly amplified polymorphic DNA analysis. Mol. Cell Probes 8:155-159.
33. Hejazi, A., C. T. Keane, and F. R. Falkiner. 1997. The use of RAPD-PCR as a typing method for Serratia marcescens. J. Med. Microbiol. 46:913-919.
34. Herman, L., M. Heyndrickx, and G. Waes. 1998. Typing of Bacillus sporothermodurans and other Bacillus species isolated from milk by repetitive element sequence based PCR. Lett. Appl. Microbiol. 26:183-188.
35. Higgins, C. F., G. F. Ames, W. M. Barnes, J. M. Clement, and M. Hofnung. 1982. A novel intercistronic regulatory element of prokaryotic operons. Nature 298:760-762.
36. Hilton, A. C., J. G. Banks, and C. W. Penn. 1996. Random amplification of polymorphic DNA (RAPD) of Salmonella: Strain differentiation and characterization of amplified sequences. J. Appl. Bacteriol. 81:575-584.
37. Hilton, A. C., J. G. Banks, and C. W. Penn. 1997. Optimization of RAPD for fingerprinting Salmonella. Lett. Appl. Microbiol. 24:243-248.
38. Hoefnagels-Schuermans, A., W. E. Peetermans, M. J. Struelens, S. Van Lierde, J.Van Eldere. 1997. Clonal analysis and identification of epidemic strains of methicillin-resistant Staphylococcus aureus by antibiotyping and determination of protein A gene and coagulase gene polymorphisms. J. Clin. Microbiol. 35:2514-2520.
39. Hughes M. S., L. A. Beck, and R. A. Skuce. 1994. Identification and elimination of DNA sequences in Taq DNA polymerase. J. Clin. Microbiol. 32:2007-2008.
40. Hulton, C. S., C. F. Higgins, and P. M. Sharp. 1991. ERIC sequences: A novel family of repetitive elements in the genomes of Escherichia coli, Salmonella typhimurium and other enterobacteria. Mol. Microbiol. 5:825-834.
41. Issaq, H. J., K. C. Chan, and G. M. Muschik. 1997. The effect of column length, applied voltage, gel type, and concentration on the capillary electrophoresis separation of DNA fragments and polymerase chain reaction products. Electrophoresis 18:1153-1158.
42. Jensen, M. A., J. A. Webster, and N. Straus. 1993. Rapid identification of bacteria on the basis of polymerase chain reaction-amplified ribosomal DNA spacer polymorphisms. Appl. Environ. Microbiol. 59:945-952.
43. Khandka, D. K., M. Tuna, M. Tal, A. Nejidat, and A. Golan-Goldhirsh. 1997. Variability in the pattern of random amplified polymorphic DNA. Electrophoresis 18:2852-2856.
44. Kobayashi, K. 1994. A simple and rapid confirmation method of the bacterial contamination using polymerase chain reaction. Kansenshogaku Zasshi 68:495-499 (in Japanese).
45. Koeuth, T., J. Versalovic, and J. R. Lupski. 1995. Differential subsequence conservation of interspersed repetitive Streptococcus pneumoniae BOX elements in diverse bacteria. Genome Res. 5:408-418.
46. Kostman, J. R., T. D. Edlind, J. J. LiPuma, and T. L. Stull. 1992. Molecular epidemiology of Pseudomonas cepacia determined by polymerase chain reaction ribotyping. J. Clin. Microbiol. 30:2084-2087.
47. Lin, A.W., M. A. Usera, T. J. Barrett, and R. A. Goldsby. 1996. Application of random amplified polymorphic DNA analysis to differentiate strains of Salmonella enteritidis. J. Clin. Microbiol. 34:870-876.
48. Louws, F. J., D. W. Fulbright, C. T. Stephens, and F. J. de Bruijn. 1994. Specific genomic fingerprints of phytopathogenic Xanthomonas and Pseudomonas pathovars and strains generated with repetitive sequences and PCR. Appl. Environ. Microbiol. 60:2286-2295.
49. Lupski, J. R., and G. M. Weinstock. 1992. Short, interspersed repetitive DNA sequences in prokaryotic genomes. J. Bacteriol. 174:4525-4529.
50. MacPherson, J. M., P. E. Eckstein, G. J. Scoles, and A. A. Gajadhar. 1993. Variability of the random amplified polymorphic DNA assay among thermal cyclers, and effects of primer and DNA concentration. Mol. Cell Probes 7:293-299.
51. Madden, R. H., L. Moran, and P. Scates. 1998. Frequency of occurrence of Campylobacter spp. in red meats and poultry in Northern Ireland and their subsequent subtyping using polymerase chain reaction-restriction fragment length polymorphism and the random amplified polymorphic DNA method. J. Appl. Microbiol. 84:703-708.
52. Madico, G., N. S. Akopyants and D. E. Berg. 1995. Arbitrarily primed PCR DNA fingerprinting of Escherichia coli O157:H7 strains by using templates from boiled cultures. J. Clin. Microbiol. 33:1534-1536.
53. Malathum, K., K. V. Singh, G. M. Weinstock, and B. E. Murray. 1998. Repetitive sequence-based PCR versus pulsed-field gel electrophoresis for typing of Enterococcus faecalis at the subspecies level. J. Clin. Microbiol. 36:211-215.
54. Martin, B., O. Humbert, M. Camara, E. Guenzi, J. Walker, T. Mitchell, et al. 1992. A highly conserved repeated DNA element located in the chromosome of Streptococcus pneumoniae. Nucleic Acids Res. 20:3479-3483.
55. Martin, F. H., M. M. Castro, F. Aboul-ela, and I. Tinoco, Jr. 1985. Base pairing involving deoxyinosine: Implications for probe design. Nucleic Acids Res. 13:8927-8938.
56. Mazurier, S. I., and K. Wernars. 1992. Typing of Listeria strains by random amplification of polymorphic DNA. Res. Microbiol. 143:499-505.
57. Mazurier, S., A van de Giessen, K Heuvelman, and K Wernars. 1992. RAPD analysis of Campylobacter isolates: DNA fingerprinting without the need to purify DNA. Lett. Appl. Microbiol. 14:260-262.
58. McClelland, M., and J. Welsh. 1994. DNA fingerprinting by arbitrarily primed PCR. PCR Meth. Appl. 4:S59-65.

59. Meier, A., Persing DH, Finken M, and E. C. Bottger. 1993. Elimination of contaminating DNA within polymerase chain reaction reagents: Implications for a general approach to detection of uncultured pathogens. J. Clin. Microbiol. 31:646-652.
60. Meunier, J. R., and P. A. Grimont. 1993. Factors affecting reproducibility of random amplified polymorphic DNA fingerprinting. Res. Microbiol. 144:373-379.
61. Micheli, M. R., R. Bova, E. Pascale, and E. D'Ambrosio. 1994. Reproducible DNA fingerprinting with the random amplified polymorphic DNA (RAPD) method. Nucleic Acids Res. 22:1921-1922.
62. Micheli, M. R., R. Bova, P. Calissano, and E. D'Ambrosio. 1993. Randomly amplified polymorphic DNA fingerprinting using combinations of oligonucleotide primers. Biotechniques 15:388-390. Microbiol Lett. 1994;115:335-9.
63. Mileham, A. J. 1995. Identification of microorganisms using random primed PCR. Meth. Mol. Biol. 46:257-267.
64. Muralidharan, K., and E. K. Wakeland. 1993. Concentration of primer and template qualitatively affects products in random-amplified polymorphic DNA PCR. Biotech. 14:362-364.
65. Nachamkin, I, K. Bohachick, and C. M. Patton. 1993. Flagellin gene typing of Campylobacter jejuni by restriction fragment length polymorphism analysis. J. Clin. Microbiol. 31:1531-1536.
66. Nakagawa, T., M. Shimada, H. Mukai, K. Asada, I. Kato, K. Fujino, and T. Sato. 1994. Detection of alcohol-tolerant hiochi bacteria by PCR. Appl. Environ. Microbiol. 60:637-640.
67. Oakey, H. J., L. F. Gibson, and A. M. George. 1998. Co-migration of RAPD-PCR amplicons from Aeromonas hydrophila. FEMS Microbiol. Lett. 164:35-38.
68. Ochman, H., A. S. Gerber, and D. L. Hartl. 1988. Genetic applications of an inverse polymerase chain reaction. Genetics 120:621-623.
69. Orita, M., H. Iwahana, H. Kanazawa, K. Hayashi, and T. Sekiya. 1989. Detection of polymorphisms of human DNA by gel electrophoresis as single-strand conformation polymorphisms. Proc. Natl. Acad. Sci. USA 86:2766-2770.
70. Park, Y. H., and R.J. Kohel. 1994. Effect of concentration of MgCl2 on random-amplified DNA polymorphism. Biotechniques 16:652-656.
71. Penner, G. A., A. Bush, R. Wise, W. Kim, L. Domier, K. Kasha, A. Laroche, G. Scoles, S. J. Molnar, G. Fedak. 1993. Reproducibility of random amplified polymorphic DNA (RAPD) analysis among laboratories. PCR Meth. Appl. 2:341-345.
72. Pitt, T. L. 1994. Bacterial typing systems: The way ahead. J. Med. Microbiol. 40:1-2.
73. Power, E. G. 1996. RAPD typing in microbiology—A technical review. J. Hosp. Infect. 34:247-265.
74. Sallen, B., A. Rajoharison, S. Desvarenne, F. Quinn, and C. Mabilat. 1996. Comparative analysis of 16S and 23S rRNA sequences of Listeria species. Int. J. Syst. Bacteriol. 46:669-674.
75. Samore, M., G. Killgore, S. Johnson, R. Goodman, J. Shim, L. Venkataraman, S. Sambol, P. DeGirolami, F. Tenover, R. Arbeit, and D. Gerding 1997. Multicenter typing comparison of sporadic and outbreak Clostridium difficile isolates from geographically diverse hospitals. J. Infect. Dis. 176:1233-1238.
76. Scheinert, P, R. Krausse, U. Ullmann, R. Soller, and G. Krupp. 1996. Molecular differentiation of bacteria by PCR amplification of the 16S-23S rRNA spacer. J. Microbiol. Meth. 26:103-117.
77. Schierwater, B., and A. Ender. 1993. Different thermostable DNA polymerases may amplify different RAPD products. Nucleic Acids Res. 21:4647-4648.
78. Schmitz, F. J., M. Steiert, H. V. Tichy, B. Hofmann, J. Verhoef, H. P. Heinz, K. Kohrer, M. E. Jones. 1998. Typing of methicillin-resistant Staphylococcus aureus isolates from Dusseldorf by six genotypic methods. J. Med. Microbiol. 47:341-351.
79. Schwarzkopf, A., and H. Karch. 1994. Genetic variation in Staphylococcus aureus coagulase genes: Potential and limits for use as epidemiological marker. J. Clin. Microbiol. 32:2407-2412.
80. Sciacchitano, C. J. 1998. DNA fingerprinting of Listeria monocytogenes using enterobacterial repetitive intergenic consensus (ERIC) motifs-polymerase chain reaction/capillary electrophoresis. Electrophoresis 19:66-70.
81. Sharples, G. J., and R. G. Lloyd. 1990. A novel repeated DNA sequence located in the intergenic regions of bacterial chromosomes. Nucleic Acids Res. 18:6503-6508.
82. Sontakke, S., and J. M. Farber 1995. The use of PCR ribotyping for typing strains of Listeria spp. Eur. J. Epidemiol. 11:665-673.
83. Stern, N. J., M. A. Myszewski, H. M. Barnhart, and D. W. Dreesen. 1997. Flagellin A gene restriction fragment length polymorphism patterns of Campylobacter spp. isolates from broiler production sources. Avian Dis. 41:899-905.
84. Stern, M. J., G. F. Ames, N. H. Smith, E. C. Robinson, and C. F. Higgins. 1984. Repetitive extragenic palindromic sequences: A major component of the bacterial genome. Cell 37:1015-1026.
85. Tenover, F. C., R. D. Arbeit, R. V. Goering, P. A. Mickelsen, B. E. Murray, D. H. Persing, B. Swaminathan. 1995. Interpreting chromosomal DNA restriction patterns produced by pulsed-field gel electrophoresis: Criteria for bacterial strain typing. J. Clin. Microbiol. 33:2233-2239.
86. Tilsala-Timisjarvi, A., and T. Alatossava. 1997. Development of oligonucleotide primers from the 16S-23S rRNA intergenic sequences for identifying different dairy and probiotic lactic acid bacteria by PCR. Int. J. Food Microbiol. 35:49-56.
87. Tyler, K. D., G. Wang, S. D. Tyler, and W. M. Johnson. 1997. Factors affecting reliability and reproducibility of amplification-based DNA fingerprinting of representative bacterial pathogens. J. Clin. Microbiol. 35:339-346.
88. Valentini, A., A. M. Timperio, I. Cappuccio, and L. Zolla. 1996. Random amplified polymorphic DNA (RAPD) interpretation requires a sensitive method for the detection of amplified DNA. Electrophoresis 17:1553-1554.
89. van Belkum, A., J. Kluytmans, W. van Leeuwen, R. Bax, W. Quint, E. Peters, et al. 1995. Multicenter evaluation of arbitrarily primed PCR for typing of Staphylococcus aureus strains. J. Clin. Microbiol. 33:1537-1547.
90. van Embden, J. D., M. D. Cave, J. T. Crawford, J. W. Dale, K. D. Eisenach, B. Gicquel, et al. 1993. Strain identification of Mycobacteriumtuberculosis by DNA fingerprinting: Recommendations for a standardized methodology. J. Clin. Microbiol. 31:406-409.
91. Van Leuven, F. 1991. The trouble with PCR machines: fill up the empty spaces! Trends Genet. 7:142.
92. Vaneechoutte, M., P. Boerlin, H. V. Tichy, E. Bannerman, B. Jager, and J. Bille. 1998. Comparison of PCR-based DNA fingerprinting techniques for the identification of Listeria species and their use for atypical Listeria isolates. Int. J. Syst. Bacteriol. 48 (Pt 1):127-139.
93. Versalovic, J., M. Schneider, F. J. De Bruijn, and J. R. Lupski. 1994. Genomic fingerprinting of bacteria using repetitive sequence-based polymerase chain reaction. Meth. Mol. Cell Biol. 5:13-24.
94. Versalovic, J., T. Koeuth, and J. R. Lupski. 1991. Distribution of repetitive DNA sequences in eubacteria and application to fingerprinting of bacterial genomes. Nucleic Acids Res. 19:6823-6831.
95. Welsh, J., and M. McClelland. 1990. Fingerprinting genomes using PCR with arbitrary primers. Nucleic Acids Res. 18:7213-7218.
96. Welsh, J., and M. McClelland. 1991. Genomic fingerprinting using arbitrarily primed PCR and a matrix of pairwise combinations of primers. Nucleic Acids Res. 19:5275-5279.
97. Wernars, K., P. Boerlin, A. Audurier, E. G. Russell, G. D. Curtis, L. Herman, N. van der Mee-Marquet. 1996. The WHO multicenter study on Listeria monocytogenes subtyping: random amplification of polymorphic DNA (RAPD). Int. J. Food Microbiol. 32:325-341.
98. Widjojoatmodjo, M. N., A. C. Fluit, and J. Verhoef. 1995. Molecular identification of bacteria by fluorescence-based PCR-single-strand conformation polymorphism analysis of the 16S rRNA gene. J. Clin. Microbiol. 33:2601-2606.
99. Widjojoatmodjo, M. N., A. C. Fluit, and J. Verhoef. 1994. Rapid identification of bacteria by PCR-single-strand onformation polymorphism. J. Clin. Microbiol. 32:3002-3007.

100. Williams, J. G., A. R. Kubelik, K. J. Livak, J. A. Rafalski, and S. V. Tingey. 1990. DNA polymorphisms amplified by arbitrary primers are useful as genetic markers. Nucleic Acids Res. 18:6531-6535.

101. Wilson, K. 1994. Preparation of genomic DNA from bacteria, p. 2.4.1-2.4.2. *In* F. M. Ausubel, R. Brent, R. E. Kingston, D. M. David, J. G. Seidman, J. A. Smith et al. (eds.), Current protocols in molecular biology. Brooklyn Green Publishing Associates, Brooklyn, N. Y.

102. Woods, C. R., J. Versalovic, T. Koeuth, and J. R. Lupski. 1993. Whole-cell repetitive element sequence-based polymerase chain reaction allows rapid assessment of clonal relationships of bacterial isolates. J. Clin. Microbiol. 31:1927-1931.

103. Wu, D. Y., L. Ugozzoli, B. K. Pal, J. Qian, and R. B. Wallace. 1991. The effect of temperature and oligonucleotide primer length on the specificity and efficiency of amplification by the polymerase chain reaction. DNA Cell Biol. 10:233-238.

104. Yoon, C. S. 1992. Examination of parameters affecting polymerase chain reaction in studying RAPD. Korean Mycol 20:315-323.

11.7 PULSED FIELD GEL ELECTROPHORESIS

11.71 Introduction

Molecular subtyping began in the 1980s when simple methods were developed to isolate bacterial DNA and cut it into fragments using restriction endonucleases. The fragments could then be separated on the basis of size using agarose gel electrophoresis, with the resulting pattern of DNA fragments being characteristic of the particular subtype. This approach to bacterial subtyping was the first of the restriction fragment length polymorphism (RFLP)-based methods. RFLP soon became a popular tool for molecular epidemiology.[9] However, restriction digestion of chromosomal DNA with frequent-cutting enzymes results in a pattern that is usually too complex to be useful in comparing large numbers of isolates. One approach to reducing the complexity of RFLP patterns is to blot the separated DNA fragments to a membrane followed by hybridization with a specific probe. This approach was very popular in the 1980s. The more popular approach today is macrorestriction analysis using enzymes which cut the DNA infrequently followed by separation of the resulting large fragments by pulsed-field gel electrophoresis (PFGE).[5,16]

11.72 Utility of PFGE in Foodborne Outbreak Investigations

The use of PFGE for molecular subtyping of *E. coli* O157:H7 was first described by Böhm and Karch in 1992.[4] A slightly modified protocol was used by Barrett et al. to investigate a large foodborne outbreak in early 1993.[1] PFGE subtyping was found to be extremely valuable in that investigation, and in several others that followed.[2,3,6,10,11,15] PFGE subtyping has also proven valuable in investigating outbreaks of other foodborne pathogens such as *Listeria monocytogenes* and *Salmonella*.[13,14,18] In most cases, PFGE has proven to be more sensitive than other subtyping methods.[1,13,15,16,18]

11.73 Interpretation of PFGE Results

It can not be emphasized enough that subtyping results, including those obtained using PFGE, must be interpreted in the context of the complete investigation including the epidemiologic, laboratory, and environmental aspects. In the absence of epidemiologic and environmental evidence, subtyping alone can neither prove nor disprove a connection between two isolates. For example, there may be multiple subtypes in a contaminated food and the isolation of a single subtype from an incriminated food that differed from the subtype of the patient isolates could lead to the erroneous conclusion that the food was not the source of the patient's infection. Likewise, mutational events could result in changes in isolates that could lead to the erroneous conclusion that isolates which actually have the same origin are not related. Clonal turnover of sequential isolates has been well documented.[8,12] This problem led Tenover et al. to suggest criteria for interpreting PFGE data.[17] While these guidelines provide an excellent theoretical framework, our experience suggests that criteria should be determined for each organism based on the genetic heterogeneity of that organism and on the prevalence of particular subtypes in each community. Unfortunately, such data are not yet available for most foodborne pathogens, and interpretation of PFGE data must be made in the context of associated epidemiologic and environmental information.

11.74 Choosing a PFGE Protocol

In order to compare patterns obtained in different laboratories, it is critically important that PFGE be performed under the same conditions. The method for preparing plugs for PFGE testing is less critical. The protocol given in the next section has been used by the Centers for Disease Control and Prevention (CDC) and 22 state health department laboratories and has proven to provide reliable, readily comparable results. A more rapid protocol for preparing DNA plugs for PFGE has been published,[7] but in our experience, this protocol is more labor intensive and less robust than longer methods. If rapid turnaround is needed, the shorter protocol of Gautom[7] may be preferred for DNA preparation, but the running conditions described below should be used to obtain results comparable to those in the CDC database. The *E. coli* O157:H7 standard strain described below (available from the Foodborne and Diarrheal Diseases Laboratory Section, CDC) should also be used if comparison with the CDC database is desired.

11.75 Standard PFGE Protocol for Subtyping E. coli O157:H7

The following protocol was developed for PFGE typing of *E. coli* O157:H7 using the restriction enzyme *Xba*I. The same conditions can be used with the restriction enzymes *Avr*II and *Spe*I when additional testing is warranted.

11.751 Day 0: Preparation of Bacterial Cultures

Inoculate blood agar plates with test cultures; streak for isolated colonies. Incubate at 37°C.

11.752 Day 1

11.7521 Preparation of gel plugs containing bacterial DNA

1. Turn on 60–65°C water bath and spectrophotometer.
2. Label two sets of 1.5-mL microcentrifuge tubes with culture numbers.
3. Add 1–1.2 mL of SE Wash Solution to the first set of labeled 1.5-mL microcentrifuge tubes. Suspend growth from overnight culture on blood plate.
4. Centrifuge 5 min at 10,000 rpm (or at another comparable time and speed) to pack cells.
5. Carefully remove supernatant, then wash cells by carefully resuspending them in 1 mL SE Wash Solution; centrifuge as described in step 4 and aspirate supernatant.
6. Make (or melt) 1.2% chromosomal grade agarose (CGA, Bio-Rad 162-0135) in Agarose Buffer during centrifugation steps.
7. Resuspend cells in 1–1.5 mL SE Wash Solution (volume will depend on specifications of spectrophotometer that is used); keep on ice if you have more than 6 cultures to process or

refrigerate if you cannot adjust the optical density (OD or absorbance) immediately.

8. Adjust OD of cell suspension to 1.4 (range 1.35—1.45) by diluting with sterile SE Wash Solution or by increasing amount of cells by centrifuging and washing more of the broth culture.
9. Transfer ~1 mL of suspension to original labeled tube and discard rest into disinfectant.
10. Label wells of PFGE plug molds with culture number. If using reusable plug molds, put strip of transparent tape on lower part of reusable plug mold before labeling wells.
11. Transfer 0.5 mL adjusted cell suspensions to second set of labeled 1.5-mL microcentrifuge tubes. If cell suspensions are cold, put in float and incubate in 37°C water bath for a few minutes, if they are at room temperature, agarose can be added directly without prewarming.
12. Add 0.5 mL melted 1.2% CGA to the 0.5-mL cell suspension; mix by gently pipetting mixture up and down a few times. Maintain temperature of agarose by keeping flask in beaker of warm water (55–65°C).
13. Dispense part of mixture into appropriate well(s) in plug mold. Do not allow bubbles to form. Allow plugs to solidify at least 15 min. They can be placed on ice or in the refrigerator to harden faster.
14. Label Falcon 2054 (or equivalent) tubes with culture numbers and date.
15. Accurately measure 2 mL of cell lysis buffer times the number of plugs (10 plugs × 2 mL = 20 mL) into the appropriate size test tube or flask. Add proteinase K to a concentration of 1 mg/mL (20 mg/20 mL).
16. Add 2 mL of cell lysis buffer with proteinase K to each labeled Falcon 2054 tube.
17. Trim excess agarose from top of plug with scalpel. Open mold and transfer plugs from mold with a 5 to 6 mm-wide spatula to the appropriately labeled tube. Be sure plug is under buffer and not on side of tube.
18. Incubate plugs for 4 h in 54°C water bath with gentle agitation.

11.7522 Washing of gel plugs after cell lysis

1. Remove tubes with plugs from water bath. Allow them to come to room temperature; they can be refrigerated for a few minutes to help harden them.
2. Carefully pour off lysis buffer into discard; plug can be held in tube with a spatula or Pasteur pipet. Add 2 mL sterile reagent grade water, mix, and pour off in 3–5 min.
3. Add 2 mL sterile TE Buffer, mix, and pour off in 3–5 min.
4. Add 2 mL sterile TE Buffer, mix, and allow to stand at room temperature for 15–20 min. Gentle agitation will enhance removal of lysis buffer.
5. Pour off TE Buffer and repeat step 4 at least three more times. If washing cannot be completed on the same day, store plugs in TE at 4°C overnight.
6. After last rinse, store plugs in 5 mL sterile TE Buffer at 4°C until used.

11.753 Day 2

11.7531 Restriction digestion of DNA in lysed gel plugs with *XbaI*

1. Label 1.5-mL microcentrifuge tubes with culture numbers; label 3 or 4 (10-well gel) or 5 (15-well gel) tubes for standards.
2. Dilute 10X H buffer (Boehringer Mannheim or equivalent) 1:10 with sterile reagent grade water.
3. Add 200 µL diluted H buffer to labeled 1.5-mL microcentrifuge tubes. Keep remaining buffer on ice.
4. Carefully remove plug from TE Buffer with narrow spatula and place in a sterile disposable petri dish.
5. Cut a 3-mm-wide slice from test samples with a #21 scalpel and transfer to tube containing diluted H buffer. Be sure plug slice is under buffer.
6. Cut three to five 3-mm-wide slices from plug of *E. coli* Standard (G5244) and transfer to tubes with diluted H buffer. Be sure plug slices are under buffer.
7. Incubate at room temperature for 15–30 min.
8. Replace rest of plug in original tube that contains TE Buffer. Store at 4°C.
9. After incubation, remove buffer from plug slice being careful not to cut plug slice with pipet tip.
10. Add 200 µL fresh restriction buffer and incubate an additional 15–30 min at room temperature.
11. Remove buffer as in step 9.
12. Dilute 10X H buffer 1:10 with sterile reagent grade water and *Xba* I restriction enzyme (~50 U/sample). It can be mixed in the same Falcon tube that was used for H buffer.
13. Add 200 µL restriction enzyme mixture to each tube; finger vortex to be sure plug is under buffer.
14. Incubate at 37°C for 4 h in a water bath.
15. Approximately 1–2 h before restriction digest reaction is finished, pour the electrophoresis gel so it has time to harden.

11.7532 Preparation of gel and electrophoresis unit for PFGE of restriction digested DNA

1. Turn on 60–65°water bath.
2. Make 0.5X Tris-Borate EDTA Buffer (TBE) by diluting: 105 mL 10X TBE to 2100 mL with reagent grade H_2O (14-cm-wide gel) or 110 mL 10X TBE to 2200 mL with reagent grade H_2O (21-cm-wide gel).
3. Make 1% Pulse Field Certified (PFC) agarose (BIO-RAD 162-0137 with 0.5X TBE as follows: for 14-cm-wide gel form (10 or 15 wells): 1.0 g agarose/100 mL 0.5X TBE.; for 21-cm-wide gel form (15 or more wells): 1.5 g agarose/150 mL 0.5X TBE. Note: < 4 mL melted 1% PFC agarose will be needed to fill wells after plugs are loaded.
4. Cool melted PFC agarose in 65°C water bath for 5–6 min; carefully pour agarose into gel form fitted with comb. Be sure there are no bubbles.
5. Turn on cooling module (14°C), power supply, and pump (setting of 60–70) approximately 30 min before gel is to be run.
6. Remove restricted plug slices from 37°C water bath; allow to come to room temperature. Remove enzyme/buffer mixture with 200 µL Microman pipet and tip (see step 9 of previous section) from each plug. Add 200 µL 0.5X TBE.
7. Remove comb from gel (gel should harden for 30-45 min).
8. Remove restricted plug slices from tubes with tapered end of spatula and load into appropriate wells. Gently push plugs to bottom and front of wells with wide end of spatula. Manipulate position with spatula and be sure that are no bubbles.
   i. Load *E. coli* standards (G5244) in lanes 1, 5, 10 or 1, 4, 7, 10 (10-well gel) or in lanes 1, 4, 8, 11, 15 (15-well gel).
   ii. Load samples in lanes 2, 3, 4, 6, 7, 8, 9 or 2, 3, 5, 6, 8, 9 (10-well gel) or in lanes 2, 3, 5, 6, 7, 9, 10, 12, 13, 14 (15-well gel).
9. Fill in wells of gel with melted 1% PFC agarose. Allow to harden for at least 5 min. Remove from form and carefully

place gel inside casting platform or corner posts in Chef electrophoresis unit chamber (BIO-RAD). Close or replace cover of chamber.

i. Select following on Chef Mapper:
Auto Algorithm
30 kb—low MW
600 kb—high MW
Select default values except where noted.
Change run time to 22 h
Initial switch time should be 2.16 s
Final switch time should be 54.17 s

ii. Set Chef DR II or III electrophoresis unit as follows:
Initial A time: 2.2 s
Final A time: 54.2 s
Start ratio: 1.0
Run time: 26 h (DR II); 24 h (DR III)
Voltage: 200 V

11.754 Day 3

11.7541 Documentation of gel

1. When electrophoresis run is over, turn off equipment and remove gel.
2. Dilute 40 µL of ethidium bromide stock solution (10 mg/mL) with 400 mL reagent grade water. Stain gel for 30 min in covered container. Ethidium bromide solution can be kept in a dark bottle and reused 4–5 times before discarding according to your institution's guidelines for hazardous waste.
3. Destain gel in reagent grade water 60–90 min; change water every 30 min. Capture image on a Gel Doc system or photograph. If background interferes with resolution, destain for an additional 30–60 min.

11.76 References

1. Barrett, T. J., H. Lior, J. H. Green, R. Khakhria, J. G. Wells, B. P. Bell, K. D. Greene, J. Lewis, and P. M. Griffin 1994. Laboratory investigation of a multistate food-borne outbreak of *Escherichia coli* O157:H7 by using pulsed-field gel electrophoresis and phage typing. J. Clin. Microbiol. 32:3013-3017.
2. Bender, J. B., C. W. Hedberg, J. M. Besser, D. J. Boxrud, K. L. MacDonald, and M. T. Osterholm. 1997. Surveillance for *Escherichia coli* O157:H7 infections in Minnesota by molecular subtyping. N. Engl. J. Med. 337:338-394.
3. Besser, R. E., S. M. Lett, J. T. Weber, M. P. Doyle, T. J. Barrett, J. G. Wells, and P. M. Griffin.1993. An outbreak of diarrhea and hemolytic uremic syndrome from *Escherichia coli* O157:H7 in fresh-pressed apple cider. JAMA 269:2217-2220.
4. Bohm, H., and H. Karch. 1992. DNA fingerprinting of *Escherichia coli* O157:H7 strains by pulsed-field gel lectrophoresis. J. Clin. Microbiol. 30:2169-2172.
5. Chu, G., D. Vollrath, and R. W. Davis. 1986. Separation of large DNA molecules by contour-clamped homogenous electric fields. Science 234:1582-1585.
6. Cieslak, P. R., T. J. Barrett, and P. M.Griffin. 1993. *Escherichia coli* O157:H7 infection from a manured garden. Lancet 342:367.
7. Gautom, R. K. 1995. Rapid pulsed-field gel electrophoresis protocol for typing of *E. coli* O157:H7 and other Gram-negative organisms in 1 day. J. Clin. Microbiol. 35:2977-2980.
8. Hartstein, A. I., P. Chetchotisakd, C. L. Phelps, and A. M. Lemonte. 1995. Typing of sequential bacterial isolates by pulsed-field gel electrophoresis. Diag. Microbiol. Infect. Dis. 22:309-314.
9. Holmberg, S. D., and K. Wachsmuth. 1989. Plasmid and chromosomal analysis in the epidemiology of bacterial diseases, p. 105-130. *In* B. Swaminathan and G. Prakash (eds.), Nucleic acid and monoclonal antibody probes: Application in diagnostic microbiology. Marcel Deker, New York.
10. Izumiya, H., J. Terajima, A. Wada, Y. Inagaki, K. I. Itoh, K. Tamura, and H. Watanabe. 1997. Molecular typing of enterohemorrhagic *E. coli* O157:H7 isolates in Japan by pulsed-field gel electrophoresis. J. Clin. Microbiol. 35:1675-1680.
11. Johnson, J. M,. S. D. Weagent, K. C. Jinneman, and J. L.Bryant. 1995. Use of pulsed-field gel electrophoresis for epidemiological study of *Escherichia coli* O157:H7 during a food-borne outbreak. Appl. Environ. Microbiol. 61:2806-2808.
12. Karch, H., H Russman, H Schmidt, A Schwarzkopf, and J. Heesemann. 1995. Long-term shedding and clonal turnover of enterohemorrhagic *Escherichia coli* O157 in diarrheal diseases. J. Clin. Microbiol. 33:1602-1605.
13. Louie, M., P. Jayaratne, I. Luchsinger, J. Devenish, J. Yao, W. Schlech, and A. Simor. 1996. Comparison of ribotyping, arbitrary primed PCR, and pulsed-field gel electrophoresis for molecular typing of Listeria monocytogenes. J. Clin. Microbiol. 34:15-19.
14. Mahon, B. E., A. Ponka, W. N. Hall, K. Komatsu, S. E. Dietrich, A. Siitonen, G. Cage, P. S. Hayes, M. A. Lambert-Fair, N. H. Bean, P. M. Griffin, and L. Slutsker 1997. An international outbreak of Salmonella infections caused by alfalfa sprout grown from contaminated seeds. J. Infect. Dis. 175:876-882.
15. Meng, J., S. Zhao, T. Zhao, and M. P. Doyle. 1995. Molecular characterization of Escherichia coli O157:H7 isolates by pulsed-field gel electrophoresis and plasmid DNA analysis. J. Med. Microbiol. 42:258-263.
16. Swaminathan, B., and G. M. Matar. 1993. Molecular typing methods, p. 26-50. *In* D. H. Persing et al. (eds.), Diagnostic molecular microbiology: Principles and applications. Mayo Foundation, Rochester, N. Y.
17. Tenover, F. C., R. D. Arbeit, R. V. Goering, P. A. Mickelsen, B. E. Murray, D. H. Persing, and B. Swaminathan. 1995. Interpreting chromosomal DNA restriction patterns produced by pulsed-field gel electrophoresis: Criteria for bacterial strain typing. J. Clin. Microbiol. 33:2233-2239.
18. Threlfall, E. J., M. D. Hampton, L. R. Ward, and B. Rowe.1996. Application of pulsed-field gel electrophoresis to an international outbreak of Salmonella agona. Emerg. Infect. Dis. 2:130-132.

Chapter 12

Labor Savings and Automation

Daniel Y. C. Fung

12.1 INTRODUCTION

Microbiological testing of foods can be both time-consuming and expensive. However, it is widely practiced as a means of improving food safety, preventing spoilage, and monitoring the quality of foods and food ingredients. It has been estimated that 420 million tests are conducted annually by food industries in the United States and Europe.[7] About 60 million of the tests are for specific organisms while the others are for groups of organisms such as aerobic mesophiles, coliforms, and fungi.

Successful microbiological analysis depends upon correct sampling plans and sample preparation. The reader is referred to Chapter 2 and to the chapters that are concerned with specific microorganisms and commodities. This chapter describes labor-saving devices that may be of use to food microbiologists.

12.2 SOLID SAMPLES

Enumeration of microorganisms in nonliquid foods often requires appropriate dilutions and homogenization of the samples. These procedures are often tedious, especially when many samples are to be analyzed.

The distribution of microorganisms on solid foods often is not uniform. When this is the situation, a larger sample size of 50 to 100 g may be required in order to obtain a reliable estimate of microbial populations.

12.21 Initial Dilutions

The Gravimetric Diluter (Spiral Biotech, Bethesda, MD) is an instrument that can be programmed to yield an initial dilution of 1:10, 1:50, 1:100 or some other factor. The food is weighed in a sterile plastic bag, the desired amount of sterile diluent is then automatically added to give the desired dilution. The unit has been found to be efficient over a wide range of dilutions.[5] Similar equipment has been marketed under the names Diluflo (Spiral Biotech) and Dilumacher (pbi, Milan, Italy).

12.22 Homogenization

The Stomacher (Tekmar Co., Cincinnati, OH) saves labor in that homogenization is performed in sterile, disposable bags and thus the cleaning and resterilization of blender jars is avoided.[6] Other advantages are that aerosols are not produced and little or no heat is generated during the 1 to 5 min treatment. With 26 categories of nonfatty foods, however, the Stomacher yielded somewhat lower aerobic plate count geometric means than those obtained with a blender.[1] The Stomacher is not applicable to foods with bones or other sharp or protruding objects since they may puncture the bag (see Chapter 2).

Instruments similar to the Stomacher are the Masticator (IUL Instruments, Erlander, KY) and the Pulsifier (Kalyx Biosciences Inc., Nepean, ONT). The latter utilizes an oval ring which houses a bag containing the sample and diluent. When activated, the ring vibrates vigorously for 30 to 60 sec to dislodge microorganisms from the food surface. In the evaluation of 96 food samples, Fung et al.[3] found that the Pulsifier and Stomacher gave comparable viable counts. Samples treated in the Pulsifier yielded clearer diluents which may be advantageous for certain subsequent analytical procedures.

12.23 Plating

12.231 Robotic Systems

Certain systems automatically dilute the foods, dispense the samples into petri dishes, pour the agar, cover and label the petri dishes, then transport them into the appropriate incubator (Zymark, Corp., Hopkinton, MA). Using similar techniques, the Colworth 2000 (A.J. Seward, London, UK) can automatically process 2000 plates per working day.

12.232 Spiral Plating

Spiral plating is discussed in Chapters 6 and 7. Its advantage is that it eliminates the need for extensive serial dilutions. A comparison between spiral plating and conventional pour plates using manual and laser counting techniques showed that the two procedures yielded comparable counts for bacteria and yeasts.[4]

12.233 Other Media and Methods

Petrifilm™, Redigel™, and Isogrids™ are other labor-saving plating procedures (see Chapters 6 and 7). The first two methods permit the culturing of foods by laboratories that have limited facilities for the preparation of agar media. On the other hand, Isogrids™ lend themselves to electronic counting and thus are labor saving when large numbers of samples are to be cultured. A comparison of the different methods against the standard plate count[2] showed high correlation coefficients for the different systems (r = 0.95+).

SimPlate™ (IDEXX, Westbrook, ME) is a new system in which a diluted sample is introduced into a plate containing 84 or 198 wells. Liquid media is then introduced and distributed evenly into the wells by swirling the SimPlate™ in a gentle, circular motion. Excess media is poured off prior to incubating the plates for 24 hr. The plates are counted under UV light in order to convert positive fluorescent wells to MPN figures. By using selective media, IDEXX has developed a procedure (Colilert™) for enumeration of coliforms and *Escherichia coli*. Other systems for coliforms have been developed by BioControl, Bothell, WA, and are marketed under the names of ColiComplete, Coliform-ColiTrak, and Colitrak Plus. The Quanti-Tray system allows MPN determinations of these organisms.

12.4 LIQUID SAMPLES

Labor saving devices for liquid foods include automated pipeting instruments such as the RapidPlate 96 Workstation (Zmark, Hopkinton, MA) that will prepare the desired dilutions. Membrane filters may be used to detect low populations of microorganisms in foods such as certain beverages that are readily filterable (see Chapter 6).

12.5 REFERENCES

1. Andrews, W. H., C. R. Wilson, P. L. Poelma, A. Romero, R. A. Rude, A. P. Duran, F. D. McClure and D. E. Gentile. 1978. Usefulness of the stomacher in a microbiological regulatory laboratory. Appl. Environ. Microbiol. 35:89.
2. Chain, V. S., and D. Y. C. Fung. 1991. Comparison of Redigel, Petrifilm, Spiral Plate System and ISOGRID and standard plate count for the aerobic count in selected foods. J. Food. Prot. 54:208.
3. Fung, D. Y. C., A. N. Sharpe, B. C. Hart, and Y. Liu. 1997. The Pulsifier: a new instrument for preparing food suspensions for microbiological analysis. J. Rapid Meth. Automat. Microbiol. 6:43.
4. Manninen, M. T., D. Y. C. Fung and R. A. Hart. 1991. Spiral system and laser counter for enumeration of microorganisms. J. Food Safety 11:177.
5. Manninen, M. T., and D. Y. C. Fung. 1992. Use of the Gravimetric diluter in microbiological work. J. Food Prot. 55:59.
6. Sharpe, A. N., and A. K. Jackson. 1975. Automation requirements in microbiological quality control of food. *In* C. G. Hedon and T. Illeni (eds.). Automation in Microbiology and Immunology. John Wiley and Sons, New York.
7. Wechsler, T. 1998. How big is pathogen testing's place in food microbiology? The Industrial Microbiology Market Review 1998. Woodstock, VT.

Chapter 13

Psychrotrophic Microorganisms

M. A. Cousin, J. M. Jay, and P. C. Vasavada

13.1 INTRODUCTION

13.11 History and Definition of Terms

In 1887 Forster observed microbial growth at 0°C, but it was not until 1902 that the term "psychrophile" was used.[58] Psychrophiles have been defined based on growth at low temperature, optimum temperature of growth, temperature of enumeration, and other criteria not related to temperature such as only including gram-negative rods or only bacteria that do not survive pasteurization.[133] Mossel and Zwart[83] and Eddy[24] proposed the term "psychrotrophs" for microorganisms that grow at low temperatures but have higher temperature optima. Morita[82] suggested that the mesophilic microorganisms that grow at 0°C be called either psychrotolerant or psychrotrophic to contrast with psychrophilic microorganisms, which have a temperature optimum of 15°C, a maximum of 20°C, and a minimum of 0°C or below. In a recent review, Brenchley [14] referred to psychrophiles as microorganisms that grow at 5°C or below regardless of the maximum growth temperature; therefore, there is still not consensus on what to call these microorganisms that grow at low temperatures.

Microorganisms that grow in foods at refrigeration temperatures (0 to 7°C) but have temperature optima above 20°C are called psychrotrophs. Psychrotrophs are defined as microorganisms that produce visible growth at 7° ± 1°C within 7 to 10 days, regardless of their optimum growth temperatures.[120] This definition honors the long-standing practice of classifying microorganisms into three temperature groups, namely thermophiles, mesophiles, and psychrophiles, with psychrotrophs being a subgroup of the mesophiles. From a practical standpoint, the microorganisms that are most commonly associated with refrigerated foods and cause food spoilage are psychrotrophs and not psychrophiles because the latter usually die at room temperature or above. Psychrotrophs grow and spoil foods that are refrigerated, but they grow better at higher temperatures in the mesophilic range.

13.12 Growth of Psychrophiles and Psychrotrophs

If a microorganism is to grow at low temperatures, then the substrate uptake, cell permeability, enzymatic systems, and synthetic pathways must all function at low temperatures. Some theories concerning the mechanism of growth of psychrophiles and psychrotrophs focus upon the generation of low activation energy for enzymes, presence of unsaturated fatty acids in the cell membranes and subsequent fluidity, conformational changes in the ribosomal proteins, and regulatory enzymes, presence of cold shock proteins, alterations in substrate uptake, and cell permeability.[19,28,41,42,50-52,82,97,99,103,115]

13.13 Psychrophiles Involved in Food Spoilage

Psychrophilic bacteria are mainly gram negative and are found in environments where temperatures are constantly below 15° to 20°C.[52] Psychrophiles grow in environments where temperatures are fairly constant; whereas, psychrotrophs grow in environments where temperatures fluctuate.[103] Most psychrophiles found in foods are species of *Aeromonas*, *Alcaligenes*, *Cytophaga*, *Flavobacterium*, *Pseudomonas*, *Serratia*, and *Vibrio*. Some gram positive genera that have been isolated from arctic waters, soils, and foods include species of *Arthrobacter*, *Bacillus*, *Clostridium*, and *Micrococcus*. Psychrophilic yeasts, molds, and algae have also been identified. *Cryptococcus*, *Leucosporidium*, and *Torulopsis* are psychrophilic genera of yeasts. Although Makarios-Laham and Levin[75,76] isolated psychrophilic *Vibrio* species from haddock, their significance in fish spoilage is unknown. Whether psychrophiles are involved in food spoilage has not been determined.

13.14 Psychrotrophs Involved in Food Spoilage

Psychrotrophic bacteria have been studied more than either yeasts or molds. These bacteria include rods and cocci; sporeformers and non-sporeformers; gram negative and gram positive bacteria; and aerobes, facultative anaerobes, and anaerobes. The major psychrotrophic bacteria found in milk and dairy products,[17,19,62,79] meats and poultry,[10,62,67,79,80] and fish and seafood[43,55,56,62,79,124] include species of *Acinetobacter*, *Aeromonas*, *Alcaligenes*, *Arthrobacter*, *Bacillus*, *Brochothrix*, *Carnobacterium*, *Chromobacterium*, *Citrobacter*, *Clostridium*, *Corynebacterium*, *Enterobacter*, *Escherichia*, *Flavobacterium*, *Klebsiella*, *Lactobacillus*, *Leuconostoc*, *Listeria*, *Microbacterium*, *Micrococcus*, *Moraxella*, *Pseudomonas*, *Psychrobacter*, *Serratia*, *Shewanella*, *Streptococcus*, and *Weissella*. In addition, species of *Alteromonas* (formerly *Pseudomonas putrefaciens*), *Photobacterium*, and *Vibrio* are also important in fish spoilage.[55,80,124] Species of *Bacillus*, *Clostridium*, *Enterobacter*, *Erwinia*, *Flavobacterium*, *Pseudomonas*, and *Yersinia* cause soft-rotting of refrigerated vegetabes.[13,15,74]

Psychrotrophic fungi have been isolated from refrigerated fresh animal and marine products and from fruits and vegetables.

Among the yeast genera involved are *Candida, Cryptococcus, Debaryomyces, Hansenula, Kluveromyces, Pichia, Saccharomyces, Rhodotorula, Torulopsis,* and *Trichosporon.*[13,19,20,57,60] Mold genera that have psychrotrophic species include *Alternaria, Aspergillus, Botrytis, Cladosporium, Colletotrichum, Fusarium, Geotrichum, Monascus, Mucor, Penicillium, Rhizopus, Sporotrichum, Thamnidium,* and *Trichothecium.*[19,57,60,67,111] Fungi predominate in refrigerated food spoilage when low water activity, high acidity or packaging conditions select for their growth over bacteria in foods, such as many fruits, jams, dried fruits, and fermented foods (cheese, sausages, yogurt, etc.).

The use of vacuum or modified atmosphere packaging of raw and processed meat, fish, and other foods favors the growth of both facultative anaerobes and true anaerobes in the oxygen-reduced environment. The major bacterial genera found in vacuum or modified atmosphere- packaged foods include psychrotrophic species and strains of *Brochothrix, Lactobacillus, Leuconostoc,* and members of the *Enterobacteriaceae* with lower populations of *Carnobacterium* spp. and *Weissella viridescens.*[10,71,100]

13.15 Psychrotrophic Pathogens

The emergence of psychrotrophic foodborne pathogens in recent years has raised new concerns about the safety of refrigerated foods. Pathogenic psychrotrophs that grow at or below 5°C include *Aeromonas hydrophila,* some strains of *Bacillus cereus, Clostridium botulinum* type E and non-proteolytic types B and F, *Listeria monocytogenes, Vibrio cholera, Yersinia enterocolitica,* and some strains of enteropathogenic *Escherichia coli.*[4,72,89,91,126] Further information on these pathogenic psychrotrophs can be obtained in their respective chapters. Several reviews on the role of psychrotrophic pathogens in vacuum or modified atmosphere-packaged foods have been published.[18,33,54,66,91,102,125,126]

Foodborne pathogens such as strains of *B. cereus, C. perfringens,* the proteolytic strains of *C. botulinum, Salmonella* serotypes, and *Staphylococcus aureus*[33,89] have minimal growth temperatures between 7° and 15°C; therefore, temperature abuse of refrigerated foods may allow these mesophilic pathogens to resume growth once temperatures rise above 10 to 15°C.

13.16 Significance of the Presence and Growth of Psychrotrophs in Foods

Psychrotrophs metabolize carbohydrates, proteins, and lipids across the range of temperatures at which foods are stored, but reaction rates are slower at temperatures at and below 7°C. Minor biochemical changes in the food may occur early during the growth phase of some psychrotrophs, but several days to weeks of refrigeration may be necessary for changes to become organoleptically apparent.[19]

Information about the spoilage of specific food commodities can be found in their respective chapters. Some general reviews are available for the spoilage of milk and dairy products,[19,26,79,109,112] meat and poultry,[7,67,79,80] fish and seafoods,[43,55,56,79] and fruits and vegetables.[13,57,74,111]

There is a trend in the United States, Europe, and Japan to market "minimally processed" refrigerated foods that range from deli-type salads to complete dinners.[15,72,106] Minimal processing uses procedures such as low heat instead of sterilization; and cleaning, peeling and cutting of fresh produce instead of leaving it whole.[1] Minimal processing includes various heat treatments, vacuum or modified atmosphere-packaging, conventional or microwave heating after product-package assembly, and strict refrigerated distribution systems.[72] Two methods, "sous-vide," in which the food is placed in an oxygen-impermeable bag, and "nouvelle carte," in which the food is packaged on a plate and placed in a vacuum pouch, involve minimal heat processing in the vacuum package and refrigerated distribution at 2° to 4°C.[72,106] These processes for refrigerated foods create new microbiological concerns for both safety and expected shelf life. Packaging in vacuum and modified gaseous atmospheres selects for facultative anaerobes and anaerobes. The minimal heating kills vegetative cells, but not spores. Since these processing and packaging methods are intended only to extend the shelf life, surviving psychrotrophic spoilage and pathogenic microorganisms can grow and dominate the microbiota of these products. To avoid this situation, proper thermal processing, vacuum packaging plus storage and distribution at 7°C or below must be maintained.

13.2 REVIEW OF METHODS USED TO ENUMERATE PSYCHROTROPHS IN FOODS

13.21 Cultural and Microscopic Methods

General reviews of the methods used to enumerate psychrotrophs have been published for dairy products,[19] fish,[56] and meats.[67] The traditional methods for enumerating psychrotrophs have involved either plate counting methods or the use of microscopy.[34,35] Examples of the time and temperatures of incubation for psychrotrophic plate counts are 10 days at 7°C for pour plates or 7 to 8 days at 7°C for spread plates and 16 hr at 17°C followed by 3 days at 7°C. Incubation conditions using shorter times and higher temperatures have included 25 hr at 21°C for milk and cream,[46,88,98,110,118] 45 hr at 18°C for milk,[86] and 24 hr at 25°C for meat.[45]

Several variations of the plate count procedure provide equivalent accuracy for the enumeration of psychrotrophs. These methods (Chapters 6 and 10) include spiral plating[59,99,109] dry rehydratable film like Petrifilm™,[3,36,78] and hydrophobic grid-membrane filter (HGMF).[16,107]

The composition of some selective media used for psychrotrophs is based on the assumption that most psychrotrophs are gram negative bacteria. In the first two editions of this book,[34,35] a selective medium was recommended for the enumeration of psychrotrophs that contained crystal violet and triphenyl tetrazolium chloride (CVT) and was incubated for either 48 hr at 30°C or 5 days at 22°C. Jay and Bue[63] found that the use of CVT agar with incubation for 48 hr at 30°C was not suitable for the enumeration of gram-negative psychrotrophs since many non-psychrotrophic mesophiles grew well under these conditions. While crystal violet at 2 ppm inhibited three gram positive bacteria but not three *Pseudomonas* species,[108] crystal violet at the recommended usage of 1 ppm[34,35] in CVT agar did not inhibit 41/44 non-psychrotrophs, nor 3/45 psychrotrophs evaluated.[68] Although some investigators have found media containing these inhibitors to be suitable for assessing gram negative psychrotrophs at temperatures that permit the growth of many non-psychrotrophic mesophiles, this is a reflection of the relatively large number of psychrotrophs in the products examined. Most non-psychrotrophic mesophiles will proliferate at temperatures between 22° and 40°C when incubated for 48 hr or more.

The direct microscopic count (DMC) where a specific sample size is placed on a defined area of a microscopic slide that is stained before counting the cells in a set number of fields and the microscopic colony count (MCC) where a specific amount of agar containing the food sample is incubated on a slide and the micro-colonies are counted after a short incubation period have been used to enumerate psychrotrophs. Zall et al.[134] used a preincubation of 5 hr at 30°C before doing a DMC and noted that this method could be used since the psychrotrophic value was about 1% of the DMC. Juffs and Babel[64] did not find very good correlation between the MCC and the psychrotrophic count done at 7°C for 10 days; however, they suggested that a slide incubated at 7°C for 48 hr may be useful for enumerating psychrotrophs.

13.22 Rapid and Automated Methods

In a recent review, Sorhaug and Stepaniak[112] emphasized the need for a rapid and sensitive method to detect psychrotrophs in milk to overcome the disadvantage of the 7- to 10-day incubation time. Because many psychrotrophic bacteria are aerobic and possess the enzyme catalase, an increase in the concentration of this enzyme has been used to estimate the number of psychrotrophic bacteria in foods. The disk flotation method using the Catalasemeter correlated well with the psychrotrophic plate count for raw poultry.[130] Dodds et al.[21] concluded that the Catalasemeter was not reliable at determining the quality of vacuum-packaged cooked turkey ham when the counts were less than 10^4 cfu/g. The feasibility of using a catalase-based method for rapid evaluation of raw and pasteurized milk quality has been studied.[31,53,96,127] Phillips and Griffiths[96] found no correlation between the catalase activity and the total count of milk; however, after a preincubation at 21°C for 25 hr in plate count agar with penicillin, crystal violet, and nisin, the detection limit was 10^5 to 10^6 cfu/mL. From these results, the catalase test would have little value for foods that have low psychrotrophic numbers or that have undergone conditions that select for a psychrotrophic spoilage microflora that is catalase negative, e.g., lactobacilli. More information on both the catalase and cytochrome c oxidase tests is given in section 13.83.

Cytochrome c oxidase has been used to estimate the numbers of psychrotrophs in milk in order to predict its keeping quality.[68-70] Kroll[68] found that more than 10^4 microbes/mL of milk were needed to detect cytochrome C oxidase. For pasteurized milk, a preincubation for 18 hr at 20°C in the presence of benzalkonium chloride to inhibit gram positive bacteria was needed for the population to reach 10^4 microbes/mL.[70] The standard plate count was a better predictor of keeping quality than the cytochrome C oxidase test; however, this method may be useful as an initial screening for the presence of $>10^4$ psychrotrophs/mL or g of food.[69] Reduction of tetrazolium salts in the presence of the gram positive inhibitor, benzalkonium chloride, has been suggested as a rapid test for psychrotrophs but they need to grow to about 10^7 cells/mL.[109]

The impedance method[22] has been used to estimate the number of bacteria in fresh fish,[87,123] raw meat,[29] raw milk,[30,117] pasteurized milk and cottage cheese.[6-8,65] The rapid estimation of psychrotrophs in cod fillets, using brain heart infusion (BHI) broth at 20°C and impedance measurements for 5 to 16 hr, correlated well with the standard psychrotrophic plate count.[123] The estimation for raw milk showed good correlation with the plate count for psychrotrophs, when the samples were analyzed after 16 to 21 hr at 20°C.[30,117] Shelf life testing of milk and cottage cheese requires a preincubation of 18 hr at 18° to 21°C before impedance is measured at 21°C.[8,65] The impedance method required only 1 to 2 days, compared to 7 to 9 days needed by the Moseley Keeping Quality Test,[132] predicted the length of shelf life better, and required less labor. When Bishop and White[6,7] compared plate counts to rapid methods for estimating the microbial shelf life of pasteurized milk and cottage cheese, both impedance and endotoxin detection were significantly correlated to shelf life. However, the impedance method produced a better predictive equation than endotoxin determination. Impedance has also been used to detect the growth of yeasts and molds in laboratory media.[105]

Gram-negative bacteria produce lipopolysaccharides (LPS) as part of their cell envelopes. A lysate produced from amoebocytes of the horseshoe crab *(Limulus polyphemus)* reacts with LPS, and the reaction can be measured by methods based on gelation, turbidity or chromogenesis.[61] Test results can be obtained in 1 hr by tube gelation or in about 30 min by the other methods. The *Limulus* amoebocyte lysate (LAL) test can be used to detect gram-negative bacteria in foods. The LAL test has been used to estimate the number of gram-negative psychrotrophs in such refrigerated foods as meats, milk, fish, and salads.[5-7,21,27,49,61,79,81,116,119] Dodds et al.[21] reported that LAL values correlated with the number of *Enterobacteriaceae* in vacuum-packaged cooked turkey with a sensitivity less than 100 cells/g. A sensitivity of 15 bacteria per test was reported for analysis of milk.[116] LAL correlated well with the bacterial count for determining the shelf life of non-acidified vegetable salads stored at refrigeration and abuse temperatures.[77] In a study of lean fish, Sullivan et al.[119] found that LAL values agreed with aerobic plate counts and total volatile bases. Using a microtiter plate method for LAL, Fallowfield and Patterson[27] were able to detect 10^2 to 10^3 *Pseudomonas* species/g in beef and pork stored at 4°C. If LAL is to be used to estimate psychrotrophs in refrigerated foods, then correlation factors for the accept-reject levels need to be established.[6,32,49,116] Further information on the use of LAL is given in section 13.82.

Since the last edition of this book, there have been a few new rapid methods researched for the detection of psychrotrophs in foods. Two methods that have been developed are the enzyme-linked immunosorbent assay (ELISA) and the polymerase chain reaction (PCR). ELISA assays have been developed experimentally to detect *Pseudomonas fluorescens* in refrigerated meat[25,38,40] and milk.[37,39] The polyclonal antibody only reacted with *Pseudomonas* species or strains[25,37,38,40] and had lower to no recognition of other psychrotrophs.[39] A monoclonal antibody recognized three *Pseudomonas* species (*P. fluorescens, P. fragi*, and *P. putida*) and *Enterobacter aerogenes*.[47] A monoclonal antibody developed to *P. fluorescens* had a sensitivity of 10^5 cfu/mL in milk[48] or 10^4 cfu/cm^2 in meat;[47] whereas, polyclonal antibodies detected 10^5 cfu/mL or cm^2 in milk or meat.[25,38,39] More research needs to be done to develop a polyclonal antibody to recognize many genera of psychrotrophs. A PCR method based on 23 S rDNA sequence from *Pseudomonas aeruginosa* was developed to detect species of *Pseudomonas, Acinetobacter, Brochothrix, Enterobacter, Flavobacterium,* and *Moraxella* in meat.[128] Much more needs to be done before a method can be commercialized for use in detecting psychrotrophs in foods.

Other rapid methods that have been studied may not always distinguish psychrotrophs from non-psychrotrophic mesophiles. These methods include HGMF,[16,107] direct epifluorescent filter technique (DEFT),[23,70,85,93-95,101] estimation of adenosine triphosphate (ATP) by bioluminescence,[2,11,12,73,90,113,114,129,131] and calorimetry.[44] Methods that are based on detection of amines[104] and aminopeptidase activity[92] still need more research before their significance in enumerating psychrotrophs can be assessed.

13.3 GENERAL RECOMMENDATIONS

13.31 Method Selection

The choice of a method for psychrotoph enumeration will depend on the intended use of the results, time and equipment available, accuracy needed, type of refrigerated food, and degree of processing. When an accurate number of psychrotrophs is needed, plate count methods must be used. It may be necessary to choose time and temperature conditions that simulate either the storage conditions of the food or the possible abuse conditions. Selection of the method for enumerating psychrotrophs must involve consideration of sublethally injured or stressed cells. Absolute conditions cannot be given for every food or every situation.

13.32 Media Selection

The selection of media for enumeration will depend on the food, types of psychrotrophs expected, recommendations of equipment manufacturers, length of incubation, reactions expected, and other relevant factors. Media other than those listed below may prove valuable.

13.4 SAMPLE PREPARATION

13.41 Sample Collection

Samples must be collected aseptically (Chapter 2) and analyzed promptly. Refrigerated storage could allow psychrotrophs to increase in numbers because some psychrotrophs have generation times as low as 6 hr. Ideally, refrigerated foods should be analyzed within 6 hr of their collection. Refrigerated samples should not be frozen because some psychrotrophs are sensitive to freezing and can be injured or killed. If samples must be frozen for shipment, then the possibility of some death must be considered when evaluating the results.

13.42 Sample Homogenization

Samples should be homogenized with diluent in a blender for 2 min or in a stomacher. Because psychrotrophs are sensitive to heat, blending for more than 2 min is discouraged to prevent the generation of heat that can result in cell injury or death. In addition, excessive blending of molds can cause fragmentation of the mycelia, depending on blade sharpness, volume, speed, and time. The use of a stomacher in preference to the blender lessens the likelihood of these problems.

13.5 EQUIPMENT, MEDIA, MATERIALS, AND REAGENTS

Refer to the specific section in recommended methods for equipment that is needed for each method. Incubators that can be maintained at 7° ± 1°C for the traditional psychrotrophic count and 17–21°C ± 1°C for rapid methods that require incubation are necessary.

13.51 Media

1. Non-selective plate media: standard methods (plate count) agar or trypticase soy agar for bacteria; and dichloran rose bengal chloramphenicol agar (DRBC) or plate count agar plus chloramphenicol or dichloran 18% glycerol (DG18) for yeasts and molds (Chapter 20).
2. Media and reagents for the rapid methods can be obtained from the manufacturers (Chapter 63).

13.6 PROCEDURES FOR ENUMERATION OF PSYCHROTROPHS

13.61 Plate Count Method

A plate count method using plate count or trypticase soy agar, or dry rehydratable film such as Petrifilm™, is recommended for general enumeration of bacteria (Chapter 7). Enumeration procedures for yeasts and molds can be found in Chapter 20. Incubate plates for 10 days at 7° ± 1°C since this is the reference definition for psychrotrophs. Alternatively, incubation for 16 hr at 17°C followed by incubation for 3 days at 7°C can be used when results are needed in less than 10 days.[120]

Count the colonies as described in Chapter 7. Record all counts as number of psychrotrophs/milliliter, gram, or square centimeter depending upon the method of sampling.

13.7 PRECAUTIONS

13.71 Incubation Temperatures

Different types of refrigerated foods are normally processed and held at temperatures of refrigeration that are specific for the food commodity. The incubation temperature used for the enumeration may not lead to adequate assessment of the psychrotrophic population that will grow in the food because microorganisms may grow in laboratory media but not in the food and vice versa. Therefore, caution must be used when interpreting results of the enumeration of psychrotrophic populations.

Sublethally injured psychrotrophs may not be detected analytically but may cause food spoilage or foodborne illness. Therefore, steps to recover sublethally injured cells should be included in enumeration and detection methods.[84] See Chapter 5 for suggestions on methods.

13.72 Pour Plate Versus Spread Plate Techniques

Psychrotrophs are especially susceptible to injury or death when agar that is held above 45°C is used for pour plating.[122] Hence, a spread/surface plate or spiral plater technique (Chapter 12) should be used whenever possible. Plates can be pre-poured and stored at 5°C for several days or weeks before use. Dry rehydratable film such as Petrifilm™ may also be used in place of the traditional plating technique.

13.8 RAPID DETECTION

Obtaining results sooner than 10 days is pragmatically desirable, and efforts must continue to find methods that are more rapid for the enumeration of psychrotrophs in foods. Rapid detection of psychrotrophs in raw ingredients, on-line quality control samples, and shelf life samples are important. Three useful rapid methods are impedance, LAL assays, and enzymatic assays.

13.81 Impedance Determination

The equipment manufacturers generally recommend media and incubation conditions that should be followed for the specific foods being analyzed because some media can interfere with the detection time. Preliminary work to determine what correlations between impedance readings and psychrotrophic counts apply to the food usually are necessary before impedance can be used for routine analysis. An example of how the method is done is as follows. Samples of solid foods (meats, poultry, fish, vegetables, etc.) are blended for 1 min in 0.1% peptone water to make a 1:10 dilution. As little as a 1.0-mL aliquot of the 1:10 dilution can be added to recommended broth in the sterile well or tube of the impedance detection instrument. Samples are incubated at 21°C and continuously monitored for up to 24 hr depending on the sample and the degree of contamination. Preliminary incubation for 15 to 24 hr at 18° to 21°C in broth may be used to obtain early detection and shelf life predictions because foods that have undergone processing normally will have lower counts than raw foods. After preincubation, 1 or more milliliters of the mixed sample is placed in the well or tube of the impedance detection instrument and incubated at 21°C; thereafter, measurements are recorded for up to 24 hr. The impedance detection time is recorded and used to estimate the number of psychrotrophs in the sample.

Liquid foods such as milk and juice can be added directly to either broth or plated media in the wells of the impedance detection instrument. Specific media developed by the instrument suppliers can also be used to detect bacteria, yeasts, and molds in liquid foods. A sample of the liquid food (amount determined by instrument used) is added to the media in the well and incubated at 21°C for up to 24 hr for bacteria and 25° to 28°C for 1 to 2 days for yeasts and molds. Preincubation in plate count broth at 18°C for 18 hr can be done to assess the keeping quality of heat-processed milk and dairy products. The impedance detection time is recorded and the estimation of microbial numbers made.

13.82 Limulus Amoebocyte Lysate (LAL) Assay

The tube gelation method has been used more extensively than either the turbidity or the chromogenic substrate methods, and its use is described below. However, the chromogenic substrate is the newer of the three basic methods, provides results in about 30 min, is more sensitive (LPS detected from 1 to < 5 pg/mL), and is becoming more widely accepted. The basic operation and automation of the chromogenic substrate has been described by Tsuji et al.[121] and reviewed by Jay.[61]

The most important considerations in the use of the tube gelation method are (a) source and sensitivity of LAL reagent, (b) whether single- or multi-test vials or reagent are to be used, and (c) the choice of endotoxin or LPS standard. Suppliers of freeze-dried LAL reagents, such as Associates of Cape Cod (Woods Hole, MA) and Sigma Chemical Co. (St. Louis, MO) provide complete instructions for the proper use of their reagents, and these should be followed carefully. LAL reagents can be obtained with different levels of sensitivity usually expressed in endotoxin units (EU).

Since LPS from different gram negative bacteria varies in its reactivity to the LAL reagent, it is essential that a standard reference endotoxin preparation be used. These are available with complete instructions for use from LAL reagent suppliers. The two reference endotoxins of choice are those prepared from *Escherichia coli* O113:H10 (EC-2) or *E. coli* O55:B5.

The tube gelation method described below is taken from the review by Jay.[61] LAL assay methods require that all utensils and glassware be pyrogen free. Glassware can be depyrogenated by heating in a dry-air oven at temperatures above 180°C for about 3 hr. Sterile pipettes and disposable tubes are generally free of pyrogens before use, and pyrogen-free water should be purchased from a firm that supplies parenteral products. Specific instructions for conducting a tube gelation test are usually provided by the LAL reagent manufacturer.

LAL reagent is supplied either in ready-to-use or single-test vials, or in multi-use vials. Follow the preparation and storage directions that come with the reagents. It is generally a good idea to cover the tubes with aluminum foil until used. The tubes should be used the same day although some manufacturers indicate that the tubes may be frozen and thawed once. Quality control procedures and both negative and positive controls are supplied by the manufacturers.

Food samples should be serially diluted, using pyrogen-free water or suitable buffer, to provide dilutions that will produce negative results. Since two-fold dilutions were not significantly different than ten-fold dilutions, use of two-fold dilutions will save on both reagents and labor.[61] Beginning with the highest dilution (lowest endotoxin concentration), the same pipette tip can be used to transfer either 0.1 or 0.2 mL of diluted sample to separate LAL reagent tubes. The tubes are vortexed gently, incubated in a waterbath at 37°C for 1 hr, and read by inverting 180° and noting gelation. Simultaneously, an endotoxin standard should be included using an appropriate reference endotoxin. The two-fold diluted endotoxin standard should be treated in the same way as the diluted test preparation; the sensitivity of the LAL preparation to LPS is determined by using this standard to define the lowest quantity that produces a gel.

The quantity of endotoxin or LPS in test samples is determined by multiplying the reciprocal of the highest sample dilution by the LAL-determined sensitivity value. For example, if the highest dilution of endotoxin standard that produces a firm gel in the LAL reagent is 0.1 ng, the sensitivity of the LAL reagent is, thus, 0.1 ng. If using the above LAL reagent and the highest dilution of food that produces a firm gel is 10^3, then the total endotoxin or LPS in food is 0.1 ng × 1000 = 100 ng/mL.

13.83 Enzymatic Methods: Catalase and Cytochrome Oxidase

Enzymatic methods are not sensitive enough to detect microbial populations below 10^4 cells/mL or g; therefore, use is restricted to foods with high microbial loads. A preincubation of 4 to 6 hr that may or may not involve selective enrichment media can improve both the selectivity and the sensitivity of these methods. Use of these methods for solid foods needs further refinement for the extraction of the enzymes from foods.

The catalase test can be done by using instruments that measure oxygen release, e.g., BioTech International (SugarLand, TX). Catalase that is naturally present in the food may need to be inactivated by heating the sample at 50°C for 10 min before doing the test for microbial catalase, which has a higher heat stability.[9] The amount of oxygen produced in the catalase test is proportional to the number of microorganisms in the food.

The oxidase test is performed by treating 4 mL of sample with 1 mL of a freshly prepared 1% *N,N,N',N'*-tetramethyl-*p*-phenylene-diamine dihydrochloride followed by incubation at 25°C for 5 min. If sample contains particulate matter, then it can be centrifuged at 7,000 × g for 10 min before taking the reading. Blue color can be evaluated either visually against reference color standards or spectrophotometrically. The intensity of the blue color is proportional to the concentration of microorganisms in the food.

13.9 INTERPRETATION

The enumeration of psychrotrophs in refrigerated foods gives an indication of the potential spoilage, keeping quality, or safety of the food. However, caution should be exercised when trying to make absolute predictions about a food based on these results. Some temperatures of refrigeration may be close to the minimal growth temperature, and the enumeration temperature may be closer to the optimum, particularly in rapid methods that require an incubation period. The temperatures used for food storage and for detection should be closely comparable to achieve meaningful results. In addition, the nature of the food is important. If the food has been refrigerated for some time, the numbers can represent a normal increase in psychrotrophs rather than a poor quality product.

Processing can kill or injure psychrotrophs, and analyzing foods immediately after processing may not allow time for injured cells to recover. If processed foods are stored in the refrigerator for extended periods, even a few cells can grow to large enough numbers to cause eventual spoilage in a few days or weeks.

13.10 REFERENCES

1. Ahvenainen, R. 1996. New approaches in improving shelf life of minimally processed fruit and vegetables. Trends Food Sci. Technol. 7:179-187.
2. Anderson, R., and F. Labell. 1988. Rapid microbial tests safeguard fresh deli foods. Food Proc. 49(12):90, 92.
3. Bailey, J. S., and N. A. Cox. 1987. Evaluation of the Petrifilm SM and VRB dry media culture plates for determining microbial quality of poultry. J Food Prot. 50:643-644.
4. Beuchat, L. R. 1996. Pathogenic microorganisms associated with fresh produce. J Food Prot. 59:204-216.
5. Bishop, J. R., and A. B. Bodine. 1986. Quality assessment of pasteurized fluid milk as related to lipopolysaccharide content. J Dairy Sci. 69:3002-3004.
6. Bishop, J. R., and C. H. White. 1985. Estimation of potential shelf life of cottage cheese utilizing bacterial numbers and metabolites. J Food Prot. 48:1054-1057,1061.
7. Bishop, J. R., and C. H. White. 1985. Estimation of potential shelf-life of pasteurized fluid milk utilizing bacterial numbers and metabolites. J Food Prot. 48:663-667.
8. Bishop, J. R., C. H. White, and R. Firstenberg. 1984. Rapid impedimetric method for determining the potential shelf-life of pasteurized whole milk. J Food Prot. 47:471-475.
9. Boismenu, D., F. Lépine, M. Gagnon, and H. Dugas. 1990. Heat inactivation of catalase from cod muscle and from some psychrophilic bacteria. J. Food Sci. 55:581-582.
10. Borch, E., M-L. Kant-Muermans, and Y. Blixt. 1996. Bacterial spoilage of meat and cured meat products. Int. J. Food Microbiol. 33:103-120.
11. Bossuyt, R. 1981. Determination of bacteriological quality of raw milk by an ATP assay technique. Milchwissenschaft 36:257-260.
12. Bossuyt, R. 1982. A 5-minute ATP platform test for judging the bacteriological quality of raw milk. Neth. Milk Dairy J. 36:355-364.
13. Brackett, R. E. 1987. Microbiological consequences of minimally processed fruits and vegetables. J. Food Qual. 10:195-206.
14. Brenchley, J. E. 1996. Psychrophilic microorganisms and their cold-active enzymes. J. Ind. Microbiol. 17: 432-437.
14. Brocklehurst, T. F., C. M. Zaman-Wong, and B. M. Lund. 1987. A note on the microbiology of retail packs of prepared salad vegetables. J. Appl. Bacteriol. 63: 409-415.
15. Brodsky, M. H., P. Entis, M. P. Entis, A. N. Sharpe, and G. A. Jarvis. 1982. Determination of aerobic plate and yeast and mold counts in foods using an automated hydrophobic grid-membrane filter technique. J. Food Prot. 45:301-304.
16. Collins, E. B. 1981. Heat resistant psychrotrophic microorganisms. J. Dairy Sci. 64: 157-160.
17. Corlett, D. A. 1989. Refrigerated foods and use of hazard analysis and critical control point principles. Food Technol. 43(2): 91-94.
18. Cousin, M. A. 1982. Presence and activity of psychrotrophic microorganisms in milk and dairy products: A review. J. Food Prot. 45: 172-207.
19. Davenport, R. R. 1980. Cold-tolerant yeasts and yeast-like organisms, p. 215-230. *In* F. A. Skinner, S. M. Passmore, and R. R. Davenport (ed.) Biology and activities of yeasts. Academic Press, New York.
20. Dodds, K. L., R. A. Holley, and A. G. Kempton. 1983. Evaluation of the catalase and *Limulus* amoebocyte lysate tests for rapid determination of the microbial quality of vacuum-packed cooked turkey. Can. Inst. Food Sci. Technol. J. 16:167-172.
21. Easter, M. C., and D. M. Gibson. 1989. Detection of microorganisms by electrical measurements. Prog. Ind. Microbiol. 26:57-100.
21. Easter, M. C., R. G. Kroll, L. Farr, and A. C. Hunter. 1987. Observations on the introduction of the DEFT for the routine assessment of bacteriological quality. J. Soc. Dairy Technol. 40:100-103.
22. Eddy, B. P. 1960. The use and meaning of the term "psychrophilic." J. Appl. Bacteriol. 23:189-190.
23. Eriksson P.V., G. N. DiPaola, M. F. Pasetti, and M. A. Manghi. 1995. Inhibition enzyme-linked immunosorbent assay for detection of *Pseudomonas fluorescens* on meat surfaces. Appl. Environ. Microbiol. 61:397-398.
24. Fairbairn, D. J., and B. A. Law. 1986. Proteinases of psychrotrophic bacteria: their production, properties, effects and control. J. Dairy Res. 53:139-177.
25. Fallowfield, H. J., and J. T. Patterson. 1985. Potential value of the *Limulus* lysate assay for the measurement of meat spoilage. J. Food Technol. 20:467-479.
26. Feller G., E. Narinx, J. L. Arpigny, M. Aittaleb, E. Baise, S. Genicot, and C. Gerday. 1996. Enzymes from psychrophilic organisms. FEMS Microbiol. Rev. 18:189-202.
27. Firstenberg-Eden, R. 1983. Rapid estimation of the number of microorganisms in raw meat by impedance measurement. Food Technol. 37(1):64-67,69-70.
28. Firstenberg-Eden, R., and M. K. Ticarico. 1983. Impedimetric determination of total, mesophilic and psychrotrophic counts in raw milk. J. Food Sci. 48:1750-1754.
29 Fischer, J. E., and P. C. Vasavada. 1987. Rapid detection of abnormal milk by the catalase test. J. Dairy Sci. 70(Suppl 1):75.
30. Forster, M. A. 1985. Factors affecting the use of the Limulus amoebocyte lysate test in the food industry. N. Z. J. Dairy Sci. Technol. 20:163-172.
31. Genigeorgis, C. A. 1985. Microbial and safety implications of the use of modified atmospheres to extend the storage life of fresh meat and fish. Int. J. Food Microbiol. 1:237-251.
32. Gilliland, S. E., H. D. Michener, and A. A. Kraft. 1976. Psychrotrophic microorganisms, p. 173-178. *In* M. L. Speck (ed.) Compendium of methods for the microbiological examination of foods, 1st ed. American Public Health Association, Washington, D. C.
33. Gilliland, S. E., H. D. Michener, and A. A. Kraft. 1984. Psychrotrophic microorganisms, p. 135-141. *In* M. L. Speck (ed.) Compendium of methods for the microbiological examination of foods, 2nd ed. American Public Health Association, Washington, D. C.
34. Ginn, R. E., V. S. Packard, and T. L. Fox. 1984. Evaluation of the 3M dry medium culture plate (Petrifilm SM) method for determining numbers of bacteria in raw milk. J. Food Prot. 47:753-755.
35. González I., R. Martín, T. Gárcia, P. Morales, B. Sanz, and P. E. Hernández. 1993. A sandwich enzyme-linked immunosorbent assay (ELISA) for detection of *Pseudomonas fluorescens* and related psychrotrophic bacteria in refrigerated milk. J. Appl. Bacteriol. 74:394-401.
36. González I., R. Martín, T. Gárcia, P. Morales, B. Sanz, and P. E. Hernández. 1994. Detection of *Pseudomonas fluorescens* and related psychrotrophic bacteria in refrigerated meat by a sandwich ELISA. J. Food Prot. 57:710-714.
37. Gonzalez I., R. Martin, T. Garcia, P. Morales, B. Sanz, and P. E. Hernandez. 1994. Polyclonal antibodies against live cells of *Pseudomonas fluorescens* for the detection of psychrotrophic bacteria in milk using a double antibody sandwich enzyme-linked immunosorbent assay. J. Dairy Sci. 77:3552-3557.
38. González I., R. Martín, T. Gárcia, P. Morales, B. Sanz, and P. E. Hernández. 1996. Polyclonal antibodies against protein F from the cell envelope of *Pseudomonas fluorescens* for detection of psychrotrophic bacteria in refrigerated meat using an indirect ELISA. Meat Sci. 42:305-313.
39. Gounot, A-M. 1986. Psychrophilic and psychrotrophic microorganisms. Experientia 42: 1192-1197.
40. Gounot, A-M. 1991. Bacterial life at low temperature: physiological aspects and biotechnical implications. J. Appl. Bacteriol. 71:386-397.
41. Gram L., and H. H. Huss. 1996. Microbiological spoilage of fish and fish products. Int. J. Food Microbiol. 33:121-137.
42. Gram, L., and H. Sogaard. 1985. Microcalorimetry as a rapid method for estimation of bacterial levels in ground meat. J. Food Prot. 48:341-345.

43. Greer, G. G. 1981. Rapid detection of psychrotrophic bacteria in relation to retail beef quality. J. Food Sci. 46:1669-1672.
44. Griffiths, M.W., J. D. Phillips, and D. D. Muir. 1980. Rapid plate counting techniques for enumeration of psychrotrophic bacteria in pasteurized double cream. J. Soc. Dairy Technol. 33:8-10.
45. Gutierrez R., T. Gárcia, I. González, B. Sanz, P. E. Hernández, and R. Martín. 1997. Monoclonal antibody detection of *Pseudomonas* spp. in refrigerated meat by an indirect ELISA. Lett. Appl. Microbiol. 24:5-8.
46. Gutierrez R., I. González, T. Gárcia, E. B. Carrera, Sanz, P. E. Hernández, and R.Martín. 1997. Monoclonal antibodies and an indirect ELISA for detection of psychrotrophic bacteria in refrigerated milk. J. Food Prot. 60:23-27.
47. Hansen, K., T. Mikkelsen, and A. Moller-Madsen. 1982. Use of the Limulus test to determine the hygienic status of milk products as characterized by levels of Gram-negative bacterial lipopolysaccharide present. J. Dairy Res. 49:323-328.
48. Hebraud M., E. Dubois, P. Potier, and J. Labadie. 1994. Effect of growth temperatures on the protein levels in a psychrotrophic bacterium, *Pseudomonas fragi*. J. Bacteriol. 176:4017-4024.
49. Herbert, R. A. 1981. A comparative study of the physiology of psychrotrophic and psychrophilic bacteria, p. 3-16. *In* T. A. Roberts, G. Hobbs, J. H. B. Christian, and N. Skovgaard (ed.), Psychrotrophic microorganisms in spoilage and pathogenicity. Academic Press, New York.
50. Herbert, R. A. 1986. The ecology and physiology of psychrophilic microorganisms. p. 1-23. *In* R. A. Herbert and G. A. Codd (ed.), Microbes in extreme environments. Academic Press, New York.
51. Hill, S. D., R. L. Richter, and C. W. Dill. 1988. Evaluation of a catalase-based method to predict the shelf-life of pasteurized milk. J. Dairy Sci. 71(Suppl 1):112.
52. Hintlian, C. B., and J. H. Hotchkiss. 1986. The safety of modified atmosphere packaging: A review. Food Technol. 40(12):70-76.
53. Hobbs, G. 1983. Microbial spoilage of fish, p. 217-229. *In* T. A. Roberts and F. A. Skinner (ed.), Food microbiology; advances and prospects. Academic Press, New York.
54. Hobbs, G., and W. Hodgkiss. 1982. The bacteriology of fish handling and processing, p. 71-117. *In* R. Davies (ed.), Developments in food microbiology–1. Applied Science Publishers, Inc., Englewood, N. J.
55. Hsu, E. J., and L. R. Beuchat. 1986. Factors affecting microflora in processed fruits, p. 129-161. *In* J. G. Woodroof and B. S. Luh (ed.), Commercial fruit processing, 2nd ed. AVI Publishing Company, Inc., Westport, Conn.
56. Ingraham, J. L., and J. L. Stokes. 1959. Psychrophilic bacteria. Bacteriol. Rev. 23: 97-108.
57. Jarvis, B., V. H. Lach, and J. M. Wood. 1977. Evaluation of the spiral plate maker for the enumeration of micro-organisms in foods. J. Appl. Bacteriol. 43:149-157.
58. Jay, J. M. 1987. Meats, poultry, and seafoods, p. 155-173. *In* L. R. Beuchat (ed.), Food and beverage mycology, 2nd ed. Van Nostrand Reinhold Co., New York.
59. Jay, J. M. 1989. The *Limulus* amoebocyte lysate (LAL) test. Prog. Ind. Microbiol. 26:101-119.
60. Jay, J. M. 1996. Modern food microbiology, 5th ed., p. 328-346. Chapman & Hall, New York.
61. Jay, J. M., and M. E. Bue. 1987. Ineffectiveness of crystal violet tetrazolium agar for determining psychrotrophic Gram-negative bacteria. J. Food Prot. 50:147-149.
62. Juffs, H.S., and F. J. Babel. 1975. Rapid enumeration of psychrotrophic bacteria in raw milk by the microscopic colony count. J. Milk Food Technol. 38:333-336.
63. Kahn, P., and R. Firstenberg-Eden. 1987. Prediction of shelf-life of pasteurized milk and other fluid dairy products in 48 hours. J. Dairy Sci. 70:1544-1150.
64. King, A. D., Jr., and H. R. Bolin. 1989. Physiological and microbiological storage stability of minimally processed fruits and vegetables. Food Technol. 43(2):132-135,139.
65. Kraft, A. A. 1986. Psychrotrophic organisms, p. 191-208. *In* A. M. Pearson and T. R. Dutson (ed.), Advances in meat research, Volume 2. Meat and poultry microbiology. AVI Publishing Company, Inc., Westport, Conn.
66. Kroll, R. G. 1985. The cytochrome *c* oxidase test for the rapid detection of psychrotrophic bacteria in milk. J. Appl. Bacteriol. 59:137-141.
67. Kroll, R. G., and U. M. Rodrigues. 1986. Prediction of the keeping quality of pasteurized milk by the detection of cytochrome *c* oxidase. J. Appl. Bacteriol. 60:21-27.
68. Kroll, R. G., and U. M. Rodrigues. 1986. The direct epifluorescent filter technique, cytochrome *c* oxidase test and plate count method for predicting the keeping quality of pasteurized cream. Food Microbiol. 3:185-194.
69. Lannelongue, M., G. Finne, M. O. Hanna, R. Nickelson II, and C. Vanderzant. 1982. Microbiological and chemical changes during storage of swordfish *(Xiphias gladius)* steaks in retail packages containing CO_2-enriched atmospheres. J. Food Prot. 45:1197-1203.
70. Lechowich, R. V. 1988. Microbiological challenges of refrigerated foods. Food Technol. 42(12):84-85,89.
71. Littel, K. J., S. Pikelis, and A. Spurgash. 1986. Bioluminescent ATP assay for rapid estimation of microbial numbers in fresh meat. J. Food Prot. 49:18-22.
72. Lund, B. M. 1983. Bacterial spoilage, p. 219-257. *In* C. Dennis (ed.), Post-harvest pathology of fruits and vegetables. Academic Press, New York.
73. Makarios-Laham, I., and R. E. Levin. 1984. Isolation from haddock tissue of psychrophilic bacteria with maximum growth temperature below 20°C. Appl. Environ. Microbiol. 48:439-440.
74. Makarios-Laham, I., and R. E. Levin. 1985. Autolysis of psychrophilic bacteria from marine fish. Appl. Environ. Microbiol. 49:997-998.
75. Manvell, P. M., and M. R. Ackland. 1986. Rapid detection of microbial growth in vegetable salads at chill and abuse temperatures. Food Microbiol. 3:59-65.
76. McGoldrick, K. F., T. L. Fox, and J. S. McAllister. 1986. Evaluation of a dry medium for detecting contamination on surfaces. Food Technol. 40(4):77-80.
77. McKellar, R.C. (ed.). 1989. Enzymes of psychrotrophs in raw food. CRC Press, Boca Raton, Fla.
78. McMeekin, T. A. 1982. Microbial spoilage of meats, p. 1-40. *In* R. Davies (ed.), Developments in food microbiology–1. Applied Science Publishers, Inc., Englewood, N. J.
79. Mikolajcik, E. M., and R. B. Brucker. 1983. Limulus amebocyte lysate assay—a rapid test for the assessment of raw and pasteurized milk quality. Dairy Food Sanit. 3:129-131.
80. Morita, R. Y. 1975. Psychrophilic bacteria. Bacteriol Rev 39:144-167.
81. Mossel, D. A. A., and H. Zwart. 1960. The rapid tentative recognition of psychrotrophic types among *Enterobacteriaceae* isolated from foods. J. Appl. Bacteriol. 23: 185-188.
82. Mossel, D. A. A., C. M. L. Marengo, and C. B. Struijk. 1994. History of and prospects for rapid and instrumental methodology for the microbiological examination of foods, p. 1-28. *In* P. D. Patel (ed.), Rapid analysis techniques in food microbiology. Blackie Academic and Professional, New York.
83. Neaves, P., D. I. Jervis, and G. A. Prentice. 1987. A comparison of DEFT clump counts obtained in eight dairy laboratories receiving replicate samples of preserved raw milk. J. Soc. Dairy Technol. 40:53-56.
84. Oehlrich, H. K., and R. C. McKellar. 1983. Evaluation of an 18°C/45-hour plate count technique for the enumeration of psychrotrophic bacteria in raw and pasteurized milk. J. Food Prot. 46:528-529.
85. Ogden, I. D. 1986. Use of conductance methods to predict bacterial counts in fish. J. Appl. Bacteriol. 61:263-268.
86. Oliveria, J. S., and C. E. Parmelee. 1976. Rapid enumeration of psychrotrophic bacteria in raw and pasteurized milk. J. Milk Food Technol. 39:269-272.

87. Palumbo, S. 1986. A. Is refrigeration enough to restrain foodborne pathogens? J. Food Prot. 49:1003-1009.
88. Patel, P. D., and A. P. Williams. 1985. A note on estimation of food spoilage yeasts by measurement of adenosine triphosphate (ATP) after growth at various temperatures. J. Appl. Bacteriol. 59:133-136.
89. Peck, M. W. 1997. *Clostridium botulinum* and the safety of refrigerated processed foods of extended durability. Trends Food Sci. Technol. 8:186-192.
90. Perez De Castro, B., M. A. Asensio, B. Sanz, and J. A. Ordoñez. 1988. A method to assess the bacterial content of refrigerated meat. Appl. Environ. Microbiol. 54:1462-1465.
91. Pettipher, G. L. 1981. Rapid methods for assessing bacterial numbers in milk. Dairy Ind. Int. 46(11):15-17, 19, 21.
92. Pettipher, G. L. 1989. The direct epifluorescent filter technique. Prog. Ind. Microbiol. 26:19-56.
93. Pettipher, G. L., R. Mansell, C. H. McKinnon, and C. M. Cousins. 1980. Rapid membrane filtration-epifluorescent microscopy technique for direct enumeration of bacteria in raw milk. Appl. Environ. Microbiol. 39:423-429.
94. Phillips, J. D., and M. W. Griffiths. 1987. A note on the use of the Catalasemetre in assessing the quality of milk. J. Appl. Bacteriol. 62:223-226.
95. Phillips, J. D., and M. W. Griffiths. 1987. The relation between temperature and growth of bacteria in dairy products. Food Microbiol. 4:173-185.
96. Philips, J. D., and M. W. Griffiths, and D. D. Muir. 1983. Accelerated detection of post-heat-treatment contamination in pasteurized double cream. J. Soc. Dairy Technol. 36:41-43.
97. Reichardt, W., and R. Y. Morita. 1982. Temperature characteristics of psychrotrophic and psychrophilic bacteria. J. Gen. Microbiol. 128: 565-568.
98. Reuter, G. 1981. Psychrotrophic lactobacilli in meat products, p. 253-258. *In* T. A. Roberts, G. Hobbs, J. H. B. Christian, and N. Skovgaard (ed.), Psychrotrophic microorganisms in spoilage and pathogenicity. Academic Press, New York.
99. Rodrigues, U. M., and R. G. Kroll. 1985. The direct epifluorescent filter technique (DEFT): increased selectivity, sensitivity and rapidity. J. Appl.. Bacteriol. 59:493-499.
100. Ronk, R. J., K. L. Carson, and P. Thompson. 1989. Processing, packaging, and regulation of minimally processed fruits and vegetables. Food Technol. 43(2):136-139.
101. Russell, N. J. 1990. Cold adaptation of microorganisms. Philos. Trans. R. Soc. London, Ser. B 326:595-611.
102. Sayem El Daher, N., and R. E. Simard. 1985. Putrefactive amine changes in relation to microbial counts of ground beef during storage. J. Food Prot. 48:54-58.
103. Schaertel, B. J., and N. Tsang, and R. Firstenberg-Eden. 1987. Impedimetric detection of yeast and mold. Food Microbiol. 4:155-163.
104. Schellekens M. 1996. New research issues in sous-vide cooking. Trends Food Sci. Technol. 7:256-262.
105. Sharpe, A. N. 1989. The hydrophobic grid-membrane filter. Prog. Ind. Microbiol. 26:169-189.
106. Smith, T. L., and L. D. Witter. 1979. Evaluation of inhibitors for rapid enumeration of psychrotrophic bacteria. J. Food Prot. 42:158-160.
107. Smithwell N., and K. Kailasapathy. 1995. Psychrotrophic bacteria in pasteurized milk: Problems with shelf life. Aust. J. Dairy Technol. 50:28-31.
108. Søgaard, H., and R. Lund. 1981. A comparison of three methods for the enumeration of psychrotrophic bacteria in raw milk, p. 109-116. *In* T. A. Roberts, G. Hobbs, J. H. B. Christian, and N. Skovgaard (ed.), Psychrotrophic microorganisms in spoilage and pathogenicity. Academic Press, New York.
109. Sommer, N. F. 1985. Strategies for control of postharvest diseases of selected commodities, p. 83-99. *In* A. A. Kader, R. F. Kasmire, F. G. Mitchell, M. S. Reid, N. F. Sommer, and J. F. Thompson (ed.) Postharvest technology of horticultural crops. Special Publication 3311. Cooperative Extension, University of California, Davis, Calif.
110. Sørhaug T., and L. Stepaniak. 1997. Psychrotrophs and their enzymes in milk and dairy products: Quality aspects. Trends Food Sci. Technol. 8:35-41.
111. Stannard, C. J. 1989. ATP estimation. Prog. Ind. Microbiol. 26:1-18.
112. Stannard, C. J., and J. M. Wood. 1983. The rapid estimation of microbial contamination of raw meat by measurement of adenosine triphosphate (ATP). J. Appl. Bacteriol. 55:429-438.
113. Stannard, C. J., A. P. Williams, and P. A. Gibbs. 1985. Temperature/growth relationships for psychrotrophic food-spoilage bacteria. Food Microbiol. 2: 115-122.
114. Südi, J., G. Suhren, W. Heeschen, and A. Tolle. 1981. Entwicklung eines miniaturisierten Limulus-Tests im Mikrotiter-System zum quantitativen Nachweis Gram-negativer Bakterien in Milch und Milchprodukten. Milchwissenschaft 36:193-198.
115. Suhren, G., and W. Heeschen. 1987. Impedance assays and the bacteriological testing of milk and milk products. Milchwissenschaft 42:619-627.
116. Suhren, G., W. Heeschen, and A. Tolle. 1982. Quantitative Bestimmung psychrotropher Mikroorganismen in Roh-und pasteurisierter Milch—ein Methodenvergleich. Milchwissenschaft 37:594-596.
117. Sullivan, J. D., Jr., P. C. Ellis, R. G. Lee, W. S. Combs Jr., and S. W. Watson. 1983. Comparison of the *Limulus* amoebocyte lysate test with plate counts and chemical analyses for assessment of the quality of lean fish. Appl. Environ. Microbiol. 45:720-722.
118. Thomas, S. B. 1969. Methods of assessing the psychrotrophic bacterial content of milk. J. Appl. Bacteriol. 32:269-296.
119. Tsuji, K., P. A. Martin, and D. M. Bussey. 1984. Automation of chromogenic substrate *Limulus* amoebocyte lysate assay method for endotoxin by robotic system. Appl. Environ. Microbiol. 48:550-555.
120. Vanderzant, C., and A. W. Matthys. 1965. Effect of temperature of the plating medium on the viable count of psychrophilic bacteria. J. Milk Food Technol. 28:383-388.
121. Van Spreekens, K. J. A., and F. K. Stekelenburg. 1986. Rapid estimation of the bacteriological quality of fresh fish by impedance measurements. Appl. Microbiol. Biotechnol. 24:95-96.
122. Van Spreekens, K. J. A., and L. Toepoel. 1981. Quality of fishery products in connection with the psychrophilic and psychrotrophic bacterial flora, p. 283-294. *In* T. A. Roberts, G. Hobbs, J. H. B. Christian, and N. Skovgaard (ed.), Psychrotrophic microorganisms in spoilage and pathogenicity. Academic Press, New York.
123. Vasavada, P. C. 1988. Low-acid foods defy liabilities. Prep. Foods 157(6):122-123, 125.
124. Vasavada, P. C. 1988. Pathogenic bacteria in milk—a review. J. Dairy Sci. 71:2809-2816.
125. Vasavada, P. C., T. A. Bon, and L. Bauman. 1988. The use of Catalasemeter in assessing abnormality in raw milk. J. Dairy Sci. 71(Suppl 1):113.
126. Venkitanarayanan, K. S., M. I. Khan, C. Faustman, and B. W. Berry. 1996. Detection of meat spoilage bacteria by using the polymerase chain reaction. J. Food Prot. 59:845-848.
127. Waes, G. M., and R. G. Bossuyt. 1982. Usefulness of the benzalkon-crystal violet-ATP method for predicting the keeping quality of pasteurized milk. J. Food Prot. 45:928-931.
128. Wang, G. I. J., and D. Y. C. Fung. 1986. Feasibility of using catalase activity as an index of microbial loads on chicken surfaces. J. Food Sci. 51:1442-1444.
129. Ward, D. R, K. A. LaRocco, and D. J. Hopson. 1986. Adenosine triphosphate bioluminescent assay to enumerate bacterial numbers on fresh fish. J. Food Prot. 49:647-650.
130. White, C. H., J. R. Bishop, and D. M. Morgan. 1992. Microbiological methods for dairy products, p. 287-308. *In* R. T. Marshall (ed.), Standard methods for the examination of dairy products, 16th ed. American Public Health Association, Washington, D. C.
131. Witter, L. D. 1961. Psychrophilic bacteria—a review. J. Dairy Sci. 44:983-1015
132. Zall, R. R., J. H. Chen, and S. C. Murphy. 1982. Estimating the number of psychrotrophs in milk using the direct microscopic method. Cult. Dairy Prod. J. 17(2):24-26, 28.

Chapter 14

Thermoduric Microorganisms and Heat Resistance Measurements

David L. Collins-Thompson and Timothy C. Jackson

14.1 INTRODUCTION

Several non-sporing bacteria exhibit a higher thermal resistance than expected given the physical or biochemical properties of the genera. These organisms are described as thermoduric, i.e., having the property of thermotolerance and hence a capacity to survive some pasteurization processes.[35] Such organisms are capable of surviving heating in a food substrate at a range of 60–80°C. This chapter deals with these organisms and related spore formers, such as *Bacillus* and *Clostridium* spp., often associated with spoilage of dairy and egg products.

While thermoduric organisms will normally grow in the mesophilic range (15–37°C), there are several examples in the literature of such organisms growing at refrigeration temperatures.[36,70] Genera or groups of bacteria reported to contain thermoduric strains are listed in Table 1.

Although not listed in Table 1, there are also reports of heat resistant coliforms such as *Enterobacter aerogenes* surviving pasteurization.[59]

Our understanding of thermoduric bacteria has been influenced by studies in milk, milk products, and liquid eggs. From these studies and others it can be concluded that the capacity for surviving pasteurization is not solely dependent upon the properties of the microorganism, but may also be influenced by factors such as initial microbial load, product composition, age of the product and method of heat treatment.

Table 1. Bacterial genera or groups reported to contain thermoduric organisms associated with foods

Genera/group	Associated foods	Reference
Actinomycetes	Milk	40
Alcaligenes	Milk	40
Arthrobacter	Milk	40, 70
Bacillus	Milk/eggs	24, 36, 62, 65
Clostridium	Milk	47
Coryneform bacteria	Eggs	55
Lactobacillus	Milk/meats/juices	27, 33, 37
Microbacterium	Milk	40
Micrococcus	Milk/eggs	24, 61, 70
Pseudomonas	Eggs	24, 61
Streptococcus/Enterococcus	Milk/meat/eggs	24, 33, 55

Studies on market milk supplies[41,42,59] indicate the importance of the *Enterococcus* spp. as thermoduric organisms (Table 2).

The sources of some thermodurics, such as *Streptococcus, Microbacterium,* and coryneform bacteria in milk supplies can be limited to milking and creamery equipment.[40] Spore formers and *Micrococcus* are often associated with soil, fodder or hay. These organisms enter the food chain because of improper handling and cleaning of processing equipment.[33]

The level of survival of many of these organisms after heat treatment is usually quite low. Survival estimates have been determined in milk for cheese making and in cream following pasteurization at 72°C for 15 sec[31] (Table 3).

Reported differences of survival data published in the literature for some thermodurics after heat treatment may simply be the result of variations in resistance or levels in the initial populations. Some of the genera may be better equipped to survive the thermal treatment.

Shafi et al.[61] demonstrated that the types of organisms surviving the pasteurization of liquid egg were similar to those surviving in milk products, with the exception of finding *Staphylococcus* as part of the heat-resistant flora. Furthermore, genera like *Bacillus* and *Micrococcus* appeared to survive pasteurization regardless of type of egg product or heat treatment. Payne et al.[55] reported that many of the organisms that survived pasteurization of whole egg were of the coryneform group, including *Microbacterium lacticum.* Two strains of coryneform bacteria survived a heat treatment of 20 and 38 min at 80°C in phosphate buffer at pH 7.1. A further characteristic of these thermoduric organisms was that none of the isolates studied grew at 5°C but all were

Table 2. Population characteristics of thermoduric isolates from market milk[41,42]

Organism	% population
Enterococcus faecalis	—
Enterococcus bovis	53
Enterococcus faecium	—
Streptococcus thermophilus	—
Micrococcus luteus	11
Corynebacteria	—
Microbacterium	7
Bacillus	29

capable of growing at 10°C. Freezing coryneforms in liquid egg at –18°C before heating had little effect on heat resistance or viability. Foegeding and Stanley[24] examined microorganisms surviving ultrapasteurization of liquid whole egg with subsequent growth at 4° or 10°C in this product. The most heat resistant isolates recovered were *Enterococcus faecalis* and *Bacillus circulans.*

In meat products, some lactobacilli and enterococci may be recovered after a pasteurization process. This survival is particularly true of the fecal enterococci in such products as canned hams where they are considered part of the normal flora.[33]

In addition to the common thermoduric non-pathogenic bacteria, there have been several thermal resistance studies and epidemiological reports of milkborne disease outbreaks, pasteurized milk and liquid egg surveys.[3,4,6,24] These have shared a concern some organisms may survive recommended pasteurization temperatures.[2,4,7,11]

The composition of a foodstuff has been reported to influence the heat resistance of thermoduric spoilage organisms. Ingredients such as pectins have been linked to the survival of *Lactobacillus fermentum* in tomato juice after heat treatment (55–60°C).[37] Factors such as water activity and pH will also influence the survival of microorganisms.

Raw milk contains proteinases from both indigenous and bacterial sources that may be involved in gelation or proteolytic activity in ultra-high-temperature (UHT) pasteurized milk. Many of these proteinases are heat resistant or may regenerate activity during storage.[1] Spoilage resulting from heat resistant enzymes may be incorrectly interpreted as having been caused by surviving thermoduric organisms.

14.2 HEAT RESISTANCE MEASUREMENTS

The type of heat resistance measurement used in an investigation will be determined by the desired use of the results. Where information is needed on the microbiological quality of a foodstuff following pasteurization, a thermoduric count may be appropriate. For this measurement, a food substrate is heated at temperatures and conditions similar to those encountered during processing, and a count is made of surviving organisms by direct plating onto a recovery medium.

Information is often required on the heat resistance of a specific organism, e.g., *Listeria monocytogenes* in a specific menstruum (such as 1% fat milk). In such circumstances, a uniform parameter is needed which would allow a comparison of heat resistance between organism or heating menstruum, and would enable an estimation of the thermal process necessary to inactivate or reduce a target population. The measurement most often used is D_T, the decimal reduction time (also known as D-value). D_T represents the time required at a specific temperature (T) for a 10-fold, or 90%, reduction of the surviving microbial population in a given menstruum (Table 4). D_T may be influenced by intrinsic characteristics of a foodstuff, or by characteristics of the organism itself, including strain, growth phase, temperature, and exposure to sublethal stresses prior to heating.

D_T may be obtained from quantitative or qualitative data. Quantitative data from successive sampling experiments, in which surviving microbial populations are enumerated at regular intervals during heating, may be used to establish a semilogarithmic survival curve (Figure 1). D_T is determined from the linear portion of the survival curve, and is calculated as the absolute reciprocal of the slope of the survival curve, where the slope = Δ number of survivors/Δ time of heating.[57,64]

Qualitative data may be obtained as positive or negative growth in enrichment broths following heating. For such data, D_T is calculated by determination of the 50% endpoint; that is, the time of heating at a constant temperature where 50% of the samples are positive for growth.[68] The holding times for such an experiment should be established to bracket the end-point of survival, yielding from all-positive to all-negative results. At least three replicate tests would be performed at each holding time, with a replicate number chosen based upon the precision desired for the estimate.[11]

The application of the Spearman-Karber estimation to determine D_T from fraction-negative data (such as 50% endpoint, t_m) has been demonstrated by Lewis [44] and Pflug and Holcomb.[57]

Correcting for the bias in the procedure, D_T can be computed as follows.[11, 44] (It is assumed that the detection method recovers a viable microorganism when present):

$$D_T = t_m/(\log_{10} N0 + 0.251)$$

t_m = the 50% time estimated from the Spearman-Karber procedure, i.e., the time when 50% of the analytical units are positive; and

$N0$ = the concentration of bacteria measured by plate count at heating time zero.

A challenge to the calculation of D_T from quantitative (successive) survivor data is that its application assumes that survivor curves will be linear, following a logarithmic order of death. While survivor curves are frequently linear, they may also be concave, convex, or sigmoidal. They may incorporate initial shoulders or declines, or demonstrate tailing in addition to a logarithmic component.[29,48,56]

Table 3. Survival rates of thermoduric organisms in milk heated at 63°C for 30 min or cream heated at 72°C for 15 sec[16]

Organism	% Occurrence among isolates
Heated Milk	
Microbacterium	100
Micrococcus	1–10
Alcaligenes	1–10
Streptococcus/Enterococcus	<1
Lactobacillus	<1
Coryneforms	<1
Pasteurized cream	
Coryneforms	70
Micrococcus	20
Bacillus	<5
Streptococcus/Enterococcus	<5

Table 4. Comparison of D values (D_T) between *S. aureus* and two thermoduric organisms (Collins-Thompson unpublished data).

Organism	Temperature of heating*°C	D Value (min)
Staphylococcus aureus	60	1.0
Enterococcus faecalis	58	3.9
Streptococcus thermophilus	58	4.2

* In phosphate buffer pH 7.00

Several strategies have been utilized for the interpretation of non-linear survivor data. Many researchers have applied linear regression analysis to the most linear portion of the survivor curve and have excluded initial declines or shoulders. The use of an intercept ratio (IR) or intercept index (II) has been suggested to account for initial shoulders or declines where present on survivor curves.[56,57] Multiple regressions have been used to analyze biphasic survivor curves,[34,54] and non-linear regression models have been applied to more complex curves.[4,45,57] Several authors have suggested that the use of thermal death point measurements, or F-values (F_T), are a more appropriate tool than D-values in situations in which survivor curves are non-logarithmic.[9,14,46,49] F-values are a measure of the thermal death time, the time necessary at a specific temperature to inactivate a microbial population in a specific menstruum. As an example, for an initial population of 5.0 $\log_{10}$ cfu/mL, $F_T = 5 \times D_T$. In practice, F is most commonly used as an indicator of process lethality in commercial sterilization systems.

Another parameter, the z-value, is an indication of the change in temperature required to change the decimal reduction time by a factor of 10. It is relatively constant for a given organism under various test conditions.[68] The z-value may be used to determine equivalent thermal processes at a range of temperatures. Where D_T is known for an organism at three temperatures, such data may be plotted as a Thermal Death Time (TDT) curve ($\log_{10}D_T$ x temperature; Figure 2). The z-value may be calculated as the absolute reciprocal of the linear portion of the slope of this curve. Where D_T is known at two temperatures (T_1, T_2), z may be calculated as[57]:

$$z = (T_2 - T_1)/(\log_{10}D_1 - \log_{10}D_2).$$

Where z for an organism is known, an established D_T (at T_1) may be used to calculate D_T at a different temperature (T_2) using the equation[64]:

$$\log_{10} D_2 = (T_1 - T_2)/z + \log_{10}D_1$$

14.3 METHODS FOR THE DETERMINATION OF HEAT RESISTANCE

The various methods for determining wet heat destruction rates for microorganisms have been categorized by Pflug and Holcomb[57] as successive or multiple-replicate-unit sampling systems.

In successive sampling systems, a small volume of a microbial suspension is inoculated into a larger volume of substrate that has been preheated to a specified heating temperature. The suspension is continuously agitated during heating. At established heating intervals, an aliquot of inoculated substrate is aseptically removed by pipette and dispensed into a sterile tube immersed in ice water for cooling. Since the substrate is already at the specified heating temperature, these systems avoid the need to correct for microbial inactivation during come-up time. Successive sampling systems may use a flask method, heating the substrate in a flask[30,32] or other container such as a canning jar[54] or multiple neck container.[43,60] A tank method has been described for the evaluation of temperatures above 100°C.[72]

In multiple-replicate-unit sampling systems, multiple sample units are prepared and heated concurrently. Heated units are removed at successive time intervals[57] and immediately exposed to ice water for cooling. Survivors may be evaluated by the presence or absence of growth or by the enumeration of survivors and the establishment of a survival curve. Sealed glass tubes, vials, ampoules and capillary tubes have been used in these sampling schemes.[5,17,21,26,28,38,50,57] The Low Temperature Long Time (LTLT) and Immersed Seal Tube (IST) methods described below are multiple-replicate unit sampling systems.

14.4 EQUIPMENT, MATERIALS, AND PROCEDURES

14.41 Preparation and Handling of Cultures

Where the survival or heat resistance of a specific microorganism is being evaluated, the preparation and handling of cultures should take into consideration factors influencing microbial heat resistance. Several reviewers have summarized the influence of environmental and physiological factors on the heat resistance of bacterial cells.[29,57,68] Characteristics of the food substrate, such as the presence of nutrients, fat content, water activity, and pH, and characteristics of the organism such as strain, growth phase, and age of culture have been demonstrated to influence heat resistance.

Relative to culture conditions, incubation and storage temperature before heating may influence a microorganism's heat resistance. For example, bacterial populations demonstrate greater heat resistance in stationary than logarithmic phase.[12,20,34,57,67,69,71] Greater heat resistance has been reported for vegetative bacteria and spores of cultures grown at higher temperatures[20,34,57] Cultures

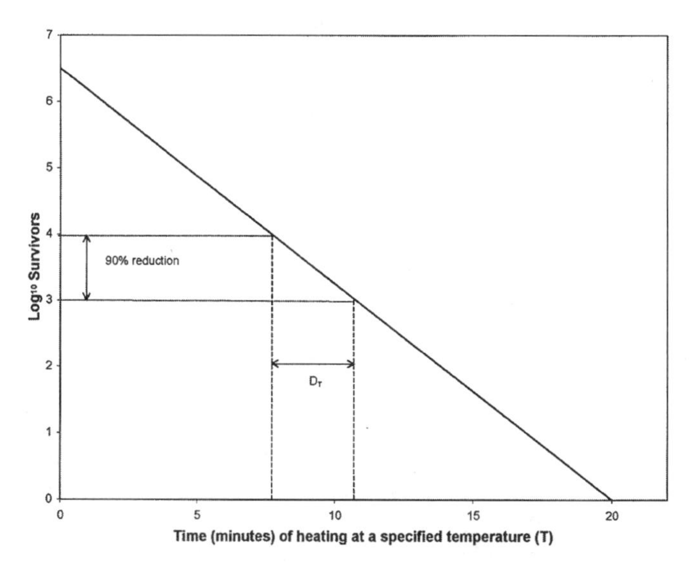

Figure 1. *$\log_{10}$ survivor curve and determination of D_T.*

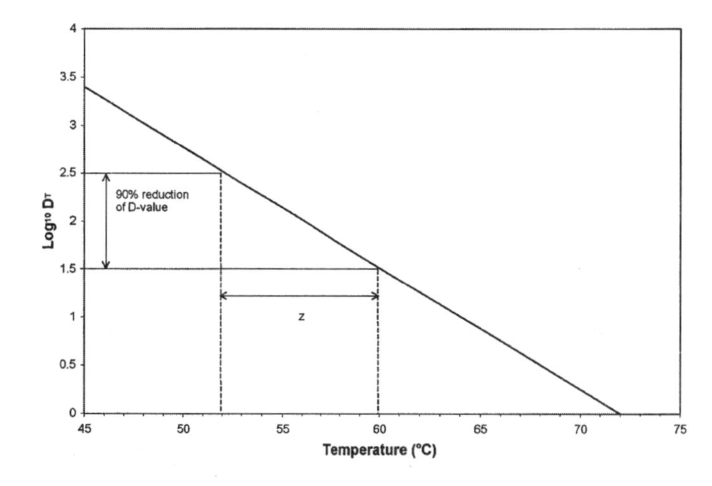

Figure 2. *Thermal death time curve and determination of z.*

held under refrigeration may exhibit greater heat sensitivity than those stored at higher temperatures.[12,32,34,39]

Since many factors inherent in a food system may influence the ability of an organism to survive heating, it is often desirable to equilibrate an inoculum to the characteristics of a menstruum before heating. Where possible, the inoculated menstruum should be equilibrated before thermal processing to a temperature similar to that which would be encountered in commercial production. In an evaluation of the heat resistance of *Listeria monocytogenes* during High Temperature Short Time (HTST) pasteurization, Farber et al.[22] resuspended a prepared inoculum pellet into 5 mL sterile whole milk that was then used to inoculate a larger volume of sterile whole milk (5 L) for evaluation of the process. The inoculated milk was stored at 4°C overnight before pasteurization to equilibrate the culture to the substrate, and to simulate commercial holding practices.

A rich, non-selective recovery medium should be used in thermal process studies, since selective media may not allow the growth of injured cells (Chapter 5). While non-selective nutrient media is suitable for the recovery of thermodurics, the use of such media to determine the survival of a specific pathogen may be limited by the presence of background flora. Selective media would allow differentiation from other organisms, but may contain components inhibitory to the recovery of injured cells. Cold enrichment schemes (4°C for 7 to 28 days) have been used to detect psychrotrophic pathogens in heated foods[12,23] but are time consuming and may produce erratic results.[11] Optimally, the heat resistance of specific bacteria should be determined by inoculation into a sterile substrate. For example, milk used in thermal resistance investigations may be UHT pasteurized or sterilized in an autoclave at 121°C for 10 to 15 min and rapidly cooled to 4°C so that caramelization does not occur.[6,7,8,13,17,18] Alternatively, milk may be heated at lower temperatures for sufficient time (such as 85°C/1 hr) to inactivate competitive organisms.[22]

For use in thermal process challenges, fresh cultures are cultivated from stock cultures (stored under refrigeration on solid media) and are generally grown to stationary phase at an optimal growth temperature. Once grown, cells are spun in a centrifuge and the supernatant decanted from the resulting pellet. The pellet is washed to remove debris and residue of the culture media which may influence heat resistance, and the culture is resuspended in the heating menstruum.

14.42 Thermoduric Measurements during Commercial Processing Procedures

A practical approach to the determination of thermoduric counts or the presence of pathogens following HTST or LTLT pasteurization would be to evaluate product during commercial processing. Such an approach would take into account the unique characteristics of a given processing system. Thermoduric counts or evaluation of pathogen survival may be accomplished by the microbiological examination of finished product produced in the commercial operation. Counts of finished product, however, may include in-process or post-processing contaminants, and may not therefore represent a true thermoduric count.

Where a definite determination of thermodurics in a raw product is desired, a more detailed sample protocol may be necessary which would evaluate samples taken at various stages in the pasteurization process.[11] For example, an evaluation of the influence of HTST pasteurization on microbial populations would include aseptic samples from several areas 1) raw milk bulk tank, 2) end of holding tube, 3) exit of pasteurizer regenerator, 4) cooler exit prior the vacuum breaker, 5) pasteurized discharge after the vacuum breaker, and 6) the finished product. Air samples should be taken from the processing environment, and from the cleaned and pasteurized equipment. Sample ports may be established in the system (e.g., TRU-TEST samplers, Food and Dairy Management, Inc., St. Paul, MN) to enable the withdrawal of sample from pipelines using a large volume syringe fitted with an 18-gauge needle.

Likewise, sampling schemes may be developed to evaluate microbial populations before, during, and following commercial LTLT vat or hold tube heating. Product taken following heating, such as at the end of the hold tube or directly from the vat, should be immediately dispensed into a sterile flask filled with glass beads previously cooled to –20°C.

14.43 Low Temperature Long Time (LTLT) Pasteurization

The following procedure has been outlined in "Standard Methods for the Examination of Dairy Products, 16th edition"[25] for the evaluation of thermoduric organisms in milk samples obtained from individual producers or for samples of finished product. The procedure may also be used to evaluate other pasteurized liquid products or well-blended 1:10 dilutions of pasteurized foods.

14.431 Equipment

Test tubes: sterile, 20 × 125 mm, screw-capped with rubber or plastic lined caps to enable proper sealing.

Pipettes: sterile, graduated for 5 or 11-mL delivery.

Thermometer with 0.1°C graduations: range of thermometer should encompass the critical temperature range for the pasteurization process. Thermometer should be checked at least biennially against a reference thermometer certified by the National Institute of Standards and Technology.

Water bath: electrically heated, thermostatically controlled to maintain a temperature of 62.8 ± 0.5°C. A sufficient volume of water should be used to absorb the cooling effect of tubes placed into the bath without the temperature drop exceeding 0.5°C.

14.432 Methodology

Thoroughly mix the liquid or diluted product to be evaluated and aseptically transfer 5.0 mL into sterile test tubes, with care taken to avoid depositing sample on the lip or upper portion of the tube. Microorganisms present on tubes above the surface of the liquid may receive uneven heating or possibly dry heating, which may influence the thermoduric count. Product samples and filled tubes should be maintained at 0–4.4°C during the filling process. Sufficient replicate tubes of each product should be prepared to enable an accounting of the variability between samples. Seal tubes containing sample with sterile rubber or plastic-lined caps.

Construct a reference (pilot) tube, containing a thermometer immersed in an equal volume (5.0 mL) of product under evaluation. The reference tube will be used to monitor product temperature during heating and cooling. Place the sample tubes and reference tube into a metal, plastic, or wire test tube rack.

Immerse the rack of tubes into a thermostatically controlled waterbath that has been preheated to 62.8°C. To assure even heating, the water level should be at least 4 cm above the level of sample in the tubes. Samples should reach the target temperature in ≤5 min. Bubbles arising from product during heating are undesirable, since contaminants may be drawn into the tubes during cooling.

Heat sample tubes for 30 min. Following heating, immediately immerse the sample rack into an ice water bath and cool to ≤10°C.

Determine the number of surviving organisms using Aerobic Plate Count procedures (Chapter 7). In the absence of regulatory issues, alternative procedures may be used. Counts may be reported as the Thermoduric Count/mL (or g)[11] or the Laboratory Pasteurization Count/mL (or g).[25]

Where products or processes other than Low Temperature Long Time Pasteurization (LTLT) are being evaluated, temperatures and times should be modified as appropriate to the process under evaluation. This method should not be used for the evaluation of High Temperature Short Time (HTST) pasteurization processes.

A disadvantage of this method is the use of relatively large volume tubes. The prescribed sample volume does not minimize the headspace in the tube above the product surface. The resulting open headspace in the tube could allow for uneven heating of organisms adhering to the sides or top of the tubes. Open headspace is also a limitation of methods that use open flask heating. For this reason, airspace heaters are required by the Pasteurized Milk Ordinance.[11,18,24,64,66]

14.44 Immersed Sealed-Tube Method

An alternative to the LTLT method in section 14.43 is the immersed sealed tube method (IST). To avoid uneven heating, the IST method utilizes small volume tubes or vials which are filled with minimal headspace, sealed, and heated by full immersion into a thermostatically controlled water bath. Using smaller sample volumes also minimizes sample "come up time," or time for samples to reach the target temperature during the initial stages of heating. This method is a preferred method for the determination of a thermoduric count or survival of an inoculum during low temperature long time pasteurization processes, or for successive sampling schemes to determine survivor curves (thermal death time curves) and corresponding D-values.[7,8,11,17,57,64]

14.441 Equipment

Sample vessels: Sterile test tubes, 7 to 10 mm diameter or 2 mL glass vials: crimp-sealable with metal caps containing metal teflon-lined seals.

Pipettes: sterile, graduated for 1, 5, 10, or 11-mL delivery.

Thermometer with 0.1 °C graduations: range of thermometer should encompass the critical temperature range for the pasteurization process. Thermometer should be checked at least biennially against a reference thermometer certified by the National Institute of Standards and Technology.

Water bath: electrically heated, thermostatically controlled to maintain a temperature of 62.8 ± 0.5°C. A sufficient volume of water should be used to absorb the cooling effect of tubes placed into the bath without the temperature drop exceeding 0.5°C.

Digital timer.

14.442 Methodology (General)

Transfer product sample or inoculated sample into small-volume vessels such as 7 to 10-mm test tubes. Transfer enough sample volume (1–4 mL) to minimize headspace in the tube. Seal the tube near its mouth in the flame of a gas burner.[52,55] Two-mL glass reaction vials may also be used, which are filled with sample and crimp-sealed using metal caps fitted with teflon-lined seals.[17,19]

Capillary tubes (0.8 mm ID × 1.5 mm OD × 7.62 cm long) have also been used in IST schemes.[46,57] A precision syringe is used to insert a 0.01-mm volume of sample into capillary tubes and the tubes are flame-sealed. The use of such tubes provides the benefit of a quick come-up time and the ability to break open tubes directly into subculture media; however, such tubes require special heating and cooling equipment, may be more difficult to fill and seal, and provide only a small volume of sample.[57]

Construct a reference or pilot tube with identical sample volume to the tubes being heated. Seal the tubes with a rubber stopper through which a thermometer or calibrated thermocouple has been inserted so that it is in the geometric center of the sample volume. The reference tube may be placed along side the sample during heating to monitor the time necessary to reach the target temperature, or to determine correction factors for heating and cooling lags under identical heating conditions as those for the heated samples. Correction factors may be calculated using the statistical method described by Anellis et al.[3] Heating and cooling lags (come up time) and consequently correction factors, will increase with increasing mass and thermal capacity of the substrate being heated.[51,52,53]

14.443 Thermoduric Count

Sufficient replicate tubes of each product should be prepared as described to enable an accounting of the variability between samples. Fully immerse sample and reference tubes in a thermostatically controlled waterbath preheated to 62.8°C ± 0.5°C. Assure that the water level is at least 4 cm above the top of sample vials. Samples should reach the target temperature in ≤5 min. Heat samples for 30 min. At the end of the heating period immediately plunge tubes or vials into an ice water bath until the temperature reaches ≤10°C.

Determine the number of surviving organisms using Aerobic Plate Count procedures (Chapter 7). In the absence of regulatory issues, alternative procedures may be used. Counts may be reported as the Thermoduric Count/mL (or g)[11] or the Laboratory Pasteurization Count/mL (or g).[25]

14.444 Multiple Replicate Sampling for Determination of Heat Resistance

The multiple replicate sampling scheme may be used to determine the heat resistance (as D_T) of a known inoculum at 62.8°C for LTLT pasteurization or other temperature lethal to the organism under investigation. The scheme involves pulling samples at regular intervals during heating to determine levels of survivors, and construction of a survivor curve. Preliminary heat inactivation runs should be performed to determine optimum heating intervals.[11]

Prepare the inoculum as described in section 14.41. A sufficient inoculum should be provided (generally 10^5 to 10^6 cfu/mL) to allow a several $\log_{10}$ reduction during heating. Prepare a sufficient number of tubes or vials containing inoculated sample to account for two to three replicate units for each heating period and for the determination of initial numbers. At least four heating periods are recommended in addition to "initial" or "time 0" samples.

Reserve a replicate set of samples for the determination of initial microbial levels. Immerse the rack containing the remaining tubes and reference tube into a thermostatically controlled waterbath preheated to the target temperature (±0.5°C). Assure that the level of the water bath is at least 4 cm above the top of the tubes. Monitor the time samples take to reach target temperature (should be ≤5 min). When temperature has been reached, immediately pull a replicate set of samples and plunge into an ice water bath to cool to ≤10°C. The mean population of these samples will represent "time 0" of heating on the survivor curve. Pull replicate sets of samples at established intervals (for example, every 1 or 5 min) and plunge into an ice water bath to cool to ≤10°C.

Plate samples onto a non-selective nutrient medium for enumeration of survivors using the Standard Plate Count procedure. The incubation period should be sufficient to recover injured microbial cells. Survivor data may be used to construct a $\log_{10}$ survivor curve and determine D_T as described in section 14.2. Where qualitative data is required, aseptically transfer the sample volume from the tubes into an appropriate non-selective subculture medium and incubate to determine growth/no growth. The experiment should be repeated to evaluate the variability inherent in the method.

14.45 High Temperature Short Time (HTST) Pasteurization

Due to the nature of the processes, thermoduric counts obtained evaluating LTLT pasteurization may not precisely correlate with survival during HTST processes.[25] It is therefore necessary to evaluate thermoduric counts for HTST pasteurization under conditions specific to this process. Evaluation of counts during and following commercial HTST processing has been discussed (section 14.42). Due to the high temperatures and short times inherent to HTST processes, the laboratory methods described for LTLT processes would not be appropriate. Laboratory techniques have been developed to simulate pasteurization conditions.[15, 63] A two-stage slug flow heat exchanger[63] is used by the U.S. Food and Drug administration to evaluate the thermal resistance of milkborne pathogens under HTST conditions.[11] The set up and operation of the apparatus is outlined by Stroup et al.[63]

14.46 Media

Media used for thermoduric/heat resistance studies should be nonselective, such as standard methods agar. The detection of fastidious organisms such as *Lactobacillus* may require APT agar or other rich medium (see Chapter 19). Foodborne pathogens that are heat stressed will require special enrichment protocols (see Chapter for specific pathogen). Incubation times and temperatures depend upon the particular application or organism. When using the aerobic plate count procedure (Chapter 7), an incubation temperature of 32°C for 48 hr is recommended.

14.5 INTERPRETATION

There are various interpretations of the significance of thermoduric microorganisms in foods. In some foods, such as milk and milk products, a link has been established between thermoduric counts and sanitary practices.[36] The number and occurrence of thermoduric microorganisms can be related to poor cleaning of equipment in milk plants.[31] These microorganisms can be responsible for flavor and spoilage problems (bitter cream and sweet curdling).[10] Thermoduric psychrotrophs in milk have been reported to invalidate the methylene blue reduction test, causing rejection of pasteurized milk of good bacterial quality.[36] Levels of thermoduric microorganisms in producer milk will vary in different situations. For example, higher levels of thermoduric organisms in milk have been associated with winter housing of cattle.[36] Counts less than 1000 cfu/mL in producer's milk are considered acceptable while levels above 1000 cfu/mL indicate a possible sanitation problem and may require investigation.

A similar situation exists with pasteurized liquid egg products. Shafi et al.[61] found that bacterial levels in pasteurized frozen whole egg samples ranged from 60–600 cfu/mL. However, bacterial levels of >1000 cfu/mL in pasteurized whole egg have been reported as unacceptable.[55]

Thermoduric organisms associated with meats and meat products have caused several problems. Studies have shown that the greening of meats may result from the survival of thermoduric lactobacilli such as *Lactobacillus viridescens*.[27] In canned hams, the survival of thermoduric enterococci and lactobacilli may give rise to problems with souring, color loss, and swollen cans. In selected dried foods, low levels of thermoduric streptococci/enterococci (<100 cfu/g) have been reported.[58] In general, thermodurics grow poorly at 5°C, but there are some psychrotrophic thermoduric microorganisms that, if present in large enough numbers, can lead to spoilage in meats within 10–15 days of refrigerated storage.[33]

14.6 REFERENCES

1. Adams, D. M., J. T. Barach, and M. L. Speck. 1976. Effect of psychrotrophic bacteria from raw milk on milk proteins and stability of milk proteins to ultrahigh temperature. J. Dairy Sci. 59:823-827.
2. Allwood, M. C., and A. D. Russell. 1970. Mechanisms of thermal injury in nonsporulating bacteria. Adv. Appl. Microbiol. 12:89-119.
3. Anellis, A., J. Lubas, and M. M. Rayman. 1954. Heat resistance in liquid eggs of some strains of the genus *Salmonella*. Food Res. 19:377-395.
4. Bhaduri, S., P. W. Smith, S. A. Palumbo, C. O. Turner-Jones, J. L. Smith, B. S. Marmer, R. L. Buchanan, L. L. Zaika, and A. C. Williams. 1991. Thermal destruction of *Listeria monocytogenes* in liver sausage slurry. Food Microbiol. 8:75-78.
5. Bigelow, W. D., and J. R. Esty. 1920. The thermal death point in relation to time of typical thermophilic organisms. J. Infect. Dis. 27:602-617.
6. Bradshaw, J. G., J. T. Peeler, J. J. Corwin, J. E. Barnett, and R. M. Twedt. 1987a. Thermal resistance of disease-associated *Salmonella typhimurium* in milk. J. Food Prot. 50:95-96.
7. Bradshaw, J. G., J. T. Peeler, J. J. Corwin, J. M. Hunt, R. M. and Twedt. 1987b. Thermal resistance of *Listeria monocytogenes* in dairy products. J. Food Prot 50:543-544.
8. Bunning, V. K., C. W. Donnelly, J. T. Peeler, E. H. Briggs, J. G. Bradshaw, R. G., Crawford, C. M. Beliveau, and J. T. Tierney. 1988. Thermal inactivation of *Listeria monocytogenes* within bovine milk phagocytes. Appl Environ. Microbiol. 54:364-370.
9. Casolari, A. 1981. A model describing microbial inactivation and growth kinetics. J. Ther. Biol. 88:1-34.
10. Collins, E. B. 1981. Heat resistant psychrotrophic microorganisms. J. Dairy Sci 64:157-160.
11. Collins-Thompson, D. L., and V. K. Bunning. 1992. Thermoduric microorganisms and heat resistance measurements, p. 169-181. *In* C. Vanderzant and D. F. Splittstoesser (ed.) Compendium of methods for the microbiological examination of foods, 3rd ed. Am. Pub. Health Assoc., Washington, D.C.
12. Condón, S., M. L. García, A. Otero, and F. J. Sala. 1992. Effect of culture age, pre-incubation at low temperature and pH on the thermal resistance of *Aeromonas hydrophila*. J. Appl. Bacteriol. 72:322-326.
13. Crawford, R. G., C. M. Beliveau, C. W. Donnelly, J. T. Peeler, and V. K. Bunning. 1988. Recovery of heat-stressed *Listeria monocytogenes* cells from pasteurized bovine milk, abstr. P46, p. 281. *In* Abstracts of the annual meeting of the American Society for Microbiology. American Society for Microbiology, Washington, D.C.
14. Dabbah, R., W. A. Moats, and V. M. Edwards. 1971. Heat survivor curves of food-borne bacteria suspended in commercially sterilized whole milk. I. Salmonellae. J. Dairy Sci. 54:1583-1588.
15. Dickerson, R. W., Jr., and R. B. Read Jr. 1968. Instrument for study of microbial thermal inactivation. Appl. Microbiol. 16:991-997.

16. Dommett, T. W. 1992. Spoilage of aseptically packaged pasteurised liquid dairy products by thermoduric psychrotrophs. Food Aust. 44:459-461.
17. Donnelly, C. W., and E. H. Briggs. 1986. Psychrotrophic growth and thermal inactivation of *Listeria monocytogenes* as a function of milk composition. J. Food Prot. 49:994-998.
18. Donnelly, C. W., E. H. Briggs, and L. S. Donnelly. 1987. Comparison of heat resistance of *Listeria monocytogenes* in milk as determined by two methods. J. Food Prot. 50:14-17,20.
19. Doyle, M. P., K. A. Glass, J. T. Beery, G. A. Garcia, D. J. Pollard, and R. D. Schultz. 1987. Survival of *Listeria monocytogenes* in milk during high temperature short time pasteurization. Appl. Environ. Microbiol. 53:1433-1438.
20. Elliker, P. R., and W. C. Frazier. 1938. Influence of time and temperature of incubation on heat resistance of *Escherichia coli*. J. Bacteriol. 36:83-98.
21. Esty, J. R., and C. C. Williams. 1924. Heat resistance studies. I. A new method for the determination of heat resistance of bacterial spores. J. Infect. Dis. 34:516-528.
22. Farber, J. M., E. Daley, F. Coates, D. B. Emmons, and R. McKellar. 1992. Factors influencing survival of *Listeria monocytogenes* in milk in a high-temperature short-time pasteurizer. J. Food Prot. 55:946-951.
23. Fernandez-Garayzabal, J. F., L. Dominguez-Rodriguez, J. A. Vazquez-Boland, E. F. Rodriguez-Ferri, V. Briones Dieste, J. L. Blanco Cancelo, and G. S. Suarez-Fernandez. 1987. Survival of *Listeria monocytogenes* in raw milk treated in a pilot plant size pasteurizer. J. Appl. Bacteriol. 63:533-537.
24. Foegeding, P. M., and N. W. Stanley. 1987. Growth and inactivation of microorganisms isolated from ultrapasteurized egg. J. Food Sci. 52:1219-1223, 1227.
25. Frank, J. F., G. L. Christen, and L. B. Bullerman (Richardson, G. H., Tech. Comm.). 1993. Tests for groups of microorganisms, p. 271–286. *In* R. T. Marshall (ed.) Standard methods for the examination of dairy products, 16th ed. Am. Public Health Assoc., Washington, D.C.
26. Goodfellow, S. J., and W. L. Brown. 1978. Fate of *Salmonella* inoculated into beef for cooking. J. Food Prot. 41:598-605.
27. Grant, G. F., A. R. McCurdy, and A. D. Osborne. 1988. Bacterial greening in cured meats. A review. Can. Inst. Food Sci. Technol. J. 21:50-56.
28. Hanna, M. O., J. C. Stewart, Z. L. Carpenter, and C. Vanderzant. 1977. Heat resistance of *Yersinia enterocolitica* in skim milk. J. Food Sci. 42:1134, 1136.
29. Hansen, N.-H., and H. Riemann. 1963. Factors affecting the heat resistance of nonsporing organisms. J. Appl. Bacteriol. 26:314-333.
30. Heather, C. D., and C. Vanderzant. 1957. Effects of temperature and time of incubating and pH of plating medium on enumerating heat-treated psychrophilic bacteria. J. Dairy Sci. 40:1079-1086.
31. Hull, R., S. Toyne, I. Haynes, and F. Lehmann. 1992. Thermoduric bacteria: A re-emerging problem in cheesemaking. Aust. J. Dairy Technol. 47:91-96.
32. Humphrey, T. J. 1990. Heat resistance in *Salmonella enteritidis* phage type 4: the influence of storage temperatures before heating. J. Appl. Bacteriol. 69:493-497.
33. ICMSF. 1980. Meat and meat products, p. 333-409. *In* Microbial ecology of foods, vol. 2. Food commodities Int. Comm. Microbiol. Spec. Foods. Academic Press, New York.
34. Jackson, T. C., M. D. Hardin, and G. Acuff. 1996. Heat resistance of *Escherichia coli* O157:H7 in a nutrient medium and in ground beef patties as influenced by storage and holding temperatures. J. Food Prot. 59:230-237.
35. Jay, J. M. 1996. Modern food microbiology, 5th ed. Chapman Hall, New York.
36. Johnston, D. W., and J. Bruce. 1982. Incidence of thermoduric psychrotrophs in milk produced in the west of Scotland. J. Appl. Bacteriol. 52:333-337.
37. Juven, B. J., N. Ben-Shalom, and H. Weisslowicz. 1983. Identification of chemical constituents of tomato juice which affect the heat resistance of *Lactobacillus fermentum*. J. Appl. Bacteriol 54:335-338.
38. Kamau, D. N., S. Doores, and K. M. Pruitt. 1990. Enhanced thermal destruction of *Listeria monocytogenes* and *Staphylococcus aureus* by the lactoperoxidase system. Appl. Environ. Microbiol. 56:2711-2716.
39. Katsui, N., T. Tsuchido, M. Takano, and I. Shibasaki. 1981. Effect of preincubation temperature on the heat resistance of *Escherichia coli* having different fatty acid compositions. J. Gen. Microbiol. 122:357-361.
40. Kikuchi, M., Y. Matsumoto, X. M. Sun, and S. Takao. 1996. Incidence and significance of thermoduric bacteria in farm milk supplies and commercial pasteurized milk. Am. Sci. Technol. Jpn. 67:265-272.
41. Koshy, C., and V. D. Padmanaban. 1989a. Studies on keeping quality of market milk with reference to thermoduric and thermophilic bacteria. Indian Vet. J. 66:138-143.
42. Koshy, C., and V. D. Padmanaban. 1989b. Characterisation of thermoduric isolates from market milk. Indian Vet. J. 66:442-448.
43. Levine, M., J. H. Buchanan, and G. Lease. 1927. Effect of concentration and temperature on germicidal efficiency of sodium hydroxide. Iowa State Coll. J. Sci. 1:379-394.
44. Lewis, J. C. 1956. The estimation of decimal reduction times. Appl. Microbiol. 4: 211-221.
45. Linton, R. H., W. H. Carter, M. D. Pierson, and C. R. Hackney. 1995. Use of a modified Gompertz equation to model nonlinear survival curves for *Listeria monocytogenes* Scott A. J. Food Prot. 58:946-954.
46. Mackey, B. M., and C. M. Derrick. 1986. Elevation of the heat resistance of *Salmonella typhimurium* by sublethal heat shock. J. Appl. Bacteriol. 61:389-393.
47. Martin, J. H. 1974. Significance of bacterial spores in milk. J. Milk Food Technol. 37:94-98.
48. Moats, W. A. 1971. Kinetics of thermal death of bacteria. J. Bacteriol. 105:165-171.
49. Moats, W. A., R. Dabbah, and V. M. Edwards. 1971. Interpretation of nonlogarithmic survivor curves of heated bacteria. J. Food Sci. 36:523-526.
50. Murano, E. A., and M. D. Pierson. 1992. Effect of heat shock and growth atmosphere on the heat resistance of *Escherichia coli* O157:H7. J. Food Prot. 55:171-175.
51. NCA Research Laboratories. 1968. Thermal death times. *In* Laboratory manual for food canners and processors, vol. 1. Microbiology and processing. The AVI Publishing Co., Inc., Westport, Conn.
52. Neidhardt, F. C., and R. A. VanBogelen. 1987. Heat shock response, p. 1334-1335. *In* F. C. Neidhardt (ed.) *Escherichia coli* and *Salmonella typhimurium*: cellular and molecular biology. Am. Soc. Microbiol., Washington, D.C.
53. Neidhardt, F. C., R. A. VanBogelen, and V. Vaughn. 1984. The genetics and regulation of heat-shock proteins. Ann. Rev. Genet. 18:295-329.
54. Palumbo, S. A., A. C. Williams, R. L. Buchanan, and J. G. Phillips. 1987. Thermal resistance of *Aeromonas hydrophila*. J. Food Prot. 50:761-764.
55. Payne, J., J. E. T. Gooch, and E. M. Barnes. 1979. Heat-resistant bacteria in pasteurized whole egg. J. Appl. Bacteriol. 46:601-613.
56. Pflug, I. J. 1990. Microbiology and engineering of sterilization processes, 7th ed. Environmental Sterilization Laboratory, Minneapolis, Minn.
57. Pflug, I. J., and R. G. Holcomb. 1983. Principles of thermal destruction of microorganisms, p. 751-810. *In* S. S. Block (ed.). Disinfection, sterilization and preservation, 3rd ed. Lea and Febiger, Philadelphia, Pa.
58. Powers, E. M., C. Ay, H. M. El-Bisi, and D. B. Rowley. 1971. Bacteriology of dehydrated space foods. Appl. Microbiol. 22:441-445.
59. Raju, V. V. R., and V. K. N. Nambudripad. 1987. Incidence and growth of heat resistant coliforms bacteria in milk and other media. Indian J. Dairy Sci. 40:354-358.
60. Read, R. B., Jr., C. Schwartz, and W. Litsky. 1961. Studies on thermal destruction of *Escherichia coli* in milk and milk products. Appl. Microbiol. 9:415-418.
61. Shafi, R., O. J. Cotterill, and M. L. Nichols. 1970. Microbial flora of commercially pasteurized egg products. Poultry Sci. 49:578-585.
62. Steward, D. B. 1975. Factors influencing the incidence of *B. cereus* spores in milk. Soc. Dairy Technol. J. 28:80-90.

63. Stroup, W. H., R. W. Dickerson Jr., and R. B. Read Jr. 1969. Two-phase slug flow heat exchanger for microbial thermal inactivation research. Appl. Microbiol. 18:889-892.
64. Stumbo, C. R. 1965. Thermobacteriology in food processing. Academic Press, New York.
65. Thomas, S. B., R. G. Druce, G. J. Peters, and D. G. Griffiths. 1967. Incidence and significance of thermoduric bacteria in farm milk supplies: a reappraisal and review. J. Appl. Bacteriol. 30:265-298.
66. Tierney, J. T., and E. P. Larkin. 1978. Potential sources of error during virus thermal inactivation. Appl. Environ. Microbiol. 36:432-437.
67. Todd, E., A. Hughes, J. MacKenzie, R. Caldiera, T. Gleeson, and B. Brown. 1993. Thermal resistance of verotoxigenic *Escherichia coli* in ground beef—initial work, p. 93-109. *In* E. C. D. Todd, and J. M. Mackenzie (ed.) *Escherichia coli* O157:H7 and other verotoxigenic *E. coli* in food. Polyscience Publications, Inc. Ottawa, Ontario.
68. Tomlins, R. I., and Z. J. Ordal. 1976. Thermal injury and inactivation in vegetative bacteria, p. 153-190. *In* Inhibition and inactivation of vegetative microbes. Academic Press, New York.
69. Verrips, C. T., R. H. Kwast, and W. de Vries. 1980. Growth conditions and heat resistance of *Citrobacter freundii*. Antonie van Leeuwenhoek 46:551-563.
70. Washam, C. J., H. C. Olson, and E. R. Vedamuthu. 1977 Heat resistant psychrotrophic bacteria isolated from pasteurized milk. J. Food Prot. 40:101-108.
71. White, H. R. 1953. The heat resistance of *Streptococcus faecalis*. J. Gen. Microbiol. 8:27-37.
72. Wilder, C. J., and H. C. Nordan. 1957. A micro-method and apparatus for the determination of rates of destruction of bacterial spores subjected to heat and bactericidal agents. Food Res. 22:462-467.

Chapter 15

Lipolytic Microorganisms

Michael J. Haas

15.1 INTRODUCTION

15.11 Enzymatic Lipid Hydrolysis (Lipolysis)

Triglycerides are tri-esters of glycerol and three fatty acids. They are categorized as either fats (lipids that are solid at room temperature) or oils (lipids that are liquid at room temperature), and are common components of foods. Other types of lipids in foods include the fatty acid mono- and di- esters of glycerol, termed monoglycerides and diglycerides, respectively. These are usually generated as intermediates in the breakdown of fats and oils. Triglycerides have very low water solubilities, while the solubilities of mono- and diglycerides can be somewhat greater. Hydrolysis of the ester bonds of the tri-, di-, and monoglycerides (lipolysis) liberates free fatty acids (FFA). In food systems, such lipolysis is usually catalyzed by enzymes, generally by the group of enzymes known as lipases.

Lipases (Enzyme Commission identification number 3.1.1.3) are defined as those enzymes capable of hydrolyzing the carboxylic acid ester bonds of water-insoluble substrates. In contrast, enzymes that hydrolyze the ester bonds of water-soluble substrates are termed esterases. The biological role of lipases is to initiate the metabolism of fats and oils by reducing them to readily metabolized FFA and glycerol. Triglycerides are not absorbed intact into the cell, and thus their initial hydrolysis must occur extracellularly. For this reason, microbial lipases are generally excreted from the cell, and are found as extracellular enzymes.

When they are part of a glyceride molecule, fatty acids contribute little to the organoleptic properties of foods. However, when freed by lipolysis they can have substantial effects. Humans readily detect the shorter chain length fatty acids, up to about 10 carbons in length, by smell or taste. In some cases, e.g., dairy products, it is often desirable that some or all sizes of these shorter FFA be released by lipolysis of endogenous lipid, or be added during processing, since they confer characteristic flavor/fragrance notes. The characteristic flavors of fermented sausages are also due, in part, to lipolytic action. However, in other situations, these same notes are distinctly undesirable, and lipolytic release of these FFA can result in rejection or disposal of the material. In addition, longer chain fatty acids, particularly those containing double bonds, will oxidize following their lipolytic release from a glyceride, resulting in the generation of compounds generally deemed organoleptically undesirable in foods and beverages.

Some hydrolysis of the fats and oils in foods is non-microbial in origin, the result of spontaneous lipid hydrolysis and the action of lipases that are naturally present in many food materials.[30] Fatty acid oxidation can also generate undesirable flavors. In some cases, oxidation and/or the actions of endogenous lipases can play a larger role in spoilage than do microbial lipases. However, lipase production is a widespread trait of bacteria, yeasts, and molds. This ability to produce a lipase does not always result in lipolytic damage, since the synthesis or activity of the enzyme may be inhibited by components of the food, or by the conditions of incubation. Nonetheless, under appropriate conditions, lipases can be significant contributors to product deterioration.

In addition to lipases, which act on triglycerides, microorganisms can produce other lipid-hydrolyzing enzymes. Chief among these are the phospholipases, which convert phospholipids, the primary components of cell membranes, to FFA, lysophosphatides, mono- and diglycerides, glycerolphosphatides and simpler materials. The detection of phospholipases will not be a subject of this chapter. It is worth noting, however, that in some cases the presence of a phospholipase can stimulate the activity of a lipase. For example, phospholipase C from *Bacillus cereus* or *Pseudomonas fluorescens* enhances the lipolytic activities of both milk lipoprotein lipase and a commercial *Rhizopus* lipase, although not that of *P. fluorescens* lipase.[13,22]

Microorganisms that produce glycosidic enzymes can, in conjunction with bacterial proteases, degrade milk membranes and thereby expose the milk lipids to lipases.[42,43] Thus, the glycosidases can contribute indirectly to lipolytic activity. The glycosidases of *P. fluorescens*, in contrast to the phospholipase C and lipase produced by that organism, are completely inactivated by milk pasteurization temperatures[41] and therefore would not be expected to play a role in potentiating lipase activity in pasteurized or otherwise similarly heated samples.

By the use of special plating media, some of which are described here, microorganisms that produce lipase can be enumerated. Such enumeration is not usually performed on a routine basis. Food manufacturers and processors analyze for lipolytic microorganisms only when a problem arises. Determination of the number of lipolytic microorganisms present in a food sample can tell the food processor whether the particular lipid-related problem is microbial or non-microbial in origin.

During the 1990s considerable information was generated regarding the basic biochemistry and molecular biology of lipases, and these enzymes were extensively developed for use in the catalysis of desired reactions *in vitro*.[37,52,53] Several microbial lipase genes were cloned, and in some cases the expression of these cloned genes was also obtained. Analysis of these cloned genes provided tremendous insight into the structure and function of lipases. Expression of cloned genes allowed the production of substantial quantities of enzyme for characterization and various applications as applied catalysts. Subsequently, rational mutagenesis was employed to investigate the basis of enzyme activity and to generate altered lipases with enhanced catalytic or physical properties.

All known microbial lipases are active as single polypeptide chains, generally with molecular masses between 30 and 65 kiloDaltons. They do not require cofactors. The eukaryotic enzymes are glycosylated to a low degree, but glycosylation is not required for activity. X-ray crystallographic methods have been employed to elucidate the three-dimensional structures of several lipases.[54] Through numerous and sophisticated studies it has become evident that there is tremendous structural homogeneity among the lipases. Despite differences in size, substrate specificity, stability, and degrees of amino acid sequence homology, these enzymes form a highly conserved structural superfamily wherein the amino acid backbone assumes a characteristic three-dimensional structure, termed an α/β hydrolase fold.[54] This generates a plane of predominantly parallel β-pleated sheets that constitutes the core of the enzyme, and is surrounded to greater or lesser degrees by α helical strands. Based on sequence similarities, the microbial lipases can be divided into four subgroups within this general structural family.[16] A key feature of α/β hydrolase fold proteins is the placement of the side chain residues of a highly conserved serine, a histidine and an acidic amino acid (aspartic or glutamic) in correct proximity, and surrounded by appropriate other amino acid side chains, to form a characteristic triad that is proficient in hydrolysis of the carboxylic acid ester bonds of lipids, as well as the hydrolysis of similar chemical structures.[54] These hydrolytic reactions are freely reversible, with hydrolysis dominating in environments such as foods where the large water concentration drives the reaction to hydrolysis. However, lipases have been skillfully adapted by many researchers for use as applied catalysts to achieve the synthesis of esters, and similar chemical structures, under conditions of limiting water.[37]

In lipase three-dimensional structures determined to date, the catalytic triad sits at the bottom or near the mouth of a hydrophobic trough or cavity that serves as the substrate binding region. In the absence of lipid substrate, this region is occluded by a peptide loop of the enzyme (termed the 'lid') that sits over it and prevents access of substrate to the catalytic triad.[15] In the presence of lipid this loop moves, freeing the active site to bind substrate.[15] This provides a mechanistic explanation for the phenomenon of interfacial activation, the feature of lipases whereby they are inactive on soluble substrates but active against insoluble ones.

15.12 Foods Involved

Virtually any fat-containing food may spoil because of the generation of free fatty acids by lipolysis. This need not be enzymatic in origin. However, lipolysis by microbial lipases can be a problematic source of product loss. Although such losses are most prevalent in fat-rich foods, such as dairy products[57] and margarine,[11] any fat-containing food is susceptible to spoilage by microbial lipolysis.

15.13 Microorganisms

Due to the competitive advantages gained by producing lipases and thereby being able to metabolize lipids present in the environment, lipase production is common in the microbial world. Some genera particularly known for lipase production are *Pseudomonas, Alcaligenes, Moraxella*, and *Staphylococcus* among the bacteria; *Rhizopus, Geotrichum, Aspergillus, Mucor*, and *Penicillium* among the molds; and *Candida, Rhodotorula*, and *Hansenula* among the yeasts. However, lipase production is widespread in nature and not limited to these organisms.

We have found some laboratory strains of *E. coli* to be acceptable lipase-negative control organisms. Others have recommended the use of *Proteus mirabilis*.[14]

15.14 Lipase Stability

Failure to detect lipase-producing microorganisms in a spoiled fatty food may indicate that the defect is non-microbial in origin. However, since microbial lipases are generally exocellular, they can be found in the absence of viable microorganisms, which may have been killed by the food processing conditions. Moreover, microbial lipases are often heat resistant or able to renature following heat treatment,[1,57] although the cells themselves are not. Lipases from some psychrotrophic *Pseudomonas* species have D values, measured at 150° or 160°C, ranging from 4.8 to 0.7 min.[1,2,8,15,17,58] Lipases from *P. fluorescens* or *Moraxella* spp., when heated for 30 sec at 100°C, lost only from 2 to 60% of their activity.[20] Thus, selected sterilization temperatures may confer microbial sterility, but leave a considerable portion of the microbial lipase intact and active. This can subsequently lead to a loss of product quality. Milk and milk products sterilized by ultra high temperature (UHT) processing at temperatures of 138° to 150°C for 2 to 6 sec are a case in point. Raw milk held under refrigerated conditions for extended periods prior to UHT processing may support the growth of psychrotrophic *Pseudomonas* species, with concomitant production of lipases that are UHT-resistant or able to renature and regain activity upon return to ambient conditions. In such a case, although UHT treatment would destroy the pseudomonads, the lipase would retain or regain activity, resulting in lipolysis and loss of the product.

Lipases may remain active, and display their activity, in foods for long periods even at low temperatures. In cell-free samples from cultures of *Pseudomonas fragi, Staphylococcus aureus, Geotrichum candidum, Candida lipolytica* and *Penicillium roquefortii*, significant amounts of free fatty acids were liberated from fats over a 2 to 4-day period at –7°C, and within one week at –18°C.[3] Some of the cell-free cultures showed significant lipase action after 3 weeks at –29°C.

Reduced metabolic activity of microorganisms in the frozen state can be partly explained by lowered water activity (a_w) caused by ice formation. Microbial lipases, however, were found active in the frozen state,[3] which means that they are active at reduced a_w. Andersson[7] studied the activity of *Pseudomonas fluorescens* lipase in full-fat milk powder as a function of a_w. Enzymatic activity remained high, and constant, over an a_w range of 0.00 to 0.54. However, at an a_w of 0.85 lipase activity was three times greater than that at 0.54. The results obtained by Alford and Pierce[3] and by Andersson[7] indicate that microbial lipases can cause drastic changes in fatty substrates during frozen or dehydrated storage even though the lipase-producing species are dormant. It has subsequently been shown that lipases retain activity over a wide range of water activities,[61] a feature that has been exploited by their use as catalysts for the conduct of reactions in organic media.

15.15 *Detection of Lipases*

Determination of the numbers and types of microorganisms that can hydrolyze fats is not always sufficient to determine the lipolytic stability of a material, especially if the microorganisms present do not survive the processing conditions. Therefore, measurement of lipase activity will give a more realistic evaluation of the spoilage potential of a food in which lipolytic microorganisms have grown. A considerable number of methods (not all of which are suitable for the detection of lipase-producing organisms, and some of which require relatively pure enzyme) have been developed for the detection and quantitation of these enzymes.[4,9,10,12,18,21,25-27,29,3133,38,40,46,48,49,51,57,62-65] These assays take a variety of approaches,[33,57] which can be generally categorized as 1) the direct detection of fatty acids released from a lipid substrate by lipolysis (e.g., by titration,[25,29,49] by extraction and determination,[63] or by the recovery and quantitation of radioactive fatty acids from a radioactive substrate); 2) production and detection of some effect that results from lipid hydrolysis or the accumulation of free fatty acids in the medium (e.g., clearing of the turbidity of an emulsified lipid substrate,[4,21,27] an increase in turbidity due to the precipitation of the calcium salts of FFA, a color change in a medium containing a pH-indicator dye as a result of the acidification that results from the ionization FFA released by lipolysis, or a color or fluorescence change of a dye in the presence of FFA[21,32,38,51,62,64]; 3) the use of non-glyceride substrates (e.g., fatty acid esters of p-nitrophenol, β-naphthol, or 4-methylumbelliferone) that are themselves not colored or fluorescent but from which colored or fluorescent products are produced by lipase action[18,31,46,48,56,65]; 4) changes in the surface tension of a lipid monolayer upon lipase-mediated hydrolysis; and 5) immunological methods that employ antibodies directed against lipases to detect these enzymes.[26] Although known lipases demonstrate a very high degree of amino acid sequence conservation around the catalytic serine, this feature has not yet been exploited by the development of nucleic acid probe technologies for lipase detection.

Only one, very straightforward and generally useful, lipase assay will be described in detail here. The assay entails making an agar-solidified gel containing triglyceride, to serve as the lipase substrate, and the dye Victoria Blue B. The assay is of the general class termed "diffusion" assays, since the enzyme diffuses through an agar-solidified liquid medium, releasing free fatty acids when it encounters substrate glycerides. Victoria Blue accumulates in areas that contain lipid, and its blue color becomes visually brighter and more vibrant in the presence of free fatty acids. The surrounding area remains its initial blue (or pink at pH values above about 9.5) or clears due to dye accumulation in the lipid-rich regions.

The chemical basis of the distinctive color change of the dye remains to be identified. It is clear that the color change is associated with the presence of FFA. However, it is not due to bulk acidification by fatty acid ionization, since the color change is relatively independent of the pH of the solution. At any rate, the net result is that lipolytic activity can be readily detected visually by the appearance of a distinctive blue signal at sites within the assay gel.

When tributyrin is the lipase substrate, Victoria Blue need not be included. Butyric acid released by enzyme action is water-soluble, and a readily detected clear halo develops in the opaque assay media at sites of lipolysis. (However, see cautions below regarding the use of tributyrin in lipase detection.) With glycerides consisting of longer chain, water-insoluble, fatty acids, no such halos develop, and the incorporation of Victoria Blue is necessary to readily detect lipolysis.

With the addition of appropriate nutrients, this method is suitable for the detection of microorganisms that produce lipases. Many lipase-producing microbes position the enzyme on the outer surface of the cell, or excrete it, and lipolytic activity can be directly detected. In other cases, including at least some of the lactic acid bacteria,[36,45,55] the lipase is intracellular. Such enzymes can be detected by the method described here, and can cause lipolysis of food lipids, at the least following their release into the food upon the death and lysis of the cells.

The disadvantages of this method are relative insensitivity and the requirement for one or more days of incubation to detect activity. However, the simplicity and general utility of the protocol argue in favor of its use.

For the direct detection of lipase activity in samples that may contain the enzyme, aliquots of the material are spotted onto the surface of the medium or into wells cut into it. For the detection of lipase-producing microorganisms, samples are spread on the surface of the assay medium or incorporated into a nutrient agar overlay. In all cases, lipase activity becomes apparent as a clearing of the opacity of the medium (when tributyrin is the substrate) or by accumulation of intense blue color in the region of the sample (for substrates containing longer chain length fatty acids). The area of the zone of clearing is proportional to the concentration of the lipase.

An alternate plate method for the detection of lipases employs agar-solidified medium containing emulsified triglyceride and rhodamine B to detect lipolysis.[38] This pink fluorescent dye changes its fluorescence maxima in the presence of FFA. Following incubation, the plates are illuminated with ultraviolet light. Visually distinctive bright pink halos appear at the location of lipolytically active samples. The logarithm of lipase activity is linearly correlated with the diameter of the halos. This method can be used either in nutrient-free media to detect free lipase, or, by using nutrient agar, for the detection of lipolytic microorganisms.[32,38] The method has been very useful in our lab not only in the identification of natural isolates that produce lipase, but also in the detection of recombinant bacteria and bacteriophage harboring and expressing cloned lipase genes,[23] and for the rapid and sensitive location of lipolytic activity in fractions generated during lipase purification.[24] The rhodamine method is more sensitive than conventional indicator dye methods, and thus gives faster responses. However, this feature can be a disadvantage since the method is also more sensitive to the presence of free fatty acids in the lipid feedstock. This can result in high background fluorescence, which precludes visual detection of enzyme-induced lipolysis.

15.2 GENERAL CONSIDERATIONS

Samples are prepared and dilutions are made as outlined in Chapters 2 and 6.

15.3 EQUIPMENT, MATERIALS, AND REAGENTS

15.31 *Reagents*

Victoria Blue B (Color Index 44045; Sigma Chemical Co., St. Louis, MO). If Victoria Blue B is not available, Victoria Blue without an alphabetic subdesignation can be used, although a deep blue zone of lipolysis may not result.[4]

On rare occasions, it is necessary to remove contaminating FFA, the products of spontaneous lipid hydrolysis, from a substrate. If high backgrounds of apparent lipolytic activity are observed in plates to which no samples have been added, the following materials will be required to purify the lipid substrate:
Petroleum ether. ACS specifications. Boiling point 3° to 6°C.
Activated alumina (commercially available).

Glass columns 2 to 3 cm in diameter × 10 to 15 cm in length, packed with 25 g activated alumina that has been prewashed with petroleum ether. A column of this size will remove the FFA from 50 g of fat containing 2% to 3% FFA.

15.32 Fat Substrates

It is convenient to use reagent grade tributyrin for the detection of lipolytic activity. However, note the cautions below concerning its lack of specificity for this class of enzymes. Also note that although the directions for making assay media call for a final lipid concentration of 5%, some researchers have cautioned against the use of tributyrin in concentrations above 1%.[21]

If fresh, or known to otherwise contain only negligible amounts of FFA, purification is not necessary. If suspected or known to contain substantial amounts of FFA, purify as follows: Dissolve tributyrin in petroleum ether (5 to 10 g per 100 mL) and pass the solution through a column of activated alumina. Collect the effluent. Remove the petroleum ether from it under vacuum or by evaporation on a steam table under a nitrogen stream, resulting in purified triglyceride.

Corn, soybean, olive, or other liquid oils can be used as the fat substrate. Any fresh commercially available cooking oil also is suitable. In cases where the growth of microorganisms will be conducted, these substrates should not contain antioxidants or other materials at concentrations that inhibit growth. Rarely, the FFA level will be sufficiently high as to necessitate their removal by passage over alumina, as described for tributyrin.

Solid fats such as lard, tallow, butter, or a solid vegetable shortening (e.g., Crisco™) also may be used, particularly for examinations of foods in which only solid fats are present. If the fat is not of high quality, remove FFAs as described above. Again, high levels of background apparent activity in media to which no putative lipase-containing sample was added are diagnostic of the presence of FFA in the substrate. Melt solid fats before emulsification.

Other triglycerides, (e.g. triolein or other defined, synthetic glyceride) may also be used if for some reason the use of these more costly lipids is appropriate. Such substrates have been used, for example, in searches for lipases exhibiting novel substrate specificities.[35]

15.33 Media

Single layer agar with or without appropriate nutrient supplementation (Chapter 63).

Base layer agar, with nutrients if appropriate, or with a nutrient agar overlay (Chapter 63).

When using media containing lipids it is important to obtain a high level of emulsification of this substrate with the aqueous portion of the media.

15.4 PROCEDURES

15.41 Single Layer Method

Single layer plating procedures with a fat-containing growth medium will generally give satisfactory counts, and can provide visual clarity superior to that of double layer methods.[27] Note that in order to foster growth of organisms to be tested, nutrient agar or other microbial growth medium must be included when formulating the substrate. Single-layer plates are appropriate for the detection of free lipase using nutrient-free, agar-solidified, lipid emulsions as substrates. It is advisable to add buffer, at concentrations of 10 to 40 mM, to these media to provide a stable background pH for the assay.

Prepare pour plates using appropriately diluted food samples as described for the agar plate count method (Chapter 7), except pour the plates with 12 to 15 mL of the single layer medium containing the fat substrate (Chapter 63). The pour plate method involves dilution of a sample of interest into molten agar-fortified media, which solidifies as it cools. With incubation, microbial colonies appear as small foci throughout the medium. Incorporation of the organisms into the medium in this manner optimizes their enumeration. An alternate approach is to simply spread or streak samples on the surface of the solidified medium and incubate. When attempting to detect free lipase, it is sufficient to simply spot aliquots of samples of interest onto the test media or into wells cut into it.

15.42 Double Layer Method

The double layer method places a nutrient agar layer between the lipid-indicator dye and the sample being tested for the presence of lipase producing microorganisms. The advantage of this is that it increases the recovery of organisms whose growth is retarded by the indicator dye, the lipid substrate, or the products of lipolysis. Double layer agar plates give the best detection of weakly lipolytic bacteria and should be particularly used when it is necessary to add carbohydrate to obtain good growth.[4,11,21,60] Pour the base agar layer as described in Chapter 63. Prepare dilutions of the food product and plate in the manner described for colony counts (Chapter 6), except that dilutions are spread on the surface of the base layer and allowed to dry, followed by the addition of 10 to 12 mL of nutrient overlay agar into which the cells are suspended by swirling of the plates before the media cools sufficiently to solidify. Another approach[4,21,60] is to pour both layers of agar and inoculate the plates by spreading or streaking samples atop the solidified upper layer.

15.43 Incubation

Incubate plates at 20° to 30°C for 3 days if tributyrin is the fat substrate and four to seven days for other fats. Positive results may be apparent as early as one day after inoculation.

15.44 Counting and Reporting Colonies

Lipolytic microorganisms on tributyrin agar without Victoria Blue B are detected by a transparent zone surrounding the colony on an opaque background. Short chain, soluble, fatty acids diffuse through the agar and give a weak blue color if Victoria Blue is present.[4] On media containing long chain fatty acid triglycerides and Victoria Blue, a dark blue zone surrounds the lipolytic colonies against an opaque, light blue background. Indistinct zones sometimes occur because of weak lipolysis, the presence of free fatty acid in the lipid substrate, or the production of acid because of sugar metabolism (see Section 15.54). Use of a stereoscopic microscope to detect these weak zones may be helpful. However, careful substrate preparation may also help minimize the appearance of weak, indistinct zones of hydrolysis. In testing for the presence of lipolytic microorganisms, count only the colonies surrounded by a lipolytic zone, and report as numbers of lipolytic microorganisms per mL (or g) of food sample.

15.45 Recommended Controls

Control plates should be run for each batch of lipid substrate and dye. *Geotrichum candidum* and *Pseudomonas fragi* are often employed as positive control organisms. A positive result can also be generated by simply spotting a few microliters of FFA (e.g., oleic acid), readily available commercially, onto a plate. Organisms that do not produce lipase, for use as a negative control, can be more difficult to obtain. Proteus mirabilis has sometimes been employed in this capacity.[14] We have shown that *Escherichia coli* JM101 lacks lipolytic activity and serves as a useful lipase-negative control. Incubate plates for three to seven days at 20° to 25°C for *G. candidum* and *P. fragi* and at 30°C for *E. coli*.

Interpretation: Tributyrin may be hydrolyzed very slowly, if at all, by *G. candidum* since the mixture of lipase isoenzymes produced by this organism, in general, preferentially hydrolyzes ester bonds where the fatty acid component contains a 9-10 cis double bond. Thus, *G. candidum* often shows best lipolytic activity against such lipids as olive, corn and soybean oil, which contain substantial proportions of oleic, linoleic or linolenic acids. The presence of inhibitory substances in the fats or oils are indicated by weak zones of hydrolysis (not the dark blue of the typical Victoria Blue B reaction) for one or both of the known lipolytic organisms. No zones of hydrolysis should be seen with *P. mirabilis* and *E. coli*.

15.5 PRECAUTIONS

15.51 Assay Conditions

The temperature of incubation is important, since lipases are not particularly thermotolerant, and also since their production is sometimes inhibited in cultures incubated at the upper limits of their growth range.[6,39] Thus it is advised that incubations be conducted below 40°C, and at least 5°C below maximum growth temperature of the culture. When the samples to be plated represent unknown microorganisms such as might be present in foods, multiple plates should be prepared and incubated at 10°, 25°, and 30°C to ensure that all potential lipase-producing colonies are detected.

Lipase activity is pH dependent. The pH optima of known lipases vary from less than 5.5 to greater than 9.5. The pH of the solid lipase detection media described here is alkaline when no nutrient agar is present, and pH 6 to 7 when a nutrient agar-like test medium is prepared. Since the majority of known lipases display optimal activity at slightly alkaline pH values, there is a high probability that any lipase present will be detected using the method recommended here. However, if it is desired to be thorough, lipase searches should also be conducted using test media adjusted to more acidic pH values (e.g., 5.5). This is particularly the case when the sample being examined for lipases, or in which lipolytic activity would cause problems, has a pH in these ranges. It is always appropriate to test the effectiveness of the assay being employed when assay parameters, such as pH, are modified. Such control tests can be conducted using microorganisms known to produce lipase, or with commercially available lipase preparations, or by direct application of FFA.

15.52 Substrate

True lipases attack only water-insoluble substrates, such as lipids. Esterases attack only water-soluble substrates. Therefore, avoid water-soluble or partially water-soluble substrates such as simple esters, monoglycerides, and Tweens, although they may be easier to prepare as substrates, and yield sharper zones of hydrolysis than when true lipids are employed. Hydrolysis of these compounds indicates esterase activity, not true lipase activity. Even the use of tributyrin, the glycerol triester of the four carbon acid butyric acid, as a substrate is controversial, and must be undertaken with caution. This is the simplest triglyceride occurring in natural fats and oils. (Although a true fat, it is hydrolyzed not only by lipases but also by esterases.) For lipase screening it is the substrate of choice[4,21,47,50] because positive results are so readily detected by formation of zones of clearing around positive cells. However, it is strongly advised that indications of lipolytic activity obtained using tributyrin be viewed as only presumptive, and that they be confirmed by repeating the determination using a lipid composed primarily of longer chain fatty acids, e.g., a vegetable oil or animal fat. As noted in Section 15.32, other substrates may give more meaningful information concerning a specific product (e.g., lard as a substrate for lipolytic microorganisms from pork). For a comparison of the use of tributyrin with other fats see Fryer et al.,[21] Lawrence,[39] and Thomas and Thomas.[59] The use of tallow as a lipase substrate in plate screening has also been described.[32]

15.53 Lipase Specificity

Most lipases are rather non-specific with respect to the natural fats or oils that they can hydrolyze. However, a number of bacterial and fungal lipases exhibit a preference for the 1-position of synthetic triglycerides.[5] Purified lipase from *P. fragi* attacked preferentially the 1-position of synthetic triglycerides and also hydrolyzed diglycerides and monoglycerides.[44] Lipases produced by members of the fungal genera *Rhizopus* and *Rhizomucor* hydrolyze primary esters (such as at the 1- and 3- positions in triglycerides), but not secondary esters, such as those at the 2-position of triglycerides. Despite these positional hydrolytic specificities, all these types of lipases would be detected by the plate methods described here. Early studies with *G. candidum* found that this organism liberated large amounts of oleic and linoleic acids during growth on media containing corn oil or lard,[34] suggesting that the lipase produced by this organism hydrolyzed the glyceride esters of unsaturated fatty acids. When the *Geotrichum* lipase was tested against synthetic triglycerides containing oleic (18:1), palmitic (16:0), and stearic acids (18:0), a preferential liberation of oleic acid occurred, regardless of its position in the triglyceride.[5,34] The structural basis for this specificity of some *Geotrichum* lipases for 9-10 cis-unsaturated fatty acids was subsequently established, and has now been elegantly defined.[28] Other lipases such as those of *Aspergillus flavus*, *Candida rugosa* (formerly *C. cylindracea*) and *Staphylococcus aureus* hydrolyze all three ester bonds of their substrates, showing no gross positional preferences.[5,39] Substantial progress has been made in the discovery and characterization of lipases, and their development as applied catalysts.[19,37] Having identified a lipase-producing strain, characterization of the enzyme should include determination of the existence of positional and/or fatty acid specificity. The results of such an analysis may suggest practical applications for the enzyme.

15.54 Growth Medium

Lipase production by some microorganisms is inhibited by the presence of readily fermentable carbohydrates,[6,39] while other microorganisms require these for growth.[11,60] The amount of carbohydrate included in a medium should be limited to that required for reasonable growth. If the microorganism will grow on nutrient agar, it is the medium of choice. If not, media such as standard methods agar or tryptic soy agar, both of which contain limited amounts of carbohydrate, may be used.

15.55 Emulsification

Simple addition of the lipid to the rest of the assay medium prior to pouring the plates is not advised. The lipid will not mix well with the aqueous phase, and there will be little of the lipid-aqueous phase interface where lipase action occurs. Thus, the sensitivity of the assay will be poor. For this reason, the directions for making up the media include emulsification by blending (e.g., with a Waring Blender, available from standard laboratory supply houses). Ultrasound, as with a common lab sonicator (Heat Systems-Ultrasonics, Inc., Farmingdale, NY) may also be employed to achieve emulsification. The goal is a stable opaque emulsion with little or no floating layer of free lipid. The agar in the media typically serves as an ample stabilizer of the emulsion once it is formed, and no further stabilizers, such as the gum arabic typically employed in liquid-phase reactions, are usually necessary. Polyvinyl alcohol 13 and polyoxyethylene-(20)-hydrogenated castor oil[50] have been recommended as emulsifiers for the preparation of tributyrin agar if needed. Gum arabic at a final concentration of 1% (w/v) is also a suitable emulsifier. Other emulsifiers such as Tweens, monoglycerides, and diglycerides should be avoided because they can serve as substrates for some esterases, leading to false positive results.

15.56 Indicator Dyes

Victoria Blue B, Spirit Blue, Nile Blue sulfate, Night Blue, and other dyes have been used as indicators of fat hydrolysis. However, toxicity to one or more microorganisms has been reported for all of them.[4,11,39] Victoria Blue B has little toxicity, and is the dye of choice when a dye indicator is necessary.

15.6 REFERENCES

1. Adams, D. M. 1979. The role of heat resistant bacterial enzymes in UHT processing, p. 89-105. *In* Proceedings of International Conference on UHT Processing and Aseptic Packaging of Milk and Milk Products, November 27-29, 1979, Dept. Food Science, N. C. State Univ., Raleigh, N. C.
2. Adams, D. M., and T. G. Brawley. 1981. Heat resistant bacterial lipases and ultra-high temperature sterilization of dairy products. J. Dairy Sci. 64:1951-1957.
3. Alford, J. A., and D. A. Pierce. 1961. Lipolytic activity of microorganisms at low and intermediate temperatures. III. Activity of microbial lipases at temperatures below 0 C. J. Food Sci. 26:518-524.
4. Alford, J. A., and E. E. Steinle. 1967. A double layered plate method for the detection of microbial lipolysis. J. Appl. Bacteriol. 30:488-494.
5. Alford, J. A., D. A. Pierce, and F. G. Suggs. 1964. Activity of microbial lipases on natural fats and synthetic triglycerides. J. Lipid Res. 5:390-394.
6. Alford, J. A., J. L. Smith, and H. D. Lilly. 1971. Relationship of microbial activity to changes in lipids of foods. J. Appl. Bacteriol. 34:133-146.
7. Andersson, R. E. 1980. Microbial lipolysis at low temperatures. Appl. Environ. Microbiol. 39:36-40.
8. Andersson, R. E., C. B. Hedlund, and U. Jonsson. 1979. Thermal inactivation of a heat resistant lipase produced by the psychrotrophic bacterium, *Pseudomonas fluorescens*. J. Dairy Sci. 62:361-367.
9. Arzoglou, P. L., A. Tavridoiu, and C. Balaska. 1989. Rapid turbidimetric determination of lipase activity in biological fluids. Anal. Lett. 22:1459-1469.
10. Blake, M. R., R. Koka, and B. C. Weimer. 1996. A semiautomated reflectance colorimetric method for the determination of lipase activity in milk. J. Dairy Sci. 79:1164-1171.
11. Bours, J., and D. A. A. Mossel. 1973. A comparison of methods for the determination of lipolytic properties of yeasts mainly isolated from margarine, moulds, and bacteria. Arch. Lebensmittelhyg. 24:197-203.
12. Brune, K. A. and F. Gotz. 1992. Degradation of lipids by bacterial lipases, p. 244-266. *In* G. Winkelmann (ed.) Microbial degradation of natural products. VCH, New York.
13. Chrisope, G. L., and R. T. Marshall. 1976. Combined action of lipase and microbial phospholipase C on a model fat globule emulsion and raw milk. J. Dairy Sci. 59:2024-2030.
14. Collins, C. H, P. M. Lyne, and J. M. Grange. 1995. Collins and Lyne's microbiological methods, p. 114. 7th ed. Butterworth-Heinemann Ltd., Linacre House, Jordan Hill, Oxford, England.
15. Cygler, M., and J. D. Schrag. 1997. Structure as basis for understanding interfacial properties of lipases, p. 3-27. *In* B. Rubin, and E. A. Dennis (ed.) Lipases, Part A: Biotechnology, methods in enzymology, vol. 284. Academic Press, New York.
16. Cygler, M., J. D. Schrag, and F. Ergan. 1992. Advances in structural understanding of lipases. Biotech. Genetic Eng. Revs. 10:143-184.
17. Driessen, F. M., and J. Stadhouders. 1974. Thermal activation and inactivation of exocellular lipases of some gram-negative bacteria common in milk. Neth. Milk Dairy J. 28:10-22.
18. Duque, M., M. Graupner, H. Stutz, I. Wicher, R. Zechner, F. Paltauf, and A. Hermetter. 1996. New fluorogenic triacylglycerol analogs as substrates for the determination and chiral discrimination of lipase activities. J. Lipid Res. 37:868-876.
19. Eigtved, P. 1992. Enzymes and lipid modification, p. 1-64. *In* F. B Padley (ed.) Advances in applied lipid research, vol. 1. JAI Press, Ltd. London.
20. Fitz-Gerald, C. H., and H. C. Deeth. 1983. Factors influencing lipolysis by skim milk cultures of some psychrotrophic microorganisms. Aust. J. Dairy Technol. 38:97-103.
21. Fryer, T. F., R. C. Lawrence, and B. Reiter. 1967. Methods for isolation and enumeration of lipolytic organisms. J. Dairy Sci. 50:477-484.
22. Griffiths, M. W. 1983. Synergistic effects of various lipases and phospholipase C on milk fat. J. Food. Technol. 18:495-505.
23. Haas, M. J., J. Allen, and T. R. Berka. 1991. Cloning, expression and characterization of a cDNA encoding a lipase from *Rhizopus delemar*. Gene 109:107-113.
24. Haas, M. J., D. J. Cichowicz, and D. G. Bailey. 1992. Purification and characterization of an extracellular lipase from the fungus *Rhizopus delemar*. Lipids 27:571-576.
25. Haas, M. J., D. Esposito, and D. J. Cichowicz. 1995. A software package to streamline the titrimetric determination of lipase activity. J. Am. Oil Chem. Soc. 72:1405-1406.
26. Hafkenscheid, J. C. M., M. Hesels, P. J. S. M. Derstens, and C. B. H. W. Lamers. 1989. An enzyme immunoassay for the determination of lipase in human duodenal fluid. Pancreas 4:90-94.
27. Harris, P. L., S. L. Cuppett, and L. B. Bullerman. 1990. A technique comparison of isolation of lipolytic bacteria, J. Food Prot. 53:176-177.
28. Holmquist, M., D. C. Tessier, and M. Cygler. 1997. Identification of residues essential for differential fatty acyl specificity of *Geotrichum candidum* lipases I and II. Biochemistry 36:15019-15025.
29. Hoppe, A., and R. R. Theimer. 1996. Titrimetric test for lipase activity using stabilized triolein emulsions, Phytochemistry 42:973-978.
30. Huis in't Veld, J. H. J. 1996. Microbial and biochemical spoilage of foods: an overview. Int. J. Food Microbiol. 33:1-18.
31. Humbert, G., M. F. Guingamp, and G. Linden. 1997. Method for the measurement of lipase activity in milk. J. Dairy Res. 64:465-469.
32. Jarvis, G. N., and J. H. Thiele. 1997. Qualitative rhodamine B assay which uses tallow as a substrate for lipolytic obligately anaerobic bacteria. J. Microbiol. Methods 29:41-47.
33. Jensen, R. G. 1983. Detection and determination of lipase (acylglycerol hydrolase) activity from various sources. Lipids 18:650-657.

34. Jensen, R. G., J. Sampugna, J. G. Quinn, D. L. Carpenter, T. A. Marks, and J. A. Alford. 1965. Specificity of a lipase from *Geotrichum candidum* for cis-octadecenoic acid. J. Am. Oil Chem. Soc. 42:1029-1032.
35. Joerger, R. D., and M. J. Haas. 1994. Alteration of chain length selectivity of a Rhizopus delemar lipase through site-directed mutagenesis. Lipids 29:377-384.
36. Kamaly, K. M., K. Takayama, and E. H. Marth. 1990. Acylglycerol acylhydrolase (lipase) activities of *Streptococcus lactis, Streptococcus cremoris,* and their mutants. J. Dairy Sci. 73:280-290.
37. Kazlauskas, R. J., and U. T. Bornscheuer. 1998. Biotransformations with lipases, p. 37-191. In Rehm, H. J. and Reed, G. (ed.) Biotechnology, 2nd ed., vol. 8, VCH Publishers, New York.
38. Kouker, G., and K. -E. Jaeger. 1987. Specific and sensitive plate assay for bacterial lipases. Appl. Environ. Microbiol. 53:211-213.
39. Lawrence, R. C. 1967. Microbial lipases and related esterases. Dairy Sci. Abstr. 29:1-8.
40. Lima, N., J. A. Teixeira, and M. Mota. 1991. Deep agar-diffusion test for preliminary screening of lipolytic activity of fungi. J. Microbiol. Methods 14:193-200.
41. Marin, A., and R. T. Marshall. 1983. Characterization of glycosidases produced by *Pseudomonas fluorescens* 26. J. Food Prot. 46:676-680.
42. Marin, A., and R. T. Marshall. 1983. Production of glycosidases by psychrotrophic bacteria. J. Food Sci. 48:570-573.
43. Marin, A., T. P. Mawhinney, and R. T. Marshall. 1984. Glycosidic activities of *Pseudomonas fluorescens* on fat-extracted skim milk, buttermilk, and milk fat globule membranes. J. Dairy Sci. 67:52-59.
44. Mencher, J. R., and J. A. Alford. 1967. Purification and characterization of the lipase of *Pseudomonas fragi*. J. Gen. Microbiol. 48:317-328.
45. Meyers, S. A., S. L. Cuppett, and R. W. Hutkins. 1996. Lipase production by lactic acid bacteria and activity on butter oil. Food Microbiol. 13:383-389.
46. Mosmuller, E. W. J., J. D. H. Van Heemst, C. J. Van Delden, M. C. R. Franssen, and J. F. J. Engbersen. 1992. A new spectrophotometric method for the detection of lipase activity using 2,4-dinitrophenyl butyrate as a substrate. Biocatalysis 1992:279-287.
47. Mourey, A., and G. Kilbertus. 1976. Simple media containing stabilized tributyrin for demonstrating lipolytic bacteria in foods and soils. J. Appl. Bacteriol. 40:47-51.
48. Negre, A., A. Dagan, and S. Gatt. 1989. Pyrene-methyl lauryl ester, a new fluorescent substrate for lipases: use for diagnosis of acid lipase deficiency in Wolman's and cholesteryl ester storage diseases. Enzyme 42:110-117.
49. Peled, N., and M. C. Krenz. 1981. A new assay of microbial lipases with emulsified trioleoyl glycerol. Anal. Biochem. 112:219-222.
50. Rapp, M. 1978. Elektive nahrmedien zurn nachweis von lipolyten. Milchwissenschaft 33:493-496.
51. Rawyler, A., and P. A. Siegenthaler. 1989. A single and continuous spectrophotometric assay for various lipolytic enzymes, using natural, non-labeled lipid substrates. Biochim. Biophys. Acta 1004:337-344.
52. Rubin, B., and E. A. Dennis. (ed.) 1997. Lipases, Part A: Biotechnology. Methods in enzymology, vol. 284. Academic Press, New York.
53. Rubin, B., and E. A. Dennis (ed.) 1997. Lipases, Part B: Enzyme characterization and utilization. Methods in enzymology, vol. 287. Academic Press, New York.
54. Schrag, J. D., and M. Cygler. 1997. Lipases and α/β fold, p. 85-106. *In* B. Rubin and E. A. Dennis (ed.) Lipases, Part A: Biotechnology, methods in enzymology, vol. 284. Academic Press, Inc., New York.
55. Sorenson, B. B. 1997. Lipolysis of pork fat by the meat starter culture Staphylococcus xylosus at various environmental conditions. Food Microbiol. 14:153-160.
56. Stead, D. 1984. A new method for measuring lipase activity from psychrotrophic bacteria in milk. Dairy Ind. Inst. 49:29.
57. Stead, D. 1986. Microbial lipases: their characteristics, role in food spoilage and industrial uses. J. Dairy Res. 53:481-505.
58. Stepaniak, L., S.-E. Birkeland, T. Sorhaug, and G. Vagias. 1987. Isolation and partial characterization of heat-stable proteinase, lipase and phospholipase C from *Pseudomonas fluorescens* P1.16. Milchwissenschaft 42:75-79.
59. Thomas, S. B., and B. F. Thomas. 1975. The bacteriological grading of bulk collected milk. Part 8. Differential and selective agar media. Dairy Ind. Int. 40:397-399.
60. Umemoto, Y. 1969. A method for the detection of weak lipolysis of dairy lactic acid bacteria on double layered agar plates. Agric. Biol. Chem. Jpn. 33:1651-1653.
61. Valivety, R. H., P. J. Halling, A. D. Peilow, and A. R. Macrae. 1992. Lipases from different sources vary widely in dependence of catalytic activity on water activity. Biochim. Biophys. Acta 1122:143-146.
62. Van Autryve, P., R. Ratomahenina, A. Riaublanc, C. Mitrani, M. Pina, J. Graille, and P. Galzy. 1991. Spectrophotometry assay of lipase activity using rhodamine 6G. Oleagineux 46:29-31.
63. Veeraragavan, K. 1990. A simple and sensitive method for the estimation of microbial lipase activity. Anal. Biochem. 186:301-305.
64. Wilton, D. C. 1990. A continuous fluorescence displacement assay for the measurement of phospholipase A2 and other lipases that release long-chain fatty acids. Biochem. J. 266:435-439.
65. Yeoh, H. H., F. M. Wong, and G. Lim. 1986. Screening for fungal lipases using chromogenic lipid substrates. Mycologia 78:298-300.

Chapter 16

Proteolytic Microorganisms

John A. Marcy and W. Payton Pruett, Jr.

16.1 INTRODUCTION

16.11 Protein Hydrolysis

Protein hydrolysis by microorganisms in foods may produce a variety of odor and flavor defects. During protein hydrolysis, proteins are degraded through proteoses, peptones, polypeptides, and dipeptides to amino acids.[3] The further dissimulation of amino acids leads to the foul odors characteristic of many spoiled foods.

Some psychrotrophic spoilage bacteria (e.g., *Acinetobacter, Flavobacterium, Pseudomonas,* and *Shewanella*) are strongly proteolytic and cause undesirable changes in dairy, meat, poultry, and/or seafood products, particularly when high populations are reached after extended refrigerated storage. Conversely, microbial proteolytic activity may be desirable in certain foods, such as in the ripening of cheese, where it contributes to the development of flavor, body, and texture.

Opinions differ about the usefulness of proteolytic counts to evaluate quality losses of refrigerated dairy, meat, poultry, and fishery products.[10,11,14] In some foods (e.g., fish fillets and fluid milk), the level of proteolytic microorganisms may be of value in predicting refrigerated storage life and assessing processing methods.[12,14]

Psychrotrophic bacteria that produce heat-stable proteases in milk were reviewed by Speck and Adams.[24] They indicated that most raw milk supplies may contain proteolytic enzymes that remain active after ultra-high temperature (UHT) sterilization, and that psychrotrophs (mainly pseudomonads) produced these proteases. Mayerhofer et al.[15] characterized a strain of *Pseudomonas fluorescens* that formed a protease of much greater heat resistance than the organism itself. While proteolysis was favored at higher temperatures (about 40°C), activity was still pronounced at refrigeration temperatures.[8]

Heat-stable proteases may adversely affect the quality of dairy products and foods that contain dairy ingredients. The onset of UHT milk spoilage by heat-stable proteases was associated with the number of *Pseudomonas* spp. present in raw milk, with higher initial counts leading to spoilage sooner.[11] As a result of proteolysis, UHT milk stored at 20°C could gel or develop sediments after a few days to a few weeks of storage. Flavor quality and textural problems in cheeses, buttermilk, and yogurt have also been attributed to heat-stable protease activity.

16.12 Microorganisms

Proteolytic species are common among the genera *Acinetobacter, Bacillus, Clostridium, Enterobacter, Flavobacterium, Micrococcus, Pseudomonas,* and *Proteus*; there are also proteolytic yeasts and molds (Chapter 20). Microorganisms that carry out protein hydrolysis and acid fermentation are called acid-proteolytic, e.g., *Enterococcus faecalis* and *Micrococcus caseolyticus.*[7]

16.13 Skim Milk Agar

The hydrolysis of casein in an opaque skim milk agar is used to determine proteolysis by microorganisms on or in agar plates[6]. Colonies of proteolytic bacteria will be surrounded by a clear zone as a result of the conversion of casein into soluble nitrogenous compounds. However, clear zones on milk agar can be created by bacteria that produce acid from fermentable carbohydrates.[6] The clear zone on common milk agar reflects only the more complete breakdown of casein, since the early stages of proteolysis cannot be detected against the opaque background. To confirm proteolysis, a chemical protein precipitant (dilute acid solution) is added to the agar surface to precipitate any undigested casein.

An improved skim milk agar was developed[14] by adding sodium caseinate, trisodium citrate, and calcium chloride to standard methods agar. Its greater sensitivity is related to the detection of the early stages of casein breakdown, the formation of a zone of precipitation (insoluble paracaseins) in a transparent medium. This medium is buffered sufficiently to reduce the occurrence of false positive zones caused by acid production. No significant differences were found between total counts of raw milk on standard methods agar and the standard methods caseinate agar. The latter medium then can be used for the simultaneous determination of total and proteolytic counts.

16.14 Gelatin Agar

Numerous methods have been used to detect gelatin hydrolysis by microorganisms, among them gelatin liquefaction (gelatin stab), and the detection of hydrolyzed gelatin in agar media with or without chemical protein precipitants. Pitt and Dey[18] developed a gelatin agar for the detection of gelatinase without the use of a chemical precipitant.

A double-layer gelatin medium with a soft agar gelatin overlay is available that detects both weak and strong gelatinolytic bacteria in a direct plating procedure.[12] Advantages claimed for the medium are [1] rapid diffusion of gelatinolytic enzymes through the soft overlay, creating large zones of clearing; [2] reduction in swarming; and [3] more rapid colony development than with pour plates. If enumeration of proteolytic bacteria is desired from dairy or seafood products, incorporation of skim milk or fish juice in the overlay will detect proteolysis without the application of a chemical precipitant. This will be advantageous if further purification or identification of colonies is needed.

A double-layer plating technique can also be applied when one or more components of the growth medium are not compatible with the protein used to check hydrolysis. A two-layer plate was developed[23] with a base indicator layer of milk agar and an upper layer of marine agar to detect proteolytic marine bacteria. Samples are placed on the upper layer by the spread plate method. No chemical precipitant is required to detect proteolysis.

16.15 Litmus Milk Assay

The litmus milk assay is a simple test that has been extensively used to biochemically characterize microorganisms. A variety of reactions, primarily involving lactose and casein, can be observed from litmus milk.[4,9] For example, acid production from sugar metabolism changes the color of litmus to pink; if sufficient acid is produced, the milk will form an acid clot or curd. Gas production will be evident by any breaks or bubbles within the curd.

Proteolytic enzyme (rennet) activity is indicated by the coagulation of casein, with the litmus dye remaining blue. This reaction is known as sweet curdling. Hydrolysis or peptonization of casein causes clearing of the milk. If reduction of litmus occurs, the leuco form of the dye will develop and the medium will become colorless.

16.16 Trinitrobenzenesulfonic Acid (TNBS) Method

Liberated amino acids from proteolysis can be detected by the TNBS method. The procedure is a spectrophotometric assay of the chromophore formed by the reaction of TNBS with primary amines.[1] The reaction is conducted under slightly alkaline conditions and is terminated by lowering the pH. Proteolysis is characterized as the increase in the concentration of trichloroacetic acid-soluble free amino groups per mL of sample.

16.17 Thin-Layer Enzyme Assay (TEA)

The TEA determines proteolytic activity by using radial diffusion in an agar gel in combination with a protein substrate absorbed to polystyrene petri dish.[26] The principle of the method relies on the fact that proteolytic digestion reduces the thickness of the protein layer on the polystyrene surface, and thus, the wettability. The wettability of the enzyme-affected surface is decreased compared to the unaffected protein surface. The amount of protein on the surface decreases in relation with an enzyme-catalyzed reaction. Proteolytic activity is easily demonstrated by condensation of water vapor on the surface after removal of the agar gel.

16.18 Hide Powder Azure (HPA) Assay

Extracellular proteinases produced by microorganisms in milk can be assayed directly using HPA as a substrate.[20,22] HPA is a general proteinase substrate prepared from collagen. The collagen is denatured by a covalently bound dye, making it more susceptible to proteolytic action. The observation of a blue color in a milk sample indicates proteolytic activity on the HPA substrate as dye-labeled peptides are released from the protein substrate. The HPA assay has been reported to detect protease produced by as few as 1.5×10^6 bacteria per mL.[5]

16.2 GENERAL CONSIDERATIONS

Prepare samples and appropriate dilutions as described for the agar plate count (Chapter 7) and for the hide powder azure (HPA) method (Section 16.47).

16.3 EQUIPMENT, MATERIALS, AND REAGENTS

16.31 Media

Skim milk agar[13]
Standard methods caseinate agar[14]
Soft agar gelatin overlay medium[12]
Soft agar basal medium[12]
Litmus milk[2]
Thin-layer enzyme assay (TEA) plate[26]

16.32 Reagents

Acetic acid, 10 %
Acetic acid, 5 %
Bovine serum albumin, 10 mg/mL
Diethyl ether
Glycine
HCl, 1 %
Hide powder azure (HPA)
Monobasic sodium phosphate, 2 M
NaOH, 1.0, 0.5, and 0.1 M
Potassium borate buffer, 1 M (pH 9.2)
Sodium sulfite, 18 mM
Trichloroacetic acid, 0.72 N
Trichloroacetic acid, 2 %
Trinitrobenzenesulfonic acid, 5 mM
Tris buffer, 0.05 M
Trypsin, 10 µg/mL

16.4 PROCEDURES

16.41 Skim Milk Agar Method

Incubate spread or pour plates of skim milk agar for 72 hr at 21°C (or use the incubation conditions recommended for the food under study). Flood the plates with 1% HCl or 10% acetic acid solution for 1 min.[13] Pour off excess acid solution, then count colonies that are surrounded by clear zones produced by proteolysis.

16.42 Standard Methods Caseinate Agar Method[14]

Place appropriate dilutions of sample in 0.1 mL quantities on the surface of standard methods caseinate agar plates and distribute evenly by spreading with a sterile bent glass rod. Allow plates to dry for 15 min. Incubate plates for 24 to 72 hr at 30°C. To enumerate proteolytic psychrotrophic bacteria, incubate plates for 10 days at 7°C. Count colonies that form white or off-white precipitate around the colony. (Organisms that are strongly proteolytic can further break down the precipitate to soluble components with the formation of an inner transparent zone.)

16.43 Soft Agar Gelatin Overlay Method[12]

Place appropriate dilutions of sample in 0.1 mL quantities on the surface of the basal medium and distribute uniformly by spreading with a sterile bent glass rod. Pour 2.5 mL of melted gelatin overlay medium and distribute evenly over surface of agar. Incubate plates for 3 days at 20°C. Flood surface of plate with 10 mL of 5% acetic acid for 15 min. Count colonies with clear zones.

16.44 *Litmus Milk Assay*[9]

Transfer a well-isolated colony to a tube containing litmus milk. Incubate the tube for up to 14 days at the microorganism's optimum temperature. If proteolysis has occurred, coagulation of the milk will be evident and the litmus dye will remain blue. Peptonization of casein is indicated by clearing of the milk.

16.45 *Trinitrobenzenesulfonic Acid Method*[16]

To duplicate 2-mL samples of milk, add 4 mL of 0.72 N trichloroacetic acid (TCA) After incubating for 20 min at 25°C, filter the samples through filter paper (Whatman #1). Mix duplicate 0.2-mL volumes of each supernatant with 2 mL of 1 M potassium borate buffer (pH 9.2) and 0.8 mL of 5 mM TNBS. Incubate the mixtures in the dark at 25°C. As standard solutions, include duplicate, unfiltered samples of glycine treated with TCA. After 30 min, add 0.8 mL of 2 M monobasic sodium phosphate containing 18 mM sodium sulfite.

Measure the absorbance of the solutions at 420 nm. Absorbances are converted to (mol of free amino acid groups per mL of milk using a standard curve. Proteolysis is defined as the increase in the concentration of trichloroacetic acid-soluble free amino groups per mL of milk.

16.46 *Thin-Layer Enzyme Assay (TEA*[26]*)*

Pour an agar medium that will support optimal growth of the test microorganism into a protein-coated polystyrene petri dish. The protein-coated surface can be prepared by mixing 1 mL of a bovine serum albumin (BSA, 10 mg/mL) solution with 9 mL of sterile distilled water on a polystyrene petri dish. After 30 min at room temperature, discard the BSA solution. Rinse the protein-coated petri dish with sterile distilled water and immediately dry the surface with sterile compressed air.

Spot inoculate with the microorganism or add culture supernatant and enzyme (e.g., trypsin) solution controls to 3 mm-diameter wells that have been aseptically drilled into the agar. After incubation at an optimal time and temperature, remove the agar, wash the bottom of the petri dish with distilled water, and blow dry with compressed air. Expose the protein-coated surface to water vapor at 50°C for 1 min. Afterwards, measure the area of water vapor condensation (proteolysis) with a ruler.

16.47 *Hide Powder Azure (HPA) Assay*[5]

Adjust the pH of suspected milk to 8.3 with 1.0, 0.5, or 0.1 M NaOH. Add 100 mg of HPA (Sigma Chemical Co.) to 5 mL of milk in a screw-capped tube. Incubate the tube for 5 hr at 37°C. Shake occasionally to keep the HPA in suspension. Terminate the reaction by cooling in ice water.

Centrifuge the tube at 800 × g for 5 min, add 1 mL of diethyl ether, and shake vigorously for 1 min. Centrifuge at 800 × g for 15 min and remove the ether layer carefully with a Pasteur pipette. Repeat the ether extraction two more times. Finally, centrifuge the aqueous phase at 150,000 × g for 60 min to remove casein. Instead of centrifugation, the casein may be precipitated with 2% TCA.[17]

Measure the blue dye released from HPA by reading the absorbance at 595 nm. If this assay is performed in 0.05 M Tris or similar buffer, the insoluble HPA may be removed by centrifugation or filtration.

16.5 PRECAUTIONS

The most serious disadvantage of the milk agar method is that the early stage of casein hydrolysis is not apparent until the medium is flooded with a chemical protein precipitant. Flooding with a chemical precipitant is required to confirm proteolysis, as clearing may result from acids produced by the fermentation of carbohydrates. Treatment with the chemical precipitant prevents the isolation of colonies for further study. However, if isolate retention is necessary, replicate plating procedures can be employed.

Simultaneous determination of total and proteolytic counts on the same plate is difficult if the ratio of the two counts varies widely. The zones of clearing may not be distinct unless the colonies are well separated. In this case, the plates that contain fewer colonies must be counted, with resulting loss of accuracy. In addition, weakly proteolytic bacteria may not be detectable unless the plate incubation period is extended, sometimes beyond the optimum length for the total count. This difficulty may be further compounded if a mixture of bacteria with widely varying proteolytic activities is present on the same plate.

The litmus milk assay can reveal substantial information about the proteolytic (rennet production or peptonization) nature of microorganisms. The assay is simple and requires few materials and little manipulation. However, reactions may not be observed for up to 14 days.

TEA is extremely sensitive because of the thinness of the protein layer. Enzyme activity quickly removes the hydrophilic film and restores the hydrophobic polystyrene surface. TEA can be used to study enzyme kinetics as well as to screen the proteolytic activity of microorganism.[25] The concentration of protein that can be absorbed on the polystyrene surface is self-limiting, thus requiring low concentrations of the substrate.

The agar concentration recommended for TEA is 2%. Beyond this level, the rate of diffusion is affected. Some media (e.g., Rogosa agar) decrease the wettability of the protein-coated surface, and thus, cannot be used with TEA.

16.6 INTERPRETATION

Ideally, the proteolytic activity of a microorganism should be measured against the specific protein(s) of the food being examined. The temperature of plate incubation should reflect that of the food during the time that microbial proteolytic activity is expected or has taken place. Convenience and the need for standardization have limited the media mainly to those containing gelatin or casein (skim milk). Gelatin is an incompletely hydrolyzed protein, and the ability to liquefy gelatin, or the lack of that ability, may not be correlated to the specific proteolytic potential being measured.[21] Ability to hydrolyze casein, on the other hand, is more closely related to the ability to hydrolyze animal protein.[11] The measurement of the level of proteolytic bacteria and the ratio of proteolytic organisms to the total microbial flora can be useful in some foods to predict their refrigerated shelf life.[12,14]

16.7 REFERENCES

1. Alder-Nissen, J. 1979. Determination of the degree of hydrolysis of food protein hydrolysates by trinitrobenzenesulfonic acid. J. Agric. Food Chem. 27:1256-1262.

2. Atlas, R. M., and L. C. Parks. 1993. Handbook of microbiological media. CRC Press, Boca Raton, Fla.
3. Banwart G. J. 1989. Basic food microbiology. 2nd ed., Von Nostrand Reinhold, New York.
4. Chan, E. C. S., M. J. Pelczar Jr., and N. R. Krieg. 1986. Laboratory exercises in microbiology. 5th ed. McGraw-Hill, New York.
5. Cliffe, A. J., and B. A. Law. 1982. A new method for the detection of microbial proteolytic enzymes in milk. J. Dairy Res. 49:209-219.
6. Frazier, W. C., and P. Rupp. 1928. Studies on the proteolytic bacteria of milk. 1. A medium for the direct isolation of caseolytic milk bacteria. J. Bacteriol. 16:57-64.
7. Frazier, W. C., and D. C. Westhoff. 1988. Food microbiology. 4th ed. McGraw-Hill, New York.
8. Gebre-Egziabher, A., E. S. Humbert, and G. Blankenagel. 1980. Heat-stable proteases from psychrotrophs in milk. J. Food Prot. 43:197-200.
9. Harrigan, W. F. 1998. Laboratory methods in food microbiology. 3rd ed. Academic Press, San Diego, Calif.
10. Jay, J. M. 1972. Mechanism and detection of microbial spoilage in meats at low temperatures: a status report. J. Milk Food Technol. 35:467-471.
11. Kazanas, N. 1968. Proteolytic activity of microorganisms isolated from fresh water fish. Appl. Microbiol. 16:128-132.
12. Levin, R. E. 1968. Detection and incidence of specific species of spoilage bacteria on fish. 1. Methodology. Appl. Microbiol. 16:1734-1737.
13. Marshall, R.T. (ed.) 1992. Standard methods for the examination of dairy products. 16th ed. American Public Health Association, Washington, D.C.
14. Martley, F. G, S. R. Jayashankar, and R. C. Lawrence. 1970. An improved agar medium for the detection of proteolytic organisms in total bacterial counts. J. Appl. Bacteriol. 33:363-370.
15. Mayerhofer, H. J., R. T. Marshall, C. H. White, and M. Lu. 1973. Characterization of a heat-stable protease of *Pseudomonas fluorescens* P 26. Appl. Microbiol. 25:44-48.
16. McKellar, R. C. 1981. Development of off-flavors in ultra-high temperature and pasteurized milk as a function of proteolysis. J. Dairy Sci. 64:2138-2145.
17. McKellar, R. C. 1984. Comparison of the hide powder azure and casein-trinitrobenzene sulfonic acid methods for determining proteolysis in skim milk. J. Food Prot. 47:476-480.
18. Pitt, T. L., and D. Dey. 1970. A method for the detection of gelatinase production by bacteria. J. Appl. Bacteriol. 33:687-691.
19. Ray, B. 1996. Fundamental food microbiology. CRC Press, Boca Raton, Fla.
20. Rinderknecht, H., M. C. Geokas, P. Silverman, and B. J. Haverback. 1968. A new ultrasensitive method for the determination of proteolytic activity. Clin. Chim. Acta. 21:197-203.
21. SAB. 1957. Manual for microbiological methods. Society of American Bacteriologists, McGraw-Hill, New York.
22. Sanjose, C. L., L. Fernandez, and P. Palacios. 1987. Compositional changes in cold raw milk supporting growth of *Pseudomonas fluorescens* NCDO 2085 before production of extra-cellular protein-ase. J. Food. Prot. 50:1004-1008.
23. Sizemore, R. K., and L. H. Stevenson. 1970. Method for the isolation of proteolytic marine bacteria. Appl. Microbiol. 20:991-992.
24. Speck, M. L., and D. M. Adams. 1976. Symposium: impact of heat stable microbial enzymes in food processing. Heat resistant proteolytic enzymes from bacterial sources. J. Dairy Sci. 59:786-789.
25. Wikström, M. B. 1983. Detection of microbial proteolytic activity by a cultivation plate assay in which different proteins absorbed to a hydrophobic surface are used as substrates. Appl. Environ. Microbiol. 45:393-400.
26. Wikström, M., H. Elwing, and A. Linde. 1981. Determination of proteolytic activity: a sensitive and simple assay utilizing substrate absorbed to a plastic surface and radial diffusion in gel. Anal. Biochem. 118:240-246.

Chapter 17

Halophilic and Osmophilic Microorganisms

John A. Baross

17.1 INTRODUCTION

Microorganisms can grow over a wide range of solute concentrations. Only a few species can grow at the high osmotic pressures characteristic of environments having supersaturated brine and sugar concentrations (i.e., reduced water activity). Inconsistency in terminology exists in the literature describing this group of microorganisms.[62] However, for practical reasons, microorganisms that require minimum concentrations of salt (NaCl and other cations and anions) are called halophiles, whereas organisms that can grow in high concentrations of organic solute, particularly sugars, have been called osmophiles. (See Tilbury[58] for a discussion of "osmophiles") The terms "osmo- and xerotolerant" have been used in place of "osmophilic" because these organisms do not have an absolute requirement for reduced water activity (a_w) or high osmotic pressure, but merely tolerate drier environments better than non osmotolerant species.[3] Osmotic pressure is usually expressed in units of megapascals (Mpa) where 0.1 Mpa equals 1 atm or 1.013 bars) and is one of the terms used to define the effects of desiccation due to increases in solute concentration.[8,47] The osmotic pressure limits the growth of the obligately halophilic *Halobacterium* spp, which grow in saturated NaCl (30%, and with an a_w of 0.80 is 41 Mpa). To minimize confusion, the name "osmophile," coined by von Richter[65] in 1912, will be adopted for discussion purposes.

In general, the requirement for salt by halophilic microorganisms is not an exclusive need for NaCl since many species require low levels of K^+, Mg^{++} and other cations and anions in addition to NaCl.[13,36,46,48,62] Furthermore, for some bacteria, the apparent requirement for NaCl is not specific but strictly osmotic, and other salts and sugars can be substituted. See Kushner[36] and Rodriquez-Valera[53] for the characteristics and microbial ecology of bacteria in hypersaline environments. The level of salt required by microorganisms varies greatly. Therefore, the microbial types associated with a particular salted food depend on the concentration and type of salt and the type of food. The most practical classification of halophilic microorganisms is based on the level of salt required.[36] *Slight halophiles* grow optimally in media containing 0.5% to 3% salt; *moderate halophiles* grow optimally in media containing 3% to 15% salt, and *extreme halophiles* in media containing 15% to 30% salt. Additionally, many halotolerant microorganisms grow without added salt as well as in salt concentrations exceeding 12%. Some halotolerant microorganisms are involved in the spoilage of foods with low a_w values, including salted foods.[62] See Chirife and Buera,[9] and Potts[48] for a discussion of a_w limitations of different microbial species. Non-halophilic bacteria grow best in a_w media containing less than 0.5% NaCl.

Among salted foods, low-salted foods (1% to 7% salt by weight) are more susceptible to microbiological spoilage and are also more likely to contain viable human pathogens. This is particularly true of untreated fresh seafoods. Heavily brined foods do not spoil easily unless maintained at elevated temperatures. A compilation of the types of halophilic, spoilage, and pathogenic microorganisms associated with various salted foods is shown in Table 1.

Yeasts are the most common osmophilic microorganisms encountered in nonionic environments of high osmolarity, such as foods containing high concentrations of sugar. Osmophilic yeasts are usually the cause of spoilage of high-sugar foods, including jams, honey, concentrated fruit juices, chocolate candy with soft centers, etc.[66] Osmophilic yeasts are of no public health significance, but are of economic importance to the food industry. For a discussion on xerophilic molds and their significance in foods, see Pitt.[46,47]

The following procedures emphasize the culturing of indigenous, spoilage and human pathogenic microorganisms from various salted foods, seafoods and foods with a low a_w. Molecular methods are rapidly being developed to detect and enumerate specific microorganisms from environmental samples without culturing.[1,2,59] Molecular probes are already available that can be used to microscopically identify and enumerate pathogens from environmental samples using fluorescent *in situ* hybridization (FISH) or PCR methods.[2] There are currently no standard molecular methods available for enumerating species of spoilage microorganisms from salted foods.

17.2 HALOPHILIC MICROORGANISMS

17.21 Equipment, Materials, and Reagents

17.211 Diluents

17.2111 Phosphate buffer with salt. Prior to sterilization, add the required amount of NaCl to pH 7, 0.1 M potassium phosphate solution. The addition of 3% NaCl in the buffer has been shown to be useful in protecting *Vibrio parahaemolyticus* against cold and heat inactivation. Under circumstances where a phosphate buffer

Table 1. Halophilic, Spoilage, and Pathogenic Microorganisms Associated with Various Salted Foods and Seafoods

Food type	Salt Associated with Food (%)	Halophilic Types	Spoilage Microorganisms	Pathogens[a]	References
I. Marine fish	1% to 4% (similar to seawater)	Slight and halotolerant types: *Listonella* spp., *Shewanella putrefaciens*, *Photobacterium phosphoreum*, *Vibrio* spp., *Pseudomonas* spp., *Acinetobacter* spp., and other gram-negative bacteria	Proteolytic *Pseudomonas* spp., *Shewanella putrefaciens*, *Photobacterium phosphoreum*, *Moraxella*, spp., and *Acinetobacter* spp.	*Vibrio parahaemolyticus*, *V. cholerae*, *V. vulnificus*, *Clostridium botulinum E*, *Clostridium perfringens*, *Staphylococcus aureus*, *Salmonella* and other pathogenic *Enterobacteriaceae*, *Erysipelothrix insidiosa*, *Listeria monocytogenes*, *Aeromonas* spp.	1, 10, 12, 21, 22, 27, 28, 31, 33, 40, 41, 43, 44, 50, 51
II. Molluscan shellfish	0.5% to 4% (similar to seawater)	Same as for fish, but greater percentage of *Vibrio* spp. and related organisms	Same as for fish at early stages. Lactic acid bacteria and yeasts in later stages of spoilage	Same as for fish plus human enteric viruses, *Gonyaulax* and other toxic algae	4, 10, 18, 27, 37, 44, 52
III. Crustaceans	0.5% to 4% (similar to seawater)	Same as for fish, but greater percentage of *Vibrio* spp. and related organisms.	Same as for fish, also yeasts and chitinoclastic microorganisms	Same as for fish except increased likelihood of chitinoclastic pathogenic *Vibrio* spp.	10, 15, 18, 22, 25, 43, 44, 61
IV. Brined meats (ham, bacon, corned beef, prepared meats, and sausage	1% to 7% brine[b]	Halotolerant molds, yeasts, gram-positive bacteria (e.g., *Micrococcus* spp., enterococci and lactic acid bacteria)	Mixed flora in aerobic packaging. Predominantly gram-positive bacteria in refrigerated meats in anaerobic packaging.	*C. botulinum*, *C. perfringens*, *S. aureus*. Pathogenic *Enterobacteriaceae* in meats containing low salts. *Listeria* spp. in sausage mix.	9, 14, 16, 19, 20, 27, 29, 34, 40, 50, 51, 55
V. Salted vegetables	1% to 15%	Moderate and halotolerant molds, yeasts, and gram-positive bacteria	Lactic acid bacteria, yeasts and molds, *Bacillus*, *Enterobacteriaceae* in foods with low salt content. *Clostridium* spp. in packaged foods.	Dependent on level of salt in food. Pathogenic members of *Enterobacteriaceae* in low salt foods. *S. aureus* in highly salted food.	24, 50
VI. Salted fish					
(A) Light salt	1% to 10%	Slight and halotolerant types	*Pseudomonas* spp. in lightly salted fish. *Clostridium* spp. in packaged fish. *Micrococcus* spp. in fish containing 5% to 10% salt.	Dependent on salt concentration, same as salted vegetables. Pathogenic *Vibrio* spp., may occur when salt is 1% to 7%. *Salmonella* spp. capable of growing in 8% NaCl at high temperatures (22° to 37°C) and surviving for up to 70 days at 5°C in curing brines. *Clostridium botulinum* Type E and *L. monocytogenes* in smoked fish.	12, 23, 26, 27, 29, 35, 50, 51, 54, 63, 64
(B) Heavy salt	20 to 25% NaCl wt/vol (10% to 15% salt in interior of fish)	Moderately and extremely halophilic types	*Halobacterium* spp., *Halococcus* spp., cause condition called "pink" in fish. Also some members of the *Micrococcaceae* and *Halomonas* spp.	*S. aureus*	26, 27, 29, 50, 51, 56, 62

[a]Refer to specific section for the isolation of specific pathogens.

[b]Brine concentration is $\frac{\text{g salt}}{\text{g salt + g water}} \times 100$

is not desirable, diluents made with 0.5% peptone supplemented with the appropriate concentrations of NaCl will also give satisfactory results.

17.2112. Synthetic sea water (SW)11 (Chapter 63). Synthetic sea water is used as a diluent and as a salt base in the preparation of sea water agar (SWA).

17.212 Media

1. Trypticase soy agar with added salt (TSA-NaCl)
   Add the required amount of NaCl to TSA prior to sterilization.
2. Sea water agar [29] (Chapter 63)
3. Halophilic agar (HA) [53] (Chapter 63)
4. Halophilic broth (HB) (Chapter 63)

Halophilic broth (HB) is used as a diluent and as an enrichment medium for the isolation of extremely halophilic bacteria. Sterilize medium by autoclaving at 121°C for 15 min.

17.22 Procedures

17.221 Slightly Halophilic Bacteria

17.2211 General. Most of the slightly halophilic bacteria originate from marine environments. Marine facultatively psychrophilic bacteria of the genera *Shewanella, Listonella, Pseudomonas, Vibrio, Moraxella, Acinetobacter,* and *Photobacterium* contribute to the spoilage of marine fish and shellfish. Gram-negative rods of terrestrial origin and other gram-negative and gram-positive facultatively psychrophilic bacteria are frequently involved.[12,27,50] Some of the organisms have complex ionic requirements and may require Mg^{++} and K^{+} in addition to NaCl for growth and proteolytic activity, whereas the salt requirement for other slightly halophilic bacteria is osmotic.[62] Growth of most of these marine bacteria is inhibited if the NaCl concentration of the growth medium is lower than 0.5% or higher than 5%. Dilution of seafood samples with distilled water may cause lysis of many representative spoilage bacteria. Holding marine foods for sustained periods at temperatures exceeding 25°C will reduce the numbers of psychrotrophic microorganisms significantly. To identify the degree of initial microbial contamination of marine origin, prechill diluent, reagents, and sampling equipment to 5°C. Avoid washing food sample with distilled water, and do not use diluents containing less than 3% NaCl.

The microbial flora associated with salted meats and vegetables are variable and dependent on many factors, including the type of food, the presence of other salts or organic preservatives, the concentration of salt, and the storage conditions of temperature and packaging.[24,51] In general, microorganisms involved in the spoilage of low salt meats and vegetables (1% to 7% brine) can be enumerated without the use of special media provided the diluents and plating media are supplemented with NaCl equivalent to the concentration in the food sample.

17.2212 Sampling. Low-salted foods to be analyzed for psychrotrophic spoilage microorganisms should be tested without delay (within 24 hr); otherwise, growth will occur. Samples should be maintained at 0° to 5°C until tested.

17.2213 Procedure.

1. **Fish (Teleosts).** Remove skin samples with a sterile cork borer (1.6 cm diameter) by punching the cork borer through the skin and then removing the disc of skin with a sterile scalpel and forceps. Collect six pieces, three from the ventral side and three from the dorsal side of the fish. Take flesh samples from just below the skin by peeling the skin with a sterile scalpel and forceps and dissecting the flesh. Add the discs of skin to 90 mL phosphate buffer-3% NaCl (Section 17.2111) or synthetic sea water with either sterile sand or glass beads (10 g). Mix sample and diluent thoroughly by shaking vigorously for 1 min. Prepare dilutions (to10^{-6}) in phosphate buffer-3% NaCl or synthetic sea water. Place 0.1-mL aliquots on either TSA plus 3% NaCl or sea water agar with the spread plate technique. Incubate plates at 7°C for 10 days for the enumeration of psychrotrophic spoilage bacteria.

   For fish flesh samples, fillets, or small whole fish (less than 6 in), cut samples aseptically into slices (2.5 cm^2) and weigh in a sterile container. Add 50 g of flesh (skin removed) to 450 mL of diluent (phosphate buffer –3 % NaCl, Section 17.2111) in a sterile blender jar and blend for 2 min. If the sample is not sufficiently homogenized, let it stand for 2 min before blending for an additional 2 min. Prepare serial dilutions and plates as described above.

   Frozen fish or other seafood blocks are sampled using an electric drill and bit. The detection and enumeration of the slightly halophilic pathogens *Vibrio parahaemolyticus, Vibrio cholerae,* and *Vibrio vulnificus* are described in detail in Chapter 40.
2. Molluscan shellfish. Collect and prepare samples for microbiological analysis as described by Hunt et al.[42] Prepare initial 1:1 dilution by blending 100 g of shellfish meat with 100 mL of sterile diluent (phosphate buffer –3% NaCl, Section 17.2111). Plate samples as described in "1" above. It is advisable to prepare two sets of plates, one to be incubated at 7°C and the other at 25°C, since shellfish generally reside in near-shore environments that are subject to wide fluctuations in temperature and, therefore, harbor both psychrotrophic and mesophilic bacteria.
3. Crustaceans. Collect and prepare samples as described by Nickelson and Finne.[43] Prepare dilutions as for fish flesh. Plate samples as described in "1" above.
4. Low-salted meats and vegetables (1% to 7% brine). Prepare dilutions by blending 50-g samples with 450 mL diluent (phosphate buffer-3% NaCl, Section 17.2111) plus NaCl equivalent to salt concentration of food sample. Plate samples on TSA supplemented with NaCl equivalent to NaCl concentration of food. If the food sample usually is not refrigerated, use 1-mL aliquot for pour plates and incubate plates for 4 days at 25°C. For refrigerated foods such as bacon, plate 0.1 mL aliquots with the spread plate technique. Incubate plates for 10 days at 7°C.

17.2214 Interpretation. In general, fish contaminated with greater than 10^8 psychrotrophic bacteria per cm^2 of skin or per g of muscle is considered spoiled.[27] The most common seafood spoilage bacteria are *Pseudomonas* species, which are psychrotrophic (Chapter 13) and actively proteolytic. Chemical and organoleptic tests (odor, particularly associated with gills) are used in conjunction with bacterial counts to assess the extent of spoilage. High total volatile nitrogen (TVN) and trimethylamine (TMA) values are used as an indication of bacterial activity.[61]

No microbiological standards have been established for low-salted meats and vegetables.

17.222 Moderately Halophilic Bacteria

17.2221 General. Most of the moderately halophilic bacteria involved in the spoilage of salted foods (5% to 20% NaCl by weight) are gram-positive species of *Bacillaceae* and *Micrococcaceae. Paracoccus halodenitrificans* and other *Micrococcus spp.* have a specific requirement for NaCl.[36,53] This is also true for moderately halophilic *Acinetobacter Moraxella* species isolated from salted herring, Vibrio costicola isolated from bacon curing brines, *Vibrio alginolyticus* isolated from seafoods and facultatively halophilic *Halomonas* spp and related organisms isolated from salted meats, fish and soy sauce mashes.[19,20,36,62] In contrast, the requirement for salt by many moderately halophilic Bacillus spp. is not specific for NaCl and many other Na^+ and K^+ salts can be substituted. *Pseudomonas* spp. isolated from sea salts and *Planococcus* halophilus can grow in the absence of added salt although they grow optimally in salt concentrations ranging from 4% to 9%.[36,62] Salted foods that can spoil because of moderately halophilic microorganisms commonly harbor high numbers of halotolerant gram-positive bacteria, yeasts, and molds. (Consult Table 1 for references on isolation and enumeration procedures of osmophilic and halotolerant pathogens.)

17.2222 Procedure. Moderately salted foods are sampled for spoilage microorganisms as described in Section 17.221.2. Prepare 1:10 dilutions by mixing 50 g of food in 450 mL sterile phosphate buffer (Section 17.211.1) with added NaCl equivalent to the salt concentration of the food sample. Plating procedures with plating media and plate incubation temperatures and times are described in Section 17.221.3d. For the isolation and enumeration of specific bacterial types, consult Table 1 and related chapters.

17.2223 Interpretation. No microbiological standards have been set for moderately salted foods. Use organoleptic tests (odor and visual evidence of spoilage such as slime or gas formation) in conjunction with total bacterial counts to determine the extent of spoilage. Staphylococcus aureus and Clostridium perfringens can grow in some moderately salted foods.

17.223 Extremely halophilic bacteria

17.2231 General. The extreme halophiles are normally found in aquatic environments of unusually high salt concentrations and in solar evaporated sea salts. These grow optimally in media containing 15% to 30% wt/vol NaCl.[13,36,53] The extreme halophiles can be divided into neutrophilic and alkaliphilic groups based on their optimum pH for growth.[53] The neutrophilic genera include *Halobacterium, Haloferax, Haloarcula,* and *Halococcus.* These generally grow optimally at pH 7.2. *Halobacterium* and *Halococcus* produce bright red or pink pigments, grow very slowly even under optimal conditions, and are readily lysed when exposed to low salt concentrations (less than 10%). The alkaliphilic group grows optimally at a pH that is approximately 9.5 and includes the genera *Natronobacterium* and *Natronococcus*. The neutrophilic genera are the most commonly encountered in the spoilage of highly salted food and hides. Extremely halophilic microorganisms are grouped with the proposed domain Archaea based on features such as 16S ribosomal RNA sequences, ether-linked membrane lipids, and their lack of a peptidoglycan cell wall.[36,67] Extremely halophilic bacteria have been incriminated in the spoilage of fish, bacon, and hides preserved in sea salts. Severe contamination of foods with *Halobacterium* or *Halococcus* will generally result in a pink discoloration on the outer surface of the sample accompanied by decomposition and putrefaction.[50]

17.2232 Procedure. Most extremely halophilic microorganisms are easy to culture and will grow on a wide variety of organic based media. Many species will grow on defined media consisting of a single carbon source, trace elements, ammonium as a source of nitrogen, and phosphate.[53] However, it is recommended that a yeast extract based medium be used for initial isolation from environmental samples. Halobacteria are very sensitive to bile salts and caution should be exercised when using a peptone in media since peptones are often contaminated with bile salts and halobacteria are very sensitive to bile salts.[53] For isolation, transfer surface slime from salted fish or bacon to an HA plate (Section 17.212.3) using a cotton or alginate swab. For the quantitative enumeration of *Halobacterium* and *Halococcus* from food samples, blend 50 g of sample with 450 mL HB. Place 0. 1 mL aliquots of each dilution (to 10^{-6}) on HA plates with the spread plate technique. For the detection of extremely halophilic bacteria from solar sea salts or brine solutions (70% to 80%), prepare serial dilutions in HB up to 10^{-6}. Prepare plates on HA. Alternately, for samples with low numbers of halophilic bacteria, place 10-mL or 10-g sample into 90 mL of HB broth and incubate at 35°C for up to 12 days. Then streak from broth onto HA plates. Incubate plates at 33° to 35°C in a humid incubator for 5 to 12 days.

17.2233 Interpretation. No microbiological criteria for heavily brined foods have been set, and usually only an organoleptic observation for presence of red or pink slime and putrefaction is performed to check for spoilage. Since *Halobacterium* and *Halococcus* are normally present in sea salts, food spoilage by these organisms can be prevented by dry heat sterilization of salt prior to use for curing. Extremely halophilic bacteria will not grow on foods stored at temperatures below 7°C.

Extreme halophiles are not pathogenic to humans, and any incidence of food poisoning associated with heavily salted foods is invariably caused by *S. aureus* (Chapter 39).

17.224 Halotolerant Microorganisms

17.2241 General. Microorganisms capable of growing in NaCl concentrations exceeding 5% as well as in media containing no NaCl are called halotolerant. Most halotolerant bacteria are gram-positive and belong to the *Micrococcaceae*, the *Bacillaceae*, and some *Corynebacterium* species. A few gram-negative species will also grow in foods cured with 10% salt. Many human pathogens, such as *S. aureus* and *C. perfringens* and some strains of *Clostridium botulinum*, are responsible for food poisoning outbreaks involving low and moderately salted foods.[50,51]

Most halotolerant microorganisms are isolated when foods are tested for slight or moderate halophiles. Specific media, however, are required to isolate halotolerant molds and yeasts from food samples (see Section 17.3).

17.2242 Procedure. Refer to Sections 17.221 and 17.222.

17.2243 Interpretation. Refer to Sections 17.221 and 17.222.

17.3 OSMOPHILIC MICROORGANISMS

17.31 General Considerations

Osmophilic yeasts can cause spoilage of honey, chocolate candy with soft centers, jams, molasses, corn syrup, flavored syrups and toppings, concentrated fruit juices, and similar products.[17,32,58,60] Many of the common spoilage yeasts in this group belong to the genus *Zygosaccharomyces*.[68]

Techniques for the enumeration of osmophilic microorganisms have been reported by numerous investigators. However, wide acceptance of standard methods has not been attained. Generally, the enumeration of osmophilic yeasts requires a consideration of both the a_w of the diluents and the plating media. Inaccurate results may be obtained if a high a_w, diluent or agar medium is used or if the agar medium has a reduced a_w but the diluent does not.[56] The type of solute used to reduce the a_w may also play an important role in enumeration techniques.[38,39]

Because of the particulate material in many food products, the pour plate technique is the method of choice for enumeration.

With products having suspected low counts, products with low viscosity, and products that may require the isolation of contaminants, membrane filtration techniques have been suggested.[30, 68] A simple presence-absence test for detection of small numbers of osmotolerant yeasts in high-sugar foods may be useful for qualitative purposes.[30]

17.32 Equipment, Materials, and Reagents

17.321 Equipment

Stainless steel filter apparatus
Type HA filters with grid marks and absorbent pads
Vacuum pump
Filter flask, 1 L
Erlenmeyer flask, 500 mL screw cap
Stomacher or blender
Petri dishes, 15 × 100 mm
Incubator, 30°C
Colony counter

17.322 Diluents

Diluents with reduced a_w are needed to detect and enumerate osmophilic microorganisms.[7,39,53] When hypotonic diluents are used, osmophilic yeasts may be destroyed. Generally, the use of 40% to 50% wt/wt of hexose in diluents is recommended. Glucose or invert sugar is most suitable. Sucrose and corn syrup in the intermediate range of dextrose equivalents are unlikely to provide a sufficiently low a_w, environment at reasonable concentrations. Additional research is needed in this area.

17.323 Media

No single medium is optimal for quantitation, though several have been suggested.[47,57] It is best to evaluate several media depending on the food being analyzed. Tests to select the best medium for each product category may be appropriate. Lenovich and Konkel[39] recommend malt extract-yeast extract 40% wt/wt glucose agar (MY40G), among others, as a general-purpose medium for most osmotolerant microorganisms. Restaino et al.[49] found that recovery of Saccharomyces rouxii from chocolate syrup was significantly improved with the addition of 60% sucrose (wt/vol) and 2% glucose to potato dextrose agar ($a_w = 0.92$). Investigators have reported the use of filtration techniques in combination with fluorescent staining to detect low numbers of osmophilic yeasts in confectionery products.[5,45]

17.33 Procedures

Several media may be used for enumeration. A more detailed discussion on appropriate media selection can be found in Pitt and Hocking.[47]

17.331 Sampling

Samples should be collected in sterile containers and stored in conditions that will not promote the growth of viable fungi. Analysis should be conducted the same day the samples are collected.

17.332 Plate Count Method

Stomach or blend samples. Prepare dilutions with sterile, phosphate-buffered water supplemented with glucose or other solute. Inoculate 1 mL of the desired dilution into Petri dishes, then pour with 15 to 20 mL of the appropriate agar medium. Incubate at 30° C for 5 to 7 days. The use of a colony counter will facilitate enumeration because many yeasts grow slowly and thus their colonies are difficult to see with the unaided eye.

17.333 Membrane Filter Technique

Enumeration of low numbers of spoilage yeasts in viscous products of low a_w has been reported using standard filtration techniques.[69]

Weigh a 25-g sample aseptically into a sterile Erlenmeyer flask. Add approximately 1 to 2 volumes of sterile distilled water. Mount a presterilized stainless steel filter apparatus containing a sterile HA filter on a filter flask. Cover the filter holder with a sterile Petri dish. Attach vacuum source to filter flask and filter sample. Rinse the walls of the sample flask and funnel with sterile water. Remove the filter from the filter base and place it, grid side up, on one of the previously discussed media. Incubate the dish for 5 days at 30° C and count the colonies.

17.34 Precautions

Analysis of products by the membrane filter technique requires that the sample be diluted with distilled water. Although dilution may facilitate sample analysis, it may cause osmotic stress to the mycoflora present by altering the a_w of the sample.

Sterilization of high-sugar media may result in compounds that are toxic to osmophiles. Steaming media containing high sugar at 100°C for 30 min will destroy all microorganisms except bacterial spores, which are unlikely to grow in media with more than 60% hexose.

Because of the high viscosity of many of these plating media, it is important to swirl the plates more than one would with standard plating media to ensure that the sample is uniformly distributed throughout the agar.

Incubation of plates with low a_w, media in a forced-draft incubator will cause plates to dry and prevent colonies from developing fully. Placing a shallow pan of distilled water in the bottom of the incubator will prevent drying.

None of the above methods account for the possibility of injury that is due to environmental stress such as heat, osmotic, or freeze injury. It may be desirable to consider methodology for recovering injured cells. However, the literature does not provide a wealth of information on the injury phenomenon in yeasts and molds.[6] One may need to develop protocols for specific food types.

17.35 Interpretation

It is important to collect baseline data on different product types in order to interpret quantitative estimates of osmophiles. Acceptable levels of osmophiles are ultimately product dependent and are based on each product's history and likelihood of spoilage. In certain circumstances, the presence of a particular osmophile at any level is unacceptable. Counts of 10 or fewer per g in products such as liquid sugars or syrups may prove to be highly significant if the yeast is a spoilage type such as *Zygosaccharomyces rouxii*. When one detects low level contamination by yeasts, identification of the species is often beneficial so that an effective course of action can be implemented.

Low counts (10 or fewer per g) may be indicative of inadequate processing that did not destroy vegetative cells originating from raw materials or post-processing contamination from equipment, air or packaging materials.

ACKNOWLEDGMENT

The author is indebted to Lawrence M. Lenovich, previous co-author of this chapter, for segments incorporated in this edition.

17.4 REFERENCES

1. Agersborg, A., R. Dahl, and I. Martinez, 1997. Sample preparation and DNA extraction procedures for polymerase chain reaction identification of Listeria monocytogenes in seafoods. Int. J. Microbiol. 35:275-280.
2. Amann, R. I., W. Ludwig, and K. H. Schleifer. 1995. Phylogenetic identification and in situ detection of individual microbial cells without cultivation. Microb. Rev. 59:143-169.
3. Anand, J. C., and A. D. Brown. 1969. Growth rate patterns of the so-called osmophilic and non-osmophilic yeasts in solutions of polyethylene glycol. J. Gen. Microbiol. 52:205-212.
4. Anderson, D. M., and D. J. Garrison (eds.). 1997. The ecology and oceanography of harmful algal blooms. Limnol. Oceanogr. 42:1009-1305.
5. Baumgart, V. J., and B. Vieregge. 1984. Rapid detection of osmo-philic yeasts in marzipan. Susswaren 28:190-193.
6. Beuchat, L. R. 1986. Consideration of media for enumerating injured fungi, p. 168-171. In A. D. King, Jr., J. I. Pitt, L. R. Beuchat, and J. E. L. Corry (eds.), Methods for the mycological examination of food. " Plenum Press, New York.
7. Beuchat, L. R. 1983. Influence of water activity on growth, metabolic activities, and survival of yeasts and molds. J. Food Prot. 46:135-141.
8. Chirife, J., and M. del Pilar Buera. 1996. Water activity, water glass dynamics, and the control of microbiological growth in foods. Crit. Rev. Food Sci. Nutr. 36:465-513.
9. Clavero, M. R. S., and L. R. Beuchat. 1996. Survival of Escherichia coli O157:H7 in broth and processed salami as influenced by pH, water activity, and temperature and suitability of media for its recovery. Appl. Environ. Microbiol. 62:2735-2740.
10. Colwell, R. R. (ed.). 1984. Vibrios in the environment, John Wiley & Sons, New York.
11. Colwell, R. R., and R. Y. Morita. 1964. Reisolation and emendation of description of *Vibrio marinus* (Russell) Ford. J. Bacteriol. 88:831-837.
12. Dalgaard, P. 1995. Qualitative and quantitative characterization of spoilage bacteria from packed fish. Int. J. Food Microbiol. 26:319-333.
13. DasSarma, S. 1995. Halophilic Archaea: An overview, p. 3-11. *In* F. T. Robb, A. R. Place, K. R. Sowers, H. J. Schreier, S. DasSarma, and E. M. Fleischmann (eds.), Archaea, A laboratory manual., Cold Spring Harbor Laboratory, Cold Spring Harbor, N. Y.
14. Draughon, F. A., C. C. Melton, and D. Maxedon. 1981. Microbial profiles of country-cured hams aged in stockinettes, barrier bags, and pariffin wax. Appl. Environ. Microbiol. 41:1078-1080.
15. Eklund, M. W., J. Spinelli, D. Miyauchi, and H. Groninger. 1965. Characterization of yeasts isolated from Pacific crab meat. Appl. Microbiol. 13:985-990.
16. Escher, F. E., P. E. Koehler, and J. C. Ayres. 1973. Production of ochratoxins A and B on country-cured hams. Appl. Microbiol. 26:27-30.
17. Fleer, G. H. 1990. *In* J. F. T. Spencer and D. M. Spencer (eds.), Food spoilage yeasts, ed. p. 124. Springer-Verlag, Berlin.
18. Fieger, E. A., and A. F. Nokva. 1961. Microbiology of shellfish deterioration, p. 561-611. *In* G. Borgstrom (ed.), Fish as food, vol. 1., Academic Press, New York.
19. Gardner, G. A. 1973. A selective medium for enumerating salt requiring *Vibrio* spp from Wiltshire bacon and curing brines. J. Appl. Bacteriol. 36:329-333.
20. Gardner, G. A., and A. G. Kitchell. 1973. The microbiological examination of cured meats, p. 11-20. *In* R. R. Board and D. N. Lovelock (eds.), Sampling-microbiological monitoring of environments. Academic Press, New York.
21. Gram, L., and H. H. Huss. 1996. Microbiological spoilage of fish and fish products. Internat. J. Food Microbiol. 33:121-137.
22. Hanninen, M. L., P. Oivanen, and V. HirvelaKoski. 1997. *Aeromonas* species in fish, fish eggs, shrimp and freshwater. Int. J. Food Microbiol. 34:17-26.
23. Hansen, L. T., R_ntved and H. H. Huss. 1998. Microbiological quality and shelf life of cold-smoked salmon from three different processing plants. Food Microbiol. 15:137-150.
24. Harrigan, W. F., and M. E. McCance. 1966. Laboratory methods in microbiology. Academic Press, New York.
25. Hatha, A. A. M., N. Paul, and B. Rao. 1998. Bacteriological quality of individually quick-frozen (IQF) raw and cooked ready-to-eat shrimp produced from farm raised black tiger shrimp (*Penaeus monodon*). Food Microbiol. 15:177-183.
26. Hennessey, J. P. 1971. Salted and dried groundfish products, p. 114-116. *In* R. Krueger (ed.), Fish inspection and quality control. Fishing News (Books) Limited, London, England.
27. Hobbs, G. 1983. Microbial spoilage of fish p. 217-229. *In* R. A. Skinner (ed.), Food microbiology: Advances and prospects. Academic Press, New York.
28. Hyytiä, E., S. Hielm, and H. Korkeala. 1998. Prevalence of *Clostridium botulinum* type E in Finnish fish and fishery products. Epidemiol. Infect. 120:245-250.
29. Ingram, M., and R. H. Dainty. 1971. Changes caused by microbes in spoilage of meats. J. Appl. Bacteriol. 34:21-39.
30. Jermini, M. F. G., O. Geiges, and W. Schmidt-Lorenz. 1987. Detection, isolation, and identification of osmotolerant yeast from high-sugar products. J. Food Prot. 50:468-472.
31. Kaper, J., H. Lockman, R. R. Colwell, and S. W. Joseph. 1979. Ecology, serology, and enterotoxin production of *Vibrio cholera* in Chesapeake Bay. Appl. Environ. Microbiol. 37:91-103.
32. Kinderlerer, J. L. 1997. *Chrysosporium* species, potential spoilage organisms of chocolate. J. Appl. Microbiol. 83:771-778.
33. Kita-Tsukamoto, K., H. Oyaizu, K. Nanba, and U. Simidu. 1993. Phylogenetic relationships of marine bacteria, mainly members of the family *Vibrionaceae*, determined on the basis of 16S rRNA sequences. Int. J. Syst. Bacteriol. 43:8-19.
34. Koelensmid, W., A. A. Blanche, and R. van Rhee. 1964. Salmonella in meat products. Ann. Inst. Pasteur de Lille 15:85-97.
35. Korkeala, H., G. Stengel, E. Hyytiä, B. Vogelsang, A. Bohl, H. Wihlman, P. Pakkala, and S. Hielm. 1998. Type E botulism associated with vacuum-packaged hot-smoked whitefish. Int. J. Food Microbiol. 43:1-5.
36. Kushner, D. J. 1978. Life in high salt and solute concentrations: Halophilic bacteria, p. 317-368. *In* D. J. Kushner (ed.), Microbial life in extreme environments. Academic Press, New York.
37. Liew, W. S., J. J. Leisner, G. Rusul, S. Radu, and A. Rassip. 1998. Survival of Vibrio spp. including inoculated *V. cholera* 0139 during heat-treatment of cockles (*Anadara granosa*). Int. J. Food. Microbiol. 42:167-173.
38. Lenovich, L. M., R. L. Buchanan, N. J. Worley, and L. Restaino. 1988. Effect of solute type on sorbate resistance in *Zygosaccharomyces rouxii*. J. Food Sci. 53:914-916.
39. Lenovich, L. M., and P. J. Konkel. 1992. Confectionary products, p. 1007. *In* C. Vanderzant and D. F. Splittstoesser (eds.), Compendium of methods for the microbiological examination of foods, 3rd ed., Am. Public Health Assoc., Washington, D. C.
40. Lindberg, A. M., A. Ljungh, S. Ahrne, S. Lofdahl, and G. Molin. 1998. Enterobacteriaceae found in high numbers in fish, minced meat and pasteurized milk or cream and the presence of toxin encoding genes. Int. J. Food Microbiol. 39:11-17.
41. Mertens, A., J. Nagler, W. Hansen, and E. Gepts-Friedenreich. 1979. Halophilic lactose-positive Vibrio in a case of fatal septicemia. J. Clin. Microbiol. 9:233-235.
42. Miescier, J. J., D. A. Hunt, J. Redman, A. Salinger, and J. P. Lucas. 1992. Molluscan shellfish: oysters, mussels, and clams, p. 897-918. *In* C. Vanderzant and D. F. Splittstoesser (eds.), Compendium of methods for the microbiological examination of foods, 3rd ed. Am. Public Health Assoc., Washington, D. C.

43. Nickelson, R., II, and G. Finne. 1992. Fish, crustaceans, and precooked seafoods, p.875-895. *In* C. Vanderzant and D. F. Splittstoes-ser (eds.), Compendium of methods for the microbiological examination of foods, 3rd ed. Am. Public Health Assoc., Washington, D. C.
44. Oliver, J. D. 1989. *Vibrio vulnificus*, p. 570-600. *In* M. P. Doyle (ed.), Food-borne bacterial pathogens. Marcel Dekker, Inc., New York.
45. Pettipher, G. L. 1987. Detection of low numbers of osmophilic yeasts in creme fondant within 25 h using a pre-incubated DEFT count. Lett. Appl. Microbiol. 4:95-98.
46. Pitt, J. I. 1975. Xerophilic fungi and the spoilage of foods of plant origin, p. 273-307. *In* R. B. Duckworth (ed.), Water relations in food. Academic Press, New York.
47. Pitt, J. I., and A. D. Hocking. 1985. Xerophiles, p. 313-333. *In* Fungi and food spoilage. Academic Press, New York.
48. Potts, M. 1994. Desiccation tolerance in prokaryotes. Microbiol. Rev. 58:755-805.
49. Restaino, L., S. Bills, and L. M. Lenovich. 1985. Growth response of an osmotolerant, sorbate-resistant *Saccharomyces rouxii* strain: Evaluation of plating medium. J. Food Prot. 48:207-209.
50. Riemann, H. 1969. Food processing and preservation effects, p. 489-541. *In* H. Rieman (ed.), Food borne infection and intoxications. Academic Press, New York.
51. Riemann, H., W. H. Lee, and C. Genigeorgis. 1972. Control of *Clostridium botulinum* and *Staphylococcus aureus* in semi-preserved meat products. J. Milk Food Technol. 35:514-523.
52. Rippey, S. R. 1994. Infectious diseases associated with molluscan shellfish consumption. Clin. Microbiol. Rev. 7:419-425.
53. Rodriquez-Valera, F. 1995. Cultivation of halophilic archaea, p. 13-16. *In* F. T. Robb, A. R. Place, K. R. Sowers, H. J. Schreier, S. DasSarma, and E. M. Fleischmann (eds.), Archaea, A laboratory manual. Cold Spring Harbor Laboratory, Cold Spring Harbor, N. Y.
54. Rorvik, L. M., E. Skjerve, B. R. Knudsen, and M.Yndestad. 1997. Risk factors for contamination of smoked salmon with *Listeria monocytogenes* during processing. Int. J. Microbiol. 37:215-219.
55. Sanz, Y., R. Vila, F. Toldra, and J. Flores. 1998. Effect of nitrate and nitrite curing salts on microbial changes and sensory quality of non-fermented sausages. Int. J. Food Microbiol. 42:213-217.
56. Seiler, D. A. L. 1986. Effect of diluent and medium water activity on recovery of yeasts from high sugar coatings and fillings, p. 162-163. *In* A. D. King, Jr., J. I. Pitt, L. R. Beuchat, and J. E. L. Corry (eds.), Methods for the mycological examination of food. Plenum Press, New York.
57. Stecchini, M., and L. R. Beuchat. 1985. Effects of sugars in growth media, diluents and enumeration media on survival and recovery of *Saccharomces cerevisiae*. Food Microbiol. 2:85-95.
58. Tilbury, R. H. 1980. Xerotolerant (osmophilic) yeasts, p. 153-179. *In* F. M. Skinner, S. M. Passmore, and R. R. Davenport (eds.), Biology and activities of yeasts Academic Press, New York.
59. van der Vossen, J. M. B. M., and H. Hofstra. 1996. DNA based typing, identification and detection systems for food spoilage microorganisms: Development and implementation. Int. J. Food Microbiol. 33:35-49.
60. van der Walt, J. P., and D. Yarrow. 1984. Methods for isolation, maintenance, classification, and identification of yeasts, p. 45-104. *In* N. J. W. Kreger-van Rij (ed.), The yeasts—A taxonomic study. Elsevier Science Publ. B. V., Amsterdam.
61. Vanderzant, C., B. F. Cobb, and C. A. Thompson, Jr. 1973. Microbial flora, chemical characteristics and shelf life of four species of pond-reared shrimp. J. Milk Food Technol. 35:443-446.
62. Ventosa, A., J. J. Nieto, and A. Oren. 1998. Biology of moderately halophilic aerobic bacteria. Microbiol. Mol. Biol. Rev. 62:504-544.
63. Vilhelmsson, O., H. Hafsteinsson, and J. K. Kristjansson. 1996. Isolation and characterization of moderately halophilic bacteria from fully cured salted cod (bachalao). J. Appl. Bacteriol. 81:95-103.
64. Villar, M., A. P. de Ruiz Holgado, J. J. Sanchez, R. E. Trucco, and G. Oliver. 1985. Isolation and characterization of *Pediococcus halophilus* from salted anchovies (Engraulis anchoita). Appl. Environ. Microbiol. 49:664-666.
65. Von Richter, A. A. 1912. Uber einem osmophilen organisms den hefepilz *Zygosaccharomyces mellis* acidi sp. n. Mykol. Zentralbl. 1:67.
66. Walker, H. W., and J. C. Ayres. 1970. Yeasts as spoilage organisms. *In* A. H. Rose and J. S. Harrison (eds.), The yeasts, vol. 3. Academic Press, New York.
67. Woese, C. R., O. Kandler, and M. L. Wheelis. 1990. Towards a natural system of organisms: Proposals for the domains Archaea, Bacteria, and Eukarya. Proc. Natl. Acad. Sci. USA 87:4576-4579.
68. Yarrow, D. 1984. *Zygosaccharomyces* Barker. chap. III, Genus 33, p. 449-465. *In* N. J. W. Kreger-van Rij (ed.), The yeasts - A taxonomic study. Elsevier Science Publ. B. V., Amsterdam.
69. Zottola, E. A., and H. W. Walker. 1984. Osmophilic microorganisms, p. 170-175. *In* M. L. Speck (ed.), Compendium of methods for the microbiological examination of foods, 2nd ed. Am. Public Health Assoc., Washington, D. C.

Chapter 18

Pectinolytic and Pectolytic Microorganisms

R. F. McFeeters, L. Hankin, and G. H. Lacy

18.1 INTRODUCTION

Pectic substances are important cell wall components of higher plants, particularly dicots. The specific functions that pectic materials perform in the cell wall are not understood. However, it appears that they are important in cementing plant cells together. BeMiller[6] has reviewed pectin structure. Pectic substances are polymers of D-galacturonic acid residues glycosidically linked alpha-1,4. Pectin molecules contain occasional rhamnose units with 1,2 linkages in the main chain. They also have side chains on both the galacturonic acid and rhamnose residues that contain mainly galactose and arabinose residues. The carboxyl groups of galacturonic acid residues are usually methylesterified to a substantial degree. This has major effects on physical properties of pectin, such as gelation. The degree of methylation in plants has been found to range from about 40% to 90% of the carboxyl groups.[2] Data on the distribution of carboxyl groups are limited, but it appears most often to be random in pectin isolated to minimize enzymatic or chemical modification.[2] In some plants, substantial numbers of the hydroxyl groups of galacturonic acid residues are acetylated. This modification inhibits pectin gelation.[6]

The nomenclature for pectic substances has been somewhat variable and confused over the years. For the purpose of this chapter, pectic substances is an inclusive term for galacturonic acid-containing polymers from plant cell walls. Pectin is used for pectic substances with a substantial fraction of the galacturonic acid carboxyl groups esterified. Pectic acid refers to polymers with a negligible amount of the carboxyl groups esterified. Polypectate or pectate refers to pectic acid with carboxyl groups in the salt form. Pectinolytic refers to the degradation of pectin and pectolytic to the degradation of pectic acid or pectate. Commercially available pectin of the type used for microbiological or enzymatic assays generally has >60% methylation. Pectic acid or polypectate is <5% esterified.

18.11 *Sources of Pectinolytic and Pectolytic Enzymes*

Most pectin-degrading organisms are associated with raw agricultural products and with soil. Up to 10% of the organisms in soil have been shown to be pectinolytic.[23] These include, but are not limited to, bacteria in the genera *Achromobacter, Aeromonas, Arthrobacter, Agrobacterium, Enterobacter, Bacillus, Clostridium, Erwinia, Flavobacterium, Pseudomonas,* and *Xanthomonas*,[54,59] as well as in yeasts, molds, protozoa, and nematodes.[3] Many of these organisms are plant pathogens.[3,17] A survey of yeasts from 13 genera[8] found pectinolytic activity produced by 69 of 207 strains tested. Pectolytic activity is not common in lactic acid bacteria. However, it has now been reported to occur in at least one strain of *Leuconostoc mesenteroides*,[30] *Lactobacillus acidophilus*,[34] *Lactobacillus casei, Lactobacillus plantarum*, and *Lactococcus lactis*.[31] The molecular biology of pectolytic enzymes from plant pathogenic erwiniae has been extensively investigated.[12,36,51] The discussion that follows refers to aerobic procedures. However, the detection of anaerobic pectinolytic bacteria also has been described.[41,46]

18.2 DETECTING PECTINOLYTIC AND PECTOLYTIC ORGANISMS

The basic method used to detect pectinolytic or pectolytic organisms has been to grow the organisms on a gel medium that contains pectin or pectate substrates, respectively. Production of enzymes by a culture is detected either by observing depressions in the gel around the colony where the substrate has been degraded or by flooding the plate with a precipitant solution. Around producer colonies a clear zone will appear where the substrate has degraded to the point that precipitation does not occur, while non-producing colonies will be surrounded by opaque gel containing the non-degraded pectin or pectate substrate. Wieringa[60] reported the first medium of this type. Over the years, many variations of this theme have been developed to address particular problems of sample handling, enzyme specificity, sensitivity, or isolation of organisms from the plates. For example, a researcher can cut holes in agar gel plates with a cork borer in order to assay liquid samples, such as culture filtrates, for enzyme activity. This is the so-called well plate or cup plate technique for enzyme assays.

18.21 *Pectate and Pectin Lyase Producers*

Considerable research has been conducted on techniques to detect pectate lyase-producing organisms. This is because these enzymes from *Erwinia* species have been cloned into *Escherichia coli.* Since lyases have alkaline pH optima, while poly-galacturonases have acidic optima, a medium pH of 7.0 or above is the main parameter used to distinguish pectate or pectin lyase producers from polygalacturonase producers. Durrands and Cooper[18] provide an example of this approach, in which media were designed to detect polygalacturonase and pectin lyase production by *Verticillium albo-atrum* mutants. Roberts et al.[52,53] and Allen et al.[1] used the pH 8.5 PEC-YA medium of Starr et al.[57] to clone pectate lyase genes from *Erwinia carotovora.*

18.211 Bacterial Pectate Lyases

Several media have been developed for the detection of bacteria that produce pectate lyases. King and Vaughn[35] developed a pectate medium with a pH of 7.0 that contained crystal violet to make it selective for gram-negative bacteria. Adding cycloheximide further inhibits the growth of yeasts and molds. Detection of pectolytic colonies is based on formation of depressions in the pectate gel because of enzymatic degradation.

Hankin et al.[24] used a mineral medium with 0.1% yeast extract, pectin, and agar to detect pectolytic colonies of *Erwinia* and *Pseudomonas.* A 1% aqueous solution of hexadecyltrimethyl-ammonium bromide[29] was used to precipitate non-degraded substrate so that pectolytic colonies showed a clear zone on a white background. The researchers emphasized that a high phosphate level in the medium was needed to observe pectolytic activity. Sands et al.[55] modified the medium of King and Vaughn[35] by using 2% pectin and 1.5% agar instead of 7% polypectate. They then added a mixture of novobiocin, penicillin G, and cycloheximide to make the medium selective for fluorescent pseudomonads. Hexadecyltrimethylammonium bromide solution was used to precipitate the pectin for visualization of clear zones around pectolytic colonies.

Cuppels and Kelman[15] did a detailed evaluation of the selectivity for and recovery of *Erwinia* from natural samples using another pectate medium containing crystal violet to prevent the growth of unwanted organisms. The medium gave an excellent recovery of pectolytic *Erwinia,* but did allow growth and enzyme production by some pseudomonads. This medium has also been used to isolate pectolytic strains of *Cytophaga johnsonae* from spoiled, fresh bell peppers and watermelon.[39] Woodward and Robinson[62] modified crystal violet medium by addition of proteose peptone to reduce the time required to detect pectolytic *Erwinia* and improve pit formation. Pierce and McCain[48] modified the medium of Miller and Schroth[43] to selectively plate for pectolytic *Erwinia.* They found it to be selective for *Erwinia* and to improve recovery compared to the Cuppels and Kelman crystal violet medium. With all of these media, it has been reported that not all commercial polypectate preparations give suitable gelation. A procedure has been described to produce polypectate from orange peel and apple pulp that will give a good gel.[13]

An essential element of work to identify and characterize pectin-degrading genes from *Erwinia chrysanthemi* was the development of plating techniques to make it possible to identify clones that contained the genes of interest. Keen et al.[33] described the isolation of *E. coli* clones that contain pectate lyase genes. They used a pectate agar at pH 8.0. After incubating samples, they detected lyase activity by flooding the plates with 1 M $CaCl_2$. A white halo formed around positive clones. Kotoujansky et al.[37] developed a technique to isolate clones of lambda-L47-1 phage to which pectate lyase genes had been transferred. They used a medium for *E. coli,* the phage host, in one layer, and a pectate medium in the bottom layer. The two gels were separated by a nylon membrane that allowed enzymes to diffuse into the pectate layer, but prevented transfer of the phage. Zones with pectate lyase activity were visualized by removing the nylon membrane and the upper gel and flooding the pectate medium with 1 M $CaC1_2$. The nylon membrane was then placed back on the plate so that phage clones could be isolated from the appropriate plaque.

18.212 Bacterial Pectin Lyases

Pectin lyases should give clearing zones on plates with pectin as the substrate. However, plates are not very sensitive for this group of pectic enzymes, and the zones produced can be indistinct.[56] Detection is accomplished by spectrophotometric assays at 235 nm on culture filtrate samples with high methyoxl pectin as substrate. Pectin lyases will give little or no measurable activity with polypectate as the substrate.

18.213 Fungal Pectate Lyases

Hankin and Anagnostakis[22] describe a plate technique with a medium that contains 1% pectin, 0.2% yeast extract, mineral salts, and 3% agar adjusted to pH 7.0. After a 3- to 5-day incubation period, plates are flooded with 1% aqueous hexadecyltrimethyl-ammonium bromide to precipitate non-degraded pectin. Clear zones occur around colonies that produce pectate lyase. If the precipitant is not allowed to remain in contact with fungal cells for more than 5 min, viable colonies can be isolated from the flooded plates. For fungal samples from natural isolations that contain bacteria, a mixture of the antibiotics neomycin and chloramphenicol provides control of bacterial growth with the least inhibition of growth or enzyme production by fungi. However, the authors emphasize that fungi should be purified and enzyme production checked in the absence of the antibiotics.

18.22 Fungal Polygalacturonase Producers

The detection of polygalacturonases by plate assays has generally been done simply by lowering the pH of a medium designed for detection of pectate lyase to 6 or below, so that poly-galacturonases will be active and pectate lyases will be inactive.[22,58]

18.23 Detecting Pectic Enzymes during Germination of Fungal Spores

Hagerman et al.[21] described a plate procedure for detection of pectolytic enzymes, protease, and cellulase activity during germination of *Botrytis cinerea* spores. The method is very sensitive for the detection of lyases and pectinesterase, but it is considerably less sensitive for polygalacturonase.

18.24 Evaluating Macerating Activity of Pectic Enzymes

Mussell and Morre[44] analyzed in detail factors affecting the maceration of cucumber tissue by commercial polygalacturonase from *Aspergillus niger,* basing their procedure on the measurement of weight loss of the cucumber tissue after enzyme treatment. They pointed out that tissue maceration assay for polygalacturonase activity was about 500 times more sensitive than viscosity assays. Ishii[28] developed a procedure to evaluate the maceration of potato, onion, and radish tissues by measuring the volume of separated cells released from tissue samples. A polygalacturonase and pectate lyase, separately and in combination with *Aspergillus japonicus,* were used. Both enzymes caused tissue maceration, but the relative activity of the enzymes varied with the plant tissue.

18.3 VISUALIZATION AND ASSAY OF PECTIC ENZYMES

18.31 Detecting Pectic Enzymes in Electrophoresis Gels

The visualization of pectic enzymes in gels following electrophoresis has been described. Cruickshank and Wade[14] incorporated citrus pectin into acrylamide slab gels. By using suitable incubation and staining procedures, they were able to differentiate pectate lyases, polygalacturonases, and pectinesterases in the gels. Bertheau et al.[7] developed a sandwich technique in which suitable buffers and substrates could diffuse into acrylamide slab gels so that a variety of enzymes, including pectate lyases and polygalacturonases, could be detected in the gels.

18.32 Detecting Pectic Enzymes on Isoelectric Focusing Gels

A sensitive and rapid method for visualizing the isoelectric profiles of pectic enzymes using activity overlays has been devised.[50] Isoelectrically focused proteins in ultra-thin (0.35 mm) 5% acrylamide gels containing broad-range (pH 3.5 to 10) and high-pH (pH 9 to 11) ampholytes bonded to plastic support (Gel Bond PAG acrylamide support, FMC BioProducts, Rockland, MD 04841-2994) are overlaid with an ultra-thin (0.35 mm) layer of 1% agarose containing 0.1% polygalacturonic acid bonded on Gel Bond agarose support and also supported on plastic. After sufficient reaction time, the overlay film is stained with ruthenium red (0.05% wt/vol). The substrate gel stains a deep pink because of reaction of the ruthenium red with polygalacturonate, while the substrate in areas over the isoelectrically focused pectic enzymes has been degraded so that a clear band shows on the pink background.

18.33 Pectinesterase

Pectinesterase can be assayed based on the release of either free carboxyl groups or methanol from pectin. The most common technique is to measure the rate of release of free carboxyl groups from pectin using a pH stat.[38] A rapid, continuous spectrophotometric assay has been developed.[20] It is based on measuring the change in absorbance of a pH indicator as the pH decreases because of the formation of free carboxyl groups in pectin. Methanol may be measured colorimetrically[61] or by gas chromatography.[42] For colorimetric analysis, the modifications described by Hudson and Buescher[26] should be used because the incubation procedure of Wood and Siddiqui[61] results in nonlinear color development.

18.34 Pectate and Pectin Lyase

Pectate lyase activity can be assayed spectrophotometrically at 235 nm because of the formation of the 4,5-double bond in the non-reducing galacturonic acid residues.[57] For pectin lyase activity, pectin may be substituted for polypectate as the substrate. A pectin lyase should show little or no activity with polypectate as the substrate.

18.35 Polygalacturonase

The release of reducing groups because of hydrolysis of polypectate is the most common method for measurement of polygalacturonase activity. Lee and MacMillan[38] used 0.5% polypectate in 0.1 M sodium acetate buffer, pH 4.5 at a temperature of 30°C for tomato polygalacturonase. These are reasonable conditions for most microbial polygalacturonases, though 0.1% polypectate is usually a saturating substrate level. For measurement of reducing groups, the procedure of Nelson[45] has been used for many years. However, this method requires the addition of one reagent solution before heating and a second after heating for color development. Also, the samples must be centrifuged prior to measurement of the absorbance because of cloudiness caused by precipitated substrate. A reducing sugar method using 2,2'-bicinchoninate[16] is highly sensitive and avoids the inconveniences of the Nelson procedure. Gross[19] demonstrated the use of 2-cyanoacetamide reagent, which also allows detection of reducing groups using a single reagent solution. Absorbance is measured at the UV region, 276 nm. Crude extracts with high levels of ultraviolet-absorbing material may cause interference with this reagent.

Sensitive, semi-quantitative assays to assess softening activity in pickle fermentation brines and spent brines that are recycled have been developed specifically for this industry. Bell et al.[4] measured loss in viscosity of pH 5 polypectate solutions after a 20-hr or 44-hr incubation period. Buescher and Burgin[9] have described a diffusion plate assay that is in common use to determine if there is polygalacturonase activity in brines before they are recycled.

Pectic enzymes are inhibited by phenolic compounds, indoleacetic acid, fatty acid, and endopolygalacturonase end-products.[3] Tannins from certain plants inhibit both pectic enzymes and cellulase.[5] Specific protein inhibitors of polygalacturonases have also been reported in various plants. Cervone et al.[11] purified an inhibitor protein from French bean hypocotyls and demonstrated that it inhibited polygalacturonases from *Colletotrichum lindemuthianum*, *Fusarium moniliforme*, and *Aspergillus niger*. Diethylpyrocarbonate, which is used to chemically modify histidyl residues in proteins, has been shown to inhibit an endo-polygalacturonase from *Aspergillus ustus*.[49]

18.4 SOURCES OF PECTIC ENZYMES

Purified endo-splitting polygalacturonases and pectin lyases from *Aspergillus niger* and *Aspergillus japonicus* are available commercially. However, the preparations each contain low levels of the other enzyme as a contaminant. The ability to clone pectate lyase genes from *E. chrysanthemi* into *E. coli* makes available clones with sequenced genes that are good producers of individual *E. chrysanthemi* pectate lyases.[32] In addition, *E. coli* clones containing *E. carotovora* genes for endo- and exo-pectate lyases are available.[52,53] Since *E. coli* does not produce pectic enzymes, these clones will not produce pectinesterase or polygalacturonase. Phaff[47] described a yeast, *Kluyveromyces fragilis*, that reliably secretes large amounts of polygalacturonase into the growth medium. The polygalacturonase was estimated to be about 95% pure in the culture filtrate. The organism does not produce either pectinesterase or pectate lyase. Three or more polygalacturonases are present in the preparation.[10,27,40] All of the enzymes appear to be endo-splitting with similar tissue-macerating properties.[40] Despite the presence of multiple enzymes, this is a source for easily produced polygalacturonases that are free of other classes of pectic enzymes. The culture can be obtained from Dr. Phaff for a nominal handling free.

18.5 EQUIPMENT, MATERIALS, AND REAGENTS

Crystal violet pectate (CVP) medium
Medium to detect pectate lyase (MP-7 medium)
Medium to detect polygalacturonase (MP-5 medium)

Polypectate gel medium
Polysaccharide precipitant
Selective medium for fluorescent pectinolytic pseudomonads (FPA medium)

18.6 PROCEDURES

18.61 Sample Preparation

Prepare a homogenous suspension and appropriate dilutions of the food (Chapter 2). In food containing a large amount of glucose, sufficient dilution may be necessary to avoid possible catabolite repression of pectic enzyme synthesis.[25] Also, in high-sugar foods such as jams and jellies, exercise care in the dilution of the sample to prevent osmotic shock to the cells by the use of an appropriate buffer or isotonic solution.

18.62 Preparation and Incubation of Plates

Place not more than 0.25 mL quantities of appropriate dilutions (to obtain 20 to 30 colonies and to avoid a wet agar surface) on the surface of prepoured plates. Distribute by a spread plate technique. Incubation of plates inoculated with a variety of plant materials has been carried out at 30°C for 48 hr. Incubation of plates inoculated with foods probably should be made at the temperature at which the food is stored. Some pectinolytic bacteria grow at 37°C, but do not produce pectate lyase until the temperature is 32°C or below. *E. carotovora* subsp. *atroseptica* often will not grow above 28°C.

18.63 Counting and Reporting

Colonies that cause depressions in polypectate gel medium or that are surrounded by a clear zone after flooding plates with a pectin precipitant are pectinolytic. Fluorescent bacteria are detected on FPA plates with long-wavelength UV light before adding pectin precipitant.

18.7 INTERPRETATION

The varieties of organisms that can degrade pectic substances are diverse. Thus, choice of the most suitable medium depends on the type of organism expected, as well as the type of food to be examined. Media and procedures are described for the detection, enumeration, and isolation of pectolytic and pectinolytic microorganisms. In addition, procedures are described for the detection of pectic enzymes. These techniques can be used to characterize degradation of fruits, vegetables, and processed foods, to screen for plant pathogens or soil organisms that degrade pectic substances, and to indicate the stability of raw plant products for storage and transport.

18.8 REFERENCES

1. Allen, C., H. George, Z. Yang, G. H. Lacy, and M. S. Mount. 1987. Molecular cloning of an *endo*-pectate lyase gene from *Erwinia carotovora* subsp. *atroseptica*. Physiol. Molec. Plant Pathol. 31:325.
2. Anger, H., and G. Dongowski. 1985. Distribution of free carboxyl groups in native pectins from fruits and vegetables. Die Nahrung 29:397.
3. Bateman, D. F., and R. L. Millar. 1966. Pectic enzymes in tissue degradation. Ann. Rev. Phytopath. 4:119.
4. Bell, T. A., J. L. Etchells, and I. D. Jones. 1955. A method for testing cucumber salt-stock brine for softening activity. U.S. Dep. Agric. Agric. Res. Serv. 72-5.
5. Bell, T. A., J. L. Etchells, and W. W. G. Smart Jr. 1965. Pectinase and cellulase enzyme inhibitor from sericea and certain other plants. Bot. Gaz. Chicago 126:40.
6. BeMiller, J. N. 1986. An introduction to pectins: Structure and properties, p. 2. *In* M. L. Fishman, J. J. Jen (ed.) Chemistry and function of pectins. Am. Chem. Soc., Washington, D. C.
7. Bertheau, Y., E. Madgidi-Hervan, A. Kotoujansky, C. Nguyen-The, T. Andro, and A. Coleno. 1984. Detection of depolymerase isoenzymes after electrophoresis or electrofocusing, or in titration curves. Anal. Biochem. 139:383.
8. Biely, P., and E. Slavikova. 1994. New search for pectolytic yeasts. Folia Microbiol. 39:485-488.
9. Buescher, R. W., and C. Burgin. 1992. Diffusion plate assay for measurement of polygalacturonase activities in pickle brines. J. Food Biochem. 16:59-68.
10. Call, H. P., and C. C. Emeis. 1983. Characterization of an "endopolygalacturonase" of the yeast *Kluyveromyces marxianus*. J. Food Biochem. 7:59.
11. Cervone, F., G. De Lorenzo, L. Degra, G. Salvi, and M. Bergami, M. 1987. Purification and characterization of a polygalacturonase-inhibiting protein from *Phaseolus vulgaris* L. Plant Physiol. 85:631.
12. Collmer, A., and N. T. Keen. 1986. The role of pectic enzymes in plant pathogenesis. Ann. Rev. Phytopath. 24:383.
13. Cother, E. J., A. B. Blakeney, and S. J. Lamb. 1980. Laboratory-scale preparation of sodium polypectate for use in selective media for pectolytic *Erwinia* spp. Plant Dis. 64:1086.
14. Cruickshank, R. H., and G. C. Wade. 1980. Detection of pectic enzymes in pectin-acrylamide gels. Anal. Biochem. 107:177.
15. Cuppels, D., and A. Kelman. 1974. Evaluation of selective media for isolation of soft-rot bacteria from soil and plant tissue. Phytopathology 64:468.
16. Doner, L. W., and P. L. Irwin. 1992. Assay of reducing end-groups in oligosaccharide homologues with 2,2′-bicinchoninate. Anal. Biochem. 202:50-53.
17. Dowson, W. J. 1957. Plant diseases due to bacteria, 2nd ed. Cambridge University Press, New York.
18. Durrands, P. K., and R. M. Cooper. 1988. Development and analysis of pectic screening media for use in the detection of pectinase mutants. Appl. Microbiol. Biotechnol. 28:463.
19. Gross, K. C. 1982. A rapid and sensitive spectrophotometric method for assaying polygalacturonase using 2-cyanoacetamide. HortScience 17:933.
20. Hagerman, A. E., and P. J. Austin. 1986. Continuous spectrophotometric assay for plant pectin methyl esterase. J. Agric. Food Chem. 34:440.
21. Hagerman, A. E., D. M. Blau, and A. L. McClure. 1985. Plate assay for determining the time of production of protease, cellulase, and pectinases by germinating fungal spores. Anal. Biochem. 151:334.
22. Hankin, L., and S. L. Anagnostakis. 1975. The use of solid media for detection of enzyme production by fungi. Mycologia 67:597.
23. Hankin, L., D. C. Sands, and D. E. Hill. 1974. Relation of land use to some degradative enzymatic activities of soil bacteria. Soil Sci. 118:38.
24. Hankin, L., M. Zucker, and D. C. Sands. 1971. Improved solid medium for the detection and enumeration of pectolytic bacteria. Appl. Microbiol. 22:205.
25. Hsu, E. J., and R. H. Vaughn. 1969. Production and catabolite repression of the constitutive polygalacturonic acid *trans*-eliminase of *Aeromonas liquefaciens*. J. Bacteriol. 98:172.
26. Hudson, J. M., and R. W. Buescher. 1986. Relationship between degree of pectin methylation and tissue firmness of cucumber pickles. J. Food Sci. 51:138.
27. Inoue, S., Y. Nagamatsu, and C. Hatanaka. 1984. Preparation of cross-linked pectate and its application to the purification of endopolygalacturonase of *Kluyveromyces fragilis*. Agric. Biol. Chem. 48:633.

28. Ishii, S. 1976. Enzymatic maceration of plant tissues by endo-pectin lyase and endo-polygalacturonase from *Aspergillus japonicus.* Phytopathology 66:281.
29. Jayasankar, N. P., and P. H. Graham. 1970. An agar plate method for screening and enumerating pectinolytic microorganisms. Can. J. Microbiol. 16:1023.
30. Juven, B. J., P. Lindner, and H. Weisslowicz. 1985. Pectin degradation in plant material by *Leuconostoc mesenteroides.* J. Appl. Bacteriol. 58:533.
31. Karam, N. E., and A. Belarbi. 1995. Detection of polygalacturonases and pectin esterases in lactic acid bacteria. World J. Microbiol. Biotech. 11:559-563.
32. Keen, N. T., and S. Tamaki. 1986. Structure of two pectate lyase genes from *Erwinia chrysanthemi* EC16 and their high-level expression in *Escherichia coli.* J. Bacteriol. 168:595.
33. Keen, N. T., D. Dahlbeck, B. Staskawicz, and W. Belser. 1984. Molecular cloning of pectate lyase genes from *Erwinia chrysanthemi* and their expression in *Escherichia coli.* J. Bacteriol. 159:825.
34. Kim, S. D., K. S. Jang, Y. A. Oh, M. J. Kim, and Y. J. Jung. 1991. Characteristics of polygalacturonases produced from *Lactobacillus acidophilus.* J. Korean Soc. Food Nutr. 20:488-493.
35. King, A. D., Jr., and R. H. Vaughn. 1961. Media for detecting pectolytic gram-negative bacteria associated with the softening of cucumbers, olives, and other plant tissues. J. Food Sci. 26:635.
36. Kotoujansky, A. 1987. Molecular genetics of pathogenesis by soft-rot erwinias. Ann. Rev. Phytopath. 25:405.
37. Kotoujansky, A., A. Diolez, M. Boccara, Y. Bertheau, T. Andro, and A. Coleno. 1985. Molecular cloning of *Erwinia chrysanthemi* pectinase and cellulase structural genes. EMB0 J. 4:781.
38. Lee, M., and J. D. MacMillan. 1968. Mode of action of pectic enzymes. I. Purification and certain properties of tomato pectinesterase. Biochemistry 7:4005.
39. Liao, C.-H., and J. M. Wells. 1986. Properties of *Cytophaga johnsonae* strains causing spoilage of fresh produce at food markets. Appl. Environ. Microbiol. 52:1261.
40. Lim, J., Y. Yamasaki, Y. Suzuki, and J. Ozawa. 1980. Multiple forms of endo-polygalacturonase from *Saccharomyces fragilis.* Agric. Biol. Chem. 44:473.
41. Lund, B. M. 1972. Isolation of pectolytic clostridia from potatoes. J. Appl. Bacteriol. 35:609.
42. McFeeters, R. F., and S. A. Armstrong. 1984. Measurement of pectin methylation in plant cell walls. Anal. Biochem. 139:212.
43. Miller, T. D., and M. N. Schroth. 1972. Monitoring the epiphytic population of *Erwinia amylovora* on pear with a selective medium. Phytopathology 62:1175-1182.
44. Mussell, H. W., and D. L. Morre. 1969. A quantitative bioassay specific for polygalacturonases. Anal. Biochem. 28:353.
45. Nelson, N. 1944. A photometric adaptation of the Somogyi method for the determination of glucose. J. Biol. Chem. 153:375.
46. Ng, H., and R. H. Vaughn. 1963. *Clostridium rubrum* sp. *N.* and other pectinolytic clostridia from soil. J. Bacteriol. 85:1104.
47. Phaff, H. J. 1966. Alpha-2,4-polygalacturonide glycanohydrolase (endo-polygalacturonase) from *Saccharomyces fragilis*, p. 636. *In* E. F. Neufeld and V. Ginsburg (ed.) Methods in enzymology, complex carbohydrates, vol. 8. Academic, New York.
48. Pierce, L., and A. H. McCain. 1992. Selective medim for isolation of pectolytic *Erwinia* sp. Plant Dis. 76:382-384.
49. Rao, M. N., A. A. Kembhavi, and A. Pant. 1996. Implication of tryptophan and histidine in the active site of endo-polygalacturonase from *Aspergillus ustus*: Elucidation of the reaction mechanism. Biochem. Biophys. Acta 1296:167-173.
50. Reid, J. L., and A. Collmer. 1985. Activity stain for rapid characterization of pectic enzymes in isoelectric focusing and sodium dodecyl sulfate-polyacrylamide gels. Appl. Environ. Microbiol. 50:615.
51. Robert-Baudouy, J. 1991. Molecular biology of *Erwinia*: From soft-rot to antileukaemics. Trends Biotechnol. 9:325-329.
52. Roberts, D. P., P. M. Berman, C. Allen, V. K. Stromberg, G. H. Lacy, and M. S. Mount. 1986. *Erwinia carotovora:* Molecular cloning of a 3.4 kilobase DNA fragment mediating production of pectate lyases. Can. J. Plant Pathol. 8:17.
53. Roberts, D. P., P. M. Berman, C. Allen, V. K. Stromberg, G. H. Lacy, and M. S. Mount. 1986. Requirement for two or more *Erwinia carotovora* subsp. *carotovora* pectolytic gene products for maceration of potato tuber tissue by *Escherichia coli.* J. Bacteriol. 167:279.
54. Rombouts, F. M. 1972. Occurrence properties of bacterial pectate lyases. Agric. Res. Report (Versl. Landbouwk. Onderz.) 779. Cent. for Agric. Publ. and Doc., Wageningen, The Netherlands.
55. Sands, D. C., L. Hankin, and M. Zucker. 1972. A selective medium for pectolytic fluorescent pseudomonads. Phytopathology 62:998.
56. Schlemmer, A. F., C. F. Ware, and N. T. Keen. 1987. Purification and characterization of a pectin lyase produced by *Pseudomonas fluorescens* W51. J. Bacteriol. 169:4493.
57. Starr, M. P., A. K. Chatterjee, P. B. Starr, and G. E. Buchanan. 1977. Enzymatic degradation of polygalacturonic acid by *Yersinia* and *Klebsiella* species in relation to clinical laboratory procedures. J. Clin. Microbiol. 6:379.
58. Vaughn, R. H., G. D. Balatsouras, G. K. York II, and C. W. Nagel. 1957. Media for detection of pectinolytic microorganisms associated with softening of cucumbers, olives, and other plant tissues. Food Res. 22:597.
59. Voragen, A. G. J. 1972. Characterization of pectin lyases on pectins and methyl oligogalacturonates. Agric. Res. Report (Versl. Land--bouwk. Onderz.) 780. Cent. for Agric. Publ. and Doc., Wageningen, The Netherlands.
60. Wieringa, K. T. 1949. A method for isolating and counting pectolytic microbes. 4th Int. Cong. Microbiol. Proc., p. 482.
61. Wood, P. J., and I. R. Siddiqui. 1971. Determination of methanol and its application to measurement of pectin ester content and pectin methyl esterase activity. Anal. Biochem. 39:418.
62. Woodward, E. J., and K. Robinson. 1990. An improved formulation and method of preparation of crystal violet pectate medium for detection of pectolytic *Erwinia.* Lett. Appl. Microbiol. 10:171-173.

Chapter 19

Acid-Producing Microorganisms

Paul A. Hall, Loralyn Ledenbach, and Russell S. Flowers

19.1 INTRODUCTION

Acid-producing microorganisms are ubiquitous in nature and associated with many raw and processed food products. The categorization of bacteria based on their ability to produce acid has traditionally been an important taxonomic tool. The production of acid via bacterial fermentation has been both a boon and bane to mankind's food supply since antiquity. Exploitation of the microbial attribute of acid production has led to a rich variety of fermented foods and beverages enjoyed around the world. On the other hand, spoilage of food and beverages through unwanted microbial fermentation leads to losses amounting to millions of dollars annually. The challenge facing the food microbiologist is to exploit the beneficial aspects of microbial fermentation while controlling the harmful impact, depending upon the food or beverage product under consideration.

One of the most industrially important groups of acid-producing bacteria is the lactic acid bacteria. This diverse group of gram-positive bacteria is generally characterized as non-spore forming, non-respiring cocci or rods, which produce lactic acid as the major end-product during the fermentation of carbohydrates. Historically, this group has included the genera *Lactobacillus, Leuconostoc, Pediococcus*, and *Streptococcus*. Currently, taxonomists generally consider the following genera to comprise this group: *Aerococcus, Alliococcus, Carnobacterium, Dolosigranulum, Enterococcus, Globicatella, Lactobacillus, Lactococcus, Lactosphaera, Leuconostoc, Oenococcus, Pediococcus, Streptococcus, Tetragenococcus, Vagococcus*, and *Weissella*.[2] Phenotypic characteristics have traditionally been used to classify lactic acid bacteria. These include cell morphology, growth temperature, type of glucose and other carbohydrate fermentation, including presence or absence of gas production, configuration of the lactic acid produced, salt tolerance, acid/alkaline tolerance, fatty acid composition, and cell wall composition. Additionally, today, taxonomists rely upon genotypic and phytogenetic characterization such as rRNA sequence determination and nucleic acid probe techniques for classifying these organisms.

Many of the genera in this group form phylogenetically distinct clusters. Additionally, some genera such as *Lactobacillus* and *Pediococcus* form phylogenetic clusters that do not correlate with classification schemes based on phenotypic characteristics. The genus *Bifidobacterium* is often grouped with the lactic acid bacteria and shares some of their features, however, they are phylogenetically unrelated.

One classically important phenotypic approach for categorizing the lactic acid bacteria is the type of sugar fermentation pathway utilized. The homofermentative lactics produce almost exclusively lactic acid as the end-product via the glycolytic Embden-Meyerhof pathway (homolactic fermentation). The heterofermentative lactics, via the 6-phosphogluconate/phosphoketolase pathway, produce significant amounts of other end-products such as ethanol, acetate, and CO_2, in addition to lactic acid (heterolactic fermentation). The metabolic pathway employed by a particular group of lactic acid bacteria can be a critical attribute impacting either the beneficial use or spoilage potential of that group in food or beverage products. Additionally, there are spore-forming bacteria such as *Sporolactobacillus* which are not phylogenetically dissimilar from some genera of vegetative lactic acid bacteria. While there is no strong scientific rationale for excluding those sporeformers which otherwise resemble lactic acid bacteria, they are, nevertheless, generally excluded from the group. This exemplifies the fact that there is not uniform agreement on what constitutes the class known as the lactic acid bacteria.

In addition to the lactic acid bacteria, many other types of acid-producing bacteria are important to the food industry from both a beneficial and spoilage perspective. Many sporeforming species such as *Bacillus* and *Clostridium* play a key role in the quality degradation and spoilage of certain foods and beverages. The role of these organisms is discussed in detail in Chapters 22 to 27 and 32 to 34.

Two other genera of acid-producing bacteria that are of importance to the food and beverage industry are *Propionibarterium* and *Acetobacter*. Several species of *Propionibacterium* are commercially important in the production of Swiss-type cheeses, imparting the characteristic flavor and eye production. *Acetobacter* has commercial importance for their use in the manufacture of vinegar.

Many of the *Enterobacteriaceae* carry out either a mixed acid or butyleneglycol fermentation. These organisms can also play a role in the spoilage of food and beverage products. The significance of specific enterics in foods, as well as methods for their detection and enumeration are detailed in Chapters 8, 35, 37, 38, and 41.

19.2 GENERAL CONSIDERATIONS

There are a number of points to consider when attempting to detect and/or enumerate acid-producing microorganisms from food and beverage products or the manufacturing environment.

Monitoring products for the presence of beneficial acid-producing organisms such as the lactic acid starter cultures requires a different approach than attempting to determine the cause of a spoilage problem. Other considerations include the particular type of food or beverage product or ingredient being examined. For example, a raw commodity with a diverse microbial flora would require different isolation/enumeration procedures than a pasteurized product which has little competitive flora. Furthermore, examination of products such as cheese and other fermented dairy products that contain relatively high levels of starter cultures requires special consideration when investigating problems. Lastly, the type of organism of interest will dictate the detection, enumeration and identification procedures employed. Often, special media and/or incubation conditions are required for a specific acid-producing organism or group of organisms of interest.

19.21 Titration and pH

Titratable acidity expressed as lactic acid may be used as an indirect measure of bacterial growth in brines and liquid foods. Measurement of pH reduction is another indicator of growth when investigating potential food spoilage or as a measure of starter culture activity under specified conditions.

19.22 Indicator Media

Complex solid media containing an indicator dye is often used to enumerate different acid-producing types of bacteria present in food products. The acid produced by the colonies will change the color of the surrounding medium and thus facilitate their identification and enumeration. Bromcresol green or bromphenol blue may be used for high acid producers while phenol red may be used for those producing moderate amounts of acid. Incorporation of indicator dyes into various liquid media containing different single sources of carbohydrates is also an important identification technique for these organisms.

19.23 Special Media

Numerous complex media are available that support the growth of various acid-producing bacteria. Some can be employed for the qualitative and quantitative differentiation of certain species such as between the lactic streptococci and propionibacteria. The majority of these media are non-selective and other microorganisms can grow on them so care must be taken in interpreting growth on these types of media. Some media are made selective through the incorporation of inhibitory compounds. For example, most fermentative organisms lack a cytochrome system and are, therefore, able to grow on media containing sodium azide. Other media may incorporate the use of a specific antibiotic to which the organisms of interest are naturally resistant. For example, most *Leuconostoc* spp. are naturally resistant to the antibiotic vancomycin.[35] Some of these organisms can be rather fastidious having special nutritional or other requirements in order to isolate them on laboratory media. For example, some of beer spoilage pediococci require media supplemented with wort extract and incubation under a reduced oxygen atmosphere in order to grow. In many instances, the isolation and numerical estimation of acid-producing bacteria depends on a number of factors. These include the recognition of certain colony characteristics such as size, shape, color, and biochemical reactions in the medium (e.g., acid production, arginine hydrolysis, and citrate utilization), cell morphology under the microscope and, often, identification of individual isolates through biochemical, genetic, or other techniques. A number of miniaturized and automated techniques are available to help in the isolation and identification of these organisms.

19.3 LIMITATIONS

A wide variety of laboratory media and isolation procedures exist for acid-producing microorganisms because of their large diversity as a group. Since many of the media employed for the isolation of these organisms are non-selective there is the potential for other microorganisms to grow. This could lead to an erroneous estimation of the number or type of organisms present. In some circumstances, certain media may only be employed where one or only a few species are present.

Another consideration is the potential for cell injury. Incorporation of selective agents into media may not allow for the recovery of sub-lethally injured or stressed cells, again leading to an erroneous interpretation. Also, many of the lactic acid bacteria are killed or injured by freezing. Therefore, samples to be examined for number of viable lactic acid bacteria should not be frozen prior to analyses. Conversely, if the food product to be examined is normally frozen, it should not be subjected to thawing and refreezing prior to microbial analyses.

Many of the lactic acid bacteria are fastidious in nature requiring special media supplements (e.g. food extracts) or incubation conditions (e.g. reduced oxygen tension) to successfully isolate and quantitate them.

When selecting colonies for further phenotypic or genotypic characterization, care must be taken to ensure purity of the isolates. Mixed isolates may lead to erroneous identification results.

Depending on the purpose of the identification or characterization scheme, reliance on one or a limited number of techniques may lead to misjudgment about an isolate. This is particularly important for the beneficial acid producing organisms such as starter cultures which often have plasmid-mediated desirable traits. While some genetic techniques give an accurate identification of the isolate, they do not necessarily indicate if a particular phenotypic trait has been lost by that isolate. While genetic characterization techniques may provide the most definitive identification of an isolate even to the sub-species level and beyond, phenotypic tests still play a role in functionally describing an isolate, for example, carbohydrate utilization, or exopolysaccharide production. The combination of both genotypic and phenotypic characterization generally provides the greatest level of information about beneficial acid-producing organisms.

19.4 EQUIPMENT, MATERIALS, AND REAGENTS

19.41 Reagents (Chapter 63)

0.01N NaOH
0.1 N NaOH
5N HCl
API 50 CH biochemical test strips (bioMerieux SA, France)
API 50 CHL medium (bioMerieux SA, France)
Bromcresol green solution, in ethanol
Bromcresol green solution, in NaOH
Bromcresol purple solution, 1.6% in ethanol
Butterfield's phophate buffer diluent
Fructose solution, 10.0% in water
0.1% peptone water diluent
RiboPrinter™ test kits
Tartaric acid solution, 10.0% in water

19.42 Media (Chapter 63)

Acetobacter agar[1]
APT agar[7,33]
Beer agar[42]
Bromcresol green agar
Dextrose tryptone agar
Fish extract medium[42]
HHD agar[28]
Hoag-Erickson medium (bioMerieux Vitek, Inc., Hazelwood, MO)
KOT medium[44]
Lactic medium for Bactometer™ (bioMerieux Vitek, Inc., Hazelwood, MO)
Modified NBB medium[21,30]
MRS broth and agar[9]
Plate count agar (Standards methods agar)
Rogosa SL broth[36]

19.43 Equipment

Anaerobic jar (Gas-Pak, BBL), AnaeroPack System™ (Mitsubishi Gas Chemical America, New York), or anaerobic chamber
Bactometer™ system (bioMerieux Vitek, Inc., Hazelwood, MO)
Blender, Stomacher™, or equivalent
Culture tubes, screw cap, for liquid media
Dilution bottles
Durham tubes
Incubators, set at various temperatures
Petri dishes, sterile, plastic or glass
Petrifilm™ Aerobic Count plates and plastic spreader (3M Microbiology Products, Minneapolis, MN)
pH meter
Quebec colony counter or equivalent
RiboPrinter system™ (Qualicon, Wilmington, DE)
Serological pipettes

19.5 PROCEDURES

19.51 Acid-producing Bacteria

A wide variety of procedures exist for the enumeration of acid-producing bacteria. The choice of method can depend upon the type of acid-producer, i.e. beneficial or desirable organisms vs. spoilage organisms, and on the food matrix involved.

Most acid producers grow slowly under aerobic conditions, but perform well if cultivated anaerobically. Because of this, solidified pour plates should be either overlaid with an appropriate agar medium or incubated in an anaerobic atmosphere using an anaerobic chamber, GasPak Anaerobic system (BBL), or other means. Lactic acid bacteria can be damaged if food products are blended with phosphate-buffered diluent, resulting in reduced counts, so sterile 0.1% peptone water diluent is recommended for dilution.[18,22,43]

After plates are incubated, individual colonies need to be gram stained, examined microscopically, and tested for catalase reaction. Gram-positive, catalase-negative cocci or rods may tentatively be considered lactic acid bacteria. Further identification can be performed using standard biochemical tests,[16,20,39] or a number of different rapid methods (Section 19.5.8).

19.511 Total acid-producer count

For a general purpose total count of acid-producers, samples can be pour-plated on plate count agar with added bromcresol purple or dextrose tryptone agar. Plated are incubated at 32 ± 1°C for 48 ± 3 hr for dairy products, or at 35 ± 1°C for 48 ± 3 hr for other products (Chapter 7). After incubation, count the colonies with a yellow halo and report as acid-forming organisms per gram of product.

19.52 Spoilage Lactic Acid-producing Bacteria

The fastidious nature of most lactic acid bacteria, especially the lactococci and lactobacilli, dictates the use of nutritionally complex media. In order to isolate spoilage organisms from a particular food matrix, the product itself is often added to a standard lactic base agar or broth, or a medium is specifically formulated for that commodity.

19.521 MRS Agar with APT Agar Overlay

This method is especially useful for salad dressing products. Stressed organisms are allowed some recovery in the MRS medium, prior to being exposed to the acidified medium. The lower pH of the final pour plate is sufficient to eliminate interference from sporeforming organisms that are often present. MRS broth[9] with added agar is used to prepare pour plates of samples as described in Chapter 7. After the plates have solidified, they are overlaid with APT agar [7,33] that has been acidified to pH 4.0 ± 0.1 with sterile 10.0% tartaric acid solution. Plates are incubated at 35 ± 1°C for 96 ± 4 hr. Identification of individual colonies is performed as in section 19.51.

19.522 Acidified MRS Agar

This medium has been shown to be useful in enumerating total lactic acid bacteria in vegetable products (Armock, personal communication). Pour plates are prepared with MRS agar acidified to pH 5.5 ± 0.1 with sterile glacial acetic acid. Plates are incubated anaerobically without an overlay at 35 ± 1°C for 72 ± 3 hours. Colonies are identified as in section 19.51.

19.523 Acidified MRS Agar with Fructose

In this method, fructose is added to MRS agar to enhance the growth of *Lactobacillus fructivorans* and *Lactobacillus plantarum*, and acidified to increase the selectivity. After MRS agar is autoclaved and tempered, sterile 10.0% fructose solution is added to achieve a final concentration of 1.0%, and the pH is adjusted to 5.4 ± 0.1 with hydrochloric acid. After pour plates have solidified, they are overlaid with additional acidified MRS agar with fructose. The plates are incubated at 30 ± 1°C for 5 days. Colonies are identified as described in section 19.51.

19.524 Modified MRS Agar

Another modification of MRS agar, specifically for the isolation of spoilage lactic acid bacteria from plant materials, is the substitution of di- instead of tri-ammonium phosphate, and the addition of 0.01% 2,3,5-triphenyltetrazolium hydrochloride (TTC).[29] This is used in conjunction with an enrichment procedure. Skim milk supplemented with 0.05% glucose and 0.1% yeast extract is blended with the sample in an MPN procedure (Chapter 6), and incubated at 30 ± 1°C for 3–5 days. Rogosa SL broth[36] with 0.04% cyclohexamide can also be used as an enrichment step,[29] but is incubated in a water bath at 45 ± 1°C. Growth from enrichment tubes is streaked onto modified MRS agar and incubated at 30 ± 1°C anaerobically for 72 ± 3 hr. Colonies are confirmed as lactic acid bacteria as in section 19.51.

19.525 HHD Agar

This medium has been shown to be effective in differentiating between homofermentative and heterofermentative lactic acid-producers in fermented vegetable products.[28] Samples are pour plated with HHD agar and after solidification are overlaid with additional HHD agar. Plates are incubated at 30 ± 1°C for 72 ± 3 hr. Homofermentative organisms appear blue to green, while heterofermentative organisms remain white. Further colony identification of the colonies is performed as in section 19.51.

19.526 MRS Broth–MPN Procedure

This method can be used as a screen for heterofermentative lactic acid-producers. MRS broth tubes containing inverted Durham tubes are inoculated in a 3-tube MPN procedure as in Chapter 6. After incubation at 35 ± 1°C for 4 days, tubes showing gas formation are counted as presumptive for heterofermentative organisms. A gram stain of the broth from suspect tubes is performed to determine if gram-positive rods or cocci are present. If only sporeforming rods are seen, those tubes are considered negative for heterofermentative lactic acid-producers. Calculation of the final count per gram using the confirmed positive tubes is performed as described in Chapter 6. Positive tubes may be streaked for isolation onto agar plates and further identification performed on individual colonies as in section 19.51.

19.527 APT Agar

Originally developed for the enumeration of heterofermentative lactic acid-producers in cured meat products, APT agar can also be used to propagate pediococci.[7,33] Pour plates are prepared as previously described and incubated at 25 ± 1°C for 72 ± 3 hr. Further identification of colonies is performed as in section 19.51.

19.528 APT Agar with Sucrose and Bromcresol Purple

APT agar with added sucrose is effective in enumerating spoilage lactic acid-producers in meat products. The addition of bromcresol purple allows for the easy differentiation between acid-producers and other organisms commonly found on meat. Pour plates of samples are prepared with APT agar to which 20.0 g of sucrose and 2.0 mL of bromcresol purple solution per L have been added prior to sterilization. Plates are incubated at 25 ± 1°C for 48–72 hours. Colonies surrounded by a yellow zone are counted and reported as acid-producers. Identification can be performed as in section 19.51.

19.529 MRS-S Agar

MRS agar modified with the addition of sorbic acid has been shown to aid in the enumeration of spoilage lactic acid bacteria in fermented meat products.[24] This medium is especially useful for preventing the growth of interfering yeasts. After the MRS agar is prepared and tempered, the pH is adjusted to 5.7 ± 0.1 with 5N hydrochloric acid, and sorbic acid dissolved in NaOH is added to give a final sorbic acid concentration of 0.1%. Pour plates are incubated anaerobically for 5 days at 20 ± 1°C, and colonies are confirmed as in section 19.51. Colonies on this medium can be examined for slime production using a sterile loop.[27]

A modification of this method has shown increased effectiveness in enumerating spoilage lactic acid bacteria by inhibiting gram-negative organisms as well as yeast.[11,47] Here 0.1% cysteine hydrochloride is added to the MRS agar. After autoclaving and tempering, it is adjusted to pH 5.7 ± 0.1 with 5N hydrochloric acid, and sorbic acid dissolved in NaOH is added to a final concentration of 0.02%. Plates are prepared and incubated as in the MRS-S method.

19.5210 APT Agar with Glucose

This medium, with the addition of 0.5% glucose, has been shown to be the most effective in isolating spoilage lactic acid bacteria from seafood products.[34] Pour plates of samples in APT agar with glucose are incubated anaerobically at 20 ± 1°C for 72 ± 3 hr, and colonies are identified as in section 19.51. To determine if an isolate is specific to spoilage in seafood, fish extract medium can be employed.[34]

19.5211 Spoilage Lactic Acid-producers in Beer

A number of media have been developed for the enumeration of spoilage lactic acid bacteria in beer, all of which include the addition of beer itself. It is recommended that the particular beer of interest be used in the medium, as this would most likely contain the nutrients needed by the particular spoilage organisms. In all cases, samples are plated on media using the spread plate technique (Chapter 7), and are incubated anaerobically at 30 ± 1°C for 7–14 days. Confirmation of isolated colonies is performed as in section 19.51. A review of detection methods for beer, along with the advantages and disadvantages of each can be found in Jespersen and Jakobsen.[19]

Beer agar. This is a good general medium for the isolation of lactic acid-producers in beer[42] and is used as described in section 19.5211.

Universal beer agar. This medium is also used in the same manner as described in section 19.5211, and is not selective specifically for lactic acid bacteria, but grows all types of beer spoilage organisms.[25]

Modified NBB medium. The spread plate technique, membrane filtration technique (Chapter 6), and an enrichment technique have all been used with this medium.[21,30,42] All methods are incubated and confirmed as in section 19.5211.

KOT medium. The addition of actidione to this medium inhibits the growth of yeasts,[44] and the presence of sodium azide prevents the outgrowth of gram-negative bacteria, so that this medium is more selective for lactic acid bacteria. It is used in the same manner as in section 19.5211, and has been shown to favor the growth of slow-growing lactic acid bacteria.

19.5212 Spoilage lactic acid producers in dairy products

The same organisms that can be beneficial in one fermented dairy product can be spoilage organisms in another dairy product. Differentiation of the various lactic acid-producing bacteria that occur in dairy products is discussed in section 19.53.

19.53 Beneficial Lactic Acid-producing Bacteria

Lactic acid bacteria have been and continue to be exploited by mankind to improve the food supply. This diverse group of organisms has been used since antiquity to improve the flavor, texture, shelf-life, and nutritional value of a wide array of food and beverage products. Lactic acid bacteria are associated with many fermented foods including dairy, meat, vegetables, fish, legumes, and cereal products. Additionally, lactic acid bacteria have been found to impart beneficial health effects in humans including control of gastrointestinal disturbances such as diarrhea, immunomodulation, and cholesterol-lowering activity.[38] These organisms also produce antimicrobial compounds which have commercial applications in the preservation of certain foods. There are at least four classes of bacteriocins (antimicrobial peptides or proteins) produced by the lactic acid bacteria including nisin produced by *Lactococcus lactis*, pediocin produced by *Pediococcus acidilactici*, and carnobacteriocin A produced by *Carnobacter pisicola*.[23] The methods used to test for these organisms will vary by commodity and type of culture. The reader is referred to Chapter 47 of this book for a description of commonly used methods for isolating and identifying dairy starter cultures. Gilliland also provides an in-depth treatment of bacterial starter cultures for foods including, milk, meat, vegetable, and bakery products.[16] In addition, Salminen and von Wright[37] published an excellent treatise on the microbiology and functional aspects of the lactic acid bacteria including their role in human health and nutrition as probiotics and prebiotics, their production of antimicrobial components, and their industrial use and applications.[37]

19.54 Spoilage Acetic Acid-producing Bacteria

While acetic acid-producing bacteria are used in the production of vinegar, they can also spoil beer, wines, and ciders. These organisms are rod-like, aerobic, and gram negative. Asai[1] describes the characterization of these organisms, mainly *Acetobacter* spp., and describes media for their enumeration.

19.55 Spoilage Butyric Acid-producing Bacteria

The bacteria that are capable of producing butyric acid are typically sporeforming organisms. Methods for the enumeration of these sporeformers are outlined in Chapter 23.

19.56 Spoilage Gluconic Acid-producing Bacteria

A single genus of bacteria involved in the spoilage of carbonated and non-carbonated beverages is *Gluconobacter* spp. This organism is an obligately aerobic, catalase-positive, gram-negative rod, and produces off-flavors associated with acid production, as well as a stringy, dextran-like material. The method for isolating this organism is a spread plate method (Chapter 6) using Bromcresol green agar. After plating the samples, plates are incubated aerobically at 25 ± 1°C for 72 ± 3 hr. *Gluconobacter* colonies are typically smooth, regular, and dark green. Individual colonies should be confirmed by gram stain and catalase reaction. For further differentiation, additional tests may be performed.[8]

19.57 Rapid Methods for the Enumeration of Acid-producing Bacteria

The difficulty in developing rapid methods for acid-producing bacteria is in achieving the required selectivity without sacrificing sensitivity. Often the selective agent is some type of acid, which can tend to reduce the numbers of stressed cells able to grow out in pre-enrichment media. Some work has been accomplished in recent years to develop rapid methods for the detection and enumeration of acid-producers, most notably in the area of spoilage *Lactobacillus* spp.

19.571 Petrifilm™ Method for Lactic Acid Bacteria

Petrifilm™ Aerobic Count plates can be used to enumerate homofermentative and heterofermentative lactic acid bacteria in certain food products (Ramos, personal communication). The method is modified by using MRS broth as the diluent and plating according to the standard Petrifilm™ procedure (Chapter 6). Plates are incubated anaerobically at 35 ± 1°C for 48 ± 3 hr. Homofermentative organisms appear as reddish-brown dots, while heterofermenters are reddish-brown dots with an adjoining gas bubble. Colonies may be isolated from the top film for further identification as in section 19.51.

19.572 Bactometer™ Method for Lactobacilli

This test has been developed for detection of spoilage lactobacilli in salad dressings, but may be used for other food products as well. It has the advantage of being able to detect very low numbers of organisms per gram of product. The Bactometer™ incubator is set at 30 ± 1°C. Samples are diluted 1:10 in Hoag-Erickson (HEM) broth and incubated at 30 ± 1°C for 24 ± 1 hr. Aliquots of 0.1 mL of pre-enriched sample are placed in the Bactometer™ module wells containing 0.5 mL of Lactic medium for Bactometer™. Modules are placed into the system and the colony count is correlated to impedence detection time (Chapter 10.6). Any growth in the Bactometer™ module well may be streaked for isolation onto agar plates and identified as in section 19.51.

19.573 DNA-based Detection Methods

The recent burst of activity involving the identification and taxonomy of lactic acid bacteria using DNA-based techniques has also given rise to a number of DNA-based direct detection methods for these organisms. Methods utilizing DNA probes as well as PCR-based methods have been developed and successfully used for detection of lactic acid bacteria in meat, milk, wine, beer, and other fermented foods. Van der Vossen, et al.[45] describes many of these methods and discusses their applications and limitations.

19.58 Identification and Characterization of Acid-producing Bacteria

In addition to the standard biochemical tests used to differentiate and identify lactic acid bacteria, there have been a number of rapid or automated methods developed to characterize these organisms. Several have been developed to the point where they are quite easy to perform, but there are some that are still quite labor intensive. These methods include SDS-PAGE,[10,14,42] cellular fatty-acid analysis,[5,13,15,46] random amplified polymorphic DNA (RAPD) analysis,[4,10,46] r-RNA targeted probe analysis,[31,32,40] and DNA-DNA hybridization.[12,26] The degree of identification and differentiation needed will often determine the choice of method. Van der Vossen, et al.[45] provides a review of some of these different methods.

19.581 API 50 CH Biochemical Test System

One of the earliest miniaturized systems developed for the biochemical characterization of *Lactobacillus* spp. and related organisms is the API 50 CH biochemical test strip (bioMerieux SA, France). Using the API 50 CHL medium (bioMerieux SA, France) as a base, the system enables the fermentation of 49 different carbohydrates to be studied. A suspension of the test organism is made in CHL medium, and each of 50 different wells in the test strips is inoculated. Strips are incubated at 35 ± 1°C, and checked at 24 and 48 hr. The biochemical profile is compared to standard references to arrive at an identification.[17,20] This system has been shown to be able to identify *Lactobacillus* spp. isolated from a number of food products.[11,34,41,42,48]

19.582 RiboPrint™ System for Identification of Lactic Acid Bacteria

Early efforts utilizing ribotyping for the identification of lactic acid bacteria have proven its effectiveness in characterizing organisms isolated from fermented vegetables[3] and dairy products.[6] The procedure used was a manual process, however, and could be difficult to duplicate. The RiboPrinter™ (Qualicon, Wilmington, DE) is an automated ribotyping system that has been recently developed for characterizing a wide variety of bacteria. Barney (personal communication) and Stogards et al.[42] have documented its use for the characterization of lactic acid bacteria isolated from brewery environments, and Ledenbach (personal communication) has also recently applied this technique to the characterization of *Lactobacillus* spp. and *Pediococcus* spp. isolated from spoiled salad dressings. All components of the RiboPrinter™ test kits are loaded into the instrument, and the rest of the procedure is completed in the instrument. Results are obtained in 8 hr from a pure colony isolate.

19.6 REFERENCES

1. Asai,T. 1968. Acetic acid bacteria classification and biochemical activities. Univ. Park Press, Baltimore, Md.
2. Axelsson, L. 1998. Lactic acid bacteria: classification and physiology, p. 1-72. *In* S. Salminen and A. von Wright (ed.), Lactic acid bacteria. Marcel Dekker, New York.
3. Breidt, F., and H. P. Fleming. 1996. Identification of lactic acid bacteria by ribotyping. J. Rapid Meth. Automatic Microbiol. 4:219-233.
4. Cocconcelli, P. S., D. Porro, S. Galandini, and L. Senini., 1995. Development of RAPD protocol for typing of strains of lactic acid bacteria and enterococci. Lett. Appl. Microbiol. 21:376-379.
5. Decallonne, J., M. Delmee, P. Wauthoz, M. El Lioui, and R. Lambert. A rapid procedure for the identification of lactic acid bacteria based on the gas chromatographic analysis of the cellular fatty acids. J. Food Prot. 54:217-224.
6. Decarvalho, A. F., M. Gauthier, and F. Grimont. 1994. Identification of dairy *Propionibacterium* species by ribosomal-RNA gene restriction patterns. Res. Microbiol. 145:667-676.
7. Deibel, R. H., J. B. Evans, and C. F. Niven. 1951. Microbiological assay for the thiamin using *Lactobacillus viridescens*. J. Bacteriol. 62:818-821.
8. de Ley, J., and J. Swings. 1984. Genus Gluconobacter, p. 275-278. *In* N. R. Kreig and J. G. Holt (ed.), Bergey's manual of systematic bacteriology, vol. 1. Williams and Wilkins, Baltimore, Md.
9. de Man, J. C., M. Rogosa, and M. E. Sharpe. 1960. A medium for the cultivation of lactobacilli. J. Appl. Bacteriol. 23:130-135.
10. Drake, M. A., C. L. Small, K. D. Spence, and B. G. Swanson. 1996. Differentiation of *Lactobacillus helveticus* strains using molecular typing methods. Food Res. Int. 29:451-455.
11. Dykes, G. A., T. J. Britz, and A. von Holy. 1994a. Numerical taxonomy and identification of lactic acid bacteria from spoiled, vacuum-packaged vienna sausages. J. App. Bacteriol. 76:246-252.
12. Dykes, G. A., T. E. Cloete, and A. von Holy. 1994b. Identification of *Leuconostoc* species associated with the spoilage of vacuum-packed vienna sausages by DNA-DNA hybridization. Food Microbiol. 11:271-274.
13. Dykes, G. A., T. E. Cloete, and A. von Holy. 1995. Taxonomy of lactic acid bacteria associated with vacuum-packaged processed meat spoilage by multivariate analysis of cellular fatty acids. Int. J. Food Microbiol. 28:89-100.
14. Faia, A. M., L. Patarata, M. S. Pimentel, B. Pot, and K. Kersters. 1994. Identification of lactic acid bacteria isolated from Portuguese wines and musts by SDS-PAGE. J. Appl. Bacteriol. 76:288-293.
15. Gilarova, R., M. Voldrich, K. Demnerova, M. Cerovsky, and J. Dobias. 1994. Cellular fatty acids analysis in the identification of lactic acid bacteria. Int. J. Food Microbiol. 24:315-319.
16. Gilliland, S. E. (ed.). 1985. Bacterial starter cultures for foods. CRC Press, Boca Raton, Fla.
17. Holzapfel, W. H., and M. E. Stiles. 1997. Lactic acid bacteria of foods and their current taxonomy. Int. J. Food Microbiol. 36:1-29.
18. Jayne-Williams, D. J., 1963. Report of a discussion on the effect of the diluent on the recovery of bacteria. J. Appl. Bacteriol. 26:398-404.
19. Jespersen, L., and M. Jakobsen. 1996. Specific spoilage organisms in breweries and laboratory media for their detection. Int. J. Food Microbiol. 33:139-155.
20. Kandler, O., and N. Weiss. 1986. The genus *Lactobacillus*, p. 1208-1234. *In* P. H. A. Sneath, N. S. Mair, M. E. Sharpe, and J. G. Holt (ed.), Bergey's manual of systematic bacteriology, vol. 2. Williams and Wilkins, Baltimore, Md.
21. Kindraka, J. A. 1987. Evaluation of NBB anaerobic medium for beer spoilage organisms. Master Brew. Am. Assoc. Technol. Quart. 24:146-151.
22. King, W. L., and A. Hurst. 1963. A note on the survival of some bacteria in different diluents. J. Appl. Bacteriol. 26: 504-506.
23. Klaenhammer, T. R., C. Ahn, C. Fremaux, and K. Milton. 1992. Molecular properties of *Lactobacillus* bacteriocins, p. 37-58. *In* R. James, C. Lazdunski, and F. Pattus (ed.), Bacteriocins, microcins and lantibiotics, Springer-Verlag, Berlin.
24. Korkeala, H., and S. Lindroth. 1987. Differences in microbial growth in the surface layer and at the center of vacuum-packed cooked ring sausages. Int. J. Food Microbiol. 4:105-110.
25. Kozulis, J. A., and H. E. Page. 1968. A new universal beer agar medium for the enumeration of wort and beer microorganisms. Proc. Am. Soc. Brew. Chem., 52-58.
26. Lonvaud-Funel, A., C. Fremaux, N. Biteau, and A. Joyeux. 1991. Speciation of lactic acid bacteria from wines by hybridization with DNA probes. Food Microbiol. 8:215-222.
27. Makela, P. M., H. J. Korkeala, and J. J. Laine. 1992. Ropy slime-producing lactic acid bacteria contamination at meat processing plants. Int. J. Food Microbiol. 17:27-35.
28. McFeeters, R. F., L. C. McDonald, M. A. Daeschel, and H. P. Fleming. 1987. A differential medium for the enumeration of homo-fermentative and heterofermentative lactic acid bacteria. Appl. Environ. Microbiol. 53:1382-1384.
29. Mundt, J. O., and J. L. Hammer. 1968. Lactobacilli on plants. Appl. Microbiol. 16:1326-1330.
30. Nishikawa, N., and M. Kohgo. 1985. Microbial control in the brewery. Master Brew. Am. Assoc. Technol. Quart. 22:61-66.
31. Nissen, H., and R. Dainty. 1995. Comparison of the use of rRNA probes and conventional methods in identifying strains of *Lactobacillus sake* and *L. curvatus* isolated from meat. Int. J. Food Microbiol. 25:311-315.
32. Nissen, H., A. Holck, and R. Dainty. 1994. Identification of *Carnobacterium* spp. and *Leuconostoc* spp. in meat by genus-specific 16S rRNA probes. Letters Appl. Microbiol. 19:165-168.
33. Niven, C. F., and J. B. Evans. 1951. Nutrition of the heterofermentative lactobacilli that cause greening of cured meat products. J. Bacteriol. 62:599.
34. Novel, G., and S. Maugin. 1994. Characterization of lactic acid bacteria isolated from seafood. J. Appl. Bacteriol. 76:616-625.
35. Orberg, P. K., and E. E. Sandine. 1984. Common occurrence of plasmid DNA and vancomycin resistance in *Leuconostoc* spp. Appl. Environ. Microbiol. 47:677.
36. Rogosa, M., J. A. Mitchell, and R. F. Wiseman. 1951. A selective medium for the isolation of oral and fecal lactobacilli. J. Bacteriol. 62:132-133.
37. Salminen, S., and A. von Wright (ed.). 1998. Lactic acid bacteria. 2nd ed. Marcel Dekker, New York.
38. Salminen, S., M. A. Deighton, Y. Benno, and S. L. Gorbach. 1998. Lactic acid bacteria in health and disease, p. 211-253. *In* S. Salminen and A. von Wright (ed.), Lactic acid bacteria. Marcel Dekker, New York.
39. Schleifer, K. H. 1986. Gram-positive cocci, p. 1208-1234. *In* P. H. A. Sneath, N. S. Mair, M. E. Sharpe, and J. G. Holt (ed.), Bergey's manual of systematic bacteriology, vol. 2. Williams and Wilkins, Baltimore, Md.

40. Schleifer, K. H., J. Ehrmann, C. Beimfohr, E. Brockmann, W. Ludwig, and R. Amann. 1995. Application of molecular methods for the classification and identification of lactic acid bacteria. Int. Dairy J. 5:1081-1094.
41. Smittle, R. B., and R. S. Flowers. 1982. Acid tolerant microorganisms involved in the spoilage of salad dressings. J. Food Prot. 45:977-983.
42. Storgards, E., M. L. Suihko, B. Pot, K. Vanhonacker, D. Janssens, P. L. E. Broomfield, and J. G. Banks. 1998. Detection and identification of *Lactobacillus linderi* from brewery environments. J. Inst. Brew. 104:47-54.
43. Straka, R. P., and J. L. Stokes. 1957. Rapid destruction of bacteria in commonly used diluents and its elimination. Appl. Microbiol. 5:21-25.
44. Taguchi, H., M. Ohkochi, H. Uehara, K. Kojima, and M. Mawatari. 1990. KOT medium, a new medium for the detection of beer spoilage lactic acid bacteria. Am. Soc. Brew. Chem. J. 48:72-75.
45. van der Vossen, J. M. B. M., and H. Hofstra. 1996. DNA based typing, identification and detection systems for food spoilage microorganisms: development and implementation. Int. J. Food Microbiol. 33:35-49.
46. Vogel, R. F., M. Muller, P. Stoiz, and M. Ehrmann. 1996. Ecology in sourdoughs produced by traditional and modern technologies. Adv. Food Sci. 18:152-159.
47. von Holy, A., T. E. Cloete, and W. H. Holzapfel. 1991. Quantification and characterization of microbial populations associated with spoiled, vacuum-packed Vienna sausages. Food Microbiol. 8:95-104.
48. Wijtzes, T., M. R. Bruggeman, M. J. R. Nout, and M. H. Zweitering. 1997. A computerised system for the identification of lactic acid bacteria. Int. J. Food. Microbiol. 38:65-70.

Chapter 20

Yeasts and Molds

L. R. Beuchat and M. A. Cousin

20.1 INTRODUCTION

Yeasts and molds constitute a large and divergent group of microorganisms consisting of several thousand species. Most can be detected in soil[37] and air. Because of their heterotrophic nature and their ability to adapt to a wide range of environmental conditions, these fungi are frequently encountered as actively growing contaminants in and on various commodities including foods, inadequately cleaned food processing equipment, and food storage facilities.

Molds and, to a lesser extent, yeasts can initiate growth over a wide pH range, from below pH 2 to above pH 9. Once growth has been established, given time and the absence of strong buffering, many fungi can change the initial pH of the substrate to one more favorable to their growth, usually a pH of 4 to 6.5. The temperature range for growth of most yeasts and molds is likewise broad (5° to 35°C), with some species capable of growth above or below this range. Many foodborne yeasts and molds, e.g., *Zygosaccharomyces*, *Eurotium*, and *Xeromyces* species, can grow in foods having a water activity (a_w) of 0.85 or below.[15,20,58] Most of the fungi of importance in food spoilage are aerobes.

Yeasts and molds can cause various degrees of food decomposition.[29,90] Invasion and growth may occur on virtually any type of food if environmental conditions are not limiting. Commodities such as corn, small grains, legumes, nuts, and fruits can be invaded before harvesting as well as during storage. Growth can also occur on processed foods and food ingredients. The detectability of fungal invasion of foods depends on the food type, the genera involved, the degree of invasion, and the methods of analysis. Appearance of contaminated food may range from no blemish to severe blemish to complete decomposition. Growth of yeasts and molds may be manifested as rot spots, pustules or scabs, slime, white or variously colored mycelia and spores, and gassing. The absence of visual fungal growth does not mean that yeasts and molds are not present; their numbers may be low or their growth may be internal. Depending on the degree of invasion, substantial economic losses may be sustained by the producer, processor, and consumer.

Numerous foodborne molds can produce mycotoxin,[17,42,75,89] (Chapter 43) and some yeasts and molds are responsible for human and animal infections.[8] Although most infectious fungi are rarely associated with foods, and their infection site is normally not the gastrointestinal tract, some foodborne fungi can cause opportunistic infections in immunocompromised humans. Certain foodborne yeasts and molds may be hazardous because of their ability to elicit allergic reactions.

20.2 GENERAL CONSIDERATIONS

This chapter is concerned with cultural procedures and biochemical techniques for the detection and enumeration of fungi. Chapter 4 describes several widely used microscopic methods.

Traditionally, acidified media have been used to enumerate yeasts and molds in foods. Such media are now recognized as inferior to antibiotic supplemented media[4,11,35,57,76-78,97] that have been formulated to suppress bacterial colony development,[60,62] enhance resuscitation of injured fungi,[12,79] and minimize precipitation of food particles.[82] Although acidified media may be appropriate for certain types of foods,[12,60,61] their use should first be carefully evaluated.

Depending on the nature of the food and the species of fungi likely to predominate, numerous media may be appropriate for viable counts.[13,14,19,29,40,49,58,63,64,77,90,97,98,99] Antibiotic-supplemented media[57,97] are recommended as general-purpose media for enumerating yeasts and molds by the dilution plating procedure. Media supplemented with solutes such as glycerol, glucose, sucrose, or sodium chloride are most appropriate for enumerating xerophilic molds and osmotolerant yeasts.[15,43,48,50,105] See Chapter 17 for methods to enumerate these fungi. Certain media are formulated to detect specific genera and species of foodborne fungi, e.g., preservative resistant yeasts,[14,41,73] *Mucor* species,[7] *Chrysosporium* species,[56] and toxigenic *Aspergillus*,[31,91] *Penicillium*[36,88] and *Fusarium*[1,2,108] species.

Yeasts and molds should be enumerated by a surface spread-plate technique rather than with pour plates. This technique provides maximal exposure of the cells to atmospheric oxygen and avoids heat stress from molten agar. Agar spread plates should be dried overnight before being inoculated.

A dry rehydratable film (Petrifilm™ yeast and mold count) has been compared to the 5-day PDA pour plate in a collaborative study using six foods inoculated with four molds and two yeasts.[59] No significant differences were noted between the means of either method for any inoculum level; however, there were some differences in repeatability as measured by the standard deviations for two foods for molds and for three foods for yeasts between the two methods. This method also has been adopted for first action by AOAC.

Most probable number (MPN) techniques may be appropriate for some foods.[65] Membrane filter techniques are applicable for beverages, and for certain solid foods.[16,33,48] Sometimes, however, entrapment of mycelia and food particles in membrane pores limits applicability.

The two-day hydrophobic grid membrane filter (HGMF) method using YM-11 agar has been evaluated in a collaborative study.[32] Naturally contaminated foods were examined by 20 laboratories using the 5-day potato dextrose agar (PDA) pour plate and the 2-day HGMF methods, with results being not significantly different for three of five products; however, for two foods, the HGMF method recovered significantly lower populations than the pour plate method. For orange juice, the HGMF method recovered statistically higher numbers of fungi than the pour plate method. Similar results for repeatability and reproducibility were noted for both methods. Based on this study, the HGMF method was recommended for first action by AOAC.

Sterile 0.1% peptone water is the most suitable diluent when culturing samples for yeasts and molds.[97] Solutes such as sucrose, glucose, or glycerol should be added to the diluent when enumerating osmotrophs in foods such as syrups and fruit juice concentrates[15,18,39] (Chapter 17).

The antibiotics most frequently employed in media of neutral pH are chlortetracycline (Aureomycin), chloramphenicol, oxytetracycline, gentamicin, streptomycin, and kanamycin (Sigma Chemical, St. Louis, MO). Chloramphenicol (100 µg/mL) or gentamicin (50 µg/mL) are recommended because of their heat stability and broad antibacterial spectra. Both antibiotics can be sterilized by autoclaving in complete media. Stock solutions of antibiotics can be stored at 5°C for up to 10 weeks without loss of activity.

Rose bengal may be added to media to restrict excessive mycelial growth by some molds.[3,47] Care should be taken, however, not to expose these media to light, since photodegradation of rose bengal yields compounds that are toxic to fungi.[5] Dichloran (2,6-dichloro-4-nitroaniline) at 2 µg/mL and surfactants[107] also restrict the spread of mold colonies.

Some degree of soaking may be beneficial for the recovery of yeasts and molds from dried or intermediate moisture foods.[49] Soaking may enhance repair of sublethally damaged cells (resuscitation) and may also soften the food and thereby facilitate release of organisms from tissues. Because mold propagules may settle out within a few minutes, it is important to mix liquid samples and diluents immediately prior to transferring or culturing.

Inoculated plates should be incubated undisturbed in an upright position at 22 to 25°C for 5 days before colonies are counted. If an incubator is not available, use room temperature and report accordingly. Plates with 15 to 150 colonies are usually counted. If the mycoflora consists primarily of molds, the lower population range is selected; if primarily yeast colonies, the upper limit is counted.

Wet mounts and gram stains of cells from at least 10 colonies per sample should be examined to confirm that bacteria are not present. Yeast cells and asexual mold spores are generally gram positive, whereas mold mycelia are gram negative.

20.3 PRECAUTIONS

Some yeasts and molds can be infectious or can elicit allergic responses, sometimes even in healthy individuals. Thus it is important to be reasonably cautious when working with fungi. Ideally, plates should be held in incubators, not in an open room. Plate lids should be removed sparingly, normally only for critical purposes such as the preparation of a slide for microscopic examination. Flamed needles should be cooled before making transfers, to avoid dispersal of conidia and other cells. Under no circumstances should a culture be smelled. Before disposal, plates and other cultures should be autoclaved. Work benches and incubators should be disinfected routinely. An acceptable quality assurance program (Chapter 1) should be established and followed.

20.4 EQUIPMENT, MATERIALS, AND REAGENTS

Methods and equipment for dilution plating are the same as for the aerobic plate count (Chapter 7)
Acetic Acid Dichloran Yeast Extract Agar (ADYS)
Asperillus Flavus/parasiticus Agar (AFPA)
Czapek Dox Iprodione Dichloran (CZID)
Dichloran Rose Bengal Chloramphenicol (DRBC)
Dichloran Rose Bengal Yeast Extract Sucrose Agar (DRYES)
Dichloran 18% Glycerol Agar (DG18)
Malt Acetic Acid Agar (MAA)
Malt Extract Agar (MEA)
Malt Extract Yeast Extract Chloramphenicol Ketoconazol Agar (MYCK)
Malt Extract Yeast Extract 50% Glucose Agar (MY50G)
Plate Count Agar (PCA) supplemented with chloramphenicol
Tryptone Glucose Yeast Extract Agar (TGY)
Tryptone Glucose Yeast Extract—Acetic Acid Agar (TGYA)

20.5 PROCEDURES

20.51 DRBC Agar (Preferred)

1. DRBC agar containing 100 µg/mL chloramphenicol is recommended. The chloramphenicol can be added to the medium before sterilization in the autoclave.
2. Petri plates, 9 cm diameter, are poured with 15 to 20 mL of the medium and then dried overnight at room temperature (21 to 25°C). Store DRBC agar plates in the dark to avoid photogeneration of inhibitors from the rose bengal dye.
3. Inoculate 0.1 mL of appropriate decimal dilutions in duplicate on the solidified agar and spread over the entire surface using a sterile bent glass rod.
4. Incubate upright plates 5 days at 22 to 25°C; do not disturb until colonies are to be counted. Count plates containing 15 to 150 colonies. Count from the underside of the plate when mold overgrowth has occurred.
5. Report the counts as cfu per g or mL of sample.

20.52 PCA Supplemented with Chloramphenicol

This method is identical to that described in Section 20.51 except that PCA supplemented with 100 µg/mL chloramphenicol is used and plates do not need to be stored in the dark.

20.53 DG18 Agar

When foods with low water activities such as cereals, flours, nuts, and spices, are to be plated, DG18 agar is best because it allows moderate xerophiles to be enumerated. The method is similar to that described for DRBC agar, but plates do not need to be stored in the dark. Consult Pitt and Hocking[90] and Samson et al.[99] for methods on direct plating of whole grains, nuts, seeds, dried fruits, etc.

20.54 Selective Media

If enumeration of specific groups of species of molds or yeasts is desired, selective agar can be used. Some of the most common selective media are given here. Further details can be found in Pitt and Hocking[90] and Samson et al.[99]

20.541 Acidophilic Molds

Plate on acetic dichloran yeast extract sucrose (ADYS) agar. This medium contains 0.5% acetic acid which selects for molds that can grow in the presence of preservatives and organic acids. Incubate at 25°C for 5–7 days.

20.452 Aflatoxigenic Molds

Plate on *Aspergillus flavus/parasiticus* agar (AFPA) and incubate at 32°C for 42–48 h. Aflatoxogenic strains produce a bright yellow-orange reverse color.

20.453 Fusarium Species

Plate on Czapek iprodione dichloran (CZID) agar and incubate at 25°C for 7 days under a black light (12 h on, 12 h off).

20.544 Heat-Resistant Fungi

See Chapter 21 for methodology.

20.545 Mucor Species

Plate on malt extract yeast chloramphenicol ketoconazol (MYCK) agar and incubate at 25°C for 5 days.

20.546 Penicillium Species

Plate on dichloran rose bengal yeast extract sucrose (DRYES) agar and incubate at 20°C for 7–8 days. Ochratoxin producers can be selectively differentiated on this agar because *P. verrucosum* and *P. viridicatum* produce a violet brown reverse color.

20.547 Preservative-resistant Yeasts

Plate on malt extract agar with 0.5% acetic acid (MAA) or tryptone glucose yeast extract (TGY) agar with 0.5% acetic acid (TGYA agar) and incubate at 25–30°C for 3–5 days.

20.548 Xerophilic Fungi

Plate on DG18 agar for most moderate xerophiles but on malt yeast 50% glucose (MY50G) agar for extreme xerophiles and incubate at 25°C for 7 days (extreme xerophiles may take up to 3 weeks to grow). Diluents for high-sugar foods should contain 40% glucose or 30% glycerol to minimize osmotic shock.

20.549 Yeasts Dominate in Foods

Plate on either malt extract agar (MEA) or tryptone glucose yeast extract (TGY) agar and incubate at 25°C for 3 days (alternately, incubate for 3 days anaerobically and 2 days aerobically). If bacteria are present in the foods, supplement agar with 100 ppm chloramphenicol or oxytetracycline. If molds are present, use DRBC agar.

20.6 INTERPRETATION

The interpretation of yeast and mold counts often is difficult because baseline data on expected and excessive levels for many foods have not been established. Criteria for fungi should take into consideration the specific food, the manner of use, the primary age group and health of the consumer, and frequency and amount of food consumed. Although admittedly difficult, determination of the identity of predominant species also is important.

20.61 Identification of Yeasts and Molds

Simple enumeration of yeasts and molds may not be sufficient to help solve some spoilage and potentially toxigenic problems associated with fungi in foods. It may be necessary to identify the genus and even the species of the fungus isolated. Yeasts are identified by biochemical and morphological techniques. Once the yeasts are isolated they can be biochemically identified by rapid diagnostic kits or conventional test tube methods. Procedures that can be used to identify yeasts to genus and species levels have been described.[6,24,29,67,90,99]

Molds are identified by macroscopic and microscopic techniques that involve cultivation on two or three suitable media at two or three incubation temperatures for 1 or 2 weeks, according to the identification scheme that is being used. The resulting mold colony is identified by size, shape, color, presence of exudate, pigmentation in the agar, colony texture, and other characteristics. Microscopically, the wet mount of the mold is examined for spore and mycelial shape, color, texture, and distinguishing marks. Techniques to identify some of the major foodborne molds have been described.[90,99] In addition, other taxonomic texts as described by Samson et al.[97] can be used. Overall, it is important to choose an identification scheme, follow the procedures closely, and use a good quality microscope to view slides.

20.7 NEW METHODS UNDER DEVELOPMENT

Methods for detection and identification of fungi in foods lag behind those for bacteria; however, in recent years alternate methods to the traditional plate count have been developed. Rapid methods that assess levels of both viable and nonviable fungi in foods are needed by the food industry, regulatory agencies, analytical consulting firms, and research laboratories.

Research is being done on rapid methods that have potential for detection and/or identification of yeasts and molds in foods. Methods that have received the most attention include: (1) microscopic analysis using selective stains or fluorescent dyes; (2) detection of a metabolite or cellular constituent produced by fungi that is absent in food; (3) use of impedimetric techniques based on conductance; (4) optically-based methods such as flow cytometry; (5) immunological detection of fungi based on antibodies that are specific for a fungus or group of fungi; and (6) molecular detection of DNA or RNA fractions that are specific to a given fungal genus or species. The development of these methods has mirrored those that have been used to detect bacteria in foods; however, there are some conditions specific to fungi that need to be considered when converting these methods for the purpose of detection and identification.

20.71 Selective Stains and Fluorescent Dyes

Staining methods have been used to rapidly detect viable fungi mainly in beverages and semi-solid foods. Yeasts have been stained with methylene blue because viable cells will reduce the dye to the colorless form while dead cells will take up the blue color.[52,69] Since some viable cells may stain blue, fluorescent dyes and optical brighteners such as acridine orange,[44] analine blue,[36,66] fluorescein-diacetate,[44,52] and viablue[45,96] have been evaluated. Fluorescent dyes have been actively researched with the direct epifluorescent filter technique (DEFT) for the detection of fungi in foods. DEFT gave an acceptable correlation with plate counts for molds[87] and yeasts[96] enumerated from foods. Distinguishing between viable and nonviable cells has not always been possible with DEFT[85,95] and has, therefore, been combined with gel microdroplets[112] and immuno-logical techniques[94] to improve detection of yeasts and molds.

20.72 Detection of Metabolites and Cellular Constituents

Any metabolite or cellular constituent that is produced by yeasts or molds but that is not present in plant or animal tissues is a potential candidate for rapid method development. Cellular components used to detect molds include chitin[22,23,70,113] and ergosterol,[9,38,92,101] whereas, mannan has been used occasionally to detect yeasts. Chitin is a linear homopolymer of β-(1→4)-N-acetyl-D-glucosamine that is present in the cell walls of most fungi but not plants and animals. Chitin or its breakdown products, chitosan, glucosamine, or N-acetylglucosamine, have been detected after acid, alkaline, or enzymatic hydrolysis by colorometric, chromatographic or near infrared spectroscopy.[22] Ergosterol is the major sterol in fungal membranes but differs from sterols in plant and animal tissues by its absorbance at 282 nm. The analysis of both chitin and ergosterol involves complex extraction procedures and instrumentation. For this reason, neither method has been adopted for rapid mold detection in foods.

20.73 Conductance

Metabolic activities of bacteria result in a resistance to the flow of an electrical current, thereby, increasing conductance.[104] However, the opposite occurs for yeasts and an indirect conductance must be measured.[100] Indirect conductance can detect as few as 10 yeast cells/mL of fruit juices within 2 to 3 days.[26,27] Conductance measurements compared favorably with plate counts on DRBC agar and acidified PDA. A lower population of 1 cell/10 mL can be detected by indirect conductance in carbonated beverages after a preliminary incubation in tryptone glucose yeast extract broth.[28] There has been less success with detecting molds in foods by measuring conductance. This method may have use for detecting yeasts in beverages; however, the need for an incubation period does not enable rapid analysis.

20.74 Flow Cytometry

Flow cytometry is an optically-based technique that has been used to detect and quantify yeasts in beverages and semi-solid foods. It has been used to measure DNA, protein, cell concentrations, enzymatic activity, and similar parameters. The flow cytometer has a laser light source, a sample chamber, an optical assembly, and photodetectors. The sample is put into a fast flowing stream that is forced through a small orifice of about 50 to 100 μm at speeds from 1 to 10 m/sec. The sample is then evaluated by a light source that measures the intensity of light scatter that is then translated into information on cell size and shape, cell viability and density, surface morphology, and other data. *Saccharomyces cerevisiae* was detected at a population of 50 cells/mL in lemonade, but in tomato juice 14,000 cells/mL were needed.[86] The detection limit is dependent on the type of juice or soft drink analyzed. Flow cytometry has been used to determine if yeasts could be detected in quarg. When products had >10 cells/g and were preincubated at 25°C for 22 h, the method was comparable to the plate count, although there were still some false negative results.[53] In lager beer, one wild yeast could be detected in a population of 10^6 cells/mL after 48 to 72 h at 25°C.[51] Flow cytometry has been combined with immunofluorescence, RNA probes, and 2D-image analysis. Detection of *S. cerevisiae* using flow cytometry with a fluorescent rRNA probe has been reported.[10] Flow cytometry shows promise for detection of yeasts in liquids or viscous foods, but more research is needed to define the conditions needed. The cost of the equipment and the need for trained personnel may be drawbacks for potential users.

20.75 Immunological Methods

Over the past 15 years, immunological methods have been developed with some even being commercialized for the detection of fungi in foods. Immunoassays are based on the binding of an antigen from food to an antibody that was produced to a species or group of fungi, and the subsequent visualization of that reaction by linking the antibody to an enzyme that will produce a colored compound upon reaction with the appropriate substrate.[21,30] These assays can be very sensitive and specific, depending upon the preparation of the antigens, antibodies, and other reagents.

Research on the detection of molds by the enzyme-linked immunosorbent assay (ELISA) began in the mid-1980's.[72,80,93] These assays have been used to detect molds in foods such as fruits,[71,74] nuts and spices,[81] and cheese and yogurt.[109,114] Very little research has been done on the use of ELISA to detect yeasts in foods, with most directed to beverages such as orange juice[116] and wine.[68] Interest in detecting toxigenic molds in foods has led to new research efforts to develop an ELISA that can detect *Aspergillus flavus* and *A. parasiticus* and distinguish them from nonaflatoxigenic *Aspergillus* species.[103,110,115]

A rapid latex agglutination assay for detecting *Penicillium* and *Aspergillus* antigens in foods has been developed.[54] This was most likely the basis for the commercial latex agglutination test kits that were available in Europe in the early 1990s.[21,30,55,111] The latex agglutination test was not suitable for testing cheese, nuts, some spices, and low moisture foods for molds and there was no correlation with mold colony counts.[55] When two latex agglutination tests were compared, both tests did not give similar results with the same foods.[111] These assays took 15 to 30 min to complete, depending on the time needed for sample preparation. Although latex agglutination tests show promise, they are no longer on the market because they were not readily adopted by the food processing industry. One commercial assay kit that can detect *Aspergillus* and *Penicillium* species is produced by a German company and marketed in the USA through Bio-Tek Instruments, Inc. Although these immunoassays have shown that molds and yeasts can be detected in foods in a few minutes, their adoption by food processing industries, regulatory agencies, and analytical testing laboratories is critical before they can be used routinely.

20.76 Molecular Methods

Molecular techniques have been increasingly used for the detection of pathogenic bacteria in foods, and it is only natural that they would be applied to the detection and identification of fungi in foods. Molecular assays are based on the identification of unique DNA or RNA profiles in fungi. Methods that have shown the most promise for detecting fungi in foods are those based on polymerase chain reaction (PCR), including random amplified polymorphic DNA (RAPD) and restriction fragment length polymorphism (RFLP), pulsed-field gel electrophoresis (PFGE), electrophoretic karyotyping, or combinations of these methods.[25,84] Peterson[84] reviewed the use of molecular methods for molds and suggested that PCR-based analyses are sensitive and reproducible, and the reagents are now commonly available. A 336 bp fragment from *Penicillium* subgenus *Penicillium* was able to recognize all isolates from this subgenus but not *Penicillium* species from other subgenera[83]; the 300 bp fragment only recognized *Penicillium roqueforti* and *P. carneum*. This method took 7 h to complete. PCR methods have also been developed to detect aflatoxigenic molds in foods.[34,102] A PCR method based on three biosynthetic genes for aflatoxin production was used to detect *A. flavus* in figs; however, the sensitivity was ten times lower than in pure culture.[34] A similar approach was used to detect *A. parasiticus* and *A. flavus* in a corn extract medium.[102] This limited research shows that PCR has potential for the detection of molds, especially specific species, in foods.

Deak[25] reviewed the use of PFGE to identify yeasts for taxonomic as well as industrial uses such as wine manufacture and the research that has been done with PCR, RAPD and RFLP for the identification of yeasts. Ten *Dekkera* and *Brettanomyces* strains can be detected in sherry by PCR in less than 10 h;[46] however, wine inhibited the reaction, indicating that more work needs to be done on detecting yeasts in food matrices. *Zygosaccharomyces bailii* could be distinguished from other *Zygosaccharomyces* species by a PCR ligase-detection reaction within one workday if the amplified product had 3×10^7 targets.[106] With more research, PCR may be a useful tool for detecting and identifying yeasts in foods.

20.77 Continued and Future Rapid Method Development

New method development for the detection and identification of fungi in foods will probably not revolve around one specific method because of the diversity of fungi. The purpose for using the method will be a driving force in future development. Also, the intended use of the results will determine what methods may hold the most promise because they may be used to estimate numbers, to determine presence or absence, to identify toxigenic fungi or specific spoilage species or strains, and other purposes. Both rapidity and ease of obtaining results will remain as critical factors in any method development. Many different types of foods from worldwide distribution will need to be analyzed before these methods can be recommended for routine use. The issues of detection of viable versus nonviable cells and correlation of these new methods to traditional plate counts need to be addressed. A reevaluation of what the fungal count means versus what information the new rapid method provides should give guidelines for rapid method development.

20.8 REFERENCES

1. Abildgren, M. P., F. Lund, U. Thrane, and S. Elmholt. 1987. Czapek-Dox agar containing iprodione and dichloran as a selective medium for the isolation of *Fursarium* species. Lett. Appl. Microbiol. 5:83-86.
2. Andrews, S., and J. I. Pitt. 1986. Selective medium for isolation of *Fursarium* species and dematiaceous hyphomycetes from cereals. Appl. Environ. Microbiol. 51:1235-1238.
3. Baggerman, W. I. 1981. A modified rose bengal medium for the enumeration of yeasts and molds from foods. Eur. J. Appl. Microbiol. Biotechnol. 12:242-244.
4. Banks, J. G., and R. G. Board. 1987. Some factors influencing the recovery of yeasts and molds from chilled foods. Int. J Food Microbiol. 4:197-206.
5. Banks, J. G., R. G. Board, and J. Paton. 1985. Illuminated rose bengal causes adenosine triphosphate (ATP) depletion and microbial death. Lett. Appl. Microbiol. 1:7-11.
6. Barnett, J. A., R. W. Payne, and D. Yarrow. 1990. Yeasts: characteristics and identification, 2nd ed. Cambridge Univ. Press, New York. 1002 pp.
7. Bartschi, C., J. Berthier, C. Guiguettaz, and G. Valla. 1991. A selective medium for the isolation and enumeration of *Mucor* species. Mycol. Res. 95:373-378.
8. Beneke, E. S., and A. L. Rogers. 1971. Medical mycology manual, 3rd ed. Burgess Publishing Co., Minneapolis, Minn.
9. Bermingham, S., L. Maltby, and R. C. Cooke. 1995. A critical assessment of the viability of ergosterol as an indicator of fungal biomass. Mycol. Res. 99:479-484.
10. Bertin, B., O. Broux, and M. van Hoegaerden. 1990. Flow cytometric detection of yeast by in situ hybridization with a fluorescent ribosomal RNA probe. J. Microbiol. Meth. 12:1-12.
11. Beuchat, L. R. 1979. Comparison and acidified and antibiotic-supplemented potato dextrose agar from three manufacturers for its capacity to recover fungi from foods. J. Food Prot. 42:427-428.
12. Beuchat, L. R. 1984. Injury and repair of yeasts and moulds, p. 293-308. *In* M. H. E. Andrew and A. D. Russell (ed.) Revival of injured microbes. Academic Press, London.
13. Beuchat, L. R. 1992. Media for detecting and enumerating yeasts and moulds. Int. J. Food Microbiol. 17:145-158.
14. Beuchat, L. R. 1993. Selective media for detecting and enumerating foodborne yeasts. Int. J. Food Microbiol. 19:1-14.
15. Beuchat, L. R., and A. D. Hocking. 1990. Some consideration when analyzing foods for the presence of xerophilic fungi. J. Food Prot. 53:984-989.
16. Brodsky, M. H., P. Entis, M. P. Entis, A. N. Sharpe, and G. A. Jarvis. 1982. Determination of aerobic plate and yeast and mold counts in foods using an automated Hydrophobic Grid-Membrane Filter technique. J. Food Prot. 45:301-304.
17. Bullerman, L. B. 1997. Fusaria and toxigenic molds other than aspergilli and penicillia, p. 409-434. *In* M. P. Doyle, L. R. Beuchat, and T. J. Montville (ed.), Food microbiology: fundamentals and frontiers. American Society for Microbiology Press, Washington, D. C.
18. Corry, J. E. L. 1976. The effect of sugars and polyols on the heat resistance and microscopic morphology of osmophilic yeasts. J. Appl. Bacteriol. 40:269-276.
19. Corry, J. E. L. 1982. Assessment of the selectivity and productivity of media used in analytical mycology. Arch. Lebensmittel. Hyg. 33:160-164.
20. Corry, J. E. L. 1987. Relationships of water activity of fungal growth, p. 51-99. *In* L. R. Beuchat (ed.), Food and beverage mycology, 2nd ed. Van Nostrand Reinhold, New York.
21. Cousin, M. A. 1990. Development of the enzyme-linked immuno≠sorbent assay for detection of molds in foods: a review. Dev. Ind. Microbiol. 31:157-163.
22. Cousin, M. A. 1996. Chitin as a measure of mold contamination of agricultural commodities and foods. J. Food Prot. 59:73-81.
23. Cousin, M. A., C. S. Zeidler, and P. E. Nelson. 1984. Chemical detection of mold in processed foods. J. Food Sci. 49:439-445.
24. Deak, T. 1993. Simplified techniques for identifying foodborne yeasts. Int. J. Food Microbiol. 19:15-26.
25. Deak, T. 1995. Methods for the rapid detection and identification of yeasts in foods. Trends Food Sci. Technol. 6:287-292.
26. Deak, T., and L. R. Beuchat. 1993. Comparison of conductimetric and traditional plating techniques for detecting yeasts in fruit juices. J. Appl. Bacteriol. 75:546-550.
27. Deak, T., and L. R. Beuchat. 1993. Evaluation of the indirect conductance method for the detection of yeasts in laboratory media and apple juice. Food Microbiol. 10:255-262.
28. Deak, T., and L. R. Beuchat. 1995. Modified indirect conductimetric technique for detecting low populations of yeasts in beverage concentrates and carbonated beverages. Food Microbiol. 2:165-172.
29. Deak, T., and L. R. Beuchat. 1996. Handbook of food spoilage yeasts. CRC Press, Boca Raton, Fla. 210 pp.
30. DeReuiter, G. A., S. H. W. Notermans, and F. M. Rombouts. 1993. New methods in food mycology. Trends Food Sci. Technol. 4:91-97.

31. Dryer, S. K., and S. McCammon. 1994. Detection of toxigenic isolates of *Aspergillus flavus* and related species on coconut cream agar. J. Appl. Bacteriol. 76:75.
32. Entis, P. 1996. Two-day hydrophobic grid membrane filter method for yeast and mold enumeration in foods using YM-11 agar: collaborative study. J. AOAC Int. 79:1069-1082.
33. Entis, P., and I. Lerner. 1996. Two-day yeast and mold enumeration using the ISO-GRID® membrane filtration system in conjunction with YM-11 agar. J. Food Prot. 59:416-419.
34. Färber, P., R. Geisen, and W. H. Holzapfel. 1997. Detection of aflatoxigenic fungi in figs by a PCR reaction. Int. J. Food Microbiol. 36:215-220.
35. Flannigan, G. 1974. The use of acidified media for enumeration of yeasts and molds. Lab. Pract. 23:633-634.
36. Frisvad, J. C. 1983. A selective and indicative medium for groups of *Penicillium viridicatum* producing different mycotoxins in cereals. J. Appl. Bacteriol. 54:409-416.
37. Gilman, J. C. 1957. A manual of soil fungi, 2nd edition. Iowa State Univ. Press, Ames.
38. Gourama, H., and L. B. Bullerman. 1995. Relationship between aflatoxin production and mold growth as measured by ergosterol and plate count. Lebensm. Wiss. Technol. 28:185-189.
39. Hernandez, P., and L. R. Beuchat. 1995. Evaluation of diluents and media for enumerating *Zygosaccharomyces rouxii* in blueberry syrup. Int. J. Food Microbiol. 25:11-18.
40. Hocking, A. D. 1981. Improved media for enumeration of fungi from foods. CSIRO Food Res. Quart. 41:7-11.
41. Hocking, A. D. 1996. Media for preservative resistant yeasts: a collaborative study. Int. J. Food Microbiol. 29:167-175.
42. Hocking, A. D. 1997. Toxigenic *Aspergillus* species, p. 393-405. *In* M. P. Doyle, L. R. Beuchat, and T. J. Montville (ed.), Food microbiology: fundamentals and frontiers. American Society for Microbiology Press, Washington, D. C.
43. Hocking, A. D., and J. I. Pitt. 1980. Dichloran-glycerol medium for enumeration of xerophilic fungi from low moisture foods. Appl. Environ. Microbiol. 39:488-492.
44. Hope, C. F. A., and R. S. Tubb. 1985. Approaches to rapid microbial monitoring in brewing. J. Inst. Brew. 91:12-15.
45. Hutcheson, T. C., T. McKay, L. Farr, and B. Seddon. 1988. Evaluation of the stain viablue for the rapid estimation of viable yeast cells. Lett. Appl. Microbiol. 6:85-88.
46. Ibeas, J. I., I. Lozano, F. Perdigones, and J. Jimenez. 1996. Detection of *Dekkera-Brettanomyces* strains in sherry by a nested PCR method. Appl. Environ. Microbiol. 62:998-1003.
47. Jarvis, B. 1973. Comparison of an improved rose bengal chlortetracycline agar with other media for the selective isolation and enumeration of molds and yeasts in foods. J. Appl. Bacteriol. 36:723-727.
48. Jarvis, B., and A. P. Williams. 1987. Methods for detecting fungi in foods and beverages, p. 599-626. *In* L. R. Beuchat (ed.), Food and beverage mycology, 2nd ed. Van Nostrand Reinhold, New York.
49. Jarvis, G., D. A. L. Seiler, A. J. L. Ould, and A. P. Williams. 1983. Observations on the enumeration of moulds in food and feedingstuffs. J. Appl. Bacteriol. 55:325-336.
50. Jermini, M. F. G., O. Geiges, and W. Schmidt-Lorenz. 1987. Detection and identification of osmotolerant yeasts from high-sugar products. J. Food Prot. 50:468-472.
51. Jespersen, L., S. Lassen, and M. Jakobsen. 1993. Flow cytometric detection of wild yeast in lager breweries. Int. J. Food Microbiol. 17:321-328.
52. Jones, R. P. 1987. Measures of yeast death and deactivation and their meaning: Part I. Proc. Biochem. 22:118-128.
53. Joosten, H. M. L. J., W. G. F. M. van Dijck, J. W. M. Spikker, and H. M. Traa. 1996. Detection of yeasts in quarg by flow cytometry. Milchwissenschaft 51:202-204.
54. Kamphuis, H. J., S. Notermans, G. H. Veeneman, J. H. Van Boom, and F. M. Rombouts. 1989. A rapid and reliable method for the detection of molds in foods: using the latex agglutination assay. J. Food Prot. 52:244-247.
55. Karman, H., and R. A. Samson. 1992. Evaluation of an immunological mould latex detection test: a collaborative study, p. 229-232. *In* R. A. Samson, A. D. Hocking, J. I. Pitt, and A. D. King (ed.), Modern methods in food mycology. Elsevier, New York.
56. Kinderlerer, J. L. 1995. Czapek casein 50% glucose (CZC50G): a new medium for the identification of foodborne *Chrysosporium* spp. Lett. Appl. Microbiol. 21:131-136.
57. King, A. D., A. D. Hocking, and J. I. Pitt. 1979. Dichloran-glycerol medium for enumeration of molds from foods. Appl. Environ. Microbiol. 39:959-964.
58. King, A. D., J. I. Pitt, L. R. Beuchat, and J. E. L. Corry (ed.). 1986. Methods for mycological examination of food. Plenum Press, New York. 315 pp.
59. Knight, M. T., M. C. Newman, M. J. Benzinger Jr, K. L. Neufang, and J. R. Agin. 1997. Comparison of the Petrifilm dry rehydratable film and conventional culture methods for enumeration of yeasts and molds in foods: collaborative study. J. AOAC Int. 80:806-823.
60. Koburger, J. A. 1970. Fungi in foods. I. Effects of inhibitor and incubation temperature on enumeration. J. Milk Food Technol. 33:433-434.
61. Koburger, J. A. 1970. Fungi in foods. II. Some observations on acidulants used to adjust media pH for yeast and mold counts. J. Milk Food Technol. 34:475-477.
62. Koburger, J. A. 1972. Fungi in foods. IV. Effect of plating medium pH on counts. J. Milk Food Technol. 35:659-660.
63. Koburger, J. A. 1986. Effect of pyruvate on recovery of fungi from foods. J. Food Prot. 49:231-232.
64. Koburger, J. A., and B. Y. Farhat. 1975. Fungi in foods. VI. A comparison of media to enumerate yeasts and molds. J. Milk Food Technol. 38:466-468.
65. Koburger, J. A., and A. R. Norden. 1975. Fungi in foods. VII. A comparison of surface, pour plate, and most probable number methods for enumeration of yeasts and molds. J. Milk Food Technol. 38:745-746.
66. Koch, H. A., R. Bandler, and R. R. Gibson. 1986. Fluorescence microscopy procedure for quantitation of yeasts in beverages. Appl. Environ. Microbiol. 52:599-601.
67. Kreger-van Rij, N. J. W. 1984. The yeasts, a taxonomic study, 3rd ed. Elsevier, Amsterdam. 1082 pp.
68. Kuniyuki, A. H., C. Rous, and J. L. Sanderson. 1984. Enzyme-linked immunosorbent assay (ELISA) detection of *Brettanomyces* contaminants in wine production. Am. J. Enol. Vitic. 35:143-145.
69. Lee, S. S., F. M. Robinson, and H. Y. Wang. 1981. Rapid determination of yeast viability. Biotechnol. Bioeng. Symp. 11:641-649.
70. Lin, H. H., and M. A. Cousin. 1985. Detection of mold in processed foods by high performance liquid chromatography. J. Food Prot. 48:671-678.
71. Lin, H. H., and M. A. Cousin. 1987. Evaluation of enzyme-linked immunosorbent assay for detection of molds in foods. J. Food Sci. 52:1089-1094, 1096.
72. Lin, H. H., R. M. Lister, and M. A. Cousin. 1986. Enzyme-linked immunosorbent assay for detection of mold in tomato puree. J. Food Sci. 51:180-183, 192.
73. Makdesi, A. K., and L. R. Beuchat. 1996. Improved selective medium for enumeration of benzoate-resistant, heat-stressed *Zygosaccharomyces bailii*. Food Microbiol. 13:281-290.
74. Marois, J. J., L. H. Kenyon, C. D. Lamison, and J. P. Smith. 1994. A rapid immunoassay method for quantifying mold in harvested wine grapes. Am. J. Enol. Vitic. 45:300-304.
75. Mislivec, P. B. 1981. Mycotoxin production by conidial fungi, p. 37. *In* G. T. Cole and B. Kendrick (ed.), Biology of conidial fungi, vol. 2. Academic Press, New York.
76. Mislivec, P. B., and V. R. Bruce. 1976. Comparison of antibiotic-amended potato dextrose agar and acidified potato dextrose agar as growth substrate for fungi. J. Assoc. Off. Anal. Chem. 59:720
77. Mossel, D. A. A., M. Visser, and W. H. Mengerink. 1962. A comparison of media for the enumeration of moulds and yeasts in foods and beverages. Lab. Pract. 11:109-112.

78. Mossel, D. A. A., C. L. Vega, and H. M. Put. 1975. Further studies on suitability of various media containing antibacterial antibiotics for the enumeration of molds in food and food environments. J. Appl. Bacteriol. 39:15-22.
79. Nelson, F. E. 1972. Plating medium pH as a factor in apparent survival of sublethally stressed yeasts. Appl. Microbiol. 24:236-239.
80. Notermans, S., and C. J. Heuvelman. 1985. Immunological detection of moulds in food by using the enzyme-linked immunosorbent assay (ELISA); preparation of antigens. Int. J. Food Microbiol. 2:247-258.
81. Notermans, S., J. Defrenne, and P. S. Soentoro. 1988. Detection of molds in nuts and spices: The mold colony count versus the enzyme-linked immunosorbent assay (ELISA). J. Food Sci. 53:1831-1833, 1843.
82. Overcast, W. W., and D. J. Weakley. 1969. An aureomycin-rose bengal agar for enumeration of yeast and mold in cottage cheese. J. Milk Food Technol; 32:442-445.
83. Pedersen, L. H., P. Skouboe, M. Boysen, J. Soule, and L. Rossen. 1997. Detection of *Penicillium* species in complex food samples using the polymerase chain reaction. Int. J. Food Microbiol. 35:169-177.
84. Peterson, S. W. 1995. Species concepts, molecular systematics and the rapid identification of toxigenic hyphomycetes, p. 3-17. *In* M. Eklund, J. L. Richard, and K. Mise (ed.), Molecular approaches to food safety. Issues involving toxic microorganisms. Alaken, Inc., Fort Collins, Colo.
85. Pettipher, G. L. 1987. Detection of low numbers of osmophilic yeasts in creme fondant within 25 h using a pre-incubated DEFT count. Lett. Appl. Microbiol. 4:95-98.
86. Pettipher, G. L. 1991. Preliminary evaluation of flow cytometry for the detection of yeasts in soft drinks. Lett. Appl. Bacteriol. 12:109-112.
87. Pettipher, G. L., R. A. Williams, and C. S. Gutteridge. 1985. An evaluation of possible alternative methods to the Howard mould count. Lett. Appl. Microbiol. 1:49-51.
88. Pitt, J. I. 1993. A modified creative sucrose medium for differentiation of species in *Penicillium*. J. Appl. Bacteriol. 75:559-563.
89. Pitt, J. I. 1997. Toxigenic *Penicillium* species, p. 406-418. *In* M. P. Doyle, L. R. Beuchat, and T. J. Montville (ed.), Food microbiology: fundamentals and frontiers. American Society for Microbiology Press, Washington, D. C.
90. Pitt, J. I., and A. D. Hocking. 1997. Fungi and food spoilage, 2nd ed. Blackie Academic and Professional, New York. 593 pp.
91. Pitt, J. I., A. D. Hocking, and D. R. Glenn. 1983. An improved medium for the detection of *Aspergillus flavus* and *A. parasiticus*. J. Appl. Bacteriol. 54:109-114.
92. Regnér, S., J. Schnürer, and A. Jonsson. 1994. Ergosterol content in relation to grain kernelweight. Cereal Chem. 71:55-58.
93. Robertson, A., D. Upadhyaya, S. Opie, and J. S. Sargeant. 1986. An immunochemical method for the measurement of mold contamination in tomato paste, p. 163-179. *In* B. A. Morris, M. N. Clifford, and R. Jackson (ed.), Immunoassays for veterinary and food analysis—1. Elsevier Applied Science, London.
94. Robertson, A., N. Patel, and J. S. Sargeant. 1988. Immunofluorescence detection of mould—an aid to the Howard mould counting technique. Food Microbiol. 5:33-42.
95. Rodriques, U. M., and R. G. Kroll. 1986. Use of the direct epifluorescent filter technique for the enumeration of yeasts. J. Appl. Bacteriol. 61:139-144.
96. Rowe, M. T., and G. J. McCann. 1991. A modified direct epifluorescent filter technique for the detection and enumeration of yeast in yoghurt. Lett. Appl. Microbiol. 11:282-285.
97. Samson, R. A., A. D. Hocking, J. I. Pitt, and A. D. King (ed.). 1992. Modern methods in food mycology. Elsevier, New York. 388 pp.
98. Samson, R. A., E. S. Hoekstra, J. C. Frisvad, and O. Filtenborg. 1995. Methods for the detection and isolation of food-borne fungi, p. 235-242. *In* R. A. Samson, E. S. Hosekstra, C. A. N. Van Vorschot (ed.), Methods for the introduction to the food-borne fungi. Centraal-bureau Schimmelcultures, Baarn.
99. Samson, R. A., E. S. Hoekstra, J. S. Frisvad, and O. Filtenborg (ed.). 1995. Introduction to food-borne fungi, 4th ed. Centraalbureau voor Schimmelcultures, Baarn. 322 pp.
100. Schaertel, B. J., N. Tsang, and R. Fistenberg-Eden. 1987. Impedimetric detection of yeast and mold. Food Microbiol. 4:155-163.
101. Schnürer, J. 1993. Comparison of methods for estimating the biomass of three food-borne fungi with different growth patterns. Appl. Environ. Microbiol. 59:552-555.
102. Shapira, R., N. Paster, O. Eyal, M. Menasherov, A. Mett, and R. Salomon. 1996. Detection of aflatoxigenic molds in grains by PCR. Appl. Environ. Microbiol. 62:3270-3273.
103. Shapira, R., N. Paster, M. Menasherov, O. Eyal, A. Mett, T. Meiron, E. Kuttin, and R. Salomon. 1997. Development of polyclonal antibodies for detection of aflatoxigenic molds involving culture filtrate and chimeric proteins expressed in *Escherichia coli*. Appl. Environ. Microbiol. 63:990-995.
104. Silley, P., and S. Forsythe. 1996. Impedance microbiology - a rapid change for microbiologists. J. Appl. Bacteriol. 80:233-243.
105. Snow, D. 1949. The germination of mould spores at controlled humidities. Ann. Appl. Biol. 36:1-13.
106. Stubbs, S., R. Hutson, S. James, and M. D. Collins. 1994. Differentiation of the spoilage yeast *Zygosaccharomyces bailii* from other *Zygosaccharomyces* species using 18S rDNA as target for a non-radioactive ligase detection reaction. Lett. Appl. Microbiol. 19:268-272.
107. Tapia de Daza, M. S., and L. R. Beuchat. 1992. Suitability of modified dichloran glycerol (DG18) agar for enumerating unstressed and stressed xerophilic molds. Food Microbiol. 9:319-333.
108. Thrane, U. 1996. Comparison of three selective media for detecting *Fusarium* species in foods: a collaborative study. Int. J. Food Microbiol. 29:149-156.
109. Tsai, G-J, and M. A. Cousin. 1990. Enzyme-linked immunosorbent assay for detection of molds in cheese and yogurt. J. Dairy Sci. 73:3366-3378.
110. Tsai, G-J, and C-S Yu. 1997. An enzyme-linked immunosorbent assay for the detection of *Aspergillus parasiticus* and *Aspergillus flavus*. J. Food Prot. 60:978-984.
111. Van der Horst, M., R. A. Samson, and R. Karman. 1992. Comparison of two commercial kits to detect molds by latex agglutination, p. 241-251. *In* R. A. Samson, A. D. Hocking, J. I. Pitt, and A. D. King (ed.), Modern methods in food mycology. Elsevier, New York.
112. Weaver, J. C., G. B. Williams, A. Klibanov, and A. L. Demain. 1988. Gel microdroplets: rapid detection and enumeration of individual microorganisms by their metabolic activity. Bio/Technol. 6:1084-1089.
113. Wing, Y. M., and J. H. Hotchkiss. 1991. Headspace gas composition and chitin content as measures of *Rhizopus stolonifer* growth — a research note. J. Food Sci. 56:274-275.
114. Yong, R. K., and M. A. Cousin. 1995. Detection of molds in foods by a nonspecific enzyme-linked immunosorbent assay. J. Food Sci. 60:1357-1363.
115. Yong, R. K., and M. A. Cousin. 1998. Detection of aflatoxigenic molds in corn and peanuts by immunoassays. Submitted for publication.
116. Yoshida, M., H. Maeda, and Y. Ifuku. 1991. Rapid detection of yeast in orange juice by enzyme-linked immunosorbent assay. Agric. Biol. Chem. 55:2951-2957.

Chapter 21

Detection and Enumeration of Heat-Resistant Molds

L. R. Beuchat and J. I. Pitt

21.1 INTRODUCTION

Spoilage of thermally processed fruits and fruit products by heat-resistant molds has been recognized in several countries.[5,8,9,10,12,15,22,24,29,30,31,39] *Byssochlamys fulva, B. nivea, Neosartorya fischeri, Talaromyces macrosporus, T. bacillisporus,* and *Eupenicillium brefeldianum* have been most frequently encountered.[10] *Byssochlamys* species have been recognized as spoilage molds in canned fruit since the early 1930s[12,13] and have been extensively studied.[2,5,14,21,29] Spoilage by other heat-resistant molds is a less serious problem recognized only in recent years. Consequently, less information is available concerning the behavior of these other genera in thermally processed fruit products.

Heat-resistant molds are characterized by the production of ascospores or similar structures with heat resistance, in some instances comparable to bacterial spores. This enables them to survive the thermal processes given to some fruit products (Table 1). Germination of ascospores may result in visible growth of mycelia on fruits and fruit products. Production of pectic enzymes by *Byssochlamys* can result in complete breakdown of texture in fruits[30] and also can result in off-flavor development.

Some *Byssochlamys* species produce patulin, byssotoxin A, and byssochlamic acid, all having toxic effects on laboratory animals.[5,20,27,32] *Neosartorya fischeri* is known to produce fumitremorgin A, B, and C, terrein, and verruculogen.[11,26] Heat-resistant molds, therefore, may constitute a public health hazard as well as a spoilage problem.

21.11 Distribution

Heat-resistant molds are widely distributed in the soil, particularly in vineyards, orchards, and fields in which fruits are grown.[8,12,13,29,36] Consequently, these molds may become contaminants on fruit and other vegetation upon contact with soil, before delivery to the processing plant. The number of ascospores on fruits is generally low, less than 1 per g.[20,36]

21.2 GENERAL CONSIDERATIONS

21.21 Samples

Because of their low incidence in fruit, ascospores are not likely to exceed 1 to 10 per 100 g or mL of processed products. Thus, for their effective detection, it is important that relatively large samples be analyzed. Centrifugation may be used to concentrate ascospores in liquid fruit products, the force and time necessary being influenced by volume, viscosity, and specific gravity of the sample.

Since the viability of ascospores is not appreciably affected by freezing and thawing, food samples can be stored frozen prior to analysis.

21.22 Enumeration Principles

A secondary but perhaps important point is that ascospores of heat-resistant molds may require heat activation before growth will occur.[3,16,18,20,38] The composition of the heating menstruum can influence the rate and extent of activation.[4,7,31,34] Maximal activation of *B. fulva* and *N. fischeri* var. *glaber* ascospores was obtained by heating at 70°C for 30 min in grape juice; in distilled water, 120 min were required for *B. fulva* while only 1% of the *N. fischeri* ascospores were activated.[37] Different strains within the same species may require different treatment times and temperatures to achieve maximal activation.

Detection and enumeration of heat-resistant ascospores rely on a selective heat treatment that inactivates vegetative cells of fungi and bacteria as well as less heat-resistant spores.

Heat-resistant molds are not fastidious in their nutrient requirements, and therefore the media listed below as well as many fruit juice agars will support germination of ascospores and subsequent vegetative growth. Since ascospores may be stressed by the heating process, highly acidic media are not recommended. However, when culturing low-acid foods that are heavily contaminated with bacterial spores, acidification or the addition of chloramphenicol to the plating medium may be required to inhibit the bacteria.

21.3 EQUIPMENT, MATERIALS, AND REAGENTS

Potato dextrose agar (PDA)
Malt extract agar (MEA)
Orange serum agar (OSA)
Czapek yeast autolysate (CYA) agar

Table 1. Tolerance of Heat-resistant Molds Isolated from Foods[a]

Mold	Heat-resistant Structure	Heating Medium	Heat Resistance
Byssochlamys fulva	Ascospores	Glucose-tartaric acid, pH 3.6	90°C, 51 min, 1000-fold[2]
		Grape juice, 26° Brix	85°C, 150 min, 100-fold[21]
Byssochlamys nivea	Ascospores	Grape juice	88°C, survived 60 min[19]
		Apple Juice	99°C, survived in juice[1]
Eupenicillium lapidosum	Ascospores	Blueberry juice	81°C, 10 min, survival; 81°C, 15 min, death[41] z = 10.3°F
	Cleistothecia	Blueberry juice	93.3°C, 9 min, growth; 93.3°C, 10 min, death[41] z = 10.6°F
Eupenicillium brefeldianum	Ascopores	Apple juice	90°C, 1 min, death[40] z = 7.2°C
	Cleistothecia	Apple juice	90°C, 220 min, death[40] z = 11.7°C
Talaromyces macrosporus	Ascospores	Apple juice	90°C, 2 min, death[40] z = 7.8°C
	(3 isolates)	Fruit-based fillings	D91°C = 2.9 to 5.4 min[3] z = 9.4 to 23.3°F
		Apple juice	D90.6°C = 1.4 min[3] z = 9.5°F
		Apple juice	D90.6°C = 2.2 min[33] z =5.2°C
	Cleistothecia	Apple juice	90°C, 80 min, death[40] z = 11.7°C
Monascus purpureus	Whole culture	Grape juice	Survival several min 100°C[9]
Humicola fuscoatra	Chlamydospores	Water	80°C, 101 min, 10-fold inactivation[23]
Phialophora sp.	Chlamydospores	Apple juice	80°C, 2.3 min, 10-fold inactivation[14]
Neosartorya fischeri	Ascospores	Water	100°C, 60 min, survival[17]
	(3 isolates)	Fruit-based fillings	D91°C = <2.0 min; D88°C = 4.2–16.2 min[3] z = 5.4 = 11°F
		Apple juice	87.8°C, 1.4 min[33] z = 5.6°C
Neosartorya fischeri var. *glaber*	Ascospores	Water	90°C, 60 min, survival[24]
		Grape juice	85°C, 10 min, 10% survival[37]
Thermoascus auranthiacum	Whole culture	Grape juice	88°C, 60 min, survival[19]

[a] Adapted from Splittstoesser and King.[35]

21.4 PROCEDURES

21.41 Petri Dish Method[25,36]

Fruits and products containing pieces of fruits must be blended or homogenized before analysis can proceed. To a tared sterile blender jar, add 100 g of fruit or fruit product plus 100 mL of sterile water and blend 5 min or until mixture appears homogeneous. After blending, enclose the jar in a polyethylene bag to safeguard against leakage through the bottom bushing and place the jar in a closed 75°C to 80°C waterbath for 1.5 hr. This holding time ensures that the homogenate will be at 75°C for at least 30 min.

Alternatively, samples (100 g plus 100 mL of sterile water) may be homogenized by a stomacher for 2 to 4 min (see Chapter 2). Two 50-mL portions of homogenate are then transferred to sterile 200 × 30 mm test tubes and placed in a closed waterbath at 75°C to 80°C for 30 min. The surface of the sample in the jars or tubes should be well below the surface of the water in the bath throughout the heat treatment.

After heating, duplicate 50-mL samples of thoroughly mixed homogenate in blender jars or the entire contents (50 mL) of duplicate tubes of sample that had been homogenized in a stomacher are combined with PDA or MEA in 150-mm diameter petri dishes. Each 50-mL sample is equally distributed in four dishes and thoroughly mixed with 10 mL of 1.5 strength agar. Petri dishes are loosely sealed in a plastic bag to prevent drying and incubated at 30°C for up to 30 days. Most viable ascospores will germinate and form visible colonies within 7 to 10 days; however, heat-injured and other debilitated ascospores may require additional time to form colonies. The 30-day incubation time also enables molds to mature and sporulate, thus aiding their identification.

Fruit juices (35° Brix or less) are analyzed in a manner identical to that described above. Fruit juice products (35° Brix or more) and fruit juice concentrates should be diluted (1:2) with sterile water and thoroughly mixed before heat treatment is applied to duplicate 50-mL samples. The general procedure is illustrated in Figure 1.

21.42 *Direct Incubation Method*[18,30]

The petri dish method is subject to error from aerial contamination (see Section 21.51). An alternative method that avoids this problem is described in this section. It is suitable for fruit pulps and homogenates.

Homogenized 50-mL samples are heated in flat-sided bottles such as 100-mL medicine flats. Bottles are heated in an upright position in a waterbath at 80°C for 30 min and then incubated on their sides, allowing as large a surface area as possible, at 30°C for up to 30 days. This procedure avoids the risk of contamination from the air and minimizes loss of moisture. Colonies develop on the surface of the homogenate.

Larger samples, such as 100-mL quantities in 200-mL bottles, can also be handled by this method. One apparent disadvantage is that colonies developing in the bottles must be picked and grown on suitable media for identification. However, cultivation on identification media is also recommended with the petri dish method.

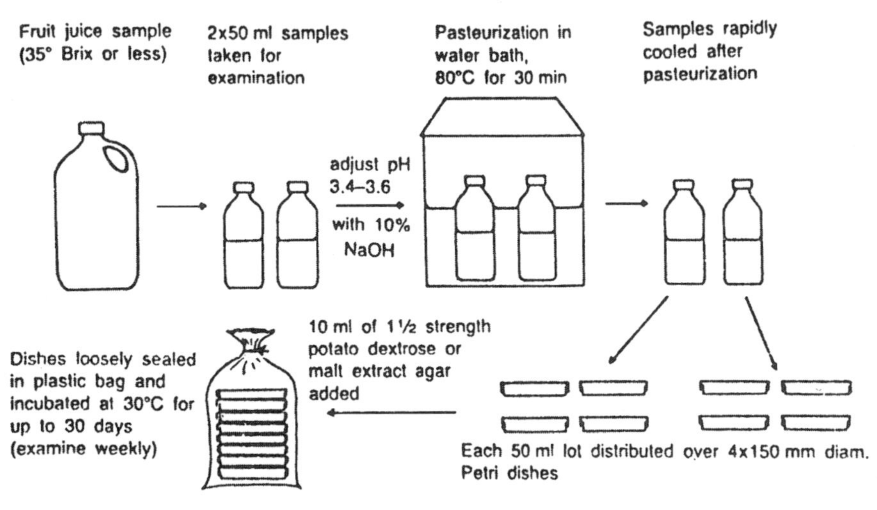

Figure 1. *Procedure for detection and enumeration of heat-resistant mold spores. (From Hocking and Pitt[10]).*

21.5 PRECAUTIONS

21.51 *Air Contamination*

With the petri dish method, aerial contamination during plating may be a problem. The appearance of green *Penicillium* colonies, or colonies of common *Aspergillus* species such as *A. flavus* and *A. niger* is a clear indication of contamination, as these molds are not heat resistant. To minimize this problem, pour plates in clean, still air or use a laminar flow hood, if possible. Alternatively, use the direct incubation method.

21.6 INTERPRETATION

Heat-resistant mold ascospores are not uncommon on fruit when it is received at the processing plant.[36] The acceptable level of contamination will depend on whether the fruit is a major or minor ingredient, whether the final product will contain a preservative such as sorbate or benzoate, and the thermal process to which the fruit is to be subjected. A count of 5 ascospores per 100 g (mL) of product at a stage just prior to the retort or heat exchanger indicates a serious problem. For ultra-high temperature (UHT) processed fruit juice blends that do not contain a preservative, even a lower level of contamination is unacceptable. One manufacturer of pasteurized fruit juices has issued a specification that calls for the absence of *Byssochlamys* from a 100-mL sample taken from each 200-L drum of raw material.[6]

21.7 TAXONOMY OF IMPORTANT HEAT-RESISTANT MOLDS

The basis for high heat resistance in these molds is the production of a teleomorph; that is, they form ascospores. Ascospores are produced, generally, in groups of eight, within a closed sac, the ascus (plural, asci); ascospores are the prime characteristic of the class of fungi called ascomycetes. In nearly all ascomycete genera, asci are in turn enclosed, in large numbers, within larger bodies. In genera of interest here, these bodies may have a solid, totally enclosed wall (a cleistothecium) or be composed of fine, interwoven hyphae (a gymnothecium). Only in *Byssochlamys* are asci borne singly and unenclosed on a layer of fine contorted hyphae.

As well as ascospores, ascomycetes generally produce an anamorph with asexual spores called conidia (singular, conidium). Conidia are not very heat resistant and are usually readily destroyed by pasteurizing heat processes or the screening techniques outlined above. The molds of interest here produce conidial states characteristic of the genera *Aspergillus, Paecilomyces,* and *Penicillium.*

21.71 *Identification of Isolates*

To identify heat-resistant mold isolates, proceed as follows. Inoculate each isolate onto media in four 90-mm diameter petri dishes, two each of CYA and MEA. Inoculate each plate at three equally spaced points. Incubate one plate each of MEA and CYA at 25°C, and the others at 30°C.

After incubation for 7 days, examine plates by eye, measuring colony diameters with a ruler, and make wet mounts to examine small pieces of mold under a compound microscope. The following key will assist in identification of common heat-resistant molds. For less common species, see Hocking and Pitt.[10,28]

21.72 *Key to Common Heat-resistant Fungi*

1. Asci produced in discrete bodies with totally enclosed walls (cleistothecia)—*Neosartorya fischeri*
   Asci produced in bodies with walls of woven hyphae (gymnothecia) or openly—2
2. Asci enclosed in gymnothecia—*Talaromyces macrosporus*
   Asci produced openly; fine hyphae may be present, but asci not enclosed—3
3. Colonies on CYA and MEA predominately buff or brown—*Byssochlamys fulva*
   Colonies of CYA and MEA persistently white or cream—*Byssochlamys nivea*

21.721 *Genus* Byssochlamys *Westling*

Byssochlamys has the distinction of being almost uniquely associated with food spoilage and in particular with the spoilage of heat-processed acid foods. Its natural habitat appears to be soils, but the genus is mentioned very seldom in lists of molds from soils other than those used for the cultivation of fruits.

Byssochlamys is an ascomycete genus characterized by the absence of cleistothecia, gymnothecia, or other bodies that in most ascomycetes envelop asci during development. Asci in *Byssochlamys* are borne in open clusters, in association with, but not surrounded by, unstructured wefts of fine white hyphae.

In our experience, the temperature range for observation of *Byssochlamys* asci and ascospores in the laboratory is sometimes very narrow. Cultures need to be incubated at 30°C as some isolates do not produce asci at 25°C or 37°C. However, presumptive evidence of the presence of *Byssochlamys* or other heat-resistant molds can be made from plates incubated at 25°C or 37°C if the isolate has come from heat-processed food or raw materials.

Byssochlamys fulva Olliver and G. Smith (Figure 2)
Anamorph: *Paecilomyces fulvus* Stolk and Samson

At 25°C, colonies on CYA and MEA are at least 60 mm diameter, often covering the whole petri dish, relatively sparse, low, or somewhat floccose, with conidial production heavy, uniformly colored olive brown, and reverse in similar colors or pale. At 30°C, colonies on CYA and MEA usually cover the entire surface, and are low to moderately deep, sparse, with moderate conidial production, colored brown, overlaid by white hyphae from which asci are produced; reverse is olive brown to deep brown.

Teleomorph, observed as single asci, which are borne from, but not enveloped by, wefts of contorted white hypae, best developed at 30°C, maturing in 7 to 12 days, occasionally formed at 25°C in fresh isolates but maturing slowly if at all, with asci spherical to subspheroidal, 9 to 12 µm diameter, and ascospores ellipsoidal, hyaline, or straw-colored, 5 to 7 µm long, and smooth-walled.

Anamorph is best observed at 25°C, consisting of penicilli borne from surface hyphae or long, trailing, aerial hyphae, stipes 10 to 30 µm long, with phialides of variable appearance, flask-shaped or narrowing gradually to the apices, 12 to 20 µm long, and conidia mostly cylindrical or barrel-shaped, narrow and 7 to 10 µm long, but sometimes longer, wider, or ellipsoidal from particular phialides, smooth-walled.

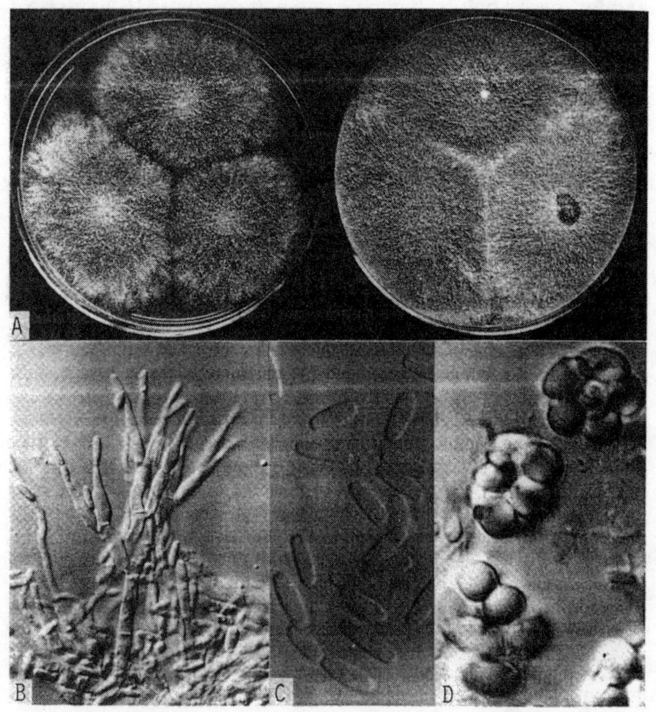

Figure 2. *Byssochlamys fulva: (A) colonies on CYA and MEA, 7 days, 25°C; (B) penicillus, x 750; (C) conidia, x 1875; (D) asci and ascospores, x 1875.*

Byssochlamys nivea Westling
Anamorph: *Paecilomyces nivea* Stolk and Samson

At 25°C, colonies on CYA are 40 to 50 mm diameter, low and quite sparse, white to slightly grey, with reverse pale to mid-brown. Colonies on MEA cover the whole petri dish, low and sparse, white to creamish, with small knots of dense hyphae, and reverse pale to brownish. At 30°C on CYA, colonies cover the whole petri dish, similar to those on MEA at 25°C, but often more dense, enveloping distinct knots of dense hyphae.

Teleomorph is similar to that of *B. fulva* except for slightly smaller asci (8 to 11 µm diam) and ascospores (4 to 6 µm diam), maturing in 10 to 14 days at 25°C and in 7 to 10 days at 30°C.

Anamorphs of two kinds are produced, aleurioconidia and penicilli; aleurioconidia borne singly, common at 30°C and 37°C, spherical to pear-shaped, 7 to 10 µm diam, with irregular penicilli sparsely produced and phialides sometimes borne solitarily from hyphae as well, phialides 12 to 20 µm long, cylindrical then gradually tapering, and conidia ellipsoidal to pear-shaped, 3 to 6 µm long, smooth-walled.

21.722 Genus Neosartorya *C. R. Benjamin*

A genus of soil fungi, *Neosartorya* is of interest in food microbiology only because its highly heat-resistant ascospores occur from time to time in heat-processed foods and it has occasionally been reported as a cause of spoilage. Although there are several species, only *N. fischeri* is commonly isolated from foods.

Neosartorya fischeri (Wehmer) Malloch & Cain (Figure 3)
Anamorph: *Aspergillus fischerianus* Samson & W. Gams

On CYA and MEA at 25°C, colonies 50 to 65 mm or more diameter, low and sparse to moderately deep, of cottony white to cream mycelium, surrounding abundant white developing cleistothecia and overlaid by scattered, usually inconspicuous blue to green conidial heads, with reverse pale to yellow. At 30°C, colonies cover the whole petri dish, similar to those at 25°C, but often are deeper and more luxuriant.

Cleistothecia white, 300 to 400 µm diameter, mature in 1 to 2 weeks at 25°C, with ascospores ellipsoidal, overall 6 to 7 × 4 to 5 µm, ornamented with two prominent, sinuous, longitudinal ridges and usually with other irregular ridges as well. Anamorph *Aspergillus*, with sparse conidiophores, 300 to 1000 µm long, terminating in small swellings, 12 to 18 µm diameter, with phialides crowded, 5 to 7 µm long, and conidia spheroidal, 2.0 to 2.5 µm diameter, with finely roughened walls.

21.723 Genus Talaromyces *C. R. Benjamin*

Talaromyces is characterized by the production of yellow or white gymnothecia in association with anamorph characteristics of *Penicillium, Paecilomyces,* or *Geosmithia.* It is a genus of about 25 species, mostly soil inhabiting. The commonly encountered species is *T. flavus*, and until recently this was believed to be the correct name for the species which sometimes occurs in heat processed foods. However, the heat resistant species is now considered to be distinct and is referred to as *T. macrosporus.*

Talaromyces macrosporus (Stolk and Samson) Frisvad et al. (Figure 4)
Anamorph: *Penicillium macrosporus* Frisvad et al.

At 25°C, colonies on CYA are usually 20 to 40 mm diameter, plane, low, and quite sparse to moderately deep and cottony, with mycelium pale to bright yellow, in most isolates concealing developing gymnothecia, also clear to reddish exudate present occasionally, and reverse sometimes yellow, more usually orange, reddish, or brown. Colonies on MEA are 30 to 50 mm diameter, generally similar to those on CYA, but gymnothecia are more abundant, with reverse usually dull orange or brown, but sometimes deep brown or deep red. At 30°C on CYA, colonies are 30 to 45 mm diameter, and generally similar to those at 25°C, but sometimes with white or brown mycelium or overlaid with grey conidia or with conspicuous red soluble pigment and reverse color. At 30°C on MEA, colonies are similar to those at 25°C, usually produce abundant gymnothecia, and in reverse sometimes produce red or olive colors also.

Gymnothecia of tightly interwoven mycelium, bright yellow, about 200 to 500 μm diameter, closely packed, mature within 2 weeks, with ascospores yellow, ellipsoidal, commonly 5.0 to 6.5 μm long, with spinose walls. Anamorph *Penicillium*, with conidiophores borne from aerial hyphae and stipes 20 to 80 μm long, bear terminal biverticillate or, less commonly, monoverticillate penicilli, and have phialides needle-shaped, 10 to 16 μm long, and conidia ellipsoidal, 2.5 to 4.0 μm long, with smooth to spinulose walls.

21.8 REFERENCES

1. Baumgart, J., and G. Stocksmeyer 1976. Heat resistance of ascospores of the genus *Byssochlamys*. Alimenta 15:67-70.
2. Bayne, H. G., and H. D. Michener. 1976. Heat resistance of *Byssochlamys* ascospores. Appl. Environ. Microbiol. 37:449-453.
3. Beuchat, L. R. 1976. Extraordinary heat resistance of *Talaromyces flavus* and *Neosartorya fischeri* ascospores in fruit products. J. Food Sci. 51:1506-1510.
4. Beuchat, L. R. 1988. Influence of organic acids on heat resistance characteristics of *Talaromyces flavus* ascospores. Int. J. Food Microbiol. 6:97-105.
5. Beuchat, L. R., and S. L. Rice. 1979. *Byssochlamys* spp. and their importance in processed fruits. Adv. Food Res. 25:237-288.
6. Cartwright, P., and A. D. Hocking. 1984. *Byssochlamys* in fruit juices. Food Technol. Aust. 36:210.
7. Conner, D. E, and L. R. Beuchat. 1987. Heat resistance of ascospores of *Neosartorya fischeri* as affected by sporulation and heating medium. Int. J. Food Microbiol. 4:303-312.
8. Fravel, D. R., and P. B. Adams. 1986. Estimation of United States and world distribution of *Talaromyces flavus*. Mycologia 78:684-686.
9. Hellinger, E. 1960. The spoilage of bottled grape juice by *Monascus purpureus* Went. Ann. Inst. Pasteur Lille 11:183-192.
10. Hocking, A. D., and J. I. Pitt. 1984. Food spoilage fungi: heat-resistant fungi. CSIRO Food Res. Q. 44:73-82.
11. Horie, Y., and M. Yamazaki. 1981. Productivity of tremorgenic mycotoxins, fumitremorgins A and B in *Aspergillus fumigatus*, and allied species. Trans. Mycol. Soc. Jpn. 22:113-119.
12. Hull, R. 1933-34. Investigation of the control of spoilage of processed fruit by *Byssochlamys fluva*, p. 64. *In* Annual report of the fruit and vegetable preservation research station. University of Bristol, Bristol, United Kingdom.
13. Hull, R. 1938. Study of *Byssochlamys fulva* and control measures in processed fruits. Ann. Appl. Biol. 26:800-822.
14. Jensen, M. 1960. Experiments on the inhibition of some thermoresistant moulds in fruit juices. Ann. Inst. Pasteur Lille; 11:179-182.
15. Jesenska, D., I. Havranekova, and I. Sajbidorova. 1984. On the problems of moulds on some products of canning industry. Cs Hyg.; 29:102-109.
16. Katan, T. 1985. Heat activation of dormant ascospores of *Talaromyces flavus*. Trans. Br. Mycol. Soc. 84:748-750.
17. Kavanagh, J., N. Larchet, and M. Stuart. 1963. Occurrence of a heat-resistant species of *Aspergillus* in canned strawberries. Nature 198:1322.

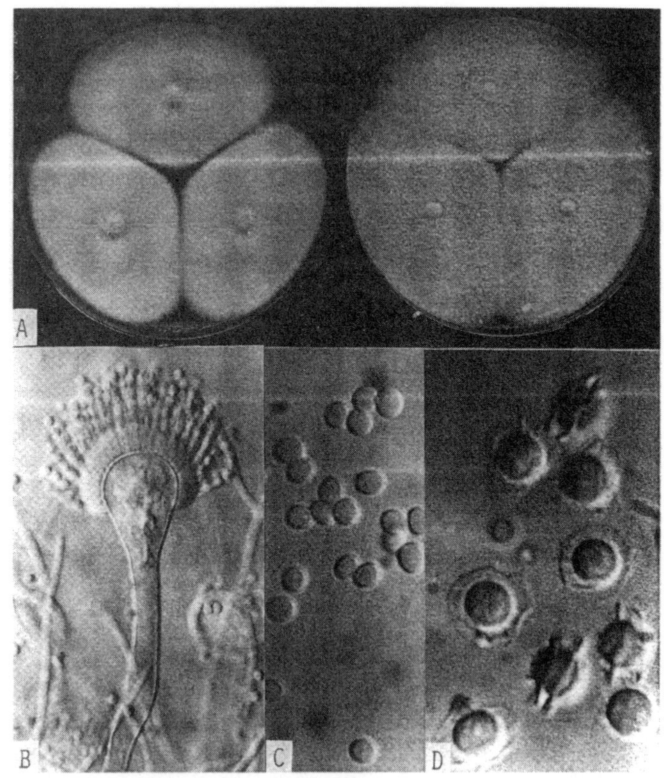

Figure 3. *Neosartorya fischeri: (A) colonies on CYA and MEA, 7 days 25°C; (B) conidiophore, x 750; (C) conidia, x 1875; (D) ascospores, x 1875.*

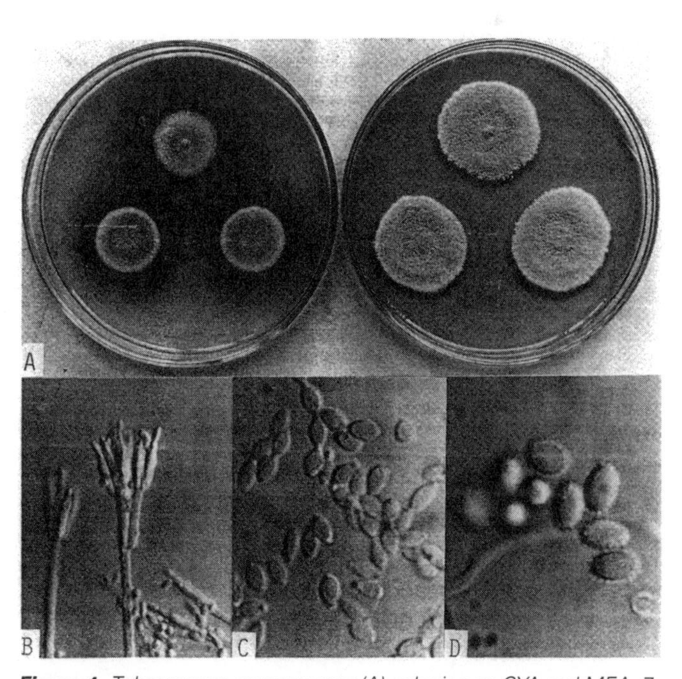

Figure 4. *Talaromyces marcosporus: (A) colonies on CYA and MEA, 7 days, 25°C; (B) penicillus, x 750; (C) conidia, x 1875; (D) ascospores, x 1875.*

18. King, A. D. 1997. Heat resistance of *Talaromyces flavus* ascospores as determined by a two phase slug flow heat exchanger. Int. J. Food Microbiol. 35:147-151.
19. King, A. D., Jr., H. G. Bayne, and G. Alderton. 1979. Nonlogarithmic death rate calculations for *Byssochlamys fulva* and other microorganisms. Appl. Environ. Microbiol. 37:596-600.
20. King, A. D., Jr., A. N. Booth, A. E. Stafford, and A. C. Waiss, Jr. 1972. *Byssochlamys fulva*, metabolite toxicity in laboratory animals. J. Food Sci. 37:86-89.
21. King, A. D., Jr., H. D. Michener, and K. A. Ito. 1969. Control of *Byssochlamys* and related heat-resistant fungi in grape products. Appl. Microbiol. 18:166-173.
22. Kotzekidou, P. 1997. Heat resistance of *Byssochlamys nivea*, *Byssochlamys fulva* and *Neosartorya fischeri* isolated from canned tomato paste. J. Food Sci; 62:410-412, 437.
23. Lubieniecki-von Schelhorn, M. 1973. Influence of relative humidity conditions on the thermal resistance of several kinds of spores of molds. Acta Aliment. 2:163-171.
24. McEvoy, I. J., and M. R. Stuart 1970. Temperature tolerance of *Aspergillus fischeri* var. *glaber* in canned strawberries. Irish J. Agric. Res. 9:59-67.
25. Murdock, D. I., and W. S. Hatcher, Jr. 1978. A simple method to screen fruit juices and concentrates for heat-resistant mold. J. Food Prot. 41:254-256.
26. Nielsen, P. V., L. R. Beuchat, and J. C. Frisvad. 1988. Growth and fumitremorgin production by *Neosartorya fischeri* as affected by temperature, light, and water activity. Appl. Environ. Microbiol. 54:1504-1510.
27. Pieckova, E., and Z. Jesenska. 1997. Toxigenicity of heat-resistant fungi detected by a bio-assay. Int. J. Food Microbiol. 36:227-229.
28. Pitt, J. I., and A. D. Hocking. 1997. Fungi and food spoilage, 2nd ed. Blackie Academic and Professional, London.
29. Put, H. M. C. 1964. A selective method for cultivating heat resistant moulds, particularly those of the genus *Byssochlamys*, and their presence in Dutch soil. J. Appl. Bacteriol. 27:59-64.
30. Put, H. M. C., and J. T. Kruiswijk. 1964. Disintegration and organoleptic deterioration of processed strawberries caused by the mould *Byssochlamys nivea*. J. Appl. Bacteriol. 27:53-58.
31. Rajashekhara, E., E. R. Suresh, and S. Ethiraj. 1996. Influence of different heating media on thermal resistance of *Neosartorya fischeri* from papaya fruit. J. Appl. Bacteriol. 81:337-340.
32. Rice, S. L. , L. R. Beuchat, and R. E. Worthington. 1977. Patulin production by *Byssochlamys* spp. in fruit juices. Appl. Environ. Microbiol. 34:791-796.
33. Scott, V. N., and D. T. Bernard. 1987. Heat resistance of *Talaromyces flavus* and *Neosartorya fischeri* isolated from commercial fruit juices. J. Food Prot. 50:18-20.
34. Splittstoesser, D. F., A. Einset, M. Wilkison, and J. Preziose. 1974. Effect of food ingredients on the heat resistance of *Byssochlamys fulva* ascospores. Proc. 4th Int. Congress Food Sci. and Technol., III; p. 79, Madrid, Spain.
35. Splittstoesser, D. F., and A. D. King, 1984. *In* M. Speck (ed.), "Compendium of methods for the microbiological examination of foods," 2nd ed. American Public Health Association, Washington, D. C.
36. Splittstoesser, D. F., F. R. Kuss, W. Harrison, and D. B. Prest. 1971. Incidence of heat resistant molds in eastern orchards and vineyards. Appl. Microbiol. 21:335-337.
37. Splittstoesser, D. F., and C. M. Splittstoesser. 1977. Ascospores of *Byssochlamys fulva* compared to those of a heat-resistant *Aspergillus*. J. Food Sci. 42:685-688.
38. Splittstoesser, D. F., M. Wilkison, and W. Harrison. 1972. Heat activation of *Byssochlamys fulva* ascospores. J. Milk Food Technol. 35:399-401.
39. Tournas, V. 1994. Heat-resistant fungi of importance to the food and beverage industry. Crit. Rev. Microbiol. 20:243-263.
40. Van der Spuy, J. E., F. N. Matthee, and D. J. A. Crafford. 1975. The heat resistance of moulds *Pencillium vermiculatum* Dangeard and *Pencillium brefeldianum* Dodge in apple juice. Phytophylactica 7:105-108.
41. Williams, C. C., E. J. Cameron, and O. B. Williams. 1941. A facultative anaerobic mold of unusual heat resistance. Food Res. 6:69-73.

Chapter 22

Mesophilic Aerobic Sporeformers

Kenneth E. Stevenson and Wayne P. Segner

22.1 INTRODUCTION

22.11 Classification

Members of the mesophilic aerobic sporeformers of significance in the spoilage of food belong to the genera *Bacillus* and *Sporolactobacillus*. Of these genera, the genus *Bacillus* is, by far, the more important.

The genus *Sporolactobacillus* includes endospore-forming rods which are mesophilic, microaerophilic, motile, catalase-negative, and homofermentative. Except for their ability to produce endospores, these organisms share many characteristics that are typical of bacteria in the genus *Lactobacillus*. Since spores produced by the type species *S. inulinus* possess a comparatively low order of heat resistance and they are apparently distributed in low numbers in food and the environment, this organism appears to have little importance in food spoilage.[6]

Mesophilic aerobic sporeformers classified in the genus *Bacillus* are aerobic to facultatively anaerobic rods that may appear as spindles, clubs, or wedges when bulged by endospores. However, swelling of the sporangium is not a prerequisite to membership in this genus.[8] While most mesophilic aerobic sporeformers are catalase-positive, catalase-negative strains culturally similar to *B. macerans* have been isolated from spoiled, insufficiently acidified, commercially canned onions and from inadequately sterilized, commercially canned cream-style corn.[20] All of the *Bacillus* species have the ability to produce refractile endospores under aerobic conditions, provided that other requirements for sporulation are met. Growth temperature is not a criterion for distinguishing *Bacillus* species that are important in food spoilage, except for *B. stearothermophilus* and *B. coagulans.* These two organisms differ in that *B. stearothermophilus* is capable of growing at 65°C[8] and will not grow on Thermoacidurans agar (pH 5.0). Overlapping temperature ranges for growth *of Bacillus* strains necessitate an arbitrary distinction between mesophiles and thermophiles. The mesophilic aerobic sporeforming bacteria are considered here as all strains of *Bacillus* species that grow better at 35°C than at 55°C, a delimitation that is compatible, but not identical, with views of others.[3,19]

The balance of this discussion is confined to the mesophilic aerobic sporeformers of the genus *Bacillus.* The facultative thermophile *B. coagulans* and the mesophile *B. cereus* are discussed separately in Chapters 24 and 32, respectively.

22.12 Sources and Growth Requirements

Bacillus species encountered in foods are generally widespread. Spores and also vegetative cells of mesophilic aerobic sporeformers are found in food, water, soil, and decomposing vegetation. Excepting the insectivorous strains, and those residing primarily in the alimentary tract of animals, the mesophilic *Bacillus* species appear to occupy no distinctive habitat.[5,8] An incubation temperature of 30° to 35°C is favorable for culture and sporulation of mesophilic sporeformers important to food microbiologists. Most strains recovered from food spoilage sporulate well aerobically on nutrient agar slants containing manganese.[4] The catalase-positive strains commonly sporulate readily. The catalase-negative strains are somewhat more demanding in their nutritional requirements; growth and sporulation are favored by the presence of yeast extract in the sporulation medium and by incubation at 35°C rather than 30°C.[10]

22.13 Nature and Characteristics

The presence of spores of the mesophilic *Bacillus* species permits their survival in mixed populations such as those found in foods. Resistance to bacteriophage, lytic factors such as bacteriocins, antibiotics produced by other organisms, lethal radiations, extremes of temperature, germicidal chemicals, and autolytic principles generally is greater in spores than in vegetative cells. Therefore, it is not surprising that spores are detected readily in many foods and ingredients such as starches, dried fruits, vegetables, cereal grains, dried milk, and spices.

22.14 Significance in Foods

Food handling equipment properly designed to eliminate niches wherein bacteria may multiply prevents the problem of the buildup of sporeforming populations during processing. Equipment with pits, crevices, dead ends, open seams, and square corners produces opportunities for bacterial buildup that may lead to deterioration of food even as it is being processed (incipient spoilage), especially when a shutdown extends the dwell time. The terminal thermal processes for low-acid foods in hermetically sealed containers generally provide lethality sufficient to inactivate spores of mesophilic bacteria. However, products that receive a mild thermal process, such as canned, cured meats, may undergo spoilage that is due to *Bacillus* species.[12] Cooling water and equipment surfaces that contact seams or lids on jars immediately following the heat process should be kept clean and as free of organisms as possible.[9]

22.15 Spoilage

Spoilage in thermally processed low-acid foods (pH>4.6) by mesophilic aerobic sporeformers is often characterized by flat sour spoilage in normal-appearing containers. Occasionally, loss of container vacuum or bulging of the container ends occurs because of growth and gas production by strains of either *B. macerans* or *B. polymyxa*. In addition, some *B. licheniformis* strains produce small amounts of gas during spoilage of thermally processed foods, although gas production is not a characteristic reported for *B. licheniformis* in "Bergey's Manual."[5]

Failure to carry out the prescribed thermal process, rather than a faulty thermal process design, is commonly responsible for spoilage of commercially processed low-acid foods, since the spores of mesophilic bacteria are usually only moderately resistant to moist heat. $D_{121°C}$ values are frequently of the order of 0.1 min or less. Spores of *B. macerans* or *B. polymyxa* strains commonly have $D_{100°C}$ values from 0.1 to 0.5 min.[21] In contrast, spores of the catalase-negative strains isolated from spoiled cream-style corn are quite resistant when heated in this product with a $D_{121°C}$ value similar to that found for spores of *Clostridium sporogenes* PA 3679.[20]

Spores of mesophilic aerobic sporeformers in low acid canned foods commonly exhibit a relatively low order of heat resistance during thermal processing if they are in fully hydrated products. However, spores of these organisms may survive and cause spoilage if they are entrapped in food particles or food ingredients that are in a dry state and not sufficiently hydrated during the thermal process. This type of spoilage has occurred, for example, in canned pasta in tomato sauce from failure to completely hydrate the dry noodles prior to thermal processing, in aseptically canned tapioca pudding from failure to adequately hydrate the tapioca kernels, and in canned chocolate flavored nutritional drink and aseptically canned chocolate pudding from failure to fully hydrate the cocoa particles prior to the thermal process. Thus, when mesophilic aerobic sporeformers are responsible for spoilage of low acid canned foods, the possibility of the spores surviving in "dry" conditions should be considered carefully.

In some instances, *B. licheniformis, B. polymyxa, B. subtilis* and, of course, *B. coagulans* (see Chapter 24) have caused spoilage of acid and acidified foods (pH<4.7). Growth of *B. licheniformis, B. polymyxa* and *B. coagulans* in acid foods, such as tomatoes, tomato juice, and acidified onions and green peppers, is of concern, because some strains can raise the product pH into the range where growth of *Clostridium botulinum* is possible.[2,7,14,15]

On occasion, mesophilic aerobic sporeformers may be recovered from, or cause spoilage of, acidified foods and fruit drinks, since such products normally receive only a hot-fill-and-hold or pasteurization heat treatment. For example, at least one report has been published of growth of spoilage isolates of *B. polymyxa* and *B. macerans* at as low a pH as 3.8 to 4.0 in sucrose nutrient broth and in canned cling peaches with a sucrose syrup.[23] However, if a food or drink shows normal appearance and odor and no evidence of growth microscopically, the significance of finding viable mesophilic aerobic sporeformers upon culturing is nil since germination of spores and vegetative cell growth in the food is prevented by its acid content. The product is commercially sterile (Chapter 61). In any instance where there is a question, spores of the isolate should be inoculated back into the original product for incubation and confirmation of growth.

22.2 SPECIAL CONSIDERATIONS

22.21 Ropy Bread Spoilage

Food products that do not receive a sporicidal terminal process, such as bakery products (Chapter 55), dried foods, and frozen foods, are not usually susceptible to spoilage by mesophilic aerobic sporeformers unless abused. However a case study[18] illustrates an exception, the problem of ropy bread, a troublesome defect that can occur in a clean, modern bakery. Breaking or cutting open the spoiled loaf releases a strong odor of decomposed or overripe watermelons.[17] The discolored, softened portions can be drawn out into long threads when teased with forceps or a glass rod.

Mucoid variants of *Bacillus subtilis,* formerly classified as *B. mesentericus* strains, are considered agents of ropy bread spoilage by virtue of endospore resistance to temperatures reached in the centers of loaves where the usual water activity (a_w) is about 0.95.[11] Almost any ingredient used in bread making may contribute "rope spores"; coarse and uncommon flours are reported likely sources.[22]

Good sanitation, modern bakery practices, and preservatives combine to keep rope spoilage under control so that incidents are rare.[11] However, the current trend toward elimination of preservatives in some bakery products increases the potential for this type of spoilage. Under any given conditions of bread production and storage, the higher the rope spore content of the ingredients, the greater the likelihood of the product becoming ropy.[13] Flour and other ingredients have been tested by a number of methods, but none of the procedures are sufficiently selective or discriminative to enumerate spores of only those organisms that produce ropy bread. A bread-baking and incubation test[22] is most reliable, but is qualitative, not quantitative. A widely used method,[1] a modified version of which is given below (Section 22.61), may be employed to determine presumptive counts of rope spores in flour, yeast, or other ingredients.

22.3 SOURCES OF ERROR

22.31 Mesophilic Aerobic Spores

The dilutions of sample material selected for culture in the given procedure permit enumeration of mesophilic spores in foods that contain relatively few spores and also in materials such as raw spices or soil samples where dense spore populations may occur. However, spore crops prepared in the laboratory will require greater dilution prior to the heating step. Certain strains of *Bacillus* that are considered facultatively thermophilic because they grow at 55° to 65°C are also capable of growth at 35°C.[10] Such organisms should not be counted since sporangia and vegetative cells are killed by the heat shock, and the incubation temperature is below that required for germination of surviving spores. Separate enumeration of (facultatively thermophilic) aciduric flat-sour sporeformers and thermophilic flat-sour sporeformers may be conducted according to methods described in Chapters 24 and 25, respectively.

22.32 Rope Spores

The enumerative techniques for rope spores all detect spores that survive a heat treatment to form colonies or pellicles in media that are neither selective nor differential. Indeterminate error stems from the assumption that all spores thus counted can produce rope. Inoculation of cells from morphologically typical colonies or pellicles into a susceptible bakery product,[22] or stepwise identification of *B. subtilis,* as described below, may be used to confirm presumptive results as occasion demands.

22.4 EQUIPMENT, MATERIALS, AND MEDIA

22.41 Equipment

- Thermostatically controlled waterbath, with stirrer, adjusted to 45°C.
- Weighted rings that fit 250-mL and 500-mL Erlenmeyer flasks to prevent bobbing and possible upset in bath.
- Waterbaths for 80° and 100°C operation.
- Incubators rated for operation in the 30° to 35°C temperature range with uniformity of temperature of ± 1°C at the extremes of the range.
- Microscope, pH meter, blenders.

22.42 Media, Dilution Menstruum (Chapter 63)

22.421 Dilution menstruum

Peptone water diluent (0.1%)

22.422 Culture media

Dextrose tryptone agar
Tryptone glucose extract (TGE) agar

22.5 PROCEDURES

22.51 Mesophilic Aerobic Spores

22.511 Sampling

Sample collection, shipment, and preparation are conducted in conformance with procedures given in Chapter 2. Ingredients need not be sampled routinely except in special cases, e.g., spices or gelatin to be used in pasteurized canned ham or ham products that receive mild heat treatment. The commonly used aerobic colony count (Chapter 7), a portion of which is contributed by aerobic spores, indicates overall "bacteriological quality" of ingredients without resorting to a separate count of mesophilic spores. A similar index of good manufacturing procedures holds true for food processing lines and equipment, where swab or surface contact platings indicate the adequacy of cleaning and other sanitation practices.

A freshly produced, thermally processed product is not likely to contain mesophilic spores or vegetative cells in numbers detectable by usual plating procedures. Consequently, about 24 containers of each code lot should be placed in incubators adjusted to 35°C, where they are held, usually for 14 days. If "flat sour" spoilage is detected, by lowered pH and/or liquefaction in a given sample set held at 35°C, the procedures listed in Chapter 62 should be followed.

22.512 Procedure

Fifty grams of ingredient or food material are weighed in a sterile, tared container and transferred to 450 mL of sterile 0.1% peptone water in a blender jar. Dispersion is accomplished by blending at high speed for 2 min. If a stomacher is used, dispersion requires 30 to 60 sec (Chapter 2).

TGE agar is prepared, 100 mL per 500-mL Erlenmeyer flask. One additional flask of medium is prepared to serve as a sterility control. Sterilization at 121°C moist heat for 15 min is followed by cooling to 45°C in a waterbath. Volumes of blended food are pipetted into a set of 3 flasks of TGE agar while they are held in the bath: 10 mL into the first, 1 mL into the second, and 0.1 mL into the third. Flasks are agitated gently to disperse the blended material throughout the medium.

Flasks are transferred without delay to a stirred water or oil bath adjusted to 80°C and held so that the liquid level is above the sample level in the flasks for 30 min. Flasks are occasionally agitated gently to assist heat distribution. Cooling is done in tepid tap water, taking care that the temperature does not fall to the point where agar congeals. Flasks are transferred to the 45°C bath following the rapid cooling step and held there for a period not to exceed 10 min. The 100-mL volume in each flask, representing test material and sterility control, is poured into a set of 5 sterile plates in approximately equal volumes, i.e., about 20 mL per plate. When the agar has solidified, plates are inverted in the 35°C incubator and allowed to incubate 48 hr.

Counts are made of surface and subsurface colonies. The sum of colonies on the set of 5 plates poured from TGE agar containing 10 mL of blended sample represents the number of aerobic mesophilic spores per gram. Similarly, the number of colonies in sets of plates receiving 1.0 and 0.1 mL of blended sample is equal to 0.1 and 0.01 of the number of spores per g and must be multiplied by 10 and 100, respectively, to calculate the count per g. The number of spores that can be enumerated by this method ranges from 1 to 150,000 spores per g.

22.52 Rope-Producing Spores

22.521 Sampling

Sampling, collection, shipment, and preparation are conducted as directed in Chapter 2.

22.522 Procedure

Fifty grams of flour or other ingredient are weighed in a tared, sterile container and transferred to 450 mL of sterile 0.1% peptone water in a blender jar. Dispersion is achieved by blending at high speed for 2 min. If a stomacher is used, mixing is done for 30 to 60 sec (Chapter 2).

Ten- and one-mL volumes of the peptone water suspension are pipetted into separate 100-mL portions of melted dextrose tryptone agar contained in 250-mL Erlenmeyer flasks and held at about 45°C. The flasks, and an additional uninoculated control flask, are submerged in a boiling waterbath so that bath water level is above the liquid level in the flasks. Contents are gently swirled from time to time as the internal temperature climbs to about 94°C at the end of 5 min heating. An additional 15 min in the boiling waterbath is required during which time the temperature should hold between 94° and 100°C.

After heating, cool the contents of the flask in tepid water to about 45°C, taking care that the agar does not lump, and pour contents of each flask into 5 sterile plates in approximately equal volumes. When the agar has solidified, invert the plates and incubate them at 35°C for 48 hr.

Count as rope-producing organisms the surface colonies that are grey-white, vesicle-like, becoming at first drier and finally wrinkled. Add to this count any subsurface colonies that display stringiness when tested. The total colonies on the set of 5 plates from the flask that received 10-mL suspension is considered as rope spores per g. The sum of colonies on plates representing the 1.0-mL volume of suspension is multiplied by 10 to get the number of rope spores per g of sample. The concentrations of spores enumerated by this method range from 1 to 15,000 per g.

When required, *B. subtilis* may be identified by conducting the following tests: catalase; acetoin production; nitrate reduction; utilization of citrate; growth in 7% NaCl; acid from glucose, arabinose, xylose, and mannitol; growth at pH 5.7; and hydrolysis of starch, casein, and gelatin (see reference #8 for media and procedures). Positive results should be obtained for each test. Additionally, the isolate should be examined microscopically for morphology, gram-stain reaction, and the presence of ellipsoidal or cylindrical spores that do not produce a distinct swelling of the sporangium[5,8].

22.6 MODIFICATIONS

22.61 Rope Spore Count

The method given is essentially that of the AACC[1] modified as follows: a) the analytical sample weight was increased from 20 g to 50 g; b) tetrazolium salts were not added to the agar medium; c) the 24-hr examination of incubating plates and "drawing" subsurface colonies to the surface were deleted; d) subsurface colonies displaying stringiness are added to the count of surface colonies displaying typical characteristics; and e) peptone water diluent was substituted for Butterfield's buffered phosphate diluent.

22.62 Mesophilic Aerobic Spore Count

The method given is basically that of the National Food Processors Association (formerly the National Canners Association) for enumeration of thermophilic spores in starch[16]. Modifications consisted of a) an increase in analytical sample weight from 20 g to 50 g; b) a reduction in heat stimulation from 10 min at 108°C (autoclave) to 10 min at 80°C, excluding equilibration time (waterbath); c) a reduction in incubation temperature to 35°C from 55°C; and, d) a change in dilution procedure to increase the enumeration limit to 150,000 spores per g.

The heating step was considered critical to reproducibility and efficiency of recovery because treatments stimulatory to some spores in a mixed, natural population are lethal to others. The method described above was used to test a variety of dry foods and soils. A family of curves reflecting the recovery of spores naturally occurring in mixed populations is presented in Figure 1. Results were similar at all temperatures in that no net stimulatory effect was observed. Heating at 90°C caused a reduction in spores recovered during the 19-min equilibration period. Somewhat greater spore inactivation was experienced during the 24-min temperature rise to 100°C.

The 80°C heat treatment was selected to provide minimum loss of endospore viability while ensuring death of vegetative cells. Fungal spores that may have been present in the samples examined did not produce colonies even after the 70°C heat treatment. While ascospores of *Byssochlamys fulva* and other molds that produce relatively heat-resistant spores (Chapter 21) are expected to survive the 80°C heat treatment, subsequent growth in TGE agar in competition with *Bacillus* species is unlikely.

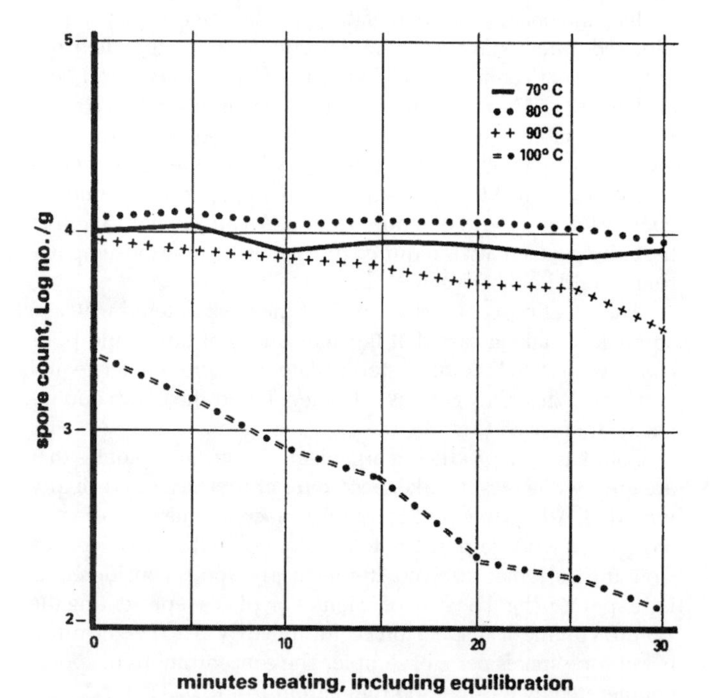

Figure 1. Survival of a mixed population of mesophilic aerobic spores occurring in Chili powder. Initial temperature = 47°C. Average equilibration time at 70°C = 18 min; 90°C = 19 min; 100°C = 24 min. Heating menstruum was tryptone glucose extract agar, which also served as recovery medium.

22.7 INTERPRETATION

22.71 Rope Spore Count

The range in which rope-producing spores are estimated in the above procedure, assuming that 300 colonies per pour plate are countable, is 1 to 15,000 per g. Additional decimal dilutions prior to the boiling water step may be introduced to extend the upper limit of enumeration.

Negative results are significant if a suitable control plating established that the recovery medium supports growth of *B. subtilis.*

Positive results should be interpreted following confirmatory inoculation tests or stepwise identification of *B. subtilis.* Counts must be interpreted in accordance with the requirements of bakeries or other food processors, whose acceptance specifications vary.

22.72 Mesophilic Aerobic Spores

Elevated counts in ingredients to be used in low-acid products (pH>4.6) that undergo a mild heat process, such as canned pasteurized luncheon meats or canned cured hams, may result in spoilage if mishandling occurs in marketing channels or by consumers. Otherwise the mesophilic spore count is of little significance to the processor. A report of spoilage in commercially canned fruit and fruit products because of *B. macerans* or *B. polymyxa*[23] left open the question of whether underprocessing or post-process recontamination was at fault. The presence of mesophilic aerobic spores in frozen or dried foods is innocuous provided mishandling does not result in large populations of *Bacillus cereus* (Chapter 32).

22.8 REFERENCES

1. American Association of Cereal Chemists. 1995. Approved methods of the American Association of Cereal Chemists, 9th ed. AOAC, St Paul, Minn.
2. Anderson, R. E. 1984. Growth and corresponding elevation of tomato juice pH by *Bacillus coagulans.* J. Food Sci. 49:647, 649.
3. Cameron, E. J., and J. R. Esty, 1926. The examination of spoiled foods II. Classification of flat sour spoilage organisms from nonacid foods. J. Infect. Dis. 39:89-105.
4. Charney, J., W. P. Fisher, and C. P. Hegarty. 1951. Manganese as an essential element for sporulation in the genus *Bacillus.* J. Bacteriol. 62:145-148.
5. Claus, D., and R. C. W. Berkeley. 1986. Genus *Bacillus* Cohn 1872, p. 1105-1139. *In* P. H. A. Sneath, N. S. Mair, and M. E. Sharpe (eds.),

Bergey's manual of systematic bacteriology, vol. 2. Williams & Wilkens, Baltimore, Md.
6. Doores, S., and D. Westhoff. 1981. Heat resistance of *Sporolactobacillus inulinus.* J. Food Sci. 46:810-812.
7. Fields, M. L., Zamora, A.F. and Bradsher, M. 1977. Microbiological analysis of homecanned tomatoes and green beans. J Food Sci. 42:931-4.
8. Gordon, R. E., W. C. Haynes, and C. H-N. Pang. 1973. The genus *Bacillus.* USDA Agriculture Handbook No 427. U.S. Department of Agriculture, Washington D.C.
9. Graves, R. R., R. S. Lesniewski, and D. E. Lake, 1977. Bacteriological quality of cannery cooling water. J. Food Sci. 42:1280-1285.
10. Harris, O., and M. L. Fields, 1972. A study of thermophilic, aerobic, sporeforming bacteria isolated from soil and water. Can. J. Microbiol. 22:917-923.
11. ICMSF. 1998. Microorganisms in foods 6: Microbial ecology of food commodities. Blackie Academic and Professional, New York.
12. Jensen, L. B. 1954. Microbiology of meats, 3rd ed. Garrard Press. Champaign, Ill.
13. Kent-Jones, D. W., and A. J. Amos. 1967. Modern cereal chemistry, 6th ed. Food Trade Press, London, England.
14. Montville, T. J. 1982. Metabolic effect of *Bacillus licheniformis* on *Clostridium botulinum:* Implications for home-canned tomatoes. Appl. Environ. Microbiol. 44:334-338.
15. Montville, T. J., and G. M. Sapers. 1981. Thermal resistance of spores from pH elevating strains of *Bacillus licheniformis.* J. Food Sci. 46:1710-1712, 1715.
16. National Canners Association. 1968. Laboratory manual for food canners and processors, vol. 1. Microbiology and processing, 3rd ed. AVI Publishing Co., Westport, Conn.
17. Pyler, E. J. 1973. Baking science and technology. Siebel Press, Chicago, Ill.
18. Raffaele, F. 1981. Case histories in sanitation, Case no. 17: Rope in bread. Dairy Food Sanit. 1:302-303.
19. Richmond, B., and M. L. Fields. 1966. Distribution of thermophilic aerobic sporeforming bacteria in food ingredients. Appl. Microbiol. 14:623-626.
20. Segner, W.P. 1979. Mesophilic aerobic sporeforming bacteria in the spoilage of low-acid canned foods. Food Technol. 33:55-59, 80.
21. Stumbo, C. R. 1973. Thermobacteriology in food processing. 2nd ed. Academic Press, New York.
22. Tanner, F. W. 1944. The microbiology of foods. 2nd ed. Garrard Press, Champaign, Ill.
23. Vaughn, R. H., I. H. Kruelevitch, and W. A. Mercer. 1952. Spoilage of canned foods caused by the *Bacillus macerans-polymyxa* group of bacteria. Food Res. 17:560-570.

Chapter 23

Mesophilic Anaerobic Sporeformers

Virginia N. Scott, Jean E. Anderson, and Guodong Wang

23.1 INTRODUCTION

The mesophilic sporeforming anaerobes all belong to the genus *Clostridium.* The various species of clostridia are basically either proteolytic (putrefactive) or nonproteolytic (saccharolytic), although there may be strain differences, and some species (e.g., *C. laramie*) may show characteristics of both groups, fermenting some carbohydrates and hydrolyzing some proteins, but not others. Since this property relates to their significance in food spoilage, they will be separated into these two groups for the purpose of further discussion.

One group consists of *C. sporogenes,* the proteolytic strains of *C. botulinum,* and other relatively heat resistant putrefactive anaerobes. The other group consists of *C. perfringens,* the nonproteolytic strains of *C. botulinum* and a variety of other similar clostridia, such as the butyric acid anaerobes (e.g., *C. butyricum, C. pasteurianum*) and some newly recognized psychrotrophic meat spoilage clostridia (e.g., *C. laramie,*[30] *C. algidicarnis*[37]), that are relatively nonresistant to heat. Only general methodology for the mesophilic sporeforming anaerobes will be considered here, since specific methods for detecting and enumerating *C. botulinum* and *C. perfringens* are provided in Chapter 33 and Chapter 34, respectively. This chapter also does not cover sporulation media for mesophilic anaerobic sporeformers, as this is beyond the scope of the chapter; consult the references provided for sporulation media appropriate for the sporeformer of interest.

Mesophilic sporeforming anaerobes of concern in foods are gram-positive, catalase-negative rods of varying size (depending in part on age and culture medium).[12] Rods occur singly, in pairs, in short chains, and occasionally as long filaments. Spores are generally subterminal to central and may or may not distend the rod. Motility is characteristic of most species (*C. perfringens* is nonmotile). Although considered to be strict anaerobes, some strains are able to grow in the presence of relatively high levels of oxygen.[12,41,56,70] Anaerobes are distributed widely in nature.[12] They are found in the soil, and therefore may be normal contaminants of vegetables at the time of harvest.[12,26,56] Some species commonly are found in the intestinal tracts and excreta of animals, and hence may become contaminants of milk and meat.[26] Others are found in fish, shellfish, and aquatic environments.[12,27]

23.11 Proteolytic (Putrefactive) Anaerobes

All mesophilic sporeforming anaerobes that digest proteins such as coagulated egg white, coagulated serum, or coagulated milk or blacken and digest brain or meat media with a putrid odor will be grouped together as proteolytic anaerobes and called by the more common term, putrefactive anaerobes.

These organisms are capable of decomposing proteins, peptides, or amino acids anaerobically with resulting foul-smelling, sulfur-containing products such as hydrogen sulfide, methyl and ethyl sulfide, and mercaptans. Ammonia and amines such as putrescine and cadaverine, usually are produced, along with indole and skatole, as well as carbon dioxide and hydrogen.[20,26] Growth occurs within an approximate temperature range of 10° to 50°C, with the exception of *C. putrefaciens,* which will grow from 0° to 30°C. This growth range covers the normal storage temperature of canned and other processed foods, including the refrigerated storage of cooked or cured meats.

The spores of putrefactive anaerobes, in general, have high heat resistance and are the organisms most often associated with underprocessing spoilage of low-acid canned foods. A number of species have been isolated from underprocessed canned foods having an original pH above 4.8. These organisms include *C. sporogenes, C. bifermentans, C. putrefaciens, C. histolyticum, C. botulinum* types A and B, and the closely related nontoxic organism identified as PA 3679.[12,20,25,26,60]

Since putrefactive anaerobes will not grow in foods with a pH of 4.6 or less, testing foods in this pH range for the presence of these organisms is not necessary.

23.12 Nonproteolytic Anaerobes

Mesophilic sporeforming anaerobes that do not digest protein and, therefore, do not produce the putrid growth end-products typical of the putrefactive anaerobes, will be grouped as nonproteolytic anaerobes. These anaerobes ferment carbohydrates and are therefore often referred to as saccharolytic.[20] Fermentation end-products include butyric acid, acetic acid, carbon dioxide and hydrogen. Some nonproteolytic anaerobes can germinate and grow at refrigeration temperatures.[8,9,10,30,31,39,54] The spores of these anaerobes have low heat resistance compared to the putrefactive anaerobes, and generally will not survive minimal retort processing. Therefore, they usually are not encountered in underprocessing spoilage of low-acid canned foods. The isolation of nonproteolytic anaerobes from spoilage in low acid canned foods is more likely to be an indication of post-processing contamination rather than underprocessing. When post-processing contamination with organisms does occur, poor sanitation of the post-processing can-handling lines is suspected, since the primary cause of this type of spoilage is physical abuse of wet containers following the process.[4,34,51,53] The quality of the can cooling water also should be questioned; however, spores of anaerobes are found infrequently in cooling water, and if found, generally are present in very low numbers.[22,47,64] At least one spoilage incident in a canned product was attributed to buildup of a psychrotrophic anaerobic sporeformer in a crabmeat picking machine.[54]

The butyric acid anaerobes are exemplified by *C. butyricum* and *C. pasteurianum*, but also include strains such as *C. tyrobutyricum*, *C. beijerinckii* and *C. acetobutylicum*. In fact, these were the first sporeformers recognized as food spoilage organisms when Pasteur investigated butyric acid fermentation in wine.[55] Butyric acid anaerobes have been isolated from canned tomatoes, pears, blueberries, pineapples, cucumbers and figs.[6,7,17,43,59,67] Spores of the butyric acid anaerobes generally do not cause spoilage unless the pH is above 4.2.[43] Although butyric acid anaerobes are not considered to be pathogenic, strains of *C. butyricum* that produce a toxin that can be neutralized by *C. botulinum* type E antitoxin have been isolated from cases of infant botulism.[3,40,61] These toxigenic strains have not been associated with any food products, and have been shown to have a lower heat resistance than spores of a nontoxigenic strain of *C. butyricum* isolated from spoiled blueberries.[44] These toxigenic strains were also unable to grow and produce toxin at pH 5.0, whereas the nontoxigenic strain from blueberries grew at pH 4.4.[44]

23.13 Spoilage in Food

With the high heat resistance of many strains of mesophilic sporeforming anaerobes, their ability to grow in the absence of oxygen and over a range that covers the temperature of the normal storage of canned and other processed foods, including refrigerated storage of cooked or cured meats and modified atmosphere packaged foods, these anaerobes are of primary importance in the spoilage of low-acid foods packed in hermetically sealed containers.

Proteolytic mesophilic anaerobes may cause spoilage in any canned food having a pH of 4.8 or above (the lowest pH at which *C. botulinum* spores have been reported to germinate and grow in commercial food products) if the spores are not destroyed by the thermal process to which the canned foods have been subjected, or if they gain access through container leakage after processing. Such spoilage is generally characterized by gas production and a putrid odor, although Cameron and Esty[11] noted that putrefactive anaerobes may show abnormal development resulting in nongaseous spoilage in low-acid products in the range of pH 4.6 to 5.0. Montville[42] has shown that cultural conditions influence gas and protease production. Irregular spoilage patterns can occur in inoculated packs of low acid foods having a pH value as low as 5.5.[57]

The butyric anaerobes are capable of germination and growth at a pH 4.2–4.4[43,68] and consequently are of spoilage significance in non-pressure-processed acid foods such as tomatoes, tomato products, and certain fruits, particularly if the pH is above 4.2. A number of factors such as fruit variety, growing conditions, etc., may cause the pH of the product to exceed 4.2. In those cases where the acidity of these products is too low, spoilage by butyric acid anaerobes may be controlled by acidification of the product, or by increasing the thermal process. Butyric acid anaerobes are also of spoilage significance in underprocessing of other acid and acidified foods, and occasionally in low-acid canned foods as well. Growth of these organisms in foods is characterized by a butyric odor and the production of hydrogen. Occasionally strains will be encountered that can grow at a pH lower than 4.2; if these strains are present in high numbers, the heat process may be inadequate and spoilage can occur. An unusual spoilage incident occurred in canned, acidified mung bean sprouts in which acid tolerant strains of *C. perfringens* and *C. barati*, as well as *C. butyricum*, were isolated.[15] The strains of *C. perfringens* and *C. barati*, which grew to high numbers during the sprouting process, were able to grow as low as pH 3.7.

Spoilage of foods by mesophilic anaerobic sporeformers is not limited to canned foods. Clostridia, particularly *C. tyrobutyricum*, and occasionally *C. butyricum* or *C. sporogenes*, are responsible for a major defect in cheese called "late blowing" or "late gas," because it occurs after the cheese has aged for several weeks.[19,20,27] Emmental, Swiss, Gouda and Edam cheeses are most often affected because of their high pH and moisture content and low interior salt content.[19] Psychrotrophic clostridia have also been associated with the spoilage of crabmeat,[54] and refrigerated meats.[8,9,14,30,31,37] The spoilage organism isolated by Kalchayanand et al.[30] germinated at 2°C, had an optimum growth temperature of 15°C, fermented some carbohydrates, digested meat but not gelatin or milk, and was destroyed by heating at 90°C for 10 min. This unique organism, named *C. laramie*, is probably more appropriately classified as a psychrophile rather than a psychrotroph, since its optimum growth temperature falls outside the mesophilic range of 20–45°C.[20] However, Broda et al.[9] found that the *Clostridium* spp. that had been isolated from spoiled vacuum-packaged beef and lamb represented two groups: one with an optimum growth temperature range of 25–30°C and one with an optimum growth temperature range of 15–20°C. Thus, these new spoilage organisms include mesophilic anaerobic sporeformers.

Because of the public health significance of their toxins, all types of *C. botulinum* (Chapter 33) are the most important anaerobes to control in food processing. However, the spores of many of the proteolytic nontoxigenic species are much more heat resistant than those of *C. botulinum*. These nontoxigenic strains are not a hazard to health, but their growth can result in severe economic losses through spoilage if not controlled through adequate thermal processing. As an illustration of their relative heat resistance, the $D_{121°C}$ value of proteolytic *C. botulinum* spores of Types A and B ranges from about 0.10 to 0.20 min, while that of PA 3679 ranges from 0.50 to 1.50 min.[60] Thus, adequate thermal processing to achieve commercial sterility in low-acid canned foods also would control the greatest health hazard.

Because of the lack of anaerobic conditions in most equipment in food processing plants, except for the dead ends of pipes, porous surfaces (such as wood), etc., conditions are not conducive to the growth and sporulation of anaerobes. Thus heavy buildup of these organisms seldom occurs. However, when this does occur, understerilization spoilage can result because of the heavy load of spores in the product. The spoilage pattern within the affected lots is often spotty and scattered, more typical of post-processing spoilage that is due to container leakage, than of the pattern expected from sporeformers that survive a thermal process. Thermal processing records and other canning parameters usually give no indications of any irregularities. In most cases the problem can be identified only by investigation at the cannery, including a bacteriological survey, plus the absence of demonstrable leakage and package defects in the spoiled containers.

The recovery in pure culture of mesophilic anaerobic sporeformers from putrid swollen cans is the classic condition encountered in underprocessing spoilage of low-acid canned foods. In rare instances spoilage of canned foods by these organisms may be the result of post-processing contamination, and the microflora observed can be a mixed culture, mesophilic anaerobic sporeformers in pure culture, or a combination of both.[34] Therefore, it is critical in the diagnosis of canned foods spoilage that cannery thermal processing data, results of container leakage tests using recommended laboratory test procedures, and any other information relative to the spoilage outbreak be evaluated to determine the cause of spoilage.[38]

It is not possible with our present knowledge to make an unqualified statement of the significance of numbers or presence of mesophilic anaerobe spores in ingredients or at any step in a food processing operation. It must be remembered also that heat processes for low-acid commercially canned foods are designed to destroy an average load of putrefactive anaerobe spores. Moreover, the process delivered is usually in excess of the designed process. Thus, to be significant, the spore count would have to be extremely large or consist of a population of extremely resistant spores.

Underprocessing spoilage is of primary concern because it presents the potential danger of toxin production by *C. botulinum*. However, three outbreaks of botulism from food canned in the United States have occurred in which defective cans resulted in postprocessing contamination of fish products with Type E *C. botulinum*.[13,28,46,65]

Although spoilage by mesophilic sporeforming anaerobes usually occurs in low-acid foods packed in hermetically sealed and thermally processed containers, these organisms must be considered a potential problem in spoilage of any low-acid food that fulfills the necessary growth requirements, particularly a sufficiently low level of oxygen. As examples, such conditions can be present in fresh produce marketed with plastic film overwrap or in plastic bags, and in modified-atmosphere-packaged foods such as fresh pasta, processed meats, poultry products, fish, vegetables, and prepared sandwiches.[16,36] The oxygen present in such packages may be insufficient to inhibit the growth of some of the clostridia, since certain strains are relatively oxygen tolerant.[52,70] Under appropriate storage temperatures, growth is possible in low-oxygen containing packages. It has been shown that *C. botulinum* is capable of reproducing and forming toxin in fresh mushrooms wrapped in nonventilated plastic film.[33,63] The use of perforated film prevents the growth by increasing the oxygen levels in the packages.[62]

Other food products, which may or may not be thermally processed, rely on either acidification or reduced water activity to prevent spoilage. Most mesophilic anaerobic sporeformers cannot germinate and grow below a pH of 4.0–4.2[5] (vegetative cells may grow at somewhat lower pH values) or a water activity of 0.93.[58] Thus, if properly prepared, these foods will not support the growth of mesophilic sporeforming anaerobes. If bacteriological examination of these foods is made, caution should be exercised in interpreting the results of anaerobic cultures if the usual culture media are employed. The cultures may show positive growth of mesophilic sporeforming anaerobes as the result of removing the spores from an inhibitory environment into one favorable to germination and growth. Consequently, isolation of bacteria in this situation is of no significance, unless they are present in large numbers or can be shown to produce spoilage when inoculated into the product. The same holds true for other foods such as cured meats that employ preservatives or other ingredients to retard spoilage.

Other potential sources of contamination with mesophilic sporeforming anaerobes include spices, cereals and cereal products, dried eggs, dried milk and milk products, and dried vegetables (onion and garlic). In general, no microbiological standards exist by which the suitability of these ingredients may be evaluated, but since they are often substantial contributors to the microbial flora and, at the same time, pose special problems in handling for analysis, specific procedures will be described for their examination.

23.2 GENERAL CONSIDERATIONS

It is sometimes desirable to have an estimate of the total number of mesophilic sporeforming anaerobes in a food material or ingredient. However, there is probably no way in which such an estimate can completely represent the total number present. Methods in use inevitably favor one group of anaerobes or the other of the total possible flora present. For example, if an estimate of the proteolytic clostridia is desired, it is usual to heat-shock for maximum germination of the spores, but the heat shock generally used is destructive of the spores of most nonproteolytic clostridia (such as those of *C. botulinum* Type E—Chapter 33). If, on the other hand, ethanol treatment[29] is used to destroy other contaminating nonsporulating organisms, a more accurate estimate of the nonproteolytic clostridia will be obtained; however, this does not provide for maximal germination and outgrowth of proteolytic strains. Thus, in either case, the total count will be in error by the failure of some clostridia to grow. Furthermore, neither method will destroy the spores of *Bacillus* spp. (Chapter 22), many of which are facultative anaerobes and inevitably will be counted along with the strict anaerobes. The same, of course, is also true with respect to both clostridia and facultatively aerobic sporeformers that are also facultative thermophiles.[1] Therefore, facultative aerobes and facultative thermophiles also may be counted by any method designed for the enumeration of mesophilic anaerobes. An additional complication arises if the total microflora capable of surviving either treatment contain bacteriocin or antibiotic-producing organisms and, at the same time, strains sensitive to these agents.[32] Given these limitations, it is best to choose a procedure most likely to give a reasonable estimate of the groups expected to be present in a particular food, since it is likely that only when spore suspensions of pure cultures are being counted can estimates be made with any degree of certainty.

Several methods are used for counting mesophilic anaerobes. Most probable number (MPN) methods (Chapter 6) seem to be the most widely used, and generally give more reproducible results with higher counts on pure cultures than other methods. Colony counts in deep agar tubes and agar plates incubated anaerobically also have had considerable use. Methods that incorporate the inoculum into melted agar with the production of deep colonies, as in pour plates or agar deep tubes, often have the handicap that gas production concomitant with the development of colonies splits the agar and renders accurate counting of colonies impossible unless a very short incubation period is used. Such counts are subject to the possible further inaccuracy that slow growing colonies may not be counted at all. Counts made by spreading the inoculum over the surface of a solid agar medium may show too much variation between replicates to be reliable, and are generally not recommended.

If the count is to represent only sporeformers, then vegetative cells must be destroyed prior to enumeration. As noted above, this is generally done by heat shocking the suspension to be counted. If the type of anaerobic sporeformer expected is known, then the heat shock can be chosen to select the appropriate type. For the proteolytic strains, which are generally more heat resistant, a heat shock of 10–13 min at 80°C is usually used; for the more heat sensitive spores a heat shock of 15–30 min at 60°C is more common, although ethanol treatment (45–60 min in an equal volume of 95% or absolute ethanol) has also been used. If the type of sporeformer is unknown, or both heat resistant and heat sensitive types are expected, the sample can be tested using two different heat shocks. If this is not possible, an alternative is to use an intermediate heat shock such as 71°C for 10 min.[35]

Although oxygen will not affect the viability of dormant spores of anaerobes, care must be taken, nevertheless, to exclude oxygen from culture media at the time of use. Vegetative cells of these organisms are especially sensitive to the lethal effects of oxygen, particularly at the stages of spore germination and outgrowth. The simplest procedure for oxygen removal is to boil or steam the medium for 10 to 20 min and quickly cool it just prior to use. Similarly, care must be taken when mixing the spore inoculum into molten agar or other liquid media to avoid the re-incorporation of more oxygen.

Stress-damaged spores may also require special consideration. If, as a consequence of heat treatment or the use of other lethal agents, it is desired to know how many spores have survived, prolonged incubation may be necessary. The recovery of viable spores injured by heat or the presence of a stressing agent may require weeks or months in a rich culture medium. Refer to Chapter 6 for recommended methods for recovery of injured spores.

23.3 EQUIPMENT, MATERIALS, AND MEDIA

23.31 Equipment and Materials

Autoclave—for heating food samples in addition to normal laboratory sterilization purposes.

Culture tubes, 25 × 200 mm, screw cap, for heat shocking plant survey samples

Culture tubes, 18 × 150 mm, either snap top or screw-cap, for liquid media

Culture tubes, 16 × 125 mm screw-cap for agar deep tubes

Petri dishes

Pipettes (serological or bacteriological)

Quebec colony counter or similar device

Anaerobic jar (BBL Gaspak system with disposable hydrogen and carbon dioxide gas generators or similar system) or anaerobic chamber

Incubator set at 30° to 35°C

Sterile swabs

Sterile knives, spatulas, spoons or other sampling tools

10-mL water or 0.1% peptone blanks in screw-cap tubes

10-mL water blanks containing 60 ppm sodium thiosulfate

9-mL and 99-mL water or 0.1% peptone dilution blanks

Scale or analytical balance

Sterile water collection bottles 250-mL screw cap

Sterile water

Sterile 250 mL Erlenmeyer flasks

Glass beads

Ethanol (95% or absolute; filter sterilized)

3% sodium thiosulfate (for neutralizing halogens in water samples)

Sterile, sealable plastic bags

Stomacher bags

Stomacher

Thermometer to determine temperature of samples during heat shocking

Water bath for heat shocking and for tempering agar

23.32 Media (Chapter 63)

2% Agar

AC agar

Andersen's pork-sea infusion broth or agar

Beef heart infusion broth—commercially available

Cooked Meat Medium (CMM)—commercially available

CMM with 0.1% soluble starch and 0.1% glucose

Dextrose tryptone agar (DTA)—commercially available

Differential Clostridial Agar (DCA)

Liver broth

Modified PA3679 Agar

Orange serum broth—commercially available

PE-2 medium

Peptone Yeast Extract Glucose Starch (PYGS) agar

Sodium Bicarbonate (0.14%)

SFP agar with egg yolk—commercially available

Sodium thioglycollate (0.1%)

Thioglycollate agar

Tomato Liver Broth

Trypticase peptone (TP) agar

Tryptose cycloserine dextrose (TCD) agar

Tryptose-Sulfite-Cycloserine (TSC) Agar

Trypticase peptone glucose yeast extract (TPGY) broth and agar

Yeast Extract Agar (YEA)

Sterile petroleum jelly or Vaspar to overlay broth media (stor age at 55°C will keep Vaspar fluid and ready to use)

23.4 PRECAUTIONS

Whenever spoiled low-acid canned foods are examined to determine the cause of spoilage, the possibility exists that this spoilage may be due to the growth of mesophilic sporeforming anaerobes. One must assume that the causative organism could be *C. botulinum.* Likewise, refrigerated spoiled foods may contain pathogens. Therefore, extreme care should be taken in handling the spoiled food if it has a putrid odor or typical clostridial forms are observed in direct smears of the food prior to culturing. In such cases the food and its container should be sufficiently heated to destroy botulinal toxin before disposal. The amount of heat will vary according to the size of the container and the nature of the product; however, autoclaving the food and the opened containers should be more than adequate to destroy botulinal toxin. It is strongly recommended that all persons who handle spoiled canned food be immunized with antibotulinal toxoid.

23.5 PROCEDURES

Chapter 62 describes the procedures to be used for the examination of canned foods for various types of bacteria, including mesophilic sporeforming anaerobes; therefore, this subject is not covered in detail here.

It may be desirable to identify the source of the mesophilic sporeforming anaerobes if they have been implicated in food spoilage. This may be determined by a bacteriological survey of the production facility and by examination of ingredients that had been used in the preparation of the food.

23.51 Plant Survey

For low-acid canned foods showing putrefactive spoilage, the first step in determining the source of the spoilage organisms is to collect canned samples at the closing machine before the heat process and culture for putrefactive anaerobes. If putrefactive anaerobic spores are found in relatively high numbers, it may be advantageous to collect product samples at key points along the canning line. Samples should be collected, using sterile sampling tools, in sterile plastic bags from the washed or blanched product, from mix tanks (formulated products), from the filler, and from closed but unprocessed containers. Equipment samples (swabs or scrapings) should be collected from hard-to-clean places that show visible buildup of food. Examples of typical problem areas are pumps, dead ends of pipes, frayed or worn conveyor belts, and wooden equipment. Similar procedures apply when looking for the source of nonproteolytic sporeformers, either at canning facilities or at facilities producing refrigerated products.

All product, swabs, scrapings, or other samples should be examined immediately after collection at the production facility. However, if they must be transported to a laboratory for examination, the samples should be transported under refrigeration. Wherever possible, these samples should be cultured the same day they are collected.

Swab samples from equipment and scrapings or splinters from wooden or other porous materials that have been placed in 10-mL water blanks should be heated for 10 min at 80°C (proteolytic sporeformers) or 30 min at 60°C (nonproteolytic sporeformers) and cooled prior to culturing.

The line (product) samples collected in sealable plastic bags are weighed and an equal weight of sterile distilled water is added. The samples are mixed thoroughly after sealing the bag securely. (Product may be aseptically weighed into stomacher bags and mixed using a stomacher, if desired.) Half-fill 25 × 200 mm screw-cap tubes with the liquid. Heat the contents for 10 min after the temperature has reached 80°C (proteolytic sporeformers) or for 30 min at 60°C (nonproteolytic sporeformers) prior to culturing. (Use a thermometer or thermocouple in a duplicate sample tube to determine when the appropriate temperature has been reached.) Alternatively, pipette portions of the unheated diluted sample into the anaerobic medium of choice, and then heat as above.

The unprocessed cans collected at the closing machine should be opened aseptically, and portions of brine or liquid transferred into sterile 25 × 200 mm screw-cap tubes for heating 10 min at 80°C (proteolytic sporeformers) or for 30 min at 60°C (nonproteolytic sporeformers) before culturing. Solids or formulated products may be cultured directly or diluted as above and pipetted into the medium of choice; these samples are heated for 10 min at 80°C (proteolytic sporeformers) or for 30 min at 60°C (nonproteolytic sporeformers).

When canned food spoilage has been determined to be caused by mesophilic anaerobic sporeformers that entered the container through leakage following thermal processing, a microbiological survey of the post-processing can-handling lines will be helpful in identifying the areas of contamination. Sterile cotton swabs moistened with diluent (10 mL sterile distilled water containing 60 ppm sodium thiosulfate to inactivate residual halogen[23,45,66]) are used to swab 8 in^2 of sample area. The swab samples are placed in 10 mL of diluent in an 18 × 150 mm screw-cap tube. A 150-mL sample of can cooling water should also be obtained for examination. Collect the sample in a sterile bottle and add 1.1 mL of sterile 3% sodium thiosulfate. All samples may be held up to 30 hr prior to culturing.[21] They should be heat activated at 71°C for 10 min[35] (alternately 60°C for 30 min can be used).

23.52 Ingredient Examination

For the preparation of foods or ingredients other than those specifically mentioned in succeeding paragraphs, some judgment will have to be used. Samples should be collected in sterile plastic bags; collection utensils, if needed, should be sterile also. Some recommended methods for ingredients are as follows.

1. **Sugar.** Weigh 20 g of granulated sugar or 30 g of liquid sugar (approximately 67° Brix) into sterile 250 mL Erlenmeyer flasks and add sterile water to make 100 mL samples. Shake the flasks until the sugar is in solution. Heat the samples rapidly to boiling and hold at boiling for 5 min before chilling rapidly to room temperature.
2. **Starch, flour, and other cereal products.** Weigh 11 g of the product into a sterile 250-mL Erlenmeyer flask and add 99 mL of sterile cool water. Shake until the sample is in suspension. Do not heat before culturing. (Samples are heated after subculturing.)
3. **Dehydrated vegetables.** Weigh 10 g of dehydrated vegetable into a sterile 250 mL Erlenmeyer flask and add 190-mL of sterile water. Rehydrate for 30 min at 4–5°C. Then shake vigorously for 2 to 3 min. Heat the sample for 10 min at 5 psi (108°C) in an autoclave and cool to room temperature. Alternatively, samples may be heated for 20 to 30 min at 100°C and then cooled.
4. **Spices.** The weight of the spice sample to be examined will depend on the type of spice to be examined. Weigh 10 g of whole spice, 2 g of crushed spice, or 1 g of ground spice into a sterile 250-mL Erlenmeyer flask; add sterile water to make 100 mL. Shake vigorously and boil for 5 min. Allow coarse particles to settle out before culturing or making dilutions for culturing.
5. **Dried Eggs, Dried Milk or Dried Milk Products** Weigh 11 g of dried eggs, dried milk, or dried milk product into dilution bottles containing 99 mL of sterile water or saline and glass beads. Shake the samples vigorously until all lumps of dried eggs, dried milk, or dried milk product have been dispersed. Do not heat the sample before culturing.

23.53 Culture Methods

Samples obtained from the plant survey, which include swab or scraping samples and product samples, are placed in broth cultures (see 23.544, Media) freshly heated and quickly cooled to exclude dissolved oxygen. Usually 1-mL and 0.1-mL portions of samples are cultured. The cultures are stratified with either sterile petroleum jelly, Vaspar, or 2% agar, depending on the medium used, before incubation. If PE-2 medium[18] or CMM is used, stratification is not necessary.

The prepared samples of heated sugar, dehydrated vegetables, and spices are cultured by taking 20-mL portions of the heated ingredients and dividing approximately equally among 6 tubes of freshly heated culture medium. These are handled in the same manner as the survey samples. The cultures are incubated at 30° to 35°C for 72 hr. Although most of the organisms will show growth within 2 or 3 days, it is sometimes desirable to incubate for 7 days, since some spores may be slow in germinating and growing out.

Twenty milliliter portions of the unheated starch, flour, cereal products and dried milk, or dried egg samples are distributed among 6 tubes of culture medium. The tubes are then heated at 100°C for 20 min. The cultures must be agitated several times during this heating period. After heating, the tubes are stratified with sterile petroleum jelly or Vaspar. The cultures are incubated for 72 hr at 30° to 35°C and examined for the growth of putrefactive anaerobes.

Procedures will need to be adjusted for mesophilic anaerobic sporeformers other than putrefactive anaerobes. Novel organisms in particular may require non-standard procedures. It may be necessary to try a variety of media and supplements, as well as different heat shock and incubation temperatures. For example, Kalchayanand et al.[31] had difficulty in isolating the *Clostridium* spp. from refrigerated beef; subsequently the organism named *C. laramie* was cultured in fluid thioglycollate or tryptic soy broth containing 0.1% hemin and 0.001% vitamin K at 10°C.[30] Broda et al.[9] determined that no single treatment could be universally recommended to maximize the recovery of psychrotrophic *Clostridium* spp. from spoiled meat or meat processing environments. They recommended both heat shock (80°C for 10 min) and ethanol treatment (60 min with equal volumes of sample and ethanol) to ensure recovery of both heat-resistant and heat-sensitive spore types.

23.54 Enumeration of Mesophilic Anaerobic Spores

If counts of mesophilic anaerobic spores are desired on a food product or ingredient, several methods are in current use. The selective differential enumeration of these spores relies on destruction of vegetative cells (generally with a heat shock, and occasionally by ethanol treatment), growth in an anaerobic environment, and, often, the ability of the organism to reduce sulfite or to produce hydrogen sulfide from sulfur-containing media components (visualized by the presence of ferric or ferrous cations resulting in a black precipitate in and around colonies or in liquid media). The identity of the black colonies should be confirmed, as some *Bacillus* spp. may occasionally produce a similar reaction.[69] Other media that do not result in black colonies are also used, especially when counting pure cultures, since there can be variability in the production of black colonies and because of occasional false positives from non-clostridia.

23.541 Enumeration by the MPN method

For the general procedure, see Chapter 6. Serial dilutions may be made in distilled water, 0.1% peptone, or culture medium. PE-2[48,64] or CMM (containing 0.1% soluble starch and 0.1% glucose for nonproteolytic strains) are useful general media for MPN enumeration, although other media listed below may also be acceptable. Preparation of dilutions in the culture medium used for detecting growth has the advantage of minimizing any effect of the inoculum on the nutrients in the final culture. Generally, 3 culture tubes for each dilution will give sufficient accuracy, but if greater precision is needed, 5 or more tubes may be used. From each dilution inoculate 1.0 mL or 0.1 mL into the number of tubes chosen. If samples were not heat-shocked before inoculation of the MPN medium, the tubes should be heat-shocked and cooled at this time. If not using PE-2 medium or CMM, overlay the media with petroleum jelly. Incubate the inoculated cultures at 30° to 35°C. Record the number of tubes of each dilution showing growth and refer to the appropriate MPN table for the number of tubes per dilution and the size of the inoculum (1.0 mL or 0.1 mL) to determine the count of viable spores per g or mL of the product. Although most mesophilic sporeforming anaerobes will show growth within 2 or 3 days, the final reading should be made in 1 to 3 weeks because of slow germination, particularly if the organisms are coming out of an inhibitory environment.

23.542 Enumeration by the Plate Count Method

Prepare heated samples and dilutions (Chapter 6) and inoculate duplicate petri dishes with each of the appropriate dilutions of the food material; 1.0 mL serological pipettes or 2.2 mL bacteriological pipettes may be used. Pour inoculated plates with TPGY agar, TCD agar, AC agar, or TP agar. Overlay the plates with a layer of thioglycollate agar. An alternate procedure using pour or spiral plating with DCA and a DCA overlay has been developed by Weenk et al.[69] Other media noted in 23.544 may also be suitable. Incubate the inoculated plates at 30 to 35°C for 2 to 5 days in an anaerobic atmosphere such as that produced by the BBL Gaspak™ or similar system or in an anaerobic chamber. Use a colony counter to count all colonies, multiply by the dilution factor, and report as anaerobic sporeformers per g or mL of food (if the samples were heated or alcohol treated).

Plating may be the least reliable method of enumerating mesophilic sporeforming anaerobes. It is not uncommon to find wide variations in the numbers of colonies developing on replicate plates of the same dilution. The production of gas bubbles by subsurface colonies frequently causes separation of the agar, reducing the accuracy in counting colonies. If gas production is a problem, reducing the amount of carbohydrate in the medium or another medium with less carbohydrate may help.

23.543 Enumeration by the Deep Agar Tube Count Method

Prepare heated samples and dilutions (Chapter 6) and inoculate duplicate sterile 16 × 125 mm screw-cap tubes with each of the appropriate dilutions of the food material, using a sterile 1.0 mL serological pipette or a 2.2 mL bacteriological pipette. Pour the inoculated tubes approximately two-thirds full with an agar medium such as TP, TCD, etc. (approximately 12 mL). Use media that are clear so that colonies may be readily detected. The action of pouring the agar into the inoculated tubes should be sufficient for adequate distribution of the inoculum throughout the tube. Gentle swirling, without incorporation of air into the medium may be necessary. Cool and solidfy the agar by immersion of the tubes in cold water. Overlay the solidified medium in the inoculated tubes with thioglycollate agar (3 mL) to serve as an oxygen barrier. Incubate the tubes at 30 to 35°C for 2 to 3 days.

Count all colonies, multiply by the dilution factor, and report as anaerobic sporeformers per g or mL of food if the samples were heated or alcohol treated. Counting colonies in the agar tubes may be facilitated by laying the tube on its side. The production of gas bubbles by the subsurface colonies frequently causes separation of the agar, thereby complicating their enumeration. A medium with reduced levels of carbohydrate may resolve the problem.

23.544 Alternative Media and Methods

For culturing ingredients for putrefactive anaerobes as described in 23.53, use Liver Broth, Beef Heart Infusion Broth, CMM or PE-2 medium. For other food samples, these media are also appropriate for detection of putrefactive anaerobes. For detection of nonproteolytic mesophilic anaerobes, TPGY broth or CMM with added 0.1% soluble starch and 0.1% glucose may be used. For culturing butyric acid anaerobes, especially those from tomato products, tomato liver broth may be the most suitable medium; some strains appear to require nutrients present in tomatoes in order to grow.

In order to ensure anaerobic conditions during incubation, broth cultures should be stratified with sterile petroleum jelly or Vaspar. If tubed solid or semisolid media are used, sterile 2% agar may be substituted. As noted previously, stratification is not necessary with PE-2 or CMM. For enumerating, rather than determining the presence of, mesophilic anaerobic spores, other media are recommended.

Several media for enumerating mesophilic anaerobic spores are available. Trypticase peptone agar (TP) and yeast extract agar (YEA) have been shown to be good media for enumeration of heat activated spores of *C. sporogenes*.[49,50] Their effectiveness is enhanced by the addition of filter sterilized sodium bicarbonate (0.14%) and sterile sodium thioglycollate (0.1%) to the tempered media just prior to use. Although modified PA3679 agar[24] was shown to produce higher counts of heat activated spores of *C. sporogenes* than either TPA or YEA, this medium is more complicated to prepare. Experience in a number of laboratories has shown that TPGY, Andersen's pork-pea infusion broth or agar,[2,49] or modified PA3679 agar[24] are the media of choice for recovering spores that have received a high heat treatment such as in thermal death time (TDT) determinations. If the medium has not been freshly sterilized, it is important to steam it at 100°C for 10 to 15 min to drive off dissolved oxygen shortly before inoculating. It is also important to cool the tubes quickly by partial immersion in tempered water to avoid re-absorption of oxygen. 15g of agar per L are added to a broth medium if it is to be used as a solid medium.

For detecting and enumerating butyric acid anaerobes, either dextrose tryptone agar (DTA)[44] or orange serum broth (the agar is unsuitable for counting because of the large amount of gas produced)[45] can be used, although PE-2 medium is also a suitable growth medium and can be used in an MPN procedure.[48,64] PE-2 was used to isolate and enumerate mesophilic anaerobic sporeformers in recycled cannery cooling water, where the most frequent isolate was identified as *C. butyricum*.[64] Tryptose cycloserine dextrose (TCD) agar has been effectively used as a selective medium for the isolation and enumeration of mesophilic anaerobic sporeformers from environmental samples collected from cannery plant surveys.[35]

For detecting and enumerating psychrotrophic clostridia from meats a variety of specialized media and selective techniques may be needed.[10] The procedures also depend on whether the organisms have a low optimum growth temperature (15–20°C) or a high one (25–30°C). Where contamination of the sample with high numbers of competing microorganisms is not expected, PYGS agar at 20°C may be used for the first type. If there are significant levels of competing microorganisms, selective media such as SFP with egg yolk may be used. This medium or Tryptose-Sulfite-Cycloserine (TSC) agar (Chapter 63) is also satisfactory for the high optimum growth temperature of psychrotrophic clostridia, with incubation at 30°C.

23.6 INTERPRETATION OF RESULTS

The diagnosis of canned food spoilage when mesophilic sporeforming anaerobes are isolated is covered in Chapter 62 and will not be discussed here.

The data obtained from a plant survey may provide the information needed to implement a corrective sanitation program in a canning plant or other processing facility. Evidence of heavy buildup of mesophilic sporeforming anaerobes on the equipment indicates that corrective action must be taken immediately to prevent a possible incident of underprocessing spoilage or to control a current one in a canned product. In the case of refrigerated products, the survey may help pinpoint the source of a spoilage organism and allow the plant to take actions to prevent product contamination with the organism from the environment.

No standards for mesophilic anaerobe spores in ingredients have been established as they have been for thermophilic spores in sugars or starches.[45] Therefore, judgment must be exercised as to whether or not contaminated ingredients should be used, taking into account the spore count level, how much of the ingredient is used in the product, the sterilizing value of the process used for the canned product, the ability of the organism to grow in the product given the storage conditions of the product, and any other criteria that may appear to be significant. Similar considerations would be given with regard to refrigerated products.

The interpretation given to positive growth in any anaerobic cultures should be made carefully, since the growth observed may be due totally or in part to facultative anaerobes (usually aerobic sporeformers) rather than clostridia. Therefore, before making a judgment, take certain other factors into account, such as sample preparation (e.g., degree of heat treatment, if any), the presence of a putrid odor, amount of gas produced in the cultures, colony morphology, cell morphology (particularly spore location), catalase production, ability to grow aerobically etc. Additional identification tests may be needed.

23.7 REFERENCES

1. Allen, M. B. 1953. The thermophilic aerobic sporeforming bacteria. Bacteriol. Rev. 17:125.
2. Andersen, A. A. 1951. A rapid plate method of counting spores of *Clostridium botulinum* J. Bacteriol. 62:425.
3. Aureli, P., L. Fenicia, B. Paolini, M. Gianfranceschi, L. M. McCroskey, and C. L. Hatheway. 1986. Two cases of type E infant botulism caused by neurotoxigenic *Clostridium butyricum* in Italy. J. Infect. Dis. 154:207.
4. Bee, G. R., and L. R. Hontz. 1980. Detection and prevention of post-processing container handling damage. J. Food Prot. 43:458.
5. Blocher, J. C., and F. F. Busta. 1983. Bacterial spore resistance to acid. Food Technol. 37(11):87.
6. Bowen, J. F., C. C. Strachan, and A. W. Moyls. 1954. Butyric acid fermentation in canned pears and tomatoes. Food Technol. 8:239.
7. Bowen, J. F., C. C. Strachan, and A. W. Moyls. 1954. Further studies of butyric fermentation in canned tomatoes. Food Technol. 8:471.
8. Broda, D. M., K. M. De Lacy, R. G. Bell, T. J. Braggins, and R. L. Cook. 1996. Psychrotrophic and cold tolerant *Clostridium* spp. associated with "blown pack" spoilage of chilled vacuum-packaged red meats and dog rolls in gas-impermeable plastic casings. Int. J. Food Microbiol. 29:335.
9. Broda, D. M., K. M. De Lacy, and R. G. Bell. 1998. Efficacy of heat and ethanol spore treatments for the isolation of psychrotrophic *Clostridium* spp. associated with the spoilage of chilled vacuum-packed meats. Int. J. Food Microbiol. 39:61.
10. Broda, D. M., K. M. De Lacy, and R. G. Bell. 1998. Influence of culture media on the recovery of psychrotrophic *Clostridium* spp. associated with the spoilage of chilled vacuum-packed meats. Int. J. Food Microbiol. 39:69.
11. Cameron, E. J., and J. R. Esty. 1940. Comments on the microbiology of spoilage in canned foods. Food Res. 5:549.
12. Cato, E. P., W. L. George, and S. M. Finegold. 1986. Genus *Clostridium* Prazmowski 1880, p. 23. *In* P. H. A. Sneath, N. S. Mair, M. E. Sharpe, and J.Holt (eds.), Bergey's manual of svstematic bacteriology, vol. 2.. Williams and Wilkens, Baltimore, Md.

13. Corwin, D. 1980. A food poisoning whodunit. FDA Consumer. November. Department of Health and Human Services, Rockville, Md.
14. Dainty, R. H., R. A. Edwards, and C. M. Hibbard. 1989. Spoilage of vacuum-packed beef by a *Clostridium* sp. J. Sci. Food Agri. 49:473.
15. De Jong, J. 1989. Spoilage of an acid food product by *Clostridium perfringens*, *C. barati*, and *C. butyricum*. Int. J. Food Microbiol. 8:121.
16. Farber, J. M. 1991. Microbiological aspects of modified-atmosphere packaging technology: A review. J. Food Prot. 54:58.
17. Fleming, H. P., M. A. Daeschel, R. F. McFeeters, and M. D. Pierson. 1989. Butyric acid spoilage of fermented cucumbers. J. Food Sci. 54:636.
18. Folinazzo, J. F., and V. S. Troy. 1954. A simple medium for the growth and isolation of spoilage organisms from canned foods. Food Technol. 8:280.
19. Frank, J. 1997. Milk and dairy products. *In* M. Doyle, L. Beuchat, and T. Montville (eds.), Food microbiology fundamentals and frontiers. ASM Press, Washington, D. C.
20. Frazier, W. C., and D. C. Westhoff. 1988. Food microbiology, 4th ed. McGraw-Hill Inc., New York.
21. Goldreich, E. E., H. D. Nash, D. J. Reasoner, and R. H. Taylor. 1972. The necessity of controlling bacterial populations in potable waters: community water supply. J. Am. Water Works Assoc. 64:596.
22. Graves, R. R., R. S. Lesniewski, and D. E. Lake. 1977. Bacteriological quality of cannery cooling water. J. Food Sci. 42:1280.
23. Green, B. L., and W. Litsky. 1974. The evaluation of sodium sulfate as a neutralizer for iodine disinfectants, p.15. *In* Abstracts of the 74th Annual Meeting of the American Society for Microbiology. American Society for Microbiology, Washington, D. C.
24. Grischy, R. O., R. V. Speck, and D. M. Adams. 1983. New media for enumeration and detection of *Clostridium sporogenes* (PA3679) spores. J. Food Sci. 48:1466.
25. Gross, C. E., E. Vinton, and C. R. Stumbo. 1946. Bacteriological studies relating to thermal processing of canned meat. V. Characteristics of putrefactive anaerobes used in thermal resistance studies. Food Res. 11:405.
26. Hersom, A. C., and E. D. Hulland. 1981. Canned foods: Thermal processing and microbiology, 7th ed. Chemical Publ. Co. Inc., New York.
27. ICMSF. 1998. Microorganisms in Foods 6: Microbial Ecology of Food Commodities. Blackie Academic & Professional, New York.
28. Johnston, R. W., J. Feldman, and R. Sullivan. 1963. Botulism from canned tuna fish. Publ. Health. Rep. 78:561.
29. Johnston, R., S. M. Harmon, and D. A. Kautter. 1964. Method to facilitate the isolation of *Clostridium botulinum* type E. J. Bacteriol. 88:1521.
30. Kalchayanand, N., B. Ray, and R. A. Field. 1993. Characteristics of psychrotrophic *Clostridium laramie* causing spoilage of vacuum-packaged refrigerated fresh and roasted beef. J. Food Prot. 56:13.
31. Kalchayanand, N., B. Ray, R. A. Field, and M. J. Johnson. 1989. Spoilage of vacuum-packaged refrigerated beef by *Clostridium*. J. Food Prot. 52:424.
32. Kautter, D. A., S. M. Harmon, R. K. Lynt, Jr., and T. Lilly, Jr. 1966. Antagonistic effects on *Clostridium botulinum* type E by organisms resembling it. Appl. Microbiol. 14:616.
33. Kautter, D. A., T. Lilly, Jr., and R. K. Lynt, Jr. 1978. Evaluation of the botulism hazard in fresh mushrooms wrapped in commercial polyvinylchloride film. J. Food Prot. 41:120.
34. Lake, D. E., R. R. Graves, R. S. Lesniewski, and J. E. Anderson. 1985. Post-processing spoilage of low acid canned foods by mesophilic anaerobic sporeformers. J. Food Prot. 48:221.
35. Lake, D. E., R. S. Lesniewski, J. E. Anderson, R. R. Graves, and J. F. Bremser. 1985. Enumeration and isolation of mesophilic anaerobic sporeformers from cannery post-processing equipment. J. Food Prot. 48:794.
36. Larson, A. E., E. A. Johnson, C. R. Barmore, and M. D. Hughes. 1997. Evaluation of the botulism hazard from vegetables in modified atmosphere packaging. J. Food Prot. 60:1208.
37. Lawson, P., R. H. Dainty, N. Kristiansen, J. Berg, and M. D. Collins. 1994. Characterization of a psychrotrophic *Clostridium* causing spoilage in vacuum-packed cook pork: description of *Clostridium algidicarnis* sp. nov. Lett. Appl. Microbiol. 19:153.
38. Lin, R. C., P. H. King, and M. R. Johnston. 1998. Chapter 22: Examination of containers for integrity. *In* Food and Drug Administration bacteriological analytical manual, 8th ed., Revision A. AOAC International, Gaithersburg, Md.
39. Lynt, R. K., D. A. Kautter, and H. M. Solomon. 1982. Differences and similarities among the proteolytic and nonproteolytic strains of *Clostridium botulinum* Types A, B, E and F: A review. J. Food Prot. 45:466.
40. McCroskey, L. M., C. L. Hathaway, L. Fenicia, B. Pasolini, and P. Aureli. 1986. Characterization of an organism that produces type E botulinal toxin but which resembles *C. butyricum* from the feces of an infant with type E botulism. J. Clin. Microbiol. 23:201.
41. Meyer, K. F. 1929. Maximum oxygen tolerance of *C. botulinum* A, B, and C, of *C. sporogenes*, and *C. welchi*. J. Infect. Dis. 44:408.
42. Montville, T. J. 1983. Dependence of *Clostridium botulinum* gas and protease production on culture conditions. Appl. Environ. Microbiol. 46:961.
43. Morton, R. D. 1998. Spoilage of acid products by butyric acid anaerobes – A review. Dairy, Food Environ. Sanit. 18:580.
44. Morton, R. D., V. N. Scott, D. T. Bernard, and R. C. Wiley. 1990. Effect of heat and pH on toxigenic *Clostridium butyricum*. J. Food Sci. 55:1725.
45. National Canners Association. 1968. Laboratory manual for food canners and processors, vol. 1. Natl. Canners Assoc. Research Labs, The AVI Publishing Co., Inc., Westport, Conn.
46. NFPA/CMI Container Integrity Task Force. 1984. Botulism risk from post-processing contamination of commercially canned foods in metal containers. J. Food Prot. 47:801.
47. Odlaug, T. E., and I. J. Pflug. 1978. Microbiological and sanitizer analysis of water used for cooling containers of food in commercial canning factories in Minnesota and Wisconsin. J. Food Sci. 43:954.
48. Ogunrinola, O. A., C. G. Edwards, and P. M. Davidson. 1997. Evaluation of four pea (*Pisum sativum*) cultivars in PE-2 medium for the MPN enumeration of anaerobic spore-forming organisms. J. Food Prot. 60:1574.
49. Polvino, D. A., and D. T. Bernard. 1982. Media comparison for the enumeration and recovery of *Clostridium sporogenes* P.A. 3679 spores. J. Food Sci. 47:59.
50. Pflug, I. J., M. Scheyer, G. M. Smith, and M. Kopelman. 1979. Evaluation of recovery media for heated *Clostridium sporogenes* spores. J. Food Prot. 42:946.
51. Put, H. M. C., H. J. Witvoet, and W. R. Warner. 1980. Mechanism of leaker spoilage of canned foods: Biophysical aspects. J. Food Prot. 43:488.
52. Rolf, R. D., D. J. Huetger, B. J. Campbell, and J. T. Barret. 1978. Factors relating to the oxygen tolerance of anaerobic bacteria. Appl. Environ. Microbiol. 36:306.
53. Segner, W. P. 1979. Mesophilic aerobic sporeforming bacteria in the spoilage of low acid canned foods. Food Technol. 33 (1):55.
54. Segner, W. P. 1992. Spoilage of pasteurized crabmeat by a nontoxigenic psychrotrophic anaerobic sporeformer. J. Food Prot. 55:176.
55. Setlow, P., and E. A. Johnson. 1997. Spores and their significance. *In* M. Doyle, L. Beuchat, and T. Montville (eds.), Food microbiology fundamentals and frontiers., ASM Press, Washington, D. C.
56. Smith, L. D. S. 1975. Common mesophilic anaerobes, including *Clostridium botulinum* and *Clostridium tetani*, in 21 soil specimens. Appl. Microbiol. 29:590.
57. Sognefest, P., G. L. Hays, E. Wheaton, and H. A. Benjamin. 1948. Effect of pH on thermal process requirements of canned foods. Food Res. 13:400.
58. Sperber, W. H. 1983. Influence of water activity on foodborne bacteria—A review. J. Food Prot. 46:142.
59. Spiegelberg, C. H. 1940. *Clostridium pasteurianum* associated with spoilage of an acid canned fruit. Food Res. 5:115
60. Stumbo, C. R. 1973. Thermobacteriology in food processing, 2nd ed. Academic Press, New York.

61. Suen, J. C., C. L. Hatheway, A. G. Steigerwalt, and D. J. Brenner. 1988. Genetic confirmation of identities of neurotoxigenic *Clostridium baratii* and *Clostridium butyricum* implicated as agents of infant botulism. J. Clin. Microbiol. 26:2191.
62. Sugiyama, H., and K. S. Rutledge. 1978. Failure of *Clostridium botutinum* to grow in fresh mushrooms packed in plastic film overwraps with holes. J. Food Prot. 41:348.
63. Sugiyama, H., and K. H. Yang. 1975. Growth potential of *Clostridium botulinum* in fresh mushrooms packaged in semipermeable plastic film. Appl. Microbiol. 30:964.
64. Thompson, P. J., and M. A. Griffith. 1983. Identity of mesophilic anaerobic sporeformers cultured from recycled cannery cooling water. J. Food Prot. 46:400.
65. Thompson, R. C. 1982. The tin of salmon had but a tiny hole. FDA Consumer. June. Department of Health and Human Services. Rockille, Md.
66. Thompson, R. E. 1944. Bacteriological examination of chlorinated water. The use of thiosulfate-treated bottles for collecting samples. Water Sewage 82(2):27.
67. Townsend, C. T. 1939. Spore-forming anaerobes causing spoilage in acid canned foods. Food Res. 4:231.
68. Townsend, C. T., and J. R. Esty. 1939. The role of microorganisms in canning. Western Canner Packer 31(8):21.
69. Weenk, G. H., J. A. van den Brink, C. B. Struijk, and D. A. A. Mossel 1995. Modified methods for the enumeration of spores of mesophilic *Clostridium* species in dried foods. Int. J. Food Microbiol. 27:185.
70. Whiting, R. C., and K. A. Naftulin. 1992. Effect of headspace oxygen concentration on growth and toxin production by proteolytic strains *of Clostridium botulinum*. J. Food Prot. 55:23.

ACKNOWLEDGMENTS

The authors wish to acknowledge the contributions of Donald E. Lake, Dane T. Bernard, and Donald A. Kautter, who authored this chapter in the 3rd edition of this book.

Chapter 24

Aciduric Flat Sour Sporeformers

George M. Evancho and Isabel Walls

24.1 INTRODUCTION

In this chapter, spoilage of acid products by two varieties of sporeforming microorganisms, *Bacillus coagulans* and acidophilic sporeformers such as *Alicyclobacillus acidoterrestris*, will be considered.

The taste of spoiled tomato juice has been described as "medicinal," "phenolic," and "fruity," and this taste is usually accompanied by a reduction of from 0.3 to 0.5 in pH.[20] Ends of cans of spoiled tomatoes remain flat; hence the term "flat sour." In 1933, while investigating off-flavor in commercially canned tomato juice, Berry isolated and described a new type of spoilage organism.[4] He determined the organism to be a sporeforming bacterium of soil origin.

In 1981, novel acidophilic sporeformers were isolated from soil. Cells formed subterminal to terminal endospores, which slightly swelled the sporangium. The pH range for growth was 2–5 over a temperature range of 22–62°C, a considerably lower pH than had been seen with typical flat sour sporeformers.[12] The first reported spoilage incident caused by these acidophilic sporeformers occurred in aseptically packed apple juice (pH 3.15) in Germany in 1982.[6] The spoilage organism was shown to be the same as Hippchen's[12] isolates from soil. Spoiled juice was reported as having a bad taste and light cloudiness. In 1990, sporeforming rod-shaped organisms were isolated from fruit juice.[16] Subsequently, similar organisms were isolated from other juice and acid products.[19,22,23,27,28] Spoiled products had a "medicinal" or "phenolic" off odor and light sediment but gas was not produced.

24.11 Classification

Berry[4] named the organism responsible for "flat sour" spoilage of tomato products *Bacillus thermoacidurans*. From comparative cultural studies, Smith *et al.*[18] concluded that *B. thermoacidurans* was identical with *Bacillus coagulans* of Hammer.[11] From their careful studies of the two species, Becker and Pederson[3] stated: "There is no justification for considering *B. thermoacidurans* as a species distinct from *B. coagulans*, and the latter name has priority."

B. coagulans is a nonpathogenic, motile, sporeforming aerobe having as many as 10 flagella per cell. The gram-stain reaction is usually positive although a few variable strains have been observed. The species is identified stepwise by the following test results: catalase positive, Voges-Proskauer positive (this property may be variable), growth in anaerobic agar (without glucose or Eh indicator), growth at 50°C, and absence of growth in 7% NaCl.[9]

Deinhard *et al.*[8] undertook characterization studies on organisms isolated from soil by Hippchen and his colleagues[12], and proposed a new name for these organisms, *Bacillus acidoterrestris*. In 1992, the creation of a new genus, *Alicyclobacillus*, was proposed[26] to comprise the species *Alicyclobacillus acidocaldarius*, *A. acidoterrestris* and *A. cycloheptanicus*. Comparative rDNA sequence analyses showed that the three strains were sufficiently different from other *Bacillus* spp. to warrant reclassification in a new genus. Also, *Alicyclobacillus* is unique in its fatty acid profiles, containing ϕ-alicyclic fatty acid as the major natural membranous lipid component. Some strains of sporeforming organisms causing spoilage of juices were determined to be *A. acidoterrestris* based on ribotyping.[25] *A. acidoterrestris* is a nonpathogenic, motile, sporeforming rod shaped organism. Central, subterminal and terminal oval spores may be seen. Free spores are 1 to 2 μM in length, and 0.7 to 0.9 μM in diameter. Colonies are round, creamy white, opaque, 3–5 mm in diameter after 5 days growth on K medium, pH 3.7, at 35°C. The Gram-stain reaction is either positive or variable. Strains described by Deinhard et al.[8] and Cerny et al.[6] were found to be aerobic, while some strains are facultatively anaerobic.[23] Strains are VP negative, most are catalase positive, all strains produce acid from D-mannitol, most produce acid from L-arabinose and D-xylose, and results are variable for D-glucose and D-trehalose. Strains are indole negative, dihydroxyacetone negative, utilize citrate but not propionate, hydrolyze starch, are negative for deamination of phenylalanine and egg yolk lecithinase and do not reduce nitrate. They do not grow in the presence of 0.001% lysozyme but most grow in the presence of 0.02 % azide. Strains do not grow in the presence of 5% NaCl.[23]

24.12 Occurrence

B. coagulans is a common soil organism. It has been isolated from canned tomato products, particularly tomato juice, and from cream, evaporated milk, cheese, and silage. Spoilage and subsequent curd formation of evaporated milk have been caused by this organism. Hammer's original studies on the coagulation of evaporated milk led to the naming of the organism.[11]

Among tomato products, *B. coagulans* has been isolated from canned whole tomatoes, tomato juice, tomato puree, tomato soup, and tomato-vegetable juice mixes.

The organism has been found to multiply in tomato-washing equipment where the volume of cold water is insufficient and the water temperature may reach 27° to 32°C. Spores have been isolated from empty cans and empty can washers, tomato product lines, conveyor belts, and filled can runways. Spores of the organism have been isolated from chipboard separators frequently used in the packaging of empty cans. They have been found in the sweepings of railcars previously used to ship grain products.

A. acidoterrestris and other acidophilic sporeformers have been isolated from soil,[12,27] the surface of unwashed and washed fruit,[27] fruit juices and juice concentrates, including apple juice, apple cranberry juice,[19] berry juice,[16] pear juice, orange juice,[27] and canned diced tomatoes.[25] Spores were also isolated from condensate water.[27] Additionally, the organism has been shown to spoil white grape juice, grapefruit juice and tomato juice.[24]

24.2 GENERAL CONSIDERATIONS

24.21 Sampling

24.211 Ingredients

Since the adequacy of a thermal process is, among other things, related to the bacterial spore load of the product to be processed, it is often advantageous to determine the load of flat sour spores in the unprocessed product as well as in the product ingredients. Pinpointing the ingredient that is contributing most to the total spore load may prove very beneficial. Depending on the product being manufactured, spore analysis of ingredients such as raw tomatoes, fresh tomato pulp, puree, concentrated or evaporated milk, cream, and nonfat dry milk may be undertaken. Sampling dairy products is conducted in accordance with the procedures given in "Standard Methods for the Examination of Dairy Products," 17th edition.[1]

As *A. acidoterrestris* is a soil-based organism, it is likely that it enters the processing plant on the surface of fruit.[27] It may be of value to test the surface of fruit, as presence of this organism will indicate the potential for spoilage in the final product. It is important that washing procedures are in place to reduce the level of contamination on the surface of the fruit. The organism has been found in juice concentrates in low numbers (<50 per mL). Testing juice concentrate may be of value as spoilage can occur in juice inoculated with less than 10 spores per mL.[24] Spoilage does not occur in juice concentrate at 30° Brix or above but spores can survive in juice concentrate for long periods of time (greater than 3 months). Spores will generally survive the heat process given to fruit juice ($D_{95°C}$ in apple juice pH 3.6 = 1.64 ± 0.24 min).

22.212 Equipment and systems

Periodic sampling of tomato wash and material from conveyor belts, pipelines, and tanks may reveal potential problem areas. A preseasonal bacteriological survey of pipelines, valves and valve bonnets, storage tanks, heaters, and other equipment surfaces that normally contact a product may indicate where cleaning and sanitation need to be improved. Sampling of such equipment may be carried out using swabbing or membrane filter techniques.

Alicyclobacillus spp. have been isolated from processing equipment;[7] therefore, sampling may be of value to determine if a buildup of spores is occurring. Additionally, spores have been isolated from condensate water[27] and filtration units.

24.213 Product in process

Canners of tomatoes and tomato products may find it advantageous to monitor flat sour spores in certain phases of manufacturing. On analysis, finished tomato products, such as juice, puree, samples of chopped tomatoes before and after the hot break, or extracted tomatoes, may indicate potential foci of spore buildup. In the canning of whole tomatoes, sampling should include juice accumulated during peeling and puree made from sound tomatoes, both of which are used frequently as the liquid portion of canned tomatoes. Aciduric flat sour spore counts of tomato products prior to processing may assist in preventing spoilage where the product is not presterilized.

24.214 Finished products

Since most processes for tomato products either completely eliminate flat sour spores or drastically reduce their number, the advantage of making spore counts of the finished product must be seriously questioned. Where spore numbers have been reduced substantially, the probability of recovering spores by subculturing 1 mL in replicates up to five or more is small. No attempt should be made to plate more than 1-mL quantities of tomato juice because of its inhibitory action on *B. coagulans*. Experience has shown that incubation of the canned finished product is much more meaningful.

Juice concentrate may be tested for the presence of acidophilic sporeformers. If present, counts are likely to be low (< 50 per mL), however, under ideal conditions, 1 spore per mL may be sufficient to cause spoilage.[24] *A. acidoterrestris* strains are unable to grow in juice concentrates at 30° Brix or above but they can survive in these products. Spoilage usually occurs in single strength juices after products have been at an elevated temperature for a period of time, for example, at 35°C for 1 week. The minimum temperature for growth ranges from 20 to 35°C, depending on juice type, pH and inoculum level. An off flavor or odor and presence of guaiacol (2-methoxy phenol) are indicators of spoilage.[14,15] 2,6 di-bromo-phenol has also been reported as a cause of the off flavor.[5] Spoiled juices may have counts as low as 100 cells per mL. Colonies should be opaque, cream colored, slightly raised. Under a microscope, organisms should be rod shaped with central, subterminal or terminal spores.

Spoiled apple juice may appear normal or have light sediment. Sediment may become darker with time. Spoiled tomato juice appears separated and certain strains will grow on the surface of the juice, producing a white film. Spoiled products have an off odor and off flavor (medicinal or phenolic). Gas is not produced.

24.22 Temperature and pH Requirements

Becker and Pederson[3] reported that *B. coagulans* was obligately thermophilic, yet they were successful in growing the organism at temperatures as low as 18°C. Optimum growth in an artificial medium occurred between 37° and 45°C. In 1949, Gordon and Smith reported that 53 of 73 cultures studied grew at 28°C; 73 at 33°, 37°, and 45°C; 72 at 50°C; 66 at 55°C; 23 at 60°C and none at 65°C.[10] In a more recent article, Thompson states the organism is generally regarded as "facultatively thermophilic," growing in artificial media at 20° to 55°C (all strains) or 60°C (16 of 22 strains).[21] The latter view is accepted more readily.

Packers of tomato products have observed spoilage development at temperatures of 21° to 38°C. *B. coagulans* will not grow in tomato products with normal pH at 55°C. Berry indicated that a temperature of 37°C appeared optimum for the production of off-flavor.[4] *B. coagulans* does not grow in tomato products at solids levels over 18%.[13]

B. coagulans grows well in artificial media at pH values between 5.0 and 7.0. Pederson and Becker[17] showed that many cultures in their vegetative form could grow at values as low as pH 4.02. In artificial media, heat-resistant spores were incapable of germinating and producing growth below pH 5.0. It has been shown that a low pH in tomato serum decreased the heat resistance of spores, and the combined effect of acidity and NaCl concentrations (1% to 3%) caused more damage to spores at the same temperature.[2]

Hippchen et al.[12] reported the pH range for growth of *A. acitoterrestris* in laboratory media (complex medium and semi-synthetic medium) as 2–5 over a temperature range of 22–62°C. Cerny et al.[6] reported the pH range for growth of the organism in semi-synthetic medium to be 2.5–5.5 over a temperature range of 26–50°C. Deinhard et al.[8] reported the pH range for growth of the organism in *Bacillus acidocaldarius* medium was 2.5–5.8 over a temperature range of 35–55°C, with an optimum growth temperature of 42–53°C. McIntyre et al.[16] reported growth on Potato Dextrose Agar over a pH range of 3–5.3 at 30–55°C. Walls[23] reported growth in Orange Serum Broth at pH 2.5–5.0 over a temperature range of 20–55°C. Yamazaki et al.[28] reported growth in *Bacillus acidocaldarius* medium[8] at pH 2.5–6.0 over a temperature range of 25–60°C.

24.3 EQUIPMENT, MATERIALS, AND REAGENTS

24.31 Culture Media, Reagents

ALI Agar
ALI Broth
Dextrose tryptone agar (with bromcresol purple)
Gum arabic
Gum tragacanth
Litmus milk
K Agar (pH 3.7)
Sodium hydroxide (0.02 N)
Thermoacidurans agar
2-methoxyphenol and phenol (Sigma, St. Louis, MO). Prepare stock standard solution of 2-methoxyphenol at 1000 mg/L. Prepare working solutions by diluting the stock solution to a range of 5-500 µg/L. Add 100 µg/L phenol to the working solutions.

24.32 Equipment

Autoclave
Blender and blender jars
Colander/sieve
Colony counter
Incubators, 43°C ±1°C, 45°C ±1°C and 55°C ± 1°C
Water baths, 80°C and 90°C
Ice bath
Stirrer with heater
4 mL vials with open end caps
Solid phase microextraction (SPME) device:
SPME fiber, 65 µm polydimethylsiloxane (PDMS)-divinyl-benzene (DVB), (Supelco, Bellefonte, PA)
SPME fiber holder and sampling stand (Supelco, Bellefonte, PA)
Gas chromatograph with mass selective detector
DB-5 chromatography column

24.33 Glassware

Blender jars and blender
Erlenmeyer flasks, 250 mL marked at 100 mL
Petri plates
Rubber stoppers
20 × 150-mm screw-cap tubes
25 × 150-mm screw-cap tubes
Thermometer (–20° to 110°C)

24.4 PROCEDURES

24.41 Isolation of B. coagulans from Whole Tomatoes, Tomato Pulp, Tomato Puree, Concentrated Milk, and Tomato Wash Water

24.411 Sample Preparation

Extract the juice from raw whole or chopped tomatoes by pressing the sample in a sterile colander or sieve. The sample also may be prepared using a sterile blender jar and suitable blender. Transfer 10 mL of the expressed juice to a sterile 20 × 150-mm screw-cap tube for heat shocking, and tighten the rubber-lined caps securely. Samples of tomato puree and products of similar consistency are handled more conveniently using 25 × 150-mm tubes. Completely immerse the tubes containing the samples in a waterbath adjusted to 90°C. Using an extra tube fitted with a slotted rubber stopper, check the rising temperature of a similar sample of ingredient or product. After the temperature in the "control" tube reaches 90°C, start timing of the heat-shock treatment. Shock for 5 min. Normally, 10 mL of water or product of a similar consistency in a 20 × 150-mm screw-cap tube requires approximately 3 min to reach 90°C. Cool the sample tubes in cold water immediately after the shock treatment, keeping the screw-cap tops well above the surface of the water. Samples of ingredient or product that, in their preparation or manufacture, have had a recent heat treatment of 82°C or higher need no further shock treatment and can be plated directly. Place a control tube containing sterile tomato juice inoculated with spores known to be *B. coagulans* in the waterbath with the tubes of samples to be shocked.

24.412 Cultural Procedure

Transfer 1 mL of the shocked sample or decimal volume thereof into each of 4 petri dishes. Add 18 to 20 mL of dextrose tryptone agar tempered to 44° to 46°C to each of two plates and add 18 to 20 mL of thermoacidurans agar tempered to 44° to 46°C to each of two plates. After solidification, invert the plates and promptly incubate at 55°C ± 1°C for 48 hr ± 3 hr. Surface colonies on dextrose tryptone agar resulting from the germination and growth of spores of *B. coagulans* will appear slightly moist, usually slightly convex, and pale yellow. Subsurface colonies on this medium are compact with fluffy edges. They are slightly yellow to orange and usually 1 mm or greater in diameter. A yellow zone caused by acid formation will surround both surface and subsurface colonies. In 48 hr, plates may turn completely yellow. *B. stearothermophilus* will also grow on dextrose tryptone agar giving pinhead-size colonies, usually brown, which are of no consequence since they do not spoil tomato juice. Suspicious colonies should be transferred to litmus milk, where they will produce coagulation if the organisms are *B. coagulans; B. stearothermophilus* will not grow on the thermoacidurans agar at pH 5.0, and therefore counts on the latter acid medium may have more significance. Typical colonies on the latter agar are large and white to cream in color.

24.42 Isolation of B. coagulans from Nonfat Dry Milk

24.421 Sample preparation

Weigh 10 g of the sample into a sterile 250-mL Erlenmeyer flask marked at 100 mL. Add 0.02 N sodium hydroxide to the mark and shake to dissolve the sample completely. Heat 10 min at 5 psi (108.4°C) steam pressure, then cool immediately. Bring volume back to mark with sterile 0.02 N sodium hydroxide.

24.422 Cultural Procedure

Transfer 2 mL of the solution to each of 10 sterile Petri plates. Add to each plate 18 to 20 mL of dextrose tryptone agar tempered to 44°–46°C. After solidification, invert the plates and promptly incubate at 55°C ±1°C for 48 hr ±3 hr. Count the typical acid flat sour colonies previously described and report on the basis of number of colonies per 10 g of sample.

24.43 Isolation of B. coagulans from Cream

24.431 Sample Preparation

Mix 2 g of gum tragacanth and 1 g of gum arabic in 100 mL of water in an Erlenmeyer flask. Sterilize in the autoclave for 20 min at 121°C. Transfer 20 mL of sample to a sterile 250-mL Erlenmeyer flask marked at 100 mL. Add the sterilized gum mixture to the 100-mL mark and carefully shake, using a sterile rubber stopper. Loosen stopper and autoclave for 5 min at 5 psi (108.4°C).

24.432 Cultural Procedure

Because of the viscosity of the mixture, first pour 5 Petri plates with dextrose tryptone agar, then immediately transfer 2 mL of the cream emulsion and swirl in the usual manner. After solidification, invert the plates and promptly incubate at 55° ±1°C for 48 hr ±3 hr. Count the typical acid flat sour colonies and report on the basis of 1 mL of sample.

24.44 Isolation of Alicyclobacillus from Fruit Juice Concentrates, Fruit Juices and Other Acid Products

24.441 Sample Preparation

For use with K Agar. Juice concentrates should be heat shocked. Transfer 10 mL of juice concentrate to a sterile 20 × 150 mm screw cap tube for heat shocking. Immerse the tube in a waterbath adjusted to 80°C to a depth that is above the level of juice in the tube. Using an extra tube fitted with a slotted rubber stopper and thermometer, check the rising temperature of a similar sample of juice or juice concentrate. After the temperature in the control tube reaches 80°C, hold the sample for 10 min. Cool the sample tube in an ice bath immediately after the heat shock treatment, keeping the screw cap top well above the surface of the water. Spoiled juices should not be heat shocked.

For use with ALI Broth. Dilute juice concentrates to 11–14 °Brix with sterile water. Place 100 mL in a 250 mL flask. Place flasks in a water bath at 90°C for 20 min. Ensure that the liquid inside the flask is at least 3 cm below the hot water level. Keep the water bath covered. After heating, cool flasks immediately in ice water and incubate at 45°C.

24.442 Cultural Procedure

Using K Agar. Transfer 1 mL of the sample into each of 3 Petri plates. Add 18 to 20 mL of K Agar pH 3.7 tempered to 44° to 46°C to each plate. After solidification, invert the plates and promptly incubate at 43°C ± 1°C for 3 days. Plates may be incubated an extra day or two if necessary to better visualize the colonies. Plates may be placed in bags to prevent drying. Colonies on the surface of the agar should be opaque, cream colored, slightly raised. Pinpoint colonies may be present under the agar surface. Count all colonies on or under the agar on each of three replicate plates, take and average and read as count per mL.

If the sample can be filtered, a more sensitive result may be obtained by filtering 10 mL or more of sample through a 0.45μ membrane, placing this on the surface of K Agar plates, and incubating as above. Juice concentrate can be diluted in sterile water to enable filtration (e.g.10 mL juice concentrate with 50 mL water). A total count should be made and reported as the count per volume of sample filtered, e.g., count per 10 mL if a 10 mL sample volume was filtered.

Using ALI Broth. When product or juice becomes turbid or an off odor is detected, streak on to ALI agar plates and incubate at 45°C for up to 7 days.

24.443 Confirmatory Tests

Isolated colonies should be examined microscopically for rod shaped organisms that produce terminal, subterminal and central spores that slightly swell the sporangium. This will indicate the presence of acidophilic sporeformers. If the organism survives the heat shock and can grow in K Agar, it has the potential to spoil juices. For spoiled juices that are not heat shocked, further confirmation to verify that the organisms are sporeformers may be needed. Isolated colonies may be streaked to Potato Dextrose Agar pH 3.5 and incubated at 43°C for 5 days to allow sporulation. A spore suspension in sterile water can be made from this, and spores should survive a heat shock of 80°C for 10 minutes. This can be verified by plating on K Agar. Identification as *A. acidoterrestris* can be accomplished by ribotyping. A chemical test for guaiacol can be made as described below.[14]

24.45 Isolation of Alicyclobacillus from Whole Fruit

24.451 Sample Preparation

Swab surfaces of fruit with a sterile sponge. Place sponge in sterile bag with 150 mL sterile peptone saline solution (PSS: 8.5 g NaCl/L and 1 g/L bacto-peptone). Knead bag by hand for 30 seconds. Aseptically transfer 5 mL to 100 mL ALI broth in a 250 mL flask. Heat shock as described in 24.441.

24.452 Cultural Procedure

Culture as outlined in 24.442.

24.46 Isolation of Alicyclobacillus from Soil

24.461 Sample Preparation

Add 5 g of soil to 100 mL ALI broth in a 250 mL flask. Heat shock as described in 24.441.

24.462 Cultural Procedure

Culture as outlined in 24.442.

24.47 Chemical test for Guaiacol

24.471 Sample Preparation

Add 100 ppb phenol to test sample. Pipette 3 mL of sample into a 4 mL vial and add 1.2 g of sodium chloride. Put a stirrer in and cap the vial. Load the vial on the vial receptacle of the SPME stand. Load the SPME fiber holder on holder cartridge, plunge the needle and pierce the septum. Push the plunger and expose the SPME fiber to the liquid sample for 60 min while stirring. Retrieve the fiber and remove the SPME fiber holder from the stand. Insert the SPME holder needle into the GC injector and push the plunger to expose the fiber for 2 min. Remove the fiber holder from the injector in reverse order.

24.472 GC-MS Analysis

Perform the analysis using a GC 5890 series II gas chromatograph (or equivalent) coupled with a 5970 series mass spectrometer (Hewlett-Packard or equivalent) at 70eV of ionization energy. Use a split/splitless injector in a splitless mode and maintain at 220°C. Use 2 min desorption time for all fiber injections. Use a DB-5 column, 0.25mmx 15m, 0.25(m (J&W Co., Folsom, CA) or equivalent. Hold the column temperature at 50°C for 5 min and then increase to 150°C at 4 °C/min, then from 150°C to 300°C at 20° C/min. Use helium as a carrier gas at a velocity of 30 cm/sec.

Collect the data in the selective ion monitoring mode (SIM), choose m/z 81, 109, and 124 for 2-methoxyphenol; m/z 66 and 94 for phenol.

The minimum detection level is 1.5 ppb 2-methoxyphenol. Detection of this level or more is indicative of spoiled juice (growth of acidophilic sporeformers). The limit of quantitation is 5.0 ppb.

24.5 MAINTENANCE OF CULTURES

Alicyclobacillus isolates may be cultured on Orange Serum Agar (OSA, DIFCO) pH 5.0, with incubation at 43°C or 55°C, and may be maintained on OSA slants at 4°C. Isolates may be stored in ALI broth with 20% glycerol at –76°C, or they may be lyophilized, or maintained as spore suspensions. Spores suspensions can be prepared by inoculating Orange Serum broth, pH 3.5, with a loopful of culture from a slant or plate and incubating at 43°C. Once the broth is turbid, the surface of Potato Dextrose Agar (pH 3.5 adjusted with tartaric acid) plates can be inoculated with 2 mL of Orange Serum culture and incubated at 43°C. Spores can be harvested after 5–7 days incubation.

24.6 PRECAUTIONS AND LIMITATIONS OF THE PROCEDURES

When plating samples of acid products such as tomato juice and tomato puree for *B. coagulans*, the tempered medium should be poured directly on the sample in a Petri plate. A minimum of 18 mL of medium per plate should be used per 1 mL of tomato product, to minimize the inhibitory activity against *B. coagulans*. Pouring should be followed immediately by gentle swirling to ensure adequate dispersion of the sample in the melted medium. In addition to providing a uniform distribution of any surviving spores, the precautions will also ensure a uniform color to the poured plate because of the acid product and the indicator in the medium. Precautions also must be taken to prevent drying out or splitting of agar in plates during incubation at 55°C. This can be accomplished by placing the plates in appropriate canisters or by providing additional moisture in the air within the incubator.

Isolation and identification procedures for *Alicyclobacillus* are considered the most appropriate methods at this time. Use of different isolation media or incubation temperatures will give different results. Ribotyping of strains is not universally available.

24.7 INTERPRETATION

True acid "flat sour" (*B. coagulans)* spores have significance in canned products in the pH range of 4.1 to 5.0. Surviving spores of *B. stearothermophilus* will not grow in this pH range, and, therefore, care must be used in distinguishing between the two organisms. The presence of *B. coagulans* can lead to serious spoilage while presence of *B. stearothermophilus* spores at this pH range is of no consequence.

Some representatives of industry show concern when counts in excess of five spores of *B. coagulans* per g are encountered in the ingredients used in canned foods in the pH range of 4.1 to 5.0.

Studies have shown that the presence of 1 spore per 10 mL tube of *A. acidoterrestris* in a susceptible product, can germinate, grow and cause spoilage under appropriate conditions of pH and incubation temperature. Inadequate washing and poor hygiene measures in concentrate production have been reported to increase the likelihood of finding *A. acidoterrestris* in raw material.[5] Filtering through a 0.45 µm membrane may be a suitable control mechanism for clear products.[5]

24.8 REFERENCES

1. APHA. 2001 (In press). Standard methods for the examination of dairy products, 17th ed., M. Wehr (ed.). Am. Public Health Assoc., Washington, D. C.
2. Alain, A. M., N. M. El-Shimi, M. M. Abd El-Magied, and A. G. Tawadrous. 1986. Heat resistance of *Bacillus coagulans* spores isolated from tomato juice. J. Food Sci. 14(2):323.
3. Becker, M. E., and C. S. Pederson. 1950. The physiological characters of *Bacillus coagulans (Bacillus thermoacidurans)*. J. Bacteriol. 59:717.
4. Berry, R. N. 1933. Some new heat resistant, acid tolerant organisms causing spoilage in tomato juice. J. Bacteriol. 25:72.
5. Borlinghaus A., and R. Engel. 1997. Fruit processing 7:262.
6. Cerny, G., W. Hennlich, and K. Poralla. 1984. Fruchtsaftverderb durch Bacillen: Isolierung und Charakterisierung des Verderb-serregers. (Spoilage of fruit juice by Bacilli: Isolation and characterisation of the spoiling microorganism). Z. Lebensmit. Unters. Forsch. 179:224.
7. De Lucca, A. J., R. A. Kitchen, M. A. Clarke, and W. R. Goynes. 1992. Mesophilic and thermophilic bacteria in a cane sugar refinery. Zuckerind, 117:237.
8. Deinhard, G., P. Blanz, K. Poralla, and E. Altan. 1987. *Bacillus acidoterrestris* sp. nov., A new thermotolerant acidophile isolated from different soils. Syst. Appl. Microbiol. 10:47.
9. Gordon, R. E., W. C. Haynes, and C. H. Pang. 1973. The genus *Bacillus*. USDA Handbook No. 427, Washington, D. C.
10. Gordon, R. E., and N. R. Smith. 1949. Aerobic spore forming bacteria capable of growth at high temperatures. J. Bacteriol. 58:327.
11. Hammer, B. W. 1915. Bacteriological studies on the coagulation of evaporated milk. Iowa Agric. Exp. Sta., Res. Bull. 19.
12. Hippchen, B., A. Roll, and K. Poralla. 1981. Occurrence in soil of thermo-acidophilic bacilli possessing ϕ-cyclohexane fatty acids and hopanoids. Arch. Microbiol. 129:53.
13. H. J. Heinz Co., Microbiology Dep., Pittsburgh, Pa. Unpublished communication. 1985.
14. Huang, C. J., I. Walls, and R. Lyon. 1998. Chemical detection of *Alicyclobacillus* spp. in apple juice. Presented at the Annual Meeting of the Institute of Food Technologists, Atlanta, Ga.
15. Krueger, D. 1997. Personal communication.
16. McIntyre, S., J.Y. Ikawa, N. Parkinson, J. Haglund, and J. Lee, 1995. Characteristics of an acidophilic *Bacillus* strain isolated from shelf stable juices. J. Food Protect. 58:319.
17. Pederson, C. S., and M. E. Becker. 1949. Flat sour spoilage of tomato juice. Tech. Bull. No. 287. N.Y. State Agricultural Experiment Station, Cornell Univ., Geneva, N. Y.
18. Smith, N. R., R. E. Gordon, and F. E. Clark. 1946. Aerobic mesophilic sporeforming bacteria. USDA Misc. Publ. No. 559, Washington, D. C.
19. Splittstoesser, D. F., J. J. Churey, and C. Y. Lee. 1994. Growth characteristics of aciduric sporeforming bacilli isolated from fruit juices. J. Food Prot. 57:1080.
20. Stern, R. M., C. P. Hagarty, and O. B. Williams. 1942. Detection of *Bacillus thermoacidurans* (Berry) in tomato juice, and successful cultivation of the organism in the laboratory. Food Res. 7:186.

21. Thompson, P. J. 1981. Thermophilic organisms involved in food spoilage: Aciduric flat sour sporeforming aerobes. J. Food Prot. 44:154.
22. Walls, I., 1994. Sporeformers that can grow in acid and acidified foods. Report of Scientific and Technical Regulatory Activities, NFPA, Washington, D. C.
23. Walls, I., and R. Chuyate. 1998. *Alicyclobacillus*—Historical perspective and preliminary characterization study. Dairy, Food and Environmental Sanitation 18:499-503.
24. Walls, I., and R. Chuyate. 1998. Growth of *Alicyclobacillus acidoterrestris* in acid products. Presented at the Annual Meeting of the International Association of Milk, Food and Environmental Sanitarians, Nashville, Tenn.
25. Webster, J. A., I. Walls, C. I. McDowell, J. J. Neubauer, and R. J., Hubner. 1996. Use of normalized ribotyping to describe acidophilic sporeformers isolated from fruits and fruit juices. Presented at the Annual Meeting of the American Society for Microbiology, New Orleans, La.
26. Wisotzkey, J. D., P. Jurtshuk, G. E. Fox, G. Deinhard, and K. Poralla. 1992. Comparative sequence analyses on the 16S rRNA (rDNA) of *Bacillus acidocaldarius, Bacillus acidoterrestris* and *Bacillus cycloheptanicus* and proposal for creation of a new genus, *Alicyclobacillus* gen. nov. Int. J. Syst. Bacteriol. 42:263.
27. Wisse, C. A. and Parish, M. E. 1998. Isolation and enumeration of thermo-acidophilic rod shaped bacteria from citrus processing environments. Dairy, Food and Environmental Sanitation, 18:504.
28. Yamazaki, K., H.Teduka, and H. Shinano. 1996. Isolation and identification of *Alicyclobacillus acidoterrestris* from acidic beverages. Biosci. Biotech. Biochem. 60:543.

Chapter 25

Thermophilic Flat Sour Sporeformers

Karl E. Olson and Kent M. Sorrells

25.1 INTRODUCTION

Canned "commercially sterile" products, e.g., tomato products, vegetables, evaporated milk can undergo thermophilic flat sour spoilage, if held at high ambient temperatures.[6]

In canned low-acid foods, particularly those having a pH no lower than about 5.3, thermophilic flat sour spoilage seldom occurs if holding temperatures are maintained below approximately 43°C. If these foods are held at a temperature above about 43°C long enough and if the food contains viable spores capable of germinating and growing out in the product, then the product may undergo flat sour spoilage.[3]

Typical thermophilic flat sour spoilage of low-acid canned foods is caused by the growth of sporeforming, thermophilic facultative aerobes in the genus *Bacillus*. *Bacillus stearothermophilus* and *Bacillus coagulans* are the typical species responsible for this type of spoilage.[2,5,6,11] These organisms characteristically ferment carbohydrates with the production of short-chain fatty acids that "sour" the product. They do not produce enough, if any, gas to change the usual "flat" appearance of the ends of the container. Although the flat sour bacteria are considered obligate thermophiles, in fact, they may grow at temperatures as low as 30°–45°C, especially if the incubated organisms are in the vegetative state and if proper environmental conditions are imposed. The group's upper temperature limit for growth is 65°–75°C.

Spores of *B. stearothermophilus* have exceptionally high thermal resistance. The D_{121}°C ranges between 4.0 and 5.0 min. with a z-value (slope of the thermal death time curve) between 7.8°–12.2°C.[10] Thus, their presence in some containers of any given lot of commercially sterile low-acid canned foods may be considered normal. Since flat sour spoilage does not develop unless the product is held at temperatures above 43°C, proper cooling after thermal processing and avoiding high temperatures during warehouse storage or distribution are essential.

Bacterial spores enter food processing plants in soil, on raw foods, and in ingredients, e.g., spices, sugar, starch, flour.[7,8,10] Populations may increase at any point where a proper environment exists. For example, food handling equipment in a processing line that is operated within the thermophilic growth range (about 43°–75°C) may serve as a focal point for the buildup of an excessive flat sour spore population.

Spores of flat sour bacteria show exceptional resistance to destruction by heat and chemicals; hence, they are difficult to destroy in a product or in the plant. Methods to minimize spore contamination include control of spore population in ingredients and products entering the plant, as well as the use of sound plant sanitation practices.

Inadequate cooling subsequent to thermal processing is a major contributor to the development of flat sour spoilage. Localized heating of sections of stacks of canned foods placed too close to heaters is another.

25.2 GENERAL CONSIDERATIONS

25.21 Sampling

25.211 Ingredients

Approximately half-pound samples from each of five bags, drums, or boxes of a shipment or lot of dry sugar, starch, flour, or similar ingredients should be collected and sealed in cans, jars, plastic bags, or other appropriate containers and transported to the laboratory.[7] Samples of ingredients used in lesser proportions in the finished product, e.g., spices, may be sampled in appropriately smaller volumes. In any case, samples should be reasonably representative of the entire lot or shipment in question. For liquid sugar, collect five separate 6- to 8-oz. (200- to 250-mL) samples from a tank or truck when it is being filled or emptied.[1]

25.212 Equipment and product in process

Only those units held at temperatures within the thermophilic growth range are of direct concern. Scrapings or swab samples of food contacting surfaces or of wet surfaces positioned directly over food materials, from which drippage may gain access to the food materials, may be cultured. Collect the samples in sterile tubes or sampling bags. Examination of food samples taken before and after passage through a particular piece of equipment, e.g., a blancher or a filler, will reveal whether a significant buildup of flat sour spores is occurring in that item of equipment. Multiple samples of a volume equivalent to the volume of the container being packed generally are taken. Unused clean metal cans and covers are convenient sample containers. (Do not use glass containers to collect samples because of the danger of breakage or of being dropped into the equipment). Solid materials in the presence of excess liquid may be collected in a sieve or similar device; this permits draining excess liquid. Chill the samples thoroughly without delay to prevent the growth of thermophilic bacteria prior to the laboratory examination. Immersion of the sample container in cold tap water usually is adequate for this purpose.

25.213 Finished product

The method of sampling depends on the objective of the examination. When spore contamination levels during production are the concern, obtain processed containers representing conditions (a) at the start of operations when the shift begins, (b) before midshift shutdown, (c) at startup after midshift shutdown, and (d) at the end of the shift. Samples of each time period should consist of at least 10 containers. The probability of finding one positive can in 10 samples if 25% of the production contains viable flat sour spores is 0.95.[7] Incubate at 55°C for 5 to 7 days.

When known or suspected insufficient cooling or storage at temperatures above 43°C is suspected, obtain containers at random from the production lot in question. Record the locations from which each container was obtained, i.e., its position on a pallet and the location of the pallet. The larger the number of samples examined, the greater will be the probability of detecting flat sour spoilage. The probability of detecting at least one spoiled container in the sample when the real spoilage level is 1% is about 95% with a 300-unit sample; 89% with a 200-unit sample, and 62% with a 100-unit sample[7] (See Chapter 2). Because growth of flat sour organisms may cause a slight loss of vacuum or a loss of consistency of a product, separation of these products sometimes is possible without destruction of normal product cans.[7]

25.22 Spore Recovery

Much of the work in the area of spore recovery deals with recovery from heat resistance tests, and these results may or may not apply in the recovery of the spores from ingredients and the recovery of vegetative cells from production equipment. Recovery of heated spores occurs best at 45° to 50°C and in neutral media. After spore heating, water is the best diluent for spore recovery[3] and best recovery occurs if heating has been done in distilled water.

For optimum vegetative growth, the cell needs adequate oxygen supply and a culture medium at pH 7. No growth of *B. stearothermophilus* occurs at pH 5.

25.3 EQUIPMENT, MATERIALS, AND REAGENTS

Equipment and supplies are needed in accordance with specifications in Chapters 2 and 63. The following additional media and apparatus are recommended.

25.31 Culture Media

Dextrose tryptone agar
Dextrose broth
Nutrient broth or dextrose tryptone broth (if needed for nutrient supplementation: see Section 25.52)
2% Agar

25.32 Equipment

Autoclave—for heat shocking food samples in addition to normal laboratory sterilization purposes.

Sanitary Can Opener (Bacti-Disc Cutter) (Figure 3 in Chapter 62)—for opening canned food samples aseptically. It may be purchased from Wilkens-Anderson Co., 4525 West Division Street, Chicago, Ill. 60651. No. 10810-01, or from Dixie Canner Equipment Co., P.O. Box 1348, Athens, Ga. 30601. No. D-173-A.

Flasks—250 mL and 300 mL Erlenmeyer flasks (for analysis of thermophilic spores ingredients)

Glassware—dilution bottles; 6 to 8 oz. for dilution blanks containing distilled water.

Incubator—temperature controlled to about 55°C

Microscope

Petri dishes—sterile: either glass or plastic may be used.

pH Meter, Electrometric—pH Color Comparator with bromcresol purple and methyl red reagents and standards may be substituted.

Pipettes, sterile—1 mL and 10 mL Mohr pipettes; sample pipettes made from either straight wall borosilicate tubing (7 to 8 mm in diameter × 35 to 40 cm). Disposable pipettes may also be used.

Special waterbath (starch)[4]

Swabs—sterile 6 in cotton or alginate swabs

25.33 Reagent

Crystal violet solution

25.4 PRECAUTIONS

Samples other than finished products must be handled so that there will be no opportunities for spore germination or spore production between the collection of the samples and the start of examination procedures.

Before making a positive judgment on a sample based on pH or microscopic examination of direct smears, be sure that these characteristics are known for "normal obtained" control products. Controls should be obtained from the same production code as the suspect samples. If such controls are not available, use a product from the same manufacturer and bearing the next closest production code. This is particularly important where formulated products are concerned, although it is not necessarily confined to such products. Incubated agar plates should not be allowed to dehydrate during incubation. Placement in oxygen-permeable bags will minimize dehydration.

25.5 PROCEDURE

Thermophilic flat sour spores possess greater heat resistance than most other organisms encountered in foods. This characteristic is advantageous to the examination of foods and ingredients because, by controlled heat treatment of samples (heat shock), it is possible to eliminate all organisms except the spore with which we are concerned. Further, heat shock, or activation, is necessary to induce germination of the maximum number of spores in a population of many species, including the flat sours.[3,7] Because the most heat-resistant spores are generally the ones of concern in food canning operations, a heat shock favoring recovery of such spores is preferable. Unless otherwise specified, i.e., in a standard procedure, 30 min at 100°C or 10 min at 110°C, followed by rapid cooling, should be used.

25.51 Sample Preparation and Examination

Sugar and starch: The National Food Processors Association (NFPA) (formerly National Canners Association) has suggested a method and standard for determining thermophilic flat sour spore contamination of sugar and starch to be used in low-acid canned foods.[4,7,10]

1. *Sugar (AOAC)*.[1] Place 20 g of dry sugar in a sterile 250-mL Erlenmeyer flask marked at 100 mL. Add sterile water to the

100-mL mark. Agitate thoroughly to dissolve the sugar. (Liquid sugar is examined by the same procedure, with this difference: a volume of liquid sugar calculated to be equivalent, based on degree Brix, to 20 g of dry sugar is added to the 250-mL flask and diluted with water to 100 mL.)

Bring the prepared sample rapidly to a boil and continue boiling for 5 min., then water cool immediately. Pipette 2 mL of the heated sugar solution into each of 5 petri plates. Add dextrose tryptone agar (Chapter 63), swirl gently to distribute the inoculum, and allow to solidify. Incubate the inverted plates at 50° to 55° for 48 to 72 hr.

2. *Starch.* Place 20 g of starch in a dry, sterile 250-mL Erlenmeyer flask and add sterile cold water to the 100-mL mark, with intermittent shaking. Shake well to obtain a uniform suspension of the starch in water. Pipette 10 mL of the suspension into a 300-mL flask containing 100 mL of sterile dextrose tryptone agar at a temperature of 55° to 60°C. Use large-bore pipettes; keep the starch suspension under constant agitation during the pipetting operation. After the starch has been added to the agar, shake the flask in boiling water for 3 min. to thicken the starch. Then place the flask in the autoclave and heat at 5 lb. pressure (108.4°C) for 10 min. After autoclaving, the flask should be gently agitated while cooling as rapidly as possible. Violent agitation will incorporate air bubbles into the medium, which subsequently may interfere with the reading of the plates. When the agar starch mixture is cooled to the proper point, (about 45°C) distribute the entire mixture about equally into 5 plates and allow to solidify. Then stratify with a thin layer of sterile plain 2% agar in water and allow to solidify. This prevents possible "spreader" interference. Incubate the inverted plates at 50° to 55°C; count colonies in 48 hr. and in 72 hr.
3. *Other ingredients.* The NFPA procedures for sugar and starch may be applied to other ingredients used in low-acid canned foods.[7] Modifications may be necessary because of physical or chemical characteristics of a particular ingredient, e.g., use of smaller sample sizes or plating smaller volumes of suspension in more than 5 plates because of colony particle size interference during counting.
4. *Calculating counts.* Flat sour colonies are round, are 2 to 5 mm in diameter, show a dark, opaque center, and usually are surrounded by a yellow halo in a field of purple. The yellow color (acid) of the indicator may be missing when low-acid-producing strains are present, or where alkaline reversion has occurred. (Subsequent colonies are compact and biconvex to pinpoint in shape. If the analyst is unfamiliar with subsurface colonies of flat sour bacteria; it is advisable to streak subsurface colonies on dextrose tryptone agar to confirm surface colony morphology).

The combined count of typical flat sour colonies from the 5 plates represents the number of flat sour spores in 2 g of the original sample (20-g sample diluted to 100 mL; 10 mL of this dilution plated). Multiply this count by 5 to express results in terms of number of spores per 10 g of sample.

The total thermophilic spore count is made by counting every colony on each of the 5 plates, then calculating in terms of number of spores per 10 g of sample.

25.52 Equipment and Product in Process

The source of excessive flat sour spores in a food processing operation may best be determined by "line samples" (Section 25.21). Use quantities of sample equivalent to the amount of the material included in a container of finished product; prepare several replicates (5 to 20); after closing, warm the containers to the initial temperature of the commercial process, subject them to the normal commercial thermal process, incubate at 55°C for 5 to 7 days; open and determine growth of flat sour bacteria by pH measurement, supplemented by microscopic examination of a direct smear if necessary. Many line samples are nutritionally complete so that water may be added to fill the container; however, some formulated product components may lack essential nutrients, in which case nutrient broth or dextrose broth (Chapter 63), for example, should be added instead of water. If in doubt, inoculate a control sample with *B. stearothermophilus* spores, heat shock, and incubate at 55° C for 48 to 72 hr to determine whether the sample material will support spore germination and outgrowth.

An alternate procedure is to make agar plate counts on serial dilutions of each heat-shocked line sample. Use dextrose tryptone agar and incubate at 55°C for 48 to 72 hr.

Swabs or scrapings from equipment surfaces should be shaken in a known volume of diluent, and the suspension heat shocked and plated on dextrose tryptone agar and incubated at 55°C for 48 to 72 hr. Then calculate the numbers of microorganisms per unit area sampled.

25.53 Finished Product

Incubate containers of processed product at 55°C for 5 to 7 days, open, and examine for flat sour spoilage.[7] Comparison of pH of incubated samples and normal unincubated controls will usually be sufficient to show the presence of flat sour spoilage. If the results are not clear, confirm the presence or absence of spoilage flora by direct microscopic examination of smears of the product from both incubated and unincubated control containers. The bacteriological condition of products whose physical characteristics provide confusing artifacts when seen in a stained smear can be examined best in a wet mount using phase optics.

Samples collected from a warehouse where insufficient cooling or storing at elevated temperatures is known or suspected are examined as above, but without preliminary incubation at 55°C.

25.6 INTERPRETATION OF RESULTS

25.61 Ingredients

NFPA standards for thermophilic flat sour spores in sugar or starch for canners' use[7] state, "For the five samples examined there shall be a maximum of not more than 75 spores and an average of not more than 50 spores per 10 g sugar (or starch)."

The total thermophilic spore count standard is, "For the five samples examined there shall be a maximum of not more than 150 spores and an average or not more than 125 spores per 10 g of sugar (or starch)."[7]

The sugar and starch standard may be used as a guide for evaluating other ingredients, keeping in mind the proportion of the other ingredients in the finished product relative to the quantity of sugar or starch used.[4,7,10]

The presence of thermophilic flat sour spores in ingredients for foods other than thermal-processed low-acid foods is probably of no significance provided those foods are not held within the thermophilic growth range for many hours. The flat sour bacteria have no public health significance.[9]

25.62 Equipment and Product in Process

Canned and processed line samples usually indicate a point or points at which a spore buildup has occurred. A high percentage of positive samples taken from one point in the line, when a low level or no positive samples were found prior to this point, shows a spore buildup in this piece of equipment. The time of day yielding positive samples may indicate whether the buildup is due to operating temperatures within the thermophilic growth range, or whether inadequate cleanup and sanitization procedures were used prior to sampling. The former condition is suggested when a majority of positive samples occurs in those taken after the line has been in operations for several hours. The latter condition is suggested when samples at the startup of the line are predominantly positive and those taken later are predominantly negative.

Plate count data for equipment line samples may be meaningful, especially when taken over a long period of time. They can show trends regarding buildups, inadequate cleanups, etc. Because counts are made in a laboratory medium that may be a better spore germination and growth medium than certain specific food products themselves, and because the sample does not receive the equivalent of the commercial thermal process, results may reflect greater than the actual potential surviving spore load in the finished product. They can, however, indicate buildup situations that are undesirable.

25.63 Finished Products

Dormant thermophilic spores are of no concern in commercially sterile canned foods destined for storage and distribution where temperatures will not exceed about 43°C. However, some canned foods are destined for exposure to temperatures above 43°C during part or most of their shelf life, i.e., those shipped to tropical areas and those intended for hot-vend service. To be considered commercially sterile, these specialized foods must not contain thermophilic spores capable of germination and outgrowth in the product.

Randomly selected warehouse samples of low-acid canned foods, some of which are found to have undergone flat sour spoilage, can reveal information about the condition contributing to the development of spoilage. Spoilage confined to product situated in the outer layers or rows of cases on a pallet suggests localized heating, e.g., from close proximity to a space heater or having been too close to the building roof during hot weather. Spoilage confined to inner cases on a pallet is indicative of insufficient cooling, i.e., stacking cases on pallets while the product was still in the thermophilic growth temperature range. Inner cases are insulated by exterior cases and may retain heat for several days.

25.7 REFERENCES

1. AOAC. 1995. Official methods of analysis of AOAC International, 16th ed., vol. II, chap. 44, P. Cunniff (ed.). AOAC International, Arlington, Va.
2. Ayres, J. C., J. O. Mundt, and W. E. Sandine. 1980. Microbiology of foods. W. M. Freeman and Company, San Francisco, Calif.
3. Cook, A. M., and R. J. Gilbert. 1968. Factors affecting the heat resistance of *Bacillus stearothermophilus* spores, J. Food Technol. 3:285.
4. Department of Defense., 1985. Military standard. Bacterial standards for starches, flours, cereals, alimentary pastes, dry milks, and sugars used in the preparation of thermostabilized foods for the Armed Forces. MIL-STD-900C. U.S. Department of Defense, Washington, D.C.
5. Gordon, R. E., W. C. Haynes, and C. H.-N. Pang. 1973. The genus *Bacillus*. Agric. Handb. No. 427. U.S. Department of Agriculture, Washington, D.C.
6. ICMSF. 1998. Microorganisms in Foods 6: Microbial ecology of food commodities. Blackie Academic and Professional, New York.
7. NCA Research Laboratories. 1968. Laboratory manual for food canners and processors, vol. I, p. 88, 102. Natl. Canners Assoc. (now Natl. Food Processors Assoc.) AVI Publ. Co. Inc., Westport, Conn.
8. Richmond, B., and M. L. Fields. 1966. Distribution of thermophilic aerobic sporeforming bacteria in food ingredients. Appl. Microbiol. 14:623.
9. Schmitt, H. P., 1966. Commercial sterility in canned foods, its meaning and determination. Q. Bull. Assoc. Food Drug Off. US. 30:141.
10. Stumbo, C. R. 1973. Thermobacteriology in food processing, 2nd ed. Academic Press, New York.
11. Walker, P. D., and J. Wolf. 1971. The taxonomy of *Bacillus stearothermophilus* p. 247. *In* A. N. Barber, G. W. Gould, and J. Wolf (ed.), England spore research. Academic Press, New York.

Chapter 26

Thermophilic Anaerobic Sporeformers

David Ashton and Dane T. Bernard

26.1 INTRODUCTION

The thermophilic anaerobes that do not produce hydrogen sulfide have been responsible for the spoilage of canned products such as spaghetti with tomato sauce, sweet potatoes, pumpkin, green beans, and asparagus.[7] They have also caused spoilage of highly acid products such as mixtures of fruit and farinaceous ingredients. The nonhydrogen sulfide-producing thermophilic anaerobic sporeformers are classified in the family *Bacillaceae*, genus *Clostridium*.[8] The type species of this group is *Clostridium thermosaccharolyticum*.[9] These organisms are obligately anaerobic and are strongly saccharolytic, producing acid and abundant gas from glucose, lactose, sucrose, salicin, and starch. Proteins are not hydrolyzed, and nitrites are not produced from nitrates.[5] Vegetative cells are long, slender, straight or slightly curved, often weakly staining, gram-negative rods. Spores are terminal and swollen. Neither toxins nor infections are produced, and therefore, the organisms are of spoilage, but not of public health, significance.

One of the noticeable characteristics of these organisms is the heat resistance exhibited by their spores. It is not unusual for the spores to have D-values at 121°C of 3 to 4 minutes or higher. Their z-value (slope of the thermal death time curve) is about 6 to 7C°. Thus, the organisms can have extreme resistance in the 105 to 113°C range.[1,10] Their survival in canned foods is therefore not unexpected, but thermophilic anaerobes are rarely found in foods processed above 121°C. Only when the finished product is improperly cooled or is held for extended periods at elevated temperatures do the thermophilic anaerobes express themselves.

The optimum growth temperature of these organisms is 55 to 60°C. They seldom grow at temperatures below 32°C but can produce spoilage in 14 days at 37°C if the spores are first germinated at a higher temperature. They have an optimum for growth of pH 6.2 to 7.2 but grow readily in products having a pH of 4.7 or higher. They have been responsible on occasion for spoilage in tomato products at pH values of 4.1 to 4.5.[7]

Ingredients such as sugar, dehydrated milk, starch, flour, cereals, and alimentary pastes have been found to be the predominant sources of thermophilic anaerobes. These organisms occur widely in the soil and therefore are found on raw materials, such as mushrooms and onion products, that have a history of contact with the soil. Excessive populations of thermophilic anaerobes can develop in ingredients such as chicken stock, beef extract, or yeast hydrolysate if an incubation period in the thermophilic temperature range is provided during concentration or hydrolysis steps. The thermophilic anaerobes do not multiply on equipment and handling systems unless an anaerobic environment containing nutrients and moisture at an elevated temperature is provided. The organism has also been observed to grow well in the cooling leg (55°C area) of hydrostatic cookers, if the water is contaminated with food. Accumulation of excessive numbers of organisms in this area, followed by improper cooling, has resulted in leaker-type spoilage of canned foods.

26.2 GENERAL CONSIDERATIONS

Methods outlined in this chapter are dictated by the fact that *C. thermosaccharolyticum* is a thermophilic, obligately anaerobic sporeformer. The primary objective is to limit the number of spores in ingredients used in canned foods.

The recommended substrate for recovery and growth of nonhydrogen sulfide-producing thermophilic anaerobes is PE-2 medium.[4] The medium should be supplemented to contain 0.3% yeast extract for detection of severely heat-stressed spores.

The AOAC International detection procedure for thermophilic anaerobes not producing hydrogen sulfide specifies liver broth as the medium of choice,[2] but experience indicates that noncommercially prepared liver broth is difficult to make and is a potential source of metabolic inhibitors, including antibiotics, without offering any increased sensitivity of detection.

26.3 EQUIPMENT, MATERIALS, AND SOLUTIONS

26.31 Equipment

Blender
Incubator that will maintain a uniform temperature of 55°C ± 2C°
Microscope with 1000X oil immersion objective
Pipettes with 10-, 1.0-, and 0.1-mL capacity
18 × 150-mm tubes with venting caps

26.32 Media

Liver broth
PE2
Vaspar
2% Agar

26.4 PRECAUTIONS

Every precaution should be taken to ensure that the ingredients of the detection medium are free from growth inhibitors. For example, try to obtain peas free of pesticides. As an added precaution, each new lot of ingredients should be incorporated into the medium and tested for growth inhibitors with a known suspension of a thermophilic anaerobe. These precautions will help to eliminate or minimize the occurrence of false negatives.

The detection procedures described in Section 26.5 are not truly quantitative. The objective in surveying ingredients is to detect spores in a known quantity of the ingredient rather than to achieve absolute quantitation.

It is important that in the preparation of PE-2 the dried peas be soaked in the peptone solution 1 hr before autoclaving to ensure the proper sterilizing effect. Repeated steaming of unused tubes of medium does not reduce its effectiveness as a substrate for the thermophilic anaerobes.

Spoiled canned food suspected of thermophilic growth should not be refrigerated because vegetative cells of thermophilic anaerobes usually die under refrigeration and spores are not generally produced in canned foods.

26.5 PROCEDURE

The following procedures apply for the detection of spores only rather than of spores and vegetative cells. If the heating step is omitted, vegetative cells can be propagated by these procedures.

26.51 Culture Medium

Unless freshly prepared medium is used, previously sterilized tubes should be subjected to flowing steam for 20 min to exhaust oxygen and cooled to 55°C before use. After inoculation, tubes are stratified with 3 mL of sterile 2% agar or Vaspar that is allowed to harden before tubes are tempered to 55°C and incubated. As a safety precaution, venting caps are recommended on tubes because of the abundant gas production by the organism of interest.

26.52 Sampling

26.521 Ingredients

Samples of dry ingredients should consist of 200 g taken aseptically from five different bags or barrels per shipment or lot-for-lot sizes of 50 or fewer containers, from 10% of the containers for lot sizes 50 to 100, and from a number of containers equal to the square root of the lot size for shipments with greater than 100 containers. Liquid sugar should be sampled by drawing five 200 g portions per tank during transfer or at the refinery during the tankfilling operation. Samples should be placed in sterile, sealed containers. If preliminary analyses indicate considerable variability in a lot, the number of samples should be increased.

26.522 Equipment and Systems

The thermophilic anaerobes will not develop on equipment unless elevated temperatures are provided in a relatively microaerophilic environment containing nutrients. Accumulated food materials in such locations should be sampled with a sterile spatula or similar device and placed in sterile, sealed containers, and the analysis should be conducted as soon as possible. Examination of food materials before and after exposure to processing equipment will help to reveal the contamination level of the equipment.

26.523 Product in process

A 200 g sample of product in process should be obtained periodically to monitor the system. Sample timing should be arranged to coincide with the introduction of a new batch of ingredients or a shutdown that may have permitted an incubation period. The samples should be cooled immediately but slowly to room temperature and the analysis conducted as soon as possible. Refrigeration is not recommended.

26.524 Finished Product

Representative containers of finished product should be obtained to reflect the condition of the entire population of containers in a production period. The need for sampling will be dictated by considerations such as the previous record of the product with respect to thermophilic spoilage and the temperature stresses to which the product is expected to be subjected during transit and storage. The number of containers sampled should be of the order of one per thousand containers produced. If immediate post-process cooling to 43°C is not achievable, monitoring of surviving thermophiles becomes extremely important. Incubate the finished product at 55°C for 5 to 7 days.

26.53 Enumerating

26.531 Dry Sugar and Powdered Milk[2]

Place 20 g of sample in a sterile flask and add sterile distilled water to a final volume of 100 mL. Aseptically stir or swirl to dissolve the sample and bring the contents of the flask to a boil rapidly. Boil for 5 min, cool by placing the flask in cold water, and bring the volume back to 100 mL with sterile distilled water. Divide 20 mL of boiled solution equally among 6 freshly exhausted tubes of PE-2 medium. Stratify each tube with 3 mL of sterile 2% agar or Vaspar, allow the agar to solidify, preheat the tubes to 55°C, and incubate at 55°C for 72 h.

26.532 Liquid Sugar

Place a sample containing 20 g of dry sugar, determined on the basis of degree Brix (29.411 g of 68° Brix liquid sugar is equivalent to 20 g of dry sugar) in a sterile flask, and proceed as for dry sugar.

26.533 Fresh Mushrooms

Homogenize 200 g of mushrooms in a sterile blender jar. Blend the diced sample until the pieces are finely chopped. Frequent shaking of the jar is essential to ensure proper blending. Place 20 g of blended sample in a sterile flask, and proceed as for dry sugar.

26.534 Starches and Flours[6]

Place 20 g of sample in a sterile flask containing a few glass beads and add sterile distilled water to a final volume of 100 mL. Shake well to obtain a uniform suspension. Divide 20 mL of the suspension equally among 6 freshly exhausted tubes of PE-2 medium. Spin 3 tubes at a time in the hands immediately after adding the sample. Place the tubes in a boiling waterbath and continue to spin the tubes for the first 5 min of heating. Continue heating for an additional 10 min; then remove the tubes and place them in cold water. Stratify the tubes with 3 mL of sterile 2% agar or Vaspar, allow to solidify, preheat the tubes to 55°C, and incubate at 55°C for 72 hr.

26.535 Cereals and Alimentary Pastes[6]

Place 50 g of well-mixed sample into a sterile blender jar and add 200 mL of sterile distilled water. Blend for 3 min to obtain a uniform suspension. Proceed as for starches and flours. For calculations assume that 10 mL of the blended materials contain 2 g of the original sample.

26.536 Product in Process

Place 100 g of product in a sterile blender jar and blend for 3 min. Distribute 20 mL or g of the blended sample equally among 6 freshly exhausted tubes of PE-2 medium, and proceed as for starches and flours.

26.537 Finished Product

Representative samples of finished product should be incubated at 55°C for 5 to 7 days and observed daily for evidence of loss of vacuum or container distortion. Samples that show signs of spoilage such as gas formation or cloudy brine should be removed from incubation and opened aseptically. Three grams of the contents should be placed in each of 2 tubes of freshly exhausted PE-2 medium by means of a wide bore pipette. Smears of the product should be made for morphological confirmation. The conditions necessary for preventing laboratory contamination when subculturing cans of finished product are detailed in the literature.[3]

26.538 Spore Suspensions

When spore suspensions are prepared for thermal inactivation studies, a greater degree of quantitation is desirable than is practiced for ingredients or finished product. In this case, 10 mL of the desired dilution of the spore suspension is placed in an 18 × 150-mm screw-cap tube and immersed in boiling water for 8 min, followed by rapid cooling in cold water. A conventional 5-tube most probable number (MPN) dilution series of the boiled suspension is prepared in freshly exhausted PE-2 medium. The inoculated tubes are treated as for dry sugar. The population of the original spore suspension is computed from MPN tables.

26.6 INTERPRETATION

Tubes of PE-2 medium positive for growth of non-hydrogen sulfide-producing thermophilic anaerobes show gas production with the peas rising to the top of the liquid medium. Thermophilic flat sour bacteria may change the color from purple to yellow without gas (see Chapter 25).

26.61 Ingredients

1. For canners' use
   Spores of non-hydrogen sulfide-producing thermophilic anaerobes should not be found in more than 60% of the samples tested or in more than 66% of the tubes for any single sample.[5] Use of ingredients meeting this standard will minimize the possibility of spoilage in the finished product. Canned foods with a pH below 4.0 are not susceptible to spoilage by thermophilic anaerobes.
2. For other use
   The presence of excessive numbers of spores of thermophilic anaerobes that do not produce hydrogen sulfide in ingredients for use other than in canned products is of little significance unless a thermophilic incubation period is provided during processing. In such a case, the number of vegetative cells present after a processing step is important and should be determined as outlined above, but omitting the boiling step.

26.62 Equipment and Systems

The presence of detectable levels of spores of non-hydrogen sulfide-producing thermophilic anaerobes on equipment and systems suggests that equipment is in need of thorough cleaning and sanitation, or growth is occurring, or both. If proper sanitation is practiced, and if the systems are properly designed, spore buildup should not occur. It is especially important to sample foams and their residues on equipment because of evaporative cooling in exposed surface areas.

26.63 Product in Process

Excessive numbers of vegetative cells or spores in the product in process, prepared from ingredients meeting the requirements of ingredients for canners' use and for other use, suggest that multiplication is occurring during one or more of the manufacturing steps. The manufacturing sequence should be sampled and the point of increase in the microbial population determined. Remedial steps should be taken immediately. The presence of vegetative cells suggests that sporulation can and will occur.

26.64 Finished Products

The presence of low numbers of spores of the non-hydrogen sulfide-producing thermophilic anaerobes in processed canned foods is not unusual. The organisms possess extreme resistance to the center-can temperatures achieved in many commercial processes.[10] An attempt to eliminate the spores by increased thermal treatments may endanger the nutritional and functional integrity of many products.

If the cooling of processed cans to a center-can temperature of 43°C or less is effected immediately and the cans are stored at temperatures below 35°C, remote from heating ducts and other sources of heat, the presence of spores of thermophilic anaerobes is of no consequence. However, with such a presence, the potential for spoilage exists if temperature abuse of the cans occurs, and therefore, this situation should be avoided through the use of meticulously selected ingredients that are carefully handled throughout the production sequence. If thermophilic anaerobes are present, the importance of efficient cooling followed by storage below 35°C cannot be overemphasized.

The presence of detectable thermophilic anaerobes in canned foods destined for hot-vend service or tropical distribution constitutes an unacceptable spoilage hazard. The situation must be overcome by the use of thermophile free ingredients or by increasing the thermal process.

26.7 REFERENCES

1. Ashton, D. H. 1981. Thermophilic organisms involved in food spoilage: Thermophilic anaerobes not producing hydrogen sulfide. J. Food Prot. 44:146.
2. AOAC. 1972. Detecting and estimating numbers of thermophilic bacterial spores in sugars. J. Assoc. Off. Anal. Chem. 55:445.
3. Evancho, G. M., D. H. Ashton, and E. J. Briskey. 1973. Conditions necessary for sterility testing of heat-processed canned foods. J. Food Sci. 38:185.
4. Folinazzo, J. F., and V. S. Troy. 1954. A simple bacteriological medium for the growth and isolation of spoilage organisms from canned foods. Food Technol. 8:280.
5. NCA Research Laboratories. 1968. Laboratory manual for food canners and processors, vol. 1, p. 104. National Canners Association

(now National Food Processors Association), Avi Publishing Co., Westport, Conn.
6. Powers, E. M. 1973. Microbiological requirements and methodology for food in military and federal specifications. Tech. Rep. 73-33-FL. U.S. Army Natick Lab., Natick, Mass.
7. Rhoads, A. T., and C. B. Denny. 1964. Spoilage potentialities of thermophilic anaerobes. Research Rep. No. 3-64. National Canners Association (now National Food Processors Association) Washington, D. C.
8. Sneath, P. H. A. 1986. Endospore-forming gram-positive rods and cocci, p. 1141. *In* P. H. A. Sneath, N. S. Mair, M. E. Sharpe, and J. G. Ho (ed.), Bergey's manual of systematic bacteriology, vol 2. Williams and Wilkins, Baltimore, Md.
9. Stumbo, C. R. 1973. Thermobacteriology in food processing, 2nd ed. Academic Press, New York.
10. Xezones, H., J. L. Segmiller, and I. J. Hutchings. 1965. Processing requirements for a heat tolerant anaerobe. Food Technol. 19:1001.

Chapter 27

Sulfide Spoilage Sporeformers

L. Scott Donnelly and Todd Hannah

27.1 INTRODUCTION

Early studies on "sulfide stinker" spoilage in canned sweet corn and other vegetables were reported by Werkman and Weaver[18] and Werkman.[17] The cans involved showed no evidence of swelling; however, upon opening, a decided odor of hydrogen sulfide was evident. The product had a blackened appearance caused by the reaction between the sulfide and the iron of the container.

Sulfide spoilage, the presently preferred designation for this type of spoilage, is not common. It is nonexistent in acid foods because of the pH requirements for the growth of the causative organism.

Products suffering this spoilage, although possessing a strong, disagreeable odor of hydrogen sulfide, exhibit no other putrefactive odor.

No evidence of pathogenicity for man or laboratory animal has been associated with *Desulfotomaculum nigrificans,* the causative organism, or with products spoiled by this organism. Evidence has indicated that *D. nigrificans* spores may possess heat resistance in excess of the $D_{120°C}$ value of 2.0 to 3.0 min.[10]

Cameron and Williams[3,4] and Cameron and Yesair[5,6] found sugar and starch to be important sources of these organisms in canneries. More recently, Donnelly and Busta[11] reported that soy isolates and carrageenan contain *D. nigrificans* spores.

27.11 Classification

The type species of the sulfide spoilage group was originally classified and named *Clostridium nigrificans* by Werkman and Weaver.[18] In 1938, Starkey isolated cultures from mud, soil, and sewage at both 30° and 55°C.[16] Those organisms growing at 55°C were large, slightly curved, sporeforming rods, while those isolated at 30°C were asporogenous, short vibrios. Cultures isolated at 30°C failed to grow when transferred directly to 55°C. Those isolated at 55°C and transferred to 30°C underwent morphological changes, eventually resulting in small vibrios resembling those originally isolated at this temperature. As a result, Starkey[16] proposed the new genus *Sporovibrio* for the anaerobic vibrio-shaped cells that produced endospores. The organism *Sporovibrio desulfuricans* was later shown by Campbell et al.[7] to be identical to *C. nigrificans.* Since the latter had taxonomic priority, the thermophilic sporeforming organism that reduced sulfate was considered properly to be named *C. nigrificans.* Campbell and Postgate later proposed the name *Desulfotomaculum nigrificans* for this organism.[8] This classification is followed by Sneath.[14]

27.12 Occurrence

Although relatively rare, sulfide spoilage may occur in canned sweet corn, peas, mushroom products, infant formulas, and other non-acid foods. Spoiled peas sometimes show no discoloration, but more frequently show blackening, with a dark-colored brine. In many instances, spangling of the enamel system of the can occurs as a result of the interaction of the dissolved hydrogen sulfide with the iron of the container. This is evident through the interior enamel because of the semitransparency of the coating.

The cause of sulfide spoilage is a combination of high spore numbers, the heat resistance of the spores, and the holding of finished product at elevated temperatures. This latter factor may be the result of inadequate cooling of processed product.

27.2 GENERAL CONSIDERATIONS

27.21 Sampling

Since there is little evidence to indicate a serious in-plant buildup potential with *D. nigrificans* in modern food processing plants, recommended sampling will be limited to frequently used ingredients.[13]

27.211 Ingredients

Half-pound samples of sugar, starch, or flour are taken from each of 5, 100 pound bags of a shipment or lot. In the case of bulk shipments of such ingredients, sampling probably will have to be carried out through a loading port or hatch at the top of the car or tank. A suitable sampler should be used so that samples can be taken from various depths of the load. Samples of liquid sugar are obtained from tank trucks with the use of a sterile, long-handled dipper.

The adequacy of sampling will vary with the size of a shipment; however, when there is any significant variability in the shipment, individual tests on 5 samples are likely to make this evident in the majority of cases.

27.22 Temperature and pH Requirements

Most isolates from sulfide spoilage achieve optimum growth at 55°C. Most of these strains will grow at 43°C but not 37°C.

Therefore, using the Cameron and Esty definition,[2] these are considered to be obligate thermophiles. Organisms resembling *D. nigrificans* have been isolated from soil, mud, and sewage, as well as from certain food ingredients. Such isolates may be classified as mesophiles, facultative thermophiles, or obligate thermophiles, using again the Cameron and Esty guidelines.

According to Breed et al., the type species, isolated from canned corn showing "sulfur stinker" spoilage, will grow between 65° and 70°C, with optimum growth at 55°C.[1] Campbell and Postgate reported that the organism can be "trained" to grow slowly at 37° or 30°C.[8]

Optimum growth of *D. nigrificans* occurs between pH 6.8 and 7.3. Scanty growth occasionally occurs as low as pH 5.6; however, pH 6.2 is considered the lower limit. Maximum pH for growth has been reported as pH 7.8. The pH values of most vegetables fall below pH 5.8, corn and peas being the exceptions. This may be responsible for the limited and relatively uncommon occurrence of sulfide spoilage.

27.3 EQUIPMENT MATERIALS, AND REAGENTS

27.31 Culture Media

Common 6-d nails or iron strips, or 5% ferric citrate
Gum arabic
Gum tragacanth
Hydrochloric acid
Sodium hydroxide (.02 N)
Sulfite agar
Vaspar (half mineral oil, half paraffin)

27.32 Equipment

Erlenmeyer flasks, 250 mL marked at 100 mL
Incubator 55°C ± 1°C
Petri dishes
Rubber stoppers
20 × 150-mm screw-cap tubes
Thermometer (–10° to 110°C)
Waterbath[9,12]

27.4 PRECAUTIONS AND LIMITATIONS OF PROCEDURE

When analyzing ingredients, thorough dispersion of the sample solution or slurry in each tube of medium is essential. More difficulty will be encountered in the analysis of starch or flour because of the thickening effect during heating. Frequent swirling of the tubes during the first 10 min of heating will assure proper dispersion.

Since tubes containing colonies of *D. nigrificans* may become completely blackened after 48 hr of incubation, a preliminary count should be made after 20 to 24 hr ± 3 hr.

When preparing tubes of sulfite agar, nails or iron strips should be cleaned in hydrochloric acid and rinsed well to remove all traces of rust before adding to the tubes of medium. The clean nails will combine with any dissolved oxygen in the medium and provide an anaerobic environment. As an alternative to the iron nails, add 10 mL of a 5% solution of ferric citrate to the sulfite agar.

Since the organism is extremely sensitive to oxygen, the inoculum should be added below the surface of the medium in the tube to obtain maximum counts.

27.5 PROCEDURE

27.51 Sugar, Starch, Flour[13]

The following method for preparation of a sample of sugar is recommended by the Association of Official Analytical Chemists (AOAC International). The similar method that follows is used for preparation of starch or flour samples.

27.511 Sample preparation

Place 20 g of dry sugar into a dry, sterile, 250-mL Erlenmeyer flask closed with a rubber stopper. Add sterile water to the 100-mL mark, and shake to dissolve. Replace the stopper with a sterile cotton plug, bring the solution rapidly to a boil, and continue boiling for 5 min. Replace the evaporated liquid with sterile water. Cool immediately in cold water.

Prepare samples of liquid sugar the same way, except the amount added to the sterile flask should be determined, depending upon the degree Brix, to be equivalent to 20 g of dry sugar.

Place 20 g of starch or flour in a dry, sterile, 250-mL Erlenmeyer flask, and add sterile cold water to the 100-mL mark, with intermittent swirling. Close the flask with the sterile rubber stopper and shake well to obtain a uniform, lump-free suspension of the sample in water. Sterile glass beads added to the sample mixture will facilitate thorough mixing during shaking.

27.512 Cultural Methods

When examining sugar, divide 20 mL of the heated solution among 6, 20 × 150-mm screw-cap tubes each containing approximately 10 mL of sulfite agar and a nail. Make the inoculations into freshly exhausted medium, and solidify rapidly by placing the tubes in cold water. Preheat the tubes to 50° to 55°C, and incubate at that temperature for 24 and 48 hr.

In the case of starch or flour, divide 20 mL of the cold suspension among 6, 20 × 150-mL screw-cap tubes, each containing approximately 10 mL of sulfite agar and a nail. The tubes should be swirled manually several times before heating and during the 15-min heating period in a boiling waterbath to ensure even dispersion of the starch and flour in the tubes of medium. Following heating, cool the tubes immediately in cold water. Preheat the tubes to 50° to 55°C and incubate at that temperature for 24 to 48 hr. *D. nigrificans* will appear as jet-black spherical areas, the color due to the formation of iron sulfide. No gas is produced. Certain thermophilic anaerobes that do not produce H_2S give rise to relatively large amounts of hydrogen, which splits the agar and, in the case of sulfite agar, reduces the sulfate, thereby causing general blackening of the medium. Count the number of colonies in the 6 tubes. Calculate and report as number of spores per 10 g of ingredient.

A more sensitive method for the detection (not enumeration) of *D. nigrificans* was recommended by Speck.[15]

27.52 Nonfat Dry Milk

27.521 Sample Preparation

Weigh 10 g of the sample into a sterile, 250-mL Erlenmeyer flask marked to 100 mL. Add .02 N sodium hydroxide to the 100-mL mark and shake to dissolve the sample completely. Heat 10 min at 5-lb steam pressure, then cool immediately.

27.522 Culturing Methods

Transfer 2 mL of the heated solution to each of 2, 20 × 150-mm screw-cap tubes of freshly exhausted sulfite agar and a nail. Gently swirl several times and solidify rapidly by placing the tubes in cold water. Preheat the tubes to 50° to 55°C, and incubate at that temperature for 24 and 48 hr ± 3 hr. Count colonies of *D. nigrificans* described earlier, and report on the basis of 10 g of sample.

27.53 Cream

27.531 Sample Preparation

Mix 2 g of gum tragacanth and 1 g of gum arabic in 100 mL of water in an Erlenmeyer flask. Sterilize in the autoclave for 20 min at 121°C. Transfer 20 mL of sample to a sterile, 250-mL Erlenmeyer flask marked for 100 mL. Add the sterilized gum mixture to the 100-mL mark and carefully shake, using a sterile rubber stopper. Loosen the stopper and autoclave for 5 min at 5-psi pressure.

27.54 Soy Protein Isolates

27.541 Sample Preparation

Prepare a 10% suspension of soy protein isolate in sterile 0.1% peptone water in milk dilution bottles (or equivalent). Adjust the pH to 7.0 ± 0.1. Steam in an autoclave (approximately 5 lb steam) for 20 min.

27.542 Culturing Methods

Following the steaming procedure, add 1.0 mL of the suspension to each of 10 tubes containing molten sulfate agar and a nail. Heat the tubes immediately before inoculation to eliminate oxygen. After inoculation, tubes are mixed, solidified in an ice waterbath, overlaid with Vaspar, and preheated to 55°C. Incubate the tubes for 14 days at 55°C. Count the jet-black spherical areas for the 10 tubes and report as the number of spores per gram of soy isolate. Preliminary counts should be made at 48 hr, 7 days, and 14 days in case tubes become completely blackened. Note that Donnelly and Busta[11] reported on an alternative medium to sulfite agar for use with soy protein isolates. Additional 10-tube sets can be used to examine a larger sample.

27.6 INTERPRETATION

A standard for sulfide spoilage applies only to ingredients (sugar, starch, flour, etc.) to be used in low-acid, heat-processed canned foods.

Sulfide spoilage spores should be present in not more than 2 (40%) of the 5 samples tested, and in any one sample to the extent of not more than 5 spores per 10 g.[12]

27.7 REFERENCES

1. Breed, R. S., E. G. D. Murray, and N. R. Smith 1957. Bergey's manual of determinative bacteriology, 7th ed. Williams and Wilkins, Baltimore, Md. p. 649.
2. Cameron, E. J., and J. R. Esty. 1926. The examination of canned spoiled foods, 2. Classification of flat sour spoilage organisms from nonacid foods. J. Infect. Dis. 39:89
3. Cameron, E. J., and C. C. Williams. 1928a. The thermophilic flora of sugar in its relation to canning. Centbl. Bakt. 76:28.
4. Cameron, E. J., and C. C. Williams. 1928b. Thermophilic flora of sugar in its relation to canning. J. Bacteriol. 15:31.
5. Cameron, E. J., and J. Yesair. 1931a. About sugar contamination: Its effect in canning corn. Canning Age 12:239.
6. Cameron, E. J., and J. Yesair. 1931b. Canning tests prove presence of thermophiles in sugar. Food Indus. 3:265.
7. Campbell, L. L., Jr., H. A. Frank, and E. R. Hall. 1957. Studies on thermophilic sulfate-reducing bacteria. I. Identification of *Sporovibrio desulfuricans as Clostridium nigrificans*. J. Bacteriol. 73:516.
8. Campbell, L. L., and J. R. Postgate. 1956. Classification of the sporeforming sulfate-reducing bacteria. Bact. Rev. 29:359.
9. Department of Defense. 1985. Military standard. Bacterial standards for starches, flours, cereals, alimentary pastes, dry milks, and sugars used in preparation of thermostabilized foods for the Armed Forces, MIL-STD-900C. U.S. Dep. of Defense, Washington, D. C.
10. Donnelly, L. S., and F. F. Busta. 1980. Heat resistance of *Desulfotomaculum nigrificans* spores in soy protein infant formula preparations. Appl. Environ. Microbiol. 40:721.
11. Donnelly, L. S., and F. F. Busta. 1981 Alternative procedures for the enumeration of *Desulfotomaculum nigrificans* spores in raw ingredients of soy protein-based products. J. Food Sci. 46:1527.
12. NCA Research Laboratories. 1968. Laboratory manual for food canners and processors, vol. 1, p. 104. Natl. Canners Assoc. (now Natl. Food Processors Assoc.) AVI Publ. Co., Inc., Westport, Conn.
13. AOAC. 1972. Official first action: Detecting and estimating numbers of thermophilic bacterial spores in sugars. J. Assoc. Off. Anal. Chem. 55:445.
14. Sneath, P. H. A. 1986. Endospore-forming gram-positive rods and cocci, p. 1200. *In* P. H. A. Sneath, N. S. Mair, M. E. Sharpe, and J. G. Holt (eds.), Bergey's manual of systematic bacteriology, vol. 2. Williams and Wilkins, Baltimore, Md.
15. Speck, R. V. 1981. Thermophilic organisms in food spoilage: Sulfide spoilage anaerobes. J. Food Prot. 44:149.
16. Starkey, R. L. 1938. A study of spore formation and other morphological characteristics of *Vibrio desulfuricans*. Arch. Mikrobiol. 9:268.
17. Werkman, C. H. 1929. Bacteriological studies on sulfide spoilage of canned vegetables. Iowa Agric. Exp. Sta. Res. Bull. 117:163.
18. Werkman, C. H., and H. J. Weaver. 1927. Studies in the bacteriology of sulfur stinker spoilage of canned sweet corn. Iowa State College J. Sci. 2:57.

Chapter 28

Investigation of Foodborne Illness Outbreaks

Irving Weitzman, Oliver D. Cook, and Jeffrey P. Massey

28.1 INTRODUCTION

Outbreaks associated with foodborne illness are a serious public health and economic problem in the United States. Current estimates suggest that each year, between 6.5 and 33 million persons become ill from foodborne diseases and up to 9,000 of these individuals will die.[4,5] The actual incidence of foodborne illness, however, may be 10 to 200 times higher than these numbers because many afflicted persons do not seek medical treatment, or reports are not made to health agencies by the victims or by the medical care providers. The estimated cost of foodborne disease is between $5 and $34.9 billion annually. These costs include direct medical care, lost wages and productivity, investigational costs, and industry losses through embargo, voluntary destruction, and recall of the products involved.[12]

The Centers for Disease Control and Prevention has defined a foodborne outbreak as the occurrence of two or more cases of a similar illness resulting from the ingestion of a common food. Between 1988 and 1992, 2,423 outbreaks of foodborne illness were reported to state or territorial public health departments. Laboratory identification of the causative agent provides invaluable assistance to the epidemiologists investigating the outbreak. Unfortunately, the etiologic agent was only identified in 41% of these outbreaks. Bacterial pathogens were responsible for 79% of the outbreaks in which a pathogen is identified. Chemical agents (14%), viruses (4%), and parasites (2%), were identified in outbreak situations to a much smaller degree.[5]

Identification of foodborne outbreaks and the etiologic agent are critical since select populations are at a greater risk for serious complications following infection with certain pathogens. Those particularly at risk include children under the age of five years, the elderly and immuno-compromised individuals. In these individuals, foodborne pathogens may lead to complications in organ systems other than the gastrointestinal tract. *Eschericia coli* O157:H7 is a leading cause of hemolyticuremic syndrome in children,[1] salmonellosis may lead to reactive arthritis or invasive disease,[2] and campylobacteriosis may result in Guillain-Barre syndrome.[17]

The epidemiology of foodborne outbreaks has changed significantly since the mid-1990's.[1] In the 1980s and early 1990s, most outbreaks were typically local in nature. The outbreaks tended to cluster both temporally and geographically.

Examples included a milkborne outbreak of salmonellosis centered in Chicago in 1985, which involved more than 16,000 confirmed cases. Since 1995, reports of multi-state outbreaks include consumption of alfalfa sprouts contaminated with *E. coli* O157:H7,[6] listeriosis associated with consumption of hot dogs,[8] and imported Guatemalan raspberries contaminated with *Cyclospora cayetanensis*.[9,24,25]

This chapter will explore the general strategies involved with investigation of food borne outbreaks. These investigations require the collaboration of epidemiologists and laboratorians who work closely together to identify both the causative agent and the mechanism of transmission. Once this information is obtained, a strategy can then be implemented to control the current outbreak and prevent similar outbreaks from occurring in the future.

28.2 PURPOSE OF INVESTIGATION

Surveillance and investigation of foodborne outbreaks have three general purposes:

1. Prevention of further illness by identifying the offending product and stopping further distribution. This step has the most direct and immediate effect on public health. Actions involved could include embargo of the product, recall, or closure of the food processing facility or food service facility.
2. Identification of the causative agent and its source. Once the causative agent is identified, public health agency can make specific recommendations for patient treatment. In addition, preventive measures, such as providing immune globulin to persons exposed to hepatitis A can be instituted. Information can also be used to prevent future outbreaks. For example, baked potatoes were not considered a hazardous food until they were involved in outbreaks of botulism; now they must be handled like any other hazardous food, with appropriate time and temperature control. Similarly, based on outbreak data, advice is given against the use of galvanized containers for storage and serving of acidic foods because of the possibility of heavy metal leaching, which may lead to poisoning.
3. Assessment of trends in prevalence of foodborne pathogens and vehicles of disease transmission. This information is critical to state and local public health departments involved in the implementation of food-protection programs.

28.3 THEORY OF INVESTIGATION

The investigation of foodborne outbreaks is based on the principles of epidemiology, the study of the factors determining the occurrence of disease in populations.[14] An understanding of these factors can lead to the development of effective methods for disease prevention.

An epidemiological investigation of a foodborne outbreak involves interviews, sample collection, and data analysis. Each of these steps can provide information vital to demonstrating the relationship between a food and a disease. Inferences can be drawn about the relationship from the following elements:

- Signs and symptoms consistent with the disease occurring in the exposed population;
- Clinical specimens to confirm the diagnosis;
- Demonstration that ill individuals were more likely to have consumed the food product than individuals who did not become ill;
- Presence of the organism or toxin in the food product in a concentration sufficient to have caused the disease; and
- Evidence of contamination and/or mishandling of the product that resulted in the presence of the organism or toxin.

Outbreaks may be divided into either one of known etiology or one of unknown etiology. When the etiologic agent has been identified, an intervention policy can be initiated based upon the known clinical feature of that organism (Table 1). When the etiologic agent is not identified, the outbreak is divided into four subgroups based on the incubation period of the illness. These subgroups are: less than one hour (probable chemical poisoning), 1 to 7 hours (e.g., *Staphylococcus aureus* food poisoning), 8 to 14 hours (e.g., *Clostridium perfringens* food poisoning), and ≥ 15 hours (e.g., *Salmonella*). Epidemiologists investigating outbreaks of unknown etiology must first attempt to determine an epidemiologic association with a food product and then implement an appropriate intervention strategy.

Although a food product may be epidemiologically associated with a disease outbreak, isolation of the suspected pathogen from the food source serves to confirm this linkage. Laboratory identification relies upon microbial analysis of samples, including clinical specimens and leftover food sample products. The development of molecular based procedures such as the polymerase chain reaction (PCR) and pulsed field gel electrophoresis (PFGE) have provided a valuable tool in the investigation of foodborne outbreaks. These techniques are described in greater detail in Chapter 11.

28.4 CONSIDERATION OF SAMPLES

Identification of the causative agent of a foodborne outbreak depends heavily on the collection of adequate samples and the appropriate handling of those samples. Although these steps are not under the direct control of the laboratory analyst, ongoing communication between the analyst and the epidemiologist or investigator can lead to a better understanding of sampling needs and, thus, a greater likelihood of identifying the causative agent. For example, it is usually helpful to collect not only implicated prepared foods, but also all remaining ingredients. The ingredient samples can provide a "baseline," or else reveal that the contamination occurred at a stage earlier than food preparation stage. Sometimes, all that remains is leftovers in the garbage, or product wrappers, but even those can be important: botulism toxin was confirmed from the wrapper of a patty melt sandwich associated with an outbreak in Peoria, IL.[8] At the time of collection, available information may be insufficient to implicate a specific food. Numerous samples must then be collected. Information that becomes available during an investigation should be then considered in determining which samples should be analyzed.

Rapid handling and appropriate temperature control of samples are important considerations in investigating foodborne outbreaks. Appropriate temperature control of the samples is essential to maintain the integrity of samples. Freezing may lead to destruction of the causative organism, while lengthy refrigeration may permit overgrowth by competing organisms. The appropriate conditions should be based on the hypothesis of the causative organism or toxin, as well as the nature of the food (e.g., ice cream would normally be handled in the frozen state).

When the laboratory receives the samples (or is notified that they are en route), decisions must be made concerning the appropriate tests to perform. The epidemiologist may have developed a hypothesis as to the cause of the outbreak and thus may request a specific test. Alternatively, knowledge of the symptoms exhibited by persons involved in the outbreak, and/or of the nature of the suspect food, can be used in the decision process. For example, sudden onset of vomiting and diarrhea 2 to 4 hr after a meal that included ham or potato salad may indicate staphylococcal intoxication. Table 1 contains information concerning incubation periods, predominant symptoms, specimens to analyze, and the organism or toxin involved.

One aspect that is sometimes overlooked is the effect of processing on the presence or absence of an organism or toxin. As an example, many heat processes will destroy *Staphylococcus aureus* but not its toxin. Thus, if the organism had the opportunity to multiply in food that was later cooked, the organism would not be identified through standard culture techniques. A test for staphylococcal toxin, however, would be appropriate.

28.5 EQUIPMENT, MATERIALS, AND SOLUTIONS

The primary tools used in investigating a foodborne illness outbreak are questionnaires, sample containers, and inspection equipment.

28.51 Questionnaires

Questionnaires are used by public health professionals to systematically collect specific information about an outbreak. Relevant data are important in developing a hypothesis concerning the source of the outbreak. Because individual state and local health departments usually develop and maintain standard questionnaires, they will not be described in detail in this chapter.

Generally, a questionnaire is used to obtain case history information from the first few individuals identified: demographic data such as address, age, and sex; signs and symptoms; other related medical information such as allergies and physician involvement; and contacts with other ill persons. A 72-hr food history is obtained to gather information to include the incubation period for most foodborne illnesses. When an illness with a long incubation period is suspected, such as listeriosis or hepatitis A, it would be necessary to attempt to obtain food preference history information for up to 2 months before onset. As the investigation expands, and there is a need to systematically collect data concerning hypothesized exposures, the investigator may develop a questionnaire specific to the outbreak to ensure that the same questions are asked of both ill and well persons who had the same opportunity for exposure (i.e., a case/control study). The questionnaire will usually include a list of specific items available at the impli-

Table 1. Guide for Laboratory Tests Indicated by Certain Symptoms and Incubation Periods[a]

Incubation Periods	Predominant symptoms	Specimens to analyze	Organism, toxin, or toxic substances
Upper gastrointestinal tract symptoms (nausea, vomiting) occur first or predominate			
5 min - 8 hrs, usually 10-45 mins	Nausea, vomiting, metallic taste, burning of mouth	Vomitus, urine, blood, stool	Heavy metals[b] (antimony, cadmium, copper, iron, tin and zinc)
1 to 2 hr	Nausea, vomiting, cyanosis, headache, dizziness, dyspnea, trembling, weakness, loss of consciousness	Blood	Nitrites[c]
30 min to 8 hrs, usually 2 to 4 hrs	Nausea, vomiting, retching, diarrhea, abdominal pain, prostration	Vomitus, stool	*Staphylococcus aureus* and its enterotoxins
1 to 6 hr	Vomiting, abdominal cramps, diarrhea, nausea, fever uncommon	Vomitus, stool	*Bacillus cereus* emetic toxin
6 to 24 hr	Diarrhea and abdominal cramps for 24 hrs followed by hepatic and renal failure	Detection of toxin in food	Amanita mushrooms[d]
Sore throat and respiratory symptoms			
1 to 4 days	Sore throat, fever, scarlet fever, upper respiratory infection	Throat swab	*Streptococcus* Group A
2 to 5 days	Inflamed throat and nose, spreading grayish exudate, fever, chills, sore throat, malaise, difficulty in swallowing, edema of cervical lymph node	Throat swabs, blood	*Corynebacterium diphtheriae*
Lower gastrointestinal tract symptoms (abdominal cramps with watery diarrhea) occur first or predominate			
4 to 30 hrs	Diarrhea, abdominal cramps	Stool	*Vibrio parahaemolyticus*
6 to 24 hrs	Diarrhea, abdominal cramps; vomiting and fever are uncommon	Stool	*Clostridium perfringens*
6 to 24 hrs	Diarrhea, abdominal cramps and vomiting in some patients, fever uncommon	Stool	*Bacillus cerus* diarrheal toxin
6 to 48 hrs	Diarrhea, abdominal crampls, and nausea; vomiting and fever are less common	Stool	Enterotoxigenic *Escherichia coli*
Variable	Diarrhea, fever, abdominal cramps	Stool	Enteropathogenic *Escherichia coli*
6 hrs to 10 days, usually 6 to 48 hrs	Diarrhea, often with fever and abdominal cramps	Stool	Nontyphoidal *Salmonella*
8 to 22 hr, usually 10 to 12 hr	Abdominal cramps, diarrhea, putrefactive diarrhea associated with *C. perfringens*	Stool	*Enterococcus faecalis* *Enterococcus faecium*
15 to 77 hours, usually 24 to 48 hrs	Nausea, vomiting, diarrhea, abdominal cramps, usually no fever	Stool	Norwalk family of viruses, small round-structured viruses (SRSV), Astrovirus Calicivirus
1 to 5 days	Watery diarrhea, often accompanied by vomiting	Stool	*Vibrio cholerae* O1 or O139
1 to 5 days	Watery diarrhea	Stool	*Vibrio cholerae* non-01 or non-O139
1 to 10 days	Diarrhea, abdominal pain (often severe)	Stool	*Yersinia enterocolitica*
1 to 11 days, median 7 days	Fatigue, protracted diarrhea, often relapsing	Stool	*Cyclospora cayetanensus*
2 to 28 days, median 7 days	Diarrhea, nausea, vomiting, fever	Stool	*Cryptosporidium parvum*
3 to 25 days, median 7 days	Diarrhea, gas, cramps, nausea, fatigue	Stool	*Giardia lamblia*
12 to 74 hr, usually 18 to 36 hr	Abdominal cramps, diarrhea, vomiting, fever, chills, malaise	Stool	*Aeromonas hydrophila*, *Enterobacteriaceae*, *Pseudomonas aeruginosa* (?), *Plesiomonas shigelloides*

Continued

Table 1. Guide for Laboratory Tests Indicated by Certain Symptoms and Incubation Periods[a]—*Continued*

Incubation Periods	Predominant symptoms	Specimens to analyze	Organism, toxin, or toxic substances
Lower gastrointestinal tract symptoms (abdominal cramps with watery diarrhea) occur first or predominate, *continued*			
3 to 6 months	Nervousness, insomnia, hunger pains, anorexia, weight loss, abdominal pain, sometimes gastroenteritis	Stool	*Taenia saginata, Taenia soliurn*
Unknown	Diarrhea, abdominal cramps, fever	Stool	*Listeria monocytogenes*
Lower gastrointestinal tract symptoms (abdominal cramps with bloody diarrhea) occur first or predominate			
12 hrs to 6 days, usually 2 to 4 days	Diarrhea (often bloody), frequently accompanied by fever and abdominal cramps	Stool	*Shigella* spp.
1 to 10 days, usually 3 to 4 days	Diarrhea (often bloody), abdominal cramps (often severe), little or no fever	Stool	Enterohemorrhagic *E. coli* (*E. coli* O157:H7 and others)
Variable	Diarrhea (may be bloody), fever, abdominal cramps	Stool	Enteroinvasive *E. coli*
2 to 10 days, usually 2 to 5 days	Diarrhea (often bloody), abdominal pain, fever	Stool	*Campylobacter jejuni* *Campylobacter coli*
1 to several weeks, usually 3 to 4 weeks	Abdominal pain, bloody diarrhea and pus, constipation, headache, drowsiness, ulcers, variable-often asymptomatic	Stool	*Entamoeba histolytica*
Neurological symptoms (visual disturbances, vertigo, tingling, paralysis)			
10 min to 3 hrs, usually 10 to 45 min	Parasthesia of lips, tongue, face, or extremities, often following numbness; loss of proportion or "floating" sensation	Implicated food	Tetraodon toxin (puffer fish poisoning)
30 min to 3 hrs	Parasthesia of lips, mouth or face, and extremities; intestinal symptoms or weakness, including respiratory difficulty	Implicated food or water	Paralytic or neurotoxic shellfish poison
Less than 1 hr	Gastroenteritis, nervousness, blurred vision, chest pain, cyanosis, twitching, convulsions	Blood, urine, fat biopsy	Organic phosphate insecticides
Less than 2 hrs	Usually vomiting and diarrhea, other symptoms differ with toxin:	Urine	Various mushrooms:
	Confusion, visual disturbance		Muscimol
	Salivation, diaphoresis		Muscarine
	Hallucinations		Psilocybin
	Disulfiram-like reaction		*Coprinus artrementaris*
	Confusion, visual disturbance		*Ibotenic acid*
1 to 6 hr	Nausea, vomiting, tingling, dizziness, weakness, anorexia, weight loss, confusion	Blood, urine, stool, gastric washings	Chlorinated hydrocarbons (insecticides)
1 to 48 hrs, usually 2 to 8 hrs	Usually gastrointestinal symptoms followed by neurological symptoms (including paresthesia of lips, tongue, throat, or extremities) and reversal of hot and cold sensation	Implicated food	Ciguatera toxin
2 hrs to 8 days, usually 12 to 48 hrs	Vertigo, double or blurred vision, loss of reflex to light, difficulty in swallowing, speaking, and breathing, dry mouth, weakness, respiratory paralysis	Blood, stool	*Clostridium botulinum* and its neurotoxins
12 to 36 hrs	Infant botulism, afebrile symmetric descending flaccid paralysis	Stool	*Clostridium botulinum* and its neurotoxins
More than 72 hrs	Numbness, weakness of legs, spastic paralysis, impairment of vision, blindness, coma	Urine, blood, stool, hair	Organic mercury

Continued

Table 1. Guide for Laboratory Tests Indicated by Certain Symptoms and Incubation Periods[a]—*Continued*

Incubation Periods	Predominant symptoms	Specimens to analyze	Organism, toxin, or toxic substances
Neurological symptoms (visual disturbances, vertigo, tingling, paralysis), *continued*			
More than 72 hrs	Gastroenteritis, leg pain, ungainly high stepping gait, foot and wrist drop		Triorthocresyl phosphate
2 to 6 weeks	Gastroenteritis, meningitis, neonatal sepsis, fever	Cerebral spinal fluid, blood	*Listeria monocytogenes*
Allergic symptoms (facial flushing, itching)			
1 min to 3 hrs, usually < 1 hr	Headache, dizziness, nausea, vomiting, peppery taste, burning of mouth and throat, facial swelling and flushing, stomach pain, itching of skin	Vomitus	Histamine[e] (Scombroid toxin)
3 mins to 2 hrs., usually < 1 hr	Burning sensation in chest, neck, abdomen, or extremities; sensation of lightness and pressure over face or heavy feeling in chest		Monosodium glutamate (usually food containing ≥1.5 g MSG)
Less than 1 hr	Flushing, sensation of warmth, itching, abdominal pain, blood, puffing of face and knees		Nicotinic acid
Generalized infection symptoms (fever, chills, malaise, prostration aches swollen lymph nodes)			
12 to 36 hrs	Primary septicemia, hypotensive shock bulbulous skin lesions	Blood	*Vibrio vulnificus*
1 to 2 days, for intestinal phase 2 to 4 weeks, for systemic phase	Gastroenteritis, fever, edema about eyes, perspiration, muscular pain, chills, prostration, labored breathing	Muscle biopsy	*Trichinella spiralis*
3 to 60 days, usually 7 to 14 days	Malaise, headache, fever, cough, nausea, vomiting, constipation, abdominal pain, chills, rose spots, bloody stools	Stool, blood	*Salmonella typhi*
10 to 13 days	Fever, headache, myalgia, rash	Lymph node biopsy, blood	*Toxoplasma gondii*
15 to 50 days, median 28 days	Jaundice, dark urine, fatigue, anorexia, nausea	Urine, blood	Hepatitis A virus
Several days to several months, usually > 30 days	Fever, chills, head- or joint ache, prostration,malaise, swollen lymph nodes, and other specific symptoms of disease in question	Blood, stool, urine, sputum, lymph node, gastric washings (one or more, depending on organism)	*Bacillus anthracis, Brucella melitensis, Brucella abortus, Brucella suis, Coxiella burnetii, Francisella tularensis, Mycobacterium spp., Pasteurella multocida, Streptobacillus moniliformis*

[a] Modified From APHA, "Compendium of Methods for the Microbiological Examination of Foods," 2nd ed. (1984) and "Guidelines for Confirmation of Foodborne-Disease Outbreaks", MMWR vol.45/No. SS-5, (1996).

[b] Consider chemical tests for such substances as zinc, copper, lead, cadmium, arsenic, or antimony.

[c] Test for discoloration of blood.

[d] Identify mushroom species eaten, test urine and blood for evidence of renal damage (SGOT, SGPT enzyme tests).

[e] Scrombroid poisoning should be considered. Examine foods for *Proteus* species or other organisms capable of decarboxylating histidine into histamine and for histamine.

cated meal, restaurant, etc., as well as symptoms, onset, time, demographic information, and contact identification.[27]

The agency may also use laboratory requisition reports for food and patient samples; these forms contain a description of the sample collected and a designation of the desired analysis. The laboratory may use the same or a different one to report the results of its analysis back to the health department personnel conducting the investigation.

28.52 Sample and Laboratory Specimen Containers

The second category of tools used for the investigation is laboratory specimen containers. For food sample specimens, these would include sterile plastic bags, jars, or other containers, as well as sterile sampling equipment such as swabs and spoons. Sterilizing agents, such as a propane torch, may be important if access to sterile equipment is difficult. Sufficient insulated containers and refrigerants for sample transport are also important. For patient specimens, stool collection kits, swabs, blood collection tubes, and equipment for transporting them is vital. Frequently, the laboratory will be responsible for providing specimen collection and transport equipment to the epidemiologist/investigator. Occasionally patients will collect specimens in their own containers. As these containers are not sterile, this should be discouraged.

28.53 Inspection Equipment

Inspection equipment is important for the follow-up at the location where the suspect food was prepared. In addition to questionnaires and sample containers, the investigator may use equipment such as a thermometer for measuring the temperature of the food at various stages of preparation and serving.

28.6 INVESTIGATIVE PROCEDURES

The procedures to investigate foodborne outbreaks are described in significant detail in *Procedures to Investigate Foodborne Illness*, 5th edition, published by the International Association of Milk, Food and Environmental Sanitarians, Des Moines, IA.[14] The major steps in conducting an investigation of a foodborne outbreak are outlined below.

28.61 Receiving Notification of Illness

A public health agency may identify the possibility of a foodborne outbreak from information developed from numerous sources, such as individuals who believe their illness was caused by a particular food or meal, a physician who treated a patient for what appears to be a foodborne illness, a laboratory analysis, a school nurse, or a poison control center. The agency evaluates the information in the context of its experience with foodborne illness, and other information at hand, such as the information on a case history form or complaint log. Whether or not a complaint or possible outbreak is investigated by the agency depends on guidelines for follow-up established by the health officer and the local government.

28.62 Case Histories

The next step in following up a foodborne outbreak is to obtain case history information for each ill individual. The investigator will usually obtain a 72-hr food history but will focus especially on the time period indicated by the predominant symptoms (e.g., most recent foods eaten if vomiting is the predominant symptom, or food eaten 6 to 20 hr before onset of cramps and diarrhea).

28.63 Verifying the Diagnosis

Since the clinical signs of many infections are often nonspecific, establishment of the identity of the causative agent is an essential component of the investigation. A variety of specimens may be obtained from individuals associated with the outbreak. These specimens include blood, stool, vomitus, throat swabs, urine or other body fluids. The presence of the causative agent can be confirmed through culture techniques, direct observation by electron microscope, serology, or polymerase chain reaction (PCR). Identification of the causative agent allows the epidemiologist investigating the outbreak to evaluate modes of transmission consistent with the pathogen.

Additional medical information is sought from the attending physician or health professional. Clinical specimens should be obtained as soon as possible to enhance the likelihood of identifying the causative organism or toxin. The specimens collected should be appropriate to the disease suspected or incubation period. For example, a blood sample would probably not be meaningful if the only symptom was diarrhea (with no fever or associated symptoms).

Clinical specimens may be collected from the first 10 to 20 people who appear to be associated with the outbreak to identify the causative organism or toxin and show that it is associated with the outbreak rather than a chance occurrence in one or two patients. The number of specimens needed will depend on whether the outbreak is typical, whether the suspect organism is easy to grow, or whether it is an unknown organism. Once the organism or toxin is positively identified, the information should be made available to physicians so that they can appropriately treat later cases.

28.64 Food samples

Leftover food samples, ingredients, and other evidence should also be collected during the initial visit to the ill person, or as soon as possible after the illness is reported. With leftovers, the sample size will be limited by the amount available, but ingredients may be available in sufficient amounts for analysis. The investigator collecting the food samples should record the conditions under which the food was being held (such as temperature), as well as the conditions for shipment to the laboratory. It is vital that these samples be analyzed as quickly and accurately as possible so that appropriate public health measures can be taken. If analysis reveals that an ingredient is contaminated, the population at risk may be much larger than if the contamination occurred on a one-time basis in a home or at one meal. The levels of contaminant in an ingredient may not be high enough to have directly caused the outbreak. However, the ingredient may have been the source, with mishandling of the product during or after processing resulting in the outbreak.

28.65 Epidemiologic Associations

Based on information obtained initially and through case interviews, the epidemiologist or investigator must determine whether an outbreak occurred and hypothesize its cause. It is frequently difficult to determine whether an illness was caused by a food, particularly if only one or two cases are involved. Common associations between people are a means of identifying outbreaks and developing a hypothesis.

There are three categories of associations: time, place, and person.

- Time refers to when the onset of illness occurred, in a common-source outbreak, onset times will cluster within a few hours or days (depending on the normal incubation period for the disease).
- Place refers to the source of the causative food, i.e., a common meal, product or food establishment.
- Person associations refer to demographic factors such as age, sex, and ethnic or religious group of the affected people.

Once identified, the association(s) can be used to develop a hypothesis concerning the illness, the food involved, and how and where the food became contaminated. Using a specific questionnaire to interview additional persons who share the association(s) helps prove or disprove the hypothesis. It is important to interview both ill and well persons who had the opportunity for the same exposures to refine the hypothesis and, for example, narrow the suspect vehicle from an entire meal to just one of the foods served. If the outbreak is large, interviews may be limited to a random sample, but generally it is helpful to interview 2 or 3 uninfected individuals (controls) for every ill individual (cases) interviewed. Clinical specimen can also be collected as described above.

28.66 Food Processing Investigation

An investigation is also made to determine how the food(s) became contaminated and/or allowed growth of pathogens. The investigation would start at the associated location and include a thorough review of how the product or meal in question was prepared. Other parameters that should be investigated include the source of ingredients, the times and temperatures involved in the preparation, opportunities for cross-contamination, employee practices and employee health status.

Frequently, differences in the process from normal procedures can provide valuable clues as to the source of the problem. It is sometimes difficult to obtain accurate information concerning an incident, particularly if mistakes were made and punishment is feared. The investigator would look for a situation in which such a food was cooked, then either cooled too slowly, stored at temperatures in the critical range (45° to 140°F), or reheated inadequately. The nature of the food and handling problems involved in the incident may indicate the type of contamination or whether the causative organism or toxin is likely to have been present in the suspect food. For example, *Bacillus cereus* foodborne illness is generally associated with starchy foods that have been heat-shocked, such as rice or dry beans, rather than with meat and poultry. Therefore, if the causative toxin in an outbreak was *B. cereus* enterotoxin, the investigation would emphasize starchy foods and/or possible cross-contaminations to other types of food. Nevertheless, the possibilities of new vehicles and/or organisms should always be kept in mind, and the investigation kept broad enough to uncover these.

28.67 Major Processing Defects

Several major causative factors occur repeatedly in foodborne outbreaks. During 1992, 407 outbreaks involing 11,015 cases of foodborne diseases were reported to the CDC.[5] The most common causative factors were:

- Improper holding temperature
- Inadequate cooling
- Inadequate cooking
- Poor personal hygiene on the part of food handlers
- Contaminated equipment

Yersinia enterocolitica, *Listeria monocytogenes*, and other bacteria which can survive and grow at refrigerator temperatures are increasingly recognized as foodborne pathogens. Thus, it is clear that refrigeration alone is not necessarily an adequate control measure for food preservation.

Poor personal hygiene is a significant contributing factor in foodborne outbreaks. In addition to their role in food processing, investigators must interview food workers concerning their personal hygiene habits. Particular attention must be paid to their hand washing habits and whether bare hands are used to process uncooked foods such as salads or sandwiches.

28.7 ANALYSIS AND CONCLUSION

28.71 Data Analysis

The data gathered from persons interviewed about the outbreak are analyzed to determine whether the hypothesis about the cause of the outbreak is true. Parameters that must be assessed include identification of the predominant symptoms, determination of the epidemic curve, determination of the incubation period, and calculation of the attack rate.

28.72 Predominant Symptoms

The predominant symptoms reported by ill persons must be identified and enumerated. The number of persons reporting a symptom is counted and compared to the total number of ill persons to obtain a percentage. The relative order of prevalence of symptoms is characteristic of a particular illness; for example, in an outbreak of *Clostridium perfringens* gastroenteritis, diarrhea would be the most prevalent symptom, with a few or no persons reporting vomiting or fever. If clinical specimens are unavailable or have not yet been analyzed, the predominant symptom data will suggest whether the illness is an infection or intoxication, which in turn will suggest possible food sources and laboratory tests to be run (Table 1).

28.73 Epidemic Curve

Data identifying the time of onset of symptoms permits the plotting of an epidemic curve. The number of persons reporting illness is plotted against time. The scale of the curve is dependent on the "time span" involved in the outbreak; if all onsets occurred within a few hours, then the scale would be hours instead of days or weeks. The shape of the curve produced is dependent on whether the outbreak occurred from a "point-source" (a single meal, for example) or from person-to-person contact. A point-source outbreak curve will rise sharply to a peak, then taper off more gradually. The length of the curve will approximate the range of the incubation period of the disease. A person-to-person outbreak will show a much slower rise and a length extending over several incubation periods.

28.74 Attack Rate

In most foodborne outbreaks, the data gathered will show that some individuals ingesting a particular food became ill, while others did not. This disparity is due to factors such as individual resistance, the amount of product consumed, and nonhomogeneity of the organism or toxin within the product. Some of those who reportedly did not eat that product will also report illness. This may be coincidental, in sympathy with those who became ill, or because the person forgot having eaten the particular food. Thus, the association of a particular food with illness usually is not 100%, and statistical methods are used to determine whether the association is significant. The starting point for this method is to determine the food-specific attack rates.

For each food served at a suspect meal, the percentage of persons who ate that food and became ill—the attack rate—is compared to the percentage of persons who did not eat the food and became ill. These two percentages are compared for each of the foods, and the food showing the largest difference in percentages becomes the suspect food in the outbreak. Another statistic used to show measurements of disease association is relative risk. Relative risk is the ratio of illness in the persons eating the food to those who did not eat the suspected food. This measurement is called the odds ratio in case control studies.

The statistical significance of attack rates, relative risk or odds ratio is determined by calculating the probability of this difference having occurred by chance alone. Either the 'chi-square' or Fisher's exact test is used to make this statistical determination. Probabilities of less than 5% ($p<0.05$) are considered statistically significant since such differences are not likely to occur by chance alone. Should more than one food turn out to be statistically significant, further data analysis must be done. This problem may occur, for example, with foods commonly eaten together, such as meat and gravy.

28.75 Reports

The final element in investigating a foodborne outbreak is to write a report. This report will itemize the pertinent information during the investigation, such as the number of exposed and affected infividuals, the clinical picture, statistical information (attack rates, relative risk, etc.), causative agent and intervention taken. In addition to documenting the investigation, a timely report can be a valuable means of disseminating knowledge about outbreaks and their causes and thus can aid in reducing or preventing future outbreaks. The report is issued by the agency which investigated the outbreak and is directed toward the health care organizations responsible for intervention strategies.

28.8 MOLECULAR SUBTYPING TECHNIQUES—BACTERIA

The National Food Safety Initiative was established by President Clinton in 1997 to serve as a nationwide early-warning system to improve the safety of the U.S. food supply. An integral part of this enhanced surveillance system for foodborne illness is a multistate system of state public health laboratories which provide DNA fingerprinting of a select group of foodborne bacterial pathogens. This system is known as PulseNet and includes establishment of a centralized national electronic database of DNA fingerprint patterns. PulseNet uses a standardized PFGE technique that allows consistency among the several state public health laboratories that are participants. The CDC serves as a repository for electronic images of DNA fingerprint patterns collected across the nation. The simultaneous emergence of identical patterns for a specific pathogen, such as *E. coli* O157:H7, from different states is an indication of food contamination from a common source prior to distribution.

28.81 Pulsed Field Gel Electrophoresis

PFGE is a DNA fingerprinting technique that allows the comparison of genetic material from different isolates of the same bacteria.[16,22] Chromosomal DNA from bacterial isolates is isolated and cut into several pieces by restriction enzymes. These pieces of DNA are separated by agarose gel electrophoresis. Electrophoresis allows separation of the DNA pieces based on size. The smallest pieces move quickly to the bottom of the gel, while the large pieces are retained at the top. The pieces are then separated as distinct bands on the gel. This resulting band pattern is referred to as a "fingerprint." Refer to Chapter 11 for a more thorough description of this technique.

The basic assumption of this technique is that genetically related organisms will share an identical DNA fingerprint, while unrelated organisms will have unique patterns. Thus, organisms derived from the same point source will share the identical pattern.

28.82 Practical Applications

PFGE has been utilized on several occasions since 1995 to support the investigation of reported outbreaks of foodborne illness. In the two outbreaks summarized below, the presence of several isolates with identical PFGE patterns allowed investigators to identify the source of the outbreak and implement appropriate intervention strategies.

28.821 *Listeria monocytogenes* Contamination of Hot Dogs and Deli Meats

Between August 1998 and March 1999, CDC identified more than 90 individuals with listeriosis in 22 states. These individuals were all infected with an identical strain of *Listeria monocytogenes*. This strain was linked to production lots of hot dogs and deli meats produced at one meat processing plant. PFGE analysis of isolates from a previously unopened package of hot dogs and from patients with listeriosis has identified that a common outbreak pattern (designated "E") is responsible for the outbreak. Cases were defined as patients from whom the outbreak pattern "E" or the outbreak sub-patterns "E1" and "E2" of *L. monocytogenes* had been isolated (Figure 1). At least 14 adult deaths and 6 fetal deaths occurred nationwide following infection with the outbreak strain of *Listeria monocytogenes*. Several of the infected individuals recalled eating implicated product(s). After the source of the contamination was identified, the meat processing plants shut down operations and the suspected food was recalled. This investigation demonstrated the value of molecular genotyping for investigations of foodborne outbreaks.[8]

28.822 *E. coli* O157:H7 Contamination of Alfalfa Sprouts

In June and July of 1997, concurrent outbreaks of *E. coli* O157:H7 were reported in Michigan and Virginia. A total of 139 individuals from both states became ill with this pathogen during this period. PFGE analysis of these isolates revealed that 85 cases shared an identical PFGE pattern. Isolates of *E. coli* O157:H7 that were not epidemiologically linked to the outbreak had distinct PFGE patterns. This indicated that the outbreak was the result of a common source, rather than an increase due to sporadic disease. Furthermore, PFGE analysis indicated that the predominant pattern observed among isolates from Virginia was identical to the pattern observed in Michigan. An ensuing epidemiologic investigation showed that outbreak resulted from ingestion of alfalfa

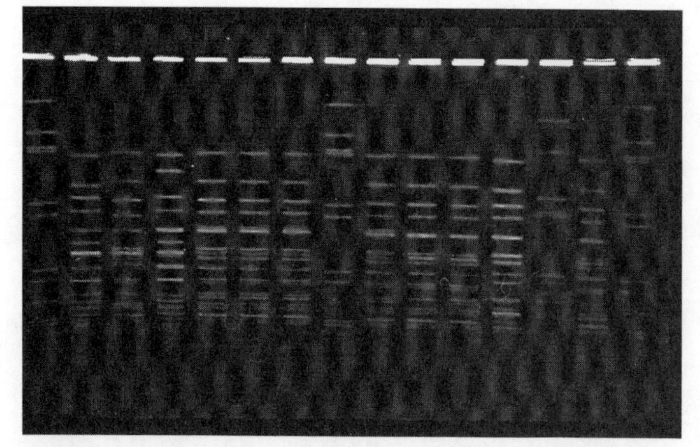

Figure 1. *Analysis of* Listeria monocytogenes *by Pulsed Field Gel Electrophoresis. Chromosomal DNA was digested with the* Apa I *restriction endonuclease. Lane M,* E. coli *strain H2446 was utilized as molecular weight size markers. Lanes 4, 5, 6, and 7 are "E" outbreak patterns, Lane 8 is the "e-1" outbreak sub-pattern, and lane 9 is the E-2" outbreak sub-pattern. Each of these isolates was obtained from individuals who reported consumption of implicated food products. Lanes 1,2,3, 10, 11, and 12 contain DNA from isolates of patients with no epidemiologic link to the implicated food products.*

sprouts grown from one common lot of seed harvested in Iowa. The use of PFGE assisted epidemiologists in confirming that the outbreaks in these two states were linked by a common strain of *E. coli* O157:H7.[6]

28.9 MOLECULAR SUBTYPING TECHNIQUES—VIRUSES

28.91 Polymerase Chain Reaction

Another molecular biology tool that can be utilized in the investigation of outbreaks of foodborne illness is the polymerase chain reaction (PCR). This technique permits the identification of either viral RNA or DNA in specimens obtained directly from the patient. This procedure, which is based upon the enzymatic amplification of a specific DNA or RNA sequence, has become an essential part of both the research and clinical microbiology laboratory. The literature contains numerous applications for viral, bacterial, parasitic, and fungal infections. For a review of these applications, refer to the references of Pershing[19,20] and Chapter 11.

PCR is of value in investigations of foodborne illnesses of a viral etiology since this technique does not require initial propogation of the virus in cell culture. The viruses associated with outbreak of foodborne illness such as Hepatitis A virus, Norwalk virus, and other enteric viruses (refer to Table 1), are not readily grown in cell culture. Instead, PCR is capable of identifying the viral agents directly from patient samples such as serum or plasma[28] or stool.[11] With the exception of shellfish,[10] PCR has not been utilized to detect enteric viruses from food.

In addition, viral genetic material amplified by PCR can be sequenced to determine the specific genetic content of the virus involved with the suspected outbreak.[21] As with PFGE, the presence of several viral isolates with identical genetic sequences can be used to support an hypothesis that a point source was responsible for the outbreak.

28.92 Practical Applications

In 1992, a multistate outbreak of hepatitis A virus was observed among school age children. Consumption of frozen strawberries processed at a single plant was shown to be the source of the infection. Cohort analysis of affected individuals showed a significant association between consumption of stawberry shortcake and development of illness. Hepatitis A virus was identified in the stool of five primary patients by PCR. Futhermore, PCR based sequence analysis showed that these patients from the two states involved all shared an identical HAV genotype. The viral genotype characteristic of this outbreak was distinct from other HAV genotypes not associated with the outbreak.[18] Another large, multistate outbreak of hepatitis A virus occurred in 1997. Large numbers of school children from Michigan and Maine became ill over a two month period. Frozen strawberries from a common food processing plant were implicated to be the source of outbreaks in schools in Michigan and Maine, as well as sporadic cases in Wisconsin, Arizona, and Louisiana. HAV isolates from suspected cases were shown to be identical by sequence analysis. The ensuing epidemiologic investigation established a significant association between illness and consumption of frozen strawberries prepared at a common food processing plant.[13]

28.10 CONCLUSION

A statistically significant association of a food with an outbreak does not "prove" that a particular food caused the outbreak. Additional information from the laboratory is critical in determining causation. The optimum situation is one in which samples of the food are available for analysis and are found to contain sufficient numbers of organisms, or amount of toxin, to have caused illness in persons consuming the food. Additionally, appropriate specimens from ill persons would have shown the presence of the organism or toxin. The confirmation of an identical genetic clone of an organism in both clinical and food specimens using the molecular subtyping techniques described above, strengthens the argument that the food caused the outbreak. The investigation at the food preparation site would have disclosed food sources and/or handling practices that would account for the presence of the organisms or toxin. Unfortunately, in many outbreaks, one or more of these elements is lacking, usually because of unavailability of samples or specimens.

28.11 REFERENCES

1. Altekruse, S. F., M. L. Cohen, and D. L. Swedlow. 1997. Emerging foodborne diseases. Emerg. Infect. Dis. 3(3):285-293.
2. Altekruse S. F., F. H. Hyman, K. C. Klontz, B. T. Timbo, and L. K. Tollefson. 1994. Foodborne bacterial infections in individuals with the human immunodeficiency virus. Souther Med. J. 87:169-173.
3. APHA. 1984. Compendium of methods for the microbiological examination of foods, 2nd ed., M. L. Speck (ed.) American Public Health Association, Washington, D. C.
4. Archer, D. L., and J. E. Kvenberg. 1985. Incidence and cost of foodborne diarrheal disease in the United States. J. Food Prot. 48:887.
5. Bean, N. H., J. S. Gould, C. Lao, and F. J. Angulo. 1996. Surveillance for Foodborne-Disease-Outbreaks-United States, 1988-1992. Morbid. Mortal. Wkly. Rep. 45/SS-5.
6. Breuer, T., D. H. Benkel, R. Shapiro, W. N. Hall, M. M. Winnett, M. J. Linn, J. Neimann, T. Barrett, S. Dietrich, F. P. Downes, D. Toney, J. Pearson, H. Rolka, L. Slutsker, P. M. Griffin, and the Investigation Team. 1999. A Multistate Outbreak of Diarrhea and Hemolytic Uremic Syndrome from *Escherichia coli* O157:H7 Infections Linked to Consumption of Alfalfa Sprouts Grown From Contaminated Seeds. Submitted for publication.
7. CDC. 1984. Foodbome botulism-Illinois. Morbid. Mortal. Wkly. Rep. 33(2):22.
8. CDC. 1999. Update: Multistate outbreak of Listeriosis - United States, 1998-1999. Morbid. Mortal. Wkly. Rep. 47(51 & 52):1117-1118.
9. Chambers, J., S. Somerfeldt, L. Mackey, S. Nichols, R. Ball, D. Roberts, N. Dufford, A. Reddick, and J. Gibson. 1996. Outbreaks of *Cyclospora cayetanensis* Infection—United States, 1996. Morb. Mortal. Wkly. Rep. 45(25):548-551.
10. Cromeans, T. L., O. V. Nainan, and H. S. Margolis. 1997. Detection of Hepatitis A virus RNA in oyster meat. Appl. Environ. Microbiol. 63(6):2460-2463.
11. De Serres, G., T. L. Cromeans, B. Levesque, N. Brassard, C. Barthe, M. Dionne, H. Prud'homme, D. Paradis, C. N. Shapiro, O. V. Nainan, and H. S. Margolis. Molecular confirmation of Hepatitis A virus from well water: epidemiology and public health implications. 1999. J. Infect. Dis. 179:37-43.
12. Food Safety from Farm to Table: A National Food Safety Initiative. Report to the President, May 1997. Food and Drug Administration, U.S. Department of Agriculture, U.S. Environmental Protection Agency, CDC, May 1997.
13. Hutfin, Y. J. F., V. Pool, E. H. Cramer, O. V. Nainan, J. Weth, I. T. Williams, S. T. Goldstein, K. F. Gensheimer, B. P. Bell, C. N. Shaprio, M. J. Alter, and M. S. Margolis. 1999. A multistate foodborne outbreak of Hepatitis A. New Engl. J. Med. 340:595-602.
14. International Association of Milk, Food and Environmental Sanitarians Inc. 1987. Procedures to investigate foodborne illness. Inter. Assoc. of Milk, Food and Environ. Sanitarians, Inc., Ames, Iowa.
15. Jaykus, L. 1997. Epidemiology and detection as options for control of viral and parasitic foodborne disease. Emerg. Infect. Dis. 3(4):529-539.

16. Maslow, J. N., A. M. Slutsky, and R. D. Arbeit. 1993. Application of pulsed field gel electrophoresis to molecular epidemiology, p. 563-572. *In* Diagnostic molecular microbiology: principles and applications, American Society for Microbiology, Washington, D. C.
17. Mishu B., J. Koehler, L. A. Lee, D. Rodrigue, F. Hickman-Brenner, and P. Blake. 1994. Outbreaks of *Salmonella enterititids* infections in the United States, 1985-1991. J. Infect. Dis. 169:547-552.
18. Niu, M. T., L. B. Polish, B. H. Robertson, B. K. Khanna, B. A. Woodruff, C. N. Shapiro, M. A. Miller, J. D. Smith, J. K. Gedrose, M. J. Alter, and H. S. Margolis. Multistate outbreak of Hepatitis A associated with frozen strawberries. 1992. J. Infect. Dis. 166:518-524.
19. Persing, D. H. (ed.) 1996. PCR protocols for emerging infectious diseases. American Society for Microbiology, Washington, D. C.
20. Persing, D. H., T. F. Smith, F. C. Tenover, and T. J. White. 1993. Diagnostic molecular microbiology: principles and applications. American Society for Microbiology, Washington, D. C.
21. Robertson, B.H., B. Khanna, O. V. Nainan, and H. S. Margolis. 1991. Epidemiologic patterns of wild-type Hepatitis A virus determined by genetic variation. J. Infect. Dis. 163:286-292.
22. Smith, C. L., S. R. Klco, and C. R. Cantor. 1988. Pulsed field gel electrophoresis and the technology of large DNA molecules. *In* Genome analysis: a practical approach. IRL Press, Oxford Press, England.
23. St. Louis, M. E., D. L. Morse, M. E. Potter, T. M. Demelfi, J. J. Guzewich, R. V. Tauxe, P. A. Blake, and the Salmonella enteriditis Working Group. 1988. The emergence of grade A eggs as a major source of Salmonella enteriditis infections. J. Am. Med. Assoc. 259(14):2103.
24. Sterling, C. R., and Y. R. Ortega. 1999. Cyclospora: an enigma worth unraveling. Emerg. Infect. Dis. 5(1).
25. Tauxe, R. V. 1997. Emerging foodborne diseases: an evolving public health challenge. Emerg. Infect. Dis. 3(4):425-434.
26. Tauxe, R. V. 1998. New approaches to surveillance and control of emerging foodborne infectious diseases. Emerg. Infect. Dis. 4(3)
27. Yang, S., M. G. Leff, D. McTaque, K. A. Horvath, J. Jackson-Thompson, T. Murayi, G.K. Boeselager, T. A. Melnik, M. C. Gildemaster, D. L. Ridings, S. F. Altekruse, F. J. Angulo. 1998. Multistate surveillance for food-handling, preparation, and consumption behaviours associated with foodborne diseases: 1995 and 1996 BRFSS Food-Safety Questions. Morb. Mortal. Wkly. Rep. 47(SS-4), 33-57.
28. Yotsuyanagi, H., S. Iino, K. Koike, K.Yasuda, K. Hino, and K. Kurokawa. 1993. Duration of viremia in human Hepatitis A viral infection as determined by polymerase chain reaction. J. Med. Virol. 40:35-38.

Chapter 29

Microbial Food Safety Risk Assessment

Anna M. Lammerding, Aamir M. Fazil, and Greg M. Paoli

29.1 INTRODUCTION

Foodborne disease arises from the consumption of microbial pathogens, and/or microbial toxins, by a susceptible individual. The risk of foodborne disease is a combination of the likelihood of exposure to the pathogen, the likelihood of infection or intoxication resulting in illness, and the severity of the illness. In a system as complex as the production and consumption of food, many factors affect both the likelihood and severity of the occurrence of foodborne disease. Many of these factors are variable, and often there are aspects for which little information is currently available. To effectively manage food safety, a systematic means of examining these factors is necessary. Risk assessment is a process that provides an estimate of the probability and impact of foodborne disease.

Risk assessment was introduced as a structured process for estimating risk during the late 1970s. It was developed as a means to standardize the basis for regulatory decision-making, specifically in areas concerning human exposure to chemical substances.[2,70,88] Risk assessment methodologies are now used routinely to evaluate risks in many diverse fields, ranging from toxicology and ecology to engineering and economic investment.[4,10,27,57,71,72,80,87,94,106]

Evaluations of the risks associated with foodborne hazards, in general or attributable to specific foods, have been predominately qualitative descriptions of the hazard, routes of exposure, handling practices, and/or consequences of exposure. Quantifying any of these elements is challenging, since many factors influence the risk of foodborne disease, complicate interpretations of data about the prevalence, numbers and behavior of microorganisms, and confound the interpretation of human health statistics. Consequently, policies, regulations and other types of decisions concerning food safety hazards have been largely based on subjective and speculative information. However, advances in our knowledge, analytical techniques and public health reporting, combined with increased consumer awareness, global trade considerations and realization of the real economic and social impacts of microbial foodborne illness have moved us toward the threshold of using quantitative risk assessment to support better prioritizing and decision-making in managing food safety.[1,17,21,22,84,86,98,100,109,115]

It is only recently that risk assessment has been applied to microbial hazards, and techniques for its application are still evolving.[41,50,53,55,56,63,64,69,74,112] Unfortunately, the entire field of risk assessment has historically been rift with assignments of different meanings to different terminologies. However, currently accepted for food safety are the definitions and processes in Table 1, defined by the Codex Alimentarius Commission, the international regulatory body for foods.[24, 29, 42] Other processes and approaches have been proposed and as the field evolves undoubtedly consensus will be achieved.

29.2 RISK ASSESSMENT: A TOOL FOR DECISION-MAKING

Risk assessment is a systematic compilation and analysis of the current data and knowledge about a risk issue. This information may be found in published research and surveillance reports, from outbreak investigations, or it may be necessary to rely on expert judgment. The outcome of the process should ideally provide a clear and balanced representation of all available information relevant to a specific situation, described in terms of the probability and impact of an adverse event. Risk assessment quantifies the magnitude of a risk to assist in the evaluation of whether or not a hazard requires increased management or regulation. This provides the basis for what is called risk analysis, an approach to managing risks through three distinct but closely linked activities: risk assessment, risk management and risk communication (Figure 1[31]).

Risk management involves synthesizing the risk assessment information into some form of action so as to manage it.[31,32,47,71,72,106] The goal of risk management is not to achieve "zero risk" but rather to chose and implement "... scientifically sound, cost-effective, integrated actions that reduce or prevent risks while taking into account social, cultural, ethical, political and legal considerations."[106] Different considerations are due when the risk management issue pertains to international trade, national policy-making, industry interventions, and/or consumer concerns. The tasks involved in risk management have been described as: the determination of what hazards present more danger than society is willing to accept; considerations of what control options are available, and deciding appropriate actions to reduce or eliminate unacceptable risks.

Risk communication pertains to the exchange of information between risk assessors, risk managers, and stakeholders in the risk issue.[33] The utility of risk assessment lies in providing appropriate information for decision-makers, and therefore it is important

Table 1. Definitions of Risk Analysis Terminology for Foodborne Hazards

Term	Definition
Hazard	A biological, chemical or physical agent in, or condition of, food with the potential to cause an adverse health effect.
Risk	A function of the probability of an adverse health effect and the severity of that effect, consequential to a hazard(s) in food.
Risk Analysis	A process consisting of three components: risk assessment, risk management and risk communication.
Risk Assessment	A scientifically based process consisting of the following steps: (i) hazard identification, (ii) hazard characterization, (iii) exposure assessment, and (iv) risk characterization.
Quantitative Risk Assessment	A Risk Assessment that provides numerical expressions of risk and indication of the attendant uncertainties.
Qualitative Risk Assessment	A Risk Assessment based on data which, while forming an inadequate basis for numerical risk estimations , nonetheless, when conditioned by prior expert knowledge and identification of attendant uncertainties permits risk ranking or separation into descriptive categories of risk.
Hazard Identification	The identification of biological, chemical, and physical agents capable of causing adverse health effects and which may be present in a particular food, or group of foods.
Hazard Characterization	The qualitative and/or quantitative evaluation of the nature of the adverse health effects associated with the hazard. For the purpose of Microbiological Risk Assessment the concerns relate to microorganisms and/or their toxins.
Dose-Response Assessment	The determination of the relationship between the magnitude of exposure (dose) to a chemical, biological or physical agent and the severity and/or frequency of associated adverse health effects (response).
Exposure Assessment	The qualitative and/or quantitative evaluation of the likely intake of biological, chemical, and physical agents via food as well as exposures from other sources if relevant.
Risk Characterization	The process of determining the qualitative and/or quantitative estimation, including attendant uncertainties, of the probability of occurrence and severity of known or potential adverse effects in a given population based on hazard identification, hazard characterization, and exposure assessment.
Risk Estimate	Output of Risk Characterization.
Transparent	Characteristics of a process where the rationale, the logic of development, constraints, assumptions, value judgements, decisions, limitations and uncertainties of the expressed determination are fully and systematically stated, documented, and accessible.
Sensitivity Analysis	A method used to examine the behavior of a model by measuring the variation in its outputs resulting from changes to its inputs.
Uncertainty Analysis	A method to estimate the uncertainty associated with model inputs, assumptions, and structure/form.
Risk Management	The process of weighing policy alternatives in the light of results of risk assessment and, if required, selecting and implementing appropriate control options, including regulatory measures.
Risk Communication	The interactive exchange of information and opinions concerning risk and risk management among risk assessors, risk managers, consumers and other interested parties.

Source: Codex Alimentarius Commission, 1999[24]

that the specific issue of concern is understood by both assessors and managers. Stakeholders in the risk issue may include, for example, producers, processors, or food handlers, the general public and/or specific subpopulations with increased risk. Increasingly, it is realized that the entire process of assessing and managing risks should be transparent and interactive. The participation of all parties during the process tends to increase the acceptability of the final outcome, as opposed to decision-making with no apparent explanation or justification.[32,33,84,106]

In summary, risk assessment is the measurement of risk and the identification of factors that influence it. Risk management is the development, selection, and implementation of strategies to control that risk if warranted, and risk communication is the exchange of information pertinent to the risk issue. Risk assessment and risk management functions should, however, be distinct, so that the assessment remains an objective scientific evaluation of the risk and is not influenced by the preferred options of the risk manager.[24] It should also be recognized that the process is not static, and risk assessments and/or management decisions may need to be reviewed and revised as new information becomes available. A framework that has evolved from the US regulatory experience, and which captures many of these considerations, is that of the PCCRARM (Figure 2[106]). This is a slightly modified form of that depicted schematically in Figure 1, and outlines more specifically the sequence of activities that may be involved in the management of risks.

29.3 APPROACHES TO MICROBIAL FOOD SAFETY RISK ASSESSMENT

Many frameworks have been developed to guide the steps involved in generating a risk assessment. Usually, there are four distinct steps: hazard identification, exposure assessment, dose-response analysis, and risk characterization.[61,70,71] The Codex Alimentarius Commission framework for food safety hazards introduced a modification that uses the term "hazard characterization" under which dose response analysis could be done if the data existed (Figure 3).

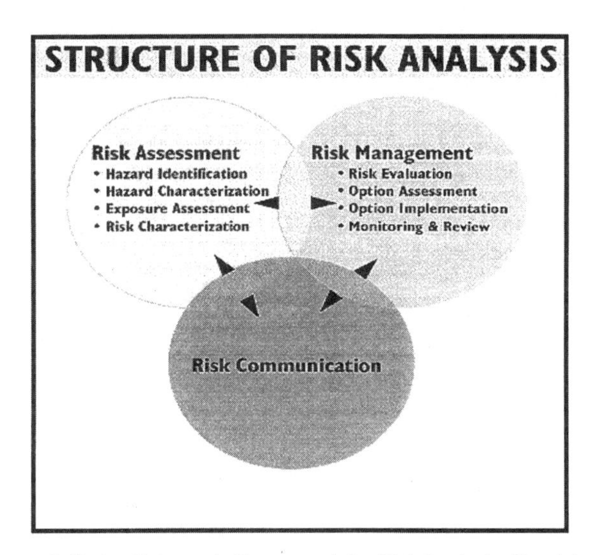

Figure 1. *Codex Schematic Framework for Risk Analysis. The risk assessment may be initiated from any source, but its conduct will typically be under the control of a risk manager who will coordinate the process and exchange of information, and turn the results of the assessment into a plan of action. The elements of risk analysis are distinct but the process is interactive and iterative.*

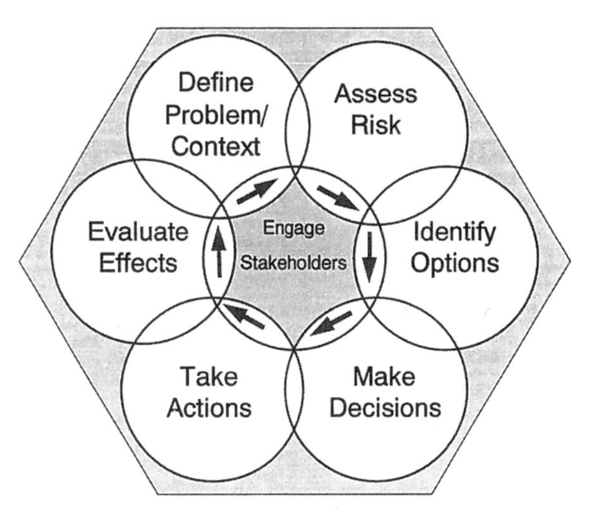

Figure 2. *Risk Management Framework Proposed by PCCRARM.*[106] *The framework encompasses the same elements as the Codex risk analysis process, but in a different format. The overall process is considered as risk management, and includes the scientific assessment of risk and the decision-making activities that include identifying options, making the decision, taking action, and evaluating effects. Re-assessment and re-evaluation may be part of the process. Again, communication among all parties is critical.*

While described as a linear process, in practice the collection and analysis of data for the different sections of a risk assessment are typically done concurrently. Information pertaining to the hazard characterization of a risk may influence the scenario and inputs considered in exposure assessment. Alternatively, insights arising during the exposure assessment may influence the considerations for hazard characterization or the dose-response relationship.[52]

The scope of the assessment is dependent on the risk management question, and the reason for doing the assessment. Traditionally, risk assessment has been viewed simply as a means to provide an estimate of risk. However, it can serve much broader applications that require the understanding of a system or process. Hence, the information required and the form the risk assessment takes will invariably be driven by the management issue or decision it is intended to aid.

There are two general approaches to risk assessment, broadly described as qualitative and quantitative. Qualitative risk assessments are descriptive or categorical treatments of information, whereas quantitative assessments rely heavily on numerical data and assumptions. If time, resource, and/or data restrictions limit the detail of the assessment, or if the issue does not warrant substantial efforts, a qualitative assessment may be appropriate as long as it provides some estimate of the magnitude of probable harm. It is also important to recognize the necessity of being precise about "qualitative" statements and measurements since descriptive characterizations of likelihood and impact can be misinterpreted. A qualitative estimate of risk may be conducted by assigning probability and impact ratings such as negligible, low, medium, or high to the risk factors. If such a system is used for rating exposure and dose-response information, specific guidelines and definitions of the assigned ranges for each rating must be clearly described and justifiable.

The outcome of a quantitative risk assessment is a numerical estimation of risk. Typically, quantitative risk assessments are derived using single "point-estimates" as inputs, e.g., values such as the mean, or "worst-case" estimates, and yield a single mean or "worst-case" estimate of the risk. This approach has limitations in producing realistic outputs, particularly for diverse and dynamic biological systems. Alternatively, risk assessments that incorporate the variation in input parameters provide more insightful assessments of the risk issue and are called probabilistic/stochastic assessments.

The conclusions of a risk assessment should include some description of the variability and uncertainty in the information used to derive the risk estimate.[24] Variability is essentially a property of nature, a result of the natural random processes. It represents diversity in a well-characterized population or parameter. Variability cannot be reduced through further study or additional measurements. An example of this could be the amounts of food people eat. Conducting surveys on eating habits will give us information on how much food people eat, however, some people will always eat more or less than others no matter how much data we collect on them. Each step in the production, processing, and marketing of a food has variability; both the microbial pathogen and human host responses are highly variable.[9,21,38,48,58]

Uncertainty is a property of the risk assessor. It results from the lack of knowledge about a phenomenon or parameter and the inability to characterize it. Uncertainty can, in many cases, be reduced through further measurement or study. An illustration of this could again be drawn from the description of how much

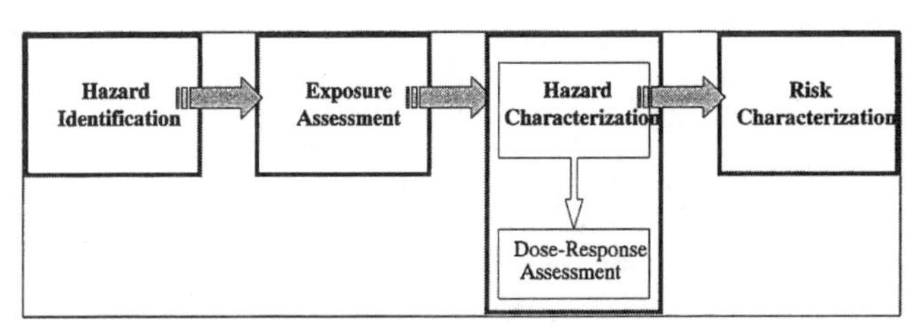

Figure 3. *General Risk Assessment Framework.*[24]

food people eat. With little information available, perhaps a minimum and maximum amount of food consumed could be estimated. By conducting additional research, the amounts of food people eat and how frequently different amounts are consumed could be determined more precisely.

Variability and uncertainty have different ramifications in the results of a risk assessment and the risk management decisions pursued. If the variability in a parameter is the driving force in the risk, then better control of the process or factor may be warranted to reduce risk. If the output of interest is influenced by the uncertainty in a parameter, the management decision may be to focus more research or data collection activities to better characterize or understand that factor, and the assessment redone. However, if a risk mitigation decision is required under circumstances where uncertainty is significant, and additional data are not readily obtainable, then a conservative decision may be warranted, with the understanding that more information would allow a better strategy. Currently, this philosophy is regarded as the precautionary principle, an action-oriented decision based on lack of sufficient information that favors assured safety.

29.4 STEPS OF RISK ASSESSMENT

29.41 Preliminary Considerations

A risk assessment is only as useful as the extent to which it answers the risk management problem. Therefore, a critical initial step is to develop a clear and unambiguous statement of the purpose and scope of the assessment. The risk manager may need to determine the exact risk, the risk compared to other risks, factors influencing this risk, factors which the decision-maker should focus on, or, acceptability of the risk. The manager may be concerned about making a decision now, identifying research needs, or where to place promising interventions. The assessment may be needed to address broad policy needs, risks associated with a specific product or process, or a commodity group. A clear understanding of the purpose will help to define and bound the assessment.[24,52,106]

The essence of microbial risk assessment is describing a system in which a microbial hazard reaches its host and causes harm. Risk assessments for microbial hazards in foods must take into consideration all the factors that affect the final prevalence and concentration of the pathogen in a food at the time of consumption, and the response of the host upon ingestion. This requires the development of models that describe the real-life scenarios of complex food systems, with consideration for the natural variability of biological systems. All data, models and assumptions used should be clearly documented. The structure of the risk assessment should ideally provide a basic framework for future updating as better and more complete information becomes available in our understanding of the behavior of foodborne microorganisms.

29.42 Hazard Identification

The first activity in risk assessment is hazard identification. Traditionally, in chemical risk assessment, this step was aimed at determining if a chemical substance could cause some form of an adverse health effect (e.g., cancer). In microbial risk assessment, the microbe is usually already identified as being a pathogen capable of causing human illness.[50,52]

Hazard identification as it applies to microbial risk assessment usually involves an evaluation of the epidemiological data linking foods and pathogens.[24,83] In addition, issues such as acute vs.

Hazard Identification • **Is there a problem?** • **How much of a problem?** • **Details of the problem?**	⇒ Evidence linking the food and pathogen to human illness. ⇒ Epidemiological investigations ⇒ National surveillance data bases ⇒ Microbiological research ⇒ Process evaluations ⇒ Clinical studies

chronic disease, the existence of specific sensitive populations, and other complications should be acknowledged. The characteristics of the organism and its action should also be recognized. Some organisms affect the host through the action of toxins, either in the food before consumption or after consumption in the intestine. Some organisms tend to cause sporadic illness as opposed to outbreaks, and this tends to diminish the attention given to them in epidemiological statistics. The hazard identification step is largely a qualitative evaluation of the available information, and considers much of the same information that is analyzed in more detail in the subsequent steps of the assessment process.

29.43 Exposure Assessment

Exposure assessment is concerned with estimating the likelihood of consumption, and the likely number, or dose of the pathogen that consumers may be exposed to in a food. In a quantitative assessment, the majority of the modeling and simulation work goes into this stage.

Exposure Assessment • **How many organisms are ingested by the consumer?** • **How often does the consumer ingest them?**	⇒ Sources of contamination: frequency and concentration, and an estimation of the probability and concentration that will be consumed. ⇒ Distribution, growth, inhibition or inactivation from primary contamination, through processing, handling at retail, consumer preparation practices. ⇒ Growth studies, predictive models ⇒ Food manufacturer data ⇒ Food surveillance data—primary process & retail ⇒ Animal/zoonotic disease data ⇒ Food composition—pH, Aw, nutrient content, presence of antimicrobial substances, competing micro flora. ⇒ Population demographics ⇒ Consumption patterns

Microbial risk assessment is faced with a much more dynamic hazard compared to traditional chemical risk assessment because of the potential for microbes to multiply in foods.[48] Assessments concerned with exposures to microbial toxins are faced with a combination of the microbe's characteristics, and the chemical-like effects of the toxin itself. This step should estimate prevalence and extent of microbial contamination of the product at the time of consumption, the likelihood that an individual consumes the certain food product in a given period of time, the circumstances under which the food was consumed (home prepared vs. institutional or food service, etc.), and the amount of the product consumed at each meal. The exposure unit should be considered as the unit that could potentially result in illness, and for most biological agents in food, that is typically considered to be a single meal serving size.

All sources of entry of the hazard into the food product should be evaluated. Since it is not possible to measure precisely the population of the pathogen present in a food at the time of consumption, models or assumptions must be developed to estimate the likely exposure. For bacteria, the growth and death of the organism must be accounted for within the food and under predicted handling and preparation practices. Temperature, time, the food chemistry, and competing microflora may affect the growth and death rates of pathogens.[49] For viral and parasitic agents that do not grow in foods, the effectiveness of decontamination and/or inactivation steps are of primary concern. Predictive microbial modelling techniques are becoming increasingly more sophisticated, and provide valuable tools for the derivation of probable exposure estimates.[13,35,62,93,108,113]

In assimilating the data for exposure assessment, consideration should be given to the verifiable effectiveness of existing control measures, and to the sensitivity, specificity, and validity of sampling and testing methods used to collect empirical information.

The scope of the risk assessment, as determined by the risk management needs, will determine the comprehensiveness and detail required for the exposure assessment. To simply derive an estimation of the health risk associated with a food, for example an estimate of the annual number of illnesses, the most direct way would be to use empirical data about contamination rates and concentrations. However, for microbial pathogens in foods, most data about contamination rates are obtained at the end of processing or at retail. Additionally, the organisms may be present, but in numbers too low to detect with the test protocol. The assessment therefore, must still consider growth/inactivation under plausible scenarios of food handling and preparation.[50]

A complete "farm-to-fork" model provides the most information relative to food safety risk management (Figure 4). This type of assessment allows consideration of a broad range of risk management options that could be implemented at one or more stages that contribute the most to an unacceptable risk outcome. Some recent examples of this approach are quantitative risk assessments for *Escherichia coli* O157:H7 in ground beef hamburgers,[18,59] *Salmonella enteritidis* in shell eggs[3] and liquid pasteurized eggs,[114] and *Listeria monocytogenes* in soft cheeses made from raw milk.[5] However, undertaking a farm-to-fork assessment is complex and information-intensive, and generally requires substantial inputs from experts in diverse fields. It may be appropriate to focus on only one part of the food chain, with the express purpose of trying to understand and reduce exposure or risk at that point, until such time as reductions at a different point in the chain can be applied.[8,73,75,81,116,117]

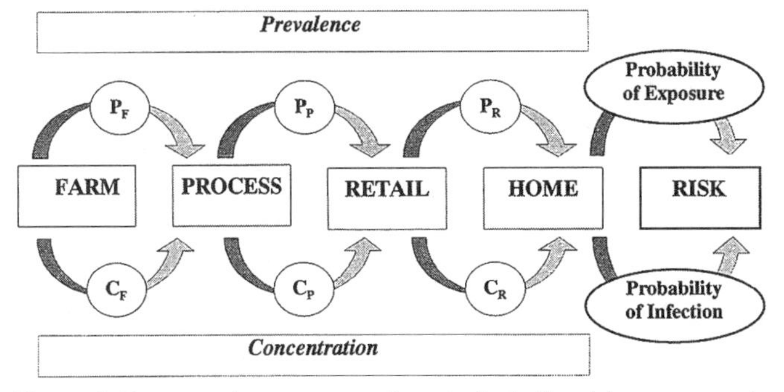

Figure 4. *Exposure Assessment – Farm to Fork. The risk assessment evaluates the influence of each stage of the process on the prevalence (P) and concentration (C) of the pathogen in the food up to the point of consumption. The hazard characterization/dose-response links the exposure assessment to the human health outcome to measure the risk.*

29.44 Hazard Characterization and Dose-Response Assessment

Hazard characterization involves providing a qualitative or quantitative description of the severity and duration of adverse effects that may result from the ingestion of a microorganism or its toxin in food. The purpose of this step is to describe the consequences of exposure to a pathogen as an estimate of the magnitude of adverse effects for an individual consumer or a population.

Hazard Characterization	
• **How serious is the illness?** • **How long does it last?** • **If possible perform dose-response analysis?**	⇒ Pathogen: Virulence parameters ⇒ Food: Factors that may protect the organism e.g. high fat content/incr. resistance to gastric acids ⇒ Host—Susceptibility/resistance factors ⇒ Population characteristics

This step also provides the measure for evaluating the value of food safety efforts; e.g, a decrease in the number of people becoming ill and/or severity of illnesses as a result of an intervention.

The likelihood of a pathogen causing illness is dependent on a) the characteristics of the organism itself, e.g., mechanism of pathogenesis, virulence factors, and resistance to host factors such as gastric acidity and immune response; b) the susceptibility of the host; e.g. immunocompetence or nutritional status; and c) the characteristics of the food in which the pathogen is carried, e.g., a food with high fat content will protect the organism from the gastric acidity of the stomach.[21,25,38]

If the data are available for microbial pathogens, a dose-response analysis should be performed. This is a mathematical relation that translates the number of organisms ingested, or dose, into a probability of infection, morbidity or mortality. It is important to specify in the dose-response analysis what response or adverse effect is being measured.

Dose-Response	
• **How likely is infection based on the amount ingested?**	⇒ Outbreak investigations ⇒ Animal studies ⇒ Human feeding trials ⇒ Severity, long term sequallae

Currently, this is a highly controversial area. Few data are available, and the paucity of dose-response data is compounded by relevance to normal human populations. Those few controlled studies that have been conducted usually involved healthy adult males.[30,101] Ethical considerations make it unlikely that many such studies will be conducted in the future. Nevertheless, the available data have provided valuable insights into the nature of the pathogen/host relationship. Information from experimental feeding trials with animal models must be cautiously interpreted when extrapolating to the human population.[76] Data collected from foodborne outbreak investigations are valuable to help determine the nature of dose-response relationships, but generally, information such as numbers of the pathogen in the food, and the proportion of people exposed and becoming ill or remaining healthy are not typically collected during an investigation.

Another, perhaps more promising approach, is to use epidemiological data on the prevalence and concentrations of a pathogen in food, and the incidence of related illness in a population exposed to that food. Buchanan et al.[12] calculated a dose-response relationship for *Listeria monocytogenes* based on the incidence and level of contamination in a specific food product, compared to the incidence of listeriosis in a defined population of consumers. However, for most pathogens, sufficient quantitative data about the levels in foods are currently not available, nor does the routine reporting of most foodborne illnesses provide accurate statistics. Finally, consensus about dose-response relationships may also be derived by elicitation of expert opinion.[51,60] The results of an expert elicitation process cannot substitute for the scientific data needed to accurately estimate dose-response relationships and their variance, but given the difficulty of collecting experimental data on harmful pathogens, the information can provide a characterization of scientific judgments to estimate relative risk to the population.[60]

Two distinct hypotheses have been proposed for the nature of the dose-response relationship for foodborne pathogens. The first is an historical notion that there is a threshold number of organisms, or minimum infectious dose, that must be ingested before any infection or adverse effects occur. The second hypothesis is that each pathogen cell has an equal capacity to cause an infection or illness.[40,95,96] Thus there is no threshold number, and the probability of causing infection increases as the levels of the biological agent increases. In essence, the latter hypothesis states that if a large enough number of people were exposed to only one cell of the pathogen, the outcome would be infection and illness in a certain proportion of those exposed. For example, it has been estimated that a single cell of *Shigella* spp., a pathogen noted for its high infectivity, has a probability of 0.005 of causing an infection.[28] Another way of expressing this concept is that if 1,000 people each consumed one *Shigella* spp. cell, five individuals in the group would become infected.

Several investigators have examined the available data and proposed non-threshold models for a number of pathogens.[34,46,101] Through comparison of predictions using these models with data from actual outbreaks, it appears that for many foodborne pathogens, the dose-response relationship is consistent with a non-threshold mechanism of infection.[28,82,90] The beta-Poisson and exponential distributions have been found useful for describing the dose-response relationship for different biological agents, particularly when low numbers of the agent are ingested. Both equations are non-threshold sigmoidal functions. Figure 5a shows an exponential dose-response model, with the log-dose plotted against probability of illness. The non-threshold character of the relationship is more evident when the probability of a illness is converted to log values, shown in Figure 5b.[50]

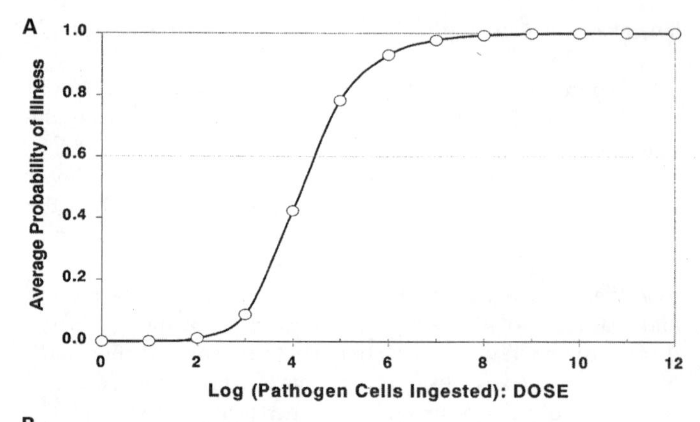

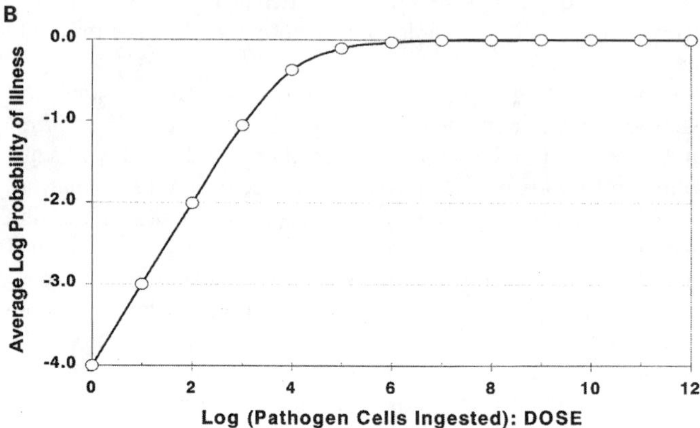

Figure 5. *Hazard Characterization (Dose-Response)*
A) Log Dose vs Average Probability of Infection;
B) Log Dose vs Average Log Probability of Infection

29.45 Risk Characterization

This is the concluding task in risk assessment. Risk characterization combines the information generated in hazard identification, exposure assessment and hazard characterization to produce a complete picture of the assessed risk. The resulting risk estimate should be considered within the context of independent epidemiological data, or other reference information, to assess the validity of the models, data, and assumptions that were chosen. The outcome of the assessment and the types of questions to be answered should be articulated by risk managers and risk assessors at the onset of the assessment process. This helps to guide the direction and consideration of inputs for the assessment. However, the assessment should at the least strive to answer the following questions in risk characterization[106]:

- What is the nature and magnitude of the risk?
- Which individuals or groups are at risk?
- How severe are the adverse impacts or effects at likely exposures?
- What is the evidence and how strong is it?
- What is uncertain about the nature of the risk?
- What is the range of informed views about the nature and probability of the risk?
- How confident are the risk assessors about their prediction?

The risk estimate should reflect a distribution of risk which represents the range of contamination of a food product, factors that might affect growth or inactivation of the pathogen, and the variability of the human response to the microbial pathogen. Risk characterization should also provide insights about the nature of the risk which are not captured by a simple qualitative or quantitative statement of risk. Such insights include, for example, a description of the most important factors contributing to the average risk, the largest contributions to the uncertainty and variability of the risk estimate, and a discussion of gaps in data and knowledge. The risk assessor may also include a comparison of the effectiveness of alternative methods of risk reduction for consideration by the risk manager.

29.5 MATHEMATICAL APPROACHES TO RISK ASSESSMENT

Quantitative risk assessments use mathematical models to estimate risk as a function of one or more inputs. Historically, many risk assessments have used point-estimates, single values such as the means or maximum values of variable data sets, to generate a single numerical value for the risk estimate. These are also referred to as deterministic assessments. The most common complaint raised against this type of assessment is that they are frequently used to determine the extremes, or "worst-case" of the risk situation, without regard for how likely these extremes are to occur. Or, if the "average" risk is calculated based on mean values, the extremes are disregarded, which may represent highly susceptible subpopulations or infrequent, but severe circumstances.

Alternatively, risk assessments can be constructed that incorporate the variability and uncertainty in the input parameters by using all the data available, described by probability distributions.

To illustrate these two approaches, the following is a hypothetical scenario for an exposure assessment which estimates the dose of a microbial pathogen that is ingested by individuals consuming a certain food product. This is a simplified example where only three input parameters are considered: concentration of the pathogen in the food before cooking, inactivation of the organism during the cooking process, and the amount of food consumed (Table 2). In this example we assume that the pathogen is reported to be present at concentrations between a minimum of 2 log CFU/g and a maximum of 5 log CFU/g, with a mean concentration of 3.5 logs. Based on thermal inactivation studies, and knowledge about typical cooking practices for the food, the pathogen population will be decreased by between 4 and 5 log due to cooking. Consumption data show that the most typical meal size is 75g of the product, although some people only eat 25g and some people actually eat as much as 175g.

The deterministic approach uses the single point estimates of the mean concentration in the food (3.5 log CFU/g), the mean log reduction from cooking (4.5 log) and the most frequently amount of food that is eaten (75 g). These point estimates are then used to calculate the "best estimate" for the number of organisms ingested:

$$\text{Dose Ingested} = 10^{([\text{Conc. In Food}] - [\text{Log Reduction}])} \times (\text{Amount Eaten})$$
$$= 10^{[3.5-4.5]} \times 75 = \text{approx. 8 cells.}$$

The analysis can be taken further by using the maximum and minimum point estimates of each of the variables to calculate possible outcomes based on different combinations of concentration, reductions due to cooking, and consumed amounts. However, as the model gets more complex, the number of possible combinations increases dramatically. In the simple example shown above there are 27 different possible combinations that could be generated, calculated as:

$$\text{No. of scenario} = (\text{No. of point estimates at each variable})^{(\text{No. of variables})}$$

Thus with one more step or variable in the model, the number of possible combinations would increase to 81. It is common practice, at best, to calculate only the bounds on the outcome. In this example, the maximum possible dose that might be ingested occurs when the concentration of the pathogen in the food is at a maximum, the log reduction during cooking is the minimum, and the maximum amount of the food is eaten. This calculation yields an estimated ingestion of 1760 pathogen cells in a meal. This is easy enough to do in this simple model, however, with a more complex model determining the combination of events that lead to a worst- or best-case scenario is not always a trivial task.

When conducting deterministic risk assessments, it has often been the practice to err on the side of conservatism. The selection of conservative point estimates has been a contentious issue: how conservative an estimate is conservative enough? Selecting a conservative value for each variable and propagating this conservatism through the model often results in an unrealistic estimate of risk. It also reduces the credibility of the assessment and results in risk management decisions based not in scientific realities or all the information but rather on regulatory guidelines or the assessor's conservatism. The two possible outcomes of using deterministic risk assessment with conservative point estimates is summarized by Burmaster[15] as follows:

- If a conservative point estimate of risk falls below some regulatory definition of maximum acceptable risk, then the risk assessor, risk manager, and members of the public can have confidence that the distribution of risk is truly acceptable. However the amount of overprotection is unknown; or,
- If a conservative point estimate of risk falls above the regulatory definition of maximum acceptable risk, then the risk assessor, risk manager, and members of the public do not know if the distribution of risk is truly unacceptable—or if the apparently unacceptable risk is merely an artifact of conservatism propagated through the analysis.

Another drawback of the deterministic approach is that the likelihood or probability of a point estimate risk actually occurring is ignored. All values between the minimum and maximum points are regarded as equally likely to occur. In reality, however, some values within the interval are more likely to occur than others. Using the illustration above, while it may be true that consumers do not eat more than 175 g or less than 25 g of the food per meal, it may be more accurate to say that there are a small proportion of people who eat the extreme amounts and that the true consumption pattern follows some probability distribution. This is an example where probabilistic techniques can be applied to provide estimates that are more accurate, provide more information to the risk manager, do not propagate conservative values through the model, and represent reality better than point estimates.

Probabilistic assessments represent all the information available for each parameter, described as a distribution of possible values. The distribution used to describe a data set is dependent on the amount of data available, and knowledge about the nature of the phenomenon. Appendix 1 gives examples of some commonly used distributions and the parameters used to describe them. When a great deal of information is available and the variable being described cannot be characterized by some theoretical distribution, then an empirical distribution fit to the data may be most appropriate. In cases where there is very little information available, the uniform distribution can be used, only a minimum and maximum value are required for this distribution. If an estimate of the value that is most likely to occur can be elicited, then the triangular distribution would offer an improvement over the uniform.

Table 2. Point-estimate Parameters for Exposure Assessment

Input Variable Description	Minimum	Most Likely	Maximum	Units
Concentration in food	2	3.5	5	$\log_{10}$ CFU/g
Log reduction from cooking	4	4.5	5	
Amount eaten of food eaten	25	75	175	g

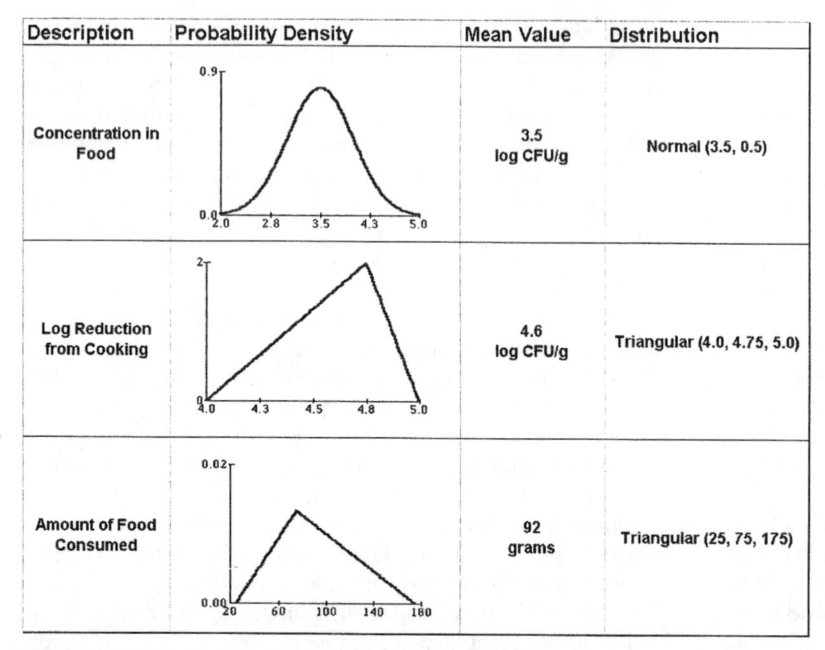

Figure 6. *Distributions for Example Probabilistic Assessment.*

For the example described, distributions can be used to replace the point estimate representations (Figure 6). In these figures, the x-axis represents the value of the parameter, and the y-axis represents the likelihood of the value occurring, between 0 and 1, or 0 and 100%.

It should be noted that these distributions are only illustrative and are not necessarily the most appropriate for describing the variables listed. The concentration of the pathogen in the food is represented by a normal distribution centered at 3.5 log CFU/g, indicating that this value will be the most frequently occurring concentration found in the food. Triangular distributions are used to describe the log reductions from cooking and the amount eaten. Cooking is expected to give a 4.5 log reduction the majority of the time, however slightly higher and lower values than this are possible, between 4 and 5 log reductions. To describe the amount eaten, approximately 75 grams of the product is consumed most frequently, however up to 175 grams is occasionally consumed by some individuals and at the other extreme, some individuals only eat 25 grams.

The outcome of the probabilistic analysis is a distribution of possible ingested doses. The results are shown in Figures 7a and 7b. For comparison, the bounds generated using point estimates are also shown. It can be seen that using point estimates to set the bounds does not provide as much information as may be necessary to make the best management decisions. By comparison with the distribution of likely exposures in Figure 7a, it is readily evident that when a "worst-case" scenario is used to derive a point estimate of maximum ingested dose, the estimated value is high (3.24 logs, or 1760 cells) but this would occur very rarely. This observation is especially evident in Figure 7b, where dose rather than log-dose is plotted on the x-axis. Without this knowledge, the risk manager may inappropriately allocate valuable resources to reducing an event that rarely occurs. It should be kept in mind however, if the outcome of ingestion of this particular pathogen is severe, it may be an appropriate management decision to ensure that this adverse outcome, however unlikely, is prevented.

A mathematical description of the production and consumption of a food using probability distributions is very difficult to calculate analytically. While some analysis is practical on very small and simple models, a compound model of food production involving pathogen growth, destruction and infection is too complex to interpret without computational tools. However, complex probabilistic risk assessments for food safety have been made feasible through the availability of commercial software and powerful desktop computers. Monte Carlo simulation* is a computational tool that aids in the analysis of models involving probability distributions. *Note: Examples of currently available software include @Risk (Palisade Corp., Newfield, NY), Crystal Ball (Decisioneering, Inc., Denver, CO) and Analytica (Lumina Decision Systems, Inc., Los Gatos, CA).

29.6 MONTE CARLO ANALYSIS

The mathematician Stansilaw Ulam (1909–1984) is most credited or associated with the development of Monte Carlo simulation. Ulam, and John von Neumann, at the hydrogen bomb superconference in 1946, realized the potential application of the method to simulate the probabilistic problem concerned with random neutron diffusion in fissile materials.[97] Despite initial development and application in the late 1940s, Monte Carlo methods were largely ignored in the risk assessment arena until recently. The capability of Monte Carlo simulations to accommodate probabilistic representations has led to its emergence as a standard tool in the field of environmental health, and in the growing field of microbial risk assessment.[15,19,64,97,105,111]

Monte Carlo analysis, as it applies to risk assessment, is a relatively straightforward procedure. The method can be applied to existing deterministic models by replacing the point estimates with probability distributions. Monte Carlo simulation involves randomly sampling each probability distribution within the model hundreds or even thousands of times, producing a new scenario at each iteration. In essence, a new "point-estimate" is generated at each iteration for each parameter within the model and the result recorded. The process is then repeated until each individual probability distribution has been sufficiently re-created.

Figure 8 illustrates how a triangular distribution with a minimum value of 1, a maximum of 8, and a most likely value of 4 is re-created as the iterations in the Monte Carlo simulation proceed. At the first iteration, only one value has been selected, which is comparable to a simple point estimate randomly selected between the limits of the distribution. After five iterations, the distribution still appears to be random point estimates between the limits. However, after 100 iterations it can be seen that values around the most likely value, 4, have been selected more frequently than those at the extremes. Finally, after 5000 iterations the triangular distribution can be observed to be sufficiently re-created with the majority of the samples selected around the most likely value, and spreading out towards the extremes which were sampled with decreasing frequency.

In this example only one parameter is shown. Typically, in a complex model there are many distributions that are sampled at each iteration, summed according to the mathematical relationships defined in the model, and the results stored. The sampling

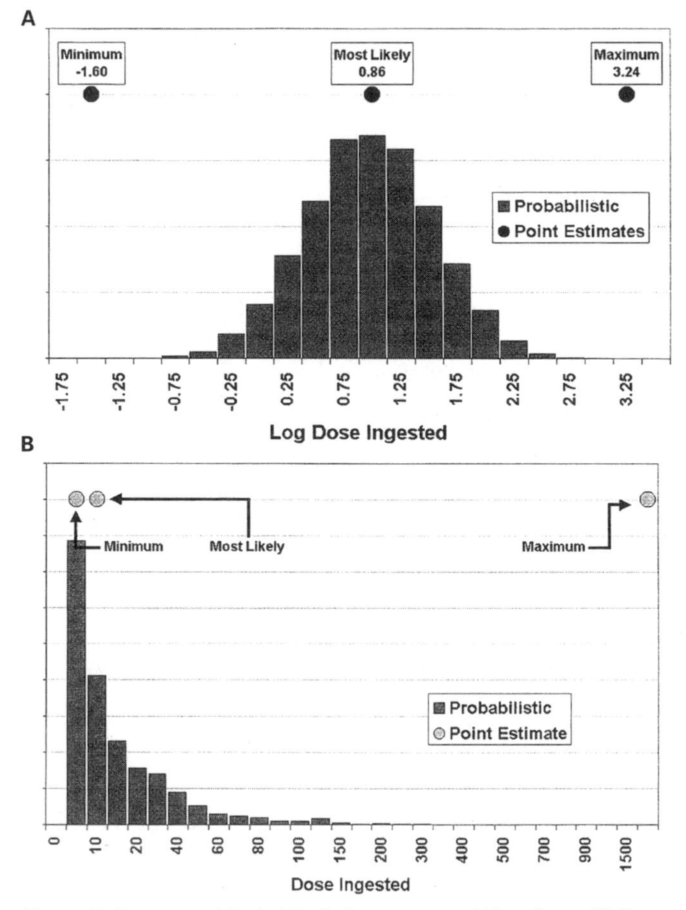

Figure 7. Outcome of Probabilistic Assessment. A) Log Dose; B) Dose.

of the input distributions in this way and the subsequent evaluation of the model then generates a single distribution for the output of interest. The output represents a result that encompasses most of the possible combinations for the inputs (Figure 9). An important characteristic of Monte Carlo analysis is the selection of a sample value at every iteration that is based on the defined probability distribution. Thus, based on the parameters of a distribution, the more likely values are selected more frequently than the less likely values, which reflects real world events much more faithfully.

The simulation of a model using Monte Carlo analysis allows the model to be used for more than just the estimation of risk. By conducting a sensitivity analysis on the model, the variables within the model which are strongly correlated with the magnitude of the risk estimate can be determined. This analysis can serve to focus research, management or modeling efforts on the most important factors influencing the risk. Variables identified as having a significant impact on the output should be first examined by the risk assessor. Inappropriate model simplifications or other errors in the construction of the model may lead to erroneous correlations. Variables that are found to be significant should be described and highlighted in the risk characterization.

The use of Monte Carlo simulation for microbial food safety risk assessment is a valuable and powerful tool. However, as with any analytical tool, these techniques have strengths and limitations, and may not be appropriate for all applications. Users should be familiar with the basis for probability statistics, and principles of good practice for use of Monte Carlo methods in risk assessment.[16,66,99,105,111,112]

29.7 UNCERTAINTY & VARIABILITY

The variability in inputs used in risk assessment may arise as a result of two influences. One is the true variability or diversity of natural phenomena, and the second the lack of knowledge, or uncertainty, about the phenomenon. Recognizing uncertainty and variability and their influence on the risk estimate has been emphasized. Techniques are available to separate and quantitatively measure the magnitude of uncertainty and variability in probabilistic risk assessments[68,85,102]; however, these may be cumbersome to apply for each input of a complex risk scenario. Such techniques might be considered for detailed analysis at significant focal points, and/or for inputs identified as having an important influence on the magnitude of the risk estimate. Characterization of variability, either quantitatively or descriptively, provides valuable insights about the extremes of distributions. For example the uppermost tail of the risk distribution curve may represent individuals or subpopulations at high risk. It is important to characterize uncertainty when a risk manager needs to know how robust the assessment is, or if the consequences of an incorrect decision based on the analysis are significant or asymmetrical. This additional knowledge allows the risk manager to

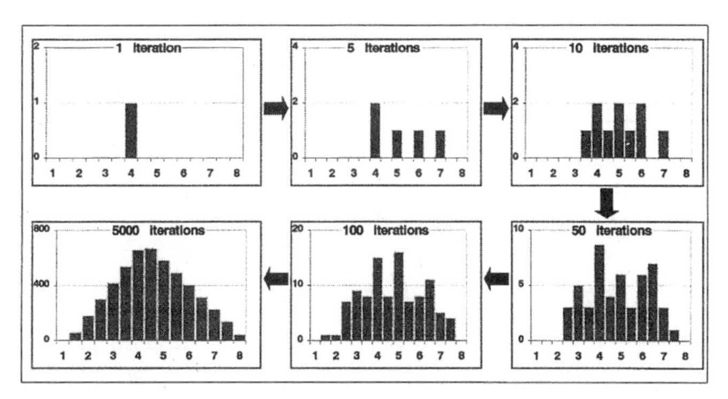

Figure 8. Re-creation of a Triangular Distribution using Monte Carlo Simulation

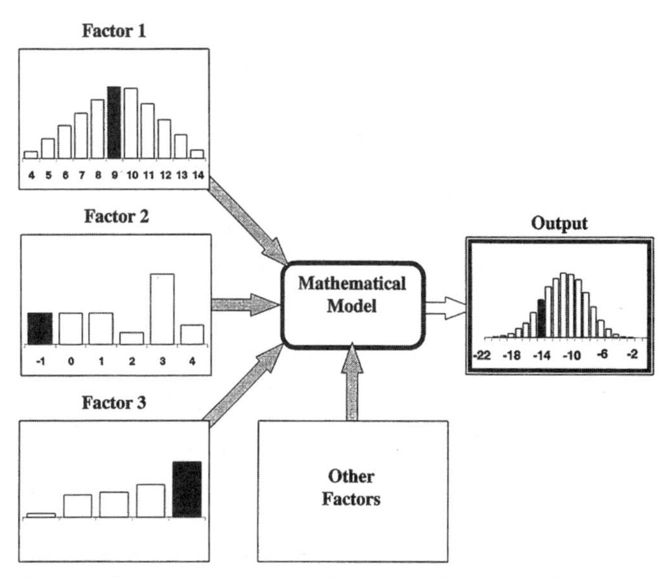

Figure 9. Example of Combining Several Distributions in a Monte Carlo Simulation.

decide to choose a more conservative option until such time more information is available.

29.8 APPLICATIONS OF RISK ASSESSMENT

A major impetus for food safety risk assessment are the provisions of the World Trade Organization's Sanitary and Phytosanitary (SPS) Agreement.[29,43,45] Essentially, these agreements state that decisions based on microbial criteria to deny the importation of products must be justified by scientific risk assessment. Thus, either quantitative or qualitative risk assessments must be used to substantiate claims of increased risk from products of importing countries, or to verify the equivalence of risk reduction activities in different countries.[42,43,44,45,50] Concerted international efforts to develop principles and guidelines for risk assessment, to ensure sound science, transparency and consistency, should lead to greater harmonization of risk assessment policies and methodologies.

National food safety and public health policies should address those risks that cause the greatest impact on human health and social resources.[21,22,54,67,88] Estimations of the magnitude of risk presented by different types of hazards, commodities, or processes provide the basis for cost/benefit analyses in decision-making. In food manufacturing, the hazard analysis critical control point (HACCP) program is a risk management approach to identifying and preventing, reducing or eliminating hazards in foods. Risk assessment quantifies the significance of hazards in terms of their impact on human health risk, and provides the means to evaluate the effectiveness of control measures and the overall HACCP program in reducing human health risks.[11,14,26,65,77,78,110] Risk assessment is a systematic approach to analyzing any food safety issue, and can be used to identify critical information gaps that may warrant focused research or data collection.

Quantitative risk assessments to assess the risk associated with infectious biological agents first focused on drinking water, reflecting an initiative by the US EPA to establish drinking water standards on a more scientific basis.[23,36,37,39,41,57,89,91] Several risk assessments were done for enteric viruses, bacteria and protozoa to use in cost/benefit analyses of pathogen control programs for drinking water. Among the first examples to systematically evaluate foodborne hazards for regulatory policy considerations were semi-quantitative assessments for cracked shell eggs and four broad food commodity groups.[103,104] Rose and Sobsey first used probabilistic modeling to quantify the risk of acquiring a viral infection from consumption of contaminated shellfish.[92] Increasingly more sophisticated models that combine predictive microbiology models with probabilistic Monte Carlo techniques have been developed to describe the complexity associated with the production, processing, distribution and consumption of foods. The entire food chain is analyzed as an inter-related system comprising unit models that are combined to present the overall risk characterization.

Cassin et al.[18] produced an example of this approach in a "farm-to-fork" process risk model for *Escherichia coli* O157:H7 in ground beef hamburgers. To support strategic decision-making, the scope of an assessment should include activities that provide relevant information for the risk manager. The main goal of the work was to develop a tool that could be used to analyze the relationship between the factors that affect the presence and behavior of microorganisms, and the probability of human illness. The basis of the assessment was a mathematical model that predicted the probability of illness as a function of multiple process parameters (Figure 10). Further application of the model was to evaluate the effects of hypothetical risk reduction strategies. By manipulating the initial data inputs, within the boundaries of realistic "what-if" scenarios, the effectiveness of proposed changes in the process, including a consumer education strategy, were estimated by comparing the change in the risk prediction as a result of the new inputs.

Other investigators have used a somewhat similar approach to the same problem, referred to as a dynamic flow tree process.[59] Risk models have been developed to assess the risk of infection with *Salmonella* Enteritidis associated with the consumption of shell eggs[3]; identify critical control points in the pasteurization of liquid whole egg[114]; estimate the risk of listeriosis associated with the consumption of soft cheeses made from raw milk[5]; and to quantitatively assess the public health protection afforded by meat inspection activities.[107] In other applications, assessors have focused only on modeling and quantifying the factors that contribute to exposure, without quantifying the ultimate human health risk. Some examples are the contamination of milk by *Listeria monocytogenes*[20,81] *Bacillus cereus*[75,118] and *Mycobacterium paratuberculosis*[73]; contamination of animal and poultry carcasses during processing,[7,8,79,116,117] and modeling the factors influencing pathogen carriage of live animals.[6] The outcomes of such evaluations could ultimately be linked with independent dose-response models to provide risk estimates.

29.9 SUMMARY

Risk assessment for microbial hazards in foods is a rapidly developing discipline. As in any application, risk assessment relies on

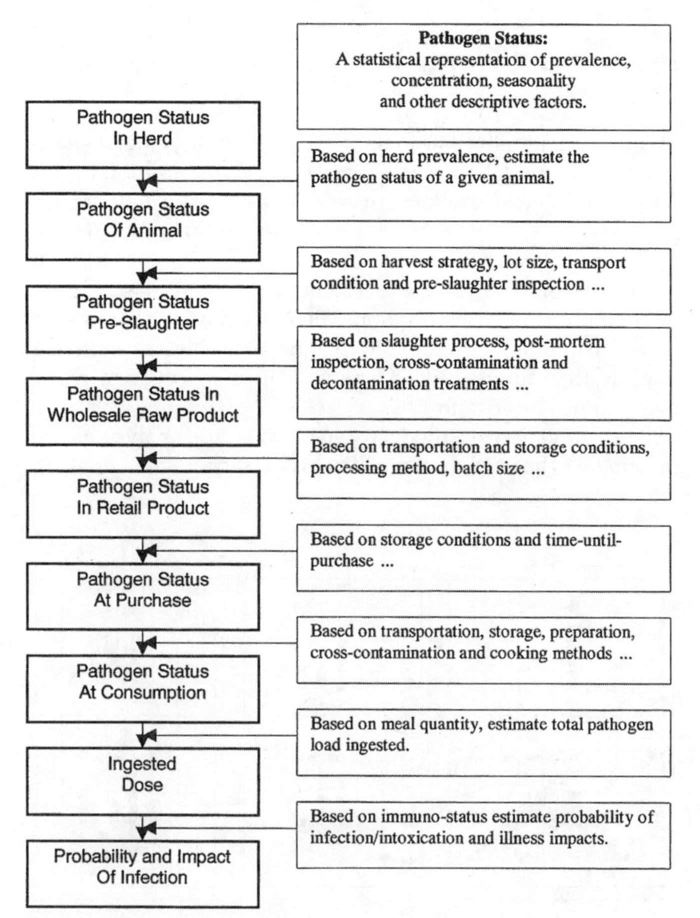

Figure 10. *Sample Elements of a Process Risk Model for* E. coli *O157:H7 in ground beef (18).*

the availability of valid data and informed knowledge. Managing microbial food safety risks encompasses integrated efforts throughout the food chain, from production to consumption. As a result, concerted efforts are needed to acquire better data and improved understanding of processes and interactions in the food chain, improved understanding of the ecology, physiology and biovariability of microbial pathogens, and the nature and variability of the pathogen/host relationship that leads to foodborne illness.

Even in the absence of complete data, the risk assessment approach is a valuable tool for gaining insights into food safety issues by encouraging the collection and analysis of information in a sequential and logical format. It is a tool that combines data and information from a multitude of sources in a structured format and presents a coherent characterization of the issue.

29.10 REFERENCES

1. Altekruse, S. F., M. L. Cohen, and D. L. Swerdlow. 1997. The future of foodborne diseases. Emerg. Infect. Dis. 3:285.
2. American Chemical Society. 1998. Understanding risk analysis. A short guide for health, safety, and environmental policy making. Am. Chem. Soc., Washington, D. C.
3. Baker, A. R, E. D. Ebel, A. T. Hogue, R. M. McDowell, R. A. Morales, W. D. Schlosser, and R. Whiting. 1998. *Salmonella* Enteritidis risk assessment: Shell eggs and egg products. U.S. Dep. of Agric. Food Safety and Inspect. Serv., Washington, D. C.
4. Barnard, R. C. 1994. Scientific method and risk assessment. Regul. Toxicol. Pharm. 19:211.
5. Bemrah, N., M. Sanaa, M. H. Cassin, M. W. Griffiths, and O. Cerf. 1998. Quantitative risk assessment of human listeriosis from consumption of soft cheese made from raw milk. Prev. Vet. Med. 37:129.
6. Berends, B. R., H. A. P. Urlings, J. M. A. Snijders, and F. Van Knapen. 1996. Identification and quantification of risk factors in animal management and transport regarding *Salmonella* spp. in pigs. Int. J. Food Microbiol. 30:37.
7. Berends, B. R., F. Van Knapen, and J. M. A. Snijders. 1996. Suggestions for the construction, analysis and use of descriptive epidemiological models for the modernization of meat inspection. Int. J. Food Microbiol. 30:27.
8. Berends, B. R., F. Van Knapen, J. M. A. Snijders, and. D. A. A. Mossel. 1997. Identification and quantification of risk factors regarding *Salmonella* spp. in pork carcasses. Int. J. Food Microbiol. 36:199.
9. Bernard, D. T., and V. N. Scott. 1995. Risk assessment and food borne microorganisms: The difficulties of biological diversity. Food Control 6:329.
10. Bernstein, P.L. 1996. Against the gods: the remarkable story of risk. John Wiley & Sons, Inc., New York
11. Buchanan, R. L. 1995. The role of microbiological criteria and risk assessment in HACCP. Food Microbiol. 12:421.
12. Buchanan, R. L., W. G. Damert, R. C. Whiting, and M. van Schothorst. 1997. Use of epidemiologic and food survey data to estimate a purposefully conservative dose-response relationship for *Listeria monocytogenes* levels and incidence of listeriosis. J. Food Prot. 60:918.
13. Buchanan, R. L., and R. C. Whiting. 1996. Risk assessment and predictive microbiology. J. Food Prot. 59 (Suppl.):31.
14. Buchanan, R. L., and R. C. Whiting. 1998. Risk assessment: A means for linking HACCP and public health. J. Food Prot. 61:1531.
15. Burmaster, D. E. 1996. Benefits and costs of using probabilistic techniques in human health risk assessments, with emphasis on site-specific risk assessments. Human Ecol. Risk Assess. 2:35-43.
16. Burmaster, D. E., and P. D. Anderson. 1994. Principles of good practice for the use of Monte Carlo techniques in human health and ecological risk assessments. Risk Anal. 14:477.
17. Buzby, J. C., T. Roberts, J. C.-T. Lin, and J. M. MacDonald. Bacterial foodborne disease, medical costs & productivity losses. U.S. Dep. Agric. Econ. Res. Serv. Agric. Econ. Rep. 741, Washington, D. C.
18. Cassin, M. H., A. M. Lammerding, E. C. D. Todd, W. Ross, and R. S. McColl. 1998. Quantitative risk assessment for *Escherichia coli* O157:H7 in ground beef hamburgers. Int. J. Food Microbiol. 41:21.
19. Cassin, M. H., G. M. Paoli, and A. M. Lammerding. 1998. Simulation modeling for microbial risk assessment. J. Food Prot. 61:1560.
20. Cassin, M. H., G. M. Paoli, R. S. McColl, and A. M. Lammerding. 1996. A comment on "Hazard assessment of *Listeria monocytogenes* in the processing of bovine milk", J. Food. Prot. 57:689 (1994). J. Food Prot. 59:341.
21. CAST. 1994. Foodborne pathogens: risk and consequences. Counc. for Agric. Sci. and Technol., Ames, Iowa. Task Force Report No. 122.
22. CAST. 1998. Foodborne pathogens: review of recommendations. Counc. for Agric. Sci. and Technol., Ames, Iowa. Special Report No. 22.
23. Clark, R. M., C. J. Hurst, and S. Regli. 1993. Costs and benefits of pathogen control in drinking water, p. 181. *In* G. F. Craun (ed.), Safety of water disinfection: Balancing chemical and microbial risks. ILSI Press, Washington, D. C.
24. Codex Committee on Food Hygiene. 1998. Principles and guidelines for the conduct of microbiological risk assessment. Draft guidelines at step 8 of procedure. Alinorm 99/13A, Appendix II, Report of the thirty-first session. Codex Alimentarius Commission, Rome.
25. Coleman, M., and H. Marks. 1998. Topics in dose-response modeling. J. Food Prot. 61:1550.
26. Corlett, D. A., and R. F. Stier. 1991. Risk assessment within the HACCP system. Food Control 2:71.
27. Covello, V. T., and M. W. Merkhofer. 1993. Risk assessment methods: Approaches for assessing health and environmental risks. Plenum Press, New York.
28. Crockett, C. S., C. N. Haas, A. Fazil, J. B. Rose, and C. P. Gerba. 1996. Prevalence of shigellosis in the U.S.: Consistency with dose-response information. Int. J. Food Microbiol. 30:87.
29. Dawson, R. J. 1995. The role of the Codex Alimentarius Commission in setting food standards and the SPS agreement implementation. Food Control 6:261.
30. Dupont, H. L., C. L. Chappel, C. R. Sterling, P. C. Okhuysen, J. B. Rose, and W. Jakubowski. 1995. The infectivity of *Cryptosporidium* in healthy volunteers. N. Engl. J. Med. 332:855.
31. FAO/WHO. 1995. Application of risk analysis to food standards issues. Report of the Joint FAO/WHO Expert Consultation. WHO, Geneva. WHO/FNU/FOS/95.3.
32. FAO/WHO. 1997. Risk management and food safety. Report of a Joint FAO/WHO Expert Consultation. FAO, Rome. FAO Food and Nutrition Paper No. 65.
33. FAO/WHO. 1998. The application of risk communication to food standards and safety matters. Report of a Joint FAO/WHO Expert Consultation. FAO, Rome. FAO Food and Nutrition Paper No. 70.
34. Farber, J. M., W. H. Ross, and J. Harwig. 1996. Health risk assessment of *Listeria monocytogenes* in Canada. Int. J. Food Microbiol. 30:145.
35. Foegeding, P. M. 1997. Driving predictive modelling on a risk assessment path for enhanced food safety. Int. J. Food Microbiol. 36:87.
36. Gale, P. 1996. Developments in microbiological risk assessment models for drinking water—A short review. J. Appl. Bacteriol. 81:403.
37. Gerba, C. P., and C. N. Haas. 1988. Assessment of risks associated with enteric viruses in contaminated drinking water, p. 489. *In* J. J. Lichtenber, J. A. Winter, C. I. Weber, and L. Fradkin (ed.), Chemical and biological characterization of sludges, sediments, dredge spoils, and drilling muds. Am. Soc. of Test. and Mater., Philadelphia, Pa.
38. Gerba, C. P., J. B. Rose, and C. N. Haas. 1996. Sensitive populations: Who is at the greatest risk? Int. J. Food Microbiol. 30:113.
39. Gerba, C. P., J. B. Rose, C. N. Haas, and K. D. Crabtree. 1996. Waterborne rotavirus: a risk assessment. Water Res. 30:2929.

40. Haas, C. N. 1983. Estimation of risk due to low doses of microorganisms: a comparison of alternative methodologies. Am. J. Epidemiol. 118:573.
41. Haas, C. N., J. B. Rose, and C. P. Gerba, C. P. 1999. Quantitative microbial risk assessment. John Wiley & Sons, Inc. New York.
42. Hathaway, S. C. 1993. Risk assessment procedures used by the Codex Alimentarius Commission and its subsidiary and advisory bodies. Food Control 4:189.
43. Hathaway, S.C. 1997. Development of food safety risk assessment guidelines for foods of animal origin in international trade. J. Food Prot. 60:1432.
44. Hathaway, S. C. 1999. The principle of equivalence. Food Control, in press.
45. Hathaway, S. C., and R. L. Cook. 1997. A regulatory perspective on the potential uses of microbial risk assessment in international trade. Intl. J. Food Microbiol. 36:127-133.
46. Havelaar, A. H., and P. F. M. Teunis. 1998. Effect modelling in quantitative microbiological risk assessment, p. 1. In A. M. Henken, and E. G. Evers (ed.), Proceedings of the 11th Annual Meeting of the Dutch Society for Veterinary Epidemiology and Economics. Bilthoven, The Netherlands.
47. Hudson, C. B. 1991. Risk assessment and risk management: Key factors in food safety decision making. Food Australia 43:S10.
48. International Commission for the Microbiological Specifications for Foods. 1996. Microorganisms in Foods 5: Characteristics of microbial pathogens. T. A. Roberts (ed.). Blackie Academic & Professional, London.
49. International Commission for the Microbiological Specifications for Foods. 1998. Microorganisms in Foods 6: Microbial ecology of food commodities. T. A. Roberts (ed.). Blackie Academic & Professional, London.
50. International Commission for the Microbiological Specifications for Foods. 1998. Potential application of risk assessment techniques to microbiological issues related to international trade in food and food products. J. Food Prot. 61:1075.
51. International Life Science Institute, Europe. 1995. A scientific basis for regulations on pathogenic microorganisms in foods. Dairy Food Environ. Sanit. 15:301.
52. International Life Science Institute, North America–Risk Science Institute Pathogen Risk Assessment Working Group. 1996. A conceptual framework to assess the risks of human disease following exposure to pathogens. Risk Anal. 16:841.
53. Jaykus, L-A. 1996. The application of quantitative risk assessment to microbial food safety risks. Crit. Rev. Microbiol. 22:279.
54. Kindred, T. P. 1996. Risk analysis and its application in FSIS. J. Food Prot. (Suppl.):24.
55. Lammerding, A. M. 1997. An overview of microbial food safety risk assessment. J. Food Prot. 60:1420.
56. Lammerding, A. M., and G. M. Paoli. 1997. Quantitative risk assessment: an emerging tool for emerging foodborne pathogens. Emerg. Infect. Dis. 3:483.
57. Macler, B. A., and S. Regli. 1993. Use of microbial risk assessment in setting US drinking water standards. Int. J. Food Microbiol. 18:245.
58. Marks, H., and M. Coleman. 1998. Estimating distributions of numbers of organisms in food products. J. Food Prot. 61:1535.
59. Marks, H. M., M. E. Coleman, J. C.-T. Lin, and T. Roberts. 1998. Topics in microbial risk assessment: dynamic flow tree process. Risk Anal. 18:309.
60. Martin, S. A., T. S. Wallsten, and N. D. Beaulieu. 1995. Assessing the risk of microbial pathogens: application of judgement-encoding methodology. J. Food Prot. 58:529.
61. McKone, T. E. 1996. Overview of risk analysis approach and terminology: The merging of science, judgement and values. Food Control 7:69.
62. McMeekin, T. A., J. Olley, T. Ross, and D. A. Ratkowsky. 1993. Predictive microbiology. Theory and application. Research Studies Press, Taunton.
63. McNab, W. B. 1997. A literature review linking microbial risk assessment, predictive microbiology, and dose-response modeling. Dairy Food Environ. Sanit.17:405.
64. McNab, W. B. 1998. A general framework illustrating an approach to quantitative microbial food safety risk assessment. J. Food Prot. 61:1216.
65. Miller, A. J., R. C. Whiting, and J. L. Smith. 1997. Use of risk assessment to reduce listeriosis incidence. Food Technol. 51:100.
66. Moore, D. R. J. 1996. Using Monte Carlo analysis to quantify uncertainty in ecological risk assessment: are we gilding the lily or bronzing the dandelion? Human Ecol. Risk Assess. 2:628.
67. Morales, R. A., and R. M. McDowell. 1998. Risk assessment and economic analysis for managing risks to human health from pathogenic microorganisms in the food supply. J. Food Prot. 61:1567.
68. Morgan, M. G., and M. Henrion. 1990. Uncertainty: A guide to dealing with uncertainty in quantitative risk and policy analysis. Cambridge University Press, New York.
69. National Advisory Committee on Microbiological Criteria for Foods. 1998. Principles of risk assessment for illnesses caused by foodborne biological agents. J. Food Prot. 61:1071.
70. National Research Council. 1983. Risk assessment in the Federal Government: Managing the process. Natl. Acad. Press, Washington, D. C.
71. National Research Council. 1994. Science and judgement in risk assessment. Natl. Acad. Press, Washington, D. C.
72. National Research Council. 1996. Understanding risk: Informing decisions in a democratic society. Natl. Acad. Press, Washington, D. C.
73. Nauta, M. J., and J. W. B. van der Giessen. 1998. Human exposure to *Mycobacterium paratuberculosis* via pasteurised milk: A modelling approach. Vet. Rec. 143:293.
74. Neumann, D. A,. and J. A. Foran. 1997. Assessing the risks associated with exposure to waterborne pathogens: An expert panel's report on risk assessment. J. Food Prot. 60:1426.
75. Notermans, S., J. Dufrenne, P. Teunis, R. Beumer, M. T. Giffel, and P. P. Weem. 1997. A risk assessment study of *Bacillus cereus* present in pasteurized milk. Food Microbiol. 14: 43.
76. Notermans, S., J. Dufrenne, P. Teunis, and T. Chackraborty. 1998. Studies on the risk assessment of *Listeria monocytogenes*. J. Food Prot. 61:244.
77. Notermans, S., G. Gallhoff, M. H. Zwietering, and G. C. Mead. 1995. The HACCP concept: specification of criteria using quantitative risk assessment. Food Microbiol. 12:81.
78. Notermans, S., and G. C. Mead. 1996. Incorporation of elements of quantitative risk analysis in the HACCP system. Int. J. Food Microbiol. 30:157.
79. Oscar, T. P. 1997. Use of computer simulation modeling to predict the microbiological safety of chicken. *In* Proceedings of 32nd National Meeting Poultry Health Processing. Georgetown, Del.
80. Pariza, M. W. 1992. Risk assessment. Crit. Rev. Food Sci. Nutr. 31:205.
81. Peeler, J. T., and V. K. Bunning. 1994. Hazard assessment of *Listeria monocytogenes* in the processing of bovine milk. J. Food Prot. 57:689.
82. Perez, J. F., F. K. Ennever, and S. M. Le Blanc. 1998. *Cryptosporidium* in tapwater. Comparison of predicted risks with observed levels of disease. Am. J. Epidemiol. 147:289.
83. Potter, M. E. 1994. The role of epidemiology in risk assessment: A CDC perspective. Dairy Food Environ. Sanit. 14:738.
84. Powell, D., and W. Leiss. 1997. Mad cows and mother's milk. The perils of poor risk communication. McGill Queen's University Press, Ontario, Canada..
85. Rai, S. N., D. Krewski, and S. Bartlett. 1996. A general framework for the analysis of uncertainty and variability in risk assessment. Hum. Ecol. Risk Assess. 2:972.
86. Roberts, T., J. Buzby, J. Lin, P. Nunnery, P. Mead, and P. I. Tarr. 1998. Economic aspects of E. coli O157:H7–Disease outcome trees, risk, uncertainty and the social cost of disease estimates, p. 155. In

B. Greenwood and K. De Cock (ed.), New and resurgent infections: Prediction, detection, and management of tomorrow's epidemics. John Wiley & Sons Inc., New York.

87. Rodricks, J. V. 1992. Calculated risks: The toxicity and human health risks of chemicals in our environment. Cambridge University Press, New York.
88. Rodricks, J. V. 1994. Risk assessment, the environment and public health. Environ. Health Perspect. 102:258.
89. Rose, J. B., and C. P. Gerba. 1991. Use of risk assessment for development of microbial standards. Water Sci. Technol. 24:29.
90. Rose, J. B., C. N. Haas, and C. P. Gerba. 1995. Linking microbiological criteria for foods with quantitative risk assessment. J. Food Safety 15:132.
91. Rose, J. B., C. N. Haas, and S. Regli. 1991. Risk assessment and control of waterborne giardiasis. Am. J. Public Health 81:709.
92. Rose, J. B., and M. D. Sobsey. 1993. Quantitative risk assessment for viral contamination of shellfish and coastal waters. J. Food Prot. 56:1043.
93. Ross, T., and T. A. McMeekin. 1994. Predictive microbiology—A review. Int. J. Food Microbiol. 23:241.
94. Royal Society, The. 1992. Risk analysis, perception and management. Report of a Royal Society Study Group. The Royal Society, London.
95. Rubin, L. G. 1987. Bacterial colonization and infection resulting from multiplication of a single organism. Rev. Infect. Dis. 9:488.
96. Rubin, L. G., and E. R. Moxon. 1984. Haemophilus influenzae type B colonization resulting from survival of a single organism. J. Infect. Dis. 149:278.
97. Rugen, P., and B. Callahan. 1996. An overview of Monte Carlo: A fifty year perspective. Hum. Ecol. Risk Assess. 2:671.
98. Soby, B. A., A. C. D. Simpson, and D. P. Ives. 1994. Managing food-related risks: integrating public and scientific judgements. Food Control 5:9.
99. Starfield, A. M., K. A. Smith, and A. L. Bleloch. 1990. How to model it: Problem solving for the computer age. McGraw-Hill, New York.
100. Tauxe, R. V. 1997. Emerging foodborne diseases: an evolving public health challenge. Dairy Food Environ. Sanit. 17:788.
101. Teunis, P. F. M., O. G. van der Heijden, J. W. B. van der Geissen, and A. H. Havelaar. 1996. The dose-response relation in human volunteers for gastro-intestinal pathogens. Report 2845500002. Natl. Inst. of Public Health and Environ., Bilthoven, The Netherlands.
102. Thompson, K. M., D. E. Burmaster, and E. A. C. Crouch. 1992. Monte Carlo techniques for quantitative uncertainty analysis in public health risk assessments. Risk Anal. 12:53.
103. Todd, E. C. D. 1996. Risk assessment of use of cracked eggs in Canada. Int. J. Food Microbiol. 30:125.
104. Todd, E. C. D., and J. Harwig. 1996. Microbial risk analysis of food in Canada. J. Food Prot. (Suppl.): 10.
105. U.S. Environmental Protection Agency. 1997. Guiding principles for Monte Carlo analysis. USEPA. Washington, D. C. EPA/630/R-97/001.
106. U.S. Presidential/Congressional Commission on Risk Assessment and Risk Management (PCCRARM). 1997. Framework for Environmental Health Risk Management. The Presidential/Congressional Commission on Risk Assessment and Risk Management, Washington, D. C.
107. van der Logt, P. B., S. C. Hathaway, and D. J. Vose. 1997. Risk assessment model for human infection with the cestode *Taenia saginata*. J. Food Prot. 60:1110.
108. van Gerwen, S. J. C., and M. H. Zwietering. 1997. Growth and inactivation models to be used in quantitative risk assessments. J. Food Prot. 61:1541.
109. van Schothorst, M. 1996. Practical approaches to risk assessment. J. Food Prot. 60:1439.
110. van Schothorst, M. 1998. Principles for the establishment of microbiological food safety objectives and related control measures. Food Control 9:379.
111. Vose, D. 1996. Quantitative risk analysis: A guide to Monte Carlo simulation modelling. John Wiley & Sons, Inc. New York.
112. Vose, D. 1998. The application of quantitative risk assessment to microbial food safety. J. Food Prot. 61:640.
113. Whiting, R. C., and R. L. Buchanan. 1994. IFT scientific status summary: Microbial modeling. Food Technol. 44:113.
114. Whiting, R. C., and R. L. Buchanan. 1997. Development of a quantitative risk assessment model for *Salmonella* Enteritidis in pasteurised liquid eggs. Int. J Food Microbiol. 36:111.
115. WHO. 1998. Food safety and globalization of trade in food, a challenge to the public health sector. WHO, Geneva. WHO/FSF/FOS/97.8 Rev. 1
116. Zwitering, M. H., and A. P. M. Hasting. 1997. Modelling the hygienic processing of foods—A global process overview. Trans. Ind. Chem. Eng. 75:159.
117. Zwitering, M. H., and A. P. M. Hasting. 1997. Modelling the hygienic processing of foods—Influence of individual process stages. Trans. Ind. Chem. Eng. 75:168.
118. Zwitering, M. H., J. C. de Wit, and S. Notermans. 1996. Application of predictive microbiology to estimate the number of *Bacillus cereus* in pasteurized milk at the point of consumption. Int. J. Food Microbiol. 30:55.

APPENDIX 1.

Common Distributions

Name	Figure	Comments
Beta	***Beta [11, 21]***	The beta distribution can be used to model the probability of the occurrence of an event. In microbial risk assessment, it is commonly used to model prevalence. With data on the number of trials conducted and the number of successes in those trials, the parameters for the distribution can be estimated, α = number positive + 1, β = number negative +1. If we are modeling the prevalence of pathogen "A" in some food and data has been collected that indicates, out of 30 samples, 10 were positive, the probability that any other sample collected will be positive can be modeled using a beta probability with α = 11 (positive +1) and β = 21 (number of samples – number positive +1). This is the distribution shown at left, which reflects the most likely prevalence of 33% (10/30).
Binomial	***Binomial [10, 0.5]*** 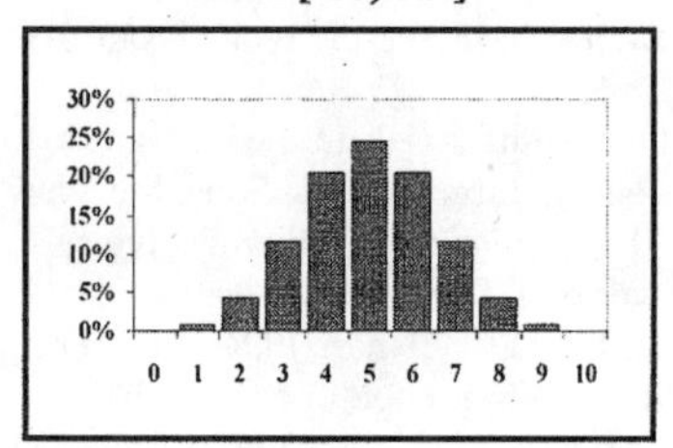	The binomial distribution can be used to model the number of successes (x) out of a number of trials (n) assuming the probability of success (p) on each trial is constant. A simple example of the application of this distribution could be estimating the number of heads we would get if we tossed a coin 10 times. The probability of success, if the coin is perfectly fair, is 50%. The distribution for the binomial distribution function with p = 0.5 and n = 10 is shown at left. The x-axis represents the number of heads and the y-axis the associated probability of getting that number of heads in 10 trials.

Name	Figure	Comments
Poisson	***Poisson [5]*** 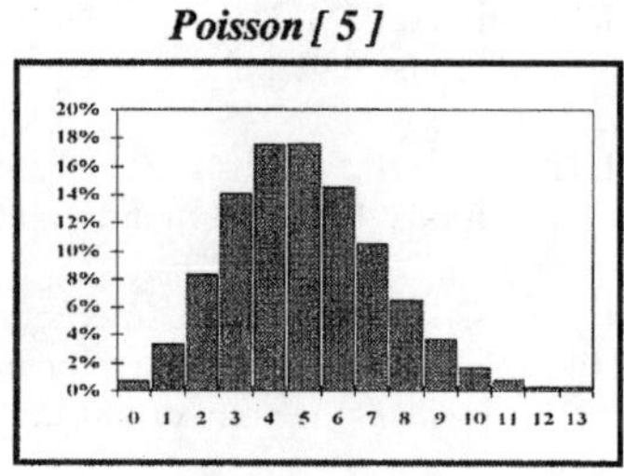	The poisson distribution can be used to model the number of occurrences of an event within an interval. The distribution can be defined with one parameter, the mean number of occurrences within a unit interval. For example, we have a large batch of some substance, and the average concentration of a pathogen in the substance is 0.5 CFU/g. We want to model how many organisms we might find if we took 10-g samples of that substance. This could be modeled using a Poisson [5] distribution (0.5 x 10). Note, it would not be correct to model this using a Poisson [0.5] and then multiplying by 10.
Exponential	***Exponential [2]*** 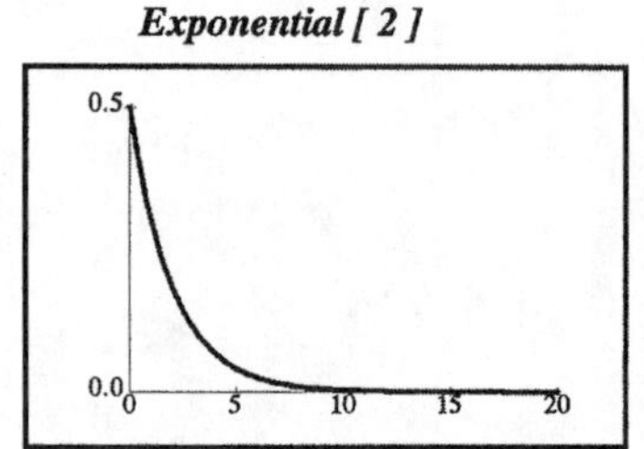	The exponential distribution is used to describe the interval (time, volume, mass etc) between the occurrence of an event. The exponential distribution is related to the poisson distribution and requires one parameter, the mean interval before the occurrence of an event. The parameter β in the exponential distribution is 1/λ as defined in the poisson distribution. Using a similar example as that of the poisson distribution, let's say we want to model the amount of the substance that we must sample before we encounter a pathogen. This can be accomplished using an Exponential [2] distribution. The parameter "2", represents the mean amount of the material that would have to be sampled before one organism would be expected to be encountered. The exponential [2] distribution is shown at left.

Name	Figure	Comments
Gamma	*Gamma [10, 0.5]* 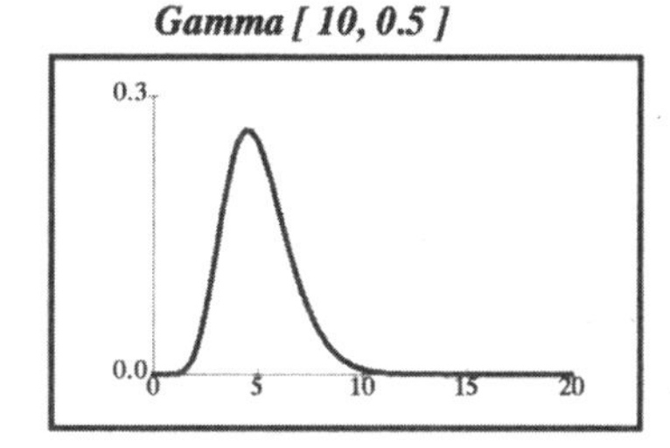	The Gamma distribution is related to the exponential and poisson distributions, and can be used to model the interval between the occurrence of an event. Thus when α, the number of events, is 1, the gamma distribution is equivalent to the exponential distribution. If for instance we want to simulate how much of the substance from the previous examples needs to be sampled in order to encounter 10 organisms, we could achieve this using a Gamma [10,0.5] distribution.
LogNormal	*Lognormal [2, 1]* 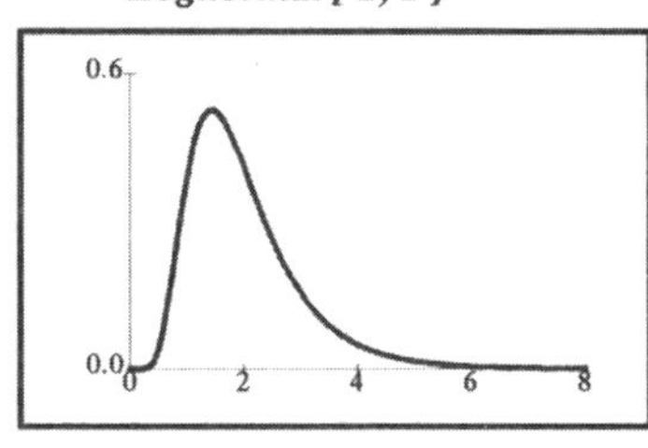	Lognormal distributions are commonly used to describe naturally occurring phenomenon. Many physical and biological processes tend to follow lognormal distributions. The products of naturally occurring random variables produce a random variable that tends to be lognormally distributed (central limit theorem). In addition, lognormal distributions often provide a good representation for physical quantities that extend from zero to positive infinity.[111] The lognormal distribution has commonly been used in microbial risk assessment to describe the concentration of microorganisms in substrates. The natural distribution of microorganisms in a lake, for instance, can be described using a lognormal distribution.
Normal	*Normal [10, 2]* 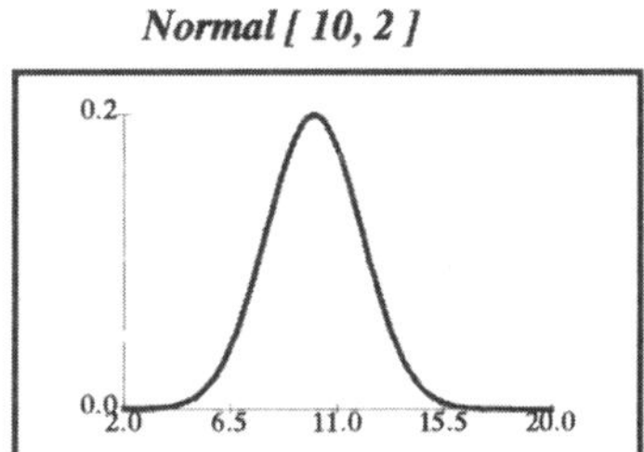	The normal distribution is perhaps one of the most recognizable and commonly used distributions in all fields of study. Like the lognormal distribution that is commonly used to describe the product of random variables, the normal distribution is commonly used to describe the sum of random variables. Its most common application has been the description of physical characteristics of populations, such as height, weight etc. Theoretically, the normal distribution stretches from negative to positive infinity, as a result, some caution should be used in applying the distribution to avoid potentially erroneous results.

Name	Figure	Comments
Triangular	*Triangular [0, 2, 8]* 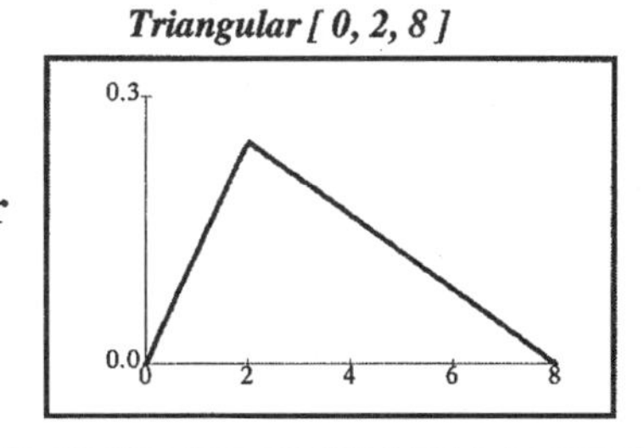	The triangular distribution is used in risk assessment as a rough modeling distribution in the absence of sufficient data. It requires only estimates of the maximum, minimum and most likely value that a variable can take. As a result, it is also commonly used in translating expert opinion into a distribution, since the parameters required for estimation are often intuitive.
Uniform	*Uniform [2, 10]* 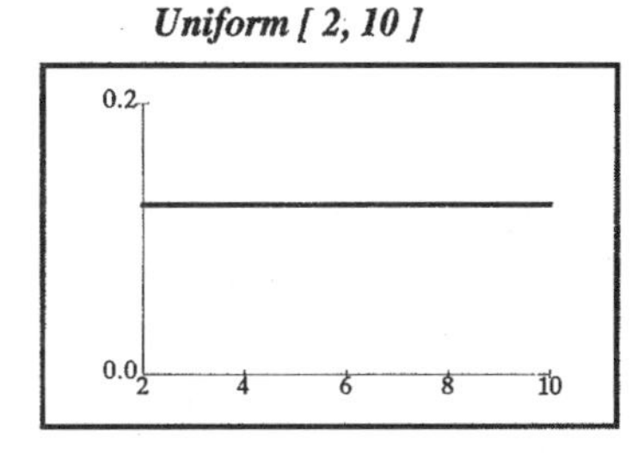	The uniform distribution is even coarser than the triangular distribution. The only estimates required are the minimum and maximum values that a variable could take. Like the triangular distribution, it is commonly used for rough modeling in the absence of data and for expert opinion. The uniform distribution is usually a poor representation of reality since it assigns the same probability for all the values within the range.

Chapter 30

Aeromonas, Arcobacter, and *Plesiomonas*

Samuel Palumbo, Carlos Abeyta, Gerard Stelma, Irene W. Wesley, Cheng-i Wei, John A. Koberger, Sharon K. Franklin, Linda Schroeder-Tucker, and Elsa A. Murano

30.1 *AEROMONAS HYDROPHILA*

30.11 General Characteristic of the Genus

Since the previous edition, major changes and improvements have occurred in the taxonomy of the genus *Aeromonas*. While a study by Colwell et al.[25] which, based on molecular techniques, indicated that *Aeromonas* should be a separate family, *Aeromonas* remains in the family *Vibrionaceae*. This is based on physiological traits rather than phylogenetic relationships. However, within the genus *Aeromonas*, individual species are now separable into 14 hybridization groups (HG). Recent work summarized by Joseph and Carnahan[42] indicated that these 14 HGs can also be separated into species and identified by a series of 11 phenotypic tests (see Table 1).

General traits of the genus *Aeromonas* include: facultative anaerobic gram-negative rod, motile by single polar flagellum (exception: *A. salmonicida*), metabolism of glucose by both fermentative and respiratory pathways, positive catalase and oxidase tests, production of exoenzymes such as amylase, protease, phosholipase, and DNase, resistance to the vibriostatic agent O/129, and lack of the enterobacterial antigen. The genus *Aeromonas* can be divided into two major groups: the mesophilic or motile aeromonads [sometimes referred to as the *A. hydrophila* group] (growth at 37°C, motility) and the *A. salmonicida* group (non-motile, psychrophilic, production of a brown pigment, and pathogenic for salmonid fish). It is the mesophilc aeromonads which are of interest and concern to human public health as a cause of extra-intestinal infections and a potential agent of diarrheal disease.

30.12 Ecology

A. hydrophila was first reported by Sandrelli,[80] who isolated the bacterium from frogs. Motile aeromonads occur widely in the aquatic environment, both fresh and saline waters. Its name translates simply as a "water-loving, gas-producing" bacterium. The bacterium also occurs widely in foods, both fresh and processed, undoubtedly as result of the food's contact with a source of water containing the bacterium. These bacteria also occur widely in sludge and sewage.

Water appears to represent a major reservoir of this group because these bacteria are able to grow at low nutrient levels and they can survive for long periods of time in water, particularly at low temperatures.[66] Motile aeromonads are present in both chlorinated and unchlorinated waters. Low levels of these bacteria have been found in chlorinated drinking water that is free of *E. coli*. Motile aeromonads isolated from chlorinated drinking water in the United States have been shown to possess virulence factors that may be related to enteric disease. Epidemiological studies suggest that *Aeromonas*-associated gastroenteritis peaks during the warmer months when the highest levels of motile aeromonads are present in the aquatic habitats.

Motile aeromonads are not considered normal inhabitants of the human gastrointestinal tract. The fecal carriage rate of *A. hydrophila* in asymptomatic hospitalized patients varies from 0 to 8%. This suggests that motile aeromonads are transient inhabitants of the human intestinal tracts and humans are not the major contributors of the bacterium to the environment, although aeromonads that are shed can multiply in sewage lines to significant numbers prior to discharge into receiving waters. In general, there is a low rate of isolation of aeromonads in the feces of food animals such as bovine, swine, and poultry and it is unlikely that feces of these animals represent a source of the bacteria in foods. Borch et al.[12] has indicated that motile *Aeromonas* spp. are an environmental contaminant in swine slaughter facilities and their presence on swine carcasses come from the processing plant rather than the animal itself.

As indicated above, water is the likely source of these bacteria in foods, with fish and seafood most often contaminated. Motile aeromonads are not known to be part of the intestinal flora or tissue of healthy fish. However, they are easily isolated from retail market fish. Shellfish, particularly oysters, represent another known reservoir of the bacterium.

Motile aeromonads are often associated with refrigerated animal products such as chicken, beef, pork, lamb, and dairy products. In addition, they can be isolated from a wide range of other foods such as vegetables, spices, and other fresh and processed foods.[64]

30.13 Epidemiology

Though motile aeromonads have long been associated with various fresh and processed foods, their definitive recognition as a causative agent of diarrhea has not been conclusively demonstrated.[58] There are many anecdotal accounts as well as specific instances of mesophilic *Aeromonas* spp. being the sole pathogen isolated from cases of diarrhea. In addition, as will be seen be-

Table 1. Phenotypic Test to Speciate Bacteria Within the Genus *Aeromonas*, from Joseph and Carnahan[42]

DNA HG	phenospecies	Test										
		esculin hydrolysis	gas from glucose	ampicillin resistance	acid from arabinose	lysine	cephalothin resistance	ornithine	acid from sucrose	Voges-Proskauer	acid from mannitol	indole
HG 1	*A. hydrophila*	+	+	+	±	+	+	–	+	+	+	+
HG 2/3	*A. hydrophila*	+	+	+	+	–	+	–	+	+	+	+
HG 4/5a	*A. caviae*	+	–	+	+	–	+	–	+	–	+	+
HG 5b	*A. media*	+	–	–	+	–	±	–	+	–	+	±
HG 6	*A. eucrenophila*	–	–	–	+	–	–	–	–	–	+	+
HG 7	*A. sobria*	–	±	±	–	+	–	–	+	–	+	+
HG 8x	*A. veronnii* bv *sobria*	–	+	+	–	+	+	–	+	–	+	+
HG 8y	*A. veronii* bv *sobria*	–	+	+	–	+	–	–	+	+	+	+
HG 9	*A. jandaei*	–	+	+	–	+	+	–	–	+	+	+
HG 10	*A. veronii* bv *veronii*	+	+	+	–	+	–	±	+	+	+	+
HG 11	*Aeromonas* sp.	–	–	–	+	–	+	+	+	–	+	+
HG 12	*A. schubertii*	–	–	±	–	+	–	–	–	–	–	–
HG 13	*A.* spp. gr. 501	–	–	+	–	–	+	–	–	+	–	+
HG 14	*A. trota*	–	+	–	–	+	+	–	±	–	+	+

low, many mesophilic *Aeromonas* cultures, including those isolated from water and various foods,[48,49,65,75] possess factors associated with virulence in other enteric pathogens

30.131 Enterotoxins

The most controversial issue related to *Aeromonas* pathogenicity concerns the relative roles of various putative enterotoxins in causing diarrhea. There are reports of both heat-stable (56°C, 10 min) and heat-labile cytotonic enterotoxins and cytotoxic enterotoxins, some related and others unrelated to cholera toxin. The controversy is not over the variety of enterotoxins reported, which is not unusual for a diverse group of organisms, but that researchers who reported evidence for one of these toxins usually were unable to observe any of the others. The various putative enterotoxins produced by *Aeromonas* species are described below.

Ljungh et al. reported partially purifying an enterotoxin with a molecular weight of 15 kDa that was stable after treatment at 56°C for 10 min.[54,55] This molecule, which was unrelated serologically to cholera toxin (CT), caused fluid accumulation in the permanently ligated rabbit ileal loop (RIL), rounding of Y-1 cells without death, and stimulation of cyclic AMP (adenosine monophosphate) synthesis. Further evidence for a heat-stable cytotonic enterotoxin was provided by Chakraborty et al., who reported cloning the cytotonic enterotoxin gene into *E. coli*.[22] Culture filtrates of the clone caused elongation of Chinese hamster ovary (CHO) cells and fluid accumulation in the RIL after treatment at 56°C for 20 min. These activities were also unrelated to CT. Both of these groups reported that the beta hemolysin was inactive in the RIL.

Potomski et al.[73] reported using affinity chromatography to isolate an *A. sobria* toxin, caused rounding of Y-1 cells, and fluid accumulation in both RILs and infant mice after heating at 56°C for 20 min. All of these activities were reportedly neutralized by antiserum to CT. Schultz and McCardell[81] also reported rounding of Y-1 cells and stimulation of cyclic AMP synthesis by *Aeromonas* culture filtrates that were heated at 56°C for 20 min. These activities were neutralized partially by antiserum to CT. They also reported that DNA from strains producing the cytotonic enterotoxin reacted with one or more synthetic oligonucleotide probes coding for CT.

Chopra and Houston[23] reported purifying a cytotonic enterotoxin that caused fluid accumulation in the RIL, stimulated cyclic AMP synthesis, and caused elongation of CHO cells. The purified toxin had a molecular weight of 44 kDa, was free of hemolytic and cytotoxic activities, and was not cross-reactive with antiserum to CT. The biological activities of the purified toxin were heat-labile at 56°C.

The first evidence for a cytotoxic enterotoxin was provided by Cumberbatch et al.,[27] who reported a correlation between cytotoxic and RIL activities. All of their cytotoxic isolates were also hemolytic. They found no evidence for a separate cytotonic activity in any of their isolates. Turnbull et al.[86] and Burke et al.[16] also observed a strong correlation between enterotoxin production and hemolytic activity. Asao et al.[6] provided the first direct evidence for a cytotoxic enterotoxin by purifying a hemolysin from *Aeromonas hydrophila* strain AH-I to electrophoretic homogeneity and observing that it was cytotoxic to Vero cells and had enterotoxic activity in both the RIL and suckling mouse assays. The hemolysin had a molecular weight of 50 kDa and was heat labile at 56°C for 5 min. Stelma et al.[84] showed that the hemolysin described by Asao was beta-hemolytic and that antiserum to the purified hemolysin completely neutralized the RIL activities of filtrates of beta-hemolytic *Aeromonas* isolates, indicating that beta hemolysin alone can cause the changes in intestinal permeability associated with diarrhea. These researchers also used polyclonal antiserum to show serological cross-reactions between the hemolysin purified from the Japanese strain, AH-1, and hemolysins produced by isolates from diverse geographic origins.

Several additional studies provided evidence that the *Aeromonas* beta-hemolysins are a family of molecules related to the "aerolysin" originally described by Bernheimer and Avigad[11]. Rose et al.[77,78] purified a 52-kDA protein toxin that possessed hemolytic, cytotoxic and enterotoxic activities as well as serological cross-reactivities to both CT and the Asao (AH-1) hemolysin. The biological activities of their toxin were neutralized by homologous antiserum and antiserum against the AH-1 hemolysin but not by antiserum against cholera toxin. Asao et al.[7] demonstrated that the hemolysin produced by *A. hydrophila* CA11, a U.S. Gulf Coast isolate, was related immunologically to AH-1 hemolysin but also possessed unique antigenic determinants. Millership et al.,[57] Kozaki et al.,[47] and Stelma et al.[85] provided evidence that hemolysins from *A. sobria* and *A. veronii* also possessed enterotoxic activities and were related serologically to the AH-1 hemolysin. Haque et al.[16] reported that some *Aeromonas* stains produce a cytotoxin that is antigenically related to *E. coli* 0157:H7 Shiga-like toxin. The relationship of that toxin to the AH-1 toxin has not been investigated.

Evidence from several surveys suggests that the cytotoxic enterotoxins are the most common *Aeromonas* enterotoxins. Cumberbatch et al.[27] found no cytotonic activity in the filtrates of 96 isolates. Likewise, Johnson and Lior found no cytotonic activity in the filtrates of 73 isolates,[41] and Seidler et al. found cytotonic activity in the filtrates of only 20 of 330 isolates (6%).[82] Stelma et al.[84] found evidence for cytotonic activity in 1 of 24 isolates, but this later was shown to be caused by sublethal doses of partially denatured cytotoxic enterotoxin after heating at 56°C.[15]

The absence of significant enterotoxic activity in the beta-hemolysin purified by Ljungh et al.[55] may be due to differences in the purification procedures. Ljungh et al. used a six-step procedure that recovered only 0.6% of the original hemolytic units.[55] In contrast, Asao et al. used a two-step procedure that recovered 65% of the original hemolytic units and probably yielded a product closer to the native molecule in all of its properties.[6] The failure of the *E. coli* clones carrying the *A. hydrophila* hemolysin gene to cause fluid accumulation in the RIL[22] was shown later to be due to the inability of the *E. coli* to release the hemolysin from the cells.[21]

Although the heat-labile cytotoxic enterotoxins appear to be the most common *Aeromonas* enterotoxins, the relative roles of these toxins and the cytotonic enterotoxins in *Aeromonas* diarrhea are not known. Determination of the relative roles of these toxins will require the development of better animal models and either the use of strains that produce only one toxin or the use of transposon mutagenesis to inactivate various toxins one at a time.

It has been agreed generally that *A. caviae* isolates were noncytotoxigenic and were not enteric pathogens.[34,40] However, Namdari and Bottone reported detecting a heat-stable cytotoxin in culture filtrates of isolates grown in double-strength trypticase soy broth (TSB).[60] They, like earlier investigators, did not detect enterotoxin in filtrates of *A. caviae* grown in single-strength TSB. young children and individuals aged 50 years or more.[50,51,61]

30.132 Other Virulence Factors

The negative results of the human feeding study performed by Morgan et al.[58] provided evidence that strains producing enterotoxin active in the RIL are not always diarrheagenic and suggests that multiple virulence factors may be involved. The strains used in that study possessed neither adhesion factors for colonization of the intestine nor the ability to invade tissues. Several studies have provided evidence that some aeromonads possess adhesion factors that correlate with the presence of pili.[24,36,41] Kirov et al.[44] observed that environmental enterotoxigenic isolates possessed numerous pili that appeared to be lost once infection was established. This suggests that in future studies environmental isolates may be more appropriate for human feedings than clinical isolates. Other possible virulence factors related to adherence include production of collagen-binding and mucin-binding proteins.[8,9]

The relationship between virulence factors and the ability of *Aeromonas* strains to cause disseminating infections has not been established. Lethality tests with both normal mice and mice immunosuppressed by X-irradiation have focused primarily on relating virulence to biotype or phenospecies.[13,38] The results of those studies and one of invasiveness to mammalian cells[88] indicated that *A. hydrophila* and *A. sobria* were inherently more virulent than *A. caviae*, but with significant strain-to-strain variation within a species.

One property linked to virulence is a surface-array protein (S layer) commonly found in human isolates from extraintestinal infections. The role of the S layer in human and animal infections is not clear, but it appears to be substantially different from the role of the S layer of *A. salmonicida* in fish disease.[39,45] Other properties of motile aeromonads that have been linked to virulence in other bacterial species include the ability to invade mammalian cells,[69,88] resistance to serum bactericidal effects,[37,65] production of a siderophore capable of removing iron from transferrin,[10,56] a mechanism for utilization of iron from haem compounds,[56] and production of proteases.[52] Loss of viability after growth in broth containing 0.5% glucose (suicide phenomenon) has been associated with both enteropathogenicity and virulence,[59] but the significance of this phenomenon is not known. The relative significance of each of these putative virulence characteristics cannot be determined until appropriate animal models are developed, preferably models in which the animals are compromised in a way that mimics the condition of the susceptible human host.

30.14 Isolation Media

Food microbiologists have traditionally depended on clinical laboratory media for the isolation of human pathogens from food. However, the requirements of food microbiologists are different from clinical microbiologists. Specifically, food microbiologists are interested in not only which bacteria may be present in the food, but also how many. In addition, bacteria in food may have been stressed or injured by various processing operations such as heating, freezing, drying, acidification etc., and thus can not grow on the selective media typically used for their isolation.

Numerous plating media have been developed for the isolation of *Aeromonas* from both clinical and environmental sources. Many plating media depend on specific physiological traits of these bacteria such as resistance to ampicillin and/or fermentation or utilization of different unique substrates. In addition, media typically used for enteric gram-negative bacteria also have been used. However, because of the requirements of the food microbiologist for quantitative recovery, many of these are not satisfactory. Further, on many of the media, differentiation of *Aeromonas* from the background microflora can be difficult. To correct these problems, Palumbo et al.[67] formulated a new medium, starch ampicillin agar (SAA), with starch hydrolysis as the differential trait (relatively few bacteria in food are capable of hydrolyzing starch) and ampicillin (final level of 10 mg/mL instead of the usual 30 used in most selective media for mesophilic *Aeromonas* spp.) to suppress the background microflora. Since its development, SAA has been used widely by food microbiologists for the isolation of *Aeromonas* from many different food products and, when compared to other media, has shown a very high percentage of presumed colonies confirmed as *Aeromonas*.[64] However, Nishikawa and Kishi,[64] when studying the recovery of motile *Aeromonas* from foods and environmental specimens, observed a problem with swarming of *Proteus* when they used SAA and formulated a new medium with starch hydrolysis as the differential system and bile salts and brilliant green as inhibitory substances.

Direct plating media can not be used to recover various pathogens from foods when low numbers are present (and particularly in the presence of a large background microflora) or when injured cells may be present. In these instances, an enrichment broth is needed. The enrichment broth allows repair of injured cells, which can then form colonies on selective media. Two enrichment broths have seen limited use for the isolation of motile *Aeromonas* from foods: a) alkaline peptone water (used for lamb, ice cream, and mixed seafoods), and b) tryptic soy broth + ampicillin (used for frozen oysters and raw and pasteurized milk). At this time however, there have not been any studies comparing their effectiveness in recovering injured cells from a wide range of foods, particularly those which might contain a high background microflora.

Membrane filter techniques have been useful for the isolation and quantitation of bacteria, including motile aeromonads from aquatic environments, especially when they occur in low numbers. These techniques may be useful in the analysis of certain foods. For water samples, Rippey and Cabelli[76] employed trehalose in a primary medium as the fermentable carbohydrate and ampicillin and ethanol as selective inhibitors. After 2 hr incubation at 37°C, an *in situ* mannitol fermentation followed by an *in situ* oxidase test was used to differentiate *A. hydrophila* from other organisms present in fresh waters. Their procedures gave recoveries that exceeded other procedures, with greater confirmation rates and fewer negative samples.

30.15 Characterization and Speciation in the Genus Aeromonas

Bacteria may be recognized as belonging to the genus *Aeromonas* by biochemical tests and thus easily separated from other closely related bacteria. In the past, motile aeromonads were often confused biochemically with the family *Enterobacteriaceae*. The oxidase test is an easy and useful test to separate *Vibrionaceae* and *Enterobacteriaceae*. Differentiation of the genera within the family *Vibrionaceae* can be accomplished by biochemical tests (Table 2). Two tests for differentiating *Vibrio* and *Aeromonas* are O/129 (vibriostatic agent) sensitivity and salt requirement. Sensitivity to the water-soluble vibriostatic agent 2, 4-diamino-6, 7-diisopropyl pteridine (O/129) phosphate is determined by disk sensitivity testing on nutrient agar containing 0.5% (wt/vol) NaCl. Disks containing either 10 µg or 150 µg of O/129 phosphate are used. Any zone of inhibition around the 150 µg disk is considered as sensitive to the vibriostatic agent. *Aeromonas* spp. are resistant to both levels. Growth at different concentrations of NaCl (e.g., 0% and 1%) is determined in 1% tryptone broth for 24 to 48 hr at 35°C. *Aeromonas* spp. will grow in the broth with and without NaCl, whereas *Vibrio* spp. require NaCl for growth.

As indicated previously, taxonomy of the genus *Aeromonas* has been clarified with the introduction of DNA hybridization techniques. Based on recent work, there are currently 14 hybridization groups (HG).[42] These authors have developed an identification scheme based on 11 phenotypic tests to speciate isolates within the genus *Aeromonas* (Table 1). While the proposed scheme at this time can be used to speciate isolates, the reactions for some of the tests are based on only a few isolates (*A. Carnahan*, personal communication). Since this scheme is fairly recent, many of the older studies which identified isolates from foods have used the biochemical tests proposed by Popoff and Veron[72] and Popoff[71] and found only *A. hydrophila*, *A. caviae*, and *A. sobria*. There have been very few studies which have identified other species.

Other identification schemes have been proposed, primarily for the identification of clinical isolates. Abbott et al.[1] devised a scheme consisting of 9 to 16 tests and permitted them to correctly assign 99% of their isolates to predicted hybridization groups. Carnahan et al.[22] developed the Aerokey II which consisted of 7 tests; they were able to correctly identify 97% of their isolates. Which of these schemes can be applied to the identification of food isolates remains to be studied.

Kaper et al.[43] described a single-tube medium for the rapid presumptive identification of *A. hydrophila*. It also offered good differentiation of *Klebsiella*, *Proteus*, and other enteric bacteria. The reactions that could be observed in the single tube were fermentation of mannitol and inositol, ornithine decarboxylase and deamination, indole production, motility, and H_2S production from sodium thiosulfate and cysteine. They found this medium useful in identifying environmental isolates of the *A. hydrophila* group.

Table 2. Differentiation of the Genus *Aeromonas* from Other Genera of the Family *Vibrionaceae* Encountered in Foods

Test	*Aeromonas*	*Vibrio*	*Plesiomonas*
Na^+ required for or stimulates growth	–	+	–
Inositol (acid)	–	–	+
Mannitol (acid)	+	+	–
Ornithine decarboxylase	–	D[1]	+
Production of exoenzymes			
Amylase	+	–	–
Gelatinase	+	+	–
Lipase	+	–	–
mol % G + C of DNA	57-63	38-51	51
Growth on thiosulfate-citrate-bile salt-sucrose agar	–	+	–
Inhibition by O/129			
10 µg	R[2]	D	D
150 µg	R	S[3]	S
Gas from glucose	+	–	–

[1]D = different biotypes [2]R = Resistant [3]S = Sensitive

Biochemical screening is usually done with the standard test methodology, but there are several rapid identification test procedures and kits available. These have been used in the identification of motile aeromonads. Ogden et al.[63] compared three rapid kits (API 20E, API 20NE, and Microbact 24E) for identification of food isolates of *Aeromonas*. They found that API 20NE and Microbact 24E correctly identified 96.3% of the food isolates studied.

The temperature at which the biochemical tests are incubated are important, especially when using one of the rapid systems. Microbiologists should use the temperature at which the cultures in the system's data base were incubated. Altwegg et al.[5] observed differences in the decarboxylase activity of motile aeromonad isolates as a function of both medium and incubation temperature, with 29°C being the best to retain two important biochemical reactions in these bacteria: ornithine negativity and arginine positivity. Ewing et al.[30,31] also reported differences in other biochemical reactions depending on temperature. Popoff and Veron[73] incubated all their test media at 30°C; because their scheme is recognized as the definitive one, this temperature should be used in determining the biochemical activity of isolates. Palumbo (unpublished data) observed that for various food isolates of motile aeromonads, many biochemical reactions are negative when incubated at 37°C compared to incubation at 28°C.

In addition to biochemical reactions, clinicians and epidemiologists often use other tests to characterize and differentiate isolates from various sources. These techniques permit tracing the epidemiological patterns of the bacteria. Some of these techniques undoubted can be applied to food isolates. These tests and techniques include phage typing (phage typing has also been used for diagnostic purposes in studies of *A. salmonicida*), serotyping,[28,33,70,79] multilocus enzyme electrophoresis, and esterase and haemagglutinination patterns.[17,26,46] The motile aeromonads are antigenically heterogeneous and this trait has been reviewed extensively.[14,29,53,73,83] The antigenic studies have dealt primarily with the serologic specificity of extracellular antigens and the diversity of somatic and flagellar antigens. Ewing and coworkers[30,31] found 12 O antigenic groups and one H antigenic group among 71 *A. hydrophila* strains tested. The presence of K antigens, partially inhibiting the O agglutination, has also been reported.

The profile of outer membrane proteins also appears as a potential means of distinguishing isolates of *A. hydrophila* from different sources. Canonica and Pisano[19] analyzed the fatty acid methyl esters from clinical isolates of motile aeromonads and observed the presence of two hydroxy fatty acid species: 3-hydroxy-12:0 and 3-hydroxy-14:0. The 3-hydrox-12:0 acid was unique to the clinical isolates and thus permitted separation of motile aeromonads of clinical original from nonclinical isolates. This approach might prove useful when applied to food isolates. Figura and Guglielmetti[32] describe testing motile *Aeromonas* from production of a CAMP-like factor. *A. hydrophila* strains were positive whether tested aerobically or anaerobically, *A. sobria* strains were positive only under aerobic conditions, and *A. caviae* strains were always negative for both conditions.

30.16 Handling of Samples

Since *A. hydrophila* group organisms are capable of ready growth in fresh foods of both animal[67] and plant origin[18] at 5°C, the food samples should be processed as soon as possible upon arrival at the laboratory. Motile aeromonads are somewhat sensitive to pH values below 5.5; therefore, acidic foods should be processed soon after arrival in the laboratory. The ability to withstand freezing has not been studied in motile aeromonads, though Abeyta et al.[2] were able to isolate *A. hydrophila* from oysters held frozen for 1.5 years at –72°C. An enrichment technique may be necessary when analyzing frozen foods for the presence of *A. hydrophila* group organisms.

30.17 Equipment, Materials, And Reagents

30.171 Equipment

Incubator, 28°C
Stomacher

30.172 Media

Alkaline peptone water
Bile salts brilliant green starch (BBGS)
Dnase test agar (Difco) or toluidine blue agar (Lachica et al. 1971. Appl. Environ. Microbiol. 24: 585–587)
Kaper's medium
Modified SA agar
Nutrient agar
Peptone (0.1%)
SA agar
Tryptic soy agar (TSA)
Tryptic soy broth + 30 mg/L ampicillin (TSBA)

30.173 Reagents

Gram stain reagents
Kovac's reagent
Lugol's iodine solution
1% N,N, N, N, tetramethyl-p-phenylenediamine dihydrochoride (for the oxidase test, necessary to differentiate members of the *Vibrionaceae* from the biochemically similar *Enteriobactriaceae*)
3% Hydrogen peroxide (H_2O_2)
Vibriostatic 0/129 discs (10 and 150 µg)

30.18 Procedures

The following procedures are suggested for isolating and quantitating motile aeromonads from foods.[3] These procedures are based on our experience in handling various foods and various media and techniques used for the *A. hydrophila* group.

30.181 Sampling

Aseptically weigh 25 g of the food into a stomacher bag, add 225 mL of sterile peptone (0.1%) water, and blend for 2 to 3 min in a stomacher laboratory blender. Further dilute peptone water as needed and surface plate 0.1 mL portions onto media described below. Use sterile bent glass rods to distribute the sample dilutions evenly over the surface of the media.

Another procedure may be used, where food samples are weighed into special stomacher filter bags (Spiral Systems, Bethesda, MD.) and the samples prepared as above. By use of the special filter bags, the food samples and dilutions can then be plated using a spiral plater and number of *A. hydrophila* can be determined quantitatively by an appropriate calculation.

30.182 Inoculation of Media

SA agar[67] and modified SA agar (Lachica, personal communication, 1990) are recommended. At this point, SA agar has seen the widest usage and, for food and many environmental investigators, upwards of 85% of the presumptive colonies have been verified as *A. hydrophila* group.[64] Investigators recommend BBGS agar for samples in which large numbers of *Proteus* spp. might be encountered.[62]

After inoculating these media with appropriate dilutions of the food samples, incubate the plates at 28°C overnight (24 hr maximum). After incubation, flood SA and BBGS agar plates with 5 mL of Lugol's iodine solution. Count typical colonies (5 mm, yellow to honey-colored on SA agar) surrounded by a clear zone of hydrolyzed starch against a black background as *A. hydrophila* groups. Suspect colonies must be picked at this step for verification since the iodine is rapidly lethal to the cells. On overcrowded plates, the growth can be scraped off with a sterile loop and zones counted to provide an estimate on the number of *A. hydrophila* colonies. On the modified SA agar, *A. hydrophila* colonies are surrounded by a light halo against a blue background; the iodine solution is not needed.

30.183 Enrichment Media

In instances when low numbers of *A. hydrophila* are anticipated or when injured cells are suspected, use an enrichment broth. The two recommended enrichment broths are APW[86] and TSBA.[2] The enrichment media can be inoculated with a 25 g sample of food or dilutions of the original 1:10 slurry prepared in the stomacher bag. Incubation at 28°C is recommended. When quantitation is desired, either of the suggested enrichment broths can be incorporated into a most probable number (MPN) procedure in which various dilutions are employed in a 3- or 5-tube method. After 24 hr, streak portions of the broths onto the above two plating media, and observe plates after 24 hr at 28°C. Score plates as positive or negative for typical *A. hydrophila* colonies, and obtain a quantitative estimate of the number of *A. hydrophila* in the original sample using MPN tables (Chapter 6).

30.184 Verification of Isolates as A. hydrophila

In our experience, amylase-producing colonies from SA agar plates have all been verified as belonging to the *A. hydrophila* group. The following procedure allows investigators to verify food isolates as the *A. hydrophila* group.[4]

Pick typical *A. hydrophila* colonies from SA agar and streak them onto plates of nutrient agar (Difco), tryptic soy agar (Difco), or any other suitable growth medium that does not contain carbohydrate; also streak a plate of Dnase test agar. Test the suspected colonies from the SA plate for amylase activity by placing a drop of Lugol's iodine on each colony. If the colonies are amylase-positive, incubate the two plating media for 24 hr at 28°C. Before incubation, place a disk of O/129 in a heavy streak area so that zones of inhibition can be seen.

Nutrient agar (or TSA) plates are used for several procedures: a) to determine if a pure culture was obtained from the initial streaking; b) to perform a Gram stain (*A. hydrophila* group organisms are short gram-negative rods); c) to perform a catalase test (a few groups of 3% H_2O_2 are added to an isolated colony; bubbles formed indicate a positive reaction); and d) to test resistance to the vibriostatic agent O/129 (*A. hydrophila* is not inhibited by the compound).

Further verification of the isolates can be easily accomplished by inoculating into a tube of Kaper's medium[43] (5-mL amounts incubated at 35°C). After 18 to 24 hr, read the tubes. A typical *A. hydrophila* group reaction is alkaline band at top, acid butt, motility +, H_2S –, and indole + (add 2 drops of Kovac's reagent to the tube and look for a red/scarlet color). The isolate also can be inoculated into the API 20E series test strip and reactions read after 24 hr. Identify by comparing the isolate's biochemical reactions with a reference table and by using a numerical key.

30.19 Interpretations

The significance of organisms of the *A. hydrophila* group in foods is currently not known. They can readily be isolated, often in high numbers, from a wide variety of fresh and processed foods and from various water supplies. Cultures of these bacteria isolated from foods and water supplies as well as clinical isolates often possess factors and traits associated with virulence in other gram-negative bacteria. These bacteria are recognized as pathogens in select groups of individuals such as immuno-compromised persons and patients with underlying malignancies, young children, and travelers in developing countries. These bacteria can grow competitively in most foods held at 5°C, a temperature formerly thought adequate to protect foods from the hazard of foodborne pathogens. They can readily be destroyed by heat, irradiation, and other food processing treatments. For the foreseeable future, they will continue to represent an enigma to both food and clinical microbiologists.

30.110 Media, Reagents, and Stains

Alkaline Peptone Water (APW)
Bile Salts Brilliant Green Starch (BBGS) Agar
Dnase Test Agar
Kaper's Medium
Modified SA Agar (Lachica's Medium
Nutrient Agar
Peptone Broth (1.0%)
SA Agar
Tryptic Soy Broth + 30 mg/L (TSBA)
Kovac's Reagent (Indole)
Lugol's Iodine Solution

30.111 References

1. Abbott, S. L., Cheung, W. K. W., Kroske-Bystrom, S., Malekzadeh, T., and Janda, J. M. 1992. Identification of *Aeromonas* strains to the genospecies level in the clinical laboratory. J. Clin. Microbiol. 30: 1262-1266.
2. Abeyta, C. Jr., Kaysner, C. A., Wekell, M. M., Sullivan, J. J., and Stelma, G. N. 1986. Recovery of *Aeromonas hydrophila* from oysters implicated in an outbreak of foodborne illness. J. Food Protect. 49: 643.
3. Abeyta, C. Jr., and Stelma, G. N. 1987. Isolation and Identification of Motile *Aeromonas* Species. In "FDA Bacteriological Analytical Manual", 6th ed., (Supplement 9/87), AOAC International, Gaithersburg, MD, p. 30.01.
4. Abeyta, C., JR. 1987. Unpublished data. U. S. Food and Drug Administration, Seafood Products Research Center, Seattle, WA
5. Altwegg, M., von Graevenitz, A. and Zollinger-Iten, J. 1987. Medium and temperature dependence of decarboxylase reactions in *Aeromonas* ssp. Curr. Micrbiol. 15: 1-4.
6. Asao, T., Kinoshita, Y., Kozaki, S., Uemura, T., and Sakaguchi, G. 1984. Purification and some properties of *Aeromonas hydrophila* hemolysin. Infect. Immun. 46: 122-127.
7. Asao, T., Kozaki, S., Kato, K., Kinoshita, Y., Otsu, K., Uemura, T., and Sakaguchi, G. 1986. Purification and characterization of an *Aeromonas hydrophila* hemolysin. J. Clin. Microbiol. 24: 228-232.
8. Ascencio, F., Martinez-Arias, W., Romero, M. J. and Wadstrom, T. 1998. Analysis of the interaction of *Aeromonas caviae, A. hydrophila* and *A. sobria* with mucins. FEMS Immunol. Med. Microbiol. 20: 219-229.
9. Ascencio, F. and Wadstrom, T. 1998. A collagen-binding protein produced by *Aeromonas hydrophila*. J. Med. Microbiol. 47: 417-425.
10. Barghouthi, S., Young, R., Olson, M. O. J., Arceneaux, J. E. L., Clem, L. W., and Byers, B. R. 1989. Amonobactin, a novel tryptophan- or phenylalamine-containing phenolate siderophore in *Aeromonas hydrophila*. J. Bacteriol. 171: 1811-1816.

11. Bernheimer, A. W., and Avigad, L. S. 1974. Partial characterization of aerolysin, a lytic exotoxin from *Aeromonas hydrophila.* Infect. Immun. 9: 1016-1021.
12. Borsh, E., Nesbakken, T, and Christensen, H. 1996. Hazard identification in swine slaughter with respect to foodborne bacteria. Intl. J. Food Microbiol. 30: 9-25.
13. Brenden, R. A., and Huizinga, H. W. 1986. Susceptibility of normal and X-irradiated animals to *Aeromonas hydrophila* infections. Curr. Microbiol. 13: 129-132.
14. Bullock, G. L. 1966. Precipitation and agglutination reactions of aeromonads isolated from fish and other sources. Bull. Off. Int. Epizoot. 68: 805-824.
15. Bunning, V. K., Crawford, R. G., Stelma, G. N., Jr., Kaylor, L. O., and Johnson, C. H. 1986. Melanogenesis in murine B16 cells exposed to *Aeromonas hydrophila* cytotoxic enterotoxin. Can. J. Microbiol. 32:815-819.
16. Burke, V., Gracey, M., Robinson, J., Peck, D., Beaman, J., and Bundell, C. 1983. The microbiology of childhood gastroenteritis: *Aeromonas* species and other infective agents. J. Infect. Dis. 148: 68-74.
17. Burke, V., Cooper, M. and Robinson, J. 1986. Haemagglutination patterns of *Aeromonas* spp. related to species and source of strains. Aust. J. Exp. Biol. Med. Sci. 64: 563-570.
18. Callister, S. M. and Agger, W. A. 1987. Enumeration and characterization of *Aeromonas hydrophila and Aeromonas* caviae isolated from grocery store produce. Appl. Environ. Microbiol. 53: 249-253.
19. Canonica, F. P. and Pisano, M. A. 1985. Identification of hydroxy fatty acids in *Aeromonas hydrophila, Aeromonas sobria, and Aeromonas* caviae. J. Clin. Microbiol. 22: 1061-1062.
20. Carnahan, A. M., Behram, S., and Joseph, S. W. 1991. Aerokey II: a flexible key for identifying clinical *Aeromonas* species. J. Clin. Microbiol. 20: 2843-2849.
21. Chakraborty, T., Huhle, B., Bergbauer, H., and Goebel, W. 1986. Cloning expression and mapping of the *Aeromonas hydrophila* aerolysin gene determinant in *Escherichia coli* K12. J. Bacteriol. 167: 368-374.
22. Chakraborty, T., Montenegro, M. A., Sanyal, S. C., Helmuth, R., Bulling, E., and Timmis, K. N. 1984. Cloning of enterotoxin gene from *Aeromonas hydrophila* provides conclusive evidence of production of a cytotonic enterotoxin. Infect. Immun. 46: 435-441.
23. Chopra, A. K., and Houston, C. W. 1989. Purification and partial characterization of a cytotonic enterotoxin produced by *Aeromonas hydrophila.* Can. J. Microbiol. 35: 719-727.
24. Clark, R. B., Knoop, F. C., Padgitt, P. J., Hu, D. H., Wong, J. D., and Janda, J. M. 1989. Attachment of mesophilic aeromonads to cultured mammalian cells. Curr. Microbiol. 19: 97-102.
25. Colwell, R. R., MacDonell, M. T., and DeLey, J. 1986. Proposal to recognize the family *Aeromonadaceae* fam. nov. Intl. J. Syst. Bacteriol. 36: 473-477.
26. Crichton, P. B. and Walker, J. W. 1985. Methods for the detection of haemagglutinins in *Aeromonas.* J. Med. Microbiol. 19: 273-277.
27. Cumberbatch, N., Gurwith, M. J., Langston, C., Sack, R. B., and Brunton, J. 1979. Cytotoxic enterotoxin produced by *Aeromonas hydrophila*:Relationship of toxigenic isolates to diarrheal disease. Infect. Immun. 23: 829-837.
28. Dooley, J. S. G., Lallier, R. and Trust, T. J. 1986. Surface antigens of virulent strains of *Aeromonas hydrophila.* Vet. Immun. Immunopathol. 12: 339-344.
29. Eddy, B. P. 1960. Cephalotrichous, fermentative gram-negative bacteria: The genus *Aeromonas.* J. Appl. Bacteriol. 23: 216-249.
30. Ewing, W. H., Hugh, R. and Johnson, J. G. 1961. Studies on the *Aeromonas* group. U. S. Department of Health, Education and Welfare, Centers for Disease Control, Atlanta, GA.
31. Ewing, W. H., Hugh, R. and Johnson, J. G. 1961. Studies on the *Aeromonas* group. In Public Health Service monograph, Centers for Disease Control, Atlanta, GA.
32. Figura, N. and Guglielmetti, P. 1987. Differentiation of motile and mesophilic *Aeromonas* strains into species by testing for a CAMP-like factor. J. Clin. Microbiol. 25: 1341-1342.
33. Fricker, C. R. 1987. Serotyping of mesophilic *Aeromonas* spp. on the basis of lipopolysaccharide antigens. Lett. Appl. Microbiol. 4: 113-116.
34. Gracey, M., Burke, V., and Robinson, J. 1982. *Aeromonas*-associated gastroenteritis. Lancet 2: 1304-1306.
35. Haque, Q. M., Sugiyama, A., Iwade, Y., Midorikawa, Y. and Yamauchi, T. 1996. Diarrheal and environmental isolates of *Aeromonas* spp. produce a toxin similar to shiga-like toxin 1. Current Microbiol. 32: 239-245.
36. Hokama, A. Honma, Y., and Nakasone, N. 1990. Pili of an *Aeromonas hydrophila* strain as a possible colonization factor. Microbiol. Immunol. 34: 901-915.
37. Janda, J. M., Brenden, R., and Bottone, E. J. 1984. Differential susceptibility to human serum by *Aeromonas* spp. Curr. Microbiol. 11: 325-328.
38. Janda, J. M., Clark, R. B., and Brenden, R. 1985. Virulence of *Aeromonas* species as assessed through mouse lethality studies. Curr. Microbiol. 12: 163-168.
39. Janda, J. M., Oshiro, L. S., Abbott, S. L., and Duffey, P. S. 1987. Virulence markers of mesophilic aeromonads: Association of the autoagglutination phenomenon with mouse pathogenicity and the presence of a peripheral cell-associated layer. Infect. Immun. 55: 3070-3077.
40. Janda, J. M., Reitano, M., and Bottone, E. J. 1984. Biotyping of *Aeromonas* isolates as correlate to delineating a species-associated disease spectrum. J. Clin. Microbiol. 19: 44-47.
41. Johnson, W. M., and Lior, H. 1981. Cytotoxicity and suckling mouse reactivity of *Aeromonas hydrophila* isolated from human sources. Can. J. Microbiol. 27: 1019-1027.
42. Joseph, S. W. and Carnahan, A. 1994. The isolation, identification, and systematics of the motile *Aeromonas* species. Ann. Rev. Fish Dis. 315-343.
43. Kaper, J., Seidler, R. J., Lockman, H. and Colwell, R. R. 1979. Medium for the presumptive identification of *Aeromonas hydrophila* and Enterobacteriaceae. Appl. Environ. Microbiol. 38: 1023-1026.
44. Kirov, S. M., Rees, B., Wellock, R. C., Goldsmid, J. M., and van Galen, A. D. 1986. Virulence characteristics of *Aeromonas* spp. in relation to source and biotype. J. Clin. Microbiol. 24: 827-834.
45. Kokka, R. P., Vedros, N. A., and Janda, J. M. 1991. Characterization of classic and atypical serogroup 0:11 *Aeromonas*: evidence that the surface array protein is not directly involved in mouse pathogenicity. Microb. Pathog. 10: 71-79.
46. Kozaki, S., Kurokawa, A., Asao, T., Kato, K., Uemura, T. and Sakaguchi, G. 1987. Enzyme-linked immunosorbent assay for *Aeromonas hgdrophila* hemolysins. FEMS Microbiol. Letters 41: 147-151.
47. Kozaki, S., Asao, T., Kamata, Y., and Sakaguchi, G. 1989. Characterization of *Aeromonas sobria* hemolysis by use of monoclonal antibodies against *Aeromonas hydrophila* hemolysins. J. Clin. Microbiol. 27: 1782-1786.
48. Krovacek, K, Faris, A., Baloda, S. J., Lindberg, T., Peterz, M., and Mansson, I. 1992. Isolation and virulence profiles of *Aeromonas* spp. from different municipal drinking water supplies in Sweden. Food Microbiol. 9: 215-222.
49. Krovacek, K, Faris, A., Baloda, S. J., Lindberg, T., Peterz, M., and Mansson, I. 1992. Prevalence and characterization of *Aeromonas* spp. isolated from foods in Uppsala, Sweden. 1992. Food Microbiol. 9: 29-36.
50. Kuijper, E. J., and Peeters, M. F. 1991. Bacteriological and clinical aspects of *Aeromonas*-associated diarrhea in the Netherlands. *Experientia* 47: 432-434.
51. Kuijper, E. J., Zanen, H. C., and Peeters, M. F. 1987. *Aeromonas*-associated diarrhea in the Netherlands. Ann. Intern. Med. 106: 640-641.
52. Leung, K.-Y., and Stevenson, R. M. W. 1988. Characteristics and distribution of extracellular proteases from *Aeromonas hydrophila*. J. Gen. Microbiol. 134: 151-160.
53. Liu, P. V. 1961. Observations on the specificities of extracellular antigens of the genus *Aeromonas and* Serratia. J. Gen. Microbiol. 24: 145-153.

54. Ljungh, A. Wretlind, B., and Mollby, R. 1981. Separation and characterization of enterotoxin and two hemolysins from *Aeromonas hydrophila*. Acta Pathol. Microbiol. Scand. Sect B 89: 387-397.
55. Ljungh, A., Enroth, P., and Wadstrom, T. 1982. Cytotonic enterotoxin from *Aeromonas hydrophila*. Toxicon. 20: 787-794.
56. Massad, G. Arceneaux, J. E. L., and Byers, B. R. 1991. Acquisition of iron from host sources by mesophilic *Aeromonas* species. J. Gen. Microbiol. 137: 237-241.
57. Millership, S. E., Barer, M. R., Mulla, R. J., and Maneck, S. 1992. Enterotoxic effects of *Aeromonas sobria* hemolysin in a rat jejunal perfusion system identified by a specific neutralization with a monoclonal antibody. J. Gen. Microbiol. 138: 216-267.
58. Morgan, D. R., Johnson, P. C., Dupont, H. L. Satterwhite, T. K., and Wood, L.V. 1985. Lack of correlation between known virulence properties of *Aeromonas hydrophila* and enteropathogenicity for humans. Infect. Immun. 50: 62-65.
59. Namdari, H., and Bottone, E. J. 1988. Correlation of the suicide phenomenon in *Aeromonas* species with virulence and enteropathogenicity. J. Clin. Microbiol. 26: 2615-2619.
60. Namdari, H., and Bottone, E. J. 1990. Cytotoxin and enterotoxin production as factors delineating enteropathogenicity of *Aeromonas caviae*. J. Clin. Microbiol. 28: 1796-1798.
61. Namdari, H., and Bottone, E. J. 1990. Microbiologic and clinical evidence supporting the role of *Aeromonas caviae* as a pediatric enteric pathogen. J. Clin. Microbiol. 28: 837-840.
62. Nishikawa, Y. and Kishi, T. 1987. A modification of bile salts brilliant green agar for isolation of motile *Aeromonas* from foods and environmental samples. Epidem. Inf. 98: 331-336.
63. Ogden, I. D., Millar, I. G., Watt, A. J., and Wood, L. 1994 A Comparison of three identification kits for the confirmation of *Aeromonas* spp. Lett. Appl. Microbiol. 18: 97-99.
64. Palumbo, S, A., Stelma, G. N., and Abeyta, C. 1999 The *Aeromonas hydrophila* Group. In the Microbiology of Foods, B. Lund et al., editors, Aspen Publishers, Gaithersburg, MD.
65. Palumbo, S. A., Bencivengo, M. M., Del Corral, F., Williams, A.C., and Buchanan, R. L. 1989. Characterization of the *Aeromonas hydrophila* group isolated from retail foods of animal origin. J. Clin. Microbiol. 27: 854-859.
66. Palumbo, S. A., Call, J. E., Huynh, B., and Fanelli, J. 1996. Survival and growth potential *Aeromonas hydrophila* in reconditioned pork-processing- plant water. J. Food Prot. 59: 881-881.
67. Palumbo, S. A., Maxino, F., Williams, A. C., Buchanan, R. L., and Thayer, D. W. 1985. Starch-ampicillin agar for the quantitative detection of *Aeromonas hydrophila*. Appl. Environ. Microbiol. 50: 1027.
68. Palumbo, S. A., Williams, A. C., Buchanan, R. L., and Phillips, J. G. 1987. Thermal resistance of Aeromonas hydrophila. J. Food Protect. 50: 761.
69. Pazzoglia, G., Sack, R. B., Bourgeois, A. L., Froehlich, J., and Eckstein, J. 1990. Diarrhea and intestinal invasiveness of *Aeromonas* strains in the removable intestinal tie rabbit model. Infect. Immun. 58: 1924-1931.
70. Peduzzi, R., De Meuron, P. A. and Grimaldi, E. 1983. Investigation of *Aeromonas* isolated from water; a serological study using Ouchterlony and immunoelectrophoresis techniques. Experientia 39: 924-926.
71. Popoff, M. 1984. Genus III. *Aeromonas* Kluyver and Van Niel 1936, 398, In N. R. Kreig and J. G. Holt (eds.), Bergey's Manual of Systematic Bacteriology, Vol. I, The Williams & Wilkins Co., Baltimore, p. 545-548.
72. Popoff, M. and M. Vernon. 1976. A taxonomic study of the *Aeromonas hydrophilia-Aeromonas* punctata group. J. Gen. Microbiol. 94:11-22.
73. Potomski, J., Burke, V., Robinson, J., Fumarola, D., and Miragliotta, G. 1987. *Aeromonas* cytotonic enterotoxin cross-reactive with cholera toxin. J. Med. Microbiol. 23: 179-186.
74. Rao, V. B. and Foster, B. G. 1977. Antigenic analysis of the genus *Aeromonas*. Texas J. Sci. 29: 85-91.
75. Rashad, F. M. and Abdel-Kareem, H. 1995. Characterization of *Aeromonas* species isolated from foods. J. Food Safety 15: 321-336.
76. Rippey, S. R. and V. J. Cabelli. 1979. Membrane filter procedure for enumeration of *Aeromonas hydrophila* in fresh waters. Appl. Environ. Microbiol. 38: 108-113.
77. Rose, J. M., Houston, C. W., and Kurosky, A. 1989. Bioactivity and immunological characterization of a cholera toxin-cross-reactive cytolytic enterotoxin from *Aeromonas hydrophila*. Infect. Immun. 57: 1170-1176.
78. Rose, J. M., Houston, C. W., Coppenhaver, D. H., Dixon, J. D., and Kurosky, A. 1989. Purification and chemical characterization of a cholera toxin-cross-reactive cytolytic enterotoxin produced by a human isolate of *Aeromonas hydrophila*. Infect. Immun. 57: 1165-1169.
79. Sakazaki, R. and T. Shimada. 1984. 0-serogrouping scheme for mesophilic *Aeromonas* strains. Japan J. Med. Biol. 37: 247-255.
80. Sandrelli, G. 1891. Uber einen neuen Mikroorganismes des Wassers, welcher fur tiere mit veraenderlicher und konstanter Temperature pathogen ist. Zentralbl. Bakteriol. Parasitenk. Infektionskr. Hyg. Abt. 192: 222-228.
81. Schultz, A. J., and McCardell, B. A. 1988. DNA homology and immunological cross-reactivity between *Aeromonas hydrophila* cytotonic enterotoxin and cholera toxin. J. Clin. Microbiol. 26: 57-61.
82. Seidler, R. J., Allen, D. A., Lockman, H. Colwell, R. R., Joseph, S. W., and Daily, O. P. 1980. Isolation enumeration and characterization of *Aeromonas* from polluted waters encountered in diving operations. Appl. Environ. Microbiol. 39: 1010-1018.
83. Shaw, D. H. and Hodder, H. J. 1978. Lipopolysaccharides of motile Aeromonads: Core oligosaccharide analysis as an aid to taxonomic classification. Can. J. Microbiol. 24: 864-868.
84. Stelma, G. N., Jr., Johnson, C. H., and Spaulding, P. 1986. Evidence for the direct involvement of $-hemolysin in *Aeromonas hydrophila* enteropathogenicity. Curr. Microbiol. 14: 71-77.
85. Stelma, G. N., Jr., Johnson, C. H., and Spaulding, P. L. 1988. Experimental evidence for enteropathogenicity in *Aeromonas veronii*. Can. J. Microbiol. 34: 877-880.
86. Turnbull, P. C. B., Lee, J. V. Miliotis, M. D., Van De Walle, S., Kornhof, H. J., Jeffrey, L., and Bryant, T. N. 1984. Enterotoxin production in relation to taxonomic grouping and source of isolation of *Aeromonas* species. J. Clin. Microbiol. 19: 175-180.
87. von Graevenitz, A. 1985. *Aeromonas* and Plesiomonas, In: E. H. Lennette, A. Balows, W. J. Hausler, Jr., and H. J. Shadomy. (eds.), Manual of Clinical Microbiology, 4th Ed., American Society for Microbiology, Washington, DC, p. 278-281.
88. Watson, I. M., Robinson, J. O., Burke, V., and Gracey, M. 1985. Invasiveness of *Aeromonas* spp. in relation to biotype virulence factors and clinical features. J. Clin. Microbiol. 22: 48-51.

30.2 *ARCOBACTER*

30.21 Introduction

The genus *Arcobacter* includes bacteria formerly designated *Campylobacter cryaerophila* (Latin; loving cold and air).[49] *Arcobacter* (Latin, arc-shaped bacterium) was first isolated from aborted bovine tissue and later from porcine fetuses.[12,13,21,37] *Arcobacter*, unlike other *Campylobacter* species, grows in the presence of atmospheric oxygen (aerotolerant) and at temperatures (15°–25°C), which are lower than those used for incubation of *Campylobacter*.[37]

Three species have been recovered from man and animals:[49] *Arcobacter butzleri*, *Arcobacter cryaerophilus*, and *Arcobacter skirrowii* (Table 3). With respect to public health importance, *A. butzleri* is perhaps the most likely species to be involved in human illness.[24] *Arcobacter cryaerophila* is recovered from livestock[49] and infrequently from humans.[24] *Arcobacter skirrowii* is infrequently isolated from livestock; it has never been recovered from humans.[49] *Arcobacter nitrofigilis* is the type strain and is found within the roots of *Spartina*, a salt-marsh plant.[33] No isolations have been made from animals.

30.211 Pathology

Arcobacter infections in animals are associated with abortions and enteritis.[21,37,53] Enteritis and occasionally septicemia occur in humans, including neonates.[19,24,28,38,47,48] Primates are naturally infected with *Arcobacter* and may develop colitis, which may provide insight into its pathogenesis in humans.[1,44]

30.212 Distribution and Epidemiology

Arcobacter species are present in livestock and poultry, and may, like *Campylobacter* exist as commensals. For cattle, *Arcobacter* organisms have been reported in the feces of clinically healthy dairy cattle, in calves with diarrhea, from the preputial swabs of bulls, and in cows with mastitis.[8,15,30,53] *Arcobacter* spp. have been cultured from 1.5% of minced beef (n = 68) samples examined.[45]

Arcobacter is present in both healthy and clinically ill pigs and in aborted porcine fetuses.[8,12,53,54] Neonatal pigs are susceptible to experimental infection with *A. butzleri*.[8,12,53,54,57] In addition, *Arcobacter* has been detected in 22% of ground pork samples (n = 290) obtained from five Iowa slaughter facilities. Significant variation was seen in the recovery of *Arcobacter* between slaughter plants.[7] In contrast, *Arcobacter* was cultured from only 0. 5% (1 of 194) of pork products purchased in The Netherlands.[45] The difference between ground pork and minimally processed pork cuts as well as isolation methods may explain the differences between the findings of the two studies.

Arcobacter, like *Campylobacter*, has been reported more frequently from poultry products than from red meats. Thus, poultry may be a significant reservoir of *A. butzleri*.[14] In France, *A. butzleri* was recovered from 81% of poultry carcasses examined (n=201). Nearly half of the poultry isolates in that study were of serogroup 1. Serogroups 1 and 5 are primarily associated with human infection.[32] In a survey of poultry products in Canada, *A. butzleri* was recovered from 97% (121 of 125) of poultry carcasses obtained from five different processing plants. In addition, *A. butzleri* was cultured from retail-purchased whole and ground poultry, chicken and turkey. As was the case in the French study, serotype 1 was the predominant serotype isolated from Canadian poultry.[27]

In the US, *Arcobacter* (77% of samples) and *A. butzleri* (57% of samples) were isolated from mechanically separated turkey meat (n=395 samples) tested from three different states.[31] In contrast, *Arcobacter* was detected in only 24% (53 of 224) of retail purchased poultry products in The Netherlands.[45] No isolations were made from poultry in a limited study conducted in Italy.[58] In the UK, a single isolation was made from a limited number of poultry intestines analyzed, but from all of 25 carcass rinse samples.[2,4] In a more extensive study in Germany, *Arcobacter* was recovered from 57% of skin but from none of caecal samples collected from 170 freshly slaughtered broilers.[18] *Arcobacter* species, including *A. skirrowii*, were detected in 80% of duck carcasses (n=10) and in a limited number of duck ceca analyzed.[43] Despite surveys which show its prevalence in poultry products, conventional outbred turkey poults and chicks appear resistant to experimental infection with *A. butzleri*.[55] Taken together these data suggest that extensive contamination occurs during processing.

Table 3. Summary of *Arcobacter* species and host distribution

Bacterium	Source	Target Organ
Arcobacter butzleri	Humans	Intestine
	Livestock	Intestine, Placenta, Fetus
Arcobacter cryaerophilus	Humans	Intestine
	Livestock	Intestine, Placenta, Fetus
Arcobacter skirrowii	Livestock	Intestine
		Reproductive tract
Arcobacter nitrofigilis	*Spartina* plant	Roots

The distribution of *Arcobacter* spp. in seafood, shellfish, and raw milk is unknown. It is probable that its distribution may parallel that of *Campylobacter*.

Transmission of *A. butzleri* may involve drinking contaminated water. *Arcobacter* spp. may be more common in developing nations with inadequate water supplies since *A. butzleri* accounted for 16% of the *Campylobacter*-like isolates made from cases of diarrhea in Thai children.[47] *Arcobacter butzleri* has been reported in drinking water reservoirs in Germany, in water treatment plants, rivers, and in the canal waters of Bangkok.[9,20,34,46] In the US, *A. butzleri* was recovered from a well serving an Idaho youth camp following an outbreak of waterborne enteritis.[42] The presence of *A. butzleri* in unchlorinated water supplies suggests this as a possible source of contamination.

It is postulated that *Arcobacter* is hardier than *Campylobacter*. By comparing D_{10} values (the irradiation dose which reduces by 10-fold the number of viable bacteria), *A. butzleri* (0.27 kGy) was found to be more resistant to irradiation than *C. jejuni* (0.18 kGy).[6] *Arcobacter* grows over a pH range of 5.5–7.5 with optimal growth of *A. butzleri* at pH 6 and optimal growth of *A. cryaerophilus* at pH 7.0–7.5. Thermal tolerance studies indicate that the thermal death time for a single strain of *A. butzleri* to be 2.5, 5 and 15 min at 60, 55 and 50°C, respectively.[10] Based on these data, *Arcobacter* will not survive minimum pasteurization treatment (63°C for 30 min or 71.7°C for 15 sec) of milk.

30.213 Species Identification

Aerotolerance and growth at 15°C–30°C are the key features, which distinguish *Arcobacter* from thermophilic *Campylobacter* species.[49,37] However, in one study, more than 50% of *A. butzleri* field strains examined were reported to grow at 42°C.[49] *Arcobacter* spp. are oxidase positive and hydrolyze indoxyl acetate, which are traits also exhibited by *Campylobacter*.

There are few reliable phenotypic traits to distinguish the four species of *Arcobacter*.[24,37,39,54] Generally, a strong positive catalase test result distinguishes *A. cryaerophilus* from *A. butzleri* (weak to negative).

The use of molecular methods, such as DNA probes and polymerase chain reaction (PCR) primers, has accelerated the identification of *Arcobacter* and its species.[5,16,17] The assays do not require pure cultures and bypass the ambiguities of biochemical identification. PCR assays to detect all members of the genus *Arcobacter* as well as PCR primers, which are specific for *A. butzleri*, *A. cryaerophilus*, and *A. skirrowii*, have been described.[5,16,17] A multiplex PCR assay to simultaneously identify *Arcobacter* and *A. butzleri* in livestock and foods, which can be conducted directly from the EMJH-P80 enrichment, is shown in Figure 1.[17]

Field strains of *A. butzleri* may be distinguished by serotyping.[14,27,29,32] At least 72 serogroups are recognized. As with *Campylobacter*, the availability of serotyping reagents limits the applicability of this method.

Molecular-based assays have been used for differentiation of *A. butzleri* strains. Methods include restriction fragment length polymorphisms (RFLP) and PCR-based DNA fingerprinting.

When the RFLP method targets the genes encoding ribosomal DNA the method is called ribotyping.[23,25,54,56] Assays targeting the enterobacterial repetitive intergenic consensus (ERIC) sequences generate fingerprints which have been used to track the source of *Arcobacter* contamination. PCR-mediated ERIC fingerprinting confirmed the identity of *A. butzleri* isolates recovered from a nursery school outbreak, suggested a single source of contamination and person-to-person transmission.[51] PCR-based ERIC fingerprinting was used to analyze over 120 *A. butzleri* field strains recovered from mechanically separated turkey meat. The resultant 86 ERIC patterns generated from the 120 strains indicated multiple contaminating sources.[31]

30.22 General Considerations

30.221 Methods of Isolation

There is no standard method for the isolation of *Arcobacter* spp., which restricts comparison of field survey data. Likewise, there has not been a comparison of the sensitivity of these protocols. Because the method of sample preparation will vary with the media used for isolation, it is advisable to refer to the original publication for specific instructions. In general, protocols used for the recovery of *Campylobacter* are adequate for *Arcobacter*, if incubation occurs at 37°C.

Collection and transport methods which have been described for *Campylobacter* are suitable for *Arcobacter*. We have found that *Arcobacter*, like *Campylobacter*, may be transported in buffered peptone water on ice.

30.23 Media and Reagents

30.231 Media

Ellinghausen, McCullough, Johnson, and Harris Polysorbate 80 (EMJH-P80) media
Sheep blood agar supplemented with cefoperzone, vancomycin, and amphotericin B (CVA plates)
Johnson-Murano (JM) semi-solid enrichment broth
Johnson-Murano (JM) agar

30.232 Reagents

Catalase test
Gram stain with carbol fuchsin counterstain
Indoxyl acetate hydrolysis
Cadmium chloride sensitivity
Oxidase test

30.24 Procedures

30.241 General Sampling Protocol

Methods for collecting and transporting fecal samples, poultry carcass rinses and livestock carcass swabs are generally as described for *Campylobacter*[7,55,57] Like *Campylobacter*, samples may be transported on ice in buffered peptone water or Cary-Blair transport media.

30.242 Selective Enrichment and Plating

Isolation protocols for *Arcobacter* species use formulations based on: 1) Ellinghausen, McCullough, Johnson, and Harris Polysorbate 80 (EMJH-P80) medium,[11,40] 2) modifications of isolation media used for *Campylobacter*,[2,19,27,28,38,46,47,48] and 3) unique formulations specific for *Arcobacter*.[3,36,45]

Since their initial isolation in *Leptospira* medium, *Arcobacter* spp. have been recovered using a two step procedure. In the first step EMJH-P80 semi-solid medium (supplemented with oleic albumin and polysorbate 80 and 0.5% agar and 5-fluorouracil) is inoculated and incubated aerobically at 30°C. Growth of the organisms as a distinct zone below the surface may occur as soon as 24 hrs or to 5 weeks. Bacterial growth from the semisolid isolation medium is subcultured on blood agar and incubated at 34°C microaerobically.[7,12,13,15,21,30-32,37,45,54,55,57]

EMJH-P80 is commercially available as Bovuminar Microbiological Media (PLM 53 5X concentrated) from Intergen Company, Purchase, NY 10577. To prepare EMJH-P80 for *Arcobacter* isolation from commercial media, make a 0.16% agar solution in water, autoclave, cool, and aseptically add 4 parts of the tempered agar to one part of Bovuminar 5X stock. Aseptically add 5-fluorouracil (Hoffmann-LaRoche, Inc, catalogue 0688) to achieve a final concentration of 100 µg/mL. Dispense (20 mL) to sterile disposable conical centrifuge tubes (50 mL) and to (9 mL) sterile glass screw-capped test tubes (20 mm × 125 mm). Media may be stored at room temperature for up to six months.[11]

For isolation from foods, place a 10 gram sample in 20 mL of EMJH-P80 and incubate aerobically (30°C, 3–7 days). After incubation, a 1 mL aliquot is subcultured to 9 mL of fresh EMJH-P80 and incubated aerobically (3 days, 30°C). The second incubation appears to be critical for achieving high titers of *Arcobacter*.[31] An aliquot (200 µL) from the second enrichment is analyzed by PCR (Figure 1).[16,17] Alternatively, inoculate foods into EMJH-P80 (10% w/v), incubate (30°C, 3–7 days), and screen by either darkfield microscopy[40] or by PCR (Figure 1).[16,17] EMJH-P80 enrichments may be refrigerated for up least two weeks without apparent loss of viability.

For isolation of *Arcobacter* from heavily contaminated water, add 1 mL of water to 9 mL of EMJH-P80, incubate aerobically (30°C, 3–5 days) and examine either microscopically[40] or by PCR.[16,17] Alternatively, filter concentrate the water sample, as described for *Campylobacter*.[42] The concentrate is placed in EMJH-P80, incubated (30°C, 3–5 days) and examined by either darkfield microscopy[40] or by PCR (Figure 1).

For isolation of *Arcobacter* from livestock, feces, cloacal swabs, or tissue samples (~1 gram) are placed in EMJH-P80 (9 mL) medium, incubated aerobically (30°C, 3–5 days), and an aliquot removed for PCR analysis (Figure 1).[54,55,57]

To screen EMJH-P80 tubes microscopically, prepare wet mounts and examine under 400X darkfield. Slender, curved rods with rapid darting motility are seen. Helical chains may also be present and move with a spiral motion.[12,13,21,37,40,45] Scan the field for agar particles. Motility is reduced when the microbes are trapped and thus concentrated in the agar.

For bacterial isolation, an aliquot from the EMJH-P-80 is streaked to the surface of either CVA (Remel. 01-270) or mCCDA (Remel) agar and incubated microaerobically in a CampyGas Pak (BBL) at 25–30°C for 2 days.[30,51] Modified CCDA agar is prepared as described for *Campylobacter*.[40] If the sample is heavily contaminated, isolate *Arcobacter* from EMJH-P80 by placing a 0.45 µm membrane filter (Millipore HAWP) onto the surface of Brain Heart Infusion agar (Oxoid CM225) supplemented with 0.6% yeast extract and 10% defibrinated bovine or ovine blood as follows.[45] With a sterile cotton plugged Pasteur pipette, remove an aliquot from the zone growth, which is below the surface of EMJH-P80. Place a drop or two of the culture fluid on the filter. Leave filter undisturbed for 1 hr at room temperature. *Arcobacter*, if present in high enough numbers, will swim through the membrane filter. Remove the filter and streak for isolation, using the area where the filter was placed as the first quadrant.[54] Incubate (30°C, 2 days) plates microaerobically (CampyGas Pak). Colonies

(1 to 2 mm) appear in color ranging from white translucent to buff colored. Colonies may appear to swarm.

Identification of suspect colonies is confirmed by either phenotypic tests (e.g. indoxyl acetate hydrolysis, oxidase, catalase, cadmium chloride sensitivity), polyacrylamide gel electrophoresis, fatty acid profile, or DNA-based methods, as described below.

Members of the genus *Arcobacter* were initially assigned to the species *Campylobacter cryaerophila*.[37,49] Thus, isolations of these

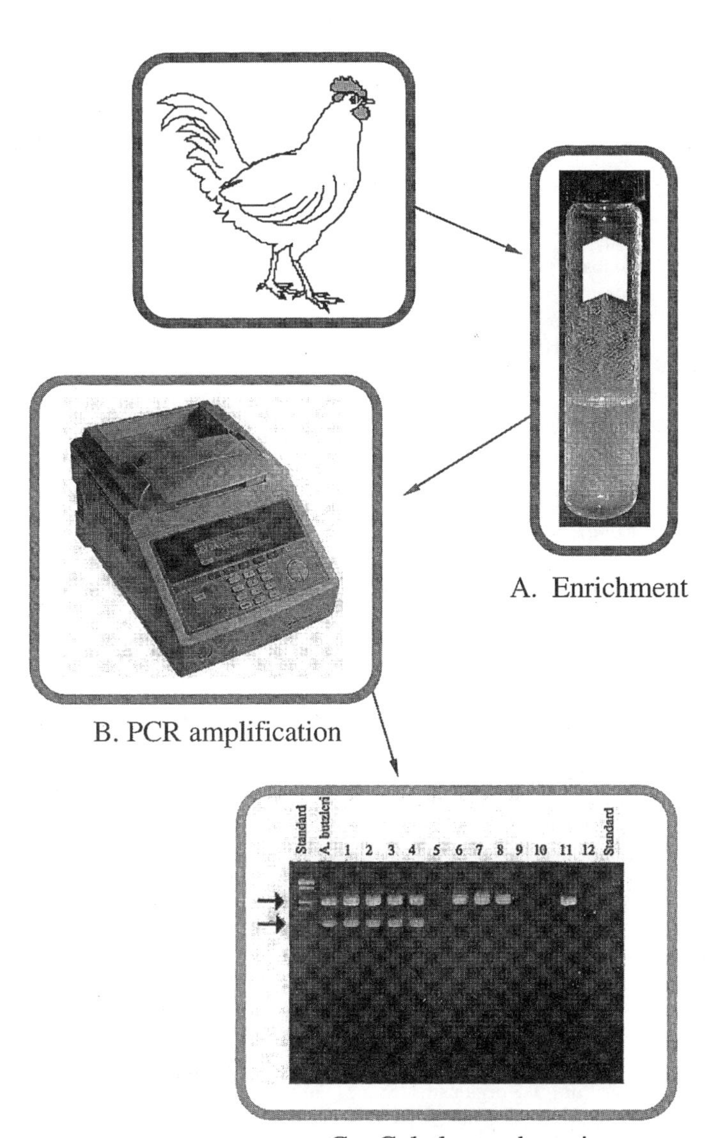

Figure 1. *Generalized scheme for PCR-based detection of Arcobacter in poultry, livestock feces, meats, or water.* ***(A).*** *Samples are enriched in EMJH-P80 media (1 week, 30C).* ***(B)*** *After incubation, cells are lysed by boiling, and template DNA with PCR reagents are placed in a thermal cycler for amplification, as described.*[17] ***(C).*** *The PCR products or amplicons are detected by gel electrophoresis. The presence of a 1223 bp PCR amplicon (upper arrow) is specific for Arcobacter. Arcobacter butzleri exhibits two PCR products: the 1223 bp amplicon characteristic of the genus and a smaller 686 bp (lower arrow) product unique to A. butzleri (lanes 1 through 4). The presence of the single 1223 bp amplicon indicates the presence of Arcobacter species other than A. butzleri (lanes 6 through 8; lane 11). The absence of a visible PCR product indicates that Arcobacter is not in the sample (lanes 5, 9,10). Positive (lane A. butzleri) and negative (lane 12) controls are shown.*

aerotolerant campylobacter-like organisms can be accomplished by using protocols designed for *Campylobacter*.[9,19,21,28,38,44,47,48,53,58]

In an attempt to simplify isolation protocols, Johnson and Murano (JM) developed a semi-solid enrichment broth followed by plating to charcoal-based agar to quench oxygen. The JM protocol was more sensitive than EMJH-P80 in detecting *Arcobacter* from heavily contaminated poultry.[36,55]

For the JM procedure, a 1:2 dilution of the sample is prepared in buffered peptone water, stomached, and a 1 mL aliquot placed in JM semi-solid enrichment medium (9 mL) and incubated (48 hr, 30°C aerobically). Following aerobic incubation, a loopful of culture is removed from 1 cm below the surface and streaked onto JM plating medium and incubated in air (48 hr, 30°C in air).[36] Presumptive colonies, characterized as gray or off-white and round with or without a surrounding red color are subsequently picked and purified. Colonies are then tested for oxidase and catalase activity and gram stained. Identification is confirmed by the multiplex PCR assay for *Arcobacter* and *Arcobacter butzleri*.[17]

This commercially available *Arcobacter* Selective Media (Oxoid CM 0965B) contains no oxygen quenching system, such as haemin, blood, activated charcoal, or ferrous sulfate, sodium metabisulfite and sodium pyruviate (FBP), to neutralize the toxic effects of atmospheric oxygen. Incubation in air (30°C , for 24 hr) and the absence of an oxygen quenching system suppress the growth of *Campylobacter*.[3] Peptones in the base medium facilitate the growth of *Arcobacter* species. The inclusion of cefoperazone, amphotericin B and teicoplanin inhibits the growth of competing flora.

Cary-Blair transport medium is recommended for transport of cultures.

30.25 Identification Tests

Species of *Arcobacter* are differentiated from *Campylobacter* by their ability to grow in air at 25°C–37°C.[37,49] For species identification, biochemical differentiation,[35,37,39] polyacrylamide-gel electrophoresis (PAGE) analysis,[2,4,49] fatty acid profiles,[19,26] DNA hybridization,[24] and PCR[5,16,17] have been reported. Of the four *Arcobacter* species, *A. butzleri* and *A. cryaerophilus* will be most frequently encountered in the laboratory. Based on their relative simplicity and comparison of methods in our laboratory, only biochemical differentiation and PCR assays are described herein. For all methods, strains of *A. butzleri* ATCC 49616, *A. cryaerophilus* 1A ATCC 43158, and *A. cryaerophilus* 1B ATCC 43157 should be included as reference standards when analyzing field strains.

30.251 Arcobacter *Test Methods*

Biochemical tests to phenotype *Arcobacter* species are limited.[24,39,49,54] All isolates hydrolyze indoxyl acetate as do *C. jejuni*, *C. coli*, and *C. upsaliensis*.[24,35,39,41,49] The most reliable tests to identify *A. butzleri* are negative or weak catalase production, growth in Campylobacter minimal medium, abundant growth on blood agar, growth on MacConkey agar, and resistance to cadmium chloride.[24,37,39,49,54] In contrast, *A. cryaerophilus* is strongly catalase positive, does not grow on *Campylobacter* minimal medium or on MacConkey agar, and is sensitive to cadmium chloride. There does not appear to be a single test or combination of biochemical assays to reliably differentiate *A. butzleri* from *A. cryaerophilus*.

Perform indoxyl acetate hydrolysis as described for *Campylobacter*.[24,41,49,54] *Arcobacter* species hydrolyze indoxyl acetate, which is indicated by the blue color. This characteristic is also shared by some species of *Campylobacter*.[41]

The catalase test is conducted as described for *Campylobacter*. Add several drops of 3% hydrogen peroxide to 24-48 hr growth on a heart infusion agar (BHI) slant. Production of any bubbles is considered positive.

A. cryaerophilus and *A. skirrowii* are catalase positive. *A. butzleri* is weak or negative for catalase production.

To perform cadmium chloride test, add cadmium chloride to filter paper disks to achieve a final concentration 2.5 μg. Allow to dry. Place on the surface of BHI supplemented with 5% defibrinated bovine blood agar which has been inoculated with the test organism and incubate (30°C, 1–2 days).[24,54]

A. cryaerophilus is sensitive to cadmium chloride (2.5 μg); *A. butzleri* is resistant.[24,54] Data are not available for *A. skirrowii*.

Perform oxidase test as described for *Campylobacter*.[24,54]

All *Arcobacter* species are oxidase positive, a characteristic which they share with *Campylobacter*.

Because of the ease and reproducibility of results, the PCR tests for *Arcobacter* identification are highly recommended. PCR may be used for the speciation of pure cultures or to screen EMJH-P80 enrichments. PCR assays targeting both the 16S and 23S rRNA genes of the genus *Arcobacter* and the individual species have been designed.[5,16,17]

After PCR amplification of the 16S rRNA gene target sequences, the appearance of a 1223 bp product indicates the presence of *Arcobacter*.[16] We have used PCR to detect *Arcobacter* sp. in 40% of fecal samples obtained from clinically healthy swine (n=~1,000) and from 11% of clinically healthy dairy cows in the US (n=~1,000).[8]

A multiplex PCR assay using two sets of primers has been described which differentiates *A. butzleri* from other species of *Arcobacter* (Figure 1).[17] It may be used to speciate pure cultures. Amplification of the 16S rRNA gene, as described above, yields the 1223 bp product whereas amplification of the 23S rRNA gene of *A. butzleri* generates the smaller 686 bp product. Thus, *A. butzleri* isolates exhibit two products (638 bp and 1223 bp) whereas all other *Arcobacter* species, such as *A. cryaerophilus* and *A. skirrowii*, yield a single amplicon (1223 bp). As a rapid screening assay, the multiplex PCR may also be conducted directly from EMJH-P80 enrichment thus bypassing the need for additional plating.[55]

30.252 DNA-fingerprinting

A pure culture is required for DNA fingerprinting, which can discriminate between field strains of the same species. At least two methods—restriction fragment length polymorphism (RFLP) and randomly amplified repetitive DNA polymorphism (RAPD) analysis—have been used for DNA fingerprinting of *A. butzleri*. The methods are similar to those detailed for *Campylobacter*.

The RFLP assay requires DNA isolation, which can be accomplished with commercially available kits, endonuclease digestion with *Pvu*II, Southern blot transfer, and hybridization with DNA probes targeting the 16S rRNA gene.[23,25,56] The 16S rRNA gene probes may be radioactively labeled with P^{32}.[56] Alternatively, oligonucleotide probes may be non-radioactively labeled with commerically available detection systems.[23,25,54] RFLP analysis is highly reproducible, which is especially important when analyzing food isolates. Microbiologists should refer to the publications in which ribotyping is detailed for specific protocols.[23,25,54,56]

DNA fingerprints to identify of *A. butzleri* strains may also be generated by PCR-based randomly RAPD targeting the enterobacterial repetitive intergenic consensus (ERIC) sequences.[31,51,52] The assay requires DNA isolation, which can be accomplished using any of the commercially available kits and ERIC primers, which can be commercially synthesized. Although, much easier to perform than ribotyping, PCR-based RAPD patterns generated in one laboratory may not be consistently reproduced in another facility even if the same field isolates are examined. For specific protocols, microbiologists should refer to the publication in which RAPD analyses targeting the ERIC sequences are described for *Arcobacter*.

30.26 References

1. Anderson, K. F., J. A. Kiehlbauch, D. C. Anderson, H. M. McClure, and I. K. Wachsmuth. 1993. *Arcobacter* (*Campylobacter*) *butzleri*-associated diarrheal illness in a nonhuman primate population. Infect. Immun. 61:2220-2223.
2. Atabay, H. I., and J. E. L. Corry. 1997. The prevalence of campylobacters and arcobacters in broiler chickens. J. Appl. Microbiol. 83:619-626.
3. Atabay H. I., and J. E. L. Corry. 1998. Evaluation of a new arcobacter enrichment medium and comparison with two media developed for enrichment of *Campylobacter* spp. Int. J. Food Microbiol. 41:53-48.
4. Atabay HI, Corry JEL and On SLW. Diversity and prevalence of *Arcobacter* spp. in broiler chickens. J Appl Microbiol 1998; 84:1007-1016.
5. Bastyns, K., D. Cartuyvels, S. Chapelle, P. Vandamme, H. Goossens, and R. de Wachter. 1995. A variable 23S rDNA region is a useful discriminating target for genus-specific and species-specific PCR amplification in *Arcobacter* species. Syst. Appl. Microbiol. 28:353-356.
6. Collins, C. I., E. A. Murano, and I. V. Wesley. 1996. Survival of *Arcobacter butzleri* and *Campylobacter jejuni* after irradiation treatment in vacuum-packaged ground pork. J. Food Prot. 11:1164-1166.
7. Collins, C. I., I. V. Wesley, and E. A. Murano. 1996. Detection of *Arcobacter* spp. in ground pork by modified plating methods. J. Food Prot. 59:448-452
8. DeBoer, E., J. J. H. C. Tilburg, D. L. Woodward, H. Lior, and W. M. Johnson. 1996. A selective medium for the isolation of *Arcobacter* from meats. Lett. Appl. Microbiol. 23:64-66.
9. Dhamaburtra, N., P. Kamol-Rathanakul, and K. Pienthaweechai. 1992. Isolation of *Campylobacter* from the canals of Bangkok metropolitan area. J. Med. Assoc. Thailand 75:350-363.
10. Dõsa E., and M. A. Harrison. 1998. Fate of *Arcobacter* spp. to environmental stresses of temperature, pH and NaCl levels. International Association of Milk, Food and Environmental Sanitarians. Abstract P57.
11. Ellinghausen, H. C., Jr, and W. G. McCullough. 1965. Nutrition of *Leptospira pomona* and growth of 13 other serotypes: Fractionation of oleic albumin complex and a medium of oleic albumin and polysorbate. Am. J. Vet. Res. 26:45-51.
12. Ellis, W. A., S. D. Neill, J. J. O'Brien, and J. Hanna. 1978. Isolation of spirillum-like organisms from pig fetuses. Vet. Rec. 102:106.
13. Ellis, W. A., S. D. Neill, J. J. O'Brien, H. A. W. Ferguson, and J. Hanna. 1977. Isolation of *Spirillum/Vibrio* like organisms from bovine fetuses. Vet. Rec. 10:451-452.
14. Festy, B. F., F. Squinazi, M. Marin, R. Derimaryu, and H. Lior. 1993. Poultry meat and water as the possible sources of *Arcobacter butzleri* associated with human disease in Paris, France. Acta Gastro-Enterol. Belg. 56(suppl 34):35.
15. Gill, K. P. W. 1983. Aerotolerant campylobacter strain isolated from a bovine preputial sheath washing. Vet. Rec. 112:459.
16. Harmon, K. M., and I. V. Wesley. 1996. Identification of *Arcobacter* isolates by PCR. Lett. Appl. Microbiol. 23:241-244.
17. Harmon, K. M., and I. V. Wesley. 1997. Multiplex PCR for the identification of *Arcobacter* and differentiation of *Arcobacter butzleri* from other arcobacters. Vet. Microbiol. 58:215-227.
18. Harrass, B., S. Schwarz, and S. Wenzel.1998. Identification and characterization of *Arcobacter* isolates from broilers by biochemical tests, antimicrobial resistance patterns and plasmid analysis. Zentralbl. Vet. 45:87-94.

19. Hsueh, P. R., L. J. Teng, P. C. Yang, S. K. Wang, S. C. Chang, S. W. Ho, W. C. Hsieh, and K. T. Luh. 1997. Bacteremia caused by *A. cryaerophilus* 1B. J. Clin. Microbiol. 35:489-491.
20. Jacob, J., H. Lior, and I. Feuerpfeill. 1993. Isolation of *Arcobacter butzleri* from a drinking water reservoir in Eastern Germany. Zbl. Hyg. 193:557-562.
21. Jahn, B. 1983. Untersuchungen uber Campylobacter-stamme aus dem genitaltrakt von schweinen. Fortsch. Vet. 37:154-157.
22. Johnson, L. G., and E. A. Murano. Comparison of three protocols for the isolation of *Arcobacter* from poultry. Submitted for publication.
23. Kiehlbauch, J. A., B. D. Plikaytis, B. Swaminathan, D. N. Cameron, and I. K. Wachsmuth. 1991. Restriction fragment length polymorphisms in the ribosomal genes for species identification and subtyping of aerotolerant *Campylobacter* species. J. Clin. Microbiol. 29:1670-1676.
24. Kiehlbauch, J. A., D. J. Brenner, M. A. Nicholson, C. N. Baker, C. M. Patton, A. G. Steigerwalt, and I. K. Wachsmuth. 1991. *Campylobacter butzleri* sp. nov. isolated from humans and animals with diarrheal illness. J. Clin. Microbiol. 29:376-385.
25. Kiehlbauch, J. A., D. N.Cameron, and I. K. Wachsmuth. 1994. Evaluation of ribotyping techniques as applied to *Arcobacter, Campylobacter* and *Helicobacter*. Mol. Cell. Probes 8:109-116.
26. Lambert, M. A., C. M. Patton, T. J. Barrett, and C. W. Moss. 1987. Differentiation of *Campylobacter* and *Campylobacter*-like organisms by cellular fatty acid composition. J. Clin. Microbiol. 25:706-713.
27. Lammerding, A. M., J. E. Harris. H. Lior, D. E. Woodward, L. Cole, and C. A. Muckle. 1996. Isolation method for the recovery of *Arcobacter butzleri* from fresh poultry and poultry products, p. 329-333. *In* D. G. Newall, J. Ketley, and R. A. Feldman (ed.), Campylobacters, helicobacters, and related organisms. Plenum, New York.
28. Lerner, J., V. Brumberger, and V. Preac-Mursic. 1994. Severe diarrhea associated with *Arcobacter butzleri*. Eur. J. Clin. Microbiol. Infect. Dis. 13: 660-662.
29. Lior, H. J., and D. L.Woodward. 1993. *Arcobacter butzleri:* A serotyping scheme. Acta Gastro-Enterol. Belgica 56 suppl.:29.
30. Logan E. F., S. D. Neill, and D. P. Mackie. 1982. Mastitis in dairy cows associated with an aerotolerant *Campylobacter*. Vet. Rec. 110:229-230.
31. Manke, T. R., I. V. Wesley, J. S. Dickson and K. M. Harmon. 1998. Prevalence and genetic variability of *Arcobacter* species in mechanically separated turkey. J. Food Prot. 61:1623-1628.
32. Marinescu, M., A. Collilgnon, F. Squinazi, D. Woodward, and H. Lior. 1996. Biotypes and serogroups of poultry strains of *Arcobacter* sp. isolated in France, p. 519-520. *In* D. G. Newall. J. H. Ketley, and R. A. Feldman (ed.), Campylobacters, helicobacters, and related organisms. Plenum, New York.
33. McClung, C. R., D. G. Patriquin, and R. E. Davis. 1983. *Campylobacter nitrofigilis* sp. Nov., a nitrogen-fixing bacterium associated with roots of *Spartina alterniflora* Loisel. Int. J. Syst. Bacteriol. 33:6905-6912.
34. Musmanno, R.A., Russi, M., Lior, H., and Figura. N. 1997. In vitro virulence factors of *Arcobacter butzleri* strains isolated from superficial water samples Microbiologica 20:63-68.
35. Nachamkin, I. 1995. Campylobacter and Arcobacter, 483-491. *In* P. R. Murray, E. J. Baron, M. A. Pfaller, F. C. Tenover, and R. H. Yolken (ed.), Manual of clinical microbiology. American Society for Microbiology Press, Washington, D. C.
36. Nash, P., and M. Krenz. 1991. Culture media, p. 1232.. *In* A. Balows, W. J. Hausler, K. L. Hermann, G. H. D. Isenberg, and H. J. Shadomy (ed.), Manual of clinical microbiology, 5th ed. American Society for Microbiology Press, Washington, D. C.
37. Neill, S. D., J. N. Campbell, J. J. O'Brien, S. T. C. Weatherup, and W. A. Ellis. 1985. Taxonomic position of *Campylobacter cryaerophila* sp. nov. Int. J. Syst. Bacteriol. 35:342-356.
38. On, S. L. W., A. Stacey, and J. Smyth. 1995. Isolation of *Arcobacter butzleri* from a neonate with bacteraemia. J. Infect. 31:225-227.
39. On, S. L. W., B. Holmes, and M. J. Sackin. 1996. A probability matrix for the identification of campylobacter, helicobacters and allied taxa. Appl. Bacteriol. 81:425-432.
40. Penner, J. 1991. *Campylobacter* , *Helicobacter* and related spiral bacteria, p. 402-409, chap. 39. *In* A. Balows, W. J. Hausler, K. L. Hermann, G. H. D. Isenberg, and H. J. Shadomy (ed.), Manual of clinical microbiology, 5th ed. American Society for Microbiology Press, Washington, D. C.
41. Popovic-Uroic, T., C. M. Patton, M. A. Nicholson, and J. A. Kiehlbauch. 1990. Evaluation of the indoxyl acetate hydrolysis test for rapid differentiation of *Campylobacter, Helicobacter* and *Wolinella* species. J. Clin. Microbiol. 28:2335-9.
42. Rice, WE, Rodgers MR, Wesley IV, Johnson CH and Tanner AS. 1998. Isolation of *Arcobacter* spp from ground water. Lett Applied Microbiology 28:31-35.
43. Ridsdale JA, Atabay HI, and Corry HEL. 1998. Prevalence of campylobacters and arcobacters in ducks at the abattoir. J Appl Microbiol 85:567-573.
44. Russell, R. G., J. A. Kiehlbauch, C. J. Gebhard, and L. J. deTolla. 1992. Uncommon *Campylobacter* species in infant *Macaca nemestrina* monkeys housed in a nursery. J. Clin. Microbiol. 30:3024-3027.
45. Schroeder-Tucker, L., I. V. Wesley, J. Kiehlbauch, D. J. Larson, L. Thomas, and G. A. Erickson. 1996. Phenotypic and ribosomal RNA characterization of *Arcobacter* species isolated form porcine aborted fetuses. J. Vet. Diagn. Invest. 8:186-195.
46. Stampi, S., O. Varoli, F. Zanetti, and G. De Luca. 1993. *Arcobacter cryaerophilus* and thermophilic campylobacters in a sewage treatment plant in Italy: Two secondary treatments compared. Epidemiol. Infect. 110:633-639.
47. Taylor, D. N., J. A. Kiehlbauch, W. Tee, C. Pitarangsi, and P. Echeverria. 1991. Isolation of group 2 aerotolerant *Campylobacter* species from Thai children with diarrhea. J. Infect. Dis. 163:1062-1067.
48. Tee, W. R., R. Baird, M. Dyall Smith, and B. Dwyer. 1988. *Campylobacter cryaerophila* isolated from a human. J. Clin. Microbiol. 267:2469-2473.
49. Vandamme, P., M. Vancanneyt, B. Pot, L. Mels, B. Hoste, D. Dewerttinck, L. Vlaes, C. Van den Borre, R. Higgins, J. Hommez, K. Kersters, J. P. Butzler and H. Goossens. 1992. Polyphasic taxonomic study of the emended genus *Arcobacter* with *Arcobacter butzleri* comb. nov. and *Arcobacter skirrowii* sp. nov., an aerotolerant bacterium isolated from veterinary specimens. Int. J. Syst. Bacteriol. 42:344-356.
50. Vandamme, P., P. Pugina, G. Baenzi, R. Van Etterijck, L. Vlaes, K. Kersters, J. P. Butzler, H. Lior, and S. Lauwers. 1992. Outbreak of recurrent abdominal cramps associated with *Arcobacter butzleri* in an Italian school. J. Clin. Microbiol. 30:2335-2337.
51. Vandamme P., B. A. J. Giesendorf, A. van Belkum, D. Pierard, S. Lauwers, K. Kersters, J. P. Butzler, H. Goossens, and W. G. V. Quint. 1993. Discrimination of epidemic and sporadic isolates of *Arcobacter butzleri* by polymerase chain reaction-mediated DNA fingerprinting. Clin. Microbiol. 31:3317-3319.
52. Versalovic, J., T. Koeuth, and J. R. Lupski. 1991. Distribution of repetitive DNA sequences in eubacteria and application to fingerprinting of bacterial genomes. Nucleic Acids Res. 19:6823-6831.
53. Wesley, I. V. 1994. *Arcobacter* infections, p. 181-90. *In* G. W. Beran and J. H. Steele (ed.), Handbook of zoonoses. CRC Press, Boca Raton, Fla.
54. Wesley, I. V. 1997. *Helicobacter* and *Arcobacter:* Potential human foodborne pathogens? Trends Food Sci. Technol. 89:293-299.
55. Wesley, I. V., and A. L. Baetz. 1999. Natural and experimental infection of birds with *Arcobacter butzleri*. Poultry Sci. 78:536-545.
56. Wesley, I. V., L. Schroeder-Tucker, A. L. Baetz, F. E. Dewhirst, and B. J. Paster. 1995. *Arcobacter*-specific and *Arcobacter butzleri*-specific 16S rRNA-based DNA probes. J. Clin. Microbiol. 33:1691-1698.
57. Wesley, I. V., A. L. Baetz, and D. J. Larson. 1996. Infection of cesarean-derived colostrum deprived 1-day old piglets with *Arcobacter butzleri, Arcobacter cryaerophilus,* and *Arcobacter skirrowii.* Infect. Immun. 64:2295-2299.
58. Zanetti, F., O. Varoli, S. Stampi, and G. de Luca. 1996. Prevalence of thermophilic *Campylobacter* and *Arcobacter butzleri* in food of animal origin. Int. J. Food Microbiol. 33:325-331.

30.3 PLESIOMONAS SHIGELLOIDES

30.31 Introduction

Plesiomonas shigelloides is an opportunistic pathogen associated with gastroenteritis ranging from mild to severe, especially in the very young, aged, and immunologically compromised.[9,17,23,32,57,58,74] Information from both foreign literature and studies in the United States supports *P. shigelloides* as a primary pathogen in diarrheal episodes,[27,30,62,76,84] although controversy still exists as to its role as an enteropathogen.[10,28,29,30,66,67] *P. shigelloides* has been reported to be the responsible cause of death in some of the infected individuals.[12,15,54] Many of the gastroenteritis cases can be associated with the consumption of uncooked or undercooked mollusks, seafood, or with foreign travel.[2,14,27,30,40,49,60,63,76,84]

Symptoms caused by *P. shigelloides* are typical of those associated with a gram-negative bacterial infection. The incubation period ranges from 1 to 2 days, and the symptoms can include diarrhea, abdominal pain, nausea, fever, headache, and vomiting and may last 10 days or longer.[30,60] The attacks are usually self-limited, and treatment involves replacement of fluids and electrolytes.[29] Studies reported that *Plesiomonas* is sensitive to cefixime, ceftriaxone, cefuroxime, cephalexin, chloramphenicol, ciprofloxacin, deoxycycline, enoxacin, gentamicin, nalidixic acid, quinolones, sulfamethoxazole, tetracycline, and trimethoprim.[60,61,62,81] *Plesiomonas*-associated gastroenteritis has been successfully treated using these antimicrobial agents.[44]

Plesiomonas shigelloides is a motile, oxidase- and catalase-positive, glucose-fermenting, facultative anaerobic, and gram-negative rod.[51] It has two to seven polar flagella and may also produce peritrichous flagella.[33] *P. shigelloides* is the only species within the genus and was previously known as *Aeromonas shigelloides*.[49] The other genera within the family Vibrionaceae, *Aeromonas* and *Vibrio,* can be separated by routine biochemical tests. *Plesiomonas* is readily differentiated from the other members of Vibrionaceae by positive inositol fermentation and its lack of extracellular hydrolases.[43] *Plesiomonas* can be misidentified as a member of the *Enterobacteriaceae* if care is not taken to conduct an oxidase test during the identification procedures.[11,30] All plesiomonads cultures ferment glucose and inositol without the production of gas, whereas sucrose, mannitol, dulcitol, adonitol, and xylose are not fermented.[43] Miller[48] demonstrated that all of the 40 strains tested ferment maltose. Some strains of *P. shigelloides* are antigenically related to *Shigella sonnei*[11,30,80] and possibly, *S. flexneri* and *S. dysenteriae*.[3] Both O and H antigens are included in a proposed international typing scheme.[4,5,6,14] Attempts at grouping the plesiomonads by biotyping appear to be of limited value.[49] Various studies have reported on the serological profile of *P. shigelloides*[6,10,64,71,78]; however, antisera are not commercially available. Rapid identification kits have been useful in the identification of *Plesiomonas*.[39,56,73]

With the increased interest in *P. shigelloides*, a number of studies have reported on the mechanism by which it may cause diarrhea in humans.[35,38,66] Saraswathi et al.[68] described the production of both heat-stable and heat-labile cytotoxic enterotoxins from all *Plesiomonas* strains tested, irrespective of origin or serotype. Gardner et al.[24] reported that 24 of the 29 strains of plesiomonads studied produce a cholera-like toxin when tested in Chinese hamster ovary (CHO) cells. Activity of sterile filtrates could be removed by heating or by preincubation of the filtrates with cholera antitoxin. Amin et al.[1] detected heat-labile toxin in 7 of 19 *P. shigelloides* strains isolated from stool samples of diarrheal children using rat ileal loop assay. Abbott et al.[8] detected a low-level of cytolysin in all the 16 strains recovered from humans with extraintestinal and intestinal illnesses, infected animals, and environmental sources, and grown in either Evan Casamino Acids-yeast extract or Penassay broth. However, Brenden et al.[14] speculated that endotoxin may play a role in *Plesiomonas* virulence as it does for other pathogens. Several reports using rabbit ileal loop and suckling mouse assays, rabbit skin permeability toxin assay, and Y1, Vero, and CHO cell culture assays reported that *Plesiomonas* did not produce toxins.[14,19,49] Janda and Abbott[36] demonstrated that more than 90% of the *P. shigelloides* strains they tested produced cell-associated beta-hemolysin at both 25°C and 35°C. Since hemolysin may play a role in iron acquisition *in vivo* via the lysis of erythrocytes, it may contribute to disease development.

Other reports indicated that the plesiomonads may be enteroinvasive.[13,18,30] Binns et al.[13] in a study with freshly isolated *P. shigelloides* from children with acute diarrhea showed that five of 16 strains exhibited invasiveness for HeLa cells, comparable to that of *S. sonnei.* However, none of the five strains gave a positive reaction in the Sereny test. Since an elevated white cell count was found in the feces of a patient with an acute *Plesiomonas* infection, Downey and Clark[18] suggested mucosal invasion by this organism. Holmberg et al.[30] showed that 23 of the 27 isolates from *Plesiomonas*-infected patients contained plasmids, and Sayeed et al.[69] observed diversity in occurrence of plasmids among different strains of *P. shigelloides*. Furthermore, the plasmid profile of *P. shigelloides* SVC 01 differed from that of *S. sonnei.* Kelly and Hain[41] showed that 40% of the 68 clinical and five environmental plesiomonads isolates tested harbored between one and seven plasmids. Small-size plasmids of approximately 2.5 kb, 3.8 kb, and 5.3 kb were found in five strains of *P. shigelloides* isolated from Louisiana blue crabs.[45] Thus, the exact mechanism of enteropathogenicity of *P. shigelloides* is uncertain at this time. To further confound the problem, in a study of 33 healthy volunteers, ingestion of 10^3 to 4×10^9 organisms did not result in any clinical manifestation of a disease.[28] Whether the plasmids play a role in *Plesiomonas* pathogenicity needs further investigation. Further insight into various aspects of this organism may be gained from a number of published reviews.[29,30,43,49,79]

30.32 Ecology

Isolation of *P. shigelloides* has been reported worldwide. Isolation from various mammals (including cats and dogs), birds (both wild and zoo birds), reptiles, crustaceans, fish (both wild and aquacultured), and environmental sites have been reported.[10,29,34,42,45,55,75,77,82,83] Isolations from humans have been reported for both symptomatic and asymptomatic individuals.[10,11,29,79] *P. shigelloides* was isolated at a very low rate (0.0078% or 3 carriers) from 38,454 food handlers and school children in Japan,[10] but a much higher rate (24%, or 12 individuals) from 51 adults in Thailand.[14] Fresh and brackish water are usually excellent sources of this organism, as are freshwater fish.[7,10,19,20,43,47,77] Freshwater fish has been implicated as one of the primary reservoirs of *P. shigelloides*, and accounts for the ease with which this organism can be isolated from the aquatic environment.[42,72] Antibiotic resistant strains of *P. shigelloides* were isolated from the intestinal tracts of catfish and from water and sediment in aquaculture ponds following the use of antimicrobial agents for treatment of fish diseases.[46] A number of studies have shown a seasonal effect on isolation rates from these sources as well as an increase in the reported cases of diarrhea during the warmer seasons.[10,75] *P. shigelloides* is also a known pathogen of fish and reptiles.[16] On at least one occasion the fish or the reptile has been implicated as the vehicle of *Plesiomonas* infection in a human handler.[26] Tropical aquaria from pet shops may constitute an infection hazard since both *Aeromonas* spp. and *P. shigelloides* were isolated from this environment.[65] It has been reported that some isolates can grow in broth containing 5% sodium chloride and at a pH as low as 4 and as high as 9. The isolates could not grow at a temperature of 5°C nor at 50°C. Pasteurization at 60°C for 30 min readily destroyed the plesiomonads tested.[51]

30.33 Isolation and Identification

Our experience has shown that *P. shigelloides* does not grow on some media (thiosulfate citrate bile salts sucrose agar, TCBS), grows slowly on others (inositol brilliant green bile salts agar, IBB), and grows well on others (Plesiomonas agar, PL). Huq et al.[31] compared the growth characteristics of *P. shigelloides* on plesiomonas differential agar, IBB agar, and modified salmonella-shigella agar at incubation temperatures of 37°C, 42°C and 44 °C and showed plesiomonas differential agar to be ideal when incubated at 44°C. *Plesiomonas* colonies could be recognized readily after 24 hr. Species differences, an injury phenomenon, nutritional demands, and lack of tolerance to certain media components complicate any attempt at recommending one simple procedure for isolating this organism.

Early isolations of *P. shigelloides* were usually accomplished with enteric agars.[10,52,75,77,79,83] Salmonella-shigella agar, eosin-methylene blue (EMB), MacConkey, and xylose lysine desoxycholate (XLD) are a few that have been used. *P. shigelloides* is quite tolerant to bile salts and brilliant green. Because lactose is generally only slowly fermented, the plesiomonads appear as lactose-negative on solid media, including triple sugar iron (TSI) agar. Because of their apparent lack of competitiveness and nutritional dependency, enrichment techniques are of only limited usefulness for the plesiomonads.[22] Direct plating of the sample should always be done, and an enrichment step can be employed. Selenite, gram-negative (GN) broth, alkaline peptone water (APW), tetrathionate, and various bile broths including bile salts brilliant green broth have been evaluated with differing success.[21,53,59,77] Success following enrichment is probably a reflection of the nature of the sample, including the presence of other organisms as well as the incubation temperature.[21,22] For routine analysis of environmental and food samples for *P. shigelloides,* Jeppesen[37] recommended spread plating on IBB and PL agars.

P. shigelloides shares many characteristics with both the *Vibrionaceae* and the *Enterobacteriaceae.* However, once an oxidase test is conducted and found to be positive, further separation and identification is fairly rapid. Classical biochemical testing is conducted to confirm identity of the organism. Table 4 summarizes some of the characteristics of the plesiomonads. Complete descriptions of the plesiomonads are reported elsewhere.[43,49,70]

Our laboratory used the following protocol when isolating the plesiomonads. Samples, depending upon consistency and expected numbers, are diluted and directly surface streaked to IBB and PL agars.[50] One to 10 grams of sample are added to 90 mL of tetrathionate broth. Plates are incubated at 35°C and the enrichment broths at 40°C. Following 24 hr of incubation, suspect colonies are picked from the plates to TSI slants and inositol gelatin medium (deep stab).[50] The enrichment broths are streaked to the isolation media and the plates incubated at 35°C for 24 hr. Isolates that are alkaline over acid without gas or hydrogen sulfide on TSI, produce acid but no gas from inositol, and do not hydrolyze gelatin are tested for oxidase. If the organism is oxidase-positive and a gram-negative rod, it is *Plesiomonas*.

30.34 Equipment, Materials, and Solutions

Materials in general are as for the Enterobacteriaceae.

Alkaline peptone water (APW) (0.1%)
Inositol brilliant green bile (IBB) salts agar
Inositol gelatin deeps
MacConkey agar
Oxidase reagent
Peptone water (0.1%)
PL agar
Salmonella shigella (SS) agar
Tetrathionate broth (without iodine)
Triple sugar iron (TSI) agar

Plesiomonads appear as yellow colonies on IBB agar, pink on PL agar, and colorless translucent on both MacConkey and SS agar. The organisms produce *Shigella*-like reaction on MacConkey agar and TSI agar slants. In inositol gelatin deeps, plesiomonads will turn the medium yellow and will not hydrolyze the gelatin.

30.35 Procedure

30.351 Selective Agar

Prepare dried plates of two of the following media: MacConkey, brilliant green lactose bile agar, PL, SS, or IBB. If the sample is fluid, transfer a 0.5 mm loopful to the surface of duplicate plates of each medium and streak for isolation. If the sample is solid,

Table 4. Some Identifying Characterstics of *Plesiomonas shigelloides*

Character		Reaction
Gram stain		– rod
Motility		lophotrichous flagella/ peritrichous flagella
Oxidase		+
Catalase		+
Glucose		acid no gas
Inositol		+/no gas
Gelatinase		–
Arginine dihydrolase		variable
Lysine decarboxylase		+
Ornithine decarboxylase		+
Histidine decarboxylase		+
Alkaline phosphatase		+
Leucine aminopeptidase		+
Nitrate to nitrite		+
Deoxyribonuclease		–
String test		–
O/129 sensitivity		sensitive
Acid from	lactose	variable
	sucrose	–
	mannitol	–
	inositol	+
	arabinose	–
	xylose	–
Urease		–
Citrate		–
Growth at 42°C		+
Halophilism	0% NaCl	+
	3% NaCl	+
	6% NaCl	–

– = negative; + = positive

prepare a 1:10 dilution in 0.1% peptone water and streak. Incubate at 35°C for 24 hr.

30.352 Enrichment Broth

Transfer a 10 g sample to 90 mL tetrathionate broth without iodine and incubate at 40°C for 24 hr. Following enrichment, streak to duplicate plates of two of the above selective media and incubate plates at 35°C for 24 hr.

30.353 Identification

Pick three typical colonies from each of the selective media into TSI and inositol gelatin deeps. Incubate at 35°C for 24 hr. *Plesiomonas* appears as alkaline over acid with no gas or hydrogen sulfide in TSI, will ferment inositol without gas, and will not hydrolyze gelatin. Conduct an oxidase test and gram stain from the TSI slant.

30.354 Precautions

A known wild-type isolate as well as an ATCC strain should be tested concurrently with the unknown isolates to ensure accuracy of media reactions. While *P. shigelloides* is not highly infectious, reasonable care should be taken when working with this organism.

30.36 References

1. Abbott, S. L., Kokka, R. P., and Janda, J. M. 1991. Laboratory investigations on the low pathogenic potential of *Plesiomonas shigelloides*. J. Clin. Microbiol. 29:148.
2. Albert, S., Weber, B., Schafer, V., Rosenthal., P., Simonsohn, M., and Doerr, H. W. 1990. Six enteropathogens isolated from a case of acute gastroenteritis. Infection 18: 381.
3. Albert, M. J., Ansaruzzaman, M., Qadri, F., Hossain, A., Kibriya, A. K., Haider, K., Nahar, S., Faruque, S. M., and Alam, A. N. 1993. Characterization of *Plesiomonas shigelloides* strains that share type-specific antigen with *Shigella flexneri* 6 and common group 1 antigen with *Shigella flexneri* spp. And *Shigella dysenteriae* 1. J. Med. Microbiol. 39:211.
4. Aldova, E. 1994. Serovars of *Plesiomonas shigelloides*. Int. J. Med. Microbiol. Virol. Parasitol. Infect. Dis. 281:38.
5. Aldova, E., Danesova, D., Postupa, J., and Shimada, T. 1994. New serovars of *Plesiomonas shigelloides* 1992. Cent. Eur. J. Public Health 1:32.
6. Aldova, E. 1995. The importance of serotyping *Plesiomonas shigelloides*. Epidemiol. Mikrobiol. Immunol. 44:147.
7. Aldova, E. and Schubert. H. W. 1996. Serotyping of *Plesiomonas shigelloides*—a tool for understanding ecological relationships. Med. Microbiol. Lett. 5:33.
8. Amin, I. I., Hossain, M. A., Hossain, M., Miah, M. R., Rahman, Z., and Rahman, K. M. 1992. Studies on virulence determinants of *Plesiomonas shigelloides*. Bangladesh Med. Res. Counc. Bull. 18:12.
9. Appelbaum, P. C., Bowen, A. J., Adhikari, M., Robins-Browne, R. M., and Koornhof, H. J. 1978. Neonatal septicemia and meningitis due to *Aeromonas shigelloides*. J. Pediatr. 92: 676.
10. Arai, T., Ikejima, N., Itoh, T., Sakai, S., Shimada, T., and Sakazaki, R. 1980. A survey of *Plesiomonas shigelloides* from aquatic environments, domestic animals, pets, and humans. Hyg. (Camb.) 84: 203.
11. Bhat P., Shanthakumari, S., and Rajan, D. 1974. The characterization and significance of *Plesiomonas shigelloides* and *Aeromonas hydrophila* isolated from an epidemic of diarrhoea. Ind. J. Med. Res. 62: 1051.
12. Billiet, J., Kuypers, S., van Lierde, S., and Verhaegen, J. 1989. *Plesiomonas shigelloides* meningitis and septicaemia in a neonate: report of a case and review of the literature. J. Infect. 19:267.
13. Binns, M. M., Vaughan, S., Sanyal, S. C., and Timmis, K. N. 1984. Invasive ability of *Plesiomonas shigelloides*. Zentralbl. Bakteriol. Mikrobiol. Hyg. A. 257:343.
14. Brenden, R. A., Miller, M. A., and Janda, J. M. 1988. Clinical disease spectrum and pathogenic factors associated with *Plesiomonas shigelloides* infections in humans. Rev. Infect. Dis. 10:303.
15. Clark, R. B., Westby, G. R., Spector, H., Soricelli, R. R., and Young, C. L. 1991. Fatal *Plesiomonas shigelloides* septicemia in a splenectomised patient. J. Infect. 23:89.
16. Cruz, J. M., Saraiva, A., Eiras, J. C., Branco, R., and Sousa, J. C. 1986. An outbreak of *Plesiomonas shigelloides* in farmed rainbow trout, *Salmo gairdneri* Richardson, in Portugal. Bulletin Eur. Assoc. Fish Pathol. 6:20.
17. Dahm, L. J. and Weinberg, A. G. 1980. *Plesiomonas (Aeromonas) shigelloides* septicemia and meningitis in a neonate. South. Med. J. 73: 393.
18. Downey, D. J. and Clark, J. N. 1984. A case of diarrhea associated with *Plesiomonas shigelloides*. N. Z. Med. J. 97 (749):92.
19. Farmer, J. J., III, Arduino, M. J., and Hickman-Brenner, F. W. 1992. The genera *Aeromonas* and *Pleasiomonas*, In "The Prokaryotes," eds. A. Balows, H. G. Truper, M. Dworkin, W. Harder, and K.-H. Schleifer, 2nd ed., p. 3012. Springer-Verlag, New York. N. Y.
20. Fernandes, C. F., Flick, G. J., Jr., Silva, J. L., and McCaskey, T. A. 1997. Comparison of quality in aquacultured catfish fillets. II. Pathogens *E. coli* O105:H7, *Campylobacter*, *Vibrio*, *Plesiomonas*, and *Klebsiella*. J. Food Prot. 60:1182.
21. Freund, S. M., Koburger, J. A., and Wei, C. 1. 1988. Enhanced recovery of *Plesiomonas shigelloides* following an enrichment technique. J. Food Prot. 51: 110.
22. Freund, S. M., Koburger, J. A., and Wei, C. I. 1988. Isolation of *Plesiomonas shigelloides* from oysters using tetrathionate broth enrichment. J. Food Prot. 51: 925.
23. Fujita, K., Shirai, M., Ishioka, T., and Kakuya, F. 1994. Neonatal *Plesiomonas shigelloides* septicemia and meningitis: a case and review. Acta Paediatr. Jpn. 36:450.
24. Gardner, S. B., Fowlston, S. B., and George, W. L. 1987. *In vitro* production of cholera toxin-like activity by *Plesiomonas shigelloides*. J. Infect. Dis. 156: 720.
25. Glunder, G. 1988. Occurrence of Plesiomonas shigelloides in wild and zoo birds. Berliner Munchener Tierarztliche Wochenschrift. 101: 334.
26. Gopal, V., and Burns, F. E. 1991. Cellulitis and compartment syndrome due to *Plesiomonas shigelloides*: A case report. Mil. Med. 156:43.
27. Haeberberger, R. L., Jr., Mikhail, I. A., Burans, J. P., Hyams, K. C., Glenn, J. C., Diniega, B. M., Sorgen, S., Mansour, N., Blacklow, N. R., and Woody, J. N. 1991. Travelers' diarrhea among United States military personnel during joint American-Egyptian armed forces exercises in Cairo, Egypt. Mil. Med. 156:27.
28. Herrington, D. A., Tzipori, S., Robins-Browne, R. M., Tall, B. D., and Levine, M. M. 1987. *In vitro* and *in vivo* pathogenicity of *Plesiomonas shigelloides*. Infect. Immun. 55: 979.
29. Holmberg, S. D. and Farmer, J. J. 1984. *Aeromonas hydrophila* and *Plesiomonas shigelloides* as causes of intestinal infections. Rev. Infect. Dis. 6: 633.
30. Holmberg, S. D., Wachsmuth, I. K., Hickman-Brenner, F. W., Blake, P. A., and Farmer, J. J. 1986. *Plesiomonas* enteric infections in the United States. Ann. Intern. Med. 105: 690.
31. Huq, A., Akhtar, A., Chowdhury, M. A., and Sack, D. A. 1991. Optimal growth temperature for the isolation of *Plesiomonas shigelloides*, using various selective and differential agars. Can. J. Microbiol. 37:800.
32. Ingram, C. W., Morrison, A. J. Jr., and Levitz, R. E. 1987. Gastroenteritis, sepsis, and osteomyelitis caused by *Plesiomonas shigelloides* in an immunocompetent host: Case report and review of the literature. J. Clin. Microbiol. 25:1791.
33. Inoue, K., Kosako, Y., Suzuki, K., and Shimada, T. 1991. Peritrichous flagellation in *Plesiomonas shigelloides* strains. Jpn. J. Med. Sci. Biol. 44:141.
34. Islam, M. S., Alam, M. J., and Khan, S. I. 1991. Distribution of *Plesiomonas shigelloides* in various components of pond ecosystems in Dhaka, Bangladesh. Microbiol. Immunol. 35:927.

35. Janda, J. M. 1987. Effect of acidity and antimicrobial agent-like compounds on viability of *Plesiomonas shigelloides.* J. Clin. Microbiol. 25: 1213.
36. Janda, J. M. and Abbott, S. L. 1993. Expression of hemolytic activity by *Plesiomonas shigelloides*. J. Clin. Microbiol. 31:1206.
37. Jeppesen, C. 1995. Media for *Aeromonas* spp., *Plesiomonas shigelloides* and *Pseudomonas* spp. From food and environment. Int. J. Food Microbiol. 26:25.
38. Johnson, W. M. and Lior, H. 1981. Cytotoxicity and suckling mouse reactivity of *Aeromonas hydrophila* isolated from human sources. Can. J. Microbiol. 27: 1019.
39. Jorgensen, J. H., Dyke, J. W., Helgeson, N. G. P., Cooper, B. H., Redding, J. S., Crawford, S. A., Andruszewski, M. T., and Prowant, S. A. 1984. Collaborative evaluation of the Abbott Advantage System for identification of frequently isolated nonfermentative or oxidase-positive gram-negative bacilli. J. Clin. Microbiol. 20:899.
40. Kain, K. C. and Kelly, M. T. 1989. Clinical features, epidemiology, and treatment of *Plesiomonas shigelloides* diarrhea. J. Clin. Microbiol. 27: 998.
41. Kelly, M. T. and Hain, K. C. 1991. Biochemical characteristics and plasmids of clinical and environmental *Plesiomonas shigelloides*. Experientia 15:439.
42. Klein, B. U., Kleingeld, D. W., and Bohm, K. H. 1993. First isolations of *Plesiomonas shigelloides* from samples of cultured fish in Germany. Bulletin Eur. Assoc. Fish Pathol. 13:70.
43. Koburger, J. A. 1988. *Plesiomonas shigelloides. In* "Foodborne Bacterial Pathogens," ed. M. P. Doyle, p. 311. Marcel Dekker, Inc. New York, N.Y.
44. Leelarasamee, A., Sukrungreang, C., Thian. G. S., and Vudhivatana, M. 1996. Therapeutic efficacy of oral ofloxacin in acute diarrhea and dysentery. J. Infect. Chemother. 2:79.
45. Marshall, D. L., Kim, J. J., and Donnelly, S. P. 1996. Antimicrobial susceptibility and plasmid-mediated streptomycin resistance of *Plesiomonas shigelloides* isolated from blue crab. J. Appl. Bacteriol. 81:195.
46. McPhearson, R. M., DePaola, A., Zywno, S. R., Motes, M. L., Jr., and Guarino, A. M. 1991. Antibiotic resistance in Gram-negative bacteria from cultured catfish and aquaculture ponds. Aquaculture 99:203.
47. Medema, G. and Schets, C. 1993. Occurrence of *Plesiomonas shigelloides* in surface water: Relationship with faecal population and trophic state. Zentralbl. Hyg. 194: 398.
48. Miller, M. L. 1985. *Plesiomonas shigelloides*: A food and waterborne pathogen. M. S. Thesis, University of Florida, Gainesville.
49. Miller, M. L. and Koburger, J. A. 1985. *Plesiomonas shigelloides:* An opportunistic food and waterborne pathogen. J. Food Prot. 48:449.
50. Miller, M. L. and Koburger, J. A. 1986. Evaluation of inositol brilliant green bile salts and Plesiomonas agars for recovery of *Plesiomonas shigelloides* from aquatic samples in a seasonal survey of the Suwannee River estuary. J. Food Prot. 49:274.
51. Miller, M. L. and Koburger, J. A. 1986. Tolerance of *Plesiomonas shigelloides* to pH, sodium chloride and temperature. J. Food Prot. 49:877.
52. Millership, S. E. and Chattopadhyay, B. 1984. Methods for the isolation *of Aeromonas hydrophila* and *Plesiomonas shigelloides* from faeces. J. Hyg. (Camb). 92:145.
53. Nair, P., and Millership, S. E. 1987. Isolation of *Plesiomonas shigelloides* from nutrient broth with brilliant green: its use in screening stool samples from an African population. J. Clin. Pathol. 40:680.
54. Nolte, F. S., Poole, R. M., Murphy, G. W., Clark, C., and Panner, B. J. 1988. Proctitis and fatal septicemia caused by *Plesiomonas shigelloides* in a bisexual man. J. Clin. Microbiol. 26:388.
55. Obi, C. L., Coker, A. Q., Epoke, J., and Ndip, R. 1995. *Aeromonas* and *Plesiomonas* species as bacterial agents of diarrhea in urban and rural areas of Nigeria: antibiogram of isolates. Cent. Afr. J. Med. 41: 397.
56. Overman, T. L. and Overley, J. K. 1986. Feasibility of same-day identification of members of the Family *Vibrionaceae* by the API 20E system. J. Clin. Microbiol. 23:715.
57. Paul, R., Siitonen, A., and Karkkainen, P. 1990. *Plesiomonas shigelloides* bacteremia in a healthy girl with mild gastroenteritis. J. Clin. Microbiol. 28:1445.
58. Penn, R. G., Giger, D. K.. Knoop, F. C., and Preheim, L. C. 1982. *Plesiomonas shigelloides* overgrowth in the small intestine. J. Clin. Microbiol. 15:869.
59. Rahim, Z. and Kay, B. A. 1988. Enrichment for *Plesiomonas shigelloides* from stools. J. Clin. Microbiol. 26:789.
60. Rautelin, H., Sivonen, A., Kuikka, A., Renkonen, O. V., Valtonen, V., and Kosunen, T. U. 1995. Enteric *Plesiomonas shigelloides* infections in Finnish patients. Scan. J. Infect. Dis. 27:495.
61. Reinhardt, J. F. and George, W. L. 1985. Comparative in vitro activities of selected anti-microbial agents against *Aeromonas* species and *Plesiomonas shigelloides.* Antimicrob. Agents Chemother. 27:643.
62. Reinhardt, J. F. and George, W. L. 1985. *Plesiomonas shigelloides*- associated diarrhea. J. Am. Med. Assoc. 253:3294.
63. Rutala, W. A., Sarubbi. F. A. Jr., Finch, C. S., MacCormack, J. N., and Steinkraus, G. E. 1982. Oyster-associated outbreak of diarrheal disease possibly caused by *Plesiomonas shigelibides.* Lancet 1:739.
64. Sack, D. A., Hoque, A. T., Huq, A., and Etheridge, M. 1994. Is protection against shigellosis induced by natural infection with *Plesiomonas shigelloides*? Lancet 343 (8910): 1413.
65. Sanyal, D., Burge, S. H., and Hutchings, P. G. 1987. Enteric pathogens in tropical aquaria. Epidemiol. Infect. 99:636.
66. Sanyal, S. C., Saraswathi. B., and Sharma, P. 1980. Enteropathogenicity of *Plesiomonas shigelloides.* J. Med. Microbiol. 13:401.
67. Sanyal, S. C., Singh. S. J., and Sen. P. C. 1975. Enteropathogenicity of *Aeromonas hydrophila* and *Plesiomonas shigelloides.* J. Med. Microbiol. 8:195.
68. Saraswathi, B., Agarwal, R. K., and Sanyal, S. C. 1983. Further studies on enteropathogenicity of *Plesiomonas shigelloides.* Indian J. Med. Res. 78:12.
69. Sayeed, S., Sack, D. A., and Qadri, F. 1992. Occurrence of a large plasmid in strain of *Plesiomonas shigelloides* with cross-reactivity against *Shigella sonnei.* Indian J. Med. Res. 95:21.
70. Schubert, R. H. W. 1984. Genus IV *Plesiomonas,* In Bergey's Manual of Systematic Bacteriology,"eds. N. R. Krieg and J. G. Holt, Vol. 1, p. 548. Williams and Wilkins, Baltimore, MD.
71. Shimada, T. and Sakazaki. R. 1978. On the serology of *Plesiomonas shigelloides*. Jap. J. Med. Sci. Biol. 31:135.
72. Sugita, H., Nakamura, T., and Deguchi, Y. 1993. Identification of *Plesiomonas shigelloides* isolated from freshwater fish with the microplate hybridization method. J. Food Prot. 56:949.
73. Taylor, P. W., Crawford, J. E., and Shotts., E. B., Jr. 1995. Comparison of two biochemical test systems with conventional methods for identification of bacteria pathogenic to warmwater fish. J. Aquatic Animal Health 7:312.
74. Terpeluk, C., Goldmann, A. Bartmann, P., and Pohlandt, F. 1992. *Plesiomonas shigelloides* sepsis and meninggoencephalitis in neonate. Eur. J. Pediatr. 151:499.
75. Tsukamoto, T., Kinoshita. Y., Shimada, T., and Sakazaki, R. 1978. Two epidemics of diarrhoeal disease possibly caused by *Plesiomonas shigelloides*. J. Hyg. (Camb.) 80:275.
76. Ueda, Y., Suzuki, N., Mori, H., Miyagi, K., Noda, K., Hirose, H., Takegaki, Y., Hashimoto., S., Oosumi, Y., Miyata Y., Taguchi, M., Ishibashi, and M., Honda, T. 1996. Bacteriological study of traveller's diarrhoea. 5) Analysis of enteropathogenic bacteria at Osaka Airport Quarantine Station from January 1992 through September 3rd, 1994. Kansenshogaku Zasshi. 70:29.
77. Van Damme, L. R. and Vandepitte, J. 1980. Frequent isolation of *Edwardsiella tarda* and *Plesiomonas shigelloides* from healthy Zairese freshwater fish: A possible source of sporatic diarrhea in the tropics. Appl. Environ. Microbiol. 39:475.
78. Van Loon, F. P., Rahim, Z., Chowdhury, K. A., Kay, B. A., and Rahman, S. A. 1989. Case report of *Plesiomonas shigelloides*-associated persistent dysentery and pseudomembranous colitis. J. Clin. Microbiol. 27:1913.

79. Vandepitte, J., Makulu. A., and Gatti, F. 1974. *Plesiomonas shigelloides* survey and possible association with diarrhoea in Zaire. Ann. Sec. Belge. Med. Trop. 54:503.
80. Viret, J. F., Cryz, S. J., Jr., Lang, A. B., and Favre, D. 1993. Molecular cloning and characterization of the genetic determinants that express the complete *Shigella* serotype D (*Shigella sonnei*) lipopolysaccharide in heterologous live attenuated vaccine strain. Mol. Microbiol. 7:239.
81. Visitsunthorn, N. and Komolpis, P. 1995. Antimicrobial therapy in *Plesiomonas shigelloides*-associated diarrhea in Thai children. Southeast Asian J. Trop. Med. Public Health 26:86.
82. Weber, J. M. and Roberts, L. 1989. A bacterial infection as a cause of abortion in the European otter, *Lutra lutra*. J. Zoology 219: 688.
83. Winsor, D. K., Bloebaum. A. P., and Mathewson, J. J. 1981. Gram-negative, aerobic, enteric pathogens among intestinal microflora of wild turkey vultures *(Cathartes aura)* in West Central Texas. Appl. Environ. Microbiol. 42:1123.
84. Yoshida, A., Noda, K., Omuram K., Miyagi, K., Mori, H., Suzuki, N., Takai, S., Matsumoto, Y., Hayashi, K., Miyata Y. et al. 1992. Bacteriological study of traveller's diarrhoea. 4) Isolation of enteropathogenic bacteria from patients with traveller's diarrhoea at Osaka Airport Quarrantine Station during 1984-1991. Kansenshogaku Zasshi. 66:1422.

Chapter 31

Campylobacter

Norman J. Stern, J. Eric Line, and Hui-Cheng Chen

31.1 INTRODUCTION

Campylobacter jejuni, C. coli, and *C. lari* are carried in the intestinal tract of warm-blooded animals and, therefore, contaminate foods of animal origin. *C. jejuni* is recognized as a leading cause of acute bacterial gastroenteritis and can lead to serious pathological sequelae. *C. coli* and *C. lari* also are recognized causes of gastroenteritis but occur less frequently than *C. jejuni.* Because these campylobacters are recognized causes of human diarrheal disease, and because eating foods of animal origin has been associated with many of these illnesses, the presence of any of these species in food represents a potential hazard to human health. For simplicity's sake, these three species will be collectively referred to as the *C. jejuni* group, or *Campylobacter*, unless otherwise noted.

31.11 Description of the Organisms

As of 1989 the genus *Campylobacter* consisted of 18 species, subspecies, and biovars, with 17 names officially recognized by the International Committee on Systematic Bacteriology and one species name proposed but not official.[102,104] The taxonomic status of the genus is in flux and is likely to continue changing for some time to come. At least nine species appear to be important in human disease. This number includes *Campylobacter pylori,* the agent of human gastritis[67] whose genetics are only distantly related to *Campylobacter*[99] and which is now classified as *Helicobacter pylori.*

Campylobacters grow between 25°C and 43°C, are gram-negative, motile, curved or spiral rods, are oxidase-positive, and do not ferment or oxidize carbohydrates. The campylobacters can be broadly placed into two groups based on the catalase test results. The catalase-positive campylobacters are most frequently associated with human disease; however, the catalase-negative species "*C. upsaliensis*" also appears related to human disease.[85] Consequently, the isolation of a catalase-negative *Campylobacter* from foods should be pursued and the strains characterized.

The campylobacters are generally inactive in many conventional biochemical tests and identification is based on only a few morphological and biochemical features. With certain strains, species identification is based on the results of only one test, and in some cases, definitive identification is not possible with routinely available laboratory tests. Confirmation of *Campylobacter* isolation may be achieved through application of immunologically based latex agglutination assays or by specific genetic tests. The need for species differentiation appears less important than is the genetic characterization of isolates.

31.12 Pathology

Symptoms of *C. jejuni* infections vary from profuse watery diarrhea (cholera-like) to bloody diarrhea containing mucus and white blood cells (dysentery-like). Although the mechanisms by which *C. jejuni* produces disease are as yet largely undefined, several possible virulence factors such as toxins, motility, invasion, and adherence have been described. It has been proposed that *C. jejuni* colonizes human intestinal mucus in a manner similar to that of the spiral bacteria of the normal microbiota.[61,62] These bacteria have a spiral morphology, are microaerophilic, and can penetrate deeply into intestinal crypts. The adaptation to the intestinal mucus niche may be an important determinant of virulence for *C. jejuni.* Motility may also be an important virulence factor for *C. jejuni.*[17,40,107] *C. jejuni* infections that result in a dysentery-like disease suggest that some strains of the species may be invasive.[48] The profuse watery diarrhea observed in some *Campylobacter* infections may indicate that these strains could be enterotoxigenic. Certain strains of *C. jejuni* produce a toxin immunologically related to cholera toxin (CT) and *Escherichia coli* heat-labile toxin (LT).[45,46,52,69] *C. jejuni* cytotoxins have been described.[82] The *Campylobacter* toxin produces cytotonic changes in tissue culture cells via stimulation of cyclic AMP production.[68,101] The role of the cytotoxin in pathogenesis has not been defined, but animals challenged with cytotoxin-positive strains have more severe diarrhea than those given cytotoxin-negative strains.[82] The presence of cytotoxin is also correlated in patients with bloody diarrhea.[53]

The lack of an easy-to-use, inexpensive animal model that mimics human infections has seriously hampered the identification of virulence factors in *C. jejuni.* Although many animal models have been proposed,[8,16,28,40,50,61,100103,107] none has gained wide acceptance.

Guillain-Barré syndrome (GBS) is a human neurological disease in which demyelination of neuronal materials occur and this process can result in acute neuromuscular paralysis.[73] Approximately 40% of patients with GBS have evidence of recent *Campylobacter* infection. Perhaps 20% of GBS patients have long term disability and approximately 5% die. Another sequel to *Campylobacter* infection is Reiter's Syndrome, a reactive arthropathy in which joint pain can be chronic. There appears to be an autoimmune component to both GBS and Reiter's syndrome in its involvement with *Campylobacter* infection.

31.13 Distribution and Epidemiology

C. jejuni has been isolated from individuals with gastroenteritis throughout the world.[6,18,21,49,60,80,89,92,94,108] A recent paper from the Centers for Disease Control indicates that Campylobacter is the most common bacterial cause of gastroenteritis in the U.S. with approximately 2.5 million cases of campylobacteriosis occurring annually.[129] *C. jejuni is* often isolated from patients with diarrhea at greater isolation rates than reported for *Salmonella* spp. The rate of isolation of *C. jejuni* infection on college campuses in the United States is 10 and 46 times more frequent than *Salmonella* and *Shigella* infection, respectively.[120] Research indicates that as few as 500 cells of *C. jejuni* can produce illness in humans.[4,95] However, the true infectious dose also depends on the buffering capacity of the challenge milieu and the immune competence of the host.

Foods of animal origin appear to be the primary vehicles involved in human infection. The sources of the other *Campylobacter* species implicated in human disease, i.e., *C. hyointestinalis, "C. upsaliensis," C. cinaedi, C. fennelliae,* and *C. pylori (H. pylori),* have not been determined. Unpasteurized milk has been by far the most commonly implicated vehicle in foodborne outbreaks of *C. jejuni* enteritis.[7,13,15,27,58,90,91,96,122,123,128] *C. jejuni* can cause a mastitis in cows,[59] and two reports suggest that the presence of *C. jejuni* in the implicated raw milk resulted from udder expression rather than from fecal contamination.[39,43]

Campylobacter is more frequently associated with sporadic cases of illness than recognized widespread outbreaks. Although raw milk is frequently reported as the vehicle of *Campylobacter* enteritis outbreaks, epidemiologic studies of sporadic cases have revealed that mishandled poultry is more important than raw milk in transmitting *C. jejuni* enteritis.[20,30,35,121] Several outbreaks also have been associated with eating poultry.[14,44,76] For these reasons public health authorities now believe that poultry is the most prominent vehicle in transmission of *Campylobacter* enteritis within the United States and in other developed countries.

31.2 GENERAL CONSIDERATIONS

31.21 Methods of Isolation

Because *C. jejuni* does not grow below 30°C and is sensitive to normal atmospheric concentrations of oxygen, only small numbers of *Campylobacter* may be present in foods. Selective enrichment may be needed to detect the few culturable cells of *C. jejuni* that may be present. In food products such as freshly processed poultry where large numbers may be expected, direct plating is appropriate. Large sample size, selective enrichment broth, suitable microaerobic conditions, and selective isolation media or filtration techniques are important for isolating campylobacters from non-poultry foods. Various selective enrichment broth systems[12,23,84,97,127] selective isolation agar media[6,11,117] and methods to produce a microaerobic atmosphere[51] have been developed for the isolation of *C. jejuni.* These systems will be discussed. Enrichment-filtration techniques with incubation at 37°C[71] have also been used to increase the isolation of *C. jejuni* and allow growth of other *Campylobacter* species of human health concern.

31.22 Treatment of Sample

Food samples must be processed rapidly to ensure optimum isolation of *C. jejuni*, which is sensitive to many environmental conditions. Analysis of samples must address the fragile nature of the organism and should take the following facts into consideration: The optimum temperature for growth is 42°C to 43°C, and thermal inactivation occurs at 48°C. Heat injury and repair of *C. jejuni* occurs at 46°C,[81] and the existence of an injured, but nonculturable form of the organism has been reported to colonize chicks[98,118] *C. jejuni* will not survive the minimum pasteurization treatment (63°C for 30 min or 71.7°C for 15 sec) for milk or milk products, hence a *Campylobacter*-free milk product should be anticipated with good manufacturing practices.[26,32] Similar observations have been made regarding heat inactivation of campylobacters in meat products.[31,55,111]

C. jejuni is not likely to grow in foods held under typical storage conditions. The organism dies more quickly at 25°C than at either 4°C or 30°C.[5,22] Reports indicate that *C. jejuni* survives best in foods held at refrigeration temperature,[19,56,111] but it is highly susceptible to freezing conditions.[1,31,34,79,111,112] Within 5 days of refrigerated storage, *Campylobacter* on processed chicken carcasses become injured and require enrichment for recovery.[119] Survival of the organism under frozen conditions is enhanced with cryoprotective agents.[111] Also, freeze-stressed *C. jejuni* may not be culturable from contaminated frozen foods,[93] a concern addressed by Park and Sanders[83] in their enrichment procedure.

C. jejuni is very sensitive to sodium chloride.[24,33] The optimum concentration for recovery or enumeration of *C. jejuni* is 0.5% NaCl. Although *C. jejuni* dies more rapidly in the presence of sodium chloride, the organism can persist for weeks in refrigerated (4°C) foods with 6.5% salt. Campylobacters are also quite sensitive to drying conditions, although they survive for up to 6 weeks at 4°C in an environment of 14% or less relative humidity.[25] In laboratory media, *C. jejuni* grows well at pH 5.5 to 8.0. Optimal growth is in the pH range of 6.5 to 7.5, while no growth occurs at pH 4.9 or lower.[22] Lactic and acetic acids have been shown to reduce the number of *Campylobacter* on chicken broiler halves.[113]

Campylobacter is both microaerophilic and capnophilic, that is, the organism grows best under conditions of reduced atmospheric concentrations of oxygen and under high carbon dioxide concentrations. The optimum atmospheric composition for growth of *C. jejuni* is 5% oxygen, 10% carbon dioxide, and 85% nitrogen.[10,51] At 4°C, the organism survives optimally in the presence of 0.01% sodium bisulfite in an atmosphere of 100% nitrogen.[56] Recovery of *C. jejuni* from inoculated unpasteurized milk is improved by the addition of reducing agents.[57] Survival of *C. jejuni* on refrigerated meat is similar irrespective of the packaging treatment.[34,114,126]

31.3 MEDIA AND REAGENTS

31.31 Media

31.311 Non-Selective Media for Checking Purity and Propagation and Testing Tolerance to Certain Chemicals and Antimicrobial Compounds

Brucella-FBP agar[29]
Brucella-FBP broth
Semisolid Brucella-FBP medium

31.312 Selective Isolation Agars

Campylobacter charcoal differential agar (CCDA)—Preston blood-free medium[42]
Campy-Cefex agar[117]
Campy-Line agar[63]

31.313 Enrichment Broths

Bolton[9]
Park and Sanders[83]
Hunt[41]

31.314 Semisolid Media for Maintenance, Storage, and Transport of Cultures

Cary-Blair transport medium
Wang's transport/storage medium[125]

31.315 Media for Taxonomic Criteria

Fermentation base for *Campylobacter*
Nitrate broth (heart infusion broth + 0.002% potassium nitrate)
Oxidation-fermentation medium for *Campylobacter* (O-F)
Semisolid (0.16% agar) brucella (albimi) medium
Trimethylamine N-oxide (TMAO) medium
Triple sugar iron (TSI) agar

31.316 Additional Media

Buffered peptone water
Brucella broth
Heart infusion agar (HIA) slant
Heart infusion agar + 5% defibrinated rabbit blood (HIA-RB)
Heart infusion broth (HIB)
Nutrient broth

31.32 Reagents

Cephalothin (30 µg/disc)
Glycine (1%)
Gram stain with carbol fuchsin counterstain
Hydrogen peroxide (3%)
Lead acetate paper strip
McFarland No. 1 turbidity standard
Nalidixic acid—30 µg/disc
Ninhydrin solution
Nitrate reduction—
  Solution A—Sulfanilic acid in acetic acid (5N)
  Solution B—Alpha-naphthylamine in acetic acid (5N)
Phenylenediamine dihydrochloride
Ryuflagella stain[54]
Saline (3.5%)
Sodium hippurate
Zinc

31.4 PROCEDURES

31.41 General Sampling Protocol

Appropriate sample handling and transport must be emphasized when attempting to isolate *C. jejuni* from foodstuffs. As indicated above, the organism is sensitive to drying, oxygen and storage at room temperature. Therefore, if the samples are improperly handled before testing, the value of any isolation procedure is compromised.

Optimally, the food sample should be stored in an oxygen-free environment (100% N_2) with 0.01% sodium bisulfite, and maintained at refrigeration temperatures. Under these conditions the organism will survive 10 times longer than when the same strain is held in a bisulfite-free medium exposed to air at 25°C.[56] Storing meat samples at 4°C with an equal amount of Cary-Blair diluent results in little change in *Campylobacter* spp. viability during 14 days of storage and may be used as a transport and storage procedure.[110]

To optimize recovery of *C. jejuni* from foods, cultures from both food samples and enrichment cultures should be streaked onto selective agar plates. CCDA[42] and Campy-Cefex[117] media are commonly used with *C. jejuni* isolation from foods. After the media have been poured into plates, they should not be excessively exposed to light[47] and should be dried to limit moisture on the agar before streaking.

31.411 Surface Rinse Technique

This procedure is useful for sampling poultry carcasses and moderately large pieces of foods. Place the undivided sample (about 1 to 2 kg) in a sterile plastic bag with 100 to 400 mL of buffered peptone water (BPW) or other appropriate liquid. Rinse the surface by shaking and massaging for approximately 1 min as desired. For enrichment culture, a 25-mL portion of the rinse may then be added to 25 mL of double strength enrichment media. We have used the Hunt enrichment technique[41] and the Bolton enrichment broth[9] (both with microaerobic incubation) with favorable results.

31.412 Enrichment Methods

Comparison of most probable number (MPN) enrichment techniques and direct Campy-Cefex plating[119] indicated no statistically significant difference in populations of *Campylobacter* isolated from freshly processed carcasses by either method.[64] For freshly processed chicken carcass samples we estimate that direct plating enumeration procedures cost approximately 5 times less than comparable MPN-enrichment procedures. The time required for direct plating for enumeration is greatly diminished relative to the MPN technique.

31.413 Swab Sample Technique

This procedure is useful for qualitatively sampling surfaces of large animal carcasses and equipment (see Chapter 2). Dip a sterile swab into enrichment broth and remove excess moisture. Take a representative sample by wiping the surface of the sample with the moistened swab. Return the swab to the enrichment broth or directly inoculate a plate of selective agar by rotating the swab over a 3 to 5-cm^2 area and streak for isolation with a sterile loop.

31.42 Direct Plating Before Enrichment

Transfer two or three loops from the above surface rinse suspension of the sample to each of two selective isolation agar plates (Section 31.312) and streak for isolated colonies. Swab samples should be directly inoculated to two selective agar plates. Twenty-five grams of foods may be added directly to enrichment broth.

Incubate the inoculated plates in a microaerobic atmosphere (5% O_2, 10% CO_2, 85% N_2) at 42°C for 18 to 48 hr. The plates can be inspected for characteristic colonies after 18 to 24 hr. but must be reincubated under the same conditions for an additional day if typical colonies are not observed. Typically on moist media, *C. jejuni* growth swarms, which is a useful diagnostic growth characteristic; however, this type of confluent growth makes it difficult to enumerate or to obtain isolated colonies. Individual colonies can be obtained by cultivating the culture on dry plates. Dry plates can be obtained by holding plates overnight at room temperature in the absence of light, or by adding 4 drops of glycerol to a filter paper placed in the vessel containing the plates.[110]

31.43 Selective Isolation Agar Media

Several selective agar media can be used for isolating *C. jejuni* from foods. Most prominent among these selective media are the modified CCDA-Preston blood-free medium,[42] and Campy-Cefex.[117] In our experience, the Preston blood-free and the Campy-Cefex media provide reasonable selectivity and allow for good quantitative recovery from poultry carcasses. A recently developed medium, Campy-Line agar (CLA), has been proposed for use in enumerating campylobacters from poultry carcass rinses: this plating medium simplifies enumeration by increasing color contrast between growing colonies and the agar.[63]

31.44 Selective Broth Enrichment and Plating

Enrichment culture is needed when small numbers of culturable *C. jejuni* are present in foods.[109] Several effective enrichment broths have been developed for this purpose. The Doyle and Roman[23] enrichment broth (90 or 100 mL) is inoculated with 10 or 25 g of food, respectively, and incubated with agitation under a microaerobic atmosphere at 42°C for 16 to 18 hr. One enrichment procedure for isolating *C. jejuni* from chicken carcasses was originally described by Park et al.[84] This enrichment has been modified, by the Food and Drug Administration[41] and now takes into account the high sensitivity of the organism to freezing. The procedure involves resuscitation of the organism for 3 to 4 hr at 31°C to 32°C in a microaerobic environment without sample agitation. Subsequently, the enrichment culture is supplemented with antibiotics, held at 35°C to 37°C for 1 to 2 hr and is further incubated at 42°C for an additional 20 hr, with agitation.

The Doyle and Roman method[23] has been reported as more effective than the original procedure described by Park et al.[84] in recovering *C. jejuni* from chicken carcasses. When comparing the Doyle, Park and Hunt enrichment procedures for recovery of campylobacters from chicken carcasses, the Hunt method was most productive.[116] In a recent comparison, the Hunt enrichment procedures were compared to methods using Bolton enrichment broth (BEB). When incubated under microaerobic conditions, the simpler BEB procedure provided similar recovery to the Hunt method.[64] However, these enrichment methods are qualitative and do not provide quantitative information. A thorough comparison between numerous enrichment procedures still is required before one method can be considered superior to another.

31.45 Microaerobic Requirements

C. jejuni is a strict microaerophile[106] and this requirement must be taken into account in using enrichment and plating procedures. Approaches to obtain microaerobic conditions include atmosphere exchange,[38] atmosphere generating systems[15] and even the use of an "Alka-Seltzer"™ gas-generating system.[88] The use of candle jars is not as effective for growing *Campylobacter* as the other approaches described above.

31.451 Enrichment Cultures

The most effective means for providing a microaerobic atmosphere in flasks with enrichment broth is by introducing a mixture of 5% O_2, 10% CO_2, 85% N_2 from a gas cylinder. The normal atmosphere can be evacuated by a standard laboratory vacuum system (about 15 mm Hg). The gas mixture may be introduced through a side arm of a flask or through a rubber stopper fitted with a short glass tube in the top of a flask. A tube and clamp are attached to the short glass tube or side arm to transfer and retain the gas in the flask. The stopper can be held in place with a screw cap that allows the glass tube to pass through. This additional cap will prevent the rubber stopper from being dislodged as the gaseous pressure increases during incubation at 42°C. Introduce the gas slowly from the cylinder into the flask, tighten the clamp, and gently swirl the flask to incorporate the microaerobic gas mixture into the enrichment broth. Release the clamp for a short interval to equalize the gaseous pressure, and repeat the entire procedure twice.

An alternative approach is to place the enrichment medium and sample in cotton-plugged flasks into an anaerobic jar. The air content of the jar is evacuated and refilled with the microaerobic gas mixture. If agitation is required for enrichment, the entire anaerobic jar can be shaken. Tissue culture flasks with vented lids also work well as sample containers and may be used in anaerobe jars or in gas charged incubators which provide the appropriate microaerobic conditions.[83] Hunt[41] used a continual flow of the microaerobic gas mixture through the enrichment broth vessel. This method requires hook-up of tubes to the vessel and uses substantial amounts of gas. The method is technically difficult to master.

31.452 Selective Agar Plates

A container to hold agar plates and a means to introduce and maintain the microaerobic atmosphere are also required. An anaerobe jar or a self-sealing bag with a portal for introducing the gas mixture can be used. As with the enrichment flasks, the microaerobic gas mixture should be introduced and evacuated from the container several times to ensure an optimal environment for growth of *C. jejuni*. Exchange of the gas mixture can be accomplished by creating a vacuum in the vessel or, when using plastic bags, by expelling the atmosphere through a small opening in the zipper of the bag. The container with the petri plates is then placed into a 42°C incubator.

31.453 Media Supplement

Enhancing the aerotolerance of *Campylobacter* spp. is also advisable to optimize recovery of the organism from foodstuffs. Supplementing the growth media with 0.025% each of ferrous sulfate, sodium metabisulfite and sodium pyruvate (FBP) increases the oxygen tolerance of *C. jejuni*.[29,36,37,115]

31.46 Isolation

Select *Campylobacter*-like colonies from each selective agar medium. Colonies on blood-supplemented or charcoal-based media are smooth, convex, and glistening with a distinct edge or flat, translucent, shiny, and spreading with an irregular edge; colorless to grayish or light cream; usually 1 to 2 mm in diameter but can range from pinpoint to 4 to 5 mm in diameter; and are nonhemolytic on blood agar. Growth may be confluent without distinct colonies.

Make a wet-mount preparation from a colony and examine for morphology and motility by dark-field or phase-contrast microscope. *C. jejuni* cells are curved, S-shaped, gull-winged or spiral rods, 0.2 to 1 µm wide, and 0.5 to 5 µm long with darting or corkscrew-like motility. Cells in older cultures may be coccoid and non-motile.

Prepare thin smears of young cultures and Gram stain by standard procedures, but use Ziehl-Neelsen's carbol-fuchsin stain as the counterstain (Chapter 63). Carbol-fuchsin stains *Campylobacter* better than safranin. Flagella can be demonstrated by staining with the method of Kodaka et al.[55] Cells have a single flagellum at one or both ends of the cell.

31.5 IDENTIFICATION TESTS

Select a single colony from the primary isolation medium, and transfer it to a plate of HIA-RB. Incubate plate in a microaerobic atmosphere at 42°C until growth is adequate, usually 24 hr. Transfer the fresh culture to 5 mL of HIB, and adjust the density of cells to match a McFarland No. 1 turbidity standard (Chapter 63). Use the cell suspension in HIB to inoculate tubes and plates of media for biochemical and growth tests.

Inoculate tubes of glucose fermentation base, nitrate, glycine, NaCl, cysteine, and HIA slant with 2 drops (approximately 0.1 mL) of cell suspension using a Pasteur pipette. Place a lead acetate paper strip over the cysteine medium for detection of H_2S. Incubate tubes in a microaerobic atmosphere at 35°C to 37°C for 3 days with the caps loose.

Inoculate two HIA-RB plates to test for growth at 25°C and 42°C by saturating a fiber-tipped swab with the broth suspension and making a single streak across each plate. Use a fresh surface of the swab for each plate by rotating the swab one-half of a turn. Four isolates can be inoculated to each plate. Incubate plates in a microaerobic atmosphere at each temperature for 3 days.

Inoculate the entire surface of an HIA-RB plate with a swab soaked with the bacterial suspension. Place discs of nalidixic acid (30 μg) and cephalothin (30 μg) on the plate and incubate in a microaerobic atmosphere at 35°C to 37°C until zones of inhibition of growth around the disks can be measured, usually within 24 hr.

Inoculate the remaining tubes of media, TSI, O-F base with and without glucose, and TMAO, with an inoculating needle with a small amount of growth from the HIA-RB plate. Streak heavily the slant of TSI medium and stab the butt one time. Stab the O-F and TMAO media three times in the upper one-third of the medium. Include in the TMAO test a tube of medium inoculated with a strain of *C. lari* (positive control) and an uninoculated tube of medium. Incubate tubes of TSI and O-F media in a microaerobic atmosphere at 35°C to 37°C for 3 days with caps loose. Incubate TMAO tubes in an anaerobic atmosphere for 7 days with caps loose.

31.51 Campylobacter Test Methodology

31.511 Catalase Test

Add several drops of 3% hydrogen peroxide to 24 to 48 hr growth on the HIA slant. Production of any bubbles is considered positive. Confirm negative results by suspending growth in a drop of 3% hydrogen peroxide on a slide, and examine for bubbles with hand lens or dissecting microscope. (Media containing blood can give false-positive reactions.)

31.512 Oxidase Test

Place a piece of filter paper in an empty petri dish and spread a loopful of growth in a line 3 to 5 mm long on the pad. Add 1 to 2 drops of oxidase reagent to the growth. A dark purple color that develops within 10 sec is considered positive.

31.513 Nitrate Reduction

To a three-day-old culture growing in nitrate broth, add an equal volume (5 drops from Pasteur pipette) of solutions A (acetic acid and sulfanilic acid) and B (acetic acid and alpha-naphthylamine). Development of a red color within 1 to 2 min is a positive result indicating production of nitrite from nitrate. If no color develops, add a small amount of powdered zinc to the tube. Development of a red color in 5 to 10 min with zinc indicates nitrates are present (negative test). No color with zinc indicates nitrates have been reduced to other compounds or to nitrogen as indicated by gas in the insert tube (positive test).

31.514 Glucose Utilization

Fermentation is indicated by a red color (acid production) in fermentation broth with Andrade's indicator. Oxidation is indicated by a yellow color (acid production) in the O-F medium with phenol red indicator. O-F medium will be yellow (acid) when removed from the microaerobic atmosphere due to absorption of CO_2. To read O-F reactions, let tubes stand in room atmosphere until O-F control becomes neutral or alkaline, usually in 1 to 2 hr.

31.515 H_2S Production

Lead acetate strip and TSI agar. Any blackening of the medium indicates H_2S production. Brownish-black coloration of the strip is a positive test. Record the degree of blackening from trace to 4+. Lead acetate strips are more sensitive than sulfide indicators in media. Some strains may be positive by the strip procedure and negative in TSI medium.

31.516 Glycine and NaCl Tolerance

Any growth, usually in the top 10 mm of the medium, is considered positive.

31.517 Temperature Tolerance

Substantive growth in the line of inoculum across the HIA-RB plate is considered positive. A trace of growth that is difficult to distinguish from initial inoculum is not considered positive. Incubate replicate plates at 42°C and 25°C under microaerobic atmosphere.

31.518 Antimicrobial Discs

Any zone of inhibition around disks indicates the organism is sensitive to the antimicrobial. Measure and record diameter of zone.

31.519 TMAO

Growth away from the stab or dispersed in the medium is considered positive. Compare with controls.

31.520 Hippurate Hydrolysis

Prepare 1% sodium hippurate in sterile distilled water, dispense in 0.4-mL amounts in 13 × 100-mm screw cap tube, and freeze at –20°C until used. Emulsify in a thawed tube of sodium hippurate a large loopful of 18 to 24 hr growth from a HIA-RB plate. Incubate suspension in a 37°C waterbath for 2 hr.

After incubation, overlay 0.2 mL of 3.5% ninhydrin solution in a 1:1 mixture of acetone and butanol. (Store ninhydrin solution in the dark at room temperature.) Re-incubate tube in a water bath at 37°C for 10 min. Development of a crystal violet color is a positive result. A colorless or light to medium purple is considered negative.

31.52 Test Interpretations

Characteristic test reactions are listed in Table 1.

31.53 Identification Criteria

Table 1 summarizes characteristics useful in identifying the catalase-positive campylobacters associated with foods. From a practical standpoint in the food microbiology laboratory, separating *C. jejuni* and *C. coli* may not be necessary. *Campylobacters* grow well at 42°C, do not grow at 25°C, and *C. lari* strains are resistant to nalidixic acid. Isolates of *C. lari* that are sensitive to nalidixic acid

Table 1. Characteristics of *Campylobacter* Species

	Growth							Biochemical Reaction						
Species	25°C	42°C	1% Glycine	3.5% NaCl	0.1% TMAO	Nalidixic Acid	Cephalothin	Oxidase	Catalase	Glucose Utilization	NO_3 Reduction	H_2S, TSI	H_2S, paper strip	Hippurate Hydrolysis
C. jejuni	–	+	+	–	–	S	R	+	+	–	+	–	+	+[a]
C. coli	–	+	+	–	–	S	R	+	+	–	+	D	+	–
C. lari	–	+	+	–	+	R	R	+	+	–	+	–	+	–
C. fetus subsp. fetus fetus	+	D	+	–	D	R	S	+	+	–	+	–	+	–
C. fetus subsp. venerealis	+	–	–	–	–	R	S	+	+	–	+	–	+	–

[a] Hippurate-negative *C. jejuni* strains have been reported.

+ = 90% or more of strains are positive; – = 90% or more of strains are negative; D = 11% to 89% of strains are positive; S = susceptible; R = resistant

have been reported. Anaerobic growth in 0.1% TMAO may be used to separate these strains from *C. jejuni* and *C. coli.* Hippurate hydrolysis is the most reliable test for separating *C. jejuni* and *C. coli. C. jejuni* strains are positive in this test. Strains with unclear or negative results by the ninhydrin method for hippurate hydrolysis can be tested by the more sensitive gas liquid chromatographic procedure.[54,75] Other methods to separate *C. jejuni* and *C. coli* have been described,[105] but hippurate hydrolysis appears to correlate best with genetic classification of the species. Hippurate-negative isolates of *C. jejuni* have been reported and probably occur infrequently.[124] *C. fetus* subsp. *fetus* grows at 25°C. This feature separates *C. fetus* subsp. *fetus* from the thermophilic campylobacters. Details of characteristics and methods to identify other *Campylobacter spp.* can be found elsewhere.[2,3,74,75]

Our laboratory uses none of the biochemical testing described above. Our approach continues to put emphasis on employing appropriate selective plating, growth in microaerobic atmosphere at elevated temperature (42°C) and careful analyses of the resulting colony formation. As indicated above, characteristic colonial morphology is requisite, followed by observing typical microscopic appearances (spiral shape) and motility. Confirmation is achieved by subjecting typical colonies to appropriate immunological assays, such as latex agglutination assay (Integrated Diagnostics, Inc., Baltimore, MD; Mercia Diagnostics, Shalford, UK; Becton Dickinson Microbiology Systems, Cincinnati, OH). These latex agglutination assays are far less cumbersome and are more reliable in characterizing isolates of *Campylobacter*. Alternatively, enzyme linked fluorescent assay systems, such as the VIDAS (BioMerieux Vitek, Hazelwood, MO), provide an automated immunological approach for detecting *Campylobacter*.

31.54 Subtyping

Subtyping was originally developed to identify common sources for disease outbreaks or, to track individual clones of a bacterium throughout a defined ecological niche. Methods for subtyping can be divided into phenotypic and genotypic discriminators. The phenotypic subtyping of *Campylobacter* that has most frequently been employed are the classical heat stable or heat labile serotyping schemes. Strains of *C. jejuni, C. coli,* and *C. lari* can be serotyped for epidemiologic studies. Two systems are in common use, one for soluble heat-stable antigens identified by an indirect hemagglutination procedure for unabsorbed antisera[87] and one for heat-labile antigens using a slide agglutination technique and absorbed antisera.[65] The heat-stable serotyping scheme employs antisera to 42 *C. jejuni* and 18 *C. coli* strains. The heat-labile serotyping system has 108 serotypes, 63 *C. jejuni,* 37 *C. coli,* and 8 *C. lari.* Commercial antisera are not readily available for either serotyping system. Serotyping can be obtained on request from the *Campylobacter* Reference Laboratory, Centers for Disease Control, for strains from outbreaks and for approved studies.

The genotypic analyses would include restriction fragment length polymorphism,[77] *flaA* gene sequencing,[72] ribotyping and pulsed field gel electrophoretic analysis among other techniques.[86] It is likely that after various comparisons, one or a combination of two techniques in the genotypic analyses will gain wide acceptance. The criteria for acceptance will inevitably be based upon the ability of the test to discriminate non-clonal isolates, consistency between laboratories, simplicity, cost, throughput, and ability to group epidemiologically relevant isolates. These genotyping procedures are still being evaluated.

31.55 Stock Culture Maintenance

The fragility of *C. jejuni* requires that special attention be given in stock culture maintenance. To prepare a stock culture, inoculate a heavy loopful of the strain into semisolid brucella broth in a screw cap test tube. Loosen the caps to allow exchange of the atmosphere. Alternatively, grow the isolate on brucella agar medium overnight. Inoculate into a semisolid brucella medium. Grow the culture to mid-log phase, i.e., for approximately 24 hr at 42°C. Cultures may be stored up to 1 month at 4°C in a microaerobic atmosphere or in a vacuum. Wang et al.[125] described a semisolid (0.15% agar) brucella medium with 10% sheep blood for storing the organism. Studies revealed the organism survived in this medium at 25°C for at least 3 weeks. Nair et al.[78] described an egg-based medium that maintained viability of *C. jejuni* for over 3 months when held at 4°C. This medium was superior to Wang's preservation medium when cultures were held at 27°C. Cary-Blair medium with 0.16% agar has also been shown to be a useful medium for storing *C. jejuni*.[66]

For long-term storage, grow *C. jejuni* on brucella agar with FBP supplement in a microaerobic atmosphere for 24 hr at 42°C, suspend in a diluent containing 15% glycerol. Small quantities (1 to 2 mL) of this thick bacterial suspension can be stored in tightly sealed vials for several years when held at –70°C. Similar preparations may also be conveniently stored using commercially prepared beads with a cryoprotectant. Cultures can also be lyophilized in skim milk and stored indefinitely at –20°C.

31.6 INTERPRETATION

C. jejuni is a fragile organism that is readily inactivated by methods used to eliminate other enteropathogens from foods. In spite of this fragility, the organism causes millions of cases of gastroenteritis in the United States each year. This discrepancy can usually be explained by the cross-contamination of either utensils or the ready-to-eat product with contaminated raw foods, especially those of animal origin. Experience has shown that bacterial disease is often transmitted when a contaminated raw food enters the food preparation area.

The best available means to protect health at this time is through consumer education and reinforcement of hygienic food handling practices. Care is required to segregate raw foods and processing utensils from any food or material that comes directly or indirectly into contact with the consumer. This approach will diminish and prevent many cases of campylobacteriosis. The paramount goal, however, should be to eliminate *C. jejuni* from animals used for production of food. Preventing the organism from colonizing animals used for food would prevent contamination of meat products, and would greatly reduce potential for foodborne transmission.

31.7 REFERENCES

1. Barrell, R. A. E. 1984. The survival of *Campylobacter jejuni* in red meats stored at different temperatures. Int. J. Food Microbiol. 1:187-196.
2. Barrett, T. J., C. M. Patton, and G. K. Morris. 1988. Differentiation of *Campylobacter* species using phenotypic characterization. Lab. Med. 19:96.
3. Benjamin, J., S. Leaper, R. J. Owen, and M. B. Skirrow. 1983. Description of *Campylobacter lanais,* a new species comprising the nalidixic acid resistant thermophilic *Campylobacter* (NARTC) group. Curr. Microbiol. 8:231.
4. Black, R. E., M. M. Levine, M. L. Clements, T. P. Hughes, and M. J. Blaser. 1988. Experimental *Campylobacter jejuni* infection in humans. J. Infect. Dis. 157:472-479.
5. Blankenship, L. C. and S. E. Craven. 1982. *Campylobacter jejuni* survival in chicken meat as a function of temperature. Appl. Environ. Microbiol. 44:88-92.
6. Blaser, M. J., I. D. Berkowitz, F. M. LaForce, J. Cravens, L. B. Reller, and W. L. L. Wang. 1979a. Campylobacter enteritis: clinical and epidemiologic features. Ann. Intern. Med. 91:179-185.
7. Blaser, M. J., J. Cravens, B. W. Powers, F. M. LaForce, and W. L. L. Wang. 1979b. *Campylobacter* enteritis associated with unpasteurized milk. Am. J. Med. 67:715-718.
8. Blaser, M. J., D. J. Duncan, G. H. Warren, and W. L. L. Wang. 1983. Experimental *Campylobacter jejuni* infection of adult mice. Infect. Immun. 39:908-916.
9. Bolton, F. J. VIDAS *Campylobacter* product insert, bioMerieux Vitek, Inc., 595 Anglum Dr., Hazelwood, MO 63042 USA.
10. Bolton, F. J., and D. Coates. 1983. A study of the oxygen and carbon dioxide requirements of thermophilic campylobacters. J. Clin. Pathol. 36:829-834.
11. Bolton, F. J., D. N. Hutchinson, and D. Coates. 1984. Blood-free selective medium for isolation of *Campylobacter jejuni* from feces. J. Clin. Microbiol. 19:169-171.
12. Bolton, F. J., and L. Robertson. 1982. A selective medium for isolating *Campylobacter jejuni/coli.* J. Clin. Pathol. 35:462-467.
13. Brieseman, M. A. 1984. Raw milk consumption as a probable cause of two outbreaks of *Campylobacter* infection. N. Z. Med. J. 97:411-413.
14. Brouwer, R., M. J. Mertens, T. H. Siem, and J. Katchaki. 1979. An explosive outbreak of Campylobacter enteritis in soldiers. Antonie van Leeuwenhoek. J. Microbiol. Serol. 45:517-519.
15. Buck, G. E., C. Fojtasek, K. Calvert, and M. T. Kelly. 1982. Evaluation of the CampyPAK II gas generator system for isolation of *Campylobacter fetus* subsp. *jejuni.* J. Clin. Microbiol. 15:41-42.
16. Caldwell, M. B., R. Walker, S. D. Stewart, and J. E. Rogers. 1983. Simple adult rabbit model for *Campylobacter jejuni* enteritis. Infect. Immun. 42:1176-1182.
17. Caldwell, M. B., P. Guerry, E. C. Lee, P. Burans, and R. Walker, R. 1985. Reversible expression of flagella in *Campylobacter jejuni.* Infect. Immun. 50:941-943.
18. Chowdhury, M. N., and E. S. Mahgoub. 1981. Gastroenteritis due to *Campylobacter jejuni* in Riyadh, Saudi Arabia. Trans. Soc. Trop. Med. Hyg. 75:359-361.
19. Christopher, F. M., G. C. Smith, and C. Vanderzant. 1982. Examination of poultry giblets, raw milk, and meat for *Campylobacter fetus* subsp. *jejuni.* J. Food Prot. 45:260.
20. Deming, M. S., R. V. Tauxe, P. A. Blake, S. E. Dixon, B. S. Fowler, T. S. Jones, E. A. Lockamy, C. M. Patton, and R. O. Sikes. 1987. *Campylobacter* enteritis at a university: Transmission from eating chicken and from cats. Am. J. Epidemiol. 126:526.
21. DeMol, P., and E. Bosmans. 1978. *Campylobacter* enteritis in central Africa. Lancet i: 604.
22. Doyle, M. P. and D. J. Roman. 1981. Growth and survival of *Campylobacter fetus* subsp. *jejuni* as a function of temperature and pH. J. Food Prot. 44:596.
23. Doyle, M. P., and D. J. Roman. 1982. Recovery of *Campylobacter jejuni* and *Campylobacter coli* from inoculated foods by selective enrichment. Appl. Environ. Microbiol. 43:1343-1353.
24. Doyle, M. P., and D. J. Roman. 1982. Response of *Campylobacter jejuni* to sodium chloride. Appl. Environ. Microbiol. 43:561-565.
25. Doyle, M. P., and D. J. Roman. 1982. Sensitivity of *Campylobacter jejuni* to drying. J Food Prot. 45:507.
26. Ehlers, J. G., M. Chapparo-Serrano, R. L. Richter, and C. Vanderzant. 1982. Survival of *Campylobacter fetus* subsp. *jejuni* in Cheddar and cottage cheese. J. Food Prot. 45:1018-1021.
27. Finch, M. J., and P. A. Blake. 1985. Foodborne outbreaks of campylobacteriosis: The United States experience, 1980-1982. Am. J. Epidemiol. 122:262-268.
28. Fitzgeorge, R. B., A. Baskerville, and K. P. Lander. 1981. Experimental infection of Rhesus monkeys with a human strain of *Campylobacter jejuni.* J. Hyg. 86:343-351.

29. George, H. A., P. S. Hoffman, R. M. Smibert, and N. R. Krieg. 1978. Improved media for growth and aerotolerance of *Campylobacter fetus*. J. Clin. Microbiol. 8:36-41.
30. Gill, C. O., and L. M. Harris. 1982. Survival and growth of *Campylobacter fetus* subsp. *jejuni* on meat and in cooked foods. Appl. Environ. Microbiol. 44:259-263.
31. Gill, C. O., and L. M. Harris. 1984. Hamburgers and broiler chickens as potential sources of human *Campylobacter* enteritis. J. Food Prot. 47:96-99.
32. Gill, K. P. W., P. G. Bates, and K. P. Lander. 1981. The effect of pasteurization on the survival of *Campylobacter* species in milk. Br. Vet. J. 137:578-584.
33. Hanninen, M. L. 1981. The effect of NaCl on *Campylobacter jejuni/coli.* Acta Vet Scand. 22:578.
34. Hanninen, M. L., H. Korkeala, and P. Pakkala. 1984. Effect of various gas atmospheres on the growth and survival of *Campylobacter jejuni* on beef. J. Appl. Bacteriol. 57:89-94.
35. Hams, N. V., D. Thompson, D. C. Martin, and C. M. Nolan. 1986. A survey of *Campylobacter* and other bacterial contaminants of premarket chicken and retail poultry and meats, King County, Washington. Am. J. Publ. Health 76:401.
36. Hoffman, P. S., N. R. Krieg, and R. M. Smibert. 1979a. Studies of the microaerophilic nature of *Campylobacter fetus* subsp. *jejuni.* 1. Physiological aspects of enhanced aerotolerance. Can. J. Microbiol. 25:1-7.
37. Hoffman, P. S., H. A. George, N. R. Krieg, and R. M. Smibert. 1979b. Studies of the microaerophilic nature of *Campylobacter fetus* subsp. *jejuni.* II. Role of exogenous superoxide anions and hydrogen peroxide. Can. J. Microbiol. 25:8-16.
38. Holdeman, L. V., E. P. Cato, and W. E. C. Moore (ed.) 1977. Anaerobe Laboratory Manual, 4th ed. Virginia Polytechnic Institute and State University, Blacksburg, Va.
39. Hudson, P. J., R. L. Vogt, J. Brondum, and C. M. Patton. 1984. Isolation of *Campylobacter jejuni* from milk during an outbreak of campylobacteriosis. J. Infect. Dis. 150:789.
40. Humphrey, C. D., D. M. Montag, and F. E. Pittman. 1985. Experimental infection of hamsters with *Campylobacter fetus jejuni.* J. Infect. Dis. 151:485-493.
41. Hunt, J. M. 1992. *Campylobacter*, pp. 77-94. *In* Food and Drug Administration Bacteriological Analytical Manual, 7th ed. Assoc. Off. Anal. Chem. Int., Arlington, Va.
42. Hutchinson, D. N., and F. J. Bolton. 1984. An improved blood-free selective medium for isolation of *Campylobacter jejuni* from fecal specimens. J. Clin. Pathol. 37: 956-957.
43. Hutchinson. D. N., F. J. Bolton, P. M. Hinchliffe, H. C. Dawkins, S. D. Horsley, E. G. Jesse, P. A. Robertshaw, and D. E. Counter. 1985. Evidence of udder excretion of *Campylobacter jejuni* as the cause of milk-borne *Campylobacter* outbreak. J. Hyg. 94:205-215.
44. Istre, G. R., M. J. Blaser, P. Shillam, and R. S. Hopkins. 1984. *Campylobacter* enteritis associated with undercooked barbecued chicken. Am. J. Publ. Health 74:1265-1267.
45. Johnson, W. M., and H. Lior, H. 1984. Toxins produced by *Campylobacter jejuni* and *Campylobacter coli.* Lancet i:229-230.
46. Johnson, W. M., and H. Lior. 1986. Cytotoxic and cytotonic factors produced by *Campylobacter jejuni, Campylobacter coli,* and *Campylobacter laridis*. J. Clin. Microbiol. 24:275-281.
47. Juven, B. J., and I. Rosenthal. 1985. Effect of free-radical and oxygen scavengers on photochemically generated oxygen toxicity and on the aerotolerance of *Campylobacter jejuni.* J. Appl. Bacteriol. 59:413-419.
48. Kapperud, G. K., and G. Bukholm. 1987. Expression of *Campylobacter jejuni* invasiveness in cell cultures coinfected with other bacteria. Infect. Immun. 55:2816.
49. Kazmi, R. R., A. Hafeez, and S. U. Kazmi. 1986. Isolation of *Campylobacter jejuni* from diarrheal stool specimens in Karachi. Abstracts Ann. Mtg. Am. Soc. Microbiol. C82, p. 341.
50. Kazmi, S. U., B. S. Roberson, and N. J. Stern. 1984. Animal-passed, virulence- enhanced *Campylobacter jejuni* causes enteritis in neonatal mice. Curr. Microbiol. 11:159.
51. Kiggins, E. M., and W. N. Plastridge. 1956. Effect of gaseous environment on growth and catalase content of *Vibrio fetus* cultures of bovine origin. J. Bacteriol. 72:397.
52. Klipstein, F. A., and R. F. Engert. 1984. Properties of crude *Campylobacter jejuni* heat-labile enterotoxin. Infect. Immun. 45:314-319.
53. Klipstein, F. A., R. F. Engert, H. B. Short, and E. A. Schenk. 1985. Pathogenic properties of *Campylobacter jejuni:* assay and correlation with clinical manifestations. Infect. Immun. 50:43-49.
54. Kodaka, H., G. L. Lombard, and V. R. Dowell, Jr. 1982. Gas-liquid chromatography technique for detection of hippurate hydrolysis and conversion of fumarate to succinate by micro-organisms. J. Clin. Microbiol. 16:962.
55. Koidis, P., and M. P. Doyle. 1983a. Survival of *Campylobacter jejuni* in fresh and heated red meat. J. Food Prot. 46:771-774.
56. Koidis, P., and M. P. Doyle. 1983b. Survival of *Campylobacter jejuni* in the presence of bisulfite and different atmospheres. Eur. J. Clin. Microbiol. 2:384-388.
57. Koidis, P., and M. P. Doyle. 1984. Procedure for increased recovery of *Campylobacter jejuni* from inoculated unpasteurized milk. Appl. Environ. Microbiol. 47:455-460.
58. Korlath, J. A., M. T. Osterholm, L. A. Judy, J. C. Forfang, and R. A. Robinson. 1985. A point-source outbreak of campylobacteriosis associated with consumption of raw milk. J. Infect. Dis. 152:592-596.
59. Lander, K. P., and K. P. W. Gill. 1980. Experimental infection of the bovine udder with *Campylobacter coli/jejuni.* J. Hyg. 84:421-428.
60. Lauwers, S., M. DeBoeck, and J. P. Butzler. 1978. *Campylobacter* enteritis in Brussels. Lancet i:604-605.
61. Lee, A. J. L. O'Rourke, P. J. Barrington, and T. J. Trust. 1986. Mucus colonization as a determinant of pathogenicity in intestinal infection by *Campylobacter jejuni*: a mouse cecal model. Infect. Immun. 51: 536 - 546.
62. Lee, A., S. M. Logan, and T. J. Trust. 1987. Demonstration of a flagella antigen shared by a diverse group of spiral-shaped bacteria that colonize intestinal mucus. Infect. Immun. 55:828-831.
63. Line, J. E. 1999. Development of a selective differential agar for isolation and enumeration of *Campylobacter* spp. 10th International Workshop on *Campylobacters, Helicobacters* and Related Organisms. Baltimore, Md. Abstract.
64. Line, J. E., C. Lattuada, and N. Stern. 1998. *Campylobacter* recovery and enumeration from broiler carcasses. 85th Ann. Meet. Int. Assoc. Milk Food Environ. Sanit. Nashville, Tenn. Abstract p. 68.
65. Lior, H., D. L. Woodward, J. A. Edgar, L. J. Laroche, and P. Gill. 1982. Serotyping of *Campylobacter jejuni* by slide agglutination based on heat-labile antigenic factors. J. Clin. Microbiol. 15:761-768.
66. Luechtefeld, N. W., W. L. L. Wang, M. J. Blaser, and L. B. Reller. 1981. Evaluation of transport and storage techniques for isolation of *Campylobacter fetus* subsp. *jejuni* from turkey cecal specimens. J. Clin. Microbiol. 13:438-443.
67. Marshall, B. 1983. Unidentified curved bacilli on gastric epithelium in active chronic gastritis. Lancet i:1273.
68. McCardell, B. A., J. M. Madden, and E. C. Lee. 1984a. *Campylobacter jejuni* and *Campylobacter* coli production of a cytotonic toxin immunologically similar to cholera toxin. J. Food Prot. 47:943-949.
69. McCardell, B. A., J. M. Madden, and E. C. Lee. 1984b. Production of cholera- like toxin by *Campylobacter jejuni/coli.* Lancet i:448-449.
70. Mead, P. S., L. Slutsker, V. Dietz, L. F. McCaig, J. S. Bresee, C. Shapiro, P. M. Griffin, and R. V.Tauxe. 1999. Food-related illness and death in the United States. Emerg. Infect. Dis. 5:1-36 wysiwig://5/ http://www.cdc.gov/ncidod/eid/vol5no5/mead.htm.
71. Megraud, F. 1987. Isolation of *Campylobacter* spp. from pigeon feces by a combined enrichment-filtration technique. Appl. Environ. Microbiol. 53:1394-1395.

72. Meinersmann, R. J., L. O. Helsel, P. I. Fields, and K. L. Hiett. 1997. Discrimination of *Campylobacter jejuni* strains by *fla* gene sequencing. J. Clin. Microbiol. 35:2810-2814.
73. Mishu-Allos, B. 1997. Association between *Campylobacter* infection and Guillain-Barré syndrome. J. Infect. Dis. 176:125-128.
74. Morris, G. K., and C. M. Patton. 1985. *Campylobacter. In* E. H. Lennette, A. Balows, and W. J. Hausler Jr. (ed.), Manual of clinical microbiology, 4th ed. Am. Soc. Microbiol., Washington, D. C.
75. Morris, G. K., M. R. El Sherbeeny, C. M. Patton, H. Kodaka, G. L. Lombard, P. Edmonds, D. G. Hollis, and D. J. Brenner. 1985. Comparison of four hippurate hydrolysis methods for identification of thermophilic *Campylobacter* sp. J. Clin. Microbiol. 22:714-718.
76. Mouton, R. P., J. J. Veltkamp, S. Lauwers, and J. P. Butzler. 1982. Analysis of a small outbreak of Campylobacter infections with high morbidity, p. 129. *In* D. G. Newell (ed.), Campylobacter-epidemiology, pathogenesis and biochemistry. MTP Press Ltd., Lancaster, England.
77. Nachamkin, I., H. Ung, and C. M. Patton. 1996. Analysis of HL and O serotypes of *Campylobacter* strains by the flagellin gene typing system. J. Clin. Microbiol. 34:277-281.
78. Nair, G. B., S. Chowderhury, P. Das, S. Pal, and S. C. Pal. 1984. Improved preservation medium for *Campylobacter jejuni*. J. Clin. Microbiol. 19:298-299.
79. Oosterom, J., G. J. A. DeWilde, E. DeBoer, L. H. DeBlaauw, and H. Karman. 1983. Survival of *Campylobacter jejuni* during poultry processing and pig slaughtering. J. Food Prot. 46:702-706, 709.
80. Pai, C. H., S. Sorger, L. Lackman, R. E. Sinai, and M. I. Marks. 1979. *Campylobacter* gastroenteritis in children. J. Pediatr. 94:589-591.
81. Palumbo, S. A. 1984. Heat injury and repair in *Campylobacter jejuni*. Appl. Environ. Microbiol. 48:477-480.
82. Pang, T., P. Y. Wong, D. Puthucheary, K. Sihotang, and W. K. Chang. 1987. In-vitro and in-vivo studies of a cytotoxin from *Campylobacter jejuni*. J. Med. Microbiol. 23:193-198.
83. Park, C. E., and G. W. Sanders. 1989. A sensitive enrichment procedure for the isolation of *Campylobacter jejuni* from frozen foods. Abstr. Fifth International Workshop on Campylobacter Infections. Puerto Vallarta, Mexico.
84. Park, C. E., Z. K. Stankiewicz, J. Lovett, J. Hunt, and D. W. Francis. 1983. Effect of temperature, duration of incubation, and pH of enrichment culture on the recovery of *Campylobacter jejuni* from eviscerated market chickens. Can. J. Microbiol. 29:803.
85. Patton, C. M., N. Shaffer, P. Edmonds, T. J. Barrett, M. A. Lambert, C. Baker, D. M. Perlman, and D. J. Brenner, D. 1989. Human disease associated with "*Campylobacter upsaliensis*" (Catalase-negative or weakly positive *Campylobacter* spp. in the United States). J. Clin. Microbiol. 27:66-73.
86. Patton, C. M., I. K. Wachsmuth, G. M. Evins, J. A. Kiehlbauch, B. D. Plikaytis, N. Troup, L. Tompkins, and H. Lior. 1991. Evaluation of 10 methods to distinguish epidemic-associated *Campylobacter* isolates. J. Clin. Microbiol. 29:680-688.
87. Penner, J. L., and J. N. Hennessey. 1980. Passive hemagglutination technique for serotyping *Campylobacter fetus* subsp. *jejuni* on the basis of soluble heat- stable antigens. J. Clin. Microbiol. 12:732-737.
88. Pennie, R. A., J. N. Zunino, C. E. Rose Jr. and R. L. Guerrant. 1984. Economical, simple method for production of the gaseous environment required for cultivation of *Campylobacter jejuni*. J. Clin. Microbiol. 20:320-322.
89. Piscoya-Hermosa, Z. A. 1985. Aislamiento y caracterizacion fenotipica de *Campylobacter jejuni* en Lima, Peru-Determinacion de biotipos. Master's thesis, Faculty of Science and Philosophy, University of Peru-Cayetano Heredia, Tingo Mana, Peru.
90. Porter, I. A., and T. M. S. Reid. 1980. A milk-borne outbreak of *Campylobacter* infection. J. Hyg. 84:415-419.
91. Potter, M. E., M. J. Blaser, R. K. Sikes, A. F. Kaufmann, and J. G. Wells. 1983. Human *Campylobacter* infection associated with certified raw milk. Am. J. Epidemiol. 117:475-483.
92. Rajan, D. P., and V. I. Mathan. 1982. Prevalence of *Campylobacter fetus* subsp. *jejuni* in healthy populations in southern India. J. Clin. Microbiol. 15:749-751.
93. Ray, B., and C. Johnson. 1984. Survival and growth of freeze-stressed *Campylobacter jejuni* cells in selective media. J. Food Safety 6:183.
94. Ringertz, S., R. C. Rockhill, O. Ringertz, and A. Sutomo. 1980. *Campylobacter fetus* subsp. *jejuni* as a cause of gastroenteritis in Jarkata, Indonesia. J. Clin. Microbiol. 12:538-540.
95. Robinson, D. A. 1981. Infective dose of *Campylobacter jejuni* in milk. Br. Med. J. 282:1584.
96. Robinson, D. A., W. J. Edgar, G. L. Gibson, A. A. Matchett, and L. Robertson, L. 1979. Campylobacter enteritis associated with consumption of unpasteurised milk. Br. Med. J. 1:1171-1173.
97. Rogol, M., B. Shpak, D. Rothman, and I. Sechter. 1985. Enrichment medium for isolation of *Campylobacter jejuni-Campylobacter coli*. Appl. Environ. Microbiol. 50:125-126.
98. Rollins, D. M., and R. Colwell. 1986. *Campylobacter* dormancy under environmental conditions. Appl. Environ. Microbiol. 52:531.
99. Romaniuk, P. J., B. Zoltowska, T. J. Trust, D. J. Lane, G. J. Olsen, N. R. Pace, and D. A. Stahl. 1987. *Campylobacter pylori*. the spiral bacterium associated with human gastritis, is not a true *Campylobacter* sp. J. Bacteriol. 169:2137-2141.
100. Ruiz-Palacios, G. M., E. Escamilla, and N. Torres. 1981. Experimental *Campylobacter* diarrhea in chickens. Infect. Immun. 34:250-255.
101. Ruiz-Palacios, G. M., J. Torres, N. I. Torres, E. Escamilla, B. R. Ruiz-Palacios, and J. Tamayo. 1983. Cholera-like enterotoxin produced by *Campylobacter jejuni*: characterisation and clinical significance. Lancet ii:250-253.
102. Sandstedt, K., and J. Ursing. 1986. *Campylobacter upsaliensis*, a new species, formerly the CNW group. Abstr. XIV International Congress of Microbiology. P.B8-17, p. 61.
103. Sanyal, S. C., K. M. N. Islam, P. K. B. Neogy, M. Islam, P. Speelman, and M. I. Huq. 1984. *Campylobacter jejuni* diarrhea model in infant chickens. Infect. Immun. 43:931-936.
104. Skerman, V. B. D., V. McGowan, and P. H. A. Sneath. 1980. Approved lists of bacterial names. Int. J. Syst. Bacteriol. 30:225.
105. Skirrow, M. B., and J. Benjamin. 1980. '1001' Campylobacters: Cultural characteristics of intestinal campylobacters from man and animals. J. Hyg. Camb. 85:427-442.
106. Smibert, R. M. 1984. *Campylobacter*, p. 11. *In* N. R. Krieg and J. G. Holt (ed.), Bergey's manual of systematic bacteriology. Williams and Wilkins, Baltimore, Md.
107. Stanfield, J. T., B. A. McCardell, and J. M. Madden. 1987. *Campylobacter* diarrhea in an adult mouse model. Microbiol. Pathogen. 3:155-165.
108. Steele, T. W., and S. McDermott. 1978. *Campylobacter* enteritis in south Australia. Med. J. Aust. 2:404-406.
109. Stern, N. J. 1981. *Campylobacter fetus* spp. *jejuni*: recovery methodology and isolation from lamb carcasses. J. Food Sci. 46:660-661, 663.
110. Stern, N. J. 1982. Selectivity and sensitivity of three media for the recovery of inoculated *Campylobacter fetus* spp. *jejuni* from ground beef. J. Food Safety 4:169-175.
111. Stern, N. J., and A. W. Kotula. 1982. Survival of *Campylobacter jejuni* inoculated into ground beef. Appl. Environ. Microbiol. 44:1150-1153.
112. Stern, N. J., S. S. Green, N. Thaker, D. J. Krout, and J. Chiu. 1984. Recovery of *Campylobacter jejuni* from fresh and frozen meat and poultry collected at slaughter. J. Food Prot. 47:372-374.
113. Stern, N. J., P. J. Rothenberg, and J. M. Stone. 1985. Enumeration and reduction of *Campylobacter jejuni* in poultry and red meats. J. Food Prot. 48:606-610.
114. Stern, N. J., M. D. Greenberg, and D. M. Kinsman. 1986. Survival of *Campylobacter jejuni* in selected gaseous environments. J. Food Sci. 51:652-654.

115. Stern, N. J., S. U. Kazmi, B. S. Roberson, K. Ono, and B. J. Juven. 1988. Response of *Campylobacter jejuni* to combinations of ferrous sulphate and cadmium chloride. J. Appl Bacteriol. 64:247-255.
116. Stern, N. J., and J. E. Line. 1992. Comparison of three methods for recovery of *Campylobacter* spp. from broiler carcasses. J. Food Prot. 55:663-666.
117. Stern, N. J., B. Wojton, and K. Kwiatek. 1992. A differential-selective medium and dry ice-generated atmosphere for recovery of *Campylobacter jejuni*. J. Food Prot. 55:514-517.
118. Stern, N. J., D. M. Jones, I. V. Wesley, and D. M. Rollins. 1994. Colonization of chicks by non-culturable *Campylobacter* spp. Lett. Appl. Microbiol. 18:333-336.
119. Stern, N. J. 1995. Influence of season and refrigerated storage on *Campylobacter* spp. Contamination of broiler carcasses. J. Appl. Poultry Res. 4:235.
120. Tauxe, R. V., M. S. Deming, and P. A. Blake. 1985. *Campylobacter jejuni* infections on college campuses: a national survey. Am. J. Publ. Health 75:659-660.
121. Tauxe, R. V., N. Hargrett-Bean, C. M. Patton, and I. K. Wachsmuth. 1988. Campylobacter isolates in the United States, 1982- 1986. Morb. Mortal. Wkly. Rep. 37(S52):1-13.
122. Taylor, D. N., B. W. Porter, C. A. Williams, H. G. Miller, C. A. Bopp, and P. A. Blake. 1982. *Campylobacter* enteritis: a large outbreak traced to commercial raw milk. West J. Med. 137:365-369.
123. Tosh, F. E., G. A. Mullen, and D. E. Wilcox. 1981. Outbreak of *Campylobacter* enteritis associated with raw milk—Kansas. Morb. Mortal. Wkly. Rep. 30:218.
124. Totten, P. A., C. M. Patton, F. C. Tenover, T. J. Barrett, W. E. Stamm, A. G. Steigerwalt, J. Y. Lin, K. K. Holmes, and D. J. Brenner. 1987. Prevalence and characterization of hippurate-negative *Campylobacter jejuni* in King County, Washington. J. Clin. Microbiol. 25:1747-1752.
125. Wang, W. L. L., N. W. Luechtefeld, L. B. Reller, and M. J. Blaser. 1980. Enriched brucella medium for storage and transport of cultures of *Campylobacter fetus* subsp. *jejuni*. J. Clin. Microbiol. 12:479-480.
126. Wesley, R. D., and W. J. Stadelman. 1985. The effect of carbon dioxide packaging on detection of *Campylobacter jejuni* from chicken carcasses. Poultry Sci. 64:763-764.
127. Wesley, R. D., B. Swaminathan, and W. J. Stadelman. 1983. Isolation and enumeration of *Campylobacter jejuni* from poultry products by a selective enrichment method. Appl. Environ. Microbiol. 46:1097-1102.
128. Wright, E. P., H. E. Tillett, J. T. Hague, F. G. Clegg, R. Darnell, J. A. Culshaw, and J. A. Sorrell. 1983. Milk-borne Campylobacter enteritis in a rural area. J. Hyg. 91:227-233.

Chapter 32

Bacillus cereus

Reginald W. Bennett and Negash Belay

32.1 INTRODUCTION

Bacillus cereus has a ubiquitous distribution in the environment and can be isolated from a variety of processed and raw foods. However, its presence in foods is not a significant health threat unless it is able to grow. Consumption of food containing more than 10^5 viable *B. cereus* cells/g has resulted in outbreaks of foodborne illnesses[12,13,23] and the establishment of specifications for various food ingredients by food manufacturers. Foods frequently incriminated in outbreaks of *B. cereus* poisoning include boiled and fried rice, cooked pasta, cooked meats, cooked vegetables, soups, salads, puddings, and vegetable sprouts. Moreover, in recent years psychrotrophic *B. cereus* strains have been isolated from foods stored at refrigeration temperatures and the observation that such strains are enterotoxigenic is of increasing concern to the food industry.[10,11,39]

Two types of illness have been attributed to the consumption of food contaminated with *B. cereus*. The "diarrheal syndrome" is characterized by abdominal pain and diarrhea. It has an incubation period of 8 to 16 hr and symptoms that last 12 to 24 hr. The "emetic syndrome" is characterized by an acute attack of nausea and vomitting within 1–5 hr after a meal. Diarrhea is not a predominant feature in this type of illness, but it does occur in some cases. The diarrheal type of illness is caused by a protein enterotoxin with a molecular weight of approximately 38,000–50,000[36,37] and there are reports that the toxin is a multicomponent complex.[18,5,6] The diarrheagenic toxin is inactivated by heating for 5 min at 56°C, although its thermostability is reported to be greater in milk than in culture supernatants.[4] The purified enterotoxin has been reported to cause increased vascular permeability in the skin of rabbits and a positive fluid accumulation response in the rabbit ligated ileal loop test[36,5,6] This toxin is antigenic and can be used to raise specific antibodies in rabbits for use in diagnostic tests. A microslide gel-diffusin assay for detection and quantitation of the diarrheagenic factor has been described.[7,8,9] Commercial immunoassays for detection of the toxin are also currently available which include a reversed passive latex agglutination (RPLA) test kit (Oxoid, UK) and an enzyme-linked immunosorbent assay (ELISA) [Tecra Diagnostics, Roseville, NSW, Australia].

The symptoms experienced by patients in emetic syndrome outbreaks are caused by a completely different toxin. The toxin, recently isolated on the basis of its ability to induce vacuolation in Hep-2 cells,[1] is a 1.2 kDa cyclic peptide that is unusually resistant to heat (withstanding 120°C for more than 1 hr), acidic pH and proteolysis.[37,15] The emetic activity of strains incriminated in emetic syndrome outbreaks has been demonstrated experimentally by monkey feeding[29] and in kittens by intravenous injection of heated culture fluids.[8] The tissue culture assay using Hep-2 cells[20,31] and a recently described spermatozoa toxicity bioassay[3] may also be useful for the detection and quantitation of the emetic toxin. Immunotechniques or other rapid and specific methods for detection of this toxin are yet to be developed.

In recent years detection and differentiation of *B. cereus* strains by means of new approaches have been reported. However, the characterization of *B. cereus* in foods by these newer approaches has not been widely evaluated. Among these newer methods are polymerase chain reaction (PCR) assays based on characterized genes,[32,2] cellular fatty acid analysis,[33,38] and molecular fingerprint analysis of cells by Fourier Transform Infrared Spectroscopy (FTIR).[28]

32.11 *Taxonomic Position of* Bacillus cereus

The work of Smith et al.[34] and Gordon et al.[14] has brought a measure of order to the diversity of strains and species in the genus *Bacillus*. *B. cereus* is classified as a large-celled species of Group 1 (species with a cell width greater than 0.9 μm and whose spores do not appreciably swell the sporangium). Whether *B. anthracis*, *B. thuringiensis*, and *B. cereus* var. *mycoides* should be accorded species status or considered varieties of *B. cereus* is arguable. Certainly they are closely related, and, for all practical purposes, the other members of this group differ from *B. cereus* by only a single characteristic that may be lost with repeated culturing, viz., pathogenicity for animals by *B. anthracis*, production of an endotoxin crystal by *B. thuringiensis*, and rhizoid growth by *B. cereus* var. *mycoides*. Consequently, absolute separation of this group into distinct species is not possible in all instances. Nevertheless, the typical characteristics of *B. cereus* strains seem to be quite stable and the other biotypes usually can be readily differentiated from them when the variant properties are evident.[16] In this chapter, each will be considered a distinct species (except the rhizoid stains) and procedures will be described for distinguishing the other biotypes from typical *B. cereus*.

32.12 *Characteristics of* B. cereus *and Similar Species*

The salient features of the four biotypes of the *B. cereus* group are summarized in Table 1. Some caution is necessary in applying and interpreting tests for identification purposes, since even within a species, variability and strain heterogeneity are common. It should also be noted that some tests are more valuable or more easily performed than others and for that reason are advocated for identification of *B. cereus* in Section 32.23. The term egg yolk reaction is used to describe the turbidity developed in egg yolk or in agar containing egg yolk. The responsible agent is an extracellular substance(s) referred to in some cases as egg yolk turbidity factor, lecithinase, or phospholipase.

Although it has been established that *B. cereus* does produce a phospholipase C, there is some evidence that the turbidity developed in egg yolk may be due to a more complex series of events than the action of a single enzyme.[25] Quantitation of *B. cereus* in a sample is obtained with a simple surface plating technique. In the United States and in many parts of Europe, use is made of a) the ability of *B. cereus* organisms to produce turbidity surrounding colonies growing on agar containing egg yolk, and b) the resistance of *B. cereus* to the antibiotic polymyxin B to create a selective and differential plating medium.

At least four such media have been described and in three of them mannitol is incorporated to enhance differentiation.[19,22,30,35] The PEMBA medium of Holbrook and Anderson[19] is apparently in wide use in the United Kingdom and a few other countries, but reports in the United States and Canada have been critical of its performance.[35,17] Blood agar overlaid with polymyxin B has also been employed by some investigators, especially for the examination of stools.[24] Colonies of appropriate morphology that show the characteristic alteration of the surrounding medium are considered to be presumptive *B. cereus*. Confirmatory tests include microscopic examination of sporulating cultures and a variety of other determinations as described in Section "Confirmatory Tests."

32.13 Treatment of Sample

1. **Collection.** The objective is to obtain a representative sample of the material to be examined (Chapter 2).
2. **Holding.** There are no data available to suggest that refrigeration of samples will cause a reduction in the number of viable *B. cereus* (Chapter 2).
3. **Homogenization and dilution**. The surface plating procedure is used. See Chapter 2 for details on dilution of fluid materials, homogenization, and dilution of solid or semisolid samples. Also see Chapter 7.
4. **Heat shocking.** The spores of many, if not most, strains of *B. cereus* germinate readily on the plating media used for enumeration. In most cases, heat shocking treatment is not needed to enhance germination. Sometimes the investigator may desire only a spore count, or for other reasons may wish to use a heat-shock procedure. In such cases, the temperature-time treatment of 70°C for 15 min is recommended.

32.14 Special Equipment and Supplies

It is sometimes necessary to identify the group to which a particular *Bacillus* sp. isolate belongs. A smear made from a sporulating culture is examined to determine whether the spore has distended the sporangium. This can be determined most easily by dark-field examination of the smear under oil immersion at 600 to 1200×. Alternatively, a regular laboratory bright-field microscope equipped with a 90 to 100× oil immersion objective can be used to examine stained preparations. Cell size is also quite important; the cell width of *B. cereus* and culturally similar species usually exceeds 0.9 μm.

32.15 Special Reagents and Media

Colbeck's egg yolk broth
Egg yolk emulsion 50%
Kim-Goepfert (KG) agar[22]
Mannitol yolk polymyxin (MYP) agar[30]
Polymyxin B sulfate

32.16 Additional Media and Reagents

Alpha-naphthol solution
BC motility medium
Basic fuchsin stain
Creatine
40% Potassium hydroxide
L-tyrosine agar
Lipid globule stain (Burdon)
Lysozyme broth
Methanol
Modified Voges-Proskauer (VP) broth
Nitrate broth
Nitrite test reagents (method 2)
Nutrient agar

Table 1. Characteristics of *Bacillus cereus* and Culturally Similar Species

Feature	*B. cereus*	*B. cereus* var. *mycoides*	*B. thuringiensis*	*B. anthracis*[a]
Gram stain	+	+	+	+
Catalase	+	+	+	+
Egg yolk reaction	–	–	–	(+)
Motility	–	–	–	–
Acid from mannitol	–	–	–	–
Hemolysis (sheep RBC)	+	(+)	+	–
Rhizoid growth	–	+	–	–
Toxin crystals produced	–	–	+	–
Anaerobic utilization of glucose	+	+	+	+
Reduction of nitrate	–	+	+	+
VP reaction	+	+	+	+
Tyrosine decomposition	+	(+)	+	(+)
Resistance to lysozyme	+	+	+	+

[a] Additional tests are required to identify this species.[27]
+ = positive; – =usually positive but occasionally may be negative; (+)= often weakly positive; – = negative

Phenol red dextrose broth
Spore stain (Ashby's)
Trypticase soy polymyxin broth
Trypticase soy sheep blood agar

32.17 Controls

It is advisable to have at least one strain each of *B. cereus*, *B. cereus* var. *mycoides*, and *B. thuringiensis* available so that the analyst can become familiar with the reactions obtained on the different media and thus learn to recognize the specific traits that differentiate them. Suitable cultures for this purpose can be obtained from the American Type Culture Collection, Manassas, VA. Prototype strains of the "emetic syndrome" and "diarrheal syndrome" types of *B. cereus* are available from the National Collection Of Type Cultures, London, England, UK, as Numbers 11143 and 11145, respectively.

32.18 Precautions and Limitations

32.181 General

Several limitations are common to the plating media described above. *B. cereus* colonies on both KG and MYP agar may show various types of colonial morphology. Those most commonly seen on KG agar are round, flat, dry, with a ground glass appearance; they may be translucent or creamy white. Less commonly, but by no means rarely, colonies may be rather amorphous with highly irregular edges. In these cases the central portion of the colony is usually white, while the perimeter is translucent. The colonies on MYP agar are similar except that the colonies and surrounding medium are pink because of the failure of *B. cereus* to ferment mannitol.

Most commonly, *B. cereus* produces a very strong reaction in egg yolk agar that is characterized by a wide zone of turbidity surrounding the individual colonies after 20 to 24 hr of incubation. Quite frequently zones from individual colonies will coalesce if many *B. cereus* are present, and estimating the true number of "zone forming colonies" is difficult. This situation often develops when more *B. cereus* colonies appear on a plate. For this reason, the countable zone forming colony range on these media is reduced to 10 to 100 rather than the 30 to 300 normally used in most quantitative plating analyses. It should be noted also that these media are not 100% selective and that other organisms (mostly *Bacillus* spp.) are often encountered. Frequently, some colonies will be moist or almost mucoid in appearance, and, when viewed from the underside of the plate, will appear to have a zone of precipitate immediatley beneath, though not extending beyond, the border of the colony. These colonies are not *B. cereus*, and the investigator should be concerned only with colonies having the typical morphology with, or infrequently without, a zone of turbidity surrounding them. Neither of these media is especially proficient for enumerating *B. cereus* in feces, particularly if there are fewer than 100,000 cells/g of feces. It has been suggested that blood agar overlaid with polymyxin B is a more appropriate plating medium for the isolation of *B. cereus* from fecal samples,[24] taking advantage of the characteristic colonial morphology and hemolytic activity of these organisms on blood agar. Although human and horse blood have been used successfully, our experience indicates that both rabbit and sheep blood are suitable alternatives.

32.182 Specific

1. **MYP agar.** This medium has been widely used in Europe and the United States for enumeration of *B. cereus*. On this medium, *B. cereus* is differentiated from most commonly occurring *Bacillus* spp. by its inability to ferment mannitol and its unusual production of lecithinase. However, acid produced by colonies other than *B. cereus* often diffuses throughout the agar, making it difficult to distinguish mannitol-fermenting from nonfermenting organisms. Suspect colonies must be transferred to a fresh medium to ascertain their true character. *B. cereus* often sporulates poorly on MYP agar, and the transfer to a medium promoting sporulation is necessary before proceeding with the identification.
2. **KG agar.** This medium is equally sensitive and selective but is used much less frequently than MYP agar. KG agar was formulated to promote free spore formation within the 20- to 24-hr incubation period. This feature allows the direct confirmation of zone-forming organisms as Group 1 bacilli by means of microscopic examination, and the immediate differentiation of *B. cereus* from *B. thuringiensis* by visualization of the endotoxin crystal in sporulated cells of the latter organism. An additional advantage to KG agar is that Group 2 bacilli, such as *B. polymyxa*, which produce lecithinase, are unable to form lecithinase under the rather nutritionally poor conditions imposed by KG agar. Because of their similarity in composition and operating principles, the PEMBA medium of Holbrook and Anderson[19] or PEMPA medium of Szabo et al.[35] may be substituted for KG agar.

32.2 PROCEDURE

32.21 Sample Collection, Preparation, and Dilution

Inoculate duplicate predried MYP or KG agar plates with suitable dilutions (usually 10^{-1} to 10^{-6}) by spreading 0.1 mL of food homogenate over the entire plate surface with a bent glass rod or "hockey stick". Use a separate inoculating rod sterilized by autoclaving for each dilution. After spreading, allow the inoculum to dry and incubate the plates at 30°C to 32°C for 20 to 24 hr.

After incubation, examine the plates for typical colonies, which are usually surrounded by a precipitate caused by lecithinase activity. Colonies on MYP agar will be pink to violet, indicating that mannitol is not fermented. The number of such colonies multiplied by the reciprocal of the dilution that the countable plate represents is the presumptive *B. cereus* count. Remember that the dilution factor is tenfold higher than the sample dilution since only 0.1 mL was dispensed onto the plating medium.

32.22 Most Probable Number Technique

A most probable number (MPN) technique is a suitable alternative to the direct plate count for examining foods that are expected to contain fewer than 1000 *B. cereus* per gram.[26] In this method, 3 tubes each of trypticase soy polymyxin broth are inoculated with 1 mL of the 1:10, 1:100, and 1:1000 dilutions of food homogenate. Incubate the tubes at 30°C for 48 hr and examine them for dense growth typical of *B. cereus*. Streak presumptive positive tubes on MYP agar, and select characteristic colonies for confirmation as *B. cereus*. The confirmed *B. cereus* count is determined using the appropriate MPN table (Chapter 6) on the basis of the number of tubes at each dilution in which *B. cereus* was detected.

32.23 Confirmatory Tests

Confirmatory tests are necessary to establish the identity of presumptive colonies as *B. cereus*. Two groups of tests are described which may be used, depending on the extent of identification desired. The first group of tests (Section 32.232) is designed to differentiate the typically reacting strains of *B. cereus* from other members of the *B. cereus* group when a clear-cut result has been obtained on MYP or KG agar. Since the results obtained on the plating media with members of the *B. cereus* group are so characteristic, other *Bacillus* species are unlikely to be mistaken for them. Therefore, the rapid confirmatory test of Holbrook and Anderson[19] can be substituted for the biochemical tests described in Section 32.233 unless the isolates are atypical or must be more definitely identified for regulatory purposes.

32.231 Rapid Confirmatory Test

Colonies from MYP agar to be tested should be subcultured on nutrient agar slants and incubated at 30°C for 24hr; those from KG agar may be tested directly. The objectives are to determine vegetative cell, sporangium, and spore morphology and to demonstrate the presence of lipid globules within the vegetative cell. A staining procedure that combines the spore stain of Ashby and the intracellular lipid stain of Burdon is used as an aid in these determinations. Smears are made from the center of 1-day-old colonies or from the edge of 2-day-old colonies, air dried, and heat fixed with minimal flaming. The slide is placed over boiling water and flooded with 5% malachite green (heating slides at least twice at 1-min intervals with a bunsen burner until steam is seen is an acceptable alternative). After 2 min, the slide is washed, blotted dry, and stained with 0.3% wt/vol Sudan black B in 70% ethanol for 20 min. The stain is poured off and the slide blotted dry and washed with reagent grade xylene for 5 to 10 sec. The slide is blotted immediately and counterstained for 20 sec with 0.5% wt/vol of safranin. The stained slides are examined microscopically under oil immersion for the presence of lipid globules within the cytoplasm (stained dark blue) and central-to-subterminal spores that do not obviously swell the sporangium. Spores are usually pale to mid-green, contrasting with the red vegetative cells. The presence of intracellular lipid globules and typical spores is a good indication that isolates from MYP or KG agar are members of the *B. cereus* group. These properties, however, are not unique to *B. cereus*; therefore, the rapid confirmatory test must always be used in conjunction with more specific tests described below before an identification can be made.

32.232 Differentiating Members of the B. cereus Group

1. **Motility tests.** Semisolid BC motility medium or direct microscopic examination may be used to determine motility. BC motility medium is inoculated by stabbing down the center with a 3 mm loopful of culture and incubating for 18 to 24 hr at 30°C. Motile strains produce diffuse growth out into the medium away from the stab. Nonmotile strains (except *B. cereus* var. *mycoides*) grow only in and along the stab. Rhizoid strains of *B. cereus* var. *mycoides* usually produce characteristic fuzzy growth in semisolid media because of the expansion of the filamentous growth, but they are not motile by means of flagella. Doubtful results should be confirmed by the alternative microscopic motility test. This test is performed by adding 0.2 mL of sterile distilled water to a nutrient agar slant, which is then inoculated with a 3 mm loopful of culture. After incubation at 30°C for 6 to 8 hr, a loopful of liquid culture from the base of the slant is suspended in a drop of water on a clean slide, covered with a cover slip, and examined immediately for motility. *B. cereus* and *B. thuringiensis* are usually actively motile, where as *B. anthracis* and the typically rhizoid strains of *B. cereus* var. *mycoides* are nonmotile.
2. **Rhizoid growth.** To test for rhizoid growth, a predried nutrient agar plate is inoculated near the center with a 2 mm loopful of culture and incubated for 24 to 48 hr at 30°C. If the culture is rhizoid, root or hairlike structures will develop up to several centimeters from the point of inoculation. This property is characteristic only of strains that are classified as *B. cereus* var. *mycoides*.
3. **Hemolytic activity.** To test for hemolytic activity, the bottom of a standard trypticase soy sheep blood agar plate is marked into six or eight equal segments. Each segment is labeled and inoculated near its center by gently touching the agar surface with a 2 mm loopful of culture. The plate is then incubated at 30° to 32°C for 24 hr and checked for hemolytic activity as indicated by a zone of complete hemolysis surrounding the growth. *B. cereus* is usually strongly hemolytic, whereas *B. thuringiensis* and *B. cereus* var. *mycoides* are often weakly hemolytic, or produce hemo-lysis only under the growth. *B. anthracis* is usually nonhemolytic. Caution: Nonmotile, nonhemolytic cultures could be *B. anthracis* and should be handled with special care. Contact the nearest public health laboratory for identification of *B. anthracis*.
4. **Detection of toxin crystals.** The endotoxin crystals of *B. thuringiensis* may be detected by phase-contrast microscopy, but are probably most conveniently detected by staining as follows: A nutrient agar slant is inoculated with a loopful of culture, incubated at 30°C for 24 hr, and then held at room temperature for 2 or 3 days to permit sporangiolysis. A smear is made on a clean slide, air dried, and lightly heat fixed. The smear is further fixed by flooding the slide with methanol. After 30 sec, the methanol is poured off and the slide is dried thoroughly by passing it through a flame. The smear is stained by flooding the slide with 0.5% aqueous basic fuchsin or TB carbol fuchsin ZN and gently heating the slide from below until steam is seen. After 1 or 2 min, the slide is heated again until steam is seen, held for 30 sec, and the stain is poured off. The slide is rinsed thoroughly in tap water, dried without blotting, and examined under oil immersion with a microscope for the presence of free spores and darkly stained tetragonal toxin crystals. Free toxin crystals are usually abundant within 3 days but will not be detectable by staining until the sporangia have lysed. *B. thuringiensis* produces endotoxin crystals that usually can be detected by staining. Other members of the *B. cereus* group do not produce such crystals. Indeed toxin crystal formation was the basis for a report implicating *B. thuringiensis* in a food poisoning outbreak.[21]
5. **Interpreting results.** On the basis of test results, those isolates that are actively motile, strongly hemolytic, and do not produce rhizoid growth or endotoxin crystals should be tentatively identified as *B. cereus*. Nonmotile or weakly hemolytic strains of *B. cereus* may occasionally be encountered. These strains can be differentiated from *B. anthracis* by their resistance to penicillin and gamma bacteriophage.[27] Noncrystalliferous variants of *B. thuringiensis* and nonrhizoid strains derived from *B. cereus* var. *mycoides* cannot be differentiated from *B. cereus* by the tests described. When implicated in food poisoning, such isolates should be tested for enterotoxigenicity by biological or serological assays.[37,7,8,20]

32.24 Biochemical tests

In some instances, identification of isolates with a greater degree of certainty may be desired. The following tests are recommended because they confirm the most salient characteristics of the *B. cereus* group and virtually eliminate the possibility of confusing strains of this group with any other *Bacillus* species.

1. **Reactions on MYP agar.** This test can usually be omitted if the isolates were picked from MYP agar and the reactions of all isolates were typical. However, the test should be included with isolates from KG agar to test for mannitol fermentation. The bottom of each plate should be marked into 6 or 8 equal segments. Each segment is labeled and inoculated near its center by touching the agar with a 2 mm loopful of culture. The plate is incubated at 30°C to 35°C for 24 hr and checked for lecithinase production as indicated by a zone of precipitate surrounding growth. Mannitol is not fermented if the growth and surrounding medium are pink. *B. cereus* and other members of the *B. cereus* group usually produce lecithinase but do not ferment mannitol.
2. **Anaerobic glucose fermentation.** The ability of isolates to metabolize glucose anaerobically is determined by inoculating phenol red dextrose broth with a loopful of culture and incubating the tube in an anaerobic jar at 35°C for 24 hr. A color change from red to yellow indicates that acid was produced. Acid is produced from glucose anaerobically by *B. cereus* and other members of the *B. cereus* group.
3. **Nitrate reduction.** Nitrates are usually reduced to nitrite by *B. cereus* and other members of the *B. cereus* group. Nitrate broth is inoculated and incubated at 35°C for 24 hr. Nitrite is detected by adding 0.25 mL each of sulfanilic acid and alpha-naphthol solutions. An orange color that develops within 10 min indicates the presence of nitrite.
4. **VP test.** Acetylmethylcarbinol is produced from glucose by members of the *B. cereus* group. Modified VP medium is inoculated and incubated at 35°C for 48 hr. The presence of acetylmethylcarbinol is determined by adding 0.2 mL of 40% potassium hydroxide and 0.6 mL of 5% alcoholic alpha-aphthol solution to 1 mL of culture in a test tube. Addition of a few crystals of creatine speeds the positive reaction, development of a purple color within 15 min.
5. **Tyrosine decomposition.** The ability of *B. cereus* and culturally similar species to decompose tyrosine is determined by inoculating a slant of nutrient agar containing 0.5% tyrosine and incubating for 48 to 72 hr at 35°C. Decomposition of tyrosine as indicated by clearing of the medium immediately under the growth, is caused by dissolution of tyrosine crystals. This clearing progresses to a depth of 3 or 4 mm after 3 days of incubation. *B. cereus* and other members of the *B. cereus* group (except *B. anthracis*) readily decompose tyrosine.
6. **Lysozyme resistance.** The ability of isolates to grow in the presence of lysozyme is determined by inoculating nutrient broth containing a final concentration of 0.001% lysozyme. Growth that develops during 48 hr of incubation at 35°C indicates that the organism is resistant to lysozyme. *B. cereus* as well as other members of the *B. cereus* group are resistant to lysozyme.

32.3 INTERPRETATION OF DATA

Colonies from MYP and KG agar that meet the criteria defined in Section 32.232 should be provisionally identified as *B. cereus*. Isolates that also fulfill all criteria listed in Section 32.24 may be definitively identified as *B. cereus* (except as noted in Section 32.232 item 5), e.g., large gram-positive bacilli that produce lecithinase, are negative for mannitol fermentation on MYP agar, grow and produce acid from glucose anaerobically, reduce nitrate to nitrite (a few strains are negative), produce acetylmethylcarbinol, decompose L-tyrosine, grow in the presence of 0.001% lysozyme, exhibit motility, are hemolytic, and do not produce endotoxin crystals or rhizoid growth. The plate count of such colonies, or the most probable number, times the dilution factor is the confirmed *B. cereus* count.

32.4 REFERENCES

1. Agata, N., M. Mori, M. Ohta, S. Suwan, I. Ohtani, and M. Isobe. 1994. A novel dodecadepsipeptide, cereulide, isolated from *Bacillus cereus* causes vacuole formation in Hep-2 cells. FEMS Microbiol. Lett. 121:31.
2. Agata, N., M. Ohta, Y. Arakawa, and M. Mori. 1995. The bceT gene of *Bacillus cereus* encodes an enterotoxic protein. Microbiology 141:983.
3. Andersson, M. A., R. Mikkola, J. Helin, M. C. Andersson, and M. Salkinoja-Salonen. 1998. A novel sensitive bioassay for detection of *Bacillus cereus* emetic toxin and related depsipeptide ionophores. Appl. Environ. Microbiol. 64:1338.
4. Baker, J. M., and Griffiths. 1995. Evidence for increased thermostability of *Bacillus cereus* enterotoxin in milk. J. Food Prot. 58:443.
5. Beecher, D. J., J. L. Schoeni, and A. C. Wong. 1995. Enterotoxin activity of hemolysin BL from *Bacillus cereus*. Infect. Immun. 63:4423.
6. Beecher, D. J., and A. C. L. Wong. 1994. Improved purification and characterization of hemolysin BL, a hemolytic dermonecrotic vascular permeability factor from *Bacillus cereus*. Infect. Immun. 62:980.
7. Bennett, R.W. 1985. Serological and biological activities of *Bacillus cereus* diarrheal toxin. Seminar on *Bacillus cereus* enterotoxins: current concepts and developments. Ann. Meet. Am. Soc. Microbiol, Las Vegas, Nev., March 30, 1985.
8. Bennett, R. W., and S. M. Harmon. 1988. *Bacillus cereus* food poisoning, p. 830. *In* A. Balows, W. J. Hausler Jr., and E. Lennette (ed.), Laboratory diagnosis of infectious diseases: principles and practices. Springer-Verlag, New York.
9. Bennett, R. W., G. Murthy, L. Kaylor, S. Cox, and S. M. Harmon. 1993. Biological characterization and serological identification of *Bacillus cereus* diarrheal factor. Neth. Milk Dairy J. 47:105.
10. Christiansson, A., A. S. Naidu, I. Nilsson, T. Wadstrom, and H.-E. Pettersson. 1989. Toxin production *by Bacillus cereus* dairy isolates in milk at low temperatures. Appl. Environ. Microbiol. 55:2595.
11. Dufrenne, J., P. Soentoro, S. Tatini, T. Day, and S. Notermans. 1994. Characteristics of *Bacillus cereus* related to safe food production. Int. J. Food Microbiol. 23:99.
12. Gilbert, R. J. 1979. *Bacillus cereus* gastroenteritis, p. 495. *In* H. Riemann and F. L. Bryan (ed.), Foodborne infections and intoxications, 2nd ed. Academic Press, New York.
13. Goepfert, I. M., W. M. Spira, and H. U. Kim. 1972. *Bacillus cereus*: Food poisoning organism. A review. J. Milk Food Technol. 35:213.
14. Gordon, R. E., W. C. Haynes, and C. Hor-Nay Pang. 1973. The genus *Bacillus*. USDA Agriculture Handbook, No. 427, Washington, D. C.
15. Granum, P. E. 1994. *Bacillus cereus* and its toxins. J. Appl. Bacteriol. 76:61S.
16. Harmon, S. M. 1982. New method for differentiating members of the *Bacillus* cereus group. J. Assoc. Off. Anal. Chem. 65:1133.
17. Harmon, S. M., D. A. Kautter, and F. D. McClure. 1984. Comparison of selective plating media for enumeration of *Bacillus cereus* in foods. J. Food Prot. 47:65.
18. Heinrichs, J. H., D. J. Beecher, J. M. MacMillan, and B. A. Zilinskas. 1993. Molecular cloning and characterization of the *hbla* gene encoding the B component of hemolysin BL from *Bacillus cereus*. J. Bacteriol. 175:6760.
19. Holbrook, R., and J. M. Anderson. 1980. An improved selective diagnostic medium for the isolation and enumeration of *Bacillus cereus* in foods. Can. J. Microbiol. 26:753.
20. Hughes, S., B. Bartholomew, J. C. Hardy, and J. M. Kramer. 1988. Potential application of a HEP-2 cell assay in the investigation of *Bacillus cereus* emetic-syndrome food poisoning. FEMS Microbiol. Lett. 52:7.

21. Jackson, S. G., R. B. Goodbrand, R. Ahmed, and S. Kasatiya. 1995. *Bacillus cereus* and *Bacillus thuringiensis* isolated in a gastroenteritis outbreak investigation. Lett. Appl. Microbiol. 21:103.
22. Kim, H. U., and J. M. Goepfert. 1971. Enumeration and identification of *Bacillus cereus* in foods. I. 24-hour presumptive test medium. Appl. Microbiol. 22:581.
23. Kramer, J. M., and R. J. Gilbert. 1989. *Bacillus cereus* and other *Bacillus* species, p. 21. *In*M. Doyle (ed.), Foodborne bacterial pathogens. Marcel Dekker, New York.
24. Kramer, J. M., P. C. B. Turnbull, G. Munshi, and R. J. Gilbert. 1982. identification and characterization of *Bacillus cereus* and other *Bacillus* species associated with food poisoning, p. 261. *In* J. E. L. Corry, F. Roberts, and F. A. Skinner (ed.), Isolation and identification methods for food poisoning organisms. Soc. Appl. Bacteriol. Tech. Ser. No. 17, Academic Press, London.
25. Kushner, D. J. 1957. An evaluation of the egg yolk reaction as a test for lecithinase activity. J. Bacteriol. 73:297.
26. Lancette, G. A., and S. M. Harmon. 1980. Enumeration and confirmation of *Bacillus cereus* in foods: collaborative study. J. Assoc. Off. Anal. Chem. 63:581.
27. Leise, J. A., C. H. Carter, H. Friedlander, and S. N. Freed. 1959. Criteria for identification of *Bacillus anthracis*. J. Bacteriol. 77:655.
28. Lin, S. F., H. Schraft, and M. W. Griffiths. 1998. Identification of Bacillus cereus by Fourier Transform Infrared Spectroscopy (FTIR). J. Food Prot. 61:921.
29. Melling, J., B. J. Capel, P. C. B. Turnbull, and R. J. Gilbert. 1976. Identification of a novel enterotoxigenic activity associated with *Bacillus cereus*. J. Clin. Pathol. 29:938.
30. Mossel, D. A. A., M. J. Koopman, and E. Jongerius. 1967. Enumeration of *Bacillus cereus* in foods. Appl. Microbiol. 15:650.
31. Nishikawa, Y., M. J. Kramer, M. Hanaoka, and A. Yasukawa. 1996. Evaluation of serotyping, biotyping, plasmid banding pattern analysis, and Hep-2 vacuolation factor assay in the epidemiological investigation of *Bacillus cereus* emetic-syndrome food poisoning. Int. J. Food Microbiol. 31:149.
32. Schraft, H., and M. W. Griffiths. 1995. Specific oligonucleotide primers for detection of lecithinase-positive *Bacillus* spp. by PCR. Appl. Environ. Microbiol. 61:98.
33. Schraft, H., M. Steele, B. McNab, J. Odumeru, and M. W. Griffiths. 1996. Epidemiological typing of Bacillus spp. isolated from food. Appl. Environ. Microbiol. 62:4229.
34. Smith, N. R., R. E. Gordon, and F. E. Clark. 1952. Aerobic sporeforming bacteria. USDA Monograph No. 16, Washington, D. C.
35. Szabo, R. A., E. C. D. Todd, and M. K. Rayman. 1984. Twenty-four hour isolation of *Bacillus cereus* in foods. J. Food Prot. 47:856.
36. Thompson, N. E., M. J. Ketterhagen, M. S. Bergdoll, and E. J. Shantz. 1984. Isolation and some properties of an enterotoxin produced by *Bacillus cereus*. Infect. Immun. 43:887.
37. Turnbull, P. C. B. 1986. *Bacillus cereus* toxins, p. 397. *In* F. Dorner and J. Drews (ed.), Pharmacology of bacterial toxins. Pergamon Press, Oxford, England.
38. Vaisanen, O. M., N. J. Mwaisumo, and M. S. Salkionja-Saonen. 1991. Differentiation of dairy strains of the Bacillus cereus group by phage typing, minimum growth temperature, and fatty acid analysis. J. Appl. Bacteriol. 70:315.
39. Van Netten, P., A. van de Moosdijk, P. van Hoensel, D. A. A. Mossel, and L. Perales. 1990. Psychrotropic strains of *Bacillus cereus* producing enterotoxin. J. Appl. Bacteriol. 69:73.

Chapter 33

Clostridium botulinum and Its Toxins

Haim M. Solomon, Eric A. Johnson, Dane T. Bernard, Stephen S. Arnon, and Joseph L. Ferreira

33.1 INTRODUCTION

Clostridium botulinum is an anaerobic, gram-positive, spore-forming, rod-shaped bacterium that produces the most potent poison known, a protein of characteristic neurotoxicity. Severe food poisoning, botulism, results from the consumption of botulinum toxin produced in food in which this organism has grown.

Antigenic types of *C. botulinum* are defined by the toxins they produce and each antigenic toxin type is neutralized completely by the homologous antitoxin only and cross-neutralization by heterologous antitoxin types is absent or minimal. The seven recognized *C. botulinum* types are designated A, B, C, D, E, F, and G. Five of these apparently produce only one type of toxin but all are given type designations corresponding to the sole or major type of toxin produced. Type C produces predominantly C1 toxin with lesser amounts of C2 or only C2, and type D produces predominantly type D toxin along with smaller amounts of C2 toxin. The production of more than one type of toxin may be a more common phenomenon than previously realized. There is a slight reciprocal cross-neutralization of types E and F, and recently strains of *C. botulinum* have been identified which produce a mixture of toxins consisting mostly of the dominant type of toxin plus small amounts of different types of toxins, e.g. Ab, Af, and Bf.

Botulism as a type of food poisoning in humans is rare, but the case fatality remains relatively high. In the United States from 1899 through 1995, 1,026 outbreaks of botulism were recorded. These involved 2,444 cases and caused 1,040 deaths.

Of outbreaks in which the toxin type was determined, 446 were due to type A, 117 to type B, 149 to type E and 6 to type F.[27] The implicated foods of two outbreaks contained both A and B toxins. The limited number of reports of C or D toxin to be the causative agent of human botulism have not received general acceptance. All except types F and G, about which little is known, are important causes of animal botulism. Human botulism also may result from wounds infected with *C. botulinum* in which the organism grows and elaborates its toxin, but this is a rare occurrence. Gastrointestinal symptoms are usually absent in such cases.

Infant botulism, first recognized as a distinct clinical entity in 1976, is now the most common form of human botulism reported in the United States. It affects infants 12 months of age or less, with 95% of cases occurring between 2 and 26 weeks of life. This form of botulism results from growth and neurotoxin production by *C. botulinum* within the intestinal tract of infants rather than from the ingestion of preformed toxin. It is usually caused by *C. botulinum* types A or B, but a few cases have been reported as being caused by other toxin types.[20] Infant botulism has been diagnosed in most states of the United States and in every populated continent except Africa. As of January 1994, 1,270 hospitalized cases of infant botulism had been reported worldwide. Of these 1,206 (95.0%) occurred in the United States.[2-4,12,13,19,20]

In infant botulism, constipation almost always precedes the characteristic signs of neuromuscular paralysis by a few days or weeks. Illness varies greatly in severity. Some infants show only mild weakness, lethargy and reduced feeding and do not require hospitalization. Severe symptoms, such as generalized muscle weakness, weakened suck and swallowing, faint cry and diminished gag reflex with a pooling of oral secretions are more commonly reported. Generalized muscle weakness and loss of head control reaches such a degree that some infants appear "floppy." Approximately half of all (hospitalized) patients require endotracheal intubation and mechanical breathing support at some point during their hospital stay. High quality intensive care is responsible for a case-fatality ratio that is <1%. The administration of the recently developed human Botulism Immune Globulin (BIG) shortens the mean hospital stay by over 50%.[4]

Definitive diagnosis of infant botulism depends on the demonstration of toxin and/or organisms in the feces. *C. botulinum* has been recovered from patients' feces for as long as 5 months after onset of illness and toxin for as long as 4 months. Although testing of serum is very useful for establishing the diagnosis of botulism in adults, it is of limited value in infants. In a recently reported study, toxin was found in the serum of only 9 of 67 (13%) culture-positive infant botulism patients.[13] Honey is a common source of C. *botulinum* spores implicated in infant botulism. In studies of honey, up to 13% of the test samples contained low numbers of *C. botulinum* spores.[14] For this reason the U.S. FDA, the U.S. CDC, the American Academy of Pediatrics, as well as several honey industry groups have all recommended that honey not be fed to infants under the age of 1 year.

The organism *C. botulinum* is distributed widely in soils and in the sediments of oceans and lakes, so that there is a diversity of sources for food contamination. The finding of type E organisms in aquatic environments by many investigators correlates with the tracing of most cases of type E botulism to contaminated fish or other sea foods. Types A and B are most commonly encountered terrestrially, and the primary vehicles of botulism caused by these two types, are foods commonly contaminated with soils. In the U S these foods have been primarily home-canned vegetables, but in Europe home canned meat products also have been important vehicles for intoxication.

C. botulinum isolates are further subdivided into four distinct groups by properties other than toxin antigenic types, with each group composed of strains of different types but having similar cultural and physiological characteristics. Group I includes all strains of type A plus the proteolytic strains of types B and F; Group II includes all strains of type E plus the non-proteolytic strains of types B and F; Group III includes all strains of types C and D; Group IV contains the proteolytic but non-saccharolytic type G.[23] A tentative fifth group containing strains of *C. butyricum and C. baratii* that produce botulinum toxins type E and F, respectively is under consideration. All type A strains and some B and F strains are proteolytic, whereas all type E strains and the remaining B and F strains are non-proteolytic. Type G shows slow proteolytic activity. Optimum temperature for growth and toxin production of the proteolytic strains is close to 35°C, while that for the non-proteolytic strains is approximately 26° to 28°C. Non-proteolytic types B, E, and F can produce toxin at refrigeration temperature (3° to 4°C). Toxins of the non-proteolytic strains do not manifest maximum potential toxicity until activated with trypsin; toxins of the proteolytic strains generally occur in fully, or close to fully, activated form. These and other differences are important in epidemiological and laboratory investigations of botulism outbreaks.

Measures to prevent botulism from foods include reduction of the microbial contamination level, acidification, reduction of moisture level, and destruction of all *C. botulinum* spores in the food. Heat is the most common method of destruction; properly processed canned foods do not contain viable *C. botulinum*. The greater incidence of botulism from home-canned foods than from commercially canned foods undoubtedly reflects the commercial canners' greater awareness and better control of the destructive heating required.

A certain food may contain viable *C. botulinum* spores and still not cause botulism. As long as the organisms do not grow, toxin is not synthesized. Many foods satisfy the nutritional requirements of *C. botulinum*, but not all provide the necessary anaerobic conditions. Many canned foods and many meat and fish products meet both nutritional and anaerobic requirements for growth of *C. botulinum*. However, growth in otherwise suitable foods is prevented if the product, naturally or by design, is acidic (pH = less than 4.6), has low water activity ($A_w < 0.9$) , a high sodium chloride concentration (5% for non-proteolytics; 10% for proteolytics), an inhibitory sodium nitrite concentration (100–200 ppm used in conjunction with other inhibitors), or two or more of these factors in combination. Unless temperature is strictly controlled and kept below 3°C, refrigeration will not prevent growth and toxin formation by non-proteolytic strains. Moreover, the usual vehicles of botulism are foods processed to prevent spoilage and are not normally refrigerated. Botulinum toxin is heat-labile, therefore, botulism can be prevented by thoroughly heating all processed foods prior to consumption, e.g. boiling for 10 minutes before serving.[6,15-18,23,24]

33.2 TREATMENT OF SPECIMENS

The diagnosis of botulism is most effectively confirmed in the laboratory by demonstrating botulinum toxin in the blood, feces, or vomitus of the patient. Specimens must be collected before the patient is treated with botulinum antitoxin. In foodborne botulism, identifying the causative food as soon as possible is important to prevent further cases.

33.21 Foods

Suspect foods should be refrigerated until tested, except for unopened canned foods, which, unless badly swollen and in danger of bursting, need not be refrigerated.

Before testing, record such identifying data as product, manufacturer or home canner, source, type of container and size, labeling, manufacturer's batch, lot, or production code, and condition of container.

Clean and mark the container with laboratory identification number or symbol, disinfect with an alcohol disinfectant, and open aseptically for sampling. Carefully avoid aerosols. (see Section 33.61).

Check for normal ingredients which, by their presence or concentration in the product, could be lethal for mice by the intraperitoneal route of administration, e.g., a high salt concentration (anchovies) or high sugar concentration (heavy syrups).[6,17,27,28]

33.211 Clinical Specimens

All clinical specimens should be collected as soon as botulism is suspected and before botulinum antitoxin is administered.[6,12,22,27]

1. **Serum (Adult patient).** Collect enough blood (approx. 50 mL) to provide at least 20 mL of serum for toxin neutralization tests. Allow blood to clot in the refrigerator; centrifuge, and remove serum to a sterile vial or test tube with a leak-proof cap. Examine immediately or refrigerate at 4°C. Examination of post-treatment (8–12 hrs) serum is also helpful to evaluate antitoxin therapy.
2. **Feces.** Collect 25 g of the patient's feces (as much as possible from infants) in a sterile, unbreakable, leak-proof container. Preferably, use a screw-cap wide-mouth plastic bottle. Seal caps with waterproof tape. Cardboard containers are not acceptable. Refrigerate specimens at 4°C until examined. A "soap-suds" enema should not be given before the feces are collected, since the soap may inactivate the toxin. If a passed stool is not available, the physician should be requested to obtain a specimen using a sterile water enema.
3. **Miscellaneous clinical specimens.** Specimens, such as vomitus, gastric aspirates, cerebrospinal fluid or tissues obtained at autopsy should be collected in sterile leak- proof containers and refrigerated at 4°C.

33.3 SPECIAL EQUIPMENT AND SUPPLIES

33.31 Culture and Isolation of C. botulinum

Anaerobic jars or anaerobic glovebox
Bunsen burner; micro-incinerator or electric incinerator
Clean dry towels
Culture tube racks
Mechanical pipetting devices (never pipette by mouth)
Microscope; phase-contrast or bright-field
Microscope slides
Refrigerator
Sanitary can opener; bacti-disc or puncture type
Sterile culture tubes; at least a few with screw-caps
Sterile forceps
Sterile mortar and pestle
Sterile 100 mm Petri dishes
Sterile pipettes

Sterile sample jars
Incubators; 35 and 28°C
Transfer loops
Millipore filters; 0.45 μm pore size

33.32 Toxin Identification and Assay

[Sterility of equipment is desirable but not absolutely necessary except as noted]
Centrifuge tubes; some sterile for separation of patient's serum from clot
Mice; Swiss-Webster, one gender, 16–24 g.
Mortar and pestle
Mouse cages, feed, water bottles
Syringes: 1 or 3 mL with 25 gauge, 5/8-inch needles for injecting mice
Refrigerated high-speed centrifuge
Sterile 25 to 50 mL syringes, with 18 to 20 gauge, 1 1/2-inch needles to obtain blood from patients
Sterile vials for storage of serum
Water-bath; 37° C
Trypsin (Difco 1:250)

33.4 MEDIA AND REAGENTS

33.41 Culture and Isolation Procedures

Ethanol (190 Proof)
Alcoholic solution of iodine (or other suitable disinfectant)
Cooked meat medium (CMM), liver or beef heart
Gram stain, crystal violet, or methylene blue solutions
Liver, veal, egg yolk agar or anaerobic egg-yolk agar
Sterile culture media (Chapter 63)
Sterile gel phosphate buffer, pH 6.2
Trypticase-peptone-glucose-yeast extract (TPGY) or with trypsin (TPGYT) broth
Use TPGYT when suspecting a nonproteolytic strain of types B, E, or F

33.46 Toxin Identification and Assay

Gel phosphate buffer, pH 6.2
Monovalent antitoxins, types A through F (obtainable from CDC, Atlanta, GA 30333)
Physiological saline
Trypsin solution (prepared from Difco 1:250 trypsin)
1N hydrochloric acid
1N sodium hydroxide

33.5 PRECAUTIONS AND LIMITATIONS OF METHODS

Botulinum toxin is heat-labile. Store test samples and cultures under refrigeration. In addition, the pH of the toxic material must be controlled to keep it slightly acidic since botulinum toxin is less stable at alkaline pH.

When performing toxicity tests in mice, take care to distinguish between symptoms of botulism and other causes of death, such as high concentrations of salt, acid, protein degradation products, or other toxic substances that may be present in the food sample.

Use "universal precautions." Be careful to avoid the creation of aerosols.

Never pipette by mouth. Use one of several mechanical pipetting devices available.

Never recap needles after injecting mice. Dispose of used syringe and needle in a puncture-resistant "sharps" container.

Autoclave all glassware and utensils coming in contact with contaminated or potentially contaminated samples before handling them.

Use CDC Biosafety Manual as reference.

Botulinum toxins are among the "most lethal" proteins known; the specific toxicity of type A toxin is $1{\times}10^8$ mouse LD_{50}/mg protein. Symptoms of botulism (paralysis of throat, eyes, respiratory musculature, etc.) are the result of toxin inhibiting the release of the acetylcholine neurotransmitter at the peripheral synapses. Laboratory personnel who expect to be exposed to the toxin on a routine basis should be immunized with toxoid (available in the USA from the CDC, Atlanta, GA 30333).

33.6 PROCEDURE FOR IDENTIFYING VIABLE *C. BOTULINUM*

33.61 Opening Canned Foods

Sanitize the uncoded end of the can with an effective alcohol disinfectant. Allow a contact time of a few minutes, then remove the disinfectant and wipe the sanitized area with a sterile, dry towel. If the can is swollen, position the can so that the side seam is away from the analyst. A container with buckled ends should be chilled before opening and flamed with extreme caution to avoid bursting the can. Flame-sterilize the sanitized can end with a Bunsen burner by directing the flame down onto the can until the visible moisture film evaporates. Avoid excessive flaming which may cause scorching and blackening of the inside enamel coating. Remove a disc of metal from the center area of the flamed end with a sterile, sanitary can opener. Remove a disc about 5 cm in diameter, except from cans which are 202 diameter where a 3-cm disc is satisfactory (see Chapter 62).[6,17,27,28]

33.62 Solid Foods

Transfer solid foods with little or no liquid aseptically to a sterile mortar. Add an equal amount of gel phosphate buffer solution and grind with a sterile pestle in preparation of media inoculation. Alternatively, small pieces of the product may be inoculated directly into the enrichment broth using sterile forceps, or placed in a stomacher bag and pummeled with an equal volume of gel phosphate buffer.

33.63 Liquid Foods

Inoculate liquid foods directly into the culture media, using sterile pipettes.

33.64 Reserve Test Sample

After culturing, aseptically remove a reserve portion of the test sample to a sterile jar for later tests.

33.65 Examining a Product for Appearance and Odor

Visually, note any evidence of decomposition, but do not taste the product under any circumstance. Record observations.

33.66 Preparation of Enrichment Cultures

1. Before inoculation, heat broth media in flowing steam or boiling water for 15 min. After heating, cool rapidly to room temperature in cold water without agitation.
2. Inoculation of enrichment media. Inoculate 1 to 2 g of solid or macerated food or 1 to 2 mL of liquid food per 15 mL of enrichment broth. Inoculate duplicate tubes of CMM, either

beef heart or chopped liver, and duplicate tubes of TPGY broth.

3. Incubate the CMM at 35°C and the TPGY broth at 26° to 28°C.
4. After 5 days of incubation, examine each culture for turbidity, the production of gas, and the digestion of the meat particles. Note the odor. Examine the cultures microscopically by a wet mount preparation under high-power, phase-contrast microscope, or a stained smear (Gram stain, crystal violet, methylene blue) with bright field illumination. Observe the morphology of the bacteria and note the presence of clostridial cells, the occurrence and relative extent of sporulation, and the location of spores within the cells.
5. Test each enrichment culture for toxin, and, if it is demonstrated, determine the toxin type according to the procedure described in Sections 33.72 and 33.73. The highest concentration of botulinum toxin is usually present after the period of active growth, generally 5 days. An enrichment culture showing no growth after 5 days should be incubated an additional 10 days to permit possible delayed germination of injured *C. botulinum* spores before the culture is discarded.
6. For pure culture isolation, gently mix and transfer 2 mL of the culture at peak sporulation to a sterile screw-cap tube and refrigerate.

33.67 Isolation of Pure Cultures

1. The possibility of isolating *C. botulinum* in pure culture from a mixed flora in the enrichment culture is greatly improved if spores are present. To 1 or 2 mL of enrichment culture showing some sporulated cells (or the retained test sample) add an equal volume of absolute ethanol in a sterile screw-cap tube. Mix the alcohol with the culture and incubate the mixture at room temperature for 1 hour, after which, this mixture is plated as described in # 3 below.
2. An alternative procedure to the alcohol method is to heat 1 to 2 mL of the enrichment culture sufficiently to destroy the vegetative cells but not the spores of *C. botulinum* present. A simple distinction, with some exception, is based on the origin of the product investigated: if the product is of aquatic origin the organism would be of the nonproteolytic types, for products of terestrial origin the organism would be of the proteolytic types. For a nonproteolytic type, do not use heat; for a proteolytic type, heat at 80°C for 10 to 15 min.
3. Streak the alcohol- or heat-treated culture on petri dishes containing either liver, veal, egg yolk agar, or anaerobic egg yolk agar (Chapter 63) in order to obtain well separated colonies. Dilution of the culture may be necessary before plating in order to select well-isolated colonies. To prevent spreading of the colonies, the plates must be well dried. Alternatively, untreated enrichment cultures or stools can be streaked directly to isolate *C. botulinum* on one of the selective differential plating media recently developed.[19,22]
4. Incubate the inoculated plates anaerobically at 35°C for about 48 hr. A Case anaerobic jar, Gaspak or other anaerobic systems are adequate to obtain anaerobiosis.
5. After anaerobic incubation, select about 10 well-separated and typical colonies from each plate. Colonies of *C. botulinum* may be raised or flat, smooth or rough; they commonly show spreading and have an irregular edge. On egg yolk medium the colonies usually exhibit a surface irridescence when examined by oblique light. This luster zone is due to lipase activity and is often referred to as a pearly layer; it usually extends beyond but follows the irregular contour of the colony. Besides the pearly layer, colonies of *C. botulinum* types C, D, and E are ordinarily surrounded by a zone (2 to 4 mm) of a yellow precipitate caused by lecithinase activity. Colonies of types A and B generally show a smaller zone of precipitation. However, considerable difficulty in selecting toxin producing colonies may be experienced since certain other members of the genus *Clostridium*, which do not elaborate toxin, produce colonies with characteristics similar to those of *C. botulinum*.

   Inoculate each colony into a tube of sterile broth with a sterile transfer loop. For nonproteolytic *C. botulinum*, inoculate TPGY or TPGYT broth; for the proteolytic types, inoculate CMM. For orientation concerning the type of organism apply the same reasoning suggested above. Incubate the inoculated tubes for 5 days as previously described; then test for toxin as described in Section 33.72. If toxin is demonstrated, determine the toxin type (Section 33.73). Restreak the toxin producing culture in duplicate on egg yolk agar medium. Incubate one plate anaerobically and the other aerobically at 35°C for 48 hours. If colonies typical of *C. botulinum* are found only on the plate incubated anaerobically, and no growth is found on the plate incubated aerobically, the culture may be considered pure. Failure to isolate *C. botulinum* from at least one of the colonies selected means that its presence in the mixed flora of the enrichment culture is at a very low level. Sometimes the numbers can be increased enough to permit isolation by repeated serial transfers through additional enrichment steps.
6. Store the pure culture in the sporulated state under refrigeration.

33.7 IDENTIFYING BOTULINUM TOXIN IN FOODS

33.71 Preparing Food

After sampling for viable *C. botulinum*, remove a portion of the food for direct toxicity testing. Store the remainder in a refrigerator. Centrifuge food samples containing suspended solids in the cold and use the supernatant fluid for the toxin assay. Test liquid foods directly.

Extract solid food with an equal volume of gel phosphate buffer. Macerate the food and buffer with a prechilled mortar and pestle or mix in a stomacher. Centrifuge to remove the solids. Wash out emptied containers suspected of containing toxic foods with a few mL of gel phosphate buffer. Do not use excessive amount of buffer as the toxin may be diluted below the detection level. To avoid or minimize nonspecific deaths of mice, filter supernatant fluid through a 0.45 μm millipore filter before injecting mice. For non-proteolytic samples or cultures, trypsinize after filtration.[6,17,27,28]

33.72 Toxicity Determinations in Foods or Culture

Toxins of nonproteolytic types, if present in the food, may need trypsin potentiation in order to be detected. Therefore, treat a portion of the food supernatant fluid, liquid food, TPGY or cooked meat culture with trypsin before testing for toxin. Do not trypsinize TPGYT cultures since they already contain trypsin. At the same time, test another portion of supernatant fluid extract, TPGY, or cooked meat culture for toxin without trypsin treatment, since the fully active toxin of a proteolytic strain, if present, may be degraded by trypsin. The same is true if TPGYT containing the fully activated toxin of a non-proteolytic strain is further trypsinized.

To trypsinize, adjust a portion of the supernatant fluid to pH 6.2 with 1-N NaOH or HCl. To 1.8 mL of each supernatant fluid add 0.2 mL of an aqueous trypsin solution and incubate at 37° C for 1 hour with occasional gentle agitation. Prepare trypsin solution by placing 1 g of trypsin (Difco,1:250) in a clean culture tube and adding 20 mL of sterile distilled water. Agitate from time to time and keep at room temperature until as much of the trypsin as possible has been dissolved. Check the pH of the trypsin solution and adjust to pH 6.0 if necessary.

Dilute a portion of the untreated fluid or culture 1:2, 1:10, and 1:100 in gel phosphate buffer. Make the same dilutions of each trypsinized test sample fluid or culture. Inject separate pairs of mice intraperitoneally (IP) with 0.5 mL of the undiluted fluid and 0.5 mL of each dilution, using a 1.0 mL or 3.0 mL syringe with a 25 gauge 5/8-inch needle. Repeat this procedure with the trypsinized test samples. Heat 2 mL of the untreated supernatant fluid or culture at 100°C for 10 min., cool, and inject each of two mice with 0.5 mL of the undiluted fluid.

Observe all of the mice periodically for symptoms of botulism and record deaths. If all of the mice die at the dilutions used, repeat, using higher dilutions to determine the end point, or the minimum lethal dose (MLD) as an estimate of the amount of toxin present. The MLD is contained in the highest dilution killing both mice (or all mice inoculated). From this, the number of MLD per mL may be calculated.

It is very important to observe the mice closely for signs of botulinum intoxication during the first 24 hr. after injection. Death of mice without clinical signs of botulism is not sufficient evidence that the material injected contained botulinum toxin.

Typical botulism symptoms in mice in sequence are ruffling of the fur, labored but not rapid breathing, weakness of the limbs, gasping for breath (opening of lower jaw), and death due to respiratory failure. Mice that die immediately after injection usually were injured on injection or react to some toxic substance other than botulinum toxin (e.g. ammonia, high salt concentration, etc.). Mice that die after 12 hours with closed, matted eyes are generally killed by infection, not toxin, and do not have the typical botulism symptoms. Either of the latter can obscure the presence of botulinum toxin. Bacteria can be removed from the contaminated specimen by filtration, although this may also lower the toxin titer. If necessary, chemical agents, including medications used in treatment of patients, may be removed from specimen by dialysis. However, if neither type of treatment reveals the presence of toxin, subcultures may have to be relied upon, the results may remain indeterminate, or the product did not contain botulinum toxin.

33.73 Typing Toxin

In determining the type of toxin, either the untreated or trypsin-treated fluid may be used, provided that it was lethal to mice. Use the preparation that gave the higher toxin titer, but, if the trypsinized fluid is to be used, prepare a freshly trypsinized one. The continued action of trypsin may destroy the toxin.

Rehydrate the lyophilized vials of antitoxin and dilute the monovalent antitoxins to types A, B, E, and F in physiological saline to contain 1 international unit per 0.5 mL. Prepare enough of this solution to inject each of 2 mice with 0.5 mL of the antitoxin for each dilution of the toxic preparation to be tested.

Protect separate groups of mice by injecting each mouse with 0.5 mL of one of the above antitoxins 30 min to 1 hr before injecting them with the suspected toxic preparation. Inject both the unprotected and protected mice with a sufficient number of dilutions to cover a range of 10, 100, and 1000 MLD below the previously determined endpoint of toxicity. Observe the mice for 48 hr for symptoms of botulism, and record deaths.

An alternative procedure is to perform the test as described in Section 33.81.

If the toxin is not neutralized, repeat the test using monovalent antitoxins to types C and D, and a pool of types A through F.

33.74 Rapid Methods

Enzyme-linked immunosorbent assays (ELISAs) have been developed for the detection of botulinum toxins A, B, E, and F. A microtiter plate indirect sandwich ELISA has been reported for types A, B, and E neurotoxins with sensitivity in the range of 10-100 pg/mL, corresponding to 1-5 mouse intraperitoneal (IP) LD_{50} doses. This test has been successfully used to detect botulinum toxins in foods. A similar ELISA plate system has been used for the detection of types A, B, E, and F in liquid culture medium.[10] Sensitivity range for this test is 50-100 pg/mL for types A, B, and E and 100-500 pg/mL for type F, corresponding to IP — LD_{50} of 1-10/mL for types A, proteolytics B and F, and E. The sensitivity for non-proteolytic types B, and F is 10-100 LD_{50}. A third ELISA system utilizes an enzyme linked coagulation assay for, the detection of types A, non-proteolytics B and E with a sensitivity range of 5-10 pg/mL.[5,21] In addition, an increasing number of rapid methods are being proposed in the form of gene probes and especially polymerase chain reaction (PCR) tests for the detection of *C. botulinum* in both culture media and foods.[1,7-9,11,25,26]

33.8 IDENTIFYING BOTULINUM TOXIN IN CLINICAL SPECIMENS

33.81 Toxin Neutralization Tests

Botulinum toxins in clinical specimens are identified by injecting mice with the suspect specimen alone, and the same specimen mixed with one or more botulinum antitoxins. If botulinum toxin is present in sufficient quantity to be determined, mice receiving the un-neutralized toxin will die, and mice receiving the toxin neutralized by the specific antitoxin will survive. Although botulinum toxin kills mice within 6 to 24 hr after injection, death may be delayed if the quantity of toxin is near the minimum lethal concentration. Therefore, the animals should be observed for 48 hr and kept another 2-3 days before recording as negative.[6,12,22,27]

33.82 Identification of Botulinum Toxin in Serum

Prepare serum-antitoxin mixtures as follows:

Patient's serum	*Normal serum or antitoxin*
1.2 mL	0.3 mL normal serum
1.2 mL	0.3 mL anti-A
1.2 mL	0.3 mL anti-B
1.2 mL	0.3 mL anti-E
1.2 mL	0.3 mL anti-F
1.2 mL	0.3 mL polyvalent (anti-A,B,C,D,E,F)

If limited volume of serum is available, only anti-A and B, together with the normal serum control, should be tested initially.

Incubate mixtures in a 37°C water-bath for 30 min.

Inject 0.5 mL of the serum-antitoxin mixtures IP into each of 2 mice (16 to 24 g) for each test mixture. Each mouse, therefore, receives 0.4 mL of the patient's serum and 0.1 mL of normal serum or antitoxin.

Observe mice at intervals for 48 hr, note any clinical signs, and record any deaths.

If no toxin is found by using 0.4 mL of patient's serum per mouse, repeat the test using larger mice (24 to 30 g) injected with 0.8 mL of patient's serum mixed with 0.1 mL of normal serum per mouse. To do this, add 2.4 mL of patient's serum to the same amount of normal serum or antitoxin used before and inject each mouse with 0.9 mL of the mixture using the same scheme as outlined above. Do not inject mice with a total volume larger than 1.0 mL since excessive amounts of normal serum can cause death. Trypsinization of the serum is not necessary for demonstration of toxin.

If all the mice injected with the test mixtures described above develop signs suggestive of botulism and die, it is possible that type G, or an unidentified toxin is involved. In this case, the patient's serum should be submitted to a reference laboratory for additional tests. At present, type G antitoxin is not available commercially.

If mice were protected only by neutralization of the toxin with the polyvalent antitoxin, repeat the test using monovalent type C, type D, and type F, the polyvalent antitoxin, and the normal test mixtures. If the mice receiving the polyvalent antitoxin mixture are the only survivors again, test the other combinations of monovalent antitoxins to determine if multiple toxin types are present. To estimate the quantity of botulinum toxin in the patient's serum, prepare dilutions (1:2, 1:4, etc.) with gel phosphate buffer and determine the MLD of toxin. Inject groups of mice with each dilution of the patient's serum, 0.5 mL per mouse.

33.83 Identifying Botulinum Toxin in Feces

Experience from the investigation of botulism outbreaks has revealed that examination of feces for botulinum toxin is a valuable diagnostic procedure. This is especially true of infant botulism.

Homogenize 10 to 15 g of feces (as much as can be obtained from infants) in an equal quantity (wt/vol) of gel phosphate buffer with a chilled mortar and pestle. Cover and refrigerate at 4°C for 6 to 8 hr to extract the toxin.

After refrigeration, centrifuge (12,000 × g) at 4°C for 20 min and remove the liquid for a toxin neutralization test. If the extract is not clear, clarify by repeating the centrifugation procedure.

Prepare the following mixtures:

Feces from adults

Extract of feces	*Treatment*
1.0 mL	none
1.0 mL	heated*
1.0 mL	0.25 mL 0.5 % trypsin
1.0 mL	0.25 mL anti-A
1.0 mL	0.25 mL anti-B
1.0 mL	0.25 mL anti-E
1.0 mL	0.25 mL anti-F
1.0 mL	0.25 mL polyvalent (anti-ABCDEF)

Feces from Infants

Extract of feces	*Treatment*
1.0 mL	none
1.0 mL	heated*
1.0 mL	0.25 mL 0.5 % trypsin
1.0 mL	0.25 mL anti-A
1.0 mL	0.25 mL anti-B
1.0 mL	0.25 mL polyvalent

*A portion of the extract is heated in a boiling water bath for 10 min; botulinum toxin is heat–labile.

Incubate the mixture in a 37°C water bath for 30 min.

Inject 0.5 mL of the extract-trypsin and extract-antitoxin mixtures and 0.4 mL of the others IP into 2 mice (16 to 24 g) for each test mixture.

If toxin is demonstrated only with the trypsinized specimen, repeat the neutralization tests using trypsinized extract. Add 1.5 mL of 0.5 % trypsin to 6.0 mL of extract, and incubate the mixture at 37°C for 30 min.

Prepare the following extract-antitoxin mixtures:

Extract	*Antitoxin*
1.25 mL	0.25 mL anti-A
1.25 mL	0.25 mL anti-B
1.25 mL	0.25 mL anti-E
1.25 mL	0.25 mL anti-F
1.25 mL	0.25 mL anti-ABCDEF

Incubate and inject 2 mice per mixture as before, but give each mouse 0.6 mL.

Observe the mice at intervals for 48 hr, note clinical symptoms of botulism and record deaths and survivors. Mice injected with the heated extract should survive. Interpret results of the tests and perform repeat tests, if required, as described above, for identification of botulinum toxin in serum.

33.84 Miscellaneous Clinical Specimens

Tests for botulinum toxin in cerebrospinal fluid, urine, and other body fluids are performed as described for serum (Section 33.82) after centrifugation, if necessary, to clarify the liquids.

Dilute vomitus and gastric washings 1:2 with gel phosphate buffer (Chapter 63) before clarification by centrifugation and perform the toxin neutralization tests.

Grind tissues with an equal volume of gel phosphate buffer using a sterile chilled mortar and pestle. Cover and refrigerate at 4°C for 12 to 18 hr, after which you centrifuge the mixture at 12,000 × g and 4°C. Remove the clear liquid and perform a toxin neutralization test as described for feces (Section 33.83). Omit the heated extract serum mixture.

33.85 Preparing Fecal Cultures

Enrichment cultures from feces are prepared as described for foods in Section 33.66, items 1 and 2, by inoculating 0.5 to 1.0 mL of the chilled homogenate from Section 33.83 into 2 tubes of CMM, one of which is cooled as described and the other equilibrated at 80°C before inoculation. Hold the latter at 80°C for 10 min after inoculation. Incubate both cultures at 35°C for 5 days.

1. Toxin testing. Test for toxin in fecal cultures and type as in Sections 33.72 and 33.73 or 33.83.
2. At the time enrichment cultures are tested for toxicity, or when sporulation occurs, streak cultures on *C. botulinum* isolation agar (CBI). To obtain well-isolated colonies, dilute enrichment culture with gel phosphate buffer, if necessary.
3. Incubate as in Section 33.67-4.
4. Select several well-isolated colonies, test for toxicity, type, and check for purity as in Section 33.67-5.

33.9 INTERPRETING DATA

Laboratory tests required in an actual or potential botulism outbreak are designed to identify botulinum toxin and/or the organism in foods and clinical specimens. Toxin in a food means that the product, if consumed without thorough heating, could cause botulism. In clinical specimens, it means that a clinical diagnosis of botulism is consistent with the laboratory findings. The presence of viable *C. botulinum* but no toxin in specimens has a different meaning. In foods, the observation by itself is not proof that the food in question caused botulism, and in clinical specimens it does not necessarily mean that the patient has botulism. The presence of toxin in the food is required for an outbreak of botulism to occur. Ingested organisms may be found in the alimentary tract, but they are considered unable to multiply and produce toxin in vivo except in infants and in a very few adults who have received antibiotics. Assuming proper handling of specimens to prevent toxin inactivation, failure to find toxin means that a discernible level is not present.

Presence of botulinum toxin and/or organisms in low-acid (i.e. above pH 4.6) canned foods means that the items were underprocessed or were contaminated through post-processing leakage. These failures occur more frequently in home processed than in commercially canned foods. Swollen containers are more likely than intact containers to contain botulinum toxin since the organism produces gas during growth. The rare occurrence of toxin in a flat can, may imply that the seams were loose enough to allow gas to escape. Botulinum toxin in canned foods is usually type A or proteolytic type B, since spores of the proteolytic strains can be among the more heat-resistant of bacterial spores. Spores of the non-proteolytics, types B, E, and F generally have low heat resistance, and do not normally survive even mild heat treatment, but can be present in canned foods through post-processing leakage.

The protection of mice from botulism and death with one of the monovalent botulinum antitoxins confirms the presence of botulinum toxin and determines the serological type of the toxin. However, there is a slight degree of cross-neutralization between types E and F. For example, type E antitoxin at a concentration of 1000 anti-MLD will neutralize 2 or 3 MLD of type F toxin. Therefore, if low levels of type F toxin are present, mice protected with monovalent type E and those protected with monovalent type F antitoxin may all survive. It may also happen, if low levels of toxin are present, that the food may be toxic on initial testing, but may be nontoxic on subsequent testing.

If the mice are not protected by one of the monovalent botulinum antitoxins, then:

1. there may be too much toxin in the food,
2. there may be more than one type of toxin,
3. death may be due to some other cause.

This situation requires retesting at a higher dilution of toxin and the use of mixtures of A, B, C, D, E, and F antitoxins in place of the monovalent antisera.

If both the heated and unheated supernatant fluids are lethal to mice, the deaths are probably not due to botulinum toxin. Yet, it is possible that a toxic substance not destroyed by heat may mask the presence of botulinum toxin, which is destroyed by heat. If the botulinum toxin is present in large quantities, further dilutions may eliminate the heat-stable toxic substance and allow the botulinum toxin to be detected.

When injecting mice to determine the presence and type of botulinum toxin, the following are helpful in interpreting results:

1. Mice injected with botulinum toxin may become hyperactive before symptoms of botulism occur.
2. The first 24 hr after injections are critical since 90% to 95% of mice will die within this time period. Typical symptoms of botulism (ruffled hair, labored breathing, weakness of limbs, paralysis of hind legs, total paralysis with gasping for breath) and death may occur within 4 to 6 hr.
3. Unless typical botulism symptoms are evident, death of mice occurring after 24 hr, or only in those receiving the 1:5 but not any higher dilution of extract, may be due to non-specific causes.

ACKNOWLEDGMENTS

The authors are indebted to Donald A. Kautter (deceased), Richard K. Lynt, Donald E. Lake, and Daniel C. Mills, past authors of this chapter, for segments incorporated in this edition. We thank James McGee of California State Health Department for valuable suggestions made in reviewing this chapter.

33.10 REFERENCES

1. Aranda, E., M. M. Rodriguez, M. A. Asensio, and J. J. Cordoba, 1997. Detection of *Clostridium botulinum* types A, B, E and F in foods by PCR and DNA probe. Lett. Appl. Microbiol. 25:186-190.
2. Arnon, S. S. 1998. Infant botulism, p. 1570-1577. *In* R. D. Feigen and J. D. Cherry (eds.), Textbook of pediatric infectious diseases, 4th ed. W.B. Saunders, Philadelphia, Pa.
3. Arnon, S. S. 1995. Botulism as an intestinal toxemia, p. 257-271. *In* M. J. Blaser, P. Smith, J. I. Ravdin, H. B. Greenberg, and R. L. Guerrant (eds.), Infections of the gastrointestinal tract. Raven Press, New York.
4. Arnon, S. S. 1993. Clinical trial of human botulism immune globulin, p. 477-482. *In* B. R. DasGupta (ed.), Botulinum and tetanus neurotoxins: Neurotransmissions and biomedical aspects. Plenum Press, New York.
5. Doellgast, G. J., M. X. Triscott, G. A. Beard, J. D. Bottoms, T. Cheng, B. H. Roh, M. G. Roman, P.A. Hall, and J. E.Brown, 1993. Sensitive enzyme-linked immunosorbent assay for detection of *Clostridium botulinum* neurotoxins A, B, and E using signal amplification via enzyme-linked coagulation assay. J. Clin. Microbiol. 31:2402-2409.
6. Dowell, V. R., and T. M. Hawkins. 1974. Laboratory methods in anaerobic bacteriology, CDC Laboratory Manual. PHS Publ. No1803, U.S. Dept. of Health, Ed., and Welfare. U.S. Public Health Serv., Washington, D. C.
7. Fach, P., D. Hauser, J. P. Guillou, and M. R. Popoff.1998. Polymerase chain reaction for the rapid identification of *Clostridium botulinum* type A strains and detection in Food samples. J. Appl. Bacteriol. 75:234-239.
8. Fach, P., M. Gilbert, R. Griffais, J. P. Guillou, and M. R. Popoff. 1995. PCR and gene probe identification of botulinum neurotoxin A, B, E, F, and G-producing Clostridium spp. and evaluation in food samples. Appl. Environ. Microbiol. 61:389-392.
9. Ferreira, J. L., M. K. Hamdy, S. G. McCay, M. Hemhill, N. Kirma, B. R. Baumstark, 1994.Detection of *Clostridium botulinum* type F using the polymerase chain reaction. Mol. Cell Probes 8:365-373.
10. Ferreira, J. L. 1997.ORA/NCTR/CDC, Initiative for Development of an ELISA Method for the Detection of type A, B, E, and F *Clostridium botulinum* Toxin. Food and Drug Administration. LIB#4093.
11. Franciosa, G., J. L. Ferreira, and C. L. Hatheway, 1994. Detection of type A,B, and E botulism neurotoxin genes in *Clostridium botulinum* and other *Clostridium* species by PCR: Evidence of unexpressed type B toxin genes in type A toxigenic organisms. J. Clin. Microbiol. 63:1911-1917.
12. Hatheway, C. L. 1979. Laboratory procedures for cases of suspected infant botulism. Rev. Infect.Dis.1:647-651.
13. Hatheway, C. L., and L. McCroskey.1987.Examination of feces and serum for diagnosis of infant botulism in 336 patients. J. Clin. Microbiol. 25:2334-2338.
14. Hauschild, A. H. W., R. Hilsheimer, K. F. Weiss, and R. B. Burke. 1988. Clostridium botulinum in honey, syrups, and dry infant cereals. J. Food Prot. 51:892-894.
15. Herzberg, M. (ed.). 1970. Toxic microorganisms: Mycotoxins, Botulism. Proceedings of the 1st U.S.-Japan Comp. Prog. *In* Natural resources. U.S. Dep. of the Interior, Washington, D. C.
16. Ingram, M., and T. A. Roberts (eds.).1967. Botulism 1966. Chapman and Hall. London, England.

17. International Commission on Microbiological Specifications for Foods. 1978. Microorganisms in Foods 1, 2nd University of Toronto Press, Toronto, Canada.
18. Lewis, K. H., and K. Cassel, Jr.(eds.). 1964. Botulism, Proceedings of a Symposium. U.S. Dept. of Health, Ed., and Welfare. Public Health Serv., Washington, D. C.
19. Mills, D. C., T. F. Midura, and S. S. Arnon. 1985. Improved selective medium for the isolation of lipase-positive *Clostridium botulinum* from feces of human infants. J. Clin. Microbiol. 21:947-950.
20. Paisley, J. W., B. A. Lauer, and S. S. Arnon. 1995. A second case of infant botulism caused by *Clostridium baratii.* Pediatr. Infect. Dis. J. 14:912-914.
21. Roman, M. G., J. Y. Humber, P. A. Hall, N. R. Reddy, H. M. Solomon, M. X. Triscott, G. A. Beard, J. D. Bottoms, T. Cheng, and G. J. Doellgast. 1994. Amplified immunoassay ELISA-ELCA for measuring *Clostridium botulinum* type E neurotoxin in fish fillets. J. Food Prot. 57:985-990.
22. Silas, J. C., J. A. Carpenter, M. K. Hamdy, and M. A. Harrison. 1985. Selective and differential medium for detecting *Clostridium botulinum.* Appl. Environ. Microbiol. 50:1110-1111.
23. Smith, L. D. S., and H. Sugiyama. 1988. Botulism: The organism, its toxin, the disease, 2nd ed. Charles C. Thomas, Springfield, Ill.
24. Stumbo, C. R.1973. Thermobacteriology in food processing, 2nd ed. Academic Press, New York.
25. Szabo, E. A., J. M. Pemberton, and P. M. Desmarchelier.1993. Detection of the genes encoding botulinum neurotoxin types A to E by the polymerase chain reaction. Appl. Environ. Microbiol. 59:3011-3020.
26. Takeshi, K., Y. Fujinaga, K. Inoue, H. Nakajima, K. Oguma, T. Ueno, H. Sunagawa, and T. Ohyama. 1996. Simple method for detection of *Clostridium botulinum* type A to F neurotoxin genes by polymerase chain reaction. Microbiol. Immunol. 40(1):5-11.
27. U.S.HEW/PHS/CDC.1979.Botulism in the United States 1899-1977. Publ. No.(CDC)74-8279, U.S. Dep. of Health, Ed., and Welfare, Public Health Serv., Washington, D. C. plus the incidence reports presented at the annual meetings of the Interagency Botulism Research Coordinating Committee (IBRCC).
28. U.S.HHS/PHS/FDA. 1995. Bacteriological analytical manual, 8th ed., Assoc. of Off. Anal. Chem., Gaithersburg, Md.

Chapter 34

Clostridium perfringens

Ronald G. Labbe

34.1 INTRODUCTION

The methods included in this chapter are useful for quantitation of *Clostridium perfringens* in food or feces and for detection and quantitation of *C. perfringens* enterotoxin in feces of food poisoning patients. The culture methods are generally accepted as the most effective for this purpose. They are in essential conformance with those adopted as official by the Association of Official Analytical Chemists (AOAC) and by the International Standards Organization. The reversed passive latex agglutination (RPLA) method for enterotoxin has been evaluated extensively in the United Kingdom[5] and in the United States,[13,18] but a full-scale collaborative study has not yet been done.

34.11 Food Poisoning and Enterotoxin Formation

Clostridium perfringens food poisoning is one of the most common types of human foodborne illness.[25] The foods usually involved are cooked meat or poultry products containing large numbers of viable bacterial cells. A heat-labile enterotoxin produced only by sporulating cells[7] induces the major symptom of diarrhea in perfringens poisoning. The enterotoxin appears to be released *in vivo* in the intestine by the sporulating organism. Although the enterotoxin is generally not preformed in the food, the foods in which conditions are favorable for sporulation may contain enterotoxin.[6,30]

34.12 Importance of Cell Numbers

Clostridium perfringens is not uncommon in raw meats, poultry, dehydrated soups and sauces, raw vegetables, and certain other foods or food ingredients. Thus, its mere presence in foods may be unavoidable. In food poisoning outbreaks, demonstration of hundreds of thousands or more organisms per gram in a suspect food supports a diagnosis of perfringens poisoning when substantiated by clinical and epidemiological evidence. The value of enumerating *C. perfringens* spores in feces of food poisoning patients as a means of confirming outbreaks has also been confirmed.[14]

Clostridium perfringens cells may lose viability if suspect foods are frozen or held under prolonged refrigeration before analysis, thereby making it difficult to incriminate the organism in food poisoning outbreaks.[38] In such cases, Gram-stained smears of food remnants of homogenates should be examined for the presence of large cell bacilli typical of *C. perfringens*. Foods responsible for food poisoning outbreaks usually contain a large enough number of cells to be readily detectable by direct microscopic examination.

Spores of different strains of *C. perfringens* may vary widely in their heat resistance. Some may withstand 100°C for an hour or more, whereas others are inactivated by a few minutes or less at the same temperature. In most environments, heat-sensitive strains outnumber heat-resistant strains. Both heat-resistant and heat-sensitive strains may cause food poisoning. Since *C. perfringens* does not sporulate in food, or does so only rarely, food should not be heated before the organism is enumerated. In feces only elevated levels of spores of *C. perfringens* are of diagnostic value. Heating a suspension of feces is recommended for detection of *C. perfringens* to eliminate competitive organisms such as members of the family *Enterobacteriaceae*.

34.13 Selective Differential Media

Several solid media have been devised for quantitation of *C. perfringens*, including neomycin blood agar,[37] sulfite polymyxin sulfadiazine (SPS) agar,[3] tryptone-sulfite neomycin (TSN) agar,[27] Shahidi-Ferguson perfringens (SFP) agar,[34] D-cycloserine-blood agar,[9] oleandomycin-polymyxin-sulfadiazine-perfringens (OPSP) agar,[8] tryptose-sulfite-cycloserine (TSC) agar,[16] egg yolk-free tryptose sulfite cycloserine (EY-free TSC) agar,[20] and trypticase-soy-sheep blood (TSB) agar.[15] The selectivity of these media is derived from the incorporation of one or more antibiotics that inhibit certain anaerobes or facultative anaerobes. With the exception of the blood agars, the media contain iron and sulfite. Clostridia reduce the sulfite to sulfide which reacts with the iron to form a black iron sulfide precipitate. Black colonies are presumptive *C. perfringens* and must be confirmed by additional tests.

The selectivity of TSN and SPS media results in the inhibition of some strains of *C. perfringens*. SPS also is unsatisfactory because many strains fail to form colonies that are characteristic black. Although the selectivity of SFP and neomycin blood agars is limited, these media may be adequate when *C. perfringens* is the predominant organism. The selectivity of OPSP agar also may be of limited use with some facultative anaerobes. D-cycloserine blood agar may be useful for the selective isolation of *C. perfringens*, although has not been tested for routine isolation of the organism from foods. TSC agar, or its modified form, EY-free TSC agar, has been documented as the most useful of the media for quantitative recovery of *C. perfringens*, with adequate suppression of the growth of practically all facultative anaerobes.[11,21] SFP, OPSP, and TSC also contain egg yolk for differential purposes. The lecithinase of *C. perfringens* hydrolyzes egg yolk lecithin and produces an opaque halo around the black colonies. However, other sulfite-reducing clostridia or other facultative anaerobes may produce a similar reaction. In some instances the egg yolk reaction of *C. perfringens* alpha-toxin may be masked by other organisms. In addition, false-negative *C. perfringens* colonies without detectable halos may occur on the plates.[20] EY-free TSC agar which is not dependent on alpha-toxin production for its differential utility is an improvement over TSC agar.[20,22]

TSC agar contains egg yolk and must be used for surface plating. EY-free TSC agar is used in pour plates. The methods described here for quantitation of *C. perfringens* use TSC agar or EY-free TSC agar. The EY-free TSC may give results as good as or better than those of TSC agar. For outbreak stool samples, the EY-free TSC and TSB agars are superior to elevated-temperature (46°C) most probable number (MPN) methods for enumeration of *C. perfringens* spores.[15]

34.14 Rapid Methods for Detecting C. perfringens Enterotoxin

A number of serological assays have been reported for the rapid detection of enterotoxin. The most rapid methods available are the enzyme-linked immunosorbent assays (ELISA[4,23,24,29,32,33,39]) and RPLA,[13] marketed by Oxoid U.S.A., Inc., Columbia, Md. 21045. A rapid and inexpensive slide latex agglutination assay has also been reported,[28] but requires the coating of latex beads with immunoglobulin. The RPLA method is easier to perform than the ELISA, which requires special test reagents and equipment which may not be generally available.[5,18]

To determine the enterotoxigenicity of *C. perfringens* from food or feces, it is necessary to induce sporulation of the organism. A number of sporulation media have been proposed. The two recommended here are the modified[14] AEA medium of Taniguti[36] and the modified[14] medium of Duncan and Strong.[8,26]

34.15 Enumeration

A MPN procedure is called for in the case of routine sampling where low numbers are expected. Iron milk medium (IMM), consisting of pasteurized whole milk with 2% iron powder or ferrous sulfate, is simple to prepare, inexpensive, and relatively sensitive.[1] Selection is based on the rapid growth of *C. perfringens* at 45°C and the typical "stormy fermentation" reaction due to the production of an acid curd (lactic acid fermentation) with subsequent disruption of the curd by gas. Similar counts of *C. perfringens* were obtained from food samples using TSC medium or IMM medium. Procedures for MPN determination are given in Chapter 6. Presumptive positives from IMM are confirmed following plating on TSC agar.[2] In the case of shellfish the IMM method has been adopted first action by AOAC International.

34.2 SAMPLING

For routine sampling of foods, *C. perfringens* spores or vegetative cells are enumerated by direct plating. In cases of outbreaks, fecal samples are examined for *C. perfringens* enterotoxin by RPLA and for *C. perfringens* spores by enumeration. Remnants of foods are examined microscopically to determine the type of analysis to be performed.

34.21 Maintaining Viability of Vegetative Cells

Generally it is recommended that outbreak food samples to be tested for *C. perfringens* be analyzed immediately, or refrigerated and tested as soon as possible, but not frozen. Loss of viability of some strains that occurs during refrigeration may be even greater when the cells are frozen. Foods that must be stored for more than 48 hr or shipped to the laboratory should be treated with buffered glycerol salt solution to give a 10% final concentration of glycerol and stored frozen at –55° to –60°C until the sample is analyzed. Treated samples shipped with dry ice show a minimal loss of viability of *C. perfringens*.[19] Fecal samples can be stored at –20°C for several months with only minimal reductions in the spore count.[17]

34.3 EQUIPMENT AND SUPPLIES

34.31 Isolation and Quantitation

Air incubator 35 to 37°C.

Anaerobic containers or anaerobic incubator with equipment and materials for obtaining anaerobic conditions. These may be anaerobic devices in which the air is replaced 3 or 4 times with 90% N_2 + 10% CO_2, or those in which oxygen is catalytically removed.

Colony counter with a piece of white tissue paper over the counting background area to facilitate counting black colonies.

Sterile blender jar, container and motor, or Stomacher, or sterile mortar and pestle and sterile sand. The blender or Stomacher is preferable.

Vortex mixer

Water bath

34.32 Commercial Tests for Quantitation of C. perfringens Enterotoxin

A kit for the detection of *C. perfringens* enterotoxin by reversed passive latex agglutination in feces and culture broths is available from Oxoid U.S.A. (Columbia, Md. 21045). An ELISA kit for the detection of enterotoxin is available from Tech Lab (Blacksburg, VA 24060). Additional materials required for performing these tests include the following:

- Microtiter plate (V-type)
- Dropper (25 µL) or micropipet
- Diluter (25 µL)
- Centrifuge

34.4 SPECIAL REAGENTS AND MEDIA

34.41 Isolation and Quantitation

Chopped liver broth

Cooked meat medium (CMM) (Difco)

Fermentation medium

Fluid thioglycollate medium (Difco)

Iron Milk Medium

Lactose gelatin

Modified AEA sporulation medium

Modified Duncan Strong (DS) medium

Motility-nitrate medium

Nitrate reduction reagents (Method 1)

0.1% peptone water diluent

Phosphate-buffered saline (PBS)

Physiological saline (0.85% sodium chloride)

TSC agar, or EY-free TSC agar (Oxoid) (Same as SFP Agar Base [Difco] plus 0.04% cycloserine)

Trypticase peptone glucose yeast extract broth (buffered)

34.5 RECOMMENDED CONTROLS

34.51 Direct Quantitation

For the selective differential media for quantitation of *C. perfringens*, use a control strain of the organism to validate the performance of the media. This control strain may also be useful when typical *C. perfringen* colonies have not been observed on the medium.

34.52 Reversed Passive Latex Agglutination

Sensitized and control latex, control enterotoxin and diluent are included in the Oxoid commercial kit.

34.6 PRECAUTIONS AND LIMITATIONS OF METHODS

34.61 Isolation and Quantitation

34.611 TSC and EY-free TSC

Both TSC and EY-free TSC appear to be suitable for enumeration of *C. perfringens*. However, some strains of *C. perfringens* may not produce distinguishable halos via the egg yolk reaction on TSC agar. Therefore, the absence of a halo around a black colony does not eliminate the possibility of the strain being *C. perfringens*. The halo of one colony also may be masked by the halo of another colony.

34.612 Contamination with Other Sulfite-Reducing Clostridia

Other sulfite-reducing clostridia that produce black colonies and are egg yolk positive can grow in TSC and EY-free TSC agar, including *Clostridium bifermentans, C. botulinum, C. paraperfringens (C. baratii), C. sardiniense (C. absonum)*, and *C. sporogenes*.

34.613 Incorporation of D-cycloserine

Enterococci may be present in high numbers in some foods. In media other than those containing D-cycloserine, the overgrowth of these organisms may interfere with or prevent the isolation of *C. perfringens*. Incorporating D-cycloserine into TSC and EY-free TSC agar effectively inhibits growth of most enterococci.

34.614 Confirmation of the Presumptive Plate Count

Presumptive sulfite-reducing colonies of *C. perfringens* from the selective differential media have often been confirmed by their nonmotility and their ability to reduce nitrate to nitrite. A variety of clostridia have these properties, including *C. celatum*,[20] *C. paraperfringens* (*C. baratii*), and *C. sardiniense* (*C. absonum*).[30] These species usually can be distinguished from *C. perfringens* by their inability to liquefy gelatin with 44 hr in lactose gelatin medium and by their inability to produce acid from raffinose within 3 days.

34.62 Reversed Passive Latex Agglutination

Because of interference from other components of the extract, it is often necessary to dilute the fecal extract 1:60 or more before enterotoxin can be quantitated.

34.7 PROCEDURE

34.71 Isolation and Quantitation

In some instances, it may be indicated to attempt to isolate *C. perfringens* from food samples contaminated with a very low number of cells. An enrichment procedure using chopped liver broth or trypticase peptone glucose yeast extract broth (buffered) may be used. Inoculate about 2 g food sample into 15 to 20 mL of medium. Incubate the samples at 35° to 37°C for 20 to 24 hr. Positive tubes show turbidity and gas production. Streak TSC agar plates containing egg yolk to obtain presumptive *C. perfringens*.

34.72 Anaerobic Total Plate and Spore Count

1. **Preparation of food and fecal homogenate.** Blend for 2 min at slow speed (or homogenize with stomacher or macerate with sterile sand) a 10 to 20 g food sample with 0.1% peptone to obtain a 1:10 dilution. For fecal specimens, homogenize 1.0 g (or 1 mL liquified stool) in 9 mL of 0.1% peptone in screw cap tubes. Homogenize on a vortex mixer. For spore counts, heat homogenates in screw cap tubes in a water bath at 75°C for 20 min before diluting and plating.
2. **Preparation of dilutions.** Prepare serial decimal dilutions (through at least 10^{-7}) using 0.1% peptone dilution blanks.
3. **Plating procedures.** Make duplicate spread platings of each dilution using 0.1 mL amounts on TSC agar. After the agar has dried slightly, overlay the surface with 5 mL (or more) of EY-free TSC agar. If EY-free TSC is used as the plating medium, make duplicate pour plates solidify, cover with an additional 5 mL (or more) of EY-free TSC agar.
4. **Incubation.** Incubate the plates upright and anaerobically for 18 to 24 hr at 35° to 37°C.
5. **Presumptive *C. perfringens* plate count.** Select plates containing, preferably, 20 to 200 black colonies, which may be surrounded by a zone of precipitate on the TSC agar but not on the EY-free TSC agar. Count all black colonies and calculate the average number of colonies in the duplicate plates.

34.73 Confirmation of C. perfringens

Select five representative black colonies (10 for official analyses) from TSC agar and stab inoculate motility-nitrate and lactose gelatin media in parallel using a stiff inoculating needle with a hook at the tip. Transfer colonies from crowded or contaminated plates to fluid thioglycollate medium. Incubate 18 to 20 hr at 35°C and streak on TSC agar to obtain pure cultures before proceeding with confirmation.

1. **Obtaining pure cultures.** Inoculate a portion of each selected black colony into a tube of buffered TPGY broth or fluid thioglycollate medium. Incubate for 4 hr in a water bath at 46°C or overnight at 35° to 37°C. After incubation examine typical colonies microscopically for the presence of large gram-positive rods typical of *C. perfringens*. Endospores are not produced in this medium. Streak the culture onto TSC agar and incubate anaerobically for 24 hr at 35°C to obtain isolated colonies. Typical colonies are yellowish gray, 1 to 2 mm in diameter, usually surrounded by an opaque zone caused by lecithinase production. These colonies then may be picked and inoculated into fluid thioglycollate medium. For longer term storage, cultures can be grown and kept frozen in Difco CMM.
2. **Motility nitrate reduction test.** Stab inoculate each fluid thioglycollate medium culture into motility nitrate medium. The medium recommended contains 0.5% each of glycerol and galactose to improve the consistency of the nitrate reduction reaction with different strains of the organism.[20] Incubate the inoculated medium at 35° to 37°C for 24 hr. Check for motility. Since *C. perfringens* is nonmotile, growth should occur only along the line of inoculum and not diffuse away

from stab. Test for reduction of nitrate to nitrite. A red or orange color indicates reduction of nitrate to nitrite. If no color develops, test for residual nitrate by addition of powdered zinc. A negative test (no violet color) after zinc dust is added indicates that nitrates were completely reduced. A positive test after addition of zinc dust indicates that the organism cannot reduce nitrate.

3. **Lactose gelatin medium.** Stab inoculate suspect colony into lactose gelatin medium. Incubate at 35° to 37°C for 24 to 44 hr. Lactose fermentation is indicated by gas bubbles and a change in color of the medium from red to yellow. Gelatin usually is liquefied by *C. perfringens* within 24 to 44 hr.[20]
4. **Carbohydrate fermentation.** Subculture isolates that do not liquify gelatin within 44 hr, or are atypical in other respects, into fluid thioglycollate medium. Incubate the cultures for 18 to 24 hr at 35° to 37°C, make gram-strained smear, and check for purity. If pure, inoculate a tube of fermentation medium containing 1% salicin and a tube of fermentation medium containing 1% raffinose with 0.15 mL of thioglycollate culture of each isolate. Incubate inoculated media at 35° to 37°C for 24 hr and check for production of acid. To test for acid, transfer 1 mL of culture to a test tube or spot plate and add 2 drops of 0.04% bromthymol blue. A yellow color indicates that acid has been produced. Reincubate cultures for an additional 48 hr and retest for the production of acid. Salicin is rapidly fermented with the production of acid by culturally similar species such as *C. paraperfringens (C. baratii), C. sardiniense (C. absonum)* and *C. celatum*, but usually not by *C. perfringens.* Acid is produced from raffinose within 3 days by *C. perfringens* but is not produced by culturally similar species. The typical cultural reactions of each species are shown in Table 1.

34.74 Detection of Enterotoxin in Culture Supernatants by Reversed Passive Latex Agglutination.

Subculture isolates to be tested for enterotoxin in CMM (Difco) and incubate for 1 or 2 days at 37°C. Mix the cooked meat culture with a Vortex mixer and transfer 2 to 3 drops to 10 mL of freshly steamed fluid thioglycollate medium. Heat the inoculated medium in a water bath at 75°C for 10 min and incubate for 18 hr at 37°C. Subculture 0.5 mL of this culture in fresh fluid thioglycollate and incubate for 4 hr at 37°C. Use the 4 hr subculture to inoculate (1% to 5% vol/vol) 15 mL of modified AEA sporulation medium or modified DS medium. Incubate the inoculated spore broth for 18 to 24 hr at 37°C. For best results, incubate AEA medium in an anaerobic jar or incubator. Check the resulting culture for spores by using a phase-contrast microscope or by examining stained smears. Fewer than 5 mature spores per field is not considered adequate sporulation. Retest such cultures in another sporulation medium. Centrifuge a portion of the sporulated culture for 15 min at 10,000 × g and test the cell-free culture supernatant for enterotoxin.

The procedures, including controls for detecting and quantitating enterotoxin by RPLA, are those specified by the manufacturer of the test kit. Note, however, that the single sporulation medium recommended by the manufacturer may not result in sporulation of all isolates. Enterotoxin levels as low as 2 ng/mL can be reliably detected using this method. It is usually indicated to screen samples for the presence of enterotoxin over a broad range of dilutions before attempting to quantitate it precisely, e.g., a tenfold dilution series to 10^{-6}. Culture supernatants are usually positive at dilutions from 10^{-3} to 10^{-6}.

To screen samples for enterotoxin, make a series of tenfold dilutions of the sample in the diluent provided with each kit. Transfer 25 µL of each dilution to two separate wells in adjacent rows of a "V" type microtiter plate. Add 25 µL of sensitized latex beads (coated with specific antibody to enterotoxin) to each well of the first row and 25 µL of control latex to the adjacent row. Mix the contents using a minishaker (Dynatech Corp., Alexandria, Va. 22314) or by stirring with clean flat toothpicks. Avoid spilling or mixing the contents of different wells. Cover the plate to minimize evaporation and incubate at room temperature (22° to 24°C) for 24 hr. If enterotoxin is present, it binds to specific antibodies on the sensitized latex to give an agglutination pattern that can be scored from 1+ to 3+. The values obtained can be used to estimate the amount of enterotoxin in the sample. Control latex beads coated with normal immunoglobins usually does not react with substances in the test sample, so the beads sediment to a compact pellet at the bottom of the test wells. A positive result with the sensitized latex and a negative result with control latex is an indication that *C. perfringens* enterotoxin is present in samples. Although not usually required for routine work, the enterotoxin can be quantitated with an accuracy of approximately 50% by assuming that the sensitivity of the RPLA test is about 2 ng/mL

Table 1. Characteristics of *Clostridium perfringens* and Phenotypically Similar Species[a]

		Motility-nitrate medium		Lactose-gelatin medium		Fermentation medium	
Species	Number of strains	Motility	Nitrite	Acid/gas	Gelatin liquefied	Salicin (24 h)	Raffinose (72 h)
C. perfringens Type A	38	–	4+	AG/T	+(48 h)	–	A
C. perfringens Type A	3	–	4+	AG/T	+(48 to 72 h)	(AG)	A
C. absonum	4	+	(+)	AG/CS	–	AG	–
C. baratti	2	–	3+	AG/CS	–	AG	–
C. celatum	2	–	2+	A/CS	–	A	–
C. paraperfringens	10	–	3+	AG/CS	–	AG	–
C. sardiniense	8	±	(+)	AG/CS	–	AG	–

[a] Determined using buffered supplemented motility-nitrate medium,[20] and Spray's fermentation medium.[35] A = acid; AG = acid and gas; CS = clear with sedimented cells; + = positive; (+) = weak; – = negative.

of sample. After determining the approximate end point of activity by screening samples, repeat the assay using a two-fold series of dilutions within the appropriate range. The reciprocal of the highest dilution that yields a positive reaction with the sensitized latex and a negative result with the control latex divided by 2000 gives the indicated amount of enterotoxin in µg/mL of undiluted extract or culture fluids. If more precise quantitation is needed, use an enterotoxin standard whose reactivity has been determined by other serological assays to determine the specific activity of the sensitized latex supplied with the test kit.

34.75 Detection of Enterotoxin in Feces by Reversed Passive Latex Agglutination

Prepare fecal extracts as described in Section 37.72, except centrifuge homogenized sample for 30 min at 15,000 × g to remove insoluble solids.

The procedures are those specified by the manufacturer of the RPLA kit. Prepare serial twofold dilutions of the test sample. However, to conserve reagents, conduct a preliminary trial using fecal extracts at dilutions of 1:10, 1:100, and 1:1000 using the diluent provided in the test kit. Also see 37.62 above.

34.8 INTERPRETATION OF DATA

34.81 Quantitation of C. perfringens *Population Based on Confirmed Anaerobic Plate Counts*

Cultures obtained from presumptive *C. perfringens* black colonies on selective, differential TSC or EY-free TSC medium are confirmed as *C. perfringens* if they are nonmotile, reduce nitrate, ferment lactose, liquefy gelatin within 44 hr, and produce acid from raffinose. Calculate the number of viable *C. perfringens* per gram of food sample as follows: multiply the presumptive plate count by the reciprocal of the dilution plated and then by the ratio of the colonies confirmed as *C. perfringens* to total colonies tested. (Note: If the surface plating method is used, the result must be multiplied by 10 since only 0.1 mL was tested.)

34.9 REFERENCES

1. Abeyta, C. Jr., and J. Wetherington. 1994. Iron milk medium method for recovering *Clostridium perfringens* from shellfish. Collaborative study. J. AOAC Int. 77:351-356.
2. Abeyta, C., Jr., M. Wekell, and J. Peeler. 1985. Comparison of media for enumeration of *Clostridium perfringens* in foods. J. Food Sci. 50:1732-1735.
3. Angelotti, R., H. E. Hall, M. J. Foster, and K. H. Lewis. 1962. Quantitations *of Clostridium perfringens* in food. Appl. Microbiol. 10:193.
4. Bartholomew, B., M. Stringer, G. Watson, and R. Gilbert. 1985. Development and application of an enzyme-linked immunosorbent assay for *Clostridium perfringens* type A enterotoxin. J. Clin. Pathol. 38:222.
5. Berry, P.R., M. F. Stringer, and T. Uemura. 1986. Comparison of latex agglutination and ELISA for the detection of *Clostridium perfringens* type A enterotoxin in feces. Lett. Appl. Microbiol. 2:101.
6. Craven, S. E., L. C. Blankenship, and J. L. McDonel. 1981. Relationship of sporulation enterotoxin formation, and spoilage during growth of *Clostridium perfringens* Type A in cooked chicken. Appl. Microbiol. 41:1184.
7. Duncan, C. L. 1973. Time of enterotoxin formation and release during sporulation of *Clostridium perfringens*, type A. J. Bacteriol. 113:932.
8. Duncan, C., and D. Strong. 1969. Improved medium for sporulation of *Clostridium perfringens*. Appl. Microbiol. 16:82.
9. Fuzi, M., and Z. Csukas. 1969. New selective medium for the isolation of *Clostridium perfringens*. Acta Microbiol. Acad. Sci. Hung. 16:273.
10. Hanford, P. M., and J. J. Cavett. 1973. A medium for the detection and enumeration of *Clostridium perfringens* (*Welchii*) in foods. J. Sci. Food Agric. 24:487.
11. Harmon, S. M. 1976. Collaborative study of an improved method for the enumeration and confirmation of *Clostridium perfringens* in foods. J. Assoc. Off. Anal. Chem. 59:606.
12. Harmon, S. M., and D. A. Kautter. 1978. Media for confirming *Clostridium perfringens* from food and feces. J. Food Prot. 41:626.
13. Harmon, S. M., and D. A. Kautter. 1986. Evaluation of a reversed passive latex agglutination test kit for *Clostridium perfringens* enterotoxin. J. Food Prot. 49:523.
14. Harmon, S. M., and D. A. Kautter. 1986. Improved media for sporulation and enterotoxin production by *Clostridium perfringens*. J. Food Prot. 49:706.
15. Harmon, S. M., and D. A. Kautter. 1987. Enumeration of *Clostridium perfringens* spores in human feces: Comparison of four culture media. J. Assoc. Off. Anal. Chem. 70:994.
16. Harmon, S. M., D. A. Kautter, and J. T. Peeler. 1971. Improved medium for enumeration of *Clostridium perfringens*. Appl. Microbiol. 22:688.
17. Harmon, S. M., D. A. Kautter, and C. L. Hatheway. 1986. Enumeration and characterization of *Clostridium perfringens* spores in the feces of food poisoning patients and normal control. J. Food Prot. 49:23.
18. Harmon, S. M., C. L. Hatheway, E. H. Peterson, G. Ransome, and J. Wimsatt. 1987. Interlaboratory study of a reversed passive latex agglutination test kit for Clostridium perfringens enterotoxin. Abstracts of the 101st Annual Meeting, Assoc. Off. Anal. Chem., p. 84.
19. Harmon, S. M., and A. M. Placecia. 1978. Method for maintaining viability of *Clostridium perfringens* in foods during shipment and storage: Collaborative study. J. Assoc. Off. Anal. Chem. 61:785.
20. Hauschild, A. H. W., and R. Hilsheimer. 1974. Evaluation and modifications of media for enumeration of *Clostridium perfringens*. Appl. Microbiol. 27:78.
21. Hauschild, A. H. W., and R. Hilsheimer. 1974. Enumeration of foodborne *Clostridium perfringens* in egg yolk-free typtose-sulfite cycloserine agar. Appl. Microbiol. 27:521.
22. Hauschild, A. H. W., R. J. Gilbert, S. M. Harmon, M. F. O'Keefe, and R. Vahefeld. 1977. ICMSF methods studies. VIII. Comparative study for the enumeration of *Clostridium perfringens* in foods. Can. J. Microbiol. 23:884.
23. Jackson, S. G., D. A. Yip-Chuck, and M. H. Brodsky. 1985. A double antibody sandwich immunoassay for *Clostridium perfringers* type A enterotoxin detection in stool specimens. J. Immunol. Meth. 83:141.
24. Jackson, S. G., D. A. Yip-Chuck, and M. H. Brodsky. 1986. Evaluation of the diagnostic application of an enzyme immunoassay for *Clostridium perfringens* type A enterotoxin. Appl. Environ. Microbiol. 52:969.
25. Labbe, R. 1989. *Clostridium perfringens*, p. 191. *In* FM. Doyle (ed.), Foodborne bacterial pathogens. Marcel Dekker, New York.
26. Labbe, R. G., and D. K.Rey. 1979. Raffinose increases sporulation and enterotoxin production by *Clostridium perfringens* type A. Appl. Microbiol. 37:1196.
27. Marshall, R. S., J. F. Steenbergen, and L. S. McClung. 1965. Rapid technique for the enumeration of *Clostridium perfringens*. Appl. Microbiol. 13:559.
28. McClane, B. A., and J. Snyder. 1987. Development and preliminary evaluation of a slide latex agglutination assay for the detection of *Clostridium perfringens* type A enterotoxin. J. Immunol. Meth. 100:131.
29. McClane, B. A., and R. J. Strouse. 1984. Rapid detection of *Clostridium perfringens* type A enterotoxin by enzyme-linked immunosorbent assay. J. Clin. Microbiol. 19:112.
30. Naik, H. S., and C. L. Duncan. 1977. Enterotoxin formation in foods by *Clostridium perfringens* type A. J. Food Safety 1:7.
31. Nakamura, S., T. Shimamura, M. Hayase, and S. Nishida. 1973. Numerical taxonomy of saccharolytic clostridia, particularly *Clostridium perfringens*-like strains: Description of *Clostridium absonum* sp. n. and *Clostridium paraperfringens*. Int. J. Syst. Bacteriol. 23:419.

32. Notermans, S., C. Heuvelman, H. Beckers, and T. Uemura. 1984. Evaluation of the ELISA as a tool in diagnosing *Clostridium perfringens* enterotoxin. Zentralbl. Bakteriol. Hyg. I. Abt. Orig. B. 179:225.
33. Olsvik, U., P. E. Granum, and B. Berdal. 1982. Detection of *Clostridium perfringens* type A enterotoxin by ELISA. Acta Pathol. Microbiol. Immunol. Scand. Sect. B 90:445.
34. Shahidi, S. A., and A. R. Ferguson. 1971. New quantitative, qualitative, and confirmatory media for rapid analysis of food for *Clostridium perfringens*. Appl. Microbiol. 21:500.
35. Spray, R. S. 1936. Semisolid media for cultivation and identification of the sporulating anaerobes. J. Bacteriol. 32:135.
36. Taniguti, T. 1969. Sporulation media for *Clostridium perfringens*: A method with a new medium (AEA medium) for sporulation of *Clostridium perfringens* and some properties of formed spores. J. Food Hyg. Soc. Jpn. 9:219.
37. Thatcher, F. S., and D. S. Clark. 1968. "Microorganisms in foods: Their significance and methods of enumeration," p. 128. Univ. of Toronto Press, Toronto, Canada.
38. Traci, P. A., and C. L. Duncan. 1974. Cold shock lethality and injury in *Clostridium perfringens*. Appl. Microbiol. 28:815.
39. Wimsatt, J., S. M. Harmon, and D. Shah. 1986. Detection of *Clostridium perfringens* enterotoxin in stool specimens and culture supernatants by enzyme-linked immunosorbent assay. Diag. Microbiol. Infect. Dis. 4:307.

Chapter 35

Pathogenic *Escherichia coli*

Jianghong Meng, Peter Feng, and Michael P. Doyle

35.1 INTRODUCTION

Escherichia coli, a member of the family Enterobacteriaceae, is a part of normal flora of the intestinal tract of humans and a variety of animals. Included in the Enterobacteriaceae family are some of the most important enteric pathogens such as *Salmonella*, *Shigella*, and *Yersinia*. Although most of *E. coli* do not cause gastrointestinal illnesses, certain groups of *E. coli* can cause life-threatening diarrhea, and severe sequelae or disability.[21]

E. coli are classified on the basis of antigenic differences (serotyping) and virulence factors (virotyping).[78] Two *E. coli* surface components form the primary basis for the serological classification system: the O antigen of the lipopolysaccharide (LPS, O) and the flagella (H, for 'hauch,' the German term for flagella). The O antigen identifies the serogroup of a strain, and the combination of O antigen and H antigen identifies its serotype. For example, two strains identified as O157:H7 and O157:H19 react with the same anti-O antibody and are of the same serogroup but react with different anti-H antibodies and are different serotypes. More than 170 different serogroups have been identified. Isolates lacking flagella are non-motile and designated NM in the serotype formula. There is some, but not a definite, correlation between serogroup and virulence. For example, antigen O86 is commonly associated with *E. coli* strains that are members of the resident colonic microflora and *E. coli* of this serogroup rarely cause disease, whereas serogroup O55 is rarely associated with the resident microflora and is frequently associated with disease. Some *E. coli* strains also possess a capsular (K) antigen that also is used for classification.

The characteristics that form the basis for the virotyping system include patterns of bacterial attachment to host cells, effects of attachment on host cells, production of toxins, and invasiveness. Both classification schemes (i.e., serotyping and virotyping) used in combination are a useful approach for grouping *E. coli* strains; however, absolute categorization of strains can not always be made on the basis of antigenic differences and virulence factors.[68] Furthermore, the genetic codes for many of these virulence traits reside on plasmids or phages or chromosomal pathogenicity islands; hence, the capacity to be a pathogen consequently may be transferred from one strain of *E. coli* to another. *E. coli* strains, that have virulence properties associated with more than one pathogenic group, have been reported.[45] Currently, diarrheagenic *E. coli* are categorized into six virotypes. A brief description of the salient features and examples of foodborne outbreaks caused by *E. coli* of the different virotypes follows.

35.11 Enterotoxigenic E. coli *(ETEC)*

ETEC strains resemble *Vibrio cholerae* in that they adhere to the small intestinal mucosa via surface fimbriae (type 1 pili and colonization factor antigens) and produce symptoms not by invading the mucosa but by elaborating one or two enterotoxins, heat-labile toxin (LT) and heat-stable toxin (ST). The enterotoxins act on intestinal mucosal cells to cause diarrhea. ETEC strains cause traveler's diarrhea and severe, often fatal, diarrhea in infants in developing countries. Epidemiologic investigations have implicated contaminated food and water as the most frequent vehicles of ETEC infection. Although ETEC is rarely considered as a possible cause of diarrhea in the United States, several foodborne outbreaks have been documented since the first such outbreak was reported in 1970s.[21,77] ETEC outbreaks also have occurred aboard cruise ships. Foods that were involved in these outbreaks include cheese, curried turkey mayonnaise, crabmeat cocktail, and drinking water.[21] More recently, two outbreaks of ETEC in Rhode Island and New Hampshire were reported.[11] The outbreak in Rhode Island had 47 cases of ETEC infection that occurred on an airline flight, whereas 121 persons became ill after eating a buffet dinner served at a mountain lodge in New Hampshire. ETEC serotype O6:NM was isolated from patients in both outbreaks. Carrots that were grown in the same state and used in salads were implicated as the vehicle. In 1994, ETEC O153:H43 was identified as the cause of two foodborne outbreaks.[77] One outbreak was associated with eating pan-fried potatoes in Milwaukee, Wisconsin, whereas corn bread dressing was implicated in another outbreak in Louisiana. Interestingly, the two ETEC strains causing these outbreaks were both serotype O153:H43, produced ST enterotoxin, and were resistant to tetracycline, ampicillin, sulfisoxazole, and streptomycin; however, they were not related based on the epidemiologic investigation and molecular subtyping data generated by pulsed-field gel electrophoresis. In June 1998, potato salad contaminated with an ETEC strain was identified as the vehicle of the largest ETEC outbreak in the United States. As many as 4,500 people were sick after attending 300 parties in Illinois. In Japan, a large outbreak of ETEC affecting more than 800 students occurred at four elementary schools in 1996.[65] Tuna paste was implicated as the vehicle food and *E. coli*

O25:NM was isolated from the food and stool specimens of the patients.

35.12 *Enteroaggregative* E. coli *(EAEC)*

EAEC strains cause persistent diarrhea, particularly in children.[15] EAEC are defined as *E. coli* strains that do not secrete enterotoxins LT or ST and that adhere to HEp-2 cells in an aggregative adherence pattern. The first description of EAEC as a cause of diarrhea among children was reported in Santiago, Chile, in the 1980s. Subsequent studies have supported the association of EAEC with diarrhea in other developing countries.[67] EAEC, however, are not strictly confined to the developing world, and have been isolated from diarrhea patients in the United Kingdom. Recently, Smith et al.[81] reported four foodborne outbreaks of EAEC infections in the United Kingdom in 1994, affecting a total of 133 patients. The symptoms included vomiting and diarrhea, with persistent diarrhea in a small number of patients. Each of the outbreaks was associated with consumption of a restaurant meal, but no single source was implicated.

35.13 *Enteropathogenic* E. coli *(EPEC)*

EPEC strains cause severe diarrhea in young children, especially infants.[20] The pathogen adheres to the intestinal mucosa and causes extensive rearrangement of host cell actin. As with many other *E. coli*, transmission of EPEC is by fecal-oral route, with contaminated hands or contaminated foods or infant formula as vehicles. EPEC once caused frequent outbreaks of infantile diarrhea in the United States, but is no longer as important a cause of diarrhea in developed countries as they were in the 1940s and 1950s. However, several outbreaks of EPEC infection have been reported in the last two decades in the United States and other developed countries. These outbreaks frequently occur in day care centers and occasionally in pediatric wards.[21] Foodborne outbreaks of EPEC also have been reported. A foodborne outbreak caused by atypical EPEC was recently documented among adults who ate a gourmet buffet in Minnesota.[45] No specific food item was identified as the vehicle in this outbreak. In 1995, two outbreaks in Northern France were associated with prawn mayonnaise vol au vents, and lettuce and gherkins, respectively. A total of 59 cases were reported, and EPEC O111 was isolated from patients.[85]

35.14 *Enterohemorrhagic* E. coli *(EHEC)*

EHEC cause illnesses, including hemolytic uremic syndrome (HUS), which are similar to those caused by *Shigella dysenteriae* predominantly in the young and elderly.[23,38] EHEC also are similar to EPEC except that EHEC produce one or two Shiga toxins. Although many serotypes of *E. coli* belong to the EHEC group, serotype O157:H7 predominates as a foodborne pathogen. *E. coli* O157:H7 was first recognized as a human pathogen in 1982 when two outbreaks of hemorrhagic colitis were associated with consumption of undercooked ground beef that had been contaminated with this organism.[76] Since then, this pathogen has caused many foodborne outbreaks of hemorrhagic colitis and HUS worldwide.[23] Large outbreaks of *E. coli* O157:H7 infections include an outbreak in 1992–1993 that affected more than 700 individuals with 4 deaths in the western United States,[5] and several outbreaks in 1996 in Japan in which more than 9,000 cases were reported.[84] The CDC estimates there are 20,000 cases of *E. coli* O157:H7 infection annually in the U.S. and 500 associated deaths.

Cattle[3,43,90] and to a less extent other ruminants[52,55] are considered to be the major reservoirs of *E. coli* O157:H7, although the organism has been isolated from other animals such as dogs, horses, and birds.[42] Many outbreaks of *E. coli* O157:H7 infection have been associated with the consumption of undercooked ground beef. Other foods also have been associated with *E. coli* O157:H7 outbreaks worldwide, including roast beef, venison jerky, salami, raw milk, pasteurized milk, yogurt, lettuce, unpasteurized apple cider/juice, cantaloupe, radish sprouts, and alfalfa sprouts.[62] An outbreak associated with handling potatoes was reported in the United Kingdom.[66] It is noteworthy that certain foods such as apple cider/juice and dry-cured salami that previously were considered safe and ready to eat because of their high acidity, and are usually not heated before consumption, have been vehicles of outbreaks. The pathogen has been shown experimentally to survive for several weeks to months in a variety of acidic foods, including mayonnaise,[88] sausages,[35] apple cider,[89] and cheddar cheese.[74] Sprouts were recently added to the spectrum of foods serving as vehicles of *E. coli* O157:H7 infection. Radish sprouts were implicated as the vehicle outbreaks in Japan that occurred in 1996 and 1997.[84] In addition, two outbreaks of EHEC infection in Michigan and Virginia were associated with consumption of alfalfa sprouts.[12] Reports of person-to-person and waterborne (particularly recreational lake water) transmission have been increasing.[13]

35.15 *Enteroinvasive* E. coli *(EIEC)*

EIEC strains are biochemically, genetically, and pathogenically related closely to *Shigella* spp., but do not produce Shiga toxin. EIEC infection presents most commonly as watery diarrhea, which can be indistinguishable from the secretory diarrhea associated with ETEC infection.[1] Only a minority of patients experience the dysentery syndrome. HUS, however, is not a complication of EIEC infection. Documented EIEC outbreaks are usually foodborne or waterborne, although person-to-person transmission has been reported.[21] The incidence of EIEC in developed countries is reportedly low, but occasional foodborne outbreaks, such as a French camembert cheese associated outbreak in 1973[61] and a restaurant-associated outbreak involving 370 people in Texas in 1994, have been reported.[37] A large outbreak of EIEC infection in Japan with 670 cases occurred in 1988.[86] High fever and watery diarrhea were characteristic of the infection. "Godofu (Sasayuki tofu)" was identified as the vehicle food, and EIEC O164:NM was isolated from patient stools and Godofu.

35.16 *Diffusely adherent* E. coli *(DAEC)*

DAEC strains have been associated with diarrhea in children, and are defined by a characteristic diffuse-adherent pattern of adherence to Hep-2 or HeLa cell lines.[67] The pathogenesis of DAEC, however, is poorly understood. These organisms generally do not elaborate heat-labile or heat-stable toxins or elevated levels of Shiga toxins, nor do they possess EPEC adherence factor plasmids or invade epithelial cells. No outbreaks associated with food to date have been reported.

35.2 CONVENTIONAL ISOLATION PROCEDURES

35.21 *General Considerations*

Pathogenic *E. coli* are phenotypically diverse, hence, there are no standard microbiological methods for isolating these pathogens. Unlike typical *E. coli*, some pathogenic strains such as EIEC do not ferment lactose[57]; hence, coliform methods that are based on lactose fermentation are not suitable for the detection of EIEC. Also, elevated incubation temperatures that are applied in fecal coliform confirmation or enrichment procedures can be inhibitory to the growth of EHEC O157:H7.[22] Elevated temperatures and sodium lauryl sulfate used in lauryl tryptose broth (LTB) in Most Probable Number (MPN) analysis can cause the loss of plasmids, which often encode virulence-associated factors in *E. coli*.[47] Hence, no standard methods are available for detecting all pathogenic *E. coli* and existing methods are either adapted from general methods for isolating *E. coli* or developed for a specific group of pathogenic *E. coli*. Regardless of methods, however, when testing for pathogenic *E. coli*, it is important that isolates are first identified biochemically as *E. coli* before they are serotyped and assayed for virulence factors associated with a respective pathogenic group.

35.22 Media and reagents

Levine's eosin-methylene blue agar (EMB)
Tryptone phosphate broth (TP)
Sorbitol MacConkey agar (SMAC)
Hemorrhagic colitis agar (HC)
Hektoen enteric agar (HE)
E. coli broth (EC)
Brain heart infusion broth (BHI)
Modified EC broth (mEC)
Trypticase soy broth (TSB)
Modified trypticase soy broth (mTSB)
EHEC enrichment broth (EEB)
Tellurite cefixime sorbitol MacConkey agar (TC SMAC)
Phenol red sorbitol-MUG agar (PSA-MUG)
Novobiocin solution (100 mg/mL)
Kovac's reagent
Butterfield Phospate Buffer
MacConkey Sorbitol 5-bromo-4-chloro-3-indoxyl-β-D-glucuronide sodium salt (MSA-BCIG)
MacConkey Agar (MAC)

35.23 Pathogenic E. coli (EIEC, ETEC, EPEC, EAEC)

Pathogenic *E. coli* strains that ferment lactose and are not affected by elevated temperatures can be distinguished from other enterics using the standard MPN method for *E. coli*. However, fairly extensive serotyping and virulence analysis of colonies from Levine's eosin-methylene blue (EMB) agar plate may be required to identify a specific pathogenic group.

Alternatively, the following method has been described in Bacteriological Analytical Manual (BAM) of Food and Drug Administration (FDA), for isolating pathogenic *E. coli* from foods.[48] Aseptically weigh a 25-g sample into 225 mL of brain heart infusion broth (BHI) and pre-enrich for 3 hr at 35°C to resuscitate damaged cells. Transfer the entire pre-enrichment to 250 mL of tryptone phosphate (TP) broth and selectively enrich at 44°C for 20 hr. From the TP enrichment, streak for isolation on EMB and MacConkey (MAC) agar plates. If high numbers of *E. coli* ($>2.5 \times 10^4$/g) are suspected, samples from BHI may be directly streaked onto EMB and MAC for differentiation and isolation. Some pathogenic *E. coli* strains may not produce colonies with the typical green sheen or may grow poorly on EMB; hence, pick 10 typical and 10 atypical colonies from EMB and MAC, and identify them biochemically as *E. coli* before serological typing and virulence analysis. For EIEC, Hektoen enteric (HE), MAC and Salmonella-Shigella agars may be used for selective plating; however, HE and MAC are least inhibitory and best suited for the isolation of EIEC.[21]

35.24 EHEC O157:H7 and other serotypes

35.241 General Method

EHEC serotypes other than O157:H7 are phenotypically diverse; hence, may be isolated from foods using the same procedure outlined above for the other pathogenic groups of *E. coli*. Serotype O157:H7 is unique in that most strains do not ferment sorbitol within 24 hr[24] and do not express β-glucuronidase which most *E. coli* posses. Hence, *E. coli* O157:H7 is typically negative in the fluorogenic, 4-methyl-umbelliferyl β-D-glucuronide (MUG) assay.[22,28] These phenotypes, especially the inability to ferment sorbitol (SOR), are used extensively to isolate O157:H7 from foods. However, relying on only a few phenotypic characteristics as the bases for selecting isolates for further characterization can yield an undesirable level of false-positive colonies. Enteric bacteria such as *Enterobacter hermanii* and *Hafnia* spp., have similar sorbitol and MUG reactions, hence, they resemble O157:H7 colonies on media that use sorbitol and MUG reagents for differentiation. Disease-causing, phenotypic variants of O157:H^- strains that produce Stx2 have been isolated from HUS patients in Germany,[40] and are fairly prevalent in Central Europe.[2] These strains genetically belong to the O157:H7 clonal group,[27] but they ferment sorbitol and are positive in the MUG assay. In the U.S. only MUG positive variants of O157:H7 have been isolated.[44] However, culturing sorbitol-negative O157:H7 strains in sorbitol-containing media has resulted in strains capable of fermenting sorbitol,[33] hence, it is possible that sorbitol- fermenting variants will be encountered in the U.S. in the future. However, presently, sorbitol-containing media are effective and economical for the preliminary differentiation of EHEC serotype O157:H7 from other *E. coli*.

A number of sorbitol-containing media are used for isolating O157:H7; some are commercially available or can be prepared from individual ingredients. Also, several specialty media that use chromogenic substrates to detect enzymatic activities unique to O157:H7 and other EHEC have become available (Table 1). These, however, can be expensive and few have been tested by collaborative studies or evaluated for analysis using food samples.

35.2411 Sorbitol MacConkey Agar (SMAC). SMAC is frequently used in clinical analysis for isolating O157:H7.[60] Prompt plating of bloody stool specimens onto SMAC has been effective in isolating O157:H7.[39] For food analysis, homogenize a 10-g sample in 90 mL of 0.01% peptone water, make ten-fold serial dilutions and plate 0.1 mL, in duplicate, onto SMAC. Incubate plates overnight at 35°C. Disregard the pink, sorbitol-fermenting colonies; pick 5 to 10 non-fermenting colonies (pale to neutral/grey) to verify a negative MUG reaction, then serotype for the O157 antigen (see Section 35.4). If O157 positive, type for the H7 antigen and examine for EHEC virulence factors (see Section 35.5). Since SMAC is not very selective, high levels of normal flora in foods may mask the presence of O157:H7 on this medium.[41]

35.2412 Hemorrhagic Colitis Agar (HC). HC is another sorbitol-containing, direct plating medium, but, was developed specifically for foods.[83] HC agar is incubated at 43°C to provide greater selectivity and it also contains MUG for additional differentiation. O157:H7 on HC appear as clear colonies that do not fluoresce under long wave UV light due to the absence of GUD activity. Like SMAC, however, high levels of normal flora in foods can overgrow and mask the presence of O157:H7.[41] To provide greater selectivity as is needed in food analysis, a 25-g sample of food is blended in 225 mL of pre-enrichment broth and incubated at 35°C overnight, before plating on SMAC and HC agars. Enrichment broths often used are modified EC (mEC) or modified

Table 1. Partial list of commercially available chromogenic media for differentiation of *Escherichia coli* strains

Target Bacteria	Commercial Name	Differentiating Characteristics	Manufacturer
O157:H7 & *E. coli*	Rainbow agar	β-galactosidase β-glucuronidase	Biolog, Hayward, CA
O157:H7 & enterics	BCMO157:H7(+)	β-glucuronidase sorbitol fermentation	BIOSYNTH, Naperville, IL
E. coli	CHROMagar	β-glucuronidase	CHROMagar, Montecino, CA

TSB (mTSB), supplemented with novobiocin. However, even with the pre-enrichment, foods having very high levels of normal microflora can be problematic with overgrowth problems on SMAC and HC agars.

35.2413 Tellurite-Cefixime Sorbitol MacConkey Agar (TC-SMAC). Due to the difficulties in isolating O157:H7 from foods with high levels of normal microflora, more selective enrichment and plating media have been developed. One enrichment broth, developed for analyzing carcass samples, uses the antibiotics vancomycin, cefixime and cefsulodin to inhibit gram-positive bacteria, aeromonads and *Proteus* spp., respectively.[14] The selectivity of SMAC plating medium also has been improved with the inclusion of tellurite-cefixime (TC), which effectively inhibits growth of other *E. coli* and other non-sorbitol fermenting bacteria. TC-SMAC agar appears to be very selective for O157:H7.[87]

Although no standardized methods are available for isolating O157:H7 from foods, the enrichment and plating media described above have been incorporated into procedures used by FDA for testing foods and methods used by the US Department of Agriculture (USDA) Food Safety and Inspection Service (FSIS) for meat analysis.

35.242 FDA Method—Bacteriological Analytical Manual (BAM)

The following is a summary of the method outlined in Chapter 4 of the BAM, 8th (rev.A), 1998, for isolating O157:H7 from foods.

1. **Enrichment.** Homogenize 25 g of sample in 225 mL of EHEC enrichment broth (EEB) and incubate at 37°C with agitation for 6 hr. Perform the "isolation" procedure outlined below, then re-incubate the enrichment broth overnight at 37°C.
2. **Isolation.** Spread-plate 0.1 mL of 6-hr enrichment culture onto a TC-SMAC plate and streak one loopful onto a second TC-SMAC plate and incubate both plates overnight at 37°C. Disregard the pink, sorbitol-fermenting colonies, and select the non-sorbitol-fermenting, neutral/grey colonies with smoky centers (1-2 mm diameter) that are typical of O157:H7. Pick 5 to 10, transfer onto slants of TSA with 0.6% yeast extract, and incubate the cultures overnight at 37°C. If no colonies typical of O157:H7 colonies grow on TC-SMAC plates, plate the overnight enrichment culture on TC-SMAC plates according to the procedure described above.

   Both EEB and TC-SMAC are highly selective and preliminary data indicate that this procedure may inhibit O157:H7 strains in foods that do not have high levels of normal microflora (S. Weagant, FDA, Bothell, WA, personal communication).
3. **Confirmation.** Use sterile toothpicks to obtain small amounts of colonies from TSA and test for indole reaction using a paper filter wetted with Kovac's reagent. If the isolate is indole-negative, it is most likely not *E. coli* and can be discarded. If isolates are indole-positive, test for the O157 antigen using commercial O157 antiserum (see Section 35.4). Latex agglutination kits such as Prolex *E. coli* O157 (PROLAB) and RIM *E. coli* O157:H7 (REMEL) give satisfactory results. Isolates that are agglutination positive with anti-O157 reagent are presumptively O157:H7. Those isolates presumptively identified as O157:H7 should be confirmed as *E. coli* by biochemical tests using miniaturized kits or conventional prodecures before performing PCR analysis for Stx1 and Stx2 (see Section 35.5).

35.243 USDA-FSIS Microbiology Laboratory Guidebook (MLG)

The following is a summary of the procedure outlined in the USFA-FSIS MLG, 3rd ed., 1998, for testing raw ground beef for O157:H7.

1. **Enrichment.** Blend 65 g of ground beef (usually 5 samples are analyzed as a composite) with 300 mL of mEC with novobiocin; add more media (about 285 mL) to bring the total volume to 650 mL. Incubate at 35°C for 18 to 24 hr.
2. **Screening.** Apply the enrichment culture to a commercial O157:H7/NM (non-motile) antibody-based test kit (Table 2) according to the manufacturer's instructions. The test kit must conform to the following performance characteristics: sensitivity ≥98%; specificity ≥90%; false-negative rate ≤2%; and false-positive rate ≤10%. Negative samples are reported as such and discarded and positive specimens are only considered presumptive positive for O157:H7, and must be confirmed by isolating the O157:H7 from the enrichment culture, and characterizing it by the biochemical and serological assays described above.
3. **Isolation.** From the presumptive-positive enrichment cultures, prepare in Butterfield's phosphate buffer ten-fold dilutions up to 10^{-7}. Spread plate 0.1 mL from each of the four highest dilutions onto MacConkey Sorbitol agar with 0.1 g of 5-bromo-4-chloro-3-indoxyl-β-D-glucuronide sodium salt (MSA-BCIG) and incubate at 42 ± 0.5°C.[70] This medium is same as SMAC, but contains BCIG, a chromogenic substrate to detect GUD activity. SOR fermenting and/or GUD positive colonies appear as red to purple. Typical O157:H7 colonies are SOR and GUD-negative and appear white. These colonies are selected and streaked onto a plate of EMB and also onto a plate of phenol red sorbitol-MUG agar (PSA-MUG) and incubated overnight at 35°C to verify the SOR and GUD reactions.[69] Disregard the yellow SOR-positive colonies on PSA-MUG that also fluoresce under long-wave UV light. Select from PSA-MUG the clear, non-fluorescent colonies and from EMB the purple colonies with a metallic green sheen and serotype the colonies for O157 antigen using a latex agglutination kit (RIM *E. coli* O157:H7 Latex Test kit or equivalent) (see Section 35.4).
4. **Confirmation.** Confirm as *E. coli* the colonies that agglutinated with anti-O157 sera, using the biochemical tests listed in Table 3. Isolates that have the biochemical profile of *E. coli* O157:H7 should be serotyped for the H7 antigen by a tube agglutination test or latex agglutination kit (RIM *E. coli*

Table 2. Partial list[*] of commercially available assays for detecting EHEC serotypes in foods.

Target	Trade name	Assay Format[a]	Manufacturer
O157	PetrifilmHEC	blot-ELISA	3M, St. Paul, MN
	EZ COLI	Tube-ELISA	Difco, Detroit, MI
	Dynabeads	Ab-beads	Dynal, Lake Success, NY
	Assurance[b]	ELISA	BioControl, Bellview, WA
	TECRA	ELISA	TECRA, Redmond, WA
	E. coli O157	ELISA	LMD Lab, Carlsbad, CA
	E. coli O157	ELISA	Meridian, Cincinnati, OH
	VIP[b]	Ab-ppt	BioControl, Bellview, CA
	Reveal	Ab-ppt	Neogen, Lansing, MI
	NOW	Ab-ppt	Binax, Portland, ME
	Immunocard	Ab-ppt	Meridian, Cincinnati, OH
	QUIX	Ab-ppt	UniversalHealthWatch, Columbia, MD
	VIDAS	ELISA[c]	bioMerieaux, St.Louis , MO
O157:H7	BAX	PCR	Qualicon, Wilmington, DE
O157:H7 & O26:H11	EHEC-TEK	ELISA	Organon-Teknika, Durham, NC

* Table modified from: Feng, P. 1998. U.S. FDA Bacteriological Analytical Manual, 8th ed. (Rev.A).

[a] ELISA, enzyme-linked immunosorbent assay; Ab-ppt, immunoprecipitation or immunochromatography; PCR, polymerase chain reaction.

[b] Approved Official first action—AOAC International.

[c] Automated ELISA

O157:H7 Latex Test kit or equivalent), using the bacterial culture grown on a blood agar plate. If the H7 test is negative, attempt to induce the motility of the isolates by growing them in motility medium and test again with anti-H7 antibody.

35.3 RAPID METHODS

Many innovative technologies have been introduced that currently are being used for the microbiological analysis of foods. Unlike conventional methods that rely on specific media to select and grow pathogens, newly developed "rapid methods" use procedures such as molecular biology techniques or immunoassay to detect in foods, gene sequences or antigens that are specific for the bacterial pathogen of interest.[26] Compared to conventional procedures, rapid methods are generally easier and more rapid to perform, with some requiring only minutes to complete. However, today's rapid methods still require that prior to analysis food samples undergo culture enrichment to grow the pathogen of interest to detectable populations. Rapid methods usually shorten analysis time by 1 to 2 days compared to conventional assays, hence they are useful for rapid screening of foods for bacterial pathogens. For screening purposes, negative results are regarded as definitive, whereas positive results are considered only as presumptive and must be confirmed by standard methods. In addition, since rapid methods use different technologies, the detection efficiencies of these assays can be dependent on the composition and microflora of the food being assayed. Hence, it is important that these methods are evaluated by collaborative studies prior to routine use in food analysis.[25] In the U.S., only methods that have been collaboratively studied and approved by the Association of Official Analytical Chemists International (AOAC) are included in the Official Methods of Analysis (OMA) and regarded as standard methods. Therefore, it is recommended to use rapid methods only for those food types specified by the OMA. Perform all rapid methods according to manufacturers' instructions provided with the kit, including the use of specified enrichment medium recommended for the assay.

Table 3. Biochemical reactions for *E. coli* O157:H7/NM

Medium	Reaction
Triple sugar iron agar	yellow butt/yellow slant, no H_2S
Sorbitol fermentation	negative
Cellobiose fermentation	negative
Tryptone broth	indole positive
MR-VP broth	MR positive; VP negative
Simmons citrate	negative (no growth)
Lysine decarboxylase	positive (purple)
Ornithine decarboxylase	positive (purple)
Decarboxylase broth base	negative (yellow)
Motility test medium	Motile (H7) or non-motile (NM)

35.31 Pathogenic E. coli *(EIEC, ETEC, EPEC, EAEC)*

Very few commercially available rapid methods have been developed for detecting in foods pathogenic *E. coli* other than EHEC, because these groups of pathogenic *E. coli* are not as frequently implicated in foodborne illness as EHEC. However, there are many antibody-, DNA probe-, and PCR-based assays reported for detecting characteristic virulence factors associated with these pathogens, including the toxins of ETEC, the invasion gene of EIEC, and the attachment and effacing gene of EPEC. Some of these assays are described in Section 35.5.

35.32 EHEC O157:H7 and other serotypes

Although EHEC serotypes O111:NM, O104:H21, O48:H21 and O26:H11 have been associated with outbreaks of foodborne disease, O157:H7 is the predominant serotype implicated in outbreaks worldwide. Hence, most commercially available assays are designed to detect the O157:H7 serotype (Table 2). With the exception of the BAX PCR assay that targets genetic sequences of O157:H7 and EHEC-TEK which detects O157:H7 and O26:H11 serotypes,[71] most of the assays target the O157 antigen and are not specific for the O157:H7 serotype. Assays that target the O157 antigen also detect O157, non-H7 serotypes that may be present in foods. Most assays use a sandwich enzyme-linked immunosorbent assay (ELISA) format or variants of ELISA that are automated or done on membranes, tubes, dipsticks, or other solid matrixes. Several, however, use immunochromatography or immunoprecipitation technology that was developed for clinical diagnostics and home pregnancy tests, and can provide a result with minimal manipulations and in less than 10 minutes after cultured enrichment. Regardless of the format, all of the assays require prior to analysis cultural enrichment in broth media such as mEC or mTSB with novobiocin or a medium specified with the kit. Also, only a few assays to date have been evaluated by AOAC collaborative studies and approved as official methods for analysis of O157:H7 in selected foods (Table 2).

35.4 SEROTYPING

Since pathogenic *E. coli* are classified based on virulence factors, serotyping does not provide a definitive identification of its pathogenic group, because some strains with the same serotype may belong to more than one pathogenic group. However, one exception, is O157:H7, which to date is recognized exclusively as an EHEC serotype. Hence, typing for the O157 and H7 antigens is useful for identification of this pathogen. A listing of the major O serogroups with their associated H antigen types for each pathogenic *E. coli* group has been published recently in a review by Nataro and Kaper.[68]

35.41 Pathogenic E. coli groups (EIEC, ETEC, EPEC, EAEC)

There are no assay kits available for serotyping these groups; however, sera for typing various O and H serotypes of these pathogenic groups are commercially available from several sources (Table 4). Slide agglutination of suspected colonies using commercial polyvalent antisera recognizing O antigens considered to represent EPEC has been a common procedure used for the diagnosis of EPEC . The advantage of this method is that it is extremely easy to perform. The major disadvantage is lack of accuracy.

35.42 EHEC and O157:H7

Sera for O157 and other EHEC serotypes are also available from several sources (Table 4). In addition, many latex agglutination kits are available for typing the O157 antigen of pure culture isolates (Table 4). These assays are rapid and easy to use, and simply involve making a cell suspension from growth on an agar plate, then mixing this suspension with a drop of O157 antibody bound to colored latex beads. In the presence of O157 bacteria, the antibody-latex agglutinates with the cells to form within a few minutes visible clumps. It is important that latex agglutination assays are performed according to manufacturers' instructions. With some kits, when a positive reaction is obtained, the isolate needs to be retested using the control latex reagent provided to rule out cases of autoagglutination.[8] Also, a single colony should be used when making the cell suspensions, because the use of heavy inoculum with more than one colony can result in false-positive reactions in latex agglutination assays.[82]

A positive reaction obtained with anti-O157 reagents is only a presumptive indication that the isolate may be of serotype O157:H7. Many *E. coli* of serogroup O157, but not H7, have been isolated from clinical and food samples.[9] These O157, non-H7, strains do not produce Shiga toxin(s) and are generally regarded as non-pathogenic for humans. Furthermore, sera to O157 can cross-react with antigens of *E. hermanii*,[58] *Salmonella* group N,[75] a few species of *Citrobacter*,[6,72] and other bacteria. Many of these cross-reacting genera, however, ferment sorbitol and can be distinguished from O157:H7 phenotypically. Because of the potential for false-positive reactions, it is important that presumptively positive O157 strains be further tested for the H7 antigen or for the characteristic EHEC virulence markers to obtain a definitive identification.

Some kits listed in Table 4 also provide latex reagents for typing the H7 antigen. However, expression of the H7 antigen can be variable, and often, O157:H7 strains isolated from clinical and environmental samples are non-motile and need to be induced for motility before H typing.[29,82] Sowers et al.[82] reported that a single colony from an agar plate is not sufficient to agglutinate the H7 latex reagent, hence, a sweep of growth must be used. They also recommended that the H7 latex reagent not be used independently of the O157 latex reagent, as a few false-positive H7 reactions were observed. The use of the H7 latex reagent alone will also detect other non-O157 *E. coli* serotypes that posses the H7 antigen.

Table 4. Partial list* of commercially available sera for serotyping of pathogenic *E. coli* strains

Target	Trade Name	Assay Format[a]	Manufacturer
O (various)		sera	Denka Seiken, Tokyo, Japan
		sera	Difco, Detroit, MI
O157	RIM	LA[a]	REMEL, Lenexa, KS
	E. coli O157	LA	Oxoid, Hampshire, England
	Prolex	LA	PRO-LAB, Austin, TX
	Ecolex O157	LA	Orion Diagnostica, Somerset, NJ
	Wellcolex	LA	Murex, Dartford, United Kingdom
	Bactrace	Serum[b]	Kirkegaad & Perry Lab, Gaithersburg, MD
		Serum	Difco
		Serum	Denka Seiken
H (various)		Sera	Denka Seiken
H7	RIM	LA	REMEL
	Wellcolex	LA	Murex
		Serum	Difco

* Table modified from: Feng, P. 1998. U.S. FDA Bacteriological Analytical Manual, 8th ed. (Rev.A).

[a] LA , latex agglutination

[b] available as unlabeled, fluorescein-labeled for fluorescent antibody assays or enzyme conjugate for ELISA.

35.5 Pathogenicity Testing

Pathogenic *E. coli* strains are defined on the basis of virulence properties, hence, pathogenicity tests are essential for the identification of pathogenic *E. coli* strains. Two approaches can be taken in the laboratory to detect pathogenic *E. coli*. These are based on phenotypic and genotypic characteristics of the pathogen. The phenotypic approach includes bioassays, tissue cultures, and immunoassays to detect production of toxins or activity of virulence factors. The genotypic methods include DNA hybridization, PCR assay, and other DNA amplification methods such as isothermal amplifications, to determine presence of virulence genes. Bioassays for identifying pathogenic *E. coli* have been described in detail in the 3rd edition of the *Compendium of Methods for the Microbiological Examination of Foods*.[49] The following section addresses immunoassays and DNA-based methods for detecting foodborne pathogenic *E. coli,* with an emphasis on EHEC. For more information on pathogenicity testing of diarrheagic *E. coli,* the reader is referred to a recent review by Nataro and Kaper.[68]

35.51 ETEC

Enterotoxins (LT and ST) and fimbrial attachment structures (colonization factor antigens) are characteristic virulence factors of ETEC. Detection of ETEC in foods has mainly relied on identifying the presence of LT and/or ST. The LTs of *E. coli* are oligomeric toxins that are closely related in structure and function to the cholera enterotoxin expressed by *Vibrio cholerae.* The LTs can be divided into two major serogroups, LT-I and LT-II, which do not cross-react immunologically. LT-I is expressed by ETEC that are pathogenic for both humans and animals, whereas LT-II is found primarily in animals and has not been associated with human or animal disease. The STs are small monomeric toxins that contain multiple cysteine residues, whose disulfide bonds account for the heat stability of these toxins. The STs can be divided into two unrelated categories, STa (ST-I) and STb (ST-II), which differ in structure and mechanism of action. STa can also be abbreviated STp or STh to indicate strains of porcine or human origin. STb is associated primarily with ETEC strains isolated from pigs.

A rabbit ligated ileal loop assay used to be performed to detect ST, but the expense and lack of standardization resulted in this test being replaced by the suckling-mouse assay, which became the standard bioassay for the presence of ST for many years.[34] The suckling-mouse assay involves the measurement of intestinal fluid in CD4 infant mice after percutaneous injection of culture supernatants. Several immunoassays also have been developed for detection of ST, including radioimmunoassay and an ELISA.[18] A competitive enzyme immunoassay is commercially available (Denka Seiken, Co Ltd., Tokyo, Japan) for detecting STa in culture fluid (Table 5). This test uses a synthetic peptide that is an analogue of STa to coat the wells and an anti-STa monoclonal antibody-enzyme conjugate. The antibody binds specifically to the solid-phase peptide. If the toxin is present in the culture filtrate, the toxin reduces binding of the conjugate to the solid phase, resulting in a reduced color intensity when compared to the negative control. These tests correlate well with results of the suckling-mouse assay.

The two standard assays for LT are the Biken test and the Y1 adrenal cell culture test. In the Biken test, LT is detected by the appearance of a precipiting line between growth of the organism and antisera against LT. In the Y1 assay, ETEC culture supernatants are added to Y1 cells and the cells are examined for rounding. Immunologic methods such ELISA, latex agglutination, and reversed passive latex agglutination test also have been developed for detecting LT (Table 5). In the reversed passive latex agglutination test, latex particles are sensitized with purified polyclonal antiserum raised in rabbits immunized with purified cholera toxin. These latex particles agglutinate in the presence of LT or cholera toxins. DNA probes and PCR assay targeting ST and LT also have been developed.[17,19] Three DNA probes are introduced in FDA BAM: the LT probe, eltA11, is a 22-mer oligonucleotide targeting the DNA sequence encoding the A subunit of LT; the STP probe is a 22-mer oligonucleotide for ETEC first isolated from pigs; and the STH probe is also a 22-mer oligonucliotide for ETEC isolated from a human patient.[46] Nataro and Kaper[68] have listed the sequences of oligonuclotides used for probing and PCR amplification of diarrheagenic *E. coli* strains. These DNA-based methods have been determined to be quite sensitive and specific when used directly on clinical samples or on isolated bacterial colonies.

35.52 EPEC

The most important virulence characteristics of EPEC strains are the attaching and effacing (A/E) histopathology, the possession of the EPEC adherence factor (EAF) plasmid, and the absence of Shiga toxins. Many EHEC strains also produce the A/E lesion, hence the determining the presence or absence of Stx is essential to differentiate EPEC from EHEC. Several assays have been developed for determining A/E activities or genes, including tissue culture tests for adherence phenotypes, fluorescence microscopy with actin probes for cytoskeletal disruption, radioactive and nonradioactive DNA probes, and PCR assays. The localized adherence phenotype is highly conserved among EPEC from the most common serotypes[20] and easily tested using cultured Hep-2 or HeLa cells and a light microscope. Transformed cell lines are incubated with bacteria, fixed, strained, and examined for adherent microcolonies of bacteria. A fluorescence actin staining (FAS) test also has been developed to identify the concentrated filamentous actin in epithelial cells beneath the sites of EPEC attachment, which serves as a surrogate marker for the A/E effect.[53] This latter test appears to correlate well with electron microscopy for attaching and effacing. The EAF probe and the *bfpA* gene probe targeting the EAF plasmid[30] and the gene probe targeting the *eae* gene that is associated with the A/E lesion,[50] are useful for the diagnosis of EPEC. Since many EHEC strains also carry an *eae* gene that is highly homologous to the EPEC *eae*, DNA probes and PCR primers derived from within the *eae* coding sequences often react with both EPEC and EHEC. Therefore, additional tests that determine the lack of Stx expression are necessary for identifying EPEC. Louie et al.[59] and Meng et. al.[64] exploited the heterogeneity of the 3′ end of oligonucleotide primers derived from the

Table 5. Partial list* of commercially available assays for detecting toxins produced by pathogenic *Escherichia coli* strains

Organism	Target	Trade Name	Assay Format[a]	Manufacturer
EHEC	Shiga toxins 1 & 2	VEROTEST	ELISA	MicroCarb, Gaithersburg, MD
		Premier EHEC	ELISA	Meridian, Cincinnati, OH
		Verotox-F	RPLA	Denka Seiken, Tokyo, Japan
		VTEC	RPLA	Oxoid, Hampshire, England
ETEC	Heat labile toxin	VET	RPLA	Denka Seiken
	Heat stable toxin	COLIST	ELISA	Denka Seiken
	Heat stable toxin	E.coli ST	ELISA	Oxoid

* Table modified from: Feng, P. 1998. U.S. FDA Bacteriological Analytical Manual, 8th ed. (rev.A).

[a] ELISA = enzyme-linked immunosorbent assay; RPLA = reverse passive latex agglutination

eae gene region to design PCR assays specific for EHEC O157:H7. The assays were quite specific, but failed to exclude EPEC O55:H7 and O55:NM strains as they are reported to be the progenitor of O157:H7.[27]

35.53 EIEC

As EIEC can be difficult to differentiate from *Shigella* spp., identification of EIEC involves demonstrating that the organism possesses the biochemical profile of *E. coli*, but has the genotypic or phenotypic characteristics of *Shigella* spp. (except the production of Shiga toxin). EIEC strains can be presumptively identified by determining their ability to invade HeLa cells and confirmed for invasive potential by the Sereny test (keratoconjunctivitis) using the eye of a guinea pig.[54] DNA probes and PCR methods have been developed to detect invasion genes, encoded on a large (220 kb) plasmid of EIEC.[31,36] Probes pMR17 (17 kb) derived from the pInv plasmid of a *Shigella flexneri* strain, and *ial* (2.5 kb) derived from the pInv plasmid of a EIEC strain have shown 100% sensitive and specific for EIEC strains that have retained their virulence. A PCR assay amplifying *ial* was also developed.[31] Several DNA probes for detecting EIEC are listed in FDA BAM.[46]

35.54 EHEC

Some of the characteristic virulence factors of EHEC include: Shiga toxins or verotoxins; the *eae* gene-encoded intimin protein that is associated with A/E lesions; and enterohemolysin encoded by a 60-mDal plasmid.[68] Of these, the Shiga toxins, i.e., Stx1 and Stx2, are closely correlated with human illness. The original assay for Stx was based on cytotoxic effect of Stx on Vero cells. The assay played an important role historically in establishing a diagnosis of EHEC infection in clinical laboratories. However, easier assays that test bacterial isolates for *stx* genes or the production of Stx have been developed. With few exceptions, most assays identify all *E. coli* serotypes that produce Stx, hence, they are not specific for serotype O157:H7. There are several commercially available antibody assays (Table 5) for detecting both Stx1 and Stx2, but only Verotox-F differentiates the two toxins. These assays should be performed according to the instructions provided by the manufacturer.

DNA probes also have been developed for detecting EHEC[46]; and most are targeted to the *eae* or *stx* genes. Samadpour et al.[79] used *stx* gene probes to detect STEC in calf fecal samples and foods, including ground beef, raw goat milk, blueberries, and surimi-based delicatessen salad. Two procedures were used: colony hybridization (colonies formed by plating of enrichment cultures were probed for STEC) and dot blotting (enrichment cultures were analyzed for STEC). A DNA probe (CVD419) based on the 60 MDa plasmid that is common among EHEC was developed by Levine et al.[56] The probe hybridized with most EHEC serotypes, isolated from patients with hemorrhagic colitis and HUS, including O157:H7, O26:H11. Using the same probe and an *eae* gene probe derived from EPEC, Barrett et al.[3] determined that both the *eae* and CVD419 sequences were significantly more common among human isolates than among cattle isolates.

There are many PCR assays for detecting *eae*, EHEC-*hlyA* encoded on the 60MDa plasmid, and *stx1* and *stx2* genes in EHEC isolates in pure culture,[32,51,63,73] but none of these assays is commercially available. Two multiplex PCR methods for *stx* genes for testing the toxigenic potential of isolates are listed in the FDA BAM. The PCR procedure detects only *stx1* and *stx2* genes, whereas the mismatch amplification mutation assay (MAMA) can simultaneously identify both toxin genes as well as the O157:H7 serotype based on a unique base mutation in the *uid*A gene.[10] The primers for these methods are shown in Table 6. The PCR parameters and conditions are described in the specific references.

Table 6. Primer sequences and expected size of products of PCR assays used in the FDA BAM for detecting Shiga-toxin genes in isolates

Assay	Target	Expected Size	Primer	Sequence
PCR	Stx1	274 bp	807F	5′-GCAATTCTGGGAAGCGTGGC-3′
			1095R	5′-CACAATCAGGCGTCGCCAGC-3′
	Stx2	364 bp	722F	5′-CTGGGGGCGAATCAGCAATG-3′
			1089R	5′-CCGCCGCCATTGCATTAACA-3′
MAMA	Stx1	348 bp	LP30	5′-CAGTTAATGTGGTGGCGAAGG-3′
			LP31	5′-CACCAGACAATGTAACCGCTG-3′
	Stx2	584 bp	LP43	5′-ATCCTATTCCCGGGAGTTTACG-3′
			LP44	5′-GCGTCATCGTATACACAGGAGC-3′
	UidA	252 bp	PT-2	5′-GCGAAAACTGTGGAATTGGG-3′
	(O157:H7)	252 bp	PT-2	5′-GCGAAAACTGTGGAATTGGG-3′
			PT-3	5′-TGATGCTCCATAACTTCCTG-3′

35.55 EAEC and DAEC

EAEC can be identified using HEp-2 assay or DNA probe or PCR assays. The characteristic aggregative adherence of EAEC on HEp-2 cells can be distinguished by prominent autoagglutination of the bacterial cells to each other, which often occurs on the surfaces of the cells, as well as on the glass coverslip free from the HEp-2 cells.[16] A DNA probe derived from a large plasmid of EAEC hybridized 56 (89%) of 63 EAEC strains examined (by HEp-2), compared to only 2 of 376 strains representing the normal flora and other diarrheagenic groups.[4] A PCR assay with primers derived from this EAEC probe also gave similar sensitivity and specificity.[80]

Methods used for identification of DAEC are similar to those used for EAEC. Presence of the diffusely adherent pattern in the HE-2 assay defines DAEC.[67] A DNA probe derived from the daaC gene necessary for expression of the F1845 fimbriae also has been used for the diagnosis of DAEC.[7]

35.6 REFERENCES

1. Acheson, D., and G. Keusch. 1995. *Shigella* and enteroinvasive *Escherichia coli*, p. 763-784. *In* M. J. Blaser, P. D. Smith, J. I. Ravdin, H. B. Greenberg, and R. L. Guerrant (eds.), Infections of gastrointestinal tract. Raven Press, New York.
2. Aleksic, S., H. Karch, and J. Bockemuhl 1992. A biotyping scheme for Shiga-like (Vero) toxin-producing *Escherichia coli* O157 and a list of serological cross-reactions between O157 and other gram-negative bacteria. Int. J. Med. Microbiol. Virol. Parasitol. Infect. Dis. 276:221-30.
3. Barrett, T., J. Kaper, A. Jerse, and I. Wachsmuth. 1992. Virulence factors in Shiga-like toxin-producing *Escherichia coli* isolated from humans and cattle. J. Infect. Dis. 165:979-980.

4. Baudry, B., S. Savarino, P. Vial, J. Kaper, and M. Levine. 1990. A sensitive and specific DNA probe to identify enteraggregative *Escherichia coli*, a recently discovered diarrheal pathogen. J. Infect. Dis. 161:1249-1251.
5. Bell, B. P., M.,Goldoft, P. M. Griffin, M. A. Davis, D. C. Gordon, P. I. Tarr, C. A. Lewis, J. H. Bartleson, T. J. Barrett, J. G. Wells, et al. 1994. A multistate outbreak of *Escherichia coli* O157:H7-associated bloody diarrhea and hemolytic uremic syndrome from hamburgers. The Washington experience. JAMA 272:1349-1353.
6. Bettelheim, K. A., H. Evangelidis, J. L. Pearce, E. Sowers, and N. A. Strockbine. 1993. Isolation of a *Citrobacter freundii* strain which carries the *Escherichia coli* O157 antigen. J. Clin. Microbiol. 31:760-761.
7. Bilge, S. S., J. M. Apostol, Jr., K. J. Fullner, and S. L. Moseley. 1993. Transcriptional organization of the F1845 fimbrial adhesin determinant of *Escherichia coli*. Mol. Microbiol. 7:993-1006.
8. Borczyk, A. A., N. Harnett, M. Lombos, and H. Lior. 1990. False-positive identification of *Escherichia coli* O157 by commercial latex agglutination tests. Lancet 336:946-947.
9. Borczyk, A. A., H. Lior, and S. Thompson. 1989. Sorbitol-negative *Escherichia coli* O157 other than H7. J. Infect .18:198-199.
10. Cebula, T. A., W. L. Payne, and P. Feng. 1995. Simultaneous identification of strains of *Escherichia coli* serotype O157:H7 and their Shiga-like toxin type by mismatch amplification mutation assay-multiplex PCR. J. Clin. Microbiol. 33:248-250.
11. Centers for Disease Control and Prevention. 1994. Foodborne outbreaks of enterotoxigenic *Escherichia coli*—Rhode Island and New Hampshire, 1993. Morb. Mortal Wkly. Rep. 43:81, 87-89.
12. Centers for Disease Control and Prevention. 1997. Outbreaks of *Escherichia coli* O157:H7 infection associated with eating alfalfa sprouts —Michigan and Virginia, June–July 1997. Morb. Mortal. Wkly. Rep. 46:741-744.
13. Centers for Disease Control and Prevention. 1997. Surveillance for outbreaks of *Escherichia coli* O157:H7 infection—preliminary summary of 1996 data. Personal communication.
14. Chapman, P. A., C. A. Siddons, D. J. Wright, P. Norman, J. Fox, and E. Crick. 1993. Cattle as a possible source of verocytotoxin-producing *Escherichia coli* O157 infections in man. Epidemiol. Infect. 111:439-447.
15. Cobeljic, M., B. Miljkovic-Selimovic, D. Paunovic-Todosijevic, Z. Velickovic, Z. Lepsanovic, N. Zec, D. Savic, R. Ilic, S. Konstantinovic, B. Jovanovic, and V. Kostic. 1996. Enteroaggregative *Escherichia coli* associated with an outbreak of diarrhoea in a neonatal nursery ward. Epidemiol. Infect. 117:11-16.
16. Cravioto, A., R. Gross, S. Scotland, and B. Rowe. 1979. An adhesive factor found in strains of *Escherichia coli* belonging to the traditional infantile enteropathogenic serotypes. Curr. Microbiol. 3:95-99.
17. Cryan, B. 1990. Comparison of the synthetic oligonucleotide gene probe and infant mouse bioassay for detection of enterotoxigenic *Escherichia coli*. Eur. J. Clin. Microbiol. Infect. Dis. 9:229-232.
18. Cryan, B. 1990. Comparison of three assay systems for detection of enterotoxigenic *Escherichia coli* heat-stable enterotoxin. J. Clin. Microbiol. 28:792-794.
19. Deng, M. Y., D. O. Cliver, S. P. Day, and P. M. Fratamico. 1996. Enterotoxigenic *Escherichia coli* detected in foods by PCR and an enzyme-linked oligonucleotide probe. Int. J. Food Microbiol. 30:217-229.
20. Donnenberg, M. 1995. Enteropathogenic *Escherichia coli*, p. 700-726. *In* M. J. Blaser, P. D. Smith, J. I. Ravdin, H. B. Greenberg, and R. L. Guerrant (eds.), Infections of gastrointestinal tract. Raven Press, New York.
21. Doyle, M. P., and V. V. Padhye. 1989. *Escherichia coli*, p. 236-282. *In* M. P. Doyle (ed.), Foodborne bacterial pathogens. Marcel Dekker, New York,
22. Doyle, M. P., and J. L. Schoeni. 1984. Survival and growth characteristics of *Escherichia coli* associated with hemorrhagic colitis. Appl. Environ. Microbiol. 48:855-856.
23. Doyle, M. P., T. Zhao, J. Meng, and S. Zhao. 1997. *Escherichia coli* O157:H7, p. 171-191. *In* M. Doyle, L. Beuchat, and T. Montville (eds.), Food microbiology—Fundamentals and frontiers. ASM Press, Washington, D. C.
24. Farmer, J. J. D., and B. R. Davis. 1985. H7 antiserum-sorbitol fermentation medium: A single tube screening medium for detecting *Escherichia coli* O157:H7 associated with hemorrhagic colitis. J. Clin. Microbiol. 22:620-625.
25. Feng, P. 1996. Emergence of rapid methods for identifying microbial pathogens in foods. J. AOAC Int. 79:809-812.
26. Feng, P. 1997. Impact of molecular biology on the detection of foodborne pathogens. Mol. Biotechnol. 7:267-278.
27. Feng, P., K. A. Lampel, H. Karch, and T. S. Whittam. 1998. Genotypic and phenotypic changes in the emergence of *Escherichia coli* O157:H7. J. Infect. Dis. 177:1750-1753.
28. Feng, P. C., and P. A. Hartman. 1982. Fluorogenic assays for immediate confirmation of *Escherichia coli*. Appl. Environ. Microbiol. 43:1320-1329.
29. Fields, P. I., K. Blom, H. J. Hughes, L. O. Helsel, P. Feng, and B. Swaminathan. 1997. Molecular characterization of the gene encoding H antigen in *Escherichia coli* and development of a PCR-restriction fragment length polymorphism test for identification of *E. coli* O157:H7 and O157:NM. J. Clin. Microbiol. 35:1066-1070.
30. Franke, J., S. Franke, H. Schmidt, A. Schwarzkopf, L. H. Wieler, G. Baljer, L. Beutin, and H. Karch. 1994. Nucleotide sequence analysis of enteropathogenic *Escherichia coli* (EPEC) adherence factor probe and development of PCR for rapid detection of EPEC harboring virulence plasmids. J. Clin. Microbiol. 32:2460-2463.
31. Frankel, G., J. A. Giron, J. Valmassoi, and G. K. Schoolnik. 1989. Multigene amplification: Simultaneous detection of three virulence genes in diarrhoeal stool. Mol. Microbiol. 3:1729-1734.
32. Fratamico, P., K. S. Solomon, M. Wiedmann, and M. Y. Deng. 1995. Detection of *Escherichia coli* O157:H7 by multiplex PCR. J. Clin. Microbiol. 33:2188-2191.
33. Fratamico, P. M., R. L. Buchanan, and P. H. Cooke. 1993. Virulence of an *Escherichia coli* O157:H7 sorbitol-positive mutant. Appl. Environ. Microbiol. 59:4245-4252.
34. Giannella, R. A. 1976. Suckling mouse model for detection of heat-stable *Escherichia coli* enterotoxin: Characteristics of the model. Infect. Immun. 14: 95-99.
35. Glass, K., J. Loeffelholz, J. Ford, and M. Doyle. 1992. Fate of *Escherichia coli* O157:H7 as affected by pH or sodium chloride in fermented, dry sausage. Appl. Environ. Microbiol. 58:2513-2516.
36. Gomes, T. A., M. R. Toledo, L. R. Trabulsi, P. K. Wood, and J. G. Morris, Jr. 1987. DNA probes for identification of enteroinvasive *Escherichia coli*. J. Clin. Microbiol. 25:2025-2027.
37. Gordillo, M. E., G. R. Reeve, J. Pappas, J. J. Mathewson, H. L. DuPont, and B. E. Murray, 1992. Molecular characterization of strains of enteroinvasive *Escherichia coli* O143, including isolates from a large outbreak in Houston, Texas. J. Clin. Microbiol. 30:889-893.
38. Griffin, P. M. 1995. *Escherichia coli* O157:H7 and other enterohemorrhagic *Escherichia coli*, p. 739-761. In M. J. Blaser, P. D. Smith, J. I. Ravdin, H. B. Greenberg, and R. L. Guerrant (eds.), Infections of gastrointestinal tract. Raven Press, New York.
39. Griffin, P. M., and R. V. Tauxe.1991. The epidemiology of infections caused by *Escherichia coli* O157:H7, other enterohemorrhagic *E. coli*, and the associated hemolytic uremic syndrome. Epidemiol. Rev. 13:60-98.
40. Gunzer, F., H. Bohm, H. Russmann, M. Bitzan, S. Aleksic, and H. Karch. 1992. Molecular detection of sorbitol-fermenting *Escherichia coli* O157 in patients with hemolytic-uremic syndrome. J. Clin. Microbiol. 30:1807-1810.
41. Hammack, T. S., P. Feng, R. M. Amaguana, G. A. June, P. S. Sherrod, and W. H. Andrews. 1997. Comparison of sorbitol MacConkey and hemorrhagic coli agars for recovery of *Escherichia coli* O157:H7 from brie, ice cream, and whole milk. J. AOAC Int. 80:335-340.
42. Hancock, D., T. Besser, and D. Rice. 1998. Ecology of *Escherichia coli* O157:H7 in cattle and impact of management practices, p. 85-91. *In* J. Kaper and A. O'Brien (eds.), *Escherichia coli* O157:H7 and other Shiga toxin-producing *E. coli* strains. ASM Press, Washington, D. C.

43. Hancock, D. D., T. E. Besser, M. L. Kinsel, P. I. Tarr, D. H. Rice, and M. G. Paros, 1994. The prevalence of *Escherichia coli* O157.H7 in dairy and beef cattle in Washington State. Epidemiol. Infect. 113:199-207.
44. Hayes, P. S., K. Blom, P. Feng, J. Lewis, N. A. Strockbine, and B. Swaminathan. 1995. Isolation and characterization of a beta-D-glucuronidase-producing strain of *Escherichia coli* serotype O157:H7 in the United States. J. Clin. Microbiol. 33:3347-3348.
45. Hedberg, C. W., S. J. Savarino, J. M. Besser, C. J. Paulus, V. M. Thelen, L. J. Myers, D. N. Cameron, T. J. Barrett, J. B. Kaper, and M. T. Osterholm. 1997. An outbreak of foodborne illness caused by *Escherichia coli* O39:NM, an agent not fitting into the existing scheme for classifying diarrheogenic *E. coli*. J. Infect. Dis. 176:1625-1628.
46. Hill, W., A. Datta, P. Feng, K. Lampel, and W. Payne. 1998. Identification of foodborne bacterial pathogens by gene probes, p. 24.01. *In* FDA bacteriological analytical manual. Association of Official Analytical Chemistry International, Gaithersburg, Md.
47. Hill, W. E., and C. L. Carlisle. 1981. Loss of plasmids during enrichment for *Escherichia coli*. Appl. Environ. Microbiol. 41:1046-1048.
48. Hitchins, A., P. Feng, W. Watkins, S. Rippey, and L. Chandler. 1998. *Escherichia coli* and the coliform bacteria, p. 4.01-4.29. *In* FDA bacteriological analytical manual. Association of Official Analytical Chemistry International, Gaithersburg, Md.
49. Hitchins, A., P. Hartman, and E. Todd. 1992. Coliforms—*Escherichia coli* and its toxins, p. 325-367. *In* C. Vanderzant and D. Splittstoesser (eds.), Compendium of methods for microbiological examination of foods. American Public Health Association, Washington, D. C.
50. Jerse, A. E., W. C. Martin, J. E. Galen, and J. B. Kaper. 1990. Oligonucleotide probe for detection of the enteropathogenic *Escherichia coli* (EPEC) adherence factor of localized adherent EPEC. J. Clin. Microbiol. 28:2842-2844.
51. Karch, H., and T. Meyer. 1989. Evaluation of oligonucleotide probes for identification of Shiga-like toxin producing *Escherichia coli*. J. Clin. Microbiol. 27:1180-1186.
52. Keene, W. E., E. Sazie, J. Kok, D. H. Rice, D. D. Hancock, V. K. Balan, T. Zhao, and M. P. Doyle. 1997. An outbreak of *Escherichia coli* O157:H7 infections traced to jerky made from deer meat. JAMA 277:1229-1231.
53. Knutton, S., T. Baldwin, P. H. Williams, and A. S. McNeish, 1989. Actin accumulation at sites of bacterial adhesion to tissue culture cells: Basis of a new diagnostic test for enteropathogenic and enterohemorrhagic *Escherichia coli*. Infect. Immun. 57:1290-1298.
54. Kopecko, D. 1994. Experimental keratoconjunctivitis (Sereny) assay, p. 39-46. *In* V. Clark and P. Bavoil (eds.), Bacterial pathogenesis, part A. Academic Press, San Diego, Calif.
55. Kudva, I. T., P. G. Hatfield, and C. J. Hovde. 1996. *Escherichia coli* O157:H7 in microbial flora of sheep. J. Clin. Microbiol. 34:431-433.
56. Levine, M., J. Xu, J. Kaper, H. Lior, V. Prado, B. Tall, J. Nataro, H. Karch, and K. Wachsmuth. 1987. A DNA probe to identify enterohemorrhagic *Escherichia coli* of O157:H7 and other serotypes that cause hemorrhagic colitis and hemolytic uremic syndrome. J. Infect. Dis. 156:175-182.
57. Levine, M. M. 1987. *Escherichia coli* that cause diarrhea: Enterotoxigenic, enteropathogenic, enteroinvasive, enterohemorrhagic, and enteroadherent. J. Infect. Dis. 155:377-389.
58. Lior, H., and A. A. Borczyk. 1987. False positive identifications of *Escherichia coli* O157. Lancet 1:333.
59. Louie, M., J. de Azavedo, R. Clarke, A. Borczyk, H. Lior, M. Richter, and J. Brunton. 1994. Sequence heterogeneity of the *eae* gene and detection of verotoxin- producing *Escherichia coli* using serotype-specific primers. Epidemiol. Infect. 112:449-461.
60. March, S. B., and S. Ratnam. 1986. Sorbitol-MacConkey medium for detection of *Escherichia coli* O157:H7 associated with hemorrhagic colitis. J. Clin. Microbiol. 23:869-872.
61. Marier, R., J. Wells, R. Swanson, W. Callahan, and I. Mehlman. 1973. An outbreak of enteropathogenic *Escherichia coli* foodborne disease traced to imported French cheese. Lancet 2:1376-1379.
62. Meng, J., and M. P. Doyle. 1998. Microbiology of Shiga toxin-producing *Escherichia coli* in foods, p. 92-111. *In* J. Kaper and A. O'Brien (eds.), *Escherichia coli* O157:H7 and other Shiga toxin-producing *E. coli* strains. ASM Press, Washington, D. C.
63. Meng, J., S. Zhao, M. Doyle, S. Mitchell, and S. Kresovich. 1997. A multiplex PCR for identifying Shiga-like toxin-producing *Escherichia coli* O157:H7. Lett. Appl. Microbiol. 24:172-176.
64. Meng, J., S. Zhao, M. P. Doyle, S. E. Mitchell, and S. Kresovich. 1996. Polymerase chain reaction for detecting *Escherichia coli* O157: H7. Int. J. Food Microbiol. 32:103-113.
65. Mitsuda, T., T. Muto, M. Yamada, N. Kobayashi, M. Toba, Y. Aihara, A. Ito, and S.Yokota. 1998. Epidemiological study of a food-borne outbreak of enterotoxigenic *Escherichia coli* O25:NM by pulsed-field gel electrophoresis and randomly amplified polymorphic DNA analysis. J. Clin. Microbiol. 36:652-656.
66. Morgan, G., C. Newman, S. Palmer, J. Allen, W. Shepherd, A. Rampling, R. Warren, R. Gross, S. Scotland, and H. Smith. 1988. First recognized community outbreak of haemorrhagic colitis due to verotoxin-producing *Escherichia coli* O157:H7 in the UK. Epidemiol. Infect. 101:83-91.
67. Nataro, J. 1995. Enteroaggregative and diffusely adherent *Escherichia coli*, p. 727-737. *In* M. J. Blaser, P. D. Smith, J. I. Ravdin, H. B. Greenberg, and R. L. Guerrant (eds.), Infections of gastrointestinal tract. Raven Press, New York.
68. Nataro, J. P., and J. B. Kaper. 1998. Diarrheagenic *Escherichia coli*. Clin. Microbiol. Rev. 11:142-201.
69. Okrend, A., B. Rose, and B. Bennett. 1990. A screening method for the isolation of *Escherichia coli* O157:H7 from ground beef. J. Food Prot. 53:249-252.
70. Okrend, A., B. Rose, and C. Lattuada. 1990. Use of 5-bromo-4-chloro-3-indoxyl-β-D-glucuronide in MacConkey sorbitol agar to aid in the isolation of *Escherichia coli* O157:H7 from ground beef. J. Food Prot. 53:941-943.
71. Padhye, N. V., and M. P. Doyle. 1991. Production and characterization of a monoclonal antibody specific for enterohemorrhagic *Escherichia coli* of serotypes O157:H7 and O26:H11. J. Clin. Microbiol. 29:99-103.
72. Park, C. H., E. A. Martin, and E. L. White. 1998. Isolation of a non-pathogenic strain of *Citrobacter sedlakii* which expresses *Escherichia coli* O157 antigen. J. Clin. Microbiol. 36:1408-1409.
73. Read, S. C., R. C. Clarke, A. Martin, S. A. De Grandis, J. Hill, S. McEwen, and C. L. Gyles. 1992. Polymerase chain reaction for detection of verocytotoxigenic *Escherichia coli* isolated from animal and food sources. Mol. Cell. Probes 6:153-161.
74. Reitsma, C., and D. Henning. 1996. Survival of enterohemorrhagic *Escherichia coli* O157:H7 during the manufacture and curing of Cheddar cheese. J. Food Prot. 59:460-464.
75. Rice, E. W., E. G. Sowers, C. H. Johnson, M. E. Dunnigan, N. A. Strockbine, and S. C. Edberg. 1992. Serological cross-reactions between *Escherichia coli* O157 and other species of the genus *Escherichia*. J. Clin. Microbiol. 30:1315-1316.
76. Riley, L. W., R. S. Remis, S. D. Helgerson, H. B. McGee, J. G. Wells, B. R. Davis, R. J. Hebert, E. S. Olcott, L. M. Johnson, N. T. Hargrett, P. A. Blake, and M. L. Cohen. 1983. Hemorrhagic colitis associated with a rare *Escherichia coli* serotype. N. Engl. J. Med. 308:681-685.
77. Roels, T. H., M. E. Proctor, L. C. Robinson, K. Hulbert, C. A. Bopp, and J. P. Davis. 1998. Clinical features of infections due to Escherichia coli producing heat- stable toxin during an outbreak in Wisconsin: A rarely suspected cause of diarrhea in the United States. Clin. Infect. Dis. 26:898-902.
78. Salyers, A., and D. Whitt. 1994. Bacterial pathogenesis—A molecular approach. ASM Press, Washington, D. C.
79. Samadpour, M., J. E. Ongerth, J. Liston, N. Tran, D. Nguyen, T. S. Whittam, R. A. Wilson, and P. I. Tarr. 1994. Occurrence of Shiga-like toxin-producing *Escherichia coli* in retail fresh seafood, beef, lamb,

pork, and poultry from grocery stores in Seattle, Washington. Appl. Environ. Microbiol. 60:1038-1040.

80. Schmidt, H., C. Knop, S. Franke, S. Aleksic, J. Heesemann, and H. Karch. 1995. Development of PCR for screening of enteroaggregative *Escherichia coli*. J. Clin. Microbiol. 33: 701-705.
81. Smith, H. R., S. M. Scotland, G. A. Willshaw, B. Rowe, A. Cravioto, and C. Eslava. 1994. Isolates of *Escherichia coli* O44:H18 of diverse origin are enteroaggregative. J. Infect. Dis. 170:1610-1613.
82. Sowers, E. G., J. G. Wells, and N. A. Strockbine. 1996. Evaluation of commercial latex reagents for identification of O157 and H7 antigens of *Escherichia coli*. J. Clin. Microbiol. 34:1286-1289.
83. Szabo, R., E. Todd, and A. Jean. 1986. Method to isolate *Escherichia coli* O157:H7 from food. J. Food. Prot. 49:768-772.
84. Takeda, Y. 1997. Enterohaemorrhagic *Escherichia coli*. World Health Stat. Q. 50:74-80.
85. Wight, J. P., P. Rhodes, P. A. Chapman, S. M. Lee, and P. Finner. 1997. Outbreaks of food poisoning in adults due to *Escherichia coli* O111 and campylobacter associated with coach trips to northern France. Epidemiol. Infect. 119:9-14.
86. Yamamura, K., N. Sumi, Y. Egashira, I. Fukuoka, S. Motomura, and R. Tsuchida. 1992. Food poisoning caused by enteroinvasive Escherichia coli (O164:H-)—A case in which the causative agent was identified. Kansenshogaku Zasshi 66:761-768.
87. Zadik, P. M., P. A. Chapman, and C. A. Siddons. 1993. Use of tellurite for the selection of verocytotoxigenic *Escherichia coli* O157. J. Med. Microbiol. 39:155-158.
88. Zhao, T., and M. Doyle. 1994. Fate of enterohemorrhagic *Escherichia coli* O157:H7 in commercial mayonnaise. J. Food Prot. 57:780-783.
89. Zhao, T., M. Doyle, and R. Besser. 1993. Fate of enterohemorrhagic *Escherichia coli* O157:H7 in apple cider with and without preservatives. Appl. Environ. Microbiol. 59:2526-2530.
90. Zhao, T., Doyle, M. P., Shere, J. and Garber, L. 1995. Prevalence of enterohemorrhagic *Escherichia coli* O157:H7 in a survey of dairy herds. Appl Environ Microbiol 61: 1290-1293.

Chapter 36

Listeria

E. T. Ryser and C. W. Donnelly

36.1 INTRODUCTION

36.11 Description and Taxonomy of the Genus

Listeria monocytogenes was first definitively described by Murray et al.[89] in 1926 in conjunction with an outbreak of disease among laboratory rabbits at Cambridge University. These researchers succeeded in both isolating the organism from the blood of infected rabbits and reinfecting healthy animals, thereby establishing the organism's pathogenicity. They named the organism *Bacterium monocytogenes* after the mononucleosis-like illness that was observed. The following year, Pirie[94] documented a similar outbreak in South Africa involving wild gerbils. This disease, termed "Tiger River Disease," was characterized by marked liver involvement with the causative agent named *Listerella hepatolytica* in honor of Lord Lister. Both isolates were soon shown to be identical and the name was changed to *Listerella monocytogenes*.[88] However, after learning that the generic name *Listerella* had been previously used to describe a slime mold and a marine protozoan,[48] the name was changed to *Listeria monocytogenes* in 1940.[95] Prior to 1940, *Listeria monocytogenes* was already recognized as an organism capable of infecting both ruminant animals and humans. In 1929, Gill[49] was first to describe an illness in sheep that he called "circling disease." This term is still used to describe listerial encephalitis, encephalomyelitis, and meningoencephalitis, which are the most common manifestations of listeriosis in sheep, cows, and goats.[107] During the same year, Nyfeldt[90] isolated *L. monocytogenes* from the blood of three human patients who had developed symptoms resembling infectious mononucleosis. By 1935, *L. monocytogenes* was well established as a cause of meningitis and perinatal septicemia in the United States.[18] However, this organism remained a relatively obscure human pathogen until the widely publicized foodborne outbreaks of the mid 1980s.

Listeria monocytogenes is a short, gram positive, non-spore-forming, rod-shaped bacterium that can appear coccoidal or filamentous in older cultures.[98] Cells are found both singly and in short chains, as well as in V forms, Y forms, and palisades. Consequently, members of the genus *Listeria* have been occasionally misidentified as *Corynebacterium* spp., *Haemophilus influenza*, *Erysipelothrix* spp., pneumococci, streptococci, or staphylococci.[15] The organism grows both aerobically and anaerobically but prefers a microaerophilic environment. Rapid growth occurs on most commonly used bacteriological media. Broth cultures typically become turbid within 8–24 hr of incubation at 35°C. When grown on nutrient agar, *Listeria* colonies are typically smooth, bluish gray and slightly raised and measure 0.2–0.8 mm in diameter after 24 hr of incubation. Cultures on clear media exhibit a characteristic blue-green iridescence when examined with a binocular microscope under obliquely transmitted light[55,75] which makes such colonies readily discernable, even among high numbers of contaminants. *Listeria* generally multiplies at temperatures ranging from 1 to 45°C[70] with optimum growth occurring at 30 to 37°C. Consequently, *L. monocytogenes* is considered a psychrotrophic foodborne pathogen with cold enrichment first used to recover this organism. Characteristic umbrella-shaped growth occurs in tubed motility media when incubated at 25°C, but not 35°C, with tumbling motility also visible in wet mounts. Although *Listeria* will grow in laboratory media over a pH range of 4.4 to 9.6 (optimal growth at pH 7), this organism is acid tolerant and can survive in foods of similar acidity for days or weeks. *Listeria* is one of the few foodborne pathogens that can grow at an a_w of 0.90.[87] Growth of this organism occurs in laboratory media containing up to 10% NaCl[120] with survival in more concentrated brine solutions also reported.[104]

Presumptive identification of *Listeria* is based on colony morphology, Gram stain reaction, tumbling motility, catalase reaction, and, for *L. monocytogenes* β-hemolysis on blood agar, which is closely linked to pathogenicity. All six currently recognized *Listeria* spp. show the following biochemical reactions: catalase positive, oxidase negative, fermentation of carbohydrates to acid without gas, hydrolysis of esculin and sodium hippurate, methyl red positive, ammonia production from arginine, and negative reactions for hydrogen sulfide production, indole, nitrate reductase, gelatin liquefaction, and starch and urea hydrolysis.[65]

For many years, the genus *Listeria* was monospecific, containing only the type species, *L. monocytogenes*. However, following several additions and subtractions, the genus *Listeria* now contains six species: *L. monocytogenes*, *L. innocua*, *L. ivanovii*, *L. welshimeri*, *L. seeligeri* and *L. grayi*, as evidenced by DNA homology values, 16S rRNA sequencing homology, chemotaxonomic properties and multilocus enzyme electrophoresis.[22,99,100,113] Although phenotypically very similar, these six *Listeria* spp. can be readily distinguished by the following five tests: hemolysin production, and acid production from D-xylose, L-rhamnose, a-methyl-D-mannoside, and mannitol, with hemolytic activity clearly the most important and most difficult characteristic to detect for identification of *L. monocytogenes*.[84]

36.12 Ecology

Listeria monocytogenes is widely distributed in the environment and has been isolated from a variety of sources including soil,[129,132] mud,[129,132] silage,[37,59] decaying vegetation,[131] water,[28] sewage,[2,128] and feces.[58,66] However, the primary habitat for *L. monocytogenes* appears to be in soil and vegetation where the bacterium leads a saprophytic existence,[38,130] with soil serving as a reservoir for later infections transmitted to animals and humans.[129] Improperly fermented or moldy silage having a pH > 5 will support the growth and extended survival of a diverse group of *Listeria* strains including *L. monocytogenes*,[36,108] with such silage formerly cited as a source of infection in cows, sheep and goats.[37,39,50] The relationship between ingestion of *Listeria*-contaminated silage, mastitis in dairy cattle and subsequent asymptomatic shedding of listeriae in milk destined for human consumption also has been documented.[7,115] However, changes in production methods have now reduced levels of *L. monocytogenes* in silage, which in turn has led to a considerable decrease in the incidence of listeriosis in silage-fed animals.[134]

Listeria is ubiquitous and has been associated with a variety of mammals, birds, fish amphibians, and insects.[38,56] Numerous animal species are susceptible to listerial infections, with many healthy animals cited as asymptomatic fecal carriers of *L. monocytogenes*. Although most infections are subclinical, listeriosis in animals can occur either sporadically or as epidemics, often leading to fatal forms of encephalitis. Virtually all domestic animals are susceptible to infection,[82] with sheep, cattle, goats, and less frequently chickens succumbing to listeriosis.[7] In humans, fecal carriage rates have reportedly varied from 0 to 77% depending on the extent of exposure, with approximately 5% of the general population assumed to be asymptomatic shedders of *L. monocytogenes*.[82]

36.13 Disease Syndrome

The disease listeriosis is a frequent cause of encephalitis, septicemia, and abortion in cattle sheep and goats and occurs most frequently during stressful climactic conditions of winter and early spring. Clinically healthy ruminants can act as asymptomatic carriers of *L. monocytogenes* and intermittently secrete the organism intracellularly within neutrophils and macrophages in their milk for months over several lactation periods.[56] Numerous surveys have shown that 2–4% of the raw milk produced in the United States can be expected to contain low levels (generally < 10 CFU/mL) of *L. monocytogenes*, with similar contamination rates likely occurring elsewhere.[103] Cattle, sheep and goats frequently shed *L. monocytogenes* in feces and manure, and these materials, along with spoiled silage and soil, appear to comprise the primary vehicles of listerial infection for ruminants. Although birds are susceptible to listeriosis, with up to 33 % of healthy chickens identified as fecal shedders of *L. monocytogenes*, clinical symptoms are uncommon in domestic fowl.[134]

Human listeriosis is most often seen as an invasive illness in certain well-defined high-risk groups, including immunocompromised adults, pregnant women and neonates, but may also occasionally occur in individuals with no predisposing conditions.[122] Widespread use of immunosuppressive medications for treating malignancies and managing organ transplantations in combination with an aging population and the continued epidemic of acquired immunodeficiency syndrome,[32] have greatly expanded the number of individuals at risk of acquiring listeriosis. Unlike most other foodborne illnesses, listeriosis may take several months to develop after initial exposure and exhibit a mortality rate of approximately 20%.[44] An active surveillance program in the United States has shown bacteremia to be the most common manifestation of listeriosis in non-pregnant adults, followed by meningitis and meningoencephalitis.[45] Patients with bacteremia most often experience fever, malaise, fatigue, and abdominal pain, while those individuals with central nervous system involvement develop fever, malaise, ataxia, seizures and altered mental status. Additional manifestations of infection have included endocarditis, endophthalmitis, septic arthritis, osteomyelitis, and peritinitis,[122] with non-bacteremic cutaneous infections and conjunctivitis primarily confined to accidentally exposed laboratory workers and individuals handling infected animals.[5,92] Among pregnant women, listeriosis is most often reported during the third trimester, with such individuals typically exhibiting only a mild flu-like illness. However, fetal infections can result from either transplacental transmission of *L. monocytogenes* to the fetus via the bloodstream or through ascending colonization of the vagina. Such intrauterine infections can result in premature delivery, spontaneous abortion, stillbirth, or early-onset neonatal listeriosis, which is characterized by sepsis, granulomatosis infantiseptica (a necrotic disease affecting the internal organs) and a mortality rate of 20 to 30%.[77] A late-onset form of neonatal listeriosis characterized by a highly fatal form of meningitis can also develop several weeks after birth as a result of infection during delivery.[45] Diagnosis of listeriosis is normally dependent on the isolation of *L. monocytogenes* from a normally sterile site such as blood or cerebrospinal fluid. Successful treatment of invasive listeriosis usually involves the prompt administration of high doses of penicillin, ampicillin, or trimethoprim-sulfamethoxazole for periods of 2 to 4 weeks.[8] Evidence from several recent foodborne outbreaks indicates that *L. monocytogenes* can produce a mild, self-limiting non-invasive form of gastrointestinal illness in normal hosts characterized by the development of fever, diarrhea, headaches and myalgia within 12 to 24 hours of exposure.[27,111]

36.14 Epidemiology

Five major outbreaks have occurred in North America. In 1981, 41 cases of listeriosis involving 7 adults and 34 infants were traced to consumption of *Listeria*-contaminated coleslaw in the Maritime provinces of Canada, making this the first confirmed outbreak of foodborne listeriosis recorded in the literature.[116] Fifteen infants (44%) and 2 adults died, giving an overall mortality rate of 41%. Coleslaw was prepared from cabbage that was harvested from fields known to have been fertilized with both raw and composted manure from sheep, two animals of which previously died from listeriosis. Contamination of the cabbage with *L. monocytogenes* serotype 4b was compounded by cold storage of the cabbage from October until early spring, which likely permitted some growth of the pathogen.

During the summer of 1983, 49 cases of listeriosis in Massachusetts involving 42 immunocompromised adults and 7 mother-infant pairs were epidemiologically linked to consumption of a specific brand of whole and 2% milk that met the legal requirements for pasteurization.[42] Fourteen patients died, giving a mortality rate of 29%. Unlike the previous outbreak involving coleslaw, *L. monocytogenes* was never isolated from the incriminated product. Although 15 samples of raw milk from the factory, as well as from several farms and a milk cooperative supplying the factory, did yield various isolates of *L. monocytogenes*, the epidemic strain of *L. monocytogenes* serotype 4b was never recovered, which in turn raises some questions as to the exact role of milk in this outbreak.[102]

In June of 1985, *L. monocytogenes* emerged as a serious foodborne pathogen when consumption of contaminated Jalisco brand Mexican-style cheese was responsible for nearly 300 listeriosis cases, including 85 deaths, among Hispanics in Southern California.[102] A total of 142 cases involving 93 pregnant women or their offspring and 49 non-pregnant immunocompromised adults were documented in Los Angeles County.[76] Forty-eight of these individuals died, giving a mortality rate of 33.8%. The patient and cheese isolates of *L. monocytogenes* serotype 4b were identical using phage typing and several DNA fingerprinting methods,[102] thereby confirming the role of cheese in this outbreak. Subsequent factory inspections suggested that the implicated cheese was illegally manufactured from a combination of raw and pasteurized milk. Additional contributing factors to this outbreak included widespread occurrence of *L. monocytogenes* in the cheese plant environment and probable growth in the final product at a pH of 6.6.

In July of 1994, Dalton et al.[27] reported that 54 of 60 (90%) previously healthy individuals developed listeriosis 9–32 hr after consuming pasteurized chocolate milk at a picnic in Illinois, with 12 additional cases also documented in Illinois, Wisconsin and Michigan. Unlike the aforementioned outbreaks, gastrointestinal symptoms (diarrhea, fever, chills, nausea, and vomiting) predominated. Additionally, only four victims required short hospitalization, with one pregnant woman delivering a healthy baby 5 days after experiencing a 6-hr bout of diarrhea. The epidemic strain of *L. monocytogenes* serotype 1/2b was recovered from unopened containers of chocolate milk at levels of 10^8 to 10^9 CFU/mL, with the product's taste and quality reportedly poor. Post-pasteurization contamination of the milk from the factory environment, inadequate and/or non-existent refrigeration during packaging and transit, and probable growth of the pathogen during this period were cited as contributing factors.

During August 1998 to February 1999, consumption of hot dogs was associated with the second worst outbreak of listeriosis in the US with 101 cases of reported in 22 states.[3,4] Over 80% of the victims were adults (median age of 70), > 60% of which suffered from an underlying illness. A total of 21 fatalities were reported (15 deaths and 6 miscarriages), giving a mortality rate of 21%. The epidemic strain of *L. monocytogenes* belonged to serotype 4b, which has been responsible for most other foodborne outbreaks. However, when further characterized by DNA fingerprinting, the organism belonged to pulsed-field gel electrophoresis type E which had seldom been seen in the United States or elsewhere. While the exact source of this epidemic strain and the route of contamination remain obscure, the start of the outbreak did coincide with the removal of a large ventilation unit near the hot dog packaging line.

Elsewhere, major outbreaks of invasive listeriosis have been traced to consumption of Vacherin Mont d'Or soft-ripened cheese in Switzerland (122 cases) during 1983–1987[16] and to paté in the United Kingdom (366 cases) during 1987–1989.[85] Three additional outbreaks also occurred in France in which jellied pork tongue (279 cases),[114] pork paté "rilletes" (39 cases)[67] and Brie de Meaux cheese prepared from raw milk (33 cases)[53] were identified as vehicles of infection.

36.15 Foods Contaminated by Listeria monocytogenes

Listeria monocytogenes is a frequent food contaminant, with this pathogen present in 2–4% of the raw milk supply and also commonly recovered from raw meat, poultry and seafood, as well as numerous varieties of processed dairy, meat, fish, and seafood and delicatessen products.[107] *Listeria* most often enters cooked/ready-to eat foods as a post-processing contaminant, with this organism typically found within the manufacturing environment. Confirmed vehicles of infection in cases of listeriosis have included coleslaw,[116] chocolate milk,[27] various soft and surface-ripened cheeses,[16,67,102] paté,[85] jellied pork tongue,[114] cooked chicken,[71] and smoked mussels.[9] Five of the nine major outbreaks described earlier were linked to dairy products. *L. monocytogenes* has the ability to survive in most cheeses during manufacture and storage[104,105,109] and attain high levels in certain soft surface-ripened varieties such as Camembert during ripening.[106] Similar multiplication of *Listeria* also can occur in paté during refrigerated storage. Although rapid growth of *L. monocytogenes* has been reported in processed meats,[51] liquid eggs,[121] various seafoods including smoked salmon,[60] growth of *L. monocytogenes* in paté and certain soft surface-ripened cheese having pH values > 6.5 appear to pose the greatest threat of listeriosis.

36.16 Response to Environmental Stress

Growth, inhibition and inactivation of *L. monocytogenes* in a food or food-processing environment is linked to the organism's ability to withstand the combined effects of various environmental stresses and food processing/preservation techniques. The unusual thermal tolerance of *L. monocytogenes* has been particularly well documented,[78] with numerous earlier studies showing that *L. monocytogenes* could survive during pasteurization of milk.[12,17,31,72] However, later work[17,81] has demonstrated that this pathogen is somewhat less heat resistant with representatives from the Centers for Disease Control and Prevention (CDC), U. S. Food and Drug Administration (FDA), and the World Health Organization (WHO)[17] now confirming that high-temperature short- time pasteurization (71.7°C/15 sec) is sufficient to completely inactivate normally occurring populations of *L. monocytogenes* in raw milk.

Listeria monocytogenes also exhibits varying degrees of resistance to acid,[1,23,91,123] salt,[78] freezing,[47] and drying.[69] According to Parish and Higgins,[93] *L. monocytogenes* survived 21 and > 90 days in orange juice of pH 3.6 and 5.8, respectively, during storage at 4°C with decreased acid resistance reported at higher storage temperatures. Given the psychrotrophic nature of *L. monocytogenes*, proper acid development is critical to minimize growth and survival of this pathogen in fermented meats and cheeses during refrigerated storage. However, recognition of acid-tolerant mutants has raised additional concerns.[91] Sublethal injury of *L. monocytogenes* through heating, freezing, drying, irradiation or exposure to various chemicals (i.e., acids, preservatives, sanitizers) is also well documented.[78,112] Therefore, the ability of this organism to recover from such sublethal injury and grow in certain foods such as pasteurized milk during storage[86] has forced a re-examination of the original *Listeria* recovery methods.

36.17 Detection and Confirmation of Listeria

Numerous selective enrichment broths and plating media have been developed for recovery of *Listeria* species from food and environmental samples. Successful isolation depends on the ability of the method to promote the growth of small numbers of potentially injured cells (e.g., < 10^2 *Listeria* CFU/mL) while at the same time minimizing the growth of non-*Listeria* background organisms. The most commonly used procedures in the United

States are those developed by the FDA[65] and U. S. Department of Agriculture (USDA),[68] which utilize one and two selective enrichment steps at 30 or 35°C followed by plating on one or more *Listeria*-selective media.

Several taxonomic tests are needed to confirm *Listeria* isolates to the genus level with additional biochemical testing required to identify *L. monocytogenes*. Traditional speciation of *Listeria* is dependent on the organism's ability to produce acid from rhamnose, mannitol, and/or xylose and hemolyze sheep blood in the CAMP test. A variety of biochemically-based test kits which simplify identification of purified isolates are commercially available[6,35,65] along with several ELISA,[6,33,34,35,73] DNA hybridization,[35,65] and PCR assays[11,35,65] that have been developed to determine the presence or absence of *Listeria* following enrichment.

36.2 GENERAL CONSIDERATIONS

36.21 Methods of isolation

Almost all of the methods developed for isolation of *L. monocytogenes* use one or both of two distinct characteristics of the bacterium. These characteristics are the abilities of *Listeria* to grow at refrigeration temperatures and to display resistance to many antibiotics. The earliest isolation methods[57] used a cold enrichment technique to allow *L. monocytogenes* to grow at the expense of non-psychrotrophic bacteria. Although this technique has contributed much to our present day knowledge concerning the epidemiology of listeriosis, the prolonged incubation period necessary to obtain positive results is a major disadvantage. Because time is often critical, the use of antimicrobial agents has replaced cold enrichment to select for *L. monocytogenes*.

The type of food along with the enrichment conditions employed will likely influence not only the populations of *L. monocytogenes* present but also the type and populations of contaminants with which one must contend. Thus, the choice of media must sometimes be tailored to the type of food being analyzed.[52,61,62,63,110] This is particularly true if one is directly plating food samples on a selective medium, as discussed below.

Three general types of methods are presently being used to analyze foods for *L. monocytogenes*. The first and least used of these is direct plating of a food suspension onto a selective solid medium.[68] This technique offers the advantage of allowing the analyst to quantify the populations of *L. monocytogenes* in the food directly. Unfortunately, direct plating can detect only ≥100 bacteria per gram *L. monocytogenes* in foods, not smaller numbers. For the detection of low numbers of *L. monocytogenes*, enrichment procedures must be employed. The second and most popular method uses one or more enrichment steps followed by plating onto a selective agar and is the basis for both the FDA[65] and USDA, Food Safety and Inspection Service (FSIS) isolation methods.[68] Foods are usually mixed with an enrichment broth and allowed to incubate for 24 to 48 hr. Following incubation, a portion of the enrichment mixture is either again mixed with an enrichment broth or plated onto the final isolation agar. Enrichment broths are usually nutritious liquid media that employ various antimicrobial agents to which *L. monocytogenes* is resistant. The most common antimicrobial agents include nalidixic acid, acriflavin, and cycloheximide. Isolation agars include those used for direct plating,[26,52,68,125] although less selective agars have also been used successfully. The third group of methods used to analyze food products for *L. monocytogenes* are rapid analytical procedures. These procedures go beyond traditional identification methods through incorporation of genetic and immunological techniques to reduce identification time. The use of enzyme immunosorbent assays and DNA probes has been reported and commercial kits are available for both[6,33,34,35,65] (Section 36.53). Rapid methods are faster than conventional methods and performance is similar. The main drawbacks to the rapid methods developed to date are cost, requirement for sufficient cell density to record positive results, and in certain instances inability to distinguish pathogenic from non-pathogenic species, along with viable versus non-viable cells.

36.22 Treatment of Sample

Food samples should be stored and shipped at 4°C, and analyzed for *L. monocytogenes* as soon as possible. *Listeria monocytogenes* can grow slowly at refrigeration temperatures. Therefore, samples should not be stored for prolonged times under refrigeration unless there is a specific need for cold enrichment.[62] Although *L. monocytogenes* is quite resistant to freezing,[52] freezing may cause reduction of *L. monocytogenes* in certain foods. For this reason, frozen storage is recommended only when immediate analysis or refrigerated handling is not possible.

The growth of *L. monocytogenes* in refrigerated foods is not uniform, and foods should be sampled where growth of this organism is concentrated. For example, the growth of *L. monocytogenes* in soft cheeses like Brie is concentrated near the subsurface. In vacuum-packaged meats, the growth of *L. monocytogenes* is concentrated on or near the surface of the packaging film where it can be detected fairly easily. The growth of *L. monocytogenes* in refrigerated foods tends to be very spotty and erratic; thus, one should sample a large number of retail samples to obtain an accurate assessment of the extent of contamination.

36.3 MEDIA, REAGENTS AND EQUIPMENT

36.31 Media

Bile Esculin Agar slants
Brain Heart Infusion (BHI) Broth
Brain Heart Infusion Agar slants
CAMP Test Agar (TSA Blood Agar)
Columbia Blood agar base
Dey-Engley Neutralizing Broth (Nasco)
FDA *Listeria* enrichment broth (LEB)
Fraser Broth (FB)
University of Vermont (UVM) broth
Horse Blood Overlay (HL) agar
Listeria Repair Broth (LRB)
Lithium-phenylethanol-moxalactam (LPM) agar
LPM plus esculin/Fe^{3+} agar
M52
Methyl Red-Voges Proskauer (MR-VP) medium
Modified McBride agar (MMA)
Modified Oxford medium (MOX)
Motility test medium (preferred) or SIM Motility Medium
Neutralizing buffer, Difco
Nitrate Broth or Nitrate Reduction Medium
Nutrient broth
Oxford Medium (OXA)
Oxidation/Fermentation (O/F) Medium
PALCAM Agar
Purple carbohydrate fermentation broth base
Triple sugar iron agar (TSI)
Trypticase soy agar (TSA)
Trypticase soy broth
Trypticase soy agar-yeast extract (TSAYE)

Trypticase soy broth-yeast extract (TSBYE)
TSA blood agar (TSA with 5% defibrinated sheep blood)
Urea broth

36.32 Reagents

Acetic Acid, 5N
Acriflavin HCl (Sigma)
b-discs (Remel, Leneka, KS)
Bile esculin
Blood Agar Base No. 2 (Unipath)
CAMP test cultures
Staphylococcus-ATCC 25923 or FDA strain ATCC 49444 (CIP 5710; NCTC 7428)
Rhodococcus equi- ATCC 6939
Columbia Blood Agar Base
CO_2
Cycloheximide (Sigma)
Defibrinated sheep blood
Fetal calf serum
Glucose
Gram stain kit
Hydrogen peroxide (H_2O_2), 3%
Isopropanol, 70%
Maltose
Mannitol
Moxalactam, Eli Lily and Co., Distra, or Sigma M-1900
MR-VP reagents
(1) a-naphthol solution, 5 g per 100 mL absolute alcohol
(2) Potassium hydroxide solution, 40 g per 100 mL distilled H_2O
(3) Methyl red indicator: 0.1 g methyl red in 300 mL 95% ethanol made up to 500 mL in distilled water
Nalidixic Acid (Sigma)
Nitrate reduction reagents
(1) Reagent A—0.8 g sulfanilic acid in 100 mL 5 N acetic acid
(2) Reagent B—0.5 g a-naphthylamine in 100 mL 5 N acetic
(3) Zinc powder
Oxidase reagents- 1 g N, N, N^1, N^1-tetramethyl-*p*-phenylenediamine•2HCl in 199 mL distilled water.
Polyvalent Antiserum (Difco)
Type 1
Type 4
Potassium phosphate buffer, 0.1 M, pH 6.0
Rhamnose
Saline, 0.85%
Xylose

36.33 Equipment

Balance-gram range
Chisel, steel, small
Cover slips, glass
Dissecting microscope or microscope, inverted
Erlenmeyer flask, 500 mL
Fermentation tubes (Durham) (optional)
Filter paper, Whatman 541
Forceps
Gauze pad, 3" × 3", sterile
Grease pencil or permanent marker
Immersion oil (for microscope)
Incubators, 20–25°C; 30°C and 35°C
Inoculating loop
Inoculating needle
Inoculating needle, pure platinum
Lamp, fluorescent desk
Lamp with AO 653 Reichert light (or Bausch and Lomb Illuminator Light)
Microscope slide, glass
Microscope, phase with 40× objective
Mirror, concave
Mixer, Vortex
Needle, 26 gauge, 3/8 inch
Pans, aluminum pie
Petri plates
pH meter
Pipettes (25 mL, 10 mL, 1 mL)
Scissors, steel
Sponge, 3" × 3", sterile with no anionic detergents
Sticks, glass hockey
Stomacher Model 400 (Tekmar, Cincinnati, Ohio) and sterile bags
Swabs, sterile cotton
Syringe, tuberculin, sterile, disposable
Tubes, 16 × 125 mm, screw capped
Twist-tie for sealing plastic bags

36.4 PRECAUTIONS AND LIMITATIONS OF THE METHODS

Various methods differ in their ability to recovery *L. monocytogenes* from foods.[62,63,96,110] In most cases, a particular medium will sacrifice either sensitivity or selectivity.

This is particularly true if cells have been sublethally injured by treatments including dehydration, heating, freezing, exposure to acids or chemical sanitizers. Foods that may contain injured *L. monocytogenes* cells and high populations of contaminants will almost certainly require pre-enrichment to allow the injured *L. monocytogenes* to recover. Resuscitation procedures are discussed in Section 36.514. Numerous studies have been conducted to compare efficacy of current *Listeria* detection procedures. Hayes et al.[62] compared use of the USDA-FSIS procedure with cold enrichment as a means of identifying *L. monocytogenes* in suspect food samples. Both procedures were able to identify *L. monocytogenes* in 28 of 51 positive samples. The USDA-FSIS procedure identified 21 samples missed by cold enrichment while the cold enrichment procedure identified two additional samples that the USDA-FSIS procedure missed. A comparison of three enrichment methods was also made by Hayes et al.[63] when examining foods from the refrigerators of patients with active clinical cases of listeriosis. Two thousand two hundred twenty-nine (2,229) foods were examined in the study and 11% were positive for *L. monocytogenes*. A comparative evaluation of three microbiological procedures was conducted on 899 of the examined foods. The USDA-FSIS,[20] FDA,[79] and Netherlands Government Food Inspection Service (NGFIS)[125] methods were not statistically different in their ability to isolate *Listeria* from the 899 samples. The FDA procedure detected *L. monocytogenes* in 65% of foods shown to be positive, while the USDA-FSIS and NGFIS procedures detected *L. monocytogenes* in 74% of foods shown to be positive. Thus, none of the widely used conventional methods proved to be highly sensitive when used independently for analysis of *Listeria* contamination in foods. It was noted, however, that use of a combination of any two methods improved detectability from 65–74% (for individual protocols) to 87–91% for combined protocols.

The highly virulent nature of *L. monocytogenes* and the high mortality rates associated with listeriosis demand strict safety precautions. Laboratory supervisors should insist that standard good laboratory safety practices be followed.[41] Laboratory personnel should be especially mindful of generating aerosols during blending and mixing procedures and should be meticulous in rinsing work areas often with bactericidal solutions. In addition, pregnant women or other immunocompromised personnel should be prohibited from entering laboratories in which *L. monocytogenes* will be analyzed. Individuals under medication should seek medical advice to determine whether their particular medications are known to compromise immunity.

36.5 PROCEDURES

Rapid progress has been made in the isolation of *L. monocytogenes* from foods and food manufacturing environments. Methodology has advanced from the long and cumbersome cold enrichment procedure to the relatively rapid regulatory procedures employed by the FDA, USDA, and others. Even more rapid procedures, promising to shorten the isolation and identification process to a few days, are employing monoclonal antibodies (ELISA methods) and DNA probes.

36.51 Enrichment and Direct Plating Methods

Detection of *L. monocytogenes* in foods is not difficult. Low numbers of the organism are commonly isolated from raw milk, meats, vegetables, seafoods and the food processing environment.[7,44,45,59] Enrichment procedures are used to isolate low numbers of *L. monocytogenes* from dairy and vegetable products (Section 36.511) and meats (Section 36.512), and food samples with ≥100/gram *L. monocytogenes* can be enumerated by direct plating on MOX agar.[68]

Two methods are widely used in the United States to isolate low numbers of *L. monocytogenes* in foods. The FDA method is used to examine milk, milk products (particularly ice cream and cheese), seafood, and vegetables.[65] The USDA method[68] is used to recover *L. monocytogenes* from meat products and environmental swabs. The main difference between the USDA and FDA methods is the type of selective enrichment and plating media used.

36.511 FDA's Enrichment Procedure for the Isolation of L. monocytogenes

For liquid samples, pipette a 25-mL sample into 225-mL enrichment broth, M52[65] in a 500-mL flask and mix well by shaking. For solid samples, weigh 25 g into a Stomacher 400 bag and add 225 mL M52. Mix the sample by stomaching for 2 min, and incubate the mixture in the plastic bag at 30°C. Samples are enriched without selective agents for 4 hr at 30°C. Following addition of selective agents (10 mg/L acriflavin, 40 mg/L nalidixic acid and 50 mg/L cycloheximide), samples are incubated for an additional 44 hr for a total incubation period of 48 hr at 30°C. This procedure is a modification of the original procedure that called for sample enrichment for 1 and 7 days at 30°C.[79] The original broth was modified by increasing its buffering capacity, thereby positioning this medium to be used successfully in conjunction with DNA probe and other methods that are more sensitive than conventional cultural procedures. After 24 and 48 hr, M52 enrichment cultures are streaked onto Oxford Medium (OXA[26]) and *Listeria* plating medium (LPM) agar[83] or LPM plus esculin/Fe^{3+} agars, both of which have replaced the previously recommended modified McBride agar (MMA) as the selective isolation media. PALCAM agar[125] may be used in place of LPM agar. This substitution brings the method into closer alliance with methodology used outside the United States and decreases the reliance on the Henry technique (Section 36.521). Incubate OXA and PALCAM plates (with optional use of a CO_2 air atmosphere) at 35°C for 24–48 hr, with LPM plates incubated at 30°C for 24–48 hr. LPM plates can be viewed using Henry illumination or alternatively, esculin and ferric iron salt may be added to LPM to eliminate the need for Henry illumination. With OXA and PALCAM, *Listeria* colonies develop a black halo. It is recommended that five or more typical colonies be picked from OXA and PALCAM or LPM and transferred to TSAYE for confirmation of purity and typical isolated colonies. The selection of five colonies insures that multiple species of *Listeria*, if present, will be identified. TSAYE plates are incubated at 30°C for 24–48 hr (35°C incubation may be utilized if colonies are not being used for wet mount motility confirmation). Purified isolates are subjected to a series of standard confirmatory tests (Gram stain, motility, fermentation of mannitol, rhamnose and xylose) with a total of 10–11 days being required to isolate and confirm the presence of *Listeria* in food samples via the FDA procedure. This procedure was specifically developed to optimize detection of *Listeria* in milk and dairy products.

36.512 USDA's Enrichment Procedure for Isolation of L. monocytogenes

The original USDA-FSIS selective enrichment protocol for isolation of *Listeria* from raw meat and poultry was developed by McClain and Lee.[83] The revised USDA procedure[24,68] differs from the original method in that a) LEB II has been replaced by Fraser Broth[43] as the secondary enrichment medium; b) LPM agar has been replaced by Modified Oxford agar (MOX); and c) the sample size has been increased to 25 g. The surfaces of meat packages are disinfected by swabbing with 70% isopropanol or 3% H_2O_2 before opening. Meat samples are cut into small pieces with sterilized steel chisels, scissors, or forceps on a sterilized aluminum pie pan. Twenty-five grams of meat sample are weighed into a sterile Stomacher 400 bag and 225 g of the primary enrichment broth (UVM) are added to the bag. The mixture is stomached for 2 min and closed with a wire twist tie with some air trapped in the bag. To guard against leakage, the sample bag is placed inside another bag or beaker during incubation for 22 ± 2 hr at 30 ± 2°C. Following incubation, the UVM primary enrichment culture is streaked onto MOX plates, which are examined for *Listeria*-like colonies after 26–28 and 48 hr of incubation at 35°C. In addition, 0.1 ± 0.02 mL of the UVM culture is pipetted into 10 ± 0.5 mL of Fraser's secondary enrichment broth (FB) and examined for blackening after 26–28 and 48 hr of incubation at 35°C. Following incubation, blackened FB cultures are streaked onto MOX plates which are incubated at 35 ± 2°C for a minimum of 24 hr. If no growth is evident on MOX plates or if the FB cultures have not blackened, FB tubes and MOX plates are reincubated until a total incubation period of 48 ± 2 hr has been achieved.

MOX plates are examined for typical rounded 1-mm colonies surrounded by a black zone. To purify suspect *Listeria* colonies, as well as to identify b-hemolytic *Listeria* colonies from MOX agar, contact a minimum of 20 typical colonies and streak to HL agar for isolation (Section 36.522). The USDA-FSIS isolation scheme is presented in Figure 1; presumptive *Listeria* isolates are identified according to the procedures in Section 36.52 (Figure 2).

36.513 Enrichment Procedure for Isolating L. monocytogenes from the Environment

L. monocytogenes is persistent on food processing machinery and in the food processing plant environment. It can be detected by swabbing surfaces with a sterile 3" × 3" sponge or gauze moistened with Neutralizing buffer or Dey-Engley neutralizing broth (Caution: Some sponges contain anionic detergents that inhibit the growth of *L. monocytogenes*). Gauzes and sponges are placed in 200 ± 5 mL of UVM or LRB, stomached for 2 min and incubated at 30 ± 2°C for 22±2 hr. Continue as outlined in Section 36.512 for isolation of *Listeria*.

36.514 Resuscitation for the Recovery of Injured L. monocytogenes

Injured *Listeria monocytogenes* are sublethally stressed as a result of exposure to heat, freezing, chemical sanitizers, or acids[25,40,52,86,112,127] These injured cells and possibly intracellular bacteria may not grow in the selective media described in this chapter. The degree to which the particular selective medium suppresses the injured bacterial cell varies depending on the medium's ingredients and on the extent of the stress or the injury. Several studies have reported on the ability of commonly used plating media to recover injured *Listeria*. Among the compounds examined, phenylethanol, acriflavin, polymyxin-acriflavin and sodium chloride were found to inhibit recovery of thermally stressed and non-stressed *Listeria*.[19,25,126,127] It should be noted that these compounds comprise media which are routinely used for detection of *Listeria*. Sublethally stressed *L. monocytogenes* require resuscitation in a non-selective medium at a temperature favoring repair of sublethal injury. Injured *L. monocytogenes* not detected by conventional enrichment procedures may resuscitate and grow to high numbers in foods stored at refrigeration temperatures, thus presenting a consumer health problem. Media developed for resuscitation of injured *Listeria* include M52[65] (See section 36.511) and *Listeria* Repair Broth.[19] Ryser et al.[110] evaluated the ability of UVM and LRB to recover different strain-specific ribotypes of *L. monocytogenes* from meat and poultry products. Forty-five paired 25-g retail samples of ground beef, pork sausage, ground turkey, and chicken were enriched in UVM and LRB (30°C/24 hr), followed by secondary enrichment in Fraser broth (35°C/24 hr) and plating on modified Oxford agar. A three-hour non-selective enrichment period in LRB at 30°C was used with tested food to enable repair of injured *Listeria* prior to addition of selective agents. Of 180 meat and poultry products tested, LRB identified 73.8% (133/180) and UVM 69.4% (124/180). Although there was not a statistically significant difference in these results, combining UVM and LRB results increased overall *Listeria* recovery rates to 83.3%. These results demonstrate that use of LRB for repair/enrichment of samples in conjunction with the USDA/FSIS method has the potential to improve recovery of *Listeria* from meat and poultry products. Continuing work on enrichment of dairy environmental sample in UVM and LRB has shown that combining aliquots from these two primary enrichment media into a single tube of Fraser broth for secondary enrichment yields a significantly higher ($p<.05$) percentage of *Listeria*-positive samples than when either UVM or LRB are used alone. Roth and Donnelly[101] assessed the survival of acid-injured *L. monocytogenes* in four different acidic food systems, as well as examined the efficacy of two different enrichment media, LRB and UVM, to re-

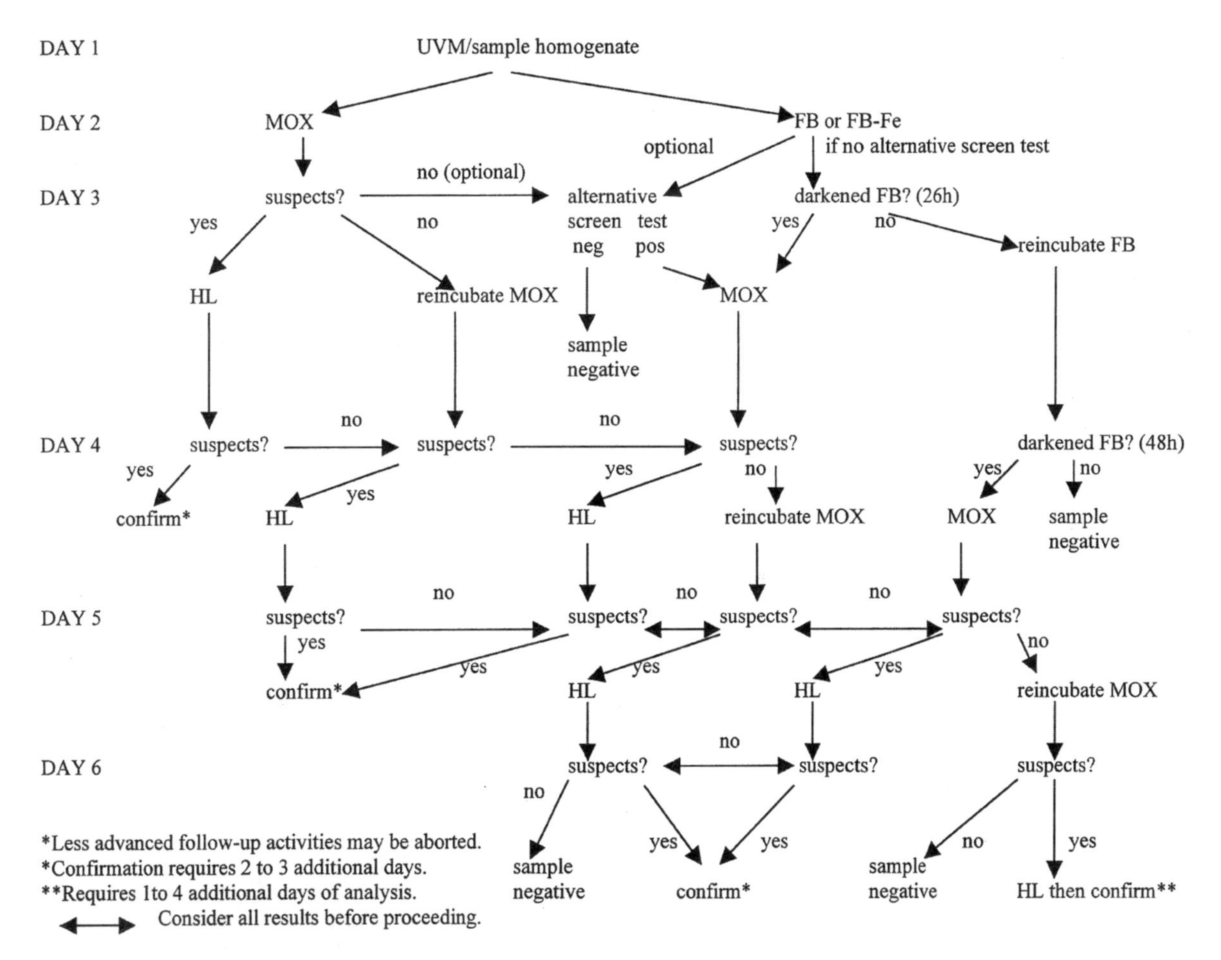

Figure 1. *Flow diagram for* Listeria monocytogenes *analysis.*

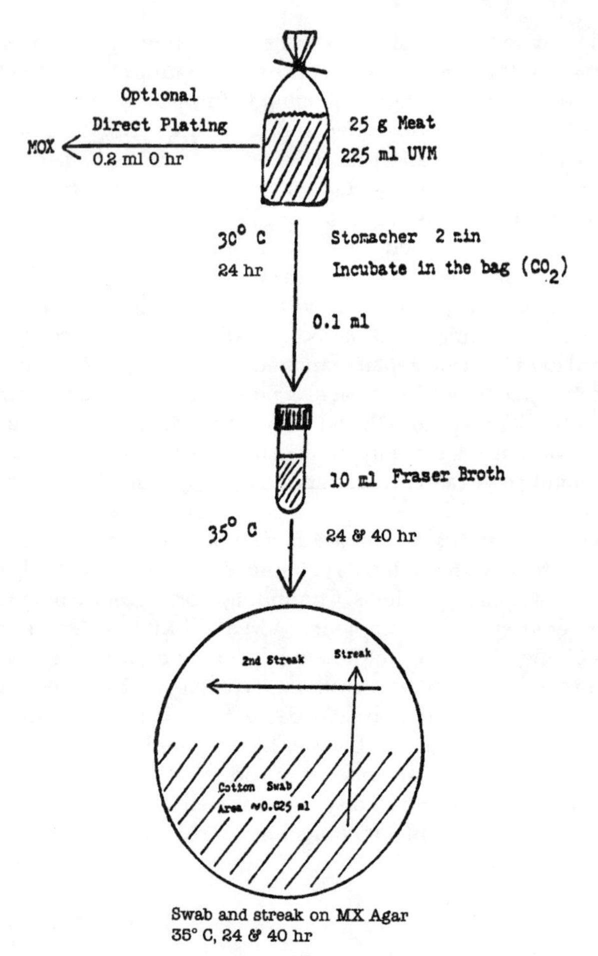

Figure 2. *Schematic presentation of the USDA enrichment method for the isolation of low numbers of* L. monocytogenes *from meat products.*

cover acid-injured *Listeria* from acidic foods which included fresh apple cider (pH 3.3); plain non-fat yogurt (pH 4.2); fresh coleslaw (pH 4.4) and fresh salsa (pH 3.9). Acid-injured *Listeria* were added to each acidic food system and monitored by selective and non-selective plating methods. Simultaneously, samples were tested using LRB and UVM enrichment procedures. Additionally, the survival of healthy *L. monocytogenes* was also monitored at 4°C (storage temperature) and 30°C (abuse temperature). Results indicated that acid-injured *Listeria* failed to repair in the foods tested, but survived in the tested foods for over a week. Storage temperature was found to affect survival, with 4°C storage having bacteriostatic effects whereas 30°C imparted bactericidal effects. Parameters involved in the survival of acid-injured *Listeria* included the degree to which the bacterial population was injured (percent injury), storage temperature, and the pH of the food. At time points where differences were detected, LRB proved to be superior, detecting 22 of 54 samples, compared to UVM which detected only three of 54 positive samples.

36.52 Identification of L. monocytogenes Colonies from Isolation Agars

36.521 Recognition of *Listeria* Colonies by 45° Transillumination

When the FDA *Listeria* protocol is used, presumptive *Listeria* colonies on LPM (prepared without esculin and Fe^{3+}) and TSAYE should be examined under a dissecting or inverted microscope arranged for 45° transillumination as first documented by Henry.[64] Not all light sources and mirrors will provide proper transillumination for *Listeria*. Best results have been obtained using the AO 653 (Reichert 653) or Bausch and Lomb illuminator light with a variable focus lens in conjunction with either a plain or concave mirror (Figure 3a). Alternatively, a more stable 45° transillumination without the use of a mirror can be achieved by using the AO 653 light and a low power (25×) invert microscope such as Leitz Diavert (Figure 3b). For proper 45° transillumination, place a piece of white filter paper on the microscope stage and focus the AO 653 light by adjusting the distance of the light as well as the focusing lens until a sharp image of the lamp filament appears on the filter paper. Check the microscope arrangement routinely using an 18-24 hr culture of *L. monocytogenes* on LPM or TSAYE.

Using this form of transillumination large *Listeria* colonies are dense white to iridescent white and exhibit a crushed glass appearance while smaller colonies tend to be blue. After some practice, *Listeria* isolates can be readily distinguished from other bac-

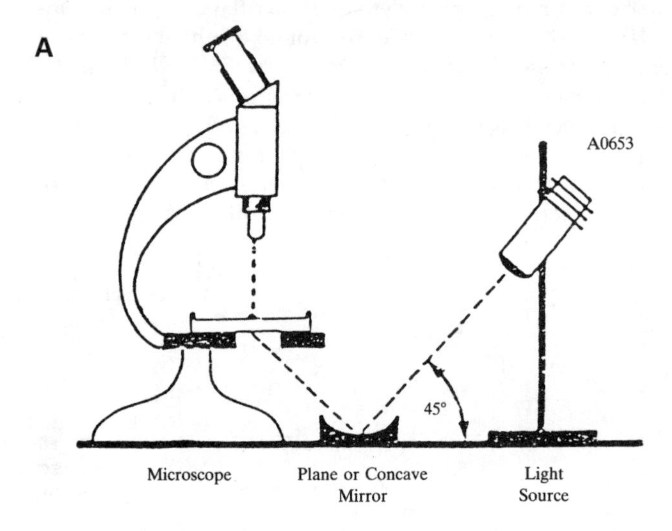

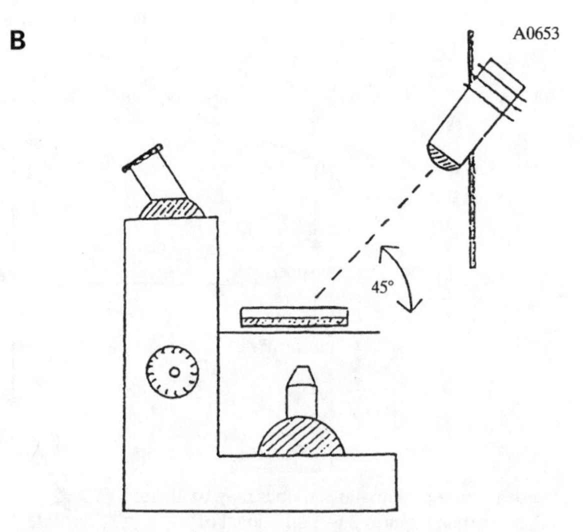

Figure 3. *Examination of TSAVE and LPM plates for* Listeria *colonies using 45° transillumination; 3A, using an illuminator light with a variable focus lens in conjunction with a plane or concave mirror; 3b, using a low-power invert microscope.*

terial colonies, which tend to be either yellowish or orange. Typical *Listeria* colonies are picked from LPM and restreaked to TSAYE for purity.

36.522 Biochemical and CAMP Test Identification of L. monocytogenes

Pick a well-isolated colony from the TSAYE plate and prepare a heavy wet mount suspension using 0.85% saline as the suspending medium. When viewed through the oil-immersion objective of a phase contrast microscope, all *Listeria* will appear as slim, short rods with slight rotating or tumbling motility. Using a known culture for comparison, *Listeria* isolates can be readily differentiated from non-*Listeria* organisms, which appear most frequently as cocci, large rods or rods with rapid swimming motility. Test a colony for Gram stain reaction and catalase activity. All *Listeria* spp. are gram positive and catalase positive, with older cells appearing coccoidal and gram variable. From the TSAYE plate, heavily inoculate a thickly poured and well-dried plate of TSA containing 5% sheep blood agar by stabbing. Draw a grid on the backside of the plate and stab one culture per grid space, reserving three spaces for known strains of *L. monocytogenes* and *L. ivanovii* (positive controls) and *L. innocua* (negative control). Following 48 hr of incubation at 35°C, examine the blood agar plate under bright light (Figure 4). *L. monocytogenes* and *L. seeligeri* produce slight clearing zones around the stab, whereas *L. innocua* and *L. ivanovii* are non-hemolytic and strongly hemolytic, respectively.

For biochemical identification, transfer a typical colony into a tube of TSBYE. After 18–24 hr of incubation at 35°C, inoculate tubes of purple carbohydrate fermentation broth containing glucose, bile esculin, maltose, rhamnose, mannitol, and xylose with or without Durham tubes. The esculin test can be omitted if obvious blackening was observed during initial isolation on OXA, MOX, or PALCAM. After 7 days of incubation at 35°C, all *Listeria* spp. should produce acid from glucose, esculin, and maltose, whereas all *Listeria* spp. except *L. grayi* should be negative for mannitol. As an alternative to microscopic observation of tumbling motility, tubes of SIM motility medium or motility test medium (preferred) can be inoculated from the TSBYE culture and observed for an umbrella-like growth pattern during 7 days of incubation at ambient temperature.

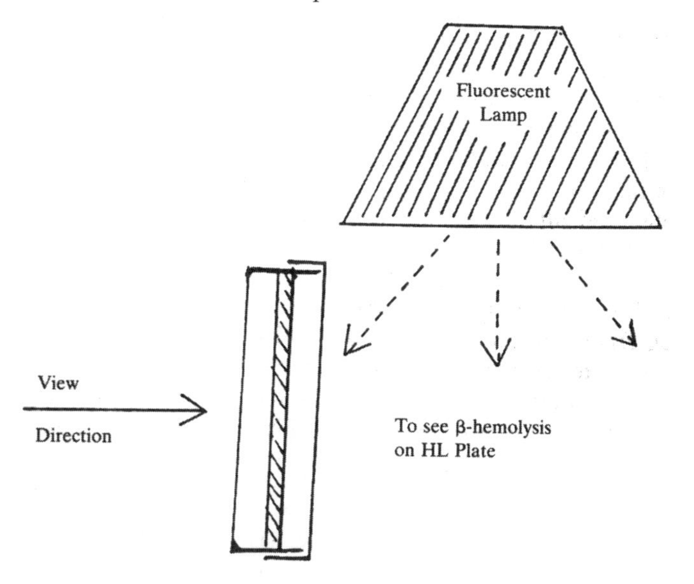

Figure 4. *Proper viewing of* Listeria *β-hemolysis on HL agar.*

The Christie-Atkins-Munch-Peterson (CAMP) test is useful for speciating *Listeria*, particularly when results from blood agar stabs are questionable. To perform this test, streak a b-hemolytic strain of *Staphylococcus aureus* (ATCC 49444 or ATCC 25923 and a culture of *Rhodococcus equi* (ATCC 6939) (both available from the American Type Culture Collection, Rockville, MD) in parallel and diametrically opposed to each other on a plate of sheep blood agar. Separate these vertical streaks so that the *Listeria* strains to be speciated along with a positive control (*L. monocytogenes*) and negative control (*L. innocua*) can be streaked close to but not touching and at a right angle to the vertical streak. After 24 to 48 hr of incubation at 35°C, examine the plate for hemolysis in the zone of interaction near the vertical streaks. Hemolytic activity of *L. monocytogenes* and *L. seeligeri* is enhanced near the *S. aureus* streak, whereas the hemolysis produced by *L. ivanovii* is stronger near the *R. equi* streak. The three remaining *Listeria* spp., *L. innocua, L. welshimeri,* and *L. grayi*, are non-hemolytic and do not react in this test. The CAMP test differentiates *L. ivanovii* from *L. seeligeri* and can differentiate a weakly hemolytic *L. seeligeri* from *L. welshimeri*. Isolates giving typical reactions for *L. monocytogenes* except for hemolysin production should be CAMP-tested before being identified as non-hemolytic *L. innocua.*

The major test results for speciating isolates of the genus *Listeria* are summarized in Table 1. All *Listeria* spp. are small, catalase-positive, gram positive rods that are motile in wet mounts and exhibit an umbrella-like growth pattern in semi-solid motility media. Glucose, esculin, and maltose are utilized by all *Listeria* with some species also producing acid from mannitol, rhamnose, and xylose. *L. grayi* produces acid from mannitol. Since *L. monocytogenes, L. ivanovii,* and *L. seeligeri* hemolyze sheep blood, all three species yield a positive CAMP reaction. Of these three, only *L. monocytogenes* fails to utilize xylose and is positive for rhamnose. The CAMP test can be used to overcome difficulties in differentiating *L. ivanovii* and *L. seeligeri*. Hemolytic activity of *L. seeligeri* is enhanced near the *S. aureus* streak, whereas *L. ivanovii* shows particularly strong hemolysis near the *R. equi* streak. Among the non-hemolytic species, *L. innocua* may yield the same rhamnose-xylose reactions as *L. monocytogenes* but is negative in the CAMP test. Occasional isolates of *L. innocua* are unable to utilize both rhamnose and xylose. *L. welshimeri*, which is rhamnose negative, may be confused with a weakly hemolytic *L. seeligeri* unless resolved by the CAMP test.

36.53 Rapid Detection Methods

As an alternative to conventional biochemical identification, a variety of miniaturized biochemical test kits are now commercially available for speciation of purified *Listeria* isolates. Biochemical test kits listed by the FDA[35] include 20 S™ (Analytical Products, Plainview, NY), API-ZYM (Analytical Products, Plainview, NY), Vitek Automicrobic Gram Positive and Gram Negative Identification Cards (BioMerieux Vitek, Hazelwood, MO), API Listeria (BioMerieux Vitek, Hazelwood, MO) and Micro-ID™ kit (Organon Teknika Corp., Durham, NC), the last of which allows speciation of *Listeria* only after the CAMP reaction is known. AOAC International, Gaithersburg, MD, has adopted Micro-ID and the Vitek Automicrobic System as official final action methods.

A variety of non-conventional methods that utilize enzyme-linked immunosorbant assays (ELISA) techniques have been developed for rapid identification of *Listeria* to the genus and species level. These methods, most of which can be used to screen enrichment cultures, have greatly accelerated the identification

Table 1. Biochemical Identification and Differentiation of *Listeria* Species[69]

	b-hemolysis	Tumbling motility	CAMP—*s. aureus*	CAMP—*Rhodococcus*	Umbrella motility	Gram reaction	Catalase	Oxidase	Urea	TSI	Glucose O/F	Bile esculin	MR-VP	NO_2	Mannitol	Xylose	Rhamnose
L. monocytogenes	+	+	+	–	+	+	+	–	–	a/a	+/+	+	+/+	–	–	–	+
L. seeligeri	+	+	+	–	+	+	+	–	–	a/a	+/+	+	+/+	–	–	+	–
L. ivanovii	+	+	+	+	+	+	+	–	–	a/a	+/+	+	+/+	–	–	+	–
L. innocua	–	+	–	–	+	+	+	–	–	a/a	+/+	+	+/+	–	–	–	+/–
L. grayi	–	+	–	–	+	+	+	–	–	a/a	+/+	+	+/+	–	+	–	–
L. murrayi	–	+	–	–	+	+	+	–	–	a/a	+/+	+	+/+	–	+	–	+/–

+ = positive reaction within 48 hr. – = negative reaction within 48 hr; a/a = acid slant/acid butt; O/F = oxidative, fermentative utilization of glucose.

of *Listeria*-positive samples. Commercially available AOAC- approved ELISA-based test kits for identifying *Listeria* to the genus level include Listeria-Tek (Organon-Teknika, Durham, NC), Assurance *Listeria* (BioControl Systems, Inc., Bothell, WA), and the Tecra *Listeria* Visual Immunoassay (Bioenterprises Pty. Ltd., Roseville, NSW, Australia). VIP™-a visual immunoprecipitate assay (BioControl Systems, Inc., Bothell, WA) and the fully automated VIDAS *Listeria* assay (BioMerieux Vitek, Hazelwood, MO) also have received AOAC approval for detecting *L. monocytogenes* and related species. Other genus-specific immunoassays awaiting AOAC approval include and Listertest, which utilizes immunomagnetic beads (VICAM, Sommerville, MA). The fully automated VIDAS *L. monocytogenes* assay (BioMerieux Vitek, Hazelwood, MO) is one of very few ELISA assays that is specific for *L. monocytogenes* and is currently awaiting AOAC approval.

Several non-radioactive DNA probe-based test kits also are commercially available for identifying *Listeria* isolates to the genus or species level. The AOAC approved Gene-Trak Colorimetric *Listeria* assay (Gene-Trak Systems, Framingham, MA) is specific for the genus *Listeria* while a similar *L. monocytogenes*-specific assay is also available from the same manufacturer and awaiting AOAC approval. AccuProbe™, a chemiluminescent genus-specific assay (Gen-Probe, San Diego, CA) is suitable for screening purified *Listeria* colonies and broth cultures. Finally, Qualicon (Wilmington, DE) has developed two additional DNA-based *Listeria* identification systems a) the automated Riboprinter Microbial Characterization System which will identify all pure *Listeria* cultures to the species level based on ribotype analysis and b) a PCR-based amplification method for specific detection of *L. monocytogenes* in enrichment cultures. These last two methods have yet to receive AOAC approval.

36.54 Serology

Isolates of *Listeria* can be further characterized based on the presence of specific heat-stable somatic (O) and heat-labile flagellar (H) antigens[29] with serotyping primarily confined to isolates obtained from outbreak investigations. Thirteen different serotypes of *L. monocytogenes* have been identified based on different combinations of 13 O and 4 H antigens[107] with serotypes 1/2a, 1/2b and 4b responsible for greater than 95% of all human infections.[44] However, since all non-pathogenic *Listeria* spp. except *L. welshimeri* share one or more somatic antigens with *L. monocytogenes*, serotyping alone, without thorough biochemical characterization, is inadequate to confirm isolates as *L. monocytogenes*.

Commercially prepared types 1, 4 and polyvalent antisera (Difco Laboratories, Detroit, MI) can be used in a slide or tube agglutination format to serotype 90% or more of all *L. monocytogenes* strains obtained from patients and the environment.[65] Specific O and H antigens and antisera must be prepared according to the methods of Bennett[13] or Seeliger and Hohne[118] for more complete serotyping of *Listeria*. Based on results from a recent collaborative study,[117] a critical need still exists for the commercial production of high quality sera and antisera from a standardized set of *Listeria* strains.

36.55 Subtyping

Characterization of *L. monocytogenes* isolates beyond the species level is primarily confined to cases of foodborne illness where investigators attempt to confirm the vehicle of infection and trace the source of contamination back to a particular food processing facility and/or environment. Methods for subtyping *Listeria* can be separated into two broad categories: (a) conventional methods, which include serotyping,[117] phage typing,[30] and bacteriocin typing[10] and (b) molecular methods, which encompass multilocus enzyme electrophoresis (MEE),[21] chromosomal DNA restriction endonuclease analysis (REA),[46] ribotyping,[124] DNA macrorestriction analysis by pulsed-field gel electrophoresis (PFGE),[14] random amplification of polymorphic DNA (RAPD) by PCR,[133] and DNA sequence-based subtyping.[54,97] In general, serotyping and phage typing are best suited as preliminary subtyping strategies, with phage typing now particularly popular in Europe for routine screening of isolates.[54] Ribotyping and MEE lack sufficient discrimination to be used alone in epidemiological investigations and, with the exception of the RiboPrinter™ (a fully automated ribotyping system developed by Qualicon, Wilmington, DE), are also fairly labor-intensive. At present, REA, PFGE, and RAPD are the preferred methods for subtyping *Listeria* because of their high discrimination and ease of use. PFGE has been used in conjunction with phage typing since the late 1980s for routine screening of *Listeria* isolates in France.[119] Furthermore, PFGE now forms the basis of the newly developed PulseNet-a nationwide foodborne disease surveillance network initiated in 1995 for subtyping *L. monocytogenes*, *Escherichia coli* O157:H7, *Salmonella typhimurium* and other foodborne pathogens in the United States.

36.56 Pathogenicity

Pathogenicity within the genus *Listeria* is generally restricted to two species, *L. monocytogenes* and *L. ivanovii*. *L. monocytogenes* is responsible for virtually all cases of human listeriosis and approximately 90% of listeriosis cases involving animals,[107] with *L. ivanovii* considered to be less virulent and of primary concern in animals.[74] The process by which *L. monocytogenes* causes disease can be divided into three stages: a) ingestion of the organism, b) penetration of the gastrointestinal lining, followed by replication after being phagocytized by macrophages and c) lysis of the macrophage, which leads to septicemia and invasion of various target organs, including the placenta and central nervous system. Consequently, *L. monocytogenes* must survive the acidity of the stomach, penetrate the intestinal lining and grow in the host before the organism can produce illness.

Like other enteroinvasive pathogens, *L. monocytogenes* produces a variety of virulence factors that damage the host's tissue, thus allowing the organism to invade the bloodstream and produce illness. Production of hemolysin (listeriolysin O) by the *hly* gene as observed in the CAMP test has been long associated with virulence in *L. monocytogenes*. Additional genes encoding for virulence factors in *L. monocytogenes* include *plcA* (a phosphatidylinositol-specific phospholipase C), *plcB* (a phosphatidylcholine-specific phospholipase C), *mpl* (a metalloprotease), *ActA* (a protein involved in actin polymerization), *inlA*, *inlB* and *inlC* (internalins A, B, and C for invasion of phagocytes) and *iap* (protein p60 for invasion of phagocytes).[74] Beginning in the 1930's, several animal models involving rabbits, guinea pigs, and chicken embryos were developed for assessing virulence of *L. monocytogenes* isolates,[107] with these tests later replaced by the mouse pathogenicity assay.[80] However, since virtually all b-hemolytic strains of *L. monocytogenes* are now considered to be virulent, the FDA and USDA have dropped their previous recommendation for mouse pathogenicity testing.

36.6 INTERPRETATION OF DATA

Several problems have been encountered in recovering *L. monocytogenes* from food and environmental samples. Most conventional and rapid *Listeria* detection procedures such as those developed by the FDA and USDA use highly selective enrichment and plating to suppress competitive background flora. However, these protocols, which are qualitative rather than quantitative, typically exhibit false negative rates of approximately 20% and also generally fail to recover sublethally injured *Listeria* cells that may be present in heated, frozen or acidic foods or within areas of food processing facilities exposed to sanitizers or other types of environmental stress.[112] Consequently, a negative test result does not necessarily ensure absence of *Listeria*. Use of multiple enrichment broths coupled with a recovery step for potentially injured cells can be used to alleviate this problem.[96, 110] Due to mixed populations of *Listeria* spp. and overgrowth by *L. innocua*, subtyping of multiple isolates is also frequently required to detect particular strains of *L. monocytogenes*.[110]

Listeria can be readily isolated from a variety of raw foods, which, after processing, are typically free of *Listeria*, provided that post-processing contamination has been prevented. Unlike many European countries which permit *L. monocytogenes* in certain cooked and "ready-to-eat" foods at levels up to 1000 CFU/g, the United States maintains a policy of "zero tolerance" for *L. monocytogenes*. Despite considerable progress in assessing relative risk, continued inability to accurately quantitate *Listeria* in foods and define the oral infective dose for *L. monocytogenes*, which varies according to host resistance, will likely continue to hamper any movement away from the present "zero tolerance" policy.

Most recalls of *Listeria*-contaminated cooked and "ready-to eat" products in the United States and elsewhere have been attributed to post-processing contamination. However, considerable progress has been made in minimizing such recalls through the implementation of good manufacturing practices, standard sanitation operating procedures and well-designed hazard analysis critical control point programs.

36.7 REFERENCES

1. Ashamed, N., and E. H. Marth. 1990. Behavior of *Listeria monocytogenes* at 7, 13, 21, and 35°C in tryptose broth acidified with acetic, lactic or citric acid. J. Food Prot. 52:688.
2. Al-Ghazali, M. R., and S. K. Al-Azawi. 1986. Detection and enumeration of *Listeria monocytogenes* in a sewage treatment plant in Iraq. J. Appl. Bacteriol. 60:251.
3. Anonymous. 1998. Multistate outbreak of listeriosis—United States, 1998. Morbid. Mortal. Wkly. Rep. 47(50):1085.
4. Anonymous. 1999. Update: Multistate outbreak of listeriosis—United States, 1998-1999. Morbid. Mortal. Wkly. Rep. 47(51):1117
5. Anspacher, R., K. A. Borchardt, M. W. Hannegan, and W. A. Boyson. 1966. Clinical investigation of *Listeria monocytogenes* as a possible cause of human fetal wastage. Am. J. Obstet. Gynecol. 94:386.
6. AOAC Official Methods of Analysis, 16th ed. 1998. Chap. 17 *Listeria*. AOAC International, Gaithersburg, Md.
7. Arimi, S. M., E. T. Ryser, and C. W. Donnelly. 1997. Dairy cattle and silage as potential sources of *Listeria* ribotypes common to dairy processing facilities. J. Food Prot. 60:811
8. Armstrong, R. W., and P. C. Fung. 1993. Brainstem encephalitis (rhomboencephalitis) due to *Listeria monocytogenes*: case report and review. Clin. Infect. Dis. 16:689.
9. Baker, M., and M. Brett. 1993. Listeriosis and mussels. Comm. Dis. New Zealand 93(1):13.
10. Bannerman, E., P. Boerlin, and J. Bille. 1996. Typing of *Listeria monocytogenes* by monocin and phage receptors. Int. J. Food Microbiol. 31:245.
11. Batt, C. A. 1999. Rapid methods for detection of Listeria, p. 261-278. *In* E. T. Ryser, and E. H. Marth (ed.) *Listeria, Listeriosis and Food Safety*, 2nd ed. Marcel Dekker, New York.
12. Bearns, R. E., and K. F. Girard. 1958. The effect of pasteurization on *Listeria monocytogenes*. Can. J. Microbiol. 4:55.
13. Bennett, R. W., and R. E. Weaver. 1995. Serodiagnosis of *Listeria monocytogenes*, p. 10.01 *In* Food and Drug Administration Bacteriological Analytical Manual, 8th ed. AOAC International, Gaithersburg, Md.
14. Brosch, R., M. Brett, B. Catimel, J. B. Luchansky, B. Ojeniyi, and J. Rocourt. 1996. Genomic fingerprinting of 80 strains from the WHO multicentre international typing study of *Listeria monocytogenes* via pulsed-field gel electrophoresis (PFGE). Int. J. Food Microbiol. 32:343.
15. Buchanan, R. E., H. G. Stahl, and D. L. Archer 1987. Improved plating media for simplified, quantitative detection of *Listeria monocytogenes* in foods. Food Microbiol. 4:269.
16. Bula, C. J., J. Bille, and M. P. Glausner. 1994. An epidemic of foodborne listeriosis in Western Switzerland: description of 57 cases involving adults. Clin. Infect. Dis. 20:66.
17. Bunning, V. K., R. G. Crawford, J. T. Tierney, and J. T. Peeler. 1992. Thermal tolerance of heat-shocked *Listeria monocytogenes* in milk

exposed to high-temperature short-time pasteurization. Appl. Environ. Microbiol. 58:2096.

18. Burn, C. G. 1936. Clinical and pathological features of an infection caused by a new pathogen of the genus *Listerella*. Am. J. Pathol. 12:341.
19. Busch, S. V., and C. W. Donnelly. 1992. Development of a repair-enrichment broth for resuscitation of heat-injured *Listeria monocytogenes* and *Listeria innocua*. Appl. Environ. Microbiol. 58:14.
20. Carnevale, R. A. and R. W. Johnston. 1989. Method for the isolation and identification of *Listeria monocytogenes* from meat and poultry products. United States Department of Agriculture Food Safety and Inspection Service, Laboratory Communication No. 57, Revised May 24. USDA-FSIS, Washington, D. C.
21. Caugant, D. A., F. E. Ashton, W. F. Bibb, P. Boerlin, W. Donachie, C. Low, A. Gilmour, J. Harvey, and B. Norrung. 1996. Multilocus enzyme electrophoresis for characterization of *Listeria monocytogenes* isolates: results of an international comparative study. Int. J. Food Microbiol. 32:301.
22. Collins, M. D., S. Wallbanks, D. J. Lane, J. Shah, R. Nietupski, J. Smida, M. Dorsch, and E. Stackebrandt. 1991. Phylogenic analysis of the genus *Listeria* based on reverse- transcriptase quenching of 16S rRNA. Int. J. Syst. Bacteriol. 41:240.
23. Conner, D. E., V. N. Scott, and D. T. Bernard. 1990. Growth, inhibition and survival of *Listeria monocytogenes* as affected by acidic conditions. J. Food Prot. 53:652.
24. Cook, L. V. 1999. Isolation and identification of *Listeria monocytogenes* from red meat, poultry, egg and environmental samples (Revision 2; 9/23/99). Chapter 8 in USDA-FSIS Microbiology Laboratory Guidebook.
25. Crawford, R. G., C. M. Beliveau, J. T. Peeler, C. W. Donnelly, and V. K. Bunning. 1989. Comparative recovery of uninjured and heat-injured *Listeria monocytogenes* cells from bovine milk. Appl. Environ. Microbiol. 55:1490.
26. Curtis, G. D. W., R. G. Mitchell, A. F. King, and E. J. Griffen. 1989. A selective differential medium for the isolation of *Listeria monocytogenes*. Lett. Appl. Microbiol. 8:95
27. Dalton, C. B., C. C. Austin, J. Sobel, P. S. Hayes, W. F. Bibb, L. M. Graves, B. Swaminathan, M. E. Proctor, and P. M. Griffin. 1997. An outbreak of gastroenteritis and fever due to *Listeria monocytogenes* in milk. N. Engl. J. Med. 336:100.
28. Dijkstra, R. G. 1982. The occurrence of *Listeria monocytogenes* in surface water of canals and lakes, in ditches of one big polder, and in the effluents of canals of a sewage treatment plant. Zentrabl. Bakteriol. Hyg. I. Abt. Orig. B. 176:202.
29. Donker-Voet, J. 1972. *Listeria monocytogenes*: Some biochemical and serological aspects. Acta Microbiol. Acad. Hung. 19:287.
30. Donnelly, C. W. 1999. Conventional methods to detect and isolate *Listeria monocytogenes*. Rapid methods for detection of Listeria, p. 225-260. *In* E. T. Ryser, and E. H. Marth, (ed.) *Listeria, Listeriosis and Food Safety*, 2nd ed. Marcel Dekker, New York.
31. Doyle, M. P., K. A. Glass, J. T. Beery, G. A. Garcia, D. J. Pollard, and R. D. Schultz. 1987. Survival of *Listeria monocytogenes* in milk during high-temperature short-time pasteurization. Appl. Environ. Microbiol. 53:1433.
32. Ewert, D. P., L. Lieb, P. S. Hayes, M. W. Reeves, and L. Mascola. 1995. *Listeria monocytogenes* infection and serotype distribution among HIV-infected persons in Los Angelos County, 1985-1992. J. Acquir. Immune Defic. Syndr. Hum. Retrovirol. 8:461.
33. Feldsine, P. T., A. H. Lienau, R. L. Forgey, and R. D. Calhoon. 1997. Assurance polyclonal enzyme immunoassay for detection of *Listeria monocytogenes* and related *Listeria* species in selected foods: Collaborative study. J. AOAC Int. 80:775.
34. Feldsine, P. T., A. H. Lienau, R. L. Forgey, and R. D. Calhoon. 1997. Visual immunoprecipitate assay (VIP) for *Listeria monocytogenes* and related *Listeria* species detection in foods: Collaborative study. J. AOAC Int. 80:791.
35. Feng, P. 1995. U. S. Food and Drug Administration. Bacteriological Analytical Manual, 8th ed., Rapid methods for detecting foodborne pathogens, App.1.01. AOAC International, Gaithersburg, Md.
36. Fenlon, D. R. 1985. Wild birds and silage as reservoirs of *Listeria* in the agricultural environment. J. Appl. Bacteriol. 59:537.
37. Fenlon, D. R. 1986. Rapid quantitative assessment of the distribution of *Listeria* in silage implicated in a suspected outbreak of listeriosis in calves. Vet. Rec. 118:240.
38. Fenlon, D. R. 1999. *Listeria monocytogenes* in the natural environment, p. 21-37. *In* E. T. Ryser, and E. H. Marth, (ed.) *Listeria, Listeriosis and Food Safety*, 2nd ed. Marcel Dekker, New York.
39. Fensterbank, R., A. Audurier, J. Godu, P. Guerrault, and N. Malo. 1984. Study of *Listeria* strains isolated from sick animals and silage consumed. Ann. Rech. Vet. 15:113.
40. Flanders, K. J., T. J. Pritchard, and C. W. Donnelly. 1995. Enhanced recovery of *Listeria* from dairy plant processing environments through combined use of repair enrichment and selective enrichment/detection procedures. J. Food Prot. 58:404.
41. Fleming, D. O., J. H. Richardson, J. J. Tullis, and D. Vesley. 1995. Laboratory safety: principles and practices, 2nd ed., ASM Press, Herndon, Va.
42. Fleming, D. W., S. L. Cochi, K. L. MacDonald, J. Brondum, P. S. Hayes, B. D. Plikaytis, M. B. Holmes, A. Audurier, C. V. Broome, and A. L. Reingold. 1985. Pasteurized milk as a vehicle of infection in an outbreak of listeriosis. N. Engl. J. Med. 312:404.
43. Fraser, J. A., and W. H. Sperber. 1988. Rapid detection of *Listeria* spp. in food and environmental samples by esculin hydrolysis. J. Food Prot. 51:762.
44. Gellin, B. G., and C. V. Broome. 1989. Listeriosis. JAMA 261:1313.
45. Gellin, B. G., C. V. Broome, W. F. Bibb, R. E. Weaver, S. Gaventa, L. Mascola, and the Listeriosis Study Group. 1991. The epidemiology of listeriosis in the United States—1986. Am. J. Epidemiol. 133:392.
46. Gerner-Smidt, P., P. Boerlin, F. Ischer, and J. Schmidt. 1996. High-frequency endonuclease (REA) typing: results from the WHO collaborative study group on subtyping of *Listeria monocytogenes*. Int. J. Food Microbiol. 32:313.
47. Gianfranceschi, M., and P. Aureli. 1996. Freezing and frozen storage on the survival of *Listeria monocytogenes* in different foods. Ital. J. Food Sci. 8:303.
48. Gibbons, N. E. 1972. *Listeria* Pirie—whom does it honor? Int. J. Syst. Bacteriol. 22:1.
49. Gill, D. A. 1931. Circling disease of sheep in New Zealand. Vet. J. 87:60.
50. Gitter, M., St. J. R. Stebbings, J. A. Morris, D. Hannam, and C. Harris. 1986. Relationship between ovine listeriosis and silage feeding. Vet. Rec. 118:207.
51. Glass, K. A., and M. P. Doyle. 1989. Fate of *Listeria monocytogenes* in processed meat products during refrigerated storage. Appl. Environ. Microbiol. 55:1565.
52. Golden, D. A., L. R. Beuchat, and R. E. Brackett. 1988. Evaluation of selective direct plating media for their suitability to recover uninjured, heat-injured, and freeze-injured *Listeria monocytogenes* from foods. Appl. Environ. Microbiol. 54:1451.
53. Goulet, V., C. Jacquet, V. Vaillant, I. Rebiere, E. Mouret, C. Lorente, E. Maillot, F. Stainer, and J. Rocourt. 1995. Listeriosis from consumption of raw milk cheese. Lancet 345:1581-1582.
54. Graves, L. M., B. Swaminathan, and S. B. Hunter. 1999. Subtyping *Listeria monocytogenes*, p. 279-297. *In* E. T. Ryser, and E. H. Marth, (ed.) *Listeria, Listeriosis and Food Safety*, 2nd ed. Marcel Dekker, New Yok.
55. Gray, M. L. 1956. A rapid method for the detection of colonies of *Listeria monocytogenes*. Zentrabl. Bakteriol. Parasit. Infekt. Hyg.I Orig. 169:373.
56. Gray, M. L., and A. H. Killinger. 1966. *Listeria monocytogenes* and listeric infections. Bacteriol. Rev. 30:308
57. Gray, M. L., H. J. Stafseth, F. Thorp Jr., L. B. Sholl, and W. F. Riley. 1948. A new technique for isolating listerellae from the bovine brain. J. Bacteriol. 55:471.
58. Gronstol, H. 1979. Listeriosis in sheep—*Listeria monocytogenes* excretion and immunological state in healthy sheep. Acta Vet Scand. 20:168.

59. Gronstol, H. 1979. Listeriosis in sheep—isolation of *Listeria monocytogenes* from grass silage. Acta Vet. Scand. 20:492.
60. Guyer, S., and T. Jemmi. 1991. Behavior of *Listeria monocytogenes* during fabrication and storage of experimentally smoked salmon. Appl. Environ. Microbiol. 57:1523.
61. Hayes, P.S., J. C. Feeley, L. M. Graves, G. W. Ajello, and D. W. Fleming, 1986. Isolation of *Listeria monocytogenes* from raw milk. Appl. Environ. Microbiol. 51:438.
62. Hayes, P.S., L. M. Graves, G. W. Ajello, B. Swaminathan, R. E. Weaver, J. D. Wenger, A. Schuchat, C. V. Broome, and the *Listeria* study group. 1991. Comparison of cold enrichment and the U.S. Department of Agriculture methods for isolating *Listeria monocytogenes* from naturally contaminated foods. Appl. Environ. Microbiol. 57:2109.
63. Hayes, P.S., L. M. Graves, B. Swaminathan, G. W. Ajello, G. B. Malcolm, R. E. Weaver, R. Ransom, K. Deaver, B. D. Plikaytis, A. Schuchat, J. D. Wenger, R. W. Pinner, C. V. Broome, and the *Listeria* Study Group. 1992. Comparison of three selective enrichment methods for the isolation of *Listeria monocytogenes* from naturally contaminated foods. J. Food Prot. 55:952.
64. Henry, B. S. 1933. Dissociation of the genus *Brucella*, p. 10.01. J. Infect. Dis. 52:374.
65. Hitchins, A. D. 1995. *Listeria monocytogenes*. *In* Food and Drug Administration bacteriological analytical manual. 8th ed. AOAC International, Gaithersburg, Md.
66. Hofer, E. 1983. Bacteriologic and epidemiologic studies on the occurrence of *Listeria monocytogenes* in healthy cattle. Zentrabl. Bakteriol. Hyg. A 256:175.
67. Jacquet, C., B. Catimel, V. Goulet, A. Lepoutre, P. Veit, P. Dehaumont, and J. Rocourt. 1995. Typing of *Listeria monocytogenes* during epidemiological investigations of the French listeriosis outbreaks in 1992, 1993 and 1995. Proceedings of XII International Symposium on Problems of Listeriosis, Perth, Western Australia, p. 161.
68. Johnson, J. L. 1998. Isolation of *Listeria monocytogenes* from meat, poultry and egg products. *In* USDA-FSIS Microbiology Laboratory Guidebook. 3rd ed., vol. 1.
69. Johnson, J. L., M. P. Doyle, R. G. Cassens, and J. L. Schoeni. 1988. Fate of *Listeria monocytogenes* in tissues of experimentally infected cattle and in hard salami. Appl. Environ. Microbiol. 54:497
70. Junttila, J. R., S. L. Niemela, and J. Hirn. 1988. Minimum growth temperatures of *Listeria* and non-haemolytic *Listeria*. J. Appl. Bacteriol. 65:321.
71. Kerr, K. G., S. F. Dealler, and R. W. Lacy. 1988. Materno-fetal listeriosis from cook-chill and refrigerated food. Lancet 2:1133.
72. Knabel, S. J., H. W. Walker, P. A. Hartman, and A. F. Mendonca. 1990. Effects of growth temperature and strictly anaerobic recovery on survival of *Listeria monocytogenes* during pasteurization. Appl. Environ. Microbiol. 56:370.
73. Knight, M. T., M. C. Newman, M. J. Benzinger Jr., and J. R. Agin. 1996. TECRA *Listeria* visual immunoassay (TLIVA) for detection of *Listeria* in foods: collaborative study. J. AOAC Int. 79:1083.
74. Kuhn, M., and W. Goebel. 1998. Pathogenesis of *Listeria monocytogenes*, p. 97-130. *In* E. T. Ryser and E. H. Marth (ed.) Listeria, listeriosis and food safety, 2nd ed. Marcel Dekker, New York.
75. Lachica, R. V. 1990. Simplified Henry technique for initial recognition of *Listeria* colonies. Appl. Environ. Microbiol. 56:1164.
76. Linnan, M. J., L. Mascola, X. D. Lou, V. Goulet, S. May, C. Salminen, D. W. Hird, M. L. Yonkura, P. Hayes, R. Weaver, A. Audurier, B. D. Plikaytis, S. L. Fannin, A. Kleks, and C. V. Broome. 1988. Epidemic listeriosis associated with Mexican-style cheese. N. Engl. J. Med. 319:823.
77. Lorber, B. 1997. Listeriosis. Clin. Infect. Dis. 24:1
78. Lou, Y., and A. E. Yousef. 1999. Characteristics of *Listeria monocytogenes* important to food processors, p. 131-224. *In* E. T. Ryser, and E. H. Marth (ed.) *Listeria, listeriosis and food safety*, 2nd ed. Marcel Dekker, New York.
79. Lovett, J. and A. D. Hitchins. 1989. *Listeria* isolation, p. 29.01. Bacteriological analytical manual, 6th ed., Supplement, Sept. 1987 (Second Printing 1989), Association of Official Analytical Chemists, Arlington, Va.
80. Lovett, J., D. W. Francis, and J. M. Hunt. 1987. *Listeria monocytogenes* in raw milk: Detection, incidence and pathogenicity. J. Food Prot. 50:188.
81. Lovett, J., I. V. Wesley, M. J. Vandermaaten, J. G. Bradshaw, D. W. Francis, R. G. Crawford, C. W. Donnelly, and J. W. Wesser. 1990. High-temperature short-time pasteurization inactivates *Listeria monocytogenes*. J. Food Prot. 53:734.
82. Low, J. C., and W. Donachie. 1997. A review of *Listeria monocytogenes* and listeriosis. Vet. J. 153:9.
83. McClain, D., and W. H. Lee. 1988. Development of a USDA-FSIS method for isolation of *Listeria monocytogenes* from raw meat and poultry. J. Assoc. Off. Anal. Chem. 71:660.
84. McKellar, R. C. 1994. Use of the CAMP test for identification of *Listeria monocytogenes*. Appl. Environ. Microbiol. 60:4219.
85. McLauchlin, J., S. M. Hall, S. K. Velani, and R. J. Gilbert. 1991. Human listeriosis and paté: a possible association. Br. Med. J. 303:773.
86. Meyer, D.H., and C. W. Donnelly. 1992. Effect of incubation temperature on repair of heat- injured *Listeria monocytogenes*. J. Food Prot. 55:579.
87. Miller, A. J. 1992. Combined water activity and solute effects on growth and survival of *Listeria monocytogenes* Scott A. J. Food Prot. 55:414.
88. Murray, E. G. D. 1963. A retrospect of listeriosis. *In* M.L. Gray (ed.). Second Symposium on *Listeria* Infection, Bozeman, Mont., Artcraft Printer.
89. Murray, E. G. D., R. A. Webb, and M. B. R. Swann. 1926. A disease of rabbits characterized by a large mononuclear leucocytosis caused by a hitherto undescribed bacillus *Bacterium monocytogenes* (n.sp.). J. Pathol. Bacteriol. 29:404.
90. Nyfeldt, A. 1929. Etiologie de la mononucleose infectieuse. C. R. Soc. Biol. 101:590.
91. O'Driscoll, B., C. G. M. Gahan, and C. Hill. 1996. Adaptive acid tolerance response in *Listeria monocytogenes*: isolation of an acid-tolerant mutant which demonstrates increased virulence. Appl. Environ. Microbiol. 62:1693.
92. Owen, C. R., A. Meis, J. W. Jackson, and H. G. Stoenner. 1960. A case of primary cutaneous listeriosis. N. Engl. J. Med. 262:1026.
93. Parish, M. E., and D. P. Higgins. 1989. Survival of *Listeria monocytogenes* in low pH model broth systems. J. Food Prot. 52:144.
94. Pirie, J. H. H. 1927. A new disease of veld rodents, "Tiger River Disease". Pub. S. Afr. Inst. Med. Res. 3:163.
95. Pirie, J. H. H. 1940. The genus *Listerella* Pirie. Science 91:383.
96. Pritchard, T. J., and C. W. Donnelly. 1999. Combined secondary enrichment of primary enrichment broths increases *Listeria* detection. J. Food. Prot. 62:532.
97. Rasmussen, O. F., P. Skouboe, L. Dons, L. Rossen, L., and J. E. Olsen. 1995. *Listeria monocytogenes* exists in at least three evolutionary lines: evidence from flagellin, invasive associated protein and listeriolysin O genes. Microbiology 141:2053.
98. Rocourt, J. 1999. The genus *Listeria* and *Listeria monocytogenes*: phylogenetic position, taxonomy, and identification, p. 1-20. *In* E. T. Ryser and E. H. Marth (ed.) *Listeria, listeriosis and food safety*, 2nd ed. Marcel Dekker, New York.
99. Rocourt, J., F. Grimont, P. A. D. Grimont, and H. P. R. Seeliger. 1982. DNA relatedness among serovars of *Listeria monocytogenes sensu lato*. Curr. Microbiol. 7:383.
100. Rocourt, J., U. Wehmeyer, P. Cossart, and E. Stackebrandt. 1987. Proposal to retain *Listeria murrayi* and *Listeria grayi* in the genus *Listeria*. Int. J. Syst. Bacteriol. 37:298.
101. Roth, T. T., and C. W. Donnelly. 1995. Injury of *Listeria monocytogenes* by acetic and lactic acids: mechanisms of repair and sites of sublethal damage. IFT Ann. Meet. Book of Abstracts 81D-1, p. 246.
102. Ryser, E. T. 1999. Foodborne listeriosis, p. 299-358. *In* E. T. Ryser and E. H. Marth (ed.), Listeria, listeriosis and food safety, 2nd ed. Marcel Dekker, New York.

103. Ryser, E. T. 1999. Incidence and behavior of *Listeria monocytogenes* in unfermented dairy products, p. 359-409. *In* E. T. Ryser and E. H. Marth (ed.), Listeria, listeriosis and food safety, 2nd ed. Marcel Dekker, New York.
104. Ryser, E. T., and E. H. Marth. 1989. Behavior of *Listeria monocytogenes* during manufacture and ripening of brick cheese. J. Dairy Sci. 72:838.
105. Ryser. E. T., and E. H. Marth. 1987. Behavior of *Listeria monocytogenes* during the manufacture and ripening of Cheddar cheese. J. Food Prot. 50:7.
106. Ryser. E. T., and E. H. Marth. 1987. Fate of *Listeria monocytogenes* during manufacture and ripening of Camembert cheese. J. Food Prot. 50:372.
107. Ryser, E. T., and E. H. Marth. 1991. Listeria, listeriosis and food safety. Marcel Dekker, New York
108. Ryser, E. T., S. M. Arimi, and C. W. Donnelly. 1997. Effects of pH on distribution of *Listeria* ribotypes in corn, hay and grass silage. Appl. Environ. Microbiol. 63:3695.
109. Ryser. E. T., E. H. Marth, and M. P. Doyle. 1985. Survival of *Listeria monocytogenes* during manufacture and storage of cottage cheese. J. Food Prot. 48:746.
110. Ryser, E. T., S. M. Arimi, M.-C. Bunduki, and C. W. Donnelly. 1996. Recovery of different *Listeria* ribotypes from naturally contaminated, raw refrigerated meat and poultry products with two primary enrichment media. Appl. Environ. Microbiol. 62:1781.
111. Salamina, G., E. D. Donne, A. Niccolini, G. Poda, D. Cesaroni, M. Bucci, R. Fini, M. Maldini, A. Schuchat, B. Swaminathan, W. Bibb, J. Rocourt, N. Binkin, and S. Salmoso. 1996. A foodborne outbreak of gastroenteritis involving *Listeria monocytogenes*. Epidemiol. Infect. 117:429-436.
112. Sallam, S. S., and C. W. Donnelly. 1992. Destruction, injury and repair of *Listeria* species exposed to sanitizing compounds. J. Food Prot. 55:771-776.
113. Sallen, B., A. Rajoharison, S. Desvarenne, F. Quinn, and C. Mabilat. 1996. Comparative analysis of 16S and 23S rRNA sequences of *Listeria* species. Int. J. Syst. Bacteriol. 46:669.
114. Salvat, G., M. T. Toquin, Y. Michel, and P. Colin. 1995. Control of *Listeria monocytogenes* in the delicatessen industries: the lessons of a listeriosis outbreak in France. Int. J. Food. Microbiol. 25:75.
115. Sanaa, M., B. Poutrel, J. L. Menard, and F. Serieys. 1993. Risk factors associated with contamination of raw milk by *Listeria monocytogenes* in dairy farms. J. Dairy Sci. 76:2891.
116. Schlech, W. F., P. M. Lavigne, R. A. Bortolussi, A. C. Allen, E. V. Haldane, A. J.Wort, A. W. Hightower, S. E. Johnson, S. H. King, E. S. Nicholls, and C. V. Broome. 1983. Epidemic listeriosis - Evidence for transmission by food. N. Engl. J. Med. 308:203.
117. Schonberg, A., E. Bannerman, A. L. Cortieu, R. Kiss, J. McLauchlin, S. Shah, and D. Wilhelms. 1996. Serotyping of 80 strains from the WHO multicentre international typing study of *Listeria monocytogenes*. Int. J. Food Microbiol. 32:279.
118. Seeliger, H. P. R., and K. Hohne. 1979. Serotyping of *Listeria monocytogenes* and related species, p. 31. *In* T. Bergen and J.R. Norris (ed.) *Methods in microbiology*, Academic Press, London.
119. Seeliger, H. P. R., J. Rocourt, A. Schrettenbrunner, P. A. D. Grimont, and D. Jones. 1984. *Listeria ivanovii* sp. *nov*. Int J. Syst. Bacteriol. 34:336.
120. Shahamat, M., A. Seaman, and M. Woodbine. 1980. Survival of *Listeria monocytogenes* in high salt concentrations. Zentralbl. Bakteriol. Hyg., I. Abt. Orig. A. 246:506.
121. Sionkowski, P. J., and L. A. Shelef. 1990. Viability of *Listeria monocytogenes* strain Brie-1 in the avian egg. J. Food Prot. 53:15-17, 25.
122. Slutsker, L., and A. Schuchat. 1999. Listeriosis in humans, p. 75-95. *In* E. T. Ryser, and E. H. Marth, (ed.) Listeria, listeriosis and food safety, 2nd ed. Marcel Dekker, New York.
123. Sorrells, K. M., D. C. Enigl, and J. R. Hatfield. 1989. Effect of pH, acidulant, time and temperature on the growth and survival of *Listeria monocytogenes*. J. Food Prot. 50:730.
124. Swaminathan, B., S. B. Hunter, P. M. Desmarchelier, P. Gerner-Smidt, L. M. Graves, S. Harlander, R. Hubner, C. Jacquet, B. Pedersen, K. Reineccius, A. Ridley, N. A. Saunders, and J. A. Webster. 1996. WHO-sponsored international collaborative study to evaluate methods for subtyping *Listeria monocytogenes*: restriction fragment length polymorphism (RFLP) analysis using ribotyping and Southern hybridization with two probes derived from *L. monocytogenes* chromosome. Int. J. Food Microbiol. 32:263.
125. Van Netten, P., I. Perales, and G. D. W. Curtis. 1989. Liquid and solid selective differential media for the detection and enumeration of *L. monocytogenes* and other *Listeria* spp. Int. J. Food Microbiol. 8:299.
126. Warburton, D. W., J. M. Farber, A. Armstrong, R. Caldeira, N. P. Tiwari, T. Babiuk, P. Lacasse, and R. Read. 1991. A Canadian comparative study of modified versions of the "FDA" and "USDA" methods for the detection of *Listeria monocytogenes*. J. Food Prot. 54:669.
127. Warburton, D.W., J. M. Farber, A. Armstrong, A. R. Caldeira, T. Hunt, S. Messier, R. Plante, N. P. Tiwari, and J. Vinet. 1991. A comparative study of the "FDA" and "USDA" methods for the detection of *Listeria monocytogenes* in foods. Int. J. Food Microbiol. 13:105.
128. Watkins, J., and K. P. Sleath. 1981. Isolation and enumeration of *Listeria monocytogenes* from sewage, sewage sludge, and river water. J. Appl. Bacteriol. 50:1.
129. Weis, J., and H. P. R. Seeliger. 1975. Incidence of *Listeria monocytogenes* in nature. Appl. Microbiol. 30:29.
130. Welshimer, H. J. 1960. Survival of *Listeria monocytogenes* in soil. J. Bacteriol. 80:316.
131. Welshimer, H. J. 1968. Isolation of *Listeria monocytogenes* from vegetation. J. Bacteriol. 95:300.
132. Welshimer, H. J., and J. Donker-Voet. 1971. *Listeria monocytogenes* in nature. Appl. Microbiol. 21:516.
133. Wernars, K., P. Boerlin, A. Audurier, E. G. Russell, G. D. W. Curtis, L. Herman, and N. van der Mee-marquet. 1996. The WHO multicentre study on *Listeria monocytogenes* subtyping: random amplification of polymorphic DNA (RAPD). Int. J. Food Microbiol. 32:325.
134. Wesley, I. V. 1999. Listeriosis in animals, p. 39-73. *In* E. T. Ryser and E. H. Marth (ed.), Listeria, listeriosis and food safety. 2nd ed. Marcel Dekker, New York.

Chapter 37

Salmonella

Wallace H. Andrews, Russell S. Flowers,
John Silliker, and J. Stan Bailey

37.1 GENERAL BASIS OF METHODS

37.11 Introduction

The incidence of salmonellosis appears to be increasing. Most recent summary data[13] from the Centers for Disease Control and Prevention (CDC) are shown in Figure 1. Large outbreaks caused by raw and processed foods continue to occur. For example, an estimated 224,000 people nationwide were estimated to be made ill by ice cream produced in a Minnesota factory.[66] New vehicles of transmission have emerged, most notably shell eggs contaminated internally with *S. enteritidis*. In fact, egg-associated illness caused by this serotype now accounts for 25% of all salmonellosis reported in humans in the United States, and *S. enteritidis* is now the most common serotype associated with illness, replacing *S. typhimurium*. Fresh produce is of increasing concern.[83] Two major outbreaks were attributed to cantaloupe. In one, due to *S. chester*, cases were reported in 30 states, and it was estimated that more than 25,000 individuals were eventually infected.[107] The second was due to *S. poona* involving 185 confirmed cases in the United States and 56 in Canada.[12] Cantaloupe on salad bars was implicated in both outbreaks. A substantial volume of produce is imported, often from under-developed areas with poor hygiene practices. Increased surveillance for enteric pathogens in imported produce is essential.

The data in Figure 1 likely underestimates the extent of the *Salmonella* problem. The U.S. Department of Agriculture (USDA) estimates 696,000–3,840,000 cases of foodborne illness due to non-typhoid *Salmonella* with 870–1,920 deaths annually.[11] The estimated costs of illness range from 0.9 to 12.2 billion dollars.

Control of salmonellosis rests upon consumer education and implementation of adequate laboratory quality control programs in the food industry. Though the application of the Hazard Analysis and Critical Control Point (HACCP) system purports to reduce the need for testing, verification that HACCP is working properly will pose a continuing need for microbiological analysis. Greater surveillance of imported products is recognized because of the poor hygienic standards in many exporting countries.

The "official" method for analysis of foods for *Salmonella* detailed in this chapter is that of the Association of Official Analytical Chemists International (AOAC).[8] This method is tedious and labor-intensive. From the pioneering fluorescent antibody (FA) technique,[8] a variety of rapid screen techniques has been developed and will be discussed elsewhere in this chapter. These procedures give presumptively positive results one to 2 days earlier than the conventional culture method. However, these techniques still require pre-enrichment of food samples, and presumptively positive results must be confirmed by the conventional method.

37.12 General Description of Methods for the Isolation of Salmonella from Foods

37.121 Introduction

The examination of foods for *Salmonella* requires methods different from those used in clinical laboratories. Generally low numbers are present in foods, necessitating the analysis of larger samples than are used with clinical materials. The introduction of large amounts of food directly into selective media tends to reduce the selectivity of the media for *Salmonella* organisms.[112] Further, the *Salmonella* organisms in foods are often in a poor physiological condition as a result of food processing and storage. These two problems are simultaneously circumvented by pre-enrichment of the sample in a non-selective medium which permits resuscitation of debilitated organisms. Inoculation of selective media from the pre-enrichment medium obviates introduction of food into the selective medium. The analysis is a qualitative test and determines the presence or absence of *Salmonella* in a given sample. However, by analyzing replicate samples at a series of dilutions, it is possible to obtain a quantitative estimate of numbers, i.e., a most probable number.

The reference procedure described hereafter for the isolation and identification of *Salmonella* (Section 37.5) has been adapted from the *Official Methods of Analysis of the AOAC International*.[8] The biochemical and serological tests used to identify *Salmonella* are described in the reference procedure and are listed in Table 1.

The classification of the *Salmonella-Arizona* group of *Enterobacteriaceae* has been a source of confusion because of different systems of nomenclature.[41,75,83] The Enteric Bacteriology Laboratories at the CDC recently introduced changes in their nomenclature of the *Salmonella-Arizona* group by incorporating the *Arizona* group[41] into the *Salmonella* genus.[19,48a,84,85] Table 2 summarizes the properties of the various subspecies within the genus *Salmonella* defined on the basis of DNA hybridization studies.[19,84,85]

Many scientific journals require the nomenclature of microorganisms to comply with *Bergey's Manual of Systematic Bacteriology*. *Bergey's Manual* has taken a different approach from CDC and has adopted the position that the use of "species" names for *Salmo-*

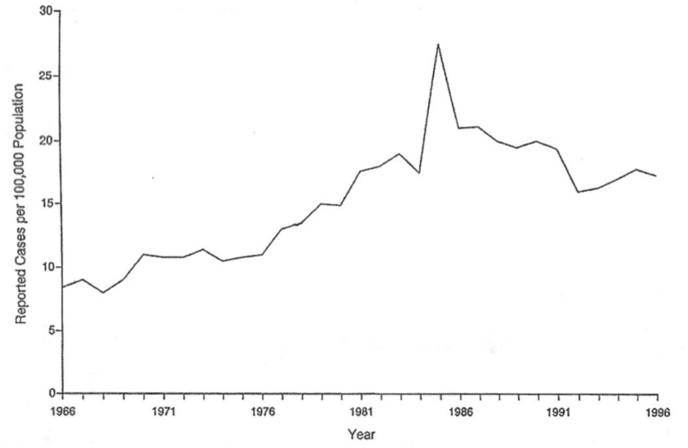

Figure 1. *Reported cases of human Salmonellosis (1966–1996), excluding typhoid fever. (Source: Centers for Disease Control and Prevention. 1997. Summary of notifiable diseases, United States. 1996. Morbid Mortal Wkly Rep. 45:54.)*

nella serovars is useful in many fields, so long as serovars' names are not taxonomically equated with species. The nomenclature in this chapter and the classic definition of the genus *Salmonella* are as described in *Bergey's Manual of Systematic Bacteriology.*[68,83]

Not all serovars within the genus *Salmonella* are equally pathogenic to humans and animals, but all are important to public health, and isolation procedures should recover all serovars of *Salmonella*. Analysts should be thoroughly familiar with the various biochemical and serological characteristics involved in differentiating *Salmonella* from the other foodborne microorganisms.

37.122 Sampling

General directions for sample collection and handling are discussed in Chapter 2. Various organizations have suggested sampling plans that can be used specifically for *Salmonella*. One of these plans has been proposed by the National Academy of Sciences (NAS)[94] through the Committee on *Salmonella* of the National Research Council (NRC). Foods are categorized on the basis of hazards presented by the particular food: 1) the food or food ingredient is a significant potential source of *Salmonella*; 2) the manufacturing process does not include a controlled step that should kill *Salmonella* microorganisms; and 3) a potential exists for microbiological growth if the food is mishandled in distribution or by consumers. Considering these hazards, foods are placed in one of five categories:

Category I—foods intended for use by infants, the aged, and the infirmed (the restricted population of high risk);

Category II—foods that present the three hazards;

Category III—food with two hazards;

Category IV—foods with one hazard;

Category V—foods with none of the hazards.

Criteria for acceptance of any particular lot of food are based on the results of analyses of a required number of 25-g analytical units, the actual portions of food analyzed. Each analytical unit is taken from a larger sample unit, usually a minimum of 100 g, the balance being set aside for possible testing. A series of random sample units make up the sample, and a sample that is representative of the lot is used to determine the acceptability of the entire lot of food. A food lot is defined as an identifiable unit of food produced and handled under similar conditions, usually determined by a limited period of time. This sampling plan is based on the premise that the distribution of the *Salmonella* organisms within the lot will be homogeneous. This premise implies that any analytical unit is as likely as any other analytical unit to contain *Salmonella.* The NAS/NRC Committee proposed criteria for the acceptance of any particular food lot in the various categories:

Category I—60 analytical units tested and found negative, indicating a 95% probability of ≤1 *Salmonella* organism per 500 g of food in the lot tested;

Table 1. Typical Biochemical and Serological Reactions of *Salmonella*

Test or Substrate	Positive	Negative	*Salmonella* Species Reaction[a]
1. Glucose (TSI)	Yellow butt	Red Butt	+
2. Lysine decarboxylase (LIA)	Purple butt	Yellow butt	+
3. H_2S (TSI and LIA)	Blackening	No blackening	+
4. Urease	Purple-red color	No color change	–
5. Lysine decarboxylase broth	Purple color	Yellow color	+
6. Phenol red dulcitol broth	Yellow color and/or gas	No gas; no color change	+[b]
7. KCN broth	Growth	No growth	–[c]
8. Malonate broth	Blue color	No color change	–[d]
9. Indole text	Red color at surface	Yellow color at surface	–
10. Polyvalent flagellar test	Agglutination	No agglutination	+
11. Polyvalent somatic test	Agglutination	No agglutination	+
12. Phenol red lactose broth	Yellow color and/or gas	No gas; no color change	–[e]
13. Phenol red sucrose broth	Yellow color and/or gas	No gas; no color change	–
14. Voges-Proskauer test	Pink-to-red color	No color change	–
15. Methyl red test	Diffuse red color	Diffuse yellow color	+
16. Simmons citrate	Growth; blue color	No growth; no color change	V

[a] +, 90% or more positive in 1 or 2 days; –, 90% or more negative in 1 or 2 days; V, variable.
[b] Majority of *Salmonella* subspecies 3a, 3b, 4, and 6 are negative (see Table 2).
[c] Majority of *Salmonella* subspecies 4 and 5 grow in KCN broth (see Table 2).
[d] Majority of *Salmonella* subspecies 2, 3a, and 3b, are positive (see Table 2).
[e] Majority of *Salmonella* subspecies 3b are positive (see Table 2).

Category II—29 analytical units found negative, indicating ≤1 *Salmonella* organism per 250 g of food;

Categories III, IV, V—13 analytical units found negative, indicating ≤ 1 *Salmonella* organism per 125 g food.

The International Commission on Microbiological Specifications for Foods (ICMSF)[71] proposed a general microbial sampling plan that also includes a sampling plan specific for *Salmonella*. Basically, any particular food is sampled according to its placement in 1 of 15 categories or "cases." Such placement depends on two factors: 1) type of microbial hazard involved and 2) anticipated conditions of handling and use of the food after sampling. The microbial hazard can range from no direct health hazard (e.g., spoilage or reduced shelf life) to severe hazard (e.g., suspected presence of *Clostridium botulinum*). Anticipated conditions of handling and use of the food can be reduced hazard (e.g., cooking prior to consumption), unchanged hazard (e.g., direct consumption of dried foods), or increased hazard (e.g., storage of thawed frozen food at ambient or high temperatures). The stringency of sampling varies directly with the case number assigned to the food.[71]

Under the ICMSF plan, food sampling for most *Salmonella* corresponds to Cases 10, 11, and 12 (decreased, unchanged, and increased hazard in use, respectively). The case number increases as the stringency of the criteria for acceptability of any particular food lot increases. However, foods suspected of containing *S. typhi*, *S. paratyphi*, or *S. cholerae suis* are assigned higher risk levels and are sampled with greater stringency (Cases 13, 14, or 15), as are any foods associated with outbreaks ("investigational sampling"). In contrast, the NAS/NRC sampling plan recognizes an equally high degree of risk for all serovars and thus calls for greater sampling stringency than the ICMSF plan under most conditions. An added feature of the ICMSF plan, not found in the NAS/NRC plan, however, is the distinction made between sampling under normal (routine) and special (investigational) conditions, such as investigation of a foodborne outbreak. The ICMSF plan compares in stringency to the NAS/NRC plan only in the investigational mode.

The two sampling plans are nevertheless similar in that both recommend increasing the number of analytical units with increased health risk. In general, the assignment of a food to a particular category or case depends on the sensitivity of the consumer group, the history of the food, whether there is a step lethal to *Salmonella* microorganisms during processing or in the home, and

Table 2. Properties of the Seven Subspecies within *Salmonella enterica* ("the *Salmonella-Arizona* Group")[a]

	Designation						
	Salmonella subspecies 1	*Salmonella* subspecies 2	*Salmonella* subspecies 3a	*Salmonella* subspecies 3b	*Salmonella* subspecies 4	*Salmonella* subspecies 5	*Salmonella* subspecies 6
DNA-hybridization group of Crosa et al.[19]	1	2	3	4	5	Not studied	
Salmonella subgenus names formerly used	I	II	III	III	IV	V	
Genus according to Ewing[46]	*Salmonella*	*Salmonella*	*Arizona*	*Arizona*	*Salmonella*	*Salmonella*	*Salmonella*
Subspecies according to Le Minor et al.[82]	*enterica*	*salamae*	*arizonae*	*diarizonae*	*houtenae*	*bongori*[d]	*indica*
Usually monophasic (Mono) or diphasic (Di) flagella	Di	Di	Mono	Di	Mono	Mono	Di
Usually isolated from humans and warm-blooded animals	+	–	–	–	–	+	+
Usually isolated from cold-blooded animals and the environment	–	+	+	+	+	+	+
Pathogenic for humans	+	+	+	+	+	+	+
TESTS TO DIFFERENTIATE THE SUBSPECIES							
Dulcitol fermentation	96[b]	90	0	1	0	100	variable
Lactose fermentation	1	1	15	85	0	0	? not given
ONPG Test	2	15	100	100	0	100	variable
Malonate utilization	1	95	95	95	0	0	0
Growth in KCN medium	1	1	1	1	95	100	0
Mucate fermentation	90	96	90	30	0	100	+
Gelatin hydrolysis[c]	–	+	+	+	+	–	+
D-Galacturonate fermentation[c]	–	+	–	+	+	–	+
Analysis by 01 bacteriophage[c]	+	+	–	+	–	V	+

[a] Adapted from Le Minor et al.[81,82,83]

[b] The numbers give the % positive for the tests after 2 days incubation at 35%C and are based on CDC data. The vast majority of the positive tests occur within 24 hr; reactions positive after 2 days are not considered.

[c] Based on the data of Le Minor et al.[81,82,83] + = 90% or more positive; – = 10% or less positive. The test for gelatin hydrolysis is the rapid "film" method at 35%C (almost all strains are negative by the tube method at 22%C within 2 days).

[d] Currently recognized as the second species of *Salmonella*, i.e., *S. enterica* and *S. bongori*.

the abuse potential of the product. The sensitivity of the consumer group and whether the food undergoes a step lethal to *Salmonella* during processing or in the home were the most important to the U.S. Food and Drug Administration (FDA) in selecting a sampling plan for *Salmonella*. Accordingly, a sampling plan has been implemented, placing foods into one of three categories:

Category I—foods that would normally be placed in Category II, except that they are intended for consumption by the aged, the infirmed, and infants (60 analytical units);

Category II—foods that would not normally be subjected to a process lethal to *Salmonella* between the time of sampling and consumption (30 analytical units);

Category III—foods that would normally be subjected to a process lethal to *Salmonella* between the time of sampling and consumption (15 analytical units).

Various studies have demonstrated that dry compositing (combining analytical units into a single test)[69,110] or wet compositing (combining pre-enrichment cultures of individual analytical units)[100,109,110] substantially reduces the analytical workload without compromising method sensitivity. In the FDA sampling plan, up to 15 × 25-g analytical units may be combined and tested as a single 375-g composite unit. The presence of *Salmonella* in foods assigned to Categories I, II, or III may be determined by the analysis of 4, 2, 1 composite unit(s), respectively.

Whether the FDA or the ICMSF sampling plan is used, the assumption is often made that negative results indicate that the product is deemed free of *Salmonella* (95% confidence level), ergo the zero tolerance connotation. In applying the NRC Category I sampling plan, negative results indicate that the sample contains no more than one *Salmonella* per 500 grams, not that it is completely free of the organism. Further, this limit assumes random distribution of the contaminant, which is certainly not always the case. To reach the same degree of confidence with non-random distribution would require negative results of a much larger number of randomly drawn samples.

37.123 Pre-enrichment

Salmonella organisms in foods are often present in low numbers and in a debilitated condition. As indicated previously, pre-enrichment in a non-selective medium obviates this problem as well as that caused by adding large amounts of food into the selective enrichment broth. North[96] first noted enhanced recovery of *Salmonella* from egg albumen as a result of pre-enrichment in lactose broth. Even though lactose is not fermented by *Salmonella* organisms and is fermented by competing coliforms, attempts to replace lactose with sugars fermented by *Salmonella* organisms proved no more effective than lactose broth. Lactose broth is commonly used in the prescribed methods for a wide variety of foods. A number of other pre-enrichment media is used in the standard analysis of specific foods (Table 3).

In connection with *Salmonella* control programs, a variety of environmental samples may be analyzed, e.g., air filters, contents of dust collectors, static material collecting on processing equipment, swabs from floor drains, etc. The analysis of these samples does not differ from that of the finished product, except that the pre-enrichment procedure must be consistent with the nature of the environmental sample. For example, an entire air filter may be examined. Pre-enrichment should take into account the mass of the filter examined. Beyond pre-enrichment, the analytical method is the same as for food samples.

37.124 Selective Enrichment

Enrichment media, which are inoculated with portions of pre-enrichment cultures, favor the growth of *Salmonella* organisms. In the past, direct enrichment of highly contaminated raw products, such as meats and sewage, was advocated. However, it has been concluded that such samples should be pre-enriched for greater recovery of the target organism.[25]

In the third edition of this *Compendium*, only selenite cystine (SC) and tetrathionate brilliant green (TBG) broths were used for selective enrichment. However, the 8th edition of the *Bacteriological Analytical Manual* (BAM)[129] specifies that a modified Rappaport medium, Rappaport-Vassiladis (RV) medium[130] be used for raw flesh foods, highly contaminated foods and animal feeds. It is the intent of the FDA to eventually replace SC broth with RV medium for the analysis of all foods because of the known toxicity of selenium which is formed from the reduction of sodium acid selenite during incubation. Moreover, disposal of the medium is expensive according to the guidelines of the U.S. Environmental Protection Agency. The use of two or more selective media provides greater method sensitivity than a single enrichment medium.

A number of studies has shown increased recovery of *Salmonella* organisms when incubating selective broths at 43°C, instead of the more frequently employed 35–37°C.[24,40,93,111] However, McCoy[89] and Aleksic et al.[1] reported that, in some instances, incubation at 43°C prevented recovery of *Salmonella*. For foods with a high microbial load, the reference method described hereafter (Section 37.2) prescribes incubation of TBG broth at 43°C and RV medium at 42°C. For foods with a low microbial load, the SC and TBG broths are incubated at 35°C.

In an attempt to provide greater analytical flexibility, D'Aoust et al.[30] investigated the potential of refrigerating pre-enrichment and selective enrichment cultures for 72 hr, thereby avoiding weekend work. Collaborative studies showed that the refrigeration (4°C) of pre-enrichment and enrichment cultures did not adversely affect the detection of *Salmonella* in a variety of low- and high-moisture foods, thereby increasing the number of weekdays on which *Salmonella* analyses could be initiated.[24]

37.125 Selective Plating Media

Selective plating media are formulated so that growth of *Salmonella* bacteria results in the formation of discrete colonies with concomitant repression of competing microflora. Separation of *Salmonella* from non-*Salmonella* bacteria is obtained through the incorporation of various dyes, bile salts, and other selective agents into the agar media. When non-*Salmonella* colonies appear on these media, they are generally distinguished by their ability or inability to produce hydrogen sulfide and metabolism of one or more discriminating carbohydrates in the media.[26] Selective plating media that have been used for the isolation of *Salmonella* include brilliant green (BG),[81] BG sulfa,[97] bismuth sulfite (BS),[133] *Salmonella Shigella*,[37] MacConkey's,[86] desoxycholate citrate,[82] Hektoen enteric (HE),[78] xylose lysine desoxycholate citrate (XLD),[109] Shanson's,[108] Rappold-Bolderdijk modified lysine iron,[102] Rambach[101] lysine-iron-cystine,[105] modified semi-solid RV (MSRV) medium[35] and novobiocin-brilliant green-glucose.[37] None of these agars are ideal for all situations, justifying the recommendation in many reference methods for the use of two or more agar media.[8,63,74,95,129]

Since their original formulation, some of these agars have been modified by the addition of antibiotics, sulfa drugs, surfactants,

Table 3. Analysis of Foods by Methods of the *Bacteriological Analytical Manual* and the Association of Official Analytical Chemists International

Food	Pre-enrichment[a]	Preparation[b]
Dried egg yolk, egg white, and whole egg, liquid milk; powdered bread and pastry mixes; infant formulas; oral or tube feedings containing egg; coconut[c]; food dyes with pH ≥ 6.0 (10% aqueous solution); gelatin[d]; guar gum[e]	Lactose broth	Mix
Pasta (noodles, macaroni, spaghetti); egg rolls; cheese; dough; prepared salads; fresh, frozen, or dried fruits and vegetables; nut meats; crustaceans; fish; meats, meat substitutes, meat by-products, animal substances, glandular products, meals (fish, meat, and bone); casein	Lactose broth	Blend
Shell eggs, liquid whole eggs (homogenized), and hard-boiled eggs	Trypticase (tryptic) soy broth[f]	Mix[f]
Food dyes with pH<6.0 (10% aqueous solution)	None[g]	Mix 25 g/225 mL TBG
Nonfat dry milk (instant and noninstant)	Brilliant green water (add 2 mL 1% brilliant green dye solution per 1,000 mL distilled water)	Soak (gently layer powder on broth; do not shake, do not adjust pH
Dry whole milk	Distilled water and add 0.45 mL 1% brilliant green solution	Mix
Dried active[h] and inactive yeast; black pepper, white pepper, celery seed and flakes, chili powder, cumin paprika, parsley flakes, rosemary, sesame seed, thyme and vegetable flakes	Trypticase (tryptic) soy broth	Mix
Onion flakes and powder, garlic flakes and powder	Trypticase (tryptic) soy broth containing 0.5% K_2SO_3 final concentration	Mix
Allspice, cinnamon, oregano	Trypticase (tryptic) soy broth	Mix, using 1:100 sample/broth ratio
Clove	Trypticase (tryptic) soy broth	Mix, using 1:1,000 sample/broth ratio
Leafy condiments[i]	Trypticase (tryptic) soy broth	Mix, using >1:10 sample/broth ratio
Candy and candy coating; chocolate	Reconstituted nonfat dry milk,[j] add 0.45 mL 1% brilliant green solution	Blend
Frosting and topping mixes	Nutrient broth	Mix
Frog legs	Lactose Broth	Immerse/rinse let pairs, (individual leg if ≥ 25 g) and examine rinsings

[a] Unless specified otherwise, 25-g samples are pre-enriched in 225 mL of indicated medium.
[b] Unless specified otherwise, 1-mL volumes of incubated pre-enriched cultures of foods are subcultured to 10-mL volumes of selenite cystine and tetrathionate brilliant green (TBG) broth. For RV enrichment, 0.1 mL of pre-enriched cultures are subcultured to 10 mL RV medium.
[c] Add up to 2.2 mL of Tergitol 7 or Trition X-100 to initiate foaming.
[d] Add 5 mL of 5% papain solution.
[e] Add 2.25 mL of filter-sterilized 1% cellulase solution.
[f] For shell eggs and hard-boiled eggs, examine egg yolk only.
[g] Pre-enrichment in lactose broth may provide improved recovery.
[h] One-mL volumes of pre-enrichment cultures of dried active yeast are transferred to 10-mL volumes lauryl tryptose broth and TBG broth.
[i] It may be necessary to examine leafy condiments at a ratio of broth-to-sample that is greater than 1:10 because of the physical difficulties encountered by absorption of broth by the dehydrated product.
[j] Reconstituted nonfat dry milk is prepared by adding 100 g of nonfat dry milk to 1 L distilled water.

and other chemical substances in an attempt to improve media selectivity and/or sensitivity. Gadberry and Highby[62] incorporated dimethylchlortetracycline hydrochloride into BG agar and reported greater inhibition of species of *Proteus, Providencia,* and *Citrobacter.* Sulfadiazine, sulfapyridine, or sulfanilamide reportedly improves BG agar selectivity.[63,91,97,135] Moats and Kinner[92] subsequently found that addition of hydrogen sulfide detection system to BG agar aided in the identification of *Salmonella* colonies, particularly in the presence of large populations of competing microorganisms. Addition of novobiocin to XLD agar or to HE agar resulted in improved media selectivity[67,106] and increased rates of *Salmonella* recovery.[106]

Variations in the preparation of the selective agars are common and can adversely affect media performance. Although Difco[38] recommends using freshly prepared plates of BS agar, Cook[15] reported that such plates were inhibitory to several *Salmonella* serovars; such inhibition decreased upon storage of prepared plates under refrigeration for up to 5 days before use. McCoy[89] found that freshly poured BS plates were too inhibitory for serovars other than *S. typhi*. D'Aoust[24] also reported that inoculated plates of freshly poured BS agar required incubation for 48 hr to yield *Salmonella* recoveries similar to those obtained on homologous refrigerated plates. When a manufacturer's directives for medium preparation differ from those described in a reference method, the reference method should be followed. Where such directives are not specified in the analytical method, the media should be prepared according to the manufacturer's instructions.

The selective plating media are inoculated by streaking a loopful of the enrichment culture onto the dry surface of prepared plates to obtain well-separated colonies. Plates are usually incubated at 35° or 37°C. Incubation at an elevated temperature (41.5°C) was reportedly advantageous in recovering different *Salmonella* serovars on BG agar and *S. typhi* on XLD and HE agars.[104,134] Wilson et al.,[132] however, reported that incubation of BS agar plates at 43°C resulted in the appearance of small atypical *Salmonella* colonies and, in several instances, reduced sensitivity. Earlier results had also shown significantly lower recoveries on BS plates incubated at 43°C than at 35°C.[61] DeSmedt and Bolderdijk[35] introduced a motility enrichment on MSRV medium incubated at 42°C. With this technique, a serologically confirmed *Salmonella* result can be obtained 48 hr after the beginning of pre-enrichment.

Inoculated plates are usually incubated 18 to 24 hr. BS agar plates should be observed after 24 hr of incubation, and re-incubated for additional 24 hr before making a final determination on the presence or absence of *Salmonella* in the test material.

37.126 Biochemical Media

Biochemical screening media are nonselective and generally establish different biochemical traits of the test culture through color change(s) in the medium. Such data generally pertain to the production of hydrogen sulfide or to the utilization of one or more fermentable carbohydrates in the medium. Additionally, inoculated differential media also provide an inoculum for subsequent biochemical and serological testing.

Two differential agar tube media are widely used. Triple sugar iron agar (TSI) measures production of hydrogen sulfide and utilization of glucose, lactose, and sucrose,[114] whereas lysine iron agar (LIA) monitors production of hydrogen sulfide and decarboxylation of lysine.[42] These two agars are usually used in combination and provide a preliminary biochemical screening of suspect cultures. It is important to recognize that some of the TSI reactions may be redundant with those observed on plating media. Cultures that appear to be contaminated on these agar tube media should be restreaked on an appropriate selective plating agar and a small portion of a single colony exhibiting typical characteristics transferred to the differential media for repeat testing. Other *Salmonella* differential media include dulcitol lactose iron agar[118]; malonate dulcitol lysine iron agar[113]; DMS agar[10]; dulcitol-malonate-phenylalanine agar[45]; and selective Padron-Dockstader agar.[98]

37.13 General Description for the Identification of Salmonella

37.131 Biochemical Tests

Many biochemical tests are available for the characterization of foodborne isolates, obtained from food products. A comprehensive list of biochemical and nutritional characters of the family *Enterobacteriaceae* is found in *Bergey's Manual of Systematic Bacteriology*,[83] and many of these tests have been described in detail by Ewing.[46] Presumptive identification of typical *Salmonella* can be obtained from a minimal number of biochemical tests that may include TSI/LIA reactions, urease, KCN, and indole. The urease, lysine decarboxylase, growth in KCN, and indole tests are generally sufficient for presumptive identification of salmonellae.[8,129] Note: Final identification rests with serological tests, not biochemical tests, because *Salmonella* strains do not always produce typical biochemical reactions.

37.132 Commercial Biochemical Multitest Systems or Kits

Many biochemical tests are routinely used to differentiate members of the family *Enterobacteriaceae* and to characterize presumptive *Salmonella* isolates. Commercially available multitest systems or diagnostic kits are more convenient than conventional tube systems and generally provide reliable test results.[16,17] The API 20E, Minitek, and Enterotube systems received AOAC approval in 1978,[99] the Micro-ID Kit in 1989,[76] and the Vitek GNI + system in 1991.[79] Other miniaturized systems have shown a high degree of correlation with conventional tube tests but have not been subjected to AOAC collaborative studies. These and other diagnostic systems and kits are thoroughly reviewed in Chapter 10. Note: Complete identification of *Salmonella* should not be based solely on biochemical tests because strains do not always produce typical biochemical reactions (Tables 1 and 2).

37.133 Confirmatory Serological Tests

The genus *Salmonella* is characterized serologically by specific antigenic components.[46] The antigens are divided into somatic (O), flagellar (H), and capsular (K) antigens. The somatic (O) antigens are composed of lipopolysaccharide complexes that are heat stable and resistant to alcohol and dilute acid. The proteinaceous flagellar (H) antigens are heat labile, whereas the surface K antigens consist of heat-sensitive polysaccharides that occur in the capsule or in the outer membrane of the bacterium. K antigens such as the Vi antigen tend to inhibit somatic (O) agglutination reactions and must be thermally denatured before undertaking somatic agglutination reactions. For further information concerning *Salmonella* antigens and antisera, see Edwards and Ewing,[41] Kauffmann,[75] and ICMSF.[72]

Since other members of the *Enterobacteriaceae* e.g., *Citrobacter* and *Providencia*, have related somatic (O) antigens, false-positive serological tests may be encountered, on occasion, upon testing unknown isolates with *Salmonella* antisera. This problem is diminished when more specific single factor or adsorbed antisera are used. Definitive serotyping of a culture should be performed by specially trained personnel working in a reference laboratory.

Concurrent serological confirmation and biochemical characterization of a single colony forming unit by some biochemical kits can be completed within 8 to 24 hr.[16,17]

37.14 Rapid Cultural Methods

Early attempts to reduce the time required for *Salmonella* analysis involved the use of short incubation times. Abbreviated incubation of pre-enrichment (≤6 hr) is desirable because it would allow presumptive identification of salmonellae one day earlier compared to standard culture methods employing 18 to 24-hr pre-enrichment. However, attempts to reduce pre-enrichment incubation to 6 hr resulted in an unacceptable number of false-negative results compared to 18 to 24-hr pre-enrichment.[25,29]

Combining pre-enrichment and enrichment steps into a single analytical procedure also has been proposed to reduce analytical time. In this approach, selective agents were added to the pre-enrichment medium, either gradually through the use of time-release capsules[115] or by direct addition after 4-hr incubation.[116] The time-release capsule method was not effective in reducing analytical time required for analysis, presumably because of the inconsistency in the dissolution of the capsules from lot to lot. In a subsequent study, Sveum and Kraft[116] added selective agents to a nonselective basal medium after 4 hr of incubation and continued incubation for a total of 24 hr at 35°C. Results with pure cultures of heat-injured and freeze-injured salmonellae and with artificially contaminated foods compared favorably with the standard culture method. However, this novel approach has not found wide application, possibly because addition of selective agents after 4-hr incubation is inconvenient and could result in a longer workday (>8 hr). The need for large quantities of selective agents to convert pre-enrichment cultures into selective enrichment broths would significantly increase analytical costs.

Reduced incubation of selective enrichment broths also has been investigated. One study utilizing a 6-hr selective step following overnight pre-enrichment yielded results similar to that obtained with a 24-hr selective step for feeds and feed ingredients.[29] However, use of a 6-hr enrichment with raw and processed foods resulted in a high percentage of false-negative results compared to that obtained after 24 hr of incubation.[34] The influence of reduced incubation time (≤8 hr) on productivity of tetrathionate and Rappaport broths for detection of *Salmonella* in naturally- and artificially-contaminated foods was examined by Rappold et al.[103] Modified Rappaport broth identified a similar number of *Salmonella*-positive samples after 8 hr and 24 hr of incubation. With tetrathionate broth, all but 3 of 27 contaminated samples were positive after 8 hr of incubation. Thus, the use of reduced incubation of selective enrichment periods may provide adequate detection under certain conditions, but is not consistently reliable and is not recommended. Some evidence suggests that extended incubation (> 24 hr) of selective enrichment may improve detection,[40,65,130] but other data indicate no benefit and reduced recovery from extended (>24 hr) selective enrichment.[25,26,28]

Although reduced incubation of enrichments may not always provide optimal conditions for isolation of salmonellae, some evidence shows that reduced enrichment can be productive when used in combination with rapid screening methods. For example, 6-hr selective enrichment procedures for dry foods are employed in three rapid methods approved by the AOAC: hydrophobic grid membrane filtration (HGMF),[43] enzyme immunoassay (EIA),[58,59] and DNA-DNA hybridization.[57]

Most of the methods proposed for the rapid detection of *Salmonella* in foods are screening tests and designed to provide rapid negative results, but requiring cultural confirmation of presumptive *Salmonella* isolates. Ideally, a rapid screening method would be performed directly on the food product without the need for enrichment of the sample. However, none of the methods currently available possesses the required sensitivity to detect the low numbers of salmonellae usually encountered in foods. Most employ one or more enrichment steps whose purpose is the same as described in standard cultural methods; i.e., pre-enrichment allows for the recovery and growth of injured cells whereas selective enrichment favors the growth of salmonellae and inhibits the growth of competitive bacteria. Several methods also employ a third nonselective enrichment broth (post-enrichment) which results in a modest increase in cell numbers.

37.15 Association, Agency, and International Methods

37.151 Association of Official Analytical Chemists International (AOAC)

The primary objective of the AOAC is ". . .to obtain, improve, develop, test, and adopt uniform, precise, and accurate methods for the analysis of foods, drugs, feeds, fertilizers, pesticides, water, or any other substances affecting public health and safety, economic protection of the consumer, or quality of the environment. . . .".[4] To be adopted as official, a method must be 1) reliable, 2) practical, 3) available to all analysts, and 4) substantiated. The proposed method of analysis must be subjected to a collaborative study and the results from that study reviewed by an AOAC committee. If approved, the study is published in the AOAC's journal and the method is adopted as an official first action method, the first level of sanction granted by AOAC. After a minimum period of two years, the first action method is adopted as a final action method, the highest level of AOAC sanction, after all adverse comments, if any, have been satisfactorily addressed. The recommended procedure for the analysis of various foods delineated in Section 37.5 is based on AOAC methods.

37.152 U.S. Food and Drug Administration (FDA)—Bacteriological Analytical Manual

This book[129] contains the microbiological methods used in FDA laboratories to analyze foods. These methods are currently considered to be the most useful to the FDA in enforcing the provisions of the Federal Food, Drug, and Cosmetics Act.[126] Methods having official AOAC status are included in this manual. The *Bacteriological Analytical Manual* (BAM) and AOAC methods of preparing various foods for the isolation of *Salmonella* are listed in Table 3. Isolation and confirmation procedures are described in Sections 37.52 and 37.53.

37.153 Centers for Disease Control and Prevention

A manual[77] used in the CDC's course on *Salmonella* isolation procedures is available. Directions for sampling, isolation, and identification of *Salmonella* are provided.

37.154 United States Department of Agriculture

The Agricultural Research Service publishes manuals for *Salmonella* analysis of poultry[18] and animal feeds.[123] Recommended test procedures are presented, followed by descriptions of serological and biochemical confirmatory test procedures.

The Poultry Division of the Agricultural Marketing Service publishes a laboratory handbook[125] that includes directions for the analysis of eggs and egg products.

The Food Safety and Inspection Service publishes a laboratory guidebook[124] that contains a comprehensive discussion of isolation and identification procedures for *Salmonella* in foods.

37.155 International Organizations

Many organizations are concerned with microbiological criteria and safety of foods involved in international trade. Specifically, these agencies are involved in the collection and assessment of microbiological criteria for foods, and the development, study, and standardization of methods for the microbiological examination of foods. Figuring prominently in this area of endeavor are the ICMSF (of the International Association of Microbiological Societies) and the Codex Alimentarius Commission, as well as temporary Expert Consultations of the Joint Food Standards Programme of the United Nations' Food and Agriculture Organization and World Health Organization, and Technical Committee 34 of the International Organization for Standardization (ISO).

37.2 TREATMENT OF SAMPLE

37.21 Collection (see Chapter 2 and Section 37.122)

37.22 Holding (see Chapter 2)

37.23 Mixing and Homogenization

If the food product is frozen, thaw a suitable portion at 2° to 5°C for 18 hr before analysis, or, if rapid thawing is desired, thaw at <45°C for ≤15 min. If the food product is powdered, ground, or comminuted, mix it with a sterile spoon or other sterile equipment before withdrawing an analytical unit. A homogeneous suspension of most powdered products can be obtained by mixing the analytical unit and broth with a sterile glass rod or other appropriate sterile instrument. Homogeneous suspensions are obtained in some cases by shaking the food-broth mixture by hand or by using a mechanical shaker. Mechanical blending may be required if the food consists of large pieces. A blending time of 2 min at 8,000 rpm is usually satisfactory for most foods.

37.3 EQUIPMENT AND SUPPLIES

37.31 Equipment and Materials

1. Blender and sterile blender jars
2. Sterile 16-oz (500-mL) wide mouthed screw-cap jars, sterile 500-mL Erlenmeyer flasks, sterile 250-mL beakers, sterile glass or paper funnels of appropriate size, and containers of appropriate capacity to accommodate composited samples
3. Sterile bent-glass spreader rods
4. Balance, with weights; 2,000-g capacity, sensitivity of 0.1 g
5. Balance, with weights; 120-g capacity, sensitivity of 5 mg
6. Incubator, 35°C
7. Waterbaths, 48°C to 50°C (polyvalent flagellar and Spicer-Edwards testing; 42 ± 0.2° C (RV medium); and 43 ± 0.2° (TBG broth); waterbaths for RV medium and TBG broth should be circulating and thermostatically-controlled
8. Sterile spoons or other appropriate instruments for transferring food specimens
9. Sterile culture dishes, 15 × 100 mm, glass or plastic
10. Sterile pipettes, 1 mL, with 0.01 mL graduations; 5 and 10 mL, with 0.1 mL graduations
11. Inoculating needle and inoculating loop (about 3 mm inner diameter), nichrome, platinum-iridium, or chromel wire
12. Sterile test or culture tubes, 16 × 150 mm and 20 × 150 mm; serological tubes, 10 × 75 mm or 13 × 100 mm
13. Test or culture tube racks
14. Vortex mixer
15. Sterile shears, large scissors, scalpel, and forceps
16. Lamp (for observing serological reactions)
17. Fisher or Bunsen burner
18. pH test paper (pH range 6 to 8) with maximum graduations of 0.4 pH units per color change
19. pH meter
20. Plastic bags, 28 × 37 cm, sterile, with resealable tape. (Items 20 to 22 are needed in the analysis of frog legs and rabbit carcasses.)
21. Plastic beakers, 4 L, autoclavable, for holding plastic bag during shaking and incubation
22. Mechanical shaker, any model that can be adjusted to give 100 excursions per min with a 4-cm (1 1/2-in) stroke, such as the Eberback shaker with additional 33- and 48-cm clamp bars

37.32 Media and Reagents

For preparation of media and reagents, refer to Section 967.25 in *Official Methods of Analysis.*[8]

1. Nonselective media for pre-enrichment and propagation
   a. Lactose broth
   b. Nonfat dry milk (reconstituted)
   c. Trypticase (BBL) or tryptic (Difco) soy broth
   d. M-broth
   e. GN-broth
   f. Nutrient broth
   g. Brain heart infusion broth
2. Selective enrichment media
   a. Selenite cystine broth
   b. Tetrathionate broth (commercial formulation from BBL or Difco) with brilliant green dye
   c. Rappaport-Vassiliadis (RV) medium
3. Selective isolation media
   a. Xylose lysine desoxycholate agar
   b. Hektoen enteric agar
   c. Bismuth sulfite agar
   d. MacConkey agar
4. Media for biochemical characterization of isolates
   a. Triple sugar iron agar
   b. Tryptone (tryptophane) broth
   c. Trypticase (tryptic) soy broth
   d. Lauryl tryptose broth
   e. Trypticase soy-tryptose broth
   f. MR-VP broth
   g. Simmons citrate agar
   h. Urea broth
   i. Urea broth (rapid)
   j. Malonate broth
   k. Lysine iron agar (Edwards and Fife)
   l. Lysine decarboxylase broth
   m. Motility test medium (semi-solid)
   n. Phenol red carbohydrate broth
   o. Potassium cyanide
   p. Bromcresol purple carbohydrate broth
5. Reagents
   a. Papain solution, 5%
   b. Cellulase solution, 1%
   c. Potassium sulfite powder, anhydrous
   d. Kovacs' reagent
   e. Voges-Proskauer test reagents
   f. Creatine phosphate crystals
   g. Potassium hydroxide solution, 40%
   h. 1 N sodium hydroxide solution
   i. 1 N hydrochloric acid
   j. Brilliant green dye solution, 1%
   k. Malachite green oxalate solution, 0.4%
   l. Bromcresol purple dye solution, 0.2%
   m. Methyl red indicator
   n. Magnesium chloride solution, 40%
   o. Sterile distilled water
   p. Tergitol Anionic 7
   q. Triton X-100

r. Physiological saline solution (sterile)
s. Formalinized physiological saline solution
t. *Salmonella* polyvalent somatic (O) antiserum
u. *Salmonella* polyvalent flagellar (H) antiserum
v. *Salmonella* somatic group (O) antisera: A, B, C_1, C_2, C_3, D_1, D_2, E_1, E_2, E_3, E_4, F, G, H, I, Vi, and other groups as appropriate
w. *Salmonella* Spicer-Edwards flagellar (H) antisera

37.33 Additional Materials and Equipment

37.331 Fluorescent Antibody Technique

1. Fluorescent microscope with exciter filter (330 to 500 nm) and barrier filter (>400 nm)
2. Multiwell slides coated with fluorocarbon material (available from Cell-Line Associates, Minotola, NJ 08341 or Clinical Sciences, Inc., 30 Troy Road, Whippany, NJ 07981); slides may also be prepared as described in the AOAC method.[8]

37.332 Hydrophobic Grid Membrane Filtration

1. HGMF filter, polysulfone membrane filter with pore size of 0.45 µm and imprinted with nontoxic hydrophobic material in a grid pattern (QA Laboratories, San Diego, CA), or equivalent
2. Filtration units for HGMF filter, equipped with 5-µm mesh prefilter to remove food particles during filtration; one unit is required for each test sample
3. Vacuum pump, water aspirator vacuum source is necessary
4. Manifold or vacuum flask

37.333 Immunodiffusion

Salmonella 1-2 TEST unit (BioControl Systems, Inc., Bellevue, WA 98005)

37.334 Enzyme Immunoassays

1. Multipipettes capable of delivering accurate amounts in ranges 50 to 250 µL
2. Waterbath, autoclave, or steamer capable of maintaining 100°C
3. Microtiter plate reader; specifications of instrument may vary with the assay and are described in the methods (Section 37.74)

37.335 DNA Hybridization

1. Micropipettes capable of accurately dispensing 0.10 mL, 0.25 mL, and 0.754 mL
2. Waterbath capable of maintaining 65°C
3. Photometer to read absorbance at 450 nm in 12 × 75-mm test tubes
4. Dipstick holders and wash basins (available from GENE TRAK Systems, Hopkington, MA 01748)

37.336 Conductance

1. Microbiological analyzer instrument operating at 35°C that measures microbial growth based on conductance changes recorded at frequency of 10 kHz (Malthus Instruments Ltd, The Manor, Manor Royal, Crawley, West Sussex, RH10 2PY, United Kingdom)
2. Automatic pipette capable of delivering 100 µL

37.4 PRECAUTIONS AND LIMITATIONS OF REFERENCE AND RAPID METHODS

37.41 Sampling

Specific sampling plans for *Salmonella* have been in place for several years.[71,94,129] To maintain the statistical significance of these plans, they should be used exactly as directed, without modification. Particular attention should be given to instructions for compositing samples. The maximal number of 25-g analytical units that may be composited is 15, resulting in composites weighing 375 g.[128]

In applying the soak method for the analysis of low-moisture, powdered foods, be aware that composited samples of certain foods such as non-instant nonfat dry milk, dry whole milk and soy flour are not readily wetted. In these cases, 25-g analytical units should not be composited but analyzed individually.

37.42 Media Preparation and Disposal

Unreliable laboratory results can arise from improper preparation of laboratory media and absence of appropriate controls that form an integral part of sound laboratory quality assurance programs.

A positive medium control ensures that no substances in the laboratory medium are inhibitory to salmonellae. This control is prepared by inoculating the medium with a suitable strain of *Salmonella* and proceeding through the entire analytical protocol used for test sampling.

A negative medium control ensures that the prepared medium is not contaminated with salmonellae. This control is prepared by carrying a flask of the uninoculated medium through the analytical procedure used for test samples.

In addition to the above, the following specific points should be considered in the preparation of media for *Salmonella* isolation:

1. Potential toxicity of various lots of brilliant green dye should be determined using appropriate laboratory strains of salmonellae.
2. Flasks of TBG broth should be shaken frequently during dispensing into culture tubes because calcium carbonate tends to settle.
3. SC broth should be dispensed in 16 × 150-mm tubes to a depth of at least 5 cm, since this medium is most efficient in an environment of reduced oxidation-reduction potential.[98]
4. Plates of BS agar should be made the day before use since freshly prepared agar may be inhibitory for *Salmonella*. Prepared plates should be stored at 4° to 5°C and tempered to room temperature before use to avoid condensation of moisture on the surface.
5. LIA slants should be prepared with a deep butt (4 cm) since lysine decarboxylation reactions are more reliable under microaerophilic conditions.
6. KCN should not be stored under refrigeration longer than 2 weeks since it may become unstable and give false-positive results.
7. TSI tubes should be capped loosely; otherwise erroneous reactions occur. By definition/convention, H_2S production by *Salmonella* is defined using TSI.
8. In many localities, the selenium formed during the incubation of SC broth is considered a toxic waste. Therefore, local environmental guidelines concerning disposal of selenite should be followed at all times.

37.43 Conventional Culture Procedure

For the analysis of a test sample, the following specific points should be considered:

1. The actual rpm speed of a food blender should be determined by a tachometer or similar device. Use of a blending speed higher than that recommended in Section 37.23 may result in injury or death of *Salmonella* organisms. To avoid spillage, initiate blending at lowest speed for a few seconds, and then gradually increase to the recommended blending speed.
2. If it is necessary to thaw a frozen sample for analysis, water-bath temperature must not exceed 45°C.
3. Modified Rappaport medium (RV medium) has replaced SC broth as a selective enrichment broth for raw flesh and highly contaminated foods and animal feeds. Because this medium is less selective than some others, it works best at an incubation temperature of 42°C.
4. Many procedures call for incubating enrichment broths at 43°C. While this elevated temperature of incubation is a valuable tool for recovery of *Salmonella*, care should be taken to assure that incubators do not operate above 43°C. A recording thermometer should be used to monitor the temperature.
5. Freshly prepared BS agar is often too inhibitory for serovars other than *S. typhi*. Storage of BS agar plates in the refrigerator for one to 5 days before use will greatly decrease this toxicity and improve the efficiency of the medium.
6. For sample types that contain many competing organisms such as *Proteus*, addition of novobiocin to HE and XLD agars at a concentration of 15 ppm will greatly improve the efficiency of these media.
7. Analysts should be aware that atypical *Salmonella* strains, e.g., *S. arizonae* and other lactose/sucrose-positive and H_2S-negative *Salmonella* biotypes, may be encountered on differential agar media. Lactose- and/or sucrose-positive cultures may resemble coliforms on HE, XLD, and TSI agars. TSI cultures with an acid slant should not be discarded as non-*Salmonella* but further screened biochemically and serologically.
8. The center of suspect *Salmonella* colonies on plating media should be picked lightly to avoid transfer of variable non-salmonellae that may lie under or adjacent to the suspect *Salmonella* colony.
9. TSI slants should be capped loosely during incubation for maintenance of aerobic conditions to prevent erroneous acid reactions on slant and excessive H_2S production in the butt of the tube. If heavy H_2S production masks the reaction in the butt, glucose utilization should be assumed.
10. Caps of inoculated lysine decarboxylase broth should be replaced tightly. Lysine decarboxylation occurs anaerobically, and exclusion of air will eliminate false-positive reactions resulting from oxidative deamination of peptones in the medium. For LIA slant and lysine decarboxylation broth reactions, only a distinct yellow color in the butt or broth should be considered negative. Cultures giving weak, indeterminate reactions in the decarboxylase broth medium should be retested by adding a few drops of 0.2% bromcresol purple dye and reading immediately.
11. If the purity of the TSI slant culture is in doubt, streak the culture to MacConkey, HE, BG sulfa, XLD or similar agars and repeat the TSI test using a well isolated colony.
12. For each type of biochemical determination, a tube of uninoculated medium (negative medium control) should be included.
13. Only the AOAC approved diagnostic kits described in Section 37.132 should be used for the rapid biochemical characterization of *Salmonella*.

37.44 Rapid Screening Methods

37.441 Fluorescent Antibody Technique

1. This method is recommended only as a screening test for the presence of *Salmonella* because the antibodies may produce false-positive reactions by cross-reacting with other members of the *Enterobacteriaceae* and through nonspecific reactions. Consequently, presumptive FA-positive enrichment cultures must be streaked onto plating media and suspect colonies confirmed by biochemical and serological techniques described in Section 37.5. Interpretation of FA results is highly subjective and requires trained analysts.
2. The method is subject to variations in the preparation of smears, conjugate, and other reagents, and misalignment of the fluorescent microscope may lead to erroneous results.

37.442 Hydrophobic Grid Membrane Filtration Described in Section 37.72

1. In this technique, TBG broth medium (10 mL) is inoculated with 0.1 mL of pre-enrichment culture instead of the more standard 1.0 mL transfer volume; larger volumes do not provide favorable *Salmonella* to non-*Salmonella* ratios in TBG broth after a 6-hr incubation.
2. Tubes of TBG broth are inoculated with a portion of pre-enrichment culture and incubated at 35° to 37°C in a water-bath, rather than in a convection incubator, because a full 6-hr incubation period is needed for *Salmonella* organisms to reach detectable levels.
3. Growth units on HGMF filters should be picked with a fully cooled sterile needle.

37.443 Immunodiffusion: 1-2 Test™ Described in Section 37.73

1. This method is a screening test for foodborne *Salmonella*.
2. Nonmotile *Salmonella* strains will not be detected because the method is based on the migration of salmonellae in a semi-solid medium.
3. The chamber plug that seals the opening between the motility chamber and the TBG broth inoculation chamber must be removed prior to inoculation of the 1-2 Test. Failure to remove this plug will prevent *Salmonella* from migrating from the inoculation site to the motility chamber.
4. The gel former tip, an interior protrusion of the cap of the motility chamber, must be cut off to prevent displacement of the added antibody solution.

37.444 Enzyme Immunoassays (Salmonella-Tek™, TECRA™, Assurance™ Described in Section 37.74)

Although commercially available kits differ in their sensitivity and specificity, the inherent limitations of these diagnostic tools are similar.

1. These methods are screening tests for the presence of *Salmonella* in foods. The antibodies employed may produce false-

positive reactions through cross-reactions with other *Enterobacteriaceae* or as a result of nonspecific reactions. Presumptively positive EIA samples must be confirmed by streaking homologous enrichment broths on appropriate plating media and identifying suspect colonies as described in Section 37.5.
2. Antibodies used in these diagnostic kits react with salmonellae in the "common" serological (O) groups, e.g., B, C, D, and E. Thus, a certain percentage of salmonellae reportedly produce false negative results.[26,27,32]
3. Commercially available kits contain standardized reagents. Use of other materials or procedures may yield erroneous results.
4. Positive and negative controls are included in test kits and must be included with each group of test samples. Controls must fall within acceptable ranges to validate test results.

37.445 DNA Hybridization (GENE TRAK™, Described in Section 37.75)

1. This method is a screening test for the presence of *Salmonella* in foods. Presumptively positive assays must be confirmed by streaking onto appropriate plating media and suspect colonies identified as described in Section 37.5.
2. Commercially available kits contain standardized reagents, and use of other materials or procedures may yield erroneous results.
3. Positive and negative controls are included in the test kits and must be included with each group of samples. Controls must fall within acceptable ranges to validate test results.
4. Reagents that contain sodium azide should be flushed down sinks with an abundant supply of water. Do not pipette reagents by mouth.

37.446 Conductance Method

1. This method is a screening test for the presence of *Salmonella* in foods. Presumptively positive assays must be confirmed by streaking onto appropriate plating media and suspect colonies identified as described in Section 37.5.
2. Samples are pre-enriched in buffered peptone water-lysine-glucose broth followed by two-tube conductance assay in two selenite-based media containing trimethylamine-N-oxide (TMAO) and dulcitol (*Salmonella* Medium 1) and lysine (*Salmonella* Medium 2).
3. *Salmonella* spp. typically give large conductance changes in these two media compared to those for non-salmonellae. Competitive microorganisms will occasionally give similar conductance changes which result in false-positive screening results. Moreover, some *Salmonella* organisms may not show sufficient changes in conductance, thereby resulting in false-negative readings.

37.5 REFERENCE METHOD

This section contains the approved methods of the *Bacteriological Analytical Manual*[129] and the AOAC International[8] for the isolation and identification of foodborne *Salmonella* sp. organisms.

Methods of pre-enrichment and preparation of sample for each food are listed in Table 3.

37.51 Pre-enrichment

Aseptically open a sample container, and continue as directed under appropriate food section below.

37.511 Dried Egg Yolk, Dried Egg Whites, Dried Whole Eggs, Liquid Milk (Skim Milk, 2% Fat Milk, Whole Milk, and Buttermilk), Prepared Powdered Mixes (Cake, Cookie, Doughnut, Biscuit, and Bread), Infant Formula, and Oral or Tube Feedings Containing Egg

1. For non-powdered foods, add 225 mL of sterile lactose broth to 25-g analytical unit. For powdered foods, gradually add 225 mL of sterile lactose broth in small increments to 25-g analytical unit. Add a small portion (15 mL) of sterile lactose broth and mix with a suitable sterile stirrer to obtain a homogeneous suspension. Repeat this procedure three times and then add the remainder of the lactose broth. Stir until a lump-free suspension is obtained.
2. Cap the container securely, and let stand at room temperature for 60 min.
3. Mix well by swirling, and determine the pH with appropriate test paper.
4. Adjust the pH, if necessary, to 6.8 ± 0.2 with 1 N sodium hydroxide or 1 N hydrochloric acid.
5. Loosen the container cap about one-quarter turn, and incubate 24 ± 2 hr at 35°C.
6. Continue as directed under Section 37.52.

37.512 Eggs

37.5121 Contents of shell eggs

1. Wash eggs with stiff brush and drain. Soak in 200 ppm Cl^- solution containing 0.1% sodium dodecyl sulfate (SDS) for 30 min. Prepare the 200 ppm Cl^-/0.1% SDS solution by adding 8 mL of commercial bleach (5.25% sodium hypochlorite) to 992 mL of distilled water containing 1 g of SDS. This disinfectant should be prepared immediately before analysis.
2. Crack eggs aseptically and, using a sterile egg separator, discard the whites.
3. Aseptically weigh 25 g egg yolk into a sterile 500 mL Erlenmeyer flask or other appropriate container.
4. Add 225 mL tryticase (tryptic) soy broth (TSB), and mix well by swirling.
5. Continue as directed under Section 37.511, 2 to 6.

37.5122 Liquid whole eggs (homogenized)

1. Aseptically weigh 25 g into a sterile 500 mL Erlenmeyer flask or other appropriate container.
2. Add 225 mL of TSB, and mix well by swirling.
3. Continue as directed under Section 37.511, 2 to 6.

37.5123 Hard-boiled eggs

1. Aseptically separate the egg yolk from the egg white.
2. Weigh 25 g pulverized egg yolk solid into a sterile 500 mL Erlenmeyer flask or other appropriate container.
3. Add 225 mL of TSB, and mix well by swirling.
4. Continue as directed under Section 37.511, 2 to 6.

37.513 Nonfat Dry Milk

37.5131 Instant

1. Aseptically weigh 25-g analytical unit into a sterile beaker (250 mL) or other appropriate container.
2. Using a sterile glass or paper funnel (made with tape to withstand autoclaving), pour the 25-g analytical unit gently and slowly over the surface of 225 mL of brilliant green water contained in a sterile 500 mL Erlenmeyer flask or other appropriate container. Alternatively, 25-g analytical units may be composited and poured over the surface of proportion-

ately larger volumes of brilliant green water. Prepare brilliant green water by adding 2 mL of 1% brilliant green dye solution per 1,000 mL of sterile distilled water.
3. Allow container to stand undisturbed for 60 ± 5 min.
4. Incubate loosely capped container, without mixing or pH adjustment, for 24 ± 2 hr at 35°C.
5. Continue as directed under Section 37.52.

37.5132 Non-instant

Examine as described for instant nonfat dry milk, except that the 25-g analytical units may not be composited.

37.514 Dry Whole Milk

1. Aseptically weigh 25-g analytical unit into a sterile, wide-mouth, screw-cap jar (500 mL) or other appropriate container.
2. Add 225 mL of sterile distilled water, and mix well by stirring.
3. Cap the jar securely, and let stand for 60 min at room temperature.
4. Mix well by swirling, and determine the pH with test paper.
5. Adjust pH, if necessary, to 6.8 ± 0.2. Cap the jar securely, and mix well before determining the final pH.
6. Add 0.45 mL of a 1% aqueous brilliant green dye solution and mix well.
7. Loosen the jar cap 1/4 turn, and incubate the jar for 24 ± 2 hr at 35°C.
8. Continue as directed under Section 37.52.

37.515 Casein

1. Aseptically weigh 25-g analytical unit into sterile blender jar.
2. Add 225 mL of sterile lactose broth to 25-g analytical unit, and blend for 2 min.
3. Aseptically transfer blended homogenate into a sterile, wide-mouth, screw-cap jar (500 mL) or other appropriate container.
4. Continue as directed under Section 37.511, 2 to 6.

37.516 Soy Flour

1. Aseptically weigh 25-g analytical unit into a sterile beaker (250 mL) or other appropriate container.
2. Using a sterile glass or paper funnel (made with tape to withstand autoclaving), pour 25-g analytical unit gently and slowly over the surface of 225 mL of lactose broth contained in a sterile 500 mL Erlenmeyer flask or other appropriate container. Analytical units (25 g) may not be composited.
3. Allow container to stand undisturbed for 60 ± 5 min.
4. Incubate loosely capped container, without mixing or pH adjustment, for 24 ± 2 hr at 35°C.
5. Continue as directed under Section 37.52.

37.517 Egg-Containing Products (Noodles, Egg Rolls, Macaroni, Spaghetti, Cheese, Dough); Prepared Salads (Ham, Egg, Chicken, Tuna, Turkey); Fresh, Frozen, or Dried Fruits and Vegetables; Nut Meats; Crustaceans (Shrimp, Crab, Crayfish, Langostinos, Lobster); and Fish

1. Preferably, do not thaw frozen foods before analysis. If frozen foods must be tempered to obtain an analytical portion, thaw below 45°C for ≤ 15 min with continuous agitation in thermostatically-controlled waterbath, or thaw within 18 hr at 2–5°C.
2. Aseptically weigh 25-g analytical unit into a sterile blending container.
3. Add 225 mL of sterile lactose broth and blend for 2 min.
4. Aseptically transfer blended homogenate into sterile, wide-mouth, screw-cap jar (500 mL) or other appropriate container.
5. Continue as directed under Section 37.511, 2 to 6.

37.518 Dried Yeast

1. Aseptically weigh 25-g analytical unit into a sterile, wide-mouth, screw-cap jar (500 mL) or other appropriate container.
2. Add 225 mL of sterile TSB, and mix well to form a smooth suspension.
3. Continue as directed under Section 37.511, 2 to 6.

37.519 Spices

37.5191 Black pepper, white pepper, celery seed or flakes, chili powder, cumin, paprika, parsley flakes, rosemary, sesame seed, thyme, and vegetable flakes

1. Aseptically weigh 25-g analytical unit into a sterile, wide-mouth, screw-cap jar (500 mL) or other appropriate container.
2. Add 225 mL of TSB, and mix well by swirling.
3. Continue as directed under Section 37.511, 2 to 6.

37.5192 Onion flakes, onion powder, garlic flakes, and garlic powder

1. Aseptically weigh 25-g analytical unit into a sterile, wide-mouth, screw-cap jar (500 mL) or other appropriate container.
2. Pre-enrich spice in TSB with added K_2SO_3 (5 g K_2SO_3 per 1,000 mL of TSB, resulting in final 0.5% K_2SO_3 concentration). Add K_2SO_3 to broth before autoclaving 225-mL volumes in 500-mL Erlenmeyer flasks at 121°C for 15 min. After autoclaving, aseptically determine and, if necessary, adjust the final volume to 225 mL with sterile broth. Add 225 mL of sterile TSB with added K_2SO_3 to the 25-g analytical unit, and mix well.
3. Continue as directed under Section 37.511, 2 to 6.

37.5193 Allspice, cinnamon, cloves, and oregano

1. Examine allspice, cinnamon, and oregano at 1:100 spice/broth ratio and cloves at 1:1000 spice/broth ratio. Examine leafy condiments at a spice/broth ratio that is greater than 1:10.
2. Examine these spices as directed under Section 37.5191.

37.5194 Candy and candy coating (including chocolate)

1. Aseptically weigh 25-g analytical unit into a sterile blending container.
2. Add 225 mL of sterile, reconstituted nonfat dry milk, and blend 2 min.
3. Aseptically transfer homogenized mixture to sterile, wide-mouth, screw-cap jar (500 mL) or other appropriate container.
4. Continue as directed under Section 37.511, 2 to 4.
5. Add 0.45 mL of a 1% aqueous brilliant green dye solution, and mix well.
6. Loosen the jar caps 1/4 turn, and incubate 24 ± 2 hr at 35 C.
7. Continue as directed under Section 37.52.

37.5195 Coconut

1. Aseptically weigh 25-g analytical unit into a sterile, wide-mouth, screw-cap jar (500 mL) or other appropriate container.

2. Add 225 mL of sterile lactose broth, and shake well.
3. Continue as directed under Section 37.511, 2 to 4.
4. Add up to 2.25 mL of steamed (15 min) Tergitol Anionic 7 (undiluted) or Triton X-100 (undiluted), and mix well.
5. Loosen the jar caps about 1/4 turn, and incubate 24 ± 2 hr at 35 C.
6. Continue as directed under Section 37.52.

37.5196 Gelatin

1. Aseptically weigh 25-g analytical unit into sterile, wide-mouth Erlenmeyer flask (500 mL) or other appropriate container.
2. Prepare a 5.0% papain solution (add 5 g of papain to 95 mL of sterile distilled water). Add 225 mL of lactose broth and 5 mL of the 5% papain solution to the 25-g analytical unit, and mix well by swirling.
3. Cap the jar securely, and incubate at 35°C for 60 min.
4. Continue as directed under Section 37.511, 3 to 6.

37.5197 Guar gum

1. Aseptically weigh a 25-g analytical unit into a sterile beaker (250 mL) or other appropriate container.
2. Prepare a filter-sterilized 1.0% cellulase solution (add 1 g of cellulase to 99 mL of distilled water). Sterilize by filtration through a 0.45-μm membrane.
3. Add 225 mL sterile lactose broth and 2.25 sterile 1% cellulase solution to sterile wide-mouth, screw-cap jar (500 mL) or other appropriate container.
4. While vigorously stirring the cellulase/lactose broth with a magnetic stirrer, pour the 25-g analytical unit quickly through a sterile glass funnel into the cellulase/lactose broth.
5. Cap jar securely, and let stand for 60 min at room temperature.
6. Incubate the loosely capped container, without pH adjustment, for 24 ± 2 hr at 35°C.
7. Continue as directed under Section 37.52.

37.5198 Meats, meat substitutes, meat by-products, animal substances, glandular products, and meals (fish, meat, and bone)

1. Aseptically weigh 25-g analytical unit into a sterile blending container.
2. Add 225 mL of sterile lactose broth and blend for 2 min, and transfer blended homogenate to sterile Erlenmeyer flask (500 mL) or other appropriate container. For foods that do not require blending (powdered, ground, or comminuted foods), add 225 mL of sterile lactose broth to a 25-g analytical unit in a sterile Erlenmeyer flask (500 mL), and swirl thoroughly.
3. Continue as directed under Section 25.511, 2 to 4.
4. Add up to 225 mL of steamed (15 min) Tergitol Anionic 7 (undiluted) or Triton X-100 (undiluted), and mix well.
5. Loosen container caps 1/4 turn, and incubate food mixtures for 24 ± 2 hr at 35°C.
6. Continue as directed under Section 37.52.

37.52 *Isolation of* Salmonella

Tighten the lid and gently shake the incubated sample mixture. For foods with a high microbial load, transfer a 0.1-mL portion of the preenrichment cultures to 10 mL of RV medium and another 1.0 mL to 10 mL of TBG broth. For foods with a low microbial load, transfer replicate 1.0-mL portions to 10 mL of SC broth and to 10 mL of TBG broth.

For foods with a high microbial load, incubate RV medium for 24 ± 2 hr at 42 ± 0.2°C (waterbath for better temperature control) and TBG broth for 24 ± 2 hr at 43 ± 0.2°C (waterbath). For foods with a low microbial load, incubate SC and TBG broths for 24 ± 2 hr at 35°C (air incubator).

Vortex enrichment culture tubes and streak 3-mm loops of SC culture on BS, HE, and XLD agars.

Repeat plate inoculations with TBG and RV (as appropriate) enrichment cultures.

Incubate plates 24 ± 2 hr at 35°C.

Examine plates for suspect *Salmonella* colonies.

1. *BS agar*. Typical *Salmonella* colonies occur as black colonies with or without a metallic sheen. Medium surrounding *Salmonella* colonies gradually changes from brown to black with increased incubation to produce the so-called halo effect.
2. *HE agar*. Salmonellae generally produce blue-green to blue colonies with or without black centers. Strong H_2S-producing strains may produce colonies with large, glossy black centers or appear as completely black colonies. Atypical lactose-positive and/or sucrose-positive *Salmonella* strains and coliforms produce salmon-colored colonies.
3. *XLD agar*. Typical salmonellae occur as pink colonies with or without black centers. Strong H_2S-producing strains may yield colonies with large, glossy black centers or appear as completely black colonies. Atypical lactose-positive and/or sucrose-positive *Salmonella* strains and coliforms produce yellow colonies with or without black centers.

Inoculate two or more suspect colonies from each selective plating medium into TSI and LIA slants. BS agar plates with no suspect colonies or with no growth should be incubated for an additional 24 hr.

Touch the center of a suspect colony with a sterile inoculating needle and inoculate a TSI slant by streaking the slant and stabbing the butt. Without flaming the needle, inoculate an LIA slant by stabbing the butt twice and then streaking the slant. Loosen cap of TSI slant and close firmly LIA slant.

Incubate TSI and LIA slants at 35°C for 24 ± 2 hr. *Salmonella* in TSI cultures typically produces alkaline (red) slant and acid (yellow) butt, with or without production of H_2S (blackening of agar). However, some lactose and/or sucrose-positive salmonellae will produce atypical reactions; i.e., acid slant and acid butt, with or without blackening. In LIA, *Salmonella* typically produces an alkaline (purple) butt from decarboxylation of lysine. Consider only a distinct yellow butt as a negative reaction. Do not eliminate cultures that produce yellow (lysine negative) discoloration of the butt solely on the strength of this reaction because lysine negative biotypes have been reported. Most *Salmonella* cultures produce H_2S in LIA.

Re-examine 48-hr BS agar plates for suspect *Salmonella* colonies. Pick two or more of these colonies and continue procedure.

Retain all presumptive-positive TSI cultures (alkaline slant and acid butt) for further biochemical and serological testing, whether corresponding LIA reaction is positive (alkaline butt) or negative (acid butt). Do not exclude a TSI culture that appears to be non-*Salmonella* if the corresponding LIA is typical (alkaline butt) for *Salmonella*. Treat these cultures as presumptive-positive and submit them to further biochemical/serological examination. LIA is useful for the detection of *S. arizonae* and atypical strains of *Salmonella* that utilize lactose and/or sucrose. Discard only cultures that appear not to be *Salmonella* on TSI (acid slant and acid butt) if corresponding LIA reactions are negative (acid butt) for *Salmonella*. Test retained presumed-positive TSI cultures as directed in Section 37.53 to determine if they are *Salmonella*. If original TSI cultures failed to indicate the presence of *Salmonella*, inoculate additional suspect colonies from appropriate selective agar plates into fresh TSI and LIA slants.

Apply biochemical and serological identification tests to three presumptive TSI cultures recovered from each set of plates streaked from SC, TBG, and RV broth cultures. Examine a minimum of six TSI cultures for each 25-g analytical unit.

37.53 Identification of Salmonella

37.531 Mixed TSI Cultures

Streak TSI cultures that appear to be mixed on MacConkey, HE, or XLD agar. Incubate plates 24 ± 2 hr at 35°C and examine plates for the presence of suspect *Salmonella* colonies.

1. *MacConkey agar*. Typical colonies appear transparent and colorless, sometimes with opaque center. *Salmonella* will clear areas of precipitated bile caused by lactose-fermenting microorganisms.
2. *HE agar*. See Section 37.52.
3. *XLD agar*. See Section 37.52.

Transfer at least two presumptive *Salmonella* colonies from the plates used to purify mixed TSI cultures to fresh TSI and LIA slants and proceed as described in Section 37.52.

37.532 Pure TSI Cultures

1. *Urease test (conventional)*. With a sterile needle, inoculate tubes of urea broth with growth from each presumptive-positive TSI culture and incubate 24 ± 2 hr at 35°C. An uninoculated negative control should be included because urea broth can produce a false-positive reaction (purple red) on standing.
2. *Optional urease test (rapid)*. Transfer two 3-mm loops of growth from each presumptive-positive TSI culture into a tube of rapid urea broth. Incubate for 2 hr in a waterbath at 37 ± 0.5°C.

Most salmonellae are urease-negative; however, urease-positive cultures have been reported. Retain all urease-negative cultures for further testing.

37.533 Serological Screening with Polyvalent Flagellar (H) Antisera or Spicer-Edwards Antisera

1. Perform the polyvalent flagellar (H) test at this point to eliminate false-positive TSI cultures, or later as described in Section 37.535. Inoculate growth from each urease-negative TSI slant into 1) brain heart infusion broth and incubate 4 to 6 hr at 35°C until visible growth occurs (to test on same day); or 2) trypticase (tryptic) soy-tryptose broth and incubate 24 ± 2 hr at 35°C (to test on following day). Add 2.5 mL formalinized physiological saline solution to 5 mL of either broth culture. Select two formalinized broth cultures from each 25-g analytical unit and test with *Salmonella* polyvalent flagellar (H) antisera. Mix 0.5 mL each of appropriately diluted *Salmonella* polyvalent flagellar (H) antiserum and formalinized antigen in a 10 × 75-mm or 13 × 10-mm serological test tube. Prepare a saline control by mixing 0.5 mL formalinized physiological saline solution with 0.5 mL formalinized test culture (autoagglutination). Incubate mixtures in a waterbath at 48 to 50°C. Observe at 15-min intervals for flocculation and record final results after 1 hr. A positive tube reaction consists of agglutination/flocculation in the test mixture and absence of agglutination in the control tube. A negative reaction corresponds to absence of agglutination in both the test and control tubes. A nonspecific autoagglutination reaction produces agglutination in both the test and control tubes. Cultures giving such nonspecific results should be tested with Spicer-Edwards antisera.
2. The Spicer-Edwards serological test can be used as an alternative to the polyvalent flagellar (H) screening test. Perform Spicer-Edwards flagellar (H) antisera test using Spicer-Edwards antisera 1, 2, 3, 4, in addition to the en, L, and 1 complexes.[8]
3. If both formalinized BHI or trypticase (tryptic) soy-tryptose broth cultures are serologically-negative, repeat serological tests on additional broth cultures to obtain a minimum of two serologically-positive cultures for additional serological and biochemical testing (Sections 37.533 and 37.536). If all TSI cultures from a given sample give negative serological flagellar (H) test results, check them for motility (Section 37.539).

37.534 Serological Somatic (O) Tests for Salmonella

(Pre-test all antisera with known *Salmonella* cultures.)

1. *Polyvalent somatic (O) test*. Using a wax pencil, mark off two sections about 1 × 2-cm each on the inside of a glass or plastic petri dish (15 × 100 mm). Commercially available sectioned slides may be used. Transfer a loopful of growth from a 24-hr TSI agar culture in the upper portion of both marked sections. Add 1 loopful of physiological saline solution to the lower portion of one section (negative control) and 1 drop of *Salmonella* polyvalent somatic (O) antiserum to the lower portion of the remaining (test) section. With a sterile loop or needle, emulsify the culture in the saline solution; repeat for the test section using a sterile loop or needle. Tilt mixtures back and forth for 1 min and observe against a dark background in good illumination. Consider any degree of agglutination (clumping) as a positive reaction. A positive sample shows agglutination in the test but not in the control section. A negative sample shows no agglutination in either the control or test section. A nonspecific reaction corresponds to agglutination in both the test and control sections.
2. *Somatic (O) group tests*. Test polyvalent (O)-positive cultures using individual group somatic (O) including Vi antisera. Cultures giving a positive Vi agglutination reaction should be examined as described in Section 967.28(B) of the *Official Methods of Analysis*.[8] Record cultures that agglutinate with individual somatic (O) antiserum as positive for that group.

37.535 Interpretation of Serological Tests

Cultures producing a positive polyvalent flagellar (H) test (Section 37.533) or polyvalent somatic (O) test (Section 37.534) should be further tested with Spicer-Edwards antisera to determine flagellar antigens (Section 37.533) and somatic (O) antigens using single grouping antisera (Section 37.534). Biochemically type using a rapid diagnostic kit (Section 37.132) or by traditional biochemical tests (Section 37.536).

If a specific somatic group (O) (Section 37.534) and flagellar antigen(s) (Section 37.533) reactions are obtained with one or more TSI cultures, the sample is positive; additional biochemical testing may be performed but is not required.

TSI cultures that produce negative, variable, or nonspecific serological reactions should be tested biochemically by a rapid diagnostic kit (Section 37.132) or traditional methods (Sections 37.536 and 37.537). Subculture isolate in broth and plate on nonselective plating medium, and then retest with antisera.

37.536 Biochemical Testing of Cultures

Following is a list of tests for the biochemical screening of *Salmonella*; kits for the biochemical characterization of *Enterobacteriaceae* may be substituted (Section 37.132).

1. *Lysine decarboxylase broth.* If the LIA test was satisfactory, it need not be repeated. Use lysine decarboxylase broth for definitive determination of lysine decarboxylase activity if culture gives doubtful LIA reaction. Inoculate broth with a small amount of growth from presumptive-positive TSI slant. Replace cap tightly and incubate 48 ± 2 hr at 35°C, but examine after 24 hr of incubation. *Salmonella* species produce an alkaline reaction (purple) throughout the medium. A negative test is indicated by yellow color throughout the medium. If the medium is discolored (neither purple or yellow), add a few drops of 0.2% bromcresol purple dye and repeat the reading.
2. *Phenol red dulcitol broth or bromcresol purple broth base with 0.5% dulcitol.* Inoculate the broth with a small amount of growth from a presumptive-positive TSI culture. Replace the cap loosely and incubate 48 ± 2 hr at 35°C, but examine after 24 hr. Most *Salmonella* species give a positive test indicated by gas formation in an inverted fermentation vial and acidity (yellow) of the medium. Production of acid should be interpreted as a positive reaction. A negative test is indicated by the absence of gas formation and a red (phenol red) or purple (bromcresol purple) color throughout the medium.
3. *Tryptone (or tryptophan) broth.* Inoculate broth with small amount of growth from TSI culture. Incubate 24 ± 2 hr at 35°C and proceed as follows:
   a. *Potassium cyanide (KCN) broth.* Transfer 3-mm loopful of 24-hr tryptophan broth culture to KCN broth. Heat rim of tube so that a good seal is formed when tube is stoppered with wax-coated cork or other material not susceptible to corrosion. Incubate 48 ±2 hr at 35°C, but examine after 24 hr. Interpret growth (indicated by turbidity) as positive. Most *Salmonella* species do not grow in this medium, as indicated by lack of turbidity.
   b. *Malonate broth.* Transfer a 3-mm loopful of 24-hr tryptone broth culture to malonate broth. Since occasional uninoculated tubes of malonate broth turn blue (positive test) on standing, include an uninoculated tube of this broth as control. Incubate 48 ±2 hr at 35°C, but examine after 24 hr. Most *Salmonella* species give a negative test (green or unchanged color) in this broth, but many members of subspecies III (*S. arizonae*) are malonate-positive.
   c. *Indole test.* Transfer 5 mL of 24-hr tryptone broth culture to a sterile test tube and add 0.2 to 0.3 mL Kovacs' reagent. Most *Salmonella* cultures give a negative reaction (no deep color at the surface of the broth). Record varying shades of orange and pink as indeterminate.
4. Classify, as "confirmed" *Salmonella,* those cultures that exhibit typical *Salmonella* reactions for test nos. 1 to 11, shown in Table 1, or produce a positive flagellar (H) agglutination test and typical biochemical profile in an acceptable biochemical identification kit (Section 37.132). If one TSI culture from a single 25-g analytical unit or from a (15 × 25 g) composite is identified as *Salmonella,* further testing of other TSI cultures from the same 25-g analytical unit is unnecessary. Cultures that contain demonstrable *Salmonella,* flagellar (H) antigens but do not have biochemical characteristics of *Salmonella* should be purified (Section 37.531) and retested, beginning as in Section 37.532.

37.537 Additional Biochemical Tests

1. *Phenol red or bromcresol purple lactose broth.* Inoculate the broth with a small amount of growth from the TSI agar slant. Examine after 24 and 48 hr of incubation at 35°C. A positive reaction consists of acid production (yellow with either phenol red or bromcresol purple) and gas production in the inverted fermentation vial. A negative reaction consists of a red (with phenol red) or purple (with bromcresol purple) color throughout the medium. Most *Salmonella* are lactose-negative.
2. *Phenol red or bromcresol purple sucrose broth.* Follow procedure described in Section 37.537, item 1. Most *Salmonella* are sucrose-negative.
3. *MR-VP broth.* Inoculate the medium with a small amount of growth from each suspect TSI slant and incubate for 48 ± 2 hr at 35°C.
   a. *Voges Proskauer test.* Transfer 1 mL broth culture to a sterile test tube and incubate the remainder of the MR-VP culture for an additional 48 hr at 35°C. Add 0.6 mL alpha-naphthol and shake well. Add 0.2 mL 40% KOH solution and shake. To intensify and accelerate the reaction, add a few crystals of creatine; read reactions after 4 hr of incubation at room temperature. Development of a pink to ruby red color throughout the medium corresponds to a positive test. Most *Salmonella* cultures are Voges Proskauer-negative.
   b. *Methyl red test.* To 5 mL or MR-VP broth (96 hr) culture, add 5 to 6 drops of methyl red indicator. Read results immediately. Most *Salmonella* cultures give a positive test, as indicated by a diffuse red color in the medium. A distinct yellow color constitutes a negative test.
4. *Simmons citrate agar.* Inoculate the agar with growth from the TSI slant, using a sterile needle. Inoculate by streaking the slant and stabbing the butt and incubate for 96 ± 2 hr at 35°C. A positive reaction consists of visible growth, usually accompanied by a color change from green to blue. Most *Salmonella* cultures are citrate-positive. A negative reaction shows little or no growth with no color change in the medium.

37.538 Identification of Salmonella cultures

Typical *Salmonella* cultures produce reactions described in Table 1 or a suitable biochemical profile in diagnostic kits and positive flagellar (H) agglutination tests. If neither of two TSI cultures carried through these tests confirms the isolates as *Salmonella,* perform tests on remaining TSI cultures from the same 25-g analytical unit.

1. Confirm, as *Salmonella,* cultures producing specific Spicer-Edwards flagellar (H) test (Section 37.533) and somatic group (O) (Section 37.534) reactions.
2. Confirm, as *Salmonella,* a culture identified as presumptive *Salmonella* according to the criteria in Table 1 or with a commercial biochemical kit and positive *Salmonella* somatic (O) and flagellar (H) tests (polyvalent or Spicer-Edwards).
3. Discard cultures classified as non-*Salmonella* with commercial biochemical kits using data base-dependent numerical profiles and that fail to give positive flagellar (H) agglutination reactions (polyvalent or Spicer-Edwards). Also discard cultures that are negative for both somatic (O) and flagellar

(H) tests (polyvalent or Spicer-Edwards) and produce atypical reactions in biochemical tests (Table 1).

4. Send presumptive positive cultures that do not conform to 1 or 2 to a reference laboratory for definitive serotyping and identification.

37.539 Treatment of Cultures Giving Negative Flagellar (H) Test

If biochemical reactions of a flagellar (H)-negative culture strongly suggest that it could be *Salmonella*, the negative flagellar agglutination may be the result of nonmotile organisms or insufficient development of flagellar antigen. Proceed as follows: Inoculate motility test medium held in a petri dish, using a small amount of growth from the TSI slant. Inoculate by lightly stabbing the medium once about 10 mm from the edge of the plate. Do not stab to the bottom of the plate or inoculate any other portion of the plate. Incubate 24 hr at 35°C. If organisms have migrated 40 mm or more, retest as follows: Inoculate a 3-mm loopful of the growth that migrated farthest into trypticase soy-tryptose broth. Repeat polyvalent flagellar (H) or Spicer-Edwards (Section 37.533) serological tests. If cultures are not motile after the first 24 hr, incubate an additional 24 hr at 35°C; if still not motile, incubate up to 5 days at 25°C. Classify a culture as nonmotile if the motility tests remain negative. If a flagellar (H)-negative culture is strongly suspected as *Salmonella* on the basis of its biochemical reactions or appearance on plating medium (i.e., BS agar), the culture should be sent to a reference laboratory for serotyping and final identification.

37.6 INTERPRETATION OF DATA—REFERENCE METHOD

If no suspect colonies are observed on isolation media (Section 37.52), or if none of the suspect colonies picked from isolation media produce reactions indicative of *Salmonella* in TSI or LIA, then the sample is negative for *Salmonella*.

If presumptive-positive cultures are obtained in either TSI or LIA, a minimum of 6 cultures must be subjected to serological and biochemical confirmation tests. However, if the first TSI culture confirms as *Salmonella*, then the remaining 5 TSI cultures do not need to be characterized biochemically and serologically. If fewer than 6 presumptive cultures are obtained, apply confirmation tests to all presumptive-positive cultures.

Confirm, as *Salmonella*, presumptive cultures that meet criteria 1 or 2 in Section 37.538. If one or more presumptive cultures meet either criterion, the sample is positive for *Salmonella*.

Discard, as non-*Salmonella*, cultures meeting criterion 3 in Section 37.538. If all presumptive cultures examined meet criterion 3, the sample is negative.

If presumptive-positive cultures cannot be classified as "confirmed" or "non-*Salmonella*," additional tests should be performed[41] until the cultures can be definitively identified.

37.7 RAPID SCREENING TESTS

Acceptance of rapid screening methods as reliable alternatives to conventional culture methods for detection of *Salmonella* in foods generally arises from comparative studies that establish the equivalence of standard culture and rapid test methods.[2] Because of the public health, regulatory, and potential legal significance of *Salmonella* in foods, validation of method performance by a widely accepted protocol is desirable. Numerous rapid methods have been proposed for detection of *Salmonella* in foods, employing a variety of technologies ranging from relatively simple rapid biochemical screening to sophisticated nucleic acid amplification and detection systems. These technologies are discussed in Chapter 10 of this book. The objective of this chapter is to introduce those rapid methods that have been validated against a standard method using a widely recognized validation protocol.[2a,2b]

Worldwide there are numerous method validation schemes in use. The most widely recognized and accepted of these schemes are the ISO and the AOAC collaborative study program. In 1992 the AOAC Research Institute (AOAC-RI) performance tested validation program was initiated. The AOAC-RI performance tested program was developed to specifically address the validation of test kits containing proprietary components. Through the mid-1980s up to 1997, several rapid methods for *Salmonella* have been evaluated via the AOAC-collaborative study and AOAC-RI programs. A number of rapid test kits has been evaluated against the ISO *Salmonella* method, but these evaluations have been performed using various validation schemes which are accepted in specific countries, but not on an international basis. Therefore, the methods described below are limited to those validated by AOAC collaborative study and AOAC-RI programs. These two validation schemes differ principally in that the AOAC-collaborative study program involves several laboratories, whereas the AOAC-RI requires independent laboratory verification of selected claims made by the manufacturer of the test kit. Both programs require that the manufacturer of the kit supply an extensive amount of data to verify performance parameters, e.g., sensitivity, specificity, range of application to products; inclusivity with regard to *Salmonella;* and exclusivity to competing organisms. The AOAC-RI program also includes verification of the manufacturer's quality assurance systems for production, shelf life of the kit, and ruggedness testing of the method across a range of conditions that might be encountered in routine use.

Most of the validated rapid methods are screening tests that require confirmation. In other cases, the rapid method may be as sensitive in detection of salmonellae as the cultural method but does not produce identical results; i.e., neither method provides 100% detection, but false-negative rates are similar. In comparing data, consider that the basis for detection of *Salmonella* by rapid methods may be different from that for isolation by culture methods. Positive results by culture methods require that three cultural conditions be met. First, *Salmonella* must grow to the level necessary to assure that cells are transferred into selective enrichments. Second, a sufficient number of salmonellae must be present in at least one of the selective enrichment cultures to ensure that sufficient *Salmonella* cells are streaked onto plates. Third, the relative proportion of salmonellae to other organisms capable of growth on the selective/differential isolation media must be such that at least one isolated colony of *Salmonella* can be obtained. Thus, the basis for obtaining positive results by culture is primarily selectivity, but the importance of pre-enrichment for resuscitation of injured cells cannot be minimized. Rapid screening methods require that a sufficient number of salmonellae/mL be present to allow detection by DNA hybridization, serological, or biochemical tests. High numbers of competitors might inhibit growth of salmonellae to the level necessary for detection by the rapid methods as well as prevent isolation by standard culture conditions. However, under certain conditions, sufficient numbers of salmonellae could develop to produce positive results by the rapid assays but not by the culture procedure because of large populations of competitive flora (e.g., swarming *Proteus*). Although it may be unrealistic to expect 100% agreement between conventional culture and rapid assay results, the high sensitivity of standard cultural methods for *Salmonella*[2] will generally provide for a high agreement between rapid and conventional culture methods. Rapid methods that have been evaluated by AOAC and/or AOAC-RI are discussed briefly, but not in the detail needed to perform the test. The published method in AOAC's *Official Methods of Analysis*[8] and the package insert for the kit (AOAC-RI) should be consulted for performance of the methods.

37.71 Fluorescent Antibody Method

Many of the rapid methods proposed for detection of salmonellae are based on immunological reactions. The FA technique was the first immunological method applied to the detection of salmonellae. Thomason et al.[120] reported that the O, Vi and H antigens of *S. typhi* could be stained with fluorescent-labeled antibodies and observed by fluorescence microscopy. With the advent of antisera of greater sensitivity and specificity, a protocol was developed for detection of salmonellae in foods.

The method was extensively evaluated and approved by AOAC in 1975. A historical summary of the development of the FA procedure has been published.[119] The AOAC-approved method employs a 24-hr pre-enrichment (for all foods but raw meat samples, which are placed directly in selective enrichment), and a 24-hr selective enrichment followed by a 4-hr post-enrichment.[47] Portions of the post-enrichment are then stained and examined for fluorescent cells. The presence of fluorescent cells morphologically typical of salmonellae constitutes a presumptive-positive result that must be confirmed by culture methods. Negative results are generally available one day earlier than with conventional culture methods. Thomason[119] compiled data from three studies and indicated a 97.6% agreement between standard culture methods and the FA method, and a false-negative and false-positive rate of 0.1% and 2.3%, respectively. However, some laboratories observed much higher false-positive rates, especially with certain products.[119] It was also established that staining and reading of stained slides is both laborious and subjective, requiring a high degree of technical training and experience. A detailed description of the FA method is not presented in this edition of the *Compendium of Methods for the Microbiological Examination of Foods* but appears in the second edition[73] and in the 16th edition of *Official Methods of Analysis*.[8]

37.72 Hydrophobic Grid Membrane Filtration[5]

The method uses a membrane filter impregnated with hydrophobic material in a grid pattern of 1,600 compartments. The hydrophobic material acts as a barrier, preventing the spread of "growth units" between adjacent compartments. An appropriate dilution of an enriched sample is passed through a HGMF filter and placed on a plate of appropriate agar medium. Following incubation, growth units are isolated within the grid.

HGMF has found application in the qualitative analysis of foods for *Salmonella*. Entis et al.[44] reported the use of an HGMF method for the detection of *Salmonella* in foods, and this method was subsequently subjected to an AOAC collaborative study. In the AOAC method,[8] originally approved for the analysis of six foods (nonfat dry milk, powdered eggs, cheese powders, pepper, chocolate, and raw poultry), the test sample was incubated in the appropriate pre-enrichment medium for 18 to 24 hr at 35° C. A 0.1-mL volume of the pre-enriched test sample was subcultured in 10 mL of TBG broth. After 6 to 8 hr of incubation, a 1-mL portion of diluted (raw poultry) or undiluted (the other five foods) selective TBG enrichment was filtered through the HGMF filter. The filters were then placed on selective agar plates. Following incubation of the selective lysine agar (SLA) (43°C) and the HE (35°C) agar plates for 24 hr, the filters were examined for suspect *Salmonella* growth units. If present, these growth units were confirmed by the conventional AOAC culture method.[8] Since the original collaborative study of the HGMF method, the method has been modified, and SLA and HE plates have been replaced by a single EF-18 agar plate. This method is now approved for all foods.

Use of the HGMF method has one limitation. The use of a single selective enrichment and a single selective plating agar may preclude the detection of strains that are more readily recovered by other selective enrichment broths and/or selective agars.

37.73 Immunodiffusion[33,54]

Detection of *Salmonella* in the self-contained and disposable *Salmonella* 1-2 TEST unit (BioControl Systems, Inc., Bellevue, Washington) is based on the formation of an opaque zone or band when *Salmonella* bacteria are immobilized by polyvalent H (flagellar) antibodies in a semi-solid motility medium. The clear, plastic device consists of two interconnected chambers: a motility chamber containing L-serine and a peptone-based, nonselective motility medium and an inoculation chamber containing L-serine and TBG broth.

Most food samples are pre-enriched for 24 ± 2 hr at 35°C as described in the culture reference method. The pre-enrichment culture is then seeded into the inoculation chamber of the 1-2 TEST system. Raw meats and other highly contaminated products are pre-enriched for 24 ± 2 hr at 35°C and then enriched in TBG broth and incubated 6–8 hours at 42 ± 0.5°C.[48,51,131] Enrichment cultures of raw meats and other highly contaminated products are then added to an empty inoculation chamber of the 1-2 TEST system.[8] Preparation of the 1–2 unit involves three steps. First, the TBG broth in the inoculation chamber is activated by the addition of one drop of iodine-iodide solution. Then one drop of a *Salmonella* antibody solution is added to the motility chamber.

Following removal of the chamber plug that seals the opening between the motility chamber and the inoculation chamber, a 0.1-mL portion of pre-enrichment (processed food) culture is added to the inoculation chamber. For raw meat and other highly contaminated products, 1.5 mL of the separately inoculated TBG broth is added to a previously emptied inoculation chamber. The 1-2 unit is incubated for a minimum 14 hr at 35°C. Motile *Salmonella* organisms, if present, migrate from the inoculation chamber into the motility chamber. A visible band of precipitation occurs where motile *Salmonella* and diffusing antibody interface. Such presumptive-positive reactions are then confirmed by streaking a loopful of TBG broth from the inoculation chamber onto selective plating media as described by the method.[8] Isolates are then confirmed using the reference method.

In addition to yielding presumptive-positive results for processed foods and raw meat samples one day earlier than with the conventional culture method, the *Salmonella* 1-2 TEST offers other advantages. It is easy to use, requires no specialized training or sophisticated equipment, and is readily adaptable for use in laboratories that handle small or large numbers of test samples.

Because the detection of *Salmonella* by the 1-2 TEST is based on immobilization, nonmotile serovars such as *S. gallinarum* and *S. pullorum* and other variants will not be detected.

In 1986, DeSmedt et al.[35] developed a motility enrichment technique based on the use of a semi-solid selective medium. The analytical procedure consisted of pre-enrichment for 20 hours, followed by motility enrichment in MSRV medium for 24 hr and serological tests on the migrated cultures. Since then, the method has undergone collaborative evaluation by comparing the method to the method of the International Office of Cocoa, Chocolate and Sugar Confectionery (IOCC) for detecting *Salmonella* in cocoa and chocolate products.[36] The rapid immunodiffusion method was at least as sensitive as the IOCC method, and, in some cases, provided superior detection of lactose-fermenting *Salmonella*. This method was subsequently compared to the AOAC reference method in a collaborative study. The method was approved by AOAC in 1992 for detection of *Salmonella* in cocoa and chocolate.[8]

37.74 Enzyme immunoassays[21,49,50,52,56,58,59]

Several EIAs have been proposed for detection of salmonellae in foods.[70,87] Several methods have been evaluated collaboratively and approved by AOAC.[49,52,56,58,59]

The first EIA method approved by AOAC was the BioEnzaBead™ method (Organon Teknika, Durham, NC, formerly Litton Bionetics). Comparative studies on the productivity of this method based on the reaction of two antibodies, MOPC 467 and 6H4, with *Salmonella* antigens were conducted by D'Aoust and Sewell,[32] Eckner et al.,[39] Flowers et al.,[61] and Todd et al.[121,122] The EIA method evaluated by Eckner et al.[39] employed a 24 ± 2-hr pre-enrichment, an 18-hr selective enrichment, and a 6-hr M-broth (35° C) post-enrichment.

Flowers et al.[52] reported on a collaborative study comparing the BioEnzaBead to the AOAC culture method. Twenty-five laboratories participated in the study, with each laboratory analyzing 12 samples of each of six food types. No significant difference appeared in the productivity of EIA and the culture procedure at the 5% probability level for any of the foods. Based on the results of this collaborative study[52] and an earlier comparative study,[39] the BioEnzaBead EIA method was approved by AOAC.[4] The approved method employed pre-enrichment in nonselective medium for 24 ± 2 hr at 35°C, selective enrichment in TBG/SC at 35°C for 18 to 24 hr, post-enrichment in M-broth for 6 hr (35°C), and a centrifugation/heating step prior to performing the EIA procedure. Subsequent work indicated that the incubation period for selective enrichment could be reduced and the centrifugation step eliminated with low-moisture foods.[59] A collaborative study was performed to compare the efficiency of the modified *Salmonella* method with the AOAC culture method.[59] The modified method required 18 to 24-hr pre-enrichment, 6 to 8-hr selective enrichment, at 35°C, and 14 to 18-hr post-enrichment at 35°C, and heating prior to EIA. The minimum time to complete the modified method was 40 hr, compared with 48 hr for the longer EIA method and 96 hr for the AOAC culture method. Based on the results of the collaborative study, which indicated no significant difference ($P > 0.05$) in detection of *Salmonella* by the modified EIA and the AOAC culture method, the modified BioEnzaBead™ method was approved for low-moisture foods by AOAC.[58]

The BioEnzaBead immunoassay originally used plastic-coated ferrous metal beads as the solid support phase. Monoclonal antibodies specific for *Salmonella* antigens were adsorbed onto the surface of the coated beads. The assay was initiated by dropping an antibody-coated metal bead into a microtiter dish well containing a heated extract of the food sample. Washing and subsequent steps in the assay were performed by moving the metal beads between plates containing wash solutions, conjugate, or substrate, using a magnetic transfer device. If *Salmonella* antigens were present in the sample, the conjugate would bind to the *Salmonella* antigens already attached to the antibody-coated bead. The beads were washed to remove any unbound conjugate and then placed in a substrate solution. Appearance of color indicated the presence of *Salmonella* antigen in the sample.

In 1988, this immunoassay was modified in favor of an antibody-coated microelisa 12-well strip-plate developed by Organon Teknika. This format (*Salmonella*-Tek™) offers several advantages over the bead assay, including the need for only one, rather than the eight, microtiter plates required by the bead assay and the use of a more sensitive substrate. Small numbers of samples can be tested by removing only the required number of strip wells. In addition, more antibody molecules are bound to the well surfaces of the strip-plate, providing for greater method sensitivity.

The *Salmonella*-Tek™ strip-plate assay was compared with the BioEnzaBead method using both inoculated and naturally contaminated samples.[23] This study demonstrated equivalent results by the BioEnzaBead assay and the *Salmonella*-Tek plate assay when compared to those obtained by the culture method. The *Salmonella*-Tek assay was more sensitive than the BioEnzaBead assay; i.e., it was capable of detecting lower levels of *Salmonella* in the post-enrichment cultures. Thus, the *Salmonella*-Tek assay can reliably replace the BioEnzaBead assay for detecting *Salmonella* in foods when the same enrichment protocol is employed. Consequently, it was recommended to AOAC that the *Salmonella*-Tek assay method replace the BioEnzaBead method and that the BioEnzaBead method be discontinued. Both assays employ the same two *Salmonella*-specific antibodies and thus should offer similar degrees of specificity. After boiling, portions of the cooled M-broth cultures are dispensed into the wells of the microelisa strips and incubated with agitation to allow capture of *Salmonella* antigens. The wells are then washed, and peroxidase-labeled conjugate binds to the captured *Salmonella* antigen (if present) to form an antibody-antigen antibody-peroxidase complex. Unbound conjugate is removed from the wells by washing, and enzyme substrate is added. If a blue color develops, the sample is presumptive-positive for *Salmonella*. The reaction is stopped with an acid solution that turns the blue color to an intense yellow. The plates are read on an EIA reader; all presumptive-positives must be confirmed culturally.

Four other EIA techniques which include enrichment protocols similar *to Salmonella*-Tek have undergone comparative and collaborative studies.

The TECRA™ EIA method (Bioenterprises Pty. Ltd., Roseville, NSW, Australia), in contrast to *Salmonella*-Tek, utilizes polyclonal, rather than monoclonal, antibodies. The assay is performed in microtiter plates or strip wells and can be read visually or spectrophotometrically.

The TECRA method is simple to perform, with no requirement for sophisticated instrumentation. Wash steps can be performed manually, and assay results can be readily obtained visually, thereby eliminating the need for an EIA spectrophotometer. Based on the results of the collaborative[55] and comparative studies,[53] the TECRA method was approved by AOAC.[6]

The Q-TROL *Salmonella* assay (Dynatech Laboratories, Inc., Chantilly, VA) utilizes monoclonal antibodies and a fluorogenic reagent. An AOAC collaborative study of the Q-TROL™ method based on 1,700 food samples, was performed in 1988.[56] Based on the results of this study, the method was approved by AOAC.[7] Subsequently, a colorimetric enzyme assay, using the substrate phenolphthalein monophosphate and the enzyme alkaline phosphatase, was developed and compared with the fluorogenic assay. The results indicated that the colorimetric assay was at least as sensitive as the previously approved fluorogenic assay.[21]

The VIDAS SLM method for detection of *Salmonella* is an automated qualitative enzyme-linked fluorescent immunoassay. The VIDAS SLM method, including culture enrichment steps and assay, requires 48 hr for completion. The enrichment sequence is similar to other EIA methods discussed above, but has the advantage that the fluorogenic assay is fully automated. The VIDAS SLM has undergone comparative and collaborative studies and was approved by AOAC in 1997.[20]

The LOCATE EIA method for *Salmonella* (Rhone-Poulenc Diagnostics, Ltd.) received AOAC approval in 1998.[9] This is a microplate assay similar to the *Salmonella*-Tek and TECRA assays, incorporating similar enrichment steps.

The extensive evaluations of the EIA methods clearly indicate that this technology is an attractive alternative to conventional culture methods for detection of salmonellae in foods. This analytical approach provides detection of salmonellae one day earlier than by conventional culture methods, with low numbers of both false-negative and false-positive results for most processed products, although relatively high false-positive rates have been reported for some raw products.[31,32,39,52] Advantages of the method include automation and a specific end-point that contrasts with the subjective interpretation associated with culture methods.

37.75 DNA Hybridization

Genetic relatedness among related bacteria can be measured by the DNA hybridization (DNAH) technique that probes for unique nucleotide sequences in target organisms. The performance of a ^{32}P DNA hybridization assay for *Salmonella* in foods (GENE TRAK Systems, Hopkington, Mass.) was compared to the AOAC culture method.[60] The results showed that the sensitivity of the DNAH method compared favorably with the culture method under all test conditions and, in some cases, DNAH appeared to provide superior detection.

A collaborative study compared the sensitivity of the ^{32}P DNA procedure to the standard culture method for detection of *Salmonella* in six artificially contaminated foods: ground pepper, soy flour, dry whole egg, milk chocolate, nonfat dry milk, and raw deboned turkey.[57] The results of this collaborative study[60] and earlier comparative studies[52] led to an AOAC approval of the DNAH method for the rapid detection of salmonellae in foods.[5] The isotopic DNAH method includes pre-enrichment (18 to 24 hr), 6-hr selective enrichment in TBG/SC, and a 12- to 16-hr post-enrichment in GN broth for processed foods and 18-hr enrichment in TBG/SC + 6-hr GN broth for raw meats. After inoculation of the post-enrichment GN broth with 1.0 mL of enrichment (TBG/SC) culture, both TBG/SC enrichments are reincubated for confirmation for up to a total of 24 ± 2 hr. A portion of the GN enrichment culture is then filtered, and the DNAH assay performed on the filters.

Positive reactions are reported when counts for the test filter are >500 cpm above the average of three negative control filters. The assay itself requires about 4 hr to perform, and the culture steps require a minimum of 42 hr. Thus, the total time required for the DNAH method is 46 hr. The DNAH assay is a screening method, and positive results are confirmed by culture methods. Both post-enrichment and reincubated selective enrichment cultures are streaked to selective agar media for confirmation of positive DNAH assays.

Subsequent to approval of the isotopic DNAH method, a non-isotopic, colorimetric DNAH method was developed.[22] The colorimetric assay employed DNA probes different from those used in the isotopic method. The target for the colorimetric assay probes is ribosomal RNA; i.e., the assay is based on hybridization of the DNA probes to target rRNA. The assay, which is carried out in an aqueous environment, involves hybridization between rRNA and *Salmonella*-specific oligonucleotide probes, capture of rRNA target:DNA probe hybrids onto a solid support, and a colorimetric end-point detection. Following pre-enrichment, selective enrichment, and post-enrichment of test samples as described above for the DNAH assay, bacteria are lysed and *Salmonella*-specific fluorescent-labeled detection and polydeoxyadenylic (poly dA) capture probes are added. The mixture is then incubated to allow hybridization of both probes to target rRNA, if present. A polydeoxythymidylic (poly dT)-coated plastic dipstick solid phase is then introduced into the hybridization solution. Base pairing between the poly dA tail sequence of the capture probe and the poly dT on the dipstick facilitates capture of the DNA probe:rRNA target hybrids onto the dipstick. Unbound labeled probe is removed by washing the dipstick, followed by incubation in a solution containing anti-fluorescent antibody conjugated to horseradish peroxidase (HRP). The peroxidase conjugate binds to the fluorescent label on the hybridized detector probe. Unbound conjugate is washed away, and the dipstick is incubated in a substrate-chromogen solution. Reaction of HRP with substrate-chromogen results in development of a blue color. The enzyme:substrate reaction is stopped by addition of acid with a concomitant change in color from blue to yellow. Absorbance at 450 nm in excess of threshold value indicates the presumptive presence of *Salmonella* in the test samples.

The colorimetric DNAH assay can be completed in 2.5 hr following approximately 44 hr of culture enrichments. Chan et al.[14] compared the productivity of the colorimetric DNAH (cDNAH) method with the AOAC culture method,[8] and an AOAC collaborative study was performed in 11 laboratories to validate the cDNAH method.[22] No significant difference appeared in the ability of the methods to detect salmonellae in any of the six artificially-contaminated foods examined in the latter study.

37.76 Conductivity

The principle of conductance microbiology is that microorganisms growing in a broth culture increase the conductance of the medium as they metabolize neutral components to charged metabolites. This change in conductance can be measured by introducing platinum electrodes and applying a low-frequency voltage. Scanning the conductance at regular intervals generates a curve that is analogous to a growth curve. Change in conductance may arise from the metabolism of general components (e.g., peptones) or specific components (TMAO or lysine). TMAO is a neutral, uncharged molecule which, when reduced to trimethylmine (TMA), produces a large change in conductance[38]; decarboxylation of lysine produces the same result.

In a collaborative study, Gibson[64] showed no significant difference between the AOAC/BAM[8,129] and the conductance method (Malthus Instruments Ltd., The Manor, Manor Royal, Crawley, West Sussex, United Kingdom) for the detection of *Salmonella* in 6 food types by 17 participating laboratories.

37.8 AOAC-RI PERFORMANCE TESTED METHODS

Ten test kit methods for rapid detection of *Salmonella* spp. in foods have certified as Performance Tested (Table 4). Prior to 1998, AOAC-RI did not require that the data upon which the certification was based be published. Therefore, the validation data for the

Table 4. AOAC-RI "Performance Tested Methods" for Detection of *Salmonella* in Foods and Related Products

Test kit	Manufacturer	Matrixes
Path-Stik	Lumac B.V.	food
Reveal	Neogen	food
Bioline ELISA	Bioline	food, feed
Salmonella Screen/ *Salmonella* Verify	Vicam L.P.	food, feed, stainless steel, concrete
Oxoid Rapid Test for *Salmonella*	Oxoid Ltd.	food, feed, environmental samples
Salmonella DLP Assay	GENE-TRAK Systems Corp.	food, feed, environmental samples
Dynabeads anti-*Salmonella*) Kit	Dynal S.A.	food, feed
BIND *Salmonella* Rapid Assay Kit	IDEXX Laboratories, Inc.	food, feed
Bax for Screening *Salmonella*	Qualicon	milk, chicken, turkey, beef, and pork
TECRA Unique *Salmonella*	TECRA Diagnostics	food and environmental samples

approval of methods listed (Table 4) are not readily available in the scientific literature. Information on the methods and their performance should be requested directly from the manufacturers.

A brief discussion of each method is provided below, along with the manufacturer's address.

37.81 Immunological Tests

Several of the "Performance Tested" methods are immunologically based. The *Salmonella* Screen/*Salmonella* Verify method (Vicam L.P., Watertown, MA, 0217) and the Dynabeads anti-*Salmonella* Kit (Dynal, Inc., Lake Success, NY, 11042) employ magnetic anti-*Salmonella* labeled beads to capture salmonellae from enrichment broths. The captured *Salmonella* are then plated directly onto selective/differential agar media for isolation and identification. The Vicam *Salmonella* Verify Method includes a step where colonies developing from the beads can be quickly screened by a latex agglutination technique to determine if they are presumptive *Salmonella* spp.

The TECRA Unique *Salmonella* screening test (TECRA Diagnostics, Roseville, NSW 2069, Australia) is a 22-hr dipstick test, which includes a 16-hr pre-enrichment step followed by: capture, wash, replication (4–5 hr at 35–37°C), conjugate binding, wash, and color development steps within a disposable device composed of a series of tubes containing the appropriate reagents. The Bioline ELISA (Bioline, DK-7100 Vejte, Denmark) is a 36-hr microtiter strip assay. The Reveal dipstick method (Neogen Corp., Lansing, MI, 48912) and Path-Stik methods (Celsis-Lumac Ltd., Cambridge Science Park, Cambridge CB4 4FX, UK) are other immunological dipstick methods that have received AOAC-RI approval for food testing. These methods include preenrichment and selective enrichment steps followed by the dipstick assay.

37.82 DNA Probe Assays

GENE-TRAK Systems (Hopkinton, MA) has received "Performance Tested" certification for the dipstick assay modification of their AOAC-OMA method.

Qualicon, a DuPont subsidiary (Wilmington, DE, 19880-0357) has developed a one-day *Salmonella* detection method, "BAX for Screening *Salmonella*," based on polymerase chain reaction (PCR) amplification of *Salmonella*-specific nucleic acid. After pre-enrichment (20 hr) and a short (3 hr) secondary enrichment step, cells are lysed and added to PCR tubes which contain the reagents required to perform the amplification. *Salmonella*-specific PCR gene product is detected using a gel electrophoresis procedure. The BAX *Salmonella* method requires approximately 28 hr including enrichment, amplification, and assay steps.

37.83 Other Methods

The Oxoid *Salmonella* Rapid Test (Unipath, Nepean, Ontario, K2E 7K3 Canada) is a biochemical screening method. Samples are pre-enriched (18–24 hr) in appropriate media and then aliquots transferred to a disposable culture vessel containing an elective medium and two tubes. Each tube contains a selective medium in the lower portion of the tube, and an indicator medium in the upper portion separated by a porous partition. *Salmonella* grow in the elective enrichment and migrate into the tubes through the lower selective media to the upper indicator media where their presence is indicated by a color change. After a 24-hr incubation of the enrichment/screening system, tubes are examined for the possible presence of *Salmonella*. Tubes producing positive results are screened using the Oxoid *Salmonella* Latex Test, and positive reactors reported as presumptively containing *Salmonella*.

The BIND (bacterial ice nucleation detection) *Salmonella* assay (Idexx Laboratories, Westbrook, ME, 04092) is a two-day *Salmonella* screening method which includes preenrichment followed by a simple assay system based on bacteriophage infection of *Salmonella* resulting in expression of ice nucleation proteins that lead to ice crystal formation which can be detected with a colorimetric assay.

37.9 REFERENCES

1. Aleksic, S., R. Rohde, and A. Quddus Khan. 1973. The isolation of *Salmonella* from human fecal specimens in selenite enrichment medium at incubation temperatures of 37°C or 43°C. Zentrabl. Bakt. Hyg., I Abt. Orig. A. 225:27.
2. Andrews, W. H. 1985. A review of culture methods and their relation to rapid methods for the detection of *Salmonella* in foods. Food Technol. 39:77.

2a. Andrews, W. H. 1996. Three validation programs for methods used in the microbiological analysis of foods. Trends Food Sci. Technol. 7:147.

2b. Andrews, W. H. 1996. Validation of modern methods in food microbiology by AOAC International collaborative study. Food Control 7:19.

3. AOAC. 1985. Changes in methods. *Salmonella* detection in foods. Hydrophobic grid membrane filter method. J. Assoc. Off. Anal. Chem. 68:405.
4. AOAC. 1986. Changes in methods. *Salmonella* in foods: enzyme immunoassay screening method. J. Assoc. Off. Anal. Chem. 69:381.
5. AOAC. 1987. Changes in methods. *Salmonella* in foods: DNA hybridization method. J. Assoc. Off. Anal. Chem. 70:394.
6. AOAC. 1989. Changes in official methods. *Salmonella* in foods: Colorimetric polyclonal EIA screening method. J. Assn. Off. Anal. Chem. 72:201.
7. AOAC. 1989. Changes in official methods. *Salmonella* in foods: Fluorogenic monoclonal EIA screening method. J. Assoc. Off. Anal. Chem. 72:203.
8. AOAC. 1995. Official methods of analysis, 16th ed. (Supplements 1996, 1997, and 1998). Assoc. Off. Anal. Chem., Gaithersburg, Md.
9. AOAC International. 1998. Changes in official methods of analysis. LOCATE enzyme-linked immunosorbent assay for detection of *Salmonella* in food. J. Assoc. Off. Anal. Chem. Int. 81:323
10. Bonev, S.I. 1976. DMS agar, a new composite tube medium for differentiation within the genus *Salmonella*. Int. J. Syst. Bacteriol. 26:79.
11. Buzby, J. C,. and T. Roberts. 1996. ERS updates foodborne disease costs for seven pathogens. Food Safety. Sept.-Dec.:20.
12. Centers for Disease Control and Prevention. 1991. Multistate outbreak of *Salmonella poona* infections—United States and Canada. Morb. Mortal. Wkly. Rep. 40:549.
13. Centers for Disease Control and Prevention. 1997. Summary of Notifiable Diseases, United States. 1996. Morb. Mortal.
14. Chan, S. W., S. G. Wilson, A. Johnson, K. Whippie, A. Shah, A. Wilby, M. Ottaviani, M. Vera-Garcia, M. Mozola, and D. Holbert. 1992. Comparative study of a colorimetric DNA hybridization assay and the conventional culture procedure for the detection of *Salmonella* in foods. J. Assoc. Off. Anal. Chem. 75:685.
15. Cook, G. T. 1952. Comparison of two modifications of bismuth-sulphite agar for the isolation and growth of *Salmonella typhi* and *Salmonella typhimurium*. J. Pathol. Bacteriol. 64:559.
16. Cox, N. A., J. S. Bailey, and J. E. Thomson. 1983. Evaluation of five miniaturized systems for identifying *Enterobacteriaceae* from stock cultures and raw foods. J. Food Prot. 46:914.
17. Cox, N. A., and A. J. Mercuri. 1979. Rapid biochemical testing procedures for *Enterobacteriaceae* in foods. Food Technol. 33:57.
18. Cox, N. A., J. E. Thomson, and J. S. Bailey. 1983. Procedure for isolation and identification of *Salmonella* from poultry carcasses. Agricultural Handbook No. 603. Agric. Res. Serv., U.S. Department of Agriculture, Washington, D. C.
19. Crosa, J. H., D. J. Brenner, W. H. Ewing, and S. Falkow. 1973. Molecular relationships among the salmonellae. J. Bacteriol. 115:307.
20. Curiale, M. S., V. Gangar, and C. Gravens. 1997. VIDAS enzyme-linked fluorescent immunoassay for detection of *Salmonella* in food: Collaborative study. J. AOAC Int. 81:419.
21. Curiale, M. S., M. J. Klatt, W. E. Gehle, and H. Chandonnet. 1990. Colorimetric and fluorometric substrate immunoassays for the detection of *Salmonella* in all foods: A comparative study. J. Assoc. Off. Anal. Chem. 73:961.
22. Curiale, M. S., M. J. Klatt, and M. A. Mozola. 1990. Colorimetric deoxyribonucleic acid hybridization for rapid screening of *Salmonella* in foods. J. Assoc. Off. Anal. Chem. 73:248.
23. Curiale, M. S., M. J. Klatt, B. J. Robison, and L. T. Beck. 1990. Comparison of colorimetric monoclonal enzyme immunoassay screening methods for detection of *Salmonella* in foods. J. Assoc. Off. Anal. Chem. 73:43.
24. D'Aoust, J.-Y. 1977. Effect of storage conditions on the performance of bismuth sulfite agar. J. Clin. Microbiol. 5:122.
25. D'Aoust, J.-Y. 1981. Update on pre-enrichment and selective enrichment conditions for detection of *Salmonella* in foods. J. Food Prot. 44:369.
26. D'Aoust, J.-Y. 1989. *Salmonella*. *In* M .P. Doyle (ed.), Foodborne bacterial pathogens, 327 pp. Marcel Dekker, New York.
27. D'Aoust, J.-Y. 1995. Methods for the detection of foodborne *Salmonella* spp.: a review. Southeast Asian J. Trop. Med. Public Health. 26:195
28. D'Aoust, J.-Y., H. J. Beckers, M. Boothroyd, A. Mates, C. R. McKee, A. B.Moran, P. Sado, G. E. Spain, W. H. Sperber, P. Vassiliadis, D. E. Wagner, and C. Wiberg. 1983. ICMSF methods studies. XIV. Comparative study on recovery of *Salmonella* from refrigerated pre-enrichment and enrichment broth cultures. J. Food Prot. 46:391.
29. D'Aoust, J.-Y., and C. Maishment. 1979. Pre-enrichment conditions for effective recovery of *Salmonella* in foods and feed ingredients. J. Food Prot. 42:153.
30. D'Aoust, J.-Y., C. Maishment, D. M. Burgener, D. R. Conley, A. Loit, M. Milling, and U. Purvis. 1980. Detection of *Salmonella* in refrigerated pre-enrichment and enrichment broth cultures. J. Food Prot. 43:343.
31. D'Aoust, J.-Y., and A. M. Sewell. 1986. Detection of *Salmonella* by the enzyme immunoassay (EIA) technique. J. Food Sci. 51:484.
32. D'Aoust, J.-Y., and A. M. Sewell. 1988. Detection of *Salmonella* with the BioEnzabead™ enzyme immunoassay technique. J. Food Prot. 51:538.
33. D'Aoust, J.-Y. and Sewell, A.M. 1988. Reliability of the immunodiffusion 1-2 Test™ system for detection of *Salmonella* in foods. J. Food Prot. 51:853.
34. D'Aoust, J.-Y., A. Sewell, and A. Jean. 1990. Limited sensitivity of short (6h) selective enrichment for detection of foodborne *Salmonella*. J. Food Prot. 53:562.
35. DeSmedt, J. M., and R. F. Bolderdijk. 1987. Dynamics of *Salmonella* isolation with modified semisolid Rappaport-Vassiliadis medium. J. Food Prot. 50:658.
36. DeSmedt, J., R. Bolderdijk, and J. Milas. 1994. *Salmonella* detection in cocoa and chocolate by motility enrichment on modified semi-solid Rappaport-Vassiliadis medium: Collaborative study. J. Assoc. Off. Anal. Chem. 77:365.
37. Devonish, J. A., B. W. Ciebin, and M. H. Brodsky. 1986. Novobiocin-brilliant green- glucose agar: new medium for isolation of salmonellae. Appl. Environ. Microbiol. 52:539.
38. Difco Laboratories. 1984. Difco manual of dehydrated culture media and reagents for microbiological and clinical laboratory procedures, 10th ed. Difco Laboratories, Detroit, Mich.
39. Eckner, K. F., R. S. Flowers, B. J. Robison, J. A. Mattingly, D. A. Gabis, and J. H. Silliker. 1987. Comparison of *Salmonella* BioEnzabead™ immunoassay method and conventional culture procedure for detection of *Salmonella* in foods. J. Food Prot. 50:379.
40. Edel, W., and E. H. Kampelmacher. 1973. Comparative studies on the isolation of "sublethally injured" salmonellae in nine European laboratories. Bull. World Health Org. 48:167.
41. Edwards, P. R., and W. H. Ewing. 1972. Identification of *Enterobacteriaceae*, 3rd ed. Burgess Publishing Co., Minneapolis, Minn.
42. Edwards, P. R., and M. A. Fife. 1961. Lysine-iron agar in the detection of *Arizona* cultures. Appl. Microbiol. 9:478.
43. Entis, P. 1985. Rapid hydrophobic grid membrane filter method for *Salmonella* detection in selected foods: Collaborative study. J. Assoc.Off. Anal. Chem. 68:555.
44. Entis, P., M. H. Brodsky, A. N. Sharpe, and G. A. Jarvis. 1982. Rapid detection of *Salmonella* spp. in food by use of the ISO-GRID hydrophobic grid membrane filter. Appl. Environ. Microbiol. 43:261.
45. Eskenazi, S., and A. M. Littell. 1978. Dulcitol-malonate-phenylalanine agar for the identification of *Salmonella* and other *Enterobacteriaceae*. Appl. Environ. Microbiol. 35:199.
46. Ewing, W. H. 1986. Edwards and Ewing's identification of *Enterobacteriaceae*, 4th ed., Elsevier Science, New York.
47. Fantasia, L. D., J. P. Schrade, J. F. Yager, and D. Debler. 1975. Fluorescent antibody method for the detection of *Salmonella*: development, evaluation, and collaborative study. J. Assoc. Off. Anal. Chem. 58:828.

48a. Farmer, J.J., McWhorter, A.C., Brennen, D.J., and G.K. Morris. 1984. The Salmonella-Arizona group of Enterobacteriaceae: Nomenclature, classification and reporting. J. Clin. Microbiol. Newsletter. 6:63.

48. Feldsine, P. T., and M. T. Falbo-Nelson. 1995. Comparison of modified immunodiffusion and *Bacteriological Analytical Method* (BAM) methods for the detection of *Salmonella* in raw flesh and highly contaminated food types. J. AOAC Int. 78:993.
49. Feldsine, P. T., M. T. Falbo-Nelson, and D. L. Hustead. 1992. Polyclonal enzyme immunoassay method for detection of motile and non-motile *Salmonella* in foods:Collaborative study. J. AOAC Int. 75:1032.
50. Feldsine, P.T., Falbo-Nelson, M.T. and Hustead, D.L. 1993. Polyclonal enzyme immunoassay method for detection of motile and non-motile *Salmonella* in foods: collaborative study. J. AOAC Int. 76:694.
51. Feldsine, P. T., M. T. Falbo-Nelson, D. L. Hustead, R. S. Flowers, and M. J. Flowers. 1995. Comparative and multilaboratory studies of two immunodiffusion method enrichment protocols and the AOAC/Bacteriological Analytical Manual culture method for the detection of *Salmonella* in all foods. J. AOAC Int. 78:987.
52. Flowers, R. S., K. F. Eckner, D. A. Gabis, B. J. Robison, J. A. Mattingly, and J. H. Silliker. 1986. Microbiological methods. Enzyme immunoassay for detection of *Salmonella* in foods: Collaborative study. J. Assoc.Off. Anal. Chem. 69:786.
53. Flowers, R. S., and J. J. Klatt. 1987. Evaluation of a visual immunoassay for detection of *Salmonella* in foods. Inst. of Food Technol. Annual Meeting, Abstract 122. Dallas, Tex.
54. Flowers, R. S., and M. J. Klatt. 1989. Immunodiffusion screening method for detection of motile *Salmonella* in foods: collaborative study. J. Assoc.Off. Anal. Chem. 72:303.
55. Flowers, R. S., M. J. Klatt, and S. L. Keelan. 1988. Visual immunoassay for detection of *Salmonella* in foods: collaborative study. J. Assoc.Off. Anal. Chem. 71:973.
56. Flowers, R. S., M. J. Klatt, S. L. Keelan, B. Swaminathan, W. D. Gehle, and H. E. Chandonnet. 1989. Fluorescent enzyme immunoassay for rapid screening of *Salmonella* in foods: collaborative study. J. Assoc.Off. Anal. Chem. 72:318.
57. Flowers, R. S., M. J. Klatt, M. A. Mozola, M. S. Curiale, D. A. Gabis, and J. H. Silliker. 1987. Microbiological methods. DNA hybridization assay for detection of *Salmonella* in foods: collaborative study. J. Assoc.Off. Anal. Chem. 70:521.
58. Flowers, R. S., M. J. Klatt, B. J. Robison, and J. A. Mattingly. 1988. Evaluation of abbreviated enzyme immunoassay method for detection of *Salmonella* in low-moisture foods. J. Assoc. Off. Anal. Chem. 71:341.
59. Flowers, R. S., M. J. Klatt, B. J. Robison, J. A. Mattingly, D. A. Gabis, and J. H. Silliker. 1987. Enzyme immunoassay for detection of *Salmonella* in low-moisture foods: collaborative study. J. Assoc. Off. Anal. Chem. 30:530.
60. Flowers, R. S., M. A. Mozola, M. S. Curiale, D. A. Gabis, and J. H. Silliker. 1987. Comparative study of a DNA hybridization method and the conventional culture procedure for detection of *Salmonella* in foods. J. Food Sci. 52:781.
61. Gabis, D. A., and J. H. Silliker. 1977. ICMSF methods studies. IX. The influence of selected enrichment broths, differential plating media, and incubation temperatures on the detection of *Salmonella* in dried foods and feed ingredients. Can. J. Microbiol. 23:1225.
62. Gadberry, J. L., and S. N. Highby. 1970. An improved medium for the isolation of species of *Salmonella* and *Arizona* from animals. Public Health Lab. 28:157.
63. Galton, M. M., W. D. Lowery, and A. V. Hardy. 1954. *Salmonella* in fresh and smoked pork sausage. J. Infect. Dis. 95:232.
64. Gibson, D. M. 1992. Automated conductance method for the detection of *Salmonella* in foods: collaborative study. J. Assoc. Off. Anal. Chem. 75:293.
65. Harvey, R. W. S., and T. H. Price. 1977. Observations on pre-enrichment for isolating salmonellas from sewage polluted natural water using Muller-Kauffman tetrathionate broth prepared with fresh and desiccated ox bile. J. Appl. Bacteriol. 43:145.
66. Hennessy, T. W., C. W. Hedberg, L. Slutsker, K. E. White, J. M. Besser-Wiek, M. E. Moen, J. Feldman, W. W. Coleman, L. M. Edmonson, K. L. MacDonald, M. T. Osterholm,and the Investigation Team. 1996. A national outbreak of *Salmonella enteritidis* infections from ice cream. N. Engl. J. Med. 334(20):1281.
67. Hoben, D. A., D. H. Ashton, and A. C. Peterson. 1973. Some observations on the incorporation of novobiocin into Hektoen enteric agar for improved *Salmonella* isolation. Appl. Microbiol. 26:126.
68. Holt, J. G., N. R. Krieg, P. H. A. Sneath, J. T. Staley, and S. T. Williams. 1994. Facultatively anaerobic gram-negative rods. *In* W. R. Hensyl (ed.), Bergeys manual of Derterminative bacteriology, 9th ed., Williams and Wilkins, Baltimore, Md.
69. Huhtanen, C. N., J. Naghski, and E. S. Dellamonica. 1972. Efficiency of *Salmonella* isolation from meat-and-bone meal of one 300-g sample versus ten 30-g samples. Appl. Microbiol. 23:688.
70. Ibrahim, G. F., and G. H. Fleet. 1985. Detection of salmonellae using accelerated methods. A review. Int. J. Food Microbiol. 2:259.
71. ICMSF. 1974. Microorganisms in foods, 2. Sampling for microbiological analysis: principles and specific applications. Int. Comm. Microbiol. Spec. Foods. Univ. of Toronto Press, Toronto, Ontario, Canada.
72. ICMSF. 1988. Microorganisms in foods, 1. Their significance and methods of enumeration, 2nd ed. Int. Comm. Microbiol. Spec. Foods. Univ. of Toronto Press, Toronto, Ontario, Canada.
73. Insalata, N. F., and R. A. Chordash. 1984. Fluorescent antibody detection of salmonellae, p. 327. *In* M. L. Speck (ed.), Compendium of methods for the microbiological examination of foods, 2nd ed., American Public Health Association, Washington, D. C.
74. ISO 1981. Microbiology—general guidance for the detection of *Salmonella*, Int. Org. for Standardization 6579, Geneva, Switzerland.
75. Kauffmann, F. 1966. The bacteriology of *Enterobacteriaceae.* Williams and Wilkins Co., Baltimore, Md.
76. Keelan, S. L., R. S. Flowers, and B. J. Robinson. 1988. Microbiological methods. Multitest system for biochemical identification of *Salmonella, E.coli,* and other *Enterobacteriaceae* isolated from foods: collaborative study. J. Assoc. Off. Anal. Chem. 71:968
77. Kent, P. T., B. M. Thomason, and G. K. Morris. 1981. Salmonellae in foods and feeds. Review of isolation methods and recommended procedures. Centers for Disease Control, Atlanta, Ga.
78. King, S., and W. I. Metzger. 1968. A new plating medium for the isolation of enteric pathogens. I. Hektoen enteric agar. Appl. Microbiol. 16:577.
79. Knight, M. T., D. W. Wood, J. F. Black, G. Gosney, R. O. Rigney, and J. R. Agin. 1990. Gram-negative identification card for identification of *Salmonella, Escherichia coli,* and other *Enterobacteriaceae* isolated from foods: Collaborative study. J. Assoc.Off. Anal. Chem. 73:729
80. Krieg, N. R., and J. G. Holt. 1984. Bergey's manual of systematic bacteriology, vol. 1. Williams and Wilkins, Baltimore, Md.
81. Kristensen, M., V. Lester, and A. Jurgens. 1925. On the use of trypsinized casein, bromthymol-blue, brom-cresol-purple, phenol-red and brilliant-green for bacteriological nutrient media. Brit. J. Exp. Pathol. 6:291.
82. Leifson, E. 1935. New culture media based on sodium desoxycholate for the isolation of intestinal pathogens and for the enumeration of colon bacilli in milk and water. J. Pathol. Bacteriol. 40:581.
83. LeMinor, L. 1984. Genus III. *Salmonella,* p. 427. *In* N. R. Krieg and J. G. Hold (ed.), Bergey's manual of systematic bacteriology, vol. 1. Williams and Wilkins, Baltimore, Md.
84. LeMinor, L., M. Vernon, and M. Popoff. 1982. Proposition pour une nomenclature des *Salmonella*. Ann. Microbiol. 133B:245.
85. LeMinor, L., M. Vernon, and M. Popoff. 1982. Taxonomie des *Salmonella*, Ann. Microbiol. 133B:223.
86. MacConkey, A. 1905. Lactose-fermenting bacteria in faeces. J. Hyg. 5:333.
87. Mattingly, J. A., B. J. Robison, A. Boehm, and W. D. Gehle. 1985. Use of monoclonal antibodies for the detection of *Salmonella* in foods. Food Technol. 39:90.
88. Madden, J. M. 1992. Microbial pathogens in fresh produce—the regulatory perspective. J. Food Prot. 55:821.

89. McCoy, J. H. 1962. The isolation of salmonellae. J. Appl. Bacteriol. 25:213.
90. Moats, W. A. 1978. Comparison of four agar plating media with and without added novobiocin for isolation of salmonellae from beef and deboned poultry meat. Appl. Environ. Microbiol. 37:747.
91. Moats, W. A., and J. A. Kinner. 1974. Factors affecting selectivity of brilliant green-phenol red agar for salmonellae. Appl. Microbiol. 27:118.
92. Moats, W. A., and J. A. Kinner. 1976. Observations on brilliant green agar with an H_2S indicator. Appl. Microbiol. 31:380.
93. Morris, G. K., and C. G. Dunn. 1970. Influence of incubation temperature and sodium heptadecyl sulfate (Tergitol No. 7) on the isolation of salmonellae from pork sausage. Appl. Microbiol. 20:192.
94. NAS/NRC. 1969. An evaluation of the *Salmonella* problem. Pub. 1693, National Academy of Science/National Research Council, Washington, D. C.
95. NAS/NRC. 1971. Reference methods for the microbiological examination of foods. Subcomm. Food Microbiol., Food Protection Committee National Academy of Science/National Research Council, Washington, D. C.
96. North, W. R., Jr. 1961. Lactose pre-enrichment method for isolation of *Salmonella* from dried egg albumen. Appl. Microbiol. 9:188.
97. Osborne, W. W., and J. L. Stokes. 1955. A modified selenite brilliant-green medium for the isolation of *Salmonella* from egg products. Appl. Microbiol. 3:295.
98. Padron, A. P., and W. B. Dockstader. 1972. Selective medium for hydrogen sulfide production by salmonellae. Appl. Microbiol. 23:1107.
99. Poelma, P. L., A. Romero, and W. H. Andrews. 1978. Comparative acccuracy of five biochemical systems for identifying *Salmonella* and related foodborne bacteria: collaborative study. J Assoc.Off. Anal. Chem. 61:1043.
100. Price, W. R., R. A. Olsen, and J. E. Hunter. 1972. *Salmonella* testing of pooled pre-enrichment broth cultures for screening multiple food samples. Appl. Microbiol. 23:679.
101. Rambach, A. 1990. New plate medium for facilated differentiation of *Salmonella* spp. from *Proteus* spp. and other enteric bacteria. Appl. Environ. Microbiol. 56:301.
102. Rappold, H., and R. F. Bolderdijk. 1979. Modified lysine iron agar for isolation of *Salmonella* from food. Appl. Environ. Microbiol. 38:162.
103. Rappold, H., R. F. Bolderdijk, and J. M. DeSmedt. 1984. Rapid cultural method to detect *Salmonella* in foods. J. Food Prot. 47:46.
104. Read, R. B., and A. L. Reyes. 1968. Variation in plating efficiency of salmonellae on eight lots of brilliant green agar. Appl. Microbiol. 16:746.
105. Reamer, R. H., R. E. Hargrove, and F. E. McDonough. 1974. A selective plating agar for direct enumeration of *Salmonella* in artificially contaminated dairy products. J. Milk Food Technol. 37:441.
106. Restaino, L., G. S. Grauman, W. A. McCall, and W. M. Hill. 1977. Effects of varying concentrations of novobiocin incorporated into two *Salmonella* plating media on the recovery of four *Enterobacteriaceae*. Appl. Environ. Microbiol. 33:585.
107. Ries, A. A., S. Zaza, C. Langkop, R. V. Tauxe, and P. A. Blake. 1990. A multistate outbreak of *Salmonella chester* linked to imported cantaloupe. Abstract. Interscience Conference on Antimicrobial Agents and Chermotherapy. p. 238.
108. Shanson, D. C. 1975. A new selective medium for the isolation of salmonellae other than *Salmonella typhi*. J. Med. Microbiol. 8:357.
109. Silliker, J. H. 1969. "Wet compositing" as an approach to control procedures for the detection of salmonellae, Appendix D, p. 206. *In* An evaluation of the *Salmonella* problem. Committee on *Salmonella*, National Academy of Science/National Research Council Washington, D. C. Publ. 1683.
110. Silliker, J. H., and D. A. Gabis. 1973. ICMSF methods studies. I. Comparison of analytical schemes for detection of *Salmonella* in dried foods. Can. J. Microbiol. 19:475.
111. Silliker, J. H., and D. A. Gabis. 1974. ICMSF Methods Studies V. The influence of selective enrichment media and incubation temperatures on the detection of salmonellae in raw frozen meats. Can. J. Microbiol. 20:813.
112. Silliker, J. H., and W. I. Taylor. 1958. Isolation of *salmonellae* from food samples. II. The effect of added food materials upon the performance of enrichment broths. J. Appl. Microbiol. 6:228.
113. Stroup, J. R. 1972. Malonate dulcitol lysine iron agar—a new differential medium for the identification of *Salmonella* subgenera I-III. J. Assoc. Off. Anal. Chem. 55:214.
114. Sulkin, S. E., and J. C. Willett. 1940. A triple sugar-ferrous sulfate medium for use in identification of enteric organisms. J. Lab. Clin. Med. 25:649.
115. Sveum, W. H., and P. A. Hartman. 1977. Timed-release capsule method for the detection of salmonellae in foods and feeds. Appl. Environ. Microbiol. 33:630.
116. Sveum, W. H., and A. A. Kraft. 1981. Recovery of salmonellae from foods using a combined enrichment technique. J. Food Sci. 46:94.
117. Taylor, W. I. 1965. Isolation of shigellae. I. Xylose lysine agars; new media for isolation of enteric pathogens. Am. J. Clin. Pathol. 44:471.
118. Taylor, W. I., and J. H. Silliker. 1958. Isolation of salmonellae from food samples. III. Dulcitol lactose iron agar, a new differential tube medium for confirmation of microorganisms of the genus *Salmonella*. Appl. Microbiol. 6:335.
119. Thomason, B. M. 1981. Current status of immunofluorescent methodology for salmonellae. J. Food Prot. 44:381.
120. Thomason, B. M., W. B. Cherry, and M. D. Moody. 1957. Staining bacterial smears with fluorescent antibody. III. Antigenic analysis of *Salmonella typhosa* by means of fluorescent antibody and agglutination reactions. J. Bacteriol. 74:525.
121. Todd, L. S., D. Roberts, B. A. Bartholomew, and R. J. Gilbert. 1986. Evaluation of an enzyme immunoassay kit for the detection of salmonellae in foods and feeds, p. 418. *In* Proceedings of. 2nd World Congress Foodborne Infection and Intoxications, vol. 1. Berlin.
122. Todd, L. S., D. Roberts, B. A. Bartholomew, and R. J. Gilbert. 1987. Assessment of an enzyme immunoassay for the detection of salmonellae in foods and animal feeding stuffs. Epidemiol. Infect. 98:301.
123. USDA. 1968. Recommended procedure for the isolation of *Salmonella* organisms from animal feeds and feed ingredients. ARS 91-68. U.S. Department of Agriculture, Animal Health Division, Agricultural Research Service, Hyattsville, Md.
124. USDA. 1974. Microbiological laboratory guidebook, Food Safety and Inspection Serv., U.S. Department of Agriculture, Washington, D. C.
125. USDA. 1984. Laboratory methods for egg products. Agriculture Marketing Service, U.S. Department of Agriculture, Washington, D. C.
126. U.S. FDA. 1971. Federal Food, Drug, and Cosmetic Act, as amended, January 1971. Stock No. 1712-0126. Food and Drug Admin., U.S. Govt Print. Off. Washington, D.C.
127. U.S. FDA. 1978. Chap. VII *In* Bacteriological analytical manual, 5th ed. p. 1, Assoc. Off. Anal. Chem., Washington, D. C.
128. U.S. FDA. 1995. Chap. 1. *In* Bacteriological analytical manual, 8th ed. Assoc. Off. Anal. Chem. Int., Gaithersburg, Md
129. U.S. FDA. 1995. Chap. 5. *In* Bacteriological analytical manual, 8th ed. Assoc. Off. Anal. Chem. Int., Gaithersburg, Md.
130. Vassiliadis, P., D. Trichopoulos, G. Papoutsakis, and E. Pallandiou. 1979. A note on the comparison of two modifications of Rappaport's medium with selenite broth in the isolation of *Salmonellas*. J. Appl. Bacteriol. 46:567.
131. Warburton, D. W., P. T. Feldsine, and M. T. Falbo-Nelson. 1995. Modified immunodiffusion method for detection of *Salmonella* in raw flesh and highly contaminated foods: Collaborative study. J. AOAC Int. 78:59.
132. Wilson, C. R., W. H. Andrews, and P. L. Poelma. 1980. Recovery of *Salmonella* from milk chocolate using a chemically defined medium and five nondefined broths. J. Food Sci. 45:310.

133. Wilson, W. J., and E. M. Blair. Mcv. Blair. 1927. Use of a glucose bismuth sulphite iron medium for the isolation of *B. typhosus* and *B. proteus*. J. Hyg. 26:374.
134. Wun, C. K., J. R. Cohen, and W. Litsky. 1972. Evaluation of plating media and temperature parameters in the isolation of selected enteric pathogens. Health Lab. Sci. 9:225.
135. Yamamoto, R., W. W. Sadler, H. E. Adler, and G. F. Stewart. 1961. Comparison of media and methods for recovering *Salmonella typhimurium* from turkeys. Appl. Microbiol. 9:76.

Chapter 38

Shigella

Keith A. Lampel

38.1 INTRODUCTION

Bacteria in the genus *Shigella* are members of the family Enterobacteriaceae, are nearly genetically identical to *Escherichia coli* and are closely related to *Salmonella* and *Citrobacter* spp. There are four species of shigellae that are divided serologically based on their somatic O antigen: *S. dysenteriae* (serogroup A), *S. flexneri* (serogroup B), *S. boydii* (serogroup C) and *S. sonnei* (serogroup D). Each serogroup except *S. sonnei* has several different serovars and subserovars. Shigellae are the causative agent of bacillary dysentery, also known as shigellosis, a debilitating diarrheal disease characterized by the production of bloody mucoid stools accompanied by abdominal cramps, fever, and tenesmus. In some cases, infected individuals experience only mild, watery diarrhea. Further complications[2] include Reiter's syndrome and hemolytic uremic syndrome.

Shigella spp. are gram-negative, nonmotile rods.[3,18] The principal mode of transmission of this pathogen is by the fecal–oral route; contaminated water and foods are also important vectors for the distribution of shigellae. Shigellosis is one of the most communicable forms of bacterial diarrheal disease. An important hallmark of bacillary dysentery is the rapid spread of *Shigella* throughout a population. This is partially due to the low infectious dose (10-200 organisms). The number of shigellae in stool samples of infected individuals varies according to the stage of illness. In the first few days of the acute phase of the disease, 10^3 to 10^9 colony forming units (CFU) per gram of stool can be isolated whereas in convalescent patients, 10^2 to 10^3 CFU are recovered.

Humans and higher primates are the only known hosts. The incubation period from the ingestion of contaminated food or water to the onset of illness for shigellosis is 12 to 50 hours. The disease is self limiting in otherwise healthy individuals and usually lasts for 4 to 7 days but may persist up to 14 days. *S. dysenteriae* type 1produces the most severe form of illness and has a fatality rate up to 20%. Although quite rare in the United States and other industrialized nations, it is more common in developing countries. Mild forms of shigellosis are caused by *S. sonnei* and this is usually the predominant isolate found in developed countries. Treatment is usually electrolyte replacement for patients suffering from dehydration. Antibiotics are not usually required for patients suffering from mild diarrhea but in the case of *S. dysenteriae* infections, therapy with antimicrobial agents may reduce the duration of the illness and carriage state of the patient and may lower mortality. Multiple-drug resistant shigellae have been reported and may pose a problem in treating these infections.

Virulence of *Shigella* spp.[11] (and enteroinvasive *E. coli*) depends upon the controlled expression of genetic factors encoded in the chromosome and on a large virulence plasmid. These pathogens survive the acidic environment of the stomach and eventually invade epithelial cells lining the large intestine. Shigellae multiply intracellularly and then spread intercellularly resulting in dissemination of this pathogen to neighboring cells without further exposure to the intestinal lumen.

Temperature is a key factor in regulating the virulence genes of shigellae. At 30°C, *Shigella* spp. are not able to invade epithelial cells *in vitro*; however, at 37°C, these cells are phenotypically invasive. Most *Shigella* virulence genes reside on a 180-220 kilobase pair (kbp) plasmid within a 37 kbp region. The 33 known genes can simply be categorized based on their function in *Shigella* pathogenesis. The immundominant antigens, the Ipa proteins, aggregate on the bacterial cell surface and are responsible for mediating the entry of the bacterium into the host cell through a bacterial–directed phagocytosis. Products of another class of genes, the *ipg* (invasion plasmid gene) and *mxi-spa* (membrane expression of invasion plasmid antigens–surface presentation of Ipa antigens) loci, are responsible for exporting these Ipa proteins from inside the bacterial cell to the extracellular milieu. Two plasmid-encoded genes (*virF* and *virB*) and one gene (*hns*) in the chromosome are involved in regulating the expression of the plasmid–encoded virulence genes. Besides temperature, other environmental stimuli, such as pH and osmolarity, modulate the expression of the regulatory gene cascade. *S. dysenteriae* type 1 produces a potent cytotoxin; the role of this toxin in *Shigella* pathogenesis is unclear. This toxin may be partially responsible for a sequella of *Shigella* infections, namely, hemolytic uremic syndrome.

38.11 Shigella *in Foods*

Shigella spp. are not indigenous to any food. Contamination of foods is primarily due to the poor personal hygiene of food handlers. Foods commonly found to be associated with foodborne outbreaks caused by shigellae are potato or tossed salads, chicken and shellfish. Establishments that were identified as the source ranged from homes to restaurants, camps, picnics, airlines, schools, nursing homes, cruise ships and military mess halls. Even when epidemiological data may implicate a source, recovery of *Shigella* spp. from incriminated foods is usually unsuccessful. And since shigellae are not associated with one particular food, routine inspection of foods for *Shigella* spp. is not performed.

Although shigellae do survive a transient passage through the stomach, it has been shown that these organisms are sensitive to acidic environments, particularly at elevated temperatures, e.g., 37°C. Isolating *Shigella* from foods with high numbers of natural microbial flora, e.g. bean sprouts with 10^8 CFU/gram, is a challenge. In many cases, the recognition of an outbreak due to contaminated foods comes first from the clinical laboratory and epidemiological findings. Meanwhile, food samples may have been discarded or compromised by storage at improper temperatures.

38.12 Survival of Shigellae in Foods

Most studies on the ability of shigellae to survive and grow in foods have been conducted with *S. flexneri* and *S. sonnei*. Temperature, pH and the presence of inorganic acids in foods greatly influence the survival and growth of shigellae. *Shigella* survival in foods has been tested over a range of temperatures, from −20°C to room temperature. In general, shigellae do not survive well in acidic environments, but in foods, such as citric juices and carbonated soft drinks, survival was recorded up to 10 days. Contaminated vegetables have been a common source for outbreaks caused by *Shigella*. In one study, an initial inoculum of 10^5–10^6 CFUs of *S. flexneri* was added to sterile and nonsterile vegetables[13] and the number of *Shigella* present in these foods remained constant for several days. A more comprehensive summary of the survival and growth of shigellae in foods has been reviewed by the ICMSF[6] and Smith.[14]

38.2 RECOMMENDED PROCEDURES

38.21 Isolation and Detection of Shigella from Foods: Bacteriological

Since *Shigella* spp. are not indigenous to any particular food, the possibility of a foodborne outbreak due to shigellae is first recognized by a clinical laboratory and later confirmed by epidemiological investigations. This process can take 7 to 10 days before a putative food source is identified. *Shigella* spp. are difficult to isolate when present in low numbers or stored for a short period of time at 4°C. Therefore, suspected foods should be processed quickly for analysis.

38.22 Culture Media and Reagents

Acetate agar
Antisera to *Shigella* O groups A, B, C, and D.
Butterfield's phosphate buffer
Christensen citrate agar
Desoxycholate agar
Eosin methylene blue (EMB) agar
GN (gram-negative) broth
Hektoen agar
Kliger Iron agar slant
Luria broth (LB)
Lysine Iron agar
MacConkey agar
Motility agar
Novobiocin
Nutrient agar slant
Salmonella–Shigella agar
Shigella broth
Tryptic soy broth
Tergitol-7 agar
Triple sugar iron agar
Xylose-lysine-desoxycholate agar (XLD)

38.23 Enrichment

Aseptically add 25 grams of food sample to 225 mL of *Shigella* broth[1] or GN broth. In some cases the addition of novobiocin (0.3 µg/mL for *S. sonnei*; 3 µg/mL for other *Shigella* spp.) to *Shigella* broth increases the chance for recovery. Samples in *Shigella* broth should be held at room temperature for a minimum of 10 min and then the broth poured into a 500 mL Erylenmeyer flask. The culture is incubated overnight at 37°C with shaking. Alternatively, the enrichment broth, such as GN, is added to a sterile stomacher bag, and the mixture blended for 3 min. In all cases, the pH of the broth should be adjusted to pH 7.0 ± 0.2. with sterile 1 N NaOH or 1N HCl. Compositing of samples, where different lots or portions of the food are analyzed to statistically evaluate the suspected contaminated food, can be done in accordance with the specifications as outlined by the ICMSF.[5] Enrichment cultures are grown for 16 to 20 hours at 37°C. It is recommended that 2 to 3 different agar plates be used with various selective media to enhance the possibility of recovering *Shigella* from food samples.

Injured cells,[16] usually found in processed and particularly in underprocessed foods, are grown in media without bile salts or desoxycholate. These selective compounds can inhibit the growth of injured cells but have no affect on uninjured cells. Food samples are blended with 100 mL tryptic soy broth and the pH adjusted to 7. Incubate at 37°C for 8 hours, add 125 mL of enrichment broth, and transfer mixture to stomacher bags and continue incubation at 37°C for 16 to 20 hours.

38.24 Plating

Cultures can be inoculated onto different types of selective media (see Table 1). On MacConkey agar plates (low selectivity), *Shigella* are lactose negative and produce colonies that are translucent and slightly pink with or without rough edges. Other al-

Table 1. Growth Characteristics of *Shigella* on Selective Media

Organism	Medium	Colony Appearance
Shigella	MacConkey	Colorless (lactose non-fermentor); *S. sonnei* colonies are flat with jagged edges.
	Hektoen	Green
	XLD	Colorless
E. coli	MacConkey	Lactose fermentor; flat, pink colonies surrounded by darker pink region (indicates sorbitol fermentors, non-sorbitol fermentors form colorless colonies).
	Hektoen	Yellow
	XLD	Yellow
Salmonella	MacConkey	Colorless (lactose non-fermentors)
	Hektoen	Green
	XLD	Red with black center

ternative lactose-containing media with low selectivity are EMB and Tergitol-7 agar. *Shigella* spp. produce colorless colonies on EMB agar plates and bluish colonies on the yellowish-green Tergitol-7 agar. Intermediate selective media, such as desoxycholate and XLD, are preferred media to isolate *Shigella* spp. In XLD agar, D-xylose is incorporated into this medium as a differentiating agent. Xylose is not fermented by most *Shigella* strains (*S. boydii* is variable), and colonies formed are translucent and red (alkaline) on XLD plates. Some *Shigella* strains do ferment xylose and will be missed on these plates; therefore plating on both XLD and MacConkey plates is recommended. On desoxycholate agar, *Shigella* colonies are reddish. *Salmonella–Shigella* agar is a highly selective medium that may be too stringent for some shigellae, such as *S. dysenteriae* type 1. On *Salmonella–Shigella* agar, *Shigella* spp. appear as colorless, translucent colonies. On Hektoen agar, *Shigella* colonies are green as are *Salmonella;* however, *E. coli* strains form yellow colonies. All presumptive isolates are inoculated onto motility agar; *Shigella* spp. are nonmotile.

38.25 Identification

Shigella spp. are gram-negative, non-motile rods. Presumptive colonies are inoculated onto Kliger iron agar or lysine iron agar. A typical reaction for shigellae is alkaline slant, acid butt and no gas. Further characterization includes biochemical and serological tests. Shigellae are negative for H_2S production, phenylalanine deaminase, sucrose fermentation, do not utilize acetate, KCN, citrate, malonate, adonitol, inositol, and salicin, lack lysine decarboxylase and are negative for Voges-Proskauer test activity (*S. sonnei* and *S. boydii* serotype 13 are positive) but all shigellae are methyl red positive and produce acid from glucose and some other carbohydrates (acid and gas production seen with *S. flexneri* serotype 6, *S. boydii* serotypes 13 and 14, and *S. dysenteriae* 3). *S. dysenteriae* type 1, unlike other *Shigella* sp., is catalase negative and has ornithine decarboxylase activity. To differentiate between *E. coli* strains and *Shigella* spp., growth on Christensen citrate agar and acetate agar can be performed as well as an assay for the ability to ferment mucate. Shigellae are negative for these reactions. See Tables 2 and 3 for key characteristic traits to differentiate *Shigella* from *E. coli* and the four species of *Shigella*.

38.26 Serological testing

Confirmation of shigellae is performed using a slide agglutination assay for serologic identification. Presumptive isolates are inoculated on nutrient agar slants and grown 16–24 hours at 37°C. Colony material is suspended in 13 mL of 0.85% saline to produce a thick suspension of bacteria. Nine 3 × 1 cm squares are made with a wax pencil on a petri dish. A drop of cell suspension and commercially available antisera (polyvalent group antisera; follow manufacturer's instructions for performance of the assay and interpretation of results) are added to each square, mixed gently with a needle to avoid any mixing of neighboring samples and the petri dish rocked 34 minutes to accelerate agglutination. A saline control should be included to monitor autoagglutination. Reaction results are as follows: 0 is no agglutination, 1+ is slight agglutination, 2+ is agglutination with 50% clearing, 3+ is agglutination with 75% clearing and 4+ is visible floc with fluid totally clear. In some cases, a false negative result can occur. Suspensions that agglutinate poorly or not at all should be boiled for 30 minutes to destroy any interfering capsular (K) antigen, and then after cooling, reexamined in the slide agglutination assay.

38.27 Isolation and Detection of Shigella from Foods: DNA probes

The application of DNA-based assays can facilitate and accelerate the identification of *Shigella* spp. from foods. DNA probes are specifically designed to hybridize with known genetic markers, usually virulence genes, either in the chromosome or on the large virulence plasmid. Presumptive colonies are grown overnight at 37°C in Luria broth in a 96-well microtiter plate. Cultures are plated onto TSA plates and grown overnight at 37°C. If radioactive probes are to be used, colonies are transferred to Whatman 541 filter paper (Hillsboro, OR), the paper (colony side up) is then saturated with 5 mL of 1.5 M NaCl in 0.5 M NaOH for 5 min and then heated in a microwave for 30 seconds at 750 W. Filters are neutralized in 5 mL 2.0 M NaCl in 1.0 M Tris buffer, pH 7.0 for 5 minutes, air dried and are ready for hybridization.

For nonradioactive hybridization analysis, colonies are lifted onto nylon membranes, (e.g., Micron Separations, Inc., Westborough, MA) and prepared according to the Boehringer Genius System User's Guide (Boehringer Mannheim Biochemicals, Mannheim, Germany). The agar plate is chilled at 4°C for 1 hour, the membrane is placed on top of the agar for 1 minute and then placed colony side up onto Whatman 3 filter paper soaked with 0.5 M NaOH in 1.5 M NaCl for 5 minutes. Membranes are neutralized on filter paper saturated with 2.0 M NaCl in 1.0 M Tris buffer, pH 7.0 for 5 min and then baked *in vacuo* at 80°C for 2 hours.

DNA probes from either isolated DNA fragments or polymerase chain reaction-generated products are labeled with α [^{32}P] dATP. Oligonucleotides are 5'end-labeled with γ [^{32}P]ATP using T4 Kinase (New England Biolabs, Beverly, MA). Nonradioactive DNA probes are labeled with digoxygenin (DIG)-dUTP using the DIG oligonucleotide tailing kit (Boehringer Mannheim) according to the protocol of the manufacturer.

The hybridization solution for radioactive probes is 5× Denhardt's solution, 6× SSC (0.9 M NaCl, 0.09 M sodium citrate), 10 mM EDTA and 20 µg/mL of sheared, boiled salmon sperm DNA. Membranes are hybridized in 10 mL of hybridization solution in petri dishes with 1–2 × 10^6 cpm of labeled probe at 60°C for 2 hours. Filters are then washed twice for 30 min each at 60°C in pre-warmed 6× SSC, then air dried and exposed to X-ray film for 4–16 hours at –70°C. Nonradioactive hybridization using DIG-labeled oligonucleotide probes (5 pmol) are performed for 6 hours. Hybridization conditions and immunological detection are described by the manufacturer (Boehringer Mannheim) using a nucleic acid detection kit.

38.28 Detection of Shigella from Foods: PCR

For most foods, e.g., vegetables, blending of samples is not required. Ten grams of food sample are placed in 10 mL of Butterfield's phosphate buffer and agitated for 10 minutes. The food material is pelleted by centrifugation at 2000 × g. The supernatant is decanted into sterile tubes and extracted twice with an equal volume of buffered phenol:chloroform. To remove any traces of phenol, the aqueous layer is extracted with chloroform and then ethanol precipitated with the addition of 1/10 volume of 3 M sodium acetate and 2× volume of 100% ethanol. Template DNA for PCR is pelleted by centrifugation at 14,000 × g for 10 minutes at 4°C followed by a 70% ethanol wash. The pellet is air dried and suspended in 25 µL of dH_2O. The protocol for PCR is as follows for a 50 µL reaction[7]: 1X PCR buffer, 200 µM of each

Table 2. Characteristics for Identifying *Shigella* and Differentiation from *E. coli*

Characteristics/Biochemical Test	Comment
Morphology, Gram stain	Gram-negative rods
Motility	Non-motile
Urease	Negative
Lysine decarboxylase	*Shigella* are negative, *E. coli* are positive
Christensen citrate	*Shigella* are negative, *E. coli* are positive
KCN	Negative
Ornithine decarboxylase	Negative except for *S. sonnei* and *S. boydii* 13
Methyl red-Voges Proskauer	Negative VP; positive MR
Oxidase	Negative
Lactose	Negative except *S. sonnei* ferments lactose slowly
Mucate	*Shigella* are negative, *E. coli* are positive
Sodium acetate	Negative[1]
Gas from glucose	Most *Shigella* are negative[2]
Polyvalent antisera	Four serogroups, A,B,C,D [3]

[1] *S. flexneri* 4a strains that are mannitol-negative may slowly utilize sodium acetate.

[2] Exceptions include *S. flexneri* 6, *S. boydii* serotypes 13 and 14, and *S. dysenteriae* 3.

[3] Several *E. coli* strains, such as enteroinvasive *E. coli*, have cross reactivity with the O-antigens of *Shigella*.

Table 3. Biochemical and Serological Tests to Differentiate *Shigella* spp. (Smith and Buchanan[15])

	Shigella spp.			
Test	*dysenteriae*	*flexneri*	*boydii*	*sonnei*
Polyvalent antisera	A[1]	B[1]	C[1]	D
Gas from glucose	–[2]	–[2]	–[2]	–
Ornithine decarboxylase	–	–	–[3]	+
Indole	±[4]	±[4]	±	–
Acid from:				
Dulcitol	–[5]	–[5]	–	–
Lactose	–	–	–[6]	+[7]
Mannitol	–	+	+	+
Raffinose	–	±	–	+[7]
Sucrose	–	–	–	+[7]
Xylose	–	–	±	–

[1] *S. dysenteriae* (15 serovars); *S. flexneri* (6 serovars); *S. boydii* (19 serovars).

[2] Gas is produced by *S. dysenteriae* 3, *S. flexneri* 6 and *S. boydii* 13 and 14 strains.

[3] *S. boydii* 13 strains are positive.

[4] *S. dysenteriae* 1 and *S. flexneri* 6 strains are negative; *S. dysenteriae* 2 are positive.

[5] *S. dysenteriae* 5 and *S. flexneri* 6 serovars may be positive.

[6] *S. boydii* 9 may be positive.

[7] Delayed positive reactions may take more than 24 hours.

Symbols: (–) indicates that less than 10% of the strains tested were positive after 24 hours of incubation; (+) indicates that greater than 90% of strains tested were positive; and (±) represents the range between 10 and 90% that are positive.

deoxynucleotide, 1 µM of each PCR primer (primers are selected to amplify a specific *Shigella* virulence gene (*ial*); 5′-TAATAC-TCCTGAACGGCG-3′ and 5′-TTAGGTGTCGGCTTTTCTG-3′), 1–5 µL of template DNA from extracted food sample, 1 unit of thermostable polymerase (Hot Tub polymerase, Amersham Corp., Arlington Heights, IL). The parameters for the thermocycler are set for 30 cycles, each cycle with three steps; 1 minute each at 94°C, 56°C and 72°C. PCR products (expected size of 760 base pairs; bp) are resolved on a 1% agarose gel using 0.5 to 1X Tris-acetate buffer (depending on the electrophoresis system). A molecular weight marker standard, preferably 100 bp ladder, is included on the gel.

Presumptive colonies from agar plates can quickly be analyzed by PCR to determine if *Shigella* are present. Colonies are lysed (colony is picked and placed in a 1.5 mL tube with 150 µL distilled water and boiled for 5 min) and 5 µL of the lysed material containing template DNA is added to the above reaction mixture. A PCR product of the correct size (760 bp), as determined by agarose gel electrophoresis, indicates the presence of *Shigella* in the food sample. False negatives are a problem; PCR-based assays may not yield a positive result because (1) the sensitivity of these assays is probably 100–200 CFUs per gram of food sample and (2) the reactions can be inhibited by interfering chemicals from foods, such as fats.

38.29 Epidemiological studies

Other techniques using molecular biology can be applied to determine the source(s) of an outbreak or if the outbreak is caused by a single source. These protocols, plasmid profiles,[8] pulse-field gel electrophoresis (PFGE),[17] restriction fragment length polymorphism,[4] colicin typing for *S. sonnei* isolates,[12] and antibiograms are useful in identifying the source of an outbreak.

38.3 PRECAUTIONS

The infectious dose of *Shigella* spp. is quite low and laboratory personnel should be made aware of proper procedures when analyzing suspected food samples and with bacteriological material.[9] The successful isolation of shigellae from foods is problematic; therefore any delay in analyzing food samples lessens the chance of recovery. Enrichment in broth medium and subsequent plating on selective agar plates should be followed by examination of sufficient numbers of suspected colonies by biochemical and serological testing. Since *Shigella* and *E. coli* are closely related genetically, the discrimination of these microbes is essential for proper identification. Application of DNA-based assays, such as PCR, may be useful in detecting the presence of *Shigella* in foods.

38.4 REFERENCES

1. Andrews, W. H., G. A. June, and P. S. Sherrod. 1995. Bacteriological analytical manual, 8th ed. p. 6.01. AOAC International, Gaithersburg, Md.
2. Bennish, M. L. 1991. Potentially lethal complications of shigellosis. Rev. Infect. Dis. 13 (suppl. 4):S319-S324.
3. Dupont, H. L. 1995. *Shigella* species (bacillary dysentery), p. 2033-2039. *In* G. L. Mandell, J. E. Bennett, and R. Dolin (ed.), Principles and practice of infectious diseases. Churchill Livingstone, New York.
4. Hinojosa-Ahumada, M., B. Swaminathan, S. B. Hunter, D. N. Cameron, J. A. Kiehlbauch, I. K. Wachsmuth, and N. A. Strockbine. 1991. Restriction fragment length polymorphism in rRNA operons for subtyping *Shigella* sonnei. J. Clin. Microbiol. 29:2380-2384.
5. International Commission on Microbiological Specifications for Foods. 1986. *In* Microorganisms in Foods 2. Sampling for microbiological analysis: Principles and specific applications. University of Toronto, Canada.
6. International Commission on Microbiological Specifications for Foods. 1996. *Shigella*. *In* Microorganisms in Foods 5. Blackwell Scientific Publications, Oxford, United Kingdom.
7. Lampel, K.A., J. J. Jagow, M. Trucksess, and W. E. Hill. 1990. Polymerase chain reaction detection of invasive *Shigella flexneri* in food. Appl. Environ. Microbiol. 56:1536-1540.
8. Litwin, C. M., A. L. Storm, S. Chipowsky, and K. J. Ryan. 1991. Molecular epidemiology of *Shigella* infections: Plasmid profiles, serotype correlation, and restriction endonuclease analysis. J. Clin. Microbiol. 29:104-108.
9. Miller, B. M. (ed.). 1986. Laboratory safety: Principles and practices. American Society for Microbiology, Washington, D. C.
10. O'Brien, A. D., and R. K. Holmes. 1987. Shiga and Shigalike toxins. Microbiol. Rev. 51:206-220.
11. Parsot, C., and P. J. Sansonetti. 1996. Invasion and the pathogenesis of *Shigella* infections. Curr. Top. Microbiol. Immunol. 209:25-42.
12. Pruneda, R. C., and J. J. Farmer III. 1977. Bacteriophage typing of *Shigella* sonnei. J. Clin. Microbiol. 5:66-74.
13. Rafii, F., M. A. Holland, W. E. Hill, and C. E. Cerniglia. 1995. Survival of *Shigella* flexneri on vegetables and detection by polymerase chain reaction. J. Food Prot. 58:727-732.
14. Smith, J. L. 1987. *Shigella* as a foodborne pathogen. J. Food Prot. 50:788-801.
15. Smith, J. L., and R. L. Buchanan. 1992. *Shigella*, p. 423-431. *In* C. Vanderzant and D. F. Splittstoesser (ed.). Compendium of methods for the microbiological examination of foods, 3rd ed. American Public Health Association, Washington, D. C.
16. Smith, J. L., and S. A. Palumbo. 1982. Microbial injury reviewed for the sanitarian. Dairy Food Sanit. 2:57-63.
17. Soldati, L., and J. C. Piffaretti. 1991. Molecular typing of *Shigella* strains using pulsed field gel electrophoresis and genome hybridizations with insertion sequences. Res. Microbiol. 142:489-498.
18. Watanabe, H., and N. Okamura. 1992. The genus *Shigella*, p. 2754-2759. *In* A. Balows, H. G. Truper, M. Dworkin, W. Harder, and K.H. Schleifer (ed.). The prokaryotes. A handbook on the biology of bacteria: Ecophysiology, isolation, identification, applications, 2nd ed. Springer-Verlag, New York.

Chapter 39

Staphylococcus aureus and Staphylococcal Enterotoxins

G.A. Lancette and R. W. Bennett

39.1 INTRODUCTION

The growth of *Staphylococcus aureus* in foods presents a potential public health hazard since many strains of *S. aureus* produce enterotoxins that cause food poisoning if ingested. Among the reasons for examining foods for *S. aureus* are: a) to confirm that this organism may be the causative agent of foodborne illness; b) to determine whether a food or food ingredient is a potential source of enterotoxigenic staphylococci, and c) to demonstrate postprocessing contamination, which usually is due to human contact with processed food or exposure of the food to inadequately sanitized food-processing surfaces. Foods subjected to postprocess contamination with enterotoxigenic types of *S. aureus* represent a significant hazard because of the absence of competitive organisms that normally restrict the growth of *S. aureus* and the production of enterotoxins.

Of the various metabolities produced by the staphylococci, the enterotoxins pose the greatest risk to consumer health. Enterotoxins are proteins produced by some strains of staphylococci,[1] which, if allowed to grow in foods, may produce enough enterotoxin to cause illness when the contaminated food is consumed. These structurally-related toxicologically similar proteins are produced primarily by *Staphylococcus aureus*, although *S. intermedius* and *S. hyicus* also have been shown to be enterotoxigenic.[2] Normally considered a veterinary pathogen,[3,4] *S. intermedius* was isolated from butter blend and margarine in a food poisoning outbreak.[5,6] A coagulase negative *S. epidermidis* was reported to have caused at least one outbreak.[7] These incidents support testing staphylococci other than *S. aureus* for enterotoxigenicity, if they are present in large numbers in a food suspected of causing a food poisoning outbreak.

Although the true incidence is not known, staphylococcal enterotoxins are among the leading cause of foodborne illness. The most common symptoms are vomiting and diarrhea, which occur 2 to 6 hours after ingestion of the toxin. The illness is relatively mild, usually lasting only a few hours to 1 day, however, in some instances the illness is severe enough to require hospitalization.

The need to identify enterotoxins in food encompasses two areas: foods that have been incriminated in foodborne illness and foods that are suspected of containing enterotoxin. In the former situation, the presence of enterotoxin in a suspect food confirms staphylococcal food poisoning. In the latter situation, the presence or absence of the enterotoxin determines the marketability of the product. The latter cannot be overemphasized because of the difficulty in preventing staphylococcal contamination of some types of foods and food ingredients. Routine testing of certain types of foods for the presence of enterotoxins, however, is not the basis for good manufacturing practices. Because toxin presence is only discernible at levels of 10^6 *Staphylococcus aureus* cells/g of product, emphasis must be more rigidly placed on preventing the contamination and subsequent outgrowth of *S. aureus* in food products.

Foods commonly associated with staphylococcal food poisoning are meat (beef, pork and poultry) and meat products (ham, salami, hotdogs), salads (ham, chicken, potato), cream filled bakery products and dairy products (cheese). Many of these items are contaminated after processing or cooking when competing microorganisms are eliminated. During preparation in homes or foodservice establishments they are subsequently mishandled (e.g. improper refrigeration) prior to consumption. In processed foods, contamination may result from human, animal or environmental sources. Therefore, the potential for enterotoxin development is greater in foods that are exposed to temperatures that permit the growth of *S. aureus*. This is especially true for fermented meat and dairy products. Though the potential is there, it is only when improper fermentation takes place that the development of staphylococcal enterotoxin occurs.

In processed foods in which *S. aureus* is destroyed by processing, the presence of *S. aureus* usually indicates contamination from the skin, mouth, or nose of food handlers. This contamination may be introduced directly into foods by process line workers with hand or arm lesions caused by *S. aureus* coming into contact with the food, or by coughing and sneezing, which is common during respiratory infections. Contamination of processed foods also may occur when deposits of contaminated food collect on or adjacent to processing surfaces to which food products are exposed. When large numbers of *S. aureus* are encountered in processed food, it may be inferred that sanitation, temperature control or both were inadequate.

In raw food, especially animal products, the presence of *S. aureus* is common and may not be related to human contamination. Staphylococcal contamination of animal hides, feathers and skins is common and may or may not result from lesions or bruised tissue. Contamination of dressed animal carcasses by

S. aureus is common and often unavoidable. Raw milk and unpasteurized dairy products may contain large numbers of *S. aureus*, usually a result of staphylococcal mastitis. Separating raw and processed foods to prevent *S. aureus* cross contamination is important for food safety.

The significance of the presence of *S. aureus* in foods should be interpreted with caution. The presence of large numbers of the organism in food is not sufficient cause to incriminate a food as the vector of food poisoning. Not all *S. aureus* strains produce enterotoxins. The potential for staphylococcal intoxication cannot be ascertained without testing the enterotoxigenicity of the *S. aureus* isolate and/or demonstrating the presence of staphylococcal enterotoxin in food. Neither the absence of *S. aureus* nor the presence of small numbers is complete assurance that a food is safe. Conditions inimical to the survival of *S. aureus* may result in a diminished population or death of viable microbial cells, while sufficient toxin remains to elicit symptoms of staphylococcal food poisoning.

The method to be used for the detection and enumeration of *S. aureus* depends, to some extent, on the reason for conducting the test. Foods suspected to be vectors of staphylococcal food poisoning frequently contain a large population of *S. aureus*, in which case a highly sensitive method will not be required. A more sensitive method may be required to demonstrate an unsanitary process or postprocess contamination, since small populations of *S. aureus* may be expected. Usually, *S. aureus* may not be the predominant species present in the food, and, therefore, selective inhibitory media are generally employed for isolation and enumeration.

The methods for identifying enterotoxins involve the use of specific antibodies (polyclonal or monoclonal).[8,9,10] The fact that there are several antigenically different enterotoxins complicates their identification because each one must be assayed separately. Another problem is that unidentified enterotoxins exist for which antibodies are not available for *in vitro* serology. These unidentified toxins, however, appear to be responsible for only a small percentage of food poisoning outbreaks.

39.2 GENERAL CONSIDERATIONS FOR ISOLATION OF *S. AUREUS*

39.21 Techniques for Isolation and Enumeration

Enrichment isolation and direct plating are the most commonly used approaches for detecting and enumerating *S. aureus* in foods. Enrichment procedures may be selective [11,12,13] or nonselective.[14] Nonselective enrichment is useful for demonstrating the presence of injured cells, whose growth is inhibited by toxic components of selective enrichment media. Enumeration by enrichment isolation, or selective enrichment isolation, may be achieved by determining either an indicated number or the most probable number (MPN) of *S. aureus* present. Common MPN procedures use three tubes or five tubes for each dilution.[14,15]

For enumeration, samples may be applied to a variety of selective media in two main ways: surface spreading, and pour plates used in direct plating procedures. Surface spreading is advantageous in that the form and appearance of surface colonies are somewhat more characteristic than the subsurface colonies encountered with pour plates. The principal advantage of pour plates is that greater sample volumes can be used.

Since the same types of selective media frequently are employed in both enrichment and direct plating, the relative sensitivity of the two procedures depends largely on the sample volumes. Larger volumes of sample normally are used in enrichment tubes, but equivalent volumes can be used in direct plating procedures by increasing the number of replicate plates. The relative precision of the two procedures for enumeration of *S. aureus* has not been established, but generally, the plate counting procedures are more precise.

39.22 Media Commonly Used for Isolation

Selective media employ various toxic chemicals, which are inhibitory for *S. aureus* to a varying extent as well as to competitive species. The adverse effect of selective agents is more acute in processed foods containing injured cells of *S. aureus*. A toxic medium may help prevent overgrowth of *S. aureus* by competing species. The two selective toxic chemicals most frequently used in staphylococcal isolation media are sodium chloride (NaCl) and potassium tellurite (K_2TeO_3). Various concentrations of these agents have been used, ranging from 5.5% to 10% NaCl and from 0.0025% to 0.05% K_2TeO_3. Other chemicals such as ammonium sulfate, sorbic acid, glycine, lithium chloride, and polymyxin frequently are combined with NaCl and K_2TeO_3. Sodium azide alone or in combination with NaCl and neomycin also has been used in selective isolation media. Additional differences in media are contributed by combinations of selective agents, pH and by the inclusion of different combinations of diagnostic features.

The principal diagnostic features of contemporary media include a) the ability of *S. aureus* to grow in the presence of 7.5% or 10% NaCl[15,16,17]; b) the ability to grow in the presence of 0.01% to 0.05% lithium chloride, and from 0.12% to 1.26% glycine,[18,19,20,21,22] or 40 mg/mL polymyxin[23,24]; c) the ability of *S. aureus* to reduce potassium tellurite (K_2TeO_3) producing black colonies,[18,22,25,26] aerobically and anaerobically[19,27]; d) the colonial form, appearance and size; e) the pigmentations of colonies; f) coagulase activity and acid production in a solid medium[28]; g) the ability of *S. aureus* to hydrolyze egg yolk[29,30]; h) the production of phosphates[21]; i) the production of thermonuclease[31,32,33,34,35]; and j) growth at 42°–43°C on selective agar.[36,37] Media used in the detection and enumeration of *S. aureus* may employ one or more of these diagnostic features.

39.23 Tests Used for Identification

Sometimes additional diagnostic features may be required to confirm *S. aureus* colonies because the inhibitors used may not completely prevent growth of other organisms, such as bacilli, micrococci, streptococci and some yeasts. Microscopic morphology helps to differentiate bacilli, streptococci and yeasts from staphylococci, which form irregular or grape like clusters of cocci. Staphylococci may be further differentiated from streptococci on the basis of the catalase test, with the former being positive. Additional features are needed to differentiate staphylococci further from micrococci. Usually, staphylococci are lysed by lysostaphin[34,38] but not by lysozyme, and they can grow in the presence of 0.4 µg/mL of erythromycin. Micrococci are not lysed by lysostaphin,[34] may be lysed by lysozyme, and will not grow in the presence of erythromycin. In a deep stab culture micrococci will grow at the surface, whereas most staphylococci grow throughout the agar. *Staphylococcus aureus* will grow and produce acid from glucose and mannitol anaerobically,[39,40] whereas micrococci do not. Staphylococcal cells contain teichoic acids in the cell wall and do not contain aliphatic hydrocarbons in the cell membrane, whereas the reverse is true with micrococci.[41] Further, the G + C content (mole percentage) of staphylococci is 30 to 40 and 66 to 75 for micrococci.[42] Testing for some of these features is difficult, time-consuming, and expensive, and usually is not required for routine detection and enumeration procedures. Several commercial miniaturized systems are available to speciate staphylococci.[43]

S. aureus is differentiated from the 32 other staphylococcal species by a combination of the following features: colonial morphology and pigmentation, production of coagulase, thermonuclease, acetone, β-galactosidase, phosphates and alpha toxin (hemolysis), acid from mannitol, maltose, xylose, sucrose and trehalose, novobiocin resistance, presence of ribitol teichoic acid, protein A and clumping factor in the cell wall.[36,39,43,44] The ultimate species identification may be established by DNA-DNA hybridization with reference strains. A non-isotopic DNA hybridization assay[45] and a polymerase chain reaction procedure[46] have been used to successfully identify *S. aureus.*

The confirmation procedure most frequently used to establish the identity of *S. aureus* is the coagulase test. Coagulase is an enzyme that clots plasma of human and other animal species. Differences in suitability among plasmas from various animal species have been demonstrated.[47] Human[48] or rabbit plasma is most frequently used for coagulase testing and is available commercially. The use of pig plasma sometimes has been found advantageous, but it is not widely available. Coagulase production by *S. aureus* may be affected adversely by physical factors, such as culture storage condition, pH of the medium and denigration. The extent to which the production of coagulase may be impaired by the toxic components of selective isolation media has not been demonstrated clearly.

Presence of clumping factor in cells is another unique feature of *S. aureus*. It can be used to distinguish tube-coagulase-positive *S. aureus* from other tube-coagulase-positive species such as *S. hyicus*. Clumping factor present in *S. aureus* cells binds to fibrinogen or fibrin present in human or rabbit plasma, resulting in agglutination of cells. This is referred to as slide coagulase, bound coagulation or agglutination. Clumping of cells in this test is very rapid (less than 2 minutes) and the results are more clearcut than 1+ or 2+ clotting in the tube coagulase test. Clumping factor can be detected using commercially available latex agglutination reagents.[49,50] Anti-protein A immunoglobulin G (IgG) and fibrinogen are used to coat polystyrene latex beads to simultaneously bind protein A and coagulase, both of which are specific cell surface components of *S. aureus*. One latex kit was successfully collaboratively studied by comparing a latex agglutination method to the coagulase test.[50]

Thermonuclease is also frequently used as a simple, rapid and practical test for routine identification of *S. aureus*.[51,52,53] Coagulase and heat stable nuclease tests are very efficient for the identification of foodborne *S. aureus* strains isolated on Baird-Parker agar.[54] However, the use of the coagulase and/or thermonuclease test may result in erroneous species designation from a taxonomic standpoint. Two species, *S. intermedius*[55] and *S. hyicus*[56] subspecies *hyicus* are both coagulase and thermonuclease positive. However, the latter species can easily be differentiated from *S. aureus* on the basis of the clumping factor test. Coagulase and/or thermonuclease negative staphylococci are being reported to be enterotoxigenic.[57,58,59]

39.3 PRECAUTIONS AND LIMITATIONS OF METHODS

Many factors affect the usefulness and reliability of *S. aureus* detection and enumeration procedures. Among the more important factors are: a) physiological state of organism; b) competitive position of *S. aureus* in the sample menstruum; and c) limitations of isolation media.

It has been demonstrated that the growth of injured cells of *S. aureus* is restricted by many of the selective isolation media used. Media satisfactory for detecting the presence of *S. aureus* in animal lesions, excretory products and nonprocessed foods may not be adequate for analyzing processed foods. Factors such as heating, freezing, desiccation, ripening and storage, which are common elements of food processing, have been shown to affect the growth of *S. aureus*[60] adversely. The extent of cellular injury inflicted during processing depends on the type or severity of treatment.

The importance of the physiological state of *S. aureus* to the selection of media for use in isolation and enumeration procedures is receiving increased attention. Frequently used staphylococcal isolation media that may restrict the growth of sublethally heated cells are mannitol salt agar, Vogel and Johnson agar, egg yolk azide agar, phenolphthalein phosphate agar containing polymyxin, milk-salt agar, tellurite glycine medium, staphylococcus medium number 110 and tellurite polymyxin egg yolk agar.[25,61,62,63,64,65,66,67,68] Selective media containing salt were more satisfactory than the media containing tellurite and tellurize azide in recovering *S. aureus* presumably injured by the ripening process of cheese.[67] Metabolically impaired cells that survive the toxic chemicals of selective media also may fail to show typical appearance.

Agents used in contemporary media to improve the recovery of stressed cells include: a) sodium pyruvate or catalase, which acts to prevent cell death from hydrogen peroxide accumulation during aerobic growth and repair[69]; b) Tween 80 for repair of damaged cell membranes where lipid and phospholipid are located[70]; c) a combined supplement of 0.05% (wt/vol) Tween 80 and 0.1% $MgCl_2 \cdot 6H_2O$, where Mg^{2+} may be required for repair of damaged ribosomes as a consequence of Mg^{2+} loss after stress[52,64,71]; and (d) phosphatidyl choline (2 mg/mL medium) or lecithin which acts similar to egg yolk for increasing enumeration of heat injured *S. aureus*.[25]

Limitations in detection and enumeration methods are generally those associated with limitations of the isolation media in supporting growth of *S. aureus* and suppressing growth of competing species. In addition to variations contributed by the competition for growth media nutrients, procedural efficiency may be affected by other factors, such as acid-base changes and production of growth limiting products, antibiotics, bacteriocins, bacteriophages and the microflora of food products. It is generally conceded that none of the staphylococcal isolation media will prevent growth of all competing species without restricting growth of some *S. aureus*. Among the sources of variation shown to affect media efficiency significantly are: a) type of food examined; b) the relative competitive position of *S. aureus;* and c) the strain of *S. aureus* involved.[24]

Diagnostic criteria used in most staphylococcal isolation media make visual colony identification of *S. aureus* impossible without further testing. The physiology of *S. aureus* is diverse, and not all strains of the species demonstrate similar activity because of their source. For example, not all biotypes have the capacity to hydrolyze egg yolk, a common diagnostic feature in many detection and enumeration procedures.[29,45,72] Considerable divergence also has been demonstrated in the response of various strains to the chemical agents used in selective isolation media. This diversity may lead to considerable confusion regarding the suitability of various isolation media.

Instability has been shown in certain of the physiological traits demonstrated by this species. Variability has been attributed to both physiological and genetic factors. The frequency with which physiological traits may change and the elements stimulating such change, have not been demonstrated clearly. In applying the customary procedures for detection and enumeration of *S. aureus*, possible variation in certain physiological traits should be considered. Some of the usual diagnostic features characteristic of the species may not always be displayed.

39.31 Treatment of Samples

Procedures for sample collection, shipment and preparation described in Chapter 2 should be observed.

Conclusions regarding the potential hazard of foods in noncommercial opened commercial containers in which the presence of *S. aureus* has been detected should be made with considerable caution. Correlation of biotypes isolated from food containers and from food poisoning victims should be established.

39.32 Handling Stock Cultures

Stock cultures of the following properly identified organisms should be maintained for testing the quality of media and reagents:

- *S. aureus* (ATCC 12600) A coagulase positive biotype with the combined characteristics of egg yolk hydrolysis and pigment production is preferable;
- *S. epidermidis* (ATCC 14990);
- *Micrococcus* spp.-*M. varians* (ATCC 15306).

Storage of stock cultures on laboratory media that results in desiccation of the media, and thus requires frequent transfer of stock cultures, should be avoided to lessen the risk of loss of certain diagnostic traits.

39.33 Recommended Controls

Each batch of medium prepared for isolation and enumeration of *S. aureus* should be tested for sterility, productivity and suitability of diagnostic criteria. To test sterility, pour melted solid media into sterile plates and incubate 45 to 48 hours at 35°–37°C. Liquid media also should be incubated 45 to 48 hours at 35°–37°C. Media productivity testing may be accomplished by determining counts of *S. aureus* obtained in 18 to 24 hour broth cultures grown in a noninhibitory medium such as brain heart infusion (BHI) broth. Enumeration should be accomplished on noninhibitory solid plating media such as BHI agar. The isolation medium being tested for productivity should give counts not significantly less (20%) than the noninhibitory medium. Each prepared batch of medium should be streaked with known cultures of *S. aureus* to test for appropriate diagnostic characteristics, such as colony size and appearance, pigmentation and egg yolk reaction. Each lot of coagulase plasma or latex reagents should be tested with known cultures of *S. aureus* and *S. epidermidis* to determine the suitability of the plasma for distinguishing positive and negative reactions.

39.4 EQUIPMENT, REAGENTS AND MEDIA

39.41 Equipment and Supplies

Glass spreading rods: sterile, fire polished, hockey or hoe shaped, approximately 3 to 4 mm diameter, 15 to 20 cm long, with an angled spreading surface 45 to 55 mm long.

Drying cabinet or incubator for drying surfaces of agar plates and for checking thermonuclease positive colonies.

39.42 Reagents

1. Coagulase plasma containing EDTA. (Plasma derived from blood for which EDTA was used as the anticoagulant, or to which is added 0.1% EDTA (wt/vol).
2. Commercial latex reagents for slide agglutination tests; and
3. Gram stain reagents.

39.43 Media

Baird-Parker agar
Baird-Parker agar containing rabbit plasma fibrinogen
Baird-Parker agar without egg yolk
Brain heart infusion agar
Brain heart infusion broth
Giolitti & Cantoni broth
Pork plasma fibrinogen overlay agar
Rabbit plasma fibrinogen agar
Toluidine blue DNA agar
Trypticase soy or tryptic soy agar
Trypticase soy or tryptic soy broth
Trypticase soy or tryptic soy broth (double strength)
Trypticase soy or tryptic soy broth containing 20% NaCl
Trypticase soy or tryptic soy broth containing 10% NaCl and 1% sodium pyruvate.

39.5 PROCEDURES

39.51 Repair-Selective Enrichment Procedure [14]

This procedure is recommended for testing processed foods likely to contain a small population of injured cells.

Prepare food samples by the procedure described in Chapter 2. Transfer 50 mL of a 1:10 dilution of the sample into 50 mL of double strength trypticase soy broth. Incubate 3 hours at 35°–37°C.

Add 100 mL of single strength trypticase soy broth containing 20% NaCl. Incubate for 24 hours ± 2 hours at 35°–37°C.

Transfer 0.1 mL aliquots of culture to duplicate plates of a Baird-Parker agar and spread inoculum to obtain isolated colonies. Incubate the plates for 46 ± 2 hours at 35°–37°C.

Select two or more colonies suspected to be *S. aureus* (Section 39.54) from each plate and subject to coagulase test (Section 39.55) or clumping factor test (Section 39.56). Report results as *S. aureus* present or absent in 5 g of food, as indicated by results of coagulase or clumping factor testing.

39.52 Selective Enrichment Procedure[11,73]

This procedure is recommended for detecting small numbers of *S. aureus* in raw food ingredients and nonprocessed foods expected to contain a large population of competing species.

Prepare food samples by the procedure described in Chapter 2.

Inoculate three tubes of trypticase soy broth containing 10% NaCl and 1% sodium pyruvate at each test dilution with 1 mL aliquots of decimal sample dilutions. Maximum dilution of sample tested must be high enough to yield a negative endpoint. Incubate 48 ± 2 hr at 35°–37°C. Subculture all tubes showing growth.

Alternatively, inoculate 3 tubes of Giolitti & Cantoni broth at each test dilution with 1 mL aliquots of decimal sample dilutions. Pour a plug of agar cooled to 45° C onto the top of the medium in each inoculated tube and allow it to solidify to form a seal. Incubate at 35°–37° C. Observe after 24 and 48 hr. Subculture any tubes developing black precipitate.[73]

Subculture using a 3 mm inoculating loop. Transfer 1 loopful from each growth-positive tube to dried Baird-Parker agar plates. Vortex-mix tubes before streaking if growth is visible only on bottom or sides of tubes. Streak plates to obtain isolated colonies. Incubate 48 ± 2 hr at 35°–37°C.

From each plate showing growth, pick at least one colony suspected to be *S. aureus* (Section 39.54) and subject to coagulase

(Section 39.55) or clumping factor (Section 39.56) testing. Report most probable number (MPN) of *S. aureus*/gram from tables of MPN values (Chapter 6).

39.53 Surface Plating Procedure[11,74]

This procedure is recommended for the detection of *S. aureus* in raw, unprocessed food. The sensitivity of this procedure may be increased by using larger volumes (>1 mL) distributed over three plates.

Prepare food samples by the procedure given in Chapter 2. Plating of two or more decimal dilutions may be required to obtain plates with the desired number of colonies per plate.

For each dilution to be plated, aseptically distribute 1 mL of sample suspension on three plates of Baird-Parker agar (e.g., 0.4, 0.3 and 0.3 mL). Spread the inoculum over the surface of the agar using sterile, bent spreading rods. Avoid the extreme edges of the plate. Keep the plates in an upright position until the inoculum is absorbed by the medium (about 10 min on properly dried plates). If the inoculum is not readily absorbed, plates may be placed in an incubator in an upright position for about 1 hr before inverting. Invert plates and incubate 45 to 48 hr at 35°–37°C.

Select a plate containing 20 to 200 colonies unless plates at only lower dilutions (>200 colonies) have colonies with the typical appearance of *S. aureus*. If several types of colonies are observed that appear to be *S. aureus*, count the number of colonies of each type and record counts separately. When plates at the lowest dilution plated contain less than 20 colonies, they may be used. If plates containing over 200 colonies have colonies with the typical appearance of *S. aureus* and typical colonies do not appear on plates at higher dilution, use these plates for enumeration of *S. aureus*, but do not count atypical colonies.

Select one or more colonies of each type counted and test for coagulase production (Section 39.55) or clumping factor (Section 39.56). Coagulase or clumping factor positive cultures may be considered *S. aureus*.

Add number of colonies on triplicate plates represented by colonies giving a positive coagulase or clumping factor test and multiply by the sample dilution factor. Report this number as number of *S. aureus*/g of product tested.

39.54 Description of S. aureus Colonies on Baird-Parker agar[18,11]

S. aureus colonies are typically circular, smooth, convex, moist, 2–3 mm in diameter on uncrowded plates, gray-black to jet-black, frequently with light-colored (off-white) margin, surrounded by opaque zone (ppt), and frequently with outer clear zone; colonies have buttery to gummy consistency when touched with inoculating needle. Occasional non-lipolytic strains my be encountered which have same appearance, except that surrounding opaque and clear zones are absent. Colonies isolated from frozen or desiccated foods which have been stored for extended periods are frequently less black than typical colonies and may have rough appearance and dry texture.

39.55 Coagulase Test[11]

With sterile needle transfer colonies to tubes containing 0.2 mL BHI broth and to trypticase or tryptic soy agar slants (TSA). Incubate culture suspensions and slants 18 to 24 hr at 35°–37°C. Keep slant cultures at ambient temperature for ancillary or repeat tests in case the coagulase test results are questionable.

Add 0.5 mL coagulase plasma with EDTA to 0.2 mL of each broth culture tube and mix thoroughly. Incubate at 35°–37°C and examine periodically during a 6 hr interval for clot formation. A 3 + or 4 + clot formation (Figure 1) is considered a positive reaction for *S. aureus*.[75,76] Small or poorly organized clots (1+ and 2+) should be confirmed by performing the ancillary tests listed below. Recheck doubtful coagulase test results on broth cultures that have been incubated at 35°–37°C for more an 18 hr but less than 48 hr. Assure culture purity before rechecking coagulase test results. Do not store rehydrated plasma longer than 5 days (at 2° to 8°C).

39.56 Clumping Factor by Latex Agglutination[50]

Transfer a loopful of growth from an 18 to 24 hr trypticase or tryptic soy agar slant or 1 or more colonies from Baird-Parker agar to a control circle and to test circle on kit reaction card. Add 1 drop control latex to control circle. Add 1 drop test latex reagent to test circle. Mix contents of control circle with inoculation loop or wooden stick and then using the same loop or stick, mix contents of test circle. Gently rock reaction card back and forth by hand for about 1 minute. Look for agglutination and significant clearing of milky background under ambient light. Test positive and negative controls simultaneously with test cultures. Test is positive if agglutination is observed in test circle and suspension in control circle remains homogeneous. Test is negative if no agglutination is observed in test circle. Test is uninterpretable if agglutination is observed in both control and test circles. Strains showing uninterpretable latex reaction must be confirmed by additional tests as in 39.6. Alternatively, use directions with test kit.

39.57 Direct Enumeration of Coagulase Positive S. aureus

This method is specifically for use with all types of cheese, milk and dairy desserts.[74] For each dilution of product, aseptically transfer 1 mL into sterile 90 mm petri dish. Pour 10 mL melted rabbit plasma fibrinogen agar into each dish. Immediately, after pouring the medium, mix the medium and inoculum by five to and fro movements followed by five clockwise movements, followed by five circular clockwise movements, followed by five to and fro movements at right angles to the first set followed by five anti-clockwise movements, then allow to set. Incubate inverted at 35°–37°C for 48 hr. On Rabbit Plasma Fibrinogen agar, *S. aureus* forms grey to black colonies surrounded by an opaque or cloudy zone indicating coagulase activity. Use plates for enumeration that have between 10 and 100 typical colonies. Average number of

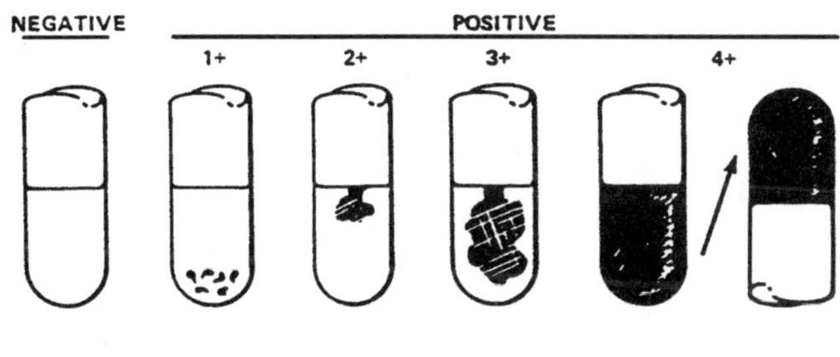

Figure 1. *Types of Coagulase Test Reactions*

typical colonies per dilution and muliply by the sample dilution factor. Report as *S. aureus* per g or mL of product tested.

39.58 Direct Enumeration of Coagulase and Thermonuclease Positive S. aureus

This procedure is recommended for raw or processed foods. The sensitivity of the procedure may be increased by plating larger inoculum volumes (>1 mL) distributed over three or more plates.

Prepare two or more decimal dilutions of food. Spread 1 mL of sample suspension of each dilution equally over three plates of Baird-Parker agar without egg yolk[77] or Baird-Parker agar containing rabbit plasma-fibrinogen tellurite[32,78] to which 0.5 mL of 20% sodium pyruvate[79] was added just prior to use and then dried by incubating at 50°C for 1 hr. Keep the plates in an upright position until the inoculum is completely absorbed at 35°–37°C.

If Baird-Parker agar containing no egg yolk is used, dispense 8 mL of tempered pork plasma-fibrinogen overlay agar[77] onto each plate. While this overlay is poured, the plates must be on a horizontal surface. After the overlay agar solidifies, invert and incubate plates for 45 to 48 hr at 35°–37°C.

Select plates containing 20 to 200 colonies and count all black colonies showing the opaque fibrin halos (coagulase-positive) surrounding the colonies.[32,77,78] Incubate these plates at 65°C for 2 hr and then overlay each plate with 10 mL of melted toluidine blue DNA agar and let solidify. After solidification, incubate plates at 35°–37°C for 4 hr.

Count all colonies showing pink halos against blue background as thermonuclease-positive. Add all colonies that showed both fibrin halos (coagulase-positive) and pink (thermonuclease-positive) halos, multiply by the sample dilution factor and report as *S. aureus* per g or mL of product tested.

39.6 ADDITIONAL TESTS

39.61 S. aureus Speciation

If anomalies are encountered during testing, additional testing may be required to establish speciation of *S. aureus*. The following tests are usually adequate:

A gram stain of *S. aureus* cultures will produce gram-positive cocci, 0.8 to 1.0 µm in diameter, occurring singly, in pairs or most typically in irregular clusters resembling clusters of grapes.

Emulsify growth from TSA slant in 1 drop 3% hydrogen peroxide on a glass slide. Immediate bubbling is a positive catalase test. Cultures of *S. aureus* are catalase-positive.

Boil a portion of culture grown in BHI broth for 15 min and use for thermonuclease test. Cut 2-mm or larger wells in toluidine blue DNA agar plates and fill with boiled culture growth using a Pasteur pipette. Touching the bottom of the well will usually draw enough liquid to fill the well to level; if not, retouch the liquid with the pipette until the well fills. It may be necessary to refill the pipette before retouching. Trial test will indicate how much liquid to fill the pipette with prior to touching the bottom of the well. Incubate plates at 35°–37°C for 4 hr or 50°C for 2 hr. Pink halos extending 1 mm beyond the well are considered positive for thermonuclease and considered to be *S. aureus*. Include positive and negative controls using *S. aureus* (ATCC 12600) and *S. epidermidis* (ATCC 14990), respectively.

Use unboiled culture growth utilized for coagulase test. Mix 0.1 mL of cell suspension with 0.1 mL of lysostaphin (dissolved in a 0.02 M phosphate buffer containing 2% NaCl) to give a final concentration of 25 µg lysostaphin/mL. To another portion of 0.1-mL cell suspension, add 0.1 mL of phosphate buffer with NaCl (negative control). Also include *S. aureus* (ATCC 12600) as a positive control and *Micrococcus varians* (ATCC 15306) as a negative control in the assay. Partial or complete clearing of cell turbidity in test and positive cultures with no clearing in negative control is considered positive. *S. aureus* is lysed (clearing) by lysostaphin.

Follow the procedures recommended by the Subcommittee on Taxonomy of Staphylococci and Micrococci. Include controls of *S. aureus* (ATCC 12600) and *M. varians* (ATCC 15306). *S. aureus* will utilize glucose and usually mannitol anaerobically; *M. varians* will not.

Rapid identification kits available commercially may also be used. However, care must be taken in interpreting results of non-typical strains.

Bacteriophage typing of *S. aureus* by the methods of Blair and Carr[83] or Blair and Williams[84] may be useful in elucidating the epidemiology of staphylococcal food poisoning outbreaks. A basic set of 22 phages for typing cultures of human origin is recommended by the Subcommittee on Phage-Typing of Staphylococci.[85]

39.7 GENERAL CONSIDERATIONS FOR DETECTION OF STAPHYLOCOCCAL ENTEROTOXINS

The amount of enterotoxin required to cause illness in humans is not known. However, information from food poisoning outbreaks[86,87] and human challenge studies[88] indicate that individuals experiencing illness probably consumed at least 100 ng of enterotoxin A, the serotype most frequently involved in foodborne staphylococcal illness.[89]

The minimum level measurable with the microslide gel double diffusion technique is 30 to 60 ng of enterotoxin per 100 g of food; chromatographic and concentration procedures must be used before serological assay.[90] The microslide test method is approved by the AOAC International[11] and is the guide for testing new methods. It should also be the guide for those using the methods presented in this chapter.

A number of methods employing specific antibodies have been used to identify and measure enterotoxins. For food extracts, the method used should be equal in sensitivity (about 0.05 µg enterotoxin per mL) to the microslide method, which is used to identify enterotoxin in extraction-concentration procedures and is the method used by the Food and Drug Administration (FDA).[80] To obtain this sensitivity, extracts must be concentrated from 100 g of food to about 0.2 mL. Any method less sensitive than the microslide is inadequate. Methods such as radioimmunoassay (RIA), agglutination, and enzyme-linked immunosorbent assay (ELISA) require less concentration or no concentration of the food extracts; thus they are less time-consuming and more sensitive.

The reversed passive hemagglutination method[91] presents two main problems: It is impossible to absorb enterotoxin antibodies from all of the antisera preparations onto red blood cells, and some food materials produce nonspecific agglutination cells. However, subsequently developed methods using latex[92] as a substitute for red blood cells appear promising as a serological tool for the identification of the staphylococcal enterotoxins.

The RIA method[93,94,95] has not been used widely in the routine microbiological laboratory because of the need for radioactively labeled purified toxins and the handling of radioactive materials; however, it may still be used in specialty and research laboratories. Several ELISA methods[96,97,98,99,100,101,102] have been proposed to identify enterotoxins in foods, although except for the polyvalent ELISA, their specificity has not been studied exhaustively. As a result, they should be used with the recommended controls. Of the ELISA methods proposed, the "sandwich" ELISA has been the method of choice, and the reagents necessary for its use are commercially available. Several of these methods are presented in this chapter.

39.71 Enterotoxigenicity of Staphylococcal Strains

Examining staphylococci for enterotoxin production is helpful for identifying enterotoxin in foods and desirable for examining strains isolated from various sources. The methods outlined here are designed to determine the minimum amount of enterotoxin produced by a strain that could cause food contamination. A relatively simple and easily performed method is the membrane-over-agar[103] technique for enterotoxin production (see section 39.911) and the optimum sensitivity plate (OSP)[104] method for serological assay. The amount of enterotoxin produced with the membrane-over-agar method is adequate for testing with the OSP, although it is not as sensitive as the microslide. This system gives results equivalent to those obtainable with the monkey feeding test in designating a strain as enterotoxigenic.[105]

An alternative method for the production of toxin is the sac culture method of Donnelly et al.[106] Although it is the best method for the production of all the staphylococcal enterotoxins, this method is somewhat more cumbersome and time consuming to set up than the membrane-over-agar method.[87] The OSP method can be used for determination of the enterotoxins with this method of production.

The semisolid agar method is an AOAC International-approved method and is used by the FDA for the production of enterotoxin. It is simple to perform and requires a minimum of items commonly found in the routine analytical laboratory. A smaller amount of enterotoxin is produced by this method than by the other two methods, and the microslide or the more sensitive rapid methods must be used to identify adequate levels of enterotoxin in order to classify a strain as enterotoxigenic. Although the microslide is more sensitive than the OSP, it is more difficult for inexperienced operators to achieve consistent results because it is a micro system. To determine the presence of enterotoxin in culture fluid, latex agglutination or ELISA methods can be applied as well, although commercial kits generally recommend broth media, which are comparable in enterotoxin production to the semisolid agar.[107] It should be remembered, however, that *S. aureus* in pure culture may occasionally produce substances that react nonspecifically with the immunoglobulin used.

The enterotoxigenicity of staphylococcal strains may also be determined by DNA hybridization techniques.[108] The nucleotide sequences of enterotoxin serotypes A, B, and C have been determined[109,110,111] therefore, oligonucleotides can be synthesized and used as DNA probes to demonstrate the potential expression of toxin production by an organism.

39.72 Enterotoxin in Foods

The major problem in identifying enterotoxin in foods is the small amount that may be present in foods incriminated in food poisoning outbreaks. Marketable foods should contain no enterotoxin. Toxins can be identified if the counts are, or at some time were, $\geq 10^6$ staphylococci cells/g. Such high counts are not acceptable; therefore, instead of routinely testing products for the presence of toxins, rules of good manufacturing practices emphasize avoiding contamination and outgrowth of *S. aureus*. An additional problem may occur with pasteurized and thermally processed foods if toxins are rendered serologically inactive during processing.[112]

Methods have been developed and evaluated to restore serological activity to heat-altered toxin in extracts of heat-processed foods.[90,113,114,115,116,117,118,119] However, some current toxin detection assays are sensitive enough to detect unaltered toxin that may persist after heat without such treatment if relatively large amounts of toxin are present. To identify small amounts of toxin, a very sensitive procedure or a satisfactory means of concentrating the food extract must be available. The sensitivity of the RIA and ELISA is such that food extracts seldom need to be concentrated; however, with the microslide, the extract must be concentrated from 100 g (or less) of food to 0.2 mL. At the same time, interfering substances must be removed from the extract.[90]

Two methods are presented for extraction and concentration of the enterotoxin in foods: the method of Casman and Bennett developed for use by the FDA[80] with minor modifications,[120] requiring 5 to 6 days to complete (including serological testing); and the method developed by the Food Research Institute to shorten the assay time. With this method, results equivalent to those of Casman and Bennett can be obtained on the third day. If latex agglutination or ELISA methods are used to identify toxins in foods, extraction procedures are more simplified.

39.8 HANDLING OF SAMPLES

Food to be analyzed should be kept refrigerated or frozen and should not be allowed to stand at ambient temperature except during use. This is particularly true of foods that contain live organisms. Foods involved in poisoning outbreaks should be collected and refrigerated as soon as possible to avoid mishandling.

39.9 EXAMINING STAPHYLOCOCCAL ISOLATES FOR ENTEROTOXIN PRODUCTION

Determining the enterotoxigenicity of *S. aureus* isolated from food, food ingredients, or the food processing environment can be a significant step in predicting the toxin serotype (A–E) in foods incriminated in foodborne intoxications. A number of methods[104,106,107] have been developed for the laboratory production of the staphylococcal enterotoxins. The membrane-over-agar,[104] sac culture,[106] and semisolid agar[107] methods are recommended for obtaining culture fluid for testing. Some commercial kits propose the use of simple broth such as brain heart infusion (BHI). The semisolid agar method is the one approved by the AOAC International[11] for the laboratory production of the staphylococcal enterotoxins. The isolates are grown, the cells are removed by centrifugation, and the culture fluid is examined to determine whether enterotoxin has been produced by the optimum sensitivity method,[104] microslide gel double diffusion, reversed passive latex agglutination, or the ELISA (Tecra Diagnostics, Roseville, N.S.W. 2069, Australia) and is distributed by International Bioproducts, Inc., Redmond, WA). The ELISA has performed well in studies evaluating enterotoxin presence in a variety of foods.[141,142,143,144] Tecra Diagnostics also has a monovalent ELISA (Figure 2) SET kit. Follow manufacturers directions for analyzing samples.

39.91 Semisolid Agar[107] Microslide Method[121] (AOAC International-Approved Method)

39.911 Enterotoxin Production (Semisolid Agar)

1. **Special equipment, supplies, and media**
   a. *Test tubes.* Test tubes (25 × 200 mm) are used to sterilize medium in 25-mL lots. Tubes containing medium may be stored until needed.
   b. *Centrifuge*

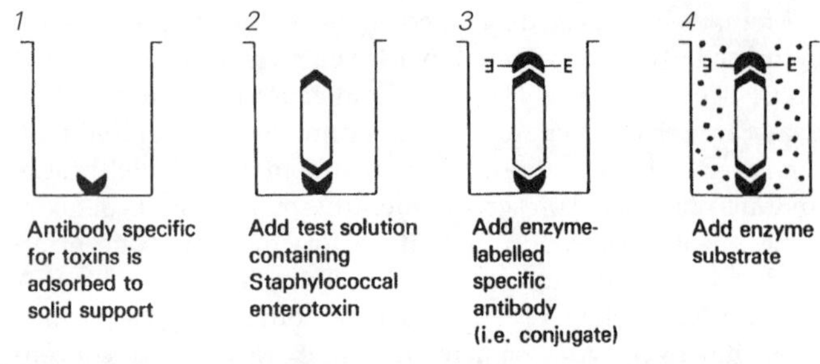

Figure 2. *Staphylococcal Enterotoxin (SET) Visual Immunoassay. ELISA performed in a double antibody "sandwich" configuration. Capture antibodies specific for SET types A-E absorbed to plastic microtiter well; in diagram 1; addition of preparation containing the suspect SET in diagram 2; addition of the enzyme-labeled specific antibodies to enterotoxins A-E in diagram 3; addition of enzyme substrate in diagram 4 with enzyme action on the substrate if toxin is present in sample.*

c. *Centrifuge tubes*
d. *Agar*
e. *Media.* The culture medium normally used is BHI broth, although other media such as 3% N-Z Amine A plus 1% yeast extract are satisfactory.
f. *Barium chloride and sulfuric acid.* $BaCl_2$ and H_2SO_4 are used to make the No. 1 McFarland standard for inoculum comparison and standardization.

2. **Preparation of materials**
   a. *Agar medium.* Add 0.7% agar to BHI broth at pH 5.3 (0.7 g/100 mL). Dissolve agar by minimal boiling. Distribute medium in 25-mL quantities in test tubes (25 × 200 mm) and autoclave at 121°C for 10 min. Store the medium in test tubes. Pour it aseptically into petri dishes (15 × 100 mm). (Alternative procedure: Dissolve agar by autoclaving batchwise and pour 25-mL quantities into petri dishes.)
   b. *Turbidity standard.* Prepare turbidity standard No. 1 of the McFarland nephelometer scale[122] by mixing 1% $BaCl_2$ with 99 parts of 1% H_2SO_4 in distilled water.
3. **Production of enterotoxin**
   a. *Inoculum.* Pick representative colonies (5 to 10 for each culture), transfer each to nutrient agar (or comparable medium) slant, and grow 18 to 24 hr at 35°–37°C. Add loopful of growth from agar slant to 3 to 5 mL of sterile distilled water or saline. Turbidity of the suspension should be approximately equivalent (by visual examination) to the turbidity standard (approximately 3×10^8 organisms/mL). Spread four drops of aqueous suspension over entire surface of agar medium with a sterile spreader.
   b. *Incubation.* Incubate plates at 35°–37° C for 48 hr (pH of culture should be approximately 8.0 or higher).
   c. *Enterotoxin recovery.* Transfer contents of petri dish to a 50-mL centrifuge tube with an applicator stick or equivalent. Centrifuge 10 min at 32,000 × g. Test the supernatant fluid for enterotoxin using the microslide method (39.912) or by other methods.

39.912 Enterotoxin Testing (microslide)

1. **Special equipment, supplies, and reagents**
   a. *Electrician's tape.* Although either Scotch Branch vinyl plastic electrical tape No. 33, 3M Company or Homart plastic tape 3/4 inch wide (Sears, Roebuck and Company) are recommended, any good-quality electrician's plastic tape should be satisfactory. The tape must be able to stick to the glass slides and not be readily removable with repeated washings.
   b. *Plexiglass template.* The templates are made from plexiglass (Figure 3).
   c. *Silicon grease.* Use silicone grease (Dow Corning), silicone lubricant spray (available at hardware stores), Lubriseal, or similar lubricant to coat the template so that it can be removed from the microslide after development without disrupting the layer of agar on which it is resting.
   d. *Petri dishes.* The 20 × 150-mm size is convenient for incubation of slides.
   e. *Platform for filling templates.* A Cordis Laboratories viewer with frosted glass over the lighted area is ideal for observing the presence of bubbles and removing them from the wells.
   f. *Capillary pipettes.* Pasteur capillary pipettes (9-inch) are essential for applying the enterotoxin reagents and food materials to microslides.
   g. *Incubator.* Any incubator or storage device that can be held constant at temperatures from 25°–37° C will suffice.
   h. *Fluorescent lamp.* Any fluorescent desk lamp to which the microslide can be held at an oblique angle is adequate.
   i. *Staining equipment.* A Wheaton horizontal staining dish with removable slide rack, or Coplin jars can be used to stain microslides.
   j. *Enterotoxins.* Crude enterotoxin preparations are adequate for the microslide test as long as only one enterotoxin precipitate line is obtained with the antiserum specific to that enterotoxin. These preparations may be obtained from Toxin Technology, Inc., Sarasota, Florida 34231.
   k. *Enterotoxin antisera.* Specific antisera to each of the enterotoxins that give only one line in the microslide with the respective crude enterotoxin are necessary. Antisera may be obtained from Toxin Technology, Inc. (See j. above).
   l. *Agar.*
   m. *Thimersol* (Merthiolate; ethylmercurithiosalicylate).

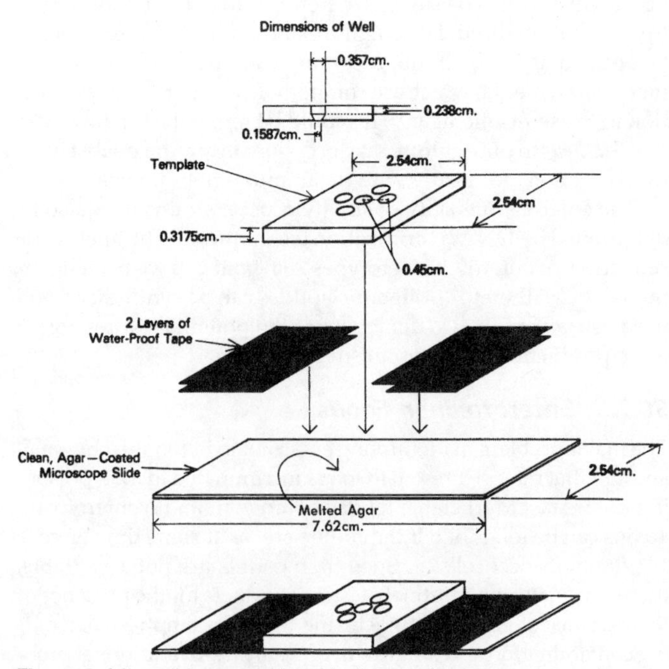

Figure 3. *Microslide assembly with diagram for preparation and specifications for plastic template.*

n. *Sodium barbital.* Any reagent-grade sodium barbital may be used.
o. *Thiazine Red-R.* This stain enhances precipitate lines in the microslide and can be obtained from a number of sources. Color index No. 14780: Matheson, Coleman and Bell, Norwood, OH 45212. An alternative stain, which some investigators prefer to Thiazine Red-R is Woolfast Pink RL: American Hoechst Corp., Bridgewater, NJ 05507.
p. *Synthetic sponge* (strips H_2O saturated)

2. **Preparation of standards**
   a. *Enterotoxins.* Dilute enterotoxin preparations according to specific instructions supplied by the manufacturer. Reagents must be balanced, i.e., the concentration of the enterotoxin must be adjusted to that of the antiserum, so that a line of precipitate will appear approximately halfway between the antigen and antiserum wells in the microslide. If they are out of balance, lines may not appear or may be difficult to observe. The solution used to dissolve and dilute the enterotoxin should be prepared according to the manufacturer's instructions.
   b. *Enterotoxin antisera*
3. **Recommended controls**
   a. Enterotoxin reagents should be used at concentrations that give a visible line halfway between the antigen and antibody wells. The test can be made more sensitive by reducing the concentration of the reagents so that a line is visible only after staining or enhancement. The amount of enterotoxin necessary to give visible lines varies with the antisera, but normally a visible line can be obtained with 0.25 to 0.50 μg enterotoxin/mL of enterotoxin solution. It is essential that, under the conditions of the test, only one line be observed with the control reagents. A set of slides should be prepared that contains the antiserum and reference enterotoxin only.
4. **Procedures**
   a. *Agar for coating slids.* Add 2 g of agar to 1 liter of boiling distilled water and heat with stirring until solution becomes clear. Pour 30 mL into 6-oz prescription bottles or other suitable containers. Store at ambient temperature. The agar can be remelted until all is used.
   b. *Gel diffusion agar for slides.* Add 12 g agar to 988 mL boiling sodium barbital saline buffer (0.9% NaCl, 0.8% sodium barbital, 1:10,000 merthiolate) adjusted to pH 7.41; continue boiling until agar dissolves, filter quickly with suction through two layers of filter paper, and store in 15- to 25-mL quantities in 4-oz prescription bottles. Remelting more than twice may break down the purified agar.
   c. *Slides.* Wash microscope slides, 3 × 1 inch, in detergent solution, rinse thoroughly in tap water followed by distilled water, air dry, and store in dust-free containers. Slides must be scrupulously clean or the agar will not adhere to the glass and will tear when the template is removed.
   d. *Taping the slides.* Wrap a 9.5–10.5-cm length (length needed will depend on how much the operator stretches the tape in applying it) of black plastic electrician's insulating tape twice around each end of the slide, about 0.5 cm from each end with a 2 cm space between the two strips. Stretch the tape slightly and press down firmly while wrapping it around the slide to avoid air bubbles. The finishing edge should end where the starting edge begins.
   e. *Rewashing slides.* Wash slides after taping as described in section 39.912, 4d.
   f. *Precoating.* Wipe area between tapes with 95% ethanol, using a small stick with piece of absorbent cotton twisted on the end. Cover surface between tapes with two drops of 0.2% agar (held at approximately 100°C with a 1-mL pipette, and rotate slide to cover surface evenly. If agar forms beads instead of spreading evenly, rewash the slide. Air dry slides on a flat surface in a dust-free atmosphere.
   g. *Petri dishes for slides.* Place two strips of synthetic sponge (approximately 1.2 × 1.2 × 6.5 cm) or two 7.5 cm strips of synthetic sponge opposite each other around the periphery of 15-cm petri dishes. Saturate strips with distilled water, pouring off any excess. Dishes will hold up to four slides and can be used repeatedly as long as the sponge or cotton is kept saturated with water.
   h. *Template.* Spread a thin uniform film of silicone grease or other suitable lubricant on bottom of template. If too little lubricant is applied, the template cannot be removed easily from the agar layer; if too much is applied, the template will not stay firmly in place and precipitate lines will be distorted. If using silicone spray, spray the silicone of a piece of cotton on the end of a stick and wipe bottom surface of template, keeping coat as thin as possible. Before reusing templates, wash them in a warm detergent solution with a piece of cheesecloth to remove silicone, being careful not to nick or scratch the template. Rinse templates thoroughly with tap and distilled water to remove any traces of detergent.
   i. *Preparation of microslide (see Figure 2).* Place 0.45 to 0.50 mL of 1.2% melted agar (held at 80°–90°C) on precoated slide area between the tape and immediately lay the silicone-coated template on the agar by placing one edge on the edge of the tape on one side and bringing it down onto the edge of the tape on the other side. After the agar solidifies, place slides in petri dishes. Temperature of agar should be such that agar does not begin to harden before template can be put into place. Take care not to force agar into bottom of wells of the template.
   j. *Stain solution.* If using Thiazine Red-R stain, dissolve 100 mg Thiazine Red-R in 100 mL of 1% acetic acid. If using Woolfast Pink RL, dissolve 1 g in 100 mL of the following solution: 5% trichloracetic acid, 1% acetic acid, and 25% ethanol in distilled water.
   k. *Preparation of record sheet.* Draw pattern of holes of template in a notebook or use a rubber stamp of the hole pattern. Indicate materials that are placed in each well.
   l. *Addition of reagents and test materials to slides.* Partially fill a capillary pipette by capillary action with the solution to be added. Remove excess liquid by touching pipette to edge of tube. Slowly lower pipette into well to be filled until it touches agar surface. (This leaves a small drop of liquid in bottom of well.) Refill capillary pipette with more reagent, lower into well, and fill well to convexity. (Do not overfill because of the danger of mixing reagents from different wells.) Place antiserum in center well of templates, standard enterotoxin reagent in upper and lower wells, and unknowns in the other two wells. Figure 4 shows the reagent arrangement for the bivalent and monovalent systems.

      Apply reagents carefully to avoid formation of air bubbles in bottom of wells. All bubbles must be removed so that reagents make proper contact with agar layer. Remove bubbles by inserting end of heat-sealed capillary

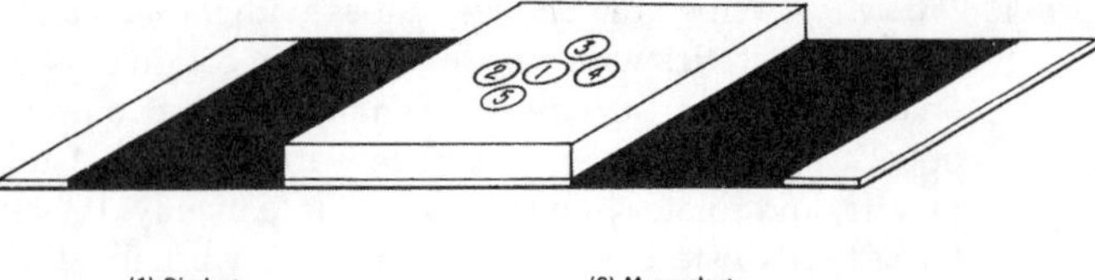

(1) Bivalent
1. Combination Antisera (e.g., Anti A and B)
2. Prepn under test
3. Ref. enterotoxin (e.g., Type A)
4. Prepn under test
5. Ref. enterotoxin (e.g., Type B)

(2) Monovalent
1. Antiserum (e.g., Anti A)
2. Dilns of prepn under test
3. Ref. enterotoxin (e.g., type A)
4. Dilns of prepn under test
5. Dilns of prepn under test

Figure 4. *Arrangement of antiserum (antisera) and homologous reference enterotoxin(s) when assaying preparation(s) under test for presence of two serologically distinct enterotoxins simultaneously (bivalent detection system) or when assaying dilutions of a preparation under test (monovalent detection system).*

pipette to bottom of well. This operation and filling of wells are best done against a lighted background. Place slides in petri dishes as soon as possible to avoid undue evaporation of liquid from wells. An alternative method is to add reagents to slides without removing them from petri dishes.

m. *Slide incubation.* Incubate slides for 48 to 72 hr at ambient temperature or for 24 hr at 35°–37°C.

n. *Reading the slide.* Remove template by sliding it to one side; clean slide with a momentary dip into water. Examine slide by holding it at an oblique angle against a fluorescent desk lamp. If precipitate lines are faint or none are visible, immerse slides in 1% cadmium acetate for 5 to 10 min, or immerse rinsed slide directly into Thiazine Red-R solution for 5 to 10 min. Alternative procedure: Place slides in a Wheaton jar filled with distilled water and extract with stirring for 30 min. Place slides in Woolfast Pink RL stain for 20 min at ambient temperature. Rinse off excess dye and destain in distilled water for 30 min or until stain is adequately washed out.[123]

o. *Interpretation of data.* Typical results (Figure 5): The control toxin should always give one precipitate line between antigen and antibody wells, as shown in Diagram 1, with toxin in wells 1 and 3, and water or unknowns containing no toxin in wells 2 and 4. The formation of a line by the unknown that joins with the control line, as illustrated in Diagrams 2 (wells 2 and 4) and 3 (well 2), shows that the unknown contains toxin at the concentration in the controls. The unknown in well 4 of Diagram 3 contains no toxin of the type present in the controls. The unknown in well 2 of Diagram 4 contains a smaller amount of enterotoxin than is present in the controls, and the unknown in well 4 contains a larger amount than is present in the controls. The unknown in well 2 of Diagram 5 contains much less toxin than is in the controls, the unknown in well 4 contains a much larger amount than is in the controls. Both unknowns in Diagram 6 contain more enterotoxin than is present in the controls, and both unknowns in Diagram 7 contain much more. In the latter case the unknown should be diluted and the slide rerun because the excess toxin has prevented the formation of the control line.

Atypical results: Occasionally results are difficult to interpret, especially for the less experienced operator (Figure 6). In Diagram 1, the reference lines do not meet or cross the lines produced by the unknown in wells 2 and 4. It is impossible to interpret these results positively; hence, the slide should be made again. Frequently extraneous lines are produced by unknowns, as shown in Diagrams 2 and 3. Lines from unknowns (wells 2 and 4) that join with the control lines (wells 1 and 3) show unknowns to be positive. The unknowns in Diagram 3 (wells 2 and 4) are negative because there are no joining lines. The extraneous lines are artifacts that are sometimes produced by unknowns. The results illustrated in Diagram 4 cannot be interpreted as positive because the lines extend beyond those produced by the controls, even though the latter lines do not cross the unknown lines. This type of slide should also be made again. The results shown in Diagrams 5 and 6 indicate channeling of the serum under the template because of improper contact between the template and the agar layer. The results shown in Diagram 5 indicate that the unknowns in wells 2 and 4 are positive because their lines join with the control lines. The results shown in Diagram 6 cannot be interpreted as positive for the unknown in well 4, and the slide should be made again. The partially double ring in Diagram 7 is a result of movement of the template during filling or development. The results indicate the presence of enterotoxin in the unknowns (wells 2 and 4), but to be certain the slide should be made again. The haze around the unknowns in Diagram 8 (wells 2 and 4) is occasionally encountered with food extracts. Such results are very dif-

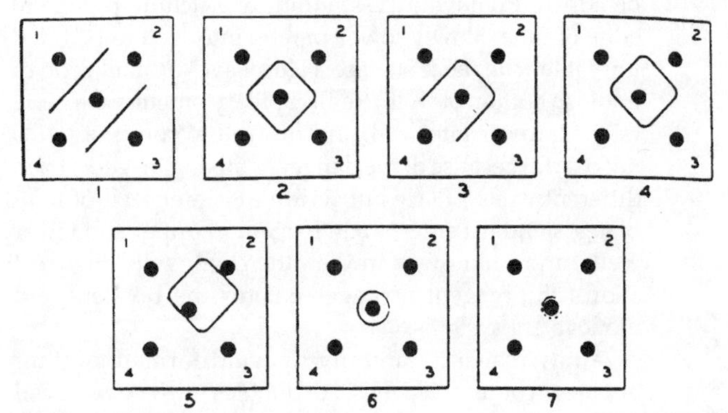

Figure 5. *Typical results obtained from the microslide test. See interpretation of data.*

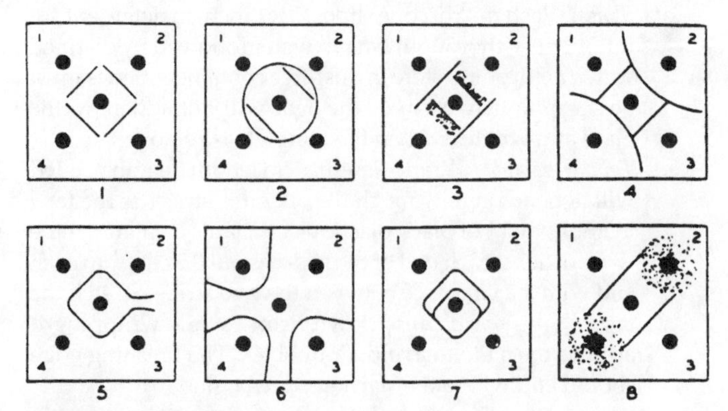

Figure 6. *Atypical results obtained with the microslide test. See interpretation of data.*

ficult, if not impossible, to interpret. The haze results from the presence of contaminating substances (usually proteins) remaining in the concentrated extract. This problem may be remedied by additional $CHCl_3$ extractions, or possibly longer treatment with trypsin.[123,124]

p. *Preserving the slide.* Thiazine Red-R staining: Rinse away any reactant liquid remaining on the slide by dipping the slide momentarily in water and immersing it for 10 min in each of the following baths: 0.1% Thiazine Red-R in 1% acetic acid, 1% acetic acid, 1% acetic acid, and 1% acetic acid containing 1% glycerol. Drain excess fluid from the slide, and dry it in a 35° C incubator if it is to be stored as a permanent record. After prolonged storage, lines of precipitation may not be visible until slide is immersed in water.

Woolfast Pink RL staining: Place slides in a Wheaton jar filled with distilled water and extract, stirring for 30 min. Place slides in Woolfast Pink RL stain for 20 min at ambient temperature. Rinse off excess dye and destain in distilled water for 30 min or until stain is adequately washed out.[124]

39.10 ENTEROTOXIN IDENTIFICATION IN FOODS

At present, the most sensitive precipitation method generally available for the identification of enterotoxin is the microslide, which is sensitive to 50 mg/mL. To identify the small amounts of enterotoxin that may be present in foods, concentrate the extract from 100 g of food to 0.2 mL. Make extractions with buffer solution and absorb the enterotoxin from the extract with ion exchangers. Elute the toxin from the exchangers and concentrate with polyethylene glycol to 0.2 mL of eluate. Two extraction-concentration methods are presented: the Reiser et al. method[105] (Section 39.101) and the Casman and Bennett method[90,120] (Section 39.102). Results obtained with the two methods are equivalent, but the Reiser et al. method requires 3 days, whereas the Casman and Bennett method requires 5 to 6 days.

Some recently developed rapid methods for identifying enterotoxins in foods described in this chapter with their own more simplified extraction procedures are the reversed passive latex agglutination; microtiter plate-ELISA, polystyrene ball-ELISA, polyvalent (A-E serotypes) visual ELISA, automated immunoanalyzer and the immunoenzymatic test.

39.101 The Reiser, Conaway, and Bergdoll Method[105]

For food extractions, add a small amount of enterotoxin (0.25 to 0.50 μg) per 100 g of uncontaminated food of the same type being examined for enterotoxin. This control will indicate whether the test is sensitive enough to detect the small amounts of enterotoxin that may be present. If known, the type of enterotoxin that may be present in the food can be used as the control. If not, then enterotoxin A is the recommended control. Crude enterotoxins can be used as the controls.

Prepare a 30% (w/v) polyethylene glycol (PEG) 20,000 MW solution by adding 3 g of PEG for each 7 mL of distilled water. Cut a piece of dialysis tubing (± inch flat width) long enough to accommodate the volume of food extract to be concentrated. Soak tubing in two changes of distilled water to remove glycerol coating. Tie one end of tubing with two knots close together. Fill tube with distilled water and test for leaks by squeezing the filled sac while holding the untied end tightly closed. Empty the sac and place it in distilled water until used.

39.1011 Special Equipment, Supplies, and Reagents

1. *pH meter.* The pH at which the extraction is done and the pH of the buffers used in the CG-50 extraction procedure are important. The pH adjustments usually are made within ± 0.1 pH unit.
2. *Omnimixer.* An Omnimixer (Sorvall) is convenient for grinding food directly in the stainless steel centrifuge tubes.
3. *Refrigerated centrifuge.* Food extracts are centrifuged at relatively high speeds at 5°C in a refrigerated centrifuge, such as a Sorvall RC-2B, which can reach speeds of 20,000 rpm. The lower the centrifuge speed, the more difficult the clarification of extracts.
4. *Centrifuge tubes.* 285-mL stainless steel centrifuge bottles (Sorvall No. 530)
5. *Magnetic stirrer.* A magnetic stirrer keeps test samples agitated during pH adjustments, dialysis, etc.
6. *Filter cloth.* At various stages in the procedures, food is filtered through several layers of coarse material, such as cheesecloth, placed in a funnel. Wetting the cheesecloth before placing it in the funnel reduces adherence of food to the cloth. The course material allows rapid flow with efficient removal of food particles, the chloroform layer, etc.
7. *Amberlite CG-50 ion exchange resin.* The extract is purified partially by absorption onto Amberlite CG-50 ion exchange resin (100 to 300 mesh) (Mallinckrodt Chemical Works, St. Louis, MO 63160).
8. *Polyethylene glycol.* Food extracts are concentrated with polyethylene glycol (Carbowax 20,000) (Union Carbide Corp., Chemical Division, Chicago, IL 60638).
9. *Lyophilizer.* The extract is finally concentrated by freeze-drying, which conveniently reduces the volume to 0.2 mL and completely recovers the extract.
10. *Dialysis tubing.* Cellulose casing of 1 inch flat width and an average pore diameter of 48 Angstrom units is used.
11. *Chloroform.* The food extract is treated with $CHCl_3$ (several times in some instances) to remove lipids and other substances that interfere with the concentration of the extract to small volumes.
12. *Trypsin.* The extract is treated with trypsin to eliminate extraneous proteins carried through the concentration procedures (crude trypsin Type II, Sigma Chemical Co., P. O. Box 14508, St. Louis, MO 63178).

39.1012 Recommended Controls[1]

For food extractions, add a small amount of enterotoxin (0.25 to 0.50 μg) per 100 g of uncontaminated food of the same type being examined for enterotoxin. This control will indicate whether the test is sensitive enough to detect the small amounts of enterotoxin that may be present. If known, the type of enterotoxin that may be present in the food can be used as the control. If not, then enterotoxin A is the recommended control. Crude enterotoxins can be used as the controls.

39.1013 Procedures

Prepare a 30% (w/v) polyethylene glycol (PEG) 20,000 MW solution by adding 3 g of PEG for each 7 mL of distilled water. Cut a piece of dialysis tubing (± inch flat width) long enough to accommodate the volume of food extract to be concentrated. Soak tubing in two changes of distilled water to remove glycerol coating. Tie one end of tubing with two knots close together. Fill tube with distilled water and test for leaks by squeezing the filled sac while holding the untied end tightly closed. Empty the sac and place it in distilled water until used.

Suspend 100 g resin in 1.5 liters of distilled water, adjust to pH 12 with 5 NaOH, and stir for 1 hr at ambient temperature. Let resin settle and decant supernatant fluid. Resuspend resin in distilled water. Resuspend resin in distilled water, adjust pH to 2.0 with 6 N HCl, and stir for 1 hr. Let resin settle and decant supernatant fluid. Resuspend resin in distilled water, let resin settle, and decant. Repeat this washing procedure until pH is that of the distilled water. Suspend resin in 0.0005 M sodium phosphate buffer at pH 5.6. If pH is below 5.4, add 5 N NaOH and stir until the pH is corrected.

39.1014 Extracting Enterotoxin from Food

1. Place a 100-g test sample of food in a 285-mL stainless steel centrifuge bottle with 140 mL distilled water and grind to even consistency (1 to 2 min) with a Sorvall Omnimixer.
2. Blend 10 mL 1.0 M HCl into the mixture (pH 4.5 to 5.0). For meat products, add only 5 mL of HCl; the pH should not drop below 4.5. If it does not, readjust with NaOH.
3. Centrifuge at 27,300 × *g* at 5° C for 20 min. Pour supernatant fluid (some of which may be quite viscous) into a beaker.
4. Adjust pH to 7.5 with 5 N NaOH. Extract with $CHCl_3$ (1 mL/ 10 mL test sample) by mixing with a magnetic stirrer for 3 to 5 min.
5. Transfer test sample to stainless steel centrifuge bottle, and centrifuge at 27,300 × *g* at 5° C for 20 min. Filter slowly through cheesecloth placed in a glass funnel to trap $CHCl_3$.
6. Adjust pH to 4.5 with 6 N HCl. Centrifuge as above to remove any precipitate. Remove supernatant fluid.
7. Adjust pH to 7.5 with 5 N NaOH and recentrifuge if any precipitate forms.
8. Add about 20 mL settled Amberlite CG-50 ion exchange resin, pH 5.4 to 5.9, per 100 g of original test sample. Measure resin by decanting buffer and transferring settled resin to a 20-mL beaker with a spatula. After enough resin has been removed from the stock supply, add decanted buffer back and store resin at 5° C.
9. Stir mixture with a magnetic stirrer at 5° C for 1 hr. Remove resin from the liquid by filtration through cheesecloth in a Buchner funnel (Coors No. 1) with suction.
10. Wash resin in funnel with 200 mL of a 1:10 dilution of 0.15 M sodium phosphate-buffered 0.9% NaCl, pH 5.9. Discard the wash.
11. Resuspend resin in 30 mL of 0.15 M Na_2HPO_4 in 0.9% NaCl. Adjust pH to 6.8 with 5 N NaOH.
12. Stir mixture with a magnetic stirrer at 5° C for 45 min to elute enterotoxin from resin. Filter with suction through filter paper in a Buchner funnel. Wash resin on filter paper with a small amount of phosphate NaCl buffer. Discard resin.
13. Into the eluate stir 1 g of purified agar and continue stirring at 5°C for 1 hr. Filter mixture through a glass fiber filter (Type E) (Gelman Instrument Co., P. O. Box 1448, Ann Arbor, MI) in a Buchner funnel. Discard agar.
14. Place eluate into a dialysis tube (3 cm flat width). Place dialysis tube into (w/v) PEG overnight.
15. Remove sacs from PEG and wash outside of sacs with warm tap water. Place sacs in warm tap water for 10 to 20 min to loosen toxin adhering to walls of sac.
16. Empty contents of sacs into 15 mL centrifuge tubes (Corex high speed, Corning No. 8441). Rinse sacs three times with distilled water, keeping final volume to 5 mL or less).
17. Add 0.5 mL of $CHCl_3$ to centrifuge tube and mix contents vigorously (a Vortex mixer serves this purpose very well.)
18. Centrifuge test sample at 34,800 × *g* at 5°C for 10 min. Decant water layer to a test tube (18 × 100 mm) and freeze dry.
19. Dissolve dried test sample in 0.2 mL of a 1% trypsin suspension in distilled water. Digest for at least 30 min at ambient temperature.
20. Test for enterotoxin by the microslide method.

39.1015 Precautions and Limitations

The food to be tested must be ground to a very fine consistency for adequate extraction of toxin. This is particularly important for foods such as cheese in which enterotoxin may be present throughout and not just on the surface. As much as possible of the nonsoluble food materials must be removed to eliminate their interference in the extraction procedures, such as in the CG-50 absorption process. The number of chloroform extractions needed to remove extraneous materials in the food depends on whether the extract can be concentrated to 0.2 mL and be of a consistency that can be applied readily to the microslide. One of the critical points is the ionic strength of the enterotoxin solution to which the CG-50 resin is added (Section 39.1011, item 7). Instructions should be followed closely; if the ionic strength is too high, the toxin will not be absorbed completely, or possibly not at all.

This method cannot be used to recover the enterotoxin quantitatively, particularly at the levels usually found in foods involved in food poisoning outbreaks. One need not be too concerned about the amount recovered as long as the quantity is sufficient to determine the presence of 0.125 to 0.250 μg enterotoxin in 100 g of food.

39.102 The Casman and Bennett Method[11,90,120] (AOAC International—Official Final Action Method)

39.1021 Special equipment, Supplies, and Reagents

1. **Refrigerated cabinet or cold ambient.** Part of the extraction procedures, such as the carboxnlmethyl cellulose (CMC) column extraction, is done at about 5°C, primarily because the column is allowed to run overnight. This type of space is useful for storing food materials and extracts, thus eliminating the need for a refrigerator.
2. **Waring blender.** The foods must be ground into a slurry for adequate extraction of enterotoxin.
3. **pH meter.** The pH at which the extraction is done is important, as is the pH of the buffers used in the CMC column. The pH adjustments usually are made within ± 0.1 pH unit.
4. **Refrigerated centrifuge** (Section 39.1011, item 3)
5. **Centrifuge tubes** (Section 39.1011, item 4)
6. **Magnetic stirrer** (Section 39.1011, item 5)
7. **Filter cloth** (Section 39.1011, item 6)
8. **Carboxymethyl cellulose.** The extract is purified partially by absorption onto CMC Whatman CM 22, 0.6 meq/g (H. Reeve Angel, Inc., Clifton, NJ). Soluble extractants are removed by this step.
9. **Polyethylene glycol** (Section 39.10.11, item 8)
10. **Lyophilizer** (Section 39.1011, item 9)
11. **Chromatographic tube.** The enterotoxin in the food is partially purified by using CMC, with elution in a chromatographic tube. For this purpose, a 19-mm inside diameter col-

umn, e.g., chromaflex plain with stopcock, size 234 (Kontes Glass Company, Vineland, NJ) is recommended.

12. **Dialysis tubing** (Section 39.1011, item 10)
13. **Separatory funnels.** Separatory funnels of various sizes are needed in the $CHCl_3$ extractions and with the chromatographic column.
14. **Glass wool.** Glass wool makes ideal plugs for use in the chromatographic columns.
15. **Chloroform.** The food extract is treated with $CHCl_3$ (several times in some instances) to remove lipids and other substances that interfere with the concentration of the extract to small volumes.

39.1022 Recommended Controls[1]

See Section 39.1012.

39.1023 Procedures

1. **Polyethylene glycol (PEG) solution** (Section 39.1011, item 8)
2. **Dialysis tubing** (Section 39.1011, item 10)
3. **Separatory funnel.** Attach a piece of latex tubing (approximately 24 inches) to the lower end of a 2-liter separatory funnel. Attach a piece of glass tubing in a No. 3 rubber stopper to the other end of the latex tubing. Place the separatory funnel in a ring stand above the chromatographic column.
4. **Carboxymethyl cellulose column.** Suspend 1 g of CMC in 100 mL of 0.005 M sodium phosphate buffer, pH 5.7, in a 250-mL beaker. Adjust pH of CMC suspension with 0.005 M H_3PO_4. Stir suspension intermittently for 15 min, recheck pH, and readjust to pH 5.7, if necessary. Pour suspension into the 1.9-cm chromatographic tube, and let CMC particles settle. Withdraw liquid from column through stopcock to within about 1 inch of surface of settled CMC. Place a loosely packed plug of glass wool on top of CMC. Pass 0.005 M sodium phosphate buffer, pH 5.7, through column until washing is clear (150 to 200 mL). Check pH of last wash. If pH is not 5.7, continue washing until the pH of wash is 5.7. Leave enough buffer in column to cover CMC and glass wool to prevent column from drying out.

39.1024 Extracting Enterotoxin from Food

1. Grind 100 g of food material in a Waring blender at a high speed for 3 min with 500 mL of 0.2 M NaCl.
2. Adjust pH to 7.5 with 1 N NaOH or HCl if food material is highly buffered, and 0.1N NaOH or HCl if food material is weakly buffered (e.g. custards). Let slurry stand for 10 to 15 min, recheck pH, and readjust if necessary.
3. Transfer slurry to two 285-mL stainless steel centrifuge bottles. Centrifuge at 27,300 × g for 20 min at 5°C. Lower speeds with longer centrifuging time can be used, but clearing of some food materials is not as effective. Separation of fatty materials is ineffective unless centrifugation is done at refrigeration temperature.
4. Decant supernatant fluid into 800-mL beaker through cheesecloth or other suitable filtering material placed in a funnel.
5. Reextract residue with 125 mL of 0.2 M NaCl by blending for 3 min. Adjust pH to 7.5 if necessary.
6. Centrifuge at 27,300 × g for 20 min at 5°C. Filter supernatant through cheesecloth, and pool filtrate with the original extract.
7. Place pooled extracts into a dialysis sac. Immerse sac into 30% PEG (W/V) at 5° C until volume is reduced to 15 to 20 mL or less (normally an overnight procedure).
8. Remove sac from PEG and wash outside thoroughly with cold tap water to remove any PEG adhering to sac. Soak in distilled water for 1 to 2 min and 0.2 M NaCl for 1–2 minutes.
9. Pour contents into small beaker. Rinse inside of sac with 2 to 3 mL of 0.2 M NaCl by running the fingers up and down the outside of the sac to remove material adhering to sides of tubing. Repeat rinsing until rinse is clear. Keep the volume as small as possible.
10. Adjust pH of extract to 7.5. Centrifuge at 32,800 × *g* for 10 min. Decant supernatant fluid into a graduated cylinder to measure volume.
11. Add extract with 1/4 to 1/2 volume of $CHCl_3$, to a separatory funnel. Shake vigorously 10 times through a 90-degree arc.
12. Centrifuge $CHCl_3$ extract mixture at 32,800 × *g* for 10 min at 5°C. Return fluid layers to separatory funnel. Draw off $CHCl_3$ layer from bottom of separatory funnel and discard.
13. Measure volume of water layer and dilute with 40 volumes of 0.005 M sodium phosphate buffer, pH 5.7. Adjust pH to 5.7 with 0.005 M H_3PO_4 or 0.005 M Na_2HPO_4. Place diluted solution in the 2-liter separatory funnel.
14. Place stopper (attached to bottom of separatory funnel) loosely into top with liquid from separatory funnel. Tighten stopper in top of tube and open stopcock of separatory funnel. Let fluid percolate through CMC column at 5° C at 1 to 2 mL per min by adjusting flow rate with stopcock at the bottom of column. Adjust flow rate so that percolation can be completed overnight. If all liquid has not passed through the column during the night, stop the flow when the liquid level reaches the glass wool layer. If all the liquid has passed through overnight, rehydrate the column with 25 mL of distilled water.
15. After percolation is complete, wash CMC column with 100 mL of 0.005 M sodium phosphate buffer (1 to 2 mL/min), stopping the flow when liquid level reaches glass wool layer. Discard wash.
16. Elute enterotoxin from CMC column with 200 mL of 0.05 M sodium phosphate buffer, pH 6.5 (0.05 M phosphate-0.05 M NaCl buffer, pH 6.5), at a flow rate of 1 to 2 mL/min at ambient temperature. Force last of liquid from CMC by applying air pressure to top of chromatographic tube.
17. Place eluate in a dialysis sac. Place sac in 30% PEG (W/V) at 5° C and concentrate almost to dryness.
18. Remove sac from PEG and wash (as in Section 39.1024, item 8). Soak sac in 0.2 M phosphate buffer, pH 7.4. Remove concentrated material from sac by rinsing five times with 2 to 3 mL of 0.01 M sodium phosphate buffer, pH 7.4 to 7.5.
19. Extract concentrated solution with $CHCl_3$ (as described in section 39.1024, item 11). Repeat $CHCl_3$ extractions until precipitate is so lacy that it falls apart in the $CHCl_3$ layer in the cheesecloth.
20. Place extract in a short dialysis sac (approximately 6 inches). Place sac in 30% PEG (W/V) and let it remain until all liquid has been removed from the sac (usually overnight).
21. Remove sac from PEG and wash outside with tap water. Place sac in distilled water for 1 to 2 min.
22. Remove contents by rinsing inside of sac with 1-mL portions of distilled water. Keep volume below 5 mL.

23. Place rinsings in a test tube (18 × 100 mm) or other suitable container and freeze dry.
24. Dissolve freeze-dried test sample in as small an amount of 0.2 M NaCl as possible (0.1 to 0.15 mL).
25. Check for enterotoxin by the microslide method.

39.1025 Precautions and Limitations

The food to be tested must be ground to a very fine consistency for adequate extractions of toxin. This is particulary important for foods such as cheese, in which enterotoxin may be present throughout and not just on the surface. As much as possible, the nonsoluble food materials must be removed to eliminate their interference later in the extraction procedures, such as in the CMC absorption process. The number of chloroform extractions needed to remove extraneous materials in the food depends on whether the extract can be concentrated to 0.2 mL and be of a consistency that can be readily applied to the microslides.

The method cannot be used to recover the enterotoxin quantitatively, particularly at the levels usually found in foods involved in food poisoning outbreaks. At the level of 1 µg or less per 100 g of food, the recovery is very low, probably 10 to 20%. One need not be too concerned about the amount recovered as long as the quantity is sufficient to determine the presence of 0.125 to 0.250 µg enterotoxin in 100 g of food.[90]

39.11 REFERENCES

1. Adesiyun, A. A., S. R. Tatini, and D. G. Hoover. 1984. Production of enterotoxin(s) by *Staphylococcus hyicus*. Vet. Microbiol. 9:487-495.
2. Anderson, J. E. 1996. Survival of the serological and biological activities of staphylococcal enterotoxin A in canned mushrooms. UMI Dissertation Services, Ann Arbor, Mich.
3. Andrews, G. P., and S. E. Martin. 1978. Modified Vogel and Johnson agar for *Staphylococcus aureus*. J. Food Prot. 41: 530.
4. Baird-Parker, A. C. 1962. An improved diagnostic and selective medium for isolating coagulase-positive staphylococci. J. Appl. Bacteriol. 25: 12.
5. Baird-Parker, A. C. and E. Davenport. 1965. The effect of recovery medium on the isolation of *Staphylococcus aureus* after heat treatment and after storage of frozen or dried cells. J. Appl. Bacteriol. 28: 390.
6. Batish, V. K., D. R. Ghodeker, and B. Ranganathan. 1978. The thermostable deoxyribonuclease (DNase) test as rapid screening method for detection of staphylococcal enterotoxin in milk and milk products. Microbiol. Immunol. 22: 437.
7. Bautista, L. and P. Taya. 1988. A quantitative study of enterotoxin production by sheep milk staphylococci. Appl. Environ. Microbiol. 54: 566.
8. Bennett, R. W. 1992. The biomolecular temperament of staphylococcal enterotoxins in thermally processed foods. J. Assoc. Off. Anal. Chem. 75:6.
9. Bennett, R. W. 1994. Urea renaturation and identification of staphylococcal enterotoxin. *In*: R. C. Spencer, E. P. Wrights and S. W. B. Newsom (eds.) RAMI-93. Rapid Methods and Automation in Microbiology and Immunology. Intercept Limited, Andover, Hampshire, England.
10. Bennett, R. W. 1996. Atypical toxigenic *Staphylococcus* and non-*Staphylococcus aureus* species on the Horizon? An update. J. Food Prot. 59:1123-1126.
11. Bennett, R. W., and V. Atrache. 1989. Applicability of visual immunoassay for simultaneous indication of staphylococcal enterotoxin serotype presence in foods. ASM Abstracts, 1989:323, p 28.
12. Bennett, R. W., and F. McClure. 1994. Visual screening with immunoassay for staphylococcal enterotoxins in foods: collaborative study. JAOAC Intern. 77:357.
13. Bennett, R. W., and M. R. Berry Jr. 1987. Serological reactivity and in vivo toxicity of *Staphylococcus aureus* enterotoxins A and D in selected canned foods. J. Food Sci. 52:416.
14. Bennett, R. W., and R. N. Matthews. 1995. Evaluation of polyvalent ELISA's for the identification of staphylococcal enterotoxin in foods. AOAC International Abstracts 1995:17-B-016.
15. Bennett, R. W., D. L. Archer, and G. A. Lancette. 1988. Modified procedure to eliminate elution of food proteins during seroassay for staphylococcal enterotoxins. J. Food Safety 9:135.
16. Bennett, R. W., K. Catherwood, L. J. Luckey, and N. Abhayaratn, N. 1993. Behavior and serological identification of staphylococcal enterotoxin in thermally processed mushrooms. *In*: S. Change, J. A. Buswell and S. Chiu (eds.). Mushroom Biology and Mushroom Products. Chapter 21 (p. 193-207). The Chinese University Press. Hong Kong.
17. Bennett, R. W., M. Ash, and V. Atrache. 1989. Visual screening with enzyme immunoassay for staphylococcal enterotoxins in foods: An interlaboratory study. AOAC Abstracts. 1989 p. 72.
18. Bergdoll, M. S. 1972. The enterotoxins, p. 301-331. *In* J. O. Cohen (ed.), The staphylococci.John Wiley and Sons, New York.
19. Bergdoll, M. S. and R. W. Bennett. 1984. Staphylococcal enterotoxins. In "Compendium of Methods for the Microbiological Examination of Foods," ed. M. L. Speck. P. 428-457. Am. Pub. Health Assn., Washington, D.C.
20. Bergdoll, M.S. 1990. Staphylococcal food poisoning. In Foodborne Disease. D.O. Cliver (Ed.) Academic Press, Inc. San Diego, CA. P. 86-106.
21. Betley, M. J., and J. J. Nekalanos. 1988. Nucleotide sequence of the type A staphylococcal enterotoxin gene. J. Bacteriol. 170:34.
22. Blair J. E., and M. Carr. 1960. The techniques and interpretation of phage typing of staphylococci. J. Lab. Clin. Med. 55:650.
23. Blair, J. E. and R. E. O. Williams. 1961. Phage typing of staphylococci. Bull. World Health Org. 24: 771.
24. Bohach, G. A., and P. M. Schlievert. 1987. Nucleotide sequence of the staphylococcal enterotoxin C1 gene and relatedness to other pyrogenic toxins. Mol. Gen. Genet. 209:15.
25. Boothby, J., C. Genigeorgis,. and M. H. Fanelli. 1979. Tandem coagulase/thermonuclease agar method for the detection of *Staphylococcus aureus*. Appl. Environ. Microbiol. 37: 298.
26. Borja, C. R. and M. S. Bergdoll. 1967. Purification and partial characterization of enterotoxin C produced by *Staphylococcus aureus* strain 137. Biochemistry 6:1467.
27. Borja, C. R., E. Fanning, I.-Y Huang, and M. S. Bergdoll. 1972. Purification and some physiochemical properties of staphylococcal enterotoxin E. J. biol. Chem. 247:2456.
28. Breckinridge, J. C., and M. S. Bergdoll. 1971. Outbreak of foodborne gastroenteritis due to a coagulase negative enterotoxin producing staphylococcus. N. Engl. J. Med. 248:541-543.
29. Brewer, D. G., S. E. Martin, and Z. J. Ordal. 1977. Beneficial effects of catalase or pyruvate in a most-probable-number technique for the detection of *Staphylococcus aureus*. Appl. Environ. Microbiol. 34: 797.
30. Brunner, K. G. and A. C. L. Wong. 1992. *Staphylococcus aureus* growth and enterotoxin production in mushrooms. J. Food Sci. 57:700.
31. Busta, F. F. and J. J. Jezeski. 1963. Effect of sodium chloride concentration in an agar medium on growth of heat-shocked *Staphylococcus aureus*. Appl. Microbiol. 11: 404.
32. Casman, E. P. and R. W. Bennett. 1963. Culture medium for the production of staphylococcal enterotoxin A. J. Bacteriol. 86:18.
33. Casman, E. P., Bennett, R. W., Dorsey, A. E., and Issa, J. A. 1967. Identification of a fourth staphylococcal enterotoxin—enterotoxin D. J. Bacteriol. 94:1875.
34. Casman, E. P., R. W. Bennett, A. E. Dorsey, and J. E. Stone. 1969. The microslide gel double diffussion test for the detection and assay of staphylococcal enterotoxins. Health Lab. Sci. 6:185.
35. Chang, T. C., and S. H. Huang. 1996. Efficacy of a latex agglutination test for rapid identification of *Staphylococcus aureus*: collaborative study. J. AOAC Int. 79:661-669.

36. Chapman, G. H. 1945. The significance of sodium chloride in studies of staphylococci. J. Bacteriol. 50: 201.
37. Chu, F. S., K. Thadhani, E. J. Schantz, and M. S. Bergdoll. 1966. Purification and characterization of staphylococcal enterotoxin A. Biochemistry 5:3281.
38. Collins, W. S., A. D. Johnson, J. F. Metzger, and R. W. Bennett. 1973. Rapid solid-phase radioimmunoassay for staphylococcal enterotoxin A. Appl. Micrbiol. 24:774.
39. Collins-Thompson, D. L., A. Hurst, and B. Aris. 1974. Comparison of selective media for the enumeration of sublethally heated food-poisoning strains of *Staphylococcus aureus*. Can. J. Microbiol. 20:1072.
40. Crisley, F. D., J. T. Peeler, and R. Angelotti. 1965. Comparative evaluation on five selective and differential media for the detection and enumeration of coagulase-positive staphylococci in foods. Appl. Micriobiol. 13:140.
41. Crisley, F. D., R. Angelotti, and M. J. Foter. 1964. Multiplication of *Staphylococcus aureus* in synthetic cream fillings and pies. Pub. Health Rep. 79:369.
42. Crowle, A. J. 1958. A simplified micro double-diffusion agar precipition technique. J. Lab. Clin. Med. 52:784.
43. Cunningham, L., B. W. Catlin, and De Garllhe, M. Privat. 1956. A deoxyribonuclease of *Micrococcus pyogenes*. J. Am. Chem. Soc. 78:4642.
44. Dangerfield, H. G. 1973. Effects of enterotoxins after ingestion by humans. Presented at the 73rd Annual Meeting of the American Society for Microbiology. May 6-11, Miami Beach, Fla.
45. Danielsson, M. L., and B. Hellberg. 1977. The biochemical activity of enterotoxin and nonenterotoxin producing staphylococci. Acta Vet. Scand. 18:266.
46. Devriese, L. A., V. Hajek, P. Oeding, S. A. Meyer, and K. H. Schleifer. 1978. *S. hyicus* (sompolinsky 1953) comb. nov. and *S. hyicus* subsp. chromogenes subsp. Nov. Ind. J. Syst. Bacteriol. 28:482.
47. DeWaart, J., D. A. A. Mossel, R. ten Broeke, and A. Van De Moosdyk. 1968. Enumeration of *Staphylococcus aureus* in foods with special reference to egg-yolk reaction and mannitol negative mutants. J. Appl. Bacteriol. 31:276.
48. Dickie, N., Y. Yano, C. Par, H. Robern, and S. Stavric. 1973. Solid-phase radio immunioassay of staphylococcal enterotoxins in food. Proceedings of Staphylococci in Foods Conference, Pennsylvania State University, University Park, Pa.
49. Donnelly, C. B., J. E. Leslie, L. A. Black, and K. H. Lewis. 1967. Serological identification of enterotoxigenic staphylococci from cheese. Appl. Microbiol. 15:1382.
50. Essink, A.W.G., S.J. M.W. Arkesteijn, and S. Notermans. 1985. Interference of lysozyme in the sandwich enzyme-linked immunosorbent assay (ELISA). J. Immunol. Methods 80:91.
51. Evenson, M. L., M. W. Hinds, R. S. Berstein, and M. S. Bergdoll. 1988. Estimation of human dose of staphylococcal enterotoxin A from a large outbreak in staphylococcal food poisoning involving chocolate milk. Int. J. Food Microbiol. 7:311.
52. Fey, H. 1983. Nachweis von Staphylokokken Enterotoxinen in Lebensmitteln. Schweiz. Gesellschaft fur Lebensmittelhygiene, schniffenreihe Heft 13:51.
53. Fey, H. and H. Pfister. 1983. A diagnostic kit for the detection of staphylococcal enterotoxins (SET) A, B, C, and D (SEA, SEB, SEC, SED). 2nd Int. Symp. Immunoenzym Tech., Cannes, 1983. *In* Avrameas et al. (ed.), Immunoezymatic techniques. Elsevier, Amsterdam.
54. Food and Drug Administration Bacteriological Analytical Manual. 1995, 8th ed., Chap. 12. AOAC Int.ernational, Gaithersburg, Md.
55. Freed, R. C., M. L. Evenson, R. F. Reiser, and M. S. Bergdoll. 1982. Enzyme-linked immunosorbent assay for detection of staphylococcal enterotoxins in foods. Appl. Environ. Microbiol. 44a:1349.
56. Giolitti, G., and C. Cantoni. 1966. A medium for the isolation of staphylococci from foodstuffs. J. Appl. Bacteriol. 29:395.
57. Gray, R. J. H., M. A. Gaske, and Z. J. Ordal. 1974. Enumeration of thermally stressed *Staphylococcus aureus* MF 31. J. Food Sci. 39:844.
58. Hajek, V. 1976. *Staphylococcus intermedius*, a new species isolated from animals. Int. J. Syst. Bacteriol. 26:401.
59. Harvey, J,. and A. Gilmour. 1985. Application of current methods for isolation and identification of staphylococci in raw bovine milk. J. Appl. Bacteriol. 59:207-221.
60. Hauschild, A. H. W., C. E. Park, and R. Hilheimer. 1979. A modified pork plasma agar for the enumeration of *Staphylococcus aureus* in foods. Can. J. Microbiol. 255:1052.
61. Heidelbaugh, N. D., D. B. Rowley, E. M. Powers, C. T. Bourland, and J. L. McQueen. 1973. Microbiological testing of skylab foods. Appl. Microbiol. 25:55.
62. Holbrook, R., J. M. Anderson, and A. C. Baird-Parker. 1969. The performance of a stable version of Baird-Parker's medium for isolating *Staphylococcus aureus*. J. Appl. Bacteriol. 32:187.
63. Hurst A. 1977. Bacterial injury:A review. Can. J. Microbiol. 23:935.
64. Hurst, A., and A. Hughes. 1978. Stability of ribosomes of *Staphylococcus aureus* S6 sublethally heated in different buffers. J. Bacteriol. 133:564.
65. Hurst, A., A. Hughes, and D. L. Collins-Thompson. 1974. The effect of sublethal heating of *Staphylococcus aureus* at different physiological ages. Can. J. Microbiol. 20:765.
66. Iandolo, J. J., and Z. J. Ordal. 1966. Repair of thermal injury of *Staphylococcus aureus*. J. Bacteriol. 91:134.
67. International Dairy Federation. 1990. Enumeration of *Staphylococcus aureus* Dried Milk Products Most probable number technique. IDF Brussels Belgium Provisional IDF Standard.60B.
68. International Dairy Federation. 1990. Enumeration of *Staphylococcus aureus* Milk and Milk-Based Products Colony Count Technique at 37 C. IDF, Brussels, Belgium. Provisional IDF Standard. 145.
69. Jackson, H., and M. Woodbine. 1963. The effect of sublethal heat treatment on the growth of *Staphylococcus aureus*. J. Appl. Bacteriol. 26:152.
70. Jarvis, A. W., and R. C. Lawrence. 1970. Production of high titers of enterotoxins for the routine testing of staphylococci. Appl. Microbiol. 19:699.
71. Johnson, H. M., J. A. Bukovic, and P. E. Kauffman. 1973. Staphylococcal enterotoxins A and B: Solid-phase radioimmunoassay in food. Appl. Microbiol. 26:309.
72. Jones, C. L., and S. A. Khan. 1986. Nucleotide sequence of the enterotoxin B gene from *Staphylococcus aureus*. J. Bacteriol. 106:29.
73. Julseth, R. M., and R. P. Dudley. 1973. Improved methods for enumerating staphylococci and detecting staphylococcal enterotoxin in meat foods. *In* 19th European Meeting of Meat Research Workers, vol. II, 511 pp. Centre Technique de la chareuteric, Paris, France.
74. Jurgens, D., B. Sterzik, and F. J. Fehrenbach. 1987. Unspecific binding of group B streptococcal cocytolysin (cAMP factor) to immunoglobulins and its possible role in pathogenicity. J. Exp. Med. 165:720.
75. Kauffman, P. E. 1980. Enzyme immunoassay for staphylococcal enterotoxin A. J. Assoc. Off. Anal. Chem. 63:1138.
76. Khambaty, F. M., R. W. Bennett, and D. B. Shah. 1994. Application of pulsed field gel electrophoresis to the epidemilogical characterization of *Staphylococcus intermedius* implicated in a food-related outbreak. Epidemiol. Infect. 113:75-81.
77. Kloos, W. E., and T. L. Bamerman. 1995. Staphylococcus, p. 282-298. *In* Manual of clinical microbiology, 6th ed., American Society of Microbiology, Washington D. C.
78. Koper, J. W., A. B. Hagenaars, and S. Notermans. 1980. Prevention of crossreactions to the enzyme-linked immunosorbent assay (ELISA) for detection of *Staphylococcus aureus* enterotoxin type B in culture filtrates and foods. J. Food Safety 2:35.
79. Koskitalo, L. D., and M. E. Milling. 1969. Lack of correlation between egg yolk reaction in staphylococcus medium 110 supplemented with egg yolk and coagulase activity of staphylococci isolated from cheddar cheese. Can. J. Microbiol. 14:132.
80. Kuo, J. K. S., and G. J. Silverman. 1980. Application of enzyme-linked immunosorbent assay for detection of staphylococcal entertotoxins in foods. J. Food Prot. 43:404.
81. Lachica, R. V. 1980. Accelerated procedure for the enumeration and identification of foodborne *Staphylococcus aureus*. Appl. Environ. Microbiol. 39:17.

82. Lachica, R. V. 1984. Egg yolk-free Baird-Parker medium for the accelerated enumeration of foodborne *Staphylococcus aureus*. Appl. Environ. Microbiol. 48:870.
83. Lachica, R. V. F., P. D. Hoeprich, and C. Genigeorgis. 1971. Nuclease production and lysostaphin susceptibility of *Staphylococcus aureus* and other catalase-positive cocci. Appl. Microbiol. 21:823.
84. Lachica, R. V. F., P. D. Hoeprich, and C. Genigeorgis. 1972. Metachromatic agar-diffusion microslide technique for detecting staphylococcal nuclease in foods. Appl. Microbiol. 23:168.
85. Lancette G. A., J. T. Peeler, and J. M. Lanier. 1986. Evaluation of an improved MPN medium for recovery of stressed and nonstressed *Staphylococcus aureus*. J. Assoc. Off. Anal. Chem. 69:44-46.
86. Lancette, G. A., and J. Lanier. 1987. Most probable number method for isolation and enumeration of *Staphylococcus aureus* in foods: collaborative study. J. Assoc. Off. Anal. Chem. 70:35.
87. Lewis, K. H., and R. Angelotti. 1964. Examination of foods for enteropathogenic and indicator bacteria. PHS Pub. No 1142, p. 98. U.S. Government Printing Office, Washington, D. C.
88. Lotter, L. P., and C. A. Genigeorgis. 1977. Isolation of coagulase-positive variants from coagulase-negative enterotoxigenic staphylococci. Zbl. Bakt. Hyg., I. Abt./Orig. A239, 18.
89. Lumpkins, E. D., Sr. 1972. Methods for staining and preserving agar gel diffusion plates. Appl. Microbiol. 24:499.
90. Macario, Vol. II, p. 23-59. Academic Press, Orlando, Fla.
91. Matthews, K. R., J. Roberson, B. E. Gillespie, D. A. Luther, and S. P. Oliver. 1997. Identification and differentiation of coagulase-negative *Staphylococcus aureus* by polymerase chain reaction. J. Food Prot. 60:686.
92. McFarland, J. 1907. The nephelometer: an instrument for estimating the number of bacteria in suspensions used for calculating the opsonic index and for vaccines. JAMA 49:1176.
93. Mintzer-Morgenstern, L., and E. Katzenelson. 1982. A simple medium for isolation of coagulase-positive staphylococci in a single step. J. Food Prot. 45:218.
94. Moore, T. D., and E. E. Nelson. 1962. The enumeration of *Staphylococcus aureus* on several tellurite-glycine media. J. Milk Food Technol. 24:124.
95. Nakane, P. K., and A. Kawaoi. 1974. Peroxidase-labeled antibody. A new method of conjugation. J. Histochem. Cyochem. 22:1084.
96. Notermans, S., and J. B. Dufrenne. 1982. A simple purification method for enterotoxin F produced by *Staphylococcus aureus* and some properties of the toxin. Antoine van Leeuwenhoek J. Microbiol. Serol. 48:447.
97. Notermans, S., H. L. Verjans, J. Bol, and M. Van Schorthorst. 1978. Enzyme-linked immunosorbent assay (ELISA) for determination of *Staphylococcus aureus* enterotoxin type B. Health Lab. Sci.14:28.
98. Notermans, S., K. J. Heuvelman, and K. Wernars. 1988. Synthetic enterotoxin B DNA probes for detection of enterotoxigenic *Staphylococcus aureus* strains. Appl. Environ. Microbiol. 54:531.
99. Notermans, S., R. Boot, and S. R. Tatini. 1987. Selection of monoclonal antibodies for detection of staphylococcal enterotoxins in heat processed food. Int. J. Food Microbiol. 5:49-55.
100. Notermans, S., R. Boot, P. D. Tips, and M. P. Denooij. 1983. Extractions of staphylococcal enterotoxins (SE) from minced meat and subsequent detection of SE with enzyme-linked immunosorbent assay (ELISA). J. Food Prot. 46:238.
101. Oda, T. 1978. Application of SP-Sephadex chromatography to the purification of staphylococcal enterotoxins A, B, C_2. Jpn. J. Bacteriol. 33:743.
102. Official Methods of Analysis (1997) 16th ed., 3rd Rev. 1997, AOAC International, Gaithersburg, Md. Methods 987.09, 975.55, 976.31, 980.32, 995.12
103. Orth, D. S., and A. W. Anderson. 1970. Polymyxin-coagulase-mannitol-agar. I. A selective isolation medium for coagulase-positive staphylococci. Appl. Microbiol. 19:73.
104. Orth, D. S., L. R. Chugg, and A. W. Anderson. 1971. Comparison of animal sera for suitability in coagulase testing. Appl. Microbiol. 21:420.
105. Park, C. E. , A. De Melo Serrano, M. Landgraf, J. C. Huang, Z. Stankiowicz, and M. K. Rayman. 1980. A survey for microorganisms for thermonuclease production. Can. J. Microbiol. 26:532.
106. Park, E. E., and R. Szabo. 1986. Evaluation of the reversed passive latex agglutination (RPLA) test kit for detection of staphylococcal enterotoxins A, B, C, and D in food. Can. J. Microbiol. 32:723.
107. Pennell, D. R., J. A. Rott-Petri, and T. A. Kurzynski. 1984. Evaluation of three commercial agglutination tests for the identification of Staphylococcus aureus. J. Clin. Microbiol. 20:614.
108. Raus, J., and D. Love. 1983. Characterization of coagulase-positive *Staphylococcus intermedius* and *Staphylococcus aureus* isolated from veterinary clinical specimens. J. Clin. Microbiol. 18:789-792.
109. Rayman, M. K., C. E. Park, J. Philport, and E. C. D. Dodd. 1975. Reassessment of the coagulase and thermostable nuclease tests as means of identifying *Staphylococcus aureus*. Appl. Microbiol. 29:451.
110. Rayman, M. K., J. J. Devoyod, U. Purvis, D. Kusch, J. Lanier, R. J. Gilbert, D. G. Till, and G. A. Jarvis. 1978. ICMSF methods studies. X. An international comparative study of four media for the enumeration of *Staphylococcus aureus* in foods. Can. J. Microbiol. 24:274.
111. Reiser, R., D. Conaway, and M. S. Bergdoll. 1974. Detection of staphylococcal enterotoxin in foods. Appl. Microbiol. 27:83.
112. Robbins, R., S. Gould, and M. S. Bergdoll. 1974. Detecting the enterotoxigenicity of *Staphylococcus aureus* strains. Appl. Microbiol. 28:946.
113. Rosec, J. P., J. P. Guraud, C. Dalet, and N. Richard. 1997. Enterotoxin production by staphylococci isolated from foods in France. Int. J. Food. Microbiol. 35:213-221.
114. Saint-Martin, M., G. Charest, and J. M. Desranleau. 1951. Bacteriophase typing in investigations of staphylococcal food-poisoning outbreaks. Can. J. Pub. Health. 42: 51.
115. Saunders, G. C., and M. L. Bartlett. 1977. Double-antibody solid-phase enzyme immunoassay for the detection of staphylococcal enterotoxin A. Appl. Environ. Microbiol. 34:518.
116. Schindler, C. A., and V. T. Schuhardt. 1964. Lysostaphin: A new bacteriolytic agent for the staphylococcus. Proc. Nat. Acad. Sci. 51:414.
117. Shinagawa, K., N. Kunita, and S. Sakaguchi. 1975. Simplified methods for purification of staphylococcal enterotoxin A and C and preparation of anti-enterotoxin sera. Jpn. J. Bacteriol. 30:683.
118. Silverman, S. J., A. R. Knott, and M. Howard. 1968. Rapid, sensitive assay for staphylococcal enterotoxin and a comparison of serological methods. Appl. Microbiol. 16:1019.
119. Simon, E., and G. Terplan. 1977. Nachweis von staphylokokken enterotoxin B Mittels ELISA-test. Zentralbl. Vesterinaermed. Reihe B. 24:842.
120. Sinell, H. J., and J. Baumgart. 1966. Selektionahrboden ru Isolierung von Staphylokokken aus Lebensinitteln. Zeptralbl. Bakteriol. Parasitenk. Infektionskr. Hug: Abt. Orig. 197:447.
121. Sperber, W. H., and S. R. Tatini. 1975. Interpretation of the tube coagulase test for identification of *Staphylococcus aureus*. Appl. Microbiol. 29:502.
122. Stiffler-Rosenberg, G., and H. Fey. 1978. Simple assay for staphylococcal enterotoxins A, B, and C. Modification of enzyme-linked immunosorbent assay. J. Clin. Microbiol. 8:473.
123. Stiles, M. E. 1977. Reliability of selective media for recovery of staphylococci from cheese. J. Food Prot. 40:11.
124. Stiles, M. E., and L. D. Witter. 1965. Thermal inactivation, heat injury and recovery of *Staphylococcus aureus*. J. Dairy Sci. 48:677.
125. Su, Y-C, and A. C. L. Wong. 1997. Current perspectives on detection of staphylococcal enterotoxins. J. Food. Prot. 609:15.
126. Subcommittee on Phage-Typing of Staphylococci. 1970. Report to the international committee on nomenclature of bacteria. Int. J. Syst. Bacteriol. 21:167.
127. Talan, D. A., D. Staatz, E.J. Goldstein, K. Singer, and G.D. Overturf. 1989. *Staphylococcus intermedius* in canine gingiva and canine-inflicted human wound infections: laboratory characterization of a newly recognized zoonotic pathogen. J. Clin. Microbiol. 27:78-81.

128. Tatini, S. R. 1976. Thermal stability of enterotoxins in food. J. Milk Food Technol. 39:432.
129. Thompson, N. E., M. Razdan, G. Kuntsman, J. M. Aschenbach, J. L. Everson, and M. S. Bergdoll. 1986. Detection of staphylococcal enterotoxins by enzyme-linked immunosorbent assays and radioimmunoassays: comparison of monoclonal and polyclonal antibody systems. Appl. Environ. Microbiol. 51:885-890.
130. Thompson, N. E., M. S. Bergdoll, R. F. Meyer, R. W. Bennett, L. Miller, and J. D. MacMillan. 1986. Monoclonal antibodies to the enterotoxins and to the toxic shock syndrome toxin produced by *Staphylococcus aureus*, p. 23-59. *In* Monoclonal antibodies, A. J. L. Macario and E. C. Macario, vol. II. Academic Press, Orlando, Fla.
131. Tirunarayanan, M. O., and H. Lunderbeck. 1968. Investigations on the enzymes and toxins of staphylococci. Acta Pathol. Microbiol. Scand. 73:429.
131. Turner, F. S., and B. S. Schwartz. 1958. The use of lyophilized human plasma standardized for blood coagulation factors in the coagulase and fibrinolytic tests. J. Lab. Clin. Med. 52:888.
132. Van Der Zee, H., and H. B. Nagel. 1993. Detection of staphylococcal enterotoxin with Vidas automated immunoanalyzer and conventional assays, p. 38. *In* 7th International Congress on Rapid Methods and Automation in Micriobiology and Immunology. RAMI-93. Conference Abstracts 1993. PI127.
133. Van Dorne, H., R. M. Baird, D. T. Hendriksz, D. M. Van Der Krock, and H. P. Pauwels. 1981. Liquid modification of Baird Parker's medium for the selective enrichment of *Staphylococcus aureus*. Antonie van Leeuwenboek. J. Microbiol. Serol. 47:267.
134. Van Dorne, H., P. Pauwels, and D. A. A. Mossel. 1982. Selective isolation and enumeration of low numbers of *Staphylococcus aureus* by a procedure that relies on elevated temperature culturing. Appl. Environ. Microbiol. 44:1459.
135. Van Schouwenberg-Van Foeden, A. W. J., J. Stadhouders, and J. A. Jans. 1978. The thermonuclease test for assessment of coagulase-positive with a normal acidity development. Neth. Milk Dairy J. 32:217.
136. Wieneke, A. A., and R. J. Gilbert. 1987. Comparison of four methods for the detection of staphylococcal enterotoxin in foods from outbreaks of food poisoning. Int. J. Food Microbiol. 4:135.
137. Williams, M. L. B. 1972. A note on the development of a polymyxin-mannitol phenolphthalein diphosphate agar for the selective enumeration of coagulase-positive staphylococci in foods. J. Appl. Bacteriol. 35:139.
138. Wilson I. G., R. Gilmour, J. E. Cooper, A. J. Bjourson, and J. Harvey. 1994. A non-isotopic DNA hybridisation assay for the identification of *Stapylococcus aureus* isolated from foods. Int. J. Food Microbiol. 22:43-54.
139. Zebovitz, E. J., J. B. Evans, and C. F. Niven Jr. 1955. Tellurite-glycine agar: A selective plating medium for the quantitative detection of coagulase-positive staphylococci. J. Bacteriol. 70:686.
140. Zehren, V. L., and V. P. Zehren. 1968. Examination of large quantities of cheese for staphylococcal enterotoxin, A. J. Dairy Sci. 51:635.

Chapter 40

Vibrio

Charles A. Kaysner and Angelo DePaola, Jr.

40.1 INTRODUCTION

40.11 Description of Genus

Members of the genus *Vibrio* are defined as gram negative, asporogenous rods that are straight or have a single, rigid curve. They are motile, most having a single polar flagellum when grown in liquid medium. Most produce oxidase and catalase, and ferment glucose without producing gas.[7] Three species, *V. cholerae, V. parahaemolyticus*, and *V. vulnificus*, are well-documented human pathogens.[53,77,78,87,97] *V. mimicus*,[23,99,107] also a recognized pathogen[99] and similar in characteristics to *V. cholerae*, can be distinguished from that species primarily by its inability to ferment sucrose. Other species within the genus, such as *V. alginolyticus*,[50] *V. fluvialis*,[70] *V. furnissii*,[16] *V. metschnikovii*,[38,69] and *V. hollisae*[39] are occasional human pathogens.[1,38,93] *Vibrio* species account for a significant number of the human foodborne infections from the consumption of raw or undercooked shellfish.[93] A Florida study of illnesses from raw shellfish consumption reported the following species in descending order of frequency: *V. parahaemolyticus*, non-O1/O139 *V. cholerae, V. vulnificus, V. hollisae, V. fluvialis*, O1 *V. cholerae*.[63,71]

40.111 V. cholerae

V. cholerae,[6] the type species of the genus *Vibrio*, is the causative agent of cholera outbreaks and epidemics.[34,53,122] Various biochemical properties and antigenic types characterize this species. It can be differentiated from other *Vibrio* species, except *V. mimicus*, because its obligate requirement for sodium ion (Na+)[6] can be satisfied by the trace amounts present in most media constituents. Cholera enterotoxin (CT) is the primary virulence factor of the disease, cholera. The gene for CT has recently been demonstrated to be regulated by a genetic pathogenicity island, VPI, which contains the other virulence genes.[54] Most *V. cholerae* strains recovered from epidemic cholera cases contain a somatic antigen in common and include serogroup O1.[53] Over 150 known somatic antigenic types have been demonstrated. Strains that are agglutinable in Inaba or Ogawa serotypes of O1 antiserum are well-documented human pathogens. Until recently, only the O1 serogroup was associated with cholera epidemics. However, in 1993 a large outbreak of cholera occurred in India/Bangladesh from a new, until then, unknown serogroup, O139.[3] Numerous cases were recorded in which patients had the typical symptoms of classical cholera, cholera gravis, previously associated with the O1 serogroup. Except for the O antigen, this serogroup is nearly identical to the seventh pandemic strain of *V. cholerae*.[10] The O139 strain has become endemic in the Bengal region and may be the cause of what will be known as the Eighth Cholera Pandemic.[34,113]

V. cholerae strains that are identical to, or closely resemble, disease-causing strains in biochemical characteristics but fail to agglutinate in either anti-O1 or -O139 sera are now referred to as *V. cholerae* non-O1/O139.[34,53] Non-O1/O139 strains are abundant in estuarine environments. Evidence indicates that non-O1/O139 strains are sporadically involved in cholera-like diarrheal disease,[23,72,80,93,101] but rarely in outbreaks. Indeed, the permeability factor produced by a non-O1/O139 strain during an investigation of a cholera outbreak has been found biologically and immunologically indistinguishable from CT. Some non-O1/O139 strains also are invasive and have caused septic infections in individuals with pre-disposing medical conditions and produce a heat stable toxin.[80,82,90,95] Most strains do not produce CT, which is the key difference between these and epidemic *V. cholerae* O1/O139.

40.112 V. mimicus

V. mimicus[24,99] has been associated with diarrhea following consumption of raw or undercooked seafood.[93] Isolated from samples during a search for *V. cholerae, V. mimicus* can be differentiated from that closely related pathogen by lack of sucrose fermentation. The organism will appear as green colonies on thiosulfate citrate bile salts sucrose (TCBS) agar and will grow in most common media without added NaCl. Virulence is poorly characterized, but some strains have been found to possess the cholera toxin gene,[107] produce CT demonstrable in a tissue culture assay, and the *ctx* gene can be detected by PCR amplification.

40.113 V. parahaemolyticus

V. parahaemolyticus,[36,78] the leading bacterial cause of diarrhea associated with seafood consumption in Florida[63] and probably the US,[93] is a halophilic estuarine organism found in coastal waters of virtually all temperate regions[26,51,97] and occasionally causes septicemia. In temperate regions, a seasonal occurrence in shellfish and in human infections have been reported, the majority in the warmer months of the year, except in Florida. All strains share a common H antigen, but to date 12 O type and 65 K (capsular) antigens have been described, but many strains are untypable.[78,97] Clinical strains of *V. parahaemolyticus* are differentiated from most environmental strains by the ability to produce a thermostable direct hemolysin (TDH), termed the Kanagawa phenomenon.[79,117] The *tdh* gene has been cloned and sequenced[83,84] and DNA probes are now available to test for the presence of this virulence marker in *V. parahaemolyticus* isolates.[40] This past decade, a second hemolysin has been identified that has been associated with strains causing gastroenteritis.[44,45] This thermostable hemolysin shares 60% homology to TDH and has subsequently been designated the thermostable related hemolysin, TRH. Presently, there is no in-vitro test to detect TRH production. After further evaluation of *V. parahaemolyticus* collections, both TDH- and TRH-producing strains predominate in clinical strains.[8,102] Taniguchi et al.[119] described a thermolabile hemolysin, TL, which was found in all *V. parahaemolyticus* strains, but not in other species. PCR procedures and gene probes have been developed to detect *tl*, *tdh*, and *trh* in *V. parahaemolyticus* strains.[8,48,74]

40.114 V. vulnificus

V. vulnificus,[33] the leading cause of death in the US related to seafood consumption and nearly always associated with raw Gulf Coast oysters,[87,100] resembles *V. parahaemolyticus* on TCBS agar, but can be differentiated by several biochemical reactions, including β-galactosidase activity.[31] Epidemiological and clinical investigations show that *V. vulnificus* causes septicemia and death following ingestion of seafood or after wound infections originating from the marine environment.[42,114,125] Recent gene probe assays,[28,129] PCR procedures,[41] fatty acid profiles,[67] and enzyme immunoassays[31,118] have been developed to detect and identify this pathogen.

40.115 Other Species

The following species have also been isolated from human stools and/or patients with gastroenteritis, with the consumption of shellfish the predominate source of infection.[93] *V. metschnikovii* differs from all other *Vibrio* species in its lack of cytochrome oxidase.[7] Some strains (biotype II) of *V. fluvialis* sp. nov. (now designated *V. furnissii*) produce gas during the fermentation of D-glucose.[16] *V. hollisae* is a halophilic species that grows poorly if at all on TCBS agar and exhibits a delayed motility pattern (>48 hr) uncharacteristic of the other vibrios.[7] Some strains of *V. hollisae* have been demonstrated to contain the *tdh* gene found in pathogenic strains of *V. parahaemolyticus*.[39]

40.116 Differentiation of Species

Table 1 presents the differential characteristics of the species most often associated with human illness related to seafood consumption. Tables can also be found in several publications, including Baumann and Schubert,[7] Elliot et al.,[31] McLaughlin,[75] and West et al.[127]

40.12 Distribution and Sources of Contamination

40.121 V. cholerae

V. cholerae O1 is excreted in great numbers in the feces of cholera patients and convalescents.[34,53] The disease is transmitted primarily by the fecal-oral route, indirectly through contaminated water supplies.[30,77,112,122,126] Direct person-to-person spread is not common. Food supplies may be contaminated by the use of human feces as fertilizer or by freshening vegetables for market with contaminated water.[30,56,60,91] Cholera outbreaks in several countries and the US are thought to have resulted from the consumption of raw, undercooked, contaminated, or recontaminated seafood. Toxigenic *V. cholerae* O1 is rarely isolated from US environments and foods and no isolations of serogroup O139 have been reported in this country. In contrast, non-O1/O139 strains are commonly isolated from estuarine water and shellfish.[52,122] Evidence suggests that *V. cholerae* O1 is a component of the autochthonous flora of brackish water, estuaries, and salt marshes of coastal areas of the temperate zone, posing an ongoing hazard to public health.[11,122] Various O1 strains have become endemic in many regions in the world, including Australia and the Gulf Coast region of the US.[20,123]

40.122 V. parahaemolyticus

This organism is frequently isolated from coastal waters and seafood in temperate zones throughout the world. It is the most frequent cause of foodborne disease in Japan,[86] where raw fish consumption is popular. A number of common-source gastroenteritis outbreaks attributed to *V. parahaemolyticus* have occurred in the US,[56] including association with oyster consumption.[85,93] Some foods implicated in the US are crab, shrimp, and lobster that, unlike fish in Japan, were typically cooked before eating. Mishandling practices, such as improper refrigeration, insufficient cooking, cross-contamination, or recontamination are suspected in these outbreaks. Recently, consumption of raw oysters was associated with large outbreaks of *V. parahaemolyticus* gastroenteritis on the West Coast in 1997[18] and in Texas and New York in 1998.[19] Clinical isolates from the West Coast were urease positive and possessed both the *tdh* and *trh* genes. The Texas and New York outbreaks were caused by an O3:K6 serotype, urease negative and possessing only *tdh*, which appears to have become pandemic and is the most prevalent strain in Asia.

40.123 V. vulnificus

The invasive species, *V. vulnificus*, the causative agent of septicemic shock,[62,87,114] is a common organism in coastal waters of some areas of the US and other countries.[59,87,118,120] It is reported to cause 20 to 40 cases each year of primary septicemia with a 50% mortality rate of individuals with liver disease and elevated serum iron levels.[100] A review of cases has determined an association between septicemia and consumption of raw oysters, nearly all from Gulf Coast waters. This species has also been responsible for wound infections in individuals that have association with skin trauma in marine environments.[87] This halophilic species will grow on or in many laboratory formulations of media that contain NaCl; a 0.5% minimum concentration is recommended. Virulence is associated with a capsule, but no reliable marker has been demonstrated, as most tests cannot distinguish between pathogenic and environmental strains.[76,104,130]

40.124 Other Halophilic Vibrios

V. alginolyticus, V. fluvialis, V. furnissii, V. metschnikovii, and *V. hollisae* are, like *V. parahaemolyticus, V. cholerae*, and *V. vulnificus*, recovered from brackish coastal waters, sediment, and sea life taken from the temperate estuarine environment.[7] These species are normal components of that environment, appear on a seasonal basis, and have been associated with human illness.[12,93]

40.13 Methods of Isolation

Vibrio species, like many other gram negative bacteria, grow in the presence of relatively high levels of bile salts. They are facultatively anaerobic and grow best in alkaline conditions. Isolation from foods is facilitated by the use of media formulated with an alkaline pH due to tolerance of this condition by these organisms. Alkaline peptone water (APW) is used commonly for isolating several of the species of concern.

The strict halophilic nature of *V. parahaemolyticus* probably accounts for the fact that illnesses caused by this organism were not documented in the US until workers began examining food and feces on appropriate media containing added salt. Media used for testing the biochemical reactions of *V. parahaemolyticus* should contain 2% to 3% NaCl. *V. vulnificus* requires NaCl for growth. A minimum of 0.5% NaCl, the concentration of most prepared media, is adequate. Diluent used for transfer of cell suspensions or dilution preparation must contain NaCl, i.e., phosphate buffered saline, PBS.[31]

TCBS agar[31] is a common medium used for the isolation of *V. cholerae, V. parahaemolyticus* and other *Vibrio* spp. from seafood. This medium supports good growth of most species while inhibiting most non-vibrios.[65] Recent formulations for selective agars for the isolation of *V. vulnificus* have also proved effective. Modified cellobiose polymyxin colistin (mCPC)[31] and CC[43] agars have been designed to differentiate *V. vulnificus* from other vibrios. *V. cholerae* strains, except the Classical biotype, will grow on mCPC agar, while most *V. parahaemolyticus* strains and other species will not. To facilitate the identification of suspect isolates, the rapid diagnostic kit API 20E can be used in lieu of the preparation of the many biochemical media needed for identification.[31,89]

40.2 GENERAL CONSIDERATIONS

40.21 Storage of Sample

The sample should be cooled immediately after collection (about 7°C to 10°C) and analyzed as soon after collection as possible. Direct contact with ice should be avoided to maximize survival and recovery of vibrios and reduce the tendency for overgrowth by indigenous marine microflora. Vibrios can be injured by rapid cooling, but grow rapidly in seafood at ambient temperatures.[21,22] Despite the recognized fragility of the vibrios to extremes of heat and cold, their survival is enhanced under mild refrigeration.[15,17,37,49,92] If frozen storage of the sample is required, a temperature of –80°C is recommended, if feasible.[14]

Shellfish samples should be handled according to recommended procedures described by the American Public Health Association.[4] Ten to twelve animals are pooled, aseptically shucked to a sterile blender jar, and blended at high speed for 90 sec. This composite is used to prepare dilutions using a NaCl-containing solution, such as PBS.

To facilitate the storage and further analysis of numerous isolates from a sample, the following procedure is recommended. This allows for the gene probe analysis of many of the isolates obtained from a sample, in contrast to the minimal number that can be feasibly handled using traditional biochemical tests. A sterile 96-well microtiter plate is filled with 100 µL/well of APW. Numerous presumptive vibrio colonies are transfered from a selective agar plate using a sterile toothpick or wood transfer stick to individual wells. The inoculation

TABLE 1. Biochemical Characteristics of the *Vibrionaceae* Commonly Encountered in Seafood*

	V. alginolyticus	*V. cholerae*	*V. fluvialis*	*V. furnissii*	*V. hollisae*	*V. metschnikovii*	*V. mimicus*	*V. parahaemolyticus*	*V. vulnificus*	*A. hydrophilia***	*P. shigelloides***
TCBS agar	Y	Y	Y	Y	NG	Y	G	G	G	Y	G
Oxidase	+	+	+	+	+	–	+	+	+	+	+
Arginine dihydrolase	–	–	+	+	–	+	–	–	–	+	+
Lysine decarboxylase	+	+	–	–	–	+	+	+	+	V	+
Growth in (w/v):											
0% NaCl	–	+	–	–	–	–	+	–	–	+	+
3% NaCl	+	+	+	+	+	+	+	+	+	+	+
6% NaCl	+	–	+	+	+	+	–	+	+	+	–
8% NaCl	+	–	V	+	–	V	–	+	–	–	–
10% NaCl	+	–	–	–	–	–	–	–	–	–	–
Growth at 42°C	+	+	V	–	nd	V	+	+	+	V	+
Acid from:											
Sucrose	+	+	+	+	–	+	–	–	–	V	–
D-Cellobiose	–	–	+	–	–	–	–	V	+	+	–
Lactose	–	–	–	–	–	–	–	–	+	V	–
Arabinose	–	–	+	+	+	–	–	+	–	V	–
D-Mannose	+	+	+	+	+	+	+	+	+	V	–
D-Mannitol	+	+	+	+	–	+	+	+	V	+	–
ONPG	–	+	+	+	–	+	+	–	+	+	–
Voges-Proskauer	+	V	–	–	–	+	–	–	–	+	–
Sensitivity to:											
10 µg 0/129	R	S	R	R	nd	S	S	R	S	R	S
150 µg 0/129	S	S	S	S	nd	S	S	S	S	R	S
Gelatinase	+	+	+	+	–	+	+	+	+	+	–
Urease	–	–	–	–	–	–	–	V	–	–	–

* Adapted from Elliot et al. (31)

** *Aeromonas hydrophila, Plesiomonas shigelloides*

Y = yellow; G = green; NG = no or poor growth; nd = not done

V = variable among strains; R = resistant; S = susceptible

pattern is recorded and the plate is incubated overnight at 35° ± 2°C. The isolates in the wells can be replicated using a 48-prong replicator to an agar plate for gene probe analysis. After replication, 100 µL of TSB-1% NaCl-24% glycerol is dispensed to each well. The plate is wrapped in a double layer of foil or plastic and placed in an ultra-low freezer, –72° to –80°C, for storage of cultures. When needed, the plate is thawed at 35°C and the cultures from the well(s) transferred, or replicated to a new microtiter plate or tubed medium. Purity of the culture can be determined by streaking to an agar medium such as T1N3.

40.22 Genetic-Based Techniques

These newer technologies have added the factor of more rapid detection and identification schemes to the analysis and are included for those laboratories that possess the proper equipment. The identification by PCR offers a one-day analysis,[5,8,9,35,41,64,66,103,115,121] while the gene probe procedures are two- to three-day analyses, including the ones presented in this chapter.[13,28,40,57,68,74,83,96,128,129,131,132] The traditional qualitative procedure and the most probable number (MPN) technique require four to seven days to complete.[31] The alkaline phosphatase-labeled probes offer the advantage of a 1-hr hybridization time, thus allowing for completion of the analysis within one working day of lifting the colonies from a plating medium. AP-labeled probes to identify the presence of *V. parahaemolyticus* and strains harboring the *tdh* gene, and detecting *V. vulnificus* in a sample are available commercially. One lot of commercially AP-labeled probe is enough to process approximately 500 filters. AP-labeled oligonucleotide probes offer the advantage of short hybridization times, usually 1 hr, completion of analyses within one day and inexpensive paper filters (Whatman #541) for colony lifts.

Digoxigenin-labeled probes[13] are also presented for the three species of concern. Advantages of the digoxigenin-labeled probe procedure are: a) the procedure can be performed inexpensively, b) more reporter groups can be included per probe molecule, i.e. signal amplification c) twice the number of copies of the probe can be prepared; complimentary to the target DNA strand, d) the probe solution can be used several times, e) the hybridization and wash temperature is the same for all digoxigenin-probes, f) the nylon membrane can be stripped of a probe and hybridized with an additional probe(s) and g) the use of a nylon membrane allows for the transfer between agar surfaces, i.e., a non-selective agar for resuscitating cells prior to moving to a selective and differential agar.

40.23 Recommended Controls

Duplicate plating media should be used for vibrios because strains may vary in their growth characteristics. Positive and negative control strains should be used for all phenotypic and genotypic assays to ensure appropriate reading of the reactions.

40.3 EQUIPMENT, MATERIALS, AND REAGENTS

40.31 Media and Reagents

Alkaline peptone water (APW)
AKI medium
Arginine glucose slants (AGS)
Blood agar (5% sheep red blood cells)
Casamino acids yeast extract (CAYE) broth
modified Cellobiose polymyxin colistin (mCPC) agar
Cellobiose colistin (CC) Agar[43]
Kligler iron agar (KIA)
Motility test medium-1% NaCl
Oxidase reagent (1% N,N,N′,N′-tetramethyl-*p*-phenylenediaminex2HCl in dH_2O)
Peptone-Tween-salt diluent (PTS)
Phosphate buffered saline (PBS)
Polymyxin B disks, 50 U (Difco or equivalent)
Saline solution—0.85% in dH_2O
2% NaCl solution
Sodium desoxycholate—0.5% in sterile dH_2O
Thiosulfate citrate bile salts sucrose (TCBS) agar
T1Nl and T1N3 agars (1% tryptone and either 1% or 3% NaCl)
T1N0, T1N3, T1N6, T1N8, T1N10
Tryptic soy agar—magnesium sulfate—3% NaCl (TSAMS)[32]
Trypticase (or tryptic) soy broth (TSB), agar (TSA) (with added NaCl, 2%)
TSB-1% NaCl-24% glycerol
Urea broth (or Christensen's urea agar) with added NaCl (2%)
V. cholerae polyvalent O1 and O139 antiserum (i.e., Difco)
VET-RPLA TD920A enterotoxin detection kit (Oxoid, Inc.)
Vibrio parahaemolyticus sucrose agar (VPSA)[32]
Vibrio vulnificus agar (VVA)
Wagatsuma agar
API 20E diagnostic strips and reagents (BioMerieux)

40.311 Probe Reagents and Materials Required

Shaking water bath(s) capable of up to 65°C. (temperatures needed, 42°, 54°, 55° and 65°C)
Shaker platform at room temperature
Microwave
Long wave UV light box or UV Crosslinker (254 nm wave length)
Heat tolerant bags (and sealer) or plastic tubs with lids (300-500 mL capacity)
96 well microtiter plates with lids
8 or 12 channel micro-pipetter
48 prong replicator
Whatman #541 filters, 85 mm (special order for this diameter, 1541-085 from Whatman)
Whatman #3 or equivalent absorbent filter or pad
Nylon membranes (positive charge), 82 mm (MSI, Westboro, MA, Magnagraph, gridded-NJOHG08250, plain-NJOHY08250)
Fiberglass mesh screens, household window screen available at hardware stores[61]
Sterile hockey sticks
Sterile toothpicks or wood applicator sticks
Glass Petri dishes, 100 mm
Lysis solution (0.5 M NaOH, 1.5 M NaCl)
Neutralizing solution (1.0 M Tris-HCl, pH 7.0 in 2.0 M NaCl) for nylon membranes
2M ammonium acetate buffer (for AP-labeled probe and #541 filters)
1X SSC, 5X SSC, 20X SSC—(standard saline citrate)
1X SSC—1% SDS (sodium dodecyl sulfate)
10% Sarkosyl solution (N-lauroyl-sarcosine, sodium salt)
10% SDS solution (sodium dodecyl sulfate)
1M Tris, pH 7.5 (Sigma T1503 Trizma base)
1M Tris, pH 9.5 (Sigma T1503 Trizma base)
3M NaCl
1M $MgCl_2$
Proteinase K stock solution (20 mg/mL)
Hybridization solution (BSA, SDS, PVP in 5X SSC) (for AP-labeled probe)

NBT/BCIP color reagent [nitro blue tetrazolium chloride/ 5-bromo-4-chloro-3-indolyl phosphate, toluidine salt] B-M cat. no. 1697471 (for colorimetric detection)
Dig buffers 1, 2, 3 and 4[13]
10 mM Tris-HCl, 1 mM EDTA, pH 8.0
Blocking reagent (B-M cat. no.1096176)
Wash solution A and B[13]
Anti-Dig AP [Anti-digoxigenin alkaline phosphatase, Fab fragments] (B-M cat. no. 1093274)
dig-11-dUTP B-M cat. no. 109308
CSPD B-M cat. no. 1755633 (for chemiluminescent detection)

40.4 PRECAUTIONS

A good selective enrichment broth has not been developed for *V. cholerae*, however due to its rapid generation time, short incubation periods are effective for isolation. APW provides suitable enrichment for incubation periods of 6 to 8 hr, but other competing microflora may overgrow *V. cholerae* during longer enrichment periods of certain types of samples. Overnight periods (16 to 18 hr), although not desirable, have been used to facilitate sample analysis during work hours. If the product has been subjected to a processing step, i.e., heating, freezing, drying, or low densities are expected, further incubation overnight is recommended to thoroughly resuscitate injured cells. The incubation of raw oyster samples in APW at 42°C for 6 to 8 hr has proved to be effective for the isolation of *V. cholerae* and is recommended.[29] DePaola and Hwang[27] found that enrichment incubation for 18 to 21 hr, instead of 6 to 8 hr, gave a higher recovery of O1 *V. cholerae* when low inocula was used. It is recommended for raw oysters to use a 1:100 ratio of oyster to APW.[27]

40.5 PROCEDURES

40.51 V. cholerae

40.511 Enrichment and Plating

Weigh 25 g of sample into a tared jar (capacity approximately 500 mL). Products such as seafood or vegetables may be blended or cut into small pieces with sterile scissors.

Add 225 mL APW to jar. Thoroughly mix the sample or blend 2 min at high speed.

Incubate APW at 35° ± 2°C for 6 to 8 hr. Re-incubate the jar overnight if the sample had been processed in some way. For analysis of raw oysters, include a second tared flask with 25 g of product plus 2475 mL APW. This flask should be incubated 18 to 21 hr at 42° ± 0.2°C in a water bath.[29,31] An enumeration technique by most probable number (MPN) may also be performed if desired (section 40.521 for procedure).

Prepare dried plates of TCBS agar. Modified-CPC or CC agars may also be included.

Transfer a 3-mm loopful from the surface pellicle of APW culture to the surface of a dried TCBS plate (and mCPC or CC), and streak in a manner that will yield isolated colonies. Incubate TCBS overnight (18 to 24 hr) at 35°± 2°C. Incubate mCPC or CC agar at 40° ± 2°C for 18 to 24 hr.

Typical colonies of *V. cholerae* on TCBS agar are large (2 to 3 mm), smooth, yellow and slightly flattened with opaque centers and translucent peripheries. Typical colonies of *V. cholerae* on mCPC or CC agar are small, smooth, opaque, and green to purple in color, with a purple background on extended incubation.

For biochemical identification, colonies from crowded plates must be streaked to a non-selective agar (T1N1, T1N3, or TSA – 2% NaCl agar) for purity. Incubate overnight at 35° ± 2°C and proceed with identification using a single isolated colony. Subculture three or more typical colonies from each plating medium to T1N1 agar slants or motility test medium stabs. Incubate slants or stabs overnight at 35° ± 2°C.

40.512 Screening and Confirmation

Arginine glucose slant (AGS). Inoculate each suspect T1N1 culture to AGS by streaking the slant and stabbing the butt. Incubate AGS with loose cap overnight at 35° ± 2°C. *V. cholerae* and *V. mimicus* cultures will have an alkaline (purple) slant and an acid (yellow) butt, as arginine is not hydrolyzed. No gas or H_2S is produced.

Kligler iron agar (KIA). Inoculate each suspect TlNl agar culture to a tube of KIA by streaking the slant and stabbing the butt. Incubate the cultures overnight at 35° ± 2°C. *V. cholerae* and *V. mimicus* cultures will have an alkaline (red) slant and an acid (yellow) butt, no gas, and no blackening (H_2S production) in the butt. Use of triple sugar iron agar (TSI) agar is less desirable since *V. cholerae* ferments sucrose, resulting in an acid slant and acid butt. *V. mimicus* cultures do not ferment sucrose and will have an alkaline slant and an acid butt in TSI agar.

Salt tolerance. From T1N1 culture, lightly inoculate one tube each of T1N0 and T1N3 broths. Incubate tubes overnight at 35° ± 2°C. *V. cholerae* and *V. mimicus* cultures will grow without NaCl.

String test. The string test[106] is a useful presumptive test for suspected *V. cholerae* as all strains are positive. Emulsify a large colony from a T1Nl agar culture in a small drop of 0.5% sodium desoxycholate in sterile dH_2O. Within 60 sec the cells lyse (loss of turbidity) and DNA strings when a loopful is lifted (up to 2 to 3 cm) from the slide.

Oxidase reaction. Add growth from an overnight T1N1 culture to a filter paper saturated with oxidase reagent (1% N,N,N',N'-tetramethyl-*p*-phenylenediaminex-2HCl). A dark purple color developing within 10 sec indicates a positive test. Transfer the growth using a platinum wire or wood applicator stick. Alternatively, add a drop of reagent to the growth on a T1N1 slant or agar plate. *V. cholerae* and *V. mimicus* are oxidase positive.

Serologic agglutination test. Serotyping of suspect *V. cholerae* cultures passing the string test using somatic or O antigens gives important epidemiological evidence. Two major serotypes of serogroup O1, Ogawa and Inaba, and one rarely encountered serotype, Hikojima, and serogroup O139 are recognized as human pathogens. The three serotypes of O1 are seen in both the classical *V. cholerae* and the El Tor biotypes. The O139 serogroup resembles only the El Tor biotype.

For each culture, mark off three sections about 1 × 2 cm on the inside of a glass petri dish or on a 2 × 3-inch glass slide and add one drop of 0.85% saline solution to the lower part of each marked section. With a sterile transfer loop or needle, emulsify the T1N1 culture in the saline solution for one section, and repeat for the other section. Check for agglutination.

Add a drop of polyvalent *V. cholerae* O1 antiserum to one section of emulsified culture and mix with a sterile loop or needle. Add a drop of anti-O139 to a separate section. Tilt the mixture back and forth for 1 min and observe against a dark background. A positive reaction is indicated by a rapid, strong agglutination in a clear background.

Test each culture with polyvalent O1 antiserum and, if positive, test separately with Ogawa and Inaba antisera. The Hikojima serotype reacts with both antisera. Antibodies to the Inaba, Ogawa, and group O1 antigen are commercially available (i.e.,

Columbia Diagnostics Inc., Springfield, VA). Similarly, O139 antiserum is commercially available.

Results of non-agglutinable cultures should be reported as non- O1/O139 *V. cholerae*.

40.513 Biochemical Tests

Table 2 presents the minimal number of characters needed to identify *V. cholerae* strains. The ability of *V. cholerae* to grow in 1% tryptone without added NaCl differentiates it from other sucrose-positive vibrios. The API 20E diagnostic strip has been used successfully for identification and confirmation of isolates.[89] The microtiter plate system for storage of suspect isolates can be used here, section 28.21.

40.514 Differentiation of El Tor and Classical Biotypes

Although the Classical biotype is currently rarely determined, the following are optional tests to differentiate them from the El Tor biotype.

Beta-hemolysis. The most common means of differentiating the biotypes of O1 *V. cholerae*, and perhaps the easiest, is to determine β-hemolytic ability on sheep blood agar. El Tor strains are β-hemolytic, while classical strains (now extremely rare) do not produce a hemolysin. Inoculate a blood agar plate with test cultures by spotting to the surface and incubate 18-24 hr at 35° ± 2°C. Beta-hemolysin can be determined by a clear zone around the growth of the culture.

Polymyxin B sensitivity. Streak a suspect culture to a dry T1N1 agar plate and place a 50 unit disc of polymyxin B on the surface. Invert the plate, incubate overnight at 35° ± 2°C, and record the result. Classical strains are sensitive (>12 mm zone); El Tor strains are resistant. If the suspect culture grows on mCPC agar, which contains polymyxin B, it is considered to be of the El Tor biotype.

40.515 Determination of Enterotoxigenicity

Most strains of *Vibrio cholerae* isolated from foods or the environment do not produce cholera toxin, CT, and are not considered virulent. Isolates identified as *V. cholerae* or *V. mimicus* should be tested for the production of CT or the CT gene.[107]

Y-1 mouse adrenal cell assay.[94] CT has been shown to stimulate the enzyme adenylate cyclase with the production of cyclic adenosine monophosphate that ultimately influences several cellular processes. In the Y-1 cell assay, CT promotes the conversion of elongated fibroblast-like cells into round refractile cells.

Inoculate test cultures from TINI slants to tubes of CAYE broth and incubate overnight at 30°± 2°C. Inoculate a 10-mL portion of CAYE broth in a 50 mL Erlenmeyer flask from each stationary culture; incubate for 18 hr with shaking. Centrifuge each test culture; filter the supernatant through a 0.22 µm filter. Refrigerated filtrates may be stored for up to 1 week.

Add aliquots of 0.025 mL from each filtrate, both unheated and heated to 80°C for 30 min, to wells of the microtiter assay plate. In addition to filtrates from known toxigenic and non-toxigenic cultures, add 0.025 mL aliquots from preparations containing 1.0 and 0.1 ng CT/mL. Suppression of cell rounding by treatment of test filtrates with anti-CT serum is an advisable control for non-specific reactions.

Immunoassay for CT. A commercially available immunoassay has been developed to detect the presence of CT in culture filtrates of *V. cholerae* and *V. mimicus* (VET-RPLA, Oxoid, Inc., Ogdensburg, NY). Inoculate test cultures into AKI medium and incubate at 35° ± 2°C 18 hr with shaking at 100 rpm. Centrifuge 5 to 7 mL of culture broth at 8,000 × g for 10 min. The supernatant can be filter sterilized through a 0.2 m filter or used as is. Test the supernatant or filtrate following the manufacturer's protocol using conical 96 well microtiter plates. Incubation of the RPLA is overnight, undisturbed at room temperature.

40.516 Other Toxins

The significance of other toxins in human pathogenicity is poorly understood and these assays are not recommended for routine analysis. Madden et al.[72] also demonstrated clinical isolates that were pathogenic for infant rabbits. A heat-labile cytolysin produced by *V. cholerae* non-O1/O139 was found by McCardell et al.[73] to be cytotoxic to Y-1 mouse adrenal and Chinese hamster ovary cells, to be rapidly fatal upon intravenous injection into adult mice, and to cause fluid accumulation in rabbit ileal loops.[108] Cultures may be tested for heat stable enterotoxin (ST)[73,116] or cytotoxin[80,96] if desired.

40.517 Genotypic Detection of the Cholera Toxin Gene by Polymerase Chain Reaction[66]

The CT gene can be present in strains of *V. cholerae* and *V. mimicus*, but not expressed under experimental conditions. Thus, a genotypic assay is recommended such polymerase chain reaction (PCR) amplification of the CT gene. This procedure offers a more rapid result and is less complicated than phenotypic assays.

Cholera toxin PCR primers, 10 pmol/µL stock solutions.
Forward 5′-tga aat aaa gca gtc agg tg-3′
Reverse 5′-ggt att ctg cac aca aat cag-3′
The PCR product is a 777 bp fragment.

APW enrichment. From sample preparation above of the 6–24 hr incubation (section 28.511), prepare a crude lysate for PCR by boiling 1 mL of APW enrichment mixture in a 1.5 mL microcentrifuge tube for 10 min. Lysate can be used immediately for PCR or stored at –20°C until use. For suspect *V. cholerae* and *V. mimicus* isolates and control cultures, inoculate 1-mL volume of APW, incubate 18 hr at 35° ± 2°C and proceed with boiling step.

To minimize cross-contamination of PCR reagents, it is recommended that a PCR master mix be prepared and aliquots stored frozen (–20°C) until use. Master mixes contain all necessary reagents except *Taq* polymerase and the lysate (template) to be amplified. The final reaction contains: 10 mM Tris-HCl, pH 8.3; 1.5 mM $MgCl_2$; 200 µM each of dATP, dTTP, dCTP, dGTP; 2 to 5% (v/v) APW lysate (template); 0.5 µM of each primer and 2.5 U *Taq* polymerase per 100 µL reaction. Volumes of 25 to 100 µL may be used. Add *Taq* polymerase to the master mix and add template upon distribution to 0.6 mL microcentrifuge tube reaction vessels. Some thermocyclers may require a mineral oil overlay (50–70 µL).

Thermocycler conditions	time (min)	temperature (°C)
Initial denaturation	3	94
Denaturation	1	94
Primer annealing	1	55
Primer extension	1	72
Final extend	3	72
No. cycles	No more than 35	

Agarose gel analysis of PCR products. Mix 10 mL PCR product with 2 µL 6X loading gel and load sample wells of 1.5 to 1.8% agarose gel containing 1 µg/mL ethidium bromide submerged in 1X TBE. Use a constant voltage of 5 to 10 V/cm. Illuminate gel with a UV transluminator and visualize bands relative to molecular weight marker migration. The primers listed amplify a 777 bp fragment of *ctxAB*. Polaroid photographs can be taken of the gel for documentation. Positive and negative culture controls and reagent control should be included with each PCR run.

Probes have been developed to also detect the presence of *ctxAB*.[128] A dig-labeled probe can also be prepared of the PCR amplification product for detection of the *ctxAB* gene using colony hybridization. The preparation of the probe, the hybridization conditions for colony blots of suspect isolates and wash protocol follow that outlined in Boehringer-Mannheim.[13]

40.518 Final Report

The final report for *V. cholerae* should include biochemical and serological identification of the isolate and enterotoxicity results. The minimal number of characters to identify the species are presented in Table 2.

40.52 V. parahaemolyticus

Three analytical schemes for enumerating *V. parahaemolyticus* are presented. The first is the MPN procedure commonly used by many laboratories.[31] In addition, this procedure is nearly identical for enumeration of *V. vulnificus*. The second is a membrane filtration procedure using a hydrophobic gridded membrane (HGMF).[32] The third is a direct plating method using DNA probes for identification of the total *V. parahaemolyticus* population[74] and pathogenic (TDH containing) strains. In addition, a TRH gene probe procedure and a PCR confirmation analysis[8] are also included.

40.521 Seafood Samples

Enrichment, isolation, and enumeration. Weigh 50 g of seafood sample into a blender. Obtain surface tissues, gills, and gut of fish. Sample the entire interior of shellfish, normally 12 animals are composited, blended at high speed for 90 sec and 50 g of composite used for analysis. For crustaceans such as shrimp, use the entire animal if possible; if it is too large, select the central portion including gill and gut.

Table 2. Minimal Number of Characters Needed to Identify *V. cholerae* and *V. parahaemolyticus* Strains

	Reaction	Percentage Positive
Gram-negative, asporogenous rod	+	100
Oxidase	+	100
Glucose, acid under a petrolatum seal	+	100
Glucose, gas	–	0
D-mannitol, acid	+	99.8
Meso-inositol, acid	–	0
Hydrogen sulfide, black butt in TSI	–	0
L-lysine decarboxylase	+	100
L-arginine dihydrolase	–	0
L-ornithine decarboxylase	+	98.9
Growth in 1% tryptone broth[a]	+	99.1[b]/0[c]

[a] No sodium chloride added.
[b] *V. cholerae* (and *V. mimicus*)
[c] *V. parahaemolyticus*
From Hugh and Sakazaki[47]

Add 450 mL PBS dilution water to the blender jar and blend for 1 min at 8,000 RPM. This constitutes the 1:10 dilution. Prepare 1:100. 1:1000, 1:10,000 dilutions or higher, if necessary, in PBS.

For shellfish, pool 10–12 animals. Blend 90 sec with an equal volume of PBS (1:2 dilution).[4] Prepare further dilutions as necessary.

For product that has been processed, i.e., heated, dried or frozen, inoculate in triplicate 10 mL portions of the 1:10 dilution into 3 tubes containing 10 mL of 2X APW. This represents the 1 g portion. Similarly, inoculate in triplicate 1 mL portions of the 1:10, 1:100, 1:1000, and 1:10,000 dilutions into 10 mL of single-strength APW. If high numbers of *V. parahaemolyticus* are expected, the examination may start at the 1:10 dilution of product.

Incubate APW overnight at 35° ± 2°C. Streak a 3-mm loopful from the top 1 cm of APW tubes containing the three highest dilutions of sample showing growth onto TCBS and mCPC or CC agars. Incubate TCBS plates at 35° ± 2°C and mCPC or CC plates at 40° ± 2°C overnight.

V. parahaemolyticus appear as round, opaque, green or bluish colonies, 2 to 3 mm in diameter on TCBS agar. Interfering, competitive *V. alginolyticus* colonies are, large, opaque, and yellow. Most strains of *V. parahaemolyticus* will not grow on mCPC or CC agar. If growth does occur, colonies will be green-purple in color due to lack of cellobiose fermentation.

Purify isolates as described previously and inoculate a microtiter plate for freezer storage, section 28.21.

40.522 Screening and Confirmation

Biochemical identification of isolates. Unless otherwise specified, all media in this section are prepared to contain 2% to 3% NaCl. The API 20E diagnostic strip can be alternatively used here. Prepare a cell suspension of the suspect cultures in 2% NaCl for the API 20E.

Screen suspect cultures of *V. parahaemolyticus*, and *V. vulnificus*, using AGS, and T1N0 and T1N3 broths as described previously. Incubate tubes at 35° ± 2°C for 18–24 hr.

Transfer two or more suspicious colonies from TCBS agar with a needle to arginine glucose slant (AGS). Streak the slant, stab the butt, and incubate with the cap loose overnight at 35° ± 2°C. Both *V. parahaemolyticus* and *V. vulnificus* produce an alkaline (purple) slant and an acid (yellow) butt (arginine dihydrolase negative), but no gas or H_2S in AGS.

For TSB and TSA slants, inoculate both media and incubate overnight at 35° ± 2°C. These cultures provide inocula for other tests as well as material for the Gram stain and for microscopic examination. Both *V. parahaemolyticus* and *V. vulnificus* are oxidase positive, gram negative, pleomorphic organisms exhibiting curved or straight rods with polar flagella.

Inoculate a tube of motility test medium by stabbing the column of the medium to a depth of approximately 5 cm. Incubate overnight at 35° ± 2°C. A circular outgrowth from the line of stab constitutes a positive test. *V. parahaemolyticus* and *V. vulnificus* are motile.

V. parahaemolyticus and *V. vulnificus* will only grow in T1N3 and not in T1N0. Only the salt-requiring cultures need to be tested further.

Only motile, gram negative rods that produce an acid butt and an alkaline slant on AGS and do not form H_2S or gas and are salt requiring are examined further.

The identifying characteristics of *V. parahaemolyticus* and *V. vulnificus* are presented in Table 1.

Biochemically, *V. parahaemolyticus* and *V. vulnificus* are phenotypically similar, but can be differentiated by differences of the ONPG, halophilism, cellobiose and lactose reactions (Table 1). By using selected biochemical traits, *V. parahaemolyticus* and *V. vulnificus* can be distinguished from most interfering marine vibrios and other marine microorganisms.

It is recommended that all *V. parahaemolyticus* isolates be tested for the presence of urease, by either using urea broth supplemented with 3% NaCl or on Christensen's urea agar supplemented with NaCl, 3% final concentration. Clinical strains from the US West Coast and from Asian countries have predominantly been urease positive and this production is correlated with the presence of the *tdh* and/or *trh* genes.[2,48,58,85,88,110,111] The urease reaction is a valuable screening test for the potentially pathogenic strains.[58]

Inoculate urea broth—3% NaCl with a heavy inoculum of culture or spot culture to surface of Christensen's-urea-NaCl agar plate or slant. Incubate 35° ± 2°C 18–24 hr. Production of urease is determined by a pink (alkaline) color to the medium. Negative cultures should be incubated an additional 24 hr for the rare, slow urease producing strains.

When the colonies are finally identified biochemically as *V. parahaemolyticus* (see below) refer to the original positive dilutions in the enrichment broth and apply the 3-tube-MPN tables for final enumeration of the organism (Chapter 6).

40.523 Hydrophobic Grid Membrane Filtration Enumeration Procedure (HGMF)[25,32]

Also see Chapter 6. The apparatus, filters, and specific instructions may be obtained from QA Laboratories, San Diego, CA.

Prepare a 1:10 dilution of seafood sample with peptone Tween-salt diluent (PTS), and blend 60 sec at high speed. Filter 1.0 mL or other volume of homogenate through HGMF using sterile diluent as a carrier. With forceps, aseptically transfer the HGMF from the filtration apparatus to the surface of a dry tryptic soy agar magnesium sulfate NaCl plate (TSAMS). Incubate 4 hr at 35° ± 2°C.

With forceps, aseptically transfer the HGMF from the TSAMS to the surface of a dry *Vibrio parahaemolyticus* sucrose agar (VPSA) plate. Invert plate and incubate at 42°C for 18 to 20 hr.

On VPSA, *V. parahaemolyticus* colonies will be green to blue filling at least one-half of the grid square. Other growth will normally be yellow due to sucrose fermentation. Count squares with green/blue colonies and calculate the MPN/g of seafood. *V. vulnificus* colonies will also be blue/green in color. DNA probes can differentiate the species.[28,57,74,129]

40.524 Serologic Typing[46,78]

V. parahaemolyticus possesses three antigenic components: H, O, and K. The H antigen is common to all strains of *V. parahaemolyticus* and is of little value in serotyping. The K, or capsular, antigen may be removed from the bacterial body by heating the isolate for 1 or 2 hr at 100°C. This process exposes the O, or somatic, antigen, which is thermostable. Since the K antigen masks the O antigen, it is necessary to remove the former by heating before performing the O agglutination tests.

There are 12 O group and 65 K antigens.[46] Five of the K antigens have been found to occur with either of two O group antigens; therefore, there are 76 recognized serotypes, Table 3. Serologic tests by themselves are not used to identify *V. parahaemolyticus* because of cross-reactions with many other marine organisms. However, during investigations of foodborne outbreaks, serologic tests become a valuable epidemiologic tool.

V. parahaemolyticus diagnostic antiserum kits are produced commercially in Japan and available from Nichimen Co., New York, NY 10036; or Accurate Chemical and Science Corp, San Diego, CA. Because the antiserum is expensive, it is not recommended for most laboratories.

40.525 Determining Pathogenicity

Kato et al.[55] showed that *V. parahaemolyticus* isolates from the stools of patients with enteric infections are hemolytic on a special high-salt human blood agar, whereas *V. parahaemolyticus* isolates from seafood and marine water usually are not. Wagatsuma[124] later modified this special agar to avoid confusion with the regular normal hemolytic activity of *V. parahaemolyticus* on conventional 5% sheep blood agar. The special agar was named Wagatsuma agar and the special hemolytic response the Kanagawa phenomenon. Freshly drawn human, dog or sheep blood is used in preparation of the agar.[31]

The correlation has been well established that *V. parahaemolyticus* strains that cause illness in humans are almost always Kanagawa-positive and isolates recovered from seafood are almost always Kanagawa-negative.[78,97,102] In addition, extensive investigation in animal models suggests that the Kanagawa hemolysin is the primary virulence factor in *V. parahaemolyticus*.[117] The Kanagawa test, or hybridization with the *tdh* gene probe provides reliable information on the presence of pathogenic strains isolated from foods.

Kanagawa test. Inoculate suspect organism into TSB-3% NaCl and incubate overnight at 35° ± 2°C.

Drop a loopful of culture on a previously well-dried Wagatsuma blood agar plate. Several drops in a circular pattern may be made on a single plate. Plates may be divided into sections to test more than one isolate per plate. Include known Kanagawa-positive and negative strains on each plate.

Incubate at 35° ± 2°C and read in less than 24 hr. A positive test consists of β-hemolysis, a zone of transparent clearing of the blood cells around the area of growth. Alpha-hemolysis is considered a negative test. No observation made beyond 24 hr is valid. A PCR procedure[8] and gene probe procedure for *tdh* detection are presented below as alternatives to the Kanagawa in-vitro test.

Table 3. Antigenic Scheme of *V. parahaemolyticus* (1986)[a]

O group	K type
1	1, 25, 26, 32, 38, 41, 56, 58, 64, 69
2	3, 28
3	4, 5, 6, 7, 27, 30, 31, 33, 37, 43, 45, 48, 54, 57, 58, 59, 65
4	4, 8, 9, 10, 11, 12, 13, 34, 42, 49, 53, 55, 63, 67
5	5, 15, 17, 30, 47, 60, 61, 68
6	6, 18, 46
7	7, 19
8	8, 20, 21, 22, 39, 70
9	9, 23, 44
10	19, 24, 52, 66, 71
11	36, 40, 50, 51, 61
12	52
Total	
12	65

[a] The antigenic scheme was first established by Sakazaki et al.[98] and later extended by the

40.526 Genotypic Detection of Hemolysin Genes of V. parahaemolyticus

Alkaline phosphatase- and digoxigenin-labeled DNA probes can be used for the identification of *V. parahaemolyticus*. A thermolabile hemolysin gene, *tl*, has been found in all strains of *V. parahaemolyticus*, but not in other species[119] and DNA probes have been used for identification. Two DNA probe procedures that have been shown to be equivalent are presented. DNA probes have also been constructed to detect the thermostable direct hemolysin, *tdh*[83] and thermostable related hemolysin, *trh*,[45] genes that are associated with pathogenic strains.

An alkaline phosphatase-labeled (AP) *tl* probe[74] is commercially prepared for use with Whatman #541 colony lifts. The hybridization and detection procedure for the AP-*tl* and AP-*tdh* probes are presented below, using a hybridization and wash temperature of 54°C. Digoxigenin-labeled probes for *tl* and *trh* were constructed of PCR amplification products using the primer sets reported by Bej et al.[8] The *tdh* probe was constructed using a primer set based on the oligonucleotide probes of Nishibuchi et al.,[83] using *tdh1* as the forward primer and the reverse compliment of *tdh4 (tdh4c)* as the reverse primer. The probes are labeled with digoxigenin during amplification according to the procedure described by Boehringer-Mannheim.[13] These amplicons are of the following sizes; 450 bp *tl*, 424 bp *tdh* and 500 bp *trh*.

Alkaline phosphatase-labeled oligonucleotide probes (AP-*tl* and AP-*tdh*)74. Probe sequences are:

Species specific thermolabile hemolysin (*tl*)

AP-labeled 5′Xaa agc gga tta tgc aga agc act g 3′ (where X is the AP-label)

Thermostable direct hemolysin (*tdh*), the Kanagawa hemolysin,

AP-labeled 5′Xgg ttc tat tcc aag taa aat gta ttt g 3′

Probes can be purchased from DNA Technology ApS, Science Park Aarhus, Gustav Wledsd Vej 10, DK-8000, Aarthus C, Denmark.

Sample preparation and dilutions are the same as with MPN procedure. In addition, the preparation of sample and the hybridization conditions are the same for the simultaneous enumeration of *V. vulnificus*, except plating to VVA and a 55°C hybridization temperature.[28] Just before use, thoroughly dry T1N3 (and VVA) agar plates inverted with lids cracked open for 1 hr at 35°C. This is a non-selective agar.

Pool 10–12 oyster meats and homogenize in equal part, by weight, with PBS for 90 sec at high speed (1:2 dilution).[4]

Weigh 0.20 g of this oyster:PBS (1:2) homogenate directly from blender (represents 0.1 g or 1:10 dilution of oyster tissue) onto T1N3 plate using balance with 0.01g sensitivity to tare plate. Pipette 100 µL of 1:10, 1:100 and 1:1000 dilutions onto T1N3 plates labeled 1:100, 1:1000 and 1:10000, respectively. For shellfish harvested between December through March plating 1:10 and 1:100 dilutions is adequate and while during May through October, summer months, 1:100, 1:1000 and 1:10000 dilutions are adequate. The initial dilution of 1:10 of other seafood products can be spotted (0.1 g) to T1N3, plus further dilutions if necessary. The detection level is 10 CFU/g.

For seafood other than oysters, the initial dilution of 1:10 should be used due to the product debris from homogenizing. Inoculate the surface of T1N3 with 100 µL of this 1:10 dilution of sample. Thus the detection level will be 100/g.

Use sterile spreading sticks to spread inoculum evenly onto T1N3 agar plates. Dry plates and uniform distribution of inoculum are essential for adequate colony isolation. Incubate plates 18–24 hr at 35° ± 2°C. All plates should be used for colony lifts and hybridization unless there is confluent growth.

Filter preparation. Overlay T1N3 (and VVA) plates with #541 Whatman filters (85 mm) for 30 min. Transfer filters with colony side up to #3 Whatman filters, or equivalent, that have been prewetted with lysis solution (4 mL). Microwave in glass Petri dishes (full power) for 30 sec/filter; rotate dishes with filters and repeat microwaving. Filters should be hot and almost completely dry but not brown. Microwave maximum of 6 filters at one time.

Wet #3 Whatman filter with ammonium acetate buffer (4 mL). Place steamed filters colony side up on #3 filter. Incubate 5 min at room temperature.

Briefly rinse #541 Whatman filter 2 times in 1X SSC buffer (10 mL/filter). Up to 10 filters can be combined in one container. (Filters can be dried and stored at this point.)

Proteinase K treatment. Incubate up to 30 filters in proteinase K solution (10 mL/filter) for 30 min at 42°C with shaking (50 rpm) in plastic container to destroy naturally occurring alkaline-phosphatase and digest bacterial protein.

Rinse filter 3 times in 1X SSC (10 mL/filter) for 10 min at room temperature with shaking, 50 rpm. (Filters can be dried and stored at this point.)

Hybridization. In a plastic bag, presoak filter in hybridization buffer for 30 min at 54°C (correct temperature is critical, 55°C for *V. vulnificus*) with shaking, 50 rpm. Use maximum of 5 filters/bag with 10-15 mL of buffer.

Pour off buffer from bag and add 10 mL fresh pre-warmed buffer/filter. Add probe (final concentration is 0.5 pmol/mL) to bag with filters and incubate 1 hr at 54°C (55°C for *V. vulnificus*) with shaking.

Rinse filter 2 times for 10 min each in SSC/SDS (10 mL/filter) at 54°C (55°C for *V. vulnificus*) with shaking. In plastic container, rinse filter 5 times for 5 min each in SSC at room temperature with shaking, 100 rpm.

Color development. In Petri dish add 20 mL of NBT/BCIP solution. Add filters (5 or less) to dish and incubate shaking at room temperature, covered to omit light. Check development of positive control every 30 min; usually complete by 1–2 hr.

Rinse in tap water (10 mL/filter) 3 times for 10 min each to stop development. Do not expose filters to light as they will continue to develop. Store in dark or under acetate holders.

***V. parahaemolyticus* enumeration by digoxigenin-labeled probes.** The primer sequences for amplicon preparation of digoxigenin-labeled probes and the PCR conditions for construction:

tl	thermolabile hemolysin[8]
L-TL	5′ aaa gcg gat tat gca gaa gca ctg 3′
R-TL	5′ gct act ttc tag cat ttt ctc tgc 3′
tdh	thermostable direct hemolysin (Kanagawa hemolysin)[83]
tdh-1	5′ cca tct gtc cct ttt cct gcc 3′
tdh-4c	5′ cca cta cca ctc tca tat gc 3′
trh	thermostable related hemolysin[8]
VPTRH-L	5′ ttg gct tcg ata ttt tca gta tct 3′
VPTRH-R	5′ cat aac aaa cat atg ccc att tcc g 3′

PCR conditions	*tl* and *trh*		*tdh*	
Hold	94°C	3 min	94°C	10 min
Denature	94°C	1 min	94°C	1 min
Anneal	60°C	1 min	58°C	1 min
Extend	72°C	2 min	72°C	1 min
Hold	72°C	3 min	72°C	10 min
Hold	8°C	indefinite	8°C	indefinite
		25 cycles		25 cycles

Triplicate membranes should be inoculated if hybridizations with the three *V. parahaemolyticus* digoxigenin-labeled probes are desired. Densities of *V. parahaemolyticus* can be determined with the *tl* probe and the densities of the pathogenic strains can be determined using *tdh* and *trh* hybridizations. The results are reported as the respective number/g.

Dilutions of a sample are prepared as previously described. One hundred µL volumes are directly plated to labeled nylon membranes on the surface of well dried T1N3 agar plates. A separate membrane is used for each of the three probes. Normally, the 1:10, 1:100, and 1:1000 dilutions are adequate. The minimum detection level is 100/g. A sterile spreading stick is used to gently spread the inoculum over the membrane surface.

Incubate the T1N3 plates for 3 hr at 35° ± 2°C. With forceps, transfer the membrane to the surface of a TCBS agar plate and incubate overnight at 35° ± 2°C.

After incubation, estimate the number of green colonies and proceed with the hybridization steps. The microtiter plate system of retention of suspect isolates can be used at this point by picking green colonies to plate wells with sterile toothpicks.

The colonies on membranes are lysed by placing them colony side up on an absorbent pad containing 4 mL lysis solution for 30 min at room temperature. A slight heating by microwaving for 20 sec may also be used.

The membranes are then transferred with forceps to an absorbent pad containing 4 mL neutralizing solution for 30 min at room temperature.

Dry membranes briefly on a paper or cloth towel, then cross-link the DNA to the membrane for 3 min under UV light source, 254 nm, or in a UV cross-linker.

Hybridization with digoxigenin-labeled probes and colorimetric or chemiluminescent detection.

Day 1. Hybridization

Warm shaker water bath to 65°C.

Place membrane(s) in heat tolerant, sealable bag or plastic container with lid. Membranes can be stacked back to back with a fiberglass mesh screen spacer between each pair.[61]

Cover the stack with pre-hybridization solution. Incubate submerged at 65°C for 1 hr.

Remove membrane(s) from bag, container and gently wipe each with a lab tissue wetted with pre-hybridization solution. This step removes excess colony material. (This step is optional depending on colony size).

Cover membrane(s) with pre-hybridization solution and incubate at 65°C submerged for 2 hr. Longer pre-hybridization times are acceptable.

Boil double stranded dig-probe for 10 min.

Pour off pre-hybridization solution and add probe while still hot. Hybridize submerged at 65°C overnight.

Day 2. Wash and Detection Steps

After hybridization, **save probe solution** in a heat resistant plastic tube in freezer. Probe can be stored up to one year.

Remove membrane(s) to plastic tray and wash twice (in a stack) by covering in Wash solution A for 5 min at room temperature on shaker, 50–100 rpm.

Wash membrane(s) twice by covering with pre-warmed Wash solution B in a bag or container at 65°C for 15 min.

Prepare Genius Dig buffer 2 by adding 0.25 g powdered blocking reagent to 50 mL of Genius Dig Buffer 1, agitate vigorously, and microwave or put in 65°C bath to dissolve blocking reagent. Agitate every 10 min until dissolved, then cool to room temperature.

For chemiluminescent detection.

Warm CSPD reagent to room temperature.

Remove membrane(s) from bag/container, cover in dish with Dig buffer 1 for 1 min at room temperature.

In a container/bag, cover membrane(s) with Dig buffer 2 and incubate at room temperature on shaker tray for 1 hr, 50 rpm. Longer blocking times are acceptable.

Drain off Dig buffer 2.

Cover membrane(s) with Dig buffer 2 with added anti-dig-alkaline phosphatase at 1:5000 (add 5 µL anti-dig for each 25 mL Dig buffer 2).

Incubate for 30 min at room temperature on shaker tray, 50 rpm.

For colorimetric detection.

Prepare color solution by adding one NBT/BCIP tablet dissolved in 10 mL of Dig buffer 3.

Pour out antibody solution. Cover membrane with Dig buffer 1 in plastic tray with shaking for 15 min at room temperature, 50 rpm. (*Use a freshly washed dish or bag, one that has not been in contact with anti-Digoxigenin*).

Pour off Dig buffer 1 and cover again in Dig buffer 1. Incubate 15 min at room temperature with shaking.

Pour off Buffer 1 and cover in tray with Dig buffer 3 for 3 min at room temperature.

For Chemiluminescent Detection.

Wipe off acetate document holder with 95% ethanol. Place membrane on acetate, add 0.5 mL CSPD/100 sq cm of membrane. Wipe off outer surfaces of acetate, put in cassette with X-ray film. Expose (usually no longer than 1 hr) and develop film as per manufacturer's specifications.

For Colorimetric Detection.

Add approximately 10 mL NBT/BCIP color substrate solution per 2 membranes. Incubate in bag or dish **in dark** at room temperature. Do not shake container while color is developing. The color precipitate starts to form within a few minutes and is usually complete after 12 hr.

Once desired spots are detected, comparing with known control spots, wash membrane with 50 mL Dig buffer 1 for 5 min to stop reaction.

The membrane can be stored, damp, in a bag after a brief rinse in Dig buffer 4 to retain color.

Probe stripping.

Nylon membranes can be stripped and re-probed if desired. The specifics are outlined in the Boehringer-Mannheim manual.[13]

Multiplex PCR identification of *V. parahaemolyticus*[8]

V. parahaemolyticus multiplex PCR analysis, an alternative confirmation step for suspect isolates.

Culture templates are prepared by growing overnight at 35° ± 2°C in TSB ± 2% NaCl. One mL of culture is centrifuged 15,000 x g for 3 min in a microcentrifuge tube. The pellet is washed twice with physiological saline. The pellet is resuspended in 1.0-mL dH_20 and boiled 10 min. Template can be stored at –20°C until used.

Three primer sets:

1. *tl* gene species specific (450 bp) same as above.

L-TL	5′ aaa gcg gat tat gca gaa gca ctg 3′
R-TL	5′ gct act ttc tag cat ttt ctc tgc 3′

2. *trh* gene (500 bp) same as above.

VPTRH-L	5′ ttg gct tcg ata ttt tca gta tct 3′
VPTRH-R	5′ cat aac aaa cat atg ccc att tcc g 3′

3. *tdh* gene (270 bp)

VPTDH-L	5′ gta aag gtc tct gac ttt tgg ac 3′
VPTDH-R	5′ tgg aat aga acc ttc atc ttc acc 3′

Reaction volume (final concentrations are the same as those in section 40.521)

dH_2O	28.2 µL
10X Buffer $MgCl_2$	5.0 µL
dNTPs	8.0 µL
primer mix (6 primers)	7.5 µL
template	1.0 µL
Taq polymerase	0.3 µL
Total volume	50.0 µL

PCR conditions:	denature	94°C 3 min
		94°C 1 min
	anneal	60°C 1 min
	extend	72°C 2 min
	final extend	72°C 3 min
	hold	8°C indefinite
		25 cycles

Agarose gel analysis of PCR products. Mix 10 µL PCR product with 2 µL 6X loading gel and load sample wells of 1.5 to 1.8% agarose gel containing 1 µg/mL ethidium bromide submerged in 1X TBE. Use a constant voltage of 5 to 10 V/cm. Illuminate gel with a UV transluminator and visualize bands relative to molecular weight marker migration. Polaroid photographs can be taken of the gel for documentation. Positive and negative culture controls and reagent control should be included with each PCR run.

40.53 V. vulnificus

40.531 Identification and Enumeration Method

Two analytical schemes for isolating and enumerating *V. vulnificus* are described. The first is a most probable number (MPN) analysis that uses an EIA procedure for identification of suspect isolates. This method also is described in the FDA BAM[31] and has been used in a number of laboratories for the analysis of oysters and environmental samples. Two direct plating methods employing hybridization with a DNA probe for colony identification have been used in several studies and have been shown to be equivalent to the MPN method.[28,129] A gas chromatographic technique that compares fatty acid profiles has also been successful for identifying *V. vulnificus*.[67]

40.532 Seafood Samples

Enrichment, isolation, and enumeration. Prepare an initial 1:10 dilution of sample in PBS following the procedure for *V. parahaemolyticus* (section 40.521). Prepare decimal dilutions in PBS. Prepare a 1:10 dilution of the oyster homogenate as follows: Weigh 20 grams of the homogenate into a sterile bottle and add PBS to dilute to a final weight of 100 g. Mix by shaking.

Inoculate 3 × 1 mL portions of the dilutions into 3 tubes containing 10 mL APW. If low numbers are expected 2-gram portions (1 g of oyster) directly from the blender can be inoculated in triplicate into tubes containing 100 mL APW. Incubate tubes 12 to 16 hr at 35° ± 2°C.

Streak a 3-mm loopful from the top 1 cm of APW tubes with growth onto mCPC or CC selective agars. Incubate mCPC and CC agars overnight at 40° ± 1°C. On either agar, colonies are round, flat, opaque, yellow, and 1 to 2 mm in diameter.

Upon identification of *V. vulnificus*, refer to the original positive dilutions of APW and apply the 3-tube-MPN tables (Chapter 6) for final enumeration of the organism.

40.533 Biochemical Identification of Isolates

Unless otherwise specified, all media in this section are prepared to contain a minimum of 0.5% NaCl.

Transfer two or more suspicious colonies with a needle from CC or mCPC agar plates to TSA and streak for isolation.

Inoculate biochemical media using a single colony. Screening reactions, AGS, oxidase reaction, motility, and salt tolerance, are used as for *V. parahaemolyticus*, section 40.522.

API 20E diagnostic strips can also be used. Prepare culture suspension in 2% NaCl solution. Biochemical reactions to differentiate *V. vulnificus* from *V. parahaemolyticus* can be found in Table 1.

40.534 Enzyme Immunoassay

This procedure is part of the FDA BAM[31] and presently the monoclonal antibody is only available to FDA laboratories. The stepwise procedure is outlined in the FDA BAM. Contact FDA, Gulf Coast Seafood Laboratory, Dauphin Island, AL 36528, for availability of the monoclonal antibody.

40.535 DNA Probe Identification of Species Specific Cytolysin Gene, vvhA[28,129]

The oligonucleotide sequence for alkaline phosphatase label:

5′ Xga gct gtc acg gca gtt gga acc a 3′

Source of this probe is the same as for *V. parahaemolyticus*, section 40.5241.

Sample preparation and dilutions are same as with MPN procedure and that presented for the AP-probe hybridization for *V. parahaemolyticus* except dry Vibrio vulnificus agar (VVA) is the plating medium.

Weigh 0.20 g of oyster:PBS homogenate directly from blender (0.1 g of oyster tissue and 1:10 dilution) onto VVA plate using balance with 0.01g sensitivity to tare plate.

Pipette 100 µL of 1:10, 1:100 and 1:1000 dilutions on VVA plates labeled 1:100, 1:1000 and 1:10000, respectively. From December through March plating 1:10 and 1:100 dilutions is adequate and May through October, summer months, 1:100, 1:1000 and 1:10000 dilutions are adequate.

Use sterile spreading sticks to spread oyster inoculum evenly onto VVA plates. Dry plates and uniform distribution of inoculum are essential for adequate colony isolation. Incubate plates 18–24 hr at 35°±2°C. Relatively large (1–2 mm) yellow opaque colonies (fried egg appearance) are typical of *V. vulnificus* on VVA.

Streaking control cultures will aid colony counting to select plates for colony lifts and to select isolates for identification and storage.

Plates with 25–250 typical colonies should be used for colony lifts and isolate selection if available. Additional dilutions can be used if uncertain. Colony lifts from plates with confluent growth or no typical colonies will probably be unproductive.

Whatman #541 filters of colony lifts are lysed and neutralized as described previously.

The microtiter plate system for multiple culture retention can be used at this point.

Filter preparation and Proteinase K treatment follow the procedures outlined in section 40.526.

40.536 Enumeration of V. vulnificus *by DNA Gene Probe*

The hybridization steps are the same as for *V. parahaemolyticus*, except the temperature for hybridization and washing of filters is 55°C. All other steps including colorimetric detection are the same.

40.537 Confirmation of V. vulnificus by Polymerase Chain Reaction[40]

Isolates obtained by the MPN procedure can be confirmed by PCR.

Primers for PCR *vvhA* or dig-labeling of probe for enumeration (519 base amplicon) are from base 785 to 1303 of the cytolysin gene.

Vvh-785F 5′ ccg cgg tac agg ttg gcg ca 3′
Vvh-1303R 5′cgc cac cca ctt tcg ggc c 3′

Reagent concentrations are the same as in section 40.517. These primers can also be used to construct a dig-labeled amplicon for identification and enumeration of the organism.

PCR conditions:		
	denature	94°C 10 min
		94°C 1 min
	anneal	62°C 1 min
	extend	72°C 1 min
	final extend	72°C 10 min
	hold	8°C indefinite
		25 cycles

Agarose gel analysis of PCR products. Mix 10 µL PCR product with 2 µL 6X loading gel and load sample wells of 1.5 to 1.8% agarose gel containing 1 µg/mL ethidium bromide submerged in 1X TBE. Use a constant voltage of 5 to 10 V/cm. Illuminate gel with a UV transluminator and visualize bands relative to molecular weight marker migration. Polaroid photographs can be taken of the gel for documentation. Positive and negative culture controls and reagent control should be included with each PCR run.

Culture templates are prepared by growing overnight at 35° ± 2°C in TSB ± 2% NaCl. One mL of culture is centrifuged 15,000 × g for 3 min. The pellet is washed twice with physiological saline. The pellet is resuspended in 1.0 mL dH_20 and boiled 10 min. Template can be stored at –20°C until used.

The gene probe enumeration of *V. vulnificus* using dig-*vvh* follows the same direct plating procedure outlined in section 28.526 for *V. parahaemolyticus*, except VVA is used. The hybridization and wash temperature are the same, 65°C.

40.6 INTERPRETATION

Contamination of food or water with enterotoxigenic *V. cholerae*, or *V. mimicus* (although rarely encountered) is a significant public health finding. The entire lot of contaminated food should be withheld from distribution until the appropriate health authorities are notified and an epidemiologic investigation can be undertaken. The serogroup, biotype and enterotoxigenicity results should be reported for each sample.

The isolation of *V. parahaemolyticus* from seafood is not unusual. *V. parahaemolyticus* is a normal saprophytic inhabitant of the coastal marine environment and multiplies during the warm summer months.[26] During this period the organism is readily recovered from most of the seafood species harvested in coastal areas. The virulent strains are separated from the avirulent strains of *V. parahaemolyticus* by means of the Kanagawa test or gene detection.[117] In most instances, *V. parahaemolyticus* Kanagawa-negative seafood strains do not cause human gastroenteritis. The presence of Kanagawa positive strains, or *tdh* and/or *trh* containing strains, constitutes a public health concern. A heat-processed product should not contain viable *V. parahaemolyticus* and if so, would indicate a significant problem in manufacturing practices or post-process contamination.

During the summer months, shellfish normally will contain *V. vulnificus* and high levels have been isolated from warm estuarine areas.[118] Most strains isolated have been demonstrated to be potentially virulent. Clinical, environmental and food isolates have been found to be highly virulent to mice[59,87,109,120], but infections are relatively rare even among those of increased risk, e.g., liver disease. However, those individuals at risk should be cautioned to consume raw shellfish during certain periods of the year when levels of *V. vulnificus* are increased, normally May through October.[81,100] As with *V. parahaemolyticus*, a heat processed product should not contain viable *V. vulnificus* and its isolation is a significant finding.

40.7 ACKNOWLEDGMENT

The authors thank colleague, Robert M. Twedt, retired, whose contribution to this chapter originated in the first editions of the *Compendium* and as a co-author in the third edition.

40.8 REFERENCES

1. Abbott, S. L., and J. M. Janda. 1994. Severe gastroenteritis associated with *Vibrio hollisae* infections: Report of two cases and review. Clin. Infect. Dis. 18:310-312.
2. Abbott, S. L., C. Powers, C. A. Kaysner, Y. Takeda, M. Ishibashi, S. W. Joseph, and J. M. Janda. 1989. Emergence of a restricted bioserovar of *Vibrio parahaemolyticus* as the predominant cause of *Vibrio*-associated gastroenteritis on the West Coast of the United States and Mexico. J. Clin. Microbiol. 27:2891-2893.
3. Albert, M. J. 1994. *Vibrio cholerae* O139 Bengal. J. Clin. Microbiol. 32:2345-2349.
4. American Public Health Association. 1970. Recommended procedures for the examination of seawater and shellfish. 4th ed. American Public Health Association, Washington, D. C.
5. Arias, C. R., E. Garay, and R. Aznar. 1995. Nested PCR method for rapid and sensitive detection of *Vibrio vulnificus* in fish, sediments, and water. Appl. Environ. Microbiol. 61:3476-3478.
6. Baumann, P., A. L. Furniss, and J. V. Lee. 1984. Genus I. *Vibrio* Pacini 1854, 411[AL]. *In* N. R. Krieg and J. G. Holt (ed.), Bergey's manual of systematic bacteriology. The Williams & Wilkins Co., Baltimore, Md.
7. Baumann, P., and R. H. W. Schubert. 1984. Family II. *Vibrionaceae*, p. 516-550. *In* N. R. Krieg, and J. G. Holt (ed.), Bergey's manual of systematic bacteriology. The Williams & Wilkins Co., Baltimore, Md.
8. Bej, A. K., D. P. Patterson, C. W. Brasher, M. C. L. Vickery, D. D. Jones, and C. A. Kaysner. 1999. Detection of total and hemolysin-producing *Vibrio parahaemolyticus* in shellfish using multiplex PCR amplification of *tl*, *tdh* and *trh*. J. Microbiol. Meth. 36:215-225.
9. Bej, A. K., A. L. Smith III, M. C. L. Vickery, D. D. Jones, and M. H. Mahbubani. 1996. Detection of viable *Vibrio cholerae* in shellfish using PCR. Food Test. Anal. Dec/Jan:16-21.
10. Berche, P., C. Poyart, E. Abachin, H. Lelievere, J. Vandepitte, A. Dodin, and J. M. Fournier. 1994. The novel epidemic strain O139 is closely related to the pandemic strain O1 of *Vibrio cholerae*. J. Infect. Dis. 170:701-704.
11. Blake, P. A., D. T. Allegra, J. D. Snyder, T. J. Barrett, L. McFarland, C. T. Caraway, J. C. Feeley, J. P. Craig, J. V. Lee, N. D. Puhr, and R. A. Feldman. 1980. Cholera—a possible endemic focus in the United States. N. Engl. J. Med 302:305-309.
12. Blake, P. A., R. E. Weaver, and D. G. Hollis. 1980. Diseases of humans (other than cholera) caused by vibrios. Annu. Rev. Microbiol. 34:341-367.
13. Boehringer Mannheim Corp. 1992. The genius system user's guide for filter hybridization, Version 2 ed. Boehringer Mannheim Corp, Indianapolis.
14. Boutin, B. K., A. L. Reyes, J. T. Peeler, and R. M. Twedt. 1985. Effect of temperature and suspending vehicle on survival of *Vibrio parahaemolyticus* and *Vibrio vulnificus*. J. Food Prot. 48:875-878.
15. Bradshaw, J. G., D. W. Francis, and R. M. Twedt. 1974. Survival of *Vibrio parahaemolyticus* in cooked seafood at refrigeration temperatures. Appl. Microbiol. 27:657-661.

16. Brenner, D. J., F. W. Hickman-Brenner, J. V. Lee, A. G. Steigerwalt, G. R. Fanning, D. G. Hollis, J. J. Farmer III, R. E. Weaver, S. W. Joseph, and R. J. Seidler. 1983. *Vibrio furnissii* (formerly aerogenic biogroup of *Vibrio fluvialis*), a new species isolated from human feces and the environment. J. Clin. Microbiol. 18:816-824.
17. Bryan, P. J., R. J. Steffan, A. DePaola, J. W. Foster, and A. K. Bej. 1999. Adaptive response to cold temperatures in *Vibrio vulnificus*. Curr. Microbiol. 38(3):168-175.
18. Centers for Disease Control. 1998. Outbreak of *Vibrio parahaemolyticus* infections associated with eating raw oysters—Pacific Northwest, 1997. Morbid. Mortal. Wkly. Rep. 47(22):457-462.
19. Centers for Disease Control. 1999. Outbreak of *Vibrio parahaemolyticus* infection associated with eating raw oysters and clams harvested from Long Island Sound—Connecticut, New Jersey, and New York, 1998. Morbid. Mortal. Wkly. Rep. 48(3):48-51.
20. Colwell, R. R., R. J. Seidler, J. Kaper, S. W. Joseph, S. Garges, H. Lockman, D. Maneval, H. Bradford, N. Roberts, E. Remmers, I. Huq, and A. Huq. 1981. Occurrence of *Vibrio cholerae* O1 in Maryland and Louisiana estuaries. Appl. Environ. Microbiol. 41:555-558.
21. Cook, D. W. 1997. Refrigeration of oyster shellstock: Conditions which minimize the outgrowth of *Vibrio vulnificus*. J. Food Prot. 60:349-352.
22. Cook, D. W., and A. D. Ruple. 1989. Indicator bacteria and *Vibrionaceae* multiplication in post-harvest shellstock oysters. J. Food. Prot. 52:343-349.
23. Craig, J. P., K. Yamamoto, Y. Takeda, and T. Miwatani. 1981. Production of a cholera-like enterotoxin by a *Vibrio cholerae* non-O1 strain isolated from the environment. Infect. Immun. 34:90-97.
24. Davis, B. R., G. R. Fanning, J. M. Madden, A. G. Steigerwalt, H. B. Bradford Jr., H. L. Smith Jr., and D. J. Brenner. 1981. Characterization of biochemically atypical *Vibrio cholerae* strains and designation of a new pathogenic species, *Vibrio mimicus*. J. Clin. Microbiol. 14:631-639.
25. DePaola, A., L. H. Hopkins, and R. M. McPhearson. 1988. Evaluation of four methods for enumeration of *Vibrio parahaemolyticus*. Appl. Environ. Microbiol. 54:617-618.
26. DePaola, A., L. H. Hopkins, J. T. Peeler, B. Wentz, and R. M. McPhearson. 1990. Incidence of *Vibrio parahaemolyticus* in U.S. coastal waters and oysters. Appl. Environ. Microbiol. 56:2299-2302.
27. DePaola, A., and G-C. Hwang. 1995. Effect of dilution, incubation time, and temperature of enrichment on cultural and PCR detection of *Vibrio cholerae* obtained from the oyster *Crassostrea virginica*. Molec. Cell. Probes. 9:75-81.
28. DePaola, A., M. L. Motes, D. W. Cook, J. Veazey, W. E. Garthright, and R. Blodgett. 1997. Evaluation of an alkaline phosphatase-labeled DNA probe for enumeration of *Vibrio vulnificus* in Gulf Coast oysters. J. Microbiol. Meth. 29:115-120.
29. DePaola, A., M. L. Motes, and R. M. McPhearson. 1988. Comparison of the APHA and elevated temperature enrichment methods for recovery of *Vibrio cholerae* from oysters: Collaborative study. J. Assoc. Off. Anal. Chem. 71:584-589.
30. Dobosh, D., A. Gomez-Zavaglia, and A. Kuljich. 1995. The role of food in cholera transmission. Medicina 55:28-32.
31. Elliot, E. L., C. A. Kaysner, L. Jackson, and M. L. Tamplin. 1995. *Vibrio cholerae, V. parahaemolyticus, V. vulnificus* and other *Vibrio* spp, p. 9.01-9.27. *In* L. A. Tomlinson (ed.), Food and Drug Administration Bacteriological Analytical Manual, 8th ed. AOAC International, Gaithersburg, Md.
32. Entis, P., and P. Boleszczuk. 1983. Overnight enumeration of *Vibrio parahaemolyticus* in seafood by hydrophobic grid membrane filtration. J. Food Prot. 46:783-786.
33. Farmer, J. J., III. 1980. Revival of the name *Vibrio vulnificus*. Int. J. Syst. Bacteriol. 30:656.
34. Faruque, S. M., M. J. Albert, and J. J. Mekalanos. 1998. Epidemiology, genetics, and ecology of toxigenic *Vibrio cholerae*. Microbiol. Molec. Biol. Rev. 62:1301-1314.
35. Fields, P. I., T. Popovic, K. Wachsmuth, and O. Ølsvik. 1992. Use of polymerase chain reaction for detection of toxigenic *Vibrio cholerae* O1 strains from the Latin American cholera epidemic. J. Clin. Microbiol. 30:2118-2121.
36. Fujino, T. 1951. Bacterial food poisoning. Saishin Igaku 6:263-271.
37. Guthrie, R. K., C. A. Makukutu, and R. W. Gibson. 1985. Recovery of *Vibrio cholerae* O1 after heating and/or cooling. Dairy Food Sanit. 5:427-430.
38. Hansen, W., J. Freney, H. Benyagoub, M. N. Letouzey, J. Gigi, and G. Wauters. 1993. Severe human infections caused by *Vibrio metschnikovii*. J. Clin. Microbiol. 31:2529-2530.
39. Hickman, F. W., J. J. Farmer III, D. G. Hollis, G. R. Fanning, A. G. Steigerwalt, R. E. Weaver, and D. J. Brenner. 1982. Identification of *Vibrio hollisae* sp. *nov.* from patients with diarrhea. J. Clin. Microbiol. 15:395-401.
40. Hill, W. E., A. R. Datta, P. Feng, K. A. Lampel, and W. L. Payne. 1995. Identification of food-borne bacterial pathogens by gene probes, p. 24.01-24.33. *In* L. A. Tomlinson (ed.), Food and Drug Administration Bacteriological Analytical Manual, 8th ed. AOAC International, Gaithersburg, Md.
41. Hill, W. E., S. P. Keasler, M. W. Trucksess, P. Feng, C. A. Kaysner, and K. A. Lampel. 1991. Polymerase chain reaction identification of *Vibrio vulnificus* in artificially contaminated oysters. Appl. Environ. Microbiol. 57:707-711.
42. Hlady, W. G., R. L. Mullen, and R. S. Hopkins. 1993. *Vibrio vulnificus* from raw oysters. Leading cause of reported deaths from foodborne illness in Florida. Fla. J. Med. Assoc. 80:536-538.
43. Hoi, L., I. Dalsgaard, and A. Dalsgaard. 1998. Improved isolation of *Vibrio vulnificus* from seawater and sediment with cellobiose-colistin agar. Appl. Environ. Microbiol. 64:1721-1724.
44. Honda, S., I. Goto, I. Minematsu, N. Ikeda, N. Asano, M. Ishibashi, Y. Kinoshita, M. Nishibuchi, T. Honda, and T. Miwatani. 1987. Gastroenteritis due to Kanagawa negative *Vibrio parahaemolyticus*. Lancet i:331-332.
45. Honda, T., Y. Ni, and T. Miwatani. 1988. Purification and characterization of a hemolysin produced by a clinical isolate of Kanagawa phenomenon-negative *Vibrio parahaemolyticus* and related to the thermostable direct hemolysin. Infect. Immun. 56:961-965.
46. Hugh, R., and J. C. Feeley. 1972. Report (1966-1970) of the subcommittee on taxonomy of vibrios to the International Committee on Nomenclature of Bacteria. Int. J. Syst. Bacteriol. 2:123-126.
47. Hugh, R., and R. Sakazaki. 1972. Minimal number of characters for the identification of *Vibrio* species, *Vibrio cholerae* and *Vibrio parahaemolyticus*. Public Health Lab. 30:133-137.
48. Iida, T., K-S. Park, O. Suthienkul, J. Kozawa, Y. Yamaichi, K. Yamamoto, and T. Honda. 1998. Close proximity of the *tdh*, *trh* and *ure* genes on the chromosome of *Vibrio parahaemolyticus*. Microbiology 144:2517-2523.
49. Jackson, H. 1974. Temperature relationships of *Vibrio parahaemolyticus*, p. 139-146. *In* T. Fujino, G. Sakaguchi, R. Sakazaki, and Y. Takeda (ed.), International Symposium on *Vibrio parahaemolyticus*. Saikon Publishing Co., Ltd., Tokyo.
50. Ji, S. P. 1989. The first isolation of *Vibrio alginolyticus* from samples which caused food poisoning. Chin. J. Prevent. Med. 23:71-73.
51. Kaneko, T., and R. R. Colwell. 1973. Ecology of *Vibrio parahaemolyticus* in Chesapeake Bay. J. Bacteriol. 113:24-32.
52. Kaper, J., H. Lockman, R. R. Colwell, and S. W. Joseph. 1979. Ecology, serology, and enterotoxin production of *Vibrio cholerae* in Chesapeake Bay. Appl. Environ. Microbiol. 37:91-103.
53. Kaper, J. B., J. G. Morris Jr., and M. M. Levine. 1995. Cholera. Clin. Microbiol. Rev. 8:48-86.
54. Karaolis, D. K., J. A. Johnson, C. C. Bailey, E. C. Boedeker, J. B. Kaper, and P. R. Reeves. 1998. A *Vibrio cholerae* pathogenicity island associated with epidemic and pandemic strains. Proc. Natl. Acad. Sci. USA 95(6):3134-3139.
55. Kato, T., Y. Obara, H. Ichinoe, K. Nagashima, A. Akiuama, K. Takitawa, A. Matsuchima, S. Yamai, and Y. Miyamoto. 1965. Grouping of *Vibrio parahaemolyticus* with a hemolysis reaction. Shokuhin Eisae Kankyu 15:83-86.

56. Kaysner, C. A. 1999. *Vibrio* species. In Press. *In* B. M. Lund, A. D. Baird-Parker, and G. W. Gould (ed.), The microbiological safety and quality of food. Aspen Publishers, Gaithersburg, Md.
57. Kaysner, C. A., C. Abeyta Jr., K. C. Jinneman, and W. E. Hill. 1994. Enumeration and differentiation of *Vibrio parahaemolyticus* and *Vibrio vulnificus* by DNA-DNA colony hybridization using the hydrophobic grid membrane filtration technique for isolation. J. Food Prot. 57:163-165.
58. Kaysner, C. A., C. Abeyta Jr., P. Trost, J. H. Wetherington, K. C. Jinneman, W. E. Hill, and M. M. Wekell. 1994. Urea hydrolysis can predict the potential pathogenicity of *Vibrio parahaemolyticus* strains isolated in the Pacific Northwest. Appl. Environ. Microbiol. 60:3020-3022.
59. Kaysner, C. A., C. A. Abeyta Jr., M. M. Wekell, A. DePaola Jr., R. F. Stott, and J. M. Leitch. 1987. Virulent strains of *Vibrio vulnificus* isolated from estuaries of the U.S. West Coast. Appl. Environ. Microbiol. 53:1349-1351.
60. Kaysner, C. A., and W. E. Hill. 1994. Toxigenic *Vibrio cholerae* O1 in food and water, p. 27-39. *In* I. K. Wachsmuth, P. A. Blake, and O. Ølsvik (ed.), *Vibrio cholerae* and Cholera: New perspectives on a resurgent disease. ASM Press, Washington, D. C.
61. Kaysner, C. A., S. D. Weagant, and W. E. Hill. 1988. Modification of the DNA colony hybridization technique for multiple filter analysis. Molec. Cell. Probes 2:255-260.
62. Kim, J. J., K. J. Yoon, H. S. Yoon, Y. Chong, S. Y. Lee, C. Y. Chon, and I. S. Park. 1986. *Vibrio vulnificus* septicemia: Report of four cases. Yonsei Med. J. 27:307-313.
63. Klontz, K. C., L. Williams, L. M. Baldy, and M. Campos. 1993. Raw oyster-associated *Vibrio* infections: Linking epidemiologic data with laboratory testing of oysters obtained from a retail outlet. J. Food Prot. 56:977-979.
64. Kobayashi, K., K. Seto, S. Akasaka, and M. Makino. 1990. Detection of toxigenic *Vibrio cholerae* O1 using polymerase chain reaction for amplifying the cholera enterotoxin gene. Kansenshogaku Zasshi 64:1323-1329.
65. Kobayashi, T., S. Enomoto, and R. Sakazaki. 1963. A new selective isolation medium for the vibrio group on modified Nakanishi's medium (TCBS agar). Jpn. J. Bacteriol. 18:387-392.
66. Koch, W. H., W. L. Payne, and T. A. Cebula. 1995. Detection of enterotoxigenic *Vibrio cholerae* in foods by the polymerase chain reaction, p. 28.01-28.09. *In* L. A. Tomlinson (ed.), Food and Drug Administration Bacteriological Analytical Manual, 8th ed. AOAC International, Gaithersburg, Md.
67. Landry, W. L. 1994. Identification of *Vibrio vulnificus* by cellular fatty acid composition using the Hewlett-Packard 5898A Microbial Identification System: Collaborative study. J. AOAC Int. 77:1492-1499.
68. Lee, C., L-H. Chen, M-L. Liu, and Y-C. Su. 1992. Use of an oligonucleotide probe to detect *Vibrio parahaemolyticus* in artificially contaminated oysters. Appl. Environ. Microbiol. 58:3419-3422.
69. Lee, J. V., T. J. Donovan, and A. L. Furniss. 1978. Characterization, taxonomy, and emended description of *Vibrio metschnikovii*. Int. J. Sys. Bacteriol. 28:99-111.
70. Lee, J. V., P. Shread, A. L. Furniss, and T. N. Bryant. 1981. Taxonomy and description of *Vibrio fluvialis* sp. *nov.* (synonym Group F Vibrios, Group EF6). J. Appl. Bacteriol. 50:73-94.
71. Levine, W. C., and P. M. Griffin. 1993. *Vibrio* infections on the Gulf Coast: results of first year of regional surveillance and the Gulf Coast *Vibrio* Working Group. J. Infect. Dis. 167:479-483.
72. Madden, J. M., W. P. Nematollahi, W. E. Hill, B. A. McCardell, and R. M. Twedt. 1981. Virulence of three clinical isolates of *Vibrio cholerae* non O-1 serogroup in experimental enteric infections in rabbits. Infect. Immun. 33:616-619.
73. McCardell, B. A., J. M. Madden, and D. B. Shah. 1985. Isolation and characterization of a cytolysin produced by *Vibrio cholerae* serogroup non-O1. Can. J. Microbiol. 31:711-720.
74. McCarthy, S. A., A. DePaola, D. W. Cook, C. A. Kaysner, and W. E. Hill. 1999. Evaluation of alkaline phosphatase- and digoxigenin-labeled probes for detection of the thermolabile hemolysin (*tlh*) gene of *Vibrio parahaemolyticus*. Lett. Appl. Microbiol. 28:66-70.
75. McLaughlin, J. C. 1995. *Vibrio*, p. 465-474. *In* P. R. Murray, E. J. Baron, M. A. Pfaller, F. C. Tenover, and R. H. Yolken (ed.), Manual of clinical microbiology, 6th ed. ASM Press, Washington, D. C.
76. McPherson, V. L., J. A. Watts, L. M. Simpson, and J. D. Oliver. 1991. Physiological effects of the lipopolysaccharide of *Vibrio vulnificus* on mice and rats. Microbios 67:141-149.
77. Mintz, E. D., T. Popovic, and P. A. Blake. 1994. Transmission of *Vibrio cholerae* O1. *In* I. K. Wachsmuth, P. A. Blake, and O. Ølsvik (ed.), *Vibrio cholerae* and cholera: molecular to global perspectives. ASM Press, Washington, D. C.
78. Miwatani, T., and Y. Takeda. 1976. *Vibrio parahaemolyticus*: A causative bacterium of food poisoning. Saikon Publishing Co., Ltd., Tokyo.
79. Miyamoto, Y., T. Kato, Y. Obara, S. Akiyama, K. Takizawa, and S. Yamai. 1969. In vitro hemolytic characteristic of *Vibrio parahaemolyticus*: its close correlation with human pathogenicity. J. Bacteriol. 100:1147-1149.
80. Morris, J. G., Jr. 1994. Non-O1 group *Vibrio cholerae* strains not associated with epidemic disease. *In* I. K. Wachsmuth, P. A. Blake, and O. Ølsvik (ed.), *Vibrio cholerae* and cholera: molecular to global perspectives. ASM Press, Washington, D. C.
81. Mouzin, E., L. Mascola, M. P. Tormey, and D. E. Dassey. 1997. Prevention of *Vibrio vulnificus* infections: assessment of regulatory educational strategies. JAMA 278:576-578.
82. Newman, C., M. Shepherd, M. D. Woodard, A. K. Chopra, and S. K. Tyring. 1993. Fatal septicemia and bullae caused by non-O1 *Vibrio cholerae*. J. Am. Acad. Dermatol. 29:909-912.
83. Nishibuchi, M., W. E. Hill, G. Zon, W. L. Payne, and J. B. Kaper. 1986. Synthetic oligodeoxyribonucleotide probes to detect Kanagawa phenomenon-positive *Vibrio parahaemolyticus*. J. Clin. Microbiol. 23:1091-1095.
84. Nishibuchi, M., and J. B. Kaper. 1985. Nucleotide sequence of the thermostable direct hemolysin gene of *Vibrio parahaemolyticus*. J. Bacteriol. 162:558-564.
85. Nolan, C. M., J. Ballard, C. A. Kaysner, J. Lilja, L. B. Williams, and F. C. Tenover. 1984. *Vibrio parahaemolyticus* gastroenteritis: An outbreak associated with raw oysters in the Pacific Northwest. Diag. Microbiol. Infect. Dis. 2:119-128.
86. Okabe, S. 1974. Statistical review of food poisoning in Japan, especially that by *Vibrio parahaemolyticus*, p. 5-8. *In* T. Fujino, G. Sakaguchi, R. Sakazaki, and Y. Takeda (ed.), International Symposium on *Vibrio parahaemolyticus*. Saikon Publishing Co., Ltd., Tokyo.
87. Oliver, J. D. 1989. *Vibrio vulnificus*, p. 569-600. *In* M. P. Doyle (ed.), FoodbornebBacterial pathogens. Marcel Dekker, Inc., New York.
88. Osawa, R., T. Okitsu, H. Morozumi, and S. Yamai. 1996. Occurrence of urease-positive *Vibrio parahaemolyticus* in Kanagawa, Japan with specific reference to the presence of thermostable direct hemolysin (TDH) and TDH-related hemolysin genes. Appl. Environ. Microbiol. 62:725-727.
89. Overman, T. L., J. F. Kessler, and J. P. Seabolt. 1985. Comparison of API20E, API Rapid E and API Rapid NFT for identification of members of the family *Vibrionaceae*. J. Clin. Microbiol. 22:778-781.
90. Pal, A., T. Ramamurthy, R. K. Bhadra, T. Takeda, T. Shimada, Y. Takeda, G. B. Nair, S. C. Pal, and S. Chakrabarti. 1992. Reassessment of the prevalence of heat-stable enterotoxin (NAG-ST) among environmental *Vibrio cholerae* non-O1 strains isolated from Calcutta, India, by using a NAG-ST DNA probe. Appl. Environ. Microbiol. 58:2485-2489.
91. Popovic, T., Ø. Olsvik, P. A. Blake, and K. Wachsmuth. 1993. Cholera in the Americas: Foodborne aspects. J. Food Prot. 56:811-821.
92. Reily, L. A., and C. R. Hackney. 1985. Survival of *Vibrio cholerae* during cold storage in artificially contaminated seafoods. J. Food Sci. 50:838-839.
93. Rippey, S. R. 1994. Infectious diseases associated with molluscan shellfish consumption. Clin. Microbiol. Rev. 7:419-425.

94. Sack, D. A., and R. B. Sack. 1975. Test for enterotoxigenic *Escherichia coli* using Y-1 adrenal cells in miniculture. Infect. Immun. 11:334-336.
95. Safrin, S., J. G. Morris Jr., M. Adams, V. Pons, R. Jacobs, and J. E. Conte Jr. 1988. Non O1 *Vibrio cholerae* bacteremia: Case report and review. Rev. Infect. Dis. 10:1012-1017.
96. Said, B., S. M. Scotland, and B. Rowe. 1994. The use of gene probes, immunoassays and tissue culture for the detection of toxin in *Vibrio cholerae* non-O1. J. Med. Microbiol. 40:31-36.
97. Sakazaki, R., S. Iwanami, and H. Fukumi. 1963. Studies on the enteropathogenic, facultatively halophilic bacteria, *Vibrio parahaemolyticus*. I. Morphological, cultural and biochemical properties and its taxonomic position. Jpn. J. Med. Sci. Biol. 16:161-188.
98. Sakazaki, R., and T. Shimada. 1986. *Vibrio* species as causative agents of food-borne infection. *In* R. K. Robinson (ed.), Developments in food microbiology. 2nd ed. Elsevier Applied Science Publishers, New York..
99. Shandera, W. X., J. M. Johnston, B. R. Davis, and P. A. Blake. 1983. Disease from infection with *Vibrio mimicus*, a newly recognized *Vibrio* species. Ann. Intern. Med. 99:169-173.
100. Shapiro, R. L., S. Altekruse, L. Hutwagner, R. Bishop, R. Hammond, S. Wilson, B. Ray, S. Thompson, R. V. Tauxe, P. M. Griffin, and the Vibrio Working Group. 1998. The role of Gulf Coast oysters harvested in warmer months in *Vibrio vulnificus* infections in the United States, 1988-1996. J. Infect. Dis. 178:752-759.
101. Sharma, C., M. Thungapathra, A. Ghosh, A. K. Mukhopadhyay, A. Basu, R. Mitra, I. Basu, S. K. Bhattacharaya, T. Shimada, T. Ramamurthy, T. Takeda, S. Yamasaki, Y. Takeda, and G. B. Nair. 1998. Molecular analysis of non-O1 non-O139 *Vibrio cholerae* associated with an unusual upsurge in the incidence of cholera-like disease in Calcutta, India. J. Clin. Microbiol. 36:756-763.
102. Shirai, H., H. Ito, T. Hirayama, Y. Nakamoto, N. Nakabayashi, K. Kumagai, Y. Takeda, and M. Nishibuchi. 1990. Molecular epidemiologic evidence for association of thermostable direct hemolysin (TDH) and TDH-related hemolysin of *Vibrio parahaemolyticus* with gastroenteritis. Infect. Immun. 58:3568-3573.
103. Shirai, H., M. Nishibuchi, T. Ramamurthy, S. K. Bhattacharya, S. C. Pal, and Y. Takeda. 1991. Polymerase chain reaction for detection of the cholera enterotoxin operon of *Vibrio cholerae*. J. Clin. Microbiol. 29:2517-2521.
104. Simpson, L. M., V. K. White, S. F. Zane, and J. D. Oliver. 1987. Correlation between virulence and colony morphology in *Vibrio vulnificus*. Infect. Immun. 55:269-272.
105. Skangkuan, Y. H., Y. S. Show, and T. M. Wang. 1995. Multiplex polymerase chain reaction to detect toxigenic *Vibrio cholerae* and to biotype *Vibrio cholerae* O1. J. Appl. Bacteriol. 79:264-273.
106. Smith, H. L., Jr. 1970. A presumptive test for vibrios: The "string" test. Bull. WHO 42:817-819.
107. Spira, W. M., and P. J. Fedorka-Cray. 1984. Purification of enterotoxins from *Vibrio mimicus* that appear to be identical to cholera toxin. Infect. Immun. 45:679-684.
108. Spira, W. M., R. Sack, and J. L. Froehlich. 1981. Simple adult rabbit model for *Vibrio cholerae* and enterotoxigenic *Escherichia coli* diarrhea. Infect. Immun. 32:739-743.
109. Stelma, G. N., Jr., A. L. Reyes, J. T. Peeler, C. H. Johnson, and P. L. Spaulding. 1992. Virulence characteristics of clinical and environmental isolates of *Vibrio vulnificus*. Appl. Environ. Microbiol. 58:2776-2782.
110. Suthienkul, O., M. Ishibashi, T. Iida, N. Nettip, S. Supavej, B. Eampokalap, M. Makino, and T. Honda. 1995. Urease production correlates with possession of the *trh* gene in *Vibrio parahaemolyticus* strains isolated in Thailand. J. Infect. Dis. 172:1405-1408.
111. Suzuki, N., Y. Ueda, H. Mori, K. Miyagi, K. Noda, H. Horise, Y. Oosumi, M. Ishibashi, M. Yoh, K. Yamamoto, and T. Honda. 1994. Serotypes of urease producing *Vibrio parahaemolyticus* and their relation to possession of *tdh* and *trh* genes. Jpn. J. Infect. Dis. 68:1068-1077.
112. Swerdlow, D. L., E. D. Mintz, M. Rodriguez, E. Tejada, C. Ocampo, L. Espejo, K. D. Greene, W. Saldana, L. Seminario, R. V. Tauxe, J. G. Wells, N. H. Bean, A. A. Ries, M. Pollack, B. Vertiz, and P. A. Blake. 1992. Waterborne transmission of epidemic cholera in Trujillo, Peru: Lessons for a continent at risk. Lancet 340:28-32.
113. Swerdlow, D. L., and A. A. Ries. 1993. *Vibrio cholerae* Non-O1 - the eighth pandemic? Lancet 342:382-383.
114. Tacket, C. O., F. Brenner, and P. A. Blake. 1984. Clinical features and an epidemiological study of *Vibrio vulnificus* infections. J. Infect. Dis. 149:558-561.
115. Tada, J., T. Ohashi, N. Nishimura, Y. Shirasaki, H. Ozaki, S. Fukushima, J. Takano, M. Nishibuchi, and Y. Takeda. 1992. Detection of the thermostable direct hemolysin gene (*tdh*) and the thermostable direct hemolysin-related hemolysin gene (*trh*) of *Vibrio parahaemolyticus* by polymerase chain reaction. Molec. Cell. Probes 6:477-487.
116. Takeda, Y., T. Peina, A. Ogawa, S. Dohi, H. Abe, G. B. Nair, and S. C. Pal. 1991. Detection of heat-stable enterotoxin in a cholera toxin gene-positive strain of *Vibrio cholerae* O1. FEMS Microbiol. Lett. 80:23-28.
117. Takeda, Y. 1983. Thermostable direct hemolysin of *Vibrio parahaemolyticus*. Pharmac. Ther. 19:123-146.
118. Tamplin, M. L. 1992. The seasonal occurrence of *Vibrio vulnificus* in shellfish, seawater and sediment of United States coastal waters and the influence of environmental factors on survival and virulence, 87 pp. *In* Final Report. Saltonstall-Kennedy Program. National Marine Fisheries Service, Washington, D. C.
119. Taniguchi, H., H. Hirano, S. Kubomura, K. Higashi, and Y. Mizuguchi. 1986. Comparison of the nucleotide sequences of the genes for the thermolabile hemolysin from *Vibrio parahaemolyticus*. Microb. Pathogenesis. 1:425-432.
120. Tison, D. L., and M. T. Kelly. 1986. Virulence of *Vibrio vulnificus* strains from marine environments. Appl. Environ. Microbiol. 51:1004-1006.
121. Varela, P., M. Rivas, N. Binsztein, M. L. Cremona, P. Herrmann, O. Burrone, R. A. Ugalde, and A. C. C. Frasch. 1993. Identification of toxigenic *Vibrio cholerae* from the Argentine outbreak by PCR for *ctxA1* and *ctxA2-B*. FEBS Lett. 315:74-76.
122. Wachsmuth, I. K., P. A. Blake, and O. Ølsvik. (ed.) 1994. *Vibrio cholerae* and cholera: Molecular to Global Perspectives. ASM Press, Washington, D. C.
123. Wachsmuth, K., O. Ølsvik, G. M. Evins, and T. Popovic. 1994. Molecular epidemiology of cholera, p. 357-370. *In* I. K. Wachsmuth, P. A. Blake, and O. Ølsvik (ed.), *Vibrio cholerae* and cholera: Molecular to global perspectives. ASM Press, Washington, D. C.
124. Wagatsuma, S. 1968. On a medium for hemolytic reaction of *Vibrio parahaemolyticus*. Media Circ. 13:159-162.
125. Warnock, E. W., III, and T. L. MacMath. 1993. Primary *Vibrio vulnificus* septicemia. J. Emerg. Med. 11:153-156.
126. Weber, J. T., E. D. Mintz, R. Canizares, A. Semiglia, I. Gomez, R. Sempertegui, A. D'avila, K. D. Greene, N. D. Puhr, D. N. Cameron, F. C. Tenover, T. J. Barrett, N. H. Bean, C. Ivey, R. V. Tauxe, and P. A. Blake. 1994. Epidemic cholera in Ecuador: Multidrug-resistance and transmission by water and seafood. Epidemiol. Infect. 112:1-11.
127. West, P. A., P. R. Brayton, T. N. Bryant, and R. R. Colwell. 1986. Numerical taxonomy of vibrios isolated from aquatic environments. Int. J. Syst. Bacteriol. 36:531-543.
128. Wright, A. C., Y. Guo, J. A. Johnson, J. P. Nataro, and J. G. Morris Jr. 1992. Development and testing of a nonradioactive DNA oligonucleotide probe that is specific for *Vibrio cholerae* cholera toxin. J. Clin. Microbiol. 30:2302-2306.
129. Wright, A. C., G. A. Miceli, W. L. Landry, J. B. Christy, W. D. Watkins, and J. G. Morris Jr. 1993. Rapid identification of *Vibrio vulnificus* on nonselective media with an alkaline phosphatase-labeled oligonucleotide probe. Appl. Environ. Microbiol. 59:541-546.
130. Wright, A. C., L. M. Simpson, and J. D. Oliver. 1981. Role of iron in the pathogenesis of *Vibrio vulnificus* infections. Infect. Immun. 34:503-507.
131. Yamamoto, K., T. Honda, T. Miwatani, S. Tamatsukuri, and S. Shibata. 1992. Enzyme-labeled oligonucleotide probes for detection

of the genes for thermostable direct hemolysin (TDH) and TDH-related hemolysin (TRH) of *Vibrio parahaemolyticus*. Can. J. Microbiol. 38:410-416.

132. Yoh, M., K. Miyagi, Y. Matsumoto, K. Hayashi, Y. Takarada, K. Yamamoto, and T. Honda. 1993. Development of an enzyme-labeled oligonucleotide probe for the cholera toxin gene. J. Clin. Microbiol. 31:1312-1314.

Chapter 41

Yersinia

Stephen D. Weagant and Peter Feng

41.1 INTRODUCTION

The genus *Yersinia* was initially separated into three species: *Y. pestis*, *Y. pseudotuberculosis* and *Y. enterocolitica*. Based on biochemical profiles and DNA relatedness, this genus was found to be a heterogeneous group and four more species were added: *Y. intermedia*, *Y. frederiksenii*, *Y. kristensenii and Y. ruckeri*.[9] Through additional revisions however, the genus has since grown to include eleven species,[2,9,10,77] three of which, *Y. pestis*, *Y. enterocolitica* and *Y. pseudotuberculosis*, are potentially pathogenic to humans. Of these, *Yersinia pestis*, the agent of the black plagues during medieval times in Europe, is probably the most infamous.[41] Today, *Y. pestis* only appears sporadically as epizootic plague in a number of rodent species; but, human infections can still occur as a result of contact with diseased animals or through flea vectors. Unlike *Y. pestis*, infections by *Y. enterocolitica* and *Y. pseudotuberculosis* primarily cause gastroenteritis and usually involve the consumption of contaminated food or water.[41,68] These two species, often referred to collectively as enteropathogenic *Yersinia*, are the focus of this chapter.

41.11 Yersinia enterocolitica

Y. enterocolitica and other closely related bacteria are ubiquitous, and are isolated frequently from the environment and a variety of foods. The reservoirs for *Y. enterocolitica* are the gastrointestinal tract of domestic and wild animals, including swine, cattle, deer, dogs and rodents as well as frogs, birds and flies. Several studies have shown that the predominant source of virulent *Y. enterocolitica* appears to be the alimentary tract of pigs.[22,41,69] However, a recent study from China, conducted over a period of 11 years, showed no serotype correlation between the animal isolates vs. those that caused infections in humans suggesting that swine, at least in China, may not be the source of *Y. enterocolitica* in humans.[82] In the environment, *Y. enterocolitica* has been found in surface waters, such as streams, lakes and estuarine water and in samples of soil and in animal waste.

The association of human illness with the consumption of *Y. enterocolitica*-contaminated food and water, such as meats (raw pork, lamb, deli luncheon meats), poultry, dairy products and tofu, and unchlorinated water, is well-documented.[5,6,15,22,24,45] Especially refrigerated foods, possibly contaminated at the manufacturing site[5,6,14,66] or in the home,[48] are potential sources of problem with *Yersinia*, because of the ability to proliferate during refrigerated storage.[15] Although unrelated to foods, the ability of *Yersinia* to grow at cold temperatures has also caused problems in clinical situations, namely with bacterial sepsis and endotoxic shock resulting from transfusion of blood or blood products contaminated with *Y. enterocolitica*.[74]

Strains of *Y. enterocolitica* and its closely related species can be grouped serologically based on their heat-stable somatic antigens. Wauters described 54 serogroups for *Y. enterocolitica* and its related species,[75] but Aleksic and Bockemuhl proposed simplifying this to 18 serogroups within species.[1] Not all *Y. enterocolitica* are pathogens. However, pathogenic strains are widely distributed among serogroups and have been identified in serogroups O:1,2a,3; O:2a,3; O:3; O:8; O:9; O:4,32; O:5,27; O:12,25; O:13a,13b; O:19; O:20; and O:21.[15] Those that predominate in human illness belong to serogroup O:3, O:8, O:9, and O:5,27. Initially, it appeared that pathogenic *Yersinia* serotypes were distributed according to geographical niches.[24] Serotype O:9 was found only in Northern Europe, while the O:8 serotype was almost exclusively isolated in the U.S. However, these geographical boundaries for serotype distribution may no longer be valid, as serotype O:9 is now found in the U.S. and isolations of serotype O:8 have also been reported from Europe and Japan.[24] The broader distribution of *Yersinia* serotypes worldwide may be attributed in part to increases in international importation and exportation of meat products.[28] Infringement into natural habitats may also be a factor, as wild rodents in Japan have been found to carry *Y. enterocolitica* serotype O:8 and suspected to be a source of this pathogen in human infections.[32] Serogroups of *Y. enterocolitica* can be further subgrouped by pulsed-field gel electrophoresis of restriction endonuclease digestion profiles. Buchrieser et al.[17] found that these profiles were well conserved within serogroups and may be a useful tool for determining genetic relatedness among strains.

Since not all *Y. enterocolitica* isolates are pathogenic, a number of virulence tests have been proposed to distinguish potentially pathogenic strains. *Yersinia* spp. that cause human yersiniosis carry a 41–48 Mdal plasmid that is closely associated with a number of virulence-related traits.[29,44,83] These include: autoagglutination in certain media at 35–37°C[7,46]; inability to grow in calcium-deficient media at 35–37°C[29] and to bind crystal violet dye at 35–37°C[13]; increased resistance to normal human sera[58]; production of a series of outer membrane proteins at 35–37°C[60]; ability to produce conjunctivitis in guinea pig or mouse (Sereny test)[71,83];

and lethality in adult and suckling mice after intraperitoneal (i.p.) injection of live organisms[7,19,62,64] The presence of the virulence-associated plasmid in *Yersinia* can be detected by agarose gel electrophoresis, by DNA colony hybridization[35,53,80] or by polymerase chain reaction (PCR). Despite the important role of this plasmid in the virulence mechanism, studies have shown that the presence of the plasmid alone is not sufficient for the full expression of virulence in *Yersinia*.[33,61,68] For example, the severity of some plasmid-mediated virulence properties such as mouse lethality and conjunctivitis vary with the presence of genes carried on the bacterial chromosome. This provides strong evidence that chromosomal genes also contributed to *Yersinia* virulence.[58-60]

One such chromosomally determined virulence trait provides enteropathogenic *Yersinia* the ability to invade mammalian cells such HeLa tissue culture cells.[49] Strains that have lost other virulence properties retain HeLa invasiveness, because the invasive phenotype is encoded by chromosomal loci. Two chromosomal genes, *inv* and *ail*, that encode the phenotype for mammalian cell invasion, have been identified in *Y. enterocolitica*.[54,55] Transfer of these genetic loci into *E. coli* confers the invasive phenotype to the *E. coli* host.[54] The *inv* gene allows *Yersinia* to invade several tissue culture cell lines.[54] However, genetic analyses show that *inv* gene sequences are present on both invasive and noninvasive strains of *Y. enterocolitica*.[55,65] Although this suggests that the presence of the *inv* gene in *Y. enterocolitica* may not be directly correlated with invasiveness, further genetic evidence shows that the *inv* gene sequences in the noninvasive isolates are nonfunctional.[59] The *ail* gene also encodes a factor for cell invasion, but shows greater specificity and invades only certain cell lines. Analysis of disease-causing, virulent *Y. enterocolitica* strains showed all these isolates to be tissue culture-invasive and to also carry the *ail* gene.[54] Therefore, the *ail* gene appears to be present only on pathogenic strains and it may be an essential chromosomal virulence factor in *Y. enterocolitica*.[54,55]

Some *Yersinia* also produce a heat-stable enterotoxin (ST), that can be detected by intragastric injection of cultural filtrates into suckling mice. The production of ST, encoded by the *yst* gene, has been observed mainly in isolates of *Y. enterocolitica* and some strains of *Y. kristensenii*; but, not in *Y. pestis* or *Y. pseudotuberculosis*.[20] The ST of *Yersinia* is very similar to the ST of enterotoxigenic *Escherichia coli*,[16] in its mode of action, and physicochemical and antigenic properties. Until recently, however, the role of ST in the disease process of *Yersinia* has been viewed with doubt. Early studies concluded that it was produced only at temperatures below 30°C; hence, not produced in the mammalian host.[16] Also, many virulent strains of *Y. enterocolitica* were found to not produce ST. More recent studies seem to suggest that ST is indeed involved in pathogenesis. Analysis of pathogenic and nonpathogenic *Y. enterocolitica* isolates revealed that the *yst* gene was present only in the pathogenic strains.[21] Studies also showed that recent isolates of pathogenic *Y. enterocolitica* tend to produce ST, but in pathogenic strains of older collections *yst* can become silent, resulting in the absence of toxin production.[20] Infection studies using animal models demonstrated that rabbits infected with ST-producing strains developed diarrhea and most of them died, while rabbits infected with isogenic bacterial strains, that did not produce ST, remained healthy as the uninfected control rabbits.[20] Finally, genetic analysis of *yst* regulation showed that even though the gene was not expressed at 37° C, the pH and increased osmolarity conditions, similar to that in the lumen, can induce *yst* to express at 37°C.[50] Hence, *Yersinia* can produce ST in the infected host and therefore, ST most likely is a virulence factor.

41.12 Yersinia pseudotuberculosis

In the United States, *Y. pseudotuberculosis* is less commonly found than *Y. enterocolitica*, and although it is frequently associated with animals (including swine, birds, rodents and hare), *Y. pseudotuberculosis* has not been implicated in foodborne illness and has only rarely been isolated from soil, water, and foods.[26,73] In Japan, on the other hand, a number of foodborne and waterborne outbreaks have been linked to *Y. pseudotuberculosis*.[27,73] In one case, children were infected after drinking water from a garden pond that was contaminated with feces from a stray cat that carried *Y. pseudotuberculosis*.[27] One study at a hospital in Ireland[3] found that 28% of patients presenting with appendicitis symptoms and 11% of patients presenting with nonspecific abdominal pain had serum titer to *Y. pseudotuberculosis* compared to 1% of controls. Strains of *Y. pseudotuberculosis* appear to be fairly homogenous and there is little or no variation in biochemical reactions, except for the fermentation of the sugars melibiose, raffinose, and salicin. The species is classified serologically into six groups, based on a heat-stable somatic antigen, and each group contains pathogenic strains. Like *Y. enterocolitica*, the presence of the 42-Mdal plasmid in *Y. pseudotuberculosis* has been correlated with virulence, and Gemski et al. have shown that strains in serogroup II and III that harbor the 42-Mdal plasmid are lethal to adult mice.[30]

Some virulence genes of *Y. pseudotuberculosis* are also present on the chromosome.[37,40]An *inv* gene that is homologous, but not identical to the invasion gene of *Y. enterocolitica*, has been identified in *Y. pseudotuberculosis*, and transfer of this *inv* locus into *E. coli* K-12 also conferred the invasive phenotype to *E. coli*.[37] The *inv* gene of *Y. pseudotuberculosis* is thermoregulated,[36,39]and it encodes for a 103 Kdal protein, known as invasin, which binds to specific receptors on mammalian cells and facilitates the entry of bacteria into tissue cells.[38] Virulence tests for *Y. pseudotuberculosis* isolates are not as abundant as for *Y. enterocolitica*. However, some of the same phenotypic tests used for evaluating pathogenicity in *Y. enterocolitica* can also so be used to test for potential virulence in *Y. pseudotuberculosis*.[24,70,79] In addition, DNA probes can be used to test for the presence of the virulence plasmid[53,80] or invasiveness,[23] and the invasion phenotype can also be tested using a HeLa cell acridine orange staining assay.[51,52]

41.2 ANALYSIS FOR *YERSINIA* IN FOODS

Traditional methods for detecting *Y. enterocolitica* and *Y. pseudotuberculosis* still rely on enrichment and selective plating to isolate *Yersinia* from foods.[24,70,79] Many of these use cold-enrichment to take advantage of the ability of Yersinia to outgrow other mesophiles at or below room temperatures[67]; but, others have incorporated antimicrobials to inhibit growth of competitive microflora that are present in foods.[67,76] For example, the inclusion of irgasan during enrichment has been found effective in selecting for *Yersinia* from a variety of foods.[12,72] However, because the *Yersinia* group is heterogeneous, some of these methods may not be effective for isolating all serotypes or biotypes of *Y. enterocolitica*.

There are several differential and selective plating media used for isolating *Yersinia*.[24,70,79] Some are media used to isolate enteric bacteria in general, but others are developed specifically to test for characteristic traits of *Yersinia*.[12] The two selective media, MacConkey[81] and CIN,[67] that are described in the procedures in this chapter, have broad acceptance for use in isolating *Yersinia* from foods. The inclusion of these two selective agars improves the sensitivity of recovery over either used separately.

To identify the isolates as *Yersinia* requires a battery of biochemical tests.[3,70,79] Many miniaturized biochemical kits have been evaluated and are generally good for presumptive identification of *Yersinia* to the genus level; but are not as reliable for speciation.[41,57] Conventional biochemical tests can be used to supplement the kits for speciating *Yersinia* isolates and must be used for complete biotyping of *Y. enterocolitica* isolates.

In addition to traditional culture-based assays, a number of molecular methods have also been developed to detect *Y. enterocolitica* and *Y. pseudotuberculosis*.[47] A DNA probe, specific for the *yadA* gene, was used to specifically detect *Y. enterocolitica* serotype O:3 in naturally contaminated pork products[56] and serotype O:8 in artificially contaminated tofu.[80] Polymerase chain reaction (PCR) amplification was used to detect *Y. enterocolitica* in blood,[25] and PCR coupled with immunomagnetic separation (IMS) was used to detect each of the common pathogenic serogroups of *Y. enterocolitica* in artificially contaminated samples of water and meat.[43] IMS-PCR assays were capable of detecting 10 to 30 CFU/g of a specific *Yersinia* in inoculated beef and pork without culture enrichment and in the presence of 10^6 cells/g of normal flora.[43] However, the IMS procedure selected for only the specific serotype targeted by the antibody linked to the magnetic beads and these purified antibodies are not yet commercially available. PCR based detection methods also have the disadvantage of not isolating the contaminating culture for further confirmatory tests. Despite the promise of these molecular methods, they have not been adopted in routine testing for *Yersinia* in foods for these reasons.

Enteropathogenic *Yersinia* are highly infective; hence, it is important to exercise cautions when working with these pathogens to prevent accidental infection of laboratory workers. Avoid creating aerosols during handling of the organism or mouth pipetting cultures and samples. Surfaces and utensils that are exposed to viable bacteria should be disinfected with the use of 500 ppm sodium hypochlorite or by autoclaving. Spent media used for growing *Yersinia* cultures should be autoclaved before being discarded.

41.21 Microbiological Procedure for Detecting for Yersinia enterocolitica *in Foods*

41.211 Equipment and Materials

1. Incubators, maintained at 10° C, 30° C and 35–37° C.
2. Blender, Oster or equivalent, 8000 rpm, with 500 mL-1 liter jar
3. Sterile glass or plastic petri dishes, 15 × 100 mm
4. Microscope, light 900X and illuminator
5. Sterile syringes, 1 mL; 26–27 gauge needle
6. Disposable borosilicate tubes, 10 × 75 mm; 13 × 100 mm
7. Wire racks to accommodate 13 × 100 mm tubes
8. Vortex mixer
9. Mouse racks and mouse cages
10. Laminar flow animal isolator
11. Anesthetizing jar
12. CO_2, compressed

41.212 Media

1. Peptone sorbitol bile broth (PSBB)
2. MacConkey agar (use presumptive mixed bile salts; DIFCO Mac CS is preferred)
3. Celfsulodin-irgasan-novobiocin (CIN) agar
4. Bromcresol purple broth supplemented individually with the following carbohydrates, each at 0.5%: mannitol, sorbitol, cellobiose, adonitol, inositol, sucrose, rhamnose, raffinose, melibiose, salicin, xylose, and trehalose
5. Christensen's urea agar (plate media or slants)
6. Phenylalanine deaminase agar (plate media or slants)
7. Motility test medium. Add 5 mL of 1% 2,3,5-triphenyl tetrazolium chloride per liter before autoclaving.
8. Tryptone broth, 1%
9. MR-VP broth
10. Simmons citrate agar
11. Veal infusion broth
12. Bile esculin agar
13. Anaerobic egg yolk agar
14. Trypticase (tryptic) soy agar (TSA)
15. Lysine arginine iron agar (LAIA)
16. Decarboxylase basal medium (Falkow) supplemented individually with 0.5% arginine, 0.5% lysine, or 0.5% ornithine
17. Congo Red-brain heart infusion agarose (CRBHO)
18. Pyrazinamidase agar slants
19. PMP broth
20. β-D-glucosidase test (*see* instructions at end of chapter)

41.213 Reagents

1. Gram stain reagents
2. Voges-Proskauer (VP) test reagents
3. Ferric chloride, 10% in distilled water
4. Oxidase test reagent
5. Saline, 0.5% (sterile)
6. Kovacs' reagent
7. 0.5% Potassium hydroxide in 0.5% NaCl, freshly prepared
8. Mineral oil, heavy grade, sterile
9. API 20E system
10. 1% Ferrous ammonium sulfate

41.214 Enrichment

The following simplified procedure is effective for isolating *Yersinia* from food, water, and environmental samples.

1. Analyze samples promptly after receipt, or refrigerate at 4°C. Although *Yersinia* have been recovered from frozen products, freezing of the samples before analysis is not recommended. Aseptically weigh 25-g sample into 225 mL PSBB enrichment broth. Homogenize for 30 s and incubate at 10°C for 10 days.
2. On day 10, remove PSBB from incubator and mix well. For alkaline treatment,[4] transfer one loop-full of enrichment to 0.1 mL of 0.5% KOH in 0.5% saline and mix for 5 s, then, streak one loop-full to a MacConkey plate and another loop-full to a CIN plate. Transfer an additional loop-full from PSBB to 0.1 mL of 0.5% saline and mix 5 s before streaking, as above. Incubate the agar plates at 30° C for 1 day.

41.215 Isolation of Yersinia

After incubation, examine MacConkey agar plates; disregard the red or mucoid colonies and select small (1–2 mm diameter), flat, colorless, or pale pink colonies. Similarly, examine CIN plates and select small (1–2 mm diameter) colonies having deep red center with sharp border surrounded by clear colorless zone with entire edge. Inoculate each selected colony by stabbing with an inoculation needle, into a LAIA slant,[78] Christensen's urea agar plate or slant, and bile esculin agar plate or slant, and incubate for 48 h at 22–26°C. Isolates which give an alkaline slant and acid butt (KA–), with no gas and no H_2S production on LAIA, and which are also urease-positive, are presumptively regarded as *Yersinia*. Discard cultures that produce H_2S and/or any gas in LAIA or are urease-negative. Give preference to typical isolates that fail to hydrolyze (blacken) esculin.

41.216 Identification

To identify presumptive cultures, use growth from LAIA slant to streak a plate of anaerobic egg yolk (AEY) agar and incubate at 22-26° C. Check the growth on AEY for culture purity, then use for lipase reaction (at 2-5 days), oxidase test, and Gram stain. From colonies on AEY plate, inoculate the following biochemical test media and incubate all at 22-26° C for 3 days. Exceptions are one tube of motility test medium and one tube of MR-VP broth, which are incubated at 35-37°C for 24 h.

1. Decarboxylase basal media (Falkow) supplemented with each of 0.5% lysine, arginine, or ornithine; overlay with sterile mineral oil
2. Phenylalanine deaminase agar
3. Two semisolid motility test medium, incubated at 22-26° C and 35-37° C
4. Tryptone broth
5. Indole test (see instructions at 41.3)
6. Two tubes of MR-VP broth. Incubate one at 22-26 °C for the autoagglutination test (see 41.217, below), then, after 48 h incubation, use it for the V-P test (see instructions at 41.3. Incubate the other MR-VP tube at 35-37° C for the autoagglutination test (see 41.217)
7. Bromcresol purple broth with 0.5% of the following filter-sterilized carbohydrates: mannitol, sorbitol, cellobiose, adonitol, inositol, sucrose, rhamnose, raffinose, melibiose, salicin, trehalose, and xylose
8. Simmons citrate agar
9. Veal infusion broth
10. Pyrazinamidase agar slants (48 h) (see instructions at 41.3)
11. β-D-glucosidase test (Incubate at 30°C for 24 h) (see instructions at 41.3)
12. Lipase test. A positive reaction is indicated by oily, iridescent, pearl-like colony surrounded by a precipitation ring and an outer zone of clearing.

Yersinia are oxidase-negative, gram-negative rods. The biochemical profiles shown in Tables 1 and 2 are used to identify *Yersinia* isolates to species and to biotype. Currently, only strains of *Y. enterocolitica* of biotypes 1B, 2, 3, 4, and 5 are known to be pathogenic. These pathogenic biotypes and *Y. enterocolitica* biotype 6 (now reclassified as *Y. molleretti* or *Y. bercovierii*) and *Y. kristensenii* do not hydrolyze esculin rapidly (within 24 h) or ferment salicin (Tables 1 and 2). However, *Y. enterocolitica* biotype 6 and *Y. kristensenii* are relatively rare and can be distinguished from the pathogenic biotypes by their failure to ferment sucrose, and they are also positive for pyrazinamidase.[42] One note of caution

Table 1. Biochemical Characteristics[a] of *Yersinia* species[2,8,9,75]

	Yersinia species										
Reaction	*Y. pestis*	*Y. pseudo-tuberculosis*	*Y. entero-colitica*	*Y. intermedia*	*Y. frederik-senii*	*Y. kristen-senii*	*Y. aldovae*	*Y. rohdei*	*Y. mollaretii*	*Y. bercovieri*	*Y. ruckeri*
Lysine	–	–	–	–	–	–	–	–	–	–	–
Arginine	–	–	–	–	–	–	–	–	–	–	–
Ornithine	–	–	+[c]	+	+	+	+	+	+	+	+
Motility at RT (22-26°C)	–	+	+	+	+	+	+	+	+	+	+
35-37°C	–	–	–	–	–	–	–	–	–	–	–
Urea	–	+	+	+	+	+	+	+	+	+	–
Phenylalanine deaminase	–	–	–	–	–	–	–	–	–	–	–
Mannitol	+	+	+	+	+	+	+	+	+	+	+
Sorbitol	+/–	–	+	+	+	+	+	+	+	+	–
Cellobiose	–	–	+	+	+	+	–	+	+	+	–
Adonitol	–	–	–	–	–	–	–	–	–	–	–
Inositol	–	–	+/–(+)	+/–(+)	+/–(+)	+/–(+)	+	–	+/–	–	
Sucrose	–	–	+[c]	+	+	–	–	+	+	+	–
Rhamnose	–	+	–	+	+	–	+	–	–	–	–
Raffinose	–	+/–	–	+	–	–	–	+/–	–	–	–
Melibiose	–	+/–	–	+	–	–	–	+/–	–	–	–
Simmons citrate	–	–	–	+/–	+/–	–	–	+	–	–	+
Voges-Proskauer	–	–	+/–(+)	+	+	–	+	–	–	–	–
Indole	–	–	+/–	+	+	+/–	–	–	–	–	–
Salicin	+/–	+/–	+/–	+	+	-(+/–)	–	–	+/–	(+)	
Esculin	+	+	+/–	+	+	–	+	–	(+)	(+)/–	–
Lipase	–	–	+/–	+/–	+/–	+/–	+/–	–	–	–	
Pyrazinamidase	–	–	+/–	+	+	+	+	+	+	+	

[a] + = positive after 3 days, (+) = positive after 7 days.

[b] Some strains of *Y. intermedia* are negative for either Simmons citrate, rhamnose, and melibiose, or raffinose and Simmons citrate.

[c] Some biotype 5 strains are negative.

Table 2. Biotype scheme[a] for *Y. enterocolitica*

Biochemical test	Reaction for biotypes[b]						
	1A	1B	2	3	4	5	6
Lipase	+	+	–	–	–	–	–
Esculin/salicin (24 h)	+/–	–	–	–	–	–	–
Indole	+	+	(+)	–	–	–	–
Xylose	+	+	+	+	–	V	+
Trehalose	+	+	+	+	+	–	+
Pyrazinamidase	+	–	–	–	–	–	+
β-D-Glucosidase	+	–	–	–	–	–	–
Voges-Proskauer	+	+	+	+/–[c]	+	(+)	-

[a] Based on Wauters.[48]

[b] ( ) = Delayed reaction; V = variable reactions.

[c] Biotype of serotype O:3 found in Japan.

however, is that although most pathogenic *Y. enterocolitica* are positive for sucrose fermentation, some fully virulent, atypical phenotypic variants, that are unable to ferment sucrose, have been isolated from clinical samples.[31] *Y. enterocolitica* isolates of biotypes 1B, 2, 3, 4, and 5 should be tested for pathogenicity.

41.217 Pathogenicity Testing

Autoagglutination test. Examine the MR-VP tube, incubated at 22-26°C for 24 h, for uniform turbidity due to bacterial growth. The tube incubated at 35-37°C however, should show clear supernatant fluid, with agglutination (clumping) of bacteria along walls and/or bottom of tube. Strains that showed this pattern of agglutination are considered presumptive positive for the presence of the virulence plasmid.

Freezing cultures. The virulence plasmids of *Yersinia* can be spontaneously lost with lengthy culture and repeated passage at temperature below 30° C or during routine laboratory culture at temperatures above 30° C. For instance, prolonged incubation of culture at 37°C for 48 h has been shown to trigger plasmid loss.[11] It is important, therefore, to immediately freeze presumptive positive cultures to prevent plasmid loss. Inoculate cultures into veal infusion broth and incubate 48 h at 22-26°C. Add sterile glycerol to a final concentration of 10% and freeze immediately. Storage at –70° C is recommended.

Low calcium response Congo red agarose virulence test. Inoculate the test organism into BHI broth and incubate overnight at 22-26° C. This should yield a broth cultures of approximately 10^9 bacterial cells per mL. Make decimal dilutions in physiologic saline to obtain a suspension of 10^3 cells/mL and spread-plate 0.1 mL dilution onto duplicate Congo red agarose plates. Incubate one plate at 35°C and the other at 25°C and examine at 24 and 48 h. Presumptive plasmid-bearing *Y. enterocolitica* isolates will appear as pinpoint, round, convex, red, opaque colonies on the plate incubated at 25°C. Plasmidless *Y. enterocolitica* isolates will appear as large, irregular, flat, translucent colonies.

Crystal violet binding test.[13] This assay may be useful as a rapid screening test to differentiate potentially virulent *Y. enterocolitica* cultures. As described in H-3 above, prepare an overnight culture grown at 22-26°C and dilute to obtain a suspension of 10^3 cells/mL. Spread-plate 0.1 mL onto duplicate BHI agar plates and incubate one at 25°C and the other at 37°C for 30 h. Gently flood each plate with 8 mL of crystal violet (CV) solution (85 mg/mL) and decant after 2 min. Compare the plates grown at the two temperatures; only plasmid-bearing colonies grown at 37°C will bind CV.

DNA colony hybridization. Several DNA probes have been developed for the detection of virulence factors in *Yersinia*.[23,34,58,80] The *inv* probe is specific only for the chromosomal invasion gene of *Y. pseudotuberculosis* and all probe-reactive isolates were verified to be invasive for HeLa cells.[23] The *ail* probe specifically detects the invasion gene of *Y. enterocolitica*. Results of colony and Southern blot analysis using the *ail* probe showed close correlation between probe reactivity and HeLa cell invasion.[23] DNA probes for the *yadA* and *virF* genes,[80] as well as an oligonucleotide sequence identified by Miliotis et al.,[53] are specific for the 41-48 Mdal virulence plasmid in both *Y. enterocolitica* and *Y. pseudotuberculosis*. The *yadA* and *virF* genes encode for the regulator operon of the outer membrane virulence proteins and for the outer membrane protein responsible for attachment and invasion phenotype, respectively, in both enteropathogenic species.

Invasiveness. An in vitro HeLa cell staining assay is available for screening *Yersinia* isolates for their invasive potential.[51,52] Acridine orange dye is used to stain infected HeLa cell monolayers, which are then examined under fluorescence microscope for the presence of intracellular cells.[51,52] This in vitro staining technique can be used to determine invasiveness in both *Y. enterocolitica* and *Y. pseudotuberculosis* isolates.[23]

Intraperitoneal infection of adult mice pretreated with iron dextran and desferrioxamine. Any isolate testing positive in the virulence tests above (H-1 to H-5) is only presumptive evidence of *Yersinia* pathogenicity and must be confirmed by a biological test for conclusive results. As described in H-3 above, culture each presumptive virulent isolate and dilute to yield a suspension of 10^3 cells/mL. Spread-plate 0.1 mL onto duplicate plates of TSAYE and CRBHO. Incubate TSAYE at 22-26°C for 48 hr and CRBHO at 35°C for 24 hr. Count the number of colonies on TSAYE to determine total cell number and inoculum level and the number of red vs. clear colonies on CRBHO to establish the ratio of plasmid- to non-plasmid-bearing cells in the inoculum. Retain the entire dilution series for mouse infection studies.

One day prior to infection, pre-treat the Swiss Webster adult mice by injecting i.p., 0.2 mL of physiologic saline solution containing 25 mg/mL each of iron dextran (Fermenta Animal Health Co., Kansas City, MO 64153) and desferrioxamine B (Desferal mesylate, Ciba Geigy, Greensboro, NC 27409). The next day, inject 0.1 mL of decimally diluted bacterial suspension i.p. into pretreated mice (use 5 mice per dilution), and observe the mice for 7 days. If possible, maintain infected mice in a laminar flow isolator to prevent cross-contamination with other animals. Deaths occurring within 7 days, especially preceded by signs of illness, is indicative of virulence and can be used to calculate the LD_{50} titer by the method of Reed and Muench[63] as outlined in Table 3. A calculated LD_{50} titer of less than 10^4 cell is typical for virulent *Y. enterocolitica*, regardless of its biotype or serotype. Alternatively, a screening test may be performed by inoculating 5 pretreated mice at the 10^{-4} dilution only. Virulent *Yersinia* will kill at least four of five mice.

41.22 Microbiological Procedures for detecting Yersinia pseudotuberculosis *in foods*

Generally, all *Y. pseudotuberculosis* strains are biochemically homogeneous except for the production of acid from melibiose, raffinose, and salicin. Species are sub-grouped based on a heat-stable somatic antigen and at present, there are six serogroups, designated by Roman numerals I-VI. Serogroups I, II, III, and IV. Subtypes have been described, but subtyping is not as easily determined, as the antiserum to the serogroup type will often

Table 3. Calculation of LD_{50} by Reed-Muench Method[63]

A	B	C	D	E	F	G	H
					Cumulative values [a]		
	Bacterial	Cells/	No. of mice		No. of mice		
Dilution	cells/ml	mouse $\log_{10}$	dead ↑	live ↓	dead	live	% Mortality
10^{0}	3×10^{8}	7.477	3	0	15	0	100
10^{-1}	3×10^{7}	6.477	3	0	12	0	100
10^{-2}	3×10^{6}	5.477	3	0	9	0	100
10^{-3}	3×10^{5}	4.477	3	0	6	0	100
10^{-4}	3×10^{4}	3.477	2	1	3	1	75
10^{-5}	3×10^{3}	2.477	1	2	1	3	25[c]
10^{-6}	3×10^{2}	1.477	0	3	0	6	0
10^{-7}	3×10^{1}	0.477	0	3	0	9	0

[a] Cumulative values in columns F and G are determined by adding values vertically in columns D and E, respectively. Arrows indicate direction of addition for accumulated values.

[b] % Mortality (H) is determined by dividing the number from column F by the total of the number from column F and column G for each dilution.

[c] The 50% endpoint lies between 10^{-4} and 10^{-5} dilutions.

$$\text{Fractional titer} = \frac{(50\%) - (\%\ \text{Mortality below } 50\%)}{(\text{Mortality above } 50\%) - (\text{mortality below } 50\%)}$$

$$\text{In the above example} = \frac{(50\% - 25\%)}{75\% - 25\%} = \frac{0.25}{0.50} = 0.5$$

$\log LD_{50}$ titer = $\log_{10}$ (cells/mouse at dilution below 50% mortality + fractional titer)

In the above example = 2.477 + 0.5 = 2.977

LD50 titer = anti log 2.977 = 948 cells

cross-react with the different subtype strains and vice versa. Strains in serogroups II and III do not elaborate lipase, but are lethal when fed to adult mice. Also, unlike *Y. enterocolitica*, HeLa cell-invasive strains are esculin-positive. Pathogenic *Y. pseudotuberculosis* associated with yersiniosis in humans also harbor the 41-48 Mdal plasmid[36]; hence, the phenotypic traits associated with the presence of the virulence plasmid in *Y. enterocolitica* are also present in *Y. pseudotuberculosis*.

41.221 Enrichment

Aseptically weigh 25 g of food sample into 225 mL PMP broth,[26] homogenize for 30 s and incubate at 4°C for 3 weeks. Analyze at the end of each week by mixing the sample well, and subsequently transferring 0.1 mL enrichment to 1 mL of 0.5% KOH in 0.5% NaCl and mix for 5-10 s. Streak a loop-full to MacConkey agar and another loop-full to CIN agar. Also, streak one loop-full directly from PMP enrichment broth to MacConkey and to CIN agar plates. Incubate all plates at 22-26°C.

41.222 Isolation and Identification

Continue as outlined in steps E to H above but noting the biochemical differences shown in Table 1. Characteristically, *Y. pseudotuberculosis* strains are negative for fermentation of ornithine, sorbitol and sucrose.

41.3 *YERSINIA* IDENTIFICATION TESTS

Phenylalanine deaminase test. Add 2-3 drops of a 10% ferric chloride solution to growth on agar slant. Development of green color is a positive test.

Indole test. Add 0.2-0.3 mL of Kovacs' reagent. Development of deep red color on the surface of the broth is a positive test.

V-P test: Add 0.6 mL of 6% a-naphthol and shake well. Add 0.2 mL of 40% KOH solution with creatine and shake. Development of pink-to-ruby red color in the medium after 4 h is a positive test.

Pyrazinamidase test. Flood 1 mL of freshly prepared 1% ferrous ammonium sulfate over the growth on a slant of pyrazinamidase agar. Development of pink color within 15 min is a positive test, which indicate the presence of pyrazinoic acid formed by the enzymatic activity of pyrazinamidase.

β-D-Glucosidase test. Add 0.1 g of 4-nitrophenyl-β-D-glucopyranoside to 100 mL of a solution of 0.666 M NaH_2PO_4 [pH 6]; dissolve and filter-sterilize the reagent. Emulsify growth from an agar or slant in physiologic saline to obtain a cell density equivalent to McFarland Turbidity Standard No. 3. Add 0.75 mL of this cell suspension to 0.25 mL of test reagent and incubate at 30°C overnight. Production of a distinct yellow color indicates a positive reaction.

41.4 REFERENCES

1. Aleksic, S., and J. Bockemuhl. 1984. Proposed revision of the Wauters et al. antigenic scheme for serotyping of *Yersinia enterocolitica*. J. Clin. Microbiol. 20:99-102.
2. Aleksic, S., A. Steigerwalt, J. Bockemuhl, G. Huntley-Carter, and D. J. Brenner. 1987. *Yersinia rohdei* sp. nov. isolated from human and dog feces and surface water. Int. J. Syst. Bacteriol. 37:327-332.
3. Attwood, S. E. A., M. T. Caffertey, A. B. West, E. Healy, K. Mealy, T. F. Buckley, N. Boyce and F. B. V. Keane. 1987. Yersinia infection and acute abdominal pain. Lancet i:529-533.
4. Aulisio, C. C. G., I. J. Mehlman, and A. C. Sanders. 1980. Alkali method for rapid recovery of *Yersinia enterocolitica* and *Yersinia pseudotuberculosis* from foods. Appl. Environ. Microbiol. 39:135-140.
5. Aulisio, C. C. G., J. M. Lanier, and M. A. Chappel. 1982. *Yersinia enterocolitica* O:13 associated with outbreaks in three southern states. J. Food Prot. 45:1263.
6. Aulisio, C. C. G., J. T. Stanfield, S. D. Weagant, and W.E. Hill. 1983. Yersiniosis associated with tofu consumption: Serological, biochemical and pathogenicity studies of *Yersinia enterocolitica* isolates. J. Food Prot. 46:226-230.
7. Aulisio, C. C. G., J. T. Stanfield, W. E. Hill, and J. A. Morris. 1983. Pathogenicity of *Yersinia enterocolitica* demonstrated in the suckling mouse. J. Food Prot. 46:856-860.

8. Aulisio, C. C. G., W.E. Hill, J. T. Stanfield, and R. L. Sellers Jr. 1983. Evaluation of virulence factor testing and characteristics of pathogenicity in *Yersinia enterocolitica*. Infect. Immun. 40:330-335.
9. Bercovier, H., D. J. Brenner, J. Ursing, A. G. Steigerwalt, G. R. Fanning, J. M. Alonso, G. P. Carter, and H. H. Mollaret. 1980. Characterization of *Yersinia enterocolitica sensu stricto*. Curr. Microbiol. 4:201-206.
10. Bercovier, H., A. G. Steigerwalt, A. Guiyoule, G. Huntley-Carter, and D. J. Brenner. 1984. *Yersinia aldovae* (Formerly *Yersinia enterocolitica*-like group X2): A new species of Enterobacteriaceae isolated from aquatic ecosystems. Int. J. Syst. Bacteriol. 34:166-172.
11. Bhaduri, S. 1997. Presence of virulence plasmid in *Yersinia enterocolitica* after its expression at 37°C. J. Rapid Meth. Automation Microbiol. 5:29-36.
12. Bhaduri, S., and B. Cottrell. 1997. Direct detection and isolation of plasmid-bearing virulent *Yersinia enterocolitica* from various foods. Appl. Environ. Microbiol. 63:4952-4955.
13. Bhaduri, S., L. K. Conway, and R. V. Lachica. 1987. Assay of crystal violet binding for rapid identification of virulent plasmid-bearing clones of *Yersinia enterocolitica*. J. Clin. Microbiol. 25:1039-1042.
14. Black, R. F., R. J. Jackson, T. Tsai, M. Medvesky, M. Shayegani, J. C. Feeley, K. I. MacLeod and A. M. Wakelee. 1978. Epidemic *Yersinia enterocolitica* infection due to contaminated chocolate milk. New Engl. J. Med. 298:76-79.
15. Bottone, E. J. 1997. *Yersinia enterocolitica*: The charisma continues. Clin. Microbiol. Rev. 10:257-276.
16. Boyce, J. M., E. J. Evans Jr., D. G. Evans, and H. L. DuPont. 1979. Production of heat-stable methanol-soluble enterotoxin by *Yersinia enterocolitica*. Infect. Immun. 25:532-537.
17. Buchrieser, C., S. D. Weagant, and C. W. Kaspar. 1994. Molecular characterization of *Yersinia enterocolitica* by pulsed-field gel electrophoresis and hybridization of DNA fragments to *ail* and pYV probes. Appl. Environ. Microbiol. 60:4371-4379.
18. Butler, T. 1994. Yersinia infections: Centennial of the discovery of the plague bacillus. Clin. Infect Dis. 19:655-663.
19. Carter, P. B., and F. M. Collins. 1974. Experimental *Yersinia enterocolitica* infection in mice: Kinetics of growth. Infect. Immun. 9:851-857.
20. Delor, I., and G. R. Cornelis. 1992. Role of *Yersinia enterocolitica* Yst toxin in experimental infection of young rabbits. Infect. Immun. 60:4269-4277.
21. Delor, I., A. Kaeckenbeeck, G. Wauters, and G. R. Cornelis. 1990. Nucleotide sequence of *yst*, the *Yersinia enterocolitica* gene encoding the heat-stable enterotoxin, and prevalence of the gene among pathogenic and nonpathogenic yersiniae. Infect. Immun. 58:2983-2988.
22. Doyle, M. P., M. B. Hugdahl, and S. L. Taylor. 1981. Isolation of virulent *Yersinia enterocolitica* from porcine tongues. Appl. Environ. Microbiol. 42:661-666.
23. Feng, P. 1992. Identification of invasive *Yersinia* species using oligonucleotide probes. Mol. Cell Probes 6:291-297.
24. Feng, P., and S. D. Weagant. 1994. *Yersinia*, p. 427-460. *In* Y. H. Hui, J. R. Gorham, K. D. Murrell, and D. O Cliver (ed.), Foodborne disease handbook. Marcel Dekker, Inc., New York.
25. Feng, P., S. P. Keasler, and W. E. Hill. 1992. Direct identification of *Yersinia enterocolitica* in blood by polymerase chain reaction amplification. Transfusion 32:850-854.
26. Fukushima, H., K. Saito, M. Tsubokura, and K. Otsuki. 1984. *Yersinia* spp. in surface water in Matsue, Japan. Zentralbl. Bakteriol. Abt. 1 Orig. B. Hyg. Kranhenhaushyg. Betreibshyg. Praev. Med. 179:235-247.
27. Fukushima, H., M. Gomyoda, S. Ishikua, T. Nishio, S. Moriki, J. Endo, S. Kaneko, and M. Tsubokura. 1989. Cat-contaminated environmental substances lead to *Yersinia pseudotuberculosis* infections in children. J. Clin. Microbiol. 27:2706-2709.
28. Fukushima, H., K. Hoshima, H. Itogawa, and M. Gomyoda. 1997. Introduction into Japan of pathogenic *Yersinia* through imported pork, beef and fowl. Int. J. Food Microbiol. 35:205-212.
29. Gemski, P., J. R. Lazere, and T. Casey. 1980. Plasmid associated with pathogenicity and calcium dependency of *Yersinia enterocolitica*. Infect. Immun. 27:682-685.
30. Gemski, P., J. R. Lazere, T. Casey, and J. A. Wohlhieter. 1980. Presence of a virulence-associated plasmid in *Yersinia pseudotuberculosis*. Infect. Immun. 28:1044-1047.
31. Guiyoule, A., F. Guinet, L. Martin, C. Benoit, N. Desplaces, and E. Carniel. 1998. Phenotypic and genotypic characterization of virulent *Yersinia enterocolitica* strain unable to ferment sucrose. J. Clin. Microbiol. 36:2732-2734.
32. Hayashidani, H., Y. Ohtomo, Y. Toyokawa, M. Saito, K. I. Kaneko, J. Kosuge, M. Kato, M. Ogawa, and G. Kapperud. 1995. Potential sources of sporadic human infection with *Yersinia enterocolitica* serovar O:8 in Aomori Prefecture, Japan. J. Clin. Microbiol. 33:1253-1257.
33. Heesemann, J., B. Algermissen, and R. Laufs. 1984. Genetically manipulated virulence of *Yersinia enterocolitica*. Infect. Immun. 46:105-110.
34. Hill, W. E., A. R. Datta, P. Feng, K. A. Lampel, and W. L. Payne. 1998. Identification of bacterial pathogens by gene probes. *In* FDA Bacteriological analytical manual, 8A ed. AOAC International, Gaithersburg, Md.
35. Hill, W. E., W. L. Payne, and C. C. G. Aulisio. 1983. Detection and enumeration of virulent *Yersinia enterocolitica* in food by DNA colony hybridization. Appl. Environ. Microbiol. 46:636-641.
36. Isberg, R. 1989. Determinants for thermoinducible cell binding and plasmid-encoded cellular penetration detected in the absence of *Yersinia pseudotuberculosis* invasin protein. Infect. Immun. 57:1998-2005.
37. Isberg, R., and S. Falkow. 1985. A single genetic locus encoded by *Yersinia pseudotuberculosis* permits invasion of cultured animal cells by *Escherichia coli* K-12. Nature 317:262-264.
38. Isberg, R., and J. M. Leong. 1988. Cultured mammalian cells attach to the invasin protein of *Yersinia pseudotuberculosis*. Proc. Natl. Acad. Sci. USA 85:6682-6686.
39. Isberg, R., A. Swain, and S. Falkow. 1988. Analysis of expression and thermoregulation of the *Yersinia pseudotuberculosis inv* gene with hybrid protein. Infect. Immun. 56:2133-2138.
40. Isberg, R., D. L. Voorhis, and S. Falkow. 1987. Identification of invasin: A protein that allows enteric bacteria to penetrate cultured mammalian cells. Cell 50:769-778.
41. Janda, J. M., and S. Abbott. 1998. The yersiniae in the enterobacteriaceae. Lippincott-Raven, Philadelphia, Pa. p. 206-244.
42. Kandolo, K., and G. Wauters. 1985. Pyrazinamidase activity in *Yersinia enterocolitica* and related organisms. J. Clin. Microbiol. 21:980-982.
43. Kapperud, G., T. Vardund, E. Skjerve, E. Hornes, and T. E. Michaelsen. 1993. Detection of pathogenic *Yersinia enterocolitica* in foods and water by immunomagnetic separation, nested polymerase chain reactions, and colorimetric detection of amplified DNA. Appl. Environ. Microbiol. 59:2938-2944.
44. Kay, B. A., K. Wachsmuth, and P. Gemski. 1982. New virulence-associated plasmid in *Yersinia enterocolitica*. J. Clin. Microbiol. 15:1161-1163.
45. Keet, E. E., 1974, *Yersinia enterocolitica* septicemia. Source of infection and incubation period identified. N. Y. State J. Med. 74:2226-2230.
46. Laird, W. J., and D. C. Cavanaugh. 1980. Correlation of autoagglutination and virulence of *Yersinia*. J. Clin. Microbiol. 11:430-432.
47. Lampel, K. A., P. Feng, and W. E. Hill. 1992. DNA probes for the identification of pathogenic foodborne bacteria and viruses, p. 151-188. *In* Bhatnagar and Leveland (ed.), Molecular approaches to improving food quality and safety. Van Nostrand Reinhold, New York.
48. Lee, L. A., A. R. Gerber, D. R. Lonsway, J. D. Smith, G. P. Carter, N. D. Pohr, C. M. Parrish, R. K. Sikes, R. J. Finton, and R. W. Tauxe. 1990. *Yersinia enterocolitica* O:3 infection in infants and children associated with the household preparation of chitterlings. New Engl. J. Med. 322:984-987.
49. Lee, W. H., P. P. McGrath, P. H. Carter, and E. L. Eide. 1977. The ability of some *Yersinia* enterocolitica strains to invade HeLa cells. Can. J. Microbiol. 23:1714-1722.

50. Mikulskis, A.V., I. Delor, V. H. Thi, and G. R. Cornelis. 1994. Regulation of the *Yersinia enterocolitica* enterotoxin Yst gene. Influence of growth phase, temperature, osmolarity, pH and bacterial host factors. Mol. Microbiol. 14:905-915.
51. Miliotis, M. D. 1991. Acridine orange stain for determining intracellular enteropathogens in HeLa cells. J. Clin. Microbiol. 29:830-832.
52. Miliotis, M. D., and P. Feng. 1992. In vitro staining technique for determining invasiveness in foodborne pathogens. FDA Laboratory Information Bulletin, March, 9(3):3754.
53. Miliotis, M. D., J. E. Galen, J. B. Kaper, and J. G. Morris. 1989. Development and testing of a synthetic oligonucleotide probe for the detection of pathogenic *Yersinia* strains. J. Clin. Microbiol. 27:1667-1670.
54. Miller, V. A., and S. Falkow. 1988. Evidence of two genetic loci in *Yersinia enterocolitica* that can promote invasion of epithelial cells. Infect. Immun. 56:1242-1248.
55. Miller, V., J. J. Farmer III, W. E. Hill, and S. Falkow. 1989. The *ail* locus is found uniquely in *Yersinia enterocolitica* serotypes commonly associated with disease. Infect. Immun. 57:121-131.
56. Nesbakken, T., G. Kapperud, K. Dommarsnes, M. Skurnik, and E. Hornes. 1991. Comparative study of a DNA hybridization method and two isolation procedures for detection of *Yersinia enterocolitica* O:3 in naturally contaminated pork products. Appl. Environ. Microbiol. 57:389-394.
57. Neubauer, H., T. Sauer, H. Becker, S. Aleksic, and H. Meyer. 1998. Comparison of systems for identification and differentiation of species within the genus *Yersinia*. J. Clin. Microbiol. 36:3366-3368.
58. Pai, C. H., and L. DeStephano. 1982. Serum resistance associated with virulence in *Yersinia enterocolitica*. Infect. Immun. 35:605-611.
59. Pierson, D. E., and S. Falkow. 1990. Nonpathogenic isolates of *Yersinia enterocolitica* do not contain functional *inv*-homologous sequences. Infect. Immun. 58:1059-1064.
60. Portnoy, D. A., S. L. Moseley, and S. Falkow. 1981. Characterization of plasmids and plasmid-associated determinants of *Yersinia enterocolitica* pathogenesis. Infect. Immun. 31:775-782.
61. Portnoy, D. A., and R. J. Martinez. 1985. Role of plasmids in the pathogenicity of *Yersinia* species. Curr. Top. Microbiol. Immunol. 118:29-51.
62. Prpic, J. K., R. M. Robins-Brown, and R. B. Davey. 1985. In vitro assessment of virulence in *Yersinia enterocolitica* and related species. J. Clin. Microbiol. 22:105-110.
63. Reed, L. J., and J. Muench. 1938. A simple method of estimating fifty per cent endpoints. Am. J. Hyg. 27:493-497.
64. Robins-Brown, R., and K. Prpic. 1985. Effects of iron and desferrioxamine on infections with *Yersinia enterocolitica*. Infect. Immun. 47:774-779.
65. Robins-Brown, R., M. D. Miliotis, S. Cianciosi, V. L. Miller, S. Falkow, and J. G. Morris. 1989. Evaluation of DNA colony hybridization and other techniques for detection of virulence in *Yersinia* species. J. Clin. Microbiol. 27:644-650.
66. Shayegani, M., D. Morse, I. DeForge, T. Root, L. Parsons, and P. S. Maupin. 1983. Microbiology of a major foodborne outbreak of gastroenteritis caused by *Yersinia enterocolitica* serogroup O:8. J. Clin. Microbiol. 17:35-40.
67. Schiemann, D. A. 1982. Development of a two-step enrichment procedure for recovery of *Yersinia enterocolitica* from food. Appl. Environ. Microbiol. 43:14-27.
68. Schiemann, D. A. 1989. *Yersinia enterocolitica* and *Yersinia pseudotuberculosis*, pp. 601-672. *In* Foodborne Bacterial Pathogens. M. Doyle (ed). Marcel Dekker, New York.
69. Schiemann, D. A., and C. A. Fleming. 1981. *Yersinia enterocolitica* isolated from the throats of swine in Eastern and Western Canada. Can. J. Microbiol. 27:1326-1333.
70. Schiemann, D. A., and G. Wauters. 1992. *Yersinia*. Compendium of methods for the microbiological examination of foods, 3rd ed. American Public Health Association, Washington, D. C.
71. Sereny, B. 1955. Experimental *Shigella* keratoconjunctivitis. Acta Microbiol. Acad. Sci. Hung. 2:293-296.
72. Toora, S., E. Budu-Amoako, R. F. Ablett and J. Smith. 1994. Isolation of *Yersinia enterocolitica* from ready-to-eat foods and pork by simple two step procedure. Food Microbiol. 11:369-374.
73. Tsubokura, M., K. Otsuki, K. Sato, M. Tanaka, T. Hongo, H. Fukushima, T. Maruyama, and M. Inoue. 1989. Special features of distribution of *Yersinia pseudotuberculosis* in Japan. J. Clin. Microbiol. 27:790-791.
74. Wagner, S. J., L. I. Friedman, and R. Y. Dodd. 1994. Transfusion-associated bacterial sepsis. Clin. Microbiol. Rev. 7:290-302.
75. Wauters, G. 1981. Antigens of *Yersinia enterocolitica*, p. 41-53. *In Yersinia enterocolitica*. E. J. Bottone (ed). CRC Press, Boca Raton, Fla.
76. Wauters, G., V. Goosens, M. Janssens, and J. Vandepitte. 1988. New enrichment method for isolation of pathogenic *Yersinia enterocolitica* serogroup O:3 from pork. Appl. Environ. Microbiol. 54:851-854.
77. Wauters, G., M. Janssens, A. G. Steigerwalt, and D. J. Brenner. 1988. *Yersinia molleretii* sp. nov. and *Yersinia bercovieri* sp. nov., formerly called *Yersinia enterocolitica* biogroups 3A and 3B. Int. J. Syst. Bacteriol. 38:424-429.
78. Weagant, S. D. 1983. A screening procedure and medium for the presumptive identification of *Yersinia enterocolitica*. Applied and Environmental Microbiol. 45:472-473
79. Weagant, S., P. Feng, and J. Stanfield. 1998. *Yersinia enterocolitica* and *Yersinia pseudotuberculosis*. *In* FDA bacteriological analytical manual, 8A ed., AOAC International, Gaithersburg, Md.
80. Weagant S. D, J. A. Jagow, K. C. Jinneman C. J. Omiecinski, C. A. Kaysner, and W. E. Hill. 1998. Development of digoxigenin labeled PCR amplicon probes for use in the detection and identification of enteropathogenic *Yersinia* and Shiga toxin-producing *Escherichia coli* from foods. J. Food Prot., in press.
81. Weissfeld, A. S., and A. C. Sonnenwirth. 1982. Rapid isolation of *Yersinia* spp. from feces. J. Clin. Microbiol. 15:508-510.
82. Zheng, X. B., and C. Xie. 1996. Note: Isolation, characterization and epidemiology of *Yersinia enterocolitica* from humans and animals. J. Appl. Microbiol. 81:681-684.
83. Zink, D. L., J. C. Feeley, J. G. Wells, C. Vanderzant, J. C. Vickery, W. D. Roof, and G. A. O'Donovan. 1980. Plasmid-mediated tissue invasiveness in *Yersinia enterocolitica*. Nature 283:224-226.

Chapter 42

Waterborne and Foodborne Parasites

Ronald Fayer, H. Ray Gamble,
J. Ralph Lichtenfels, and Jeffrey W. Bier

42.1 INTRODUCTION

Morphologic identification of protozoa and helminths can be quite difficult. Brightfield, interference-contrast, or phase-contrast microscopy and a variety of specialized stains or live specimens are used to aid identification. In addition to specific references cited in the text, several atlases are recommended for assistance in identification.[4-7,34 46,57,71,73] To further aid in the identification of some parasites new techniques have been developed utilizing biochemical, immunological, and molecular methods that increase both the sensitivity and specificity of detection.

42.11 Parasites Involved in Foodborne and Waterborne Diseases

The prevalence of diseases caused by parasitic animals in humans is relatively low for most modern countries as a result of hygienic standards. Acquisition of parasites from contaminated food or drink is minimized by water-treatment systems and quality control in the production and inspection of meat from food animals. Nevertheless, many species of parasites persist in all human populations. Among the parasites that humans acquire through ingestion of food and water are protozoa, flukes (trematodes), tapeworms (cestodes), roundworms (nematodes), and spiny-headed worms (acanthocephalids). A detailed list of the parasites, even those uncommonly or rarely found in humans, is given in Table 1. This list includes the source of infection, the name of the parasite, and the infective stage of 118 parasites. Some parasites of public health importance are not listed because they are not normally considered contaminants of food or drink. For example, *Dientamoeba fragilis* is considered nonpathogenic by many workers but is thought by others to cause diarrhea and abdominal discomfort in some people. The route of transmission is unknown. There is no known cyst stage and some workers have suggested that the organisms might be transmitted within the eggs of the pinworm *Enterobius vermicularis* or the whipworm *Trichuris trichiura.*

42.12 Sources of Parasites

All waterborne parasites originate from human or animal fecal contamination of the environment. Numerous surface waters such as rivers, lakes, and ponds have been found to harbor encysted stages of parasites. Water-treatment facilities have reduced the opportunities for transmission of parasites by water. However, outbreaks of giardiasis (see Section 42.23) and cryptosporidiosis (see Section 42.25) following flooding or other events that result in contamination of the treated water with untreated water or debris emphasize the cosmopolitan distribution of these organisms.

Vegetables can be contaminated with numerous species of parasites through the use of fertilizers consisting of animal or human feces. Vegetables can also become contaminated during food preparation if rinsed with water containing parasites. Because of such contamination it is important to avoid drinking water and eating raw fruit or vegetables in endemic areas. Even washing of contaminated fruits and vegetables does not completely remove oocysts[59] or other encysted stages. Drink only boiled or bottled beverages and eat only thoroughly cooked vegetables.

A few species of parasites have intermediate stages, found in molluscs, that encyst on vegetation (Table 1). Shellfish, fish, poultry, beef, goat meat, horse meat, lamb, pork, or game meats such as bear, deer, elk, moose, walrus, pheasant, quail, and others sometimes contain encysted, encapsulated, or free stages of parasites. These stages are not merely accidental contaminants but specific developmental life-cycle stages in the animal hosts. All are temperature-sensitive and are killed by thorough cooking. Many, but not all, are killed by freezing. Perhaps the best known of these foodborne parasites is *Trichinella spiralis,* the roundworm (nematode) that causes trichinellosis (see Section 42.3212). Although the nematode is well known, fewer than 100 cases of human trichinellosis are currently diagnosed annually in the United States. The foodborne parasite, *Toxoplasma gondii,* causes toxoplasmosis in humans (see Section 42.3211) and has been detected serologically in 30% or more of the population of the United States.[23,29] The major source of infection has not been unequivocally established, but oocysts from cat feces and encysted bradyzoites in undercooked meat (especially pork) are the two most likely sources.

42.13 Inspection of Fish and Meat

The National Marine Fisheries Service (NMFS) in the National Oceanographic and Atmospheric Administration (NOAA) of the United States Department of Commerce offers a voluntary seafood program on a fee-for-service basis that focuses on quality.

Table 1. Parasites Transmitted to Humans by Water and Food

Source	Parasite	Stage in water or food
Water	Protozoa	
	Entamoeba histolytica	Cyst
	Entamoeba chattoni	*Cyst*
	Entamoeba coli	Cyst
	Endolimax nana	Cyst
	Iodamoeba buetschlii	Cyst
	Retortamonas intestinalis	Cyst
	Retortamonas sinensis	Cyst
	Chilomastix mesnili	Cyst
	Enteromonas hominis	Cyst
	Pentatrichomonas hominis	Trophozoite
	Giardia duodenalis	Cyst
	Isospora belli	Oocyst
	Balantidium coli	Cyst
	Toxoplasma gondii	Oocyst
	Cryptosporidium	*parvum* Oocyst
	Cyclospora cayetanenesis	Oocyst
	Cestoda	
	Spirometra sp.	Procercoid in *Cyclops*
	Taenia sp.	Cyst
	Echinococcus granulosus	Cyst
	Echinococcus multilocularis	Cyst
	Echinococcus vogeli	Cyst
	Nematoda	
	Capillaria hepatica	Cyst
	Trichuris trichiura	Cyst
	Rhabditis sp.	Larva in molluscs
	Mammomonogamus sp.	Cyst
	Enterobius vermicularis	Cyst
	Syphacia obvelata	Cyst
	Ascaris lumbricoides	Cyst
	Toxocara cati	Cyst
	Toxocara canis	Cyst
Vegetables or Fruits	Protozoa	
	Cryptosporidium parvum	Oocyst
	Cyclospora cayetanensis	Oocyst
	Trematoda	
	Watsonius watsoni	Metacercaria
	Gastrodiscoides hominis	Metacercaria
	Fasciola hepatica	Metacercaria
	Fasciola gigantica	Metacercaria
	Fasciolopsis buski	Metacercaria
	Fischoederius elongatus	Metacercaria
	Nematoda	
	Ascaris lumbricoides	Larva in egg
	Angiostrongylus costaricensis	Larva
	Angiostrongylus cantonensis	Larva
	Trichuris trichiura	Larva
Molluscs		
Oysters	Protozoa	
	Cryptosporidium parvum	Oocyst
	Nematoda	
	Echinocephalus sp.	Larva
Snails	*Angiostrongylus cantonensis*	Larva
Snails, slugs	*Angiostrongylus costaricensis*	Larva
	Trematoda	
Snails, oysters	*Echinostoma* sp.	Metacercaria
Clams	*Himasthla muehlensi*	Metacercaria
Crustacea		
	Cestoda	
	Mesocestoides sp.	Larva
	Trematoda	
	Paragonimus westermani	Metacercaria
Fish (F)		
Frog legs (FL),		
Snakes (S)		
	Nematoda	
F	*Capillaria philippinensis*	Larva
F	*Dioctophyme renale*	Larva
F, FL	*Gnathostoma* sp.	Larva
F	*Pseudoterranova* sp.	Larva
F	*Anisakis* sp.	Larva
F	*Porrocaecum* sp.	Larva
F	*Contracaecum* sp.	Larva
F	*Gnathostoma spinigerum*	Larva
F	*Eustrongylides* sp.	Larva
	Cestoda	
F	*Diphyllobothrium latum*†	Plerocercoid
F	*Diplogonoporus grandis*	Plerocercoid
F	*Digramma brauni*	Plerocercoid
F	*Ligula intestinalis*	Plerocercoid
F(probably)	*Braunia jasseyensis*	Plerocercoid
	Spirometra sp.	Plerocercoid
F	*Mesocestoides* sp.	Pleurocercoid
	Trematoda	
F	*Artyfechinostomum mehrai*	Metacercaria
F,S	*Echinostoma* sp.	Metacercaria
F	*Echinochasmus* sp.	Metacercaria
F	*Echinoparyphium recurvatum*	Metacercaria
F	*Episthmium caninum*	Metacercaria
F	*Euparyphium melis*	Metacercaria
F	*Hypoderaeum conoideum*	Metacercaria
F	*Paryphostomum surfartyfex*	Metacercaria
F	*Opisthorchis felineus*	Metacercaria
F	*Opisthorchis viverrini*	Metacercaria
F	*Clonorchis sinensis*	Metacercaria
F	*Appophalus donicus*	Metacercaria
F	*Centrocestus* sp.	Metacercaria
F, FL	*Centrocestus formosanus*	Metacercaria
F	*Cryptocotyle lingua*	Metacercaria
F	*Diorchitrema* sp.	Metacercaria
F	*Haplorchis* sp.	Metacercaria
F	*Heterophyes nocens*	Metacercaria
F	*Heterophyes hyterophyes*	Metacercaria
F	*Heterophyopsis continua*	Metacercaria
F	*Metagonimus yokogawai*	Metacercaria
F	*Metagonimus minutus*	Metacercaria
F	*Phagicola* sp.	Metacercaria
F	*Procerovum* sp.	Metacercaria
F	*Pygidiopsis summa*	Metacercaria
F	*Stellantchasmus falcatus*	Metacercaria
F	*Stictodora fuscata*	Metacercaria
F	*Diorchitrema pseudocirratum*	Metacercaria
F	*Diorchitrema formosenum*	Metacercaria
F	*Nanophyetus salmincola*	Metacercaria

Continued

Table 1. Parasites Transmitted to Humans by Water and Food—*Continued*

Source	Parasite	Stage in water or food
F	*Isoparorchis hypselobagri*	Metacercaria
FL	*Alaria americana*	Metacercaria
F	*Neodiplostomum seoulense*	Metacercaria
F	*Paralecithodendrium* sp.	Metacercaria
F	*Phaneropsolus* spp.	Metacercaria
F	*Prosthodendrium molenkampi*	Metacercaria
F	*Gymnophalloides seoi*	Metacercaria
F	*Fischoederius elongatus*	Metacercaria
F	*Watsonius watsoni*	Metacercaria
F	*Spleotrema brevicaeca*	Metacercaria
F	*Plagiorchis* sp.	Metacercaria
	Acanthocephala	
F	*Corynosoma strumosum*	Juvenile
F	*Bulbosoma* sp.	Juvenile
	Trematoda	
S	*Paragonomus* sp.	Mesocerceria
S	*Neodiplostomum seoulense*	Metacercaria
	Nematoda	
S	*Gnathostoma* sp.	Larva
Poultry	Protozoa	
	Toxoplasma gondii	Tissue cyst
	Nematoda	
	Gnathostoma spinigerum	Larva
Pork	Protozoa	
	Sarcocystis suihominis	Tissue cyst
	Toxoplasma gondii	Tissue cyst
	Nematoda	
	Trichinella spiralis	Larva
	Cestoda	
	Taenia solium	Cysticerci
	Taenia asiatica	Cysticerci
	Trematoda	
	Paragonimus sp.	Metacercaria
Beef	Protozoa	
	Sarcocystis hominis	Tissue cyst
	Toxoplasma gondii	Tissue cyst
	Cestoda	
	Taenia saginata	Cysticercis
Goat meat	Protozoa	
	Toxoplasma gondii	Tissue cyst
Lamb	Protozoa	
	Toxoplasma gondii	Tissue cyst
Horse meat	Nematoda	
	Trichinella spiralis	Larva
Venison/meat from other wild ruminants		
	Protozoa	
	Toxoplasma gondii	Tissue cyst
Bear meat	Protozoa	
	Toxoplasma gondii	Tissue cyst
Nematoda		
	Trichinella spiralis‡	Larva

† There have been approximately 13 species of *Diphyllobrothrium* reported to infect humans

‡ In arctic areas *Trichinella* is referred to as *Trichinella nativa*

Any type of product, from whole fish to reconstructed products, can be inspected and certified. Labels for the various inspection procedures include "Packed Under Federal Inspection," indicating that plant sanitation and processing methods, from raw material to final product, have been inspected. The inspection can include product grading. NOAA also offers inspection of specific lots or shipments of seafood (including imported lots) for compliance with labeling requirements, wholesomeness, safety, and suitability for human consumption.

The Food and Drug Administration (FDA) of the United States Department of Health and Human Services conducts Hazard Analysis and Critical Control Point (HACCP) inspections for domestic seafood processors and imports. Under these new regulations, the FDA focuses on food safety, including pathogenic bacteria, parasites in products to be consumed raw, parasites as filth in products intended to be heated, antibiotics, toxic chemicals, marine toxins and decomposition, as well as labeling claims.

The Food Safety Inspection Service (FSIS) of the United States Department of Agriculture (USDA) inspects all interstate food processors of poultry and red meats. The inspection checks for product adulteration, sanitation practices, and misbranding. There is no mandatory inspection of meat for *Toxoplasma gondii* or *Trichinella spiralis* in the United States. Voluntary inspection of pork by the pooled sample digestion method for *T. spiralis* is allowed under the Code of Federal Regulations (CFR) Title 9, Chap. 3, 318.10.[20]

Inspection for *Trichinella* in pork exported from the United States is required by European Union (EU) member countries and Russia. Hog carcasses are inspected using 1 g of tissue in a pooled digestion test as described in EU directives[26,27] and Russian meat inspection regulations.[65] Horse carcasses intended for export to the EU member countries must be inspected by testing a 5 g. sample in a pooled digestion test using EU methods.[26,27] Both pork and horse meat testing programs are administered by the Agricultural Marketing Service of the USDA.

Beef and pork carcasses are inspected by the USDA for the presence of tapeworm cysts of the genus *Taenia* (CFR, Title 9, Chap. 3, 311.23, 311.24).[18,19] Cysts, also called cysticerci, are small, fluid filled, oval, white bladderworms about 7–10 × 4-6 mm. Methods of inspection are organoleptic, involving only the visual examination of the carcass. The sensitivity of organoleptic inspection for cattle is relatively low; hence only some of the infected carcasses are detected. Infested areas of meat are excised; in a case of heavy infection the entire carcass may be condemned.

42.2 WATERBORNE PARASITES: PREVALENCE, DISEASE, TRANSMISSION, AND METHODS OF IDENTIFICATION

The Office of Groundwater and Drinking Water of the United States Environmental Protection Agency (US EPA) provides guidance and regulations that effect the purity and safety of the fresh water supply in the United States. Water treatment plants providing potable water adhere to US EPA regulations on filtration, chemical flocculants and disinfectants, turbidity, and microbial contaminants.

42.21 Amoebae

Of the intestinal amoebae parasitic in humans, only *Entamoeba histolytica* is consistently considered a pathogen.[47] Between 1976 and 1985 in the United States, approximately 3,000 to 7,000 cases were reported annually to the Centers for Disease Control. *Enta-*

moeba histolytica is transmitted without known animal reservoir hosts. The trophozoite stage, a motile form resembling a macrophage, is found most often in the intestine where it causes ulcers, acute colitis, diarrhea, or dysentery. Trophozoites can be tissue invasive from the gut resulting in abscesses in organs such as the liver and lungs, and ulcerations of the anus, perianal skin, and skin of the vulva. One to 5% of patients develop an ameboma, or pseudotumor of the colon. Because the trophozoite is easily destroyed outside the body, transmission usually involves passage of the resistant cyst stage in the stool. This stage is round and surrounded by a tough outer wall. Ingestion of cysts in contaminated food or water or direct passage from person to person are the usual methods of transmission. Cysts can be transmitted from soiled fingers of infected food handlers who have poor personal hygiene. Metronidazole, chloroquine, and iodoquinol are used in the treatment of intestinal amebiasis.[2]

Identification of *E. histolytica* from water is based on microscopic examination of sediment obtained either from back-flushing filters through which water had passed or in the pellet of particulate matter from centrifuged water (see Section 42.28). Trophozoites are 20 to 30 µm, have a thick, clear ectoplasm and granular endoplasm; pseudopods might be visible. The nucleus, unclear in fresh specimens, is distinct when stained with hematoxylin. It has a ring of small peripheral granules and a central dark body (endosome). Cysts are 10 to 20 µm in diameter, often have four nuclei but sometimes have one, two, or eight nuclei, and sometimes contain rod-like bodies (chromatoidal bodies) with rounded ends; they are morphologically distinguishable from *E. coli* but indistinguishable from nonpathogenic *E. dispar* and *E. moshkovski*, requiring biochemical, immunological or molecular techniques to confirm species identity.[28]

42.22 Balantidium coli

Balantidium coli, a ciliated protozoa of medical importance, is the only species in the genus *Balantidium*. Although distributed worldwide, it is not a prevalent infection in humans. The trophozoite stage in the large intestine causes ulcerative colitis and diarrhea. The cyst is thought to be the infective stage, capable of survival in moist environments for several weeks. *B. coli* is a commensal widely distributed in pigs and indistinguishable organisms have been found in monkeys, dogs, rats, guinea pigs, and buffalo. Water obtained from drainage area contaminated by human or pig feces is thought to be the major source of human infection. The treatment of choice is tetracycline, but diiodohydroxyquin or paromomycin are suitable substitutes.[67]

The *B. coli* cyst can be identified in suspect water after concentration by filtration or centrifugation (see Section 42.28). The cyst is large (40 to 65 µm) and surrounded by a distinct wall, and it contains a large macronucleus that is often bean-shaped and a smaller micronucleus.[55] Both cysts and trophozoites can be identified by microscopic examination of suspect stool specimens mixed with physiological saline. The freshness of the stool sample is important because trophozoites can disintegrate within a few hours after passage. Ciliated trophozoites are motile and rotate as they move. In formalin-fixed samples trophozoites often resemble debris, artifacts, or eggs.[67]

42.23 Giardia

Giardia, the first protozoa observed with a microscope, was described by Leeuwenhoek (1632–1723) from his own stool. *Giardia duodenalis* are flagellated protozoa of the intestine found worldwide in humans and animals.[50] Of the 40 or more so-called species from animals and humans, it is not known for certain how many can cause human infection. Infection is characterized primarily by diarrhea. Cysts and trophozoites can harbor viruses, bacteria, and fungi but the human health significance of these is unknown. It is suspected that many animals serve as sources for human infection but the occurrence of animal to human transmission remains controversial. The life cycle of *Giardia* spp. includes a binucleated, flagellated, motile stage called a trophozoite, which is responsible for the enteritis and diarrhea, and multinucleated cyst stage, which is responsible for survival outside the host and transmission to another host. Cysts are transmitted by the fecal-oral route. Soiled fingers, contaminated drinking water and fecal contamination of food by food handlers are the most common methods of infection. Sand filtration is needed to eliminate cysts from community drinking water because *Giardia* spp. can survive normal chlorination. Quinacrine, metronidazole, furazolidone, and paromomycin have been used for treatment.[21]

Giardia spp. cysts are identified by bright-field or phase-contrast microscopy in suspect water after concentration by filtration or centrifugation (see Section 42.28). *Giardia* spp. can be identified as trophozoites or cysts in stool specimens mixed with physiological saline. Trophozoites are motile in specimens less than a few hours old. They are 9 to 21 µm long, by 5 to 15 µm wide but only about 2 µm thick. When fixed and stained with iron hematoxylin, the trophozoites are pear shaped and when viewed with the narrow end down they look like a human face with nuclei for eyes and a median bar for the mouth.[6] The cyst stage, with a visible but not prominent wall (often of uneven thickness), is ellipsoid and contains two or four nuclei.

An enzyme-linked immunosorbent assay is available to detect a *Giardia* spp. antigen excreted in the feces (Alexon, Santa Clara, Calif.). At least two fluorescent antibody test kits are available to detect *Giardia:* an indirect fluorescent kit for feces (Meridian Diagnostics, Cincinnati, Ohio) and a direct fluorescent antibody kit for water samples (Biovir Laboratories, Benicia, Calif.). Biovir also markets a polyclonal fluorescein-linked antibody to *Giardia* spp., of both human and animal origin, and a monoclonal fluorescein linked antibody, which is reportedly specific for human *Giardia* spp. When no *Giardia* has been found in stools of suspected cases, duodenal or jejunal fluid can be obtained and examined by duodenal tube, endoscope, or Enterotest (capsule swallowed by patient and retrieved by an attached nylon string).[68] A cDNA probe has been used to detect *Giardia* in water samples.[54]

42.24 Isospora *(Coccidia)*

The genus *Isospora* is part of a large group of protozoa collectively known as coccidia. *Isospora belli* is the only known species infectious for humans. Relatively few cases of the infection have been reported. The most prominent clinical sign is diarrhea with dehydration in severe cases. It is thought to be transmitted from person to person because there are no other known hosts. Even the route of transmission is not known with certainty, but a fecal–oral route similar to that of other coccidia is suspected. Severe diarrhea, sometimes of long duration, produces stools that contain the oocyst, an egg-like form surrounded by a tough outer wall, which can remain infectious in the environment for months under moist conditions.

Although there are no reports of *I. belli* being found in water, oocysts of other coccidia have been identified with bright-field or phase-contrast microscopy in water after concentration by centrifugation (see Section 42.28).

42.25 *Cryptosporidium spp.*

Several species of *Cryptosporidium*, coccidian protozoa, have been named from various animal hosts but only *C. parvum* has been found infectious for humans and other mammals.[30] Unrecognized as a human pathogen until the late 1970s, it is now known to be widespread, prevalent, and highly pathogenic under certain conditions. Infection results in enteritis with clinical signs marked by diarrhea, abdominal discomfort, and dehydration in severe cases. Of several waterborne outbreaks, one in Georgia involved about 36% of a population of nearly 36,000 persons[36] and another in Milwaukee, WI affected over 400,000 persons.[40,48] *Cryptosporidium parvum* is transmitted by the oocyst stage in the stool. Hand-to-mouth transmission may account for its spread in day care centers. However, the finding of oocysts in numerous surface waters and in tap water indicates the likelihood of spread to the general population through the water supply. Of 107 surface water samples collected in six western states, 72% were positive for the oocyst stage of a *Cryptosporidium* sp.[45]

Identification of the oocyst stage of *Cryptosporidium* spp. is based on microscopic examination of sediment from filtered or centrifuged water (see Section 42.28). The oocyst stage is most easily identified after flotation in sucrose solution (see Section 42.28) and observation by phase-contrast microscopy. Oocysts resemble yeasts without buds but appear very light, some with what appears to be a dark central granule. In acid-fast stained smears examined by bright-field microscopy, oocysts stain bright red. An indirect fluorescent antibody test kit for the detection of *Cryptosporidium* spp. in feces is commercially available (Meridian Diagnostics, Cincinnati, Ohio). A kit to detect *Giardia* spp. and/or *Cryptosporidium* using an indirect fluorescent antibody in water samples is also available (Meridian Diagnostics, Cincinnati, Ohio). The Polymerase Chain Reaction (PCR) technique and other molecular tools are in various stages of development and application for detection of *C. parvum* in clinical and environmental specimens.[69] DNA analysis has shown distinct genotypes within the species *C. parvum*; they may have different pathogenicities for humans.

42.26 *Cyclospora*

Originally described as a cyanobacterium-like organism, *Cyclospora cayetanensis* is a newly emerging disease organism and is the only member of the genus *Cyclospora* reported to infect humans. Other species have been described from snakes, moles and rodents.[58] Oocysts similar to *C. cayetanensis* were identified in dogs.[70] *Cyclospora* has been found in monkeys and apes but the species may be different from *C. cayetanensis*.

The life cycle has not been thoroughly illucidated but appears similar to other coccidian parasites. The oocyst is excreted unsporulated in the feces and is not infectious. After 1–2 weeks in moisture at 20–30°C the oocysts sporulate, containing 2 sporocysts each with 2 sporozoites. Sporulated oocysts are infectious. After ingestion of fecal contaminated food or water, oocysts release sporozoites which penetrate duodenal or jejunal epithelial cells and develop asexually into schizonts. The number of asexual generations is not known and sexual stages have not been described.

C. cayetanensis has been reported in human feces worldwide.[28] Contaminated drinking water is considered a source of infection. In the United States an outbreak occurred in a hospital dormitory among 21 staff and employees. It was suggested that infection may have resulted from drinking water contaminated in a rooftop holding tank.[41] The possibility of other sources of infection, such as food, was not extensively examined. Foodborne cyclosporiasis has been reported. Vegetables were collected in markets in an endemic area of Peru and were found contaminated with *Cyclospora* oocysts.

In 1996 an outbreak in North America was reported in 20 states, the District of Columbia, and 2 Canadian provinces affecting 1465 persons who ate Guatemalan raspberries.[37] In 1997, 510 cases were reported with raspberries from Guatemala as the probable vehicle of transmission in most cases and mesclun lettuce (a mixture of various baby leaves of lettuce) as the other.[16] In 1997 at least 20 cluster cases involving approximately 185 adults were associated with foods containing fresh basil.[17] In the summer of 1998 when raspberries from Guatemala were banned from import into the United States, no cases of infection were reported. In contrast, raspberries were imported into Canada where an outbreak was reported.

Visualizing of oocysts in feces is the diagnostic method of choice. Oocysts are spheroidal 8.6± 0.6 µm (7.7–9.9 µm) in diameter with a bilayered wall.[58] Unsporulated oocysts contain a granular cytoplasm; sporulated oocysts contain 2 sporocysts each with 2 sporozoites. Oocysts auto-fluoresce when viewed with fluorescence microscopy using an excitation filter at wavelengths of 340–380 nm[24] that pass through a long-pass filter of 400 or 420 nm. Oocysts usually appear red when stained by acid-fast methods (Kinyoun and Ziehl-Neelson staining procedures) but are variably acid-fast and sometimes refractory to staining.

Trimethoprim-sulfamethoxazole (co-trimoxazole) has been effective for treatment of children, immunocompetent and immunocompromised adults.[39,49]

42.27 *Toxoplasma*

(See Section 42.32.)

42.28 *Methods of Concentrating Protozoa in Suspect Water*

Recovery rates of parasites from water are influenced by sample volume, number of parasites present, and water quality. Large volume, small numbers of parasites, and high turbidity decrease recovery. Other methods might be useful for recovery of a variety of protozoa, but two filtration methods found effective for recovery of *Cryptosporidium* spp. are provided as examples.[62]

The first uses 293-mm diameter polycarbonate membrane filters with a 5-µm pore size.[56] Up to 20-L of water are prefiltered under vacuum. The filtrate is collected and filtered again through a 1.0-µm pore-size filter. Oocysts retained on the 1.0-µm filter are recovered by vibrating the inverted filter in 200-mL of distilled water for 3-min at the medium setting, on a Toothmaster Investment Vibrator (Whaledent International, Division of IPCO Corp.), concentrating the sediment by centrifugation, clarifying it in a potassium citrate (1.19-g/mL) density gradient, and filtering the suspension through a 1.2-µm pore-size cellulose nitrate filter. Oocysts stained with fluorescein isothiocyanate-labeled polyclonal antibody are detected by fluorescence microscopy. The second method uses a Micro Wynd II 250-µm diameter polypropylene cartridge filter (AMF/CUNO, Meriden, Conn.) with a 1.0-µm pore size, which can be transported to a sampling site where large volumes (100-gal; 378-L) of water are to be examined.[52] When such a large volume is filtered, 6-L of 0.1% Tween 80 is needed to backflush and rinse the cut-up filter. The 6-L is concentrated by centrifugation and the pellet clarified by flotation on sucrose solution (1.24-g/mL). High recovery rates can be obtained when 0.1% Tween 80 and 1% sodium dodecyl sulfate are mixed with the sample and sonication is used to disaggregate oocysts from sediment. Oocysts recovered from the surface of the sucrose are identified on a glass slide by indirect immunofluorescence.

The oocyst stage of virtually all coccidia can be identified by microscopic examination of suspect stool samples or water contaminated with feces. Usually, a differential density flotation-concentration method is used to separate fecal or other debris from oocysts. Zinc sulfate, sodium chloride, cesium chloride, Percoll™ (Pharmacia, Uppsala, Sweden), sucrose (0.4–0.2 M), and other solutions with specific gravities between 1.02 and 1.12 are useful for flotation of the oocysts. They constitute the lower layer in a centrifuge tube. The upper layer consists of contaminated water or a mixture of feces and water. The tube is centrifuged for 5 min at 325–750 × *g*. A portion of the surface layer of the higher density solution is removed by aspiration and a drop is placed on a glass slide for microscopic examination.

42.3 FOODBORNE PARASITES: PREVALENCE, DISEASE, TRANSMISSION, AND METHODS OF IDENTIFICATION

42.31 Parasites of Molluscs, Crustaceans, Fish, and Frog Legs

Parasites that can be transmitted to humans by eating fish (marine and freshwater), molluscs (including clams, oysters, and snails), and crustaceans (crabs or crayfish) are helminths (worms), including nematodes, cestodes, and trematodes (Table 1).[43] Only a few helminths, the tapeworms *(Diphyllobothrium latum* and related species), and the anisakids, *Anisakis* spp. and *Pseudoterranova (Phocanema* spp.) pose important threats to humans in North America.[38] Four other helminths transmitted only rarely to humans in North America include *Eustrongylides* spp., *Paragonimus* spp., *Nanophyetus* spp., and *Alaria* spp.

42.311 Anisakids

Anisakiasis in humans results from ingestion of the third-stage larva of nematodes of the family Anisakidae [principally of the genera *Anisakis* and *Pseudoterranova (Phocanema)*] in raw, semi-raw, or pickled fish. More than 4,000 cases from Japan and 200 from Scandinavia have been reported since the disease was first recognized in 1955.[13] In the United States, relatively few cases of anisakiasis have been described in detail, but many more have been documented. Larvae of *Anisakis* are found in pelagic fishes such as herring, mackerel, and salmon.[10] Principal definitive hosts are whales and porpoises. Larvae of *Pseudoterranova* are found in inshore fish such as cod, flounder, and flukes. Principal definitive hosts are seals. Anisakid larvae are white or clear (some *Pseudoterranova* have a reddish tint), measure 15 to 37 mm long, and are 1 to 2 mm wide. Morphological features have been described.[1,53] Surveys in fish markets in the United States indicate a high prevalence of anisakidae. Clinical symptoms in humans range from a tickle in the throat before "coughing up" a larva of *Pseudoterranova* (which usually does not invade human tissues), to severe epigastric pain requiring surgery if the stomach wall is punctured. Treatment may include removal of the larva with the aid of gastroenteroscopy or more traditional surgery.[1]

The detection of larval anisakids in fish muscle by candling has been adopted as an official procedure by the Association of Official Analytical Chemists.[66] The candling table consists of a framework to hold a light source below a rigid acrylic plastic or other suitable work surface with 45% to 60% translucency. The surface must be large enough to accommodate an entire fillet. The light source must be "cool white" with a color temperature of 4,200°K; at least two 20-watt fluorescent bulbs are recommended. The tubes and electrical connection should be constructed to prevent overheating of the work space. The light intensity must average 1500 to 1800 lux when measured from a distance of 30 cm. Distribution of the illumination directly above the light source should be three times greater than that of the outer field. Indirect light in the vicinity of the candling table should be about 500 lux. Compression candling increases the sensitivity of this method[9] in scallops and clams.

Another study compared recovery of nematodes from digestion of fish muscle and viscera in 2% pepsin (1:10,000) adjusted to pH 2 with HCl at 36 ± 0.5° C for 24 hr with elution into saline over a 16 to 18 hr period. The concentrates from both procedures were examined for parasites with the aid of a dissecting microscope. Digestion was more sensitive for recovering potentially pathogenic nematodes, whereas elution was more sensitive for recovering non-pathogens.[42] Pepsin-HCl digestion[64] may be the most complete and accurate method for collecting larvae from fish flesh, but a method using a food blender to free the larvae from the tissue of fresh and frozen fish and UV light to detect them was regarded as accurate, quick, and inexpensive.[11] Freezing (–20°C for 60 hr) or cooking (60°C for 5 min) kills larvae in fish.[10]

42.312 Eustrongylides *spp.*

A rare zoonotic disease is caused by larvae of the genus *Eustrongylides* which normally mature in fish-eating birds. Human infections result when freshwater fish harboring encysted larvae are eaten raw.[63] Eggs, from the feces of infected birds, are deposited in freshwater streams, and are infective for segmented worms (oligochaetes) in which larvae of the parasites develop and become infective for fish. Numerous species of freshwater fish ingest the oligochaetes and harbor fourth-stage larvae, which are infective for birds. Prevalence in small fish such as killifish may be 50% or higher in some areas. Only six human cases are known. All are from North America. Five Chesapeake Bay fishermen swallowed their bait fish raw (and whole) and experienced abdominal pain within 24 hr. Two of the symptomatic fishermen underwent surgery, and large (8 to 12 cm long) bright red nematodes were removed from their peritoneal cavity. Another man experienced symptoms resembling appendicitis one day after eating sashimi. During surgery a bright red *Eustrongylides* sp. was found. Larval nematodes are detected by examination of fish flesh. The large red nematodes, usually within a tough white oval cyst, are easily seen.

42.313 Angiostrongylus *spp.*

Angiostrongyliasis, or eosinophilic meningoencephalitis, results when nematodes of the genus *Angiostrongylus* enter blood vessels of the brain. The nematode is normally parasitic in rats. Humans are infected by ingesting the larvae in invertebrate intermediate hosts or by eating food contaminated with infective larvae from an invertebrate host. Molluscs such as slugs and aquatic and terrestrial snails, as well as planarians, freshwater shrimp, land crabs, and frogs serve as either intermediate or transport hosts. The disease, widespread in the South Pacific and Southeast Asia including Hawaii, has recently been reported in rats and humans in Cuba[22] and in rats in Louisiana.[14] Another form of angiostrongyliasis has been reported in Central America. It is caused by *A. costaricensis,* which resides in the mesenteric arteries of rats and humans.[51] Clinical symptoms of *A. cantonensis* infection include high eosinophil counts in blood and spinal fluid and elevated lymphocytes in cerebrospinal fluid. When nematodes invade the brain symptoms include headache, paralysis, stiff neck, coma, and death. Control or prevention of infection is difficult in endemic areas because larvae survive in drinking water and on vegetables. Boiling infected snails for 2 min or freezing at –15°C for 24 hr kills all infective larvae.

Detection is difficult because of the small size of the larvae (less than 0.5 mm long).[4] Larvae from suspected intermediate hosts are recovered by Baermann elution of minced tissue or by artificial digestion techniques. In Baermann elution,[61] samples wrapped in gauze are placed in a funnel fitted with a short rubber hose, which is clamped shut. Lukewarm water is poured into the funnel until it wets the bottom of the sample. Invertebrates migrate into the warm water where most species settle to the bottom of the funnel and can be collected by opening the clamp. Liquid collected from the funnel is examined microscopically for larvae.[5]

42.314 Diphyllobothrium *spp.*

Diphyllobothrium latum, the broad fish tapeworm of humans and related species, causes diphyllobothriasis in much of the world, with greatest prevalence in northern Europe, Asia, and North America. Of an estimated 100,000 human cases in North America, most were in the Great Lakes region. In Alaska and other arctic regions, where infections with other species of *Diphyllobothrium* have been reported, the prevalence is unknown. Species of this genus also infect carnivores, such as dogs, cats, bears, seals, and sea lions. The infective stage for humans and other carnivores is the wormlike, slightly flattened, flesh-colored plerocercoid (5 to 40 mm long) found in the flesh of fish. In North America most cases result from eating infected freshwater or anadromous fish but marine fish are also infected. Prevalence in fish varies greatly; 50% to 70% of pikes and walleyes are infected in some small lakes in North America. Infections result primarily from cultural practices, the ingestion of raw or inadequately cooked infected fish. Some diphyllobothriid tapeworms can become quite large. *Diphyllobothrium latum* and *D. ursi* can grow to10 feet in length but *D. dendriticum*, the most common diphyllobothriid in humans, usually reaches only about 1 meter in length. Clinical symptoms of diphyllobothriasis include epigastric fullness, pressure or pain, diarrhea, nausea, anorexia, vomiting, and, rarely, anemia.[12] Treatment with niclosamide is usually successful in removing the tapeworm from the intestine.[15] Other species of *Diphyllobothrium* have been described and are known to infect humans in more restricted geographic regions.

Plerocercoids are difficult to detect in fish flesh because they are small and flesh-colored (creamy white). Plerocercoids can be recovered by shredding the tissue and inspecting it with a dissection microscope or hand lens or by artificial digestion. The latter has not been critically evaluated for efficiency. Plerocercoids are rendered noninfectious by freezing, 18°C for at least 24 hr) or by heating (56°C for 5 min).[12]

42.315 Nanophyetus

Human infection with the intestinal fluke *Nanophyetus salmincola* was first reported from the northwestern United States in 1987. The fluke is also called the salmon poisoning fluke because it transmits a rickettsia fatal to dogs that eat infected salmon. *Nanophyetus schikhobalowi* was found in up to 98% of the population in some Russian villages in eastern Siberia. The stage transmitted to humans is an ovoid metacercaria about 0.5 by 0.25 mm found on the skin or in the flesh or viscera of salmonids and to a lesser extent other species of fish and amphibians. Symptoms include right lower quadrant abdominal discomfort, bloating, episodic diarrhea of several weeks duration, nausea, vomiting, decreased appetite, cramping, and weight loss. Anthelmintic treatment with bithionol or niclosamide has been effective in eliminating parasites.[25]

Metacercariae are detected by homogenizing the fish flesh in saline and examining the sediment after screening. In one method, the whole animal or a selected part was homogenized in 200 mL of saline and washed through a 212-µm mesh screen into a fingerbowl and allowed to settle for 1 min. The supernatant is decanted, and the sediment is diluted with saline to 1-L, transferred to a 1-L pharmaceutical flask, and allowed to settle for 1 min. The supernatant is then decanted. The wash is repeated and the sediment is examined for cysts.[35] To our knowledge this method has not been confirmed. Cooking is recommended for killing metacercariae in fish; the effect of freezing is unknown.

42.316 Paragonimus *spp.*

Paragonimiasis is caused by lung flukes of the genus *Paragonimus*, which includes more than 30 species. Human infections are known in Asia, the South Pacific Islands, and the Americas. The species that caused the five clinical cases reported from North America are unidentified. Endemic cases in Central and South America and Africa, where the prevalence may be high, were attributable to *P. westermanni*.

Adult *Paragonimus* spp. infect cats, rats, mice, mongooses, monkeys, dogs, pigs, and cattle. Infection results from ingestion of globular metacercariae (0.3–0.5 mm in diameter) in various tissues of freshwater crustaceans (crabs, prawns, or crayfish). Metacercariae have been recovered from crayfish collected from the eastern and midwestern United States. Human cases involving the lungs are characterized by fever and malaise at onset, followed by dyspnea, chest pain, and eosinophilia, with accumulation of pleural fluid. Several months later a productive cough develops and operculate ova are detectable in the sputum. The infection can spread to other organs. Cerebral paragonimiasis usually takes a year after the onset of pulmonary symptoms and can result in an array of severe central nervous system symptoms. Abdominal paragonimiasis may be more common but with milder or no symptoms. Anthelmintic treatment with praziquantel or niclofolan may be effective in eliminating symptoms and detectable ova.[72] Detection in crayfish is by removal of the carapace and examination of the heart. The metacercariae are easily recognized without magnification in living specimens because the white excretory bladder contrasts sharply with the yellowish host tissue.[3]

Data are lacking on the ability of the *Paragonimus* spp. metacercariae to survive freezing. Infections can be prevented by cooking freshwater crustaceans to an internal temperature of 75°C. Care should also be exercised in preparing raw crayfish for cooking by avoiding inadvertent ingestion of uncooked cysts.

42.317 Alaria *spp.*

Alaria americana, *Alaria* spp., and a trematode identified only as a member of the subfamily Alarinae were each reported once in humans in North America. The mature adult stages of these flukes are found in the bobcat, coyote, fox, lynx, martin, skunk, and wolf throughout North America. The mesocercarial stage in muscles of the frog (second intermediate host) is infective for a variety of hosts, including humans. In the most severe human case reported the autopsy revealed thousands of larvae migrating through tissues of the lungs, liver, spleen, stomach, heart, brain, and spinal cord.[31]

Digestion of suspected infected flesh will recover the small (0.5 mm long) motile larvae (see Section 42.312). They can also be recovered by processing in a kitchen blender and sedimentation or by dissection and sedimentation alone. Extreme care should be exercised when handling the flesh of frogs.

42.32 Parasites of Meat

42.321 Parasites in Pork

42.3211 *Toxoplasma.* *Toxoplasma gondii* is a coccidian protozoa that has been studied extensively because of its economic and public health significance.[23] It is estimated that 30%–40% of the population of the United States has serum antibodies to *T. gondii*, indicating current or past infection with the organism. Members of the cat family are the only known definitive hosts (i.e., those capable of contaminating the environment with the oocyst stage in feces). All other vertebrates, including humans, are potential intermediate hosts (i.e., they can acquire the infection either by ingesting the oocysts or by eating animal tissues that contain the cyst stage). Infectious stages of *T. gondii* have been isolated from muscles and various organs such as the heart, brain, and liver of virtually all food animals including chickens, pigs, cattle, goats and sheep. Cooking meat at 60°C or irradiation to 30 krad renders the encysted organisms noninfectious. The acquired postnatal form of the infection is often associated with mild influenza-like symptoms; however, potentially life-threatening infections are well documented in the congenital or neonatal form of the infection, especially if the mother becomes infected during the first or second trimester of pregnancy. The disease can also be serious in immunosuppressed patients and often is fatal in AIDS patients. Acute toxoplasmosis can usually be treated with drugs such as pyrimethamine and sulfas or with clindamycin or spiramycin, but these drugs are relatively toxic and are not prescribed during pregnancy. Toxoplasmosis can be prevented by avoiding direct and indirect exposure to cat feces and by thoroughly cooking meat.[29] The extent or importance of *Toxoplasma* in drinking water has not been evaluated although outbreaks related to waterborne *Toxoplasma* oocysts have been reported.

The presence of oocysts in cat feces or water can be determined as described in Section 42.28. Because oocysts are 10 µm in diameter or slightly greater, the pore size of the filter should be adjusted accordingly. Oocysts can be detected in cat feces by mixing feces in water followed by flotation of oocysts in sucrose solution and examination by light microscopy. However, oocysts of *T. gondii* are identical in size and shape with those of two other species of coccidia and can be differentiated only by oral inoculation and testing of mice for antibodies to *T. gondii*. Cysts are not detected during routine organoleptic meat inspection.

Diagnosis of *Toxoplasma* in meat requires a biological assay as follows: a 50-g sample of meat passed through a grinder is placed in 625 mL of an acid-pepsin digestion solution (10 g pepsin with 1:10,000 activity and 10 mL of 37% HCI/L of water). The meat-digestion solution is incubated at 37°C for 90 min on a shaker. The digested sample is filtered through two layers of gauze and centrifuged at 400 × g for 10 min. The sediment is resuspended in 50 mL of saline, centrifuged again, and the sediment is resuspended in 6 mL of saline containing 1,000 IU penicillin/mL and 100 µg dihydrostreptomycin/mL. Usually, 1 mL of the sample is inoculated into each of six 25-g *Toxoplasma*-free mice. After 30 days the blood serum of mice is tested for antibodies to *Toxoplasma* and brain smears are examined for cysts of *T. gondii.*

42.3212 *Trichinella.* Human trichinellosis results from eating raw or undercooked meat, most often pork, containing the infective muscle larval stage of the nematode *Trichinella spiralis*. Human trichinellosis is worldwide, but clinical cases are rare. Fewer than 100 human cases have been diagnosed per year in the United States over the last decade, although sporadic postmortem surveys have indicated a much higher subclinical or undiagnosed rate (2.2%). In countries that practice meat inspection for trichinae, human infection from pork has virtually disappeared. *T. spiralis* can infect virtually any warm-blooded carnivore. Pork, horse meat, bear meat, and game meats containing the infective muscle larvae of *T. spiralis* are the most common source of human infection. The current infection rate of *T. spiralis* in hogs in the United States is about 1 per 10,000. Hogs acquire the infection by eating raw or undercooked garbage containing infected meat, infected rodents, wildlife, or infected hog carcasses. Human infection results in many symptoms as the worms develop in the intestine and produce larvae that migrate to and encyst in striated muscles. During the larval migration and encystment, symptoms include fever, myalgia, and periorbital edema (the general trichinellosis syndrome). The diagnosis of human trichinellosis depends on a combination of factors, including a history of consumption of infected meat, symptoms, and laboratory findings of which eosinophilia is most notable. Definitive diagnosis relies on the presence of larvae in a muscle biopsy. Infections detected at the intestinal level can be treated with mebendazole or thiabendazole; mebendazole is less effective against the larvae in the muscle. Corticosteroids (e.g., prednisone) are often given during the muscle phase to reduce tissue inflammation. *T. spiralis* is detected in meat by one of two direct methods.[20,26,27,44] Serological methods for the antemortem or postmortem detection of trichinellosis are available but are currently used for epidemiologic studies only.[32,33]

1. *The Compression Method.* In this method, a sample of muscle tissue is obtained from a suspected animal carcass or meat. A carcass sample should be collected from one of the areas where the larvae accumulate in highest density: the tongue, diaphragm, or masseter tissues. The sample is cut into small pieces (thin slices are best) and squeezed between two glass microscope slides until the tissue becomes translucent. The sample is then examined with the aid of a compound microscope (at low magnification) or a dissecting microscope for the presence of encysted larvae. Muscle larvae in tissue are coiled within a muscle nurse cell or "cyst."
2. *The Digestion Method.* The level of infection of *T. spiralis* larvae in muscle tissue is often too low to be detected by the compression method; therefore, a second method involving tissue digestion is used. In this method, a larger sample of meat (one to several hundred grams) is obtained. The tissue to be examined is cleaned of adhering fat and ground in a meat grinder. (Alternatively meat may be minced with scissors.) Ground meat is subjected to digestion in a solution of 1% pepsin (1:10,000 activity) and 1% HCl; digestion solution should be used at a ratio of 1 L for 100 g of tissue. Digestion is carried out for 3 hr at 37°C with continuous agitation, either by stirring or shaking. The digested material is then allowed to settle for 20 min, and approximately 3/4 of the supernatant fluid is decanted. The sediment is resuspended and passed through a 400-µm mesh screen to remove undigested material. The filtered digestion fluid is then allowed to settle for 20 min in a conical glass. The upper 3/4 of the supernatant is decanted and the sediment is resuspended in several volumes of tap water. This process is repeated until the supernatant is clear. The final sediment is examined under a microscope for the presence of motile larvae.

42.3213 *Taenia solium.* Human taeniasis (tapeworm disease) results from ingestion of raw or undercooked pork containing the infective cysticercus larvae of *Taenia solium. T. solium* is distributed worldwide, with higher prevalence rates of human infection in less developed countries; it is not prevalent in the United States.

Pigs acquire larval stages by ingesting *T. solium* eggs from the environment. The eggs hatch, liberating the oncosphere that penetrates the intestine and travels via the circulatory system to muscle tissue. The oncosphere develops to the infective cysticercus stage in 10 to 12 weeks. Humans acquire taeniid infection by ingesting raw or undercooked pork containing cysticerci. Human taeniasis is a relatively mild disease which can be asymptomatic or cause vague symptoms of nausea, abdominal cramps, weight loss, or headache. Human taeniasis is diagnosed by the presence of eggs or proglottids (segments) in the stool. Effective drugs are available for treatment of tapeworms; the drug of choice is praziquantel. Humans can also serve as intermediate hosts, harboring the cysticercus stage *(Cysticercus cellulosae)* of *T. solium.* Human cysticercosis caused by ingestion of eggs of *T. solium* is a widespread and serious disease involving damage primarily to the central nervous system.[60]

Hog carcasses are inspected for the presence of *C. cellulosae* at slaughter. Inspection involves palpation and incision of tongue and other muscular tissues including the masseters, esophagus, heart, and diaphragm. Because of the random distribution and low levels of infection, the presence of one cysticercus in a portion of muscle would not necessarily indicate infection in other cuts from the same carcass.

42.322 Parasites in Beef

42.3221 *Toxoplasma.* (See Section 42.3211.)

42.3222 *Taenia saginata.* Human taeniasis (tapeworm disease) results from the ingestion of raw or undercooked beef containing the infective cysticercus *(Cysticercus bovis)* stage of *Taenia saginata*. This parasite is distributed worldwide with higher prevalence of human infection in less developed countries. Its prevalence varies regionally but is more common than *T. solium* in the United States. The larval cysticercus stage is found in cattle. The life cycle, transmission, and stages of infection are similar to those of *T. solium* (see Section 42.3213), except that the larval stage of *T. saginata* does not infect humans.[60] Methods of detection are similar to those for *T. solium* (See Section 42.3213). Because infection levels in beef carcasses are quite low, antemortem inspection procedures are relatively insensitive.

42.323 Parasites in Goat Meat and in Lamb

42.3231 *Toxoplasma.* (See Section 42.3211.)

42.4 EMERGING METHODS

Enzyme-linked immunosorbent assays to detect specific secreted parasitic antigens are currently being developed. As indirect fluorescent antibody test kits become available for more species, the level of detection in foods will increase. Gene probes and genetic tests based on the polymerase chain reaction may increase the sensitivity of parasite detection to that available for other microbial pathogens.

42.5 REFERENCES

1. Andersen, E. M., and J. R. Lichtenfels. 1998. Anisakiasis. *In* C. H. Binford and D. H. Connor (eds.), Pathology of tropical and extraordinary diseases, vol. 2. Armed Forces Institute of Pathology, Washington, D. C.
2. Anonymous. 1998. Physicians' Desk Reference, 52nd ed. Medical Economics Company, Montvale, N. J.
3. Ameel, D. J. 1934. Paragonimus, its life history and distribution in North America and its taxonomy (Trematoda:Troglotrematidae). Am. J. Hyg. 19:279-317.
4. Ash, L. R. 1970. Diagnostic morphology of the third-stage larvae of *Angiostrongylus cantonensis, Angiostrongylus vasorum, Aeluorostrongylus abstrusus and Anafilaroides rostratus* (Nematoda:Metastrongyloidae). J. Parasitol. 56:249-253.
5. Ash, L. R., and T. C. Orihel. 1987. Parasites: A guide to laboratory procedures and identification. American Society of Clinical Pathologists. Chicago, Ill.
6. Ash, L. R., and T. C. Orihel. 1997. Atlas of human parasitology, 4th ed. American Society of Clinical Pathologists, Chicago, Ill.
7. Beaver, P. C., R. C. Jung, and E. W. Cupp, 1984. Clinical parasitology. Lea and Febiger, Philadelphia, Pa.
8. Bhabulaya, M. 1982. Angiostrongylosis, p. 25. *In* M. G. Shultz (ed.), Handbook series on zoonoses, Section C: Parasitic zoonoses, vol. 11. CRC Press, Boca Raton, Fla.
9. Bier, J. W., G. J. Jackson, R. L. Sellers, and R. A. Rude 1984. Parasitic animals in foods, p. 21.01. *In* FDA bacteriological analytical manual, 6th ed. Assoc. Off. Anal. Chem., Arlington, Va.
10. Bier, J. W. 1988. Anisakiasis, p. 768. *In* W. Balows, W. J. Hausler, Jr., and E. H. Lennette (eds.), Laboratory diagnosis of infectious diseases-Principles and practice. Springer-Verlag, New York.
11. Brattey, J. 1988. A simple technique for recovering larval ascaridoid nematodes from the flesh of marine fish. J. Parasitol. 74:735-737.
12. Byland, B. G. 1982. Diphyllobothrium, p. 217-243. *In* M. G. Schultz (ed.), Handbook series on zoonoses, Section C: Parasitic zoonoses, vol. 11. CRC Press, Boca Raton, Fla.
13. Cheng, T. C. 1982. Anisakiasis, p. 37. *In* M. G. Schultz (ed.), Handbook series on zoonoses, Section C: Parasitic zoonoses, vol. 11. CRC Press, Boca Raton, Fla.
14. Campbell, B. G., and M. D. Little. 1988. The finding of *Angiostrongylus cantonensis* in rats in New Orleans, Louisiana, USA. Am. J. Trop. Med. Hyg. 38:568-570.
15. CDC. 1981. Diphyllobothriasis associated with salmon-United States. Morb. Mortal Wkly Rep. 30:331-334.
16. CDC. 1997. Update: Outbreaks of cyclosporiasis-United States and Canada, 1997. Morb. Mortal Wkly. Rep. 46:521-523.
17. CDC. 1997. Update: Outbreak of cyclosporiasis—Northern Virginia–Washington, D. C.–Baltimore, Maryland metropolitan area. Morb. Mortal Wkly. Rep. 46:689-691.
18. CFR. 1987. Code of Fed. Reg., Title 9-Animals and Animal Products, chap. 111, 311.23, p.142. U.S. Government Printing Office, Washington, D. C.
19. CFR. 1987. Code of Fed. Reg., Title 9-Animals and Animal Products, chap. 111, 311.24, p.142. U.S. Government Printing Office, Washington, D. C.
20. CFR. 1987. Code of Fed. Regs., Title 9-Animals and Animal Products, Chapter 111, 318. 10, p. 207. U.S. Government Printing Office, Washington, D. C.
21. Davidson, R. A. 1990. Treatment of giardiasis: The North American perspective. *In* E. A. Meyer (ed.), Giardiasis. Elsevier, New York.
22. Dorta Contreras, A. J., M. Ferra Valdes, R. Plana Bouiv, A. G. Diaz Martinez, N. Gonzalez Garcia, R. Escoba, and X. Perez. 1987. Eosinophilic meningoencephalitis caused by *Angiostrongylus cantonensis* Chen, 1935; immunological study. Rev. Esp. Pediatr. 43:379-382.
23. Dubey, J. P., and C. P. Beattie. 1988. Toxoplasmosis of animals and man. CRC Press, Boca Raton, Fla.
24. Dytrych, J. K., and R. P. Cooke 1995. Autofluorescence of *Cyclospora*. Br. J. Biomed. Sci. 52:76.
25. Eastburn, R. L., T. R. Fritsche, and C. A. Terhune. 1987. Human intestinal infection with *Nanophyetus salmincola* from salmonid fishes. Am. J. Trop. Med. Hyg. 36:586-591.
26. EEC. 1977. Commission directive 77/96/EEC. Off. J. Eur. Commun. 26: 67.
27. EEC. 1984. Commission directive 84/319/EEC. Off. J. Eur. Commun. 167: 34.

28. Fayer, R. 1998. Waterborne and foodborne protozoa. *In* Y. H. Hui, J. R. Gorham, K. D. Murrell, and D. O. Cliver (eds.), Foodborne disease handbook: Diseases caused by viruses parasites and fungi, vol 2. Marcel Dekker, New York.
29. Fayer, R., and J. P. Dubey. 1985. Methods for controlling transmission of protozoan parasites from meat to man. Food Technol. 39(3):57-63.
30. Fayer, R., C. A. Speer, and J. P. Dubey. 1997. Biology. *In* R. Fayer (ed.), Cryptosporidium and cryptosporidiosis. CRC Press, Boca Raton, Fla.
31. Freeman, R. S., P. F. Stuart, J. B. Cullen, A. C. Ritchie, A. Mildon, B. J. Fernandes, and R. Bonin. 1976. Fatal human infection with mesocercariae of the trematode *Alaria Americana.* Am. J. Trop. Med. Hyg. 25:803-807.
32. Gamble, H. R., W. R. Anderson, C. E. Graham, and K. D. Murrell 1983. Serodiagnosis of swine trichinosis using an excretory-secretory antigen. Vet. Parasitol. 13:349-361.
33. Gamble, H. R., and K. D. Murrell. 1988. *Trichinellosis*, p. 1017. *In* W. Balows, W. J. Hausler, Jr., and E. H. Lennette (eds.), Laboratory diagnosis of infectious diseases-Principles and practice. Springer-Verlag, New York.
34. Gardiner, C. H., R. Fayer, and J. P. Dubey 1988. An atlas of protozoan parasites in tissues. U.S. Government Printing Office, Washington, D. C.
35. Gebhart, G. A., R. E. Milleman, S. E. Knapp, and P. A. Wyberg. 1966. "Salmon poisoning" diseases second intermediate host susceptibility. J. Parasitol. 52:54-60.
36. Hayes, E. R., T. D. Matte, T. R. O'Brien, G. S. Logsdon, Rose JB, Ungar BL, Word DM, Pinsky PF, Cummings ML, et alet al. 1989. Large community outbreak of cryptosporidiosis due to contamination of a filtered public water supply. N. Eng. J. Med. 3201:1372-1376.
37. Herwaldt, B. R., and M. L. Ackers. 1997. An outbreak in 1996 of cyclosporiasis associated with imported raspberries. N. Eng. J. Med. 336:1548-1556.
38. Higashi, G. I. 1985. Foodborne parasites transmitted to man from fish and aquatic foods. Food Technol. 39(3):69-72.
39. Hoge, C. W., D. R. Shlim, M. Ghimire, J. G. Rabold, A. Welch, R. Rajah, P. Gaudio, and P. Escheverria. 1995. Placebo controlled trial of co-trimoxazole for Cyclospora infections among travelers and foreign residents in Nepal. Lancet 345:691-693.
40. Hoxie, N. J., J. P. Davis, J. M. Vergemont, R. D. Nashold, and K. A. Blair. 1997. Cryptosporidiosis associated mortality following a massive waterborne outbreak in Milwaukee, Wisconsin. Am. J. Public Health 87:2032-2035.
41. Huang, P., J. T. Weber, D. M. Sosin, P. M. Griffin, E. G. Long, J. J. Murphy, F. Kocka, C. Peters, and C. Kallick. 1995. the first reported outbreak of diarrheal disease associated with *Cyclospora* in the United States. Ann. Intern. Med. 123:409-414.
42. Jackson, G. J., J. W. Bier, W. L. Payne, and F. D. McClure. 1981. Recovery of parasitic nematodes from fish by digestion or elution. Appl. Environ. Microbiol. 41:912-914.
43. Ko, R. C. 1995. Fishborne parasitic zoonoses. *In* P. T. K. Woo (ed.), Fish diseases and disorders. Commonwealth Agricultural Bureau International, University Press, Cambridge, England.
44. Kohler, G., and E. J. Ruitenberg. 1974. Comparison of three methods for the detection of *Trichinella spiralis*infections in pigs by five European laboratories. Bull. WHO 50:413.
45. LeChevalier, M. W., W. D. Norton, and R. G. Lee. 1991. *Giardia* and *Cryptosporidium* spp. in filtered drinking water supplies. Appl. Environ. Microbiol. 57:2617-2621.
46. Lee, J. J., S. H. Hutner, and E. C. Bovee. 1985. An illustrated guide to the protozoa. Society of Protozoologists, Lawrence, Kans.
47. Lushbaugh, W. B., and F. E. Pittman. 1982. Amebiasis, p. 5. *In* L. Jacobs and P. Arambulo (eds.), Handbook series on zoonoses, Section C: Parasitic zoonoses, vol. 11. CRC Press, Boca Raton, Fla.
48. MacKenzie, W., N. Hoxie, M. Proctor, M. S. Gradus, K. A. Blair, D. E. Peterson, J. J. Kazmierczak, D. G. Addiss, K. R. Fox, , J. B. Rose, et al. 1994. A massive outbreak in Milwaukee of *Cryptosporidium* infection transmitted through the public water supply. N. Eng. J. Med. 331:161-167.
49. Madico, G., R. H. Gilman, E. Miranda, L. Cabrera, and C. R. Sterling. 1993. Treatment of Cyclospora infections with co-trimoxazole. Lancet 342:122-123.
50. Meyer, E. A., and E. L. Jarroll. 1982. Giardiasis, p. 25. *In* L. Jacobs and P. Arambulo (eds.), Handbook series on zoonoses, Section parasitic zoonoses, vol. 11. CRC Press, Boca Raton, Fla.
51. Morera, P. 1973. Life history and redescription of *Angiostrongylus costaricensis* Morera and Despedes 1971. Am. J. Trop. Med. Hyg. 22:613-621.
52. Musial, C. E., M. J. Arrowood, C. R. Sterling, and C. P. Gerba. 1987. Detection of *Cryptosporidium* in water by using polypropylene cartridge filters. Appl. Environ. Microbiol. 53:687-692.
53. Myers, B. J. 1975. The nematodes that cause anisakiasis. J. Milk Food Technol. 38:774.
54. Nakhforoosh, M., and J. B. Rose. 1989. Detection of *Giardia* with a gene probe, p. 239. *In* Abstracts of the 89th Annual Meet. American Society for Microbiology. American Society for Microbiology, Washington, D. C.
55. Neafie, R. C. 1976. Balantidiasis, p. 325-327. *In* C H. Binford and D. H. Connor (eds.), Pathology of tropical and extraordinary diseases, vol. 1. Armed Forces Institute of Pathology, Washington, D. C.
56. Ongerth, J. E., and H. H. Stibbs. 1987. Identification of *Cryptosporidium* oocysts in river water. Appl. Environ. Microbiol. 53:672-676.
57. Orihel, T. C., and L. R. Ash. 1995. Parasites in human tissues. American Society of Clinical Pathologists, Chicago, Ill.
58. Ortega, Y. R., R. H. Gilman, and C. R. Sterling. 1994. A new parasite (Apicomplexa; Eimeriidae) from humans. J. Parasitol. 80:625-629.
59. Ortega, Y. R., C. R. Roxas, R. H. Gilman, N. J. Miller, L. Cabrera, C. Taquiri, and C. R. Sterling. 1997. Isolation of *Cryptosporidium parvum* and *Cyclospora cayetanensis* from vegetables collected in markets of an endemic region in Peru. Am. J. Trop. Med. Hyg. 57:683-686.
60. Pawlowski, Z. S. 1982. Taeniasis and cysticercosis, p. 313. *In* L. Jacobs and P. Arambulo (eds.), Handbook series on zoonoses, Section C: Parasitic zoonoses, vol.. 11. CRC Press, Boca Raton, Fla.
61. Pritchard, M. H., and G. O. W. Kruse. 1982. The collection and preservation of animal parasites. Tech. Bull. No.1. University of Nebraska Press, Lincoln, Nebr.
62. Rose, J. B. 1988. Occurrence and significance of *Cryptosporidium* in water. J. Am. Water Works Assoc. 80:53-56.
63. Shirazian, D., E. L. Schiller, C. A. Glasser, and S. L. Vonderfecht 1984. Pathology of larval *Eustrongylides* in the rabbit. J. Parasitol. 70:803.
64. Stern, J. A., D. Chakravarti, J. R. Uzmann, and M. N. Hesselhold. 1958. Rapid counting of Nematoda in salmon by pepsin digestion. U.S. Fish and Wildlife Special Scientific Report No. 25S, p. 1.
65. USSR. 1972. Veterinary statute of the USSR, vol. 2. Kolos, Moscow.
66. Valdimarsson, G., H. Einarsson, and F. J. King. 1985. Detection of parasites in fish muscle by candling techniques. J. Assoc. Off. Anal. Chem. 68:549-550.
67. Walzer, P. D., and G. R. Healy. 1982. Balantidiasis, p. 15. *In* L. Jacobs and P. Arambulo (eds.), Handbook series on zoonoses, Section C: Parasitic zoonoses, vol. 2. CRC Press, Boca Raton, Fla.
68. Wolfe, M. S. 1990. Clinical symptoms and diagnosis by traditional methods. *In* E. A. Meyer (ed.), Giardiasis. Elsevier, New York.
69. Xiao, L., I. Sulaiman, R. Fayer, and A. A. Lal. 1998. Species and strain specific typing of *Cryptosporidium* parasites in clinical and environmental samples. Mem. Inst. Oswaldo Cruz, Rio de Janeiro, in press.
70. Yai, L. E., A. R. Bauab, M. P. Hirschfeld, M. L. de Oliveira, and J. T. Dmaceno. 1997. The first two cases of *Cyclospora* in dogs, Sao Paulo, Brazil. Rev. Inst. Med. Trop. Sao Paulo 39:177-179.
71. Yamaguchi, T. 1981. Color atlas of clinical parasitology. Wolfe Medical Publ., London, England.
72. Yokogawa, S., W. W. Court, and M. Yokogawa. 1960. *Paragonimus* and paragonimiasis. Exp. Parasitol. 10:81-92.
73. Zaman, V. 1980. Atlas of medical parasitology. Lea and Febiger, Philadelphia, Pa.

Chapter 43

Toxigenic Fungi and Fungal Toxins

Lloyd B. Bullerman and Hassan Gourama

43.1 INTRODUCTION

The beneficial effects of certain molds have been known for years. Molds have not only served to synthesize antibiotics, but also to produce some foods. Fermented foods, such as some cheeses, soy sauce, miso, tempeh and other delicacies are all prepared with the help of molds in combination with lactic acid bacteria and/or yeasts. However, it is also well documented that some molds produce toxic substances. These toxic substances are known as mycotoxins, from the Greek word "mykes," which means fungus and the Latin word "toxicum," which means toxin, or poison.[16] Toxigenic molds have caused food safety problems for as long as foods have been harvested and stored. Some of the plagues mentioned in the Bible may have been caused by mycotoxins.[31] However, the intensive study of toxigenic fungi and fungal toxins has only taken place since 1960. At that time there was a severe outbreak of disease among turkey poults and other young farm animals in England in which 100,000 young turkeys died.[5] Because of the mysterious nature of the disease and the high death loss of turkeys, the disease was first called "Turkey X Disease."[12] The study of Turkey X Disease led to the discovery of aflatoxin, which had been produced by *Aspergillus flavus* (*parasiticus*) in peanuts that had been used to make peanut meal.[38] The peanut meal contained acutely toxic amounts of aflatoxin, which caused the poisonings when the peanut meal was used as a feed ingredient. The discovery of aflatoxins led to the realization that mold metabolites could cause disease and death in animals, and possibly humans, and stimulated intensive and extensive research on toxigenic molds and mycotoxins.

43.2 MYCOTOXINS

Since the discovery of aflatoxins, numerous molds have been tested in the laboratory for the production of toxic metabolites. Of the hundreds of mycotoxins produced under laboratory conditions, only about 20 are known to naturally occur in foods and feeds with sufficient frequency and in potentially toxic amounts to be of concern to food safety. The molds that produce the mycotoxins of most potential concern are found in five taxonomic genera, *Aspergillus, Penicillium, Fusarium, Alternaria* and *Claviceps*.[16,31]

Aspergillus species produce aflatoxins B_1, B_2, G_1, G_2, M_1 and M_2, ochratoxin A, sterigmatocystin and cyclopiazonic acid among other lessor known toxic substances.[2,16,27,28,31,39] *Penicillium* species produce patulin, citrinin, citreoviridin, penitrem A, PR Toxin, roquefortine C, secalonic acid, rubratoxin and a number of other toxic substances as well as ochratoxin A and cyclopiazonic acid.[36,39] *Fusarium* species produce the trichothecenes: deoxynivalenol (DON, vomitoxin), 3-acetyl deoxynivalenol, 15-acetyl deoxynivalenol, nivalenol, diacetoxyscirpenol and T-2 toxin; zearalenone, fumonisins and moniliformin as well as other potentially toxic and possibly unknown toxic substances.[15,16,31] *Alternaria* species produce a number of biologically active compounds of questionable mamalian toxicity including tenuazonic acid, alternariol and alternariol methyl ether. *Alternaria alternata* f. sp. *lycopersici* (AAL), which is pathogenic to tomatoes, produces AAL toxin, which is structurally and toxicologically similar to fumonisins.[15] *Claviceps* toxins are primarily the ergot alkaloids that can be found in ergot parasitized grasses and small grains.[16,31] A summary of potentially toxic molds isolated from various food and agricultural commodities is given in Table 1.

Of the 20 or so naturally occurring mycotoxins mentioned above, there are five toxins, or groups of related compounds, that are of greatest concern. These are the aflatoxins, ochratoxin, zearalenone, deoxynivalenol (or nivalenol in some regions) and fumonisins.[31] Toxins of growing concern that may be added to that list are patulin, cyclopiazonic acid and moniliformin. The mycotoxigenic molds of greatest concern and the toxins they produce are summarized in Table 2.

Analytical methods for detection and quantitation of mycotoxins are influenced by many factors. No single method can readily be used for all of the mycotoxins, though some multi-toxin detection methods have been published.[42] Adequate product sampling, in the nature of obtaining a true representative sample, is critical to achieving accurate analytical results.[42] Once a representative sample is obtained the mycotoxin in question must be extracted from the food or commodity. Extraction solvents for many mycotoxins are either organic solvents or combinations of organic solvents and water, and for a few mycotoxins, water alone may suffice.[42] Frequently mycotoxin extracts must be purified or "cleaned up" to remove substances, such as lipids and pigments from the substrate, that will interfere with analytical techniques. Cleanup steps may involve partitioning between solvents to remove interferences, or using various types of columns or cartridges packed with different absorbent materials to separate the mycotoxins from the interfering materials.[42] Analytical quantitation of mycotoxins is most often done by some type of chromatography, either thin-layer chromatography (TLC), high perfor-

Table 1. Summary of Selected Reports of Isolations of Potentially Toxic Molds from Various Food or Agricultural Commodities[14,17]

Commodity	Potentially toxic genera/species found		Potential mycotoxins
Wheat, flour, bread, cornmeal, popcorn	*Aspergillus flavus* *ochraceus* *versicolor* *Cladosporium* *Fusarium* *moniliforme, proliferatum* *subglutinans, graminearum*	*Penicillium citrinum,* *citreonigrum,* *aurantiogriseum,* *griseafulvum,* *commune*	Aflatoxin, ochratoxin, sterigmatocystin, patulin, penicillic acid, fumonisins moniliformin, deoxynivalenol, zearalenone
Peanut, in-shell pecans	*Aspergillus flavus,* *parasiticus, ochraceus,* *versicolor* *Fusarium, Rhizopus,* *Chaetomium*	*Penicillium* *aurantiogriseum* *expansum* *citrinum*	Aflatoxins, ochratoxin, patulin, sterigmatocystin
Apples and apple products	*Penicillium expansum*		Patulin
Meat pies, cooked meats, cocoa powder, hops, cheese	*Aspergillus flavus* *Cladosporium*	*Penicillium verrucosum,* *viridicatum, roqueforti* *griseofulvum, commune*	Aflatoxins, ochratoxin, patulin, penicillic acid
Aged salami and sausage, country cured ham, moldy meats, cheese	*Aspergillus flavus,* *ochraceus, versicolor*	*Penicillium verrucosum* *viridicatum,* *aurantiogriseum*	Aflatoxins, ochratoxin, patulin, penicillic acid, sterigmatocystin
Black and red pepper, macaroni	*Aspergillus flavus,* *ochraceus*	*Pencillium* species	Aflatoxins, ochratoxin
Dry beans, soybeans, corn, sorghum, barley	*Aspergillus flavus,* *ochraceus, versicolor* *Alternaria, Cladosporium*	*Penicillium aurantiogriseum* *verrucosum, viridicatum,* *citrinum, expansum* *islandicum, griseofulvum*	Aflatoxins, ochratoxin, sterigmatocystin, penicillic acid, patulin, citrinin, griseofulvin alternariol, altenuene, altertoxin
Refrigerated and frozen pastries	*Aspergillus flavus,* *versicolor*	*Penicillium aurantiogriseum,* *citrinum, viridicatum* *commune, roqueforti* *griseofulvum, viridicatum*	Aflatoxins, sterigmatocystin, ochratoxin, citrinin, patulin, penicillic acid
Moldy supermarket foods	*Penicillium cyclopium* *Fusarium oxysporum* *solani*	*Aspergillus* species	Penicillic acid, possibly other *Penicillium* toxins, T-2 toxin
Foods stored in homes, both refrigerated and non-refrigerated	*Penicillium* species	*Aspergillus* species	Aflatoxin, kojic acid, ochratoxin A, patulin, penicillic acid

mance liquid chromatography (HPLC), gas chromatography (GC) and gas chromatography linked to mass spectrometry (GCMS). Numerous detection methods are used including u.v. absorbance, fluorescence, electrochemical, flame ionization and others. In recent years enzyme-linked immunosorbant assays (ELISA) have been developed for a number of mycotoxins (Tables 3 and 4) and are commercially available.[4,11,34,42] The Technical Services Division of the Grain Inspection, Packers and Stockyards Administration (GIPSA) of USDA has published methods for sampling, chemical analyses and a list of approved test kits for use in testing grain.[25]

Development and validation of the necessary official chemical methods for determining the presence and level of mycotoxins in foods is an undertaking of international scope. Coordination is achieved through a Joint Mycotoxin Committee representing the AOAC International, the American Oil Chemists' Society (AOCS), the American Association of Cereal Chemists (AACC), and the International Union of Pure and Applied Chemists (IUPAC). All the methods validated to date under the aegis of the Joint Mycotoxin Committee have been adopted by the AOAC and published in Chapter 49 of the "Official Methods of Analysis."[40] The AOCS[20] and AACC[1] have each included methods related to their specific activities in their compendia of official methods. IUPAC has recommended a number of the methods, and these various adoptions are noted for each method in the separate AOAC chapter on Natural Toxins.

Table 2. Mycotoxigenic Molds of Greatest Concern and the Mycotoxins They Produce[2,16,27,31]

Molds	Mycotoxins
Aspergillus flavus	Aflatoxins B_1 and B_2
Aspergillus parasiticus *Aspergillus nomius*	Aflatoxins B_1, B_2, G_1 and G_2
Aspergillus ochraceus *Aspergillus niger var. niger* *Penicillium verrucosum*	Ochratoxin
Penicillium expansum, other *Penicillium* species, some *Aspergillus* species	Patulin
Aspergillus flavus	Cyclopiazonic Acid
Fusarium graminearum, *F. culmorum*, *F. crookwellense*	Zearalenone
Fusarium graminearum, *F. culmorum*, *F. crookwellense*	Deoxynivalenol 3-acetyldoxynivalenol 15-acetydeoxynivalenol nivalenol, trichothecenes
Fusarium moniliforme, *F. proliferatum*, *F. subglutinans*	Fumonisins
Fusarium proliferatum, *F. subglutinans*	Moniliformin

43.3 MYCOTOXICOSES

A mycotoxicosis (plural mycotoxicoses) is the disease or adverse effect of a mycotoxin in an animal. Mycotoxins produce a number of adverse effects in a range of biological systems, including microorganisms, plants, animals and humans. The toxic effects of mycotoxins in humans and animals, depending upon dose, may include a) acute toxicity and death as a result of exposure to high amounts of a mycotoxin, b) reduced milk and egg production and lack of weight gain in food producing animals from subchronic exposure, c) impairment or suppression of immune functions and reduced resistance to infections from chronic exposures to low levels of toxins, and d) tumor formation, cancers and other chronic diseases from prolonged exposure to very low levels of a toxin. In addition, mycotoxins may be mutagenic, capable of inducing mutations in susceptible cells and organisms, and teratogenic, capable of causing deformities in developing embryos. Other manifestations of mycotoxins that can affect the food supply in economic terms are reduced growth rates and increased reproductive problems in food producing animals and livestock.[16]

In acute doses, aflatoxins cause severe liver damage and death in animals. Swine, young calves and young poultry are quite susceptible, whereas mature ruminants and chickens are more resistant. Mature sheep seem to be particularly resistant. Sub-acute and chronic exposures to aflatoxin cause liver damage, decreased milk production, decreased egg production, lack of weight gain and immune suppression. The young of all species are more susceptible, but older individuals will also be affected. Clinical signs of sub-acute or chronic exposures of animals to aflatoxins include gastrointestinal problems, decreased feed intake and efficiency, reproductive problems, anemia and jaundice.[16,19]

Ochratoxin A causes kidney damage in many animals and is most commonly associated with mycotoxic nephropathy of swine. In addition, ochratoxin may be immunosuppressive and is now classified as a carcinogen. In high doses ochratoxin can also cause liver damage, intestinal necrosis and hemorrhage. While swine are very susceptible to ochratoxin, ruminants are more resistant, presumably due to degradation in the rumen.[16,36]

The trichothecenes include deoxynivalenol. These toxins generally cause gastroenteritis, feed refusal, necrosis and hemorrhage in the digestive tract, destruction of bone marrow, suppression of blood cell formation, and suppression of the immune system. Clinically, animals show signs of gastrointestinal problems, vomiting, loss of appetite, poor feed utilization and efficiency, bloody diarrhea, reproductive problems, abortions and death. Poultry frequently develop mouth lesions and extensive hemorrhaging in the intestines.[15,16]

Fumonisins cause several diseases in animals, including equine leukoencephalomalacia, porcine pulmonary edema and liver cancer in rats. Fumonisins are not as potent liver carcinogens as aflatoxins. The lifetime carcinogenic potential of fumonisin B_1 in the rat is somewhere between carbon tetrachloride and dimethynitrosamine, however fumonisins, and the organisms that produce them, commonly occur in corn and have been associated with higher incidences of human esophageal cancer in areas where corn is a dietary staple.[15,16]

Zearalenone causes reproductive problems in animals, especially swine, where it disrupts estrus cycles and causes vulvovaginitis in females and feminization of males. In high concentrations it can interfere with conception, ovulation, implantation, fetal development and viability of newborn animals. Ruminants are more resistant to zearalenone than monogastric animals, presumably due again to degradation in the rumen.[15,16]

The range of adverse effects caused by mycotoxins in animals include embryonic death, inhibition of fetal development, abortions and teratogenicity (deformities) in developing embryos. Nervous system dysfunctions also are observed, including such signs as tremors, weakness of limbs, uncoordinated movement, staggering, sudden muscular collapse and loss of comprehension, due to brain tissue destruction. Other symptoms include seizures, profuse salivation and gangrene of limbs, ears and tails. Several mycotoxins also cause cancers in liver, kidney, urinary tract, digestive tract and lungs.[15,16]

The effects of mycotoxins on human health are harder to document than effects on animals. Numerous cases of poisoning from consumption of bread made from moldy grain along with deaths of children from consuming moldy cassava and rice have been reported in the literature.[16] Also, toxins produced by fungi in the genera *Claviceps* and *Fusarium* have caused the human diseases ergotism and alimentary toxic aleukia, respectively. These two diseases have been well documented and described in detail because large numbers of people contracted them and their effects could be readily observed.[15,16]

Human exposure to acute dosages of aflatoxins has resulted in edema, liver damage and death. Aflatoxins have also been associated with human liver cancer in regions where liver cancer and hepatitis B virus are endemic. Ochratoxin A has been associated with the human kidney disease known as Balkan Endemic Nepropathy. Zearalenone was implicated in an outbreak of precocious pubertal changes in thousands of young children in Puerto Rico. Zearalenone is now considered to be an endocrine disrupter and speculation has suggested that it could play a role in human breast cancer. T-2 Toxin is believed to be the cause of

Table 3. Commercially Available Test Kits for Mycotoxin Testing[11]

Mycotoxin	Product Name	Format	Analysis Time	Range of Detection	Description
International Diagnostics Systems (St. Joseph, MI)					
Aflatoxin	Afla-Cup	Cup ELISA	5 minutes	5, 10, 20 ppb	Visual Qualitative
Editek (Medtox Laboratories, Burlington, NC)					
Aflatoxin	EZ Screen	Card ELISA	10 minutes	20 ppb	Visual Qualitative
Zearalenone	EZ Screen	Card ELISA	10 minutes	50 ppb	Visual Qualitative
T-2	EZ Screen	Card ELISA	10 minutes	12.5 ppb	Visual Qualitative
Ochratoxin	EZ Screen	Card ELISA	10 minutes	5 ppb	Visual Qualitative
Neogen Corporation (Lansing, MI)					
Aflatoxin	Veratox	Microwell ELISA	5 minutes	5-50 ppb	Quantitative
Aflatoxin	Veratox AST	Microwell ELISA	10 minutes	5-320 ppb	Quantitative
Aflatoxin	Veratox HS	Microwell ELISA	20 minutes	1-8 ppb	Quantitative
Aflatoxin	Agri-Screen	Microwell ELISA	5 minutes	20 ppb	Visual Qualitative
Aflatoxin	NPC	Immunoaffinity Column	7 minutes	2-100 ppb	Quantitative
Vomitoxin	Veratox	Microwell ELISA	20 minutes	0.5-6.0 ppm	Quantitative
Vomitoxin	Agri-Screen	Microwell ELISA	20 minutes	1.0 ppm	Visual Qualitative
Fumonisin	Veratox	Microwell ELISA	20 minutes	1.0-6.0 ppm	Quantitative
Fumonisin	Agri-Screen	Microwell ELISA	20 minutes	5.0 ppm	Visual Qualitative
T-2 Toxin	Veratox	Microwell ELISA	20 minutes	50-1500 ppb	Quantitative
Zearalenone	Veratox	Microwell ELISA	20 minutes	125-1000 ppb	Quantitative
Ochratoxin	Veratox	Microwell ELISA	20 minutes	10-50 ppb	Quantitative
R-Biopharm USA (Marshall, MI)					
Aflatoxin	Ridascreen	Microwell ELISA	15 minutes	5-135 ppb	Quantitative
Vomitoxin	Ridascreen	Microwell ELISA	15 minutes	0.5-6.0 ppm	Quantitative
Fumonisin	Ridascreen	Microwell ELISA	15 minutes	1.0-6.0 ppm	Quantitative
Zearalenone	Ridascreen	Microwell ELISA	15 minutes	100-1000 ppb	Quantitative
T-2 Toxin	Ridascreen	Microwell ELISA	15 minutes	50-1000 ppb	Quantitative
Ochratoxin	Ridascreen	Immunoaffinity Column	15 minutes	2-100 ppb	HPLC Clean-up
Rhone Diagnostics Technologies Ltd. (Glasgow, Scotland)					
Aflatoxin	AflaPrep	Immunoaffinity Column	10 minutes	2-100 ppb	HPLC Clean-up
Aflatoxin	AflaPlate	Microwell ELISA	35 minutes	2-50 ppb	Quantitative
Ochratoxin	OchraPrep	Immunoaffinity Column	10 minutes	2-100 ppb	HPLC Clean-up
Romer Laboratories (Union, MO)					
Aflatoxin	FluoroQuant	Clean-up column (Fluorometer)	15 minutes	5-100 ppb	Quantitative
Vomitoxin	FluoroQuant	Clean-up column (Fluorometer)	15 minutes	0.5-100 ppm	Quantitative
Vomitoxin	Accu-Tox	Tube ELISA	25 minutes	0.5-5.0 ppm	Quantitative
VICAM LP (Watertown, MA)					
Aflatoxin	Aflatest	Immunoaffinity Column	7 minutes	2-500 ppb	Quantitative/HPLC Clean-up
Vomitoxin	DonTest	Immunoaffinity Column	15 minutes	0.5-10 ppm	Quantitative/HPLC Clean-up
Fumonisin	Fumonitest	Immunoaffinity Column	15 minutes	0.2-2.0 ppm	Quantitative/HPLC Clean-up
Zearalenone	Zearaletest	Immunoaffinity Column	7 minutes	100-1000 ppb	Quantitative/HPLC Clean-up
Ochratoxin	Ochratest	Immunoaffinity Column	7 minutes	2-100 ppb	Quantitative/HPLC Clean-up

Alimentary Toxic Aleukia (ATA), the severe human disease that occurred in Russia during World War II as a result of food shortages that forced people to eat overwintered moldy cereal grains. The disease was manifested by destruction of bone marrow, damage to the hematopoietic system and loss of blood-making capacity, severe hemorrhaging, anemia and death. Deoxynivalenol is believed to be the cause of a number of gastrointestinal syndromes reported in different parts of the world, including the former Soviet Union, China, Korea, Japan and India. *Fusarium moniliforme* and possibly fumonisins have been linked with high rates of

Table 4. Sources of Information on Commercially Available Mycotoxin Test Kits and Analytical Testing Supplies in the U.S.

Manufacturer/Distributor	Mycotoxins
Editek (Medtox Laboratories) 1238 Anthony Road Burlington, NC 27215 Telephone: 800-334-1116 Fax: 336-229-4471 Web Site: www.medtox.com	Aflatoxins, Ochratoxin T-2 Toxin, Zearalenone
International Diagnostic Systems Corp. P. O. Box 799 St. Joseph, MI 49085 Telephone: 616-428-8400 Fax: 616-428-0093	Aflatoxins
R-Biopharm Inc. 7950 Old US27 South Marshall, MI 49068 Telephone: 877-789-3033 Fax: 616-789-3070 Web Site: www.rbiopharm.com	Aflatoxins, Aflatoxin M, Ochratoxin, Zearalenone, Deoxynivalenol (DON, Vomitoxin), T-2 Toxin, Fumonisins
Neogen Corporation 620 Lesher Place Lansing, MI 48912 U.S.A. Telephone: 517-372-9200 Fax: 517-372-0108 Web Site: www.neogen.com	Aflatoxins, Ochratoxin, Deoxynivalenol (DON, Vomitoxin), T-2 Toxin, Zearalenone, Fumonisins
Romer Labs 1301 Stylemasster Drive Union, MO 63084 U.S.A. Telephone: 314-583-8600 Fax: 314-583-6553 Web Site: www.romerlabs.com	Aflatoxins, Ochratoxin, Patulin, Sterigmatocysin, Deoxynivalenol, (DON, Vomitoxin), T-2 Toxin, Zearalenone, Fumonisins, Other Fusarium Toxins
Vicam 313 Pleasant Street Watertown, MA 02472 U.S.A. Telephone: 617-926-7045 Fax: 617-923-8055 Web Site: www.vicam.com	Aflatoxins, Ochratoxin, Deoxynivalenol, Zearalenone, Fumonisins

esophogeal cancer in the Transkei region of South Africa, Northeastern Italy and Northern China. Moniliformin has been suggested by Chinese scientists as a possible cause of a degenerative heart disease known as "Keshan Disease," occurring in regions of China where corn contaminated with moniliformin is consumed by the human population. The disease is a human myocardiopathy involving myocardial necrosis. Immunotoxicity in humans may be another effect of mycotoxins, particularly *Fusarium* mycotoxins. T-2 Toxin is reported to be highly immunosuppressive and there are reports of so-called "sick houses" where residents have developed diseases, such as leukemia and where *Fusarium* species have been found in the houses. Deoxynivalenol can cause elevated immunoglobin A levels in mice resulting in kidney damage that is very similar to a human kidney disease known as glomerulonephritis or immunoglobulin A (IgA) nephropathy. The involvement of mycotoxins in human disease is less clear than their involvement in animals diseases, but there is growing evidence that these toxins are also causative factors in human diseases.[15,16]

43.4 MYCOTOXIN PRODUCING MOLDS

The molds that produce the mycotoxins of greatest concern are species of *Aspergilllus, Penicillium* and *Fusarium. Aspergillus* and *Penicillium* species tend to be saprophytic and often attack commodities, such as cereal grains and nuts while in storage, though some aspergilli can also invade in the field. *Fusarium* species may be plant pathogenic as well as saprophytic types, and tend to invade grain in the field. A number of factors affect mycotoxin production by molds, including moisture, relative humidity, temperature, substrate, pH, competitive and associative growth of other fungi and microorganisms, and stress on plants, such as drought and damage to seed coats from hail, insects and mechanical harvesting equipment. The major commodities that are susceptible to contamination with mycotoxins include corn (maize), peanuts, cottonseed and some tree nuts. Wheat and barley are also susceptible to contamination, primarily with deoxynivalenol.[16,31]

The mycotoxins of greatest concern are produced by the following mold species. Aflatoxins are produced by *Aspergillus flavus, Aspergillus parasiticus* and *Aspergillus nomius*.[28,30] *Aspergillus flavus* can also produce cyclopiazonic acid. Ochratoxin is produced by *Aspergillus ochraceus, Aspergillus niger* and *Penicillium verrucosum*.[2,27,36] *Penicillium expansum,* as well as other *Penicillium* species and some *Aspergillus* species can produce patulin.[36] Zearalenone is produced by *Fusarium graminearum, Fusarium culmorum* and *Fusarium crookwellense,* deoxynivalenol or nivalenol are produced by the same three species, depending on the geographic origin of the producing strain. These same species may also produce the related toxins 3-acetyldeoxynivalenol and 15-acetyldeoxynivalenol, again depending on the geographic origin of the organism.[15,16,31] Fumonisins are produced by *Fusarium moniliforme, Fusarium proliferatum* and *Fusarium subglutinans*.[15,16,31] *Fusarium proliferatum* and *F. subglutinans* are also capable of producing moniliformin.[15,16]

Mold contamination of raw farm commodities is not completely preventable. Thus, it becomes necessary to determine the extent of contamination. To do so, methods must be available for detecting and enumerating the molds capable of producing the various mycotoxins, i.e., the "toxigenic fungi." Further, there are situations where it may be more feasible and convenient to test for the toxigenic fungus than for the mycotoxin itself. The presence of toxigenic fungi can be used as a means of assessing the quality and acceptability of a commodity before actual toxin production has occurred. An awareness of the types of molds present in a food or feed can be an early indicator of the types of mycotoxins that could be encountered.

Media selective for specific fungi are becoming more widely available, but compared to the development of selective media for specific spoilage and pathogenic bacteria, the menu of selective media for toxigenic fungi is limited. However, several media are available for the detection of toxigenic fungi. These include media for *A. flavus, A. parasiticus, P. verrucosum, P. viridicatum* and *Fusarium* species.

Bell and Crawford[9] first developed a medium designed to isolate *A. flavus* from peanuts. The medium contained rose bengal, dichloran, and streptomycin. Bothast and Fennell[13] developed the next medium, called *Aspergillus* differential medium (ADM). It contained 1.0% yeast extract, 1.5% tryptone, and 0.05% ferric acid and required incubation at 28°C for 3 days. On this medium *A. flavus* and *A. parasiticus* form insoluble bright organge-yellow pigments on the colony reverse, an easily recognized phenomenon.

Hamsa and Ayres[26] modified ADM by adding streptomycin and dichloran and recommend incubating cultures at 28°C for 5 days. Later Assante et al.[7] showed that the orange-yellow pigment formation was due to reaction of ferric ions from the ferric citrate with aspergillic acid molecules forming a colored complex of three aspergillic acid molecules with ferric ions. Pitt et al.,[35] further modified these media and developed a medium called *Aspergillus flavus-parasiticus* agar or AFPA. Incubation of AFPA is done at 30°C for 42 to 48 hr, resulting in sufficient color development to recognize *A. flavus* and *A. parasiticus* before the development of the typical olive green spores (conidia).

Pitt et al.,[35] reported that few other mold species produce the orange-yellow color reaction on AFPA. Occasional false positives can be obtained with some strains of *A. niger* and *A. ochraceus*. Questionable species can readily be distinguished with an additional incubation of 24 to 48 hr. *Penicillium* species apparently do not give false positives and only a few (2.9%) *A. flavus* and *A. parasiticus* isolates gave false negatives. AFPA has been recommended for detection of *A. flavus* and *A. parasiticus* in spices, nuts, peanuts, oilseeds, grains, corn meal, cowpeas and livestock feeds.[10,29]

Frisvad[21,22,23] has reported a selective medium for detecting nephrotoxin-producing strains of *P. verrucosum* and related species. The medium was later modified with the addition of chlortetracycline.[24] The medium consists of yeast-extract sucrose (YES) agar, as the basal medium, supplemented with chloramphenicol (50 mg/L), chlortetracycline (50 mg/L), dichloran (2 mg/L), and rose bengal (25 mg/L). This medium, referred to as DRYES agar, reportedly distinguishes between producers of ochratoxin A and citrinin and producers of xanthomegnin and viomellein. The ochratoxin A- and citrinin-producing strains of *P. verrucosum* are reported to be consistently indicated by a violet brown pigment produced on the colony reverse. The xanthomegnin- and viomellein-producing strains of *P. viridicatum* and *P. aurantiogriseum (P. cyclopium)* are indicated by a yellow reverse and obverse.

Several attempts have been made to develop selective media to detect and isolate *Fusarium* species. These include Nash-Snyder medium,[32] modified Czapek-Dox agar,[33] Czapek iprodione-dichloran (CZID) agar,[3] potato-dextrose-iprodione-dichloran (PDID) agar,[41] and dichloran-chloramphenicol-peptone agar (DCPA).[6,37] The Nash-Snyder medium and modified Czapek-Dox agar contain pentachloronitrobenzene, a known carcinogen, and are not favored for routine use in food microbiology laboratories. However, these media can be useful for evaluating samples that are heavily contaminated with bacteria and other fungi. CZID agar is becoming a regularly used medium for isolating *Fusarium* species from foods, but rapid identification of *Fusarium* isolates to the species level is difficult if not impossible on this medium. Isolates must be subcultured on other media, such as carnation leaf agar (CLA) for identification.[33] However, CZID agar is a good selective medium for *Fusarium* species. While most molds are completely inhibited on CZID agar, *Fusarium* species can be readily distinguished. Thrane et al.[41] reported that PDID agar is as selective as CZID agar for *Fusarium* species, with the advantage that it supports *Fusarium* growth with morphological and cultural characteristics that are the same as on PDA, which facilitates more rapid identification, since various monographs and manuals for *Fusarium* identification describe characteristics of colonies grown on PDA. Thrane et al.[41] compared several media for their suitability to support colony development by *Fusarium* species and found that PDID and CZID agars were better than DCPA. Growth rates were much higher on DCPA, making colony counts more difficult. Conner,[18] however, modified DCPA by adding 0.5 µg of crystal violet per mL and reported increased selectivity by inhibiting *Aspergillus* and *Penicillium* species but not *Fusarium* species.

Other considerations for the detection of toxigenic fungi include such things as direct vs. dilution plating and adjustment of the water activity of media. For the detection of mycotoxic fungi in stored grains or seeds, direct plating of surface sanitized kernels may be preferable to grinding the sample and using dilution plating. The direct plating method permits the detection of internal mycoflora, which is usually of greater concern. The direct plating method is described and has been reported to be more efficient than the dilution plating method for detecting individual mold species.[8] Combining this technique with a selective medium such as AFPA, DRYES, CZID or DCPA provides an additional technique for detecting toxigenic species. While most known toxigenic species will grow on media of unadjusted water activity (a_w), it may occasionally be desirable to reduce the a_w of a medium. The most common way of doing this is to add 7.5% NaCl. The a_w of medium can also be lowered by increasing concentrations of sucrose (20%) or glucose (40%), or with glycerol.

43.5 EQUIPMENT, MATERIAL, AND REAGENTS

For the dilution plating technique, preparation of the sample is essentially the same as for the aerobic plate count (Chapter 7). For direct plating of foods additional equipment will be required. The direct plating technique can be applied to intact particles and whole foods, such as grain kernels, seeds, dried beans, nuts, coffee beans, cocoa beans and whole spices.[8] Before plating, samples should be held in a freezer for 72 hr (–20°C) to kill insects and mites and their eggs.

43.6 PRECAUTIONS

Foods containing toxigenic mold species may also contain mycotoxins. Samples should be handled accordingly. Some toxigenic mold species are also potential pathogens to humans and have spores that are allergenic. Also, spores of *A. flavus* and *A. parasiticus* can contain low levels of aflatoxins. Therefore, precautions should be taken to prevent inhalation of airborne spores and aerosolized sample dusts.

43.7 MEDIA AND PROCEDURES

***Aspergillus flavus-parasiticus* agar (AFPA) for aflatoxin-producing molds.** Do not invert the plates for incubation. Plate counts on this medium should be incubated at 30 ± 1°C for 42 hr. Plates may be incubated longer if necessary. Colonies exhibiting a bright yellow-orange reverse should be counted as *A. flavus* and *A. parasiticus* colonies.

Dichloran rose bengal yeast extract sucrose agar (DRYES) for nephrotoxin-producing *Penicillium* strains. Do not invert plates. Incubate at 20° to 25°C for 5 to 7 days or until colonies develop. Colonies exhibiting a violet brown pigment on the reverse are counted as potential ochratoxin- and citrinin-producing strains of *P. verrucosum*. Colonies exhibiting a yellow reverse and obverse are counted as potential xanthomegnin- and viomellelin-producing strains of *P. viridicatium* and *P. aurantiogriseum (P. cyclopium)*.

Czapek iprodione-dichloran agar (CZID) for *Fusarium* species.

Dichloran-chloramphenicol peptone agar (DCPA) for *Fusarium* species.

43.8 REVIEWS AND COMPENDIA

1. Betina, V., ed. 1984. "Mycotoxins, production, isolation, separation and purification." Elsevier Science Publishing Co., Inc., New York.
2. Betina, V. ed. 1989. Mycotoxins: Chemical, biological and environmental aspects. Elsevier Scientific Publishing Co., New York. 438 pp.
3. Bullerman, L. B. 1997. Fusaria and toxigenic molds other than aspergilli and penicillia. In "Food Microbiology Fundamentals and Frontiers". eds. M. P. Boyle, L. R. Beuchat and T. J. Montville. ASM Press, Washington, D.C.
4. Bullerman, L. B. 1979. Significance of mycotoxins to food safety and human health. Journal of Food Protection 42:65-86.
5. Bullerman, L. B. 1986. Mycotoxins and food safety. A scientific status summary by the Institute of Food Technologists' Expert Panel on Food Safety and Nutrition. Institute of Food Technologists, Chicago.
6. CAST. 1989. Mycotoxins. Economic and health risks. Council for Agricultural Science and Technology, Task Force Report No. 116. Ames, Iowa.
7. Cole, R. J. and Cox, R. H. 1981. Handbook of toxic fungal metabolites. Academic Press, New York.
8. Eaton, D. L. and Groopman, J. D. (eds.). 1994. The toxicology of aflatoxins, human health, veterinary and agricultural significance. Academic Press, Inc., New York.
9. GIPSA. 1999. Grain fungal diseases & mycotoxin reference. Grain Inspection, Packers and Stockyards Administration. Technical Services Division. United States Department of Agriculture. Kansas City, MO.
10. Gourama, H. and Bullerman, L. B. 1995. *Aspergillus flavus* and *Aspergillus parasiticus*: Aflatoxigenic fungi of concern in foods and feeds: A review. J. Food Protection 58:1395-1404.
11. Hocking, A. D. 1997. Toxigenic *Aspergillus* species. In "Food Microbiology Fundamentals and Frontiers." eds. M. P. Doyle, L. R. Beuchat and T. J. Montville. ASM Press, Washington, D.C.
12. International Agency for Research on Cancer. 1993. Mycotoxins. In: "IARC Monographs on the Evaluation of Carcinogenic Risks to Humans." Vol. 56. "Some naturally occurring substances: Food Items and Constituents, Heterocyclic Aromatic Amines and Mycotoxins." pp. 245-521. IARC, Lyon, France.
13. International Agency for Research on Cancer. 1987. Aflatoxins. In: "IARC Monograph on the Evaluation of Carcinogenic Risks to Humans," Supplement 7, pp. 83-87. IARC, Lyon.
14. International Agency for Research on Cancer. 1982. Environmental Carcinogens, Selected Methods of Analysis, Vol. 5. "Analysis of Mycotoxins in Foods," IARC Sci. Publ. 44. IARC, Lyon, France.
15. Kurata, H. and Ueno, Y., eds. 1984. Toxigenic Fungi. Their toxins and health hazard. Developments in Food Science 7. Elsevier Science Publishing Co., Inc., New York.
16. Marasas, W. F. O., Nelson, P. E., and Toussoun, T. A. 1984. Toxigenic *Fusarium* species. Identity and Mycotoxicology. The Pennsylvania State U. Press, University Park, PA.
17. Marasas, W. F. O. and Nelson, P. E. 1987. Mycotoxicology. The Pennsylvania State U. Press, University park, PA.
18. Miller, J. D. and Trenholm, H. L. eds. 1994. Mycotoxins in grain. Compounds other than aflatoxins. Eagan Press, St. Paul, MN.
19. Moreau, C. 1979. Moulds, Toxins and Food. (English trans. with additions by M. Moss). John Wiley and Sons, Chichester, England.
20. Pitt, J. I. 1997. Toxigenic *Penicillium* species. In "Food Microbiology Fundamentals and Frontiers". eds. M. P. Doyle, L. R. Beuchat and T. J. Montville. ASM Press, Washington, D.C.
21. Pitt, J. I. and Hocking, A.D. 1997. "Fungi and food spoilage." Second Edition. Blackie Academic and Professional (Chapman & Hall), New York.
22. Sharma, R. P. and Salunkhe, D. K. eds. 1991. Mycotoxins and phytoalexins. CRC Press, Inc., Boca Raton, FL.
23. Sinha, K. K. and Bhatnagar, D. eds. 1998. Mycotoxins in Agriculture and Food Safety. Marcel Dekker, Inc., New York.
24. Smith, J. E. and Henderson, R. S. eds. 1991. Mycotoxins and animal foods. CRC Press, Inc., Boca Raton, FL.
25. United States Department of Agriculture. 1990. Aflatoxin Handbook. Federal Grain Inspection Service. U.S.D.A., Washington, D.C.

43.9 REFERENCES

1. American Association of Cereal Chemists. 1990. Approved methods of the American Association of Cereal Chemists. 8th ed. AACC, St. Paul, Minn.
2. Abarca, M. L., M. R. Bragulat, G. Castella, and F. Cabanes. 1994. Ochratoxin A production by strains of *Aspergillus niger* var. *niger*. Appl. Environ. Microbiol. 60:2650-2652.
3. Abildgren, M. P., F. Lund, U. Thrane, and S. Elmholt. 1987. Czapek-Dox agar containing iprodione and dichloran as a selective medium for the isolation of *Fusarium* species. Lett. Appl. Microbiol. 5:83-86.
4. Abouzied, M. M., J. I. Azcona, W. E. Braselton, and J. J. Pestka. 1991. Immunochemical assessment of mycotoxins in 1989 grain foods: Evidence for deoxynivalenol (vomitoxin) contamination. Appl. Environ. Microbiol. 57:672-677.
5. Allcroft, R., R. B. A. Carnaghan, K. Sargeant, and J. O'Kelly. 1961. A toxic factor in Brazilian groundnut meal. Vet. Rec. 73:428-429.
6. Andrews, S., and J. L. Pitt. 1986. Selective medium for isolation of *Fusarium* species and dematiaceous hyphomycetes from cereals. Appl. Environ. Microbiol. 51:1235-1238.
7. Assante, G., L. Camarda, R. Locci, L. Merlilni, G. Nasini, and E. Papadopoulus. 1981. Isolation and structure of red pigments from *Aspergillus flavus* and related species grown on differential medium. J. Agric. Food Chem. 29:785.
8. Bandler, R., M. E. Stack, H. A. Koch, V. Tournas, and P. B. Mislivec. 1995. Yeasts, molds and mycotoxins. *In* Bacteriological analytical manual. AOAC International, Gaithersburg, Md.
9. Bell, D. K., and J. L. Crawford. 1967. A Botran-amended medium for isolating *Aspergillus flavus* from peanuts and soil. Phytopathology 57:939.
10. Beuchat, L. R. 1986. Evaluation of media for simultaneously enumerating total fungi and *Aspergillus flavus* and *A. parasiticus* in peanuts, corn meal and cowpeas, p. 129. *In* A. D. King, J. I. Pitt, L. R. Beauchat, and J. E. L. Corry (ed.), Methods for the mycological examination of food. Plenum Press, New York.
11. Bird, C. 1999. Personal communication.
12. Blount, W. P. 1961. Turkey "X" disease. J. Br. Turkey Fed. 9(2):52, 55-58, 61-71. Mar/Apr.
13. Bothast, R. J., and D. I. Fennell. 1974. A medium for rapid identification and enumeration of *Aspergillus flavus* and related organisms. Mycologia 66:365.
14. Bullerman, L. B. 1986. Mycotoxins and food safety. A Scientific Status Summary by the Institute of Food Technologists' Expert Panel on Food Safety and Nutrition. Institute of Food Technologists, Chicago, Ill.
15. Bullerman, L. B. 1997. Fusaria and toxigenic molds other than Aspergilli and Penicillia, Chapter 23, p. 393-405. *In* M. P. Doyle, L. R. Beuchat, and T. J. Montville (ed.) Food microbiology, fundamentals and frontiers. American Society of Microbiology Press, Washington, D. C.
16. Bullerman, L. B. 1999. Mycotoxins classification. *In* R. Robinson, C. Batt, and P. Patel (ed.), Encyclopedia of food microbiology. Academic Press, London.
17. CAST. 1989. Mycotoxins. Economic and health risks. Council for Agricultural Science and Technology, Task Force Report No. 116. Ames, Iowa.
18. Conner, D. E. 1992. Evaluation of methods for selective enumeration of *Fusarium* species in feedstuffs, p. 299-302. *In* R. A. Samson, A. D. Hocking, J. I. Pitt, and A. D. King (ed.), Modern Methods in Food Mycology. Elsevier Scientific Publishers, Amsterdam.
19. Cullen, J. M., and P. M. Newberne. 1994. Acute hepatotoxicity of aflatoxins p. 3-26. Chapter 1. *In* D. L. Eaton and J. D. Groopman (ed.), The toxicology of aflatoxins: Human health, veterinary and agricultural significance. Academic Press, New York.

20. Firestone, D. 1989. Official methods and recommended practices of the American Oil Chemists' Society, 4th ed. AOCS Press, Champaign, Ill.
21. Frisvad, J. C. 1981. Physiological criteria and mycotoxin production as aids in identification of common asymmetric penicillia. Appl. Environ. Microbiol. 41:568.
22. Frisvad, J. C. 1983. A selective and indicative medium for groups of *Penicillium viridicatum* producing different mycotoxins in cereals. J. Appl. Bacteriol. 54:409.
23. Frisvad, J. C. 1986. Selective medium for *Penicillium viridicatum* in cereals, p. 132. *In* A. D. King, J. I. Pitt, L. R. Beuchat, and J. E. L. Corry (ed.), Methods for the mycological examination of food. Plenum Press, New York.
24. Frisvad, J. C., O. Filtenborg, U. Thrane, and P. V. Nielsen. 1992. Collaborative study on media for detecting and enumerating toxigenic *Penicillium* and *Aspergillus* species, p. 255-261. *In* R. A. Samson, A. D. Hocking, J. I. Pitt, and A. D. King (ed.), Modern methods in food mycology. Elsevier Science Publishers B. V., New York.
25. GIPSA. 1999. Grain fungal diseases and mycotoxin reference. Grain Inspection, Packers and Stockyards Administration. Technical Services Division. U.S. Department of Agriculture. Kansas City, Mo.
26. Hamsa, T. A. P., and J. C. Ayres. 1997. A differential medium for the isolation of *Aspergillus flavus* from cottonseed. J. Food Sci. 42:449.
27. Heenan, C. N., K. J. Shaw, and J. I. Pitt. 1998. Ochratoxin A production by *Aspergillus carbonarius* and *A. niger* isolates and detection using coconut cream agar. J. Food Mycol. 1:67-72
28. Hocking, A. D. 1977. Toxigenic *Aspergillus* species, Chapter 21, p. 393-405. *In* M. P. Doyle, L. R. Beuchat, and T. J. Montville (ed.), Food microbiology, fundamentals and frontiers. ASM Press, Washington, D. C.
29. Hocking, A. D., and J. I. Pitt. 1986. A selective medium for rapid detection of *Aspergillus flavus*, p. 127. *In* A. D. King, J. I. Pitt, L. R. Beuchat, and J. E. I. Corry (ed.), Methods for the mycological examination of food. Plenum Press, New York.
30. Kurtzman, C. P., B. W. Horn, and C. W. Hesseltine. 1987. *Aspergillus nomius,* a new aflatoxin producing species related to *Aspergillus flavus* and *Aspergillus tamari.* Antonie van Leewenhoek 53:147-158.
31. Miller, J. D. 1995. Fungi and mycotoxins in grain: Implications for stored product research. J. Stored Prod. Res. 31:1-16.
32. Nash, S. M., and W. C. Snyder. 1962. Quantitative estimations by plate counts of propagules of the bean root rot *Fusarium* in field soils. Phytopathology 52:567-572.
33. Nelson, P. E., T. A. Tousoun, and W. F. O. Marasas. 1983. *Fusarium species:* An illustrated manual for identification. Pennsylvania State University Press, University Park, Pa.
34. Pestka, J. J., M. N. Abouzied, and Sutikno. 1995. Immunological assays for mycotoxin detection. Food Technol. 49(2):120-128.
35. Pitt, J. I., A. D. Hocking, and D. K. Glenn. 1983. An improved medium for the detection of *Aspergillus flavus* and *A. parasiticus*. J. Appl. Bacteriol. 54:109.
36. Pitt, J. I. 1997. Toxigenic *Penicillium* species, Chapter 22, p. 406-418. *In* M. P. Doyle, L. R. Beuchat and T. J. Montville (ed.), Food microbiology, fundamentals and frontiers. American Society of Microbiology Press, Washington, D. C.
37. Pitt, J. I., and A. D. Hocking. 1997. Fungi and food spoilage, 2nd ed. Blackie Academic and Professional (Chapman & Hall), London.
38. Sargeant, K., A. Sheridan, J. O'Kelly, and R. B. A. Carnaghan. 1961. Toxicology associated with certain samples of groundnuts. Nature 192:1096.
39. Scott, P. M. 1994. *Penicillium* and *Aspergillus* toxins, Chapter 5, p. 261-285. *In* J. D. Miller and H. L. Trenholm (ed.). Mycotoxins in grain: Compounds other than aflatoxins. Eagan Press, St. Paul, Minn.
40. Scott, P. M. 1995. Natural toxins, Chapter 49, p. 49-1. *In* P. Cunniff (ed.), Official methods of analysis of AOAC International, 16th ed. AOAC International, Arlington, Va.
41. Thrane, U., O. Filtenborg, J. C. Frisvad, and F. Lund. 1992. Improved methods for the detection and identification of toxigenic *Fusarium* species, p. 285-291. *In* R. A. Samson, A. D. Hocking, J. I. Pitt, and A. D. King (ed.), Modern methods in food mycology. Elsevier Science Publishers B.V., New York.
42. Wilson, D. M., E. W. Sydenham, G. A. Lombaert, M. W. Trucksess, D. Abramson, and G. A. Bennett. 1998. Mycotoxin analytical techniques, Chapter 5, p. 135-182. *In* K. K. Sinha and D. Bhatnagar (ed.), Mycotoxins in agriculture and food safety. Marcel Dekker, Inc., New York.

Chapter 44

Foodborne Viruses

Gary P. Richards and Dean O. Cliver

44.1 INTRODUCTION

44.11 Etiologic Agents of Foodborne Viral Illnesses

Viral agents of foodborne illness may be characterized by their ability to elicit human infection, tolerate the proteolytic secretions of the stomach and bile secretions in the duodenum, and spread via the fecal-oral route. Viruses most noted for foodborne illness in the United States are hepatitis A virus and human caliciviruses. Human caliciviruses may be broken down into three genogroups: Norwalk virus, Snow Mountain agent and Sapporo virus.[61,92] Norwalk-like viruses include the Hawaii[136] and Montgomery County agents.[64] Snow Mountain virus[9,39] is becoming a more frequently reported cause of foodborne illness. Less commonly associated with foodborne illness is the hepatitis E virus, which is also classified as a calicivirus, although it appears more closely related to the alphavirus superfamily based on nucleotide structure.[112] A novel genetic variant of human calicivirus, known as the Parkville virus, was responsible for an outbreak of foodborne illness.[94] Small round viruses implicated in foodborne illness include the Cockle, Ditchling, "W," Parramatta and Wollan agents, the mini-rotaviruses, the mini-reoviruses, and the Otofuke agent.[65,114] Astroviruses, morphologically distinct from the Norwalk virus, infect children and include the Marin County[96] and United Kingdom strains.[114]

Other viruses have been associated with diarrheal illness and may occasionally be transmitted via food. Included in this group are rotaviruses (primarily group A), enteric adenoviruses (types 40 and 41), enteric coronaviruses, parvovirus-like agents, and other small round viruses. Rotaviruses are the most commonly known cause of gastroenteritis among infants and young children, with adenoviruses the second most likely cause of gastrointestinal illness.

There is a broad spectrum of rotaviruses and adenoviruses that infect birds and mammals. These viruses do not typically infect across species lines,[21,142] although group B rotaviruses can occasionally spread from animals to humans. Group A rotaviruses cause most human illness, however group B rotaviruses derived from cows, pigs, and other animals have caused epidemics, primarily in China.

Humans are the natural hosts for Norwalk and hepatitis A virus infections, although some non-human primates are susceptible to experimental infection. Laboratory chimpanzees have been infected experimentally with hepatitis A and shown clinical symptoms similar to those in man. Norwalk virus, on the other hand, elicited an immune response in chimpanzees, but the animals did not exhibit clinical illness.[149]

Enterically transmitted non-A, non-B hepatitis virus has been responsible for large outbreaks of waterborne illness in Algeria, Burma, India, Mexico, Nepal, Pakistan, Somalia, and the Soviet Union.[13,14] Polluted water and person-to-person spread of the virus appear to be important means of disease transmission although direct foodborne transmission appears likely. Non-A, non-B hepatitis is an old term replaced today with hepatitis C or hepatitis E. Hepatitis C virus is not considered an enterically transmitted virus, as its primary means of spread is percutaneous, through blood products or through injections with contaminated needles. Outbreaks of foodborne hepatitis type E have been identified in several countries and may be associated with a variety of foods, particularly shellfish.[17,29,132] Animal caliciviruses from ocean reservoirs have been associated with human illness and could pose another group of potential foodborne pathogens.[126]

Poliovirus, once a serious foodborne pathogen in the United States, may still be a cause of foodborne illness in developing countries where vaccines are not administered routinely, although the World Health Assembly has adopted the goal of eradicating poliomyelitis from the world. Accordingly, substantial progress toward this goal has been made, however there remains reservoirs of wild-type poliovirus in some parts of the world.[16] Poliomyelitis was for many years prevalent in some underdeveloped countries and may have been spread through fecally contaminated milk and other food products. Immunity to poliovirus, conferred through childhood immunizations, and improved hygienic practices has essentially eliminated poliomyelitis spread by foods. Continued use of attenuated poliovirus vaccines suggests that vaccine viruses may be found in the environment for the foreseeable future and that they may be used as indicators for the possible presence of human viruses that are difficult to detect. Other human enteroviruses (coxsackieviruses and echoviruses) have rarely been reported to be transmitted via foods.[24]

Reports indicate that tick-borne encephalitis from flaviviruses has been responsible for illnesses in Central Europe associated with raw milk from goats, sheep, and cows.[51,128] Ticks become infected when they bite flavivirus-infected rodents. The virus is subsequently transmitted through tick bites to goats, sheep, and cows. Humans of all ages may become ill when they drink raw milk from infected animals. Many other viruses cause animal diseases; these are generally not transmissible to humans, especially via foods. Human viruses transmitted via food and water are largely produced only in humans, however foods of both animal and vegetable origin can serve as vehicles for these viruses.

44.12 Seasonality of Virus Infections

The seasonality of foodborne virus infection in humans varies with the particular virus and the vehicle. Some foodborne viral infections occur predominantly during certain times of the year, while others may have no seasonal predilection at all. Norwalk illness may be transmitted during any season, however summer outbreaks of Norwalk are often associated with raw shellfish consumption. The reasons for seasonal predilection are not fully understood, but seasonality of food harvesting and consumption patterns for specific foods may play a role. Environmental factors are influenced by season and may affect virus inactivation and distribution in the environment. Since the source of foodborne virus contamination is usually the human, human activities, which are often seasonal, may also affect the levels of virus shed directly or indirectly into the environment. Farm practices, like irrigation of crops with effluent from sewage treatment plants, may contribute to a higher incidence of viral illness from consuming some freshly harvested fruits and vegetables.[137] Virus infectivity may decrease in stored produce consumed during the non-growing season, thus reducing the risk of illness.[73]

44.13 Routes of Food Contamination

Foods destined for human consumption may become contaminated with enteric viruses by several means. In each case, the viruses are ultimately transmitted by the fecal-oral route. A major source of virus contamination of foods in the United States and abroad is poor personal hygiene where improperly sanitized hands of food handlers and consumers contaminate foods and food-contact surfaces. Many outbreaks have been caused by food handlers working while they are incubating, sick with, or recovering from an enteric viral illness.[97,98,148] Outbreaks of illness often lead to secondary spread of the virus from one infected individual to another.

Food handlers in developing countries are often immune to hepatitis A, because hepatitis A virus (HAV) infections are common in childhood and confer permanent immunity. It is noteworthy that the risk of hepatitis A spreading in adults within these countries may be lower than in developed countries where the majority of food handlers and consumers are still susceptible to infection and virus shedding.

Effluent from sewage treatment and septic tank discharge are sources of virus transmission to lake, river, coastal, and estuarine waters. Molluscan shellfish, like oysters, clams, and mussels, biologically concentrate viruses and other impurities within their edible tissues during normal feeding activities. Boat waste discharged into shellfish beds is also a potential source of seafood contamination, as are improper handling, shucking, and food preparation practices. Polluted ground- or municipal-water may contaminate foods and beverages with enteric viruses. Water used in preparing and washing foods, especially those served raw, is not always free from fecal pollution and can readily contaminate foods. Polluted water may also contaminate food contact surfaces and utensils.

Drinking water is an important source of hepatitis A virus and probably most other enteric viruses. Foods and beverages may also become contaminated from unsanitary ice used in drinks or to chill foods. Many commercial operations use ice to cover or immerse perishable foods before, during, or after storage. In all cases, ice should be obtained from approved sources.

Irrigation of crops with effluent water and fertilization with sewage sludge pose some degree of contamination risk.[116] It is obvious that such practices may externally contaminate vegetables, fruits, and other produce with enteric viruses. Vegetables contaminated in such a manner have been shown to harbor viable viruses for several weeks.[137] Although the extent to which crop irrigation and fertilization contribute to foodborne illness is uncertain, fruits such as strawberries have recently led to outbreaks of hepatitis A virus infection.[15,93]

Insects may transfer sewage-borne pathogens to food products, and cross contamination of foods is another means of contaminant transfer. Work surfaces used for processing and handling raw materials should be properly sanitized before placing foods in contact with them. An outbreak of Norwalk illness associated with the cross contamination of salad by raw seafood[52] might have been averted if a few simple precautions had been observed.

44.14 Virus Reservoirs and Persistence

Viruses may be harbored in and on a variety of animate and inanimate objects which serve as vectors or fomites of infectious virus particles. Reservoirs include humans, animals, sewage, fresh and marine waters, soils, and sediments. Humans are the natural reservoir of human enteric viruses. Replication outside the human host does not naturally occur with human pathogenic viruses. Fecal shedding of viruses may lead directly or indirectly to food contamination and these foods may retain contaminants for extended periods. Viruses in sewage, soils, fresh and marine waters, and sediments may persist for days, weeks, and even months, depending on temperature and other factors.

Molluscan shellfish, such as oysters, clams, and mussels, biologically concentrate viruses from the water column during their regular feeding activities. Although virus replication does not take place within the shellfish, bioaccumulated viruses may be maintained in high numbers and retain infectivity for up to 3 months.[87] Shellfish, contaminated in their native environment, are one of the most common sources of foodborne viral illness.

Immunodeficient individuals appear to be at increased risk of infection from enteric viruses. Immunosuppressed patients with HIV infection or AIDS-related opportunistic infections showed high levels of gastrointestinal viruses especially rotavirus and adenovirus.[30] Enteric viruses, including the adenoviruses, astroviruses, caliciviruses, rotaviruses and small round structured viruses, have been associated with severe or persistent diarrhea in persons with impaired immune function.[125] Children with immunodeficiencies, particularly those with severe combined immunodeficiency syndrome (SCIDS), can chronically shed viruses associated with diarrhea.[74] The extent of foodborne viral infection among the immunosuppressed at large is not known.

44.15 Incidence of Foodborne Viral Illnesses

A survey of foodborne illnesses in the United States showed that, of the reported 2,423 outbreaks involving 77,373 cases reported to the Centers for Disease Control and Prevention between 1988 and 1992, 4.5% of the outbreaks and 6.5% of the cases whose etiology was determined were attributed to enteric viruses.[6] Sporadic foodborne outbreaks were not included in this report, but likely involve far more individuals than those reported.[6] This is attributed to the facts that many of the illnesses are mild and of short duration and persons afflicted do not report the illness or seek medical treatment. In addition, the etiologic agent can not be determined for many of the outbreaks that are likely caused by viral agents, since laboratory diagnosis of the illnesses and detection of the viruses in foods are more difficult than for their bacterial counterparts; 59% of the outbreaks and 52% of the cases in the total for the period of this summary were of undetermined etiology.[6] Only large outbreaks or those that have severe consequences can justify the expense of an extensive epidemiological investigation. Further, enteric virus illnesses may be difficult to attribute to specific foods unless an epidemiological investigation is performed. In the case of hepatitis A virus with its nearly 1-month average incubation period, individuals are unlikely to remember what they may have eaten a month or more preceding the illness.

Molluscan shellfish are an important vehicle of foodborne outbreaks of hepatitis A and Norwalk or Norwalk-like virus illness in the United States.[37,91,104,105] The true incidence of such illness remains a mystery because many, including hepatitis A, are often subclinical (unrecognized even by the infected individual) or are not distinctive. Outbreaks have a potential to become extremely serious, as demonstrated in a 1988 outbreak of clam-associated hepatitis A in China. During a 2-month period, nearly 300,000 cases of hepatitis A were reported from the consumption of contaminated clams in and around Shanghai, China.[55,150] Such outbreaks serve as a reminder to maintain vigilance over food safety.

44.2 FOODS TO BE TESTED

44.21 Molluscan Shellfish

Oysters, clams, and mussels are filter feeders capable of bioconcentrating viruses from their surrounding waters. Viruses associated with fecal waste and attached to particulate matter may be readily bioconcentrated by molluscan shellfish. A portion of the viruses generally pass through the digestive tract and may be excreted in the feces, however a portion of the viral contamination may persist within the digestive epithelial cells or be transported to more internal tissues, where they may persist for extended periods. Shellfish obtained from waters approved by state health authorities for the harvest of shellfish may contain some level of viral contamination, as regulatory agencies do not routinely survey shellfish for viruses.[44] Viruses may be infectious or inactivated due to sewage treatment or other processes. The detection of infectious viruses is of primary concern and may be accomplished for some viruses by classic cell culture techniques or by procedures involving cell culture propagation of the virus to levels detectable by molecular biological techniques. Infectivity can not be determined for viruses that do not replicate in cell cultures, with the rare exception that tests may be performed in human volunteers.[53]

44.22 Fruits and Vegetables

Fruits and vegetables may be contaminated by irrigation water, fertilization practices, contact surfaces or the hands of infected workers. Contamination is generally believed to be limited to the external surface, however cutting or processing of products can lead to internal contamination by the hands of an infected worker or a contaminated knife or contact surface. Leafy vegetables need only be tested on the surfaces for viruses.[66,67] Other fruits and vegetables may contain only surface contamination and may be tested by surface rinsing.[25] Efforts should be made to maximize and quantify the amount of surface area tested. Cut fruits and vegetables must be more extensively sampled and require more effort to extract enteric virus contamination.[57,69] Frozen products may be sampled with a sterile drill bit. Fruits and vegetables should be tested before freezing as viruses are more difficult to extract from frozen products. Likewise, the virus titer may be reduced by freezing or drying. Products should be transported or stored at 2°C to 4°C in a manner to avoid desiccation.

44.23 Animal Products

Human pathogenic viruses of animal origin are not generally present in the United States, therefore animal products are generally contaminated through human contact. Poor plant sanitation, contaminated processing water, or ill workers can lead to meat, poultry or fish contamination during processing. Comminuted products, like hamburger, can contain contamination throughout the product,[57,69] whereas steaks, roasts, fish fillets and chicken parts would be likely to contain primarily surface contamination. Fortunately, such vehicles seldom cause illness due to the fact that these products are usually cooked before serving.[41]

44.24 Processed Foods

Processed foods may become contaminated by the food handlers or contaminated packaging materials or by ingredients added during processing. Infectious foodborne viruses are not usually a problem in foods subjected to cooking or pasteurization unless the product is recontaminated after heating. In two instances, bread caused outbreaks of hepatitis A virus because it was handled by unsanitized hands after baking.[143,145] Products contaminated before entry to a processing facility may retain the viruses or spread them to yet other products if the processing steps are ineffective for their inactivation. The hazard analysis-critical control point (HACCP) food safety system is intended to minimize opportunities for events such as these.

44.3 DETECTING VIRUSES IN FOODS

44.31 General Considerations

Testing foods for enteric viruses involves virus extraction, concentration and assay, however not all of the needed equipment is likely to be present in most food microbiology laboratories. Further, many of the procedures are costly, time consuming, and require a level of technical expertise not commonly found in most laboratories. In many cases, virus extraction and concentration can be performed concurrently. Until the advent of molecular biology, about a decade ago, foodborne viruses were detected using cell culture propagation techniques such as plaque assays and various cytopathogenicity assays. Difficult-to-culture viruses were detected using enzyme linked immunosorbent assays, radioimmunoassays, radioimmunofocus assays, and other immunological procedures. Some laboratories detected viruses visually, by electron microscopy. The development of polymerase chain reaction techniques has engendered many innovative approaches to detecting foodborne viruses. However, molecular biological procedures have limitations that reduce their practicality for detection of enteric viruses in the foods that are most likely to be contaminated.[108] Some of the technical limitations are discussed below. There are also limitations from a practical standpoint. They include the fact that both cell culture and molecular biology-based assays are costly, and cell culture assays may require a week or two to complete. Molecular biological techniques are faster, however they require a level of technical expertise not usually available in a general food microbiology laboratory. Many food microbiology laboratories also do not have the basic equipment required to perform cell culture or molecular biology-based assays, and in the case of cell culture facilities, the equipment is of little use for other purposes.

44.32 Classical Cell Culture Assays

Cell culture-based assays to detect infectious virus particles in foods are generally preferred over molecular biological methods, when such assays are available. Cell culture assays are useful for the detection of the enteroviruses, but not for hepatitis A, caliciviruses like the Norwalk and Snow Mountain viruses, astroviruses such as the Hawaii agent, or small round structured viruses (SRSV) that may be present in contaminated foods. Wild-type hepatitis A virus can often be propagated in cell culture after prolonged incubation periods, however quantitative cell culture-based assays are unavailable. Although poliovirus may not be detectable in all sewage, most municipal sewage treatment systems contain vaccine strains of poliovirus, strains that may serve as an indicator of the possible presence of other enteric viruses. These strains originate from infants and children who shed viruses at high levels for weeks to months after oral administration of the vaccine. Cell culture methods for the detection of poliovirus include the plaque assay and cytopathogenicity assay.

44.321 Limitations of Cell Culture-Based Assays

Assays involving the cell culture propagation of enteric viruses are useful for viruses, like poliovirus, that readily replicate in cell or tissue cultures, but are ineffective in detecting hepatitis A or Norwalk-like viruses. Hepatitis A virus may be propagated in cell cultures, but its replication is very slow. Methods also are not quantitative, as virus infectivity of cells appears to be quite variable. Hepatitis A virus may infect an individual cell with little or no spread to adjoining cells, making its visual detection by the plaque assay technique impossible. Small round structured viruses like Norwalk virus have not been propagated in cell or tissue culture and cannot be assayed by cell culture techniques. Rotaviruses may be assayed in cell culture, however their replication is slow and results are not very quantitative. There are also no cell culture-based assays for many astroviruses.

44.322 Cell Cultures for Food Virology

There are a variety of cells and cell lines available from commercial suppliers. Since viruses are host specific, it is essential to select cells capable of supporting the replication of the viruses of interest. Some cell lines are more susceptible than others to infection by particular viruses, so a careful screening of cells is essential to achieve maximal assay sensitivity. Buffalo green monkey kidney (BGM) cells derived from the African green monkey, are often used for the detection of enteroviruses.[5,31,32,120] BGM cells have been shown to be highly sensitive for polio-, echo-, and coxsackie- viruses.[11,12,20] A human rhabdomyosarcoma (RD) cell line[119,121] and the MA-104 cell line derived from fetal rhesus monkey kidneys,[95] are also useful for the isolation of enteroviruses. Efforts to propagate hepatitis A virus have achieved only limited success because the viruses reproduce very slowly in vitro and seldom kill the infected cells. Nevertheless, cells used for the propagation and adaptation of hepatitis A virus to growth in cell culture include: fetal rhesus monkey kidney (FRhK-4) cells,[26,28,33,42,48,49,50,86,103,113,122,146,154] BS-C-1 cells from the African green monkey,[33,152] and human embryonic kidney (HEK) and human embryonic fibroblast (HFS) cells.[43,48,50] To date, there have been no reports of the successful *in vitro* replication of Norwalk virus. From among the cells used for virus propagation, BGM, RD, MA-104, BS-C-1, and FRhK-4 cells are continuous cell lines readily adaptable for long-term use in the laboratory. HEK and HFS cells are used as primary to tertiary cells and are more costly to use than continuous cell lines. However, primary to tertiary cells from African green, cynomolgus or rhesus monkey kidneys are considered among the most sensitive cells for propagating enteroviruses.

44.33 Molecular Biological Techniques

Molecular biology is a rapidly expanding field which has produced an array of novel methods for detecting viral nucleic acids in food, environmental, and clinical samples.[115] Nucleic acid sequencing of the genomes of DNA and RNA viruses has led to the development of primers for use in the polymerase chain reaction (PCR) and probes needed for confirmation of virus presence.[1,2,3,27,46,60,75,76,80,151] Methods include reverse transcription (RT) multiplex PCR,[40] nested and semi-nested PCR[54,79,124] and the use of genomic probes.[78] In addition, methods have been developed employing antibody or antigen techniques, such as immunocapture PCR,[7] magnetic immunoseparation PCR,[63,85] and antigen capture PCR.[36]

The use of molecular biological techniques for detecting viral nucleic acids, either by PCR or RT-PCR, includes several steps. In many methods for detecting RNA viruses, the RNA must first be extracted and semi-purified. The RNA is then copied into a complimentary strand (cDNA) by reverse transcription (RT) followed by amplification of the cDNA product by polymerase chain reaction (PCR). If the virus is a DNA virus, the DNA may be extracted and amplified directly by PCR. The PCR products are usually run on an electrophoretic gel to determine the size and quantity of cDNA present. The nucleic acids may then be transferred to a membrane and the membrane reacted with specific, labeled virus probes. Probes are frequently radiolabeled, however non-radiometric procedures are becoming more popular and may be more practical for food testing laboratories. Non-radiometric procedures include chemiluminescence detection,[133,147] colorimetric tests,[19] and didoxigenin-labeled oligonucleotide probes.[34,77] The RT-PCR procedure has been used for the detection of RNA viruses, like poliovirus, hepatitis A virus, caliciviruses and astroviruses. To date, there have been numerous advances in molecular techniques, however none of the methods have received the scrutiny that cell culture-based assays has received. In addition, molecular biology-based approaches are limited in their applications to determining food safety as discussed in section 44.331, below. Because interlaboratory trials have not been performed and no consensus approach has yet emerged, a detailed comparison of these latest methods will not be attempted here. Collaborative testing or long-term usage of specific methods is needed to define their utility in food virology.

44.331 Limitations of Molecular Biological Techniques

Depending on the products tested and the information desired, molecular biological techniques may be inappropriate for use as an assay procedure. PCR techniques can determine the presence of nucleic acids, but not the infectivity of the virus particles. Infectivity must be based on culturing procedures. Whereas infectivity assays can produce quantitative results, molecular biological techniques can only show presence or absence and are considered quantal in nature. Quantal assays have some practicality when applied appropriately to address the right questions or issues. To determine the utility of molecular biological versus culturing methods, one should determine into which of five categories the food product and data requirements fall (Table 1).

Category I , Sanitary Survey, is appropriate to determine the presence or absence of food contamination during growth, harvesting, processing, packaging or distribution of a food commodity. In this situation, it would be assumed that products may occasionally contain enteric virus and that the detection of viruses during any stage of a process would constitute a failure in the system and the need for remedial steps. Such testing could be an asset in HACCP programs seeking to identify critical control points in a process and might be undertaken anywhere from the farm to the table. An example could be strawberries or celery where one might want to determine whether fertilization or irrigation practices, or picking and handling, contributed to the transfer of enteric viruses to products previously shown to be free from contamination. For processed foods, care must be exercised to ensure that changes in product form, pH, salt concentration, etc. do not alter either virus extraction or assay efficiencies. To date, no methods have been found to be universally acceptable for the extraction or assay of viruses from food products.

Category II, End-point Testing, involves the evaluation of foods that are customarily free from enteric viral contamination. Molecular biological assays may be useful in determining presence or absence in a food traditionally or historically found to be free from viral contamination. Category II testing differs from category I in that category II foods are not considered likely candidates for virus transmission and would only be spot checked, at most, under normal operating or processing conditions. They also differ in that category II testing would usually be performed on finished products or raw and finished products, rather than at incremental steps during processing.

Category III, Process Inactivation, involves an evaluation of processing effectiveness in eliminating viral and other contaminants. It is characterized by the incorporation of techniques designed to inactivate pathogens or spoilage microorganisms such as cooking, treatment with disinfectants like chlorine or ozone or with ultraviolet irradiation from disinfection lamps, elevated temperatures or extended exposure at ambient or sub-optimal temperatures, etc. Milk pasteurization would be an example where product may be tested at distinct stages during the process or at the end to determine the efficiency of the treatment process in inactivating viruses. Another example would be shellfish depuration, a process where shellfish are disinfected in tanks of seawater that is recirculated through an ultraviolet light, ozone or other system.[106,107] Assays should be performed on raw materials as well as on processed products to obtain a valid comparison. Foods in category III differ from foods in category II in that category III foods are processed using steps designed to inactivate microbial contaminants. Foods under category III should not be monitored by molecular biological techniques. In a HACCP system, such testing would be done only to validate the virucidal processing step, after which the process, rather than the product, would be monitored continuously.

Category IV, Suspect Products Testing, relates to foods known or suspected to contain inactivated viral contaminants. It applies to products that may contain infectious viruses as well as some portion of inactivated virus particles. Examples of products under this category include molluscan shellfish and some fruits and vegetables. Shellfish are included because they may be grown in waters that contain chlorinated sewage effluent expected to contain principally inactivated enteric viruses. Some fruits and vegetables may also be included if they were fertilized with composted or processed sewage sludge or irrigated with treated wastewater. Category IV applies to products subjected to viruses which have undergone partial or complete inactivation through natural processes, rather than through processing steps designed to inactivate viral contaminants as in category III, above. Foods under category IV should not be tested by molecular biological procedures since such methods can not differentiate between infectious and inactivated viruses. Plaque or cytopathogenicity assays can provide quantitative measurements of the level of infectious virus particles for some of the enteric viruses.

Category V, Outbreak Investigation, involves products suspected of being involved in specific outbreaks of enteric virus illness. Unlike foods listed under categories III and IV, products suspected to be involved in outbreaks may be assayed by either cell culture or molecular biological techniques. The

Table 1. Categorization of Testing Situations for Determining the Appropriateness of Molecular Biological Versus Cell Culture-Based Assays for Monitoring the Virological Safety of Foods

Category	Description	Molecular Biological Assay	Cell Culture-based Assay[a]
I	Sanitary Survey—Comparison for virus presence at incremental points in a process	Yes[b]	Yes
II	End Point Testing—Testing for virus presence of final product and possible comparison with raw materials	Yes[b]	Yes
III	Process Inactivation—Testing of products subjected to processing for the express purpose of preservation or disinfection.	No	Yes
IV	Suspect Products Testing—Testing of of products suspected to contain inactivated viruses	No	Yes
V	Outbreak Investigation—Testing of suspect foods in outbreaks of enteric virus illness	Yes	Yes

[a] Applicable to viruses which can be amplified in cell culture systems and can produce cytopathic effects

[b] Applicable as long as raw materials or early stage products are adequately characterized as having no enteric contamination using similar methods.

reason for this distinction becomes evident in the following example. If an outbreak of hepatitis A were to occur, then any indication of the presence of hepatitis A virus in a food, whether determined by cell culture or molecular methods, would implicate the food in the outbreak. This information could then be compared with epidemiological data to narrow down the suspect products and ultimately identify the source of the virus.

Table 1 is intended to identify situations where cell culture-based assays are preferable over molecular-biologically-based assays. Given the current state-of-the-art, it is clear that many of the enteric viruses can not be propagated in cell or tissue culture. It is those viruses that may require continued use of molecular biological testing. The limitations of such testing must be recognized when interpreting the results. As research scientists provide new and innovative approaches for the propagation of Norwalk and Snow Mountain viruses, Hawaii agent, small round structured viruses and other enteric viral pathogens, cell culture-based assays may become more readily available to identify the presence of infectious virus particles in foods.

There are applications for molecular biological assays in combination with cell culture-based assays to ascertain the presence or absence of viruses propagated in a cell culture system. This is applicable to viruses that do not produce cytopathology in cell or tissue cultures but are amplified sufficiently to convert an initially negative RT-PCR test of the cells and virus to a positive test after propagation. It must be demonstrated that only after cell culture amplification would there be sufficient viral nucleic acids to be detected by PCR or RT-PCR. This again would be a quantal assay, unable to measure viral contamination quantitatively. It would, however, show the presence of infectious virus particles and might be useful in risk assessments.

RT-PCR can support nucleic acid amplification from only small volumes of extract, and the extract must be relatively free from components that may interfere with the reactions. Interfering compounds carried over into the extracts from foods may include lipopolysaccharides, glycogen, and proteins. Steps are under development or evaluation to eliminate some of these interfering compounds.

Food Sample (25 to 100g)[a]	Purpose of Step
↓	
Dilute sample 1:5 to 1:10 with 50-90 mM glycine-NaOH plus 0.8% NaCl, pH8.8-9.5 Homogenize or mix for 1 to 2 min at 0-22°C	To elute viruses from foods at high pH
↓	
Add Cat-Floc to 0.1 to 0.5% Let settle for 20 min Centrifuge at 10,000×g for 15 min Retain supernatant	To remove extraneous food solids by flocculation
↓	
Add beef extract, meat extract, or skim milk Adjust pH to 3.5 to 4.5 Mix slowly for 15 min Centrifuge at 10,000×g for 15 min	To add carrier protein and acidify to precipitate protein and absorbed viruses
↓	
Resuspend pellet in 15 to 20 ml of 0.1-0.15M Na_2HPO_4, pH 9.0	To resuspend viruses at alkaline pH
↓	
Treat with antibiotics or filter	To remove non-viral contaminants
↓	
Assay for viruses	

Figure 1. *Elution–Precipitation Method for Extraction of Viruses from Foods.*

[a] Experimentally determine the optimal concentration and conditions within the indicated range for each food/virus combination.

44.4 EQUIPMENT, MATERIALS AND REAGENTS

Equipment needs for the virus testing laboratory vary with the methods used. The types of equipment required can be dictated by: the virus to be sought, type of food, procedure for extracting the virus particles, and the analytical techniques to be employed. Equipment for virus extraction often includes blenders or other liquefaction apparatus, and centrifuges. Cell culture-based assays usually require a CO_2 incubator, a rocker platform, a biohazard cabinet, a vacuum pump and filter units, appropriate cell cultures and sera, tissue culture flasks or multi-well plates, and various reagents including antibiotics. Molecular biology-based assays would require a thermal cycler, oligonucleotide primers, probes and enzymes, a microcentrifuge, electrophoresis equipment, photographic capabilities, and precision pipetting instrumentation. Disposable labware should be used whenever possible.

44.5 SAMPLE PROCESSING PROCEDURES

There are a host of published methods for extracting and assaying viruses from some foods and little or no information available for others. None of the procedures is officially approved. The greatest amount of work on the development of extraction methods has been devoted to molluscan shellfish, and reviews or laboratory comparisons of the various methods have been published.[10,45] Methods are available for the extraction of viruses from Eastern oysters, *Crassostrea virginica*[8,45,68,89,90,110,130,131,135,138,139]; Pacific oysters, *Crassostrea gigas*[72,141]; hard-shell clams, *Mercenaria mercenaria*[8,89,90,135,138]; mussels, *Mytilus edulis*[8,144]; soft-shell clams, *Mya arenaria*[8,89,90,138]; and the Japanese clam, *Tapes japonica*.[62] These methods are designed to produce extracts for assay in cell culture. A few other studies have been useful in providing guidance for extracting enteric viruses from the surfaces of fruits, vegetables, and mushrooms.[4,25,57,66,67,69,135] Foods with internal viral contaminants (such as shellfish or hamburger) would normally be subjected to homogenization in a blender followed by clarification by differential centrifugation. Products suspected to contain only external contamination may be washed with a Stomacher® (Seward Medical Limited, London, England) followed by concentration of the resulting rinse. The Stomacher® can also be used to recover viruses from particulate foods (including raw ground beef)—its advantage over homogenization is that it leaves the food solids in a form that facilitates clarification of the suspension. Methods also exist for the recovery of enteric viruses in milk.[56,134]

Two approaches have been widely used for the extraction of enteric viruses from foods for the purpose of cell culture assays: the elution-precipitation method (Figure 1) and the adsorption-elution-precipitation method (Figure 2). The elution-precipitation approach involves the liquefaction of the food at alkaline pH, which causes elution of the viruses from the solids. Undissolved food particles are removed by clarification (see Section 44.53) and the remaining suspensions are concentrated (see Section 44.54) and assayed in cell culture or by molecular biological techniques (see Section 44.6). The adsorption-elution-precipitation approach relies on the adsorption of viruses to homogenized food solids under acidic conditions followed by centrifugation to pellet the virus-solids mixture. Viruses may then be eluted from the particulates by resuspension of the pellet in an alkaline pH buffer. Undissolved food components are removed by clarification (see Section 44.53) and the resulting fluids concentrated (see Section 44.54) and assayed (see Section 44.6).

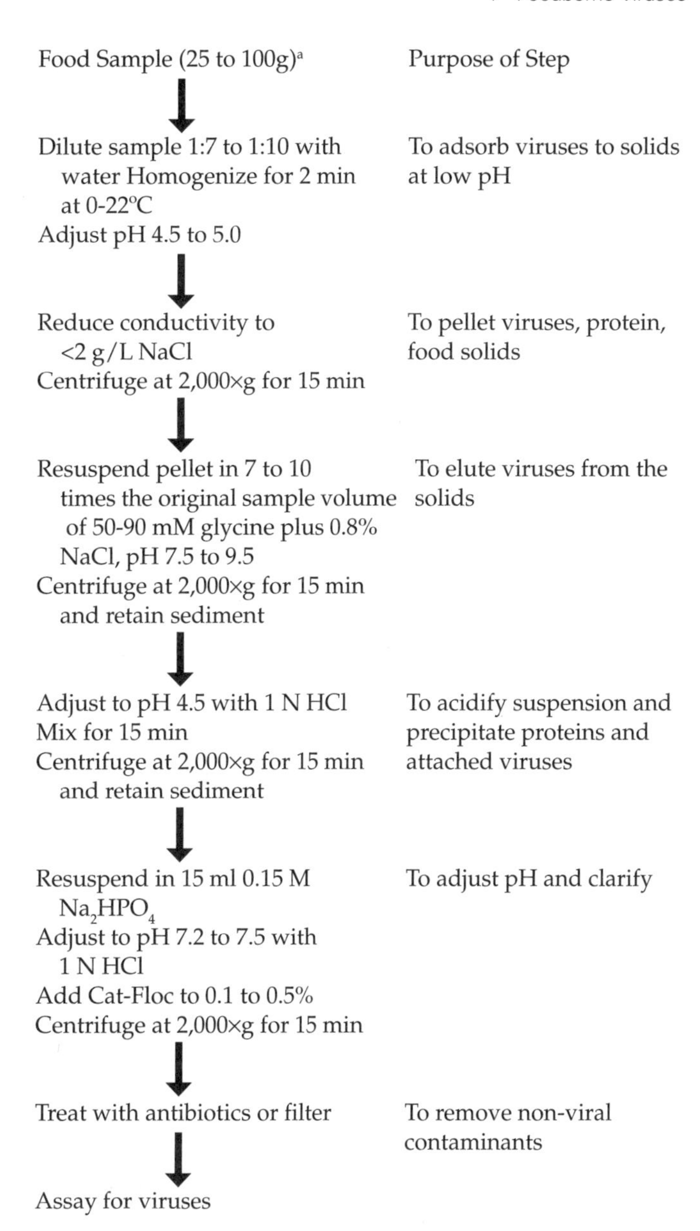

Figure 2. *Adsorption–Elution–Precipitation Method for Extraction of Viruses from Foods.*

[a] Experimentally determine the optimal concentration within the indicated range for each food/virus combination.

44.51 Sample Size and Sampling

Sample size is important if a true representation of the level of contamination is desired. Samples usually comprise 25 to 100 g of food. In the case of surface-contaminated foods, analysts should attempt to standardize the surface area of the sample analyzed. It should be decided in advance whether a single, pooled sample of multiple units (e.g., multiple shellfish) or multiple samples of individual units are to be tested. For practical purposes, it is generally more efficient and cost effective to process fewer samples containing multiple units. If products are co-mingled during processing, and low level contamination is expected, then larger amounts of sample may need to be evaluated. The limits on sample size are dictated by the ability to manipulate large volumes and the sensitivity desired. Virus recoveries may be less efficient for samples in excess of 200 g.[90] For category I studies involving sanitary surveys (Table 1), it is necessary to standardize the assays for comparison of results between different points in the growth, harvesting, processing, packaging, or distribution of the food item, in order to permit process verification or to identify critical control points along the farm to fork continuum. For category II and III studies, standardization of sample size is critical for accurate comparisons between the raw and finished product or for comparative studies to determine the relative levels of contamination between different processing batches. Suspect products testing, category IV, may not require as large a sample size, since the levels of contamination may be typically higher for foods in this category. When an outbreak is being investigated (Category V), any representative sample that is available may be tested. Quantities are often limited, but if a large quantity is available, several subsamples from various portions of the food mass should be tested, in that uniform distribution of the virus can not be assumed.

The consistency of the sample and complexity of processing must also be considered when deciding on a sampling strategy. Efforts should be made to standardize, to the extent possible, the sampling strategies with those used in other, similar facilities, so that comparative studies between like facilities may be made. Such inter-facility comparisons may afford a basis for virological standards for foods when official testing methods are adopted.

44.52 Liquefaction of Solid Foods

Solid food samples must first be converted to a liquid or a semi-liquid form. Means should be chosen that will minimize problems in clarification, by not reducing the food solids to smaller suspended particles than necessary. The first step for the extraction of viruses contained within foods often involves homogenization of the food, usually in a buffer solution, at a ratio of 5 mL to 10 mL of buffer per gram of sample. Water may be substituted for buffer in some cases. Precautions must be taken to prevent heat buildup during the homogenization process, as heat may inactivate the viruses. Sonication may also be used, provided it is of brief duration, without heat buildup, and does not cause suspension of too much of the food product, which could complicate subsequent clarification. Precautions must also be exercised to prevent aerosolization of potential contaminants into the laboratory. Liquefaction of the food is not appropriate for samples containing only surface contamination, like salad greens, where simple shaking or Stomacher® processing may be sufficient to dislodge attached virus particles.

44.53 Clarification of Food Suspensions

Clarification is the process by which viruses are separated from food solids and from potentially cytotoxic or inhibitory substances, by means of chemical or physical treatment of the liquid suspension. Clarified extracts should be relatively clear and sufficiently free of cytotoxic materials to be able to assay in cell culture. Clarification may involve multiple processes and treatments. Separation of viruses from soluble proteins may be accomplished by the use of the polycation sewage flocculent, Cat-Floc® (Calgon Corp., Pittsburg, PA), at a final concentration of 0.1–0.5%.[68,69,71] Cat-Floc may be added to homogenates or partially clarified extracts to further reduce the presence of materials potentially cytotoxic to cell cultures or that inhibit molecular biological amplification of viral nucleic acids by PCR or RT-PCR. After Cat-Floc coagulation, the suspended solids are removed by centrifugation or with a fiberglass prefilter. Other compounds and methods found to be effective in removing potentially cytotoxic materials during virus extraction are described below (see Section 44.55).

44.54 Concentration of Food Extracts

Concentration of food extracts is required to reduce the volume to a manageable size and to reduce the costs of the overall assay. Concentration is particularly important for molecular biological-based assays which require PCR or RT-PCR as these amplification procedures can test only 50 to 100 µL of sample per amplification. Concentration may be accomplished by acid precipitation, a process where the mixture is acidified, the viruses become bound to the proteins and the proteins precipitate out of solution. The precipitate is then collected after centrifugation and the supernatant fluids are discarded. As a result, the viruses are concentrated and partially purified in a single step. At times, a carrier protein must be added to enhance virus attachment and subsequent acid precipitation. Beef extract V (BBL Microbiology Systems, Cockeysville, MD) or Lab-lemco Meat Extract Powder (Oxoid USA, Columbia, MD)[110] and skim milk[131] may also serve as carrier proteins to facilitate organic flocculation and subsequent precipitation of the protein-virus complex. Lots of beef extract must be evaluated before use as some will not flocculate[58] while others may carry potentially cytotoxic or RT-PCR/PCR inhibitory substances into the final virus-containing extract. The effects of carryover of the proteins into the final extract must be evaluated as beef extract and skim milk at elevated levels were found to significantly ($P < 0.05$ and $P < 0.01$, respectively) reduce plaque counts.[109] However, meat extract powder and nonfat dry milk did not significantly reduce plaque numbers compared to controls.[109]

Other methods of concentration include ultrafiltration, ultracentrifugation and hydroextraction. Hydroextraction involves the withdrawal of fluids through dialysis membranes using polyethylene glycol or some other hydrophilic compound to produce an increasingly concentrated sample within the dialysis membrane. Unfortunately, this method tends to concentrate potentially cytotoxic materials or compounds inhibitory to RT-PCR reactions. Both the ultrafiltration and ultracentrifugation methods are costly and of limited value in the development of practical food testing procedures.

44.55 Removal of Contaminants

Most foods contain constituents that may be cytotoxic to cell cultures or inhibitory to PCR or RT-PCR reactions. It is important to remove such contaminating materials before assay. Compounds and methods found to be effective in removing potentially cytotoxic materials during various stages of virus extraction include polyethylene glycol precipitation, Freon TF® (1,1,2-trichloro-1,2,2-trifluoroethane) (E. I. DuPont de Nemours, Wilmington, DE) extraction, Pro-Cipitate (Affinity Technology, Brunswick, NJ) precipitation, and the use of cationic detergents. Polyethylene glycol precipitation techniques for viruses have been around for many years, however they have only recently assumed an integral role in concentrating enteric viruses in food extracts.[22,38,60,75,82,140,153] Organic extraction of shellfish tissues with Freon® has been effective in eliminating lipids and nonpolar compounds from the extracts.[57] Such methods have been widely used, however bans on the production and distribution of this product, in the interest of preserving the earth's ozone layer, indicate the need to develop alternative approaches for eliminating potential inhibitors.

Several procedures have been reported to further concentrate extracts and to eliminate substances potentially inhibitory to the RT-PCR reaction. Shellfish polysaccharides can be removed, at least partially, by the incorporation of the cationic detergent cetyltrimethylammonium bromide (CTAB) to the extraction procedure.[1,47,118,153] Likewise, Pro-Cipitate, a protein precipitating agent, has also been used successfully to remove inhibitors of RT-PCR.[22,38,60,123]

For cell culture-based assays, shellfish extracts may be diluted to reduce potentially cytotoxic materials and tested in additional flasks of cell culture at minimum cost. Cell monolayers may also be drained of inoculum after the adsorption period or rinsed with isotonic medium to reduce cytotoxic materials. Dilution is only appropriate for molecular biological-based assays if the volume of the extract is very low or the concentration of viruses is expected to be very high, otherwise, the number of individual assays required to test the extract becomes cost prohibitive.

It is essential that the final extracts be free of contaminating microbes, if they are to be assayed in cell culture systems. Molecular biology-based assays may not require sterility, as is necessary for cell culture assays, however, excessive microbial contamination may affect pH or other factors important to the RT or PCR reactions. In addition, microbial contamination of samples destined for molecular biological assays may lead to unintended microbial products being amplified during PCR and subsequent ambiguous results. For cell culture-based assays, microbial contamination can often be eliminated by the incorporation of antibiotics into the extract followed by a short period of incubation (perhaps 20 min) at room temperature for the antibiotics to act on the microbes. Antibiotics may be added not only to the extract, but to the cell culture medium (for assays involving cytopathogenicity in liquid cultures) or to the agar (for plaque assay overlays). Antibiotics in common use include gentamicin (50–100 µg/mL), penicillin (500 units/mL), streptomycin (500 µg/mL), kanamycin (5–10 µg/mL) and nystatin (mycostatin, 50 units/mL). The use of antibiotics may not be useful in molecular biological-based assays, since portions of the nucleic acids of bacterial and other contaminants would still be expected to amplify after antibiotic treatment if nonspecific priming occurred.

Filtration of extracts can also be used to remove undesirable bacteria, yeasts, and molds. Effective filtration relies on the passage of relatively clean extracts through 0.45 or 0.22 µm filters. Viruses may be bound to these filters, particularly if they are attached to particulate matter or maintain a charge opposite that of the filter. Pre-treatment of filters with sterile serum or selection of filters made from non-adsorptive polymers is usually effective in reducing the binding of enteric viruses.[23] Other methods to destroy microbial contaminants include treatment of the extracts with chloroform or ether followed by a period for the organic solvent to dissipate into the air. This work must be performed under a sterile hood, properly vented, however either treatment may result in residues that could affect subsequent assays.

44.6 CELL CULTURE-BASED VIRUS ASSAYS

There are a variety of assays available for detecting poliovirus, related enteroviruses and rotavirus in foods. Most of these assays were developed before the advent of molecular biology; as molecular techniques, in spite of some of their limitations in food virology (see Section 44.331), have now captured the attention of many researchers. As a result, substantially less effort is being devoted to the development of new cell culture methods for detection of hepatitis A, Norwalk and related viruses. Cell culture-based enumerative assays include the plaque and radioimmunofocus assays (RIFA). The plaque assay is a quantitative procedure that allows for the enumeration of individual foci of virus infection in cell culture monolayers. There are published RIFA procedures for the detection of hepatitis A[81] and rotavirus.[84] The RIFA assay can be used in certain research settings, but is not practical for routine food screening. Quantal assays differ from enumerative assays because they are either non-quantitative or only semi-quantitative. Quantal assays rely on the scoring of cultures as positive or negative for cytopathic effects (CPE), a condition where cells become rounded and change in their optical properties as determined microscopically. In theory, a single infectious virus could cause CPE in a culture and the destruction of most of the cells. Whether one or a thousand infectious viruses are present, the result would be the same—positive CPE. If multiple cultures are inoculated with each of several serial dilutions, the quantal results can be interpreted to estimate the most probable number of cytopathic units (MPNCU) or the number of median infectious doses (TCD_{50}) in the original sample.[18,129]

44.61 Preparation of Cell Cultures for Virus Inoculation

The preparation of cell cultures requires utmost care to ensure the sterility and sensitivity of the cells. General guidance is available on cell culture techniques.[99] In essence, cells are grown in commercially prepared culture medium either obtained in liquid form or rehydrated from powder. Rehydrated medium must be filter sterilized. The most common medium is minimum essential medium (MEM) with Earle's balanced salts. Depending on the cells used, it may be necessary to supplement the commercial medium with L-glutamine, sodium bicarbonate, fetal bovine serum and antibiotics; non-essential amino acids and vitamins are seldom required. Antibiotics may be and often are used to eliminate microbial contamination (see Section 44.55). Culture media prepared in the laboratory should be passed through 0.45 or 0.22 µm filters. Pressure filtration through a membrane is preferable to vacuum filtration, particularly if serum is present in the medium or if large volumes of media must be prepared.

The preparation of cell cultures leading up to an assay must be tightly controlled. Monolayers must not be overly confluent (too old or stressed or dense) or under-confluent (leaving gaps in a plaque assay flask—gaps that might be mistaken for plaques at the end of the assay). The medium in the flasks prior to the assay should be at the correct pH, not too acidic or alkaline. This can often be determined by the color, since most commercial media contain a pH indicator. Plaque assays are often performed with young cells that have just reached confluency and are actively growing. Quantal assays, based on cytopathic effect and often performed in liquid cultures, should be conducted in healthy and actively growing cultures maintained at the proper pH, and under optimal growth conditions (temperature, CO_2 concentration, etc.) for the cell line used.

44.62 Inoculation of Cell Cultures

Plaque assays are not performed to any standard methods, and the basic procedures may vary. Generally, a virus-containing extract is pipetted onto a confluent monolayer of susceptible cells and viruses are allowed to adsorb to the cells. Adsorption may take place under static conditions on a level table with manual redistribution of the inoculum every 15 to 20 min or with uniform spreading of the inoculum over the surface of the monolayer by means of a rocker platform. Very different results may be obtained depending on the amount of virus inoculum added; the size of the tissue culture flask; the condition, age and density of the cells; and the length of time the extract is in the flask. Rocking duration has been shown to significantly influence the plaque counts; 2 hours is considered optimal for poliovirus.[109] Rocking rates from 1 to 5 back and forth oscillations per minute did not cause significant differences in the plaque counts.[109] An inoculum volume of ≤ 0.02 mL/cm^2 is appropriate for most plaque assays. For larger volumes of unconcentrated food extracts, volumes as high as 1.4 mL/cm^2 may be used in quantal assays if the adsorption period is extended to 16 to 20 hr.[70] Food extract may contain cytotoxic materials and is often removed after the adsorption period to prevent damage to the cell culture monolayer. Cell lines may differ in their susceptibility to wild-type viruses, so it is preferable to split the extract into two portions, one to be assayed in one cell line and the other to be assayed in a second cell line.[88] Alternatively, after the adsorption period in the first cell line, the inoculum may be removed and placed in a second cell line for adsorption and subsequent agar overlay.[70,89,90]

For plaque assays, an agar overlay, consisting of sterile and tempered agar (about 42°C) is combined with tempered cell culture medium, serum, antibiotics and any supplements required to sustain the cell cultures, and added to the cells.[117] Agar is usually at a final concentration of 0.75% to 1%. Neutral red may be added to the overlay to stain living cells (plaques will appear clear), but the flasks must be maintained in the dark. After solidification of the agar, the flasks are inverted, cell side up, and incubated at 37°C for several days. For more protracted incubation periods, a second agar overlay may be added to provide nutrients necessary to support the cells and to maintain the pH. When a second overlay is used, the neutral red is omitted from the first overlay medium and is added to the second agar to reduce the adverse effects of neutral red on the cell monolayer.

Plaque assays may be performed for rotaviruses, but trypsin or other enzymes are usually added to the agar to enhance virus uptake by the cells.[59,100-102,127] Rotavirus may also be assayed in liquid culture (quantal assay) with the incorporation of trypsin.[102]

There are two forms of quantal assays. In one, viruses are adsorbed to cells followed by the addition of liquid medium. Another is where the viruses are added to freshly split cells in medium and the cells are allowed to attach to the bottom of the container as viruses adsorb to the cells in suspension as well as during and after attachment. In either case, the cells are observed microscopically for signs of cytopathic effects at various intervals over several days. Quantal assays may be semi-quantitative if they are performed using multiple plates, flasks, or wells. This works particularly well in most probable number (MPN) assays when virus counts are high and where virus may be diluted down to extinction. Again, results may be expressed as MPNCU or TCD_{50}.[18,129]

44.63 Maintenance and Handling of Cultures during Assay

Plaque assays are frequently performed in tissue culture flasks and the flasks are incubated at 37°C. Many cells require the presence of 5% CO_2 and incubators should be monitored for both temperature and CO_2 level regularly. Other cell lines may grow perfectly well in tightly capped flasks without added CO_2. In such cases, the CO_2 can be obtained from a bicarbonate buffering system derived from ingredients in the media. If extended incubation is anticipated, it may be necessary to humidify the air in the incubator by adding a pan of distilled water to the bottom shelf. Since neutral red can damage cells in the presence of light, plaque assay flasks may be placed in a cardboard box with a loose fitting lid or paper or foil cover to exclude excessive exposure to light during opening and closing of the incubator. Sufficient air holes should be present to allow air circulation and CO_2 entry, if applicable. Flasks should be observed for plaques every few days for at least 2 weeks and for up to 4 weeks if the cells remain viable. Plaques should be marked on each flask with a marking pen as they are observed and the marks should be tallied at the completion of the assay. The agar and cells within some of the plaques should be picked, added to a small volume of culture medium, frozen at –70°C and thawed to release cell-bound viruses. A portion of the culture medium should be tested to confirm the presence of viruses by quantal assay. Cultures that exhibit CPE in the quantal assay should be frozen at –70°C for virus identification.

Cells in use for quantal assays must be maintained at the proper temperature and CO_2 concentration. If assays are performed in small volumes of medium (as in a 96-well plate), precautions must be in place to prevent dessication of the cells. 24-well plates are more appropriate for this type of assay and are especially useful for MPN-type analyses with up to six serial dilutions plated in quadruplicate. Wells, tubes or flasks exhibiting CPE should be subcultured in freshly prepared cells to confirm virus presence. This may be performed by transferring a small drop of the medium to a fresh culture, incubating and observing for CPE. Alternatively, the medium and cells may be frozen at –70°C to release cell-bound viruses, thawed, and assayed in fresh cells.

CPE and microbial contamination appear similar, especially to the untrained eye. CPE is usually recognized by a rounding and glistening of the cells as their light refractivity changes with virus infection. One sign of microbial contamination is the early clouding of the cell culture medium. Clouding of medium later into the assay may be normal as infected and lysed cells will detach from the surface and enter the media. Rapid changes in the color of the medium may also be an indication of microbial presence. Bacterial, fungal and yeast contamination of the medium often causes rapid acidification of the medium, often within a day, changing the medium from red to yellow. Cell death may also be the result of cytotoxic materials present in the inoculum. Confirmation of CPE by transfer of a drop of culture fluid from a seemingly infected culture to a new culture will permit the differentiation between cytotoxicity and true infection.

44.64 Controls

Foods suspected of being contaminated should be extracted along with uncontaminated food of the same type. The latter serves as a negative control for the extraction. Food extracts often contain cytotoxic materials that interfere with plaque or quantal assays, and a negative control can often alert the analyst to possible cytotoxicity problems. Food samples received for testing may be partially decomposed—it may be difficult to improvise a negative control that is in a comparable state. A positive control for the extraction procedure serves as an indicator of the extraction efficiency. It can be performed using known quantities of viruses, preferably at levels suspected to be present in the test samples, and added during the earliest stages of the extraction.

Positive and negative controls must also be prepared for plaque and quantal assays. Positive controls would include preparations of viruses suspected to be in the samples. They are included as a test of the cell line's sensitivity toward the particular virus sought. Negative controls could include cells inoculated with: a) cell culture medium, serum and medium supplements (as a test for medium sterility) or b) negative control extract (as a test for cytotoxicity or failure to obtain sterile extract). Cytotoxicity can lead to false negative results if the cells are sufficiently altered to prevent virus replication. More often, cytotoxicity leads to patches or spots of dead cells (occasionally associated with visible particulates in an extract) which may be registered as presumptive plaques. For this reason, plaques must be confirmed by subculture.

44.65 Precautions

Safe handling of contaminated foods, extracts and virus propagation systems is important to the well-being of the analyst as well as the surrounding environment. As much precaution must be exercised in handling potential virus-infected materials as is required to assure sterility of cell culture systems. Viruses propagated in many research laboratories are often attenuated strains, not generally associated with disease. Wild-type viruses found in contaminated foods, whether originating from the hands of infected workers or environmental sources, are potentially pathogenic, or even lethal. Work must be performed only under the proper supervision in laboratories with adequate equipment and containment capabilities. Used cultures and virus extracts must be autoclaved or properly disinfected and discarded in an approved manner. Work surfaces must be disinfected with chlorine, iodine, ultraviolet irradiation, or other means found to be successful in eliminating a broad range of viral pathogens. Ultraviolet disinfection can only be accomplished with lamps operating at the proper wavelength; the older the lamp, the less effective it becomes. Lamps will continue to light long after they have lost their germicidal effectiveness, so they should be replaced regularly or monitored for their level of germicidal emissions.

44.7 VIRUS IDENTIFICATION

Some of the methods described are so specific that they will detect only a single type of virus, either based on antigen-antibody interaction or a unique nucleotide sequence in the viral genome. When such methods yield a positive result, the identity of the detected virus is immediately known. However, no matter how sensitive these methods are to their homologous virus, their specificity causes them to have no sensitivity for other or unknown viruses. Cell culture methods generally afford the only means of detecting viruses that were not previously known and where there are no antisera, probes, or primers available. Other broad-spectrum methods are in short supply, but a nonspecific ELISA method has been described[35]; and a primer set that reacts with most, if not all, human enteroviruses is available.[75]

Conventional methods for virus identification have relied on the use of various antisera, such as the Lim-Benyesh-Melnick antiserum pools for serotyping the enteroviruses.[83] Antiserum is also necessary for the identification of viruses by immune electron microscopy. This method has been used rather extensively in research and clinical settings, but has major cost and technical requirements prohibiting its direct use in food testing laboratories. Other immunoassays used in the identification of enteric viruses include the immunofluorescence assay, radioimmunoassay, radioimmunofocus assays, enzyme linked immunosorbent assay, and immunoenzymmatic assays. Of these, only the immunofluorescence assay[111] and the radioimmunofocus assay[81] can detect viruses in infected cells that do not exhibit cytopathic effects.

Virus identification has been complicated in the past, due in large part to the need for numerous antisera, costly reagents or expensive instrumentation beyond the reach of small laboratories. With the advent of molecular biology, many exciting and innovative means are under development for the identification of enteric viruses. PCR and RT-PCR can be employed to identify polioviruses, hepatitis A virus, Norwalk and related viruses, rotavirus, astrovirus and other enteric pathogens. Although molecular techniques have limitations in their application to direct testing of food extracts, their utility for detecting and identifying specific viruses propagated in cell culture systems is, without doubt, a major breakthrough. As improved primers and probes become available, the identification of viruses will become much easier. Unlike the problems associated with inhibitors of PCR and RT-PCR in food extracts, there are few if any anticipated inhibitory substances in cell culture-enriched viruses. Therefore, molecular biological identification of viruses should become popular. Sequencing of a portion of the viral genome may also be used as a method of "fingerprinting," to determine the relationship of multiple isolates with a single outbreak of viral illness.[93]

44.8 INTERPRETATION OF RESULTS

44.81 Negative Results

Negative results obtained from use of the above methods will likely occur and may be indicative of food products that are free from viral pathogens. However, it should be recognized that virus contamination is unlikely to be uniform throughout a lot of food, so the small sample that yielded the negative test result may not be representative of the entire lot from which it was taken. There is always the possibility of false negative results from inefficient extraction of the viruses from the food; virus inactivation by heat, storage, mishandling practices, etc.; and the use of cell cultures not susceptible to infection by the particular virus strain. Likewise, the assays may not have been performed under optimal conditions, a problem that can lead to reduced counts or no counts. False negative results are often caused by cytotoxic materials in the extracts which lead to inefficient plaque assays or inhibit PCR or RT-PCR reactions. False negatives can be guarded against by the incorporation of positive controls as indicated in Section 44.64.

44.82 Positive Results

The identification of plaques or cytopathic effects is subjective, in some cases, and requires a trained eye and adequate controls. Presumptive positive results must be confirmed by secondary passage and identified by serological or molecular biological testing. False positive results may be associated with cytotoxic materials in the extracts—materials which create spots in cell culture monolayers used for plaque assays or generalized cell death in quantal assays. False positive results may also occur if the cells were mishandled before or during plaque assays. Damaged cells may begin to slough off the flask, causing holes in the monolayer. Likewise, if plaque assays are performed in sub-confluent cells, then gaps in the monolayer may be misinterpreted as plaques. A sign of plaques is that they increase in size and are generally regular (round), whereas gaps in monolayers may be more irregular and fixed in size. There is always the possibility that the sample to be tested was contaminated in the laboratory, and repeat analyses of extracts may be warranted. If the detection was done entirely by molecular biological methods, the identity of the PCR product can be verified with an oligonucleotide or cDNA probe specific to the amplified product, but there will be no way to demonstrate that the virus was infectious when the sample was taken.

44.9 REFERENCES

1. Atmar, R. L., Metcalf, T. G., Neill, F. H., and Estes, M. K. 1993. Detection of enteric viruses in oysters by using the polymerase chain reaction. Appl. Environ. Microbiol. 59: 631.
2. Atmar, R. L., Neill, F. H., Romalde, J. L., Le Guyader, F., Woodley, C. W., Metcalf, T.G., and Estes, M.K. 1995. Detection of Norwalk virus and hepatitis A virus in shellfish tissues with the PCR. Appl. Environ. Microbiol. 61: 3014.
3. Atmar, R. L., Neill, F. H., Woodley, C. M., Manger, R., Fout, G. S., Burkhardt, W., Leja, L., McGovern, E. R., Le Guyader, F., Metcalf, T. G., and Estes, M. K. 1996. Collaborative evaluation of a method for the detection of Norwalk virus in shellfish tissues by PCR. Appl. Environ. Microbiol. 62: 254.
4. Badawy, A.S., Gerba, C. P. and Kelly, L. M. 1985. Development of a method for recovery of rotavirus from the surface of vegetables. J. Food Prot. 48: 261.
5. Barron, A.L., Olshevsky, C. and Cohen, M. M. 1970. Characteristics of the BGM line of cells from African green monkey kidney. Brief report. Arch. Ges. Virusforsch. 32: 389.
6. Bean, N.H., Goulding, J. S., Lao, C. and Angulo, F. J. 1996. Surveillance for foodborne-disease outbreaks—United States, 1988-1992. Morbid. Mortal. Weekly Rept. 45(SS-5): 1.
7. Beaulieux, F., See, D. M., Leparc-Goffart, I., Aymard, M., and Lina, B. 1997. Use of magnetic beads versus guanidium thiocyanate-phenol-chloroform RNA extraction followed by polymerase chain reaction for the rapid, sensitive detection of enterovirus RNA. Res. Virol. 148: 11.
8. Bemiss, J.A., Logan, M.M., Sample, J.D. and Richards, G.P. 1989. A method for the enumeration of poliovirus in selected molluscan shellfish. J. Virol. Methods 26: 209.
9. Benedict, I. T., Madore, H. P., Menegus, M. A., Nitzkin, J. L. and Dolin, R. 1987. Snow Mountain agent of gastroenteritis from clams. Am. J. Epidemiol. 126: 516.
10. Bouchriti, N. and Goyal, S. M. 1993. Methods for the concentration and detection of human enteric viruses in shellfish: a review. New Microbiol. 16: 105.
11. Bryden, A. S. 1992. Isolation of enteroviruses and adenoviruses in continuous simian cell lines. Med. Lab. Sci. 49: 60.
12. Cao, Y., Walen, K. H. and Schnurr, D. 1988. Coxsackievirus B-3 selection of virus resistant Buffalo green monkey kidney cells and chromosome analysis of parental and resistant strains. Arch. Virol. 101: 209.
13. CDC 1987a. Enterically transmitted non-A, non-B hepatitis—East Africa. MMWR 36: 241.
14. CDC 1987b. Enterically transmitted non-A, non-B hepatitis—Mexico. MMWR 36: 597.
15. CDC 1997. Hepatitis A associated with consumption of frozen strawberries—Michigan, March 1997. MMWR 46: 288.
16. CDC 1998. Progress toward poliomyelitis eradication—West Africa, 1997–September 1998. MMWR 47:882.

17. Chan, T. Y. 1995. Shellfish-borne illnesses. A Hong Kong perspective. Trop. Geogr. Med. 47: 305.
18. Chang, S. L., Berg, G., Busch, K. A., Stevenson, R. E., Clarke, N. E. and Kabler, P. W. 1958. Application of the most probable number method for estimating concentrations of animal viruses by the tissue culture technique. Virology 6: 27.
19. Chenal, V. and Griffais, R. 1994. Chemiluminescent and colorimetric detection of a fluoroscein-labeled probe and a digoxigenin-labeled probe after a single hybridization step. Mol. Cell. Probes 8: 401.
20. Chonmaitree, T., Ford, C., Sanders, C. and Lucia, H. L. 1988. Comparison of cell cultures for rapid isolation of enteroviruses. J. Clin. Microbiol. 26: 2576.
21. Christensen, M. L. 1989. Human viral gastroenteritis. Clin. Microbiol. Rev. 2: 51.
22. Chung, H., Jaykus, L.A., and Sobsey, M. D. 1996. Detection of human enteric viruses in oysters by in vivo and in vitro amplification of nucleic acids. Appl. Environ. Microbiol. 62: 3772.
23. Cliver, D. O. 1968. Virus interactions with membrane filters. Biotechnol. Bioeng. 10: 877.
24. Cliver, D. O. 1997. Virus transmission via foods, an IFT Scientific Status Summary. Food Technol. 51: 71.
25. Cliver, D. O. and Grindrod, J. 1969. Surveillance methods for viruses in foods. J. Milk Food Technol. 32: 421.
26. Cromeans, T., Fields, H.A., and Sobsey, M.D. 1989. Replication kinetics and cytopathic effect of hepatitis A virus. J. Gen. Virol. 70:2051.
27. Cromeans, T. L., Nainan, O.V. and Margolis, H. S. 1997. Detection of hepatitis A virus RNA in oyster meat. Appl. Environ. Microbiol. 63: 2460.
28. Cromeans, T., Sobsey, M.D. and Fields, H.A. 1987. Development of a plaque assay for a cytopathic, rapidly replicating isolate of hepatitis A virus. J. Med. Virol. 22:45.
29. Cubitt, W. D. 1990. Human, small round structured viruses, caliciviruses and astroviruses. Baillieres Clin. Gastroenterol. 4: 643.
30. Cunningham, A. L., Grohmann, G. S., Harkness, J., Law, C., Marriott, D., Tindall, B. and Cooper, D. A. 1988. Gastrointestinal viral infections in homosexual men who were symptomatic and seropositive for human immunodeficiency virus. J. Infect. Dis. 158: 386.
31. Dahling, D. R. and Wright, B. A. 1986. Optimization of the BGM cell line culture and viral assay procedures for monitoring viruses in the environment. Appl. Environ. Microbiol. 51: 790.
32. Dahling, D. R., Berg, G. and Berman, D. 1974. BGM, a continuous cell line more sensitive than primary rhesus and African green kidney cells for the recovery of viruses from water. Health Lab. Sci. 11: 275.
33. Day, S. P., Murphy, P., Brown, E. A., and Lemon, S. M. 1992. Mutations within the 5′ nontranslated region of hepatitis A virus RNA which enhances replication in BS-C-1 cells. J. Virol. 66: 6533.
34. De Leon, R., Matsui, S. M., Barric, R. S., Herrmann, J. E., Blacklow, N. R., Greenberg, H. B., and Sobsey, M. D. 1992. Detection of Norwalk virus in stool specimens by reverse transcriptase-polymerase chain reaction and nonradioactive oligoprobes. J. Clin. Microbiol. 30: 3151.
35. Deng, M., and Cliver, D. O. 1984. A broadspectrum enzymelinked immunosorbent assay for the detection of human enteric viruses. J. Virol. Meth. 8: 87.
36. Deng, M. Y., Day, S. P., and Cliver, D.O. 1994. Detection of hepatitis A virus in environmental samples by antigen-capture PCR. Appl. Environ. Microbiol. 60: 1927.
37. Desenclos, J. C., Klontz, K. C., Wilder, M. H., Nainan, O. V., Margolis, H. S., and Gunn, R. A. 1991. Am. J. Pub. Health 81: 1268.
38. Dix , A. B., and Jaykus, L. A. 1998. Virion concentration method for the detection of human enteric viruses in extracts of hard-shelled clams. J. Food Prot. 61:458.
39. Dolin, R., Reichman, R. C., Roessner, K. D., Tralka, T. S., Schooley, R. T., Gary, W., and Morens, D. 1982. Detection by immune electron microscopy of the Snow Mountain agent of acute viral gastroenteritis. J. Infect. Dis. 146: 184.
40. Egger, D., Pasamontes, L., Ostermayer, M., and Bienz, K. 1995. Reverse transcription multiplex PCR for differentiation between polio- and enteroviruses from clinical and environmental samples. J. Clin. Microbiol. 33: 1442.
41. Filppi, J. A., and Banwart, G. J. 1974. Effect of the fat content of ground beef on the heat inactivation of poliovirus. J. Food Sci. 39: 865.
42. Flehmig, B. 1981. Hepatitis A virus in cell culture. II. Growth characteristics of hepatitis A virus in FRhK-4/R cells. Med. Microbiol. Immunol. (Berl.) 170: 73.
43. Flehmig, B., Vallbracht, A., and Wurster, G. 1981. Hepatitis A virus in cell culture. III. Propagation of hepatitis A virus in human embryo kidney cells and human embryo fibroblast strains. Med. Microbiol. Immunol. (Berl.) 170: 83.
44. Fugate, K. J., Cliver, D. O., and Hatch, M. T. 1975. Enteroviruses and potential bacterial indicators in Gulf Coast oysters. J. Milk Food Technol. 38: 100.
45. Gerba, C. P., and Goyal, S. M. 1978. Detection and occurrence of enteric viruses in shellfish: a review. J. Food Prot. 41:,743.
46. Goswami, B. B., Koch, W. H., and Cebula, T. A. 1993. Detection of hepatitis A virus in *Mercenaria mercenaria* by coupled reverse transcription and polymerase chain reaction. Appl. Environ. Microbiol. 59: 2756.
47. Gouvea, V., Santos, N., Timenetsky, M. C., and Estes, M. K. 1994. Identification of Norwalk virus in artificially seeded shellfish and selected foods. J. Virol. Methods 48: 177.
48. Graff, J., Kasang, C., Normann, A., Pfisterer-Hunt, M., Feinstone, S.M., and Flehmig, B. 1994. Mutational events in consecutive passages of hepatitis A virus strain GBM during cell culture adaptation. Virology 204: 60.
49. Graff, J., Normann, A., Feinstone, S. M., and Flehmig, B. 1994. Nucleotide sequence of wild-type hepatitis A virus GBM in comparison with two cell culture-adapted variants. J. Virol. 68: 558.
50. Graff, J., Normann, A. and Flehmig, B. 1997. Influence of the 5' noncoding region of hepatitis A virus strain GBM on its growth in different cell lines. J. Gen. Virol. 78: 1841.
51. Grešíková, M. 1994. Tickborne encephalitis. In "Foodborne Disease Handbook: Vol. 2. Diseases Caused by Viruses, Parasites, and Fungi," eds. Y. H. Hui, J. R. Gorham, K. D. Murrell and D. O. Cliver, p. 113. Marcel Dekker, New York.
52. Griffin, M. R., Surowiec, J. J., McCloskey, D. I., Capuano, B., Pierzynski, B., Quinn, M., Wojnarski, R., Parkin, W. E., Greenberg, H., and Gary, G. W. 1982. Foodborne Norwalk virus. Am. J. Epidemiol. 115: 178.
53. Grohmann, G. S., Murphy, A. M., Christopher, P. J., Auty, E., and Greenberg, H. B. 1981. Norwalk virus gastroenteritis in volunteers consuming depurated oysters. Austral. J. Exper. Biol. Med. 59 (Pt. 2): 219.
54. Hafliger, D., Gilgen, M., Luthy, J., and Hubner, P. 1997. Seminested RT-PCR systems for small round structured viruses and detection of enteric viruses in seafoods. Int. J. Food. Microbiol. 37: 27.
55. Halliday, M. L., Kang, L. Y., Zhou, T. K., Hu, M. D., Pan, Q. C., Fu, T. Y, Huang, Y. S., and Hu, S. L. 1991. An epidemic of hepatitis A attributable to the ingestion of raw clams in Shanghai, China. J. Infect. Dis. 164: 852.
56. Hassen, A., Hachicha, R., Jedidi, N., Agbalika, F., and Harteman, P. 1991. A method for recovery of enteroviruses from milk. Arch. Inst. Pasteur Tunis 68: 261.
57. Herrmann, J. E., and Cliver, D. O. 1968. Methods for detecting foodborne enteroviruses. Appl. Microbiol. 16: 1564.
58. Hurst, C. J., Dahling, D. R., Saffermen, R. S., and Goyke, T. 1984. Comparison of commercial beef extracts and similar materials for recovering viruses from environmental samples. Can. J. Microbiol. 30: 1253.
59. Imamura, Y., Yamamoto, S., and Shingu, M. 1991. Improvement of the plaque technique for human rotaviruses: effect of fetal bovine serum, acetyltrypsin and agar. Kurume Med. J. 38: 251.

60. Jaykus, L. A., De Leon, R., and Sobsey, M. D. 1996. A virion concentration method for detection of human enteric viruses in oysters by PCR and oligoprobe hybridization. Appl. Environ. Microbiol. 62: 2074.
61. Jiang, X., Matson, D. O., Cubitt, W. D., and Estes, M. K. 1996. Genetic and antigenic diversity of human caliciviruses (HuCVs) using RT-PCR and new EIAs. Arch. Virol. Suppl. 12: 251.
62. Johnson, K. M., Cooper, R. C., and Straube, D. C. 1981. Procedure for recovery of enteroviruses from the Japanese cockle *Tapes japonica*. Appl. Environ. Microbiol. 41: 932
63. Jothikumar, N., Cliver, D. O., and Mariam, T. W. 1998. Immunomagnetic capture PCR for rapid concentration and detection of hepatitis A virus from environmental samples. Appl. Environ. Microbiol. 64: 504.
64. Kapikian, A. Z., Wyatt, R. G., Dolin, R., Thornhill, T. S., Kalica, A. R., and Chanock, R. M. 1972. Visualization by immune electron microscopy of a 27-nm particle associated with acute infectious nonbacterial gastroenteritis. J. Virol. 10: 1075.
65. Kogasaka, R., Nakamura, S., Chiba, C., Sakuma, Y., Terashima, H., Yokoyama, T., and Nakao, T. 1981. The 33- to 39-nm virus-like particles, tentatively designated Sapporo agent, associated with an outbreak of acute gastroenteritis. J. Med. Virol. 8:187.
66. Konowalchuk, J., and Speirs, J. I. 1975. Survival of enteric viruses on fresh vegetables. J. Milk Food Technol. 38: 469.
67. Konowalchuk, J., and Speirs, J. I. 1977. Virus detection on grapes. Can. J. Microbiol. 23: 1301.
68. Kostenbader, K. D. Jr., and Cliver, D. O. 1972. Polyelectrolyte flocculation as an aid to recovery of enteroviruses from oysters. Appl. Microbiol. 24: 540.
69. Kostenbader, K. D. Jr., and Cliver, D. O. 1973. Filtration methods for recovering enteroviruses from foods. Appl. Microbiol. 26: 149.
70. Kostenbader, K. D. Jr., and Cliver, D. O. 1977. Quest for viruses associated with our food supply. J. Food Sci. 42: 1253.
71. Kostenbader, K. D. Jr., and Cliver, D. O. 1981. Flocculants for recovery of food-borne viruses. Appl. Environ. Microbiol. 41: 318.
72. Landry, E. F., Vaughn, J. M., and Vicale, T. J. 1980. Modified procedure for extraction of poliovirus from naturally-infected oysters using Cat-Floc and beef extract. J. Food Prot. 43: 91.
73. Larkin, E. P. 1982. Viruses in wastewater sludges and in effluents used for irrigation. Environ. Internat. 2:29.
74. LeBaron, C. W., Furutan, N. P., Lew, J. F., Allen, J. R., Gouvea, V., Moe, C., and Monroe, S. S. 1990. Viral agents of gastroenteritis: Public health importance and outbreak management. Morbid. Mortal. Wkly Rept. 39 (No. RR-5): 1.
75. Lees, D. N., Henshilwood, K., and Dore, W. J. 1994. Development of a method for detection of enteroviruses in shellfish by PCR with poliovirus as a model. Appl. Environ. Microbiol. 60:2999.
76. Lees, D. N., Henshilwood, K., Green, J., Gallimore, C. I., and Brown, D. W. 1995. Detection of small round structured viruses in shellfish by reverse transcription-PCR. Appl. Environ. Microbiol. 61: 4418.
77. Le Gall-Recule, G., and Jestin, V. 1995. Production of digoxigenin-labeled DNA probes for detection of Muscovy duck parvovirus. Mol. Cell. Probes 9: 39.
78. Le Guyader, F., Aparie-Marchais, V., Brillet, J., and Billaudel, S. 1993. Use of genomic probes to detect hepatitis A virus and enterovirus RNAs in wild shellfish and relationship of viral contamination to bacterial contamination. Appl. Environ. Microbiol. 59: 3963.
79. Le Guyader, F., Dubois, E., Menard, D., and Pommepuy, M. 1994. Detection of hepatitis A virus, rotavirus, and enterovirus in naturally contaminated shellfish and sediment by reverse transcription-seminested PCR. Appl. Environ. Microbiol. 60: 3665.
80. Le Guyader, F., Neill, F. H., Estes, M. K., Monroe, S. S., Ando, T., and Atmar, R. L. 1996. Detection and analysis of a small round-structured virus strain in oysters implicated in an outbreak of acute gastroenteritis. Appl. Environ. Microbiol. 62: 4268.
81. Lemon, S.M., Binn, L.N., and Marchwicki, R.H. 1983. Radioimmunofocus assay for quantitation of hepatitis A virus in cell cultures. J. Clin. Microbiol. 17: 834.
82. Lewis, G. D., and Metcalf, T. G. 1988. Polyethylene glycol precipitation for recovery of pathogenic viruses, including hepatitis A virus and human rotavirus, from oyster, water, and sediment samples. Appl. Environ. Microbiol. 54: 1983.
83. Lim, K. A., and Benyesh-Melnick. 1960. Typing of viruses by combinations of antiserum pools. Application to typing of enteroviruses (coxsackie and echo). J. Immunol. 84: 309.
84. Liu, S., Birch, C., Coulepis, A., and Gust, I. 1984. Radioimmunofocus assay for detection and quantitation of human rotavirus. J. Clin. Microbiol. 20: 347.
85. López-Sabater, E. L., Deng, M. Y., and Cliver, D. O. 1997. Magnetic immunoseparation PCR assay (MIPA) for detection of hepatitis A virus (HAV) in American oyster (*Crassostrea virginica*). Lett. Appl. Microbiol. 24: 101.
86. Maidaniuk, A.G., Tiunnikov, G.I., Konakova, V.E., Bondarenko, E.P., Nemtsov, I.V., Karpovich, L.G., Kalashnikova, T.V., Mamaeva, N.V., Sosnovtsev, S.V., and Chizhikov, V.E. 1993. The isolation and study of the characteristics of a cytopathic strain of the hepatitis A virus. Vopr. Virusol. 38: 101.
87. Melnick, J. L., and Gerba, C. P. 1980. The ecology of enteroviruses in natural waters. CRC Critical Rev. Environ.Control 10: 65.
88. Metcalf, T. G. 1978. Indicators of viruses in shellfish. In "Indicators of viruses in water and food," ed. G. Berg, p. 383. Ann Arbor Science Publishers, Ann Arbor, MI.
89. Metcalf, T. G., Eckerson, D., and Moulton, E. 1980a. A method for recovery of viruses from oysters and hard and soft shell clams. J. Food Prot. 43: 89.
90. Metcalf, T. G., Moulton, E., and Eckerson, D. 1980b. Improved method and test strategy for recovery of enteric viruses from shellfish. Appl. Environ. Microbiol. 39: 141.
91. Morse, D. L., Guzewich, J. J., Hanrahan, J. P., Stricof, R., Shayegani, M., Deibel, R., Grabau, J. C., Nowak, N. A., Herrmann, J. E., Cukor, G., and Blacklow, N. R. 1986. Widespread outbreaks of clam- and oyster-associated gastroenteritis. Role of Norwalk virus. N. Engl. J. Med. 314: 678.
92. Nakata, S., Kogawa, K., Numata, K., Ukae, S., Adachi, N., Matson, D. O., Estes, M. K., and Chiba, S. 1996. The epidemiology of human calicivirus/Sapporo/82/Japan. Arch. Virol. Suppl. 12: 263.
93. Niu, M. T., Polish, L. B., Robertson, B. H., Khanna, B. K., Woodruff, B. A., Shapiro, C. N., Miller, M. A., Smith, J. D., Gedrose, J. K., and Alter, M. J. 1992. Multistate outbreak of hepatitis A associated with frozen strawberries. J. Infect. Dis. 166: 518.
94. Noel, J.S., Liu, B.L., Humphrey, C.D., Rodriguez, E.M., Lambden, P.R., Clarke, I.N., Dwyer, D.M., Ando, T., Glass, R.I., and Monroe, S.S. 1997. Parkville virus: a novel genetic variant of human calicivirus in the Sapporo virus clade, associated with an outbreak of gastroenteritis in adults. J. Med. Virol. 52: 173.
95. Oglesbee, S. E., Wait, D. A., and Meinhold, A. F. 1981. MA104: A continuous cell line for the isolation of enteric viruses from environmental samples. Ann. Mtg. Am. Soc. Microbiol. Abstr. A113.
96. Oshiro, L. S., Haley, C. E., Roberto, R. R., Riggs, J. L., Croughlan, M., Greenberg, H., and Kapikian, A. Z. 1981. A 27-nm virus isolated during an outbreak of acute infectious nonbacterial gastroenteritis in a convalescent hospital: a possible new serotype. J. Infect. Dis. 143: 791.
97. Patterson, W., Haswell, P., Fryers, P. T., and Green, J. 1997. Outbreak of small round structured virus gastroenteritis arose after kitchen worker vomited. Commun. Dis. Rpt. CDC Rev. 7: R101.
98. Patterson, T., Hutchings, P., and Palmer, S. 1997. Outbreak of SRSV gastroenteritis at an international conference traced to food handling by a post-symptomatic caterer. Epidemiol. Infect. 111: 157.
99. Pollard, J.W. and Walker, J.M. 1997. Methods in Molecular Biology. Basic Cell Culture Protocols. 2nd ed, eds. J.W. Pollard and J.M. Walker, 489 pp., Humana Press, Totowa, New Jersey.
100. Ramia, S., and Sattar, S. A. 1979. Simian rotavirus SA-11 plaque formation in the presence of trypsin. J. Clin. Microbiol. 10: 609.
101. Ramia, S., and Sattar, S. A. 1980a. Proteolytic enzymes and rotavirus SA-11 plaque formation. Can. J. Comp. Med. 44: 232.

102. Ramia, S. and Sattar, S. A. 1980b. The role of trypsin in plaque formation by simian rotavirus SA-11. Can. J. Comp. Med. 44: 433.
103. Reiner, P., Reinerova, M. and Veselovska, Z. 1992. Comparison of two defective hepatitis A virus strains adapted to cell culture. Acta. Virol. 36: 245.
104. Richards, G. P. 1985. Outbreaks of shellfish-associated enteric virus illness in the United States: requisite for development of viral guidelines. J. Food. Prot. 48: 815.
105. Richards, G. P. 1987. Shellfish-associated enteric virus illness in the United States, 1934-1984. Estuaries 10: 84.
106. Richards, G. P. 1988. Microbial purification of shellfish: a review of depuration and relaying. J. Food. Prot. 51: 218.
107. Richards, G. P. 1991. Shellfish depuration. In "Microbiology of Marine Food Products," ed. D. R. Ward and C. Hackney. p. 395. Van Nostrand Reinhold, New York, NY.
108. Richards, G. P. 1999. Limitations of molecular biological techniques for assessing the virological safety of foods. J. Food Prot. (In Press)
109. Richards, G. P., and Weinheimer, D. A. 1985. Influence of adsorption time, rocking and soluble proteins on the plaque assay of monodispersed poliovirus. Appl. Environ. Microbiol. 49: 744.
110. Richards, G. P., Goldmintz, D., Green, D. L., and Babinchak, J. A. 1982. Rapid methods for the extraction and concentration of poliovirus from oyster tissues. J. Virol. Meth. 5: 285.
111. Riggs, J. L. 1979. Immunofluorescent staining. In "Diagnostic Procedures for Viral, Rickettsial and Chlamydial Infections," 5th ed., ed. E. H. Lennette and N. J. Schmidt., p. 141. Amer. Pub. Health Assn., Washington, DC.
112. Robertson, B. H., and Lemon, S. M. 1998. Hepatitis A and E, In "Topley and Wilson's Microbiology and Microbial Infections, Vol.1, Virology," 9th ed., eds. B. W. J. Mahy and L. Collier, p. 693, Oxford University Press, New York.
113. Robertson, B.H., Khanna, B., Brown, V.K., and Margolis, H.S. 1988. Large scale production of hepatitis A virus in cell culture: effect of type of infection on virus yield and cell integrity. J. Gen. Virol. 69: 2129.
114. Rodriguez, W. J. 1989. Viral enteritis in the 1980s: perspective diagnosis and outlook for prevention. Pediatr. Infect. Dis. J. 8: 570.
115. Romalde, J. L. 1996. New molecular methods for the detection of hepatitis A and Norwalk viruses in shellfish. Microbiologia 12: 547.
116. Sadovski, A. Y., Fattal, B., Goldberg, D., Katzenelson, E., and Shuval, H. I. 1978. High levels of microbial contamination of vegetables irrigated with wastewater by the drip method. Appl. Environ. Microbiol. 36: 824.
117. Safferman, R. S., ed. 1993. Chapter 10. Cell culture procedures for assaying plaqueforming viruses, and Chapter 11. Virus plaque confirmation procedures. *In* Manual of Methods for Virology—EPA publication EPA/600/484/013, U.S. Environmental Protection Agency, Washington, DC.
118. Santos, N., and Gouvea, V. 1994. Improved method for purification of viral RNA from fecal specimens for rotavirus detection. J. Virol. Methods 46:11.
119. Schmidt, N. J., Ho, H. H., and Lennette, E. H. 1975. Propagation and isolation of group A coxsackievirus in RD cells. J. Clin Microbiol. 2: 183.
120. Schmidt, N. J., Ho, H. H., and Lennette, E. H. 1976. Comparative sensitivity of the BGM cell line for the isolation of enteric viruses. Health Lab. Sci. 13: 115.
121. Schmidt, N. J., Ho, H. H., Riggs, J. L., and Lennette, E. H. 1978. Comparative sensitivity of various cell culture systems for isolation of viruses from wastewater and fecal samples. Appl. Environ. Microbiol. 36: 480.
122. Schultz, D. E., Honda, M., Whetter, L. E., McKnight, K. L., and Lemon, S. M. 1996. Mutations within the 5' nontranslated RNA of cell culture-adapted hepatitis A virus which enhance cap-independent translation in cultured African green monkey kidney cells. J. Virol. 70: 1041.
123. Schwab, K. J., De Leon, R., and Sobsey, M. D. 1995. Concentration and purification of beef extract mock eluates from water samples for the detection of enteroviruses, hepatitis A virus, and Norwalk virus by reverse transcription-PCR. Appl. Environ. Microbiol. 61: 531.
124. Severini, GM., Mestroni, L., Falaschi, A., Camerini, F., and Giacca, M. 1993. Nested polymerase chain reaction for high-sensitivity detection of enteroviral RNA in biological samples. J. Clin. Microbiol. 31: 1345.
125. Simmons, A. 1998. "Virus infections in immunocompromised patients, In "Topley and Wilson's Microbiology and Microbial Infections, Vol. 1, Virology, 9th ed," eds. B. J. W. Mahy and L. Collier, p. 917, Oxford University Press, New York.
126. Smith, A. W., Skilling, D. E., Cherry, N., Mead, J. H., and Matson, D. O. 1998. Calicivirus emergence from ocean reservoirs: zoonotic and interspecies movements. Emerg. Infect. Dis. 4: 13.
127. Smith, E. M., Estes, M. K., Graham, D. Y., and Gerba, C. P. 1979. A plaque assay for simian rotavirus SA11. J. Gen. Virol. 43: 513.
128. Smorodintsev, A. A. 1958. Tick-borne spring-summer encephalitis. Progr. Med. Virol. 1: 210.
129. Sobsey, M. D. 1976. Field monitoring techniques and data analysis. In "Virus Aspects of Applying Municipal Wastes to Land." eds. L. B. Baldwin, J. M. Davidson and J. F. Gerber, p. 87. University of Florida, Gainesville.
130. Sobsey, M. D., Carrick, R. L., and Jensen, R. H. 1978. Improved methods for detecting enteric viruses in oysters. Appl. Environ. Microbiol. 36: 121.
131. Speirs, J. I., Pontefract, R. D., and Harwig, J. 1987. Methods for recovering poliovirus and rotavirus from oysters. Appl. Environ. Microbiol. 53: 2666.
132. Stolle, A., and Sperner, B. 1997. Viral infections transmitted by food of animal origin: the present situation in the European Union. Arch. Virol. Suppl. 13: 219.
133. Stone, T., and Durrant, I. 1996. Hybridization of horseradish peroxidase-labeled probes and detection by enhanced chemiluminescence. Mol. Biotechnol. 6: 69.
134. Sullivan, R., and Read, R. B., Jr. 1968. Method for recovery of viruses from milk and milk products. J. Dairy Sci. 51: 1748.
135. Sullivan, R., Peeler, J. T., and Larkin, E. P. 1986. A method for recovery of poliovirus 1 from a variety of foods. J. Food Prot. 49: 226.
136. Thornhill, T. S., Wyatt, R. G., Kalica, A. R., Dolin, R., Chanock, R. M., and Kapikian, A. Z. 1977. Detection by immune electron microscopy of 26- to 27-nm virus-like particles associated with two family outbreaks of gastroenteritis. J. Infect. Dis. 135: 20.
137. Tierney, J. T., Sullivan, R., and Larkin, E. P. 1977. Persistence of poliovirus 1 in soil and on vegetables grown in soil previously flooded with inoculated sewage sludge or effluent. Appl. Environ. Microbiol. 33: 109.
138. Tierney, J. T., Fassolitis, A. C., Van Donsel, D., Rao, V. C., Sullivan, R., and Larkin, E. P. 1980. Glass wool-hydroextraction method for recovery of human enteroviruses from shellfish. J. Food Prot. 43: 102.
139. Tierney, J. T., Sullivan, P., Peeler, J. T., and Larkin, E. P. 1985. Detection of low numbers of poliovirus 1 in oysters: Collaborative study. J. Assn. Off. Anal. Chem. 68: 884.
140. Traore, O., Arnal, C., Mignotte, B., Maul, A., Laveran, H., Billaudel, S., and Schwartzbrod, L. 1998. Reverse transcriptase PCR detection of astrovirus, hepatitis A virus, and poliovirus in experimentally contaminated mussels: comparison of several extraction and concentration methods. Appl. Environ. Microbiol. 64: 3118.
141. Vaughn, J. M., Landry, E. F., Vicale, T. J., and Dahl, M. C. 1979. Modified procedure for the recovery of naturally accummulated poliovirus from oysters. Appl. Environ. Microbiol. 38: 594.
142. Wadell, G. 1984. Molecular epidemiology of human adenoviruses. Curr. Top. Microbiol. Immunol. 110: 191.
143. Warburton, A. R., Wreghitt, T. G., Ramplng, A., Buttery, R., Ward, K. N., Perry, K. R., and Parry, J. V. 1991. Hepatitis A virus outbreak involving bread. Epidemiol. Infect. 106: 199.
144. Watson, P. G., Inglis, J. M., and Anderson, K. J. 1980. Viral content of a sewage-polluted intertidal zone. J. Infect. 2: 237.
145. Weltman, A. C., Bennett, N. M., Ackman, D. A., Misage, J. H., Campana, J. J., Fine, L. S., Doniger, A. S., Balzano, G. J., and Birkhead,

G. S. 1996. An outbreak of hepatitis A associated with a bakery, New York, 1994: the 1968 "West Branch, Michigan" outbreak repeated. Epidemiol. Infect. 117: 333.

146. Wheeler, C. M., Fields, H. A., Schable, C. A., Meinke, W. J., and Maynard, J. E. 1986. Adsorption, purification, and growth characteristics of hepatitis A virus strain HAS-15 propagated in fetal rhesus monkey kidney cells. J. Clin. Microbiol. 23: 434.
147. Whitby, K., and Garson, J. A. 1995. Optimisation and evaluation of a quantitative chemiluminescent polymerase chain reaction assay for hepatitis A virus RNA. J. Virol. Methods 51: 75.
148. White, K. E., Osterholm, M. T., Mariotti, J. A., Korlath, J. A., Lawrence, D. H., Ristinen, T. L., and Greenberg, H. B. 1986. A foodborne outbreak of Norwalk virus gastroenteritis: evidence for post recovery transmission. Am. J. Epidemiol. 124: 120.
149. Wyatt, R. G., Greenberg, H. B., Dalgard, D. W., Allen, W. P., Sly, D. L., Thornhill, T. S., Chanock, R. M., and Kapikian, A. Z. 1978. Experimental infection of chimpanzees with the Norwalk agent of epidemic viral gastroenteritis. J. Med. Virol. 2: 89.
150. Xu, Z. Y., Li, Z. H., Wang, J. X, Xiao, Z. P., and Dong, D. X. 1992. Ecology and prevention of a shellfish-associated hepatitis A epidemic in Shanghai, China. Vaccine 10 Suppl 1: S67.
151. Yang, F., and Xu, X. 1993. A new method of RNA preparation for detection of hepatitis A virus in environmental samples by the polymerase chain reaction. J. Virol. Methods 43: 77.
152. Zhang, H., Chao, S.F., Ping, L.H., Grace, K., Clarke, B., and Lemon, S.M. 1995. An infectious cDNA clone of a cytopathic hepatitis A virus: genomic regions associated with rapid replication and cytopathic effect. Virol. 212: 686.
153. Zhou, Y. J., Estes, M. K., Jiang, X., and Metcalf, T. G. 1991. Concentration and detection of hepatitis A virus and rotavirus from shellfish by hybridization tests. Appl. Environ. Microbiol. 57: 2963.
154. Zou, S., and Chaudhary, R. K. 1991. Kinetic study of the replication of a cell-culture-adapted hepatitis A virus. Res. Virol. 142: 381.

Chapter 45

Meat and Poultry Products

R.B. Tompkin, A.M. McNamara, and G.R. Acuff

45.1 INTRODUCTION

Raw red meats and poultry are derived from warm-blooded animals. Their microbial flora is heterogeneous and consists of mesophilic and psychrotrophic bacteria from the animal itself, soil and water bacteria from their environment, and bacterial species introduced by man and equipment during processing.[45,50,51]

The surface flora on freshly slaughtered carcasses, usually about 10^2 to 10^3 bacteria per cm^2, is primarily mesophilic, having originated from the gastrointestinal tract and external surfaces of the live animal. Contamination from the slaughtering environment is also largely mesophilic in nature since this process occurs in rooms ambient in the summer and heated in the winter. Psychrotrophic organisms originating from soil and water are present but usually only to about 10^1 per cm^2. Mesophiles are important because they indicate the degree of sanitation during the slaughtering process. However, since some growth of these bacteria can be expected during refrigerated storage, their usefulness as sanitation indicators diminishes rapidly over time.

The existence of viable bacteria within the deep tissues of freshly slaughtered livestock has been a debated issue.[40,67] The development of sour rounds of beef during slow chilling and spoilage near the bone of improperly processed country hams suggest that viable bacteria can be present within the deep tissue of freshly slaughtered carcasses. However, since sterile muscle tissue can be obtained easily,[41] it is probable that if bacteria are present within muscle tissues of healthy live animals, initial numbers are exceedingly low. Increasing use of mechanical tenderization, and other processes that may compromise the sealed nature of intact cuts of meat, may introduce microorganisms from the surface into the interior.

When red meats and poultry are cooked and subsequently refrigerated to deter spoilage, bacteria on the raw tissue are greatly reduced, leaving only sporeformers and, occasionally, small numbers of thermoduric bacteria, notably the enterococci, micrococci, and some lactobacilli. Because the post-processing environment frequently is refrigerated in federally inspected establishments, a low-level recontamination with psychrotrophic bacteria almost always occurs. Psychrotrophs are important because of their ability to grow even when products are stored at proper refrigeration temperatures. They ultimately cause spoilage and thus determine the shelf life of the product. There is also some concern about the growth of psychrotrophic pathogens, such as *Listeria monocytogenes* in refrigerated cooked products.

The level of coliforms has been commonly used as an indicator for hygienic conditions and the microbiological quality of processed meat and poultry products. For example, the presence of coliforms on the surface of properly cooked products indicates post-processing contamination. However, as cooked products are subsequently held in storage, the interpretation of coliform counts changes because the natural flora of meat and poultry plants may include certain coliforms capable of slow growth at refrigeration temperatures (e.g., 2° to 5°C).[9,10,49,68,71,78] Some contamination of cooked products will occur between cooking and packaging. Since multiplication of these psychrotrophic coliforms will occur during refrigerated storage, the coliform counts lose their significance as an indicator of the hygienic conditions existing during production. This is one of several factors that prevent coliform counts from being an effective indicator of the safety of refrigerated meat and poultry products.

Numerous factors influence the type of microbial spoilage that occur in fresh and processed meat and poultry products. These factors include the inherent pH of the meat; the addition of salt, nitrite, sugar, smoke (liquid or natural), or acidulants; and the state of the meat (heated, fermented, or dried). After processing, the type and rate of spoilage are influenced by the type of packaging, temperature of storage, final composition of the product, and surviving or contaminating microorganisms. The same basic principles apply equally to red meat and poultry products.

Historically, numerous distinct products evolved as attempts were made to prolong the quality of meat and poultry for future use and to add variety to the diet. This process has led to a wider variety of products than exists for most other commodities. In the following text, meat and poultry products are categorized as to whether they are raw or ready-to-eat, and then further subdivided into process categories. Please note that the term "cured" is used only when nitrite or nitrate is added to the products.

This chapter will address meat and poultry products commercially prepared under federal inspection and distributed to retail stores, or for use in food service establishments. Products subject to state inspection in food service establishments, retail outlets, homes, or on the farm also may be encountered by the analyst. Approximate microbial levels are mentioned throughout the text to introduce the reader to the subject. More specific information on microbial populations and microbial criteria is available.[28-31,33,34,50-52] The normal flora, spoilage flora, and microorganisms of public health concern will be discussed within each product category.

45.2 RAW MEAT AND POULTRY PRODUCTS

45.21 Raw Meat

The initial microflora on freshly slaughtered red meat carcasses is largely mesophilic in nature and cannot multiply at the temperatures used for carcass chilling and holding. Bacterial presence is limited almost exclusively to the exterior surface of carcasses unless the surface has been penetrated by utensils allowing transport of bacteria to interior muscle tissues. During refrigerated holding, the carcass flora begins to shift toward gram-negative psychrotrophs of the *Pseudomonas-Moraxella-Acinetobacter* group. Continued growth of this group of organisms in an aerobic environment can eventually spoil the meat.[8] Usually, carcasses are cut into smaller portions in refrigerated fabrication rooms. Most bacteria on processing equipment in refrigerated rooms are psychrotrophic in nature, further assuring the presence of these bacteria on the meat surfaces. Cuts are generally vacuum packaged in a high-oxygen barrier film before distribution, which results in the development of a facultative anaerobic or microaerophilic gram-positive bacterial population consisting primarily of lactic acid bacteria.[39]

Spoilage of whole cuts of meat at refrigeration temperatures is primarily a surface phenomenon resulting in formation of slime, off-odor, and possibly, some gas production. Shelf life of raw chilled meat is prolonged by those factors affecting the growth rate of the psychrotrophs: dry surface, low initial numbers of psychrotrophs, the inherent pH of the meat, oxygen limitation, and temperature.[20,42] Carcass decontamination technologies implemented to reduce the presence of pathogenic bacteria also are expected to lower initial numbers of spoilage bacteria and result in an overall increase in shelf life.[48] The microbial level of chilled meats after transportation and storage at the retail level may have little or no relationship to that at the processing level because bacterial growth will have continued.

The production of hydrogen sulfide, sometimes with green discoloration, is a defect that can occur in vacuum packed fresh meats. A variety of bacteria (e.g., *Lactobacillus sake, Pseudomonas mephitica , Enterobacter liquifaciens*) have been associated with this type of spoilage.[24,26,47,64,72,80]

Psychrotrophic clostridia (e.g., *C. laramie, C. estertheticum, C. algidicarnis*) capable of germination and/or growth at 2°C or below have been reported to cause spoilage of vacuum-packaged raw and cooked meats. A variety of offensive odors may be present upon opening the packages with the predominant aromas being dependent upon the clostridial species. The packages may or may not be distended with gas depending on the type of clostridia. Two groups of clostridia have been proposed on the basis of their optimum growth temperature: group 1 (15 to 20°C), group 2 (25 to 30°C).[12] Isolation and recovery of the clostridia may require the use of special methods. The reader should refer to the literature for guidance.[11,12,19,21,56-58,63]

Pathogenic *Escherichia coli, Campylobacter jejuni/coli, Staphylococcus aureus, C. perfringens, L. monocytogenes,* and *Salmonella* may be present on fresh red meat tissues because the slaughtering process does not include a bactericidal step sufficient for assured elimination. The frequency and levels of these bacteria on freshly slaughtered animal carcasses will vary, depending upon climatic, farm, livestock transport, stockyard, and processing conditions. Data collected by the U.S. Food Safety and Inspection Service (FSIS)[28] in a nationwide microbiological baseline data collection program for carcasses produced from steers and heifers reported carcasses positive for *C. perfringens* at 2.6%, *S. aureus* at 4.2%, *L. monocytogenes* at 4.1%, *C. jejuni/coli* at 4.0%, *E. coli* O157:H7 at 0.2%, and *Salmonella* at 1.0%. Data collected for cows and bulls[29] were similar, reporting low percentages of pathogens on carcass surfaces. Nationwide data collected on market hog carcasses[32] reported carcasses positive for *C. perfringens* at 10.4%, *S. aureus* at 16.0%, *L. monocytogenes* at 7.4%, *C. jejuni/coli* at 31.5%, *E. coli* O157:H7 at 0.0%, and *Salmonella* at 8.7%. FSIS also collected raw ground beef samples[31] from facilities subject to Federal inspection and reported *C. perfringens* in 53.5%, *S. aureus* in 30%, *L. monocytogenes* in 11.7%, *C. jejuni/coli* in 0.002%, *E. coli* O157:H7 in 0.0%, and *Salmonella* in 7.5% of samples examined.

The presence of pathogens in raw meat is an obvious public health concern. Although efforts are being made to reduce this contamination, none of the currently available procedures can assure the provision of pathogen-free raw meat. Thus, for the foreseeable future, proper handling of raw meat is essential to prevent foodborne illness caused by indigenous pathogenic bacteria. Critical control points for preventing foodborne illness include eliminating cross-contamination from raw products to ready-to-eat foods, using adequate cooking times and temperatures, avoiding recontamination after cooking by disinfecting surfaces previously contaminated by raw meat, and properly chilling and storing meat after cooking.

45.22 Raw Poultry

The internal muscle tissues of healthy poultry are essentially free of bacteria and contamination is limited to exterior surfaces. After evisceration, poultry carcasses are chilled quickly by a variety of procedures, including immersion chilling, spray chilling, air chilling, slush ice chilling, and carbon dioxide chilling systems. Initial contamination consists primarily of mesophilic bacteria and has been reported to include *Achromobacter, Aerobacter, Alcaligenes, Bacillus, Corynebacterium, Escherichia, Flavobacterium, Lactobacillus, Micrococcus, Proteus, Pseudomonas, Staphylococcus, Streptococcus, Listeria, Campylobacter* and *Salmonella*.[6,46,65,66] Initial numbers range from 10^1 to 10^4 bacteria per cm^2 in various reports. The microflora eventually is predominated during aerobic refrigerated storage by a gram-negative psychrotrophic microflora consisting of the *Pseudomonas-Moraxella-Acinetobacter* group. Packaging of the product in high-oxygen barrier film results in a microflora dominated by lactic acid bacteria.

Storage of fresh poultry products and transport to retail markets is conducted under refrigerated temperatures that can allow surfaces to freeze. This results in very little growth of spoilage bacteria or pathogens. Growth of spoilage bacteria occurs throughout display and subsequent handling and storage by the consumer. Cold storage warehouses normally store frozen poultry at temperatures around –20°C or lower. Shelf life may be maintained easily for six months to a year at these temperatures as growth of spoilage bacteria is halted.[69]

Data collected by FSIS in a nationwide microbiological baseline data collection program for broiler chicken carcasses[30] reported carcasses positive for *C. perfringens* at 42.9%, *S. aureus* at 64.0%, *L. monocytogenes* at 15.0%, *C. jejuni/coli* at 88.2%, *E. coli* O157:H7 at 0.0%, and *Salmonella* at 20.0%. FSIS also collected raw ground chicken samples from plants under federal inspection and reported estimated nationwide levels of *C. perfringens* of 50.6%, *S. aureus* of 90%, *L. monocytogenes* of 41.1%, *C. jejuni/coli* of 59.8%, *E. coli* O157:H7 of 0.0%, and *Salmonella* of 44.6%.[33] Nationwide data collected on raw, young turkey carcasses reported carcasses positive for *C. perfringens* at 29.2%, *S. aureus* at 66.7%, *L. monocytogenes* at 5.9%, *C. jejuni/coli* at 90.3%, *E. coli* O157:H7 at 0.0%, and *Salmo-*

nella at 18.6%.[37] Raw ground turkey samples collected from plants under federal inspection yielded estimated nationwide levels of *C. perfringens* at 28.1%, *S. aureus* at 57.3%, *L. monocytogenes* at 30.5%, *C. jejuni/coli* at 25.4%, *E. coli* O157:H7 at 0.0%, and *Salmonella* at 49.9%.[34]

The presence of pathogens in raw poultry continues to be a public health concern. Irradiation is the only process available, at the present time, that can ensure raw poultry free of enteric pathogens such as salmonellae and *C. jejuni/coli*. Thus, in order to prevent foodborne illness, proper handling of raw poultry should include eliminating cross-contamination from raw products to ready-to-eat foods, using adequate cooking times and temperatures, avoiding recontamination after cooking by disinfecting surfaces previously contaminated with raw tissue, and properly chilling and storing poultry after cooking.

45.23 Shelf-Stable Raw Salted and Salted-Cured Products

This category of meat products was, of necessity, extremely common in the early history of the U.S. before refrigeration was available. The salting of beef, mutton, and pork in barrels is no longer practiced in the U.S., but many raw salted products are still prepared and sold, particularly in regions with predominantly agricultural histories. Salt pork, dry-cured bacon, and country-cured hams continue to be produced in volume today. Basic manufacturing steps are similar, although innumerable variations exist in processing and spicing. Salt pork, salt bacon, and salt hams are prepared by coating the meat with salt only and storing below 10°C in bins. During this period, water is drawn from the product by the salt. At intervals, the pieces of meat are recoated with salt. At the end of the salting period, the product contains high levels of salt, and is racked and held at ambient temperatures until the surface dries. It is then rubbed with a thin coating of salt and spices (where used), netted, and sold. The salt content of these products is sufficiently high that subsequent refrigeration is not necessary. Dry-cured hams and bacon frequently sold as country-cured are processed similarly except that nitrite and/or nitrate is used along with a lower salt content. The product is hung to dry in ambient temperature rooms for 35 to 140 days before preparation for sale as shelf-stable meat.

During the initial refrigerated treatment of these processes, the salt and curing agents penetrate and equilibrate in the tissue. Nonsporeforming bacteria of public health concern are subjected to stress or rendered nonviable. During the subsequent drying period, salt-tolerant micrococci and enterococci grow and appear to render the products even more refractile to the growth of microorganisms of public health concern (e.g., *S. aureus*). The nitrite- or nitrate-cured products must be held in the dry room for a specified period to destroy trichinellae. Although almost all U.S. consumers cook these products before eating them, dry-cured hams have been consumed raw in Europe for hundreds of years. Since it is likely that a small segment of the population will not cook these products, the trichinellae control requirement is necessary.

Commercially prepared, raw salted and salted-cured meats have an enviable public health record in the U.S. Local health officials frequently are concerned about these products being sold at ambient temperatures. However, if the products do not bear the "Keep Refrigerated" label, it is likely that they already have been held at elevated temperatures during drying for 3 to 6 months, in which case, the concern is unjustified. It is unlikely that a microbiological analysis of raw salted meats will yield bacteria of public health concern. Bone taint, a type of deep tissue spoilage occurring near the bone, is attributed to improper initial salt equalization.[70] Massive levels of micrococci and enterococci are usually encountered and have been confused with staphylococci when inappropriate methods of examination were used. Mold may develop on these products, particularly during storage in humid conditions.

45.24 Perishable Raw Salted and Salted-Cured Products

Fresh pork sausage, fresh turkey sausage, chorizo, bratwurst, Polish sausage, and Italian sausage are the perishable raw salted meat products encountered most commonly. These products are also sold as perishable cooked products. The method of packaging is the major factor determining the predominant spoilage flora. Fresh sausage sold in bulk form on trays or stuffed into edible casings has a relatively short shelf life (e.g., 7–21 days). The spoilage flora consists of a variety of psychrotrophs, including pseudomonads. Fresh sausage sold in oxygen-impermeable chubs has a longer refrigerated shelf life (a few weeks). The restriction of oxygen leads to a predominant flora of lactic acid-producing bacteria, which impart a tangy flavor. When prepared using hot hog sausage technology (pre-rigor meat), chubs of pork sausage packaged in an oxygen impermeable film can keep for about 2 months.

Large quantities of perishable red meat and poultry products are cured with solutions of nitrite and/or nitrate salts, sodium chloride, and cure accelerators such as sodium ascorbate, sugar, and flavoring materials. More cured meats are cooked before shipping from processing firms; however, some are sold with little or no heat treatment and must be cooked before eating. Examples of some raw cured meats sold in volume include uncooked hams, bacon, and corned beef. The initial microbial content of these products is identical to the microflora of fresh meat until the curing salts are applied. During refrigerated storage, often under vacuum, the predominant normal flora of the raw meat remains constant or decreases as a consequence of the curing ingredients. Psychrotrophic lactic acid bacteria, enterococci, micrococci, and yeasts grow slowly because they are less inhibited by the salt and curing compounds. During cold storage they become the predominant microbial flora of these meats and, in time, will cause the product to spoil.

Some raw cured meats are heated or smoked to produce dry surfaces and a smoked flavor; these procedures may extend shelf-life. Heat treatments to 54.5°C, with or without smoke deposition on surfaces, are common for bacon. Both the heat treatment and smoking procedures cause microbial reductions. Following vacuum packaging, psychrotrophic lactic acid bacteria predominate as the cause of spoilage during cold storage. If bacon is packaged loosely in bulk quantities without vacuum or if leakers occur in vacuum packages, a more heterogeneous flora of bacteria, yeasts, and molds develops and causes spoilage.

Perishable raw salted and salted-cured meats are infrequently associated with foodborne illnesses, probably because consumers understand that these products require refrigeration. Traditionally, they are cooked well before serving.

45.25 Marinated Products

Marinade is a sauce or mixture of water, spices, salt, tenderizers, and often acid, used to improve the flavor and/or tenderize raw meat or poultry. If an acidic or alkaline solution is used for marinade, the pH of the tissue is altered and may positively affect the shelf life of the product.[14] The addition of certain spices with natural antimicrobial activity may improve shelf life.

Marinated products for retail sale are usually packaged in high-oxygen barrier packaging and stored refrigerated. The lack of oxygen and addition of acid to the marinade selects for domination of the microflora by lactic acid bacteria. Spoilage is accompanied occasionally by gas production and, in some cases, sulfide odors produced by *Hafnia alvei* and *Enterobacter liquefaciens* (formerly *Serratia liquifaciens*) have been reported.[55]

45.26 Raw Breaded Products

Battered and breaded meat and poultry products have a long history of popularity with consumers, and coating seafood, poultry, meat, and vegetables with batter and/or breading is a common practice in homes and food service operations. Recently, an increase in the production of raw breaded meat and poultry products has been noted in the processing industry. Batter commonly consists of a thick mixture of flour, milk, water and eggs, and may also contain seasonings. Breading usually consists of a mixture of dry, coarse ingredients, typically including flour, starch and seasonings. Fresh or restructured pieces of meat are dipped in or coated with batter, then covered with breading. Breaded products may be frozen immediately or given a short initial cook to set the batter, ensuring good adhesion to the piece of meat, followed by freezing. This breaded, frozen product is either raw or slightly cooked on the exterior surfaces, and raw in the interior. Bacteria will be present on the raw meat products and also will be introduced by the batter and breading. However, since these products will most likely be stored frozen, no bacterial growth is expected. Cooking by various methods will occur before consumption, so there should be no significant health hazard unless the product is temperature-abused. Obviously, inadequate heat treatment or mishandling of the product before consumption may lead to a hazardous product.

45.3 READY-TO-EAT MEAT AND POULTRY PRODUCTS

45.31 Perishable Cooked Uncured Products

The microbiology of cooked meats begins with the raw materials and the cooking process. Most cooked pork and poultry products are cooked to sufficiently high temperatures so that only spores survive. Beef products are cooked to lower temperatures which destroy nonsporeforming pathogens, but not necessarily some of the thermoduric bacteria, such as the enterococci and spores. Internal microbial levels in cooked meats depend on the microbial levels and types before heating, the thermal process, and the subsequent holding time and temperatures. Freshly prepared cooked uncured meats normally have 35°C plate counts of 10^2 or less per gram. During subsequent handling, packaging, or serving of cooked products contamination occurs on the surface of the products from equipment and food handlers. In addition, some cooked products have their surfaces coated with spices, herbs, and other flavorings before final packaging. Contamination may add coliforms at levels of 10^1 to 10^2 per gram to the surface of the product. The presence of *E. coli* in these products indicates unsanitary conditions and warrants investigation of the manufacturing facility to determine the source of *E. coli* and recommend corrective measures. Human contact with cooked food, as in slicing or deboning, can add *S. aureus*. Such levels are harmless, but sufficient for growth to hazardous levels if subsequent conditions of time-temperature abuse occur. Thus, for example, proper refrigeration of salads containing cooked boned meat and poultry is essential to prevent the growth of *S. aureus* and enterotoxin production.

Cooked uncured meat products are ideal menstrua for microbial growth for they are highly nutritious, have a favorable pH, and are normally low in salt content. Given time and favorable temperatures, contaminants, including pathogens, will grow rapidly. Recognizing this fact, most regulations prohibit holding precooked meats between 5 and 60°C except during necessary periods of preparation, heating, or chilling. Many such foods are frozen for shipment and distribution. If the product does not thaw in commerce, microbial numbers at retail will relate fairly well to numbers present shortly after the product was frozen. Such foods, held too long above the freezing point, will spoil from growth of a wide variety of organisms, including enterococci, pseudomonads, lactic acid bacteria, psychrotrophic coliforms, and yeasts. The type of spoilage flora will be influenced by factors such as packaging and temperature.

Adding uncooked ingredients such as celery, spices, cheese, sauces, or gravies to a cooked meat product changes the microbial composition so that the above estimates do not apply. A full description of formulation and processing is essential for proper interpretation of laboratory data.

Foods in which the level of *S. aureus* or *C. perfringens* has reached 10^6 per gram may cause illness. Cooked uncured meat, poultry products, and gravy have historically been associated with *C. perfringens* outbreaks. The mere presence of salmonellae is potentially hazardous. Botulism rarely has been a problem in commercially uncured precooked meats in North America.[16,18,81] Products of this category have been involved in foodborne illness when there has been a serious departure from good practices in preparing, holding, and serving in homes, restaurants, and institutions.

Certain cooked products (e.g., roast beef, cooked pork, or turkey breast) have been associated with spoilage from psychrotrophic clostridia. The products usually have been cooked in plastic film and subsequently stored at normal refrigeration temperatures with no evidence of temperature abuse. The spoilage is proteolytic in nature and may not be detected until the packaging is removed. More than one type of *Clostridium* appears to be involved. Attempts to detect toxin production have been negative. The source of the clostridia is likely the raw meat or poultry, where similar spoilage has been reported.[11,12,21,56-58,63]

45.32 Perishable Cooked Cured Products

Precooked cured meats include franks, bologna, ham, and a wide variety of luncheon meats made from red and/or poultry meat. The heating step applied to these products destroys the normal raw meat microflora with the exception of spores and, possibly, thermoduric bacteria. During chilling and preparation for packaging, some contamination will occur on exposed surfaces. Salt, nitrite, and other inhibitors (e.g., sodium diacetate, sodium lactate, buffered citric acid) in the meat limit the growth of survivors and contaminants somewhat selectively. Upon prolonged refrigeration, lactic acid bacteria, micrococci, enterococci, molds, and yeasts may grow and cause spoilage. When packaged in an oxygen-impermeable package, the spoilage flora at 5°C and below consists primarily of lactic acid bacteria.[1,60] The presence of loose or gassy vacuum packages is normally due to heterofermentative lactic acid bacteria. Products of bacterial action sometimes react with meat pigments to form a green color.[44] Other forms of spoilage that may lead to laboratory examination include small areas of yellow discoloration on luncheon meat caused by a heat resistant strain of *Streptococcus*[87] and the formation of slime by *Leuconostoc* spp on products formulated with sucrose.

After packaging, these products normally have counts of 10^3 or less per gram; higher levels in products from retail cases reflect the time-temperature history of the product. Under proper refrigeration, such meats do not support the growth of mesophilic pathogens, so high aerobic plate counts are unrelated to health hazard. Coliforms, present as unavoidable contaminants at low levels ($\pm 10^1$ per gram), can grow slowly in refrigerated products if the coliforms are psychrotrophic. Human contact could introduce pathogens, but these products rarely have been implicated as vehicles of foodborne illness.

The risk of staphylococcal food poisoning from commercially packaged products is very low because *S. aureus* does not grow as well anaerobically in the presence of salt and nitrite (luncheon meats usually are packaged under vacuum or in a modified atmosphere having very low oxygen content). More importantly, *S. aureus* competes poorly with the lactic acid bacteria that dominate in commercially packaged cured meats and will not grow below 6.7°C (luncheon meats are usually well refrigerated).

Ham has been frequently implicated in outbreaks of staphylococcal foodborne illness. Most commonly, the ham was cooked or re-heated and then subjected to contamination by a food handler (e.g., during slicing). Subsequent holding at warm temperatures for a sufficient length of time provided the conditions for *S. aureus* to multiply and produce enterotoxin. Cooking (which destroys the normal lactic flora) and the combination of warmth, oxygen, and salt in the ham (which favors staphylococcal growth over other chance contaminants) are underlying factors.

Cooked cured meats seldom cause other types of foodborne illness. Commercial processes destroy salmonellae and other nonsporeforming pathogens. If these products become contaminated after heating, salmonellae can survive and, if the temperature is favorable, multiply. The low incidence and numbers of *C. botulinum* coupled with the presence of nitrite and salt, the growth of lactic acid bacteria, and the prevalence of refrigeration contribute to the control of these bacteria. Similar factors probably apply to *C. perfringens*, but there has been very little research using cured meats to confirm this commonly held opinion. Despite the frequency with which *L. monocytogenes* occurs in these products throughout the world, relatively few outbreaks have been reported. Only two cases of human listeriosis in North America have been linked to the consumption of these and other ready-to-eat meat and poultry products.

45.33 Canned Cured Products

45.331 Shelf-Stable Canned Cured Products

Canned shelf-stable cured meats include a) canned viennas, corned beef, frankfurters, meat spreads, and chicken or turkey luncheon meat given a botulinal cook ($F_0 \geq 2.78$); b) 12-oz. canned luncheon meat and small canned hams made from pork given less than a botulinal cook; c) canned sausages that may be covered with hot oil in the final container and have a water activity of ≤ 0.92, and sliced dried beef or prefried bacon in vacuum-sealed or modified-atmosphere packaging that rely on low water activity (≤ 0.86) for stability; and d) vinegar-pickled meats such as sausages or pigs feet.

Products in category (a) are similar in their microbiology to other low-acid canned foods and should be examined as per Chapters 61 or 62. The shelf-stable canned cured pork products in category (b) are limited to 3 pounds or less in size because the thermal process ($F_0 = 0.1–0.7$) required for stability will cause an unacceptable, mushy texture in products of a larger size. Although these products are given less than a botulinal cook, they have an excellent, but not unblemished, record of safety and stability. Their stability is dependent on four interrelated factors: nitrite, salt, low indigenous spore levels in the raw materials, and thermal processing that destroys many of the spores. Those heat-injured spores that survive are inhibited by the presence of sufficient nitrite and salt. Since it is possible to recover low levels of mesophilic sporeformers from products in category (b), "commercial sterility," as defined in the Code of Federal Regulations (21CFR Part 113.3) for low-acid foods in hermetically sealed containers does not apply to these products. Spoilage does not occur unless the curing process, retorting process, or can seal is faulty. Thermophilic spoilage can occur if the products are stored at abnormally high temperatures for a sufficient length of time. A faulty cure with insufficient levels of salt or nitrite can permit the growth of sporeformers that survive the thermal process. Slight underprocessing can allow survival of a higher than normal level of mesophilic spores and, thereby, increase the chance that subsequent growth will occur. Gross underprocessing results in additional survival and growth of enterococci and other nonsporeformers. Faulty can seals may allow a variety of bacteria to enter the can and cause spoilage of the contents.

Sausages, sliced dried beef, and prefried bacon in category (c) do not spoil unless the water activity is higher than recommended or the seal of the package or container is broken. These products may be packaged under vacuum or with a modified atmosphere having a low residual level of oxygen. Products that exceed a water activity of 0.86 support the growth of more salt-tolerant bacteria such as micrococci or enterococci, particularly if the products have been sealed under a low vacuum or nitrate has been added during curing. Theoretically, it may be possible for *S. aureus* to grow in these products at water activities near 0.86, but such growth has not been observed in commercial practice. This fact may be due to a low level of *S. aureus,* more rapid growth of the other flora, a combined effect of pH and water activity, or more importantly, that anaerobic growth of *S. aureus* is prevented below 0.90 water activity.[77] Overt spoilage does not always accompany bacterial growth in these products when the water activity is too high (e.g., 0.86 to 0.90).

Pickled pigs feet, pickled sausages, and similar products in category (d) are immersed in vinegar brine and owe their stability to low pH, acetic acid, little or no fermentable sugar remaining in the tissue, and/or an airtight package.[73] Lactic acid bacteria and spores in moderate numbers may be present. Aerobic plate counts are variable and unpredictable; coliforms and *S. aureus* are rarely present. Lactic acid bacteria in numbers exceeding 10^7 per gram may make the brine cloudy.[74] In hot weather the container may build up internal pressure. Foodborne disease organisms cannot survive, but mold can grow if the container is not properly sealed.

Hermetically sealed containers of pickled bone-in meat may develop gas and even explode from the action of the vinegar on the bone. This nonmicrobial problem should be considered by microbiologists investigating the cause of swelling.

45.332 Perishable Canned Cured Products

Perishable canned cured meats made from pork can be up to 22 pounds in size and must be labeled "Perishable, Keep Refrigerated." They contain levels of nitrite and salt similar to those in shelf-stable products but are heated insufficiently to control the indigenous spore population. Hence, these products require refrigerated storage for both safety and stability. These products also include ham and other cured meats that are thermally processed in hermetically sealed oxygen-impermeable films and distributed as "cook-in-bag" products.

These products can retain their acceptable quality for 1 to 3 years when properly processed and refrigerated. Spoilage of these products at refrigeration temperature usually is due to the survival and growth of psychrotrophic, thermoduric, nonsporeforming bacteria (e.g., enterococci, *Lactobacillus viridescens*) that were present in abnormally high levels in the raw materials or that survived when a substandard thermal process was used. This spoilage often is characterized by sourness or off-odor, and a loss of vacuum or swelling of the cans. A green discoloration may be associated with spoilage by *L. viridescens*. Perishable canned cured meats normally contain low levels ($\leq 10^2$ per gram) of viable mesophilic aerobic and anaerobic sporeformers that survive processing, but remain dormant at refrigeration temperatures. The presence of abnormally high levels (e.g., $\geq 10^3$ per gram) of mesophilic anaerobic sporeformers that are incapable of growth below 10°C is evidence of temperature abuse and a potential botulinal hazard must be considered. An initial indication of this risk is a putrid aroma on opening the can, however one must be familiar with the putrid aroma produced by putrefactive anaerobes. The presence of spores (beware of fat droplets from the product) in a wet mount preparation under phase contrast microscopy is additional, readily attainable evidence of temperature abuse.

Spoilage by psychrotrophic clostridia is rare, but about six different episodes have been observed in perishable canned cured meat products.[54,82] Experience indicates that this form of spoilage is more common in products that have lower brine levels (e.g., ≤3%) and have been in refrigerated storage for several months to a year. Bacterial isolates from spoiled product (i.e., gas production or loss of vacuum, and a putrid aroma) have been identified as *C. putrefaciens*.[83]

45.34 Shelf Stable Canned Uncured Products

Typical products include a) roast beef with gravy, beef stew, chili con carne, tamales, and canned whole chicken, which are low-acid canned foods and given a botulinal cook or greater, and b) sloppy joe and spaghetti sauce with ground meat, which are high-acid canned foods and given a milder heat treatment. Unstable or otherwise suspect products in categories (a) and (b) should be analyzed by the procedures described in Chapter 62.

45.35 Fermented and Acidulated Sausages

Dry fermented products such as German and Italian-style salamis, pepperoni, Lebanon bologna, and summer sausage depend upon a lactic fermentation and relatively low water activity for preservation. These sausages are produced by stuffing a blend of ground or chopped meat and other ingredients into a casing, holding at controlled temperatures (e.g., 20° to 45°C) and humidity to facilitate acid production, and then drying at reduced temperatures (e.g., 10° to 15°C) and humidity. Most manufacturers add commercial starter cultures during formulation for more rapid controlled processing, consistent quality, and to reduce the risk of developing excessive levels of *S. aureus*. After fermentation, the lactic acid-producing microflora can exceed 10^8 per gram, while *Enterobacteriaceae* have not increased beyond their initial levels and *S. aureus* remains at less than 10^3 per gram. During drying and subsequent storage, organic acids (primarily lactic acid) and a relatively high salt content may result in an overall decrease in the microbial population, although 10^6 per gram of lactic acid bacteria surviving in some products at the retail level is not uncommon. Some products are smoked or heated, thus greatly reducing the bacterial levels in the final product. Fermentation acids, salt, heating, and drying influence the survival of pathogens that could cause foodborne illness.

Concern about *Trichinella spiralis*, *S. aureus*, salmonellae, and more recently, *E. coli* O157:H7 has resulted in significant changes in processing conditions over the past 30 years, thus changing the surviving microbial flora in modern sausage products. The risk of *T. spiralis* is controlled by specific regulatory requirements designed to ensure inactivation of the parasite.[86] Since 1995, the USDA has required products produced under federal inspection be produced with processes validated to cause a 5D kill of *E. coli* O157:H7, or its equivalent, before the products are released for distribution. This should be more than adequate to ensure the absence of nonsporeforming enteric pathogens in products found at retail stores and elsewhere.

The risk of staphylococcal food poisoning is limited to the fermentation phase of processing. The increased use of starter cultures since 1970 and improvements in process control have been effective in reducing this risk.[5] While there have been no reported outbreaks of staphylococcal foodborne illness from commercial products in the past 25 years, it is important for processors to be aware that under certain circumstances, toxigenic staphylococci can multiply during fermentation and produce toxin in the periphery of the sausage.[7] While the staphylococcal population may decline during heating, drying and storage, the toxin will remain. Thus, the level of viable *S. aureus* in the final product is not an accurate measure of the risk of enterotoxin.

When testing fermented sausages for their potential in causing staphylococcal food poisoning, the outer 3 mm layer of the sausage should be sampled for viable *S. aureus* at the end of the fermentation cycle, before the product is heated and/or dried. Since *S. aureus* death may have occurred in these products before distribution from the manufacturer, it may be necessary to test the outer 3 mm of suspect product for enterotoxin (Chapter 39).

Yeasts and molds may grow on the surface of the casing during production and subsequent storage. The major cause of microbial spoilage of vacuum-packaged fermented sausages consists of a lactic flora and occasionally, yeasts. Spoilage may appear as a milky fluid between the product and the film, and perhaps, swelling of the package due to the production of gas. Spoilage from mold is limited to product packaged in bulk or leaking vacuum packages. As in the case of dried meats, the growth of mold on fermented products has not been shown to have public health significance.[75] Some processors cultivate a nontoxigenic surface mold to impart a characteristic flavor and appearance. Information about the formulation and processing procedure is helpful for the proper interpretation of laboratory data from fermented sausages. Further information on the microbiology of fermented meat and poultry products is available.[4]

Semi-dry sausages can be produced by fermentation or by adding an acidulant (citric acid, glucono-delta-lactone, lactic acid) to the meat during formulation. Acidulants are usually encapsulated to facilitate a more controlled, rapid pH drop during subsequent heating. As the product is heated, the encapsulating material breaks down and the acidulant is released. Since these products may be heated to higher temperatures than those used for making dry sausage, very low microbial levels are normally detected after processing. Acidulated products can be differentiated from fermented products by reading the ingredient phrase on the label or by contacting the producer. The spoilage pattern for these products is similar to that of dry fermented products. For product that is in chub form (i.e., not sliced), the growth of the spoilage flora is limited to the surface of the product. Susceptibility to spoilage is influenced by a wide variety of factors (e.g., storage temperature). Dry and semi-dry sausages normally are stored and distributed at refrigeration temperatures to maintain optimum quality and avoid the risk of fat rendering at warmer storage temperatures. Pathogen growth on these products after release from a commercial processor has not been involved in foodborne illness.

45.36 Dried Meat Products

Most commercial procedures for the preparation of dried meat (e.g., jerky, beer sticks, basturma, tosajo, dehydrated meat or poultry) include a cooking step that destroys the normal vegetative flora of the raw meat or poultry, and a rapid drying procedure that reduces the water activity to inhibitory levels to restrict microbial growth. Molds, and sometimes yeasts, subsequently may grow on the surface of dried meats. Drying to a lower moisture level, using sorbic acid and vacuum packaging, and avoiding storage in a humid atmosphere will control such growth. Spoilage of dried meat or poultry by molds has no known public health significance.[50,76] Occasionally, dried meats and certain perishable cured meats will develop white deposits on the surface that appear to be mold growth. A crystalline structure on microscopic examination will exclude the possibility of mold. Deposits of this type may be due to crystallized inorganic salts such as phosphate,[2,3] hard fat, or lactate salts. Consumers and analysts might confuse these crystals with broken glass or mold.

Microbial levels on dried meats are highly variable and depend on the nature of the product, its ingredients, and method of production. Commercial products rarely have been involved in foodborne illness. Salmonellae not destroyed by heating, or resulting from recontamination of the dried product, can lead to illness.[15,17] Dried meats improperly prepared in the home may permit survival, if not growth, of pathogens during processing or lead to contamination.[59]

45.4 PATHOGENS

S. aureus, C. perfringens, Campylobacter, L. monocytogenes, and *Salmonella* occasionally are present in low numbers on raw meat surfaces. These species are most hazardous when they grow without competition in cooked products. *S. aureus* causes a foodborne intoxication from the ingestion of one or more enterotoxins. Humans and animals can be sources of contamination, and human cases have been traced to food preparers who inadvertently inoculate food during preparation.[53] Outbreaks attributed to *C. perfringens* appear to be limited to cooked noncured meats, such as roast beef or turkey, as opposed to ham and other salted or cured meats. Since 1996, *Campylobacter* has been identified as the major cause of sporadic cases of human diarrhea in the U.S., despite its fragile nature and microaerophilic growth requirements.[35] *L. monocytogenes* is of concern in cooked packaged meat and poultry products held under refrigeration for prolonged periods, particularly when refrigeration is marginal (10°C or higher). Salmonellae are ubiquitous in raw meat and poultry products. New drug resistant strains such as *S. typhimurium* DT104 are becoming major causes of human illness in Europe and the U.S.[43]

Other animal pathogens can cause human disease with close animal-to-man contact, such as among farmers and persons working in slaughtering plants. Some of these include *Brucella, Mycobacterium, Leptospira, Coxiella burnetii, C. tetani,* and *Chlamydia psittaci.* Theoretically, some of these pathogens could be transmitted to consumers by contact with raw meat, but epidemiological evidence is lacking to support this possibility. The relative significance of meat and poultry as vehicles for the transmission of foodborne illness from *C. jejuni/coli, L. monocytogenes, Yersinia enterocolitica, Arcobacter,* and enteropathogenic *E. coli* continues to be investigated. The literature should be surveyed for current information on these pathogens.

45.5 RECOMMENDED METHODS

A wide variety of methods are used for sampling meat and poultry products. The sampling procedure depends on the product and the reasons for analysis. Poultry carcasses can be sampled using a whole bird rinse technique,[38,79,84,85] skin snips;[62] or sponging.[38,85] Excision samples[13,22] or sponges[22,23,38,84] are often used for larger carcasses, such as cattle and swine. Additional suggestions for sampling other types of meat and poultry products are available in other reference materials.[25,27,36,61] Ground products can be sampled by aseptically removing random portions to form a composite sample of the weight to be analyzed.[25,36,38,84] Cooked meat products can be sliced, diced, or cored aseptically to form composite samples[25,36] for analysis. In addition, the free juice or liquid in the package or a rinse of the product with sterile diluent can suffice for some purposes (e.g., determining the cause of spoilage).

After samples are collected and prepared for analysis, the microbiological methods can be performed as described in this book. The selection of the analyses depends on the sample and the information desired. Some of the tests that can be performed and their appropriate chapters are as follows:

Aerobic Plate Count (mesophilic)—Chapter 7
Aerobic Plate Count (psychrotrophic)—Chapter 13
Anaerobic plate count—Chapter 23
Lactic acid bacteria—Chapter 19
Campylobacter—Chapter 31
Clostridium botulinum—Chapter 33
Clostridium perfringens—Chapter 34
Coliforms, *E. coli*, and Fecal Coliforms—Chapter 8
Salmonella—Chapter 37
S. aureus—Chapter 39
Staphylococcal enterotoxins—Chapter 39
Yeasts and Molds—Chapter 20
Enterococci—Chapter 9
L. monocytogenes—Chapter 36
Yersinia—Chapter 41
Water Activity—Chapter 64

In certain specific investigations, other analyses may be required. Many of these are discussed elsewhere, while others, such as protein, salt, water, and fat, are not included in this book. To perform these tests, the analyst should refer to other publications.

45.6 REFERENCES

1. Allen, J. R., and E. M. Foster. 1960. Spoilage of vacuum-packaged sliced processed meats during refrigerated storage. Food Res. 25:19-25.
2. Arnau, J., E. Maneja, L. Guerrero, and J. M. Monfort. 1993. Phosphate crystals in raw cured ham. Fleischwirtschaft 73:859-860.

3. Arnau, J., L. Guerrero, and P. Gou. 1998. The precipitation of phosphates in meat products. Fleischwirtschaft 78:701-702.
4. Bacus, J. N. 1984. Utilization of microorganisms in meat processing: a handbook for meat plant operators. John Wiley & Sons, New York.
5. Bacus, J. N. 1986. Fermented meat and poultry products, p. 123-164. *In* A. M. Pearson and T. R. Dutson (eds.), Advances in meat research. Vol 2. Meat and poultry microbiology. AVI Publishing, Westport, Conn.
6. Bailey, J. S., J. E. Thomson, and N. A. Cox. 1987. Contamination of poultry during processing, p. 193-211. *In* F. E. Cunningham and N. A. Cox (eds.), The microbiology of poultry meat products. Academic Press, Orlando, Fla.
7. Barber, L. E., and R. H. Deibel. 1972. Effect of pH and oxygen tension on staphylococcal growth and enterotoxin formation in fermented sausage. Appl. Microbiol. 24:891-898.
8. Barnes, E. M. 1976. Microbiological problems of poultry at refrigerator temperatures—A review. J. Sci. Food Agric. 27:777-782.
9. Bersani, C., P. Cattaneo, C. Balzaretti, and C. Cantoni. 1984. Psychrotrophic Enterobacteriaceae occurring in refrigerated meat products. Ind. Aliment. 23:112-118.
10. Beyer, K., and H. J. Sinell. 1981. Psychrotrophic microorganisms in vacuum packaged chilled beef trimmings, p. 191-198. *In* T. A. Roberts, G. Hobbs, J. H. B. Christian, and N. Skovgaard, (eds.), Psychrotrophic microorganisms in spoilage and pathogenicity. Academic Press, New York, NY.
11. Broda, D. M., K. M. De Lacy, and R. G. Bell. 1998. Efficacy of heat and ethanol spore treatments for the isolation of psychrotrophic *Clostridium* spp. associated with the spoilage of chilled vacuum-packed meats. Int. J. Food Microbiol. 39:61-68.
12. Broda, D. M., K. M. De Lacy, and R. G. Bell. 1998. Influence of culture media on the recovery of psychrotrophic *Clostridium* spp. associated with the spoilage of vacuum-packed chilled meats. Int. J. Food Microbiol. 39:69-78.
13. Brown, M. H., and A. C. Baird-Parker. 1982. The microbiological examination of meat, p. 423-520. *In* M. H. Brown (eds.), Meat microbiology. Applied Science Publishers, New York.
14. Cannon, J. E., F. K. McKeith, S. E. Martin, J. Novakofski, and T. R. Carr. 1993. Acceptability and shelf-life of marinated fresh and precooked pork. J. Food Sci. 58:1249-1253.
15. CDC. 1985. Salmonellosis associated with carne seca—New Mexico. Morb. Mortal Wkly. Rep. 34:645-646.
16. CDC. 1990. Foodborne disease outbreaks—Five year summary, 1983-1987. *In* CDC Surveillance Summaries. Morb. Mortal Wkly. Rep. 39 (SS-1):15-57.
17. CDC. 1995. Outbreak of salmonellosis associated with beef jerky—New Mexico, 1995. Morb. Mortal Wkly. Rep. 44:785-788.
18. CDC. 1996. Surveillance for foodborne-disease outbreaks—United States, 1988-1992. *In* CDC Surveillance Summaries. Morb. Mortal Wkly. Rep. 45 (SS-5):1-66.
19. Collins, M. D., U. M. Rodrigues, R. H. Dainty, R. A. Edwards, and T. A. Roberts. 1992. Taxonomic studies on a psychrophilic *Clostridium* from vacuum-packed beef: Description of *Clostridium estertheticum* sp. nov. FEMS Microbiol. Lett. 96:235-240.
20. Dainty, R. H., B. G. Shaw, and T. A. Roberts. 1983. Microbial and chemical changes in chill-stored red meats, p. 151-178. *In* T. A. Roberts and F. A. Skinner (eds.), Food microbiology: Advances and prospects. Academic Press, New York.
21. Dainty, R. H., R. A. Edwards, and C. M. Hibbard. 1989. Spoilage of vacuum-packed beef by a *Clostridium* sp. J. Sci. Food Agric. 49:473-486.
22. Dorsa, W. J., C. N. Cutter, and G. R. Siragusa. 1996. Evaluation of six sampling methods for recovery of bacteria from beef carcass surfaces. Lett. Appl. Microbiol. 22:39-41.
23. Dorsa, W. J., G. R. Siragusa, C. N. Cutter, E. D. Berry, and M. Koohmaraie. 1997. Efficacy of using a sponge sampling method to recover low levels of *Escherichia coli* O157:H7, *Salmonella typhimurium*, and aerobic bacteria from beef carcass surface tissue. Food Microbiol. 14:63-69.
24. Egan, A. F., B. J. Shay, and P. J. Rogers. 1989. Factors affecting the production of hydrogen sulfide by *Lactobacillus sake* L13 growing on vacuum-packed beef. J. Appl. Bacteriol. 67:255-262.
25. FDA. 1995. Bacteriological analytical manual, 8th ed. AOAC International, Gaithersburg, Md.
26. Fernandez-Coll, F., and M. D. Pierson. 1985. Enumeration of hydrogen sulfide-producing bacteria from anaerobically packaged pork. J. Food Prot. 48:982-986.
27. FSIS. 1975. Microbiology laboratory guidebook. U.S. Dept. of Agric., Washington, D. C.
28. FSIS. 1994. Nationwide beef microbiological baseline data collection program: Steers and heifers. U.S. Dep. of Agric., Washington, D. C.
29. FSIS. 1996. Nationwide beef microbiological baseline data collection program: Cows and bulls. U.S. Dep. of Agric., Washington, D. C.
30. FSIS. 1996. Nationwide broiler chicken microbiological baseline data collection program. U.S. Dep. of Agric., Washington, D. C.
31. FSIS. 1996. Nationwide federal plant raw ground beef microbiological survey. U.S. Dep. of Agric., Washington, D. C.
32. FSIS. 1996. Nationwide pork microbiological baseline collection program: market hogs. U.S. Dep. of Agric., Washington, D. C.
33. FSIS. 1996. Nationwide raw ground chicken microbiological survey. U.S. Dep. of Agric., Washington, D. C.
34. FSIS. 1996. Nationwide raw ground turkey microbiological survey. U.S. Dep. of Agric., Washington, D. C.
35. FSIS. 1998. FoodNet: An active surveillance system for bacterial foodborne diseases in the United States. Report to Congress. U.S. Dep. of Agric., Washington, D.C.
36. FSIS. 1998. Microbiology Laboratory Guidebook, 3rd ed. U.S. Dep. of Agric., Washington, D. C.
37. FSIS. 1998. Nationwide young turkey microbiological baseline data collection program. U.S. Dep. of Agric., Washington, D. C.
38. FSIS. 1998. Self-instructional guide for collecting raw meat and poultry product samples for *Salmonella* analysis. FSIS Directive 10,230.5.
39. Gardner, G. A. 1981. *Brochothrix thermosphacta (Microbacterium thermosphactum)* in the spoilage of meats—A review, p. 139-173. *In* T. A. Roberts, G. Hobbs, J. H. B. Christian, and H. Skovgaard (eds.), Psychrotrophic microorganisms in spoilage and pathogenicity. Academic Press, New York.
40. Gill, C. O. 1979. Intrinsic bacteria in meat. J. Appl. Bacteriol. 47:367-378.
41. Gill, C. O. 1982. Microbial interaction with meats, p. 225-264. *In* M. H. Brown (ed.), Meat microbiology. Applied Science Publishers, New York.
42. Gill, C. O. 1986. The control of microbial spoilage in fresh meats, p. 49-88. *In* A. M. Pearson and T. R. Dutson, (eds.), Advances in meat research. Vol 2. Meat and poultry microbiology. AVI Publishing, Westport, Conn.
43. Glynn, M. K., C. Bopp, W. Dewitt, P. Dabney, M. Mokhtar, and F. J. Angulo. 1998. Emergence of multidrug-resistant *Salmonella enterica* serotype typhimurium DT104 infections in the United States. N. Engl. J. Med. 338:1333-1338.
44. Grant, G. F., A. R. McCurdy, and A. D. Osborne. 1988. Bacterial greening in cured meats: a review. Can. Inst. Food Sci. Technol. J. 21:50-56.
45. Grau, F. H. 1986. Microbial ecology of meat and poultry, p. 1-47. *In* A. M. Pearson and T. R. Dutson (eds.), Advances in meat research. Vol. 2. Meat and poultry microbiology. AVI Publishing, Westport, Conn.
46. Gunderson, M. F., H. W. McFadden, and T. S. Kyle. 1954. Bacteriology of commercial poultry processing. Burgess Publishing, Minneapolis, Minn. p. 59-63.
47. Hanna, M. O., G. C. Smith, L. C. Hall, and C. Vanderzant. 1979. Role of *Hafnia alvei* and a *Lactobacillus* species in the spoilage of vacuum-packaged strip loin steaks. J. Food Prot. 42:569-571.
48. Hardin, M. D., G. R. Acuff, L. M. Lucia, J. S. Oman, and J. W. Savell. 1995. Comparison of methods for contamination removal from beef carcass surfaces. J. Food Prot. 58:368-374.

49. Hechelmann, H., Z. Bem, K. Uchida, and L. Leistner. 1974. Occurrence of the tribe Klebsielleae in refrigerated meats and meat products. Fleischwirtschaft 54:1515-1517.
50. ICMSF. 1980. Meats and meat products, p. 333-409. *In* ICMSF, Microbial ecology of foods. Vol. 2. Food commodities. Academic Press, New York.
51. ICMSF. 1980. Poultry and poultry meat products, p. 410-458. *In* ICMSF, Microbial ecology of foods. Vol. 2. Food commodities. Academic Press, New York.
52. ICMSF. 1986. Microorganisms in foods. Vol 2. Sampling for microbiological analysis: Principles and specific applications, 2nd ed. University of Toronto Press, Toronto, Canada.
53. Jablonski, L. M., and G. A. Bohach. 1997. *Staphylococcus aureus*, p. 353-375. *In* M. P. Doyle, L. R. Beuchat, and T. J. Montville (eds.), Food microbiology fundamentals and frontiers. ASM Press, Washington, D. C.
54. Johnston, R. W., and G. W. Krumm. 1980. The microbiological safety of canned, cured, perishable meat products, p. 295-299. Proceedings of the 26th European Meeting of Meat Research Workers, Vol 2. Colorado Springs, Colo.
55. Jones, D. K., J. W. Savell, G. R. Acuff, and C. Vanderzant. 1988. Retail case-life and microbial quality of pre-marinated, vacuum packaged beef and chicken fajitas. J. Food Prot. 51:260-262.
56. Kalchayanand, N., B. Ray, and R. A. Field. 1993. Characteristics of psychrotrophic *Clostridium laramie* causing spoilage of vacuum-packaged refrigerated fresh and roasted beef. J. Food Prot. 56:13-17.
57. Kalchayanand, N., B. Ray, R. A. Field, and M. C. Johnson. 1989. Spoilage of vacuum-packaged refrigerated beef by *Clostridium*. J. Food Prot. 52:424-426.
58. Kalinowski, R. M. 1996. M.S. thesis. Psychrotrophic clostridia causing spoilage in cooked meat products. Illinois Institute of Technology, Chicago, Ill.
59. Keene, W. E., E. Sazie, J. Kok, D. H. Rice, D. D. Hancock, V. K. Balan, T. Zhao, and M. P. Doyle. 1997. An outbreak of *Escherichia coli* O157:H7 infections traced to jerky made from deer meat. JAMA 277:1229-1231.
60. Kempton, A. G., and S. R. Bobier. 1970. Bacterial growth in refrigerated, vacuum-packed luncheon meats. Can. J. Microbiol. 16:287-297.
61. Kotula, A. W., J. C. Ayres, C. N. Huhtanen, N. J. Stern, W. C. Stringer, and R. B. Tompkin. 1980. Guidelines for microbiological evaluation of meat, p. 65-70. *In* Proceedings of the American Meat Science Association's 33rd Annual Reciprocal Meat Conference. West Lafayette, Ind.
62. Kotula, K. L., and Y. Pandya. 1995. Bacteriological contamination of broiler chickens before scalding. J. Food Prot. 58:1326-1329.
63. Lawson, P., R. H. Dainty, N. Kristiansen, J. Berg, and M.D. Collins. 1994. Characterization of a psychrotrophic *Clostridium* causing spoilage in vacuum-packed cooked pork: Description of *Clostridium algidicarnis* sp. nov. Lett. Appl. Microbiol. 19:153-157.
64. Leisner, J. J., G. G. Greer, and M. E. Stiles. 1996. Control of beef spoilage by a sulfide-producing *Lactobacillus sake* strain with bacteriocinogenic *Leuconostoc gelidum* UAL187 during anaerobic storage at 2°C. Appl. Environ. Microbiol. 62:2610-2614.
65. Lillard, H. S. 1989. Incidence and recovery of salmonellae and other bacteria from commercially processed poultry carcasses at pre- and post-evisceration steps. J. Food Prot. 52:88-91.
66. Lillard, H. S. 1990. The impact of commercial processing procedures on the bacterial contamination and cross-contamination of broiler carcasses. J. Food Prot. 53:202-204.
67. Mackey, B. M., and C. M. Derrick. 1979. Contamination of the deep tissues of carcasses by bacteria present on the slaughter instruments or in the gut. J. Appl. Bacteriol. 46:355-366.
68. Mossel, D. A. A., and H. Zwart. 1960. The rapid tentative recognition of psychrotrophic types among Enterobacteriaceae isolated from foods. J. Appl. Bacteriol. 23:185-188.
69. Mountney, G. J. 1976. Poultry products technology, 2nd ed. AVI Publishing, Westport, Conn. p. 63-66.
70. Mundt, J. O., and H. M. Kitchen. 1951. Taint in southern country-style hams. Food Res. 16:233-238.
71. Newton, K. G. 1979. Value of coliform tests for assessing meat quality. J. Appl. Bacteriol. 47:303-307.
72. Nicol, D. J., M. K. Shaw, and D. A. Ledward. 1970. Hydrogen sulfide production by bacteria and sulfmyoglobin formation in prepacked chilled beef. Appl. Microbiol. 19:937-939.
73. Niven, C. F. 1952. Significance of the lactic acid bacteria in the meat industry, p. 31. *In* Proceedings of the American Meat Institute's Fourth Research Conference. AMI, Chicago, Ill.
74. Niven, C. F. 1956. Vinegar pickled meats: A discussion of bacterial and curing problems encountered in processing. Am. Meat Found. Bull. No. 27.
75. NRC. 1985. An evaluation of the role of microbiological criteria for foods and food ingredients. National Academy Press, Washington, D. C.
76. Rojas, F. J., M. Jodral, F. Gosalvez, and R. Pozo. 1991. Mycoflora and toxigenic *Aspergillus flavus* in Spanish dry-cured ham. Int. J. Food Microbiol. 13:249-256.
77. Scott, W. J. 1953. Water relations of *Staphylococcus aureus* at 30°C. Aust. J. Bio. Sci. 6:549.
78. Stiles, M. E., and L-K. Ng. 1981. *Enterobacteriaceae* associated with meats and meat handling. Appl. Environ. Microbiol. 41:867-872.
79. Surkiewicz, B. F., R. W. Johnston, A. B. Moran, and G. W. Krumm. 1969. A bacteriological survey of chicken eviscerating plants. Food Technol. 23:1066-1069.
80. Taylor, A. A., and B. G. Shaw. 1977. The effect of meat pH and package permeability on putrefaction and greening in vacuum packed beef. J. Food Technol. 12:515-521.
81. Tompkin, R. B. 1980. Botulism from meat and poultry products—A historical perspective. Food Technol. 34:229-236, 257.
82. Tompkin, R. B. 1986. Microbiology of ready-to-eat meat and poultry products, p. 89-121. *In* A. M. Pearson and T. R. Dutson (eds.), Advances in meat research. Vol 2. Meat and poultry microbiology. AVI Publishing, Westport, Conn.
83. Tompkin, R. B., A. B. Shaparis, and L. N. Christiansen. 1982. Unpublished data.
84. U.S. Government. 1996 (July 25). Pathogen reduction; hazard analysis and critical control point (HACCP) systems. Fed. Reg. 61(144): 38806-38906.
85. U.S. Government. 1997 (May 13). Pathogen reduction; hazard analysis and critical control point (HACCP) systems; technical corrections and amendments. Fed. Reg. 62(92):26211-26219.
86. U. S. Government. National Archives and Records Administration. 1997. CFR. Title 9. Chapter III. Part 318. Section 10—Prescribed treatment of pork and products containing pork to destroy trichinae, p. 258-268.
87. Whiteley, A. M., and M. D. D'Souza. 1989. A yellow discoloration of cooked cured meat products-isolation and characterization of the causative organism. J. Food Prot. 52:392-395.

Chapter 46

Eggs and Egg Products

Steven C. Ricke, Sarah G. Birkhold, and Richard K. Gast

46.1 INTRODUCTION

In 1997, the number of eggs consumed in the United States was 240 per person. Of this, 71% were consumed as shell eggs and 29% as egg products in the form of liquid, frozen, and dried products, as well as specialties such as frozen precooked omelettes. An increasing proportion of eggs is consumed as egg products. In 1987, only 17% of all eggs consumed in the United States were consumed as egg products.

Shell eggs have many uses in the home, in restaurants, and in institutions, either alone or as ingredients in other foods. Egg products generally are used in the food industry to provide desirable functional properties in products such as mixes, bakery foods, noodles, mayonnaise, salad dressings, candies, and ice cream. There are also several other uses of eggs including pet foods.[17] The widespread use of eggs as ingredients of many foods makes them prime suspects in foodborne disease outbreaks because of their excellent nutritional environment for supporting bacterial growth. This potential hazard requires careful microbiological control in production and usage.

Commercial shell egg production has undergone considerable change during the past 40 years. In the 1940s, most eggs were produced on family farms by small flocks of a few hundred birds or less. Today, most eggs are produced by flocks of 30,000 or more, with some commercial operations reaching millions of birds. Many of these operations are automated with mechanical feeding, watering, egg collection, and cartoning systems. Over the years, shell egg quality has been improved by genetic selection of birds, better production practices, refrigeration of the eggs, and more rapid marketing systems.

Shell eggs are graded according to exterior and interior quality factors and sized according to weight. The standards, grades, and weight classes are established by the U.S. Department of Agriculture (USDA). USDA operates a voluntary grading program. Before marketing, shell eggs go through a mechanized process of sorting, washing, drying, candling (grading), weighing, and packaging.

Shell eggs have been designated by the FDA Model Food Code as "potentially hazardous foods." Although the Food Code requires that such foods be received and stored at 41°F or less, an exemption from this rule exists for eggs because they are often received at higher temperatures. Instead, current recommendations are that eggs be held in units capable of maintaining 41°F or less. There is, however, continued concern regarding this practice. USDA and the Department of Health and Human Services have recently jointly proposed regulations requiring shell eggs to be stored and transported at an ambient temperature of no more than 45°F.[97] Recent research has focused on using cryogenic gases to rapidly cool eggs before placing them into cartons.[24] The length of egg holding time before consumption is an important factor in limiting the growth of pathogenic bacteria. Research has shown that age-related changes occur in eggs that permit the growth of *Salmonella enteritidis* (See Chapter 37 for explanation of nomenclature).[50] Therefore, there is increasing interest in establishing appropriate guidelines for implementing a shell egg dating system that will be useful to consumers.

Undergrade eggs (with poor shell condition or poor interior quality) are usually sent to an egg breaking firm for the production of egg products. Graded, as well as nest-run eggs (not sized or graded), are also commonly used in producing egg products. Federal law prohibits the use of "leakers" (with shell and shell membranes broken) and eggs showing evidence of spoilage for producing products for human consumption.

The quality of egg products has greatly improved in recent years. USDA is responsible for mandatory inspection of egg products plants. On July 1, 1971, federal legislation went into effect requiring mandatory USDA inspection of egg products operations (Egg Products Inspection Act, Public Law 91-5971). Regulations established under this law specify minimum standards for breaking stock, the sanitary conditions of facilities and equipment, pasteurization conditions, holding conditions, and other parts of the operations. This law also provides that egg products shall be *Salmonella* negative, using a specified sampling and testing program. Over the years the industry has applied new technologies to improve the quality of egg products, and companies purchasing egg products have specified tight production and microbiological requirements.

Current regulations and quality-control procedures have continued to evolve from work performed in the 1950s and 1960s as a cooperative effort between industry, the Food and Drug Administration (FDA), and USDA. Before implementation of continuous USDA inspection in 1971, numerous outbreaks of salmonellosis in humans were attributed to egg products. As a result of industry-government cooperation, new processes and control procedures were developed, and thus since 1971, commercially produced egg products have not been a major source of human salmonellosis.

46.2 GENERAL CONSIDERATIONS

46.21 Composition

As in other foods, the composition of the egg influences the types of organisms that will develop. Since the parts of the egg differ considerably in composition,[16] susceptibility to spoilage or the growth of pathogens differs considerably in each part.

Shell eggs consist of approximately 9.5% shell, 63% white, and 27.5% yolk.[21] The shell is relatively porous, containing mostly calcium carbonate crystals; a keratin-type protein coats the shell and fills the pores. Two membranes separate the egg white from the shell. As the egg cools, an air cell forms generally at the large end of the egg where the two membranes separate. During storage, this air cell increases in size as water evaporates through the shell. A membrane also surrounds the yolk. The thick portion of the egg white, known as the chalaza, holds the yolk near the center of the egg.

Egg white contains approximately 10.5% to 11.5% solids, of which 86.0% is protein, 9.0% total reducing sugars (3.2% free glucose), and 5.0% ash. Only a trace of lipids is present. Several of the egg white proteins have biological activity that retards bacterial growth either through lysis of bacteria or by tying up certain nutrients: lysozyme lyses many gram-positive microorganisms; conalbumin binds iron and other metals, retarding growth of certain bacteria; ovomucoid inhibits trypsin activity; avidin binds biotin; and riboflavin is bound to protein. The pH of egg white is less than 8.0 at oviposition but increases rapidly to pH 9.0 during holding in air, then slowly levels off at about pH 9.3. The increase in pH is due to the loss of CO_2 from the egg white.

Egg yolk contains 52% solids, of which 31.0% is protein, 64.0% is lipids (41.9% triglycerides, 18.8% phospholipids, and 3.3% cholesterol), 2.0% is total carbohydrates (0.4% free glucose), and 3.0% is ash. The pH of the yolk in freshly laid eggs is 6.0 and slowly increases to between 6.4 and 6.9. The pH of the blended yolk and white varies between pH 7.0 and pH 7.6.

46.22 Functional Properties

Egg yolk and egg white possess unique functional properties that make them useful, not only when prepared by themselves, but also when used as ingredients in other foods.[16] Coagulation, emulsifying power, and the ability to form heat stable foams are the most important of these functional properties. In the manufacture of egg products, the processing and holding conditions must not damage these properties. For example, pasteurization times and temperatures have been established to destroy pathogens while not substantially damaging functional properties.

46.23 Pasteurization

USDA prescribes minimum conditions of pasteurization for egg products sufficient to destroy harmful viable microorganisms. Liquid whole eggs are required to be held at 60°C for 3.5 min. Recently, research has been conducted on in-shell pasteurization of eggs using either heat[90] or irradiation.[79]

Many factors affect the thermal resistance of *Salmonella* in eggs. Salt, pH, sugar, type of acid used to adjust the pH, and total solids are all important. In liquid whole egg, *Salmonella* species have maximum heat resistance at about pH 5.5. The actual pasteurizing condition required is determined experimentally for each product. Cotterill et al.[22] constructed thermal destruction curves for several egg products. These curves are useful in estimating pasteurization time and temperatures.

USDA permits dried egg whites to be pasteurized in the dry state (hot room method) at a minimum temperature of 54.4°C for not less than 7 days in lieu of pasteurization of the liquid egg white before drying.[23] Spray-dried whites having approximately 6% moisture should be filled into bulk packages as they discharge from the dryer and the packages are moved immediately to the "hot room." Otherwise, it may take 3 or 4 days to heat the center of the package to the hot-room temperature.[93]

46.3 NORMAL FLORA

46.31 Shell Eggs

The shell and the contents of the egg at the time of oviposition are generally sterile or harbor very few microorganisms.[34,35,89] Contamination of the shell occurs afterwards from nesting material, floor litter, and avian fecal matter. Due to the presence of pores, the shell is a poor barrier to microorganisms, whereas the shell membrane appears to act as a fair barrier.[89] The contaminating flora are mainly gram-positive cocci, but gram-negative rods are also present in low numbers[11]; however, the gram-negative organisms generally cause spoilage of shell eggs. The egg contents may become contaminated by improper washing and storage methods. The physical and chemical barrier provided by the egg shell, shell membranes[36,44] and antimicrobial substances in the albumen favor the penetration and multiplication of the gram-negative bacteria.[10] The egg yolk offers an excellent medium for growth of microorganisms.[89]

46.32 Egg Products

The number of bacteria in liquid egg products before pasteurization depends on several factors, including the condition and quality of the shell eggs used for breaking, the method of washing and sanitizing the shell eggs, the sanitation of the equipment, and the time and temperature at which the liquid is held. The most common types of bacteria found in liquid eggs before pasteurizing are gram-negative rods.

The total aerobic plate count of commercial egg products is usually relatively low, less than 25,000 per gram, because of pasteurization or other heat treatment.[20,81] Yolk containing 10% salt need not be pasteurized if used in mayonnaise or salad dressing having a pH of 4.1 or less in accordance with the FDA requirement. Bacteria that survive pasteurization are usually heat resistant *Bacillus*, enterococci, and micrococci.[75] After pasteurization, care must be taken to avoid recontamination.

46.4 FLORA CHANGES IN SPOILAGE

46.41 Shell Eggs for Breaking

As discussed in Section 46.21, eggs possess several natural barriers that prevent bacterial invasion including the shell, membranes, and antibacterial factors present in the egg white.[12,84,86] Almost all eggs in the United States are washed before they are packed.[6] Bacterial contamination can occur through improper washing.[68] For example, if the wash solution temperature is less than the temperature of the egg, bacteria can be drawn into the egg through shell pores when the contents contract. Also, eggs washed with water contaminated with iron are more susceptible to spoilage, apparently because excessive iron overcomes the ability of conalbumin to inhibit bacterial growth. Egg washers that are not properly designed, maintained, and cleaned can increase contamination of the shell.[64] Cracks in the shell also present an avenue of contamination.

The most common spoilage organisms of shell eggs are gram-negative, such as *Pseudomonas, Serratia, Proteus, Alcaligenes,* and *Citrobacter.* For example, *Pseudomonas* spoilage is commonly found in eggs that have been improperly cleaned and stored for extended periods. They usually show a green fluorescence of the whites under ultraviolet light or other discolorations and off-odors.[31] If sufficiently advanced, spoilage is evident by odor and liquefaction of the yolk as well as by a thinning of the whites.

The manufacture of egg products begins with egg-breaking and separating operations. An automated system consists of a loader, a washer, and a breaker-separator unit. Three persons are required to operate this system. The person who places the shell eggs on the automatic loader also removes eggs not suitable for use in edible products. Another person checks the eggs coming from the washer and removes eggs that are not clean, as well as eggs that have become leakers or are otherwise unsuitable. The operator of the eggbreaking machine controls its speed to permit proper inspection of the contents. This person removes cracker heads and separating cups when an egg is found unsuitable (contents showing off-odor or liquefaction must be removed and not allowed to contaminate other eggs). With such a system, the liquid egg products should have relatively good microbiological quality. Any bacteria in the products will come from the exterior of the shell, or from the interior where insufficient growth of the organisms had not been detected by the operator. Since the contents of individual eggs must be examined, centrifugal egg separators are not allowed for edible eggs in breaking plants. Centrifugal separators have been used recently by the food service industry. Many states have banned their use in restaurants because of microbiological concerns. These devices also are routinely used for separating inedible eggs in many breaking plants. USDA has specified the maximum temperatures and times for holding liquid egg products after preparation to minimize bacterial growth.

46.42 New Egg Products and Designer Eggs

Many new value-added egg products have been developed and marketed in recent years. Dehydrated, pasteurized egg whites, previously available only to food service, are successfully marketed to consumers in grocery stores. Eggs are frequently featured as hand-held breakfast items such as breakfast burritos and tacos. Eggs are also found as toppings for breakfast pizzas. Ready-to-cook quiche and scrambled egg mixes are available in many grocery stores. Hard-cooked eggs continue to be popular in salad bars and in various pickled products. Commercial hard-cooked eggs are packed in containers with an organic acid and mold inhibitors such as 0.1% sodium benzoate or potassium sorbate.[41] To maintain low bacterial loads, hard-cooked eggs must be cooled rapidly after cooking, and strict sanitation must be maintained during peeling, handling, and packaging.[88] It has been recommended that hard-cooked eggs packed in an acid-benzoate solution be given a final heat treatment in the package.[89] Eggs packed in a citric acid-sodium benzoate solution should be stored at –1.0°C for maximum shelf life.

One particular area of egg product development that has received considerable attention is the production of designer eggs for health conscious consumers. Extensive reviews have been written on this subject.[43,65] Designer eggs are eggs whose nutritional profile have been altered by feeding hens a diet rich in specific nutrients that are passed on to the egg. Components of both the albumen and yolk can be enhanced to provide a more healthful nutritional profile. Currently eggs from hens fed diets rich in omega-3 and omega-6 fatty acids can be found in many large supermarkets. Eggs whose albumen solids have been altered to increase specific proteins may prove useful to companies that break eggs for use as food ingredients.[73,80] Designer eggs with elevated Vitamin E content can also be produced.[52] Designer eggs taste and cook like any other egg, therefore they can be easily used in egg recipes.

Shell eggs are also used for the production of ethnic foods. Only fresh, clean eggs with sound shells should be selected for these methods of production. Baluts, also called vit lon, are duck eggs that have been incubated 14–18 days; usually 16 days is preferred. They are considered ethnic delicacies and are exempt from the Egg Products Inspection Act. At this stage of incubation the embryo has soft bones and minimal feather development. They are boiled before consumption. Once the egg is removed from the incubator they must be either refrigerated or immediately cooked. These products must be labeled with the word "embryo" or "balut" preceded by the name of the kind of poultry, or labeled as "Incubated Fertile Eggs" or with similar wording. The label must include the complete name and address of the hatchery.[9] Baluts are considered potentially hazardous foods, and therefore must be kept refrigerated. FDA and individual states regulate the sale of baluts in the United States.

Both salt and alkali solutions are used to preserve eggs. Typically duck eggs are used to make these products. Coating or burying shell eggs in a strong alkali solution will cause the white to turn to a dark, firm gel and the yolk will turn green. When properly preserved these eggs have a pH of 9.0–10.0 and will keep 60–90 days without refrigeration. These eggs are commonly referred to as either "Century Eggs," "Black Eggs," or "Blue Eggs." They are not considered potentially hazardous foods. Salted duck eggs are produced by soaking shell eggs in highly concentrated salt solution for about 60 days. Salted eggs are cooked prior to consumption. When properly salted, the egg should have a natural odor. They are not considered potentially hazardous foods.

46.5 PATHOGENS OF CONCERN

46.51 Salmonella

Salmonella is the principal pathogen currently associated with eggs and egg products. Although other pathogenic microorganisms have been isolated from eggs, they have not been frequently implicated in foodborne disease outbreaks. Outbreaks caused by diverse *Salmonella* serotypes were attributed to eggs and egg products with considerable frequency before the 1970 Egg Products Inspection Act, but the institution of mandatory USDA inspection to remove cracked and dirty eggs and the implementation of effective pasteurization standards for egg products substantially reduced the association of eggs with human illness.

Beginning in the late 1980's, the incidence of human *S. enteritidis* infections began to increase dramatically in the United States and elsewhere.[74] Outbreaks of *S. enteritidis* have been associated with eggs more than any other food source. Unlike previous experience with other serotypes, *S. enteritidis* outbreaks have typically involved clean, intact, Grade A shell eggs.[76] This highly invasive serotype can cause systemic (although often inapparent) infections in laying hens, with the colonization of reproductive organs leading to internal contamination of eggs before oviposition.[38] However, *S. enteritidis* contamination of eggs has been shown to occur very infrequently in infected flocks, and contaminated eggs usually contain very small numbers of bacterial cells.[49] As a consequence, most egg-associated *S. enteritidis* outbreaks have also involved temperature abuse that allowed the pathogen to multiply to more dangerous levels before cooking, cross-contamination within commercial or institutional kitchens, and cooking methods that were inadequate to destroy all bacteria present.

By the mid 1990s, *S. enteritidis* was responsible for more reported cases of human salmonellosis in the United States than any other serotype. Accordingly, control programs have been developed and implemented with the aim of reducing both the incidence of *S. enteritidis* infections in laying chickens and the likelihood that consumers will be exposed to infectious doses of *S. enteritidis* in eggs.[46] Risk reduction programs in laying flocks that emphasize sanitation and biosecurity practices are complemented by renewed emphasis on proper refrigeration and handling of eggs, thorough cooking practices, and the use of pasteurized products for highly susceptible populations of consumers.

The number of *Salmonella* organisms isolated from unpasteurized liquid egg products is usually less than 1 per gram. The specified pasteurization procedures produce a 6 to 8 log reduction of *Salmonella* (including *S. enteritidis*) in liquid egg products.[20,81,87] The likelihood of finding *Salmonella* in a pasteurized product is thus, very low in the absence of opportunities for recontamination after processing. The isolation of salmonellae from pasteurized egg products indicates a need for all processing procedures and equipment to be carefully reviewed and inspected.

46.52 Other Egg Pathogens

Pathogenic organisms of emerging concern in eggs and egg products include *Campylobacter jejuni, Listeria monocytogenes,* and *Yersinia enterocolitica.* All three are resistant to lysozyme.[48] *Campylobacter* can survive (under anaerobic conditions) or grow (under aerobic conditions) in inoculated yolk, in albumen and yolk mixtures, in liquid whole egg, on the surface of the shell, and in the inner shell and shell membranes.[19,27,42,51,70] The organism is not sensitive to the high pH of the albumen but is sensitive to conalbumin.[19] Therefore, *Campylobacter* cannot survive for extended time periods (> 48 hr) in albumen.[19,42,51] The killing effect in albumen is temperature dependent, with the longest survival time at 4°C.[19,51] *Campylobacter* could not be recovered from commercially pasteurized yolk, albumen, liquid whole egg, or scrambled egg mix, or from chopped whole eggs or egg and cheese omelets.[51] It has been reported that 8.1% of laying hens are chronic shedders (positive > 30% of sampling times) of the organism, but the organism has not been recovered from the contents of eggs produced by infected hens and it is believed to be unlikely that vertical transmission occurs under commercial conditions.[7,27,82,88] Identical serotypes were recovered from infected hens' feces and human patients in an outbreak in which eggs were uncooked.[30]

Listeria monocytogenes and *Y. enterocolitica* both grow at refrigeration temperatures.[8,13,63] It has been estimated that 10% to 30% of personnel who work in egg product plants and slaughterhouses harbor *L. monocytogenes*.[54] Although listeriosis occurs frequently in the hen, producing lesions in the oviduct, there is no indication of transovarian transmission.[40,53,78] However, there is growing evidence that *L. monocytogenes* represents a potential hazard in egg products.[62] *Listeria* can grow in liquid egg or reconstituted dried egg at room or refrigerated temperatures, is stable in frozen egg products and can persist for up to 180 days in powdered egg products.[14,56,62,92] Although *L. monocytogenes* apparently can not survive in egg white it can grow in whole egg or yolk and can survive commercial egg wash water conditions.[14,28,29,32,60,62,83] In addition, *L. monocytogenes* has been detected in liquid egg products and appears to be resistant to at least minimal pasteurization.[32,33,61,62,69] The organism has also been recovered from pasteurized milk (61.7°C for 3.5 min, and 72.2°C for 16.4 sec) when pre-pasteurization levels were greater than 5×10^4/mL and 1×10^5, respectively.[8,15] Present egg product pasteurization requirements[23] may not be adequate for destruction of the organism if high levels are present. However, when thermal resistance studies have been done on specific strains there is an indication that minimal pasteurization requirements for liquid egg yolk would be more lethal to *L. monocytogenes* than the corresponding prescribed thermal processing conditions for egg whites and may also be a function of other factors such as pH.[71,77] *Y. enterocolitica* has been isolated from domestic fowl, game birds, wild birds, processed poultry, and egg products.[25,26,45,63] The organism can grow in commercially pasteurized unsalted eggs[28] and is capable of survival under extremely alkaline conditions.[4,57,91] Certain strains of *Y. enterocolitica* can survive in egg wash water under simulated commercial conditions of pH 10 and 38°C.[85] *Y. enterocolitica* has the potential to penetrate egg shells and contaminate egg contents due to improper washing, storage, or handling procedures.[1] Southam et al.[85] recommended that wash water temperatures be maintained at pH ≥10 and a temperature ≥40°C. On peeled hard-cooked eggs, the organism can survive for 8 to 9 weeks at –20°C[55] and can grow at 4°C in a 0.5% citric acid solution.[5]

46.6 INDICATORS OF LACK OF SANITATION

Coliform, as well as yeast and mold counts, are generally used as indicators of unsanitary conditions. *E. coli* counts may be a more important indicator than coliform counts.

Chemical methods also indicate poor egg quality or poor storage conditions. For example, the amount of lactic acid formed during bacterial growth has been used to show improper handling and poor sanitation.[59]

46.7 RECOMMENDED METHODS[3,98]

46.71 Sampling and Preparation of the Sample

46.711 Shell Eggs

Select shell eggs at random from cases or cartons representative of the lot. Transfer the eggs to clean cartons or cases for transport to the laboratory, and maintain the eggs at temperatures below 10°C until analyzed. Avoid egg sweating that can result when shell eggs are transferred from cold storage temperatures (around 4.4°C) to room temperature without a tempering period, especially when humidity is high. Returning sweating eggs to cold storage can promote the penetration of bacteria through the eggshell. Moisture on the shell surface can also increase opportunities for the egg contents to become contaminated when the shell is broken for sampling, so eggs should not be handled until they have been warmed to room temperature.

The interior contents of shell eggs can be sampled as follows:

a. Wash each egg with a brush, using soap and water 11 degrees warmer than the egg (generally at least 32.2°C).
b. Drain off excess moisture and immerse the egg in 70% alcohol for 10 min.
c. Remove from alcohol, drain, and flame.
d. Handle the alcohol-flamed eggs using sterile gloves, crack the egg against a sterile sharp-edged surface (or using a sterile breaking knife), and aseptically pour out the contents into a sterile container (blender jars, mason jars, disposable bea-

kers, or stomacher bags are all satisfactory). Egg contents are then mixed using sterile glass beads, a sterile spoon or similar utensil, a sterile electric mixer, or a stomacher until the sample is homogeneous. For separate examination of egg yolk and albumen, use a sterile egg separator or spoon.

e. For bacterial enumeration, weigh 11 g of the homogenous sample into a 99-mL phosphate buffered saline or peptone water dilution blank containing glass beads and shake 25 times by hand through a 1 ft arc for 7 sec. Prepare appropriate serial dilutions using similar dilution blanks.

Because of the extremely low incidence at which *S. enteritidis* contamination has been found to occur, the contents of 20–30 shell eggs are often pooled together so that a satisfactory detection sensitivity can be achieved using a manageable number of samples. Such pooled samples must be pre-incubated before further bacteriological analysis to allow *S. enteritidis* contaminants to reach consistently detectable levels.[37] Iron supplementation of incubating egg pools can increase the rate of growth of many *S. enteritidis* isolates.[39] Sampling methods to isolate and identify *Salmonella* are described in detail elsewhere.[3,96]

46.712 Liquid and Frozen Egg Products

Although testing for *Salmonella* is the principal concern for eggs in the United States, the same sampling methods are applicable for other microbiological analyses. Three general types of samples are taken for *Salmonella* determination: surveillance, confirmation, and certification. Surveillance samples, the routine samples taken to determine the presence or absence of *Salmonella*, are collected and analyzed by the processor. Confirmation samples are taken by USDA inspectors for verification of the surveillance sampling and testing programs and are analyzed by a USDA laboratory. Certificates are issued by USDA for *Salmonella*-negative products usually when requested by a customer of the processor. For example, some customers in foreign countries request USDA certification. Certificates are issued only if a certified sample has been drawn by USDA, analyzed by its laboratory, and found to be negative for *Salmonella*. Where egg products are to be used in Category I products (intended for consumption by infants and aged, infirm, or immunocompromised individuals), more intensive sampling and testing are usually required. Some customers, for example, require using the sampling plan described by Foster[36] and the testing of 1,500 g per lot.

Obtain samples of liquid eggs from vats or tanks at the plant or from containers. Make sure that the product has been thoroughly mixed. Use a sterilized dipper or sampling tube to withdraw the sample. Sterile pint mason jars or friction top cans are satisfactory for holding the samples. Obtain about three-fourths of a pint of sample and hold it below 4.4°C for no more than 4 hr if possible. Avoid freezing, which will destroy many of the bacteria present. If there is doubt as to the homogeneity of the liquid in a vat, take several three-fourths-pint samples and composite them in a 2-quart jar to give a more representative sample.

Record the temperature of the containers from which the samples were taken. Often temperature may be the key to abnormal bacterial populations because temperatures above 7.2°C indicate improper handling.

Select cans of frozen egg representative of the lot. Open the cans, and with a sterile spoon remove any ice or frost on the frozen egg. The area selected for drilling should not be humped or peaked. With a high-speed electric drill, puncture the egg about 1 inch from the edge of the can using the following steps:

a. Slant the bit so it will go through the center of the frozen egg to within an inch or two of the opposite lower edge of the can.
b. Transfer the shavings to a sample container with a sterile spoon. Keep the shavings frozen at all times.
c. Pack in an insulated box with dry ice for transport to the laboratory.

Frequency of sampling of liquid and frozen egg products is based on the performance and history of a plant's capability to produce a *Salmonella*-negative product. Frequency of sampling is increased or decreased according to a flow chart issued by USDA[94,95] and updated in 1996 (Figure 1; FSIS Directive 10, 230.4). One 4-oz sample is collected per lot (one lot may represent a day's production) and from this amount 100 g are used for *Salmonella* analysis. For a plant without a history of *Salmonella*-negative product, all lots must be sampled until 60 consecutive lots show negative *Salmonella* results. Sampling then is decreased to 1 in every 2 lots. If another 60 consecutive lots are negative, sampling is decreased to 1 in every 4 lots. After this period, if 60 or more consecutive lots are negative, sampling is decreased to 1 in every 8 lots. If *Salmonella*-positive lots occur at any time, sampling must then be increased according to the flow chart.

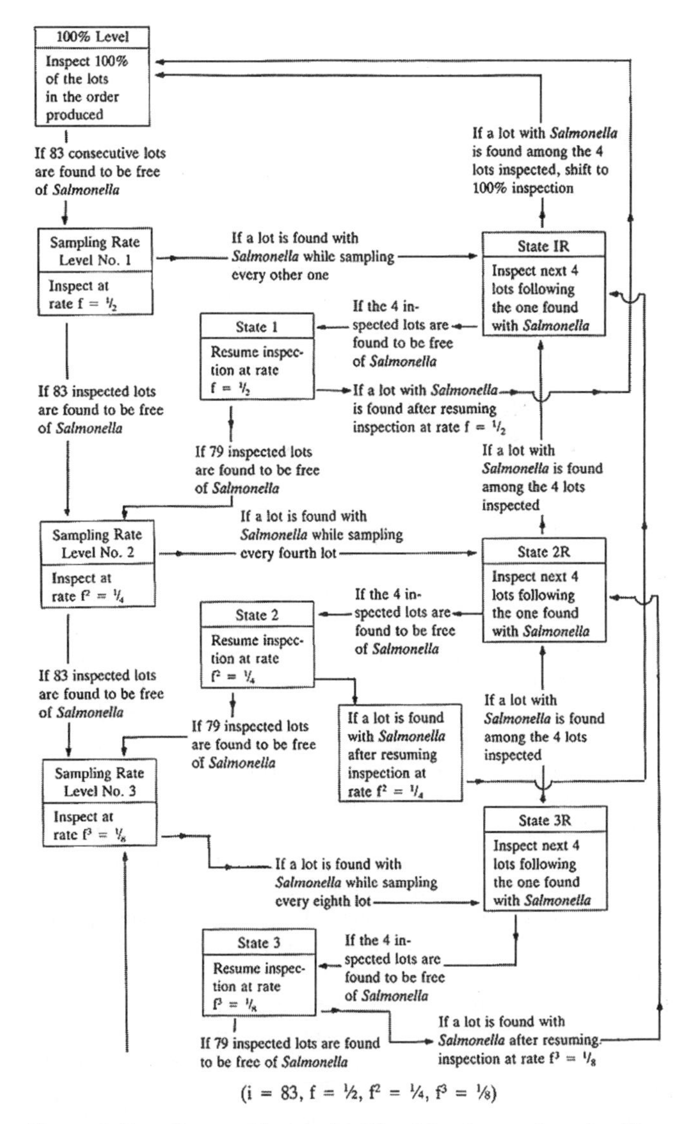

Figure 1. *Flow Process Chart for Multilevel Continuous Sampling Plans*

46.713 Dried Egg Products

Sample dried egg products with a sterilized spoon or trier. Transfer sample to a sterile jar or plastic bag. Do not fill the container more than two-thirds full to leave room for mixing.

If the product is in small packages, select several unopened packages. Open them at the laboratory under aseptic conditions and transfer a liberal quantity to a sterile can or beaker. Thoroughly mix with a sterile spoon to obtain a homogeneous mixture.

As required by USDA, each lot of dried egg product must be sampled and tested for *Salmonella.*[95] Usually a lot is considered to be a day's production of each product from each dryer.

For surveillance sampling, the company is required to have analyzed 3 samples drawn from each lot of each product produced from each dryer. Alternatively it may elect to have surveillance samples drawn and analyzed at the same level as that prescribed for certification. For certification, the number of samples drawn and the amount of product tested from each lot are given as follows:

Number of containers in lot	Number of individual samples to be drawn	Number of analyses and amount of sample to be analyzed per analysis
50 or less	4	(1) 100 g
51 to 150	8	(2) 100 g
151 to 500	12	(3) 100 g
501 to 1500	16	(4) 100 g
Over 1500	20	(5) 100 g

Twenty-five grams of each sample are removed, and 4 of these samples are combined to make the 100 g composite used for the *Salmonella* test.

46.714 Tests to Be Done on Each Lot

Aerobic Plate Count 35°C—Chapter 7.
Aerobic Plate Count 32°C—Chapter 7.
Coliform Group—Chapter 8.
Enterobacteriaceae—Chapter 8.
E. coli—Chapter 35.
Yeast and Mold Count—Chapters 5, 6, and 20.
Direct Microscopic Count—Chapter 4.
Salmonella—Chapters 5, 6, and 37.

In addition to the culture method for *Salmonella,*[2] several rapid screening methods for the isolation of *Salmonella* have been adopted as official methods by the Association of Official Analytical Chemists.[3] These methods are also recognized as acceptable substitutes for conventional culturing by the FDA.[2] While negative results using these rapid methods are considered definitive, positive results are regarded as presumptive and confirmation must be done by the conventional culture method.[2,3] Positive results are confirmed by plating, biochemical testing, and serotyping according to the culture method.[2] Technological advancements for faster, more specific and more sensitive detection systems for salmonellae and other foodborne pathogens continue to be developed and commercially marketed.[72] However, in order for these methods to supplant current methods, approaches must be emphasized that avoid required preenrichment steps to achieve sufficient numbers of organisms prior to application of the rapid assay. Otherwise such "rapid" methods will fail to yield conclusive confirmation during the time required to process the egg or egg product. Two avenues have been explored to circumvent the problem of detecting low numbers of foodborne pathogenic organisms such as salmonellae in eggs. One approach involves concentrating low numbers of *S. enteritidis* in pooled egg contents by a centrifugation method that involves enzymatic digestion and chemical reduction of egg albumen[66] or using a magnetic bead-ELISA system to bind the organism in pooled liquid egg contents.[47] The other approach has been to use polymerase chain amplification to detect *S. enteritidis* DNA[67] or luminescent recombinant bacteriophages specific for *Salmonella* spp.[18] These approaches once standardized should reduce detection times to hours rather than days by virtue of their ability to allow detection of far fewer organisms. Additional information on rapid methods is presented in Chapter 12.

46.8 INTERPRETATION OF DATA

The aerobic plate count of the interior contents of shell eggs should be less than 10 microorganisms per gram. A count of more than 100 microorganisms per gram usually indicates bacterial invasion through the shell, possibly due to improper washing and sanitizing of the shell surface, followed by a period of storage or excessive handling of "sweated" eggs. The storing of eggs in a cooler at a relative humidity above 85% encourages mold growth on the surface of shell eggs.

Aerobic plate counts of unpasteurized liquid egg from commercial eggbreaking operations are generally in the range of 10^3 to 10^6 per gram. Aerobic plate counts exceeding 10^7 per gram usually indicate the use of poor-quality eggs for breaking but may also be due to poor sanitation or improper storage of the liquid eggs. Coliform counts in raw liquid eggs can be expected to be from 10^2 to 10^5 per gram, yeast and mold less than 10 per gram, and *Salmonella* less than 1 per gram.

All egg products, liquid, frozen, or dried, should meet the following specifications:

Aerobic Plate Count	< 25,000 per g
Coliform Group	< 10 per g
Yeast and Mold	< 10 per g
Salmonella	Negative by prescribed sampling and testing procedures

The reasons that egg products do not meet the above specifications include poor microbiological quality of the unpasteurized liquid eggs, improper pasteurization, and recontamination after pasteurization. Liquid or frozen eggs subject to temperature abuse also will be of poor microbiological quality.[58]

46.81 Potential Hazards

Generally egg products are used in foods that are cooked or baked in such a way that *Salmonella* and other pathogens are destroyed, or used in foods such as mayonnaise and salad dressing, where the pH is sufficiently low to inhibit bacterial growth. However, if a contaminated egg product is brought into a food plant, there is always the possibility of contamination of other foods. Egg products are also used in some foods where pathogens, such as *Salmonella,* can survive, such as in meringue for pies. In any case, it is important that egg products be microbiologically safe.

46.82 Action Taken when Egg Products Are below Specification

If *Salmonella*-positive samples of egg products are found, USDA allows the following procedures:

1. If *Salmonella*-positive results are obtained in either liquid or frozen egg products, the egg processor must test the next 4

consecutive lots and go through the frequency of sampling outlined in the USDA-specified flow diagram.

2. For frozen egg products, containers are divided into sublots of 100 or less and the number of samples taken is the square root of the number in the sublot. The samples from each sublot are composited, and a 100-g aliquot is analyzed from each composite sample.
3. For dried egg whites the entire lot must be heat-treated again, resampled, and tested. Alternately, the product may be resampled and tested as prescribed for yellow products below.
4. For dried yellow products (made from whole egg and/or yolk), the entire lot must be reconstituted, repasteurized, and redried. Alternatively, yellow products may be resampled and tested as follows:
   a. For containers of 100 lbs (net) or over, containers must be divided into sublot groups of 4 consecutively produced containers and samples drawn from each container in the sublot.
   b. For containers of 50 lbs to 99 lbs (net), containers must be divided into sublot groups of 8 consecutively produced containers and samples taken from every other container in the sublot.
   c. For containers of less than 50 lbs (net), containers must be divided into sublot groups of 12 consecutively produced containers and samples drawn from every third container in the sublot.
   d. Twenty-five grams from each of the 4 consecutively numbered samples are combined into a 100-g composite for testing.

If the sublots of frozen or dried egg products are still found to be *Salmonella*-positive, they must be reprocessed and repasteurized.

46.9 REFERENCES

1. Amin, M. K., and F. A. Draughon. 1990. Infection of shell eggs with *Yersinia enterocolitica*. J. Food Prot. 53:826-830.
2. Andrews, W. H., G. A. June, P. S. Sherrod, T. S. Hammack, and R. M. Amaguana. 1995. *Salmonella*, Chapter 5. *In* FDA, Bacteriological analytical manual, 8th ed. AOAC International, Gaithersburg, Md.
3. AOAC. 1997. Official methods of analysis of the AOAC, 16th ed. AOAC International, Gaithersburg, Md.
4. Aulisio, C. C. G., I. J. Mehlman, and A. C. Sanders. 1980. Alkali method for rapid recovery *of Yersinia enterocolitica* and *Yersinia pseudotuberculosis* from foods. Appl. Environ. Microbiol. 39:135-140.
5. Bailey, J. S., D. L. Fletcher, and N. A. Cox. 1987. The influence of added egg yolk on the microbiological quality of hard-cooked eggs stored in a citric acid/sodium benzoate solution. Poult. Sci. 66:861-865.
6. Baker, R.C., and C. Bruce. 1994. Effects of processing on the microbiology of eggs, p. 153-173. In R. G. Board and R. Fuller (ed.), Microbiology of the avian egg. Chapman and Hall, London.
7. Baker, R. C., M. D. C. Paredes, and R. A. Qureshi. 1987. Prevalence of *Campylobacter jejuni* in eggs and poultry meat in New York State. Poult. Sci. 66:1766-1770.
8. Bearns, R. E., and K. F. Girard. 1958. The effect of pasteurization on *Listeria monocytogenes*. Can. J. Microbiol. 4:55-61.
9. Birkhold, S. Personal communication with USDA, Agriculture Marketing Service, Poultry Division.
10. Board, R. G. 1964. The growth of gram-negative bacteria in the hen's egg. J. Appl. Bacteriol. 27:350.
11. Board, R. G., and H. S. Tranter. 1995. The microbiology of eggs, p. 81-104. *In* W. J. Stadelman and O. J. Cotterill (ed.), Egg science and technology, 8th ed. Haworth Food Products Press, Inc., New York.
12. Board, R. G., C. Clay, J. Lock, and J. Dolman. 1994. The egg: A compartmentalized, aseptically packaged food, p. 43-61. *In* R. G. Board and R. Fuller (ed.), Microbiology of the avian egg. Chapman and Hall, London.
13. Bojsen-Møller, J. 1972. Human listeriosis. Diagnostic, epidemiological, and clinical studies. Acta Path. Microbiol. Scand., Sect. B. Suppl. 229:13-157.
14. Brackett, R. E., and L. R. Beuchat. 1991. Survival of *Listeria monocytogenes* in whole egg and egg yolk powders and in liquid whole eggs. Food Microbiol. 8:331-337.
15. Bradshaw, J. G., J. T. Peeler, J. J. Corwin, J. M. Hunt, and R. M. Twedt. 1987. Thermal resistance of *Listeria monocytogenes* in dairy products. J. Food Prot. 50:543-544, 556.
16. same question as for 6.
17. Burley, R. W., and D. V. Vadehra. 1989. The avian egg - Chemistry and biology. John Wiley and Sons, Inc., New York.
18. Case, L. P., D. P. Carey, and D. A. Hirakawa. 1995. Canine and feline nutrition - A resource for companion animal professionals. Mosby-Year Book, Inc., St. Louis, Mo.
19. Chen, J., and M. W. Griffiths. 1996. *Salmonella* detection in eggs using *Lux*+ bacteriophages. J. Food Prot. 59:908-914.
20. Clark, A. G., and D. H. Bueschkens. 1986. Survival and growth of *Campylobacter jejuni* in egg yolk and albumen. J. Food Prot. 49:135-141.
21. Cotterill, O. J. 1968. Equivalent pasteurization temperatures to kill salmonellae in liquid egg white at various pH levels. Poult. Sci. 47:354-365.
22. Cotterill, O. J., and J. L. Glauert. 1979. Nutrient values for shell, liquid/frozen, and dehydrated eggs derived by linear regression analysis and conversion factors. Poult. Sci. 58:131-134.
23. Cotterill, O. J., J. L. Glauert, and G. F. Krause. 1973. Thermal destruction curves for *Salmonella oranienburg* in egg products. Poult. Sci. 52:568-577.
24. Cunningham, F. E. 1995. Egg-product pasteurization, p. 289. *In* W. J. Stadelman and O. J. Cotterill (ed.). Egg science and technology, 8th ed. Haworth Food Products Press, Inc., New York.
25. Curtis, P. A., K. E. Anderson, and F. T. Jones. 1995. Cryogenic gas for rapid cooling of commercially processed shell eggs before packaging. J. Food Prot. 58:389-394.
26. De Boer, E., B. J. Hartog, and J. Oosterom. 1982. Occurrence of *Yersinia enterocolitica* in poultry products. J. Food Prot. 45:322-325.
27. De Boer, E., W. M. Seldam, and J. Oosterom. 1986. Characterization of *Yersinia enterocolitica* and related species isolated from foods and porcine tonsils in the Netherlands. Int. J. Food Microbiol. 3:217-224.
28. Doyle, M. P. 1984. Association of *Campylobacter jejuni* with laying hens and eggs. Appl. Environ. Microbiol. 47:533-536.
29. Erickson, J. P., and P. Jenkins. 1992. Behavior of psychrotrophic pathogens *Listeria monocytogenes*, *Yersinia enterocolitica* and *Aeromonas hydrophila* in commercially pasteurized eggs held at 2, 6.7 and 12.8°C. J. Food Prot. 55:8-12.
30. Farber, J. M., E. Daley, and F. Coates. 1992. Presence of *Listeria* spp. in whole eggs and wash water samples from Ontario and Quebec. Food Res. Int. 25:143-145.
31. Finch, M. J., and P. A. Blake. 1985. Foodborne outbreaks of campylobacteriosis: The United States experience, 1980-1982. Am. J. Epidemiol. 122:262-268.
32. Florian, M. L. E., and P. C. Trussell. 1957. Bacterial spoilage of shell eggs. IV. Identification of spoilage organisms. Food Technol. 11:56-60.
33. Foegeding, P. M., and N. W. Stanley. 1990. *Listeria monocytogenes* F5069 thermal death times in liquid whole egg. J. Food Prot. 53:6-8, 25.
34. Foegeding, P. M., and S. B. Leasor. 1990. Heat resistance and growth of *Listeria monocytogenes* in liquid whole egg. J. Food Prot. 53:9-14.
35. Food Investigation Board. 1955. Eggs and egg products. Rep. Food Invest., No. 60, H.M.S.O. London, England.
36. Forsythe, R. H., J. C. Ayres, and J. L. Radlo. 1953. Factors affecting the microbiological populations of shell eggs. Food Technol. 7:49-56.

37. Foster, E. M. 1971. General Session Address: The control of Salmonellae in processed foods: A classification system and sampling plan. J. AOAC 54:259-266.
38. Gast, R. K. 1993. Recovery of *Salmonella enteritidis* from inoculated pools of egg contents. J. Food Prot. 56:21-24.
39. Gast, R. K., and C. W. Beard. 1990. Production of Salmonella enteritidis-contaminated eggs by experimentally infected hens. Avian Dis. 34:438-446.
40. Gast, R. K., and P. S. Holt. 1995. Iron supplementation to enhance the recovery of *Salmonella enteritidis* from pools of egg contents. J. Food Prot. 58:268-272.
41. Gray, M. L. 1958. Listeriosis in fowls—A review. Avian Dis. 2:296-314.
42. Hale, K. K., L. M. Potter, and R. B. Martin. 1981. Firmness and microbial quality of hard cooked eggs stored in citric acid. Poult. Sci. 60:1664.
43. Hänninen, M. L., H. Korkeala, and P. Pakkala. 1984. Growth and survival characteristics of *Campylobacter jejuni* in liquid egg. J. Hyg. Camb. 92:53-58.
44. Hargis, P. S., and M. E. Van Elswyk. 1993. Manipulating the fatty acid composition of poultry meat and eggs for the health conscious consumer. World's Poult. Sci. J. 49:251-264.
45. Hartung, T. E. and W. J. Stadelman. 1962. The influence of metallic cations on the penetration of the egg shell membranes by *Pseudomonas fluorescens*. Poult. Sci. 41:1590-1596.
46. Heddleston, K. L. 1972. Fowl cholera, p. 219-241. *In* M. S. Hofstad, B. W. Calnek, C. F. Helmboldt, W. M. Reid, and H. W. Yoder, Jr. (ed.), Diseases of Poultry, 6th ed. Iowa State Univ. Press, Ames, Iowa.
47. Hogue, A., P. White, J. Guard-Petter, W. Schlosser, R. Gast, E. Ebel, J. Farrar, T. Gomez, J. Madden, M. Madison, A. M. McNamara, R. Morales, D. Parham, P. Sparling, W. Sutherlin, and D. Swerdlow. 1998. Epidemiology and control of egg-associated *Salmonella* Enteritidis in the United States of America. Rev. Sci. Tech. Off. Int. Epiz. 16:542-553.
48. Holt, P. S., R. K. Gast, and C. R. Greene. 1995. Rapid detection of *Salmonella enteritidis* in pooled liquid egg samples using a magnetic bead-ELISA system. J. Food Prot. 58:967-972.
49. Hughey, V. L., and E. A. Johnson. 1987. Antimicrobial activity of lysozyme against bacteria involved in food spoilage and food-borne disease. Appl. Environ. Microbiol. 53:2165.
50. Humphrey, T. J. 1994. Contamination of egg shell and contents with *Salmonella enteritidis*: A review. Int. J. Food Microbiol. 21:31-40.
51. Humphrey, T. J., and A. Whitehead. 1993. Egg age and the growth of *Salmonella enteritidis* PT4 in egg contents. Epidemiol. Infect. 111:209-219.
52. Izat, A. L., and F. A. Gardner. 1988. The incidence of *Campylobacter jejuni* in processed egg products. Poult. Sci. 67:1431-1435.
53. Jiang, Y. H., R. B. McGeachin, and C. A. Bailey. 1994. a-Tocopherol, b-Carotene, and retinol enrichment of chicken eggs. Poult. Sci. 73:1137-1143.
54. Kampelmacher, E. H. 1962. Animal products as a source of listeric infection in man, p. 145-151. *In* M. L. Gray (ed.), Second Symposium on Listeric Infection. Montana State College, Bozeman, Mont.
55. Kampelmacher, E. H., and L. M. van Noorle Jansen. 1969. Isolation of *Listeria monocytogenes* from faeces of clinically healthy humans and animals. Zentralbl. Bakteriol. Parasitenkd. Infektionskr. I. Abt. Orig., Reine A 211:353-359.
56. Kendall, M. R., and J. Gilbert. 1980. Survival and growth of *Yersinia enterocolitica* in broth media and in food. Tech. Ser., Soc. Appl. Bacteriol. 15:215-226.
57. Khan, M. A., I. A. Newton, A. Seaman, and M. Woodbine. 1975. The survival of *Listeria monocytogenes* inside and outside its host. *In* Problems of Listeriosis. Lecture: Microbiol. Unit, Department of Applied Biochemistry & Nutrition, University of Nottingham, Nottingham, U.K.
58. Kinner, J. A., and W. A. Moats. 1981. Effect of temperature, pH, and detergent on survival of bacteria associated with shell eggs. Poult. Sci. 60:761-767.
59. Kraft, A. A., L. E. Elliott, and A. W. Brant. 1958. The shell membrane as a barrier to bacterial penetration of eggs. Poult. Sci. 37:238-240.
60. Kraft, A. A., G. S. Torrey, J. C. Ayres, and R. H. Forsythe. 1967. Factors influencing bacterial contamination of commercially produced liquid egg. Poult. Sci. 46:1204-1210.
61. Laird, J. M., F. M. Bartlett, and R. C. McKellar. 1991. Survival of Listeria monocytogenes in egg washwater. Int. J. Food Microbiol. 12:115-122.
62. Leasor, S. B., and P. M. Foegeding. 1989. *Listeria* species in commercially broken raw liquid whole egg. J. Food Prot. 52:777-780.
63. Leclair, K., H. Heggart, M. Oggel, F. M. Bartlett, and R. C. McKellar. 1994. Modelling the inactivation of *Listeria monocytogenes* and *Salmonella typhimurium* in simulated egg wash water. Food Microbiol. 11:345-353.
64. Leistner, L., H. Heckelmann, M. Kashiuazaki, and R. Albertz, R. 1975. Nachweis von *Yersinia enterocolitica* in Faeces und Fleisch von Schweinen, Rindern und Geflügel. Fleischwirtschaft 55:1599-1602.
65. Lepper, H. A., M. T. Bartram, and F. Hillig. 1944. Detection of decomposition in liquid, frozen, and dried eggs. J. AOAC 27:204-223.
66. Leskanich, C. O., and R. C. Noble. 1997. Manipulation of the *n*-3 polyunsaturated fatty acid composition of avian eggs and meat. World's Poult. Sci. J. 53:155-183.
67. McElroy, A. P., N. D. Cohen, and B. M. Hargis. 1995. Evaluation of a centrifugation method for the detection of *Salmonella enteritidis* in experimentally contaminated chicken eggs. J. Food Prot. 58:931-933.
68. McElroy, A. P., N. D. Cohen, and B. M. Hargis. 1996. Evaluation of polymerase chain reaction for the detection of *Salmonella enteritidis* in experimentally inoculated eggs and eggs from experimentally challenged hens. J. Food Prot. 59:1273-1278.
69. Moats, W. A. 1981. Factors affecting bacterial loads on shells of commercially washed eggs. Poult. Sci. 60:2084-2090.
70. Moore, J., and R. H. Madden. 1993. Detection and incidence of *Listeria* species in blended raw egg. J. Food Prot. 56:652-654, 660.
71. Morishige, M., T. Kinjo, and N. Minamoto. 1984. Growth and survival of *Campylobacter jejuni* in yolk. Res. Bull. Faculty of Agric., Gifu Univ., Gifu-ken, Japan.
72. Palumbo, M. S., S. M. Beers, S. Bhaduri, and S. A. Palumbo. 1996. Thermal resistance of *Listeria monocytogenes* and *Salmonella* spp. in liquid egg white. J. Food Prot. 59:1182-1186.
73. Pillai, S. D., and S. C. Ricke. 1995. Strategies to accelerate the applicability of gene amplification protocols for pathogen detection in meat and meat products. Crit. Rev. Microbiol. 21:239-261.
74. Prochaska, J. F., J. B. Carey, and D. J. Shafer. 1996. The effect of L-lysine intake on egg component yield and composition in laying hens. Poult. Sci. 75:1268-1277.
75. Rodrigue, D. C., R. V. Tauxe, and B. Rowe. 1990. International increase in *Salmonella enteritidis*: A new pandemic? Epidemiol. Infect. 105:21-27
76. Rosser, F. T. 1942. Preservation of eggs. II. Surface contamination on egg shell in relation to spoilage. Can. J. Res. 20D:291-296.
77. St. Louis, M. E., D. L. Morse, M. E. Potter, T. M. DeMelfi, J. J. Guzewich, R. V. Tauxe, P. A. Blake, and the *Salmonella enteritidis* Working Group. 1988. The emergence of grade A eggs as a major source of *Salmonella enteritidis* infections: New implications for the control of salmonellosis. JAMA 259:2103-2107.
78. Schuman, J. D., and B. W. Sheldon. 1997. Thermal resistance of *Salmonella* spp. and *Listeria monocytogenes* in liquid egg yolk and egg white. J. Food Prot. 60:634-638.
79. Seeliger, H. P. R. 1961. Listeriosis. Hafner Publishing Co., New York.
80. Serrano, L. E., E. A. Murano, K. Shenoy, and D. G. Olson. 1997. D values of *Salmonella enteritidis* isolates and quality attributes of shell eggs and liquid whole eggs treated with irradiation. Poult. Sci. 76:202-206.
81. Shafer, D. J., J. B. Carey, and J. F. Prochaska. 1996. Effect of dietary methionine intake on egg component yield and composition. Poult. Sci. 75:1080-1085.
82. Shafi, R., O. J. Cotterill, and M. L. Nichols. 1970. Microbial flora of commercially pasteurized egg products. Poult. Sci. 49:578-585.

83. Shane, S. M., D. H. Gifford, and K. Yogasundram. 1986. *Campylobacter jejuni* contamination of eggs. Vet. Res. Comm. 10:487-492.
84. Sionkowski, P. J., and L. A. Shelef. 1990. Viability of *Listeria monocytogenes* strain Brie-1 in the avian egg. J. Food Prot. 53:15-17, 25.
85. Solomon, S. E., M. M. Bain, S. Cranstoun, and V. Nascimento. 1994. Hen's egg shell structure and function, p. 1-24. *In* R. G. Board and R. Fuller (ed.), Microbiology of the avian egg. Chapman and Hall, London.
86. Southam, G., J. Pearson, and R. A. Holley. 1987. Survival and growth of *Yersinia enterocolitica* in egg washwater. J. Food Prot. 50:103-107.
87. Sparks, N. H. C. 1994. Shell accessory materials: structure and function, p. 25-42. *In* R. G. Board and R. Fuller (ed.), Microbiology of the avian egg. Chapman and Hall, London.
88. Speck, M. L., and F. R. Tarver Jr. 1967. Microbiological populations in blended eggs before and after commercial pasteurization. Poult. Sci. 46 (Abstr.):1321.
89. Stadelman, W. J., and O. J. Cotterill (ed.). 1995. Egg science and technology, 4th ed. Haworth Food Products Press, Inc., New York.
90. Stadelman, W. J., A. I. Ikeme, R. A. Roop, and S. E. Simmons. 1982. Thermally processed hard cooked eggs. Poult. Sci. 61:388-391.
91. Stadelman, W. J., R. K. Singh, P. M. Muriana, and H. Hou. 1996. Pasteurization of eggs in the shell. Poult. Sci. 75:1122-1125.
92. Stern, N. J., M. D. Pierson, and A. W. Kotula. 1980. Effects of pH and sodium chloride on *Yersinia enterocolitica* growth at room and refrigeration temperatures. J. Food Sci. 45:64-67.
93. Urbach, H., and G. Schabinski. 1955. Zur Listeriose des Menschen. Z. Hyg. Infer. 141:239-248.
94. U.S. Department of Agriculture. 1969. Egg pasteurization manual. ARS 74-48. USDA, Agricultural Research Service, Washington, D. C.
95. U.S. Department of Agriculture. 1972. Egg products inspection handbook. Dried egg products instructions. AMSPY Instruc. 910 (Egg Products)—5. Revision 2. USDA, Washington, D. C.
96. U.S. Department of Agriculture. 1975. Egg products inspection handbook. Section 8—Sampling for bacteriological, chemical, and physical testing. AMS-PY Instruc. 910 (Egg Products)-l. USDA, Washington, D. C.
97. U.S. Department of Agriculture. 1993. Laboratory methods for egg products. Science Division, USDA, Washington, D. C.
98. U.S. Government. 1998 (May 19). Salmonella Enteritidis in eggs. Fed. Reg. 63(96):27502-27511.
99. Vanderzant, C., and D. F. Splittstoesser (ed.). 1992. Compendium of methods for the microbiological examination of foods. American Public Health Association, Washington, D. C.

Chapter 47

Milk and Milk Products

R. L. Richter and E. R. Vedamuthu

47.1 INTRODUCTION

Microorganisms are important in dairy products. The proper selection and balance of starter cultures is critical for the manufacture of fermented products of desirable rheological, texture, and flavor characteristics. Undesirable microorganisms are responsible for the spoilage of dairy products, and pathogens introduced into milk and milk products by unsatisfactory milk production practices, failures in processing systems, or unsanitary practices are of primary concern. The importance of the desirable and undesirable microorganisms in milk and milk products has resulted in the development of methods to enumerate them and in the establishment of standards to reflect the safety or quality of milk and milk products.

The microbiological quality of milk and milk products is influenced by the initial flora of the raw milk, the processing conditions, and post-pasteurization contamination. High moisture products such as fluid milks, cottage cheese, concentrated milk, ice cream mixes, and cultured milks have a maximum shelf life of 2 to 3 weeks because of microbial growth even under good processing and refrigerated storage conditions. The shelf life of cultured milk products is somewhat longer than that of unfermented products because of their high acidity. Other dairy products such as ultrapasteurized milk products, sweetened condensed and evaporated milk, ripened cheeses, and butter have an extended shelf life because of reduced water activity, pH, temperature of storage, and/or heat treatment applied during processing.

Recognition by public health authorities and the dairy industry in the early 1900s that raw milk was a major public health concern resulted in the development of recommendations that defined proper conditions for the production and handling of milk. These recommendations evolved into the Milk Ordinance of 1924 and an interpretation of these regulations in the Code in 1927.[30] The currently used document is known as the Pasteurized Milk Ordinance, 1995 Revision.[22] It defines the conditions under which Grade A milk products must be produced and processed. Methods to evaluate milk and milk products for quality and safety have been standardized and are published by the American Public Health Association as the *Standard Methods for the Examination of Dairy Products* (SMEDP).[48]

The Food and Drug Administration (FDA) and the U.S. Department of Agriculture (USDA) both define the microbiological limits for products (Tables 1 and 2). These microbiological criteria can be met consistently provided good manufacturing practices are employed, but they do not necessarily provide guidelines that ensure maximum shelf life or stability of product quality. The dairy industry has accepted the responsibility for processing safe, wholesome products that retain desirable qualities throughout the shelf life of the product. Acknowledgment of this responsibility has resulted in the establishment of quality criteria that are more demanding than those established by regulatory agencies.

47.2 RAW MILK

Raw milk, as it leaves the udders of healthy animals, normally contains very low numbers of microorganisms. Total counts usually are less than 10^3 per mL. *Micrococcus*, *Staphylococcus*, *Streptococcus*, and *Corynebacterium* spp. are the most common bacteria usually present.[15,40,42,62] If lactating cows have mastitis, large numbers of the infectious organisms might be shed into the milk and increase the total counts of bulk milk, if the milk from infected cows is not kept separate.[5,36] *Staphylococcus aureus* and *Streptococcus agalactiae* are commonly associated with contagious mastitis while coliforms, *Pseudomonas*, and other *Streptococcus* spp. are more related to environmental mastitis.[59] Mastitic infections are generally accompanied by a rise in the somatic cell count of milk. Most bacteria in milk as it leaves the cow generally do not grow well under refrigeration, thus their numbers in bulk milk usually do not increase significantly.

After it leaves the udder, milk may become contaminated with microorganisms from the surfaces of the cow, the environment, and unclean milking systems.[62,76,77,78,79] Contamination is generally bacterial; yeasts and molds occur rarely or in low numbers. Unclean udders and teats can contribute organisms from a variety of sources (e.g., manure, soil, feed, and water). They include lactic acid bacteria, coliform and other gram-negative bacteria, *Bacillus*, and *Clostridium* spp., as well as the normal surface flora (e.g. *Micrococcus* and *Staphylococcus* spp.). Improperly cleaned milk contact surfaces can add substantial numbers of microorganisms to subsequent milkings by providing conditions (nutrients) for growth of contaminating microorganisms. Large numbers of thermoduric bacteria (*Micrococcus*, *Microbacterium*, *Streptococcus*, *Lactobacillus*, *Enterococcus*, *Bacillus* spp.) are associated with persistent poor cleaning of milking machines, pipelines, bulk storage

Table 1. Microbiological Standards for Raw and Fluid Products

Product	Standard	Reference
Grade A raw milk for pasteurization		22
Individual producer samples	100,000/mL	
Commingled samples	300,000/mL	
USDA raw milk for manufacturing		93
Grade no. 1	500,000/mL	
Grade no. 2	1,000,000/mL	
Undergrade	>1,000,000/mL	
Grade A pasteurized fluid products		22
Bacteria	20,000/mL	
Coliforms		
Packaged	10/mL	
Bulk for transport	100/mL	
Grade A aseptically processed		22
Bacteria—No growth tests specified in Section 6 of the Grade A Pasteurized Ordinance		

tanks, and transfer hoses while higher numbers of gram-negative bacteria or lactococci may occur from occasional neglect.[62] Thermoduric bacteria have a direct influence on the total bacterial counts of freshly pasteurized milk, but their growth in the farm bulk tank is minimal. In summary, the types of organisms that prevail in raw milk depend on the initial microbial population, the extent of cleaning and sanitizing of milking equipment and utensils, and the time and temperature of storage.

Because raw milk is cooled immediately and held at refrigeration temperatures, increases in the microbial load of milk in the bulk tank are usually caused by psychrotrophic bacteria. *Pseudomonas, Flavobacterium*, and *Alcaligenes* spp. as well as some of the coliform bacteria tend to be the predominant psychrotrophic bacteria of raw milk in storage.[2,10,74,77,78] These bacteria, if allowed to grow to large numbers, can cause bitter, fruity, rancid, and unclean flavors that persist throughout further processing. Some gram-negative psychrotrophic bacteria, particularly *Pseudomonas* species, are also capable of producing heat-stable enzymes (proteases and lipases) in raw milk that can cause defects in the final product.[2,10,73] Some gram-positive organisms capable of growth in refrigerated milk are species of *Bacillus, Micrococcus, Enterococcus*, and *Arthrobacter*.[10] They are usually of less significance as defect producers than are the gram-negative bacteria. The lactococci grow and produce acid in poorly refrigerated raw milk. A malty flavor defect may also occur under these conditions. However, the use of refrigerated bulk tanks has reduced the frequency of milk spoilage by lactic streptococci. Other flavor defects of raw milk that are non-microbial in nature must be considered in evaluating raw milk quality. These include such flavors as rancid(lipolized), oxidized, feed flavors, and defects associated with high somatic cell count milk (mastitic milk).

Microorganisms associated with foodborne illness may enter the raw milk supply through infected animals, milking personnel, or the environment. Recently, the consumption of raw milk (on the farm or as certified raw milk) has been implicated in outbreaks of foodborne illness involving *Salmonella, Campylobacter jejuni*, and *Yersinia enterocolitica*.[6,7] The contamination of pasteurized product with raw milk containing *Listeria monocytogenes* and *Salmonella* has been implicated in recent major outbreaks.[37,69]

Efforts should be made to minimize microbial numbers in raw milk supplies because processing does not eliminate quality defects. Standard Plate Counts (SPC) are routinely obtained to assess the overall sanitation and storage conditions of dairy farms. Counts of less than 10,000 CFU/mL are easily maintained. Some believe the Preliminary Incubation method for raw milk to be a more reliable indicator of farm sanitation than the SPC because it enumerates those organisms that may occur through poor dairy hygiene.[48] Thermoduric counts are useful in determining the number of microorganisms that will survive pasteurization. Raw milk received by a dairy plant is generally processed before results of the plating methods are available. The direct microscopic count (DMC) and direct microscopic somatic cell count (DMSCC) methods can be used for rapid estimates of total numbers of microorganisms and somatic cells before the raw milk is further processed. However, these counts are most reliable when numbers of cells are high. They are of limited value for normal Grade A milk.

47.21 Recommended Methods

Standard Plate Count—Chapter 6, SMEDP[48]
Thermoduric—Chapter 8, SMEDP[48]
Direct Microscopic Count—Chapter 10, SMEDP[48]
Direct Microscopic Somatic Cell Count—SMEDP[48]
Electronic Somatic Cell Count—SMEDP[48]

Other Methods for Specific Determinations

Psychrotrophic Plate Count—Chapter 8, SMEDP[48]
Coliform Count—Chapter 7, SMEDP[48]
Preliminary Incubation—SMEDP[48]
For individual pathogens such as *Salmonella, Campylobacter, Listeria*, and *Yersinia* consult appropriate chapters of this book.
Mastitis Pathogens—Laboratory Handbook on Bovine Mastitis; National Mastitis Council.[59]
Consult Chapter 10, for rapid methods.

47.3 PASTEURIZED MILK

The initial microflora of freshly pasteurized milk consists primarily of thermoduric bacteria and spores. The types and numbers of thermoduric bacteria are dependent on the microbial population of the raw milk before pasteurization. Gram-positive *Bacillus, Micrococcus, Lactobacillus, Microbacterium, Corynebacterium, Streptococcus, Enterococcus*, and *Arthrobacter* spp. are among the more common thermoduric organisms.[10] Large numbers of these bacteria in the raw milk supply can contribute significantly to the SPC of pasteurized products and in some cases cause the SPC to exceed regulatory standards. Most thermoduric bacteria grow slowly in refrigerated milk and are generally outgrown by gram-negative psychrotrophic species that gain entry primarily as post-pasteurization contaminants.[10,105] However, in the absence of psychrotrophic bacteria or if large numbers of thermoduric bacteria survive pasteurization, certain thermodurics, particularly psychrotrophic sporeforming *Bacillus* spp., can grow and cause spoilage (e.g., sweet-curdling).[4,63]

The predominant gram-negative psychrotrophic bacteria in pasteurized milk are species of *Pseudomonas, Flavobacterium*, and *Alcaligenes*, as well as some members of the coliform group.[10,62] Spoilage of pasteurized milk by gram-negative psychrotrophic bacteria results in fruity, rancid, bitter, and unclean flavors. Gen-

Table 2. Microbiological Standards for Manufactured Milk Products

Product	Standard		Reference
Dry whole milk			
U.S. extra	50,000/g	SPC	82
	10/g	Coliform	82
U.S. standard	100,000/g	SPC	82
	10/g	Coliform	82
U.S. grade not assigned	>100 x 10^6/g	DMC	82
Instant dry whole milk			
	30,000/g	SPC	86
	<40 x 10^6/g	DMC	86
	10/g	Coliform	86
Nonfat dry milk	(Spray) (Roller)		
U.S. extra	40,000/g; 50,000/g	SPC	83, 90
U.S. standard	75,000/g; 100,000/g	SPC	83, 90
U.S. grade not assigned	>100 × 10^6/g	DMC	83, 90
Instant nonfat dry milk			
U.S. extra	30,000/g	SPC	91
	10/g	Coliform	91
U.S. grade not assigned	>75 × 10^6/g	DMC	91
Dry buttermilk			
U.S. extra	30,000/g	SPC	85
U.S. standard	200,000/g	SPC	85
Dry whey			
U.S. extra	50,000/g	SPC	84
	10/g	Coliform	84
Edible dry casein (acid)			
U.S. extra	30,000/g	SPC	80
	Negative/0.1g	Coliform	80
	<5/0.1g	Yeast and molds[a]	80
	<5,000/g	Thermophiles[a]	80
	Negative	Staphylococcus[a]	80
	Negative/100g	Salmonella[a]	80
U.S. standard	100,000/g	SPC	80
	2/0.1g	Coliform	80
	<5/0.1g	Yeast and molds[a]	80
	<5,000/g	Thermophiles[a]	80
	Negative	Staphylococcus[a]	80
	Negative/100g	Salmonella[a]	80
Plastic and frozen cream			
	30,000/g	SPC	96, 97
	10/g	Coliform	96, 97
	20/g	Yeast and mold	96, 97
Ice Cream			
	50,000/g	SPC	99
Unflavored	10/g	Coliform	99
Flavored (nuts, berries, etc.)	20/g	Coliform	99
Sherbet			
	50,000/g	SPC	100
	10/g	Coliform	100
Sweetened condensed milk			
	1,000/g	SPC	101
	<10/g	Coliform	101
	<5/g	Yeasts	101
	<5/g	Molds	101
Butter			
	100/g	Proteolytic	94
	20/g	Yeast and mold	94
	10/g	Coliform	94
	10/g	Enterococci[a]	94
Whipped butter			
	50/g	Proteolytic	88, 95
	10/g	Yeast and mold	88, 95
	10/g	Coliform	88, 95
	10/g	Enterococci[a]	95
Light whipped butter			
	1,000/g	SPC	89
	10/g	Coliform	89
	Negative	*E. coli*	89
	10/g	Yeast and mold	89
Cottage cheese			
	10/g	Coliform	98
	100/g	Psychrotrophic	98
	10/g	Yeast and mold	98
Cream cheese products			
	25,000/g	SPC	87
	10/g	Coliform	87
	Negative	*E. coli*	87
	10/g	Yeast and mold	87

[a] Optional except when required or requested

erally, populations in excess of 10^6 per mL are required before flavor defects are detectable flavor or odor. The rate of microbial growth and of quality deterioration of the product is influenced by the numbers and types of bacteria in the freshly pasteurized product and the storage temperature. For optimum shelf life, refrigerated storage should be below 4°C. Under ideal processing and handling conditions, the shelf life of pasteurized milk should well exceed 14 days. Spoilage caused by gram-positive bacteria will occur in the absence of gram-negative bacteria, although at a much slower rate.

Microbiological tests most commonly used to evaluate freshly pasteurized milk are the SPC and the coliform count. Although these methods are used routinely to determine compliance with regulatory standards, they are of limited value in predicting the keeping quality (shelf life) of milk, which is considered to be a function of the extent of post-pasteurization contamination. The presence of coliforms in freshly pasteurized milk is normally an indication of post-pasteurization contamination. However, the numbers of coliforms and other post-pasteurization contaminants in freshly processed milk may be below detection levels by conventional methods. The Moseley Keeping Quality test has been used as an indication of the shelf life of the product by determining the extent of microbial growth under typical storage conditions. The primary disadvantage of the test is that 7 to 9 days are required to obtain results. Other methods designed to determine, within a shorter time period, the presence of low levels of coliforms and other gram-negative psychrotrophic bacteria that presumably do not survive pasteurization have been suggested.[2] Most involve preliminary incubation of the samples at temperatures selective for psychrotrophic bacteria, with or without reagents inhibitory to gram-positive bacteria, followed by selective plating procedures and/or the detection of microbial metabolites.[2] The usefulness of these methods in detecting post-pasteurization contamination and predicting shelf life has not been fully determined. However, the reader should be aware that there are procedures that may be useful in a given situation.[2]

Bacteria related to foodborne illness are destroyed by proper pasteurization.[6,16,37,62] Recent outbreaks of salmonellosis and listeriosis in pasteurized milk have been linked to post-pasteurization contamination. Contamination of pasteurized milk through the addition of ingredients was implicated in an outbreak caused by *Y. enterocolitica* in chocolate milk.[6] Post-pasteurization contamination with *L. monocytogenes* and *Y. enterocolitica* is of major concern to the dairy industry since these organisms grow at refrigeration temperatures.

47.31 Recommended Methods

Standard Plate Count—Chapter 6, SMEDP[48]
Coliform Count—Chapter 7, SMEDP[48]

Other Methods for Specific Determinations
Psychrotrophic Count—Chapter 8, SMEDP[48]
Moseley Keeping Quality Test—SMEDP[48]
Preliminary Incubation—SMEDP[48]
For individual pathogens such as *Salmonella*, *Campylobacter*, *Listeria*, and *Yersinia* consult appropriate chapters of this book.
Consult Chapter 10, for rapid methods.

47.4 DRIED PRODUCTS

Dairy products that are dried include milk, skimmed milk (nonfat dry milk), buttermilk, whey, cheese, and some fermented milk products. The manufacture of these products usually involves the preparation of a concentrate before drying by spray, roller, or foam processes. Milk is separated or standardized, pasteurized, and preheated at selected time and temperature combinations to produce low-heat, medium-heat, or high-heat products. Preheated milk is concentrated in an evaporator with the temperature of the concentration process selected to produce the desired characteristics in the product. Increasing temperatures of concentration are used to produce low-heat, medium-heat, and high-heat products. Concentrated milk is briefly exposed to high temperatures during drying with the amount of heat applied depending on the type of product being manufactured and the method of drying. The microflora of dried milks is affected by the time and temperature combinations used during preheating, concentration, and drying. Failure to achieve satisfactory bacteria counts in dried milk products is usually due to the use of poor-quality raw milk. Psychrotrophic bacteria, coliform bacteria, and yeasts and molds are reduced to very low levels during preheating[43] and their presence in dried milk products indicates contamination from equipment or the environment during or after manufacture. The typical microflora of dried milk consists of thermoduric micrococci, thermoduric streptococci, corynebacteria, and aerobic sporeformers.[24]

Most dried dairy products are often used as ingredients of other foods and are subject to further processing. Yet, dried milks must be considered sensitive products from a public health aspect because they are often consumed after reconstitution without additional heating. It is well established that dried milk can be a source of foodborne illness because of contamination with *Salmonella* and *Staphylococcus*.[25,49] Food poisoning caused by *S. aureus* is not currently a serious problem. However, poor temperature control during storage of raw milk before processing or following heat treatments during processing can lead to growth of *S. aureus* and the production of enterotoxin that can survive subsequent manufacturing processes. Growth of *S. aureus* in milk after preheat treatment and before drying has been identified as the probable cause of several outbreaks of *Staphylococcus* intoxication.[49] Proper storage temperatures and process controls will eliminate the hazard of food poisoning from *S. aureus* in dried milk products. Environmental contamination of dried milk products with *Salmonella* is a major concern for manufacturers of dried milk products. The history of foodborne illness in dried milk caused by *Salmonella* has been reviewed by Marth.[49] The seriousness of *Salmonella* as a pathogen and the probability that dried milk might be consumed without further processing caused the institution of process control measures and an extensive monitoring program that has significantly reduced the incidence of *Salmonella* in dried milk products.[62] The strengths and weaknesses of the regulatory *Salmonella* surveillance program used in the United States to routinely monitor dried milk products for *Salmonella* have been identified by the National Research Council.[61]

Because of their low water activity, dried products rarely spoil or deteriorate because of microbial growth. Standards specifying SPC, coliform counts, and yeast and mold counts of dried milk products reflecting quality grades and product processes have been established by the USDA and the American Dairy Product Institute (Tables 1 and 2).[1,92] The sanitary quality of the processing conditions and environment can be monitored by evaluating the dried milk products for post-pasteurization contamination through indicator organisms such as yeasts, molds, and coliforms. These microorganisms are destroyed by the heat treatments used during the manufacture of dried milk products and their presence in these products reflects contamination from unsanitary equipment or the environment following manufacture. Total counts, thermoduric counts, and aerobic mesophilic spore counts, in increasing order of sensitivity, reflect the effect of changing conditions during the manufacture of dried milk.[43] These counts can be used as indicators of the quality of the raw milk used to manufacture the dried product. The DMC, which measures both viable and dead bacteria, can also be used to provide information relative to the original bacterial population of the raw milk. When there is reason to suspect possible public health concerns because of *Staphylococcus*, it would be advisable to test for staphylococcal toxin. The *Salmonella* surveillance program is not a substitute for microbiological control by the manufacturer; therefore manufacturers should test products for *Salmonella* in accordance with the sampling plans recommended in the NAS/NRC *Salmonella* report.[60] Improper storage of dry milk bags in moist areas, ingress of water in storage areas, or loss of integrity of packaging will promote the growth of molds on dried milk products. Molds can cause discoloration, musty flavors, breakdown of milk components resulting in off-flavors, and possible production of mycotoxins.

47.41 Recommended Methods

Standard Plate Count—Chapter 6, SMEDP[48]
Coliform Count—Chapter 7, SMEDP[48]
Yeast and Mold Count—Chapter 8, SMEDP[48]
Direct Microscopic Clump Count—Chapter 10, SMEDP[48]
Thermoduric Count—Chapter 8, SMEDP[48]
Psychrotrophic Count—Chapter 8, SMEDP[48]
Salmonella—Chapter 37
Staphylococcal Enterotoxins—Chapter 39
Aerobic Spore Count—Chapter 10, SMEDP[48]

47.5 BUTTER

Butter, one of the few foods defined by law, must contain at least 80% milkfat.[30] It can be salted or unsalted, and it may or may not contain added starter cultures consisting of *Lactococcus lactis subsp. lactis*, *Lactococcus lactis subsp. cremoris*, or *Leuconostoc* spp. for additional flavor. The composition of butter is approximately 80.5% fat, 16% to 16.5% moisture, 1.0% to 1.5% curd, and 1.75% to 2.0% salt. Butter is manufactured by creating a water-in-oil emulsion by churning cream, causing a phase inversion of cream in a continuous process or in a conventional churn. During the churning process, moisture from the buttermilk or from water added to adjust the moisture content of the butter is worked into the lipid structure. The moisture is dispersed as fine droplets throughout butter. In properly worked butter, the droplets are uniform and of small size. This composition is critical to stability because poorly worked butter will have an uneven distribution of moisture; areas of high moisture will permit microbial growth if the butter is contaminated during or after manufacture. Additional storage stability is provided by the addition of salt, which can result in a salt-in-water concentration as high as 16%.

The microflora of butter reflects the quality of the cream, the sanitary conditions of the equipment used to manufacture the butter, and the environmental and sanitary conditions during packaging and handling. Pasteurization of cream causes a significant reduction in the number of all but the most heat-resistant bacteria. Palatability problems caused by bacteria include rancid, putrid, malty, and fishy tastes. The development of yeasts and molds on the surface of butter can cause surface discoloration and flavor problems.[23] These flavor and surface growth problems are usually caused by species of *Pseudomonas*, *Lactococcus*, *Geotrichum*, and *Candida*. Butter has been implicated in an outbreak of food poisoning caused by staphylococci.[9] Minor and Marth[54] studied the growth and survival of *S. aureus* in butter and cream. Survival and growth of this organism in butter were affected by storage temperature of the cream and salt content of the butter. Poor-quality cream could be a source of food poisoning if conditions were sufficient to permit growth and enterotoxin production by *Staphylococcus*.

Various tests can be used to determine the quality and safety of butter or to determine sources of contamination during the manufacturing process. SPCs, coliform counts, and yeast and mold counts can be useful to detect sources of contamination and to assess the degree of processing sanitation. Because of the longer viability and relatively higher salt tolerance of *Enterococcus* species, they are better than coliforms as indicators of production sanitation. Lipolytic and proteolytic counts are useful for evaluating the keeping quality of butter since lipolysis and proteolysis are directly related to the development of specific flavor defects. In cases where potential public health problems related to staphylococcal growth and enterotoxin production are suspected, staphylococcal counts might be desirable. However, these determinations would be of limited value if growth of *S. aureus* and enterotoxin production preceded pasteurization.

47.51 Recommended Methods

Standard Plate Count—Chapter 6, SMEDP[48]
Coliform Count—Chapter 7, SMEDP[48]
Lipolytic Count—Chapter 8, SMEDP[48]
Proteolytic Count—SMEDP[48]
Yeast and Mold Count—Chapter 8, SMEDP[48]
Psychrotrophic Count—Chapter 8, SMEDP[48]
Staphylococcus aureus Count—Chapter 39

47.6 FROZEN DAIRY PRODUCTS

Frozen dairy products include ice cream, sherbet, novelties, and frozen yogurt. The microbial content of frozen products largely reflects the quality of the ingredients used for their manufacture: milk, cream, nonfat milk solids, sugar, chocolate, fruits and nuts, egg products, emulsifiers, and stabilizers. Milk, cream, and the soluble components are normally blended and pasteurized. Therefore, microbial counts of the pasteurized mix are generally low (< 100 per mL). Sporeformers (*Bacillus* spp.) and some of the hardier thermoduric bacteria originating from the fluid or dry components are usually the only survivors. Flavors, coloring agents, and ingredients such as fruits, nuts, and chocolate chips added to the mix after pasteurization can be a source of contaminants. In addition, post-pasteurization contamination can occur from poorly cleaned equipment, air incorporation, poor use of product rerun, and personnel.[29] The presence of coliforms in pasteurized mixes and frozen products is an indication of post-pasteurization contamination, although false positives are possible because of the ability of certain noncoliforms to ferment sucrose.[45]

The quality of soft-serve mixes (ice cream and frozen yogurt) suffers when these products are stored for extended periods, particularly at the upper range of refrigeration temperatures, before freezing. The numbers and types of spoilage organisms will depend on the extent of post-pasteurization contamination and the time-temperature exposure.

Although there is no growth in frozen dairy products, many types of bacteria (including pathogens, if present) can survive in frozen products.[7] *L. monocytogenes* has been isolated from ice cream and novelties.[69] Ice cream has also been implicated as a vehicle for *Salmonella* infection, although most of the outbreaks associated with this organism occurred as a result of using raw eggs in the manufacture of homemade ice cream.[6] However, *Salmonella* infections can result from ice cream if they are introduced at the retail level by the improper handling of serving utensils. Failure to store ice cream scoops in flowing water allows the accumulation of ice cream solids which provides a rich environment for bacterial growth. Staphylococcal intoxications involving soft-serve ice creams have been reported.[6]

Normally, SPC and coliform counts of the finished product are used as indicators of microbial contamination usually caused by poor quality ingredients or inadequate plant sanitation.[48] Products and ingredients may be tested at various stages during processing to determine the source of microbial contamination; SPC, coliform counts, yeast and mold counts, and thermoduric counts are useful. All bulk ingredients should be evaluated before use in the manufacture of frozen dairy foods.

47.61 Recommended Methods

Standard Plate Count—Chapter 6, SMEDP[48]
Coliform Count—Chapter 7, SMEDP[48]
Thermoduric Count—Chapter 8, SMEDP[48]
Yeast and Mold Counts—Chapter 8, SMEDP[48]

For the microbial evaluation of a non-dairy ingredient (fruits, nuts, cocoa, sugars, and egg and egg products) consult the appropriate chapter in this book. For individual pathogens such as *Salmonella*, *Campylobacter*, *Listeria*, and *Yersinia* consult the appropriate chapters of this book.

47.7 CONCENTRATED PRODUCTS

Concentrated milk products include evaporated milk, concentrated milk, sweetened condensed milk, and, to a limited extent, condensed sour products. They are manufactured by the removal of moisture by heat and differ in the amount of heat treatment given during processing, the degree of concentration, and in the ingredients added to increase shelf stability or to influence the final characteristics of the product. The procedure for manufacturing concentrated milk or skim milk and condensed sour products consists of pasteurization, preheating, evaporation, and cooling. Concentration of milk for the preparation of low-heat products can be in single or multiple effect evaporators that operate at temperatures from 37° to 52°C under reduced pressure.[8] The reduced temperatures employed in multiple effect evaporators can provide an opportunity for the growth of thermoduric and thermophilic bacteria. Plain concentrated milk products are usually concentrated 3:1, contain no added ingredients to inhibit bacterial growth, and will support microbial growth. They are susceptible to microbial spoilage and must be stored under refrigerated conditions. The microflora of plain concentrated milk products will consist of gram-positive thermoduric bacteria that survive pasteurization and the heat treatment applied during concentration, and bacteria associated with unsanitary equipment or environmental contamination. *Bacillus, Micrococcus, Lactobacillus, Microbacterium*, coryneform bacteria, *Streptococcus, Enterococcus*, and *Arthrobacter* spp. are common thermoduric bacteria that survive these heat treatments.[23] The number and types of these bacteria in concentrated milk reflect the quality of the milk used to manufacture the concentrated milk. The presence of psychrotrophic bacteria, coliform bacteria, and yeasts and molds in concentrated milk indicates contamination of the product during or after manufacture. If milk is concentrated at temperatures below 60°C, growth of thermophilic bacteria might occur and reduce the quality of the product. The SPC can be used to evaluate the quality of the raw milk used to manufacture concentrated milk and will provide some indication of the sanitary conditions of the manufacturing environment. However, if more specific information regarding the quality of the raw milk is desired, thermophilic counts or spore counts might be useful.

The sanitary condition of the processing equipment and the environment can be evaluated by coliform and psychrotrophic counts of plain concentrated milk. Concentrated fermented products are preserved by the high concentration of acid and low pH of the products. Spoilage may occur if sufficient air is available for mold growth. Growth of mold on these products can cause the pH to increase near the mold growth permitting the growth of bacteria that can lead to rapid spoilage of the product.[62]

Sweetened condensed milk must contain a minimum of 8.5% milkfat and 28% total milk solids. The minimum concentration of sugar is not specified other than to be of sufficient quantity to prevent spoilage.[80] A sucrose concentration of 42% to 43% is usually sufficient to produce a sucrose/water ratio of 60% to 66%. Sugar concentrations at the higher limit are necessary for maximum shelf life. The major differences between concentrated milk and sweetened condensed milk are that the latter is usually sold in cans and has sugar added; lactose is also crystallized in sweetened condensed milk to prevent the formation of large lactose crystals that would cause the product to be "grainy" or "sandy."[8] However, the type of spoilage that might occur is considerably different.

The main types of spoilage in sweetened condensed milk are osmophilic sucrose-fermenting yeasts such as *Torula* spp. and molds. Species of *Aspergillus* and *Penicillium* have been implicated in the production of mold on the surface of sweetened condensed milk when sufficient air is available for their growth.[24] This problem can be eliminated by filling cans to a level that eliminates air for growth, by using practices that reduce the probability of yeast and mold contamination after processing, and by using sucrose with low yeast and mold counts. Spoilage by other types of microorganisms is rare and indicates a problem related to the composition of the product, especially the concentration of sugar. Coliform counts and yeast and mold counts are useful to evaluate sanitary conditions during processing and afterward. High yeast and mold counts can also indicate contamination from sucrose. Consequently, sucrose should be checked for these organisms. Standard Plate Counts, thermoduric counts, thermophilic counts, and spore counts can provide information about the quality of the raw milk used to produce these products.

Evaporated milk production differs from the production of plain condensed milk in these ways: a more intensive preheat treatment is given to evaporated milk to provide storage stability; a stabilizing mineral mixture may be added to reduce gelation during storage; and the product is sterilized in a can by batch or continuous retort procedures. Consequently, the product can be considered a commercially sterile, low-acid food. Federal regulations specify a minimum fat and total milk solids content of 7.5% and 25%, respectively.[81] Spoilage results from inadequate heat treatment or from leakage of the cans after packaging. Spoilage by *Bacillus stearothermophilus* may occur if the product is stored at abnormally high ambient temperatures.[63] Methods for determining commercial sterility and for identifying causes for the absence of sterility in low-acid foods can be found in Chapters 61 and 62.

47.71 Recommended Methods

Standard Plate Count—Chapter 6, SMEDP[48]
Coliform Count—Chapter 7, SMEDP[48]
Yeast and Mold Count—Chapter 8, SMEDP[48]
Thermoduric Count—Chapter 8, SMEDP[48]
Thermophilic Count—SMEDP[48]

47.8 FERMENTED DAIRY PRODUCTS

Cheeses, yogurts, and fermented milks represent a diverse class of dairy products derived from the alterations of milk by microbial and enzymatic activities. A typical dairy fermentation is initiated by the growth of lactic acid bacteria responsible for the production of lactic acid from lactose. The compositional, structural, and flavor characteristics of the fermented product are determined by the processing conditions, such as type of starter culture used, type of enzyme addition, incubation and ripening temperature, milk composition and handling, salt addition, processing and aging conditions, and ripening microflora. These parameters also determine the types of micro-organisms capable of survival and/or growth in the product.

47.81 Cheeses

Cheeses are fresh or unripened (cottage cheese, cream cheese, mozzarella, or Neufchatel) or ripened (aged). Ripened cheeses can be further categorized by moisture content as soft surface-ripened (Camembert, Brie), semisoft (Muenster, Gouda, Edam, Roquefort, or Blue), hard (Cheddar, Swiss, Emmentaler, or Gruyère), or hard-grating cheese (Romano or Parmesan).[62] In cheese production, the initial microflora of the raw or heated milk (subpasteurized to pasteurized) is rapidly overshadowed by the active starter organism(s). Subnormal activity of the starter culture may allow growth of undesirable microorganisms. The primary functions of starter cultures (Table 3) include production of lactic acid, which promotes curd formation in conjunction with coagulating enzymes such as rennet, lowering of the redox potential, and destruction or prevention of the growth of pathogens and spoilage organisms. They also contribute to the flavor characteristics of the product. Starter populations in excess of 10^9 per gram of milk or curd are common in the initial stage of the fermentation.

Ripening of cheeses under controlled conditions of temperature and humidity determines the final flavor and body characteristics of the product. The development of these characteristics involves enzymes of the starter organism, inherent milk enzymes, and the activity of the secondary flora (Table 4). During ripening of Cheddar cheese, the number of the starter organisms generally declines as the secondary flora increases (>10^8 per gram). Gram-negative rods, micrococci, and other non-lactic gram-positive bacteria also tend to decrease[74] during the ripening of cheese.

Table 3. Lactic Starter Cultures and Related Products

Culture	Product
Lactococcus lactis subsp *Lactis* *Lactococcus lactis* subsp. *cremoris*	Cottage cheese, buttermilk, sour cream, Cheddar, soft and semisoft cheeses, Gouda, blue-vein cheese, other cheeses
Lactococcus lactis subsp. *lactis biovar diacetylactis* *Leuconostoc cremoris*	Cottage cheese, buttermilk, sour cream, semisoft cheese, Cheddar
Streptococcus thermophilus	Yogurt, mozzarella, Emmentaler, Gruyère, Swiss, hard Italian cheese
Lactobacillus delbrueckii subsp. bulgaricus *Lactobacillus delbrueckii subsp. lactis* *Lactobacillus lactis*	Yogurt, mozzarella, Emmentaler, Gruyère, Swiss, hard Italian cheese, kefir, koumis
Lactobacillus acidophilus	Acidophilus milk, yogurt
Propionibacterium shermanii	Emmentaler, Gruyère, Swiss, Gouda
Bifidobacterium ssp.	Probiotic milk and cultured milk products

Adapted from Kosikowski,[41] Law,[45] and Olson and Mocquot[62]

Microbial spoilage in cheese is generally limited because of the combined effect of acid and salt, and is less likely in the lower moisture cheeses. Spores of *Clostridium tyrobutyricum* in the milk used for the manufacture of Emmentaler, Edam, and Gouda can survive the heat treatment used for cheese milk and cause late gas formation (blowing defect) and related off-flavors during ripening. Thermoduric *Streptococcus thermophilus* can cause flavor defects in Gouda cheese.[45] The presence of heat-stable enzymes (from psychrotrophic bacteria) can be detrimental to the quality of both fresh and ripened cheese[11,12,45] by causing bitter or rancid flavors and by impairing the coagulation properties of the milk. Fresh cheeses, such as cottage cheese and other high-moisture cheeses, may be subject to spoilage by gram-negative psychrotrophic bacteria (*Pseudomonas*, *Flavobacterium*, or *Alcaligenes*), coliforms, and yeast and molds that enter as post-pasteurization contaminants.[10] Most hard-ripened cheeses are not subject to gram-negative spoilage though coliform contamination has been associated with the gassy defect in cheese making (for example, Cheddar).[41] Ripened cheese is prone to surface growth of yeast and molds, particularly if exposed to atmospheric oxygen. Defects in cheese have also been attributed to the starter cultures, culture failure, and undesirable secondary flora.[45]

Table 4. Secondary Flora of Ripened Cheese

Cheese	Secondary flora
Soft (surface ripened)	
Camembert	Yeasts
Brie	*Penicillium caseiolum*
Semisoft	
Caephilly	Lactobacilli
Limburger (surface ripened)	Yeasts, *Brevibacterium linens*
Blue-vein	
Roquefort	*Penicillium roqueforti*, Yeasts, micrococci
Gorgonzola	
Stilton	
Hard	
Cheddar	Lactobacilli, pediococci
Emmentaler	*Propionibacterium shermanii*, Group D streptococci
Gruyère	*Proprionibacterium shermanii*, Group D streptococci, Yeasts, coryneforms, *B. linens*

Adapted from Law[45]

Microbial competition, reduced water activity, organic acids, and a low pH generally limit the growth of pathogens in cheese. A slow starter culture (due to bacteriophage, antibiotics, etc.) can allow growth of bacteria related to foodborne illnesses such as *Staphylococcus*, *Salmonella*, *Listeria*, and enteropathogenic *E. coli*,[6,37,52] which enter with raw milk or as post-pasteurization contaminants. Numbers of *S. aureus* will normally decline during the ripening stage, but if sufficient numbers (>10^7 per mL) are reached during cheese making, enterotoxin may persist in the cheese. *Salmonella* spp. survive beyond the ripening period[6] with the potential to cause infection at relatively low doses.[13] Most, but not all, enteropathogenic strains of *E. coli* are inactivated at pH < 5.0,[52] although in low-acid, semisoft, surface-ripened cheese, fecal coliforms are commonly found.[6] Post-heating contamination or the use of contaminated raw milk can be a source of *L. monocytogenes*, as was implicated in an outbreak involving a low-acid Mexican-style cheese.[16] *Listeria* is capable of surviving in Cheddar, Camembert, and cottage cheese[69] although growth is limited because of the low pH of most cheeses. Biogenic amine formation in fermented dairy products has been reported by many workers. Edwards and Sandine[19] reviewed the occurrence of biogenic amines in cheeses and discussed the public health significance of these amines. Histamines produced by the decarboxylation of histidine, and other biogenic amines such as tyramine and tryptamine cause abdominal cramps, diarrhea, nausea, headache, palpations, tingling, and flushing, etc. Medical intervention is unnecessary, with most symptoms disappearing a few hours after onset. Biogenic amines also affect some antidepressant drugs by counteracting monoamine oxidase inhibitors. Enterococci and certain other lactic acid bacteria were implicated in the production of biogenic amines in fermented dairy products.

47.82 Yogurt and Fermented Milks

Yogurt, fermented milks (buttermilk), and cultured cream (sour cream) are unripened, cultured dairy products. They are generally ready for consumption with minimum processing after development of the desired acidity through a lactic acid fermentation. Yogurt fermentation involves a mixed culture of *S. thermophilus* and *Lactobacillus delbrueckii* subsp. *bulgaricus*, which are thermophilic in nature, while *Lactococcus* spp. and *Leuconostoc* spp. are normally used in cultured milks and sour cream. The more exotic cultured milks (e.g., kefir, koumis) are derived from mixed fermentations involving yeasts, *Lactobacillus* spp., *Lactococcus* spp., and *Leuconostoc* spp. *Lactobacillus acidophilus* may be used in the production of both yogurts and fermented milks.[41]

In a normal fermentation, a final pH of < 4.5 is developed in cultured milk products. This low pH generally prevents the growth of most spoilage and pathogenic organisms, although interference with acid development may allow growth of undesirable microorganisms. Microorganisms that cause deterioration of fermented milk products can enter products through poor sanitation techniques or can be introduced by the addition of flavoring materials such as fruit, nuts, and other flavoring materials. Yeasts and molds, which tolerate the lower pH, are the more predominant organisms involved in the spoilage of cultured milks. *Bacillus subtilis* and *B. cereus* can cause bitter flavors if large numbers survive pasteurization. Coliforms, if present, decline rapidly after manufacture of yogurt, although they may survive in cultured buttermilk and sour cream.[52]

Fermented dairy products normally contain high numbers of starter microorganisms or secondary ripening flora, making total counts insignificant except in products that are heated to inactivate added cultures. Yeast and mold counts and coliform counts may be used as indicators of adequacy of processing sanitation with some fermented products. However, it needs to be determined if the source of the contamination is poor sanitation or contaminated flavoring materials that might have been added to the products. Enterococci are probably better indicators of improper sanitation than coliforms because coliforms are quite sensitive to high acid conditions prevalent in fermented products and might not be recovered on selective media because of acid injury. Counts of psychrotrophic bacteria are useful for cottage cheese and similar products that are subject to spoilage by psychrotrophic bacteria.

47.83 Recommended Methods

Yeast and Mold Counts—Chapter 8, SMEDP[48]
Coliform Counts—Chapter 7, SMEDP[48]
Psychrotrophic Counts—Chapter 8, SMEDP[48]
(Cottage cheese)—SMEDP[48]

Other Methods for Specific Determinations
Staphylococcus Enterotoxin/Thermonuclease—Chapter 39

For individual pathogens such as *Salmonella, Campylobacter, Listeria,* and *Yersinia* consult appropriate chapters of this book.

Consult Chapter 10, for rapid methods.

47.9 ACID-PRODUCING BACTERIA IN DAIRY FOODS

One of the most important groups of acid-producing bacteria in the food industry is the lactic acid bacteria. Members of this group are gram-positive, nonsporulating cocci or rods, dividing in one plane only, with the exception of the pediococci. They are catalase negative (with the exception of some pediococci, which either form a pseudo-catalase or incorporate preformed hemin, supplied exogenously, into a catalase molecule). The organisms are usually nonmotile and are obligate fermenters, producing mainly lactic acid and sometimes also volatile acids and CO_2. They are subdivided into the genera *Lactococcus, Leuconostoc, Pediococcus,* and *Lactobacillus.* The homofermentative species produce lactic acid from available sugars, while the heterofermentative types produce, in addition to lactic acid, mainly acetic acid, ethanol, CO_2, and other components in trace amounts. Lactic acid bacteria are widespread in nature and are best known for their activities in major foods such as dairy, meat, and vegetable products. *Propionibacterium* also produce acid and several species are important in the development of the characteristic flavor and eye production in Swiss-type cheese.

47.91 Lactic Acid-Producing Bacteria

The large number of media proposed for lactic acid bacteria, particularly for streptococci and lactobacilli, is indicative of the difficulties encountered in growing some strains of these organisms. The choice of medium is governed to some extent by the particular strains under study and, therefore, by product or habitat. The media listed below have merit in the support of colony development of lactic acid bacteria, but are not highly selective, and some, such as Eugon agar, are not selective at all. Hence, organisms other than lactic acid bacteria may develop on these media and produce acid.

While the lactic acid bacteria in general are tolerant of low pH, they can be very sensitive to other adverse conditions. Samples to be examined for numbers of viable lactic acid bacteria should not be frozen prior to analyses. Many of the lactic acid bacteria are easily killed or injured by freezing. If the product to be examined is normally frozen, it should not be thawed and refrozen prior to microbial analyses since this would tend to increase damage caused by freezing.

Dilution of products with phosphate-buffered diluent for plating can damage lactic acid bacteria in samples to the point that reduced counts are obtained. Thus, it is best to use sterile 0.1% peptone water (Chapter 7) as the diluent since it protects bacteria during the dilution process.[31,35,39]

Depending on the product, it may be advantageous to blend the initial dilution for the plate count to disrupt chains of lactic acid bacteria. This is especially true for many freshly prepared cultured food products. Blending the initial dilution can produce a more accurate count of the number of bacteria actually present. Chilled diluent and a chilled blender cup should be used. These bacteria do not grow well aerobically, although most of them are considered to be facultative. Thus, it is usually important to pour overlays of the appropriate agar medium onto the surface of the solidified agar in plates containing the lactic acid bacteria. An alternative is to incubate the plates in an environment containing little or no oxygen.

The fastidious nature of lactic acid bacteria restricts them in the environment to wherever carbohydrate, protein breakdown products, vitamins, and minerals occur in ample quantity and proportion. Therefore, the greatest natural reservoir for these bacteria is growing green plants. Enrichment culture of blended plant material added to skim milk supplemented with 0.05% glucose and 0.1% yeast extract, with subsequent plating on appropriate media is a common isolation procedure. Mundt,[56] Mundt et al.,[58] Mundt and Hammer[57] and Sandine et al.[70] are often consulted for the isolation of lactococci and lactobacilli from plants, including vegetables.

47.911 Selective Media for Lactic Acid-Producing Bacteria

47.9111 Lactic agar[20] (Chapter 63). This medium was developed to support colony development of lactococci (lactic streptococci) and lactobacilli. Prepare pour plates of samples as described in Chapter 7 using lactic agar. After incubation of plates, prepare gram stains of individual colonies, examine these microscopically, and test for catalase reaction. Gram-positive, catalase-negative cocci or rods may be tentatively considered to be lactic acid bacteria. If further identification is needed, consult Sharpe.[72]

47.9112 MRS media (Chapter 63). MRS broth was developed by deMan, Rogosa, and Sharpe[14] to support the growth of various lactobacilli, particularly of dairy origin. Pediococci and leuconostocs grow luxuriously in this medium. MRS broth with added agar may be used to prepare pour plates of samples as described in Section 47.9111 using MRS agar instead of lactic agar. Identification of individual colonies is described in Section 47.9111.

47.9113 RMW agar[68] (Rogosa SL agar) (Chapter 63). This is a selective medium for the cultivation of oral and fecal lactobacilli. Pediococci can also be isolated using this medium. For the preparation of agar plates, proceed as in Section 47.9111, but substitute RMW agar for lactic agar. Since some lactobacilli will not grow on this medium if incubated aerobically, plates should be incubated in a CO_2-enriched atmosphere. The plates can be placed in plastic bags, flushed 1 min with CO_2, sealed, and incubated at the desired temperature.

47.9114 M 16 agar[47] (Chapter 63). This medium was developed to support growth of lactococci (lactic streptococci) used in cheddar cheese manufacturing in New Zealand. Prepare agar plates as in Section 47.9111, but substitute M 16 agar for lactic agar.

47.9115 M 17 agar[75] (Chapter 63). This medium was developed by Terzaghi and Sandine[75] to support the growth of lactococci (lactic streptococci). It is buffered with β-disodium glycerophosphate and also is useful for plaque assay of lactic bacteriophages. Prepare plates as in Section 47.9111, but substitute M 17 agar for lactic agar.

47.9116 Eugon agar[102] (Chapter 63). This medium is reported to support the surface growth of lactic acid bacteria.[103] Prepare agar plates as in Section 47.9111.

47.9117 APT agar[21] (Chapter 63). APT agar was developed for the cultivation and enumeration of heterofermentative lactic acid bacteria of discolored, cured meat products.[21] This medium is also commonly used for the propagation of pediococci. Prepare agar plates as in Section 47.9111.

47.9118 LBS oxgall agar[27] (Chapter 63). LBS oxgall agar is made selective for bile-resistant lactobacilli[28] by incorporating 0.15% oxgall into its formulation. More consistent results often are obtained by preparing the LBS plus oxgall from individual ingredients rather than by using commercially available premixed media. Agar plates are prepared as in Section 47.9111 using LBS agar plus oxgall in place of lactic agar. The plates should be placed in plastic bags, flushed with CO_2 for 1 min, sealed, and incubated at 37°C for 48 hr. Incubation for dried products should be increased to 72 hr.[27]

47.912 Differential Media for Lactic Acid-Producing Bacteria

This section presents methods for the qualitative and quantitative differentiation of some lactic acid bacteria employed in the dairy industry. The media are not selective and must be used only in pure culture studies or as recommended in this section. The broth and agar media for lactococci (lactic streptococci) can be used to identify members of the lactic group streptococci when commercial starter cultures or products are first plated on a general-purpose agar, or on media more efficient for the detection of lactic acid bacteria, such as lactic agar or eugon agar.

47.9121 Separation of lactic streptococci—*Lactococcus lactis* subsp. *lactis, Lactococcus lactis* subsp. *cremoris,* and *Lactococcus lactis* subsp. *diacetylactis.* The most common microorganisms in starter cultures used in dairy products are streptococci. *Lactococcus lactis* and its subspecies *cremoris* and *diacetylactis* belong to this group. Their separation can be made by biochemical tests, the major criteria being arginine hydrolysis and tests for diacetyl and acetoin. Reddy et al.[66] developed differential broth for lactic streptococci to separate these species. In addition to direct inoculation with a loopful of an active (pure) milk culture, the broth is suitable for the qualitative differentiation of individual colonies of lactic streptococci developing on agar plates containing dilutions of a commercial starter culture. After inoculation, close the test tube caps tightly to prevent escape of liberated NH_3 and CO_2. Incubate the tubes at 30°C for 24 to 72 hr and observe the indicator color reactions and CO_2 accumulation at 24 hr intervals. Subspecies *cremoris* produces a deep yellow color (acid) in the broth. *L. lactis* initially turns the broth yellow (acid), but later the violet hue returns because of the liberation of NH_3 from arginine. *L. lactis* subsp. *diacetylactis* yields a violet color and produces copious amounts of CO_2 (from citrate) in the fermentation tubes within 48 hr. *L. lactis* subsp. *diacetylactis* produces a more intense purple than *L. lactis. Leuconostoc* starter strains cause no appreciable color change in the violet differential broth, and only minute amounts of gas are observed with some *Leuconostoc mesenteroides* subsp. *dextranicum* strains. Arginine hydrolysis in the differential broth can be further checked by testing a portion of the broth with Nessler's reagent on a porcelain spot plate. A deep red precipitate indicates arginine hydrolysis.

47.9122 Differential enumeration of lactococci. Agar medium for differential enumeration of lactic streptococci[67] can be used for the qualitative and quantitative differentiation of a mixture of *L. lactis* and its subspecies *cremoris* and *diacetylactis.* This medium contains arginine and calcium citrate as specific substrates, diffusible (K_2HPO_4) and non-diffusible ($CaCO_3$) buffer systems, and bromcresol purple as the pH indicator. Milk is added to provide carbohydrate (lactose) and growth-stimulating factors. Production of acid from lactose causes yellow bacterial colonies. Subsequent arginine utilization by *L. lactis* and *L. lactis* subsp. *diacetylactis* liberates NH_3 and results in a localized pH change back to neutrality, with a return of the purple indicator hue. *L. lactis* subsp. *diacetylactis* utilizes suspended calcium citrate, and after 6 days of incubation, the citrate degrading colonies exhibit clear zones against a turbid background. The buffering capacity of $CaCO_3$ limits the effects of acid and NH_3 production around individual colonies. From a practical standpoint this medium can be used a) to study associative growth relationships in starter mixtures of these species; b) to verify the composition of mixed starter cultures; and c) to screen single strains for compatibility in mixed cultures.

Prepare decimal dilutions of the culture with sterile 0.1% peptone solution (Chapter 63) and spread 0.1-mL quantities of the dilution evenly over the surface of agar plates (Chapter 7) with a sterile bent glass rod. Incubate the plates in a candle-oats jar (Section 47.922) at 32°C and examine plates after 36 to 40 hr and after day 6 of incubation.

After 36 to 40 hr, count all colonies and then count the yellow subspecies *cremoris* colonies separately. Return the plates to the candle-oats jar for an additional 4 days. After this period, expose the plates to the air for 1 hr. First determine the total count, and then count all colonies showing zones of clearing of the turbid suspension of calcium citrate (*L. lactis* subsp. *diacetylactis*). Subtract the subspecies *cremoris* (after 36 to 40 hr) and *diacetylactis* counts from the total count to obtain the *L. lactis* population in the mixture. Slow arginine hydrolyzing or nonhydrolyzing strains of *L. lactis* subsp. *diacetylactis* in cultural mixtures sometimes produce yellow colonies similar to subspecies *cremoris* after 36 to 40 hr. In such an instance, mark the yellow colonies (after 36 to 40 hr) with an indelible felt pen. When the final count is taken, count the marked colonies that show clearing as *L. lactis* subsp. *diacetylactis* and subtract their number from the original yellow colony count to obtain the accurate value for subspecies *cremoris*.

Maximum differential efficiency is obtained only when the counts on the individual plates do not exceed 250 colonies and when fresh medium is employed. This medium is not selective and must be used in pure culture studies only. Mullan and Walker[55] described lactic streak agar for the differentiation of lactic streptococci. The basic principles applied in this medium for differentiation are the same as those of Reddy et al.[67] Mullan and Walker[55] reported that results were available faster by their technique than by that of Reddy et al.[67] and that, with the use of their agar, there was no need for incubation of petri plates in a CO_2-enriched atmosphere. In addition to these two media, Kempler and McKay[38] described KM agar for the differentiation of citrate-fermenting lactic streptococci (*L. lactis* subsp. *diacetylactis*) from the noncitrate-fermenting *L. lactis* and subsp. *cremoris*. Recently, Vogensen et al.[104] described another differential agar medium for separating lactic streptococci and *Leuconostoc mesenteroides* subsp. *cremoris* found in mixed cultures [Modified Nickels and Leesment agar in association with X-Gal solution (5-Bromo-4-chloro-3-indolyl-β-D-galactopyranoside dissolved in dimethylsulfoxide at 1 mg/mL and filter-sterilized)]. Samples (0.5 to 1.0 mL of suitable dilutions) are mixed with 6 to 8 mL of modified Nickels and Leesment agar and allowed to solidify. An overlayer of 4 to 5 mL of the same medium is then poured. The plates are incubated at 25°C for 3 days. One-half mL of X-Gal solution is added to the plate, evenly distributed over the entire agar surface, reincubated for another day at 25°C, and examined. Lactococcus colonies appear white, while *Leuconostoc mesenteroides* subsp. *cremoris* colonies appear blue.

47.9123 Ratio of *Streptococcus thermophilus* and *Lactobacillus delbrueckii* subsp. *bulgaricus* in yogurt. Yogurt is a fermented milk product in which *S. thermophilus* and *L. delbrueckii* subsp. *bulgaricus* are the essential microbial species and are active in a symbiotic relationship. To obtain optimum consistency, flavor, and odor, many investigators claim that the two species should be present in about equal numbers in the culture. Dominance by either species can cause defects. Because of the emphasis on maintaining a balance between coccus and rod, techniques are needed to determine the relative proportions of *S. thermophilus* and *L. delbrueckii* subsp. *bulgaricus* when grown together in milk cultures. A microscopic examination to determine the ratio of coccus to rod is inadequate because dead cells cannot be distinguished from viable ones by this technique.

An agar medium (Lee's agar) for differential enumeration of yogurt starter bacteria has been described by Lee et al.[46] This medium contains sucrose, which most *L. delbrueckii* subsp. *bulgaricus* strains will not ferment, but *S. thermophilus*, will, and lactose, which both species utilize. With a suitable combination of sucrose and lactose, the rate of acid production by *S. thermophilus* is enhanced and that of *L. delbrueckii* subsp. *bulgaricus* restricted. Sufficient lactose is provided to obtain adequate colony formation of *L. delbrueckii* subsp. *bulgaricus* on the agar. Directions for the preparation of the agar plates should be followed very carefully.

Dilute culture in 0.1% peptone solution to 1×10^{-6} and spread 0.1 mL volumes of the dilutions over the agar surface with a sterile bent glass rod. Incubate plates for 48 hr at 37°C in a CO_2 incubator. *S. thermophilus* will form yellow colonies and *L. delbrueckii* subsp. *bulgaricus* will form white colonies. For satisfactory differentiation, the total number of colonies on the plates should not exceed 250. A preponderance of either species in a mixture prevents the distinction of colony types on this medium. This is because differentiation on this medium is based on acid-producing activity, the restriction of acid diffusion within a small area, and its detection with a pH indicator. Obviously, this is not a selective medium, and many other microorganisms can be expected to grow on it. In addition, some strains of *L. delbrueckii* subsp. *bulgaricus* can form yellow colonies indistinguishable from those of *S. thermophilus*. This difficulty may be eliminated by the use of pretested strains in the culture mixtures.

Porubcan and Sellars[65] described a medium (HYA agar) on which *L. delbrueckii* subsp. *bulgaricus* grows as diffuse, low-mass colonies (2 to 10 mm in diameter) and *S. thermophilus* as discrete, high-mass colonies (1 to 3 mm in diameter). Differentiation is achieved in this medium by adding an appropriate sugar or sugar mixture to the melted agar base before plating. The limitation of this method, particularly when used by personnel with limited training, is that differentiation is based on colony morphology.

Shankar and Davies[71] found that β-glycerophosphate, when incorporated into growth media generally, was inhibitory for *L. delbrueckii* subsp. *bulgaricus* strains, but did not affect *S. thermophilus* strains. They suggested that this principle could be used to get a differential count of rod-coccus cultures. A total count is obtained on lactic agar, and a differential *S. thermophilus* count on β-glycerophosphate-containing medium such as M-17 agar.[75] The *L. delbrueckii* subsp. *bulgaricus* population is calculated by subtracting the *S. thermophilus* count from the total count. Driessen et al.[18] reported two separate media to count cocci and rods respectively in mixed cultures. The medium used for *S. thermophilus* is called ST agar, and that for *L. delbrueckii* subsp. *bulgaricus* is designated LB agar. Matalon and Sandine[51] described modifications of lactic agar that allowed good differentiation of *S. thermophilus* and *L. delbrueckii* subsp. *bulgaricus*. The media are easy to make and give good differentiation with various strains of the two species in mixtures. The basal medium is lactic agar with added 0.1% Tween 80. One modification involved the addition of a sufficient amount of filter-sterilized 1% aqueous solution of 2,3,5-triphenyltetrazolium chloride to give a final concentration of 50 µg per mL of the dye in the agar. On this agar, *L. delbrueckii* subsp. *bulgaricus* appeared as white, large, smooth colonies with entire edges, while *S. thermophilus* formed smaller red entire colonies. The second modification involved the addition of 7.0% (vol/vol) reconstituted nonfat dry milk (11% solids, sterilized by autoclaving at 121°C for 15 min). Plates are poured and surface dried. Suitable dilutions are surface-plated and incubated at 35 to 37°C for 48 to 72 hr. On this agar, *L. delbrueckii* subsp. *bulgaricus* appeared as large, white, smooth, slightly raised entire colonies surrounded by a distinctive cloudy halo. *S. thermophilus* formed small, white, smooth colonies with no halo. Millard et al.[53] described the use of a hydrophobic grid membrane system for the simultaneous enumeration of *S. thermophilus* and *L. delbrueckii* subsp. *bulgaricus*. The membrane system was used in conjunction with erioglaucine supplemented tryptone-phytone-yeast extract agar. They reported that the procedure allowed single plate determinations of cocci to rods ratios from 20:1 to 1:5. Ibrahim and Yamani[34] used a whey based medium for the differential enumeration of *L. delbrueckii* subsp. *bulgaricus* and *S. thermophilus*. Bromcresol green whey was prepared by mixing one part of a solution of sterile agar containing 0.004% bromcresol green, 1.2% K_2HPO_4, and 3% yeast extract with two parts of whey which contained 4% agar. *L. delbrueckii* colonies were light colored and had an irregular mass while the *S. thermophilus* colonies were green, lenticular, and had smooth edges.

The introduction of *L. acidophilus* and *Bifidobacterium* into yogurt has caused the development of media to enumerate these bacteria in yogurt. Lankaputhra and Shah[44] determined the effects of carbon source and antibiotics on the growth of *S. thermophilus, L. delbrueckii* subsp. *bulgaricus, L. acidophilus,* and *Bifidobacterium* spp. They successfully enumerated *L. acidophilus* in pure culture and in yogurt using a minimal nutrient base with salicin as the sole source of carbon. This medium suppressed the growth of all of the bacteria except *L. acidophilus*. Ghoddusi and Robinson[26] evaluated several media available for the enumeration of starter cultures. They reported that tryptone proteose peptone yeast extract agar containing Prussian blue made visual separation of *L. delbrueckii* subsp. *bulgaricus, S. thermophilus, L. bulgaricus,* and *Bifidobacterium* spp. possible in bio-yogurts. Trypticase phytone yeast agar with a mixture of antibiotics was suitable for the discrete enumeration of *Bifidobacterium* spp. in mixed cultures.

47.92 Propionic Acid-Producing Bacteria

The propionibacteria can be difficult to isolate from foods and other natural sources. They grow slowly on solid media and prefer anaerobic or microaerophilic conditions. For samples in which different types of competing microorganisms are present in equal or higher concentrations, the propionibacteria may be the last colonies to appear on agar plates and may be difficult to pick out. Selective media designed for the propionibacteria have been based on their ability to metabolize lactic acid under anaerobic conditions. Complex media do not completely suppress the growth of competing organisms, especially those present on agricultural materials and plant surfaces. Defined media are more selective, but may not support the growth of all propionibacteria present in natural sources.

47.921 Media

Sodium lactate agar is a complex medium containing protein digests, yeast extract, various salts, and sodium lactate as a carbon source. Various versions of this medium[32,33,64] have been used since it was described by Vedamuthu and Reinbold.[102] If trypticase soy broth is used as an ingredient (modified sodium lactate agar), dextrose is also present, and the medium loses some selectivity. Pour plates or spread plates may be prepared. These are incubated at 30° to 32°C for 5 to 7 days under anaerobic or microaerophilic conditions (see below). Colonies that appear within 2 days are probably not propionibacteria. Individual colonies may be confirmed as propionibacteria by microscopic examination for typical pleomorphic rod shape and by detection of propionic acid production by gas chromatography or HPLC.

A defined selective medium (sodium lactate agar) has been described by Peberdy and Fryer[64] for the isolation of propionibacteria from cheese. This medium contains several salts, four vitamins, cysteine, ammonium sulfate as nitrogen source, and sodium lactate as carbon source. Several known bacterial species commonly found in cheese did not grow on this medium; 16 of 22 *Propionibacterium* strains tested did grow. Incubation is at 30° to 32°C under anaerobic or microaerophilic conditions for 11 to 14 days. Colonies may be confirmed as propionibacteria as described above. Competing organisms are less likely to grow on this medium than on sodium lactate agar.

47.922 Anaerobic Conditions

These methods to obtain anaerobic or microaerophilic conditions are listed in increasing order of convenience.

Vedamuthu and Reinbold[102] described the use of the candle-oats jar to obtain a CO_2-enriched atmosphere. Plates are incubated in a desiccator with moistened oats in the desiccant chamber. A candle is placed on the platform in the chamber and lit just before closing the lid. An atmosphere high in humidity and CO_2 is obtained.

Hettinga et al.[32] adapted the pouch method, first described by Bladel and Greenberg[3] for enumeration of clostridia, to the cultivation of propionibacteria. Pouches are prepared by sealing two sheets of plastic film together with a pouch-shaped sealing iron. A 1-mL aliquot of sample is pipetted into the pouch through the neck, and 30 mL of tempered growth medium (containing 2% agar) is added. The sample and medium are mixed by gently working the pouch between the fingers, and the pouch is placed in a holder until the medium solidifies. The agar seal in the neck of the pouch excludes oxygen sufficiently. Colonies can be easily seen and counted through the sides of the pouch.

Pour plates or spread plates prepared in standard petri plates can be incubated in an anaerobic or CO_2-enriched atmosphere by placing them in sealed plastic jars. The jars can be flushed repeatedly with CO_2 through a vent that can be closed, or GasPak (BBL) hydrogen + CO_2 generator envelopes can be used in the jars. Colonies appear sooner and grow better on a defined medium if the GasPak envelopes are used.

47.10 ADDITIONAL READING

For more extensive information on the processing, microbiology, and spoilage of dairy products and their possible role in the transmission of disease, the reader should refer to Doyle, Beuchat, and Montville[17] and Marth and Steele.[50]

47.11 REFERENCES

1. ADPI. 1965. Standards for grades of dry milks, including methods of analysis. Am. Dairy Products Inst., Chicago, Ill.
2. Bishop, J. R., and C. H. White. 1986. Assessment of dairy product quality and potential shelf-life-A review. J. Food Prot. 49:739-753.
3. Bladel, B. O., and R. A. Greenberg. 1965. Pouch method for the isolation and enumeration of clostridia. Appl. Microbiol. 13:281-285.
4. Bodyfelt, F. 1980. Quality assurance: Heat resistant psychrotrophs affect quality of fluid milk. Dairy Record (March):97.
5. Bramley, A. J., C. H. McKinnon, R. T. Staker, and D. L. Simpkin. 1984. The effect of udder infection on the bacterial flora of the bulk milk of ten dairy herds. J. Appl. Bacteriol. 57:317-323.
6. Bryan, F. L. 1983. Epidemiology of milk-borne diseases. J. Food Prot. 146:637-649.
7. Bryan, F. L. 1988. Risks associated with vehicles of foodborne pathogens and toxins. J. Food Prot. 51:498-508.
8. Campbell, J. R., and R. T. Marshall. 1975. The science of providing milk for man. McGraw-Hill, New York.
9. Centers for Disease Control. 1970. Staphylococcal food poisoning traced to butter-Alabama. MMWR Morb. Mortal Wkly. Rep. 19:271.
10. Cousin, M. A. 1982. Presence and activity of psychrotrophic microorganisms in milk and dairy products: A review. J. Food Prot. 45:172-207.
11. Cousin, M. A., and E. H. Marth. 1977. Cheddar cheese made from milk that was precultured with psychrotrophic bacteria. J. Dairy Sci. 60:1048-1056.
12. Cousins, C. M., M. E. Sharpe, and B. A. Law. 1977. The bacteriological quality of milk for cheddar cheese making. Dairy Ind. Int. 42:12-13, 15, 17.
13. D'Aoust, J. Y., D. W. Warburton, and A. M. Sewell. 1985. *Salmonella typhimurium* phage-type 10 from cheddar cheese implicated in a major Canadian foodborne outbreak. J. Food Prot. 48:1062-1066.

14. DeMan, J. C., M. Rogosa, and M. E. Sharpe. 1960. A medium for the cultivation of lactobacilli. J. Appl. Bacteriol. 23:130-135.
15. de Vries, T. 1975. Primary infection of milk. 1. Bacterial infection inside the udder and its relation with the cell count in milk. Neth. Milk Dairy J. 29:127-134.
16. Donnelly, C. W. 1988. Listeria and U.S. dairy products: The issues in perspective. Dairy Food Sanit. 8:297-299.
17. Doyle, M. P., L. R. Beuchat, and T. J. Montville. 1997. Food microbiology-fundamentals and frontiers. ASM Press, Washington, D. C.
18. Driessen, F. M., J. Ubbels, and J. Stadhouders. 1977. Continuous manufacture of yogurt I. Optimal conditions and kinetics of the prefermentation process. Biotechnol. Bioeng. 19:821-839.
19. Edwards, S. T., and W. E. Sandine. 1991. Public health significance of amines in cheese. J. Dairy Sci. 64:2431-2438.
20. Elliker, P. R., A. W. Anderson, and G. Hannesson. 1956. An agar culture medium for lactic acid streptococci and lactobacilli. J. Dairy Sci. 39:1611-1612.
21. Evans, J. B., and C. F. Niven. 1951. Nutrition of the heterofermentative lactobacilli that cause greening of curing products. J. Bacteriol. 62:599-603.
22. FDA. 1995. Grade A pasteurized milk ordinance (rev.). Publ. No. 229. HHS/PHS/FDA, Washington D. C.
23. Foster, E. M., F. E. Nelson, M. L. Speck, R. N. Doetsch, and J. C. Olson. 1957. Dairy microbiology. Prentice-Hall, Englewood Cliffs, N. J.
24. Frazier, W. C. 1958. Food microbiology. McGraw-Hill, New York.
25. George, E., J. C. Olson, J. I. Jezeski, and S. T. Coulter. 1959. The growth of staphylococci in condensed skim milk. J. Dairy Sci. 42:816-823.
26. Ghoddusi, H. B., and R. K. Robinson. 1996. Enumeration of starter cultures in fermented milks. J. Dairy Res. 63:151-158.
27. Gilliland, S. E., and M. L. Speck. 1977. Enumeration and identity of lactobacilli in dietary products. J. Food Prot. 40:760-762.
28. Gilliland, S. E., M. L. Speck, and C. G. Morgan. 1975. Detection of *Lactobacillus acidophilus* in feces of humans, pigs, and chickens. Appl. Microbiol. 30:541-545.
29. Goff, H. D. 1988. Hazard analysis and critical control point identification in ice cream plants. Dairy Food Sanit. 8:131-135.
30. Gunderson, F. L., H. W. Gunderson, and E. R. Ferguson. 1963. Food standards and definitions in the United States-A guide book. Academic Press, New York.
31. Hartman, P. A., and D. V. Huntsberger. 1961. Influence of subtle differences in plating procedure on bacterial counts of prepared frozen foods. Appl. Microbiol. 9:32-38.
32. Hettinga, D. H., E. R. Vedamuthu, and G. W. Reinbold. 1968. Pouch method for isolating and enumerating propionibacteria. J. Dairy Sci. 51:1707-1709.
33. Hofherr, L. A., B. A. Glatz, and E. G. Hammond. 1983. Mutagenesis of strains of propionibacterium to produce coldsensitive mutants. J. Dairy Sci. 66:2482-2487.
34. Ibrahim, S. A., and M. I. Yamani. 1998. An elective whey based medium for the differential enumeration of *Lactobacillus delbrueckii* subspecies *bulgaricus* and *Streptococcus salivarius* subspecies *thermophilus* in yogurt. J. Dairy Sci. 81(Suppl. 1):27.
35. JayneWilliams, D. J. 1963. Report of a discussion on the effect of the diluent on the recovery of bacteria. J. Appl. Bacteriol. 26:398-404.
36. Jeffrey, D. C., and J. Wilson. 1987. Effect of mastitis related bacteria on total bacterial count of bulk milk supplies. J. Soc. Dairy Tech. 40:23-26.
37. Jervis, D. I. 1988. Behaviour of pathogens in dairy products. Dairy Ind. Int. 53:15-19.
38. Kempler, G. M., and L. L. McKay. 1980. Improved medium for detection of citratefermenting *Streptococcus lactis* subsp. *diacetylactis*. Appl. Environ. Microbiol. 39:926-927.
39. King, W. L., and A. Hurst. 1963. A note on the survival of some bacteria in different diluents. J. Appl. Bacteriol. 26:504-506.
40. Kleter, G. 1975. The bacterial flora in aseptically drawn milk. Neth. Milk Dairy J. 28:220-237.
41. Kosikowski, F. W. 1982. Cheese and fermented milk foods, 2nd ed. F. V. Kosikowski and Assoc., Brooktondale, NY.
42. Kurzweil, R., and M. Busse. 1973. Total count and microflora of freshly drawn milk. Milchwissenschaft 28:427-431.
43. Kwee, W. S, T. W. Dommett, J. E. Giles, R. Roberts, and R. A. D. Smith. 1986. Microbiological parameters during powdered milk manufacture. 1. Variation between processes and stages. Aust. J. Dairy Tech. 41:3-6.
44. Lankaputhra, W. E. V., and N. P. Shah. 1996. A simple method for selective enumeration of *Lactobacillus acidophilus* in yogurt supplemented with *Lactobacillius acidophilus* and *Bifidobacterium* spp. Milchwissenschaft 51:446-451.
45. Law, B. A. 1984. Microorganisms and their enzymes in the maturation of cheeses, p.245-283. *In* Progress in industrial microbiology, vol. 19. Modern applications of traditional biotechnologies. Elsevier Scientific Publishing, New York.
46. Lee, S. Y., E. R. Vedamuthu, C. J. Washam, and G. W. Reinbold. 1974. An agar medium for the differential enumeration of yogurt starter bacteria. J. Milk Food Technol. 37:272-276.
47. Lowrie, R. J., and L. E. Pearce. 1971. The plating efficiency of bacteriophages of lactic streptococci. N. Z. J. Dairy Sci. Technol. 6:166-177.
48. Marshall, R. T. (ed.). 2001 (In press). Standard methods for the examination of dairy products, 17th ed. Am. Public Health Assoc., Washington, D. C.
49. Marth, E. H. 1985. Pathogens in milk and milk products, p. 43-87. *In* G. H. Richardson (ed.), Standard methods for the examination of dairy products, 15th ed. Am. Public Health Assoc., Washington, D. C.
50. Marth, E. H., and J. L. Steele. 1998. Applied dairy microbiology. Marcel Dekker, New York.
51. Matalon, M. E., and W. E. Sandine. 1986. Improved media for differentiation of rods and cocci in yogurt. J. Dairy Sci. 69:2569-2576.
52. Mikolajcik, E. M. 1980. Psychrotrophic bacteria and dairy product quality. 3. Organisms of public health importance in fermented dairy foods. Cult. Dairy Prod. J. 15:14-17.
53. Millard, G. E., R. C. McKellar, and R. A. Holley. 1990. Simultaneous enumeration of the characteristic microorganisms in yogurt using the hydrophobic grid membrane filter system. J. Food Prot. 53:64-66.
54. Minor, T. E. and E. H. Marth. 1972. *Staphylococcus aureus* and enterotoxin A in cream and butter. J. Dairy Sci. 55:1410-1414.
55. Mullan, M. A., and A. L. Walker. 1979. An agar medium and simple streaking technique for the differentiation of lactic streptococci. Dairy Ind. Int. 44:13, 17.
56. Mundt, J. O. 1973. Litmus milk reaction as a distinguishing feature between *Streptococcus faecalis* of human and nonhuman origins. J. Milk Food Technol. 36:364-367.
57. Mundt, J. O., and J. L. Hammer. 1968. Lactobacilli on plants. Appl. Microbiol. 16:1326-1330.
58. Mundt, J. O., A. H. Johnson, and R. Khatchikian. 1958. Incidence and nature of enterococci on plant materials. Food Res. 23:186-193.
59. National Mastitis Council, Inc. 1998. Laboratory handbook on bovine mastitis. Natl. Mastitis Council, Inc., Madison, Wis.
60. National Research Council. 1969. An evaluation of the Salmonella problem. Natl. Acad. Sci./Natl. Res. Council Comm. on Salmonella, Washington, D. C.
61. National Research Council. 1985. Application of microbiological criteria to foods and food ingredients, p. 184-307. *In* An evaluation of the role of microbiological criteria for foods and food ingredients. National Academic Press, Washington, D. C.
62. Olson, J. C., and G. Mocquot. 1980. Milk and milk products, p. 470-520. *In* J. H. Silliker, R. P. Elliott, A. C. Baird-Parker, F. L. Bryan, J. H. Christian, D. S. Clark, J. C. Olson, and T. A. Roberts (ed.), Microbial ecology of foods. Academic Press, New York.

63. Overcast, W. W., and K. Atmaram. 1974. The role of *Bacillus cereus* in sweet curdling of fluid milk. J. Milk Food Technol. 37:233-236.
64. Peberdy, M. F., and T. F. Fryer. 1976. Improved selective media for the enumeration of propionibacteria from cheese. N. Z. J. Dairy Sci. Technol. 11:10-15.
65. Porubcan, R. S., and R. L. Sellar. 1973. Agar medium for differentiation of *Lactobacillus bulgaricus* from *Streptococcus thermophilus*. J. Dairy Sci. 56:634.
66. Reddy, M. S., E. R. Vedamuthu, and G. W. Reinbold. 1971. A differential broth for separating the lactic streptococci. J. Milk Food Technol. 34:43-45.
67. Reddy, M. S., E. R. Vedamuthu, C. J. Washam, and G. W. Reinbold. 1972. Agar medium for differential enumeration of lactic streptococci. Appl. Microbiol. 24:947-952.
68. Rogosa, M., J. A. Mitchell, and R. F. Wiseman. 1951. A selective medium for the isolation and enumeration of oral and fecal lactobacilli. J. Bacteriol. 62:132-133.
69. Rosenow, E. M., and E. H. Marth. 1987. Listeria, listeriosis and dairy foods: A review. Cult. Dairy Prod. J. 22:13-17.
70. Sandine, W. E., P. C. Radich, and P. R. Elliker. 1972. Ecology of lactic streptococci. A review. J. Milk Food Technol. 35:176-185.
71. Shankar, P. A., and F. L. Davies. 1977. A note on the suppression of *Lactobacillus bulgaricus* in media containing betaglycerolphosphate and application of such media to selective isolation of *Streptococcus thermophilus* from yogurt. J. Soc. Dairy Technol. 30:28-30.
72. Sharpe, M. E. 1979. Identification of the lactic acid bacteria, p. 65-79. *In* R. A. Skinner and D. W. Lovelock (ed.), Identification methods for microbiologists, 2nd ed. Academic Press, New York.
73. Speck, M. L., and D. M. Adams. 1976. Symposium: Impact of heat stable microbial enzymes in food processing. Heat resistant proteolytic enzymes from bacterial sources. J. Dairy Sci. 59:786-789.
74. Stadhouders, J. 1975. Microbes in milk and dairy products: An ecological approach. Neth. Milk Dairy J. 29:104-126.
75. Terzaghi, B. E., and W. E. Sandine. 1975. Improved medium for lactic streptococci and their bacteriophages. Appl. Microbiol. 29:807-813.
76. Thomas, S. B. 1972. The significance of thermoduric bacteria in refrigerated bulk collected milk. Dairy Ind. Int. 37 475-476, 478-480.
77. Thomas, S. B. 1974a. The microflora of bulk collected milk-Part 1. Dairy Ind. Int. 39:237-240.
78. Thomas, S. B. 1974b. The microflora of bulk collected milk-Part 2. Dairy Ind. Int. 39:279-282.
79. Thomas, S. B., and R. G. Druce. 1972. The incidence and significance of coli-aerogenes bacteria in refrigerated bulk collected milk. Dairy Ind. Int. 37:583-593.
80. U.S. Department of Agriculture. 1968. United States standards for grades of edible casein (acid). USDA Agr. Marketing Serv., Dairy Division, Washington, D. C.
81. U.S. Department of Agriculture. 1977. Federal and state standards for the composition of milk products (and certain non-milkfat products). USDA Agricultural Handbook No. 51, Washington, D. C.
82. U.S. Department of Agriculture. 1983. United States standards for grades of dry whole milk. USDA Agric. Market. Serv., Dairy Div., Washington, D. C.
83. U.S. Department of Agriculture. 1984. United States standards for grades of nonfat dry milk (roller process). USDA Agric. Market. Serv., Dairy Div., Washington, D. C.
84. U.S. Department of Agriculture. 1990. United States standards for grades of dry whey. USDA Agr. Market. Serv., Dairy Div., Washington, D. C.
85. U.S. Department of Agriculture. 1991. United States standards for grades of dry buttermilk and dry buttermilk product. USDA Agric. Market. Serv., Dairy Division, Washington, D. C.
86. U.S. Department of Agriculture. 1993. United States specifications for instant dry whole milk. USDA Agric. Market. Serv., Dairy Div., Washington, D. C.
87. U.S. Department of Agriculture. 1994. United States specifications for cream cheese, cream cheese with other foods, and related products. USDA Agric. Market. Serv., Dairy Div., Washington, D. C.
88. U.S. Department of Agriculture. 1994. United States standards for grades of whipped butter. USDA Agric. Market. Serv., Dairy Division, Washington, D.C.
89. U.S. Department of Agriculture. 1995. United States specifications for light butter. USDA Agric. Market. Serv., Dairy Div., Washington, D. C.
90. U.S. Department of Agriculture. 1996. United States standards for grades of nonfat dry milk (spray process). USDA Agric. Market. Serv., Dairy Div., Washington, D. C.
91. U.S. Department of Agriculture. 1996. United States standards for instant nonfat dry milk. USDA Agr. Market. Serv., Dairy Div., Washington, D. C.
92. U.S. Government. National Archives and Records Administration. 1998. CFR. Title 7. Part 58—Grading and inspection, general specifications for approved plants and standards for grades of dairy products, p. 67-143.
93. U.S. Government. National Archives and Records Administration. 1998. CFR. Title 7. Part 58. Subpart B. Section 135—Bacterial estimate, p. 98.
94. U.S. Government. National Archives and Records Administration. 1998. CFR. Title 7. Part 58. Subpart B. Section 345—Butter, p. 115-116.
95. U.S. Government. National Archives and Records Administration. 1998. CFR. Title 7. Part 58. Subpart B. Section 346—Whipped butter, p. 116.
96. U.S. Government. National Archives and Records Administration. 1998. CFR. Title 7. Part 58. Subpart B. Section 348—Plastic cream, p. 116.
97. U.S. Government. National Archives and Records Administration. 1998. CFR. Title 7. Part 58. Subpart B. Section 349—Frozen cream, p. 116-117.
98. U.S. Government. National Archives and Records Administration. 1998. CFR. Title 7. Part 58. Subpart B. Section 528—Microbiological requirements, p. 126.
99. U.S. Government. National Archives and Records Administration. 1998. CFR. Title 7. Part 58. Subpart B. Section 648—Microbiological requirements for ice cream, p. 130.
100. U.S. Government. National Archives and Records Administration. 1998. CFR. Title 7. Part 58. Subpart B. Section 653—Microbiological requirements for sherbet, p. 130.
101. U.S. Government. National Archives and Records Administration. 1998. CFR. Title 7. Part 58. Subpart B. Section 938—Physical requirements and microbiological limits for sweetened condensed milk, p. 141-142.
102. Vedamuthu, E. R., and G. W. Reinbold. 1967. The use of candleoats jar incubation for the enumeration, characterization, and taxonomic study of propionibacteria. Milchwissenschaft 22:428-431.
103. Vera, H. D. 1947. The ability of peptones to support surface growth of lactobacilli. J. Bacteriol. 54:14.
104. Vogensen, F. K., T. Karst, J. J. Larsen, B. Kringelum, D. Ellekjaer, and E. W. Nielsen. 1987. Improved direct differentiation between *Leuconostoc cremoris, Streptococcus lactis,* subsp. *diacetylactis*, and *Streptococcus cremoris/ Streptococcus lactis* on agar. Milchwissenschaft 42:646-648.
105. Washam, C. J., H. C. Olson, and E. R. Vedamuthu. 1977. Heat-resistant psychrotrophic bacteria isolated from pasteurized milk. J. Food Prot. 40:101-108.

Chapter 48

Fish, Crustaceans, and Precooked Seafoods

Ranzell Nickelson II, Susan McCarthy, and Gunnar Finne

48.1 INTRODUCTION

Seafoods are more perishable than other high-protein foods. In seafood products, changes in flavor, odor, texture, and color reflect the level of freshness vs. decomposition, and are caused primarily by microbial activity. The rate of decomposition is influenced by the initial number and types of bacteria, and storage conditions, such as temperature, humidity, and gaseous atmosphere. Certain seafoods contain high levels of osmoregulators in the form of nonprotein nitrogen (e.g., amino acids, trimethylamine oxide, or urea) that are readily available to bacteria. Large quantities of various fish and crustaceans are harvested from cold water; therefore, the microflora is not inhibited as effectively by refrigeration as is the normal microflora of warm-blooded animals or fish caught in warm water regions. The place and method of processing seafood species and warm-blooded animals (i.e., aboard ship vs. a slaughter plant) also affect perishability.

Concern with the presence of bacterial contamination in seafood products is usually related to the potential for foodborne illness. However, poor quality products (spoiled or decomposed) are rarely responsible for food poisoning because they usually are discarded before consumption. Food poisoning from seafood products, as with other foods, is normally the result of mishandling during or after preparation (except in instances such as scombroid or histamine poisoning).

During 1988–1992, shellfish and other fish were responsible for 3% and 14%, respectively, of the 1,072 foodborne disease outbreaks of confirmed etiology. Confirmed outbreaks attributed to shellfish were due to microbial origin (67%) and paralytic shellfish poisoning (29%), while those caused by other fish were due to bacterial origin (13%), ciguatera (31%), and scombrotoxin (55%).[9]

This chapter, devoted to fish, crustaceans, and precooked seafood, focuses on many different seafood species and products. Within the limits of this text, it would be difficult to describe the bacteriology of all seafood species because of the differing areas of the world and times of year for seafood harvest, the variety of product forms, and the numerous testing procedures and conditions utilized. The information presented here should permit a practical approach for the microbiological analysis of products and for interpretation of test results.

48.2 FRESH AND FROZEN FISH AND CRUSTACEANS

48.21 Natural Flora

The subsurface flesh of live, healthy fish is considered bacteriologically sterile. The largest concentrations of microorganisms are found in the intestine, gills, and surface slime. The numbers and types of microorganisms found on freshly caught fish are influenced by the geographical location of the catch, the season, and the method of harvest.[56] The microflora of fish and crustaceans reflect the microbial population of their surrounding waters.[44,57]

Differences to be expected include variations in salinity, temperature, organic matter, and water quality found in each harvest area. For example, the average number of bacteria found on freshly caught Gulf of Mexico shrimp is reported to range from 10^3 to 10^4/g, while freshly caught inshore bay shrimp have counts from 10^4 to 10^5/g.[29,69] The area of catch also can influence the types of microorganisms found on fish. The incidence of *Salmonella* has been shown to be higher in fish from inshore waters impacted by human and animal pollution than in fish from the open ocean.[30]

The microflora of water, sediments, and marine organisms of the South Atlantic Ocean and the Gulf of Mexico can change considerably from summer to winter, particularly for organisms such as *Vibrio parahaemolyticus*, which have been shown to cycle with zooplankton blooms.[12] When water temperatures increase, both plankton production and the incidence of *V. parahaemolyticus* increase. The vibrios cannot be detected in the water during the winter months, but they may be found in the sediment. Results of an investigation of the microbial flora of pond-reared brown shrimp *(Penaeus aztecus)* show the changes that can occur in shallow waters along the western Gulf Coast.[71] The flora of shrimp and pond water were monitored from June through October. During the initial months of the experimental period, coryneform bacteria were predominant. Lowest counts occurred in August when water temperatures and salinities were highest. Coryneform bacteria and species of *Vibrio, Flavobacterium, Moraxella,* and *Bacillus* were common isolates. At harvest, a decrease in coryneform bacteria and subsequent increases in *Vibrio, Flavobacterium,* and *Moraxella* species were noted.

The microflora of shrimp harvested from the Gulf of Mexico consisted predominantly of coryneform bacteria, *Pseudomonas,*

Moraxella, and *Micrococcus*. *Pseudomonas*, the most common spoilage organism, was not isolated from pond water or from shrimp, possibly indicating a difference in the harvesting and handling techniques.[69] The harvest method also has been shown to influence the number of bacteria on fish.[56] Trawled fish usually exhibit higher numbers of bacteria than do line-caught fish because the trawled fish are dragged along the bottom (exposing them to mud) and their intestinal contents may be forced out as the trawl is hauled in.

In an examination of wholesale/frozen and retail/previously frozen raw imported shrimp products, Berry et al.[4] reported samples contaminated with *Listeria* spp. (16.7%) and *L. monocytogenes* (6.7%). *Vibrio* spp. were recovered from 63.3% of the samples. Most of these isolates were *V. parahaemolyticus* (36.7%), *V. alginolyticus* (26.7%), or *V. vulnificus* (16.7%).

Before any conclusions can be drawn about the natural microflora of seafood, certain nonintrinsic factors, such as type of organism, initial isolation media used, and incubation temperatures, must be considered. Plating medium can affect the number and the types of bacteria isolated because of differences in nutrient and salt requirements of the various microorganisms. Differences in types of microflora reported for shrimp[7,31,69,75] may have been influenced by the type of media used. Many of the coryneform bacteria appear on various plating media as small colonies only after 2 to 3 days incubation and fail to multiply after initial transfer.[71] The nutrient requirements for some of these organisms are quite strict. Salinity of the medium also can affect the numbers and types, especially if halophilic organisms are present. *V. parahaemolyticus* can best be isolated on appropriate media prepared with added salt.[70] It is now generally accepted that a plate incubation temperature of 25°C will produce significantly higher numbers than incubation at 35°C (which was the accepted standard for many years). Vanderzant et al.[72] found counts on breaded shrimp ranging from 1.1×10^4 to 6.8×10^6/g at 35°C. In contrast, the same samples yielded 6.0×10^4 to 2.7×10^7/g when plates were incubated at 25°C.

A survey of fresh retail products from Seattle, WA showed the overall quality of seafoods to be high.[1] Most probable number (MPN) coliforms averaged 199/g, MPN *Escherichia coli* 21/g, MPN *Staphylococcus aureus* 66/g, enterococci 9,121/g, *Clostridium perfringens* 18/g, *Bacillus cereus* 100/g, and *V. parahaemolyticus* 3.7/g. *Vibrio cholerae, Clostridium botulinum, Salmonella*, and *Shigella* spp. were not detected. Aerobic plate counts (APCs) exhibited a mean of 2.0×10^5/g at 22°C, which indicates an historical improvement in quality. Foster et al.[27] reported only 39% of products surveyed met the International Commission on Microbiological Specifications for Foods (ICMSF)[33] criteria ($n = 5, c = 3, M = 10^6$/g, $M = 10^7$/g), whereas the more recent survey shows that 98% met these criteria.

48.22 Spoilage Microflora

Shelf life, the time of storage before microbial spoilage of a fish is evident, is determined by the number and types of bacteria and the storage temperature. The numbers and types of microorganisms are determined by the natural microflora and the manner in which the fish was handled between harvesting and storage. The true storage temperature is not only the final market box temperature, but the temperature history of the product, which includes ambient temperature at harvest, delays in refrigerated storage, and fluctuations in storage temperature.

It is recognized generally that the predominant spoilage organisms belong to the genus *Pseudomonas*. These organisms are capable of causing spoilage because of two important characteristics. First, they are psychrotrophic and thus, multiply at refrigeration temperatures. Secondly, they metabolize various substances in the fish tissue which results in metabolic by-products associated with off-flavors and off-odors. These compounds are reported to be methyl mercaptan, dimethyl disulfide, dimethyl trisulfide, 3-methyl-l-butanal, trimethylamine, and ethyl esters of acetate, butyrate, and hexanoate.[45,46,55]

As noted in the previous section, the natural microflora of tropical-subtropical products can be expected to contain *Pseudomonas* as an insignificant proportion of the total population; nonetheless, pseudomonads predominate at spoilage. There are several possible reasons for subsequent spoilage by *Pseudomonas*, involving one or more of the following: (1) shorter generation time than other organisms, (2) antagonistic or synergistic reactions, (3) the ability to metabolize large protein molecules, and (4) overall biochemical activity. Regardless of the mode of action, the time from catch to spoilage is based to a great extent on the storage temperature, and on the initial numbers of *Pseudomonas* and related organisms present.

Since few are present in the natural microflora, it may be assumed that extensive numbers and subsequent spoilage are due to abuses in handling practices and/or storage.

48.23 Indigenous Bacterial Pathogens

48.231 Vibrio parahaemolyticus

V. parahaemolyticus is one of the major etiologic agents of gastroenteritis and is responsible for about 40% of foodborne illness in Japan.[77] It has been isolated from clinical and nonclinical sources including estuarine and marine environments.[36] Foodborne illness caused by *V. parahaemolyticus* may result from consumption of insufficiently heated or raw seafood, or properly cooked seafood contaminated after cooking. Implicated foods include shrimp, crab, lobster, and oysters.[10]

48.232 Vibrio vulnificus

V. vulnificus has been isolated from seawater, sediment, finfish, shellfish, crustacea, and plankton, and is found along the Atlantic, Pacific, and Gulf coasts of the U.S.[13] *V. vulnificus* causes primary septicemia, wound infections, and gastroenteritis. Mortality rates as high as 55% can result from primary septicemia which is associated with raw oyster consumption and liver disease.[37] Isolation of *V. vulnificus* from marine samples is highest when water temperature exceeds 20°C and salinities are between 5‰ and 20‰.[59]

48.233 Vibrio cholerae

V. cholerae non-O1 is often isolated from shellfish from U.S. coastal waters and may cause infection in humans following consumption of raw, improperly cooked, or recontaminated cooked shellfish. Frequent sporadic cases usually are associated with consumption of raw oysters during warmer months.[23]

V. cholerae O1 is found in the temperate estuarine and marine environments along the U.S. coast.[24] This organism is responsible for epidemic cholera. Sporadic cholera cases in the U.S. between 1973 and 1991 were associated with consumption of raw, improperly cooked, or recontaminated cooked shellfish. An outbreak in 1991, involving eight people in New Jersey, was due to consumption of noncommercial crabmeat that was mishandled and illegally transported into the U.S.[8,26] Previous cases acquired in the U.S. were associated with undercooked crabs or raw oysters from the Gulf of Mexico.

48.234 Aeromonas hydrophila

A. hydrophila has been found in freshwater, brackish water, fish, and shellfish.[32] Due to the ubiquity of the organism, it is thought that not all strains are pathogenic. Most cases have been sporadic rather than associated with large outbreaks.[20]

48.235 Clostridium botulinum *type E*

Isolated almost exclusively from marine environments, *C. botulinum* type E is of special concern because it is more psychrotrophic than other types within the species, is capable of reproducing during refrigeration, and is non-proteolytic. Class II, non-proteolytic, types E and B are capable of growth in a water phase salt concentration of 5% to 7% at storage temperatures as low as 3.3°C.[15,16,58]

Kapchunka, a ready-to-eat, air-dried, salt-cured, uneviscerated whitefish, presents a potentially life-threatening health hazard from botulinum toxin that is much greater than that of cleaned fish. *C. botulinum* spores are likely to be in the fish gut as a result of ingestion during feeding. It is also difficult to attain sufficiently high salt levels in all portions of the fish during curing. In 1985 and 1987, *C. botulinum* type E in kapchunka was responsible for 10 botulism cases, with three deaths in the U.S. and Israel.[64] Both hot- and cold-smoked fish also have caused outbreaks of type E botulism.[21] Other concerns center around the use of modified- or controlled-atmosphere packaging, during which oxygen levels are decreased or other gases are increased, to reduce the normal spoilage microflora and extend the shelf life of the product.[39]

48.236 Plesiomonas shigelloides

Most human *P. shigelloides* infections occur in the summer months and correlate with environmental contamination of freshwater; it also has been isolated from freshwater fish and shellfish.[47] Most strains associated with human gastrointestinal disease have been from patients living in tropical and sub-tropical areas. Infections are rarely reported in the U.S. or Europe due to their self-limiting nature.[22]

48.237 Yersinia enterocolitica

Y. enterocolitica has been isolated from raw seafood products and can grow at refrigeration temperatures. Pathogenic strains have been isolated from crabs harvested from cold waters.[19] There are no documented outbreaks from consumption of seafood products caused by this organism.

48.24 *Nonindigenous Bacterial Pathogens*

48.241 Clostridium perfringens

C. perfringens is found normally in the intestinal tract of man or in soil, and usually is not associated with freshly harvested seafood products. It has the potential for causing problems in sauces or soups, such as gumbo. Although it may contribute to histamine production in scombroid fish held at high temperatures (>30°C), it is not a problem in properly prepared and stored seafood products.

48.242 Listeria monocytogenes

The literature contains information on *Listeria* isolations from soil, birds, sewage, silage, stream water, estuarine environments, and mud. *Listeria monocytogenes* has been recovered from frozen raw and cooked fish or crustaceans, hot- and cold-smoked fish, fermented fish, and squid.[35,73,76] Listeriosis has been implied from raw shellfish or raw finfish in an outbreak in New Zealand.[41] Contaminated water may infect fish and shellfish in freshwater and marine environments. Agricultural runoff, fecal contamination by animals, and sewage effluents contribute to contamination of the aquatic environment with *Listeria*.[6,48] *L. monocytogenes* differs from other foodborne pathogens in its psychrotrophic nature while having some resistance to heat, high salt concentrations, and acidic conditions.[51]

48.243 Staphylococcus aureus

Although seldom isolated from freshly harvested seafood products, *S. aureus* can be found in products that involve extensive human handling, like picked crab meat.

48.244 Salmonella *spp.*

Although not normally isolated from fish and shellfish from open seas, these organisms can be isolated from seafoods harvested from contaminated inland waters. *Salmonella* presents no direct health hazard in raw products expected to be cooked before consumption because normal cooking destroys them. They are of concern, however, in products consumed raw and in products ready for consumption without further heat processing. The presence of *Salmonella* also is of concern because it can be transmitted to other foods via cross-contamination, from raw to cooked foods. *Salmonella* is a normal inhabitant of the intestinal tract of reptiles and amphibians, and has been of concern in imported processed frog legs. It has not been a problem in other properly handled, prepared, and stored seafood products. Its presence indicates direct or indirect fecal contamination from man or animal.

48.245 Shigella *spp.*

Shigella spp. is an incidental foodborne contaminant directly linked to man. Cases of shigellosis have been reported from finfish, shellfish, shrimp or tuna salad, cooked shrimp, and raw oysters.[38]

48.246 Escherichia coli

E. coli has been used traditionally as an indicator of fecal contamination and for the potential presence of *Salmonella. E. coli* is not considered a good indicator of fecal pollution for seafood from cold water because of a rapid decline of *E. coli* in seawater at low temperatures.[19] It has been established, however, that the isolation of *E. coli* from blue crab meat, in conjunction with inspections that show a source of fecal material, does constitute "filth" in the form of fecal material.

For the role of viruses and parasites in seafoods, please consult Chapters 44 and 42 of this book.

48.3 COOKED CRUSTACEAN PRODUCTS

Crab, shrimp, lobster, and langostino products are precooked to extend shelf life, to impart a desirable characteristic flavor, or to facilitate picking or peeling. In addition, crab meat and langostino may be pasteurized at 85°C for 1 min after picking and peeling to enhance refrigerated storage life.

Blue crabs are steamed under pressure or boiled before picking to produce what is termed fresh-picked crab meat. Survival of the natural microflora, multiplication of organisms during storage before picking, and post-cooking contamination contribute collectively to the number and types of bacteria found in crab meat. One survey indicated that 93% of the plants operating under good sanitary conditions produced crab meat with MPN coliforms of less than 20/g, MPN *E. coli* <3g, MPN coagulase-positive staphylococci less than 30/g, and APCs (at 35°C) of less than 10^5/g.[53] A survey of retail samples of blue crab meat complicates the issue of microbiological standards and where they should be applied. Blue crab meat with an APC (geometric mean) of 5.2×10^5 at 30°C had counts of 3.6×10^9 in the upper 10% of samples examined.[74] These large differences in count may reflect a wide range of sanitary practices, but differences in time and temperature profiles during distribution most likely are involved.

Dungeness, snow, and king crab (cooked whole or cooked portions) and cooked lobsters fall into a unique category because little edible flesh actually is exposed to recontamination after cooking.

Exposed areas, flesh and shell, are useful in determining post-cooking contamination, whereas meat extracted from beneath the shell is a better indication of cooking effectiveness. APCs for cooked frozen products should be conducted at 35°C and 25°C for public health and shelf-life indicators, respectively. Distribution patterns of the microbial flora in Dungeness crab meat revealed the presence of three types of microorganisms.[40] *Moraxella, Pseudomonas, Acinetobacter*, and *Flavobacterium-Cytophaga* spp. originated from the raw crab and became dominant during refrigerated storage. *Arthrobacter* and *Bacillus* spp. also originated from the raw crab, but did not grow in the meat. *Micrococcus, Staphylococcus*, and *Proteus* were introduced during processing and did not grow. A high incidence of false-positive fecal coliforms as detected in EC broth has been reported.[54] Fecal coliforms were detected in only 10% of the samples at an MPN of 40/100g.

Geometric means for APCs incubated at 35°C for cooked peeled shrimp and raw peeled shrimp have been reported at 7.3×10^3/g and 3.0×10^6/g, respectively.[62]

Pasteurization of shrimp, crabmeat, and langostinos is intended to extend refrigerated shelf life and is not intended to render the product commercially sterile.[11] Since *C. botulinum* is normally found in the marine environment and in marine food animals, its presence raises important public health questions concerning pasteurized products. The recommended pasteurization process of 85°C for 3 min at the coldest point in a can of crabmeat is adequate to kill 10^6 spores/g of *C. botulinum* type E that may have been present in the fresh meat. Type E spores can be introduced into the crabmeat post-pasteurization during the cooling process if the can seams are defective; however, refrigeration at 2.2°C or below is required and protects the product from becoming toxic.

Pasteurization (83°C for 1 min) for blanched, peeled langostinos caused the following changes in microbial counts: Before pasteurization APC 1.3×10^5/g (25°C); MPN coliforms 9.1/g; MPN *E. coli* <3/g; and MPN coagulase-positive staphylococci >1100/g; and after pasteurization APC $<1.0 \times 10^3$–1.2×10^4/g; coliforms <3/g; *E. coli* <3/g; and MPN coagulase-positive staphylococci <3-23/g.[63]

48.4 BREADED AND PREPARED SEAFOOD PRODUCTS

Breaded and prepared seafood products can vary greatly in numbers and types of microorganisms because of the addition of non-marine ingredients. The weight of breaded products may include 25% to 65% flour, seasonings, nonfat dry milk, and dried eggs. Prepared products such as fish cakes and crab cakes contain spices, onions, and celery. The degree of cooking employed (i.e., raw, partial, or complete) further affects the number and types of organisms.

Surveys of frozen raw breaded shrimp processing plants have indicated that plants with good quality control can produce consistently a finished product with bacterial loads lower than the original incoming raw shrimp.[60] APCs on finished product samples from plants operating under good sanitary conditions ranged from 1.1×10^4 to 6.8×10^6/g at 35°C and 6.0×10^4 to 2.7×10^7/g at 25°C.[72] The initial flora consisted primarily of *Pseudomonas, Achromobacter, Aeromonas, Bacillus, Moraxella, Microbacterium, Micrococcus*, or coryneform bacteria. *Bacillus, Microbacterium, Micrococcus*, and coryneform bacteria were predominant in retail samples. The bacterial flora of batter and breading, which consisted primarily of *Bacillus* and *Microbacterium* species, probably contributed to this condition.

In a survey of fish breading operations, it was demonstrated that raw fish from plants operating under good conditions had average APC values of less than 10^5/g, MPN coliform less than 100/g, and no more than 20% of the units positive for *E. coli* or staphylococci.[3] Fried breaded fish had APC values of less than 2.5×10^4/g, MPN coliform was less than 10/g, and no more than 10% of the units were positive for *E. coli* or staphylococci.[61]

In a survey of plants by Duran et al.[14] in which all were operating under good manufacturing practices, aerobic plate counts ranged from 3.3×10^5 to 2.1×10^6 at 35°C and 7.6×10^5 to 7.8×10^6 at 30°C. Coliforms ranged from 148–160/g; MPN *E. coli* was <3/g; *S. aureus* was <10/g; and *Salmonella* was isolated from only 1 of 188 samples. The latter was traced back to the dry batter containing egg products. Again the peeling, batter, and breading operations produced a finished product of lower APCs than the incoming raw shrimp.

Products such as crab stuffing or crab cakes contain fresh or dehydrated vegetables and normally have higher coliform counts than other products. This fact was reflected in a proposed guideline MPN for frozen fish sticks, frozen fish cakes, and frozen crab cakes of 230, 1500, and 4300/g, respectively.[65,66]

48.5 SALTED AND SMOKED PRODUCTS

The main feature of salting fish involves removing some of the water from the fish tissue and partially replacing it with salt. Salting or brining may be done either as a hard cure (stacking fish with layers of dry salt), as a light dry cure (where limited dry salt is used), or in a brine solution (pickling). Depending upon the method used and the time of treatment, salted fish or seafood products may range from 1% salt to a fully salted dry fish containing in excess of 20% salt. The heavy or hard-cured fish is most commonly dried and stored as a stable product under dry storage conditions; less salted fish may be used for smoking or consumed as a salted or pickled product.

Smoking fish and fish products serves a number of different functions. Although traditionally used as a method of preservation, fish is smoked to give it a distinct flavor. In addition, smoking contributes to an increase in the shelf life of the product because of microbial destruction and lowering of the water activity. Several types of smoked fish products are available. These can be classified according to the temperature maintained during smoking, cold or hot smoked, the salt content, and the duration of smoking and/or drying. For cold smoked fish, the smoke is applied at a low temperature. Cold smoking may continue for a short period to produce lightly smoked fish, such as lox, or over long periods to produce kippered products. Hot smoking is conducted at much higher temperatures and results in a barbecued or cooked product. Smoked products are dried to varying degrees, depending on the smoking procedures, which result in a wide range of free water content. The result is a great variety in microbial distribution. For hard-smoked products with high heat input, the more heat-stable organisms such as *Bacillus, Micrococcus,* and yeasts will be predominant. For milder smoked fish, a number of gram-negative organisms will survive; *Pseudomonas* and *Moraxella-Acinetobacter* are common in the microflora of such products.

Smoked seafood products are perishable and should be treated as such. Storage stability will be dependent primarily on the water activity, salt content, total heat input during smoking, and storage conditions. For lightly brined and lightly smoked products, the refrigerated spoilage pattern will be similar to that of fresh refrigerated fish; *Pseudomonas* will be the dominant organism. For heavier smoked and more salted products, gram-positive organisms, together with halophiles, will prevail and molds will be seen more frequently. Hard-smoked products with a low water activity are spoiled primarily by molds.

The potential for growth and toxin production by *C. botulinum* type E is the concern related to the safety of smoked seafood products. This organism can be a part of the normal microflora of fish and marine sediments, and can grow and produce toxin at temperatures as low as 3.3°C. In products where the heat input during smoking is limited, the spores may survive and, since surface areas of smoked fish tissue have an oxidation-reduction potential sufficiently low to permit sporulation and growth, *C. botulinum* may present a health hazard. The elimination of the heat-sensitive spoilage microflora and temperature abuse of the final product encourage toxin production. Eklund et al.[16] reported that a hot-smoked fish process coupled with a vacuum-packaged, post-processing, heat-pasteurization step is capable of inactivating the Class II, non-proteolytic, type E and B spores, but not the more heat-resistant proteolytic *C. botulinum* spores. Their recommendation is that such a product be labeled "Keep refrigerated. Store below 38°F (3.3°C)."

48.6 OTHER SEAFOOD PRODUCTS

48.61 *Minced Fish Flesh, Surimi, and Seafood Analogs*

The mechanical separation (deboning) of fish flesh from dressed fish, filleting waste, frames, or V-cuts, has become an established practice in larger filleting and freezing plants. Minced fish flesh, the resulting product, can be processed directly into consumer products, frozen in blocks or prepared into surimi, which is a food-processing intermediate material.

The deboning process adds concerns relative to microbial contamination as compared to whole or filleted fresh or frozen fish. In addition to the increased number of possible contamination points during the deboning process, the nature of the product gives ample opportunity for an increased microbial population to develop. During deboning, tissue masceration not only increases the surface area, but it also allows for the release of cellular fluids rich in free amino acids and other substrates ideal for microbial growth. It is essential, therefore, that equipment be kept scrupulously clean and that the minced flesh be kept as cold as possible during processing. Blackwood[5] demonstrated the importance of good sanitary practices during deboning of fish flesh. When comparing two plants using similar raw materials, facilities, and processing equipment, one plant produced minced fish flesh with an average APC of 6.7×10^5/g with 1 out of 56 samples positive for fecal coliforms. The APC for minced flesh produced in the second plant was 1.5×10^7/g with 60% being positive for fecal coliforms.

The types of organisms present in minced flesh were shown by Nickelson et al.[50] to be very similar to the types of organisms present on the raw material. Ten processing lots representing six species of fish showed *Moraxella-Acinetobacter* spp. to be the most prevalent microbial type in the fish before, during, and after processing into minced fish flesh. Increases of up to one log during the production of minced flesh from whole fish also were reported. During freezing, marked decreases in the total microbial counts of the minced flesh were noted. Licciardello and Hill[42] examined 208 imported frozen minced fish samples and found all to be in compliance with the proposed standards of the ICMSF.[33]

Surimi is a food-processing intermediate material made from heavily washed minced fish flesh normally produced from pollock-type species. To the washed minced flesh are added cryoprotectants such as sorbitol, other sugars, and polyphosphates. The cryoprotectants allow surimi to be frozen for long periods of time without any substantial loss in the gel-forming capacity of the myofibrillar proteins. Because washing and addition of cryoprotectants dramatically alter the composition of the minced flesh, Ingham and Potter[34] compared the microbiological properties of minced flesh and surimi produced from Atlantic pollock. The APC and psychrotrophic count of the minced flesh and surimi samples stored at 5°C and 13°C were initially similar, but reached higher levels in surimi with time. They also pointed out that careful handling and storage are equally important for both products to maintain microbiological quality.

Because of its gel-forming ability, surimi can be processed into a variety of different seafood analogs. These analogs are prepared from chopped surimi with added salt, starch, and polyphosphates, together with flavor and aroma components. The final product can be flavored and shaped into imitation crab, shrimp, lobster, scallops, or fish portions. Amano[2] reported that the initial load of bacteria on the raw fish plays an insignificant role in relation to the storage life of the final product. The number of organisms associated with the raw fish is dramatically reduced during the extensive washing of the mince and additional organisms are destroyed during heat setting of the gel. This fact was also demonstrated by Yoon et al.[79] who investigated the microbiological and chemical changes in imitation crab during storage. They showed that because of the heat processing and added ingredients, crab analogs contain lower microbial numbers and a different microflora from uncooked fish. During storage at 0 and 5°C, psychrotrophic gram-negative organisms that survived the heat processing and initially were present in low numbers became the major spoilage organisms. At higher storage temperatures, *Bacillus* spp. dominated the spoilage population. The source of *Bacillus* spores was shown to be the starch and other ingredients added to the surimi during processing.

In general, fresh fish are seldom the cause of foodborne illness because fish spoils before pathogenic microorganisms proliferate and form toxin. However, heat used in the production of analogs reduces competing organisms. If these products are subsequently contaminated with pathogens and subjected to conditions where these pathogens can grow, they may become a health hazard. Yoon and Matches[78] inoculated samples of imitation crab legs and flaked crab meat with *S. aureus, Salmonella, Y. enterocolitica,* and *A. hydrophila* and stored them at 0°C, 5°C, 10°C, and 15°C. *S. aureus, Salmonella,* and *Aeromonas* grew at 10°C and 15°C, but not at the two lower temperatures. *Yersinia* grew at all four temperatures tested. The rapid growth at these temperatures indicates the potential hazard if seafood analogs should become contaminated and subjected to temperature abuse during distribution.

48.62 Fish Protein Concentrate

During the production of fish protein concentrate (FPC), the combination of heat and isopropyl alcohol extraction will remove most lipid material and also lower the water activity to an extent that microorganisms will not be able to multiply. Heat, together with the organic solvent, will destroy most of the nonsporeforming microorganisms present on the raw material.[28,52] Post-processing contamination is the major source of bacteria in FPC. According to FDA regulations,[68] FPC shall be free of *E. coli* and pathogenic organisms, including *Salmonella,* and have a total bacterial plate count of not more than 10,000 organisms/g. The organisms most frequently found in FPC are *Bacillus, Micrococcus,* and molds.

48.63 Pickled Products

Pickle-curing uses salt, vinegar, and spices to preserve and enhance the flavor of herring or similar fish. Pickled fish are presalted, washed to remove excess salt, and then repacked into smaller containers. A hot pickle solution consisting of vinegar, sugar, and spices is added. During the presalting, the most common microflora are *Micrococcus* and *Bacillus* spp. The same types of organisms are found most frequently in freshly pickled products as well. Secondary microbial contamination from handling may occur during repacking and additional organisms can be added with the pickling ingredients. During the storage of pickled products, the microflora undergoes changes with a selective proliferation of halotolerant microaerophiles.

Erichsen[17] reported that three to four types of microorganisms were found to dominate successively in pickled seafood during storage. Initially, most strains belonged to *Micrococcus* spp. with a lower level of *Staphylococcus* spp.; both are found commonly in the raw material used. No coagulase-positive staphylococci could be isolated, either in the raw material or in the pickled product. The next group of organisms to appear during the storage of pickled fish belonged to the genus *Pediococcus.* The contribution of these organisms to the flavor and quality of the final product is not known. They are homo-fermentative and known to lower the pH of the product. The last group of organisms predominating in pickled fish at the time of visual gas production belonged to the genera *Lactobacillus* and *Leuconostoc.* These organisms were hetero-fermentative and haloduric, and they preferred the lower pH for carbohydrate fermentation. The organisms that ultimately cause product deterioration were atypical for fresh fish. During prolonged storage, pickled fish display some softening of the tissue and cloudiness of the pickling solution, indicating proteolysis of the product.

48.64 Fermented Fish

Fermentation of fish is a common type of preservation used in southeast Asia. Many fish sauces and pastes are produced through natural fermentation. The process is quite simple: the fish are heavily salted and allowed to stand at natural atmospheric conditions for several months. Since the high salt content retards microbial growth, hydrolysis of the fish protein is thought to occur by natural tissue enzymes, cathepsins.

For low salt products, halophilic or halotolerant populations consisting of gram-positive organisms will prevail. One of these products, "i-sushi," which is fermented by various strains of *Lactobacillus,* has been associated with food poisoning outbreaks from toxin production by *C. botulinum* type E.

48.7 MODIFIED-ATMOSPHERE PACKAGING AND STORAGE

In general, there are three different ways in which modification of the atmosphere surrounding a fresh food material can be utilized as a method of preservation: 1) atmospheric modification through a packaging technique such as vacuum packaging; 2) atmospheric pressure reduction within the container to accomplish a hypobaric condition; and 3) atmospheric modification through enrichment of the container atmosphere by addition of different gas blends. Although no commercial feasibility for the use of hypobaric preservation of fishery products has been demonstrated, the interest in both vacuum and modified-atmosphere packaging of seafood products has increased greatly over the last years.

To prevent growth of aerobic spoilage organisms, fresh seafoods can be packaged under vacuum in gas-impermeable packages or containers. Under such conditions, residual oxygen is used by the resident microflora and tissue enzymes to produce carbon dioxide, resulting in a lowering of the surface oxidation-reduction potential. These conditions will suppress the growth of common aerobic spoilage organisms and favor the growth of facultative anaerobic organisms including lactic acid bacteria. The delay in spoilage is achieved due to the slower growth rates and the less extensive organoleptic changes that are characteristic of these organisms.

The mechanism of shelf-life extension for products packaged in modified atmospheres is similar to that of vacuum-packaging. High levels of carbon dioxide during storage at refrigeration temperatures will selectively inhibit the growth of gram-negative pseudomonads and other gram-negative psychrotrophs, which normally grow rapidly and are responsible for the characteristic off-odors and off-flavors of seafood products. The gram-positive micrococci, streptococci, and lactobacilli are more tolerant of high concentrations of carbon dioxide and often become the dominant microflora.

Temperature abuse of vacuum- and controlled-atmosphere-packaged fresh foods may result in rapid growth of both spoilage and pathogenic bacteria. With respect to fishery products preserved short of sterilization, a large amount of research has centered around the significance of *C. botulinum.* Two factors contribute to a higher risk of *C. botulinum* toxigenesis in seafoods than other foods: 1) *C. botulinum* type E and non-proteolytic types B and F are able to grow at temperatures of 3.3°C and 5°C, respectively, and 2) the prevalence of *C. botulinum* spores in fresh and salt water fish is relatively high. Eyles and Warth[18] have reported spore loads of 17 spores/100 g fish in haddock fillets.

With existing refrigeration equipment and distribution practices, there is no guarantee that temperatures of vacuum- and modified-atmosphere-packaged seafood products never exceed 3°C. An important question is whether organoleptic spoilage will precede toxin production and thus, warn the consumer. A few studies have addressed this question with conflicting results. However, there is a general indication that as the storage temperature is increased from 0°C to 10°C, the time interval between unacceptable spoilage and detectable toxin production in fish stored under modified atmospheres shortens, decreasing the safety margin. The conflicting data relative to rate of spoilage versus rate of toxigenesis can best be explained by the large number of variable factors that will affect both of these rates. Some of these factors are initial numbers and types of organisms present, species of fish used, number and quality of spores used as inoculum, packaging methodology, and criteria to evaluate spoilage. Even though raw fish normally are cooked adequately before consumption, which ordinarily assures destruction of any botulinum toxin, modified atmosphere technology has not been widely adopted by the seafood industry because of the potential for botulinum toxicity.

48.8 RECOMMENDED METHODS

48.81 Sample Shipment

All fish and crustacean samples should be collected and transported in such a manner as to represent closely the original product. They should be collected in the original package and shipped at a temperature near that of the product at the time of collection. Severe changes in temperature, either higher (thawing) or lower (direct contact with dry ice), can greatly influence the results of a bacteriological analysis.

Frozen seafood products can be packaged with dry ice in an insulated container for 24 to 30 hr with no significant change in APC. The coliform population can be reduced by the extreme cold (–30°C within 1 hr); prolonged frozen storage should be avoided. After 30 hr, 10 to 12 pounds of dry ice will vaporize and result in subsequent increases in temperature and APCs.

Packaged chemical gels are gaining popularity and, if previously frozen for at least 36 hr, will maintain an insulated container for 24 to 30 hr. Once melting begins, however, the temperature rises rapidly, and the microbial population increases at the rate of one log/24 hr. Coliform counts do not appear to be reduced by the cold of the ice gel; these tend to increase with rising temperature.

Wet ice is effective for 24 hr, but the "washing" action of the melting ice can produce inaccurate bacterial counts or destroy the product if it has not been wrapped properly.

A frozen product may be shipped in an insulated container without additional cooling, but the temperature will increase gradually after 6 hr. Bacterial counts are relatively stable after 12 hr, but they show a dramatic increase after 24 hr.

When freezing a fresh product is unavoidable because of delays in shipping or analysis, destruction or injury of microorganisms can be expected. Special precautions should be taken with products to be examined for *V. parahaemolyticus* or *C. perfringens* because both are sensitive to refrigeration.

48.82 Sample Method

48.821 Rinse Samples

Since most contamination is surface related, the rinse technique offers a rapid, reliable, and nondestructive means of sampling. The product is placed into a sterile plastic bag and weighed. An appropriate volume of sterile diluent, 0.1% peptone, phosphate buffer, or 0.5% NaCl is added, and the bag is massaged by hand for 1 to 2 min. Examples of products best suited for this technique include green headless (shell-on tails) shrimp,[49] small dressed fish, fillets, peeled and deveined shrimp, frog legs, and shell-on crab legs.

48.822 Swab Samples

The usefulness of swab samples is limited generally to comparing surface areas of large fish, such as belly cavity vs. head, or to sampling food contact surfaces. (See Chapter 2.)

48.823 Tissue Samples

Tissue samples in this context refers to the blending or "stomaching" of a product or part of a product. Seafoods best suited for this technique include crab meat, fillets, breaded portions, breaded shrimp, squid, crab cakes, and minced flesh. Frozen samples can be taken with a sterile drill bit. (See Chapter 2).

48.824 Skin Samples

Removal of a known area of skin by template or bore is applicable to larger fish. After incisions are made, the skin is removed from the muscle with sterile forceps and scalpel. With packaged fish, there is a need to include microorganisms on the inside of the packaging film. After treatment of the exterior of the film with a suitable bactericidal solution, sampling of film and fish can be carried out as described above.

48.83 Microbiological Procedures

48.831 Aerobic Plate Count

Use the method described in Chapter 7, modified by the incorporation of 0.5% NaCl in the medium (some nonselective media already contain 0.5%, e.g., trypticase soy agar). For routine assessment of quality, plate incubations at 25°C for fresh and frozen, and 35°C for cooked products are recommended. Occasionally, both temperatures are used to assess the quality as well as the safety of a product.

48.832 Coliform Group

Use the method described in Chapter 8. The MPN technique for breaded products should be used with some caution because materials from the breading-batter may contribute to false counts in lower dilutions.

48.833 S. aureus

The MPN technique has limited value in the examination of fresh and fresh frozen seafood products because of the high number of halotolerant organisms exhibiting growth in media containing 10% NaCl. It is recommended that the enumeration of *S. aureus* from fresh and fresh frozen seafoods be made by direct plating onto Baird-Parker agar plates. Cooked products can be examined according to the MPN technique. Both procedures are described in Chapter 39.

48.834 Additional tests

E. coli (See Chapter 35)
V. parahaemolyticus and V. cholerae (see Chapter 40)
C. botulinum (see Chapter 33)
L. monocytogenes (see Chapter 36)
Y. enterocolitica (see Chapter 41)
Salmonella (see Chapter 37)

48.835 Direct Microscopic Count (Chapter 4)

This is not a substitute for an APC, but may be used as a rapid, simple means of assessing the quality of seafood products that are subject to surface contamination. Counts normally will be 1 to 2 logs higher.

48.9 INTERPRETATION OF RESULTS

No uniform guidelines can be used to interpret the result of bacteriological testing of seafood products. Instead, each product must be evaluated on the basis of its own characteristics and guidelines must be established for practical good manufacturing procedures. Several studies [25,33,43,65,66,67] may be helpful in evaluating the acceptability of seafood and seafood products.

48.10 REFERENCES

1. Abeyta, C., Jr. 1983. Bacteriological quality of fresh seafood products from Seattle retail markets. J. Food Prot. 46:901-909.
2. Amano, K. 1962. The influence of fermentation on the nutritive value of fish with special reference to fermented fish products of South-east Asia, p. 180-200. *In* R. Kreuzer and E. Heen (ed.), Fish in nutrition. Fishing News Ltd., London.
3. Baer, E. F., A. P. Duran, H. V. Leininger, R. B. Read, Jr., A. H. Schwab, and A. Swartzentruber. 1976. Microbiological quality of frozen breaded fish and shellfish products. Appl. Environ. Microbiol. 31:337-341.
4. Berry, T. M., D. L. Park, and D. V. Lightner. 1994. Comparison of the microbial quality of raw shrimp from China, Ecuador, or Mexico at both wholesale and retail levels. J. Food Prot. 57:150-153.
5. Blackwood, C. M. 1974. Utilization of mechanically separated fish flesh-Canadian experience, p. 325-329. *In* R. Kreuzer (ed.), Fishery products. Fishing News Ltd., Surrey, U. K.
6. Brackett, R.E. 1988. Presence and persistence of *Listeria monocytogenes* in food and water. Food. Technol. 42:162-164.
7. Campbell, L. L., Jr., and O. B. Williams. 1952. The bacteriology of Gulf Coast shrimp. IV. Bacteriological, chemical and organoleptic changes with ice storage. Food Technol. 6:125-126.
8. Centers for Disease Control. 1991. Cholera—New Jersey and Florida. Morb. Mortal Wkly. Rep. 40:287-289.
9. Centers for Disease Control. 1996. Surveillance for foodborne-disease outbreaks—United States, 1988-1992. Morb. Mortal Wkly. Rep. 45(SS-5):1-66.
10. Chen, C. H., and T. C. Chang. 1995. An enzyme-linked immunosorbent assay for the rapid detection of *Vibrio parahaemolyticus*. J. Food Prot. 58:873-878.
11. Cockney, R. R., and T.-J. Chai. 1991. Microbiology of crustacea processing: crabs, p. 41-63. *In* D. R. Ward and C. R. Hackney (ed.), Microbiology of marine food products. Van Nostrand Reinhold, New York.
12. Colwell, R. R. 1974. *Vibrio parahaemolyticus*-Taxonomy, ecology and pathogenicity. *In* Proceedings of the 2nd U.S.-Japan Conference of Toxic Microorganisms. International Symposium *of Vibrio parahaemolyticus*. Saikon Publ. Co., Tokyo, Japan.
13. DePaola, A., G. M. Capers, and D. Alexander. 1994. Densities of *Vibrio vulnificus* in the intestines of fish from the U.S. Gulf Coast. Appl. Environ. Microbiol. 60:984-988.
14. Duran, A. P., B. A. Wentz, J. M. Lanier, F. D. McClure, A. B. Schwab, A. Swartzentruber, R. J. Barnard, and R. B. Read, Jr. 1983. Microbiological quality of breaded shrimp during processing. J. Food Prot. 46:974-977.
15. Eklund, M. W. 1982. Significance of *Clostridium botulinum* in fishery products preserved short of sterilization. Food Technol. 36(12):107-112.
16. Eklund, M. W., M. E. Peterson, R. Paranjpye, and G. A. Pelroy. 1988. Feasibility of a heat-pasteurization process for the inactivation of nonproteolytic *Clostridium botulinum* types B and E in vacuum-packaged, hot-process (smoked) fish. J. Food Prot. 51:720-726.
17. Erichsen, I. 1967. The microflora of semi-preserved fish products. III. Principal groups of bacteria occurring in titbits. Antonie van Leeuwenhoek. J. Microbiol. Ser. 33:107-112.
18. Eyles, M. J., and A. D. Warth. 1981. Assessment of the risk of botulism from vacuum-packaged raw fish: A review. Food Technol. Aust. 33:574-580.
19. Faghri, M. A., C. L. Pennington, L. S. Cronholm, and R. M. Atlas. 1984. Bacteria associated with crabs from cold waters with emphasis on the occurrence of potential human pathogens. Appl. Environ. Microbiol. 47:1054-1061.
20. Food and Drug Administration. 1992. *Aeromonas hydrophila* and other spp. Foodborne pathogenic microorganisms and natural toxins handbook. USFDA, Center for Food Safety and Applied Nutrition. Washington, D. C. http://vm.cfsan.fda.gov/~mow/chap17. html.
21. Food and Drug Administration. 1992. *Clostridium botulinum*. Foodborne pathogenic microorganisms and natural toxins handbook. USFDA, Center for Food Safety and Applied Nutrition. Washington, D. C. http://vm.cfsan.fda.gov/~mow/chap2.html.
22. Food and Drug Administration. 1992. *Plesiomonas shigelloides*. Foodborne pathogenic microorganisms and natural toxins handbook. U.S. Food and Drug Administration, Center for Food Safety and Applied Nutrition. Washington, D. C. http://vm.cfsan.fda. gov/~mow/chap18.html.
23. Food and Drug Administration. 1992. *Vibrio cholerae* Serogroup Non-O1. Foodborne pathogenic microorganisms and natural toxins handbook. USFDA, Center for Food Safety and Applied Nutrition. Washington, D. C. http://vm.cfsan.fda.gov/~mow/chap8. html.
24. Food and Drug Administration. 1992. *Vibrio cholerae* Serogroup O1. Foodborne pathogenic microorganisms and natural toxins handbook. USFDA, Center for Food Safety and Applied Nutrition. Washington, D. C. http://vm.cfsan.fda.gov/~mow/chap7.html.
25. Food and Drug Administration. 1996. Fish & fisheries products hazards & controls guide, 1st ed. USHHS/FDA, Washington, D. C.
26. Finelli, L., D. Swerdlow, K. Mertz, H. Ragazzoni, and K. Spitalny. 1992. Outbreak of cholera associated with crab brought from an area with epidemic disease. J. Infec. Dis. 166:1433-1435.
27. Foster, J.F., J. L. Fowler, and J. Dacey. 1977. A microbial survey of various fresh and frozen seafood products. J. Food Prot. 40:300-303.
28. Goldmintz, D., and J. C. Hull. 1970. Bacteriological aspects of fish protein concentrate production. Dev. Ind. Microbiol. II:335.
29. Green, M. 1949. Bacteriology of shrimp. II. Quantitative studies on freshly caught and iced shrimp. Food Res. 14:372-383.
30. Gulasekharam, J., T. Velaudapillai, and G. R. Niles. 1956. The isolation of *Salmoitellci* organisms from fresh fish sold in a Colombo fish market. J. Hyg. 54:581.
31. Harrison, J. M., and J. S. Lee. 1968. Microbiological evaluation of Pacific shrimp processing. Appl. Microbiol. 18:188-192.
32. Hird, D. W., S. L. Diesch, R. G. McKinnell, E. Gorham, F. B. Martin, C. A. Meadows, and M. Gasiorowski. 1983. *Enterobacteriaceae* and *Aeromonas hydrophilia* in Minnesota frogs and tadpoles, (*Rana pipiens*). Appl. Environ. Microbiol. 46:1423-1425.
33. ICMSF. 1974. Microorganisms in foods. 2. Sampling for microbiological analysis: Principles and specific applications. International Committee on Microbiological Specifications for Foods. University of Toronto Press, Canada.
34. Ingham, S. C., and N. N. Potter. 1987 Microbial growth in surimi and mince made from Atlantic pollock. J. Food Prot. 50:312-315.
35. Jemmi, T. 1990. Actual knowledge of *Listeria* in meat and fish products. Mitt. Gebsch. Lebensm. Hyg. 31:144-157.
36. Kishishita, M., N. Matsuoka, K. Kumagai, S. Yamasaki, Y. Takeda, and M. Nishibuchi. 1992. Sequence variation in the thermostable direct hemolysin-related hemolysin (*trh*) gene of *Vibrio parahaemolyticus*. Appl. Environ. Microbiol. 58:2449-2457.
37. Klontz, K.C., S. Lieb, M. Schreiber, H. T. Janowski, L. M. Baldy, and R. A. Gunn. 1988. Syndromes of *Vibrio vulnificus* infections. Ann. Int. Med. 109:318-323.
38. Kvenberg, J. 1991. Nonindigenous bacterial pathogens, p. 267-284. *In* D. R. Ward and C. R. Hackney (ed.), Microbiology of marine food products, Van Nostrand Reinhold, New York.
39. Lee, D. A., and M. Solberg. 1983. Time to toxin detection and organoleptic determination in *Clostridium botulinum* incubated fresh fish fillets during modified atmosphere storage, abstr. 483. *In* Abstracts of the 43rd Annual Meeting of the Institute for Food Technologists.
40. Lee, J. S., and D. K. Pfeifer. 1975. Microbiological characteristics of Dungeness Crab *(Cancer magister).* Appl. Microbiol. 30:72-78.
41. Lennon, D., B. Lewis, C. Mantell, D. Becroft, K. Farmer, S. Tonkin, N. Yeates, R. Stamp, and K. Mickleson. 1984. Epidemic perinatal Listeriosis. Pediatr. Inf. Dis. 3:30.
42. Licciardello, J. J., and W. S. Hill. 1978. Microbiological quality of commercial frozen minced fish blocks. J. Food Prot. 41:948-952.
43. Martin, R. E., and G. T. Pitts. 1981. Handbook of state and federal microbiological standards and guidelines. National Fisheries Institute, Inc., Washington, D. C.

44. Miget, R. J. 1991. Microbiology of crustacean processing: Shrimp, crawfish, and prawns, p. 65-87. *In* D. R Ward and C. R. Hackney (ed.), Microbiology of marine food products. Van Nostrand Reinhold, New York.
45. Miller, A. III, R. A. Scanlan, J. S. Lee, and L. M. Libbey. 1973a. Identification of the volatile compounds produced in sterile fish muscle *(Sebastes melonops)* by *Pseudomonas fragii*. Appl. Microbiol. 25:952-955.
46. Miller, A. III, R. A. Scanlan, J. S. Lee, and L. M. Libbey. 1973b. Volatile compounds produced in sterile fish muscle *(Sebastes melonops)* by *Pseudomonas putrefaciens, Pseudomonas fluorescens,* and an *Achromobacter* species. Appl. Microbiol. 26:18-21.
47. Miller, M. L., and J. A. Koburger. 1986. Tolerance of *Plesiomonas shigelloides* to pH, sodium chloride and temperature. J. Food Prot. 49:877-879.
48. Motes, M. L., Jr. 1991. Incidence of *Listeria* spp. in shrimp, oysters, and estuarine waters. J. Food Prot. 54:170-173.
49. Nickelson, R., II, J. Hosch, and L. E. Wyatt. 1975. A direct microscopic count procedure for the rapid estimation of bacterial numbers on green-headless shrimp. J. Milk Food Technol. 38:76-77.
50. Nickelson, R., II, G. Finne, M. O. Hanna, and C. Vanderzant. 1980. Minced fish flesh from nontraditional Gulf of Mexico finfish species: Bacteriology. J. Food Sci. 45:1321-1326.
51. Pace, J., C. Y. Wu, and T. Chai. 1988. Bacterial flora in pasteurized oysters after refrigerated storage. J. Food Sci. 53:325-327, 348.
52. Paskell, S. L., and D. Goldmintz. 1973. Bacteriological aspects of fish protein concentrate production from a large-scale experiment and demonstration plant. Dev. Ind. Microbiol. 14:302.
53. Phillips, F. A., and J. T. Peeler. 1972. Bacteriological survey of the blue crab industry. Appl. Microbiol. 24:958-966.
54. Powell, J. C., A. R. Moore, and J. A. Gow. 1979. Comparison of EC broth and medium A-1 for the recovery of' *Escherichia coli* from frozen shucked snow crab. Appl. Environ. Microbiol. 37:836-840.
55. Ryser, E. T., E. H. Marth, and S. L. Taylor. 1984. Histamine production by psychrotrophic pseudomonads isolated from tuna fish. J. Food Prot. 47:378-380.
56. Shewan, J. M. 1971. The microbiology of fish and fishery products—A progress report. J. Appl. Bacteriol. 34:299-315.
57. Shewan, J. M. 1977. The bacteriology of fresh and spoiling fish and the biochemical changes induced by biochemical action, p. 51. *In* Proceedings of the Conference of Handling, Processing, and Marketing of Tropical Fish. Tropical Products Institute, London, U.K. *and* Miget, R. J. 1991. Microbiology of crustacean processing: Shrimp, crawfish, and prawns, p. 65-87. *In* D.R Ward and C.R. Hackney (ed.), Microbiology of marine food products. Van Nostrand Reinhold, New York.
58. Solomon, H. M., R. K. Lynt, T. Lilly Jr., and D. A. Kautter. 1977. Effect of low temperatures on growth of *Clostridium botulinum* spores in meat of the blue crab. J. Food Prot. 40:5-7.
59. Stelma, G. N., Jr., P. L. Spaulding, A. L. Reyes, and C. H. Johnson. 1988. Production of enterotoxin by *Vibrio vulnificus* isolates. J. Food Prot. 51:192-196.
60. Surkiewicz, B.F., J. B. Hyndman, and M. V. Yancey. 1967. Bacteriological survey of the frozen prepared foods industry. II. Frozen breaded raw shrimp. Appl. Microbiol. 15:1-9.
61. Surkiewicz, B.F., R. J. Groomes, and L. R. Shelton. 1968. Bacteriological survey of the frozen prepared foods industry. IV. Frozen breaded fish. Appl. Microbiol. 16:147-150.
62. Swartzentruber, A., A. H. Schwab, A. P. Duran, B. A. Wentz, and R. B. Read, Jr. 1980. Microbiological quality of frozen shrimp and lobster tail in the retail market. Appl. Environ. 40:765-769.
63. Tillman, R. E., R. Nickelson, and G. Finne. 1981. The bacteriological quality and safety of pasteurized langostino tails, p. 161-168. *In* Proceedings of the Sixth Tropical and Subtropical Fish. Tech. Conference of the Americas. Sea Grant College Program, Texas A & M University, College Station, Tex.
64. U.S. Government. 1988. Salt-cured, air-dried, uneviscerated fish; Compliance policy guide; Availability. Fed. Reg. 53(215):44,949-44,951.
65. U.S. Government. National Archives and Records Administration. 1980. Federal Register 45(108):37,524-37,526—Frozen fish sticks, frozen fish cakes, and frozen crab cakes: Recommended microbiological quality standards.
66. U.S. Government. National Archives and Records Administration. 1981. Federal Register 46(113):31,067-31,068—Frozen fish sticks, frozen fish cakes, and frozen crab cakes: Recommended microbiological quality standards.
67. U.S. Government. National Archives and Records Administration. 1998. CFR. Title 21. Part 123.6 —Hazard analysis and hazard analysis critical control point (HACCP) plan, pp. 258-268.
68. U.S. Government. National Archives and Records Administration. 1998. CFR. Title 21. Part 172.385—Whole fish protein concentrate, p. 47-48.
69. Vanderzant, C., E. Mroz, and R. Nickelson. 1970. Microbial flora of Gulf of Mexico and pond shrimp. J. Milk Food Technol. 33:346-350.
70. Vanderzant, C., R. Nickelson, and J. C. Parker. 1970. Isolation of *Vibro parahaemolyticus* from Gulf Coast shrimp. J. Milk Food Technol. 33:161-162.
71. Vanderzant, C., P. W. Judkins, R. Nickelson, and H. A. Fitzhugh. 1972. Numerical taxonomy of coryneform bacteria isolated from pond-reared shrimp *(Penaeus aztecus)* and pond water. Appl. Microbiol. 23:38-45.
72. Vanderzant, C., A. W. Matthys, and B. F. Cobb, III. 1973. Microbiological, chemical, and organoleptic characteristics of frozen breaded raw shrimp. J. Milk Food Technol. 36:253.
73. Weagent, S. D., P. N. Sado, K. G. Colburn, J. D. Torkelson, F. A. Stanley, M. H. Krane, S. C. Sheilds, and C. F. Thayer. 1988. The incidence of *Listeria* species in frozen seafood products. J. Food Prot. 51:655-657.
74. Wentz, B. A., A. P. Duran, A. Swartzentruber, A. H. Schwab, and R. B. Read. 1983. Microbiological quality of fresh blue crabmeat, clams and oysters. J. Food Prot. 46:978-981.
75. Williams, O. B., L. L. Campbell, Jr., and H. B. Rees, Jr. 1952. The bacteriology of Gulf Coast shrimp. II. Qualitative observations on the external flora. Tex. J. Sci. 4: 53-54.
76. Wong, H.-C., W.-L. Chao, and S.-J. Lee. 1990. Incidence and characterization of *Listeria monocytogenes* in foods available in Taiwan. Appl. Environ. Microbiol. 56:3101-3104.
77. Yamamoto, K., T. Honda, and T. Miwatani. 1992. Enzyme-labeled oligonucleotide probes for detection of the genes for thermostable direct hemolysin (TDH) and TDH-related hemolysin (TRH) of *Vibrio parahaemolyticus*. Can. J. Microbiol. 38:410-416.
78. Yoon, I. H., and J. R. Matches. 1988. Growth of pathogenic bacteria on imitation crab. J. Food Sci. 53:688-690. Erratum J. Food Sci. 53:1582.
79. Yoon, I. H., J. R. Matches, and B. Rasco. 1988. Microbiological and chemical changes of surimi-based imitation crab during storage. J. Food Sci. 53:1343-1346.

Chapter 49

Molluscan Shellfish: Oysters, Mussels, and Clams

David W. Cook, William Burkhardt, III, Angelo DePaola, Susan A. McCarthy, and Kevin R. Calci

49.1 INTRODUCTION

Bivalved molluscs such as oysters, mussels, and clams are economically important marine shellfish species found in abundance in estuarine and marine waters. These animals feed on plankton and other microflora by filtering large quantities of the surrounding water. If this environment is fecally polluted, there is a potential that pathogenic bacteria or viruses could become concentrated and thus pose a health concern to the consumer.

Historically, *Salmonella typhi* was the most significant microbial contaminant involved in shellfishborne epidemics in the U.S.[47] During the 1960s and 1970s however, hepatitis A became the major public health problem associated with the consumption of raw shellfish.[59] More recently, viral gastroenteritis caused by Norwalk and Norwalk-like viruses has become the most commonly reported illness associated with shellfish consumption.[12,44,60]

Worldwide, the largest shellfish-associated epidemic occurred in Shanghai, China, in 1988.[34] More than 300,000 cases of hepatitis A were reported. The epidemic was caused by the consumption of raw shellfish harvested from a harbor receiving untreated domestic sewage.

Since the mid-1970s, bacteria of the genus *Vibrio* have been recognized as a significant public health concern related to the consumption of raw molluscan shellfish. *Vibrio* illnesses have been most prevalent with shellfish harvested from the Gulf Coast and include *V. vulnificus, V. cholerae* O1, *V. cholerae* non-O1, *V. parahaemolyticus, V. hollisae, V. mimicus, V. fluvialis, V. alginolyticus* and *V. damsela*.[19,35,46] These vibrios are not generally associated with human pollution and some have been shown to multiply in shellfish before[65] and after harvest.[14]

Primary septicemia from *V. vulnificus* is a leading cause of death associated with consumption of seafood in the U.S. It nearly always occurs in the chronically ill, especially those with liver disorders, who consume raw Gulf Coast oysters harvested between April and November.[56,60] Fifteen to 30 cases are reported annually with about 50% resulting in death.[68] Additionally, healthy individuals can develop gastroenteritis from this organism.[42] Control efforts are hindered by the inability to distinguish pathogenic strains, an incompletely defined at-risk population and unknown range of infectious doses. There are no tolerance levels established for this bacterium. Typically, Gulf Coast oysters harvested from May to October contain approximately 1000 *V. vulnificus* per g,[54] but the numbers can increase more than 100-fold if oysters are not refrigerated.[15]

V. cholerae O1 and non-O1 have been linked to sporadic cases of diarrhea; non-O1 strains have occasionally caused primary septicemia. Approximately 70 cases caused by the Gulf Coast *V. cholerae* O1 strain have been reported in the U.S. since 1970, but few cases have been associated with consumption of shellfish.[7] The Latin American strain of *V. cholerae* was found in oysters from Mobile Bay, AL, in 1991 and 1992, but no cases were linked to consumption of any foods produced in the U.S.[53] In Florida, *V. cholerae* non-O1 accounted for 22% of the shellfish associated vibrio infections over a 13-year period.[35] It is prevalent in the environment and in molluscan shellfish.[18,35]

Consumption of raw shellfish containing pathogenic strains of *V. parahaemolyticus* can cause gastrointestinal illness and in a few cases has resulted in primary septicemia.[36] In Florida, the incidence of illness does not appear to be seasonal,[35] but in the Pacific Northwest, most illnesses occur between June and September.[63] Less is known about its epidemiology along the Atlantic seaboard.

V. hollisae, V. mimicus, V. fluvialis, V. alginolyticus and *V. damsela* have been implicated in illness associated with consumption of shellfish.[35] With the exception of *V. hollisae*, they can be isolated from shellfish using standard methods for *V. vulnificus, V. cholerae* and *V. parahaemolyticus*, but there are no official methods for the detection or enumeration of these vibrios.

Campylobacter species are gram-negative bacteria that occur in the intestines of mammalian and avian species and in waters contaminated with their feces. *Campylobacter jejuni* is currently recognized as one of the principal bacterial causes of human gastroenteritis. Raw shellfish have been implicated in outbreaks of campylobacteriosis.[1,32]

Listeria monocytogenes, a gram-positive coccoid rod, is widely distributed in nature and is resistant to salt, alkali, desiccation, and cold temperatures.[24,25,64] The organism has been recovered from silage, sewage, soil, vegetation, mammals, birds, and freshwater and estuarine environments.[13,48,52] It has been associated with food products, including milk, cheese, meats, fruits, vegetables, and ready-to-eat seafood such as cooked shrimp, crabmeat, crawfish, lobster, imitation fish products, and hot or cold smoked fish.[20,21,67] *L. monocytogenes* has also been reported as the causative agent in three incidences of seafood-borne listeriosis, the most recent occurred in New Zealand and was due to contamination of smoked mussels.[49] The organism was also implicated as the causative agent in an epidemic of perinatal infections in New Zealand that involved consumption of shellfish or raw fish.[45]

The sanitary control of shellfish in the United States is based primarily on the classification and control of harvest areas through comprehensive sanitary surveys of the shoreline, bacteriological monitoring of growing area waters, and prohibition of harvesting from areas not meeting "approved" growing area criteria. Routine control procedures are based on the guidelines of the National Shellfish Sanitation Program (NSSP).[27]

Historically, it appears that strict adherence to NSSP guidelines for the classification and control of shellfish growing areas has resulted in the production of safe shellfish when the hazard has been directly associated with sewer outfalls. Shellfish-associated outbreaks have occurred that have been linked to improper adherence to the NSSP guidelines involving wet storage, misclassification of shellfish growing areas coupled with shellfish purification, and more recently, the illegal overboard dumping of human fecal wastes into shellfish growing areas.[12] With the possible exception of gastroenteritis caused by naturally occurring marine vibrios, shellfish-borne outbreaks have been usually associated with a breakdown in growing area control procedures resulting in the harvesting of contaminated shellfish. However, the question of the effectiveness of NSSP guidelines in preventing shellfish-borne illness caused by *Vibrio sp.*, which grow in estuarine waters long distances from sewer outfalls, remains to be resolved.[17]

The safety of shellfish is predicated on the densities of indicator organisms, primarily the coliform group, present in the growing waters and on the direct relationship of these organisms to known sources of pollution. In 1964, the National Shellfish Sanitation Workshop adopted the fecal coliform criterion for the wholesale market standard for shucked oysters[37] and, in 1968, for all species of fresh and fresh frozen shellfish.[51] The validity of the NSSP shellfish meat and growing water bacteriological standards has been challenged throughout the history of the program.[38,39]

The only microbiological standard for the shellfish meats developed by the NSSP is for product at the time of receipt at the wholesale market. Fresh or fresh frozen shellfish are generally considered to be satisfactory at the wholesale market if the fecal coliform MPN does not exceed 230/100 g and the Standard Plate Count (SPC) is not more than 500,000/g of sample. It is recognized, however, that the level of accumulation of indicator organisms under a given set of environmental conditions may vary according to the type of shellfish and individual species.[8,10,38] Further, fecal coliforms may multiply in shellstock unless they are cooled immediately after harvest.[16]

The National Advisory Committee on Microbiological Criteria for Foods[55] recommended that the SPC procedure[5] be replaced by a 25°C spread plate count procedure and that the plate count agar be supplemented with 1% NaCl. Cook[15] found that the procedure measured a greater number of bacteria and more closely approximated the rate of multiplication of *V. vulnificus* in temperature abused shellstock than did the APHA[5] procedure.

Bacteria counts of shellfish meats may be reduced in connection with a process known as microbial depuration. According to the NSSP, depuration (or controlled purification) is intended to reduce the number of pathogenic microorganisms that may be present in shellfish harvested from moderately polluted (restricted) waters to such levels that the shellfish will be acceptable for human consumption without further processing.[27] Depuration is not intended for shellfish from heavily polluted (prohibited) waters nor is it intended to reduce the levels of poisonous or deleterious substances such as marine biotoxins that the shellfish may have accumulated from their environment. The acceptability of depuration process is contingent on the state shellfish control authority exercising very stringent supervision over all phases of the process.

Controlled purification studies of hardshell clams and oysters show that a processing time of 48 hr will reduce initial fecal coliform densities by 99%, but under certain conditions longer depuration periods may be required for adequate shellfish purification.[9] Shellfish intended for the depuration process typically harbor less than 2000 fecal coliform/100 g and it is expected that after 48 hours of depuration, fecal coliform densities would be reduced to less than 20/100 g. Efficacy of the depuration process, as outlined by the NSSP, is based upon an endpoint criterion of fecal coliform densities which are shellfish species specific.

When shellfish are examined for regulatory purposes, the procedures accepted by the NSSP[35] should be used. In addition to microbiological methods, procedures for the collection, handling and preparation of samples are recommended. Information in sections 49.2 and 49.3 of this chapter is taken in part from the APHA.[5] Additional information has been added to facilitate processing and to address changes in the procedure necessary to insure detection or enumeration of vibrios.

49.2 EXAMINATION OF SHELLFISH

49.21 Collection and Transportation of Samples

Samples of shellstock (shellfish in the shell) and shucked unfrozen shellfish should be examined within 6 hr after collection; in no case should they be examined if they have been held more than 24 hr after collection. The report of the examination should include a record of the time elapsed between collection and examination.

Individual containers of shellfish samples should be marked for identification; the same mark should be put in its proper place on the descriptive form that accompanies the sample.

A history and description of the shellfish should accompany the sample to the laboratory, including 1) date, time, and place of collection, 2) the area from which the shellfish were harvested, 3) the date and time of harvesting, 4) the conditions of storage between harvesting and collection, and 5) the initials of sample collector. When shellfish are collected from beds, the temperature and salinity of the overlying water should be recorded.

Not all of this information may be obtainable for shellfish samples collected in market areas. In such cases, the identification of the shipper, the date of shipment, and the harvesting area should be determined, as well as the date, time, and place of collection.

49.22 Shellstock

Samples of shellstock should be collected in clean containers. The containers should be waterproof and durable enough to withstand the cutting action of the shellstock and abrasion during transportation. Waterproof paper bags, paraffined cardboard cups, and plastic bags are suitable types of containers. A tin can with a tight lid is also suitable.

Shellstock samples should be kept in dry storage at a temperature above freezing but lower than 10°C until examined. Shellstock must not be allowed to come in contact with ice because the shellfish may die, allowing their shells to gape and microorganisms from the melt water to contaminate the meat. Sample integrity is of utmost importance.

In general, a minimum of 12 shellfish must be taken in order to obtain a representative sample and to allow for the selection of sound animals suitable for shucking. With most species, allowing for the necessary culls, approximately 200 g of shell liquor and meats will be obtained. (Figure 1)

Because of their larger size, certain species such as the Pacific oyster, *Crassostrea gigas*, the surf clam, *Spisula solidissima*, and the hardshell clam, *Mercenaria mercenaria*, may require using blender jars of a larger size than usual for the designated 10 to 12 specimens. Certain blenders will accept ordinary mason jars and thus, a two-quart container may be used for blending these species.

When two-quart containers are not available, the 10 to 12 shellfish should be ground for 30 sec. Then 200 g of this meat homogenate should be blended with 200 g of sterile diluent for an additional 60 sec.

On the other hand, 10 or 12 specimens of other species, such as the Olympia oyster, *Ostrea lurida*, and small sizes of the Pacific little neck clams, *Protothaca staminea* and *Tapes japonica*, may produce much less than 100 g of shell liquor and meats. Blender containers of smaller size are indicated, but even when pint or half-pint jars are used, as many as 20 to 30 specimens of these species may be required to produce an adequate volume for proper blending.

49.23 Shucked Shellfish

A sterile wide-mouth jar of a suitable capacity with a watertight closure is an acceptable container for samples of shucked shellfish taken in shucking houses, repacking establishments, or bulk shipments in the market. The shellfish may be transferred to the jar with sterile forceps or spoon. Samples of the final product of shucking houses or repacking establishments may be taken in the final packing cans or containers. The comments pertaining to species of various sizes in the section on shellstock applies to shucked shellfish. Consumer-size packages are acceptable for examination provided they contain an adequate number of specimens.

Samples of shucked shellfish should be refrigerated immediately after collection by packing in crushed ice; they should be so kept until examined.

Figure 1. Several commercially important shellfish from the United States (top left clockwise): Hardshell clam (Mercenaria mercenaria), Blue mussel (Mytilus edulis), Eastern oyster (Crassostrea virginica), Softshell clam (Mya arenaria).

49.24 Frozen Shucked Shellfish

If the package contains an adequate number of specimens (10 to 12), packages may be taken as a sample. Samples from larger blocks may be taken by coring or drilling with a suitable instrument, or by quartering, using sterile technique. Samples should be transferred to sterile wide-mouth jars for transportation to the laboratory.

It is desirable to keep samples of frozen shucked shellfish in the frozen state at temperatures close to those at which the commercial stock was maintained. When such storage is not possible, containers of frozen shucked shellfish should be packed in crushed ice and kept so until examined.

49.3 PREPARATION OF SAMPLE FOR EXAMINING SHELLFISH IN THE SHELL

49.31 Cleaning the Shells

The hands of the examiner must be scrubbed thoroughly with soap and potable water. New, clean, latex or PVC gloves can be worn to protect the hands.

Scrape off all growth and loose material from the shell, and scrub the shellstock with a sterile stiff brush under running water (Figure 2) of drinking quality, paying particular attention to the crevices at the junctions of the shells. Place the cleaned shellstock in clean containers, on clean towels or absorbent paper, and allow the shellfish to drain.

49.32 Removal of Shell Contents

Before starting the removal of shell contents, the hands of the examiner must be thoroughly scrubbed with soap and water and rinsed with 70% ethanol or donned in new, clean, latex or PVC gloves. A protective mail glove may be worn under the latex or PVC glove to prevent accidental injury. Open the shellfish as directed below, collecting the appropriate quantities of shell liquor and meats in a sterile blender or other suitable sterile container (Figures 3 a-d).

49.321 Oysters

Hold the oyster in the hand or on a fresh clean paper towel on the bench with the deep shell on the bottom. Using a sterile oyster knife, insert the point between the shells on the ventral side, (at the right when the hinge is pointed away from the examiner), about one-fourth the distance from the hinge to the bill. Entry also

Figure 2. Removal of surface debris with sterile brush.

may be made at the bill after making a small opening with a sterile instrument similar to bonecutting forceps.

Cut the adductor muscle from the upper flat shell and pry the shell wide enough to drain the shell liquor into a sterile tared beaker, widemouthed jar, or blender jar. The upper shell then may be pried loose at the hinge and discarded, and the meats transferred to the beaker or jar after severing the muscle attachment to the lower shell (Figure 4).

49.322 Hardshell Clams

Entry into the hardshell clam, *Mercenaria mercenaria*, or the Pacific little neck clam, *Protothaca sp.*, is best done with a sterile, thin-bladed knife such as a paring knife. To open the clam, hold it in the hand, place the edge of the knife at the junction of the bills, force it between the shells with a squeezing motion, and drain the shell liquor into the sample container. Cut the adductor muscles from the shells and transfer the body of the animal to the sample container. An alternative method of entry is to nibble a small hole in the bill with sterile bonecutting forceps and with the knife sever the two adductor muscles.

49.323 Other Clams

The softshell clam, *Mya arenaria*, the Pacific butter clam, *Saxidomus giganteus*, the surf clam, *Spisula solidissima*, and similar species may be shucked with a sterile paring knife, entering at the siphon end and cutting the adductor muscles first from the top valve and then from the bottom valve.

Mussels, *Modiolus* and *Mytilus* species, may be entered at the byssal opening. The byssal threads should be removed during the cleansing of the shell. The knife may be inserted and the shells spread apart with a twisting motion, allowing the draining of the shell liquor. Cut away the many attachments from the shell.

49.324 Shucked Shellfish

Transfer a suitable quantity from a sample jar to a sterile tared blender jar or other container using a sterile spoon.

49.33 Dilution and Grinding

Weigh the sample to the nearest gram. Transfer the weighed sample to a sterile blender jar and add an equal amount, by weight, of sterile phosphate buffered dilution water or 0.5% sterile peptone water. When samples are to be examined for vibrios, phosphate buffered saline[26] should be used as the diluent. Grind for 60 to 120 sec in a laboratory blender operating at approximately 14,000 RPM (Figure 5). Two mL of this mixture contain 1 g of shellfish meat. The optimum grinding time will vary with make and condition of the machine, species of shellfish, and probably, the physical state of the meats. In general, a grinding time of 60 to 90 sec will be optimum for all species. Excessive grinding in small containers should be avoided to prevent overheating. (See Chapter 2 on Sampling and Sample Preparation.)

A dilution of equal amounts by weight of certain species of shellfish such as the hardshell clam, surf clam, and butter clam, often results in a mixture that after grinding is of too heavy a consistency for pipetting and transferring to culture tubes. In these cases, using a greater proportion of dilution water is permissible; the addition of 3 parts by weight of dilution water to 1 part of the weighed sample is suggested. With such dilutions,

A

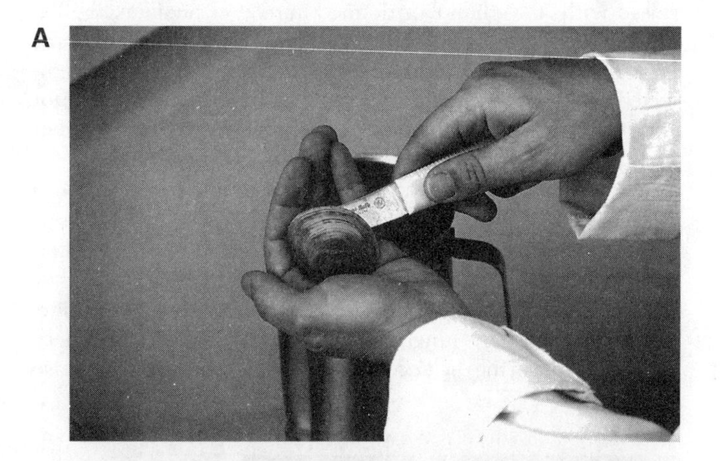

B

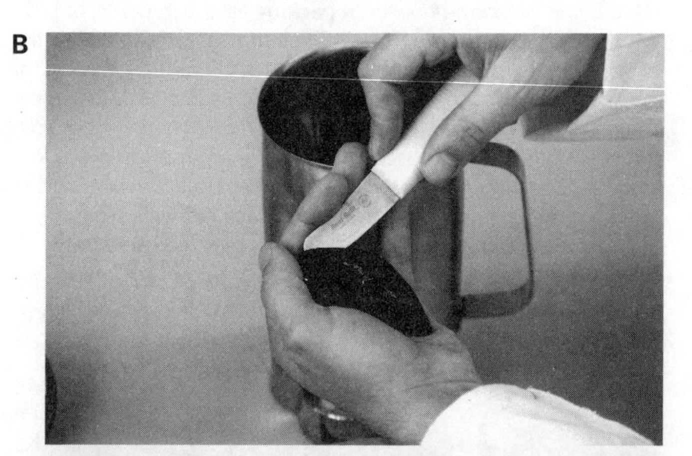

C

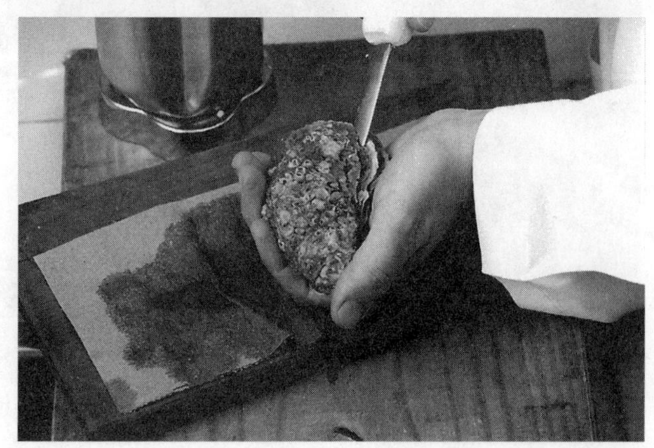

D

Figure 3. Proper entry location of sterile knife into scrubbed shellfish. (a) Hardshell clam (b) Blue mussel (c) Eastern oyster (d) Softshell clam.

4 mL of the ground sample will be equal to a 1 g portion of shellfish. If the 1:4 dilution method is used, adjustment in the concentration of presumptive broth in the tubes receiving the 1 g portions should be made accordingly.

49.4 MICROBIOLOGICAL PROCEDURES

49.41 Tests for Members of Coliform Group

The procedures accepted by the NSSP for the microbiological examination of shellfish are described in "Recommended Procedures for the Examination of Seawater and Shellfish"[5] and "Accumulation of *Escherichia coli* in the Northern Quahaug."[11] They include methods for coliforms, fecal coliforms, SPC, and Elevated Temperature Coliform Plate Count (acceptable for controlled purification of soft- and hard-shell clams only). For methods in conformance with these recommendations, refer to Chapter 7 (Aerobic Plate Count) and Chapter 8 (Coliform, *E. coli* and other Enterobacteriaceae).

49.42 Salmonella

Hackney and Potter[33] provide a comprehensive review of salmonellae in shellfish including methodology for salmonellae. Methods for detection of salmonellae in food are presented in Chapter 37 of this book and the BAM.[26]

49.43 Vibrios

The Seventh National Shellfish Sanitation Workshop[58] recommended that the BAM procedure be accepted as the official reference of the NSSP for the examination of shellfish for *V. parahaemolyticus*. It was recommended further that state laboratories conduct the test for this organism when routine tests of marine foods suspected in foodborne outbreaks fail to demonstrate other enteric pathogens or bacterial toxins. There are no official NSSP procedures for the analysis of shellfish for *V. cholerae* or *V. vulnificus*. Refer to the current edition of the BAM[26] and to Chapter 40 of this book for the detection of vibrios.

49.44 Listeria monocytogenes

There are no official NSSP procedures for analyzing shellfish for *L. monocytogenes*. Refer to the methodology in Chapter 36 of this book, the BAM[26] and AOAC.[4]

49.5 INTERPRETATION OF DATA

The NSSP microbiological quality standard for fresh and fresh frozen oysters, mussels, and clams was developed as a guideline of acceptable quality for shellfish harvested, processed, and shipped according to recommended practices, and sampled upon receipt at the wholesale market, retail markets, or restaurants, if shipped directly to the receiver. The standard, a maximum fecal coliform MPN of 230/100 g of sample and a SPC not exceeding 500,000/g for shellfish, can be used by control receiver agency programs to determine product quality as delivered by the shipper, or, in case of excessive counts, forwarded to the producer state control agency for investigative and corrective purposes.

Bacteriological water-quality standards are one of the critical control points used by the NSSP. The coliform standard for approved growing areas, a median MPN value of 70, with not more than 10% of samples exceeding an MPN of 230 for a 5 tube, 3 dilution method (or an MPN of 330 for a 3-tube, 3-dilution method) has been unchanged essentially since 1946. To develop a growing area standard more indicative of fecal pollution, the FDA proposed using fecal coliforms at the 1974 National Workshop.[69] The proposal stated, "The median fecal coliform MPN value for a sampling station shall not exceed 14 per 100 mL of sample and not more than 10 percent of the samples shall exceed 43 for a 5 tube, 3 dilution test, or 49 for a 3 tube, 3 dilution test." The fecal coliform criterion and standard are now used by most states participant in the NSSP to verify classification of shellfish waters.

Shelf life and bacteriological quality, as measured by the SPC, will vary according to species, salinity, and bacteriological quality of growing area waters, climate, processing controls, refrigeration, and other conditions. For example, the northern quahaug (*Mercenaria mercenaria*), harvested from New England waters, normally will have a lower bacteriological count and longer shelf life than a softshell clam (*Mya arenaria*) harvested from the warmer, less saline waters of the Mid-Atlantic States. At the time of harvest, shellfish in approved areas generally have fecal coliform levels of <50/100 g and rarely exceed 230/100 g. However, fecal coliform counts in Gulf Coast oysters frequently exceed 230/100 g when received at processing plants. These high counts are attributed to post harvest multiplication in unrefrigerated shellstock.[16,62] Fecal coliform counts in excess of the NSSP quality standard may raise concern that the shellfish had come from sewage contaminated areas. Therefore, the proper interpretation of

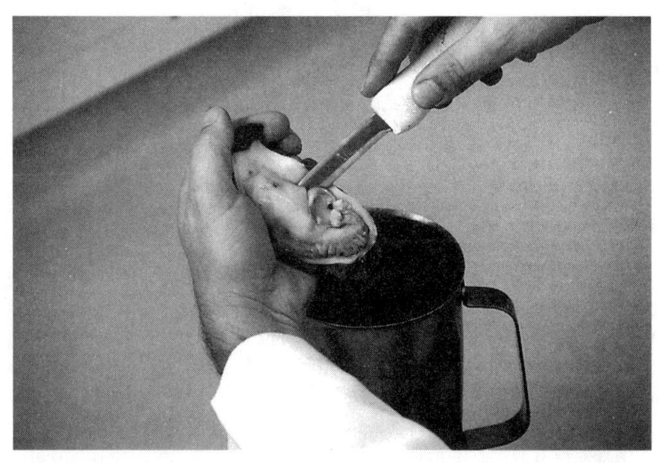

Figure 4. Shucking oyster meat and liquor into a sterile, tared blender jar.

Figure 5. Weighing shellfish meat and liquor followed by the addition of an equal amount of diluent.

wholesale market data requires knowledge of the sanitary control conditions of the product from growing area to market.

The FDA has a zero tolerance for *L. monocytogenes*, *Salmonella* species and *V. cholerae* (toxin producing O1 and non-O1 organisms) in ready-to-eat fishery products including raw shellfish. The action level for *V. parahaemolyticus* is 10,000/g (Kanagawa positive or negative).[28]

Experience with a shellfish product is valuable when interpreting results, and since this is not always easily attainable, the analyst is advised to consult with the appropriate agency of the state or country in which the product originated.

49.6 RAPID METHODS

There is considerable interest in the development of rapid methods to determine the sanitary quality of shellfish from controlled purification (depuration) systems and for interstate shellfish monitoring programs. Cabelli and Heffernan[11] developed a 24 hr plate method for the enumeration of elevated temperature coliforms. This method was shown comparable with the standard MPN procedure and accepted as an official procedure for monitoring the depuration process for softshell (*Mya arenaria)* and the hardshell clam (*Mercenaria mercenaria*). It is important to note that this procedure has not proven equivalent for analyses of oysters (*Crassostrea virginica).*

Using a variety of frozen foods including seafoods, Fishbein et al.[29] effectively recovered *Escherichia coli* using LST broth incubated at 44°C in a waterbath for 24 hr. Andrews et al.[3] and Andrews and Presnell [2] evaluated a multiple tube test for the recovery of *E. coli* from shellfish and shellfish waters using A-1 medium incubated at 44.5°C for 24 hr. The APHA MPN method[5] for fecal coliforms was compared by Hunt and Springer[40] to the A-1 and a modification of the A-1 method (A-1-M) for recovery of fecal coliforms and *E. coli* from sea water. In a broader study involving 16 state, federal, and Canadian provincial laboratories, Hunt et al.[41] compared the modified A-1 method with the APHA MPN method for recovery of fecal coliforms and *E. coli* in shellfish meats. These studies verified that the 24hr modified A-1 method produced fecal coliform counts for shellfish growing waters comparable to those produced by the APHA method which requires 48 to 72 hr.

Qadri et al.[57] developed two rapid methods for determining fecal contamination in oysters reported to be as sensitive and accurate as the standard APHA method. One involves incubating MacConkey broth for 2 hr at 37°C, then for 22 to 24 hr at 44°C. The second method uses the same incubation system but includes the inoculation of peptone broth as well as MacConkey broth.

Rippey et al.[61] developed a fluorometric method for enumerating *E. coli* in molluscan shellfish. This method, a modification of the traditional APHA MPN method[5] for fecal coliform bacteria, incorporates a fluorogenic molecule, 4-methylumbelliferyl-β-D-glucuronide (MUG), into the confirmatory broth medium (EC broth) at a concentration of 50 μg/mL. To avoid natural fluorescence associated with some glass formulations, media should be prepared in borosilicate glass test tubes. After incubation at 44.5°C for 24 ± 2 hr in a water bath, inoculated tubes are examined for fluorescence under long-wave length UV light. The presence of *E. coli* is indicated by both the production of gas and fluorescence. The MUG method, which requires up to a maximum of 72 hr for enumerating *E. coli* in shellfish, represents considerable savings in time and labor over the standard *E. coli* MPN procedure[5] which takes up to 10 days to complete. Attempts to incorporate MUG into the presumptive media (Lauryl Tryptose Broth) and into spread plate preparations have been without success due to endogenous glucuronidase activity present in shellfish tissue resulting in false positive results.[43] Incorporation of MUG into spread plate media has also been attempted; however, diffusion of the β-glucuronidase activity from producing colonies interfered with accurate enumeration of *E. coli*.[30]

Watkins et al.[66] and Frampton et al.[30] independently characterized the specificity of a chromogenic chemical compound, 5-bromo-4-chloro-3-indoxyl-β-D-glucuronic acid cyclohexylammonium salt (BCIG) as an alternative to MUG to detect *E. coli* from secondary wastewater and shellfish meats using a spread plate technique. In the former study, BCIG was successfully incorporated into a Lauryl Tryptose Agar (3%) at a final concentration of 100 μg/ mL. Results indicated a false negative rate of 1% and a false positive rate of 5%. The benefit of such a compound was the development of a rapid (24 hr), direct method for the enumeration and identification of *E. coli* from shellfish; a drawback was the cost of BCIG.

Other approaches for direct, rapid enumeration of indicator microorganisms from shellfish tissue involve the use of membrane filtration technology in conjunction with selective growth media and incubation conditions. Progress has been made using this approach for the enumeration of *E. coli, Enterococci*, and *Clostridium perfringens*[22,31] from oyster and clam meats. Development of these methods has focused on the enzymatic digestion of mechanically homogenized shellfish tissue by proteases such as trypsin and pronase. Following digestion, these protocols incorporate course prefiltration with the filtrate being re-filtered through a final recovery filter. Filters are incubated on selective media for a prescribed time and incubation conditions. This approach has produced variable results possibly due to the stress on microorganisms caused by the enzymatic digestion or loss of microorganisms in the prefiltration.

Further research is warranted on rapid, direct, and cost effective methods to enumerate microbial indicators of sanitary quality from shellfish meat. Consideration must be given to both the indicator organisms and the shellfish species.

49.7 INTERNATIONAL MICROBIOLOGICAL STANDARDS

The FDA has established international memoranda of understanding (MOU's) with official agencies in countries that wish to export shellfish into the U.S. MOU countries include Canada, Chile, Mexico, New Zealand, and Republic of Korea; all use the official NSSP microbial criteria, standards and methods to assess the sanitary quality of shellfish.

Countries in the European Union (EU) use bacterial quality of shellfish meats rather than water quality of the growing area to measure sanitary quality of the shellfish. Generally, all countries use an MPN procedure however, there is not a standard analytical method; some countries enumerate fecal coliforms and others *E. coli*. The United Kingdom[50] uses Modified Glutamate Broth as a presumptive medium followed by confirmation in Brilliant Green Lactose Bile (BGLB) broth and 1% tryptone water incubated at 44°C. Tubes of BGLB showing gas production are scored as fecal coliforms and those showing gas in BGLB and indole production in tryptone water are scored as *E. coli*.

The methods used by France[23] and Spain[6] are similar in that both use BGLB incubated at 35°C as the presumptive medium and BGLB incubated at 44°C along with 1% tryptone water for confirmation. However, the French use a 3-tube MPN and report results as *E. coli* and the Spanish use a 5-tube MPN and report results as fecal coliforms.

Other countries use various microbiological criteria, standards, and methods to assess the sanitary quality of molluscan shellfish. As microbiological data from different countries is interpreted, care should be taken to insure that respective methods produce comparable data.[6]

49.8 REFERENCES

1. Abeyta, C., F. G. Deeter, C. A. Kaysner, R. F. Stott, and M. M. Wekell. 1993. *Campylobacter jejuni* in a Washington state shellfish growing bed associated with illness. J. Food Prot. 56:323-325.
2. Andrews, W. H., and M. W. Presnell. 1972. Rapid recovery of *Escherichia coli* from estuarine water. Appl. Microbiol. 23:521-523.
3. Andrews, W. H., C. D. Diggs, and C. R. Wilson. 1975. Evaluation of a medium for the rapid recovery of *Escherichia coli* from shellfish. Appl. Microbiol. 29:130-131.
4. AOAC. 1995. *Listeria monocytogenes* in milk and dairy products: Selective enrichment and isolation method, p. 94a. *In* P. A. Cunniff (ed.), Official methods of analysis of AOAC International, 16th ed. AOAC International, Gaithersburg, Md.
5. APHA. 1970. Recommended procedures for the examination of sea water and shellfish, 4th ed. Am. Public Health Assoc., Washington, D. C.
6. Araujo, M., R. A. Sueiro, A. Amezaga, and M. J. Garrido. 1995. Underestimation of fecal coliform counts in shellfish-growing waters by the Spanish official method. J. Food Prot. 58:791-795.
7. Blake, P. A. 1994. Endemic cholera in Australia and the United States, p. 309-319. *In* I. K. Wachsmuth, P. A. Blake, and O. Olsvik, (ed.), Vibrio cholerae and cholera: Molecular to global perspectives. ASM Press, Washington, D. C.
8. Burkhardt, W., and K. R. Calci. 1998. Temperature mediated bio-accumulation of indicator organisms by Gulf coast oysters. *In* Proceedings of the 98th Annual Meeting of the American Society for Microbiology. ASM Press, Washington, D. C.
9. Burkhardt, W., S. R. Rippey, and W. D. Watkins. 1992. Depuration rates of northern quahogs, Mercenaria mercenaria (Linnaeus, 1758) and eastern oysters, Crassostrea virginica (Gmelin, 1791) in ozone- and ultraviolet light-disinfected seawater systems. J. Shellfish Res. 11:105-109.
10. Burkhardt, W., W. D. Watkins, and S. R. Rippey. 1992. Seasonal effects on accumulation of microbial indicator organisms by *Mercenaria mercenaria*. Appl. Environ. Microbiol. 58:826-831.
11. Cabelli, V. J., and W. P. Heffernan. 1970. Accumulation of *Escherichia coli* by the northern quahaug. Appl. Microbiol. 19:239-244.
12. CDC. 1997. Viral gastroenteritis associated with eating oysters—Louisiana, December 1996-January 1997. Morb. Mortal Wkly. Rep. 46(47):1109-1112.
13. Colburn, K. G., C. A. Kaysner, C. Abeyta, Jr., and M. M. Wekell. 1990. *Listeria* species in a California coast estuarine environment. Appl. Environ. Microbiol. 56:2007-2011.
14. Cook, D. W. 1994. Effect of time and temperature on multiplication of *Vibrio vulnificus* in postharvest Gulf coast shellstock oysters. Appl. Environ. Microbiol. 60:3483-3484.
15. Cook, D. W. 1997. Refrigeration of oyster shellstock: conditions which minimize the outgrowth of *Vibrio vulnificus*. J. Food Prot. 60:349-352.
16. Cook, D. W., and A. D. Ruple. 1989. Indicator bacteria and *Vibrionaceae* multiplication in post-harvest shellstock oysters. J. Food Prot. 52:343-349.
17. DePaola, A. 1981. *Vibrio cholerae* in marine foods and environmental waters: A literature review. J. Food Sci. 46:66-70.
18. DePaola, A., M. W. Presnell, M. L. Motes, R. M. McPhearson, R. M. Twedt, R. E. Becker, and S. Zywno. 1983. Non-O1 *Vibrio cholerae* in shellfish, sediment and waters of the U.S. Gulf Coast. J. Food Prot. 46:802-806.
19. Desenclos, J. A., K. C. Klontz, L. E. Wolfe, and S. Hoecheri. 1991. The risk of *Vibrio* illness in the Florida raw oyster eating population, 1981-1988. Am. J. Epidemiol. 134:290-297.
20. Dillon, R. M., and T. R. Patel. 1992. *Listeria* in seafoods: a review. J. Food Prot. 55:1009-1015.
21. Dorsa, W. J, D. L. Marshall, and M. Semien. 1993. Effect of potassium sorbate and citric acid sprays on growth of *Listeria monocytogenes* on cooked crawfish (*Procambaris clarkii*) tail meat at 4°C. Lebensm. Wiss. Technol. 26:480-482.
22. Dufour, A., and L. Hopkins. 1986. A membrane filter procedure for enumerating bacteria in oysters. *In* Abstracts of the 86th Annual Meeting for the American Society of Microbiology. ASM Press, Washington, D. C.
23. Dupont, J., D. Menard, C. Herve, F. Chevalier, B. Beliaeff, and B. Minier. 1996. Rapid estimation of *Escherichia coli* in live marine bivalve shellfish using automated conductance measurement. J. Appl. Bact. 80:81-90.
24. Farber, J. M., F. Coates, and E. Daley. 1992. Minimum water activity requirements for the growth of *Listeria monocytogenes*. Lett. Appl. Microbiol. 15:103-105.
25. FDA. 1992. Preventing foodborne listeriosis. US Dep. of Health and Human Serv., Food and Drug Admin., USDA/FSIS. Available from URL: http://vm.cfsan.fda.gov/~fsis/fsislist.html.
26. FDA. 1998. Bacteriological analytical manual, 8th ed., Revision A, AOAC International, Gaithersburg, Md.
27. FDA. 1997. National shellfish sanitation program. Guide for the control of molluscan shellfish (rev.). U.S. Dept. of Health and Human Serv., Public Health Serv., FDA, Washington, D. C.
28. FDA. 1998. Fish & fisheries products hazard & controls guide, 2nd ed. U.S. Food and Drug Admin., Washington, D. C.
29. Fishbein, M., B. F. Surkiewicz, E. F. Brown, H. M. Oxley, A. P. Padron, and R. J. Groomes. 1967. Coliform behavior in frozen foods. I. Rapid test for the recovery of *Escherichia coli* from frozen foods. Appl. Microbiol. 15:233-238.
30. Frampton, E. W., L. Restaino, and N. Blaszko. 1988. Evaluation of the BBB-glucuronidase substrate 5-bromo-4-chloro-3-indolyl-BBB-D-glucuronide (X-gluc) in a 24-hour direct plating method for *Escherichia coli*. J. Food Prot. 51:402-404.
31. Grabow, W. O. K., J. C. deVilliers, and N. Prinsloo. 1991. An assessment of methods of the microbiological analysis of shellfish. Wat. Sci. Tech. 24:413-416.
32. Griffin, M. R., E. Dalley, M. Fitzpatrick, and S. H. Austin. 1983. *Campylobacter* gastroenteritis associated with raw clams. J. Med. Soc. N. J. 80:607-609.
33. Hackney, C. R., and M. E. Potter. 1994. Animal-associated and terrestrial bacterial pathogens, p. 172-209. *In* C. R. Hackney and M. D. Pierson (eds.), Environmental indicators and shellfish safety. Chapman and Hall, New York.
34. Halliday, M. L., L.-Y. Kang, T.-K. Zhou, M.-D. Hu, Q.-C. Pan, T.-Y. Fu, Y.-S. Huang, and S.-L. Hu. 1991. An epidemic of hepatitis A attributable to the ingestion of raw clams in Shanghai, China. J. Infect. Dis. 164:852-859.
35. Hlady, W. G. 1997. *Vibrio* infections associated with raw oyster consumption in Florida.1981-1994. J. Food Prot. 60:353-357.
36. Hlady, W. G., and K. C. Klontz. 1996. The epidemiology of *Vibrio* infections in Florida, 1981-1993. J. Infect. Dis. 173:1176-1183.
37. Houser, L. S. (ed.). 1964. Proceedings of Fifth National Shellfish Sanitation Workshop of the U.S. Dept of Health, Education, and Welfare. US DHEW, Public Health Serv., Washington, D. C.
38. Hunt, D. A. 1977. Indicators of quality for shellfish waters, p.337. *In* A. W. Hoadley and B. J. Dutka (eds.), Bacterial indicators/health hazards associated with water. Amer. Soc. for Testing and Materials, Philadelphia, PA.
39. Hunt, D. A. 1979. Microbiological standards for shellfish growing waters—Past, present and future utilization. Proc. Natl. Shellfish Assoc. 69:142-146.

40. Hunt, D. A., and J. Springer. 1978. Comparison of two rapid test procedures with the standard EC test for recovery of fecal coliform bacteria from shellfishgrowing waters. J. AOAC 61:1317-1323.
41. Hunt, D. A., J. P. Lucas, F. D. McClure, J. Springer, and R. Newell. 1981. Comparison of modified A1 method with standard EC test for recovery of fecal coliform bacteria form shellfish. J. AOAC 64:607-610.
42. Klontz, K. C., S. Lieb, M. Schreiber, H. T. Janowski, L. M. Baldy, and R. A. Gunn. 1988. Syndromes of *Vibrio vulnificus* infections: clinical and epidemiological features in Florida cases, 1981-1987. Ann. Intern. Med. 109:318-323.
43. Koburger, J. A., and M. L. Miller. 1985. Evaluation of a fluorogenic MPN procedure for determining *Escherichia coli* in oysters. J. Food Prot. 48:244-245.
44. Kohn, M. A., T. Farley, T. A. Ando, M. Curtis, S.A. Wilson, Q. Jin, S. S. Monroe, R. C. Baron, L. M. McFarland, and R. I. Glass. 1995. An outbreak of Norwalk virus gastroenteritis associated with eating raw oysters. JAMA 273:466-471.
45. Lennon, D., B. Lewis, C. Mantell, D. Becroft, B. Dove, K. Farmer, S. Toukin, N. Yeates, R. Stamp, and K. Mickleson. 1984. Epidemic perinatal listeriosis. Pediatr. Infect. Dis. 3:30-34.
46. Levine, W. C., P. M. Griffin, and Gulf Coast *Vibrio* Working Group. 1993. *Vibrio* infections on the Gulf Coast: Results of first year regional surveillance. J. Infect. Dis. 167:479-483.
47. Lumsden, L. L., H. E. Hasseltine, J. P. Leake, and M. V. Veldee. 1925. A typhoid fever epidemic caused by oysterborne infection (192425). Public Health Reports Suppl. 50. U.S. Public Health Serv., Washington, D. C.
48. McCarthy, S. A. 1990. *Listeria* in the environment, p. 25-29. *In* A. J. Miller, J. L. Smith, and G. A. Somkuti, (eds.), Foodborne listeriosis. Elsevier, New York.
49. McCarthy, S. A. 1997. Incidence and survival of *Listeria monocytogenes* in ready-to-eat seafood products. J. Food Prot. 60:372-376.
50. Ministry of Agriculture, Fisheries and Food; Department of Health, Public Health Laboratory Service Working Group. 1992. Bacteriological examination of shellfish. PHLS Microbiol. Dig. 9:76-82.
51. Morrison, G. (ed.). 1968. Proceedings of Sixth National Shellfish Sanitation Workshop of the U.S. Dep. of Health, Education, and Welfare. U.S. DHEW, Public Health Serv., Washington, D. C.
52. Motes, M. L. 1991. Incidence of *Listeria* spp. in shrimp, oysters, and estuarine waters. J. Food Prot. 54:170-173.
53. Motes, M., A. DePaola, S. Zywno-Van Ginkel, and M. McPhearson. 1994. Occurrence of toxigenic *Vibrio cholerae* O1 in oysters in Mobile Bay, Alabama: An ecological investigation. J. Food Prot. 57:975-980.
54. Motes, M. L., A. DePaola, D. W. Cook, J. E. Veazey, J. C. Hunsucker, W. E. Garthright, R. J. Blodgett, and S. J. Chirtel. 1998. Influence of water temperature and salinity on *Vibrio vulnificus* in Northern Gulf and Atlantic Coast oysters (*Crassostrea virginica*). Appl. Environ. Microbiol. 64:1459-1465.
55. NACMCF. 1992. Microbiological criteria for raw molluscan shellfish. J. Food Prot. 55:463-480.
56. Oliver, J. D. 1989. *Vibrio vulnificus*, p. 570-600. *In* M. P. Doyle (ed.), Foodborne bacterial pathogens, Marcel Dekker, Inc., New York.
57. Qadri, R. B., K. A. Buckle, and R. A. Edwards. 1974. Rapid methods for the determination of faecal contamination in oysters. J. Appl. Bacteriol. 37:7-14.
58. Ratcliffe, S. D., and D. S. Wilt (eds.). 1971. Proceedings of Seventh National Shellfish Sanitation Workshop of the U.S. Dept of Health, Education, and Welfare. U.S. DHEW, Public Health Serv., Food and Drug Admin., Washington, D. C.
59. Richards, G. P. 1985. Outbreaks of shellfish-associated enteric virus illness in the United States: Request for development of viral guidelines. J. Food Prot. 48:815-823.
60. Rippey, S. R. 1994. Infectious diseases associated with molluscan shellfish consumption. Clin. Microbiol. Rev. 7:419-425.
61. Rippey, S. R., L. A. Chandler, and W. D. Watkins. 1987. Fluorometric method for enumeration of *Escherichia coli* in molluscan shellfish. J. Food Prot. 50:685-690, 710.
62. Ruple, A. D., and D. W. Cook. 1994. Factors affecting indicator and pathogen populations during handling of shellstock, p. 274-281. *In* C. R. Hackney and M. D. Pierson (ed.). Environmental indicators and shellfish safety. Chapman and Hall, New York.
63. Sample, T., and C. Swarson. 1997. Minutes of the *Vibrio parahaemolyticus* workshop of the U.S. Food and Drug Administration, Oct. 14. Seattle, Wash.
64. Seeliger, H. P. R., and D. Jones. 1986. Genus *Listeria*, p. 1235-1245. *In* P. H. R. Sneath, N. S. Mair, M. E. Sharpe, and J. G. Holt (eds.), Bergey's manual of systematic bacteriology, vol. 2. Williams and Wilkins, Baltimore Md.
65. Tamplin, M. L., and G. M. Capers. 1992. Persistence of *Vibrio vulnificus* in tissues of Gulf Coast oysters, *Crassostrea virginica*, exposed to seawater disinfected with UV light. Appl. Environ. Microbiol. 58:1506-1510.
66. Watkins, W. D., S. R. Rippey, C. R. Clavet, D. J. Kelley-Reitz, and W. Burkhardt. 1988. Novel compound for identifying *Escherichia coli*. Appl. Environ. Microbiol. 54:1874-1875.
67. Wehr, H. M. 1987. *Listeria monocytogenes*—A current dilemma. J. AOAC 70:769-772.
68. Whitman, C. 1994. Overview of the important clinical and epidemiologic aspects of *Vibrio vulnificus* infections. *In* W. Watkins and S. McCarthy (eds.), Proceedings of the 1994 *Vibrio vulnificus* Workshop. U.S. Dept. of Health and Human Serv., Public Health Serv., Food and Drug Admin., Washington, D. C.
69. Wilt, D. S. (ed.). 1974. Proceedings of Eighth National Shellfish Sanitation Workshop of the U.S. Dept of Health, Education, and Welfare. U.S. DHEW, Public Health Serv., Food and Drug Admin., Washington, D. C.

Chapter 50

Fruits and Vegetables

R. E. Brackett and D. F. Splittstoesser

50.1 INTRODUCTION

This chapter concerns fresh produce and nonsterile-processed fruits and vegetables. Examples of the latter are products preserved by freezing or dehydration. Some of these foods will be cooked before consumption, and others will be eaten raw.

Products that might be classified as both fresh and processed vegetables include minimally processed and fresh-cut products such as precut salad mixes, salad ingredients, e.g., shredded lettuce and carrots that are sold in grocery stores and to institutions. Contamination during processing and changes in microbial growth patterns during storage of the minimally processed products may alter the microflora of these foods quantitatively and qualitatively.[8,9,17,43,58]

50.2 FRESH PRODUCE

50.21 Normal Flora

1. *Sources.* There is great variation in the types of microorganisms on fruits and vegetables received from orchards, vineyards, and growing fields. All green plants possess a resident microflora that normally subsists on the slight traces of carbohydrates, protein, and inorganic salts that dissolve in the water exuding from, or condensing on, the epidermis of the host.[59] Other important factors include contamination from soil, water, dust, wild and domestic animals and other natural sources and the extent of contact during harvest with the soiled surfaces of harvesters and containers. In addition, intrinsic properties of the fruit or vegetable, such as a thick skin, could protect against surface damage and subsequent growth of saprophytic organisms. Moreover, it is also possible for viable bacteria to exist on internal tissues of even undamaged fruits and vegetables.[24,30]
2. *Predominant organisms.* Fruits and vegetables differ in a variety of ways and are often defined differently. However, fruits are usually defined as the seed-baring organs of plants whereas vegetables typically include all other types of produce. Some fruits, most notably tomatoes, are often thought of as vegetables because they are often used in meals as vegetables. Although similar types of microorganisms can be present on both fruits and vegetables, the intrinsic characteristics of the specific commodity can affect the ultimate residential microflora that develops on each.[44] The most notable difference between fruits and vegetables is pH. Although exceptions exist (e.g. melons), vegetables typically possess near neutral pH, whereas fruits are typically more acidic.

   Both fungi and bacteria grow well on vegetables. However, bacteria typically grow faster than yeasts and molds in the neutral environment of most vegetables. Hence, vegetables tend to select for bacteria, which are then present in greater abundance. In contrast, vegetables normally tend to harbor several orders of magnitude fewer fungi.[18]

   The specific types of microorganisms predominating on freshly harvested vegetables is quite variable. Because these foods are subjected to so many environmental conditions, it is conceivable that almost any type of microorganisms could be present. In general, however, the typical bacterial microflora of freshly harvested vegetables consists of gram-negative genera. The specific genera comprising this group include *Enterobacter cloacae, E. agglomerans,* and various *Pseudomonas* species.[9,44,46,63] However, products grown in or in close association with the soil can also contain various gram-positive bacteria including *Bacillus* and coryneform bacteria.

   Unlike vegetables, most fruits select for fungi. Fruits typically contain sufficient acidity to effectively restrict growth of many bacteria. In addition, most fruits contain organic acids that possess antibacterial properties. Hence, it is not uncommon to find a greater diversity of yeasts and molds present on the surface of fruits as compared to bacteria.
3. *Populations.* Populations of microorganisms on fresh produce can vary dramatically. Aerobic plate counts can be as high as 10^9 per g but are more typically in the range of 10^4–10^6.[14,15,27,35] A myriad of factors are responsible for this large degree of variability. Environmental factors such as weather conditions, the presence and abundance of animals or insects, or physical damage can potentially influence microbial populations. Webb and Mundt[61] for instance, observed that population of molds isolated from vegetables were influenced by the duration of the growing season, proximity of the vegetable to the ground, and rainfall. Moreover, populations of microorganisms are often not uniformly distributed even on or in an individual product. For ex-

ample, Maxie[29] isolated over 10^4 mesophilic aerobic bacteria/g on external leaves of lettuce but only about 32 cfu/g on innermost leaves. Hence, great caution should be exercised in attempting to specify an "acceptable" population of microorganism on freshly harvested produce.

4. *Indicator bacteria.* The presence or populations of coliform bacteria and enterococci in foods is frequently used as indicators of fecal contamination. In the case of produce, however, there is little or no relationship between these organisms and fecal contamination. Unlike some other types of foods, such as pasteurized milk, coliform bacteria and other organisms classified as fecal coliforms are part of the naturally occurring microflora of plants.[23,37,41,42,46] Some species of *Klebsiella* and *Enterobacter* in particular give a positive fecal coliform test.[63] *Escherichia coli* are not part of the normal microflora of fresh produce. Therefore, their presence can be related to the use of polluted water for irrigation[19] or washing, the presence of animal feces, unclean hands, or contaminated surfaces of harvesters and containers.

50.22 Floral Changes and Spoilage

Once fresh produce is harvested, it immediately begins the process of senescence, making it more susceptible to microbial decay.[21,38] Populations of microorganisms will normally increase dramatically during storage. The degree and rate of this increase will depend on the form of the product and storage conditions. For example, microbes will usually grow faster on cut vegetables than on whole vegetables.[8,17] Temperature, humidity, and the use of packaging or a modified atmosphere will also influence the microflora.[5,24] Although high populations of microorganisms are associated with spoilage, in many cases high populations can be present in the absence of spoilage.[5] Ultimate populations of bacteria often reach 10^6–10^7 cells per g before the product appears spoiled.[5,36] Gram-negative bacteria will usually predominate on stored fresh vegetables, although storage and packaging conditions may result in changes in types of microorganisms that develop.[5]

Market or storage diseases are the terms given the rots produced in fresh fruits and vegetables. Many of the diseases are named after their appearance, e.g., gray, brown, cottony, and stem-end rots. Molds are generally responsible for spoilage, particularly in fruits. More than 20 mold genera, including *Alternaria, Botrytis, Penicillium,* and *Phytophthora,* have been recognized as causes of spoilage. Some of these molds are opportunistic plant pathogens, whereas others are true plant pathogens that can invade healthy plant tissue.[1,4,47]

Certain bacteria can also cause fruit and vegetable spoilage. The bacterium of greatest concern is *Erwinia carotovora,* the cause of bacterial soft rot in a great variety of vegetables. However, some members of *Pseudomonas, Bacillus,* and *Clostridium* are also important spoilage organisms.[28]

Various types of vegetable sprouts have become popular in recent years and present a unique situation. Optimum conditions for producing sprouts are often the same as those supporting microbial growth.[26] Hence, these products can often develop very large populations of a variety of microorganisms, including human pathogens, during production of the sprouts.

50.23 Human Disease Microorganisms

Fresh fruits and vegetables have previously been thought to constitute only minimal risk for microbial foodborne illness. However, several outbreaks of human illness associated with produce (Table 1)[34] in recent years has shown that these foods can constitute a significant risk in transmitting foodborne illness. Virtually every type of pathogen can potentially be present on fresh produce but highly virulent gram-negative bacteria, such as *E. coli* O157:H7, are of most concern. Similarly, several parasites previously thought to be problems only in developing countries have caused produce-related illness in the U.S.[39] For example, *Cyclospora cayetanensis* infections have been traced to raspberries[34] and mesclun lettuce salad mix.[34] Foodborne viruses are likewise also increasingly being associated with fresh produce and have been associated with both fruits and vegetables (Table 1).

The fact that vegetables are often in contact with soil means that they commonly harbor spores of *Clostridium botulinum.* Unexpected outbreaks of botulism have occurred in vegetables that had been processed in a nontraditional manner. An outbreak of botulism resulting from the consumption of potato salad was believed to have been caused by toxin formation in foil-wrapped baked potatoes that had been held at room temperature.[57] In addition, *C. botulinum* can grow and produce toxin in products, such as garlic, which are stored in oil.[45]

Pathogenic microorganisms gain access to fresh produce in a number of ways. Produce can become contaminated before harvest through contaminated irrigation water or exposure to flood waters, use of raw manure as fertilizer, contact with wild or domestic animals, or workers.[10,60] After harvest, pathogens can be transmitted to foods via soiled processing equipment or transportation equipment, poor quality wash or cooling water, workers, or insects.[16,25,40]

50.3 PROCESSED FRUITS AND VEGETABLES

50.31 Factors Affecting Flora

As discussed previously, fruits and vegetables may harbor large numbers of microorganisms at the time of harvest. Further pro-

Table 1. Pathogens Causing Outbreaks Associated with Fresh or Frozen Produce[a]

Pathogen	Produce
Shigella spp.	Lettuce
	Green onions
Salmonella spp.	Sliced tomatoes
	Sprouts
	Sliced watermelon
	Sliced cantaloupe
	Unpasteurized orange juice
Escherichia coli O157:H7	Unpasteurized apple cider/juice
	Lettuce varieties
	Alfalfa sprouts
Enterotoxigenic *E. coli* (ETEC)	Carrots
Vibrio cholerae	Coconut milk
Listeria monocytogenes	Cabbage
Bacillus cereus	Sprouts
Hepatitis A virus	Lettuce
	Raspberries
	Frozen strawberries
	Sliced tomatoes
Norwalk/Norwalk-like virus	Sliced melon
	Green salad
	Celery
Cyclospora cayetanensis	Raspberries
	Mesclun lettuce
	Basil/Basil-containing products
Cryptosporidium parvum	Unpasteurized apple cider

[a] Adapted from Ref. 34.

cedures to prepare these products for canning, freezing, or distribution can affect both populations and types of microorganisms present.

1. *Washing.* Washing is among the first procedures to which fruits and vegetables are exposed after harvest. Water sprays are often used to dislodge field soil as well as cool the product. In doing so, washing removes some of the surface microorganisms, i.e., those that are not protected by the native mucilaginous material of the plant surface. In general, washing will remove about 1–2 logs of microorganisms.[64] However, washing produce in soiled or recycled water can result in increases in populations of microorganisms or contamination by pathogens. In addition, temperature of the water can affect the distribution of microorganisms within the produce. For example, bacteria can be drawn into internal tissues when wash water is cooler than the product being washed.[12,64]
2. *Heat.* Fruits and vegetables are subjected to heat during blanching and thermal processing (canning). Blanching is typically done to products that are to be frozen and usually results in reduction or elimination of most vegetative organisms.[52,55] In contrast, canning eliminates virtually all microorganisms present, the exception being some very heat resistant spores. Hence, microorganisms will only be present in canned fruits and vegetables as a result of post-process contamination, usually when a can seam fails.
3. *Sanitizers and other antimicrobials.* Sanitizers or other antimicrobial agents are commonly used to control populations of microorganisms in wash and processing waters. Unfortunately, the use of most common sanitizers approved for use with foods is of limited value and rarely reduces populations of microorganisms on produce by more than about 2 logs.[9,34,62,64]

   The relatively high concentrations of sulfur dioxide used to treat apple slices before freezing and various fruits and vegetables prior to dehydration destroy most vegetative microorganisms. In winemaking, the addition of 100 ppm sulfur dioxide reduces the number of wild yeasts by 99.9% or more.
4. *Cutting and chopping.* Cutting, chopping, and slicing of produce allows greater opportunity for microorganisms to invade internal tissues of plants and also releases nutrients which microorganisms can use for growth. Consequently, these practices usually result in higher populations of microorganisms and more rapid spoilage.[8] In addition, improperly cleaned and sanitized cutting, chopping and slicing equipment can sometimes contaminate produce, resulting in increases in populations.[17]
5. *Freezing and dehydration.* Although many organisms are destroyed by these processes, large numbers, including many vegetative forms, manage to survive. A further decrease in viable numbers occurs during the storage of frozen or dried fruits and vegetables.[20, 50] The rate of this decrease is influenced by many factors, such as storage conditions, the food type, and the predominant microflora.
6. *Irradiation.* Although gamma irradiation is not widely used to process fruits or vegetables, it has been used for disinfestation and control of fungal spoilage in some especially perishable commodities. It may also have use for reducing populations of human pathogens in fruit and vegetable products.

   The effects of gamma irradiation differ depending on the type of organism. Gram-negative bacteria are the most sensitive microorganisms whereas bacterial endospores and fungi are the most resistant.[5] Hence, treatment of fresh products by low doses of irradiation will tend to eliminate bacteria in favor of fungi. An irradiation dose of 1.8 kGy is sufficient to achieve a 5D inactivation of *E. coli* O157:H7 in apple juice.[11]
7. *Modified atmosphere packaging and storage.* Many fruits and vegetables marketed in recent years have been subjected to some type of storage or packaging in modified atmospheres. The primary purpose of using modified atmosphere is to extend the shelf life of produce by maintaining physiological quality and slowing senescence.[38] However, the use of modified atmosphere packaging can also affect the microflora. Extending shelf life allows more time for microorganisms present to grow, resulting in greater ultimate populations.[2] Modified atmospheres containing greatly elevated CO_2 or reduced O_2 can select for certain groups of microorganisms, specifically gram-positive facultative or anaerobic bacteria.[7] Unlike meats and poultry, however, fruits and vegetables are at least somewhat sensitive to and suffer damage from such gas concentrations. Because of this sensitivity, processors do not intentionally expose produce to gas concentrations to significantly repress growth of aerobic bacteria or encourage growth of anaerobic bacteria. However, respiratory activity of improperly packaged or stored plant tissues can reduce O_2 concentrations sufficiently to allow growth of anaerobic bacteria.[9]

50.32 Floral Changes—Frozen Vegetables

1. *Sources of contamination.* Blanching is the critical control step in the processing of frozen vegetables. Since it destroys most of the contaminating organisms, the microflora of the packaged product reflect mainly recontamination after blanching.

   The major source of organisms on frozen vegetables is contaminated equipment.[49] Units that have been especially troublesome are choppers, slicers, conveyor and inspection belts, and filling machines. The surfaces of some of these units are difficult to reach for proper cleaning. Belts, which generally are quite accessible, may present problems because of the tenacity with which microorganisms adhere to certain surfaces. In addition, some fabrics absorb liquids and thus permit microbial biofilms to accumulate within the belt interior. These biofilms later become a source of contamination when fractures develop in the belt surface.

   The degree of difficulty in controlling post-blanch contamination is related to vegetable type. With corn, large amounts of starch are released onto equipment surfaces and into flume water, whereas minimal quantities of soluble solids are leached from the vegetable during the processing of green beans and peas. Chopped leafy vegetables usually have higher microbial counts than the non-chopped products.[48]
2. *Predominant flora.* A wide variety of saprophytic bacteria can be isolated from frozen vegetables. The predominant flora is influenced by vegetable type and perhaps even geographic location. Gram-negative rods predominate on certain vegetables such as greens; on others, Lactococci and *Leuconostoc* species are most numerous. The proportion of lactic acid bacteria on frozen blanched vegetables has been

found to increase as the processing season progresses; with corn, for example, the percentage increased from 30% of the aerobic count on the first day of processing to more than 90% by the 15th day.[49] This is due to development of a characteristic microflora on processing lines. Many of the streptococci isolated from vegetables differ in some characteristics from described species.[32, 33, 54]

Low numbers of *Geotrichum candidum* filaments can be recovered on vegetables at various processing stages, but most are removed by blanching and subsequent hydro-cooling.[56] As a result, frozen vegetables are often negative for this mold. Other molds also may be present in low numbers.

3. *Indicator organisms.* Coliforms and enterococci are common contaminants of frozen vegetables and may be present in relatively large numbers, sometimes thousands per g.[51] As is the case with fresh produce, their presence usually does not indicate fecal contamination. It appears that they are introduced onto equipment surfaces, perhaps via the air, and become a part of the microflora of the processing line, along with the more numerous contaminants such as lactic acid-producing cocci. *Escherichia coli* is a relatively rare contaminant of blanched vegetables, and thus its recovery may indicate fecal contamination. A majority of frozen vegetables that yield positive tests for fecal coliforms are negative for *E. coli.*[45] Enterococci resembling *Streptococcus faecalis* (see Chapter 9) differ from strains of human origin.[31]

50.33 Floral Changes—Frozen Fruits

1. *Sources of contamination.* Although fruits generally are not blanched, much of the microflora acquired in the orchard and during the harvest is removed by various steps in the washing and peeling. Thus, as with vegetables, many of the contaminants on processed fruits originate from equipment within the plant.
2. *Predominant flora.* Because of the low pH of most fruits, aciduric yeasts and molds predominate, with yeasts usually the more numerous. *Geotrichum candidum* has been termed "machinery mold" because it may accumulate on fruit-processing equipment.[13] Of the acid-tolerant bacteria, the lactic acid group is most common, although species of *Acetobacter, Gluconobacter,* and *Zymomonas* may also develop in the acid environment of fruit-processing lines.
3. *Indicator organisms.* Coliforms can be recovered from various fruits even though the pH may be too low to support growth of these organisms. As with vegetables, the presence of indicator bacteria on frozen fruits and fruit products usually does not indicate a public health concern.[61]

50.34 Human Disease Microorganisms

Frozen blanched vegetables appear to present few problems with respect to organisms that cause foodborne illnesses. However, frozen fruits and vegetables that are not subject to a blanch can potentially be contaminated with foodborne pathogens. One well-publicized outbreak of hepatitis A was traced to contaminated frozen strawberries.[34] Contamination of blanched products would be expected to be a problem of post-blanch contamination whereas unblanched products could be contaminated at any point from growing through packaging.

Fruits are generally too acidic for growth of the more common foodborne pathogens. Many organisms do not survive in a low-pH environment, e.g., *Salmonella* and *Shigella* die off rapidly in citrus juices.[20] In contrast, *L. monocytogenes* can survive well on both chopped and whole tomatoes. Toxigenic fungi are a potential problem. Patulin, a mycotoxin produced by *Penicillium expansum* and a number of other molds, has been found in apple juice. It appears that the most effective control is to exclude moldy fruit from the product.[22]

50.4 RECOMMENDED METHODS

Standard Plate Count (Chapter 7)
Geotrichum Count (Chapter 4)
Yeasts and Molds (Chapter 20)
Coliform Bacteria (Chapter 8)
Listeria Monocytogenes (Chapter 36)
Escherichia coli (Chapter 35)
Salmonella (Chapter 37)

50.5 INTERPRETATION OF DATA

The microbiology of fresh fruits and vegetables often has little relationship to their quality or safety. Sound vegetables, for example, may yield extremely high aerobic plate counts because of high resident populations, or because of contamination from soil and other natural sources. The routine microbiological examination of most fresh fruits and vegetables, therefore, is not recommended. However, it might be prudent to do routine microbiological examination of fruits and vegetables destined for chronically ill or immunocompromised persons. Some pathogens, such as *L. monocytogenes,* can be particularly hazardous for such people.

Because most of the organisms on frozen and other processed fruits and vegetables originate from equipment surfaces, aerobic plate counts provide a means of assessing sanitation of the processing line. Problems in controlling contamination may differ with the type of fruit or vegetable. For example, a given population of bacteria on green beans may signify poor sanitation, while the same count on chopped broccoli may indicate excellent conditions.[48] Many frozen fruits and vegetables have aerobic plate counts below 50×10^3 per g.

Coliforms and enterococci are part of the normal flora of plant products, and populations of 10^2 or 10^3 per g of processed product are not uncommon. Hence, the presence of coliforms or fecal coliforms should not be interpreted as reflecting the sanitary quality or safety of fresh produce without the further analysis for more specific indicators of fecal contamination, most notably *E. coli.*

Coagulase-positive *Staphylococcus aureus* may be present on vegetables,[53] but usually in low numbers, under 10 per g. The routine culturing of fruits and vegetables for staphylococci is not justified.

50.6 REFERENCES

1. Beneke, E. S., L. S. White, and F. W. Fabian. 1954. The incidence and pectolytic activity of fungi isolated from Michigan strawberry fields. Appl. Microbiol. 2:253.
2. Berrang, M. E, R. E. Brackett, and L. R. Beuchat. 1990. Microbial, color and textural qualities of fresh asparagus, broccoli, and cauliflower stored under controlled atmospheres. J Food Prot. 53:391-395.
3. Beuchat, L. R., B. V. Nail, B. B. Adler, and M. R. S. Clavero. 1998. Efficacy of spray application of chlorinated water in killing pathogenic bacteria on raw apples, tomatoes, and lettuce. J. Food Prot. 61:1372-1374.
4. Brackett, R. E. 1986. Vegetables and related products, p. 129. *In* J. R. Beuchat (ed.), Food and beverage mycology, 2nd ed. Van Norstead, New York.
5. Brackett, R. E. 1987. Microbiological consequences of minimally processed fruits and vegetables. J. Food Qual. 10:195-206.

6. Brackett, R. E. 1988. Changes in the microflora of packaged tomatoes. J. Food Qual. 11:89-105.
7. Brackett, R. E. 1990. Influence of modified atmosphere packaging on the microflora and quality of fresh bell peppers. J. Food Prot. 53:255-257.
8. Brackett, R. E. 1993. Microbial quality, p.125-148. *In* R. L. Shewfelt and S. E. Prussia (eds.), Postharvest handling: A systems approach. Academic Press, New York.
9. Brackett, R. E. 1998. Fruits, vegetables, and grains, p.117-126. *In* M. P. Doyle, L. R. Beuchat, and T. J. Montville (eds.), Food microbiology: fundamentals and frontiers. ASM Press, Washington, D. C.
10. Brackett, R. E. 1998. Safe handling of fruits and vegetables. *In* J. M. Farber and E. C. D. Todd (eds.), Safe handling of foods, in press. Marcell Dekker, New York.
11. Buchanan, R. L., S. G. Edelson, K. Snipes, and G. Boyd. 1998. Inactivation of *Escherichia coli* O157:H7 in apple juice by irradiation. Appl. Environ. Microbiol. 64:4533-4535.
12. Buchanan, R. L., S. G. Edelson, R. L. Miller, and G. M. Sapers. 1999. Contamination of intact apples after immersion in an aqueous environment containing *Escherichia coli* O157:H7, in press. J. Food Prot.
13. Eisenberg, W. V., and S. M. Cichowicz. 1977. Machinery mold-Indicator organisms in food. Food Technol. 31:52-56.
14. Ercolani, G. L. 1976. Bacteriological quality assessment of fresh marketed lettuce and fennel. Appl. Environ. Microbiol. 31:847-852.
15. Etchells, J. L., R. N. Costilow, T. A. Bell, and H. A. Rutherford. 1961. Influence of gamma radiation on the microflora of cucumber fruit and blossoms. Appl. Microbiol. 9:145.
16. Fisher, T. L., and D. A. Golden. 1998. Fate of *Escherichia coli* O157:H7 in ground apples used in cider production. J. Food Prot. 61:1372-1374
17. Garg, N., J. J. Churey, and D. F. Splittstoesser. 1990. Effect of processing conditions on the microflora of fresh-cut vegetables. J. Food Prot. 53:701,702,703.
18. Geeson, J. D. 1979. The fungal and bacterial flora of stored white cabbage. J. Appl. Bacteriol. 46:189-193.
19. Geldreich, E. E., and R. H. Bordner. 1971. Fecal contamination of fruits and vegetables during cultivation and processing for market. A review. J. Milk Food Technol. 34:184-195.
20. Hahn, S. S., and M. D. Appleman. 1952. Microbiology of' frozen orange concentrate. I. Survival of enteric organisms in frozen orange concentrate. Food Technol. 6:156.
21. Hao, D. Y. Y., and R. E. Brackett . 1994. Pectinase activity of vegetable spoilage bacteria in modified atmosphere. J. Food Sci. 59:175-178
22. Harwig, J., Y. K. Chen, B. P. C. Kennedy, and P. M. Scott. 1973. Occurrence of patulin and patulin-producing strains of *Penicillium expansum* in natural rots of apple in Canada. Can. Inst. Food Sci. J. 6:22-25.
23. Hayward, A. C. 1974. Latent infections by bacteria. Ann. Rev. Phytopath.12:87-97.
24. Hobbs, G. 1986. Ecology of food microorganisms. Microbiol. Ecol. 12:15-30.
25. Janisiewicz, W. J., W. S. Conway, M. W. Brown, G. M. Sapers, P. Fratamico, and R. L. Buchanan. 1998. Fate of *Escherichia coli* O157:H7 on fresh cut apple tissue and its transmission by fruit flies, in press. Appl. Environ. Microbiol.
26. Jaquette, C. B., L. R. Beuchat, and B. E. Mahon. 1996. Efficacy of chlorine and heat treatment in killing Salmonella stanley inoculated onto alfalfa seeds and growth and survival of the pathogen during sprouting and storage. Appl. Environ. Microbiol. 62:2212-2215.
27. Kaferstein, F. K. 1976. The microflora of parsley. J. Milk Food Technol. 39:837-840.
28. Lund, B. M. 1983. Bacterial spoilage, p. 219. *In* C. Dennis (ed.), Postharvest pathology of fruits and vegetables. Academic Press, New York.
29. Maxie, R. B. 1978. Lettuce salad as a carrier of microorganisms of public health significance. J. Food Prot. 41:435-438.
30. Meneley, J. C, and M. E. Stanghellini. 1974. Detection of enteric bacteria within jocular tissue of healthy cucumbers. J. Food Sci. 39:1267-1268.
31. Mundt, J. O. 1973. Litmus milk reaction as a distinguishing feature between *Streptococcus faecalis* of human and nonhuman origins. J. Milk Food Technol. 36:364-367.
32. Mundt, J. O. 1975. Unidentified streptococci from plants. Int. J. Syst. Bacteriol. 25:281.
33. Mundt, J. O., W. F. Graham, and L. E. McCarty. 1967. Spherical lactic acid-producing bacteria of southern-grown raw and processed vegetables. Appl. Microbiol. 15:1303-1308.
34. National Advisory Committee on Microbiological Criteria for Foods. 1998. Microbiological safety evaluations and recommendations for fresh produce. Food Control 9 (Pt 6):321-347.
35. Nguyen-the, C., and F. Carlin. 1994. The microbiology of minimally processed fresh fruits and vegetables. Crit. Rev. Food Sci. Nutr. 34(4):371-401.
36. Priepke, P. E., L. S. Wei, and A. L. Nelson. 1976. Refrigerated storage of prepackaged salad vegetables. J. Food Sci. 41:379-382.
37. Riser, E. C., J. Grabowski, and E. P. Glenn. 1984. Microbiology of hydroponically-grown lettuce. J. Food Prot. 47:765-769.
38. Rolle, R. S., and G. W. Chism. 1987. Physiological consequences of minimally processed fruit and vegetables. J. Food Qual. 10:157-177.
39. Rude, R. A., G. J. Jackson, J. W. Bier, T. K. Sawyer, and N. G. Risty. 1984. Survey of fresh vegetables for nematodes, amoebae, and *Salmonella*. J. Assoc. Off. Anal. Chem. 47:613-615.
40. Ruiz, B. G-V., R. Galvez Vargas, and R. Garcia-Villanova. 1987. Contamination on fresh vegetables during cultivation and marketing. Int. J. Food Microbiol. 4:285-291.
41. Senter, S. D., N. A. Cox, J. S. Bailey, and F. I. Meredith. 1984. Effects of harvesting, transportation, and cryogenic processing on the microflora of southern peas. J. Food Sci. 49:1410-1437.
42. Senter, S. D., N. A. Cox, J. S. Bailey, and W. R. Forbus, Jr. 1985. Microbiological changes in fresh market tomatoes during packing operations. J. Food Sci. 50:254-255.
43. Shapiro, J. E., and I. A. Holder. 1960. Effect of antibiotic and chemical dips on the microflora of packaged salad mix. Appl Microbiol. 8: 341.
44. Skovgaard, N. 1984. Vegetables as an ecological environment for microbes, p. 27-33. *In* I. Kiss, T. Deak, and K. Incze (eds.), Microbial associations and interactions in food. Akademia Kiado, Budapest.
45. Solomon, H. M., and D. A. Kautter. 1988. Outgrowth and toxin production by *Clostridium botulinum* in bottled chopped garlic. J. Food Prot. 51:862.
46. Splittstoesser, D. F. 1970. Predominant microorganisms on raw plant foods. J. Milk Food Technol. 33:500-505.
47. Splittstoesser, D. F. 1986. Fruits and fruit products, p. 101-128. *In* L. R. Beuchat (ed.), Food and beverage mycology, 2nd ed. Van Norstead, New York.
48. Splittstoesser, D. F., and D. A. Corlett, Jr. 1980. Aerobic plate counts of frozen blanched vegetables processed in the United States. J. Food Prot. 43:717.
49. Splittstoesser, D. F., and L. Gadjo. 1966. The groups of microorganisms composing the "total" count population in frozen vegetables. J. Food Sci. 31:234-239.
50. Splittstoesser, D. F., and B. Segen. 1970. Examination of frozen vegetables for salmonellae. J Milk Food Technol; 33: 111-113.
51. Splittstoesser, D. F., and W. P. Wettergreen. 1964. The significance of coliforms in frozen vegetables. Food Technol. 18:134-136.
52. Splittstoesser, D. F., W. P. Wcttergreen, and C. S. Pederson. 1961. Control of microorganisms during preparation of vegetables for freezing. II. Peas and corn. Food Technol. 15:332-334.
53. Splittstoesser, D. F., G. E. R. Hervey II, and W. P. Wettergreen. 1965. Contamination of frozen vegetables by coagulase-positive staphylococci. J. Milk Food Technol. 28:148.
54. Splittstoesser, D. F., J. Mautz, and R. R. Colwell. 1968. Numerical taxonomy of catalase negative cocci isolated from frozen vegetables. Appl. Microbiol. 16:1024.
55. Splittstoesser, D. F., D. T. Queale, J. L. Bowers, and M. Wilkison. 1980a. Coliform content of frozen blanched vegetables packed in the United States. J. Food Safety 2:1-11.

56. Splittstoesser, D. F., J. Bowers, L. Kerschner, and M. Wilkison. 1980b. Detection and incidence of *Geotrichum candidum* in frozen blanched vegetables. J. Food Sci. 45:511-513.
57. Sugiyama, H., M. Woodburn, K. H. Yang, and C. Movroydis. 1981. Production of botulinum toxin in inoculated pack studies of foil-wrapped baked potatoes. J. Food Prot. 44:896-898.
58. Terry, R. C., and W. W. Overcast. 1976. A microbiological profile of commercially prepared salads. J. Food Sci. 41:211-213.
59. Thaysen, A. C., and L. D. Galloway. 1930. The microbiology of starch and sugars. Oxford University Press, New York. p. 191.
60. Tierney, J. T., R. Sullivan, and E. P. Larkin. 1997. Persistence of poliovirus 1 in soil and on vegetables grown in soil previously flooded with inoculated sewage sludge or effluent. Appl. Environ. Microbiol. 33:109-113.
61. Webb, T. A., and J. O. Mundt. 1978. Molds on vegetables at the time of harvest. Appl. Environ. Microbiol. 35:655-658.
62. Zhang, S., and J. M. Farber. 1996. The effects of various disinfectants against *Listeria monocytogenes* on fresh-cut vegetables. Food Microbiol. 13:311-321.
63. Zhao, T., M. R. S. Clavero, M. P. Doyle, and L. R. Beuchat. 1997. Health relevance of the presence of fecal coliforms in iced tea and leaf tea. J. Food Prot. 60:215-218.
64. Zhuang, R. Y., L. R. Beuchat, and F. J. Angulo. 1995. Fate of Salmonella montevideo on and in raw tomatoes as affected by temperature and treatment with chlorine. Appl. Environ. Microbiol. 61:2127-2131.

Chapter 51

Fermented and Acidified Vegetables

H. P. Fleming, R. F. McFeeters, and F. Breidt

51.1 INTRODUCTION

Vegetables may be preserved by fermentation, direct acidification, or a combination of these along with other processing conditions and additives to yield products that are referred to as pickles. Pasteurization and refrigeration are used to assure stability of certain of these products. Organic acids and salt (sodium chloride) are primary preservatives for most types of pickles. Lactic acid is produced naturally in fermented products. Acetic acid (or vinegar) is the usual acid added to pasteurized, unfermented (fresh-pack) pickles. Acetic acid also is added to many products made from fermented (salt-stock) cucumbers. Other preservatives such as sodium benzoate, potassium sorbate, and sulfur dioxide may be added to finished products. Although the term "pickles" in the United States generally refers to pickled cucumbers, the term is used herein in a broader sense to refer to all vegetables that are preserved by fermentation or direct acidification. Cucumbers, cabbage, olives, and peppers account for the largest volume of vegetables and fruits commercially pickled. Lesser quantities of onions, tomatoes, cauliflower, carrots, melon rinds, okra, artichokes, beans, and other produce also are pickled.

The fermentation of vegetables is due primarily to the lactic acid bacteria, although yeasts and other microorganisms may be involved, depending on the salt concentration and other factors. Salt serves two primary roles in the preservation of fermented vegetables: It influences the type and extent of microbial activity, and it helps prevent softening of the vegetable tissue. Some vegetables are brined at such high salt concentrations as to greatly retard or preclude fermentation. Salt may be added in the dry form, as with cabbage, or as a brine solution, as with most other vegetables. The concentration of salt used varies widely among vegetables, depending on tendency of the vegetable to soften during brine storage. Softening of brined cucumbers can be reduced or prevented by adjusting the level of salt to inhibit pectinolytic enzymes.[3,4] Fermentation is an economical means for temporary preservation of produce such as cucumbers, cabbage, and olives. The produce is fermented and stored in large tanks until it is needed for further processing. After removal from brine storage, brined cucumbers may be desalted if needed before being finished into various products such as dills, sweets, sours, hamburger dill chips, mixed vegetables, and relishes.[50] Finished salt-stock dill cucumber pickles contain a minimum of 0.6% lactic acid, according to USDA grade standards.[137] The products may or may not be pasteurized, depending on the addition of sugar and other preservatives. Extensive reviews are available on the brining and fermentation of cabbage,[118,131] cucumbers,[48,53] and olives.[52,68,139,140,141]

Direct acidification with acetic acid (without pasteurization) has been a primary method for many years of preserving various pickles and sauces in the United Kingdom, where the products are referred to as acetic acid preserves. British researchers have determined that the minimum acetic acid concentration necessary to achieve satisfactory preservation of all pickles and sauces is 3.6%, calculated as a percentage of the volatile constituents of the product.[9] The high concentration of acid needed for preservation results in such a strong acid flavor, however, that the relative importance of this method of preservation has diminished. Milder acidic flavors are more in demand today, and use of acidification in combination with pasteurization has become more important. Nevertheless, some specialty products such as hot pepper sauce and sliced peppers still are preserved principally by high concentrations of acetic acid without pasteurization.

Fresh-pack cucumber pickles are preserved by mild acidification (0.5% to 1.1% acetic acid[137]) of fresh cucumbers, followed by heating to an internal product temperature of 74°C and holding for 15 minutes, according to the original recommendations of Etchells et al.[30,37,109] Such products are effectively pasteurized, since they are heated enough to inactivate microbial vegetative cells, and sufficient acid has been added to prevent outgrowth of bacterial spores. Although some packers still use this heat process, others now vary the times and temperatures, depending on product type and risk factors (for spoilage problems of non-health significance) acceptable to the packer. Fermented pickles, such as whole genuine dills and hamburger dill chips, may or may not be heated. If pasteurized, these products may be given a milder heat treatment than fresh-pack pickles, such as an internal product temperature of 71°C with no holding time. The fresh-pack process has been applied to peppers and other vegetables. Fresh-pack pickles are considered acidified foods for regulatory purposes. According to the U.S. Food and Drug Administration (FDA), "'Acidified foods' means low-acid foods to which acid(s) or acid food(s) are added; these foods include, but are not limited to, beans, cucumbers, cabbage, artichokes, cauliflower, puddings, peppers, tropical fruits, and fish, singly or in any combination. They have a water activity (a_w) greater than 0.85 and have a finished equilibrium pH of 4.6 or below. These foods may be called, or may purport to be, 'pickles' or 'pickled'...."[136]

Refrigerated pickles may or may not be fermented before refrigeration. Also, they may or may not be acidified, although mild acidification is highly recommended.[49] Most commercially prepared and distributed refrigerated pickles sold today are not fermented, but are acidified and contain a preservative such as sodium benzoate.

Increasing environmental concerns related to waste disposal are influencing methods for preservation of pickled vegetables, particularly those involving use of salt for bulk storage. The U.S. Environmental Protection Agency (EPA) has proposed a maximum of 230 ppm of chloride in fresh waters,[29] a limit that may not be readily achievable by many vegetable briners who discharge chloride wastes into freshwater streams. Organic acids (lactic and acetic) in combination with calcium chloride and preservatives (e.g., sodium benzoate) are now used instead of sodium chloride for bulk storage of olives for "green-ripe" processing into canned black olives in California.[142] Salt is still used, however, for fermented olives. The use of calcium salts (chloride or acetate) has led to reduced levels of sodium chloride for bulk fermentation and storage of cucumbers. Calcium salts have been found to enhance firmness retention of cucumbers at reduced concentrations of sodium chloride.[14,61,65,84,135] Studies have revealed, however, that spoilage microorganisms may present a serious problem in fermented cucumbers if the salt concentration is too low.[68] Recently, the use of sulfite was proposed as a way to store cucumbers in the absence of salt, and the sulfite removed by reaction with hydrogen peroxide after storage and before conversion into finished products.[100]

51.2 NORMAL FLORA

Fresh produce contains a varied epiphytic microflora (Chapter 50). Pickling cucumbers were found to contain as high as 5.3×10^7 total aerobes, 1.9×10^4 aerobic spores, 9.8×10^5 total anaerobes, 5.4×10^2 anaerobic spores, 6.1×10^6 coliforms, 5.1×10^4 total acid formers, 4.6×10^3 molds, and 6.6×10^3 yeasts per g of fresh cucumber.[46] The numbers increased during storage at higher temperatures (21°C) and humidity (>70% relative humidity). Although some investigators have held that the interior of sound, fresh cucumbers is sterile, others have found microorganisms, mostly gram-negative rods, within the healthy fruit.[106,127] In cucumbers, bacteria were more often near the skin and less often in the central core; in tomatoes, their frequency was highest near the stem-scar and central core and decreased toward the skin.[127] Cabbage contains the greatest number of bacteria on the outer leaves and lower numbers toward the center of the head.[118]

The floral changes during natural fermentation of brined vegetables may be characterized into four stages: initiation, primary fermentation, secondary fermentation, and post-fermentation.[53] During initiation, the various gram-positive and gram-negative bacteria that were on the fresh vegetable compete for predominance. *Enterobacteriaceae*, aerobic spore-formers, lactic acid bacteria, and other bacteria may be active. Eventually, the lactic acid bacteria gain predominance by lowering the pH, and primary lactic fermentation occurs. During primary fermentation, five species of lactic acid-producing bacteria are active, listed in approximate order of their occurrence: *Streptococcus* (*Enterococcus*) *faecalis, Leuconostoc mesenteroides, Pediococcus cerevisiae* (probably *P. pentosaceous* and/or *P. acidilactici*, according to recent classification[130]), *Lactobacillus brevis*, and *Lactobacillus plantarum*. Although all five species are active during fermentation of sauerkraut,[118] which contains relatively low concentrations of salt (ca. 2.25%), only the latter three species predominate in fermentation of cucumbers, which contain higher concentrations of salt (ca. 5% to 8%).[48] *Lactobacillus plantarum* characteristically terminates the lactic fermentation, apparently because of its greater acid tolerance.[121]

During fermentation of brined cucumbers, lactic acid bacteria may grow within the cucumber tissue as well as the brine.[22] Gas composition of the cucumbers at the time of brining greatly influences the ratio of bacterial growth in the cucumbers and the brine.[25] Yeasts were found not to grow within the cucumber tissue, presumably because of their larger size, which prevented their entry through stomata of the cucumber skin.

Green olives contain inhibitors of lactic acid bacteria,[55,57,87] which are thought to influence fermentation of Spanish-type green olives.[44,88] Yeasts are not inhibited and predominate in the fermentation when the olives are neither properly lye treated nor heat shocked before brining.[44]

Various species of fermentative yeasts also are active during primary fermentation. If fermentable sugars remain after primary fermentation, these sugars may give rise to secondary fermentation dominated essentially by yeasts. Fermentative yeasts grow as long as fermentable sugars are available; this may result in severe gaseous spoilage (bloater formation).[32,39,88] During post-fermentation, growth of oxidative yeasts, molds, and bacteria may occur on brine surfaces of open tanks that are not exposed to ultraviolet radiation or sunlight.[33, 112] Vegetable brining tanks are typically uncovered and are held outdoors to allow sunlight to reduce or prevent surface growth. Surface growth does not occur in fermented and anaerobically stored green olives.[140] Attempts have been made to develop a suitable anaerobic tank for the cucumber-brining industry.[63,66,85]

Lactic starter cultures have been used commercially on a limited basis in sauerkraut, olives, cucumbers, and other products.[54] *Pediococcus cerevisiae* and *L. plantarum* have been used in pure culture or controlled fermentations of cucumbers,[43,47] and olives.[44] Although starter cultures have been used on a limited commercial scale for fermenting cucumbers over the past 10 years, they are not widely used.

51.3 FLORA CHANGES IN SPOILAGE

51.31 Fermented Vegetables

Production of CO_2 in the cover brine of fermenting cucumbers by various bacteria, including heterofermentative lactic acid bacteria and fermentative species of yeasts, is associated with bloater spoilage. Even homofermentative lactic acid bacteria such as *L. plantarum* and *P. cerevisiae* produce sufficient CO_2, when combined with CO_2 from cucumber tissue, to cause bloater formation in brined cucumbers.[56] The major source of CO_2 production by homofermentative lactic acid bacteria is decarboxylation of malic acid, a natural constituent of pickling cucumbers.[103] It has been demonstrated that cultures that do not degrade malic acid will ferment cucumbers with reduced bloater damage.[104] Procedures have been developed to produce and isolate non-malate-decarboxylating mutants of *L. plantarum*.[24,26] Purging fermenting cucumber brines with nitrogen has been shown to be effective in preventing bloater formation.[20,47,56,60] Purging is now widely used by the pickle industry. Air purging also is effective in preventing bloater formation,[20,60] but can result in cucumber softening due to mold growth,[21,60,75] reduced brine acidity due to yeast growth,[123] and off-colors and -flavors unless the purging regimen is carefully controlled. In fact, air is most commonly used today by the pickle industry, with intermittent purging regimes and addition of potassium sorbate to prevent growth and spoilage by yeasts and molds. Bloater formation has been attributed to growth of gas-forming microorganisms in the brine surrounding the cucumbers[45] or within the cucumber.[22,126]

Softening of brined vegetables may be caused by pectinolytic enzymes of plant or microbial origin. Growth of film yeasts on brine surfaces may occur and result in loss of brine acidity. Accompanying mold growth on the brine surface can cause softening of sauerkraut, cucumbers, or olives. Heavy scum yeast and/or mold growth is usually the result of neglecting brined material during extended storage. It is important for the brine surface of cucumber tanks to be exposed to direct sunlight to inhibit yeast and mold growth. Sauerkraut tanks are usually held indoors, and a properly seated plastic cover weighted down with water or brine is necessary to prohibit aerobiosis and thereby prevent surface growth of yeasts and molds. Softening of brined cucumbers may also result from mold polygalacturonases that accompany the cucumbers, especially cucumbers with flowers attached,[6,42] into the brine tank. This problem was reduced in former times by draining and rebrining of the tank ca. 36 hours after initial brining,[41] but this solution is not normally used today because of environmental concerns with salt disposal. Recycled brine may be treated to inactivate softening enzymes.[76,102] Adding calcium chloride can slow down the rate of enzymatic softening of fermenting cucumbers.[15] However, this should not be relied upon to eliminate enzymatic softening problems. Care must be taken to minimize contamination of cucumbers, particularly the small fruit, with flowers and plant debris, which can be a source of contamination by pectinolytic molds.

Butyric acid spoilage of brined olives has been attributed to two distinct types of microbial action. In one type, *Clostridium butyricum* and a closely related group of clostridia produce butyric acid from sugars during the primary stage of fermentation.[78] In a second type of malodorous olive fermentation, "zapatera" spoilage results from decomposition of organic acids at a time when little or no sugar is present and the lactic acid fermentation stops before the pH has decreased below pH 4.5.[90] Propionibacteria were isolated from brined olives with indications of "zapatera" spoilage and were hypothesized to grow and cause a rise in pH because of degradation of lactic acid, thus permitting subsequent growth by *Clostridium* species.[10,80,122]

Butyric acid spoilage of brined cucumbers was found to occur after an apparently normal primary fermentation by lactic acid bacteria.[68] *Clostridium tertium* was identified as contributing to the spoilage. Evidence from end products indicated that unidentified bacteria, possibly propionibacteria species, degraded lactic acid, causing a rise in pH that allowed C. *tertium* to grow. This problem, particularly acute when cucumbers are stored at relatively low salt concentration, can be reduced/avoided by assuring a final brine pH of 3.5 or lower after fermentation.[70]

51.32 Finished Pickle Products from Salt-stock Vegetables (Not Pasteurized)

Fully cured salt-stock vegetables are made into various types of finished pickle products by a series of operations involving leaching out most of the salt, souring with vinegar, and then sweetening with sugar if desired. Preservation of these products depends on sufficient amounts of vinegar alone (for sour pickles), or a combination of vinegar and sugar with or without sodium benzoate (for sweet pickles).[2] If the concentration of these ingredients is inadequate and the product is not pasteurized, fermentation usually takes place. Osmotolerant yeasts are the principal spoilage organisms in such products.[2] Molds and film yeasts may grow on the surface of the liquid chiefly as the result of faulty jar closure. Lactic acid bacteria, propionibacteria and butyric acid bacteria also may cause spoilage in unpasteurized fermented vegetables that do not contain adequate concentrations of acetic acid or other preservatives.

51.33 Acetic Acid Preserves

The 3.6% acetic acid (as a percentage of volatile constituents) required for preservation of pickles and sauces[9] is similar to, but slightly lower than the concentration found necessary by Bell and Etchells[2] to prevent yeast spoilage in finished pickles from salt-stock cucumbers. Microbial spoilage apparently occurs in acetic acid preserves when the concentration of acid is marginal. Spoilage microorganisms include yeasts[27] and lactobacilli, particularly the heterofermentative *Lactobacillus fructivorans*.[28]

51.34 Pasteurized Pickle Products

Spoilage usually occurs in these products when they are improperly pasteurized or improperly acidified so that an equilibrated brine product of pH 3.8 to 4.0 is not achieved. Spoilage is due chiefly to acid-forming bacteria and, to a lesser extent, yeasts that survive faulty heat treatment, or butyric acid bacteria when the product is not acidified adequately at the outset. Molds and film yeasts are factors in cases of poor jar closure.

51.35 Refrigerated Pickle Products

51.351 Fermented

A wide array of fermented, refrigerated cucumber pickle products are prepared as specialty products.[49] Examples of such products include overnight dills, half-sour dills, genuine kosher dills, kosher new dills, sour garlic pickles, half-sour new pickles, fresh-packed half-sour pickles, new half-sours, home-style new pickles, half-sour kosher new dills, and the like. The cover brine may or may not be acidified. The products are held in barrels for a few days or longer at room temperature and then refrigerated at 2° to 5°C. They may be distributed in bulk or in consumer-size glass containers. In some cases, they may be initially brined, held, and distributed in consumer-size containers. Under such conditions and at equilibrated brine strengths of 10 to 12 salometer (1 salometer = 0.264% salt by weight), microbial growth (chiefly coliforms, gas-forming and non-gas-forming lactics, and fermentative yeasts) and enzymatic activity (pectinolytic and cellulolytic) together with the curing process continue at a slow rate.[45] Gaseous spoilage of the product is caused chiefly by the gas-forming microbial groups mentioned earlier. Gas production may be sufficient to reach 15 psi pressure within the container.

Softening problems may be even greater than for salt-stock cucumbers since these products are held at much lower concentrations of salt. Fresh, whole garlic cloves and other spices are normally added to such products. These spices may contain high activities of softening enzymes that increase softening problems. In a few months, the stored pickles may have lost much of their characteristic flavor, texture, and color and also may be bloated because of gaseous fermentation by the principal gas-forming microbial groups present.

Whether these pickles are made in bulk or in the retail jar, the very nature of the product makes it difficult to maintain good quality for any reasonable length of time. The barreled product reaches the good manufacturing practices (GMP)-recommended brine pH of 4.6 or below for acidified foods, usually before refrigeration or shortly thereafter, and then slowly continues acid development. This recommended condition for brine-product pH cannot be assured for the product made in the retail jar because there is no uniform process accepted by packers wherein the product is acidified at the outset or where it is deliberately incubated for development of natural lactic acid fermentation.

Sauerkraut marketed in plastic bags in refrigerated display cases is preserved by the addition of sodium benzoate and bisulfite.[133] The shelf life of such products is influenced by chemical changes that may result in discoloration (browning) and objectionable flavor formation.

51.352 Not Fermented

Most of these products for national distribution are acidified with vinegar to an equilibrium pH well below 4.6, contain 2% to 3% NaCl, and are immediately refrigerated upon packing.[49] They may contain sodium benzoate or other preservatives. Like the fermented refrigerated product, the cucumbers are not heated either before or after packing. If properly acidified, refrigerated, and preserved, the products will maintain acceptable quality for several months and do not present a public health concern. Recipes that do not contain vinegar or other acid in the initial cover liquor, however, should be viewed with caution.

51.4 PATHOGENIC MICROORGANISMS

We know of no authenticated reports of pathogenic microorganisms associated with standard commercial pickle products prepared under "good manufacturing practices" of acid, salt, and sugar content (and combinations thereof) from brined, salted, and pickled vegetable brine-stock, including cucumbers. The Commissioner of the FDA stated that "No instances of illness as the result of contamination of commercially processed fermented foods with *Clostridium botulinum* have been reported in the United States."[136] Even so, certain types of microorganisms that may cause spoilage of the product may, at times, be encountered, such as molds, yeasts, and acid-tolerant lactic acid bacteria. These organisms, usually under conditions associated with neglect, may reduce the quality of the texture and flavor of the product—whether prepared in bulk or retail container—and render it unusable. However, these organisms are not considered human pathogens.

Essentially the same pattern of consumer safety applies to fresh-pack (pasteurized) pickle products. These products have continued to increase in popularity until they now use over 40% of the annual cucumber crop in the United States. These pickles usually are prepared from raw cucumbers, but may include other vegetables in a mixture; also, vegetables other than cucumbers may be packed, such as various types of peppers, okra, carrots, green beans, and tomatoes. The process calls for the packed product to be acidified at the outset with a sufficient amount of food-grade organic acid, e. g., vinegar, acetic acid, or lactic acid, to result in an equilibrated brine product pH of 4.0 or below (preferably 3.8). Vinegar (acetic acid) is usually the acidulant of industry choice for cucumber pickle products. The basic pasteurization procedure, with acidified product heated to an internal temperature of 74°C and held for 15 minutes, has been used successfully by industry since ca. 1940.[30,35,37] Insufficient acidification of pasteurized pickles can result in butyric acid-type spoilage, possibly involving public health concerns.

Listeria monocytogenes, a food-borne pathogen, has become a major concern to the food industry over the past 15 years. The bacterium is commonly found in the environment and has been isolated from various plant materials, including silage,[51] soybeans, corn,[144,145] and cabbage.[128] Beuchat et al.[8] showed that *L. monocytogenes* was able to grow on raw cabbage and in cabbage juice. Conner et al.[19] found death of *L. monocytogenes* (one strain tested, LCDC 81-861, is a pathogen isolated from coleslaw) to occur in cabbage juice adjusted to pH ≤ 4.6 with lactic acid and incubated at 30°C; at 5°C, the death rate was slower than at 30°C. However, two strains tested grew well at pH values of 5.0 to 6.1. In a nutrient medium acidified with hydrochloric acid, the minimum pH values at which growth of *L. monocytogenes* was detected at 30°, 20°, 10°, 7°, and 4°C were, respectively, 4.39, 4.39, 4.62, 4.62, and 5.23.[77] Johnson et al.[86] demonstrated that *Listeria* could be recovered from fermented sausage made with beef intentionally contaminated with the bacterium. To our knowledge, *Listeria* has not been reported in fermented vegetables. However, the observation that *Listeria* can be isolated from fermented materials (silage and sausage), coupled with the bacterium's ability to tolerate moderately low pH[51,19,77,86] and high salt concentrations (growth in complex media at 10% salt[129]) suggests that *Listeria* may pose a concern for mildly acidified or fermented vegetables.

While pathogenic bacteria have not been reported in commercially fermented vegetable products, the potential for the survival and growth of some pathogens in acidic environments has been investigated.[83,124] These studies suggest that *Escherichia coli*, *L. monocytogenes*, *Salmonella* species, and others may potentially grow in mildly acidified or fermented vegetables. Outbreaks of *E. coli* O157:H7 in unpasteurized apple juice or cider, which typically has a pH between 3.5 and 4, have resulted in over 100 reported illnesses and at least 1 death.[7,16,17] Researchers have found that adaptation to acidic conditions can be induced in some pathogenic bacteria,[72,89,92,94] and may even increase virulence.[113] Acid adaptation, or acid tolerance response, in bacteria typically involves an initial sub-lethal acid shock, which results in changes in gene expression analogous to the response observed with heat shock,[96] and the resulting physiological changes allow the treated cells to survive for extended periods in normally lethal acid conditions. A number of acid shock proteins and pH regulated genes in *Listeria, Salmonella,* and other bacteria, have been identified.[1,71,81,82,114,116] It has been shown that acid adaptation can enhance the survival of *Salmonella* and *Listeria* in fermented dairy products,[74,95] and may be part of a more general phenomenon called stress hardening, which results in increased resistance of bacteria to a variety of environmental stresses.[97,115,124] The significance of acid tolerance/adaptation of food pathogens in fermented and acidified vegetables is yet to be revealed.

51.5 RECOMMENDED METHODS

51.51 Collection and Storage of Brine Samples

In examination of pickle products, brine or pickle liquor covering the vegetable material is required. The size of container to be sampled may range from a small jar of pickles to a 1,000-bu tank of fermented brine stock. Brine samples from containers such as tanks and barrels should be taken for bacteriological analysis as follows:

For large brine tanks, insert a suitable length of 3/16-in stainless steel tubing, sealed at one end with lead or solder and perforated with several 1/16-in holes for a distance of 6 to 8 in from the sealed end, through an opening between the wooden boards composing the false head down into the brine toward the mid-depth of the vegetable material. Withdraw brine through a sanitized, attached piece of flexible tubing into a sample container. The length of the steel sampling tube is governed by the depth of the container to be sampled. Withdraw and discard approximately 100 mL of brine before taking the final sample, about 10 mL, into a sterile test tube. If microbial changes during the fermentation are to be followed, start sampling at the time the material is salted or brined and continue at regular intervals of 1 to 2 days during active fermentation. After each sampling, wash the whole assembly thoroughly.

For tightly headed barrels such as those used for genuine dills and salted vegetables for non-pickle use, take the sample through the top or side bung. For smaller containers, such as jars or cans of pickle products, shake thoroughly and take the sample from the center of the material by means of a sterile pipette. Wash the tops of the metal cans with alcohol, flame, and puncture. If the containers show evidence of gas pressure, carefully release gas by puncturing the sanitized top with a flamed ice pick. Containers under heavy gas pressure may be refrigerated overnight to reduce the gas pressure prior to sampling.

Brine samples from actively fermenting material should be examined as promptly as possible after collection to prevent changes in the microbial flora. The same is true for samples of packaged pickle products. If it is necessary to ship or store samples, this should be done so under refrigerated conditions; the elapsed time from collection to examination should not exceed 24 hours. When shipment by air is required, samples are collected in sterile, 16 × 105-mm tubes fitted with plastic screw caps having rubber liners. Pulp and oil liners, or plastic liners such as Teflon, may leak because of changes in air pressure.

Brine samples may be preserved for subsequent chemical determinations by the addition of one to two drops of toluene or Merthiolate (1% aqueous solution) per 10 mL of sample. Samples preserved with the above chemicals are unfit for human consumption and should be so marked.

Many techniques have been developed for sample preparation and storage for ascorbic acid analyses. For fermented and acidified vegetables, quickly mixing a sample with at least four volumes of 3% wt/vol metaphosphoric acid is a good sample preparation procedure. Little or no ascorbic acid is lost after 24 hours of storage in the refrigerator. Metaphosphoric acid stabilizes ascorbic acid much better than sulfuric acid or oxalic acid.

51.52 Comminution of Whole Vegetables and Particulates

To enumerate microflora of whole or particulate vegetables, approximately 300 grams of tissue are homogenized aseptically with an equal weight of sterile saline (0.85% NaCl). The samples are homogenized in a heavy duty commercial blender (e.g., Waring Blendor model 31BL46, Waring Products, New Hartford, CT). With a 1 liter blender jar homogenize for 1 minute at maximum RPM. To initiate blending, it may be necessary to cut whole or large pieces of vegetables in the jar using a sterile knife. Approximately 100 mL of the vegetable slurry is immediately removed after blending for further processing in a stomacher apparatus (e.g., Stomacher 400 homogenizer, Spiral Biotech, Inc., Bethesda, MD). The slurry is dispensed into a stomacher bag containing a filter (Stomacher 400 filter bags, Spiral Biotech) and processed using the maximum force setting for 1 minute. The filtrate removed from these bags should contain particles approximately 40 microns in diameter or less and can be used in a spiral plater or plated directly on agar petri plates. For sauerkraut in particular, as well as vegetables in general, the Robot Coupe blender model RSI 2YI (Robot Coupe USA, Inc., Ridgeland, MS) has proven to be very useful for preparing samples for chemical analyses. This model will accommodate up to 1000 grams of tissue and does not require additional liquid for homogenization. The sample is blended for 3–4 minutes, which results in a homogenous slurry. This slurry may then be filtered or centrifuged prior to chemical analyses.

51.53 Microscopic Examination

Microscopic examination of brine and vegetable samples for bacteria and yeasts is helpful at times, particularly when carried out in conjunction with plate count observations.

51.531 Bacteria

Make direct counts for bacteria according to the following procedures:

1. Place 0.01 mL amounts of liquid on slides using a calibrated pipette or loop and spread evenly over a 1 cm^2 area; fix with heat.
2. Stain according to the Kopeloff and Cohen modification of the gram stain.[91] Count according to the Wang[143] modification of the Breed[11] technique.
3. Report results as "numbers of different morphological types of gram-positive and gram-negative bacterial cells per mL of brine."
4. To determine the number of bacteria within brined vegetable tissue, blend the tissue to a homogeneous slurry and filter through coarse filter paper (Reeve Angel 202, Whatman Laboratory Products, Inc., Clifton, NJ). Bacteria within the filtrate are then enumerated with a Petroff-Hauser counting chamber at a magnification of about 500×.[25]

51.532 Yeasts

Use the microscopic technique for determining yeast populations in fermenting vegetable brines and various types of finished pickle products undergoing gaseous spoilage by the organisms, particularly where populations are in excess of 10^4 cells per mL of sample and where yeast colonies are not required for isolation and study. The use of a vital stain permits differentiation of yeast population into viable and non-viable cells and increases the usefulness of the direct counting technique.

The counting procedure is essentially the method of Mills[108] as modified by Bell and Etchells[2] for counting yeasts in high salt content brines and in high sugar content liquors:

1. Add 1 mL of brine or pickle liquor sample to 1 mL of 1:5,000 (0.02%) erythrosin stain.
2. Shake the sample stain mixture to obtain an even suspension.
3. Using a 3-mm diameter platinum loop, transfer enough of the mixture to the area under the cover glass of an improved Neubauer double-ruled hemacytometer to fill the chamber in one operation.
4. Allow cells to settle for approximately 5 minutes, and count the yeast cells using a microscope equipped with a 4-mm objective and 15× oculars.
5. Record cells stained pink as "dead yeast cells" and unstained cells as "live yeast cells."

The number of yeast cells per mL of brine or pickle liquor may be calculated thus:

$$\frac{\text{Number of yeast cells counted} \times \text{dilutions} \times 250{,}000}{\text{number of large squares counted}} = \text{Numbers per mL}$$

If only one side of the hemacytometer counting chamber is used (25 large squares), the lowest yeast count obtainable is 20,000 per mL, while, if both sides are counted (50 large squares), a population as low as 10,000 per mL can be counted.

Report yeast count as "total yeast cells," "live yeast cells," and "dead yeast cells" per mL of sample.

51.54 Enumeration of Viable Microorganisms

In addition to the procedures described in this section, we have found certain alternative methods such as use of Petri Film (3M Microbiology Products, St. Paul, MN) to be particularly useful for field studies (see Chapter 10 for an array of rapid methods). Also, the use of a spiral plater (Spiral Biotech, Inc., Bethesda, MD) has become highly useful in our laboratory for microbial enumeration (see Chapter 12 for automated methods).

51.541 Aerobic Plate Count

Use plate count agar or nutrient agar and incubate at 30°C for 18 to 24 hours. For longer incubation, overlay the solidified, plated samples with about 8 to 10 mL of the same medium to prevent or minimize spreaders.

51.542 Lactic Acid Bacteria

Selective enumeration of lactic acid bacteria may be carried out using MRS agar containing 0.02% sodium azide and incubating for 1 to 4 days at 30°C.[23] Lactic acid bacteria of the genus *Lactobacillus* or *Pediococcus* may be enumerated with *Lactobacillus* selective medium (LBS), supplemented with 1% fructose[132] and 200 ppm cycloheximide to inhibit yeasts. Bromcresol green (or brilliant green, as in Chapter 63), 0.0075%, may be added to aid in colony counting, but may restrict the growth of some lactic acid bacteria. To differentially enumerate all species of lactic acid bacteria associated with vegetable fermentations, plate fermenting samples on a non-selective medium such as tryptone-glucose-yeast extract agar.[120] After incubation at 30°C for 24–48 hours, isolate colonies for later identification on the basis of cell morphology, acid and gas production, and mucoid growth;[119] other reactions may be used. For incubation of agar plates longer than 24 hours, place the petri plates in a sealed plastic bag (it is convenient to use the sterile plastic sleeve in which the petri plates come packaged), or in a humidified incubator to prevent desiccation of the medium.

Several kinds of differential and selective media are available for the characterization of lactic acid bacteria. HHD medium (see Chapter 63 for composition) is used for the differential enumeration of homofermentative and heterofermentative lactic acid bacteria.[98] This medium incorporates fructose, which is reduced to mannitol by heterofermentative but not homofermentative lactic acid bacteria. In agar medium, homofermentative colonies of lactic acid bacteria are blue to green, while heterofermentative colonies are white. MD medium (see Chapter 63 for composition) may be used to differentiate malate-decarboxylating (MDC^+) and malate-non-decarboxylating (MDC^-) lactic acid bacteria. The decarboxylation of malic acid is undesirable in cucumber fermentations because of the CO_2 produced. The differential reaction is based upon pH changes in the medium caused by malate decarboxylation. A pH decline (MDC^-) is shown by a color change from blue to green, whereas no color change indicates an MDC^+ reaction. HHD and MD media are non-selective for lactic acid bacteria. MS agar medium may be used to select for MDC^+ bacteria.[12] This medium contains malate and has a low initial pH 4.0. MDC^+ bacteria are capable of raising the pH via the malolactic fermentation and initiating growth, while MDC^- bacteria cannot initiate growth.

51.543 Total Enterobacteriaceae and Coliform Bacteria

Add 1% glucose to violet red bile agar, which is referred to as MacConkey glucose agar[110] or VRBG agar. Incubate for 18–24 hours at 30°C. For coliform bacteria, use violet red bile agar without added glucose, and count all purplish red colonies surrounded by a reddish zone of precipitated bile, 0.5 mm in diameter or larger. In addition to these methods, a number of rapid methods for enumeration of *Enterobacteriaceae*, such as *E. coli* O157:H7 or *Salmonella* species, are available (see Chapters 8 and 10).

51.544 Yeasts and Molds

Yeast and mold populations may be enumerated on YM agar (Difco Laboratories, Detroit, MI or see Chapter 20) supplemented with 20 mL/L of an antibiotic solution containing 0.5% chlortetracycline and 0.5% chloramphenicol. We have found it desirable to use both antibiotics to preclude growth of bacteria. Alternatively, acidify sterile tempered molten dextrose agar (at 45°C) with 10% tartaric acid, usually 5% by volume, to achieve a final pH 3.5. Incubate YM or acidified dextrose agar for 24 to 48 hours at 30°C. Small colonies of lactic acid bacteria may appear on the acidified medium, but are suppressed in the antibiotic medium.

Mold colonies are filamentous and, thus, are distinguished readily from yeasts on acidified dextrose agar. Differentiation of subsurface yeasts and film yeasts presents more difficulty. Surface colonies of the common film-forming yeasts associated with pickle products and vegetable brines, i.e., species of *Debaryomyces, Endomycopsis, Candida,* and *Pichia,*[33,111] are generally dull and very rough, as contrasted to the usual round, raised, white, glistening colonies of the fermentative, subsurface yeasts, i.e., species of *Torulopsis, Brettanomyces, Hansenula, Saccharomyces,* and *Torulaspora.*[36,39,40] However, even when distinguishing colony characteristics of the two yeast groups exist, they are not considered sufficiently clear-cut for separation. Because of this, the procedure outlined under Section 51.545 should be used. Film yeasts rapidly form a heavy wrinkled surface film at one or both salt concentrations. Certain species, such as *Saccharomyces halomembranis,* form heavier films at 10% salt than at 5%.[31,36,38,111]

51.545 Film Yeasts

For an estimate, pick representative filamentous colonies from the yeast plates into tubes of dextrose broth containing 5% and 10% salt. Incubate 3 to 5 days at 32°C and look for heavy surface film. Two salt concentrations are suggested because some species develop heavier films at the lower salt strength (5%), whereas, with other species, the reverse is true.

51.546 Obligate Halophiles

Use tubes of liver broth plus salt (Chapter 63). Prepare decimal dilutions, seal with sterilized, melted petroleum jelly, and incubate 7 days at 32°C. Record positive tubes daily by noting the raising of the petroleum seal caused by gas production and the absence of any distinctive odor.

This medium has proved satisfactory for detecting obligate halophiles sometimes found in brined and dry-salted vegetables. The salt content of the medium should approximate that of the sample. No growth of coliforms or yeasts has been encountered in this medium. This is probably due to the inability of either group to initiate satisfactory early growth in laboratory media even at moderately high salt concentrations in competition with the very fast-growing obligate halophiles.

51.547 Butyric Acid-forming Bacteria

Neutralize the brine sample with an excess of sterile calcium carbonate. Heat a 50- to 100-mL sample in a water bath for 20 minutes at 80°C to kill vegetative cells. Prepare decimal dilutions and inoculate previously heated and cooled tubes of liver broth medium. Seal with melted petroleum jelly and incubate 7 days at 32°C. Examine tubes daily for production of gas and a strong butyric acid odor.

51.55 Chemical Analyses

51.551 Titratable Acidity and pH

Determine titratable acidity of a 10-mL sample of the fermentation brine or finished pickle liquor (liquid of the final product) by diluting the sample with 30 to 50 mL of distilled water; titrate with 0.1N NaOH using phenolphthalein as the indicator. Alternatively, samples may be titrated to pH 8.2 with a pH meter. Report values for fermented, brined samples as g of lactic acid per 100 mL of sample, and for finished pickle liquor samples as g of acetic acid per 100 mL of sample.

For a 10-mL sample, use the following calculations:

a. mL of 0.1N alkali used $\times$ 0.090 = g of lactic acid per 100 mL.
b. mL of 0.1N alkali used $\times$ 0.060 = g of acetic acid per 100 mL.

When only a small amount of the original sample is available, use a 2-mL amount for titration purposes. Such small samples are not recommended. For the 2-mL sample, multiply the mL of 0.1N alkali by 5, then by the above number for lactic or acetic acid.

Carry out pH determinations of the samples with a pH meter, checking the instrument frequently with a standard buffer in the pH range of the sample under test.

51.552 Chloride and Calcium

It is often helpful to know the approximate salt content in performing microbiological examination of brines. Use a salometer, and test about 200 mL of brine. A chemical test for salt is required for small amounts of sample or when a higher degree of accuracy is desired than that obtainable with the salometer.

The following method is recommended. Transfer 1 mL of sample to a flask, and dilute with 15 to 20 mL of distilled water. Titrate with 0.171N silver nitrate solution, 29.063 g per liter, using 3 to 5 drops of 0.5% dichlorofluorescein as the indicator. Agitate to keep the precipitate broken up until a light salmon pink color develops. Report as "g of sodium chloride per 100 mL of the sample." When 1 mL of sample is titrated, each mL of silver nitrate solution is equal to 1 g of sodium chloride per 100 mL. A chloride test strip is commercially available from Environmental Test Systems (http://www.etsstrips.com/water.html).

A rapid colorimetric procedure based upon calcium binding by methylthymol blue can be used to measure the calcium content of brines or blended tissue.[79] Samples are mixed with an equal volume of 4% wt/vol trichloroacetic acid solution. Acidified solution, containing 50 to 600 µg calcium, is added to a test tube, and reagent solution is added. An immediate absorbance change at 612 nm occurs. The relationship between calcium concentration and absorbance is hyperbolic. A standard curve can be constructed by hand, or the data may be fitted to a hyperbola using nonlinear regression. There are also test strips available for calcium and some other inorganic components of brines such as sulfite and nitrate. One supplier of test strips is EM Science (http://www.emscience.com).

51.553 Fermentation Substrates and Products

To determine whether the intended fermentation occurred or to determine the nature of an off-fermentation, it is important to measure both the substrates and products of a fermentation process. High-performance liquid chromatography (HPLC) procedures to measure all major substrates and products of both heterolactic acid and homolactic acid fermentations have been developed using refractive index detection.[105] Reversed phase C_{18} columns gradually lose resolution of organic acids and need to be replaced periodically. More recently, a procedure to determine all these compounds in a single HPLC injection has been developed. For this method, a conductivity detector is used to measure organic acids and an electrochemical detector connected in series is used to detect sugars and alcohols.[99] Resin columns in the H^+ form are extremely stable and reproducible in their separations of compounds in fermented vegetable samples. We have used a single column (Bio-Rad HPX-87H) for over 3 years without loss of performance. Sucrose degrades during chromatography on resin columns in the H^+ form and, if mannitol is present as occurs in heterolactic fermentations such as sauerkraut, it will not separate adequately from fructose to analyze either compound. All major sugars and sugar alcohols involved in vegetable fermentation can be separated using a Dionex Carbopak PA1 column with dilute NaOH as the eluant.[117] Due to low analyte capacity on columns of this type, electrochemical detection of sugars is preferred over a refractive index detector. For the analysis of organic acids, reversed-phase columns gradually lose resolution and need to be replaced periodically. Also, not all C_{18} columns adequately separate malic acid, lactic acid, acetic acid, and ethanol. Resin columns in the H^+ form are extremely stable and reproducible for organic acid analysis using 0.02N sulfuric acid as the eluant. One problem with this procedure for cucumber pickle analysis is that fructose and malic acid coelute. The coelution problem has been solved in two ways using the same column as cited above. Lazaro et al.[93] developed equations to quantify fructose and malic acid differentially based on peak heights obtained from ultraviolet and refractive index detectors connected in series. Frayne[73] actually resolved malic acid and fructose by connecting two of the HPLC columns in series.

Though HPLC is today the method of choice for analysis of fermentations, it has the disadvantage that an expensive instrument is required. An alternative approach for analysis of many fermentation substrates and products is enzymatic analysis using commercially available kits. The analysis can be done manually with an inexpensive visible colorimeter or spectrophotometer. The main disadvantage is that only a single compound can be analyzed at a time. Compounds for which kits are available include glucose, fructose, malic acid, L-lactic acid, acetic acid, ethanol, and CO_2. Commercial sources for such kits include Boehringer Mannheim Biochemicals (www://biochem.boehringer.com) and Sigma Chemical Company (http://www.sigma-aldrich.com). Enzymatic analysis is the only routine way to measure the L-isomer of lactic acid specifically.

It is important for the stability of fermented vegetables that all fermentable sugars be metabolized by the end of the fermentation process. For products like cucumbers that have little or no sucrose, colorimetric measurement of reducing sugars is a simple, rapid quality control procedure to assess the completion of sugar utilization. The dinitrosalicylic acid (DNS) procedure[107,134] is recommended. The reagent is stable at room temperature for many months. The assay can be reliably performed with an inexpensive colorimeter. A fermentation can be considered complete if the brine contains less than 0.05% reducing sugar and acid shows no increase for several days.

51.554 Softening Enzyme Activity

Softening enzymes in brines of fermenting cucumbers and other vegetables may be determined by the highly sensitive viscometric method of Bell et al.[5] The procedure, which has been widely used in the pickle industry for many years, is based on viscosity loss of a buffered polypectate solution. Brine samples, 25 mL, are dialyzed in running water for 3 hours and distilled water for 1 hour. One mL of the dialyzed sample is added to 5 mL of 1.2% sodium polypectate, which is dissolved in 0.018 M, pH 5.0 citrate buffer in an Ostwald-Fenske no. 300 viscometer. A drop of toluene is added to the sample to prevent microbial growth during incubation. The flow time of the pectate solution is measured after sample addition and at 20 hours. The viscosity loss is calculated according to the following equation:

$$\text{Percent loss in viscosity} = \frac{A - B}{A - W} \times 100 \quad (1)$$

where A is the initial flow time in seconds, B the flow time at 20 hours, and W is the flow time for water. Bell et al.[5] provide a table that relates loss in viscosity to the units of pectate depolymerizing activity. A less than 9% loss of viscosity in 20 hours is considered to represent weak to negative activity in brine samples. Buescher and Burgin[13] described a diffusion plate assay that is in common use in the pickle industry to determine if there is polygalacturonase activity in brines before they are recycled.

Refer to Chapter 18 for isolation of pectinolytic organisms and characterization of pectinolytic enzymes.

51.555 Dissolved Carbon Dioxide

The advent of purging to remove CO_2 from fermenting cucumber brines and, thereby preventing bloater formation, has created a need to determine the concentration of dissolved CO_2 in the brine. For the highly accurate determinations that may be required for research purposes, dissolved CO_2 is determined by the micro distillation procedure.[59] A 10-mL brine sample is injected by syringe into a capped jar containing a phosphoric acid solution. A small vial containing 5 mL of 0.200N NaOH placed inside the jar traps the CO_2 as it distills from the acidified solution. After 24 hours at 37°C, the vial is removed, 5 mL of 0.2 M $BaC1_2$ is added, and the remaining base is titrated to the phenolphthalein end-point with 0.100N HCL. Values are expressed as mg CO_2 per 100 mL brine. When exact amounts and concentrations of acid and base specified in the method are adhered to, the following equation may be used:

$$\text{mg}CO_2/100 \text{ mL brine} = (1.000 - \text{mL } 0.1\text{N HCl}) \times 220$$

For quick estimates that may be required for quality control tank monitoring, dissolved CO_2 is determined with a micro CO_2 apparatus. Adaptation of this instrument for the determination of CO_2 in fermenting cucumber brines has been described.[58] This is a gasometric method based on the classical Van Slyke procedure. A 1-mL brine sample is placed in the instrument vial, a volumetric syringe is clamped into place, an acid solution is added, the apparatus and sample vial are shaken, and the gas volume displacement is read on the calibrated syringe scale. Carbon dioxide in the brine sample is calculated from scale readings of the brine compared to a CO_2 solution of known concentration and is expressed as mg CO_2 per 100 mL brine. It is suggested that brine samples be taken from brine tanks through a siphon tube (see Section 50.51) and 8.5 mL injected by syringe through a needle into a Vacutainer tube (10 mL draw, Becton-Dickinson, containing 0.5 mL of ca. 3N NaOH) to minimize CO_2 1oss. The samples are then equilibrated to the same temperature as the known solution before analysis.

In both methods, the total CO_2 content of the solution is determined and is expressed as mg CO_2 per 100 mL brine, or as percent saturation.[62]

51.556 Ascorbic Acid

Brine samples should be stored for ascorbic acid analysis as described in section 51.51. The HPLC procedure of Vanderslice and Higgs,[138] which involves separation of ascorbic acid on a reversed phase polymer column with pH 2.14 phosphate buffer as the eluant solution followed by fluorometric detection, has been shown to give accurate ascorbic acid and dehydroascorbic acid analysis in a wide variety of food products. It has been used in this laboratory to measure ascorbic acid in fermented cucumber samples using an electrochemical detector instead of a fluorometric detector (Zhou et al., unpublished).

51.6 INTERPRETATION OF DATA

51.61 Fermented Vegetables

Proper record-keeping of salting procedures and chemical and microbiological data can greatly aid the commercial briner in assessing causes for success or failure in preserving the quality of brined vegetables. Records of chemical determinations of salt, titratable acidity, pH, fermentable sugars, dissolved CO_2, and softening enzyme activity are very useful in such assessments, depending on the particular commodity. A recordkeeping system for brined cucumbers has been published.[34]

In fermented vegetables, it is important that the lactic acid fermentation become established early to preclude growth by spoilage bacteria. Acidity and pH data provide this information. Salt concentrations above 8% for cucumbers and olives or above 2.5% for cabbage may prevent or retard a desirable lactic fermentation. Unusually low salt concentrations may result in softening of the brined vegetables.

If the dissolved CO_2 concentration in the brine of fermenting cucumbers is allowed to exceed about 50% saturation (equals 54 mg per 100 mL at 21°C and 6.6% NaCl) at any time during brine storage, bloater damage may result. Maintaining the brine CO_2 concentration below 50% saturation will greatly aid in reducing bloater damage.[62] Sporadic bloater damage may occur even in effectively purged brine-stock cucumbers. Such damage may be due to growth of bacteria within the brined fruit.[22] Since brines must be purged as long as fermentation occurs, it is important to monitor the level of fermentable sugars in the brine. When fermentable sugars are not detected and acid development has ceased, the fermentation is considered to be complete, and purging can be safely discontinued.

Microbial softening enzyme activity of brines may indicate the cause of soft brine-stock pickles, especially if the cucumbers are held at relatively low brine strengths (5% to 8% NaCl). Higher salt concentrations will prevent softening by these enzymes,[3] but high salt levels present disposal problems, in addition to affecting the lactic fermentation adversely. Studies have indicated that calcium chloride, ca. 0.2% to 0.4%, and other salts of calcium may inhibit the action of softening enzymes.[14] Calcium chloride is now being added to commercial cucumber brines. The extent of protection against softening offered by calcium has not been fully assessed.

The absence of softening enzyme activity in older brine-stock pickles does not necessarily mean that such activity did not cause the softening. Softening enzymes that accompany the cucumbers and attached flowers into the brine tank may exert their influence early in brine storage and then be dissipated or inactivated so as not to be detectable later.

Softening in the seed area of large cucumbers, commonly termed "soft centers," is thought to be due to natural polygalacturonase of overly mature cucumbers,[141] not to microorganisms.

The advent of reliable HPLC procedures to measure changes in substrates and products of fermentations has made it practical to assess the balance between substrate utilization and product formation in complex food fermentations. Carbon recovery of <100% indicates that some fermentation products have been missed in the analysis, while recovery of >100% suggests that unknown substrates have been fermented. Examples of fermentation balances done on complex fermentations have been published.[18,66,64,67,101] The determination of fermentation end products was extremely useful in recent efforts to determine microorganisms responsible for spoilage of fermented cucumbers.[68]

51.62 *Finished Pickle Products from Salt-Stock Vegetables*

These products normally contain a few thousand microorganisms per mL. These counts may be composed chiefly of spores of aerobic bacteria that remain inactive in the acid medium and tend to decrease during storage. Fermentative yeasts and lactic acid bacteria may cause vigorous gas production, which causes the pickle liquor to become highly charged with gas and to possess a tang when tasted. Viable microorganisms, normally latent in properly fermented and preserved products, may cause gaseous spoilage in improperly finished products. Gaseous spoilage and cloudy cover brine may be the result in hamburger dill chips, genuine dill pickles, Spanish-style green olives, and similar products, if residual sugar remains.

51.63 *Pasteurized Pickle Products*

Properly acidified, packaged, and pasteurized pickle products are not subject to microbial spoilage. When spoilage occurs, it is usually due to underpasteurization. Some commercial packers minimize heat processing in order to maintain greater product quality. Minimal processing is done at the risk of spoilage. Spoilage results in recall of the product at the packer's expense. No public health problem exists in pasteurized pickle products that have been properly acidified. After spoilage occurs, however, as evidenced by gas pressure and brine turbidity, there is no way to ensure that the product was properly acidified initially. Lactic acid bacteria are normally found in such products. The spoiled product usually contains acid, but it is not known at that point if the original product, particularly if it was fresh produce, was sufficiently acidified to prevent growth of *Clostridium* before growth of the lactic acid bacteria.

Improper acidification can also be a source of spoilage with potential public health significance, as discussed in Section 51.4. Improper closure can result in growth of aerobic microorganisms on the surface of the brine and a reduction in acidity.

51.7 REFERENCES

1. Bearson, S. M. D., W. H. Benjamin, Jr., W. E. Swords, and J. W. Foster. 1996. Acid shock induction of RpoS is mediated by the mouse virulence gene *mivA* of *Salmonella typhimurium*. J. Bacteriol. 178:2572-2579.
2. Bell, T. A., and J. L. Etchells. 1952. Sugar and acid tolerance of spoilage yeasts from sweet-cucumber pickles. Food Technol. 6:468.
3. Bell, T. A., and J. L. Etchells. 1961. Influence of salt (NaCl) on pectinolytic softening of cucumbers. J. Food Sci. 26:84.
4. Bell, T. A., J. L. Etchells, and I. D. Jones. 1950. Softening of commercial cucumber salt-stock in relation to polygalacturonase activity. Food Technol. 4:157.
5. Bell, T. A., J. L. Etchells, and I. D. Jones. 1955. A method for testing cucumber salt-stock brine for softening activity. USDA-ARS Publ. 72-5, 18 p.
6. Bell, T. A., J. L. Etchells, and R. N. Costilow. 1958. Softening enzyme activity of cucumber flowers from northern production areas. Food Res. 23:198.
7. Besser, R. E., S. M. Lett, J. T. Weber, M. P. Doyle, T. J. Barrett, J. G. Wells, and P. M. Griffin. 1993. An outbreak of diarrhea and hemolytic uremic syndrome from *Escherichia coli* O157:H7 in fresh-pressed apple cider. JAMA 269:2217-2220
8. Beuchat, L. R., R. E. Brackett, D. Y. Hao, and D. E. Conner. 1986. Growth and thermal inactivation of *Listeria monocytogenes* in cabbage and cabbage juice. Can. J. Microbiol. 32:791.
9. Binsted, R., J. D. Devey, and J. C. Dakin,. 1971. Pickle and sauce making, 3rd ed. Food Trade Press Ltd., London.
10. Borbolla y Alcala, J. M. R., and L. Rejano Navarro. 1981. On the preparation of Sevillian style olives. The fermentation. II. Grasas Aceites; 32:103.
11. Breed, R. S. 1911. The determination of the number of bacteria in milk by direct microscopic examination. Zentralt. Bakteriol. Parasitenkd. 11(30):337.
12. Breidt, F., and H. P. Fleming. 1992. Competitive growth of genetically marked malolactic-deficient *Lactobacillus plantarum* in cucumber fermentations. Appl. Environ. Microbiol. 58:3845-3849.
13. Buescher, R. W., and C. Burgin. 1992. Difffusion plate assay for measurement of polygalacturonase activities in pickle brines. J. Food Biochem. 16:59-68.
14. Buescher, R. W., J. M. Hudson, and J. R. Adams. 1979. Inhibition of polygalacturonase softening of cucumber pickles by calcium chloride. J. Food Sci. 44:1786.
15. Buescher, R. W., J. M. Hudson, and J. R. Adams. 1981. Utilization of calcium to reduce pectinolytic softening of cucumber pickles in low salt conditions. Leben. Wiss. Technol. 14:65.
16. Centers for Disease Control. 1996. Outbreak of *Escherichia coli* O157:H7 infections associated with drinking unpasteurized commercial apple juice — British Columbia, California, Colorado, and Washington. Morb. Mortal Wkly. Rep. 45:975.
17. Centers for Disease Control. 1997. Outbreaks of *Escherichia coli* O157:H7 infection and crytosporidiosis associated with drinking unpasteurized apple cider—Connecticut and New York. Morb. Mortal Wkly. Rep. 46:4-8.
18. Chen, K. H., R. F. McFeeters, and H. P. Fleming. 1983. Fermentation characteristics of heterolactic acid bacteria in green bean juice. J. Food Sci. 48:962.
19. Conner, D. E., R. E. Brackett, and L. R. Beuchat. 1986. Effect of temperature, sodium chloride and pH on growth of *Listeria monocytogenes* in cabbage juice. Appl. Environ. Microbiol. 52:59.

20. Costilow, R. N., C. L. Bedford, D. Mingus, and D. Black. 1977. Purging of natural salt-stock pickle fermentations to reduce bloater damage. J. Food Sci. 42:234.
21. Costilow, R. N., K. Gates, and M. L. Lacy. 1980. Molds in brined cucumbers: Cause of softening during air purging of fermentations. Appl. Environ. Microbiol. 40:417.
22. Daeschel, M. A., and H. P. Fleming. 1981. Entrance and growth of lactic acid bacteria in gas-exchanged, brined cucumbers. Appl. Environ. Microbiol. 42:1111.
23. Daeschel, M. A., J. O. Mundt, and I. E. McCarty. 1981. Microbial changes in sweet sorghum *(Sorghum bicolor)* juices. Appl. Environ. Microbiol. 42:381.
24. Daeschel, M. A., R. F. McFeeters, H. P. Fleming, T. R. Klaenhammer, and R. B. Sanozky. 1984. Mutation and selection of *Lactobacillus plantarum* strains that do not produce carbon dioxide from malate. Appl. Environ. Microbiol. 47:419.
25. Daeschel, M. A., H. P. Fleming, and E. A. Potts. 1985. Compartmentalization of lactic acid bacteria and yeasts in the fermentation of brined cucumbers. Food Microbiol. 2:77.
26. Daeschel, M. A., R. F. McFeeters, H. P. Fleming, T. R. Klaenhammer, and R. B. Sanozky. 1987. Lactic acid bacteria which do not decarboxylate malic acid and fermentation therewith. U.S. Patent 4,666,849.
27. Dakin, J. C., and P. M. Day. 1958. Yeasts causing spoilage in acetic acid preserves. J. Appl. Bact. 21:94.
28. Dakin, J. C., and J. Y. Radwell. 1971. Lactobacilli causing spoilage of acetic acid preserves. J. Appl. Bact. 34:541.
29. Environmental Protection Agency. 1987. Water quality criteria; availability of document. Fed. Reg. 52: 37,655-37,656. EPA, Washington, D. C.
30. Etchells, J. L. 1938. Rate of heat penetration during the pasteurization of cucumber pickles. Fruit Prod. J. 18:68.
31. Etchells, J. L. 1941. Incidence of yeasts in cucumber fermentations. Food Res. 6:95.
32. Etchells, J. L., and T. A. Bell. 1950a. Classification of yeasts from the fermentation of commercially brined cucumbers. Farlowia 4:87.
33. Etchells, J. L., and T. A. Bell. 1950b. Film yeasts on commercial cucumber brines. Food Technol. 4:77.
34. Etchells, J. L., and L. H. Hontz. 1973. Information on the nature and use of an improved system for recording quality control data during the brining of cucumbers (Quality control report for brining cucumbers). Published and distributed by Pickle Packers International, Inc., St. Charles, Ill.
35. Etchells, J. L., and I. D. Jones. 1942. Pasteurization of pickle products. Fruit Prod. J. 21:330.
36. Etchells, J. L., and J. D. Jones. 1943. Bacteriological changes in cucumber fermentation. Food Indust. 15:54.
37. Etchells, J. L., and I. D. Jones. 1944. Procedure for pasteurizing pickle products. Glass Packer 23:519.
38. Etchells, J. L., F. W. Fabian, and I. D. Jones. 1945. The *Aerobacter* fermentation of cucumbers during salting. Mich. Agric. Exp. Sta. Tech. Bull. No. 200.
39. Etchells, J. L., R. N. Costilow, and T. A. Bell. 1952. Identification of yeasts from commercial fermentations in northern brining areas. Farlowia 4:249.
40. Etchells, J. L., T. A. Bell, and I. D. Jones. 1953. Morphology and pigmentation of certain yeasts from brines and the cucumber plant. Farlowia 4:265.
41. Etchells, J. L., T. A. Bell, and J. D. Jones. 1955. Cucumber blossoms in salt stock mean soft pickles. Res. Farm 13:14-15.
42. Etchells, J. L., T. A. Bell, R. J. Monroe, P. M. Masley, and A. L. Demain. 1958. Populations and softening enzyme activity of filamentous fungi on flowers, ovaries and fruit of pickling cucumbers. Appl. Microbiol. 6:427.
43. Etchells, J. L., R. N. Costilow, T. E. Anderson, and T. A. Bell. 1964. Pure culture fermentation of brined cucumbers. Appl. Microbiol. 12:523.
44. Etchells, J. L., A. F. Borg, I. D. Kittel, T. A. Bell, and H. P. Fleming. 1966. Pure culture fermentation of green olives. Appl. Microbiol. 14:1027.
45. Etchells, J. L., A. F. Borg, and T. A. Bell. 1968. Bloater formation by gas-forming lactic acid bacteria in cucumber fermentations. Appl. Microbiol. 16:1029.
46. Etchells, J. L., T. A. Bell, R. N. Costilow, C. E. Hood, and T. E. Anderson. 1973a. Influence of temperature and humidity on microbial, enzymatic and physical changes of stored, pickled cucumbers. Appl. Microbiol. 26:943.
47. Etchells, J. L., T. A. Bell, H. P. Fleming, R. E. Kelling, and R. L. Thompson. 1973b. Suggested procedure for the controlled fermentation of commercially brined pickling cucumbers—The use of starter cultures and reduction of carbon dioxide accumulation. Pickle Pak Sci. 3:4.
48. Etchells, J. L., H. P. Fleming, and T. A. Bell. 1975. Factors influencing the growth of lactic acid bacteria during brine fermentation of cucumbers, p. 281. *In* Lactic acid bacteria in beverages and food. Academic Press, New York.
49. Etchells, J. L., T. A. Bell, and W. R. Moore Jr. 1976. Refrigerated dill pickles—Questions and answers. Pickle Pak Sci. 5:1.
50. Fabian, F. W., and R. G. Switzer. 1941. Classification of pickles. Fruit Prod. J. 20:136.
51. Fenlon, D. R. 1985. Wild birds and silage as reservoirs of listeria in the agricultural environment. J. Appl. Bacteriol. 59:537.
52. Fernandez-Diez, M. J. 1971. The olive, p. 255. *In* A. C. Hulm (ed.), The biochemistry of fruits and their products, vol. 2. Academic Press, New York.
53. Fleming, H. P. 1982. Fermented vegetables. *In* A. H. Rose (ed.), Economic microbiology. Fermented foods, vol. 7. Academic Press, New York.
54. Fleming, H. P., and R. F. McFeeters. 1981. Use of microbial cultures: Vegetable products. Food Technol. 35:84.
55. Fleming, H. P., W. M. Walter Jr., and J. L. Etchells. 1969. Isolation of a bacterial inhibitor from green olives. Appl. Microbiol. 18:856.
56. Fleming, H. P., R. L. Thompson, J. L. Etchells, R. E. Kelling, and T. A. Bell. 1973a . Bloater formation in brined cucumbers fermented by *Lactobacillus plantarum.* J. Food Sci. 38:499.
57. Fleming, H. P., W. M. Walter Jr., and J. L. Etchells. 1973b. Antimicrobial properties of oleuropein and products of its hydrolysis from green olives. Appl. Microbiol. 26:777.
58. Fleming, H. P., R. L. Thompson, and T. A. Bell. 1974a. Quick method for estimating CO_2 in cucumber brines. Advisory statement published and distributed by Pickle Packers International, Inc., St. Charles, Ill.
59. Fleming, H. P., R. L. Thompson, and J. L. Etchells. 1974b. Determination of carbon dioxide in cucumber brines. J. Assoc. Off. Anal. Chem. 57:130.
60. Fleming, H. P., J. L. Etchells, R. L. Thompson, and T. A. Bell. 1975. Purging of CO_2 from cucumber brines to reduce bloater damage. J. Food Sci. 40:1304.
61. Fleming, H. P., R. L. Thompson, T. A. Bell, and L. H. Hontz. 1978a. Controlled fermentation of sliced cucumbers. J. Food Sci. 43:888.
62. Fleming, H. P., R. L. Thompson, and R. J. Monroe. 1978b. Susceptibility of pickling cucumbers to bloater damage by carbonation. J. Food Sci. 43:892.
63. Fleming, H. P., E. G. Humphries, and J. A. Macon. 1983a. Progress on development of an anaerobic tank for brining of cucumbers. Pickle Pak Sci. VII:3-15.
64. Fleming, H. P., R. F. McFeeters, R. L. Thompson, and D. C. Sanders. 1983b. Storage stability of vegetables fermented with pH control. J. Food Sci. 48:975.
65. Fleming, H. P., R. F. McFeeters, and R. L. Thompson. 1987. Effects of sodium chloride concentration on firmness retention of cucumbers fermented and stored with calcium chloride. J. Food Sci. 52:653.
66. Fleming, H. P., R. F. McFeeters, M. A. Daeschel, E. G. Humphries, and R. L. Thompson. 1988a. Fermentation of cucumbers in anaerobic tanks. J. Food Sci. 53:127.

67. Fleming, H. P., R. F. McFeeters, and E. G. Humphries. 1988b. A fermentor for study of sauerkraut fermentation. Biotech. Bioeng. 31:189.
68. Fleming, H. P., M. A. Daeschel, R. F. McFeeters, and M. D. Pierson. 1989. Butyric acid spoilage of fermented cucumbers. J. Food Sci. 54:636.
69. Fleming, H. P., K. H. Kyung, and F. Breidt. 1995. Vegetable fermentations, p. 629-661. *In* H. J. Rehm and G. Reed (ed.), Biotechnology, vol. 9. Enzymes, biomass, food and feed, 2nd ed. VCH Publishers, Inc., New York.
70. Fleming, H. P., R. L. Thompson, and R. F. McFeeters. 1996. Assuring microbial and textural stability of fermented cucumbers by pH adjustment and sodium benzoate addition. J. Food Sci. 61:832-836.
71. Foster, J. W. 1991. *Salmonella* acid shock proteins are required for the adaptive acid tolerance response. J. Bacteriol. 173:6896-6902.
72. Foster, J. W., and H. K. Hall. 1991. Inducible pH homeostasis and the acid tolerance response of *Salmonella typhimurium*. J. Bacteriol. 173:5129-5135.
73. Frayne, R. F. 1986. Direct analysis of the major organic components in grape must and wine using high performance liquid chromatography. Am. J. Enol. Vitic. 37:281.
74. Gahan, C. G. M., B. O'Driscoll, and C. Hill. 1996. Acid adaptation of *Listeria monocytogenes* can enhance survival in acidic foods and during milk fermentation. Appl. Environ. Microbiol. 62:3128-3132.
75. Gates, K., and R. N. Costilow. 1981. Factors influencing softening of salt-stock pickles in air-purged fermentations. J. Food Sci. 46:274.
76. Geisman, J. R., and R. E. Henne. 1973. Recycling food brine eliminates pollution. Food Eng. 45:119.
77. George, S. M., B. M. Lund, and T. F. Brocklehurst. 1988. The effect of pH and temperature on initiation of growth of *Listeria monocytogenes*. Lett. Appl. Microbiol. 6:153.
78. Gililland, J. R., and R. H. Vaughn. 1943. Characteristics of butyric acid bacteria from olives. J. Bacteriol. 46:315.
79. Gindler, E. M., and J. D. King. 1972. Rapid colorimetric determination of calcium in biologic fluids with methylthymol blue. Am. J. Clin. Pathol. 58:376.
80. Gonzalez Cancho, F., L. Rejano Navarro, and J. M. R. Borbolla y Alcala. 1980. Formation of propionic acid during the conservation of table green olives. III. Responsible microorganisms. Grasas Aceites 31:245.
81. Hall, H. K., and J. W. Foster. 1996. The role of fur in the acid tolerance response of *Salmonella typhimurium* is physiologically and genetically separable from its role in iron acquisition. J. Bacteriol. 178:5683-5691.
82. Hartke, A., S. Broche, J-C Giard, A. Benachour, P. Boutibonnes, and Y. Auffray. 1996. The lactic acid stress response of *Lactococcus lactis* subsp. *lactis*. Curr. Microbiol. 33:194-199.
83. Hill, C., B. O'Driscoll, and I. Booth. 1995. Acid adaptation and food poisoning microorganisms. Intern. J. Food Microbiol. 28:245-254.
84. Hudson, J. M., and R. W. Buescher. 1985. Pectic substances and firmness of cucumber pickles as influenced by CaCl, NaCl and brine storage. J. Food Biochem. 9:211.
85. Humphries, E. G., and H. P. Fleming. 1989. Anaerobic tanks for cucumber fermentation and storage. J. Agric. Eng. Res. 44:133-140.
86. Johnson, J. L., M. D. Doyle, R. G. Cassens, and J. L. Schoeni. 1988. Fate of *Listeria monocytogenes* in tissues of experimentally infected cattle and hard salami. Appl. Environ. Microbio.; 54:497.
87. Juven, B., and Y. Henis. 1970. Studies on the antimicrobial activity of olive phenolic compounds. J. Appl. Bacteriol. 33:721.
88. Juven, B., Z. Samish, Y. Henis, and B. Jacoby. 1968. Mechanism of enhancement of lactic acid fermentation of green olives by alkali and heat treatments. J. Appl. Bacteriol. 31:200.
89. Karem, K. L., J. W. Foster, and A. K. Bej. 1994. Adaptive acid tolerance response (ATR) in *Aeromonas hydrophilia*. Microbiology 140:1731-1736.
90. Kawatomari, T., and R. H. Vaughn. 1956. Species of *Clostridium* associated with zapatera spoilage of olives. Food Res. 21:481.
91. Kopeloff, N., and P. Cohen. 1928. Further studies on a modification of the Gram stain. Stain Technol. 3:64.
92. Kroll, R. G., and R. A. Patchett. 1992. Induced acid tolerence in *Listeria monocytogenes*. Lett. Appl. Microbiol. 14:224-227.
93. Lazaro, M. J., E. Carbonell, M. C. Aristoy, J. Safon, and M. Rodrigo. 1989. Liquid chromatographic determination of acids and sugars in homolactic cucumber fermentations. J. Assoc. Off. Anal. Chem. 72:52.
94. Leyer, G. H., L-L Wang, and E. A. Johnson. 1995. Acid adaptation of *Escherichia coli* O157:H7 increases survival in acidic foods. Appl. Environ. Microbiol. 61:3752-3755.
95. Leyer, G. J., and E. A. Johnson. 1992. Acid adaptation promotes survival of *Salmonella* spp. in cheese. Appl. Environ. Microbiol. 58:2075-2080.
96. Lindquist, S. 1986. The heat shock response. Ann. Rev. Biochem. 55:1151-1191.
97. Lou, Y., and A. E. Yousef. 1997. Adaptation to sublethal environmental stresses protects *Listeria monocytogenes* against lethal preservation factors. Appl. Environ. Microbiol. 63:1252-1255.
98. McDonald, L. C., R. F. McFeeters, M.A. Daeschel, and H. P. Fleming. 1987. A differential medium for the enumeration of homofermentative and heterofermentative lactic acid bacteria. Appl. Environ. Microbiol. 53:1382.
99. McFeeters, R. F. 1993. Single-injection HPLC analysis of acids, sugars, and alcohols in cucumber fermentations. J. Agric. Food Chem. 41:1439-1443.
100. McFeeters, R. F. 1998. Use and removal of sulfite by conversion to sulfate in the preservation of salt-free cucumbers. J. Food Prot. 61:885-890.
101. McFeeters, R. F., and K-H Chen. 1986. Utilization of electron acceptors for anaerobic mannitol metabolism by *Lactobacillus plantarum*. Compounds which serve as electron acceptors. Food Microbiol. 3:73.
102. McFeeters, R. F., W. Coon, M. P. Palnitkar, M. Velting, and N. Fehringer. 1978. Reuse of fermentation brines in the cucumber pickling industry. EPA 600/2-78-207.
103. McFeeters, R. F., H. P. Fleming, and R. L. Thompson. 1982. Malic acid as a source of carbon dioxide in cucumber juice fermentations. J. Food Sci. 47:1862.
104. McFeeters, R. F., H. P. Fleming, and M. A. Daeschel. 1984a. Malic acid degradation and brined cucumber bloating. J. Food Sci. 49:999.
105. McFeeters, R. F., R. L. Thompson, and H. P. Fleming. 1984b. Liquid chromatographic analysis of sugars, acids, and ethanol in lactic acid vegetable fermentations. J. Assoc. Off. Anal. Chem. 67:710.
106. Meneley, J. C., and M. E. Stanghellini. 1974. Detection of enteric bacteria within locular tissue of healthy cucumbers. J. Food Sci. 39:1267.
107. Miller, G. L. 1959. Use of dinitrosalicylic acid reagent for determination of reducing sugar. Anal. Chem. 31:426.
108. Mills, D. R. 1941. Differential staining of living and dead yeast cells. Food Res. 6:361.
109. Monroe, R. J., J. L. Etchells, J. C. Pacilio, A. F. Borg, D. H. Wallace, M. P. Rogers, L. J. Turney, and E. S. Schoene. 1969. Influence of various acidities and pasteurizing temperatures on the keeping quality of fresh-pack dill pickles. Food Technol. 23:71.
110. Mossel, D. A. A., W. H. J. Mengerink, and H. H. Scholts. 1962. Use of a modified MacConkey agar medium for the selective growth and enumeration of *Enterobacteriaceae*. J. Bacteriol. 34:381.
111. Mrak, E. M., and L. Bonar. 1939. Film yeasts from pickle brines. Zentralbl. Bakteriol. Parasitenkd. Abt. 2 100:289.
112. Mrak, E. M., R. H. Vaughn, M. W. Miller, and H. J. Phaff. 1956. Yeasts occurring in brines during the fermentation and storage of green olives. Food Technol. 10:416.
113. O'Driscoll, B., C. G. M. Gahan, and C. Hill. 1996. Adaptive acid tolerance response in *Listeria monocytogenes*: Isolation of an acid-tolerant mutant which demonstrates increased virulence. Appl. Environ. Microbiol. 62:1693-1698.
114. O'Driscoll, B., C. G. M. Gahan, and C. Hill. 1997. Two-dimensional polyacrylamide gel elecvtrophoresis analysis of the acid tolerance response in *Listeria monocytogenes* LO28. Appl. Environ. Microbiol. 63:2679-2685.

115. Okereke, A., and S. S. Thompson. 1996. Induced acid-tolerance response confers limited nisin resistance on *Listeria monocytogenes* Scott A. J. Food Prot. 59:1003-1006.
116. Olson, E. R. 1993. Influence of pH on bacterial gene expression. Mol. Microbiol. 8:5-14.
117. Pakach, T. J., H-P Lieker, P. J. Reilly, and K. Thielecke. 1991. High-performance anion-exchange chromatography of sugars and sugar alcohols on quaternary ammonium resin under alkaline conditions.
118. Pederson, C. S., and M. N. Albury. 1969. The sauerkraut fermentation. N.Y. Agric. Exp. Sta. Bull. 824.
119. Pederson, C. S., and M. N. Albury. 1950. Effect of temperature upon bacteriological and chemical changes in fermenting cucumbers. N.Y. Agric. Exp. Sta. Bull. 744.
120. Pederson, C. S., and M. N. Albury. 1954. The influence of salt and temperature on the microflora of sauerkraut fermentation. Food Technol. 8:1.
121. Pederson, C. S., and M. N. Albury. 1961. The effect of pure culture inoculation on fermentation of cucumbers. Food Technol. 15:351.
122. Plastourgos, S., and R. H. Vaughn. 1957. Species of *Propionibacterium* associated with zapatera spoilage of olives. Appl. Microbiol. 5:267.
123. Potts, E. A., and H. P. Fleming. 1979. Changes in dissolved oxygen and microflora during fermentation of aerated, brined cucumbers. J. Food Sci. 44:429.
124. Rowbury, R. J. 1995. An assesment of environmental factors influencing acid tolerence and sensitivity in *Escherichia coli*, *Salmonella* spp. and other enterobacteria. Lett. Appl. Microbiol. 20:333-337.
125. Saltveit, M. E., Jr., and R. F. McFeeters. 1980. Polygalacturonase activity and ethylene synthesis during cucumber fruit development and maturation. Plant Physiol. 66:1019.
126. Samish, Z., D. Dimant, and T. Marani. 1957. Hollowness in cucumber pickles. Food Manuf. 32:501.
127. Samish, Z., R. Etinger-Tulczynska, and M. Bick. 1963. The microflora within the tissue of fruits and vegetables. J. Food Sci. 28:259.
128. Schlech, W. F., III, P. M. Lavigne, R. A. Bortolussi, A. C. Allen, E. V. Haldane, A. J. Wort, A. W. Hightower, S. E. Johnson, S. H. King, E. S. Nicholls, and C. V. Broome. 1983. Epidemic listeriosis—Evidence for transmission by food. N. Engl. J Med. 308:203.
129. Seelinger, H. P. R., and D. Jones. 1986. Genus *Listeria*, p. 1235. *In* P. H. A. Sneath, N. S. Mair, M. E. Sharpe, and J. G. Holt (ed.), Bergey's manual of systematic bacteriology, vol. 2. Williams and Wilkins, Baltimore, Md.
130. Sneath, P. H. A., N. S. Mair, M. E. Sharpe, and J. G. Holt (ed.). 1986. Bergey's manual of systematic bacteriology, vol. 2. Williams and Wilkins, Baltimore, Md.
131. Stamer, J. R. 1975. Recent developments in the fermentation of sauerkraut, p. 267. *In* J. G. Carr , C. V. Cutting, and C. G. Whiting (ed.). Lactic acid bacteria in beverages and food. Academic Press, New York.
132. Stamer, J. R., and B. O. Stoyla. 1967. Growth response of *Lactobacillus brevis* to aeration and organic catalysts. Appl. Microbiol. 15:1025.
133. Stamer, J. R., and B. O. Stoyla. 1978. Stability of sauerkraut packaged in plastic bags. J. Food Prot. 41:525.
134. Sumner, J. B., and E. B. Sisler. 1944. A simple method for blood sugar. Arch. Biochem. 4:333.
135. Tang, H-CL., and R. F. McFeeters. 1983. Relationships among cell wall constituents, calcium, and texture during cucumber fermentation and storage. J. Food Sci. 48:66.
136. US HEW/PHS. 1979. Acidified foods and low-acid canned foods in hermetically sealed containers. Public Health Serv., Food and Drug Admin. Fed. Reg. 44:16,204-16,238.
137. U.S. Department of Agriculture. 1966. U.S. standards for grades of pickles. USDA Fed. Reg. 31:10,231-10,305.
138. Vanderslice, J. T., and D. J. Higgs. 1993. Quantitative determination of ascorbic, dehydroascorbic, isoascorbic, and dehydro-isoascorbic acids by HPLC in foods and other matrices. J. Nutr. Biochem. 4:184-190.
139. Vaughn, R. H. 1954. Lactic acid fermentation of cucumbers, sauerkraut, and olives, p. 417. *In* L. A. Underkofler and R. J. Hickey (ed.), Industrial fermentations, vol. 2. Chem. Publ., New York.
140. Vaughn, R. H. 1975. Lactic acid fermentation of olives with special reference to California conditions, p. 307. *In* J. G. Carr, C. V. Cutting, and G. C. Whiting (ed.), Lactic acid bacteria in beverages and food. Academic Press, New York.
141. Vaughn, R. H. 1981. Lactic acid fermentation of cabbage, cucumbers, olives, and other products, p. 185. *In* G. Reed (ed.), Prescott and Dunn's industrial microbiology, 4th ed. Avi Publ., Westport, Conn.
142. Vaughn, R.H., M. H. Martin, K. E. Stevenson, M. G. Johnson, and V. M. Crampton. 1969. Salt-free storage of olives and other produce for future processing. Food Technol. 23:124.
143. Wang, S. H. 1941. A direct smear method for counting microscopic particles in fluid suspension. J. Bacteriol. 42:297
144. Welshimer, H. J. 1968. Isolation of *Listeria monocytogenes* from vegetation. J. Bacteriol. 95:300.
145. Welshimer, H. J., and J. Donker-Voet. 1971. *Listeria monocytogenes* in nature. Appl. Microbiol. 21:516.

Chapter 52

Gums and Spices

Rodney J. H. Gray and Joan M. Pinkas

52.1 INTRODUCTION

52.11 Gums

The terms gums, hydrophilic colloids, hydrocolloids, and water-soluble polymers have been used interchangeably to refer to a wide range of useful plant and microbial polysaccharides, or their derivatives, that hydrate in cold or hot water to form viscous solutions, dispersions or gels.[16,31,105] Those of plant origin[32,44,91,104,105] include acacia (gum arabic), gum tragacanth, karaya, and ghatti. In addition to these exudates, natural gums include seaweed extracts[32,44,91,104,105] (agar, alginates, carrageenan), gums from seed[33,44,91,104,105] (locust bean or carob, guar) gum from tubers[38,44,91,104] (Konjac), microbial gums[13,31,39,44,53,91,104,105] (gellan and xanthan), and fruit extracts[31,33,44,91,104,105] (pectins). Modified gums include amidated pectins, propylene glycol alginate, and cellulose derivatives such as sodium carboxymethyl cellulose, hydroxypropyl cellulose, and methyl cellulose.[5,16,30,33,44,91,55,104,105,106]

Industrial importance stems from the functional properties of aqueous solutions or dispersions of these hydrocolloids. Stabilization, dispersion, or suspension are the functional properties usually exhibited. In addition, hydrocolloids may function as gelling agents or serve as emulsifiers, binders, flocculating agents, film formers, foam stabilizers, mold release agents, or lubricants.[4,5,16,31,105] Gums therefore have potential for application in a wide variety of foods. Among these are meat and dairy products, sauces, pie fillings, salad dressing, whips, soups, emulsions, puddings and jellies.

52.12 Spices

The term "spice" originated from the Latin words "species aromatacea" meaning fruits of the earth and was later shortened to "species" meaning a commodity of special value or distinction.[27] While current definitions vary among trade organizations[93] and governments,[101] the term "spice" is generally used to describe the whole family of dried plant seasonings, including spices, herbs, blends and dehydrated vegetables.[93] The International Organization for Standardization (ISO) defines spices and condiments as "whole or ground vegetable products or mixtures thereof, without extraneous matter, that are used for flavoring, seasoning and imparting aroma in foods".[45] The ISO officially recognizes over 100 spices grown throughout the world.[45] Included are leafy herbs such as oregano, basil and sage that are cultivated in temperate climates, the tropical aromatics such as pepper, cassia (cinnamon) and cloves and certain dehydrated vegetables like onion and garlic. The main centers of production are India, Indonesia, China, Brazil and Central America, among others. Major suppliers to the United States include Canada (mustard seed), Mexico (sesame seed, capsicums, oregano), Indonesia (black pepper, cassia) and India (cumin seed), however, the sources can change rapidly as a result of political or natural disasters.[37] Within the United States, California supplies dehydrated onion and garlic, paprika, chili peppers and more than a dozen different herbs.[37]

Spices are obtained from various parts of botanically diverse aromatic and herbaceous plants; for example, berries (black pepper and allspice), seeds (mustard and poppy), bark (cassia and cinnamon), flower parts (cloves and saffron), rhizomes (ginger and turmeric), leaves (marjoram and thyme) and bulbs (onion and garlic).

52.2 GENERAL CONSIDERATIONS

52.21 Gums

As an item of commerce, gums are predominantly marketed in the dehydrated form and as such are microbiologically stable. Most gums, however, will support microbiological growth in the presence of sufficient moisture. Rehydrated gums should therefore be used promptly following preparation. Enzymes excreted by bacteria, primarily *Bacillus* spp.,[15,89] can degrade unused gel or gum solutions. Most susceptible to bacterial action are the gums tragacanth, acacia, karaya, guar, locust bean (carob), carrageenan, and sodium alginates.[88,89] Kappa carrageenan has been suggested as a low-cost substitute for agar in media intended for the culture of yeasts and molds.[62]

Most gums produce high-viscosity solutions when dissolved in water at very low concentrations, usually less than 1%. This property confounds the classical analytical procedures used in microbiology that center around the use of decimal dilutions.

In the preparation of gums for microbiological analysis, a 1:10 dilution is so viscous as to be unworkable and often a 1:50 dilution will be marginal. Hence, the detection limit of the normal aerobic plate count on these materials may be relegated to ~2000 per g (i.e., where a minimum of 20 colony-forming units per plate of a 10^{-2} dilution is required). Multiple plating, depositing larger volumes per plate, the use of large Petri plates and enzyme pre-treatment[2] are all approaches to consider in addressing this viscosity-based problem.

With the exception of the viscosity property of gums, their microbiological analysis may follow the standard procedures used in the microbiological examination of foods.

52.22 Spices

The microbiological analysis of spices can, at times, be challenging due to the presence of natural antimicrobial compounds. Of the recognized inhibitory components in spices, many are phenolic constituents of the essential oil fraction.[17,10,83] For example, the antimicrobial action of cloves and allspice is due to eugenol[17] while thymol and carvacrol are the antimicrobial agents in both oregano[83] and thyme.[51] Cinnamaldahyde and allyl isothiocyanate are responsible for the inhibitory properties of cinnamon and mustard seed, respectively.[83] The sulfur compound, allicin, is the principal antibacterial and antifungal agent in onion and garlic.[17]

The *in vitro* antimicrobial effects of spices and herbs are well documented.[9,32,34,40,46,48,49,51,65,76,84] Comparing the results of the studies is difficult, however, because of the differences in test designs. Some investigations involved pure spices while others tested essential oils or alcoholic extracts. Variations also exist in the spectrum of microorganisms and spices studied. What can be stated is that the minimum inhibitory spice concentration varies with the microorganism and the spice. There is consensus that gram positive bacteria are more sensitive to spices than gram negative bacteria.[24,25,65,84] Extensive antimicrobial activity has been noted with onion,[48,49] garlic,[34,48,76] oregano,[9,46,49] cassia (cinnamon),[46,49,70] cloves[24,46,70] and allspice,[49] among others.[9,84] One study demonstrated complete inhibition of *Staphylococcus aureus* by 0.5% alcoholic extracts of mace, nutmeg, cinnamon, thyme, oregano, turmeric, white pepper, black pepper, rosemary and cloves in culture media and inhibition with 0.2% extracts of bay leaf and sage.[65] In the same study, the antibacterial activity of most of the 22 spice and herb extracts tested against *Bacillus stearothermophilus* was generally high in that a 0.1% extract inhibited growth.[65] *Bacillus cereus var mycoides* was comparatively sensitive to mace and sage at 0.1% and to thyme, black pepper, rosemary and cloves at 0.2%.[65] In contrast, *Escherichia coli* and *Salmonella typhimurium* showed no sensitivity to most of the 0.5% extracts with the exception of cloves and cinnamon.[65] Other studies demonstrated that 1% (w/v)[48] to 10% (w/v)[49] of onion, 5% (w/v) of garlic[48] and 10% (w/v) of allspice,[49] cassia[49] and oregano[49] were toxic to various *Salmonella spp.* in culture media. In addition to exhibiting antibacterial activity, the essential oils of some spices also have antifungal properties. For example, cinnamon and clove oils have been shown to inhibit mold growth and aflatoxin production, *in vitro*.[70]

Numerous factors determine the extent of the antimicrobial action of spices in foods. In addition to spice concentration and the nature of the microorganisms, food composition plays a major role. Typically, foods with complex compositions require increased concentrations of preservatives to inhibit growth.[83] One study demonstrated that fat and/or protein are more effective than carbohydrates in shielding bacteria from the action of sage inhibitors.[83] The concentration of spices often required to inhibit bacterial and fungal growth in foods is generally higher than in culture media and may exceed normal usage levels.[17,83] For this reason, the use of spices as antimicrobial agents in foods may be limited.[65]

Some spices have also been shown to exert a stimulatory effect on certain microorganisms. Enhanced growth of *Listeria monocytogenes* strain Scott A has been noted in a liquid medium with 1% white pepper.[40] At low concentrations, oregano has been observed to accelerate acid production by lactic acid bacteria without increases in cell number.[108]

52.3 NORMAL FLORA

52.31 Gums

The microbiology of the major botanical gums is the subject of very few articles in the scientific literature. Some work in the 1970s provides one of the few sources of published information on this topic.[85,86,87,88,89] These authors reported high numbers of organisms on gums (10^8 per g), but the data referred primarily to the raw agricultural product or unprocessed starting materials. Gum arabic, for example, carries low levels of microflora in comparison to other exudates. Typically, bacterial counts are in the 1000 cfu/g range and pathogens are absent.[12] Much lower counts were found in the finished products (alginate, carrageenan, locust bean gum, and guar flour). Aerobic sporeformers, primarily *Bacillus* sp., predominated. Even lower levels of microflora can be expected in pectin and cellulose derivatives.

Streptococcus faecalis occurred in some of the gum tragacanth, carrageenan, and guar samples.[85] *Staphylococcus aureus* (coagulase-positive strains) was also detected in some of the gum tragacanth and locust bean flour samples at levels up to 3×10^5 per g. The latter findings, however, have not been confirmed by laboratories in the United States on routine analysis of gums. *Clostridium perfringens* is found occasionally in tragacanth, acacia, and locust bean flour,[86] but not in gum karaya and guar flour. Anaerobic sporeformers were not detected in alginates and carrageenan. *Escherichia coli* was not found in any of the samples representing the 1970 study[87]; however, laboratories in the United States have found this organism occasionally in raw samples.

52.32 Spices

As raw agricultural commodities, spices and herbs commonly harbor large numbers of bacteria and fungi including potential spoilage organisms and, occasionally, organisms of sanitary and public health significance.[35] The manner and environment in which they are grown, harvested and handled, as well as the chemical nature of the spice, directly impacts its microbiological quality. In general, roots, berries and herbs carry a greater microbiological load than the bark and seed items.[35] While a considerable number of vegetative cells are killed during the drying process, many bacteria and molds may survive.[57,94] Bacterial multiplication is not a concern if the products are sufficiently dried, stored and shipped under normal, dry conditions, however, fungal spoilage may occur if the spices are subjected to improper storage. After harvest, various types of cleaning processes are used to progressively reduce the number and types of microorganisms.[74,92] Additional means for effectively reducing the microbial populations are treatment with ethylene oxide,[26,60,69,102,103] high temperature steam[59] or irradiation.[26,102]

Large variations in total aerobic plate counts are found between spices, between samples of a spice from different origins and in spice samples from the same source.[8,54,49,67] Of all the spices, untreated black pepper typically has the highest aerobic plate counts, usually in excess of 10^6 cfu/g.[8,49,69,79,94] Untreated paprika,[35,49,94] celery seed,[35] coriander,[8] white pepper,[8] turmeric,[29] thyme, basil and others can also have aerobic plate counts in the millions per gram. Moderately low counts have been noted in cassia and nutmeg.[69] The lowest aerobic plate counts are found in spices, such as cloves,[35,69] whose essential oils exhibit antimicrobial effects. Aerobic sporeformers are the predominant microorganisms in the aerobic plate count.[8,58,59] One study has shown that they account for 50% to 95% of all organisms isolated from 15 different spices.[8] The majority of the aerobic sporeformers detected in samples of black pepper, white pepper, paprika, marjoram, coriander, pimento (allspice) and onion powder were proteolytic organisms.[8] High numbers of amylolytic organisms were also noted in black pepper, white pepper, pimento, onion powder and cinnamon.[8]

Much like the variation in aerobic plate counts, it is not unusual to find a wide range of mold counts in spices and herbs. Mold counts generally do not correlate with total aerobic bacterial counts.[49] Cinnamon,[67,69] thyme,[35,68] rosemary,[69] celery,[69] sage,[69] white pepper[35,69,94] and cardamom[69] have been reported to harbor molds in excess of 10^5 cfu/g while other spices[49,67,69] contain negligible levels. The molds most frequently isolated are from the *Aspergillus* genus and include *A. niger*,[1,11,21,29,28,50] *A. flavus*,[1,11,21,28,29,50] *A. glaucus* group,[67,68,94] *A. parasiticus*[11,29] and *A. nidulans*.[28] *Penicillium* spp.[21,28,50,94] and *Absidia* spp.[21,94] are also represented. Although *A. flavus* is a prominent species in spices, aflatoxin contamination is generally low. One study indicated the absence of aflatoxins in 20 spice and herb samples containing *A. flavus*.[21] Other studies have demonstrated the presence of low concentrations of aflatoxins in capsicums,[1,80] turmeric,[81] ginger[81] and nutmeg.[81] Cinnamon[66] and cloves,[66] and to some extent oregano,[66] mustard,[66] black pepper[47] and white pepper[47] are considered poor substrates for aflatoxin production. In general, yeasts are seldom detected.[35] However, moderate to high levels of yeasts can be found in herbs such as dill and basil, primarily because of the moderate climate in which they are grown.[69]

Low concentrations of sporeforming bacteria that are capable of causing gastroenteritis, such as *Bacillus cereus* and *Clostridium perfringens*, are present in spices. They are, however, of no public health significance unless the food to which they have been introduced has been mishandled and the microorganisms have multiplied to a concentration of 10^5 to 10^6 cfu/g of food. In a survey of assorted spices and herbs at import, *B. cereus* was isolated from all products tested except cloves, cayenne pepper and chives.[69] The counts were generally less than 10^4 cfu/g with a few exceptions.[69] Approximately one-half of the processed spices tested in other studies contained *B. cereus* with counts rarely over 5000 cfu/g.[71,78] Likewise, *C. perfringens* has been isolated from a wide variety of spices and herbs[20] including garlic powder,[77] black pepper,[78,77] cumin seed,[77] bay leaves,[72] cayenne pepper,[72] cinnamon[72] and oregano.[77,72] While the reported incidence of this sporeformer varies, the counts are generally less than 500 cfu/g.[20,72,77]

Nonsporeforming bacteria, including indicator organisms and those of a potentially more serious nature, may also be present. Coliforms are often detected in spices and herbs at import[69,79] with *Citrobacter freundii*,[79] *Enterobacter cloacae*,[79] *Klebsiella pneumoniae*[79] and *Enterobacter aerogenes* frequently represented. *Escherichia coli* has been detected at import[69,79] and in samples from various retail markets,[7,54] but usually in low numbers.[69,79,94] Salmonellae have occasionally been isolated from a variety of products including paprika,[8,14] black pepper,[8,58,69,78,79] fenugreek,[69] white pepper,[69] cumin and oregano. While the public health safety record in the use of spices is generally good, a few incidents of human salmonellosis have been traced to the consumption of foods seasoned with black pepper,[36,61] white pepper[82] and paprika.[63]

In a study of 160 samples of 55 different retail spices and herbs, *Listeria spp*. was not detected.[58] Coagulase positive staphylococci are also not characteristically found.[8,21,29,35,58,72,107] Thermophilic bacteria typically comprise less than 1 to 10% of the bacterial population, with higher percentages occasionally noted.[52] Other microorganisms detected sporadically, usually in small numbers, include anaerobic sporeformers[58,67] and lactic acid bacteria.[58]

52.4 POST-HARVEST CHANGES

52.41 Gums

Although processing methods for many gums are proprietary, patents and other literature provide some insight. In general, tree gums secreted at wounds and incisions are colorless, but they darken on drying and aging. Most tree gums are collected by hand, sorted, graded, packed, and shipped from the growing area. The processor further grades, cleans, mills, and blends the gum. Tragacanth gum is sold as ribbons, flakes, granules, or powder. Gum ghatti exudates are sun-dried and pulverized. Since the production of these gums is largely physical in nature, e.g., collection, grinding, blending, etc., there may be great variation in the microbial content and in some instances high levels of microorganisms may be present.[85,86] The use of propylene oxide has utility for these gums as an antimicrobial treatment. It should be noted that residues are limited to 300 ppm by regulation,[99] and that permission for the use of propylene oxide may well be revoked in the future.

Carrageenan is extracted at high temperature under alkaline conditions from red algae, filtered and concentrated, precipitated with alcohol, dried, and then milled. A similar process under acidic conditions is used to derive pectin from citrus peels. The extremes of temperature and pH utilized in the initial stages of production of these gums effect a marked reduction in their bioburden. Care must be exercised in subsequent process steps to capitalize on this effect and to avoid re-contamination. Gums such as carrageenan and pectin are commonly standardized to provide the user with constant gelling and thickening properties. Standardization is often achieved by addition of sugar or buffer salts. The contribution of standardizing agents to the microbial status of gums should not be ignored.

52.42 Spices

After harvest, spices are subjected to different types of processes such as washing, peeling, curing, drying, fumigation, cleaning, grading and milling.[74] While the processes vary in accordance with the spice, drying is, by far, the most important step for all products. It is critical that they be dried quickly to an a_w that will prevent mold growth. For economic reasons, this is typically accomplished by natural means. Dried spices undergo extensive cleaning to remove extraneous matter such as dirt, stones, stalks, leaves, insects and metallic contamination. During the milling process, a considerable reduction in the total bacterial load has been noted for cumin, coriander, chilies and black pepper as a result of the increase in product temperature.[73,75] Cryogenic milling is used for some spices to retain more of the flavor components that may be lost during regular ambient grinding.[74,92]

Ethylene oxide fumigation is a very effective treatment that is used in some countries to reduce the microbiological populations of spices. While the term "sterilization" is frequently used to describe this process, it is inaccurate. A more appropriate term is "pasteurization" because it more closely reflects the results of the doses used in the industry. Vegetative cells, including coliforms,[60,69] *E. coli*[69] and *Salmonella*,[69] are eliminated, with low to moderate concentrations of bacterial spores typically remaining.[69]

Many factors affect the overall reduction in counts, including the initial microbial type and load, the temperature and relative humidity in the chamber, the concentration of ethylene oxide, the physical and chemical nature of the spice and its moisture content.[68] In a study of 136 ground, powdered and granulated spice samples, there was destruction of at least 99.9% of the aerobic plate count bacteria in approximately one-half of the samples.[94] In the same study, the decline in mold counts following treatment was generally from 99 to 99.9+%.[94] Fumigation with propylene oxide is used in some countries in place of ethylene oxide to reduce microbial contaminants. On a weight basis, it is less bactericidal than ethylene oxide.[94] Current U.S. Environmental Protection Agency regulations limit the ethylene oxide and propylene oxide residues to a maximum of 50 ppm[97,98] and 300 ppm,[99] respectively.

Irradiation of spices with gamma rays has been shown to be equivalent to or better than ethylene oxide treatment from the point of view of microbiological efficacy.[26,57,102] Studies indicate that less volatile oils are lost from irradiated spices than from those treated with ethylene oxide.[26,102] In addition to being simple, safe and efficient, gamma irradiation allows the treatment of products in their final packaging, which eliminates recontamination issues.[57] Microwave treatment has also been investigated with inconsistent results reported.[23,102] Treatment with high temperature steam is another method for decontaminating whole spices. It is a safe, highly economical and efficient process for significantly reducing microbial populations of some spices without sacrificing appearance and flavor levels.[43]

52.5 SAMPLING AND PREPARATION FOR ANALYSIS

The stringency of a sampling plan for food depends on the spoilage and health hazards associated with the microorganisms of concern and how the food will be handled and consumed after it is sampled.[95] A three-class attribute sampling plan with five samples taken at random from each lot of material, as described by the International Commission on Microbiological Specifications for Foods (ICMSF), is appropriate for routine microbiological examinations of gums and spices for aerobic plate count bacteria, yeasts, molds, coliforms and *E. coli*.[95] A two-class attribute plan is recommended when testing for pathogens that are moderately to seriously hazardous, such as *Salmonella*.[95] When a potential problem has been identified, investigational sampling should be conducted.[95] Other sampling plans may also be suitable.

As with any type of food sampling, aseptic technique should always be employed. Most gums and spices can easily be sampled with sterile three zone powder samplers, needle point samplers, triers, scoops or spoons. The size and shape of some whole spices such as turmeric, ginger, nutmeg and cassia limits the type of device that can be used. Sterile scoops and spoons are the most appropriate implements for removing these types of spices from containers. The samples (approximately 200 g each) should be placed in sterile, polyethylene sample bags that are clearly labeled, submitted to the laboratory and tested individually. Store gum and spice samples in a cool (≤20°C), dry (≤60% relative humidity) area if they can not be tested promptly.[93]

52.51 Procedures for Gums

The initial dilution procedures described below for gums are applicable to assays including the aerobic plate count, total coliform bacteria, *Escherichia coli* and aerobic sporeformers. One exception is the assay for yeasts and molds which may be performed using a direct plating technique. This may be performed on Dichloran 18% glycerol (DG18) agar,[100] although where both yeasts and molds must be enumerated Dichloran rose bengal chloramphenicol (DRBC) agar should be used.[100]

52.511 Initial Dilution

Unless some pretreatment[2] is used, preparation of the classical 1:10 dilution is not feasible for gums due to viscosity development in these substances upon hydration. In most instances the maximum gum concentration that can be manipulated using standard methods is 2%; in a few instances even 2% is too viscous and use of a 1% concentration is recommended.

52.5111 Carrageenan and Locust Bean Gum (Carob). Prepare a 1:100 initial dilution by aseptically weighing a 5 g gum sample into 495 mL sterile Butterfield's phosphate-buffered diluent.

52.5112 Other Gums. For other gums a 1:50 initial dilution may be prepared by aseptically weighing a 10 g sample into 490 mL sterile buffered diluent.

Following stomaching of the initial dilution for 2 minutes the typical 10-fold serial dilution may then be performed. Multiple plating of 2.5 mL of each dilution into each of 4 plates, and using the sum of the plates for the dilution in the 25–250 colony range, will allow determination of, e.g., the aerobic plate count. (Note 1:50 dilution ×5 = cfu/g gum and for the 1:100 dilution ×10 = cfu/g gum).

52.52 Procedures for Spices

The following sample preparation procedure is applicable to all analyses. The initial dilution procedures are appropriate for examinations such as aerobic plate counts, yeasts, molds, coliform bacteria, *Escherichia coli*, and others. Pre-enrichment dilutions for *Salmonella* analysis are described in the Methods section (52.62). Procedures for sporeformer analyses are presented in the respective chapters in this text.

Sample preparations and their initial dilution vary according to the nature of the material being examined. Whole berries, roots, bark and large seeds should be reduced to a moderate particle size before testing. Aseptically weigh 100 g of the sample into a sterile, dry blender jar. Blend the sample at the lowest speed for 30 seconds or more. Take special care not to generate excessive heat during the blending step for this may injure or destroy the microorganisms.

52.521 Initial Dilution

52.5211 Ground spices, herbs, seasonings and small whole seeds. Prepare a 1:10 dilution by aseptically weighing 11 ± 0.1g of the sample into a sterile filter stomacher bag and adding 99 ± 2 mL of sterile 0.1% peptone water. Stomach for 30 to 60 seconds. Some seeds, even though small, may have sharp edges that will puncture the stomacher bag. Therefore, stomaching may not be appropriate for all products. Alternately, weigh 11 ± 0.1g of the material into a sterile 250 mL widemouth polypropylene or glass bottle containing 99 ± 2 mL of sterile 0.1% peptone water and shake at least 25 times in a 1 foot arc or aseptically weigh the sample into a sterile blender jar, add sterile, chilled diluent and blend at the lowest speed for 2 minutes.

52.5212 Whole and coarsely ground leafy herbs and ground cassia (cinnamon). Whole and coarsely ground leafy herbs require an initial 1:20 dilution because of their low density and high absorption properties. Ground cassia should also be diluted 1:20 because it contains mucilaginous substances[22,90] that become highly gelatinous and stringy when suspended in water. Aseptically weigh 11 ± 0.1 g of the product into a sterile filter stomacher bag, polypropylene bottle or blender jar. Add 209 ± 2 mL of sterile 0.1% peptone water or phosphate buffered diluent (chilled, if blending) and either stomach for 30 to 60 seconds, shake at least 25 times in a 1 foot arc or blend at low speed for 2 minutes, depending on the type of container.

52.5213 Dehydrated vegetables. Prepare a 1:20 dilution by aseptically weighing 11 ± 0.1 g of the sample into a filter stomacher bag and adding 209 ± 2 mL of chilled 0.1% peptone water or phosphate buffered diluent. Gently swirl the stomacher bag to wet the vegetables. Allow the sample to rehydrate in the refrigerator for 30 minutes. Stomach for 30 to 60 seconds. Alternately, the sample can be diluted in a sterile blender jar and rehydrated as above before blending for 2 minutes at low speed.

52.6 METHODS

In general, gums, spices and herbs are tested with standard methods for aerobic plate count bacteria, yeasts, molds, total coliform bacteria, *E. coli* and *Salmonella*. Depending on the application, certain food processors may require additional analyses such as enumeration of aerobic and anaerobic sporeformers (mesophilic and thermophilic), lactic acid bacteria, *B. cereus*, *C. perfringens* and others.

Several agars have been used successfully to enumerate yeasts and molds from dried food ingredients. Dichloran Rose Bengal Chloramphenicol (DRBC) agar produces discrete, non-spreading mold colonies that are easy to count.[18,56,64] Dichloran 18% Glycerol (DG18) agar is a low a_w medium designed to maximize the enumeration of common xerophilic fungi.[41,42,100] Phytone Yeast Extract (PYE) agar has also been used to assess the quality of consumer spices and herbs.[58] It is important to note that some molds spread rapidly on DG18[64] and PYE agars which can make enumeration challenging if the plates are crowded.

52.61 Gums

Hydrocolloids have a unique property of developing viscosity upon hydration. In conducting microbial assays on gums this viscosity development, however, only impacts the initial dilution phase, as previously described (Sec. 52.511). Methods for the detection and enumeration of micro-organisms in gums otherwise follow standard methodology.

52.62 Spices

It is important to note that some spices and herbs contain essential oils that are inhibitory to bacteria[9,34,40,46,48,65,65,84] and fungi[70] and may produce low counts on lower dilution plates and moderately high counts on higher dilution plates. This can be attributed to the transfer of antimicrobial compounds with the inoculum. When testing certain spices, prepare a sufficient number of serial dilutions to overcome this natural inhibitory effect and prevent the reporting of erroneous low counts. This is particularly important for cloves, allspice, cassia (cinnamon), mustard seed, oregano, onion and garlic. Results are generally not reported from the first dilution plates and tubes for these seven products. Inhibition may also be noted with nutmeg, sage, rosemary, marjoram and others.

The antimicrobial effects of the essential oils must also be taken into account when analyzing a spice for the presence of *Salmonella* and other pathogens that require a pre-enrichment step. Increasing the conventional 1:9 spice to pre-enrichment medium ratio is recommended for some products. Toxicity studies with regard to *Salmonella* have demonstrated that spice to Trypticase Soy Broth (TSB) pre-enrichment ratios of 1:1000 are necessary for cloves, pimento (allspice), cinnamon, oregano and mustard seed to ensure the detection of this microorganism of public health significance.[69] To neutralize the inhibitory properties of onion and garlic, prepare a 1:9 ratio of sample to TSB containing 0.5% K_2SO_3.[6] Seasoning blends containing a large portion of one or more of the above mentioned spices should be pre-enriched like the pure spice. To overcome the physical difficulties encountered by their low density and high absorption properties, dilute whole and coarsely ground leafy herbs 1:20 in TSB.

52.7 INTERPRETATION OF DATA

Gums and spices are used in a wide variety of foods prepared in assorted ways. For this reason, it is inappropriate to have a single microbiological limit. The International Commission on Microbiological Specifications for Foods (ICMSF) recommends that gums and spices be treated as raw agricultural commodities, and as such, the ultimate use of the products will dictate the specifications.[95] For example, black pepper that contains a high concentration of sporeforming bacteria may be suitable as a table condiment for seasoning cooked foods that will be eaten immediately, but may be unacceptable to a canned food processor.[100] Likewise, an herb that contains a large number of yeasts may be unsatisfactory to a salad dressing manufacturer but may be acceptable to a canned soup processor. More stringent standards are typically required for spices added to cooked foods that are not reheated before eating and foods that are only minimally processed. Because of their formulation, many low fat, low sodium and natural foods also require the use of low microbial count ingredients.

The ICMSF recommendation[95] recognized the nature of gums and spices and supported the use of the in-process controls inherent in a Hazard Analysis and Critical Control Point (HACCP) system rather than relying on examination of end-product against a single figure microbiological limit. Nothwithstanding, several official monographs on gums indeed contain single figure microbiological limits.

Standard specifications for most gums are published, in the form of monographs, in Food Chemicals Codex, the U.S. Pharmacopoeia/National Formulary, the Official Journal of the European Communities, and the Compendium of Food Additive Specifications (Joint FAO/WHO Expert Committee on Food Additives). As mentioned above, several of these monographs contain single figure microbial limits, e.g., in the USP/NF,[96] microbial limits for gums range from a requirement for gelatin which reads, "The total bacterial count does not exceed 1000 per g, and the tests for *Salmonella* species and *Escherichia coli* are negative." Several gums (for example, alginic acid, sodium alginate, propylene glycol alginate, and carrageenan) have a requirement for a 200/g maximum " total bacterial count" with tests for *Salmonella* and *Escherichia coli* negative. Still others (for example, acacia, pectin, tragacanth, and xanthan) have no "total bacterial count" requirement alongwith the *Salmonella* and *Escherichia coli* negative requirement while in the case of the guar gum USP/NF monograph there are no microbial limits.

Government agencies, academia, and industrial microbiologists continue to advocate the establishment of in-process controls to assure product integrity, rather than reliance on end-product testing for compliance with specifications. In the gums and spice industries the strengthening of HACCP systems that encompass all stages of production, processing and distribution will serve to further enhance the microbial safety of these products.[3,19] In addition, educating consumers as to the importance of correct food handling practices will also help to prevent spoilage and illness incidents.

52.8 REFERENCES

1. Adegoke, G. O., A. E. Allamu, J. O. Akingbala, and A. O. Akanni. 1996. Influence of sundrying on the chemical composition, aflatoxin content and fungal counts of two pepper varieties- *Capsicum annum* and *Capsicum frutescens*. Plant Foods Human Nutr. 49:113-117.
2. Amaguana, R. M., P. S. Sherrod, T. S. Hammack, G. A. June, and W. H. Andrews.1996. Usefulness of Cellulase in recovery of Salmonella spp. from Guar Gum., J. Assn. Offic. Anal. Chem. 79(4):853.
3. American Spice Trade Association, Inc. 1994. "Spice microbiology hazard background review." American Spice Trade Association, Inc., Englewood Cliffs, N. J.: Technical bulletin No. 941024.
4. Andres, C. Stabilizers 1. 1975. Gums. Food Proc. 36 (12):31.
5. Andres, D. Stabilizers 2. 1976. Gums. Food Proc. 37(1):83.
6. Andrews, W. H., D. Wagner, and M. J. Roetting. 1979. Detection of *Salmonella* in onion and garlic powders: collaborative study. J. Assoc. Off. Anal. Chem. 62(3):499-502.
7. Arias, M. L., D. Utzinger, and R. Monge. 1997. Microbiological quality of some powder spices of common use in Costa Rica. Rev. Biol. Trop. 44(3)/45(1):692-694.
8. Baxter, R., and W. H. Holzapfel. 1982. A microbial investigation of selected spices, herbs, and additives in South Africa. J. Food Sci. 47:570-578.
9. Beuchat, L. R. 1976. Sensitivity of *Vibrio parahaemolyticus* to spices and organic acids. J. Food Sci. 41:899-902.
10. Beuchat, L. R., and D. A. Golden. 1989. Antimicrobials occurring naturally in foods. Food Technol. 43:134-142.
11. Bhat, R., H. Geeta, and P. R. Kulkarni. Microbial profile of cumin seeds and chili powder sold in retail shops in the city of Bombay. J. Food Prot. 50(5):418-419, 433.
12. Blake, S. M., D. J. Dreble, G. O. Philips, and A. DuPlessey. 1988. The effect of sterilizing doses of γ-irradiation on the molecular weight and emulsifying properties of gum arabic. Food Hydrocolloids 2(5):407-415.
13. Blanshard, J. M. V., and J. R. Mitchell. 1979. "Polysaccharides in food." Butterworth, Inc., Woburn, Mass.
14. Bruchmann, M. 1995. Salmonella contamination of spices. Results of studies carried out in Brandenburg in 1993. Arch Lebensmittelhygiene 46(1):17-19.
15. Cadmus, M. C., L. K. Jackson, K. A. Burton, R. D. Plattner, and M. E. Slodki. 1982. Biodegradation of xanthan gum by *Bacillus* sp. Appl. Environ. Microbiol. 44:5.
16. Cottrell, I. W., and J. K. Baird. 1980. Gums, *In* "Kirk-Othmeer encyclopedia of chemical technology," 3rd ed. Vol. 12, p. 45, John Wiley & Sons, Inc., New York.
17. Council for Agricultural Science and Technology. 1998. Naturally occurring antimicrobials in food. Task Force Report No. 132, 1998 April. CAST, Iowa.
18. Cousin, M. A., and H. H. Lin. 1986. Comparison of DRBC medium with PDA containing antibiotics for enumerating fungi in dried food ingredients. *In* A. D. King, Jr., J. I. Pitt, L. R. Beuchat, and J. E. L. Corry (eds.), Methods for the mycological examination of food. NATO ASI series A: Life Sciences 122, p. 97-101. Plenum Press, New York.
19. D'Aoust, J. Y. 1994. Salmonella and the international food trade. Int. J. Food Microbiol. 24:11-31.
20. DeBoer, E., and E. M. Boot. 1983. Comparison of methods for isolation and confirmation of *Clostridium perfringens* from spices and herbs. J. Food Prot. 46(6):533-536.
21. DeBoer, E., W. M. Spiegelenberg, and F. W. Janssen. 1985. Microbiology of spices and herbs. Antonie van Leeuwenhoek; 51:435-438.
22. Dutta, A. B. 1961. A chemical method for distinguishing cinnamon from cassia. J. Assoc. Off. Anal. Chem. 44(4):639-640.
23. Emam, O. A., S. A. Farag, and N. H. Aziz. 1995. Comparative effects of gamma and microwave irradiation on the quality of black pepper. Z. Lebensm Unters Forsch. 201:557-561.
24. Farag, R. S., Z. Y. Daw, F. M. Hewedi, G. S. A. El-Baroty. 1989. Antimicrobial activity of some Egyptian spice essential oils. J. Food Prot. 52(9):665-667.
25. Farbood, M. I., J. H. MacNeil, and K. Ostovar. 1976. Effect of rosemary spice extractive on growth of microorganisms in meats. J. Milk Food Technol. 39(10):675-679.
26. Farkas, J., and E. Andrassy. 1988. Comparative analysis of spices decontaminated by ethylene oxide or gamma radiation. Acta Aliment. 17(1):77-94.
27. Farrell, K. T. 1985. "Spices, condiments, and seasonings." Van Nostrand Reinhold Company Inc., New York.
28. Garrido, D., M. Jodral, and R. Pozo. 1992. Mold flora and aflatoxin-producing strains of *Aspergillus flavus* in spices and herbs. J. Food Prot. 55(6):451-452.
29. Geeta, H., and P. R. Kulkarni. 1987. Survey of the microbiological quality of whole, black pepper and turmeric powder sold in retail shops in Bombay. J. Food Prot. 50(5):401-403.
30. Glicksman, M. 1969. "Gum technology in the food industries." Academic Press, New York.
31. Glicksman, M. 1982. "Food hydrocolloids," Vol. I, CRC Press, Boca Raton, Fla.
32. Glicksman, M. 1986. "Food hydrocolloids," Vol. II, CRC Press, Boca Raton, Fla.
33. Glicksman, M. 1986. "Food hydrocolloids," Vol. III, CRC Press, Boca Raton, Fla.
34. Gonzalez-Fandos, E., M. L. Garcia-Lopez, M. L. Sierra, and A. Otero. 1994. Staphylococcal growth and enterotoxins (A-D) and thermonuclease synthesis in the presence of dehydrated garlic. J. Appl. Bacteriol. 77:549-552.
35. Guarino, P. A. 1973. Microbiology of spices, herbs, and related materials. Spec. Rep. No 13, Proceedings of the 7th Annual Symposium, Fungi and Foods, 1972 Oct 19; New York State Agricultural Experiment Station, Geneva, N. Y.
36. Gustavsen, S., and O. Breen. 1984. Investigation of an outbreak of *Salmonella oranienburg* infections in Norway, caused by contaminated black pepper. Am. J. Epidemiol. 119(5):806-812.
37. Hannigan, K. J. 1980. Spices: changes ahead. Food Eng. June:47-50.
38. Harris, P. 1990. "Food gels." Elsevier Science Publishing Co., Inc., New York.
39. Hauschild, A. H. W. 1973. Food poisoning by *Clostridium perfringens*. Can. Inst. Food Sci. Technol. J. 62:106.
40. Hefnawy, Y. A., S. I. Moustafa, and E. H. Marth, 1993. Sensitivity of *Listeria monocytogenes* to selected spices. J. Food Prot. 56(10):876-878.
41. Hocking, A. D. 1992. Collaborative study on media for enumeration of xerophilic fungi. *In* R. A. Samson, A. D. Hocking, J. I. Pitt, A. D. King (eds.). "Modern methods in food mycology," p. 121-125. Elsevier, New York.
42. Hocking, A. D., and J. I. Pitt. 1980. Dichloran-glycerol medium for enumeration of xerophilic fungi from low-moisture foods. Appl. Environ. Microbiol. 39:488-492.
43. Hsieh, R. C., S. M. Johnson, D. H. Dudek, inventors. 1989 July 4. McCormick & Company, Inc, assignee. Process for sterilization of spices and leafy herbs. U.S. patent 4,844,933.
44. Imeon, A. 1997. "Thickening and gelling agents for food," 2nd ed. Chapman & Hall, New York.
45. International Organization for Standardization. 1995. ISO 676 Spices and condiments-botanical nomenclature. ISO, Switzerland.
46. Ismaiel, A., and M. D. Pierson. 1990. Inhibition of growth and germination of *C. botulinum* 33A, 40B, and 1623E by essential oil of spices. J. Food Sci. 55(6):1676-1678.
47. Ito, H., H. Chen, and J. Bunnak. 1994. Aflatoxin production by microorganisms of the *Aspergillus flavus* group in spices and the effect of irradiation. J. Sci. Food Agric. 65:141-142.

48. Johnson, M. G., and R. H. Vaughn. 1969. Death of *Salmonella typhimurium* and *Escherichia coli* in the presence of freshly reconstituted dehydrated garlic and onion. Appl. Microbiol. 17(6):903-905.
49. Julseth, R. M., and R. H. Deibel. 1974. Microbial profile of selected spices and herbs at import. J. Milk Food Technol. 37(8):414-419.
50. Juri, M. L., H. Ito, H. Watanabe, and N. Tamura. 1986. Distribution of microorganisms in spices and their decontamination by gamma-irradiation. Agric. Biol. Chem. 50(2):347-355.
51. Juven, B. J., J. Kanner, F. Schved, and H. Weissiowicz. 1994. Factors that interact with the antibacterial action of thyme essential oil and its active constituents. J. Appl. Bacteriol. 76:626-631.
52. Kadis, V. W., D. A. Hill, and K. S. Pennifold. 1971. Bacterial content of gravy bases and gravies obtained in restaurants. Can. Inst. Food Technol. J. 4(3):130-132.
53. Kang, K. S., and I. W. Cottrell. 1979. Polysaccharides. *In* H. J. Peppler and D. Perlman (eds.), "Microbial technology," 2nd ed., Vol. 1, p. 417. Academic Press, New York.
54. Kaul, M., and N. Taneja. 1989. A note on the microbial quality of selected spices. J. Food Sci. Technol. 26(3):169-170.
55. Kennedy, J. F., G. O. Phillips, D. J. Wedlock, and P. A. Williams. 1985. "Cellulose and its derivatives." John Wiley & Sons, New York.
56. King, A. D., Jr., A. D. Hocking, J. I. Pitt. 1979. Dichloran-rose bengal medium for enumeration and isolation of molds from foods. Appl. Environ. Microbiol. 37(5):959-964.
57. Kiss, I., and J. Farkas. 1988. Irradiation as a method for decontamination of spices. Food Rev. Int. 4(1):77-92.
58. Kneifel, W., and E. Berger. 1994. Microbiological criteria of random samples of spices and herbs retailed on the Austrian market. J. Food Prot. 57:893-901.
59. Kovacs-Domjan, H. 1988. Microbiological investigations of paprika and pepper with special regard to spore formers including *B. cereus*. Acta Alimentaria; 17(3): 257-64.
60. Krishnaswamy, M. A., J. D. Patel, K. K. K. S. Nair, and M. Muthu. 1974. Microbiological quality of certain spices. Indian Spices 11(1,2):6-11.
61. Laidley, R., S. Handzel, D. Severs, and R. Butler. 1974. *Salmonella weltevreden* outbreak associated with contaminated pepper. Epidemiol. Bull. Canada 18(4):62.
62. Laserna, E. C., F. Uyenco, E. Epifanio, R. L. Veroy, and G. J. B. Cajipe. 1981. Carrageenan from Euche*uma striatum* (Schmitz) in media for fungal and yeast culture. Appl. Environ. Microbiol. 42:174.
63. Lehmacher, A., J. Bockemuhl, and S. Aleksic. 1995. Nationwide outbreak of human salmonellosis in Germany due to contaminated paprika and paprika-powdered potato chips. Epidemiol. Infect. 115:501-511.
64. Lenovich, L. M., J. L. Walters, and D. M. Reed. 1986. Comparison of Media for the Enumeration of fungi from dried foods. *In* A. D. King, Jr, J. I. Pitt, L. R. Beuchat, and J. E. L. Corry (eds.), "Methods for the mycological examination of food." NATO ASI series A: Life Sciences 122, p. 76-83. Plenum Press, New York.
65. Liu, Z. H., and H. Nakano.. 1996. Antibacterial activity of spice extracts against food-related bacteria. J. Fac. Appl. Biol. Sci. 35:181-190.
66. Llewellyn, G. C., M. L. Burkett, and T. Eadie. 1981. Potential mold growth, aflatoxin production and antimycotic activity of selected natural spices and herbs. J. Assoc. Off. Anal. Chem. 64(4):955-960.
67. Malmsten, T., K. Paakkonen, and L. Hyvonen. 1991. Packaging and storage effects on microbiological quality of dried herbs. J. Food Sci. 56(3):873-875.
68. Michael, G. T., and C. R. Stumbo. 1970. Ethylene oxide sterilization of *Salmonella senftenberg* and *Escherichia coli*: death kinetics and mode of action. J. Food Sci. 35:631-634.
69. Pafumi, J. 1986. Assessment of the microbiological quality of spices and herbs. J. Food Prot. 49(12):958-963.
70. Patkar, K. L., C. M. Usha, H. S. Shetty, N. Paster, and J. Lacey. 1993. Effect of spice essential oils on growth and aflatoxin B_1 production by *Aspergillus flavus*. Lett. Appl. Microbiol. 17:49-51.
71. Powers, E. M., T. G. Latt, and T. Brown. 1976. Incidence and levels of *Bacillus cereus* in processed spices. J. Milk Food Technol. 39(10):668-670.
72. Powers, E. M., R. Lawyer, and Y. Masuoka. 1975. Microbiology of processed spices. J. Milk Food Technol. 38(11):683-687.
73. Pruthi, J. S. 1964. Chemistry, microbiology and technology of curry powders. Spices Bull. 3(6):7.
74. Pruthi, J. S. 1980. "Spices and condiments: chemistry, microbiology, technology." Advances in Food Research Supplement 4. Academic Press, New York.
75. Pruthi, J. S., and B. D. Misra. 1963. Physico-chemical and micro-biological changes in curry powders during drying, milling and mixing operations. Spices Bull. C 3(3-5):8.
76. Rees, L. P., S. F. Minney, N. T. Plummer, J. H. Slater, and D. A. Skyrme. 1993. A quantitative assessment of the antimicrobial activity of garlic (*Allium sativum*). World J. Microbiol. Biotechnol. 9:303-307.
77. Rodriguez-Romo, L. A., N. L. Heredia, R. G. Labbe, and J. S. Garcia-Alvarado. 1998. Detection of enterotoxigenic *Clostridium perfringens* in spices used in Mexico by dot blotting using a DNA probe. J. Food Prot. 61(2):201-204.
78. Salmeron, J., R. Jordano, G. Ros, and R. Pozo-Lora. 1987. Microbiological quality of pepper (*Piper nigrum*) II. Food poisoning bacteria. Microbiol. Aliments Nutr. 5:83-86.
79. Satchell, F. B., V. R. Bruce, G. Allen, and W. H. Andrews. 1989. Microbiological survey of selected imported spices and associated fecal pellet specimens. J. Assoc. Off. Anal. Chem. 72(4):632-637.
80. Scott, P. M., and B. P. C. Kennedy. 1973. Analysis and survey of ground black, white, and capsicum peppers for aflatoxins. J. Assoc. Off. Anal. Chem. 56(6):1452-1457.
81. Scott, P. M., and B. P. C. Kennedy. 1975. The analysis of spices and herbs for aflatoxins. Can. Inst. Food Sci. Technol. J. 8(2):124-125.
82. Severs, D. 1974. *Salmonella* food poisoning from contaminated white pepper. Epidemiol, Bull. (Canada) 18:80.
83. Shelef, L. A. 1983. Antimicrobial effects of spices. J. Food Safety 6:29-44.
84. Shelef, L. A., O. A. Naglik, and D. W. Bogen. 1980. Sensitivity of some common food-borne bacteria to the spices sage, rosemary, and allspice. J. Food Sci. 45:1042-1044.
85. Souw, P., and N. J. Rehm. 1973. Investigations on microorganisms in thickening agents. I. Cell counts of aerobic microorganisms. Chem. Mikrobiol. Technol. Lebensm. 2(6):187.
86. Souw, P., and H. J. Rehm. 1975a. II. Cell counts of anaerobic sporeformers. Chem. Mikrobiol. Technol. Lebensm. 4(3):71.
87. Souw, P., and N. J. Reh 1975b. III. Survival of *Escherichia coli*, *Streptococcus faecalis*, and *Staphylococcus aureus* in dried thickening agents. Chem. Microbiol. Technol. Lebensm. 4:97.
88. Souw, P., and N. J. Rehm. 1975c. IV. Microbial degradation of three plant exudates and two seaweed extracts. Z. Lebensm. Unters. Forsch. 159(5):297.
89. Souw, P., and N. J. Rehm. 1976. V. Degradation of the galactomannans guar gum and locust bean gum by different bacilli. Eur. J. Appl. Microbiol. 2: 47.
90. Stahl, W. H., J. N. Skarzynski, and W. A. Voelker. 1969. Differentiation of certain types of cassias and cinnamons by measurement of mucilaginous character. J. Assoc. Off. Anal. Chem. 52(4):741-744.
91. Stephen, A. M. 1995. "Food polysaccharides and their applications." Marcel Dekker Inc., New York.
92. Tainter, D. R., and A. T. Grenis. 1993. "Spices and seasonings: A food technology handbook." VCH Publishers, Inc., New York.
93. The American Spice Trade Association, Inc. 1990. "The foodservice & industrial spice manual." ASTA, Englewood Cliffs, N. J.
94. The International Commission on Microbiological Specifications for Foods. 1980. "Microbial ecology of foods, Vol. 6, Food commodities. Blackie Academic & Professional, New York.
95. The International Commission on Microbiological Specifications for Foods. 1986. "Micro-organisms in foods 2, Sampling for microbiological analysis: principles and specific applications," 2nd ed. University of Toronto Press, Toronto.
96. The United States Pharmacopoeia/National Formulary. 1995. Rand McNally, Tauton, Mass.

97. U.S. Environmental Protection Agency. 1994. Ethylene oxide, tolerances for residues. 40CFR 180.151. U.S. Government Printing Office, Washington, D. C.
98. U.S. Environmental Protection Agency. 1997. Ethylene oxide. 40 CFR 185.2850. U.S. Government Printing Office, Washington, D. C.
99. U.S. Environmental Protection Agency. 1997. Propylene oxide. 40 CFR 185.5150. U.S. Government Printing Office, Washington, D. C.
100. U.S. Food and Drug Administration. 1998. "Bacteriological analytical manual," 8th ed. Revision A. AOAC International, Gaithersburg, Md.
101. U.S. Government National Archives and Records Administration. 1997. 21CFR101.22. Foods; Labeling of spices, flavorings, colorings and chemical preservatives. U.S. Government Printing Office, Washington, D. C.
102. Vajdi, M., and R. R. Pereira. 1973. Comparative effects of ethylene oxide, gamma irradiation and microwave treatments on selected spices. J. Food Sci. 38:893-895.
103. Weber, F. E. 1980. Controlling microorganisms in spices. Cereal Foods World 25(6):319-321.
104. Whistler, R. L., and J. N. BeMiller, JN. 1993. "Industrial gums," Academic Press, New York.
105. Whistler, R. L., and J. N. BeMiller. 1997. "Carbohydrate chemistry for food scientists." Eagan Press, St. Paul, Minn.
106. Whistler, R. L., and J. R. Zysk. 1978. Carbohydrates. *In* Kirk-Othmer encyclopedia of chemical technology, 3rd ed. vol. 4, p. 535. John Wiley & Sons, New York.
107. Zaied, S. E. A., N. H. Aziz, A. M. Ali. 1996. Comparing effects of washing, thermal treatments and gamma irradiation on quality of spices. Nahrung 40(1):32-36.
108. Zaika, L. L., and J. C. Kissinger. 1981. Inhibitory and stimulatory effects of oregano on *Lactobacillus plantarum* and *Pediococcus cerevisiae*. J. Food Sci. 46:1205-1210.

Chapter 53

Salad Dressings

Richard B. Smittle and Michael C. Cirigliano

53.1 INTRODUCTION

Commercially manufactured salad dressings began to appear on the American market around 1912.[44] Mayonnaise, cooked starch-based dressings resembling mayonnaise, and pourable dressings are the types of salad dressings most commonly marketed. Sufficient cooking to sterilize would destroy the physical integrity of these products; thus, preservation usually depends on the vinegar (acetic acid) or lemon juice present. Acetic acid concentration in excess of 1.5% makes a product unpalatable, but a concentration much below this level may permit spoilage. Worrell[44] suggests that the shelf life of properly prepared mayonnaise is between 3 and 6 months. However, it is common in the salad dressing and mayonnaise industry to have a 9- to 12-month shelf life. The shelf life of commercial mayonnaise and salad dressing is largely dictated by the physical and chemical characteristics of the product as opposed to the microbiological stability.

53.2 GENERAL CONSIDERATIONS

Mayonnaise is a creamy, pale yellow food with a mild to tangy flavor. The pH range is usually from 3.6 to 4.0, which is mainly due to the product's containing 0.5% to 1.2% acetic acid. Acetic acid, and occasionally other organic acids, are important for flavoring and are also the primary preservative agents. However, sodium benzoate and potassium sorbate are also frequently included as preservatives. The oil content (U.S.A.) ranges from 65% to 80%.[36,37] In the aqueous phase, the salt (NaCl) content is about 9.0% to 11.0%, and sugar usually represents 7.0% to 10.1%. The water activity (a_w) of a typical dressing with 12% salt is around 0.925. Traditionally, mayonnaise and salad dressings have been manufactured with oil, vinegar, water, and various flavor ingredients such as tomato, spices, sugar, and vegetable pieces. Today, salad dressings range from the traditional Italian type (two-phase high acetic acid and high oil levels) to the refrigerated spoonable type containing fresh dairy ingredients. In the 1980s, a trend has developed toward the manufacture of less tart (reduced vinegar), low calorie (reduced oil) products. Two significant microbiological consequences occur with both of these reductions. First, as the acetic acid content decreases, the pH of the product is raised. Higher pH levels create a more microbiologically unstable product. Second, as the oil content decreases, the water phase is increased, which in turn decreases the salt and the organic acid concentration in the water phase. This water phase is the critical microbiological concern.

53.3 NORMAL FLORA

The microorganisms in salad dressings come from the ingredients, from manufacturing equipment, and from the air. Few species are able to survive the low pH of salad dressings, and these few generally appear in low numbers.[11,23] Bachmann[2] isolated *Bacillus subtilis, B. mesentericus (B. pumilis* and *B. subtilis),*[13] micrococci, a diplococcus, and a mold from several types of unspoiled dressings. Fabian and Wethington[11] found no thermophiles, coliforms, or lipolytic bacteria, and only a few yeasts in 103 samples of unspoiled dressings. Some dressings contained a few molds. Of 10 unspoiled dressings examined by Kurtzman et al.,[23] nine appeared sterile, and one contained *B. subtilis* and *B. licheniformis,* but with fewer than 50 organisms per g.

53.4 FLORAL CHANGE IN SPOILAGE

The microflora responsible for the spoilage of salad dressings seems quite restricted and ordinarily consists of a few species of *Lactobacillus, Saccharomyces,* and *Zygosaccharomyces.* Typically, the spoilage organisms of mayonnaise have been a few species of *Hansenula, Pichia, Geotrichum, Saccharomyces,* and *Lactobacillus.*[32] The source of spoilage organisms generally can be traced to unsanitary equipment such as infrequently cleaned mixing, pumping, and filling machines. Perhaps such equipment is contaminated initially by low levels of spoilage organisms from ingredients and, rarely, from aerosols created during cleaning. Surface spoilage of dressings may also result from airborne contaminants. Mold and film yeasts have on occasion been encountered on dressings with large headspaces or improperly torqued caps, where increased amounts of air were available for their growth.[32]

Mayonnaise and other salad dressings spoil for a variety of reasons: separation of emulsion, oxidation, and hydrolysis of the oils by strictly chemical processes and from the growth of microorganisms.[12,16,39]

Microbiological spoilage is frequently manifested by gas formation that forces out the dressing when the container is opened. Other indicators of spoilage, such as off-flavor and change in color, odor, or texture, may occur. Iszard,[17,18,19] one of the first to report microbiological spoilage of mayonnaise, demonstrated *B. petasites (B. megaterium)*[13] as the cause. Spoilage of a Thousand Island dressing was caused by *B. vulgatus (B. subtilis)*,[26] and the source of contamination was found to be the pepper and paprika used in the formulation. Mayonnaise and dressings, as they are now produced, would probably not be spoiled by *Bacillus* spp. because of their generally low pH and high acetic acid contents.

Lactobacillus fructivorans was first isolated from spoiled salad dressing,[4] and later proved to be a common spoilage organism[23] that required special isolation media. Cirigliano[5] and Smittle[32] have reported that salad dressings are frequently spoiled by *L. plantarum, L. buchneri, L. fermentum, L. brevis, L. fructivorans*, and on occasion by *L. cellobiosus.*

Yeasts frequently cause spoilage in a variety of dressings. Fabian and Wethington[10] found a species of *Saccharomyces* in spoiled French dressing and mayonnaise, and Williams and Mrak[43], showed that a yeast similar to *Saccharomyces globiformis (S. bailii?)* caused spoilage of a starch-based dressing. Two-thirds of the spoiled dressing samples examined by Kurtzman et al.[23] contained *S. bailii (Zygosaccharomyces bailii).* Since the samples came from widely separated areas of the United States, this suggests *S. bailii* to be the yeast primarily responsible for dressing spoilage. Appleman et al.[1] found a mixture of *B. subtilis* and *Saccharomyces sp.* responsible in one instance of mayonnaise spoilage. *S. bailii* and *L. plantarum* were present in high numbers in a blue cheese dressing.[23]

53.5 HUMAN DISEASE MICROORGANISMS

The survival of pathogenic microorganisms in salad dressings and mayonnaise has been investigated by numerous workers. When Wethington and Fabian[41] inoculated salad dressing with *Salmonella* and *Staphylococcus,* they found that survival time depended on product pH. At pH 5.0, one strain of *Staphylococcus* survived 168 hr, but at pH 3.2 survival was limited to 30 hr. The longest survival time for *Salmonella* in dressing was 144 hr at pH 5.0, and 6 hr at pH 3.2. Similar data have been reported by other investigators.[8,12,15,21,31] Erickson and Jenkins[8] showed that *Salmonella* spp. including *Salmonella* phage type 4 were rapidly inactivated in commercial reduced calorie mayonnaise dressings at the same rate as real mayonnaise regardless of the aqueous phase acetic acid, salt, sucrose, and other compositional factors. Vladimirov and Nefedieva[38] reported *Escherichia coli* was able to survive one day in mayonnaise, and the data of Bachmann[2] suggested survival of no more than 10 days. More recent work has reported that *E. coli* 0157:H7 does not grow in commercial mayonnaise and mayonnaise-based dressings[9,28,40,45] *E. coli* 0157:H7 rapidly died-off at normal storage temperatures (22–25°C). However, at lower temperatures they can survive for longer periods of time, for example at 20°C they can survive for as long as 8–21 days,[45] at 7°C they can survive for as long as 35 days,[40] and at 5°C they can survive for as long as 55 days.[45] There are no reports that *E. coli* 0157:H7 grew in commercial salad dressings. Smittle[31] reviewed the literature concerning the microbiology of these products and concluded that mayonnaise and salad dressing prepared according to the U.S. Food and Drug Administration (FDA) Standard of Identity[36,37] are bactericidal to the vegetative cells of *Salmonella, Staphylococcus aureus, Clostridium botulinum, Clostridium perfringens, Streptococcus viridans, Shigella flexneri,* and *Bacillus cereus.* Furthermore, he concluded that mayonnaise of pH 4.1 or less (0.25% acetic acid) ensures a bacteriologically safe product.[31] Codex Alimentarius Commission[6] recommendations for standards of identity, if accepted by the national government, become the law of the accepting country.

53.6 MICROBIAL INDICATORS OF POOR SANITATION

Microorganisms tolerant of the conditions existing in salad dressings are used as indicators of poor sanitation. Not only are a few selected organisms capable of growing in the undiluted product, but some can grow to large populations in the diluted product as encountered in improperly cleaned and sanitized equipment. Three groups of organisms are commonly used as microbial indicators. They are the yeasts and molds, lactobacilli, and aerobic bacilli. High numbers of these groups are indicative of poor sanitation and potential spoilage problems. Specifically, the presence of any spoilage organism at any level is unacceptable. Since most yeasts and molds, and lactobacilli introduced in low numbers, die quickly, salad dressings and mayonnaise usually contain <10 per g of these organisms. Any persistence or increase in numbers to >10 per g should be a warning of a sanitation problem.

When salad dressings become diluted, bacteria can grow. However, when diluted material containing bacteria is mixed with undiluted product, the vegetative cells of most bacteria usually die quickly unless acid or preservative adaptation has taken place. Usually, only bacterial spores survive, and these are detected using standard plating procedures. These spores remain viable almost indefinitely. The presence of aerobic bacilli in excess of 50 per g should be considered indicative of a sanitation problem. The exception to this discussion is salad dressings containing unprocessed (unheated) cheese or other cultured dairy products (e.g., cream and buttermilk). Cheeses and other fermented dairy products containing live molds and cultured bacteria must have their own set of standards, which need to be individually determined. All of these indicators are minimized by using microbiologically acceptable ingredients and by cleaning and sanitizing equipment properly. Areas of particular concern are pumps, mixing and filling equipment, and product transfer lines. The implementation of frequent CIP and CPD cleaning regimens, as well as the periodic breakdown of all equipment and replacement of worn gaskets and O-rings, is recommended, particularly where the manufacture of sensitive low-acid salad dressings is involved.

53.7 EQUIPMENT, MATERIALS, AND SOLUTIONS

53.71 Equipment

Balance
Durham tubes
Glass slides and coverglasses
Glass rod
Incubator, 22° to 28°C
Incubator, 35°C
Stomacher bags
Metal spatulas
Microscope
Petri dishes
pH meter
Pipettes
Quebec Colony Counter

53.72 Materials and Solutions (Chapter 63)

Carbon dioxide

Crystal violet, 0.5%
Fructose (for MRS agar) (0.5%)
Gram stain reagents
Lactobacillus heterofermentative screen broth
MRS agar
0.1% peptone water
Plate count agar (PCA)
Phosphate buffer (Butterfield's)
Potato dextrose agar (PDA), pH 3.5
PDA with antibiotics
Chlortetracycline HCL (100 mg/L)
Chloramephenicol (100 mg/1L)
Tartaric acid (10%)
Cycloheximide (100 mg/mL)
0.1N Sodium hydroxide

53.8 RECOMMENDED METHODS

53.81 *Sample Preparation*

Stir sample with a sterile glass rod or a sterile metal spatula; place 50 g into a sterile stomacher bag. Add enough diluent to prepare a 1:10 dilution using sterile phosphate buffer or 0.1% peptone water and treat for 2 min. Make subsequent dilutions to 10^6 with phosphate buffer or peptone water.[23,35]

53.82 *Yeasts and Molds*

For isolation, use PDA agar with 100 mg/L of chlortetracycline HCl and chloramphenicol to inhibit bacteria. Incubate plates at 25° ± 5°C and examine at 3 and 5 days. As an alternative, use PDA acidified with 10% tartaric acid to at least pH 3.5.

53.83 *Aerobic Bacteria*

For isolation, use PCA containing 100 µg/mL of cycloheximide for inhibition of fungi. Some yeasts, not likely to be found in salad dressings, are resistant to this level of cycloheximide. Incubate plates at 28° to 35°C, and examine at 2 and 5 days.[23]

53.84 *Lactobacilli*

Fastidious lactobacilli such as *Lactobacillus fructivorans* and *L. brevis* cannot be detected easily on PCA, but they are readily isolated on *Lactobacillus*-selective MRS agar.[3,24,27] Incubate plates of MRS agar supplemented with 0.5% fructose at 20° to 28°C in a carbon dioxide-enriched atmosphere and examine at 3, 5, and 14 days.[34] The use of a carbon dioxide-enriched atmosphere generally shortens the incubation time.[33] The CO_2 necessary for incubation can be obtained by flushing the incubation container with CO_2 from a gas cylinder or generating CO_2 with commercially available devices. Alternatively, overlay and incubate MRS agar plates at 20° to 28°C, and examine at 5, 7, and 14 days.

53.85 *Heterofermentative Screen*

To screen salad dressings and ingredients for acetophilic bacteria, pipette 1 mL of 1:10 dilution into each of 3 *Lactobacillus* heterofermentative screen broth tubes. Incubate tubes at 32°C for 72 hr ± 2 hr. Positive tubes have trapped CO_2 in the Durham tube or bubbles of CO_2 clinging to the inside of the tube and a color change from green to yellow indicating acid production.

Perform a gram stain or wet mount to verify the presence or absence of yeasts. This procedure can also be used as a yeast screen by omitting the actidione from the formula followed by staining and isolation on PDA for confirmation.

53.86 *Coliforms, fecal coliforms,* E. coli, Salmonella, Staphylococcus aureus, Listeria monocytogenes, *and* Yersinia enterocolitica

If mayonnaise and salad dressings are suspected of causing illness, they should be examined for coliforms, including *E. coli* (Chapter 35), *Salmonella* spp. (Chapter 37), *Staphylococcus aureus* (Chapter 39), *Listeria monocytogenes* (Chapter 36), and *Yersinia enterocolitica* (Chapter 41). Because of their acidity, the products should first be neutralized. Product pH and total acidity must also be measured (See Chapter 64).

53.87 *Microscopic Observation of Dressings*

Yeasts and bacteria are readily stained for microscopic observation by using crystal violet. A small drop of 0.5% crystal violet may be mixed directly with the product on a microscope slide or mixed with a drop of the product diluted 1:10 with distilled water.

53.88 *Identification of Spoilage Microorganisms*

Yeasts can be identified by the culture techniques of Wickerham[42] and the classification systems found in "The Yeasts."[22] Criteria for identification of Lactobacillus can be found in "Bergey's Manual"[20] and in papers by Rogosa and Sharpe,[30] Charlton et al.,[4] Kurtzman et al.,[23] and Smittle and Flowers.[33] Species of *Bacillus* can be identified on the basis of the scheme presented by Gordon et al.[13]

53.9 INTERPRETATION OF DATA

Sometimes, obviously spoiled dressings contain few viable microorganisms (fewer than 10^2 per g) or none at all. In these instances, the microorganisms have probably died after the nutrients were exhausted or after the accumulation of metabolic by-products. A direct microscopic examination will usually reveal the dead cells. If even a few cells are seen per field, this sighting indicates that large populations of viable organisms were present in the dressing at one time and were probably responsible for the spoilage observed.

Gaseous fermentation may not be evident in spoiled dressings until several weeks after manufacture. *Lactobacillus fructivorans* and other lactobacilli grow slowly, and considerable time is needed for microbial population increase and visible gas buildup. Some spoilage may be observed only by an increase in acid or a flavor change, especially spoilage from the homofermentative lactobacilli, such as *L. casei* and *L. plantarum*. This slow growth is particularly evident on isolation plates; 10 to 14 days may elapse before colonies are observed.[23] It seems likely that this slow growth, as well as failure to use the proper isolation medium, may account for the relatively few reports of *L. fructivorans* in spoiled salad dressings.

Zygosaccharomyces bailii and certain other haploid species of *Zygosaccharomyces* ferment fructose and glucose quickly, but may give a delayed fermentation of sucrose.[23,25] This explains the long delay between manufacture and spoilage of products contaminated with these yeasts when sucrose is the sweetener. Kurtzman et al.[23] showed that 9 out of 13 strains of *S. bailii (Z. bailii)*[22] from spoiled dressings fermented sucrose vigorously, but fermentation did not begin until 12 to 56 days after inoculation.

Salad dressings average approximately pH 4 (3.0 to 4.6)[11,23] because of their acetic acid content, and this acidity accounts for the absence of food poisoning microorganisms. Dressings formulated at a significantly higher pH and stored below 20° C should be examined for the presence of microorganisms causing foodborne illness.

Mixing of dressings into meat, potato, and similar salads dilutes the acetic acid so that its inhibitory properties may be diminished.[14] Nevertheless, mayonnaise and salad dressings frequently retain much of their inhibitory characteristic when mixed with meat and vegetables [7,14,29] and Smittle[31] recommended that when these ingredients are to be used in salads, they should be mixed with the salads as soon as possible to retard microbial growth.

53.10 REFERENCES

1. Appleman, M. D., E. P. Hess, and S. C. Rittenberg. 1949. An investigation of a mayonnaise spoilage. *Food Technol.* 3:201.
2. Bachmann, F. M. 1928. A bacteriological study of salad dressings. *Wisc. Acad. Sci. Arts Lett. Trans.* 23: 529.
3. Carr, J. G. 1975. VI.3 Lactics of the world unite, p. 369. *In* J. G. Carr, C. V. Cutting, and G. C. Whiting (eds.), Lactic acid beverages and food. Academic Press, New York.
4. Charlton, D. B., M. E. Nelson, and C. H. Werkman. 1934. Physiology of *Lactobacillus fructivorans sp.* nov. isolated from spoiled salad dressings. *Iowa State J. Sci.* 9:1.
5. Cirigliano, M. C. 1985. Microbiological hazards to processing. Presented at Sanitation Seminar, Sept. 1985. Assoc. for Dressings and Sauces, Atlanta, Ga.
6. Codex Alimentarius Commission. 1998. Codex Alementaries Commission Standards for Mayonnaise CAC/Vol 11. 2nd ed. World Health Organization, Rome.
7. Douglas, H. C., and J. C. M. Fornachon. 1949. The taxonomy of *Lactobacillus hilgardii* and related heterofermentative lactobacilli. *Hilgardia* 19:133.
8. Erickson, J. P., and P. Jenkins. 1991. Comparative *Salmonella* spp. and *Listeria monocytogenes* inactivation rates in four commercial mayonnaise products. *J. Food Prot.* 54:913.
9. Erickson, J. P., J. W. Stamer, M. Hayes, D. N. McKenna, and L. A. Van Alstine. 1995. An assessment of *Escherichia coli*: 0157:H7 contaminaion risks in commercial mayonnaise from pasteurized eggs and environmental sources, and behavior in low-pH dressings. *J. Food Prot.* 58:1059.
10. Fabian, F. W., and M. C. Wethington. 1950a. Spoilage in salad and French, dressing due to yeasts. *Food Res.* 15:135.
11. Fabian, F. W., and M. C. Wethington. 1950b. Bacterial and chemical analyses of mayonnaise, salad dressing, and related products. *Food Res.* 15:138.
12. Frazier, W. C. 1967. Food microbiology, p. 537. McGraw-Hill, New York.
13. Gordon, R. E., W. C. Haynes, and C. H-N. Pang, 1973. The genus *Bacillus*. Agric. Handb. No. 427. U.S. Dep. Agric., Washington, D. C.
14. Gould, S., A. Woolford, H. Rappaport, and J. M. Goepfert. 1976. Factors affecting the behavior of *Salmonella* and *Staphylococci* in meat salad. Annual Report 1976. Food Res. Inst., Madison, Wisc.
15. Gram, H. G. 1957. Abtotung von *Salmonellen, Staphylococcus aureus, B. proteus,* und *B. alkaligenes* durch mayonnaise. *Fleischwirtsch* 9:111.
16. Gray, H. G. 1927. Bacterial spoilage in mayonnaise, relishes, and spreads. *Canning Age* 8:643.
17. Iszard, M. S. 1927a. The value of lactic acid in the preservation of mayonnaise dressing and other dressings. *Canning Age* 8:434.
18. Iszard, M. S. 1927b. The value of lactic acid in the preservation of mayonnaise dressing and other products. *J. Bacteriol.* 13:57.
19. Iszard, M. S. 1927c. Supplementary report on the use of lactic acid as a preservative in mayonnaise and allied products. *Spice Mill* 50:2426.
20 Kandler, O., and N. Weiss. 1986. Regular, nonsporing gram-positive rods, p. 1208. *In* J. G. Holt (ed.), Bergey's manual of systemic bacteriology. Williams and Wilkens, Baltimore, Md.
21. Kintner, T. C., and M. Mangel. 1953. Survival of staphylococci and salmonellae experimentally inoculated into salad dressing prepared with dried eggs. *Food Res.* 18:6.
22. Kreger-van Rij, N. J. W. The yeasts, A taxonomic study, 3rd ed. North-Holland Publishing Co., Amsterdam, Holland.
23. Kurtzman, C. P., R. Rogers, and C. W. Hesseltine. 1971. Microbiological spoilage of mayonnaise and salad dressings. *Appl. Microbiol.* 21:870.
24. Lawrence, D. R., and P. A. Leedham. 1979. Detection of lactic acid bacteria. *J. Inst. Brew.* 85:119.
25. Pappagianis, D., and H. J. Phaff. 1956. Delayed fermentation of sucrose by certain haploid species of *Saccharomyces*. Antonie van Leeuwenhoek *J. Microbiol. Serol.* 22:353.
26. Pederson, C. S. 1930. Bacterial spoilage of a Thousand Island dressing. *J. Bacteriol.* 20:99.
27. Peladan, F., D. Erbs, and M. Moll. 1986. Practical aspects of detection of lactic bacteria in beer. *Food Microbiol.* 3:281.
28. Raghubeer, E.V., J. S. Ke, M. L. Campbell, and R. S. Meyer. 1995. Fate of *Escherichia Coli* 0157:H7 and other coliforms in commercial mayonnaise and refrigerated salad dressings. *J. Food Prot.* 58:13.
29. Rappaport, H., and J. M. Goepfert. 1975. Behavior of *Salmonella* and *Staphylococcus aureus*. Annual Report 1975. Food Res. Inst., Madison, Wisc.
30. Rogosa, M., and M. E. Sharpe. 1959. An approach to the classification of the lactobacilli. *J. Appl. Bacteriol.* 22:329.
31. Smittle, R. B. 1977. Microbiology of mayonnaise and salad dressing: A review. *J. Food Prot.* 40:415.
32. Smittle, R. B. 1987. The microbiology of dressings and sauces. Presented at Microbiology Quality Assurance Seminar, Sept. 1987. Assoc. for Dressings and Sauces, Atlanta, Ga.
33. Smittle, R. B., and R. M. Flowers. 1982. Acid tolerant microorganisms involved in the spoilage of salad dressings. *J. Food Prot.* 45:977.
34. Splittstoesser, D. F., L. L. Lienk, M. Wilkinson, and J. R. Stamer. 1975. Influence of wine composition on the heat resistance of potential spoilage organisms. *Appl. Microbiol.* 30:369.
35. Straka, R. P., and J. L. Stokes. 1957. Rapid destruction of bacteria in commonly used diluents and its elimination. *Appl. Microbiol.* 5:21.
36. U.S. Government National Archives and Records Administration. 1998. Code of Federal Regulations Title 21 Part 169.140 Mayonnaise p. 529. U.S. Government Printing Office, Washington, D. C.
37. U.S. Government National Archives and Records Administration. 1998. Code of Federal Regulations Title 21 Part 169. 150 Salad Dressing p. 529-530. U.S. Government Printing Office, Washington, D.C.
38. Vladimirov, B. D., and N. P. Nefedieva. 1937. Mayonnaise as a culture medium for microorganisms. *Vopr. Pitan.* 6:85.
39. Walker, H. W., and J. C. Ayres. 1970. Yeasts as spoilage organisms, p. 463. *In* A. H. Rose and J. S. Harrison (eds.), The yeasts, vol. 3, ed. Academic Press, New York.
40. Weagant, S. D., J. L. Bryant, and D. H. Bark. 1994. Survival of *Escherichia coli* 0157:H7 in mayonnaise and mayonnaise- based sauces at room and refrigerated temperatures. *J. Food Prot.* 57:629.
41. Wethington, M. C., and F. W. Fabian. 1950. Viability of food-poisoning staphylococci and salmonellae in salad dressing and mayonnaise. *Food Res.* 15:125.
42. Wickerham, L. J. 1951. Taxonomy of yeasts. U.S. Dep. Agric. Tech. Bull. No. 1029 1.
43. Williams, O. B., and E. M. Mrak. 1949. An interesting outbreak of yeast spoilage in salad dressing. *Fruit Prod. J.* 28: 41.
44. Worrell, L. 1951. Flavors, spices, condiments, p. 1706. *In* M. B. Jacobs (ed.), The chemistry and technology of food and food products, vol. 2. Interscience Publ., New York.
45. Zhao, T., and M. P. Doyle. 1995. Fate of enterohemorrhagic *Escherichia coli* 0157:H7 in commercial mayonnaise. *J. Food Prot.* 57:780.

Chapter 54

Sweeteners and Starches

Richard B. Smittle and John P. Erickson

54.1 INTRODUCTION

Natural sweeteners are derived from plant material and are processed to an extent that their physical appearance is completely unlike the source material. They come in a variety of sugar compositions, depending on their origin and processing steps, and are sold as liquid syrups or in dry crystalline or powder forms. In many cases, the source material is highly vulnerable to microbial spoilage during harvesting and processing. However, the final products are usually microbiologically stable, depending on the water activity (a_w).[17]

The most commonly used sweetener, sucrose, is extracted and purified from cane, *Saccharum officinarum*, and beets, *Beta vulgaris*.[17,24] Other sweeteners are derived from the enzymatic and acid hydrolysis of corn starch to yield corn syrups with various concentrations of reducing sugars. The glucose resulting from hydrolysis can be converted to fructose using isomerase to form a glucose-fructose mixture called high-fructose corn syrup. Maple syrup, honey, and molasses are other natural sweeteners used in many foods as flavoring agents.

Next to cellulose, starch is the most prevalent naturally occurring biological substance on earth.[32] Of the 18 million tons of starch purified annually, half is used primarily for food.[2] Corn is the most common source of starch in the United States; potatoes and manioc are common sources of starch in Europe and Asia.

Starch molecules are comprised of polymers of D-anhydroglucose occurring in linear or branched forms. Amylose, the linear form of starch, is the primary component responsible for the swelling action (viscosity increase) when the starch granules are hydrated.[20] Amylopectin, the other polymer in starch, is a highly branched structure that resists gelatinization and is characterized by good clarity and stability. With the advent of modern processing techniques and genetic control of the corn plant resulting in chemical and physical modifications of starch, its use has increased dramatically in modern food product development. Examples of products that routinely contain starch or modified starch include thermally processed low-acid foods such as pork and beans, puddings, and gravies; acid foods such as salad dressings, barbecue sauces, and other condiments; baked goods and snack foods; dry blended food such as instant puddings, seasoning mixes, and soups; and dry coatings on candies and chewing gums.[32] Recently, starch-based polymers have been developed for use as biodegradable plastics in which the starch is the substrate for microbial action that helps to degrade the synthetic polymer.

In the United States, starch production and its use in the food industry are regulated by the Food and Drug Administration (FDA). Chemical modification of starch is defined in 21 CFR (Code of Federal Regulations) 172.892[36] and in the Food Chemicals Codex.[25] In other countries, recommendations of the Codex Alementarius Commission[10] if accepted by a national government, becomes laws of the accepting country. The process and chemicals used to modify starch vary but usually fall into one of four classes: bleached, converted, crossed linked, and stabilized.[11,31,32]

54.2 GENERAL CONSIDERATIONS

Low water activity is largely responsible for the microbial stability of starches and natural sweeteners.[34] Crystalline or powdered starches and sugars will remain microbially stable if they are kept dry.[17] On the other hand, liquid syrups, which are solutions of sweeteners and water, may spoil because of their higher water content.[17,29,34,37] Maple syrup, honey, molasses, mixtures of sucrose and invert, glucose, and fructose or glucose with fructose are in this category. Industrially prepared syrups range from 67° to 86° Brix, depending on the sweetener. In general, small sugar molecules exert greater osmotic pressure than large molecules. The preservatives sodium benzoate and potassium sorbate may be added to syrups to prevent spoilage, but their efficacy may be limited if the pH is not low enough. In addition, these preservatives cannot be added to some consumer products because of undesirable organoleptic changes.

The general steps of corn starch processing involve steeping, wet milling, washing and purification, and modification and drying.[28,31] Steeping is a controlled enzymatic degradation in which corn is soaked in 45° to 50°C water containing 0.1% to 0.2% sulfur dioxide[28] for 24 to 48 hr at a pH of *ca.* 4.0.[38] These conditions are critical since the high moisture state of the product makes it vulnerable to microbial attack, which can result in alcoholic or butyric acid byproducts.[31] After steeping, the kernels are cracked open by liquid cyclone, releasing the germ. The germ is pressed to remove oil, and the remaining aqueous starch-gluten mixture is separated, washed, chemically and/or physically modified, and dried to 10% to 17% moisture using flash, belt, or drum dryers.[38]

Tapioca starch is produced from the roots of the manioc plant, which grows in equatorial regions.[38] After harvest, the roots are washed prior to starch separation. Sulfur dioxide and low pH are used as described for corn starch to help control microbial contamination during the steeping process. After drying, the raw starch is often exported to processing plants where it is further processed and modified to meet specific needs of the food industry.

54.3 NORMAL AND INDICATOR FLORA

Microorganisms in starches and sweeteners come from the raw source materials or from the manufacturing processes.[18] High processing temperatures and/or low water activity of starches and natural sweeteners afford little opportunity for microbial survival and growth, but recontamination may occur after heat treatment.

The steeping process is a critical control point in the manufacture of corn starch. Here, the starch grain is subjected to high moisture during the enzymatic degradation of the corn. Sulfur dioxide and low pH are used to inhibit gram-negative bacteria, but lactic acid bacteria and flat sour sporeformers have been reported to proliferate during the steeping process.[16,19] Although aflatoxin can be present on corn used to produce starch, there is little if any carryover of aflatoxin to the edible product of the wet milling process (starch). However, the residuals (steepwater, gluten, germ) may contain the toxin.[4]

Raw tapioca starch is produced in areas such as Thailand, Brazil, and some African countries. Since the manioc roots are harvested from the soil, various soil microorganisms are initially present on the product and often proliferate during storage of the root prior to processing. Sodium hypochlorite and hydrogen peroxide are frequently used to reduce the microbial load of starch slurries,[31] and propylene oxide has been used as a package fumigant for bulk quantities of starch.[3]

Plant and equipment sanitation as well as water quality have been identified as factors affecting microbial contamination.[38] As a rule, there are usually $<10^2$ microorganisms per gram of sweetener.[30] There is usually <1 yeast per gram.[34] Organisms that are likely to survive are sporeforming mesophilic aerobic and anaerobic bacilli. Further, after heating, especially in syrups, aerobic mesophilic bacteria, yeasts, and molds from the environment may contaminate the product.

Microorganisms such as *Bacillus stearothermophilus, B. coag-ulans, Clostridium thermosaccharolyticum, C. nigrificans*, certain mesophilic bacteria, yeasts, and molds[7,9] may be present, but not grow in starches, sugars, and syrups. However, these organisms can cause spoilage of products when they are present in starches and sweeteners used as ingredients for other foods.[17,34]

The following is a summary of standards for these organisms for application of starches and sweeteners in food canning and beverage manufacturing.

1. National Food Processors Association's Bacterial Standards for Sugar:[14,26]
   Five samples are examined after heating.
   a. Total thermophilic spore counts—average of not more than 125 spores/10 g of sugar.
   b. Flat sour spores—average of not more than 50 spores/10 g of sugar.
   c. Thermophilic anaerobic (TA) spores—may be present in up to three of the five samples, but in any one sample, not more than four of six tubes inoculated by the standard procedure should contain TA spores.
   d. Sulfide spoilage spores—may be present in up to two of the five samples.
2. Bottlers' Standards for Dry, Granulated Sugar.[27]
   a. <200 mesophilic bacteria/10 g; <10 yeasts/10 g; <10 molds/10 g.
3. Bottlers' Standards for Liquid Sugar (Sugar Syrup) in 10 g of Dry Sugar Equivalent (DSE).[27]
   a. <100 mesophiles; <10 yeasts; <10 molds.

Food processors typically set microbiological specifications for starch according to the risk of spoilage that the starch may pose for the product and also according to the industry's ability to produce the ingredient with a certain microbial load. In general, starch contains <50,000 SPC/g; <0.3 MPN coliform/g; <0.3 MPN *E. coli*/g; <100 yeasts and molds/g; and is negative for *Salmonella* and *Staphylococcus aureus*. Liquid sweeteners select for microorganisms that tolerate low a_w, and contamination by osmotolerant organisms, especially yeasts, presents the greater risk of spoilage.

Because of their importance as spoilage organisms, yeasts capable of growth at <0.85 a_w have been investigated thoroughly and found to be part of the normal flora in the raw intermediate products of honey, maple syrup, molasses, corn syrup, and cane and beet sugar.[34,37] The most prominent osmotolerant yeasts appear to be in the genus *Zygosaccharomyces* (*Saccharomyces*).[37] Some other yeasts that appear as part of the normal flora belong to the genera *Pichia, Candida, Torula, and Schizosaccharomyces*.[37]

54.4 SPOILAGE ORGANISMS

The most common agents of undesirable fermentation in liquid syrups such as honey, maple syrup, and corn syrups are osmophilic yeasts, especially *Zygosaccharomyces rouxii*.[17,35,37] In addition to *Z. rouxii, Saccharomyces cerevisiae* and *S. mellis* may grow in liquid sucrose. It has been reported that honey is particularly prone to spoilage by *Z. japonicus, Z. barkeri, Z. mellis, Z. prioriano, Z. nussbaumeri*, and *Z. richteri*, all of which have been found in normal and fermented honeys.[37] The hygroscopicity and viscosity of honey may allow the development of a water-sugar gradient where yeasts can grow in sectors of the gradient with sufficiently high a_w.[35] *Torulopsis apicola* has spoiled white crystalline sucrose that had become contaminated with moisture.[34] Raw sugar that was stored in the country of origin has been shown to be spoiled by *S. rouxii* and *T. candida*.[34] Molasses has been spoiled by *S. heterogenicus* and *T. holnii*.[34] Brown and white sugar syrups have been spoiled by *Z. rouxii, Z. bailii* var. *osmophilus, Saccharomyces* spp., *Candida valida, Hansenula anomola* var. *anomola, Kloeckera apiculata, Candida* spp., *Torulopsis* spp., and *S. cerevisiae*.[34]

Saccharomyces aceris-sacchari, S. behrensianus, S. monocensis, Z. mellis, Z. barkeri, Z. japonicus, and *Z. nussbaumeri* have been isolated from spoiled maple syrup. *Z. bailii* and *Z. rouxii* have been routinely isolated from corn syrup (R. B. Smittle, personal communication). *S. zsopfi* has been isolated from canned fermented cane syrup.[37]

Microbial spoilage of liquid sugars and corn syrups may be prevented by destruction or removal of contaminating microorganisms, use of sanitizing agents on processing and syrup storage equipment, and prevention of water vapor condensates in storage vessels. To prevent condensation, filtered air treated by ultraviolet irradiation is forced over the surface of the liquid. Prompt use of syrups also helps to reduce the incidence of spoilage.

Under aerobic conditions, molds can grow and cause visible spoilage of syrups.[37] For example, in table syrups, especially maple syrup without an added preservative, molds may grow on the surface of the syrup, particularly after the consumer opens the container. Syrups preserved by potassium sorbate may be spoiled by sorbate-resistant molds, which produce a solvent-like odor.[35]

The presence of thermophilic spores is the most important issue in the microbiology of starch. Since starch is a product of soil-grown crops, it is not uncommon to find various soilborne *Bacillus* species in the finished product. In addition to the presence of microorganisms themselves, the presence of heat-stable amylases, which can remain after the cells have been killed, may also cause spoilage of products made with enzyme-contaminated starches. Often the microbial load may be reduced during the manufacturing process only to leave active enzymes that may continue to degrade the product post-production or when the raw ingredient is incorporated into a finished product. For this reason, not only should the number of microbial contaminants be determined during routine quality testing, but the types of organisms should also be identified.

54.5 PATHOGENS

The refining processes of crystalline sweeteners and liquid syrups derived directly from plant material destroy vegetative cells of pathogens. With the exception of honey, commercially produced starches and sweeteners are not involved in outbreaks of foodborne illness. Honey has been implicated as the source of botulinal spores in cases of infant botulism. Spores of *Clostridium botulinum* can grow and produce toxin in the intestines of infants, but preformed botulinal toxins have not been found in honey.[1,6,8,21,22] In the United States, surveys of honey not related to illness suggest the presence of about 1 to 10 spores/kg[15,23,33]; in Europe, surveys of honey not associated with infant botulism indicate the absence of spores of C. *botulinum*.[12,13]

The gram-positive pathogen *Bacillus cereus* and related aerobic sporeformers may be present in starches used as food ingredients. These organisms have the potential to grow in such foods as puddings and sauces that have been prepared to a high a_w with contaminated ingredients and then temperature abused.[18] Foodborne illness could then result from consumption of such temperature-abused foods.

54.6 RECOMMENDED METHODS

The following chapters in this compendium discuss specific procedures required for sample preparation and analyses for particular microorganisms.

54.61 Methods for Osmophilic/Xerotolerant Yeasts and Molds

Halophilic and osmophilic organisms: Chapter 17
Yeasts and molds: Chapter 20
Confectionery products: Chapter 56

54.62 Methods for Nonosmophilic/Xerotolerant Microorganisms

Sample preparation: Chapter 2
Aerobic plate count: Chapter 7
Yeasts and molds: Chapter 20
Coliforms, *E. coli*: Chapter 8
Salmonella: Chapter 37
Staphylococcus aureus: Chapter 39
Bacillus cereus: Chapter 32
Microscopic examination: Chapter 4
Mesophilic aerobic sporeformers: Chapter 22
Thermophilic flat sour sporeformers: Chapter 25
Aciduric flat sour sporeformers: Chapter 24
Thermophilic anaerobic sporeformers: Chapter 26
Sulfide spoilage sporeformers: Chapter 27
Clostridium botulinum spores in honey: References 12, 15, 22, 23, 33

The standard heat-shocking procedures, i.e., heating for 10 min under 5 pounds per square inch of steam pressure, used in the enumeration of thermophilic spores in starches and sweeteners, must be followed strictly to minimize the inaccuracy and unreliability of the method. For example, substitute heat-shocking treatments of samples in a hot water bath for longer times assumes a z-value that may be a source of error relative to the standard heat-shock procedure.[5]

54.63 Special Procedures

Water activity: Chapter 64

54.7 INTERPRETATION OF DATA

Water activity is the most important ecological factor for controlling microbial growth in starches and sweeteners.[34] Dry starch, crystalline, granulated, and powdered sweeteners are microbiologically stable and need only be kept dry to prevent microbial degradation. However, syrups with an a_w range of 0.65 to 0.70 are subject to spoilage by yeasts.[35] The presence of only a few viable osmophilic yeasts is a more important spoilage risk for syrups than is the presence of large numbers of other microorganisms that cannot grow at the low a_w of these products. Thus, in conducting tests, evaluating analytical data, and making decisions for possible remedial actions, it is important to identify the types of organisms that can grow in the product. The amount of sugar in solution, storage temperature, available oxygen, pH, added preservatives, and storage time will select for the type of organisms, if any, that can cause spoilage.[34]

Normally, syrups and dry sweeteners contain few microorganisms; most contain <100/g[30] and few yeasts.[34] Any syrup or granulated sugar in excess of these guidelines or those set by various trade associations must be treated with care, depending on the use of the product, e.g., in confectionery manufacturing, canning, or bottling.[14,26,27]

54.8 REFERENCES

1. Arnon, S. S., T. F. Midura, K. Damus, B. Thompson, R. M. Wood, and J. Chin. 1979. Honey and other environmental risk factors for infant botulism. *J. Pediatr.* 94:331-336.
2. Aspinall, G. O. 1985. The polysaccharides, vol. 3, p. 210. Academic Press, Orlando, Fla.
3. Banwart, G. J. 1989. Basic food microbiology, 2nd ed., p. 305-306. Van Nostrand Reinhold, New York.
4. Bennett, G. A., and R. A. Anderson. 1978. Distribution of aflatoxin and/or zearalenone in wet-milled corn products: A review. *J. Agric. Food Chem.* 26:1055-1060.
5. Bernard, D. 1980. NFPA position in regards to changes in the thermophile enumeration procedure as it applies to the use of NFPA standards for thermophiles in starches. NFPA Laboratory Memorandum. May 28, 1980.
6. Brown, L. W. 1979. Commentary: Infant botulism and the honey connection. *J. Pediatr.* 94:337-338.
7. Cameron, E. J., and C. C. Williams. 1928. The thermophilic flora of sugar in its relation to canning. *Zentralbl. Bakteriol. Parasitenkd. Infektionskr. Abt.* 176:28-37.

8. Chin, J., S. S. Arnon, and T. F. Midura. 1979. Food and environmental aspects of infant botulism in California. *Rev. Infect. Dis.* 1:693-696.
9. Clark, F. M., and F. W. Tanner. 1937. Thermophilic canned food spoilage organisms in sugar and starch. *Food Res.* 2:27-39.
10. Codex Almentarius Commission. 1994. Codex Almentarius Commission Standards for Sugars, Standard For Honey, Standards for Edible Cassava Flour CAC/Vol 11, 2nd ed. World Health Organization, Rome.
11. Corn Refiners Association. 1986. Corn starch, 7th ed. Corn Refiners Assoc., Washington, D. C.
12. Fleming, R., and V. Stojanowic. 1980. Examination of honeys for spores of *Clostridium botulinum. Arch. Lebensmittelhyg.* 31:179-180.
13. Hartgen, V. H. 1980. Examination of honeys for botulinum toxin. *Arch. Lebensmittelhyg.* 31:177-178.
14. Horwitz, W. (ed.). 1975. Thermophilic bacterial spores in sugars: Official first action, p. 920-921. *In* Official methods of analysis of the Association of Official Analytical Chemists. AOAC, Washington, D. C.
15. Huhtanen, C. N., D. Knox, and H. Shimanuki. 1981. Incidence and origin of *Clostridium botulinum* spores in honey. *J. Food Prot.* 44:812-814.
16. ICMSF. 1980. Microbial ecology of foods, vol. I, Factors affecting life and death of microorganisms, p. 2428. Academic Press, New York.
17. ICMSF. 1998. Microorganisms in Foods 6: Microbial ecology of food commodities. Sugar, cocoa, chocolate, and confectioneries. Blackie Academic and Professional, London.
18. Kramer, J. M., and R. J. Gilbert. 1989. Bacillus *cereus* and other Bacillus species. *In* M. P. Doyle (ed.), Foodborne bacterial pathogens. Marcel Dekker, New York.
19. Liggett, R. W., and H. Koffler. 1948. Corn steep liquor in microbiology. *Bacteriol. Rev.* 12:297-311.
20. Luallen, T. E. 1985. Starch as a functional ingredient. *Food Technol.* 39:59-63.
21. Marx, J. L. 1978. Botulism in infants: A cause of sudden death? *Science* 201:799-801.
22. Midura, T. F., S. Snowden, R. M. Wood, and S. S. Arnon. 1979. Isolation of *Clostridium botulinum* from honey. *J. Clin. Microbiol.* 9:282-283.
23. Mitamura, H., K. Kameyama, and Y. Amdo. 1979. The contamination of spore-forming bacteria in honey. *Rep. Hokkaido Inst. Public Health.* 29:16-19.
24. Muller, E. G. 1986. The sugar industry. *In* S. M. Herschdoerfer (ed.), Quality control in the food industry, vol. 3, 2nd ed.. Academic Press, New York.
25. NAS. 1981. Food Chemical Codex, 3rd ed. National Academy Press, Washington, D. C.
26. National Food Processors Association. 1972. Bacterial standards for sugar, revised. Natl. Food Processors Assoc., Washington, D. C.
27. National Soft Drink Association. 1975. Quality specifications and test procedures for bottlers' granulated and liquid sugar. Natl. Soft Drink Assoc., Washington, D. C.
28. Petersen, N. B. 1975. Edible starches and starch derived syrups. Noyes Data Corp. Park Ridge, N. J.
29. Pitt, J. L. 1975. Xerophilic fungi and the spoilage of foods of plant origin. *In* R. B. Duckworth (ed.), Water relations of foods. Academic Press, New York.
30. Scarr, M. P. 1968. Symposium on growth of microorganisms at extremes of temperature: Thermophiles in sugar. *J. Appl. Bacteriol.* 31:66-74.
31. Smith, R. 1981. Quality control in corn refining, p. 24-28.. *In* Corn annual. Corn Refiners Assoc., Washington, D. C.
32. Smith, P. S., 1983. Food starches and their uses, p. 34-42. *In* D. L. Downing (ed.), Gum and starch technology Eighteenth Annual Symposium, vol. 17., Cornell University Cooperative Extension, Ithaca, N. Y.
33. Sugiyama, H., D. C. Mills, and L. J. Cathy Kvo. 1978. Number of *Clostridium botulinum* spores in honey. *J. Food Prot.* 41:848-850.
34. Tilbury, R. H. 1976. The microbial stability of intermediate moisture foods with respect to yeasts. *In* R. Davies, G. G. Birch, and K. J. Parker (eds.), Intermediate moisture foods. Applied Science Publishers, London.
35. Troller, J. A. 1979. Food spoilage by microorganisms tolerating low a, environments. *Food Technol.* 33:72-75.
36. U.S. Government National Archieves of Records Administration. 1986. Code of Fed. Regs., Title 21, Part. 172.892. Food Starch—Modified p. 108-109. U.S. Government Printing Office, Washington, D. C.
37. Walker, H. W., and J. C. Ayres. 1970. Yeasts as spoilage organisms. *In* A. H. Rose and J. S. Harrison (eds.), The yeasts, vol. 3, Yeast technology. Academic Press, New York.
38. Whistler, R. L., and E. F. Paschall. 1967. Starch: Chemistry and technology," vol. 11. Academic Press, New York.

Chapter 55

Cereal and Cereal Products

Kurt E. Deibel and Katherine M. J. Swanson

55.1 INTRODUCTION

Cereal and cereal products constitute a significant food resource for people throughout the world. Cereal grains include wheat, oats, corn, rye, barley, millet, sorghum and rice. Soybeans, which are pulses, not grains, are included in this chapter because numerous soy products are similar to those produced from cereals. Cereal products include flour(s), breakfast cereals, snack foods, corn meal, doughs, pasta, and dry mixes for cakes, pastry, and breads. Many cereal products are used in the formulation and manufacturing of other products, e.g., sausages, cold cuts, confectioneries and baby food.

Various spoilage microorganisms can proliferate on cereal grains and finished products held under improper storage conditions. Such spoilage may manifest itself as visually undesirable grains or organoleptically unpalatable products. The majority of these spoilage microorganisms represent the normal flora of cereal grains. They include yeasts, molds, psychrotrophic, thermophilic and thermoduric bacteria, lactic acid bacteria and the "rope bacteria." Pathogens of concern vary with product commodity and are discussed below.

Good manufacturing practices can serve to control, and even reduce, the levels of pathogenic and spoilage microorganisms in cereal grains during processing and storage. The aerobic plate and yeast and mold counts are important indices of good sanitation, handling, processing and storage practices. For details that are beyond the scope of this chapter, see the reviews and published studies on the microbiology of cereal and cereal products.[6,17,19,30,33]

55.2 FACTORS AFFECTING BIOLOAD

Cereal and cereal products may be arbitrarily divided into the eight general categories described in Table 1. The "bioload," as defined here, represents the total microflora of the agricultural commodity or product. The data presented in Table 1 are based on routine quality control tests performed on various items of the specified categories. The microbiological procedures used for these tests are described in this compendium. They represent industry-wide experience and are presented for illustrative purposes; they are not intended to denote microbiological acceptance criteria.

55.21 Cereal Grains

The microflora of cereal grains are generally representative of the environment in which they are grown.[36] A multitude of environmental factors influences the composition of these microflora. The rainfall, sunlight, temperature, and soil conditions during the growing season and at harvest are all important in determining the number and type of microorganisms that are present. Agricultural practices (e.g. "organic" farming, types of chemicals used, and harvesting equipment and methods) also influence the cereal grain microflora. Bird, insect, and rodent activity in the field additionally contribute to the heterogeneity of the microflora. Maintenance of Good Manufacturing Practices (GMP) is also important for pest management in grains.

The diversity of the microflora found on cereal grains at harvest is compounded by further contamination during transport and storage.[13] Abusive conditions during either of these periods may permit water uptake with subsequent microbial growth.

Different varieties of grains often do not differ markedly from each other with respect to microbial populations. Mold, yeasts, and most of the aerobic mesophiles present on cereal grains are indigenous to the plants themselves. Some grains are routinely contaminated with *Cladosporium* molds, while others contain *Aspergillus, Fusarium, Alternaria,* and other types.[2,10] As noted, external contaminants (coliforms, *E. coli,* and enterococci) may be contributed by birds, insects, and rodents, all of which are ecologically associated with cereal grains. Storage conditions such as the moisture content of the grain, the temperature, and the time of storage are critical factors in controlling the growth of microorganisms.[10,18,23,25,29,32,38]

Bacterial populations in grains normally reach levels of 10^6 per g. The wide variety of species present may include aerobic mesophilic sporeformers, lactic acid bacteria, coliforms, and pseudomonads.[17] Low numbers of pathogenic organisms have also been recovered from cereal grains. These include *B. cereus, C. perfringens, C. botulinum,* and *Salmonella* spp; however, cereal grains and their milled products have seldom been implicated in foodborne disease.[26] Generally, the low water activity (a_w) of cereal grains prevents bacterial growth. These microorganisms can survive the milling process and thus contaminate flours and the resultant products.

Yeasts and molds may contaminate cereal grains to levels of 10^4 per g.[17] Field fungi will slowly die off in grains properly dried to less than 13% moisture. Inadequate drying or improper storage in wet conditions will allow some molds to grow and spoil the grains. *Aspergillus flavus* and *A. parasiticus* are common contaminants of certain cereal grains.[28] These molds, like others, can

Table 1. Normal Microbiological Profile of Cereal Grains and Cereal Products

	Product Category	Normal microflora	Quantitative range	Remarks
I.	Cereal grains	Molds	10^2–10^4/g	a) Counts represent "normal" grains in commercial channels; "mildewed" or "musty" or "spoiled" grain would be beyond these ranges.
		Yeasts and yeast-like fungi	10^2–10^4/g	
		Bacteria		
		Aerobic plate count	10^2–10^6/g	
		Coliform group	10^2–10^4/g	
		E. coli	$<10^2$–10^3/g	
		Actinomycetes	10^3–10^6/g	
II.	Flour(s), Corn meal, Corn grits, Semolina	Molds	$<10^2$–10^4/g	a) Microbial counts in flour can vary from one storage period to another depending on moisture content and storage conditions. b) Soy flours sometimes contain salmonellae.
		Yeasts and yeast-like fungi	<10–10^2/g	
		Bacteria		
		Aerobic plate count	10^2–10^6/g	
		Coliform group	$<1 - 10^2$/g	
		"Rope" spores	$<1 - 10^2$/g	
III.	Breakfast cereals and snack foods	Molds	$<1 - 10^3$/g	a) Cereals are additionally tested for *E. coli* and salmonellae. b) Snacks are routinely tested for salmonellae and coagulase-positive staphylococci.
		Yeasts and yeast-like fungi	$<1 - 10^2$/g	
		Bacteria		
		Aerobic plate count	$<1 - 10^2$/g	
		Coliform group	$<1 - 10^2$/g	
IV.	Frozen and refrigerated dough (chemically leavened)	Molds	<10–10^4/g	a) Refrigerated (chemically leavened) have <10 yeast per g. Yeast counts represent inoculum intentionally added to frozen doughs as part of the of the formulation. b) Routinely tested for salmonellae, *E. coli*, and coagulase-positive staphylococci.
		Yeasts and yeast-like fungi	10^5–10^6/g	
		Bacteria		
		Aerobic plate count	10^2–10^7/g	
		Coliform group	$<10 - 10^2$/g	
		Psychrotrophs	$<10 - 10^3$/g	
V.	Baked goods	Molds	$<10^2$–10^3/g	a) The presence of rope-producing spores can result in spoilage of baked goods; control of ingredients, good baking practices and the use of preservatives can minimize rope
		Yeast and yeast-like fungi	$<10 - 10^3$/g	
		Bacteria		
		Aerobic plate count	$<10^2$–10^3/g	
		Coliform group	$<10 - 10^2$/g	
VI.	Soy Protein	Bacteria		a) Quantitative ranges reflect both original contamination and growth during storage of intermediate moisture products. b) Molds, yeast, salmonellae, and staphylococci are routinely tested. c) Soy protein products intended for anaerobic storage (e.g., canning) should be routinely tested for thermophilic sporeformers, flat sour organisms, putrefactive sporeformers, and sulfide spoilage organisms.
		Aerobic plate count	10^2–10^5/g	
		Coliform group	10^2–10^3/g	
		E. coli	$<10 - 10^2$/g	
		Psychrotrophs	10^2–10^4/g	
		C. perfringens	$<1 - 10^2$/g	
VII.	Pasta products	Bacteria		a) Wide ranges in bioloads in these products reflect the difference between egg-based & macaroni-type products b) Routinely tested for salmonellae and staphylococci.
		Aerobic plate count	10^3–10^5/g	
		Coliform group	$<10 - 10^3$/g	
		Molds and yeasts	$10^2 - 10^5$/g	
VIII.	Dry mixes	Molds	10^2–10^5/g	a) Routinely tested for salmonellae and *E. coli*.
		Yeasts and yeast-like fungi	10^2–10^5/g	
		Bacteria		
		Aerobic plate count	10^2–10^6/g	
		Coliform	$<1 - 10^4$/g	

Table based on "routine" quality control tests normally performed on various items of specified category; data represent industry-wide experience; data presented as "orders of magnitude" for illustrative purposes only.

produce mycotoxins, thereby presenting a potential health hazard to consumers. Fluorescence detection methods (e.g. ultraviolet light scan) are used as a crude screening test to identify potential mycotoxigenic grains. The use of rapid immunoassays for mycotoxins are more reliable than UV screening of grains and typically require limited technical ability. However, rigid application of Hazard Analysis Critical Control Point (HACCP) principles to control moisture during harvest, shipment, and storage would be a prudent measure to prevent the spoilage of grain.

55.22 Flours(s), Corn Meal, Corn Grits, Semolina

Most of the microorganisms found in flour and other milled products originate on the raw materials from which they are milled.[13,30] Other sources of potential contamination are transportation facilities, mill unloading devices, conveyors, processing equipment, the milling sequence, and exposure to moisture during the milling process. Grains (e.g., wheat, rye, and barley) are tempered by spraying with water and holding in bins for varying periods of time. This procedure may permit microorganisms to proliferate. The use of chlorine in the spray water is a proven means to reduce the microbiological bioload on the grain. Bleaching of flour can also reduce the microbial population, although spores are unaffected.[35] Corn meal, corn grits, and some corn flours are traditionally produced by a "dry milling" process that avoids the tempering steps.

Soy flour is manufactured by a different process. The soybeans are moistened, dehulled, flaked, extracted with organic solvent to remove the oil, then "caked" and ground into flour. *Salmonella* spp. has been detected in soy flour. Salmonellosis and yersiniosis in humans and animals have been traced to soybeans and soy products.[19]

The microbial levels in properly handled and processed flours will be no greater than the levels in raw grains. Bacterial populations may reach a level near 10^6 per g. Yeasts and molds may be recovered at 10^4 per g. The presence of psychrotrophs, flat-sour organisms, and thermophilic sporeformers may be of particular interest to processors of canned or chilled foods. The presence of "rope bacteria" is of interest to manufacturers of baked goods.

55.23 Breakfast Cereals and Snack Foods

The three basic breakfast cereal manufacturing processes are "flaking," "puffing," and "extrusion." In each case, moisture is introduced into the formulation, thus providing an opportunity for microbial growth. Controlling the time that the formulation is held at this moisture level will limit the amount of microbial growth. The heat applied during the cooking process when the product is moist will reduce the microbial levels. A potential for post-heat contamination may appear during an enrichment application or "enrobing" operation, to add vitamins, minerals, sweeteners, or colorings to cereals. If the additives are contaminated, or the process or equipment is unsanitary, the finished product may be contaminated. However, the low moisture levels of breakfast cereals and snack foods will prevent any further microbial growth.

55.24 Refrigerated and Frozen Doughs

The ingredients used in the formulation of doughs provide the primary source of their indigenous microflora. These include flour, dry milk, eggs, sugar, spices, flavorings, and water. The equipment and environment of manufacture also play an important role in the microbiology of the finished product. These commodities are dispatched in an unbaked state by the manufacturer. The final heat treatment is provided by the consumer.

In frozen doughs, no microbial growth will occur during distribution and storage if the products are held at the prescribed temperature. Microbial counts in refrigerated doughs will increase during storage. However, the rate of increase will depend on the types and numbers of microorganisms initially present and the storage temperature. Most refrigerated doughs are formulated with low a_w and pH in an attempt to retard bacterial growth. Refrigerated doughs are generally chemically leavened rather than formulated to contain a yeast starter culture; slow yeast growth during refrigerated storage would eventually burst the container. Wild-type yeasts may be present in low numbers in refrigerated doughs. Lactic acid bacteria are of special concern in the spoilage of refrigerated doughs.[9,33] Heterofermentative lactics are regarded as the primary spoilage agent of refrigerated doughs. Gas production by these organisms will burst the container.

55.25 Baked Goods

"Baked goods" refers to breads, cakes, pie shells, and pastries, as well as fried dough products such as donuts. The baking or frying process destroys most of the microorganisms in baked goods.[28] Post-baking contamination can result in the spoilage of certain baked goods if they are stored at ambient temperatures for prolonged periods. Mold is one of the most common forms of spoilage for breads and other baked goods. Preservatives are commonly added to reduce the potential for mold spoilage.[20]

Sporeformers that survive the baking process may also cause spoilage of baked products. Breads and other baked goods are subject to spoilage by the "rope bacteria," mucoid variants of *Bacillus subtilus*.[7] This bacterium causes a condition known as "rope" in these products, so called because of the ropy and stringy texture of the product interior.

The microbiology of filled baked goods varies considerably, based primarily on the microbiology of the filling rather than that of the cereal portion. Many fillings will not support the growth of pathogenic microorganisms; they are formulated with a pH <4.5, or an a_w <0.85. However, some custard and cream fillings are outside this range and thus support the growth of certain pathogens. Appropriately designed challenge studies are needed to define the critical storage capabilities of certain products. The use of the hurdles technology[4] is encouraged when formulating pies that are in the low acid range (pH > 4.6; water activity > 0.85). These must be considered potentially hazardous and treated as perishable foods, i.e., kept refrigerated.

55.26 Soy Protein Products

The ingredients of these products are essentially soy flour and a variety of additives (color, flavoring, and vitamins). Each additive may contribute to the microbial population of the finished commodity. Some steps in the process may also add to the contamination. Finished soy protein product range in moisture content from 2% to 64%. The higher moisture content along with the nutritional quality of some of these products may be conducive to microbiological growth. The range of counts and the variety of microbiological types encountered (see Table 1) suggest that the problems may be those of storage and sanitation.

The microorganisms of concern in soy protein are somewhat dependent on its end use. If used as an ingredient in a retorted, canned product, then thermophilic sporeformers, flat sour organisms, putrefactive sporeformers, and sulfide spoilage organisms would be of high importance. If used in a shelf stable or perishable product, which has received minimal heat during processing, then *Salmonella* spp. would be of importance. Because of the frequent isolation of *Salmonella* from soy flour, it should be treated as a sensitive ingredient of such products.[39]

55.27 Pasta Products

Pasta products (usually manufactured from durum wheat flour) essentially fall into two categories: egg-based pasta such as noodles and macaroni-type pasta such as macaroni, spaghetti, and vermicelli. The former, as its name implies, contains flour, water enrichment nutrients, and pasteurized dried frozen or fresh eggs. The latter contains only flour, water, and enrichment nutrients. Both products are manufactured in much the same fashion: mixed, extruded, shaped, cut, and dried. The initial microbiological profile of the mixed dough is directly related to the quality of the ingredients. During its manufacture, the product is semisoft, unheated dough of approximately 30% moisture that can support microbial growth. Microorganisms can also grow during the slow, low-temperature drying process. Not until the moisture drops below 13% during the drying process (several hours) would the microbiological activity be inhibited.

Microbiological counts in freshly dried pasta may be as high as 10^7 per g, but will decrease during storage because of its low moisture content. Improper drying of the pasta may allow the growth of some molds.[19] *S. aureus* has been a problem in the manufacture of pasta products.[24] Enterotoxin production is a potential hazard because of the ideal growth conditions during mixing and initial drying. Proper equipment, cleaning, and sanitation can minimize this threat. *Salmonella* spp. may also occur in pasta if contaminated eggs have been used as an ingredient. Boiling the pasta will kill vegetative cells during normal preparation, but staphylococcal enterotoxin will remain.[14,21,24]

55.28 Dry Mixes

The manufacture of dry mixes is a dry blending of such ingredients as flour, dried eggs, flavorings, sugar, and dried dairy products. The finished product will reflect the microbiological profile of the individual ingredients and the cleanliness of the mixing equipment. Usually the microbial load is not reduced in the dry mix process. Microbiological control can be affected by quality control of the ingredients, use of clean equipment, and maintenance of low, microbiologically inhibitory moisture levels in the finished product.

Sanitation of equipment in dry mix manufacturing is of primary importance. Wet cleaning should be avoided whenever possible since this usually causes more problems than it solves.

55.3 METHODS

The following summary suggests routine and special analyses that are employed to determine the microbiological condition of the eight product categories discussed above. Refer to the appropriate chapters in this text for specific procedures.

55.31 Routine Analyses

1. Mold and yeast determinations
   The mold and yeast count is an indication of the sanitary history of the product as well as a prediction of potential future spoilage during storage. Recommended sample size is 22 g diluted 1:10 in 198 mL of diluent.
2. Aerobic plate count. Sample size: 22 g in 198 mL diluent
3. Coliform organisms and *Escherichia coli*
   The finding of coliform bacteria in cereal grains and flour is common and Richter and coworkers[30] reported isolation of *E. coli* in an average of 12.8% of flour samples. Therefore, finding *E. coli* in flour containing raw products does not necessarily imply mishandling. Finding *E. coli* in a finished, ready-to-eat product, however, may be a public health concern, as it may imply post process recontamination.
4. *Staphylococci*
   Enterotoxin-producing staphylococci represent a potential hazard to pasta manufacturers. Generally, staphylococci contaminate the mix during processing rather than entering with the flour. Therefore, testing of the raw mix or the finished product is a better determination of a potential problem.
5. *Salmonella*
   Samples should be composited as discussed in the chapter on *Salmonella*.

55.32 Special Analyses

Special analyses are to be performed on certain products under specified circumstances.

1. "Rope" spores.
   Analysis for rope spores should be performed on cereal grains and flours where the prospect of "ropy" dough, from the action of *B. subtilus*, is of concern. (Chapter 22 for additional information.)
2. Mycotoxins.
   Mycotoxins are toxins elaborated by certain fungi that may grow on moist grains. Epidemiological evidence[30] suggests that humans can be affected by ingestion of these toxins. Refer to Chapter 43 for additional information on mycotoxins and methods.
3. Staphylococcal enterotoxins.
   In addition to assaying a cereal product for the presence of coagulase-positive staphylococci, assaying for enterotoxin is also sometimes advisable, as when the *S. aureus* count suggests the presence of some hazard. If the history of the product reveals that growth of staphylococci could have taken place, as with temperature abuse, it is useful to assay the product for toxin.
4. *Clostridium perfringens.*
   Clostridium perfringens food poisoning is often associated with meat and poultry. Soy protein products are often designed as substitutes for meats, and their physicochemical properties may be conducive to the growth of *C. perfringens*. Such products should be analyzed for the presence of this organism.
5. *Bacillus cereus.*
   Rice has repeatedly been incriminated in foodborne disease outbreaks because of *B. cereus*.[16] In food poisonings where rice is the suspected vehicle, analyses should be done for *B. cereus*.

55.33 Other Special Tests

Cereal products may be tested for psychrotrophs, thermophiles, anaerobes, flat sour spores, hydrogen sulfide producers, lactic acid producers, nitrate-utilizing gas producers, and sulfide spoilage spores depending on the specific product needs and applications.

55.4 REFERENCES

1. AACC. 1983. Approved methods of the American Association of Cereal Chemists, 8th ed. American Association of Cereal Chemists, St. Paul, Minn.
2. Abramson, D., R. N. Sinha, and J. T. Mills. 1984. Quality changes in granary-stored wheat at 15 and 19% moisture content. Mycopathologia 87:115.
3. AOAC. 1984. Official methods of analyses of the Association of Official Analytical Chemists,14th ed. Association of Official Anaytical Chemists, Washington, D. C.
4. Barbosa-Ca'novas, G. V., U. R. Pothakamuny, E. Palou, and B. G. Swanson. 1998. Hurdle technology, p. 235. *In* Nonthermal preservation of foods. Marcel Dekker, Inc., New York.
5. Bothast, R. J., R. F. Rogers, and C. W. Hesseltine. 1973. Microbial survey of corn in 1970-71. Cereal Sci. Today 18:18.
6. Bothast, R. J., F. R. Rogers, and C. W. Hesseltine. 1973. Microbiology of corn and milled corn products. Northern Regional Research Laboratory, Peoria, Ill.
7. Brackett, R.E. 1997. Fruits, vegetables, and grains, p. 327. *In* M. P. Doyle, L. R. Beuchat, and T. J. Montville (eds.), Food microbiology fundamentals and frontiers. ASM Press, Washington, D. C.
8. Busby, W. F., and G. N. Wogan. 1979. Foodborne mycotoxins and alimentary mycotoxicoses. *In* H. Reimann and F. L. Bryan (ed.), Foodborne infections and intoxications, 2nd ed. Academic Press, New York.
9. Chen, R. W. 1979. Refrigerated doughs. Cereal Foods World 24:46.
10. Christensen, C. M. 1968. Influence of moisture content, temperature, and time of storage upon invasion of rough rice by storage fungi. Phytopathyology 59:145.
11. Cohen, H., and M. Lapointe. 1986. Determination of ochratoxin A in animal feed and cereal grains by liquid chromatography. J. Assoc. Off. Anal. Chem. 69:957.
12. Davis, N. D., J. W. Dickens, R. L.Free, P. B. Halmilton, O. L. Shotwell, and T. D. Wyllie. 1980. Protocols for surveys, sampling, post-collection handling, and analysis of grain samples involved in mycotoxin problems. J. Assoc. Off. Anal. Chem. 63:95.
13. Dehoff, T. W., R. Stroshine, J. Tuite, and K. Baker. 1984. Corn quality during barge shipment. Trans. ASAE, p. 259.
14. Denny, C. B., P. L. Tan, and C. W. Bohrer. 1966. Heat inactivation of staphylococcal enterotoxin A. J. Food Sci. 31:762.
15. Flannigan, B. 1986. *Aspergillus clavatus* — An allergenic, toxigenic deteriogen of cereals and cereal products. Internat. Biodeteriorat. 22:79.
16. Granum, P. E. 1997. *Bacillus cereus*, p. 327. *In* M. P. Doyle, L. R. Beuchat, and T. J. Montville (ed.), Food Microbiology Fundamentals and Frontiers. ASM Press, Washington, D. C.
17. Hesseltine, C. W., and R. R.Graves. 1966. Microbiological research on wheat and flour. Econ. Bot. 20:156.
18. Hill, R. A., and J. Lacey. 1983. Factors determining the microflora of stored barley grain. Ann. Appl. Biol. 102:467.
19. ICMSF. 1998. Microorganisms in Foods 6. Microbial ecology of food commodities, chap. 8, Cereals and cereal products, p. 313-355. Blackie Academic & Professional, London.
20. Jackel, S. 1980. Natural breads may cause microbiological problems. Bakery Prod. Market.15:138.
21. Jay, J. M. 1998. Staphylococcal gasteroenteritis, p. 437. *In* Modern food microbiology, 5th ed. Aspen Publishers, Gaithersburg, Md.
22. Knight, R. A., and E. M. Menlove. 1961. Effect of the bread-baking process on destruction of certain mold spores. J. Sci. Food Agric. 12:653.
23. Kuiper, J., and G. M. Murray. 1978. Spoilage of grain by fungi. Agric. Gazette of New South Wales. Oct., p. 39.
24. Lee, W. H., C. L. Staples, and J. C. Olson, Jr. 1975. *Staphylococcus aureus* and survival in macaroni dough and the persistence of enterotoxins in the dried products. J. Food Sci. 40:119.
25. McMahon, M. E., P. A. Hartman, R. A. Saul, and L. H. Tiffany. 1975. Deterioration of high-moisture corn. Appl. Microbiol. 30:103.
26. NRC. 1985. An evaluation of the role of microbiological criteria for food and food ingredients, p. 272. National Research Council. National Academy Press, Washington, D. C.
27. Ostovar, K., and K. Ward. 1976. Detection of *Staphylococcus aureus* from frozen and thawed convenience pasta products. Lebensm Wiss. Technol. 9:218.
28. Pitt, J. I., and A. D.Hocking. 1985. Fungi and food spoilage. Academic Press, New York.
29. Prasad, D. C., W. E. Muir, and H. A. H. Wallace. 1978. Characteristics of freshly harvested wheat and rapeseed. Trans. ASAE 21:782.
30. Richter, K. S., E. Dorneanu, K. M. Eskiridge, and C. S. Rao. 1993. Microbiological quality of flours. Cereal Foods World 38(5):367-369.
31. Sadek, M. A., F. M. M. El-Zayet, M. G. ABD El-Fadel, and R. A. Taha. 1985. Isolation and enumeration of *B. cereus* and some other microorganisms from balady bread. Z. Gesamte Hyg. Ihre Grenzgeb. 31:623.
32. Sinha, R. N., W. E. Muir, and D. B. Sanderson. 1985. Quality assessment of stored wheat during drying with near-ambient temperature air. Can J. Plant Sci. 65:849.
33. Slocum, G. G. 1963. Let's look at some microbiological problems associated with cereal foods. Cereal Sci. Today 8:313.
34. Stoloff, L., M. Trecksess, P. W. Anderson, E. F. Glabe, and J. G. Aldridge. 1978. Determination of the potential for mycotoxin contamination of pasta products. J. Food Sci. 43:228.
35. Thatcher, F. S., C. Coutu, and F. Stevens. 1953. The sanitation of Canadian flour mills and its relationship to the microbial content of flour. Cereal Chem. 30:71.
36. Thomas, P. M. 1971. Role of microflora in the deterioration of agricultural commodities in warehouses. Allahabad Farmer 45:463.
37. U.S. Food and Drug Administration. 1995. Bacteriological analytical manual, 8th ed. AOAC International, Gaithersburg, Md.
38. Wallace, H., A. H., R. N. Sinha, and J. T. Mills. 1976. Fungi associated with small wheat bulks during prolonged storage in Manitoba. Can. J. Bot. 54:1332.
39. Wilson, C. R., W. H. Andrews, P. L. Poelman, and D. E. Wagner. 1985. Recovery of *Salmonella* species from dried foods rehydrated by the soak method. J. Food Prot. 48:505.

Chapter 56

Confectionery Products

Patrick J. Konkel

56.1 INTRODUCTION

Confectionery is what many people think of as "candy" or "sweets." Confectioneries are sweet, shelf-stable products with low water activity (a_w). Most confectioneries, and all of those considered in the context of this chapter, have a_w values below 0.85. Products, such as certain pastries, which do not fall within this classification, will have different microbiological issues. Confectionery belongs to either of two groups, sugar confectionery or chocolate confectionery, although broader classifications sometimes are used.[27] These products contain sugar, syrups, honey, or other sweeteners. In addition, confectionery products may contain cocoa or chocolate products; dried milk or other dairy products; nuts, coconut or other fruits; cereal grain products, including crisped rice; starch, gelatin or other thickeners; egg albumen; spices, colors, flavors or acidulants; or other ingredients.

Their low a_w makes confectioneries resistant to bacterial growth; however, bacterial survival in these products may be enhanced by low a_w.[1,26] Except for the concern about the survival of contaminant pathogenic bacteria, e.g., *Salmonella*, in these products, microbiological concerns are limited to spoilage by osmophilic yeasts or molds in those products with relatively higher a_w.

In the United States, confectionery manufacturing and standards of identity are regulated by the Food and Drug Administration (FDA).[49,50] Otherwise, the recommendations of the Codex Alimentarius Commission,[5] if accepted by a national government, become the laws of the accepting country.

Table 1. Water Activity of Some Confectioneries

Product	Water Activity
Fondant cream	0.75–0.84[15]
Mints	0.75–0.80[38]
Fruit jellies	0.59–0.76[15]
Fudge	0.65–0.75[38]
Marshmallow	0.60–0.75[38]
Marzipan	0.65–0.70[15]
Turkish delight	0.60–0.70[15]
Nougat	0.40–0.70[38]
Licorice	0.53–0.66[15]
Gums and pastilles	0.51–0.64[15]
Chocolate	0.40–0.50[38]
Caramel	0.40–0.50[38]
Hard Candy	0.20–0.35[38]

56.2 GENERAL CONSIDERATIONS

56.21 Water Activity

Water activity is the intrinsic product characteristic that most influences the microbial ecology of confectioneries (Table 1). High concentrations of sugars, especially those of low molecular weights,[4] afford low a_w. Most bacteria cannot grow at a_w below 0.85, and growth of spoilage-causing yeasts and molds is unlikely at a_w levels below 0.61 (See Chapter 64).

56.22 Sugar Confectionery

Sugar confectioneries include hard candies, toffee, caramel, fondants, creams, and pastes. Hard candies and toffee are not subject to microbial degradation because of their extremely low a_w. On occasion, improperly formulated (too high a_w) caramel used as centers for chocolate will undergo yeast spoilage. Only fondants, creams, and pastes have a significant history of microbial spoilage.[22] Sometimes, for additional product stability and subject to regulatory compliance, acidulants and preservatives are incorporated.

Fondants, creams, and pastes are often used as base materials to formulate other confections. Other components such as colors, flavors, fruits, and nuts can be added for variety.

56.23 Chocolate Confectionery

Products manufactured with cocoa and chocolate make up a large segment of the confectionery industry. Chocolate confectionery includes chocolate bars, blocks and bonbons; products with inclusions like nuts, raisins, coconut or crisped rice; and chocolate-coated sugar confectionery, nuts, fruits, or jellies.

Chocolate-coated centers have a continuous coating of chocolate, which serves as a moisture barrier against the absorption of atmospheric moisture. If moisture is taken up through the cracks or discontinuities in the coating, weeping (formation of syrup on the chocolate) may occur because absorbed water dilutes the sugar in the center.[23] This physical defect can be mistakenly attributed to yeast spoilage.

Although not directly associated with most aspects of the microbiological quality of chocolate confections, the microbiology of cocoa beans fermentation[20,26] may be of interest to some readers.

56.3 SPOILAGE OF CONFECTIONERY

56.31 Bacteria

The a_w (less than 0.85) of confectionery precludes bacterial growth (See Chapter 64).

56.32 Yeasts

Yeasts are the principal cause of spoilage in confectionery products,[2,45,51] Spoilage characteristics include gassing, slime, off-odors, off-flavors and liquefaction. The limiting a_w for osmophilic yeasts in nutrient rich confectioneries such as pralines and caramel is 0.60,[47] but in less nutrient-rich confectionery the a_w required to limit osmophilic yeasts is higher.[35]

Zygosaccharmoyces rouxii[33] and, less frequently, *Brettanomyces bruxellensis* have been identified as causes of spoilage. *Torulopsis etchellsii, Torulopsis versatilis,* and *Candida pelliculosa* have caused gassy spoilage of chocolate syrup with 75° Brix.[45]

Marzipan is spoiled by yeasts when the population reaches 10^5 to 10^7.[53] *Z. rouxii* is the common spoilage yeast and causes gassy bursting of the marzipan, accompanied by the development of weak aromatic, yeasty, or yeasty-bitter odors. The growth of osmophilic yeasts may be so slow in marzipan and coated fondants that evidence of spoilage may not be seen for several months.[24] The doubling times for *Z. rouxii* in persipan and marzipan raw masses are about 1 and 0.5 days, respectively.[2] Yeast spoilage of chocolate-covered creams and marzipan is evidenced by cracking of the coating and leaking of the fondant and syrup. Drying of the syrup as a result of cracking may decrease the a_w and thereby stop growth.

Spoilage of confectionery by osmophilic yeasts must be studied carefully because the causative organisms may lyse or may not be detected if precautions are not taken to prevent osmotic shock during sample dilution and in the plating medium (see Chapter 17). Direct microscopic observation of a spoiled product is often a valuable diagnostic tool.

56.33 Molds

Molds may spoil confectionery through development of visible mold mycelia on the surfaces of products or packaging materials. Their growth may produce a musty odor and taste. A soapy taste may develop in high fat-content products because of enzymatic hydrolysis of lipids.[24] Molds of the genera *Aspergillus, Verticillium, Penicillium, Mucor, Rhizopus,* and *Tricothecium*[17,52] have been isolated from confectionery. Visible molds usually occur on the surface or at the interface between product and packaging,[14,30] but development may also occur in the interior.[24,52]

56.34 Miscellaneous Defects

Fat blooms and sugar blooms on chocolate are physical defects that are misdiagnosed frequently as mold growth. Fat bloom makes chocolate appear gray-white. Fat bloom has a greasy appearance and under the microscope minute fat crystals are seen. Fat bloom is associated with improper temperature control at one or more stages in processing or storage.[27]

Sugar bloom is similar to fat bloom but not greasy. In severe cases sugar bloom has a crystalline appearance, is rough to the touch, and has small sugar crystals that can be seen under the microscope. Sugar bloom is associated with storage of chocolate products exposed to temperature changes sufficient to cause condensation of moisture on the surface. It is also caused by storage under conditions of high (78% or higher) relative humidity.[27]

Lipases from microorganisms can cause hydrolytic rancidity in chocolate products.[27,40] Hydrolytic rancidity occurs when cocoa butter replacers such as palm kernel oil are used. As an example, hydrogenated palm kernel oil contains about 47% lauric acid and is especially useful as a cocoa butter replacer. Active lipases separate lauric acid from the fat as a free fatty acid and this results in a distinctly soapy taste.[39] Capric and myristic acids also have a soapy taste.

Residual lipases remaining after processes have destroyed or otherwise removed microorganisms from ingredients may cause hydrolytic rancidity. For example, a soapy taste in "white chocolate" was attributed to lipolyzed milk in which a *Bacillus* had grown prior to drying.[54]

Free fatty acids may be oxidatively metabolized to ketones; this results in a defect known as perfume rancidity.[29] Esterified fatty acids in hydrogenated liquid oils and nuts in some confections may be oxidized.

Off-flavors, other than from lipolytic and oxidative rancidity, may occur in confectionery and be mistaken for microbiological or lipolytic spoilage. Absorption of odors from plastic wrappers, from inks used in printing,[27] and from storage near detergents, disinfectants, oils, or tobacco may impart off-odors to confectionery products.

56.4 PATHOGENS

Confectioneries, by the nature of their composition, processing, physical and chemical characteristics, are rarely associated with foodborne illness. However, between 1970 and 1995, five incidents were reported of salmonellosis caused by consumption of contaminated chocolate products.[8,11,13,16,21] *Salmonella* has been recovered from chocolate samples, apparently in the absence of any salmonellosis outbreak.[9,46] Besides salmonellosis, other confectionery-borne microbial infections have not been reported. Because confectioneries usually do not support the growth of bacteria, these products have not been associated with bacteriologically induced intoxications, e.g., by *Staphylococcus aureus, Bacillus cereus,* or *Clostridium perfringens.*

Bacilli are usually the predominant microorganisms in many confectionery products, with *B. cereus* and *B. subtilis* often present.[10,44] However, these products have not been associated with illness caused by *B. cereus.*

Mycotoxicoses from this product group have not been reported, although occasional product spoilage by molds has been observed, and the introduction of mycotoxins from ingredients, e.g., nuts, has been of concern.[20]

Although bacteria do not grow in most confectioneries, the water activity and/or fat contents generally afford marked microbial survival.[12,27] When a confectionery contains a pathogenic organism, it is likely to survive in the product for several months after manufacturing.[1,43]

56.5 MICROBIAL CONTROL PROCEDURES

Confectionery microbiological safety is best addressed through Hazard Analysis Critical Control Print (HACCP) procedures.[7,19] For chocolate confectioneries, cocoa bean roasting is a critical control point. For all confectioneries, raw materials and the environment are additional control points. Otherwise, confectionery microbiological quality and the prevention of spoilage are accomplished by proper formulation with respect to a_w.

56.51 Raw Materials

Pathogens and spoilage microorganisms may enter the confectionery processing environment or product through ingredients like eggs, gelatin, cereal grain products, nuts, coconut, spices and colors[34] as well as the cocoa beans and raw milk (See Section 56.53). If *Salmonella* is present, the level is usually low. Therefore it is essential to use sampling plans adequate for the detection of small numbers of microorganisms.[18] Also, refer to Chapter 2 of this book and the Bacteriological Analytical Manual.[48]

Segregation of raw materials and unit operations that process these ingredients is important in preventing cross-contamination between raw and finished goods. For the manufacture of chocolate products, it is absolutely necessary to handle cocoa beans as a contaminated raw material. Therefore operations that clean, roast, and winnow beans should be physically separated from subsequent downstream processing.

56.52 Product Formulation

Appropriate control of a_w during product formulation will prevent yeast and mold spoilage. Control of a_w in the formulation of sugar confectioneries is achieved through increasing osmotic pressure by adding glucose syrup, corn syrup with high dextrose equivalent, high fructose syrup, or invert sugar with sucrose. Formulations with a dissolved solids concentration in the syrup phase of 75% wt/wt may be near the minimum to prevent fermentation by yeasts.[32] The use of sucrose alone is not satisfactory because a saturated solution of sucrose has an a_w (0.84) suitable for yeast and mold growth. The correct ratio of sucrose, glucose, and invert sugar[3,4] will afford a product of sufficiently low a_w so as to be microbiologically stable.

Addition of 2% to 4% glycerol or sorbitol retards crystallization. Inhibition of yeasts is enhanced by the addition of a small amount of lactic or acetic acid. Where legally permitted, the addition of combinations of acidulants and preservatives, e.g., citric acid and sorbic acid, offers protection against yeast growth.[37]

56.53 Thermal Processing

Pathogenic microorganisms such as *Salmonella* may be introduced into the confectionery processing environment through raw materials, including commodities like cocoa beans and raw milk. Raw milk or other dairy products need to undergo heat treatment equivalent to pasteurization. Roasting of cocoa beans for 15 minutes to 2 hours at 105°C to 150°C should destroy non-sporeforming pathogenic bacteria such as salmonellae[12] and substantially reduce the overall microbial content of the beans. Additional cocoa bean treatments such as micronizing (infrared heating) or steam treatment,[28] or "liquor roasting" (where a small amount of water is added to cocoa liquor that is then heated before the additional water is removed),[25] reduce the overall microbial content of chocolate. Heat-resistant sporeforming bacteria such as *Bacillus cereus*, *Bacillus licheniformis*, and *Bacillus subtilis* survive roasting and will carry over into finished goods.[1,31]

The use of a thermal process during certain sugar confectionery operations presents an opportunity to control spoilage fungi. The amount of heat applied for destruction of these microorganisms varies somewhat depending on the a_w of the product. But on a practical basis, yeasts in fondants, for example, are normally destroyed within 20 min at 60°C[12] or in syrups by 15 to 20 min at 80°C.[41] Other sugar confectionery heat treatments, such as boiling, can allow for the complete destruction of vegetative bacteria.[42]

56.54 Plant Environment

Effective traffic, air, pest, and especially water control systems can prevent pathogenic contaminants from establishing residences, reproductive pockets, and subsequent cross-contamination within the confectionery processing environment.[9] Air, dust, and moisture provide means of microbial transmission in these plant environments. Dust generated by raw material handling can be conveyed throughout a plant if air handling systems are not properly installed. This may lead to cross-contamination of in-process or finished goods. Uncontrolled moisture may provide opportunities for microbes to proliferate and establish themselves permanently in a plant that produces low a_w products. Sources of uncontrolled moisture may be water and steam leaks, condensation, leaking roofs, and improper "wet cleaning" methods.

Proper cleaning and sanitation procedures should be selected based upon type of equipment, material composition, and location of moisture sensitive processes within the plant. Judicious use of water for cleaning should be established and complement the plant's moisture control plan.

Special care should be given to the control of osmophilic yeasts in processing equipment. Confectionery residues in equipment are naturally selective for osmophilic yeasts. Process equipment in which water is used for production should be washed well with aqueous cleaning solutions and disinfected with chemical sanitizers or steam. Removal of food residues from difficult-to-clean equipment is also necessary to prevent the establishment of preservative-resistant yeast populations.[36]

56.6 RECOMMENDED METHODS

56.61 Routine Methods

Aerobic plate count—Chapter 7
Coliform organisms and *Escherichia coli*—Chapter 8
Salmonella—Chapter 37 [For cocoa and chocolate products, the *Salmonella* test pre-enrichment medium should be formulated to attenuate the naturally occurring microbial inhibitors in these products.]
Yeasts and molds—Chapter 20

56.62 Supplemental Recommendations

Mycotoxins—Chapter 43
Osmophilic yeasts—Chapter 17
Water activity—Chapter 64

56.63 Rapid Alternatives

For various specific applications, some of the more "rapid" methods occasionally are useful. However, they are not broadly applicable to monitoring all raw materials, environmental control, in-process and finished product testing, or for all product types. Careful thought is necessary to identify the exact purpose of an evaluation, its urgency, the required precision, and often the correlation between the rapid test results and those from another more conventional or standard test procedure.

56.7 INTERPRETATION OF DATA

Aerobic plate counts generally are less than 10^4 per g for sugar confectioneries and 10^3 per g to 10^6 per g (often dependent upon the nature of the cocoa bean roasting) for cocoa and chocolate products.[6]

Coliform organisms frequently are undetectable and rarely are present at levels above 100 per g in confectionery. *E. coli* usually is not found—<1 per g by direct plating on violet red bile (VRB) (See Chapter 8); <3 per g by most probable number (MPN).

Molds rarely are present at levels above 100 per g in sugar confectioneries and occasionally may be found in the range of 10^2 per g to 10^3 per g in cocoa and chocolate products.

Yeast counts usually do not exceed 100 per g. The absence of osmophilic yeasts is critical for some products with a_w values above 0.62 (see Chapter 17).

Occasionally low levels, below 10^3 per g, of *Staphylococcus aureus* or *Bacillus cereus* may be present. They do not seem to indicate any health hazard at these levels in confectionery.

Any level of *Salmonella* is unacceptable in confectionery.

56.8 REFERENCES

1. Barrile, J. C., J. F. Cone, and P. G. Keeney. 1970. A study of salmonella survival in milk chocolate. Manuf. Confect. 50(9):34.
2. Blaschke-Hellmessen, R., and G. Teuschel. 1970 *Saccharomyces rouxii* Boutroux als Urasche von Garungserscheinungen in geformten Marzipan- and Persipanartikeln und deren Verhutung in Herstellerbetrieb. Nahrung 14(4):249.
3. Cakebread, S. H. 1969. Chemistry of candy: Shelf-life of candy. Manuf. Confect. 49(2):38.
4. Cakebread, S. H. 1971. Chemistry of candy: Factors in microbiological deterioration. Manuf. Confect. 51(4):45.
5. Codex Alimentarius Commission. 1994. Codex Alimentarius Commission Standards for cocoa products and chocolate. CAC/vol. 11, 2nd ed. World Health Organization, Rome.
6. Collins-Thompson, D. L., K. F. Weiss, G. W. Riedel, and C. B. Cushing. 1981. Survey of and microbiological guidelines for chocolate products in Canada. Can. Inst. Food Sci. Technol. J. 14(3):203.
7. Cordier, J. L. 1994. HACCP in the chocolate industry. Food Control 5:171.
8. Craven, P. C., W. B. Baine, D. C. Mackel, W. H. Barker, E. J. Gangarosa, M. Goldfield, H. Rosenfeld, R. Altman, G. Lachapelle, J. W. Davies, and R. C. Swanson. 1975. International outbreak of *Salmonella eastbourne* infection traced to contaminated chocolate. Lancet 1:788.
9. D'Aoust, J. Y. 1977. *Salmonella* and the chocolate industry. A review. J. Food Prot. 40(10):718.
10. Gabis, D. A., B. E. Langlois, and A. W. Rudnick. 1970. Microbiological examination of cocoa powder. Appl. Microbiol. 27:66.
11. Gastrin, B., A. Kampe, K. Nystrom, B. Oden-Johanson, G. Wessel, and B. Zetterberg. 1972. An epidemic of *Salmonella durham* caused by contaminated cocoa. Lakartidningen 69(46):5335. (Original in Swedish)
12. Gibson, B. 1973. The effect of high sugar concentrations on the heat resistance of vegetative microorganism. J. Appl. Bacteriol. 36:365.
13. Gill, O. N., C. Bartlett, P. Sockett, M. Vaile, B. Rowe, R. Gilbert, C. Dulake, H. Murrell, and S. Salmosa. 1983. Outbreak of *Salmonella napoli* infection caused by contaminated chocolate bars. Lancet 1:574.
14. Gondar, I. 1980. Marzipan production in the Duna chocolate factory. Edesipar 31(2):44. (Original in Hungarian) Food Sci. Tech. Abstr. 13(05):5K31, 1981.
15. Hilker, J. S. 1976. Confectionery products, p. 608. *In* M. L. Speck (ed.), Compendium of methods for the microbiological examination of foods, 1st ed. Am. Public Health Assoc., Washington, D. C.
16. Hockin, J. C., J. J. D'Aoust, D. Bowering, J. H. Jessop, B. Khama, H. Lior, and M. E. Milling. 1989. An international outbreak of *Salmonella nima* from imported chocolate. J. Food Prot. 52:51.
17. Hopko, I. 1979. Food hygienic aspects of the confectionery industry. Edesipar 30(1):8. (Original in Hungarian)
18. ICMSF. 1986. Microorganisms in foods, vol. 2, Sampling for microbiological analysis; principles and specific applications, 2nd ed. Int. Comm. on Microbiol. Spec. for Foods. Univ. of Toronto Press, Toronto, Canada.
19. ICMSF. 1988. Microorganisms in Foods 4. Application of the Hazard Analysis Critical Control Point (HACCP) System to ensure microbiological safety and quality. Blackwell Scientific Publications, Oxford.
20. ICMSF. 1998. Microorganisms in Foods 6. Microbial ecology of food commodities. Int. Comm. on Microbiol. Spec. for Foods. Blackie Academic & Professional, London.
21. Kapperud, G., S. Gustavsen, I. Hellesnes, A. H. Hansen, J. Lassen, J. Hirn, M. Jahkola, M. A. Montenegro, and R. Helmuth. 1990. Outbreak of *Salmonella typhimirium* infection traced to contaminated chocolate and caused by a strain lacking 60-megadalton virulence plasmid. J. Clin. Microbiol. 28(12):2597.
22. Lenovich, L., and P. Konkel. 1992. Confectionery products, p. 1007. *In* C. Vanderzant and D. F. Splittsoesser (eds.), Compendium of methods for the microbiological examination of foods, 3rd ed. Am. Public Health Assoc., Washington, D. C.
23. Lindley, P. 1972. Chocolates and sugar confectionery, jams, and jellies. *In* S. M. Herschodoerfer (ed.), Quality control in the food industry, vol. 3. Academic Press, New York.
24. Mansvelt, J. W. 1964. Microbiological spoilage in the confectionery industry. Confect. Prod. 30(1):33.
25. Martin, R. A., Jr. 1987. Chocolate. *In* C. D. Chichester (ed.), Advances in food research. Academic Press, New York.
26. Mazigh, D. 1994. Microbiology of chocolate. *In* Industrial chocolate manufacture and use. Blackie Academic & Professional, Glasgow.
27. Minifie, B. W. 1989. Chocolate, cocoa, and confectionery: Science and technology, 3rd ed. AVI Publ., Westport, Conn.
28. Minson, E. 1992. Chocolate manufacture—beans through liquor production. Manuf. Conf. 72:61.
29. Mossel, D. A. A., and F. E. M. J. Sand. 1968. Occurrence and prevention of microbial deterioration of confectionery products. Conserva 17(2):23.
30. Ogunmoyela, O. A., and G. G. Birch. 1984. Effect of sweetener type and lecithin on hygroscopicity and mould growth in dark chocolate. J. Food Sci. 49:1088.
31. Ostovar, K., and P. G. Keeney. 1973. Isolation and characterization of microorganisms involved in the fermentation of Trinidad's cacao beans. J. Food Sci. 38:611.
32. Pitt, J. I. 1975. Xerophilic fungi and the spoilage of foods of plant origin. *In* R. B. Duckworth (ed.), Water relations of foods. Academic Press, New York.
33. Pitt, J. I., and A. D. Hocking. 1985. Fungi and food spoilage. Academic Press, New York.
34. Pivnick, H., and D. A. Gabis. 1984. Confectionery products, p. 700. *In* M. L. Speck (ed.), Compendium of methods for the microbiological examination of foods, 2nd ed., Am. Public Health Assoc., Washington, D. C.
35. Pouncy, A. E., and B. C. L. Summers. 1939. The micromeasurement of relative humidity for the control of osmophilic yeasts in confectionery products. J. Soc. Chem. Ind. Engl. Trans. Commun. 58:162.
36. Restaino, L., L. M. Lenovich, and S. Bills. 1982. Effect of acids and sorbate combinations on the growth of four osmophilic yeasts. J. Food Prot. 45:1138.
37. Restaino, L., S. Bills, K. Tscherneff, and L. M. Lenovich. 1983. Growth characteristics of *Saccharomyces rouxii* isolated from chocolate syrup. Appl. Environ. Microbiol. 45(5):1614.
38. Richardson, T. 1987. ERH of confectionery food products. Manuf. Confect. 67:65.
39. Rossell, J. B. 1983. Measurements of rancidity, p. 259. *In* J. Allen and R. J. Hamilton (eds.), Rancidity in foods,Proceedings of an SCI Symposium. Applied Science Publ. Ltd., London, England.
40. Shahani, K. M. 1975. Lipases and esterases, p. 184. *In* G. Reed (ed.), Enzymes in food processing, 2nd ed. Academic Press, New York.
41. Silliker, J. H. 1977. Bacterial contaminants in confections. Presented to the 94th Annual Conv. of the Natl. Confectioners Assoc., Chicago, Ill., June 25.

42. Slater, C. A. 1986. Chocolate and sugar confectionery, jams and jellies. *In* S. M. Herschdoerfer (ed.), Quality control in the food industry, vol. 3, 2nd ed. Academic Press, New York.
43. Tamminga, S. K. 1979. The longevity of *Salmonella* in chocolate. Antonie van Leeuwenhoek J. Microbiol. Serol. 45(1):153.
44. Te Giffel, M. C., R. R. Beumer, S. Leijendekkers, and F. M. Rombouts. 1996. Incidence of *Bacillus cereus* and *Bacillus subtilis* in foods in the Netherlands. Food Microbiol. 13:53.
45. Tilbury, R. H. 1976. The stability of intermediate moisture foods with respect to yeasts, p. 138. *In* R. Davies, G. G. Birch, and K. J. Parker (eds.), Intermediate moisture foods. Applied Science Publ., London.
46. Torres-Vitela, M. R., E. F. Escartin, and A. Castillo. 1995. Risk of salmonellosis associated with consumption of chocolate in Mexico. J. Food Prot. 58(5):478.
47. Troller, J. 1979. Food spoilage by microorganisms tolerating low-a_w environments. Food Technol. 33(1):72.
48. U.S. Food and Drug Administration. 1995. Bacteriological analytical manual, 8th ed., AOAC International, Gaithersburg, Md.
49. U.S. Government, National Archives and Records Administration. 1987a. Code of Federal Regulation Title 21 Part 110—Current Good Manufacturing Practice in Manufacturing, Packing, or Holding Human Food. p. 218-223. U.S. Government Printing Office, Washington, D.C.
50. U.S. Government National Archives and Records Administration. 1987b. Code of Federal Regulation Title 21 Part 163—Cacao products. p. 496-503. U.S. Government Printing Office, Washington, D.C.
51. Windisch, S. 1977. Nachweis und Wirkung von Hefen in zuckerhaltigen. Lebensmittel. Aliment. 23.
52. Windisch, S., and I. Neumann. 1965. Uber die "Wasserflecken" des Marzipans und ihre Entsehung. Z. Lebensm. Unters. Forsch. 129:9.
53. Windisch, S., and I. Neumann. 1965. Zur mikrobiologischen Untersuchung von Marzipan. 3. Mitteilung: Erfahrungen aus der Betriebskontrolle bei der Marzipanherstellung. Susswaren 9(10):540.
54. Witlin, B., and R. D. Smyth. 1957. "Soapiness" in "white" chocolate candies. Am. J. Pharm. 129:135.

Chapter 57

Nut Meats

A. Douglas King, Jr. and Thomas Jones

57.1 INTRODUCTION

Nut meats are derived from processed nuts harvested from trees, shrubs, or plants. They are consumed raw, salted, or dry- or oil-roasted. Nut meats are sold primarily to the food processing industry and are used in baked goods, cake mixes, candy, cereals, ice cream, etc. Nut meats are also sold shelled or in-shell for direct consumer use.

The nut meat industry has shown a steady growth over the past few decades to in excess of $2.5 billion worldwide.[34] The largest nut meat production worldwide and in the United States is peanuts (groundnut). Almond, pecan, and walnut are the three largest U.S. tree nut crops. Worldwide, filbert production is second to peanuts. About one-third of the U.S. tree nut and one-fifth of the shelled peanut production is exported. Half of shelled peanut production is processed into peanut butter.

With the exception of cashew and macadamia nuts, United States Standards for grades of the various nut meats are published by the United States Department of Agriculture (USDA) Marketing Service. Macadamia nut standards are set by the Hawaii Department of Agriculture, and cashew standards follow those of The Cashew Export Promotion Council (Cochin, India). Frequently buyers specify quality parameters for nut meats for purchase. Specification covenants are made between customer and vendor.

57.2 GENERAL CONSIDERATIONS

Nut meat processing is normally a dry process so bacterial and yeast growth is not important but molds may grow on damp nuts. Nut meats with field dust deposited on them during growth and harvesting bring contaminating microorganisms into the processing plant. This is especially true of nuts that have ground contact or with soft-shells like almonds or peanuts and nuts with broken shells. Air lift (aspiration) sorting to remove lightweight pieces such as shell and shriveled nut meats, also removes some field dirt.

Peanuts are dried after digging but before storage. Peanuts are usually dry-roasted before grinding into peanut butter. Thus, peanut butter has a low water activity and microbial growth is prevented. Peanut butter has been recalled because of the presence of *Salmonella*. The presence of such pathogens in peanut butter is usually a result of mishandling after dry roasting. Aflatoxin is a concern because of the chance of mold growth and mycotoxin formation in the field or during storage if the nuts contain more than about 9% moisture.

The shells of pecans and certain other nuts are softened by humidifying to avoid breaking the nut meats during cracking.[3,11] Pistachio processing includes a flotation step to remove immature fruits. Pecan shell fragments are separated from nut pieces by flotation. Blanching in water or steam loosens the pellicle from almonds and peanuts. A salt water dip or spray is used to apply salt to nut meats. These water treatments can cause microbial contamination of the nut meat. Water treatments can also moisten the nut meats and increase their water activity. Processes, such as blanching, are often followed by drying (71°C/3 minutes) to reduce water activity as quickly as possible. Good sanitation and frequent changes of water are needed to control microbial buildup where water baths are used.

Nuts that are not dry following harvest and processing must be dried to prevent mold growth during storage. Heat is sometimes used to dry nut meats and to reduce microbial counts, but darkening of the nut meats may occur. Condensate can form on nut meats removed from refrigeration, particularly during periods of humid conditions. The condensate will cause the water activity to rise, encouraging growth of mold. Tempering the product to avoid radical temperature change, therefore, is a good practice. The moisture content of normal tree nuts ranges from 3.8% to 6.7%, which gives a water activity of 0.7 or lower where microbial growth is prevented[4] (Chapter 64).

Most of the microbial flora on nut meats die during dry storage. Survival depends on temperature, water activity, composition and time. Moisture contents that are higher, but below the level where growth can occur, and higher storage temperature accelerates the microbial death rate.[21,23] With oil-roasted nuts the oil and dry nature of the nuts causes a slower decline in bacterial numbers.

Gas sterilization of nut meats with propylene oxide reduces the microbial population.[2] Use of propylene oxide is permitted on tree nut meats but not on peanuts.[35] United States regulations limit residual content of propylene oxide in the processed nut meat to 300 ppm. A pending tolerance would limit propylene oxide residues to 150 ppm in all raw nut meats except peanuts; residues are already limited to 150 ppm in many whole almond, Brazil nut, filbert, pecan, pistachio and walnut.[8] Other countries have zero tolerance for propylene oxide residues. To avoid high residuals, propylene oxide should only be applied once, in retorts, for a maximum of 4 hr exposure, at a temperature no higher than 52°C.

For further discussion of the microbiology of nut meats see ICMSF.[18]

57.3 MICROFLORA

Under normal conditions, bacteria that contaminate nut meats do not present a spoilage problem because they will not grow at the low water activity of nuts. Molds are of somewhat greater concern because of their ability to grow in relatively dry environments (lower water activity).[31]

The number and types of microorganisms present on nut meats will depend primarily on conditions of harvest, processing, and storage. Genera of bacteria isolated from almonds and pecans (tree nuts) include *Enterobacter, Escherichia, Bacillus, Xanthomonas, Clostridium, Pseudomonas, Leuconostoc, Streptococcus, Micrococcus*, and certain members of the coryneform group (i.e., *Brevibacterium* and *Corynebacterium*).

Nut meats from hard-shelled varieties have lower microbial counts than those from soft-shelled varieties. The nut meats of almonds with intact shells have lower counts than shelled nuts, and almonds harvested onto canvas have lower counts than those knocked to the ground. Nuts with the least foreign material have the lowest counts. Insect-damaged nut meats are more heavily contaminated than those that have not been infested.[20] The microbial counts on nut meats are often several thousand per g or less. Coliforms are not uncommon, but *Escherichia coli* is present on 4–6% or less of the samples.[20,25]

The U.S. Food and Drug Administration (FDA) considers nuts to be adulterated when moldy, rancid, infested with insects, tainted with aflatoxin or positive for *E. coli*.[9] *E. coli* was not found in pecans with unbroken shells.[12,28] Also, meats of whole pecans with no visible breaks or cracks did not become contaminated when soaked 24 hr in lactose broth containing *E. coli*. Longer soaking of pecans, e.g., 48 hr in water, resulted in the opening of 24% of the shell sutures; most failed to close completely when the nuts were redried. Similar findings have been made with walnuts.[24] Nuts harvested from orchards where farm animals have grazed are more likely to be contaminated with *E. coli* than those from orchards where grazing has not occurred.[25,26] Since *E. coli* may be present on nut meats before processing, its presence on processed nuts does not necessarily indicate poor processing plant practices.[20,26,28,29] Microorganisms present on almond, Brazil nuts, filbert, peanut, pecan, and walnut are the result of contamination with orchard soil combined with a) damaged or cracked nuts, b) insect infestation, c) diseased nuts, d) contamination within the processing environment and e) time in contact with the soil.[27]

Nut meats are frequently contaminated with the molds especially the storage molds *Penicillium, Aspergillus*, and *Fusarium*.[6,7,13,14,16,17,19,20,27,30,32,36] The mold composition changes from "field fungi" to "storage fungi" from harvest through processing to storage.[19,32]

57.4 PATHOGENS

Although tree nuts and ground nuts often are subject to microbial contamination, they are seldom vehicles in food poisoning outbreaks. Rarely they may contain *Salmonella*.[25] Usual thermal treatments that have been applied to in-shell pecans and pecan halves heavily contaminated with *Salmonella senftenberg, Salmonella anatum*, and *Salmonella typhimurium* did not destroy the salmonellae consistently.[5]

57.5 MOLDS AND MYCOTOXINS

Molds are common on nut meats. If the water activity of nut meats is high enough, mold spores will germinate, grow and may form mycotoxins. After germination, the hyphae can penetrate pecan tissue.[7] Nut meats are sometimes added to other foods, resulting in an environment where molds on the nut meats can grow, causing spoilage. Propylene oxide is sometimes used to reduce mold counts.[37] Contamination, especially of peanuts, with aflatoxin produced by the mold Aspergillus flavus is particularly a health concern. For further discussion of mycotoxins see Chapter 43. In the United States, FDA guidelines permit 20 ppb of total aflatoxin in nut meats. Aflatoxin limits vary between nations and can be either above or below the United States standard. Some countries also place a limit on the amount of aflatoxin B1, the most toxic form of the contaminant.

57.6 LABORATORY TESTS

Chapters 7, 8, and 20 describe the detection and enumeration of aerobic bacteria (APC), coliforms and *E. coli*, and molds and yeasts, respectively. Microorganisms may be removed from the nut meat surface by stomaching (Chapter 2). Alternatively, serial dilutions can be prepared by the protocol used in the FDA/AOAC Bacteriological Analytical Manual.[10] In this procedure, 50 grams of nut meats (halves or larger pieces) are combined with 50 mL of Butterfield's phosphate buffer to create a 1:2 dilution. This mixture is shaken 50 times, allowed to stand for 3–5 minutes, then shaken again 5 times to resuspend any particulates before serial dilutions and inoculations are made. For nut meal, 10 g of product are combined with 90 mL of Butterfield's phosphate buffer to create a 1:10 dilution; this 1:10 dilution is shaken and treated the same way as the 1:2 dilution listed above. Nut meats with a pellicle that does not come off easily are placed in a flask with the rinse solution on a rotary shaker for 10 min before removing the aliquot for microbial analysis. Avoid undue exposure to the inhibiting and destructive effects of tannin from the nuts by making serial dilutions promptly. No more than 15 min should elapse from adding diluent until completing the dilutions and inoculations.

FDA has specified that if an *E. coli* most probable number (MPN) of at least 0.36 per g, IMVIC confirmed, is found in two or more sub-samples when less than 10 sub-samples are examined, or in 20% or more of the sub-samples where more than 10 are examined, the product is in noncompliance.[9] All nut samples tested for *E. coli* must include a 1:2 dilution. This is done to achieve the necessary level of sensitivity (0.3 per gram) for detecting samples that are out of compliance. Nut meals are sometimes difficult to manipulate when prepared as a 1:2 dilution; the meal can be added directly to the MPN tubes to overcome this problem, strained through a filtered stomacher bag or pipetted with a large-bore pipette.

Methods for detection of *Salmonella* are in Chapter 37. Compositing of analytical sample units is a time-saving measure. The presence of Salmonella in nut meats not intended for further processing is unacceptable.

Mold evaluation on nut meats can be made by dilution techniques and counting colonies. Another approach is to examine the fungal flora by sterilizing the surface of the nut meat in 10% household chlorine bleach for 1 min, then plant 50 pieces (six to ten per plate) directly on the surface of previously poured petri plates. Molds are identified preferably to species, after 5 days of incubation at 25°C. The data are reported as the percentage of nutmeats contaminated with the identified molds.

Several media are available for counting of molds and yeasts on nut meats. For general purpose enumeration, or if Mucoraceous fungi such as *Mucor* or *Rhizopus* cause overgrowth on fungal petri plates, dichloran rose bengal chloramphenicol agar (DRBC) is recommended.[22, 33] DRBC allows growth for fruiting body development and easier identification of the molds present, a requirement to assess the potential hazard of the mold population. Otherwise oxytetracycline glucose yeast extract agar (OGY) or potato dextrose agar (PDA) are suitable media.[22,31,33] Phytone yeast extract agar (PYE) as well as OGY and PDA can give faster but unidentifiable counts in as little as 3 days but are prone to fungal overgrowth. DRBC takes 3 to 5 days for counts. Dichloran 18% glycerol agar (DG18) is useful for xerophilic fungi.[15] Additional information on xerophilic molds and media can be found in Chapter 20. *Aspergillus flavus* and parasiticus agar (AFPA) is a good medium for detection and enumeration of these two molds.[22,31,33] AFPA plates are examined after 42–48 hr incubation at 30°C for a characteristic bright orange-yellow reverse color indicating aflatoxigenic colonies. Dichloran Chloramphenicol Peptone Agar (DCPA) is used to detect Fusarium and dematiaceous Hyphomycetes such as Alternaria.[1,33] Fungal media containing antibiotics to control bacterial growth is preferred to acidified media (Chapter 20). References to appropriate methods for the determination of aflatoxins and other mycotoxins are given in Chapter 43.

57.7 REFERENCES

1. Andrews, S., and J. I. Pitt. 1986. Selective medium for the isolation of Fusarium species and dematiaceous Hyphomycetes from cereals. *Appl. Environ. Microbiol.* 51:1235.
2. Beuchat, L. R. 1973. Escherichia coli on pecans: Survival under various storage conditions and disinfection with propylene oxide. *J. Food Sci.* 38:1063.
3. Beuchat, L. R. 1975. Incidence of mold on pecan nuts at different points during harvesting. *Appl. Microbiol.* 29:852.
4. Beuchat, L.R. 1978. Relationship of water activity to moisture content in tree nuts. *J. Food Sci.* 43:754.
5. Beuchat, L. R., and E. K. Heaton. 1975. Salmonella survival on pecans as influenced by processing and storage conditions. *Appl. Microbiol.* 29:795.
6. Beuchat, L. R., and E. K. Heaton. 1980. Factors influencing fungal quality of pecans stored at refrigeration temperatures. *J. Food Sci.* 45:251.
7. Chipley, J. R. and E. K. Heaton. 1971. Microbial flora of pecan meat. *Appl. Microbiol.* 22:252.
8. Duggan, P., E. D. Welsh, and M. M. Duggan (eds.). 1998. *Pesticide chemical news guide.* CRC Press LLC, Washington D. C.
9. FDA. 1996. Compliance policy guides, Sub Chapter 570. Nuts. Washington, D. C.
10. FDA. 1998. Food sampling and preparation of sample homogenate. *In* Bacteriological analytical manual, 8th ed., Revision A. AOAC International, Gaithersburg, Md.
11. Forbus, W. R., Jr., B. L. Tyson, and J. L. Ayres. 1979. Commercial feasibility of an in-line steam process for conditioning pecans to improve shelling efficiency and maintain product (nut meat) quality. *J. Food Sci.* 44:988.
12. Hall, H. E. 1971. The significance of Escherichia coli associated with nut meats. *Food Technol.* 25:230.
13. Hanlin, R. T. The distribution of peanut fungi in the southeastern United States. *Mycopath. Mycol. Appl.* 49:227.
14. Heperkan, D., N. N. Aran, and M. Ayfer. 1994. Mycoflora and aflatoxin contamination in shelled pistachio nuts. *J. Sci. Food Agric.* 66:273.
15. Hocking, A. D., and J. I. Pitt. 1980. Dichloran-glycerol medium for enumeration of xerophilic fungi from low moisture foods. *Appl. Environ. Microbiol.* 39:488.
16. Hocking, A. D., and J. I. Pitt. 1996. Fungi and mycotoxins in foods. *Fungi Aust.* 1B:315.
17. Huang, L. H., and R. T. Hanlin. 1975. Fungi occurring in freshly harvested and in-market pecans. *Mycologia* 4:689.
18. International Commission on Microbial Standards for Foods. 1980. Chapter 21, p. 635-642. *In* Microbial ecology of foods, vol. 2. Academic Press, New York.
19. Joffe, A. Z. 1969. The mycoflora of fresh and stored groundnut kernels in Israel. Mycopathol. *Mycol. Appl.* 39:255.
20. King, A. D., Jr., M. Miller, and L. C. Eldridge. 1970. Almond harvesting, processing and microbial flora. *Appl. Microbiol.* 20:208.
21. King, A. D., Jr., W. U. Halbrook, G. Fuller, and L. C. Whitehand. 1983. Almond nut meat moisture and water activity and its influence on fungal flora and seed composition. *J. Food Sci.* 48:615.
22. King, A., Jr., J. I. Pitt, L. R. Beuchat, and J. E. L. Corry (eds.). 1986. *Methods for the mycological examination of food.* Plenum Press, New York.
23. King, A. D., Jr., and J. E. Schade. 1986. Influence of almond harvesting, processing, and storage on fungal population and flora. *J. Food Sci.* 51:202.
24. Kokal, D. 1965. Viability of Escherichia coli on English walnut meats (Juglans regia). *J. Food Sci.* 30:325.
25. Kokal, D., and D. W. Thorpe. 1969. Occurrence of Escherichia coli in almonds of Nonpareil variety. *Food Technol.* 3:277.
26. Marcus, K. A., and H. J. Amling. 1973. Escherichia coli field contamination of pecan nuts. *Appl. Microbiol.* 16:279.
27. McDonald, D. 1970. Fungal infection of groundnut fruit after maturity and during drying. *Trans. Br. Mycol. Soc.* 54:461.
28. Ostrolenk, M., and A. C. Hunter. 1939. Bacteria of the colon-aerogenes group on nut meats. *Food Res.* 4:453.
29. Ostrolenk, M., and H. Welch. 1941. Incidence and significance of the colon-aerogenes group on pecan meats. *Food Res.* 6:117.
30. Pitt, J. I., A.D. Hocking, K. Bhudhasami, B. Miscramble, K. A. Wheeler, and P. Tanboon-Ek. 1993. The normal mycoflora of commodities from Thailand. 1. Nuts and oilseeds. *Int. J. Food Microbiol.* 20:211.
31. Pitt, J. I., and A. D. Hocking. 1997. Fungi and food spoilage, 2nd ed. Aspen Publishers, Gaithersburg, Md.
32. Porter, D. M., and K. H. Garren. 1970. Endocarpic microorganisms of two types of windrow-dried peanut fruit (Arachis hypogaea L.). *Appl. Microbiol.* 20:133.
33. Samson, R. A., A. D. Hocking, J. I. Pitt, and A. D. King (eds.). 1992. *Modern methods in food mycology.* Elsevier, Amsterdam.
34. USDA. 1997. Agricultural statistics. Government Printing Office, Washington, D. C.
35. U. S. Government. National Archives and Records Administration. 1978. Code of Federal Regulations. 21 Part 193, p. 380. Propylene oxide. Washington, D. C.
36. Wehner, F. C., and C. J. Rabie. 1970. The micro-organisms in nuts and dried fruits. *Phytophylactica.* 2:165.
37. Wilson-Kakashita, G. W., D. L. Gerdes, and W. R. Hall. 1995. The effect of gamma irradiation on the quality of English walnuts (Juglans regia). *Lebensm. Wiss. Technol.* 28:17.

Chapter 58

Fruit Beverages

W. S. Hatcher, Jr., M. E. Parish, J. L. Weihe,
D. F. Splittstoesser, and B. B. Woodward

58.1 INTRODUCTION

Federal standards of identity have been established for some fruit juices, but not for most drinks called ades, nectars, cocktails, and other terms. These names do not signify absolute values as to the percentage of fruit or fruit juice in the beverage. Fruit juices are the unconcentrated liquid extracted from pure mature fruit.

Fruit juices may be squeezed directly from the fruit as in citrus processing, or they may be prepared from macerated or crushed material as in the processing of grapes, cherries, berries, and apples. Juices may be highly clarified, or they may contain considerable amounts of suspended solids. They may be marketed at their natural strength or as concentrates prepared by freeze concentration or evaporation. Preservation can be accomplished by thermal processing including aseptic packaging, refrigeration, freezing, ultra-filtration, or the addition of microbial inhibitors. Fruit juice drinks may contain 5% to 20% or more of juice, often combined with acids, natural or artificial colors and flavors, and other additives.

58.2 NORMAL MICROFLORA

58.21 Grove, Orchard, and Vineyard

The microorganisms found on sound fruit surfaces, especially those with a thick rind or skin, may be any of the various genera associated with soil, air, irrigation water, and insect pests in fruit-growing areas. Populations may be relatively low, e.g., 10^4 per apple.[22] Succulent fruits with thin skins such as Concord grapes, on the other hand, commonly possess yeast populations of 10^8 to 10^9 per g.[44]

Unsound, decomposed fruits are heavily contaminated, and a small percentage of unwholesome fruit can "seed" operating equipment with spoilage microorganisms.[17,30]

58.22 Processing Effects

Certain processing operations reduce the number of viable microorganisms, while others may serve as significant sources of contamination.[21] Washing fruits may reduce counts by more than 90%. Other procedures that remove or destroy microorganisms are hot pressing, lye peeling, fining, centrifugation, sanitization, and pasteurization.

Current emphasis on water conservation and reuse can lead to increased microbial populations and necessitates appropriate water treatment and disinfection.

Equipment used in the preparation of fruit juices is frequently a significant source of contamination. Unit operations such as presses, extractors, finishers, mills, pipelines, and conveyors are areas conducive to the formation of biofilms.[40] Growth may take place in the juice itself when it is held too long at ambient temperature.

58.23 Predominant Organisms

Fruit juices are generally acidic, with pH values ranging from approximately 2.4 for lemon juice to 4.2 for tomato juice. All contain sugars with amounts varying from 2% in lemon to more than 20% in some varieties of grape.[43] The low pH of these foods selects for yeasts, molds, and a few groups of aciduric bacteria.

The microorganisms of greatest significance in processing citrus juices are the lactic acid bacteria, primarily species of *Lactobacillus* and *Leuconostoc*, yeasts, and molds. In single-strength juice the lactic acid bacteria generally outgrow the fungi; however, multiplication of these bacteria is inhibited by the high sugar content of products concentrated above 30° Brix.[6,19] Refrigerated temperatures (≤5°C) also select for the fungi, even in single-strength juices.

After concentration and freezing, the microbial population is greatly reduced. Most of the surviving organisms are yeasts that may grow if temperatures are elevated above freezing. The most frequently found yeasts in commercial orange juice concentrate belong to the genera *Candida*, *Saccharomyces*, and *Rhodotorula*.[47]

Yeasts are usually the most important group in apple and grape juices. Mills and presses can be significant sources of contamination. The yeasts most commonly isolated from these products and from carbonated fruit-containing beverages belong to the genera *Saccharomyces*, *Candida*, and *Torulopsis*. Psychrotropic species may grow in Concord grape juice during the period it is stored at 2°C to precipitate tartrates.[41]

High levels of mold contamination are generally attributed to unsound fruit entering the processing facility. Machinery mold, *Geotrichum candidum*, may be introduced into fruit products from unsanitary equipment. Low numbers of heat-resistant molds such as *Byssochlamys* spp. and *Neosartorya fisheri* often are present on the raw fruit[45] and may survive the processing steps (Chapter 21).

Other fungi frequently associated with fruit products include *Alternaria, Botrytis, Colletotrichum, Diplodia, Fusarium, Penicillium,* and *Phomopsis*.[34,43]

58.24 Indicator Bacteria

Coliforms as well as enterococci are frequently isolated from citrus and other fruit products,[37,38,39] but their presence is not necessarily indicative of avoidable insanitary conditions during harvesting, processing or packaging.[3] They may become part of the normal processing plant flora; however, their presence does not indicate fecal contamination. When examining juice concentrates for coliforms, some yeasts may give false-positive presumptive tests.[23] Total coliforms and *E. coli* have been isolated from fresh (non-pasteurized) orange juice samples, using enrichment procedures for isolation and enumeration.[31,47]

58.3 PATHOGENS

Pathogenic bacteria are usually not a problem in fruit juices or juice-based drinks, with most being heat inactivated during thermal processing, or negatively affected by the high-acid, low pH, juice environment. However, several juice-borne disease outbreaks occurred in the 20th century as detailed in a recent review[31] and the assumption of safety of fruit juices must be reexamined.

Disease syndromes attributed to juice-borne outbreaks include salmonellosis, typhoid fever, hepatitis, hemorrhagic colitis, hemolytic uremia, cryptosporidiosis, and gastroenteritis. Specific microorganisms involved in the various outbreaks were *Salmonella typhi* and other salmonellae, *E. coli* O157:H7, and *Cryptosporidium parvum*. Products associated with these illnesses were either "fresh" unpasteurized juices that were probably contaminated during juice processing, or heat treated juices contaminated by food handlers in a retail or institutional setting.

The ability of pathogenic bacteria to adapt for survival under acidic conditions is well documented.[2,9,24] Survival of *S. typhimurium* cells pre-exposed to pH 5.8 was 100 to 1000 times greater than cells not habituated to this pH prior to incubation at pH 3.3.[8] Although growth of pathogens is not expected at pH levels common to most fruit juices, extended survival of cells, especially at refrigerated temperatures, has been demonstrated. Results of several studies earlier this century suggest that certain coliforms, salmonellae and other enterics survive in low pH products for very substantial time periods.[25] One study indicates that between 1 and 49 days are needed to cause a 4-log reduction of enteric bacteria (including salmonellae and *Shigella*) inoculated into apple, orange or tomato juices.[26] After a 1974 outbreak of salmonellosis from apple cider,[1] subsequent studies showed that *Salmonella typhimurium* could survive at least 30 days in pH 3.6 apple juice.[11] Salmonellae isolated from a citrus juice processing facility survived between 14 and 73 days at 0° and 4°C when inoculated into orange juice.[36] Population death rates in juice were enhanced at lower pH values.

Toxigenic molds (Chapter 43) may grow on fruits that are processed into beverages. Traces of patulin have been found in apple juice.[20] Clinically significant yeasts were found in commercial orange concentrate,[47] and hepatitis was probably spread among hospital employees by a worker who may have contaminated orange juice during its preparation.[5] Cancer patients undergoing chemotherapy or radiation treatment and others with immunosuppresive diseases may be at risk from opportunistic bacteria and yeasts such as *Klebsiella, Enterobacter, Escherichia, Candida, Torulopis*, and related organisms that are frequently found in small numbers in fruit juices.

Listeriosis has not been associated with fruit juices. *Listeria* were not detected in 100 samples of orange juice from processing facilities in various parts of North America.[35] When inoculated into a juice beverage system, *L. monocytogenes* survive for substantial time periods under refrigeration.[33] As with other pathogens, death of *Listeria* is enhanced by lower pH and higher storage temperatures.

58.4 SPOILAGE

In the processing of citrus concentrates, single-strength juice may be held in stainless steel tanks for 30 to 120 minutes prior to high-temperature evaporation. During this period the product is most susceptible to microbial degradation. Growth of lactic acid bacteria may result in the production of acetylmethylcarbinol and diacetyl which give an off-flavor similar to buttermilk.[7,15,16] Above 30° Brix fruit juice concentrates are subject to spoilage by osmophilic yeasts.[46] Citrus juice concentrates are generally stored at 0° to 10°F, temperatures which minimize growth of yeasts and molds.[28,29]

Heat-resistant molds (Chapter 21) may survive the thermal process given canned juice beverages. Some evidence indicates that juice drinks are more susceptible to this type of spoilage than are single-strength fruit juices.

Some strains of sporeforming bacteria, all of which were once thought unable to grow in the high-acid environment of most fruit products, are now recognized as spoilage agents. Acidophilic sporeforming bacilli, belonging to the genera *Alicyclobacillus* and *Sulfobacillus*, can grow down to a pH of 3.0.[4,10,42,50] Their spores can easily withstand the heat treatment given most fruit juices and drinks. Growth in pasteurized juices, most notably apple, produces a phenolic odor and taste.

The successful use of benzoic and sorbic acids to preserve juice drinks depends on the elimination of most yeasts and aciduric bacteria from the various ingredients.

Improved flavor parameters associated with nonthermally treated juices have created a growing market for "fresh," unpasteurized fruit juices. The increased microbial instability of these products requires greater attention to process sanitation and storage conditions. Microbial spoilage of unpasteurized fruit juices is most commonly due to aciduric microbes such as lactic acid bacteria and yeasts. A qualitative survey of the bacterial flora from unpasteurized orange juice indicated the presence of *Lactobacillus fermentum, L. plantarum, Leuconostoc mesenteroides* subsp. *dextranicum, L. mesenteroides ssp. mesenteroides,* and *L. paramesenteroides*.[32]

58.5 RECOMMENDED METHODS

58.51 Aerobic Plate Count (Chapter 7)

Plate on orange serum agar. Incubate 48 hr. at 30°C.

58.52 Alicyclobacillus (Chapter 24)

58.53 Diacetyl Test[14,18]

58.54 Direct Microscopic Count (Chapter 4)

For citrus products, stain 5 mL of 12° Brix juice with an equal volume of 0.075% aqueous crystal violet, color index 681. After mixing thoroughly, distribute 0.01 mL over 1 cm^2 of a clean slide. Dry under a heat lamp.[19]

58.55 *Experimental Procedures*

Recently a number of rapid methods for detection of microorganisms have been applied to citrus juices. They include radiometry,[13] impedance,[48] bioluminescence,[12] and the plate loop method.[27] Additionally, a gas chromatographic method for the detection of diacetyl and acetoin in citrus juice has been investigated.[49] This method can separately quantify diacetyl and acetoin down to 0.05 and 10 ppm, respectively.

58.56 *Geotrichum Count (Chapter 4)*

58.57 *Heat-Resistant Molds (Chapter 21)*

58.58 *Toxigenic Fungi and Fungal Toxins (Chapter 43)*

58.59 *Yeasts and Molds (Chapter 20)*

58.6 INTERPRETATION OF DATA

The microbiology of fruit beverages will vary greatly depending on the nature of the fruit, the methods of processing, and the means of preservation. High microbial populations often indicate poor fruit quality, unsanitary equipment, or the opportunity for growth in the food at some stage in the process. However, since each product and process is different, one cannot apply the criteria developed for one to another.

Heat-processed beverages can be free of viable nonsporeforming bacteria but yield low numbers of sporeforming bacteria when cultured on nonselective media. Many of these sporeformers cannot grow in the high-acid environment of fruit products, and thus their presence has no bearing on shelf stability. However, a few strains are acidophilic and in time may spoil the product. The presence of sporeforming bacteria may mean that a high ratio of unwholesome fruit was used to manufacture the juice or that the fruit was not washed adequately before processing.

Direct microscopic counts for yeasts, bacteria, or molds may provide a clue to conditions of sanitation during processing; however, most of these organisms will have been removed in the process of creating a highly clarified beverage. A marker used increasingly by industry to assess the quality of apples used to manufacture juice is patulin. This mycotoxin is produced by several genera of molds,[20] and a level in excess of 50 ppb (in single strength juice) may mean a high percentage of unwholesome apples were used to manufacture the product.

Frozen concentrated juices generally contain microbial populations of 10^2 to 10^5 CFU per milliliter of reconstituted product. Although high temperature evaporators, operated above 90°C, destroy most microorganisms, opportunity exists for recontamination of the concentrated product. Furthermore, retail orange juice concentrates are generally prepared by diluting 60° Brix concentrate to 40° to 45° Brix using water or freshly extracted, single-strength juice. This juice may serve as a source of contamination if processors fail to pasteurize it in the belief that nonpasteurized juice provides a concentrate with the freshest, most natural flavor.

Heat-resistant mold spores may be present in low numbers in fruit juice concentrates. They can present a problem if the juice is to be used as a constituent of a beverage that will be preserved with a heat process.

Nonsterile fruit beverages may contain low numbers of coliforms and enterococci, but their presence is not a reliable indication of fecal contamination.

The low pH of most fruit beverages prevents the growth of enteric pathogens. Products that have received a heat treatment at some stage of the process should be free of these organisms. In rare instances salmonellae have been introduced into fruit juices and have survived for an extended period of time.

58.7 REFERENCES

1. Anon. 1975. *Salmonella typhimurium* outbreak traced to a commercial apple cider. Morb. Mortal Wkly. Rep. 24:87.
2. Belli, W. A., and R. E. Marquis. 1991. Adaptation of *Streptococcus mutans* and *Enterococcus hirae* to acid stress in continuous culture. Appl. Environ. Microbiol. 57:1134.
3. Dack, G. M. 1955. Significance of enteric bacteria in foods. Am. J. Public Health 45:115I.
4. Dufresne, S., J. Bousquet, M. Boissinot, and R. Guay. 1996. *Sulfobacillus disulfidooxidans* sp. nov., a new acidophilic, disulfide-oxidizing, gram-positive, spore-forming bacterium. Int. J. Syst. Bacteriol. 46:1056.
5. Eisenstein, A. B., R. D. Aach, W. Jacobson, and A. Goldman. 1963. An epidemic of infectious hepatitis in a general hospital - Probable transmission by contaminated orange juice. JAMA 185:101.
6. Faville, L. W., and E. C. Hill. 1951. Acid-tolerant bacteria in citrus juices. Food Res. 17(3):281.
7. Fields, M. L. 1964. Acetylmethylcarbinol and diacetyl as chemical indexes of microbial quality of apple juice. Food Technol. 18:114.
8. Foster, J. W., and H. K. Hall. 1990. Adaptive acidification tolerance response of *Salmonella typhimurium*. J. Bacteriol. 172:771.
9. Foster, J. W., and M. P. Spector. 1995. How *Salmonella* survive against the odds. Annu. Rev. Microbiol. 49:145.
10. Golovacheva, R. S., and G. I. Karavaiko. 1978. A new genus of thermophilic spore-forming bacteria, *Sulfobacillus*. Microbiology 47:658.
11. Goverd, K. A., F. W. Beech, R. P. Hobbs, and R. Shannon. 1979. The occurrence and survival of coliforms and salmonellae in apple juice and cider. J. Appl. Bacteriol. 46:521.
12. Graumlich, T. R. 1985. Estimation of microbial populations in orange juice by bioluminescence. J. Food Sci. 50:116
13. Hatcher, W. S., S. DiBenedetto, L. E. Taylor, and D. I. Murdock. 1997. Radiometric analysis of frozen concentrated orange juice for total viable microorganisms. J. Food Sci. 42:636.
14. Hatcher, W. S., J. L. Weihe, D. F. Splittstoesser, E. C. Hill, and M. E. Parish. 1992. Fruit beverages. D. F. Splittstoesser (ed.), Am. Public Health Assoc., Washington, D. C.
15. Hays, G. L. 1951. The isolation, cultivation and identification of organisms which have caused spoilage in frozen orange juice. Proc. Fla. State Hort. Soc. 64:135.
16. Hays, G. L. and D. W. Riester. 1952. The control of "off odor" spoilage in frozen concentrated orange juice. Food Technol. 6:386.
17. Hill, E. C., and L. W. Faville. 1951. Studies on the artificial infection of oranges with acid tolerant bacteria. Proc. Fla. State Hort. Soc. 64:174.
18. Hill, E. C., and F. W. Wenzel. 1957. The diacetyl test as an aid for quality control of citrus products. 1. Detection of bacterial growth in orange juice during concentration. Food Technol. 11:240.
19. Hill, E. C., F. W. Wenzel, and A. Barreto. 1954. Colorimetric method for detection of microbiological spoilage in citrus juices. Food Technol. 8:168.
20. Lindroth, S., and A. Niskanen. 1978. Comparison of potential patulin hazard in homemade and commercial apple products. J. Food Sci. 43:446.
21. Luthi, H. 1959. Microorganisms in noncitrus juices. Adv. Food Res. 9:221.
22. Marshall, C. R., and V. T. Walkley. 1951. Some aspects of microbiology applicable to commercial apple juice production. I. Distribution of microorganisms on fruit. Food Res. 16:448.
23. Martinez, N. B., and M. D. Appleman. 1949. Certain inaccuracies in the determination of coliforms in frozen orange juice. Food Technol. 3:392.
24. Miller, L. G., and C. W. Kaspar. 1994. *Escherichia coli* O157:H7 acid tolerance and survival in apple cider. J. Food Prot. 57:460.

25. Mitscherlich, E., and E. H. Marth. 1984. Microbial survival in the environment. Springer-Verlag, New York.
26. Mossell, D. A. A., and A. S. deBruin. 1960. The survival of *Enterobacteriaceae* in acid liquid foods stored at different temperatures. Ann. Inst. Pasteur Lille 11:65.
27. Murdock D. I., and W. S. Hatcher, Jr. 1976. Plate loop method for determining total viable count of orange juice. J. Milk Food Technol. 39:470.
28. Murdock, D. I., and W. S. Hatcher, Jr. 1975. Growth of microorganisms in chilled orange juice. J. Milk Food Technol. 38:470
29. Murdock, D. I., and W. S. Hatcher, Jr. 1978. Effect of temperature on survival of yeast in 45° and 65° Brix orange concentrate. J. Food Protect. 41:689.
30. Murdock, D. I., J. F. Folinazzo, and C. H. Brokaw. 1953. Some observations of gumforming organisms found on fruit surfaces. Proc. Fla. State Hort. Soc. 66:278.
31. Parish, M. E. 1997. Public health and nonpasteurized fruit juices. Crit. Rev. Microbiol. 23(2):109.
32. Parish, M. E., and D. P. Higgins. 1988. Isolation and identification of lactic acid bacteria from samples of citrus molasses and unpasteurized orange juice. J. Food Sci. 53:645.
33. Parish, M. E., and D. P. Higgins, 1989a. Survival *of Listeria monocytogenes* in low pH model broth systems. J. Food Prot. 52:144.
34. Parish, M. E., and D. P. Higgins, 1989b. Yeasts and molds isolated from spoiling citrus products and by-products. J. Food Prot. 52:261.
35. Parish, M. E., and D. P. Higgins. 1989c. Extinction of *Listeria monocytogenes* in a single-strength orange juice: Comparison methods for detection in mixed populations. J. Food Safety 9:267.
36. Parish, M. E., J. A. Narciso, and L. M. Friedrich. 1997 Survival of salmonellae in orange juice. J. Food Safety 17:273.
37. Patrick, R. 1951. Sources of coliform bacteria in citrus juice for concentrates. Proc. Fla. State Hort. Soc. 64:178.
38. Patrick, R. 1953. Coliform bacteria from orange concentrate and damaged oranges. Food Technol. 7:157.
39. Patrick, R., and E. C. Hill. 1958. Enterococcus-like organisms in citrus concentrates. Food Technol. 12:337.
40. Patrick, R., and E. C. Hill. 1959. Microbiology of citrus fruit processing. Res. Bull. 618. Univ. Florida Agric. Exp. Sta., Gainesville, Fla.
41. Pederson, C. S., M. N. Albury, D. C. Wilson, and N. L. Lawrence. 1959. The growth of yeasts in grape juice stored at low temperatures. I. Control of yeast growth in commercial operation. Appl. Microbiol. 7:1.
42. Previdi, M. P., F. Colla, and E. Vicini 1995. Characterization of *Alicyclobacillus*, a spore-forming thermophilic acidophilic bacterium. Ind. Conserv. 70:128.
43. Splittstoesser, D. F. 1978. Fruits and fruit products. *In* L. R. Beuchat (ed.), Food and beverage mycology. AVI Pub. Co., Westport, Conn.
44. Splittstoesser, D. F., and L. R. Mattick, 1981. The storage life of refrigerated grape juice containing various levels of sulfur dioxide. Am. J. Enol. Viticult. 32:171.
45. Splittstoesser, D. F., F. R. Kuss, W. Harrison, and D. B. Preston. 1971. Incidence of heat resistant molds in eastern orchards and vineyards. Appl. Microbiol. 21:335.
46. Troller, J. A., and J. H. B. Christian. 1978. Water activity and food. Academic Press, New York.
47. Weihe, J. L. 1986. Citrus and beverage microbiology. *In* R. R. Matthews (ed.), Proceedings of the 26th Annual short course for the food industry. Univ. of Florida, Inst. of Food and Agric. Sci., Gainesville, Fla.
48. Weihe, J. L., S. L. Siebt, and W. S. Hatcher, Jr. 1984. Estimation of microbial populations in frozen concentrated orange juice using automated impedance measurements. J. Food Sci. 49:243.
49. Wicker, L., M.E. Parish, and R. J. Braddock. 1988. Analysis of diacetyl and acetylmethylcarbinol by GLC and application to citrus concentrate quality control. Presented at 85th Annual Meeting of the South Assoc. of Agric. Sci., New Orleans, La.
50. Wisotzkey, J. D., P. Jurtshuk, Jr., G. E. Fox, G. Deinhard, and K. Poralla. 1992. Comparative sequence analyses on the 16S rRNA (rDNA) of *Bacillus acidocaldarius*, *Bacillus acidoterrestris*, and *Bacillus cycloheptanicus* and proposal for creation of a new genus, *Alicyclobacillus* gen. nov. Int. J. Syst. Bacteriol. 42:263.

Chapter 59

Soft Drinks

Ralph DiGiacomo and Phyllis Gallagher

59.1 INTRODUCTION

Soft drinks are non-alcoholic beverages, that contain carbon dioxide, nutritive or non-nutritive sweeteners, natural or synthetic flavors, colors, acidification agents, chemical preservatives, foaming and emulsifying agents, in addition to various other functional ingredients. Soft drinks may also contain various fruit juices, tea powder and be fortified with various vitamins.

Soft drinks have a pH range of 2.5–4.0, with colas between 2.5 and 3.1. Most of the soft drinks consumed in North America are carbonated with a range of 2.0 to 4.0 volumes of carbon dioxide.

59.2 MICROBIOLOGY

Due to the high acid content, and resulting low pH of soft drinks, only aciduric bacteria, yeast and mold are of significance. Microbiological problems associated with soft drinks are primarily those of product spoilage and can be a serious economic problem, but rarely a public health issue. Pathogenic bacteria cannot survive in the acidic and carbonated environment of these beverages.[15]

59.21 Types of Microorganisms Found in Soft Drinks

Yeast accounts for most of the spoilage problems in the soft drink industry due to their high acid tolerance and ability to grow anaerobically. The types of yeast found include: *Zygosaccharomyces, Brettanomyces, Saccharomyces, Candida, Toruplopsis, Pichia, Hansenula* and *Rhodotorula*.[1,7]

The highly preservative resistant *Zygosaccharomyces* are by far the most significant spoilage yeast, with *Z. bailii* documented as the most notorious spoilage yeast in soft drinks. Spoilage by this yeast results in pronounced off-odors, off-tastes, visible sediment and increased package pressure/failure due to production of carbon dioxide.[5,12]

Brettanomyces spp. are sensitive to benzoic and sorbic acids but are highly resistant to carbonation. This yeast has been implicated in the spoilage of both low and non-preserved diet beverages, flavored carbonated water as well as sugar sweetened products. *B. naardenensis* is most commonly associated with spoilage of soft drinks.[1,10]

Mold spores can survive in carbonated beverages but can not grow due to the lack of oxygen and preservation effect of carbon dioxide. Mold can cause spoilage of non-carbonated beverages or when carbonation is lost due to loss of package integrity. *Aspergillus, Penicillium* and *Rhizopus* are common types of mold present in soft drink environments.[8]

Aciduric bacteria of concern in the soft drink industry include the acetic acid bacteria (*Acetobacter* and *Gluconobacter*) and lactic acid bacteria. *Acetobacter* and *Gluconobacter* are strict aerobes and are of concern in non-carbonated fruit drinks, juices, tea, and also flavor concentrates. They are very resistant to benzoic and sorbic acids and are capable of growth at low pH.

Due to their fastidious nature, lactic acid bacteria are of concern in soft drinks that contain juice, tea or other components that contribute available nutrients such as amino nitrogen and vitamins. Some species of *Lactobacillus* are also resistant to benzoic and sorbic acids.

Bacterial spoilage typically results in off-flavors, sediment, turbidity, "ropiness" (from dextrans) and gas production (heterofermentative lactics) in the beverage.[2,9,13]

59.22 Sources of Microorganisms

The ingredients used in the production of soft drinks such as sugar syrup, flavor concentrates, water and various dry ingredients, are rarely the source of spoilage yeast and bacteria. Flavor components typically contain high levels of alcohol, propylene glycol or benzoic and sorbic acids. The finished syrups rarely present any source of microbes because they are highly acidic and typically contain high levels of preservatives (>1000 ppm). Juice components used in soft drink production are either aseptically packaged or preserved with benzoic/sorbic acids to ensure they are not a source of contamination to the bottling/canning facility.

By far, the most significant source of spoilage yeast and bacteria is the bottling plant environment and equipment. The majority of microbial contamination occurs from the proportioner/blender and all equipment down stream through the filler; the most common source is the bottle or can filler itself. Fillers vary in size and may contain from 45 to 120 filler valves and are very difficult to clean and sanitize.

Other areas where build-up of microbes is common are proportioners/blenders, carbonators, and deaerators. Improper storage of liquid sugar can lead to condensation, dilution, and subsequent growth of osmophilic yeasts. Disposable packaging materials are rarely a source of microbial contamination. Returnable (refillable) packaging, however, can be a source of microorganisms if not properly cleaned prior to filling.[3,7]

59.23 Principles of Preservation

The factors that determine the microbiological stability of soft drinks are both intrinsic and extrinsic. Intrinsic factors are based on the nature of the formulation and include acidity, carbonation, nutrients, preservatives (benzoic/sorbic acids), chelators, and other natural inhibitors found in flavors. Extrinsic factors include processing, packaging and storage, however, factors of primary concern are types and levels of contaminating microorganisms that may be present on product contact surfaces.

Preservation of soft drinks relies primarily on the formulation (intrinsic factors), coupled with effective equipment sanitation and good manufacturing practices. A multiple barrier approach is applied that balances all of these factors to result in the production of microbiologically stable drinks.

Colas have inherent microbiological stability due to the high carbonation, low pH (2.5) and natural inhibitors in the flavor systems (terpenes). Other beverages are formulated with preservatives such as benzoic acid, sorbic acid and combinations of the two. More recently, the addition of less traditional preservatives such as chelating agents (EDTA, polyphosphates), have been employed as additional barriers in soft drink preservation systems. Various acidulants and combinations also can enhance the preservation system.

Soft drinks that contain fruit juices, tea or other sources of nitrogenous compounds are particularly susceptible to microbial spoilage.

Although soft drink preservation systems are designed to be relatively robust, this should not be a substitute for good manufacturing practices. Consistently poor sanitation practices can lead to increased levels of spoilage organisms and also the development of resistant spoilage yeast and bacteria.[3,4,11]

59.3 SANITATION

Since soft drinks are typically not thermally processed and are cold-filled, sanitation is an important factor in their successful production. With the development of more uncarbonated drinks containing ingredients such as juice, tea, and vitamins, sanitation is even more critical and is the focus of much of the microbiological testing that occurs in soft drink plants.

59.4 EQUIPMENT, MEDIA AND REAGENTS

59.41 Equipment

Incubators — 25°C and 35°C

59.42 Culture Media (Chapter 63)

Tryptone glucose extract media
Potato dextrose agar (acidified)
M-Green Yeast and Mold Medium (broth)
Preservative Resistant Yeast Medium (PRY, broth or agar)
Brettanomyces Selective Medium (BSM, broth or agar)
MRS (broth or agar)
M-Endo Medium (broth)
Colilert Presence/ Absence (broth)

59.5 RECOMMENDED METHODS

59.51 Sampling

There are several areas in bottling and canning plants where microbiological sampling is important:

Water processing system
Sweetener system
Syrup process and system
Filling process
Empty packaging (returnable packaging)
Finished products
Plant environment

The only sampling and testing of raw materials include sugar syrups and water. Flavors, concentrates and dry ingredients are rarely tested. These materials are routinely evaluated by the concentrate production facilities that produce and deliver these ingredients to the bottling and canning operations. Most flavor concentrate components are highly acidic, contain very high levels of preservatives or contain alcohol or propylene glycol and thus are rarely a source of spoilage microbes.

The major focus in microbiological sampling in soft drink plants is on measuring the effectiveness of the sanitation program. The most effective way to accomplish this is with the collection of sanitation rinse water samples (final water rinse after sanitizing). Sanitation swab samples are also used in areas where rinse water sampling is not practical. Sanitation rinse water samples collected in beverage containers at the filler are the most important indicators of sanitation effectiveness since they represent product flow during the filling process. Filler valve rinse water samples must be taken from a pressurized filler (nitrogen or carbon dioxide, 15–25 psi) in beverage containers to ensure full activation of the filler valve. This ensures rinse water contact with all internal valve components. Sanitation rinse water samples are also collected from proportioners, blending pumps, syrup tanks, carbonaters, the water treatment system and other areas within the filling process.

59.52 Membrane Filtration (Chapter 6)

Membrane filtration is the standard method employed in the soft drink industry. This method is simple and the most appropriate due to the need to detect low levels of yeast, bacteria, and mold in both sanitation rinse water and product samples. It is used for water, sugar, beverage syrups, and finished product samples in addition to sanitation rinse water and swab sampling. Sample sizes typically range from 10–100 mL, depending on the sample and specification. Viscous samples such as sugar syrups or beverage syrups are diluted first before filtering.

59.53 Yeasts and Molds (Chapter 20)

Total yeast counts are the single most important indicator of the sanitation effectiveness in soft drink plants. Total yeast and mold counts in sanitation rinse water samples, sugar, and finished products are obtained by membrane filtration using m-Green yeast and mold media. Preservative Resistant Yeast medium (PRY) is used to selectively isolate and enumerate *Zygosaccharomyces,spp.* while *Brettanomyces* Selective Medium (BSM) is employed for the detection of *Brettanomyces spp.*[14,16] These selective media are typically used when trouble shooting contamination and spoilage problems.

59.54 *Acid-Producing Bacteria (Chapter 19)*

Acetic acid bacteria can be detected and enumerated with acidified Potato Dextrose medium and will also grow on m-Green yeast and mold medium. Lactic acid bacteria are more difficult to detect due to their fastidious nature. Lactobacillus MRS agar or broth is typically used for isolation and enumeration. A modification of MRS supplemented with 10 – 20% apple juice can enhance recovery of lactics from soft drink environments.

59.55 *Aerobic Plate Count (Chapter 7)*

Aerobic plate counts of water or product samples employ membrane filtration using Tryptose Glucose Extract media.

59.56 *Coliforms (Chapter 8)*

Coliform testing is routinely performed on the incoming water supply to the plant. Two methods are typically used, either membrane filtration with m-Endo medium or presence/absence (P/A) testing using Colilert (IDEXX Labs).

59.57 *Bioluminescence Assays*

Since rinse water testing remains the major focus of sanitation measurement, bioluminescence (ATP photometry) has limited application. Simple and cost effective bioluminescent assays that have the required sensitivity for rinse water testing are not currently available. In soft drink plants, it can be applied where swab testing is appropriate and as a troubleshooting tool.

59.6 INTERPRETATION OF DATA

The microbiological stability of a product will determine the level of sanitation required. Beverages can be divided into two broad categories. Those which are sensitive to microbiological spoilage and require more stringent sanitation measures, and those which are robust due to the intrinsic product parameters and are not prone to the same spoilage risks (colas).

Total yeast counts are the single most important indicator of sanitation effectiveness in a soft drink environment.

All the microbiological and sanitation process data (time/temperature/concentration/flow rates) generated should be analyzed and tracked in order to identify trends. This will enable a preventative approach, rather than a reactive one in preventing spoilage of final beverage.

Successful production of microbiologically sensitive products such as uncarbonated juice beverages or tea products require a stringent sanitation program. Filler valve rinse water total yeast levels, for example, should be less than 15 yeast colony forming units (CFU) per 100 mL. The more robust products (colas) can tolerate less stringent standards, however, to comply with good manufacturing practices, rinse water counts should not exceed 100 yeast CFU per 100 mL. Filler valve rinse water testing usually ranges from 10 to 25% of the filler valves. Since soft drink fillers typically have from 45 to 120 filling valves, variation among the valves tested can occur. Therefore, the application of a 3 class sampling plan concept when testing filler valve rinse waters may be appropriate. An example for a 120 valve can filler would be as follows: n = 30, m = 15 yeast CFU/ 100 mL, M = 50 yeast CFU/ 100 mL, c = 3 (n = number of samples, m = counts below are acceptable, M = counts above are unacceptable, counts between m and M are marginal, c = number of marginal samples permitted).[6]

Qualitative or quantitative analysis of samples with selective media is frequently applied to identify sources of spoilage organisms and assess product spoilage risk.

59.7 REFERENCES

1. Baird-Parker, A. C., and W. J. Kooiman. 1980. Soft drinks, fruit juices, concentrates, and fruit preserves, *In* J. H. Silliker, R. P. Elliott, A. C. Baird-Parker, F. L. Bryan, J. H. B. Christian, D. S. Clark, J. C. Olson, Jr., T. A. Roberts (eds.), Microbial ecology of foods, vol. 2, Food commodities. Academic Press, New York.
2. Batchelor, V. J., 1984. Further microbiology of soft drinks, *In* H. W. Houghton (ed.), Developments in soft drink technology—3. Elsevier Applied Science Publishers, New York.
3. Berry, J. M. 1979. Yeast problems in the food and beverage industry. *In* M. E. Rhodes (ed.), Food mycology. G. K. Hall, Boston, Mass.
4. Bray, D. F., D. A. Lyon, and E. Burr. 1973. Three-class attributes plans in acceptance sampling. Technometrics 15:575.
5. Davenport, R. R., 1980. An outline guide to media and methods for studying yeasts and yeast-like organisms. *In* F. A. Skinner, S. M. Passmore, and R. R. Davenport (eds.), The biology and activities of yeasts. Academic Press, New York.
6. Davenport, R. R., and D. S. Thomas. 1985. Zygosaccharomyces Bailii —A profile of characteristics and spoilage activities. Food Microbiol. 2:157.
7. Deak, T., and L. R. Beuchat. 1996. Handbook of food spoilage yeasts. CRC Press, New York.
8. Eagon, R. G., and C. R. Green. 1973. Effect of carbonated beverages on bacteria. Food Res. 22:687.
9. Eyles, M. J., and A. D. Warth. 1989. The response of Gluconobacter oxydans to sorbic and benzoic acids. Int. J. Food Microbiol. 8(4): 335-342.
10. Insalata, N. F. 1952. Carbon dioxide versus beverage bacteria. Food Eng. 24:84.
11. Juven, B. J., and I. Shomer. 1985. Spoilage of soft drinks caused by bacterial flocculation. J. Food Protect. 48:52.
12. Kolfschoten G. A., and D. Yarrow. 1970. Brettanomyces naardenensis, a new yeast from soft drinks. Antonie Van Leeuwenhoek. 36(3):458.
13. Krakowski, E., 1993. Development of Brettanomyces Selective Media (BSM), Pepsi Cola Research and Development, Valhalla, New York.
14. Pitt, J. I., and K. C. Richardson. 1973. Spoilage by preservative resistant yeast. CSIRO Food Res. Q. 33:80.
15. Sand, F. E. M. J. 1977. Gluconobacter, still drinks and plastic containers. Soft Drinks Trade J. Oct:371.
16. Witter, L. D. , J. M. Berry, and J. F. Folinazzo. 1958. The viability of *Escherichia coli* and a spoilage yeast in carbonated beverages. Food Res. 23:133.

Chapter 60

Bottled Water

Henry Kim and Peter Feng

60.1 INTRODUCTION

Bottled water is considered, legally, a food in many individual nations and by the Codex Alimentarius of the United Nations. In the United States, it is regulated by the Food and Drug Administration (FDA), which defines bottled water as "water that is intended for human consumption and that is sealed in bottles or other containers with no added ingredients except that it may optionally contain safe and suitable antimicrobial agents.[4] Optionally and within limitations, fluoride may also be added under the bottled water quality standard.[3]

Bottled water may be used as a beverage by itself or as an ingredient in other beverages (e.g., diluted juices, flavored bottled waters). In addition to "bottled water" or "drinking water," FDA also defines various types of bottled water under the bottled water identity standard.

Artesian water or artesian well water is water from a well that taps a confined aquifer in which the water level stands at some height above the top of the aquifer. Artesian water may be collected with the assistance of an external force to enhance the natural underground pressure.

Ground water is water from a subsurface saturated zone that is under a pressure equal to or greater than atmospheric pressure. Ground water must not be under the direct influence of surface water as defined by the U.S. Environmental Protection Agency (EPA).[6] This definition differs from the Codex Alimentarius definition in which only "protected ground water" is not under the direct influence of surface water.

Mineral water is water coming from a source tapped at one or more bore holes or springs, originating from a geologically and physically protected underground water source. Additionally, it must contain at least 250 parts per million (ppm) total dissolved solids (TDS) and no minerals may be added. If the TDS content is between 250 ppm and 500 ppm or is above 1500 ppm, the statement "low mineral content" or "high mineral content" must appear on the label, respectively.

Purified water or *demineralized water* is water that has been produced by distillation, deionization, reverse osmosis, or other suitable processes and that meets the definition of "purified water" in the 23d Revision of the United States Pharmacopeia.[8] In addition, purified water may be called "distilled water" if it has been produced by distillation, "deionized water" if it has been processed by deionization, and "reverse osmosis water" if it has been processed by reverse osmosis.

Sparkling bottled water is water that, after treatment and possible replacement of carbon dioxide, contains the same amount of carbon dioxide that it had at emergence from the source.

Spring water is water derived from an underground formation from which water flows naturally to the surface of the earth. Spring water shall be collected only at the spring or through a bore hole that taps the underground formation feeding the spring.

Sterile water is water that meets the requirements under the "Sterility Test" <71> in the 23d Revision of the United States Pharmacopeia.[9]

Well water is water from a hole bored, drilled, or otherwise constructed in the ground which taps the water of an aquifer.

When bottled water comes from a community water system, the statement "from a community water system" or "from a municipal source" must appear on the label, except when it has been treated to meet the definitions for "purified water" or "sterile water" and is labeled as such.

When the label or labeling of a bottled water states or implies (e.g., through label statements or vignettes with reference to infants) that the bottled water is for use in feeding infants, and the product is not commercially sterile, the product's label must bear the statement "Not sterile. Use as directed by physician or by labeling directions for use of infant formula."

Under the bottled water quality standard, FDA has established a microbiological quality requirement that is based on coliform detection levels. When tested by the membrane filter method, not more than one of the ten analytical units (portions from ten subsamples of a sample that is representative of a given lot of bottled water) in the sample shall have 4.0 or more coliform organisms per 100 mL and the arithmetic mean of the ten analytical units in the sample shall not exceed one coliform organism per 100 mL. When the multiple-tube fermentation method is used, not more than one of the ten analytical units in the sample shall have a most probable number (MPN) of 2.2 or more coliform organisms per 100 mL and no analytical unit shall have an MPN of 9.2 or more coliform organisms per 100 mL. No heterotrophic plate count (HPC) (See Chapter 7) limit is specified in the FDA's microbiological quality standard for bottled water.

The FDA's current good manufacturing practice (CGMP) regulations for bottled water[7] require weekly coliform analyses for each type of bottled water produced in a plant and, additionally, require that any source water obtained from other than a public water system be analyzed for coliforms at a minimum of once a week. Microbiological tests on containers and closures are also required under the CGMP regulations for bottled water. At a minimum of once every 3 months, at least four containers and closures are selected just prior to filling and sealing and analyzed for HPC and coliforms. All samples must be free of coliform organisms, and no more than one of the four samples may exceed the following HPC requirement: one colony per milliliter capacity of the container, or one colony per square centimeter of the surface area of the closure.

60.2 TYPES OF MICROORGANISMS

Ozone is typically used as a disinfectant in the bottled water industry. The FDA has affirmed ozone as generally recognized as safe (GRAS) for use at maximum residual level at the time of bottling of 0.4 milligrams ozone per liter of bottled water that, prior to ozonation, meets the microbiological, physical, chemical, and radiological quality standards for bottled water.[5] As a result of this practice, bottled water products are typically free of coliforms and have very low HPC (less than 100/mL) at the time of bottling. Nevertheless, even when bottled water products are produced under strict adherence to the CGMP requirements, a few hardy organisms naturally present in the water can survive. During storage and after ozone has dissipated, these organisms follow a cyclic pattern of growth before dying off, without causing microbial spoilage.

60.21 Predominant Flora

The predominant organisms found in bottled water are Gram-negative rods, which can grow in distilled or other mineral-free water as well as in waters that contain minerals. Included in the group are organisms of the genera *Pseudomonas, Flavobacterium,* and *Moraxella/Acinetobacter*.

60.22 Indicator Organisms

Although coliform organisms are not themselves pathogens and are rarely found in bottled water, samples are routinely tested for coliforms because they serve as an indicator of possible contamination. The presence of coliforms, which can occur naturally in soil, water, and vegetation, indicates possible contamination from airborne sources or from product contact surfaces that have not been effectively disinfected. No coliform organisms should be found in bottled water produced in accordance with the CGMP regulations. The presence of fecal coliforms, on the other hand, indicates a possibility of sewage contamination, with the associated possibility of the presence of pathogenic microorganisms. However, if the required weekly tests for coliforms are negative, it is not necessary to specifically test for the presence of fecal coliforms in bottled water. Therefore, testing for fecal coliforms is not usually performed. However, regulations regarding testing for fecal coliforms vary with other countries and certain organizations.

Although coliforms and fecal coliforms are considered adequate indicators of poor sanitation or indirect indicators for the presence of bacterial pathogens, they may not be effective indicators for the presence of parasitic protozoa or mammalian viruses. Cysts of some parasites have been shown to be more resistant to disinfection than coliform organisms.[10,11] Free-living protozoa have been found in bottled mineral water in some countries,[10] but parasitic protozoa and viruses are seldom detected in bottled water.

60.23 Escherichia coli

The Codex Alimentarius Committee on Food Hygiene (CACFH) is developing an international Code of Hygienic Practice for Bottled/Packaged Drinking Water (Other than Natural Mineral Water), which will contain provisions concerning the microbiological criteria for bottled water. In considering these provisions for the microbiological criteria, it is recognized that in tropical countries, testing for coliforms in bottled water is highly susceptible to obtaining positive results for coliforms that are not of fecal origin or obtaining false positive results for some species of bacteria that are not coliforms. This is due to climate conditions and indigenous bacterial populations in the water source in tropical countries. Consequently, a coliform standard for bottled water produced in tropical countries may not be practical. Therefore, CACFH is considering a microbiological criterion that no *E. coli* be detected in bottled water and that testing for *E. coli* in bottled water should be conducted using internationally recognized or validated methods (e.g., ISO methods, AOAC methods, APHA/AWWA/WEF Standard Methods for the Examination of Water and Wastewater).

60.3 EQUIPMENT, MEDIA, AND REAGENTS

60.31 Equipment

Incubator 35°C, 44.5°C, 20–28°C
Membrane filter apparatus
Sterile swabs

60.32 Culture Media (Chapter 63)

Brilliant green lactose bile broth (BGB)
Lauryl tryptose broth (LTB)
LES Endo agar
MacConkey agar
M-Endo Medium (broth or agar)
m-HPC agar
Nutrient agar
R2A agar
Tryptone glucose extract agar (TGEA)
Tryptone glucose yeast agar (Plate count agar)

60.33 Reagents (Chapter 63)

Gram stain (Hucker)
Phosphate buffer (Butterfield)

60.4 RECOMMENDED METHODS

60.41 Sampling (Chapter 2)

Whenever possible, sample directly from the original sealed container, which does not need to be refrigerated.

When it is necessary to use a secondary container, collect sample aseptically in a suitable, presterilized bottle and process the sample as soon as possible. If testing cannot be completed within 8 hr of collection, refrigerate the sample but under no circumstances should the time between collection and analysis exceed 24 hr. Record time and temperature of storage of all samples.[1]

60.42 Heterotrophic Plate Count (Chapter 7)

Use the pour plate, spread plate, or the membrane filtration method described in Chapter 7. Suitable media for pour or spread plate methods include tryptone glucose extract agar (TGEA), plate count agar (tryptone glucose yeast agar), or R2A agar. For membrane filtration method, use R2A agar or m-HPC agar.

For most bottled waters, plates suitable for counting will be obtained by plating 1 mL or 0.1 mL of undiluted test sample. The membrane filtration method (Chapter 7.71) allows for the use of 100 mL (or greater) samples and, therefore, is much more sensitive for bottled waters suspected to have very low counts.

Incubate all plated tests, except those on R2A agar, at 35°C ± 0.5°C for at least 72 hr. Bacteria found in bottled water often demonstrate a prolonged lag phase during adaptation to growth on tryptone glucose extract agar or plate count agar; hence countable colonies may not form in 48 hr incubation.[1] Bottled waters plated on R2A should be incubated at 35°C for 5 to 7 days, or at 20°C to 28°C for 7 days.

60.43 MPN for Members of the Coliform Group (Chapter 8)

For routine examination of bottled water, use 10 lauryl tryptose broth (LTB) fermentation tubes, each inoculated with 10 mL of undiluted test sample. Perform the confirmed test for coliforms by using brilliant green lactose bile (BGB) broth as specified in Chapter 8. Estimate the most probable number (MPN) of coliforms in the sample from the number of positive, confirmed BGB tubes by using the following table.

MPN Index and 95% Confidence Limits for Various Combinations of Positive and Negative Results When Ten 10-mL Test Portions Are Used

Number		95% confidence limit	
of tubes	MPN/g	lower	higher
0	< 1.1	0	3.0
1	1.1	0.03	5.9
2	2.2	0.26	8.1
3	3.6	0.69	10.6
4	5.1	1.3	13.4
5	6.9	2.1	16.8
6	9.2	3.1	21.1
7	12.0	4.3	27.1
8	16.1	5.9	36.8
9	23.0	8.1	59.5
10	>23.0	13.5	Infinite

For more precision, or for bottled waters suspected of having high bacterial load, use the 5-tube MPN method where 5 tubes each of LTB are inoculated with 10 mL, 1.0 mL, and 0.1 mL of undiluted test samples (see Chapter 6 for the 5 tube MPN table).

60.44 Completed Test for Coliforms

Submit at least 10% of all positive, confirmed tubes for analysis as follows, and use the completed test to definitively establish the presence of coliforms and to provide quality-control data.[1] With an inoculating loop, streak a LES Endo or MacConkey agar plate from each gas-positive BGB tube as soon as possible after gas formation. Incubate the plate inverted, at 24 ± 2 hr at 35°C. Be careful not to disturb the surface integrity of the agar plates, as this will make the subsequent picking of the individual colonies difficult. Pick a typical, well isolated (no colonies within 0.5 cm) colony from the surface of the plates by using a transferring needle and inoculate a nutrient agar slant and a tube of LTB. Typical colonies on LES Endo agar are pink to dark red with a green metallic surface sheen.[1] Typical lactose-fermenting colonies on MacConkey agar are red and may be surrounded by an opaque zone of precipitated bile.[1] Incubate the nutrient agar slant and the LTB at 35 ± 0.5°C for 24 ±2 hr. If no gas is observed in LTB, incubate the LTB tube for an additional 24 ± 2 hr.

For the LTB tubes that showed gas, use the corresponding nutrient agar slants to prepare a slide for gram-stain (Hucker modification). Demonstration of gram-negative, non-sporeforming, rod-shaped bacteria constitutes a positive completed test and that a member of the coliform group is present in the sample. The gram stain may be omitted from the completed test for potable water such as bottled water.

60.45 Membrane Filter Method for Coliforms (Chapter 8)

Filter 100 mL of test sample and transfer the filter to LES Endo agar or M-Endo medium (broth or agar) for incubation at 35 ± 0.5°C for 22 to 24 hr. Count colonies that are pink to dark red with a green metallic surface sheen. The sheen may vary from a small pinhead to complete coverage of the colony. Regardless of the extent of sheen, when the number of sheen colonies is 5 to 10, confirm by inoculating growth from each sheen colony into tubes of LTB and BGB and incubate at 35 ± 0.5°C for 48 hr. When the number of sheen colonies exceed 10, randomly select and confirm 10 colonies that are representative of all sheen colonies. Gas production in both LTB and BGB in 48 hr confirms the colony as coliform. If only the LTB shows gas, use a loopful of this LTB culture to inoculate a second BGB tube. Gas production within 48 hr is a confirmed coliform test. Report results as number of coliform colonies per 100 mL.

60.46 Rapid Methods for Coliforms and E. coli

Several rapid test procedures are available to detect coliforms and *E. coli* in water samples (see Chapter on Rapid Methods). The AC (Autoanalysis Colilert) test is one of these.[2]

60.47 Test Method for Containers and Closures

Containers (Chapter 3)—Pour 100 mL of sterile phosphate buffer into the container to be tested, and swirl to ensure contact with the entire inside surface area. Perform HPC and coliform test on the exposed buffer solution. Report any coliform present and the HPC as per mL of capacity of the container.

Closures—Wet a swab in sterile phosphate buffer solution and swab the entire inside surface of the closure twice. Insert the swab in a tube of 10 mL sterile phosphate buffer and break off stick below the area touched by the fingers. Mix well. Use 1 mL of suspension and perform HPC and a coliform test. Multiply the count by 10 to give the number of cells per closure or calculate the count per square centimeter of surface area of the closure.

60.5 INTERPRETATION OF DATA

The organisms that grow in bottled water as measured by the HPC are not usually of sanitary significance and do not result in microbial spoilage or flavor impairment of the product. Production of bottled water in accordance with the bottled water CGMP regulations will typically result in a low HPC (less than 100/mL) in treated source water at the time of bottling. The presence of coliform organisms in the bottled water product, however, would indicate a lack of adherence to the bottled water CGMP regulations (e.g., possible poor sanitary conditions at the plant) and potential public health problems. The presence of *E. coli* in the bottled water product indicates fecal contamination that could pose public health problems.

60.6 ACKNOWLEDGMENTS

The authors acknowledge Sarah Cowman and Robert Kelsey, authors of the Bottled Water chapter in the previous edition of the *Compendium of Methods for the Microbiological Examination of Foods*. Some of the material from the Bottled Water chapter in the previous edition has been incorporated into this chapter.

60.7 REFERENCES

1. APHA/AWWA/WEF. 1995. Standard methods for the examination of water and wastewater, 19th ed. Part 9000, Am. Public Health Assoc., Am. Water Works Assoc., Water Environ. Fed., Washington, D. C.
2. Edberg, S. C., M. J. Allen, D. B. Smith, and The National Collaborative Study. 1988. National field evaluation of a defined substrate method for the simultaneous enumeration of total coliforms and *Escherichia coli* from drinking water: Comparison with the standard multiple tube fermentation method. Appl. Environ. Microbiol. 54:1595.
3. U.S Government National Archives and Records Administration. 1997. Code of Federal Regulations, Title 21, Part 165.110(a).Bottled Water, Identity.U.S. Government Printing Office, Washington, D. C.
4. U.S Government National Archives and Records Administration. 1997. Code of Federal Regulations, Title 21, Part 165.110(b) Bottled Water, Quality. U.S. Government Printing Office, Washington, D. C.
5. U.S Government National Archives and Records Administration. 1997. Code of Federal Regulations, Title 21, Part 184. Direct food substances affirmed as generally recognized as safe. U.S. Government Printing Office, Washington, D. C.
6. U.S Government National Archives and Records Administration. 1997. Code of Federal Regulations, Title 40, Part 141. National Primary Drinking Water Regulations. U.S. Government Printing Off., Washington, D. C.
7. U.S Government National Archives and Records Administration. 1997. Title 21, Part 129. Processing and Bottling of Bottled Drinking Water. Code of Federal Regulations. U.S. Government Printing Office, Washington, D. C.
8. U.S. Pharmacopeia, 23d Revision, 1995. Purified Water, P 1637, U.S. Pharmacopeia Convention, Inc., Rockville, Md.
9. U.S. Pharmacopeia. 1995. Sterility test, p. 1686, 23rd Revision. U.S. Pharmacopeia Convention, Inc., Rockville, Md.
10. Warburton, D. W. 1993. A review of the microbiological quality of bottled water sold in Canada. Part 2. The need for more stringent standards and regulations. Can. J. Microbiol. 39:158.
11. Warburton, D. W., K. L. Dodds, R. Burke, M. A. Johnston, and P. J. Laffey. 1992. A review of the microbiological quality of bottled water sold in Canada between 1981 and 1989. Can. J. Microbiol. 38:12.

Chapter 61

Canned Foods—Tests For Commercial Sterility

Kurt E. Deibel and Michael Jantschke

61.1 INTRODUCTION

When spoilage occurs in a closed container of canned food, it manifests itself by obvious gas production, swelling the lid of the container, changes in either product consistency, odor, or pH, and/or by an increase in the number of microorganisms seen in the microscopic examination of the food. The commercial sterility test is not designed for detailed diagnosis of spoilage, and, therefore, when any of the above spoilage criteria are met, the analyst should go directly to Chapter 62 for procedures to determine the exact cause of spoilage.

Canned foods in the context of this chapter are those that have been preserved by heat in hermetically sealed containers. These foods are typically packaged in cans, glass or flexible packages and have been processed sufficiently to achieve commercial sterility. A hermetically sealed container is one that prevents the entry of microbes, thereby preventing spoilage flora and/or pathogens from entering from external sources after the container has been sealed. The canned food should remain unspoiled indefinitely if sufficiently heat processed, stored and handled properly to ensure that the seal remains intact. With the exception of foods such as canned cured meats, most canned foods do not rely on preservatives or inhibitory agents other than acids to ensure stability. The heat treatment given canned foods is enough to produce commercial sterility, not complete sterility. Complete sterility is defined here as a state completely free of all viable microorganisms. The heat process required to achieve complete sterility would destroy the product consistency and nutritional values that currently exist in commercially canned foods. The highly heat-resistant and nontoxic thermophilic sporeformers (Chapters 24, 25, 26, and 27), under proper handling and storage conditions, remain dormant in commercially canned foods and present no problems; therefore, canned foods are called "commercially sterile."

Commercial sterility of thermally processed food is defined in Title 21 of the Code of Federal Regulations, Part 113.3 (e)[11] as the condition achieved 1) by the application of heat, which renders the food free of a) microorganisms capable of reproducing in the food under normal nonrefrigerated conditions of storage and distribution and b) viable microorganisms (including spores) of public health significance; or 2) by the control of water activity and the application of heat, which renders the food free of microorganisms capable of reproducing in the food under normal nonrefrigerated conditions of storage and distribution.

Under this definition, a commercially sterile, heat-processed food may contain viable microorganisms. The microorganisms, however, do not increase in numbers and consequently do not cause physical changes in the food product. Likewise, pathogens either are not present or are incapable of reproducing in the product.

Viable microorganisms normally may be recovered from commercially sterile, heat-processed foods under three general conditions: 1) The microorganism is an obligate thermophilic, sporeforming bacterium, and the normal storage temperature is below the thermophilic range. 2) The heat-processed food has a pH of less than 4.6. Acid-intolerant microorganisms may be present, but are incapable of growth because of the acidic condition. 3) Mesophilic or thermophilic sporeformers may be recovered from canned foods that use a combined process of heat and water activity to prevent outgrowth and spoilage. Finding microorganisms in these three instances is normal, and the product is considered commercially sterile.

Food processors are able to heat-process food to achieve absolute sterility, but this is not done for several reasons. Overcooking (excessive time or temperature) encourages the development of off-flavors, color and consistency changes and nutrient losses. "Commercial sterility heat processing" enables the manufacturer to pack a microbiologically safe, shelf-stable food without undue impairment of flavor, color, consistency, or nutrient content.

Food canners have a history of cooperatively sharing technical advances for the improvement of the entire industry. In addition, numerous safeguards are employed (both voluntarily and as enforced by regulatory agencies) to ensure the adequacy of commercial sterilization cooking procedures. Requirements for processing low-acid foods and acidified foods, and appropriate records to be kept, are found in parts 108, 113, and 114 of Title 21 of the Code of Federal Regulation.[11]

Commercial sterility testing of canned foods should be conducted on normal-appearing canned food by visually examining the incubated container and the product, measuring the vacuum, odor, and pH, and, if necessary, making a microscopic examination to detect large numbers of bacteria. Subculturing rarely is done on a routine basis because it is time-consuming and expensive and has a high risk of laboratory contamination or of faulty interpretations.[5] Subculturing may be desired or even necessary when running preliminary test packs using a new processing method; therefore, procedures for this purpose will be described in this chapter. In addition, subculturing can be used in certain situations as part of a Partial Quality Control (PQC) program to reduce the time of product incubation.

"Sterility (Commercial) of Foods (Canned Low-Acid)" has "final action" status in the *AOAC Official Methods of Analysis, AOAC International (1996).*[1] That method should be *only* used for low-acid foods, those other than alcoholic beverages with a finished equilibrium pH greater than 4.6 and a water activity greater than 0.85. The methods listed in this chapter closely follow the AOAC procedure for low-acid foods; in addition, methods designed for acidified and acid foods (those with a pH of 4.6 or below) are given.

61.2 GENERAL CONSIDERATIONS

61.21 Treatment of Sample

The food manufacturer selects test samples that are representative of the production lot and incubates them for 10 days at 30°C to 35°C to determine if the product meets the criteria for commercial sterility. If conventional retorts are used for sterilization, each retort load is examined with at least one sample. If sterilization is accomplished by continuous means, samples are drawn at periodic intervals during the time of pack. Products intended to be stored at temperatures above 40°C (i.e. vending machines) would require that extra samples be drawn for 5 to 7 days of incubation at 55°C.

After incubation, record the product identity and the manufacturer's code. The container should be examined critically for abnormal conditions such as leakage, swells, flippers, prior opening, etc. (See Chapter 62 for a thorough explanation of container appearances.) If spoilage is evident, use the procedures described in Chapter 62. If the container is defective or damaged, the sample is not suitable for commercial sterility testing.

61.22 Sources of Error

Sources of error in tests for commercial sterility derive in large measure from misinterpretation of findings; therefore, it is recommended that the reader refer to sections dealing with confirmation and interpretation of data where bases of judgement are fully expanded. Misinterpretations can derive from laboratory contamination, insufficient numbers of containers sampled, and the presence of heat-resistant enzymes that may digest food components, thereby giving the impression of microbial spoilage.

61.23 Special Equipment and Supplies

1. *Work areas: Preferred method.* A laminar flow work station meeting ultra-clean environment specifications, Class 100 is preferred[7] in sterility evaluations of hermetically sealed containers. This equipment will provide a work environment free of particles 0.3 microns or larger at an efficiency of 99.99% with an air flow of 100 feet per min. Primary disinfection of the unit can be obtained by a thorough washing of the interior surfaces (being careful to avoid contact with filtration media) with a bactericidal solution. After chemical disinfection, the blower should be operated for at least 1 hr prior to performing analyses within the unit.
2. *Work areas: Alternate method.* If the equipment described in the preferred method is not available, the samples may be opened in a room secure from drafts and other sources of contamination. The counter surface should be scrubbed thoroughly with soap and water and then disinfected with an appropriate bactericidal agent such as 100-ppm chlorine solution.
3. *Bacteriological can opener.* A special can opener designed for the aseptic opening of metal containers for bacteriological sampling without distorting the can seams is the Sanitary Can Opener (Bacti-Disc Cutter) (Wilkens-Anderson Company, Chicago, IL. 60651). Under no circumstances should a common kitchen-type can opener be employed because sample contamination and distortion of the double seam will occur.

61.24 Recommended Controls

1. *Glassware.* All glassware should be autoclaved for a minimum of 20 min at 121°C. Equipment, wrapped in kraft paper, should be placed in a sterilizing oven at 170° to 180°C for 2 hr. Heat-sensitive sterilization indicators may be affixed to each autoclave load to identify readily the status of a given unit of equipment. If equipment is to be sterilized well in advance of use, double-layer aluminum foil dust covers should be placed over flasks, dilution bottles, etc.
2. *Laminar flow work station.* Efficiency of the unit should be monitored through the use of open, uninoculated control media exposed to the work station environment during the entire transfer period. Three controls should be used, one placed to the right and one to the left of the samples undergoing analysis, and one in front of the samples roughly in the middle of the work area.
3. *Media.* Media sterility checks should be performed on common liquid media by incubating for 48 hr at appropriate temperatures and then examining for the absence of growth. Uninoculated control plates should be prepared and incubated for every lot of solid media employed in routine test studies.
4. *Personnel requirements.* Prior to working with samples, the hands and face of personnel should be scrubbed thoroughly with an appropriate germicidal hand soap.[6] Personnel should wear clean lab coats, and items of personal clothing such as neckties should be removed or contained. Persons known to have colds, boils, or similar health problems should not perform this evaluation. Personnel with shoulder-length hair, sideburns below ear lobe, mustaches or beards must wear sterile protective snoods and sterile full-head coverings.[3]

61.3 Equipment, Materials, and Reagents

1. *Pipettes.* Straight walled pipettes should be used. Either glass or disposable plastic is acceptable. If non-disposable pipettes are used, they should be cotton-plugged before sterilization then wrapped not more than five to a package in heavy kraft paper, steam-sterilized at 121°C for 20 min, then dried in vacuo. Alternatively, the packages may be placed in a sterilizing drying oven at 170° to 180°C for 2 hr.
2. *pH meter.* Electronic pH meters should be used to determine pH of the product in question. Accuracy must be within 0.1 pH unit of a known buffer solution. A pH indicator paper with a suitable range may be used for pH determinations on a large number of homogeneous samples. However, any sample showing variation from the normal should be checked electrometrically (Chapter 64).
3. *Forceps.* Nonserrated forceps,[4] at least 8 in. in length, should be available to handle large particles and other nonpipettable products.

4. *Microscope.* A suitable bacteriological microscope fitted with an oil immersion lens or a phase system is needed. Microscopic examination of a food product for bacterial contamination should be made at magnification of not less than 930X.
5. *Culture tubes.* All tubes should have a screw cap closure and be manufactured of borosilicate or Pyrex™ glass or suitable plastic. If plastic is used, it must be nontoxic and capable of withstanding normal glassware sterilization temperatures. Cotton plugs are not acceptable as closures on either type of tube.
6. *Petri dishes.* Sterile dishes should be 100 × 20- or 100 × 15-mm plastic or glass. Only new, sealed sleeves of disposable plastic dishes should be used in sterility test work. If glass is used, standard petri storage cans should be employed for storage of the glassware. Only freshly sterilized containers should be used in sterility test work. Disposable plastic petri dishes are desired.
7. *Culture media and reagents.* When media are commercially available in the dehydrated form, they should be used in preference to media formulated in individual laboratories. See Chapter 63 for formulation and preparation instructions for media and reagents.

Laboratory media (see Chapter 63) suitable for canned food examination includes the following:

For low-acid foods:

Dextrose tryptone broth (DTB; for the growth of aerobes)

PE-2 medium in screw-cap tubes or Liver broth (for the growth of anaerobes)

For acid foods:

Orange serum broth or Thermoacidurans broth (TAB) (for acid-tolerant microorganisms).

Potato dextrose agar—acidified or Sabouraud dextrose agar (used for yeast and mold)

TAB, MRS or All-purpose medium with Tween (APT) (for growth of *Lactobacillus*)

Reagents for simple staining and examination.

Ziehl-Neelsen's carbol fuchsin stain

Crystal Violet Stain (0.5% to1% solution).

61.4 PRECAUTIONS

61.41 Safety Procedures

Never taste the contents of any low-acid food container that are suspected of having undergone spoilage because of the potential presence of botulism toxin!

When opening swollen containers for examination, use caution and wear personal protective equipment (safety glasses, goggles, protective face masks, etc.).

61.42 Disposal of Spoiled Containers

Refer to Chapter 62 for details of the disposal of contaminated material.

61.5 PROCEDURE

The most reliable test for determining commercial sterility of a container of product is to incubate that container at an appropriate temperature long enough to allow any significant microorganisms contained therein to grow and to manifest their presence.[8] If microorganisms proliferate under the proper storage conditions imposed, the product is not commercially sterile and should be examined as described in Chapter 62.

61.51 Incubation

Incubation conditions are governed partly by the purpose of the commercial sterility test:

1. *Routine production monitoring.* For low-acid products destined for storage at temperatures above about 40°C (hot vending), containers from each sampling period or retort load should be incubated at 55°C for 5 to 7 days. For all other low-acid products, incubate at 30° to 35°C for 10 days, except certain meat products packed under continuous regulatory agency inspection.[10] For acid or acidified foods, incubate at 25° to 30°C for 10 days.
2. *Examination of a production lot or lots because of suspected non-commercial sterility.* If possible, incubate the entire lot or lots; otherwise, incubate a statistically randomized sample of the lot or of each lot. Use incubation temperatures as in No. 1, but increase the time to 30 days in the case of mesophiles.

61.52 Examination

Containers may be removed from the incubator whenever outward manifestations of microbial growth appear (e.g., swells, or with transparent containers, noticeable product change). At the end of the incubation period, some containers should be opened to detect possible flat sour spoilage.

Weigh each suspect container to the nearest gram and record the weight on the data sheet. Subtract the average tare weight of the empty container from each unit weight to determine the approximate net weight of each sample; this information will be of value in diagnostic tests mentioned in Chapter 62.

If the only purpose of the test procedure is to determine commercial sterility of the product (and provided the analyst is thoroughly familiar with the particular product and with potential types of microbial spoilage of that product), containers may be examined without employing aseptic techniques. Open the containers carefully; note abnormal odors, consistency changes, and frothiness; measure pH electrometrically or colorimetrically; if results are not conclusive at this point, prepare a smear for microscopic examination. If the product is controlled by water activity (a_w), perform an a_w examination (Chapter 64). The observation of abnormalities compared to containers of normal control product means that the containers of the products are not commercially sterile. In that case, follow the procedures in Chapter 62 on the remaining unopened abnormal product.

If the analyst is unfamiliar with the particular product and its potential spoilage characteristics, it is necessary to use aseptic procedures in opening and examining each container.

61.53 Opening Container Aseptically[1,6,8,9]

Remove the label from the container if possible. Examine the container for defects and record all observations. The container must be clean. If obvious soil, oil, etc., are present, wash with detergent and water, rinse, and wipe dry with clean paper towels.

1. *Cans.* Hold the clean, uncoded end of the can over a large Meker burner, just above the blue portion of the flame, continuing until the visible moisture film evaporates. (If the can swells, keep side seam directly away from analyst and flame cautiously.) Flame a clean Sanitary Can Opener, then aseptically cut and remove a disc of metal from the end. (With cans having an "easy-open" feature on one end, open the opposite end. Do not disturb the "easy-open" end in the event that it is subsequently necessary to examine the score). Subculture immediately.

2. *Glass jars.* Clean and flame the closure as described for cans. An index line should be scribed or otherwise permanently marked on both the cap and the glass. This procedure will allow for the quantitative measurement of the closure security by a qualified container examiner if such an examination is deemed necessary. Open with a clean, flamed Sanitary Can Opener (avoid unseating or moving the closure with respect to the jar; subsequent seal examination may be desirable). Subculture immediately.

   If complete removal of the jar closure is necessary, the jar should be cleaned as above, then inverted in a disinfecting solution for an appropriate period of time (e.g., 100 ppm chlorine for 10 min). A sterile cotton pad then is placed on the closure. A flame-sterilized Sanitary Can Opener point is used to puncture the closure and thus relieve the vacuum. The closure then may be removed with less danger of admitting microbial contaminants over the glass-sealing surface.
3. *Nonrigid containers (e.g., pouches).* Clean and sanitize the container surface with a detergent sanitizing agent. Dry with a sterile, fresh towel. If the product is maneuverable, push it away from one end. With clean, lightly flamed scissors, cut the end off just under the container seal. (On large containers, a 2-to 3-in opening may be made by cutting off the corner of the container with a diagonal cut). Open the container without touching the cut ends. Subculture immediately.

61.54 Subculture Media—Selection

The best subculture medium for use in examining spoiled canned foods is a portion of the normal product that has been tubed and sterilized. Its use eliminates a question sometimes raised with the use of artificial laboratory medium as to whether or not organisms recovered by the laboratory medium actually grow in the product. Liquid products may be tubed without modification. Products consisting of solids and liquids should be blended mechanically, retaining the normal ratio of solids to liquids. Products lacking free moisture should be blended mechanically, with sterile water added as necessary to make a slurry capable of being tubed. Low-acid product tubed media should be sterilized by steaming 30 min at 100°C.

Laboratory media (Chapter 63) suitable for canned foods examination may also be used.

61.55 Subculture Product Samples, Observe Product Characteristics

Transfer approximately 2 g of product from each container with a straight-walled pipette to duplicate tubes of subculture media. For low-acid products, inoculate two tubes of sterile PE-2 medium. Since these tubes are intended to allow the outgrowth of anaerobes, they should be exhausted in flowing steam for 20 min. and cooled to 55°C *prior* to inoculation if not freshly prepared and autoclaved. Be sure to inoculate the lower portion of the tube and tighten the caps after inoculation. Then inoculate duplicate tubes of the aerobic medium (DTB). Incubate at 30° to 35°C. If normal temperature for storage or handling of the product is higher than 40°C, also inoculate duplicate tubes of appropriate medium used for incubation at 55°C; otherwise, do not incubate at 55°C. For acid products, inoculate two tubes of sterile product, or of the acid medium or media of choice, and incubate at 25°C to 30°C. If molds are suspected, prepare potato dextrose agar pour plates and incubate at 30°C. Incubate all subcultures for at least 5 days before declaring them negative. At the time subcultures are prepared, transfer an additional 10 g of product to a sterile culture tube and refrigerate; this will serve as a reference sample for repeat subcultures if necessary. Measure pH of the product, and observe product odor and appearance relative to a control sample.

61.56 Mass Culture Technique for Aseptically Packed Tomato or Fruit Concentrates

APT has been developed as a culture medium for the selective cultivation of *Lactobacillus*, *Leuconostoc*, and yeasts capable of causing spoilage in acid product concentrates such as tomato paste and fruit pastes. It is an excellent medium for the recovery of minimal contamination and, therefore, the utmost care must be taken by workers to ensure that the medium is not contaminated during handling. Since the broth is intended primarily as a mass culture medium for detecting low-level contaminants or localized contaminants in aseptically packed acid products, approximately 100 g of product under tests should be inoculated aseptically into 300 mL of sterile medium contained in a 500 mL screw-cap flasks. Care must be taken to disperse the product adequately through the broth; a sterile stirring rod should be used.

A minimum of three flasks per sample should be inoculated. An extra aseptic sample retained from each container should be incubated with the flasks. In addition, a retained sample should be held at refrigeration temperatures for use in microscopic comparisons with the incubated product or for repeating the test. Incubation of the cultures should be at 30°C for 5 days with visual examination for fermentation or biological surface growth daily. The extra retained incubation samples should be incubated for 10 days at 30°C. At the end of the incubation period, all samples should be examined microscopically for evidence of bacterial or yeast contamination.

61.57 Microscopic Examination of Product

Prepare a wet mount of product liquid (or of surface scrapings of product not having excess free liquid) for examination with a phase contrast-equipped microscope, or make a smear for simple staining and examination. Crystal violet or Ziehl-Neelsen carbol fuchsin stains are recommended (see Chapter 63). Products having a high solids content in the liquid portion often resist adhering to a slide, and they may be diluted advantageously with approximately equal volumes of sterile distilled water on the slide. A flamed bacteriological loop may be used for making smears. Examine both the product being tested and a normal control product to enable judgement regarding the presence of abnormal levels of microorganisms in the former.

61.58 Confirmation of Positive Laboratory Medium Subcultures

Inherent characteristics of acid, acidified, and water activity-controlled food products preclude the germination and outgrowth of some viable bacterial spores present. Laboratory media may not exhibit similar bacteriostatic properties, resulting in positive subcultures, which may lead to the erroneous conclusion that the product in question is not commercially sterile. When the true significance of such positive subcultures is in doubt, use them to inoculate the sterile product, either tubed or canned, followed by incubation. Organisms of significance will grow and manifest their presence in the product. Again, use of the sterile product as the primary subculture medium, together with direct microscopic examination and assessment of product abnormalities compared to normal control, usually will obviate a confirmatory procedure.

Laboratory contamination is indicated when an organism producing gas in an anaerobic medium at 35°C is isolated from low-acid canned foods having an obvious vacuum. Aseptically inoculate growing organisms into another normal can, close the hole with solder, and incubate for 14 days at 35°C. Swelling of the container indicates that the organism was not in the original sample and is due to laboratory contamination.[1]

61.6 INTERPRETATION OF DATA

Serious consequences can result if commercial sterility testing information is based on faulty laboratory techniques or if errors are made in evaluating data. Strict observance of the preceding methods will lessen the chances of designating a product contaminated when, in actuality, it is commercially sterile. Likewise, the instances of incorrectly claiming a product to be commercially sterile will be decreased. A similar test has been adopted by the AOAC International as "final action."[1]

The development of swelled containers (any degree) may indicate microbial activity. The presence of a swell condition by itself, particularly at the flipper, springer, or soft swell level, however, does not indicate positively that the product is not commercially sterile. Growth must be confirmed by demonstrating excessive microorganisms by direct smear, or an abnormal product (pH, consistency, odor, etc.) as compared to the normal product. Some causes of low-degree swells, which cannot be confirmed microbiologically in a commercial sterility test, are overfilling, low filling temperatures, improper vacuum closing procedures, inadequate vacuum mixing (certain comminuted products), incipient spoilage, and chemical swells ("hydrogen springers").[2] At a high elevation in such areas as Denver, Colorado, cans that would appear normal at sea level often show swells.

Nonsterility of flat containers is shown if direct smears reveal excessive microorganisms and if one or more product characteristics are abnormal. Subculture may be necessary to support these conclusions.

Duplicate cultures should show comparable biochemical reactions and microbial flora. These, in turn, should be similar to those in the original product (e.g., predominating flora in direct smears of product will usually appear in subcultures, and biochemical reactions will be comparable).

If only one tube of a duplicate set is positive, adventitious (laboratory) contamination should be suspected. Comparison of the positive culture with the original product, especially in direct smears, often will clarify the situation. If not, new subcultures should be prepared from the retained sample.

A stained smear or a wet mount examined by phase microscopy that shows only an occasional cell in some fields usually does not suggest a spoiled product. Representative coverage of a smear is necessary to form an accurate evaluation; examination of 10 to 20 fields is necessary to avoid being misled. Comparison with the normal product is mandatory.

It is sometimes difficult to differentiate bacterial cells from food particles when microscopically examining a food product. On occasion, food-grade yeast is an ingredient of the processing food. Check the list of ingredients on the label for yeast. Caution must be taken not to assume that these yeast cells are contaminating microorganisms. Fermented products (e.g., sauerkraut) will show a high normal microbial population by direct smear.

If a recovery medium has a near-neutral pH and the food being tested is an acid food, microbial growth will probably take place. This is normal and expected and does not indicate the lack of commercial sterility. When testing an acid food, an acid pH recovery medium should be used.

61.61 Confirmation Program

If, when using proper commercial sterility testing procedures, microbial growth is recovered from a container of food that displays no evidence of spoilage, the following program should be carried out:

1. The bacterial isolate, or isolates, should be grown in pure culture.
2. An unopened container of food exhibiting the same manufacturer's code as the one previously tested should be selected.
3. By aseptic techniques, a small puncture hole should be made through the can end or jar closure.
4. The product should be inoculated (under the surface) with the microbial isolate.
5. The puncture hole should be flamed to create a vacuum in the headspace and aseptically sealed with solder or similar material. (It is sometimes necessary to warm the closed container to 40°C to 45°C before puncturing to ensure that the seal will be under vacuum).
6. The inoculated container should be incubated at 30°C to 35°C for 10 days.
7. The container should be opened and the product examined.
8. If the previously tested container did contain microorganisms (i.e., was not commercially sterile), spoilage indications (or lack thereof) should be identical for both containers. If the first container exhibited a normal-appearing product but the analyst recovers growth upon subculture and if, after inoculating the second container with the microbial isolate, the analyst observes spoilage (gas liquefaction, etc.), it should be concluded that the first container was commercially sterile; growth was the result of contamination due to faulty technique.
9. Open containers should be emptied, washed, and dried. Keep them properly identified in the event subsequent container examination is deemed desirable (see Chapter 62).

61.62 Defects Caused by Heat-Resistant Enzymes

The enzymatic liquefaction of starch-based products, such as high temperature-short time (HTST)-processed puddings and sauces, is caused by heat-stable microbial enzymes such as alpha-amylase. A severe thinning of the product consistency occurs, sometimes coupled with the formation of a clear, gel-like substance. Extremely small amounts of active enzyme can cause this phenomenon and can be present in ingredient material that, by either chemical or microbiological analysis would be considered suitable for use. This defect does not result in a physical distortion of the container and is evident only after opening the container.

Direct microscopic smear of the product may now show unusual numbers of microorganisms. Bacteriological culturing of the suspect sample, assuming proper handling techniques are used will be negative for viable microorganisms normally associated with either microleakage or underprocessing.

When interpreting direct microscopic smears, be careful to examine a normal control in order to arrive at a proper judgment regarding levels of microorganisms in the suspect product.

Product that has undergone enzymatic deterioration will usually have a slightly reduced pH as measured against a normal sample. No off-odor will be apparent, but, because of the severe consistency change and possible gel formation, the product will be obviously unfit for consumption.

Enzymatic liquefaction is an economic problem for the packer and has no public health significance with regard to safety of the product.

61.7 REFERENCES

1. AOAC International, 1996. Official methods of analysis of the Association of Official Analytical Chemists International, 16th ed. Assoc. of Off. Anal. Chem., Arlington, Va. Section 17.6.01, p. 42-42a.
2. APHA. 1966. Recommended methods for the microbiological examination of foods, 2nd ed., J. M. Scharf (ed.). Am. Public Health Assoc., New York. p. 35.
3. Barbeito, M. S., C. T. Mathews, and L. A. Taylor. 1967. Microbiological laboratory hazard of bearded men. Appl. Microbiol. 15:899.
4. Corson, L. M., G. M. Evancho, and D. H. Ashton. 1973. Use of forceps in sterility testing: A possible source of contamination. J. Food Sci. 38:1267.
5 Denny, C. B. 1970. Collaborative study of procedure for determining commercial sterility for low-acid canned foods. J. Assn. Off. Anal. Chem. 53:713.
6. Denny, C. B. 1972. Collaborative study of a method for determining commercial sterility of low-acid canned foods. J. Assn. Off. Anal. Chem. 55:613.
7. Federal Standard. 209a. 1967. No. 252-522/280. U.S. Government Printing Office, Washington, D. C.
8. Hersom, A. C., and E. D. Hulland. 1981. Canned foods, thermal processing, and microbiology, 7th ed. Chemical Publ. Co., New York. p. 259.
9. Schmitt, H. P. 1966. Commercial sterility in canned foods, its meaning and determination. Q. Bill. Assoc. Food Drug Off. U.S. 30:141.
10. Governmant National Archives and Records Administration. 1995. Code of Federal Regulation Title 9 Part 381. Mandatory poultry products inspection, p 579; Part 318. Mandatory meat products inspection, entry into official establishments; reinspection and inspection of products. U.S. Government Printing Office, Washington, D.C. p 282.
11. Government National Archives and Records Administration. 1998. Code of Federal Regulation Title 21 Part 108. Emergency permit control, Part 113. Thermally processed low acid foods packaged in thermatically sealed containers. Part 114. Acidified foods. U.S. Government Printing Office, Washington, D.C. p. 198-208, 224-250, 250-256.

Chapter 62

Canned Foods—Tests for Cause of Spoilage

Cleve B. Denny and Nina Gritzai Parkinson

62.1 INTRODUCTION

Spoilage of canned foods is defined as the outgrowth of microorganisms resulting in changes in the product (pH change, texture change, gas production, etc.) The methods discussed in this chapter are designed to identify the types of microorganisms responsible for product spoilage. There are specific characteristics in the product, container and/or microorganisms recovered which give an indication of the cause of spoilage. For example, post-process contamination usually results in a mixed microbiological flora of vegetative cells and occasionally sporeformers, that may have little or no heat-resistance. Insufficient processing is characterized by pure cultures of heat-resistant sporeformers and may have serious public health implications because of the potential presence of *Clostridium botulinum* and its toxins.

The methods described in this chapter are intended for thermally processed foods packaged in hermetically sealed containers and will be referred to as "canned" foods in this chapter, despite the container type. The types of containers typically used include metal and plastic cans; glass; metal and plastic retort trays, cups, and bowls; and flexible pouches. The types of products not covered in this chapter include perishable, "pasteurized" canned cured meats (e.g. refrigerated canned hams, Chapter 45), and other refrigerated products, dried foods, bottled and carbonated beverages (Chapter 59) and alcoholic beverages. Commercial sterility determination methods for canned foods are different from spoilage analysis, and are described in Chapter 61.

A comprehensive examination procedure is required for expedient and accurate diagnosis of spoilage. It must include the history of the defective product, the structural integrity of the containers, the physical and chemical state of the food and other factors that may be related to the presence of viable microorganisms. This information is used with the microbiological examination to determine the cause of spoilage. Failure to include all information may lead to an inaccurate diagnosis, which might have public health and economic implications.

In order to test a product, one must have a basic understanding of the differences between low-acid and acid (or acidified foods), the types of microorganisms which may affect them and how these effects are manifested.

62.2 MICROBIOLOGY OF CANNED FOODS

The thermal process given to a canned food depends on many factors, including the type of product, its normal pH, its consistency, the size of the container and the type of processing equipment to be used.[4,21] Product pH is the most important factor that determines the degree of thermal processing needed to achieve product stability because of the inhibitory effect of acid on survival and outgrowth of indigenous microorganisms. For this reason commercially processed foods are divided into two major pH categories:

- low-acid foods, which are foods that have a pH above 4.6 and a water activity (a_w) greater than 0.85[37]; and
- acid (or acidified) foods, which are foods that naturally have a pH of 4.6 or lower, or are acidified to a pH of 4.6 or lower.[7,8,35]

Normal pH ranges for many foods are listed in Table 5 at the end of this chapter.

62.21 Processing of Low-Acid Foods (pH Above 4.6)

Low-acid foods are thermally processed to produce a condition known as commercial sterility. Commercial sterility is defined as the condition in which all *C. botulinum* spores and all other pathogenic bacteria have been destroyed as well as more heat-resistant organisms, which, if present, could produce spoilage under normal conditions of storage and distribution.[32]

Commercially sterile foods, however, are not necessarily sterile in the classical sense. Occasionally, "canned" low-acid foods and acidified foods (pH 4.0–4.6) may contain low numbers of certain thermophilic spores that will not cause spoilage unless the food is stored at temperatures above 43°C. To sterilize a canned product completely usually would degrade product quality.

62.22 Processing of Acid or Acidified Foods (pH 4.6 or Lower)

A pH of 4.6 or lower is sufficient to inhibit the growth of *C. botulinum* in foods, thus permitting the application of a less severe thermal process to the food.

Since there is a significant advantage in quality retention in a lower thermal process, some low-acid foods are acidified to pH 4.6 or less (thus the term acidification).[35] When "controlled acidification" is used, the control of the product pH is critical to insure against possible outgrowth of *C. botulinum* in the product. Acid foods, in the higher pH ranges (4.0 to 4.6), may contain low numbers of thermophilic sporeformers that do not become a problem unless the food is stored at temperatures above 43°C. Acid or acidified foods also may contain dormant mesophilic spores that are inhibited from growth by the acid.

62.23 Factors Responsible for Spoilage in Canned Foods in General

Microbiological spoilage in canned foods may be indicated by an abnormal odor, appearance and/or pH of the product. In addition, the container is usually, but not always swollen. The causes of these general conditions usually are related to one of the following factors:

- Insufficient processing, permitting survival of mesophilic microorganisms.
- Inadequate cooling after processing or storage and distribution conditions at elevated temperatures (above 43°C), permitting growth of various thermophilic microorganisms.
- Post-process contamination permitting microorganisms to contaminate the product after the thermal process (sometimes referred to as "leaker" spoilage). Typically, this results from microorganisms gaining access to the product via container defects, rough container handling or from contaminants in the cooling water or soiled contact surfaces.

Other conditions may occur in canned foods that result in slight to severe swelling of the container and/or abnormal product appearance and may confuse the investigator. These conditions include:

- "Incipient spoilage" which occurs before the product or ingredient(s) are thermally processed and may be caused by microbial growth or enzymatic action. Microbial growth before the thermal process may result in carbon dioxide production, changes in pH and the presence of excessive numbers of dead microbial cells. Enzymatic action also may result in evolution of carbon dioxide as well as off-odors (particularly in vegetables).
- "Hydrogen swells" which result from a chemical reaction of the food with the metal container, thereby producing hydrogen gas.[27] When advanced corrosion has occurred, "pinholes" in cans and lids of glass jars may be present. Hydrogen swells are most common in overage merchandise.
- Non-enzymatic browning (Maillard reaction) sometimes occurs in canned products having high levels of sugar, amino acids and/or acid. Carbon dioxide may be produced in sufficient amounts to bulge the container, particularly during storage of the product at elevated temperatures. This problem is occasionally encountered in canned fruit concentrates.
- Product formulation errors and/or mishandling (e.g., freezing of a canned food).
- Enzymatic changes such as liquefaction, off-flavors, curdling, and discoloration sometimes occur in foods packed by the ultra-high temperature (UHT) or high-temperature/short-time (HTST) heat processes such as those used in aseptic-type systems. Containers appear normal and high numbers of microbial cells are not seen by direct microscopic examination. Enzymatic activity creates an economic problem for the packer, but has no significance in regard to safety of the product.

62.24 Microbiological Groups Associated with Spoilage in Low-Acid Foods (pH Above 4.6)

62.241 Insufficient Processing

Insufficient processing is indicated chiefly by the survival of bacterial spores, particularly those of the *Clostridium* species and sometimes of the *Bacillus* species, which subsequently spoil the product. From the public health standpoint, this is a most serious situation because of the potential growth of *C. botulinum* and development of its toxin.

Generally, anaerobic mesophilic spoilage is associated with putrid odors and the presence of rod-shaped microorganisms in pure culture with or without typical clostridial sporangia or spores. If putrid odors and clostridial-type spores are found during examination, the product should be tested for botulinum toxin by the mouse bioassay (Chapter 33), regardless of the percentage of the pack exhibiting spoilage.[18] If there is no odor but clostridial-type spores are found, trypsin treatment may be needed to activate the toxin. If a low-acid, canned food misses retorting completely, it may or may not swell the container, but the pH is typically lower than normal and the product will have a non-cooked texture.

62.242 Thermophilic Spoilage

Low-acid foods may spoil during storage above 43°C because of growth of extremely heat-resistant sporeforming thermophilic microorganisms. Most of these bacteria grow at 55°C, some as high as 70°C. Some are facultative thermophiles and may also grow at 35°C or lower. Thermophilic bacteria are not pathogenic (Chapters 24, 25, 26 and 27).

Thermophiles occur naturally in agricultural soils and their spores frequently are present in low numbers in commercially sterile products. Certain ingredients (i.e. sugar, starch and some spices) may also contribute to the thermophile load. Since the thermal process is not designed to destroy these spores, they may survive in a canned product, but are not likely to grow unless product is held at elevated temperatures. This frequently occurs in pre-processing equipment where product may be allowed to stagnate, in cans that are not cooled properly after the thermal process, or if warehouse or other storage or transport temperatures are in excess of 43°C.

It is very important during microbiological examination of low-acid foods to avoid confusing thermophilic and mesophilic sporeforming bacterial isolates. They may be confused because in the vegetative state the cells will grow over a much wider temperature range than when they germinate and grow out from the spore state. For this reason it is essential to produce spores at the temperature of isolation (30–35°C or 55°C), then heat-shock the suspension to destroy vegetative cell forms and subculture at both 30–35°C and 55°C. The temperature of outgrowth from the spore state indicates whether the isolate is an obligate thermophile (growth only at 55°C), facultative thermophile (growth at both 30–35°C and at 55°C) or a true mesophile (optimum growth at 30–35°C).

It should be emphasized that heat-shocking for the above purposes must be conducted on subcultures. Heat-shock of the microorganisms in the food sample may unintentionally destroy sporeformers if they are all in the vegetative cell form or when acidic conditions prevail in the food caused by acid producing spoilage organisms.

In low-acid foods the most common forms of thermophilic spoilage and the causative microorganisms are categorized as follows:

1. "Flat sour" spoilage is indicated when the container is not swollen but the pH of the product is significantly lowered. The causative microorganisms are sporeformers, such as *Bacillus stearothermophilus*, a facultative aerobe. Spore germination occurs at thermophilic temperatures only, but vegeta-

tive cell growth may occur at mesophilic temperatures as well (Chapter 25).

2. "Thermophilic anaerobe" (TA) spoilage is typically characterized by swelling and sometimes even by bursting of the container. The condition is caused by obligately thermophilic, sporeforming anaerobes such as *Clostridium thermosaccharolyticum* that produce large quantities of hydrogen and carbon dioxide. The product usually has a "cheesy" odor. Vegetative cells of TA's may grow at temperatures below the thermophilic growth range (Chapter 26).
3. "Sulfide stinker" spoilage is characterized by a flat container in which the contents are darkened and have the odor of rotten eggs. This type of spoilage is caused by the sporeforming anaerobic, obligately thermophilic microorganism *Desulfotomaculum nigrificans*, which produces hydrogen sulfide. The container does not swell because hydrogen sulfide is soluble in the product, but reacts with any iron present to form black iron sulfide (Chapter 27).

62.243 Container Leakage

Environmental sources such as air, water or dirty contact surfaces may contribute to microbial contamination of processed foods through rough handling of filled containers, or because of defective or damaged containers and closures.[22] Subsequent spoilage often results in swelling of the containers when the pathway of the original leak becomes blocked, preventing the escape of gas. If the opening is large enough, leakage of the product may occur. Spoilage in flat containers, with or without vacuum, may result from post-process contamination of the product with microorganisms that produce little or no gas.

Numerous groups of microorganisms may be found in instances of container leakage, including cocci, cocco-bacilli, short and long rods (lactic acid bacteria), yeast and molds, aerobic sporeformers, and mixtures of some or all of these (which is very common). All of these organisms typically exhibit little or no heat-resistance.

Aerobic sporeformers are sometimes present in pure or mixed culture as a result of post-process contamination. The predominance of sporeformers is becoming more common because of the widespread chlorination (0.5 ppm or more of free residual chlorine) of cannery cooling water, which destroys most vegetative microorganisms.

Post-processing can handling equipment may be a source of seam or closure contamination before the seaming compound has had a chance to "set up" and seal the closure. This may occur immediately following thermal processing when containers come in contact with wet, dirty roll tracks or receive rough handling in the presence of contaminated surfaces. Rough handling sufficient to cause leakage does not necessarily result in permanent dents in the can and permanent dents do not necessarily result in leakage.

It is essential to conduct thorough leakage tests and detailed structural examinations of all containers, although failure to demonstrate leakage sites does not rule out post-process contamination spoilage. This fact cannot be overemphasized when a decision must be made to differentiate spoilage that is due to post-process contamination from spoilage that is caused by insufficient processing, especially when the microbial flora found in the product do not necessarily indicate either situation. The original container closure records are required under 21CFR113.60 and 113.100(c)[38,36] and should be examined, particularly if post-process contamination is suspected.

62.25 Microbiological Groups Associated with Spoilage of Acid or Acidified Foods (pH 4.6 or Lower)

62.251 Insufficient Processing

Since foods with a pH of 4.6 or lower do not require a severe thermal process, as compared to low acid foods, a variety of acid-tolerant sporeforming microorganisms may survive the thermal process. Their survival is typically a result of excessive pre-processing contamination. The microorganisms, which typically spoil acid or acidified foods, fall into the following groups:

1. "Butyric acid anaerobes" such as the mesophilic, sporeforming anaerobe *Clostridium pasteurianum*, which produces butyric acid as well as carbon dioxide and hydrogen (Chapter 23).
2. Aciduric "flat sours," particularly *Bacillus coagulans*[34] in tomato products. This acid-tolerant, facultative anaerobic sporeformer grows both at 30–35°C and as high as 61°C (Chapter 24). *Alicyclobacillus* spp., recently isolated in fruit juice and fruit juice concentrates, gives product an off-flavor, but seldom produces gas or acid (Chapter 24).
3. Heat-resistant molds often contaminate juice concentrates and fruits prior to processing and are able to survive the mild heat treatment given to these products. The causative microorganisms are usually *Byssochlamys fulva, Neosartorya fischeri or Talaromyces flavus,* fungi that produce very heat-resistant ascospores. Spoilage is evidenced by a moldy taste and odor, color fading, the presence of mold mycelia in the product and sometimes by slight swelling of the container (Chapter 21).
4. Yeast and asporogenous bacteria are especially seen in cases of gross underprocessing. This type of spoilage may be indistinguishable from leaker spoilage unless the containers are thoroughly examined for leakage and structural defects and post-processing handling conditions are known to be satisfactory (Chapter 20).

62.252 Thermophilic Spoilage

Thermophilic spoilage may occur in acid foods, particularly in tomato products. Cultural tests should include "cross temperature" incubation of isolates at 30–35°C and at 55°C to differentiate mesophiles from thermophiles.

62.253 Container Leakage

Pure or mixed cultures of acid-tolerant bacteria, yeast and molds are commonly found in leaker spoilage of acid products. The containers may be swollen or flat, depending on the gas-producing abilities of the spoilage microorganisms.

Gas and swelling of the can are commonly produced by bacteria (rods and cocci) or yeast and sometimes by molds. There are also bacteria (rods and cocci) that do not produce gas, therefore the container does not exhibit signs of swelling. The contents usually show a slight lowering of pH (generally less than 0.2 pH unit) except where mold may have increased it, usually with obvious organoleptic changes in the product. Mold spoilage usually is evidenced by the presence of mycelia and fungal spores in flat containers having a leak large enough to permit entrance of oxygen. Sometimes a mat of mycelial growth is present on the surface of the product. A heavily etched ring on the sidewall at the product surface level, inside the freshly opened can often indicates that oxygen has leaked into the can. In bottles and jars, mold may be growing just inside the lid.

62.3 EXAMINATION METHODS FOR DIAGNOSING SPOILAGE CONDITIONS IN CANNED FOODS

The following methods are designed to guide the investigator in conducting a thorough and carefully documented examination. Results obtained by these methods may be recorded on the sample data collection forms and compared to the "Keys to Probable Cause of Spoilage" discussed in Section 62.4.

The diagnosis of spoilage in canned foods should be conducted by trained microbiologists or trained technicians under supervision. If difficulties are encountered in the examination of samples or in the interpretation of results, assistance is available from recognized authorities in the field, such as the National Food Processors Association; the container supplier; various universities; the Food Microbiology Methods Development Branch, CFSAN, U.S. Food and Drug Administration; or the Microbiology Division, Office of Public Health Science, Food Safety and Inspection Service, U.S. Department of Agriculture.

62.31 Gathering and Interpreting Background Information

Information regarding the circumstances of production and storage and the incidence of defects in canned foods suspected of spoilage is invaluable for diagnosis.[29] In some cases, the cause and nature of spoilage may be evident after a thorough review of background information.

To illustrate this point, aseptically canned foods (those where the container and product are sterilized separately and then filled and closed) such as juices, puddings and milk do not lend themselves to the procedures listed in this chapter. Contamination may occur in the container before filling, in the cooling of the liquid product, in a diversion valve area, or in the filling or closing area. Microorganisms resulting from this cold recontamination before closing may resemble leakage types. Therefore, subculturing usually does not help in the diagnosis. Instead, a history of the spoilage often will pinpoint the cause. For instance, spoilage that starts early in the day and builds up to 100% by the end of the day usually indicates recontamination of the product in the cooling, diversion or filling equipment areas. Low numbers of spoiled food containers scattered throughout a day or period usually indicate contamination of container or lid, filtered air or sealing machine.

A suggested "Request for Spoilage Analysis" form is provided (Figure 1) for gathering background information. The best way to use the form is to request persons submitting samples to fill out the questionnaire at the time the collected samples are submitted for examination. Some detailed information may not be available initially and may need to be obtained later. Once this preliminary information is gathered, some determinations can be made as to possible problems or causes of spoilage, which may be confirmed with the microbiological analyses and testing of the containers.

For canned foods using terminal heat processing, the suggested guide for interpretation of background information is shown in Table 1. This may guide the investigator in obtaining additional information and/or records to support the laboratory findings.

62.32 Procedures for Examination of Containers and Contents

Conduct the following steps in the sequence shown. Record data on the "Spoilage Examination Data" sheet included for this procedure (Figure 2). The first line of the table refers to the section in this chapter where the item is discussed.

62.321 Sampling

The number of units available for analysis will vary with the circumstances. When the item has been in distribution for a long time or if spoilage is very sporadic, only a few sample units may be available. Items with recent codes or those in plant or wholesale warehouses are usually available in sufficient quantity.

1. When sufficient samples are available:

- Randomly select 12 sample units that are representative of an apparent or suspected spoilage condition (i.e., swollen

REQUEST FOR SPOILAGE ANALYSIS

No. Date

Type of Product and Style Pack:

A. Information from Point of Collection of Sample(s):
   1. Source
   2. Number of samples
      Sampling Instructions: Please submit 12 suspect units, if available, and 12 units of sound product from a similar code as controls. Submit all suspect containers (and an equal number of controls of similar code) if less than 12 suspect units are available.
   3. Date of collection
   4. Code(s)
   5. Container size

B. Information Concerning Code Lot of Submitted Sample(s):
   1. Location of spoiled goods
   2. Temperature of storage
      - Plant: - Warehouse: - Retail:
   3. Number of spoiled and normal containers of each size involved
   4. Percent spoilage (number spoiled divided by number in lot × 100)
   5. Packing dates, periods and lines involved
   6. Date when spoilage first appeared or noted
   7. Are any cans burst, hard swells, soft swells, flippers or low vacuum; and are any flat cans spoiled (complaints of spoilage in flat cans)
   8. Were any irregularities noted in product during:
      a. Preparation
      b. Thermal processing
      c. Cooling temperature
      d. Chlorination or sanitizer in cooling water
      e. Sanitation of post-process handling (can lines, labeling area, etc.)
      f. Container integrity
      g. Other (damage, mishandling)
   9. Type of cooker or processor
   10. Process used—I.T.: Time: Temp.:
   11. Were any heat-sensitive inks used on the containers
   12. Were color checks used on the retort baskets

Please submit appropriate records for any items noted in B.

***Figure 1.** "Request for Spoilage Analysis" Form*

Table 1. Guide for Interpretation of Background Information

Information	Interpretations
1. Number of spoiled containers	i. An isolated container is usually a random "leaker" (although insufficient processing must be considered).
	ii. More than one container, especially one or more than one per case, may indicate defective containers, rough handling or insufficient processing.
	iii. Excessive spoilage may indicate insufficient processing.
2. Age of product and storage conditions	i. Excessive age and/or excessively high storage temperatures may produce detinning and hydrogen swells in metal cans or jars with metal lids.
	ii. Perforation because of corrosion or damage of the container may produce leaking and/or mixed culture post-process contamination spoilage, with swells in some cases where the perforations are blocked.
	iii. Thermophilic spoilage may result from high storage temperature (in excess of 43°C).
3. Location of spoilage in stacks, temperature of the warehouse	i. Spoilage in center of stacks or near ceiling may indicate failure to cool product sufficiently, resulting in thermophilic spoilage.
	ii. Scattered spoilage may indicate insufficient processing or post-process contamination spoilage.
	iii. Excessive spoilage may indicate insufficient processing.
4. Processing records, including retort charts, "cook check tags" or other thermal process records	i. Irregular cooks may be correlated with spoilage from insufficient processing.
	ii. Low vacuums or flippers, change in pH of product and/or high numbers of cells in direct micro exam (albeit not viable) are indications of incipient spoilage.

containers or reports of abnormal appearance of contents) from each production code involved.

- Also select 12 control sample units of sound product from the same or a related code (same day, week) for comparison with spoiled product. Analyze suspect and control sample units. Save 2 suspect samples, if feasible, for other testing in case microbiological tests are inconclusive (i.e., headspace, organoleptic testing, etc. See Section 62.324).
- Avoid testing very hard swells or buckled cans if a sufficient number of softer swells are available. Very hard swells or buckles may burst during examination and structural damage to cans may render the containers unsuitable for microleak testing and detailed container examination.

2. When sufficient samples are not available:

- If fewer than 12 sample units are available from each production code involved, analyze all sample units representative of an apparent or suspected spoilage condition.
- Analyze an equal number of sound control units from the same or a related code.
- If only one or two sample units are available, it is advisable to request additional units while proceeding with the analysis of those on hand.
- Avoid testing very hard swells or buckled cans if a sufficient number of softer swells are available. Very hard swells or buckles may burst during examination and structural damage to cans may render the containers unsuitable for microleak testing and detailed container examination.

Segregate samples by production code (if more than one code is present). Examine the label of each container for leakage stains, and if present, circle with a marking pen. Make a mark with waterproof ink on the bottom of the label and container to indicate the alignment of the label to the container. Carefully remove labels from containers and attach to the back of the data sheet, for reference during the external examination of containers

Consecutively number each container on the side (body) with waterproof ink, diamond pen, or copper sulfate solution,[31] take care not to write on the can side seam. Record container numbers, code(s) and can size(s) on the data sheet.

62.322 External Examination

Examine the external condition of the containers for swelling of the ends, lid or body in the case of flexible pouches. Examine different types of containers for structural defects or damage to the following areas and record all observations on the data sheet:

1. Cans[20,23]
   a. End seams for any defects including crooked, excessively wide or narrow seams, "cutovers," and "sharp" seams, cracked or ruptured seams, "cable burns" (where the metal is abraded through the seam), false seams, "vees," seam bumps and any evidence of leakage.
   b. Side seam for possible weld voids, pitting, rusting or leakage.
   c. The embossed code on lid for possible fractures or perforations.
   d. The can body for perforations, imperfections (particularly "inclusions" in the tinplate) or leakage.
   e. Circle defects, when they are found, with waterproof ink. Also, mark an "X" on the end of the can exhibiting a defect(s) to show the end that should not be opened during sampling of the container. If no defects are found on either can end, save the end with the embossed code (usually the "processors" end) and mark it with an "X." Never open the lid end of a two-piece aluminum or steel can unless there is evidence of "leaker damage" on the body shell end.

Date: ______________________ Analyst: ______________ Product: ______________ Container Size: __________

Identification: ______________ Source: ______________ Lab Nos: ______________

62.321		62.322	62.323		62.324	62.328	62.329	62.330	62.333	
Sample No.	Container Code	Container Condition Exterior	Gross Gm.	Oz.	Vacuum or Gas (Hydrogen)	Odor, Appearance	Microscopic Examination	Ph	Container Exam and Microleak Testing	Container Condition Interior

Figure 2. *"Spoilage Examination Data" Form*

f. Some processors apply a spot or stripe of heat-sensitive paint or "thermal-sensitive fluid" to the body of the container as a retort check. The paint contains suspended material and has the consistency of "house paint," whereas the fluid is a solution and has the consistency of coding ink. The choice of color may vary with different processors and with different types of products. For example, paint may change from red or blue before retorting to white after retorting. For this reason, it may be necessary to confirm the specific color system used on the sample(s).

No color change indicates that the can was not retorted. A color change indicates only that the product was retorted and does not show that the correct process temperature or time was achieved. Therefore, the possibility of insufficient processing cannot be discounted even though the color check conforms to the specified post-retort color.

2. Glass containers[24]
   a. The lid for possible misalignment, damage or leakage.
   b. The glass container for "hairline" or small impact cracks or improper finish.
   c. The contents for gas bubbles and abnormal conditions (underfill, digestion, turbidity, etc.)
3. Plastic retort trays[33,25]
   a. The lid seal area for channel leaks, incomplete seals, uneven impression, evidence of severe delamination, severe seal width variation or product inclusion involving loss of hermetic integrity.
   b. The lid and body for punctures, cuts or evidence of leakage.
   c. The plastic container for swells, punctures, fractures, crush damage, cuts or cracks involving evidence of leakage or loss of hermetic integrity.
   d. The contents (if visible) for abnormal conditions including turbidity, digestion, underfill or lack of brine.
4. Flexible pouches[33,25]
   a. The seal area for blisters, severe wrinkles, channel leaks, contaminated seal, severely misaligned seal, notch leaks, pronounced seal creep, nonbonding, delamination in the sealed areas, product inclusion involving evidence of leakage or loss of hermetic integrity.
   b. The pouch body for swells, cuts, punctures or evidence of fractures or malformation involving leakage or loss of hermetic integrity.
   c. The contents (if visible) for abnormal conditions including digestion or turbidity.

62.323 Weight

Weigh each container to the nearest gram or ounce and record the weight on the data sheet. Subtract the average known tare weight of the empty container from each unit weight to determine the approximate net weight of each sample. Compare to the established net weight, (21CFR Parts 130-169)[40] as well as the maximum and minimum "permissible" net weights of the product. Following the disposition of contents and cleaning of container (Section 62.332), when the empty dried containers are available, weigh them to ensure that they fall within the average tare weight for the type of container.

If the net weights of the suspect sample units exceed the net weights of the controls and exceed the maximum permissible net weight, the suspect containers may be overfilled, causing reduced headspace and reduced or zero vacuum. In extreme cases of overfilling, the bulging of the lid gives the external appearance of a swell. Seafood, such as sardines, because of their size and their packing in short-height cans, are often inadvertently overfilled.

If the suspect units have a lower net weight than controls and are below the minimum permissible net weight for the product, they may be underfilled or may have leaked. Leakage also may be evidenced by stains or residues on the label or on the exterior areas of the container.

62.324 Vacuum Measurement or Gas Analysis

Select a portion of the suspect flat containers (if any) for measurement of internal vacuum and swollen containers for analyses of headspace gas composition. Also, select a portion of the control sample units for measurement of internal vacuum. Determine the number of units for vacuum or gas test as follows:

Number of suspect or control containers available	Number to select for vacuum or gas tests
6 to 12	2
2 to 5	1
1	None

When suspect samples consist of both flat and swollen containers, and only two to five are available, select a swollen container for gas analysis. If only one suspect unit is available, save it for aseptic sampling. A container punctured for gas or vacuum test should not be used for culturing because of the possibility of contamination.

1. **Vacuum measurement (normal appearing cans and jars).** Bring to room temperature any container that has been removed from incubation or refrigeration before measuring the vacuum. Use a puncture gauge with a rubber seal around

the puncture needle to obtain the vacuum, in inches of mercury. (VWR Scientific, P.O. Box 7900, San Francisco, CA 94120). Measure the vacuum through the lid, about halfway between the double seam and the center of the lid to avoid clogging the puncture needle. Record vacuum measurements on data sheet.

Flexible plastic retort trays are flushed with an inert gas and following processing contain a certain amount of headspace gas. Vacuum measurement, therefore, cannot be made on these packages. Some plastic containers for aseptically processed product are filled without having a vacuum and, therefore, also cannot be tested for vacuum.

2. **Gas collection and analysis (swollen containers).**
   a. *Qualitative test for hydrogen.* When there is sufficient headspace gas in a swollen metal can, the lid may be punctured with a "Sanitary Can Opener" or Bacti Disc Cutter (Wilkens-Anderson Company (WACO), Chicago, IL 60651, tel. 800-847-2222. Figure 3).[31] While puncturing, collect escaping gasses in an inverted test tube that is held at an angle directly over the point of puncture. Immediately following collection of the gas, place the open end of the tube over a burner flame. This will result in a loud "pop" if the gas is mainly hydrogen. Usually hydrogen results from detinning of metal cans or is produced by butyric acid anaerobes or thermophilic anaerobes.

      Sampling of plastic retort containers is often more difficult because of their flexibility. The plastic lid may be punctured with a sterile hypodermic needle and the gas immediately collected in an inverted tube. However, puncturing a plastic tray lid may make subsequent tests for microleaks difficult to perform. Hydrogen accumulation in plastic containers usually results from spoilage caused by thermophilic anaerobes.
   b. *Quantitative test for hydrogen—headspace gas analysis (hydrogen, carbon dioxide, nitrogen and oxygen).* A gas sample from a swollen metal container can be collected by use of a rubber gasketed puncture needle connected to a syringe or a manometric device having a "gas-holding solution." In collecting a gas sample from a swollen plastic tray, care must be exercised because of the flexibility of the container and the lid; puncture the lid near its center.

      Analyze the gas sample on a chemical gas analyzer or on a gas chromatograph fitted to a thermoconductivity detector (TCD).[41,19]

      Gas from a sound product should contain mostly nitrogen with very small amounts (generally less than 1%) of hydrogen, carbon dioxide, and oxygen. A large percentage of carbon dioxide usually indicates microbial growth, or in some products such as vegetables it may indicate preprocessing product respiration. In high sugar products, carbon dioxide production may be due to the Maillard reaction. A large percentage of hydrogen may indicate metal corrosion or detinning (hydrogen swell), or thermophilic anaerobic spoilage. A high concentration of both carbon dioxide and hydrogen may also be indicative of microbial growth, particularly of thermophilic anaerobes. Record results on the data sheet.

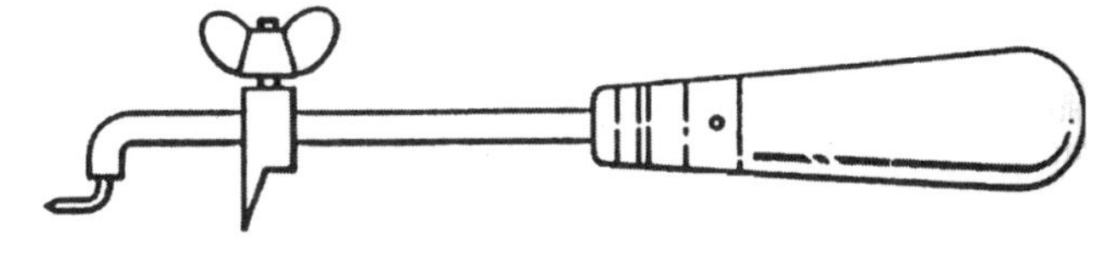

Figure 3. *Sanitary Can Opener*

62.325 Preparation of Area for Tests

The site where the samples are prepared for testing should be clean, dust free and located in an area where samples may be taken aseptically for microbiological examinations. A laminar air flow work station is ideal for providing an ultraclean working environment.[26] Analysts should wear clean protective laboratory garments. All sampling utensils and pipettes should be sterilized and ready for use. The surface of the workbench used during examination procedures should be washed, sanitized with an effective disinfectant, such as 100 ppm chlorine solution and wiped dry with clean paper towels.

These conditions are essential to minimize incidental microbial contamination during aseptic sampling, a problem which is most acute when no viable microorganisms are present in the product.[12,14]

62.326 Preparation of Containers

Wash all sample containers in warm soap and water to remove soil and grease. Rinse in clean running water and dry with clean paper towels. Place the unmarked end of the can to be opened in the upward position. Do not open containers.

62.327 Aseptic Sampling

For each sample unit, label two containers, such as sterile large screw cap test tubes (24 mm) or other sterile container with the sample unit number, product lot code and date of examination. After aseptic sampling by the following procedure, retain each set of food samples obtained from each container for cultural tests. Note: Sample units used for vacuum measurement or gas collection must not be used for aseptic sampling because of possible contamination from puncture devices.

1. **Sterilization of lid.** Arrange all containers in numerical order, positioning the end to be opened in the upward position. The end to be saved should have been marked with an "X."
   a. *Flat to moderately hard swells.* Flood the lid with a disinfecting agent, such as 100 ppm chlorine solution.[12] Do not use alcohol because it may not completely sterilize the lid surface.[13] Allow to stand 10 to 15 minutes and pour off excess sanitizing agent. Heat with a burner flame over the lid until visible moisture has evaporated. Do not overheat the lid if a canned sample is swollen (i.e. under positive pressure). Place side seam away from you during the flaming.
   b. *Hard swells, very hard swells and buckled cans.* It is advisable to chill sample containers having high positive pressure in a refrigerator for several hours prior to lid sterilization. This will minimize possible bursting of the container during flaming of the lid. If thermophilic anaerobes are suspected, do not chill the containers, because chilling often kills the vegetative cells of thermophilic anaerobes (TA's) and spores are rarely recoverable in thermophilic spoilage.

      Note: Since it takes 10 to 15 min to disinfect a lid with the sanitizing agent, flood the lids of the next containers and let them stand while proceeding with opening and sampling. Once the lid has been flamed, aseptic opening and sampling must be conducted in one continuous operation for each sample unit.

c. *Plastic retort trays.* Soak for 10 to 15 minutes in 100 ppm chlorine solution. Drain to remove moisture. Do not flame.

d. *Pouches.* Soak for 10 to 15 minutes in 100 ppm chlorine solution. Drain to remove moisture. Do not flame.

2. **Aseptic opening of lid.**

a. *Flat container.* Cut a hole slightly off-center in the lid using a Sanitary Can Opener or Bacti Disc Cutter (Wilkens-Anderson Company (WACO), Chicago, IL 60651, tel. 800-847-2222. Figure 3).[31] A hole 2 inches in diameter is usually sufficient for sampling; however, for a small can or jar, a smaller hole may be necessary. Always leave 1/4 to 1/2 inch of lid remaining in contact with the double seam for subsequent microleak testing and container examination.

Open plastic retort trays and foil pouches from the bottom or side to prevent damage to the seal area and lid. Following disinfection, lightly flame dry to avoid damage to the plastic container and cut a hole, large enough to remove a sample aseptically, using a small electric soldering iron with a sharp tip (or flamed scissors).

b. *Swollen containers.*[19] Place the container on a large, clean tray in case the contents under pressure spill over during opening.

Invert a large sterilized plastic or metal funnel over the can, insert a sterile-tipped puncture rod through the funnel and puncture the lid. Alternately, cover the area to be punctured with sterile paper towels in case product erupts. After venting has occurred, remove the funnel and open the container by the procedure described previously for a flat container (Section 62.327.2a).

Plastic containers should be punctured at the bottom, using a sterile sharp-tipped tool such as the tip of a knife

3. *Aseptic sampling of contents.* Use sterile pipettes or utensils to transfer approximately 20 g of food to each of two labeled sterile tubes or other sterile container. Always use cotton-plugged pipettes fitted with a mechanical suction device such as a rubber bulb or other mechanical pipettor. The following procedures are suggested for transfer of various types of materials to the tubes:
   - Liquids: Transfer 20 mL of liquid with sterile 20 mL pipettes.
   - Semi-liquids (with or without small particulate material): Transfer 20 mL of product with sterile pipettes having large bore tips.
   - Solids in liquids: Pipette about 20 mL of liquid per tube with a sterile 20 mL pipette having a large bore tip. Transfer about 10 g of solids with sterile forceps (non-serrated tip type), spoon or spatula to a sterile container.[16] (Do not re-use utensils without resterilizing in an autoclave.)
   - Semi-solid or solid materials: Transfer about 20 g material with sterile forceps (non-serrated tip type), spoon or spatula to a sterile container.

All transfer utensils must be wrapped and sterilized prior to use (or use sterile disposable ustensils). Alcohol flaming of metal utensils must not be used because it is not an effective means of sterilization.[13] Retain all aseptically drawn samples for procedures in Section 62.34. Store tubes in the refrigerator unless microscopic examination reveals long rods indicating possible thermophilic anaerobes, since vegetative cells of TA's often die under refrigeration.

62.328 Odor and Appearance

It is important to note the odor and appearance of a product, since these characteristics may provide valuable information. However, never taste the contents of any container during a spoilage examination because of the potential presence of botulinum toxin or the presence of other pathogens.

1. **Odor.** Compare the odor of the suspect samples to the odor of the control. This procedure aids in the detection of an off-odor. Determine the odor characteristics of each opened sample unit. The following terms are commonly used in spoilage examinations:

acidic	fermented	rotten egg
butyric	medicinal	sharp
cheesy	metallic	sour
fecal	putrid	sweet
yeasty		

As a guideline, putrid odors suggest the presence of mesophilic, anaerobic sporeformers, and, if found, the samples should be tested for botulinum toxin. However, non-proteolytic, less heat-resistant *C. botulinum* strains may be encountered that do not produce putrefactive odor. (For a reference of the putrefactive odor characteristic of anaerobic spoilage, smell a laboratory culture of non-toxic Putrefactive Anaerobe #3679 (*C. sporogenes-like)* which may be obtained from ATCC.) Fecal odors are generally from coliform bacteria and are indicative of post-process contamination. Sharp, sour odors are usually caused by acid production by bacteria. "Rotten egg" odor is caused by "sulfide stinker" spoilage *(D. nigrificans*), usually accompanied by dark discoloration of the product. These odors do not necessarily indicate the presence of toxin.

2. **Appearance.** Compare the appearance of the suspect samples to that of the normal controls. Categorize appearance according to the following characteristics:

Appearance	Characteristic
Tough, hard, etc.	Cooked vs. uncooked
Off-color	Light/dark, etc.
Texture	Softened, digested, slimy
Consistency	Fluid, viscous, ropy, liquefied, coagulated

Record the odor and characteristic on the data sheet.

62.329 Microscopic Examination

Examine the contents of each suspect and each control sample microscopically. If using a bright field microscope (not phase contrast), using a flamed loop, make a smear of the contents from each container on clean microscope slides. Let the smear air dry, flame fix, stain with crystal violet (0.5% to 1% solution) and dry. The staining will allow spores to be visible, if present. If a smear will not adhere to the slide during staining (sometimes encountered with fatty foods), rinse a second smear with xylol and allow to dry before staining with crystal violet. If a phase contrast microscope is available, examine a wet amount prepared by transferring a small drop of food slurry to a slide and overlaying it with a cover slip. Examine slides at 1000× under the oil immersion objective. Endospores appear brightly refractile under phase lighting.

Never attempt to gram stain a product because a gram stain result depends on the age of the culture. A gram stain should be made on an 18- to 24-hr culture only.

Record observations on the data sheet. Common results and interpretations are as follows:

1. A normal product may exhibit a few microbial cells in every microscopic field examined. When examining a suspect sample, the observation of a higher number of microbial cells per field, when compared to the normal product, is usually an indication of microbial activity. Fermented foods, such as sauerkraut, are an exception because numerous dead cells will be present. Cultural recovery of viable microorganisms in pasteurized, fermented foods is necessary to confirm spoilage of microbial origin.
2. Mixtures of rods, cocci, cocco-bacilli, yeast or molds usually indicate post-process contamination, but also may indicate that the product was underprocessed.
3. Pure cultures of medium to long rods with or without detectable spores (directly or indirectly in subculture) may indicate post-process contamination, especially if the morphology varies from sample to sample. If subsequent tests indicate that the container is structurally sound and no post-processing rough handling of wet cans was observed in the plant, insufficient processing may have occurred.
4. Pure cultures of medium to large rods having sporangia or free spores may indicate insufficient processing or leakage. If typical "clostridial" rods containing subterminal spores, or only free spores are present, or if the odor of the product is putrid or "putrefactive," toxin tests by mouse inoculation (bioassay) should be done immediately (Chapter 33).
5. When mixed or pure cultures of cells are observed microscopically, but viable microorganisms are not recovered in subsequent cultural tests, "incipient spoilage" may be indicated (except for fermented foods).

Record all observations on the data sheet.

62.33 pH Determination

Determine the pH of each sample unit by insertion of the electrode(s) of a previously standardized pH meter, directly into the product in the container. Record pH on the data sheet. The procedure is described in 21 CFR 114.9.[10]

After measuring the pH of each sample, clean the electrode(s) thoroughly. Use separate electrodes to measure the pH of foods that may be eaten and foods to be tested for spoilage to avoid cross contamination. Compare pH measurements of suspect and control samples. In addition, compare all results to the normal pH range given for the particular food provided in Table 5 or from product specifications.

62.331 Disposal of Contents

1. Samples suspected of having toxic spoilage: when one or more sample units are suspected of containing mesophilic, anaerobic spoilage and/or toxin, transfer the contents to another container (e.g. large beaker) and autoclave them. Using rubber gloves, rinse the sample container with 100 ppm chlorine solution and air dry at 35°C. Do not autoclave the sample container since this may change the container's lid sealing compound.
2. Low-acid foods (pH above 4.6) exhibiting no signs of toxic spoilage: empty contents into a metal tray and dispose of the contents appropriately.
3. Acid food (pH 4.6 or lower): empty contents into an appropriate disposal container and dispose of the contents appropriately.

62.332 Cleaning of Container

Wash the interior of the container with warm water and soap, using a bottle brush if necessary to remove food residues from interior seam surfaces. Rinse, soak cans in hot water for 30 minutes, rinse with hot water and drain. If available, an automatic dishwasher may be used to clean empty cans and bottles, however, do not run through the dry cycle, because heat may be too high and may change the container's seaming compound.

Dry the containers by shaking out excess water from the interior of the can and placing the container in an air incubator and drying for 24 hr at 35°C. (Higher temperatures may cause changes in the seaming compound). Containers must be dry before additional testing.

62.333 Container Examination

1. **Superficial examination.** Examine the interior of the can for patches of black or dark gray discoloration, especially along the side and end seams. Patches of discoloration indicate localized detinning and pitting that are due to chemical reaction of the can contents with the container, usually producing a hydrogen swell. Detinning may be due to leakage of oxygen into the can, faulty tin plate, abnormally corrosive contents or excessive age of the product.

   In plain (unenameled) cans, the tin may be completely corroded from the inner surface of the can and internal surfaces will be dark gray. Hydrogen swells are not dependent on concurrent microbiological spoilage, but post-process contamination spoilage may follow when corrosion has caused the container to perforate.

   For plastic trays and pouches, examine for seal imperfections, pinholes, and cracks or cuts. Flex cracks in the foil layer usually do not leak through the plastic.[33,39]
2. **Detailed examination.** Detailed container examination is necessary when containers have not exhibited obvious defects, especially when post-process contamination spoilage must be clearly differentiated from that caused by insufficient processing.

   The detailed examination requires trained personnel and adequate facilities to conduct leakage testing, either using the vacuum microleak[1] or air pressure method, preferably both methods should be used. Once leak-tested, the double seams and side seams should be evaluated and torn down.[20,23]

   When the expertise for conducting detailed container examinations is not available, it is advisable to seek assistance.

62.34 Materials

Dextrose tryptone agar
Dextrose tryptone broth
PE-2 (peptone yeast extract broth containing peas)
Cooked meat medium (consisting of liver or beef heart infusion broth)
Nutrient agar containing manganese
Liver veal agar
Orange serum broth (acidified to pH 5.5)
Thermoacidurans agar (acidified to pH 5.0) or deep tubes
Potato dextrose agar (acidified to pH 3.5)
Phytone yeast extract agar with streptomycin and chroamphenicol
APT agar
K medium
3% Hydrogen peroxide

62.341 Microbiological Culture Procedures

Microbiological culture procedures are different for the examination of low-acid and acid and acidified foods. For this reason two separate examination sequences are provided: Section 62.342 for low-acid food samples having a normal product pH above 4.6; and Section 62.343 for acid and acidified food samples having a normal product pH of 4.6 or lower.

Each section contains a data sheet for recording results during the course of the microbiological examination. When microbiological tests are completed, compare the results, along with information from Sections 62.31 and 62.32, to the Spoilage Diagnosis Keys provided in Section 62.4 (Tables 3 and 4).

Isolates obtained from microbiological tests, along with the reserve sample of product, should be saved in case further work is necessary.

62.342 Cultural Tests for Low-Acid Foods (normal pH above 4.6)

1. **Description of method.** This method employs primary media for the recovery of spoilage microorganisms and subculture media to identify growth characteristics and further characterize them into spoilage groups. Differentiation into spoilage groups is fundamentally dependent on the growth of isolates under aerobic or anaerobic conditions, growth response to 30-35°C (mesophilic temperatures) and at 55°C (thermophilic temperatures) and morphological differentiation of vegetative and sporeforming cultures. Other determinative factors, including acid and gas production, odor of the culture and specific cell morphology (cocci, cocco-bacilli, rods, yeast, etc.) are also utilized. Note that for mesophilic incubation temperatures, 30 or 35°C can be used. Different laboratories have incubators set at their preferred temperature.

   The purpose, description and abbreviated designation for each medium employed is given in the following tables. In several instances, a choice of media commonly used is given. These may be used interchangeably at the option of the investigator.

2. **Primary recovery media.**

Purpose	Primary recovery media	Abbreviation
Aerobic recovery media	Dextrose tryptone (bromcresol purple) agar in poured plates[14]	DTA[a]
	Alternate: Dextrose tryptone (bromcresol purple) broth in tubes	DTB[b]
Anaerobic recovery media	Peptone yeast extract broth containing bromcresol purple indicator and whole peas in tubes[13]	PE-2[c]
	Alternate: Cooked meat medium, consisting of liver or beef heart infusion broth in tubes	CMM[c]

[a] Prior to inoculation the surface of the agar should be free from excess moisture. Plates should be taped to prevent loss of moisture. Product with low pH will also cause media to change from purple to yellow, but this is usually noted immediately after inoculation. Acid production turns media from purple to yellow and may indicate a flat-sour thermophilic isolate, if growth occurs at 55°C. Isolates are characterized morphologically by microscopic examination of colonies.

[b] Anaerobes may grow in the lower portion of the tube. Use of a larger food sample is the chief advantage of DTB over DTA.

[c] All "CMM" media must be exhausted prior to use by heating to 100°C for 20 min to remove oxygen, and rapidly tempered (without agitation) to the intended incubation temperature (30–35°C or 55°C). After inoculation, the surface of the medium must be layered (stratified) with sterile agar or vaspar to maintain anaerobic conditions or incubated anaerobically. Gas production is evidenced by bubbles in the medium or gas accumulation under the agar or vaspar seal. PE-2 may be used directly, without exhausting.

3. **Subculture media.**

Purpose	Subculture media	Abbreviation
Growth and sporulation of aerobes	Nutrient agar containing manganese in poured plates[28]	NAMn[d]
Growth and sporulation of anaerobes (under strict anaerobic conditions)	Liver-veal agar in poured plates[15]	LVA[e]

[d] NAMn supports growth and enhances spore production by aerobic sporeformers[28,9,11] and is used primarily to differentiate mesophilic from thermophilic *Bacillus spp.* When rod-shaped aerobes in pure culture are isolated on DTA (or DTB) and sporulation is not evident, but there is reason to believe sporeformers are involved in the spoilage, the isolates should be subcultured on NAMn at the temperature of initial isolation. After incubation up to 10 days, if spore production has taken place, the spores are heat shocked to destroy all vegetative cells and cultured again on NAMn at both 30°C and 55°C. The temperature at which outgrowth occurs from the spore state indicates whether the isolate is an obligate mesophile (growth at 30°C), an obligate thermophile (growth at 55°C) or a facultative thermophile (growth at 30°C and at 55°C).

[e] LVA plates are used to differentiate anaerobes from aerobes. Isolates from PE-2 or CMM are streaked on LVA and incubated aerobically and anaerobically (in anaerobic jar or incubator). The medium supports sporulation of the anaerobic *Clostridium spp.*, but it is a good growth medium and under aerobic conditions will grow a great variety of aerobes as well. Prior to inoculation, the surface of the agar should be free from moisture.

   All results should be recorded on the Worksheet for Cultural Test Results for Low-Acid Foods (Figure 4).

4. **Cultural examination sequence.** Figures 5 and 6 summarize the aerobic and anaerobic examination sequences for low-acid food samples. Review of these sequences prior to utilization of the stepwise testing procedure will aid in determining the extent of the examination. Often negative cultural results will preclude the need for further testing.

   It should be noted that all tests starting with DTA or DTB media are designed to identify groups of aerobic microorganisms. All tests starting with PE-2 or CMM media are for identifying anaerobic microorganisms.

   Maximum incubation periods are shown in the diagrams; usually growth will take place in a much shorter time and continued incubation is unnecessary (exception, NAMn).

5. **Procedure: Recovery methods.**

   a. Select one of the two retained samples (taken aseptically in Section 62.327) for cultural testing. Choose samples that are representative of the various initial examination findings. Be sure to test all samples with putrid odor or elevated product pH. Solid and semi-solid samples may need to be diluted prior to subculturing. Save the other retained sample in the refrigerator in case additional cultural tests are necessary. For each retained sample selected for cultural tests, prepare a set of the following media. Note that duplicate media are listed for each test.

      Label four DTA plates or DTB aerobic media tubes with sample number and date. Mark two plates or tubes "30 (or 35)°C" and two "55°C."

      Label four PE-2 or CMM anaerobic media tubes with sample number and date. Mark two tubes "30 (or 35)°C" and two tubes "55°C." CMM tubes must be exhausted by heating at 100°C for 20 minutes and immediately tempered (without agitation) to the intended incubation temperature before use.

   b. Inoculate each DTA plate by transferring a small quantity of the food, via a flame-sterilized loop, to the surface of the plate. Streak to dilute out the inoculum across the

Sample No.:__________ Code: ________________ Date Received: ______________

Culture stage (Reference section)	Medium	Number of tubes or plates per test	Incubation temperature (°C)	Incubation Time (days)	Results of cultural tests[a]								
								Cell morphology					
					Growth	Acid	Gas[b]	Rods	Spores	Cocci	Yeast	Mold	Additional comments
Aerobic Recovery (62.341 2 & 6)	DTA (or DTB)	2	30-35	4									
		2	55	4[c]									
Subculture #1 (62.342 3 & 6)	NAMn	2	30-35	10									
		2	55	10									
Subculture #2 Heat Shock (62.342 3 & 6a)	NAMn	2	30-35	4									
		2	55	4									
Anaerobic Recovery (62.342 2 & 5)	PE-2 (or CMM)	2	30-35	10									
		2	55	4[c]									
Subculture (62.342 3 and 6b)	LVA (Aerobic)	2	30-35	4									
	LVA (Anaerobic)	2	30-35	4									

[a] "-" = Negative result, e.g., no growth, no acid, etc. "+" = Positive result, e.g. growth, acid, etc.
[b] Note odor of gas or culture in "Additional comments," particularly from PE-2/CMM and LVA cultures
[c] Preheat tubes prior to incubating

Figure 4. *Worksheet for cultural test results for low-acid foods (pH above 4.6)*

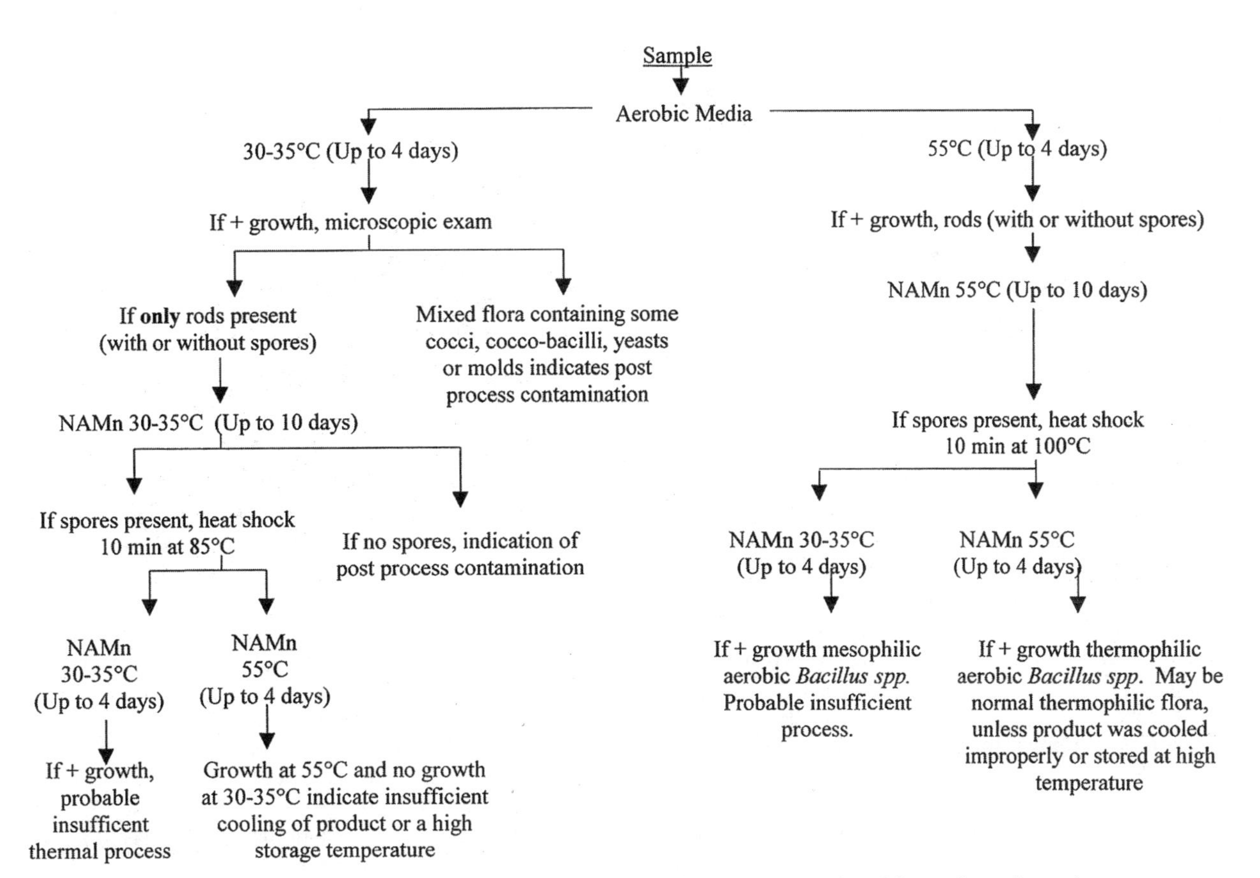

Figure 5. *Flow diagram for the aerobic cultural examination of low-acid canned foods (pH 4.6 or above) for spoilage diagnosis*

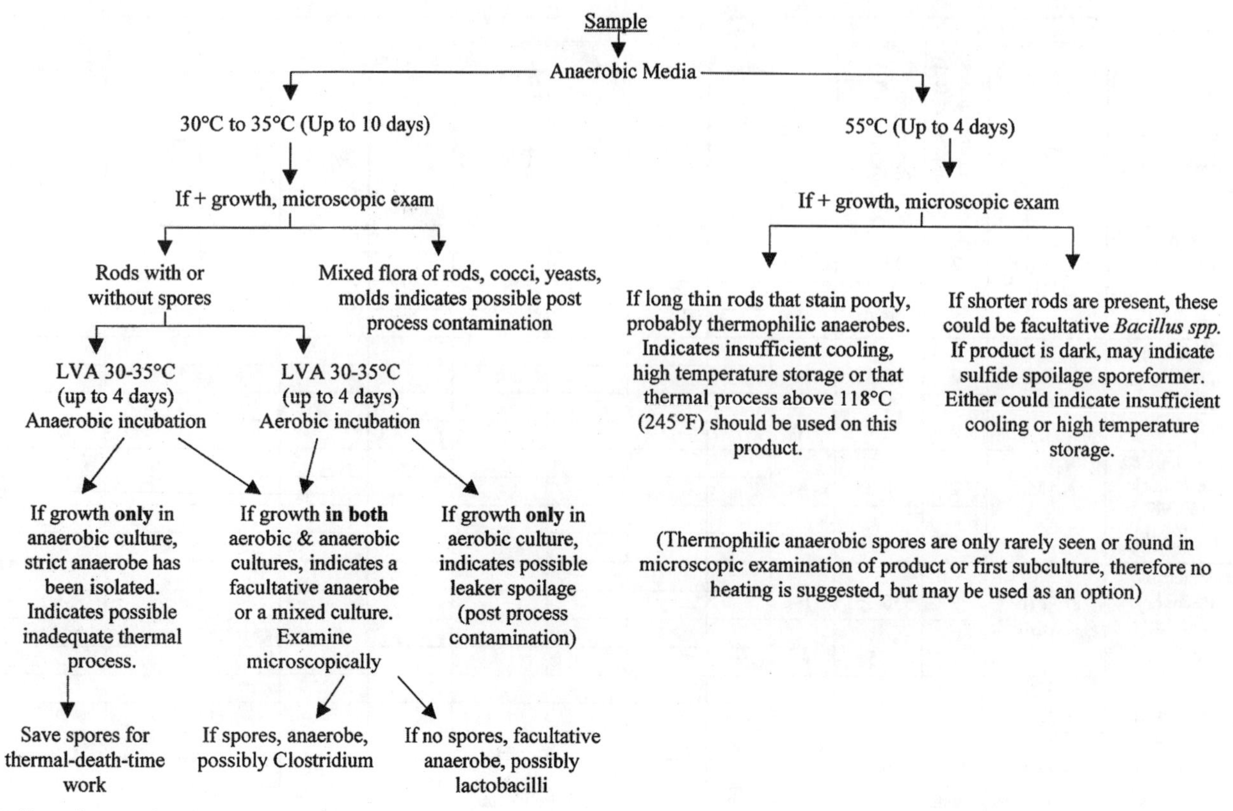

Figure 6. *Flow diagram for the anaerobic cultural examination of low-acid (pH above 4.6) canned foods for spoilage diagnosis.*

surface of the plate. If DTB tubes are used, aseptically transfer about 1 mL or 1 g of food into the broth. Inoculate each tube of PE-2 or CMM anaerobic media by aseptically transferring about 1 mL or 1 g of food into the broth below its surface. Layer the surface of the four inoculated tubes of CMM with 1/2 inch of sterile vaspar or agar to provide anaerobic conditions.

c. Incubate media as follows: Place DTA (or DTB) and PE-2 (or CMM) media, labeled "30 (or 35)°C" in an air incubator having a temperature of 30 or 35°C. Incubate DTA (or DTB) aerobic media up to 4 days; and PE-2 (or CMM) anaerobic media up to 10 days. Place DTA plates labeled "55°C" in a 55°C incubator for 4 days.
d. Preheat all tubes to be incubated at 55°C in a water bath prior to placing in incubator. Pre-warming is necessary to avoid growth and gas production by rapidly growing mesophiles before the tube contents reach the thermophilic temperatures in the incubator. Incubate thermophilic media up to 4 days.
e. Record detailed results on the data sheet in spaces provided for the recovery media (Figure 5). If high loads of organisms were observed during microscopic examination of the product, and few cells are seen in the culture, this may result from carryover and not growth, and should be noted.

Growth in DTA (or DTB) aerobic media: Record growth at the different temperatures based on acid production as indicated by bromcresol purple dye turning yellow or by visible growth when examined microscopically under oil immersion at about 1000× magnification (crystal violet stain or phase contrast microscopy). A gram stain may be made on 18- to 24-hr cultures only.

If growth consists of mixed cultures of rods, cocci, coccobacilli, yeast or molds, then leakage or gross underprocessing is indicated and further culturing is unnecessary. Cocci, yeast or molds in pure culture may also indicate post-process contamination or gross underprocessing. Very short (tiny) rods (cocco-bacilli) are also an indication of post-process contamination. Generally, rods of various widths also suggest post-process contamination, but additional testing is necessary to confirm this. If only rod-shaped bacteria are present, with or without spores, continue with subculture sequence given in Section 6a below.

Growth in PE-2 (or CMM) anaerobic media: Record presence and temperature(s) of growth, acid production (noted in PE-2 medium only, because of color change of bromcresol purple indicator) and presence of gas under the agar or vaspar seal. Note the odor. Examine growth microscopically under oil immersion at about 1000× (crystal violet stain or phase microscopy). During microscopic examination observe for clostridial morphology. Spores are brightly refractile, especially under phase microscopy.

If rod-shaped bacteria, with or without spores, are observed in PE-2 (or CMM) anaerobic media incubated at 30°C, continue with the subculture sequence given in Section 6b. Save the PE-2 (or CMM) subcultures, because if anaerobic growth is observed in the subculture from the PE-2 (or CMM) tubes, it must be tested for the presence of botulinum toxin (Chapter 33). See Figure 6.

6. **Procedure: Subculture methods.**
   a. The presence of mesophilic or thermophilic aerobic sporeformers may be established as follows (see also Figure 5):

- If only rod-shaped bacteria with or without spores were observed on DTA or DTB aerobic media, they should be subcultured on NAMn plates as follows:
- Streak cultures recovered at 30–35°C on DTA or DTB aerobic media onto two NAMn plates. Incubate the NAMn subculture at 30–35°C only.
- Streak cultures recovered at 55°C on DTA or DTB aerobic media onto two NAMn plates. Incubate the NAMn subculture at 55°C only.
- Allow NAMn subcultures to incubate up to 10 days to ensure sporulation. Verify spore formation by microscopic examination of all cultures before proceeding.
- Place two loopfuls of a colony containing spores from each NAMn plate incubated at 30–35°C and/or 55°C into separate sterile 16 mm glass tubes containing 10 mL of sterile distilled water. Avoid touching the sides of the tubes with the loop. Swirl tubes gently to disperse culture. Place the tubes in a 85°C water-bath for 10 min to heat-shock mesophilic spores and to destroy vegetative cells. Heat-shocking thermophilic spores will require higher temperatures (Chapters 25, 26), usually 100°C for 10 minutes. The level of the spore suspension in the tube should be 1 inch below the level of the water in the water-bath.
- Following heat-shock of the suspensions, streak a loopful from each tube onto each of four fresh NAMn plates. Incubate two plates at 30–35°C and two at 55°C. Incubate up to 4 days at each temperature and observe for presence or absence of growth. Examine microscopically for spores and record findings on data sheet.

b. The presence of mesophilic, obligately anaerobic sporeformers (clostridia) may be established as follows (see also Figure 6):

- If rod-shaped bacteria occur in anaerobic media incubated at 30–35°C, streak four LVA plates from each positive tube in a manner to dilute the inoculum. Incubate two of the LVA plates aerobically, in a 30–35°C incubator for up to 4 days. Incubate the other two plates in an anaerobic incubator at 30–35°C for up to 4 days. In the latter case, ensure that anaerobic conditions are maintained by use of a suitable oxidation-reduction indicator.
- After 2 to 4 days incubation, examine the aerobic and anaerobic plates for growth. Examine the cultures microscopically and record cell morphology and presence or absence of spores. Note and record the odor of the anaerobic culture (describe the odor as in Section 62.328.1). If the odor is described as putrid, clostridia may be suspected and the mouse bioassay should be done (Chapter 33).
- The following interpretations and tests are dependent on the results of subculture growth reactions in LVA plates incubated at 30–35°C under aerobic or anaerobic conditions.
- When growth occurs only on the aerobic LVA plates, an aerobe has been isolated. No further testing is necessary since anaerobic isolates were not recovered.
- When growth occurs only on the anaerobic LVA plates, a strict anaerobe has been isolated. A *Clostridium* species is indicated if microbiological examination reveals the presence of sporeforming cells having clostridial morphology (swollen rod-shaped cells containing subterminal spores and/or free spores). Growth on the anaerobic LVA plate, particularly with a putrid odor, necessitates testing the 30–35°C anaerobic isolation media for the presence of *C. botulinum* toxin by the mouse inoculation bioassay procedure (Chapter 33).
- Growth on both the aerobic and anaerobic LVA plates indicates the presence of an aerobe and an anaerobe, a facultative anaerobe or a mixture of all three. Since toxic *C. botulinum* may be present on the anaerobic LVA plate, examine the culture microscopically and, if appropriate, conduct tests for botulinum toxin (Chapter 33). Growth on the aerobic and anaerobic LVA plates may also consist of a pure culture of asporogenous, facultatively anaerobic lactobacilli. Microscopically, these organisms appear as slender rods. They will be present in instances of post-process contamination and are non-pathogenic.

62.343 Cultural Tests for Acid or Acidified Foods (normal pH of 4.6 or lower)

1. **Method.** Aciduric microorganisms are the only types capable of growing in a product having a pH of 4.6 or lower. For this reason, the subculture media must be selective for the spoilage type involved, i.e., must approach the normal pH of the food to eliminate the possibility of isolating viable organisms that are dormant in the product, but are capable of growing in a more neutral medium.[2]

   The general media of choice for recovering most acid-tolerant microorganisms are orange serum broth acidified to pH 5.5[3] or thermoacidurans agar plates acidified to pH 5.0.[30] It should be noted, however, that because the pH of these media are somewhat higher than 4.6, microorganisms may be recovered occasionally that are normally suppressed in the acid product. It is common to recover spoilage microorganisms from specific acid products when cultured in media consisting of the normal product that has been dispensed into culture tubes and sterilized. Hence, when the investigator is dealing exclusively with spoilage diagnosis of a specific acid product, it is recommended that re-inoculation of spoilage microorganisms in sterilized, unspoiled product be considered to confirm its ability to grow in the product.

   The following media are also useful to recover specific types of spoilage organisms or to differentiate microorganisms in acid foods (see also Table 2).

   Potato dextrose agar acidified to pH 3.5 or phytone yeast extract agar with streptomycin and chloramphenicol[5] are suitable for the recovery of yeast and molds without the interference of bacteria.

   Thermoacidurans agar acidified to pH 5.0 is useful for recovery of *B. coagulans* and butyric anaerobes.
   Unacidified dextrose tryptone agar with bromcresol purple indicator may be employed to demonstrate acid producers and unacidified APT agar may be used to support growth of fastidious lactic acid bacteria. These media also support the growth of acid-intolerant microorganisms and should be used in conjunction with acidified media.
2. **Procedure.**
   a. Select one of the retained samples (taken aseptically in Section 62.327) for cultural testing. Save the other sample in the refrigerator in case specific cultural tests become necessary. For each retained sample selected for cultural tests, assemble the following media (includes duplicate media per test):
   - Four orange serum broth (OSB) tubes or thermoacidurans agar (TAA) plates

Table 2. Cultural Media for Acid or Acidified Foods (pH of 4.6 or Lower)

Purpose	Medium	Abbreviation	Incubation temperature (°C)
General recovery	Orange serum broth[21] pH 5.5 (tubes)[a]	OSB	30–35
	Alternative: Thermoacidurans agar, pH 5.0 (plates)[37]	TAA	
Facultative thermophilic aerobes; aciduric flat sours *B. coagulans*	Orange serum broth[21] pH 5.5 (tubes)	OSB	55
	Thermoacidurans agar[37] pH 5.0 (plates)	TAA	30–35
Butyric acid anaerobes	Thermoacidurans agar[37] pH 5.0 (deep tubes)	TAD	30–35
Yeasts and molds. Especially heat-resistant mold *Byssochlamys fulva*	Potato dextrose agar, pH 3.5 (plates)	PDA	25–30
	Alternative: Phytone yeast extract agar with streptomycin and chloramphenicol[39] (plates)	PYE	
Aerobes and acid formers	Dextrose tryptone agar with bromcresol purple indicator (plates)	DTA	30–35
Fastidious lactic acid bacteria	Unacidified APT agar[8] (plates)[b]	APT	30–35
Alicyclobacillus spp.	K medium[c]	K medium	43

[a] If butyric acid anaerobes are indicated by a butyric odor of the product, or if spoilage by thermophilic anaerobes is indicated by high temperature storage conditions, carbon dioxide and hydrogen, and presence of sporeforming rods, layer the OSB with sterile vaspar or acidified agar to produce anaerobic conditions, or use thermoacidurans agar (pH 5.0) deep tubes.

[b] The test for catalase consists of flooding growth on this medium with a 3% solution of hydrogen peroxide. Catalase negative isolates belong to the lactic group (lactobacilli, streptococci, pediococci). This test must be performed on fresh cultures.

[c] See Chapter 24 for additional information.

- Two thermoacidurans agar (TAD) deep tubes if butyric anaerobe spoilage is suspected
- Two potato dextrose agar (PDA) and phytone yeast extract agar (PYE) plates if yeast and/or mold spoilage is suspected
- Use PDA if spoilage from heat-resistant mold *(B. fulva)* is suspected in fruit and fruit products (Chapter 21)
- Two DTA plates if aerobic acid formers are suspected. Two APT plates if fastidious lactobacilli are suspected, especially in flat containers
- Use K medium if spoilage from *Alicyclobacillus spp.* is suspected in fruit juice or fruit juice concentrates (Chapter 24)
- Label all tubes and plates with sample number and date.

b. Inoculate OSB tubes with about 1 mL or 1 g of food sample. Note and record if the medium becomes turbid from the food at time of inoculation. If TAA plates are used, streak each with a loopful of food as described.

c. Inoculate optional media by transferring a small amount of food to the plate using a flame-sterilized loop. Streak across media to dilute out the inoculum. Use a pipette for inoculation of TAD tubes; place 1 mL in bottom of melted 47° to 50°C agar, and cool rapidly.

d. Incubate all samples at temperatures indicated on the data collection form. Heat all tubed media intended for 55°C incubation at 55°C in a water-bath immediately after inoculation and before placing in the incubator.

e. If growth occurs after incubation, note the following conditions:

- Presence of acid in media containing bromcresol purple indicator (purple to yellow). Note that the sample may have already turned it yellow.
- Presence of gas in liquid media (bubbles) or in layered (OSB, TAD deep tubes) anaerobic media (gas under plug).
- Prepare a crystal violet stain or a wet mount for phase microscopy as soon as growth is apparent. Look for the presence of refractile spores in sporangia and/or free spores, as well as the morphological forms of nonsporeforming microorganisms.
- If DTA or APT medium was used, flood with a 3% solution of hydrogen peroxide. A positive catalase test is indicated by gas bubbles around the colonies; a negative catalase reaction is indicated by the absence of bubbles. Lactic acid bacteria are indicated by negative catalase reaction. The lactic acid group of bacteria includes lactobacilli, streptococci and pediococci.
- Enter results on Worksheet for Cultural Test Results for Acid or Acidified Foods (Figure 7) and compare to diagnostic keys in Section 62.4 (Table 5).

62.4 KEYS TO PROBABLE CAUSE OF SPOILAGE

Refer to the spoilage keys shown in Tables 3 and 4 to diagnose a cause of spoilage for each type of product. Each key categorizes possible spoilage conditions for food within a given pH range. The normal pH of the product being examined must be used when referring to the charts (not the spoiled product pH). Normal pH ranges of various commercially canned foods are shown in Table 5.

62.5 ADDRESSES FOR SOME FIRMS LISTED IN REFERENCES (SECTION 62.6)

AOAC International, 481 North Frederick Avenue, Suite 500, Gaithersburg, MD 20877

Food Processors Institute (FPI), 1350 "I" Street, NW, Suite 300, Washington, DC 20005

National Food Processors Association (NFPA), (previously NCA) 1350 "I" Street, NW Suite 300, Washington, DC 20005-3005

General Services Administration, Specifications Section, Room 6654, 7th and D Sts. SW, Washington, DC 20407

Crown Cork and Seal Company Inc., One Crown Way, Philadelphia, PA 19114

62.6 REFERENCES

1. AOAC/NFPA/FDA. 1989. "Classification of Visible Flexible Package Defects (Exterior). Association of Official Analytical Chemists/National Food Processors Association/Food and Drug Administration. AOAC International, Gaithersburg, MD.
2. APHA. 1966. Canned Foods. In: Recommended Methods for the Microbiological Examination of Foods. 2nd Edition. Ed. J. M. Scharf. American Public Health Association., NY. 1966. p. 40.
3. APHA. 1966. Appendix A, Culture—Media, #48. Orange serum agar. In: Recommended Methods for the Microbiological Examina-

Sample No: ________	Code: ________	Date: ________								
		Results								
	Incubation Temperature (°C) (Incubate up to 5 days)						Bacterial			
Medium		Growth	Acid	Gas	Mold	Yeast	Cocci	Rods	Spores	Catalase + or -
OSB[a] or TAA	30 – 35									
OSB[a] or TAA	55									
	30 – 35									
TAD	30 – 55									
PDA or PYE	25 – 35									
DTA[b] (opt)	30 – 35									
APT[b] (opt)	30 – 35									

[a]Note if OSB is layered to obtain anaerobic conditions.
[b]Incubation of DTA and/or APT in an anaerobic jar is useful when very poor growth is produced by aerobic incubation procedures. Permit plates to stand exposed to air for a minimum of 30 min before testing for catalase.

Figure 7. *Worksheet for Cultural test results for acid or acidified foods (pH 4.6 or lower)*

tion of Foods. 2nd Ed., Ed. J. M. Scharf, American Public Health Association, NY. p. 178.

4. Ball, C. O. And Olson, F. C. W. 1957. "Sterilization in Food Technology: Theory, Practice and Calculations." McGraw-Hill Book Co., NY.
5. BBL. Phytone yeast extract agar. In: Manual of Products and Laboratory Procedures. 6th Edition. Becton Dickinson Microbiol. Systems, Cockeysville, MD. 1988. p. 224.
6. Bee, G. R., DeCamp, R. A., and Denny, C. B. 1972. "Construction and Use of a Vacuum Micro-leak Detector for Metal and Glass Containers." National Canners Association/National Food Processors Association, Washington, D.C.
7. Cameron, E. J. 1940. Report on canned vegetables. J. Assn. Off. Anal. Chem. 23:607.
8. Cameron, E. J. and Esty, J. R. 1940. Comments on the microbiology of spoilage in canned foods. Food Res. 5:549.
9. Charney, J., Fisher, W. P., and Hepny, C. P, 1951. Manganese as an essential element for sporulation in the genus Bacillus. J. Bacteriol. 62:145.
10. Corson, L. M., Evancho, G. M., and Ashton, D. H., 1973, Use of forceps in sterility testing: A possible source of contamination. J. Food Sci. 38:1267.
11. Curran, H. R. and Evans, F. R. 1954, The influence of iron or manganese upon the formation of spores by mesophilic aerobes in fluid organic media. J. Bacteriol. 67:489.
12. Denny, C. B. 1970. Collaborative study of procedure for determining commercial sterility of low-acid canned foods. J. Assn. Off. Anal. Chem. 53:713.
13. Denny, C. B. 1972. Collaborative study of a method for the determination of commercial sterility of low-acid canned foods. J. Assn. Off. Anal. Chem. 61:613.
14. Difco. 1984. Dextrose tryptone broth (agar). In: Difco Manual of Dehydrated Culture Media and Reagents for Microbiological and Clinical Laboratory Procedures. 10th Edition. Difco Laboratories, Inc., Detroit, MI. p. 284.
15. Difco. 1984. Liver veal agar. In: Difco Manual of Dehydrated Culture Media and Reagents for Microbiological and Clinical Laboratory Procedures. 10th Edition. Difco Laboratories, Inc., Detroit, MI. p. 528.
16. Doyle, J. E. and Ernst, R. E. 1969. Alcohol flaming-a possible cause of contamination in sterility testing. Amer. Clin. Pathol. 51:507.
17. Evans, J. B. and Niven, C. F. Jr. 1951. Nutrition of the heterofermentative lactobacilli that cause greening of cured meat products. J. Bacteriol. 62:599.
18. FDA, 1995. *Clostridium botulinum*. In: Bacteriological Analytical Manual. 8th Edition. AOAC International, Gaithersburg, MD. 1995. Chapter 17.
19. FDA, 1995. Examination of Canned Foods. In: Bacteriological Analytical Manual. 8th Edition. AOAC International, Gaithersburg, MD. 1995. Chapter 21A.
20. FDA, 1995. Examination of Containers for Integrity. In: Bacteriological Analytical Manual. 8th Edition. AOAC International, Gaithersburg, MD. 1995. Chapter 22.
21. Food Processors Institute. Microbiology of Thermally Processed Foods. In: Canned Foods: Principles of Thermal Process Control, Acidification, and Container Closure Evaluation. 6th Edition. Food Processors Institute, Washington, D.C. Chapter 2.
22. Food Processors Institute. Food Container Handling. In: Canned Foods: Principles of Thermal Process Control, Acidification, and Container Closure Evaluation. Food Processors Institute, Washington, D.C. p. 27.
23. Food Processors Institute. 1995. Closures for Metal Containers. In: Canned Foods: Principles of Thermal Process Control, Acidification, and Container Closure Evaluation. 6th Edition. Food Processors Institute, Washington, D.C. p. 117.
24. Food Processors Institute. 1995. Closures for Glass Containers. In: Canned Foods: Principles of Thermal Process Control, Acidification, and Container Closure Evaluation. 6th Edition. Food Processors Institute, Washington, D.C. p. 141.

Table 3. Keys to Probable Cause of Spoilage in Low-Acid Foods (pH above 4.6)[31]

Condition of cans	Characteristics of material in cans						Diagnosis
	Odor	Appearance	Gas (CO_2 & H_2)	pH	Smear	Cultures	
Swells	Normal to "metallic"	Normal to frothy (cans usually etched or corroded)	More than 20% H_2	Normal occasional microorganisms	Negative to	Negative	Hydrogen swells
Swells	Sour	Frothy; possibly ropy brine	Mostly CO_2	Below normal	Pure or mixed cultures of rods, cocco-bacilli, cocci or yeast	Growth, aerobically and/or anaerobically at 30°C and possibly growth at 55°C	Leakage
Swells	Sour	Frothy; possibly ropy brine; food particles firm with uncooked appearance	Mostly CO_2	Below normal	Pure or mixed cultures of rods, cocco-bacilli, cocci or yeast	Growth, aerobically and/or anaerobically at 30°C and possibly growth at 55°C (If product received high exhaust, only sporeformers may be recovered.)	No process given
Swells	Normal to sour-cheesy	Frothy	H_2 and CO_2	Slightly to definitely below normal	Rods, usually granular; spores seldom seen	Gas anaerobically at 55°C and possibly slowly at 30°C Negative: Thermophilic anaerobes often autosterilize	Inadequate cooling or storage at elevated temps.—hermophilic anaerobes
Swells	Normal to cheesy to	Normal to frothy with disintegration of solid particles	Mostly CO_2 possibly some H_2	Normal to slightlybelow normal	Rods; possibly spores present	Gas anaerobically at 30°C. Putrid odor.	Insufficient processing—mesophilic anaerobes (possibly *C. botulinum*)
Swells	Slightly off —possibly ammoniacal	Normal to frothy	CO_2	Slightly to definitely below normal	Rods; occasionally pores observed	Growth, aerobically and/or anaerobically with gas at 30°C and possibly at 55°C. Pellicle in aerobic broth tubes. Spores formed on agar and in pellicle.	Insufficient processing or leakage (*B. subtillis* type)
Swells	Butyric acid volume gas	Frothy; large	H_2 and CO_2	Definitely below normal	Rods, bipolar staining; possibly spores	Gas anaerobically at 30°C. Butyric acid odor.	Insufficient processing (butyric acid anaerobes)
No vacuum and/or cans buckled	Normal	Normal	No H_2	Normal to slightly below normal	Negative to moderate number of micro organisms	Negative	Insufficient vacuum caused by: • incipient spoilage • insufficient exhaust • insufficient blanch • improper retort cooling procedures • overfill
Flat cans (0 to normal vacuum)	Sour	Normal to cloudy brine	No H_2	Slightly to definitely below normal	Rods, possibly granular in appearance	Growth without gas at 55°C and possibly at 30°C	Inadequate cooling or storage at elevated temps (thermophilic flat sours)
Flat cans (0 to normal vacuum)	Normal to sour	Normal to cloudy brine; possibly moldy	No H_2	Slightly to definitely below normal or mold	Pure or mixed cultures of rods cocco-bacilli, cocci	Growth, aerobically and/or anaerobically at 30°C and possibly at 55°C	Post-process contamination

Table 4. Keys to Probable Cause of Spoilage in Acid or Acidified Foods (pH 4.6 or lower)[31]

Condition of cans	Characteristics of material in cans						
	Odor	**Appearance**	**Gas (CO_2 & H_2)**	**pH**	**Smear**	**Cultures**	**Diagnosis**
Swells	Normal to "metallic"	Normal to frothy (cans usually etched or corroded)	More than 20% H_2	Normal micro-organisms	Negative to occasional	Negative	Hydrogen swells
Swells	Sour	Frothy; possibly ropy brine	Mostly CO_2	Below normal	Pure or mixed culturesof rods, cocco-bacillli, cocci or yeast	Growth, aerobically and/or anaerobically at 30°C and possibly at 55°C	Leakage or grossly insufficient processing
Swells	Sour	Frothy; possibly ropy brine; food particles firm	Mostly CO_2	Below normal	Pure or mixed cultures of rods, cocco-bacillli, cocci or yeast	Growth, aerobically and/or anaerobically at 30°C and possibly at 55°C. (If product received high exhaust, only sporeformers may be recovered)	No process given
Swells	Normal to sour-cheesy	Frothy	H_2 and CO_2	Normal to slightly below normal	Rods, medium short to medium long, usually granular; spores seldom seen	Gas anerobically at 55°C and possibly slowly at 30°C	Inadequate cooling or storage at elevated temperatures (thermo-philic anaerobes)
Swells	Butyric acid	Frothy; large volume gas	H_2 and CO_2	Below normal	Rods, bipolar staining; possibly spores	Gas anaerobically at 30°C. Butyric acid odor.	Insufficient processing, (butyric acid anaerobes)
Swells	Sour	Frothy	Mostly CO_2	Below normal	Short to long rods	Gas anaerobically; acid and possibly gas aerobically in broth tubes at 30°C. Possible growth at 55°C.	Grossly insufficient processing (lactobacilli) or post-process contamination
No vacuum and/or cans buckled	Normal	Normal	No H_2	Normal to slightly below normal	Negative to moderate number of microorganisms	Negative	Insufficient vacuum caused by: • Incipient spoilage • Insufficient exhaust • Insufficient blanch • Improper retort cooling procedure • Overfill
Flat cans (0 to normal vacuum)	Sour to "medicinal"	Normal	No H_2	Slightly to definitely below normal	Rods, possibly granular in appearance	Growth without gas at 55°C, and possibly at 30°C. Growth on thermoacidurans agar (pH 5.0)	Insufficient processing (*B. coagulans.*) Spoilage of this type usually limited to tomato juice.
Flat cans (0 to normal vacuum)	Normal to sour	Normal to cloudy brine; possibly moldy	No H_2	Slightly to definitely below normal	Pure or mixed cultures of rods, cocco-bacilli, cocci or mold	Growth aerobically and/or anaerobically at 30°C possibly at 55°C	Post process contamination or no process given

Table 5. Normal pH Ranges of Selected Commercially Canned Foods[24]

Kind of food	pH Range	Kind of food	pH Range
Apples, whole	3.4-3.5	Mackerel	5.9-6.2
Apples, juice	3.3-3.5	Milk	
Asparagus, green	5.0-5.8	Cow	6.4-6.8
Beans		evaporated	5.9-6.3
Baked	4.8-5.5	Molasses	5.0-5.4
Green	4.9-5.5	Mushrooms	6.0-6.5
Lima	5.4-6.3	Olives, ripe	5.9-7.3
Soy	6.0-6.6	Orange juice	3.0-4.0
Beans, with pork	5.1-5.8	Oysters	6.3-6.7
Beef, corned, hash	5.5-6.0	Peaches	3.4-4.2
Beets, whole	4.9-5.8	Pears (Bartlett)	3.8-4.6
Blackberries	3.0-3.3	Peas	5.6-6.5
Blueberries	3.2-3.6	Pickles	
Boysenberries	3.0-3.3	dill	2.6-3.8
Bread		sour	3.0-3.5
White	5.0-6.0	sweet	2.5-3.0
date and nut	5.1-5.6	Pimento	4.3-4.9
Broccoli	5.2-6.0	Pineapple	
Carrots, chopped	5.3-5.6	crushed	3.2-4.0
Carrot juice	5.2-5.8	sliced	3.5-4.1
Cheese		juice	3.4-3.7
Parmesan	5.2-5.3	Plums	2.8-3.0
Roquefort	4.7-4.8	Potatoes	
Cherry juice	3.4-3.6	White	5.4-5.9
Chicken	6.2-6.4	Mashed	5.1
Chicken with noodles	6.2-6.7	Potato salad	3.9-4.6
Chop suey	5.4-5.6	Prune juice	3.7-4.3
Cider	2.9-3.3	Pumpkin	5.2-5.5
Clams	5.9-7.1	Raspberries	2.9-3.7
Codfish	6.0-6.1	Rhubarb	2.9-3.3
Corn		Salmon	6.1-6.5
on the cob	6.1-6.8	Sardines	5.7-6.6
cream style	5.9-6.5	Sauerkraut	3.1-3.7
Whole grain		Juice	3.3-3.4
Brine packed	5.8-6.5	Shrimp	6.8-7.0
Vacuum packed	6.0-6.4	Soups	
Crab apples, spiced	3.3-3.7	bean	5.7-5.8
Cranberry		beef broth	6.0-6.2
Juice	2.5-2.7	chicken noodle	5.5-6.5
Sauce	2.3	clam chowder	5.6-5.9
Currant juice	3.0	duck	5.0-5.7
Dates	6.2-6.4	mushroom	6.3-6.7
Duck	6.0-6.1	noodle	5.6-5.8
Figs	4.9-5.0	oyster	6.5-6-9
Frankfurters	6.2	pea	5.7-6.2
Fruit cocktail	3.6-4.0	spinach	4.8-5.8
Gooseberries	2.8-3.1	tomato	4.2-5.2
Grapefruit		turtle	5.2-5.3
Juice	2.9-3.4	vegetable	4.7-5.6
Pulp	3.4	Squash	5.0-5.3
Sections	3.0-3.5	Strawberries	3.0-3.9
Grapes	3.5-4.5	Sweet potatoes	5.3-5.6
Hams, spiced	6.0-6.3	Tomatoes	4.1-4.4
Hominy, lye	6.9-7.9	juice	3.9-4.4
Huckleberries	2.8-2.9	Tuna	5.9-6.1
Jam, fruit	3.5-4.0	Turnip greens	5.4-5.6
Jellies, fruit	3.0-3.5	Vegetable juice	3.9-4.3
Lemons	2.2-2.4	Mixed vegetables	5.4-5.6
Juice	2.2-2.6	Vinegar	2.4-3.4
Lime juice	2.2-2.4	Youngberries	3.0-3.7
Loganberries	2.7-3.5		

25. Food Processors Institute. 1995. Closures for Semirigid and Flexible Containers. In: Canned Foods: Principles of Thermal Process Control, Acidification, and Container Closure Evaluation. 6th Edition. Food Processors Institute, Washington, D.C. p. 151.
26. General Services Administration, 1988. Clean room and work station requirements. Fed. Std. No. 209D, Specifications Section, General Services Administration, Washington, D.C.
27. Hartwell, R. R. 1951. Certain aspects of internal corrosion in tin plate containers. Adv. Food Res. 3:327.
28. Maunder, D. T. 1970. "Examination of Canned Foods for Microbial Spoilage." Microbiology, Metal Div. R. and D., Continental Can Co., Inc., Oak Brook, IL.
29. National Canners Association. February 1962. Diagnosis of Canned Food Spoilage. NCA Research Information. National Canners Association. Washington, D.C.
30. National Canners Association. 1968. Preparation of media for microorganisms. Thermoacidurans agar. In: Laboratory Manual for Food Canners and Processors. Vol. 1. Microbiology and Processing. National Canners Association, AVI, Westport, CT. p. 20.
31. National Canners Association. 1968. Investigating spoilage problems. In: Laboratory Manual for Food Canners and Processors. National Canners Association, AVI, Westport, CT. p. 45.
32. National Canners Association. 1968. Process calculations. In: Laboratory Manual for Food Canners and Processors. National Canners Association, AVI, Westport, CT. p. 220.
33. National Food Processors Association. 1989. Flexible Package Integrity Bulletin. Bulletin 41-L. NFPA, Washington, D.C.
34. Stern, R. M., Hegarty, C. P., Williams, 0. B. 1942. Detection of *Bacillus thermoacidurans* (Berry) in tomato juice and successful cultivation of the organism in the laboratory. Food Res. 7: 186.
35. Townsend, C. T., Yee, L., and Mercer, W. A. 1954. Inhibition of the growth of *Clostridium botulinum* by acidification. Food Res. 19:536.
36. U. S. Government, National Archives and Records Administration, 1998. Code of Federal Regulations, Title 21, Part 113.100(c), Thermally Processed Low Acid Foods Packaged in Hermetically Sealed Containers.
37. U.S. Government. National Archives and Records Administration 1998 Code of Federal Regulations, Title 21, Part 113.3(n). Thermally Processed Low Acid Foods Packaged in Hermetically Sealed Containers.
38. U. S. Government, National Archives and Records Administration, 1998. Code of Federal Regulations, Title 21, Part 113.60. Thermally Processed Low Acid Foods Packaged in Hermetically Sealed Containers. Acidified Foods.
39. U. S. Government, National Archives and Records Administration, 1998. Code of Federal Regulations, Title 21, Part 114.90. Food Standards.
40. U. S. Government, National Archives and Records Administration, 1998. Code of Federal Regulations, Title 21, Parts 130-169.
41. Vosti, D. C., Hernandez, H. H., and Strand, J. B. 1961. Analysis of headspace gases in canned foods by gas chromatography. Food Technol, 15:29.

Chapter 63

Media, Reagents, and Stains

Elsa A. Murano and Jennifer A. Hudnall

63.1 CULTURE MEDIA, INTRODUCTION

This chapter contains the formulae for all non-commercially available culture media mentioned in the text of the *Compendium* and the directions for making and using media. Commercially available media are listed and designated as commercially available by an asterisk (*) after the media names. Additional information is supplied for these media when requested by chapter authors. The names of the various media are those by which they are generally known in the literature, with a few exceptions where the name used was given by the author of the chapter referring to the medium. In the text, the recommended medium is referred to by name and listed alphabetically in Section 63.3 of this chapter.

63.11 Types of Media

In the study of microorganisms, several types of media are employed, depending on the intended purpose. *Maintenance media* are used to maintain pure cultures that are to be used to conduct further studies, or to be stored for prolonged periods of time. Such media are simple in terms of components, and do not contain compounds that will encourage luxurious growth. If glycerol is added, maintenance media can be used to freeze cultures at –70°C.

Selective media are used to select for certain microorganism(s) to the disadvantage of others in the same sample. Compounds are added to inhibit the growth of the undesirable microorganisms while not inhibiting the desirable ones. There is no medium that is universally selective for all pathogenic microorganisms, thus several kinds of selective media are commonly used to increase the chances of isolating the organism(s) of interest. *Differential media* contain an indicator, such as a dye, that will help distinguish between various organisms. This differentiation is based on the result of biochemical reactions carried out during growth by the microorganisms, which can change the color of the dye.

Selective media can be used to develop *enrichment media*. Enrichment media are liquid media into which samples are inoculated, and which require incubation periods during which the organisms of interest are selected for, while those that are not desirable are inhibited. The purpose of the media is to selectively augment the number of a certain microorganism in a mixed culture while suppressing the growth of others. Enrichment is usually necessary when the number of cells of a particular microorganism in a sample are too low to be detected by direct plating onto solid media.

Selective as well as differential media can be used in developing *isolation media*. In this case, a sample containing a mixed culture is plated onto solid media in order to detect the presence of a desired organism from among all organisms present. The medium is made with compounds that encourage the growth of wanted organisms, while also containing a dye or similar component to differentiate the organism of interest from other isolated colonies.

63.12 Forms of Media

In most cases, recommended media can be obtained in dehydrated form. The equivalent product of any manufacturer may be used and should be made in accordance with the manufacturer's instructions. Catalogs of media manufacturers contain descriptions of products with their formulations. Since it is difficult to define equivalency in objective terms, the user must test products in the laboratory as part of a routine laboratory quality assurance program to determine that the medium functions as expected. Dehydrated media should be employed unless the author specifically recommends that the user prepare the media from individual ingredients. In such cases, the laboratory should test the medium with sufficient cultures to determine that the performance of the preparation is satisfactory. In some instances, the name of a manufacturer is given to provide a source of the material mentioned in the formulation when that material is available from a single company or is not commonly employed in the laboratory.

63.121 Dehydrated Media

Dehydrated media should be stored in sealed containers in a cool, dry place, protected from light, or as instructed by the manufacturer. If the environment is hot and humid, media may be stored in a refrigerator or freezer, as preferred. Properly stored, most dehydrated media should be stable for at least three years; however, purchases should be planned to permit a complete turnover within a year or two.

After the first usage, when the seal has been broken, quality of the dehydrated medium may depend upon the storage environment. A suitable package size should be purchased to minimize repeated use of material from an unsealed container. Air and moisture entering an unsealed container can initiate reactions resulting in reduced productivity by the medium.

Care should also be taken to avoid chemical or microbiological contamination that may take place if unclean spatulas are used to transfer the material.

63.122 Prepared Media

The "life" of any prepared medium, whether in tubes, bottles, or plates, depends on the type of medium and the conditions of storage. Prepared media should not be stored unless protected against water loss. Cottonplugged containers do not adequately protect against water loss. Prepared plates should be stored in moisture-proof containers, such as plastic bags, to minimize moisture loss. Because of the instability of prepared culture media, it is advisable in most cases to use prepared plates no more than one week old, and media in screwcapped tubes no more than six months old. For exceptions (e.g., bismuth sulfite agar), the reader should refer to the formula and directions for making and using each medium found in this chapter and in catalogs and reference materials available from media manufacturers. If dehydration has occurred, media should be discarded. All prepared media, whether in plates or tubes, should be stored between 2 to 8°C unless recommended otherwise.

63.13 Common Media Constituents

Most media used in the maintenance, enrichment, and isolation of microorganisms contain one or more of the following components:

63.131 Agar

Obtained as flakes, granules, or powder, agar is made from seaweed. It has gelling properties that make it very useful in bacteriological work. It does not go into solution when mixed with water, thus it can be easily washed to remove impurities. In addition, once it is mixed with water by heating, it solidifies at room temperature, making it very easy to use in a variety of settings. It should be noted that repeated melting of medium containing agar, or prolonged sterilization by heating, will decrease gel strength.

63.132 Peptone

Enzyme or acid hydrolysis of animal or vegetable protein from muscle, liver, blood, milk, casein, or soya bean results in production of peptone. No two batches of peptone are exactly the same, which usually is not a factor for the growth of most microorganisms. However, certain batches are unsuitable for particular purposes, so peptone should be tested before it is used. For instance, gelatin hydrolysate of casein produces peptone that is deficient in tryptophan, an important component for carrying out the indole test.

63.133 Meat and Yeast Extracts

Beef heart, muscle, liver and brain are commonly used extracts. These contain soluble organic bases, products of protein degradation, vitamins, and minerals. Yeast extract is made from baker's or brewer's yeast. It is a rich source of amino acids and B-complex vitamins. It is used to supplement or even replace meat extracts in media. Typically, meat extracts are used at the 1% level, while yeast extract is used at 0.3% level without significant change in the capacity of the medium to support growth.

63.134 Blood

Horse blood, as well as that from cows, goats, rabbit, and sheep is used, both for the nutrients inherent in the material, as well as for differential reactions. For instance, sheep blood agar is used for detecting the different hemolysins produced by microorganisms. Blood must be stored in a refrigerator and should not be allowed to freeze. It is often used defibrinated so as to inhibit clotting in mixing with other medium components.

63.135 Bile Salts

Bile salts are prepared by extracting dried pig or ox bile by treatment with ethanol. The extract is then decolorized with charcoal, and the bile salts are precipitated with ether to form a yellowish-brown powder. Bile acid conjugates can be further hydrolysed by treatment with alkali to prepare sodium desoxycholate. These compounds are inhibitory to Gram positive bacteria, making them useful as selective agents in isolation media.

63.136 Carbohydrates

These compounds are added to media to promote growth and/or the development of colors, formation of gas, and production of acid in differentiation reactions. Commonly used carbohydrates are: pentoses such as arabinose and xylose; hexoses such as fructose, galactose, glucose, and mannose; disaccharides such as lactose, maltose, and sucrose; polysaccharides such as glycogen and starch; glycosides such as esculin and salicin; and sugar-alcohols such as adonitol, dulcitol, glycerol, mannitol, and sorbitol.

63.14 Basic Steps in Medium Preparation

1. Carefully weigh the proper amount of the dehydrated base medium or the correct proportion of constituent ingredients.
2. Place the requisite amount of microbiologically suitable (distilled, deionized or otherwise appropriately treated) water into a suitable container (e.g., borosilicate glass or stainless steel).
3. Add the weighed material(s) to part of the water. Mix with a stirring rod. Add remaining water, and mix again.
4. Heat, if necessary, to effect complete solution by boiling while stirring on a hot plate/magnetic stirrer. Alternatively, the medium may be heated on an asbestos-centered wire gauze over a free flame or over an electric hot plate, agitating frequently to prevent burning of medium at the bottom of the container. A nonpressurized freeflowing steam unit may also be used. Media containing agar should be boiled to ensure dissolution of the agar. Prolonged boiling may cause undesirable foaming; this can be reduced by holding the flask in cold water for a few seconds after initial boiling has been accomplished. Restore water, if necessary, to compensate for loss by evaporation.
5. Determine the pH of the medium, and adjust if necessary.
6. Distribute the medium into suitable containers. This can be most easily accomplished for liquid media by dispensing with automatic pipettors available through laboratory supply houses. A hand-held adjustable pipettor is most suitable for small numbers of test tubes; an electrically operated machine is more satisfactory for larger numbers. In each instance, before use, machines must be flushed with microbiologically suitable water, followed by medium until the water has been completely removed from the system. After use, the apparatus must be flushed with warm water, detergent solution, tap water, and distilled water. If residues accumulate, the apparatus should be disassembled and scrubbed. If the apparatus employs a syringe, a small amount of water

should be left in the syringe to prevent drying out and subsequent "freezing" of the plunger and barrel.

7. The amount of medium distributed per container should be limited so that no point within the volume of medium is more than 2.5 cm from the top surface or walls of the container (to ensure rapid equilibration of temperature when placed in a waterbath).
8. Autoclave at 121°C for 15 minutes or according to the recommended procedures for each medium. For volumes larger than 500 mL, sterilization times should be increased to 20 to 30 min or more, as needed. When sterilizing media containing heat-stable carbohydrates, temperatures should not exceed 116 or 118°C. Heat-labile carbohydrates should be filter-sterilized, and then added aseptically to cooled, autoclaved, base medium. Recheck the pH before using the medium (see Section 63.15).
9. Overloading the autoclave or improper spacing of containers will greatly reduce autoclaving efficiency. For good results, separate containers by at least one-half inch in all directions, and do not use volumes of media in excess of 3 to 4% of the volume of the autoclave (see Section 63.171).
10. Melt only enough solid media necessary for use within 3 hr. Hold solid media at 44 to 46°C. If a precipitate forms, discard media. Do not remelt media.

Note a: Chemicals and substrates such as carbohydrates must be reagent grade unless otherwise specified. Follow the manufacturer's instructions for storing stock reagents. Discontinue use of chemicals showing any evidence (e.g., color change) of contamination, decomposition, or hydration.

Note b: Automatic agar makers with a complete sterilization cycle are available and may be of value to some laboratories.

63.15 Adjusting Reaction (pH)

Determine the hydrogen ion concentration (pH) of culture media electrometrically at 25°C. This temperature is used in commercial production of media and should be used in the laboratory to determine the pH of media before use. Determinations made at 45°C to take advantage of the fluid state of agar are not accurate, differing significantly from those obtained at the recommended temperature.

Electrometric procedure (potentiometer): Allow electrodes of the instrument to equilibrate (some models may require 30 min) to the temperature at which the determination is to be made, and adjust the buffer solution to the same temperature before testing the instrument. Only buffers referenced to the National Bureau of Standards (NBS) are recommended. The pH of the buffer solution should be in pH range of the medium to be tested. The temperature of test solution and buffer should be the same; room temperature (25°C) is generally used for convenience. For solid media, macerate a suitable aliquot thoroughly with a glass rod before inserting the electrodes. Surface electrodes, if available, are convenient for checking the pH of plated media. Be sure that the temperature is maintained until the reading is complete. If in doubt, repeat the determination. Do not dilute test solutions or buffers.

Note: Plating media are not highly buffered. This can cause confusion. Meters are designed only to show a difference in pH between two solutions at the same temperature. Completely anomalous observations will be obtained if the meter is used under other conditions. Temperature compensators of pH meters do not permit correction of a temperature difference between the test solution and reference buffer. The reason for this is that the H^+ concentration of the buffer changes with temperature. Temperature compensators should be set at the temperature at which measurement is made. Since the emf/pH ratio is less at 25°C than at 45°C, for example, a correction factor must be used; this is built into the instrument and makes it possible to determine the correct pH directly.

63.16 Redox Potential (Eh)

The growth of microorganisms in a biological medium is dependent on the redox potential (Eh) of that medium. Eh is of particular concern to those working with anaerobic microorganisms, since it provides the most useful scale for measuring the degree of anaerobiosis. It can be defined for our purpose as the possibility of a medium giving up or taking electrons, in other words, to be oxidized or reduced.

The Eh of a medium can be accurately measured electrometrically with most pH meters using the millivolt mode of the meter. Consult the manufacturer's pH meter instruction manual for the step-by-step procedure for taking Eh measurements on their meter. A measuring electrode normally made of platinum and a calomel reference electrode are used for determining Eh. The electrodes, after being placed at the bottom of the medium, are allowed to reach equilibrium, and the Eh value is then read in millivolts. Because oxygen plays an important role in Eh measurement, steps must be taken to avoid the introduction of oxygen into the medium. This can be done by passing nitrogen over the medium. Caution: Commercially available nitrogen should be purified before use because it is not always oxygen-free. The nitrogen is purified by passing it through an alkaline solution of pyrogallol or an alkaline solution of sodium hydrosulfite. To remove oxygen from the medium, purified nitrogen can be passed through the medium; however, consider the effect of the Eh measurement if this is done.

Since redox potentials are dependent on pH, the pH values at which they are determined should be indicated. A more common practice in anaerobic microbiology is to measure Eh by estimating its value with the use of redox dyes. Methylene blue and resazurin are two of the most widely used of these dyes for estimating Eh. Where accurate Eh determinations are not required, the use of redox dyes are suggested; however, since some dyes are toxic to some bacteria, they should be tested before use.

63.17 Sterilization and Storage

Before sterilization, bring any medium containing agar to boiling temperature, stirring frequently. Restore lost water if necessary (see Section 63.14(4)), and autoclave. Because the pH of a medium may change during sterilization and because of possible browning reactions, it is important not to exceed the recommended temperature and time. Reduce pressure with reasonable promptness (but in no less than 15 min) to prevent undue changes in the nutritional properties of the medium and remove from the autoclave when atmospheric pressure is obtained. For medium containing agar, autoclave the medium in flasks or bottles from which melted medium may be poured into plates. Optionally, use test tubes containing 15- to 20-mL amounts for pouring melted medium into plates.

Prepare medium in such quantities that, if stored, it will be used before loss of moisture through evaporation becomes evident. To prevent contamination and excessive evaporation of moisture from a medium in flasks and bottles during storage, optionally fit pliable aluminum foil, rubber, plastic, parchment or heavy kraft paper, or viscose caps securely over closures before autoclaving. Use of screwcap or crown (cork-and-seal type) closures on containers appreciably reduces contamination and evaporation. If tubed media are used within a short time, commercially available polypropylene or stainless steel closures may be used. Media should be stored at 2 to 8°C in a dry, dustfree area and should not be exposed to direct sunlight.

63.171 Steam Sterilization

Steam-sterilize media, water and materials such as rubber, cork, cotton, paper, heatstable plastic tubes, and closures by autoclaving at 121°C for not less than 15 min. Autoclave media and dilution blanks within an hour of preparation. Slightly loosen closures to allow passage of steam into and air from closed containers when autoclaved. Place one or more spore controls in the center of the load, preferably within a container similar to those being processed. Make certain that the load is loosely packed (see Section 63.14(9)). Before allowing steam pressure to rise, automatically or manually expel all air from the sterilizer through an appropriate steam trap exhaust valve of suitable size. If expelling air manually, make sure airfree steam is being exhausted before pressurization begins. Because temperature obtained at a constant pressure of pressure-saturated steam varies according to atmospheric pressure, rely only on a properly operating and calibrated mercury thermometer for temperature readings.

Avoid overloading autoclaves so that the rate of air exhaust and heating is not appreciably delayed (see Section 63.14(9)). The autoclave should reach 121°C slowly but within 10 min after starting the air-exhaust operation. A rapid come-up time in an autoclave does not result in more efficient autoclaving. Inefficiency results from the failure of steam to replace air when steam enters the autoclave too quickly. A steam flow that is too slow also results in decreased efficiency because air-steam mixtures are forming.

Where nonliquid materials with slow heat conductance are to be sterilized, or where the packing arrangement or volume of materials otherwise retards the penetration of heat, allow extra time for materials to reach 121°C before beginning to time the sterilization period. If necessary, use longer sterilization periods to ensure sterility.

After sterilization, gradually reduce the pressure within the autoclave (using no less than 15 min) since liquids may be at temperatures above their boiling point at atmospheric pressure. If the pressure is lowered too rapidly, liquids may boil over and be lost. When sterilizing dry materials, such as sampling equipment or empty sample bottles, pressure may be released rapidly through exhaust valves at the end of the 15-min holding period at 121°C. Rapid exhaust through exhaust valves prevents collection of condensate and speeds drying of paper-wrapped equipment.

Used plates, pipettes, tubes, etc., should be routinely decontaminated in all microbiological laboratories. Dairy products, especially raw milk, may harbor potential pathogens such as *Staphylococcus* and *Streptococcus*. For decontamination, the same principles apply for loading the sterilizer and for sterilizing as for preparing media. Plastic Petri dishes are most conveniently sterilized by placing them in heat-resistant autoclavable bags. When large numbers of Petri dishes are autoclaved, sterilization time should be at least 30 min. Insufficient sterilizing of plastic Petri dishes is indicated when they retain some degree of their original shape; properly autoclaved plates become amorphous. Amorphousness by itself, however, is not proof of sterility.

63.172 Hot-Air Sterilization

Sterilize equipment with dry heat in hot-air sterilizers so that materials at the center of the load are heated to not less than 170°C for not less than 1 hr (this usually requires exposure for about 2 hrs at 170°C). To ensure sterility, do not crowd the oven. When the oven is loaded to capacity, it may be necessary to use a longer time period or slightly higher temperature. Spore packs or thermocouples should be used to establish effective sterilization procedures.

63.173 Filter-Sterilization

Some carbohydrates and other liquid preparations that may be denatured by heat should be sterilized by filtration. Membrane filters are commonly used, which are made of cellulose esters, nylon, or polytetrafluoroethylene. These filters may themselves be sterilized by autoclaving, or may be obtained commercially already sterile. To sterilize heat-labile solutions, inject the liquid through a syringe containing a membrane filter with pores no larger than 0.2 or 0.45 μm in diameter. Most bacteria become trapped in the filter, since their size typically ranges from 0.8 to >1.0 μm.

63.18 Quality Control

This chapter generally constitutes a discussion of day-to-day quality control practices; however, specific mention should be made of the need to use simple quality control measures in standardized media preparation procedures. An extensive discussion of laboratory quality control that includes media and media preparation is provided in Chapter 1, specifically Section 1.8. The following are minimal quality control procedures used by laboratories meeting standard method requirements to analyze dairy products. Many of the procedures are mentioned elsewhere in this chapter, but they are summarized here to emphasize their contribution to quality control in media production. Some of the items were taken from a CDC publication.[3]

1. Date the label of each bottle of dehydrated medium to indicate when it was received and when it was first opened.
2. Store dehydrated media in tightly capped bottles in a cool, dry place protected from light, or in a refrigerator if necessary. Keep no more than a six- to twelve- month supply on hand, being sure to use older stocks first. Dehydrated media should be free-flowing powders; if a change is noted in this property or in color, discard.
3. Whenever possible, use commercial dehydrated media.
4. Complete mixing of medium to form a homogeneous solution in water is necessary before sterilizing and dispensing. Avoid unnecessary stirring, which causes foaming.
5. The performance of an autoclave should be monitored by either a continuous temperature-recording device calibrated with standards traceable to NBS in combination with properly placed indicator strips or discs, or spore strips or suspensions. The record for each "run" should be dated, numbered, and filed.
6. Limit heating of the medium to the minimum necessary to ensure sterilization. Open the autoclave immediately following a sterilization cycle (when atmospheric pressure has been reached and the medium is below boiling temperature). Avoid prolonged storage of the medium in a water bath.
7. Check the final pH of each lot of medium after cooling to room temperature. The pH of agar is obtained using a slurry of the medium in a beaker.

8. Follow strict aseptic technique while dispensing sterilized material. Hands should not touch any part of the dispensing tubing which may come in contact with sterile material. During interruptions in a dispensing cycle, the spout of a dispenser train should be placed in a sterile glass container. If possible, a dispensing cycle should not be interrupted.
9. Moving parts of any dispensing apparatus should be oiled or greased as recommended by the manufacturer. Any leaks should be repaired immediately. The accuracy of the medium dispensed should be checked with a graduated container at the beginning and end of each batch dispensed.
10. Media containing dyes should be protected from light by storage in a dark room, in a dark glass bottle, or wrapped with brown paper.
11. Each container of autoclaved medium should be labeled with the name of the medium and the autoclave "run number."
12. Inclusive dates during which each lot and container of dehydrated medium were used should be recorded for possible future troubleshooting.
13. Each lot should be inspected visually before use for volume, tightness of closures, clarity, color, consistency, and completeness of label.
14. Balances should be checked with certified weights on a routine basis.
15. pH meters should be standardized before each use with standard buffers, and electrodes should be checked to see that they are properly filled and not cracked.

63.19 Suitability of Water for Microbiological Applications

Only water that has been treated to free it from traces of dissolved metals and bactericidal and inhibitory compounds should be used to prepare culture media, reagents, and dilution blanks. However, for routine use in dairy product analysis, the growth medium or dilution system prepared with the water is of concern, not the water itself. Culture media provide considerable protection from toxic agents present in water; therefore, the primary interest in quality of water for routine applications is in its use as a dilution fluid.

If special circumstances arise suggesting that the water itself should be examined, specific resistance in ohms may be determined as a rough indication of distilled water quality. Generally, between 400,000 and 500,000 ohms specific resistance is the breakpoint between acceptable and unacceptable distilled water. If organic contamination is suspected, an experienced chemist should make the determination.

Distillate from chlorinated water occasionally contains significant amounts of free chlorine, even when passed through an ion-exchange resin column before use. If distilled or treated water gives color with an orthotolidine test[1], the chlorine should be neutralized with sodium thiosulfate before use for milk dilution blanks.

63.110 Testing of Dilution Water for Toxicity

The fifteenth edition of "Standard Methods for the Examination of Dairy Products" describes procedures for toxicity testing of dilution water in section 5.4.[2] Tests should be conducted on a regular basis to confirm the quality of laboratory diluents.

63.111 Cleaning Glassware and Testing for Detergent Residues

Modern detergents are very effective for cleaning laboratory glassware. Most of these are of the anionic type, usually with alkaline builders such as phosphates, carbonates, or silicates. Some detergents, especially the cationic type (quaternary ammonium compounds or "quats"), are highly bactericidal, and great care must be exercised to ensure their removal. Detergents and soaps have a great affinity for all surfaces, and because of this characteristic they displace dirt and allow it to be easily washed away. However, because of this same characteristic, they are difficult to remove completely.

63.1111 Cleaning Glassware

Most common detergents in laboratory use are satisfactory for general purposes; however, deposits of "milk stone" or calcium salts that resist cleaning by ordinary means are encountered occasionally. These salts must be removed by rinsing glassware in acid solutions for several minutes before effective cleansing can be achieved. A suitable solution for removing milk stone contains dichromate and sulfuric acid. This is prepared by dissolving 40.0 g of finely ground potassium dichromate in 150.0 mL of treated water. Place in a Pyrex vessel and add 330.0 mL of sulfuric acid very slowly with continuous stirring. During the addition of acid the vessel containing the dichromate solution should be cooled by placing it in a cold water bath (a stoppered sink will suffice). The laboratory worker should wear eye protection and protective clothing when making the cleaning solution because it is extremely corrosive. Each laboratory should devise safety procedures for preparing and handling this preparation.

The detergent wash is best done with hot water after preliminary rinsing with warm water to remove most of the debris. Soaking aids removal of stubborn residues. Glassware having residues that are not removed by the detergent action should be immersed for 24 hr in the acid-dichromate cleaning solution and then rinsed thoroughly in the tap and distilled water. Traces of acid on glassware can be detected with pH indicator paper after the last rinse. Six to 12 rinses with running tap water followed by several rinses with distilled water are necessary for complete removal of detergent.

63.1112 Detergent Residues

If doubt remains as to the effectiveness of rinsing, especially when quaternary ammonium compounds were used, the following procedure may be used to detect bactericidal residues:

Prepare glass Petri dishes in three ways: wash one set of three and sterilize by the usual method; wash a second set of three in the acid cleaning solution, then in castile soap, then rinse four times in tap water and six times in distilled water before sterilizing; dip a third set in the presently used detergent solution and sterilize without rinsing.

Place a sample of milk in triplicate in these dishes, and count colonies after two days' incubation at 32°C. No significant difference in counts should appear between the first and second sets of plates. A reduction in bacterial count or diminished size of colonies may be apparent in plates of group three if the detergent is bactericidal or bacteriostatic and residues remain.

63.2 EQUIPMENT

The following equipment and supplies are those most frequently used for the preparation of media, reagents, and stains. Their specifications, listed below, meet the specifications described in "Standard Methods for the Examination of Water and Wastewater,"[1] and "Standard Methods for the Examination of Dairy Products."[2]

63.21 Autoclave

The autoclave shall be large enough to prevent crowding of the interior and shall be constructed to provide uniform temperatures within the chamber up to and including the sterilizing temperature of 121°C. Additionally, it must be equipped with an accurate mercury-filled thermometer or bimetallic helix dial thermometer properly located to register minimum temperatures within the sterilizing chamber, a pressure gauge, and a properly adjusted safety valve. A temperature recording instrument is optional. A small, 5- to 10-quart pressure cooker, equipped with a thermometer for temperature control, may be substituted for an autoclave when proper temperatures are maintained and recorded and satisfactory results are obtained. The autoclave, and especially strainers to traps or vents, should be cleaned frequently and immediately after spills or malfunction of the equipment. The laboratory should have a maintenance program designed specifically to keep the equipment in good repair and working order, with checks at least annually by reputable repairmen. These checks should include thermocouple readings to determine temperatures throughout the autoclave.

63.22 Automatic Pipettors

Pipettors shall be capable of delivering a volume with an accuracy of ±1%.

63.23 Balance

The balance shall be sensitive to 0.1 g with 200-g load; a 2-kg capacity single pan balance is preferred.

63.24 Dilution Bottles, Tubes

Bottles or tubes should be resistant glass, preferably borosilicate. Dilution bottles should have a capacity of about 150 mL and be closed with Escher type rubber stoppers or screw caps. Use friction-fit liners in screw caps, as required to make the closure leakproof. Liners that do not produce toxic or bacteriostatic compounds must be used. Be sure that each batch of dilution blanks is filled properly. Dilution bottles should be marked indelibly at 99 ± 1 mL graduation level. Plastic caps for bottles or tubes and plastic closures for sample containers must be treated when new to remove toxic residues if the manufacturer's information indicates them to be toxic or if no information is available. Treatment may be accomplished by autoclaving them twice while they are submerged in water or exposing them to two successive washings in water containing a suitable detergent at 82°C.

63.25 Hot-Air Sterilizing Oven

The oven should be large enough to prevent crowding of the interior, and should have vents suitably located to ensure prompt and uniform heat distribution for adequate sterilization. It should be equipped with a thermometer having a range of 0 to 220°C located to register the minimum temperature in the oven. In addition to the thermometer, a temperature recorder is desirable.

63.26 pH Meter

An electrometric pH meter with an accuracy of 0.1 pH unit shall be used.

63.27 Petri Dishes or Plates

Petri dishes or plates shall have bottoms of at least 85 mm inside diameter and shall be 12 mm deep, with interior and exterior surfaces of bottoms free from bubbles, scratches, or other defects. Petri dishes may be made of glass or of plastic, and should have flat bottoms.

63.28 Petri Dish Containers

Containers should be of stainless steel or aluminum, with covers. Char-resistant sacks of high-quality sulfate pulp kraft paper may be substituted. Disposable petri dishes may be stored in their original containers.

63.29 Pipettes, Glass and Plastic

Pipettes should be nontoxic, straight-walled, calibrated for bacteriological use, conforming to APHA specifications, with tips ground or fire-polished (glass). Pipettes having graduations other than those listed in the APHA specifications may be used provided they meet all the other APHA specifications: volume delivered in 4 sec maximum, last drop of undiluted milk blown out, or last drop of diluted milk touched out, 1 mL tolerance ± 0.025 mL. To allow for residual milk and dilutions on walls and in the tip of the glass pipette under the specific technique hereinafter described for rapid transfers, such pipettes shall be graduated to contain 1.075 mL of water at 20°C. In the case of styrene plastic, the pipette should be calibrated to contain 1.055 mL of water at 20°C. Use only pipettes with unbroken tips and having graduations distinctly marked so as to contrast sharply with the color of the fluid and dilutions being employed. Discard pipettes which are damaged in any way. **Note:** Do not pipette volumes greater than 10% of the pipette's delivery capacity.

63.210 Pipette Containers

Stainless steel or aluminum containers are preferred. Copper containers shall not be used. Char-resistant, high-quality sulfate-pulp kraft paper may be substituted.

63.211 Refrigerator

A refrigerator is required to cool and maintain the temperature of media between 0 and 4.4°C until used.

63.212 Thermometers

Use thermometers of appropriate range, mercury-filled, or having a distinctively colored fluid, with a freezing point lower than 1°C; or an adjustable type, with a graduation interval not to exceed 1°C unless otherwise specified and accuracy checked at least once a year with a thermometer certified by the National Bureau of Standards (NBS) or to a secondary standard traceable to the NBS thermometer. Where a record is desired of temperatures in refrigerators, autoclaves, hot-air ovens, or incubators, automatic temperature-recording instruments may be used. Two general types of mercury thermometers are in use in laboratories: those calibrated for total immersion, and those that are designed to be partially immersed. Partial immersion thermometers have a line all the way around the stem of the thermometer at the point to which they should be immersed. If this line is not indicated, then the thermometer is designed for total immersion. As examples, a partial immersion thermometer should be used in a water bath because only part of the thermometer is immersed in the medium (water, in this case) whose temperature is being measured. Conversely, a thermometer in an incubator or refrigerator is immersed totally in the warm or cold environment and should be of the complete immersion type.

The easiest way to check the calibration of laboratory thermometers is to put them in a water bath, either partially or totally immersed in the water, according to the way they will be used in the laboratory. Also place in the water bath a thermometer certified by the NBS. Most, but not all, of these thermometers are calibrated for total immersion, so that they typically are immersed totally in the water of the water bath. Vigorous stirring of the water in the bath is essential to ensure uniform temperature during thermometer calibration.

Check the calibration by comparing the temperature reading on the certified thermometer with that of the laboratory thermometer at or very near the temperature that the thermometer will be used to measure (e.g., an incubator thermometer should be checked at 32°C if this is the temperature of interest). If the thermometer is to be used for several purposes, it should be checked at three different temperatures at least. If there is a difference in the temperature reading of the laboratory thermometer and the certified thermometer after the reading of the certified thermometer has been corrected as indicated by the certificate, attach a tag to the laboratory thermometer to show the amount of correction that should be applied to obtain an accurate determination of temperature.

63.213 Water Bath

A water bath of appropriate size and thermostatically controlled is needed for holding melted media at 44 to 46°C.

63.3 ALPHABETICAL LISTING OF CULTURE MEDIA

A-1 MEDIUM

Tryptone .... 20.0 g
Lactose .... 5.0 g
Sodium chloride .... 5.0 g
Triton X-100 .... 1.0 mL
Salicin .... 0.5 g
Distilled water .... 1.0 L

Dissolve the ingredients listed above in 1.0 L of distilled water. Adjust the pH to 6.9 ± 0.1, and dispense 10-mL portions of single-strength broth into 18- × 150-mm tubes containing inverted fermentation tubes. For double-strength broth, use 22- × 175-mm tubes containing inverted fermentation tubes. The medium may have a cloudy appearance before sterilization. Autoclave 10 min at 121°C. Store in the dark for up to 7 days.

AC Agar

Prepare commercially available AC broth; add 15g agar/L.

*Acetate Differential Agar**

Acetobacter Agar

Autolyzed yeast .... 10.0 g
$CaCO_3$ .... 10.0 g
Agar .... 15.0 g
Distilled water .... 1.0 L
Glucose .... 3.0 g

Suspend the ingredients (except glucose) in distilled water and dissolve by heating to 100°C. Add 3.0 g of glucose, and then autoclave for 15 min at 121°C.

Note: When distributing the medium into tubes, distribute the $CaCO_3$ evenly among tubes. After autoclaving, the tubes should be shaken, then slanted and cooled quickly to keep the $CaCO_3$ in suspension.

Acetic Acid Dichloran Yeast Extract Agar (ADYS)

Yeast extract .... 20 g
Sucrose .... 150 g
$MgSO_4 \bullet 7H_2O$ .... 0.5 g
Dichloran .... 0.002 g
Agar .... 20 g
Distilled water .... 1.0 L

Mix all ingredients and boil to dissolve. Autoclave at 121°C for 15 min. Temper to 45-50°C and add glacial acetic acid to make a 0.5% final concentration before pouring plates.

AGAR—2%

Agar .... 20 g
Distilled water .... 1.0 L

Heat to boiling to dissolve. Dispense into media bottles and sterilize for 15 min at 250°F.

Agar Medium for Differential Enumeration of Lactic Streptococci

Tryptone .... 5.0 g
Yeast extract .... 5.0 g
Casamino acids .... 2.5 g
L-Arginine hydrochloride .... 5.0 g
Dipotassium phosphate .... 1.25 g
Calcium citrate .... 10.0 g
Carboxymethyl cellulose† .... 15.0 g
Agar .... 15.0 g

† Dupont CMC Grade P-754 at the rate of 0.6% may be substituted for CMC Cekol MV, Uddeholm, Sweden.

Suspend 15.0 g of agar in 500.0 mL of distilled water, and steam until dissolved. In another glass container suspend 10.0 g of calcium citrate and 15.0 g of CMC in 500.0 g of distilled water. Stir while heating until a homogeneous, white, turbid suspension is formed. Mix the two portions in a stainless steel vessel containing the required amounts of tryptone, yeast extract, dipotassium phosphate, casamino acids, and arginine. Cover the mixture, and steam for 15 min. After steaming, adjust the pH to 5.6 ± 0.2 with 6 N hydrochloric acid. Dispense the agar medium into bottles in 100-mL quantities and autoclave at 121°C for 15 minutes.

Just before pouring plates, add 5.0 mL of sterile, reconstituted nonfat milk (11% solids), 10.0 mL of sterile 3% (w/v) calcium carbonate in distilled water, and 2.0 mL of sterile 0.1% bromocresol purple in distilled water to 100 mL of the sterile agar medium (melted and tempered to 55°C). Mix to obtain homogeneity. Final pH should be 5.9 ± 0.2. Pour the mixture into previously chilled, sterile Petri dishes to obtain a layer of medium 4 to 5 mm thick. After solidification, dry the plates for 18 to 24 hr in an incubator at 37°C.

AKI Medium

(for Vibrio*)*

Peptone .... 15.0 g
Yeast extract .... 4.0 g
Sodium chloride .... 5.0 g
Distilled water .... 1.0 L

On day of use, dissolve basal medium. Autoclave for 15 min at 121°C. Cool to room temperature. Add 30 mL of freshly pre-

*Commercially available.

pared 10% $NaHCO_3$ solution and mix. Dispense aseptically into tubes or flasks. Adjust pH to 7.4 ± 0.2.

ALI Broth/Agar

$(NH_2)SO_4$ .... 0.20 g
$MgSO_4 \cdot H_2O$ .... 0.50 g
$CaCl \cdot H_2O$ .... 0.25 g
KH_2PO_4 .... 3.00 g
Glucose .... 1.00 g
Soluble starch .... 2.00 g
Yeast extract .... 2.00 g
Distilled water .... 1.0 L

To prepare ALI broth, mix all ingredients and adjust pH to 3.5 with 1N H_2SO_4 prior to autoclaving.

To make ALI agar, prepare ALI broth as described above with twice the concentration of all components. Adjust the pH as before. Prepare an equal volume of 3.5% agar solution in distilled water. Autoclave the broth and the agar solutions separately. Temper at 50°C and mix the two solutions using aseptic technique. Pour plates immediately after mixing.

Anaerobic Egg Yolk Agar

Fresh eggs (antibiotic free) 3.0
Yeast extract .... 5.0 g
Tryptone .... 5.0 g
Proteose peptone .... 20.0 g
Sodium chloride .... 5.0 g
Agar .... 20.0 g
Distilled water .... 1.0 L

Wash eggs with a stiff brush and drain. Soak in 70% alcohol for 10 to 15 min. Remove eggs and allow to air dry. Crack eggs aseptically, separate, and discard the whites. Add the yolks to an equal volume of sterile saline and mix thoroughly.

Combine the remaining ingredients, dissolve by heating, and adjust the pH to 7.0. Dispense the medium into appropriate containers and sterilize at 121°C for 15 min. Melt 1.0 L of the sterile agar, cool to 45 to 50°C, and add 80.0 mL of the egg yolk emulsion. Mix thoroughly and pour plates immediately.

Andersen's Pork Pea Agar

Pork infusion .... 800.0 mL
Pea infusion .... 200.0 mL
Peptone .... 5.0 g
Tryptone .... 1.6 g
Dipotassium phosphate .... 1.25 g
Soluble starch .... 1.0 g
Sodium thioglycollate .... 0.5 g
Agar .... 16.0 g
Distilled water .... 1.0 L

Prepare the pork infusion by adding 1 lb of fresh, lean, ground pork to 1.0 L of distilled water and steaming in flowing steam for 1 hr. Filter out the meat through cheesecloth. Chill to solidify, remove the remaining fat, centrifuge to remove the remaining solids, and use the infusion in the above formula.

Prepare the pea infusion by blending 1 lb of good quality, fresh green peas or frozen peas with 450.0 mL of water and steam for 1 hr in flowing steam. Remove the solids by centrifugation and clarify with celite (diatomaceous earth). Use the filtrate in the above formula.

Mix the pork infusion, pea infusion, and other ingredients, and adjust the pH to 7.2 ± 0.2. Autoclave the mixture for 5 min at 121°C, and clarify while hot by adding 25.0 g of celite. Filter the mixture through Whatman No. 4 filter paper with suction. Distribute into tubes, and store at 4°C. As needed, thaw and autoclave at 121°C for 12 min.

Prepare a 5% solution of sodium bicarbonate, sterilize by filtration, and keep refrigerated. Place 0.4 mL of bicarbonate into each plate, pour molten agar, mix thoroughly, and allow to solidify. When solid, pour a layer of thioglycollate agar over the surface and allow to harden.

APT Agar/Broth*

Modifications:

Add 20.0 g sucrose and 2.0 mL bromcresol purple solution per L to APT agar prior to sterilization to aid in the enumeration of lactic acid bacteria in meats. Adjust the pH of sterile, tempered APT agar to 4.0 ± 0.2 with sterile 10% tartaric acid for use as an overlay on MRS agar for the enumeration of lactic acid bacteria in salad dressings. Add 0.5% glucose to APT agar prior to sterilization to aid in the enumeration of lactic acid bacteria in seafood.

Arcobacter Selective Agar/Broth*

Arginine Glucose Slants (AGS)

Peptone .... 5.0 g
Yeast extract .... 3.0 g
Tryptone .... 10.0 g
NaCl .... 20.0 g
Glucose .... 1.0 g
L-arginine (hydrochloride) .... 5.0 g
Ferric ammonium citrate .... 0.5 g
Sodium thiosulfate .... 0.3 g
Bromcresol purple .... 0.02 g
Agar .... 13.5 g
Distilled water .... 1.0 L

Boil to dissolve ingredients. Dispense into tubes (5 mL into 13 × 100 mm tubes). Autoclave for 10-12 min at 121°C. After sterilization, solidify as slants. Adjust final pH to 6.8-7.2.

Aspergillus Flavus *and* Parasiticus *Agar (AFPA)*

Commercially available from Oxoid as AFPA Base (CM731) with a chloramphenicol supplement (SR78).

Yeast extract .... 20.0 g
Bacteriological peptone .... 10.0 g
Ferric ammonium citrate .... 0.5 g
Chloramphenicol .... 0.1 g
Agar .... 15.0 g
Distilled water to make .... 1.0 L
Dichloran (2,6-dichloro-4-nitroanaline), stock solution (0.2% in ethanol) .... 1.0 mL

Combine the listed ingredients and autoclave at 121°C for 15 min. The final pH should be 6.0 to 6.5. The medium may be used for pour plates or prepoured into sterile petri dishes and allowed to solidify for use in surface plating. Do not invert the plates for incubation. Plate counts on this medium should be incubated at 30 ± 1°C for 42 hr. Plates may be incubated longer if necessary. Colonies exhibiting a bright yellow-orange reverse should be counted as *A. flavus* and *A. parasiticus* colonies.

*Commercially available.

Baird-Parker Agar*

Baird-Parker Agar Containing Rabbit Plasma-Fibrinogen

Prepare Baird-Parker agar by suspending in 900.0 mL of distilled water and boiling to dissolve completely. Dispense 90.0-mL portions in screw-capped bottles and autoclave for 15 min at 121°C. Final pH should be 6.8-7.2 at 25°C.

Rabbit Plasma Fibrinogen Solution: Mix 50.0 mL of rehydrated rabbit plasma-EDTA with 50.0 mL of 15% bovine fibrinogen (citrated, fraction I, type I, Sigma Chemical Co.). Add 20.0 mg of salt-free soybean trypsin inhibitor (Schwartz/Mann), and dissolve by mixing. Filter-sterilize the mixture using sterile 1-mm millipore filters.

Preparation of Complete Medium: Prepare filter-sterilized solutions of 1% potassium tellurite and 20% sodium pyruvate. Mix 10.0 mL of prewarmed rabbit plasma-fibrinogen (45 to 50°C), 90.0 mL of melted and cooled (45 to 50°C) Baird-Parker agar medium, and 1.0 mL of 1% potassium tellurite. Avoid entrapping air bubbles. Pour 15.0 to 18.0 mL into sterile Petri dishes. After solidification store plates in plastic bags at 2 to 8°C. Plates can be stored for 4 weeks under these conditions. Just prior to use, spread 0.5 mL of sterile 20% sodium pyruvate solution on each plate, and allow to dry at 50°C or 35°C until dry (2 or 4 hr, respectively) with agar surface upwards.

Baird-Parker Agar without Egg Yolk*

Base Layer Agar with Nutrient Overlay Agar

Fat .... 50.0 g
Victoria Blue B (1:1500 aqueous solution) .... 200.0 mL
Agar .... 15.0 g
Distilled water .... 800.0 mL

Victoria Blue solution is not necessary if tributyrin is used as the fat substrate. If Victoria Blue is not used, add 200.0 mL of distilled water to the base layer medium. The fat used in this medium may consist of tributyrin, corn oil, soybean oil, any available cooking oil, lard, tallow, or triglycerides that do not contain antioxidant or other inhibitory substances. FFA (free fatty acids) in the fat substrates are removed as described in 11.32. Sterilize the fat in the autoclave for 30 min at 121°C. Sterilize the Victoria Blue solution by membrane filtration (0.45 mm). Dissolve the agar in 800.0 mL of distilled water, and sterilize for 15 min at 121°C. Cool the sterile ingredients to 50°C and mix together in a warm sterile blender for 1 min. Pour 3.0 to 4.0 mL of medium into the bottom of sterile Petri dishes (plastic Petri dishes may require a larger volume of medium because of the difficulty in wetting the surface of the plate with the fat-containing medium). Plates may be used immediately or stored for 3 to 4 days in the refrigerator.

The plates containing the base agar should be dried before use in a bacteriological safety hood (laminar flow) by partially opening the Petri dish lids (10 to 15 minutes). Dilutions of the food sample suspected of containing lipolytic microorganisms should be prepared, and 0.1 mL of each dilution should be spread onto the surface of plates containing the base agar. When the inoculum is dry (a few minutes), 10 to 12 mL of a nutrient overlay agar (nutrient agar, standard methods agar, or tryptic soy agar, listed in order of preference) should be poured over the base agar.

BC (Bacillus Cereus) Motility Medium

Trypticase .... 10.0 g
Yeast extract .... 2.5 g
Glucose .... 5.0 g
Disodium hydrogen phosphate .... 2.5 g
Agar .... 3.0 g
Distilled water .... 1.0 L

Dissolve the ingredients in distilled water, and heat to boiling to completely dissolve the agar. Mix thoroughly, and dispense 2.0 mL into 13- × 100-mm tubes. Autoclave for 15 min at 121°C. Final pH should be 7.4 ± 0.2. Allow the medium to solidify, and store at room temperature for 2 or 3 days for best results.

Beef Extract V* (BBL)

Beef Heart Infusion Broth

Beef heart infusion .... 500.0 g
Tryptose .... 10.0 g
Sodium chloride .... 5.0 g
Distilled water .... 1.0 L

Dissolve the ingredients in distilled water, distribute into tubes, and sterilize by autoclaving for 15 min at 121°C. The final pH of the medium should be adjusted to 7.4.

For best results, the medium should be freshly prepared. If it is not used the same day as sterilized, heat in boiling water or flowing steam for several minutes to remove absorbed oxygen, and cool quickly, without agitation, just prior to inoculation.

Beer Agar

Yeast extract .... 5.0 g
Fructose .... 10.0 g
Agar .... 10.0 g
Beer .... 1.0 L

Combine ingredients in beer, and heat to boiling while stirring to dissolve the agar completely. Autoclave for 15 min at 121°C.

Bile Esculin Agar*

Bile Salts Brilliant Green Starch (BBGS) Agar

Proteose peptone .... 10.0 g
Lab-lemco powder (Oxoid) .... 5.0 g
Bile salts (Oxoid L55) .... 5.0 g
Soluble starch .... 10.0 g
Agar .... 15.0 g
Brilliant green (0.05% w/v solution) .... 1.0 mL
Distilled water .... 1.0 L

Suspend the ingredients in distilled water and dissolve by heating while stirring. Autoclave for 15 min at 121°C. Final pH should be 7.2.

Bismuth Sulfite (BS) Agar*

Blood Agar

Blood agar base (commercial, dehydrated) .... 40.0 g
Distilled water .... 1.0 L

Prepare the base according to the manufacturer's instructions. Temper the sterilized base at 45 to 50°C, and aseptically add 5% sterile, defibrinated, room temperature sheep blood. Mix thoroughly, avoiding incorporation of air bubbles, and pour into plates.

Blood Agar Base*

*Commercially available.

*Brain Heart Infusion (BHI) Agar/Broth**

Brettanomyces Selective Media (BSM)

Yeast extract .... 3.0 g
Malt extract .... 3.0 g
Peptone .... 5.0 g
Dextrose .... 10.0 g
Distilled water .... 1.0 L

Prepare base broth. Autoclave at 121°C for 15 min. After cooling, add the following filter-sterilized ingredients:

Thiamine .... 1.0 mL (1% aqueous stock solution)
Chloramphenicol .... 10.0 mL (1% stock solution in ethanol)
Cycloheximide .... 10.0 mL (1% stock solution in ethanol)
Chlortetracycline .... 20.0 mL (0.5% aqueous stock solution)
Gentamicin .... 5.0 mL (1% aqueous stock solution)

All stock solutions must be refrigerated and filter-sterilized except for Chloramphenicol, which can be added prior to autoclaving. Chlortetracycline stock solution should be protected from light.

*Brilliant Green Bile (BGB) Agar/Broth**

Or

*Brilliant Green Lactose Bile Broth**

*Brilliant Green Sulfa Agar * (BBL)*

Bromcresol Green Agar

Glucose .... 25.0 g
Peptone .... 3.0 g
Yeast extract .... 5.0 g
Agar .... 15.0 g
Distilled water .... 1.0 L

Combine ingredients in 1.0 L distilled water, and heat to boiling while stirring to dissolve the agar completely. Autoclave for 15 min at 121°C. After autoclaving, add 1.0 mL Bromcresol green solution per 100.0 mL of tempered medium.

Bromocresol Purple Carbohydrate Broth Base with Carbohydrates

Basal medium:
Peptone .... 10.0 g
Beef extract (optional) .... 3.0 g
Sodium chloride .... 5.0 g
Bromocresol purple .... 0.04 g
Distilled water .... 1.0 L

Combine the ingredients above with 5 g per L of the desired carbohydrate (glucose, adonitol, arabinose, mannitol, maltose, sucrose, lactose, sorbitol, cellobiose, salicin, or trehalose). Adjust the pH to 7.0 ± 0.2. Dispense 8-mL aliquots into 16- × 150-mm tubes containing inverted 12- × 75-mm fermentation tubes. Autoclave for 10 min at 121°C. Final pH should be 6.8 to 7.0.

*Brucella Broth**

Brucella-FBP Agar/Broth

Brucella broth .... 28.0 g
$FeSO_4$ .... 0.25 g
Sodium metabisulfite (anhydrous) .... 0.25 g
Sodium pyruvate (anhydrous) .... 0.25 g
Distilled water .... 970.0 mL

Suspend the Brucella broth in 970.0 mL of distilled water, and boil for approximately 1 min to dissolve completely. Autoclave at 121°C for 15 min, and cool to 50°C. Prepare the FBP solution by dissolving $FeSO_4$, sodium metabisulfite, and sodium pyruvate together in 30 mL of distilled water, and sterilize by filtering through a 0.22-mm filter. Add the filter-sterilized FBP solution (ca. 30 mL) to the cooled medium, and distribute into tubes. Brucella-FBP agar and semi-solid medium is made in the same manner, but 15 g or 1.6 g of agar, respectively, is added to the Brucella broth before autoclaving.

Campylobacter *Charcoal Differential Agar (CCDA)–Preston Blood-Free Medium*

Nutrient broth No. 2 (Oxoid) .... 25.0 g
Bacteriological charcoal .... 4.0 g
Casein hydrolysate .... 3.0 g
Sodium deoxycholate .... 1.0 g
Ferrous sulfate .... 0.25 g
Sodium pyruvate .... 0.25 g
Agar .... 12.0 g
Sodium cefoperazone .... 32.0 mg
Distilled water .... 1.0 L

Dissolve all of the medium components except cefoperazone in 1.0 L of distilled water, and boil to dissolve completely. Autoclave at 121°C for 15 min. Cool to 50°C, and add filter-sterilized (through a 0.22-mm filter) cefoperazone. Mix well and pour into sterile Petri plates.

Campy-Bap (Blaser's Agar)

Brucella agar .... 43.0 g
Distilled water .... 925.0 mL
Lysed horse blood .... 50.0 mL
Vancomycin .... 10.0 mg
Polymyxin B .... 2,500 IU
Trimethoprim lactate .... 5.0 mg
Amphotericin B .... 2.0 mg
Cephalothin .... 15.0 mg

Dissolve the brucella agar in distilled water, heat while stirring, and boil for approximately 1 min to dissolve the agar completely. Autoclave at 121°C for 15 min. Cool the medium to 50°C, and add the blood and filter-sterilized solutions of antibiotics. Defibrinated sheep blood may be used in lieu of horse blood; use 100.0 mL of sheep blood and only 875.0 mL of distilled water to prepare the Brucella agar base medium. Difficulties in dissolving the specified antibiotics are resolved by making the water alkali. Prepare solutions of each antimicrobial compound for 1.0 L of medium in 5.0 mL of Brucella broth; dissolve the amounts of each compound required for 10.0 liters of medium in 50.0 mL of brucella broth, filter through a 0.22-mm filter, dispense 5-mL portions in vials, and freeze until use. The combination of antimicrobials is also available commercially.

Campy-Cefex Agar

Ingredients:
Brucella agar .... 44.0 g
Ferrous sulfate, heptahydrate .... 0.5 g
Sodium bisulfite (metabisulfite) .... 0.2 g
Pyruvic acid (sodium salt) .... 0.5 g
Distilled water .... 1.0 L

*Commercially available.

Supplements:
Cefoperazone (sodium salt) .... 33.0 mg
Cycloheximide .... 200.0 mg
Lysed horse blood .... 50 mL

Mix ingredients (excluding supplements) in a large Erlenmeyer flask and heat to boiling. Autoclave for 15 min at 121°C and 15 psi pressure; cool to 50°C. Add supplements after cooling to 50°C. Dispense into Petri dishes.

To prepare the supplements:

Cefoperazone: dissolve 1.0 g of cefoperazone in 10 mL dionized water. Filter-sterilize using a 0.22 µm filter. Dispense into sterile cryovial tubes and store at –80°C. Use 0.33 mL/L for 33 mg/L.

Cycloheximide: Dissolve 2.0 g of cycloheximide into 10 mL 50% methanol. Filter-sterilize using a 0.22 µm filter. Add 1.0 mL/L. Prepare fresh before use.

Lysed Horse Blood: Use only fresh blood and lyse upon receiving. If lysed when purchased, be sure it was prepared from fresh blood and shipped promptly. If not used upon receipt. Freeze within 1-2 days. To lyse and/or store, gently mix and pour about 100 mL per sterile polypropylene bottle. Place in –20°C freezer. Thaw and refreeze once more to lyse. Freezing inhibits adsorption of oxygen by hemoglobin. Use blood up to 5 months. Can be thawed and refrozen several times.

Campy Line Agar

Ingredients:
Brucella agar .... 43.0 g
Ferrous sulfate, heptahydrate .... 0.5 g
Sodium bisulfite (metabisulfite) .... 0.2 g
Pyruvic acid (sodium salt) .... 0.5 g
Distilled water .... 1000.0 mL
Ketoglutaric acid .... 1.0 g
Sodium carbonate .... 0.6 g
Yeast extract .... 3.0 g
Supplements:
Cefoperazone (sodium salt) .... 33.0 mg
Cycloheximide .... 100.0 mg
Vancomycin (hydrochloride) .... 10.0 mg
Polymyxin B (sulfate) .... 0.35 mg
Trimethoprim .... 5.0 mg
Triphenyltetrazolium choride-TTC .... 200.0 mg
Hemin (chloride, bovine) .... 10.0 mg

Mix ingredients (excluding supplements) in a large Erlenmeyer flask and heat to boiling. Autoclave for 15 min at 121°C and 15 psi pressure; cool to 50°C. Add supplements after cooling to 50°C. Dispense into Petri dishes.

To prepare the supplements:

Cefoperazone: dissolve 1.0 g of cefoperazone in 10 mL dionized water. Filter-sterilize using a 0.22 µm filter. Dispense into sterile cryovial tubes and store at –80°C. Use 0.33 mL/L for 33 mg/L.

Cycloheximide: Dissolve 2.0 g of cycloheximide into 10 mL 50% methanol. Filter-sterilize using a 0.22 µm filter. Add 0.5 mL/L. Prepare fresh before use.

Trimethoprim/Vancomycin/Polymyxin B: In a 100-mL volumetric flask, dissolve 0.1 g of trimethoprim in 20 mL of 100% ethanol. Add 0.2 g of vancomycin and 0.007 g of polymyxin B. Fill to 100 mL with deionized water. Some gentle heating may be required to comple the dissolution. Filter-sterilize using a 0.22 µm filter. Dispense into cryovials and freeze at –80°C. Use 5 mL/L of agar. Prepare a 20% stock solution of TTC in deionized water (20 g/100 mL). Slight heating will be necessary to complete dissolution. Filter-sterilize using a 0.22 µm filter. Use 1 mL/L. Stock solution may be stored at 4°C indefinitely. The solution will crystallize at refrigeration temperatures, but gentle heating at 50°C should get it back into solution.

Hemin: Prepare a stock solution of hemin in a 100-mL volumetric flask. Dissolve 0.5 g hemin in 10 mL of 1N sodium hydroxide. Add 90 mL of distilled water to a final volume of 100 mL. Autoclave for 15 min at 121°C. Use 2 mL/L.

Camp Test Agar

(same as Trypticase Soy Sheep Blood Agar)

Prepare trypticase (tryptic) soy agar, and cool to 48°C in a water bath. Add 5.0 mL of sterile defibrinated sheep blood for each 100.0 mL of medium, mix well, and dispense 18.0 mL into 15- × 100-mm culture dishes. Allow plates to dry for 24 to 48 hr at room temperature before use.

Carbohydrate Fermentation Broth

Peptone .... 10.0 g
Meat extract .... 3.0 g
NaCl .... 5.0 g
Andrade's indicator .... 10.0 mL
Distilled water .... 1.0 L

Dissolve the ingredients in distilled water, and adjust the pH to 7.1 to 7.2. Dispense the medium into tubes, and sterilize by autoclaving at 121°C for 15 min. If detection of gas formation is desired, include an inverted vial (Durham tube) before sterilization. The vial will fill with liquid upon slow cooling after autoclaving. Add each desired carbohydrate at a final concentration of either 0.5% or 1.0%. It is desirable to add the carbohydrates as filter-sterilized (0.22 mm pore size) solutions although some can be autoclaved without deleterious results. The filter-sterilized carbohydrate can be added as a 10% solution (e.g. 5.0 mL of basal broth plus 0.5 mL of 10% carbohydrate). A more accurate method is to prepare the basal broth as a 2X solution and add an equal volume of a 2X carbohydrate solution, which also eliminates problems of solubility that may be encountered at 10% concentrations.

*Cary-Blair Transport Medium**

Casamino Acids-Yeast Extract (CAYE) Broth

Casamino acids .... 30.0 g
Yeast extract .... 4.0 g
Potassium phosphate (dibasic) .... 0.5 g
Glucose .... 2.0 g
Distilled water .... 1.0 L

Dissolve all ingredients except glucose in distilled water. Adjust the pH to 7.2 ± 0.2 and sterilize at 121°C for 15 min. Cool and aseptically add 10 mL of a filter-sterilized 20% glucose solution in distilled water. After mixing, aseptically dispense into sterile tubes or bottles.

Cefoperazone Vancomycin Amphotericin B Blood Agar (CVA)

Ingredients:
Tryptone .... 10.0 g
Peptamin (Difco #0905-01-5) .... 10.0 g
Dextrose .... 1.0 g

*Commercially available.

Yeast extract .... 2.0 g
Sodium chloride .... 5.0 g
Sodium bisulfite .... 0.1 g
Agar .... 15.0 g
Distilled water .... 950 mL

CVA Antibiotic Stock Solution

Cefoperazone .... 2.0 g
Vancomycin .... 1.0 g
Amphotericin B .... 0.2 g
Defibrinated bovine or ovine blood 5%

Bring to 100 mL with sterile water. Filter-sterilize and dispense in 3.0 mL aliquots. Stable at –20°C for 3 months.

Instructions: Prepare the agar in distilled water, autoclave (121°C for 20 min) and cool at 47°C. Add 5% defibrinated bovine or ovine blood. Add 1.0 mL of the CVA antibiotic stock solution to 1 L of CVA agar to achieve a final concentration of cefoperazone (20 mg/L), vancomycin (10 mg/L), and amphotericin B (2 mg/L). Dispense to Petri plates (100 mm × 15 mm). CVA agar is also available commercially (Remel 01-270).

Cefsulodin-Irgasan-Novobiocin (CIN) Agar

Special peptone (Oxoid) .... 20.0 g
Yeast extract .... 2.0 g
Mannitol .... 20.0 g
Pyruvic acid (Na salt) .... 2.0 g
NaCl .... 1.0 g
Magnesium sulfate • $7H_2O$ .... 0.01 g
Sodium desoxycholate .... 0.5 g
Agar (Oxoid L11) .... 12.0 g
Distilled water .... 949.0 mL

Suspend all ingredients in distilled water, and bring to a boil with stirring to completely dissolve. Cool to approximately 80°C, and add 1.0 mL of 0.4% Irgasan (Ciba-Geigy) in 95% ethanol. Mix well, and cool to about 50-55°C. Add the following filter-sterilized (except NaOH, 0.22 mm pore size) solutions:

NaOH (5 N) .... 1.0 mL
Neutral red (3 mg/mL) .... 10.0 mL
Crystal violet (0.1 mg/mL) .... 10.0 mL
Cefsulodin (1.5 mg/mL) .... 10.0 mL
Novobiocin (0.25 mg/mL) .... 10.0 mL

Add 10.0 mL of a filter-sterilized 10% solution of strontium chloride slowly with stirring. Adjust the pH to 7.40, and pour the medium into sterile Petri plates. NOTE: Comparable formulations are available commercially as "*Yersinia* Selective Agar."

Cell Growth Medium

Mix the following sterile solutions aseptically:
Minimal essential medium with Hank's salts (MEMH) ..... 1.0 L
L-15 medium (Leibovitz), containing
100 IU penicillin G,
100 mg streptomycin and
50 mg gentamicin per mL .... 1.0 L

Mix the solutions on a magnetic stirrer, filter through a 0.20 mm membrane, and dispense into a sterile 2-L Erlenmeyer flask. Adjust the pH to 7.5, and cap and store at 5°C.

Just before use add:
Fetal bovine serum .... 200.0 mL
$NaHCO_3$ (7.5%) .... 50.0 mL

Cellobiose-Colistin Agar (CC)

[Same as Modified Cellobiose-Polymyxin Colistin Agar (mCPC), except no Polymyxin B is added]

Chopped Liver Broth

Ground beef liver .... 500.0 g
Soluble starch .... 1.0 g
Peptone .... 10.0 g
Dipotassium phosphate .... 1.0 g
Distilled water .... 1.0 L

Add finely ground beef liver to the distilled water, and boil for 1 hour. Adjust the pH of the broth to 7.0 ± 0.2 and boil for another 10 min. Press through cheesecloth, and bring the broth to 1.0 L with distilled water. Add the peptone and dipotassium phosphate, and adjust the pH to 7.0 ± 0.2. Place liver particles from the pressed cake in the bottom of culture tubes (about 1 cm deep), cover with 8 to10 mL of broth and sterilize by autoclaving for 20 min at 121°C. Before use, exhaust the medium for 20 min in flowing steam.

Christensen's Citrate Agar

Sodium citrate .... 3.0 g
Glucose .... 0.2 g
Yeast extract .... 0.5 g
Cysteine monohydrochloride .... 0.1 g
Ferric ammonium citrate .... 0.4 g
KH_2PO_4 .... 1.0 g
NaCl .... 5.0 g
Sodium thiosulfate .... 0.08 g
Phenol red .... 0.012 g
Agar .... 15.0 g
Distilled water .... 1.0 L

Suspend ingredients, mix thoroughly and heat with agitation. Bring to a boil for about 1 min to dissolve ingredients. Fill 16 × 150 mm tubes 1/3 full and cap to maintain aerobic conditions. Autoclave for 15 min at 121°C. Final pH should be 6.9 ± 0.2. Before the media solidifies, incline tubes to obtain a 4-5 cm slant and 2-3 cm butt. Note that the Difco formulation does not include ferric ammonium citrate and sodium thiosulfate.

Christensen's Urea Agar*

Colbeck's Egg Yolk Broth*

Columbia Blood Agar Base*

Congo Red Brain Heart Infusion Agarose (CRBHO)

Brain heart infusion .... 37 g
$MgCl_2$ .... 1 g
Agarose .... 12 g
Congo Red dye (375 mg/100 mL distilled water) .... 20 mL

Dissolve, autoclave at 121°C for 15 min and pour 20 mL per plate. Formula as per S. Bhaduri, USDA, Philadelphia, PA. ASM Abstracts, 1989, No. P26. US Patent applicationn Serial No. 07/493,662.

Cooked Meat Medium (CMM)*

*Commercially available.

Cooked Meat Medium with Starch and Glucose

Prepare commercially available cooked meat medium according to instructions, except to include 0.1% starch and 0.1% glucose in the distilled water used to suspend the meat particles.

CVP (Crystal Violet Pectate) Medium

To 500.0 mL of boiling distilled water, blending at very low speed in a preheated blender, add:

Crystal violet solution (0.075% w/v aqueous solution) .. 1.0 mL
Sodium hydroxide (1N aqueous solution) .... 4.5 mL
Calcium chloride (fresh 10% w/v $CaCl_2 \bullet H_2O$ aqueous solution) .... 3.0 mL
Agar .... 2.0 g
Sodium nitrate .... 1.0 g

Blend at high speed for 15 sec. Continue blending at low speed and slowly add:

Sodium polypectate† .... 9.0 g

† For best results use polypectate from Ral Tech Sci. Services, Inc., P.O. Box 7545, Madison, WI 53707.

Pour the incomplete medium into a 2-L flask, and add:

Sodium lauryl sulfate (10% w/v aqueous solution) .... 0.5 mL

Cap the flask with aluminum foil rather than cotton plug, and autoclave for 25 min at 121°C. Avoid foaming, and pour plates as soon as possible since the medium solidifies quickly and cannot be remelted. Since it is essential that the surface of the agar be free of water, allow the plates to dry for 48 hr at room temperature before use. The final pH of the medium is 7.2.

Czapek Dox Iprodione Dichloran (CZID)

Czapek Dox broth (Difco) .... 35.0 g
$CuSO_4 \bullet 5H_2O$ .... 0.005 g
$ZnSO_4 \bullet 7H_2O$ .... 0.01 g
Chloramphenicol .... 0.05 g
Dichloran (0.2% in ethanol) .... 1 mL
Agar .... 20.0 g
Distilled water .... 1.0 L

Mix all ingredients together and steam to dissolve. Autoclave at 121°C for 15 min. Cool to 50°C and add 10 mL of 0.5% aqueous chlortetracycline solution and 1 mL of iprodione (Royal 50WP from Rhone-Poulenc) from a sterile solution (0.3 iprodione in 50 mL distilled water) before pouring plates.

Czapek Yeast Autolysate (CYA) Agar

Yeast extract .... 5.0 g
Sucrose .... 30.0 g
$NaNO_3$ .... 3.0 g
KCl .... 0.5 g
K_2PO_4 .... 1.0 g
$MgSO_4 \bullet 7H_2O$ .... 0.5 g
$FeSO_4 \bullet 5H_2O$ .... 0.01 g
Agar .... 15.0 g
Distilled water .... 1.0 L

Suspend the ingredients in distilled water, and heat the mixture to boiling while stirring to dissolve completely. Autoclave at 121°C for 15 min.

Decarboxylase Basal Medium with Arginine, Lysine (Falkow), or Ornithine

Base

Peptone or gelysate .... 5.0 g
Yeast extract .... 3.0 g
Glucose .... 1.0 g
Bromcresol purple .... 0.02 g
Distilled water .... 1.0 L

For arginine broth, add 5 g L-arginine to 1 L base; for lysine (Falkow) broth, add 5 g L-lysine to 1 L base; for ornithine, add 5 g L-ornithine to 1 L base. Adjust pH so that value after sterilization is 6.5 ± 0.2. Dispense 5 mL portions into 16×125 mm screw-cap tubes. Autoclave loosely capped tubes 10 min at 121°C. Screw the caps on tightly for storage and after inoculation. For control, use unsupplemented base.

Dextrose Agar/Broth*

Dextrose Tryptone Agar/Broth*

Dextrose Tryptone Bromcresol Purple Agar (DTA)/ Broth (DTB)

See Dextrose Tryptone Agar/Broth.

Dey-Engley Neutralizing Medium*

Dichloran Chloramphenicol Peptone Agar (DCPA)

Peptone .... 15.0 g
KH_2PO_4 .... 1.0 g
$MgSO_4 \bullet 7H_2O$ .... 0.5 g
Chloramphenicol .... 0.1 g
Dichloran (0.2% in ethanol) .... 1 mL
Agar .... 15.0 g
Distilled water .... 1.0 L

After adding all ingredients, sterilize by autoclaving at 121°C for 15 min. The final pH of this medium should be 5.5 to 6.0.

Dichloran Rose Bengal Chloramphenicol (DRBC) Agar*

Commercially available from Oxoid as DRBC Agar Base (CM727) with a chloramphenicol supplement (SR78).

Dichloran Rose Bengal Yeast Extract Sucrose Agar (DRYES)

Yeast extract .... 20.0 g
Sucrose .... 150.0 g
Dichloran (0.2% in ethanol) .... 1 mL
Rose Bengal (5% w/v) .... 0.5 mL
Chloramphenicol .... 0.1 g
Agar .... 20.0 g
Distilled water .... 1.0 L

Mix all ingredients together and steam to dissolve. Autoclave at 121°C for 15 min. Adjust pH to 5.6 after autoclaving.

Dichloran 18% Glycerol Agar (DG18)

Peptone .... 5.0 g
Glucose .... 10.0 g
KH_2PO_4 .... 1.0 g
$MgSO_4 \bullet 7H_2O$ .... 0.5 g
Chloramphenicol .... 0.1 g
Dichloran (0.2% in ethanol) .... 1 mL
Glycerol .... 220.0 g
Agar .... 20.0 g

*Commercially available.

Distilled water .... 1.0 L

Add all ingredients except glycerol to 800 mL distilled water. Steam to dissolve before adding water to 1 L. Add glycerol and autoclave at 121°C for 15 min. The final pH is 5.6 ± 0.2 and the final water activity is 0.955.

Differential Broth For Lactic Streptococci

Tryptone .... 5.0 g
Yeast extract .... 5.0 g
Dipotassium phosphate .... 1.0 g
Arginine .... 5.0 g
Sodium citrate .... 20.0 g
Bromcresol purple .... 0.02 g
Distilled water to make .... 1.0 L

Suspend the ingredients in 800.0 mL of distilled water, add 35 mL of 11% reconstituted skim milk, and bring the total volume to 1.0 L. Steam the medium for 15 min, cool to 25°C, and adjust the pH to 6.2. Dispense 7.0-mL quantities into 10- × 126-mm screw-capped tubes containing Durham fermentation tubes. Autoclave at 121°C for 15 minutes. The final pH should be 6.2 ± 0.2. Do not open the autoclave door until the temperature has dropped below 75°C.

Differential Clostridial Agar (DCA)

Meat extract powder .... 8.0 g
Casein peptone .... 5.0 g
Meat peptone .... 5.0 g
Starch .... 1.0 g
Yeast extract .... 1.0 g
Resazurin .... 0.002 g
Cysteine HCl .... 0.5 g
D-glucose .... 1.0 g
Agar .... 20.0 g
Distilled water .... 1.0 L

Autoclave at 121°C for 15 min. Add 7.5 mL of a freshly prepared solution of 10% sodium sulfite • $7H_2O$ (filter-sterilized) and 5 mL of a solution of 20% ferric ammonium citrate (sterilized at 121°C for 15 min—stored refrigerated for no more than one month).

DNase Test Agar*

After 24 hr of incubation, the DNase medium is flooded with 0.1N HCl. DNase-positive colonies are surrounded by a clear zone. At 24 hr, DNase-positive colonies on the Toluidine Blue agar are surrounded by a purplish-pink precipitation formed by the reaction of hydrolyzed DNA with the dye.

Doyle and Roman Enrichment Medium

Brucella broth (Gibco) .... 29.0 g
Distilled water .... 910.0 mL
Sodium succinate .... 3.0 g
Cysteine hydrochloride .... 0.1 g
Lysed horse blood .... 70.0 mL
Vancomycin .... 15.0 mg
Trimethoprim lactate .... 5.0 mg
Polymyxin B .... 20,000 IU
Cycloheximide .... 50.0 mg

Dissolve by swirling the Brucella broth medium, sodium succinate, and cysteine hydrochloride in distilled water. Autoclave at 121°C for 15 min. Cool the medium to room temperature, and add the lysed horse blood and filter-sterilized (through a 0.22-mm filter) antibiotics. Dispense 90 or 100 mL into sterile side-arm flasks.

Eagle's Minimal Essential Medium (MEME)

(with Earl's salts and nonessential amino acids)

L-Alanine .... 8.9 mg
L-Arginine HCl .... 126.0 mg
L-Asparagine•H_2O .... 150.0 mg
L-Aspartic acid .... 13.3 mg
L-Cystine•2HCl .... 31.29 mg
L-Glutamic acid .... 14.7 mg
L-Glutamine .... 292.0 mg
L-Glycine .... 7.5 mg
L-Histidine HCl•H_2O .... 42.0 mg
L-Isoleucine .... 52.0 mg
L-Leucine .... 52.0 mg
L-Lysine HCl .... 72.5 mg
L-Methionine .... 15.0 mg
D-Phenylalanine .... 32.0 mg
L-Proline .... 11.5 mg
L-Serine .... 10.5 mg
L-Threonine .... 48.0 mg
L-Tryptophan .... 10.0 mg
L-Tyrosine (disodium salt) .... 52.1 mg
L-Valine .... 46.0 mg
D-Calcium pantothenate .... 1.0 mg
Choline chloride .... 1.0 mg
Folic acid .... 1.0 mg
Isoinositol .... 2.0 mg
Nicotinamide .... 1.0 mg
Pyridoxal HCl .... 1.0 mg
Riboflavin .... 0.1 mg
Thiamine HCl .... 1.0 mg
Glucose .... 1000.0 mg
$CaCl_2 \bullet 2H_2O$ .... 265.0 mg
KCl .... 400.0 mg
$MgSO_4 \bullet 7H_2O$ .... 200.0 mg
NaCl .... 6800.0 mg
$NaHCO_3$ .... 2200.0 mg
$NaH_2PO_4 \bullet H_2O$ .... 140.0 mg
Phenol red .... 10.0 mg
Distilled water .... 1.0 L

Dissolve ingredients in 1.0 L of distilled water. Sterilize by filtration, and adjust the pH to 7.2 ± 0.2. Store at 5°C.

Eagle's Minimum Essential Medium (MEM)*

Earle's Balanced Salts (ES) Solution (Phenol Red-Free)

NaCl .... 6.8 g
KCl .... 400.0 mg
$CaCl_2 \bullet 2H_2O$ .... 265.0 mg
$MgSO_4 \bullet 7H_2O$ .... 200.0 mg
$NaH_2PO_4 \bullet H_2O$ .... 140.0 mg
Glucose .... 1.0 g
$NaHCO_3$ .... 2.2 g
Distilled water .... 1.0 L

Dissolve ingredients in 1.0 L of distilled water. Sterilize by filtration, and adjust the pH to 7.2 ± 0.2.

*Commercially available.

EC (Escherichia coli) *Broth**

*Egg Yolk Emulsion 50%**

Egg Yolk-Free Tryptose-Sulfite Cycloserine (EY-Free TSC) Agar

(Same as Tryptose-Sulfite Cycloserine Agar)

EHEC Enrichment Broth (EEB)

Trypticase soy broth .... 30.0 g
Bile salts No. 3 .... 1.12 g
K_2HPO_4 .... 1.5 g
Distilled water .... 1.0 L

Dissolve ingredients in distilled water. Autoclave for 15 min at 121°C and let cool at room temperature. Prepare and filter-sterilize the final antibiotic solutions in distilled water:

Cefixime .... 1 mg/mL
Cefsulodin .... 100 mg/mL
Vancomycin .... 100 mg/mL

Add each antibiotic solution to the autoclaved broth to achieve the following final concentrations:

Cefixime .... 0.05 mg/L
Cefsulodin .... 10.0 mg/L
Vancomycin .... 8.0 mg/L

Ellinghausen McCullough Johnson Harris Polysorbate 80 (EMJH-P80)

Solution A:
Sterile distilled water .... 700 mL
25X phosphate buffered solution .... 40 mL
20X salt solution .... 50 mL
Copper solution .... 1 mL
Zinc solution .... 10 mL
Iron solution .... 20 mL
L-cystine .... 0.2 g
Solution B:
Sterile distilled water .... 120 mL
B-12 vitamin solution .... 20 mL
Thiamine HCl B-1 solution .... 0.1 mL
Tween 80 .... 1.2 mL

Autoclave number of tubes needed. Make Solutions A and B in separate beakers. Stir Solution A for 5 min (all of cystine will not dissolve). Filter Solution a through triple thickness Whatman #1 filter paper and repeat until solution is clear. Add Solution A to Solution B and stir. Add distilled water to bring final volume to 1 L. Dispense 800 mL of this final product into a 2-L Erlenmeyer flask. Add 1.3 g of agar and adjust pH to 7.2-7.4. Autoclave at 121°C for 15 min. Cool to 50°C in water bath. Add 200 mL of a filter-sterilized 5% albumin solution. Add 200 mg of 5-Fluorouracil and dispense into sterile tubes, 10 mL/tube.

Enrichment Broth (M52)

(for *Listeria*)
TSBYE supplemented with:
Monopotassium phosphate (anhydrous) .... 1.35 g/L
Disodium phosphate (anhydrous) .... 9.6 g/L
Acriflavin HCl .... 10 mg/L
Nalidixic acid (sodium salt) .... 40 mg/L
Cycloheximide .... 50 mg/L
Pyruvic acid (sodium salt) (Sigma), 10% (w/v) aqueous solution .... 11.1 mL/L

Sterilize enrichment broth without the 3 selective agents (last 3 ingredients) by autoclaving at 121°C for 15 min. Then add 2.5 mL 10% (w/v) filter-sterilized sodium pyruvate. Add the 3 selective ingredients aseptically to 225 mL enrichment broth plus the 25 g food sample after 4 h incubation at 30°C. Prepare acriflavin and nalidixic acid supplements as 0.5% (w/v) stock solutions in distilled water. Prepare cycloheximide supplement as 1.0% (w/v) stock solution in 40% (v/v) solution of ethanol in water. Filter-sterilize the 3 selective ingredients. Add stock solutions: 0.455 mL acriflavin, 1.8 mL nalidixic, and 1.15 mL cycloheximide to 225 mL enrichment broth plus the 25 g food sample.

*Eosin Methylene Blue (EMB) Agar (Levine)**

*Eugon Agar**

Fermentation Base for Campylobacter

Enteric fermentation base (Difco) .... 18.0 g
Andrade's indicator .... 10.0 mL
Glucose .... 10.0 g
Distilled water .... 1.0 L

Dissolve the enteric fermentation base and glucose in 950.0 mL of distilled water. Add Andrade's indicator, and adjust the pH to 7.2. Add distilled water to a total volume of 1000.0 mL, and dispense in 3.0-mL amounts in 15- × 125-mm tubes with inverted vials. Autoclave at 121°C for 15 min.

Fermentation Medium (Spray)

(for *Clostridium perfringens*)
Trypticase .... 10.0 g
Neopeptone .... 10.0 g
Agar .... 2.0 g
Sodium thioglycollate .... 0.25 g
Distilled water .... 1.0 L

Dissolve the ingredients except agar in distilled water, and adjust the pH to 7.4. Add the agar, and heat while stirring until the agar is dissolved. Mix well, and dispense 9.0-mL portions into 16- × 125-mm tubes. Autoclave for 15 min at 121°C. Before use, heat the tubed medium in boiling water or flowing steam for 10 min, and add 1.0 mL of 10% sterile carbohydrate solution to each tube.

Fish Extract Medium

Minced pollock fillets/distilled water .... 1:5 ratio

Centrifuge pollock mixture for 10 min at 11,000 × g. Filter the supernatant fluid by ultrafiltration through polysulphone membrane filters with a nominal limit molecular weight of 10,000 (Minitan system, Millipore). Extract may be stored lyophilized at 5°C.

*Fluid Thioglycollate Medium**

FPA (Fluorescent Pectolytic Agar) Medium

Proteose peptone No. 3 .... 20.0 g
Dipotassium phosphate .... 1.5 g
Magnesium sulfate • $7H_2O$ .... 0.73 g
Pectin .... 5.0 g
Agar .... 15.0 g

*Commercially available.

Distilled water .... 1.0 L

Dissolve ingredients in distilled water by boiling. Adjust to pH 7.1, and sterilize at 121°C for 15 min.

Add 1.0 mL ethanol to the following mixture of dry antibiotics, and allow to stand 30 min.

Penicillin G .... 75,000.0 IU
Novobiocin .... 45.0 mg
Cycloheximide .... 75.0 mg

Dilute antibiotic solution with 9 mL sterile distilled water, and add 1.0 mL of the dilute solution to 100.0 mL of sterile and cooled basal medium before pouring into sterile Petri plates. After the plates are inoculated by a spread plate technique and incubated, the fluorescent bacteria are detected under long-wavelength ultraviolet light. Plates are then treated with the polysaccharide precipitant as described under MP-7 medium.

Fraser Secondary Enrichment Broth

Proteose peptone .... 5.0 g
Tryptone .... 5.0 g
Lab Lemco powder (Oxoid) .... 5.0 g
Yeast extract .... 5.0 g
NaCl .... 20.0 g
KH_2PO_4 .... 1.35 g
Na_2HPO_4 .... 12.0 g
Esculin .... 1.0 g
Nalidixic acid (2% in 0.1 M NaOH) .... 1.0 mL
Lithium chloride .... 3.0 g
Distilled water .... 1.0 L

Combine the ingredients in distilled water, mix well and dispense in 10-mL aliquots in 20- × 150-mm test tubes. Autoclave at 121°C for 12 min. Note: Do not overheat; cool at once after removal from the autoclave. Store the sterilized medium under refrigeration. Immediately before use, to each 10-mL tube of medium add 0.1 mL of filter-sterilized acriflavin solution (2.5 mg/mL, Sigma) and 0.1 mL of filter-sterilized ferric ammonium citrate solution (5% stock solution in distilled water, Sigma).

Gelatin Agar

Gelatin .... 30.0 g
Agar .... 15.0 g
Sodium chloride .... 10.0 g
Trypticase (pancreatic digest of casein) .... 10.0 g
Distilled water .... 1.0 L

Dissolve ingredients in distilled water by bringing to a boil. Dispense in tubes or bottles, and sterilize at 121°C for 15 min. Final pH should be 7.2 ± 0.2.

Giolitti and Cantoni Broth

Tryptone .... 10.0 g
Beef extract .... 5.0 g
Yeast extract .... 5.0 g
Mannitol .... 20.0 g
Sodium chloride .... 5.0 g
Lithium chloride .... 5.0 g
Glycine .... 1.2 g
Sodium pyruvate .... 3.0 g
Distilled water .... 1.0 L

Dissolve ingredients in distilled water, warming the solution gently in order to dissolve it. Dispense 19 mL amounts into 20 mm × 200 mm test tubes. Sterilize by autoclaving at 121°C for 15 min. Cool at room temperature. Aseptically add 0.3 mL of a 3.5% potassium tellurite solution to each tube and swirl tubes gently to disperse throughout the medium.

GN-Broth

(Same as Gram-Negative Broth, Hajna)

*Gram-Negative (GN) Broth, Hajna**

Gum Tragacanth-Arabic

Gum tragacanth .... 2.0 g
Gum arabic .... 1.0 g
Distilled water .... 100.0 mL

Dissolve the ingredients in distilled water, heating slightly. Autoclave at 121°C for 15 min.

Halophilic Agar (HA)/Broth (HB)

Casamino acids .... 10.0 g
Yeast extract .... 10.0 g
Proteose peptone .... 5.0 g
Trisodium citrate .... 3.0 g
KCl .... 2.0 g
$MgSO_4•7H_2O$ .... 25.0 g
NaCl .... 250.0 g
Agar .... 20.0 g
Distilled water .... 1.0 L

Combine the ingredients with distilled water, and heat to boiling to dissolve completely. Autoclave at 121°C for 15 min. Final pH should be 7.2.

To prepare HB, omit the agar.

Ham's F-10 Medium (with Glutamine and $NaHCO_3$)

L-Alanine .... 8.91 mg
L-Arginine-HCl .... 211.0 mg
L-Asparagine•H_2O .... 15.0 mg
L-Aspartic acid .... 13.3 mg
L-Cysteine-HCl .... 35.12 mg
L-Glutamine .... 146.2 mg
L-Glutamic acid .... 14.7 mg
Glycine .... 7.51 mg
L-Histidine HCl•H_2O .... 21.0 mg
L-Isoleucine .... 2.6 mg
L-Leucine .... 13.1 mg
L-Lysine HCl .... 29.3 mg
L-Methionine .... 4.48 mg
L-Phenylalanine .... 4.96 mg
L-Proline .... 11.5 mg
L-Serine .... 10.5 mg
L-Threonine .... 3.57 mg
L-Tryptophan .... 0.60 mg
L-Tyrosine .... 1.81 mg
L-Valine .... 3.5 mg
Glucose .... 1100.0 mg
Hypoxanthine .... 4.08 mg
Lipoic acid .... 0.2 mg
Phenol red .... 1.2 mg
Sodium pyruvate .... 110.0 mg
Thymidine .... 0.727 mg
Biotin .... 0.024 mg
Choline chloride .... 0.698 mg
Folic acid .... 1.320 mg

*Commercially available.

Isoinositol .... 0.541 mg
Niacinamide .... 0.615 mg
D-calcium pantothenate .... 0.715 mg
Pyridoxine-HCl .... 0.206 mg
Riboflavin .... 0.376 mg
Thiamine HCl .... 1.010 mg
Vitamin B_{12} .... 1.360 mg
$CaCl_2 \bullet 2H_2O$ .... 44.10 mg
$CuSO_4 \bullet 5H_2O$ .... 0.0025 mg
$FeSO_4 \bullet 7H_2O$ .... 0.83 mg
KCl .... 285.0 mg
KH_2PO_4 .... 83.0 mg
$MgSO_4 \bullet 7H_2O$ .... 152.8 mg
NaCl .... 7400.0 mg
$NaHCO_3$ .... 1200.0 mg
$Na_2HPO_4 \bullet 7H_2O$ .... 290.0 mg
$ZnSO_4 \bullet 7H_2O$ .... 0.028 mg
Distilled water .... 1.0 L

Dissolve ingredients in 1.0 L of distilled water and sterilize by filtration. Adjust the final pH to 7.0 ± 0.2. Check sterility before use. Store at 2 to 8°C.

HC (Hemorrhagic Coli) Agar

Tryptone .... 20.0 g
Bile salts No. 3 .... 1.12 g
Sodium chloride .... 5.0 g
Sorbitol .... 20.0 g
4-Methylumbelliferyl-β-D glucuronide (MUG) .... 0.1 g
Bromcresol purple .... 0.015 g
Agar .... 15.0 g
Distilled water .... 1.0 L

Dissolve the ingredients listed above in distilled water by heating with stirring. Autoclave the medium for 15 min at 121°C. The final pH should be 7.2 ± 0.2.

Note: MUG is not essential for the enzyme-labeled monoclonal antibody procedure.

Heart Infusion Agar/Broth*

Hektoen Enteric (HE) Agar*

HHD Medium

Fructose .... 2.5 g
KH_2PO_4 .... 2.5 g
Trypticase peptone .... 10.0 g
Phytone peptone .... 1.5 g
Casamino acids .... 3.0 g
Yeast extract .... 1.0 g
Tween 80 .... 1.0 g
Bromcresol green solution .... 20.0 mL
Agar (when desired) .... 20.0 g
Distilled water .... 1.0 L

Combine ingredients with 1.0 L of distilled water, and heat to boiling while stirring to dissolve the agar completely. Adjust the pH to 7.0 ± 0.2, and autoclave for 15 min at 121°C.

Bromcresol green solution is prepared by dissolving 0.1 g of bromcresol green in 30.0 mL of 0.01 N NaOH.

HL Agar (Horse Blood Overlay Agar)

Base Layer: Prepare Columbia Blood Agar Base, and autoclave at 121°C for 15 min. Place 10.0 mL into Petri dishes as the base.

Top Layer: Prepare Columbia Blood Agar Base, and autoclave at 121°C for 15 min. Add 5.0% horse blood to the tempered (46°C) medium, and mix thoroughly. Pour 5.0 mL of blood agar over the base layer while the base is still warm. HL Agar may be stored under refrigeration in plastic bags; discard any plates that become discolored or show hemolysis.

Hoag-Erickson Medium

MRS broth (commercially available) .... 25.00 g
Potassium phosphate, monobasic .... 5.00 g
Potassium phosphate, dibasic .... 1.00 g
D-mannitol .... 12.50 g
Ferrous sulfate heptahydrate 0.01g
Dextrose .... 10.00 g
Distilled water .... 1.0 L

Combine ingredients with 1.0 L distilled water, and heat to boiling while stirring continuously. Adjust the pH to 6.0 ± 0.1, and autoclave for 15 min at 121°C.

Hunt Broth

Nutrient Broth No. 2:
Lab-Lemco powder (Oxoid L29) .... 10.0 g
Peptone .... 10.0 g
NaCl .... 5.0 g
Yeast extract .... 6.0 g
Distilled water .... 1.0 L

Mix the above ingredients to a final pH of 7.5 ± 0.2. Autoclave for 15 min at 121°C in graduated bottles. Use broth within one month of preparation (preferably less than 2 weeks). Media adsorbs oxygen during storage, which can inhibit recover of microaerophilic organisms. Keep bottles very tightly closed. Before use, add 50 mL frozen and thawed lysed horse blood (5%), 4 mL FBP, and 4 mL of appropriate antibiotic concentrates (solutions made separately as described below). Store powdered media tightly closed in cool, dry area to reduce oxygen infusion and peroxide formation.

Lysed Horse Blood: Use only fresh blood and lyse upon receiving. If lysed when purchased, be sure it was prepared from fresh blood and shipped promptly. If not used upon receipt. Freeze within 1-2 days. To lyse and/or store, gently mix and pour about 100 mL per sterile polypropylene bottle. Place in –20°C freezer. Thaw and refreeze once more to lyse. Freezing inhibits adsorption of oxygen by hemoglobin. Use blood up to 5 months. Can be thawed and refrozen several times.

Prepare the following solutions in volumetric flasks and filter-sterilize with 0.45 µm filters. Prepare chemicals and antibiotics with short shelf-lives (FBP, cefoperazone, vancomycin, pyruvate, and hemin) in small amounts and sterilize using 0.45 µm syringe filters. Syringes need not be sterile. Filter into sterile containers and dispense into sterile plastic tubes if storing frozen.

FBP:
Ferrous sulfate .... 6.25 g
Sodium metabisulfite .... 6.25 g
Sodium pyruvate (weigh using larger weighin boat) ..... 6.25 g
Distilled water to make .... 100 mL

Combine ingredients in 100 mL volumetric flask and bring to line with distilled water. Dissolve pyruvate in about 10 mL distilled water to facilitate pouring into volumetric flask. Filter through 0.45 µm filter and store in 15-25 mL portions in tightly

*Commercially available.

capped sterile polypropylene containers for up to one month at –20°C. Freezing inhibits reaction with oxygen. Can be refrozen one time. Protect from light.

Antibiotic Formula No. 1: (prepare each solution separately)

Cefoperazone: Antibiotic is prepared 1/2 strength to be added twice to broth (4 mL/L each time). Weigh 0.4 g into 100 mL distilled water in volumetric flask. Dissolve and filter through a 0.45 µm filter. Solution may be stored 5 days at 4°C, 14 days at –20°C, and 5 months at –70°C. Freeze in sterile plastic tubes or bottles. May be prepared in smaller amount and syringe filter-sterilized. Powder may be purchased from Sigma or obtained free from Pfizer by writing Dr. Sydney Jacobson, Roering Division, Pfizer, 235 E. 42nd St., New York, NY 10017. Request 2-4 g for in vitro use. Final concentration is 32 mg/L.

Trimethoprim Lactate (Sigma): Dissolve 0.357 g in 100 mL distilled water, and filter-sterilize. It will precipitate. May be stored 1 year at 4°C. Add 4 mL/L for final concentration of 15 mg/L.

Vancomycin (Sigma): Dissolve 0.25 g in 100 mL distilled water and filter-sterilize. Store up to 2 months at 4°C. Because of short shelf-life, prepare in smaller amounts. Add 4 mL/L for final concentration of 10 mg/L.

Amphotericin B: Dissolve 50 mg in a volumetric flask and bring to line with water. Filter-sterilize. Store at –20°C up to 1 year. Add 4 mL for final concentration of 2 mg/L.

Antibiotic Formula No. 2: (prepare each solution separately)

Rifampicin (10 mg/L final concentration): Rifampicin is substituted for vancomycin in Antibiotic Formula No. 1. It is to be added twice to broth at 4 mL/L each time. Weigh 0.125 g slowly into 30 mL alcohol, swirling repeatedly. When powder is dissolved completely, bring to 100 mL with distilled water. May be stored for 1 year at –20°C. Final concentration is 10 mg/L.

Cefoperazone, Trimethoprim lactate, and Amphotericin B: Use instructions for Antibiotic Formula No. 1.

HYA Agar

Beef extract .... 1.0 g
Proteose peptone No. 3 .... 10.0 g
Glucose .... 2.5 g
Galactose .... 2.5 g
Lactose .... 5.0 g
Agar .... 15.0 g
Distilled water .... 1.0 L

Dissolve the ingredients except for the glucose, galactose, and lactose in distilled water with heat. Adjust the pH to 6.8 ± 0.2 before autoclaving in 90.0-mL aliquots at 121°C for 20 min. Add each of the sugars as cold, sterilized, 10% solutions (filter-sterilized) just prior to plating samples.

Inositol Brilliant Green Bile Salts (IBB) Agar

Peptone .... 10.0 g
Meat extract .... 5.0 g
Sodium chloride .... 5.0 g
Bile salts No. 3 .... 8.5 g
Brilliant green .... 0.33 mg
Neutral red .... 25.0 mg
Inositol .... 10.0 g
Agar .... 15.0 g
Distilled water .... 1.0 L

Dissolve ingredients in distilled water, and adjust the pH to 7.2. Autoclave at 115°C for 15 min.

Inositol Gelatin Deeps

Gelatin .... 120.0 g
Sodium phosphate (dibasic) .... 5.0 g
Yeast extract .... 5.0 g
Inositol .... 10.0 g
Phenol red .... 0.05 g
Distilled water .... 1.0 L

Heat to dissolve ingredients in distilled water, and adjust the pH to 7.4. Distribute the medium into tubes (5 mL per tube), and autoclave at 115°C for 15 min.

Iron Milk Medium

Fresh whole milk .... 1.0 L
Ferrous sulfate•$7H_2O$* .... 1.0 g
Distilled water .... 50.0 mL

Dissolve ferrous sulfate in the distilled water. Add slowly to 1 L milk and mix. Dispense 10 mL into 16 × 150 mm culture tubes. Autoclave for 12 min at 118°C. Prepare fresh each time.

*Or add 0.2 g of iron powder to each 10-mL tube of milk.

Johnson Murano (JM) Semi-Solid Enrichment Broth

Special peptone (Oxoid L72) .... 10.0 g
Yeast extract (BRL) .... 5.0 g
Beef extract (Difco) .... 5.0 g
NaCl .... 4.0 g
Potassium phosphate monobasic .... 1.5 g
Sodium phosphate dibasic .... 3.5 g
Pyruvate .... 0.5 g
Thioglycollate .... 0.5 g
Charcoal .... 0.5 g
Bile salts No. 3 (Difco 0130-15) .... 2.0 g
Agar .... 2.0 g
Distilled water .... 1.0 L

Antimicrobials (final concentrations in broth):

5-Fluorouracil .... 200 mg/L
Cefoperazone .... 32 mg/L

5-Fluorouracil is commercially available (Hoffman LaRoche, Cat. No. 0688, Nutley, NJ). Dissolve cefoperazone (320 mg) in 10 mL distilled water. Filter sterilize (0.45 um). Store in screw-cap test tubes (20 mm × 125 mm).

Mix ingredients with distilled water (except for antimicrobials) to achieve a final volume of 1.0 L (Ph 6.8). Heat to boiling and cool to 55-60°C. Stir the mixture to keep the charcoal in suspension since the charcoal will not completely dissolve. Autoclave at 121°C for 20 min and cool to 55-60°C. Stir and add antimicrobials to achieve the desired concentrations.

Johnson Murano (JM) Agar

Special peptone (Oxoid L72) .... 10.0 g
Yeast extract (BRL) .... 5.0 g
Beef extract (Difco) .... 5.0 g
NaCl .... 4.0 g
Potassium phosphate monobasic .... 1.5 g
Sodium phosphate dibasic .... 3.5 g
Pyruvate .... 0.5 g
Thioglycollate .... 0.5 g
Agar .... 20.0 g
Distilled water .... 1.0 L

*Commercially available.

Antimicrobials (final concentrations in broth):
Cefoperazone .... 32 mg/L
Defibrinated Sheep Blood .... 50 mL

Dissolve cefoperazone (320 mg) in 10 mL of distilled water. Filter-sterilize (0.45 um pore size filter). Add ingredients to (except for antimicrobial and blood) to distilled water to achive a final volume of 1.0 L. Heat to boiling. Autoclave at 121°C for 20 min and cool to 55-60°C. Add 1.0 mL of the stock antibiotic solution and 5% defibrinated ovine blood (Lampire Biological Laboratories, Pipersville, PA). Dispense aseptically to Petri dishes (100 x 15 mm). Plates may be refrigerated for up to 1 month.

K Agar (Difco)*

Kaper's Medium

Proteose peptone (Difco) .... 5.0 g
Yeast extract (Difco) .... 3.0 g
Tryptone (Difco) .... 10.0 g
L-ornithine•HCl .... 5.0 g
Mannitol .... 1.0 g
Inositol .... 10.0 g
Sodium thiosulfate .... 0.4 g
Ferric ammonium citrate .... 0.5 g
Bromcresol purple .... 0.02 g
Agar .... 3.0 g
Distilled water .... 1.0 L

Heat to dissolve all ingredients completely in distilled water and adjust the pH to 6.7. Dispense the medium into tubes (5 mL per tube) and autoclave at 121°C for 12 minutes.

Kim-Goepfert (KG) Agar

Preparation A:
Peptone .... 1.0 g
Yeast extract .... 0.5 g
Phenol red .... 0.025 g
Agar .... 18.0 g
Distilled water .... 900.0 mL

Preparation B: Colbeck's Egg Yolk Broth or Concentrated Egg Yolk Emulsion:
Concentrated Egg Yolk Emulsion is available commercially from Oxoid Limited or as Egg Yolk Enrichment 50% from Difco Laboratories.

Preparation C: Polymyxin B Sulfate:
This selective agent is obtainable in sterile powdered form (500,000 units, i.e., 50 mg per vial) from Pfizer Inc. To use, aseptically add 5.0 mL of sterile distilled water with a sterile syringe. Mix to dissolve the powder.

Dissolve Preparation A ingredients in distilled water, heat to boiling to completely dissolve the agar, and adjust the pH to 6.8. Autoclave at 121°C for 20 min, cool to 50°C and add 100.0 mL of Preparation B and 1.0 mL of Preparation C. Mix well, pour into Petri dishes, allow to solidify, and store in a manner to eliminate excess surface moisture. Plates may be stored at 4°C for up to 7 days.

Kligler Iron Agar (KIA)*

KM Agar

Nonfat milk .... 10.0 g
Milk protein hydrolysate .... 2.5 g
Glucose .... 5.0 g
Agar .... 15.0 g
Distilled water .... 1.0 L

Combine the ingredients in distilled water, and heat while stirring to dissolve the agar completely. Adjust the pH to 6.6 before autoclaving for 12 min at 115°C (10 lb/in^2). Cool the medium to 45°C in a water bath.

To each L of KM agar at 45°C add 10.0 mL of each of the following solutions which have been steamed (110°C) for 30 min:

Solution 1: 10% potassium ferricyanide

Solution 2: 1.0 g of ferric citrate and 1.0 g of sodium citrate in 40.0 mL of distilled water.

Gently swirl the medium and pour plates. Dry the plates in the dark for 24 hr at 30°C.

Koser's Citrate Medium*

Kot Medium

Trypticase peptone .... 5.00 g
Malt extract .... 2.50 g
Yeast extract .... 2.50 g
Liver concentrate .... 1.00 g
Maltose .... 5.00 g
Glucose .... 5.00 g
L-malic acid .... 0.50 g
Tween 80 .... 1.0 mL
K_2HPO_4 .... 0.50 g
$MgSO_4$•$7H_2O$ .... 0.125 g
$MnSO_4$•$5H_2O$ .... 0.025 g
Cysteine monohydrochloride .... 0.50 g
Cytidine .... 0.20 g
Thymidine .... 0.20 g
Actidione .... 0.10 g
NaN_3 .... 0.05 g
Agar .... 20.00 g
Beer .... 800 mL
Distilled water .... adjust to 1.0 L

Combine ingredients and heat to boiling while stirring to dissolve the agar completely. Adjust the pH to 6.3 ± 0.02, and autoclave for 15 min at 121°C.

Lab-Lemco Meat Extract Powder* (Oxoid)

Lactic Acid Medium*

(Commercially available from bioMerieux SA, France for Bactometer)

Lactic (Elliker) Agar

Add 15.0 g of agar per L of commercially available Elliker broth.

Lactic Streak Agar

Sodium carboxymethylcellulose .... 10.0 g
Yeast extract .... 5.0 g
Phytone peptone .... 5.0 g
Beef extract .... 5.0 g
Lactose .... 1.5 g
Calcium citrate .... 10.0 g
L-arginine hydrochloride .... 1.5 g
Polypeptone .... 5.0 g
Agar .... 15.0 g

*Commercially available.

Bromocresol purple (0.1%) .... 2.0 mL
Distilled water .... 1.0 L

Suspend the agar in 800.0 mL of distilled water, and heat to dissolve completely. Add the other ingredients, with the exception of the sodium carboxymethylcellulose, calcium citrate, and bromcresol purple. Suspend the sodium carboxymethylcellulose and calcium citrate in 200.0 mL of water, disperse using a blender, and then add the mixture to the molten medium. Adjust the pH to 6.0, dispense in 100.0-mL amounts, and autoclave at 115°C for 10 min. Cool the sterilized medium to 50°C and add 2.0 mL of a sterile 0.1% aqueous bromcresol purple solution. Pour plates and dry at 37°C for one hr. Discard plates if not used that day. Store bottles of agar at 5°C.

Note: If higher temperatures are used in autoclaving, the lactose will have to be sterilized separately to avoid sugar-amino acid reactions.

Lactobacillus *Heteroferm Screen Broth*

(Same as Modified MRS)
Bacto proteose peptone No. 3 .... 10.0 g
Bacto yeast extract .... 5.0 g
Tween 80 .... 1.0 mL
Ammonium citrate .... 2.0 g
Sodium acetate .... 5.0 g
Magnesium sulfate .... 0.1 g
Manganese sulfate .... 0.05 g
Dipotassium phosphate .... 2.0 g
Dextrose (Bacto) .... 20.0 g
Bromcresol green (Bacto) .... 0.04 g
2-phenylethyl alcohol (to inhibit gram-negatives) ..... 3.0 g
Actidione (to inhibit yeasts) .... 4.0 mg
Distilled water .... 1.0 L

Combine the ingredients with distilled water, and stir to dissolve. Adjust the pH to 4.3 ± 0.01 (pH adjustment is critical for excluding CO_2-producing starter cultures such as *Leuconostoc* spp.) with concentrated HCl, and then dispense the medium into 16- × 125-mm tubes containing inverted Durham fermentation tubes. Autoclave for 15 min at 121°C.

Modifications for Salad Dressings: MRS agar is prepared by adding 15.0 g of agar per L of medium. For lactobacilli detection in salad dressings, adjust the medium pH to 5.5 with glacial acetic acid after sterilization.

Lactobacillus *Selection (LBS) Agar*

Note: 1.32 mL of glacial acetic acid must be added to the commercially available BBL product (LBS agar).

Modifications: Add 1% fructose to LBS medium to ensure greater enumeration of certain lactobacilli. Add 200 ppm cycloheximide if needed to inhibit yeasts. To prepare LBS Modified adjust the pH to 5.4 ± 0.2 using acetic acid and add 0.0075% sterile brilliant green prior to cooling. Mix, cool and pour plates.

*Lactose Broth**

Lactose Gelatin Medium

Tryptose .... 15.0 g
Yeast extract .... 10.0 g
Lactose .... 10.0 g
Gelatin .... 120.0 g
Phenol red (as solution) .... 0.05 g
Distilled water .... 1.0 L

Suspend the ingredients except the gelatin and phenol red in 400.0 mL of distilled water, and dissolve by heating gently while stirring. Suspend the gelatin in 600.0 mL of cold distilled water, and dissolve by heating in a water bath at 50 to 60°C with frequent stirring. When the gelatin is dissolved, combine with the other dissolved ingredients and adjust the pH to 7.5 with 1 N sodium hydroxide. Add the phenol red, mix well, and dispense 10.0-mL portions into 16- × 125-mm screw-capped tubes. Sterilize by autoclaving for 10 min at 121°C. If the medium is not used within 8 hr, deaerate by holding in a water bath at 50° to 70°C for 2 to 3 hr before use.

*Lauryl Tryptose Broth or Lauryl Sulfate Tryptose (LST) Broth**

LB (Lactobacillus bulgaricus) *Agar*

Tryptone .... 10.0 g
Yeast extract .... 5.0 g
Glucose .... 20.0 g
Dipotassium phosphate .... 2.0 g
Beef extract .... 10.0 g
Filtered tomato juice .... 40.0 mL
Tween 80 .... 1.0 g
Distilled water .... 780.0 mL
Agar .... 20.0 g

Combine the ingredients with distilled water, and heat to dissolve. Adjust the pH to 6.8, and then add 80.0 mL of acetate buffer (113.55 g of sodium acetate and 9.90 g of acetic acid per L). Autoclave at 121°C for 15 min.

LBS (Lactobacillus *Selection) Oxgall Agar*

Prepare LBS agar and add 0.15% oxgall (BBL) before adjusting the pH to 5.4 ± 0.2 and heating.

Lee's Agar

Tryptone .... 10.0 g
Yeast extract .... 10.0 g
Lactose .... 5.0 g
Sucrose .... 5.0 g
Calcium carbonate .... 3.0 g
Dipotassium phosphate .... 0.5 g
Bromcresol purple solution (0.2%) .... 10.0 mL
Agar .... 18.0 g
Distilled water to make .... 1.0 L

Dissolve the ingredients except for the bromocresol purple in distilled water with gentle heating. Adjust the pH of the medium to 7.0 ± 0.2 before autoclaving 20 min at 121°C. Carefully mix the melted medium to suspend the calcium carbonate evenly. Just before pouring plates, add 10.0 mL of a sterile (121°C for 15 min) 0.2% bromocresol purple solution. Pour the medium into previously chilled sterile Petri dishes to obtain a layer of medium 4- to 5-mm thick. After solidification, dry the plates in a 30°C incubator for 18 to 24 hr.

*LES Endo Agar**

*Levine Eosine Methylene Blue (L-EMB) Agar**

*Commercially available.

Listeria *Enrichment Broth (LEB)—FDA*

Acriflavin HCl (Sigma) .... 15.0 mg
Nalidixic acid (sodium salt, Sigma) .... 40.0 mg
Cycloheximide (Sigma) .... 50.0 mg
Trypticase soy broth-yeast extract (sterile) .... 1.0 L

Add the three supplemental ingredients aseptically to Trypticase soy broth-yeast extract (TSB-YE) after autoclaving and just before use. Make acriflavin and nalidixic acid supplements as 0.5% stock solutions in distilled water. Make the cycloheximide supplement as 1.0% stock in 40% ethanol in water. Filter-sterilize all three supplementary ingredients. Add 0.68 mL of acriflavin solution, 1.8 mL of nalidixic acid solution, and 1.15 mL of cycloheximide solution to 225.0 mL of TSB-YE to achieve the correct concentrations.

Listeria *Repair Broth (LRB)*

Trypticase soy broth .... 30.0 g
Glucose .... 5.0 g
Yeast extract .... 6.0 g
MOPS (free acid) .... 8.5 g
MOPS (sodium salt) .... 13.7 g
Ferrous sulfate .... 0.3 g
Magnesium sulfate .... 2.46 g
Pyruvate .... 10.0 g
Distilled water .... 1.0 L

Add ferrous sulfate to distilled water and let dissolve. Then add TSB, glucose, yeast extract and MOPS. With heat and mixing, allow to dissolve completely. After all ingredients are completely dissolved, add magnesium sulfate. Boil. Remove from heat and add pyruvate, mix. Solution should be clear and a deep gold to slightly reddish color. Dispense as desired. Autoclave 15 min.

*Litmus Milk**

Liver Broth

Fresh beef liver .... 500.0 g
Distilled water .... 1.0 L
Tryptone .... 10.0 g
Soluble starch .... 1.0 g
Dipotassium phosphate .... 1.0 g

Remove the fat from 1 lb of fresh beef liver. Grind the liver, mix with 1.0 L of distilled water, and boil slowly for 1 hr. Adjust the pH to 7.6, and remove the liver particles by straining through cheesecloth. Bring the volume of the broth back to 1.0 L with distilled water, and add the tryptone, dipotassium phosphate, and soluble starch. Refilter the broth. Dispense 15.0 mL of the broth into 20- × 150-mm tubes and add previously removed liver particles to a depth of 1.0 in. in each tube. Autoclave 20 min at 121°C.

Liver Broth For Mesophilic Anaerobic Sporeformers

Double-strength liver infusion .... 500 mL
Distilled water .... 500 mL
Tryptone .... 10 g
K_2HPO_4 .... 1 g
Soluble starch .... 1 g

To make liver infusion, add 1.0 L distilled water to 1.0 L ground beef liver. Boil slowly for 1 hr. Add distilled water to make to volume. Press boiled material through cheese cloth; this yields a double-strength infusion. Store frozen until needed. Retain liver particles and dry thoroughly by spreading a thin layer and allowing to air-dry at 35-57°C. Alternatively, the boiled liver particles can be frozen. To make broth, mix all ingredients and adjust pH to 8.5 using 5N NaOH. Place a small pinch of dried liver particles or 1/2 inch of frozen liver particles in the bottom of each test tube and add 9 mL of broth. Autoclave 20 min at 121°C.

Liver Broth Plus Salt

Prepare liver broth, and add 15% salt (or closely approximate the salt content of the sample to be examined).

*Liver Veal Agar (LVA)**

Liver Veal Egg Yolk Agar

Fresh eggs (antibiotic free) 3.0
Liver veal agar .... 1.0 L

Wash eggs with a stiff brush, and drain. Soak eggs in 0.1% mercuric chloride solution for 1 hr. Pour off the mercuric chloride solution, and replace with 70% ethyl alcohol. Soak in 70% ethyl alcohol for 30 min. Crack the eggs aseptically, and discard the whites. Remove the yolk with a sterile 50-mL Luerlok syringe. Place in a sterile container, and add an equal volume of sterile saline (0.85% sodium chloride). Mix thoroughly. To each 500.0 mL of melted liver veal agar tempered to 50°C add 40.0 mL of the egg yolk-saline solution. Mix thoroughly and pour plates. Dry plates at room temperature for 2 days or at 35°C for 24 hours. Discard contaminated plates, and store sterile plates under refrigeration.

LPM (Lithium-Phenylethanol-Moxalactam) Agar

Phenylethanol agar .... 35.5 g
Glycine anhydride .... 10.0 g
Lithium chloride .... 5.0 g
Distilled water .... 1.0 L

Prepare as directed by the manufacturer, and autoclave at 121°C for 12 min. Cool the medium in a 46°C waterbath, and then add 2.0 mL of 1.0% filter-sterilized moxalactam per L of medium. Mix well, and pour 12.0 mL into 100-mm plates. Refrigerate LPM plates in plastic bags. The 1% moxalactam solution is prepared by dissolving 1.0 g of sodium or ammonia moxalactam in 100.0 mL of 0.1 M potassium buffer (0.1 M KH_2PO_4 into 0.1 M K_2HPO_4 until the pH reaches 6.0). Filter-sterilized moxalactam can be frozen until needed at –60°C in 4.0-mL quantities.

LPM (Lithium-Phenylethanol-Moxalactam) Agar Plus Esculin And Ferric Iron

Phenylethanol agar .... 35.5 g
Glycine anhydride .... 10.0 g
Lithium chloride .... 5.0 g
Esculin .... 1.0 g
Ferric ammonium citrate .... 0.5 g
Distilled water .... 1.0 L

Mix ingredients and autoclave at 121°C for 12 min. Cool the medium in a 46°C waterbath, and then add 2.0 mL of 1.0% filter-sterilized moxalactam per L of medium. Mix well, and pour 12.0 mL into 100-mm plates. Refrigerate plates in plastic bags. The 1% moxalactam solution is prepared by dissolving 1.0 g of sodium or ammonia moxalactam in 100.0 mL of 0.1 M potassium buffer (0.1 M KH_2PO_4 into 0.1 M K_2HPO_4 until the pH reaches 6.0). Filter-sterilized moxalactam can be frozen until needed at –60°C in 4.0-mL quantities.

*Commercially available.

Luria Broth

Bacto tryptone .... 10.0 g
Bacto yeast extract .... 5.0 g
NaCl .... 0.5 g
1M NaOH .... 2.0 mL
Distilled water .... 1.0 L

Dissolve all ingredients in distilled water and adjust pH to 7.0 with 1M NaOH. Autoclave at 121°C for 15 min. Add 10 mL of a 20% filter-sterilized glucose solution after autoclaving.

Lysine Artginine Iron Agar (LAIA)

Peptone .... 5.0 g
Yeast extract .... 3.0 g
Glucose .... 1.0 g
L-lysine .... 10.0 g
L-arginine .... 10.0 g
Ferric ammonium citrate .... 0.5 g
Sodium thiosulfate .... 0.04 g
Bromcresol purple .... 0.02 g
Agar .... 15.0 g
Distilled water .... 1.0 L

Mix all ingredients and adjust pH to 6.8. Heat to boiling and dispense 5 mL into 13 × 100mm screw-cap tubes. Autoclave at 121°C for 12 min. Cool tubes in slanted position. This medium may also be prepared by supplementing Difco lysine iron agar with 10 g L-arginine per L.

Lysine Decarboxylase Broth*

Lysine Iron Agar (LIA)*

(Edwards And Fife)

Lysozyme Broth

Nutrient Broth: Prepare nutrient broth, and dispense 99.0-mL amounts in bottles or flasks. Autoclave for 15 min at 121°C.

Lysozyme Solution: Dissolve 0.1 g of lysozyme in 65.0 mL of sterile 0.01 N hydrochloric acid. Heat to boiling for 20 min, and dilute to 100.0 mL with sterile 0.01 N hydrochloric acid. Alternatively, dissolve 0.1 g of lysozyme chloride in 100.0 mL of distilled water, and sterilize by filtration. Test solution for sterility before use.

Add 1.0 mL of sterile 0.1% lysozyme solution to each 99.0 mL of nutrient broth. Mix thoroughly, and aseptically dispense 2.5 mL of the complete medium into sterile 13- × 100-mm tubes.

M 16 Agar

(A modification of Rogosa SL agar)

Beef extract .... 5.0 g
Yeast extract .... 2.5 g
Ascorbic acid .... 0.5 g
Phytone or soy tone .... 5.0 g
Polypeptone or tryptose .... 5.0 g
Sodium acetate trihydrate .... 3.0 g
Lactose or dextrose .... 5.0 g
Agar .... 10.0 g
Distilled water .... 1.0 L

Dissolve all of the ingredients except for the lactose or dextrose in 950.0 mL of distilled water, and boil while stirring to dissolve the agar completely. Adjust the pH to 7.2 ± 0.2 with 2 N NaOH. Autoclave at 121°C for 15 min and cool to 50°C. Add a sterile solution of lactose or dextrose (5.0 g in 50.0 mL of distilled water, autoclaved at 121°C for 15 min) to the medium.

M 17 Agar

Phytone peptone .... 5.0 g
Polypeptone .... 5.0 g
Yeast extract .... 2.5 g
Beef extract .... 5.0 g
Lactose .... 5.0 g
Ascorbic acid .... 0.5 g
Disodium glycerophosphate .... 19.0 g
1.0 M $MgSO_4 \bullet 7H_2O$ .... 1.0 mL
Agar .... 10.0 g
Distilled water .... 1.0 L

Add 10.0 g of agar to 950.0 mL of distilled water, and heat to boiling to dissolve the agar completely. The remaining ingredients, except for the lactose, should be added to the dissolved agar, and the mixture should be autoclaved at 121°C for 15 min. After cooling the sterile medium to 45°C in a temperature-controlled water bath, add a sterile solution of lactose (5.0 g in 50.0 mL of distilled water, sterilized at 121°C for 15 min). Final pH should be 7.15.

M-Broth*

MacConkey (MAC) Agar*

MacConkey Glucose Agar

(Same as Violet Red Bile Agar with 1% glucose)

Malonate Broth*

Malonate Broth Modified*

(With added glucose and yeast extract)

Malt Acetic Acid Agar (MAA)

Add glacial acetic acid to make a 0.5% final concentration to molten malt extract agar (45-50°C) before pouring plates. Pour plates immediately after adding acid.

Malt Extract Agar (MEA)*

Malt extract .... 20.0 g
Dextrose .... 20.0 g
Peptone (bacteriological) .... 1.0 g
Agar .... 20.0 g
Distilled water .... 1.0 L

Suspend the ingredients in distilled water and heat the mixture to boiling while stirring to dissolve completely. Autoclave at 121°C for 15 min. Formulations of this medium are commercially available.

Malt Extract Yeast Extract Chloramphenicol Ketoconazol Agar (MYCK)

Malt extract .... 20.0 g
Yeast extract .... 2.0 g
Chloramphenicol .... 0.5 g
Agar .... 15.0 g
Distilled water .... 1.0 L

Mix together all ingredients and steam until dissolved. Autoclave at 121°C for 15 min. The final pH is 5.6. Temper to 45-50°C

*Commercially available.

and add 50 mL of a 1% (w/v in ethanol) filter-sterilized solution of ketoconazol per L of agar.

Malt Extract Yeast Extract 40% Glucose (MY40G) Agar

Malt extract powder .... 12.0 g
Yeast extract .... 3.0 g
Agar .... 12.0 g
Distilled water to make .... 600.0 g
Glucose, AR .... 400.0 g

Combine the ingredients except glucose with 550.0 mL of distilled water, and steam to dissolve the agar. Immediately make up to 600.0 g with distilled water. While the solution is still hot, add the glucose all at once, and stir rapidly to prevent the formation of hard lumps of glucose monohydrate. If lumps form, dissolve them by steaming for a few minutes. Steam the medium for 30 min. This medium is of sufficiently low a_w not to require autoclaving. The final pH of this medium is about 5.5, and the a_w is near 0.92.

Malt Extract Yeast Extract 50% Glucose Agar (MY50G)

Malt extract powder .... 10.0 g
Yeast extract .... 2.5 g
Agar .... 10.0 g
Distilled water .... 0.5 L
Glucose monohydrate .... 500.0 g

Combine first three ingredients with 450 mL distilled water and steam to dissolve. Make up to 500 mL with distilled water. Add glucose and stir rapidly to prevent lump formation. Steam for 30 min. The final water activity is 0.89; therefore, autoclaving is not necessary.

Mannitol Yolk Polymyxin (MYP) Agar

Preparation A:
Beef extract .... 1.0 g
Peptone .... 10.0 g
D-Mannitol .... 10.0 g
NaCl .... 10.0 g
Phenol red .... 0.025 g
Agar .... 15.0 g
Distilled water .... 900.0 mL

Preparation B: Colbeck's Egg Yolk Broth or Concentrated Egg Yolk Emulsion:

Concentrated Egg Yolk Emulsion is available commercially from Oxoid Limited or as Egg Yolk Enrichment 50% from Difco Laboratories.

Preparation C: Polymyxin B Sulfate:

Dissolve 500,000 units of sterile polymyxin B sulfate (Burroughs Wellcome Co, Research Triangle Park, NC 27709) in 50.0 mL of sterile distilled water.

Mix the ingredients in distilled water, adjust the pH to 7.2 ± 0.1, heat to boiling to dissolve, and dispense 225.0-mL portions into 500-mL flasks. Autoclave at 121°C for 20 min, cool to 50°C in a waterbath and add 12.5 mL of Preparation B and 2.5 mL of Preparation C to each flask containing 225.0 mL of medium. Mix well, pour into Petri dishes, allow to solidify and dry for 24 hr at room temperature. Plates may be stored at 4°C for 7 days.

MD Medium

L-malic acid (Sigma) .... 20.0 g
Trypticase .... 10.0 g
D-(+)-Glucose (Sigma) .... 5.0 g
Casamino acids .... 3.0 g
Phytone .... 1.5 g
Yeast extract .... 1.0 g
Tween 80 (Atlas Chemical Ind.) .... 1.0 g
Bromcresol green solution .... 20.0 mL
Agar (when desired) .... 20.0 g
Distilled water .... 1.0 L

Combine ingredients with 1.0 L of distilled water. Heat to boiling while stirring to dissolve agar completely. Adjust the pH to 7.0 with 10 N KOH and autoclave for 15 min at 121°C. Medium may be stored at room temperature.

Bromcresol green solution is prepared by dissolving 0.1 g of bromcresol green in 30.0 mL of 0.01 N NaOH.

M-Endo Medium (Agar Or Broth)*

MF Endo Broth*

m-Green Yeast and Mold Medium*

m-HPC Agar*

Minerals Modified Glutamate Agar

Lactose .... 10.0 g
Sodium formate .... 0.25 g
L(-)-Cystine .... 0.02 g
L(-)-Aspartic acid .... 0.024 g
K_2HPO_4 .... 0.9 g
Thiamin .... 0.001 g
Nicotinic acid .... 0.001 g
Pantothenic acid .... 0.001 g
$MgSO_4 \cdot 7H_2O$ .... 0.1 g
Ferric ammonium citrate .... 0.01 g
$CaCl_2 \cdot 2H_2O$ .... 0.01 g
L(+)-Arginine .... 0.02 g
Bromcresol purple .... 0.01 g
Ammonium chloride .... 2.5 g
Sodium glutamate .... 6.35 g
Agar .... 15.0 g
Distilled water .... 1.0 L

Combine all ingredients, and heat to boiling to dissolve agar. Autoclave for 10 min at 116°C. The final pH should be 6.7 ± 0.2. Dispense 20-mL portions into 15- × 100-mm Petri dishes.

Minimal Essential Medium with Earle's Salts (MEME)*

Minimal Essential Medium with Hank's Salts (MEMH)

L-Arginine HCl .... 126.4 mg
L-Cystine Na .... 28.42 mg
L-Glutamine .... 292.3 mg
L-Histidine $HCl \cdot H_2O$ .... 41.90 mg

*Commercially available.

L-Isoleucine .... 52.50 mg
L-Leucine .... 52.50 mg
L-Lysine HCl .... 73.06 mg
L-Methionine .... 14.90 mg
L-Phenylalanine .... 33.02 mg
L-Threonine .... 47.64 mg
L-Tryptophan .... 40.20 mg
L-Tyrosine .... 36.22 mg
L-Valine .... 46.90 mg
D-Calcium pantothenate .... 1.0 mg
Choline chloride .... 1.0 mg
Folic acid .... 1.0 mg
i-Inositol .... 2.0 mg
Nicotinamide .... 1.0 mg
Pyridoxal HCl .... 1.0 mg
Riboflavin .... 0.1 mg
Thiamin HCl .... 1.0 mg
$CaCl_2 \bullet 2H_2O$ .... 140.0 mg
KCl .... 400.0 mg
KH_2PO_4 .... 60.0 mg
$MgSO_4$ .... 97.7 mg
NaCl .... 8000.0 mg
Na_2HPO_4 .... 47.5 mg
Dextrose .... 1000.0 mg
Phenol red, Na† .... 17.0 mg
$NaHCO_3$ (filter-sterilized and added to final medium)† .... 1875.0 mg
Distilled water .... 1.0 L

† MEMH used for extraction lacks phenol red and $NaHCO_3$.

Combine all ingredients, and sterilize by filtration through a 0.2 μm filter. Dispense into a 2-L Erlenmeyer flask. The final pH should be 7.2 ± 0.2. This medium is available in a commercial formulation.

Modified (AEA) Sporulation Medium

Polypeptone .... 10.0 g
Yeast extract .... 10.0 g
Sodium phosphate (dibasic) .... 4.36 g
Potassium phosphate (monobasic) .... 0.25 g
Ammonium acetate .... 1.5 g
Magnesium sulfate (heptahydrate) .... 0.2 g
Distilled water .... 1.0 L

Dissolve the ingredients in distilled water, and adjust the pH to 7.5 ± 0.1 with 2 M sodium carbonate. Dispense the medium in 15-mL portions in 20- × 150-mm screw-capped tubes, and sterilize by autoclaving for 15 min at 121°C. To prepare the final medium, add 0.6 mL of separately sterilized 10% raffinose and 0.2 mL each of filter-sterilized 0.66 M sodium carbonate and 0.32% cobalt chloride ($CoCl_2 \bullet 6H_2O$) dropwise to each 15 mL of base medium. Check the pH of one or two tubes. The pH should be near 7.8 ± 0.1. Just before use, steam the medium for 10 min and, after cooling, add 0.2 mL of filter-sterilized 1.5% sodium ascorbate (prepared that day) to each 15-mL tube of medium.

Modified Campylobacter Blood-Free Selective Agar (mCCDA)

CCDA agar base (Oxoid) .... 45.5 g
Yeast extract .... 2.0 g
Distilled water .... 1.0 L

Autoclave for 15 min at 121°C. Final pH should be 7.4 ± 0.2. Cool and add the following filter-sterilized reagents: 4.0 mL of sodium cefoperazone, 4.0 mL of rifampicin, and 4.0 mL of amphotericin B.

Modified Cellobiose-Polymyxin-Colistin Agar (mCPC)

Solution 1:
Peptone .... 10 g
Beef extract .... 5 g
Sodium chloride .... 20 g
Dye solution (1000X)* .... 1 mL
Agar .... 15 g
Distilled water .... 1.0 L

Mix ingredients and adjust pH to 7.6. Boil to dissolve but **do not** autoclave. Temper to 48-55°C.

*Dye solution (1000X):
Bromthymol blue .... 4.0 g
Cresol red .... 4.0 g
Ethanol (95%) .... 100 mL

For consistent medium color, prepare a stock solution. Dissolve dyes in ethanol, store in refrigerator. Add 1.0 mL of stock solution to each liter of basal medium.

Solution 2:
Cellobiose .... 10 g
Colistin .... 400,000 U
Polymxyin B .... 100,000 U
Distilled water .... 100 mL

Dissolve cellobiose in distilled water with gentle heating (avoid boiling). Cool to room temperature. Aseptically add filter-sterilized antibiotics.

Add Solution 2 to cooled Solution 1. Mix and dispense to sterile Petri dishes. Final color should be dark green to green-brown.

Modified Duncan Strong (MDS) Medium

Yeast extract .... 4.0 g
Proteose peptone .... 15.0 g
Sodium thioglycollate .... 1.0 g
Sodium phosphate (dibasic heptahydrate) .... 10.0 g
Raffinose .... 4.0 g
Distilled water .... 1.0 L

Combine the ingredients with distilled water, and mix thoroughly. Sterilize by autoclaving for 15 min at 121°C. Adjust the pH to 7.8 using filter-sterilized 0.66 M sodium carbonate, and then dispense into tubes. Check the pH of one or two of the tubes; it should be near 7.8 ± 0.1.

Modified EC Broth (mEC)

Tryptone .... 20.0 g
Bile salts No. 3 .... 1.12 g
Lactose .... 5.0 g
K_2HPO_4 .... 4.0 g
KH_2PO_4 .... 1.5 g
NaCl .... 5.0 g
Novobiocin solution .... 0.2 mL
Distilled water .... 1.0 L

Prepare without novobiocin. Dissolve ingredients in distilled water. Autoclave 15 min at 121°C and let cool to room temperature. Add novobiocin solution just before analysis. Final pH is 6.9 ± 0.2.

*Commercially available.

Modified Fecal Coliform (mFC) Agar*

Tryptose .... 10.0 g
Proteose peptone No. 3 .... 5.0 g
Yeast extract .... 3.0 g
Sodium chloride .... 5.0 g
Lactose .... 12.5 g
Bile salts No. 3 .... 1.5 g
Aniline blue (water blue) .... 0.1 g
Agar .... 15.0 g
Distilled water .... 1.0 L

Combine all ingredients, and dissolve by heating to boiling. Do not autoclave. Temper to 50°C and adjust the pH to 7.4. Pour the medium into sterile Petri plates in 20-mL aliquots, and allow to dry thoroughly. The medium may be stored for 4 weeks at 4°C.

Modified Lactic Agar for Yogurt Bacteria

Aseptically add 7.0% (v/v) sterile (121°C for 12 min) reconstituted nonfat dry milk (NFDM, 11% solids, w/w) to melted Lactic (Elliker) Agar. To avoid partial solidification of the agar, the milk should be prewarmed to 47°C before the addition. After pouring into sterile Petri dishes, dry the plates for 18 to 24 hr at 28° to 30°C.

Modified McBride Agar

Phenylethanol agar (Difco) .... 35.5 g
Glycine anhydride .... 10.0 g
Lithium chloride .... 0.5 g
Distilled water .... 1.0 L

Combine the ingredients listed, and dissolve with heat while stirring. Autoclave at 121°C for 15 min and cool to 46°C. Add filter-sterilized cycloheximide to obtain a concentration of 200.0 mg per L of medium.

Modified NBB Medium

Casein peptone .... 5.0 g
Yeast extract .... 5.0 g
Meat extract .... 2.0 g
Tween 80 .... 0.5 mL
Potassium acetate .... 6.0 g
Sodium phosphate, dibasic .... 2.0 g
L-cysteine monohydrochloride .... 0.2 mg
Chlorophenol red .... 70.0 mg
Glucose .... 15.0 g
Maltose .... 15.0 g
L-malic acid .... 0.5 g
Agar .... 15.0 g
Beer/distilled water (1:1) .... adjust to 1.0 L

Combine ingredients with 1.0 L distilled water, and heat to boiling while stirring to dissolve the agar completely. Adjust the pH to 5.8 ± 0.02, and autoclave for 15 min at 121°C.

Modified Nickels and Leesment Agar

Part 1:
Tryptone .... 20.0 g
Yeast extract .... 5.0 g
Gelatin .... 2.5 g
Lactose .... 10.0 g
NaCl .... 4.0 g
Sodium citrate .... 2.0 g
Distilled water .... 750.0 mL

Dissolve the ingredients in distilled water and adjust the pH to 6.65.

Part 2: Incubate 10% reconstituted non-fat milk inoculated with lactic starter at 20°C for 24 hr. Centrifuge the incubated milk to remove the precipitated curd, and collect the clear supernatant. Autoclave the supernatant at 121°C for 15 min and store at 4°C.

Part 3: Grind 13.3 g of calcium citrate and 0.8 g of carboxymethyl-cellulose, and slowly add to 100.0 mL of hot water. Mix well, and filter through a cotton cloth.

Part 4: Dissolve 8.0 g of calcium lactate in 50.0 mL of distilled water with gentle heating.

Prepare each of the four parts separately and autoclave at 121°C for 15 min. After autoclaving, mix the four parts in the following amounts to make up 1.0 L of medium:

Part 1 .... 750.0 mL
Part 2 .... 100.0 mL
Part 3 .... 100.0 mL
Part 4 .... 50.0 mL

Pour plates with periodic mixing of the agar.

Modified Oxford (OX) Agar*

Commercially available from Oxoid (CM856); however, commercial formulations may deviate from the original formula.

Modified PA 3679 Agar

Tryptone .... 10.0 g
Soluble starch .... 2.0 g
K_2HPO_4 .... 2.0 g
Yeast extract .... 2.0 g
Agar .... 15.0 g
Distilled water .... 1.0 L

Combine all ingredients and autoclave at 121°C for 15 min. Add 2 mL of sterile TBL solution/100 mL tempered agar before plating.

TBL Solution

3.3 g sodium thioglycolate in 20 mL distilled water, sterilized at 121°C for 15 min.

2.5 g sodium bicarbonate in 27.5 mL distilled water, filter-sterilized.

2.5 mL of 0.1 mg/mL lysozyme (78,000 units/mg) in distilled water, filter-sterilized.

Modified Rappaport (RAP) Broth

Solution A:
Tryptone .... 10.0 g
Distilled water .... 1.0 L

Solution B:
Sodium phosphate dibasic .... 9.5 g
Distilled water .... 1.0 L

Solution C:
Magnesium chloride•$6H_2O$ .... 40.0 g
Distilled water .... 100.0 mL

Sterilize by autoclaving at 121°C for 15 minutes.

Solution D:
Malachite green .... 0.2 g
Distilled water .... 100.0 mL

Do not sterilize Solution D.

Solution E:
Carbenicillin .... 10.0 mg
Distilled water .... 10.0 mL

Combine carbenicillin with distilled water and filter-sterilize.

*Commercially available.

To make 250.0 mL of medium, mix 155.0 mL of Solution A and 40.0 mL of Solution B. Sterilize the mixture by autoclaving at 121°C for 15 min. Cool to 50°C, and add 53.0 mL of Solution C. Add 1.6 mL of Solution D and 0.6 mL of Solution E, and mix thoroughly. No final pH adjustment is necessary.

Modified SA Agar (Lachica's Medium)

Heart infusion agar (Difco) .... 40.0 g
Amylose azure[†] .... 3.0 g
Ampicillin .... 0.01 mg
Distilled water .... 1.0 L

† Amylose azure is available from Calbiochem, San Diego, CA.

Suspend the ingredients in distilled water, and dissolve by heating while stirring. Autoclave for 15 min at 121°C. Pour sterilized and tempered medium into sterile Petri dishes.

Colonies of the *Aeromonas hydrophila* group appear surrounded by a light halo on a light blue background.

Modified Sodium Lactate Agar

Trypticase soy broth .... 10.0 g
Yeast extract .... 10.0 g
Sodium lactate (60% syrup) .... 20.0 mL
Agar .... 15.0 g
Distilled water .... 1.0 L

Dissolve the ingredients in distilled water with gentle heating. Adjust the pH to 7.0 before autoclaving at 121°C for 15 to 20 min. This formulation contains dextrose in the trypticase soy broth and thus is not as selective as sodium lactate agar.

Modified Trypticase Soy Broth (mTSB)

Trypticase soy broth .... 30.0 g
Bile salts No. 3 .... 1.12 g
K_2HPO_4 .... 1.5 g
Novobiocin solution .... 0.2 mL
Distilled water .... 1.0 L

Prepare without novobiocin. Dissolve ingredients in distilled water. Autoclave 15 min at 121°C and let cool to room temperature. Add novobiocin solution just before analysis. Final pH is 7.3 ± 0.2.

Novobiocin Solution:

Novobiocin (sodium salt) .... 100.0 mg
Distilled water .... 1.0 mL

Dissolve novobiocin in the water. Filter-sterilize using a 0.2 µm filter and syringe. May be stored for several months in a dark bottle at 4°C.

Modified V-P Broth

(Smith, Gordon and Clark)

Proteose peptone .... 7.0 g
Glucose .... 5.0 g
NaCl .... 5.0 g
Distilled water .... 1.0 L

Dissolve the ingredients in distilled water, and dispense 5.0 mL into 20-mm test tubes. Autoclave for 15 min at 121°C. This is a modified medium and must be formulated in the laboratory.

Motility-Nitrate Medium (Buffered)

(for *Clostridium perfringens*)

Beef extract .... 3.0 g
Peptone .... 5.0 g
Potassium nitrate .... 1.0 g
Disodium phosphate .... 2.5 g
Agar .... 3.0 g
Galactose .... 5.0 g
Glycerol .... 5.0 g
Distilled water .... 1.0 L

Dissolve the ingredients, except agar, in distilled water, and adjust the pH to 7.4. Add the agar, and heat to boiling with stirring to completely dissolve. Dispense 11-mL portions into 16- × 125-mm tubes. Sterilize the dispensed medium by autoclaving for 15 min at 121°C, and cool quickly in cold water. If the medium is not used within 4 hr after preparation, heat for 10 min in boiling water or flowing steam and chill in cold water before use.

Motility Test Medium*

Motility Test Medium with TTC

Add 5 mL of a 1% solution of 2,3,5-triphenyl tetrazolium chloride per L to Motility Test Medium before autoclaving.

MP-5 (Mineral Pectin 5) Medium

Prepare 500.0 mL of double strength (without agar) MP-7 medium, and adjust the pH to 5 to 6 with 1N hydrochloric acid. Also prepare 500.0 mL of 3% agar solution and heat to dissolve. Sterilize the double strength MP-7 medium and the 3% agar solution, at 121°C for 15 min, and then cool both to 48°C. Mix the two sterile solutions, and pour plates immediately to prevent hydrolysis.

MP-7 (Mineral Pectin 7) Medium

Basal medium:

Pectin (citrus or apple) .... 5.0 g
Monopotassium phosphate .... 4.0 g
Disodium phosphate .... 6.0 g
Yeast extract .... 1.0 g
Ammonium sulfate .... 2.0 g
Agar .... 15.0 g

Mix the dry ingredients prior to placing in liquid for better dispersion, then add 500.0 mL distilled water.

Prepare separate mineral solutions as described below.

Ferrous sulfate .... 0.2 g plus 200.0 mL distilled water
Magnesium sulfate .... 40.0 g plus 200.0 mL distilled water
Calcium chloride .... 0.2 g plus 200.0 mL distilled water
Boric acid .... 0.002 g plus 200.0 mL distilled water
Manganese sulfate .... 0.002 g plus 200.0 mL distilled water
Zinc sulfate .... 0.014 g plus 200.0 mL distilled water
Cupric sulfate .... 0.010 g plus 200.0 mL distilled water
Molybdenum trioxide .... 0.002 g plus 200.0 mL distilled water

Mix 1.0 mL of each of the above salt solutions in 492.0 mL of distilled water, producing 500.0 mL of mineral solution. Add the 500.0 mL of the mineral salt solution to the 500.0 mL of basal medium to bring the volume of the mixture to 1.0 L. Adjust the pH to 7.2 ± 0.2. Dissolve the ingredients with gentle heating, and sterilize at 121°C for 15 min. Temper medium to 48°C, and pour into sterile Petri plates.

After colonies have grown, preferably on spread plates, pour the polysaccharide precipitant (described in the Reagents Section) over the surface of the plate, taking care not to dislodge the colonies. Zones of pectin hydrolysis will appear quickly, usually within a few minutes, and can best be viewed against a black

*Commercially available.

background. The reagent precipitates intact pectin, and pectinolytic colonies are seen surrounded by a halo in an otherwise opaque medium.

MRS Agar/Broth*

Commercially available as lactobacilli MRS broth. For agar, add 15.0 g of agar per L of commercially available lactobacilli MRS broth. May be supplemented with 10-20% apple juice to enhance recovery of lactics from soft drinks.

Modifications for acid-producing microorganisms: Adjust the pH of sterile, tempered MRS agar to 5.4 ± 0.2 with sterile glacial acetic acid to inhibit the growth of sporeforming organisms.

Add 10.0% fructose solution to sterile, tempered, acidified MRS agar to a final fructose concentration of 1.0 % to enhance the growth of *Lactobacillus fructivorans*.

Adjust sterile, tempered MRS agar to a pH of 5.7 ± 0.1 with HCl, and add sorbic acid solution (sorbic acid dissolved in NaOH) to a final sorbic acid concentration of 0.1% to inhibit growth of yeasts.

Add 0.1% cysteine hydrochloride to MRS agar, and after autoclaving and tempering, adjust the agar to a pH of 5.7 ± 0.1 with HCl. Then add sorbic acid solution (sorbic acid dissolved in NaOH) to a final sorbic acid concentration of 0.02% to inhibit the growth of yeasts and gram negative organisms.

Add 0.01% 2,3,5-triphenyltetrazolium HCl (TTC) to MRS agar to favor the enumeration of lactic acid bacteria commonly associated with plants.

MR-VP Broth

(Same as Buffered Glucose Broth)*

MS Agar

Glucose .... 5 g
L-malate .... 20 g
Casamino acids .... 6 g
Yeast extract .... 6 g
Tween 80 .... 1 mL
Mineral salts supplement* .... 100 mL
Agar .... 16 g
Distilled water .... 1.0 L
*Mineral Salts Supplement:
KH_2PO_4 .... 0.6 g
KCl .... 0.45 g
$CaCl_2$ .... 0.13 g
$MgSO_4$ .... 0.13 g
$MnSO_4$ .... 0.01 g
Distilled water .... 100 mL

All of the medium components except the agar and mineral salts are prepared in volumes of 500 mL (1/2 the final volume), and the pH adjusted to 4.0. The resulting basal medium broth is then filtered through a 0.22 µm filter. The agar is mixed in 400 mL of distilled water and 100 mL of mineral salts supplement and autoclaved (final volume of 500 mL). The filter-sterilized basal medium broth and the autoclaved agar-mineral salts solution are tempered at 50°C and mixed together, and poured into petri plates.

Neutralizing Buffer*

Nitrate Broth*

(same as Nitrate Reduction Medium)

Nitrate Broth

(for *Campylobacter*)
Heart infusion broth (Difco) .... 25.0 g
Potassium nitrate .... 2.0 g
Distilled water .... 1.0 L

Dissolve the heart infusion broth and potassium nitrate in the distilled water, and adjust the pH to 7.0. Dispense 4.0-mL amounts of the medium into 15- × 125- mm tubes with inverted vials, and autoclave at 121°C for 15 min.

Nitrate Reduction Medium

(same as Nitrate Broth)
Nutrient Broth (Difco) .... 8.0 g
Potassium nitrate .... 1.0 g
Distilled water .... 1.0 L

Dissolve the nutrient broth and potassium nitrate in the distilled water, and dispense into tubes. Autoclave at 121°C for 15 min.

Nonfat Dry Milk (Reconstituted)

Nonfat dry milk .... 100.0 g
Distilled water .... 1.0 L

Combine 100.0 g of dehydrated nonfat dry milk with 1.0 L of distilled water, and swirl until dissolved. Dispense 225-mL portions into 500-mL Erlenmeyer flasks, and autoclave at 121°C for 15 min. Aseptically adjust the final volume to 225 mL just before use. 0.45 mL of a 1.0% aqueous brilliant green dye solution should be added to the medium-sample mixture after adjusting the final pH.

Nutrient Agar*

If this medium is to be used as a blood agar base, add 8.0 g of sodium chloride per L to make the medium isotonic so that red cells will not rupture, and adjust the pH to 7.3 ± 0.2.

Nutrient Agar With Manganese (NAMn)

Preparation A: Prepare commercially available nutrient agar as indicated.

Preparation B: Dissolve 3.08 g of manganese sulfate in 100.0 mL of distilled water. Add 1.0 mL of Preparation B to Preparation A (nutrient agar) and autoclave at 121°C for 15 min.

Nutrient Broth*

Nutrient Overlay Agar

(Same as Base Layer Agar with Nutrient Overlay Agar)

Orange Serum Agar (OSA)/Broth (OSB)*

Oxford Medium (OXA)

Columbia blood agar base .... 39.0 g
Esculin .... 1.0 g
Ferric ammonium citrate .... 0.5 g
Lithium chloride .... 15.0 g
Cycloheximide .... 0.4 g
Colistin sulfate .... 0.02 g
Acriflavin .... 0.005 g
Cefotetan .... 0.002 g
Fosfomycin .... 0.010 g
Distilled water .... 1.0 L

*Commercially available.

Add the 55.5 g of the first 4 components (basal medium) to 1 L distilled water. Bring gently to boil to dissolve completely. Sterilize by autoclaving at 121°C for 15 min. Cool to 50°C and aseptically add supplement, mix, and pour into sterile petri dishes. To prepare supplement, dissolve cycloheximide, colistin sulfate, acrifalvin, cefotetan, and fosfomycin in 10 mL of 1:1 mixture of ethanol and distilled water. Filter-sterilize supplement before use. Oxford basal medium and supplement mixture are available commercially.

Oxidation-Fermentation (OF) Test Medium

Basal medium without carbohydrate:

Peptone	2.0 g
NaCl	5.0 g
K_2HPO_4	0.3 g
Bromthymol blue	0.03 g
Agar	2.5 g
Distilled water	1.0 L

Heat with agitation to dissolve agar. Dispense 3-mL portions into 13 × 100 mm tubes. Autoclave for 15 min at 121°C. Cool to 50°C; pH 7.1.

Carbohydrate stock solution: Dissolve 10 g carbohydrate in 90 mL distilled water. Sterilize by filtration through 0.22 µm membrane. Add 0.3 mL stock solution to 2.7 mL base in tube. Mix gently and cool at room temperature. Inoculate tubes in duplicate. Layer one tube with sterile mineral oil. Incubate for 48 h at 35°C.

Oxidation-Fermentation Medium for Campylobacter

Basal medium without carbohydrate:

Bacto casitone	0.2 g
Agar	0.3 g
Phenol red, 1.0% aqueous	0.3 mL
Distilled water	100.0 mL

Dissolve the casitone in the distilled water and add the phenol red. Adjust the pH to 7.4, add the agar and heat to boiling for 1 min to dissolve completely. Autoclave at 121°C for 15 min and dispense 6 mL of the basal medium into 15- × 125-mm tubes. Cool tubes in an upright position.

Basal medium with carbohydrate: Carbohydrate (glucose) - 1% final concentration

10% glucose in distilled water	10.0 mL

Filter-sterilize (0.22-µm filter) the carbohydrate. When the sterile basal medium is tempered, add the carbohydrate aseptically (10 mL of 10% carbohydrate to 100 mL of basal medium) and dispense.

Oxytetracycline Glucose Yeast Extract (OGY) Agar*

Available commercially from Oxoid as Oxytetracycline-Glucose-Yeast Extract Agar (O.G.Y.E. Agar, CM545) with an oxytetracycline supplement (SR73).

Palcam Selective Agar

Basal medium:

Peptone	11.5 g
Starch	0.5 g
NaCl	2.5 g
Columbia agar	6.5 g
Mannitol	5.0 g
Ferric ammonium citrate	0.25 g
Esculin	0.4 g
Glucose	0.25 g
Lithium chloride	7.5 g
Phenol red	0.04 g
Distilled water	500.0 mL

Selective agents:

Polymyxin B sulfate	10 mg
Acriflavin	5 mg
Ceftazidine	20 mg

Weigh 34.4 g basal medium powder (all ingredients except the selective agents) and suspend in 500 mL distilled water. Sterilize by autoclaving at 121°C for 15 min. Dissolve the selective agent supplement mixture in sterile distilled water at a rate of 17.5 mg/mL and filter-sterilize. Add 1.0 mL selective agent supplement solution to 500 mL sterile basal medium which has been cooled to 50°C. Mix gently and pour plates. Final pH, 7.2 ± 0.1.

Park and Sanders Enrichment Broth

Basal Medium:

Brucella broth (dehydrated)	29.0 g
Sodium pyruvate	0.25 g
Distilled water	950.0 mL
Lysed horse blood	50.0 mL

Supplement A:

Vancomycin	0.01 g
Trimethoprim lactate	0.01 g

Supplement B:

Cefoperazone	0.032 g
Cycloheximide	0.1 g

Dissolve the brucella broth and sodium pyruvate in distilled water, and autoclave at 121°C for 15 min. Cool to room temperature, add the blood and Supplement A. Dispense 50 or 100-mL portions in a cotton-plugged 250-mL Erlenmeyer flask. After a resuscitation period of approximately 4 hr at 31° to 32°C (to recover injured cells), add Supplement B to each flask, and incubate at 37°C for 2 hr, then at 42°C under a microaerobic atmosphere for an additional 40 to 42 hr with agitation at 100 rpm. Both the resuscitation and enrichment culture must be performed in a microaerobic environment. Solutions of each antimicrobial compound for 1 L of medium preparation may be dissolved in 5 mL of brucella broth; dissolve the amounts of each compound required for 10 L of medium in 50 mL of brucella broth, filter through a 0.22-µm filter, dispense in 5 mL portions in vials, and freeze until use.

Peptone Sorbitol Bile Broth

Na_2HPO_4	8.23 g
$NaH_2PO_4 \bullet H_2O$	1.2 g
Bile salts No. 3	1.5 g
NaCl	5.0 g
Sorbitol	10.0 g
Peptone	5.0 g
Distilled water	1.0 L

Mix all ingredients and autoclave at 121°C for 15 min. Final pH is 7.6 ± 0.2.

Pe-2 Medium

Yeast extract	3.0 g
Peptone	20.0 g
Bromcresol purple (2% ethanol solution)	2.0 mL
Distilled water	1.0 L

*Commercially available.

Dissolve the ingredients in distilled water by heating, if necessary, and dispense 19-mL portions into 18- × 150-mm screw capped culture tubes. Add 8 to 10 untreated Alaska seed peas (Rogers Bros. Co., Seed Division, P.O. Box 2188, Idaho Falls, ID 83401, catalog No. 423, or Northrup King Seed Co., Minneapolis, MN 55413, or W. Atlee Burpee Co., Warminster, PA 18974). Allow ingredients to stand one hour to permit hydration. Autoclave at 121°C for 15 min. Prepare bromcresol purple 2% ethanol solution by adding 2.0 g of the dye to 10.0 mL of ethanol and dilute to 100.0 mL with distilled water. Add 2.0 mL of the solution to 1.0 L of medium.

Peptone Broth (1.0%)

Peptone .... 10.0 g
Distilled water .... 1.0 L

Dissolve peptone in distilled water by stirring. Autoclave at 121°C for 15 min.

*Phenol Red Carbohydrate Broth**

*Phenol Red Dextrose Broth**

(Difco)

Phenol Red Sorbitol-Mug Agar (PSA-MUG)

Phenol red broth base (Difco) .... as specified
Agar 2%
D-sorbitol 0.5%
MUG 0.005%

Dissolve ingredients in distilled water by heating with stirring. Autoclave for 15 min at 121°C. Final pH is 6.8-6.9.

Phenyl Alanine Deaminase Agar

Yeast extract .... 3.0 g
L-phenylalanine or .... 1.0 g
DL-phenylalanine .... 2.0 g
Na_2HPO_4 .... 1.0 g
NaCl .... 5.0 g
Agar .... 12.0 g
Distilled water .... 1.0 L

Heat gently to dissolve agar. Tube and autoclave 10 min at 121°C. Incline tubes to obtain a long slant or pour as plates. Final pH is 7.3 ± 0.2.

Phytone Yeast Extract (PYE) Agar (BBL)*

Phytone Yeast Extract Agar with Streptomycin and Chloramphenicol

Phytone or Soytone .... 10.0 g
Yeast extract .... 5.0 g
Dextrose .... 40.0 g
Streptomycin .... 0.03 g
Chloramphenicol .... 0.05 g
Agar .... 17.0 g
Distilled water .... 1.0 L

Suspend the ingredients listed above in distilled water, and dissolve by boiling gently. Dispense the medium into suitable containers, and autoclave at 118°C for 15 min. Final pH should be 6.6 ± 0.2.

Peptone Yeast Extract Glucose Starch Agar (PYGS)

Proteose peptone .... 5.0 g
Tryptone .... 5.0 g
Yeast extract .... 10.0 g
Meat extract (Oxoid Lab Lemco L29) .... 10.0 g
Soluble starch .... 1.0 g
Resazurin .... 0.001 g
Cysteine HCl .... 0.5 g
Salts solution A .... 40 mL
Salts solution B .... 40 mL
Distilled water .... 1.0 L

Combine ingredients and sterilize at 121°C for 15 min.

Salts Solution A

$CaCl_2 \cdot 2H_2O$ .... 0.265 g
$MgSO_4 \cdot 7H_2O$ .... 0.48 g
NaCl .... 2.0 g
Distilled water .... 1.0 L

Salts Solution B

KH_2PO_4 .... 1.0 g
$K_2HPO_4 \cdot 3H_2O$ .... 1.3 g
$NaHCO_3$ .... 10.0 g
Distilled water .... 1.0 L

PI Agar

Peptone .... 5.0 g
Sodium chloride .... 5.0 g
Yeast extract .... 2.0 g
Mannitol .... 7.5 g
L-Arabinose .... 5.0 g
Inositol .... 1.0 g
Lysine .... 2.0 g
Bile salts No. 2 .... 1.0 g
Phenol red .... 0.08 g
Agar .... 15.0 g
Distilled water .... 1.0 L

Dissolve ingredients in distilled water, and adjust the pH to 7.4. Autoclave at 115°C for 15 min.

*Plate Count Agar (PCA)**

Modifications: Add 2.0 mL bromcresol purple solution per L of Plate Count Agar prior to sterilization for the enumeration of acid-producing organisms.

Plate Count Agar with Antibiotic

Plate count agar .... 1.0 L
Chloramphenicol .... 100.0 mg

To 1.0 liter of commercially available plate count agar add 100.0 mg of chloramphenicol. Mix thoroughly, and autoclave at 121°C for 15 min.

PMP Broth

Na_2HPO_4 .... 7.9 g
NaH_2PO_4 .... 1.1 g
Peptone .... 2.5 g
D-Mannitol .... 2.5 g
Distilled water .... 1.0 L

Mix all ingredients and adjust pH to 7.6. Autoclave at 121°C for 15 min.

*Commercially available.

Polypectate Gel Medium

Sodium polypectate .... 70.0 g
Peptone .... 5.0 g
Dipotassium phosphate .... 5.0 g
Monopotassium phosphate .... 1.0 g
Calcium chloride•$2H_2O$ .... 0.6 g
Distilled water .... 1.0 L

Source of polypectate. Raltech, the previous source for sodium polypectate, has been sold. The remaining polypectate stock that Raltech owned was sold to someone else, but it was destroyed by mistake over a year ago. Currently, there is no known source of suitable commercial polypectate for plate assays and viscosity assays.

Heat 500.0 mL of distilled water, and place all ingredients except the polypectate in a blender with the heated water to dissolve. Add the polypectate last, in small amounts per addition, with slow stirring to diminish occluded air. Adjust the pH to 7.0 and then add the remainder of the distilled water. Sterilize by autoclaving at 121°C for 15 min, temper the medium to 48°C and then pour into sterile Petri plates.

Pork Plasma-Fibrinogen Overlay Agar

Grind Bovine Fibrinogen (BFG) fraction I (Calbiochem, San Diego) in a mortar to a fine powder, and dissolve (8.0 mg/mL) in 0.05 M sodium phosphate buffer of pH 7.0 by stirring for 30 min on a magnetic stirrer. Filter the solution through Whatman No. 41 paper and filter-sterilize.

Dissolve Trypsin inhibitor (Soya; Sigma Chemical Co., St. Louis, MO) in 0.05 M sodium phosphate buffer of pH 7.0 at a concentration of 3 mg/mL and filter-sterilize.

Obtain fresh or rehydrated commercial pork plasma-EDTA and filter-sterilize.

Prepare sterile 1.4% agar in distilled water. Melt and temper the agar to 45°C prior to use.

Mix gently and thoroughly the following amounts of prewarmed (45°C) solutions to prepare the overlay agar:

Pork plasma .... 2.5 mL
BFG .... 47.5 mL
Trypsin inhibitor .... 0.5 mL
Agar .... 50.0 mL

This mixture should be kept at 45° to 50°C and used within 1 hr. Use 8.0 mL of the overlay mixture is needed per plate of Baird-Parker agar without egg yolk but containing potassium tellurite.

Potato Dextrose Agar (PDA)*

Potato Dextrose Agar (Acidified)

As plating medium for yeasts and molds, melt previously sterilized potato dextrose agar (commercially available) in flowing steam or boiling water, cool and acidify to pH 3.5 with sterile 10% tartaric acid solution. Mix thoroughly and pour into plates. To preserve the solidifying properties of the agar, do not heat the medium after the addition of tartaric acid.

Potato Dextrose Agar (with Antibiotics)

Prepare an antibiotic solution containing 500.0 mg each of chlortetracycline HCl and chloramphenicol in 100.0 mL of sterile phosphate buffered distilled water and mix. (Not all material dissolves; therefore the suspension must be evenly dispersed before pipetting into the medium.) Add 2.0 mL of this solution per 100.0 mL of tempered potato dextrose agar, giving a final concentration in the medium of 100.0 mg/l of each of the antibiotics. After being swirled to mix, the medium is ready to use.

Potato Dextrose Agar 2% Glucose 60% Sucrose

Potato Dextrose Agar .... 39.0 g
Glucose, AR .... 20.0 g
Sucrose .... 600.0 g
Distilled water to make .... 1.0 L

Add potato dextrose agar to approximately 400.0 mL of distilled water, and heat while stirring until completely dissolved. Further heat the solution, and add the glucose and sucrose until all the sugar is dissolved. Make up to a total volume of 1.0 L with distilled water. Dispense the medium into smaller containers and autoclave at 121°C for 15 min. The final a_w of this medium is 0.92.

Preservative Resistant Yeast Medium (PRY)

Mannitol .... 10.0 g
Yeast extract .... 10.0 g
Glacial acetic acid .... 10.0 mL
Agar .... 15.0 g
Distilled water .... 1.0 L

Prepare base broth or agar containing all ingredients except the glacial acetic acid. Autoclave at 121°C for 15 min. After cooling, if necessary, replace volume lost during autoclaving with sterile water. Add glacial acetic acid.

Proteose Extract Agar

Proteose peptone No. 3 .... 15.0 g
Yeast extract .... 7.5 g
Casamino acids .... 5.0 g
Starch, soluble .... 1.0 g
K_2HPO_4 .... 5.0 g
$(NH_4)_2SO_4$ .... 1.5 g
Agar .... 15.0 g
Distilled water .... 1.0 L

Combine all ingredients, and heat with agitation to dissolve agar. Dispense 10-mL portions into 16- × 150-mm test tubes. Autoclave for 15 min at 121°C. The final pH should be 9.0 ± 0.2. Incline while cooling to obtain a long slant.

Purple Carbohydrate Broth*

Pyrazinamidase Agar

Tryptic soy agar .... 30.0 g
Yeast extract .... 3.0 g
Pyrazine-carboxamide .... 1.0 g
0.2 M Tris-maleate, pH 6.0 .... 1.0 L

Heat to boiling; dispense 5 mL in 16 × 125 mm tubes. Autoclave at 121°C for 15 min. Cool slanted.

Rappaport-Vassiliadis Medium

Broth Base

Tryptone .... 5.0 g
NaCl .... 8.0 g
Distilled water .... 1.0 L

Magnesium Chloride Solution

$MgCl_2$•$6H_2$ .... 400 g
Distilled water .... 1.0 L

*Commercially available.

Malachite Green Oxalate Solution

Malachite green oxalate .... 0.4 g
Distilled water .... 100 mL

Combine 1.0 L broth base, 100 mL magnesium chloride solution, and 10 mL malachite green oxalate solution (total volume of complete medium is 1.11 L). Broth base must be prepared on same day that components are combined to make complete medium. Dispense 10 mL volumes of medium into 16 × 150 mm test tubes. Autoclave for 15 min at 115°C. Final pH should be 5.5 ± 0.2. Store in refrigerator and use within 1 month. Note: use of commercially available media is not recommended.

*R2A Agar**

*RMW Broth/Agar**

(Commercially available as Rogosa SL Broth/Agar)

Modifications for vegetable products: Add 0.04% cyclohexamide to Rogosa SL Broth to aid in the enrichment of vegetable products for lactic acid bacteria.

Rogosa SL Broth/Agar

(Same as RMW Broth/Agar*)

SA Agar

Phenol red agar base .... 31.0 g
Soluble starch, reagent grade .... 10.0 g
Distilled water .... 1.0 L

Suspend the ingredients in distilled water, and dissolve by heating while stirring. Autoclave for 15 min at 121°C. After sterilization and tempering, 10 mg of ampicillin (A-9393, Sigma) should be dissolved in a very small quantity of sterile distilled water and added to 1 L of the medium before pouring into sterile Petri dishes.

The presence of *A. hydrophila* group organisms can be verified by the addition of approximately 5 mL of Lugol's iodine solution to each plate.

*Sabouraud Dextrose Agar**

Salmonella Shigella *(SS) Agar**

Sea Water Agar (SWA)

Yeast extract .... 5.0 g
Peptone .... 5.0 g
Beef extract .... 3.0 g
Agar .... 15.0 g
Synthetic Sea Water .... 1.0 L

Dissolve the ingredients in Synthetic Sea Water with heat while stirring. Autoclave the medium at 121°C for 15 min. The final pH should be 7.5.

*Selenite Cystine (SC) Broth**

Semisolid Brucella-FBP Medium

(Same as *Brucella*-FBP Agar/Broth)

Semisolid Brucella *(Albimi) Medium*

(With cysteine, glycine or NaCl for *Campylobacter*)
Brucella broth (Gibco) .... 29.0 g
Agar .... 1.6 g
Distilled water .... 1.0 L

The following are added individually before sterilization:
L-Cysteine hydrochloride .... 0.2 g
Glycine .... 10.0 g
Sodium chloride .... 3.5 g

Dissolve the ingredients completely, adjust the pH to 7.0, and dispense 10 mL of the medium in 16- × 150-mm screw capped tubes. Autoclave at 121°C for 15 min. Cool tubes in an upright position.

*SFP Agar with Egg Yolk**

Shigella Broth

Tryptone .... 20 g/L
K_2HPO_4 .... 2 g/L
KH_2PO_4 .... 2 g/L
NaCl .... 5 g/L
Glucose .... 1 g/L
Tween 80 .... 1.5 mL
Distilled water .... 1.0 L

Mix all ingredients and autoclave for 15 min at 121°C. Final pH should be 7.0 ± 0.2.

*Simmons Citrate Agar**

Skim Milk Agar

Standard Methods Agar .... 1.0 L
Reconstituted skim milk (10% solids) .... 100.0 mL

Melt standard methods agar, cool to 50°C, and add 100.0 mL of sterile skim milk. Mix well, and pour into sterile Petri dishes.

Sodium Lactate Agar

Trypticase .... 10.0 g
Yeast extract .... 10.0 g
Sodium lactate .... 10.0 g
Dipotassium phosphate .... 0.25 g
Agar .... 15.0 g
Distilled water to make .... 1.0 L

Dissolve the ingredients in distilled water with gentle heating. Adjust the medium to pH 7.0 ± 0.2 before autoclaving for 20 min at 121°C.

*Soft Agar Gelatin Overlay**

(Basal Medium)
Peptone .... 5.0 g
Beef extract .... 3.0 g
Sodium chloride .... 5.0 g
Manganese sulfate .... 0.05 g
Agar .... 15.0 g
Distilled water .... 1.0 L

Dissolve the ingredients in distilled water with gentle heating. Adjust the pH to 7.0 and autoclave at 121°C for 15 min. The soft overlay is the same as the Basal Medium except that 0.8% agar and 1.5% gelatin are used for the overlay.

*Sorbitol-MacConkey Agar**

(Difco)

SS Agar

(Same as *Salmonella Shigella* Agar)

*Commercially available.

SS-Deoxycholate (SSDC) Agar

SS Agar .... 60.0 g
Sodium desoxycholate .... 10.0 g
Calcium chloride .... 1.0 g
Distilled water .... 1.0 L

Combine the ingredients with distilled water and boil while stirring for 2 to 3 min to completely dissolve the agar. Do not autoclave.

This medium is available from E. Merck AG, Darmstadt, Germany, as "*Yersinia*-Agar."

ST (Streptococcus Thermophilus) *Agar*

Tryptone .... 10.0 g
Yeast extract .... 5.0 g
Sucrose .... 10.0 g
Dipotassium phosphate .... 2.0 g
Agar .... 15.0 g
Distilled water .... 1.0 L

Add the ingredients to distilled water, heat to dissolve the agar and autoclave at 121°C for 15 min. The final pH should be 6.8.

Standard Methods Agar*

(Same as Plate Count Agar)

Standard Methods Caseinate Agar

Pancreatic digest of casein (Tryptone or Trypticase) .. 5.0 g
Yeast extract .... 2.5 g
Glucose .... 1.0 g
Agar .... 15.0 g
Sodium caseinate .... 10.0 g
Trisodium citrate hydrated (0.015 M solution) .... 1.0 L
Calcium chloride (1.0 M solution) .... 20.0 mL

Prepare a 0.015 M solution of trisodium citrate by placing 4.41 g of trisodium citrate (dihydrate) into a volumetric flask, bring to 1.0 L volume with distilled water, and mix thoroughly. Dissolve the pancreatic digest of casein, yeast extract, glucose and agar in 500.0 mL of the 0.015 M trisodium citrate with gentle heating.

Sulfite Agar*

Synthetic Sea Water (SW)

NaCl .... 24.0 g
KCl .... 0.7 g
$MgCl_2 \cdot 6H_2O$ .... 5.3 g
$MgSO_4 \cdot 7H_2O$ .... 7.0 g
$CaCl_2$ .... 0.1 g
Distilled water .... 1.0 L

Suspend the ingredients in distilled water, and dissolve. Adjust the pH to 7.5 with 1 N NaOH. Dispense, and autoclave at 121°C for 15 min.

Tellurite Cefixime Sorbitol MacConkey Agar (TC SMAC)

Prepare it the same way as Sorbitol MacConkey agar, but with the following filter-sterilized additives added after autoclaving, to achieve a final concentration indicated.

Potassium tellurite .... 2.5 mg/L
Cefixime .... 0.05 mg/L

Both additives are also available as SMAC Media Cefixime Tellurite (CT) Supplement, from Dynal Inc., Lake Success, NY.

Tergitol 7 Agar*

(Difco)

Tetrathionate Broth*

Tetrathionate Broth (without Iodine)*

Thermoacidurans Agar (TAA)*

Thin-Layer Enzyme Assay (TEA) Plate

Dissolve 10 mg of bovine serum albumin (BSA) or another protein substrate in 1.0 mL of distilled water. Mix 1.0 mL of this solution with 9.0 mL of distilled water on a polystyrene Petri dish that has been previously cleaned with ethanol and dried with sterile compressed air. After 1 min, decant the protein solution, wash the surface with distilled water, and blow dry. The plate may be stored at 4°C for up to 6 months. Prepare agar medium with the protein-coated Petri dish in the usual manner.

Thioglycollate Agar

Fluid Thioglycollate medium (commercial, dehydrated) ... 29.5 g
Agar .... 20.0 g
Distilled water .... 1.0 L

Suspend the ingredients in distilled water, and heat to boiling while stirring to dissolve completely. Distribute the medium into tubes or flasks, and autoclave at 121°C for 15 min. This medium should be used within one week of preparation.

Thiosulfate Citrate Bile Salts Sucrose (TCBS) Agar*

T_1N_1 Agar/Broth

Trypticase (pancreatic digest of casein) .... 10.0 g
NaCl .... 10.0 g
Agar .... 15.0 g
Distilled water .... 1.0 L

Dissolve the ingredients in distilled water by bringing to a boil. Dispense in tubes, and sterilize at 121°C for 15 min. Allow to solidify in an inclined position (long slant). Final pH should be 7.2 ± 0.2. To prepare T_1N_1 broth, omit the agar.

Note: To prepare T_1N_3 medium, use 30.0 g of NaCl; to prepare T_1N_6 medium, use 60.0 g NaCl; to prepare T_1N_8 medium, use 80.0 g of NaCl; to prepare T_1N_{10} medium, use 100.0 g of NaCl.

Toluidine Blue Agar

Nutrient broth .... 8.0 g
Sodium chloride .... 5.0 g
Toluidine blue .... 0.083 g
DNA .... 0.3 g
Agar .... 20.0 g
Distilled water .... 1.0 L

Add the DNA to cold distilled water, and heat slowly while stirring. When the DNA is in solution, add the rest of the components and sterilize by autoclaving at 121°C for 15 min.

Toluidine Blue DNA Agar

Deoxyribonucleic acid .... 0.30 g
Calcium chloride (anhydrous) .... 5.5 mg

*Commercially available.

NaCl ..... 10.0 g
Agar ..... 10.0 g
Tris (hydroxymethyl) aminomethane ..... 6.1 g
Toluidine blue. ..... 0.083 g
Distilled water ..... 1.0 L

Suspend the ingredients in 1.0 L of distilled water, adjust the pH to 9.0, and boil to dissolve completely. Cool to 45° to 50°C and add 0.083 g of Toluidine Blue-O. Mix the medium, and dispense in small portions into screw-capped storage bottles. This medium need not be sterilized and can be stored at room temperature for 4 months even with several melting cycles. Prepare plates by dispensing 10-mL portions into 15- × 100-mm Petri dishes and store melted agar at 2° to 8°C. Prior to use cut 2-mm wells in the agar plates with a metal cannula, and remove the agar plugs by aspiration.

Tomato Liver Broth

Liver broth (see Liver Broth formulation) ..... 1.0 L
Tomato juice ..... 1.0 L

Adjust to pH 5.0. Dispense 9 mL into each test tube with a small pinch of dried liver particles (or 1/2 inch of fresh or frozen liver particles—see Liver Broth formulation). Autoclave for 20 min at 115.5°C.

Trimethylamine N-Oxide (TMAO) Medium

Nutrient broth No. 2 (Oxoid CM67) ..... 25.0 g
Yeast extract (Difco) ..... 1.0 g
New Zealand agar (Oxoid) ..... 2.0 g
Trimethylamine N-oxide (Sigma) ..... 1.0 g
Distilled water ..... 1.0 L

Dissolve the nutrient broth, yeast extract, and TMAO in the distilled water. Add the agar, and heat to dissolve. Dispense 4 mL in 13- × 100-mm screw capped tubes, and autoclave at 121°C for 15 min. Cool the tubes in an upright position, and store under refrigeration.

Triple Sugar Iron (TSI) Agar*

Trypticase Peptone (TP) Agar

Trypticase ..... 50.0 g
Peptone ..... 5.0 g
Sodium thioglycollate ..... 1.0 g
Agar ..... 15.0 g
Distilled water ..... 1.0 L

Sterilize at 121°C for 15 min. Add 14 mL of 10% sodium bicarbonate (filter-sterilized) to tempered agar just prior to pouring plates.

Trypticase Peptone Glucose Yeast Extract (TPGY) Broth and TPGY with Trypsin (TPGYT)

Trypticase or Tryptone ..... 50.0 g
Peptone ..... 5.0 g
Yeast extract ..... 20.0 g
Glucose ..... 4.0 g
Sodium thioglycollate ..... 1.0 g
Distilled water ..... 1.0 L

Dissolve solid ingredients in distilled water, adjust the pH to 7.0, and dispense into tubes in appropriate volumes. Sterilize the dispensed medium at 121°C for 8 min (15 min for large volumes), and refrigerate until used.

If trypsin is to be added, prepare a 1.5% aqueous solution of trypsin. Sterilize by filtration through a Millipore (or compatible) 0.45 μm filter, and refrigerate until needed. After steaming TPGY broth to drive off oxygen and cooling, add the trypsin to the TPGY broth immediately before inoculating to give a final concentration of 0.1%.

Trypticase Soy Agar/Broth (TSA/TSB)*

Trypticase Soy Agar Plus 3% NACL (TSA-NACL)

Trypticase peptone ..... 15.0 g
Phytone peptone ..... 5.0 g
NaCl ..... 5.0 g
Agar ..... 15.0 g
Distilled water ..... 1.0 L

Combine the ingredients, and heat with agitation to dissolve the agar. Boil the medium for 1 min and then dispense into suitable tubes or flasks. Autoclave for 15 min at 121°C. The final pH should be 7.3 ± 0.2.

Trypticase Soy Agar-Yeast Extract (TSA-YE)

Trypticase soy agar (dehydrated, TSA, BBL) ..... 40.0 g
Yeast extract ..... 6.0 g
Distilled water ..... 1.0 L

Suspend the dehydrated TSA and yeast extract in 1.0 L of distilled water, and stir while heating to dissolve completely. Autoclave at 121°C for 15 min, then temper the medium to 45° to 50°C before pouring into sterile Petri dishes.

Trypticase Soy Broth-Yeast Extract (TSB-YE)

Trypticase soy broth (dehydrated, TSB, BBL) ..... 30.0 g
Yeast extract ..... 6.0 g
Distilled water ..... 1.0 L

Suspend the dehydrated TSB and yeast extract in 1.0 L of distilled water and stir while heating to dissolve completely. Dispense into tubes, and autoclave at 121°C for 15 min.

Trypticase Soy Polymyxin Broth

Preparation A:
Trypticase peptone ..... 17.0 g
Phytone peptone ..... 3.0 g
Sodium chloride ..... 5.0 g
Dipotassium phosphate ..... 2.5 g
Dextrose ..... 2.5 g
Distilled water ..... 1.0 L

Preparation B: Polymyxin B sulfate:

This selective agent is obtainable in sterile powdered form (500,000 units, i.e., 50 mg per vial) from Pfizer, Inc. To use, aseptically add 33.3 mL of sterile distilled water with a sterile syringe to give a 0.15% solution. Mix to dissolve the powder, and store at 4°C until used.

Suspend preparation A ingredients in water, and mix thoroughly. Warm the mixture slightly if necessary to complete the solution. Dispense 15.0 mL into 20- × 150-mm culture tubes, and sterilize by autoclaving for 15 min at 121°C. Just before use, add 0.1 mL of sterile 0.15% polymyxin B sulfate solution (preparation B) to each tube and mix thoroughly.

*Commercially available.

Trypticase-Soy-Sheep Blood Agar

(Same as CAMP Test Agar)

Prepare trypticase (tryptic) soy agar, and cool to 48°C in a water bath. Add 5.0 mL of sterile defibrinated sheep blood for each 100.0 mL of medium, mix well, and dispense 18.0 mL into 15- × 100-mm culture dishes. Allow plates to dry for 24 to 48 hr at room temperature before use.

Trypticase-Soy-Sheep Blood (TSB) Agar

(For *Clostridium perfringens*)

Trypticase peptone .... 15.0 g
Phytone peptone .... 5.0 g
NaCl .... 5.0 g
Agar .... 15.0 g
Distilled water .... 1.0 L

Suspend the ingredients in distilled water, mix thoroughly, and heat to boiling while stirring for 1 min. Autoclave the medium for 15 min at 118° to 121°C and cool to approximately 45°C in a water bath. Final pH should be 7.3 ± 0.1. Add 50.0 mL of defibrinated sheep blood to each L of base medium. Thoroughly mix and dispense 18 to 20 mL in 15- × 100-mm Petri dishes. Commercially prepared plates are satisfactory.

Trypticase Soy-Tryptose Broth

Trypticase soy broth (commercial, dehydrated) .... 15.0 g
Tryptose broth (commercial, dehydrated) .... 13.5 g
Yeast extract .... 3.0 g
Distilled water .... 1.0 L

Dissolve ingredients in 1 L water. Heat gently to dissolve. Dispense 5 mL portions into 16 × 150 mm test tubes. Autoclave 15 min at 121°C. Final pH, 7.2 ± 0.2.

Tryptic Soy Agar/Broth*

Tryptic Soy Agar-Magnesium Sulfate-NACL (TSAMS)

(For *Vibrio*)

Tryptic Soy agar .... 40.0 g
Sodium chloride .... 15.0 g
$MgSO_4 \cdot 7H_2O$ .... 1.5 g
Distilled water .... 1.0 L

Adjust pH to 7.4. Dissolve by boiling and autoclave at 121°C for 15 min. The medium may be stored for 4 weeks at 4°C.

Tryptic Soy-Fast Green Agar (TSFA)

Trypticase Soy Agar (commercial, dehydrated) .... 40.0 g
Fast Green FCF (CI No. 42053) .... 0.25 g
Distilled water .... 1.0 L

Suspend the ingredients in distilled water, and heat to boiling while stirring to dissolve completely. Dispense the medium into flasks and autoclave for 15 min at 121°C. Temper the sterile medium to 50°C and adjust the pH to 7.3. Pour 20 mL of medium per plate, and allow to dry thoroughly. This medium may be stored for 4 weeks at 4°C. Fast Green FCF is available from Sigma (F7252).

Tryptone Bile Agar*

Tryptone Broth

Tryptone or trypticase .... 10.0 g
Dextrose .... 5.0 g
Dipotassium phosphate .... 1.25 g
Yeast extract .... 1.0 g
Bromcresol purple (2% alcoholic solution) .... 2.0 mL
Distilled water .... 1.0 L

Combine the ingredients with distilled water, and dissolve with gentle heat if necessary. To prepare a 2% bromcresol purple solution, add 2.0 g of bromcresol purple to 10.0 mL of ethyl alcohol, and dilute to 100.0 mL with distilled water. Add 2.0 mL of the solution to each L of medium. Dispense 10-mL portions into 20- × 150-mm screw cap culture tubes and autoclave for 20 min at 121°C. Do not exhaust before using.

Tryptone (1%) Broth For Indole*

Tryptone Glucose Extract Agar (TGEA)*

Tryptone Glucose Yeast Agar*

(Same as Plate Count Agar)

Tryptone-Glucose-Yeast Extract (TGY) Agar*

(Same as Plate Count Agar)

Tryptone Glucose Yeast Extract Acetic Acid Agar (TGYA)

Add glacial acetic agar to final concentration of 0.5% to molten (50°C) TGY agar. Pour plates immediately.

Tryptone Phosphate Broth

Tryptone .... 20.0 g
Tween 80 .... 1.5 mL
K_2HPO_4 .... 2.0 g
KH_2PO_4 .... 2.0 g
NaCl .... 5.0 g
Distilled water .... 1.0 L

Dissolve the ingredients in distilled water, and sterilize by autoclaving for 15 min at 121°C. Adjust the final pH to 7.0 ± 0.2.

Tryptophan Broth

(Same as Tryptone Broth)

Tryptose Agar*

Tryptose Cycloserine Dextrose (TCD) Agar

Tryptose .... 15.0 g
Soytone .... 5.0 g
Yeast extract .... 5.0 g
Ferric ammonium citrate .... 1.0 g
Agar .... 20.0 g
Distilled water .... 1.0 L

Dissolve the ingredients in distilled water, and adjust the pH to 7.6 ± 0.2. Autoclave the medium for 10 min at 121°C, and cool to approximately 50°C in a water bath. To each L of cooled medium add 10.0 mL of a filter-sterilized solution of 4.0% D-cycloserine to give a final concentration of 400 mg/mL.

Tryptose-Sulfite Cycloserine (TSC) Agar

Tryptose .... 15.0 g
Soytone .... 5.0 g
Yeast extract .... 5.0 g

*Commercially available.

Sodium bisulfite (meta) .... 1.0 g
Ferric ammonium citrate .... 1.0 g
Agar .... 20.0 g
Distilled water .... 1.0 L

Dissolve the ingredients in distilled water, adjust the pH to 7.6 ± 0.2, and autoclave the medium for 10 min at 121°C. To each L of autoclaved medium cooled to 50°C, add 10.0 mL of a 4.0% filter-sterilized solution of D-cycloserine to give a final concentration of approximately 400 mg per mL. Also add 80.0 mL of a sterile 50% egg yolk in saline emulsion per 1.0 L of medium, with the exception of that medium to be used to overlay the plates. Egg yolk enrichment 50% may be obtained from Difco Laboratories, Detroit, MI. Dispense the medium in standard Petri dishes for surface plating. Before use, air dry the plates at room temperature for 24 hr or until the surface of the agar is somewhat dry. Prepare plates fresh each time they are to be used. NOTE: SFP agar base available commercially from Difco Laboratories is the same as the above basal medium.

Tyrosine Agar

Preparation A: Prepare nutrient agar, and dispense 100.0 mL into bottles. Autoclave 15 min at 121°C, and cool to 48°C in a water bath.

Preparation B: Add 0.5 g of L-tyrosine to a 20 × 150 mm culture tube and suspend in 10.0 mL of distilled water using a Vortex mixer. Sterilize the suspension by autoclaving for 15 min at 121°C.

Mix Preparation A (100.0 mL) with sterile Preparation B (10.0 mL), and aseptically dispense 3.5 mL of complete medium into sterile 13- × 100-mm tubes. Slant tubes, and cool rapidly to prevent separation of the tyrosine.

Universal Beer Agar

Tomato juice broth .... 25.0 g
Peptonized milk .... 15.0 g
Dextrose .... 10.0 g
Agar .... 12.0 g
Distilled water .... 750 mL
Beer .... 250 mL

Combine ingredients and heat to boiling while stirring to dissolve the agar completely. Adjust the pH to 6.3 ± 0.2, and autoclave for 15 min at 121°C.

*Urea Broth**

*Urea $R_{(apid)}$ Broth**

*UVM (University of Vermont) Broth**

Commercially available from BBL; however, the commercial formulation may deviate from the original formula.

*Veal Infusion Agar and Broth**

Vibrio Parahaemolyticus *Sucrose Agar (VPSA)*

Tryptose .... 5.0 g
Tryptone .... 5.0 g
Yeast extract .... 7.0 g
Sucrose .... 10.0 g
Sodium chloride .... 30.0 g
Bile salts No. 3 .... 1.5 g
Bromthymol blue .... 0.025 g
Agar .... 15.0 g
Distilled water .... 1.0 L

Dissolve ingredients and sterilize by boiling. Temper the medium to 50°C, and adjust the pH to 8.6. Pour 20 mL per plate, and allow to dry thoroughly. The medium may be stored for 4 weeks at 4°C.

Vibrio Vulnificus *Agar (VVA)*

Solution 1:

Peptone .... 20 g
NaCl .... 30 g
Dye solution* (100X) .... 10 mL
Agar .... 25 g
Distilled water .... 1.0 L

Adjust pH to 8.2. Boil to dissolve agar. Autoclave for 15 min at 121°C. Temper to 50°C.

*Dye solution: Dissolve 0.6 g of bromthymol blue in 100 mL of distilled water.

Solution 2:

Cellobiose .... 10 g
Distilled water .... 100 mL

Dissolve cellobiose in water by gentle heating (avoid boiling). Filter-sterilize through a 0.2 µm filter, and add to tempered Solution 1. Prepare as pour plates.

*Violet Red Bile Agar (VRBA)**

Violet Red Bile Agar-Glucose (VRBG)

Add 1% glucose for pour-plate enumeration of total *Enterobacteriaceae* in brined vegetables.

Violet Red Bile Agar-Mug (VRB-MUG)

Add 4-methylumbelliferyl-β-D-glucuronic acid (MUG) reagent to VRBA at a level of 100 mg/mL.

Wagatsuma Agar

Peptone .... 10.0 g
Yeast extract .... 3.0 g
Dipotassium phosphate .... 5.0 g
Sodium chloride .... 70.0 g
Mannitol .... 10.0 g
Crystal violet .... 0.001 g
Agar .... 15.0 g
Distilled water .... 1.0 L

Suspend ingredients in distilled water, and dissolve by boiling gently. Adjust the pH to 8.0. Do not autoclave. Wash rabbit or human erythrocytes three times in physiological saline, and reconstitute to original blood volume. Add 2.0 mL of washed erythrocytes to 100.0 mL of agar cooled to 50°C just prior to pouring into sterile petri dishes.

Wang's Transport/Storage Medium

Brucella broth .... 28.0 g
Agar .... 4.0 g
Defibrinated sheep blood .... 100.0 mL
Distilled water .... 900.0 mL

Dissolve the Brucella broth and agar in distilled water, and boil for 1 min to dissolve the agar completely. Autoclave at 121°C for

*Commercially available.

15 min and cool to 50°C before adding the blood. Lysed horse blood may be used in lieu of sheep blood; 50 mL of horse blood and an additional 50 mL of sterile distilled water is used to replace the sheep blood. Dispense in 4.0-mL portions in sterile 15- × 125-mm screw-capped test tubes, and solidify in an upright position.

Xylose-Lysine-Desoxycholate Agar (XLD)*

Yeast Extract Agar (YEA)

Proteose peptone .... 10.0 g
Yeast extract .... 3.0 g
NaCl .... 5.0 g
Agar .... 15.0 g
Distilled water .... 1.0 L

Adjust pH to 7.2-7.4. Autoclave at 121°C for 15 min.

Yersinia Agar*

(Same as SS-desoxycholate (SSDC) agar)

YM Agar*

(Difco)

63.4 ALPHABETICAL LISTING OF REAGENTS, DILUENTS, AND INDICATORS

This section contains the formulae for preparation of all reagents, diluents and indicators used throughout this book, with the exception of a few complex preparations listed in specific chapters. Reagents, diluents, and indicators that are commercially available are indicated by asterisks. Reagents should be prepared using chemicals of the highest purity only, and double distilled water. The reagents may be heat sterilized or sterilized by membrane filtration when necessary.

Acid Phosphatase (ACP)

Sodium acetate (0.05 M solution, pH 5.0) .... 50 mL
α-naphthyl acid phosphate, monosodium .... 50 mg
β-naphthyl acid phosphate, monosodium .... 50 mg

Just before use add 20 mg of Black K salt.

Aconitase (ACO)

Tris-HCl, 0.2 M solution, pH 8.0 .... 15 mL
$MgCl_2$, 0.1 M solution .... 10 mL
Cis-aconitic acid .... 20 mg
Isocitrate dehydrogenase .... 5 U
Nicotinamide adenine dinucleotide phosphate (NADP)
1% solution .... 1 mL
Phenazine methosulfate (PMS), 1% solution .... 0.5 mL
Dimethyldiazol tetrazolium (MTT), 1.25% solution ... 0.5 mL

Use agar overlay.

Adenylate Kinase (ADK)

Tris-HCl, 0.2 M solution, pH 8.0 .... 25 mL
Glucose .... 100 mg
Adenosine diphosphate (ADP) .... 25 mg
Hexokinase (Baker's yeast type IV) .... 40 U
Glucose-6-phosphate dehydrogenase .... 15 U
$MgCl_2$, 0.1 M solution .... 1 mL
Nicotinamide adenine dinucleotide phosphate (NADP)
1% solution .... 0.5 mL
Phenazine methosulfate (PMS), 1% solution .... 0.5 mL
Dimethyldiazol tetrazolium (MTT), 1.25% solution ... 0.5 mL

Use agar overlay.

Agarose Buffer

(for PFGE)
Tris .... 10 mM solution
EDTA .... 0.1 mM

Adjust pH to 8.0

Alanine Dehydrogenase (ALDH)

Phosphate, 10 mM solution, pH 7.0 .... 50 mL
DL-alanine .... 50 mg
Nicotinamide adenine dinucleotide (NAD)
1% solution .... 2.0 mL
Phenazine methosulfate (PMS), 1% solution .... 0.5 mL
Dimethyldiazol tetrazolium (MTT), 1.25% solution ... 1.0 mL

Alcohol Dehydrogenase (ADH)

Tris-HCl, 0.2 M solution, pH 8.0 .... 50.0 mL
Ethanol .... 3.0 mL
Isopropanol .... 2.0 mL
Nicotinamide adenine dinucleotide (NAD)
1% solution .... 2.0 mL
Phenazine methosulfate (PMS), 1% solution .... 0.5 mL
Dimethyldiazol tetrazolium (MTT), 1.25% solution ... 1.0 mL

Aldolase (ALD)

Tris-acetate, 0.1 M solution, pH 7.5 .... 50.0 mL
Arsenic acid disodium salt heptahydrate .... 100 mg
Fructose-1,6-diphosphate .... 100 mg
Glyceraldehyde-3-phosphate dehydrogenase .... 50 U
Triosephosphate isomerase .... 100 U
Nicotinamide adenine dinucleotide (NAD)
1% solution .... 2.0 mL
Phenazine methosulfate (PMS), 1% solution .... 0.5 mL
Dimethyldiazol tetrazolium (MTT), 1.25% solution ... 1.0 mL

Alkaline Peptone Water (APW)

Peptone .... 10 g
Sodium chloride .... 20 g
Distilled water .... 1.0 L

Adjust pH to 8.5 ± 0.2.

Alkaline Phosphatase (ALP)

Tris-HCl, 0.05 M solution, pH 8.5 .... 50 mL
NaCl .... 1 g
$MgCl_2$, 0.1 M solution .... 2 mL
$MnCl_2$, 0.25 M solution .... 2 mL
β-naphthyl acid phosphate, monosodium .... 50 mg
Polyvinylpyrrolidone .... 100 mg

Just before use, add 50 mg Fast Blue BB salt.

Alpha-Naphthol Solution

Dissolve 5.0 g of fresh a-naphthol in 100.0 mL of 99% isopropyl alcohol. Store in brown bottles under refrigeration.

*Commercially available.

Ammonium Acetate Buffer

(for AP-probe hybridization)
Ammonium acetate .... 154 g
Distilled water .... 1.0 L
Mix ingredients, autoclave at 121°C for 15 min, and store at room temperature.

Assay Buffer

(for ribotyping)
Sodium bicarbonate/carbonate .... 50 mM
$MgCl_2$ .... 1 mM
Adjust pH to 9.5

Base Wash

(for cellular fatty acid analysis)
NaCl .... 120 g
NaOH .... 5.4 g
Distilled water .... 450 mL
Prepare and store in brown bottle. **Caution:** solution is caustic.

0.1 M Bicarbonate Buffer, pH 9.6

Na_2CO_3 .... 1.59 g
$NaHCO_3$ .... 2.93 g
Distilled water .... 1.0 L
Dissolve ingredients in distilled water, and store at room temperature for not more than 2 weeks.

Blocking Buffer

(for ribotyping)
Skim milk powder (Difco) .... 30% (w/v)
NaCl .... 25 mM
EDTA .... 1 mM
Tween 20 .... 0.3% (w/v)
Tris-HCl, pH 7.5 .... 50 mM

Brilliant Green Dye Solution, 1.0%

Brilliant green dye .... 1.0 g
Distilled water (sterile) .... 100.0 mL
Dissolve 1.0 g of brilliant green dye in sterile distilled water, and then dilute to 100.0 mL. Before use, test all batches of the dye for toxicity with known positive and negative test microorganisms.

Bromcresol Green Solution

(for Bromcresol Green Agar)
Bromcresol green dye .... 0.26 g
Ethanol (95%) .... 100.0 mL
Dissolve dye in a portion of the ethanol, then add the rest to achieve a final volume of 100.0 mL.

Bromcresol Green Solution

(for HHD Medium)
Bromcresol green dye .... 0.1 g
0.01N sodium hydroxide .... 30.0 mL
Dissolve the dye in the sodium hydroxide.

Bromcresol Purple Dye Solution, 1.6%

Bromcresol purple dye .... 1.6 g
Distilled water (sterile) .... 100.0 mL
Dissolve the bromcresol purple dye in a portion of the sterile distilled water, and then add remaining water to achieve a final volume of 100.0 mL.

Bromcresol Purple Dye Solution, 0.2%

Bromcresol purple dye .... 0.2 g
Distilled water (sterile) .... 100.0 mL
Dissolve the bromcresol purple dye in a portion of the sterile distilled water, and then add remaining water to achieve a final volume of 100.0 mL.

Butterfield's Phosphate Buffered Dilution Water

(Same as Phosphate Buffer)

Catalase Test

Flood plates with 3.0% hydrogen peroxide solution. Observe for bubble formation using a hand lens or wide field binocular microscope. Colonies exhibiting no evidence of gas formation are catalase negative. Alternative method: Transfer a loopful of colony to a slide, mix with 2.0 to 5.0% hydrogen peroxide, and observe as above.

Catalase (CAT) Gel Staining Reagents

Solution A:
Hydrogen peroxide (30% solution) .... 60 mL
Distilled water .... 50 mL
Solution B:
Potassium ferricyanide (2% solution) .... 25 mL
Ferric chlorate hexahydrate (2% solution) .... 25 mL
Incubate the gel for 15 min in Solution A. Discard solution, briefly rinse gel in water, and discard water. Add freshly made Solution B. Shake gently until light bands appear on dark blue background.

Coagulase Plasma Containing EDTA*

Cytochrome Oxidase Reagent

N,N,N,N-tetramethyl-p-phenylenediamine .... 5.0 g
Distilled water .... 1.0 L
Combine ingredients and store in dark glass bottle at 5° to 10°C. Storage life is 14 days. To perform test, add 0.3 mL of the reagent to an 18-hr blood agar base slant. Positive reaction is development of a blue color within 1 min.

Denaturing Solution

(for ribotyping)
NaOH .... 0.2 M
NaCl .... 1.5 M

Denhardt's Solution

Ficoll (M.W. ~ 400,000) .... 5.0 g
Polyvinyl pyrrolidone (average M.W. 360,000) .... 5.0 g
Bovine serum albumin (nuclease-free) .... 5.0 g
Distilled water to make .... 100.0 mL
Combine all ingredients, and filter-sterilize. Store in small aliquots at –20°C.

Dulbecco's Phosphate-Buffered Saline (DPBS)

NaCl .... 8.0 g
KCl .... 200.0 mg
Na_2HPO_4 .... 1.15 g

*Commercially available.

KH_2PO_4 .... 200.0 mg
$CaCl_2$ .... 100.0 mg
$MgCl_2 \cdot 6H_2O$ .... 100.0 mg
Distilled water .... 1.0 L

Combine ingredients and dissolve in distilled water. Sterilize by filtration. Final pH should be 7.2.

Dulbecco's Phosphate-Buffered Saline (Calcium- and Magnesium-Free)

NaCl .... 8.0 g
KCl .... 200.0 mg
Na_2HPO_4 .... 1.15 g
KH_2PO_4 .... 200.0 mg
Distilled water .... 1.0 L

Combine ingredients, and dissolve in distilled water. Sterilize by filtration.

0.5 M EDTA

Na_2EDTA .... 186.12 g
Distilled water .... 1.0 L

Dissolve in 800 to 900 mL of distilled water. Adjust the pH to 8.0 with 10 N NaOH, and bring the volume to 1.0 L with distilled water.

ECOR1 Buffer

(for ribotyping)
Tris-HCl, pH 7.5 .... 500 mM
NaOH .... 250 mM
$MgCl_2$ .... 50 mM
Bovine serum albumin .... 1 µg/µL

ELISA (Enzyme-Linked Immunosorbent Assay) Buffer

Bovine serum albumin .... 1.0 g
Sodium chloride .... 8.0 g
KH_2PO_4 .... 0.2 g
Na_2HPO_4 .... 2.9 g
Potassium chloride .... 0.2 g
Distilled water .... 1.0 L

Dissolve ingredients in distilled water, adjust pH to 7.4, and add 0.5 mL of Tween 20. Store frozen and thaw before use.

Esterase (EST)

Phosphate (10 mM solution, pH 7.0) .... 50 mL
Substrate (1% solution in acetone) .... 2 mL
Substrates: α-naphthyl acetate
β-naphthyl acetate
α-naphthyl propionate
β-naphthyl propionate
α-naphthyl butyrate
β-naphthyl butyrate
Fast Blue RR salt .... 25 mg

Prepare just before use.

Extraction Solution

(for fatty acid analysis)
Hexane .... 200 mL
Methyl tert-butyl ether .... 200 mL

Prepare and store in brown bottle. **Caution:** reagents are flammable.

Formalinized Physiological Saline Solution

Formaldehyde solution (36-38%) .... 6.0 mL
NaCl .... 8.5 g
Distilled water .... 1.0 L

Dissolve 8.5 g of NaCl in 1.0 L of distilled water and autoclave for 15 min at 121°C. Cool to room temperature, and add 6.0 mL of formaldehyde solution. Do not autoclave after the addition of formaldehyde.

Fructose Solution, 10%

Fructose .... 10.0 g
Distilled water .... 100.0 mL

Dissolve 10.0 g fructose in distilled water, then dilute to 100.0 mL. Sterilize by filtration through a Millipore (or compatible) 0.45 µm filter, and refrigerate until needed.

Fumarase (FUM)

Tris-HCl (0.2 M solution, pH 8.0) .... 25 mL
Fumaric acid (potassium salt) .... 50 mg
Malate dehydrogenase .... 50 U
Nicotinamide adenine dinucleotide (NAD)
1% solution .... 2.0 mL
Phenazine methosulfate (PMS), 1% solution .... 0.5 mL
Dimethyldiazol tetrazolium (MTT), 1.25% solution ... 1.0 mL

Use agar overlay.

Gel Loading Solution

(for ribotyping)
Solution A: Bromophenol blue .... 0.25% (w/v)
Xylene cyanol .... 0.25% (w/v)
Ficoll .... 25% (w/v)
Solution B: EDTA, 0.5 M .... pH 8.0

Mix 2.75 mL of Solution A with 0.55 mL of Solution B.

Gel Phosphate Buffer

Gelatin .... 2.0 g
Na_2HPO_4 .... 4.0 g
Distilled water .... 1.0 L

Dissolve ingredients by heating gently in 800.0 mL of distilled water. Adjust the pH to 6.2 with HCl. Add the remaining water (200.0 mL minus the volume of HCl added) and sterilize at 121°C for 20 min.

Genius DIG Buffers

(for DIG probe hybridization)
Buffer 1:
Tris-HCl (1 M, pH 7.5) .... 100 mL
NaCl (3 M) .... 50 mL
Distilled water .... 1.0 L

Store at room temperature.

Buffer 2:
Buffer 1 .... 50 mL
Blocking reagent (B-M cat. No. 1096176) .... 0.25 g

Agitate vigorously and microwave or heat gently in a 65°C water bath to dissolve blocking reagent. Agitate every 10 min until dissolved, then cool to room temperature. Store at room temperature.

*Commercially available.

Buffer 3:
Tris-HCl (1 M, pH 9.5) .... 100 mL
NaCl (3 M) .... 33.3 mL
$MgCl_2$ (1 M) .... 50 mL
Distilled water .... 1.0 L

Store at room temperature.

Buffer 4:
Tris base (10 mM) .... 1.21 g
EDTA (1 mM) .... 0.372 g
Distilled water .... 1.0 L

Dissolve first in 700 mL of distilled water, adjust pH to 8.0. Add remaining water to achieve a final volume of 1.0 L. Store at room temperature.

Glucose-6-Phosphate Dehydrogenase (G6P)

Tris-HCl (0.2 M solution, pH 8.0) .... 50.0 mL
Glucose-6-phosphate disodium salt hydrate .... 100 mg
$MgCl_2$ (0.1 M solution) .... 1.0 mL
Nicotinamide adenine dinucleotide phosphate (NADP) 1% solution .... 1.0 mL
Phenazine methosulfate (PMS), 1% solution .... 0.5 mL
Dimethyldiazol tetrazolium (MTT), 1.25% solution ... 0.5 mL

Glucose Dehydrogenase (GDH)

Phosphate (0.05 M solution, pH 7.5) .... 50.0 mL
D-glucose .... 9 g
Nicotinamide adenine dinucleotide (NAD) 1% solution .... 2.0 mL
Phenazine methosulfate (PMS), 1% solution .... 0.5 mL
Dimethyldiazol tetrazolium (MTT), 1.25% solution ... 1.0 mL

Glutamate Dehydrogenase (GD1 or GD2)

Tris-HCl (0.2 M solution, pH 8.0) .... 50.0 mL
L-glutamic acid .... 200 mg
Phenazine methosulfate (PMS), 1% solution .... 0.5 mL
Dimethyldiazol tetrazolium (MTT), 1.25% solution ... 1.0 mL

For GD1, add 2 mL nicotinamide adenine dinucleotide (NAD), 1% solution.

For GD2, add 1 mL nicotinamide adenine dinucleotide phosphate (NADP), 1% solution.

Glutamic-Oxalacetic Transaminase (GOT)

Tris-HCl (0.2 M solution, pH 8.0) .... 50.0 mL
Aspartic acid monosodium salt .... 200 mg
α-ketoglutaric acid, disodium salt .... 140 mg

Just before use, add 200 mg of Fast Blue BB salt.

Glyceraldehyde-Phosphate Dehydrogenase (GP1 or GP2)

Tris-HCl (0.2 M solution, pH 8.0) .... 40.0 mL
Fructose-1,6-diphosphate tetra(cyclohexylammonium) salt .... 100 mg
Arsenic acid sodium salt heptahydrate .... 50 mg
Aldolase .... 10 U
Phenazine methosulfate (PMS), 1% solution .... 0.5 mL
Dimethyldiazol tetrazolium (MTT), 1.25% solution ... 1.0 mL

For GP1, add 2 mL nicotinamide adenine dinucleotide (NAD), 1% solution.

For GP2, add 1 mL nicotinamide adenine dinucleotide phosphate (NADP), 1% solution.

H Buffer

(for PFGE)
Tris-HCl .... 500 mM
$MgCl_2$ .... 100 mM
NaCl 1M
Dithioerythritol .... 10 mM

Dilute 1:9 (v/v) with water to make 1X. Adjust pH to 7.5

Hank's Phosphate-Buffered Saline

(Calcium- and Magnesium-free)
NaCl .... 8.0 g
KCl .... 400.0 mg
$Na_2HPO_4 \cdot 7H_2O$ .... 90.0 mg
KH_2PO_4 .... 60.0 mg
Glucose .... 1.0 g
$NaHCO_3$ .... 350.0 mg
Distilled water .... 1.0 liter

Dissolve all ingredients in distilled water, and sterilize by filtration.

Hexokinase (HEX)

Glycine (0.1 M solution, pH 7.5 with KOH) .... 50.0 mL
$MgCl_2$ (0.1 M solution) .... 2.0 mL
D-glucose .... 200 mg
Adenosine triphosphate (ATP) .... 50 mg
Glucose-6-phosphate dehydrogenase .... 10 U
Nicotinamide adenine dinucleotide phosphate (NADP) 1% solution .... 1.0 mL
Phenazine methosulfate (PMS), 1% solution .... 0.5 mL
Dimethyldiazol tetrazolium (MTT), 1.25% solution ... 1.0 mL

Horseradish Peroxidase (HRP)

(Color Development Solution)

Solution A:
4-chloro-1-naphthol .... 60.0 mg
Methanol (ice cold) .... 20.0 mL

Solution A should be stored in the dark.

Solution B:
Tris buffered saline (TBS) .... 100.0 mL
30% H_2O_2 .... 60.0 mL

Solution B should be prepared just before use.

Combine solutions A (ice cold) and B (room temperature), and use within 10 min on antibody-soaked filters (10 mL/HGMF).

Hybridization Solution

(for ribotyping)
Salmon sperm DNA, denatured, sonicated .... 125 μg/mL
NaCl .... 0.5 M
Sodium dodecyl sulfate .... 1% (w/v)

Hybridization Solution

(for AP-probe hybridization)
Bovine serum albumin (BSA) .... 0.5 g
Sodium dodecyl sulfate (SDS) .... 1.0 g
Polyvinyl-pyrrolidone (PVP) .... 0.5 g
Standard Saline Citrate Buffer (SSC), 5X .... 100 mL

*Commercially available.

Store at 4°C no longer than 1 week. Pre-warm to hybridization temperature before use.

Hybridization Wash Buffer

(for ribotyping)
NaCl, 0.5 M
Sodium dodecyl sulfate, 1% (w/v)

Hydrochloric Acid (1 N)

Hydrochloric acid (concentrated) .... 86.0 mL
Distilled water to make .... 1.0 L

Indole Oxidant-Accelerated Reagent

Solution A:
4-dimethylaminobenzaldehyde .... 2.5 g
Ethanol (95%) .... 90.0 mL
Concentrated HCl .... 10.0 mL

Solution A can be stored for 4 weeks at 4°C if protected from light.

Solution B:
Potassium persulfate .... 1.0 g
Distilled water .... 100.0 mL

Solution B can be stored for 3 months at 4°C.

Mix equal volumes of Solution A and Solution B just before use.

Indophenol Oxidase (IPO)

Tris-HCl (0.2 M solution, pH 8.0) .... 25.0 mL
$MgCl_2$ (0.1 M solution) .... 2.0 mL
Nicotinamide adenine dinucleotide (NAD)
1% solution .... 2.0 mL
Phenazine methosulfate (PMS), 1% solution .... 0.5 mL
Dimethyldiazol tetrazolium (MTT), 1.25% solution ... 1.0 mL

Use agar overlay and do not incubate in the dark. IPO activity appears as white bands on a light blue background.

Isocitrate Dehydrogenase (IDH)

Tris-HCl (0.2 M solution, pH 8.0) .... 50.0 mL
DL-isocitric acid trisodium salt (1 M solution) .... 2.0 mL
$MgCl_2$ (0.1 M solution) .... 2.0 mL
Nicotinamide adenine dinucleotide phosphate (NADP)
1% solution .... 1.0 mL
Phenazine methosulfate (PMS), 1% solution .... 0.5 mL
Dimethyldiazol tetrazolium (MTT), 1.25% solution ... 1.0 mL

(KOH) For Voges-Proskauer (VP) Reaction*

Kovac's Reagent (Indole)

r-dimethylaminobenzaldehyde .... 5.0 g
Amyl alcohol .... 75.0 mL
Hydrochloric acid (concentrated) .... 25.0 mL

Dissolve r-dimethylaminobenzaldehyde in the amyl alcohol, and then slowly add the hydrochloric acid. To test for indole, add 0.2 to 0.3 mL of reagent to 5.0 mL of a 48-hour culture of bacteria in tryptone broth. A dark red color in the surface layer constitutes a positive test for indole.

Lactate Dehydrogenase (LDH)

Glycine (0.1 M solution, pH 7.5 with KOH) .... 50.0 mL
DL-lactic acid lithium salt .... 330 mg
Fructose-1,6-diphosphate tetra (cyclohexylammonium)
salt .... 10 mg
Nicotinamide adenine dinucleotide (NAD)
1% solution .... 2.0 mL
Phenazine methosulfate (PMS), 1% solution .... 0.5 mL
Dimethyldiazol tetrazolium (MTT), 1.25% solution ... 1.0 mL

Leucine Aminopeptidase (LAP)

KH_2PO_4 (0.1 M solution, pH 5.5) .... 50.0 mL
L-leucine-b-naphthylamide hydrochloride .... 30 mg
$MgCl_2$ (0.1 M solution) .... 1.0 mL

Just before use, add 30 mg Black K salt.

Leucine Dehydrogenase (LED)

Phosphate (10 mM solution, pH 7.0) .... 50.0 mL
L-leucine .... 50 mg
Nicotinamide adenine dinucleotide (NAD)
1% solution .... 2.0 mL
Phenazine methosulfate (PMS), 1% solution .... 0.5 mL
Dimethyldiazol tetrazolium (MTT), 1.25% solution ... 1.0 mL

Lugol's Iodine Solution

Iodine .... 1.0 g
Potassium iodide .... 2.0 g
Distilled water .... 300.0 mL

MacFarland Standards

Make suspensions of barium sulfate as follows:

1. Prepare v/v 1.0% solution of sulfuric acid.
2. Prepare w/v 1.0% solution of barium chloride.
3. Prepare 10 standards as follows:

McFarland Standard No.	1% Barium Chloride (mL)	1% Sulfuric Acid (mL)	Approximate Bacterial Suspension $\times 10^6$/mL
1	0.1	9.9	300
2	0.2	9.8	600
3	0.3	9.7	900
4	0.4	9.6	1200
5	0.5	9.5	1500
6	0.6	9.4	1800
7	0.7	9.3	2100
8	0.8	9.2	2400
9	0.9	9.1	2700
10	1.0	9.0	3000

4. Seal about 3 mL of each standard in a small test tube. Select tubes carefully for uniformity of absorbance.

Malate Dehydrogenase (MDH)

Solution A:
DL-malic acid .... 26.8 g
NaOH .... 16.0 g
Distilled water .... 100 mL

Caution: Potentially explosive reaction.

Solution B:
Tris HCl (0.2 M solution, pH 8.0) .... 40 mL
Nicotinamide adenine dinucleotide (NAD)
1% solution .... 2.0 mL

*Commercially available.

Phenazine methosulfate (PMS), 1% solution .... 0.5 mL
Dimethyldiazol tetrazolium (MTT), 1.25% solution ... 1.0 mL
Add 6 mL of Solution A to Solution B.

Malic Enzyme (ME)

Solution A:
DL-malic acid .... 26.8 g
NaOH .... 16.0 g
Distilled water .... 100 mL
Caution: Potentially explosive reaction.
Solution B:
Tris HCl (0.2 M solution, pH 8.0) .... 40.0 mL
$MgCl_2$ (0.1 M solution) .... 2.0 mL
Nicotinamide adenine dinucleotide phosphate (NADP)
1% solution .... 1.0 mL
Phenazine methosulfate (PMS), 1% solution .... 0.5 mL
Dimethyldiazol tetrazolium (MTT), 1.25% solution ... 1.0 mL
Add 6 mL of Solution A to Solution B.

Mannitol-1-Phosphate Dehydrogenase (MIP)

Tris HCl (0.2 M solution, pH 8.0) .... 50.0 mL
Mannitol-1-phosphate .... 5 mg
Nicotinamide adenine dinucleotide (NAD)
1% solution .... 2.0 mL
Phenazine methosulfate (PMS), 1% solution .... 0.5 mL
Dimethyldiazol tetrazolium (MTT), 1.25% solution ... 1.0 mL

Mannose Phosphate Isomerase (MPI)

Tris HCl (0.2 M solution, pH 8.0) .... 25.0 mL
Mannose-6-phosphate barium salt .... 10 mg
Glucose-6-phosphate dehydrogenase .... 10 U
Phosphoglucose isomerase .... 50 U
$MgCl_2$ (0.1 M solution) .... 1.0 mL
Nicotinamide adenine dinucleotide (NAD)
1% solution .... 2.0 mL
Nicotinamide adenine dinucleotide phosphate (NADP)
1% solution .... 1.0 mL
Phenazine methosulfate (PMS), 1% solution .... 0.5 mL
Dimethyldiazol tetrazolium (MTT), 1.25% solution ... 0.5 mL
Use agar overlay.

Membrane Washing Solution

(for ribotyping)
NaCl .... 0.5 M
Tween 20 .... 0.3% (w/v)

Methylation Solution

(for cellular fatty acid analysis)
Methanol .... 275 mL
HCl (6 N solution) .... 325 mL
Prepare and store in brown bottle. **Caution:** solution is acidic.

Methyl Red Indicator for MR Test

Methyl red .... 0.10 g
Alcohol, 95% (ethanol) .... 300.0 mL
Distilled water to bring volume to .... 500.0 mL

Dissolve methyl red in 300.0 mL of ethanol and add distilled water to make a total volume of 500.0 mL. Incubate test cultures 5 days at 30°C. Alternatively, incubate at 37°C for 48 hr. Add 5 or 6 drops of reagent to cultures. Do not perform tests on cultures incubated for less than 48 hr. If results are equivocal, repeat tests on cultures incubated for 4 or 5 days. Duplicated tests should be incubated at 22 to 25°C.

4-Methylumbelliferyl-β-D-Glucuronic Acid (MUG)*

Multilocus Enzyme Electrophoresis Buffer Systems

System A (Tris Citrate):
Electrode Buffer
Tris base .... 83.2 g
Citric acid monohydrate .... 33.09 g
Distilled water .... 1.0 L
Adjust pH to 8.0.
Gel Buffer
Dilute electrode buffer 1:29 (v/v) in water. Adjust pH to 8.0
System B (Tris Citrate):
Electrode Buffer
Tris base .... 27 g
Citric acid monohydrate .... 18.07 g
Distilled water .... 1.0 L
Adjust pH to 6.3.
Gel Buffer
Tris base .... 0.97 g
Citric acid monohydrate .... 0.63 g
Distilled water .... 1.0 L
Adjust pH to 6.7.
System C (Borate):
Electrode Buffer
Boric acid .... 18.5 g
NaOH .... 2.4 g
Distilled water .... 1.0 L
Adjust pH to 8.2.
Gel Buffer
Tris base .... 9.21 g
Citric acid monohydrate .... 1.05 g
Distilled water .... 1.0 L
Adjust pH to 8.7.
System D (Lithium Hydroxide):
Electrode Buffer
Lithium hydroxide monohydrate .... 1.2 g
Boric acid .... 11.89 g
Distilled water .... 1.0 L
Adjust pH to 8.1.
Gel Buffer
Tris base .... 5.58 g
Citric acid monohydrate .... 1.44 g
Distilled water .... 900 mL
Electrode buffer .... 100 mL
Adjust pH to 8.3.
System F (Tris Maleate):
Electrode Buffer
Tris base .... 12.1 g
Maleic acid .... 11.6 g
Disodium EDTA .... 3.72 g
$MgCl_2$, hexahydrate .... 2.03 g
Distilled water .... 1.0 L
Adjust pH to 8.2.
Gel Buffer
Dilute electrode buffer 1:9 (v/v) in water. Adjust pH to 8.2.

*Commercially available.

System G (Potassium Phosphate):

Electrode Buffer

Potassium phosphate monobasic .... 18.14 g
Distilled water .... 1.0 L
Adjust pH to 6.7.

Gel Buffer

Potassium phosphate dibasic .... 1.06 g
Citric acid monohydrate .... 0.25 g
Distilled water .... 1.0 L
Adjust pH to 7.0.

NBT/BCIP Color Reagent

(for AP-probe hybridization)

Add one tablet of NBT/BCIP color reagent to 10 mL of distilled water.

NBT/BCIP Color Reagent

(for DIG probe hybridization)

Add one tablet of NBT/BCIP color reagent to 10 mL of DIG buffer 3.

Nessler's Reagent*

Nitrate Reduction Reagents

Method 1

Solution A:

Sulfanilic acid .... 0.5 g
Glacial acetic acid .... 30.0 mL
Distilled water .... 120.0 mL

Solution B:

N (1-naphthyl) ethylenediamine dihydrochloride
(Marshal's reagent)† .... 0.2 g
Glacial acetic acid .... 30.0 mL
Distilled water .... 120.0 mL

† Cleve's acid (5-amino-2 naphthalene sulfonic acid) may be substituted for Marshal's reagent.

To 3.0 mL of an 18-hour culture in indole-nitrate broth, add 2 drops of Solution A and 2 drops of Solution B. A red violet color which develops within 10 min indicates that nitrate has been reduced to nitrite. If the reaction is negative, examine for residual nitrate since conceivably the nitrite may have been reduced to another state. Add a few grains of powdered zinc. If a red violet color does not develop, nitrate has been reduced. Perform tests on uninoculated medium as a control.

Method 2

Solution A:

Sulfanilic acid .... 2.0 g
Glacial acetic acid .... 60.0 mL
Distilled water .... 150.0 mL

Solution B:

α-naphthol .... 1.0 g
Absolute alcohol .... 200.0 mL

Add 0.2 mL of Solution A followed by 0.2 mL of Solution B to a culture incubated 24 to 48 hr in nitrate broth. Development of an orange color within 10 min indicates that nitrate has been reduced to nitrite.

Nitrate Reduction Reagents

(For *Campylobacter*)

Solution A:

Glacial acetic acid .... 286.0 mL
Distilled water .... 714.0 mL
Sulfanilic acid .... 7.01 g

Slowly add the glacial acetic acid to the distilled water. Mix carefully. Add the sulfanilic acid, and mix well to dissolve. Store under refrigeration.

Solution B:

Glacial acetic acid .... 286.0 mL
Distilled water .... 714.0 mL
Dimethyl-a-naphthylamine .... 5.01 mL

Slowly add the glacial acetic acid to the distilled water. Mix carefully. Add the dimethyl-a-naphthylamine, and mix well to dissolve. Store under refrigeration.

Novobiocin Solution

(for *E. coli*)

Novobiocin solution (100 mg/mL) .... 100 mg
Distilled water .... 1.0 mL

Dissolve novobiocin. Filter-sterilize using 0.2 µm filter and syringe. May be stored several months in dark bottle at 4°C.

Nucleoside Phosphorylase (NSP)

Phosphate (10 mM solution, pH 7.0) .... 25 mL
Inosine .... 20 mg
Xanthine oxidase .... 2 U
Phenazine methosulfate (PMS), 1% solution .... 0.5 mL
Dimethyldiazol tetrazolium (MTT), 1.25 % solution .... 1 mL

Use agar overlay.

O157 Monoclonal Antibody Solution

(*E. coli* O157)

E. coli O157 ascitic fluid .... 10.0 mL
Tris buffered saline (TBS) with 1% gelatin .... 11.0 mL
Horseradish peroxidase (HRP)-protein A conjugate ...... 3.0 mL

Dispense 10 mL of *E. coli* O157 ascitic fluid (National Research Council, Ottawa, Ontario, Canada) into 1.0 mL of TBS 1% gelatin. Add 3.0 mL HRP-protein A conjugate, and stir the mixture at 4°C for 1 hour. Dilute to 10.0 mL with TBS 1% gelatin. This amount is sufficient for one hydrophobic grid membrane filter (HGMF).

Oxidase Reagent

(Kovac's Modification)

Tetramethyl-r-phenylenediamine dihydrochloride ...... 50.0 mg
Distilled water .... 5.0 mL

Dissolve powder in distilled water. Store up to 1 week at 4°C in dark brown container or covered tube to protect from light.

PCR Buffer without $MgCl_2$ (Working Solution)

Tris HCl .... 100 mM
KCl .... 500 mM

Dilute 1:9 (v/v) with water to make 1X. Adjust pH to 8.3.

PCR Buffer with $MgCl_2$ (Working Solution)

Tris HCl .... 10 mM
KCl .... 50 mM
$MgCl_2$ .... 2 mM

Adjust pH to 8.3.

*Commercially available.

Peptidase (PEP)

Tris HCl (0.2 M solution, pH 8.0) .... 25 mL
Oligopeptide .... 20 mg
Horseradish peroxidase (type I) .... 800 U
Snake venom (*Crotalus atrox*) .... 10 mg
Just before use, add:
$MnCl_2$ (0.25 M solution) .... 0.5 mL
O-dianisidine dihydrochloride .... 10 mg
Use agar overlay.

Peptone/Tween 80 Diluent (PT)

Peptone .... 1.0 g
Tween 80 .... 10.0 g
Distilled water .... 1.0 L

Suspend the ingredients in distilled water, and heat if necessary to dissolve completely. Dispense the diluent into bottles in desired volumes, and autoclave 15 min at 121°C.

Peptone Tween Salt Diluent (PTS)

Peptone .... 1.0 g
Tween 80 .... 10.0 g
Sodium chloride .... 30.0 g
Distilled water .... 1.0 L

Dissolve ingredients in distilled water, and dispense into bottles of desired volume. Autoclave 15 min at 121°C.

Peptone Water Diluent (0.1%)

Peptone .... 1.0 g
Distilled water .... 1.0 L

Dissolve peptone in distilled water. Adjust pH to 7.0 ± 0.1. Prepare dilution blanks with this solution, dispensing a sufficient quantity to allow for loss during autoclaving. Autoclave at 121°C for 15 min.

Phosphate Buffer (Butterfield)

Stock solution:
KH_2PO_4 .... 34.0 g
Distilled water .... 500.0 mL

Combine the above ingredients, and adjust the pH to 7.2 with about 175 mL 1 N sodium hydroxide solution; dilute to 1.0 L. Sterilize at 121°C for 15 min and store in refrigerator.

Diluent: Dilute 1.25 mL of stock solution to 1.0 L with distilled water. Prepare dilution blanks in suitable containers. Sterilize at 121°C for 15 min.

Phosphate Buffered Saline (PBS)*

Phosphate Buffered Saline (PBS)

(For *Clostridium perfringens*)
Sodium phosphate (dibasic) .... 8.09 g
Potassium phosphate (monobasic) .... 2.44 g
NaCl .... 4.25 g
Double distilled water to make .... 1.0 L

Combine the ingredients, and adjust the pH to 7.2. Store at 4°C. The addition of 0.1 g of thiomersal (merthiolate) per L is recommended unless sediment from feces is to be cultured.

Phosphoglucose Isomerase (PGI)

Tris HCl (0.2 M solution, pH 8.0) .... 25 mL
Fructose-6-phosphate disodium salt .... 10 mg
Glucose-6-phosphate dehydrogenase .... 3 U
$MgCl_2$ (0.1 M solution) .... 0.3 mL
Nicotinamide adenine dinucleotide phosphate (NADP), 1% solution .... 1 mL
Phenazine methosulfate (PMS), 1% solution .... 0.5 mL
Dimethyldiazol tetrazolium (MTT), 1.25% solution .... 0.5 mL
Use agar overlay.

Phosphoglucomutase (PGM)

Tris HCl (0.2 M solution, pH 8.0) .... 5 mL
Distilled water .... 25 mL
$MgCl_2$ (0.1 M solution) .... 5 mL
α-D-glucose-1-phosphate disodium salt hydrate (Grade VI or approx. 95% pure preparation containing contaminating α-D-glucose-1,6-diphosphate) .... 5 mg
Glucose-6-phosphate dehydrogenase .... 50 U
Nicotinamide adenine dinucleotide phosphate (NADP), 1% solution .... 1.0 mL
Phenazine methosulfate (PMS), 1% solution .... 0.5 mL
Dimethyldiazol tetrazolium (MTT), 1.25% solution .... 0.5 mL
Use agar overlay.

6-Phosphogluconate Dehydrogenase (6PG)

Tris HCl (0.2 M solution, pH 8.0) .... 20 mL
6-phosphogluconic acid barium salt .... 10 mg
$MgCl_2$ (0.1 M solution) .... 10 mL
Nicotinamide adenine dinucleotide phosphate (NADP), 1% solution .... 1.0 mL
Phenazine methosulfate (PMS), 1% solution .... 0.5 mL
Dimethyldiazol tetrazolium (MTT), 1.25% solution .... 1.0 mL

Physiological Saline Solution 0.85% (Sterile)

NaCl .... 8.5 g
Distilled water .... 1.0 L

Dissolve NaCl in distilled water, and autoclave for 15 min at 121°C. Cool to room temperature.

Polysaccharide Precipitant

Dissolve 1.0 g hexadecyltrimethylammoniumbromide in 100.0 mL of water. Solution may be sterilized by autoclaving if desired. Flood plates with the solution. Clear zones indicative of pectic enzymes should be visible within 15 min.

Potassium Hydroxide Solution, 40%

Potassium hydroxide .... 40.0 g
Distilled water to make .... 100.0 mL

Combine the ingredients, and stir until dissolved.

Potassium Hydroxide 40% + Creatine

Potassium hydroxide .... 40.0 g
Creatine (reagent grade) .... 0.3 g
Distilled water to make .... 100.0 mL

Combine the ingredients, and stir until dissolved. Do not store over 3 days at 2° to 8°C nor more than 21 days at –17.8°C.

Pre-Hybridization/Hybridization Buffer

(for DIG probe hybridization)
Standard Saline Citrate buffer (SSC), 20X .... 25.0 mL
Sarkosyl solution (10%) .... 1.0 mL
Sodium dodecyl sulfate (10%) .... 200 µL

*Commercially available.

Blocking reagent .... 1.0 g
Distilled water .... 100 mL
Gently heat, microwave to dissolve. Store at 4°C for up to one week.

Protein A Peroxidase Conjugate*

Proteinase K Solution

(for AP-probe hybridization)
Proteinase K (Difco) .... 200 µL
Standard Saline Citrate Buffer (SSC), 1X .... 100 mL
Prepare fresh before use. Proteinase K stock is 20 mg/mL in distilled water (store at –20°C). For each filter, dilute 20 µL stock solution of Proteinase K in 10 mL of 1X SSC.

Saline Solution, 0.5% Aqueous Solution

NaCl .... 5.0 g
Distilled water .... 1.0 L
Dissolve NaCl in distilled water, and autoclave for 15 min at 121°C. Final pH should be 7.0.

Saponification Solution

(for cellular fatty acid analysis)
NaOH .... 45 g
Methanol .... 150 mL
Distilled water .... 150 mL
Prepare and store in brown bottle. **Caution:** solution is caustic.

SE Wash Solution

(for PFGE)
NaCl .... 75 mM
EDTA .... 25 mM
Adjust pH to 8.0.

Sodium Carbonate 10%

$NaHCO_3$ .... 30.0 g
Distilled water .... 1.0 L
Dissolve $NaHCO_3$ in distilled water, and filter-sterilize.

Sodium Chloride 3%

NaCl .... 30.0 g
Distilled water .... 1.0 L
Dissolve NaCl in distilled water, dispense into suitable containers and autoclave at 121°C for 15 min.

Sodium Chloride, Physiological (0.85% NaCl)

(Same as Physiological Saline Solution 0.85%)

Sodium Hydroxide (0.02 N)

Sodium hydroxide .... 0.8 g
Distilled water to make .... 1.0 L

Sodium Hydroxide Solution (1 N)

Sodium hydroxide .... 40.0 g
Distilled water to make .... 1.0 L

Spicer-Edwards EN Complex Antibody Solution

Spicer-Edwards EN complex (Difco) .... 0.1 mL
Horseradish peroxidase (HRP)-protein A conjugate ...... 0.07 mL
Tris buffered saline (TBS) 1% gelatin .... 41.0 mL
Add 0.1 mL of Spicer-Edwards EN Complex to 0.07 mL of HRP-protein A conjugate in 1.0 mL of TBS 1% gelatin, and stir at 4°C for 1 hr. Dilute to 40.0 mL with TBS 1% gelatin. Use within a few hours. This amount is sufficient for four HGMF's.

SSC/SDS (1%)

(for AP-probe hybridization)
Sodium dodecyl sulfate .... 10 g
Standard Saline Citrate Buffer (SSC), 1X .... 1.0 L
Mix ingredients, autoclave at 121°C for 15 min, and store at room temperature.

Stabilizer Solution

(for *Geotrichum* mold count)
Sodium carboxymethylcellulose .... 25 g
Formaldehyde .... 10 mL
Distilled water .... 500 mL
Mix sodium carboxymethylcellulose in 10 mL formaldehyde and 500 mL water.

Standard Saline Citrate (SSC) Solution (20X)

NaCl .... 175.4 g
Sodium citrate .... 88.2 g
Distilled water to make .... 1.0 L
Dissolve ingredients in distilled water to make a total volume of 1.0 L. Prepare 5X and 2X SSC by diluting this stock solution with distilled water.

Storage Buffer

(for ribotyping)
Tris-HCl, pH 8.0 .... 50 mM
NaCl .... 150 mM
Bovine serum albumin .... 1% (w/v)
$MgCl_2$ .... 2 mM
$ZnCl_2$ .... mM

Tartaric Acid Solution, 10.0%

Tartaric acid .... 10.0 g
Distilled water .... 100.0 mL
Dissolve 10.0 g tartaric acid in distilled water, then dilute to 100.0 mL. Autoclave for 15 min at 121°C.

TBE Buffer

(for PCR)
Tris .... 108 g/L
Boric acid .... 55 g/L
EDTA .... 9.3 g/L
Dilute 1:9 (v/v) with water to make 1X TBE buffer. Adjust pH to 8.3

TBE Buffer

(for PFGE)
Tris .... 108 g/L
Boric acid .... 55 g/L
EDTA .... 9.3 g/L
Dilute 1:9 (v/v) with water to make 1X TBE buffer. Adjust pH to 8.3

*Commercially available.

TE Buffer

(for PCR)
Tris-HCl .... 10 mM
EDTA .... 1 mM
Adjust pH to 8.0.

TE Buffer

(for PFGE)
Tris .... 10 mM
EDTA .... 1 mM
Adjust pH to 8.0.

Tergitol™ Anionic 7

This reagent is a sodium sulfate derivative of 3,9-diethyl tridecanol-6. Tergitol-7 is an anionic wetting agent manufactured by Union Carbide Corp., Chemicals and Plastics, New York, NY 10017.

Threonine Dehydrogenase (THD)

Phosphate (10 mM solution, pH 7.0) .... 50 mL
L-threonine .... 50 mg
Nicotinamide adenine dinucleotide (NAD), 1% solution .... 2.0 mL
Phenazine methosulfate (PMS), 1% solution .... 0.5 mL
Dimethyldiazol tetrazolium (MTT), 1.25% solution ... 1.0 mL

Triphenyltetrazolium Chloride (TTC)

Triphenyltetrazolium chloride .... 1.0 g
Add distilled water to bring the total volume to 100.0 mL and dissolve. Filter-sterilize before use.

Tris Buffer 1.0 M

Tris (hydroxymethyl) aminomethane .... 121.1 g
Distilled water .... 1.0 L
Dissolve the Tris in 500 mL of distilled water and adjust to the desired pH with 5 N hydrochloric acid. Bring the total volume to 1.0 L with distilled water, and store at 4 to 6°C.

Tris Buffered Saline (TBS) 1% Gelatin

Tris .... 2.42 g
NaCl .... 29.24 g
Gelatin .... 10.0 g
Distilled water .... 1.0 L
Dissolve ingredients in distilled water by heating and stirring. Autoclave for 15 min at 121°C. Final pH should be 7.5 ± 0.2.

Tris Buffered Saline (TBS) 3% Gelatin

Tris .... 2.42 g
NaCl .... 29.24 g
Gelatin .... 30.0 g
Distilled water .... 1.0 L
Dissolve ingredients in distilled water by heating and stirring. Autoclave for 15 min at 121°C. Final pH should be 7.5 ± 0.2.

Tris Buffered Saline (TBS) Tween

Tris .... 2.42 g
NaCl .... 29.24 g
Tween 20 .... 0.5 mL
Distilled water .... 1.0 L
Dissolve ingredients in distilled water by heating and stirring. Autoclave for 15 min at 121°C. Final pH should be 7.5 ± 0.2.

Tris EDTA

Tris-HCl, pH 8.0 .... 10 mM
EDTA, pH 8.0 .... 1 mM

Triton X-100*

This reagent is the registered trademark for octylphenoxy polyethoxy ethanol. Triton X-100 is a nonionic preparation manufactured by Rohm and Haas Company, Independence Mall West, Philadelphia, PA 19105. It is also sold by Fisher Scientific Company.

Trypsin Stock Solution

Trypsin (Difco No. 0152 or equivalent) .... 10.0 g
Tris buffer 1.0 M, pH 7.0 .... 100.0 mL
Warm the Tris buffer to room temperature, and then dissolve the enzyme in the buffer, warming to 35°C if necessary. Filter-sterilize the solution using a 0.45 µm membrane filter, and dispense into test tubes in 1- or 2-mL volumes. Store up to one week at 4° to 6°C or up to three months at –18°C.

TTNE Buffer

(for ribotyping)
Tricine .... 20 mM
Tris base .... 50 mM
Sodium acetate .... 5 mM
EDTA .... 10 mM
Adjust to pH 8.65.

Vaspar

Combine one part mineral oil with two parts petroleum jelly, and sterilize in an oven at 191°C for 3 hr.

Vibriostatic Agent 0/129*

(Sigma #D0656)

Voges-Proskauer (V-P) Test Reagents

Solution A:
α-naphthol .... 5.0 g
ethanol (absolute) .... 100.0 mL
Solution B:
Potassium hydroxide .... 40.0 g
Distilled water to make .... 100.0 mL
Perform Voges-Proskauer (V-P) test at room temperature by transferring 1 mL of a 48-hr culture to a test tube and adding 0.6 mL of α-naphthol (Solution A) and 0.2 mL of 40% potassium hydroxide (Solution B). Shake after addition of each solution. To intensify and speed reactions, add a few crystals of creatine to the test medium. Read results 4 hr after adding reagents. Positive V-P test is the development of an eosin pink color.

Vracko-Sherris Indole Reagent

r-dimethylaminobenzaldehyde (DABA) .... 5.0 g
1 N HCl to make .... 100.0 mL

*Commercially available.

Bring volume to 100 mL with 1 N HCl. Keep from light. DABA may take overnight to go into solution. The solution will keep for 1 week.

Wash Solutions

(for DIG probe hybridization)
Wash Solution A:
Standard Saline Citrate buffer (SSC), 20X .... 100 mL
Sodium dodecyl sulfate (10%) .... 10 mL
Distilled water .... 1.0 L
Store at room temperature.
Wash Solution B:
Standard Saline Citrate buffer (SSC), 20X .... 50 mL
Sodium dodecyl sulfate (10%) .... 10 mL
Distilled water .... 1.0 L
Store at room temperature.

63.5 STAINS, INTRODUCTION

This section contains directions for preparation of all stains used throughout this book. Dyes employed should be from batches certified by the Biological Stain Commission or dyes of equal purity (No batch is approved by the Commission unless it meets chemical and physical tests and has been found to produce satisfactory results in the procedures for which it is normally used).

Control organisms should be employed frequently, preferably with every batch of slides stained, to assure the analyst that the completed preparations and technique employed are producing appropriate results.

63.51 Types of Stains

Stains are classified as either *simple*, or *differential*. Simple stains consist of the addition of one dye that serves to show the morphology of an organism when viewed under the microscope, with all structures having the same hue. Differential stains consist of more than one dye which are added in several steps. The stained cell structures are differentiated by color, as well as by shape.

63.6 ALPHABETICAL LISTING OF STAINS

Acridine Orange

Acridine orange .... 0.025 g
Tinopal AN (Ciba-Geigy Ltd.) .... 0.025 g
Citrate-NaOH buffer, pH 6.6 .... 0.01M
Mix acridine orange and Tinopal AN to dissolve in buffer. Store protected from light.

Basic Fuchsin Stain

Dissolve 0.5 g of basic fuchsin in 20.0 mL of 95% ethanol. Dilute to 100.0 mL with distilled water. Filter if necessary to remove any excess dye particles. **Note:** Carbol-fuchsin ZN stain may be substituted.

Carbol Fuchsin

Basic fuchsin .... 0.3 g
Ethyl alcohol .... 10 mL
Phenol .... 5 mL
Distilled water .... 95 mL
Dissolve basic fuchsin in alcohol. Heat phenol crystals to 45°C to melt. Add 5 mL phenol to the water and mix. Mix solutions and allow to stand for several days. Filter.

Crystal Violet Stain

(0.5 to 1% solution)
Crystal violet (90% dye content) .... 0.5 to 1.0 g
Distilled water .... 100.0 mL
Dissolve the crystal violet in distilled water, and filter through coarse filter paper. Prepare smear, air-dry, heat-fix and stain for 20 to 30 sec. Rinse with tap water, air-dry, and examine.

Crystal Violet Stain

(for *Geotrichum* mold count)
Crystal violet (90% dye content) .... 10 g
Ethyl alcohol (95%) .... 100 mL
Dissolve crystal violet in the alcohol.

Ethidium Bromide Stain

Ethidium bromide .... 0.5 µg/mL.

Erythrosin Stain

Stock Solution:
Erythrosin B (certified) .... 1.0 g
Distilled water .... 100.0 mL
Dissolve the dye in the distilled water, and store under refrigeration.

Buffer Solution: Dissolve equal parts of 0.2 M disodium phosphate and 0.2 M monosodium phosphate in distilled water.

Make a 1:5000 concentration of the stain by diluting 1.0 mL of the stock solution with 50.0 mL of the buffer solution. To 1.0 mL of liquid sample add 1.0 mL of the 1:5000 erythrosin solution in a small serum tube. Shake to obtain an even suspension of organisms, and transfer a drop of solution to the hemacytometer with a 3 mm platinum loop.

Gel Loading Dye

Ficoll 400-DL .... 25% (w/v)
Bromphenol blue .... 0.5 mg/mL
Xylene cyanole FF .... 0.5 mg/mL
The dye can be prepared in bulk and small aliquots stored at –20°C.

Gram Stain (Typical)

Crystal Violet:
Crystal violet .... 2.0 g
Ethyl alcohol (95%) .... 20 mL
Ammonium oxalate .... 0.8 g
Distilled water .... 80 mL
Dissolve crystal violet in alcohol. Heat ammonium oxalate gently to dissolve in water. Mix crystal violet and ammonium oxalate solutions and filter.
Gram's Iodine:
Iodine crystals .... 1.0 g
Potassium iodide .... 2.0 g
Distilled water .... 300 mL
Grind iodine and potassium iodide together in a mortar. Add water slowly (in 10-mL volumes), grinding continuously until crystals are dissolved. Add remaining water and store protected from light.
Decolorizer:
Ethyl alcohol, 95%.
Safranin:
Safranin O .... 2.5 g

Ethyl alcohol (95%) .... 100 mL
Distilled water .... 900 mL

Dissolve safranin in ethyl alcohol. Add water and mix.

Gram Stain (Hucker)

Hucker Crystal Violet

Solution A:

Crystal violet (85% dye content) .... 2.0 g
Ethyl alcohol 95% .... 20.0 mL

Solution B:

Ammonium oxalate monohydrate .... 0.2 g
Distilled water .... 20.0 mL

Prepare the following four solutions:

a. Ammonium oxalate-crystal violet solution: Mix equal parts of Solutions A and B. (Sometimes the crystal violet is so concentrated that gram-negative organisms do not properly decolorize. To avoid this difficulty, the crystal violet solution may be diluted as much as tenfold prior to mixing with equal parts of Solution B.)
b. Lugol's solution, Gram's modification: Dissolve 1.0 g of iodine crystals and 2.0 g potassium iodide in 300.0 mL of distilled water.
c. Counterstain: Dissolve 2.5 g safranin dye in 100.0 mL of 95% ethyl alcohol. Add 10.0 mL of the alcohol solution of safranin to 100.0 mL of distilled water.
d. Ethyl alcohol: 95%.

Staining Procedure: Stain the heat-fixed smear for 1 min with the ammonium oxalate-crystal violet solution. Wash the slide in water; immerse in Lugol's solution for 1 min.

Wash the stained slide in water; blot dry. Decolorize with ethyl alcohol for 30 sec, using gentle agitation. Wash with water. Bolt and cover with counterstain for 10 sec, then wash, dry, and examine as usual.

Cells which decolorize and accept the safranin stain are gram-negative. Cells which do not decolorize but retain the crystal violet stain are gram-positive.

Preferably stain vigorously growing 24-hour cultures from nutrient or other agar free of added carbohydrates. Use positive and negative culture controls.

Gram Stain (Hucker Modification) for Campylobacter

Crystal Violet:

Solution A:

Crystal violet (90% dye content) .... 2.0 g
95% ethyl alcohol .... 200.0 mL

Solution B:

Ammonium oxalate .... 8.0 g
Distilled water .... 800.0 mL

Prepare solutions A and B, and then mix. Let the solution stand overnight or for several days until the dye goes into solution. Filter through coarse filter paper.

Gram's Iodine:

Iodine crystals .... 1.0 g
Potassium iodide .... 2.0 g
Distilled water .... 300.0 mL

Decolorizer:

95% ethyl alcohol

0.3% Carbol Fuchsin Counterstain:

Solution A:

Basic fuchsin .... 0.3 g
Ethyl alcohol .... 10.0 mL

Solution B:

Phenol (melted crystal) .... 5.0 mL
Distilled water .... 95.0 mL

Prepare solutions A and B and mix.

Staining Procedure:

1. Prepare a thin smear in a drop of water on a clean glass slide.
2. Air dry, and gently fix with heat.
3. Stain for 1 min with crystal violet.
4. Wash the slide with tap water.
5. Add iodine for 1 min.
6. Wash the slide with tap water.
7. Decolorize with ethyl alcohol until alcohol wash is clear.
8. Wash the slide with tap water.
9. Counterstain with carbol fuchsin for 10 to 20 sec.
10. Wash the slide with tap water, dry, and examine.

Gram Stain (Kopeloff and Cohen Modification)

Methyl Violet:

Crystal violet or methyl violet 6B (1% aqueous) .... 30.0 mL
Sodium bicarbonate solution (5%) .... 8.0 mL

Allow the solution to stand at room temperature for 5 min or more before using.

Iodine:

Iodine .... 2.0 g
Sodium bicarbonate solution (1 N) .... 10.0 mL
Distilled water .... 90.0 mL

Dissolve 2.0 g of iodine in 10 mL of the sodium hydroxide solution, and then add 90.0 mL of distilled water.

Decolorizer:

Ethyl alcohol (95%) .... 50.0 mL
Acetone .... 50.0 mL

Counterstain:

Basic fuchsin .... 0.1 g
Distilled water .... 100.0 mL

Staining Procedure:

1. Prepare a thin smear in a drop of water on a clean glass slide.
2. Air dry, and gently fix with the least amount of heat necessary.
3. Flood the smear with crystal violet or methyl violet for 5 min.
4. Flush the smear with iodine solution for 2 min.
5. Drain without blotting but do not allow the smear to dry.
6. Add the decolorizer dropwise until the runoff is colorless (10 sec or less).
7. Air-dry the slide.
8. Counterstain with basic fuchsin for 20 sec.
9. Wash the excess stain from the slide by short exposure to tap water and then air-dry. If the slide is not clear, immersion in xylol is recommended.
10. Cells which decolorize and accept the basic fuchsin stain are gram-negative. Cells which do not decolorize and retain the crystal violet or methyl violet stain are gram-positive.

Kinyoun Stain

Basic fuchsin .... 4.0 g
Phenol .... 8 mL
Ethyl alcohol (95%) .... 20 mL
Distilled water .... 100 mL

Dissolve the basic fuchsin in the alcohol, and add the water slowly while shaking. Melt the phenol in a 56°C water bath, and add 8 mL to the stain, using a pipet with a rubber bulb.

Lipid Globule Stain

(Burdon's Method)

Solution A:

Sudan black B .... 0.3 g
Ethyl alcohol 70% .... 100.0 mL

To prepare Solution A, dissolve 0.3 g of Sudan black B (C.I. 26150) in 70% ethanol. After the bulk of the dye has dissolved, shake solution at intervals during the day and allow to stand overnight. Filter if necessary to remove undissolved dye. Store in a well-stoppered bottle.

Solution B:

Safranin O .... 0.5 g
Distilled water .... 100.0 mL

Procedure:

1. Prepare smear and let dry thoroughly in air. Heat-fix with minimal flaming.
2. Flood entire slide with Solution A and leave undisturbed for 10 to 20 min.
3. Drain off the excess stain, and blot dry.
4. Wash the slide for 5 to 10 sec with chemically pure xylene, and blot dry.
5. Counterstain with Solution B for 10 to 20 sec.
6. Wash slide with tap water, blot dry, and examine.

Malachite Green

Malachite green oxalate .... 5.0 g
Distilled water .... 100 mL

Methylene Blue (Loeffler's)

Methylene blue (90% dye content) .... 0.3 g
Ethyl alcohol (95%) .... 30 mL
Potassium hydroxide .... 0.01 g
Distilled water .... 100 mL

Dissolve methylene blue in the alcohol. Dissolve potassium hydroxide in water. Mix the two solutions together.

North Aniline Oil-Methylene Blue Stain

Mix 3.0 mL of aniline oil with 10.0 mL of 95% ethanol, and then slowly add 1.5 mL of HCl with constant agitation. Add 30.0 mL of saturated alcoholic methylene blue solution, and then dilute to 100.0 mL with distilled water. Filter before use.

RYU Flagella Stain

(Kodaka Modification)

Solution 1:

5% phenol .... 10.0 mL
Powdered tannic acid .... 2.0 g
Saturated aluminum potassium sulfate 12-hydrate (crystal)
(14 g potassium alum in 100 mL distilled water) ...... 10.0 mL

Solution 2:

Saturated alcoholic solution of crystal violet
(12 g crystal violet in 100 mL 95% ethyl alcohol)

Mix 10 parts of Solution 1 (mordant) with 1 part of Solution 2 (stain). Store at room temperature indefinitely. Does not require filtration before use.

Procedure:

1. With an inoculating needle, pick a colony and lightly touch the needle in the center of 2 drops of distilled water on a new, precleaned slide.
2. Air dry the slide at room temperature.
3. Cover the smear with the staining solution, and stain for 5 min. Stain time may vary with each lot of stain.
4. Thoroughly wash the slide in running tap water.
5. Dry the slide, and examine.

The stain is available commercially from Carr Scarborough Microbiologics, Inc., P.O. Box 1328, Stone Mountain, GA 30086.

Spore Stain (Ashby's)

Solution A:

Malachite green .... 5.0 g
Distilled water .... 100.0 mL

Solution B:

Safranin O .... 0.5 g
Distilled water .... 100.0 mL

Procedure:

1. Prepare smear, and heat-fix with minimal flaming.
2. Place slide over boiling water bath until definite drops of water collect on the bottom of the slide.
3. Flood the slide with Solution A.
4. After 1 or 2 min, wash the slide thoroughly in cool tap water.
5. Counterstain with Solution B for 20 to 30 sec.
6. Rinse slide thoroughly in cool water, dry without blotting, and examine.

Ziehl-Neelsen's Carbol Fuchsin Stain* (TB Carbol Fuchsin ZN)

63.7 REFERENCES

1. American Public Health Association. Standard methods for the examination of water and wastewater. 14th ed. Washington (DC): The Association; 1975.
2. American Public Health Association. Standard methods for the examination of dairy products. 15th ed. Washington (DC): The Association; 1985.
3. Centers for Disease Control. Manual of quality control procedures for microbiological laboratories. Atlanta (GA): The Centers; 1974.
4. Stamer JR, Stoyla BO. Growth response of *Lactobacillus brevis* to aeration and organic catalysts. Appl Microbiol 1967;15:1025.

63.8 ACKNOWLEDGMENTS

This chapter has been revised from earlier editions and contains information previously provided by A.H. Schwab, H.V. Leininger, E.M. Powers, and G.R. Acuff.

Chapter 64

Measurement of Water Activity (a_w), Acidity, and Brix

Virginia N. Scott, Rocelle S. Clavero, and John A. Troller

64.1 MEASUREMENT OF A_W

Water activity or a_w is a measure of the free moisture in a product and is defined as the ratio of the vapor pressure of a substance to that of pure water at a specified temperature. This can be expressed mathematically as a function of Raoult's law, which states

$$a_W = \frac{P}{P_O} = \frac{n_1}{n_1 + n_2} \quad (1)$$

where P is the vapor pressure of a solution, P_O is the vapor pressure of pure water, n_1 is the number of moles of solvent, and n_2 is the number of moles of solute. This term is related to equilibrium relative humidity, ERH, as follows:

$$a_W = \frac{ERH}{100} \quad (2)$$

Relative humidity (RH) usually is used to characterize ambient atmospheric conditions such as might occur in a food warehouse or processing plant.

The a_w of a food describes the degree to which water is "bound" in the food, and its availability to participate in chemical/biochemical reactions and growth of microorganisms. Hence, texture, non-enzymatic browning reactions, enzymatic activity, lipid oxidation, and other aspects of foods may be influenced by manipulation of a_w levels. Microbial growth, and, in some cases, the production of microbial metabolites, may be particularly sensitive to alterations in a_w. Microorganisms generally have optimum and minimum levels of a_w for growth depending on other growth factors in their environments. Minimum levels permitting growth of a number of microorganisms are shown in Table 1. These and other aspects of the influence of a_w on microorganisms have been reviewed.[4,26,31]

The amount of water removed from or added to a food depends on the nature and amount of water-soluble substances (water binding capacity) present in the product. The amount or percent of water present often does not adequately characterize the capacity to bind water or to limit its escaping tendency. A more useful parameter is the weight of water desorbed from the food into the vapor phase, which can be defined as the partial pressure of water vapor [P in Equation (1)] at equilibrium in the food. P is considered to be the concentration of desorbable water that is present,[12] and its ratio to the vapor pressure of pure water at the same temperature describes its escaping "tendency" or fugacity ratio. The a_w level of a food actually is a measure or characterization of water that is available to participate in various chemical and physical reactions. The approximate a_w levels of a number of foods are shown in Table 2.

Because of the strong influence of a_w on the wholesomeness and safety of many foods, regulations specifying a_w levels have been established. An example is the exclusion of any food with an a_w of 0.85 or less from the acidified and low-acid canned food regulations (21 CFR Parts 113 and 114). Another regulation (9 CFR Part 319.106) specifies a brine salt content of 10% or an a_w level of 0.92 or less in hams not preserved with nitrates, nitrites, or a combination of these two. USDA specifies that semi-dry, shelf-stable sausage may be fermented to a pH of 4.6 if the a_w does not exceed 0.91. Other a_w-related standards have been proposed in the U.S. and in other countries.

A number of methods have been described for the measurement of a_w. Some involve direct measurement of vapor pressure, which is then converted to a_w by reference to appropriate tables and applying Equation (1). A vapor pressure manometer was described originally by Makower and Meyers[16]; however, subsequent articles[15,30] defined a number of improvements in this instrument. Other instruments, such as electric hygrometers, measure a_w indirectly through the measurement of crystal hydration, electrical resistance or capacitance changes. A psychrometric method for measuring a_w ranges above 0.5 has been described[21] as well as isopiestic equilibration techniques,[8,14,17] bi-thermal equilibration procedures,[28] hair hygrometers, and dew point instruments.[1] These and other methods have been summarized in Troller and Christian,[31] Prior,[22] Guilbert and Morin[11] and Diaz.[7] Comparisons of the various techniques for measuring a_w levels have been conducted in several collaborative studies.[12,13,29]

Because of their ease of use and commercial availability, electric hygrometers and dew point instruments are most frequently employed for the measurement of a_w by the U.S. food industry. Only methods relating to these two types of instruments will be discussed in this chapter. This, however, should not imply that these are the only acceptable methods. In fact, Rödel et al.[27] described a thermometric technique based on freezing point that had several advantages over hygrometers for measuring the a_w of meat products. We recognize that the range of instrumentation in this field is expanding, and a single chapter containing operational procedures for each and every instrument is not feasible.

Table 1. Approximate Minimum Levels of a_w Permitting Growth of Microorganisms at Temperatures near Optimal

Molds	a_w
Alternaria citri	0.84
Aspergillus candidus	0.75
A. flavus	0.78
A. fumigatus	0.82
A. niger	0.77
A. ochraceous	0.77
A. restrictus	0.75
A. sydowii	0.78
A. tamarii	0.78
A. terreus	0.78
A. versicolor	0.78
A. wentii	0.83
Botrytis cinerea	0.90
Byssochlamys nivea	0.89
Chrysosporium fastidium	0.69
C. xerophilum	0.71
Emericella (Aspergillus) nidulans	0.80
Eremascus albus	0.70
E. fertilis	0.77
Eurotium (Aspergillus) amstelodami	0.70
E. chevalieri	0.71
E. herbariorum	0.74
E. repens	0.71
E. rubrum	0.70
Fusarium graminearum	0.90
F. moniliforme	0.87
Monascus (Xeromyces) bisporus	0.61
Mucor plumbeus	0.93
Paecilomyces variotii	0.84
Penicillium brevicompactum	0.78
P. chrysogenum	0.78
P. citrinum	0.80
P. aurantiogriseum	0.81
P. expansum	0.83
P. fellutanum	0.78
P. glabrum	0.81

Molds *(continued)*	a_w
P. islandicum	0.83
P. griseofulvum	0.81
P. spinulosum	0.80
P. viridicatum	0.81
Rhizopus stolonifer	0.84
Wallemia sebi (Sporendonema epizoum)	0.75
Yeasts	
Debaryomyces hansenii	0.83
Pichia membranaefaciens	0.90
Rhodotorula mucilaginosa	0.92
Saccharomyces cerevisiae	0.90
Zygosaccharomyces bailii	0.80
Z. rouxii	0.62
Bacteria	**a_w adjusted with salts**
Enterobacter aerogenes	0.94
Bacillus cereus	0.95
B. megaterium	0.95
B. stearothermophilus	0.93
B. subtilis	0.90
Campylobacter jejuni	0.97
Clostridium botulinum type A	0.95
C. botulinum type B (proteolytic)	0.94
C. botulinum type E	0.97
C. perfringens	0.95
Escherichia coli	0.95
Halobacterium halobium	0.75
Lactobacillus viridescens	0.95
Listeria monocytogenes	0.93
Microbacterium spp.	0.94
Micrococcus halodenitrificans	0.86
M. lysodeikticus	0.93
Pseudomonas fluorescens	0.97
Salmonella spp.	0.95
Staphylococcus aureus	0.86
V. parahaemolyticus	0.94

We have chosen to approach this problem through specific recommendations concerning performance criteria for water activity measurement while leaving most of the specific operational procedures to the individual equipment manufacturer.

The level of precision required in a_w measurement is dictated by how close the a_w of a specific product is to a specified critical value. It is felt generally that a laboratory routinely engaged in a_w determinations should be capable of measuring the a_w of a food with a precision of ±0.01 a_w unit regardless of the method used. Since most commercial instruments, when properly calibrated and operated, are capable of achieving a standard deviation of ±0.005 or less, this level of precision should not be difficult to attain. The accuracy of these methods should be within 1% to 3% of solutions of known a_w values.

64.2 EQUIPMENT

64.21 Electric Hygrometers

Electric hygrometers consist of a potentiometer, a sample/sensor holder, and a sensor. Measurement consists of placing the product in a sample cup that is placed in the sample holder and exposed to a sensing probe. The probe monitors the humidity changes and measures temperature. The sensing stations may be placed in an incubator if the temperature is critical to the measurement and the system lacks temperature control (see Section 64.23), however most recent models offer temperature control features. Some systems have water-jacketed sample chambers for temperature control. Rotronic and Novasina are the brands of electric hygrometers most commonly in use at present. Measurements typically take 30–90 minutes or longer, depending on the product, although Rotronic has introduced a model that reaches equilibrium in 4 to 6 minutes. Many electric hygrometers have difficulty in measuring samples that contain compounds such as glycerol, propylene glycol, and volatile acids (see Section 64.5); however, Rotronic claims there are no restrictions on the types of products that can be measured with its A2101 (AwQuick) system and Novasina is introducing an improved cell that is more resistant to volatiles. An audible signal and/or a visual (LCD or LED) display (read as percent equilibrium relative humidity) are typically used to indicate an equilibrium condition within the sample holder, which is essential to all measurements. This is easily converted to a_w by means of Equation (2). Recorders may be either built-in or added-on, and newer models may interface with dataloggers or computers. Both Rotronic and Novasina offer multichannel systems for simultaneous measurement of more than one sample at a time. Portable, battery-operated instruments are also available.

Adjustable potentiometers used for the calibration of the sensor around a specific set point or points are generally built into these instruments. Nevertheless, use of a standardized curve is recommended for these instruments. Calibration and use of standard curves will be covered in detail in Section 64.3.

Table 2. Approximate a_w Values of Some Foods and Sodium Chloride and Sucrose Solutions

a_w	NaCl (%)	Sucrose (%)	Foods
1.00 to 0.95	0 to 8	0 to 44	Fresh meat, fruit, vegetables, canned fruit in syrup, canned vegetables in brine, frankfurters, liver sausage, margarine, butter, low-salt bacon
0.95 to 0.90	8 to 14	44 to 59	Processed cheese, bakery goods, high-moisture prunes, raw ham, dry sausage, high-salt bacon, orange juice concentrate
0.90 to 0.80	14 to 19	59 to saturation (0.86 a_w)	Aged cheddar cheese, sweetened condensed milk, salami, jams, candied peel, margarine
0.80 to 0.70	19 to saturation (0.75 a_w)		Molasses, soft dried figs, heavily salted fish
0.70 to 0.60			Parmesan cheese, dried fruit, corn syrup, licorice
0.60 to 0.50			Chocolate, confectionery, honey, noodles
0.40			Dried egg, cocoa
0.30			Dried potato flakes, potato chips, crackers, cake mixes, pecan halves, peanut butter
0.20			Dried milk, dried vegetables, chopped walnuts

By permission of Academic Press.[31]

1. *Sample holders.* Sample holders should be vapor-tight and of sufficient size to provide space for a representative sample, yet sufficiently small to permit equilibration of the sample within a reasonable amount of time. Samples can be stored in covered cups; however, some loss of moisture and consequent change in a_w level may occur. It is recommended that cups used for storage, for example, those containing calibration solutions, be sealed with a strip of electrical tape. Most instruments can be ordered with a sample holder and a supply of spare sample cups.
2. *Sensors.* Although the exact details of sensor composition are proprietary, these devices usually contain an immobilized electrolyte (a hygroscopic polymer, frequently lithium chloride) and associated circuitry that gives a signal relative to the ERH. Changes in the ERH within the closed sample chamber (sample and airspace are at equilibrium) are reflected in changes in the conductance of an electrical current through the sensor and across the electrolyte. This is detected electrically and read as described above. Accurate measurement requires good temperature control. The ERH is only equal to the sample a_w if the sample and sensor temperature are equal.

64.22 Dew Point Instruments

In a dew point instrument, such as the Decagon AquaLab (CX-1, CX-2 and Series 3), a sample is equilibrated with the headspace in a sealed chamber containing a mirror and a means of detecting condensation on the mirror. Cooling of the mirror is usually achieved by a Peltier system that is electronically linked to a photocell into which light is reflected from the condensing mirror. A fan agitates the air and moves it across the sample surface toward the mirror, which is cooled until condensation occurs on it. The temperature at which condensation begins is the dew point. A photodetector senses the change in reflectance when condensation occurs on the mirror, and a thermocouple attached to the mirror records the temperature at which condensation occurs; sample temperature is measured simultaneously by infrared thermometry. The a_w is computed by converting the mirror and sample temperatures to vapor pressures and calculating the ratio. Because these instruments use a primary measurement of a_w, the manufacturer claims that routine calibration prior to sample analysis is generally not necessary, although there is a linear offset adjustment that should be adjusted if a verification standard (a saturated salt solution) indicates this is necessary. The use of standard curves is recommended (see Section 64.3).[24, 25] The AquaLab has a range of 0.1 to 1.00 a_w, with an option to read as low as 0.03. One model is equipped for use with an external water bath for temperature control; another model has user-selectable internal temperature control. After a 30 minute warm-up period, equilibration time is generally less than 5 minutes, (usually less than 2 minutes, except for very dry samples [24,25]) and is signaled by audible and visual (LED) indicators. As with many of these instruments, the Decagon dew point instruments can be interfaced with a computer and/or printer. Verification standards are available and can be purchased in four water activity levels: 0.98, 0.76, 0.50 and 0.25; alternatively, standardized salt solutions can be used (see Section 64.31).

64.23 Constant Temperature Requirement for Hygrometers

Reference to standard tables relating temperature to the vapor pressure of water clearly shows that very small changes in temperature can produce disproportionate changes in vapor pressure. In saturated salt solutions, temperature affects solubility and a_w can change with temperature.[10] In addition to the effect of temperature on solubility, temperature also has effects on the state of the food matrix; these effects may be negligible in some foods, but not in others. Thus, very precise control of temperature is important when measuring a_w levels, and for products where a_w levels are near critical values, temperature control is essential. There are several methods of achieving this.

The temperature employed should be the same as that used for calibration of the instrument. A temperature of 30°C is recommended because of its relative ease of temperature maintenance; however, most workers continue to use 25°C as their working temperature. Depending on the a_w range to be measured and the precision required, the temperature of the sensor should not fluctuate more than ± 0.2°C. Most commercially available hygrometers, in addition to relative humidity capability, also are able to measure and, in some cases, record temperatures within the sample headspace.

1. *Integral thermal control.* Some instruments, e.g., certain models of the Decagon AquaLab and Novasina, have temperature control systems that are basically integral with the instrument. Usually temperature maintenance is achieved with a heating and cooling Peltier element, and temperature fluctuations are minimized by use of large masses of metal surrounding the sample holder that serve as heat sinks, thus assuring greater accuracy and minimal temperature fluctuations.
2. *Circulating water sample holders.* Both Rotronic and Decagon offer sample holders that permit the circulation of water or other fluid within it. The fluid is obtained from a constant temperature bath, adjusted to the required temperature.
3. *Other devices.* Some hygrometers, particularly older models, have no provision for controlling sample temperatures. Such control is most often attempted by placing the sensors and sample holders in a commercially available constant temperature box or incubator capable of holding the desired measurement temperature (usually 25°C or 30°C). In some instances, particularly with older models, it may be necessary to modify units with sensitive thermo-controllers and other electronic devices to achieve the level of temperature control desired.

Uniform temperatures can be obtained within the cabinet by a forced draft, such as a small duct fan, placed within the chamber. It is advisable to operate this fan continuously to avoid temperature changes associated with heat from the fan motor should it be operated intermittently. Placing the sensor and sample holder in a styrofoam box or cooler chest also helps prevent temperature fluctuations. Immersion of sample holders within a waterbath is an alternative; however, care must be exercised to assure that water or water vapor does not contact the sensors. Incubators are also useful for equilibrating the sample to the test temperature prior to measuring a_w.

64.3 CALIBRATION

Saturated salts should be used to calibrate instruments used for the measurement of a_w. The exception to this relates to measurements in the range of 0.92 to 0.97 and above. In this range, Flom et al.[9] found that calibrating with solutions of varying NaCl content were superior to calibrating with saturated solutions. Here, care must be exercised to be sure that calibration errors do not occur as a result of moisture absorption from the environment.[5]

64.31 Preparation of Saturated Salts

A list of a_w values of a number of saturated salts suitable for the calibration of electric hygrometers is shown in Table 3. Standard salt solutions are available commercially, or they may be prepared in the laboratory. Reagent grade salts and deionized distilled water should be used for the preparation of solutions.[6] The salt is added to a sealable container and water is added in small increments, followed by stirring, until free liquid is observed. The volume of undissolved salt (slush) should be at least 25% of the total volume in the container. Slushes should be permitted to equilibrate to the measurement temperature prior to use and should be stored in the measurement cabinet or other site that approximates the measurement temperature. Slush containers must be sealed tightly. Some slushes such as NaBr may solidify gradually by crystal coalescence. This has no effect on a_w.

Some compounds are not suitable for calibration. Salts that decompose, hydrolyze, or undergo reactions with the components of air do not always give reliable, reproducible results and should not be used for this purpose. Hence use of compounds such as KOH, KI, chromates and dichromates, and ammonium compounds such as NH_4Cl should be avoided.

64.32 Calibration Procedure

Depending on the type of instrument used, calibration procedures may involve either the establishment of a standard curve or the adjustment of a set point potentiometer; some systems have a series of set point potentiometers, each one corresponding to a specific a_w standard. Generally, an instrument with three or fewer set points should be calibrated by means of a standard curve that has been obtained as recommended in the instructions accompanying the instrument or as outlined below.

1. *Standard curve.* A calibration curve can be constructed that relates instrument readout to standard a_w values for particular salts. This standard curve is constructed using at least five salts covering the range to be measured. It is recommended that calibration be conducted from highest to lowest a_w values, since some sensors may show a hysteresis effect (an a_w measurement approached from a lower level does not agree with the value when the same measurement is approached from a higher level). Readings are obtained after equilibrium has been obtained and plotted against values presented in Table 3, and a line is drawn through the points. This calibration plot will then be used for referral during the measurement of unknowns. All measurements must be within the range of calibration points; plots may not be extrapolated.

   Where available, the use of a computer may be helpful in referring readout values to actual a_w levels. Usually a regression line of five or more points is plotted, and a_w levels along this plot can be programmed. In some cases, the a_w instrument may be connected directly to the computer.
2. *Adjustable potentiometers.* Some instruments are equipped with potentiometers that can be adjusted to a specific set point. The measurement of standard salt solutions is used to calibrate the potentiometers to specific a_w values according to the manufacturer's instructions.

Table 3. Equilibrium Relative Humidity (ERH) of Some Saturated Salts[a]

Salt	20°C	25°C	30°C
LiBr	6.61 ± 0.58	6.37 ± 0.52	6.16 ± 0.47
LiCl	11.31 ± 0.31	11.30 ± 0.27	11.28 ± 0.24
KAc	23.11 ± 0.25	22.51 ± 0.32	21.61 ± 0.53
$MgC1_2$	33.07 ± 0.18	32.78 ± 0.16	32.44 ± 0.14
K_2CO_3	43.16 ± 0.33	43.16 ± 0.39	43.17 ± 0.50
$Mg(NO_3)_2$	54.38 ± 0.23	52.89 ± 0.22	51.40 ± 0.24
NaBr	59.14 ± 0.44	57.57 ± 0.40	56.03 ± 0.38
$NaNO_3$	75.36 ± 0.35	74.25 ± 0.32	73.14 ± 0.31
NaCl	75.47 ± 0.14	75.29 ± 0.12	75.09 ± 0.11
KBr	81.67 ± 0.21	80.89 ± 0.21	80.27 ± 0.21
KCl	85.11 ± 0.29	84.34 ± 0.26	83.62 ± 0.25
KNO_3	94.62 ± 0.66	93.58 ± 0.55[b]	92.31 ± 0.60
K_2SO_4	97.59 ± 0.53	97.30 ± 0.45	97.00 ± 0.40

[a] See reference 10.

[b] The value of 92.7 is considered to be more correct.[23]

3. *Combination.* If the instrument to be used has a single set point potentiometer, the set points supplied are not adequate to cover the desired measurement range, or it is necessary to precisely evaluate a_w over a narrow range, a combination of calibration procedures may be used. The set point(s) is (are) first calibrated and set according to the manufacturer's instructions. A calibration plot covering the range to be measured is then prepared as described above.

64.33 Equilibration

During calibration, care must be exercised to assure that equilibrium has been achieved within the sensor holder. For electric hygrometers, equilibration time increases directly with a_w; for dew point instruments, very low a_w samples equilibrate more slowly. With analog or digital read-out devices on an electric hygrometer, two consecutive readings differing by less than 0.01 a_w unit and obtained 10 minutes apart will usually indicate adequate equilibration when dealing with standard salts. It should be noted that some samples, particularly those of relatively high a_w, may equilibrate slowly and these criteria may be difficult to achieve. Alternatively, successive dial readings can be plotted on a scale of a_w versus time. However, most models of various water activity instruments are equipped with an audible and/or an LED signal which indicates when equilibrium is achieved. These indicators are more accurate than with older models.

For most samples, equilibrium will be reached within 5 minutes with dew point instruments or 30–90 minutes for electric hygrometers. In rare cases, equilibration may take as long as 24 hours. Experience with the specific equipment and test samples may be the best guide to equilibration time.

64.34 Frequency of Calibration

The frequency with which instruments must be calibrated depends on the number of samples measured, desired precision of the instruments, type of foods measured, and age of the sensor. A standard salt with an a_w value close to that of the sample should be used to check calibration with sufficient frequency to ensure the degree of precision desired. This will permit mathematical compensation for a small amount of drift. If drift greater than 2% to 3% has occurred, the instrument should be recalibrated. It is advisable to recalibrate at least monthly; experience with each instrument usually will dictate calibration frequencies. Failure to maintain calibration for more than 1 or 2 days or the inability to bring the sensor within calibration limits usually indicates that it should be returned to the manufacturer for repair. These problems also may be the result of contamination (see Section 64.52). The instruction manual should be consulted in such cases.

64.4 MEASUREMENT OF SAMPLES

The sample to be measured must be uniform, if possible; some products will need to be comminuted, although care should be exercised if grinding a material to prevent excessive heat buildup with consequent change in a_w. The a_w of emulsions, especially water-in-oil emulsions, may be difficult to measure accurately, primarily because occlusion by the lipid portion prevents the "display" of water vapor pressure. In these cases, the emulsion may be broken through centrifugation or repeated freeze-thaw cycles and the resulting water phase measured. However, the water phase alone may not be representative of the a_w in emulsified water droplets, especially if the emulsifier itself has some capacity to bind water. Processed meat products can be especially troublesome in this regard. Some currently available instruments are automatically optimized, hence measurement of "problem" samples can be obtained.

To obtain measurements of unknown materials, the sample is placed in the sample dish, which is then placed in the holder and sealed. The sample is allowed to equilibrate and the a_w is read from the digital readout, recorder plot, or calibration curve (see Section 64.33). The a_w should be determined as the average of duplicate samples. Greater accuracy and precision may be achieved with additional replication. Roa and Tapia de Daza[25] found with a dew point instrument that triplicate samples resulted in precision of ±0.005 a_w.

64.5 PRECAUTIONS AND LIMITATIONS

64.51 Sample Considerations

The material to be sampled should be placed immediately in the sample container to minimize moisture exchange with ambient air. In the case of hygrometers, the sample volume should be as large as possible to reduce the amount of time required for equilibration. Dew point devices normally require that the sample cup be partially filled to prevent spilling and to avoid pickup of powdery materials by the circulating air stream.

Sample chambers must have properly fitted gaskets to prevent vapor exchanges with the atmosphere. Should leakage occur, it will manifest itself as erratic or continuously drifting readings or as excessively long equilibration periods. Sample holders should not absorb or contribute moisture to the sample void space.

64.52 Sensor Contamination

Mists, condensate, splashing, and all other forms of moisture must not contact the sensor. Mechanical filters and splash guards are integral with the sensor head in most instruments and should prevent this type of contamination.

A list of potential chemical contaminants is shown in Table 4. Should contamination occur, the sensor, in some cases, may be restored to normal functioning following a period of non-use and recalibration. Storage of the sensor at approximately 30°C under a slight vacuum may also be useful. In some situations, however, the damage is permanent, and the sensor must be discarded or returned to the manufacturer for repair. Some manufacturers provide a chemical filter, which is usually a fiber disc impregnated with charcoal. It should be used whenever the sensor is exposed to the substances listed in Table 4. Equilibration time will be sig-

Table 4. Potential Chemical Contaminants of Hygrometric Sensors

Type of Contamination	Compound
Ammonia	T
Amines	T
Alcohols	T
Glycols	T
Mercury vapor	P
Ketones	P
Esters	P
Halogen gases	P
Hydrogen sulfide	P
Sulfur dioxide	P
Volatile acids (e.g., acetic)	T

T = Temporary
P = Permanent

nificantly increased when a chemical filter is used. Some sensors, according to their manufacturers, are inherently resistant to chemical contamination.

64.53 Sensor Care and Storage

Sensors are very sensitive to shock and vibrations and should be treated with care. When not in use, they should be protected from dust and other contaminants at a relative humidity of less than 70% and a temperature between 20° and 50°C.

64.54 Other

The error inherent in measuring a_w by electric hygrometers is such that reporting results to three decimal places may give a false impression of the precision and accuracy of the method. It is suggested, therefore, that a_w readings be reported to only two decimal places.

64.6 MEASUREMENT OF ACIDITY

The term pH is defined as the logarithm of the reciprocal of the hydrogen ion concentration in solution. It measures the effective acidity—the degree or intensity of acidity. It is determined by measuring the difference in potential between two electrodes immersed in a sample solution. A suitable system consists of a potentiometer, a glass electrode, and a reference electrode. A precise pH determination can be made by taking an electromotive force (emf) measurement of a standard buffer solution of known pH and then comparing that measurement with an emf measurement of a sample of the solution to be tested.

Potentiometric procedures probably represent the bulk of pH determinations in the examination of food, although other methods, such as calorimetric and pH paper measurements, are available and are occasionally used. Equipment used for in-line and process determinations may differ and calibration procedures specific to the application are required. These systems are usually unique to the use intended and are beyond the scope of this chapter.

64.7 pH MEASUREMENTS

64.71 pH Meter

The primary instrument for pH determination is the pH meter or potentiometer. For most work, an instrument with a direct-reading pH scale is necessary. Batteries should be checked frequently to assure proper operation of battery-operated instruments. An instrument using an expanded unit scale or a digital read-out system is preferred, since it allows more precise measurements.

The accuracy of most pH meters is stated to be approximately 0.1 pH unit, and reproducibility is usually ±0.05 pH unit or less. Some meters permit the expansion of any pH unit range to cover the entire scale and have an accuracy of approximately 0.01 pH unit and reproducibility of ±0.005 pH unit. A pH meter should be operated in accordance with the manufacturer's instructions using National Institute of Standards and Technology (NIST) standard buffers as the primary reference.

64.72 Electrodes

The typical pH meter is equipped with a glass membrane electrode. The most commonly used reference electrode is the calomel electrode, which incorporates a salt bridge filled with saturated potassium chloride solution. Glass and reference electrodes may both be housed in a single combination electrode.

1. *Care and use of electrodes.* Calomel electrodes should be kept filled with saturated potassium chloride solution or other solution specified by the manufacturer because they may become damaged if they are allowed to dry out. For best results, electrodes should be soaked in buffer solution, distilled or deionized water, or other liquid specified by the manufacturer for several hours before using and should be kept ready by storing with tips immersed in distilled water or in buffer solution used for standardization. Electrodes should be rinsed with water before being immersed in the standard buffers and between sample determinations should be rinsed with water or the solution to be measured next.

   Drifting of combination electrodes may be encountered, which causes difficulty in obtaining accurate readings. When this occurs, the source of the problem should be identified using troubleshooting techniques provided in the manufacturer's operating manual. Often, a faulty reference electrode junction is responsible.
2. *Temperature.* To obtain accurate results, the same temperature should be used for the electrodes, the standard buffer solution, and the samples for the standardization of the meter and pH determinations. Tests should be made at a temperature in the range from 20° to 30°C (68° to 86°F).

64.73 Standardization

When operating an instrument, the manufacturer's instructions should be followed and these techniques for pH determinations used:

1. Switch the instrument on and allow the electronic components to warm up and stabilize before proceeding.
2. Standardize the instrument and electrodes with standardization buffers as close to the sample pH value as practical. Select two standard buffer solutions such that the expected pH of the test sample falls within their range. Note the temperature of the buffer solution and set the temperature compensator control at the observed temperature, if necessary.
3. Immerse the electrode tips in the buffer solution and take the pH reading, allowing 1 min for the meter to stabilize. Adjust the standardization control so that the meter reading corresponds to the pH of the known buffer (for example, 4.0) for the temperature observed.

   Rinse the electrodes with water and blot (do not wipe) with soft tissue. Repeat the procedure with fresh portions of buffer solution until the instrument remains in balance on two successive trials. To check the operation of the pH meter, check the pH reading using another standard buffer.

   Indicating electrodes may be checked for proper operation by using first an acid buffer, then a base buffer. First, standardize the electrodes using a pH 4.0 buffer at or near 25°C. The standardization control should be adjusted so that the meter reads exactly 4.0. Electrodes should be rinsed with water, then blotted and immersed in a pH 9.18 borax buffer. The pH reading should be within ±0.03 unit of the 9.18 value.

64.74 Measurement of Samples

The following procedures should be used to determine the pH of samples:

1. Adjust the temperature of the sample to room temperature, and set the temperature compensator control to the observed

temperature, if necessary. (Generally this will already be set from the standardization procedure. Note, with some expanded-scale instruments, the sample temperature *must* be the same as the temperature of the buffer solution used for the standardization.)

2. Rinse and blot the electrodes. Immerse the electrodes in the sample and take the pH reading, allowing 1 min for the meter to stabilize. Rinse and blot the electrodes and repeat on a fresh portion of sample. The two readings should be in agreement with one another to indicate that the sample is homogeneous. Report values to 2 decimal places. Oil and grease from the samples may coat the electrodes; therefore, it is advisable to clean and standardize the instrument frequently. When oily samples cause fouling problems, it may become necessary to rinse the electrode with ethyl ether.

64.75 Preparation of Samples

Some food products may consist of a mixture of liquid and solid components that differ in acidity. Other food products may be semisolid in character. The following are examples of preparation procedures for determining the pH for each of these categories:

Liquid and solid component mixtures. Drain the contents of the container for 2 min on a U.S. standard No. 8 sieve (preferably stainless steel) or equivalent inclined at a 17° to 20° angle.

1. If the drained liquid contains sufficient oil to cause electrode fouling, separate the layers with a separatory funnel and retain the aqueous layer. Adjust the temperature of the aqueous layer to 25°C and determine its pH.
2. Remove the drained solids from the sieve. Blend to a uniform paste, adjust the temperature of the paste to 25°C and determine its pH.
3. Mix aliquots of solid and liquid fractions in the same ratio as found in the original container and blend to a uniform consistency. Adjust the temperature of the blend to 25°C and determine the equilibrated pH. Alternatively, blend the entire contents of the container to a uniform paste, adjust the temperature of the paste to 25°C and determine the equilibrated pH.

Marinated oil products. Separate the oil from the solid product. Blend the solid in a blender to a paste consistency; it may become necessary to add a small amount of distilled water (10–20 mL per 100 g of product) to some samples to facilitate the blending. A small amount of added water will not alter the pH of most food products, but caution must be exercised concerning poorly buffered foods. Determine the pH by immersing electrodes in the prepared paste after adjusting the temperature to 25°C.

Semisolid products. Food products of semisolid consistency, such as puddings, potato salad, etc., may be blended to a paste consistency, and the pH may be determined on the prepared paste. Where more fluidity is required, 10 to 20 mL of distilled water may be added to 10 g of product. Adjust the temperature of the prepared paste to 25°C and determine its pH. Large amounts of water may alter pH levels; consequently, the amount of water addition should be minimized.

Large solid components. The internal pH level should be checked with spear electrodes as near as possible to the geometric center.

Emulsions. The pH of emulsions, such as margarine, peanut butter, meat products, etc., may be particularly difficult to obtain as a result of electrode contamination by lipids. In these situations, the water phase should be separated from the lipid phase and the pH of the former should be taken. This separation is most easily accomplished by centrifugation or alternating freeze-thaw cycles.

Wines.[2] Calibrate the pH meter with a freshly prepared, saturated, aqueous solution of potassium bitartrate and adjust the meter to read 3.55 at 20°C, 3.56 at 25°C, or 3.55 at 30°C. Rinse the electrode by dipping in distilled water and then in the sample before placing the electrode in a fresh sample to be tested.

Beer.[2] Check the pH meter before and after use against a standard potassium acid phthalate buffer. Rinse the electrode with distilled water and sample.

Flour.[2] Disperse 10 g of sample in 100 mL of distilled water and digest for 30 min with frequent shaking. Standardize the pH meter with buffers with pH of 4.01 and 9.18. Allow the suspension to stand for 10 min, decant the supernate and determine the pH immediately.

Cacao products other than butter.[2] Suspend 10 g of product in 90 mL boiling water with constant stirring to avoid lumps. Filter and cool filtrate to 20–25°C before determining pH. The pH meter should be standardized to pH 4.00 and 6.86. For cacao butter, the sample should be melted and dispersed in water at 50°C by mechanically stirring for 5 min. Separate the aqueous layer, cool to 20–25°C, filter and determine pH.

Water.[2] The sample bottle should not be opened before analysis; analyze the sample within a few hours of receipt. Wash the electrode 6–8 times with portions of the sample before pH determination. Equilibrium, indicated by absence of drift, must be established before readings are accepted.

64.76 Precautions and Limitations

Temperature. Temperature can affect both the electrode potential and the hydrogen ion activity of a sample. Automatic temperature compensation adjusts for the effect of temperature change on the response of pH electrodes; however, it cannot compensate for pH changes in the sample that are the result of alterations in temperature. Thus, to obtain accurate results, standardization and pH determination should always be done at the same temperature and within the range of 20° to 30°C (68° to 86°F).

Buffers. Standard buffers may be commercially purchased or may be freshly prepared as outlined in "Official Methods of Analysis of AOAC International."[2] Do not use buffers after their expiration date or if there is precipitation or microbial growth; CO_2 absorption and microbial growth or other contamination cause changes in the buffer pH. Do not pour used buffer back into the bottle. Buffers should be tightly capped during storage to reduce CO_2 absorption. Since pH buffer values change with temperature, the temperature of the buffer should be determined and the pH meter standardized accordingly.

64.77 pH Paper Measurements

Although not used in applications requiring a high level of sensitivity, pH determinations using pH paper continue to be useful in some circumstances. The use of pH paper is acceptable for products with a pH < 4.0 (21 CFR 114.90). The paper is impregnated with an indicator dye that changes color over a limited pH range; there are different pH papers for different pH ranges. The normal procedure is to tear off a 3-in strip of pH paper from a roll specified for the expected pH range. The strip should be held only at one end and should not come in contact with other materials before use. The untouched end of the strip should be dipped in the sample to be measured and the color compared with that of various pH levels on the pH paper container label. As noted above, this method should not be considered in situations requiring sensitivity.

64.8 TITRATABLE ACIDITY

Titratable acidity is a measure of the total amount of acid present and is determined by adding a standard alkali solution to a known weight or volume of sample until the acidity is completely neutralized. Measurement of the amount of acid present may be accomplished by titration of a suitable quantity of the sample with standardized 0.1N NaOH to pH 8.3 using a pH meter or to a phenolphthalein end-point.[2,18] Calibration of the pH meter with appropriate standard buffers should be done and constant stirring must be provided to obtain accurate results. For end-point determinations, phenolphthalein indicators should be freshly prepared. Measurement of titratable acidity may be affected by dilution of samples, speed of titration, amount of indicator, and the temperature of the sample.[3] All samples should be titrated as quickly as possible at room temperature.

Titratable acidity is often expressed in terms of the predominant acid present, i.e, citric acid for orange juice and most tomato products, malic acid for wines, lactic acid and/or acetic acid for fermented products. Conversion factors to use to determine titratable acidity for the most common acid present in food products are listed in Table 5.[18] The percent of acid present is calculated as follows:

$$\% \text{ acid} = \frac{(\text{mL NaOH})\ (\text{N NaOH})\ (\text{milliequivalent weight of acid})\ (100)}{\text{Weight of sample in grams}}$$

64.81 Measurement of Titratable Acidity in Food Samples

Use of titratable acidity values to measure acidity is largely dependent on the type of food analyzed. Food samples with a characteristic pigmentation/dark color or precipitates may not permit a determination of the phenolphthalein end point with the necessary precision, even when diluted. Listed below are examples of preparation procedures for titratable acidity measurements in some foods.

Milk.[3] Shake or stir milk gently; if lumpy or frozen, heat it to 37° ± 1°C before mixing. Measure a 9- or 18-g sample into a 100-mL beaker using a pipet. If the sample is measured volumetrically, rinse the pipet with two volumes of distilled water. Add the rinse water to the sample in the beaker and mix gently but thoroughly. If the sample is weighed, use a clean pipet and add twice the sample weight in distilled water, and mix thoroughly. (For greatest analytical accuracy it is generally recommended to weigh samples.) Add exactly 0.5 mL of phenolphthalein indicator and titrate with 0.1N sodium hydroxide to the first permanent color change to pink. If the sample is dark in color, titrate with a pH meter to pH 8.3.

Table 5. Conversion Factors Used for Obtaining Titratable Acidity with 0.1 N NaOH[18]

Type of acid	Factor (milliequivalent weight)
Anhydrous citric acid	0.0640
Hydrous citric acid	0.0700
Acetic acid	0.0601
Malic acid	0.0670
Lactic acid	0.0901
Tartaric acid	0.0750

Ice cream, sherbets, and ices.[3] Warm samples to room temperature and mix thoroughly. Measure a 9- or 18-g sample and proceed as described for milk.

Dry whole milk, non-fat dry milk, malted milk products.[3] Weigh 10 g of nonfat dry milk or malted milk, or 13 g of dry whole milk, into a 250-mL beaker. Mix with 100 mL of distilled water for up to 1 min with an electric mixer. Avoid formation of excessive foam. Let the sample stand for 1 hr at room temperature. Gently mix the sample and transfer 18 g into an Erlenmeyer flask. Rinse the pipet with an equal amount of distilled water, and add the rinse to the sample in the flask. Titrate the sample as described for milk.

Food dressings.[2] Weigh a 15 g sample and add 200 mL distilled water. Shake the suspension until lumps are thoroughly dispersed. Titrate with 0.1 N NaOH using phenolphthalein and express the result as % citric acid. To recognize the endpoint, have a duplicate sample available so that, by comparison, the first change in color may be noted.

Corn syrups and sugars.[2] Weigh 50 g sample and dissolve in 200 mL water. Titrate with 0.1 N NaOH to a faint pink endpoint (pH 8.3). Calculate acidity as HCl.

Cheese.[2] Add water at 40°C to 10 g prepared cheese sample to a total combined volume of 105 mL. Shake vigorously and filter. Measure 25 g filtrate (equal to 2.5 g of sample) and titrate with 0.1 N NaOH. Express the result as lactic acid (1 mL 0.1 N NaOH = 0.0090 g lactic acid) or mL 0.1 N NaOH/100 g.

Wines.[2] If CO_2 is present, remove either by 1) placing a 25 mL sample in an Erlenmeyer flask connected to a water aspirator and agitating for 1 min under vacuum or 2) heating the 25 mL sample to incipient boiling and holding for 30 s with constant swirling until cooled. Add 1 mL phenolphthalein indicator solution to 200 mL hot, boiled water and neutralize to a distinct pink. Combine with 5 mL degassed sample and titrate with 0.1 N (or 0.0667 N) standardized NaOH using a well-illuminated white background. Express as % tartaric acid, or g tartaric acid/100 mL; if 0.0667 N alkali is used, g tartaric acid/100 mL = mL NaOH/10. To calculate as malic or citric (g/100 mL wine), multiply result by 0.893 or 0.933, respectively.

64.9 BRIX

The Brix scale refers to the percentage by weight of sucrose in a pure solution of this sugar and is defined as grams of sucrose in 100 grams of solution at 60°F (20°C).[19] Technically, the term Brix should not be used when designating solids concentrations in other nutritive sweetner solutions such as invert syrups and corn sweeteners.[19] When referring to concentrations in solutions other than pure sucrose, "percent solids" or other designations are thus more appropriate. However, in practice, Brix has been used, particularly in the citrus industry, as a measure of the total soluble solids in juice.[20] These soluble solids are primarily sugars: sucrose, fructose, and glucose. Soluble components such as minerals and other dissolved substances may also contribute to the soluble solids. Brix is reported as "degrees Brix" and is equivalent to the percentage of sugars present in the product. For example, a juice which is 12° Brix has 12% total soluble solids.

Measurement of Brix may be accomplished by using a refractometer or a hydrometer. The refractometer measures the ability of solutions to bend or refract a light beam, which is proportional to the solution's concentration. A hydrometer is a weighted spindle with a graduated neck which measures the weight or specific gravity of a liquid in relation to the weight of water. The specific gravity of water is 1.000; as sugars are added or dissolved, the hydrometer will float and the specific gravity (sp. gr.) increases relative to the amount of sugars added. Specific gravity can be then converted to Brix.

64.10 REFERENCES

1. Anagnostopoulos, G. D. 1973. Water activity in biological systems: A dew-point method for its determination. J. Gen. Microbiol 77:233.
2. AOAC International. 1998. Official methods of analysis of AOAC International, 16th ed., 4th Revision, P. Cunniff (ed.). AOAC International, Gaithersburg, Md.
3. APHA. 1992. Standard methods for the examination of dairy products, 16th ed., R. T. Marshall (ed.). Am. Public Health Assoc., Washington, D.C.
4. Beuchat, L. R. 1983. Influence of water activity on growth, metabolic activities, and survival of yeasts and molds. J. Food Prot. 46:135.
5. Chirife, J., and S. Resnik. 1984. Unsaturated solutions of sodium chloride as reference sources of water activity at various temperatures. J. Food Sci. 49:1486.
6. Chirife, J., G. Favetto, C. Ferro Fontan, and S. Resnik. 1983. The water activity of standard saturated salt solutions in the range of intermediate moisture foods. Lebensm. Wiss. Technol. 16:36.
7. Diaz, R. G. 1992. Actividad del agua de los alimentos. Métodos de determinación. Alimentaria Jan.-Feb:77.
8. Fett, H. 1973. Water activity determination of foods in the range 0.80 to 0.99. J. Food Sci. 38:1097.
9. Flom, W. D., N. Tanaka, S. K. Kayots, and L. M. Finn. 1986. Improved procedure for determining water activity in a high range. J. Assoc. Off. Anal. Chem. 69:952.
10. Greenspan, L. 1977. Humidity fixed points of binary saturated aqueous solutions. J. Res. Natl. Bur. Stand. A, Phys. Chem. 81A:89.
11. Guilbert, S., and P. Morin. 1986. Definition and measurement of the activity of water: Review of usual methods and theoretical critiques. Lebensm. Wiss. Technol. 19:395.
12. Hardman, T. M. 1976. Measurement of water activity. A critical appraisal of methods. *In* R. Davies, G. G. Birch, and K. V. Parker (eds.). Intermediate moisture foods. Applied Sciences Publ., London.
13. Labuza, T. P., K. Acott, S. R. Tatini, and R. Y. Lee. 1976. Water activity determination: A collaborative study of different methods. J. Food Sci. 41:910.
14. Lang, K. W., T. D. McCune, and M. P. Steinberg. 1981. A proximity equilibration cell for rapid determination of sorption isotherms. J. Food Sci. 46:936.
15. Lewicki, P. P., G. C. Busk, P. L. Peterson, and T. P. Labuza. 1978. Determination of factors controlling accurate measurement of a_w by the vapor pressure manometer. J. Food Sci. 43:244.
16. Makower, B., and S. Meyers. 1943. A new method for the determination of moisture in dehydrated vegetables. Proc. Inst. Food Technol. p. 156.
17. McCune, T. D., K. W. Lang, and M. P. Steinberg. 1981. Water activity determination with the proximity equilibration cell. J. Food Sci. 46:1978.
18. NFPA. 1997. Tomato products bulletin 27-L, 7th ed., C. Denny (ed.). Natl. Food Proc. Assoc., Washington, D. C.
19. Pancoast, H. M., and W. R. Junk. 1980. Handbook of sugars, 2nd ed. AVI Publ. Co., Inc., Westport, Conn.
20. Potter, N. N., and J. H. Hotchkiss. 1995. Food science, 5th ed. Chapman and Hall, New York.
21. Prior, B. A. 1977. Psychrometric determination of water activity in the high a_w range. J. Food Prot. 40:537.
22. Prior, B. A. 1979. Measurement of water activity in foods: A review. J. Food Prot. 42:668.
23. Resnik, S. L., G. Favetto, J. Chirife, and C. Ferro Fontan. 1984. A world survey of water activity of selected saturated salt solutions used as standards at 25°C. J. Food Sci. 49:510.
24. Richard, J., and T. P. Labuza. 1990. Rapid determination of the water activity of some reference solutions, culture media, and cheese using a new instrument based on the dew-point method. Sci. Aliments 10:57.
25. Roa, V., and M. S. Tapia de Daza. 1991. Evaluation of water activity measurements with a dew point electric humidity meter. Lebensm. Wiss. Technol. 24:208
26. Rockland, L. B., and L. R. Beuchat. 1987. Water activity: Theory and applications to food. IFT Basic Symposium Series. Marcel Dekker, Inc., New York.
27. Rödel, W., R. Scheuer, and H. Wagner. 1990. A new method of determining water activity in meat products. Fleischwirtsch. 70:905.
28. Stokes, R. H. 1947. The measurement of vapor pressures of aqueous solutions by bi-thermal equilibration through the vapor phase. J. Am. Chem. Soc. 69:1291.
29. Stoloff, L. 1978. Calibration of water activity measuring instruments: Collaborative study. J. Assoc. Off. Anal. Chem. 61:1166.
30. Troller, J. A. 1983. Water activity measurements with a capacitance manometer. J. Food Sci. 48:73.
31. Troller, J. A., and J. H. B. Christian. 1978. Water activity and food. Academic Press, New York.

Index